D1560697

The Making, Shaping and Treating of Steel

Edited by William T. Lankford, Jr.
      Norman L. Samways
      Robert F. Craven
      Harold E. McGannon (Editor Emeritus)

United States Steel

Tenth Edition

Library of Congress Catalog Card Number: 84-81539

ISBN 0-930767-00-4

PRINTED IN U.S.A.
By
Herbick & Held
Pittsburgh, Pennsylvania

PREFACE

This, the Tenth Edition of "The Making, Shaping and Treating of Steel," has been prepared following a long tradition of providing up-to-date and complete information relating to the latest technology and current practices used in the making and processing of steel. The First Edition was distributed in 1919 in the form of hectographed notes. These notes were intended for use in training courses for salesmen and other non-technical employees of the Carnegie Steel Company. In the following year, two regularly printed and bound editions, enlarged by extensive revisions, were published. The authors of these early editions were J. M. Camp and C. B. Francis. They co-authored the book until 1925. The Fifth Edition, published in 1940, was rewritten by Mr. Francis.

Beginning with the Sixth Edition, published in 1951, writing the book became a group effort with many individuals from various departments of United States Steel Corporation making significant contributions. This work was carried out under the editorial direction of the Research Department. Three subsequent editions were published in the same manner.

"The Making, Shaping and Treating of Steel" was, and is, one of a kind and has been frequently referred to as "the bible of the steel industry." During its 65 year history, more than 230,000 copies were distributed to steel users, steel-plant employees, trainees in industry, students in technical schools and colleges, writers, librarians, and others here and abroad who had an interest in the steel industry.

Today's steel industry faces many challenges. Conditions have changed dramatically since the Ninth Edition was published in 1971. Industrial developments in the Third World, energy problems beginning with the oil embargo of 1973 and the resultant energy price increases, additional concern for the environment, and worldwide competition for steel markets have led to numerous technological changes. All of these factors were considered in the rewriting of this book. New chapters have been written on Continuous Casting and Ladle Metallurgy. The entire book has been revised and expanded.

This edition was written by United States Steel Corporation personnel with assistance from the Association of Iron and Steel Engineers. In the future, the AISE will assume total responsibility for the book.

ACKNOWLEDGEMENTS

All of the original manuscripts and most of the revisions for this Tenth Edition were written by U.S. Steel employees. Editorial support in the form of rewriting, proofreading, and editing was performed by the Association of Iron and Steel Engineers. Assistance with the rewriting of several chapters was provided by the Communications Design Center at Carnegie-Mellon University. The index was prepared by Western Reserve Publishing Services of Kent, Ohio. All coordination during the 12 months preceding publication was handled for AISE by Robert F. Craven and Norman L. Samways.

Many individuals and organizations within U.S. Steel contributed to the completion of this edition. Special credit goes to Keith K. Kappmeyer, Vice President—Technology, Steel and Related Resources, and the current and former U.S. Steel Technical Center employees listed below. Without their diligence and hard work, completion of this book would not have been possible.

H. M. Alworth	J. M. Holt	P. E. Repas
O. F. Angeles	D. H. Hubble	W. L. Roberts
R. G. Auger	R. M. Hudson	A. J. Rohrer
W. P. Benter	W. R. Johnson	R. J. Rygiel
R. J. Bentz	R. T. Jones	J. M. Sheehan
C. F. Bieniosek	R. R. Judd	D. E. Sonon
A. J. Bonfiglio	B. M. Kapadia	C. E. Spaeder
W. E. Brayton	R. J. King	R. L. Stephenson
K. G. Brickner	G. J. W. Kor	J. W. Stewart
C. G. Carson	R. W. Leonard	C. D. Stricker
L. J. Cuddy	S. J. Manganello	R. P. Stripay
D. S. Dabkowski	D. L. McBride	H. J. Tata
J. D. Defilippi	C. Mitchell	J. R. Tiskus
W. D. Doty	M. R. Moore	E. T. Turkdogan
J. Feinman	P. R. Mould	R. W. Vanderbeck
R. W. Fullerton	M. A. Orehoski	D. N. Volk
T. M. Garvey	M. J. Papinchak	J. M. Vrable
S. Gilbert	E. F. Petras	S. Waslo
E. B. Henry	H. R. Pratt	E. A. Yorkgitis
J. M. Hensler	H. M. Reichhold	R. J. Zaranek

The work of former U.S. Steel employees W. T. Lankford, Jr. and H. E. McGannon was extremely valuable. Dr. Lankford reviewed and evaluated the manuscripts for the entire book. Mr. McGannon, who has been associated with this book for over 30 years, provided continuity and perspective.

Individuals from other U.S. Steel groups, including Engineering, Resource Development, and the American Bridge Division, also made meaningful contributions. The work of J. J. Trageser was particularly helpful.

The contribution of Norman L. Samways, Technical Editor of the *Iron and Steel Engineer*, deserves special mention.

The Association of Iron and Steel Engineers is proud to be associated with this industry classic, which was acquired through the efforts of the AISE Board of Directors and 1983 AISE President Joseph G. Dickinson.

Herschel B. Poole
Managing Director and Publisher
ASSOCIATION OF IRON AND STEEL ENGINEERS

Pittsburgh, Pennsylvania
November, 1984

The Association of Iron and Steel Engineers is an international organization whose objective is the advancement of the technical and engineering phases of the production and processing of iron and steel. It maintains a well balanced program of activities to keep members informed of developments in engineering and operating practices. They include presentation of papers; research and development projects; establishing design criteria for equipment and processes; conventions and expositions; plant tours; publishing books and periodicals including *Iron And Steel Engineer, Directory Iron and Steel Plants;* and other activities that benefit its membership and industry.

CONTENTS

Association of Iron and Steel Engineers
Suite 2350, Three Gateway Center
Pittsburgh, Pa. 15222

THE MAKING, SHAPING
AND
TREATING OF STEEL

CHAPTER 1

Evolution of Iron- And Steelmaking

SECTION 1

PRINCIPLES OF MODERN STEELMAKING

Practically all steel products are made at the present time by the sequence of steps shown in Figure 1—1.

Iron-bearing materials containing principally iron oxides (iron ore, pellets, sinter, etc.) are reduced to molten iron (**pig iron**) in the blast furnace, using the carbon of coke as the reducing agent. In the process, the iron absorbs from 3.0 to 4.5 per cent of carbon. Iron containing 3.0 to 4.0 per cent of carbon can be used to make iron castings (**cast iron**). However, most modern **steel** contains considerably less than 1.0 per cent of carbon, and the excess carbon must be removed from the product of the blast furnace to convert it into steel.

The excess carbon is removed by controlled oxidation of mixtures of molten pig iron and melted iron and steel scrap in **steelmaking furnaces** to produce **carbon steels** of the desired carbon content. In this country, the principal steelmaking furnaces include the basic-oxygen furnace, open-hearth furnace, and electric-arc furnace. Various elements—chromium, manganese, nickel, molybdenum, etc—may be added singly or in combination to the molten steel during or after the carbon-removal process to produce **alloy steels**. All of the modern steelmaking processes produce molten (liquid) steel.

After the molten steel has attained the desired chemical composition in the steelmaking process, it is **tapped** (poured) from the furnace into a ladle from whence most steel is **teemed** (poured) into tall, usually rectangular molds where it solidifies to form **ingots**. Most ingots, after being removed from the molds, are reheated to a proper uniform temperature and rolled (in some cases, forged) into shapes known as **blooms, billets,** and **slabs** which are referred to as **semifinished steel:** only a relatively small number are rolled or forged directly into finished forms. Increasing quantities of semifinished steel are being produced by pouring liquid steel into the top of open-bottomed molds in continuous-casting machines where it solidifies and is continuously withdrawn from the bottom of the molds in long lengths of the desired shapes.

Blooms, slabs, and billets are referred to as semifinished steel because they form the starting material for the production by **mechanical treatment** (hot rolling, cold rolling, forging, extruding, drawing, etc.) of finished steel products that include bars, plates, structural shapes, rails, wire, tubular products, and coated and uncoated sheet steel, all in the many forms required by users of steel. Many of these products require some form of **heat treatment** at the steel mill to give them the best properties for their intended use.

The above oversimplified sequence might make it appear that the steelmaking process is simple. Actually, the manufacture of steel in its many product forms involves a complicated series of interrelated operations that will be discussed in detail in succeeding chapters.

SECTION 2

PREHISTORIC AND ANCIENT FERROUS METALS

FERROUS METALS IN ANTIQUITY

Ferrous metals are those that consist principally of iron. There were three sources of ferrous metals available to early man: meteoric iron, telluric (native) iron, and man-made ferrous metals. The relative rarity of the first two sources suggests that ferrous metals could only have come into common use after man had learned to extract iron from its ores.

The antiquity of man's use of ferrous metals is attested by references in fragmentary writings and in inscriptions on monuments, palaces and tombs that survived the collapse of such ancient civilizations as those of Assyria, Babylonia, Egypt, Persia, China and India and, later, Greece and Rome. In addition to these written records, archeologists have unearthed actual ferrous-metal tools, weapons and ornaments used by many of these historic ancient peoples, as well as some implements and jewelry in sites in many parts of the world that were occupied by prehistoric peoples who left no written records.

Gold, silver, copper and some other metals known to

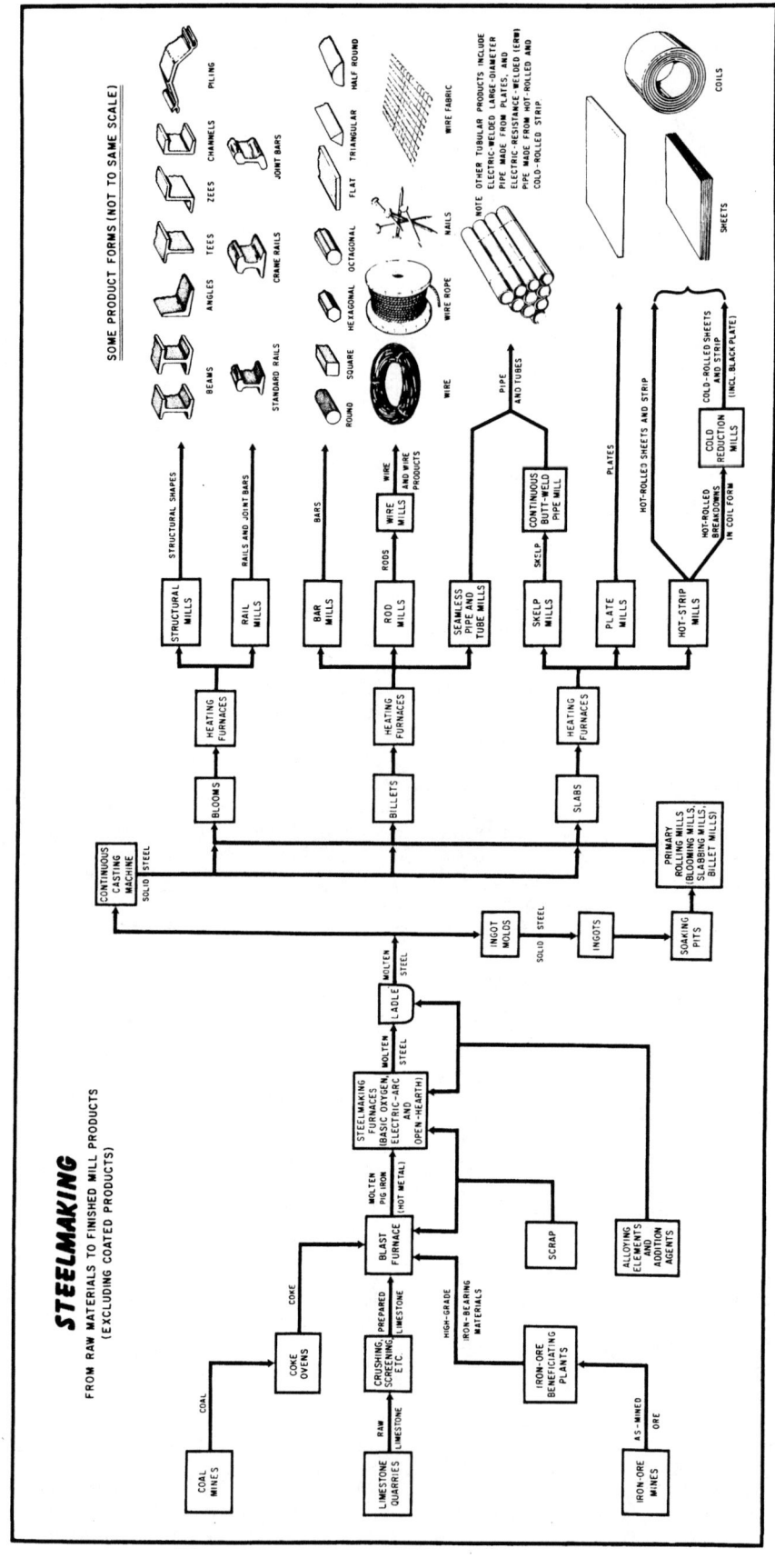

Fig. 1—1. Flow diagram showing the principal process steps involved in converting raw materials into the major product forms, excluding coated products.

the ancients often are found on or near the surface of the ground in a fairly pure metallic condition, in the form of nuggets or rough masses. Being bright in appearance, such native metals are noticed readily and, as they are capable of being shaped by hammering without prior heating, they were put to eventual use by primitive peoples. The softness of gold and silver made them useless for ordinary tools and weapons, and their ultimate chief use was for vessels and ornament. The metal copper, however, can be hardened appreciably by hammering it without previous heating, and the very hammering required to shape a tool from native copper makes it sufficiently hard for many purposes.

Archeological evidence seems to indicate that a knowledge of how to obtain copper from its ores existed before iron was intentionally made by man. Copper obtained from the reduction of its ores could be obtained in the fluid (molten) state, and was quite obviously a desirable product that could be shaped either by casting directly into molds or by hammering a solidified lump. Mixtures (alloys) of copper and tin that formed bronze, and of copper and zinc that formed brass, provided the ancients with metals that found widespread usage because they also could be melted in furnaces then available. The high melting temperature of iron, however, made it difficult if not impossible to melt in primitive furnaces, as discussed later in this section under "Man-Made Ferrous Metals."

It is not known definitely when man first began deliberately to make ferrous metals by reducing iron from its ores, or whether the knowledge of how to do so spread from a single point of discovery or if it was developed independently in several widely separated localities.

References made to ferrous metals in very ancient writings from China and India suggest that they were used in those areas at least as early as 2000 B.C., although it cannot be established if the metals were man-made. Some authorities ascribe the original discovery of practical ferrous-metal manufacture to peoples in India at a very early date.

The deliberate reduction of iron ore to produce ferrous metals seems to have spread to an ever-increasing extent over a relatively wide geographic area in the ancient world between 1350 B.C. and 1100 B.C. After that date, the production of ferrous metals began to be practiced generally, at least by the more advanced peoples. However, the art of making ferrous metals did not advance to the same degree in all areas over the same period of time: in fact, isolated peoples until quite recently employed methods not unlike those used in very ancient times.

METEORIC IRON

There is evidence that the first ferrous metal used by man may have been obtained from fragments of meteorites. Three facts form part of this evidence. In the first instance, practially all of the earliest names for iron, when translated, mean "stone (or hard substance or metal) from heaven," "star metal," or have similar meanings that suggest that the metal came from outside the earth. Secondly, chemical analysis of numerous archeological specimens has established that they contain considerable quantities of nickel, which likewise is found in similar quantities (usually 7 to 15 per cent, but sometimes as high as 30 per cent) in the iron of meteorites, whereas man-made iron, produced by any of the methods available to the ancients, would be essentially nickel-free. The third instance supporting this theory is that primitive peoples of relatively recent times used iron from meteorites to make useful implements and the main masses of the large meteorites from which bits of metal had been severed were still in the places where they had fallen and still served as sources of supply.

The foregoing evidence is not conclusive. In the first instance, ferrous metals may have been made by peoples who left no written records, before literate peoples used meteoric iron and gave it a name. Illiterate producers of ferrous metals eventually may have supplied some of their product to the literate peoples who, since there was no means for recognizing the difference in chemical composition of meteoric and man-made iron, may have called the different metals by the same name. The second and third instances only prove that useful implements can be (and were) made of meteoric iron.

TELLURIC (NATIVE) IRON

Iron is very rarely found in the native state. One of the few known occurrences is in northwestern Greenland, where iron occurs as grains or nodules in basalt (an iron-bearing igneous rock) that erupted through beds of coal. Two very rare natural nickel-iron alloys, given the mineralogical names of **awaruite** ($FeNi_2$) and **josephinite** (Fe_3Ni_5), have been found in the form of granules and small bean-shaped pebbles.

It is improbable, therefore, that early man could have found any useful quantity of naturally occurring metallic iron, certainly not enough to account for the widespread distribution of ferrous-metal artifacts that have been discovered by archeologists.

MAN-MADE FERROUS METALS

It has been known for many centuries that iron ore, embedded in burning charcoal, can be reduced to metallic iron. Charcoal consists almost entirely of carbon and the iron in ore is present chiefly in the form of iron oxides. In modern terminology, oxygen from air admitted near the bottom of the bed of charcoal combines with carbon to generate heat by forming carbon dioxide which, in the presence of the remaining hot carbon in the combustion zone, is converted almost immediately to carbon monoxide. The hot carbon monoxide reacts with oxygen in the iron oxides to form carbon dioxide that escapes through the overlying bed of charcoal and ore, leaving metallic iron. Some of the carbon also reacts directly with iron oxide to make iron and carbon monoxide. The conversion of carbon dioxide in the combustion zone, and the reaction of carbon with the iron oxides, both absorb heat that must be supplied by the combustion of additional carbon of the charcoal.

The foregoing process was carried out by early metalworkers in many types of furnaces, some of which relied on natural draft to supply air for combustion,

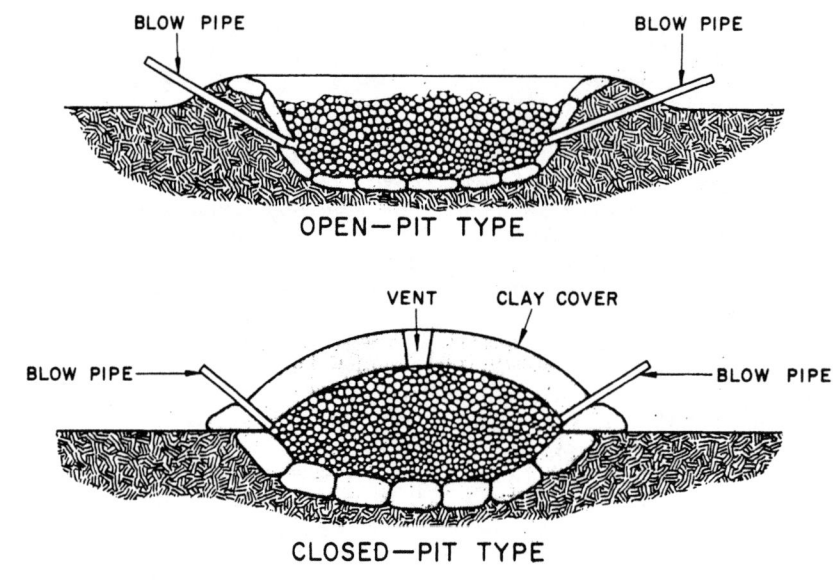

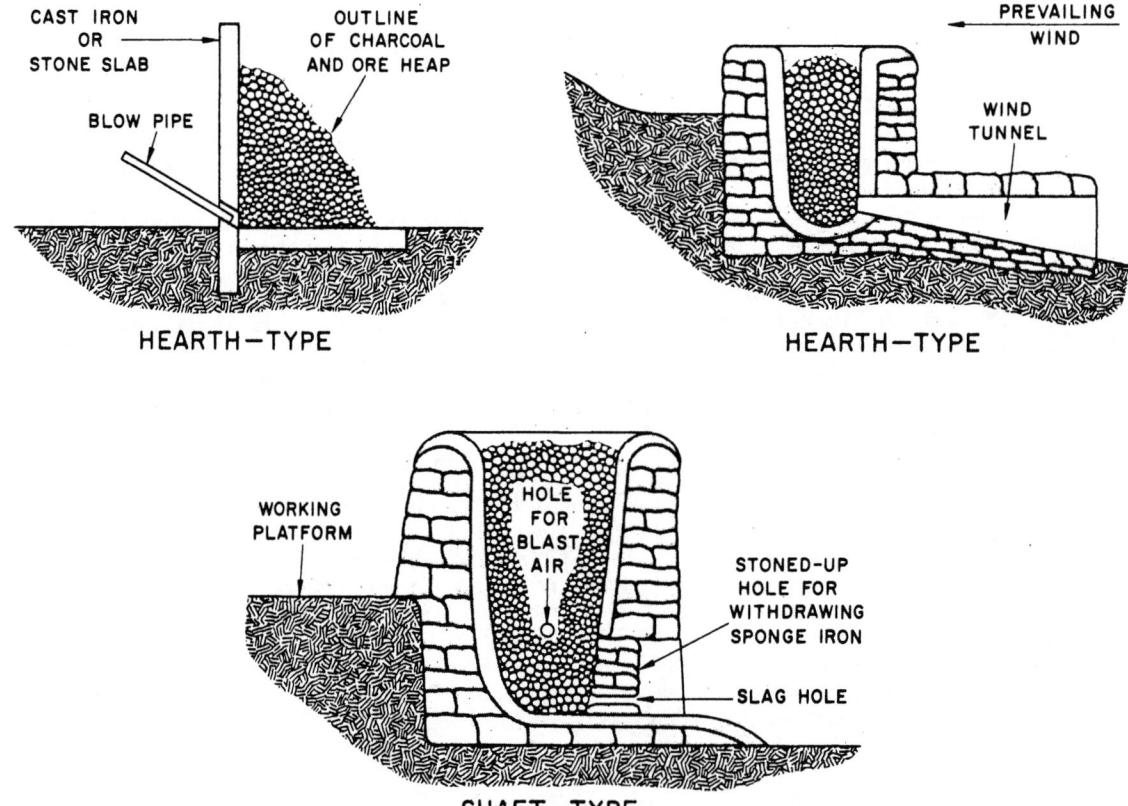

FIG. 1—2. Schematic cross-sections (not to same scale) of some early types of furnaces for reducing iron ore, using charcoal as fuel.

while others were equipped with some means for blowing the air into the fuel bed to obtain higher temperatures and better control of the process. Schematic cross-sections of some early types of furnaces are shown in Figure 1—2. Natural draft alone probably supplied the air for combustion in the first hearths or furnaces. Later, it seems to have been the practice to construct a hearth on a hillside or at the base of a cliff facing in the direction of a prevailing wind. A wind of suitable direction and velocity could be led through a trench or tunnel into the burning fuel bed through an opening in the hearth wall to provide the air blast required for intense combustion of the charcoal fuel. Still later, devices for blowing air into the fuel bed were developed to make the process independent of wind and weather. Through the centuries, these devices ranged all the way from mouth-blown hollow reeds, through foot-operated bladders of animal skins, foot-operated bellows, hand-operated bellows, and air-blowing devices operated by treadmills, water wheels and, eventually (in modern times), steam engines, internal-combustion gas engines, steam-operated turbo blowers and gas turbines. Another device somewhat widely used was the **trompe**, which utilized the aspirating effect of a falling column of water inside a tube, to draw in air near the top of the tube, expelling it into a closed chamber at the bottom; the air was piped from this chamber to the furnace (see Figure 1—3).

In all of these furnaces or hearths, the hottest zone in the bed of burning charcoal was adjacent to the opening where air entered the furnace. Most of the iron oxide was reduced to metallic iron before it reached the hottest zone. In the pit-type and hearth-type furnaces, the reduced iron was in the form of porous granules that descended gradually toward the hotter regions of the fuel bed as charcoal was consumed and iron oxides were reduced. Eventually, under the influence of the higher temperatures near the bottom, these granules became pasty and agglomerated to form a loosely coherent mass now known as **sponge iron**. The air blown into the furnace entered at some distance above the bottom. The space below the air entrance provided a place for the sponge to form where it would not be exposed to oxygen from the air which, otherwise, would combine with the reduced iron to form iron oxides. In the pit- and hearth-type furnaces and in the shaft-type furnaces of intermediate height, it would

have been difficult, if not impossible, to attain temperatures high enough to melt the sponge iron.

Impurities in the iron ore, such as silica, combined with some of the iron oxide of the ore to form a liquid siliceous slag of high iron content, at least part of which permeated the pores of the sponge iron, while the remainder either collected in the bottom of the furnace or ran out of a suitable opening.

After a sponge of sufficient size had formed, it was pried out of the fuel bed and hammered, while still hot, to compact it into a reasonably solid and uniform lump or bar of iron by expelling most of the slag while closing the pores of the sponge.

If the reduced iron were kept in contact with hot charcoal in the absence of air, it might absorb some of the carbon. Unavoidably, some of the manganese, silicon, phosphorus and sulphur that might be present as impurities in the ore also might be absorbed by the iron during the reduction process, but their effects will not be considered in relation to early processes since it was the variation in carbon content that chiefly influenced the properties of the reduced iron that were of interest to artisans.

The amount of carbon absorbed would depend upon the temperature and the time during which the iron was in contact with carbon: both of these factors could be controlled to some extent to control the amount of carbon absorbed. The type of furnace employed in the reduction process also would influence the degree to which carbon absorption could be controlled.

Iron reduced in pit-type and hearth-type furnaces was almost pure iron, and contained very little carbon because of the relatively low temperatures that could be attained and the short time the sponge was exposed to conditions favoring the absorption of carbon.

Shaft-type furnaces, on the other hand, provided conditions more favorable to carbon absorption, especially when adequate devices for blowing air were employed. Higher temperatures could be attained, and reduction of iron oxides began a considerable distance above the combustion zone so that any reduced iron would be in contact with hot carbon for a longer time at higher temperatures than in the pit- or hearth-type furnaces.

The preceding descriptions have shown that carbon from the charcoal used almost exclusively as fuel in early furnaces had several important roles:

(1) Its combustion supplied the heat necessary for the iron-reducing process.

(2) It provided the chemical agent required to reduce iron from its oxides.

(3) It protected the reduced iron from being reoxidized by oxygen from the air so long as the iron was surrounded by hot carbon.

(4) It could be absorbed by the hot reduced iron to form iron-carbon alloys of several types, characteristics of which depended upon their carbon content, as discussed in the following descriptions of products of early iron-reduction processes.

Products of Early Iron-Reduction Processes—One of the products of early iron-reduction processes was a low-carbon metal that was relatively soft, ductile, malleable, and could be hammer-welded at forging temperatures: it corresponded generally to modern

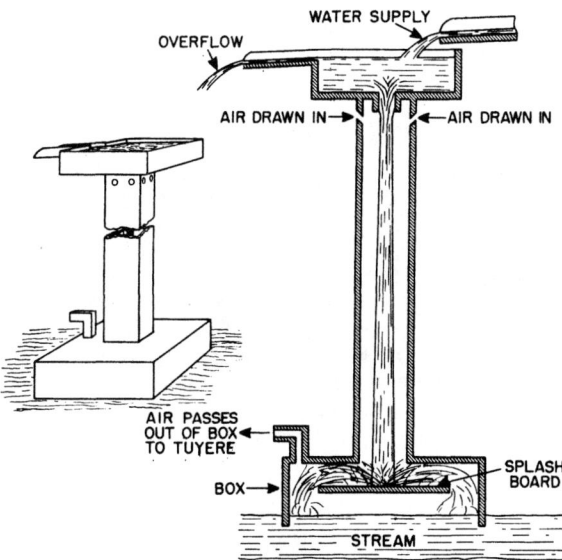

FIG. 1—3. Schematic representation of the trompe for utilizing the principal of aspiration to provide air blast for furnaces.

wrought iron (see Sections 3 and 5 of this chapter).

The type of furnace used or the manner in which a given furnace was operated could result in the absorption by the iron of moderate amounts of carbon (up to, say, one per cent). The iron-carbon alloy so produced was malleable, ductile and weldable but, if cooled slowly from forging temperature, was inherently somewhat harder than low-carbon wrought iron. More importantly, the higher carbon content made the iron-carbon alloy capable of being made very hard by quenching (cooling very rapidly from a high temperature by immersing in water or some other liquid), in a manner comparable to modern medium-carbon and high-carbon steel. The effect of quenching was evidently known at a very early date. The quenched metal was very hard and somewhat brittle. It could be reheated for a short time to a relatively low temperature to make it less brittle without too drastically lessening the hardness obtained by quenching: this process is now known as tempering. Other early steelmaking methods are discussed in Section 6 of this chapter.

The more carbon an iron-carbon alloy contains (up to something over 4 per cent), the lower the melting point of the alloy will be. Especially in shaft-type furnaces, it was possible for the iron to absorb enough carbon (say, 2.5 to 3.0 per cent) and molten high-carbon iron was produced. It is probable that the lumps that formed when the liquid metal solidified were thrown away by the early metalworkers, because they could not be shaped by hammering, being too brittle even when heated to forging temperature. Eventually, it was learned how to convert the high-carbon iron to wrought iron during remelting under controlled conditions in charcoal-fired hearths. In some areas at an early date, molten high-carbon iron was produced deliberately (at least as early as 200 B.C. in China) and poured into molds of the desired shapes where it solidified to form useful articles of cast iron. Molten high-carbon iron has come to be the chief product of the reduction of iron ore (see Section 4 of this chapter and Section 1 of Chapter 15). Processes by which molten metal is produced by the reduction of ore are called smelting.

Judgment and skill of the operator were the only means of control of any of the early iron-reducing or hardening methods, since nothing was known of the metallurgical principles that governed them. In reducing iron from its ores, the metal obtained was not predictably uniform from one operation to the next, and a large part of the iron in the ore was lost to the slag. This lack of uniformity in the product was probably not too important when the wrought-iron type of metal was made but, in the case of the steel-like iron-carbon alloys, it made it impossible to use other than "rule-of-thumb" methods for carrying out the hardening operation, and failure to obtain the desired results was probably common. In some areas, the iron ores also contained some other metals beneficial to the properties of ferrous metals which became alloyed with the iron during reduction (manganese, for instance), and the metals produced from such ores became justly celebrated, although the reason for the superiority was not then known.

It will be seen in succeeding sections of this chapter that some of the ancient methods described above survived, with modifications, until modern times.

DIRECT PROCESSES FOR MAKING WROUGHT IRON

HISTORICAL BACKGROUND OF DIRECT PROCESSES

From the earliest times to about 1300 A.D., all wrought iron was produced directly from ore. Reduction of iron ore was carried out in a relatively simple manner using charcoal as fuel, as described in Section 2. Because wrought iron was produced by the single step of reducing iron from its ores, the various processes are classified as direct processes.

In view of the many centuries in which direct processes were used, it is to be expected that many different methods and types of apparatus would have been developed. Little is known of the furnaces in use prior to the eighteenth century, but the majority probably were of the hearth type, while the remainder were of the shaft type and may be compared to small blast furnaces, as will be discussed later. While these furnaces might, and did, differ widely as to form, size, and materials of construction, the fundamental metallurgical principles were the same for all. Two variations of direct charcoal-hearth processes for making wrought iron were the Catalan process of considerable antiquity and the American bloomery process of more recent times that are described to illustrate how the same principles were employed in simple and sophisticated forms of the same furnace.

THE CATALAN PROCESS

The Catalan hearth, as the furnace used in this process that originated about the Thirteenth Century was called, appeared in various forms. A common late type was anywhere from about 0.5 metre (20 inches) square and 0.4 metre (16 inches) deep to around 0.75 metre (30 inches) or 1 metre (40 inches) square and something over 0.6 metre (24 inches) deep. The nozzle or tuyere, through which the blast was blown into the furnace, was placed about 230 millimeters (9 inches) from the bottom in the smaller hearths and about 380 millimeters (15 inches) from the bottom in the larger hearths. The hearth was filled to the level of the tuyere with charcoal, on which was piled lump ore together with charcoal. These materials were placed to form two separate columns, the charcoal against the tuyere side of the hearth, and the ore against the other side (Figure 1—4). A gentle blast of air was applied at first and carbon monoxide, formed by combustion of the charcoal,

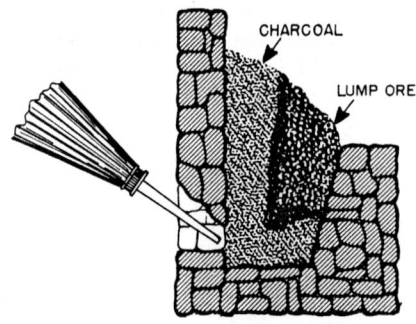

FIG. 1—4. Representation of a Catalan hearth or forge used for reducing iron ore up until relatively recent times, showing method of charging fuel and ore, and approximate position of the nozzle supplied with air by a bellows.

passed preferentially through the open pile of lump ore. The ore was reduced to metallic iron when the oxygen in the iron oxide of the ore combined with some of the carbon monoxide to form carbon dioxide. The waste gases escaped at the top of the charge. Charcoal (along with fine ore) was added at regular intervals to replace that consumed in combustion. After about two hours, the lump-ore column was gradually pushed downwards and the temperature of the hearth was raised by increasing the blast. As successive portions of the ore became reduced, they were pushed nearer the tuyere where the hearth was hottest. By the time the ore had reached the hotter regions, it was largely reduced to the metallic state.

The unreduced portion of the lump ore, along with part of the fine ore added periodically with charcoal, formed a siliceous slag of high iron content with the **gangue** (waste material). The metallic iron resulting from reduction of the ore became pasty at the temper-

atures existing near the tuyere, to form a coherent **loup** or **bloom**. After as much as possible of the ore had been reduced, the bloom was pried out of the hearth and hammered into bar form.

THE AMERICAN BLOOMERY

The American bloomery process was very similar to the Catalan process, differing from it chiefly in the fact that ore in a fine state, instead of in lumps, was mixed with charcoal to form the charge. The American bloomery represented the highest development in the simple hearth type of furnace for producing wrought iron. The bellows supplying the blast was operated by a water wheel or steam engine. The hearth was provided with a water-cooled metal bottom-plate, and cast-iron plates lined the sides. These hearths, rectangular in shape, were about 0.6 metre (2 feet) deep and 0.9 metre (3 feet) wide and were surmounted by a tall chimney in the form of a truncated pyramid for carrying off the hot waste gases. The blast was heated (to save fuel) by passing the air through cast iron pipes around which the hot waste gases passed on their way from the furnace to the opening of the stack. Usually bloomeries were open in front like an open fireplace, with the tuyere placed either at one side or at the back, about 0.5 metre (20 inches) above the bottom. Charcoal was first put into the hearth, the blast turned on, and when the fire was burning well, some ore was spread on the charcoal. Thereafter, charcoal and ore were added alternately until a sufficient amount of metal had collected upon the bottom. Then the iron, in a pasty mass and mixed with much slag, was removed from beneath the fuel bed with bars and tongs and hammered into a bloom. The last wrought iron to be produced by the bloomeries in this country was made in 1901.

SECTION 4

DEVELOPMENT OF THE BLAST FURNACE

It may be said in general that the blast furnace for producing molten high-carbon iron developed gradually from the early hearths in which chiefly wrought iron was produced. The development consisted in gradually increasing the height of the furnace, introducing the charge at intervals through the top, and providing an adequate blast of air. These higher furnaces, distinguished as a class from the Catalan type of hearth or bloomery, were termed **shaft furnaces**. Originally developed by ironmakers of Central Europe, the new type of furnace was built of masonry that enclosed a vertical chamber in the form of two truncated cones placed base to base—in a crude way resembling the lines of a modern blast furnace (see Figure 1—5). The iron ore, flux and charcoal were charged into the top of the shaft, while air under relatively low pressure was blown into the furnace through a **tuyere** or tuyeres near the bottom of the structure.

EARLY SHAFT-TYPE FURNACES

The **stuckofen** or **old high bloomery**, variations of which appear to have been called **salamander furnace, wolf furnace, wolf oven, wulf's oven** and **luppenofen** or **loup furnace**, evolved as described above from the Catalan type of hearth furnace. The earliest recorded sites of such shaft-type furnaces were in territories included in Germany (in Nassau, Siegen, and Saxony) and in parts of Austria, Belgium, and the Netherlands.

The stuckofen, in the state of development described around 1350 A.D., was a furnace roughly 3 to 5 metres (10 to 16 feet) high having a round, elliptical or rectangular shaft cross-section (greatest cross-sectional dimension about 0.9 to 1 metre or about 3 to 4 feet).

One or two tuyeres supplied the blast, which entered the stuckofen somewhat over a foot above the hearth. Fuel and ore were charged into the top of the furnace, being replenished from time to time as smelting pro-

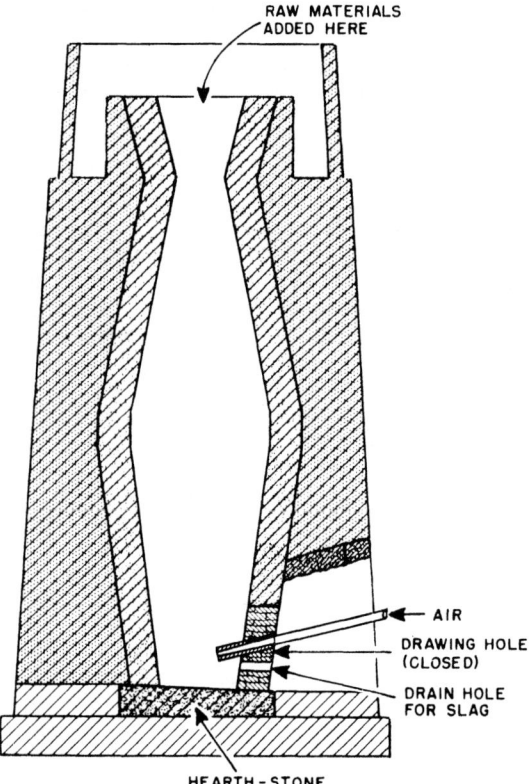

FIG. 1—5. Schematic cross-section of a stuckofen, equipped with a drawing hole for the extraction of the blooms. (After Percy.)

ceeded. A drawing hole was provided in the wall at the bottom of the shaft for extracting the blooms. This hole was closed by brick or stone work that was torn out each time a bloom was removed, after which the hole was again closed. Charcoal was the only fuel used.

The furnace called the wolf oven has been described as lower than the stuckofen, perhaps 1.8 to 2.1 metres (6 to 7 feet) in average height. Intermediate in size, between the wolf oven and the stuckofen, are the **blasofen** and **bauernofen**. The bauernofen corresponds to the **osmund furnace** (about 2.4 metres or about 8 feet high) used in Sweden (a similar type was used in India). A type of furnace which originally resembled and was operated like the stuckofen, and was later adapted to produce either blooms of low-carbon wrought iron or molten, high-carbon iron, resembled a crude blast furnace and was termed **blauofen, blau furnace** or **blue furnace**.

The stuckofen may be considered as the forerunner of the modern blast furnace which produces only liquid, high-carbon iron. Liquid high-carbon iron often was produced in the stuckofen, intentionally or otherwise. This occurred when the reduced iron was in contact with hot fuel away from the blast long enough to absorb sufficient carbon to reduce its melting point to where it would become liquid. The height of the furnace made this possible, especially if the operating temperature were high enough. A **flussofen** was strictly a primitive blast furnace intended only to produce mol-

ten, high-carbon iron. The modern blast furnace, then, is a shaft furnace, gradually evolved from the stuckofen and flussofen. In its early days it was called a **high furnace**, from its German name, **hochofen** (French: **haut fourneaux**). It is designed solely to produce molten iron and operates continuously, in that the solid raw materials (ore, coke and limestone) are charged at the top at regular short intervals, and the molten iron and slag which collect in the hearth are tapped out at longer intervals.

THE BLAST FURNACE AFTER 1500 A.D.

The blast furnace was introduced into England about 1500 A.D. Coke was first used as a blast-furnace fuel in England in 1619, but its use was not adopted generally until about 1730. In the early 1800's—again in England —the principle of heating the air before it was blown into the furnace was introduced: air so preheated is referred to as **hot blast**.

Early American Furnaces—In America, an iron works was established in Virginia on the James River about 1619; this was destroyed in an Indian raid in 1622 and never rebuilt. The Hammersmith (now Saugus), Massachusetts iron works was begun in 1645 and was the first successful iron works in what is now the United States, not being abandoned until 1675.

It is an interesting fact that the principal development connected with the blast furnace for over 400 years after its inception was the spread of its use to new localities. There was a strong family resemblance among all of the furnaces built during this period, although there were variations in size and in the design of machinery for supplying the blast, etc. For this reason, the Hammersmith furnace shown schematically in Figure 1—6 can serve as typical of American blast furnaces of as recently as 100 years ago.

The American furnaces of the middle Nineteenth Century now would be called very crude affairs. They were usually in the form of a truncated cone or pyramid, 6 to 9 metres (20 to 30 feet) high, and constructed of stonework which enclosed a shaft about 1.2 metres (4 feet) across at the top and 2.4 metres (8 feet) at the bosh. The hearth was either round or square in cross section. The capacity ranged from one to six tons a day. In 1850, for example, the production of iron in the United States was reported to be 511 438 metric tons (563 755 net tons), produced by 377 establishments.

Cupola Blast Furnaces—Rods and bands of wrought iron were employed in the construction of some of the larger furnaces to increase the stability of the stack, but the expansive forces present burst even the strongest practicable ties. The obvious answer was to completely enclose the stack in a "shell" constructed of wrought-iron plates; a furnace built at Port Henry, New York in 1854 was said to be the first to be enclosed completely in an iron shell. The shell type of construction gave to such furnaces the name of "cupola blast furnaces" (Figure 1—7).

Top Improvements—The top of the early furnaces was open and the escaping gases burned in the air above the furnace. Eventually, attempts were made to use the heat of the burning gases to preheat the blast air. The first devices for heating air were mounted on top of the stacks. A later development was the adoption

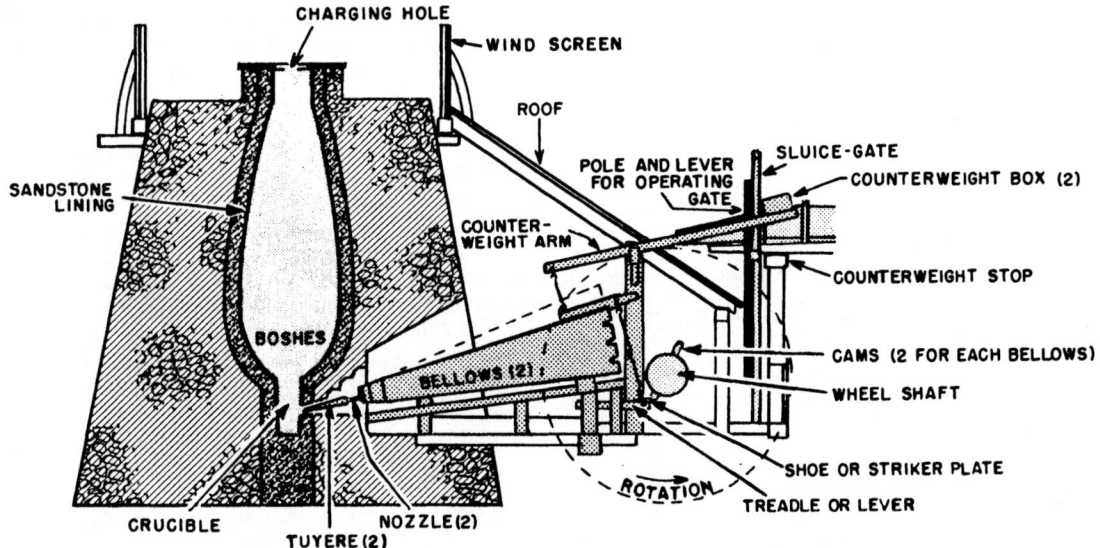

FIG. 1—6. Schematic cross-section of the Hammersmith furnace near Saugus, Massachusetts, restored by the American Iron and Steel Institute. Water from the sluice turned the overshot water wheel. Cams on the axle of the wheel engaged the treadle or lever and exerted a squeezing force on the bellows that compressed the air for the blast. The raw materials were dumped into the charge hole at the top of the stack, and molten iron was run from the furnace through an opening in the wall of the crucible. This opening was near the bottom of the crucible on the side facing the reader, and was kept plugged except when molten iron was run.

of the closed top, that involved the invention of a bell-and-hopper arrangement that kept the top closed except when the bell was lowered to charge materials into the furnace. One of the first American furnaces to adopt the closed top was the Fletcherville charcoal blast furnace near Mineville, New York about 1870. This principle was later extended to the use of a double bell and hopper (1883) that made it possible to charge materials without ever completely opening the furnace

top (Figure 1—8): this is the present usual closure, although some furnaces designed to operate at higher top pressures have been equipped with three bells or with the Paul Wurth-type bell-less top.

As early as 1859 in this country (earlier abroad) attempts were made to collect the gases at the top of the furnace before they burned, and lead them through suitable piping to ground level, where they could be burned in special structures called "stoves" in which

FIG. 1—7. The Isabella Furnaces, constructed in 1871-72. These were both open-top furnaces originally. A third furnace was added to the plant a few years later. These furnaces are typical of the design referred to as cupola-type blast furnaces.

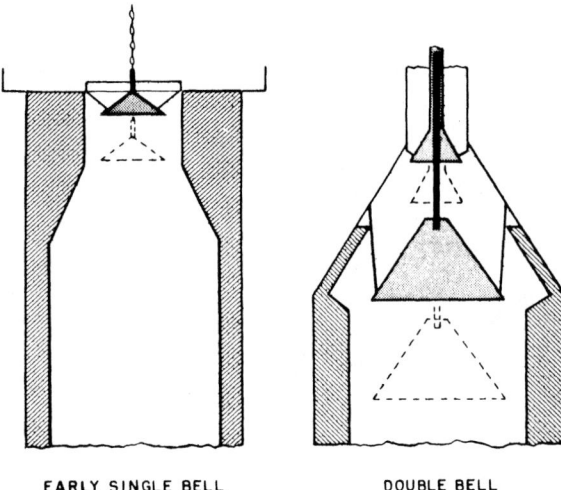

EARLY SINGLE BELL   DOUBLE BELL

Fig. 1—8. Schematic representation of the principle of operation of the single and double bell methods for effecting closure of a blast-furnace top. The distance the bells are lowered (as indicated by the dotted outlines) has been exaggerated for clarity.

many of the early furnaces were brought to the charging hole atop the stack in barrows or wheeled carts that passed over a bridge leading from an adjacent elevation to the furnace. As furnaces grew taller, vertical hoists similar to elevators were employed to raise the barrows loaded at ground level to the top of the furnace. Men called "top fillers" then wheeled the materials to the charging position and dumped them onto the bell. As furnace capacity increased, it was impractical to handle the huge quantities of raw materials by manual methods. In 1883, the first inclined skip hoist (in conjunction with the first double bell and hopper) was installed on an American furnace to raise the raw materials in a skip car to the top of the furnace and automatically dump the contents of the car into a hopper above the small bell. The fact that a skip car always dumped its load in the same location interfered with the proper distribution of materials that was essential to smooth furnace operation, so that various mechanical means had to be developed to distribute the charge over the top bell; one of these consisted of a rotating hopper over the small bell that, with modifications, is generally employed on modern furnaces. Another method employed a charging bucket that was rotated after filling in the stockhouse before delivering to the furnace top.

Stocking Methods—Increased production rates also forced the adoption of mechanical handling methods for handling and stocking raw materials before they were charged into the furnace skip car or bucket. Prior to 1890, raw materials were dumped from railroad cars run onto an elevated trestle, and manually moved to the stockhouse where the skip car or bucket was filled. In 1895, construction of the blast-furnace plant at Duquesne, Pennsylvania, included an ore yard with a stocking-bridge system similar to that employed in present-day blast-furnace plants; this was such a radically new principle that it was referred to as the "Duquesne revolution." The success of the new method led to its general adoption by the industry. This brief resume describes only some of the early principal ideas and inventions that led, step-by-step, to the designs employed in contemporary blast-furnace plants. A detailed description of a modern plant is given in Chapter 15.

the blast air could be heated before it was blown into the furnace through the tuyeres. Stoves of both recuperative and regenerative types were developed: only the regenerative type is employed at present.

Blast Improvements—The development of better machinery for compressing the blast air kept pace with —or even preceded—the construction of taller and larger furnaces. The water wheel that operated bellows or wooden cylinder-type blowing tubs was first replaced by the steam engine. Soon, steam-driven cylinder-type blowing engines of high capacity were developed and became standard for the larger furnaces about 1880. In the United States, the first gas-driven blowing engines were installed in 1903; these were internal-combustion engines that used cleaned blast-furnace gas as fuel. An important development for generation of the air blast was the turboblower, first used in 1910 and now the accepted means for the purpose.

Charging Improvements—The ore, fuel and flux for

SECTION 5

INDIRECT PROCESSES FOR MAKING WROUGHT IRON

After furnaces which produced molten high-carbon iron became commonly employed in Europe, part of their product was used to produce iron castings by pouring the liquid metal into molds of the desired shape. Such cast iron had limited usefulness, since it was inherently hard and brittle due to its high-carbon content and the presence of other elements that entered the iron during reduction of the iron ore. It was not malleable, that is, it could not be shaped at any temperature below its melting point by either hammering or rolling.

In order to utilize the high-carbon product of these

furnaces for making forged or wrought articles, it was necessary to develop purifying processes that would remove the excess carbon, manganese, silicon, etc., from the impure iron to produce relatively soft, malleable wrought iron that would have the same general composition and characteristics as the iron formerly produced directly from the ore in the Catalan and similar processes. As might be expected, a very great number of methods were developed in different localities. Two types of processes eventually became prominent: **the charcoal-hearth processes** and the **puddling process.** Since the production of wrought iron from ore by any of these processes involves two separate steps: (1) reducing the ore to make pig iron and (2) remelting and purifying the pig iron to make wrought iron, they are referred to as **indirect processes**.

Some of the most widely used charcoal-hearth processes for purifying pig iron are described below (the Walloon, South Wales and Lancashire processes).

CHARCOAL HEARTH PROCESSES

Walloon Process—Just how, when, where, and by whom wrought iron was first produced from pig iron is unknown, though it is probable the process originated in Belgium. The first attempts were, no doubt, made in the forge or on a hearth such as those already described for the production of iron directly from the ore. Here the action of the air from the blast (by that time in general use) would, if the iron were handled properly during melting, result in the oxidation of the silicon and the greater portion of the manganese and carbon, giving a ductile and workable product. The first reference to the process in written records appeared about 1620, but by that time the process had reached a stage of considerable development. Previous to that date, the Walloons of Flanders had gone to Sweden, where they had introduced the process, since known as the Walloon process. In this process a rather deep hearth with one or two tuyeres was used (Figure 1—9). With the hearth filled with charcoal and heated to a high temperature, the pig iron, in the form of long pigs, was fed into the fire so that the lower end of the pig would be gradually melted, and the molten metal would trickle to the bottom directly in front of the blast. The metal, desiliconized and decarburized by the oxygen in the air blast, would collect as a pasty mass upon the bottom, being worked vigorously as it collected. The ball of pasty metal was then separated into lumps that were raised above tuyere level and remelted. The new ball formed on the bottom was then removed from the hearth and hammered into a bloom. The second melt-

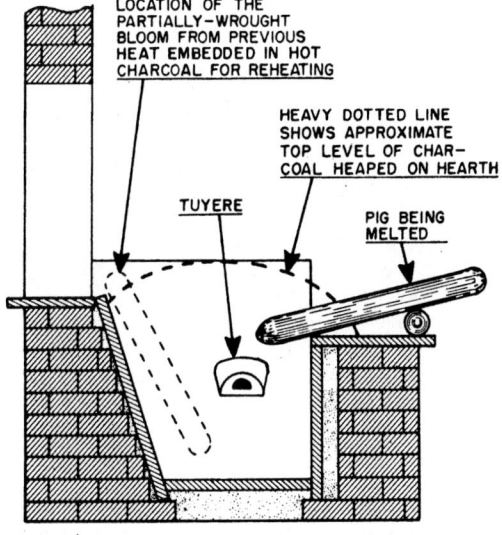

FIG. 1—9. General arrangement of a Walloon hearth used for purifying pig iron to make wrought iron.

ing freed the metal from much of its entrapped slag. The pig used in Sweden, since it was reduced from the famous Dannemora ore in charcoal furnaces, was exceptionally low in silicon, sulphur and phosphorus, hence was especially adapted to this process.

South Wales Process—Few districts outside of Dannemora are favored with ore so free from phosphorus, or were able to continue using charcoal for fuel for such a period of many years. The use of coke in the blast furnace leaves no alternative but the production of high-silicon iron, if the sulphur content is to be kept suitably within limits. At the lower temperatures necessary to produce low-silicon iron, more of the sulphur content of the coke will be picked up by the iron. Such iron, high in sulphur and silicon, could not be purified in a single operation, as in the Walloon process, where the purification was carried on in the combustion chamber with the metal and slag in contact with the fuel. It was found, however, that this iron could be purified and converted into wrought iron very readily in two stages. The South Wales process was a typical two-stage process. For the first stage, a small, rectangular, water-cooled hearth, surmounted by a stack and provided with a number of tuyeres, was used. In some cases, this hearth was a separate structure; in other cases this hearth for melting the pig iron formed part of a two-hearth furnace in which the melting hearth was slightly above the second hearth where final refining took place. When a separate melting hearth was employed, coke was used as fuel to melt the pig iron; for refining the melted iron, the second hearth was fired with charcoal. When the two hearths were incorporated into one furnace structure, charcoal was used as fuel in both. Sometimes two charcoal hearths were served by one melting hearth that tapped directly into them. The hearths were known by various names. The melting hearth, when separate, was called the **refinery**, or **refinery fire**, if the metal tapped were allowed to partially or completely solidify before being transferred to the second hearth. If the metal were allowed to flow directly from the refinery into the second hearth or hearths, as in the two-hearth furnace, the melting hearth was known as the **melting finery** or **running-out fire**. In both cases, the second hearth was known as the **finery, charcoal finery** or, more often, **knobbling fire**.

With a good fire burning upon the melting hearth, alternate charges of coke and pig iron were made upon it. As the metal melted, it would collect upon the bottom of the melting hearth where the blast from the tuyeres impinged upon it, oxidizing the silicon and some of the phosphorus along with a part of the iron. Assuming that the melting hearth in this case was a separate unit, when a sufficient quantity of partially purified and partially solidified metal had collected, it was transferred to the second hearth from the melting hearth, being piled in front of the tuyere and completely remelted while exposed to the blast. During the remelting, the metal was worked constantly and repeatedly raised slightly off the bottom, which treatment promoted the oxidation of carbon. As the carbon was removed, the metal gradually assumed a pasty condition. When it was assumed that the iron was sufficiently decarbonized, it was worked into a ball, taken from the furnace, and hammered.

Lancashire Process—The Lancashire process differed essentially from the South Wales process in that the pig iron was both melted and refined in a single hearth using charcoal as fuel. With some of the slag left from the previous operation to cover the bottom, the hearth was piled with charcoal up to above the tuyeres. The pig iron in lumps, was placed on top of the charcoal pile, covered with more charcoal, and the blast turned on. The pig iron melted in drops which became partially decarburized in passing through the tuyere area and collected on the bottom. When all of the pig was melted, it was worked with bars to mix it with the slag and become thoroughly purified. As purification proceeded the metal became stiff and pasty, and when purification was completed the pasty mass was raised above the tuyeres and melted down again to free it from the intermingled slag. The pasty lump resulting from the remelting process was then taken from the furnace and hammered into a bloom.

These three are only a few of the many types of charcoal-hearth indirect processes developed for purification of pig iron to produce wrought iron.

HAND PUDDLING PROCESSES

About 1613, Rovenson invented the reverberatory furnace which is a furnace with a shallow hearth and a roof that deflects the flame and radiates heat toward the hearth or surface of the charge. Rovenson described his furnace as a bloomery, finery or chaffery "in which the material to be melted or wrought may be kept divided from the touch of the fuel," but it was not employed for purifying pig iron until 1766, when the Cranege brothers received a British patent on a process which later came to be known as **puddling.** With careful manipulation of a reverberatory furnace, they were able to convert "white iron," or pig iron from which most of the silicon and phosphorus had been removed in a refinery, as described under the "South Wales Process," into a good malleable form of iron by the use of raw coal alone for fuel. In 1784, Henry Cort hollowed out the bottom of the furnace to contain the metal in the molten state, then by agitating this "puddle" or bath of metal with an iron bar or paddle he was able to convert white or partially-refined pig iron into a malleable form (wrought iron), the carbon being burned out by the oxidizing gases of the furnace atmosphere.

As the furnace bottom was made up of sand, it was rapidly fluxed away by the iron oxide formed. Besides, the process consumed much time and was wasteful of iron, the yield being less than 70 per cent of the metal charged. These objectionable features were largely overcome by Joseph Hall, who, in 1830, substituted old bottom material for the sand, thus introducing the iron oxide bottom, which adapted the process to any grade of iron, shortened the time of the heats, and increased the yield to about 90 per cent. The introduction of the iron oxide type bottom shortened the heat times appreciably because carbon was removed from the bath more rapidly as a result of a vigorous boiling action generated by the oxidation of carbon in the melt by the

oxides on the bottom of the furnace. Hall's process came to be known as the **pig boiling** process. Later, this process became the leading method for the production of wrought iron.

The original method was designated as **dry puddling** because of the small quantity of slag formed, the slag-forming impurities having been removed in the refinery (see "South Wales Process," above). Hall, or his associates, also introduced the use of air-cooled iron plates for supporting the bottom and sides, which materially increased the life of the furnace. During the next 30 years, few changes were made in the process, for the new process was so far superior to previous ones that there was left little incentive for improvement. This attitude was changed, however, with the introduction of the pneumatic, or Bessemer, process in 1856. Then, in order to overcome competition of the Bessemer steel, and incidentally lessen the labor of puddling, which, like all its predecessors, was very arduous, hundreds of attempts were made to improve and cheapen the process (as shown under "Mechanical Puddling"). Few of these attempts were successful, and even the most promising of the successful ones, for various reasons, failed of universal adoption.

Between the years 1920 and 1930, however, hand puddling was almost entirely abandoned to be supplanted by the Ely mechanical puddler and the Aston process, the former duplicating conditions of hand puddling as closely as possible and the latter employing radically different principles and methods to obtain a similar but more uniform product.

Construction of the Hand Puddling Furnace—Although various modifications were introduced in the construction of puddling furnaces, affecting both size and design, the tendency was to adhere to the smaller and simpler types, such as the one shown in Figure 1—10. This type was known as a single furnace, had a capacity of 227 kg (500 pounds) per heat, and was coal-fired. The furnace was made up of the following parts: the **grate,** or fireplace, located at one end of the furnace; the **neck,** at the opposite end, leading to the **flue** that connected to the **stack;** and the **hearth,** or **puddling basin,** centrally located between the grate and the neck. The furnace was constructed entirely of brick, but was encased on the sides by a shell of iron plates held in place by tie rods. As the furnace was of the reverberatory type, all these parts were covered by an arched roof which sloped down from the fireplace to the uptake flue. The roof over the fireplace was built of firebrick, but usually silica brick were used over the hearth and neck. The fireplace was enclosed on each side and at the rear by firebrick walls. To support the fire bed the space over the ash pit was bridged with iron bars. Above the bars a square hole in the firebox on the front side of the furnace was provided for firing. The neck, at the other end of the furnace, was an inclined firebrick flue, frequently lined with a course of best-quality silica brick. The neck terminated in a short uptake, or vertical flue that led to the stack, which was independently supported upon a **mantle.** At the base of the uptake, directly opposite the neck, was an opening or door, called the **floss hole,** which was provided primarily for the removal of the cinder that was carried, or overflowed, from the puddling basin.

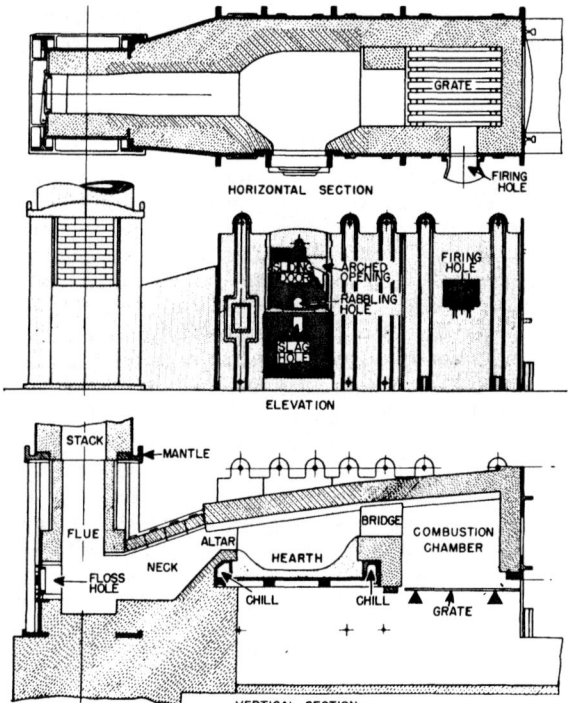

FIG. 1—10. Diagrammatic sections of a hand puddling furnace of the simple design known as a single furnace.

Externally, the bottom of the hearth or puddling basin consisted of iron plates supported on bearer bars laid transversely across the space between the side walls of the furnace. This construction provided all the benefits of air cooling. A low brick wall, laid across the furnace and known as the **bridge,** separated the hearth from the fireplace and also served as a backing for one end of the basin. At the opposite end of the hearth, a somewhat lighter and lower wall, known as the **breast wall** or **altar,** separated the basin from the neck. Imbedded in each of these walls next to the lining, was a hollow iron casting, called a **chill,** through which air or water was circulated to keep these parts cool. The other two sides of the basin were supported by the walls of the furnace itself, and were similarly air-cooled. The back wall was built up solid to the roof, but the front wall contained the arched opening to the hearth. The sides of this opening were made of specially formed silica brick, known as the **jambs,** while its bottom was made of a heavy iron plate called the **fore plate.** This opening was closed by a brick-lined sliding door, in the bottom of which was a small U-shaped opening, the **rabbling hole,** through which tools for working the heat could be inserted without raising the door.

In general, the one-piece bottom was composed of a refractory fettling consisting mainly of the ferrosoferric oxide of iron (Fe_3O_4). Certain grades of ore or of heating-furnace cinder were frequently used, but more often the bottom was made by applying, oxidizing, and fritting in successive layers of fine iron cuttings (such as thread cuttings from a pipe mill) known as swarf.

Operation of the Hand Puddling Furnace —With the hearth properly built up or repaired and the furnace in good working order and at a proper temperature, about 227 kilograms (500 pounds) of pig iron were charged by hand through the door. Following this operation, occupying 2 to 3 minutes, the purification of the pig iron and the process of puddling advanced by stages, known as **melting, clearing, boiling, balling** and **drawing.** To achieve quick melting, the door and other openings were closed, the furnace was fired vigorously, and the pigs turned once or twice by the puddler or his helper. In this way the charge was usually melted within 20 to 25 minutes after charging The molten metal, covered with a thin layer of slag, was then stirred or rabbled by the puddler to hasten the oxidation of the silicon, the manganese, and a part of the phosphorus, an operation known as clearing and requiring 8 to 10 minutes. As soon as the metal had cleared, as revealed by a change in its appearance, the puddler endeavored to bring on the boil by raising the temperature of the furnace, charging some dry roll scale (iron-oxide scale detached from bars in rolling), and stirring the bath vigorously. After some 8 to 10 minutes of strenuous effort, the oxidation of the silicon and manganese was brought to a point where the carbon could also be oxidized.

As the product of this reaction was carbon-monoxide gas, CO, and since the slag was somewhat viscous, the action caused the latter to foam or boil and rise in the furnace. At this point the slag was permitted to flow from the furnace freely unless it was desired to hold the phosphorus high in the iron, when a little coal was added from time to time and as much of the slag as possible was held in the furnace. As the elimination of carbon became more rapid, the gas would escape in larger bubbles and burst into flame at the surface of the slag to form small flames called **puddler's candles.** With the disappearance of the candles, the puddler increased the stirring of the bath during the **lowering of the heat** until the metal in terms of the puddler, would **come to nature.** In this phenomenon, most characteristic of puddling, the metal appeared in small globules, like butter in churned milk, each globule representing a portion of the iron that had become decarburized. As this reaction neared completion, the bath became pasty and very hard to work. This change occurred because the high-carbon pig iron, which was molten at that temperature, was converted to low-carbon iron, which is solid (though pasty) at the same temperature. The change progressed rapidly, lasting only 6 to 8 minutes, so that in some 30 to 35 minutes after clearing the metal was ready for balling.

The globules agglomerated by the rabbling tended to collect in sponge-like clusters on the bottom; these clusters had to be raised constantly and exposed to the heat to prevent them from freezing to the bottom. So, the temperature of the furnace was raised as high as possible, and the metal was worked into a mass which was next separated into three parts or balls of about 70 kilograms (150 pounds) each, a size convenient for handling with tongs. This operation required about 15 minutes. Each ball in turn was then grasped by tongs supported from a trolley and drawn through the door. After the last ball was removed, the furnace was permitted to cool to some extent, and the bridge and breast were covered with a special ore mix. Any necessary patching of the bottom was done, and another charge of metal was placed on the hearth for the next heat. These operations required about 2½ hours from heat to heat.

Rolling of Hand-Puddled Wrought Iron —At one time the balls were worked into the form of a rough bloom with a hammer, an operation called **shingling.** Later, hammering was superseded by the use of a device known as a **squeezer,** of which there were different types, all of which squeezed most of the excess slag out of the ball and compressed it into a form more suitable for rolling. As soon as the ball was delivered by the squeezer, it was grasped with tongs and at once delivered to the first pass of the rolling mill.

Rolling the Squeezed Ball —The squeezed ball was rolled into flats called **muck bars,** the size of which was regulated by the product to be made from them and the manner or system used in forming the product. For ordinary bar iron, which was the chief product, the muck bar was usually approximately 20 millimeters (¾-inch) thick and 65 to 205 millimeters (2½ to 8 inches) wide. Bars of these sizes required from 5 to 9 passes. On account of the rapid cooling of the bar in the rolls, it was not practicable to attempt to roll sizes smaller than these, as the slag was no longer fluid enough at this stage to be worked out of the bar. As slag was squeezed out of the bar at all passes in the mill, the muck bar had a very rough surface with some torn edges and was otherwise unfit to do service as a finished bar. Having been rolled to the size required, the muck bar was allowed to cool before being subjected to further treatment.

Variables in the Muck Bar —Owing to irregularities in the pig iron used, differences in manipulation by different puddlers, and in different plants, the small quantity of metal refined with each heat, and the fact that the metal solidified before purification was complete, muck bar was an exceedingly variable product. Since the retention of the characteristics of wrought iron did not permit melting, this variation had to be overcome through heat and mechanical treatments. To effect the necessary refinement, two methods were used, known as busheling and piling, followed by rolling.

Busheling—Obviously, the surest way to obtain a thorough mixing of the iron, was to shear the muck bars into small pieces—the smaller the better. These small pieces from the different muck bars were allowed to collect in a pile, or piles, from which portions weighing approximately 80 to 270 kilograms (180 to 600 pounds) were removed with a scoop or fork and charged into a reheating furnace, called a **balling furnace,** where they were heated "white hot," or to a self-welding temperature. With a paddle, these pieces were then collected into a ball, similar to a puddle ball, which was squeezed or shingled, then rolled or hammered into a bloom, which was then reheated and worked into the form desired. This process, known as **busheling,** was used for working up muck bar only when iron of the highest quality was desired. The process was also used in working up small scrap.

Piling—The more common practice was to shear the muck bar into lengths of from 0.6 to 0.9 metres (2 to 3 feet), then arrange these pieces in piles of from 5 to 7 or more each and bind the pieces together with wire or bands. The piles were carefully charged into a furnace and heated white hot. The high temperature caused the different bars to weld together, so that they could then be removed and rolled into bars. The first 2 or 3 passes squeezed out more and more of the liquid slag, but in the last passes the bar had cooled to a point where the slag was merely plastic and would not flow. Thus, a fairly smooth and uniform bar was produced, which would be sold as **merchant bar, single-rolled iron, single-refined iron,** or **No. 2 iron.** To attain the highest degree of uniformity, particularly with respect to distribution of slag fiber, the once-piled or so-called single-refined bars were in turn cut into short lengths, repiled and rerolled.

Double Refining—To further improve uniformity, then, merchant bar was cut up into short lengths, fagoted, reheated, and rerolled to produce the products known as **double-rolled iron, double-refined iron, best bar,** or **No. 3 bar.** The manner of **fagoting** (binding together into a bundle) or piling these bars varied in numerous ways, and depended only in part upon the use to which the iron was to be applied. Therefore, each manufacturer generally had his own methods of fagoting, which imparted to his iron an individuality detected by etching. Some of the more common methods of fagoting are illustrated by the accompanying sketches (Figure 1—11). When these fagots were heated and rolled into bars, more slag was expelled, the bar was made more uniform in composition, and the fibers were much elongated and reduced in size. As a result, the bars showed an improvement in mechanical

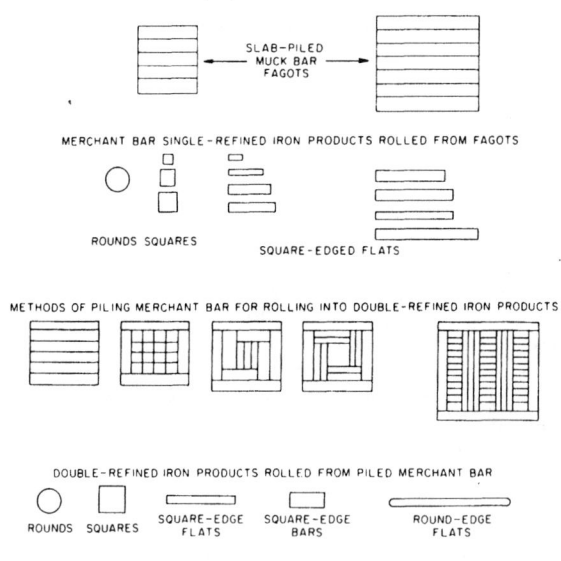

FIG. 1—11. Some of the more common methods of fagoting or binding together piles of single-refined wrought iron bars prior to heating for re-rolling to produce double-refined wrought iron bars. Approximate fps equivalents of the indicated dimensions are 3½ in. and 6 in. wide and ¾ in. in thickness.

properties, including both strength and ductility. There was a limit, in addition to the factor of cost, however, to the number of times the iron could be worked to advantage. After five or six workings its mechanical properties began to be lowered, and the bars decreased in strength and were less ductile. The cause, or causes, of this change was questionable; probably, it was due to the elimination of much of the silicate slag or possibly too much reheating and rerolling caused the ferrous silicate fibers to become oxidized to the ferric condition, thus destroying some of the characteristics of wrought iron. Wrought iron in this condition was often referred to as **dry iron.**

Reactions and Process Losses in Hand Puddling—The changes in composition of the pig iron during puddling involved the elimination, or oxidation, of silicon, manganese, phosphorus, and carbon about in the order mentioned. In these reactions, the oxidizing agents were FeO and Fe_3O_4. That Fe_2O_3 played little, if any, part in the reactions was evident from the fact that Fe_2O_3 decomposed at temperatures about $1100°$ C $(2010°F)$ to form Fe_3O_4. Also, it had been found that if hematite ore (Fe_2O_3) was used as the oxidizing medium, the boil came on very slowly but if roll scale $(Fe_3O_4$ or $FeO \cdot Fe_3O_4)$ were used, the reactions proceeded with much greater speed. Nearly all of the silicon and manganese and a part of the phosphorus were oxidized before the boil began, and at some period after melting and after the elimination of some of the silicon, the oxidation of all four elements, including carbon, might have proceeded simultaneously.

A study of the probable reactions that occurred in puddling indicates that a gain in weight of the puddled iron over the pig iron used could be expected, because iron was formed by reduction from the slag in the elimination of practically all impurities. By careful manipulation, furnaces could be operated to show a slight gain or a very slight loss. Nevertheless, in ordinary working, there was a loss of from 3 to 6 per cent, which was sometimes a little more, at other times a little less, than the total of the impurities present. If the heat were properly handled, the loss was largely due to oxidation of iron after solidification had begun in the after-part of the boil and during the balling stage. If the heat were not skillfully handled, a variable part of the loss may have been due to the escape of the metallic granules with the "boilings" before the "heat was lowered." In reheating and rolling the muck bar, there was a variable loss of from 10 to 20 per cent for each time the iron was worked depending upon the number of times it was worked, the manner of the working, and other factors incidental to the operations of heating and rolling. In general, these losses were due to surface oxidation of the metal in heating and rolling, expulsion of the slag, and cropping. Slag expulsion was the smallest item of loss, except in the case of muck bar, and depended mainly upon the number of times the iron was worked, but was also affected by the temperature at which the iron was worked, and the nature of the incorporated slag itself. The loss was a little greater on iron with a high phosphorus content than on iron low in phosphorus.

The never-ending competition in the iron and steel industry has been a constant spur to improve methods or lower costs of production. Just as the puddling process virtually eliminated the more primitive direct-reduction methods for the production of iron, so the Bessemer and open-hearth processes for the production of steel threatened the life of the wrought-iron industry. Even before the invention of these steelmaking processes, much attention was given to improving the puddling process, because the process was costly and the labor arduous, and the furnace was wasteful of heat. With a reverberatory puddling furnace of the type described earlier, from about 1000 to 2000 kilograms (2000 to 4000 pounds) of coal were required to produce one ton of muck bar. This consumption of fuel was reduced somewhat by the use of double furnaces with enlarged hearths, but, since the application of regenerative and recuperative furnaces appears to have been impracticable, efforts along this line failed to achieve much in the way of lowering costs. The installation of waste-heat boilers in the stacks effected marked economies, and their use became general. To overcome the high labor cost, many attempts were made to carry out the puddling operations mechanically. About 1870, the Danks mechanical puddling furnace appeared. From about 1870 to 1925, wrought iron lost ground to steel in spite of several efforts to revive it. In the meantime, however, it came to be recognized as a product with characteristic properties unlike those of steel, and in 1925 attempts to revive the industry were made. These endeavors advanced along two lines, the one mechanical and the other metallurgical, the former aiming to duplicate the process of hand puddling as closely as possible and the latter aiming to produce a material having all the characteristics of wrought iron through the use of the same metallurgical principles but applied in a manner entirely different from that of hand puddling. These two lines of effort are described below under the headings of mechanical puddling and the Aston process, the latter representing a method of producing wrought iron by A. M. Byers Company.

MECHANICAL PUDDLING

Principles of Mechanical Puddling—At first, mechanical puddlers took the form of stirring or rabbling appliances that could be attached to the top of the ordinary furnace. Because of the great variety of motions necessary in the different operations of charging, raising and stirring the heat, balling the iron, and drawing the balls, none of these were successful. The more successful attempts at mechanical puddling have involved a complete change in the design of the furnace, and some changes also in the process. These attempts have been too numerous to warrant description here. The furnaces themselves may be classified as follows:

1. The rectangular furnace that oscillated about a horizontal axis of rotation.
2. The circular flat-bottom furnace that revolved about an axis slightly inclined to the vertical.
3. The circular furnace with flat or troughlike bottom that oscillated about a horizontal axis of rotation.
4. The cylindrical furnace that rotated about a horizontal axis coincident with the center.

5. The cylindrical furnace that oscillated about a horizontal axis coincident with the center, or both oscillated and rotated about such an axis.

Furnaces built on any of these plans were made to puddle iron successfully, but those of the fourth and fifth types were most successful, partly on account of the facilities they afforded for controlling the agitation of the metal, and partly because of the simplicity of their construction. The **Danks furnace,** somewhat widely used in this country from 1868 to 1885, was of the fourth type, while the **Roe furnace,** built and successfully operated in 1905, was of the third type. **H. D. Hibbard's furnace,** first operated on a commercial scale in 1921, was somewhat similar to the Danks furnace. The **Ely furnace,** patented by W. C. Ely, was first designed for busheling scrap but was later (about 1920) applied to the puddling of iron: it represented the fifth type.

THE ASTON PROCESS

From the descriptions given in the preceding sections, it is apparent that the basis of wrought-iron manufacture consisted in refining the base metal to a close approach to pure iron, and incorporating therein an iron-silicate slag of desirable chemical composition in proper amount and distribution. Several correlated steps were involved, distinct in nature and capable of separation, but carried out in the usual methods for hand or mechanical puddling as one interconnected operation. Departing radically from these former methods, the Aston process was developed and put into large scale operation in 1930 by the A. M. Byers Company of Pittsburgh, Pennsylvania. In this process, metal refining, slag melting, and processing to form the slag-impregnated sponge ball were carried out as separate stages, each stage in a separate furnace or kind of equipment. The last stage of processing was the crux of the process, and was based upon the change in gas solubility from a very high amount in molten iron to an amount practically negligible on solidification. This stage of the Aston process was carried out by pouring the refined metal in a continuous stream into a large volume of molten slag. The slag acted as a heat absorbing agent which effected solidification of the metal with accompanying liberation of its dissolved gases, at a steady rate and with a force sufficient to disintegrate the plastic metal into a spongy mass, conforming in all particulars to the characteristics of high-quality wrought iron.

The essential features of the operation of the process are illustrated in Figures 1—12 to 1—15.

The iron-silicate slag was melted to exacting chemical requirements in special furnaces (see Figure 1—12), and then transferred to the processing cups (also called "thimbles").

Cupolas produced molten iron (hot metal) of Bessemer grade which was desulphurized in the ladle with caustic soda.

The molten desulphurized hot metal from the cupolas was "full-blown" in an acid-lined converter and the highly refined, deoxidized metal was poured at a controlled uniform rate into a thimble holding molten iron-silicate slag (Figure 1—13). After the metal from the converter had been poured, the surplus slag was

FIG. 1—12. Molten iron silicate slag being tapped from rotary furnaces and transferred by ladle to replenish the slag in the processing ladles (process cups) shown at lower level in Figure 1—13. (Courtesy, A. M. Byers Company.)

FIG. 1—13. At the processing floor, molten refined iron at 1545°C (2900°F) was poured by processing machines into molten iron silicate at 1380°C (2500°F) to form an iron sponge ball characteristic of the Aston process for making wrought iron. Processing machines, scale mounted, had traverse, oscillating, and tilting motions insuring distribution of the metal into the slag. (Courtesy, A. M. Byers Company.)

FIG. 1—14. Empty processing ladle received surplus decanted processing slag, was replenished if necessary, and returned by rail to processing floor for reuse. The welding-hot iron sponge ball, "wet" with molten iron silicate, was moved to the press by overhead traveling crane. (Courtesy, A. M. Byers Company.)

FIG. 1—15. The welding-hot sponge ball, "wet" with molten iron silicate retained in the processing ladle in which it was produced, was immediately transferred to and dumped on the press table, lowered between side and end rams, and pressed into a rectangular bloom prior to rolling. (Courtesy, A. M. Byers Company.)

poured from the thimble, leaving the white-hot sponge ball of wrought iron (Figure 1—14). This sponge iron was first pressed into a bloom of rectangular section (Figure 1—15) and was then rolled into slabs or billets.

Since the mass of the pressed bloom was 2.7 to 3.6 metric tons (3 to 4 net tons), most of the product—skelp, plate, etc.—was rolled from reheated solid sections, in contrast to the older wrought-iron practice of building muck bar piles.

COMPOSITION, STRUCTURE AND PROPERTIES OF WROUGHT IRON

Chemical Composition of Wrought Iron—Chemical composition had a relation to wrought-iron quality comparable with its importance in the steel industry. However, the constituents might have been, in greater or lesser degree, alloyed with the base metal or associated as oxidized constituents with the intermingled slag. The commonly reported composition of wrought iron listed the carbon, manganese, phosphorus, sulphur and silicon of the composite mass. On the basis of analyses as commonly made, the following statements applied to wrought iron of high quality: Carbon seldom, if ever, exceeded 0.035 per cent in quality wrought iron. Silicon content was 0.075 to 0.150 per cent, normally almost negligible in alloyed association with the metal, and existing almost entirely as silicates in the slag. Sulphur was always undesirable, and in well-made wrought iron, it was kept under 0.02 per cent. A sulphur content of 0.010 per cent or under was quite common in quality wrought iron. Phosphorus was almost invariably higher in wrought iron than in steel. It must be borne in mind that phosphorus was in part dissolved in the base metal, and in part associated with the slag. Good wrought iron might have a phosphorus content of from 0.10 per cent or less to 0.25 per cent or more, according to manufacturer's preference, nature of raw materials, or adaptability to service conditions. The lower order was advisable for materials subjected to shock, high temperature, or requiring higher ductility. Traditionally, the manganese content of hand-puddled or processed (Aston process) wrought iron was less than 0.06 per cent; British specifications generally had a 0.10 per cent maximum and in the United States specifications carried a limit of 0.09 per cent. Low manganese in wrought iron was usually an earmark of quality, although there was no logical ground for condemning an otherwise well-made product because of a relatively high manganese content.

Macroscopic Structure of Wrought Iron—In view of the composite nature of wrought iron, its quality was obviously affected by the nature of the association of base metal and slag. Methods of disclosing this internal structure had an importance greater even than the prominent place which the metallurgist assigns to them in the study of steel. Wrought iron exhibited a well-recognized fibrous fracture. The fracture test was a good over-all means for determining the general characteristics of wrought iron, but it could not be relied upon solely.

Macroscopic etching would reveal gross structure, reflecting such features of manufacture as methods of piling, and general slag distribution. Deep etching had

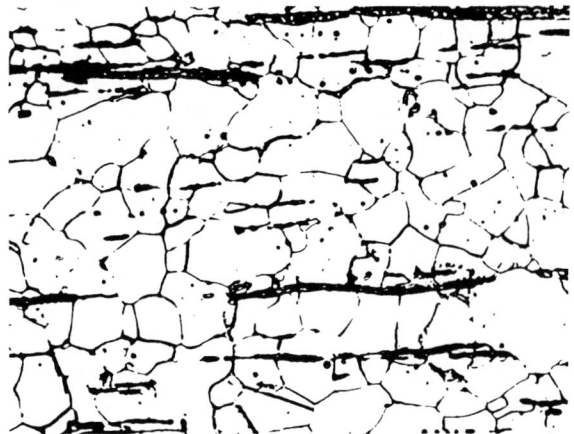

FIG. 1—16. Photomicrograph at 100X showing typical structure of wrought iron produced by the Aston process parallel to the direction of rolling. White areas are the highly refined iron matrix. Dark gray elongated lines are iron-silicate slag filaments. (Courtesy, A. M. Byers Company.)

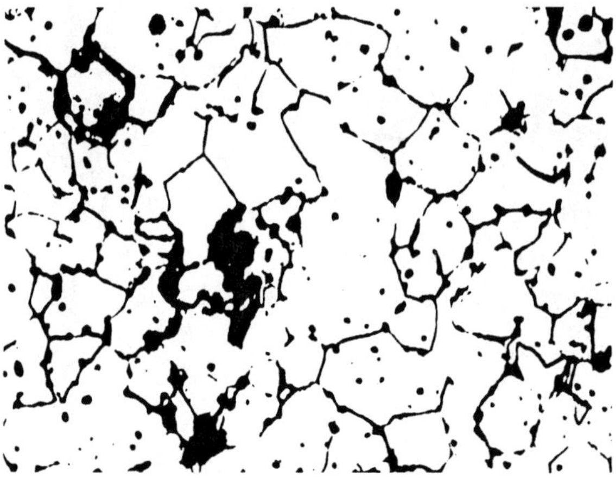

FIG. 1—17. Photomicrograph at 100X showing typical structure of wrought iron produced by the Aston process perpendicular to the direction of rolling. White areas are the highly-refined iron matrix. Dark areas are cross-sections of iron-silicate slag filaments. (Courtesy, A. M. Byers Company.)

a useful place but, like the fracture test, gave only a limited amount of information pertaining to the finer points of quality.

Microscopic Structure of Wrought Iron—Wrought iron consisted essentially of a ferrite matrix, through which the slag was uniformly disseminated in the form of several hundred thousand filaments per square inch. Figures 1—16 and 1—17 illustrate typical microstructures of wrought iron. Microstructural details related to the quality of wrought iron were:

a. **Grain Size**—Coarse grain, distortion, or lack of uniformity had a bearing upon quality in relation to mill history and use of product.

b. **Pearlitic areas** indicated the quantity and nature of distribution of the carbon, generally practically negligible or quite small in amount in real wrought iron.

c. **Slag—Type and Distribution**—Coarse, pocketed slag was undesirable. Finer textures resulted from progressive rolling reductions, and promoted better mechanical properties, especially ductility.

Mechanical Properties of Wrought Iron—The mechanical properties of wrought iron were essentially those of pure iron, modified only slightly in general practice by other chemical constituents of the base metal and profoundly by the quantity and distribution of the incorporated slag. Up to certain limits, the ductility was increased by greater work in forge or mill, which caused a finer and more threadlike distribution of the slag. This was reflected in the common practice of the puddle mill of once or twice piling in products designated as "single" or "double-refined" iron. Obviously, a similar result would be achieved by rolling relatively large initial blooms into small final sections.

In comparison with steel, the longitudinal ductility of wrought iron was somewhat lowered, due to slag incorporation, while the transverse strength and ductility were markedly reduced. However, the aforementioned specific mechanical properties had a dependence upon and were a reflection of rolling history—the

relative amounts and direction of the mechanical work imposed as they affected the ultimate orientation of the incorporated slag in the base metal—and could be altered by suitable rolling practices to obtain optimum directional values.

The values below were representative of tensile properties for various wrought-iron products, compiled from standards of the American Society for Testing and Materials. Because of the physical size of the products listed, only the longitudinal properties are reported, except for plate for which both longitudinal and transverse properties are given.

BAR IRON—SINGLE-REFINED

Under 41 mm (1⅝ in.) diameter or thickness

Tensile strength
MPa . . . . 331 (minimum)
psi . . . . 48 000 (minimum)
Yield point, tensile strength factor   0.6
Elongation in 200 mm (8 in.), per
cent . . . . 25 (minimum)
Reduction of area, per cent . . . . 40 (minimum)

BAR IRON—DOUBLE-REFINED

Under 41 mm (1⅝ in.) diameter or thickness

Tensile strength
MPa 331-372
psi . . . . 48 000 to 54 000
Yield point, tensile strength factor   0.6
Elongation in 200 mm (8 in.), per
cent . . . . 28 (minimum)
Reduction of area, per cent . . . . 45 (minimum)

The higher ductility accompanying greater work is reflected in the figures for double-refined material. For heavy sections, such as large diameter bars and forgings, strength and ductility requirements were somewhat lowered.

WELDED PIPE

Tensile strength
    MPa . . . . 290 (minimum)
    psi . . . . 42 000 (minimum)
Yield point
    MPa . . . . 172 (minimum)
    psi . . . . 25 000 (minimum)
Elongation in 200 mm (8 in.), per
    cent . . . . 12 (minimum)

Herein are reflected the effects of high temperatures in welding, and the lessened elongation in testing tubular sections. However, where special precautions were taken in the making of pipe for bending purposes, the ductility figures were bettered in practice by several per cent in the elongation obtained.

PLATE

Under normal rolling practice, plate exhibited the maximum of difference in longitudinal and transverse properties. The ASTM specifications for plates made by conventional rolling practice required that plates have the following tensile properties:

Tensile strength, longitudinal
    MPa . . . . 331 (minimum)
    psi . . . . 48 000 (minimum)
Yield point, longitudinal
    MPa . . . . 186 (minimum)
    psi . . . . 27 000 (minimum)
Elongation in 200 mm (8 in.), per
    cent:
    Longitudinal . . . . 14 (minimum)
    Transverse . . . . 2 (minimum)

By proper attention to rolling practice, it was feasible to equalize the properties, so that a specification requirement of a tensile strength of 269 MPa (39 000 psi) minimum and an elongation in 200 mm (8 in.) of 8 per cent (minimum) in either direction might be met: this was of value for producing plate for flanging and other forming purposes.

SECTION 6

EARLY PROCESSES FOR MAKING STEEL

GENERAL FORMS OF PROCESSES

It has been known since very early times that some ferrous metals would become very hard if cooled rapidly from a high temperature, as by immersing (quenching) in water or some other liquid, while others would not. For the present purpose, the ferrous metals that could be hardened in this way will be referred to as steel, while those that could not will be referred to as wrought iron.

The ability of steel to harden by quenching is now known to be due to the fact that it contains a considerable amount of carbon, whereas wrought iron, which contains very little carbon, cannot be hardened in this manner. The effective amount of carbon in hardenable steels may be considered to range between 0.30 and 1.00 per cent, in this discussion. Steelmakers in early times were not even aware of the existence of carbon as a chemical element, and therefore could not have recognized it as the essential alloying element in steel. Nevertheless, by trial and error, they developed empirical methods that produced ferrous metals having the proper carbon content, insofar as the ability of the metals to respond to the hardening treatment was concerned.

Control of carbon content is still important to the steelmaker, but is only one of the many factors involved in the production of the numerous types of steel made in modern times.

Over the centuries, the steelmaking methods that were developed took one of four general forms:

(1) Direct production of steel by reducing iron ore with carbon in the presence of an excess of the carbon.
(2) Carburizing (increasing the carbon content) of wrought iron of originally low carbon content (without melting the iron) by heating it for long periods in contact with carbon and out of contact with air.
(3) Increasing the carbon content of originally low-carbon wrought iron by melting it and adding carbon to the molten metal.
(4) Decreasing the carbon content of the high-carbon product of the stuckofen or blast furnace (pig iron) by various processes in which the excess carbon was removed by oxidation.

These methods are discussed in greater detail below.

Direct Production of Steel—Section 2 of this chapter outlined how sponge iron could absorb carbon during the process of the reduction of iron ore in charcoal-fired hearth-type and shaft-type furnaces, providing certain conditions relating to time and temperature were met. Enough carbon could be absorbed to convert the sponge iron to steel, in the sense that articles forged from the converted sponge could be hardened by quenching. It is probable that the carbon content of various parts of the sponge differed, perhaps considerably. However, during repeated heating and hammering in the processes of consolidating the sponge and ultimately shaping the metal into a useful form, carbon could diffuse from the higher-carbon portions to adjacent lower-carbon portions and the final distribution of carbon might have become quite uniform.

Another direct method for the production of steel is typified by a process used for many centuries in India to make the steel called **wootz**. Its method of manufacture has been variously described. In one description, the first step consisted of heating pure ore with carbonaceous material such as charcoal or finely-chopped wood in closed crucibles. After heating at a high temperature for several hours, the ore was reduced to metallic iron and absorbed sufficient carbon from the excess of charcoal to have a low enough melting point to become fluid. The crucibles were allowed to cool and, when broken open, a small "button" of high-carbon steel was found at the bottom. Two methods have been recorded for lowering the carbon content of the buttons to give steel having the desired carbon content or "temper." One method consisted of repeatedly heating the buttons while they were covered with a layer of iron-oxide paste. The other recorded method comprised heating the buttons for several hours in a charcoal fire to a temperature not much below their melting point and turning them over in the path of the blast, so that the metal would be partially decarburized. In both cases, the partially decarburized buttons would then be heated to be welded together by hammering to form bars.

The foregoing are only two of the types of early direct processes for making steel. It will be noted that there is some similarity between these two processes and other early processes for making steel. For example, carburization of sponge iron in the furnace in which it was reduced is related in principle to the carburization of solid wrought iron as described below, and the process for making wootz bears some resemblance to the crucible process that also is described below.

It is an interesting fact that methods for producing steel directly from iron ore are still being investigated, and, in some cases applied on at least a semi-commercial scale, as will be seen elsewhere in this book.

Carburizing of Solid Wrought Iron—If a bar or forged object of wrought iron were embedded in hot carbon that completely surrounded it and protected it from oxidation, the iron could absorb some of the carbon. As carbon was absorbed at the surface, it would begin to diffuse inward. The amount of carbon absorbed and the depth to which it would penetrate would depend upon the temperature and duration of the process. If properly performed, this process, now called **carburizing**, would produce a layer or "case" of high-carbon content on the exterior of the wrought iron bar or object without appreciably changing the shape or dimensions of the carburized piece. The high-carbon layer could be hardened by quenching, and the piece would now be described as **case-hardened**; the low carbon content of the iron in the interior of the piece would prevent hardening of the metal there. Archeological specimens from 1000 B.C. exhibit evidences of having been deliberately case-hardened to produce durable points and edges. The principle of carburizing is still in use for case-hardening (see Chapter 41).

The carburizing principle was also employed in the **cementation process**, a formerly important method for making steel that is described in detail later in this section.

A variation of the carburizing principle used until at least the late Middle Ages consisted of immersing the hot low-carbon sponge iron derived from a hearth-type furnace in the high-carbon molten iron from a shaft-type furnace. Carbon would diffuse from the molten iron into parts of the sponge, giving those parts the hardening characteristic of steel.

Melting Processes for Carburizing Wrought Iron—Another early method for converting wrought iron to steel consisted of placing small pieces of wrought iron in a closed pot or crucible, along with chopped wood, charcoal or other carbonaceous material, and heating to a high temperature for several hours. The iron absorbed carbon and melted. After the crucible had cooled, it was opened to remove the "cake" or "button" of steel that had solidified in the bottom of the crucible. The procedure just described has been used for many centuries, and is considered by some to have been one of the ways in which wootz (see above) was made in India, although its use was not limited to that geographic area.

This method evolved gradually into the **crucible process** for steelmaking that is described later in this section.

Decarburization of Pig Iron—With the advent of shaft-type furnaces that produced only pig iron of high carbon content, processes were developed for removing practically all of the carbon from pig iron to convert it to wrought iron, as described earlier (see Section 5). It is conceivable that variations of the processes for making low-carbon wrought iron by simultaneous melting of the pig iron and oxidation of carbon made it possible to produce a steel-like metal by controlling the refining process so that only part of the carbon was removed and enough was retained to produce a hardenable metal.

Present-day steelmaking processes are based largely upon the principle of the decarburization of pig iron, the exceptions being mostly in the field of electric-furnace steelmaking, as will be discussed in Section 7 of this chapter and in Chapters 16 through 18. However, prior to the invention of the Bessemer process for steelmaking in 1856 (see "The Pneumatic Steelmaking Processes" in Section 7 of this chapter), only two of the numerous methods for steelmaking outlined above had attained any considerable industrial importance. One was the process of increasing the carbon content of wrought iron by heating it in contact with hot carbon away from air; this came to be called the cementation process. The other method, the **crucible process**, consisted of melting wrought iron in clay crucibles in which carbon had been added for the express purpose of increasing the carbon content of the iron. Both of these processes were certainly known to and practiced by the ancients.

During the Middle Ages both the cementation and crucible processes appear to have been lost to civilization. The cementation process was revived in Belgium about the year 1600 A.D. while the crucible process was rediscovered in England by Benjamin Huntsman in 1742. Both processes were practiced in secret for some time after their revival. Hence, little is known of their early history. The following brief descriptions will, therefore, be confined to practices in later years.

THE CEMENTATION PROCESS

The cementation process was highly developed and flourished in England during the 18th and 19th centuries, but it has practically been replaced by other methods. The process depended upon the fact that when a low-carbon ferrous product, such as wrought iron, was heated to a red heat in contact with charcoal or other carbonaceous material, the metal absorbed carbon, which, up to the saturation point of about 2.00 per cent, varied in amount according to the time the metal was in contact with the carbon and the temperature at which the process was conducted. For carrying on the process, a type of muffle furnace or pot furnace was used, and the iron and charcoal were packed in the pots in alternate layers.

The iron used was usually in the form of bars approximately 65 to 75 millimeters wide, 15 to 20 millimeters thick and 1.83 to 3.66 metres long (2½ to 3 inches wide, ⅝ to ¾ inch thick, and 6 to 12 feet long) For the best grades of steel, only the best wrought iron was supposed to be used, though low-carbon steel made by the open-hearth process was later substituted. Charcoal, which had been passed over about a 6-millimetre (¼-inch) screen to remove fines, represented the favorite carburizing agent, though various other substances and mixtures had been tried and used. In charging the pots, their bottoms were first covered with a layer of charcoal about 50 to 75 millimeters (2 or 3 inches) thick, then, in alternation, layers of bars and charcoal were added to each until the pots were full. The bars were laid flatwise and about 12.5 millimeters (½ inch) apart, between edges, so that each bar was completely surrounded by charcoal. After applying the final layer of charcoal the charge was covered with wheel swarf (refuse from the grindstones) and all openings to the pots were closed and made as nearly airtight as possible. A fire was next lighted in the furnace, and the charge, for the next 3 or 4 days, gradually heated to a full red heat—the actual temperature varied from 800° C to 1100° C (1470° F to 2010° F) and this temperature was then maintained for 7 to 12 days, depending upon the size of the bars used, the carbon content desired, and the temperatures attained and maintained. The degree of carburization, or the **temper** of the bars, was determined by fracture tests on test bars that were withdrawn from the pots through small openings or holes provided for the purpose. When the bars had reached the desired temper, i.e., had absorbed the desired amount of carbon, the fire was banked and the furnace allowed to cool slowly. When the contents of the furnace were cool enough to handle, the bars were removed from the pots.

If the original bars were of wrought iron, their surfaces were found to be covered with irregular elevations, known as **blisters** or **beads**, which resulted from the expansive force of the carbon monoxide formed by carbon reacting with the oxides of the incorporated slag. Hence, these bars were known as **blister steel.** If the original bars were of mild steel or remelted wrought iron, these blisters were absent. Both products were frequently referred to as **converted** or **cement steel.** If air had gained access to the bars during the process, their surfaces were covered with scale and almost decarburized. Such bars were known as **aired bars.** After sorting and reheating, the cement-steel bars were hammered or rolled into what in England was called **spring plate** or **bar steel,** which formerly was used for springs, but later was used as raw material for crucible steel or for the production of shear steel. If for

the latter, the bars were broken or sheared into short lengths, fagoted, sprinkled with a little borax, covered with clay, reheated to a welding temperature, and hammered into a bar, known as **single shear.** For purposes requiring a high grade, uniform steel, the single shear bars were again broken at their centers, the two halves of each laid together, and the fagot thus formed was reheated and hammered down to the required size. These bars were known as **double shear steel.** This steel was formerly used exclusively for the manufacture of cutlery, hence the name, shear steel.

This working was necessary to obtain the highly desirable homogeneity of the steel, and, prior to the revival of the crucible process, was the only method available to attain that end. Formerly, about seven grades or tempers of shear steel were produced. These varied in average carbon content from 0.50 per cent to 1.50 per cent. But as the carbon was absorbed from the surfaces of the bars, the carbon content of the blister steel bars progressively decreased to the center. In the softer grades the center portions of each bar remained unaltered, and this core was known as **sap.** In the harder grades, the outside of the bar might show a carbon content of 1.50 to 2.00 per cent with a center of 0.85 to 1.10 per cent. It was this characteristic of cement steel that led Huntsman, who was a clockmaker, to seek a method for making a more uniform steel for his springs.

THE CRUCIBLE PROCESS

Realizing that the shortcomings of cement steel for springs lay in lack of homogeneity, Huntsman conceived the idea of melting it in crucibles, which melting he thought should make the steel perfectly homogeneous. Briefly, his method was as follows: First, cement bars were carefully selected that would give the exact temper, or carbon content, desired in the finished steel. These were cut or broken into small pieces. Then this steel was charged into large clay crucibles, which, after covering and luting the lids, were placed in a coke fire and heated until the contents were thoroughly fused. At this point the crucibles were withdrawn from the fire, and upon removing the cover and skimming off the small amount of slag formed on top, the molten metal was poured into a cast-iron mold, where it remained until solid. Finally, by the usual method of reheating and hammering then in use, the ingot was worked into the form desired. This method gave a steel that was not only homogeneous throughout, but was free from occluded slag and dirt, hence was so much superior to cement steel for many purposes that the method at once became the leader for the production of the finest steels. This position the process held for almost two hundred years, but it has been superseded for the making of special alloy steels and carbon tool steels by the electric-furnace process, including the high-frequency induction furnace. The electric-furnace process is cheaper, is capable of giving as good steel, and possesses many metallurgical advantages over the crucible method.

Although the principles and the general method of procedure remained the same, the crucible process was the subject of considerable experimentation after the time of Huntsman, and some changes in the material of the charge, in the manufacture of the crucibles, and in the furnace for melting the crucible charges were introduced, so that the details of standard practice for this now practically extinct process in different countries and localities varied somewhat.

Manufacture of Crucibles—The **crucibles**, or **pots**, a very important part of the equipment required in the manufacture of crucible steel, were costly to make and had a comparatively short life. In England, they were frequently made of clay by the steel manufacturers, but in this country so-called graphite crucibles purchased from crucible manufacturers were almost universally used. These crucibles, with a capacity of about 36 to 56 kilograms (80 to 124 pounds), varied somewhat in form and thickness of wall. In general, they were barrel-like in shape, and varied in size from about 330 to 460 millimeters (13 to 18 inches) in height and from 200 to 320 millimeters (8 to 12½ inches) in outside diameter at the bilge. As to wall thickness, they were somewhat thicker at the bottom than at the top, usually about 38 millimeters (1½ inches) at the bottom and about 19 millimeters (¾ inch) at the top.

The Crucible Melting Furnace—In this country, the crucible furnace was generally of the gas-fired regenerative type (Figure 1—18). It consisted of a number of combustion chambers or **melting holes** built in a row between two sets of checker chambers. Each melting hole, about three feet in depth and otherwise large enough to hold 4 or 6 crucibles, was connected with the two sets of checkers by short flues leading upward from its bottom. If producer gas was used as fuel, the gas checkers were placed next to the line of melting holes with the air checkers extending along the outer walls of the furnace. This plan of construction readily permitted introducing the gas into the ports below the air, which was deflected downward by the roof, thus holding the flame near the bottom of the crucibles as it swept across the melting hole. In case natural gas, fuel oil, or powdered coal was used for fuel, both checkers would be used for air, or the furnace would be constructed with but one checker chamber on each side. The shop floor was on a level with the top of the melt-

CRUCIBLES

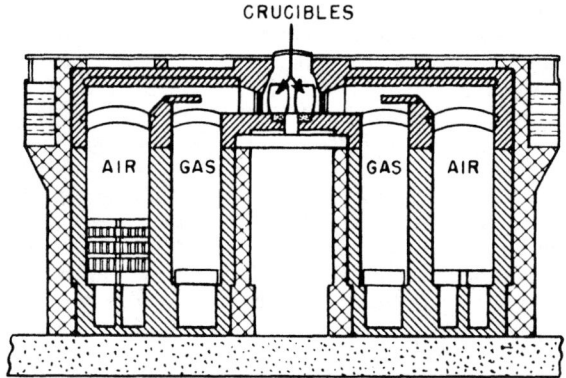

AIR   GAS   GAS   AIR

FIG. 1—18. Vertical section through checker chamber and one of the melting holes of a regenerative crucible furnace. The regenerator chambers labeled "gas" and "air" contained brickwork called checkers as indicated in the left-hand "air" chamber where the first few courses of checkerwork are shown schematically.

ing hole, so that only the covers were visible from above. After moving these covers aside, the crucibles were lowered into and lifted out of the hole by tongs with broad jaws curved to fit the crucibles.

The bottoms of the melting holes were built of first-quality firebrick laid on plates, which were supported by beams, or cross bars, spanning the space between the two sets of checkers, thus forming a continuous cellar, or vault, that extended the full length of the furnace beneath the melting holes. A small hole was provided in the center of the bottom of each melting hole, so that any refuse on the bottom of the melting hole could be poked into the vault below for cleaning. To protect the bottom in case of breakouts, it was covered to a depth of 150 to 200 millimeters (6 to 8 inches) with coke dust or crushed coke. This coke also gave some protection to the crucibles, as it helped to consume free oxygen not needed for combustion and to maintain a reducing atmosphere in the hole.

Charging the Crucibles—From what has already been said, it will be evident that, aside from the possible elimination of gases from the steel and the separation of slag and other non-metallic inclusions, the crucible process was not a purification process. Consequently, any elements capable of alloying with iron charged into the crucible were present also in the steel. On this account the greatest of care in the selection of the raw materials had to be exercised. And it was characteristic of the history of the process that, once a charge that produced the kind or grade of steel desired had been hit upon by any shop, that charge was strictly adhered to and could not be varied. For many years after the introduction of the process by Huntsman, cement steel or blister steel was the only raw material used. Eventually, attempts were made to eliminate the cementation step by the use of wrought iron with charcoal; and, provided Swedish iron, which was the purest wrought iron then made, was used, the steel produced was as good as that made from cement bars. In 1801 David Mushet introduced the use of manganese, by adding oxide of manganese with the charge, which practice was superseded some years later by the use of spiegel. It is interesting to note that the use of spiegel in Bessemer steel was patented in 1857 by a namesake, Robert Mushet (this method of adding manganese was an important factor in the early development of Bessemer steel). Then followed a great deal of unsuccessful experimental work, involving so many different substances and mixtures that the stock house came to be known as the medicine house. Later, with the introduction of the Bessemer and acid open-hearth processes, efforts were made to reduce the cost by the substitution of soft-steel scrap for the Swedish wrought iron, but owing, perhaps, to the fact that miscellaneous scrap was used, which carried a higher content of manganese, phosphorus, and sulphur and was more variable in composition than the Swedish wrought iron, the steel was found to be of inferior quality. More recently, washed metal, that is semi-purified pig iron, freed from silicon, manganese, and phosphorus, had been used with excellent results to bring up the carbon of the wrought iron. Still more recently, owing to the introduction of the cheaper electric process and the difficulty of securing good wrought iron, recourse was had to the use of steel scrap, in part at least. With good basic or acid open-hearth mill scrap, the composition of which is more uniform than miscellaneous scrap and could be more readily controlled, steel of good quality could be made, provided proper care was taken to adjust the composition with the best wrought iron and washed metal.

All the practices just described had to do with carbon steels. Since the process had always been used in the production of the finest steels, especially tool steel, the introduction of alloy steels created the greatest diversity of practices. In making alloy steels, the old rule-of-thumb methods of charging had to be abandoned for other plans in which the composition of the charge could be accurately determined by chemical analysis. This applied particularly to high-speed tool steels, which, prior to the introduction of the electric process, could be made successfully only by the crucible process. At first, these steels were made from high-grade wrought iron or muck bar by the addition of the desired elements in the form of iron alloys, and the necessary recarburizer in the form of charcoal or washed metal. Later, tool scrap was used to make up a part of the charge, and much inspecting and many analyses were required to select the scrap and determine the proportions of the charge.

Stages of the Crucible Process—The materials for the charge, after they had been inspected and analyzed, were sheared or broken into small pieces, then carefully weighed out in proper proportions and amounts in the mixing house. These were carefully placed in the crucibles with the wrought iron or steel scrap on the bottom. The crucibles were then closed by snugly fitted covers and taken to the melting hole, into which they were lowered vertically by tongs in the hands of a workman. The gas and air were then regulated to melt the charge as rapidly as possible without injuring the crucibles. The time of melting varied from 2½ to 4 hours, according to the composition of the charge and the heating conditions of the furnace. Low-carbon heats required more time than high-carbon, and high-speed steel required the most time of all grades. As the charge melted, and for some time after melting, the metal evolved gases, which finally collected in fairly large bubbles and broke slowly on the surface of the metal, producing the spectacle known as **cat's eyes**. The melter, by temporarily removing the crucible cover, now watched each pot closely, and when the steel appeared to be **dead**, that is, free from gases and in a tranquil state, he signaled the puller-out, who grasped the crucible with the long broad tongs, lifted it out of the hole, and set it on the floor. The killing of the heat usually required 30 to 40 minutes, sometimes longer, and was an important part of the process, because the practice was necessary to give sound ingots. On the other hand, if the steel was kept in the furnace too long after it had reached the dead-melt stage, it would be damaged and might be ruined by absorption of too much carbon and silicon from the crucible. The temperature was also important; if the heat were pulled too cold, the steel would begin to solidify before it could be cast, and if it were cast too hot, the resulting ingot would not be sound. In some cases the pots were pulled before the steel was completely killed, and a

little aluminum (not over 0.05 per cent) was added to the pot just before the steel was poured into the ingot mold. These methods of finishing affected the hardening properties of many of the steels made by this process.

Casting—The cover was then removed from the crucible and most of the slag was mopped or swabbed up by means of a ball of slag on one end of an iron rod. The crucible then was grasped horizontally with a pair of tongs, and, by tilting the crucible, the steel was poured into a cast-iron mold. This pouring required strength and skill, for it had to be continuous and the stream of metal could not be permitted to impinge upon the walls of the mold. Any slag that might remain on the surface of the liquid metal was held back, by a small iron bar, and prevented from flowing into the mold. The molds were closed at the bottom and to facilitate the removal of the ingots after casting, split molds were used. These molds were cast in two pieces or halves which, fitted together by rings and wedges, formed a mold of square section with the joint passing longitudinally through diagonally opposite corners. Usually, the capacity of the molds was but a single potful of steel, and their cross section was but 75 to 100 millimeters (3 or 4 in.) square; but occasionally larger ingots were required, in which case two pots might be poured at once by the use of a funnel or spout made from a worn-out crucible, or several pots might be poured into a steel ladle, and the steel teemed into the molds from it. The last plan was followed in the production of fine steel castings of a size requiring more than two pots of metal. As soon as the steel had been poured, the crucible was cleaned and carefully inspected for cracks, when, if found sound and in good condition, it was charged, as before, for another heat.

Stripping and Inspecting the Ingots—As soon as the steel had solidified in the molds, the wedges or keys were loosened, and the rings holding the two parts of the mold together were removed. In the case of lower carbon steels, the mold might be removed at once and the ingot cooled in air, but ingots of high-carbon and alloy steels had to be cooled slowly to prevent the formation of **clinks** or tiny cracks, due to non-uniform distribution of contraction and expansion forces. Such ingots were allowed to stand in their molds until solid, then removed and placed in dry lime or covered with ashes or hot, dry sand until cold. The ingots were then topped, that is, pieces of their tops were broken off until the fracture showed perfectly sound steel. From the fractures thus exposed, the temper or carbon content (except in the cases of certain alloy steels) was determined. Then they were inspected for surface defects, and carefully and slowly reheated to a forging temperature. They were next forged, or cogged, under a hammer and any cracks or defects that appeared during the forging were ground or chipped out. After being forged to the required section, usually about 25 to 64 millimeters (1 to 2½ inches) square, the billets were inspected again. Any surface defects, such as rough spots, seams, or tiny cracks were ground out with emery wheels, and these billets were reheated and rolled or forged into the sections desired. All this work was done with the greatest care, and was so meticulous as compared with the attention given ordinary steel that it was termed **crucible practice.**

Without doubt, crucible practice played as important a part in the high quality of crucible steel as crucible melting.

Chemistry of the Crucible Process—From a chemical standpoint, at least, the crucible process was the simplest of all for making steel. In the beginning, the charge carried a small quantity of iron oxides in the form of rust and scale and an almost negligible quantity of free oxygen in the air trapped in the crucible. As the charge melted, these oxides at first formed an oxidizing and very basic slag, which was soon reduced to the ferrous condition by the carbon in the charge or the crucible. Some ferrous oxide dissolved in the metal, and the remainder reacted with clay of the crucible to form ferrous silicates and ferrous aluminum silicates, which, after the charge had all melted and the temperature had been raised, absorbed more and more silica until the slag became very acid. In the meantime, the metal had been absorbing carbon, either from the crucible or the charcoal charged, and the conditions within the crucible now became strongly reducing. Any oxides dissolved or otherwise incorporated with the metal now reacted with carbon, forming carbon monoxide gas; and since the slag was very acid, silica was readily split off from the slag molecules and was at once re-duced, forming silicon, which alloyed with the iron and entered the metal. The absorption of carbon and the reduction and absorption of silicon both progressed with time and temperature, and were more rapid in new crucibles than in crucibles that had been used. The presence of silicon in the metal had the effect of cleansing it of oxides, and, without doubt, it was a large factor in "killing" the steel; but after a content of 0.50 per cent was exceeded, it began to have a noticeable embrittling effect upon most steels. Silicon also affected the welding and certain other properties adversely, hence was positively undesirable in some steels, a difficulty overcome by partial substitution of aluminum. This change in the deoxidizer used sometimes was made for the purpose of controlling the hardening properties of the steel. Hence, the time of killing as well as the temperature had to be closely watched, but it is evident that close control of either of these factors was impossible, with the result that both the carbon and the silicon contents might vary over relatively wide limits.

To produce a large quantity of steel of uniform composition, say up to one ton, it was necessary to pour a number of these 45-kilogram (100-pound) crucible melts into a ladle in such a manner that the metal would be thoroughly mixed.

SECTION 7

MODERN STEELMAKING PROCESSES

All of the steelmaking processes discussed in Section 6 of this chapter were destined eventually to be supplanted by entirely new methods. The first of the new techniques was the pneumatic or Bessemer process (1856). Closely following the invention of the pneumatic process was the development of the regenerative-type furnace that, now known as the open-hearth furnace, became adapted to steelmaking and evolved into the principal means for producing steel throughout the world.

The open-hearth process was supplanted as the world's leading steelmaking method by a pneumatic process that involves blowing high-purity oxygen onto the surface of a bath of molten pig iron; a method known in this country as the basic oxygen steelmaking process. (For brevity, this process is often referred to as BOP, and the furnace in which the process is carried out is referred to as the BOF.) A variation of the basic oxygen process involves blowing high-purity oxygen through a bath of molten pig iron; this is called the Q-BOP process. In addition, several processes have been developed that use various combinations of top and bottom blowing.

Electric-furnace processes are finding more and more applications in the quantity production of quality steels. There has been a recent revival of interest in processes for producing iron and steel by direct methods from ore, without the necessity for first reducing the ore in the blast furnace to make pig iron, and then purifying the pig iron in a second step. No one of these processes, generally referred to as "direct-reduction processes," has yet attained general acceptance although some have been successful in certain localities where a combination of favorable conditions make them practical. Some of these processes are described in Chapter 14.

In this section, some of the historical highlights will be discussed for the pneumatic, open-hearth and electric-furnace steelmaking processes, followed by a brief survey of the status of each process in the United States in recent years.

THE PNEUMATIC STEELMAKING PROCESSES

The Bottom-Blown Acid Process—The original pneumatic process, developed independently by William Kelly of Eddyville, Kentucky and Henry Bessemer of England, involving blowing air *through* a bath of molten pig iron contained in a bottom-blown vessel lined with acid (siliceous) refractories. The process was the first to provide a large-scale method whereby pig iron could rapidly and cheaply be refined and converted into liquid steel. Bessemer's American patent was issued in 1856; although Kelly did not apply for a patent until 1857, he was able to prove that he had worked on the idea as early as 1847. Thus, both men held rights to the process in this country; this led to considerable litigation and delay, as discussed later. Lacking financial means, Kelly was unable to perfect his invention and Bessemer, in the face of great difficulties and many failures, developed the process to a high degree of perfection and it came to be known as the acid Bessemer process.

The fundamental principle proposed by Bessemer and Kelly was that the oxidation of the major impurities in liquid blast-furnace iron (silicon, manganese and carbon) was preferential and occurred before the major oxidation of iron; the actual mechanism differs from this simple explanation, as outlined in the discussion of the physical chemistry of steelmaking in Chapter 13. Further, they discovered that sufficient heat was generated in the vessel by the chemical oxidation of the above elements in most types of pig iron to permit the simple blowing of cold air through molten pig iron to produce liquid steel without the need for an external source of heat. Because the process converted pig iron to steel, the vessel in which the operation was carried out came to be known as a converter. The principle of the bottom-blown converter is shown schematically in Figure 1—19.

At first, Bessemer produced satisfactory steel in a converter lined with siliceous (acid) refractories by refining pig iron that, smelted from Swedish ores, was low in phosphorus, high in manganese, and contained enough silicon to meet the thermal needs of the process. But, when applied to irons which were higher in phosphorus and low in silicon and manganese, the process did not produce satisfactory steel. In order to save his process in the face of opposition among steelmakers, Bessemer built a steel works at Sheffield, England, and began to operate in 1860. Even when low-phosphorus Swedish pig iron was employed, the steels first produced there contained much more than the admissible amounts of oxygen, which made the steel "wild" in the molds. Difficulty also was experienced with sulphur which, introduced from the coke used as fuel in melting the iron in cupolas, contributed to "hot shortness" of the steel. These objections finally were overcome by the addition of manganese in the form of spiegeleisen to the steel after blowing was completed.

FIG. 1—19. Principle of the bottom-blown converter. The blast enters the wind box beneath the vessel through the pipe indicated by the arrow and passes into the vessel through tuyeres set in the bottom of the converter.

The beneficial effects of manganese were disclosed in a patent by R. Mushet in 1856. The carbon and manganese in the spiegeleisen served the purpose of partially deoxidizing the steel, while part of the manganese combined chemically with some of the sulphur to form compounds that either floated out of the metal into the slag, or were comparatively harmless if they remained in the steel.

Because of trouble with tuyeres in the bottom, some early converters had tuyeres located in the side of the vessel but below the metal-bath surface. Many of the early converters were stationary and had to be tapped in a manner similar to the cupola or the open-hearth furnace; such converters were used for a considerable period in Sweden and Germany. However, the rotating or tilting type was favored in England and in the United States.

As stated earlier, Bessemer had obtained patents in England and in this country previous to Kelly's application; therefore, both men held rights to the process in the United States.

The Kelly Pneumatic Process Company had been formed in 1863 in an arrangement with William Kelly for the commercial production of steel by the new process. This association included the Cambria Iron Company; E. B. Ward; Park Brothers and Company; Lyon, Shord and Company; Z. S. Durfee and, later, Chouteau, Harrison and Vale. This company, in 1864, built the first commercial Bessemer plant in this country, consisting of 2¼-metric ton (2½-net ton) acid-lined vessel erected at the Wyandotte Iron Works, Wyandotte, Michigan, owned by Captain E. B. Ward. It may be mentioned that a Kelly converter was used experimentally at the Cambria Works, Johnstown, Pennsylvania as early as 1861.

As a result of the dual rights to the process a second group consisting of Messrs. John A. Griswold and John F. Winslow of Troy, New York and A. L. Holley formed another company under an arrangement with Bessemer in 1864. This group erected an experimental 2¼-metric ton (2-½-net ton) vessel at Troy, New York which commenced operations on February 16, 1865. The rival organizations, after much litigation had failed to gain for either sole control of the patents for the pneumatic process in America, decided to combine their respective interests early in 1866. This larger organization was then able to combine the best features covered by the Kelly and Bessemer patents, and the application of the process advanced rapidly.

By 1871, annual Bessemer steel production in this country had increased to approximately 40 800 metric tons (45 000 net tons), about 55 per cent of the total steel production of the country, which was produced by seven Bessemer plants.

Bessemer-steel production in the United States over an extended period of years remained significant; however, raw steel is no longer being produced by the acid Bessemer process in the United States. The last completely new plant for the production of acid-Bessemer steel ingots in the United States was built in 1949.

A. L. Holley's contributions to the early development of the process were exceedingly important. He acted as consulting engineer on most of the plants erected in the first ten to fifteen years of the development of the process in the United States. Holley also greatly improved the design of the detachable bottom, originally developed by Bessemer in 1863; this permitted increased tonnage output because it permitted replacement of the bottom when repairs were necessary without excessive loss of production.

Another notable contribution to the development of the pneumatic processes was made at a later date (1889) by William R. Jones of the Edgar Thomson Works of the Carnegie Steel Co., now a part of United States Steel Corporation, who conceived the idea of the hot-metal mixer.

As already stated, the bottom-blown acid process known generally as the Bessemer process was the original pneumatic steelmaking process. Many millions of tons of steel were produced by this method. From 1870 to 1910, the acid Bessemer process produced the majority of the world's supply of steel.

The success of acid Bessemer steelmaking was dependent upon the quality of pig iron available which, in turn, demanded reliable supplies of iron ore and metallurgical coke of relatively high purity. At the time of the invention of the process, large quantities of suitable ores were available, both abroad and in the United States. With the gradual depletion of high-quality ores abroad (particularly low-phosphorus ores) and the rapid expansion of the use of the bottom-blown basic pneumatic, basic open-hearth and basic oxygen steelmaking processes over the years, acid Bessemer steel production has essentially ceased in the United Kingdom and on the Continent of Europe.

In the United States, the Mesabi Range provided a source of relatively high grade ore for making iron for the acid Bessemer process for many years. In spite of this, the acid Bessemer process declined from a major to a minor steelmaking method in the United States and eventually was abandoned.

The early use of acid Bessemer steel in this country involved production of a considerable quantity of rail steel, and for many years (from its introduction in 1864 until 1908) this process was the principal steelmaking process. Until relatively recently, the acid-Bessemer process was used principally in the production of steel for buttwelded pipe, seamless pipe, free-machining bars, flat-rolled products, wire, steel castings, and blown metal for the duplex process.

Fully-killed acid Bessemer steel was used for the first time commercially by United States Steel Corporation in the production of seamless pipe. In addition, dephosphorized acid Bessemer steel was used extensively in the production of welded pipe and galvanized sheets.

The Basic Bessemer or Thomas Process—The bottom-blown basic pneumatic process known by the several names of the **Thomas, Thomas-Gilchrist** or **basic Bessemer process,** was patented in 1879 by Sidney G. Thomas in England. The process, involving the use of a basic lining and a basic flux in the converter, made it possible to use the pneumatic method for refining pig irons smelted from the high-phosphorus ores common to many sections of Europe. The process (never adopted in the United States) developed much more rapidly on the Continent than in Great Britain and, in

1890, Continental production was over 1.8 million metric tons (2 million net tons) as compared with 0.36 million metric tons (400 000 net tons) made in Great Britain.

The simultaneous development of the basic open-hearth process resulted in a decline of production of steel by the bottom-blown basic pneumatic process in Europe and, by 1904, production of basic open-hearth steel there exceeded that of basic pneumatic steel. From 1910 on, the bottom-blown basic pneumatic process declined more or less continuously percentagewise except for the period covering World War II, after which the decline resumed.

The Surface Side-Blown Acid Process—This pneumatic process found its principal use in foundries making steel castings. The acid electric-arc-furnace process replaced it in most steel foundries.

The side-blown acid pneumatic converter (Figure 1—20) was sometimes referred to as a surface-blown converter, or as the Tropenas converter. There were numerous variations of design of converters using the side-blowing technique, but all were characterized by having all of the tuyeres above the liquid level of the bath and entering through the side of the vessel. The majority of side-blown acid converters were of one-half to two ton capacity, although some larger vessels were built.

Oxygen-Blown Pneumatic Processes—Although the use of gaseous oxygen (rather than air) as the agent for refining molten pig iron and scrap mixtures to produce steel by pneumatic processes received the attention of numerous investigators from Bessemer onward, it was not until after World War II that commercial success was attained.

Blowing with oxygen was investigated by R. Durrer and C. V. Schwarz in Germany and by Durrer and Hellbrugge in Switzerland. Bottom-blown basic-lined vessels of the designs they used proved unsuitable because the high temperature attained caused rapid deterioration of the refractory tuyere bottom; blowing pressurized oxygen downwardly against the top surface of the molten metal bath, however, was found to convert the charge to steel with a high degree of thermal and chemical efficiency.

Plants utilizing top blowing with oxygen have been in operation since 1952-53 at Linz and Donawitz in Austria. These operations, sometimes referred to as the Linz-Donawitz or L-D process, were designed to employ pig iron produced from local ores that are high in manganese and low in phosphorus; such iron is not suitable for either the acid or basic bottom-blown pneumatic processes utilizing air for blowing. The top-blown process, however, is adapted readily to the processing of blast-furnace metal of medium- and high-phosphorus contents and is particularly attractive where it is desirable to employ a steelmaking process requiring large amounts of hot metal as the principal source of metallics. This adaptability has led to the development of numerous variations in application of the top-blown principle. In its most widely used form, which also is the form used in the United States, the top-blown oxygen process is called the basic oxygen steelmaking process (BOP for short).

The basic oxygen process consists essentially of

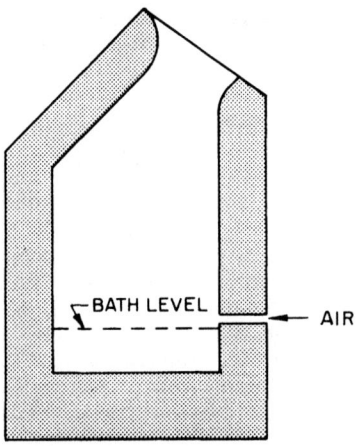

FIG. 1—20. Principle of the side-blown converter. The blast entered the vessel through tuyeres in its side, indicated by arrow. Angle at which centerline of tuyeres intersected horizontal surface of bath could be varied by tilting vessel.

blowing oxygen of high purity onto the surface of the bath in a basic-lined vessel by a water-cooled vertical pipe or lance inserted through the mouth of the vessel (Figure 1—21). Annual production of raw steel in the United States by the basic oxygen process since its introduction here in 1955 is given in Appendix E at the end of this book.

A successful bottom-blown oxygen steelmaking process has been developed in recent years. Based on developments in Germany and Canada and known as the Q-BOP process (or as the OBM process in Europe), the new method has eliminated the problem of rapid bottom deterioration encountered in earlier attempts to bottom-blow with oxygen. The tuyeres (Figure 1—22), mounted in a removable bottom, are designed in such a way that the stream of gaseous oxygen passing through a tuyere into the vessel is surrounded by a sheath of another gas. The sheathing gas is normally a hydrocarbon gas such as propane or natural gas. The first vessels for the Q-BOP process were relatively small (about 30 tons), but the process has been scaled up successfully to provide vessel capacities of 200 tons and over, comparable to the capacities of typical top-blown BOP vessels.

The desire to improve control of the oxygen pneumatic steelmaking process has led to the development of various combination-blowing processes. In these processes, 60 to 100 per cent of the oxygen required to refine the steel is blown through a top-mounted lance (as in the conventional BOP) while additional gas (such as oxygen, argon, nitrogen, carbon dioxide or air) is blown through bottom-mounted tuyeres or permeable brick elements. The bottom-blown gas results in improved mixing of the metal bath, the degree of bath mixing increasing with increasing bottom gas-flow rate. By varying the type and flow rate of the bottom gas, both during and after the oxygen blow, specific metallurgical reactions can be controlled to attain desired steel compositions and temperatures. There are, at present, many different combination-blowing processes, which differ in the type of bottom gas used, the flow rates of bottom gas that can be attained, and the equip-

ment used to introduce the bottom gas into the furnace. All of the processes, to some degree, have similar advantages. The existing combination-blowing furnaces are converted conventional BOP furnaces and range in capacity from about 60 tons to more than 300 tons. The conversion to combination blowing began in the late 1970's and has continued into the 1980's at an accelerated rate. Further details of these processes are given in Chapter 16.

Two other oxygen-blown steelmaking, the Stora—Kaldo process and the Rotor process, did not gain wide acceptance.

In 1980, there were 81 commercial basic oxygen furnaces in the United States and 2 Kaldo process furnaces; the latter are no longer in operation. The num-

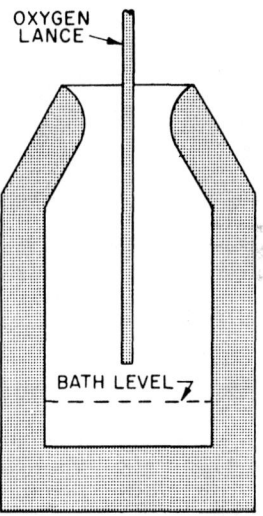

FIG. 1—21. Principle of the top-blown converter. Oxygen of commercial purity, at high pressure and velocity is blown downward vertically onto surface of bath through a single water-cooled pipe or lance, indicated by arrow.

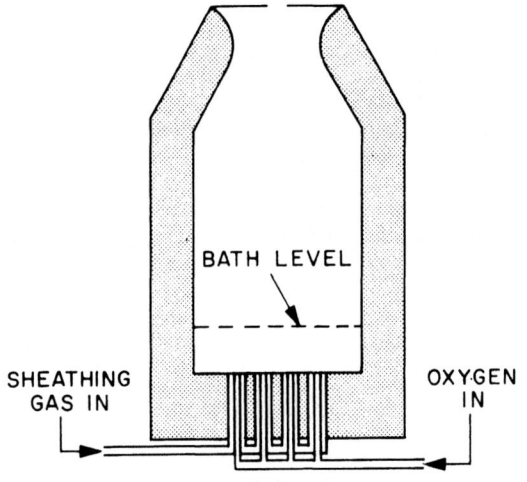

FIG. 1—22. Schematic cross-section of a Q-BOP vessel, showing how a suitable gas is introduced into the tuyeres to completely surround the stream of gaseous oxygen passing through the tuyeres into the molten metal bath.

Table 1—I. Oxygen Process Steelmaking Furnaces in the
United States, Classified into Capacity Ranges.[a]

**BASIC OXYGEN PROCESS
STEELMAKING FURNACES[d]**

Capacity Range		Number of Furnaces With Rated Capacity Within Range
(Metric Tons)[b]	(Net Tons)[c]	
69-91	76-100	6
92-113	101-125	7
114-136	126-150	7
137-159	151-175	—
160-181	176-200	15
182-204	201-225	22
205-227	226-250	8
228-250	251-275	6
251-272	276-300	7
273-295	301-325	3
	Total	81

[a] As of January 1, 1980. Based on information in "Iron and Steel Works Directory of the United States and Canada" published by the American Iron and Steel Institute, Washington, D.C., July 1980, which listed facilities expected to be in operation at year's end.
[b] Calculated to nearest metric ton.
[c] Nominal capacities.
[d] Includes two 204-metric ton (225-net ton) Q-BOP furnaces.

ber of basic oxygen furnaces in various capacity ranges is given in Table 1—I.

Mention should be made here of the argon-oxygen-decarburization (AOD) process, a duplex process in which scrap is melted in an electric furnace and refined in an AOD furnace by side blowing with a mixture of gaseous oxygen and argon (or nitrogen). This process is described in Chapter 19.

The 1982 "World Steel Industry Data Handbook, Vol. 5, USA" of 33 Metal Producing magazine published by McGraw-Hill (New York), listed 43 argon-oxygen-decarburization furnaces in steel plants in the United States as of 1981, as follows: (all ranges are maximum capacity ranges) 1 in the range less than 4.5 metric tons (5 net tons), 4 in the 10.0 to 13.5 metric tons (11 to 15 net tons) range, 5 in the 14.5 to 18.0 metric tons (16 to 20 net tons) range, 5 in the 19.0 to 22.5 metric tons (21 to 25 net tons) range, 6 in the 23.5 to 27.0 metric tons (26 to 30 net tons) range, 4 in the 32.5 to 36.0 metric tons (31 to 35 net tons) range, 10 in the 41.5 to 45.0 metric tons (46 to 50 net tons) range, 7 in the 87.1 to 90.7 metric tons (96 to 100 net tons) range, 7 in the 87.1 to 90.7 metric tons (96 to 100 net tons) range and 1 furnace with a maximum capacity of 160 metric tons (175 net tons).

THE OPEN-HEARTH STEELMAKING PROCESSES

Early History of the Processes—The phenomenal success of the Bessemer process, coupled with the ever-increasing demand for steel, attracted many other inventors to the study of new and improved methods of steelmaking. Of all the methods which received attention, the only one which was destined to become a rival

of the pneumatic process was developed through the invention of the regenerative principle by Karl Wilhelm Siemens who, although German-born, was a naturalized British citizen.

Siemens' early work with the regenerative principle as applied to steam engines showed that a very great saving of fuel and very high temperatures could be obtained by its use and, at the suggestion of his brother, Frederick, he turned his attention to the application of the principle for producing high temperatures in furnaces. The first experimental furnace was built in 1858, when it was discovered that, especially with large furnaces using solid fuel, many difficulties were to be overcome if the full efficiency which the principle promised was to be realized.

After two years or more of experimentation, Siemens fell upon the plan of gasifying the fuel prior to burning it in the furnace, whereby he found that most of the difficulties could be overcome. Thus, a correlative development of the regenerative furnace was the **gas producer.** Siemens' patent covering the successful furnace design specifically mentioned the possibility of using the furnace for the production of steel. An early design of a Siemens steelmaking furnace, with gas producer, is shown schematically in Figure 1—23.

In the ferrous industry, some of the early uses Siemens made of the furnace were for puddling, reheating iron and steel for forging and rolling, and for melting crucible steel. Siemens next turned his attention to the manufacture of steel in his furnace and, though many trials were made at different works, he met with only indifferent success. Finally, like Bessemer, he found it necessary to erect a steel works of his own. These works were located in Birmingham, England, and were employed at first in a remelting process by which steel of the best quality was obtained from processing such scrap as old rails, plates, and so on.

In the meantime, Siemens was busy developing the idea that steel could be made from pig iron by oxidizing the carbon content of the latter with iron ore and, by the year 1868, proved that this method, which came to be known as the **pig and ore process,** could be employed successfully. He next turned his attention to evolving a method whereby steel could be produced directly from iron ore, thus dispensing with the blast furnace. In this feat he actually succeeded, but the cost of production was many times that of producing steel from pig iron. Subsequent events have proved that it is more practical and economical to employ the blast furnace for the primary reduction of the iron ore to pig iron, and from this to make steel in the furnace which has been evolved from Siemens' original designs.

Principles of Siemens' Pig and Ore Process—Briefly, the method of Siemens was as follows. A rectangular covered hearth was used to contain the charge of pig iron or pig iron and scrap (see Figure 1—23). Most of the heat required to promote the chemical reactions necessary for purification of the charge was provided by passing burning fuel gas over the top of the materials. The fuel gas, with a quantity of air more than sufficient to burn it, was introduced through ports at each end of the furnace, alternately at one end and then the other. The products of combustion passed out of the port temporarily not used for entrance of gas and

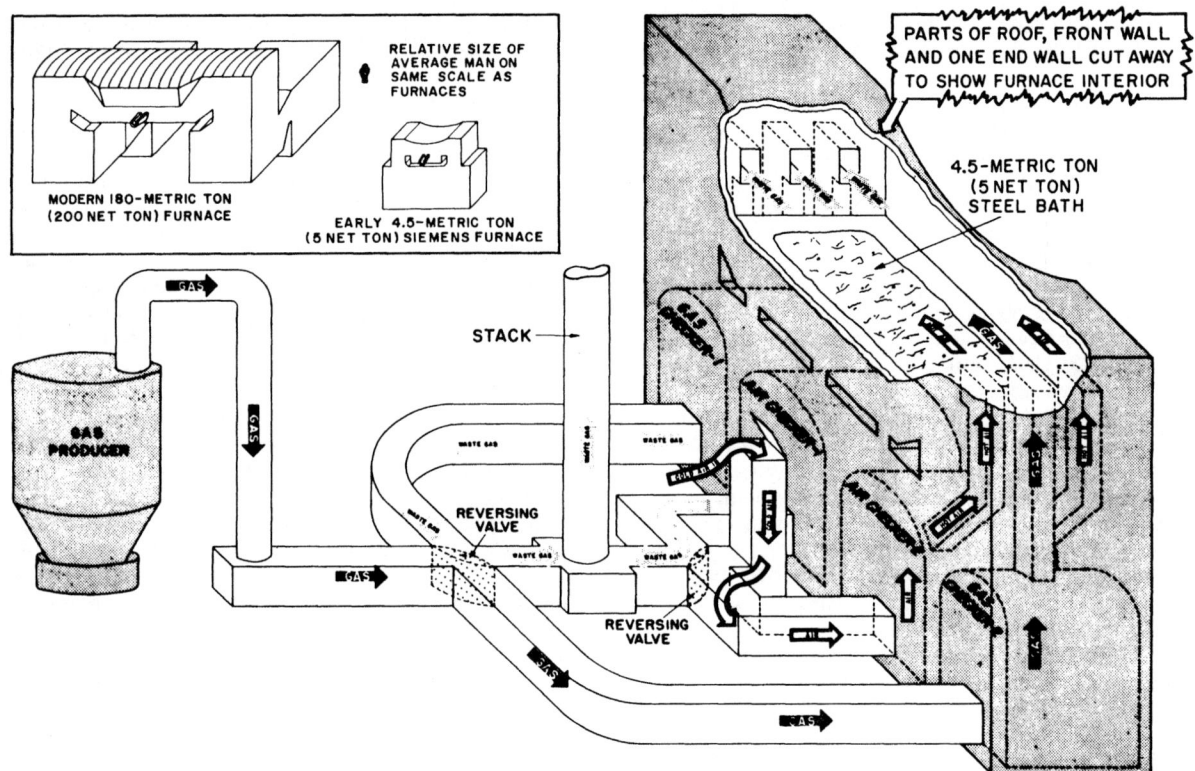

FIG. 1—23. Schematic arrangement of an early type of Siemens furnace with about a 4.5 metric ton (5 net ton) capacity. The roof of this design (which was soon abandoned) dipped from the ends toward the center of the furnace to force the flame downward on the bath. Various different arrangements of gas and air ports were used in later furnaces. Note that in this design, the furnace proper was supported on the regenerator arches. Flow of gas, air and waste gases were reversed by reversing the position of the two reversing valves. The inset at the upper left compares the size of one of these early furnaces with that of a modern 180 metric ton (200 net ton) open hearth.

air, and entered chambers partly filled with brick checkerwork. This checkerwork, commonly called **checkers,** provided a multitude of passageways for the exit of the gases to the stack. During their passage through the checkers, the gases gave up a large part of their heat to the brickwork. After a short time, the gas and air were shut off at the one end and introduced into the furnace through the preheated checkers, absorbing some of the heat stored in these checkers. The gas and air were thus preheated to a somewhat elevated temperature, and consequently developed a higher temperature in combustion than could be obtained without preheating. In about twenty minutes, the flow of the gas and air was again reversed so that they entered the furnace through the checkers and port used first; and a series of such reversals, occurring every fifteen or twenty minutes was continued until the heat was finished. The elements in the bath which were oxidized both by the oxygen of the air in the furnace atmosphere and that contained in the iron ore fed to the bath, were carbon, silicon and manganese, all three of which could be reduced to as low a limit as was possible in the Bessemer process. Of course, a small amount of iron is oxidized and enters the slag.

Thus, as in all other processes for purifying pig iron, the basic principle of the Siemens process was that of oxidation. However, in other respects, it was unlike any other process. True, it resembled the puddling process in both the method and the agencies employed, but the high temperatures attainable in the Siemens furnace made it possible to keep the final product molten and free of entrapped slag. The same primary result was obtained as in the Bessemer process, but by a different method and through different agencies, both of which imparted to steel made by the new process properties somewhat different from Bessemer steel, and gave the process itself certain metallurgical advantages over the older pneumatic process, as discussed later in this section.

Mechanical Changes and Improvements in the Siemens Process—As would be expected, many variations of the process, both mechanical and metallurgical, have been worked out since its original conception. Along mechanical lines, various improvements in the design, the size, and the arrangement of the parts of the furnace have been made. Early furnaces had capacities of only about 3.5 to 4.5 metric tons (4 or 5 net tons), while modern furnaces range from about 35 to 544 metric tons (40 to 600 net tons) in capacity, with the majority having capacities between about 180 and 270 metric tons (200 and 300 net tons).

The early development of the Siemens steelmaking process was retarded by the lack of refractory materials capable of withstanding the high temperatures and the chemical reactions of the process. Even in modern advanced stages of development of the process, it was recognized that additional production and improved steel quality might be possible of attainment if new or improved refractories eventually made feasible a sustained operation at higher temperatures.

The Siemens process became known more generally, at least in this country, as the **open-hearth process.** The name "open hearth" was derived, probably, from the fact that the steel, while melted on a hearth under a roof, was accessible through the furnace doors for inspection, sampling, and testing. Modern open-hearth steelmaking practices are described in Chapter 17.

Metallurgical Improvements—The hearth of Siemens' furnace was of acid brick construction, on top of which the bottom was made up of sand—essentially as in the **acid process** of today. Later, to permit the charging of limestone and use of a basic slag for removal of phosphorus, the hearth was constructed with a lining of magnesite brick, covered with a layer of burned dolomite or magnesite, replacing the siliceous bottom of the acid furnace. These furnaces, therefore, were designated as basic furnaces, and the process carried out in them was called the **basic process.**

The **pig and scrap process** was originated by the Martin brothers, in France, who, by substituting scrap for the ore in Siemens' **pig and ore process,** found it possible to dilute the charge with steel scrap to such an extent that less oxidation was necessary. Since the time of the Martins, these processes have undergone various modifications, chief of which are those known as the **Talbot,** the **Campbell,** the **Bertrand-Thiel,** and the **Monell** process. These latter processes have become obsolete, and they did little to determine modern practice, except that Talbot and Campbell did exert an influence which was reflected until recently in tilting-furnace practice, as discussed in Chapter 17.

Advantages of the Siemens Process—The advantages offered by the Siemens process may be summarized briefly as follows: (1) By the use of iron ore as an oxidizing agent and by the external application of heat, the temperature of the bath was made independent of the purifying reactions, and the elimination of impurities could be made to take place gradually, so that both the temperature and composition of the bath were under much better control than in the Bessemer process. (2) For the same reasons, a greater variety of raw materials could be used (particularly scrap, not greatly consumable in the Bessemer converter) and a greater variety of products could be made by the open-hearth process than by the Bessemer process. (3) A very important advantage was the increased yield of finished steel from a given quantity of pig iron as compared to the Bessemer process, because of lower inherent sources of iron loss in the former, as well as because of recovery of the iron content of the ore used for oxidation in the open hearth. (4) Finally, with the development of the basic open-hearth process, the greatest advantage of Siemens' over the acid Bessemer process was made apparent, since the basic open-hearth process is capable of eliminating phosphorus from the bath. While this element can be removed also in the basic Bessemer (Thomas-Gilchrist) process, it is to be noted that, due to the different temperature conditions, phosphorus is eliminated before the carbon in the basic open-hearth process, whereas the major proportion of the phosphorus is not oxidized in the basic Bessemer process until after the carbon, in the period termed the afterblow. Hence, while the basic Bessemer process requires a pig iron with a phosphorus content of 2.00 per cent or more in order to maintain the temperature high enough for the afterblow, the basic open-hearth process permits the economical use of iron of any phosphorus content up to 1.00 per cent. In the United States, this fact was of importance since it made available immense iron ore deposits which could not be utilized otherwise because of their phosphorus content, which was too high to permit their use in the acid Bessemer or acid open-hearth process and too low for use in the basic Bessemer process. This was one of the important reasons why the basic open-hearth process became the leading steelmaking method in this country: other reasons are discussed later.

The Open-Hearth Process in the United States—As early as 1868, a small open-hearth furnace was built at Trenton, New Jersey, but satisfactory steel at a reasonable cost did not result and the furnace was abandoned. Later, at Boston, Massachusetts, a successful furnace was designed and operated, beginning in 1870. Following this success, similar furnaces were built at Nashua, New Hampshire, and in Pittsburgh, Pennsylvania, the latter by Singer, Nimick and Company, in 1871. The Otis Iron and Steel Company constructed two 6.3 metric ton (7 net ton) furnaces at their Lakeside plant in 1874. Two 13.5 metric ton (15 net ton) furnaces were added to this plant in 1878, two more of the same size in 1881, and two more in 1887. All of these furnaces had acid linings, using a sand bottom for the hearths.

A furnace with a basic bottom, rammed from Austrian magnesite, produced basic steel at the Otis plant in January, 1886. Production rate on the basic furnace was so low compared to that normally achieved with an acid bottom that the basic bottom was torn out after four months, and replaced by an acid bottom. Following the Otis installation, the following companies also installed open-hearth furnaces: The Cleveland Rolling Mills (later a part of United States Steel Corporation); The Pennsylvania Steel Company (later Bethlehem Steel Corporation); the Schoenberger Works (now dismantled, but formerly part of the United States Steel Corporation; and Carnegie, Phipps and Company (later a part of Homestead Works of United States Steel Corporation).

The commercial production of steel by the basic process was achieved first at Homestead. The initial heat was tapped March 28, 1888. By the close of 1890, there were 16 basic open-hearth furnaces operating. From 1890 to 1900, magnesite for the bottom began to be imported regularly and the manufacture of silica refractories for the roof was begun in American plants. For these last two reasons, the construction of basic furnaces advanced rapidly and, by 1900, furnaces larger than 45 metric tons (50 net tons) were being planned.

While the Bessemer process could produce steel at a possibly lower cost above the cost of materials, it was restricted to ores of a limited phosphorus content and

Table 1—II. Open-Hearth Steelmaking Furnaces in the United States as of January 1, 1980[a]

BASIC OPEN-HEARTH FURNACES

Number of Furnaces	Rated Capacity	
	(Metric Tons)[b]	(Net Tons)[c]
2	136	150
6	163	180
5	236	260
6	281	310
8	286	315
11	290	320
7	304	335
10	308	340
9	363	400
7	381	420
71		

ACID OPEN-HEARTH FURNACES

Number of Furnaces	Rated Capacity	
	(Metric Tons)[b]	(Net Tons)[c]
1	36	40
1	45	50
1	113	125
3		

[a] Based on information in the "Iron and Steel Works Directory of the United States and Canada" published by the American Iron and Steel Institute, Washington, D.C., July 1980. Only those furnaces expected to be in operation at least until the end of 1980 were reported.
[b] Calculated to nearest metric ton.
[c] Nominal capacities.

its use of scrap was also limited. The open hearth was not subject to these restrictions, so that the annual production of steel by the open-hearth process increased rapidly, and, in 1908, passed the total tonnage produced yearly by the Bessemer process. Total annual production of Bessemer steel ingots decreased rather steadily after 1908, and has ceased entirely in the United States. In addition to the ability of the basic open-hearth furnace to utilize irons made from American ores, as discussed earlier, the main reasons were the flexibility of the open-hearth process, with respect to its ability to produce steels of many compositions, and its ability to use a large proportion of steel or iron scrap, if necessary. Also steels made by any of the pneumatic processes that utilize air for blowing contain more nitrogen than open-hearth steels; this higher nitrogen content made Bessmer steel less desirable than open-hearth steel in some important applications.

Open-hearth steel production in the United States over an extended period of time is given in Appendix E at the end of this book, and the number of open-hearth furnaces in the United States in selected years is given in Table 1—II.

ELECTRIC-FURNACE STEELMAKING

It has been said that arc-type furnaces had their beginning in the discovery of the carbon arc by Sir Humphrey Davy in 1800, but it is more proper to say that their practical application began with the work of Sir William Siemens, who in 1878 constructed, operated and patented furnaces operating on both the direct-arc and indirect-arc principles.

At this early date, the availability of electric power was limited and its cost high; also, carbon electrodes of the quality required to carry sufficient current for steel melting had not been developed. Thus, the development of the electric melting furnace awaited the expansion of the electric-power industry and improvements in carbon electrodes.

The amount of steel produced annually by electric-furnace processes in the United States is given in Appendix E at the end of this book.

Direct-Arc Furnaces—The first successful commercial direct-arc steelmaking furnace was placed in operation by Heroult in 1899, and the first shipment of electric steel was a carload of bars from Heroult's plant at La Praz to the firm of Schneider and Company at Creusot, France, on December 28, 1900. The Heroult patent, stated in simple terms, covered single-phase or multi-phase furnaces with the arcs in series through the metal bath. This type of furnace, utilizing three-phase power, has been the most successful of the electric furnaces in the production of steel. The design and operation of modern electric-arc furnaces are discussed in Chapter 18.

Simultaneous with and subsequent to Heroult's success with the direct-arc principle, many investigators abroad directed their attention to developing furnaces employing the same principle. However, in this country, there were no developments along arc-furnace lines until the first Heroult furnace was installed in the plant of the Halcomb Steel Company, Syracuse, New York, which made its first heat on April 5, 1906. This was a single-phase, two-electrode, rectangular furnace of 3.6 metric tons (4 net tons) capacity. Two years later a similar but smaller furnace was installed at the Firth-Sterling Steel Company, McKeesport, Pennsylvania, and in 1909, a 13.5 metric ton (15 net ton) three-phase furnace was installed in the South Works of the Illinois Steel Company, now a part of the United States Steel Corporation, in Chicago, Illinois. The latter was, at that time, the largest electric steelmaking furnace in the world, and was the first round (instead of rectangular) furnace. It operated on 25-cycle power at 2200 volts and tapped the first heat on May 10, 1909.

The foregoing furnaces all were for making steel for ingots. The first electric furnace for the production of steel for commercial castings was that of the Treadwell Engineering Company, Easton, Pennsylvania. It was a single-phase, two-electrode furnace of 1.8 metric tons (2 net tons) capacity, designed along the lines of the Halcomb and Firth-Sterling furnaces. It was operated first in August, 1911.

About the same time, the General Electric Company began to experiment with the design of direct-arc electric furnaces, with the view of developing a market for electrical equipment required for their operation, and built three or four units. Other furnace designs followed: the Snyder, Ludlum, Vom Baur, Booth-Hall, Moore, Green, Swindell, and Volta, the last-named being of Canadian origin.

In recent years, the direct-arc principle has been applied to vacuum arc remelting (VAR) furnaces (also called vacuum consumable-electrode furnaces) and electroslag remelting (ESR) furnaces that are described in Chapter 18.

There were in the steel industry in 1981 a total of 87 vacuum arc remelting furnaces, according to the 'World Steel Industry Data Handbook, vol. 5, USA' published by 33 Metal Producing magazine (McGraw-Hill, New York, 1982). The numbers of furnaces that could produce ingots in various maximum weight ranges were as follows: 25 in the 4.5 metric ton (5 net ton) or less range, 17 in the 5.0 to 9.0 metric ton (6 to 10 net ton) range, 10 in the 10.0 to 13.5 metric ton (11 to 15 net ton) range, 19 in the 14.5 to 18.0 metric ton (16 to 20 net ton) range, 13 in the 19.0 to 22.5 metric ton (21 to 25 net ton) range, and 1 in the 23.5 to 27.0 metric ton (26 to 30 net ton) range. In addition, there was one furnace producing 9.5 metric ton (10.5 net ton) ingots and one producing ingots of unspecified weight with a maximum diameter of 762 millimeters (30 inches).

According to the same source, there were in 1981 a total of 33 electroslag remelting units in this country. Twenty-six furnaces were identified as capable of producing round ingots in the following maximum weight ranges:

Number of Furnaces	Maximum Ingot Weight Range	
	Metric Tons	Net Tons
3	4.5 or less	5 or less
7	5.0-9.0	6-10
8	10.0-13.5	11-15
3	14.5-18.0	16-20
2	19.0-22.5	21-25
1	23.5-27.0	26-30
2	64.0-68.0	71-75

In addition to these 26 furnaces, there were 7 other units listed for which the maximum ingot weight was not given. Of these, one was described as having a capacity of 7.3 metric tons (8 net tons), 3 were stated to be of 13.6 metric ton (15 net ton) capacity, 1 was stated to have an 18.1 metric ton (20 net ton) capacity, and 2 were listed without any operational data.

Indirect-Arc Furnaces—The first work on indirect-arc furnaces was done by Stassano, in Italy. His design consisted of a vertical, cylindrical shell, with three electrodes spaced 120 degrees apart and entering the furnace just above the bath. A furnace of this design was installed in the plant of the Clark Equipment Company, Buchanan, Michigan, in 1911.

About the beginning of World War I, Rennerfelt of Sweden developed an indirect-arc furnace with two horizontal electrodes and one vertical electrode, so connected electrically that the arc was "blown" down on the charge or bath, and the horizontal electrodes were arranged to tilt downward so that they could arc directly on the bath.

None of the indirect-arc furnaces came into very great use because maintenance was difficult and power consumption was high.

The Induction Furnace—Another type of electric melting furnace, used to a certain extent for melting high-grade alloys, is the high-frequency coreless induction furnace described in Chapter 18, which gradually replaced the crucible process in the production of complex, high-quality alloys used as tool steels. It is used also for remelting scrap from fine steels produced in arc furnaces, melting chrome-nickel alloys, and high-manganese scrap, and, more recently, has been applied to vacuum steelmaking processes.

The induction furnace had its inception abroad and first was patented by Ferranti in Italy in 1877. This was a low-frequency furnace. It had no commercial application until Kjellin installed and operated one in Sweden. The first large installation of this type was made in 1914 at the plant of the American Iron and Steel Company in Lebanon, Pennsylvania, but was not successful. Some other low-frequency furnaces, however, have operated successfully, especially in making stainless steel.

A successful development using higher-frequency current is the coreless high-frequency induction furnace. The first coreless induction furnaces were built and installed by the Ajax Electrothermic Corporation, who also initiated the original researches by E. F. Northrup leading to the development of the furnace. For this reason, the furnace is often referred to as the Ajax-Northrup furnace.

The first coreless induction furnaces for the production of steel on a commercial scale were installed at Sheffield, England, and began the regular production of steel in October, 1927.

The first commercial steel furnaces of this type in the United States were installed by the Heppenstall Forge and Knife Company, Pittsburgh, Pennsylvania, and were producing steel regularly in November, 1928. Each furnace had a capacity of 272 kilograms (600 pounds) and was served by a 150-kVA motor-generator set transforming 60-hertz current to 860 hertz. The crucibles were monolithic, about 150 millimeters (6 inches) in diameter and about 900 millimeters (36 inches) in depth. Time of an ordinary heat was 55 minutes. The coreless type of induction furnace normally was operated on alternating current at a frequency of approximately 1000 hertz produced by a motor-generator set of special design with the power transmitted over co-axial cables to the primary coil of the furnace. More recently, mercury-arc rectifiers, which have the advantage of higher efficiencies over motor-generator sets, are used to supply the high-frequency power.

In plants listed in the 1980 'Iron and Steel Works Directory of the United States and Canada' published by the American Iron and Steel Institute, there were 14 electric induction melting furnaces. These included 1 furnace of 45-kg (100-lb) capacity, 1 of 113-kg (250-lb) capacity, 1 of 272-kg (600-lb) capacity, 2 of 454-kg (1000-lb) capacity, 3 of 907-kg (2000-lb) capacity, 1 of 1814-kg (4000-lb) capacity, 2 of 2268-kg (5000-lb) capacity, and 3 of 63 503-kg (140 000-lb) capacity.

There were 15 vacuum induction melting furnaces listed in the same directory. Four of these had a capacity of less than 3.5 metric tons (5 net tons), 5 had capacities ranging between 4.5 and 9 metric tons (5 to 10 net tons), 2 had capacities ranging between 10.0 and 13.5 metric tons (11 and 15 net tons), 2 had capacities ranging between 23.5 and 27.0 metric tons (26 and 30 net tons), and 2 were listed without any stated capacity. In addition, one plant was stated to use vacuum induction melting but no details were given about the number or capacities of the furnaces used.

Electric Reduction Furnaces—Electric reduction furnaces are used for smelting ores in the production of ferroallys, such as ferromanganese, ferrosilicon, and ferrochromium. These furnaces differ from the steelmaking furnaces in that production is continuous, as in a blast furnace; the charge is placed in the furnace at the top and the molten product tapped near the bottom. They are either single-phase or three-phase units. Electric furnaces are also used in the smelting of iron ore, as discussed in Chapter 14.

In the single-phase units, a single carbon electrode is suspended in the center of the furnace, with the carbon bottom of the furnace serving as the second electrode. This is an example of a furnace with a conducting bottom. In the three-phase furnaces, the three electrodes are spaced as in the arc furnace and are buried in the charge. These units normally have an open top, and the charge either is shoveled in or introduced by chutes. The electrical load on this type of furnace differs from that on the arc furnace used for steelmaking. Since charging is continuous, the load is fairly steady, and the power factor can be maintained around 90 per cent. The voltages used are lower than in the arc furnaces, and the current density greater. These requirements necessitate a careful design of the secondary circuits to keep reactance to a minimum. For some products, and on 25-hertz systems, units have electrical capacities up to 18 000 kVA.

Electric Furnaces of Special Design—There have been other types of furnaces designed that are worthy of mention. The Hering furnace was a conventional unit except that the bath was deeper than usual and the hearth contained one or more resistance tubes at the bottom which served as water-cooled electrodes. Molten metal was required for the first charge, and only two-thirds of the heat was tapped. The effect of the current flowing through the molten metal in the resistor tubes was to pinch off the column of metal and eject it from the tube. The furnace had no commercial application.

Another furnace of European design was a combination of open-hearth and electric-arc furnace principles. Still another was a combination of low-frequency induction furnace and pneumatic converter, the tuyeres of the latter being equipped so that a reducing gas might be introduced into the furnace.

COMPARATIVE STATUS OF MODERN STEELMAKING PROCESSES IN THE U. S.

Appendix E at the end of this book shows the total annual raw-steel production by the principal steelmaking methods in the United States over an extended period. The growing importance of the basic oxygen and electric-arc furnace steelmaking processes and the declining importance of the basic open-hearth steelmaking process are clearly evident.

As was stated earlier in this section, the last completely new plant for the production of acid-Bessemer steel ingots went into operation in 1949. For reasons that have been discussed, the use of the acid-Bessemer process for ingot production has been abandoned here. The basic-Bessemer or Thomas process was never adopted in this country.

The acid open-hearth process is used to make only a very small proportion of the raw steel made in the United States.

While the basic open-hearth process dominated steelmaking in the United States for many years, the last completely new basic open-hearth steelmaking plant built in the United States began full-scale operations in 1958, three years after the introduction of the basic oxygen process here. The basic oxygen steelmaking process has since become the leading steelmaking process in the United States (see Figure 1—24). The numbers of basic oxygen process and open-

Table 1—III. Electric Direct-Arc Steelmaking Furnaces in the United States as of January 1, 1980[a]

BASIC-LINED FURNACES

Capacity Range		Number of Furnaces
(Metric Tons)[b]	(Net Tons)[c]	
0.9-4.5	1-5	4
5.0-9.0	6-10	10
10.0-13.5	11-15	10
14.5-18.0	16-20	19
19.0-22.5	21-25	15
23.5-27.0	26-30	8
28.0-31.5	31-35	18
32.5-36.0	36-40	15
37.0-41.0	41-45	20
41.5-45.0	46-50	14
46.0-54.5	51-60	10
55.5-63.5	61-70	7
64.5-72.5	71-80	16
73.5-81.5	81-90	14
82.5-90.5	91-100	19
91.5-109.0	101-120	12
110.0-127.0	121-140	5
128.0-145.0	141-160	14
146.0-163.5	161-180	13
164.0-181.5	181-200	16
182.5-204.0	201-225	7
205.0-227.0	226-250	–
227.5-272.0	251-300	–
273.0-317.5	301-350	2
318.5-363.0	351-400	3
	Total	271

ACID-LINED FURNACES

Capacity Range		Number of Furnaces
(Metric Tons)[b]	(Net Tons)[c]	
27.0-36.0	30-40	1
37.0-45.5	41-50	1
46.0-54.5	51-60	–
55.5-63.5	61-70	1
	Total	3

[a] Based on information in "Iron and Steel Works Directory of the United States and Canada" published by the American Iron and Steel Institute, Washington, D.C., July 1980.
[b] Calculated to nearest one-half metric ton
[c] Nominal capacities

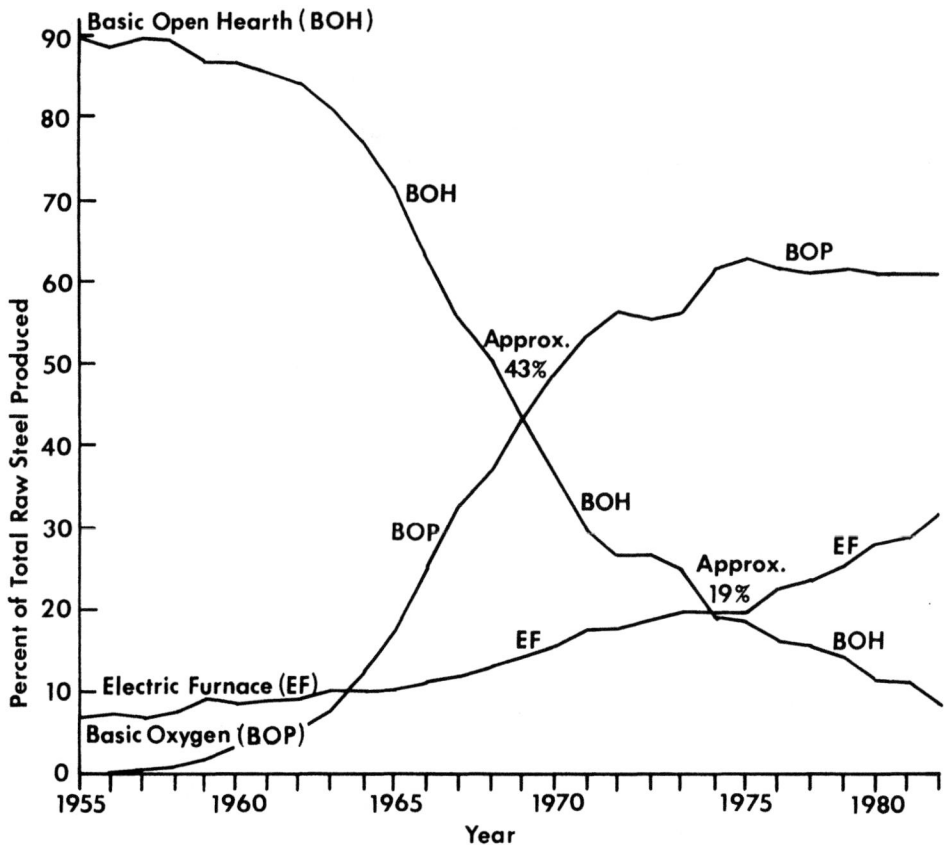

FIG. 1—24. Proportions of total annual raw-steel production by the three principal steelmaking processes in the United States, since the introduction of the basic oxygen process in 1955.

hearth furnaces in the United States in 1980, classified into the capacity ranges into which the capacities of individual furnaces fall are given in Table 1—I and 1—II, respectively.

Table 1—III lists the number of electric direct-arc steelmaking furnaces in the United States in 1980.

Raw-steel production by the three principal steelmaking processes in the United States in 1981 was as follows: the basic oxygen process 66.4 million metric tons (73.2 million net tons), the basic open-hearth process 12.2 million metric tons (13.5 million net tons), and the electric-furnace processes 30.9 million metric tons (34.1 million net tons).

The largest basic oxygen process furnaces can produce heats of steel ranging up to about 270 metric tons (300 net tons) of raw steel in about three-quarters of an hour. The basic open-hearth process can produce heats of steel of up to about 545 metric tons (600 net tons) in about 10 hours, and smaller heats in about 5 hours.

The basic open-hearth process and the basic oxygen process produce carbon and alloy steels of the same general grades. The basic oxygen process has also been used to make stainless steels, which cannot be made in the open-hearth. Electric furnaces are used to produce stainless- and alloy-steel grades which cannot be produced in basic open-hearth furnaces and to produce carbon steels where there is no supply of hot metal (liquid pig iron from blast furnaces). Although the acid-Bessemer process was the least expensive method for

producing steel, it fell into disuse because it could produce only a very limited variety of marketable steels.

The basic oxygen process has the same advantage of speed as did the acid-Bessemer process, with none of the chemical limitations of the latter. It is also capable of producing the same general grades of steel as the basic open-hearth process. The basic oxygen process does not possess the flexibility of the open-hearth or electric-furnace processes with regard to the amount of scrap that can be charged and melted. Between 1977 and 1981, inclusive, scrap made up an average of about 27.5 per cent of the charge of the basic oxygen furnaces operating in the United States, according to the Annual Statistical Reports of the American Iron and Steel Institute for those years. The open-hearth charge usually contains 45 to 55 per cent of scrap, but larger amounts (up to 100 per cent) can be used. The electric furnace charge usually consists entirely of scrap. In the basic oxygen process, preheating of the scrap charge either outside of the furnace before charging or inside the furnace before adding molten pig iron, preheating of the lime addition, or the use of liquid or gaseous fuel injected during at least part of the oxygen-blowing operation through lances of special design which also carry the oxygen supply through separate orifices, can increase the overall heat supply and permit the use of larger amounts of scrap. Exothermic substances such as calcium carbide can be added to generate additional heat during blowing with oxygen. However, any of the methods mentioned increase blowing time or cost or both.

The production of heats of steel at regular, comparatively short intervals is an advantage of the basic oxygen process when it is used to supply molten steel for the continuous-casting process.

The speed of the reactions in the basic oxygen process creates problems in control that are more critical than in the slower open-hearth and electric-furnace processes. It has been found necessary to make automatic as many operational functions as possible, and to develop accurate and very fast-acting instruments for measuring temperatures and for analyzing gases.

As was the historical case with both open-hearth and electric furnaces, the trend in basic oxygen steelmaking has been generally toward the use of larger furnaces in the newer installations.

Electric-arc furnaces until relatively recently were considered chiefly for the production of alloy steels. However, since the middle 1960's, something over 60 per cent of all electric furnace production has been carbon steel. Total steel production in electric furnaces more than quadrupled between 1960 and 1981 (about 7.7 million metric tons or 8.5 million net tons in 1960, compared with 30.9 million metric tons or 34.1 million net tons in 1981). Electric-arc furnaces possess the advantages of relatively low investment cost, and the ability to (1) produce steels of a wide range of compositions (carbon, alloy and stainless), (2) make heats smaller than the full capacity of the furnace, (3) take advantage of changing costs of scrap and pig iron, and (4) produce quality steels without a source of hot metal. Improvements in furnace design and auxiliary equipment have led to the use of larger furnaces with high output rates. Oxy-fuel burners to preheat the scrap charge in the furnace and the use of oxygen for rapid decarburization of the melted bath have decreased power consumption and increased output by decreasing the amount of time required to produce a heat of steel. Electric-furnace steelmaking can be combined readily with vacuum-degassing and continuous-casting operations.

Bibliography

Agricola, G., De re metallica. Translated from the first Latin edition of 1556 by Herbert Clark Hoover and Lou Henry Hoover, N. Y., Dover, 1950.

Aitchison, Leslie, A history of metals. MacDonald & Evans, Ltd., Great Britain; published in the U.S. by Interscience Publishers, Inc., N. Y., 1960.

Biringuccio, V., Pirotechnia. Translated from the Italian by C. S. Smith and M. T. Gnudi (first edition 1540). N.Y., Am. Institute of Mining and Metallurgical Engineers, 1943.

Beck, L., Geschichte des Eisens, Braunschweig, Germany, Vieweg, 1891-1903 (5 vols.)

Dilley, D. R. and McBride, D. L., Oxygen steelmaking—fact vs. folklore. Iron and Steel Engineer 44, October 1967, 131-152.

Forbes, R. J., Metallurgy in antiquity. Leiden, Netherlands, Brill, 1950.

Lightner, M. W., Developments in ironmaking and steelmaking. Blast Furnace and Steel Plant 54, November 1966, 1023-1027.

McBride, D. L., Progress in open-hearth steelmaking. Blast Furnace and Steel Plant 54, October 1966, 943-951.

McBride, D. L., Changes in ironmaking and steelmaking. Iron and Steel Engineer 43, January, 1966, 109-112.

Percy, John, Metallurgy: Iron and Steel. London, England, John Murray, 1864.

Rickard, T. A., Iron in antiquity. Journal of the Iron and Steel Institute (London) 120, 323-342 (1929).

Smith, Cyril Stanley, The discovery of carbon in steel. Technology and Culture 5, No. 2, Spring 1964, 149-175. Published by Wayne State University Press, Detroit, Mich., for the Society for the History of Technology.

Speer, Edgar B., The changing open-hearth. Iron and Steel Engineer 39, March 1962, 71-79.

CHAPTER 2

Steel-Plant Refractories

SECTION 1

CLASSIFICATION OF REFRACTORIES

Refractories are the primary materials used by the steel industry in the internal linings of furnaces for making iron and steel, in vessels for holding and transporting metal and slag, in furnaces for heating steel before further processing, and in the flues or stacks through which hot gases are conducted. At the risk of over-simplification, they may, therefore, be said to be materials of construction that are able to withstand temperatures from 260° to 1760°C (500° to 3200°F).

Refractories are expensive, and any failure in the refractories results in a great loss of production time, equipment, and sometimes the product itself. The type of refractories also will influence energy consumption and product quality. Therefore, the problem of obtaining refractories best suited to each application is of supreme importance. Economics greatly influence these problems, and the refractory best suited for an application is not necessarily the one that lasts the longest, but rather the one which provides the best balance between initial installed cost and service performance. This balance is never fixed, but is constantly shifting as a result of the introduction of new processes or new types of refractories. History reveals that refractory developments have occurred largely as the result of the pressure for improvement caused by the persistent search for superior metallurgical processes. The rapidity with which these ever-recurring refractory problems have been solved has been a large factor in the rate of advancement of the iron and steel industry. To discuss the many factors involved in these problems and to provide information helpful to their solution are the objectives of this chapter.

Refractories may be classified in a number of ways, no one of which is completely satisfactory. From the chemical standpoint, refractory substances, in common with matter in general, are of three classes; namely, acid, basic, and neutral. Theoretically, acid refractories should not be used in contact with basic slags, gases or fumes, whereas basic refractories can be best used in contact with a basic chemical environment. Actually, for various reasons, these rules are continually violated. Hence, the time-honored chemical classification is largely academic, and of little value as a guide to actual application. Also, the existence of a truly neutral refractory may be doubted. Classifications by use, such as blast-furnace refractories or refractories for oxygen steelmaking, are generally too broad and are constantly subject to revision.

For our purposes, refractories will be classified with reference to the raw materials used in their preparation and to the minerals predominating after processing for use. This classification is believed to offer the best possibility for a clear understanding of the origin and nature of steelplant refractories.

A. MAGNESIA OR MAGNESIA-LIME GROUP

This group includes all refractories made from natural or synthetic magnesites, brucite, and dolomite. These constitute the most important group of refractories for the basic steelmaking processes. All these materials are used as a source of magnesia (MgO).

Synthetic Magnesia—Magnesia (periclase) derived synthetically from sea water or brine represents the single most important refractory raw material used in modern steelmaking facilities. Several steps are necessary in the production of dense synthetic magnesia as briefly outlined below:

(1) $\quad MgCl_2 \;+\; Ca,Mg(OH)_2 \;\rightarrow\; Mg(OH)_2 \;+\; CaCl_2$
sea water   slaked   magnesium   salt
or brine   dolomite   hydroxide   waste

(2) $\quad Mg(OH)_2 \quad \overset{\triangle}{\rightarrow} \quad MgO$ (light weight)
heat to
982°C
(1800°F)

(3) $\quad MgO \quad \overset{\triangle}{\rightarrow} \quad MgO$ (dense)
heat to
1926°C
(3500°F)

The dense magnesia produced generally has a purity of 95 to 99 per cent MgO depending on the specifics of processing and the desired end-use. As shown above, magnesia (MgO) is recovered both from the sea-water and slaked dolomite. The important density of the final product is achieved by high-temperature firing in special shaft kilns and by mechanical briquetting of a calcine of high surface area prior to briquetting. The extreme importance of using prefired refractory raw materials from which essentially all the permanent shrinkage or expansion has been removed is obvious as materials with excessive in-service shrinkage or expansion could not be expected to contain molten metals or slags to a suitable degree. Large plants for the production of synthetic magnesia (periclase) are operated worldwide and in the United States from brine wells in Michigan and on sea-water locations in Florida, Texas, California, and Maryland.

Natural Magnesite—Some of the magnesia used is derived from natural sources such as brucite [$Mg(OH)_2$] sources in Greece and Nevada. These sources are limited in quantity and often require mineral beneficiation to remove undesirable impurities such as silica (SiO_2). Other less pure deposits of natural magnesium carbonate ($MgCO_3$) are also employed to produce lower quality grades of magnesia for refractories. In either case, high-temperature firing is again required to produce materials with suitable density for refractory applications.

Dolomite—The natural double carbonate ($CaCO_3 \cdot MgCO_3$) dolomite occurs in relative abundance in the United States in Illinois, Ohio, and Pennsylvania and can be converted to a useful refractory raw material by high-temperature firing in rotary or shaft kilns. Selective mining and beneficiation and mechanical densification after calcining can be used to produce a dense dolomite material containing 98 to 99 per cent basic oxide (CaO plus MgO) with minimum flux contents (primarily Fe_2O_3, Al_2O_3, and SiO_2). As will be later described in more detail, the dolomite is always subject to some degree of atmospheric deterioration from hydration of the lime constituent. For this reason, special manufacturing methods and packaging for shipment and storage must be employed in the use of dolomite refractories.

B. MAGNESIA-CHROME GROUP

Naturally occurring chrome ores consist of a highly refractory spinel ($RO \cdot R_2O_3$) composed of MgO, FeO, Al_2O_3, Cr_2O_3, and Fe_2O_3 in various proportions, in association with less refractory silicates. Chrome ores of widely different compositions are suitable for refractory purposes. Most of the suitable refractory grades are obtained from the Philippines and South Africa. Certain chrome ores must be beneficiated to decrease the gangue (primarily silica) content before use. Chrome ore is used primarily in combinations with magnesia in refractory products to combine the best characteristics of each material. Chrome ore requires no prefiring before use.

C. SILICEOUS GROUP

Quartzite—Quartzite or ganister is the most commonly used, and the purest, of the siliceous raw materials. Massive rock forms analyzing over 98 per cent SiO_2, such as have long been used for silica-brick manufacture, are found in Pennsylvania, Wisconsin, Alabama, Utah, and California. For coke-oven use, silica brick still are made largely from quartzite ganister. The technology is available to produce higher purity silica by washing quartz pebbles and pebble conglomerate.

Sandstone—Sandstone, or firestone, is a sedimentary rock consisting essentially of bonded sand grains, and usually containing 90 to 96 per cent SiO_2, 3 to 5 per cent Al_2O_3, and some iron oxide and lime. It is relatively soft and often striated, thus permitting easy cutting or splitting into blocks or shapes for use in the raw state.

Fused Silica—High-purity silica may be electrically fused to produce non-crystalline or cryptocrystalline fused silica as an aggregate with special characteristics for relatively low-temperature refractories.

Zircon and Zirconia—Zircon refractories ($ZrO_2 \cdot SiO_2$) are produced from special zircon sands from Australia or Florida following flotation and magnetic concentration. Stabilized zirconia (ZrO_2) is produced from the same zircon sand by electrical fusion with additions to reduce out the silica and other impurities.

D. FIRECLAY GROUP

Chemically, clays are all hydrous silicates of alumina and occur in widely scattered geographical areas. They are identified as being plastic when in a wet and finely divided state (as obtained by pulverizing, wetting, and mixing), rigid when dried, and vitreous when heated to a sufficiently high temperature. Clays may be residual or sedimentary, and have been formed by the natural decomposition or weathering of feldspathic rock. Ordinary clays contain high proportions of combined water and impurities that render them unfit for use as high-temperature refractories. Impurities include alkalies, titania, compounds of iron, calcium, and magnesium, and organic matter from various sources. Even in small quantities the alkalies, and the compounds of iron, calcium, and magnesium are important fluxes and have a pronounced effect on the refractory properties of the clay. The species used as a refractory is known as fireclay. Fireclays are of sedimentary origin, are usually associated with coal deposits, and contain limited percentages of fluxing impurities. There are several varieties, namely:

Siliceous Fireclays—While the term siliceous fireclays might properly refer to clays having a rather wide range in silica content, reference here is to those clays with a minimum of 75 per cent SiO_2 which are employed in the manufacture of semi-silica brick, and are characterized by a very low percentage of impurities such as alkalies, alkaline earths, and iron oxides.

Plastic Fireclay—A fireclay of sufficient natural plasticity to bond nonplastic materials.

Flint Fireclay—A hard or flint-like fireclay occurring as an unstratified massive rock, practically devoid of natural plasticity and showing a conchoidal fracture.

Nodular Fireclay—Also called burley or burley flint clay, nodular fireclay occurs in the form of a rock containing aluminous or ferruginous nodules, or both, bonded by fireclay.

Kaolins—While not fireclays, certain kaolins are highly refractory and are being increasingly employed in the manufacture of firebrick. Kaolins are both sedimentary and residual, and quite pure, generally closely approaching the theoretical clay composition represented by the formula $Al_2O_3 \cdot 2SiO_2 \cdot 2H_2O$.

As will be later described, fireclay refractories are usually made from a combination of pre-fired (calcined) clays and raw or unfired clays.

E. HIGH-ALUMINA GROUP

This group includes those materials for the production of refractories of higher alumina content than the maximum that can be provided by fireclays, namely, more than about 44 per cent Al_2O_3. There are several such sources with varying alumina content as listed below.

Bauxitic Kaolins—Certain domestic kaolin deposits in Georgia and Alabama contain materials from which 50 to 70 per cent Al_2O_3 raw materials may be produced by selective mining and beneficiation. These products are very low in undesirable impurities (alkali and iron oxide) and are used widely in refractories. In recent years, modern facilities have been provided for calcining these bauxitic kaolins to dense, stable materials.

Bauxite—Bauxite, which is composed primarily of the minerals diaspore ($Al_2O_3 \cdot 3H_2O$) and gibbsite ($Al_2O_3 \cdot H_2O$), is found in Arkansas, Alabama, and Georgia, but most of the calcined bauxite used in refractories comes from South American sources with new sources in China recently being used. These materials are again used in calcined form and contain 86 to 90 per cent Al_2O_3.

Sillimanite, Andalusite, and Kyanite—These minerals all have the chemical formula $Al_2O_3 \cdot SiO_2$, and theoretically contain 62.9 per cent Al_2O_3 and 37.1 per cent SiO_2. On heating, all form mullite ($3Al_2O_3 \cdot 2SiO_2$) and a siliceous glass, but differ in the ease with which this decomposition takes place, kyanite being the easiest to convert at about 1325°C (2415°F) and sillimanite the hardest at about 1530°C (2785°F). Kyanite from Virginia and North Carolina has been used extensively in domestic refractories in recent years in either the raw or calcined form.

High-Purity Alumina—Essentially pure alumina (Al_2O_3) in dense form may be produced by sintering or fusing calcined alumina made from aluminum nitrate obtained from bauxite by the Bayer process. Although very expensive, alumina materials can impart special properties to refractories when used in pure form or in combination with the aforementioned clays, bauxites, or other refractory raw materials. Alumina may be pre-reacted with pure silica to produce mullite aggregates or in-situ in the brick during processing.

F. CARBON GROUP

This group includes natural and artificial graphites, and various types of coal, coke, silicon carbide, and silicon nitride. Graphite deposits occur widely distributed, both in this country and abroad. Because it often occurs mixed with calcareous or siliceous rock, it is expensive to purify. Flake graphite, as found in Ceylon and Madagascar, is preferred for crucible and stopper-head manufacture, in which the graphite is bonded with large fractions of clay. Both amorphous and flake graphites are used to enhance the slag-resistant properties of many types of refractories in combination with other refractory materials.

Carbon brick and block are used extensively as a refractory and may be made from foundry coke, petroleum coke, or calcined anthracite coal. Pitches can also be used in such refractories as binders. Silicon carbide is made by fusing petroleum cokes and silica sands in high-temperature electric furnaces. Silicon carbide can be used in pure form or as additions to impart specific properties to other refractories (with fireclay, high-alumina, or carbon refractories).

SECTION 2

PREPARATION OF REFRACTORIES

Each of the refractory types previously described is used in a variety of forms produced by several different manufacturing methods. Included are the solid brick or block forms and the granular monolithic materials or moldable forms. Figure 2—1 is a greatly simplified flowsheet illustrating the various methods of refractory manufacture and the resultant products. This figure will be used to broadly describe this subject.

A. RAW MATERIALS

Refractory raw materials have been described in the previous section. Figure 2—1 classifies raw materials as either calcined, uncalcined (raw), or binders. Calcined materials have been fired to remove moisture and volatiles and to densify the material to minimize subsequent in-service shrinkage and reaction. The calcining temperature will range from 1093° to 1925°C (2000° to 3500°F). Raw or uncalcined materials are cheaper to use than calcined materials and are used to impart desirable characteristics such as plasticity or volume expansion (bloating) to certain refractories. Binders are used to impart strength to the refractory during manufacture or in service. Binder types include:

(a) Temporary binders such as paper by-products, sugar, or certain clays, to improve handling strength during manufacture.

(b) Chemical binders to impart strength during manufacture, after manufacture, or on installation as a monolithic material. Examples include sodium silicate, phosphoric acid, glass phosphates, chromic acid, boric acid, and magnesium sulphate.

(c) Cement binders which set hydraulically when mixed with water. The primary binders of this type used in refractories are the calcium-aluminate cements which set rapidly and are able to retain some of their bonding strength to intermediate temperatures.

(d) Organic binders such as tars, pitches, or resins for use in reducing atmospheres where the carbon residuals impart bonding strength or act to inhibit alteration.

The raw-material processing before refractory manufacture has a very important influence on the composition and properties of the final product. Although some raw materials need only be mined and sized be-

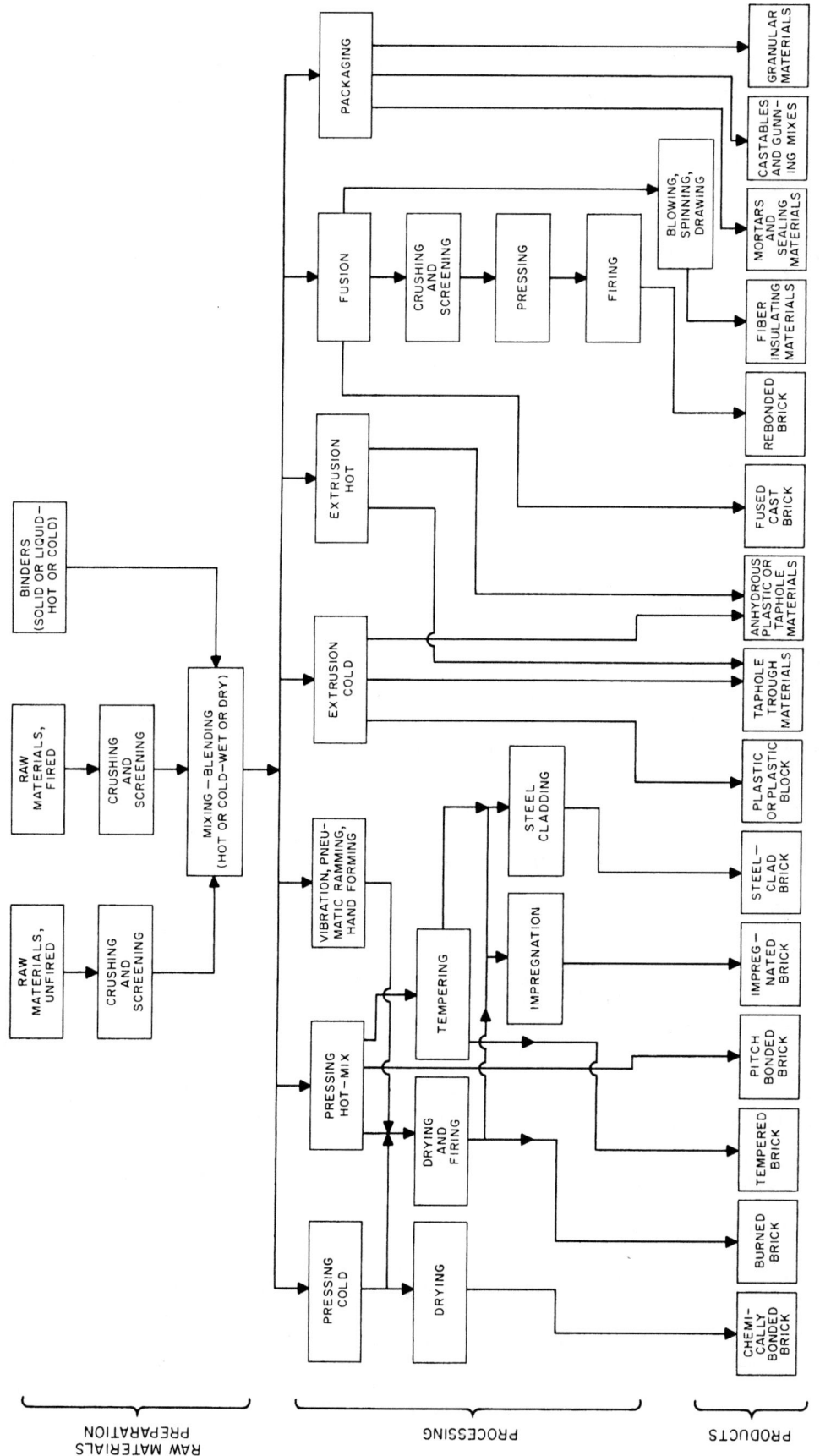

FIG. 2—1. Simplified flowsheet for refractory materials.

fore use, modern refractories use an increasing proportion of raw materials which involve selective mining, beneficiation, and various forms of thermal pretreatment as essential steps to providing desirable levels of purity and density.

B. PROCESSING

A knowledge of the various steps in refractory processing is very important in understanding the behavior of steelplant refractories. As shown in Figure 2—1, all refractories use crushing, sizing, and mixing or blending as the first steps in their manufacture. These steps are necessary to produce the proper particle sizings necessary for the desired product density and strength. In sizing a mix to produce dense brick, for example, the raw materials are crushed and screened to produce some desired particle size range as illustrated below:

Screen Size		Weight Per Cent of Mix
mm	Tyler Mesh Size	
on 4.6 through 1.65	on 4 through 10	20
on 1.65 through 1.17	on 10 through 14	30
on 1.17 through 0.83	on 14 through 20	5
on 0.83 through 0.30	on 20 through 48	5
on 0.30 through 0.15	on 48 through 100	5
on 0.15 through 0.07	on 100 through 200	5
on 0.07 through 0.04	on 200 through 325	10
Below 0.04	Below 325	20

The crushing and screening techniques used are increasingly complex, including vibratory screening equipment and air-classification techniques.

Mixing and blending steps range from the simple addition of water to clay, to hot mixing of preheated aggregates with selected pitch or other anhydrous binders. Special sequences of combining raw materials, the time of mixing, and the use of high-energy mixing equipment are used to obtain uniform mixing and equal distribution of additives.

Figure 2—1 illustrates some (but not all) of the forming methods that are used after mixing and blending. The most widely used manufacturing method involves cold pressing of the grain-sized and blended mix to produce a dense refractory shape. Power or "dry" pressing of the mix into a shape is done on hydraulic or mechanical presses capable of forming the moist material (2 to 5 per cent water) at pressures of 34.5 to 103.4 MPa (5 to 15 ksi). The degree of compaction obtained in this pressure range depends on plasticity and particle sizing, but most high-quality brick are pressed to the point where further pressure would produce laminations or internal cracking. The pressing chamber or mold may be evacuated or deaired to increase density and prevent laminations from entrapped air. Dry pressing lends itself to a wide variety of materials and can produce a wide range of properties. Certain products may be pressed hot and these materials are usually plasticized with liquid pitch.

Shapes may also be formed by applying pressure by other means such as vibration, pneumatic ramming, hand molding, or isostatic pressing. Many brick and special shapes in fire clay compositions are also formed by extrusion followed by low-pressure pressing (the stiff-mud-repress process). In this process, more-plastic mixes (10 to 15 per cent water) are forced through a die by a power-driven auger, cut into slugs, and then pressed to shape. This process usually involves deairing during extrusion.

Hot extrusion of pitch or other anhydrous bonded materials may also be used; however, this is mainly for monolithic materials.

In limited cases, raw materials are fused in very high-temperature electric furnaces and cast into larger ingots in graphite molds. These ingots can subsequently be cut to the desired shape, or may be broken and crushed into a refractory raw material for use in conventional power-pressed brick or for use in monoliths. In still another process, molten refractory may be blown, drawn, or spun into fibers for subsequent use in forming mats, blankets, or boards.

Many refractory materials are used in bulk form. Sized, granular refractories may be used in dry form or mixed with water at the plant site before installation by casting or gunning. Wet extruded material may be packaged to avoid drying and shipped to the plant site ready for application by ramming into place as a large monolithic structure. Wet bonding mortars may be shipped in sealed containers ready for use.

As shown in Figure 2—1, many products are prefired before shipment. The purpose of firing is to produce dimensionally stable products having specific properties. Firing in modern refractory plants is accomplished in continuous or tunnel kilns.

In tunnel-kiln firing, which is usually preceded by tunnel drying, the unfired brick loaded on small cars are passed slowly through a long tunnel-shaped refractory-lined structure, divided successively into preheating, firing, and cooling zones, generally taking 3 to 5 days for the trip. This time will vary widely, however, with the product being fired. Products of combustion from the fuel burned in the firing zone pass into the preheating zone (countercurrent to the direction of travel of the cars onto which the brick are stacked) and give up their heat to the oncoming loads of brick. Some refractories are also fired in batch or periodic kilns where 2 to 4 week cycles are used for heating, cooling, and loading and unloading kilns.

Temperatures of firing are important regardless of the type of kiln used, because both the quality and properties of the brick may be affected. The final properties and behavior of most brick can be modified by firing them in an oxidizing or a reducing atmosphere. By controlling the rate of heating and the maximum soaking temperature and soaking time, change in the crystalline structure can be effected, which in turn can also affect the service performance of the brick. In general, the objectives in firing are to (a) drive off hygroscopic, combined water, and CO_2; (b) bring about desired chemical changes such as oxidizing iron and sulphur compounds, and organic matter, etc.; (c) effect transformations of the mineral constituents and convert them to the most stable forms; and (d) effect necessary combinations and vitrification of bonding agents. Firing temperatures vary from as low as 1093°C (2000°F) for certain fireclay materials to over 1770°C (3200°F) for some basic products.

Certain refractories with carbon binders or containing oxidizable constituents may be indirectly fired in-

Table 2—I. Common Types of Refractories Used in Iron- and Steelmaking.

Refractory Type	Common Refractory Types in Indicated Raw Material Groups					
	Magnesia or Magnesia-Lime	Magnesia-Chrome	Siliceous	Fireclay	High Alumina	Carbon
Chemically bonded brick	x	x	—	—	x	—
Pitch-bonded brick	x	—	—	—	—	—
Burned brick	x	x	x	x	x	x
Impregnated brick....................	x	—	—	—	x	x
Steelclad brick	x	x	—	—	—	—
Plastic or plastic block................	—	—	x	x	x	x
Taphole and trough materials	—	—	x	x	x	x
Anhydrous plastic or taphole materials ..	—	—	x	x	x	x
Fused-cast brick	—	x	—	—	x	—
Rebonded brick	—	x	—	—	—	—
Fiber insulation materials.............	—	—	—	x	x	—
Granular materials	x	—	—	x	x	—
Castables and gunning mixes	x	x	x	x	x	—
Mortars and sealing materials	x	—	x	x	x	x

side muffles to prevent oxidation or may be packed in coke or graphite during firing for the same purpose. One grade of carbon refractory is hot pressed by electrically heating it during pressing. This accomplishes the forming and thermal treatment of the refractory in a single step. Nitrogen or other special atmospheres may be used to impart special binding phases such as silicon nitride.

Manufacturing processes for making lightweight or insulating brick aim for high porosity, preferably with fine pore structure. This is accomplished by mixing a bulky combustible substance, like sawdust or ground cork, or a volatile solid, such as napthalene, with the wet batch, by forcing air into the wet plastic mass, or by mixing into the batch reagents which will react chemically to form a gas and a product not injurious to the brick. In firing such brick, the combustible or volatile material is eliminated and the remaining refractory structure is rigidized. Low density, preexpanded aggregate may also be used to make products by conventional brick-making methods.

Some processing after the fired brick are produced may also be performed. For example, the brick may be steelcased for use in applications where oxidization of the steel case between brick serves to "weld" or hold

the brick together. Many brick types are also impregnated by placing the fired brick into vacuum tanks and introducing liquid pitch or resins into the brick pore structure. This treatment results in formation of a carbon phase in service which has highly beneficial effects in some applications.

C. PRODUCTS

Figure 2—1 also illustrates some of the typical products made by the procedures described. The refractory types described generally reflect the bonding system (chemically bonded, burned, pitch-bonded) of the final refractory or its final form (brick, plastic, gunning mix, etc.). These terms will be further explained under the examples of materials used later in this chapter. Each of the refractory types in Figure 2—1 might be further broken down according to the raw materials used in manufacture. Table 2—1 illustrates the most common types of each product which might be produced from the previously described raw-materials groups. Note that some refractory types (burned brick, castables, and gunning mixes) are produced in all compositions, whereas others are produced in only a few compositions.

SECTION 3

CHEMICAL AND PHYSICAL CHARACTERISTICS OF REFRACTORIES AND THEIR RELATION TO SERVICE CONDITIONS

The foregoing discussions have indicated that there is a wide variety of refractories from the standpoint of raw materials, overall composition, and method of manufacture. The requirements for refractories are equally diverse. Analysis of service conditions in iron- and steelmaking in general shows that refractories are required to withstand:

(1) A wide range of temperature, up to 1760°C (3200°F).

(2) Sudden changes in temperature; high tensile stresses accompanying these rapid temperature changes cause "thermal shock" and result in cracking or fracturing.

(3) Low levels of compressive stresses at both high

and low temperatures.

(4) Abrasive forces at both high and low temperatures.

(5) The corrosive action of slags, ranging from acid to basic in character.

(6) The action of molten metals, always at high temperatures and capable of exerting great pressures and buoyant forces.

(7) The action of gases, including CO, SO_2, Cl, CH_4, H_2O, and volatile oxides and salts of metals. All are capable of penetrating and reacting with the refractory.

(8) As a refractory is being subjected to one or more of the previously stated conditions, it usually functions as a highly effective insulator, or may also be required to be a conductor or absorber of heat depending on its application.

Since any particular service environment usually involves more than one of the above factors, predetermining the life of a refractory is a complex process involving information on physical and thermal properties as determined by laboratory testing, analysis of the effect of service or process conditions and media on the refractory, and a knowledge of the fundamental reactions between refractories and the various contaminants encountered in service. In this section the physical and chemical characteristics of selected refractories as measured in a variety of laboratory tests will be described as a general guide to understanding the complex nature of these materials in relation to their service environments.

Chemical Composition—As described in the section on refractory classification, the raw materials used in making refractories differ appreciably and result in materials with a wide range of compositions as illustrated in Table 2—II. It must be emphasized that these are unaltered refractories before use and not refractories that have been chemically changed in service. Refractories have the unique ability to withstand alteration by penetration, contamination, and/or reaction in service and still function as reliable engineering materials. The next section will describe the reactions between refractories and their environments. As a general rule, however, recent trends in refractory development require refractories with the minimum content of impurities, and these impurities are deliberately decreased during raw material or product processing. The following describes the undesirable constituents (originally present or from contamination) in several types of refractories. It should be noted that many other impurities (such as PbO, ZnO, B_2O_3, etc.) which are undesirable in all refractories because of their low melting points have not been shown.

Refractory Type	Undesirable Impurities
Silica	Al_2O_3, alkali, TiO_2
Fireclay—all types	alkali, iron oxide, CaO, MgO
High-alumina—all types	alkali, iron oxide, CaO, MgO
Magnesia-chrome—all types	SiO_2, iron oxide
Magnesia—all types	SiO_2, Al_2O_3, iron oxide
Carbon	alkali, iron oxide

Refractoriness—The maximum usable temperature for refractory materials may be measured by several techniques. The fusion temperature of most refractory materials is generally not unique, but represents a more or less gradual transition from solid to liquid. The amount of liquid that can be tolerated in a refractory as it is heated, without rendering it useless, is largely governed by the viscosity of the liquid and whether or not the liquid is primarily responsible for bonding the more-refractory particles together. For example, fireclay refractories may develop liquid and actually start to soften as low as 980°C (1800°F), but due to the high viscosity of the liquid their limiting service temperature may be several hundred degrees higher. An arbitrary procedure has, therefore, been established for gaging the refractoriness of such materials, and is called the Pyrometric Cone Equivalent, or PCE test, in which the softening behavior of small cones of the refractory is compared with that of standard pyrometric cones with a known softening behavior for a given time-temperature heating rate. The PCE is reported as the number of the standard cone that has a softening behavior like that of the cone of the refractory being investigated when tested in accordance with the Standard Method of Test for Pyrometric Cone Equivalent (ASTM designation: C 24). The composition of the standard cones and the temperature range covered make the PCE test most applicable to measuring the refractoriness of alumina-silica refractories. Figure 2—2 gives the PCE of various refractories and the corresponding softening temperatures when heated under the test conditions. It must be recognized that the end points of pyrometric cones primarily reflect the influence of time and temperature, and hence are reproducible only under identical conditions. A cone used in a kiln fired for a week may soften completely at a tempera-

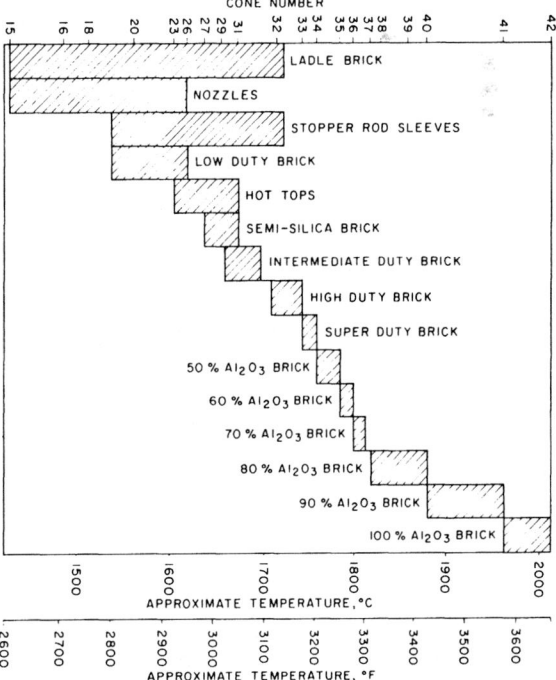

FIG. 2—2. Pyrometric cone equivalent (PCE) of various alumina-silica refractories.

ture which would leave it unaffected in the short PCE test. Fireclay refractories behave similarly, the influence of time being strikingly shown in the case of ladle brick. Normally, maximum service temperatures of fireclay brick are considerably below PCE temperatures, but low-PCE ladle brick are successfully used at temperatures 150°C (300°F) above their PCE temperatures since exposure is seldom for more than an hour at a time.

A more realistic measurement of refractoriness for most refractory materials can be obtained by placing the refractory under load at temperature and determining its behavior. This test may consist of a simple procedure where a refractory is loaded in compression and its deformation or temperature of failure measured, or it may be a more complex procedure involving the measurement of creep during long-time exposure of the refractory at some lower temperature. Figure 2—3 shows a typical creep-type test in which two 60 per cent Al_2O_3 refractories have been held at 1315°C (2400°F) for 100 hours under a 172-kilopascal (25 psi) load. Using such data it would be possible to extrapolate the long-time creep resistance of these materials. and the lower-alkali brick would obviously be preferred if the application required minimum deformation. In applying such information, the refractory engi

neer must also consider the presence of any temperature gradient through the sample that may exist in service when he uses these results. For example, these results would have considerably less meaning where only the hot face of the refractory is at 1315°C (2400°F) with a gradient to ambient conditions than where the refractory was entirely at a temperature of 1315°C (2400°F). The behavior of alumina-silica refractories under load is influenced primarily by the viscosity of the liquids in the bonding or matrix phase and is therefore very sensitive to impurity level in the brick and to factors controlling the amount and distribution of this glassy phase (microstructure, degree of burn or prefiring, etc.).

Still another widely used technique to determine the useful application temperature of a refractory is to measure the fracture strength versus temperature relationship. Figure 2—4 shows the strength of test specimens in three-point loading for two high-alumina refractories over a range of temperature. The hot-strength advantage of the low-alkali type product is again obvious. The hot strength of refractories measured in this manner is highly dependent on the bonding phase as illustrated in Figure 2—5 for various basic refractories. This figure shows the strength-temperature relationship for various periclase-chrome brick.

Table 2—II. Typical Chemical Compositions of Steelplant Refractories.

Refractory Type	Chemical Composition (Per Cent)								
	SiO₂	Al₂O₃	Cr₂O₃	MgO	CaO	Fe₂O₃	TiO₂	Alkali	C
Silica Brick									
High-Purity	96.4 to 97.4	0.1 to 0.4	—	0.1 to 0.2	2.2 to 3.4	0.3 to 2.1	0.1	0.1	—
Coke-Oven	95.5 to 96.5	0.6 to 1.2	—	0.1 to 0.2	2.0 to 3.5	0.4 to 0.7	0.1	0.1 to 0.4	—
Sandstone	89.6 to 96.1	1.8 to 2.8	—	—	0.1 to 1.4	0.5 to 1.6	0.2 to 0.5	0.6 to 1.2	—
Fused-Silica	99.6 to 99.8	0.1 to 0.2	—	—	0.0 to 0.2	0.0 to 0.2	—	0.1 to 0.2	—
Semi-Silica Brick	69.6 to 79.0	18.0 to 26.9	—	0.1 to 0.4	0.1 to 0.4	0.6 to 2.0	0.8 to 1.6	0.2 to 0.4	—
Fireclay Brick or Shapes									
Stopper-Rod Sleeves	58.1 to 61.6	30.0 to 34.0	—	0.2 to 0.6	0.3 to 0.4	2.7 to 3.1	1.7 to 1.8	2.4 to 2.6	—
Steel-Teeming Nozzles	52.4 to 62.3	29.0 to 36.9	—	0.9 to 1.0	0.3	2.8 to 4.3	1.6 to 1.9	3.0 to 3.3	—
Ladle Brick; Bloating Type	60.0 to 61.3	28.0 to 30.8	—	0.5 to 0.8	0.2 to 0.5	2.3 to 5.3	1.3 to 1.5	3.6 to 4.2	—
Ladle Brick; Volume-Stable Type .	53.4 to 61.0	30.6 to 40.6	—	0.4 to 0.6	0.2 to 0.4	1.1 to 2.7	1.4 to 2.4	0.9 to 4.2	—
Low Duty	53.0 to 69.0	25.0 to 34.0	—	0.4 to 0.6	0.3 to 0.6	2.3 to 3.4	1.0 to 2.0	1.8 to 2.9	—
Intermediate Duty	56.0 to 70.0	25.0 to 36.0	—	0.5 to 0.6	0.2 to 0.4	1.8 to 3.4	1.3 to 1.9	1.0 to 2.7	—
High Duty	51.0 to 59.0	35.0 to 40.0	—	—	0.3 to 0.5	1.6 to 2.5	2.0 to 3.0	1.5 to 2.6	—
Super Duty	50.2 to 54.0	40.0 to 46.0	—	—	0.1 to 0.5	0.8 to 2.3	2.1 to 2.5	0.2 to 1.4	—
High-Alumina Brick									
50% Al₂O₃	43.0 to 47.0	47.0 to 51.0	—	0.5 to 0.6	0.5 to 0.6	0.9 to 1.6	2.2 to 2.4	0.8 to 1.3	
60% Al₂O₃	27.7 to 37.0	58.0 to 67.0	—	0.1 to 0.6	0.1 to 0.3	0.9 to 2.7	1.7 to 3.0	0.2 to 1.2	—
70% Al₂O₃	19.4 to 28.0	68.0 to 76.7	—	0.1 to 0.2	0.1 to 0.3	0.9 to 2.2	2.0 to 3.3	0.2 to 1.2	—
80% Al₂O₃	8.5 to 17.1	78.0 to 86.5	—	0.1 to 0.2	0.1 to 0.4	0.7 to 1.7	2.5 to 3.2	0.1 to 0.6	—
90% Al₂O₃	3.0 to 10.0	87.5 to 95.8	—	0.0 to 0.2	0.1 to 1.9	0.2 to 1.1	0.1 to 2.6	0.2 to 0.9	—
100% Al₂O₃	0.4 to 1.1	97.7 to 99.0	—	0.0 to 0.1	0.1 to 0.2	0.1 to 0.3	0.0 to 0.3	0.1 to 0.3	—
Mullite	12.4 to 29.3	67.5 to 86.5	—	0.0 to 0.1	0.1 to 0.4	0.1 to 1.4	0.1 to 2.8	0.2 to 0.6	—
Fused-Cast	0.1 to 1.3	93.5 to 99.5	—	—	0.1 to 0.3	0.1 to 0.2	—	0.3 to 4.0	—
Fireclay or High Alumina Monoliths									
Fireclay Plastics	50.0 to 55.0	40.0 to 45.0	—	—	0.2 to 0.5	1.0 to 2.5	2.0 to 3 0	0.5 to 1.6	—
1095°C (2000°F) Castables	44.0 to 74.0	20.0 to 40.0	—	—	5.4 to 16.5	3.1 to 6.8	1.4 to 2.8	0.6 to 2.0	—
1230°C (2250°F) Castables	38.0 to 46.0	30.0 to 40.0	—	—	7.6 to 14.0	2.0 to 5.2	1.5 to 2.1	0.6 to 1.5	—
1480°C (2700°F) Castables	39.0 to 45.0	40.0 to 52.0	—	—	4.2 to 6.5	0.5 to 3.6	1.6 to 2.6	0.6 to 1.2	—
1650°C (3000°F) Castables	2.0 to 32.0	54.0 to 96.0	—	—	2.5 to 5.9	0.1 to 1.3	0.1 to 2.6	0.1 to 0.4	—
60% AlO, Plastics or									
Ramming Mixes	28.1 to 41.3	54.1 to 66.0	—	0.1 to 0.4	0.1 to 0.3	0.3 to 2.1	0.7 to 2.8	0.2 to 1.7	

(Continued on next page)

Table 2—II. Typical Chemical Compositions of Steelplant Refractories (Continued)

Refractory Type	Chemical Composition (Per Cent)								
	SiO_2	Al_2O_3	Cr_2O_3	MgO	CaO	Fe_2O_3	TiO_2	Alkali	C
70% Al_2O_3 Plastics or Ramming Mixes	5.8 to 26.2	66.0 to 75.5	—	0.1 to 0.3	0.1 to 0.4	0.9 to 2.4	1.6 to 4.1	0.2 to 2.6	—
80% Al_2O_3 Plastics or Ramming Mixes	7.2 to 17.2	75.8 to 85.1	—	0.0 to 1.3	0.3 to 2.8	0.3 to 2.8	1.6 to 3.1	0.0 to 1.2	—
90% Al_2O_3 Plastics or Ramming Mixes	0.0 to 11.0	72.3 to 95.0	2 to 8	0.0 to 0.2	0.0 to 5.4	0.0 to 0.9	0.2 to 2.5	0.1 to 0.8	
100% Al_2O_3 Plastics or Ramming Mixes	0.0 to 0.3	97.7 to 99.8	—	—	—	0.1 to 0.5	—	0.0 to 0.3	—
Mullite Plastics or Ramming Mixes	23.8 to 36.4	59.7 to 72.9	—	0.1 to 0.3	0.1 to 0.2	0.5 to 1.2	0.2 to 2.4	0.2 to 1.7	—
Basic Brick									
Magnesite; Unburned or Fired	0.4 to 4.5	0.1 to 1.0	0.1 to 0.9	91.7 to 98.3	0.6 to 3.8	0.1 to 2.3	—	—	—
Magnesite-Chrome; Unburned	1.8 to 8.3	5.5 to 14.1	6.1 to 15.0	49.5 to 82.0	0.8 to 1.8	2.3 to 8.4	—	—	—
Magnesite-Chrome; Fired	1.8 to 4.4	2.5 to 10.0	6.7 to 17.1	60.8 to 80.7	1.0 to 1.5	2.3 to 11.5	—	—	—
Magnesite-Chrome; Direct Bonded	1.0 to 2.5	6.5 to 10.5	10.3 to 18.1	58.8 to 65.5	0.6 to 1.1	4.3 to 10.5	—	—	—
Magnesite-Chrome; Rebonded	1.5 to 3.0	6.6 to 8.4	16.5 to 18.5	60.0 to 64.2	0.6 to 1.1	9.6 to 10.7	—	—	—
Magnesite-Chrome; Fused-Cast	2.0 to 3.0	6.9 to 8.1	17.5 to 18.5	58.8 to 62.0	0.8 to 1.1	9.5 to 11.9	—	—	—
Chrome-Magnesite; Unburned	3.5 to 5.0	17.5 to 22.4	18.8 to 24.4	37.0 to 49.0	0.7 to 1.4	8.8 to 11.5	—	—	—
Chrome-Magnesite; Fired	3.4 to 5.1	8.8 to 26.7	18.1 to 24.5	29.5 to 52.5	0.7 to 1.4	8.5 to 14.6	—	—	—
Chrome; Fired	4.9 to 8.3	27.3 to 29.1	29.9 to 32.8	18.4 to 22.5	0.5 to 0.7	12.0 to 14.6	—	—	—
Spinel Bonded; Unburned or Fired	1.7 to 1.9	10.0 to 15.9	0.0 to 0.9	76.5 to 87.0	0.9 to 1.4	0.6 to 9.3	—	—	—
Basic Monolithic Materials									
Dead-Burned Natural Magnesite	4.4 to 8.1	1.0 to 2.8	0.1 to 0.4	72.5 to 87.5	2.1 to 7.2	3.8 to 14.3	—	—	—
Dead-Burned Dolomite	0.3 to 6.7	0.1 to 2.3	—	32.5 to 43.0	45.5 to 56.9	0.3 to 10.2	—	—	—
Bottom Ramming Materials	0.6 to 3.2	0.2 to 1.4	0.0 to 0.8	92.1 to 98.4	0.7 to 1.2	0.2 to 0.9	—	—	—
Bottom Patching Materials	4.3 to 13.2	0.5 to 8.4	0.1 to 7.8	47.0 to 88.5	0.8 to 42.4	1.3 to 13.0	—	—	—
Plastic Chrome Ore	5.6 to 14.8	11.9 to 35.8	24.0 to 42.4	11.2 to 23.5	12.0 to 21.1	0.1 to 0.6	—	—	—
Pitch-Bearing Basic Refractories									
Low-Flux Dolomite Type	0.3 to 1.0	0.1 to 0.6	0.1 to 0.3	38.0 to 43.0	50.0 to 57.0	0.2 to 1.1	—	—	4 to 5
Dolomite; Magnesite Type	0.5 to 1.5	0.2 to 0.6	0.0 to 0.3	60.0 to 66.0	35.0 to 38.0	0.3 to 1.5	—	—	4 to 5
Magnesite Type	0.5 to 3.8	0.2 to 1.0	0.0 to 0.4	88.1 to 98.1	0.9 to 6.2	0.1 to 3.0	—	—	2 to 5**
Magnesia—Carbon Type	0.5 to 2.0	0.2 to 0.3	0.0 to 0.1	94.0 to 99.0	0.7 to 2.2	0.1 to 0.8	—	—	5 to 20***
Carbon Brick									
Graphite Base	1.0 to 2.5*		—	—	—	—	—	—	96 to 98
Anthracite Base	5.0 to 10.0*		—	—	—	—	—	—	88 to 94
Anthracite Base, Hot Pressed	15.0 to 20.0*		—	—	—	—	—	—	79 to 84
Zircon+		38.0 to 42.0	0.8 to 1.0	—	—	0.1 to 0.4	—	—	
Zirconia+		0.3 to 0.4	—	—	2.0 to 3.0	0.1 to 0.2	—	—	

*Al_2O_3 plus SiO_2.

**Burned-impregnated type 2 to 3 per cent carbon; pitch-bonded or tempered type 4 to 5 per cent carbon.

+ZrO_2 in Zircon is 50 to 52 per cent and in Zirconia is 95 to 97 per cent.

***Type of carbon ranging from amorphous to higher crystalline flake types.

The brick fired at a higher temperature (1720°C or 3100°F) produce higher levels of hot strength than those burned at lower temperature (1550°C or 2800°F). The reasons for these differences in strength will be described in more detail later in this chapter.

Figure 2—6 shows the hot-strength relation at 1480°C (2700°F) for two general types of periclase brick. The brick with lime-to-silica ratios between 2 and 3 to 1 and low levels of Fe_2O_3 + Al_2O_3 have high hot strengths. By proper control of raw materials and processing, brick with the desired level of hot strength may be produced.

Historically, the useful temperature limit of refractories has also been determined by measuring the perma-

nent growth or reheat expansion. Although it is not currently used for most refractories, this technique is still of importance for evaluating "bloating" type brick which expand in service due to the entrapment of sulphur gases in a pyro-plastic clay body. While normally indicative of overheating and melting, bloating in certain applications is desirable to form tight linings for handling metal or where controlled erosion may be desirable. Figure 2—7 illustrates the effect of bloating by a ladle brick on closing an open joint through the volume expansion of the brick and the development of a vesicular hot-face structure. This bloating originates at 1150° to 1200°C (2100° to 2200°F). It should be noted that any load on bloating materials at temperature will

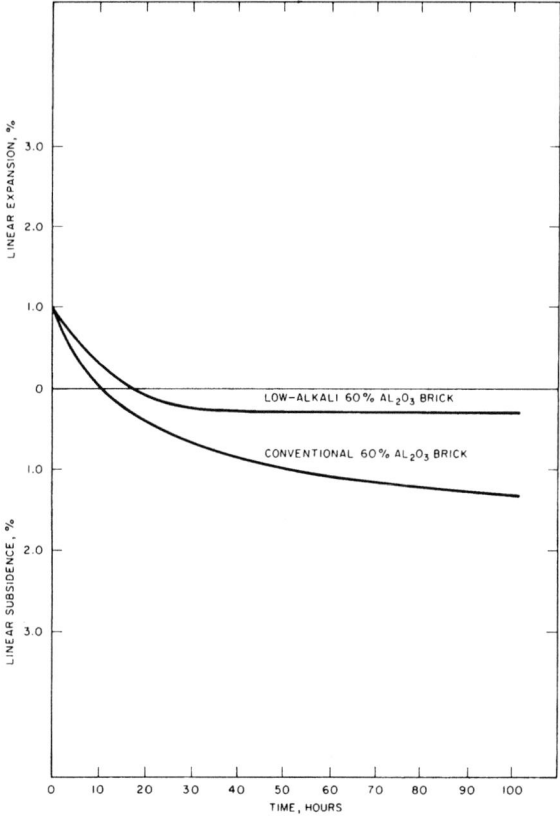

FIG. 2—3. 100-hour load tests at 1315°C (2400°F) and 172 kPa (25 psi) loading on indicated refractories.

cause them to deform plastically.

Thermal Shock—Thermal-shock resistance is perhaps the most misunderstood property of refractories and the most difficult to measure and predict. A severe thermal shock creates mechanical stresses which can cause refractories to fracture catastrophically. Thermal shock often occurs where a high heat flux creates a sharp temperature gradient within the refractory. As thermal expansion of any solid is a temperature-dependent property, sharp temperature gradients cause large thermal expansion differences to occur. Stress develops because of these differences in thermal expansion, and when the thermal stress exceeds the tensile strength of the refractory, failure or spalling usually occurs. The existence of a temperature gradient in a multiphase refractory is especially critical because each phase has unique thermal-expansion properties For example, the aggregate particles (large particles) and matrix of fireclay brick differ greatly in thermal expansion. The mullite aggregate grain has relatively low thermal expansion, whereas the glassy matrix is a relatively high-expansion material. This thermal-expansion mismatch causes the characteristic micro-cracks in the matrix of a fireclay refractory. In service, contaminants from the environment are absorbed and variably alter the microstructure, creating composition and property gradients in the refractory. This altered microstructure with composition and properties that vary with depth can cause structural-type spalling

which will be later described in more detail.

Materials with good resistance to thermally induced fracture generally have high strength, high thermal conductivity, and high thermal diffusivity, and low Young's modulus of elasticity, low Poisson's ratio, low coefficient of thermal expansion, and low emissivity. As steelplant refractories are frequently used under severe temperature gradients, the inherently low tensile strength of refractories may result in thermal cracking and spalling. Considerable interest, therefore, exists in predicting and improving thermal-shock resistance of steelplant refractories.

The classical method of measuring thermal-shock resistance of alumina-silica refractories is the panel-spalling test in which preheated panels of refractories are alternately heated and cooled with water sprays and the amount of weight loss by cracking and spalling measured. More recent empirical methods of predicting the spalling resistance of a refractory involve the measurement of the change in some measurable physical property (strength, sonic modulus) after samples are cycled many times through a temperature cycle designed to reflect specific service conditions. These same techniques are also used to develop refractory materials with improved levels of shock resistance for field trials.

Newer methods of predicting the thermal-shock resistance of refractories have been developed wherein the resistance to crack propagation rather than initiation is measured. These measurements, called "work of fracture", quantify the energy needed to form new surfaces during the fracture process in a constant-strain-rate three-point-bending test. Figure 2—8 illustrates the modes of refractory failure which may be obtained in such tests; the stable type behavior is preferred to avoid catastrophic failures due to shock. Measurements of work of fracture under specific tem-

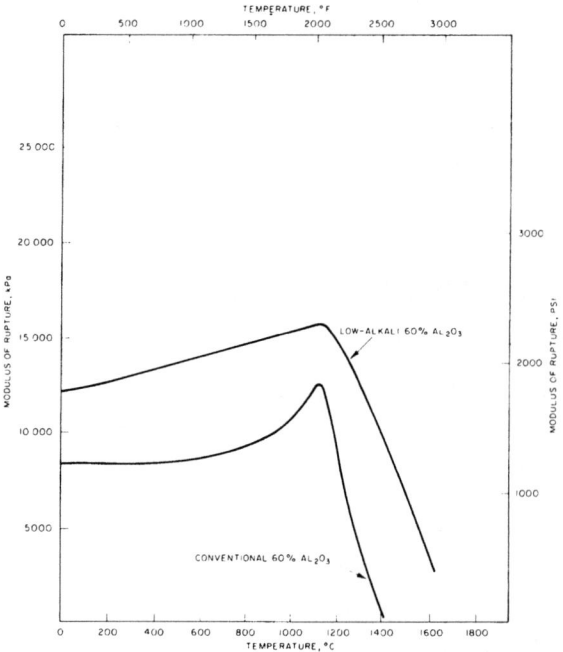

FIG. 2—4. Hot modulus of rupture of indicated refractories.

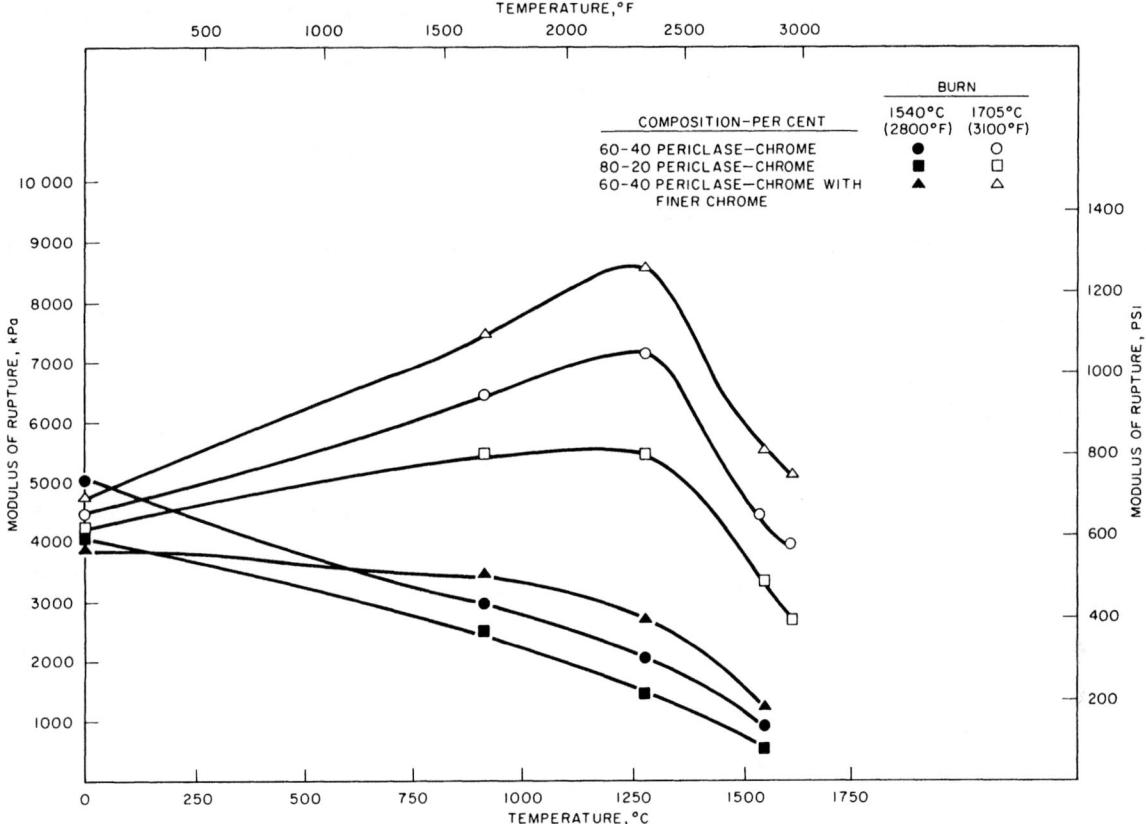

FIG. 2—5. Strength-temperature relationship for refractories of the indicated compositions.

perature and atmospheric conditions will eventually lead to new microstructures designed to resist crack propagation.

Strength—Refractories are subject to static loading,

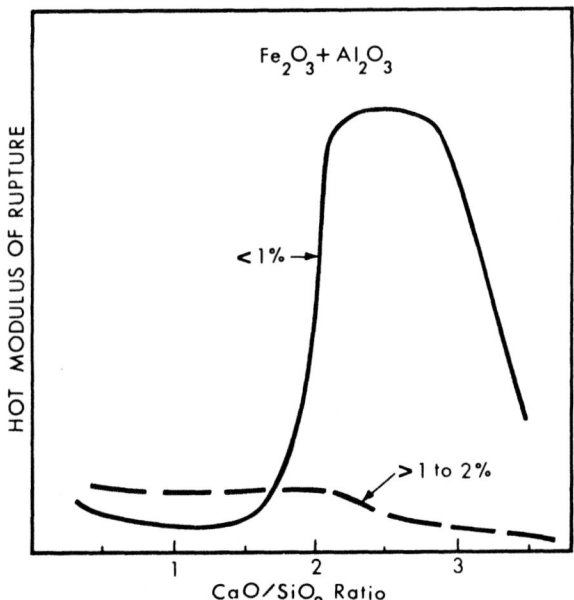

FIG. 2—6. Relation of strength at 1480°C (2700°F) to composition for burned periclase brick.

usually in compression, in any application. Because most refractories are handled a great deal before installation, they must also have sufficient strength to avoid breakage and corner loss. In service, monolithic refractories must develop sufficient strength to withstand static loads and mechanical abuse. In certain applications, refractories may also withstand impact loading. The strength of refractories is usually specified in any refractory specification for the above reasons; furthermore, because strength relates significantly to other properties, it is often used as a quality-control measure. As a rule, the cold strength of refractories is determined in compression or by the three-point loading for modulus of rupture in flexure.

Table 2—III lists typical cold and hot strengths for various refractories along with porosity and bulk-density data. In general, those refractories of a given type with higher density (lower porosity) also have the highest room-temperature strengths as illustrated in Figure 2—9.

Porosity and density measurements are another convenient method for characterizing the quality of refractories and are frequently used in refractory specifications and quality-control programs. The apparent porosity of a material indicates the fraction of the total volume which is open pore space, and hence is an indicator of the total area of surface available for reaction with slags and gases. The total porosity is the fraction of the total volume consisting of voids, both open and closed. Thus, depending on the nature of the material, method of manufacture, and degree of burn, it may slightly exceed the apparent porosity or it may be more than twice as great. The quantity, size, and distribution of pores in the microstructure of refractories play very important roles in establishing final engineering properties. The effect of porosity on specific properties will be discussed in more detail later, but, in general, increasing the amount of porosity adversely affects resistance to deformation under hot load and resistance to attack by both gases and slags. On the other hand, higher porosity generally decreases internal thermal conductivity.

Strength data on refractories must be interpreted carefully because there is no common relationship between refractory hot and cold strength. A classic case of the lack of any such relationship is illustrated in Figure 2—10 for refractory concretes (castables). In this case, the cold and hot strength values are inversely related as the liquid and glass formation which contributes to a high room-temperature strength actually causes a decrease in the strength at elevated temperatures. The reactions involved between refractory components and their effects on refractory properties will be covered more completely in the following section.

The type of loading to which a refractory is exposed must also be considered when using strength information. For example, loads of relatively low magnitude are experienced in a blast-furnace-stove checker setting, and temperature and load information can be used with hot-strength or creep information to select the refractories for any position in the checkers, Figure 2—11. In certain other applications, impact loading such as that which occurs during scrap charging must be considered and the exact strength required is largely unknown. In these cases, refractories are usu-

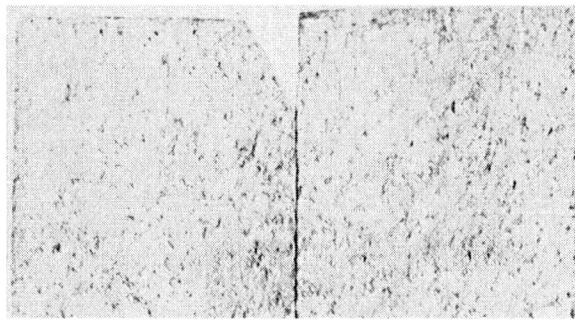

Appearance before test.

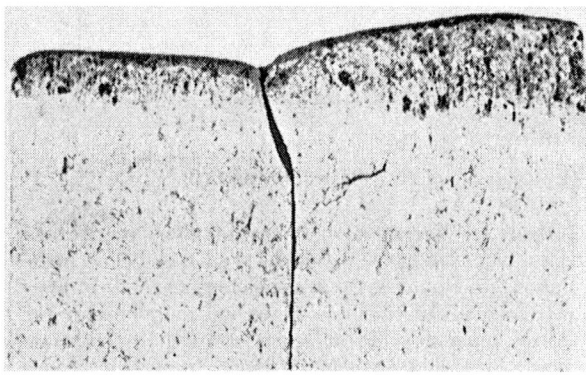

An all-clay bloating-type brick after the hot face was exposed to 1400°C (2550°F).

A brick with added bauxite (48% Al₂O₃) after the hot face was exposed to 1400°C (2550°F).

FIG. 2—7. Ladle brick with a large reheat expansion will bloat sufficiently to close the joints in a ladle lining, forming a monolithic surface even if relatively open joints or broken brick are present. The brick with less reheat expansion result in partially closed joints.

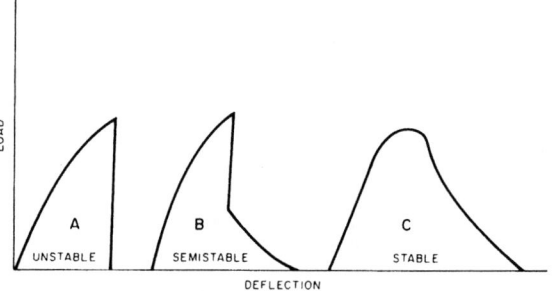

FIG. 2—8. Definition of stable fracture.

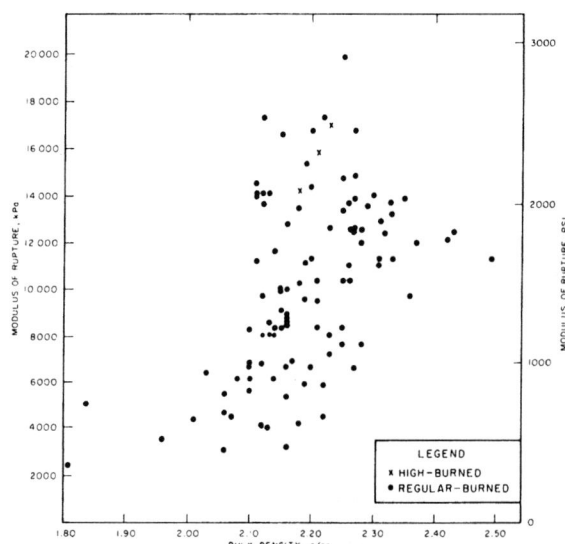

FIG. 2—9. Relation of modulus of rupture and bulk density of high-duty fireclay brick.

Table 2—III. Some Typical Physical Properties of Steelplant Refractories

Refractory Type	Apparent Porosity %	Bulk Density (g/cm³)	Bulk Density (lb/ft³)	Modulus of Rupture — At Room Temperature (kPa)***	At Room Temperature (psi)	At 1260°C (2300°F) (kPa)***	At 1260°C (2300°F) (psi)	At 1480°C (2700°F) (kPa)***	At 1480°C (2700°F) (psi)
Silica Brick									
High-Purity	19.2 to 26.2	1.72 to 1.89	107 to 118	3 210 to 7 480	465 to 1085	6 600 to 11 000	900 to 1600	5500 to 7600	800 to 1200
Coke Oven	21.0 to 30.0	1.70 to 1.86	106 to 116	2 760 to 10 340	400 to 1500	6 600 to 11 000	900 to 1600	6200 to 7500	900 to 1100
Sandstone	5.5 to 22.9	2.01 to 2.48	127 to 155	4 140 to 12 270	600 to 1780	1 030 to 12 270	150 to 1780	6200 to 7500	900 to 1100
Fused-Silica	12.0 to 21.5	1.60 to 2.27	100 to 141	2 620 to 10 690	380 to 1550	ND*	ND*	—	—
Semi-Silica Brick	23.4 to 30.8	1.77 to 2.05	110 to 128	2 210 to 4 140	320 to 600			—	—
Fireclay Brick or Shapes									
Stopper-Rod Sleeves	15.7 to 21.5	2.06 to 2.20	129 to 137	10 310 to 60 430	1495 to 8765**	—	—		
Steel-Teeming Nozzles	12.3 to 17.4	2.14 to 2.23	134 to 139	40 330 to 110 320	5850 to 16000**	—	—		
Ladle Brick; Bloating Type	9.1 to 17.7	2.14 to 2.39	134 to 149	7 270 to 18 960	1055 to 2750	—	—		
Ladle Brick; Volume-Stable Type	9.3 to 19.9	2.13 to 2.27	133 to 142	6 240 to 21 990	905 to 3190	—	—		
Low Duty	18.0 to 21.0	—	126 to 145	6 210 to 10 340	900 to 1500	—	—		
Intermediate Duty	18.0 to 21.0	—	128 to 140	6 210 to 11 030	900 to 1600	—	—		
High Duty	4.2 to 30.4	1.81 to 2.49	113 to 155	2 480 to 20 130	360 to 2920	—	—		
Super Duty	5.3 to 21.5	2.06 to 2.46	129 to 155	2 380 to 23 170	345 to 3360	5 520 to 12 960	800 to 1880		
High-Alumina Brick									
50% Al₂O₃	20.0 to 24.0	2.00 to 2.16	125 to 135	6 890 to 10 340	1000 to 1500				
60% Al₂O₃	13.0 to 28.4	2.07 to 2.56	129 to 160	3 930 to 20 240	570 to 2935	1 450 to 9 510	210 to 1380		
70% Al₂O₃	14.6 to 28.6	2.25 to 2.79	140 to 174	5 830 to 25 100	845 to 3610	2 480 to 9 930	360 to 1440		
80% Al₂O₃	14.3 to 28.7	2.45 to 2.97	153 to 185	4 650 to 20 510	675 to 4425	2 170 to 37 920	315 to 5500		
90% Al₂O₃	15.5 to 26.8	2.67 to 3.00	167 to 187	6 890 to 26 130	1000 to 3790	3 100 to 24 480	450 to 3550		
100% Al₂O₃	19.0 to 27.6	2.84 to 3.07	177 to 192	8 170 to 24 410	1185 to 3540	3 000 to 17 580	435 to 2550		
Mullite	12.0 to 24.1	2.31 to 2.66	146 to 166	8 170 to 25 170	1185 to 3650	2 450 to 27 580	355 to 4000		
Fused-Cast	0.9 to 3.1	3.18 to 3.50	194 to 218	24 130 to 41 370	3500 to 6000	16 550 to 26 200	2400 to 3800		
Fireclay or High-Alumina Monoliths									
Fireclay Plastics	20.1 to 26.8	1.96 to 2.22	122 to 139	210 to 4 170	30 to 605	—	100 to 500		
1095°C (2000°F) Castables	29.9 to 40.4	1.18 to 2.59	74 to 162	930 to 8 890	135 to 1290	—	—		
1230°C (2250°F) Castables	27.6 to 37.5	1.66 to 2.06	104 to 129	1 450 to 6 380	210 to 925	—	—		
1480°C (2700°F) Castables	28.4 to 37.8	1.71 to 2.15	107 to 134	1 240 to 10 100	180 to 1465	690 to 2 100	100 to 300		
1650°C (3000°F) Castables	26.9 to 34.7	1.87 to 2.77	117 to 173	720 to 9 070	105 to 1315	1 150 to 3 450	150 to 500		
High-Alumina Plastics or Ramming Mixes									
60% Al₂O₃	15.8 to 30.9	2.07 to 2.93	129 to 183	0 to 7 450	0 to 1080	1 690 to 7 340	245 to 1065		
70% Al₂O₃	17.4 to 30.3	2.33 to 2.65	145 to 165	170 to 8 100	25 to 1175	380 to 9 580	55 to 1390		
80% Al₂O₃	17.3 to 30.7	2.40 to 2.96	150 to 185	760 to 5 900	110 to 855	1 550 to 9 580	225 to 1390		
90% Al₂O₃	12.7 to 30.4	2.68 to 3.09	167 to 193	450 to 13 790	65 to 2000	2 210 to 24 130	320 to 2050		
100% Al₂O₃	14.8 to 23.8	2.14 to 3.02	134 to 189	0 to 15 240	0 to 2210	790 to 10 720	115 to 1555		

(Continued on next page)

Table 2—III. Some Typical Physical Properties of Steelplant Refractories (Continued)

Refractory Type	Apparent Porosity %	Bulk Density (g/cm³)	Bulk Density (lb/ft³)	Modulus of Rupture — At Room Temperature (kPa)***	(psi)	At 1260°C (2300°F) (kPa)***	(psi)	At 1480°C (2700°F) (kPa)***	(psi)
Basic Brick									
Magnesite; Unburned or Fired	15.2 to 23.8	2.70 to 2.95	168 to 185	10 860 to 24 170	1575 to 3505	590 to 18 270††	85 to 2650††	690 to 13900††	100 to 2000††
Magnesite-Chrome; Unburned	17.5 to 22.0	2.80 to 3.02	175 to 188	6 380 to 16 100	925 to 2335	960 to 3 030	140 to 440	690 to 2800	100 to 400
Magnesite-Chrome; Fired	15.9 to 19.6	2.88 to 3.16	180 to 197	3 930 to 6 070	570 to 880	760 to 2 070	110 to 300	690 to 1050	100 to 150
Magnesite-Chrome; Direct Bonded	15.5 to 19.0	2.90 to 3.10	182 to 193	4 140 to 8 960	600 to 1300	8 270 to 17 240	1200 to 2500	2100 to 6900	300 to 1000
Magnesite-Chrome; Rebonded	13.0 to 15.2	2.95 to 3.16	185 to 197	6 210 to 13 790	900 to 2000	13 790 to 17 930	2000 to 2600	2800 to 6900	400 to 1000
Magnesite-Chrome; Fused-Cast	12.0 to 16.2	3.03 to 3.21	190 to 201	13 790 to 17 240	2000 to 2500	10 340 to 20 680	1500 to 3000	3400 to 9500	500 to 9500
Chrome-Magnesite; Unburned	18.8 to 21.0	2.97 to 3.15	185 to 197	9 930 to 14 170	1440 to 2055	1 690 to 2 070	245 to 300	690 to 2100	100 to 300
Chrome-Magnesite; Fired	18.3 to 26.5	2.81 to 3.15	175 to 197	6 140 to 10 000	890 to 1450	2 900 to 11 070	420 to 1605	2800 to 5500	400 to 800
Chrome; Fired	16.6 to 18.5	3.16 to 3.40	197 to 212	9 650 to 16 720	1400 to 2425	450 to 1100	65 to 160	350 to 690	50 to 100
Spinel Bonded; Unburned or Fired	15.3 to 21.5	2.74 to 2.97	171 to 185	9 510 to 21 170	1380 to 3070	2 280 to 6 340	330 to 920	2100 to 2800	300 to 400
Basic Monolithic Materials									
Bottom Ramming Materials	3.7 to 21.7	2.69 to 2.83	168 to 177	7 620 to 12 480	1105 to 1810	720 to 2 480	105 to 360	—	—
Bottom Patching Materials	—	2.72 to 3.08	170 to 192	—	—	—	—	—	—
Plastic Chrome Ore	17.8 to 24.1	2.75 to 3.13	172 to 195	970 to 9 410	140 to 1365	690 to 2 070	100 to 300	—	—
Pitch-Bearing Basic Refractories									
Low-Flux Dolomite	—	—	178 to 185	6 210 to 10 340	900 to 1500	—	—	—	—
Dolomite; Magnesite Type	—	—	175 to 186	6 890 to 11 030	1000 to 1600	—	—	—	—
Magnesite Type	—	—	185 to 197	8 960 to 10 690	1300 to 1550	10 340 to 20 680†	1500 to 3000†	690 to 13900†	100 to 2000†
Carbon Brick									
Graphite Base	15.0 to 17.0	1.65 to 1.72	103 to 107	24 130 to 27 580	3500 to 4000	—	—	—	—
Anthracite Base	19.0 to 23.0	1.50 to 1.55	94 to 97	12 410 to 21 370	1800 to 3100	—	—	—	—
Anthracite Base; Hot Pressed	20.0 to 22.0	1.64 to 1.68	102 to 105	8 230 to 10 340	1200 to 1500	—	—	—	—

*ND = Not determined.

**Compressive strength.

***To convert to kg/cm², multiply by 0.010197.

†Values on burned-impregnated type only.

††High-strength brick are fired magnesite brick with high CaO/SiO₂ ratios.

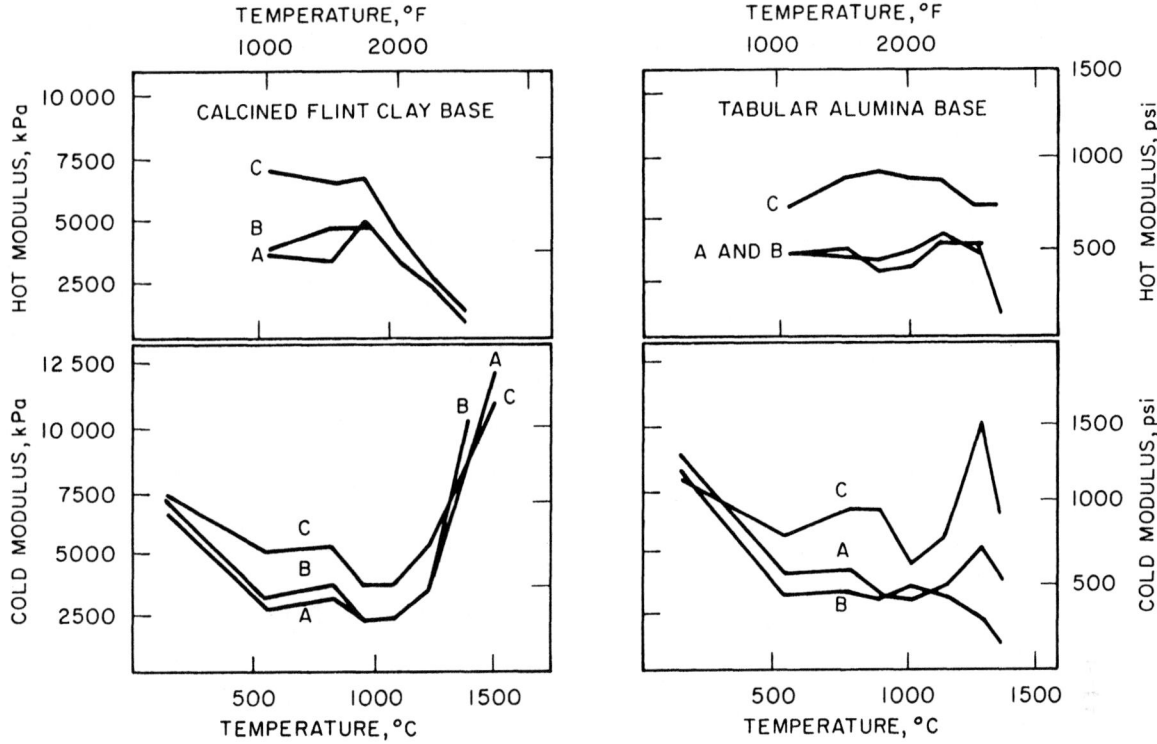

FIG. 2—10. Effect of temperature on properties of refractory castables. (Curves A, B, and C are each for a different cement.)

ally selected for service trials empirically based on measured levels of strength, and trial and error methods prevail. In still other approaches, required strengths may be estimated based on computer stress-analysis techniques. This is illustrated in Figure 2—12, where tensile stresses have been computed using finite element analysis for a special coke-oven-door shape exposed to a known temperature gradient. Such analysis can indicate if calculated stresses exceed material strengths and provide direction in decreasing in-

process stresses by changes in materials and/or designs.

Abrasion—Refractories are often subject to abrasion by moving solids, or by solids entrained in gases. It is well established that abrasion resistance as measured by a variety of different techniques can be related to cold strength of a material up to the point where significant liquids develop in the refractory. Most truly abrasive environments are encountered at a relatively low temperature (the abrasive media at higher temperatures becoming part of the process slag). Figure 2—13 illustrates the relationship of resistance to weight loss caused by abrasion in a grit-blast test to refractory strength for the type of castables often used to line duct

FIG. 2—11. Temperature and load distribution in checker setting for 1315°C (2400°F) top temperature.

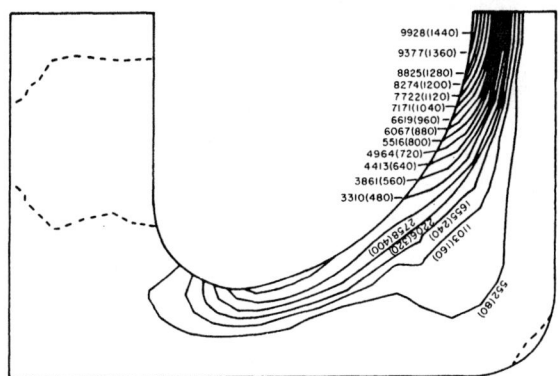

FIG. 2—12. Maximum tensile stresses in special coke-oven-door hole design for coefficient of expansion equal to 2.16 x $10^{-6} \cdot K^{-1}$ (1.2 x 10^{-6} in./in./°F) Stresses are expressed in kPa, followed by their equivalents in psi in parentheses.

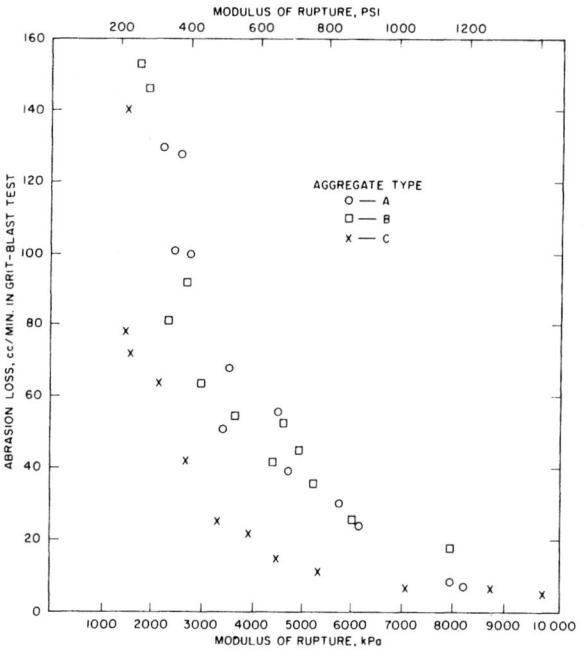

FIG. 2—13. Relation of modulus of rupture and abrasion loss for castables made from indicated aggregates and calcium-aluminate cement.

work carrying abrasive particles. The type of aggregate employed also has some lesser effect on abrasion resistance. In general, refractory materials have an extremely wide range of resistance to abrasion, as illustrated in Table 2—IV. In almost all cases, abrasion resistance is inversely related to resistance to thermal shock and the selection of materials for abrasion resist-

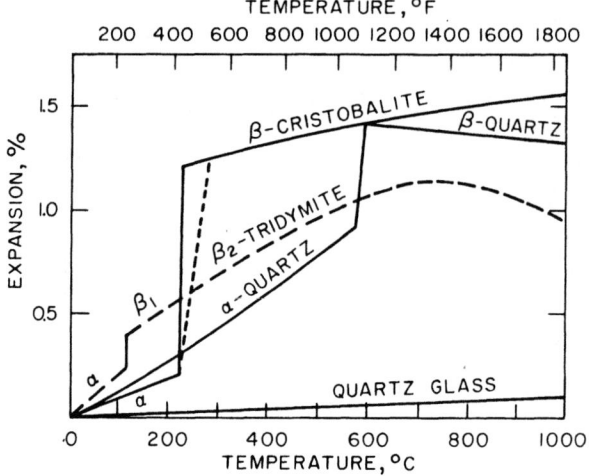

FIG. 2—15. Thermal expansion of silica minerals.

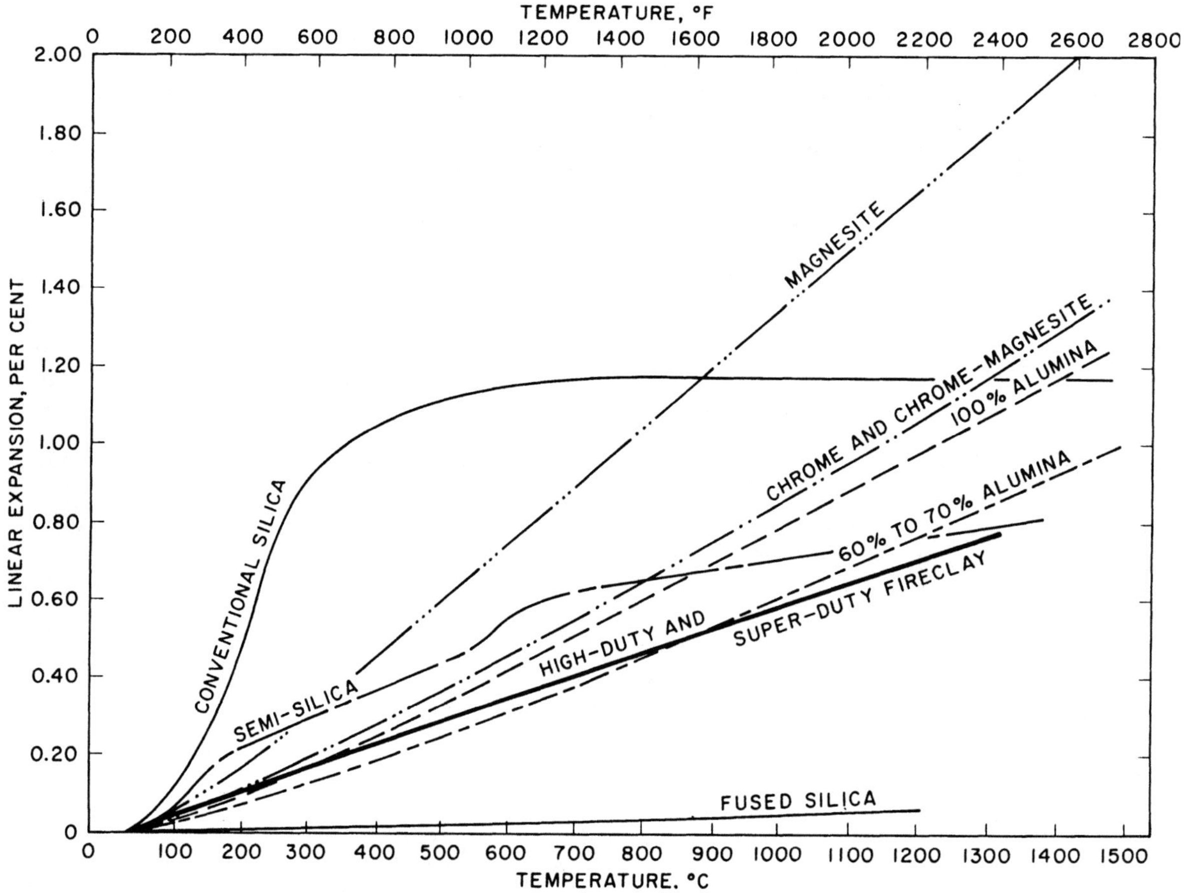

FIG. 2—14. Typical curves of linear expansion of various types of refractories.

Thermal Expansion—All refractories expand on heating in a manner related to their compositions. Figure 2—14 shows classic examples of the expansion of divergent refractory types. It is obvious that the shape of the expansion curve with temperature and the maximum magnitude of the expansions differ appreciably. Compressible materials or voids are used for relieving such thermal expansion in any refractory construction because growth of several inches can obviously be experienced. Refractories with linear or near linear thermal expansion generally require considerably less care

in heat-up than those that experience sudden expansion or contraction due to polymorphic inversions. A classic example of a refractory that requires care during heat-up is silica brick which after firing consists of a carefully controlled balance of various mineral forms of the compound SiO_2. Figure 2—15 shows the thermal expansion of the various silica mineral forms which make slow heat-up through various critical ranges necessary. As a result of the complex mineral makeup of silica brick, large constructions using these brick (such as coke ovens) are heated and cooled at very slow rates and may require several weeks to reach operating temperatures. Once the critical temperature ranges are

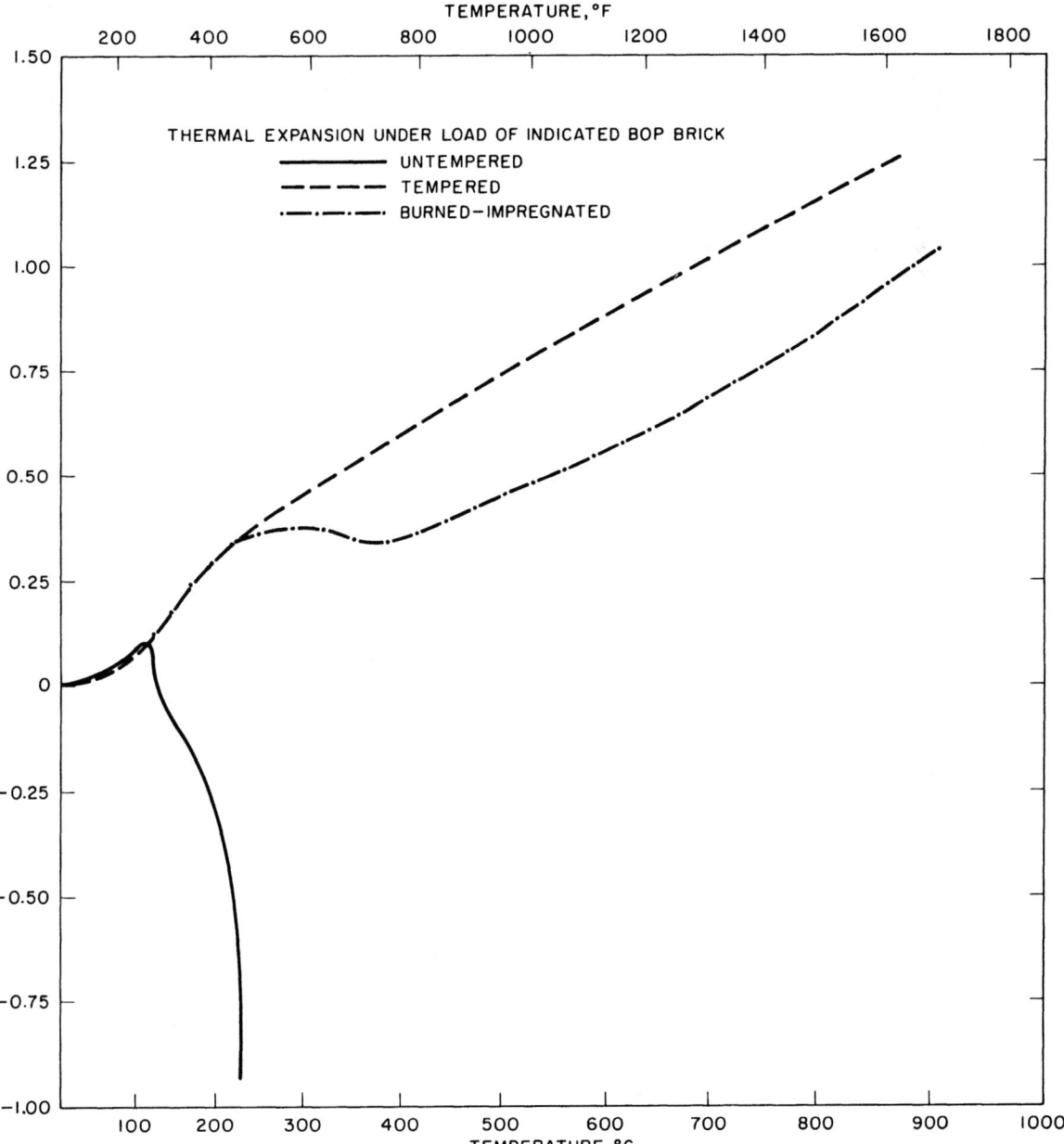

FIG. 2—16. Lineal dimensional changes with change in temperature for various BOP refractories under constant load.

passed, however, silica brick exhibit low and uniform thermal expansion and may be thermal-cycled without damage between 1000°F to 1538°C (1800° and 2800°F).

The physical makeup of the refractory being used is also critical in determining safe heating rates and expansion allowances. Figure 2—16 shows expansion versus temperature curves determined for various basic oxygen process (BOP) refractories under a constant load. Although these refractories have the same overall chemical composition, changes in processing result in considerably different yield behavior as the pitch and pitch-like binders coke during burn-in. Tempering (low-temperature baking) can reduce failure of the untempered brick and result in only a slight yield. A prefired pitch-impregnated brick with ceramic bonding shows no yield as the pitch in the brick pores cokes during heating. By properly controlling properties of BOP brick, it is possible to heat BOP linings from ambient conditions to steelmaking temperatures in 2 to 3 hours without lining damage.

Thermal Conductivity and Heat Transfer—As insulators, refractory materials have always been used to conserve heat, and their resistance to heat flow is a prime selection factor in many applications. Figure 2—17 shows thermal-conductivity curves for several refractory types ranging from dense refractories to insulating brick. Some refractories (for example, carbon or silicon-carbide) have appreciably higher conductivities (up to 43.26 watts per metre-kelvin or 300 Btu/h/ft²/°F/inch) whereas others are available with conductivities lower than 0.14 watt per metre-kelvin or 1 Btu/h/ft²/°F/inch (for example, block insulation or refractory fiber forms).

Using measured conductivity values, heat-transfer losses through single- or multiple-component refractory walls can be calculated using the general formula:

$$Q/A = \frac{t_1 - t_2}{L_1/K_1 + L_2/K_2 \ldots L_N/K_N}$$

where Q/A = heat loss expressed in watts per square metre (Btu/ft²/h in fps units)

t_1 = temperature of the hotter surface, °C (°F with fps units)

t_2 = temperature of the cooler surface, °C (°F with fps units)

$L_1, L_2 \ldots L_N$ = thickness of each material, metres (in. in fps units)

$K_1, K_2 \ldots K_N$ = thermal conductivity of each material, watts per metre-kelvin (Btu in. per square foot per hour per °F in fps units)

Such calculations are now rapidly made using computer simulations where calculated heat-transfer rates can be balanced with loss from the outer refractory surface by radiation and natural or forced convection. Moreover, where once it was a slow mathematical process to determine even the steady-state heat times, the advent of computers has made the rapid determination of the heat-transfer data for even transient conditions routine. Although it might seem that every construction should be designed for minimum heat losses, this is not always the case, and care must be taken in

Table 2—IV. Abrasion Resistance of Indicated Refractories in Grit-Blast Tests

Material	Abrasion Loss cc/min.
Abrasion-resistant castables	5–100
Normal fireclay brick	2–25
Acid-resistant brick	0.5–1.0
Dense fireclay brick	0.5–1.0
Silicon-carbide brick	0.1–0.3
Phosphate-bonded burned high-alumina brick	0.1–1.0
Fused slag or rock	<0.1–0.3
Fine-grained sintered alumina	<0.1

some situations including the following:

(1) The hot-face refractories in a particular application must be able to withstand the higher temperatures that will result when layers of highly insulating backup materials are added.

(2) The other refractory properties must be suitable for the environment. For example, most insulating ma-

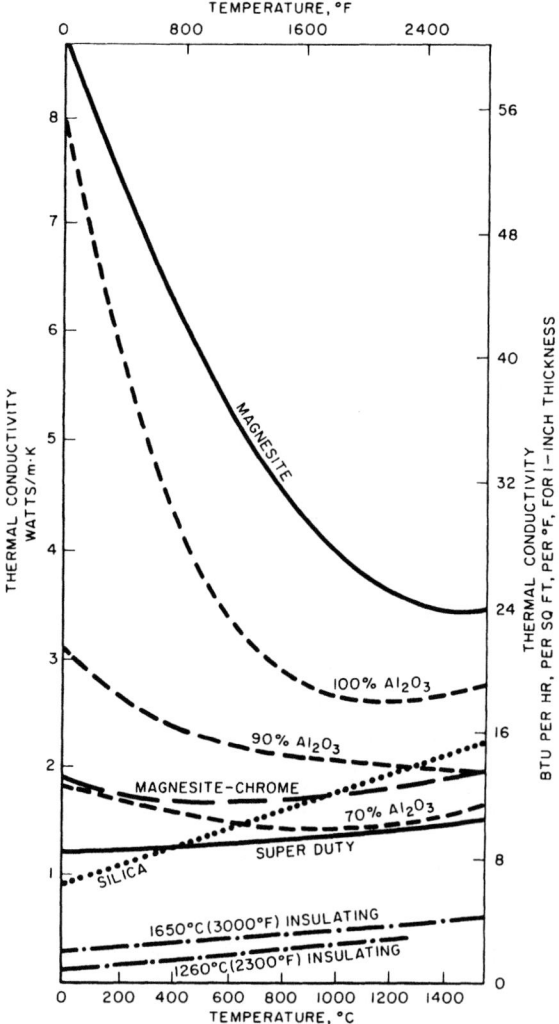

FIG. 2—17. Typical thermal conductivity curves for various refractory brick.

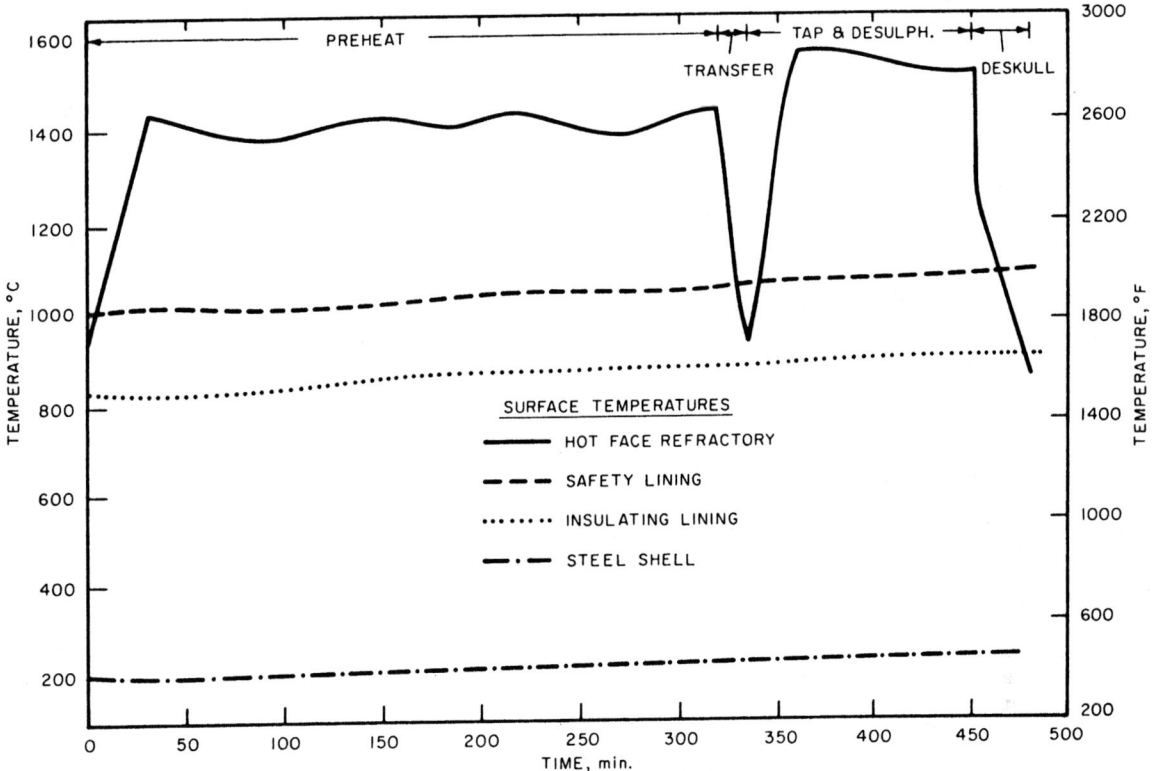

FIG. 2—18. Temperature profile of a steel ladle as a function of time.

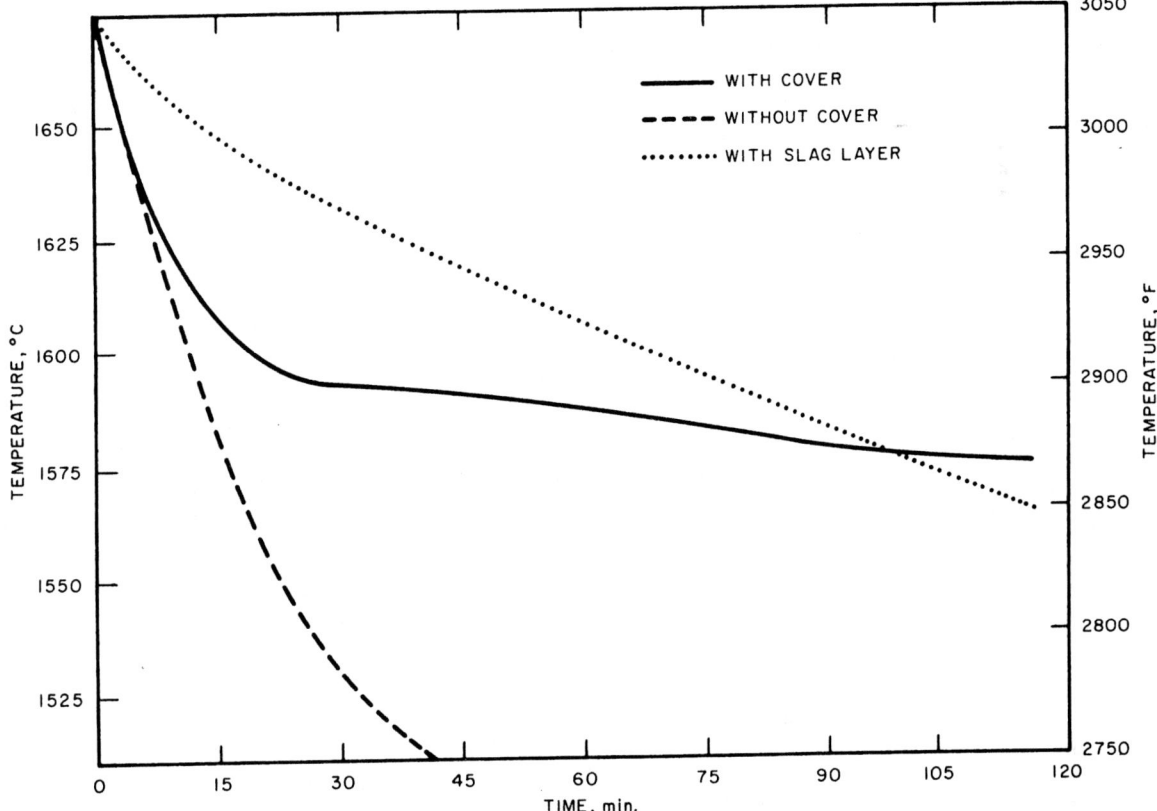

FIG. 2—19. Liquid-steel temperature as a function of time for ladles with cover, without cover, and with slag layer.

Table 2—V. Comparison of Insulative Properties of Refractory Fiber Forms with Other Materials.

Material Form	Bulk Density (kg/m³)	Bulk Density (lb/ft³)	Thermal Conductivity at Indicated Mean Temperature (watts/metre·kelvin) 260°C	540°C	815°C	1095°C	(Btu in./ft²·°F·h) 500°F	1000°F	1500°F	2000°F
Refractory Fibers	64 to 128	4 to 8	0.048 to 0.079	0.087 to 0.149	0.144 to 0.288	0.205 to 0.459	0.33 to 0.55	0.60 to 1.03	1.00 to 2.00	1.42 to 3.18
Calcium Silicate Fibers (870° to 1040°C or 1600° to 1900°F)	336 to 384	21 to 24	0.092 to 0.102	0.107 to 0.118	—	—	0.64 to 0.71	0.74 to 0.82	—	—
Insulating Firebrick 1095°C (2000°F)	561	35	0.140	0.176	0.212	—	0.97	1.22	1.47	—
1260°C (2300°F)	673	42	0.218	0.276	0.333	0.390	1.51	1.91	2.31	2.70
1425°C (2600°F)	769	48	0.277	0.320	0.364	0.407	1.92	2.22	2.52	2.82
1540°C (2800°F)	929	58	0.289	0.361	0.433	0.505	2.00	2.50	3.00	3.50
1650°C (3000°F)	1009 to 1073	63 to 67	0.447	0.462	0.483	0.519	3.10	3.20	3.35	3.60
Diatomaceous Material (870° to 1370°C or 1600° to 2500°F)	529 to 641	33 to 40	0.133 to 0.245	0.154 to 0.281	0.176 to 0.316	0.353	0.92 to 1.70	1.07 to 1.95	1.22 to 2.19	2.45
Lightweight Castables (1095° to 1425°C or 2000° to 2600°F)	465 to 977	29 to 61	0.130 to 0.231	0.159 to 0.252	0.216 to 0.289	0.260 to 0.375	0.9 to 1.6	1.1 to 1.75	1.5 to 2.0	1.8 to 2.6
Medium-Density Castables	1249 to 1410	78 to 88	0.346 to 0.462	0.303 to 0.779	0.390 to 0.736	0.433 to 0.664	2.4 to 3.2	2.1 to 5.4	2.7 to 5.1	3.0 to 4.6
High-Density Castables and Plastics	1874 to 2146	117 to 134	0.923 to 1.270	0.952 to 1.241	0.952 to 1.212	0.996 to 1.226	6.4 to 8.8	6.6 to 8.6	6.6 to 8.4	6.9 to 8.5
High-Density, High-Alumina Castables and Plastics	2227 to 2723	139 to 170	1.154 to 1.890	1.154 to 1.717	1.154 to 1.573	1.226 to 1.515	8.0 to 13.1	8.0 to 11.9	8.0 to 10.9	8.5 to 10.5
Superduty Firebrick	2082 to 2483	130 to 155	1.356	1.327	1.371	1.515	9.4	9.2	9.5	10.5
High-Alumina Firebrick	2643 to 2963	165 to 185	2.597 to 3.275	2.193 to 2.785	1.977 to 2.539	1.904 to 2.554	18.0 to 22.7	15.2 to 19.3	13.7 to 17.6	13.2 to 17.7

terials will not stand direct exposure to metal or slag, and backup materials may be subject to attack by vaporized process components (alkali, sulphur compounds, acids) or their condensates. Gas channeling through permeable materials must also be considered to prevent hot spots on shells.

(3) Insulation increases the depth of penetration and chemical attack on the hot-face layer.

Furnaces operating under larger temperature cycles may benefit further from the use of low-density materials, and the heat required to maintain equilibrium conditions can be considerably decreased.

Many steelplant refractory applications also never reach thermal equilibrium, and dynamic heat-transfer calculations must be employed in the analysis of such applications. In a normal steel ladle, for example, heat losses from the steel and shell temperatures will de-

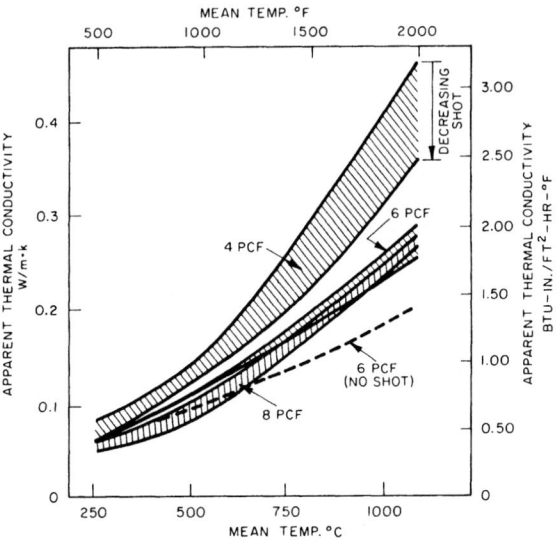

FIG. 2—20. Thermal conductivity as a function of temperature for various fiber materials.

pend greatly on ladle lining preheat and cycle time as well as the properties of the refractories used. Figures 2—18 and 2—19 show the results of one such simulation study on a special ladle lined with a basic refractory. There is a large effect of ladle preheat on lining and steel temperatures and of the ladle cover material (or slag layer) on steel-heat loss by radiation to the surroundings.

In certain severe environments, water- or air-cooled refractory constructions are used to provide added resistance to slag, metal, abrasion, shock, or other severe wear mechanisms. In general, these cooling mechanisms are designed to freeze or solidify severe chemical reactions and provide a "stable" refractory thickness which can be expressed in fps units as follows:

$$X = K\left(\frac{T}{7380} - 0.0605\right)$$

X = stable refractory thickness, in.

T = minimum refractory-process reaction temperature, °F.

K = thermal conductivity, (Btu in. per square foot per hour per °F)

In this case, high-conductivity materials have the obvious advantage of greater stable thicknesses unless their nature lowers the minimum refractory-process reaction temperature. Thus, the highest conductivity and compatible refractories are preferable for such environments. Cooling methods and materials of this type will be described in the section on blast-furnace refractories.

Table 2—V shows the conductivity of insulating fibers in comparison to more conventional refractory materials. Fibers are used because of their very low conductivity and heat-storage characteristics in comparison to other ceramic materials. Figure 2—20 shows the conductivity and temperature relationships for various fiber forms. The materials are used where heat savings and rapid product throughput are the primary consideration.

SECTION 4

REACTIONS AT ELEVATED TEMPERATURES

The foregoing discussion of the high-temperature behavior of refractories emphasizes physical factors, but refractory behavior also depends greatly on high-temperature reactions occurring not only within the refractories themselves but between refractories and contaminants encountered in service.

Phase-equilibrium diagrams have proved to be invaluable guides to understanding service reactions and the influence of composition on refractory properties. As excellent compilations of diagrams have been published (see references at end of chapter), only a few are reproduced here. However, it should be recognized that these diagrams are not without limitations in their use to predict or explain refractory behavior. For example, the various systems have been explored using simple combinations of pure oxides and represent equilib-

rium conditions, while refractories are rarely pure and seldom in equilibrium, either as manufactured or in service. Because of this complex chemical nature of refractories, information is often needed on reactions involving so many oxides that the usefulness of phase equilibrium information on systems involving three or even four oxides is minimal. The diagrams also give no information on such significant matters as viscosity of the liquids formed or the rates at which reactions proceed.

Figure 2—21 shows the Al_2O_3-SiO_2 system which applies to silica, fireclay, and high-alumina refractories. It will be noted that the lowest temperature at which any liquid is developed in the system is 1590°C (2894°F), while those compositions more aluminous than mullite ($3Al_2O_3 \cdot 2SiO_2$) or above 71.8 per cent Al_2O_3 develop

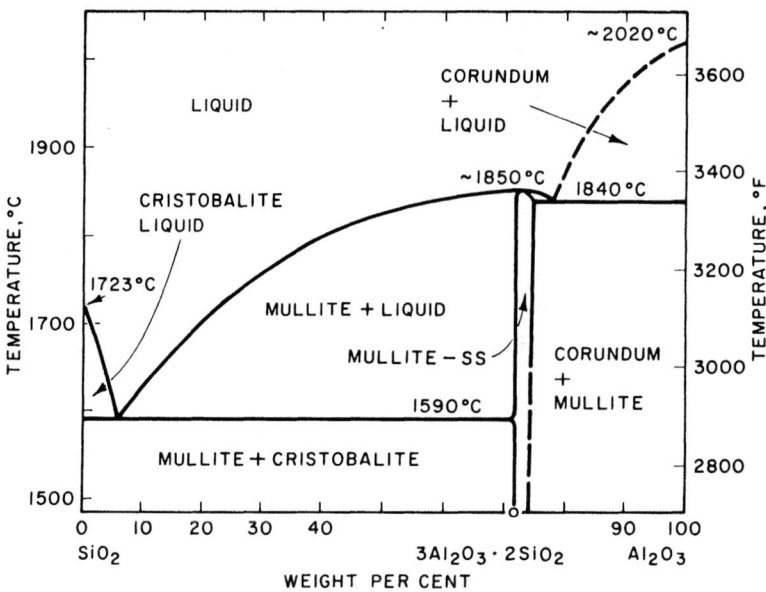

FIG. 2—21. Phase diagram of the Al₂O₃-SiO₂ system.

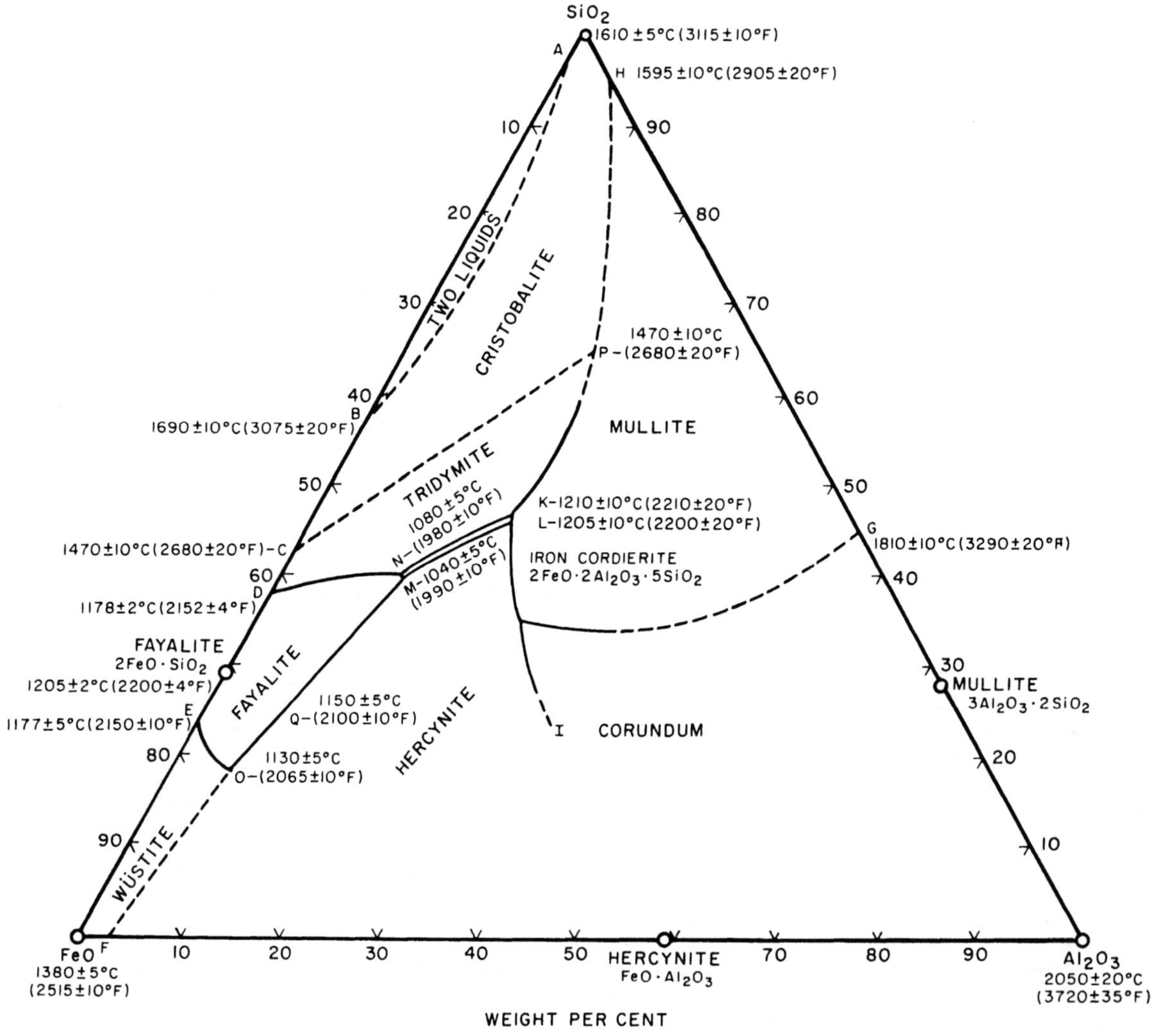

FIG. 2—22. The FeO-Al₂O₃-SiO₂ system.

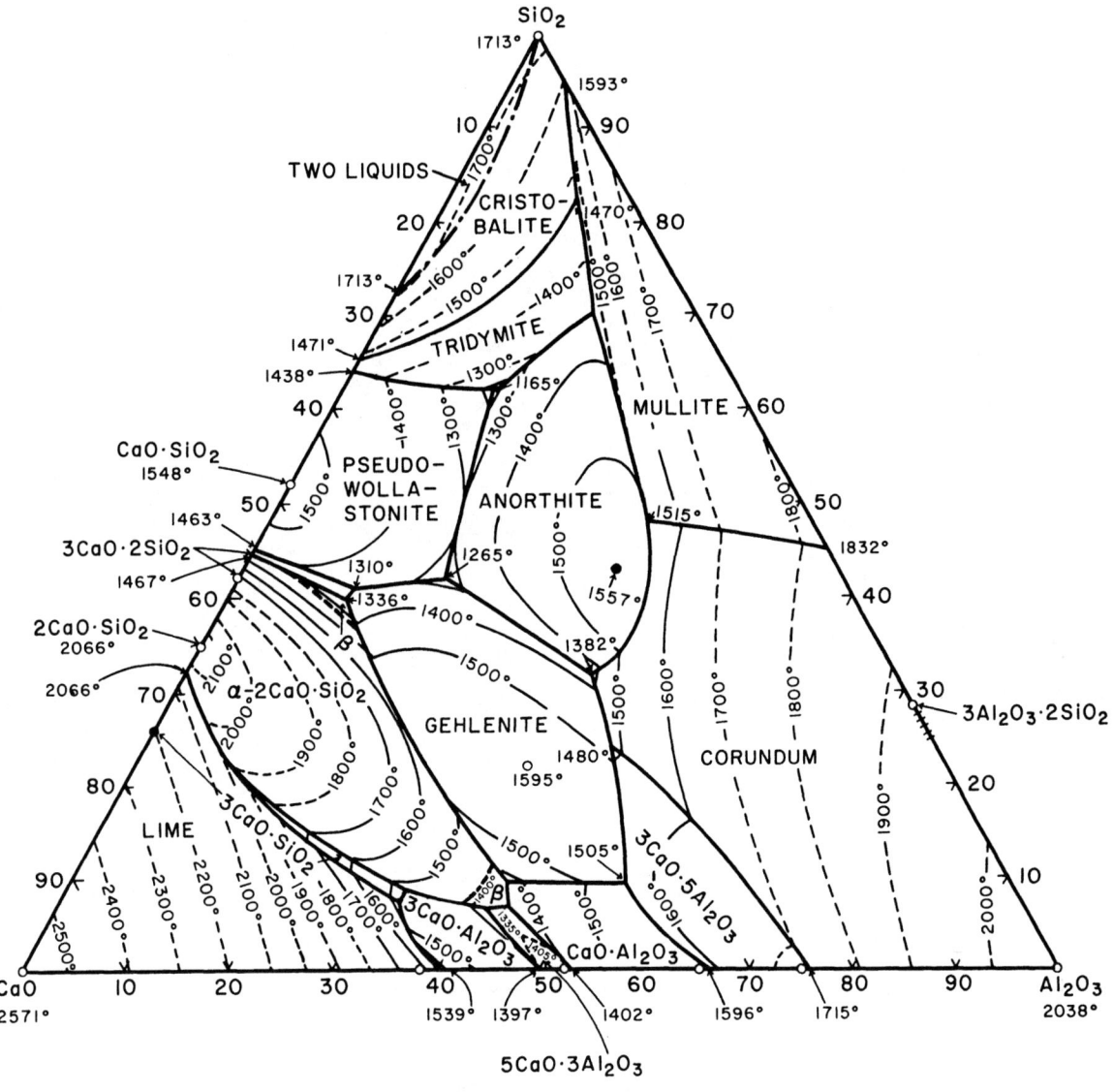

FIG. 2—23. The CaO-Al₂O₃-SiO₂ system. Temperatures are in degrees Celsius.

liquids at quite high temperatures. It is obvious that useful refractories can be made of the pure oxides Al₂O₃ or SiO₂. The pronounced effect of the impurities present in most commercial refractories in this system can be appreciated by comparing the temperatures at which they deform initially under load, 1150°-1200°C (2100°-2200°F), with the initial liquid temperatures of 1540° or 1840°C (2874° or 3344°F) indicated on the diagram. The oxide impurities largely responsible for lowering the refractoriness of fireclay refractories are CaO, MgO, FeO, Na₂O, and K₂O. As previously described, mining and beneficiation techniques are now widely used to obtain raw materials with minimum impurity levels.

Figure 2—22 shows the FeO-Al₂O₃-SiO₂ system. Here it is seen that the formation of some liquid can be expected even below about 1095°C (2000°F) with Al₂O₃-SiO₂ refractories and that very damaging amounts will be formed at the higher temperatures common to iron and steel processes. This is particularly true as iron-oxide-bearing liquids are characteristically very fluid.

Figure 2—23 is the diagram of the CaO-Al₂O₃-SiO₂ system, which is most applicable to reactions of fireclay refractories with blast-furnace slags and indicates superior resistance for higher-Al₂O₃ products in such environments. This system has also been useful in predicting behavior of silica brick, which will be discussed later.

The ternary phase equilibrium diagrams for K₂O or Na₂O reactions with Al₂O₃ and SiO₂ are reproduced in Figures 2—24 and 2—25. It is evident that the refractoriness of alumina-silica refractories will be seriously affected by very small amounts of Na₂O, less than 1 per cent being sufficient to lower the temperature of initial liquid formation to less than about 1095°C (2000°F), while approximately 10 per cent is sufficient to completely liquefy the more siliceous alumina-silica

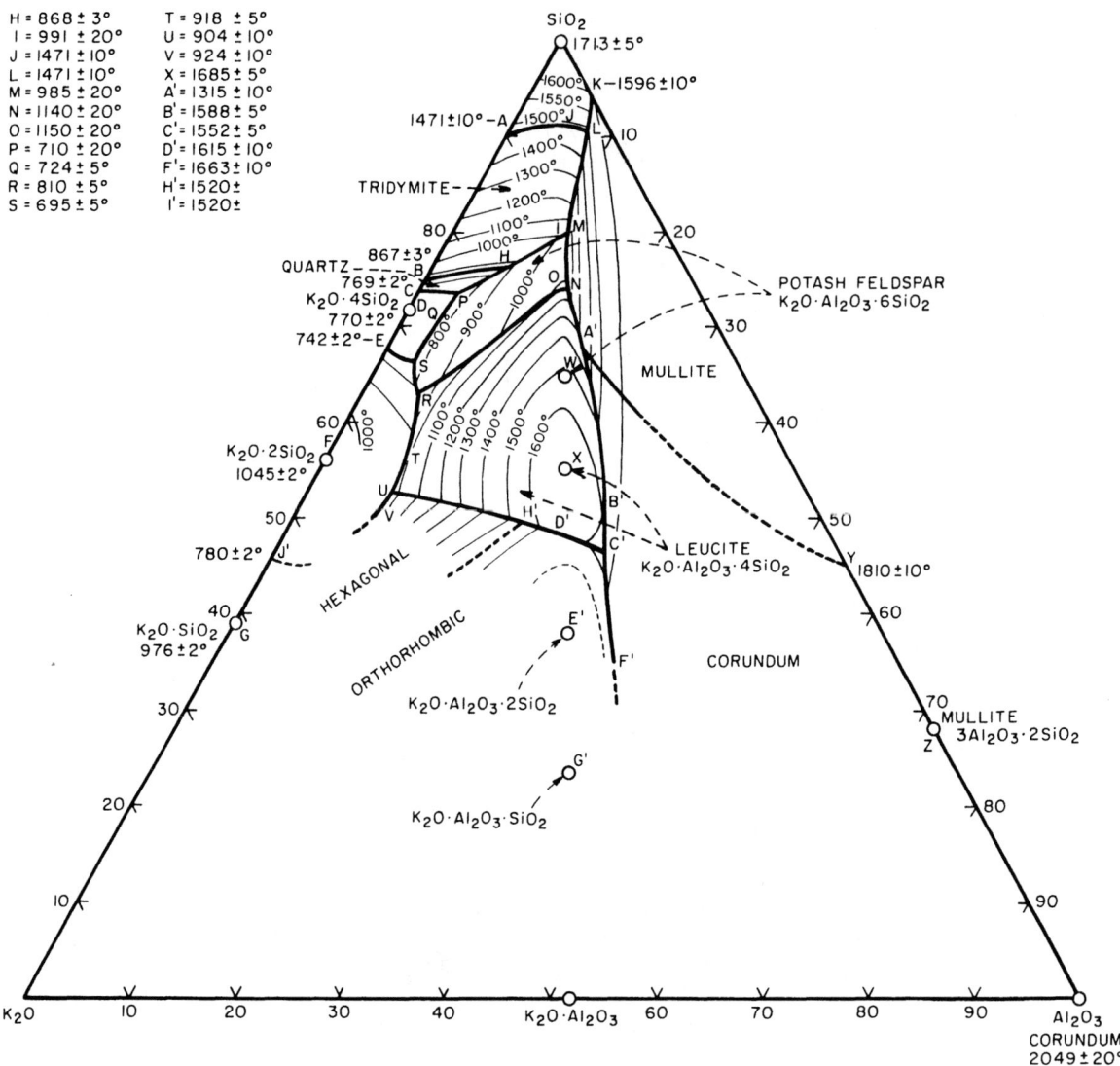

FIG. 2—24. The K_2O-Al_2O_3-SiO_2 system. Temperatures are in degrees Celsius.

compositions at about 1205°C (2200°F). Potassium oxide (K_2O) has a similar effect in amounts up to 10 per cent.

Low levels of impurities are far more critical in silica brick than fireclay brick, and alkalies are the worst offenders. The amounts of Na_2O, K_2O, and Al_2O_3 required to lower the melting point of pure silica from about 1725°C to about 1675°C (about 3140°F to about 3050°F) are, respectively 1.4, 1.9, and 3.1 per cent.

Another very deleterious fluxing agent for fireclay refractories is MnO, as can be seen in the diagram of the MnO-Al_2O_3-SiO_2 system in Figure 2—26. Note that this system is quite similar to the FeO-Al_2O_3-SiO_2 system.

Because the raw materials for silica brick lack both a natural bond and a high melting point, care must be taken that the required bonding addition has the minimum effect on refractoriness. Figure 2—27 shows the CaO-SiO_2 system and explains why lime is universally used for this purpose. With additions of CaO to SiO_2, the melting temperature remains unchanged between

1 and 27.5 per cent CaO, due to the formation of two immiscible liquids. No such phenomenon occurs in the Al_2O_3-SiO_2 system, and by referring again to the CaO-Al_2O_3-SiO_2 system, it is found that only a small amount of Al_2O_3 is required to destroy the CaO-SiO_2 immiscibility. In fact, the effect of minor increments of Al_2O_3 on the liquid development of silica brick is such that the temperature of failure under a load of 172 kilopascals (25 pounds per square inch) will decrease approximately 5°C (10°F) for each 0.1 per cent increase in Al_2O_3 in the 0.3 to 1.2 per cent range of Al_2O_3 between super duty and conventional silica brick. Figure 2—28 of the FeO-SiO_2 system shows that FeO, like CaO, also forms two immiscible liquids when added to SiO_2, thus greatly increasing the tolerance of silica brick for FeO. Furthermore, as with CaO, a small amount of Al_2O_3 can eliminate this immiscibility.

As atmospheric conditions in steelplant furnaces may range from highly reducing to highly oxidizing, the form of iron oxides present may vary from FeO to

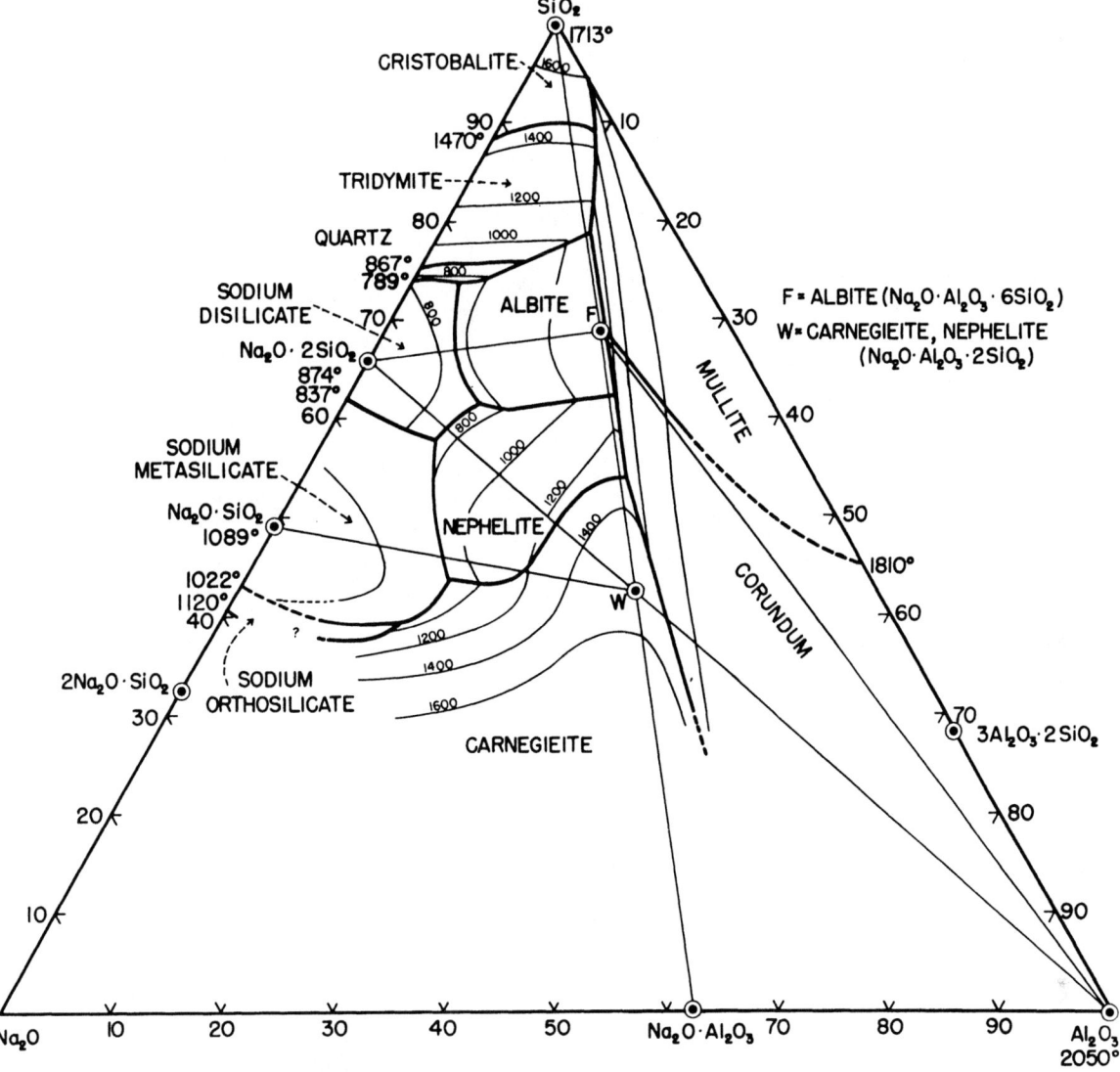

FIG. 2—25. The Na₂O-Al₂O₃-SiO₂ system. Temperatures are in degrees Celsius.

Fe₂O₃. Accordingly, Figure 2—29 shows the system FeO-Fe₂O₃-SiO₂, and is of considerable importance in understanding the behavior of silica brick in service. Thus, it is seen that the lowest-melting liquids occur from reaction of FeO and SiO₂, and that at temperatures in the range of about 1455° to 1665°C (about 2650° to 3030°F) less liquid, and a less siliceous liquid, will be produced with either FeO·Fe₂O₃, or Fe₂O₃ than with FeO, due to the greater extent of the two-liquid region under oxidizing conditions.

The two principal refractory oxides considered to be basic are magnesia (MgO) and calcia (CaO). Magnesia is noted for its tolerance to iron oxides. As shown in Figure 2—30, MgO and FeO form a continuous series of solid solutions which have high refractoriness even with very high FeO contents. Under oxidizing conditions, magnesia is even more tolerant to iron oxide. Magnesia and iron oxide form the refractory compound magnesioferrite (MgO·Fe₂O₃) which contains 80

weight per cent Fe₂O₃. Magnesioferrite forms solid solutions with magnetite (FeO·Fe₂O₃) at higher iron-oxide contents and with magnesia at lower iron-oxide contents. On the other hand, calcia is more reactive with iron oxide, forming low-melting calcium ferrites such as dicalcium-ferrite (2CaO·Fe₂O₃) that melts incongruently at about 1440°C (about 2620°F). Also, calcia is subject to hydration and disruptive disintegration on exposure to atmospheric conditions and cannot be used in refractory shapes made by conventional procedures. It is evident, therefore, that magnesia is the more useful basic refractory oxide and forms the base for all types of basic refractories including those made from magnesite, olivine, dead-burned dolomite, and magnesite and chrome ore.

Magnesia-bearing refractories, regardless of type, contain accessory refractory oxides and encounter other refractory oxides in service which exert an important influence on their performance. Figure 2—31

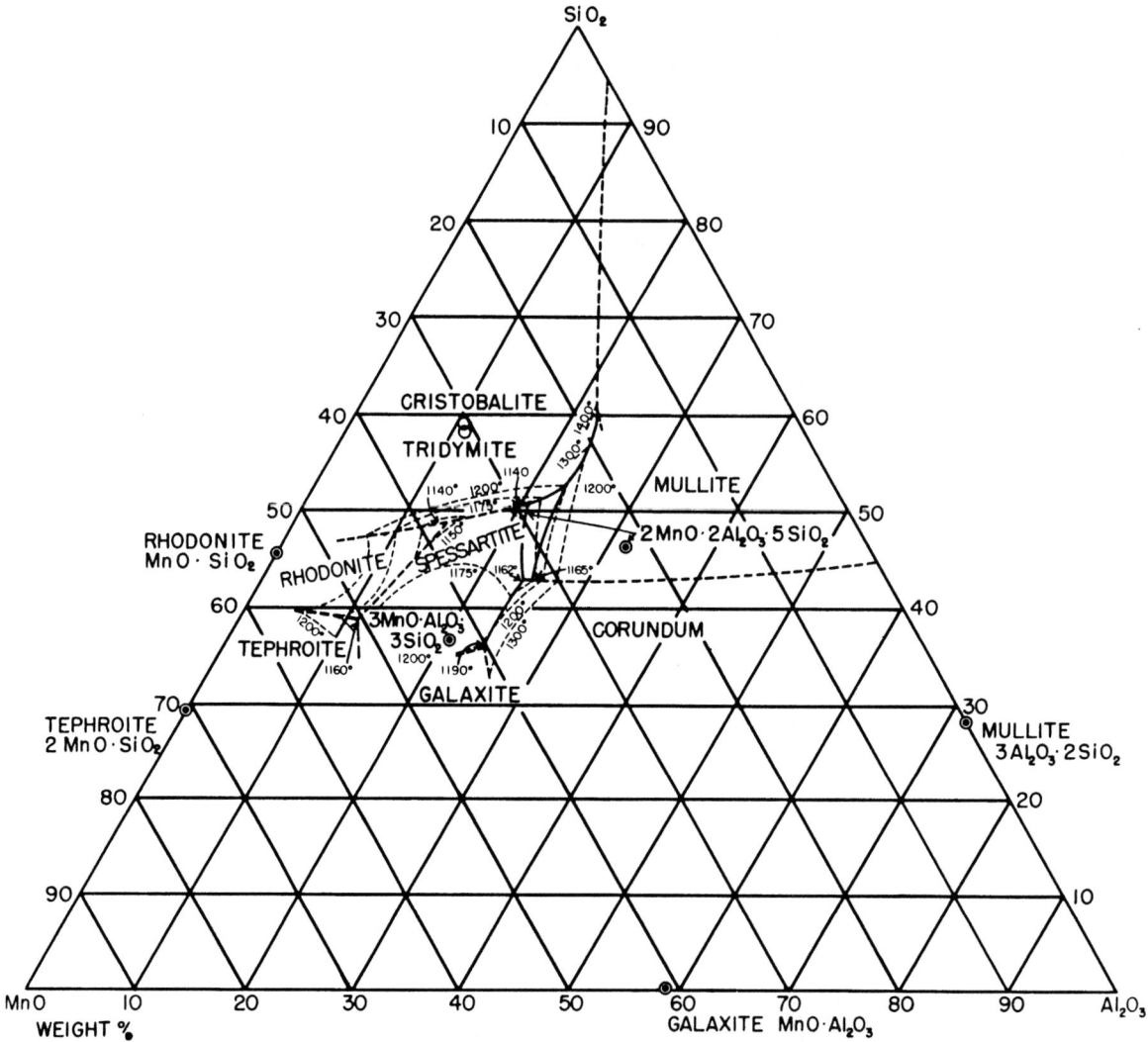

FIG. 2—26. The MnO-Al₂O₃-SiO₂ system. Temperatures are in degrees Celsius.

shows the reactions and phase assemblages in the MgO-CaO-SiO₂ system. In the high-MgO portion of this system, the principal mineral is of course always periclase. The accessory silicate bonding minerals, however, will vary considerably depending on the ratio of CaO to SiO₂. Table 2—VI presents a summary of the compounds present with periclase as affected by the CaO/SiO₂ weight ratio and their approximate melting points. The type of refractory bonding preferred will depend on the intended application. The forsterite bond, which occurs at CaO/SiO₂ ratios less than 0.93/1.00, is desired in some basic brick to prevent excessive formation of monticellite and merwinite. These minerals form low-temperature liquids and therefore have poor high-temperature load-carrying ability. Refractory bonding phases with high melting points can also be obtained at CaO/SiO₂ ratios above 1.86/1.00 where refractory dicalcium or tricalcium silicates are present. In recent years, in addition to the control of CaO/SiO₂ ratio, increased emphasis has been placed on decreasing the amounts of lime and silica

present to obtain the maximum advantage of the properties of nearly-pure magnesia. This has been accomplished largely through the use of the improved synthetic magnesites.

The high-temperature reactions of refractories made from magnesia and chrome ore are under constant study. The properties of refractories made from these two raw materials are excellent because of the tendency of each to minimize the major weaknesses of the other constituent. Chrome ore consists of a solid solution of chrome spinels (Mg, Fe)O · (Cr, Al, Fe)₂O₃ with appreciable amounts of gangue silicates. At high temperatures, the gangue silicate in chrome ores is responsible for poor resistance to deformation under load and the iron oxides, when alternately oxidized and reduced, cause expansion and contraction, often causing disintegration. Also, chrome spinel shows considerable growth or bursting when reacted with iron oxide at high temperatures as a result of the formation of solid solutions of magnetite (FeO · Fe₂O₃) and other spinels. With additions of magnesia to chrome ore, however, the gangue silicates are converted on firing during manufacture or in service to the more refractory phases such as forsterite or dicalcium silicate, and the iron oxide to the spinel $MgO \cdot Fe_2O_3$ by co-diffusion of Fe_2O_3 and MgO between the magnesia and chrome spinel. $MgO \cdot Fe_2O_3$ is more resistant to deterioration in cyclic oxidizing-reducing conditions than the iron oxides in the original chrome ore. The addition of still greater amounts of magnesia to chrome ore improves significantly the resistance of the refractory to iron-oxide bursting because of the greater affinity of magnesia for iron oxide as compared with Cr_2O_3. The addition of chrome ore to magnesia on the other hand improves the resistance of magnesia to thermal spalling through an apparent stress relief in an otherwise rigid structure.

As in magnesia refractories, the CaO/SiO_2 ratio exerts an important influence on the phases present in composite refractories of magnesia and chrome ore. At CaO/SiO_2 ratios less than 1.86/1.00, the primary phases between MgO, CaO, and SiO_2 are the same as that previously discussed with the sesquioxides Cr_2O_3, Al_2O_3, and Fe_2O_3 combined with MgO and FeO to form spinel solid solutions. At higher CaO/SiO_2 ratios, the sesquioxides form low-melting compounds with CaO.

As in magnesia refractories, efforts have been made

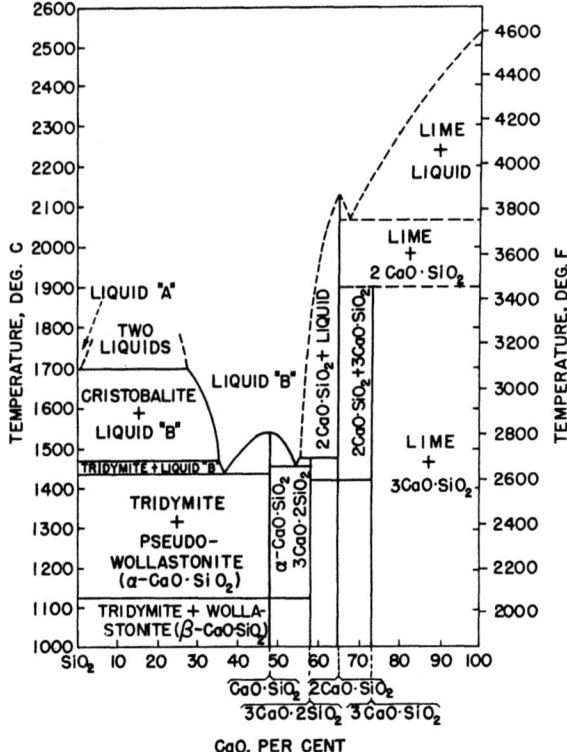

FIG. 2—27. The CaO-SiO₂ system.

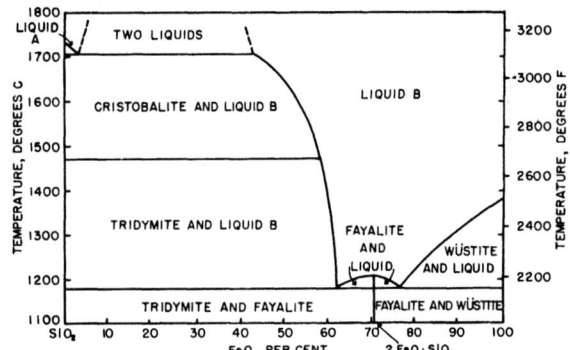

FIG. 2—28. The FeO-SiO₂ system.

Table 2—VI. Mineral Phases in Equilibrium with Periclase (MgO) in the MgO-CaO-SiO₂ System

Weight Ratio CaO/SiO₂	Minerals Present	Composition	Approximate Melting Temperature	
			°C	°F
less than 0.93	Forsterite	$2MgO \cdot SiO_2$	1900	3450
	Monticellite	$CaO \cdot MgO \cdot SiO_2$	1490*	2710*
0.93	Monticellite	$CaO \cdot MgO \cdot SiO_2$	1490*	2710*
0.93 to 1.40	Monticellite	$CaO \cdot MgO \cdot SiO_2$	1490*	2710*
	Merwinite	$3CaO \cdot MgO \cdot 2SiO_2$	1575*	2870*
1.40	Merwinite	$3CaO \cdot MgO \cdot 2SiO_2$	1575*	2870*
1.40 to 1.86	Merwinite	$3CaO \cdot MgO \cdot 2SiO_2$	1575*	2870*
	Dicalcium silicate	$2CaO \cdot SiO_2$	2130	3865
1.86	Dicalcium silicate	$2CaO \cdot SiO_2$	2130	3865
1.86 to 2.80	Dicalcium silicate	$2CaO \cdot SiO_2$	2130	3865
	Tricalcium silicate	$3CaO \cdot SiO_2$	1900**	3450**
2.80	Tricalcium silicate	$3CaO \cdot SiO_2$	1900**	3450**
More than 2.80	Tricalcium silicate	$3CaO \cdot SiO_2$	1900**	3450**
	Lime	CaO	2565	4650

*Incongruent melting.
**Stable only between 1900° and 1250°C (3450° and 2280°F).
Dissociation below and above these temperature into $2CaO \cdot SiO_2$ and CaO.

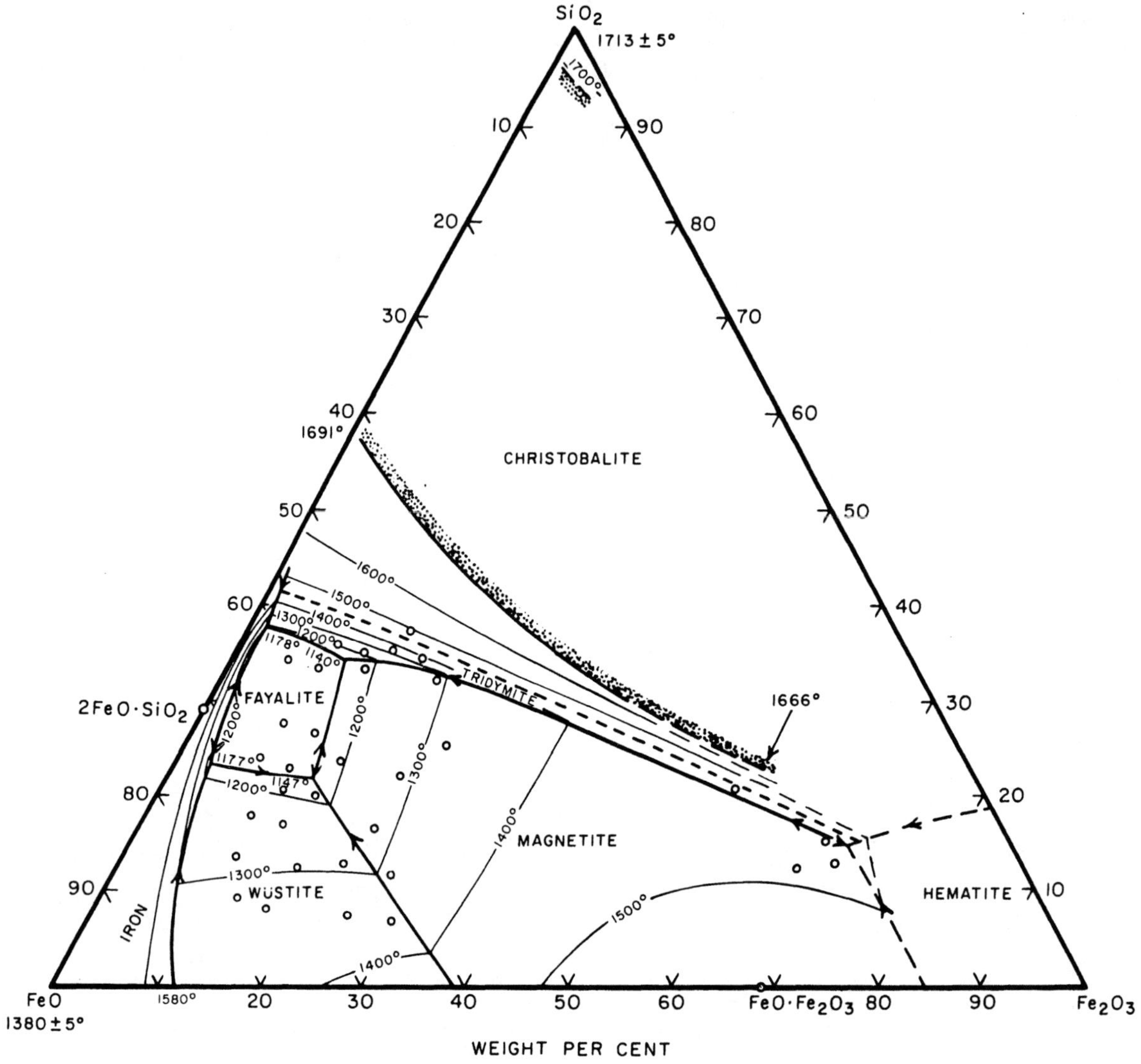

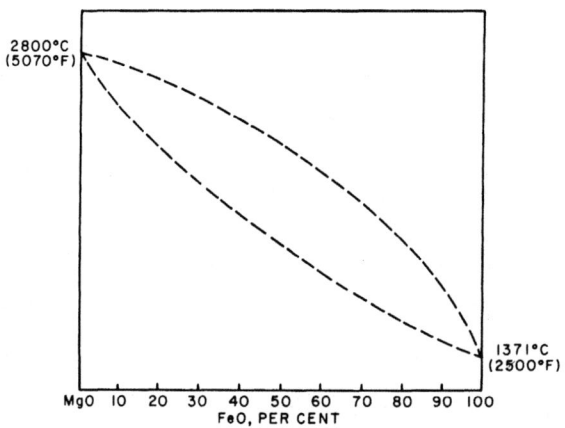

SiO2
1713 ± 5°
1700°
10
90
20
80
30
70
40
60
1691°
CHRISTOBALITE
50
50
1600°
1500°
60
40
1300°
1400°
1200°
1178°
1140°
TRIDYMITE
2FeO·SiO2
FAYALITE
1666°
30
1200°
1200°
1300°
1177°
1147°
1200°
80
20
1400°
MAGNETITE
1300°
90
WUSTITE
HEMATITE
10
1500°
IRON
1400°
FeO
1380 ± 5°
1580°
20
30
40
50
60
FeO·Fe2O3
80
90
Fe2O3
WEIGHT PER CENT

FIG. 2—29. The FeO-Fe_2O_3-SiO_2 system. Temperatures are in degrees Celsius.

2800°C
(5070°F)
1371°C
(2500°F)
MgO 10 20 30 40 50 60 70 80 90 100
FeO, PER CENT

FIG. 2—30. The MgO-FeO system.

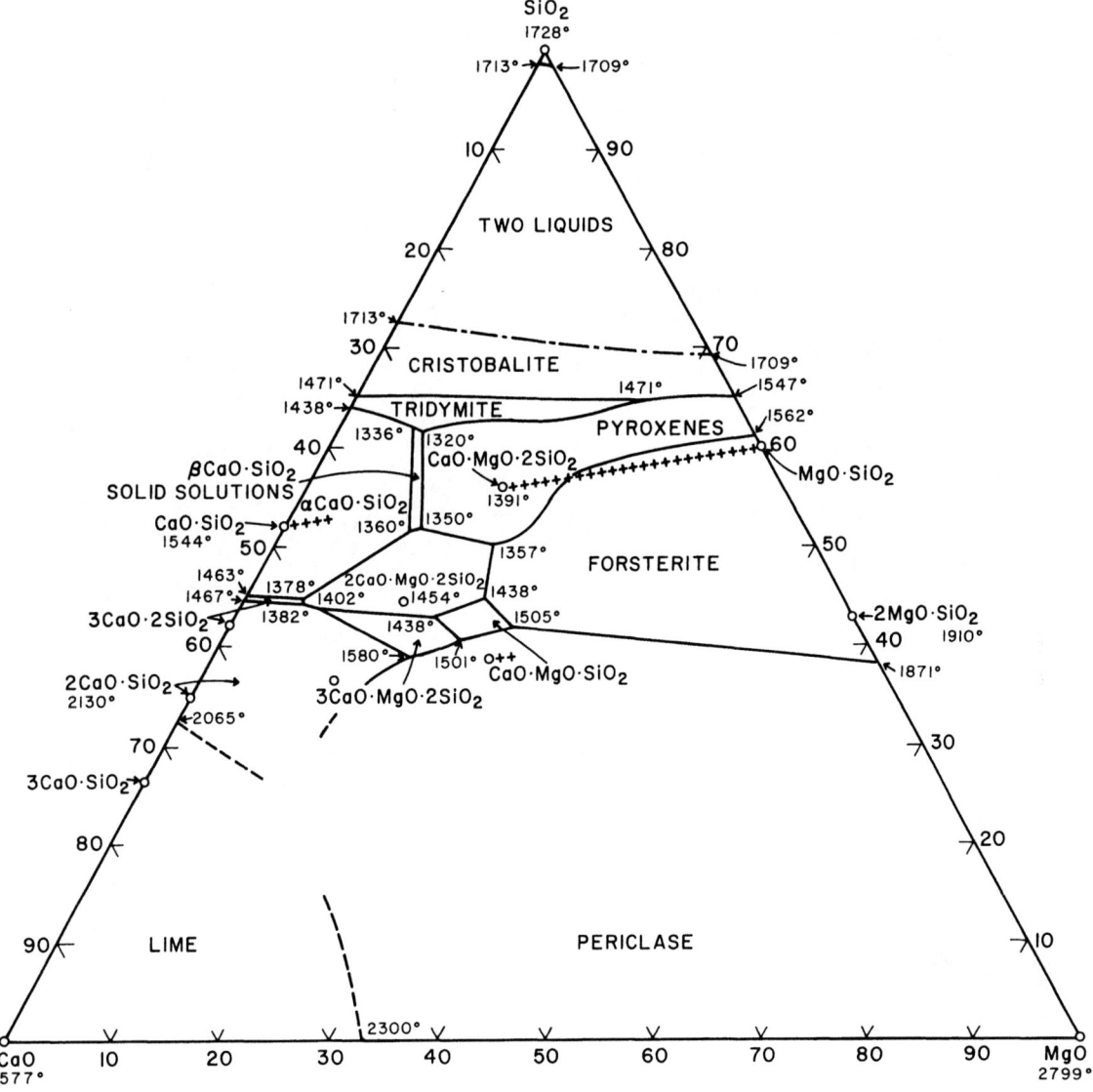

FIG. 2—31. The MgO-CaO-SiO₂ system. Temperatures are in degrees Celsius.

to replace the silicate bond in refractories made from magnesia and chrome ore. The most noteworthy of these efforts to date is the use of firing temperatures above those normally employed to produce a low-silica burned basic refractory with so-called "direct" bonding of the magnesia to the chrome spinel, and magnesia to magnesia. Figure 2—32 shows photomicrographs of the conventional silicate and these "direct" bonds, respectively. Direct-bonded brick have unusually high hot strengths, several times that of conventional brick of similar composition.

Although the phase assemblages previously discussed indicate the general combinations of iron oxides in basic refractories, it is equally important to consider the influence of furnace temperature and atmosphere on the oxidation state of the iron present in various compositions including the oxides MgO, Cr_2O_3, Al_2O_3, FeO, and Fe_2O_3. It has been shown, for example, that

in mixtures consisting originally of Fe_2O_3 and $MgO \cdot Fe_2O_3$, dissociation of the Fe_2O_3 is accompanied by solution of the dissociation product magnetite ($FeO \cdot Fe_2O_3$) in $MgO \cdot Fe_2O_3$ until a single spinel phase is formed. In this same region, it has been shown that the spinel-to-sesquioxide transition temperature decreases as the magnesia content increases, but at the same time magnesia stabilizes Fe_2O_3 at higher temperatures. Chrome oxide (Cr_2O_3), on the other hand, has an opposite effect to magnesia in that it increases the sesquioxide-to-spinel transition temperature and lowers the degree of dissociation of Fe_2O_3 at higher temperatures. In mixtures consisting initially of $MgO \cdot Fe_2O_3$ and MgO, dissociation of Fe_2O_3 proceeds with a decrease in the amount of spinel and solution of iron oxide in periclase.

Atmospheric changes may also exert an important influence on carbon-bearing refractories at tempera-

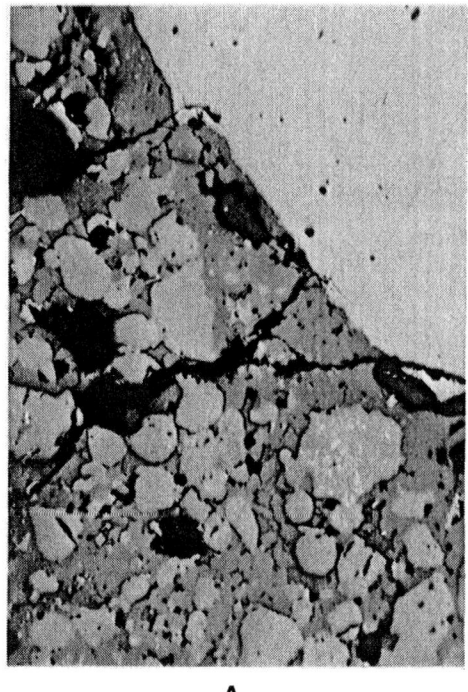
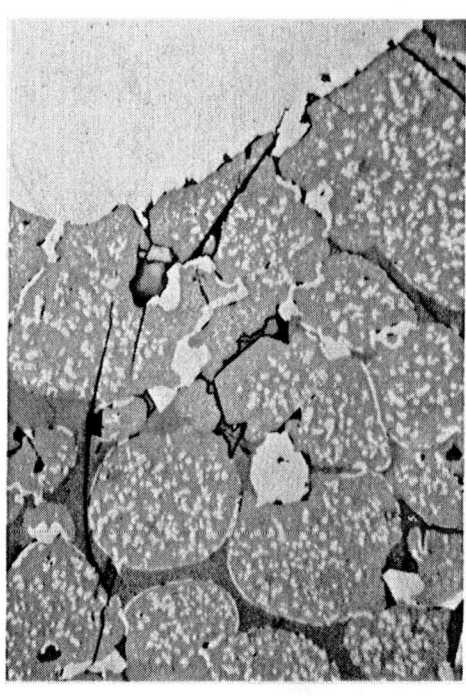

A   B

FIG. 2—32. Photomicrographs showing the structure of basic brick with different bonds. "A" is a magnesite-chrome ore composition in which the bond between the large white chrome spinel grains and rounded magnesia crystals is principally the dark silicate. "B" shows a magnesite-chrome composition having less silicate and a direct bond between white chrome spinel and periclase (MgO) containing multiple white magnesioferrite inclusions. Reflected light, X200.

ture. Studies have shown that partial vaporization and condensation of certain oxides in contact with carbon at high temperatures can cause desirable effects such as the development of a dense impervious layer which limits slag and metal penetration. On the other hand, atmospheres involving low partial oxygen pressures (such as those present in the AOD process) may cause undesirable effects in many refractories.

SECTION 5

TESTING AND SELECTION OF REFRACTORIES

Refractory testing and selection procedures can be broadly grouped into three categories: (1) physical and chemical tests on unused brick, (2) simulated service environment tests, and (3) post-mortem examinations on used refractories.

Physical and Mechanical Tests—Section 3 previously discussed several simple tests for hot strength, resistance to deformation under load, porosity, permeability, abrasion resistance, and thermal-expansion and thermal-conductivity characteristics which are performed by accepted procedures. In recent years, the trend has been toward making such tests more realistic as previously described (hot-strength rather than cold, expansion tests under load, and so on). Such tests form the basis for quality-control judgments on a given type of refractory or manufacturer in today's steel plant.

Simulated Service Tests—A larger number of diverse tests have been devised to simulate service conditions because most refractories wear by a complex combination of mechanisms. One simple example includes fail-ure of refractories in a special gaseous environment such as carbon monoxide. Carbon-monoxide disintegration occurs as carbon from the CO gas deposits around iron concentrations in a refractory:

$$2CO \rightarrow CO_2 + C$$

The iron concentrations catalyze this reaction, producing carbon buildups and growth on iron locations which may fracture the refractory behind the hot face (the above reaction is highly temperature sensitive). Simulative tests have been designed to predict resistance to carbon-monoxide disintegration in all types of refractories by heating the refractory for long periods in CO gas. Figure 2—33 shows refractory samples after such tests Although refractories without iron will not be affected by this reaction, the total removal of iron from most natural raw material is not economically practical. A more practical solution involves proper handling of raw materials and processes to remove ter-nary iron and minimize iron contamination from grind-

UNAFFECTED   SURFACE POPOUTS

LIGHT CRACKING   HEAVY CRACKING

VERY HEAVY CRACKING   DISINTEGRATED

FIG. 2—33. These specimens were rated after the carbon monoxide disintegration test.

ing. Control of the form of iron in the fired aggregate in monolithic materials or fired products has led to the development of special low-iron refractories for blast-furnace linings. Grinding-equipment improvements and magnetic separation have further been used to minimize contamination. Examples of other simple simulated service tests include testing for exposure to alkalis, hydration, sulphidation, or oxidation.

Many more complex simulative tests exist where various failure mechanisms may be simulated at one time. For example, refractory samples may be exposed to slags in a rotating furnace heated with an oxygen-gas torch as shown in Figure 2—34 A, B, and C. In this test, refractory wear from chemical reaction with the molten slag, erosion by slag, thermal cycling, and other effects can be measured for various times, temperatures, and atmospheres. The results obtained are related to refractory chemical composition (Section 4) and physical properties (Section 3). Figures 2—35 and 2—36 illustrate results using such a test on basic refractory products made from periclase and chrome ore. Lower erosion by a slag with a basicity of 2.0 is obtained on the brick with higher MgO contents, but lower brick porosity also has a significant effect on decreasing slag attack. Tests of this type are very complex and care must be taken to properly interpret the results in relation to true service environments.

Another example of a simulated service test involves testing the fused-silica tubes (shrouds) used between the tundish and mold in continuous casting (see Section 6). In the simulative tests, tubes are rotated rapidly in a steel bath of controlled composition to simulate molten steel flowing through the steel exit ports. Figure 2—37 shows the effect of rotation speed on tubes of a given type. The results obtained are also highly dependent on tube density and strength and on steel composition (particularly Mn content). As in the rotary slag tests, rotating samples in steel may produce wear by metal

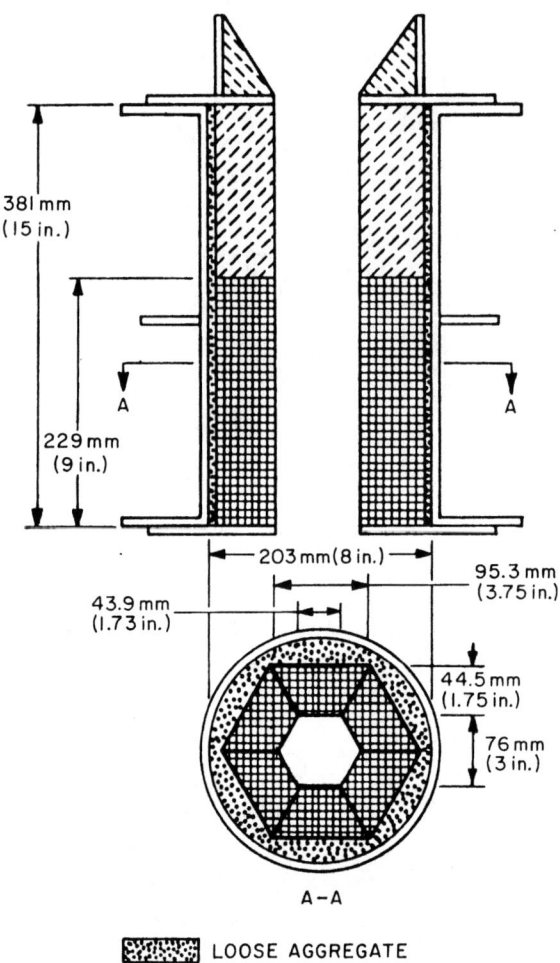

LOOSE AGGREGATE
TEST SPECIMEN
CASTABLE
REFRACTORY LINERS

Fig. 2—34B. Cross-section of rotary slag-furnace lining.

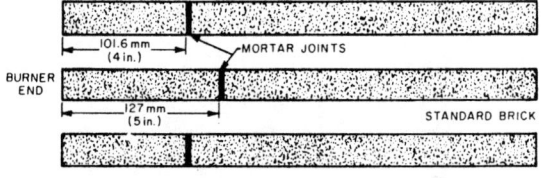

STAGGERING OF MORTAR JOINTS BETWEEN ADJACENT BRICK

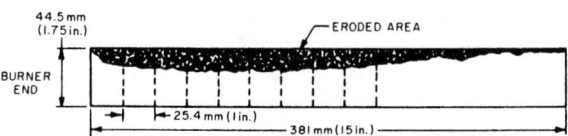

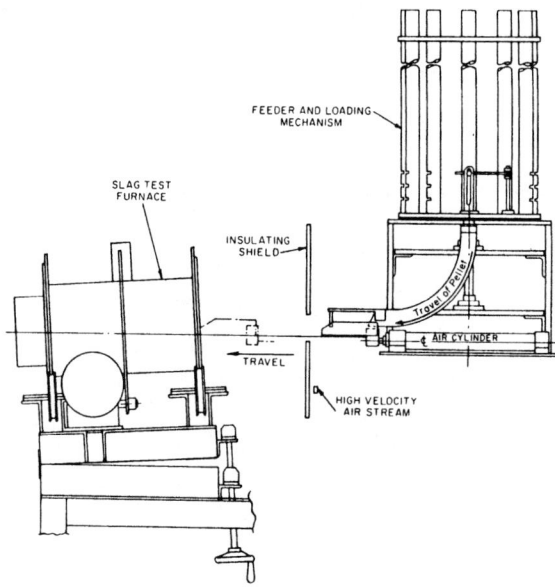

Fig. 2—34A. Rotary apparatus for determining relative slag erosion.

Fig. 2—34C. Positions at which measurements shall be taken on section specimen after slag test.

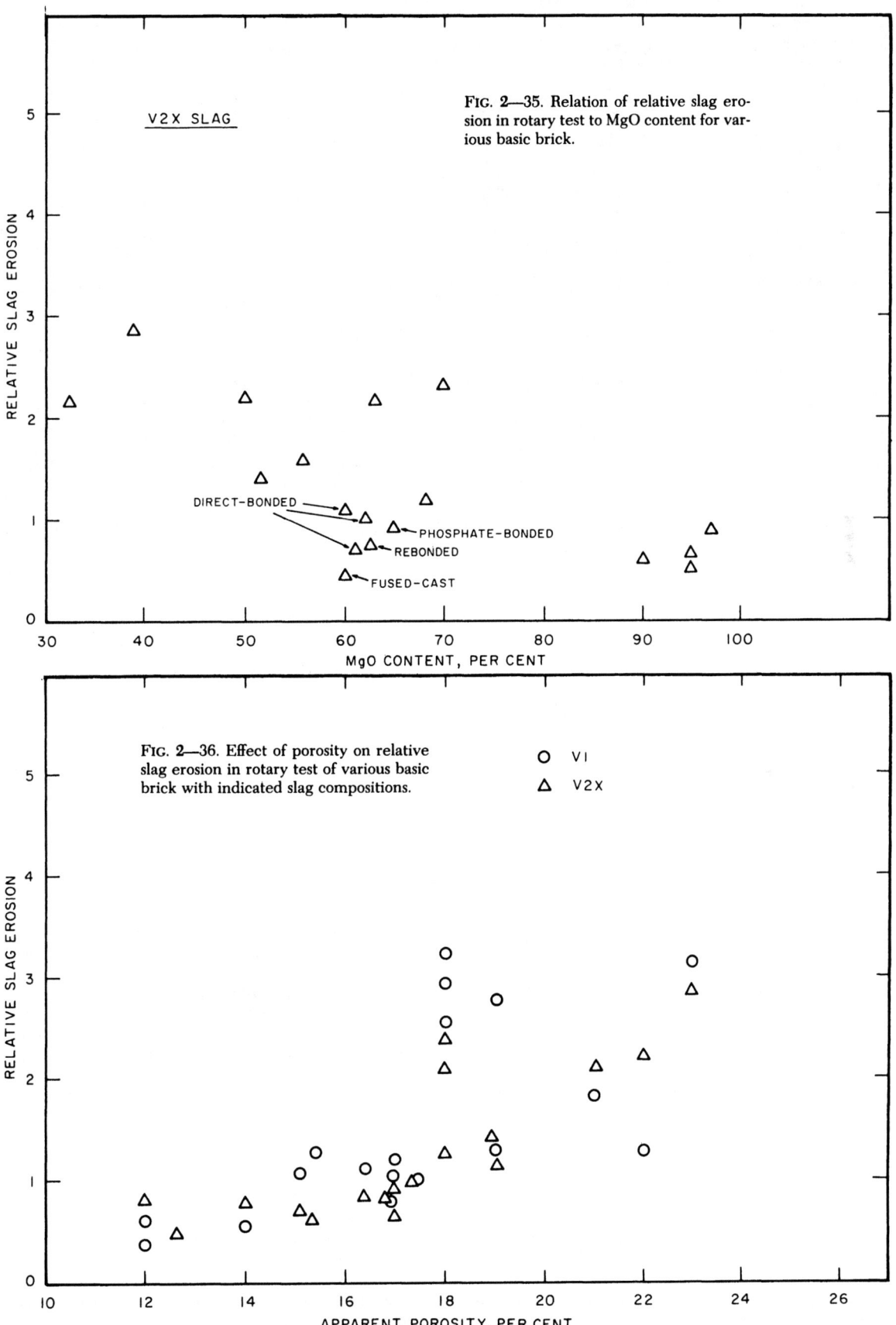

FIG. 2—35. Relation of relative slag erosion in rotary test to MgO content for various basic brick.

FIG. 2—36. Effect of porosity on relative slag erosion in rotary test of various basic brick with indicated slag compositions.

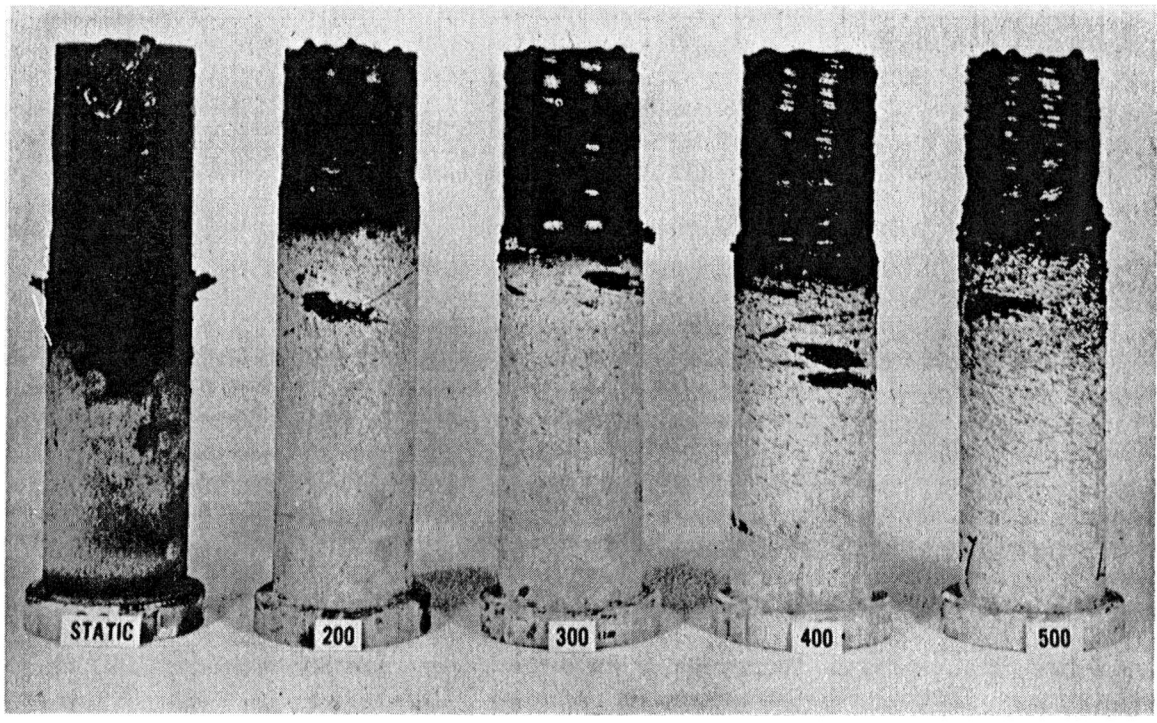

FIG. 2—37. One brand of pouring tubes after testing at various speeds for one hour in G 13450 (AISI 1345) steel.

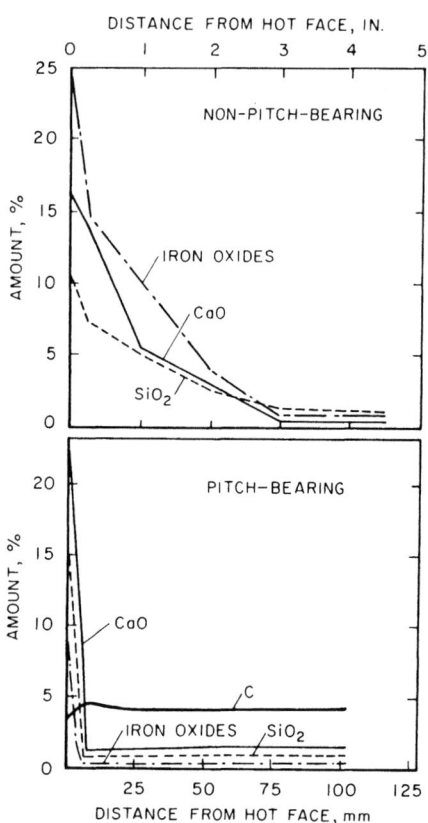

FIG. 2—38. Influence of carbon on slag penetration into magnesite brick in a basic oxygen furnace lining.

erosion, thermal shock, or a combination of mechanisms.

Post-Mortem Studies—To improve refractory life, the mechanisms by which the refractories are consumed must be fully understood. A considerable and continuing effort has therefore been made to thoroughly analyze used refractories and to integrate this information into improved refractory products. In its simplest form, used refractories are examined by chemical analyses and by microscopic methods. Figure 2—38 shows results using this approach on two types of brick taken from a BOP furnace after service. In the magnesite brick without pitch (burned product), CaO, FeO, and SiO_2 from the slag have penetrated the brick to a depth of 25 to 75 mm (1 to 3 inches) and altered the microstructure and properties of the refractory. Such a refractory may not wear uniformly due to structural spalling as well as hot-face corrosion and erosion. In a similar magnesite brick made with pitch or having retained carbon, penetration has been restricted to an area only a few millimetres thick by the presence of carbon in the brick. The carbon minimizes wetting and causes the formation of a dense impervious layer due to limited MgO volatilization. A carbon-bearing refractory of this type wears uniformly by corrosion and erosion. The brick without carbon has been severely altered by slag penetration, diffusion, and reaction with the brick components, whereas penetration has been held to 2 millimetres in the carbon-bearing brick. Unfortunately, carbon-penetration inhibitors cannot be used for all applications because the carbon is easily oxidized in some oxidizing steelmaking processes (for example, open hearths). A beneficial effect can be obtained from the carbon derived from pitch in both pitch-bonded and pitch-impregnated types of refractories.

Many more sophisticated techniques can be used as part of post-mortem examinations including x-ray, scanning electron microscope, or electron-probe techniques. Figure 2—39 illustrates the use of the electron probe to indicate the distribution of impurities in a zirconia refractory. The concentration of impurities in the grain-boundary area is typical of refractory material and helps to explain why small quantities of such impurities often produce significant detrimental effects including a significant loss in refractoriness.

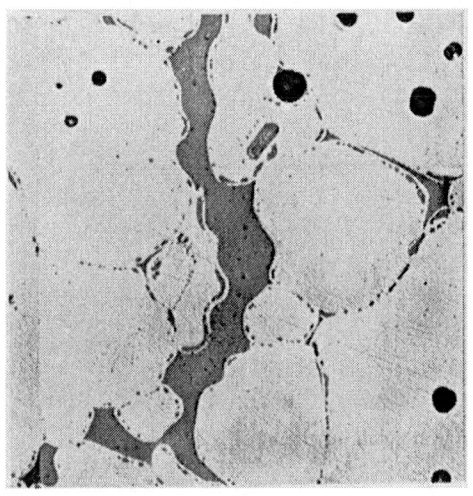

A) PHOTOMICROGRAPH

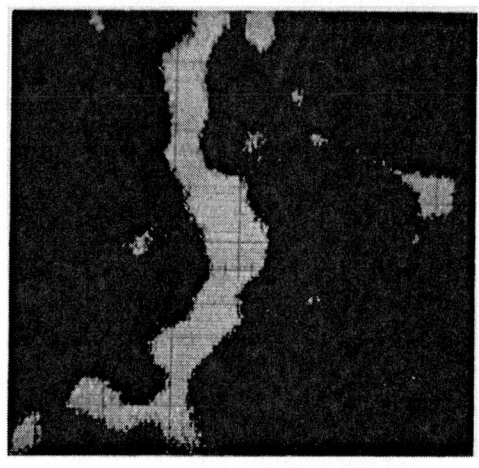

B) SILICON CHARACTER- ISTIC X-RAY IMAGE

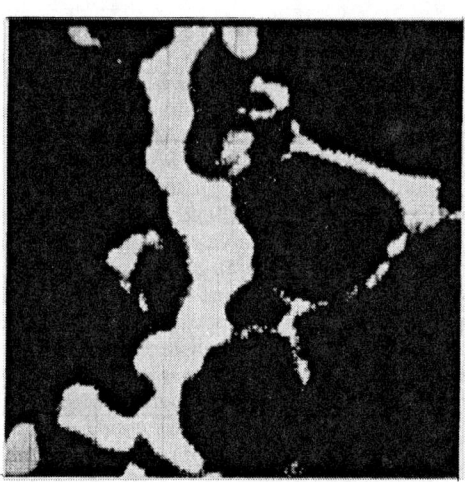

C) CALCIUM CHARACTER- ISTIC X-RAY IMAGE

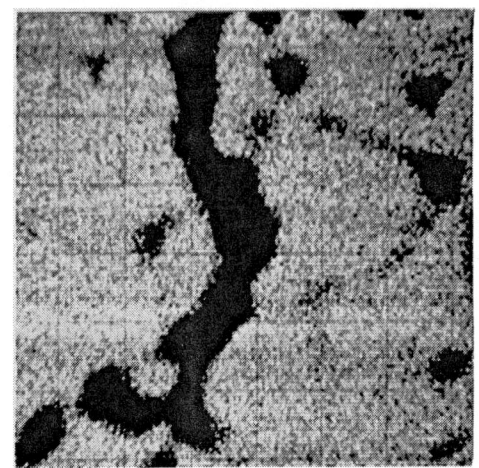

D) ZIRCONIUM CHARACTER- ISTIC X-RAY IMAGE

FIG. 2—39. Photomicrograph of center of cross-sectioned sensor button showing predominantly large grains with glassy material at grain boundaries and characteristic x-ray images (600X) showing distribution of elements.

SECTION 6

SPECIFIC USES OF REFRACTORIES

The following sections will briefly describe the use of refractories in the various areas of iron- and steelmaking, metal handling and casting, and finishing. References to other chapters should be made for further details.

REFRACTORIES IN COKE OVENS

This area is adequately covered in Chapter 4 on the manufacture of metallurgical coke.

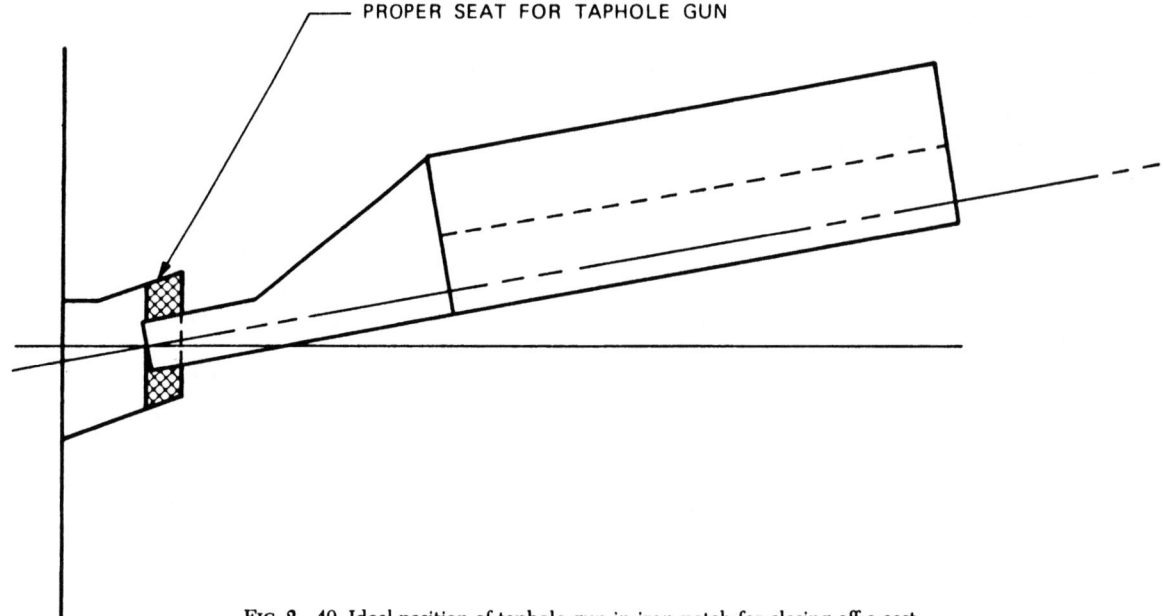

— PROPER SEAT FOR TAPHOLE GUN

FIG. 2—40. Ideal position of taphole gun in iron notch for closing off a cast.

REFRACTORIES IN THE BLAST FURNACE AND AUXILIARIES

Chapter 15 contains much useful information regarding refractory design and use in blast furnaces, and this discussion is intended to supplement that information. For convenience, the blast-furnace refractory areas shall be subdivided into refractories for the casthouse, refractories in the furnace proper, and stove and auxiliary refractories.

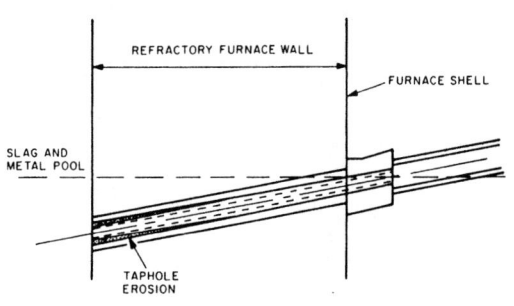

REFRACTORY FURNACE WALL

FURNACE SHELL

SLAG AND
METAL POOL

TAPHOLE
EROSION

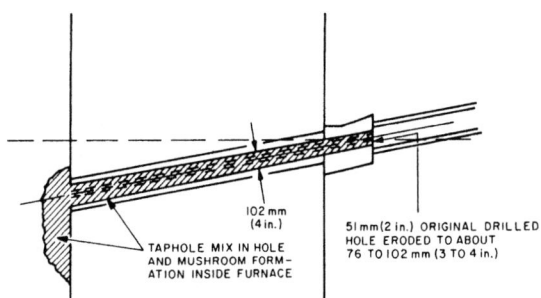

102 mm
(4 in.)

51 mm (2 in.) ORIGINAL DRILLED
HOLE ERODED TO ABOUT
76 TO 102 mm (3 TO 4 in.)

TAPHOLE MIX IN HOLE
AND MUSHROOM FORM-
ATION INSIDE FURNACE

FIG. 2—41. Idealized taphole stoppage.

Casthouse Refractories—Taphole and trough refractories exert an important influence on furnace behavior, production rate, and overall costs. In tapholes, special materials are required to stop the flow of metal-slag against pressure following a cast as ideally illustrated in Figures 2—40 and 2—41. As shown, the taphole is eroded as the materials exit the furnace and the flow is stopped by extruding a column of refractory taphole mix into the furnace. Ideally, the extruded mix will build up on the inner furnace wall to maintain a relatively constant taphole length. Excessive erosion or failure to properly stop the hole can cause serious and dangerous problems in a casthouse. The taphole filler materials are drilled to initiate a cast, and the properties of any taphole mix must be a compromise to provide proper erosion resistance and taphole length without being overly difficult to drill.

Smaller blast furnaces generally use mixtures of clays, coke, and pitch tempered with water for extrudability. The more severe conditions in larger high-production furnaces require the use of anhydrous taphole materials tempered with special tars and containing aggregates that promote erosion resistance (high-alumina aggregates, silicon carbide, silicon nitride, and so on). The nature of the anhydrous materials requires that the taphole gun remain in place for a short period while the material initially softens and flows into place and then rigidizes with heat. The amount of anhydrous material required after each cast is less than for water-tempered materials, and the hot strength of the anhydrous materials is significantly higher than for water-tempered products.

The refractory design of troughs is also dependent on furnace size. In small furnaces with single tapholes, troughs are designed to be tapped periodically, drained, and maintained by frequent gunning or tamping with low-cost materials. On larger and/or multiple-taphole furnaces, troughs are exposed to long periods of tapping and holding of the hot metal and are periodically relined with expensive high-alumina plastic and ramming materials containing carbon or silicon carbide. Figure 2—42 shows the different design for those two trough types using thin or thick monolithic layers. Figure 2—43 indicates the factors which give longer trough life, which is strongly dependent on design parameters such as the depth of metal in the area where the exiting stream contacts the trough surface. Trough lives on large furnaces may range from 40 000

to 20 000 metric tons (44 000 to 22 000 net tons) before replacement.

Blast Furnace Proper—Conditions in a blast furnace vary widely and the refractories may be subject to several wear mechanisms as generally illustrated in Figure 2—44. Figure 2—45 shows a very general description of the types of refractories used in the older standard design furnaces and in new designs now being service tested. The general trends in new furnaces are toward high-alumina products (60 to 99 per cent alumina), high thermal-conductivity carbon products, or special silicon-carbide refractories. The success of any blast-furnace lining depends strongly on the design of the cooling system used and the ability to maintain effective cooling over an extended furnace campaign. Although longer refractory life is usually the result of using more expensive refractories in modern furnaces, a major part of the improvement in life undoubtedly results from more effective cooling (twice as many stack plates, for example) and the consistent operating

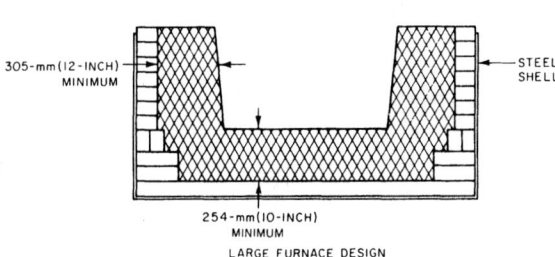

FIG. 2—42. Profile of indicated trough designs.

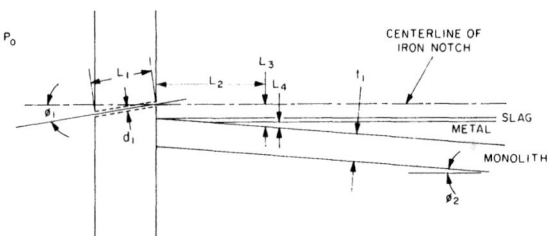

FACTORS GIVING LONGER TROUGH LIFE

P_o Lower pressure in furnace.
ϕ_1 Shallow taphole angle.
L_1 Longer taphole length.
L_3 More difference in elevation of taphole and trough surface.
L_4 More metal depth in trough in impact area.
t_1 Increased trough monolith thickness.
ϕ_2 Less trough angle for pooling trough.

FIG. 2—43. Taphole and trough parameters for maximum trough performance.

Table 2—VII. Some of the Refractory Designs Successfully Used in Bosh and Lower Stack

Area	Refractory Design		
	Type	Thickness	Cooling
Bosh	High-duty fireclay	Full	Internal plate
	Dense 90% alumina	Full	Internal plate
	High-conductivity carbon	Full	Spray or stave
	Composite high-conductivity carbon and special silicon-carbide	Silicon carbide hot-face	Spray or stave
Lower stack	High-duty fireclay	Full	Internal plate
	Impregnated 60% alumina	Full	Internal plate
	Composite dense 90% alumina and fireclay	90% alumina next to shell	Internal plate
	Composite special silicon-carbide and fireclay	Silicon-carbide next to shell	Internal plate or stave

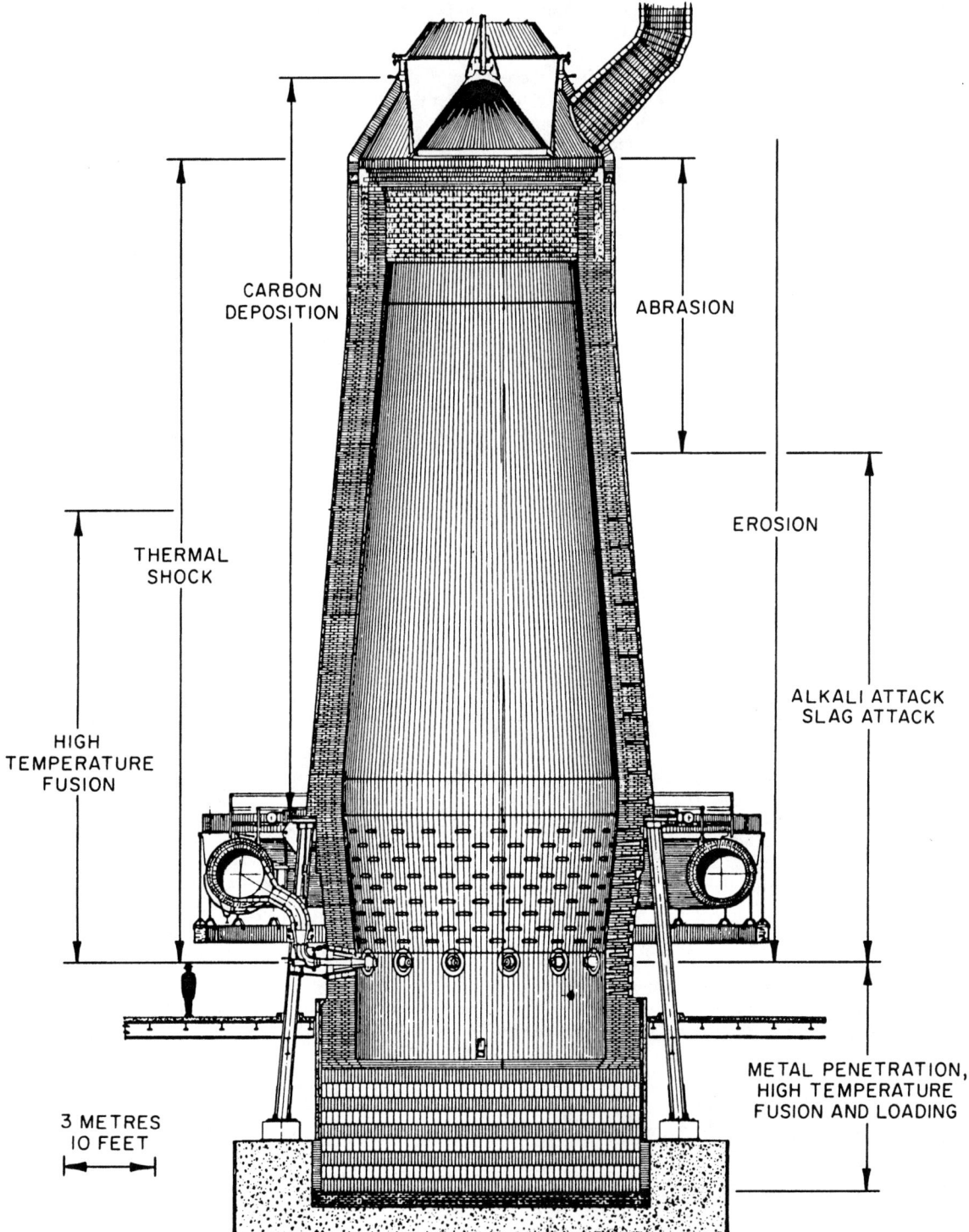

CARBON
DEPOSITION

ABRASION

THERMAL
SHOCK

EROSION

HIGH
TEMPERATURE
FUSION

ALKALI ATTACK
SLAG ATTACK

METAL PENETRATION,
HIGH TEMPERATURE
FUSION AND LOADING

3 METRES
10 FEET

Fig. 2—44. Mechanisms of refractory wear in the blast furnace.

conditions resulting from ideal furnace burdening, charging, and operation. Table 2—VII lists some of the successful bosh designs which have been employed using external spray or stave cooling or internal plate cooling. The higher thermal-conductivity type brick are generally used with external cooling, whereas lower conductivity materials use plate cooling to provide greater stable lining thicknesses. In addition to high thermal conductivity, brick in the bosh area must have resistance to the other wear parameters previ-

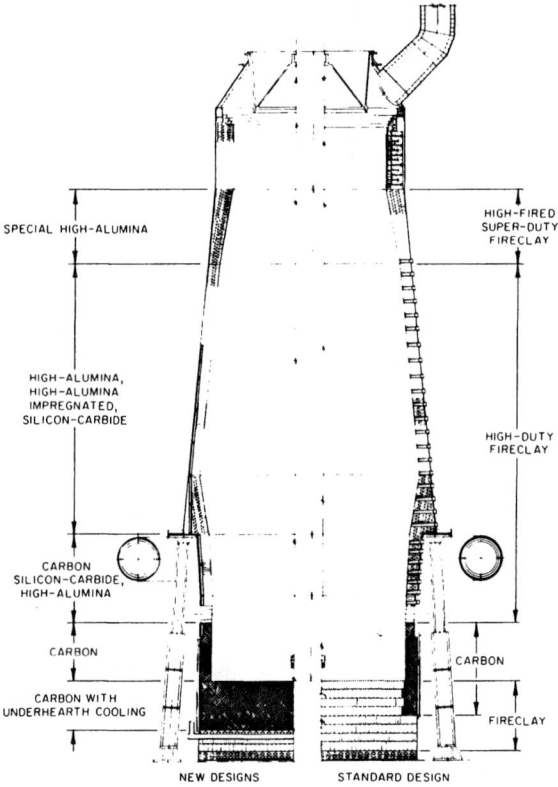

FIG. 2—45. Typical standard and new blast-furnace constructions.

ously described, and only the best grades of those refractories can perform adequately.

Recent use has also been made of high-alumina or silicon-carbide products in the lower stack in a variety of designs similar to those in the bosh. The necessity for more expensive refractories in such designs depends on

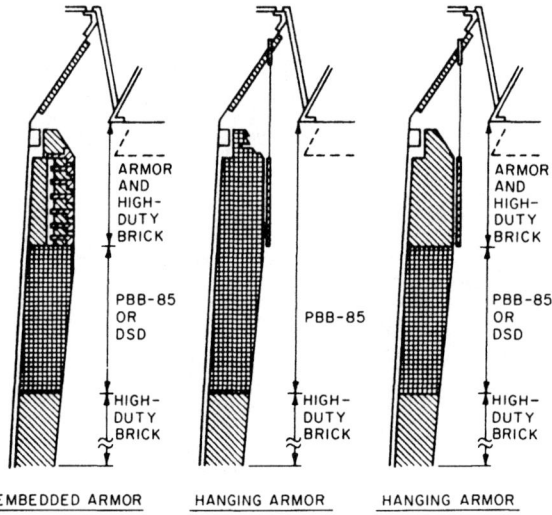

NOTE: PBB-85 = PHOSPHATE BONDED BURNED 85% Al₂O₃
DSD = DENSE SUPER-DUTY

FIG. 2—46. Typical stockline designs.

individual furnace conditions, but again refractories are selected to best resist the particular environment. While lower-stack conditions are similar to those in the bosh, the potential for oxidation normally precludes the use of carbon refractories in this area. Expensive lower-stack refractories are often faced with burn-in linings of fireclay which are lost during the early days of the campaign when unstable operating conditions may exist.

Figure 2—46 shows several armor-refractory designs used in the furnace stockline area. The refractory in this area must withstand severe abrasion and also thermal cycling from maximum-temperature operating conditions to contact with wet or frozen burdens. Trials are being conducted with a variety of materials in this area of the furnace, and several special top-charging devices have been developed which minimize the damaging effects on refractories.

In the furnace hearth, thick carbon designs with or without underneath cooling are providing increasingly longer campaigns as the metal freeze line may remain in a stable position for many years of operation.

The life of a blast-furnace lining is so strongly dependent on original design and operating conditions that it is difficult to compare refractory performance among furnaces of widely different sizes and operating practices. Lining life varies from 3 to over 10 years for tonnages of 3 to over 20 million tons per campaign. Intermediate refractory repairs are often made by gunning the furnace with special cement-bonded castables to provide additional service with a short outage period. Additional service of 1 to 3 years may be obtained on a distressed furnace using this gunning procedure. More recently, special anhydrous grout materials have been developed to build up worn bosh and lower-stack areas when the grout is injected under pressure in thin areas of the lining.

Furnace Auxiliaries—The refractory design in blast-furnace stoves is also changing rapidly as higher blast temperatures are employed to reduce coke consumption. The refractories in stoves producing higher blast temperatures obviously require higher overall stove operating temperatures, with the maximum temperature in the stove dome. Table 2—VIII illustrates the change in refractories in internal-combustion-chamber stoves designed to operate at several temperatures. In all cases, the refractories are selected to have specific creep-resistance values based on stove-temperature profiles. As shown, higher alumina-content refractories can be used to safely raise stove dome temperature to 1315°C (2400°F), whereas further increases require the use of special creep-resistant silica refractories. Refractories in stoves must also resist thermal cycling and the fluxing effects of alkali and iron oxide. The life of refractories in stoves may be quite long, but is strongly dependent on the extent of contamination of the stove by impurities in the gas (alkalis) and the extent of practices where furnace gas containing significant quantities of iron or alkali is backdrafted through the stoves.

The refractory-lined railcars used to transport molten iron to the steelmaking facilities (torpedo ladles) are usually lined with dense fireclay or high-alumina brick, Figure 2—47. More recently, improvements in the life of torpedo ladles have been achieved using pitch-impregnated linings to better resist slag and

Table 2—VIII. Change in General Refractory Construction to Meet Higher Hot-Blast Temperatures in Internal-Combustion-Chamber Stoves

Maximum Dome Temperature		Dome Lining	Combustion Chamber Lining			Checkers			Ring-Wall Lining	
(°C)*	(°F)*		Upper	Lower	Target	Top	Middle	Lower	Upper	Lower
1260	2300	Semi-silica	Semi-silica	High-duty	60% Alumina or Mullite	Super-duty or 60% Alumina	High-duty	High-duty	Semi-silica	High-duty
1315	2400	60% Alumina	60% Alumina	60% Alumina	60% Alumina or Mullite**	60% Alumina	High-duty	High-duty	60% Alumina	High-duty
1425	2600	Mullite or silica	Silica	70% Alumina	Mullite**	Silica	60% Alumina	High-duty	Mullite or Silica-60% Alumina	High-duty
1535	2800	Silica	Silica	70% Alumina	Mullite**	Silica	60% Alumina	High-duty	Silica-60% Alumina	High-duty

*Dome controls set 65° to 93°C (150°-200°F) lower.
**Ceramic burner so no target wall.

metal erosion. The life of torpedo ladles depends strongly on inherent factors in their use (travel distances, number of uses per day) and on the amount of slag in the ladle and the frequency of slag dumping. Lives of 100 000 to 200 000 metric tons (110 000 to 220 000 net tons) are common between brick relining, and the time between relining may be extended to 300 000 to 600 000 metric tons using practices where the ladles are periodically removed from service, cleaned, and gunned when cold with refractory cement-bonded materials. The overall costs per ton of the brick or the brick-gunning practices are similar and

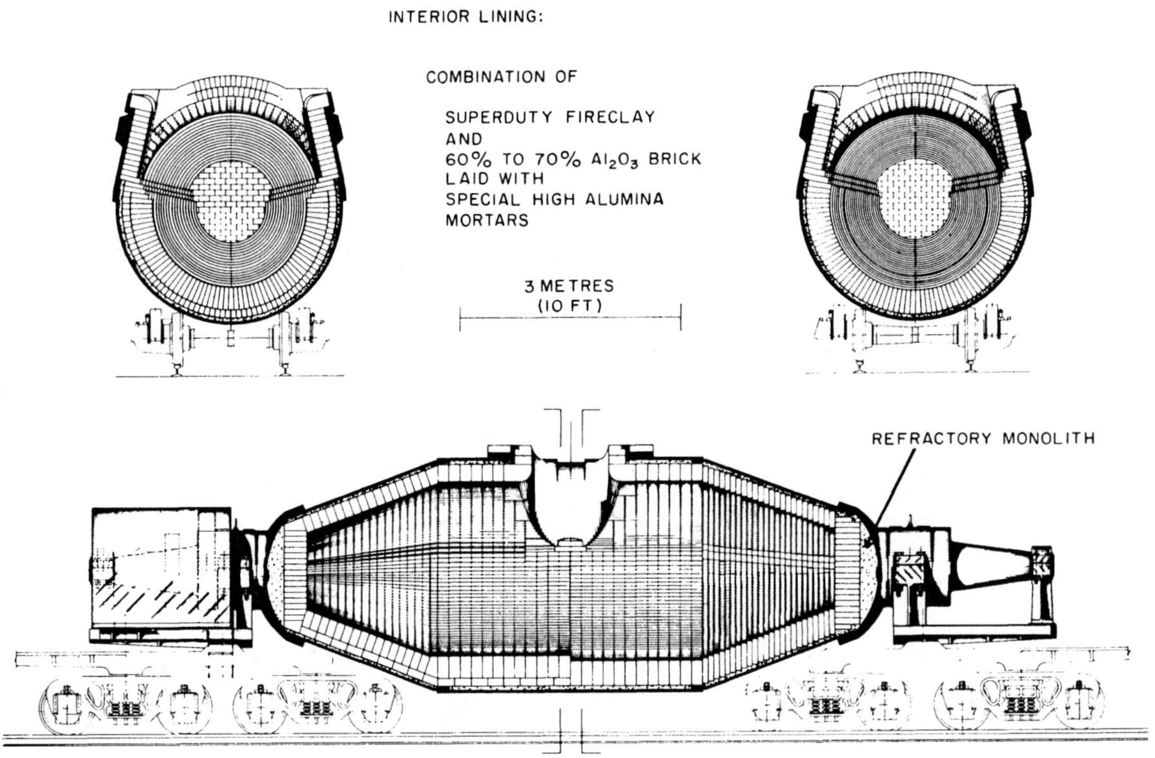

INTERIOR LINING:

COMBINATION OF

SUPERDUTY FIRECLAY
AND
60% TO 70% Al₂O₃ BRICK
LAID WITH
SPECIAL HIGH ALUMINA
MORTARS

3 METRES
(10 FT)

REFRACTORY MONOLITH

FIG. 2—47. Typical torpedo ladle.

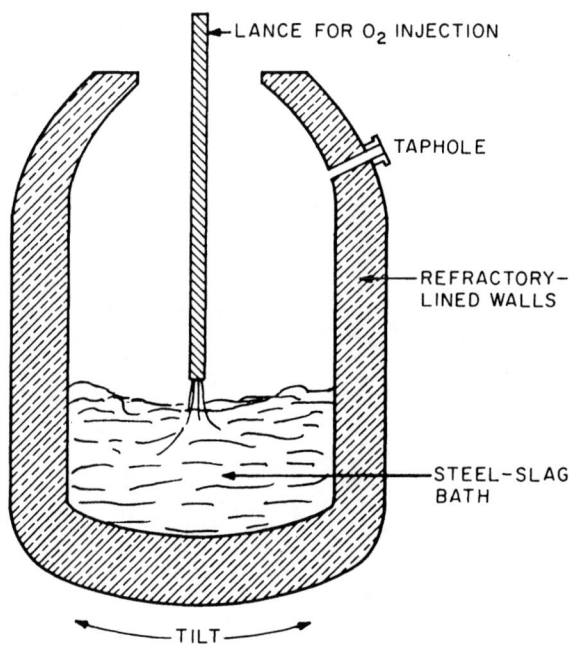

FIG. 2—48. Generalized vertical section of a BOP furnace.

the choice of practice depends on local plant preference. Desulphurization of hot metal in torpedo ladles can reduce the refractory life obtained and necessitates the use of more expensive refractories.

REFRACTORIES IN THE BASIC OXYGEN STEELMAKING PROCESS

The largest quantities of steel today are produced by top- or bottom-blown basic oxygen steelmaking furnaces as illustrated in Figures 2—48 and 2—49. In the top-blown process (BOP), high-purity oxygen is introduced through water-cooled copper lances, whereas the oxygen is introduced through special hydrocarbon-cooled tuyeres in the bottom-blown Q-BOP process. The refractories in either process are subject to a variety of wear mechanisms. Figure 2—50 presents a conceptual idea of the conditions to which a BOP lining is exposed including impact and erosion as scrap and hot-metal are charged (A), high-temperature erosion and corrosion during the oxygen blow (B), and localized slag corrosion during turndowns for steel testing (C) or tap (D). During each heat, the refractories are also subjected to changes in temperature and slag composition as generally shown in Figures 2—51 and 2—52. As the exothermic reactions involved in oxidation of carbon, silicon, manganese, and so on, proceed, the overall bath temperatures increase from 1260° to 1315°C (2300° to 2400°F) to 1590° to 1640°C (2900° to 3000°F). At the same time, some of the oxidized elements enter the slag and combine with added lime to produce slags ranging in CaO to SiO_2 ratio from 1 to 3, and iron oxidizes and enters the slags, making them increasingly more corrosive. Lime may be added during the blow

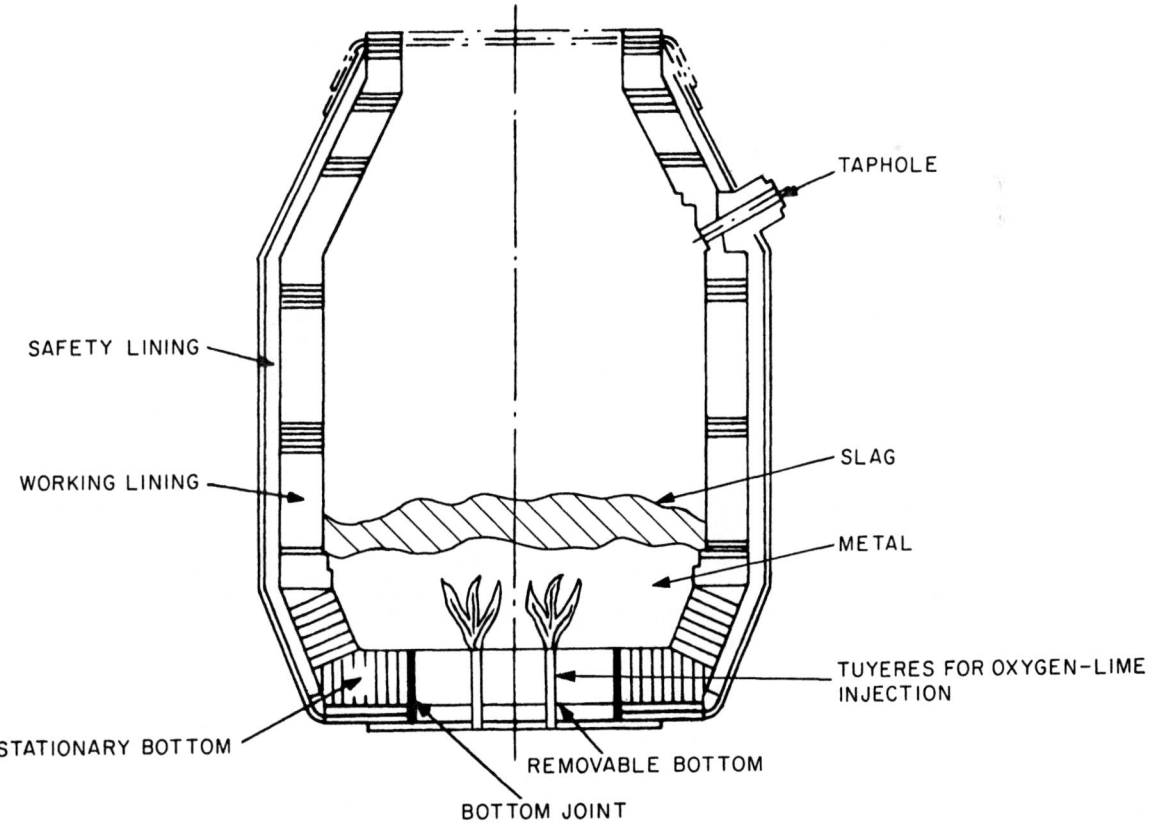

FIG. 2—49. Schematic cross-section of a Q-BOP furnace.

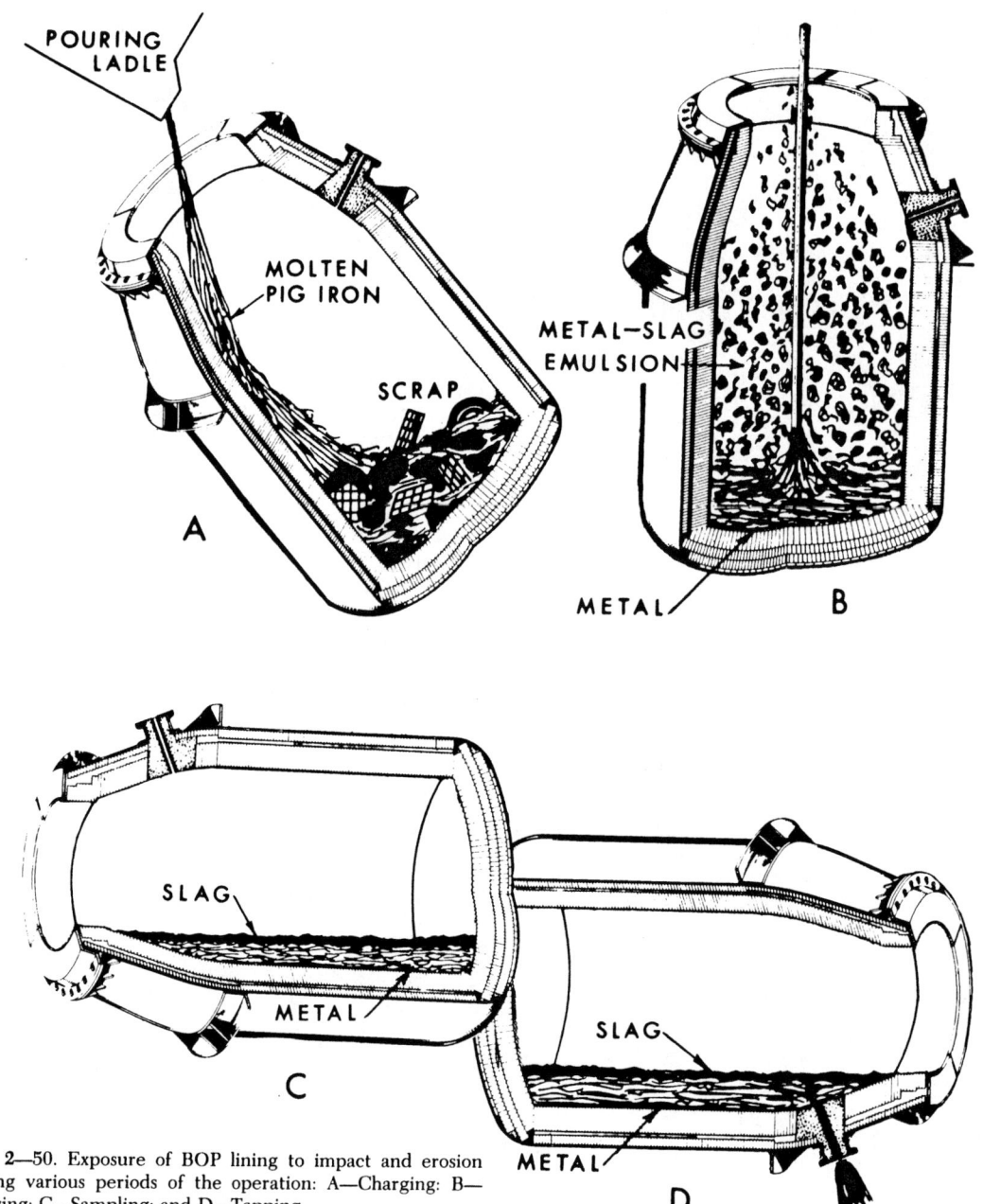

FIG. 2—50. Exposure of BOP lining to impact and erosion during various periods of the operation: A—Charging: B—Blowing: C—Sampling: and D—Tapping.

by top-charging relatively coarse lime as in the BOP or bottom-injecting finer lime as in the Q-BOP. Similar conditions and reactions are encountered in the Q-BOP process, although the exact levels of reaction and the location of the reacting zones are different with bottom blowing. Thus, basic oxygen steelmaking furnace-lining materials must resist deterioration by impact and abrasion of scrap, by metal erosion, by a variety of slag corrosion conditions, and by rapid thermal changes.

BOP or Q-BOP linings are predominantly composed of magnesite brick bonded or impregnated with certain types of pitches as described in other sections of this chapter. Figure 2—53 shows simple and more com-

plex BOP linings wherein refractories are installed in differing types and thicknesses in a zoned configuration to meet the specific demands of each area of the furnace in the most economical fashion. Ideally, such linings wear in a manner such that a minimum of refractory is left in all locations after a campaign.

While similar principles are used in designing Q-BOP linings, the thickness and grades of refractories are selected to allow for the more severe wear in Q-BOP bottoms. The removable bottoms of Q-BOP furnaces are also normally changed one or more times during a furnace campaign to obtain full use of the upper furnace lining, Figure 2—54. Special pitch-magnesite mixes which can be poured or vibrated are used to fill and seal the gap (joint) between the removable and stationary bottoms.

Special structural towers or similar delivery systems, such as illustrated in Figure 2—55 for a Q-BOP furnace, are used to facilitate the relining of basic oxygen steelmaking furnaces, but brick are manually placed. The downtime to cool, tear out, and reline a furnace is typically 4 to 5 days with 2 to 3 days being required for the brick laying alone.

Pitch-bearing refractories are installed in linings without using mortar and generally without expansion allowance because the expansion is accommodated by the yielding of tempered brick during coking or by the melting of the pitch coating on the surface of impregnated brick. Linings are rapidly brought to temperature by burning hot coke with oxygen from the lance in the BOP or by burning natural gas through the Q-BOP tuyeres. Figure 2—56 shows a typical BOP heat-up

schedule where linings are brought into service in only 2 to 4 hours. Longer burn-in times are not only wasteful but also cause loss of desirable residual carbon in the brick structure.

The life of BOP and Q-BOP furnace linings depends greatly on (1) the local hot-metal compositions (Si, P, S, and so on) which determine the amount and composition of the slag, (2) the frequency of reblowing (to correct for missed grade), and (3) operating parameters unique to each plant such as the quantity of heats made with higher tap temperatures for degassing and/or con-

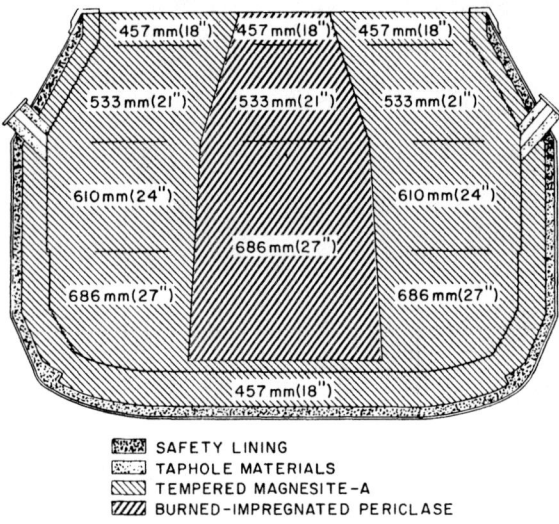

SAFETY LINING
TAPHOLE MATERIALS
TEMPERED MAGNESITE-A
BURNED-IMPREGNATED PERICLASE

NOTE:
NUMBERS INDICATE APPROXIMATE LINING THICKNESSES.
LINING SHOWN AS FOLDOUT THROUGH TAPHOLE.

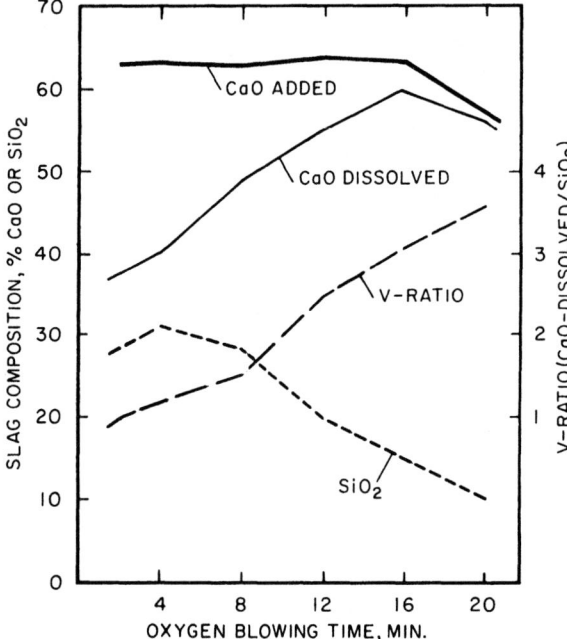

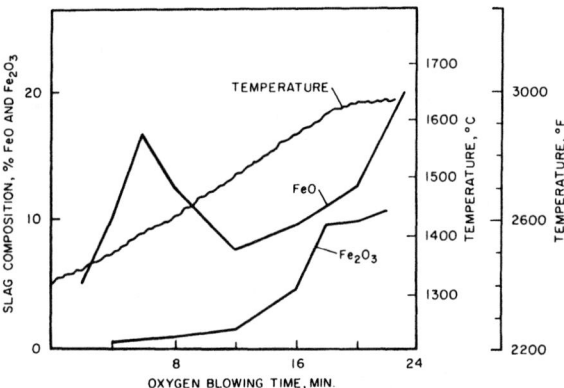

FIGS. 2—51 (Above) and 2—52 (Below). Changes in slag composition and bath temperature during a typical BOP blow.

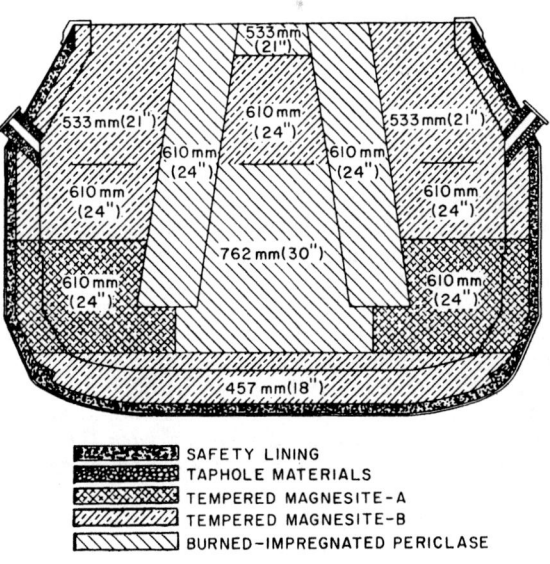

SAFETY LINING
TAPHOLE MATERIALS
TEMPERED MAGNESITE-A
TEMPERED MAGNESITE-B
BURNED-IMPREGNATED PERICLASE

NOTE:
NUMBERS INDICATE APPROXIMATE LINING THICKNESSES.
LINING SHOWN AS FOLDOUT THROUGH TAPHOLE.

FIG. 2—53. (Above) Simple zoned refractory lining for typical BOP furnaces. (Below) More complex zoned refractory lining for typical BOP furnaces.

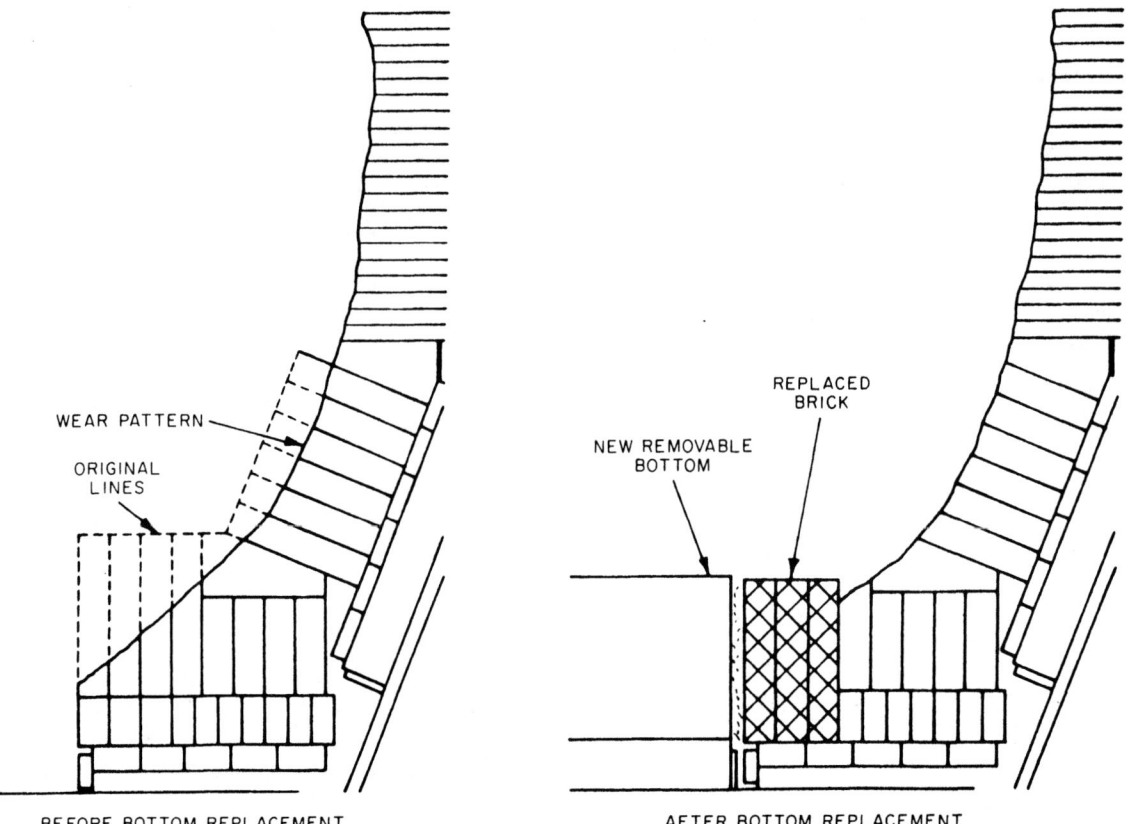

WEAR PATTERN

ORIGINAL LINES

BEFORE BOTTOM REPLACEMENT

REPLACED BRICK

NEW REMOVABLE BOTTOM

AFTER BOTTOM REPLACEMENT

FIG. 2—54. Appearance of stationary bottom area of Q-BOP furnace bottom and a bottom replacement indicating degree of brick patching.

tinuous casting. A lining life of 1000 to 3000 heats is commonly obtained and some shops have obtained over 10 000 heats. The longer life in basic oxygen steelmaking furnaces is related to a great extent to slag-conditioning practices for lining protection. Such practices involve addition of a low-cost source of magnesia to satisfy (saturate) the erosive character of the slags. For example, additions of 10 to 40 kilograms per metric ton (20 to 80 lb per net ton) of light-burned dolomite (dolomitic lime) as part of the slag-forming flux charge result in magnesia levels in the slag sufficient to minimize slag attack.

The amount of MgO in the slag necessary to prevent lining wear depends on hot-metal composition. Although it is impossible to operate without some lining wear, MgO additions to the slag within the restraints of metallurgical requirements (sulphur removal, and so on) can subsequently decrease lining wear. Inherent factors such as the silicon content of the hot metal and iron oxide content of the slag will establish the amount of MgO required for lining protection, which may vary from 4 to above 10 per cent.

Many plants also add dolomitic lime or raw dolomite to slag remaining in the furnace after tap and rotate the furnace so that this more refractory synthetic slag can coat the furnace lining. Using such techniques, it is not uncommon to build refractory slag layers several inches thick on the furnace lining which obviously re-

tard slag attack of the covered brick. Because certain areas of the furnace cannot be effectively coated in this manner, it is also common to use gunning maintenance techniques to apply dolomite or magnesite-based materials in such areas. Table 2—IX illustrates a planned gunning schedule for a typical BOP furnace campaign where the amount of gunning is increased as the furnace lining becomes older. Such gunning practices are accomplished with a minimum of lost production because the long life obtained permits cyclic operations of two or more furnaces.

Refractory consumption in the BOP and Q-BOP obviously varies widely, with the following data providing typical information:

Material	Consumption	
	kg per metric ton	lb per net ton
Brick	1 to 2.5	2 to 5
Gunning Mix	0.25 to 1.25	0.5 to 2.5
MgO slag additive	2 to 12.5	4 to 25

BASIC ELECTRIC-FURNACE REFRACTORIES

The refractories in the basic electric-arc furnace are subject to many of the same wear mechanisms as in basic oxygen steelmaking furnaces, but they are also subject to wear caused by unbalanced electrical power input. This unbalanced power input occurs adjacent to

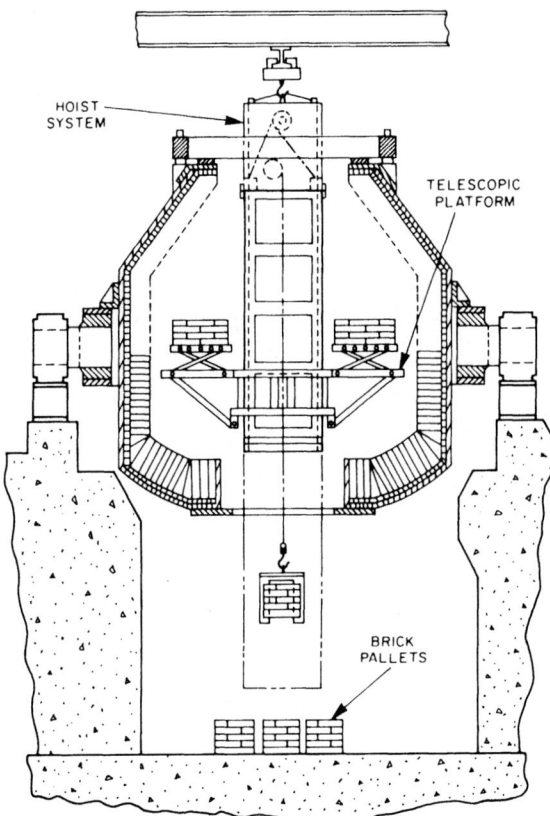

FIG. 2—55. Furnace relining tower.

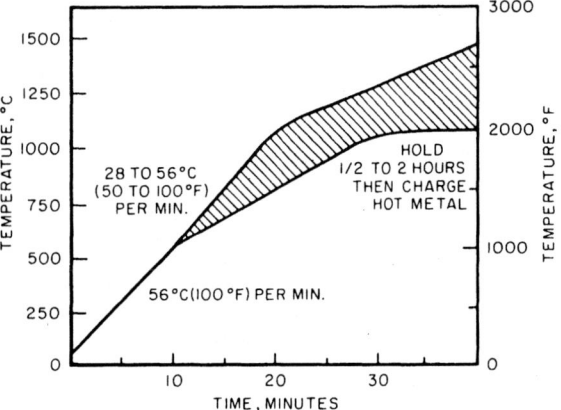

FIG. 2—56. Typical burn-in curve for a BOP lining.

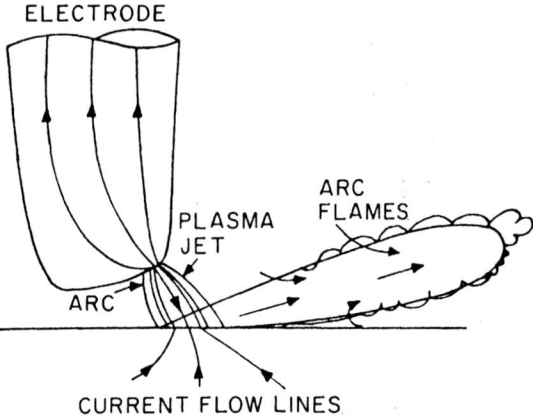

FIG. 2—57. Impingement of arc flames on electric-furnace sidewall.

the electrodes and results in very rapid temperature changes in portions of the lining, causing hot spots. Although not completely understood, this rapid temperature change apparently results from the impingement of a "flare" (luminous gases) on the sidewall. This flare is associated with the high-temperature plasma region between the electrode and steel bath, Figure 2—57. The rapid and localized temperature change associated with this phenomenon results in localized wear of the sidewalls and is conventionally the controlling factor in electric-furnace refractory life.

Table 2—IX. Typical BOP Gunning Schedule

Heats	Gunning Frequency, heats	Gunning Material Consumption kg per metric ton	lb per net ton
0 to 200	None	0	0
200 to 400	every 10	0.25	0.5
400 to 1000	every 6 to 8	0.5	1.0
Over 1000	every 4 to 6	1.25	2.5

Basic electric-arc-furnace refractories are also subjected to frequent and severe temperature cycles when cold scrap is charged on the hot lining and to the fluxing action of iron-oxide fumes and slags. The slag volume and composition may differ widely in practices such as single- or double-slag heats; however, slags are basic and require basic lining materials in all areas.

Table 2—X and Figures 2—58 and 2—59 show typical refractory constructions in the basic electric-arc-furnace hearth and sidewall. In all cases, zoned linings are employed to balance wear and minimize cost. The three concepts described were selected to represent a range of practices which differ appreciably in the sidewall configuration. In the conventional lining, various types of magnesite-chrome brick are used to meet the variable conditions at the sidewall. More resistant fused-cast or direct-bonded brick are used in the severe-wear areas. These refractories may vary in thicknesses from about 305 to 460 mm (12 to 18 inches). Refractories in this type of lining fail by the "hot-spot" mechanism previously described and by cracking from thermal cycling. The cracking is accelerated by absorption of slag components at and above the slag line and by iron-oxide fume absorption throughout the furnace.

The second lining construction shown uses special refractories made from mixtures of magnesite and various graphites bonded with pitch or resin (magnesia-carbon brick). The unique microstructures of these refractories are illustrated in Figure 2—60. The magnesia-carbon brick are less susceptible to thermal

Table 2—X. Refractory Construction in Zoned Electric Furnace

Type	Typical Refractory in Indicated Zones (Figures 2—58 and 2—59)				
	1 Metal-bath area	2 Slag-line and tap-hole	3 Hot-spots	4 Lower walls	5 Upper walls
A. Conventional lining	Magnesite brick or ramming mix Impregnated magnesite brick	Fused-cast magnesite-chrome brick or magnesite-brick	Fused-cast magnesite-chrome brick	Direct-bonded magnesite-chrome brick—clad	Unburned magnesite-chrome brick—clad. Direct-bonded—clad
B. Lining using magnesia-carbon brick	Same as A	Magnesia-carbon brick	Magnesia-carbon brick—clad	Magnesia-carbon brick or direct-bonded brick—clad	Same as A
C. Lining using water-cooled panels	Same as A	Same as A or B	WC* panels	WC* panels	Same as A or B. WC* panels

*WC = water-cooled

cracking than other electric-furnace refractories because the carbon minimizes absorption of iron-oxide fume or slag, and because of their higher thermal conductivity. Steel cladding is used on these brick to minimize the oxidation loss of the important carbon constituents. Sealing materials are also used between the brick and wall to decrease air flow and minimize cold-face oxidation. Although such magnesia-carbon brick are a recent innovation, their use is expanding in modern high-capacity furnaces.

An even more recent development involves the use of water-cooled panels in all or part of the furnace side wall as illustrated in Figure 2—59. Such panels are particularly suitable for use in the hot-spot areas. These panels are designed to provide proper water velocities in welded or cast steel, iron, or copper bodies. In most cases, systems for recycling water are provided to minimize blockage of the cooling passages. Care must be taken to avoid direct contact of the metal bath with the panels and damage of the panels by direct scrap impact. Heat losses to the panels are decreased by slag coatings on the panels, and many panels have surfaces designed to hold such coatings.

As might be expected from the foregoing, the life of sidewall refractories varies widely from a situation where the hot spots are patched at 20 to 40 heats with

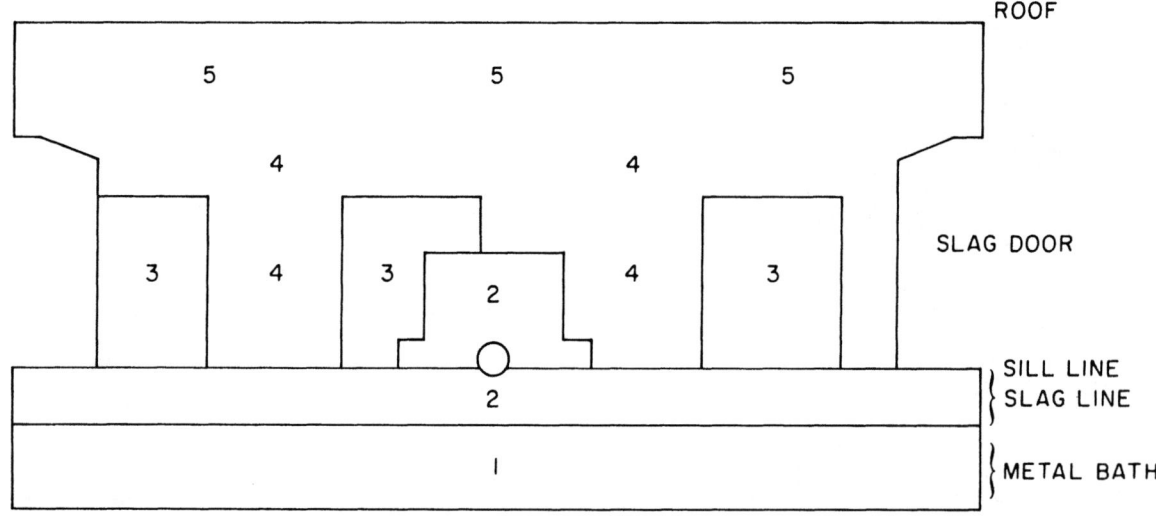

FIG. 2—58. Various areas of zoned electric-furnace refractory lining shown as foldout through slag door. See Table 2—X for the refractory identification in each zone.

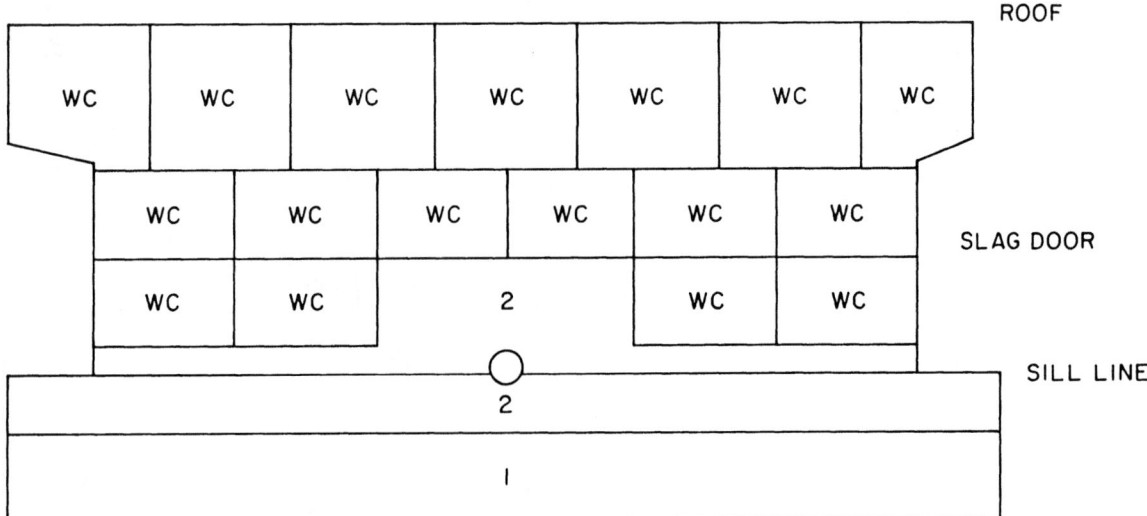

FIG. 2—59. Various areas of zoned electric-furnace refractory and water-cooled lining shown as a foldout through slag door. See Table 2—X for the refractory identification in each zone.

total sidewall life of 80 to 100 heats, to a situation where the life is 150 to 200 heats without patching. Water panels can last 1000 to more than 3000 heats without replacement, but patching of the refractories

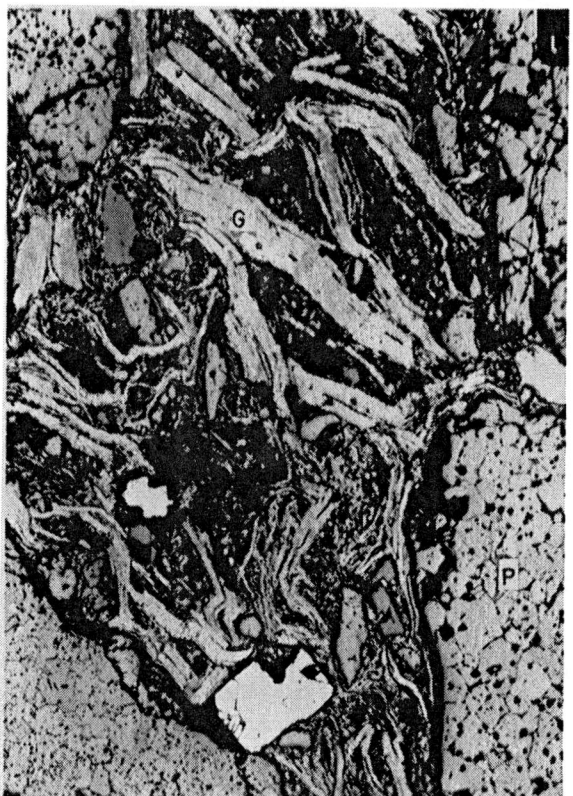

FIG. 2—60. Photomicrograph of magnesia-carbon brick for electric-furnace hot-spots. Coarse periclase (P) and flake graphite (G) particles are in a matrix of fine periclase particles. The white particles are silicon antioxidant. Reflected light: original magnification 65X (reduced slightly in reproduction).

underneath, beside, or above the panels is required at more frequent intervals. In all three constructions, magnesite or dolomite gunning materials are used to patch worn refractories, to help coat the water-cooled-panel surfaces not covered by slag, and to patch areas around panels.

In the furnace hearth, magnesite monolithic or brick materials are almost universally used, as generally illustrated in Figure 2—61. The life of hearth refractories is quite long (1000 to 5000 heats) without reline although monolithic dolomite or magnesite maintenance materials are used on parts of the hearth.

A typical roof construction is shown in Figure 2—62. The roofs generally consist of special double-tapered dome shapes installed against a water-cooled skewback ring with ramming mix around the electrode and off-gas port rings. Although a variety of materials are used, most electric-furnace roofs use 70 per cent alumina brick with 85 to 90 per cent alumina ramming mixes. Roofs last 50 to 150 heats depending on operating conditions and physical parameters such as roof height above the bath. Electric-furnace-roof refractories on modern top-charge furnaces wear by a combination of chemical reaction with iron-oxide fume and absorbed dust and structural spalling or peeling.

When operating conditions dictate, "checkerboard" roofs with alternate direct-bonded basic and high-alumina brick are used in severe wear areas. Roofs exposed to the most severe conditions may have all basic roofs in a suspended type construction or water-cooled panels in all areas except the center section.

REFRACTORIES FOR STEEL HANDLING

Special refractory materials are used to hold and transport steel between various steelmaking processes and the ingot or continuous-casting steps which result in solidified product. During this intermediate period, the steel must be protected from excessive heat loss, oxidation, or contamination and will usually undergo further metallurgical treatment in the same containers.

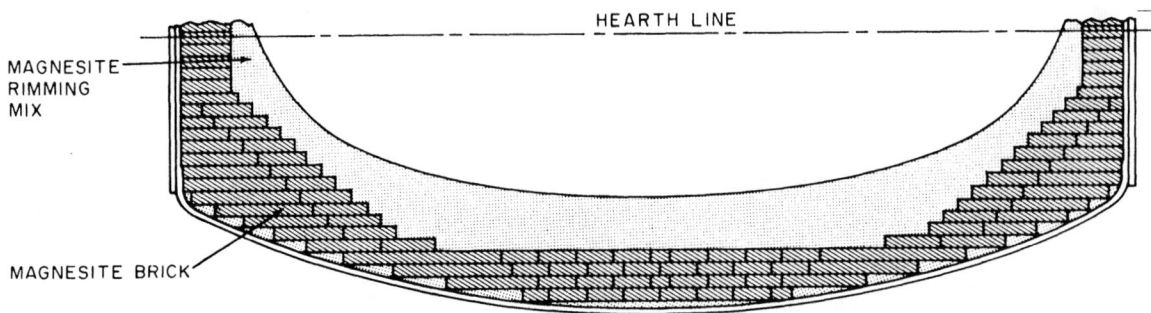

MONOLITHIC STADIUM-TYPE SUBHEARTH CONSTRUCTION

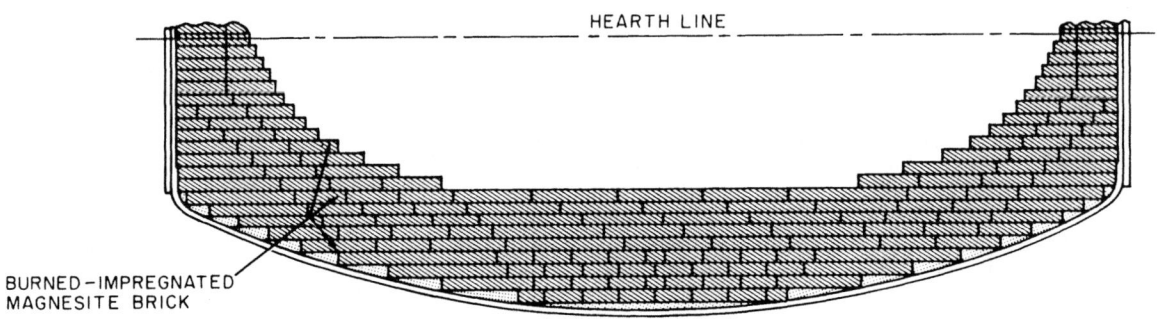

BRICK STADIUM-TYPE SUBHEARTH CONSTRUCTION

FIG. 2—61. Hearth construction in electric-arc furnaces.

Those metallurgical treatments may include one or more of the following: alloying, deoxidation, degassing, desulphurization, and gas stirring. Figure 2—63 shows a block diagram illustrating some possible alternative steps between the steelmaking process and steel solidification process and also a general description of the refractory used to contain the steel (linings), control flow (stopper rods or sliding gates), protect a steel stream (pouring tubes), or inject (desulphurizing agents or stirring gases through lances). The steel grade to be produced naturally dictates the complexity of the particular process involved, ranging from the simplest case of direct pouring into ingots to processing involving all of the aforementioned steps. The following will briefly describe the refractory use in all of these areas.

Steel Ladles—Steel ladles are generally of a size to handle one heat of steel from the steelmaking process. Figure 2—64 shows a typical ladle using about a 125- to 230-mm (5- to 9-inch) working lining and a 63.5-mm (2½-inch) safety lining. Most such ladles are lined with bloating-type fireclay brick and may use higher alumina brick materials for slag lines or bottoms if required by local conditions. The working lining of such ladles is replaced every 15 to 100 heats depending mostly on the severity of conditions such as the steel-holding time in the ladle (½ to 3 hours) and the rapidity of recycling the ladle to minimize thermal shock effects. Many steelmaking shops also practice some form of ladle maintenance by hot or cold gunning low-cost siliceous materials onto a worn lining using simple

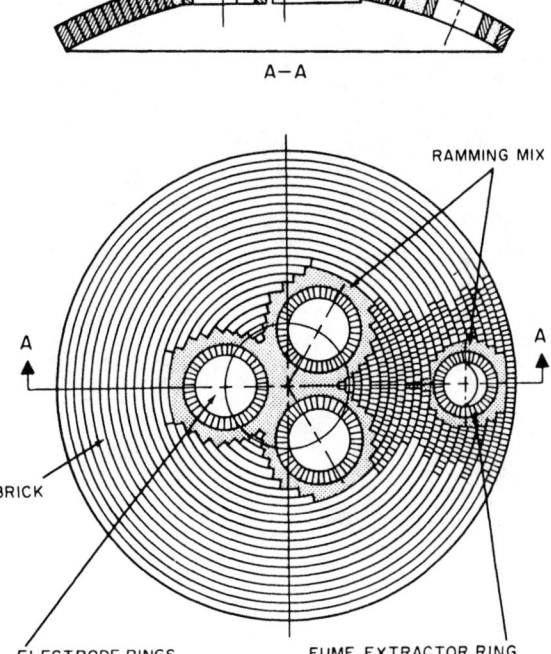

FIG. 2—62. Typical construction in a high-alumina electric-furnace roof.

manual gunning techniques or automatic rotating spray heads. Although longer life can be obtained with more expensive refractory materials (zircon, basic, high alumina) the cost per ton of steel for such materials has generally been higher than that obtained with low-cost fireclay materials. Some special linings are used where required by metallurgical conditions, such as in the desulphurization ladle shown in Figure 2—65 where a basic lining is employed to maintain a low oxygen potential in order to effectively desulphurize the metal. In this case, the ladle is also insulated to minimize heat losses to the lining, prevent skulling, and protect the

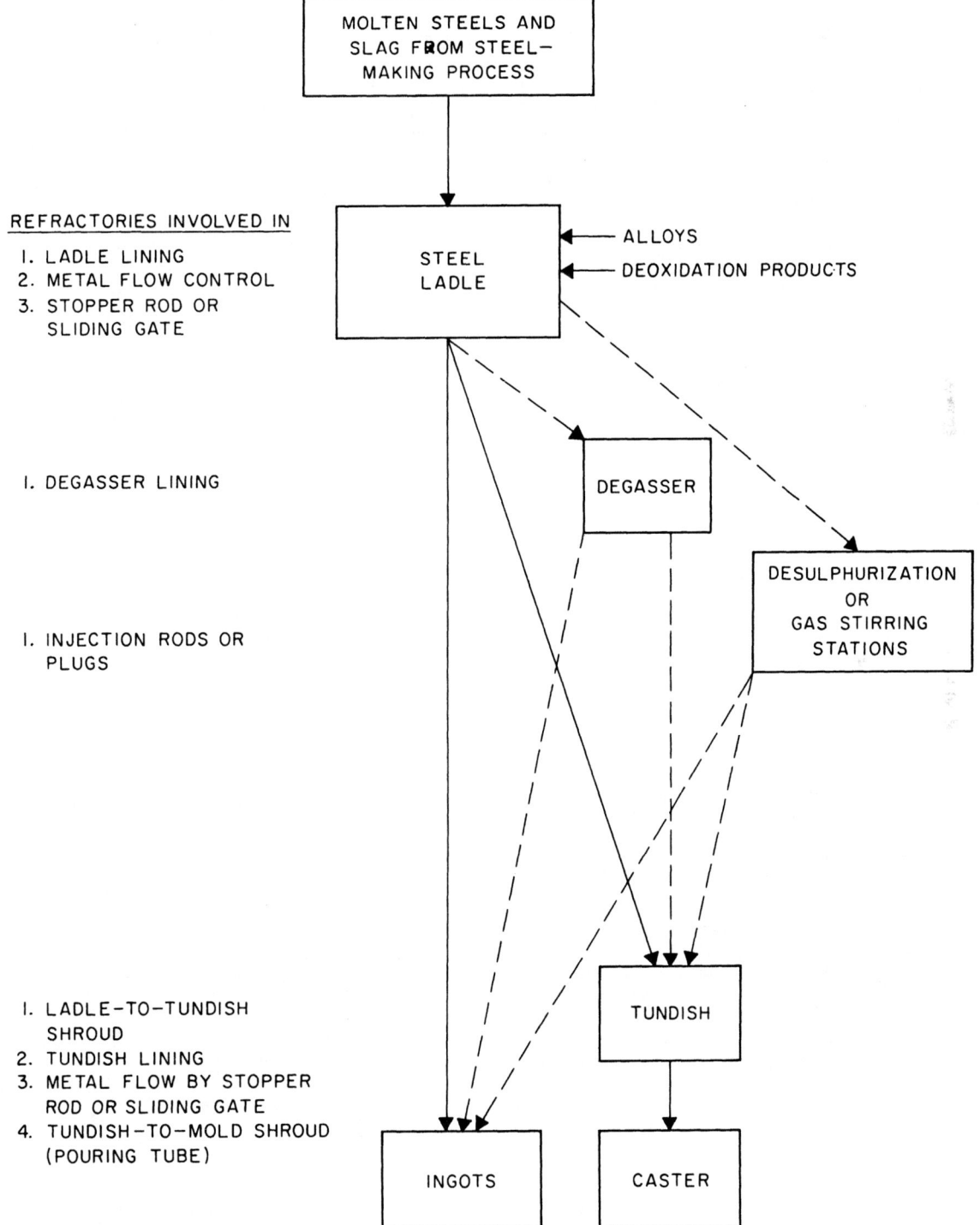

FIG. 2—63. Schematic flowsheet of steel handling from steelmaking process to solidified metal and associated refractory systems.

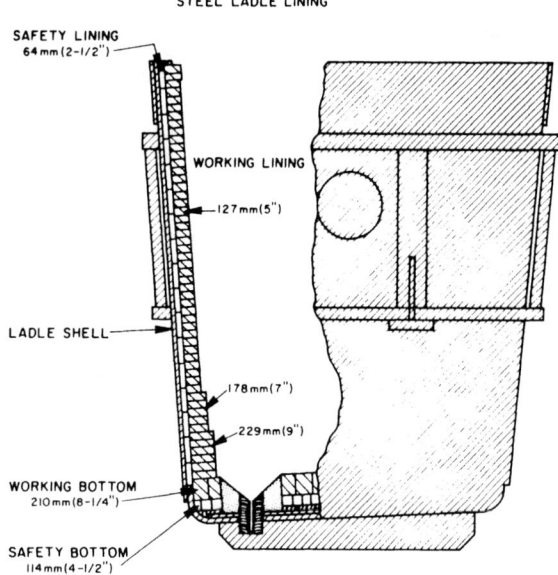

FIG. 2—64. Typical steel ladle construction.

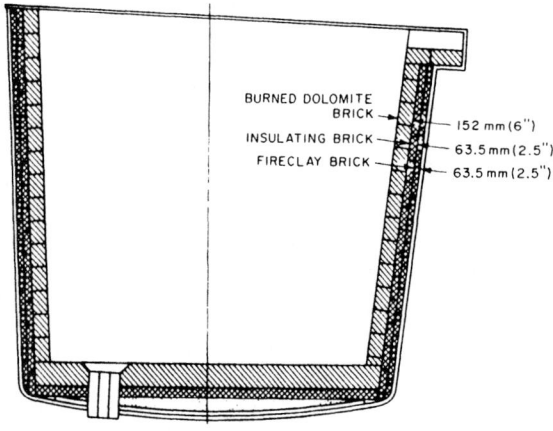

FIG. 2—65. Ladle used for special desulphurized heats.

mal shock involved in instantly cycling from ambient to molten-steel temperatures. The refractory plates in the ladle gates must be sufficiently resistant to sliding mechanical abrasion to prevent the formation of undesirable metal fins or even metal leakage in any abraded areas. Because the valve is frequently used in the throttling or partly open position, these plates must also be reasonably resistant to erosion by molten steel. Plates with low porosity and small pores slow the penetration of metal and oxides, and the denser, higher strength plates are more resistant to abrasion and erosion. Recent trends in manufacturing technology also indicate that impregnation of plates with pitch during the manufacturing process will slow liquid-steel penetration and extend plate life. Properly used plates will

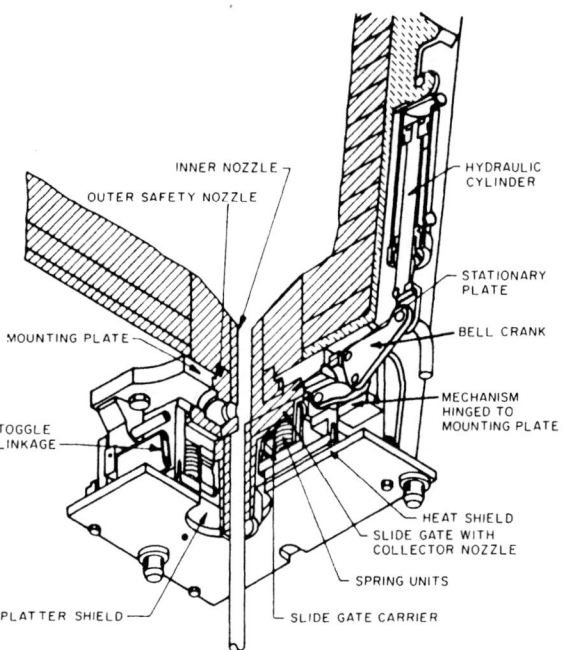

FIG. 2—66A. Sectional view of a slide gate for controlling flow of molten steel from a ladle.

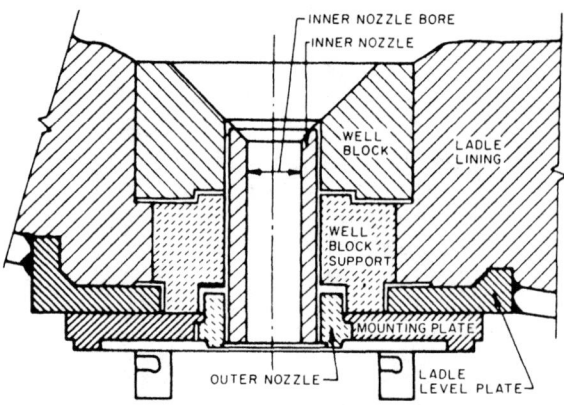

FIG. 2—66B. Refractory construction in the well area of a ladle equipped with a slide gate.

steel shell from overheating.

Metal-Flow Control in Ladle—Flow of the metal from the ladle may be controlled by stopper rods or sliding gates. More recently, more ladles have been equipped with a sliding-gate system when the extended holding times or other factors associated with continuous casting or in ladle processing have required external flow-control systems. Figure 2—66A illustrates the concept of the sliding ladle gate on a steel ladle using air-cooled stainless-steel springs to maintain a constant sealing pressure on the slide-gate interface. The valve mechanism is attached to the ladle bottom with non-adjustable linkages to permit rapid opening and closing of the valve for replacing refractories. Except for the inner nozzle, the well refractories should last the life of the ladle lining. Figure 2—66B illustrates the refractory construction in the well area. The top plate and sliding valve gate are usually made of special high-alumina refractories capable of resisting the ther-

often last several heats, and new refractory developments are extending the life of plates.

TUNDISHES

Tundishes between the ladle and continuous caster are large shallow containers for molten steel that provide a stable flow of metal into the mold by maintaining a constant ferrostatic head (independent of the diminishing steel head effect in ladles as they empty). The stable stream of steel prevents its oxidation. In addition, the tundish helps separate nonmetallics from the steel before it enters the mold. The type and design of tundish used will vary widely depending primarily on the machine design (single or multiple strand) and the casting sequence (number of consecutive heats). Table 2-XI lists some of the parameters considered in designing a tundish which dictate its desirable refractory life between maintenance steps. Figures 2—67 through 2—72 illustrate some of the tundish designs for different types of casters and operating sequences.

The location and number of strands generally dictate the overall geometry of the tundish. A trough arrangement is generally required for a multiple-strand machine, Figure 2—67, while a rectangular box is more

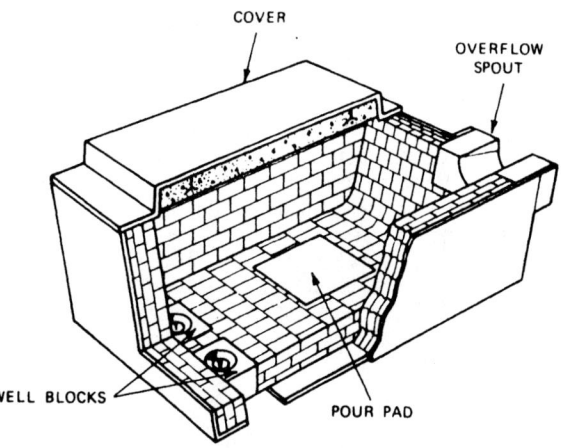

FIG. 2—68. Rectangular-shaped tundish for single-strand casting.

suitable for a single-strand machine, Figure 2—68. In addition, the transfer-ladle positioning device (rotor or ladle car) and the emergency system for handling loss of pouring control, overflow of the tundish, or unscheduled termination of a cast further determine tundish-shell geometry and the location of overflow spouts. To a lesser degree, the trunnion location for tundish handling, and structural members for shell rigidity and support, will have some influence on the final shell geometry.

The required volume or capacity of a tundish, which is essentially a metering device, is predetermined by the size of the strands and the desired casting rate. The tundish capacity may also be influenced by the desire to sequence-cast. In this type of casting, there must be sufficient capacity to allow time to change ladles while continuing to cast. Some additional tundish capacity may be required for sequence casting to accommodate the variability of the tapping schedule of the steel-producing process.

The factors discussed above determine the length, width, depth, and general geometric configuration of

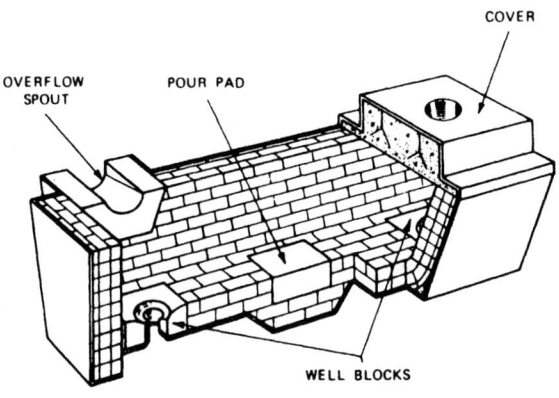

FIG. 2—67. Trough-shaped tundish for multistrand casting.

Table 2—XI. Some Operating Parameters Affecting Tundish Refractory Design

Parameter	Batch-Type Caster Operation	Short-Sequence Caster Operation	Long-Sequence Caster Operation
Product mix	Broad mix	Moderate mix	Narrow mix
Consecutive heats to be cast	1	2 to 3	> 20
Maintenance cycle on lining	1 to 3 casts	Option of 1 to 3 casts or reline after full life of 30 to 60 heats	Reline after full life of 20 to 30 heats
Degree of preheat	None to slight	Slight to significant	Significant
Special tundish features required	Stationary with no insulation or flow control	Tilting or stationary with internal or external flow control and some insulation	Stationary with full insulation and external flow control

the tundish shell. The following discussion will cover those factors that determine the lining contour. The operating practices in starting a cast and the methods of slag and skull removal influence the internal shape of the lining. For those casters where a running start (flow control out of the tundish is open while filling the tundish) or where draining of slag and residual metal through the nozzles and wells after a cast is common practice, a bottom sloping from the ladle stream to the wells is required, Figure 2—69. Tundishes that are taken off-line and cooled for deskulling or are tilted and oxygen-lanced for cleaning require smooth lining contours and sidewalls that slope away from the vertical, and in the latter case, toward the overflow spout, Figure 2—70.

The final design parameter determining shape and size is that of the flow characteristics of the steel into and out of the tundish. Water flow-model studies which simulate liquid steel in an actual tundish have been beneficial in determining critical areas of design. The bath depth and well locations at the ladle stream are critical in dissipating the fluid force of the ladle stream

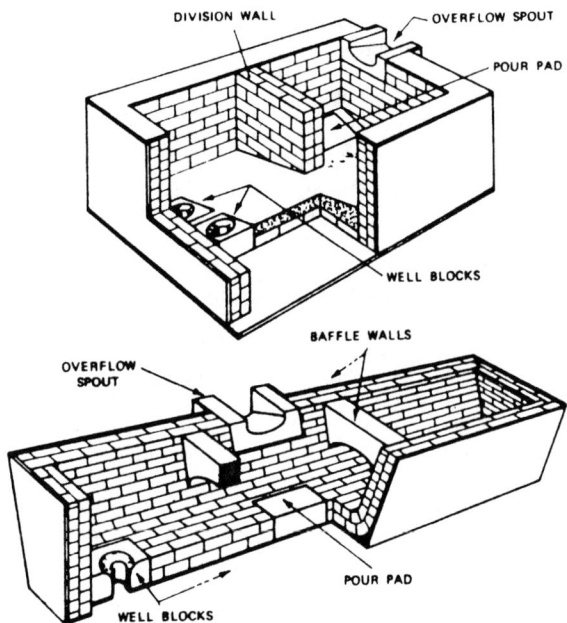

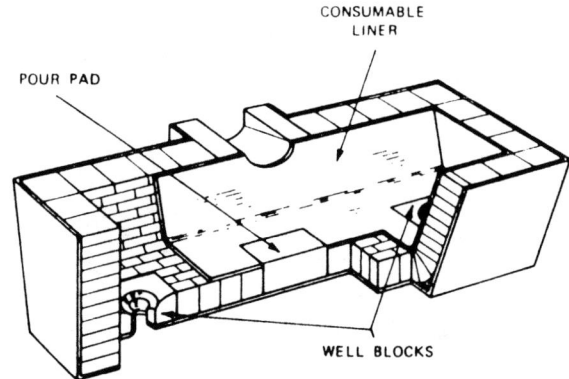

FIG. 2—71. Tundishes with division and baffle walls.

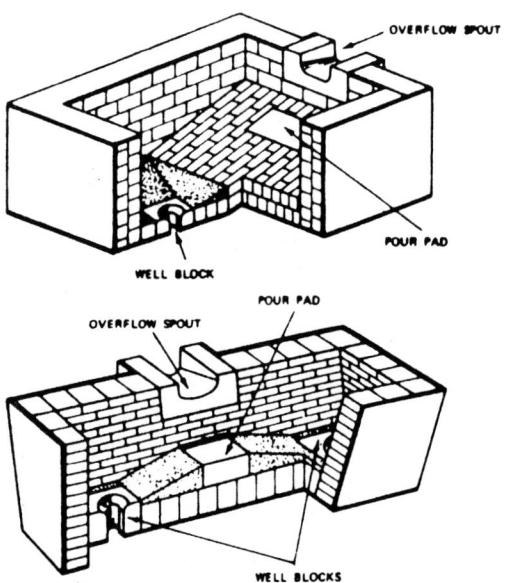

FIG. 2—69. Sloping-bottom tundish designs.

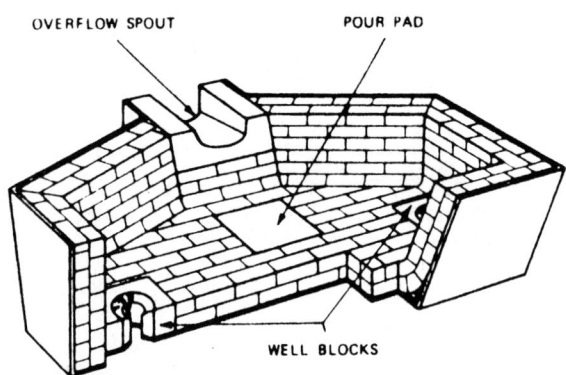

FIG. 2—70. Tundish with back walls sloping toward overflow spout.

FIG. 2—72. Insulating liner in tundish to minimize preheat and facilitate skull removal.

to minimize turbulence (which entrains slag and oxidation products into the steel) while minimizing erosive wear of refractories. Well locations are also important in helping to minimize the entrainment of nonmetallics into the tundish stream. The distance from the ladle stream to the wells is important in providing sufficient residence time for the flotation of nonmetallics. This final factor may require the use of baffles or division walls in some tundishes, Figure 2—71.

As a container for liquid steel, the tundish is of necessity lined with refractories. Some degree of thermal insulation is required to prevent skulling, to protect the shell from overheating, and to provide comfortable, safe working conditions for the operators. The operating parameters generally affecting insulation requirements are:

(1) Length of preheating time, and temperature to which the tundish is preheated.

(2) Time required to cast a heat of steel.

(3) Number of heats cast sequentially.

(4) Deskulling practice.

(5) Operating cycle (turn-around time and practices used in preparing the tundish for the next heat). Shell temperature and propensity for skulling are used as the criteria in selecting the optimum lining design.

Selection of Tundish Refractories—Once the operating parameters and degree of insulation necessary for a particular tundish have been established, considerable choice exists as to the refractories for the tundish. As noted previously, tundish linings may be allowed to wear without maintenance or maintained periodically with special patching materials. These tundish-maintenance materials are applied either by troweling or gunning and serve to provide a surface from which skull is readily released as well as acting as the working lining. The selection of the particular materials to be used for working linings is made on the basis of economic considerations for a particular shop and adjusted continually to develop the lowest-cost practices. Recently, use has also been made of insulating-type boards as the working lining of batch caster tundishes to decrease refractory cost and eliminate preheat, Figure 2—72.

Pour pads, which receive the impact of the incoming ladle stream, require special materials with high strength at operating temperature. Large refractory blocks made from alumina or alumina-chrome are usually employed to minimize erosion, and isostatically pressed alumina block have been used successfully in some tundishes.

The purpose of the tundish cover is to provide thermal insulation and protection for the operators against the splash of steel while it is being poured from the ladle. Some batch casters use steel slabs for covers when no insulation is required.

Three types of refractory materials have been used for tundish covers: insulating brick, castables, and ramming materials.

Tundish Nozzle and Flow-Control Devices—The tundish nozzle is a critical link in the continuous-casting system because it must deliver a constant and controlled rate of flow of steel to the mold with minimum stream flare to minimize spatter, spray, and atmospheric oxidation. Because the ferrostatic head remains substantially constant in the tundish throughout casting of much of each heat, the bore of the nozzle must remain at a constant diameter throughout the cast. High purity, stabilized zirconia, having low thermal conductivity and good erosion resistance, has demonstrated the most desirable overall properties of any type of nozzle used to date.

Stopper-rod flow control for tundishes is accomplished in a manner similar to that previously described for ladles. Similar refractories are used in both applications, except that the tundish nozzles are generally made of materials such as zirconia. Tundish stopper rods are generally smaller, however, and air- or inert-gas-cooled. Cooling is particularly necessary when several heats are cast back to back. In certain cases, the stopper-rod and sleeve assembly has been replaced by

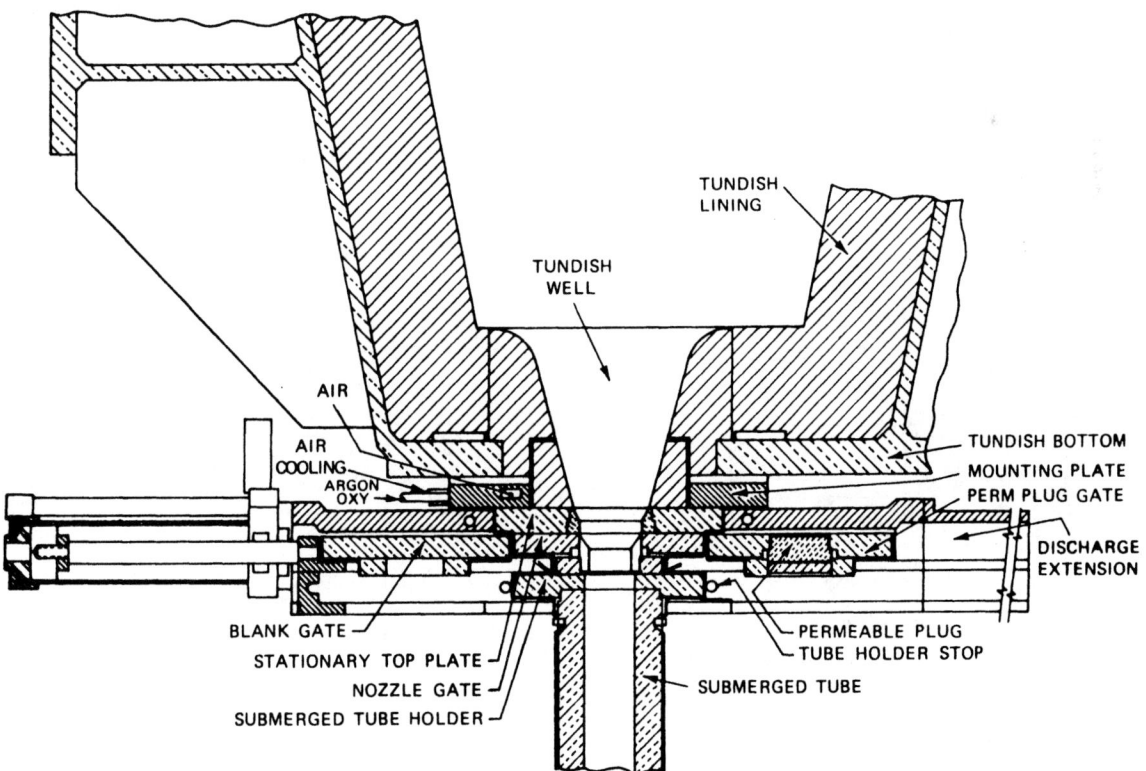

FIG. 2—73. Tundish valve showing submerged tube in position.

a one-piece unit of essentially the same outer configuration as the rod made from isostatic pressed alumina-graphite. Because this system contains no steel support rod, it can be used without cooling.

Tundish sliding gates have advantages over stopper rods. The nozzles in the tundish slide gate can be changed during casting if they become worn or clogged; this capability allows a tundish to remain in use continuously during the sequential casting of multiple heats. The nozzle size or controlling orifice can also be changed on demand if a change in a pouring rate is required. The use of a slide gate also ensures an instantaneous positive shutoff.

Figure 2—73 illustrates the general design concept of the tundish gate with the capability of changing the controlling orifice. There are also tundish gates which operate on the same general principles as the ladle gates. Sliding-gate valves on tundishes can be converted for use with a pouring tube that extends below the tundish and is submerged in the metal in the casting mold, thereby protecting the stream from oxidation. With this unit, it is possible to support a pouring tube independently and to maintain a seal between the sliding gate and the tundish and between the tube and the gate.

When a submerged pouring tube is used it is impossible to open a clogged nozzle using a conventional oxygen lance; therefore, special permeable-plug start-up gates are employed. The use of a permeable-plug gate permits an inert gas, such as argon, to be injected and thereby stir the metal in the well area while the tundish is being filled. A short burst of oxygen is generally used to open the nozzle and clear the well of possible skull just before an operating nozzle is put into place. This system has also been adopted in some casting shops as the best way to permit filling of a tundish at start-up, thus minimizing poor quality starting

streams with the usual attendant spraying of the mold cover, lubricating ring, and mold wall.

Sections through tundish slide-gate valves, Figures 2—73 and 2—74, show the refractories that are required. The slide gates, top plates, and tube holders are made of either fireclay or high-alumina refractories, depending on their duty requirements. The fireclay plates are used for shutoff gates and drainage nozzles and may be used for the nozzle gates when low tonnages are cast. High-alumina plates, usually 85 per cent Al_2O_3, are used in nozzle gates when high tonnages are to be cast continuously. Both plates and gates used with high steel-throughput systems are usually made with zirconia nozzle inserts. If inserts are used, the plates surrounding these inserts, as well as the tube-holder plates, need to be no higher than 70 to 85 per cent alumina, since they are not exposed to molten steel for extended periods of time. Erosion-resistant refractory inserts permit precise control of the pouring stream for long periods of time. Furthermore, the used zirconia insert can be removed from an undamaged high-alumina gate plate and another insert cemented in place, thereby permitting the reuse of gates and an attendant reduction in cost.

Pouring-Tube Refractories—Demands for improved steel quality have led to the use of various protective systems for minimizing the oxidation of the steel stream as it leaves the tundish nozzle and enters the continuous-casting mold. Requirements for such systems depend upon the composition of the steel and its susceptibility to oxidation but may be influenced by many other economic and operational factors. Generally, significant quantities of deoxidizers and alloying elements such as aluminum, chromium, silicon, titanium, zirconium, calcium, and manganese in steel increase the need for shrouding of the stream to achieve maximum cleanliness in the steel.

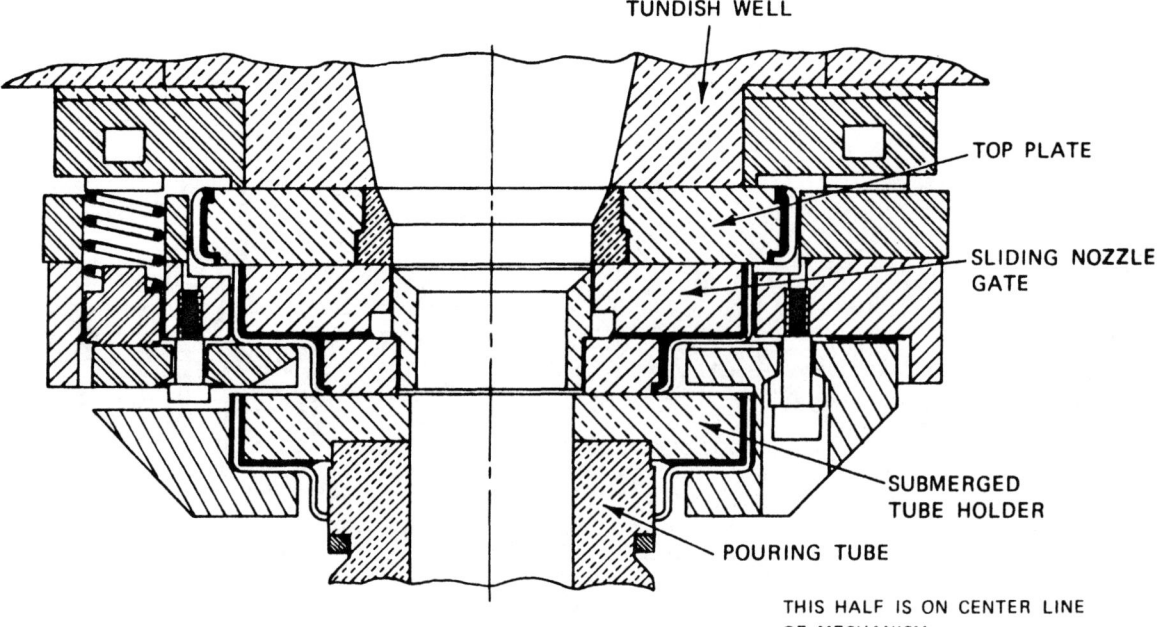

FIG. 2—74. Cross-section of sliding gate with immersed-tube attachment.

The shrouding system that has been most widely used consists of a replaceable refractory tube that is firmly attached to the bottom of the tundish nozzle and extends into the continuous-casting mold and several inches below the molten steel level or meniscus. Such tubes can be either open at the bottom or have a closed bottom containing side ports. The internal diameter of these tubes is determined by the flow rate required to achieve optimum casting rates, and their outside diameter is limited by the mold size. The tube bore should be slightly larger than that of the largest tundish nozzle that is used in casting, whereas the outside diameter will be approximately 50 to 75 mm (2 to 3 in.) less than the smallest diameter of the mold, thus allowing 25 to 38 mm (1 to 1½ in.) clearance between the side of the mold and the side of the tube. This clearance minimizes the possibility of the freezing of steel between the tube and mold wall. The number, configuration, and location of side ports in closed-bottom tubes varies, but usually they are designed to give the most desirable flow characteristics in the mold.

Many types of refractory materials have been tried as pouring tubes. The combination of severe thermal conditions, erosion by high-velocity steel streams, need for simplicity in use, and need for low-cost materials has resulted in the wide commercial acceptance of rebonded fused silica. Alumina-graphite pouring tubes have also gained a significant measure of commercial use, particularly for casting alloy steels.

The rebonded fused-silica pouring tubes now being used generally have a density of 1.88 to 1.98 g/cc, an apparent porosity of 8 to 13 per cent, and a cristobalite content no greater than 1 to 2 per cent. These tubes have excellent thermal-shock resistance so that they can be brought into contact with molten steel without preheating. Furthermore, their low thermal conductivity minimizes the possibility of steel freezing inside the tube. Consistent quality and internal soundness of pouring tubes are important to obtain the full advantage of shrouding. Recently developed nondestructive inspection techniques have significantly increased the reliability and performance of fused-silica pouring tubes.

A major limitation on the use of fused-silica shrouds is their rapid erosion in steels having high concentrations of certain elements, particularly manganese. In certain grades of steel, particularly those that are aluminum-killed, all types of tubes tend to experience a buildup in the bore that is essentially a mixture of oxides, predominantly aluminum oxide, and solidified steel. In this case, a new shrouding system is needed that will prevent the buildup and resulting restriction of the steel stream and decreased casting rate.

SOAKING PITS AND REHEATING FURNACES

Refractories in soaking pits and reheating furnaces are generally of the alumina-silica type and are se-

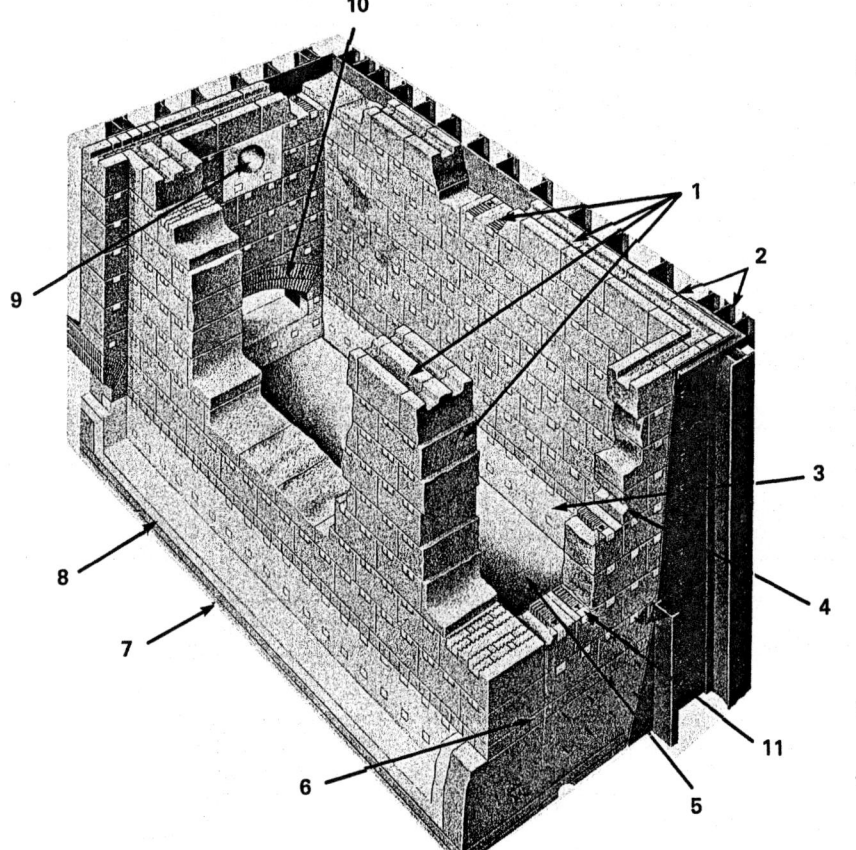

DETAIL OF ITEMS:

1. FIRECLAY PLASTIC WALL BLOCKS INCLUDING DESIGN FOR COPINGS, AND COMMON WALL CONSTRUCTION.
2. STRUCTURAL STEEL CONSTRUCTION.
3. HIGH–ALUMINA (85-90%) PLASTIC SLAG LINE.
4. THERMOCOUPLE PORT.
5. HIGH–ALUMINA PLASTIC OR CASTABLE FLOOR.
6. FIRECLAY PLASTIC BLOCK.
7. INSULATING CASTABLE.
8. FIRECLAY CASTABLE.
9. HIGH–ALUMINA PLASTIC BURNER PORT.
10. SUPERDUTY FIREBRICK RECUPERATOR ARCHES.
11. SUPERDUTY FIRECLAY REFRACTORY ANCHORS AND STAINLESS-STEEL HANGERS.

FIG. 2—75. Modern soaking-pit construction showing plastic block wall construction.

lected to minimize heat losses and to resist the localized conditions in a particular area of the furnace (slag
attack, abrasion, heat shock, and so on). Many of the
refractories today are selected to minimize furnace
downtime and labor costs. Figure 2—75 shows a modern pit construction made from plastic block. Note that
various types of monolithic block refractories have
been used in the various pit areas to resist local conditions such as the slag-resistant high-alumina block in
the slag line. Figure 2—76 shows a detail of the wall
construction used in both pits and reheating furnaces
where superduty plastic (or preformed block) is used
with insulation block and fastened to a steel shell or
superstructure with refractory-metal anchors in a variety of designs. In many cases, preassembled refractory
modules are used to decrease pit downtime. The modulus may be made of brick, plastic, or castable materials.

Details of refractory and support design exert an important influence on the performance of refractories in
pits and reheating furnaces. In pit covers, for example,
anchor design and spacing and coping design and

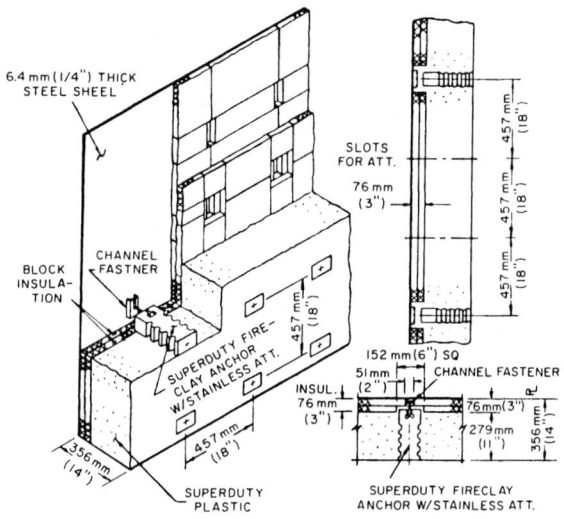

FIG. 2—76. Typical soaking-pit two-component wall construction.

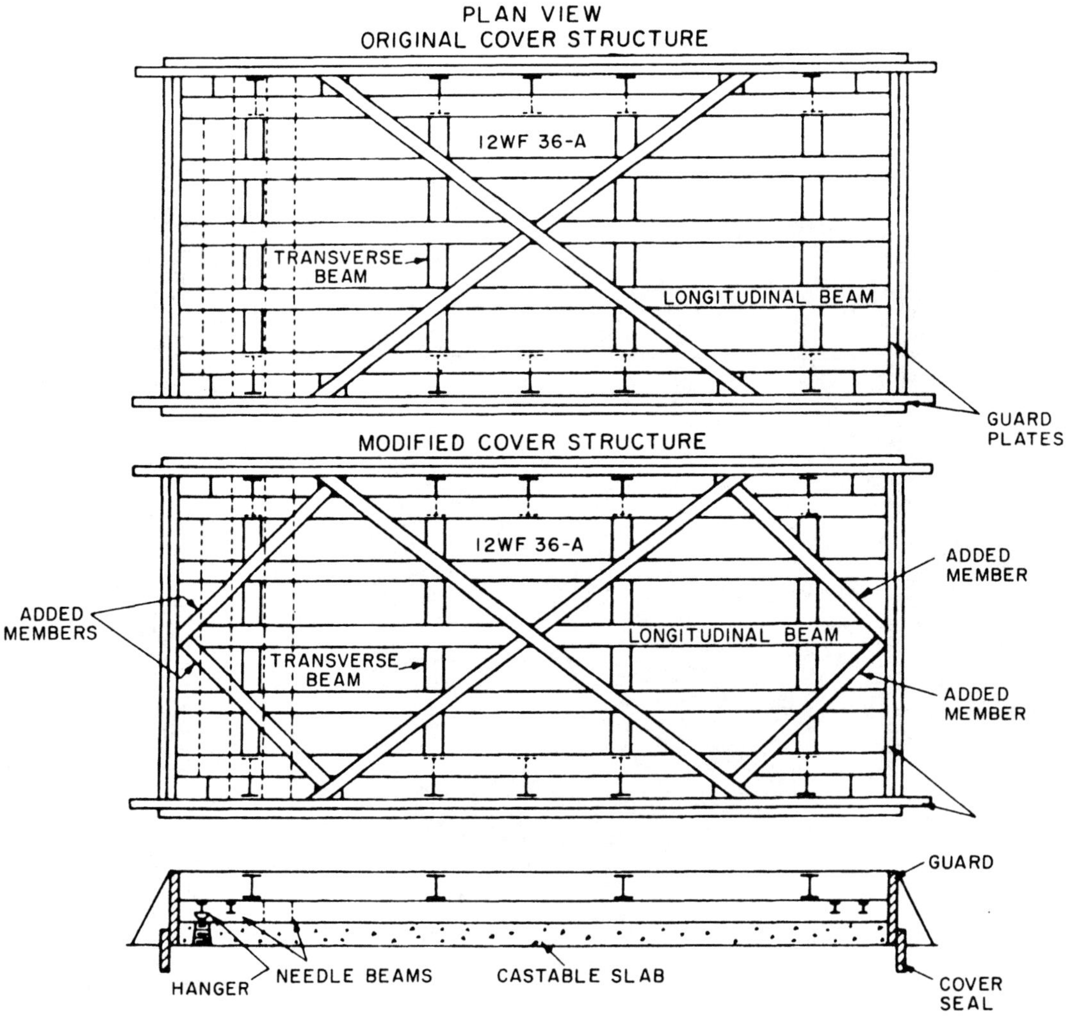

PLAN VIEW
ORIGINAL COVER STRUCTURE

MODIFIED COVER STRUCTURE

FIG. 2—77A. Improvement of soaking-pit-cover stiffness by modification of structure.

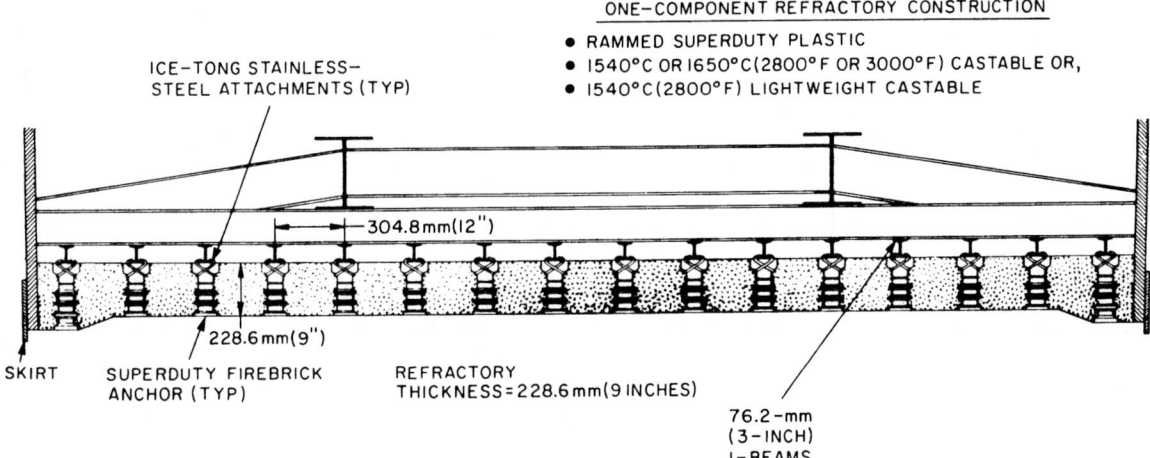

FIG. 2—77B. Cross-section of a typical soaking-pit cover showing structural members and refractory.

maintenance are equally important to refractory material selection. Figures 2—77A and B show a typical cover construction. By rigidizing the cover design properly, multiple improvements in life can be obtained when covers are failing by corner cracking. Where insulated pit covers (like any insulated refractory application) are used, the change in thermal gradient and refractory properties with insulation must be considered, as illustrated in Figure 2—78. Note, for example, that the average temperature at an insulated cover has increased from 800°C to 1150°C (1470°F to 2100°F) and the average hot strength has decreased from about 12 410 to 7580 kPa (1800 to 1100 psi). Figure 2—79 shows the proper method of anchor construction in an insulated cover to avoid anchor failures.

In slab-reheating furnaces, refractories are used over the areas of the skid-pipe system not in contact with the slab to decrease energy usage, Figure 2—80. Special refractory-metal anchoring systems or interlocking refractory shapes are used in this application to resist vibration and facilitate skid-insulation replacement. Figure 2—81 shows some examples of methods of holding refractory tile on to a skid-pipe system.

In recent years, increased use has been made of various forms of fiber refractory materials in many types of reheating furnaces because of their good insulating properties and light weight. Such materials are not suitable in contact with slag or in areas requiring abrasion resistance, but are suitable for many less-severe areas. Figure 2—82 shows a plate-mill continuous-annealing furnace lined with fiber blanket held in place with metallic anchors. Refractory-type anchors may be used in such a system to extend the useful temperature of fibers. Fibers may also be used in stackbonded or

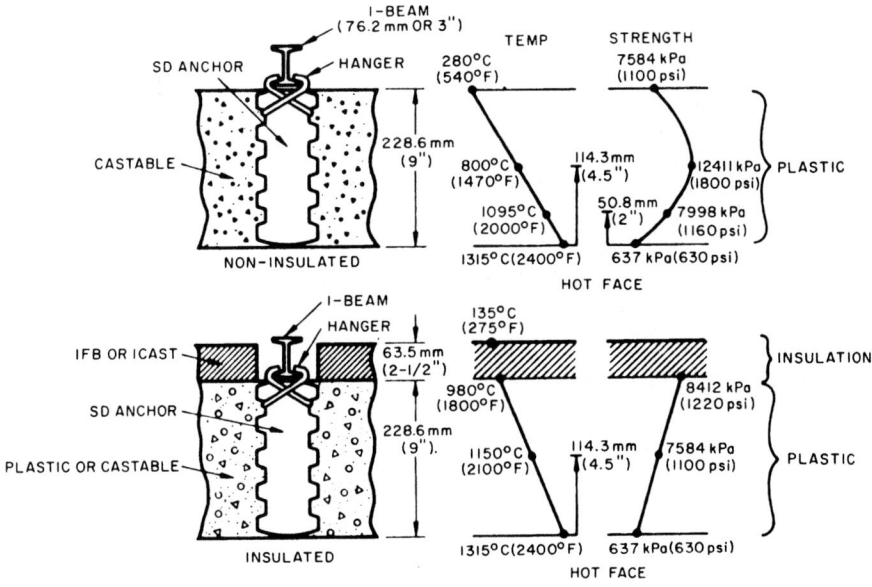

FIG. 2—78. Change in hot strength as a result of applying insulation to the cold face of soaking-pit-cover refractory.

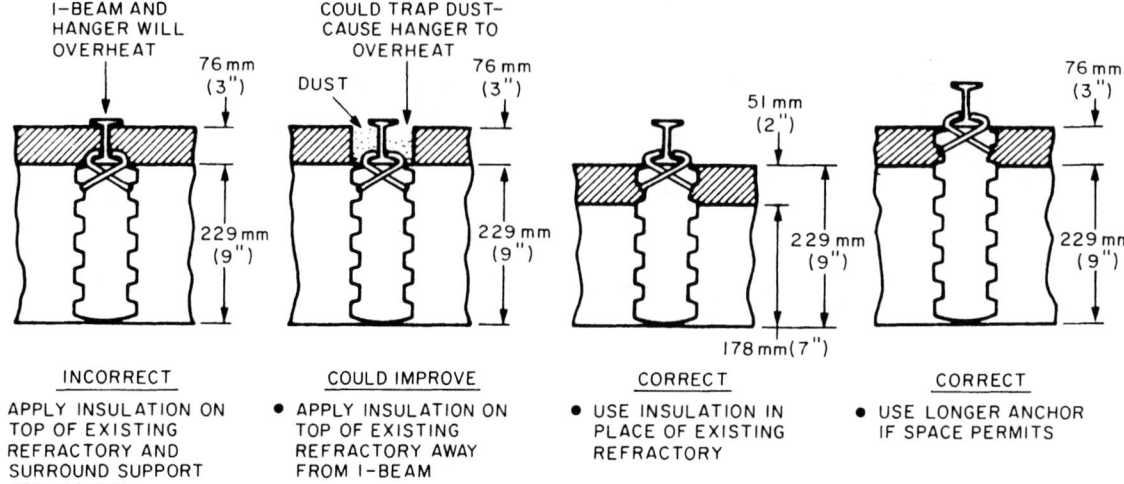

FIG. 2—79. Proper methods of integrating insulation and existing dense refractories for soaking-pit covers.

modular form where the fibers are folded or fastened to a steel backup plate attached directly to the steel super-structure. Veneers of fibers are also being evaluated over existing refractory linings to reduce energy consumption. In all cases, the economics of fiber materials must be carefully evaluated in comparison to other insulating materials including insulating brick, insulating castables, and mineral wool or other insulating blocks.

Refractories are also used in recuperative or regenerative systems. Historically, recuperators have used fireclay recuperative tubes and regenerators have used fireclay checkers. Recently, the top course of recuperator tubes has been replaced with silicon-carbide tubes to improve their resistance to cracking and to provide increased resistance to contamination by iron-oxide scale or slag, coke breeze, and other contaminants.

The refractory life in soaking pits is quite dependent on operating parameters and varies widely. For example, the life of pit covers may range from 2 months to several years, depending on cover and seal design and

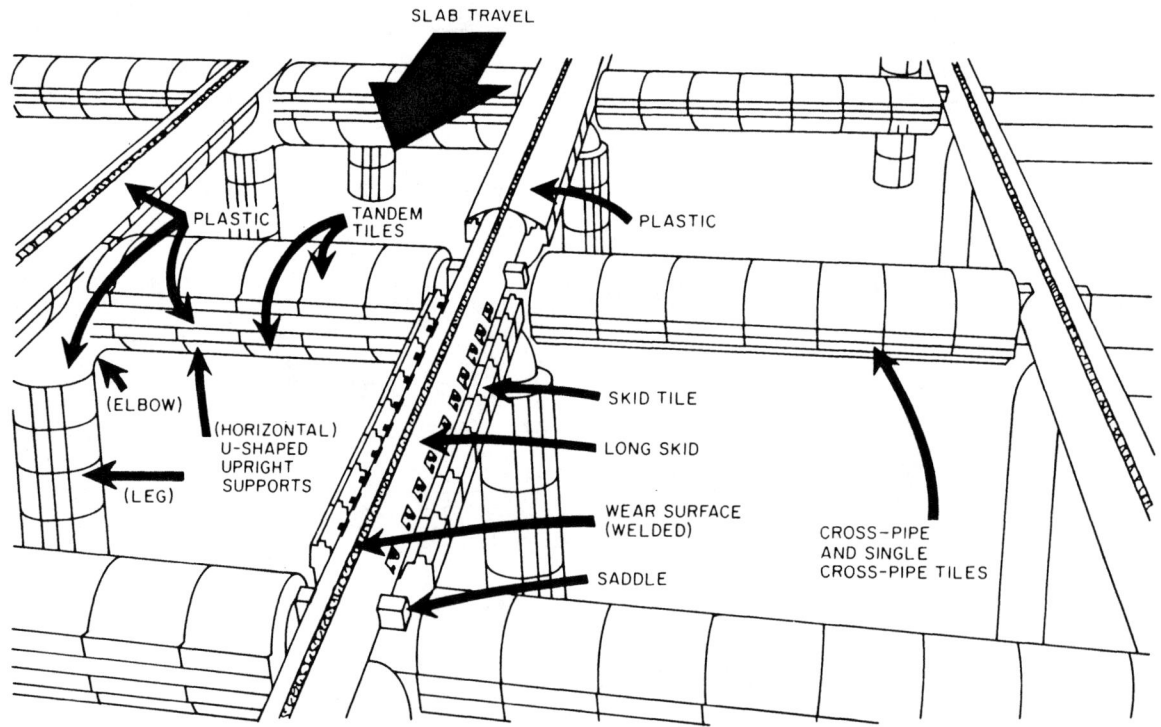

FIG. 2—80. Identification of the major components of a skid-pipe system.

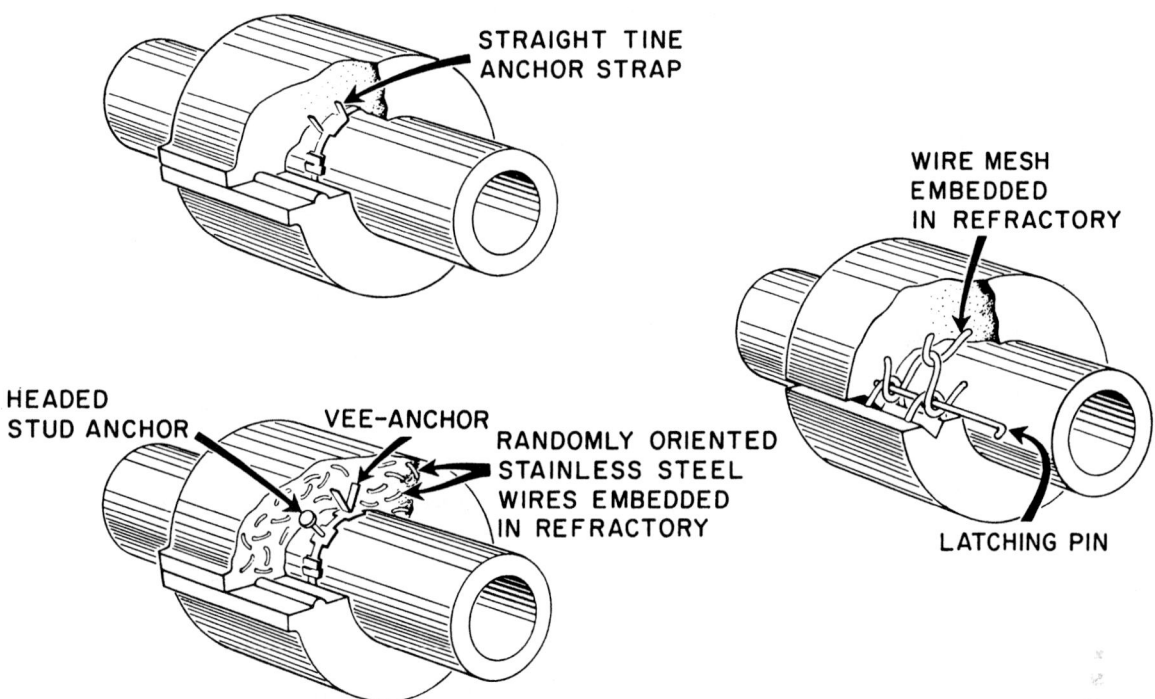

FIG. 2—81. Various methods of using metallic enhancement with anchors in skid tiles.

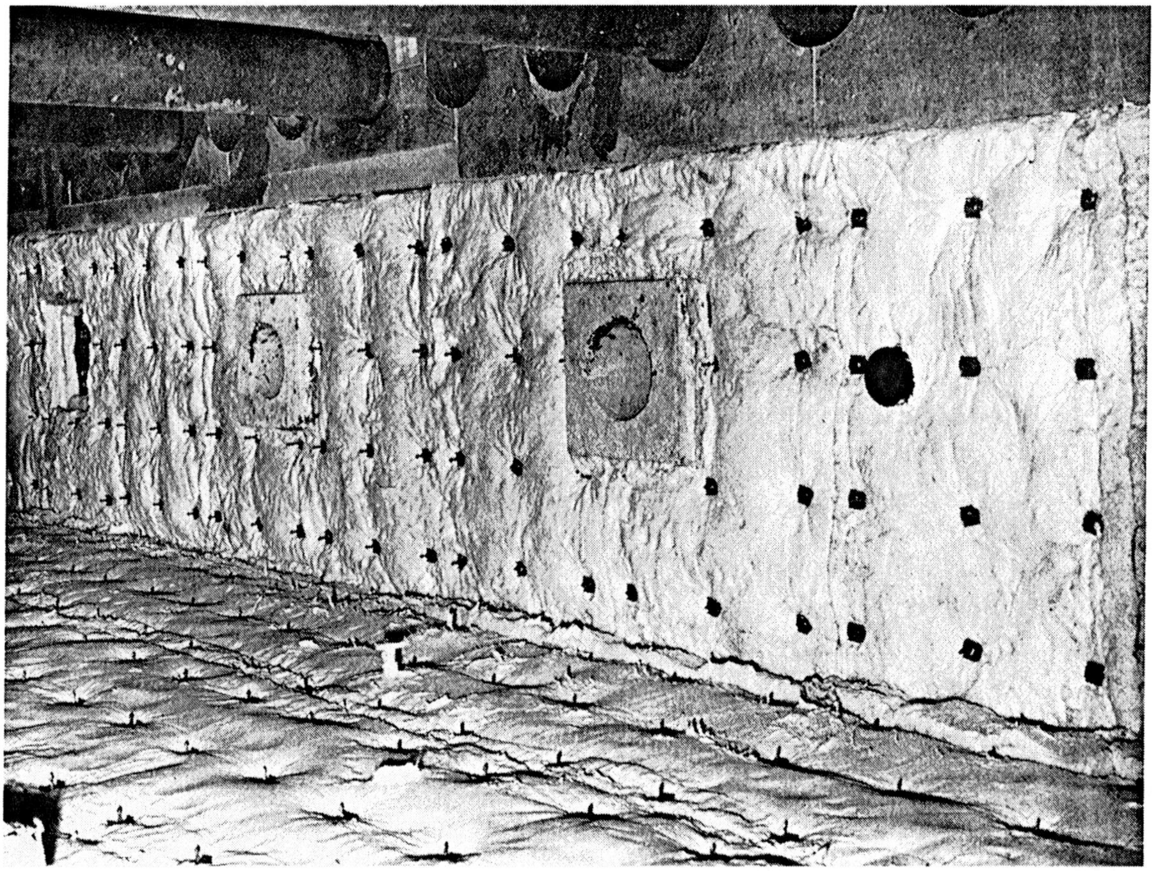

FIG. 2—82. Plate-mill continuous-annealing furnace lined with fiber blanket wallpaper construction.

maintenance. The performance of pit slag lines will also vary widely, depending on the pit operating practices (wet or dry bottom and the frequency of maintenance) and maintenance techniques such as gunning. Reheating-furnace lives can be quite long, but again are related primarily to operating practices such as mechanical damage or severe slagging or scaling conditions.

Bibliography

Alper, A. M. (ed.), High-temperature oxides, Part I, Magnesia, lime, and chrome refractories. Academic Press, 1970.

American Ceramic Society, Phase diagrams for ceramists, 1974, 1969 Supplement.

American Ceramic Society, Bulletins, 1955-83.

AIME, Proceedings of open-hearth and basic oxygen steel conferences, 1965 to 1983. The Society.

American Society for Testing and Materials, ASTM Standards, Part 17 (1983), The Society, Philadelphia, Pennsylvania.

Chesters, J. H., Refractories for iron and steelmaking. Iron and Steel Institute, London, 1973.

Chesters, J. H., Refractories, production and properties. Iron and Steel Institute, London, 1973.

Harbison-Walker Refractories, Modern refractory practice. 4th Edition, 1961.

Norton, F. H., Refractories. McGraw Hill, 3rd Edition, 1949.

CHAPTER 3

Steel-Plant Fuels and Fuel Economy

SECTION 1

FUELS, COMBUSTION AND HEAT FLOW

Any substance capable of producing heat by combustion may be termed a **fuel.** However, it is customary to rank as fuels only those which include carbon and hydrogen and their compounds. Wood was the earliest fuel used by man. Coal was known to exist in the fourth century B.C., and petroleum was used by the Persians in the days of Alexander. Prehistoric records of China and Japan are said to contain references to the use of natural gas for lighting and heating.

Heat generated by the combustion of fuel is utilized in industry directly as heat or is converted into mechanical or electrical energy. Fuel has become the major source of energy for manufacturing enterprises.

Fuel enters significantly into manufacturing costs, and in some industries represents one of the largest items of expense. The steel industry is one of the major consumers of metallurgical coal. A modern fully integrated steel plant consumes the equivalent of approximately a ton of coal for each ton of raw steel produced. The steel industry also consumes large quantities of natural gas and petroleum.

Energy conservation efforts and technological improvements have combined to decrease domestic steel industry energy consumption from 46.21 kilojoules per net ton (39.73 million Btu/ton) of shipments in 1976 to 34.40 kilojoules per net ton (29.58 million Btu/ton) of shipments in 1980.

CLASSIFICATION OF FUELS

There are four general classes of fuels; namely, **fossil, by-product, chemical** and **nuclear.** Of these classes, the first three listed achieve energy release by combustion of carbon and/or hydrogen with an oxidant, usually oxygen: the process involves electron exchange to form products of a lower energy state as a consequence of energy release in an exothermic reaction. The fourth class liberates energy by fission of the nucleus of the atom and converting mass into energy.

Fossil fuels are hydrocarbon or polynuclear aromatic compounds composed principally of carbon and hydrogen and are derived from fossil remains of plant and animal life. These fossil remains have been transformed by biochemical and geological metamorphoses into such fuels as coal, natural gas, petroleum, and so on.

By-product and **waste fuels** are derived from a main product and are of a secondary nature. Examples of these fuels are coke breeze, coke-oven gas, blast-furnace gas, wood wastes, etc.

Chemical fuels are primarily of an exotic nature and normally are not used in conventional processes. Examples of these fuels are ammonium nitrate and fluorine.

Nuclear fuels are obtained from fissionable materials. The three basic fissionable materials are uranium-235, uranium-233 and plutonium-239.

Fossil and by-product fuels currently used in the steel industry are classified further into three general divisions; namely, **solid, liquid** and **gaseous fuels.** Fuels in each general division can be classified further as **natural, manufactured** or **by-product.** Fuels found in nature sometimes are called **primary fuels;** those manufactured for a specific purpose or market, together with those that are the unavoidable by-product of some regular manufacturing process, are called **secondary fuels.** The primary fuels serve as the principal raw materials for the secondary fuels. Table 3—I gives a classified list of the important fossil fuels. It also lists some interesting by-product fuels, many of which have been utilized by industry to conserve primary fuel.

Importance of Each Class—Coal is the major fuel of public utilities for the generation of power and is essential to the steel industry for the manufacture of coke. In recent years, the development of a practical method for fission of the atom and the release of nuclear energy in controlled chain reactions has given rise to a different type of power-generation system. In this system, the boiler is used as a convenient way of transferring the heat generated in a nuclear reactor to a suitable working fluid. Currently, the major application of nuclear power systems is in the public utility field.

Coal has been supplanted almost entirely by liquid fuels for the generation of motive power by railroads in North America. However, coal continues as a major raw material for many chemical plants as a source of carbon, hydrogen, and their compounds.

The growth of petroleum consumption has been

Table 3—I. Classification of Fuels[a]

General Division	Primary Fuels	Secondary Fuels	
	Natural	Manufactured	By-Product
Solid	Anthracite coal Bituminous coal Lignite Peat Wood	Semi-coke (low-temperature carbonization residue) Coke Charcoal Briquettes { Coal slack and culm / Lignite / Peat / Sawdust / Petroleum-refining residue Pulverized coal	Charcoal—low-temperature distillation of wood Wood refuse—chips, shavings, trimmings, tan bark, sawdust, etc. Bagasse—refuse sugar cane Anthracite culm—silt refuse of anthracite screening Coke breeze { By-product coke—screenings / Petroleum coke—petroleum-refining residue Waste materials from grain { Corn / Barley / Wheat / Buckwheat / Sorghum
Liquid	Petroleum	Gasoline Kerosene Alcohol Colloidal fuels Fuel oil { Residual oils / Distillate oils / Crude petroleum Naphtha Vegetable oils { Palm / Cottonseed	Coal distillates { Tar / Naphthalene / Pitch / Benzol }—coke manufacture Acid sludge—petroleum-refining residue Pulp-mill waste
Gaseous	Natural Gas	Producer gas Water gas Carburetted water gas Coal gas Oil gas Reformed natural gas Butane[b] Propane[b] Acetylene Hydrogen	Blast-furnace gas—pig-iron manufacture Coke-oven gas[c]—coke manufacture Oil-refinery gas Sewage gas—sewage sludge

[a] Excluding chemical and nuclear fuels.
[b] Liquefiable heavier constituents of natural gas.
[c] Considered by-product of coke manufacture in steel industry but a manufactured fuel in the gas industry.

phenomenal in the past 60 years, due to the increasing demand for its distillation products. Gasoline, the most important product, is used as a motor fuel. Diesel-engine fuel is a distillate of crude oil. Distillate and residual fuel oils, and some crude petroleums of too low commercial value for distillation, are used for industrial and domestic heating. Crude and refined petroleum of various grades are used for lubrication of all types of machinery and prime movers. Petroleum and natural gas are raw materials for the petrochemical industry.

Natural gas has replaced coal to a considerable extent for domestic and industrial heating due to the installation of very large pipe lines from producing to consuming centers, the importation of LNG (liquefied natural gas), the rise in price of solid fuel, the relative level in the price of natural gas over the intervening time, and its convenience, cleanliness, controllability and versatility as a fuel. The by-product gaseous fuels—coke-oven gas and blast-furnace gas—are major iron and steel industry fuels.

PRINCIPLES OF COMBUSTION

Fossil and by-product fuels consist essentially of one, or a mixture of two or more, of four combustible constituents: (1) solid carbon, (2) hydrocarbons, (3) carbon monoxide, and (4) hydrogen. In addition to these combustible constituents, nearly all commercial fuels contain inert material, such as ash, nitrogen, carbon dioxide, and water. Bituminous coal is an example of a fuel which contains all four of the combustible constituents named above, and coke is an example of a fuel containing only one (solid carbon). The constituents which make up liquid fuels and many coals are quite complex, but since these complex constituents decompose or volatilize into the four simpler constituents named above before actual combustion takes place, a knowledge of the combustion characteristics of these constituents is sufficient for nearly all practical applications. All of these four constituents of fuels except carbon are gases at the temperatures where combustion occurs. Combustion takes place by combining oxygen, a gas present in air, with the combustible constituents of a fuel. The complete combustion of all fuels generates gases. It is apparent, therefore, that a review of the properties, thermal values and chemical reactions of gases is necessary for an understanding of any class of fuel.

Since fuels are used to develop heat, a knowledge of heat terms and the principles of heat flow are essential for the efficient utilization of this heat. The combustion of fuels involves, besides combustion reactions, the factors and principles which influence speed of combustion, ignition temperature, flame luminosity, flame development, flame temperature and limits of flammability. The ensuing divisions of this section deal generally with these subjects. Sections 2, 3 and 4, respectively, deal specifically with the combustion of solid, liquid and gaseous fuels.

Units for Measuring Heat—Heat is a form of energy and is measured in absolute joules in SI units.

In the centimetre-gram-second (cgs) system, the unit for measuring heat was the **calorie** (abbreviated cal), defined as the amount of heat required to raise the temperature of one gram of pure, air-free water 1°C in the temperature interval of 3.5° to 4.5°C at normal atmospheric pressure: this unit was the **gram-calorie** or **small calorie**, identified in the table of values below as $cal_{4°C}$. The temperature interval chosen for this definition was selected because the density and, therefore, the heat capacity of water varies slightly with temperature and the temperature of maximum density of water is very nearly 4°C. A larger heat unit in the cgs system was the **kilocalorie (kilogram-calorie** or **large calorie)**, equal to 1000 gram-calories and abbreviated kcal.

Other values for the calorie were obtained by selecting other temperature intervals, resulting, for example, in the $cal_{15°C}$ and the $cal_{20°C}$ listed below. Yet another variation was the **mean calorie** (abbreviated cal_{mean}), defined as 1/100 of the amount of heat required to raise the temperature of one gram of water from 0°C (the ice point) to 100°C (the boiling point).

None of the foregoing definitions of the calorie were completely satisfactory because of the variation of the heat capacity of water with temperature. Consequently, on the recommendation of the Ninth International Conference of Weights and Measures (Paris, 1948), the calorie came to be defined in energy units in ways that made its value independent of temperature. The **thermochemical calorie** (abbreviated $cal_{thermochem}$) was defined first in international electrical-energy units and later (1948) in terms of mechanical-energy units. The calorie used in the present International Tables, identified as cal_{IT}, was adopted in 1956 at the International Conference on Properties of Steam in Paris, and is expressed in mechanical-energy units.

As stated above, the SI unit used to define the calorie in terms of mechanical-energy units is the absolute joule: the word "absolute" differentiates the SI joule based on mechanical-energy units from the international joule formerly used which was based on international electrical units.

The presently accepted values in absolute joules of the various calories discussed above are as follows:

1 $cal_{4°C}$	= 4.204 5 joules
1 $cal_{15°C}$	= 4.185 80 joules
1 $cal_{20°C}$	= 4.181 90 joules
1 cal_{mean}	= 4.190 02 joules
1 cal_{IT}	= 4.186 8 joules (exactly)
1 $cal_{thermochem}$	= 4.184 joules (exactly)

In the foot-pound-second (fps) system, the principal unit adopted for measuring heat was the **British thermal unit** (Btu). Defined as the amount of heat required to raise by 1°F the temperature of one pound of pure, air-free water, its value depended upon the temperature interval chosen for its complete definition. As in the case of the calorie, several values came into use, notably the $Btu_{39°F}$ based on a 1°F rise in temperature at or near the temperature of maximum density of water (39.2°F), the $Btu_{59°F}$ based on the temperature interval of 58.5°F to 59.5°F corresponding nearly to the 14.5°C to 15.5°C interval of the $cal_{15°C}$, the $Btu_{60°F}$ based on the temperature interval from 60°F to 61°F, and the **mean Btu** (Btu_{mean}) that represented 1/180 of the heat required to raise the temperature of a pound of water from 32°F (the freezing point) to 212°F (the boiling point). Other values for the Btu were adopted, based on definitions that made the unit independent of the properties of water: these included the Btu of the International Tables referred to above (abbreviated Btu_{IT}), and the **thermochemical Btu** (designated $Btu_{thermochem}$). Following the recommendations of the 1948 International Conference on Weights and Measures, all of the foregoing values for the Btu came eventually to be expressed in SI mechanical-energy units, and now have been assigned the following values in absolute joules:

$Btu_{39°F}$	= 1059.67 joules
$Btu_{59°F}$	= 1054.80 joules
$Btu_{60°F}$	= 1054.68 joules
Btu_{mean}	= 1055.87 joules
Btu_{IT}	= 1055.056 joules
$Btu_{thermochem}$	= 1054.350 joules

When both are determined for the same temperature interval, 1 Btu equals very nearly 252 calories, and 1 kilocalorie (1000 cal) very nearly equals 3.9683 Btu.

Calorific Value of Fuels—The heat given up or absorbed by a body between two temperatures, provided no change of state or of allotropic form is involved, is known as **sensible heat**. The heat given up or absorbed by a body when a change of state or of allotropic form takes place and no temperature change is involved is known as **latent heat**. For example, 1 kilogram of water absorbs 418.68 kilojoules of sensible heat when being heated from 0°C to 100°C, and absorbs 2257.1 kilojoules of latent heat when converted to steam at 100°C. Likewise, 1 pound of water absorbs 189.9 kilojoules (180 Btu) of sensible heat on being heated from 32°F to 212°F, and absorbs 1023.8 kilojoules (970.4 Btu) of latent heat when converted to steam at 212°F.

Sensible heat and latent heat are used frequently in combustion calculations, particularly in problems dealing with heat losses in flue gases. Their significance is indicated in describing gross and net heating values.

The **gross heating value** of a fuel is the total heat developed by the combustion of a fuel at constant pressure after the products of combustion are cooled back to the starting temperature, assuming that all of the water vapor produced is condensed; that is, the gross heating value includes both sensible and latent heat. The **net heating value** of a fuel is defined as the heat developed by the combustion of a fuel at constant pressure after the products of combustion are cooled back to the starting temperature, assuming that all of the water vapor remains uncondensed. Accordingly, the net heating value includes only the sensible heat.

Where combustion calculations in this chapter are in SI units, the starting point is 273.15 K (0°C) at 101.325 kPa (760 mm Hg) absolute pressure*. Where the calculations are in cgs units, the starting point is 0°C at 760 mm absolute pressure. The starting point for calculations in the fps system in this chapter has been taken as 60°F at 30 in. of Hg absolute pressure; this has generally been the base for combustion calculations in the American steel industry.

When a fuel contains neither hydrogen nor hydrocarbons, no water vapor is produced by combustion and the gross and net heating value will be the same, as in the case of burning carbon or carbon monoxide. The heating value or calorific value of a fuel may be determined on a dry or wet basis. The determination may be made by laboratory tests employing **calorimeters,** or by calculation. The process of determining the calorific value of solid and liquid fuels by a calorimeter consists in completely oxidizing the fuel in a space enclosed by a metal jacket (called the **bomb**) so immersed that the heat evolved is absorbed by a weighed portion of water contained in an insulated vessel. From the rise in temperature of the water, the heat liberated by one gram of the fuel is calculated. The best types of calorimeters for solid and liquid fuels are those called oxygen-bomb calorimeters in which the fuel is burned in the presence of compressed oxygen. Gas calorimeters are of different construction to permit volumetric measurement of the gas and its complete combustion under non-explosive conditions, as well as absorption of the heat produced in a water jacket. The Junkers-type continuous-flow calorimeter is a common type.

A **saturated gas** is one which contains the maximum amount of water vapor it can hold without any condensation of water taking place. The usual basis for reporting the calorific value of a saturated fuel gas in SI units is in gross kilojoules per cubic metre measured at 273 K and 101 kPa absolute pressure. In the cgs system, calorific value usually has been reported in gross kilocalories per cubic metre measured at 0°C and 760 mm Hg absolute pressure. In fps units in the American steel industry, calorific value usually has been reported in gross Btu per cubic foot of saturated gas measured at 60°F and 30 in. Hg absolute pressure.

The heating value of a given fuel can be obtained by multiplying the calorific value of each gas by its percentage of the total fuel volume, and then totaling the individual values of the separate constituents. The heat of combustion for various dry elementary gases may be found in Table 3—II. For instance, the gross heating value of dry blast-furnace gas is 3633 kilojoules per cubic metre (92.5 Btu per cubic foot) for the composition used in the following calculations.

In the calculation of the heating value of gases saturated with water vapor, the volume of water vapor must be deducted from the unit volume of the gas. For instance, a cubic metre of dry carbon monoxide gas has

Constit- uents of Gas	Content by Volume (%)	Gross Heating Value* of Each Component		Gross Heating Value* of Each Fraction	
		Btu/ft³	kJ/m³	Btu/ft³	kJ/m³
CO₂	12.76	0	0	0	0
CO	25.69	321.4	12 623	82.6	3244
H₂	3.05	325.0	12 766	9.9	389
N₂	58.50	0	0	0	0
			Total	92.5	3633

*Btu/ft³ at 60°F, 30 in. Hg: kJ/m³ at 273 K (0°C) and 101 kPa (760 mm Hg). Conversion Factor - 39.28

a heating value of 12 623 kilojoules, but when saturated with water vapor at 273 K (0°C) and 101 kPa (760 mm Hg), a cubic metre has a heating value of only 12 405 kilojoules (see Table 3—II). Likewise, a cubic foot of dry carbon monoxide gas has a heating value of 321.4 Btu, but when saturated with water vapor at 60°F and 30 in. Hg absolute pressure, a cubic foot has a heating value of only 315.8 Btu. The amount of water vapor present in saturated mixtures can be calculated from data in Table 3—III, as discussed under "Gas Laws."

Thermal Capacity, Heat Capacity and Specific Heat—The **thermal capacity** or **heat capacity** of a substance is expressed as the amount of heat required to raise the temperature of a unit weight of the substance one degree in temperature. In SI, it is expressed in joules per kilogram kelvin (J/kg·K). The fps system has used Btu per pound per degree Fahrenheit (Btu/lbm·deg F), while the cgs system has used calories per gram per degree Celsius or Centigrade (cal/g·deg C). The **specific heat** is always a ratio, expressed as a number; for example, the specific heat of wrought iron is 0.115. There is no further designation, as this means that if it takes a certain number of joules to heat a certain number of kilograms of water a certain number of kelvins, it will take only 0.115 times as many joules to heat the same number of kilograms of wrought iron the same number of kelvins, and the same figure, 0.115, obviously applies if the centimetre-gram-second or foot-pound-second system were used.

The amount of heat required to raise the temperature of equal masses of different substances to the same temperature level varies greatly; that is to say, the specific heat varies greatly; also the specific heat of the same substance varies at different temperatures. Usually, it is necessary to know the amount of heat required to raise the temperature of a substance some appreciable amount. For that purpose, formulae and tables are usually accessible in handbooks for supplying the mean specific heat between various temperature levels. Two values of specific heat for gases are usually given: (1) specific heat at constant pressure, and (2) specific heat at constant volume. The difference is due to the heat equivalent of the work of expansion caused by an increase of volume resulting from a temperature rise. Normal combustion practice with gases in steel plants deals with a constant pressure condition (or nearly so), and for this reason specific heat at constant

Table 3—II. Essential Gas Combustion Constants [1]

| Gas | Formula | Molecular Weight | Specific Gravity (Air=1) | Heat of Combustion [2] | | | | | | | | Unit Volumes per Unit Volume of Dry Combustible (m^3 or ft^3) | | | | | |
| | | | | Btu per ft^3[3] | | Btu per lb[3] | | kJ per m^3[4] | | kJ per kg[4] | | Required for Combustion | | | Flue Products | | |
				Gross	Net	Gross	Net	Gross	Net	Gross	Net	O_2	N_2[5]	Air	CO_2	H_2O	N_2
Carbon (Graphite)	C	12.01	—	—	—	14 093	14 093	—	—	32 780	32 780	—	—	—	—	—	—
Hydrogen	H_2	2.016	0.06959	325.02	274.58	60 991	51 605	12 767	10 786	141 865	120 033	0.5	1.882	2.382	—	1.0	1.882
Oxygen	O_2	32.00	1.1053	—	—	—	—	—	—	—	—	—	—	—	—	—	—
Nitrogen	N_2	28.016	0.9718	—	—	—	—	—	—	—	—	—	—	—	—	—	—
Carbon Monoxide	CO	28.01	0.9672	321.37	321.37	4 347	4 347	12 623	12 623	10 111	10 111	0.5	1.882	2.382	1.0	—	1.882
Carbon Dioxide	CO_2	44.01	1.5282	—	—	—	—	—	—	—	—	—	—	—	—	—	—
Methane	CH_4	16.042	0.5543	1 012.32	911.45	23 875	21 495	39 764	35 802	55 533	49 997	2.0	7.528	9.528	1.0	2.0	7.528
Ethane	C_2H_6	30.068	1.0488	1 773.42	1 622.10	22 323	20 418	69 660	63 716	51 923	47 492	3.5	13.175	16.675	2.0	3.0	13.175
Ethylene	C_2H_4	28.052	0.9740	1 603.75	1 502.87	21 636	20 275	62 995	59 033	50 325	47 160	3.0	11.293	14.293	2.0	2.0	11.293
Propylene	C_3H_6	42.078	1.4504	2 339.70	2 188.40	21 048	19 687	91 903	85 960	48 958	45 792	4.5	16.939	21.439	3.0	3.0	16.939
Acetylene	C_2H_2	26.036	0.9107	1 476.55	1 426.17	21 502	20 769	57 999	56 020	50 014	48 309	2.5	9.411	11.911	2.0	1.0	9.411
Benzene	C_6H_6	78.108	2.6920	3 751.68	3 600.52	18 184	17 451	147 366	141 428	42 296	40 591	7.5	28.232	35.732	6.0	3.0	28.232
Hydrogen Sulphide	H_2S	34.076	1.1898	646	595	7 097	6 537	25 375	23 372	16 508	15 205	1.5	5.646	7.146	$SO_2 = 1.0$	1.0	5.646
Sulphur Dioxide	SO_2	64.06	2.264	—	—	—	—	—	—	—	—	—	—	—	—	—	—

[1] Adapted from "Gas Engineers Handbook." (See Segeler listing in bibliography at end of chapter).
[2] Based on perfect combustion.
[3] Based on dry gases at 60°F and 30 in. Hg. For gases saturated with water, 1.74 per cent of the heating value must be deducted.
[4] Based on dry gases at 273 K (0°C) and approx. 101 kPa (760 mm Hg). For gases saturated with water at 273 K (0°C), 0.60 per cent of the heating value must be deducted. (To convert kJ to kcal, multiply by 0.239.)
[5] N_2 in air accompanies O_2 (N_2 not required for combustion).
Note: Conversion Factors: Btu/ft^3 to kJ/m^3 = 39.28; Btu/lb to kJ/kg = 2.326.

Table 3—III. Water Vapor Pressure[1]

Temp. (°C)	Pressure (mm Hg)	Pa[2]	Temp. (°F)	Pressure (in. Hg)	Pa[3]
0	4.579	610.5	32	0.1803	608.8
2	5.294	705.8			
4	6.101	813.4	35	0.2035	687.2
6	7.013	935.0	40	0.2478	836.8
8	8.045	1072.6			
			45	0.3004	1014
10	9.209	1227.8	50	0.3626	1224
12	10.518	1402.3			
14	11.987	1598.1	55	0.4359	1472
16	13.634	1817.7	60	0.5218	1762
18	15.477	2063.4			
			65	0.6222	2101
20	17.535	2337.8	70	0.7393	2497
22	19.827	2643.4			
24	22.377	2983.4	75	0.8750	2955
26	25.209	3360.9	80	1.032	3485
28	28.349	3779.6			
			85	1.213	4096
30	31.824	4242.9	90	1.422	4802
32	35.663	4754.7			
34	39.898	5319.3	95	1.660	5606
36	44.563	5941.2	100	1.933	6527
38	49.692	6625.1			
			105	2.243	7574
40	55.324	7375.9	110	2.596	8766
42	61.50	8199			
44	68.26	9101	115	2.995	10 110
46	75.65	10 086	120	3.446	11 640
48	83.71	11 160			
			125	3.956	13 360
50	92.51	12 334	130	4.525	15 280
52	102.09	13 611			
54	112.51	15 000	135	5.165	17 440
56	123.80	16 505	140	5.881	19 860
58	136.08	18 143			
			145	6.682	22 560
60	149.38	19 916	150	7.572	25 570
62	163.77	21 834			
64	179.31	23 906	155	8.556	28 890
66	196.09	26 143	160	9.649	32 580
68	214.17	28 554			
			165	10.86	36 670
70	233.7	31 160	170	12.20	41 200
72	254.6	33 940			
74	277.2	36 960	175	13.67	46 160
76	301.4	40 180	180	15.29	51 630
78	327.3	43 640			
			185	17.07	57 640
80	355.1	47 340	190	19.02	64 230
82	384.9	51 320			
84	416.8	55 570	195	21.15	71 420
86	450.9	60 120	200	23.46	79 220
88	487.1	64 940			
			205	25.99	87 760
90	525.76	70 100	210	28.75	97 080
92	566.99	75 590			
94	610.90	81 450	212	29.92	101 035
96	657.62	87 680			
98	707.27	94 290			
100	760.00	101 325			

[1] Values for °C and mm Hg from "International Critical Tables" (E. W. Washburn, ed. in chief), published for National Research Council by McGraw-Hill Book Company, New York, 1928. Values for °F and in. Hg from "Gas Engineers Handbook" (see Segeler listing in bibliography at end of chapter.

[2] Calculated, using factor 133.3224 to convert mm Hg (0°C) to Pa.

[3] Calculated, using factor 3376.85 to convert in. Hg (60°F) to Pa. Factor for converting in. Hg (32°F) to Pa is 3386.38.

pressure is used. The **mean specific heat** is the average value of the specific heat between two temperature levels. It is obtained by integrating the equations for instantaneous specific heat over the temperature limits desired, and dividing this quantity by the difference between the temperature limits.

The **heat content** is the heat contained at a specified temperature above some fixed temperature. It is calculated by multiplying the weight of a substance by the mean specific heat times the temperature difference, or H_t = weight $\times$ mean specific heat $\times$ ($T_2 - T_1$). For convenience in calculations with gases, the unit weight of the volume of a cubic metre or a cubic foot of gas is often used.

Gas Laws—Calculations based on the gas laws to be discussed involve the concepts of **absolute zero** and **absolute temperature**. Absolute zero in SI is 0 K, in the cgs system it is −273.15°C, and in the fps system it is −459.67°F; for practical purposes to facilitate calculations, −273°C and −460°F can be taken as absolute zero in the cgs and fps systems, respectively.

Absolute temperature in SI is the temperature expressed in kelvins above 0 K at an absolute pressure of 101.325 kPa (one standard atmosphere). In the cgs system, absolute temperature (°C$_{abs}$) has been the temperature in degrees Celsius (formerly called degrees centigrade) above −273.15°C at 760 mm Hg (millimetres of mercury) at one standard atmosphere absolute pressure. Absolute temperature in the cgs system also has been expressed according to the Kelvin scale of temperature, using K instead of °C. 1°C = 1K. Absolute zero on the Kelvin scale is 0 K, and on the Celsius scale absolute zero, as stated above, is −273.15°C. Temperatures on the Kelvin and Celsius scales have the following relation:

$$T_K = T_{°C} + 273.15$$

or, to facilitate calculations,

$$T_K = T_{°C} + 273$$

Absolute temperature in the fps system is the temperature in degrees Fahrenheit (°F$_{abs}$) above −459.67°F at 29.921 inches of mercury (one standard atmosphere) absolute pressure. Also in the fps system, absolute temperatures have been expressed according to the **Rankine** temperature scale, using degrees Rankine (°R) instead of degrees Fahrenheit (°F). 1°R = 1°F. Absolute zero on the Rankine scale is 0°R, and on the Fahrenheit scale absolute zero, as stated above, is −459.67°F. Temperatures on these two scales have the following relation:

$$T_{°R} = T_{°F} + 459.67$$

or, to facilitate calculations,

$$T_{°R} = T_{°F} + 460$$

Again to facilitate calculations, rounded values for an absolute pressure of one atmosphere (101 kPa in SI and 30 in. Hg in the fps system) can be used instead of the

(Continued)

more precise values in the definitions.

The volume of an ideal gas varies in direct proportion to its absolute temperature (**Charles' Law**) and inversely as its absolute pressure (**Boyle's Law**).

For example, in SI units, the volume of 1000 m³ of a gas measured at 288 K (15°C) and 101 kPa absolute pressure, when heated to 1253 K (980°C) and 101 kPa absolute pressure, is equal to:

$$1000 \times \frac{1253}{288} = 4351 \text{ m}^3$$

and the volume of 1000 m³ of fuel gas measured at 288 K(15°C) and 101 kPa absolute pressure is equal to 802 m³ when compressed to 25 kPa gage pressure at 288 K, calculated as follows:

$$1000 \times \frac{101}{101 + 25} = 802 \text{ m}^3$$

Similarly, in the fps system, the volume of 40 000 ft³ of gas measured at 60°F and 30 in. Hg absolute pressure, when heated to 1800°F and 30 in. Hg absolute pressure, is equal to:

$$40\ 000 \times \frac{460 + 1800}{460 + 60} = 174\ 000 \text{ ft}^3$$

and the volume of 40 000 ft³ of fuel gas measured at 60° F and 30 in. Hg (standard conditions) is equal to 31 579 ft³ when compressed to 8 in. Hg gage pressure at 60° F, calculated as follows:

$$40\ 000 \times \frac{30}{30 + 8} = 31\ 579 \text{ ft}^3$$

The total pressure of any gas mixture is equal to the sum of the pressures of each component. Each component produces a partial pressure proportional to its concentration in the mixture. Therefore, in a mixture of water vapor and any other gas, each exerts a pressure proportional to its percentage by volume, and since water has a definite vapor pressure at various temperatures, as shown in Table 3—III, the concentration of water vapor in a gas is limited. When this limit of water vapor is reached, the gas is said to be **saturated.** Any drop in temperature or increase in pressure from that point will cause condensation of some of the water vapor; for instance, the water vapor in 1000 m³ of saturated fuel gas at 293 K (20°C) and 101 kPa would equal:

$$1000 \times \frac{2.3378}{101} = 23.1 \text{ m}^3$$

(2.3378 kPa is the partial pressure of water vapor in a saturated mixture at 293 K (20°C) and 101 kPa, as shown in Table 3—III.) In fps units, water vapor in 1000 ft³ of saturated fuel gas measured at 60°F and 30 in. Hg is calculated as follows:

$$1000 \times \frac{0.522}{30} = 17.40 \text{ ft}^3$$

(0.522 is the partial pressure of water vapor in a saturated mixture at 60° F and 30 in. Hg—Table 3—III). The amount of water vapor which will condense at various temperatures may be ascertained by the use of Table 3—IV.

In some combustion calculations, it is necessary to convert volumes to weights and vice versa. Such conversions can be made very conveniently by using molar units; namely, the mole (abbreviated mol and expressed in kilograms) in SI; the gram-mole (abbreviated gm-mol and expressed in grams) in the cgs system; and the pound-mole (abbreviated lb-mol and expressed in pounds) in the fps system. A mol, gm-mol or lb-mol of a substance is that quantity whose mass expressed in the proper units stated above is the same number as the number of the molecular weight. Thus, the molecular weight of oxygen is 32, so that the mol in SI is 32 kg of oxygen, the gm-mol in the cgs system is 32 grams of oxygen, and the lb-mol in the fps system is 32 lb of oxygen.

In SI, a mol of any gas (its molecular weight in kilograms) theoretically occupies 22.414 m³ at 273.15 K and 101.325 kPa absolute pressure. (Values of 22.4 m³, 273 K and 101 kPa are close enough for most calculations.) In the cgs system, a gm-mol of any gas (its molecular weight in grams) theoretically occupies 22.414 dm³ at 0°C and 760 mm Hg absolute pressure. In the fps system, a lb-mol of any gas (its molecular weight in pounds) theoretically occupies 359 ft³ at 32°F and 29.921 in. Hg absolute pressure; or, at 60°F and 30 in. Hg absolute pressure (the usual reference points for combustion problems in the steel industry) a lb-mol occupies 378.4 ft³. The simplicity of using molar units in combustion calculations is shown by the following examples:

The weight of a cubic metre of dry air is calculated in SI as follows:

0.21 (% vol. of O_2 in air) ×
    32 (mol. wt. of O_2) =   6.72
0.79 (% vol. of N_2 in air) ×
    28 (mol. wt. of N_2) = 22.12
Weight in kg of a mol of dry air = 28.84

$$\frac{28.84}{22.4} = 1.29 \text{ kg}$$ (weight per m³ of dry air at 273 K and 101 kPa absolute pressure)

The volume of 1 kg of dry air at 273 K and 101 kPa absolute pressure is equal to:

$$\frac{22.4}{28.84} = 0.78 \text{ m}^3$$

In the cgs system, the calculations would be similar to those in SI, except that the weight of a gram-mol of dry air would be determined to be 28.84 grams, and the weight of a cubic decimetre (litre) of dry air would be found to be 1.29 grams: the volume of 1 gram of dry air at 0°C and 760 mm Hg would be found to be 0.78 dm³.

In the fps system, the weight of a lb-mol of dry air would be determined to be 28.84 lbm,* and the weight of 1 ft³ of dry air at 60°F and 30 in Hg would be:

$$\frac{28.84}{378.4} = 0.076 \text{ lbm}$$ (weight per ft³ of dry air at 60°F and 30 in. Hg absolute pressure)

---

*Note: lbm denotes pounds-mass and replaces the previous designation lb(s).

Also, the volume of 1 lbm of dry air at 60°F and 30 in. Hg would be

$$\frac{378.4}{28.84} = 13.1 \text{ ft}^3$$

The relation of an ideal gas to its volume and pressure is expressed by the formula:

$$PV = NRT$$

where:
R = gas constant
P = absolute pressure
V = volume
N = number of mols.
T = absolute temperature of gas

The numerical value of R in the above equation depends upon what units (SI, cgs or fps) are used to measure P, V, N and T. Values of R for various combinations of units for measuring the other quantities are as follows:

T	P	V	N	R

SI (INTERNATIONAL SYSTEM)
(Standard conditions: 273.15 K, 101.325 kPa)

K	kilopascals (kPa)	m³	mol	8.3144 kJ/mol·K

CGS SYSTEM
(Standard conditions: 273.15°K, 760 mm Hg)

°K^a	dynes/cm²	cm³	g-mol	8.3144 J/g-mol·°K
°K^a	mm Hg	dm³	g-mol	62.37 mm Hg-dm³/g-mol·°K
°K^a	atmospheres	dm³	g-mol	0.08206 dm³-atm/g-mol·°K

FPS SYSTEM
(Standard conditions: 519.67°R, 30 in. Hg)

°R^b	psi	ft³	lb-mol	10.703 (lbf) (ft³) (in.^{52})/ lb-mol · °R
°R^b	in. of Hg	ft³	lb-mol	21.83 in. Hg-ft³/lb-mol·°R

^{a}°K = °C$_{abs}$ = °C + 273.15
^{b}°R = °F$_{abs}$ = °F + 459.67

In SI units, an example of the use of the foregoing formula would be to calculate the volume occupied by 100 kg of natural gas having a composition of 80 per cent CH_4, 18 per cent C_2H_6 and 2 per cent N_2 by volume at a gage pressure of 27 kPa and a temperature of 38°C, (using the data from Tables 3—II).
The weight of a mol of the gas is:

$$CH_4 = 0.80 \times 16 = 12.8$$
$$C_2H_6 = 0.18 \times 30 = 5.4$$
$$N_2 = 0.02 \times 28 = \underline{0.56}$$
$$18.76 \text{ kg}$$
$$P = 27 + 101 = 128 \text{ kPa absolute}$$
$$N = \frac{100}{18.76} = 5.33$$
$$R = 8.3144$$
$$T = 273 + 38 = 311$$

Substituting these values in the equation for a perfect gas (PV = NRT):

$$128V = 5.33 \times 8.3144 \times 311$$
$$V = 107.7 \text{ m}^3$$

In fps units, a similar application of the formula would be to calculate the volume occupied by 100 lbm of natural gas of the same composition as above at 8 in. Hg gage pressure and a temperature of 100°F.

$$P = 30 + 8 = 38 \text{ in. Hg absolute}$$
$$N = \frac{100}{18.76} = 5.33 \text{ lb-mols}$$

Table 3—IV. Properties of Dry Air[1]

SI UNITS[2]

Temp. (°C)	Volume of 1 kg (m³)	Mass of 1 m³ (kg)	Mass of Water Vapor to Saturate 100 kg of Dry Air at 100 Per Cent Humidity[3] (kg)
0	0.7735	1.2928	0.3774
5	0.7874	1.2700	0.5403
10	0.8019	1.2471	0.7638
15	0.8160	1.2255	1.0649
20	0.8302	1.2046	1.4702
25	0.8443	1.1844	2.0082
30	0.8584	1.1649	2.7194
35	0.8726	1.1460	3.6586
40	0.8868	1.1277	4.8872
45	0.9010	1.1099	6.5279
50	0.9151	1.0928	8.6686
55	0.9293	1.0761	11.5070
60	0.9434	1.0600	15.3233
65	0.9576	1.0443	20.5315
70	0.9717	1.0291	27.8395
75	0.9859	1.0143	38.5230
80	1.0001	0.9999	55.1492
85	1.0142	0.9860	83.6207
90	1.0284	0.9724	141.5052
95	1.0425	0.9592	312.9940

FPS UNITS[4]

Temp. (°F)	Volume of 1 lbm (ft³)	Mass of 1 ft³ (lbm)	Mass of Water Vapor to Saturate 100 lbm of Dry Air at 100 Per Cent Humidity[3] (lbm)
32	12.360	0.080906	0.3767
40	12.561	0.079612	0.5155
50	12.812	0.078050	0.7613
60	13.063	0.076550	1.1022
70	13.315	0.075103	1.5726
80	13.567	0.073710	2.2184
90	13.818	0.072370	3.0998
100	14.069	0.071077	4.2979
110	14.321	0.069829	5.0913
120	14.571	0.068627	8.0981
130	14.823	0.067463	11.0935
140	15.074	0.066338	15.2441
150	15.327	0.065244	21.1155
175	15.954	0.062679	52.6416
200	16.584	0.060298	226.384

[1]At an absolute pressure of 101.325 kPa (760 mm Hg) in SI units, and 30 in. of Hg in fps units.

[2]Calculated on basis of density of dry air at 273.15 K (0°C) and 101.325 kPa (760 mm Hg) equal to 1.2928 kg per m³.

[3]Mass of water vapor at lower humidities is approximately proportional to the humidity: e.g., at 50 per cent humidity, the mass will be one-half that at 100 per cent humidity for a given temperature.

[4]From "Gas Engineers Handbook" (see Segeler listing in bibliography at end of chapter).

$$R = 21.83$$
$$T = 460 + 100 = 560°R$$
$$38V = 5.33 \times 21.83 \times 560$$
$$V = 1715 \text{ ft}^3$$

Combustion Calculations—The combustion of fuels is carried out by chemical reaction with air, and occasionally with air enriched with oxygen, or with pure oxygen. **Dry air** is a mixture of the following gas volumes under average conditions:

$$N_2 = 78.03\%$$
$$O_2 = 20.99\%$$
$$\text{Argon} = 0.94\%$$
$$CO_2 = 0.03\%$$
$$H_2 = 0.01\%$$
$$\text{Total} = 100.00\%$$

In combustion calculations it is customary to include all elements in dry air (other than oxygen) with the nitrogen, as shown below:

	% by Volume	% by Weight
Oxygen	20.99	23.11
Nitrogen	79.01	76.89

Only the oxygen in the air reacts with a fuel in combustion processes. The nitrogen acts as a diluent which must be heated up by the heat of the reaction between the oxygen and the fuel. It, therefore, reduces, the temperature of the flame and reduces the velocity of combustion.

Water vapor which is present in air also acts as a diluent. The amount of moisture present in air is generally stated in terms of humidity. Air is capable of being saturated with water vapor the same as other gases as described under "Gas Laws." Air which is saturated completely with water vapor has a humidity of 100 per cent; if only 50 per cent saturated, it has a humidity of 50 per cent (Table 3—IV).

The principal **combustion reactions** are:

$$C + O_2 = CO_2$$
$$2CO + O_2 = 2CO_2$$
$$2H_2 + O_2 = 2H_2O$$
$$CH_4 + 2O_2 = CO_2 + 2H_2O$$
$$2C_2H_6 + 7O_2 = 4CO_2 + 6H_2O$$
$$C_2H_4 + 3O_2 = 2CO_2 + 2H_2O$$
$$2C_3H_6 + 9O_2 = 6CO_2 + 6H_2O$$
$$2C_2H_2 + 5O_2 = 4CO_2 + 2H_2O$$
$$2C_6H_6 + 15O_2 = 12CO_2 + 6H_2O$$
$$2H_2S + 3O_2 = 2SO_2 + 2H_2O$$

The amount of oxygen required and consequently air, together with the amount of the resultant products of combustion, may be calculated in SI by the use of mols and the proper chemical equation. For instance, it will require $(32 \div 12)$ or 2.667 kg of O_2 to burn 1 kg of C, and since dry air contains 23.11 per cent by weight of O_2, the weight of dry air required to burn one kilogram of carbon will be $(2.667 \div 0.2311)$ or 11.540 kg. The product of combustion, CO_2, will amount to $[(12 + 32) \div 12] = 3.667$ kg.

Combustion calculations using gases are more conveniently made in volumetric units. For instance, to burn a cubic metre of CO completely to CO_2 requires ½ m³ of O_2 in accordance with the molecular relationship in the equation. The dry air required would be $(0.5 \div 0.209)$ or 2.382 m³. For burning a cubic metre of methane, CH_4, to CO_2 and H_2O, the air required would be $(2.0 \div 0.209)$ or 9.528 m³.

The foregoing calculations may be performed with fps units by substituting pounds-mass (lbm) for kilograms and cubic feet for cubic metres.

Combustion calculations are necessary to determine the air requirements and the products of combustion for burning fuels of various compositions. The per cent of air used above theoretical requirements is called per cent **excess air**; the per cent below, the per cent **deficiency of air**. Typical combustion data on a dry basis for burning gaseous fuels of the compositions stated are shown in Table 3—V. In making calculations to include the water vapor which may be present in a saturated or partially saturated gas and in air, the same general method may be used by adding water vapor to the fuel-gas composition, and by adding the volume of water vapor which is introduced through air in the products of combustion column, headed H_2O.

In order to maintain combustion, a fuel must, after it has been ignited, be able to impart sufficient heat to its air-gas mixture so that it will not drop below **ignition temperature**, the minimum point of self-ignition. Too lean or too rich a mixture of a fuel with air is unable to support combustion. An **upper and lower limit of flammability** exists for all gases. The limits of flammability, as well as ignition temperatures, for a number of gases are shown in Table 3—VI.

In the design of burners or in the selection of fuel for a specific purpose, consideration of **velocity of combustion** is of major importance. Since gaseous fuels are composed usually of a mixture of combustible gases, a knowledge of the relative combustion speed of each elementary gas will provide means for evaluating this factor in any gaseous mixture. The **velocity of combustion**, or **rate of flame propagation**, of a given fuel, is influenced by three factors: (1) degree to which the air and gas are mixed, (2) temperature of the air-gas mixture, and (3) contact of the air-gas mixture with a hot surface (catalyst). By intimately mixing air and gas, combustion may be accelerated and a shorter, sharper flame developed. In the case of a gas containing large amounts of hydrogen, intimate mixing will provide a combustion reaction of explosive velocity relative to that of a gas containing large amounts of methane. Inert gases, such as carbon dioxide and nitrogen, present in fuel gases or in a gas-air mixture, reduce combustion velocity. The proportion of nitrogen in a fuel gas-air mixture may be reduced by oxygen enrichment of air for combustion, and combustion speed may, by this means, be accelerated many fold. Such measures also will raise the flame temperature. The use of preheated air for combustion also accelerates combustion of gases. In order to burn large volumes of fuel in a small space, a mixture of air and gas is sometimes directed against a hot, incandescent surface. By increasing the velocity of combustion, higher temperatures are localized close to the burner point. This condition is desirable for some processes and highly undesirable for others. For instance, the scarfing process requires a highly intensive localized heat, while the heating of steel for rolling requires a lower intensity distribution of heat over the

Table 3—V. Combustion Data* for Blast Furnace, Coke-Oven and Natural Gas

BLAST FURNACE GAS (ALL VOLUMES AT 0°C and 101.325 kPa).

Gas Comp.	% by Volume	Air per m³	Air Each Component	No Excess Air — CO₂	H₂O	SO₂	O₂	N₂	10% Excess Air — CO₂	H₂O	SO₂	O₂	N₂	50% Excess Air — CO₂	H₂O	SO₂	O₂	N₂
CO_2	11.5	...	...	.115	...	...	...	...	.115	...	...	.014†	.054†	.115	...	...	.071†	.269†
N_2	60.0	...	...	...	...	...	...	.600	...	...	...	...	.600	...	...	...	...	.600
CO	27.5	2.382	.655	.275	...	...	...	.517	.275	...	...	...	.517	.275	...	...	...	.517
H_2	1.0	2.382	.0238	...	.010	...	...	.019	...	.010	...	...	.019	...	.010	...	...	.019
Total	100.0	...	.679	.390	.010	...	...	1.136	.390	.010	...	.014	1.190	.390	.010	...	.071	1.405

COKE OVEN GAS (ALL VOLUMES AT 0°C and 101.325 kPa).

Gas Comp.	% by Volume	Air per m³	Air Each Component	No Excess Air — CO₂	H₂O	SO₂	O₂	N₂	10% Excess Air — CO₂	H₂O	SO₂	O₂	N₂	50% Excess Air — CO₂	H₂O	SO₂	O₂	N₂
CO_2	1.4	...	...	.014	...	...	...	...	.014	...	...	.101†	.382†	.014	...	...	.507†	1.909†
H_2S	0.6	7.146	.0429	...	.006	.006	...	.034	...	.006	.006	...	.034	...	.006	.006	...	.034
O_2	0.4	...	-.0190	...	...	...	...	-.015	...	...	...	...	-.015	...	...	...	...	-.015
N_2	4.3	...	...	...	...	...	...	.043	...	...	...	...	.043	...	...	...	...	.043
CO	5.6	2.382	.1334	.056	...	...	...	.105	.056	...	...	...	.105	.056	...	...	...	.105
H_2	55.4	2.382	1.3196	...	.554	...	...	1.042	...	.554	...	...	1.042	...	.554	...	...	1.042
CH_4	28.4	9.528	2.7060	.284	.568	...	...	2.138	.284	.568	...	...	2.138	.284	.568	...	...	2.138
C_2H_4	2.5	14.293	.3573	.050	.050	...	...	.282	.050	.050	...	...	.282	.050	.050	...	...	.282
C_2H_6	0.8	16.675	.1334	.016	.024	...	...	.105	.016	.024	...	...	.105	.016	.024	...	...	.105
Ill.	0.6	26.208	.1572	.018	.018	...	...	.124	.018	.018	...	...	.124	.018	.018	...	...	.124
Total	100.0	...	4.831	.438	1.220	.006	...	3.858	.438	1.220	.006	.101	4.240	.438	1.220	.006	.507	5.767

NATURAL GAS (ALL VOLUMES AT 0°C and 101.325 kPa).

Gas Comp.	% by Volume	Air per m³	Air Each Component	No Excess Air — CO₂	H₂O	SO₂	O₂	N₂	10% Excess Air — CO₂	H₂O	SO₂	O₂	N₂	50% Excess Air — CO₂	H₂O	SO₂	O₂	N₂
CO_2	0.08	...	...	.001	...	...	...	...	.001	...	...	.223†	.838†	.001	...	...	1.114†	4.191†
O_2	0.17	...	-.002	...	...	...	...	-.002	...	...	...	...	-.002	...	...	...	...	-.002
N_2	1.02	...	...	...	...	...	...	.010	...	...	...	...	.010	...	...	...	...	.010
CH_4	81.88	9.528	7.802	.819	1.638	...	...	6.164	.819	1.638	...	...	6.164	.819	1.638	...	...	6.164
C_2H_6	16.85	16.675	2.810	.337	.506	...	...	2.220	.337	.506	...	...	2.220	.337	.506	...	...	2.220
Total	100.00	...	10.610	1.157	2.144	...	...	8.392	1.157	2.144	...	.223	9.230	1.157	2.144	...	1.114	12.583

*The same numerical values apply if all volumes are expressed in cubic feet at 60°F and 30 in. Hg absolute pressure.

†From excess air.

Table 3—VI. Limits of Flammability and Ignition Temperature for Simple Gases and Compounds[1]

| Simple Gases and Compounds | Limits of Flammability | | Ignition Temperature (In Air) | |
	Lower % by Volume Gas in Air	Upper % by Volume Gas in Air	(°C)	(°F)
H	4.0	75	520	968
CO	12.5	74	644-658	1191-1216
CH_4	5.0	15.0	705	1301
C_2H_6	3.0	12.5	520-630	968-1166
C_3H_8	2.1	10.1	466	871
C_2H_4	2.75	28.6	542-548	1008-1018
C_3H_6	2.00	11.1	458	856
C_4H_8	1.98	9.65	443	829
C_2H_2	2.50	81	406-440	763-824
C_6H_6	1.35	6.75	562	1044
C_7H_8	1.27	6.75	536	997

[1] From U.S. Bureau of Mines Bulletin 503 (1952); see also U.S. bureau of Mines Bulletin 627 (1965); also "Gas Engineers Handbook" under Segeler listing at end of chapter.

full surface of the pieces being heated. In order to reduce combustion speed of a gaseous fuel, the air and gas streams may be stratified to produce slow mixing. Such a method creates a **diffusion flame,** a long flame of relatively uniform temperature with a relative higher degree of cracking of the hydrocarbon components.

Theoretical flame temperature is the temperature which would be attained by the products of combustion if the combustion of a fuel took place instantaneously, and there were no loss of heat to the surroundings. Such a condition never exists, but theoretical flame temperature represents another measure for comparing fuels. Fuels which develop a high flame temperature by combustion are more capable of producing a higher thermal efficiency in practice than those which develop low flame temperatures. The efficiency of heat utilization is the relation of the total heat absorbed by a substance to the heat supplied. Since the temperature level at which waste gases leave a furnace is usually fixed within a relatively narrow range, the higher the flame temperature the higher the potentiality for heat absorption by the substance to be heated. The theoretical flame temperature of a fuel may be calculated by balancing the sum of the net heating value of a given quantity of fuel and the sensible heat of the air-gas mixture against the heat content of the products of combustion. Theoretical flame temperature so calculated should be corrected for **dissociation** of CO_2 and H_2O at temperatures in excess of 1650°C (3000°F). The theoretical flame temperatures for a number of important gaseous fuels are given in Table 3—XIX (see "Combustion of Various Gaseous Fuels", Section 4). The reader is referred to The "Gas Engineers Handbook" and the books by Lewis and von Elbe and by Hougen et al. and Trinks and Mawhinney and others listed in the bibliography at the end of this chapter for a full explanation of combustion stoichiometry and fuel economy calculations and the calculation of theoretical flame temperatures and the dissociation of gases at elevated temperatures.

There are a number of factors which determine the

character, size and shape of a gas flame. Gases burned at very high combustion velocity will produce very little or no luminosity regardless of the kind of gas. The velocity and volume with which the air-gas stream leaves a burner or furnace port, the fuel-air ratio, and the amount of non-combustible material in the fuel will influence the length and shape of a flame. The kind of gas to be burned has a very great effect upon the character of the flame. Carbon monoxide and hydrogen burn with an invisible to a clear blue flame, while the hydrocarbon gases, methane, ethane, etc, are capable of developing highly luminous flames. The principal reason that these gases burn with a luminous flame is due to the thermal breakdown of the hydrocarbons into carbon and hydrogen, and under combustion conditions which permit this, the carbon particles are heated to incandescence thereby giving the flame its luminous appearance. The luminosity of a flame may be decreased or increased by varying the supply of air. A deficiency of air below theoretical requirements will increase luminosity and it also usually will lengthen the flame. An excess of air will decrease luminosity and shorten the flame with most burners or furnace ports. Increasing the temperature of preheat of the air for combustion will reduce luminosity, as is also the case when water vapor (steam), which may be introduced with the gas, air, or for atomization of liquid fuels, is increased. A luminous flame has a number of desirable qualities, the principal one being its greater ability to transfer heat by radiation from a fixed temperature level. However, it should be noted that a luminous flame is obtained usually at a lower temperature level than when the same fuel is burned with a lower degree of luminosity.

HEAT FLOW

Heat flow is caused by a difference in temperature, and heat is transmitted in three ways, namely, by conduction, by convection, and by radiation.

Conduction is the transmission of heat through a solid body without visible motion of the body, as through a steel bar. The amount of heat transferred through a homogeneous solid by conduction is expressed by the formula

$$q = \frac{kA\triangle T}{x}$$

where, in SI, the quantities are expressed in the following units:

q = watts transmitted (1 W = 1 J/s)
k = conductivity factor in W/m·K
A = area in m^2
$\triangle T$ = temperature difference in K
x = length of heat-transfer path in metres
and, in fps units

q = Btu transmitted per hour
k = conductivity factor in Btu·in./ft²·h·°F
A = area in square feet
$\triangle T$ = temperature difference in °F
x = length of heat-transfer path in inches

The flow of heat through a non-homogeneous solid body by conduction is expressed by the formula

$$q = \cfrac{\triangle T}{\cfrac{x_1}{k_1 A_1} + \cfrac{x_2}{k_2 A_2} + \cdots \cfrac{x_n}{k_n A_n}}$$

where, in SI units,

x_1, x_2 and x_n = the respective lengths of heat-transfer path through the various resistances in metres

k_1, k_2 and k_n = the corresponding conductivity factors of the various resistances expressed in $W/m \cdot K$

A_1, A_2 and A_n = the corresponding areas expressed in m^2.

Convection—When heat is transmitted by the mechanical motion of gas or water currents in contact with a solid, or by gas currents in contact with a liquid, the transfer of heat is by **convection**. In the transfer of heat by convection, it is necessary to conduct heat through the relatively stationary film between the moving and stationary bodies. This film becomes thinner as the velocity of the currents parallel to its surface increases. The transfer of heat by convection is expressed by the formula:

$$q = UA\triangle t$$

where, in SI, the quantities are expressed in the following units:

q = watts transmitted
U = film coefficient, expressed in $W/m^2 \cdot K$, dependent upon the velocity, specific gravity and viscosity of the moving fluid and the conductivity of the film
A = area in m^2
$\triangle t$ = temperature difference in °K

and, in fps units,

q = Btu transmitted per hour
U = film coefficient (Btu per ft^2 per °F per h) dependent upon the velocity, specific gravity and viscosity of the moving fluid and the conductivity of the film
A = area in ft^2
$\triangle t$ = temperature difference in °F

Radiation refers to the transmission of heat through space without the help or intervention of matter. This is the means by which the heat of the sun reaches the earth, and by which much of the heat of combustion of fuels is utilized in high-temperature processes in the steel industry. When radiant energy strikes any body a certain proportion of the total is reflected, while that absorbed is reconverted to heat energy. A perfectly **black body** is one that will not reflect radiations falling upon it but absorbs them all. The **coefficient of reflectivity** of a body receiving radiation is equal to one minus its **black body coefficient**. Emissivity refers to the rate at which a body radiates heat in relation to a black body of equal area, and this rate depends upon the temperature of the body and the nature of its surface. **Kirchoff's Law** shows that the **absorptivity** and **emissivity** of a given surface are numerically equal at the same temperature. The **Stefan-Boltzmann Law** states that the total energy of a black body is proportional to the fourth power of its absolute temperature, that is (in SI units):

$$W = \sigma T^4$$

where W equals the total emissive power of a black body, expressed in watts per square metre (W/m^2), σ is the **Stefan-Boltzmann constant** equal to 5.71 x 10^{-5} ergs/$cm^2 \cdot s \cdot K^4$ or 5.71 x 10^{-8} watts per $m^2 \cdot K^4$, and T is the absolute temperature in kelvins (°K).

In fps units, W is expressed in Btu/$ft^2 \cdot h$, and equals 0.173 x 10^{-8} Btu/$ft^2 \cdot h \cdot °R^4$ with T representing the absolute temperature in °R.

The net effect of heat transfer between two bodies, neither of which can be considered a black body, must take into account the **emissivity** factor E, which is the ratio of the emissive power of an actual surface to that of a black body; this results in the following equation in SI units:

$$q = 5.71\epsilon A\left[\left(\frac{T_1}{100}\right)^4 - \left(\frac{T_2}{100}\right)^4\right]$$

where

q = watts transmitted
5.71 = Stefan-Boltzmann constant expressed in $W/m^2 \cdot K^4$
ϵ = emissivity factor
A = surface area in m^2
T_1 = absolute temperature of body giving off heat, in kelvins (°K)
T_2 = absolute temperature of body receiving heat, in kelvins (°K)

In fps units, the above equation becomes:

$$q = 0.173\epsilon A\left[\left(\frac{T_1}{100}\right)^4 - \left(\frac{T_2}{100}\right)^4\right]$$

Table 3—VII. Emissivity Factors (A Pefect Absorber or Radiator = 1)

Material	E
Polished aluminum at 230°C (445°F)	0.039
Polished aluminum at 580°C (1075°F)	0.056
Polished brass at 300°C (570°F)	0.031
Polished nickel at 380°C (715°F)	0.086
Polished nickel-plated steel at 22°C (72°F)	0.052
Bright tinned steel plate at 24°C (75°F)	0.071
Polished mild steel	0.288
Cast iron—machined—at 22°C (72°F)	0.437
Cast iron—liquid—at 1330°C (2425°F)	0.282
Cast iron—rough oxidized	0.97
Mild steel—dull oxidized—from 26° to 355°C (79° to 672°F)	0.96
Firebrick glazed through use at 1000°C (1830°F)	0.75
Silica brick (rough)	0.81

where

q = Btu transmitted per hour

0.173 = Stefan-Boltzmann constant expressed in Btu/ft$^2 \cdot$h$\cdot$°R^4

ϵ = emissivity factor

A = surface area in ft^2

T_1 = absolute temperature of body giving off heat, in °R

T_2 = absolute temperature of body receiving heat, in °R

The emissivity factors for various materials at specified temperatures are presented in Table 3—VII. Emissivities vary from almost zero to slightly less than one, depending on the nature of the material, its surface finish, and its temperature. Polished metal surfaces have low emissivities, whereas those of oxidized surfaces and non-metals generally approach a value of one.

In the generation of heat from fuels, the character of the flame and its proximity to the receptor of heat is particularly significant in the transfer of heat by radiation. The amount of heat transferred from a flame varies widely and in proportion to its degree of luminosity. The transfer of heat by radiation varies inversely with the square of the distance between the transmitter and receptor of radiant energy. For that reason, flames should be kept close to the substance to be heated where high heat transfer rates are desirable.

SECTION 2

SOLID FUELS AND THEIR UTILIZATION

The solid fuels have played a significant role in the evolution of our modern, industrial civilization. Coal in particular has been of far-reaching importance in that it has provided the prodigious amount of energy essential to the development of the iron and steel industries. Vast quantities of this energy source remain to be exploited but the rate of utilization far exceeds the rate at which coal is being formed. It follows that the efficient use of the remaining supply is desirable. Toward this end, modern coal research is directed.

Geologically, the earliest-formed coal thus far encountered occurs in the Silurian strata of Bohemia. It is not until Lower Carboniferous time (see Table 3—VIII), however, that the source materials of coal be-

Table 3—VIII. Geologic Time Divisions.

ERAS	PERIODS		EPOCHS	MILLIONS OF YEARS
Cenozoic	Quaternary		Recent	70
			Pleistocene	
	Tertiary		Pliocene	200
			Miocene	
			Oligocene	
			Eocene	
Mesozoic	Cretaceous			500
	Jurassic			
	Triassic			
Paleozoic	Permian			3000 +
	Carboniferous	Pennsylvanian (Upper Carboniferous)		
		Mississippian (Lower Carboniferous)		
	Devonian			
	Silurian			
	Ordovician			
	Cambrian			
Proterozoic	Algonkian	Keweenawan		
		Huronian		
Archeozoic	Archean	Timiskamian		
		Keewatin		

gan to accumulate in significant quantities. Every continent, including Antarctica, contains some coal and no system of rocks younger than the Silurian is devoid of this important substance. In North America major concentrations of source materials were accumulated during the Carboniferous, Cretaceous and Tertiary periods. A smiliar statement can be made for Europe but, in contrast, some of the most important Asiatic coals occur in Triassic and Jurassic rocks of the Mesozoic Era.

Coal Resources—The known coal deposits in the United States are greater than those of any other country. Based on U.S. Geological Survey data, the remaining identified coal resources as of January 1, 1974, were 1570 billion metric tons (1731 billion net tons) of coal of all ranks under 915 meters (3000 feet) or less of overburden. The Energy Information Administration has published the demonstrated reserve base of coal in the United States as of January 1982, at 438 128 million metric tons (482 954 million net tons). This figure represents coal of all ranks in the ground at depths and bed thicknesses generally considered mineable under current economic conditions. This would be enough to supply requirements for a long period in the future if all present coal reserves were available economically

and of acceptable quality. A considerable quantity of the reserves of better quality coking coal has been utilized in the past and it is apparent that in the future it will be necessary to use coals requiring efficient extraction, cleaning, and other processing to assure proper utilization.

For obvious reasons, the steel industry has been striving to use coals which would produce metallurgical coke of optimum quality with a minimum of processing. Concentrations of coals of this class are found chiefly in the Appalachian area, although isolated deposits also exist in some Central and Western states. The preponderance of total coal reserve in the United States is in the form of lower-rank coals in the Great Plains, the Rocky Mountains, the Pacific Coast states and the Gulf region (see Figure 3—I). The manner in which this material can be used most effectively remains to be determined.

Origin and Composition of Coal—Coal is known to be a complex mixture of plant substances which have been altered in varying degrees by physical and chemical processes. Ordinarily, plant material, upon death, completely decomposes because of the action of microorganisms. Under certain circumstances, notably those associated with forested fresh-water swamps, this

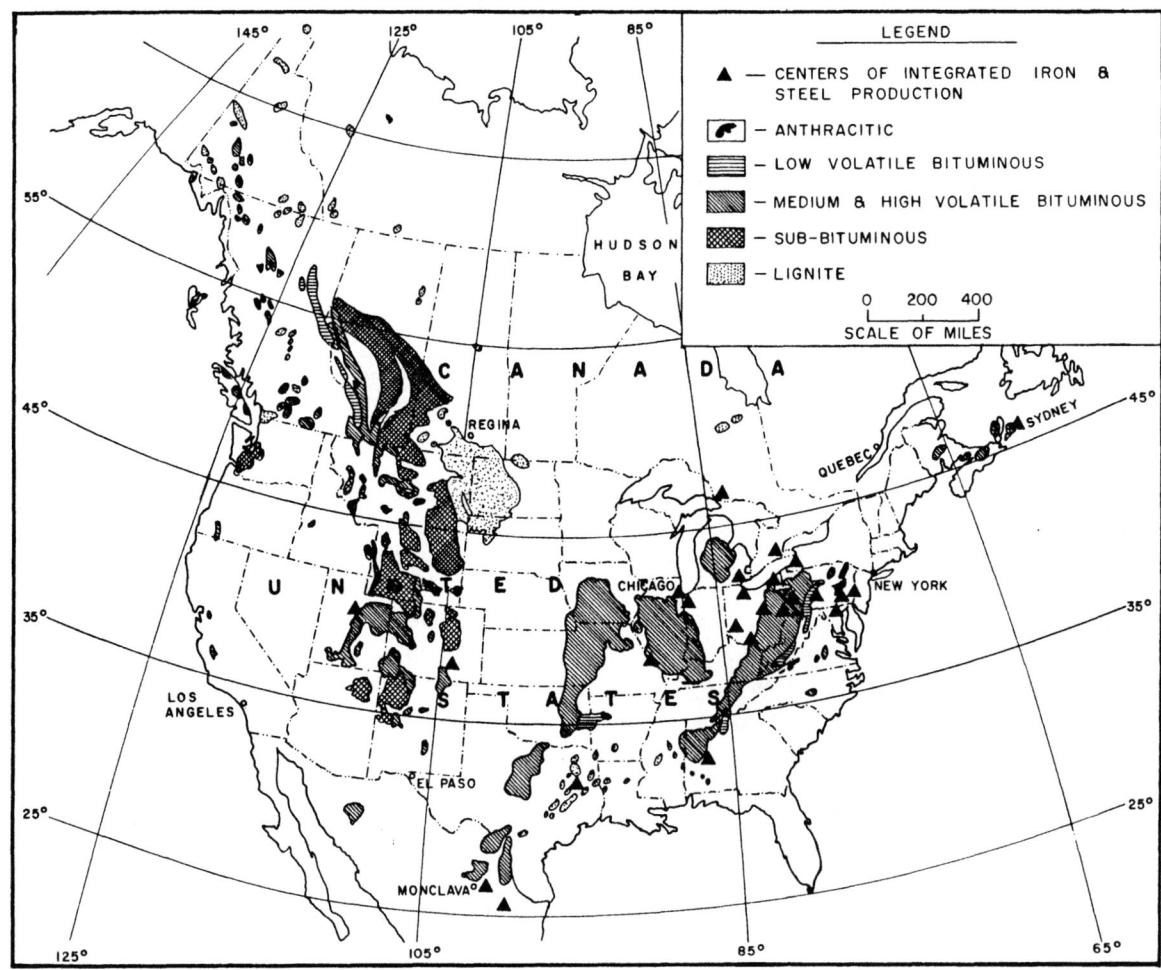

FIG. 3—1. Map showing general location and extent of the important coal fields of North America and centers of integrated iron and steel production. (Map prepared by the Canadian Department of Energy, Mines and Resources, Ottawa, Canada.)

action is inhibited by antibiotic solutions which are common in this type of environment. As a result the rate of accumulation of the plant material exceeds that of its decomposition and dispersion. Under such conditions a brown fibrous deposit known as **peat** is formed. Peat is the first step in the formation of coal.

Peat deposits formed millions of years ago subsequently were submerged through vertical movements of the earth's crust, in which position they became covered by deposits of sedimentary rocks. Later movements of the earth's crust raised many of these deposits to various heights above sea level. In the meantime, the peat had been changed, through agencies of biological action, pressure, and heat, into coal. The better ranks of coal in this country were formed during the Carboniferous period, the geologic period when conditions were most favorable for plant accumulation and decomposition. Included in the present deposits that originated in that period are the coal fields of the Appalachian and Central states.

The rate at which peat forms depends upon the rapidity of plant growth and the manner in which tissue increment is related to the rate of decomposition. It has been estimated that approximately one century is required to form a deposit of mature, compacted peat about one-third metre (one foot) in thickness. Certain studies of volatile matter relationships suggest that about a one-metre thick (a three-foot-thick) deposit of mature peat is required to produce a one-third-metre thick (one-foot-thick) layer of bituminous coal. These and other data indicate that a coal seam which is a metre or more (several feet) thick may require a time span of thousands of years for its formation. If, in the course of time, the peat is subjected to the necessary conditions it becomes modified to **brown coal** and, when adequately consolidated, to **lignite**. From the lignitic stage, the material passes progressively through the **sub-bituminous, bituminous, semi-anthracite** and **anthracite** stages with a gradual change in the composi-

tion of the individual components of the complex mass. The proximate and ultimate compositions of coal, defined later under "Chemical Composition and Coal Classification" and shown in Tables 3—IX, 3—X and 3—XI, illustrate the gradual concentration of carbon and loss of oxygen in the various stages of coal formation.

Peat varies in appearance from a light, brown-colored, fibrous material to a very black and dense, muck-like sediment. Lignite is usually brown in color and commonly shows a woody texture. It contains a large amount of moisture and usually disintegrates, or slacks, into small pieces as it dries on exposure to air. Sub-bituminous coal varies in color from very dark brown to black and fractures irregularly. Bituminous coal is black in color and usually exhibits a banded

Table 3—IX. Typical Moisture and Ash Content of Raw Solid Fuels (Per Cent).[1]

Fuel	Moisture Content	Ash Content
Peat	65-90	(a)
Lignite (North Dakota)	35-40	7.6
Sub-Bituminous (Wyoming)	15-25	3.3
Bituminous (Low-Volatile B)	2.5	11.4
Anthracite (Northeastern Pa.)	5.5	9.6

[1] For additional coal analyses, see Bureau of Mines R.I. 7104 (1968), "Analyses of Tipple and Delivered Samples of Coal" and previous reports in the same series; also "Combustion Engineering" and "Steam, Its Generation And Use" published by Combustion Engineering, Inc. and the Babcock & Wilcox Co., respectively, and cited in the references.

(a) Highly variable, from 2 to 15 per cent or higher.

Table 3—X. Typical Compositions of Peat and Coals of Different Ranks (Dry Basis)

Group (ASTM Designation D 388)	Proximate Analyses (Per Cent)			Ultimate Analyses (Per Cent)					Gross Heating Value	
	Volatile Matter	Fixed Carbon	Ash	Carbon	Hydrogen	Nitrogen	Sulphur	Oxygen	kJ/kg[a] (dry)	Btu/lbm[b] (dry)
Meta anthracite	1.2	90.7	8.1	86.8	1.6	0.6	0.9	2.0	31 797	13 682
Anthracite	3.4	87.2	9.4	84.2	2.8	0.8	0.6	2.2	32 094	13 810
Semianthracite	13.0	74.6	12.4	78.3	3.6	1.4	2.0	2.3	31 560	13 580
Bituminous										
Low-Volatile	16.0	79.1	4.9	85.4	4.8	1.5	0.8	2.6	34 860	15 000
Medium-Volatile ...	22.2	74.9	2.9	86.4	4.9	1.6	0.6	3.6	35 274	15 178
High-Volatile A	34.3	59.2	6.5	79.5	5.2	1.4	1.3	6.1	33 456	14 396
High-Volatile B	39.2	55.4	5.4	78.3	5.2	1.5	1.4	8.2	32 787	14 108
High-Volatile C	36.4	54.5	9.1	73.1	4.8	1.5	2.6	8.9	31 302	13 469
Subbituminous										
A	38.9	56.4	4.7	75.1	5.0	1.4	1.0	12.8	31 595	13 595
B	42.8	54.4	2.8	75.0	4.9	1.3	0.5	15.5	30 788	13 248
C	39.4	47.4	13.2	64.2	4.4	1.2	0.4	16.6	25 796	11 100
Lignite A and B	41.8	49.4	8.8	64.4	4.2	1.1	0.8	20.7	25 643	11 034
Peat	67.3	22.7	10.0	52.2	5.3	1.8	0.4	30.3	21 048	9 057

(a) To convert to kilocalories per kilogram, multiply by 0.2390.
(b) To convert to kJ/kg from Btu/lbm, multiply by 2.326.

Table 3—XI. Approximate Range of Moisture Contents for Peat and for Coals of Different Ranks (ASTM Designation D 388).

Fuel	Moisture Content (Per Cent)
Meta anthracite	3-10
Anthracite	1-8
Semianthracite	1-10
Low-Volatile Bituminous	2-4
Medium-Volatile Bituminous	1-4
High-Volatile A Bituminous	2-11
High-Volatile B Bituminous	4-15
High-Volatile C Bituminous	7-17
Subbituminous A	10-20
Subbituminous B	14-25
Subbituminous C	16-34
Lignite A and B	23-60
Peat	55-90

structure due to the alternate dull and vitreous layers of varying thickness. Coals of the high-volatile bituminous rank commonly burn with a smoky, yellow flame. Anthracite is black, hard and brittle and has a high luster. It ignites less easily than bituminous coal and burns with a short, bluish, yellow-tipped flame producing very little or no smoke. The characteristics of semianthracite coal are intermediate between those of bituminous coal and anthracite.

All of the solid natural fuels contain both combustible and non-combustible materials. The combustible material is composed mainly of carbon, hydrogen and, to a lesser extent, sulphur. The non-combustible constituents are water, nitrogen and oxygen, and a variety of mineral materials usually referred to as **ash.**

The bituminous coals are of greatest interest to the steel industry in view of the fact that essentially all coking coals fall in this category. The lustrous black bands which are conspicuous in a lump of bituminous coal are generally referred to as **vitrain** although some American coal petrographers employ the term **anthraxylon** in preference. Following U.S. Bureau of Mines terminology, the anthraxylon is derived from woody plant tissues and is surrounded by a dull **ground mass** made up of **translucent attritus, opaque attritus** and **fusain.** The attrital portion is composed of finely comminuted fragments of altered plant materials. Fusain is a friable, charcoal-like substance derived from woody tissues and is a term used universally without modification.

In addition to the readily recognizable bands of vitrain and fusain, European and Asiatic coal investigators have found it useful to identify silky, minutely striated layers within a coal as **clarain.** Layers of dull, compact coal are called **durain.** Thus, coal seams can be thought of as being composed, usually, of various mixtures of **vitrain, fusain, clarain** and **durain,** each occurring in the form of layers which are visually observable. Coals made up largely of vitrain and clarain are spoken of as **bright coals** whereas coals containing a high percentage of durain are called **splint coals.** Bright coals are generally better coking coals than splint coals, vitrain apparently playing an important

part in the carbonization process. Fusain will not coke, but in *small* percentages it may actually increase coke strength provided the particle size is fine enough. The fixed carbon content is higher and the volatile matter content is lower in fusain than in the other "banded ingredients."

Microscopic study has shown the banded components to be composed of identifiable plant entities called **phyterals,** but of greater significance is the fact that the vitrain, fusain, clarain and durain are made up of numerous components or **macerals** which can be defined by their physical and chemical properties. Durain, for example, may include several macerals (vitrinite, semifusinite, micrinite, cutinite, etc.) which are easily distinguished by their differing optical properties. Additional information regarding the nature and variability of these individual coal components and their contribution to the effective and efficient utilization of all types of coal appears in references on applied coal petrography at the end of this chapter.

Chemical Composition and Coal Classification —There are two methods commonly employed to determine the chemical composition of coal; namely **ultimate analysis** and **proximate analysis** (Table 3—X). An ultimate analysis determines the quantities of carbon, hydrogen, oxygen, nitrogen, sulphur, chlorine and ash in dry coal; a proximate analysis determines the fixed carbon, volatile matter, moisture and ash contents. The proximate analysis is used most commonly, since it furnishes most of the data required for normal commercial evaluations.

The analysis of coal can be made in the laboratory on an "air-dry" basis. The coal sample is delivered to the laboratory in sealed containers. In the laboratory, the coal is weighed and then exposed to the air of the laboratory for a period of time and then weighed again. The per cent loss in weight is the "air-dry" loss. However, since the "air-dry" analysis is of little value to the user, the analysis is converted to the "as-received" basis by combining air-dry loss and final moisture content.

$$1.0 - \frac{\% \text{ air-dry loss}}{100}.$$

The heating value of coal is reported on as-received basis and dry basis. To convert the analysis to "dry basis" from the as-received basis, each constituent value (except the moisture since that is being eliminated) is divided by the factor

$$1.0 - \frac{\% \text{ moisture in as-received analysis}}{100}.$$

Typical ranges of moisture are listed for various coals in Table 3—IX and Table 3—XI.

Using data provided by chemical, physical or petrographic analyses, coals are classified according to **rank, grade,** and **type.** Classification according to rank is based upon the degree of metamorphism within the coal series from the level of lignite to that of anthracite coal. The American Society for Testing and Materials ranks coals according to their fixed-carbon content on a dry basis, and the lower rank coals according to Btu content on a moist basis. The classification of coals by rank adopted by the American Society for Testing and

Table 3—XII. Classification of Coals by Rank.[a]

Class	Group	Fixed Carbon Limits, per cent (Dry, Mineral-Matter-Free Basis) Equal or Greater Than	Less Than	Volatile Matter Limits, per cent (Dry, Mineral-Matter-Free Basis) Greater Than	Equal or Less Than	Calorific Value Limits (Moist,[b] Mineral-Matter-Free Basis)[g] Equal or Greater Than Btu/lb	kJ/kg	Less Than Btu/lb	kJ/kg	Agglomerating Character
I. Anthracitic	1. Meta-anthracite	98	...	...	2	...	...	...	...	Nonagglomerating
	2. Anthracite	92	98	2	8	...	...	...	...	
	3. Semianthracite	86	92	8	14	...	...	...	...	
II. Bituminous	1. Low volatile bituminous coal	78	86	14	22	...	...	...	...	
	2. Medium volatile bituminous coal	69	78	22	31	...	...	...	...	
	3. High volatile A bituminous coal	...	69	31	...	14 000[d]	32 500[d]	...	...	Commonly agglomerating[e]
	4. High volatile B bituminous coal	...	...	...	...	13 000[d]	30 200[d]	14 000	32 500	
	5. High volatile C bituminous coal	...	...	...	...	11 500	26 700	13 000	30 200	
		...	...	...	...	10 500	24 400	11 500	26 700	Agglomerating
III. Subbituminous	1. Subbituminous A coal	...	...	...	...	10 500	24 400	11 500	26 700	Nonagglomerating
	2. Subbituminous B coal	...	...	...	...	9 500	22 100	10 500	24 400	
	3. Subbituminous C coal	...	...	...	...	8 300	19 300	9 500	22 100	
IV. Lignitic	1. Lignite A	...	...	...	...	6 300	14 600	8 300	19 300	
	2. Lignite B	...	...	...	...	...	...	6 300	14 600	

[a] This classification does not include a few coals, principally nonbanded varieties, which have unusual physical and chemical properties and which come within the limits of fixed carbon or calorific value of the high-volatile bituminous and subbituminous ranks. All of these coals either contain less than 48 per cent dry, mineral-matter-free fixed carbon or have more than 15 500 moist, mineral-matter-free British thermal units per pound (36 100 kJ per kg).

[b] Moist refers to coal containing its natural inherent moisture but not including visible water on the surface of the coal.

[c] If agglomerating, classify in low-volatile group of the bituminous class.

[d] Coals having 69 per cent or more fixed carbon on the dry, mineral-matter-free basis shall be classified according to fixed carbon, regardless of calorific value.

[e] It is recognized that there may be nonagglomerating varieties in these groups of the bituminous class, and there are notable exceptions in high volatile C bituminous group.

[f] From ASTM Designation D-388-66 in "ASTM Standards 1975," Part 26, page 215, to which reference may be made for method of calculation to mineral-matter-free basis and other information. Reproduced by permission of the American Society for Testing and Materials. The complete specification is obtainable from the society.

[g] Rounded values for kJ/kg, obtained by calculation, are not part of the original specification.

Materials (ASTM Specification D388), is shown in Table 3—XII.

In the United States, coals are also classified into **types** and such terms as **bright, semi-splint, splint, cannel** and **boghead coal** are applied. The data required are obtained from microscopic studies. The United States Bureau of Mines defines bright coal as containing less than 20 per cent opaque matter, semi-splint must have between 20 and 30 per cent, and splint coal must be made up of more than 30 per cent of this ingredient. Cannel and boghead coals are **non-banded** and are characterized by a small percentage of anthraxylon (vitrain). Boghead possesses a high percentage of volatile oils and gases, and contains an abundance of algal material. Cannel, or candle, coal is so named because it can be ignited with a match or a candle flame and it burns with unusual brilliance. Cannel coal is non-coking, often contains large quantities of spore and pollen materials, and like boghead, has a high content of volatile oil and gas.

Coals are classified to grade by their ash and sulphur contents. The mineral constituents of the ash are also important because they influence fouling and slagging in the furnace.

MINING OF COAL

It is found that seams of coals vary in thickness throughout the world from a few millimetres to over 75 metres (a fraction of an inch to over 250 feet). In this country the thickest seams are found in the sub-bituminous coals of the West, one of which approaches about 30 metres (100 feet). In the East, the Mammoth bed in the anthracite fields of Pennsylvania attains a thickness of 15 to 18 metres (50 to 60 feet) but is found to be quite variable when traced laterally. The Pittsburgh seam at the base of the Monongahela series in the Appalachian area is noteworthy because of its exceptionally uniform thickness (approximately 2 metres or 7 feet) over thousands of square miles. Figure 3—2 shows the western portion of Pennsylvania in such a manner as to make clear the areal extent as well as the sub-surface relations of the coal-bearing formations of this region. Data are provided as to thickness of seams and distance between coals.

Coal seams may dip gently as shown in Figure 3—2, or they may be horizontal, or they may exist almost vertical with respect to the Earth's surface. Mining problems are often complicated by the fact that seams seldom remain in the same plane throughout their extent. Under present conditions, a coal bed must be at least 0.75 to 0.90 metres (30 inches to 36 inches) thick to be profitable for mining. Figure 3—2 shows, also, that coal seams vary in their distance below the Earth's surface. U. S. Geological Survey estimates of coal reserves do not include coal seams deeper than about 910 metres (3000 feet) from the surface, although in Great Britain and Europe coal seams at greater depth are being mined.

The mining of coal is performed by either one of two methods: (1) **Open pit** or **stripping**, also called **contour mining,** or (2) **underground** or **deep mine.** The first method involves removing the formation (over-burden) above the seam by stripping with scrapers, bull-dozers,

or mechanically operated shovels, followed by removing the exposed coal. Stripping is applied to coal seams which are relatively close to the surface, particularly to thick seams underlying overburden about 25 to 50 metres (80 to 150 feet) deep, although the development or larger equipment and improved techniques in recent years has justified removal of layers of overburden thicker than this. Auger mining is being used extensively to recover coal where the overburden is too great for strip-mining practices to be employed. A large-diameter auger or drill with cutting bits on its end is propelled into the exposed edge of a coal seam. As the auger progresses into and along the seam, the broken coal is conveyed away from the face through the tube to the outside for transport away from the auger. Production by strip mining has increased greatly since World War I due to reduced labor and material costs and a quicker return on capital investment compared to underground mining. In the United States, strip mining accounted for slightly over 25 per cent of the coal produced in 1957. In 1982, strip and auger mining combined produced about 62 per cent of the total coal mined.

Underground mining is performed by either the **room-and-pillar** or the **longwall** method. The room-and-pillar method is in more common use in the United States, accounting for approximately 90 per cent of present underground mining. The longwall method is particularly adaptable to mining seams up to about 4.6 metres (15 feet) thick under conditions where the roof may be permitted to cave. It is used more extensively in the mines of the eastern United States. Of 118 mines using this method in 1983, 93 were in the Appalachian region. There are a number of modifications applicable to each method. The room-and-pillar system consists essentially of working out rooms, chambers, or breasts in the coal seam from passages (entries) driven from the mine entrance. Entrance to an underground mine is by drift, shaft or slope. The rooms vary in width from about 3.5 to 12 metres (12 feet to 40 feet), 6.1 metres (20 ft.) being the most common, and from about 45 to 90 metres (150 feet to 300 feet) in length, depending on such factors as weight and character of the overlying and underlying structure and thickness of seam. Pillars separating the rooms vary in width from about 2 to 30 metres (6 feet to 100 feet), depending on conditions and mining practice. These pillars are sometimes removed by retreat mining and the coal recovered.

In the longwall method, a continuous mining face is maintained in the coal seam. After mining, the roof is permitted to cave, about 5 to 9 metres (15 or 30 feet) from the mine working face.

Prior to the advent of mechanical mining, undercutting of the coal seam preparatory to blasting was done manually. Production per man was low by this method and required a number of working faces in the mine to produce high mine tonnage. Hand loading of coal into mule-drawn cars was the prevailing practice for many years until development of machinery for cutting, loading and haulage. Electric trolley-type locomotives capable of hauling longer underground trains of cars of increased capacity displaced mule-drawn trains as mine capacity increased.

In some modern underground mines, the coal is car-

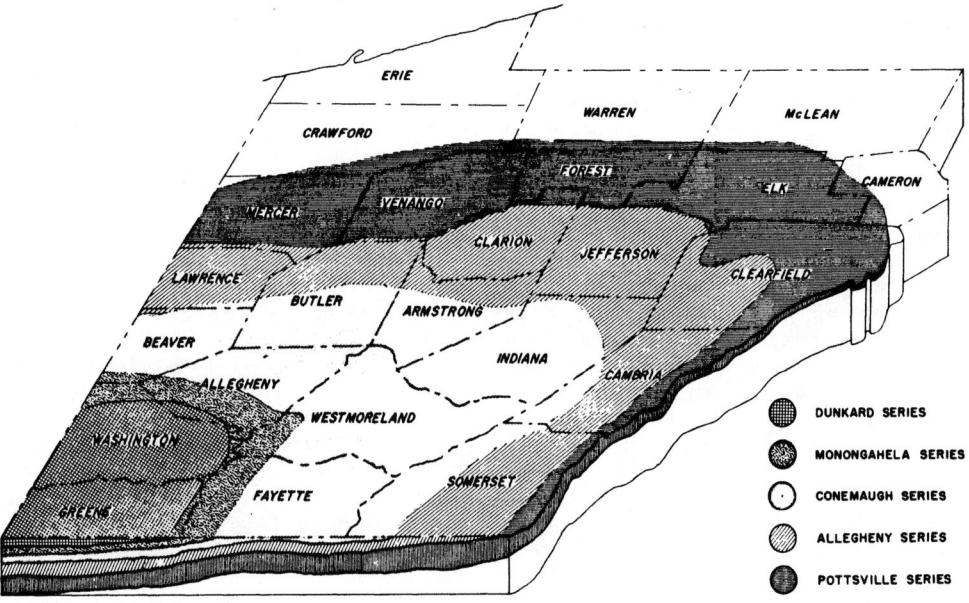

Series	Strata	Thickness		Totals From Top	
		m	ft	m	ft
DUNKARD SERIES	PROCTOR SANDSTONE	12.2-12.2	40-40	12.2	40
	WINDY GAP COAL	0.15-0.30	½-1	54.9	180
	GILMORE COAL	0.15-0.30	½-1	88.4	290
	NINEVEH "A" COAL	0.076-0.30	¼-1	138.7	455
	NINEVEH COAL	0.15-0.30	½-1	153.9	505
	HOSTETTER COAL	0.15-0.30	½-1	176.8	580
	FISH CREEK COAL	0.15-0.30	½-1	205.7	675
	DUNKARD COAL	0.15-0.30	½-1	225.6	740
	JOLLYTOWN COAL	0.15-0.30	½-1	240.8	790
	HUNDRED COAL	0.076-0.30	¼-1	254.5	835
	WASHINGTON "A" COAL	0.15-0.30	½-1	280.4	920
	WASHINGTON COAL	0.61-1.52	2-5	313.9	1030
	LITTLE WASHINGTON COAL	0.38-0.61	1¼-2	321.3	1054
MONONGAHELA SERIES	WAYNESBURG COAL	0.91-1.52	3-5	374.6	1229
	UNIONTOWN COAL	0.30-0.91	1-3	391.7	1285
	LOWER UNIONTOWN COAL	0-0.30	0-1	413.0	1355
	SEWICKLEY COAL	0.91-1.52	3-5	452.6	1485
	REDSTONE COAL	0.91-1.52	3-5	470.9	1545
	PITTSBURGH COAL	1.52-2.43	5-8	484.6	1590
CONEMAUGH SERIES	MORGANTOWN COAL	0.30-1.52	1-5	490.7	1610
	LITTLE PITTSBURGH COAL	0.30-1.83	1-6	507.5	1665
	LITTLE CLARKSBURG COAL	0.61-2.13	2-7	530.4	1740
	NORMANTOWN COAL	0-0.30	0-1	545.6	1790
	CLARYSVILLE COAL	0-0.30	0-1	560.8	1840
	ELK LICK COAL	0.61-1.22	2-4	591.3	1940
	HARLEM COAL	0.15-0.61	½-2	621.8	2040
	UPPER BAKERSTOWN COAL	0.15-0.30	½-1	650.7	2135
	BAKERSTOWN COAL	0.61-1.52	2-5	652.3	2140
	BRUSH CREEK COAL	0-1.83	0-6	714.8	2345
	MAHONING COAL	0.30-1.83	1-6	722.1	2369
ALLEGHENY SERIES	UPPER FREEPORT COAL	0.61-1.52	2-5	730.0	2395
	LOWER FREEPORT COAL	0.61-0.91	1-3	751.3	2465
	UPPER KITTANNING COAL	0-0.30	0-1	768.1	2520
	MIDDLE KITTANNING COAL	0.91-1.52	3-5	775.7	2545
	LOWER KITTANNING COAL	0-2.74	0-9	797.1	2615
	CLARION COAL	0.30-1.22	1-4	829.1	2720
POTTSVILLE SERIES	BROOKVILLE COAL	0.30-0.91	1-3	836.7	2745
	MERCER COAL	0-0.61	0-2	853.4	2800

FIG. 3—2. (Above) Distribution of coal-bearing strata in Western Pennsylvania. (Below) Important coals of the northern portion of the Appalachian coal region.

ried out of the mine by a system of conveyor belts to a shipping station or cleaning plant. In other mines, the coal is carried in mine cars to a rotary dumper which may have a capacity of as many as 37 cars. From the dumper the coal is fed by way of a conveyor or elevator to a shipping station or cleaning plant.

Continuous Mining—The cutting machines and loading machines characteristic of mechanical mining in the past were single-purpose units, and each performed essentially a single function of mining at the working face. After either unit completed its work it was withdrawn from the face to allow other units of the production setup to move up to the face to carry out succeeding functions. To keep all operating units working at full efficiency, it was necessary to have additional working places near at hand so that the single-purpose machines could enter the places in rotation and carry out their functions without interference.

To eliminate some of the difficulties attendant upon the addition of extra working places, multi-purpose machines known as **continuous miners** have been developed and the operation carried out by such machines has been given the name **continuous mining**. Continuous miners combine in a single unit the actions of dislodging the coal from the solid seam and loading it into some unit of a transportation system. Such machines, therefore, combine in one operation the separate steps of cutting, drilling, blasting and loading common to earlier mechanical mining methods. Coal planers and shearers with long-wall mining achieve these combined objectives.

There are several types of continuous miners in operation, one of which is a ripper-type miner that has cutting bits mounted in the rims of multiple wheels that are rotated to rip the coal out of the seam while the ripper wheels are propelled into and up or down in the coal seam. The coal that is ripped loose from the seam falls into the gathering head of the loader, which has dual gathering arms that sweep the broken coal into the conveyor section of the machine for loading into shuttle cars or other suitable conveying equipment.

Continuous miners of some other types employ auger-type cutters that bore into the face, the cut coal in both cases being carried by a conveyor on the machine to a shuttle car or other means of transportation.

COAL PREPARATION

As one phase of coal preparation, the objective of **coal cleaning** (often called washing) is removal of solid foreign matter, such as rock and slate, from the coal prior to its use. Reduction of ash and sulphur contents; control of ash fusibility; increase of heating (calorific) value; and improvement of coking properties of the coal can be achieved by this practice. From a coal-cleaning standpoint, the impurities in coal are of two types; namely, those which cannot be separated from the coal, usually called **fixed** or **inherent impurities;** and those which can be removed, herein referred to as **free impurities**. Altogether, these impurities are of eight types, named as follows: (1) residual inorganic matter of the coal-forming plants from which the coal was derived; (2) mineral matter washed or blown into the coal-forming mass during the periods of its formation;

(3) pyrites (FeS_2) formed by bacterial reaction of iron and sulphur in the coal-forming matter; (4) sedimentary deposits during the coal-forming periods which appear as partings, sometimes called "bone", that usually must be mined with the coal; (5) massive deposits formed through deposition on bedding planes; (6) saline deposits, somewhat rare in coal beds of the United States; (7) slate, shale, clay, etc. from the underlying and overlying strata accidentally included in mining; and (8) water or moisture, which includes that naturally carried by the coal in air-dry condition, and excess moisture producing a condition of wetness. Items (1), (2) and, for the most part, (3) form **fixed** ash, while (4), (5), (6), and (7) are partly **free** ash-forming materials that can be removed by hand-picking and suitable mechanical cleaning treatments. Item (8) involves drying operations differing from those required to separate mineral impurities, which is the primary objective of cleaning. Mechanical cleaning is possible because of the difference in specific gravity between the major impurities and the coal, the density of the former being 1.7 to 4.9, while pure coal has a density of about 1.3. Sulphur is present as pyrites, organic compounds, and sulphates, and only part of the pyrites can be removed by cleaning. Phosphorus is usually associated more with bony and impure coal than with clean coal and is, therefore, reduced by washing. Salts, particularly the alkali chlorides, lower the fusion point of the ash, affect coke-oven linings and are troublesome in waste liquors from coking operations.

The advent of full-seam mechanical coal mining and the increasing need for metallurgical coke of low and uniform ash and sulphur contents has focused attention on the needs for the most efficient types of washers.

The preparation of coal starts at the production face in the mine. If loading is done by hand, the miner is required to discard all rock and slate over 75-mm (3-in.) size. Since practically all loading is done mechanically in the United States little attempt is made to prepare the coal at the face other than to control the tonnage from various sections of the mine if sulphur content of the coal is high or variable.

The cleaning qualities of a particular coal are determined by the **float and sink test, commonly referred to as a washability test**. Fundamentally, this test effects a fractionation of the coal by size and specific gravity. This test consists in crushing coal to proper size and floating individual sizes of it on liquids having densities of 1.30, 1.40, 1.50, 1.60, etc., to determine the weight and character of the material that floats and sinks in *each* liquid. The proportion of coal, and the ash and sulphur content of the different fractions, provides reasonably complete data on the washability characteristics of a tested coal. Extreme fines may be evaluated by froth flotation.

Coal preparation is accomplished by a combination of crushing, sizing, cleaning, and dewatering operations:

    A. Crushing
       1. Mine breakers
       2. Bradford breakers
       3. Roll crushers
       4. Impact crushers

B. Sizing
   1. Grizzlies
   2. Vibrating screens
   3. Classifiers
   4. Cyclones
C. Cleaning
   1. Jigs
   2. Dense media processes
   3. Cyclone processes
   4. Tables
   5. Froth flotation
D. Dewatering
   1. Screens
   2. Centrifuges
   3. Vacuum filters
   4. Thermal dryers

By far the largest percentage of coal is cleaned by wet methods.

A complete description of each of the foregoing processes would be too lengthy for inclusion herein; hence only a brief review will be given of the principles of some of the more important types of cleaning processes in use at present. A reference list for further study of this subject is appended to this chapter.

Jigs were probably the earliest type of machine used in the mineral industry to separate materials of different densities. They consist essentially of a box with a perforated base into which the material is placed, and by alternate surges of water upward and downward through the perforations, materials of different specific gravities stratify. Materials having the highest specific gravities remain at the bottom while the lighter material rises. With proper mechanical facilities, a continuous separation is achieved. While jigs are not very efficient in cleaning a mixture of various sizes, they are capable of satisfying some market requirements, and capacities up to about 450 metric tons (500 net tons) per hour have been obtained.

In dense-media processes, only a part of the power for separating coal and refuse is supplied by an upward flow of liquid, this separating power being supplemented by using a liquid medium which is heavier than water. The medium employed is a mixture of water and some finely divided solid material, such as sand, magnetite, or barite, which can be separated readily from the washed coal and reused. In the high-density suspension process, the upward flow is discarded entirely, the liquid medium consisting of a mixture which is just dense enough so that the coal floats in it, and the impurities sink. The size of coal has less significance in the efficiency of this process than of those previously described, and material ranging from 1.6 to 254 mm ($\frac{1}{16}$ inch to 10 inches) can be cleaned in one operation. However, difficulty is encountered in separating the solid material from coal of fine size. Capacities up to about 545 metric tons (600 net tons) per hour have been obtained with bituminous coal. The Chance cone method, which uses a mixture of sand and water, is also widely used in the United States. The Tromp and Barvoys processes, using magnetite and barite respectively as the solid material in the mixture, are used extensively in Europe and in the United States. In these processes, the specific gravity of the mixture of solid material and water can be varied by changing their proportions to suit the optimum conditions in cleaning. Agitation in the separating cone is supplied by an upward current of water and by mechanical stirring.

Cyclones, which have an inverted conical shape and are fed tangentially at the widest part of the cone, are used with the dense medium process to increase the separating rate of particles in the medium. This device is particularly effective on sizes between ¾ inch (17 mm) and 100 mesh (0.15 mm). Cyclones can also be used in the absence of a specific medium, but the efficiency is lower and they must often be combined in stages. In this practice they are referred to as hydroclones or water-only cyclones.

Coal is also cleaned on table concentrators. Essentially these tables consist of a slightly inclined rectangular surface having a series of parallel grooves or cleats. The tables are mechanically agitated to permit stratification of the light and heavy material and to cause the heavy material to move with the long axis of the table. A current of water is introduced at the top edge of the table to wash the coal which has settled above the refuse to the discharge edge of the table. The refuse which settles underneath the coal moves longitudinally down the table and is discharged at the end. Tables have been used principally for cleaning coal of the smaller sizes, from about 0.3 to 12.7 mm (48-mesh up to about ½-inch).

Froth flotation of coal involves agitating fine coal with a mixture of water and a relatively small quantity of some frothing agent. In this process, coal is buoyed to the surface by the froth and removed while refuse settles. It is widely used to recover coal, finer than 100 mesh(150 mm).

With practically all wet-washing systems the water is recirculated. When the water passes through the dewatering screens it contains a considerable amount of small-size coal solids which must be recovered for efficiency reasons. Also, the circulating water and effluent from flotation must be clarified before it is returned to the cleaning unit. This clarification is accomplished in various ways, the most important being by the use of hydraulic cyclones followed by a Dorr-type thickener. Settling cones and settling tanks are also used for this purpose. Where the Dorr-type thickener and settling tanks are used, it is customary to draw off the settlings in the form of a slurry containing 40 per cent to 60 per cent of solids and to further separate the slurry in a vacuum-type filter. The filters deliver a product with approximately 25 per cent moisture. For a more complete discussion of dewatering and waste disposal, the reader is referred to a standard book on coal preparation.

CARBONIZATION OF COAL

The most important use of coal in the modern steel industry is in the manufacture of metallurgical coke, which is discussed in detail in Chapter 4.

The carbonization of coal in by-product ovens entails the production of large amounts of coke-oven gas and tar, important fuels in the steel industry, as well as light oils and various coal chemicals. The yields of gas and tar are largely a matter of the kind of coal used and the temperatures employed in coke manufacture.

COMBUSTION OF SOLID FUELS

The principal combustion reactions of solid fuels have been given in Section 1 of this chapter, under "Principles of Combustion," and this present discussion will deal with operating factors pertinent to the combustion of solid fuels in steel plants.

The combustion of **coke** in blast furnaces has been studied by a number of investigators, each of whom has found that combustion takes place in a relatively small space directly in front of each tuyere, as discussed in Chapter 15 on "The Manufacture of Pig Iron."

Coke breeze, produced by screening coke at both the coke plant and blast furnaces, is utilized as a fuel in steel-plant boiler houses to generate steam and in ore-agglomerating plants. When used as boiler fuel, coke breeze is burned on **chain-grate stokers.** Of importance in the combustion of coke breeze on chain-grate stokers is the maintenance of a relatively uniform fuel bed on the grate, about 200 to 300 mm (8 to 12 inches) thick, to prevent blowholes, and a balanced or slight positive pressure in the furnace at fuel-bed level. The operation of the grate should permit the normal combustion of about 146 kilograms per square metre (30 pounds of coke breeze per square foot) of effective grate area per hour. Chain-grate stokers are particularly adaptable to solid fuels with an ash of low fusion point. The design of front and back arches must take into consideration the fuel to be burned on chain-grate stokers. The arches are utilized to reflect heat and thereby aid ignition on the fuel bed. Use of coke breeze in ore agglomeration is discussed in Chapter 5 under "Sintering."

Stokers for firing coal have generally been used in steel-plant boilers on units whose capacity is under about 45 000 kilograms (100 000 pounds) of steam per hour and for units using exclusively a solid fuel. They are often used on boilers to provide flexibility for the adjustment of boiler output to the steam load in plants where there is an insufficient or fluctuating supply of gaseous by-product fuels. The advantage of stokers lies in their ability to control easily the rate of combustion of a solid fuel with efficient use of air. The combustion process on stoker-fired boilers consists essentially in first driving the volatile matter from a continuous supply of fuel, and then oxidizing the carbon in the residue on the stoker. The combustion of the coke-like residue on the grate produces CO_2 and CO. The CO and volatile matter are burned over the grate by secondary air admitted over the fuel bed. The temperature of the fuel bed is affected by the rate of firing and, at the top or hottest part of the bed, varies from about 1230°C (2250°F) at low to 1510°C (2750°F) at high rates. The amount of primary air supplied determines the capacity of stoker-fired furnaces and the effective use of secondary air determines the efficiency of combustion. In well operated and carefully sealed boilers, approximately 20 to 30 per cent excess air will permit combustion of the gases within seven or eight feet above the grate.

Stokers are classified in general according to the travel of the fuel. In an **overfired stoker** the fuel is fed on top of the bed, and in an **underfired** or a **retort** stoker the fuel is fed at the bottom or side of the bed. A **traveling-grate** or **chain-grate stoker** carries the bed horizontally on the flat upper surface of a conveyor as in a chain-grate stoker. There are a number of modifications of these stoker types. The **spreader stoker** projects the coal into the furnace above the fuel bed and the fuel is burned both in suspension and on the fuel bed. While the fuel bed of a stoker-fired boiler is relatively thin, usually from 100 to 300 mm (4 to 12 inches), compared to a gas-producer bed, similar zones of reaction occur. In over-fired stokers the ash zone is immediately above the grate, followed by the oxidation, reduction and distillation zones. In underfired or retort stokers the distillation of the volatile matter takes place in an oxidizing atmosphere and the volatile products pass through the incandescent residue from combustion rather than through green coal, as in the case of overfired stokers. The normal combustion rates on coal-fired stokers amounts to about 150 to 300 kilograms of coal per square metre (30 to 60 pounds per square foot) of effective grate area per hour.

Pulverized Coal—The cement industry was the first to use pulverized ("powdered") coal extensively as a fuel. Public utilities and the steel industry began applying pulverized coal on an experimental basis as a boiler fuel about 1917, and by 1935 practically all large boilers (above about 45 000 kilograms or 100 000 pounds of steam per hour) in public-utility power stations used this fuel, except for those stations located in the vicinity of oil and natural gas fields where local fuels were more competitive than coal. Large modern boiler installations in integrated steel plants generally use pulverized coal, either as a standby or as an auxiliary fuel in conjunction with blast-furnace gas for steam or power generation. Although pulverized coal has been used as a fuel for metallurgical purposes in steel plants, such as in open-hearth, reheating, forge and annealing furnaces, applications have been limited generally to periods of national fuel shortages, such as existed during the first and second World Wars, when the more desirable liquid or gaseous fuels were diverted to other uses and not available.

Pulverized-coal firing offers important combustion advantages over grate firing and an economic advantage over gaseous and liquid fuels in most sections of the country. Boiler capacities are not limited as is the case with boilers equipped with stokers. Fine particles of coal burned in suspension are capable of developing a highly luminous high-temperature flame. Coal in this form may be burned normally with less excess of air above theoretical requirements than with a solid fuel, and the rate of heat release from the combustion of pulverized coal is greater than that accomplished with the solid fuel. Coal, when pulverized to the degree common for boiler uses (70 per cent through a 0.074-mm or 200-mesh screen), has the control flexibility of gaseous and liquid fuels. Practically all ranks of coal, from anthracite to lignite, can be pulverized for combustion and each possesses specific combustion characteristics which largely influence the extent of pulverization. Pulverized-coal firing in modern boilers has certain inherent problems. Excessive fly-ash discharge from the stack, high operating power-consumption rates, excessive pulverizer-maintenance cost, erosion of induced-draft-fan blades and other boiler components, and requirements for large furnace volumes impose practical limitations in selecting this type of firing for low-capacity boilers. Dust collectors are required to control stack emissions.

The ash-disposal problem has been one of the principal deterrents to a more extended use of pulverized coal. In the cement kiln, coal ash is no problem as it is absorbed by the cement in the kiln without adverse effect on the final product. In boilers, the principal difficulty of clogged boiler tubes and deterioration of furnace walls has been overcome by the use of slagging-type furnaces in which the ash in molten form is granulated by water jets at the bottom of the furnace well. The introduction of the "cyclone furnace", which offered the removal of the ash as liquid slag, further increased the application of pulverized-coal firing for steam and power generation. This equipment was developed to solve two major problems that beset the power engineer: (1) the increasing necessity to use low-quality, high-ash fuels for steam generation; and (2) the requirement that as much of the coal ash as possible be kept in the furnace and not permitted to go through the furnace and out of the stack. However, the removal of ash as a liquid slag requires the use of coal having ash of low fusion point and such special coal is sometimes difficult to obtain.

The problem of ash contamination resulting from burning fine particles of coal in suspension above a metallic liquid bath or mass of hot steel, damage to refractories from the chemical or physical action of ash, and the clogging of furnace checkers or recuperators from ash accumulation, as well as the normal availability of other fuels, has prevented widespread use of the fuel for metallurgical purposes in steel plants.

Pulverized coal for firing boilers is relatively more modern than stokers. This fuel is used generally on boiler units having a capacity in excess of about 45 000 kilograms or 100 000 pounds of steam per hour, or on practically any size of boiler using combination firing with oil or gas. Pulverized coal offers high boiler efficiency, and means for quick regulation of boiler load. The rank of coal pulverized and the extent of pulverization particularly determine the speed of combustion. A high-volatile coal will burn faster than anthracite coal, also one with a lower ash content will burn faster. The process of combustion with pulverized coal is similar to that of lump coal but is of much higher velocity due to the introduction of the particle in suspension in a high-temperature chamber, and the greater surface exposure relative to weight. The release of volatile matter in pulverized coal is practically instantaneous when blown into the furnace, and the speed of combustion of the resulting carbonized particle and volatile gases depends upon the thoroughness with which the pulverized coal has been mixed with air. High combustion temperatures, low ash losses, and low excess air needs (10-20 per cent), with resultant high boiler efficiencies (85 to 90 per cent with good practice), make pulverized coal an ideal boiler fuel.

Air for combustion of pulverized fuel is generally preheated, with 10 to 50 per cent of that required introduced ahead of the pulverizer and the balance made up at a point near the burner. This method of introducing the air helps dry the coal and maintains a non-explosive mix in the pulverized-coal transmission system.

The combustion-chamber size for pulverized coal is generally proportioned for a heat-release range of from about 207 000 to 1 035 000 watts per cubic metre or 745 000 ') 3 725 000 kilojoules per cubic metre per hour (20 000 to 100 000 Btu per cubic foot per hour) of combustion space. However, the cyclone furnace has heat-release rates of about 5 175 000 to 9 315 000 watts per cubic metre or 18 625 000 to 33 525 000 kilojoules per cubic metre per hour (500 000 to 900 000 Btu per cubic foot per hour) within the cyclone chamber and the boiler furnace is used only for extracting heat from the flue gases. The difference in requirements is dependent upon whether pulverized coal is the sole fuel to be used in the chamber, the size of the coal particles, the rank of coal to be pulverized, the ash-slagging temperature of the coal, and the desired temperature for the combustion chamber. Spreader-stoker installations offer low first cost for smaller-size boilers, and the fly-ash emission from the boiler is not as severe as with the pulverized-fuel boilers.

Another technology is fluidized bed combustion which has been used to provide emission control for high sulphur coals. In this type of firing, the fuel limestone sorbent particles are kept suspended and bubbling or fluidized in the lower section of the furnace through the action of air under pressure through a series of orifices in a lower distribution plate. The fluidization promotes the turbulent mixing required for good combustion, which in turn promotes the three required parameters for efficient combustion; time, turbulence, and temperature. The limestone captures the freed sulphur from the coal products of combustion to form calcium sulfate. The mixture of ash and sulfated limestone sorbent discharged is a relatively inert material, disposal of which presents little hazard. This type firing also achieves a measure of NOx control in that it burns at a lower temperature, somewhat below maximum NOx formation.

SECTION 3

LIQUID FUELS AND THEIR UTILIZATION

Liquid fuels are essential to practically all parts of the American transportation system. The movement of passengers and freight by highway and air is dependent upon gasoline and other products of petroleum. The railways of the country have nearly all been equipped with diesel locomotives powered by fuel oil. Nearly all ocean-going ships are driven by oil, as are the majority of lake and river craft. Liquid fuels have also become of major importance as a source of heat and power in manufacturing plants. The particular advantages of petroleum as a source of energy and the available supplies have brought about a phenomenal growth in the petroleum industry.

Traumatic price increases initiated by the Organization of Petroleum Exporting Countries (OPEC) during the 1970's resulted in conservation efforts that have reduced U.S. dependence on foreign oil. After the initial shock, demand dropped slightly then increased to a peak of 3.0 million cubic metres (18.8 million barrels) per day. After the second price shock, demand gradually dropped to 2.4 million cubic metres (15.25 million barrels) per day in 1982. Imports crested at 1.4 million cubic metres (8.8 million barrels) per day in 1977 dropping to 0.7 million cubic metres (4.6 million barrels) per day in 1982. The U.S. has established a Strategic Petroleum Reserve (SPR) to minimize problems associated with supply interruptions. Non-OPEC sources are providing an increasing portion of U.S. supplies. For the first 10 months of 1983, Mexico (23%), the United Kingdom (10.4%), Indonesia (9.8%), Nigeria (9.7%), Saudi Arabia (7.7%) and Canada (7.6%) were the principal suppliers. In 1982 the U.S. produced about 16 percent of the world supply of crude petroleum. In addition to development of the Alaskan oil fields, large new reserves were discovered off the California coast in the Santa Maria Basin and the Santa Barbara Channel. Mexico has increased its exports and Canada has large reserves in the Beaufort Sea area and the Hibernia discovery off the coast of Newfoundland.

The reserves of crude petroleum in this country were estimated at 4 415 million cubic metres (27 858 million barrels) as of December 31, 1982, and during 1982, the annual production was about 500 million cubic metres (3 157 million barrels).

Fuel oil, tar, pitch and pitch-tar are the principal liquid fuels used in the steel industry. Table 3—XIII shows the consumption of fuel oil, tar and pitch for the year 1982.

Tar and pitch are by-products of the manufacture of coke. The virgin tar as it comes from the ovens contains valuable tar-liquor oils which can be extracted and the residue pitch used as a fuel. It is customary to mix virgin tar with this highly viscous residue to provide fluidity for facilitating handling and burning, or to utilize

tar in which only the lighter products have been removed by a topping process by which sufficient fluidity is retained. Pitch-tar mixtures and topped tar make available for use as a fuel 78 to 83 per cent of the heat in the crude tar recovered in the distillation process.

Origin, Composition and Distribution of Petroleum—Classified according to their origins, three main types of rocks make up the outer crust of the earth: igneous, sedimentary and metamorphic rocks. **Igneous rocks** are formed from **magma**, a molten (liquid or pasty) rock material originating at high pressures and temperatures within the earth. Lava is magma that reaches the surface in the liquid or pasty state. Very commonly, the magma cools and solidifies before reaching the surface. In any case, when the molten material cools sufficiently to become solid, igneous rocks are the result. If cooling is slow, the rocks will be crystalline (granite, for example); but if the cooling is rapid, the rocks will not be crystalline but glassy in nature (obsidian, for example). Because of the nature of their origin and their usually dense, nonporous structures, igneous rocks are never hosts to petroleum deposits.

Sedimentary rocks are formed from eroded particles of rocks and soil, carried away by wind or water (and sometimes glacial action) and deposited in seas, lakes, valleys and deltas in relatively even, sometimes very thick, beds or strata (sandstones and shales are formed from deposits of this type). Other types of stratified deposits may be formed by evaporation of land-locked seas (beds of rock salt), by accumulation of the mineral remains of animals (composed chiefly of calcium carbonate, which is the principal constituent of limestone), or by chemical precipitation (gypsum and some limestones originate in this manner). The beds of sand, silt, clay, calcium carbonate or whatever eventually are covered by other sedimentary deposits, sometimes to very great depths. With the passage of long periods of time, pressure of the overlying strata, heat, cementation by chemical means, earth movements, or a combination of these or other agencies, the strata are consolidated into sedimentary rocks, typified by the few mentioned parenthetically earlier. *Petroleum occurs almost entirely in sedimentary rock formations*, principally sandstones and limestones, under certain ideal conditions to be described later.

Table 3—XIII. Consumption of Fuel Oil, Tar and Pitch in the Steel Industry (1982)[1]

Purpose	Fuel Oil m³	Fuel Oil Thousands of Gallons	Tar and Pitch m³	Tar and Pitch Thousands of Gallons	Raw Steel Produced (Thousands of Tons) Metric	Net
Blast-Furnace Area	460 000	121 480[3]	80 685	21 317	—	—
Steel-Melting Furnaces	225 000	59 484	74 784	19 758	67 656	74 577
Heating and Annealing Furnaces	357 000	94 93	[2]	[2]	—	—
Other	375 000	99 122	13 191	3 485	—	—
Total	1 417 000	374 479	168 660	44 560	67 656	74 577

[1] From: Annual Statistical Report—1982; American Iron and Steel Institute. Calculated rounded values for cubic metres and metric tons were not part of the original report. (1000 gallons = 3.785 cubic metres.)

[2] Only a small amount included with other miscellaneous uses.

[3] Includes coke-oven underfiring.

Metamorphic rocks originally were sedimentary or igneous rocks. Their composition, constitution or structure have been changed through the single or combined action of natural forces such as heat, pressure, or other agencies. Marble, for example, is metamorphosed limestone.

The organic theory of the origin of petroleum, generally accepted by geologists, is that petroleum has been derived from either animal or vegetable matter, or both, by a process of slow distillation, after its burial under beds of sediments. There is evidence to indicate that the animal and vegetable matter was of marine origin; such evidence includes the association of brines with oil, the visible oily coating on seaweeds found in certain localities, and the optical phenomenon of light polarization by oils similar to that of substances found in certain plants and animals and which is not shown by inorganically synthesized petroleum. The accumulation of the matter from which petroleum has been derived, its burial by sedimentary material, and the action of pressure and heat to cause distillation, has resulted in petroleum formation in many parts of the world. Geological studies indicate that petroleum was not formed in the pools in which it is found, but that the action of water pressing against oil formations caused the petroleum to flow, over a period of many years, through porous beds or strata to points of accumulation. Pools of oil occur in "traps" in sedimentary rocks such as sandstone or limestone. These traps may be formed in various ways, a few of which are illustrated schematically in Figure 3—3. Essentially, such traps are formed by an impervious layer which prevents upward migration of the petroleum to any further extent. The oil is obtained by drilling wells into these zones of accumulation. The well is encased in a steel pipe through which it is often customary to run a number of smaller pipes to bring the product to the surface.

Crude petroleum is a liquid containing a complex mixture of solid, liquid and gaseous hydrocarbons. The solid hydrocarbons are in solution and the liquid is at least partly saturated with gases (methane, ethane, etc.). The elementary composition of American crude oils from representative fields covers the following ranges (in per cent):

Carbon	84 to 87
Hydrogen	11.5 to 14.0
Sulphur	0.05 to 3.0
Nitrogen	0.01 to 1.70

Ordinary crude petroleum is brownish-green to black in color with a specific gravity from about 0.810 to 0.981, and an ash content of 0.01 to 0.05 per cent.

The principal constituents in crude oil are the **paraffin** (C_nH_{2n+2}), **naphthene** (C_nH_{2n}), and **aromatic** (C_nH_{2n-x}) series of hydrocarbons, and **asphaltic compounds**. In **paraffin-base crudes**, such as found in Pennsylvania, the asphaltic content is low, only traces of sulphur and nitrogen are found, and the specific gravity averages about 0.810. In **mixed-base crudes** which have a lower content of paraffins and a higher content of naphthenes than the paraffin-base crudes, the content of asphaltic compounds is higher, the sulphur content usually is under 0.4 per cent and the paraffin-wax content is generally high. Mixed-base crudes occur in the mid-continent region. The **naphthene-base crudes** contain a high percentage of naphthenes and very little paraffin wax. They occur in the central, south-central and south-western areas of the United States. Light naphthene-base crudes contain a low proportion of asphalt, compared to reverse proportions in heavy naphthene-base crudes. The sulphur content varies widely. The **aromatic crudes**, which occur chiefly in California, generally have a high asphaltic-compound content, sulphur content varying from 0.1 to 4.13 per cent and a relatively high nitrogen content. The presence of wax is often widespread, although some crudes of this class are free of wax.

Crude oil is delivered by rail, ocean tankers, inland and intercoastal waterways, in specially constructed tanks, and by pipelines, including the Alaskan pipeline.

Grades of Petroleum Used as Fuels—Fuel oils may be classified generally as: (1) **raw** or **natural crude petroleums,** (2) **distillate fuel oils,** (3) **residual fuel oils,** and (4) **blended oils.** By the older methods of refining, the products from many of the oil refineries west of the Mississippi River were gasoline, naphtha, kerosene and fuel oil, while eastern refineries usually carried the fractionation of oil much further, their output being such products as gasoline, benzene, naphtha, kerosene, light machine oil, automobile oils, cylinder oils, paraffin wax and tar, pitch, or coke. Recent improvements in thermal cracking at both high and low pressure and the use of catalytic conversion processes have enabled refiners to convert more of the petroleum to gasoline and to produce lubricants from western petroleum relatively high in asphalt.

Distillate fuel oils consist of the fractions distilled intermediate between kerosene and lubricating oils. Residual fuel oils are the viscous residual products remaining after the more volatile hydrocarbons have been driven off in the refining process. Blended oils are mixtures of any or all of the three classes of fuel oils. The distribution of products obtained from crude oil in the United States in 1982 is listed in Table 3—XIV.

Properties and Specifications of Liquid Fuels—Before discussing the more important properties and specifications of fuel oil, some of the common terms will be reviewed. Further details may be found in the ASTM Standards for Petroleum (Parts 23, 24 and 25) cited in the references at the end of this chapter.

Specific gravity is the ratio of the weight of a volume of a body to the weight of an equal volume of some standard substance. In the case of liquids, the standard is water. **Baumé gravity** is an arbitrary scale for measuring the density of a liquid, the unit being called "Baumé degree." Its relation to specific gravity is shown by the formula:

$$Be^{\circ} = \frac{140}{Sp.\ Gr.} - 130 \text{ (for liquids lighter than water)}$$

For example, the Baumé hydrometer will read 10° Bé in pure water, when the specific gravity scale reads 1.00.

The **American Petroleum Institute (API) Gravity** is a modification of the Baumé scale for light liquids. API

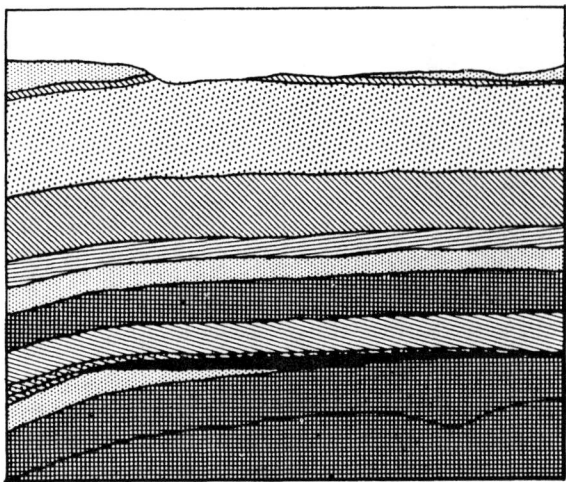

A. STRATIGRAPHIC TRAP.—In the stratigraphic trap, the producing formation gradually pinches out and disappears up the structure. An impervious layer is deposited on top of the sand, thus forming a cap rock. The solid black section represents petroleum accumulated below the cap rock.

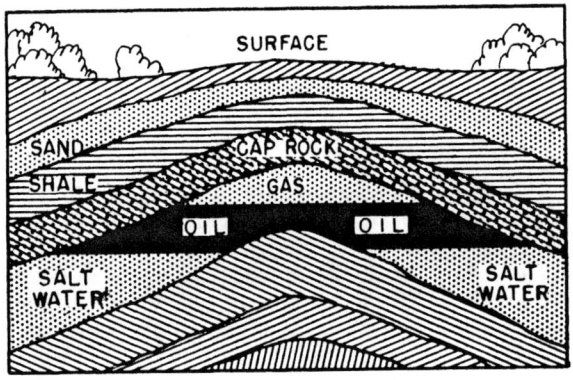

B. ANTICLINES.—In an anticlinal structure, the rocks comprising the crust of the earth are folded upward. The oil and gas are usually found on the crest of an anticlinal structure. An impervious cap rock must be present to seal the reservoir and prevent the escape of the gas and oil into higher layers. This cap rock, in one form or another, must be present in all reservoirs to contain the oil and gas within the structure.

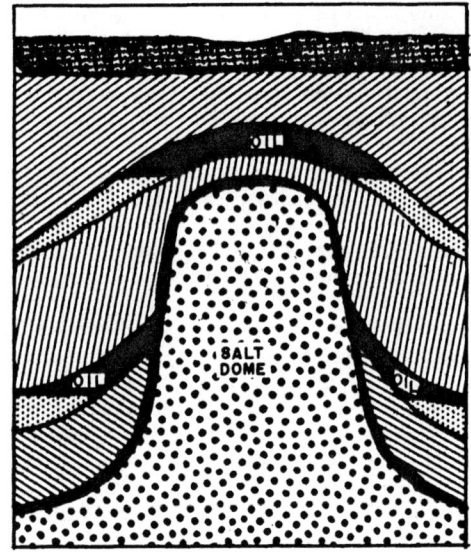

C. SALT DOMES.—The salt dome is believed to be the result of the intrusion of large masses of salt into the sediments where they are found. This intrusion creates an upward pressure and results in the doming of the overlying sedimentary rocks. In this type of structure, petroleum accumulates within the upturned porous beds about the summit and flanks of the salt core, as indicated by the solid black sections.

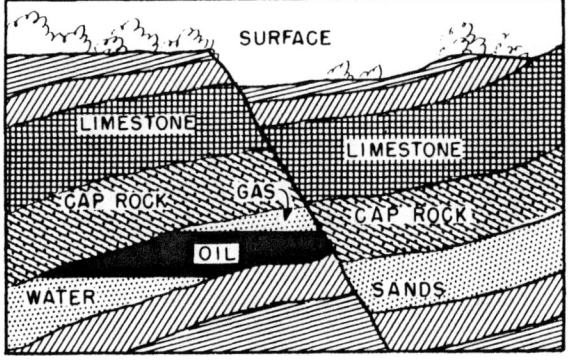

D. FAULTS.—A fault is a structural closure caused by the fracturing of the crustal rocks during earth movements. In the process of folding, a reservoir for oil may be formed when a porous rock is brought into contact with an impervious layer, thus forming a trap.

FIG. 3—3. Schematic representation of four geologic structures associated with the underground accumulation of petroleum through natural agencies. From "Fundamentals of Petroleum," NAVPERS 10883, Superintendent of Documents, U.S. Government Printing Office, Washington, D.C.

---

gravities are always reported at 15°C (60°F). The relation between API gravity and specific gravity is:

$$°API = \frac{141.5}{Sp. Gr.} - 131.5$$

The greater the degrees Baumé or API, the lighter or lower in density the fluid. There are about 90 API degrees between the heaviest and lightest oils which, therefore, make this scale valuable for determining differences between the density of various oils.

Flash point is the lowest temperature at which, under specified conditions, a liquid fuel vaporizes rapidly enough to form above its surface an air and vapor mixture which gives a flash or slight explosion when ignited by a small flame. It is an indication of the ease of combustion or of the fire hazard in handling or using oil.

Pour point is the lowest temperature at which oil will pour or flow when chilled without disturbance under specified conditions.

Viscosity is the property of liquids that causes them to resist instantaneous change of shape or rearrangement of their parts due to internal friction. Since this property has a direct relation to resistance of flow in fuel-oil pipe systems and to atomization, it is an important specification.

Absolute viscosity is a measure of internal fluid friction. It is defined as the tangential force on unit area of either of two parallel planes at unit distance apart when the space between the two planes is filled with liquid and one of the planes moves relative to the other with unit velocity in its own plane, and is referred to as **dynamic viscosity**.

The unit for expressing absolute viscosity in SI is the pascal-second (Pa·s): since by definition the pascal equals one newton per square metre (N/m²) and the newton has the formula kg·m/s², the dimensions for the pascal-second are kilograms per second per metre (kg/s/m). One Pa·s is equal to 0.1 poise or 10 centipoises (the poise and centipoise are defined below).

The cgs unit of absolute viscosity is the **poise,** which has the dimensions, grams per centimetre per second (g/s/cm). The **centipoise** is 1/100 of a poise and is the unit of absolute viscosity most commonly used in cgs. One poise equals 10 Pa·s (the SI unit) and one centipoise equals 0.1 Pa·s.

Absolute viscosity in the fps system is expressed in pounds per second per foot (lbf/s/ft)

The absolute viscosity of water at 20°C (68°F) in SI equals 0.1 Pa·s; in the cgs system it equals 1 centipoise; and in the fps system it equals 0.00209 lbf/s/ft.

Relative viscosity of a fluid is defined as the ratio of the absolute viscosity of the fluid to the absolute viscosity of water, with both the fluid and water at the same temperature and their viscosities measured in the same units.

Kinematic viscosity relates to the time for a fixed amount of a fluid to flow through a capillary tube under the force of gravity: it may be defined as the quotient of the absolute viscosity in centipoises divided by the specific gravity of a fluid, both at the same temperature. The unit of kinematic viscosity is the **stoke** or **centistoke** (0.01 stoke), derived as follows:

$$\frac{centipoises}{specific\ gravity} = centistokes$$

The viscosity of all liquids decreases with increasing temperature and ASTM viscosity determinations are made at oil temperatures of 37.8°, 50.0°, 54.4° and 99.2°C (100°, 122°, 130° and 210°F, respectively), and are often expressed as **Saybolt Universal** at 37.8°C (100°F) or **Saybolt Furol** at 50°C (122°F). The terms "Saybolt Universal" and "Saybolt Furol" represent the type of instrument used in making the viscosity determinations. Viscosity measurements made by either method may be interconverted by the use of tables.

Reid Vapor Pressure is a test for the vapor pressure of gasoline at 37.8°C (100°F). It shows the tendency of gasoline to generate vapor bubbles and is expressed in SI in kilopascals, in the cgs system in kilograms per square centimetre, and in the fps system in pounds per square inch absolute.

Octane number is the anti-knock rating of gasoline. The rating is made by matching the fuel in a test engine with a mixture of normal heptane, which detonates very easily and has an octane rating of zero, and iso-octane, which has exceptionally high anti-knock characteristics and is rated at 100. A fuel knock that matches a mixture of say 60 percent octane and 40 per cent heptane would have an octane rating or number of 60. **Cetane number** is used to show the ignition quality of diesel oils. The rating is based on a scale resembling those of octane numbers by matching the ignition delay of the fuel against blends of cetane, a fast-burning paraffin, and methyl naphthalene, a slow-burning aromatic material.

The **ASTM** has developed a table for grading fuel oils, consisting of six grades. According to this classification, heating oils generally used for domestic and small industrial heating furnaces comprise Grades 1, 2 and 3. Grades 5 and 6, formerly known as **Bunker "B"** and **Bunker "C"** fuel oils, respectively, are used extensively in the steel industry. Grade 5 fuel oil is usually cracking-still tar and Grade 6 fuel oil, a straight-run or cracked residual, or a mixture of residual and cracking-still tar blended to reduce the viscosity as required by the consumer.

All grades of fuel oil are normally sold to meet specifications mutually satisfactory to buyer and seller. A typical specification of Grade 6 fuel oil is as follows.

	Minimum	Desired	Maximum
Gravity—API at 15.6°C (60°F) ...	10	12	14
Viscosity—SSF at 50°C (122°F)	280	300	310
Pour Point		0°C (32° F)	10°C (50°F)
Flash Point	107°C (225°F)	121°C (250°F)	149°C (300°F)
Sulphur (Per Cent)	0	0.6	1.0[1]
Sodium Chloride ..	0	0	0.719 kg/m³ (0.006 lb/gal)
Ash (Per Cent)	0	0	0.15
Bottom Sediment and Water (Per Cent)	0	0	2.0

[1] For blast-furnace use, 1.5 per cent maximum.

Some purchasers may also specify the desired calorific value of Grade 6 fuel oil (for example, 41 805 kilojoules per cubic decimetre or 150 000 Btu per gallon, minimum) and the fire point of the oil (for example, 205°C (400°F) minimum and 232°C (450°F) maximum).

The yield of tar produced in by-product coke ovens by high-temperature distillation between 1000° and 1100°C (1832° and 2012°F) differs within very wide limits according to the kind of bituminous coal coked, and to the temperature, coking time, and design of oven employed in the process. Virgin tar as produced in the by-product ovens consists essentially of tar acids, neutral oils which are principally aromatic hydrocarbons, and a residue pitch.

The residue pitch from the distillation of tar is highly viscous or brittle. Pitch contains a substantial percentage of free carbon and some high-boiling and complex organic chemicals. The composition and properties of a typical pitch-tar mix and Grade 6 fuel oil are shown in Table 3—XV.

The viscosity of liquid fuels such as virgin tar, pitch-tar mixtures and topped tar decreases with temperature increase as shown in Table 3—XVI.

Combustion of Liquid Fuels—The combustion of liq-uid fuel usually is obtained by atomizing the fuel. Atomization breaks up the fuel into fine, mist-like globules, thus permitting an increased area for intimate contact between the air supplied for combustion and the fuel. The chemistry of combustion of liquid fuels is complex. The small particles of fuel either vaporize to form gaseous hydrocarbons which burn to CO_2 and H_2O through a chain of reactions, or the fuel cracks to form carbon (soot) and hydrogen which also burn with complete combustion to CO_2 and H_2O. Both of these conditions normally occur in the combustion of liquid fuels. The first condition predominates with good atomization and proper mixing with sufficient air. A deficiency of air or poor atomization will cause smoke. For large furnaces, such as open hearths and heating furnaces, the atomizing agent is usually steam at a pressure anywhere between about 415 and 860 kPa (60 and 125 pounds per square inch) gage. The steam consumed in atomization varies from 0.3 to 0.7 kilograms per kilogram of fuel. When liquid fuels are used in smaller furnaces, atomization usually is achieved by compressed air or by mechanical action. The character of a liquid-fuel flame, that is, its shape, size and luminosity, may be altered with a fixed burner design by changing the degree of atomization which is controlled by the steam pressure. Liquid fuels normally are burned in steel plants to produce a highly luminous flame at an intensity of flame propagation intermediate between that generally secured with coke-oven gas and that with natural gas.

Liquid fuels are often preferred because they permit better control of flame direction and, because of their high calorific value, control of flame temperature and luminosity.

Table 3—XIV. Percentages of Yield of Refined Petroleum Products from Crude Oil[1] For the Year 1982

Finished Motor Gasoline	46.2%
Finished Aviation Gasoline	0.2
Liquid Refinery Gases and Ethane	2.2
Naptha Type Jet Fuel	1.7
Kerosene Type Jet Fuel	6.4
Kerosene	0.9
Distillate Fuel Oil	21.5
Residual Fuel Oil	8.8
Naptha, Under 400°F For Petrochemical Feedstock	1.2
Other Oils Over 400°F For Petrochemical Feedstock	2.2
Special Napthas	0.4
Lubricants	1.2
Wax	0.1
Petroleum Coke	3.4
Asphalt	2.7
Road Oil	<0.1
Still Gas For Petrochemical Feedstocks	4.4
Still Gas For Other Use	0.7
Processing Gain	−4.4
	100.0%

Note:—Total may not equal the sum because of individual round off.

[1] From United States Department of Energy, Petroleum Supply Annual—June 1983.

Table 3—XV. Composition and Properties of Typical Liquid Fuels

Fuel	Ultimate Analysis of Fuel (%)						
	H_2O	C	H_2	N_2	O_2	S	Ash
Pitch-Tar (Dry)	...	90.78	5.35	1.39	1.65	0.61	0.22
Pitch-Tar (Natural Basis)	1.33	89.57	5.28	1.37	1.63	0.60	0.22
Grade 6 Fuel Oil (Dry)[1]	...	88.60	10.50	0.30	0.00	0.55	0.05

Fuel	Specific Gravity at 15.6°C (60°F)	Mass		Dry Air Required for Combustion		Theoretical Flame Temperature	
		kg/dm³	lb/gal	m³kg	ft°/lbm	°C	°F
Pitch-Tar (Dry)	—	—	—	—	—	—	—
Pitch-Tar (Natural Basis)	1.199	1.196	9.9855	9.41	158.13	1924	3495
Grade 6 Fuel Oil (Dry)	0.9529	0.951	7.935	10.64	180	2093	3800

Fuel	Calorific Value			
	Gross		Net	
	kJ/kg	Btu/lbm	kJ/kg	Btu/lbm
Pitch-Tar (Dry)	—	—	—	—
Pitch-Tar (Natural Basis)	37 577	16 155	36 458	15 674
Grade 6 Fuel Oil (Dry)	43 938	18 890	41 449	17 820

[1]Courtesy of Sun Oil Co.—Typical composition.

Table 3—XVI. Effect of Temperature on Viscosity of Various Tars and Tar Mixtures.

Fuel	Test Temp.		Viscosity in Sec., Saybolt Universal
	(°C)	(°F)	
Virgin Tar	79.4	175	189.4 max 73.3 min 109.4 avg
Pitch-Tar Mix	79.4	175	1940 max 181 min. 946.1 avg
Pitch-Tar Mix	98.9	210	687 max 97 min 561.7 avg
Topped Tar	93.3	200	700 max 550 min 600 avg

The amount of air required to burn liquid fuels depends upon the chemical composition of the particular fuel. Grade 6 fuel oil requires approximately 10.64 m^3 per kilogram (180 ft^3 per lbm) of dry air for perfect combustion, and tar-pitch approximately 9.34 m^3 (158 ft^3). From the ultimate analysis of a liquid fuel, the theoretical air requirements and products of combustion may be calculated, as explained in Section 1 under "Combustion Calculations."

Liquid-Fuel Burners—There are many different designs of burners for liquid fuels. Burners designed for atomization by steam or air may be classified into two general types, the **inside mixing** and the **outside mixing**. In the inside-mixing type the fuel and atomizing agent are mixed inside the burner or burner system, while in the outside-mixing type the two fluids meet immediately outside the burner tip. In open-hearth furnaces and large reheating furnaces the inside-mixing type is used. The inside-mixing type is sometimes classified as an **emulsion type** or a **nozzle-mix type** of burner. In the emulsion type the mixing is performed at a point several feet from the burner tip, while in the nozzle-mix type the two fluids meet inside the burner but very close to the burner tip. In the latter type, mixing probably takes place both inside the burner and as the stream enters the furnace.

Liquid-fuel burners used in open-hearth furnaces are water cooled due to port end design; those in reheating, forge and annealing furnaces seldom require water cooling.

The handling of liquid fuels at consuming plants requires a system for their transportation, storage and conditioning. Where liquid fuels are received by tank car, a system of receiving basins, unloading pumps, strainers and storage tanks generally is required. The storage tanks must be of ample size to meet fuel demands between deliveries and should be provided with heaters to maintain proper fluidity for flow through pipe lines to the system pressure pumps. Pressure pumps are used to deliver the liquid fuel through a pipe system to the point of consumption. Where there are a number of consuming units being served from a common fuel-storage system, the pipe feeder line is designed in the form of a loop through which the fuel flows at constant pressure and temperature. The various units tap into this loop. The fuel-oil lines are lagged and provided with tracer steam lines to maintain uniform fluidity throughout the system and to provide fuel at the burners at the proper viscosity for atomization. The temperature at which liquid fuel is delivered to the burners varies with the character of the fuel and burner design. Where pitch is used, a temperature as high as 150°C (about 300°F) in the lines is sometimes required. A temperature level usually somewhere between 95° and 120°C (about 200° and 250°F) is maintained for pitch-tar mixtures, and 65° and 95°C (about 150° and 200°F) for Grade 6 fuel oil. Additional details on burners and firing practices for liquid fuels appear in several references cited at the end of this chapter.

SECTION 4

GASEOUS FUELS AND THEIR UTILIZATION

The availability of **natural gas** in so many sections of this country has had a profound influence upon our industrial progress. It was first used as an illuminating gas at Fredonia, New York, in 1821. The discovery of new fields and the installation of pipe lines to consuming centers led to increasing demands, as the convenience, cleanliness, and general utility of this form of fuel became better known. The initial use of natural gas for steel manufacture was at a rolling mill plant at Leechburg, Pa., in 1874. A well in this area permitted exclusive use of natural gas for puddling, heating, and steam generation for a period of six months. Since 1932 there has been an accelerated demand for natural gas, which thirty years later, by 1962 for example, had grown to 0.391 trillion cubic metres (13.8 trillion cubic feet) of gas produced annually. By the close of 1982, annual production of natural gas was 0.50 trillion cubic metres (17.5 trillion cubic feet). To deliver this supply, the gas industry has installed some 1.6 million kilometers (over 1.0 million miles) of natural-gas pipe lines. In addition to pipe lines for transport, the availability of vast quantities of natural gas at economically competitive prices has been heavily dependent upon the development and use of underground gas-storage areas near large consuming centers. These storage areas permit winter peak demands to be met by storage of natural gas during off-seasons by year around pumping at high load levels.

As of November 1, 1983, there were 409 underground gas storage fields being operated or developed by 104 companies. This included 398 fields considered active and another 11 fields under development. The maximum deliverability of the 398 fields is estimated at 1.71 billion cubic metres (60.48 billion cubic feet) per day in a 24-hour period. To that could be added an estimated 1.91 billion cubic metres (67.6 billion cubic feet) of regularly scheduled pipeline deliveries per day and 0.42 billion cubic metres (14.65 billion cubic feet) from other sources for a total maximum delivery in one day that could reach 4.04 billion cubic metres (142.73 billion cubic feet).

In the steel industry, the increasing use of gaseous oxygen in steelmaking is reducing the fuel requirement for melting and refining. However, in the production of tonnage oxygen for steel production, large quantities of energy are required. It is estimated that for each ton of steel produced by the basic oxygen process, it requires the equivalent of between about 10 and 17 cubic metres (350 and 600 cubic feet) of natural gas to produce the oxygen required.

Producer gas was the first gaseous fuel successfully utilized by the iron and steel industry. This gas permitted the early experimentation in regeneration, and the utilization of this principle started a new era of steel manufacturing. The advantages of preheated gas and air were so clearly indicated in 1861 that producer gas rapidly became the major fuel utilized by open-hearth furnaces and maintained its position for almost sixty years, or until about 1920, when by-product coke plants, supplying coke-oven gas and tar, began to challenge this leadership.

Blast-furnace gas utilization by the iron and steel industry probably should rank first historically, although its adoption by the industry was slower than in the case of producer gas. The sensible heat in the blast-furnace top gases was first utilized in 1832 to transfer heat to the cold blast. Originally, this heat exchanger was mounted on the furnace top. In 1845, the first attempts were made to make use of its heat of combustion, but history indicates that the burning of blast-furnace gas was not successful until 1857. It is probable that progress in the utilization of blast-furnace gas was delayed by its dust content, the problems of cleaning and handling, and the low cost of solid fuel. Increasing cost of other fuels and competition forced its use, and by the turn of this century, blast-furnace gas had become one of the major fuels of the iron and steel industry. In 1982, the steel industry used about 58 billion cubic metres (2065 billion cubic feet) of blast-furnace gas (based on 3535 kJ/m^3 or 90 Btu/ft^3) for blast-furnace stove heating and other uses including the firing of melting, heating and annealing furnaces and coke-oven underfiring: this quantity of blast-furnace gas represents 7.1 million metric tons or 7.8 million net tons of coal equivalent.

The initial use of by-product **coke-oven gas** in the iron and steel industry was at the Cambria Steel Company, Johnstown, Pa., in 1894. This installation was followed by only a few by-product coke-plant additions until shortage of transportation facilities and the rising price of coal and natural gas during the first World War accelerated installations throughout the steel industry. The utilization of coke-oven gas has been very profitable as it reduced the purchase of outside fuels. It is estimated that plants operating steelmaking furnaces in the United States used about 11 629 million cubic metres (410 683 million cubic feet) of coke-oven gas as fuel in 1982 (based on 19 640 kJ/m^3 or 500 Btu/ft^3).

NATURAL GAS

Natural gas and petroleum are related closely to each other in their chemical composition and in geographical distribution. Both are made up predominantly of hydrocarbons. Petroleum rarely is free of natural gas, and the same fields usually produce both fuels. When natural gas exists indigenous to an oil stratum and its production is incidental to that of oil, it is called **casinghead gas.** Gas found in a field is usually under pressure which diminishes with extended use or, sometimes, from the presence of too many other wells. The life of a well varies from a few months to twenty years. Rocks bearing gas are sandstones, limestones, conglomerates, and shales—never igneous rocks. Natural gas is derived from the remains of marine animal and plant life—in theory, the same as described previously for petroleum.

Natural gas as found is usually of singular purity and is composed principally of the lower gaseous hydrocarbons of the paraffin series, methane and ethane, some of the heavier liquefiable hydrocarbons (which are recovered as **casinghead gasoline** or sold in bottled form as butane, propane, pentane, etc.) and a small amount of nitrogen or carbon dioxide. Some natural gases contain small quantities of helium. Occasionally, wells are found in which the gas contains hydrogen sulphide and organic sulphur vapors. **Sour gas** is defined as a natural gas which contains in excess of 0.0343 grams of hydrogen sulphide or 0.686 grams of total sulphur per cubic metre, equivalent to 1½ grains of hydrogen sulphide or 30 grains of total sulphur per 100 cubic feet. It is fortunate, however, that by far the greater part of natural gas available in this country is practically sulphur-free.

There are a number of great gas fields in the United States and Canada. Major fields, based on remaining reserves, are: Panhandle-Hugoton in Kansas, Oklahoma and Texas; San Juan in New Mexico and Colorado; Puckett and Chocolate Bayou in West Texas; Katy and Old Ocean on the Texas Coast; Jalmat-Eumont in New Mexico; Big Piney in Wyoming; Peace River in British Columbia; and the North Slope in Alaska.

The principal constituent of natural gas is methane, CH_4. Since natural gas contains from 60 to 100 per cent of CH_4 by volume, the characteristics of methane gas, which were shown in Section 1, largely dominate the parent gas. Comparing methane with the other principal combustible gases, it will be noted that it has a low rate of flame propagation, a high ignition temperature, and a narrow explosive range. Methane, as well as all other hydrocarbons (of which it is the lowest member), burns with a luminous flame. Typical compositions of natural gas are presented in Table 3—XVII.

The iron and steel industry consumed a total of 11 583 million cubic metres (409 064 million cubic feet) of natural gas in 1982 in blast furnaces and other uses in the blast-furnace area, steel-melting furnaces, heating and annealing furnaces, heating ovens for wire rods, and other uses. Heating and annealing furnaces consumed 54 per cent of the total gas used.

Table 3—XVII. Typical Composition of Natural Gas in Various Cities in the United States (Based on American Gas Association Survey, Fall of 1962).***

Constituents (Per Cent by Volume)	Cities				
	Birmingham, AL	Pittsburgh, PA	Los Angeles, CA	Kansas City, MO	Detroit, MI
Carbon Dioxide	1.06	0.80	0.50	0.22	0.59
Oxygen	....	0.01	....	....	....
Helium	....	....	....	....	0.20
Nitrogen	2.14	0.40	2.60	17.10	3.30
Butanes and Higher Hydrocarbons..............	0.49	0.39	0.50	0.56	0.44
Propane.....................	0.67	0.79	1.90	2.91	1.34
Methane	93.14	94.03	86.50	72.79	89.92
Ethane......................	2.50	3.58	8.00	6.42	4.21
Specific Gravity	0.599	0.595	0.638	0.695	0.616
Gross Heating Value—					
Btu/ft³*	1024	1051	1084	945	1016
kJ/m³**	40 222	41 280	42 580	37 120	39 900

*At 30 in. Hg, 60° F, dry.

**At 760 mm Hg, 0°C, dry.

***Source: "Gas Engineers Handbook" (see Segeler listing at end of chapter). Additional information is available in "Analyses of Natural Gas, 1966" by B. J. Moore and R. D. Shrewsbury in U. S. Bureau of Mines Information Circular 8356 (1967).

MANUFACTURED GASES

The four most important of the commercially used manufactured gases are **producer gas, water gas, oil gas,** and **liquefied petroleum gases.** Since none of these gases except liquefied petroleum gases are used presently in steel manufacturing or processing in the United States, only a brief description of their manufacture and characteristics will be given here.

Manufacture of Producer Gas—Producer gas is manufactured by blowing an insufficient supply of air for complete combustion, with or without the admixture of steam, through a thick, hot, solid-fuel bed. A large proportion of the original heating value of the solid fuel is recovered in the potential heat of carbon monoxide, hydrogen, tarry vapors, and some hydrocarbons, and in the sensible heat of the composite gas which also contains carbon dioxide and nitrogen. When the gas is cleaned, the sensible heat of the gases and the potential heat of the tar vapors is lost.

Table 3—XVIII gives the composition of clean producer gas made from various fuels in a well-operated updraft producer.

The gross heating value of raw producer gas, including tar, made from a high-volatile coal, 8 per cent ash, is about 6678 to 7463 kilojoules per cubic metre (170 to 190 Btu per cubic foot).

Producer gas has a very low rate of flame propagation due to the relatively large amount of inert gases, N_2 and CO_2 it contains. The hot gas, containing tar, burns with a luminous flame; the cold gas is only slightly luminous, while it is non-luminous if made from anthracite coal or coke. Producer gas is a relatively heavy gas and has a wide explosive range. The theoretical flame temperature is low, approximately 1750°C (3180°F), and the gas generally was preheated when

Table 3—XVIII. Composition of Clean Producer Gas[1]

Constituent	Solid Fuel Feed				
	Anthracite Coal	Coke		Bituminous Coal	
		100 to 125 mm (4 to 5") Lump	Breeze	A	B
CO_2...........................	6.3%	9.2%	8.7%	3.4%	9.2%
Illuminants	0.0	0.1	0.0	0.8	0.4
O_2...........................	0.0	0.0	0.0	0.0	0.0
CO	25.0	21.9	23.3	25.3	20.9
H_2..........................	14.2	11.1	12.8	9.2	15.6
CH_4..........................	0.5	0.2	0.4	3.1	1.9
N_2...........................	54.0	57.5	54.8	58.2	52.0
Total	100.0%	100.0%	100.0%	100.0%	100.0%
Gross Heating Value,					
kJ/m³[a]	5185	4753	5146	6088	6128
Btu/ft³[b]	132	121	131	155	156

[1] U.S. Bureau of Mines, Bulletin 301.

[a] At 760 mm Hg, 0°C, dry.

[b] At 30 in. Hg, 60°F, dry.

utilized in steel-plant processes.

Manufacture of Water Gas—Water gas or **blue gas** is generated by blowing steam through an incandescent bed of carbon. The gas-forming reactions are primarily:

$$C + H_2O = CO + H_2$$
$$C + 2H_2O = CO_2 + 2H_2$$

Fuel oil may be "cracked" in a separate heating chamber to form gases that are added to water gas to enrich it when **carburetted water gas** is made.

While coke generally is used as the fuel in the production of water gas because of its high carbon content and cleanliness, anthracite and bituminous coal and mixtures of coal and coke also have been used successfully, but with some sacrifice in over-all operating efficiency.

Water gas burns with a clear blue flame; hence, the name "blue gas." It is used in a number of chemical processes to supply a basic gas for synthetic processes, but it is not suitable for distribution as a domestic fuel unless it has been enriched with cracked fuel oil, when it is called carburetted water gas.

Water gas made from coke burns with a non-luminous flame. Carburetted water gas burns with a highly luminous flame. Both gases have a high rate of flame propagation. The speed of combustion for water gas exceeds that of any other extensively used fuel gas; that for carburetted water gas is practically the same as for coke-oven gas. Water gas has a slightly lower specific gravity than natural gas, but is somewhat heavier than coke-oven gas. Carburetted water gas is heavier than natural gas but lighter than producer gas. The theoretical flame temperature of both blue and carburetted water gas is very high, respectively about 2020°C and 2050°C (3670° and 3725°F) exceeding that of all other industrial fuel gases commonly used. Both gases have a relatively wide explosive range.

Oil gas is a combination of cracked petroleum and water gas made by passing oil and steam through hot refractory checker work. Oil gas is commercially important in localities where coal or coke is expensive and oil cheap.

Liquefied petroleum gases (LPG), sometimes referred to as **bottled gases,** have become commercially important because of the concentration of fuel energy in liquid form which may be converted easily into a gas. They are distributed for household use in steel cylinders called "bottles" and in tank cars or trucks for industrial purposes. They are sometimes sold under various trade names but are composed mainly of butane, propane, and pentane. The iron and steel industry in 1982 consumed about 15 000 cubic metres (about 4 million gallons) of liquefied petroleum gases, most of which were used in heating and annealing furnaces. A steel cylinder of propane as sold for domestic purposes contains approximately 50 300 kilojoules per kilogram (21 640 Btu per pound) of liquid gas.

Alternative Fuel Sources—Drastic escalation of oil prices by the Organization of Petroleum Exporting Countries (OPEC) in the 1970's stimulated interest in alternative sources of fuels to supplement existing supplies. A number of projects began operation during the early 1980's. Tennessee Eastman built a coal gasification plant to provide synthetic gas for conversion to

chemicals. The Cool Water Project in California integrates coal gasification with the generation of electricity by a combination of gas and steam turbines. Union Oil Company began operation of a shale oil facility at Parachute, Colorado, and a consortium built the Great Plains Coal gasification plant to produce synthetic natural gas.

Gas is also being recovered from sanitary landfills. The Gas Research Institute and others (including U.S. Steel) are involved with projects for recovering methane from coal seams. More detailed information is contained in the technical literature and publications such as the Proceedings of the Annual Energy Technology Conference.

BY-PRODUCT GASEOUS FUELS

The two major by-product gaseous fuels are blast-furnace and coke-oven gases. A number of other unavoidable gaseous fuels are created by regular manufacturing processes. Some of these are of minor economic consequence, but the majority are useful and generally utilized at the plant where they are produced. An exception is oil-refinery gas which is sometimes piped and marketed to industries adjacent to refineries. The calorific value and flame characteristics of by-product gases have wide ranges. Blast-furnace gas has probably the lowest heat content of any, and oil-refinery gas the highest, respectively about 3535 and 72 668 kilojoules per cubic metre (90 and 1850 Btu per cubic foot), although both vary from these values.

Blast-furnace gas is a by-product of the iron blast furnace. The paramount objective in blast-furnace operation is to produce iron of a specified quality, economically; the fact that usable gas issues from the top of the furnace is merely a fortunate attendant circumstance. When air enters the tuyeres (see Chapter 15 on "The Manufacture of Pig Iron") its oxygen reacts with the coke. The resulting gas passes up through the shaft of the furnace which has been charged with coke, ore, and limestone, and after a number of chemical reactions and a travel of some 25 metres (80 feet), issues as a heated, dust-laden, lean, combustible gas. The annual volume production of this gas is greater than that of any other gaseous fuel. Two and one-half to three and one-half tons of blast-furnace gas are generated per ton of pig iron produced. While the purpose of the gases generated by the partial combustion of carbon is to reduce iron ore, the value of a blast furnace as a gas producer is evident from the relation just noted. The essential reactions by which blast-furnace gas is produced are shown in Chapter 15.

The percentage of CO and CO_2 in blast-furnace gas is directly related to the amount of carbon in the coke and the amount of CO_2 in the limestone charged per ton of iron produced. The rate of carbon consumption depends principally upon the kind of iron to be made, the physical and chemical characteristics of the charged material, the distribution of the material in the furnace stack, the furnace lines, and the temperature of the hot blast. The total CO + CO_2 content of the top gas is about 40 per cent by volume, and when producing ordinary grades of iron the ratio of CO to CO_2 will vary from 1.25 to 2.5, to 1. The hydrogen content of the gas varies from 3 to 5 per cent depending on the type and amount of tuyere-injected fuels. The remaining percentage is made up of nitrogen, except for about 0.2 per cent CH_4.

Blast-furnace gas leaves the furnace at a temperature of about 120° to 370°C (about 250° to 700°F), and at a pressure of 345 to 1380 mm Hg gage pressure (15 inches w.g. to 14.5 psig), carrying with it 22 to 114 grams of water vapor per cubic metre (10 to 50 grains per cubic foot) and 18 to 34 grams of dust per cubic metre (8 to 15 grains per cubic foot). The particles of dust vary from 6.4 to 0.000254 mm in diameter. In early days of blast-furnace operation, the gas was used as it came from the furnace without cleaning, causing a great deal of trouble with flues, combustion chambers, and stoves due to clogging. The gas now is cleaned almost universally, the degree depending upon the use.

The outstanding characteristics of blast-furnace gas as a fuel are: (1) very low calorific value—about 2946 to 3535 kilojoules per cubic metre (75 to 90 Btu per cubic foot) depending on blast-furnace coke rate, (2) low theoretical flame temperature—approximately 1455°C (2650°F), (3) low rate of flame propagation—relatively lower than any other common gaseous fuel, (4) high specific gravity—highest of all common gaseous fuels, and (5) burns with a non-luminous flame.

Coke-Oven Gas—The steel industry, which uses nearly 90 percent of the total coke-oven gas generated in the United States, generally classifies coke-oven gas as a by-product of coke manufacture. This undoubtedly is due to the former waste of coke-oven gas and other coal products for so many years in the beehive-coke process. Actually, the production of coke-oven gas and other coal chemicals is a part of an important manufacturing process in which large sums have been expended for their recovery, as they have a value almost equal to that of the coke. Coke-oven gas is produced during the carbonization or destructive distillation of bituminous coal in the absence of air, as described in Chapter 4. Approximately 310 cubic metres of 19 640 kJ/m³ gas are produced per metric ton of coal coked (about 11 000 cubic feet of 500-Btu gas per net ton of coal coked) in conventional high-temperature coking processes.

The composition of coke-oven gas varies in accordance with grade and density of coal and operating practices. Typical percentage ranges for constituents of dry coke-oven gas by volume are as follows:

CO_2*	1.3	- 2.4
O_2	0.2	- 0.9
N_2	2.0	- 9.6
CO	4.5	- 6.9
H_2	46.5	- 57.9
CH_4	26.7	- 32.1
Illuminates	3.1	- 4.0
Specific Gravity	0.36	- 0.44

Heating Value (Gross)
    kJ/m³ ....21 093 - 22 782
    Btu/ft³ 537-580
Heating Value (Net)
    kJ/m³ ....18 854 - 20 543
    Btu/ft³ 480-523

*Includes H_2S

Coke-oven gas contains hydrogen sulphide, H_2S.

About 40 per cent of the sulphur in coal, not removed in the washing process, is evolved with the distillation products. Much of this remains in the gas. Carbonization of coals containing 1.20 per cent sulphur evolves a gas containing about 9.7 grams of sulphur per cubic metre (424 grains per 100 cubic feet), and those containing 1.60 per cent sulphur about 14 grams of sulphur per cubic metre (600 grains per 100 cubic feet). Commercial coals in the eastern part of the United States usually run from 0.5 to 1.5 per cent sulphur. Gases high in sulphur content are very undesirable for metallurgical purposes.

Coke-oven gas normally is saturated with water vapor. In distribution systems, means must be provided for draining off the condensation due to any temperature change.

Coke-oven gas burns with a non-luminous to semiluminous flame, depending upon the degree of mixing air and gas. Its rate of flame propagation is high—considerably higher than natural, producer, or blast-furnace gas. It has a low specific gravity—lowest of any of the gaseous fuels commonly utilized by the steel industry. It has a high theoretical flame temperature—about 1980°C (3600°F), a little higher than that of natural gas. The explosive range is about twice that of natural gas.

Based upon a heating value of 19 640 kJ/m³ (500 Btu/ft³), the steel industry used 11 629 million cubic metres (410 683 million cubic feet) of coke-oven gas in 1982; of this total, 30 per cent was used in heating and annealing furnaces, 31.2 per cent in coke-oven underfiring, 12.1 per cent in steel-melting furnaces, blast furnaces and other uses in the blast-furnace area and the remainder (26.7 per cent) in all other uses, according to 1982 Annual Statistics of the American Iron and Steel Institute.

USES FOR VARIOUS GASEOUS FUELS IN THE STEEL INDUSTRY

Gaseous fuels are ideal for many steel-plant applications. Below are the more important applications where gaseous fuels either must be used on account of the nature of the work or facility, or where they are preferred over liquid or solid fuel:

Coke-Oven Heating
Blast-Furnace Stoves
Gas Turbines for Power Generation
Gas Engines for Blowing or Power Generation
Soaking Pits
Reheating Furnaces
Forge and Blacksmith Furnaces
Normalizing and Annealing Furnaces
Controlled-Cooling Pits
Foundry Core Ovens
Blast Furnace and Steel Ladle Drying
Drying of Blast-Furnace Runners and Open-Hearth
    Tapping Spouts
Hot-Top Drying
Ladle Preheating
Oxy-Fuel Burners

The choice of the most desirable fuel for each of the many facilities in a steel plant is not always possible, but by judicious planning the most efficient fuel or combination can be selected from those available. The general characteristics of each gas govern, wherever possible, its selection for a specific purpose in a steel plant. An outline of the important applications of the major gaseous fuels follows.

Uses for Blast-Furnace Gas—For many years the use of blast-furnace gas for purposes other than for the firing of stoves and boilers was not economical. A number of factors have contributed, however, to the enlarged use of blast-furnace gas, the more important of which are: (1) rising cost of purchased fuel; (2) technical progress in gas cleaning, in the use of regeneration and recuperation, and in the mixing of gaseous fuels; (3) the economic advantage of using pulverized coal in boiler houses to substitute for blast-furnace gas, thereby permitting its substitution elsewhere for the more expensive liquid and gaseous fuels; and (4) seasonal shortages in the availability of purchased liquid and gaseous fuels.

In certain applications, in addition to preheating the air, the gas itself may be preheated to provide higher temperature potential. For the facilities listed below, blast-furnace gas may be utilized successfully without preheat:

Blast Furnace Stoves
Soaking Pits
Normalizing and Annealing Furnaces
Foundry Core Ovens
Gas Engines for Blowing or Power Generation
Gas Turbines for Power Generation
Boilers

The thermal advantage of using blast-furnace gas in gas engines for blowing and for electric-power generation must overcome the heavy investment and maintenance expense of this equipment. The modern boiler house utilizing high steam pressure and temperature with efficient turboblowers and generators has sufficiently reduced the thermal advantage of gas engines so that their use is difficult to justify. A relatively recent successful development has been the use of direct-connected gas turbines for driving generators, and jet engines for driving compressors.

Preheated blast-furnace gas burned with preheated air has been used successfully in the following:

Coke-Oven Heating
Soaking Pits
Reheating Furnaces

When blast-furnace gas is preheated, it should have a minimum cleanliness of 0.023 grams per cubic metre (0.01 grains per cubic foot); and in all cases where this gas is used, extra precautions must be taken to prevent the escape of fuel or unburned gas into attendable surroundings since it contains a large percentage of toxic CO gas. Blast-furnace gas is used for many applications in the steel plant and, in addition, is used frequently for heating coke ovens and sometimes is mixed with other gases as an open-hearth fuel. In blast furnace operations where the blast furnace gas has a heating value approaching a low value of 2946 kilojoules per cubic metre (75-Btu per cubic foot), it is necessary to switch the gas with other fuels to obtain very high hot-blast temperature from the stove.

Use of Coke-Oven Gas—Coke-oven gas has had a more extended use than blast-furnace gas because of: (1) relatively low distribution costs due to its low specific gravity and high calorific value; (2) its ability to develop extremely high temperatures by combustion; and (3) the high rate at which it can release heat, thereby eliminating excessively large combustion chambers. Important applications for coke-oven gas include open-hearth furnaces in addition to those previously listed for gaseous fuels. The low specific gravity of coke-oven gas is a disadvantage in the open hearth, and for this reason, it is supplemented wherever possible with a driven liquid fuel in this service. In addition, the sulphur (in the form of H_2S) present in raw (not desulphurized) coke-oven gas is a distinct disadvantage, particularly when used in making low-sulphur heats in the open hearth and in heating certain grades of alloy steel for rolling. Its presence also requires the use of materials resistant to sulphur attack in pipe lines, valves, and burners.

There are a number of fuel applications in a steel plant where neither blast-furnace gas nor coke-oven gas, when burned alone, develop the desired flame characteristics or temperature level for optimum results. By mixing two fuels of such great variance in characteristics, a more ideal fuel can be obtained for specific applications.

The speed of combustion is very high for coke-oven gas and very low for blast-furnace gas. The desired speed can be attained through the proper proportioning of the two fuels. The speed also can be modified to a limited extent when necessary by suitable combustion technique. Mixed blast-furnace and coke-oven gas is particulary suitable for application to soaking pits and reheating furnaces.

Use of Natural Gas—Due to plant balances requiring the purchase of outside gaseous fuels, mixtures of coke-oven gas and natural gas are often utilized. While the temperature-developing characteristics of these two gases are nearly identical, they have differences in other characteristics, notably in the rate of flame propagation and in luminosity. By proper proportioning, the advantage of a short, intensive cutting flame or a long, luminous, soft flame may be had to suit the applications. Use of natural gas for flame cutting and scarfing has been increasing steadily.

Use of Producer Gas—Raw, hot, producer gas was used extensively in the past in steel-plant operations for open-hearth furnaces, soaking pits, and reheating furnaces. It was customary to preheat this gas regeneratively when used in open hearths and soaking pits, and also in batch-type reheating furnaces. In continuous-type reheating furnaces, the fuel seldom was preheated. With good gas making, producer gas develops a soft, heavy, long, luminous flame desirable for reheating steel. The use of this gas largely has been superseded in many plants by natural gas and by-product gaseous and liquid fuels.

COMBUSTION OF VARIOUS GASEOUS FUELS

The major combustion reactions of the components of gaseous fuels with air and a table of essential gas combustion constants were given in Section 1 of this chapter. From chemical equations, the quantity of air

Table 3—XIX. Properties of Typical Gaseous Fuels

Fuel Gas	Constituents of Fuel Gas Per Cent by Volume (Dry Basis)							Illuminants		Specific Gravity	Unit Vols. of Air Required for Combustion of Unit Vol. of Gas (m³ or ft³)	Heating Value per Unit Volume of Gas[1]			
	CO_2	O_2	N_2	CO	H_2	CH_4	C_2H_6	C_2H_4	C_6H_6			kJ/m³ Gross	kJ/m³ Net	Btu/ft³ Gross	Btu/ft³ Net
Natural Gas (Pittsburgh)	...	...	0.8	...	...	83.4	15.8	...	...	0.61	10.58	44 347	40 105	1129	1021
Reformed Natural Gas	1.4	0.2	2.9	9.7	46.6	37.1	...	1.3	C_3H_6 0.8	0.41	5.22	23 529	21 054	599	536
Coke-Oven Gas	2.2	0.8	8.1	6.3	46.5	32.1	...	3.5	0.5	0.44	4.99	22 547	20 190	574	514
Water Gas (Coke)	5.4	0.7	8.3	37.0	47.3	1.3	...	...	...	0.57	2.10	11 273	10 291	287	262
Carburetted Water Gas	3.0	0.5	2.9	34.0	40.5	10.2	...	6.1	2.8	0.63	4.60	21 604	19 954	550	508
Oil Gas (Pacific Coast)	4.7	0.3	3.6	12.7	48.6	26.3	...	2.7	1.1	0.47	4.73	21 643	19 483	551	496
Producer Gas (Bituminous Coal)	4.5	0.6	50.9	27.0	14.0	3.0	...	...	...	0.86	1.23	6403	6010	163	153
Blast Furnace Gas	11.5	...	60.0	27.5	1.0	...	...	...	...	1.02	0.68	3614	3614	92	92
Butane (Commercial)	...	...	...	(C_4H_{10}—93.0) (C_3H_8—7.0)				...	...	1.95	30.47	126 678	116 937	3225	2977
Propane (Commercial)	...	...	...	(C_3H_8—100.00)				...	...	1.52	23.82	101 028	93 133	2572	2371

(Continued On Next Page)

Table 3—XIX. (Continued)

Fuel Gas	Products of Combustion in Unit Vol. per Unit Vol. of Fuel (m³ or ft³)				Ultimate % CO_2	Net Heat Content per Unit Volume of Products of Combustion[2]		Theor. Flame Temp. No Excess Air	
	H_2O	CO_2	N_2	Total		kJ/m³	Btu/ft³	°C	°F
Natural Gas (Pittsburgh)	2.22	1.15	8.37	11.73	12.1	3417	87.0	1961	3562
Reformed Natural Gas	1.30	0.53	4.16	5.99	11.3	3519	89.6	1991	3615
Coke-Oven Gas	1.25	0.51	4.02	5.78	11.2	3417	87.0	1988	3610
Water Gas (Coke)	0.53	0.44	1.74	2.71	20.1	3794	96.6	2021	3670
Carburetted Water Gas	0.87	0.76	3.66	5.29	17.2	3779	96.2	2038	3700
Oil Gas (Pacific Coast)	1.15	0.56	3.77	5.48	12.9	3555	90.5	1999	3630
Producer Gas (Bituminous Coal)	0.23	0.35	1.48	2.06	18.9	2930	74.6	1746	3175
Blast Furnace Gas	0.02	0.39	1.14	1.54	25.5	2337	59.5	1454	2650
Butane (Commercial)	4.93	3.93	24.07	32.93	14.0	3555	90.5	2004	3640
Propane (Commercial)	4.0	3.0	18.82	25.82	13.75	3582	91.2	1967	3573

[1] From: "Combustion," American Gas Association (Third Edition); "Gaseous Fuels," American Gas Association (1948); and "Gas Engineers Handbook" (see bibliography at end of chapter).

[2] kJ/m³ at 0°C, 760 mm Hg (dry); Btu/ft³ at 60°F, 30 in. Hg (dry).

required to provide perfect combustion and the resultant products may be calculated for any given gaseous fuel. Table 3—XIX shows the air requirements, products of combustion, and pertinent characteristics of several gaseous fuels. The degree of mixing of air with a gaseous fuel, and the degree of excess or deficiency of air to the theoretical requirements are pertinent combustion problems. The degree of mixing is controlled by burner design. Burners have been developed to produce short, intense flames or long, slow-burning flames. The short, intense flame is usually nonluminous or semi-luminous, while the long flame is luminous. This relation is not always the case, however, since a gas must contain hydrocarbons to develop luminosity. Burners capable of producing short, intense flames will liberate a large amount of heat in a small space. Some gases, due primarily to the constituents of which they are composed, are capable of a high rate of heat release; others, of a very low rate of heat release. The two extremes are evident in two common steel-plant fuels—coke-oven gas and blast-furnace gas which give high and low rates of heat release, respectively. There is also a limit to the length of flame which can be produced. It is determined by the ability of the flame to provide enough heat to propagate itself. If the short, intense-flame type burner is used with coke-oven or natural gas, combustion will be so intense that no flame

will be visible, and heat can be liberated at rates over 40 million watts per cubic metre per hour (up to several million Btu per cubic foot per hour) of combustion space; while the long, slow-burning-flame burner firing the same gases is capable of developing a visible flame 6 to 9 metres (20 to 30 feet) long with a heat liberation of 155 000 to 210 000 watts per cubic metre (15 500 to 21 000 Btu per cubic foot per hour). Both types of flames are desirable for specific steel-plant applications. It is obvious that burner selection based on degree of mixing is important. Carrying an excess or deficiency of air for combustion is practiced usually to control scale formation, but this is done sometimes to control flame characteristics. An excess of air tends to shorten, while a deficiency lengthens, a flame. An excess of air above theoretical requirements causes higher heat losses as any extra air absorbs its share of the heat of combustion. Fuels containing hydrogen must have the exit stack temperature high enough to avoid condensation of water within the furnace system and subsequent corrosion of furnace parts. In some cases, it is necessary to burn the fuel with excess air to maintain the flue gases above their dew point. When there is a deficiency of air, potential heat is lost. In problems of design and fuel conservation, the air requirements and volume and constituents of the products of combustion must be known to effect a practical solution.

SECTION 5

FUEL ECONOMY

Because fuel represents the largest single item of raw-material expense for the manufacture of iron and steel, the subject of fuel economy is of consequence to both the producer and consumer of steel products. The steel industry consumes annually during normal times over 90 million metric tons (100 million net tons) of primary fuels in coal equivalent. The efficient utilization of this large quantity of fuel is also pertinent to the conservation of our fuel resources. The history of the steel industry shows great progress has been made in reducing the amount of fuel required to produce a ton of steel. During the Revolutionary War, iron making required large quantities of charcoal, as the source of carbon, to reduce the ore. If a substitute had not been found for charcoal, our forests would have disappeared many years ago and our industrial progress arrested. In the past one hundred years, which really represents the modern era of steelmaking, a number of important developments have taken place to reduce the fuel requirements in producing steel. Some of these developments could be listed by historical sequence, while others are of such a nature that they cannot be designated by any period of time.

The major contributions to fuel economy in ironmaking and steelmaking plants have been:

(1) Development of the Bessemer converter.
(2) Development of the Siemens-Martin regenerator.
(3) Development of the hot blast.
(4) Utilization of blast-furnace gas.
(5) Installation of by-product coke plants and utilization of by-product fuels.
(6) Integration of steel plants.
(7) Electric drives for rolling mills.
(8) Improved efficiency of steam-generating equipment and steam prime movers.
(9) Large producing units.
(10) Balancing of producing units.
(11) Use of raw materials with improved chemical and physical quality.
(12) Recovery of waste heat by recuperators, boilers and other forms of heat exchangers.
(13) Development and utilization of instruments and control equipment.
(14) Insulation of high-temperature facilities.
(15) Utilization of the optimum fuel for specific facilities.
(16) Improvements in manufacturing technique and production control.
(17) More highly skilled operators.
(18) Development of oxygen-blown steelmaking processes.
(19) Development of continuous casting of steel.
(20) Recovery of sensible heat by hot charging and direct rolling.

The results of the above contributions now have made it possible to produce a ton of raw steel (by ingots or by continuous casting) utilizing less than one ton of coal (equivalent), instead of several tons as required a hundred years ago. The consumption of primary fuels in the iron and steel industry for 1982 is shown in Table

3—XX.

In addition to these outstanding contributions to fuel economy in steel mills, the importance of the effect that the rate of operations has on fuel economy should be stressed. Historically, the iron and steel industry follows the general business level maintained in the country, but its rate of operations often fluctuates more than that of many other industries. During peak production, optimum fuel economy is the natural result of operating the facilities which require fuel under the conditions for which they were designed to operate most economically. During periods of low production, fuel consumption undergoes a severe increase per unit of output; careful scheduling of production and facilities are required during this period to maintain minimum fuel losses.

The effectiveness with which by-product fuels are used in steel plants is of major significance in reducing the quantity of primary or purchased fuel required to produce a ton of steel. The consumption of by-product fuels (blast-furnace gas, coke-oven gas, pitch-tar and coke breeze) for 1982 is shown in Table 3—XXI.

The efficiency of heat utilization by blast furnaces and their auxiliaries, by steelmaking furnaces, by soaking pits and by reheating furnaces is discussed in the chapters dealing with the design and operation of these units (Chapters 15 through 18 and Chapter 22).

MEANS EMPLOYED FOR HEAT CONSERVATION

The heat lost from the combustion of fuel in steel-plant metallurgical and service facilities represents an appreciable part of the total supply. The amount lost differs among the various processes. In general, those processes having the higher temperature levels have the greatest thermal losses and, therefore, offer the best opportunity for heat recovery. For a specific facility, the amount and causes for these losses can be determined quantitatively by conducting a heat balance, and the results of this heat balance can then be used for planning a program aimed at heat conservation. Usually, the largest losses are contained in the waste flue gases and in radiation from the furnace walls. Additional losses are associated with combustion control (providing insufficient air for combustion at the burners and/or inadequate mixing of fuel and air), heating practice and air infiltration. Most important common denominators underlining all successful programs of heat conservation are proper maintenance of the facility and its instrumentation and proper scheduling of operations so that the facility is, as closely as practical, fully utilized for its intended purpose.

Recovery of Waste Heat—The recovery of heat from waste flue gases of high-temperature processes has been practiced for nearly 100 years. These high-temperature flue gases contain both sensible heat and the latent heat of vaporization of water and sometimes potential heat (unburned flue gases). The recovery of the heat of vaporization of water is not practical; however, the recovery of the sensible heat is accomplished by one or a combination of several methods, including the use of regenerators, recuperators, waste-heat boilers, or utilization of a furnace design in which the waste heat is used to preheat the product.

Regenerators are used alternately to absorb heat from one fluid and then transfer it to another fluid; recuperators are used to transfer heat continuously from one fluid to another. The fluids referred to in these two definitions are: (1) hot gaseous products of combustion which give up heat during passage through the regenerator or recuperator and (2) fuel gases or air for combustion which undergo heating while passing through the regenerator or recuperator. The importance of recuperators or regenerators for preheating the air to be used for combustion is shown in Figure 3—4. In this figure, the amount of fuel saving (in per cent) is plotted vertically, and the temperature of the flue gas at exit is plotted horizontally. The temperature of the preheated air is shown on the curves, and from these an estimate can be obtained of the amount of fuel saving to be gained from using preheated air.

Regenerators are applied usually to furnaces which can be fired alternately from the ends, the flow of gases through the furnace and regenerators being reversed according to predetermined time and/or temperature

Table 3—XX. Primary Fuels Consumed by the Iron and Steel Industry in 1982[1]

	Thousands of Metric Tons	Thousands of Net Tons
Coal[2]	33 209	36 606
Coal Equivalent from:		
Fuel Oil[3]	2 038	2 247
Natural Gas[4]	14 845	16 363
Purchased Electric Power[5]	13 458	14 835
Total Coal Equivalent Consumed	63 550	70 051
Raw Steel Production	67 656	74 577
Product Shipments	55 854	61 567
Coal Equivalent per Ton of:		
Raw Steel Produced	0.939	0.939
Product Shipped	1.138	1.138

[1] From: "Annual Statistical Reports (1982)," American Iron and Steel Institute.
[2] One metric ton of coal equivalent to 29 million kilojoules. One net ton of coal equivalent to 25 million Btu.
[3] One cubic metre of fuel oil equivalent to 1.44 metric tons of coal. One gallon of fuel oil equivalent to 0.006 net ton of coal.
[4] Twenty-five thousand cubic metres of gas equivalent to 36.29 metric tons of coal. One million cubic feet of gas equivalent to 40 net tons of coal.
[5] One kilowatt-hour equivalent to 0.395 kilograms or 0.87 pounds of coal.

cycles. Open-hearth furnaces, coke ovens, and many batch-type reheating furnaces and soaking pits, are equipped with regenerators. Blast-furnace stoves also use the regenerative principle but operate over a much longer cycle and in a somewhat different manner than that practiced in other installations. In a blast-furnace stove, checker brick of the regenerator are heated by

Table 3—XXI. By-Product Fuels Consumed by the Iron and Steel Industry in 1982 (Thousands of Tons)[1].

	Coal Equivalent	
	Metric Tons	Net Tons
Blast-Furnace Gas[2]	7 120	7 848
Coke-Oven Gas[3]	7 452	8 214
Pitch-Tar[4]	259	285
Coke Breeze[5]		
Iron-Ore Agglomeration	750	827
Steam Generation and Other Uses	152	168
Total By-Product Fuels	15 733	17 342

[1] Based on "Annual Statistical Reports (1982)," American Iron and Steel Institute and "Quarterly Coal Reports (1982)," Energy Information Administration.

[2] Based on reported consumption of 3731 kJ/m³ (95 Btu per ft³) blast-furnace gas.

[3] 100 000 m³ of gas equivalent to 64 metric tons of coal (1 million ft³ of gas equivalent to 20 net tons of coal).

[4] 1 m³ of pitch-tar equivalent to 1.53 metric tons of coal (1 gallon of pitch-tar equivalent to 0.0064 net ton of coal).

[5] Based on 1 ton of breeze equivalent to 0.840 ton of coal and including only breeze reported used in steam generation and ore agglomeration.

burning of fuel exclusively for the purpose of heating the regenerator brick, while in the open-hearth and other furnace installations the checker brick are heated by waste gases. In both cases, the heat stored in the regenerators is used to preheat air for the combustion of fuel in the furnace they serve.

Recuperators have been applied to many modern pit-batch and continuous-type reheating furnaces, and to steam boilers. When applied to steam boilers, they commonly are called "air preheaters." Recuperators are of three general types classified according to the direction of flow of the waste gases and air, as follows:

1. Counter-flow
2. Parallel or co-current flow
3. Cross-flow

Counter-flow is used to attain maximum air-preheat temperatures, and cross-current flow is used to secure optimum heat-transfer rates (kilojoules per square metre of recuperator surface per degree Celsius temperature difference per hour, or Btu per square foot of recuperator surface per degree Fahrenheit temperature difference per hour). Parallel flow is used where it is necessary, such as in metallic recuperators, to maintain the temperature difference of the division wall between the two fluids as uniform as possible throughout its length and to keep the temperature of the hot end below a maximum so that the metallic elements will not be overheated. Generally, a combination of counter-flow and cross-flow is applied to many steel-mill furnace applications where a refractory material is used to divide the two fluids. A combination of the two types is accomplished by baffling the flow of one of the

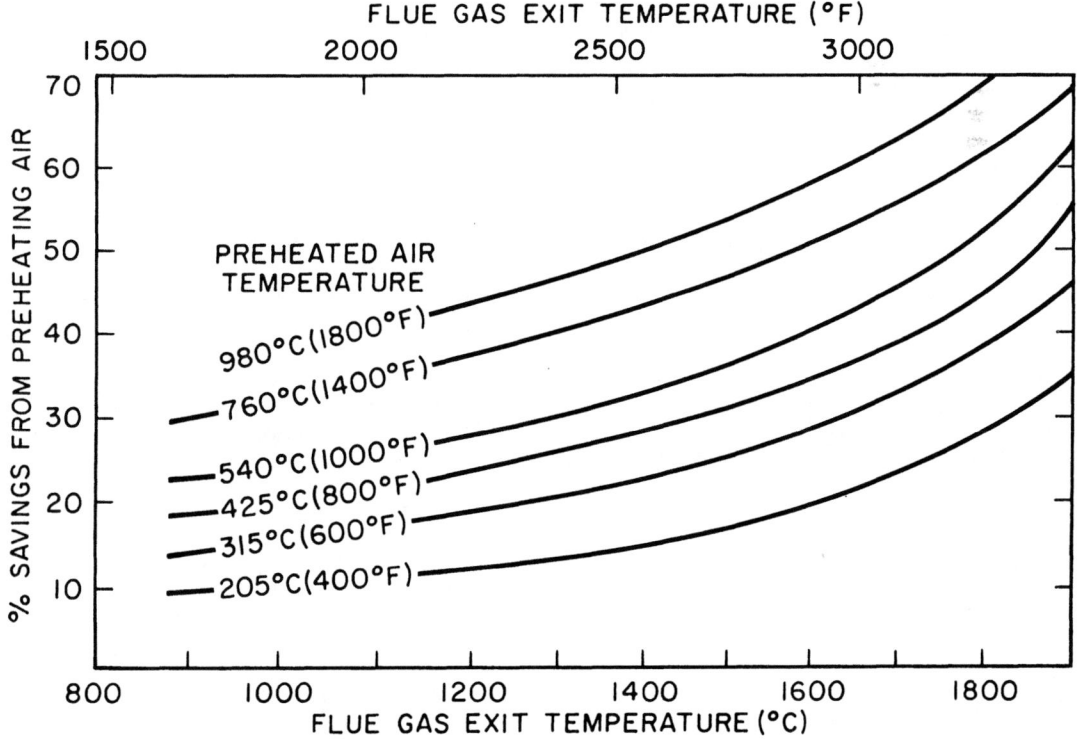

FIG. 3—4. Fuel savings resulting from use of preheated air.

fluids. In such designs, the general direction of flow of fluids exchanging heat is counter-current and the flow in each baffle section is cross-current. When the temperature of waste gas from which heat is to be extracted is relatively low, say under 980°C (about 1800°F), metallic tubes (e.g., stainless steel) are generally used because they possess an advantage against leakage. Higher-temperature recuperators are generally constructed of clay or silicon-carbide materials: often, however, these recuperators suffer the disadvantage of air leakage.

Waste-heat boilers are used to obtain heat recovery when a practical limit of recovery has been obtained by regenerators or recuperators and there is still sufficient heat left in the waste gases to justify expenditures for the waste-heat boilers. Boilers sometimes are used in place of regenerators or recuperators, depending upon conditions such as where preheated air is undesirable or where the generation of steam solves the problem of fuel conservation more satisfactorily. Waste-heat boilers are most applicable to high-temperature, continuous processes and have been used principally in the steel plant in conjunction with basic oxygen vessels, open-hearth furnaces, gas turbines and, to a lesser degree, with reheating furnaces and soaking pits. Firetube and water-tube boiler types have been installed, the former being the preferred type and generally of horizontal single-pass design. Waste-heat boilers usually are provided with superheaters and sometimes with economizers.

Minimizing Radiation Losses—Radiation losses from the walls and roofs of the furnaces can be minimized by the selection and use of an insulating system. Thermal-insulating materials have been used in steel plants for a great many years. There are many different kinds of insulating materials, each being most suitable for a specific temperature level and for the degree of insulation desired and improvements in insulating materials are continuing. A detailed discussion of insulating systems for steel-mill furnaces is given in Chapter 2.

Combustion Control—As mentioned previously, one of the large losses in a heating process is the heat contained in the waste flue gases. The amount of sensible heat in the waste flue gases is the product of the heat content per unit volume (m^3 or ft^3) or unit weight (kg or lbm) of gas multiplied by the total volume or total weight of the gas respectively, in the appropriate units. The total loss of heat in waste flue gases can be minimized by providing the proper amount of air for combustion. If too much air is provided, the volume of the waste gases is increased. In addition, an excess of air over that required for combustion lowers the flame temperature and increases the time necessary to heat the product. In a similar manner, an excess of fuel (or a deficiency of air) also decreases the flame temperature and prolongs the heating time and, in addition, results in unburned fuel being carried into the waste flue gases. This results in unnecessary usage of fuel and can be dangerous because this unburned fuel can burn in the flue and cause damage to the stack and/or waste-heat-recovery systems. In other words, fuel economy is optimized by the use of proper combustion conditions.

In modern steel-plant furnaces, and in boilers, the amount of air supplied for combustion is maintained only a little above theoretical requirements by special pressure regulators and valves which accurately proportion the amounts of fuel and air fed to the burners. For most burners, an amount of air about 10 per cent above theoretical is considered to be optimum. The use of instruments to continuously monitor the products of combustion and automatically control the air-fuel ratio to achieve this desired level of excess air has led to improved fuel economy and heating efficiency.

The amount of waste flue gases can be minimized, and the heating rate of the unit can be increased, by the oxygen enrichment of combustion air. Air is composed of only about 20.9 per cent oxygen (by volume), with the remainder consisting of inert nitrogen plus a small amount of several other inert gases. When combustion takes place, the oxygen combines with the carbon and hydrogen of the fuel and liberates heat. The inert gases of the air absorb heat from the combustion and carry it out of the furnace, and it is lost so far as the furnace process is concerned. These gases reduce flame temperature by absorbing heat, thus reducing the rate of transfer of heat to the work.

Obviously, if the inert content of the air could be diminished, more efficient combustion could be obtained. Recent technical developments that have lowered the cost for producing oxygen of commercial purity have made large-scale use of this gas economical for some industrial processes. Consequently, many plants have experimented with the addition of oxygen to ordinary air used for combustion, with generally good results. In effect, increasing the oxygen content lowers the inert content of the air; consequently, when a given amount of fuel is burned with oxygen-enriched air, the volume of the waste gas is less than if ordinary air were used. If the temperature of the waste gas is not increased, the sensible-heat loss in the flue gas will be decreased, due to the smaller heat capacity of the smaller volume. In furnaces operated at high thermal head, a decrease in the amount of the inert gases usually results in a decrease in the waste-gas temperature. With the same fuel input, enriched air for combustion raises the flame temperature of a given fuel, thereby improving heat-transfer rate and increasing production; alternatively, the fuel input may be decreased when enriched air is used to maintain the same production rate as obtained with fuel using ordinary air. Increased production rates almost always reduce the heat losses per ton of product in any furnace employing a high thermal head.

Air Infiltration—If the pressure of the gases in the heating chamber of the furnace is below atmospheric, cold outside air will be drawn into the furnace through any openings that exist. If the interior pressure is above that of the outside air, the hot gases will be forced out of the furnace through these same openings, and if too much higher will, in addition, tend to penetrate the refractories and overheat the furnace bindings with, in some cases, damaging effect. Generally, it is desirable to operate a furnace with a slight positive pressure in the heating chamber (i.e., furnace pressure slightly higher than atmospheric). It should be noted that the pressure from top to bottom of the heating chamber is not uniform, due to the "stack effect" of the hot gases. Control, therefore, is aimed at maintaining the desired pressure at the hearth level. Air drawn into a furnace operating under negative pressure upsets the fuel-air ratio which is controlled automatically or by valve settings. In some units such as reheating or heat-treating furnaces, this air aggravates the problem of oxidation (or scaling) of the work because of the oxygen present.

If the pressure in the furnace at the hearth level is equal to atmospheric or slightly positive, better heating conditions are obtained. This is especially so in furnaces where most of the heating of the work takes place through heat transfer by radiation from the flame to the bath or work. The positive pressure must be controlled to prevent excessive "sting-out" of flame from furnace openings (a small pressure imparts a high velocity to hot gases), as well as to avoid the build-up of excessive back pressure that would interfere with proper flow of fuel (if gaseous) and combustion air. Positive pressures maintained at hearth level in practical work are quite low, ranging only up to about one millimetre (a few hundredths of an inch) of water. Furnace pressure is controlled by adjusting the opening in the stack damper.

Positioning of the damper can be done manually, using the flame sting-out as an indication of the existence of positive pressure, but it is difficult to adjust the opening for the frequent changes in furnace conditions. The development, about the year 1928, of industrial-type instruments with sufficient sensitivity to measure differential gas pressures with an accuracy of plus or minus 0.06 mm (0.0025 inch) of water made possible the use of automatic control of furnace pressure.

Automatic furnace-pressure control has been provided for a majority of steel-plant furnaces and has been a principal factor in the improvements in fuel economy and efficiency of melting and reheating furnaces during the years following its adoption.

Heating Practice—The primary objective of any furnace operation is to heat the product (steel, as in the case of most furnaces, or air as in the case of blast-furnace stoves) to the desired temperature at the desired heating rate. The actual practice used to achieve the desired temperature and the desired rate will vary with each specific installation, and in each installation will vary with the level of operation. Consequently, it is beyond the scope of this discussion to provide such specific details. However, several general principles apply to all facilities.

The use of excessive amounts of fuel (high heating rate) is wasteful of the fuel itself and results in high furnace exit-gas temperatures and damage to refractories. In some processes, high fuel rates not only do not hasten transfer of heat to the material being processed, but also may cause it actual damage. On the other hand, the use of insufficient fuel reduces the rate of heat transfer and prolongs process time, thereby increasing thermal losses. The optimum rate for protection either of the material being heated or the furnace refractories, and often for control of heating or production rate, is maintained in most furnaces by automatic temperature-measuring instruments which control the fuel rate through a system of electrical relays or other units which control the operation of motors, hydraulic systems or other means for regulating fuel valves. Many modern heating installations have adopted computerized control for the heating operations.

A basic requirement for all heating operations is good temperature measurement. One problem involved in temperature measurement is the difficulty and almost impossibility of measuring the temperature inside a solid piece of steel, and in most cases the difficulty is encountered in measuring the surface temperature in one spot. Because of this inability to measure the temperature of solid steel directly, measurement is made of some other temperature that is closely related to the steel temperature. Depending upon the specific furnace installation, this involves measuring the temperature of the roof or wall of the furnace or a measurement of the temperature of the gases in the furnace. These measurements are achieved in the steel industry by instruments operating on four main principles: (1) by measuring the intensity of radiation emitted by the hot furnace or object being heated, (2) by measuring the minute electrical current generated in a circuit composed of two wires of dissimilar metals, joined end to end, when one of the joints is heated (this is the principle of the thermocouple), (3) by measuring the change in the electrical resistance of conductors when heated to the temperature in question, and (4) by measuring the change in the ratio of two separate wavelengths of radiation emitted from the hot object or furnace.

Bibliography

GENERAL

American Iron and Steel Institute, "Steel at the Crossroads". Washington D.C., 1980.

Baumeister, Theodore (ed.), Marks Standard Handbook for Mechanical Engineers (8th ed., section on fuels and furnaces), New York, McGraw-Hill, 1978.

Brame, J. S. S. and J. G. King, Fuel—Solid, Liquid and Gaseous. London, Edward Arnold, 1967.

Chapman, A. J., Heat Transfer, 4th edition. New York, MacMillan, 1984.

Considine, D. M. (editor), Energy Technology Handbook. New York, McGraw-Hill, 1977.

Fristrom, R. M. and A. A. Westenberg, Flame Structure. New York, McGraw-Hill, 1965.

Gebhart, B., Heat Transfer. New York, McGraw-Hill, 1961.

Giedt, W. H., Principles of Engineering Heat Transfer. New York, Van Nostrand, 1957.

Griswold, John, Fuels, Combustion and Furnaces. New York, McGraw-Hill, 1946.

Hauck Manufacturing Company, Hauck Industrial Combustion Data Handbook, 3rd ed. The Company, Lebanon, Pa., 1953.

Hill, R. F. (editor), A Decade of Progress, Energy Technology X, Government Institutes, Inc., Annual Publication, June, 1983.

Hottel, H. C. and A. F. Sarofim, Radiative Transfer. New York, McGraw-Hill, 1967.

Hougen, O. A., et al., Chemical Process Principles. Part I (2nd. ed., 1954), Material and Energy Balances; Part II (2nd ed., 1959), Thermodynamics; Part III (1947) Kinetics; Charts (3rd ed., 1964). New York, John Wiley and Sons.

Institute of Fuel, Waste-Heat Recovery. Proceedings of 1961 Conference. London, Chapman and Hall, Ltd. (1963).

Jakob, M., Elements of Heat Transfer (3rd. ed.). New York, John Wiley and Sons, 1957.

Johnston, A. J., and G. H. Auth, Fuels and Combustion Handbook. New York, McGraw-Hill, 1951.

Lewis, B., and R. N. Pease, Combustion Processes. Princeton University Press, 1956.

Lewis, B., and G. von Elbe, Combustion, Flames and Explosions of Gases. New York, Academic Press, 1951.

Lewis, G. N., et al., Thermodynamics (2nd ed.). New York, McGraw-Hill, 1961.

Mackey, C. O., et al. Engineering Thermodynamics. New York, John Wiley and Sons, 1957.

McAdams, W. H., Heat Transmission (3rd. ed.). New York, McGraw-Hill, 1954.

North American Manufacturing Company, North American Combustion Handbook (2nd ed.), The Company, 1978.

O'Loughlin, J. R., Generalized Equations for Furnace Combustion Calculations. Combustion 34, May 1963, 23-24.

Perry, R. H., et al., (eds.), Chemical Engineers' Handbook (6th ed.). New York, McGraw-Hill, 1984.

Proceedings of the Energy Technology Conference (Annual publication) Government Institutes Inc. Rockville, MD. 20856.

Schack, Alfred, Industrial Heat Transfer (translated from 6th German ed.). New York, John Wiley and Sons, 1965.

Smith, M. L., and K. W. Stinson, Fuels and Combustion. New York, McGraw-Hill, 1952.

Strehlow, R. A., Fundamentals of Combustion. Scranton, Pa., International Textbook Company, 1967.

"Temperature—Its Measurement and Control in Science and Industry," Volume 5, 1982, American Institute of Physics.

Trinks, W., and M. H. Mawhinney, Industrial Furnaces, Vol. I (5th ed., 1961) and Vol. II (4th ed., 1967). New York, John Wiley and Sons.

U.S. Department of Energy (DOE) Research, Development and Demonstration for Energy Conservation: Preliminary Identification of Opportunities in Iron and Steelmaking, Arthur D. Little, 1978.

STATISTICS

Averett, P., Coal Resources of the United States—January 1, 1974, U.S. Department of the Interior-Geological Survey, Bulletin No. 1412, Washington, D.C., 1975.

Coal Production—1982, U.S. Department of Energy-Energy Information Administration, Document No. DOE/EIA-0118(82), September, 1983.

1982 International Energy Annual, U.S. Department of Energy-Energy Information Administration, Document No. DOE/EIA-0219(82), September 1983.

Keystone Coal Industry Manual. New York, McGraw-Hill, 1984.

Lowie, R. L., Recovery Percentage of Bituminous Coal Deposits in the United States. United States Bureau of Mines R. I. 7109, April 1968.

Metric Practice Guide, American Iron and Steel Institute, 1978.

Petroleum Supply Annual—1982, U.S. Department of Energy-Energy Information Administration, Document No. DOE/EIA-0340(82)/1, June, 1983.

Pfleider, E. P. (ed.), Surface Mining. New York, American Institute of Mining, Metallurgical and Petroleum Engineers, 1968.

Spiers, H. M. (ed.), Technical Data on Fuels (6th ed.). London, British National Committee, World Power Conference, 1961.

United States Bureau of Mines. Mineral Facts and Problems. (chapters on bituminous coal, anthracite and petroleum and natural gas). Superintendent of Documents, Washington, D.C.

United States Crude Oil, Natural Gas, and Natural Gas Liquids Reserves—1982, U.S. Department of Energy-Energy Infor-

mation Administration, Document No. DOE/EIA-0216(82), August, 1983.

COAL AND COKE—GENERAL

Donnelley, F. J., and L. T. Barbour, Delayed coke—A Valuable Fuel. Hydrocarbon Processing 45, Nov. 1966, 221-224.

Elliot, M.A. (Editor), Chemistry of Coal Utilization. 2nd Supplementary Volume. New York, John Wiley and Sons, 1981.

Glenn, R. A. and H. J. Rose, The Metallurgical, Chemical and Other Process Uses of Coal. Pittsburgh, Bituminous Coal Research, Inc., 1958.

Machin, R. E., Science in a Coalfield (2nd ed.). New York, Pitman, 1952.

Leonard, J. W., (ed.), Coal Preparation (4th ed.). New York, American Institute of Mining, Metallurgical and Petroleum Engineers, 1979.

Raistrick, A., and C. E. Marshall, Nature and Origin of Coal and Coal Seams. New York, British Book Centre, 1952.

Van Krevelen, D. W., Coal. New York, Elsevier, 1961.

Van Krevelen, D. W., and J. Schuyer, Coal Science. New York, Elsevier, 1957.

Wilson, P. J. and J. H. Wells, Coal, Coke and Coal Chemicals. New York, McGraw-Hill, 1950.

COAL AND COKE—TESTING

American Society for Testing and Materials, ASTM Standards, Volume 06.05, Gaseous Fuels, Coal and Coke. Philadelphia, The Society, 1983.

Rees, O. W., Chemistry, Uses and Limitations of Coal Analyses. Illinois State Geological Survey, R. I. 220(1966).

United States Bureau of Mines, Bulletin 638, Methods of Analyzing and Testing Coal and Coke. Superintendent of Documents, Washington, D.C., 1967.

COAL PETROGRAPHY

Gray, R. J., et al., Distribution and Forms of Sulphur in a High-Volatile Pittsburgh Seam Coal. Transactions, American Institute of Mining, Metallurgical and Petroleum Engineers 226, 1963, 113-121.

Harrison, J. A., Coal Petrography Applied to Coking Problems. Proceedings Illinois Mining Institute, 69th Year (1961), 17-43.

Harrison, J. A. Application of Coal Petrography to Coal Preparation. Transactions of SME (American Institute of Mining, Metallurgical and Petroleum Engineers) 226 (1963), 346-357.

Kaye, N., The Application of Coal Petrography in the Production of Metallurgical Coke. The Coke Oven Managers' Yearbook (1967). London, The Coke Oven Managers' Association.

Mackowsky, M. Th., Practical Possibilities of Coal Petrography. Compterendu 31e Congres Intern. de Chemie Indust., Liege, Sept. 1958. Special Libraries Association translation 59-17595.

Marshall, C. E., Coal Petrology. Economic Geology, 50th Anniversary Volume, 1955.

Mason, D. M., and F. C. Schora, Jr., Coal and Char Transformation in Hydrogasification. Fuel Gasification, Advances in Chemistry Series No. 69, American Chemical Society, 1967, 18-30.

Parks, B. C., and H. J. O'Donnell, Petrography of American Coals. United States Bureau of Mines Bulletin 550 (1957). Superintendent of Documents, Washington, D.C.

Schapiro, N., and R. J. Gray, The Use of Coal Petrography in Coke Making. Journal Institute of Fuel 37, June 1964, 234-242.

Thomas, C. M., Coal Petrology and Its Application to Coal Preparation. Coal Preparation, March-April 1968, 50-59.

GASEOUS AND LIQUID FUELS

American Chemical Society, Fuel Gasification. Advances in Chemistry Series, The Society.

American Gas Association, Proceedings Synthetic Pipeline Gas Symposium. (Annual proceedings)

American Society for Testing and Materials, ASTM Standards, Petroleum Products, etc. Section 5, Volumes 05.01 to 05.04. Philadelphia, The Society, 1983.

Brush, S. G. (ed.) Kinetic Theory, Vol. I-1965, Vol. II-1966, and Vol. III-1972. Pergaman.

Chapman, S., and T. G. Cowling, Mathematical Theory of Non-Uniform Gases; An Account of the Kinetic Theory of Viscosity, Thermal Conduction, and Diffusion in Gases (2nd ed.). London, Cambridge, 1952.

Eilers, M. G., et al., Producer gas. Gas Engineers Handbook (loc. cit., see Segeler, C. George), 3/37-3/46.

Glover, C. B., Blue Gas and Carburetted Water Gas. Gas Engineers Handbook (loc. cit., see Segeler, C. George), 3/47-3/60.

Gumz, W., Gas Producers and Blast Furnaces. New York, John Wiley and Sons, 1950.

Hale, D., Capacity Increases Slightly, Industry Ready for Winter. Pipeline and Gas Journal, November, 1983, 28-34.

Jost, W., Diffusion in Solids, Liquids, Gases. Academic Press, 1952.

Moore, B. J. Analyses of Natural Gases, 1917-1980, United States Bureau of Mines I. C. 8870 (1982). Superintendent of Documents, Washington, D. C.

Morgan, J. J., Water Gas. Chemistry of Coal Utilization, Vol. II (H. H. Lowry, ed.). New York, John Wiley and Sons, 1945.

Segeler, C. George (ed. in chief), Gas Engineers Handbook. New York. The Industrial Press, 1977.

The Gas Research Institute (GRI), 8600 West Bryn Mawr Avenue, Chicago, Illinois 60631.

van der Hoeven, B. J. C., Producers and Producer Gas. Chemistry of Coal Utilization Vol. II (H. H. Lowry, ed.), New York, John Wiley and Sons, 1945.

Weil, S. A., et al., Fundamentals of Combustion of Gaseous Fuels. Research Bulletin 15 (1957), Institute of Gas Technology, Chicago.

POWER AND STEAM GENERATION

Babcock and Wilcox Company, Steam, Its Generation and Use (39th ed.). New York, The Company, 1978.

Bender, R. J., Low Excess Air and Sonic Atomization Team Up. Power 110, July 1966, 71.

Cizmadia, L., and F. J. Fendler, Design and Operation of Fairless Package Boiler. Iron and Steel Engineer 45, April 1968, 109-114.

Freyling, Glenn R. (ed.), Combustion Engineering (revised ed.). New York, Combustion Engineering, Inc., 1967.

Illinois Institute of Technology, American Power Conference Proceedings (annual).

Rochford, R. S., Considerations in Converting Multiple Fuel Fired Industrial Boilers. Iron and Steel Engineer 42, December 1965, 169-170, 173.

Technical publications by The Electric Power Research Institute (EPRI), Palo Alto, California 94303.

CHAPTER 4

Manufacture of Metallurgical Coke and Recovery of Coal Chemicals

SECTION 1

INTRODUCTION

Carbon As A Reducing Agent—Although the oxides of iron may be reduced to metallic iron by many agents, carbon (directly or indirectly) is the reducing agent found to be best suited for the economical production of iron. Carbon of suitable reactivity and physical strength was at one time produced from wood by distillation, yielding wood charcoal; but for the operation of a modern large blast furnace the carbon required for the smelting of iron is obtained from the destructive distillation of selected coking coals at temperatures in the range of about 900°C to 1095°C (1650°F to 2000°F).

Chemical Effects of Coking—Coal is made up principally of the remains of vegetable matter which has been partially decomposed in the presence of moisture and the absence of air and subjected to variations in temperature and pressure by geologic action (see Chapter 3). It is a complex mixture of organic compounds, the principal elements of which are carbon and hydrogen with smaller amounts of oxygen, nitrogen, and sulphur. It also contains some noncombustible components called ash. The ash consists primarily of inorganic compounds which became imbedded in the coal matrix during the coalification process.

The chemical compounds making up coal, like most of those in animal and vegetable life, are unstable when subjected to a high degree of heat or thermal treatment. When heated to high temperatures, in the absence of air, the complex organic molecules break down to yield gases, together with liquid and solid organic compounds of lower molecular weight and a relatively non-volatile carbonaceous residue (coke).

Coke, then, is the residue from the destructive distillation of coal. Structurally, it is a cellular, porous substance which is heterogeneous in both physical and chemical properties. The physical properties of metallurgical coke, as well as its composition, depend largely upon the coal used and the temperature at which it is carbonized. Not all coals will form coke, and not all coking coals will give the same firm, cellular mass characteristic of coke suitable for metallurgical purposes. Some coals will produce an acceptable coke without blending with other coals, while others are usable only as constituents of blends. The type and method of operation of coking facilities also exert a profound influence on the quality and yield of coke for the blast furnaces.

Kinds of Coke—There are three principal kinds of coke, classified according to the methods by which they are manufactured; **low-, medium-** and **high-temperature coke.** Coke used for metallurgical purposes must be carbonized in the higher ranges of temperature (between 900° and 1095°C) (1650° and 2000°F) if the product is to have satisfactory physical properties. Even with good coking coal, the product obtained by low-temperature carbonization between 480° and 760°C (900° and 1400°F) is unacceptable for good blast furnace operation.

Important Properties of Metallurgical Coke—Probably the most important physical property of metallurgical coke is its strength to withstand breakage and abrasion during handling and its use in the blast furnace. In the United States, the standard ASTM tests used to evaluate these properties are the **Stability Index** for breakage and the **Hardness Index** for abrasion. Both of these tests involve tumbling coke of selected size in a standard drum rotated for a specific time at a specific rate. The Stability Index and the Hardness Index are the percentages of coke remaining on 1-inch and ¼-inch screens, respectively, when the coke is screened after tumbling.

In modern blast-furnace practice, the trend is toward use of iron-bearing burden materials of controlled size such as sinter and pellets; thus, the size of the coke used in the burden assumes more importance than in the past when only crude ore was used. The size of coke produced in by-product ovens is somewhat dependent upon the type of coal, heating rate, width of the ovens, and the bulk density of the coal charge. Greater amounts of low-volatile coal, wider ovens and greater bulk density of the coal charge generally tend to produce larger coke while faster heating rates tend to produce smaller coke. Because relatively uniform size is desired, crushing and screening of the coke must be resorted to when controlled size is desired. Most blast-furnace operators prefer coke sized between about 18.5 and 76 mm (¾ inch and 3 inches) for optimum furnace performance. Other physical properties of the coke such as porosity, density, and combustibility are controllable only to a small extent, and their importance in affecting blast-furnace operation has not been definitely established.

Methods of Manufacturing Metallurgical Coke— There are two proven processes for manufacturing metallurgical coke, known as the **beehive** process and the **by-product process.** In the beehive process, air is admitted to the coking chamber in controlled amounts for the purpose of burning therein the volatile products distilled from coal to generate heat for further distillation. In the by-product method, air is excluded from the coking chambers, and the necessary heat for distillation is supplied from external combustion of some of the gas recovered from the coking process (or, in some instances, cleaned blast-furnace gas or a mixture of coke-oven and blast-furnace gas). With modern by-product ovens, properly operated, all the volatile products liberated during coking are recovered as gas and coal chemicals, and, when coke-oven gas alone is used as fuel, about 40 per cent of the gas produced is returned to the ovens for heating purposes. While the beehive process was the leading method for the manufacture of coke up to 1918, it now has been replaced largely by the by-product process as discussed later in this chapter. There is a difference of temperature of coking in the two processes, that of the by-product being somewhat lower than the beehive. Beehive coke is usually larger, though not as uniform in size. In general, properly carbonized beehive coke and by-product coke both are silvery gray in appearance.

Other processes for producing metallurgical coke are known as **continuous processes;** many variations have been proposed but none has been adopted on a commercial scale. In one continuous process, finely pulverized coking or non-coking coal is dried and partially oxidized with steam or air in fluidized-bed reactors to prevent agglomeration when coking coal is used. The reactor product is carbonized in two stages at successively higher temperatures to obtain a **char.** Using a binder produced from tar obtained in the carbonization stages, the char is briquetted in roll presses. The "green" briquettes are cured at low temperatures, carbonized at high temperatures, and finally cooled in an inert atmosphere to produce a metallurgical coke of low volatile content. This type of coke often is referred to as **formcoke.** Briquetting will be discussed again later in this chapter.

Products of Coal Carbonization—The reactions occurring during the carbonization of coal for the production of metallurgical coke are complex. The process can be considered as taking place in three steps: (a) Primary breakdown of coal at temperatures below 700°C (1296°F) yields decomposition products some of which are water, oxides of carbon, hydrogen sulphide, hydroaromatic compounds, paraffins, olefins, phenolic,

and nitrogen-containing compounds. (b) Secondary thermal reactions among these liberated primary products as they pass through hot coke, along hot oven walls and through highly heated free space in the oven involve both synthesis and degradation. A large evolution of hydrogen and the formation of aromatic hydrocarbons and methane occur in the stage above 700°C (1296°F). Decomposition of the complex nitrogen-containing compounds produces ammonia, hydrogen cyanide, pyridine bases and nitrogen. (c) Progressive removal of hydrogen from the residue in the oven produces hard coke.

During carbonization, from twenty to thirty-five per cent by weight of the initial charge of coal is evolved as mixed gases and vapors which pass from the ovens into the collecting mains and are processed through the coal-chemical recovery section of the coke plant to produce coal chemicals. When the production of coke is accomplished in modern by-product coke ovens with equipment for recovering the coal chemicals, one ton of coking coal in typical American practice yields about the following proportions of the coke and coal chemicals, depending upon the type of coal carbonized, carbonization temperature and method of coal-chemical recovery:

	Per Metric Ton	Per Net Ton
Blast-Furnace Coke . .	600-800 kg	1200-1400 lb
Coke Breeze	50-100 kg	100-200 lb
Coke-Oven Gas	296-358 m³	9500-11500 ft³
Tar	30.3-45.4 litres	8-12 gal
Ammonium Sulphate	10-13.8 kg	20-28 lb
Ammonia Liquor	56.8-132.5 litres	15-35 gal
Light Oil	9.5-15.1 litres	2.5-4 gal

The **coke-oven gas** contains the **fixed gases** so classified because they are gases at 760 mm (29.92 in.) pressure and 15.5°C (60°F). They are hydrogen, H_2; methane, CH_4; ethane, C_2H_6; carbon monoxide, CO; carbon dioxide, CO_2; illuminants which are essentially unsaturated hydrocarbons, such as ethylene, C_2H_4; propylene, C_3H_6; butylene, C_4H_8; and acetylene, C_2H_2. Other fixed gases present are hydrogen sulphide, H_2S; ammonia, NH_3; oxygen, O_2; and nitrogen, N_2.

Other substances in the raw gases and vapors leaving the ovens, which are **liquids** at ordinary temperatures and pressures, are:

(a) **Ammonia Liquor** (primarily the water condensing from the gas), an aqueous solution of ammonium salts of which there are two kinds—**free and fixed.** The free salts are those which are decomposed on boiling to liberate ammonia. The fixed salts are those which require boiling with an alkali such as lime to liberate the ammonia.

(b) **Tar,** the organic matter that separates by condensation from the gas in the collector mains. It is a black, viscous liquid, a little heavier than water. The following general classes of compounds may be recovered from tar: pyridine, tar acids, naphthalene, creosote oil and coal-tar pitch.

(c) **Light Oil,** a clear, yellow-brown oil somewhat lighter than water. It contains varying amounts of coal-gas products with boiling points from about 40°C to 200°C, and benzene, toluene, xylene and solvent naphthas are the principal products recovered from it.

In the recovery of coal chemicals, the first step is the recovery of the basic crude materials (coke-oven gas, ammonia liquor, tar and light oil) as a primary operation in accordance with commercial practice. Secondary operations consist of the processing of these primary products to separate them into their components as discussed in Section 7 of this chapter.

SECTION 2

COALS FOR METALLURGICAL-COKE PRODUCTION

In modern by-product coke-making operations, high-volatile coal usually is blended with either or both medium- and low-volatile coals to provide the charge for the coke ovens. These coals should contain as small amounts of sulphur and ash as is economically feasible, because the amount of these components present in the coal mixture directly affects the quality of the coke produced and its performance in the blast furnace. For this reason, most coals used for the production of metallurgical coke are beneficiated prior to use in the ovens, as already discussed in Chapter 3. Maintaining a uniform amount of these components in the coals on a day-to-day basis is of great importance. Approximately 80 per cent of the sulphur and all of the ash in coal remain in the coke produced from it. Most coals also contain small amounts of chlorides and phosphorus which, when present in the usual quantities, do not present serious problems. Excessive amounts of chlorides can cause problems with corrosion in the distillation equipment used in the recovery of coal chemicals. Phosphorus in the coal remains in the coke.

High-volatile coal usually becomes highly plastic or fluid upon heating, and when coked alone produces a highly fissured, porous and weak coke which will break up and abrade during handling and use in the blast furnace. To produce coke of acceptable quality, low-volatile coal, which usually has a low fluidity when heated, is blended with high-volatile coal with high fluidity, in amounts up to 40 per cent of the blend. In recent years, procedures have been developed to predict by microscopic techniques the strength of coke (Stability Index) which can be produced by carbonization of coals or coal blends. Such procedures permit evaluation of the coking properties of coal in extensive beds on the basis of small samples (such as drill cores) obtained in exploration programs.

The use of low-volatile coal in coal blends to improve coke properties complicates the selection of coals, in that low-volatile coals have a tendency to swell on heating and produce greater pressure on the oven walls and contract less after carbonization. The pressure and contraction properties of blends, therefore, must be ascertained in laboratory-scale test ovens prior to the use of the blends in commercial ovens to prevent possible oven damage and loss of production. Automated coal petrography in concert with computer programs is also becoming a useful tool in predicting pressure and contraction properties of coal blends, as well as the resultant coke strength.

PREPARATION OF COAL CHARGE FOR BY-PRODUCT OVENS

In the production of metallurgical coke, the selection of coals is the most important single factor in establishing coke quality. The best coals are low in ash and sulphur contents and, when blended with other compatible coals, produce a strong coke. The selection of coals in most instances is made by individuals not directly connected with the coke plant and is influenced by economics, coal availability and other criteria not necessarily connected with coke quality.

The importance of coal preparation cannot be overemphasized. It is the most important step in the coke-making process in terms of coke quality and uniformity. The proper preparation of the coal blend affects both the smoothness of operation and the productivity of the coke battery.

A simplified flow diagram of a typical coal-handling and preparation facility is shown in Figure 4—1; a brief description of the elements of this system and its operation follows.

Coal Unloading—Coals are received at coke plants by rail, river barge or, in some cases, by motor truck. Railroad cars usually are unloaded by rotary dumpers or bottom dumping. Rotary car dumpers usually are preferred. Gary Works of United States Steel Corporation uses a typical rotary-dumper system in which up to 130 railroad cars are lined up on a track with a slight grade and moved to the dumper by a series of retarders and car pullers. Twelve cars per hour (1088 metric tons or 1200 net tons of coal) can be unloaded with only one or two operators for this system.

Since coal freezes to the railroad cars during the winter months, a car-thawing facility is provided to thaw the coal prior to dumping. Some coal mines spray de-icing compounds on the coal as it is loaded at the mines to prevent freezing.

At the Clairton plant of United States Steel, all coal is received in river barges that transport about 815 to 1360 metric tons (900 to 1500 net tons) each. In this operation, there are two coal hoists each with a capacity of about 910 metric tons (1000 net tons) per hour and a bucket-elevator type unloader which can unload up to about 1815 metric tons (2000 net tons) per hour.

Throughout the unloading operation, coal identity must be maintained so that selected coals always are unloaded to the proper bin or stockpile. Any coal misplaced to the wrong storage bins can cause serious operating problems at the coke batteries in addition to having a negative affect on coke quality.

Bed Blending—The coals as received can vary in sulphur and ash content. One of the methods used to reduce this variability is to bed blend about 10 days supply of coal in horizontal layers and reclaim it vertically. Individual coals, according to classification, can be bed blended prior to coal preparation. A minimum of two stockpiles for each classification of coal is required; one

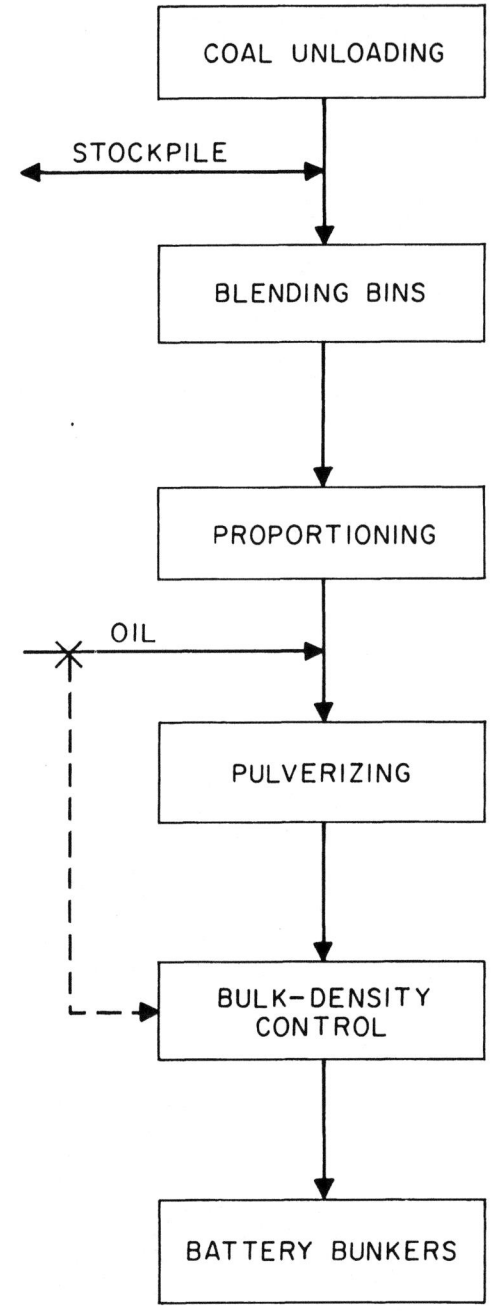

FIG. 4—1. Schematic diagram of a typical coal-preparation facility.

being stockpiled and one being reclaimed. Usually, a more uniform quality coke is produced from coals that have been bed blended, which has a beneficial effect on blast furnaces operating at high production rates. The benefits of bed blending in the form of improved blast-furnace coke can often justify the added cost of coal blending. Coal bed blending can be accomplished either by use of stacker/reclaimers or by utilizing mobile equipment such as carry-alls, trucks and front-end

loaders. Mobile equipment has a lower initial cost but the system selected will depend upon the size of the plant and the results of an economic evaluation of both systems.

Primary Crushing—Coal received at the coke plant varies in size from run-of-mine, which includes large lumps, to crushed coal which could be essentially 100 per cent minus 19 mm (¾ inch). In addition, large chunks of frozen coal are not uncommon during the winter months, and contaminants such as wood, rock and metal are usually present.

Primary crushers are used to break up the large coal with a minimum of fines generation. Not all plants are equipped with primary crushers, but plants that are usually employ roll-type crushers or Bradford breakers (Figure 4—2). The Bradford breaker is a large rotating drum over 6 metres (20 feet) long and about 4.2 metres (14 feet) in diameter, with holes about 63 mm (2½ inches) in diameter spaced uniformly over the entire shell. Baffle plates inside the drum lift and drop the coal as the drum rotates. This repeated impact crushes the coal, which then passes through the holes in the shell to a conveyor. The rock, slate and extraneous refuse are discharged from the end of the drum. Plants not equipped with breakers of this type usually use magnetic separators and/or scalping screens to separate foreign material.

Blending Bins and Weigh Feeders—Blending bins and weigh feeders are used for blending the individual coals in proportion to the final coal blend which will be charged to the coke ovens from the bunker located above the coke batteries. This operation can be performed either before or after pulverization. Some plants have raw-coal bins ahead of the pulverizers plus coal bins for storing pulverized coal prior to blending. Other plants pulverize the coal as it is unloaded and accumulate the pulverized coal in bins prior to blending.

In a well-designed plant, blending bins are provided for each coal or groups of coals. The number and size of the bins is determined by the number of coals to be blended and the daily throughput of coal. The bins can be either round or rectangular with conical bottoms constructed of concrete or steel. However, they should be designed with mass-flow bottom discharge. Bins usually are grouped together to facilitate distribution of coal to the top of the bins and to shorten the conveyor lengths required at the bottom discharge.

Quality analysis proves that a consistently well-proportioned coal blend is essential to the production of the highest quality, most uniform coke possible from the coals supplied. In any given situation, regardless of the design or age of the facility, every effort should be made to get the maximum performance from the combined mixing-bin and weigh-feeder systems. Coal proportioning is too important to take lightly or to ignore simply because the facility is too old to maintain properly or because repair parts are difficult to obtain. The benefits in uniform coke quality and in the resultant blast-furnace performance justify a concerted operating effort and an adequate maintenance program.

Each bin is provided with a discharge weigh belt for controlling the weight of the coal delivered. Coal-weight control is accomplished by scales and control

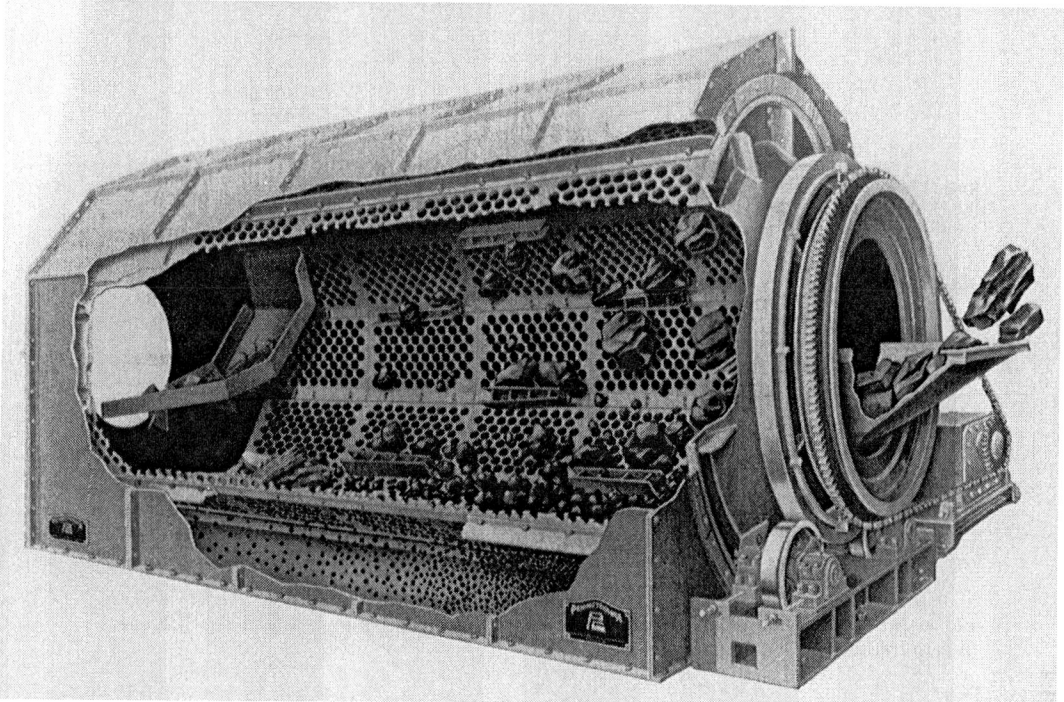

FIG. 4—2. Cut-away view of a rotary Bradford breaker. (Courtesy, Pennsylvania Crusher Division of Bath Iron Works Corporation.)

circuitry which regulate either the speed of the belt carrying a constant depth of coal or the depth of the coal on a constant-speed belt. There is probably little difference in the performance of these two types of weigh feeders provided the machines are of modern design and well maintained. Usually, if the weigh feeders and control circuits are in good condition and an error exists, it is probably due to the bin above the feeder. If the flow from the bin is not constant because of poor bin design or because of wet coal or extraneous material, the weigh feeder cannot deliver the required flow rate. The subject of bin design and mass flow of materials from bins is too extensive to discuss in detail here. Suffice it to say that the bin and feeder must be treated as an integral unit if accurate proportioning is to be accomplished. The best feeder is useless without a proper bin above it.

Coal Screening—Most of the coals received at the coke plants contain large amounts of minus 3-mm (minus ⅛ inch) coal. Screens are sometimes installed in the pulverizers (described below) to remove and by-pass this fine coal around the pulverizers. This reduces pulverizing costs, reduces plugging problems caused by wet fines, reduces the amount of very fine coal (minus 147-micron or minus 100-mesh) coal produced. These screens, however, tend to plug or blind, reducing efficiency especially when the coal is wet. Without constant cleaning, this blinding permits a large portion of the fine coal to pass over the screens and through the pulverizer. Because of this problem, larger screen openings are sometimes used and this allows larger coal to by-pass the pulverizers. This negates the effectiveness of the pulverizers, resulting in an over-sized coal blend and lower coke strength.

Secondary Crushing (Pulverizing)—Many different types of pulverizers are available for crushing coal, but only three types find wide application in coke plants: these are the hammermill (Figure 4—3), the impact mill and cage mill.

The **hammermill** until recent years was the most commonly used type. In this machine, hammers made of flat steel bars or cast-steel sections swing freely on 8 to 10 steel shafts which rotate in the mill at speeds of up to 100 rpm. The hammers first impact and grind the coal against heavy plates and finally force it through a bar screen. The degree of pulverization is determined by the speed of the hammers and the openings in the bar screen. The hammermill is an excellent heavy-duty machine which can withstand the abuse of over-size material and tramp metal, usually without serious damage. One important disadvantage, however, is its tendency to generate a large amount of fine coal (minus 147-micron or minus 100-mesh) by the grinding action of the hammers forcing the coal through the bar screen. This fine coal is undesirable in cokemaking as much of it is carried out of the ovens into the gas-collector mains during charging, adversely affecting tar quality.

Impact mills are essentially hammermills with the bar screen removed and fitted with special impact plates (Figure 4—4). Coal enters the top of the machine where the coal particles are struck by the rapidly moving hammers and thrown against the impact plates which are located at various positions outside the path of the hammers. The impact shatters the coal particles, usually along natural cleavage lines. The degree of pulverization depends on the hammer speed, the number of hammers and impact plates, and the hammer-to-

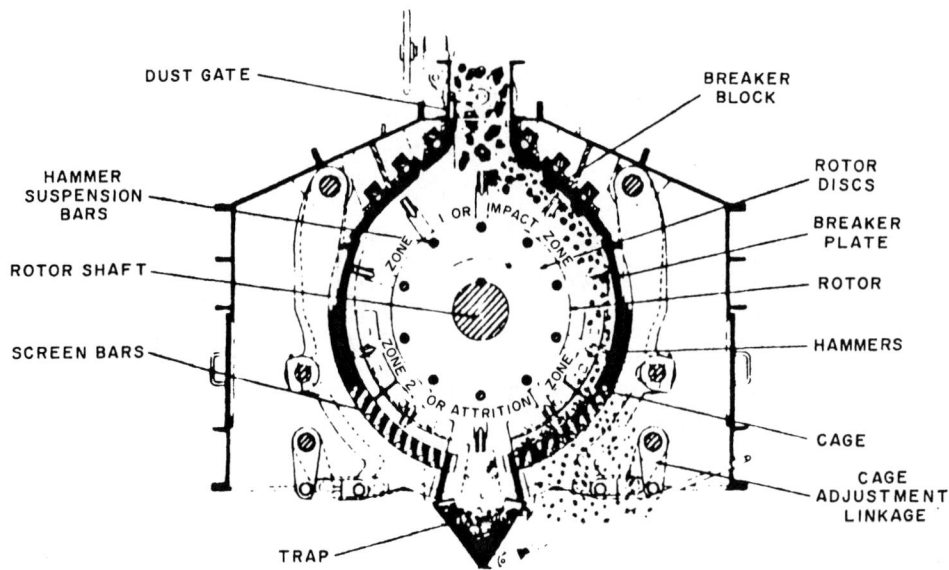

FIG. 4—3. Sectional view of a reversible hammer mill. (Courtesy, Pennsylvania Crusher Division of Bath Iron Works Corporation.)

plate spacing. Impact mills are as rugged as hammer-mills but have an added advantage of providing a high degree of pulverization with a lower level of fine-coal generation. Because of this, impact mills are the most commonly used pulverizers in coke plants at this time.

The third type of pulverizer commonly used in coal pulverization is the cage mill (Figure 4—5). In this machine, two cages, one inside the other, are rotated in opposite directions at speeds of up to 500 rpm. Coal enters the center of the machine, is struck by the impact plates of the inside cage and thrown into the plates of the outside cage, which is rotating in the opposite direction, and is discharged out of the machine.

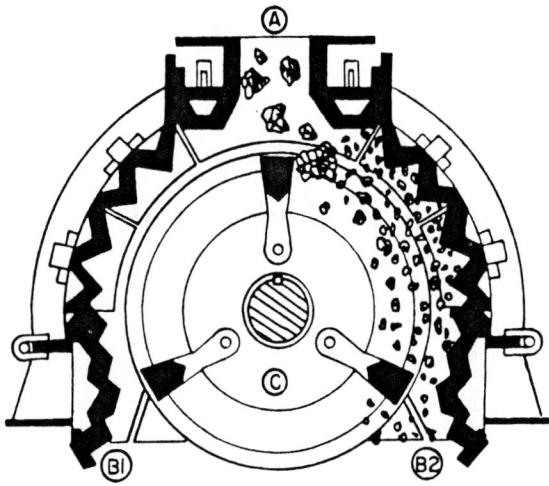

FIG. 4—4. Reversible impactor for pulverizing coal. Legend: (A) Feed chute. (B1, B2) Breaker plates. (C) Rotor. (Courtesy, Pennsylvania Crusher Company Division of Bath Iron Works Corporation.)

The degree of pulverization is controlled by the rotational speed and the spacing between the cages. The advantage of this machine is that it provides a high level of pulverization and good control of top size, with low levels of fine-coal generation. The disadvantages are its susceptibility to damage by tramp metal and plugging by softer material such as wood and rags. Each cage-mill installation must be protected by installing both magnets and metal alarms on the conveyors ahead of the pulverizers. With this protection, the cage mill is probably the best pulverizer for coal preparation in coke plants.

Constant monitoring of the performance of the crushers by screen analysis of the crushed-coal product is essential if consistently high quality coke is to be produced. When the pulverization level falls, immediate corrective action must be taken. At one United States Steel coke plant with the pulverization level target of 82 per cent minus 3-mm (minus ⅛-inch), coke stability (ASTM index) is 55. If pulverization is allowed to fall to 70 per cent minus 3-mm (minus ⅛ inch) a loss of 2 stability points to 53 would result.

Bulk-Density Control—The last step in the process of preparing coal for coking is adjustment of the bulk density of the coal blend. If some form of control is not practiced, the bulk density of the coal blend charged to the ovens could easily vary as much as 64 or 80 kilograms per cubic metre (4 or 5 pounds per cubic foot) over short periods of time (minutes) and even more over longer periods as weather conditions might affect moisture content of the coal. Additionally, without control, the average bulk density of the coal blend would usually be significantly lower than desired. This is particularly true considering the high coal-pulverization levels now considered to be standard practice at all coke plants.

A coke battery cannot be operated smoothly and efficiently without bulk-density control. It would be im-

FIG. 4—5. Cut-away view of a cage mill for pulverizing coal. (Courtesy, T. J. Gundlach Machine Division.)

possible to heat the battery properly, coal level in the ovens would be variable, stickers and oven damage by high coking pressure and insufficient contraction could occur and coke strength would be variable.

Bulk density of the coal blend is affected by moisture content, as shown in Figure 4—6. For example, coal with a bulk density of 700 kilograms per cubic metre (43.7 pounds per cubic foot) containing 8 per cent moisture actually contains only 644 kilograms of dry coal per cubic metre (40.3 pounds per cubic foot). Consequently, it is desirable to maintain a constant moisture content in the coal blend by adding water if necessary.

Most bulk-density-control systems are based on the use of No. 2 or diesel grade oil, or other similar oils, to increase bulk density of the coal blend. In a manual system, bulk-density measurements are made regularly (usually every hour) by hand in an ASTM bulk-density box, and the oil-flow rate is adjusted manually as required to hold bulk density at the control point.

There are two systems for automatic control of bulk density; namely, mechanical weigh belts and gamma-ray units. The bulk density of the coal as it leaves the pulverizer or coal mixer is measured continuously by either of these systems and the amount of oil needed to attain the control point is automatically adjusted. The oil is sprayed onto the coal stream just ahead of the pulverizers or mixers where it is distributed and mixed with the coal blend.

The gamma-ray unit when applied to measuring bulk density of coal is a very complex system in which many factors influence the measurement. Some of the more important factors are: (1) the depth of the coal on

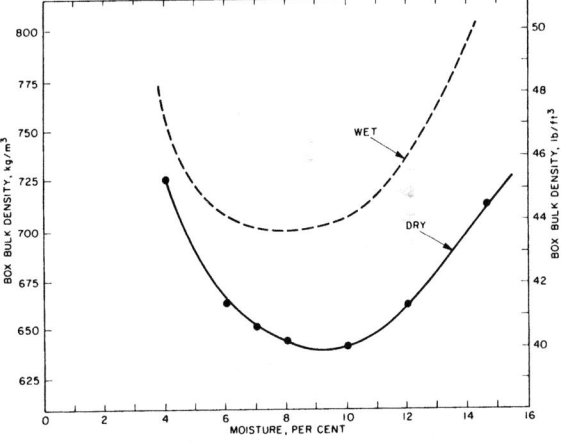

FIG. 4—6. Effect of moisture content on the bulk density of a selected (Concord) coal.

the belt as it passes under the radioactive source; (2) changes in the radiation-absorption coefficient of the coal; (3) dust or other material in the signal path; (4) temperature of the detector; (5) thickness and tension of the conveyor belt; and (6) size consist, moisture content and temperature of the coal. The advantages of the gamma-ray system are that it is less expensive than the weigh-belt system and is less expensive to operate and maintain. Both systems, properly maintained, can perform with an accuracy of at least plus or minus 16 kilograms per cubic metre (one pound per cubic foot).

Figure 4—7A shows the effect of oil addition on coal bulk density for one coal blend used by United States Steel and Figure 4—7B shows the changes in coke stability which result from changes in coal pulverization, bulk density and coking rate for the same coal blend.

The use of oil is not the only means available to increase and/or control coal-blend bulk density. In Japan, many coke plants add water to the coal in the stockpiles as it is being conveyed into the plant to control moisture content to about 9 per cent. This is done primarily to control dust, but the operators also aim for uniform moisture content. This uniform moisture content, in combination with a uniform coal-blend pulverization level, results in a uniform bulk density for the coal blend. In most cases, the bulk density is quite low, averaging only about 688 kilograms per cubic metre (43 pounds per cubic foot). Some Japanese plants briquette about one-third of their coal blend; this increases bulk densities and also improves coke quality (see Section 5). In the Saar region of West Germany, a process called "stamp charging" is employed to increase the bulk density of the fine, high-moisture coals used in that area (see Section 5). In this latter process, a large compacting device located on the pusher side of the battery is used to compress the coal blend into a rectangular mass slightly smaller in size than the coking chamber of an oven. This "stamped charge" is then

"inserted" into the oven from the pusher side. Bulk densities as high as 1040 kilograms per cubic metre (65 pounds per cubic foot) are not uncommon when using this practice. Coal preheating can also be considered as a means of controlling bulk density since the moisture content (zero per cent) and the pulverization level are controlled (see Section 5).

Coal Mixing—The use of the pulverizer for mixing blended coals can be considered adequate provided the coals are layered properly on the conveyor feeding the pulverizer. However, hammer-type pulverizers provide little or no mixing from side to side; so, if one of the coals is loaded into one side of the conveyor, it will come out of the pulverizer in the same location and will not be mixed with the rest of the coals. If this situation exists because of the arrangement of the coal conveyors and weigh feeders, it can be partially corrected by the use of belt plows or belt mixers.

Some plants install continuous mechanical mixers after the pulverizers and prior to conveying the coal to the coal bunker to ensure thorough mixing of the

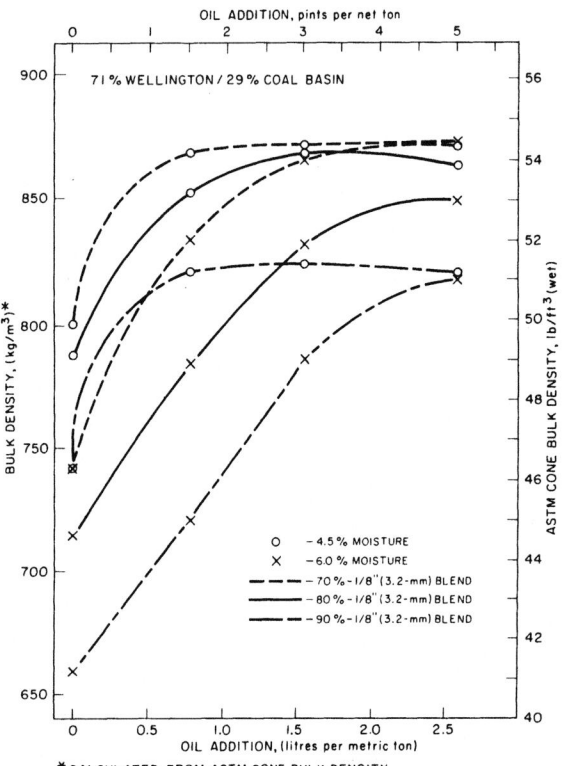

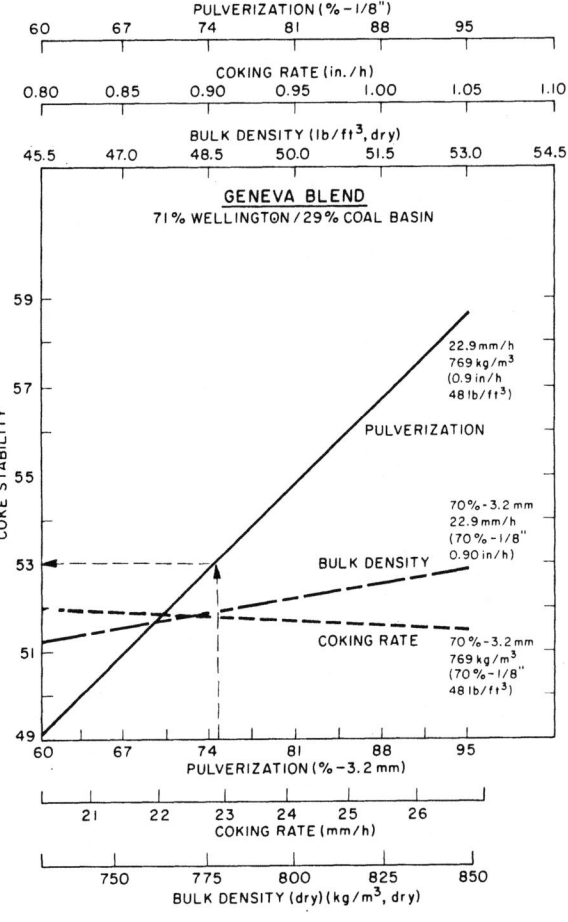

FIG. 4—7A. Bulk density response to oil additions for a coal blend coked at the Geneva Works of United States Steel Corporation.

FIG. 4—7B. Changes in coke stability resulting from changes in coal pulverization, bulk density and coking rate for a coal blend used at the Geneva Works of United States Steel Corporation.

blended coals. There are numerous types of solid-solid continuous mixers such as ribbon, paddle and zig-zag blenders. The choice of mixers for a specific installation is a function of many variables such as throughput, coal size and moisture content, retention time required, space availability, cost, and so on, and the choice is usually made after an engineering evaluation.

SECTION 3

METALLURGICAL-COKE PRODUCTION PROCESSES

A. THE BEEHIVE PROCESS FOR CARBONIZING COAL

Until 1918, most of the coke produced in the United States was made in beehive coke ovens. Beehive ovens usually were located close to coal mines so that freshly mined coal could be charged directly into the ovens; no provision was made for stocking or mixing coal at the oven site.

A beehive coke oven was a circular, domed, firebrick structure with a flat floor that sloped toward the door through which coke was removed when the coking process was completed. An opening called the **trunnel head** in the center of the domed roof permitted coal to be charged into the oven and allowed the gases generated during the coking process (called **foul gas**) to escape into the atmosphere. A typical oven was about 3.7 metres (12 feet) in diameter and processed about 4.5 to 6.3 metric tons (5 to 7 net tons) of coal per charge. From 48 to 72 hours were required to complete a coking cycle.

Beehive-coke-oven plants were constructed according to three general arrangements: (1) the **bank system**, in which the ovens were built in a single row against a bank of earth, natural or artificial, with a retaining wall in front of the ovens; (2) The **single-block system**, which consisted of a single row of ovens, all facing the same way, with retaining walls at both the front and back of the row (Figure 4—8); and (3) the **double-block system**, in which the ovens, in a double row, were built back to back or staggered with a retaining wall extending along the front of each row. In all cases, the space above the ovens between the bank and the retaining wall or between the two retaining walls was filled with earth. Tracks were laid on top of the rows to carry **larry cars** from which coal could be charged directly into the ovens; other tracks at yard level in front of the ovens carried cars which received the coke as it was removed from the ovens.

Charging—Beehive ovens were charged as soon as practicable after drawing the coke from the oven at the end of the previous cycle, in order that stored-up heat from the previous charge would be sufficient to start the coking process. New ovens had to be heated up gradually to the coking temperature by wood and coal fires, after which small charges of coal were coked until the ovens reached normal working conditions. With the oven in readiness for charging, the door was partially bricked up and the coal charge was dropped through the trunnel head from the larry car above, leaving the coal in a cone-shaped pile in the oven. In order to secure uniform coking of the coal, this pile had to be leveled so that the coal would lie in a bed of uniform depth of about 460 to 610 mm (18 to 24 inches) over the entire floor of the oven. This leveling was done by machine or by hand. In works not equipped with a machine, the leveling was accomplished with a long-handled scraper, operated through the door of the oven, which was purposely bricked up to only two-thirds of its height at the time of charging. After leveling the coal, the door opening was bricked up to within about 38 mm (1½ inches) from the top.

Coking Process—The coking process began very soon after leveling was completed, as the oven retained enough heat in the brick of the walls and the earth fill to start liberation of volatile matter from the coal. As the temperature of the coal charge increased, the temperature of the combustible volatile gases soon reached the "kindling" (ignition) point and, in the presence of the air admitted to the oven, they ignited with a slight explosion at first and then continued to burn quietly in the crown of the oven or as small candlelike flames at the surface of the coking mass, thus supplying heat to continue the process. Coking proceeded from the top of the coal downward, so that the coking time depended upon the depth of the coal. The generation of gas from the coal rapidly approached a maximum which was maintained for a period and then declined to practically nothing. The burning of the volatile matter during this period was regulated by gradually closing up the opening at the top of the door for the admission of air. This regulation was necessary to maintain the temperature at a maximum and conserve coke, as an excess of air at the beginning of the coking period tended to cool the oven and later consumed some of the carbon of the coke. The yield was also reduced by improper leveling. If the coal was not of uniform depth, the thin portions coked through before the thick, and some of the thin sections was consumed while the coking of the thick sections was being completed. On the other hand, if the process was stopped when the thin areas had coked through, there was a loss due to uncoked butts on the thick areas. Coking proceeded downward from the top of the charge, in which the coal at increasing depths passed through a plastic stage as the temperature rose. This produced expansion and contraction of the charge with the result that the coke was ramified by a great number of irregular vertical fissures that divided the coke mass into very irregular columnar pieces that extended from the top to the bottom of the mass.

Watering and Drawing—At the end of the coking period, the brickwork closing the door was torn out and the coke was **watered out** by spraying with a stream of water directed through the door of the oven or, in later ovens, by the use of a spraying device inserted through the door. After watering, the coke was removed from

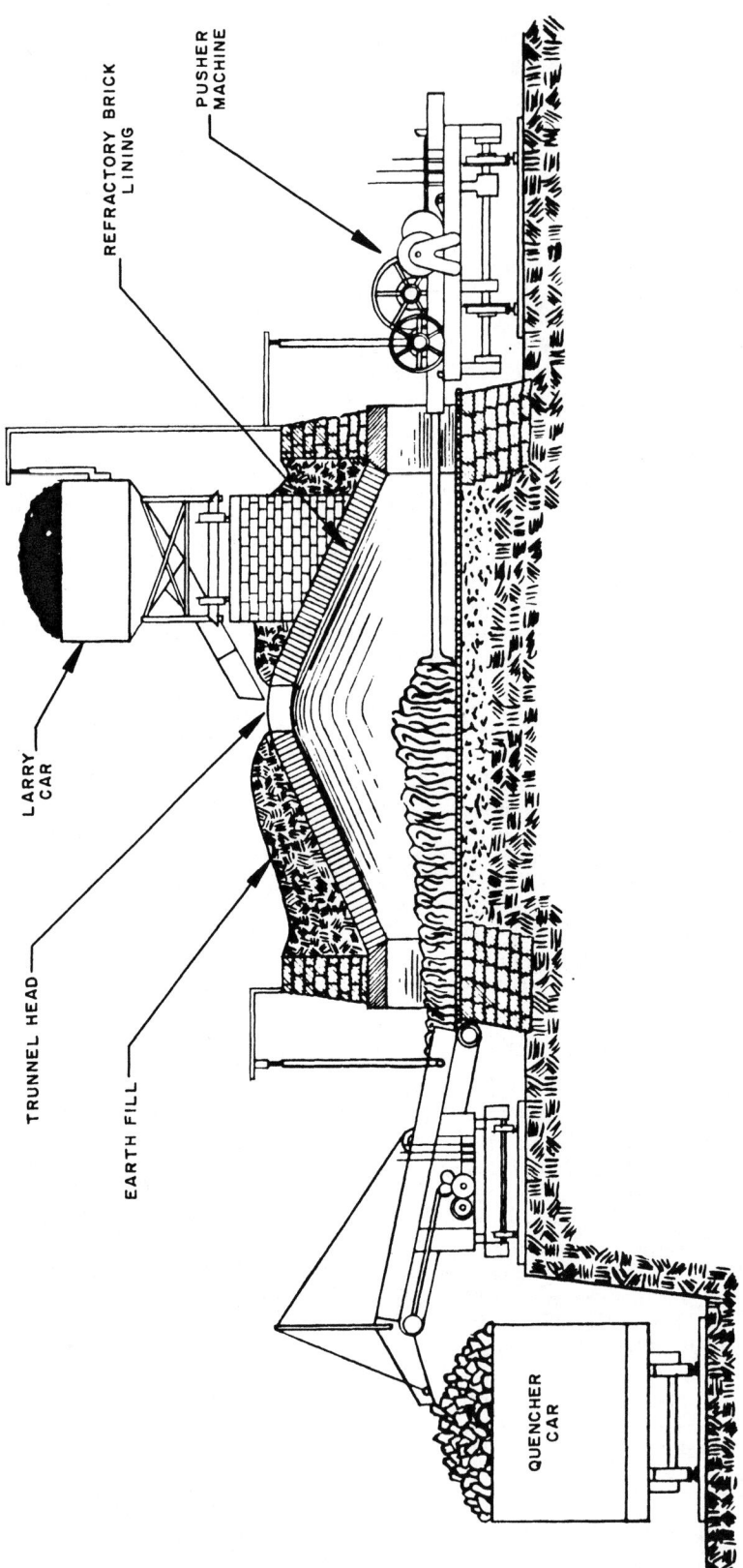

FIG. 4—8. Ideal section of a beehive coke oven in a single-block battery, showing refractory lining, earth fill, larry car, pusher, quencher car and other details, including the trunnel head at the top through which the coal was charged and the volatile products escaped. Vertical lines in the coke bed indicate the fissures that developed during coking, giving beehive coke its characteristic columnar structure.

the oven by hand or by machine. Figure 4—8 shows the use of a pusher to remove the coke.

The beehive process has virtually been phased out of production in the United States. Lack of coals suitable for the process and more stringent rules regarding air pollution were major factors causing decline of the process.

B. THE BY-PRODUCT PROCESS FOR CARBONIZING COAL

Introductory—By-product coke ovens are those designed and operated to permit collection of the volatile material evolved from coal during the coking process, as opposed to beehive and other types of ovens that allow the volatile products to escape. An overall view of a by-product coke-oven battery is shown in Figure 4—9. A variety of valuable substances are recovered from the volatile material by suitable chemical processes, as discussed in Section 7 of this chapter.

Because of the physical dimensions of the coking chamber of by-product ovens (narrow, long and tall) they sometimes are referred to as **slot ovens**. They are called **recovery-type ovens** also, because their design permits recovery of the volatile products of the coking operation.

In this section, reference is made to "furnace plants" and "merchant plants" for by-product coke making. Furnace plants generally are owned by or affiliated with iron- and steelmaking companies and produce coke mainly for use in the blast furnaces of the compa-

nies. Merchant plants include those that manufacture metallurgical and industrial grades of coke for sale on the open market; those associated with chemical companies or gas utilities; and those that supply only a small part of their output to blast furnaces of local iron works with which they are affiliated.

FOUNDRY-COKE PRODUCTION

While the discussion in this section will be primarily about plants producing blast-furnace coke, it would not be complete without some reference to facilities and procedures for making foundry coke.

In 1981, there were 14 merchant coke producers in the United States who made foundry coke for melting iron in ironmaking cupolas. Sixteen operating coke plants involving 33 coke-oven batteries were available for this service. Annual capacity for producing foundry coke from these batteries was estimated to be 3.2 million metric tons (3.5 million net tons).

Annual consumption of foundry-grade coke has been in the 2.9-million metric ton (3.2 million net ton) range; however, requirements have declined since 1977. A consensus exists that annual foundry-coke requirements will average approximately 2.3 million metric tons (2.5 million net tons) in the near future.

The production of coke for use in foundry cupolas generally involves the use of a blend of three to five coking coals in which the volatile content of the blend is in the range of 20 to 24 per cent. Coking tempera-

FIG. 4—9. Overall view of the pusher side of a typical by-product coke-oven battery.

tures are significantly lower than in the production of blast-furnace coke and seldom exceed 1150°C (2102°F) in the heating flues. Normal coking cycles are longer than in blast-furnace-coke production, usually in the 28 to 30 hour range. However, it is not uncommon to exceed 32 hours as a maximum and 23 hours as a minimum coking time.

Coke size requirements for use in cupolas are a function of the cupola diameter, and a "rule of thumb" generally gives a 10-to-1 ratio of cupola diameter to nominal coke size. Nominal sizes for foundry-coke shipments are usually: (1) 228-mm (9-inch) top size to 152-mm (6-inch) bottom size; (2) 152-mm (6-inch) top size by 101-mm (4-inch) bottom size; and (3) 101-mm (4-inch) top size by 76-mm (3-inch) bottom size. A combined size of 228-mm (9-inch) by 101-mm (4-inch) is frequently shipped.

Proximate analysis of foundry coke normally shows less than 1 per cent volatile matter, less than 0.72 per cent sulphur, less than 6.5 per cent ash, a low moisture content of about 1 per cent, and a fixed-carbon content of 92 per cent or better.

Most foundry coke is shipped in railroad hopper cars; however, a significant amount is trucked to users, resulting in less than one day delivery and the possibility of maintaining minimum coke inventories at the foundries.

EVOLUTION OF BY-PRODUCT OVENS FOR CARBONIZING COAL

Early Developments—Construction and operation of by-product coke ovens in the United States lagged behind European development. The production of manufactured gas from coal in retorts was started on a commercial basis in both England and the United States in the early 1800's. Tar, ammonia liquor and gas were recovered in this operation. These ventures, however, were unrelated to the manufacture of coke.

Prior to and for fully 50 years after industrial installations for making manufactured gas, repeated efforts were made to develop a single facility that could produce coke similar in quality to that made in beehive ovens and also provide for recovery of products such as those obtained in gas-producing retorts. Attention was being directed similarly during this period to the design of an oven that utilized a principle of coking different from those currently practiced. These developments were confined largely to Europe because of the inadequacy of the methods then used in producing coke of suitable quality from some of the more weakly coking coals known at that time to exist in various foreign countries but not in the United States.

In 1856, in Commentry, France, **Knab** built a group of retort ovens with the principal purpose of producing gas, tar and ammonia. The gas from the ovens was freed of tar and ammonia in suitable chemical apparatus and then burned in heating flues under the ovens to provide the heat for carbonizing the coal. The Knab ovens were 7 m (23 ft) long; 2 m (6 ft, 6¾ in.) high; and 1 m (39½ in.) wide.

Prior to this time, however, an oven that embodied vertical flues in its walls was developed in France and used both in that country and in the Ruhr district of Germany for producing coke. The purpose of that design was to exclude air from the coking chamber to the greatest degree possible during the carbonization of coal. The volatile gases were drawn by stack draft through orifices in the top of the coking chamber into vertical flues in the oven walls. Gas from the ovens and air were mixed and burned in a downward direction in these flues and passed to the stack through sole flues under the oven. Known as the **Francois** oven when built in France and as the **Rexroth** oven when used in Germany (hence the combination name, **Francois-Rexroth**), this so-called **waste-heat oven** was 8 m (26 ft) long, 1.5 m (5 ft) high, and 0.9 m (35 in.) wide.

In the earliest development of the industry, therefore, various facilities were developed for different purposes. From the standpoint that the Francois-Rexroth oven was a rectangular, vertical coking chamber, which embodied vertical heating flues, it could lay claim to the distinction of being the forerunner of the modern coking oven, notwithstanding the fact that it was not of the recovery type and that it was designed primarily to overcome some of the objections to the then prevailing coking method. The Knab oven similarly could be placed in the category of a pioneer among recovery-type ovens, despite its shortcomings in heating facilities. Both the Francois-Rexroth and the Knab ovens were departures from the then contemporary practice in that air was excluded from the coking chamber and external heat was supplied for carbonization.

Development of Vertical-Flue Ovens—The course that was pursued later in improving oven performance led to two separate and distinct developments.

Carves of England, using the Knab principle, developed an oven in 1862 (**Knab-Carves** design) that embodied horizontal heating flues in the oven walls. This oven was an improvement over the Knab design, providing for greater yield of coke and requiring less time for carbonization. The principal contribution by Carves, however, was the addition of an exhauster in the chemical-recovery apparatus. In the Carves oven system, gas was drawn from the oven by the exhauster, pumped through the chemical-recovery apparatus, and returned to the ovens for use as fuel, the present method of recovering coal chemicals being based on this same principle. Because of the inclusion of horizontal flues in the oven walls, the Carves system can claim the distinction of being the originator of the horizontal-flue system of oven heating.

Almost simultaneously with the Carves development, **Coppee** in Belgium, using the basic principles of the Francois-Rexroth oven, introduced the innovation of reducing the width of the oven and providing additional vertical heating flues in the oven walls. The ovens built by Coppee at that time were about 9 m (30 ft) long; 1 m (3 ft, 7 in.) high; and 0.5 m (18 in.) wide. In these ovens, stack draft was used to draw the gases of combustion from the vertical downdraft flues through waste-heat boilers, recovering the sensible heat as steam. The ovens built by Coppee embodied 28 vertical flues, and one of the chief contributions of this builder was the control of combustion by dampering the air admitted to the flues. Since the ovens built by Coppee conformed more nearly to present-day conventional-width ovens, and also contained a substantial number of vertical flues in the oven walls, this oven may, in some respects, be considered the true predecessor of the modern coke oven.

Up to that time, all these developments took place in Europe, although ovens of the Coppee or Belgian design were built and operated in the United States. The installations were made at both the Johnstown plant of the Cambria Iron Company and the Hollidaysburg plant of the Blair Iron and Coal Company, both in Pennsylvania.

Ovens of the Coppee design were built in the Ruhr district of Germany by **Dr. C. Otto and Company,** who added further refinements to the earlier Francois-Coppee design. Attempts at converting these ovens from the so-called waste-heat type to recovery-type by-product ovens have been credited to this builder. In the ovens of Otto design, the volatile products from the coal were removed from an opening in the top of the oven and the gas, after passing through the chemical-recovery apparatus, was returned to the flues in the walls of the ovens. The flow of combustion gases in this oven was different from that of the early Coppee design, with products of combustion rising and descending in alternate flues, thus forming the basis for the later ovens of "hairpin flue" design.

Heating of these ovens was rather crude and inefficient. Nevertheless, this early **Otto-Coppee** development is significant because it was the first vertical-flue oven of the recovery type built with flues completely surrounding the oven chamber.

Horizontal-Flue Ovens—Evolution of the horizontal-flue oven was also proceeding, but improvements were largely in the direction of by-product recovery rather than in coke-oven design. The basic design of the Carves-Knab oven was improved later by **Heussner** in Germany and **Simon** in England. Heussner, in 1881, developed the **Heussner-Carves** oven along more conventional lines by reducing the width and improving the heating system, making available a small quantity of surplus gas. Following this development, recovery-type ovens were established more firmly in Germany.

Simon, in England, also revised the Carves-oven design and, in 1883, erected a plant in Durham, England embodying recuperators for the transfer of heat from waste gases to incoming air. A reduction of coking time to 48 hours was the principal contribution of this **Simon-Carves** oven.

In 1885, **Louis Semet,** chief engineer of the Brussels Gas Works, and his brother-in-law, **Ernest Solvay,** built twenty-five 4-ton ovens near Mons, Belgium. The **Semet-Solvay** design (Figure 4—10) was based on the Carves principle of horizontal flues. Solvay's interest involved supplying the lime kilns with coke and the soda process with ammonia. It is of interest to recall that the ovens of horizontal-flue design did not embody regenerators, such facilities being included sometime later.

The **Rothberg** oven, a half-divided horizontal-flue oven, was developed in 1902; the first of these ovens in the United States began operation in 1903.

Improvements in Heating Facilities—Despite the fact that ovens of Coppee, Knab, Carves, Heussner, Semet-Solvay, Otto and others represented appreciable progress compared with the simple beehive system,

such facilties were lacking in heat economy. Temperatures in the ovens were not high enough to produce an acceptable metallurgical coke, notwithstanding return to the oven for heating purposes of all or nearly all of the volatile gases from the coal. These ovens left much to be desired in uniformity of temperature. Their coke product did not compare favorably in quality with that produced in beehive ovens. As a result of inferior quality and lack of uniformity of the produced coke, the use of recovery-type by-product ovens for producing metallurgical coke was viewed with skepticism, until further improvements were made in both design and operation.

The criticism leveled against this system of coking resulted in greater efforts to improve the facilities. In the ovens of vertical-flue design, attention was directed to the improvement of the heating system. Following the invention of the Siemens regenerator for preheating combustion air used in steelmaking furnaces, a patent embodying the use of this principle on coke ovens was obtained by **Gustave Hoffmann** in Germany. This application brought about far-reaching improvements in by-product ovens.

The Otto firm purchased the patents from Hoffmann and in 1883 added these facilities to the **Otto-Coppee** oven. This type, known as the **Otto-Hoffmann** oven, (Figure 4—10) was equipped with regenerators, parallel to the length of the battery, for preheating both combustion air and fuel gas. The deposit of carbon from the cracking of preheated coke-oven gas in the regenerators soon dictated abandonment of fuel preheating. Other structural and operational disadvantages required an early revision of the basic Otto-Hoffmann design but the basic principles of the oven were of unquestionable value as witnessed by their widespread use in later years.

In order to provide further regularity of temperature and to increase the capacity of recovery-type ovens, **G. Hilgenstock** in 1895, using the Otto-Hoffmann design as a base, fundamentally altered the oven originally built by Otto. The changes made by Hilgenstock consisted of providing gas inlets to the combustion flues beneath the oven base. Air for combustion was added through small flues provided in the side of the brickwork. The upper terminus of the gas ducts was in the base of the vertical flues at the approximate level of the oven floor. The use of the under-oven principle for controlling admission of fuel gas eliminated the gas flues, which in previous designs were a source of difficulty. This was a marked advance, although the underjet principle of fuel-gas control was not exploited until much later.

To obviate some of the criticism of the fairly excessive initial investment cost of the recovery-type oven, which in some respects resulted from the inclusion of regenerators, this feature was eliminated from the Otto-Hilgenstock design. This phase in the evolution of the recovery-type oven is significant because it provides a clue to the rather slow adoption of this procedure by the iron and steel industry. From an economic standpoint, the beehive oven was cheap to build, could be operated with a minimum of help, and was capable of being shut down or restarted with little or no difficulty. The recovery-type oven, however, was costly to

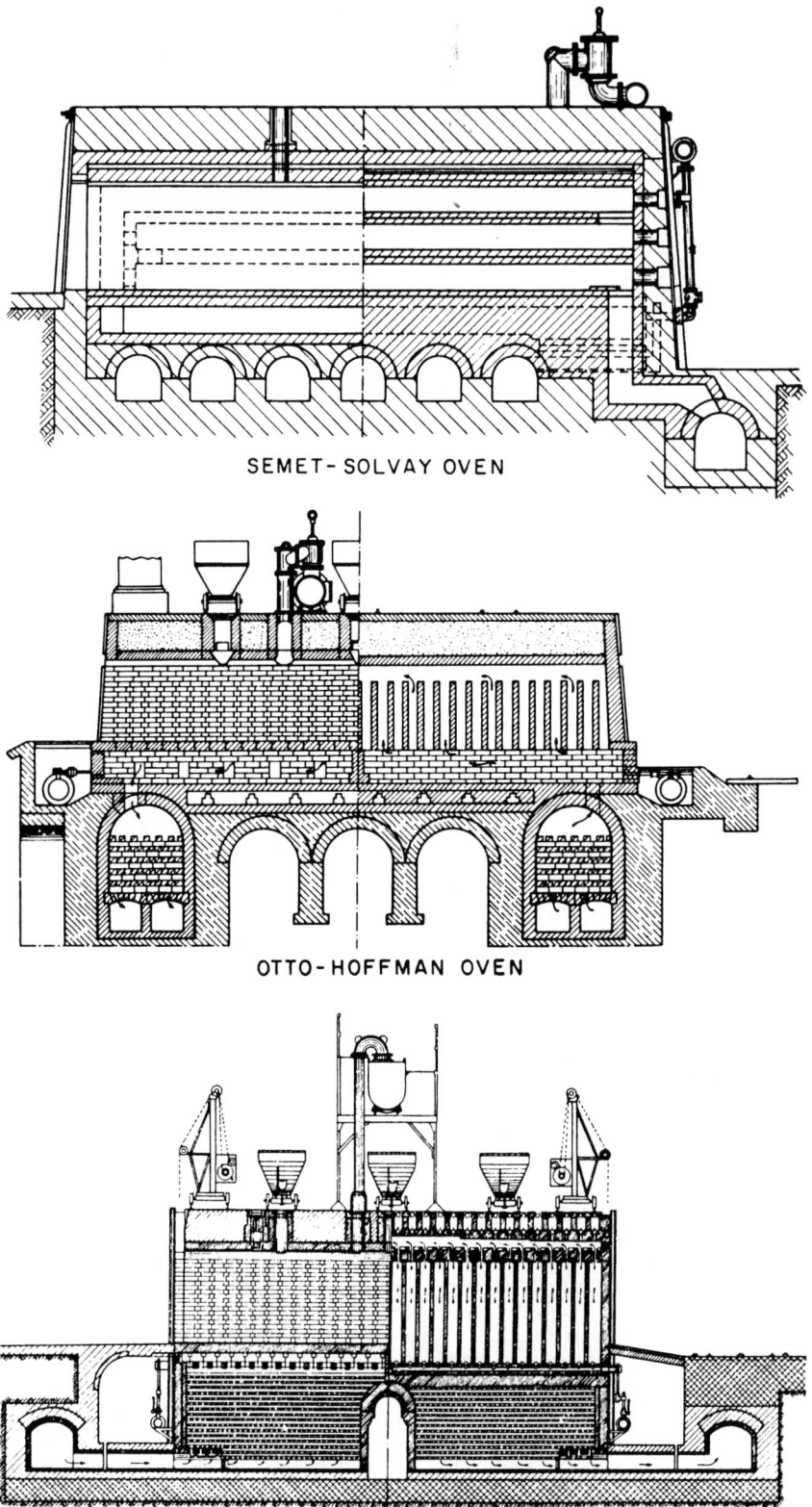

SEMET-SOLVAY OVEN

OTTO-HOFFMAN OVEN

KOPPERS ORIGINAL CROSS-REGENERATIVE OVEN

FIG. 4—10. Early designs of recovery-type coke ovens (not to same scale). (Courtesy, Koppers Company, Inc.)

build and difficult as well as expensive to operate, because of the size of the units and the need for skilled personnel. Consequently, the incentives for building recovery-type ovens were lacking in the steel industry until the economic advantages of the ability of the integrated steel plant to use the excess gas had been clearly established. The industry then began widespread adoption of this process.

Later installations of the Otto-Hilgenstock oven (also variously known as the **United-Otto** or **Schniewind** oven) were equipped with regenerators. A still further modification of this oven included Cowper regenerative stoves, where special requirements indicated their use.

Until the end of the Nineteenth Century, all the ovens equipped with regenerators were built with these facilities parallel to the length of the battery. The operational difficulties in regulation and control of air to the individual vertical heating flues presented many problems. An oven design by **Heinrich Koppers** (1904) was developed to overcome such difficulties. In the Koppers design (Figure 4—10) the regenerators were placed directly underneath the heating flues and parallel with the length of the oven; called **cross regenerators,** this type of regenerator became standard in the coke-oven industry following the expiration of the Koppers patents in 1918. The first oven of this type was built by the Illinois Steel Company (a subsidiary of United States Steel Corporation) at Joliet, Illinois and the first coke was pushed in September, 1908. The plant consisted of four batteries of 70 ovens each with coal-chemical recovery facilities.

Although silica brick as a coke-oven refractory was first used in this country by the Cambria Steel Company at Johnstown, Pennsylvania in the 1890's, a controversy existed between European and American designers as to its use for coke-oven wall construction. In order to obtain an answer to this problem, the various batteries at Joliet were constructed of different kinds of brick; No. 1 battery was constructed of silica brick and No. 2 battery was built of St. Louis quartzite, the raw material for which was composed mostly of Missouri clays. Quartzite brick imported from Germany was used in erecting Nos. 3 and 4 batteries. Results demonstrated the superiority of the silica brick and its use eventually became general.

Aside from the various constructional features, an innovation provided by Koppers was the inclusion of the gas gun flue, by which gas was fed to the base of each vertical flue through a ceramic flue and nozzles. The nozzles were removable and made with orifices of various sizes. This permitted control of the volume of gas to individual flues and provided for the replacement of nozzles when required.

The next significant development in by-product coke ovens was the erection of Koppers ovens for the Coal Products Company at Joliet in 1912. These ovens were arranged with combination cross regenerators to provide for simultaneous preheating of air and producer gas for coke-oven underfiring. By this procedure an additional amount of the richer coke-oven gas was released for sale. This type was the predecessor of the **combination oven,** so-called because the regenerators are arranged to preheat both air and fuel gas in separate chambers.

Difficulties were experienced with the single horizontal flue (bus flue) above the vertical heating flues through which the products of combustion passed from the heating flues to the downflow flues of the early cross-regenerative ovens because it had to be quite large to handle the volume of combustion gases generated.

This condition was aggravated as ovens of greater capacity were built and increasing use was made of lean fuels such as blast-furnace gas and producer gas. This problem was solved in various ways by different designers during the 1920's.

One solution to the horizontal-bus-flue problem was the **Koppers-Becker** design of oven which incorporated an entirely new system of heating, as shown later in Figure 4—42. In these ovens, first installed as an experimental battery in Chicago in 1922, gas is burned in all of the flues on one wall of an oven at the same time. The products of combustion from groups of two or more adjacent vertical flues of the "on" wall in which fuel gas is burning enter short horizontal bus flues and are thence conducted over the top of the oven coking chamber through crossover flues to a companion series of bus flues, whereby the vertical flues of the entire "off" wall are simultaneously conducting waste gases to the regenerator. On reversal, the opposite conditions obtain.

Another solution to the horizontal-flue problem is exemplified by the Wilputte double-divided oven (see Figure 4—42). The size of the flue is limited in these ovens by dividing it at the center of the oven. Each half of the flue then serves an inner and an outer zone of the heating system. Gas is burned alternately in the two outer zones with the products of combustion being carried by the horizontal flues to the two inner zones and thence to the regenerators. On reversal, gas is burned in the two inner zones and the products of combustion pass to the regenerators through the two outer zones; in effect, these designs constitute two short cross-regenerative ovens placed end-to-end.

Various improvements in the design of regenerative systems were made by different builders, and the systems employed in some modern by-product ovens are discussed in Section 6 of this chapter.

As vertical-flue ovens of greater height came into use, the uniform heating of the flues in the high walls from top to bottom became a problem, especially when heating with rich gas that has a characteristically short flame. Some of the solutions for this problem included: (1) dilution of fuel gas with waste gases to produce a longer flame (Koppers); (2) introducing combustion air in successive vertical stages along the height of the heating flue (Carl Still); and (3) using high and low burners in alternate heating flues (Wilputte). Application of these and other combustion-control principles to some modern by-product coke ovens also is discussed in Section 4 of this chapter.

The By-Product Oven in the United States— Although reasonable success had been achieved abroad with the by-product oven, it was not used in the United States prior to 1890. It is pertinent to recall that the first ovens of the by-product type installed in the United States were not for production of metallurgical coke. The distinction for launching the industry must be accredited to the Solvay Process Company, which was interested in establishing the soda-ash industry in this country. The plant that was built in Syracuse, New York and placed in operation in 1893 consisted of 12 Semet-Solvay ovens.

It was not until two years later that recovery-type (by-product) ovens were installed in the United States for producing blast-furnace coke. In 1895, the Cambria Steel Company at Johnstown, Pennsylvania built a battery of 60 Otto-Hoffmann ovens.

Until 1900, only nine plants with 1081 by-product ovens had been constructed in the United States. From 1900 through 1910, however, there was a period of rapid industrial expansion, particularly in the iron and steel industries, and the number of by-product coke plants more than tripled while the number of by-product ovens increased four-fold.

The need for coal-tar dyes and explosives in World War I added impetus to this expansion, and by 1918 the by-product coke industry had 59 plants with over 9000 ovens.

The construction of by-product ovens for gas utilities in the 1920's further expanded the industry and, by 1930, there were 89 plants with 12 771 ovens in existence.

Due to a world-wide economic depression, construction of by-product ovens was halted during the 1930's, and by 1940 there were fewer ovens in existence than during the early 1930's. However, World War II started a new construction era, and between 1940 and 1950, 2359 ovens were added to the industry total for ovens. The increase in total ovens, however, only partially indicates the vast construction program of this period because many of the ovens constructed were replacements or rebuilds of ovens originally built during World War I. Actually, over 5000 ovens were constructed during this period.

The construction of by-product ovens continued after 1950, and in 1958 the number of ovens reached an all-time peak of 16 244. Although ovens increased in number because of the expansion of furnace plants, the number of by-product coke plants in existence decreased to 77 during this period, because 7 gas-utility plants were abandoned as markets for coke-oven gas were lost to natural gas which became available in many areas through the extension of gas pipelines after World War II.

During the 1960's, 1970's, and early 1980's, the total number of by-product coke ovens in the United States started to decline, due to a steady reduction in steelmaking output, environmental controls, more efficient coke-to-hot metal ratios achieved in blast furnaces, and ovens with larger-capacity coking chambers. The total number of ovens in the United States as of 1983 had declined to 6350. Total cokemaking capacity in 1983 was 32 million metric tons (35.2 million net tons). Ten years earlier, cokemaking capacity had been 61.8 million metric tons (68 million net tons).

SECTION 4

GENERAL DESIGN AND OPERATING PRINCIPLES OF MODERN BY-PRODUCT OVENS

Principal Oven Components—The by-product coking process, being a true distillation process, involves the use of retort ovens. While there are many modifications, these ovens consist essentially of three main parts, namely: the **coking chambers,** the **heating flues,** and the **regenerative chambers**—all constructed of refractory brick. The following discussion of components applies generally to all types of ovens, but is related principally to those used in the United States.

Ovens are constructed in batteries that have contained from as few as 10 to over 100 ovens. In the United States, large batteries of 45 or more ovens generally have been preferred, while batteries of fewer ovens have been more common elsewhere.

Coking chambers in a battery alternate with heating chambers so that, in effect, there is a heating chamber on each side of a coking chamber. The regenerative chambers are underneath the heating and coking chambers. Separating walls between regenerators also serve as foundation walls for the heating and coking chambers. The entire structure is supported either from the ground or by columns under a reinforced-concrete or structural-steel base.

The coal is charged through openings in the top of the oven and, after the coal has become coke, the coke is pushed out from one end by a power-driven ram, or pusher (Figure 4—11). During the coking period, the ends of the coking chamber are closed by refractory-lined doors, which are constructed to completely seal the ends of the ovens. The ovens first constructed in the industry provided a space between the door and the jamb which was filled with a special luting mixture to seal the oven prior to charging. Later, several types of self-sealing doors were developed, which seal the opening when put in place and require no luting; several of these are described later in this section.

To permit the escape of volatile matter driven from the coal during coking, an opening is provided at the top of the oven at either one or both ends of the coking chamber. Each such opening is fitted with an offtake pipe, which connects the oven with the gas-collecting main for the battery.

The combustion chambers consist of a large number of flues which permit uniform heating of the entire length of the coking chamber. Ovens have been built with either horizontal heating flues or vertical heating flues, but vertical flues are used almost exclusively in present installations. Some of the older ovens employed the recuperative principle for preheating combustion air. Modern practice utilizes the regenerative principle to achieve higher thermal efficiency whereby less gas is required to heat the ovens. In all modern oven batteries, individual regenerators are provided for each heating wall and are located under each oven.

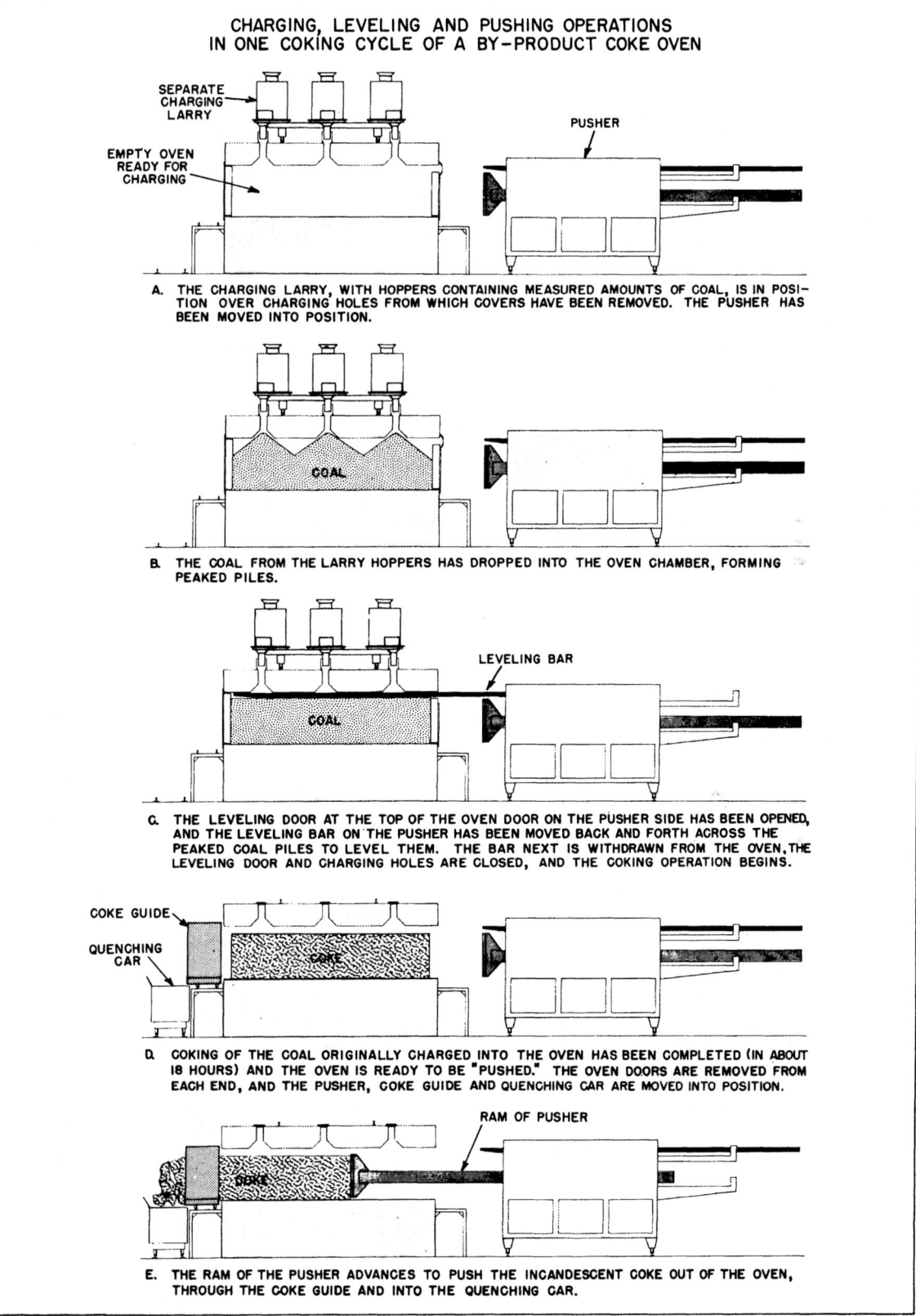

CHARGING, LEVELING AND PUSHING OPERATIONS
IN ONE COKING CYCLE OF A BY-PRODUCT COKE OVEN

SEPARATE
CHARGING
LARRY

PUSHER

EMPTY OVEN
READY FOR
CHARGING

A. THE CHARGING LARRY, WITH HOPPERS CONTAINING MEASURED AMOUNTS OF COAL, IS IN POSI-
TION OVER CHARGING HOLES FROM WHICH COVERS HAVE BEEN REMOVED. THE PUSHER HAS
BEEN MOVED INTO POSITION.

COAL

B. THE COAL FROM THE LARRY HOPPERS HAS DROPPED INTO THE OVEN CHAMBER, FORMING
PEAKED PILES.

LEVELING BAR

COAL

C. THE LEVELING DOOR AT THE TOP OF THE OVEN DOOR ON THE PUSHER SIDE HAS BEEN OPENED,
AND THE LEVELING BAR ON THE PUSHER HAS BEEN MOVED BACK AND FORTH ACROSS THE
PEAKED COAL PILES TO LEVEL THEM. THE BAR NEXT IS WITHDRAWN FROM THE OVEN, THE
LEVELING DOOR AND CHARGING HOLES ARE CLOSED, AND THE COKING OPERATION BEGINS.

COKE GUIDE

QUENCHING
CAR

COKE

D. COKING OF THE COAL ORIGINALLY CHARGED INTO THE OVEN HAS BEEN COMPLETED (IN ABOUT
18 HOURS) AND THE OVEN IS READY TO BE "PUSHED." THE OVEN DOORS ARE REMOVED FROM
EACH END, AND THE PUSHER, COKE GUIDE AND QUENCHING CAR ARE MOVED INTO POSITION.

RAM OF PUSHER

COKE

E. THE RAM OF THE PUSHER ADVANCES TO PUSH THE INCANDESCENT COKE OUT OF THE OVEN,
THROUGH THE COKE GUIDE AND INTO THE QUENCHING CAR.

FIG. 4—11. Schematic representation of the sequence of operations involved in charging,
leveling and pushing in one coking cycle of a by-product coke oven.

This permits separate control of the flow of preheated air for combustion to individual vertical-heating-flue walls, and allows close control of heating. An advantage of individual regenerators is that the control of heating for each oven is relatively independent of the operation of the remainder of the battery.

Coking-Chamber Design—The dimensions of the coking chamber are usually a compromise of many interrelated variables that will best suit the expected production requirements while making the best quality of coke within practical limits. Past experience with coals of similar properties is the best guide, as there is no well defined theoretical method of arriving at definite oven dimensions for specific coals.

In general, the dimensions of coking chambers have ranged from 1.82 metres (6 feet) to 6.7 metres (22 feet) in height, 9.14 metres (30 feet) to nearly 15.5 metres (52 feet) in length, and 304 millimetres (12 inches) to 550 millimetres (22 inches) in average width. Some recent ovens have coking chambers 15.25 metres (50 feet) in length and 6.5 metres (21 feet) in height. Generally, the preferred width of coking chambers in this country has been 460 to 488 millimetres (18 to 19 inches) average because (1) coking chambers of these widths normally produce coke of a size similar to that desired by blast-furnace operators, and (2) the time required to coke coal in chambers in this width range is acceptable. Narrower chambers (e.g., 356 millimetres or 14 inches wide) permit a more rapid rate of coking and are used to produce an acceptable blast-furnace coke from western United States coals that make poor blast-furnace coke when processed in wider chambers. However, the lower productivity of the narrower-width ovens precludes their use except where some such special conditions exist.

The coal is charged into the coking chamber through charging holes provided in the roof of the oven. The oven retort (coking chamber) and the heating system are designed to process a coal charge of definite volume, having a level top surface about one foot below the oven roof. The number of charging holes and the flow characteristics of the coal control the time required to charge an oven with coal.

To prevent escape of gases from the oven during charging in most plants, a steam-jet aspirator is used to draw gases from the space above the charged coal into the collecting main; this practice is called "charging on the main." Since this puts the oven under a slight negative pressure, air is drawn into the oven through the charging holes and leveler door. It is impossible to prevent the introduction of some air into the gas-recovery equipment; this is one of the reasons for keeping the time required for charging to a minimum. Excessive leveling of the charged coal not only extends the time during which the leveler door is opened, but also tends to pack the coal at the top of the charge, particularly under the charging holes, and may cause localized erosion of the oven wall.

The coking chambers are narrower on the pusher side and taper from 50 to 100 millimetres (2 to 4 inches) toward the coke side. The extent of taper depends upon the expanding properties of the coals to be coked. The heating walls on either side of the coking chambers are built of specially designed shapes of high-

quality silica brick, set in silica mortar that forms a ceramic bond at the higher operating temperatures. The coking-chamber floor may be of either first-quality fireclay blocks, or silica bricks.

Effectively, all of the heat for coking the coal is conducted through the coking-chamber liners. The coking, therefore, starts in the coal at the walls and progresses from both walls to the center of the oven. Since coking proceeds in this manner, the coke in the center is not fused together (Figure 4—12). This limits the length of any piece of coke to half the width of the coking chamber minus any shrinkage that may have occurred. The structure of the coke mass at the end of the coking period is that of two parallel slabs of irregularly interlaced pieces. These have sufficient strength to be pushed from the oven by the pushing ram and exert very little lateral pressure on the oven side walls. However, the oven walls must be designed with sufficient structural strength to resist a high lateral pressure should the structure of the coke mass be broken for any reason during the pushing. Additionally, these walls must be gas-tight to prevent any leakage of gases between the oven and the heating flues.

Heating-System Design—Practically all heating systems may be grouped into two general classes: the gun-flue type shown later in Figures 4—44, 4—45 and 4—46, and the underjet type shown in Figures 4—47, 4—48 and 4—49. In the gun-flue type the gas is introduced through a horizontal gas duct extending the length of each wall a little below the oven floor line. Short connecting ducts lead vertically upward to a replaceable nozzle brick at the bottom of each of the vertical heating flues. In the underjet type, the fuel gas

FIG. 4—12. View of coke being pushed from oven, showing central line of cleavage and block-like structure of the coke.

is introduced into each heating flue from the gas-distribution piping in the basement of the battery through gas ducts built integrally into the regenerator division and flue-supporting walls. Each of these separate gas risers is equipped with a regulating nozzle to control the flow of gas to each flue.

There are many variations of both general types. All attempt to heat the coal being coked at a controlled rate and temperature, uniformly from end to end of the oven, and from top to bottom of the charge (with the exception of the top few centimetres or inches which may be held slightly lower for the better control of coal-chemical yields) and at the lowest rate of heat consumption per kilogram or pound of coal carbonized.

In recent years, larger, faster operating ovens have been constructed. Designs have been keyed to higher production rates at lower operating costs, with the maximum use of automation and mechanization. The higher production rates have been achieved with larger ovens and denser brick, and with special design features described in Section 6 that deals with the more recent designs of by-product coke ovens.

Heating of individual ovens is controlled so that the temperature at the base of the heating flues does not exceed about 1425° to 1535°C (2600° to 2800°F). This is considered the maximum safe temperature range to which silica coke-oven refractories should be heated. With the flues operating within this temperature range, time required to coke the coal depends primarily upon the width of the coking chamber and the nature of the coals being coked, although a few other factors are involved. In general, a coking time is selected that will produce a uniform "skin temperature" of the block of coke in the coking chamber of from 1040° to 1095°C (1900° to 2000°F) at the time the charge has been coked all the way through to the center. The "skin temperature" referred to above applies to the coke adjacent to the walls of the coking chamber. The time required for coking coal under the above operating conditions generally varies from 16 to 20 hours. Normally, the average time is 17 to 18 hours.

Since the flue temperature should not exceed 1535°C (2800°F), the heating system must be reversed on a regular basis. The frequency with which the firing of by-product coke ovens is reversed generally has been established on a 30-minute cycle. Batteries which require a higher average temperature level to sustain high production rates can be reversed on either a 15-minute or 20-minute cycle without exceeding maximum permissible temperatures.

Of particular importance in flue design is the position of the gas inlet in the vertical flue in relation to the air port, and the manner in which the gas and preheated air are mixed at the base of the flue. Excessive turbulence at this point will result in quick combustion, and a short, intense flame will cause local over-heating at the base of the vertical flue.

Control of the rate of flame propagation and flame length in the vertical flues is desirable. This has been accomplished in several ways. In some batteries, blast-furnace gas is mixed with rich fuel gas to lower the heating value of the latter and alter its combustion properties. One system recirculates waste gas from the flues containing waste gas to the nozzle where fuel gas

is being burned. This feature is contained in the Koppers-Becker design (Figure 4—46). By this recirculation, the flame length can be controlled without cooling or purification of the waste gas and without having to heat the diluent as would be required with external mixing. The ratio of recirculated waste gas to fuel gas is controlled by the orifice size and the fuel-gas pressure, and normally approximates one volume of waste gas to one volume of fuel gas. Waste-gas recirculation also prevents the accumulation of carbon in the underjet gas ducts, as the contained carbon dioxide and water vapor both tend to inhibit carbon deposition. In ovens not recirculating waste gas, air must be introduced into the gas ducts during the "gas off" periods for decarbonization of the nozzles.

A feature of the Wilputte double-divided oven shown in Figures 4—48 and 4—49, especially for those over 10 feet in height, is the high-low burner construction. Low burners in flues alternating with high burners in adjacent flues prevent overheating at the bottom of the flues and improve the vertical heat distribution.

Gas for Heating—When blast-furnace gas or other gases having a low calorie (Btu) content (lean gases) are used for oven heating, supplementary heating with gas of a higher calorific value may be needed in order to maintain coke production at as high a rate per oven per operating hour as when firing with a straight high-calorie (rich) gas. It is not practical to add more than a limited amount of rich gas to the lean gas before the mixture enters the regenerators, as the rich component of the mixed gas can decompose (crack) partially while passing through the regenerators and cause an objectionable deposit of carbon. Because of this, any additional rich gas needed is fed through its own gas system for burning in the flues at rates needed to achieve the required temperature.

The rich fuel gas is heated to above its dew point. The fuel-gas mains located in the alleys on both the pusher and coke side (or in the basement in the case of an underjet battery) are insulated to limit condensation in the headers and to keep the gas at a uniform temperature throughout the length of the header.

Air for Combustion—In the gun-flue type of oven, the air required for combustion is taken into the sole flue at the base of the regenerator chambers through an air box equipped with small slats (finger bars) that can be adjusted to regulate the amount of air taken from the alleys. The temperature, the velocity and the direction of the wind may have a noticeable effect on the heating and should be compensated for. In the underjet type, the air for combustion may be taken from the enclosed basement and is therefore independent of wind velocity, direction and temperature.

In the Wilputte design of underjet ovens, the basements are sealed and kept at a constant air pressure with only the air required for combustion entering the basement through a wind tunnel extending along the entire length of the battery. When operating on a fast coking time with lean gas, a fan is used to deliver sufficient air to the wind tunnel. Spaced along the length of the basement are suitable openings equipped with regulating louvers to distribute the air uniformly throughout the basement. In this design, only the air required for combustion is available for basement cooling.

In Koppers-Becker underjet ovens, the air for combustion is introduced in much the same manner as that just described, with the exception that air in excess of the amount required for combustion is forced into the basement, the excess finding its way out through suitable openings around the buckstays on the pusher side (the wind tunnel being on the coke side of the basement). In this arrangement, air in addition to that required for combustion may be circulated through the basement for cooling.

Importance of Heat Control—Faulty heating affects not only the quality and quantity of the coke and coal chemicals produced, but also the ultimate life of the ovens. The most serious damage to the ovens is caused by fluxing or slagging of exposed brick surfaces due to local overheating beyond the critical temperature of the brick. This may occur in zones that are not readily accessible for repair. The advantages of an even and controlled heat throughout the oven cannot be overemphasized and remains a constant challenge.

When a new battery, or an old battery that has been allowed to go cold, is to be put into operation, great care must be taken in bringing the battery up to operating temperature. The major high-temperature portion of the battery is high-grade silica brick which has a high coefficient of thermal expansion at lower than operating temperatures (see Chapter 2 on "Refractories"). Therefore, the rate of heating must be slow enough to ensure maximum temperature equalization throughout the entire battery structure. In practice, the heating from cold to operating temperature takes from five to seven weeks. This same practice obtains in reverse when a battery of ovens is taken out of operation for repairs or extended shutdown and allowed to go cold. The cooling of a battery in preparation for refractory repairs has gained popularity in recent years.

The usual method of heating up a battery begins by burning gas in the coking chambers, using burners inserted through an opening in the door at each end of each oven. The products of combustion are allowed to enter the heating system at the horizontal- or bus-flue elevation through suitable openings provided for this purpose, which later are plugged and sealed. The hot gases during the initial or drying-out period are vented down through the vertical flues and regenerators to the stack flue and stack. When the flues become hot enough to ignite fuel gas, the gas is introduced through the normal heating ducts. Where other gas is not available for heating up, liquid butane or propane may be vaporized and used. In the earlier period, coal or coke has been used for heating up by substituting a brick bulkhead for the oven door, leaving openings for firing and ash removal. The same type of false hearth is used to protect the oven liner brick. After heating up, the false hearth and bulkhead are removed and the oven door is installed.

Oven Doors—As has been mentioned, the ends of the oven are equipped with removable, refractory-lined doors. After a coal charge is fully coked, suitable machinery located on both the pusher and coke sides of the battery is used to remove the doors and hold them during the pushing operation. After pushing, the doors are replaced and sealed preparatory to recharging the oven.

The original method of sealing the doors was to trowel and smooth ground "mud" into a V-shaped opening between the door and the door jamb. In recent years there have been developed self-sealing doors that do not require luting (sealing with "mud"). In principle, the self-sealing door has finally developed into a spring-loaded door that depends on a metal-to-metal contact between the door and the continuous machined-surfaced cast-iron or ductile-iron jamb.

Three popular types of coke-oven doors are:

(1) The Koppers door with a U-shaped sealing ring on a cast-iron door, with a series of spring-loaded plungers that force the sealing ring to conform to the oven-door jamb (Figures 4—13 and 4—14).

(2) The Wilputte fabricated steel door incorporating a heavy, reinforced channel frame with a sealing diaphragm with a renewable flexible stainless-steel sealing edge. Heavy-duty springs mounted at the locking bars and acting through a series of heat-resistant spring-loaded plungers provide a constant force to conform the sealing edge to the oven door jamb. The Wilputte leveler door is constructed with a heavy ductile-iron casting with insulation and a stainless-steel sealing edge that is adjusted by set screws. The leveler-door latch spring is made from a heat-resistant stainless steel (Figures 4—15 and 4—16).

(3) The Otto type self-sealing door with spring-loaded sealing strip (Figure 4—17).

Oven-door expense is a large factor in over-all oven repair and maintenance costs. This expense can be controlled by careful design of door-handling equipment and strict adherence to good operating practice. The lining of the door is usually sectionalized and made up of clay-brick shapes. In some plants the doors have a monolithic lining of lumnite cement made with an aggregate of various grades of crushed brick and ganister which give good service. The thickness of the lining and the position of the inside face of the lining relative to the end vertical flue of the oven are important as they influence the heating of the ends of the coal charge.

Gas-Collecting System—The oven may be equipped with one or two offtakes to carry off the volatile products liberated in the coking process. Where one offtake is provided, it is through the roof of the oven at one end of the oven, and where two are provided, there is one at each end of the oven. In either case the volatile products pass through the duct or ducts in the oven top and enter a refractory-lined standpipe (sometimes referred to as an "ascension pipe") which in turn is connected to a collecting main through a damper valve (Figures 4—18 and 4—19). Between the damper valve and the oven the standpipe is equipped with a cap valve, or "elbow cover," which, when open, vents the oven to the atmosphere.

The damper valve is usually a liquid-sealed valve, so designed that the cooling spray furnishes the seal when the damper valve is in the closed position, the excess spray overflowing into the collecting main. With this arrangement, the cooling spray is always on. The liquid, called **flushing liquor,** used in damper and collecting-main flushing, is the condensate from the volatile products driven off in the coking process.

A standpipe and valve arrangement used by Dr. C. Otto is shown in Figure 4—18.

In the Wilputte design illustrated in Figure 4—19, the standpipe is fabricated steel and the standpipe elbow is cast steel. The elbow is equipped with a spring-loaded locking cover and fitted with a renewable stainless-steel sealing edge that provides a gas-tight seal. The elbow can be fitted with a renewable flanged sealing surface.

The flushing-liquor spray nozzle is recessed in the standpipe elbow and protected by a shroud. The nozzle provides 360° spray coverage of the elbow interior, to maintain maximum cooling and cleaning. Easy access to the spray nozzles is provided.

The standpipe is lined with refractory brick or castable with a layer of insulation. The elbow is lined with anchored castable refractory. The lined standpipe and elbow maintain the oven offtake gas at a temperature high enough to minimize deposition on the walls until the spray in the elbow can cool the gas and flush any deposits into the collecting main.

A gas-tight floating elbow connection to the collecting main is accomplished by the use of a flushing-liquor seal. Additional sprays located in the collecting main provide a flushing spray to the damper-valve housing to maintain trouble-free valve operation. The height of the elbow cover permits inspection of the inside of the elbow and the spray pattern when the oven is "off the main."

All ovens of a battery thus are connected to a single or multiple pressure-equalized collecting main on either or both sides of the battery. The function of either the single or double collecting main is not only to collect gas from the ovens but also to maintain at all times an accurately controlled pressure in the oven during the coking process. Pressure in the oven during coking has a pronounced effect on the coke and coal chemicals. The pressure in the collecting main usually is kept at a point that will give about 2 mm (0.078 inch) water-gage pressure at the bottom of the oven at the end of the coking period. The pressure is regulated by a controller located in the connection between the collecting main and the suction main which carries the gas to the coal-chemical-recovery units. The collecting main also serves to transport to the tar decanters the products condensed from the gas by the flushing liquor.

The gas passing from the ascension pipe to the collecting main is shock-cooled with a flushing-liquor spray which causes (1) precipitation of tar from the gas, and (2) cooling of the gas to the desired temperature.

ACCESSORY OVEN EQUIPMENT

Coal-Storage Bins and Charging Larries—To provide coal for the ovens, every modern plant has an overhead coal bin at the ovens of sufficient surge capacity to permit flexibility in coal preparation without interference to the scheduled uniform operation of the ovens. The number and size of bins required are determined for each individual plant.

All modern by-product coke ovens are designed to take a definite volume of coal per charge, and are charged from a larry car operating between the over-

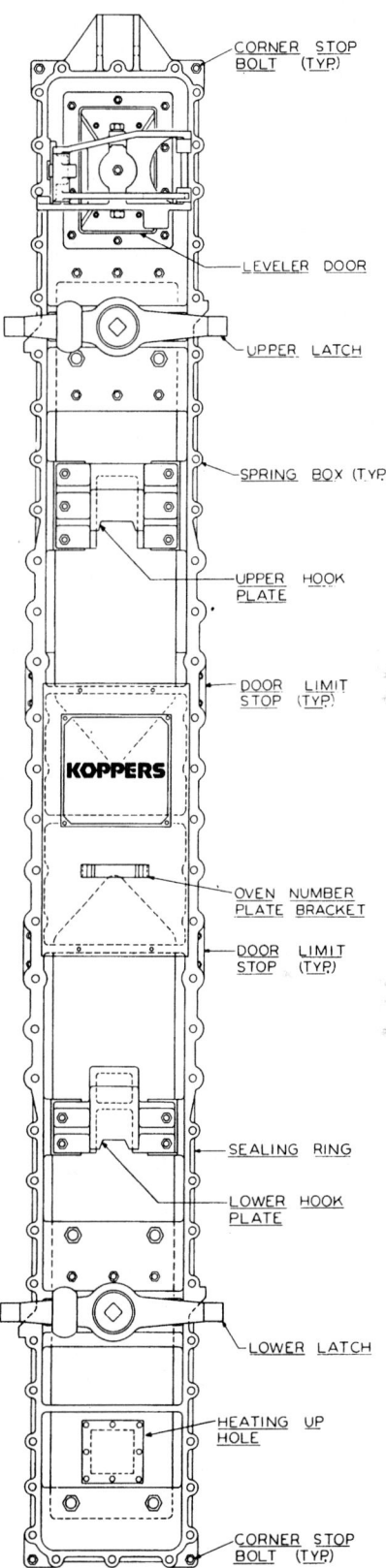

FIG. 4—13. Face view of a Koppers coke-oven door of the self-sealing type (see also Figure 4—14.) (Courtesy, Koppers Company, Inc.)

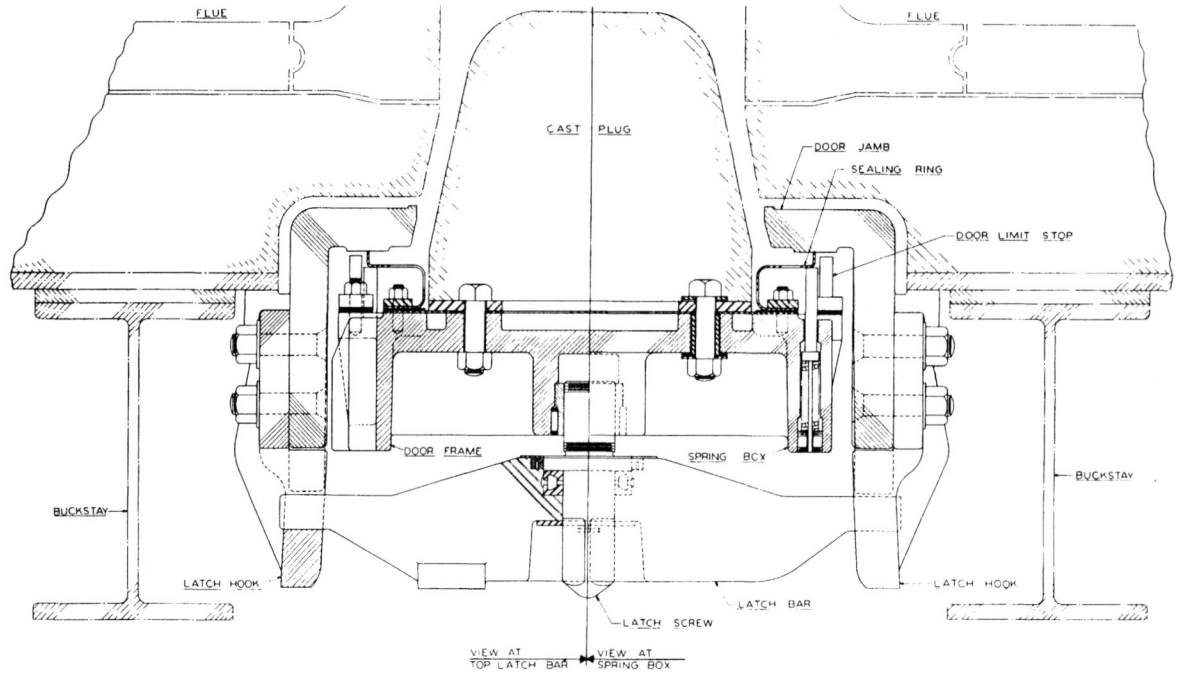

FIG. 4—14. Horizontal section of the self-sealing coke-oven door shown in Figure 4—13. (Courtesy, Koppers Company, Inc.)

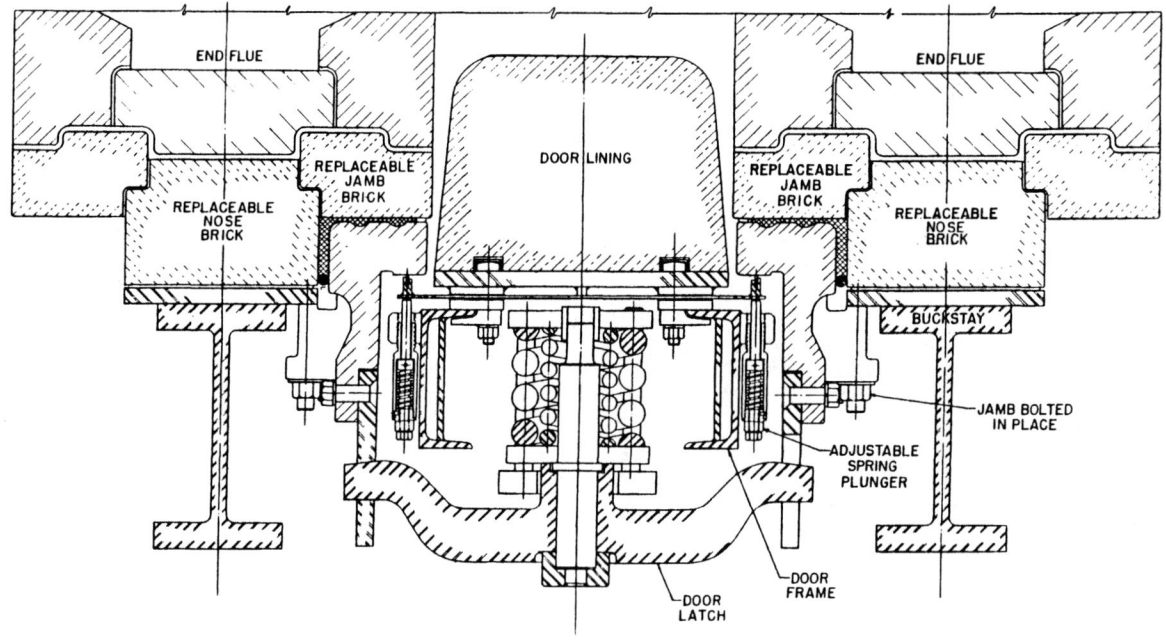

FIG. 4—15. Cross-section of a Wilputte door and door jamb arrangement. (Courtesy, Wilputte Corporation, a Salem Company.)

FIG. 4—16. Wilputte design of self-sealing coke-oven doors, showing spring-loaded bars that maintain metal-to-metal contact between the door and jamb. (Courtesy, Wilputte Corporation, a Salem Company.)

head coal-storage bin and the ovens on a track supported by the battery top (see Figures 4—20 and 4—21).

The desired amount of coal for an oven charge is drawn from the storage bin and is measured, usually by using either the track scales at the loading station, or by volumetric sleeves (choke boxes) on the larry car. When the separate hoppers of the larry, one for each charging hole, are equipped with volumetric sleeves, the gates usually are linked together and power operated as all gates may be left open until the hoppers are full and the flow of coal is stopped by the sleeves.

The larry car in principle is designed in connection with the number of charging holes per oven so that a predetermined quantity of coal is charged into the oven through each charging hole, the mechanism for controlling discharge from each hopper being independently operated by powered drives on new installations and by gravity on older batteries.

Improvements in larry cars, particularly the method of discharging coal, have been directed toward making possible better charging practices. The aim has been to reduce the charging time; to reduce the number of passes of the leveling bar necessary for leveling; to make a smokeless charge; to prevent hanging up of the coal in the larry hoppers; and to make a uniform charge, regardless of coal bulk density.

On the more modern installations (Figure 4—22), automatic lid lifters are provided to dispense with manual removal of the charging-hole covers.

The gravity-discharge larry and the mechanically unloaded larry are the two main types in use at the present time. The gravity-discharge larry is equipped with conical-shaped hoppers, shear gates, and drop-sleeve mechanisms. With this type, the coal charge flows by gravity from the hoppers into the ovens. There are two designs of mechanically unloaded larry cars in use at present; namely, the "turntable" and "screw-discharge" types. The turntable larry is equipped with a revolving table serving as the bottom of each hopper. The revolving table forces the coal through an opening in the side of the hopper leading to the shear gate and drop sleeve and thence to the oven. The screw-discharge larry is equipped with rectangular-shaped hoppers with the lower section tapering to a small opening directly over the screw trough. The trough contains the screw conveyor that forces the coal horizontally to the vertical drop sleeve and shear gate section.

The modern trend for oven charging in order to reduce emissions to the atmosphere has been the utilization of stage charging (see Figure 4—56). With this concept, the two outer hoppers of the charging car, each of which contains approximately 40 per cent of the coal charge, discharges coal into the oven while the charging-hole lids of the inner charging holes are still in place. When coal discharge from the outer hoppers is completed, the outer lids are replaced and the lids from the inner charging holes are sequentially removed and the remainder of the coal is charged into the oven. At a predetermined time during the discharge into the inner charging holes, the leveler bar levels the coal in the oven. With stage charging, the oven charging takes between 3 and 8 minutes as compared to approximately 2 minutes required when coal was discharged from all hoppers simultaneously.

The screw-discharge larry gives a slightly better performance on wet or fine coal but difficulty may be encountered with any extraneous foreign matter in the coal, whereas the gravity-discharge larry and turntable larry have less trouble with foreign matter. With fine or wet coal, the gravity-discharge larry gives more trouble than either of the other two types. With present-day methods for preparation of coal for coking, the trend is toward use of the turntable larry.

Pusher-Side Equipment—The pusher-side equipment shown in Figures 4—23 and 4—24 is generally similar on all types of ovens. The pusher is, in general, a combination of three machines: a pusher, a leveler, and a door extractor. It is designed to operate on a track parallel to but independent of the battery.

The function of the door-extracting element of this machine is to remove and hold the pusher-side door during the pushing operation. It is either electrically and/or hydraulically operated from the elevated cab which contains all of the machine controls. With self-sealing doors, an important feature of the door-extractor design is that its speed be relatively slow and easily controllable and the alignment be accurate to avoid damage to the sealing edges of the door. Contained in the head of the extractor is a mechanism for latching the door.

The function of the pushing element (Figure 4—11, Diagram E and Figure 4—24) is to push the coke cake from the oven. This is done by an electrically powered rack-and-pinion-operated ram, equipped in front with a suitable head that, when spotted immediately in front

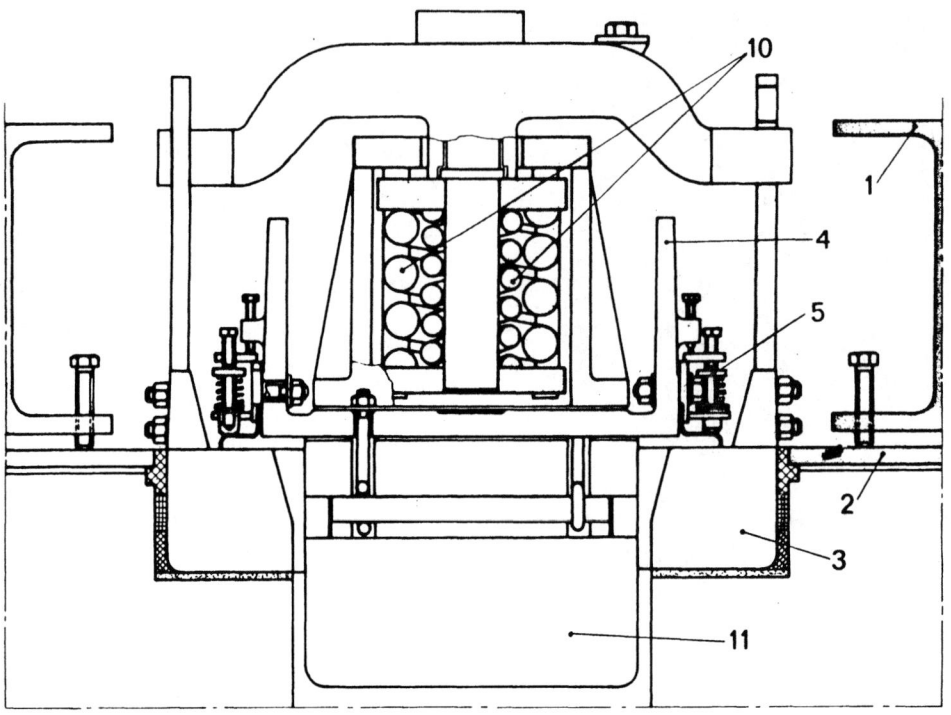

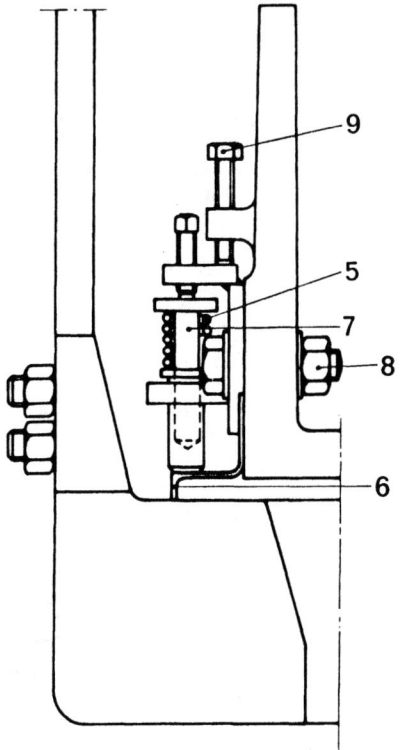

FIG. 4—17. Otto self-sealing coke-oven door with spring-loaded sealing strips. (1) Buckstay. (2) Wall-protection plates. (3) Door frame. (4) Door body. (5) Spring element. (6) Z-shaped sealing strip. (7) Spring-loaded pressure and guide pins. (8) Retaining bolts for spring elements. (9) Adjusting bolts. (10) Door latching with compression-spring loading. (11) Door plug. (Courtesy, Dr. C. Otto and Company.)

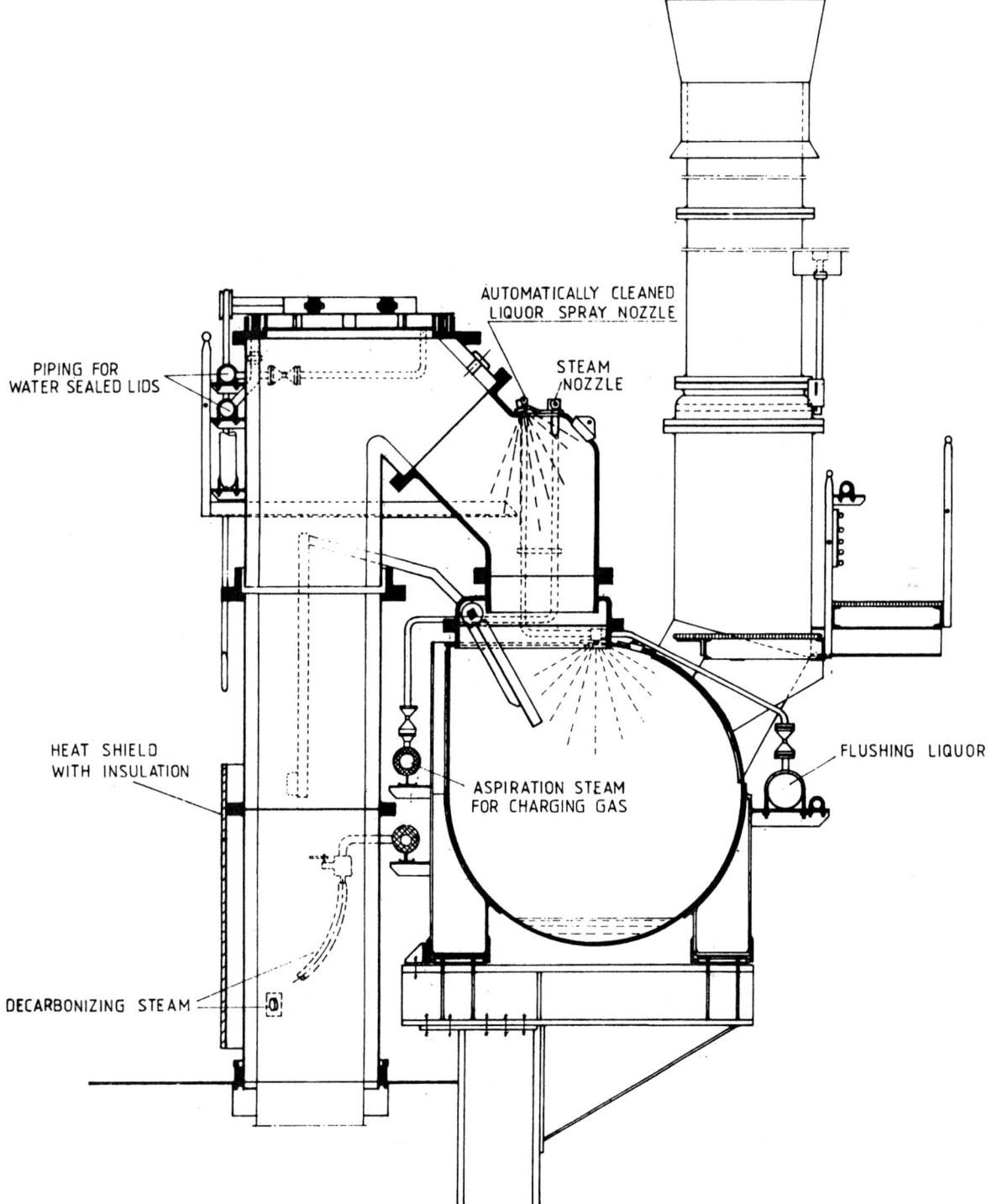

AUTOMATICALLY CLEANED
LIQUOR SPRAY NOZZLE

STEAM
NOZZLE

PIPING FOR
WATER SEALED LIDS

FLUSHING LIQUOR

HEAT SHIELD
WITH INSULATION

ASPIRATION STEAM
FOR CHARGING GAS

DECARBONIZING STEAM

Fig. 4—18. Dr. C. Otto collecting main and ascension pipe with oval water-sealed cap for automatic gooseneck cleaning. (U.S. patent). (Courtesy, Dr. C. Otto and Company, G.m.b.H.)

of the oven to be pushed, may be moved forward until all the coke has been pushed from the oven and through the coke guide into the quenching car.

The ram is equipped with a rider shoe located a short distance behind the pushing head to support the ram during its passage through the oven. This rider shoe is easily replaceable, as it is subject to considerable abrasion sliding over the brick floor of the oven, especially as there is always considerable coke breeze on the oven floor during the pushing and return. The ram is either a built-up box girder, "H"-beam, or electrically welded open-lattice structure. The open-lattice construction seems to be currently in favor as it is more easily repaired and resists the tendency to warp, due to more even cooling regardless of wind direction. With the box-girder-type ram, it is necessary to have wind-and-rain guards for the ram in its retracted position. The pushing speed of the ram is about 15 to 20 metres (50 to 60 feet) per minute, and the maximum pressure exerted is controlled by overload relays to prevent damage to the oven brickwork.

The function of the leveling element (Figure 4—11, Diagrams B and C) is to level the coal charge in the oven, leaving a free-gas space below the roof of the charged oven. This is done by an electrically operated leveling bar carried by the pusher-machine structure in such a position that it may be introduced through a suitable opening in the top of the pusher-side door. The leveling bar is a fabricated section consisting of two side plates held apart by vertical plates spaced at from about 0.6 to 1.2 metres (2 to 4 feet) which also serve as scrapers. When this bar is moved into and out of the oven, the scraper plates level the peaks of coal beneath the charging holes into the valleys and, on removal

from the oven, drag out all excess coal into a chute discharging into a receiving bin carried by the pusher machine. This excess coal periodically is dropped into a ground-level hopper or container for return to the coal bunker (bin).

The various platforms, control rooms and operating cab of the pusher machine should be designed to facilitate comfortable operation and accessibility for lubrication, adjustment and repair.

The modern pusher machine is equipped with devices to remove the carbon build-up from the doors and jambs so that the door can seat tightly in place against the jambs to prevent emission of gases during the coking cycle (Figure 4—25).

The battery has a bench serving as a walkway along its entire length and as a working platform for servicing the doors and jambs. The level of this bench is about 0.76 to 0.91 metre (2½ to 3 feet) below the oven-floor level so that the pusher ram may pass over the top of the bench railing.

Coke-Side Equipment—The coke side of the battery is equipped with a door-extracting machine and a coke guide. The coke-side equipment operates on a track integral with the coke-side bench, as shown in Figures

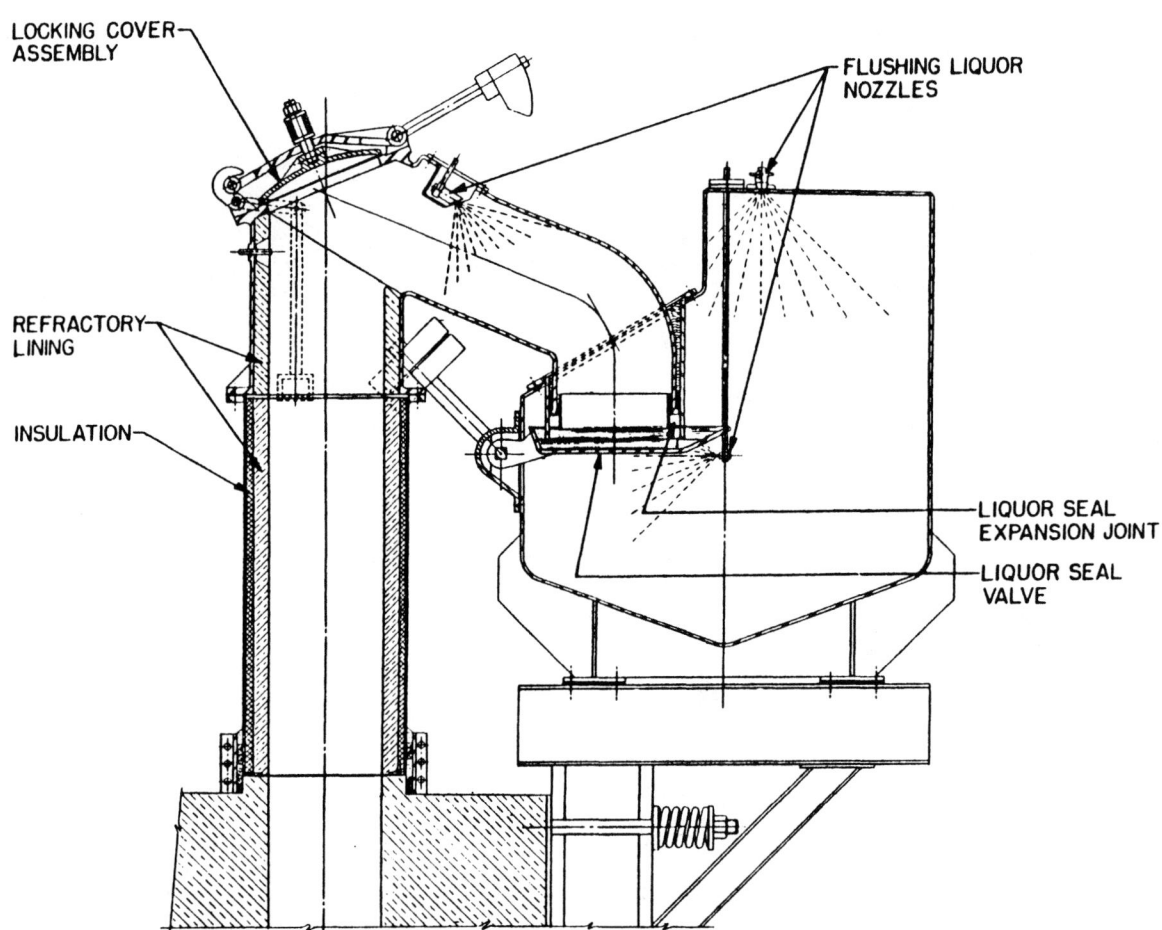

FIG. 4—19. Wilputte arrangement of standpipe, liquid-sealed damper valve, collecting main and automatic elbow cleaner. (Courtesy, Wilputte Corporation, a Salem Company.)

FIG. 4—20. View of a coal-storage bin and larry car on top of a modern coke-oven battery equipped with a double collecting main. (Courtesy, Koppers Company, Incorporated.)

4—26 and 4—27.

The function of the door machine is to remove and hold the coke-side door during the pushing of an oven, and to place an attached coke guide in the proper position to conduct the coke across the coke-side bench into a quenching car operating on a ground-level track parallel to the battery; these operations are shown in Figure 4—26.

The design and operation of the door extractor of the door machine is generally similar to the door extractor of the pusher machine. Recent developments place the extracted door behind a heat shield and have details of design that facilitate door cleaning.

The coke guide is attached to the door-extractor machine by a disconnecting coupler. At modern plants, the coke guide is equipped with a movable lattice framework that is power-operated from the door machine. When spotted at the oven to be pushed this movable framework is moved into the space between the buckstays and against the door jamb, thus preventing spillage of coke at this point during pushing. The door-extracting machine is equipped similar to the pusher machine with mechanical devices for cleaning doors and jambs.

Generally, most coke-oven machinery in the United States operates on direct-current power; however, on many new installations oven machinery is powered by alternating-current motors instead of direct-current motors and alternating current generated at the plant can be used without rectification.

Use of alternating-current motors is made possible by silicon-controlled rectifiers. There are several advantages in this method of control. With silicon-controlled rectifiers (SCR), stepless control is obtained and the movement of machines is fast and smooth with reduced stress on equipment.

COKE QUENCHING

There are two methods for quenching the hot coke pushed from the ovens, namely, wet quenching and dry quenching, the latter being used mainly in Japan and Russia.

Wet Quenching of Coke—Wet quenching in most modern plants is accomplished by receiving the charge of hot coke from the ovens in the quenching car, which is conducted to the quenching station by a locomotive, where the coke is quenched by water. The car is then taken to a coke wharf where the coke is discharged. The handling of the coke from the wharf will be discussed later.

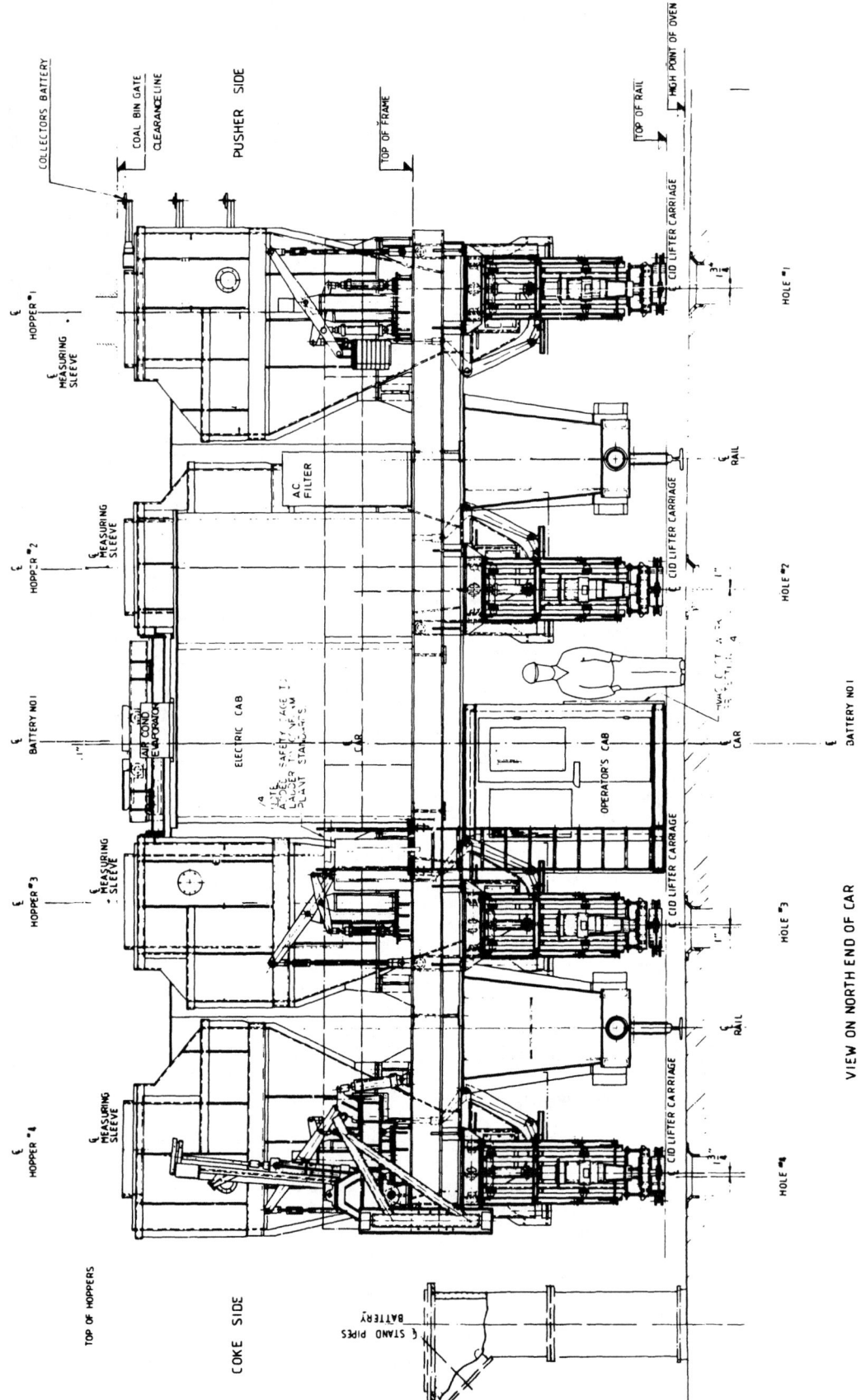

Fig. 4–21. Front elevation of a coal-charging larry for a modern by-product coke oven. (Courtesy, Dr. C. Otto and Company, G.m.b.H.)

The quenching car is designed so that by moving the car during the pushing operation the coke is caught in a relatively uniform bed about two feet thick on the sloping bottom of the car (Figure 4—26). The quenching car has gates at the low side of the sloping bottom, operated by electric or air power, so designed and arranged that the quenching water not evaporated may drain readily from the car. When the charge of hot coke has arrived at the quenching station, it is spotted under a system of stationary sprays located in the quenching tower that is operated by remote control. The purpose of quenching, obviously, is to cool the coke rapidly to stop any further combustion. However, it must be recognized that coke with a low moisture content is desired. This is accomplished by so arranging the sprays and the time of quenching that sufficient heat will remain in the center of the individual coke lumps to evaporate excess surface water. The usual practice is to aim at an average moisture content of 2½ per cent in the metallurgical coke after screening.

Modern quenching stations are of the recirculating type in which the water draining from the quench car discharges into a settling basin where the coke fines collected with the excess quench water settle out. The clear water is then reused for quenching. Because of environmental regulations, the make-up water to the quench stations must be fresh water; the discharge of contaminated water into the quench basin is no longer permitted.

Baffles have been installed near the top of some quenching-station stacks to minimize carryover of entrained dust and water droplets out of the top of the stacks by the steam generated in quenching. These eliminators consist of wooden-grid-type baffles, provided with auxiliary water sprays for periodic flushing to remove accumulations of dust.

Dry Quenching of Coke—When pushed from the oven chamber, the hot coke has a temperature of about 1000°C (1800°F) and has to be cooled before it can be further handled and used. The usual cooling method in the United States is wet quenching accomplished by spraying the coke with water in a quenching tower. This results in heat loss to the atmosphere. With dry quenching (cooling), however, the sensible heat of the hot coke is recovered and utilized for the generation of steam.

Dry quenching is a proven process but it was not much used in the past because of high investment costs and technical problems and because it was uneconomical. There is a growing interest in dry quenching due to changes in the economic sector in particular, and the more stringent regulations regarding pollution control and the reduction of emissions from coke-oven plants.

Since 1965, more than 50 dry-quenching units, some rated at 57 metric tons (63 net tons) per hour, have

FIG. 4—22. View of an automatic charging-hole lid-lifter assembly as mounted on a coal-charging car. (Courtesy, Koppers Company, Incorporated.)

been installed in the USSR. NKK of Japan has installed eight dry-quenching units, rated at about 70 metric tons (77 net tons) per hour, at their Ohgishima Works.

Figure 4—28 illustrates a method of dry quenching in which hot coke is pushed from an oven into a hot-coke bucket mounted on a car running on railway tracks. The car transfers the loaded bucket to a hoisting tower which lifts the bucket to the top of the quench tower where the hoist trolley moves the bucket into position over one of the quenching chambers. The top cover of the chamber is removed, the hot coke is discharged from the bottom of the bucket into the chamber, and the cover is replaced. The coke is cooled as it descends through the refractory-lined chamber, by recirculating inert gases passing upward through the chamber, from 1000° to 200°C (1832° to 392°F) in 2½ hours. The cooled coke is discharged in 2-metric-ton (2.2-net-ton) batches through lock hoppers under the chamber onto the conveyor belt that carries it to a screening station. The inert gas which cools the coke contains 12 per cent CO_2, 8.5 per cent CO, 3.0 per cent H_2 and 76.5 per cent N_2. The gas enters the bottom of the quenching chamber at 175°C (347°F) and exits at about 800°C (1472°F). This inert gas flows from the top

of the chamber through a drop-out chamber to remove coke particles and then enters the waste-heat boiler that serves as a steam generator. Cooled gas at 175°C (347°F) flows through cyclones for dust removal prior to entering the recirculating fan. Depending upon the intended application, as much as 0.46 metric ton (0.5 net ton) of steam at 3920 kPa (60 kg/cm² or 570 psi), and 440°C (824°F) can be generated per ton of coke.

CHARGING AND PUSHING SCHEDULES

A coke-oven battery contains many ovens which, obviously, cannot all be charged or pushed at the same time. The amount of heat required at various stages of the conversion of coal to coke varies during the time a charge remains in the oven. For these reasons and the necessity of balancing the heat, operating and combustion requirements for the many ovens contained in a single battery, specific sequences for servicing the ovens have been developed. The sequence that has the most widespread use in the United States, developed by Koppers Company, Incorporated, numbers all of the ovens in a battery in order, using a base nine series of numbers; that is, no zero, 10, 20, etc., appears. The ovens then are serviced by pushing and charging in

FIG. 4—23. Pusher-side equipment of a battery of by-product coke ovens, showing the machine which is equipped to level the coal charge in the oven and also push the finished coke out of the oven, as well as remove and replace oven doors before and after the pushing operation. (Courtesy, Nippon Steel Corporation.)

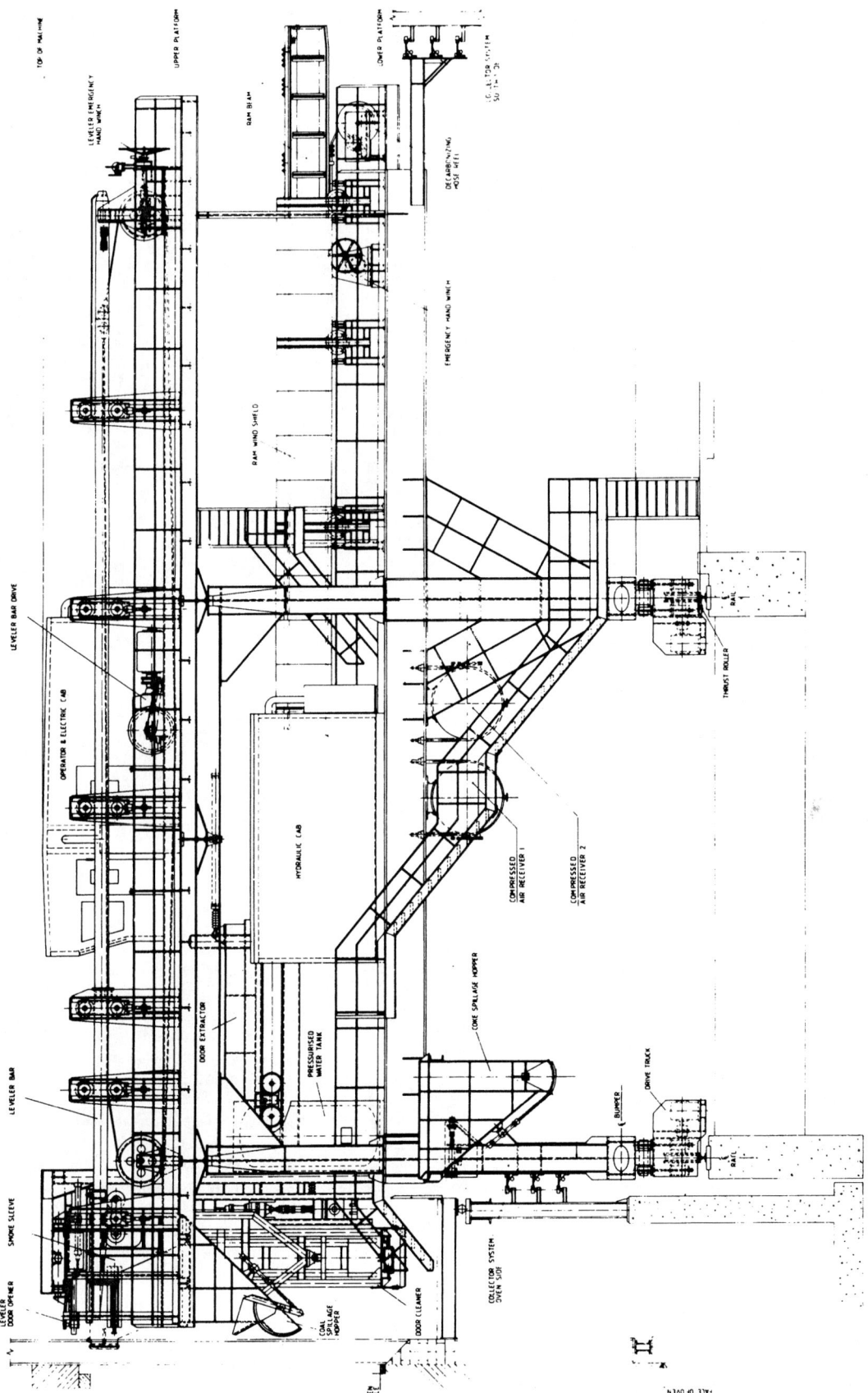

FIG. 4—24. By-product coke-oven pusher machine, equipped to level the coal charge in the oven, push the finished coke out of the oven and to remove and replace oven door before and after the pushing operation. (Courtesy, Dr. C. Otto and Company, G.m.b.H.)

sequence all the ovens whose numbers end with the numeral 1 (1, 11, 21, 31, etc.); next those ovens whose numbers end with 3 (3, 13, 23, 33, etc.) and so on until all of the odd-numbered ovens have been serviced. Next, all the even-numbered ovens beginning with No. 2 are similarly serviced. This results in a favorable local heating balance with respect to the particular oven to be serviced. At the time any given oven is scheduled to be pushed according to this system, the oven next to it on its left is 55 per cent of the way through its alloted coking time and the coking cycle of the one on its right is 45 per cent complete.

In another system, termed the Marquard system after its inventor, the ovens in a battery are divided into three groups identified as A, B and C. The ovens within each group are then numbered consecutively (including those numbers ending with zero). The order of servicing the ovens then proceeds with pushing and charging all the odd-numbered ovens starting with A-1 (A-1, B-1, C-1; A-3, B-3, C-3; and so on). The even-numbered ovens are then pushed in a similar sequence, beginning with A-2. With this system, when an oven is to be pushed, the coking cycle of the oven next to it on its left is about 52 per cent complete and that of the adjacent oven to its right is about 48 per cent complete. This system requires that two odd-numbered ovens be next to each other where the sections adjoin.

INSTRUMENTATION AND CONTROL

Most instrumentation on coke-oven batteries is confined to the heating, pushing and foul-gas-handling facilities. As has been mentioned, the variable controls such as orifices in the air boxes and restriction dampers in waste-heat boxes are so adjusted that master controls may be installed for each battery. In this manner, single machines, meters, gages and other devices control the various functions of the battery system after the individual-oven controls have been adjusted. Thus, the gas for underfiring is metered in the battery header, but the correct distribution of the gas to the individual heating walls is controlled by a fixed orifice in the gas system at each wall entry. Within each heating wall, downstream from the control orifice, nozzles of graduated orifice size control distribution of the gas to each heating flue. The air and waste gases are likewise controlled by individually adjustable orifices in the air and waste-heat boxes. Finger bars are used to control the air-box openings and butterfly valves control the outlet of waste gas to the stack canal.

A draft regulator operates a butterfly valve to maintain a preset stack draft on the battery.

The choice of a flowmeter for the underfiring gas will be influenced in large part by its accuracy over a wide range of flow. The meter should be recording, preferably one adapted to a uniform graduated chart for reasons of legibility and computation facility. It is also necessary to control accurately the pressure in the fuel-gas main with an instrument capable of close control at relatively low pressures. It may be desired to have a recording of these pressures. A recording thermometer is installed also in the fuel-gas header so that the gas quantities may be computed to standard conditions. The battery also is equipped with recording draft gages, waste-heat recording thermometers, and various

FIG. 4—25. Jamb cleaner mounted on a pusher machine, in position to remove carbon deposits. (Courtesy, Nippon Steel Corporation.)

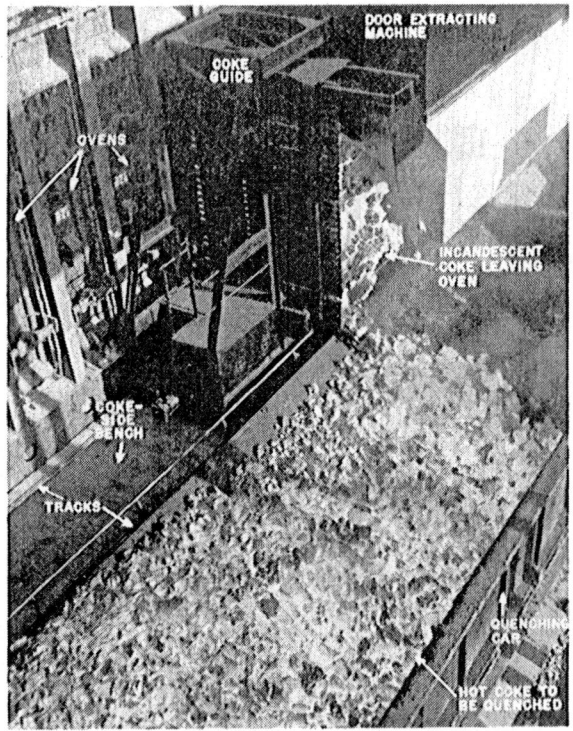

FIG. 4—26. Coke side of a battery of by-product coke ovens during the process of pushing the coke out of one of the ovens. The quenching car is self-propelled, and carries the coke to a quenching station where it is sprayed with water before being dumped on the coke wharf.

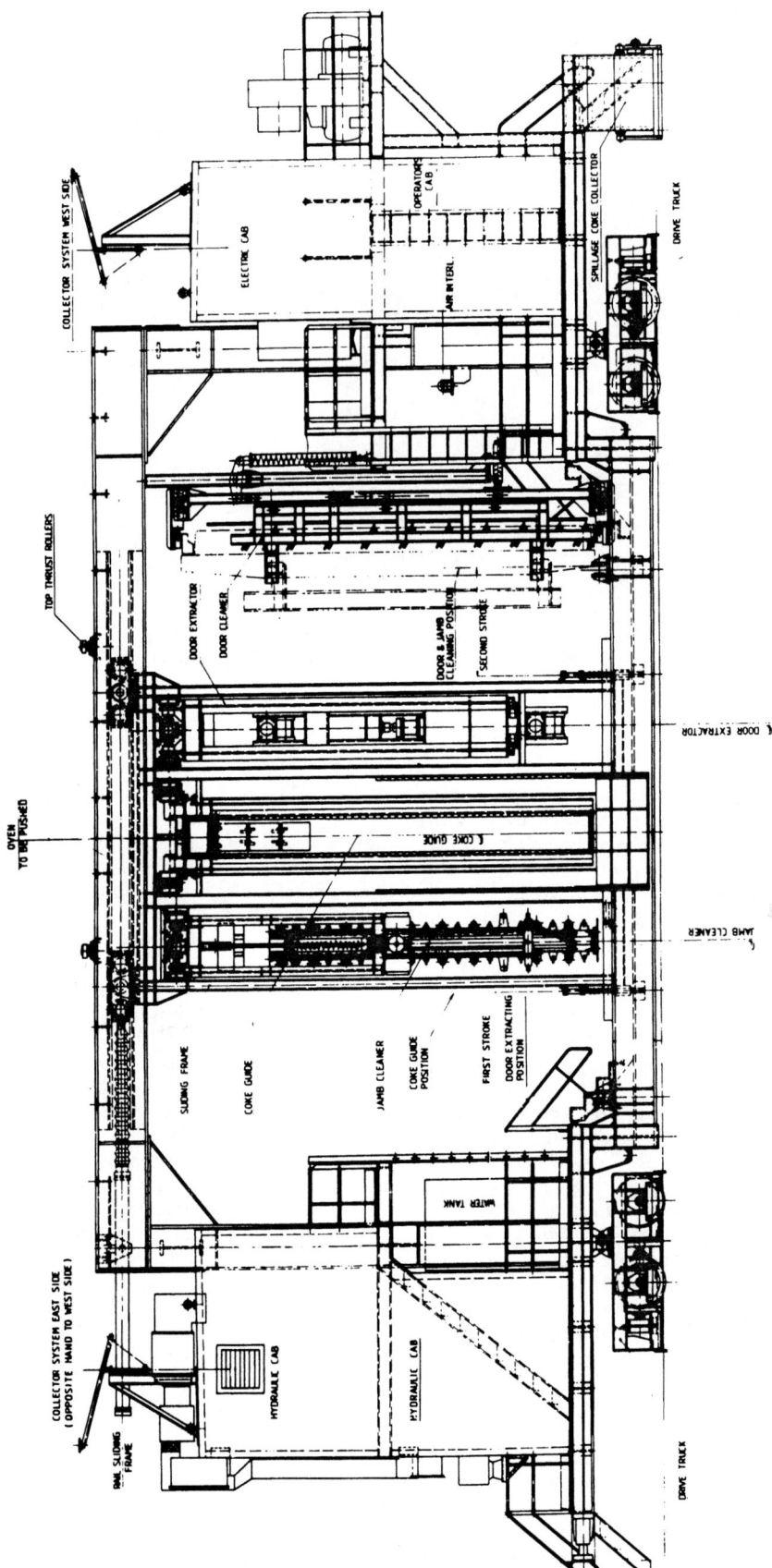

FIG. 4—27. One-spot door machine which permits extraction and replacement of an oven door, positioning of the coke guide and door-jamb cleaning by shifting the extractor, guide and cleaner on an overhead rail sliding frame without moving the entire machine. (Courtesy, Dr. C. Otto and Company, G.m.b.H.)

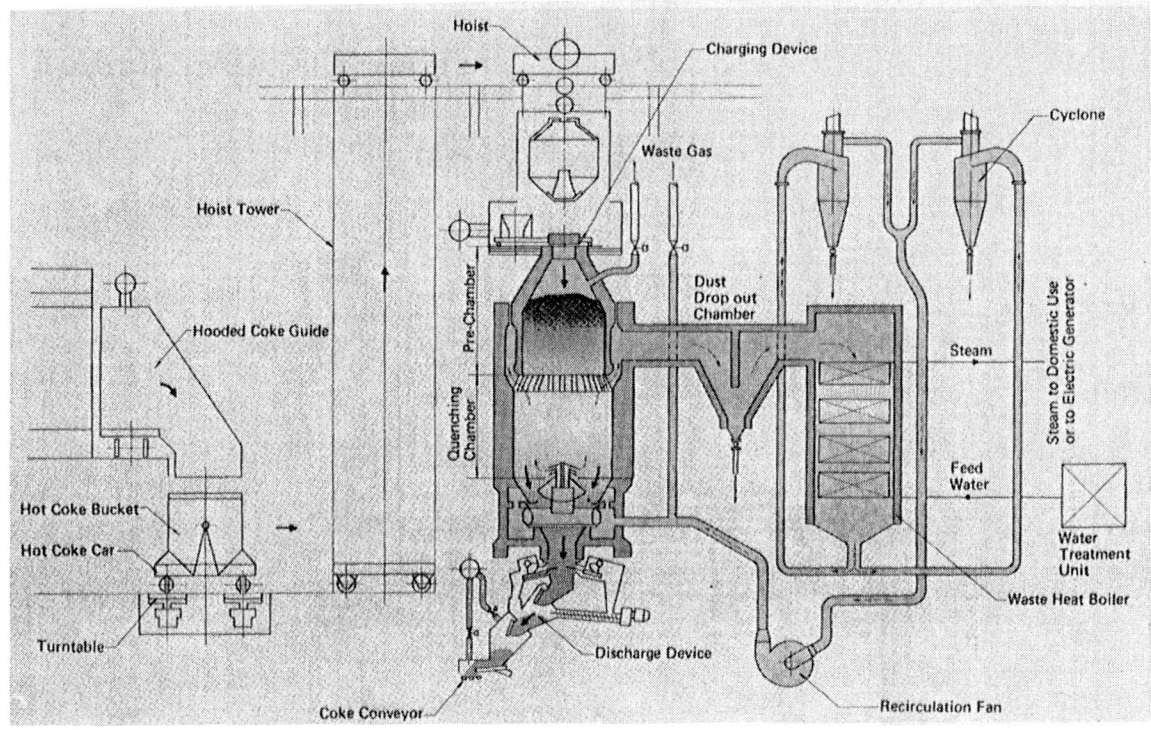

FIG. 4—28. Flow diagram of a dry-quenching system.

indicating gages. Portable pressure gages, thermometers and pyrometers are used in the setting of individual oven controls. Another instrument used by the heaters is a portable pyrometer with which periodic temperature readings are taken of the flues and emerging coke. Probably the most widely accepted pyrometer for the latter purposes is of the incandescent-filament type. Recording pressure and temperature gages also are installed in the gas-collecting mains.

The pressure in the collecting main is controlled by a back-pressure regulator located at the junction of the collecting main with the suction main; its function is to maintain a closely controlled pressure in the ovens during coking.

Many plants record the regularity of oven operation by some related activity, such as the time of pushing, by the peak-load recording of the pusher-power circuit, or by recording the time of quenching of each charge. All utilities have the usual types of instrumentation.

Computer-Controlled Battery Heating—In general, the control of coke battery heating today still depends on human experience and skill using the meters, gages, and manual control valves described above. Because of the complexity of the heating systems, this function requires relatively large numbers of skilled operators making numerous manual temperature measurements and adjustments. Difficulties with battery heating control can result in low fuel efficiency, environmental problems, and poor quality coke. Therefore, to improve control of this critical battery function, computer-controlled battery heating systems are now being developed and adapted to selected coke batteries both domestically and abroad. These new systems, al-

though still in the development phase, are proving to be quite successful in optimizing battery heating in terms of both fuel efficiency and heating uniformity. Japanese and German coke producers have expended considerable effort in battery automation, and United States Steel is also moving into this area of technology with the installation of two somewhat different systems, one at its Clairton plant and one at its Fairfield Works. The Clairton system is based on feedforward control developed in-house and the one at Fairfield is based on a feedback control system purchased from Nippon Kokan K.K.

The United States Steel Feedforward Heating System is a predictive system requiring specific knowledge of coal properties prior to charging, fuel gas prior to burning, and periodic final coke temperatures. The computer control system shown schematically in Figure 4—29, comprises four, essentially independent, subsystems: (1) coal tracking and calculation of battery heat requirements, (2) an underfiring gas flow control loop, (3) flue gas oxygen cascade control, and (4) a coke temperature feedback system. One element, coal heat of carbonization, has yet to be defined and measured with the necessary accuracy. In the present system, as shown, feedback of final coke temperature is necessary to compensate for changes in heat of carbonization and for changes in battery efficiency.

The coal proportioning runs are tracked through the coal handling system to the battery. Coal bulk density and moisture are entered into a computer for each coal run. The battery heat requirement is computed based on the coal in the battery at any time, operating rate, and battery efficiency. Oven push count inputs from

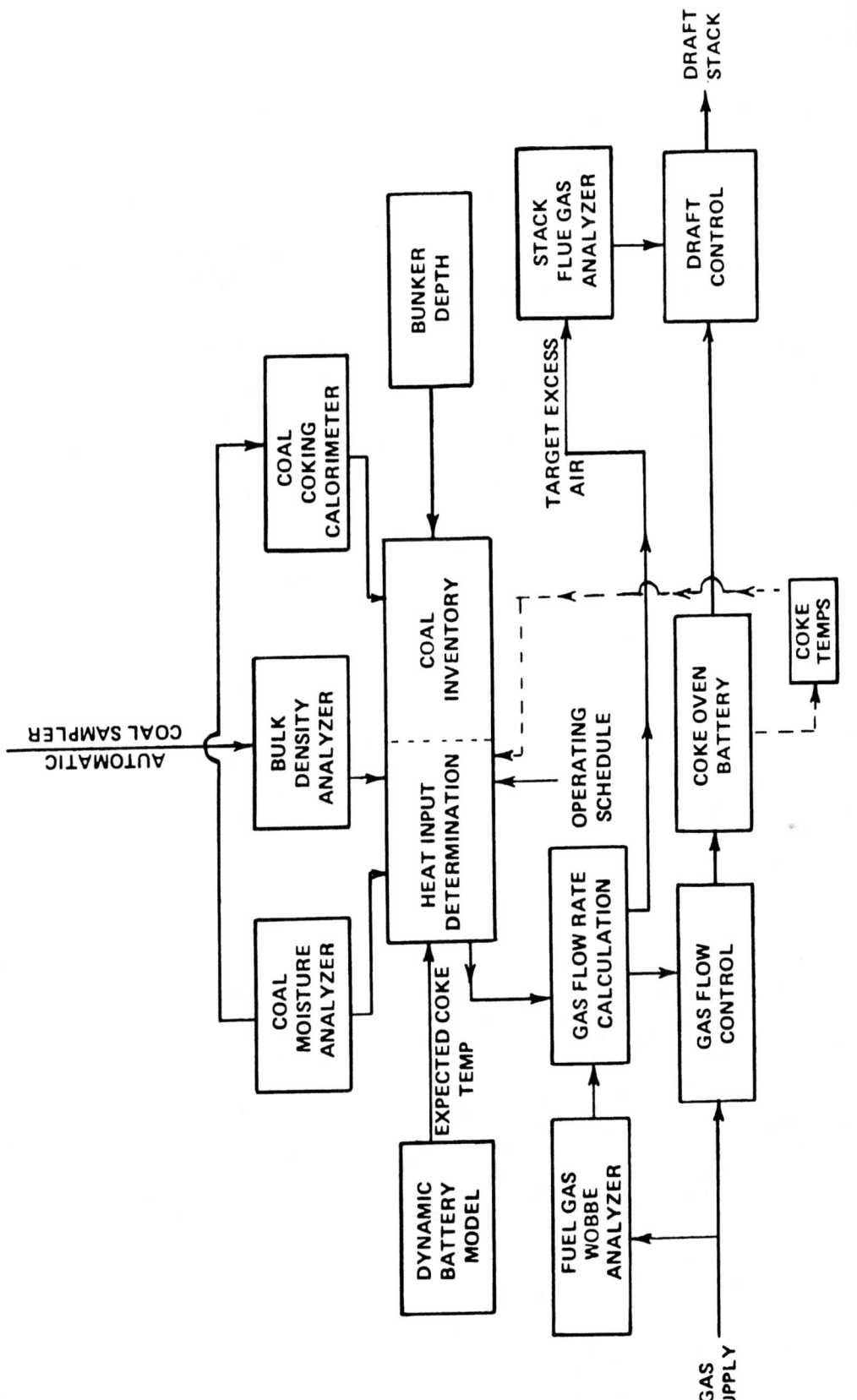

FIG. 4—29. Feedforward coke-oven-battery heating-control system.

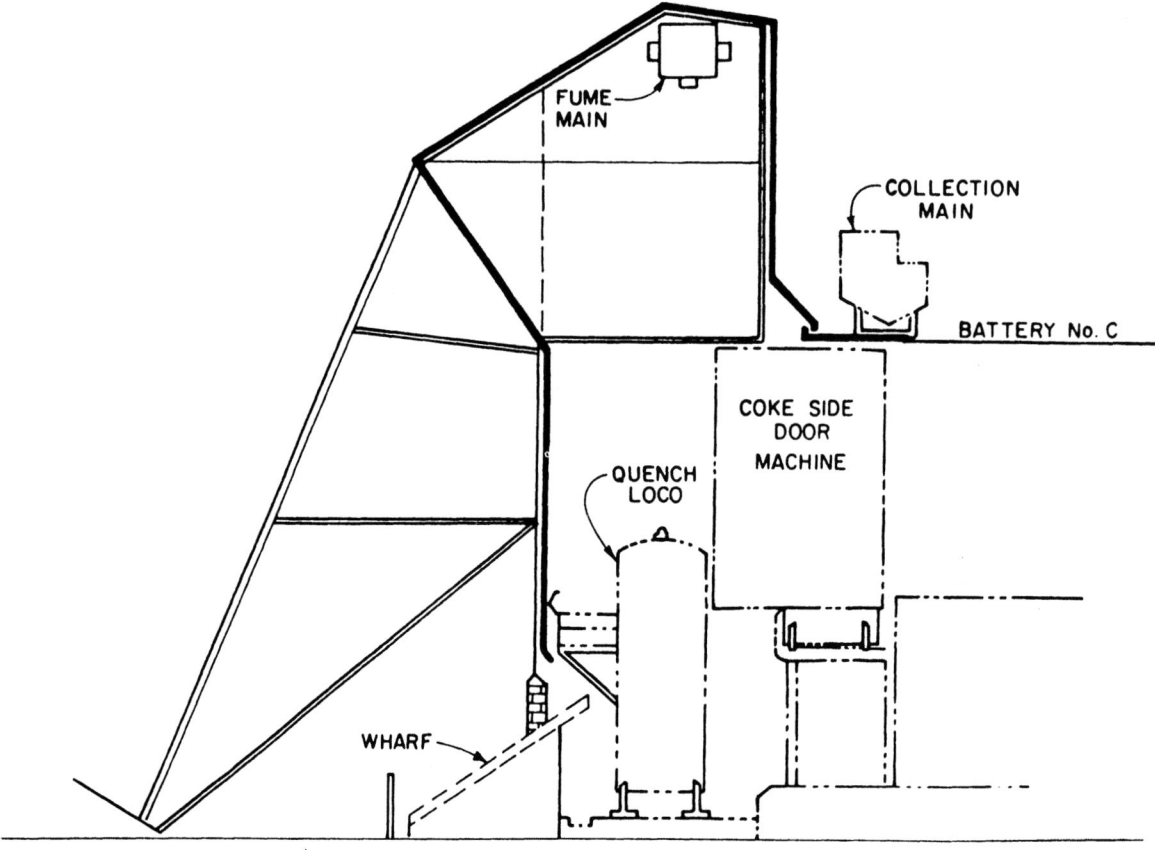

FIG. 4—30. Coke-side shed for collecting emissions from a by-product coke oven during the pushing operation.

the pusher machine are used to monitor the operation and 'charge' the coal into the computer model.

The calculated battery heat requirement is used as the setpoint for the underfiring gas flow control loop which uses the fuel gas Wobbe Index to control the gas flow to maintain the desired heat input.

The excess oxygen cascade control adjusts the stack draft to maintain a constant stack oxygen for efficient combustion and proper vertical temperature distribution in the flues. The cascade control is independent of the computer but the stack draft setpoint, consistent with desired heat input, is adjusted by the computer to keep within the range of the cascade control.

To account for changes in battery efficiency and heat of carbonization, a coke temperature feedback system was developed to close the control loop. Coke temperature is sensed using an infra-red pyrometer mounted on the coke guide and is transmitted to the computer which adjusts the heat requirement computation to maintain the desired final coke temperature.

The computer hardware provided for this control system is sized to handle all 14 coke batteries at Clairton. A Hewlett Packard HP-1000 minicomputer with 256-K bytes memory and two 50-megabyte disc-memory units is used for data handling, logging, and control. IPAC series 1500 remote multiplexers are located at each battery, at coal-side sample stations, and with Serial Line Controllers where they interface the computer with field inputs and control points. This high capacity computer not only handles the requirements of the feedforward heating system, but also is used for coke plant data storage and analysis, and as a management information system. The computer stores critical operating data and prints out daily reports useful in evaluating battery operations.

Battery heating control systems such as the one just described are now being implemented on selected batteries throughout the world. The use of the computer to manage the very complex overall battery heating strategy and to make continuous overall underfiring adjustments allows operators the time to maintain underfiring components, such as gas nozzles, and to fine-tune individual oven temperature profiles which can be done only by manual adjustments.

PUSHING-EMISSIONS CONTROL SYSTEMS

With the advent of the Environmental Protection Agency and the Clean Air Act of 1970 significant environmental pressures were placed on the steel industry and in particular coke-making facilities. Coke pushing emissions were singled out as the main source of air pollution. Subsequently considerable emphasis was given to reducing pushing emissions through the development of pushing-emissions control systems.

The four alternative types of systems considered by the industry are: (1) sheds, (2) bench-mounted hoods with fixed ducts, (3) enclosed hot-coke cars with mobile scrubbing systems and (4) water-spray systems.

In contrast to bench-mounted hoods and water-spray systems, both enclosed cars and sheds have demonstrated effective capture of emissions from pushing. Capture efficiency of one shed has been estimated at about 90 per cent, and enclosed cars have been judged to achieve about the same degree of capture. Thus, both enclosed cars and sheds are judged better for capture of pushing emissions than the other control systems.

A comparative analysis of the overall emission reduction that enclosed cars and sheds can achieve shows the performance is essentially the same for both. Emissions for an average-size plant were estimated at 2.4 and 2.9 kilograms (5.3 and 6.4 pounds) per hour for the enclosed car and shed, respectively. Performance of the two alternative types of control devices, scrubbers and wet electrostatic precipitators, were shown to be equivalent.

Many of the new systems use bag houses rather than venturi scrubbers to control particulates because of the lower pressure drop and reduced power requirement.

Coke-Side Shed—The basic design concept of a shed is to completely capture all of the emissions generated during the pushing cycle, as well as smoke emitting from coke-oven doors on the coke side of the battery. The coke-oven shed is constructed to span the entire coke side of the battery including coke-side bench, hot-car tracks, and coke-side doors. The internal support structure, geometric shape, and method of fume transmission and scrubbing vary with respect to the manufacturer's design and customer's local requirements.

The shed is simply a large structure with enough cubic capacity to hold all the emissions for a single push. These emissions are held in the shed's upper portion until system exhaust fans have had sufficient time to direct all captured gases through an appropriate cleaning system and return clean air to the atmosphere.

The first shed in the United States was erected and in operation during the first quarter of 1972. Since then, many sheds have been installed at various steel plants throughout the United States and Canada. Each of these installations is unique in that they all incorporate modifications to the first prototype, i.e., different shed dimensions, changes due to local plant conditions, and unlike fume-cleaning systems. (see Figure 4—30). Outside of the basic concept, each shed has its own inherent design and working parameters.

Traveling Hoods—The movable-hood concept incorporates the addition of a large hood over the hot car from which the fumes generated during pushing are trapped and conducted to the selected gas-cleaning system. This system was first developed by Mitsubishi and has been installed on many coke-oven batteries in Japan since 1970. Several versions of this Mitsubishi system are in operation within the United States.

In this system, an enclosure hood is suspended from the door machine and is fabricated over the coke guide and extends to cover the hot car. The emissions from the push are pulled by fan into a large duct extending the full length of the battery. This duct is stationary but allows for the interconnection between the hood transi-

tion piece and the duct at each oven by designing an appropriate opening at each oven. As the gases are collected in the hood during the push, the hot car moves under the hood and collects the coke. The effectiveness of the traveling-hood design depends mainly on the designer being able to keep all interface openings between moving parts to an absolute minimum; that is, the interface points between the coke guide, oven jamb, enclosure hood and the hot car.

Another similar hooded system is the Minister-Stein system designed by Hartung, Kuhn and Company (Figure 4—31).

The basic differences of the Minister-Stein system are:

a. The fume-collecting duct is located outside the hot-car track on heavy structural supports.
b. The duct is covered with a rubber belt. A mobile tripper carried with the hood is used for introducing the pushing emissions under the belt at the location of the push.
c. A large heat exchanger is carried with the hood to control the emissions and protect the rubber belt. Atmospheric air cools the heat exchanger between pushes.
d. A third rail mounted on the duct bearing structure supports the hood and heat exchanger.

The fumes from either system are conveyed to a wet scrubber and exhausted into the atmosphere through a short stack.

Enclosed Quench Cars With Mobile Gas-Scrubbing Systems—There are a variety of systems in this category. In principle, these systems enclose the hot coke from the time the oven is pushed until the coke is sprayed at the quench station. Therefore, emissions are captured during the pushing cycle and throughout the travel to the quench station. The coke guide is completely enclosed as the guide is in position prior to the pushing operation. Generally, the coke-guide cover is in two sections, one fixed and the other movable so that it can telescope into a fixed position over the hot-car hood during pushing.

The coke guide is equipped with a fixed hood for containing emissions. This hood has a telescoping extension which is lowered down over the hot car with sufficient clearance to permit travel of the hot car.

The traveling covered hot cars require an opening for introducing the hot coke as it discharges from the coke guide.

There are two distinct modes of operation involving quench cars. In one system, the covered hot car travels during the pushing operation. In the other method, the covered hot car is stationary (single spot) throughout the pushing cycle. With the single-spot car, the coke-guide hood extension can be lowered to a fixed position on the hot-car cover. This is a distinct advantage because it greatly increases capture of emissions.

The stationary or "one spot" quench car remains in a fixed position during the pushing operation. This car is also completely enclosed; however, the car is deeper in bed depth than the conventional hot car since the coke is not spread out in a thin layer. The increase in bed depth allows for the complete containment of the coke within the hot car.

Suction is provided through a duct on the side of the

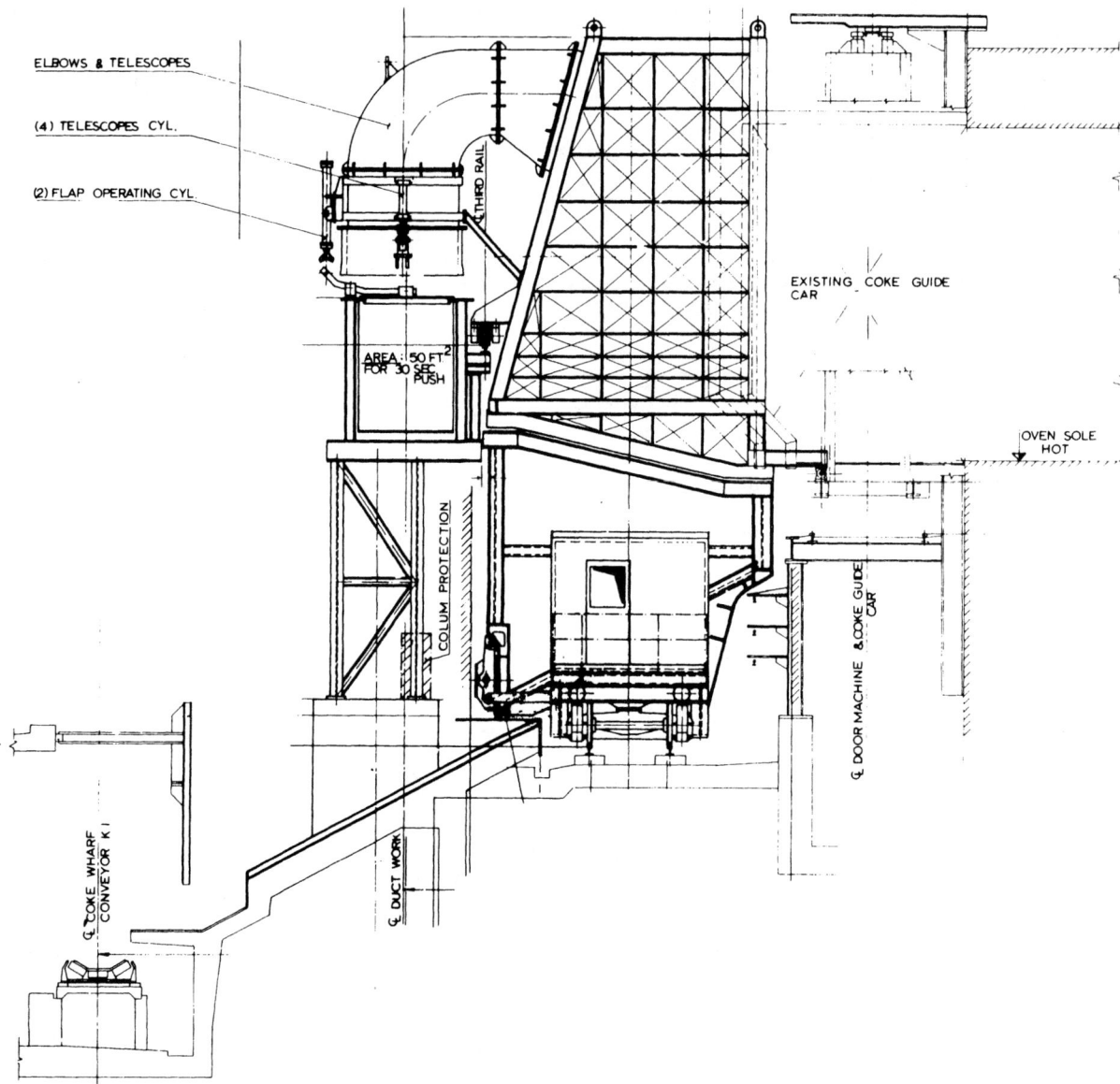

FIG. 4—31. One type of coke-side emission-control system. (Courtesy, Hartung, Kuhn and Company.)

hot car. Suction rates (cfm) during pushing, and while the hot car is traveling, are varied by adjusting louvers on the inlet to the exhaust fan. The emissions are exhausted through a duct to the exhaust fan and control device located on a separate rail car connected to the hot car. Another type of exhaust system employs a Hydro-Sonic compressed-air system where a compressed-air-powered jet exhauster creates a negative pressure in the hood enclosure and scrubs pollutants from the gas stream (see Figure 4—32).

Wet-Spray Pushing-Emission-Control Systems—Wet-spray systems are not literally a capture system, but an attempt to prevent the escape of emissions from the coke battery. These systems consist of a spray header with a series of spray nozzles which are directed at the coke as it is pushed from the oven.

Various systems for wet spraying coke have been tried with little or no success. Wet spraying has been judged by trained observers to be strictly cosmetic and not effective with regard to capture of particulate matter generated during the pushing operation.

COKE SCREENING AND HANDLING

Coke Wharf—A coke wharf is substantially a long, narrow inclined platform with the shorter dimension sloping away from the quenching-car track toward a conveyor belt that runs along the lower side of the structure. A properly designed wharf should be of such size that it will serve as surge storage for quenched coke ahead of the screens so that short delays incidental to screening and loading operations will not interrupt the desired regularity in oven-pushing sequence.

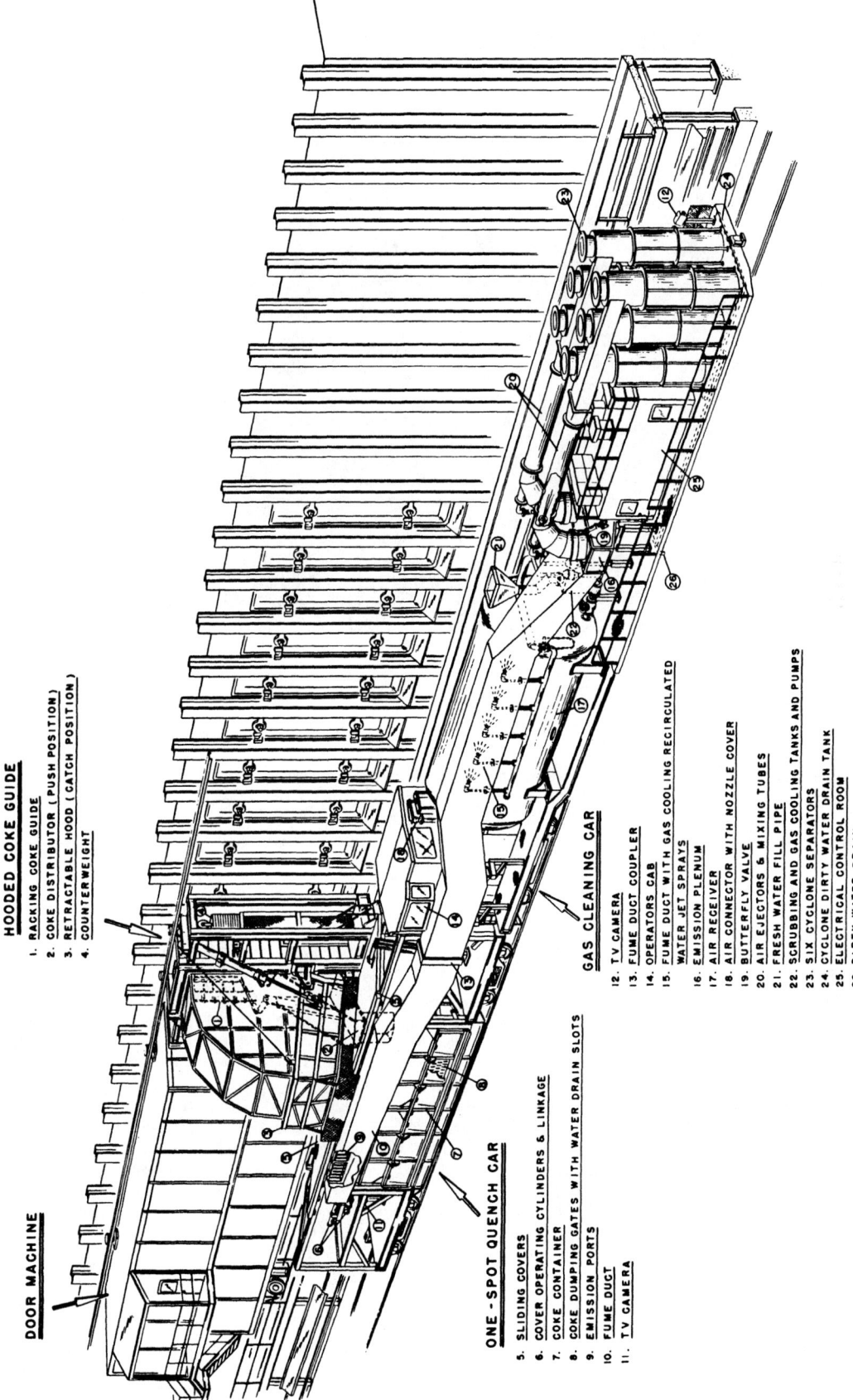

DOOR MACHINE

HOODED COKE GUIDE

1. RACKING COKE GUIDE
2. COKE DISTRIBUTOR (PUSH POSITION)
3. RETRACTABLE HOOD (CATCH POSITION)
4. COUNTERWEIGHT

ONE-SPOT QUENCH CAR

5. SLIDING COVERS
6. COVER OPERATING CYLINDERS & LINKAGE
7. COKE CONTAINER
8. COKE DUMPING GATES WITH WATER DRAIN SLOTS
9. EMISSION PORTS
10. FUME DUCT
11. TV CAMERA

GAS CLEANING CAR

12. TV CAMERA
13. FUME DUCT COUPLER
14. OPERATORS CAB
15. FUME DUCT WITH GAS COOLING RECIRCULATED
16. WATER JET SPRAYS
16. EMISSION PLENUM
17. AIR RECEIVER
18. AIR CONNECTOR WITH NOZZLE COVER
19. BUTTERFLY VALVE
20. AIR EJECTORS & MIXING TUBES
21. FRESH WATER FILL PIPE
22. SCRUBBING AND GAS COOLING TANKS AND PUMPS
23. SIX CYCLONE SEPARATORS
24. CYCLONE DIRTY WATER DRAIN TANK
25. ELECTRICAL CONTROL ROOM
26. DIRTY WATER DRAIN

FIG. 4—32. Schematic arrangement of the Koppers Hydro-Sonic gas-cleaning system. (Courtesy, Koppers Company, Incorporated.)

FIG. 4—33. General view of a manually operated coke wharf showing freshly quenched coke being discharged from the gates of the quenching car onto the brick-paved sloping wharf. When the wharf gates are opened, coke slides down the wharf as at the lower left onto a conveyor belt that carries it to the screening station. An automatic (mechanically operated) coke wharf is shown in Figure 4—34.

Most modern wharves are paved with hard-burned clay brick.

The coke wharf receives the quenched coke from the quenching car. The coke is spread out on the wharf in a thin bed for quick drying and visual inspection for the detection of unquenched coke. Manually directed streams of water are used to quench any still-hot coke detected on the wharf before the coke is conveyed on rubber-belt conveyors to the screens.

The coke is retained on the wharf by a series of gates which, when opened, permit the coke to slide down the wharf and onto the belt conveyor to be delivered to the screening station. Figure 4—33 shows a manually operated coke wharf. The present trend is to use automatic mechanically operated wharf gates (Figure 4—34) with automatic sequence and memory circuits to maintain proper utilization of the wharf for surge capacity for the screens and the quenching car. In either case, whether operated automatically or manually, the wharf design and construction is the same.

The more modern coke wharves are provided with rotating coke plows. The coke wharf is designed with a flat shelf at the bottom of the sloping wharf. The coke plow consists of a multi-bladed variable-speed rotating unit mounted on a frame which automatically travels back and forth over the length of the wharf. During travel in both directions, the rotating unit discharges coke from the wharf shelf onto the wharf belt. As the coke belt leaves the wharf, a temperature-sensing device monitors the temperature of the coke and activates water sprays to cool any hot portions of coke.

Conveyor System for Coke—The transfer of coke from the wharf to the screening station now is accomplished almost universally by a system of rubber-belt conveyors and chutes.

Screening and Crushing—The purpose of coke screening and crushing is to provide coke of a controlled size for blast-furnace use from which fines and, in some cases, pieces over a set maximum size are removed. The latter are crushed and screened before use. The very small pieces, commonly called coke breeze, usually are used in the coke plants as a boiler fuel, in sintering plants as sinter fuel, or screened for use in chemical or metallurgical processes.

Storage and Shipping—In the loading of blast-furnace coke and the subsequent transportation to the blast furnaces, care must be taken to prevent additional breakage. For this reason, the coke is loaded into railroad cars and, where practical, is consigned directly to the blast furnaces, as additional handling into and out of stock results inevitably in coke degradation. When the blast furnaces are located close to the coke plant, belt conveyors often are used to conduct the coke from the screening station to the blast-furnace stock-house bins.

REFRACTORY MATERIALS USED IN COKE-BATTERY CONSTRUCTION

A coke battery is basically a structure made of refractory materials held together by a steel exo-skeleton. While constructed at ambient temperature, it is designed to operate at maximum temperatures up to 1455°C (2650°F). During operation, the maximum temperature can be expected to cycle downward by as

FIG. 4—34. General view of part of a modern coke wharf with automatic mechanically operated gates.

much as 110°C (200°F) as part of normal battery operation. Consequently, the battery must be constructed of refractory materials that (1) can withstand the maximum as well as the cyclic temperatures, and (2) have known and predictable properties related to thermal expansion, strength, and creep (see Chapter 2).

Silica Brick—The most abundant refractory used in the construction of a coke battery is silica brick. Silica brick is manufactured primarily from the mineral quartz in finely crystalline form, and having the proper characteristics for conversion to cristobalite and tridymite, which are high-temperature crystalline forms of silica. The quartz is obtained primarily from crushed quartzite rock, which is washed to remove natural impurities. The crushed and washed quartzite is ground further and sized into specific fractions which are then reblended in specific proportions, along with 2.0 to 3.5 per cent lime (CaO), water and organic binders to achieve the desired properties in the brick.

Brick shapes are produced by power and impact pressing, and those that either cannot be mechanically formed or are only required in small numbers are hand made. Immediately after forming, the green brick,

held together by the organic binders, is dried. The two manufacturers of coke-oven silica shapes in the United States fire the silica brick in periodic kilns. The firing temperature is high enough to convert the quartz to cristobalite and tridymite, the stable high-temperature forms of silica, which in turn renders the brick volume stable for service. It is necessary to maintain a carefully planned time-temperature cycle during firing because there are critical temperature ranges through which the silica brick must pass so that a strong, well-bonded brick results. The normal permanent expansion of silica brick during firing is 12 to 15 per cent.

Silica brick has a relatively high melting temperature, 1695° to 1710°C (3080° to 3110°F), and it has the ability to withstand a 172 to 345 kPa (25 to 50 pound per square inch) load to within 28° to 56°C (50° to 100°F) of the ultimate melting point; therefore, it has excellent creep properties. The purity of the brick is important; for example, if the sum of alumina, titania, and alkalies content in a brick is 1.0 per cent, the load to failure will be 28° to 50°C (50° to 90°F) lower than in another brick in which this sum is only 0.5 per cent. At temperatures about 593°C (1100°F), silica brick is nearly volume stable and virtually free from thermal spalling, while at temperatures below 593°C (1100°F) silica brick is highly susceptible to thermal spalling.

Since the late 1950's there had been a general trend in the industry to use high-bulk-density silica brick (greater than 1858 kilograms per cubic metre or 116 pounds per cubic foot) in coke-oven battery construction, because increasing bulk density is accompanied by corresponding increases in **cold** strength and in thermal conductivity. Therefore, it was presumed that higher heat-transfer rates from flue to oven would be realized because of the higher thermal conductivity.

Measurement of flue to oven temperature differentials on operating batteries conducted by United States Steel showed that higher-bulk-density silica brick did not result in improved heat transfer over regular-bulk-density 1740 to 1804 kPa (108 to 112 pounds per cubic foot) silica brick. In addition, a series of tests on silica brick of varying bulk density showed that the **hot** strength in the range of 649° to 1316°C (1200° to 2400°F) did not appreciably change as a function of bulk density, and that the lower-bulk-density brick were less brittle and less susceptible to failure from thermal cycling. As a result of these measurements and tests, the United States Steel specification for coke-oven silica brick was revised in 1979. In addition, the battery design specification was changed so that only regular-density silica brick are to be used in battery construction.

Fireclay Brick—Fireclay brick is used throughout the cooler parts of the battery, including the regenerator chambers, checker, battery roof, in the pinion walls, and in the coke wharf. According to ASTM standards, there are five general classes of fireclay brick: super duty, high duty, semi-silicas, medium duty and low duty. Fireclay brick used in coke-oven battery construction is generally of the high-duty class, whether specified regular or high duty.

Blends of five or more ground and sized clays are used to make fireclay brick; some brick, especially those in the low-duty class, may be made from one clay. Mixes for high-duty brick commonly contain raw flint and bond clays, possibly with calcined clays. In the high-duty class, a large proportion of the mix is precalcined to control firing shrinkage, as well as to stabilize the volume and control mineral composition of the final product.

Fireclay shapes are formed by power pressing, extrusion and repressing, air ramming and hand ramming. The method of forming is selected on the basis of shape complexity as well as the properties of the fireclay that are required; however, as with silica refractories, small lots of a given shape may be hand formed. Fireclay shapes formed by all processes are dried on hot floors or in tunnel or humidity driers. After drying, the shapes are usually fired in continuous, car-type tunnel kilns; periodic-type, down-draft kilns are rarely used. The firing temperature is dependent on the nature of the clays used and on the intended service of the brick. Free and combined water are lost during firing, and iron and sulphur compounds as well as organic matter are oxidized. The particles of clay are ceramically bonded together to form a strong refractory. High-duty fireclay shapes are more resistant to spalling than medium- and low-duty products, and they are burned hard enough so that they are highly resistant to carbon-monoxide disintegration.

High-duty fireclay brick for coke-oven battery construction generally is refractory to about 1704°C (3100°F); however it does not have the creep resistance of silica brick. High-duty fireclay can be expected to deform 0.5 to as much as 4.0 per cent under a 172 kPa (25 psi) load at 1349°C (2460°F). Fireclay brick has nearly a linear thermal expansion from ambient temperature to battery operating temperatures, as compared to silica brick, which has nearly all of its expansion taking place below 593°C (1100°F). As a result, high-duty fireclay brick can be repeatedly cycled through low temperature ranges without spalling failure. As mentioned previously, medium- and low-duty fireclay refractories, being more dense, are more susceptible to thermal shock.

Mortars—There are two primary types of mortars used in coke-oven batttery construction, silica coke-oven mortar and fireclay coke-oven mortar. The silica mortar is further divided into a heat-set and an air-set type; normally the silica brick in a battery is laid up with the heat-set type.

Silica mortars are made from a sized silica sand for aggregate along with some small amount of bond clay. The air-set silica mortar is the same composition, but it has a small amount of dry sodium silicate added. The fireclay mortar is made from a sized calcined kaolin with a small amount of bond clay. No mortar used in battery construction contains any type of cementitious bond. Silica mortars are refractory up to about 1682°C (3060°F) and clay mortars up to about 1604°C (2920°F). Shrinkage on drying from a workable or trowelable consistency is an important parameter, and is specified as less than 2.5 and 4.0 per cent for silica and clay mortars, respectively.

Mortars used in coke-oven battery construction **do not** develop a coherent bond with the refractory to which they are applied at normal battery operating temperatures. Mortars therefore do not contribute to the structural integrity of a battery. The mortar used in a battery serves two primary functions. It is used to compensate for inconsistencies in brick size, and it acts as a sealant between brick to prevent gas leakage among the combustion-air, fuel-gas, and foul-gas systems within the battery.

Castables—Refractory concretes, made with calcium aluminate (CA) cements and various refractory aggregates, are used in various locations in a coke-oven battery. The service capability of a castable is dependent on the purity level of the CA cement as well as of the aggregate. There are three purity levels of CA cement, low-purity (Lumnite and Fondue), medium-purity (Refcon, Secar 50) and high-purity (CA25 and Secar 250). The aggregates can range from calcined clays to high-purity tabular alumina. Size grading of castable aggregates can also vary considerably.

From a practical standpoint, castables used in coke-oven battery construction should be limited to the low- and medium-purity types, because these materials are of adequate refractoriness for the specific area of application.

Low-purity castables are used to line the waste-heat

tunnels and flues; it can be formed up and cast in place or it can be gunited. The gunite application is faster and lower cost. Low-purity castables are also applicable to areas in the battery top, for example as a filler material in the parapet area. There is no specification on low-purity castables used in battery construction or repair.

Medium-purity castables are used for the higher temperature areas in a battery to which castables are applicable, for example door-plug refractories and stand-pipe lining.

Expansion-Joint Fillers—Expansion joints may be filled with vermiculite, an expanded mica, or with ceramic-fiber bats of adequate density and refractoriness.

Packing Materials—Packing materials are used to seal between metal components and refractory brick, for example between the jamb and jamb brick, in stand-pipe split rings and slip joints and around air and waste-heat boxes. Asbestos rope is the preferred and most commonly used packing material. Synthetic-fiber ropes have also been used with varying degrees of success.

Insulating Materials—Various types of insulating materials are used in battery construction, from insulating fire brick to low-refractoriness insulating concrete made of an expanded-clay or shale aggregate (e.g. Haydite) and Lumnite calcium aluminate cement. Insulating fireclay bricks which are available from an 871°C (1600°F) class to a 1760°C (3200°F) class are used to insulate the battery top, areas behind the buck stays, regenerator walls and pinion walls. Block insulation (e.g. Johns-Manville Superex 2000) can be used in non-load-bearing areas. Insulating concretes can be employed as fillers for irregular areas throughout lower temperature areas of the battery.

Acid-Resistant Refractory Construction—The battery-stack lining is of acid-resistant construction, which consists of high-density acid-resistant fireclay brick laid up with acid-resistant mortars. Acid-resistant brick are usually made by the stiff-mud process and are high fired so that low porosity (6 to 11 per cent) and permeability result. Acid-resistant mortars are generally sodium-silicate-bonded fireclay aggregate. Ambient temperature levels are critical for good acid-resistant construction. Mortars generally will not cure at temperatures below 13°C (55°F).

Fired Fused Silica—Silica (quartz) that is electrically fused, cooled, reground, then formed and refired is referred to as a fired fused-silica shape. Fused silica is essentially amorphous (noncrystalline), containing a maximum of 0.5 per cent quartz and 4.0 per cent cristobalite. Fired fused silica has a nearly flat thermal expansion curve, expansion being on the order of 0.72 K^{-1} (0.4 × 10^{-6} in./in./°F); therefore, it has excellent resistance to thermal shock. A fairly common use of fired fused silica is for door-plug refractories; however, refractory cost is 2 to 3 times that of a castable or brick door plug. Fired fused silica has also been used as jamb brick with varying degrees of success.

Fused silica grain along with calcium-aluminate cement is used to make fused-silica castables. The castables, however, do not have the thermal-expansion properties of the fired bodies. In general, the castables made with fused-silica aggregate are subject to similar temperature-volume relationships as normal silica-alumina castables in that aggregate sizing and cement content can have marked effects on volume stability.

Common Brick—Common red brick is used in a number of non-critical areas in battery construction, primarily in areas of low temperature and minimal exposure to thermal cycling. Common brick is a dense-fired clay body. It is used as a filler in the battery roof, backup material in the pinion walls and often as an underlying course on the pad.

COKE-OVEN REPAIR WORK

Coke-oven batteries will have an operating life of twenty to forty years, depending upon operating conditions and battery maintenance. Usually, after twenty to thirty years of service, a battery will require specific repairs to the refractories, steelwork or machinery. These repairs, if properly performed, will extend the life of the battery.

Since coke-oven repair work requires considerable expertise, only a few companies are qualified to make battery repairs. The complexity of coke-oven repairs will not permit a detailed description of the procedures; however, Wilputte Corporation, one of the recognized leaders in coke-oven repairing, offers the following general description of coke-oven repair work.

Brickwork repairs usually undertaken are the replacement of end flues, the replacement of oven walls between oven floor and oven roof, and emergency repairs inside the oven chamber.

Depending on how well a battery has been maintained, an end-flue repair (usually 2 flues or more) is undertaken after about 20 to 25 years of service. This repair can be performed while the battery is in the hot condition making coke or after the battery has been completely cooled to atmospheric temperature. During the repair the buckstays may be replaced and work may be done on the oven doors, collecting mains and oven top. An end-flue repair is estimated to last for approximately 5 years, after which additional repairs may be required.

Oven-wall brickwork repair may consist of the replacment of through walls (pusher side to coke side) above the oven floor and below the oven roof. At the same time brickwork repairs may be done in the corbel area, and face brickwork and individual oven roofs may be replaced.

If only a few oven walls are to be replaced, the repair is done with the battery in the hot condition with one or two buffer ovens (empty ovens) on each side of the wall being repaired. The other ovens of the battery may be producing coke. Before demolition of the damaged wall, roof-support beams must be added in the adjacent ovens, insulating panels must be installed on the adjacent hot walls, charging-car rail bridges must be installed on the oven top, and support provided for the oven roof under which the wall will be replaced. Care must be taken to maintain the heat in the regenerators and on the adjacent walls. On individual oven-wall replacement, the heatup period for the repaired oven is approximately 15 days.

If an entire battery of through walls is to be replaced from the oven floor to the underside of the oven roof, the repair may be performed with the battery in the idle hot condition or with the battery refractory cooled to atmospheric temperature.

The idle hot condition means that the ovens are maintained hot while the repairs are done. Pairs of walls are replaced together with the adjacent ovens maintained in the hot condition. Each pair of walls being replaced is called an increment and work may proceed on several increments at one time. The heating of each increment of the repaired walls takes approximately 15 days. Ovens may be placed back in service in increments before all the walls have been replaced.

The replacement of the through walls of an entire battery may also be accomplished by permitting the battery to be cooled down to the cold condition under controlled conditions before the repairs are made. The cool down may take up to 21 days during which time pressure is maintained on the oven walls by adjusting

the upper and lower tie rods. The oven roofs must be supported as described for the hot-repair procedure. After the walls are rebuilt and before heatup, all the checkers must be removed from the regenerator and any cracks in the pillar walls must be cleaned and packed with ceramic floss before the checkers are replaced so as not to hinder brickwork expansion during heatup. The repaired battery is heated up in the same manner as a new battery.

A repaired battery with all new through walls, with proper maintenance can be expected to have a life of approximately 15 years at a fraction of the cost of a new coke-oven battery.

With through-wall replacement of an entire battery, with the repairs done in either the hot or cold condition, additional items usually replaced include the side buckstays above the bench, collecting mains, doors and jambs, oven-top paving, and tie rods and springs.

SECTION 5

PREHEATING, STAMP-CHARGING AND BRIQUETTING OF COALS

The growing world-wide shortage of high-quality coking coal necessitates developments and measures which will enable the present range of coking coal to be augmented by coals which are less suitable for cokemaking. Therefore, in addition to the conventional method of preparing coal and charging it into by-product coke ovens, several other methods of coal preparation and charging exist. Some of these latter methods have gained wide acceptance in the coking

industry and will now be discussed.

COAL PREHEATING

Coal drying and preheating prior to charging the coal to the battery ovens is not a novel idea. Thermal treatment of coals has been investigated since the 1920's. Earlier work involved heating the coal to reduce the moisture content to 1 or 2 per cent and then charging the coal at ambient temperature into the coke

FIG. 4—35. Overall view of a coal-preheating facility. (Courtesy, Dr. C. Otto and Company.)

ovens. In the 1950's studies were made of preheating the coal to elevated temperatures, up to 300°C (570°F), in an essentially inert atmosphere so as not to affect the coking characteristics of the coal by softening, devolatization or oxidation. The major problem encountered with the preheated coal was the potential hazard in charging the ovens. In the 1960's and 1970's, serious interest in pilot-plant-scale activity became strong in Europe and the United States, culminating in the construction of commercial coal-preheating facilities in England, South Africa, France, the United States and Japan (see Figure 4—35).

The major advantages of carbonizing preheated coal are listed below:

(1) Coal preheating improves the strength and hardness of coke and allows the use of poorer quality coals. The coke produced from preheated coal is of closer size range, has smaller pores, higher true and apparent densities and a more homogeneous structure.

(2) Preheated coal increases oven throughput because of reduction in coking time. The full extent of the production increases has not been realized because of lower than anticipated bulk density of coal in the oven associated with preheated coal charging.

(3) The drying and preheating of coal in a flash dryer or fluidized bed is more efficient than in a battery oven. The higher thermal efficiency of coal-preheating systems allows reduction in overall fuel requirements.

(4) Charging preheated coal of fairly uniform moisture content results in more uniform heating of the batteries and less thermal shock to the refractory brickwork when compared to wet-coal charging.

(5) Closed oven charging via pipeline or chain

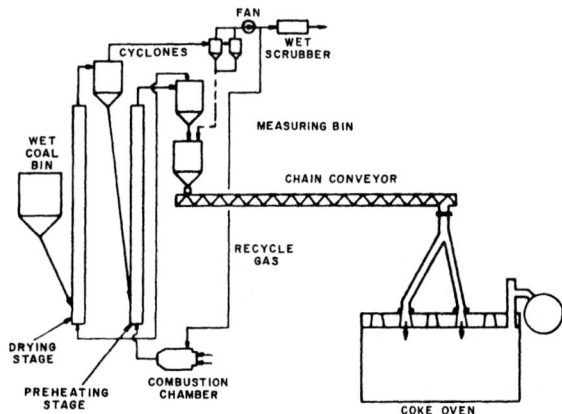

FIG. 4—36. Schematic arrangement of a Rosin-Buttner type of coal-preheating unit.

conveyors has a potential of eliminating emissions during the coal charging phase.

The disadvantages of coal preheating are the potential safety and operational problems associated with preheating and handling of very fine hot coal, overfilling of battery ovens, higher initial evolution of gases during the charging step and increased amount of coal carryover to the by-product area.

Preheating Systems—Basically two different coal-preheating systems are used in drying and preheating of the coal: the flash-drying/preheating system such as Rosin-Buttner shown in Figure 4—36 and a fluidized-bed and entrainment system such as Cerchar shown in Figure 4—37. Other processes use combinations of different preheaters, conveying and charging systems.

The **Rosin-Buttner type** preheating unit is a two-stage entrainment dryer-preheater. Wet coal is fed by

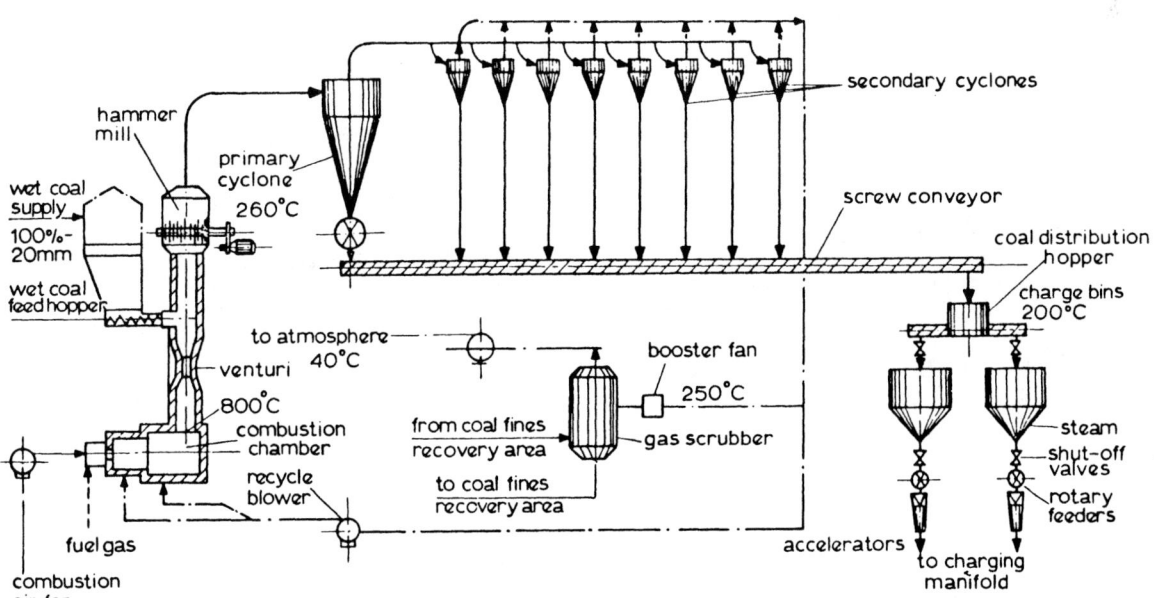

FIG. 4—37. Schematic arrangement of a Cerchar-type coal-preheating plant.

a centrifugal feeder into the first stage where it is flash-dried and preheated by concurrent flow of hot gases from the second-stage unit to approximately 80°C (175°F) and 2 to 3 per cent moisture. The entrained coal is separated from the gas in a cyclone and is conveyed to the second-stage unit in which the coal is preheated to between 200° and 230°C (390° and 445°F). The hot gases flowing concurrently are supplied in part by burning natural gas or coke-oven gas in a combustion chamber supplemented by a recycle stream of inert gas from the first stage. The preheated coal is separated from the gas stream by a series of cyclones and is conveyed to a set of storage bins to be charged to the battery either by drag conveyors or a hot-coal larry car. Part of the process gas from the first stage is scrubbed or cleaned in the electrostatic precipitator and the rest is recycled back to the combustion chamber.

The **Cerchar preheater** is a single-stage fluidized-bed and entrainment unit. The wet coal is fed by a double screw conveyor into a flash-drying section. The velocity of gases is high enough to transport all the coal upward. The coal is dried and preheated in a fluidized bed in which a crusher disperser decreases the size of coal until it is removed from the preheater by entrainment. Coal is separated from the gas stream in a series of cyclones and is conveyed to a storage bin. The off gas is split, part of it returning back to the combustion chamber and part of it being cleaned in a wet scrubber.

The most common coal-preheating processes operating in the world today are listed below:

The **Coaltek system** is a combination of a Cerchar single-stage fluidized-bed entrainment preheater coupled with a pipeline charging system developed by Allied Chemical. The preheated coal is discharged from the storage bin to weigh bins. The bins are pressurized with steam after the proper amount of coal is loaded, and coal is then conveyed via a pipeline aided by additional steam jets to the battery oven. The coal and

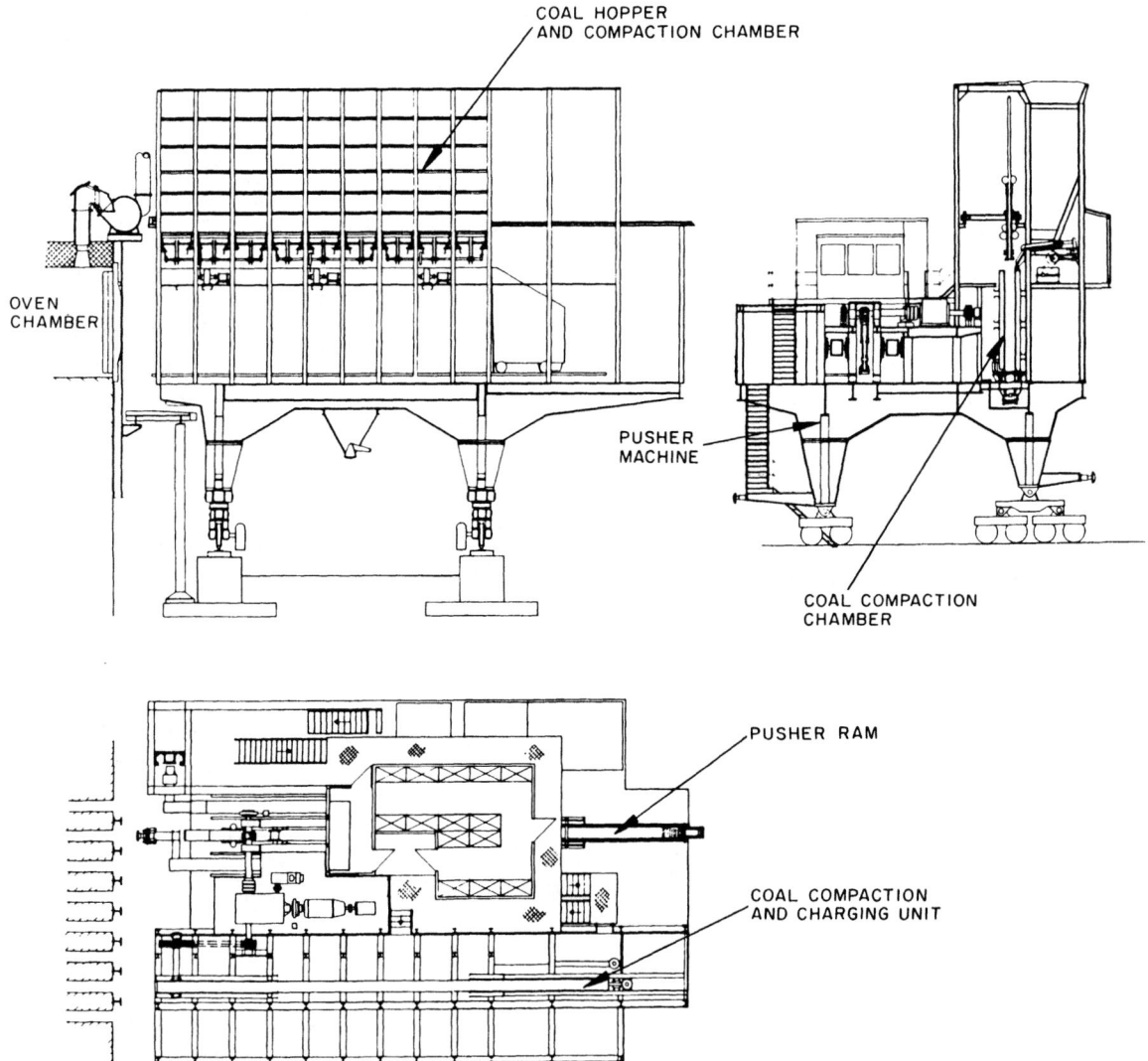

FIG. 4—38. Top, side and end elevations of a coke-oven pushing machine incorporating a stamp-charging unit.

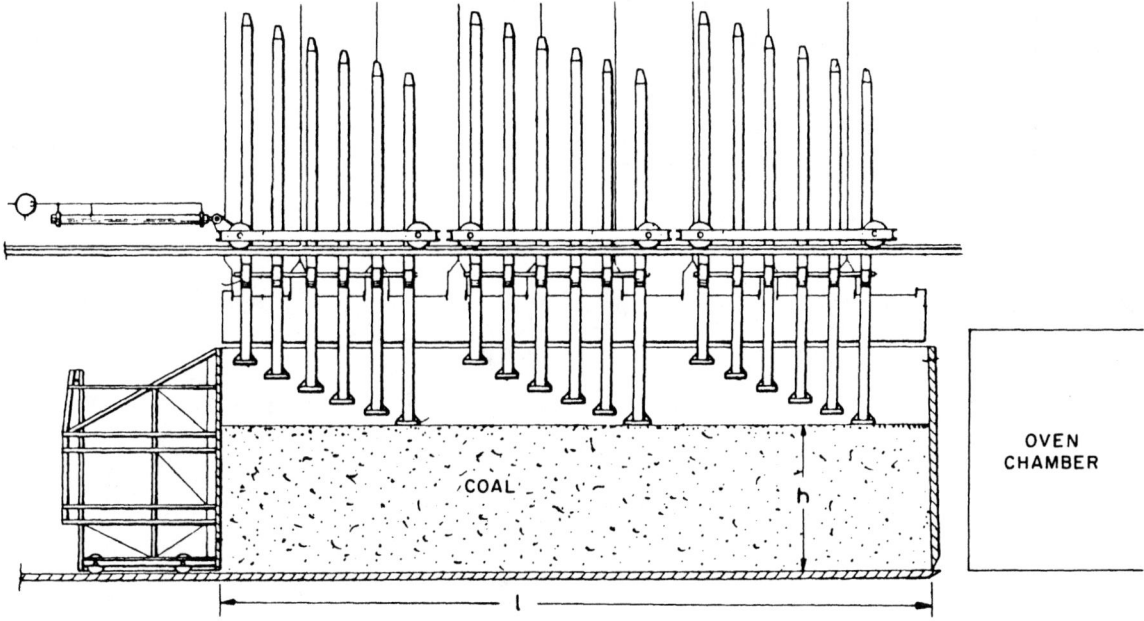

FIG. 4—39. Schematic section showing side view of a stamp-charging unit.

steam entering the oven are then separated by a specially designed baffle, and the gases emitted during charging are channeled to a specially designed collecting main.

The **Precarbon process** developed by Bergbau-Forshung and Didier Engineering of West Germany utilizes a Rosin-Buttner type of double-stage entrainment dryer-preheater coupled with sealed drag-conveyor charging of the preheated coal by way of a charging-chute system to the battery oven. The charging is accomplished by specially designed pneumatic level probes which automatically stop charging preheated coal when predetermined coal lines are obtained. Two of the three 6-metre batteries being operated by United States Steel at its Gary Works are equipped with Precarbon processes.

The **Otto-Simon Carves system** is a combination of a Rosin-Buttner type double-stage entrainment preheating system and a hot-larry charging system. This system was one of the first ones commercially installed in South Africa and in England.

A preheater developed by **Buttner-Schilde-Haas** of West Germany was installed by United States Steel in 1978. The BSH process utilizes a single-stage entrainment-type preheater which is equipped with a solids-recycle system. The classifier at the exit of the column allows recirculating of a variable amount of coarser material back to the preheat column. The preheat system is coupled with a hot-larry-car charging system.

The world-wide steel industry is challenged with providing facilities to produce high-strength coke from increasingly poorer coking coals. Coal preheating combined with charging of preheated coal is an economically viable process that offers the steel industry a method for attaining this goal, as demonstrated by general acceptance of the practice of coal preheating.

STAMP CHARGING

Stamp charging is a process where the entire coal charge to the coke oven is stamped, or compressed, and then pushed into the oven for coking. It is an old process with the greatest amount of activity in the Saar Basin area near the France-Germany border, in Eastern Germany near the Polish border, and in Poland. Other areas of use include Czechoslovakia and India. In the Saar basin, stamp charging has been in use for over a hundred years. This process has never been used in the United States due to an abundance of good coking coals.

In the stamping process, the coal blend is charged into the stamping chamber, Figures 4—38 and 4—39, which is part of the pusher machine and has slightly smaller dimensions than the coke oven. The charging to the stamping machine is done in layers, with each layer approximately 460 to 500 mm (18 to 20 inches) in thickness. Each layer is compressed using two stampers which are lifted mechanically and fall by their own weight on the coal. When enough layers to fill the oven have been made, the side walls of the chamber are withdrawn and the coal is pushed into the oven on the movable bottom. The front panel is removed on the coke side and the bottom is withdrawn. The stamping operation takes about 20 minutes, depending on the height of the oven.

The main advantage of stamp charging is the increased bulk density of the charge. Bulk densities of up to about 1.15 metric tons per cubic metre (72 lb/ft³) have been achieved using this process. Excessive coking pressure is avoided by the addition of coke breeze to the blend, which cuts down on blend expansion, and by using less pressure-generating coals in the blend. The stamping process brings the coal particles into more intimate contact with each other, which en-

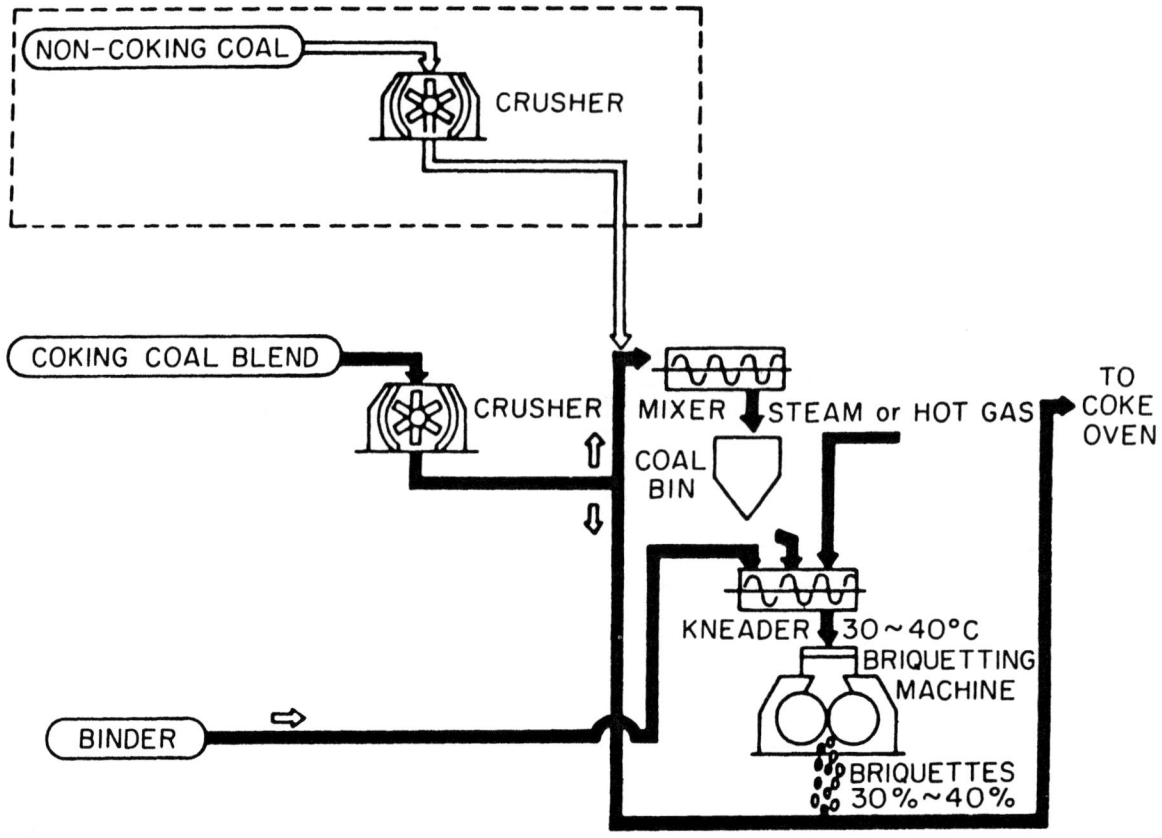

FIG. 4—40. Simplified flow diagram of Sumi-Coal system.

hances the coking properties. This process was developed as a way to make high-strength coke from the relatively poor-quality coking coals found in these areas. The coals are either too low in rank and/or too high in organic inert contents to produce strong coke by conventional means without bringing in expensive, better quality coking coals.

Stamp charging produces a more abrasion resistant coke, and the yield of blast-furnace-size coke is higher than with conventional charging. In addition, even though some coke-oven volume is lost due to the thickness of the charging chamber floor and the oven taper, and longer coking times are required, the amount of coke throughput is increased about 5 per cent over conventional charging.

The initial capital costs are high due to the expense of the combination stamping-pushing machine and the complex charging procedure. There is no need for a larry car, however. Gas-emission problems used to occur during charging, but new procedures such as charging with suction have helped in this area. The need for better coke quality and higher coal throughputs means that stamp-charging techniques have to be improved and further developed, especially in view of the fact that chamber heights are continually being increased.

BRIQUETTE BLENDING

Sumitomo Metal Industries and Nippon Steel Corpo-
ration in Japan have been engaged for many years in producing high-strength coke from partially briquetted coal blends. This enables them to use up to about 20 per cent of what they consider to be noncoking or poorly coking coals. This was out of necessity due to the increasingly large amounts of Australian, Canadian and Soviet Union coals they import. These coals have low fluidities and high organic inert contents which make them much poorer in coking quality than American medium-and low-volatile coals, but they are much less expensive than American coals, and the supply is more dependable. The processes described in this section involve partial briquetting of the coal blend with a binder, and charging the briquettes into the coke oven along with the unbriquetted coal blend.

a. **Sumi-Coal System**—In this system, about 30 per cent of the coal blend is briquetted, with the briquettes containing about 7 per cent tar or pitch as a binder. This process enables the use of high-inert, low-fluidity coals in the blend, while still producing high-strength coke.

In the briquetting process, (Figure 4—40) coking coal is drawn from the fine-coal circuit and crushed to about 85 per cent minus 3 mm (minus ⅛-inch). The non-coking coals, which are used only in the briquettes, are crushed through a special hammermill also to about 85 per cent minus 3 mm (⅛ inch). The coals are mixed in a ratio of approximately 60 to 65 per cent non-coking coal to 35 to 40 per cent coking coal, and

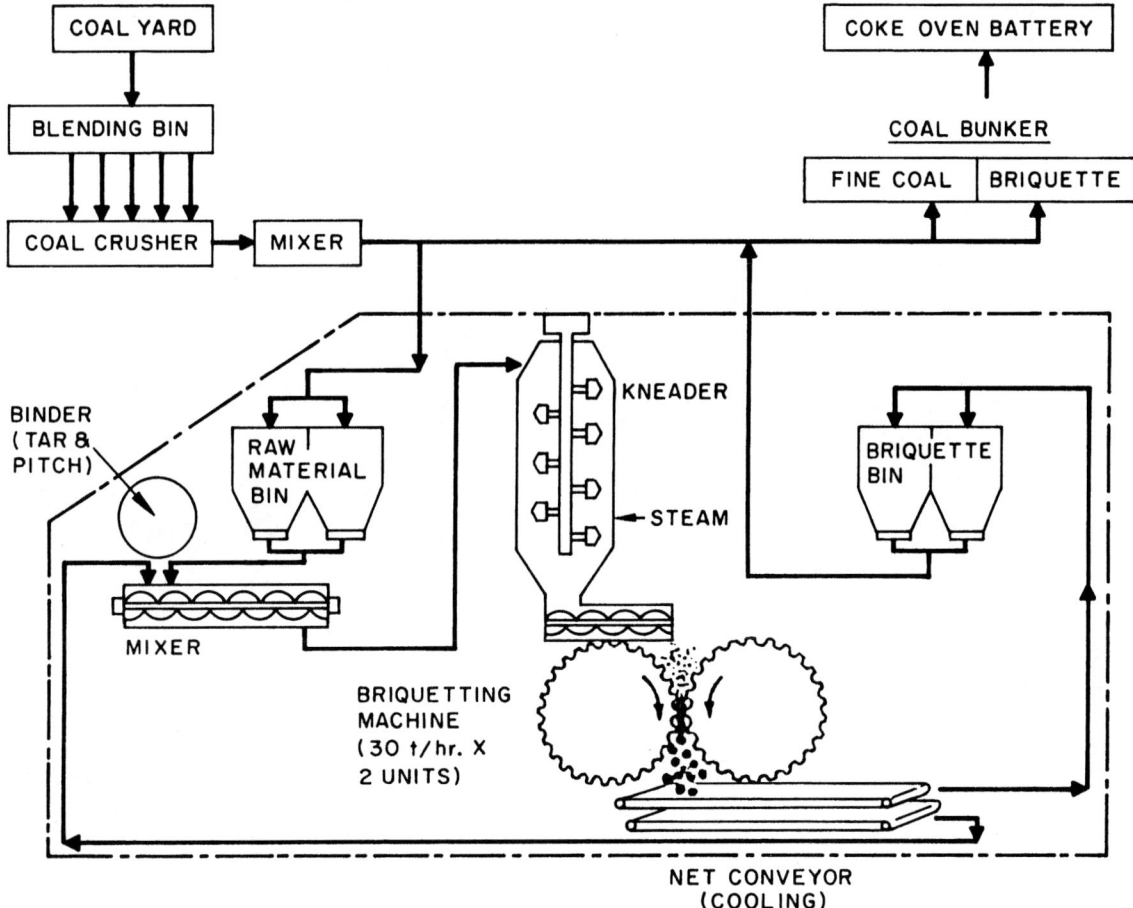

FIG. 4—41. Schematic flow of diagram of Nippon Steel's briquette-blending system.

brought by chain conveyor to three briquetting-line hoppers. Each hopper feeds two externally-heated kneaders where the binder, which has been heated to above its softening point, is added, and the materials are well mixed. The heated mixture is conveyed by screw feeder to the briquetting machine where briquettes, approximately 19.665 cm³ (1.2 cubic inches) in size, are made in a double-roll briquetting machine. Previously the unbriquetted material was recycled through the briquetting machine, but now it is fed to the fine-coal conveyor belt. The briquettes are then mixed directly into the regular blend for charging.

The briquette blending in the Sumi-Coal system enables the use of up to about 20 per cent non- or weakly coking coals in the blend without adversely affecting coke strength. Use of less than 20 per cent results in improved coke strength. The improvement is believed to be due to the more intimate contact in the briquettes between coking and non-coking coal particles, the improvement in fluidity and dilatation due to the binder, and the approximately 48 kg/m³ (3 lb/ft³) increase in the bulk density of the charge. The bulk-density increase, though, is offset by an approximately 10 per cent longer coking time required, so that productivity is basically the same at the 30 per cent briquette level.

Sumitomo feels some of the merits of the process are that it generates less coke breeze due to the increased strength, it provides stable blast-furnace operation and a reduction in coke consumption in the blast furnace, and, therefore, a correspondingly lower amount of imported coal is needed.

In addition, Sumitomo lists advantages of the briquette blending over coal preheating to be that more of the cheaper non- or poorly coking coal can be incorporated into the coal blend, operating costs are cheaper, simpler and more stable equipment is used, and installation into existing plants is easy.

b. **Nippon Steel System**—The briquette-blending system employed by Nippon Steel (Figure 4—41) is similar to that employed by Sumitomo. The principle of bringing the coal particles in more intimate contact in the briquettes and the effect of the binder are basically the same. The major differences are that:

(1) The non-coking coal is used to replace approximately 10 per cent of the American medium- and low-volatile coals, and the blend is adjusted accordingly.

(2) The non-coking coal is **not** kept separately from the coking coal, but is in the blend; therefore, some of the non-coking coal is not briquetted.

(3) In the mixing and kneading procedure a higher

temperature is used, so that the briquettes are cooled before storage.

(4) The briquettes are stored separately and not blended with the regular charge blend until just before entering the larry car for charging.

As with the Sumitomo system, Nippon Steel uses approximately 30 per cent briquettes in their coal blends. By 1976, about 27 per cent of Nippon Steel's

cokemaking capacity had briquette-blending facilities. Their policy has been to use the briquette-blending process on existing batteries, and to use coal preheating on new battery construction.

Nippon Steel began work on the briquette-blending process in the late 1950's and early 1960's, with the first commercial use on-line in 1971.

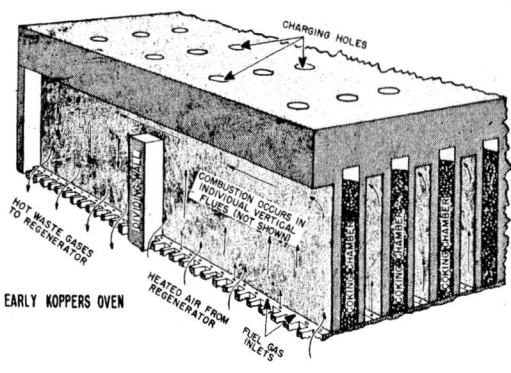

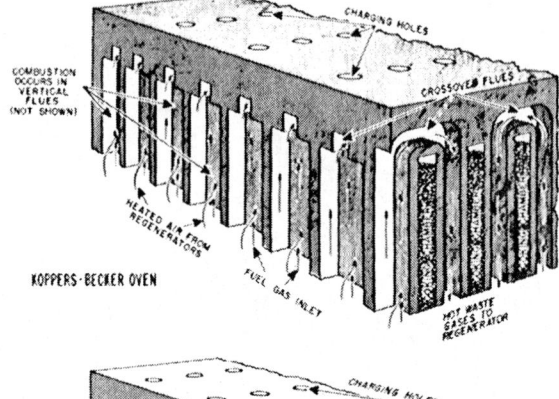

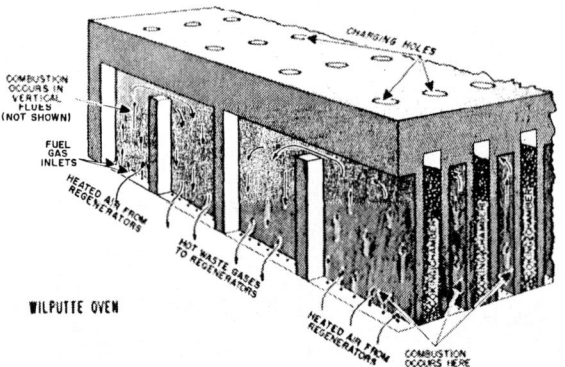

FIG. 4—42. Schematic representation of the differences in firing methods employed in the three most common types of by-product coke ovens used in the United States. Individual flues are not shown. The firing procedures shown are for a single phase of heating which is reversed at the end of a specified period, at which time the flow directions indicated by the arrows become reversed.

SECTION 6

SOME PROPRIETARY DESIGNS OF MODERN BY-PRODUCT COKE OVENS

Introductory—The most prevalent designs of by-product coke ovens in the United States have been the Koppers and Koppers-Becker ovens designed and built by Koppers Company, Inc., and the double-divided (also called four-divided) oven design of Wilputte Corporation, a Salem Company. Ovens of these designs comprise a majority of the by-product coking capability of this country. However, the Dr. C. Otto, Carl Still and Didier designs have gained popularity recently. Metallurgical coke is the primary product, since the coal chemicals recoverable in the by-product coking process are meeting competition from their counterparts derived from petroleum (petrochemicals).

The chief differences in design of these three types are in the heating system. Figure 4—42 illustrates, by simplified sketches, how each of these types is heated. A fourth general type, the hairpin- or twin-flue oven, is discussed later in this chapter; Figure 4—59 illustrates one design of twin-flue oven.

The Koppers oven, more technically referred to as a regenerative, single-divided oven, was the most promi-

nent from about 1916 to 1928, and many are still in operation. In a typical oven of this design (Figures 4—42 and 4—43) all parts except the regenerator checkers and battery top are constructed almost entirely of high-quality silica brick.

The majority of ovens of this type initially constructed in the United States had a coking volume of approximately 14 cubic metres (500 cubic feet). The general dimensions of these ovens were: length, 11.28 metres (37 feet) from face to face of the doors; height, 3 metres (9 feet, 10 inches) from floor to roof; and width, tapering from 430 millimetres (17 inches) at the pusher end to about 495 millimetres (19½ inches) at the discharge end. Usually, four charging holes were provided in the top of each oven for admitting the coal charge, while a separate opening in the top at one end provided an outlet for volatile matter. The oven is of the vertical-flue type with individual regenerative chambers (Figure 4—43). The heating chamber has a total of about thirty vertical flues per flue wall.

The flues are provided with openings to the regenerator chambers, the fuel-gas mains, and to a large horizontal flue on a level a little below the top of the coking chamber. A dividing wall near the middle of the regenerators separates the heating system, except the horizontal flue, into two parts with (in the case of the thirty-flue wall) sixteen vertical flues on the narrower end of the oven and fourteen on the wider end. Each end, approximating half of the oven, thus may be heated alternately, and in practice the reversals are made at regular intervals for each battery of ovens by a reversing mechanism controlled by a timing device. Two large underground flues, one on each side, extending along both sides in front of and parallel to the battery, and connected to the checker chambers (regenerators) by cast-iron air boxes, provide for escape of the products of combustion. These flues lead to a stack about 61 metres (about 200 feet) high at one end of each battery to furnish the draft necessary to draw the gases through the heating system.

The Koppers-Becker oven (Figures 4—42, 4—44, 4—45, 4—46 and 4—47) employs a different flue arrangement whereby gas is burned on an entire wall simultaneously (both pusher and coke sides). The products of combustion from two or more vertical flues of the "on" walls in which fuel gas is burning enter short bus flues and thence are conducted over the top of the oven through cross-over flues to a companion series of bus flues whereby the vertical flues of the entire "off" wall simultaneously are conducting waste gas to the regenerators. On reversal, the opposite conditions obtain. This design, by eliminating the large horizontal flue, is well adapted to ovens of greater size, and to

heating with blast-furnace gas of low calorific value. Since the flues in each wall (coke side to pusher side) are connected only to the flues in its companion wall, there are cross-over flues over only every other oven, and the battery thus is limited to an uneven number of ovens.

The Wilputte by-product coke oven is offered with various divided zone heating systems, each depending on the particular application. The double-divided or four-divided oven (Figures 4—42, 4—48 and 4—49) has two outer zones and one double inner zone. Larger ovens with coking chambers of 39.64 cubic metres (1400 cubic feet) and over will be of similar design, but they will be six- or eight-divided, especially when required to operate on lean gas. Twin-flue or hairpin-flue ovens of Wilputte design are discussed later in this section.

In recent years, larger and faster operating by-product coke ovens have been constructed. Designs have been keyed to higher production rates and improved efficiency, with extensive use of automatic control and mechanization. Modern coke-oven batteries planned and constructed with these features will be described in the remainder of this section. The designs to be described first originated in the United States, and represent those of Koppers Company, Incorporated and the Wilputte Corporation, a Salem Company. These will be followed by descriptions of oven designs developed by the following companies in the Federal Republic of Germany (West Germany), in the order named: Didier-Werke A.G. (Essen); Dr. C. Otto and Company G.m.b.H. (Bochum); Firma Carl Still (Reck-

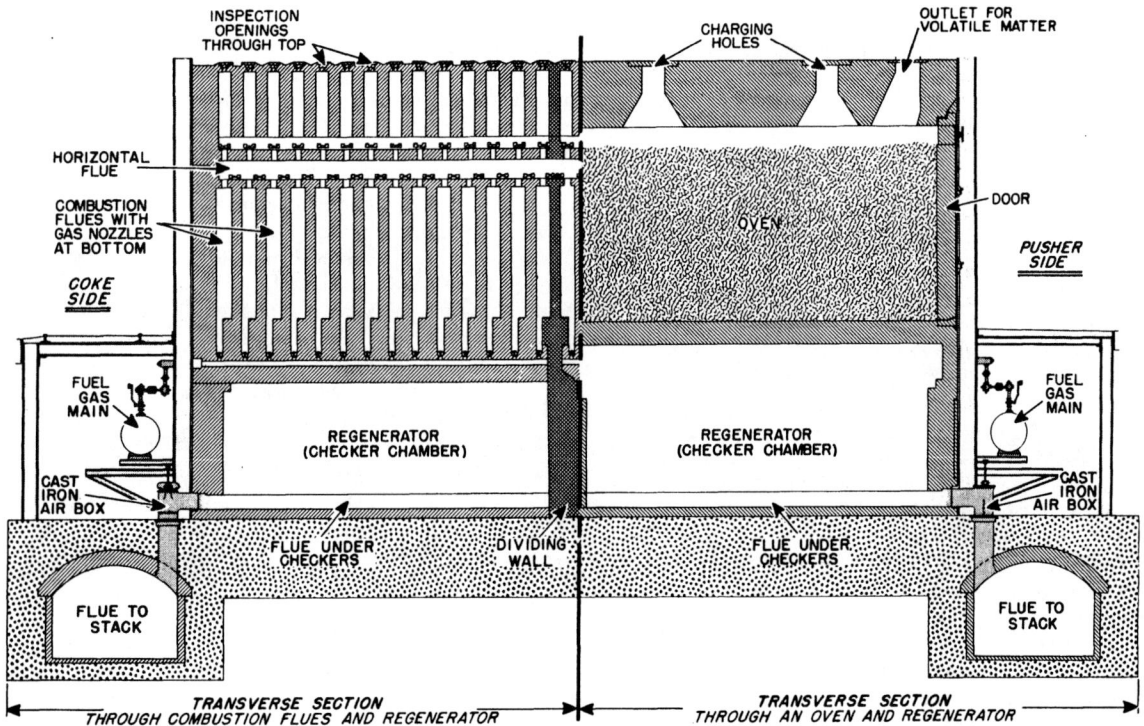

FIG. 4—43. Transverse sections of Koppers regenerative single-divided by-product coke-oven battery. Section at left is through combustion chambers (flues), that at right is through oven chamber. See also Figure 4—42. (Courtesy, Koppers Company, Inc.)

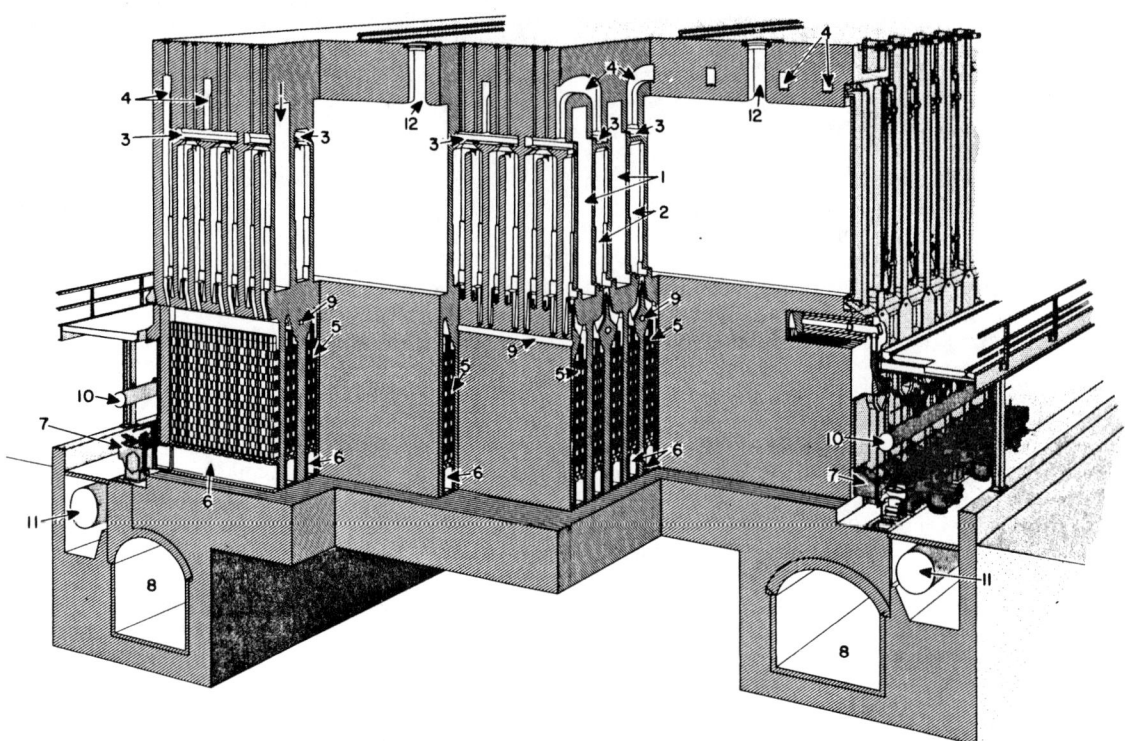

FIG. 4—44. General perspective cut-away section of Koppers-Becker combination ovens with gun-flue heating facilities (see Figure 4—47. for underjet-fired type). 1. Oven chamber. 2. Vertical combustion flues. 3. Horizontal flues. 4. Cross-over flues. 5. Regenerators. 6. Oven sole flues. 7. Gas and air connections to waste-gas flue. 8. Waste-gas flues. 9. Gas ducts for coke-oven. 10. Oven gas main. 11. Blast-furnace gas main. 12. Charging holes. (Courtesy, Koppers Company, Inc.)

linghausen); and Nippon Steel Corporation of Japan.

The descriptions of modern by-product coke ovens in this section were prepared especially for this book by the respective designers and builders, whose generous co-operation is hereby acknowledged.

KOPPERS COMPANY, INCORPORATED

Modern by-product coke-oven batteries designed and built by Koppers Company, Incorporated include a number of different types and variations which can be tailored to the requirements of each individual installation. The three types of ovens offered by Koppers are (1) the Koppers single-divided oven; (2) the Koppers-Becker oven; and (3) the Koppers twin-flue oven.

These oven types are built as either underjet or gun-flue ovens. The heating systems are designed as combinations for heating with either lean or rich gas, such as blast-furnace gas or coke-oven gas; or alternatively, for heating with rich gas only.

Oven dimensions vary considerably, depending upon whether the new ovens must meet some dimensions or clearances set by existing ovens or oven machinery. Where no such restrictions exist, the trend has been toward higher and longer ovens. Oven heights range from 3 metres (9 ft, 10 in.) to 6.5 metres (21 ft, 4 in.). Oven length ranges from about 11.3 metres (37 ft) to about 16 metres (52 ft, 6 in.). Oven average width usually is between 432 mm (17 in.) and 467 mm (18⅜ in.), although some narrow ovens of 362 mm (14¼ in.) average width have been built to handle poorly coking

coal.

Any of the three oven types can be charged with either wet coal or dry preheated coal. Koppers has built three large batteries of 6.15 metre (20 ft, 2 in.) ovens to produce high-quality blast-furnace coke from preheated coal. Two of these batteries are of the Koppers twin-flue type heated with coke-oven gas, while the third is of the Koppers-Becker combination type heated normally with blast-furnace gas.

Koppers Single-Divided Oven—The Koppers single-divided oven was the original oven type developed by Koppers. Over the years the design details have been improved and the oven dimensions increased to the point where this design provides a rugged, simple battery where limited oven height (under 4.27 metres or 14 ft) is required.

While most of the Koppers single-divided ovens have been built as gun-flue ovens, this design can be built as an underjet type. It is usually designed for heating with rich gas only.

The distinguishing features of the Koppers single-divided oven are the single division wall near the center of the regenerator chamber and the long horizontal flue connecting the tops of all the vertical heating flues in each flue wall. This design is illustrated in Figure 4—43.

The Koppers single-divided oven is designed so that air flows upward on one reverse through all the regenerators and vertical combustion flues on the pusher side of the battery, while the products of com-

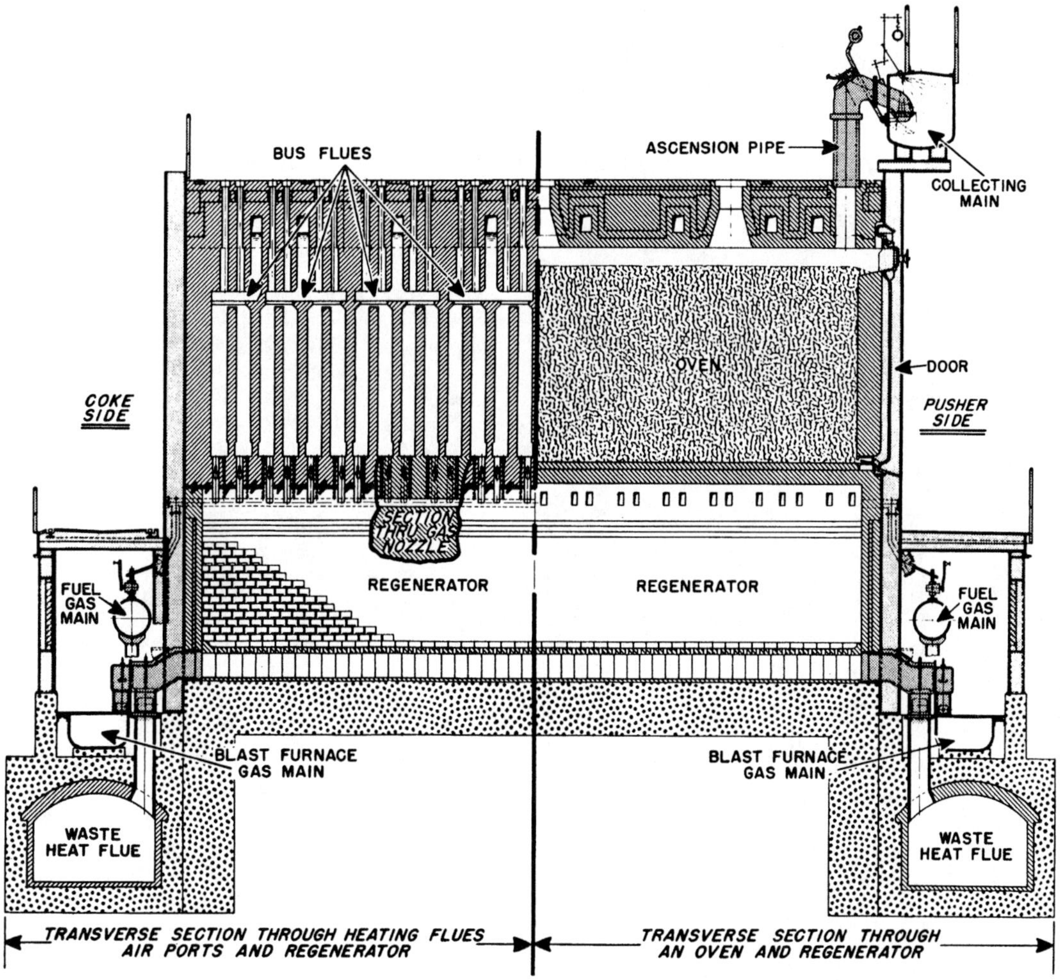

FIG. 4—45a. Transverse section through a Koppers-Becker combination gun-flue type by-product coke-oven battery. (Courtesy, Koppers Company, Inc.)

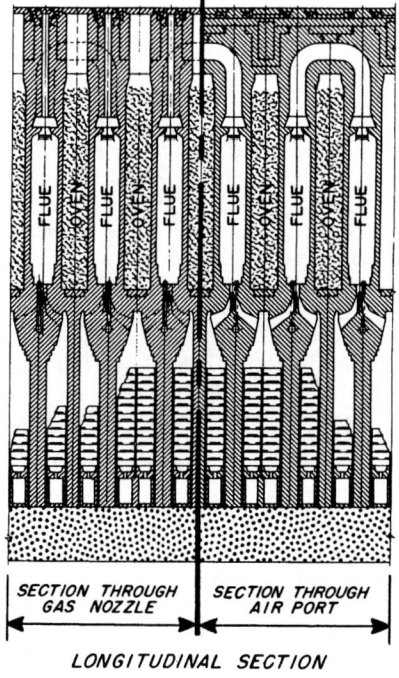

LONGITUDINAL SECTION

FIG. 4—45b. Longitudinal section (left) through gas nozzles and (right) through air ports of a portion of a Koppers-Becker combination gun-flue type by-product coke-oven battery. (Courtesy, Koppers Company, Inc.)

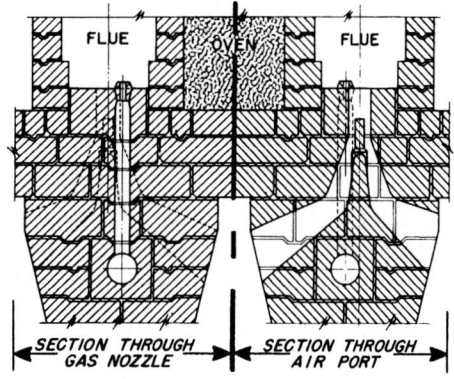

FIG. 4—46. Enlarged sections through gas nozzles and air ports of a Koppers-Becker combination gun-flue type of by-product coke oven, showing detail of part of Figure 4—45b. (Courtesy, Koppers Company, Inc.)

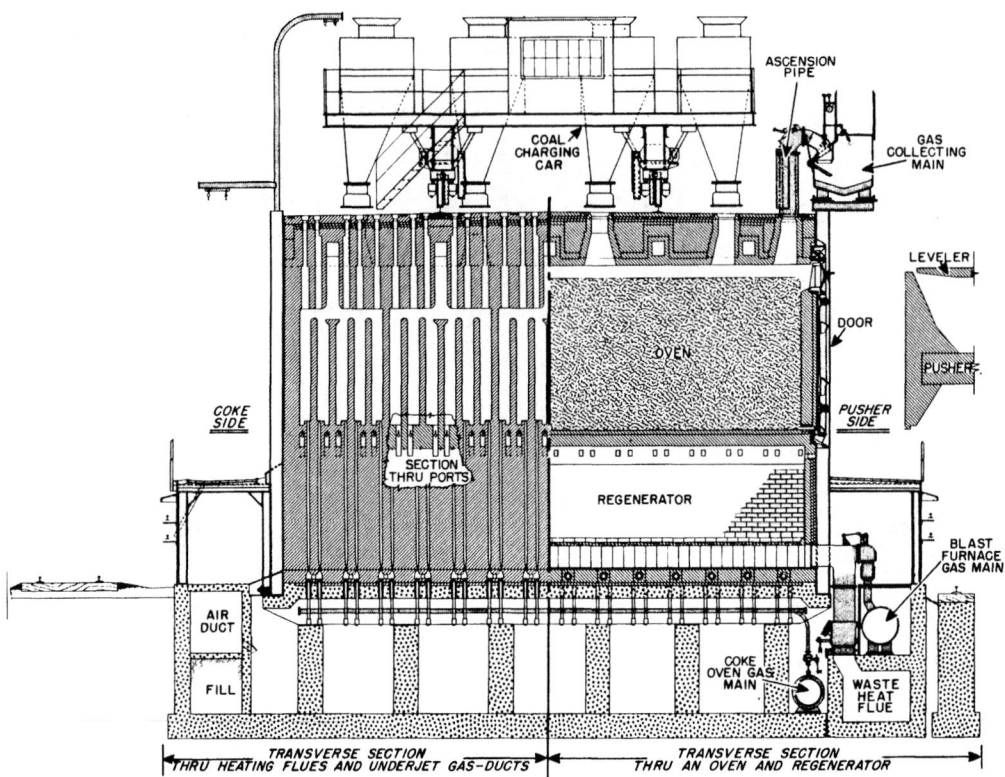

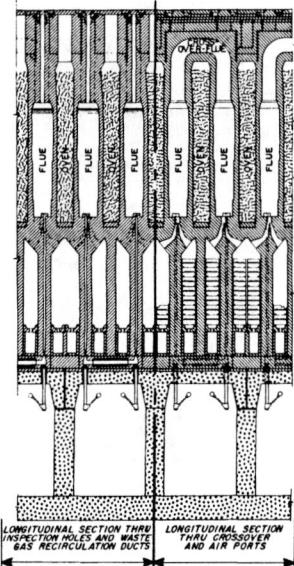

Fig. 4—47. Transverse and longitudinal sections through Koppers-Becker underjet-fired low-differential combination by-product coke oven. See Figure 4—52 for enlarged section of waste-gas recirculation ducts applied to a later oven. (Courtesy, Koppers Company, Inc.)

The proper flow of gas to each flue is controlled by calibrated nozzles, while the flow of air is controlled by setting sliding damper bricks in the horizontal flue at the top of each vertical flue. A single wide regenerator beneath each oven provides air to the flue wall adjacent to that oven.

High-quality silica brick is used throughout the entire regenerator walls and heating-flue walls, eliminating the need for any sliding joints (a serious cause of potential leakage) that would be required if clay or other dissimilar refractory materials were used in these important areas. The heating-flue walls are constructed of vertically and horizontally tongued-and-grooved hammerhead and liner brick, spaced with sturdy flue partition shapes (see Figure 4—50). The oven ends are firmly supported by strong buckstays, face plates, and rigid metal jamb castings bolted to the face plates. The jamb brick are provided with a vertical cleavage joint to permit expansion independently of the hotter interior heating-flue brickwork. These features are all provided in all three types of ovens built by Koppers Company, Incorporated.

Uniform distribution of up-flow air (and lean gas when used), as well as down-flow waste gas, between the sole flue and the regenerator chamber is accomplished by specially designed sole-flue ports. The size and taper of these ports are varied as required across

bustion flow across the horizontal flue and down through the coke-side vertical flues and regenerators into the coke-side waste-heat flue. On the next reverse, the flow is upward through all the coke-side regenerators and vertical combustion flues, across the horizontal flue and down through the pusher-side vertical flues and regenerators into the pusher-side waste-heat flue. This arrangement is known as a cross-regenerative system, and is free of any possible cross-leakage between up-flow fuel gas and down-flow waste gas.

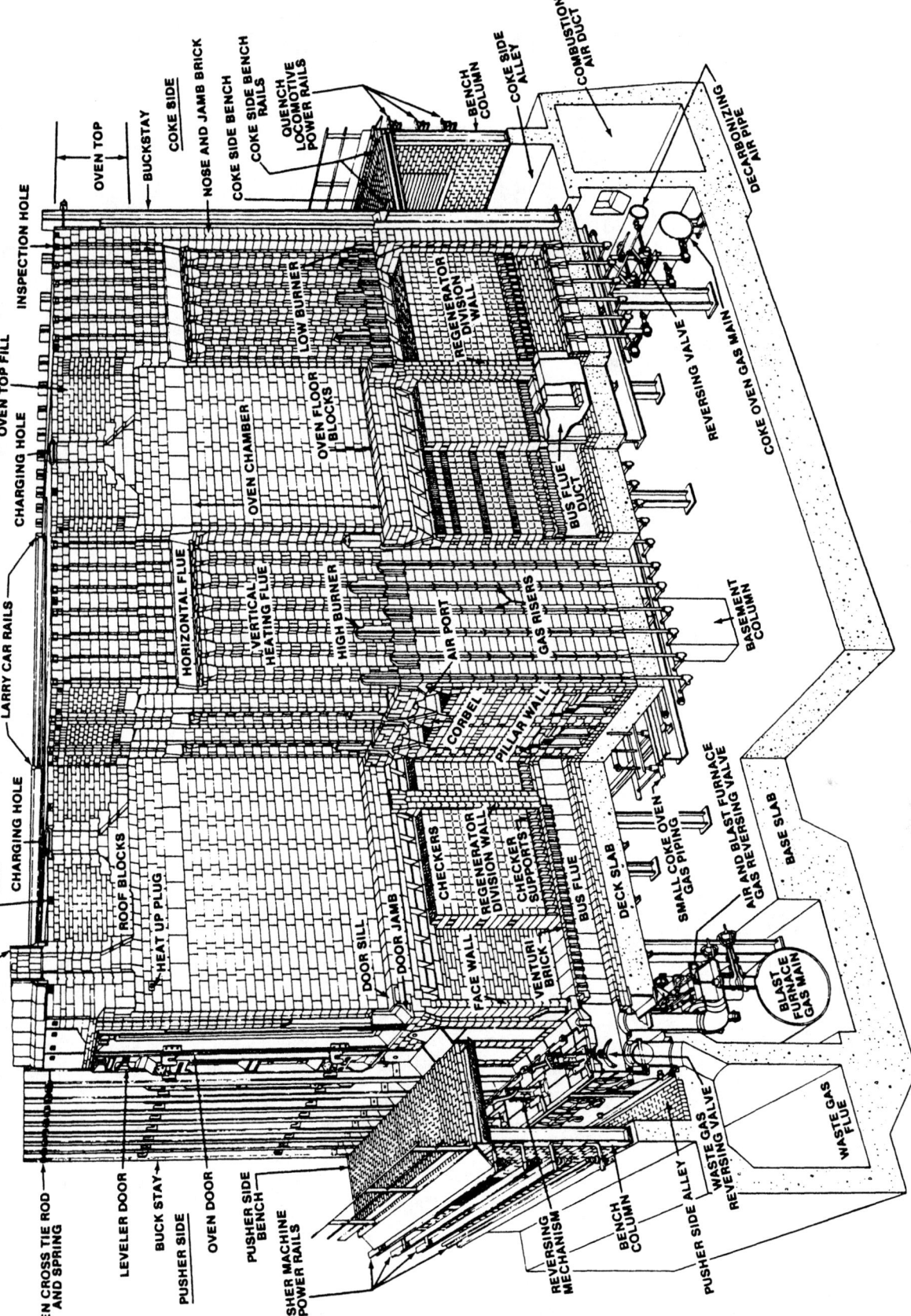

FIG. 4—48. General perspective "cut-away" drawing of a Wilputte four-divided combination underjet-fired by-product coke oven. The "rich" fuel gas referred to is coke-oven gas; the "lean" gas is blast-furnace gas. (Courtesy, Wilputte Corporation, a Salem Company).

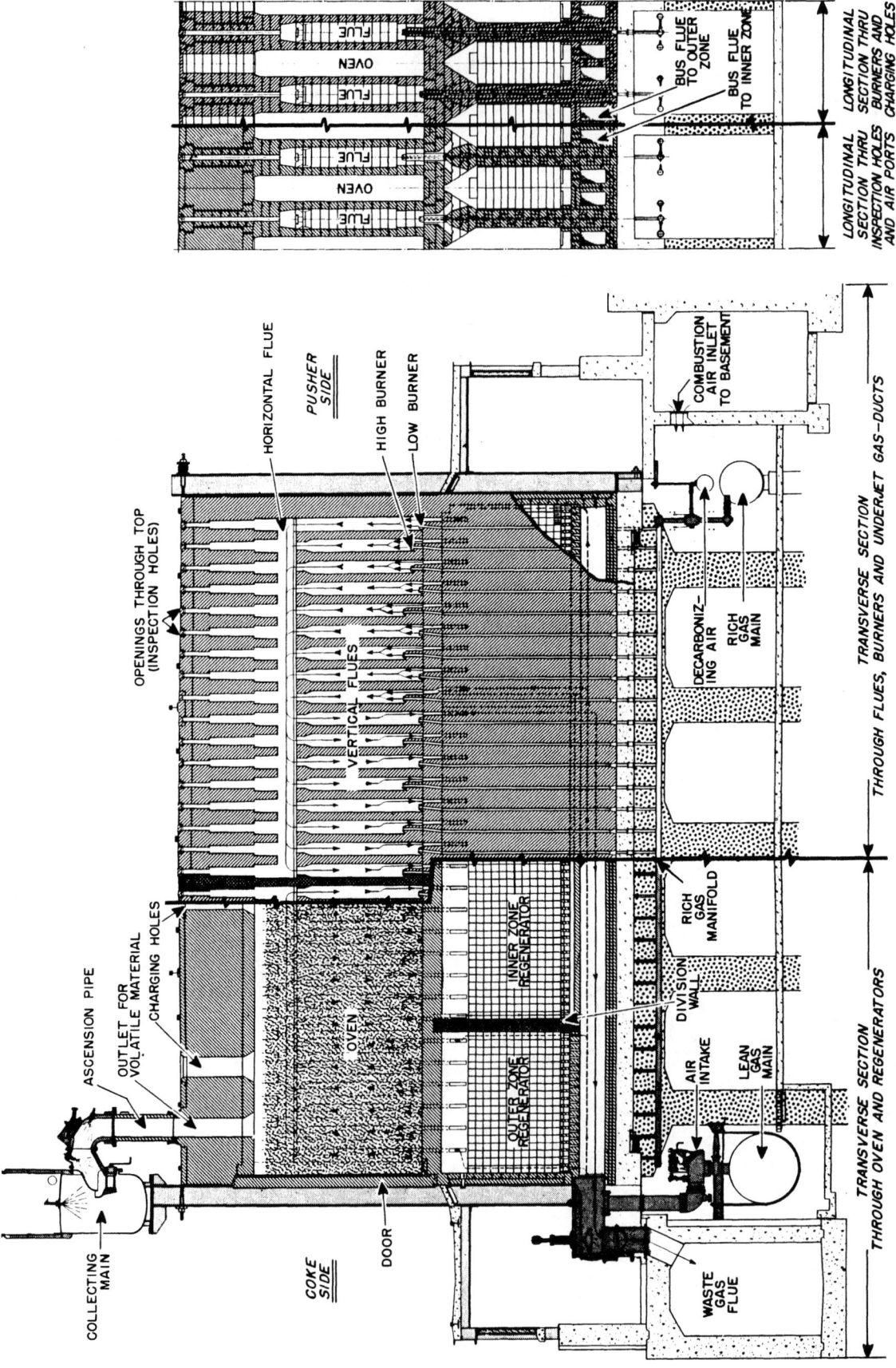

FIG. 4—49. Sections through a battery of Wilputte underjet, combination, by-product ovens, designated as "double-divided" ovens, having two outer zones and one double inner zone in the heating system. The rich gas and lean gas referred to on the drawing are coke-oven gas and blast-furnace gas, respectively. (Courtesy, Wilputte Corporation, a Salem Company.)

the length of the regenerators and sole flues to force the desired distribution of gases. A patented system is also provided to supply more heat to the checkers at the ends of the regenerators to compensate for heat losses through the regenerator faces. A high-efficiency extended-surface checker brick is used in the regenerators to secure efficient heat recovery. These features also are provided in all three types of ovens built by Koppers Company, Incorporated.

Koppers-Becker Oven—The Koppers-Becker oven was first developed and built in the 1920's to provide a system for heating taller ovens with either blast-furnace gas or coke-oven gas. This design has been refined and extended to permit uniform heating of modern tall ovens 6.5 metres (21 ft, 4 in.) or higher. It provides a number of advantages that are unique to this design.

Many Koppers-Becker ovens have been built as either underjet or gun-flue ovens. The underjet design is preferred for the modern tall ovens to provide ease of access to the fuel-gas nozzles for better combustion control. The Koppers-Becker oven is most advantageous for combination ovens to be heated with blast-furnace gas, although a great many have been built for heating only with coke-oven gas.

The distinguishing features of the Koppers-Becker oven are the grouping of the vertical flues with short, relatively small horizontal flues connecting the tops of the vertical flues in each group, and with a connection

between each horizontal flue section and the corresponding section in the adjacent companion heating flue wall by means of crossover flues extending up and over the intervening oven chamber. The regenerator chambers are arranged in groups of three, with a wide regenerator flanked by two narrow regenerators in each group. This design is illustrated in Figure 4—51.

This arrangement provides for upward flow of air (and blast-furnace gas, if used) through half the regenerator groups of three, and through the adjacent heating flue walls to which they are connected. These regenerators and flue walls burn upwardly on one reverse for their entire length from pusher side to coke side. The products of combustion flow out through the short horizontal flue sections and over the tops of the intervening ovens through the cross-over flues, passing down through all the vertical flues in the companion heating flue walls, down through the companion regenerator groups of three, and into the waste-heat flue. Each group of three regenerators feeds air (and blast-furnace gas, if used) to two adjacent heating-flue walls, so that every other pair of adjacent flue walls burns on one reverse, while the alternate pairs are "off" and receive downflow combustion products. On the next reverse the flow of combustion media and products of combustion is reversed. Waste-heat flues may be provided on one side or both sides as required.

The regenerator chambers are arranged in groups of three. In each group the flow of gases in all three cham-

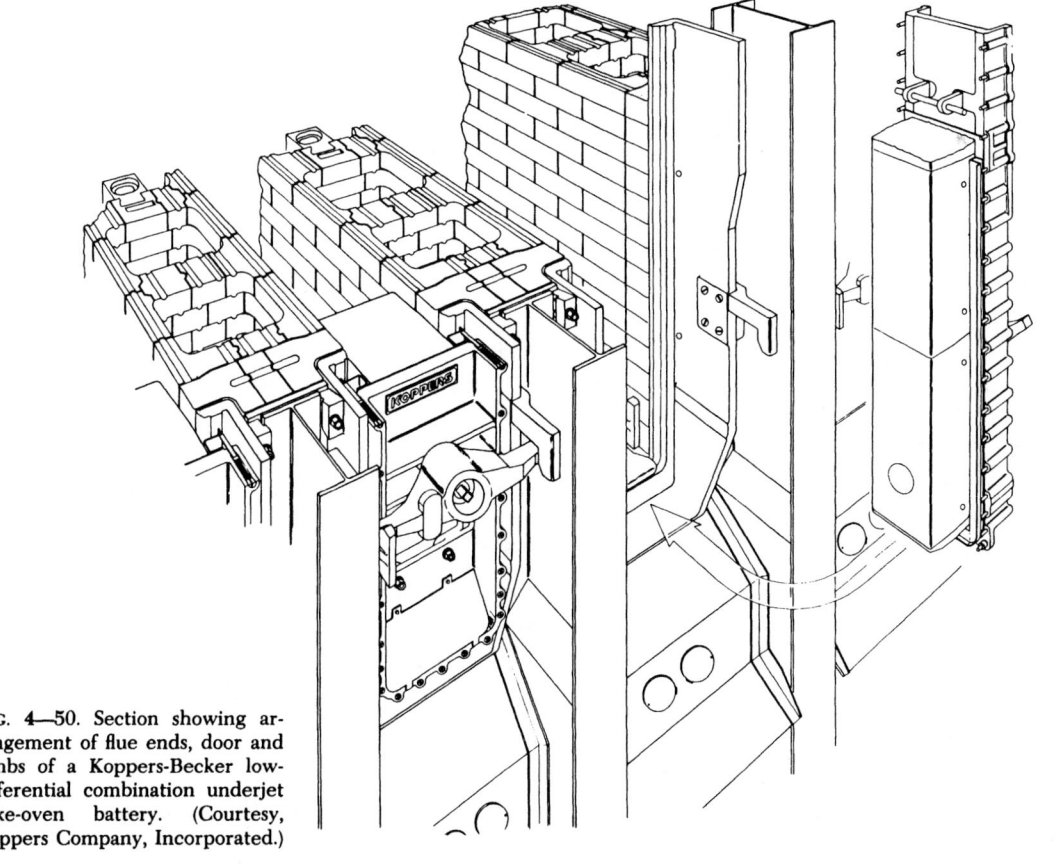

FIG. 4—50. Section showing arrangement of flue ends, door and jambs of a Koppers-Becker low-differential combination underjet coke-oven battery. (Courtesy, Koppers Company, Incorporated.)

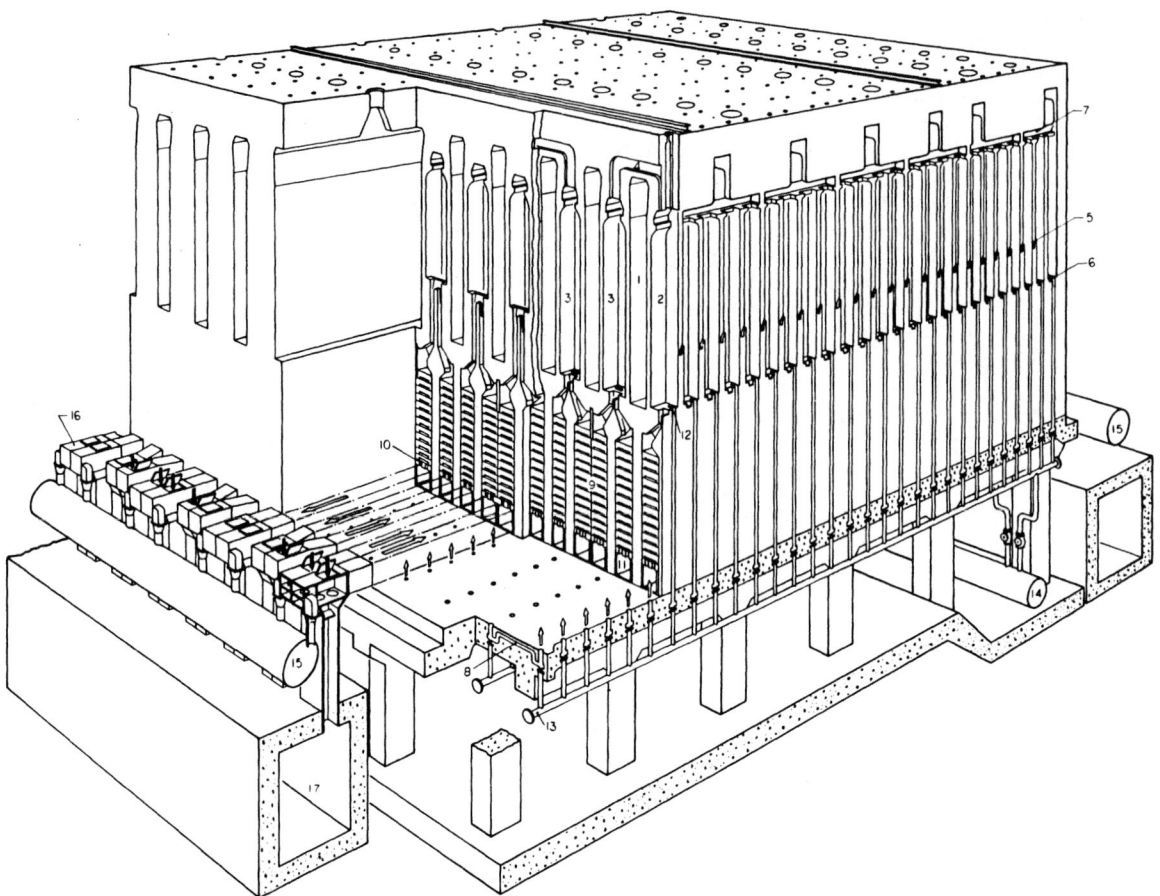

Fig. 4—51. Sectional view of a Koppers-Becker low-differential combination underjet coke-oven battery operating on rich gas. Legend: 1, Oven; 2, Burning flue; 3, Waste-gas flue; 4, Crossover flue; 5, High air port; 6, Low air port; 7, Horizontal collecting flue; 8, Recirculating duct; 9, Checker brick; 10, Sole-flue parts; 11, Sole flue; 12, Gas port; 13, Rich-gas manifold; 14, Rich-gas main; 15, Blast-furnace-gas main; 16, Gas or air and waste-heat boxes; 17, Waste-heat flue. (Courtesy, Koppers Company, Incorporated.)

bers is in the same direction with substantially equal pressure (draft) conditions. The up-flow blast-furnace gas regenerator chamber is always located between two up-flow air regenerator chambers, so that there is little or no tendency for any blast-furnace gas leakage to occur. There is no possibility of dangerous leakage of gas into the down-flow waste gas regenerators or sole flues. It is only at every third regenerator wall that an up-flow regenerator chamber is located next to a down-flow one, but here the up-flow regenerator chamber always contains air, so that if leakage would occur the fuel gas for heating would not be lost, and no damage would result.

The corbel area between the regenerators and the ovens is strong and simple. Symmetrical, short and direct ducts connect with the regenerators immediately beneath the combustion flues, resulting in great strength and tightness.

The regenerators contain no transverse partition walls, but are open for their entire length. Blast-furnace gas and air are metered to each vertical flue by ports at the base of the flues. The open regenerators supply gas and air at equalized pressure and temperature at the point of metering to each vertical flue. The

open space above the checkers at the top of the regenerators acts as an equalizing chamber permitting continued good distribution of blast-furnace gas and air to the vertical flues as the battery ages, even if portions of the regenerators become locally restricted as a result of blast-furnace gas dust deposits.

The wide regenerators are divided longitudinally by specially designed checker brick, and above these by a baffle wall extending the length of the regenerator, providing for effective control of the distribution of blast-furnace gas or combustion air to each individual flue wall. Pressure (draft) conditions at the top of the regenerators are easily controlled at the reversing boxes.

The vertical heating flues of the oven walls are arranged in small groups. Each group has its own crossover flue. Regardless of the length or height of the oven, each branch horizontal collecting flue remains small in cross-section, and short in length, giving the walls extra strength where it is most needed. Low pressure drop through the horizontal flues and crossover flues results in a low overall pressure drop through the entire combustion system. A "normal" height stack can be used.

Air and blast-furnace gas flows are metered to each

flue by the air and blast-furnace gas ports, where pressures and temperatures are equalized across the battery, and distribution does not change with coking rate changes. For tall ovens, in order to assure uniform heating for the full height of the oven, air and blast-furnace gas are supplied through ports located part way up the flue, as well as at the bottom of the flue.

When underfiring with rich gas (coke-oven gas), all of the fuel is supplied at the base of the flues. For tall ovens the combustion is "staged" by supplying a portion of the air required at the base of the flue and the remainder at a higher elevation. This elongates the flame, creating uniform vertical heating and eliminating low temperatures near the oven top.

The up-flow rich-gas risers are always at substantially the same pressure as the two adjacent regenerator chambers which contain up-flow air. There is no possibility of leakage of rich fuel gas into the down-flow regenerator chambers or sole flues.

To further lengthen the flame, assuring uniform vertical heating when underfiring with rich gas, an automatic waste-gas recirculation system is used. Recirculating ducts connect the bases of the down-flow and up-flow rich gas risers which are provided with venturi throats (see Figure 4—52). The rich fuel gas flows through jet-type nozzles into the venturi sections, serving as eductors to aspirate a portion of the products of combustion from the down-flow flues, through the down-flow risers and horizontal recirculating ducts,

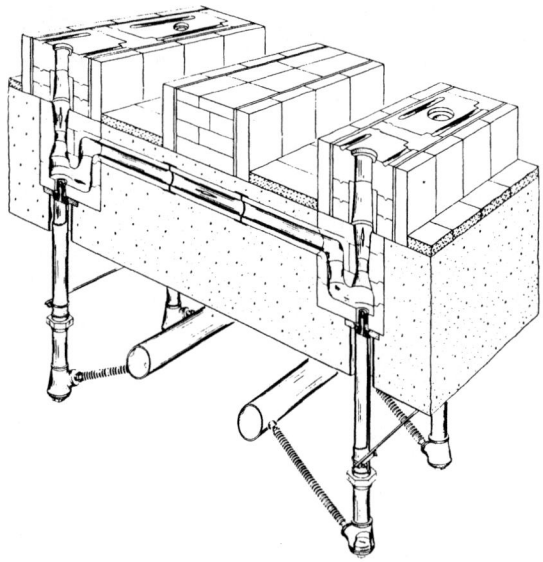

FIG. 4—52. Waste-gas recirculating-duct arrangement for a Koppers-Becker low-differential combination underjet by-product coke-oven battery. (Courtesy, Koppers Company, Incorporated.)

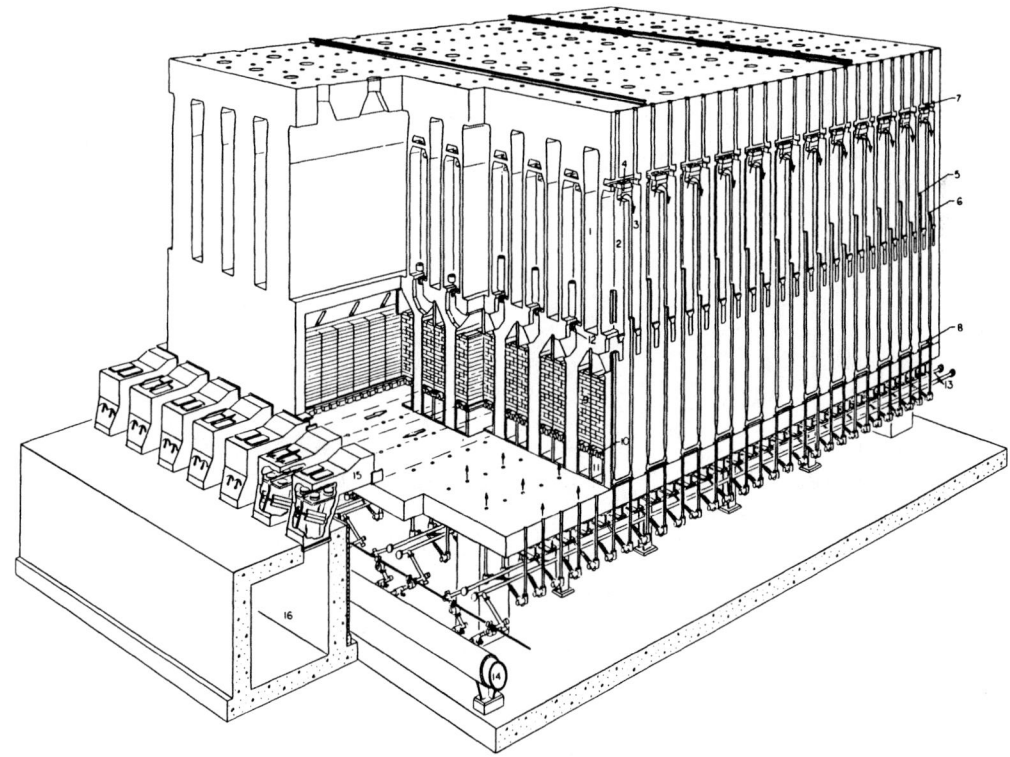

FIG. 4—53. Sectional view of a Koppers twin-flue underjet by-product coke-oven battery. Legend: 1, Oven; 2, Burning flue; 3, Waste-gas flue; 4, Turnaround; 5, High gas port; 6, Low gas port; 7, Differential heating flue; 8, Recirculating duct; 9, Checker brick; 10, Sole-flue ports; 11, Sole flue; 12, Air port; 13, Rich-gas manifolds; 14, Rich-gas main; 15, Air and waste-heat boxes; 16, Waste-heat flue. (Courtesy, Koppers Company, Incorporated.)

mixing this recirculated portion of waste gas with the rich fuel gas in the up-flow risers to the burning flues. This results in dilution of the rich fuel gas with inert products of combustion. This mixture burns more slowly, producing a longer flame of lower luminosity, similar to a blast-furnace gas flame.

The Koppers waste-gas recirculation system suppresses the tendency to form carbon near the top of the up-flow fuel-gas risers. Any carbon that may be deposited on the up-flow cycle is automatically consumed during the down-flow cycle by reaction with the excess oxygen, water vapor and carbon dioxide in the recirculated waste gas. This eliminates the need for a separate decarbonizing air system.

Koppers Twin-Flue Oven—The Koppers twin-flue oven was first built in the early 1970's. This design was developed to provide a tall oven with a simplified refractory-brick and oven-top construction. It has been built as an underjet battery, primarily for heating with rich gas, but is also adaptable to heating with blast-furnace gas.

The distinguishing features of the Koppers twin-flue oven are the arrangement of the vertical flues of each heating wall in pairs connected at the top, and the use of regenerators which are unpartitioned for their entire length (see Figure 4—53). No horizontal flue is required, providing added strength to the walls in this critical area.

There is one wide regenerator (two narrow regenerators for combination ovens) under each oven, extending the full width of the battery, separated from the regenerators on both sides by substantial silica walls of tongued-and-grooved brick. During a given reverse, every other regenerator handles up-flow air, while the alternate regenerators handle down-flow products of combustion. Air passing up one regenerator supplies air to alternate up-flow flues in each of two adjacent flue walls. Down-flow waste gases from the corresponding down-flow twin flues pass to the down-flow regenerators on each side, and from there to the waste-heat flue.

This twin-flue design incorporates most of the structural and design features that have been proven successful over the years in the Koppers single-divided and Koppers-Becker ovens, which have been described in the preceding sections. Automatic waste-gas recirculation is achieved by the use of recirculating ducts connecting the base of the rich-gas risers of each pair of twin flues. The recirculating ducts and venturi throats at the base of each riser are built into the silica brickwork of each heating-flue wall. Alternating high and low coke-oven gas burners are provided in alternate flues across each heating-flue wall to assure uniform vertical heating of the modern tall twin-flue ovens.

Auxiliary Equipment—Some auxiliary equipment

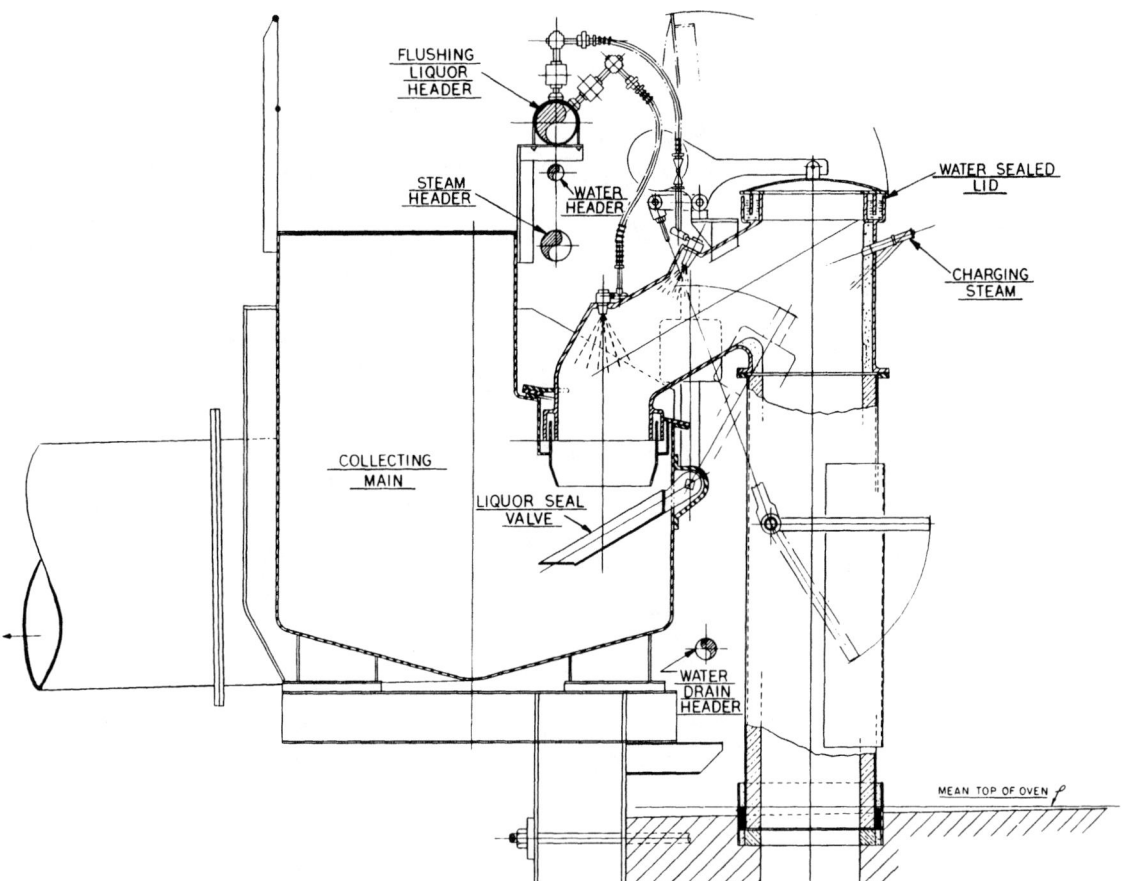

FIG. 4—54. Cross-section through a collecting main and ascension pipe on a by-product coke oven charged with wet coal. (Courtesy, Koppers Company, Incorporated.)

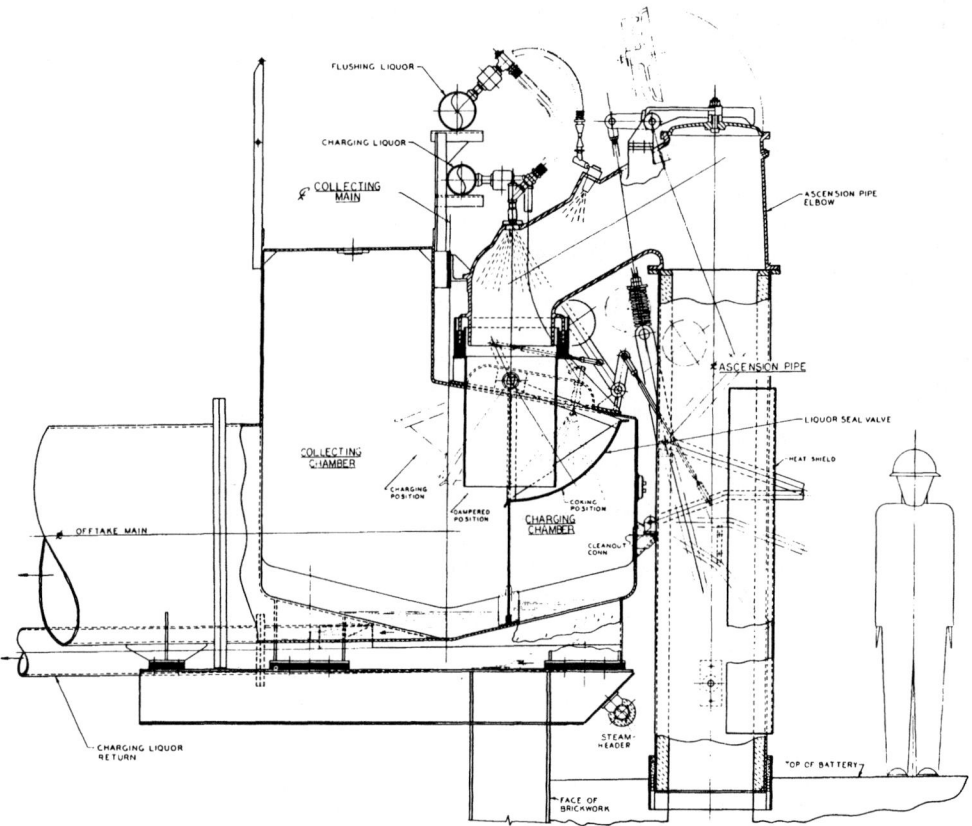

FIG. 4—55. Patented divided collecting main for by-product coke ovens charged with preheated coal. (Courtesy, Koppers Company, Incorporated.)

which can be used with any of the three types of Koppers oven is described in the following paragraphs.

The Koppers doors are of ductile iron or fabricated steel, and have the desired amount of flexibility to conform to the general shape of the jambs (see Figure 4—50). A U-shaped sealing diaphragm is provided which is backed up by a series of spring-loaded plungers. This arrangement provides added flexibility to permit the doors to conform automatically to minor differences in contour between the door and the sealing surface of the jamb. This feature permits door interchanging from oven to oven with automatic adjustment to the difference in contours. The two latches may be either of the screw type or spring-loaded type.

The door-extractor ram swings the door 90° and holds it in position in front of the cleaning station. The patented cleaner frame then moves forward to contact the door with the reciprocating scraping tools, which automatically clean the door-sealing channel both vertically and horizontally, and also clean the bottom and sides of the door plug. The number of scraping motions may be adjusted to suit varying conditions.

The patented jamb cleaner is mounted on a separate ram which positions the cleaner against the door jamb. Variable settings automatically provide a programmed number of strokes for the scraping tools which effectively remove deposits from the jamb surfaces. An additional tool enters the oven at the floor, and as the cleaning head is retracted from the oven the bottom

tool removes any floor deposits which would prevent the door from being properly replaced.

In the case of wet-coal charging, gas is removed from each oven through single or double ascension pipes at the oven ends. The collecting main is shaped to allow the ascension pipes to be as short as possible (see Figure 4—54). Each ductile-iron return-bend section contains two flushing-liquor sprays and is connected to the collecting main by a sealed expansion joint. A liquor-sealed valve at the base of the return bend allows the oven to be isolated from the main. Charging steam is directed through the return-bend section to pull the gas and smoke generated during charging into the collecting main. A return-bend cover is provided for venting the oven to the atmosphere for decarbonizing purposes.

Mechanical-cleaning devices are provided on the coal-charging car for removal of carbonaceous deposits from the return bend and ascension pipe. Other mechanical devices are available to manipulate the covers and dampers from the coal-charging car.

In the case of preheated-coal charging, a patented system allows gas to be removed from each oven through a single ascension pipe with the collecting main divided into two parts (see Figure 4—55). Since the carry-over of fine-coal particles during charging of preheated coal would be detrimental to coke-oven operation, this portion of the evolved dust and flushing liquor is collected and treated separately. After charging has been completed the "swing" damper is positioned to route the gases and flushing liquor to the opposite side of the collecting main to be processed by the normal by-product system. The collecting main is shaped to allow the ascension pipes to be as short as possible. Each ductile-iron return-bend section contains two flushing-liquor sprays and is connected to the collecting main by a sealed expansion joint. A return-bend cover with latching mechanism is provided for venting the oven to the atmosphere for decarbonizing purposes. Mechanical-cleaning devices are provided on the battery-top service car for removal of carbonaceous deposits from the return bend and ascension pipe. Other mechanical devices are available to manipulate the covers and dampers from the service car.

Pollution Control—Koppers Company offers facilities and systems to control environmental pollution from coke-oven batteries. Equipment to control pollution is an accepted part of all proposals for new coke ovens whether on new or existing sites throughout the world, and generally for existing batteries in the United States. The problems of making a coke-oven battery essentially gas-tight are being resolved with improved engineering designs, consistent operating practices and modernized maintenance practices.

To assist the coking industry in this emission control effort, Koppers Company has available the following facilities and equipment for these major sources of coke oven emissions:

(1) **Charging Emissions**—An overall charging concept, commonly called "staged charging" or "sequenced charging", is available. This system (Figure 4—56) involves the ability to discharge the coal from the charging-car hoppers in the proper sequence and at the proper times so that blockages of the gas passageways from the oven are prevented. Double collector mains or jumper-pipe systems are available to provide a gas outlet at each end of the oven chamber. Suction is provided by aspirating steam at both gas outlets during a charge. Successful smokeless charging by this method is greatly aided by certain auxiliary control devices which include mechanical feeders to control the flow of coal from the hoppers to the ovens; automatic lid-lifters with slide gates; automatic leveler-door operators with leveler-bar smoke seals; mechanical standpipe and gooseneck cleaners; and self-cleaning steam aspirating nozzles.

Another system for reducing charging emissions includes the feeding of preheated coal into the ovens through pipes using steam as a carrier. This system is attractive where poorer quality coking coals are used for producing coke.

(2) **Pushing Emissions**—Koppers Company has available a number of systems to reduce the emissions generated during the pushing operation. A coke-side enclosure or shed has been provided to contain and collect these emissions. Wet scrubbers and baghouses are being used to clean the captured emissions from the push and also the gases evolved from the coke-side oven doors.

Another system for control of pushing emissions is the Koppers-Ford system which consists of an enclosed coke guide with an attached hood that covers a quench car. The hood is connected to a fume main and a land-based gas-scrubbing system to remove and clean the emissions.

Various mobile gas-cleaning cars are available and being operated with one-spot and two-spot quenching cars. Conventional gas-cleaning equipment being provided consists of wet scrubbers and fans, as well as the Koppers HydroSonic air or steam ejector system. Refer to Figure 4—32.

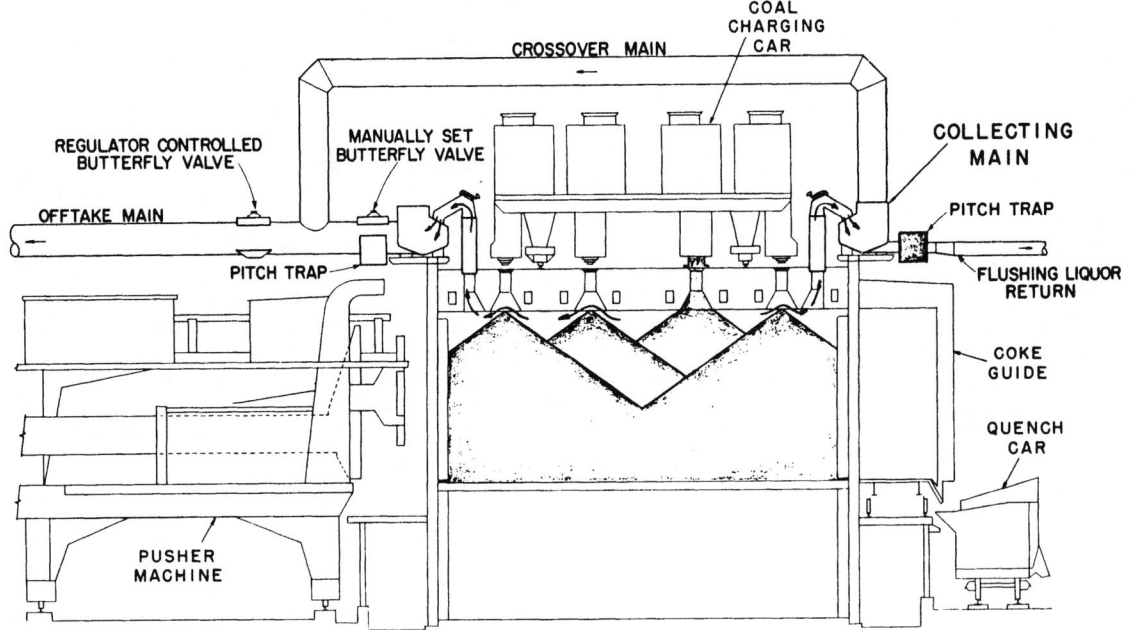

FIG. 4—56. Schematic arrangement of a coke oven equipped for "staged charging" or "sequenced charging." (Courtesy, Koppers Company, Incorporated.)

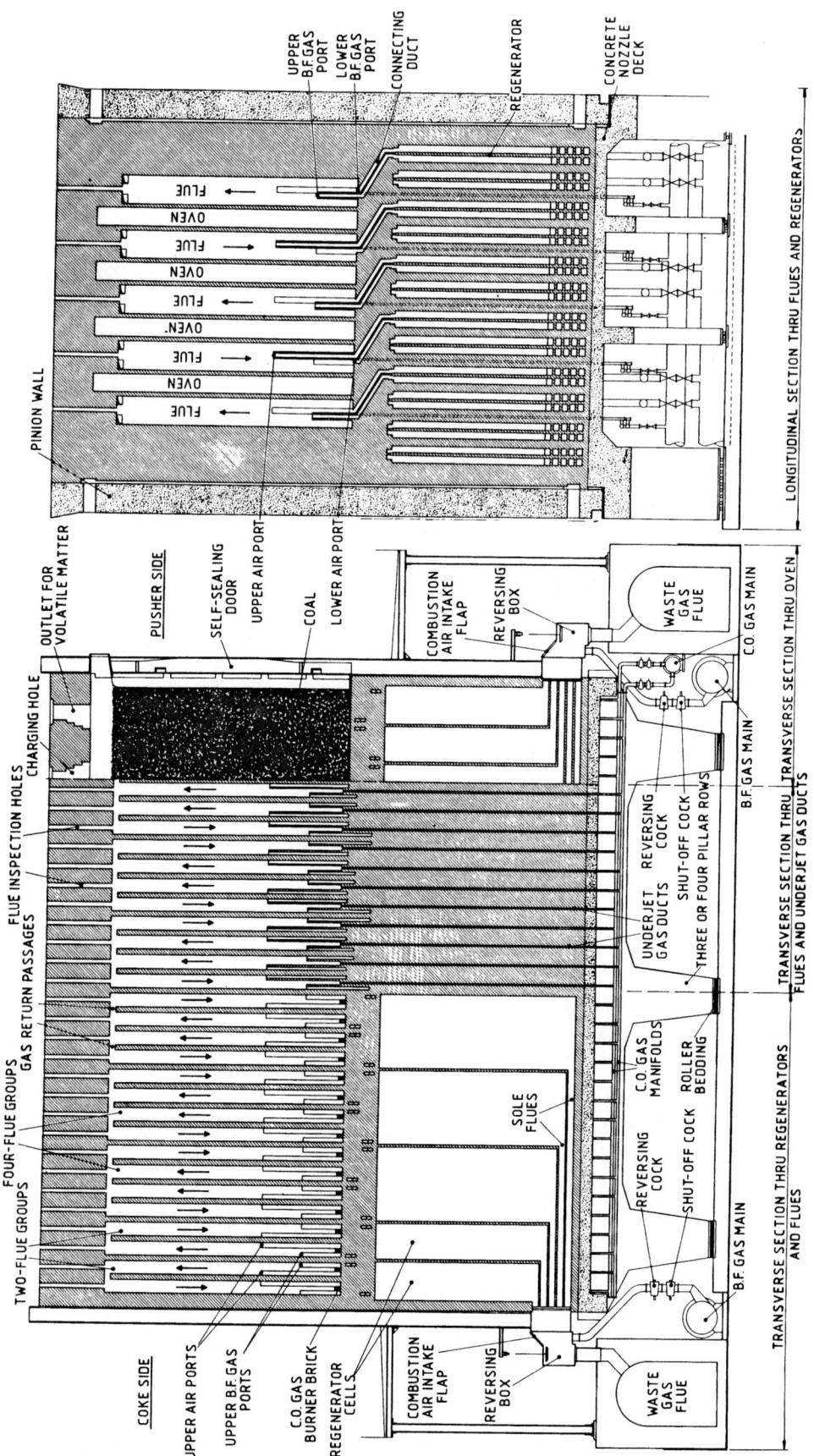

FIG. 4—57. Longitudinal and transverse sections through a modern Didier flow-controlled coke oven with heating-flue groups of the underjet compound type. (Courtesy, Didier Engineering, G.m.b.H.)

A new rotary-drum quench car has been put into operation at Ford Motor Company and has been incorporated into the previously discussed Koppers-Ford system.

(3) **Quenching Emissions**—Particulate matter entrained with steam from the quenching operation can be controlled with the use of mist suppressors installed in the quench-station stacks. Usually these are wooden baffles arranged to cover the area of the stack and are installed with cleaning sprays to wash down the collected particulates.

Dry coke cooling is another method of controlling quenching emissions. In this process, hot coke is dumped into a chamber and inert gas is circulated through it to cool the coke. This gas is then passed through a waste-heat boiler where the heat from the coke is recovered and used to generate steam. Although this system does not emit a plume of steam with particulates, control devices are still used at coke-transfer points and in the gas stream to collect and clean emissions.

WILPUTTE CORPORATION, A SALEM COMPANY

Since the early 1960's, Wilputte has been engaged actively in the development of larger capacity ovens with increased coking rates.

The first advance in this development program was the introduction of high-density silica oven liner brick. This was followed by the increase in height of the four-divided and twin-flue ovens to 5.2 metres (17 feet) with a capacity of 31.16 cubic metres (1100 cubic feet). By 1978, 1150 Wilputte ovens, each with a capacity of 31.15 cubic metres (1100 cubic feet), had been constructed, with a designed coking rate of 16 hours. Experience gained with the operation of these ovens at a fast coking rate, using both coke-oven and blast-furnace gas firing, was combined with research and development work on yet higher and larger capacity ovens. This resulted in the construction of 6.21-metre (20 ft, 4½ in.) high ovens with a capacity of 39.64 cubic metres (1400 cubic feet).

Depending upon the particular application, the 31.15-cubic-metre (1100 cubic foot) capacity ovens were either twin-flue with coke-oven gas firing, or four-divided with combination-gas firing. Gas-gun or underjet firing systems were employed, depending upon client's preference.

The 39.64-cubic-metre (1400-cubic-foot) capacity ovens and 44.46-cubic-metre (1570-cubic-foot) capacity ovens constructed up to 1969 were of the twin-flue design for coke-oven-gas firing. An underjet firing system is used exclusively for these largest ovens.

Dimensions of the coking chamber of a 39.64-cubic-metre (1400-cubic-foot) oven as compared with those of a 31.15-cubic-metre (1100-cubic foot) oven are, respectively: heights, 6.21 metres and 5.18 metres (20 feet, 4½ inches and 17 feet); lengths, 15.42 metres and 14.94 metres (50 feet, 7¼ inches and 49 feet). Average width for each is about 457 millimetres (18 inches).

The new generation of large-capacity Wilputte ovens accounted in 1969 for an installed carbonization capacity of 14.5 million metric tons (16 million net tons).

Much of Wilputte's effort has been directed also toward the more efficient handling of coal and coke, control of the environmental discharges associated with normal coke-oven operation, and the application of alternating-current drives; these features are described later.

Twin-Flue Ovens—Twin-flue ovens are comprised of multiple vertical heating flues arranged in pairs and designed to permit burning of gas simultaneously in alternate flues. Every 20 to 30 minutes the supply of fuel gas is reversed to allow burning in the flues which were carrying waste gas.

Four-Divided Ovens—The heating wall of the four-divided oven (Figures 4—48 and 4—49) consists of 33 vertical flues; located between adjacent coking chambers are 17 outer and 16 inner flues. Nine of the outer flues are located on the coke side, and eight are located on the pusher side. The outer flues are arranged to burn alternately with the 16 inner flues, reversing every 20 to 30 minutes.

Pillar Walls—The pillar walls of all sizes of Wilputte ovens are of all-silica construction from the deck slab up to provide a uniform material of construction without slip joints that could cause potential leaks in the regenerator. This design also eliminates the complex buckstay bracing system in the bench and below the oven floor.

Wall Construction with High-Density Liners—The wall liner brick in contact with the coal is high-density silica brick. This brick has a minimum density of about 1.84 g/cm³ (115 lbm/ft³) as compared to standard liners with a density of about 1.71 g/cm³ (107 lbm/ft³) The increased heat-transfer rate and reduced thickness of this brick permit a lower average flue temperature than required for 114.3-millimetre (4½-inch) thick liners of standard density, or an increased coking rate with the same temperature.

The heating walls are of interlocked header and stretcher brick construction with headers interlocked to each other and to the liners as well. Courses are not less than 152.4 millimetres (6 inches) in height, thereby reducing the number of joints substantially.

The ends of the heating walls can be constructed with removable nose and jamb brick made of silica, clay or fused silica. The construction is such that all nose and jamb brick can be replaced without disturbing the end flues.

Heating—Wilputte coke-oven batteries can burn either coke-oven or blast-furnace gas, or a combination of these. The changeover may be made from a central control room.

Alternating high and low burners are installed in the heating flues. This arrangement lengthens the initial vertical combustion zone when heating with rich gas and, by enabling the use of wall liners of uniform thickness from floor to roof, materially improves the vertical distribution of temperatures when underfiring with lean gas.

The regenerator distribution system using venturi ports which are graduated in size along the bus flue achieves uniform distribution of flow through the regenerator, thereby improving its efficiency. Also, this graduated venturi-port design, since it compensates for adverse manifold effects in the bus flue, permits the use of a single side waste-gas flue.

Coke-Oven-Gas Firing—In Wilputte ovens, coke-oven gas may be fired either through an underjet or gas-gun system.

When firing through an underjet system, coke-oven gas under regulated pressure is delivered through the main coke-oven-gas manifold to the underjet system located in the battery basement. The underjet system comprises the individual wall manifolds with reversing cocks and connections to the individual flues, each having an orifice metering device. These metering devices are sized to accommodate the progressive increase in gas requirements toward the coke side of the oven to compensate for oven taper and to meet the additional gas requirements of the outer end flues. The rich fuel gas upon leaving the metering orifices travels upward through a duct in the pillar wall to a high or low burner.

Ovens with capacities up to 31.15 cubic metres (1100 cubic feet) may be equipped with a gun-type fuel-gas system. In this system, gas guns deliver fuel gas through horizontal ducts from which it is distributed through conventional clay nozzles to the individual heating flues.

Blast-Furnace-Gas Firing—The combustion of low-Btu blast-furnace gas must be initiated at the bottom of the heating flue to avoid a cool bottom condition. This is achieved in the four-divided oven by mixing the preheated lean gas and air at a point below the oven floor to promote proper flame propagation, and through the use of uniformly thick wall liners from wall to roof, which materially improves the vertical distribution of flue temperature.

Double Collecting Mains—In order to reduce atmospheric emissions during coal charging, most batteries being built in the United States are being provided with collecting mains on both sides of the battery so as to provide a free exit for handling the gas evolved during charging. Double collecting mains with stage charging can provide for smokeless charging as the gas-escape path is unobstructed along the top of the oven to the standpipes.

With the double-collecting-main system, the space above the coal tends to be cooler in the latter portions of the coking cycle, thus reducing the tendency to form roof carbon. However, care must be exercised to close off the oven from one of the collecting mains in the latter part of the coking cycle to prevent the possibility of cool, wet, raw gas being drawn into the oven chamber and possibly causing damge to the silica brickwork.

Figure 4—19 shown earlier depicts the arrangement of the standpipe, damper valve and collecting main at one end of an oven.

Coal Charging—The Wilputte charging car is equipped with magnetic lid lifters, drop-sleeve assemblies, automatic gooseneck cleaners, coal-bin operating mechanism, smoke-abatement units and operator's cab.

With one spotting of the car, the following functions are performed: (a) clean the standpipe elbows; (b) lift lids, drop the sleeves and charge the oven; and (c) lift the sleeves and replace the lids. With the automated charging car, the travel, loading and weighing of the coal and spotting are programmed and automatic, but the discrete functions (a), (b) and (c) are performed upon initiating signals from an operator.

A rotating-table-type charging larry simultaneously discharges the coal contents evenly at a controlled rate into all charging holes. This results in an oven charge of uniform bulk density that normally requires only a single stroke for leveling.

Pusher Machine—The Wilputte portal-type pusher is designed for door removal, oven pushing and door replacement with one spotting of the machine. The hydraulically operated door cleaner locks into the retracted door, automatically cleaning the sealing edge and the door lining. Simultaneously, a hydraulically operated jamb cleaner, held in place by the door-jamb latch keeper, moves into position for cleaning the face and inside of the door jamb.

The leveler bar, provided with a smoke sleeve for sealing the leveler door opening during operation, discharges the coal withdrawn by the retracting bar into a hopper located on the pusher.

A system of induction-coil interlocks prevents pushing the oven before removal of the coke-side door and positioning the coke guide.

Hydraulic Door Machinery—The hydraulically operated door removal and replacement machine has an inherent ruggedness and reliability superior to the electric model with its multiplicity of gear reducers, brakes, torque-limiting devices, limit switches and interlocks.

The hydraulically operated door and jamb cleaner on the coke-side door machine is similar to the pusher-side machine and also locks onto the retracted door. The jamb cleaner is positioned by respotting the door machine during the door-cleaning cycle and is held in place by the door jamb's latch keeper.

Tilting Quench Car—The quench car has a tilting floor designed for uniform distribution of coke, reduction of the overall moisture content after quenching, and elimination of hot spots that might otherwise be found in coke after discharge onto the wharf.

Self-Sealing Doors—The sealing edge of the spring-loaded self-sealing doors (Figure 4—16), shown earlier, is made of a renewable flexible stainless-steel strip that is mounted on a flexible diaphragm and will adjust readily to the contour of the door-frame sealing face. The leveler door also is self-sealing.

The Wilputte door jambs can be either the ell-shaped contour or the block contour, securely bolted to a steel nose plate behind the buckstay so that the jamb and steel nose plate can be held tightly against the brickwork despite any tendency of the buckstay to distort. This design also permits replacement of door jambs without removing buckstays while providing maintenance access to the nose brick.

Automation of Coke Wharf—The coke-wharf gates are actuated pneumatically and are provided with a sequencing device for controlling the various movement patterns. Gates are interlocked with the air compressor and close automatically when the coke-wharf conveyor belt stops. An automatic temperature and water-spray system provides protection for the belt.

Another type of automatic wharf has a motor-operated plow running parallel with the coke conveyor to unload a run of coke from the wharf. This method of transferring coke onto the wharf belt promotes uniform screen loading and efficient screening.

Electrical Equipment—Alternating-current drives have been adopted for coke-oven machinery in modern Wilputte installations. Operational advantages of alternating-current over direct-current power are: (a) control consists mostly of solid-state components where the number of moving parts is minimized; (b) highly accurate, stepless and simple control is obtained; and (c) smooth starting, accelerating and decelerating are achieved.

DIDIER ENGINEERING G.m.b.H.

Since 1865, Didier has been active in the engineering and construction of cokemaking facilities. Through extensive research and development, the organization has become one of the leaders in cokemaking technology. The first Didier 6-metre and 7-metre (about 19-foot, 8-inch and 23-foot) large-capacity coke batteries were built in Germany during the 1970's. Tall Didier-type ovens were first built in the United States during the late 1970's.

The modern Didier-type coke oven is known as the Didier flow-controlled coke oven. The typical design of such an oven is shown in Figure 4—57. This oven has a heating system consisting of independent heating units with special characteristics, as follows:

(1) Grouping of the heating flues at each end of the oven in two-flue groups.

(2) Grouping of the remaining heating flues in four-flue groups.

(3) Individual sole flue for each regenerator cell.

(4) Blast-furnace-gas and air ports at two heating-flue levels.

(5) Coke-oven-gas burner brick at the heating-flue bottom only.

(6) Individual adjustment of blast-furnace-gas and air streams entering the heating flue.

(7) Heat-control adjustment for blast-furnace gas and air within the reversing box.

(8) Mechanical registers in the reversing box for immediate change-over of blast-furnace-gas and coke-oven-gas heating systems.

Adjustability of the temperature distribution is the main feature of the Didier-type heating system.

Oven Chamber—The average width of the coking chamber ranges from about 350 mm (13¾ inches) to 463 mm (18¼ inches); the height exceeds 7 m from floor to roof; the longest oven is 17.1 m (56 feet, 1¼ inches) inside buckstays. Table 4—I shows the main dimensions of some Didier type ovens.

Since the shrinkage of coal may vary with changing coal characteristics or with changing of the coking time, the ovens have a variable height of the coal line. The leveler door is designed to suit an adjustable coke-charging level.

Thickness of the oven liner walls is constant from pusher to coke side and from oven floor to oven top. Heat-transfer control from the heating flue into the coal charge is entirely effected by the combustion pattern in the heating flues. This reduces the number of different shapes required for the liner walls.

The oven chamber width is vertically constant. Temperature of the free space above the coal charge is controlled by the combustion pattern in the heating flues and by coal level adjustment.

Heating Wall—The binder walls (or "header walls") which separate the individual heating flues from each other are of uniform thickness from oven floor to top.

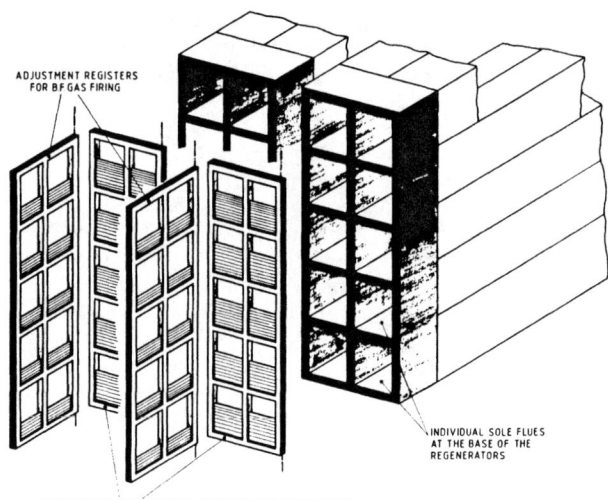

FIG. 4—58. Adjustable registers for controlling flow rates of fuel gas streams into sole flues. (Courtesy, Didier Engineering, G.m.b.H.)

The opening at the top of the binder wall, where the combustion products pass from the "on" flue to the adjacent "off" flue, is small to ensure maximum strength in this area, considering the load imposed by the charging car traveling on top of the battery and the bending and shear forces arising from the coal charge and from coke-pushing operations.

The heating-wall brick bond is of the Didier type hammer-head construction to provide structural interlocking between binder and liner wall to obtain maximum wall strength.

Liner and binder walls are both constructed of bricks with tongue and groove, which makes for strong interconnection between the individual bricks and protects against raw-gas leakage from the oven chamber into the heating flues.

The width of the heating wall is very important for the resistance of the wall to bending forces acting from the oven chamber on the wall. Wall width therefore has to increase with height, which in turn requires an increase in the center-to-center distance of the ovens.

The bricks forming the vertical ducts within the heating flue, which supply air and blast-furnace gas to the higher ports, are locked into the binder walls in order to ensure their strict vertical alignment.

Air and Gas Supply—There is a total of four ports in each heating flue to supply four streams of combustion air in vertical and parallel flows into each flue. Two of these ports are located at the bottom of the flue, and the remaining two at a higher elevation. One short rich-gas burner brick at the base of each heating flue supplies the coke-oven gas into the flue, also in vertical and parallel flow with the air. The rich-gas burner brick has a large inside diameter to reduce the velocity of the coke-oven gas flow as it enters the heating flue. This reduces friction between the air and the coke-oven gas

Table 4—I. Oven Dimensions of Typical Large-Capacity Ovens Built by Didier Engineering, G.m.b.H.

Location	West Germany	United States	Australia
Length inside Buckstays	16.53 m (54 ft, 2¾ in.)	15.6 m (51 ft, 2¼ in.)	17.1 m (56 ft, 1¼ in.)
Height of Oven Chamber	7.0 m (23 ft)	6.1 m (20 ft)	7.125 m (23 ft, 4½ in.)
Average Width of Oven Chamber	450 mm (17¾ in.)	413 mm (16¼ in.)	463 mm (18¼ in.)

flow, which minimizes turbulence and consequently delays the mixing of the gas and air.

The time required for oxygen and gas to come into contact once they have separately entered the flue is much longer than the time required for the chemical reaction which follows their contact. The time required for flame propagation (or rate of ignition) is negligible. Consequently, flow-pattern control in the heating flue is the key to flame adjustability.

The parallel flows in the Didier-type heating flue provide for a combustion zone extending over the full oven height, while the individual adjustability of each of the four air streams provides the means for changing the combustion pattern and thus for readjusting or altering the vertical temperature profile in the flues, if required.

In this arrangement, the coke-oven-gas burner brick can be very short; the subsequent short residence time prevents overheating and thermal decomposition of the coke-oven gas as it passes through the hot burner brick. This prevents operational problems resulting from build-up of carbon deposits in the burner brick even when the coke-oven gas used as fuel has a high benzol content whose decomposition products might otherwise block up the gas passage.

When heating with blast-furnace gas, some of the air is supplied through one of the ports at the flue bottom and the balance of the air through the highest port. The blast-furnace gas enters the flue through the second port at the flue base and also through an elevated port. Each of the two air streams and each of the two blast-furnace gas streams can be rate-adjusted individually and without touching the setting of the heating adjustment facilities for firing with coke-oven gas. These adjustment facilities consist of separate registers in the reversing boxes of coke-oven-gas and blast-furnace-gas operation.

This system ensures an optimum setting of the streams for coke-oven-gas firing and an independent setting for blast-furnace-gas firing. Once the two registers have been adjusted to the best setting for each gas during the commissioning of the battery, the oven setting can instantaneously be switched back and forth between these settings, whenever there is a change-over from coke-oven-gas to blast-furnace-gas operation or vice versa, as frequently may happen in an integrated steel plant. This unique system eliminates the usual compromise which must be made between the optimum coke-oven-gas and the optimum blast-furnace-gas flow-rate setting, as is commonly the case in coke-oven operating practice.

Heating-Flue Grouping—The heating wall of the Didier-type oven is divided into small groups of vertical flues, each flue group consisting of up-flow and down-flow flues. These flue groups are individually supplied with combustion air during coke-oven-gas operation or with blast-furnace gas and air upon fuel change-over. Thus, these groups form individually controllable heating units enabling readjustment or alteration of the temperature profile horizontally across the battery.

In the interior of the oven, local heat requirements depend chiefly upon the width of the coking chamber which increases from the pusher to the coke side. The resultant increase in heat requirements between two adjacent flues is very minor. Here, each group is comprised of four flues with the two inner flues on up-flow and the two outer flues on down-flow.

At each end of the oven, there is a higher heat demand per flue than in the interior of the wall; this is due to increased surface losses. In order to provide maximum flexibility in the heating adjustment capabilities at the oven ends, the four-flue groups are split up into two separate pairs of flues with one down-flow and one up-flow in each pair. Across the wall, this arrangement results in a pattern of "off" and "on" flues where one down-flow flue at the oven end is followed by two up-flow flues then two down-flow flues.

Looking longitudinally along the battery, "off" flues alternate with "on" flues from wall to wall. This provides a chessboard pattern arrangement of "off" and "on" flues.

Upon heating reversal, the direction of flow in all heating flues reverses.

Connecting Ducts—Each heating flue requires only two connecting ducts to the regenerators, one for air and one for blast-furnace-gas supply. A wall divides the total cross-section of each duct into a top and bottom section for separate flow to the top and bottom ports respectively in the heating flues.

Regenerative System—There are two regenerator chambers under each oven. Each regenerator is occupied by a single medium from pusher side to coke side over the full oven length, i.e., it is either uniformly on air or on blast-furnace gas or on waste gas.

The wall which separates the two regenerators from each other provides an additional support for the oven chamber.

Each regenerator is divided into two halves by a thin divider wall which runs parallel to the main regenerator walls. One regenerator half serves the heating-flue bottom ports, the other half serves the top ports.

Thin transverse divider walls subdivide the regenerator halves into pairs of cells for the individual supply of the connected heating-flue groups. There are as many cell pairs as there are heating-flue groups in a heating wall.

The regenerator walls are constructed of brick shapes with tongue and groove in both horizontal and vertical directions to provide for strong interlocking between the individual bricks. The bricks within the walls are arranged in bonded construction for added strength. Both tongue-and-groove design and staggering of the joints produce a labyrinth joint arrangement for maximum protection against cross leakage between the regenerators.

The lower part of the regenerator walls is of fireclay material. Since silica expands more than fireclay during the battery heatup, a slip joint is provided at the fireclay-silica interface in the regenerator walls, so that the silica will slide over the less expansive fireclay. Provisions are made for pressure to be applied to the face of the fireclay part of the regenerator walls to prevent the fireclay brickwork from being dragged by the more expansive silica in the slip joint.

The thin regenerator dividers in both longitudinal and transverse directions are also constructed of shapes, tongued and grooved in the horizontal and vertical joints. These dividers function as a flow guide only, splitting up a homogeneous flow so that any leakages across these dividers have consequently no effect on battery heating performance.

Sole Flues—The common sole flue at the base of the regenerator is divided to form a multi-sole-flue design. Each of the individual sole flues (often referred to as 'single sole flues') is connected to one cell only to ensure individual flow-rate control from the entrance of the sole flue in the reversing box. Flow-rate control is achieved by adjusting the cross-sectional area of the entrance of the sole flues.

From the sole flue, gas and air enter the cell over the full width and length of the cell bottom.

Adjustment Registers—Two adjustable registers in front of the individual sole-flue inlets control the flow rates of streams entering the sole flues (Figure 4—58). One register is set for blast-furnace-gas heating and one for coke-oven-gas heating. This arrangement provides for the adjustment of both the vertical and the horizontal heat distribution to the oven walls.

Upon fuel-gas change-over, the register with the first gas setting is swung away from the sole flue inlets and the second register with the other setting is swung into place.

The two registers for each regenerator are rotated by vertical shafts either manually from outside the reversing box or by a mechanized change-over system similar to the common heating reversal mechanism. Screw-latched doors at the side of the reversing-box body provide for access to the registers for flow-rate adjustment.

Reversing Boxes—In the Didier-type oven, air and blast-furnace gas are fed into, and waste gases are evacuated from, the regenerators on both the coke and pusher sides. Consequently, the reversing boxes on both oven sides have a double function. During one reversing semi-cycle they conduct air or blast-furnace gas into the regenerators and, after reversal, they remove the waste gases from the regenerators for evacuation into the waste-gas flues.

A double-disc mushroom valve at the bottom of the reversing box isolates the waste-gas duct from the reversing box when the box is on air or blast-furnace gas. The discs are self-aligning due to a convex sealing surface.

A manual adjustable damper between the reversing box and the waste-gas duct permits adjustment of the draft for the regenerator.

At the regenerator face, the reversing box is mounted on a cast-iron frame which incorporates the individual sole-flue entrances; these are installed during refractory installation.

Door Jamb Casting—The typical Didier-type door-jamb casting has an almost square cross section in order to minimize temperature differentials across the section which might cause warpage. The jamb castings are held in position by being clamped to the face-plate castings which are behind the buckstays. This arrangement minimizes movement of the joint between the cast-iron door jamb and the oven refractory. The clamps allow easy replacement of the jambs without cutting the buckstays. The hold-down bolt of the clamps has a hook end which reaches into a pocket in the face-plate casting and is thus easily replaceable. Projections cast on the heavy face-plate castings reach into pockets cast into the jambs for structural interlocking between jamb and face plate to prevent waisting or "hour glassing" of the jamb.

This door-jamb design can be customized to suit any

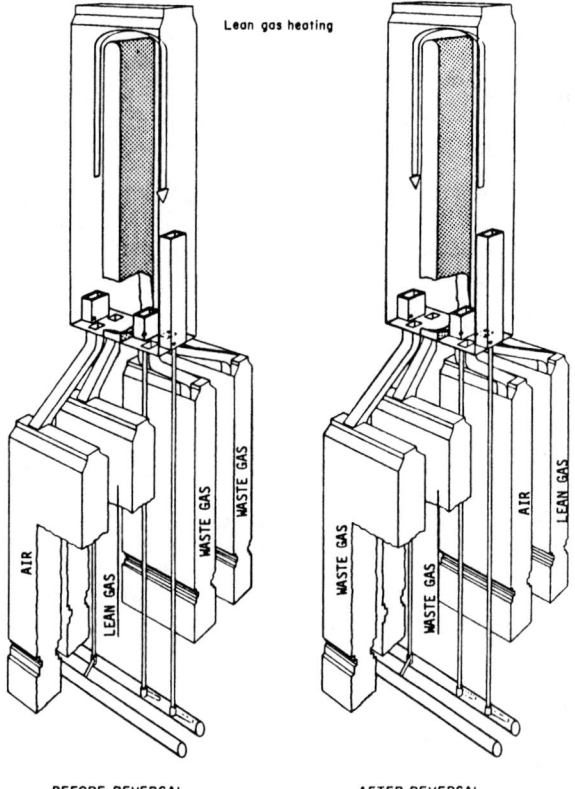

FIG. 4—59. Heating unit of the Otto twin-flue high-capacity oven with controlled two-stage heating. (Courtesy, Dr. C. Otto and Company.)

type of oven door, whether of Didier or other design, irrespective of whether it is a fabricated or a cast-iron door, with spring-adjusted sealing strips or with any other type of knife-edge adjustment.

Face Plate Castings—For heating-wall head protection, heavy face-plate castings are provided behind the buckstays with stiffening ribs at the oven face side to provide for maximum strength against warpage. The castings are lined with castable heat-insulating material at the oven-face side to protect the casting and the buckstay against the heat of the oven and to reduce the oven wall head heat losses. The castings are supported by the buckstays which also carry the weight of the door jambs.

Nozzle Deck—The nozzle deck which carries the regenerators is of reinforced concrete construction. It is supported on three or four rows of concrete pillars. In order to ensure free thermal expansion of the nozzle deck and to avoid the development of cracks during battery heat-up, which could cause air to infiltrate the regenerators, the concrete pillars are supported on steel rollers as in a typical bridge construction. A number of pillars in the center of the battery are rigidly fixed to the concrete foundation slab to provide a fixed point from which the nozzle deck can freely expand in all directions. On top of the nozzle deck, there is an extra layer of heat-insulating concrete provided to keep the nozzle-deck temperature down.

Pinion Walls—The pinion walls, at both battery ends, are vertical slabs of reinforced-concrete construction and are supported on the concrete nozzle deck. They are held by spring-loaded tie rods at ther top and by anchor bolts at the base. The anchor bolts are cast into the nozzle deck. The purpose of the spring loading at the top and the base is to achieve full pressure controllability of the refractory block.

TYPES OF DIDIER OVENS

Combination Blast-Furnace Gas and Coke-Oven Gas Heating—The compound-type Didier batteries are heated either with coke-oven gas or with blast-furnace gas. With these batteries, however, it is also possible to operate some of the battery ovens on coke-oven gas and, simultaneously, the remainder of the ovens on blast-furnace gas.

With the fully controllable gas system, it is not necessary to introduce coke-oven gas as a supplementary fuel through the coke-oven-gas heating system in order to compensate for deficiencies in the vertical temperature profile of any flues.

Rich-Gas Firing—The Didier-type oven has also been constructed for firing with coke-oven-gas only. In this case, the blast-furnace fuel gas system is eliminated.

Lean-Gas Firing—If the compound oven is operated on blast-furnace gas, it could be advantageous to install a blast-furnace-gas enrichment system in order to equalize the fluctuations in the blast-furnace-gas calorific value by adding controlled amounts of coke-oven gas to the blast-furnace gas. The resultant constant calorific value of the enriched gas will save energy since the combustion air can in this case be adjusted to a fuel of constant calorific value.

Gun-Flue Design—In the gun-flue oven, the coke-oven gas is introduced through a horizontal gas duct extending the length of each wall below the level of the oven floor. Short connecting ducts lead vertically upward to a replaceable nozzle brick at the base of each vertical flue. Individual adjustment of the coke-oven-gas rate for each flue is effected by replacing the nozzle brick through the inspection hole in the top of the battery. There is no need for this inconvenient procedure if the ovens are operated exclusively on blast-furnace gas. The Didier-type oven can be built either as a gun-flue or as an underjet oven depending upon the client's preference. In both types adjustability of air and blast-furnace gas by the Didier flow-control system is maintained.

In order to prevent leakages from the horizontal gun-flue duct, the Didier gun-flue ovens employ an interlock arrangement of the channel bricks in combination with embedding of the whole arrangement into a refractory mass within the brickwork.

Underjet Design—Existing Didier-type large-capacity ovens are of the underjet type which is attributable to clients' preferences regarding coke-oven-gas rate control for each heating flue. This control is achieved by readily exchangeable orifice plates in the underjet piping.

DIDIER HEATING SYSTEM

Fuel-Gas Piping—With the underjet-type Didier oven, there are two blast-furnace-gas mains in the battery basement extending both sides of and parallel to the battery for symmetrical supply of blast-furnace gas through the reversing boxes into the regenerators.

The single coke-oven fuel gas distribution main is arranged either on one side or in the middle of the basement depending on the size of the main.

There are two coke-oven-gas manifolds under each heating wall which alternate in step with heating reversal in supplying coke-oven fuel gas to the heating flues. Readily exchangeable orifice plates in the coke-oven-gas risers provide for fuel-gas rate control for each flue.

Reversing Operation—Modern Didier-type batteries are usually equipped with hydraulic reversing equipment for reversing the combustion-air intake flaps at the reversing boxes, the fuel-gas cocks and waste-gas valves. The equipment includes hydraulic cylinders, a pressure accumulator with reserve capacity for several reversals and stand-by facilities in case of power failure. Additional cylinders are installed if a mechanized fuel-gas change-over system is used for instantaneous switch-over of the adjustment registers in the reversing box from coke-oven-gas to blast-furnace-gas heating.

Waste-Gas System—For the underjet-type Didier battery, the two waste-gas flues, one on each battery side, extending along both sides in front of and parallel to the battery, provide for symmetrical draft on both the pusher and coke sides of the battery. The flues have half the cross section of a common waste-gas flue, since both flues simultaneously service the battery at any given moment.

Foul-Gas Collecting System—Either a single or a double collector-main system is installed, chiefly depending on the oven charging method and the charging gas-collecting system employed. For maximum leakage protection, standpipe caps can be provided with a water seal.

GENERAL OPERATING CRITERIA FOR DIDIER OVENS

Heating Control—Vertical and cross-wall temperature profiles in the Didier-type oven are exclusively adjusted by devices installed in the reversing box which is situated outside the hot refractory block and in an easily accessible area.

Cleaning Operations—Air and blast-furnace gas enter the regenerator cell via the (intentionally) fully open width and length of the cell base and without passing through flow-control ports, which would increase their velocity. This considerable velocity decrease causes dust entrained with the blast-furnace gas or air to settle within the sole flue rather than in the regenerator checkers. The sole flues are totally accessible for dust removal, unlike the checkers, and are therefore a preferred area for dust removal. Dust accumulations can be scraped out, or blown back into the reversing box by a bent-back air lance, or evacuated by an industrial-type vacuum cleaner. This avoids the necessity of removing, cleaning and replacing the checkers and maintains the efficiency of the checkers.

DR. C. OTTO AND COMPANY

Founded in 1872, Dr. C. Otto and Company is the oldest coke-oven construction company in Germany, having been engaged in the building of coke ovens and complete coke-oven plants since 1876. The Company built the first battery of nonrecovery ovens in 1876, adopting its own design and making its own refractory bricks.

In 1881, the Company introduced the by-product recovery oven in Germany, the first of these ovens already incorporating the twin-flue-principle. Further development in the by-product recovery sector led in 1883 to the emergence of the Otto-Hoffmann regenerative oven. For the first time, waste gas from the coke ovens was not used to generate steam, but to preheat combustion air in regenerators filled with refractory checker bricks. Apart from high yields of tar and ammonia, the Otto-Hoffmann oven was the first coke oven to produce a gas surplus of about 10 per cent.

The Otto-Hilgenstock underjet oven developed in 1896 was the outcome of an evaluation of practical experience and exhaustive tests carried out by Otto.

The underjet oven, designed and introduced by Dr. C. Otto, represented a substantial improvement of heating methods and an immense increase of oven capacity.

Since 1922, the twin-flue underjet oven has been the standard type of oven constructed by Otto. In 1926, the first twin-flue compound oven for rich- and lean-gas underfiring was built at the Otto test coking plant in Bochum-Dahlhausen. Of 4.20 metres (about 13 ft. 10 in.) in height, these ovens were the tallest in the world at that time. Their adaptability to varying capacity requirements has repeatedly been demonstrated in the past, the shortest coking time being 11¾ hours and the longest, during periods of stagnation, around 336 hours.

Since 1967, numerous Otto large-capacity ovens with chamber heights of up to 7.5 metres (about 24½ feet) have been built. Dimensions and performance figures for some of these ovens are given in Table 4—II.

Heating-Flue System—The heating wall consists of a series of twin flues. A typical Otto coke oven of 16.46 metres (54 ft) in length has 17 such twin flues or 34 heating flues. Combustion of fuel gas and air takes place in every second heating flue, with the combustion products flowing downward in the adjacent heating flue to the regenerators. The direction of flow is changed at predetermined intervals by the reversing system. This type of heating with fuel gas and air burning in every second heating flue creates a more uniform distribution of heat input as against the simultaneous combustion in groups of adjacent heating flues. The twin-flue design permits every second header wall to extend into the top brickwork. Therefore the stability of the twin-flue type heating wall is higher than that of any other type of coke-oven heating wall. A twin-flue heating wall not only represents maximum stability, but also offers flow conditions which result in lower and more uniform differential pressure over the length of the heating wall.

Regenerator Arrangement—The regenerators are located under the oven chambers and heating walls and are continuous from pusher to coke side. When heating with rich gas, two adjacent regenerators are supplied with combustion air and two with combustion products. When heating with lean gas, one regenerator is supplied with combustion air and one with lean gas. The regenerators exchange duties after reversal. They are subdivided lengthwise, from pusher to coke side, into as many cells as there are heating flues. This results in a controlled and uniform flow of lean gas, air and combustion products, with optimum heat recovery in the regenerators. Each regenerator cell is connected by short ducts to the heating flues; one duct to a heating flue on one side of the oven chamber and one to a heating flue on the other side.

The regenerator arrangement, in combination with the twin-flue heating system, provides numerous heating units, each consisting of a twin heating flue connected to four regenerator cells (Figure 4—59).

Supply of Fuel Gas and Air—The Otto large-capacity oven is of the underjet type, all the regulating facilities for fuel gas (coke oven or rich gas) and combustion air being located in the basement of the oven. Here the gases to be controlled are cold and the regulating system is easily accessible for cleaning, adjustment or inspection.

The rich-gas distribution system is located in the battery basement. The rich gas flows through the rich-gas ducts in the regenerator division walls under the heating flues directly to the burners, bypassing the regenerator checkers. The quantity of rich gas for each individual burner is controlled by metal nozzles located in the connections between the distribution headers and rich-gas ducts.

In the compound oven, designed for heating with either rich or lean gas, the lean gas is introduced at the bus-flue level of the oven and distributed to the individual regenerator cells through steel orifice plates with calibrated openings. These orifice plates are located between the bus flue and the regenerator and are eas-

Table 4—II. Dimensions and Performance Figures of OTTO High-Capacity Coke Ovens

	4.5-m Oven		6-m Oven		7.5-m Oven	
Coking-chamber height	4.500 m	(14 ft 9¼ in.)	6.280 m	(20 ft 7¼ in.)	7.500 m	(24 ft 7½ in.)
Length, face to face	13.520 m	(44 ft 4¼ in.)	15.480 m	(50 ft 9½ in.)	16.460 m	(54 ft —)
Length between doors	12.750 m	(41 ft 10 in.)	14.718 m	(48 ft 3⅛ in.)	15.600 m	(51 ft. 2⁵⁄₃₂ in.)
Coking-chamber width, coke side	0.430 m	(1 ft 5 in.)	0.505 m	(1 ft 7⅞ in.)	0.488 m	(1 ft 7¼ in.)
Coking-chamber width, pusher side	0.370 m	(1 ft 2⅝ in.)	0.429 m	(1 ft 4⅞ in.)	0.412 m	(1 ft 4¼ in.)
Oven taper, end to end	0.060 m	(2⅜ in.)	0.076 m	(3 in.)	0.076 m	(3 in.)
Oven center distances	1.100 m	(3 ft 7⅜ in.)	1.302 m	(4 ft 3¼ in)	1.400 m	(4 ft 7⅛ in.)
Thickness of wall liners	0.090 m	(3½ in.)	0.100 m	(4 in.)	0.100 m	(4 in.)
Thickness of headers	0.130 m	(5⅛ in.)	0.150 m	(6 in.)	0.150 m	(6 in.)
Oven-roof thickness	1.000 m	(3 ft 3⅜ in.)	1.391 m	(4 ft 6¾ in.)	1.700 m	(5 ft 7 in.)
Number of charging holes	4	4	4	4	4	4
Number of twin flues	14	14	16	16	17	17
Height of rich-gas burners	0.074 m 1.125 m	(3 in.) (3 ft 8¼ in.)	40.236 m 0.629 m 1.022 m 1.677 m	(9¼ in.) (2 ft ¾ in.) (3 ft 4¼ in.) (5 ft 6 in.)	0.248 m 0.375 m 1.621 m 1.822 m	(9¾ in.) (1 ft 2¾ in.) (5 ft 3¾ in.) (5 ft 11¾ in.)
Height of regenerators	2.550 m	(7 ft 4½ in.)	3.149 m	(10 ft 4 in.)	3.450 m	(11 ft 4 in.)
Working volume of oven	21 330 m^3	(754 cu ft)	39 670 m^3	(1402 cu ft)	48 230 m^3	(1703 cu ft)
Weight of coal charge (bulk density wet = 800 kg/m^3 — 50 lb per cu ft)	17.060 metr. tons	(18.80 net tons)	31.800 metr. tons	(35.05 net tons)	38.580 metr. tons	(42.50 net tons)
Coal throughput per oven per day	31.500 metr. tons	(34.71 net tons)	42.400 metr. tons	(46.73 net tons)	53.520 metr. tons	(59.00 net tons)
Coking rate	30.12 mm/h	(1,18 in. per hr)	25.40 mm/h	(1 in. per hr)	25.40 mm/h	(1 in. per hr)
Coking time	13 h	(13 hours)	18 h	(18 hours)	17 h	(17 hours)
Heat consumption:						
Rich-gas underfiring (calorific value of gas 2 × 10^4 kJ/m^3n — 500 Btu per standard cu ft)	2.280 kJ/kg	(980 Btu per lb)	2.260 kJ/kg	(970 Btu per lb)	2.272 kJ/kg	(975 Btu per lb)
Lean-gas underfiring (calorific value of gas 3,12 × 10^3 kJ/m^3n — 78 Btu per standard cu ft)	2.474 kJ/kg	(1062 Btu per lb)	2.448 kJ/kg	(1052 Btu per lb)	2.462 kJ/kg	(1057 Btu per lb)

ily slid into position. They are also readily accessible through sealed boxes located in the regenerator faces.

Combustion air is supplied and distributed in a manner similar to lean gas. Air is drawn into the bus flue through air-intake boxes located at the end of the bus flue and distributed through adjustable orifice plates to the individual regenerator cells. The ducts for the combustion air and lean gas from the regenerator cells extend to the heating flue sole.

Otto High-Capacity Rich-Gas Oven—On Otto large-capacity ovens designed for rich-gas heating only, the lean-gas distribution system is eliminated and there is one regenerator for combustion air and one for combustion products for each heating wall. There are no connecting ducts from each regenerator cell to the adjacent heating wall; therefore, all the heat-control features described for the compound type of oven are also incorporated in the Otto large-capacity oven for rich-gas heating.

Operation with Rich- and Lean-Gas Heating—Figure 4—60 illustrates the arrangement of the compound oven and the flow of the fuel gases, air and combustion products.

Coke-oven gas under regulated pressure is delivered through the rich-gas main (1) to the underjet system (2) which comprises two individual manifolds for each heating wall. One manifold connects to the burner in the even-numbered heating flues and one to those in the odd-numbered heating flues. One manifold supplies gas to the low-level burners and the other manifold to the high-level burners. The gas is supplied to the burners through the vertical ducts built integrally into the pillar walls of the regenerators, each having a regulating nozzle in the short connection between the manifold and the riser pipe. Each manifold is provided with a shut-off and reversing cock located between the coke-oven-gas main (1) and distribution manifolds.

Combustion air enters the oven on one side of the battery through the air-intake boxes (6, 7) and sole flue (8). The air is distributed through the adjustable orifice plates (10) to the individual regenerator cells (11) and preheated. It then passes through the short ducts to the heating flues (13). Combustion of gas and air in the upflowing flue (13) takes place, the hot products of combustion flowing down the adjacent flue (13a) and entering the regenerators (11a and 12a). The cooled combustion products are collected in the sole flue (8a) and pass through the waste-gas valve (14a) and waste-gas flue (15) to the stack.

The change-over of the fuel-gas supply from the odd-numbered heating flues to the even-numbered heating flues, or vice versa, is usually performed every 30 minutes by the hydraulically operated reversing system. The regenerators also exchange duties by the alternate opening and closing of the flaps on the air-intake boxes and by the reversal of the waste-gas valves. The two regenerators heated by waste-gas during the previous heating period are now changed over for preheating air. In order to eliminate any carbon deposits, the burners not on gas are supplied with a certain amount of decarbonizing air via the decarbonizing-air manifolds (18), the respective gas-distribution headers and rich-gas ducts. This practically eliminates carbon deposits even when heating with benzolized or partly debenzolized coke-oven gas.

When underfiring with lean gas, the gas is delivered via the main (16) located in the battery basement and through the reversing valves to the bus flue (8) and distributed by the adjustable orifice plates to the individual regenerator cells. It is preheated and supplied to the bottom of the heating flues.

Combustion air is admitted in the same manner as with rich-gas heating, through the air intake (6, 7) and bus flue (8) to the appropriate regenerator cells. When heating with lean gas, only one regenerator is used for preheating air and the adjacent regenerator for preheating lean gas, with combustion starting at the very bottom of the heating flue. The flow of combustion products and the change-over are the same as with rich-gas heating.

The Otto large-capacity oven referred to above is designed to heat with coke-oven gas or blast-furnace gas or any other lean gases, such as producer gas. The change from one fuel gas to the other is quickly accomplished either by changing the entire battery or any selected number of heating walls.

Temperature Distribution and Flow Conditions—As already explained, the entire heating system of the Otto large-capacity oven consists of numerous heating units, each separately controlled for the supply of fuel gas and air. Each heating flue is served by two regenerator cells and supplied with a controlled amount of fuel gas and air at each combustion point. The combustion products leaving the regenerator cells are controlled by the calibrated orifice plates located between the regenerator and the bus flue.

The combination of underjet system and individually controlled heating units ensures that the horizontal heat distribution is adapted to the requirements of the oven taper.

Owing to the combination of continuous regenerators with independently controlled twin-flue heating units, the Otto large-capacity oven requires only one row of waste-gas-reversing valves and one waste-gas flue. The lean gas and combustion air are introduced at the opposite side of the battery. The flow in the bus flues is uniform, irrespective of reversals, and so there is no need for additional facilities to control the incoming or outflowing media, since the Otto large-capacity oven is in this respect self-regulating.

For control of vertical heating the Otto large-capacity coke oven is provided with controlled stage heating. Each heating flue is served by separately adjustable, rich-gas underjets, and has burners at different levels. The vertical temperature distribution is influenced by locating the combustion points in the twin flue at different heights; therefore the hottest areas are at different levels in the heating flues, before and after reversal.

With the underjet system, the fuel-gas supply to the combustion points at different levels in the heating flues is separately and accurately controlled with the measuring facilities located in the cold battery basement. It allows control of the required vertical temperature gradient and adjustment if required during operation.

When heating with lean gas, preheated gas and air enter the bottom of the heating flue. The lean-gas

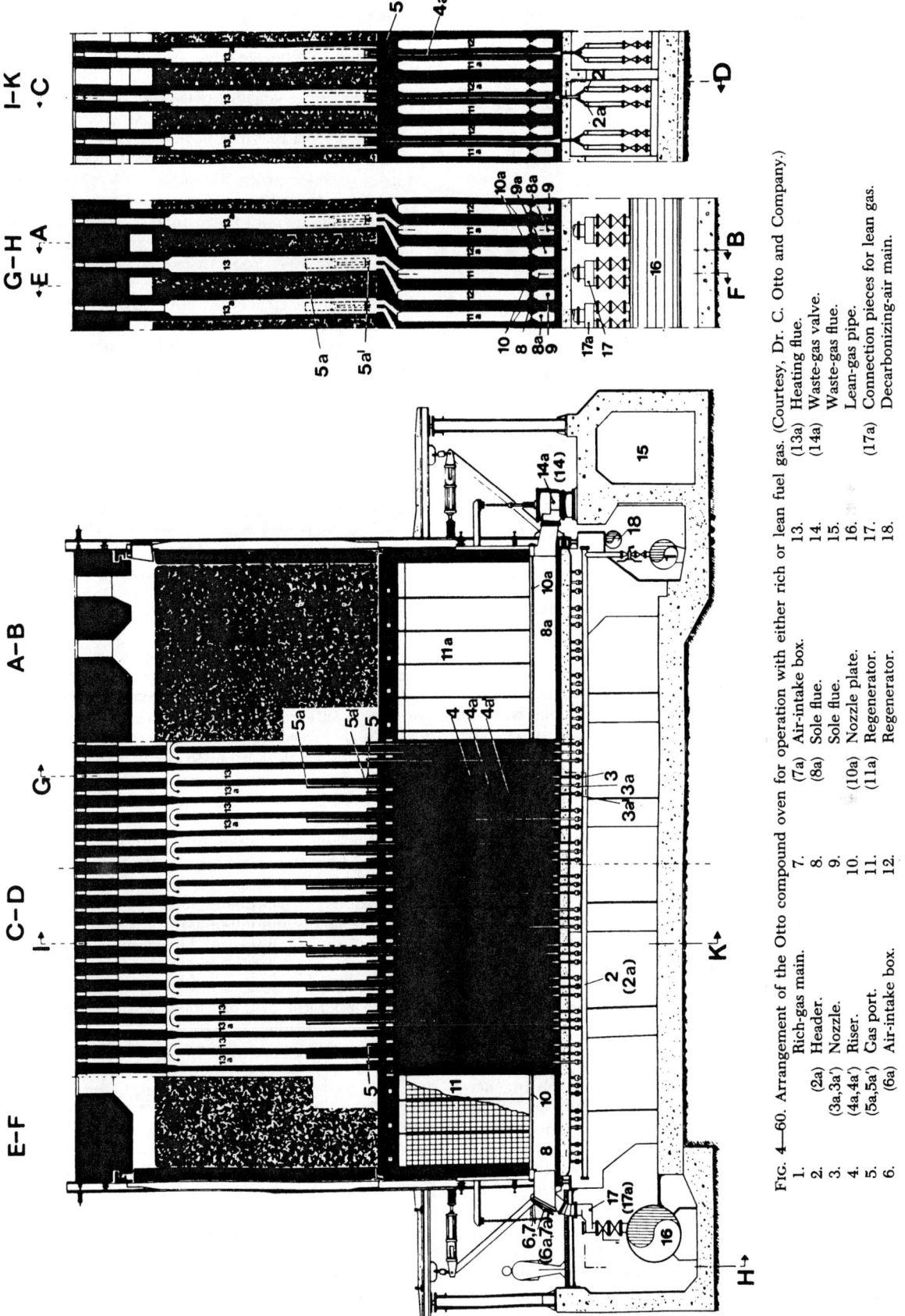

FIG. 4—60. Arrangement of the Otto compound oven for operation with either rich or lean fuel gas. (Courtesy, Dr. C. Otto and Company.)

1.	Rich-gas main.	7.	Air-intake box.	13.	(13a) Heating flue.
2.	(2a) Header.	8.	(8a) Sole flue.	14.	(14a) Waste-gas valve.
3.	(3a,3a') Nozzle.	9.	Sole flue.	15.	Waste-gas flue.
4.	(4a,4a') Riser.	10.	(10a) Nozzle plate.	16.	Lean-gas pipe.
5.	(5a,5a') Gas port.	11.	(11a) Regenerator.	17.	(17a) Connection pieces for lean gas.
6.	(6a) Air-intake box.	12.	Regenerator.	18.	Decarbonizing-air main.

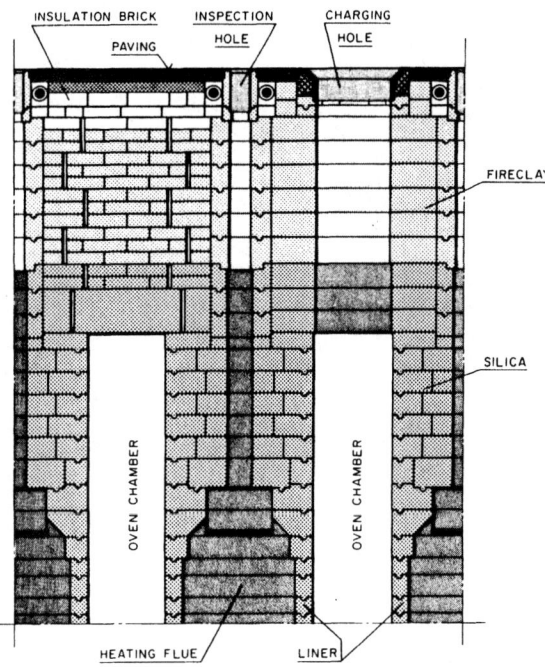

FIG. 4—61. Cross-section of battery brickwork in an Otto underjet oven. (Courtesy Dr. C. Otto and Company.)

flame is long, due to its slow rate of combustion, and no secondary combustion is required to obtain good vertical heat distribution in large ovens.

Design and Materials—The Otto underjet oven is supported on a reinforced-concrete substructure consisting of a base slab and nozzle decking rigidly interconnected by columns, thus forming a girder-type structure of high stability. As such, the concrete substructure, which serves as a basement for the fuel-gas distribution system, can largely withstand mining subsidences, earthquakes and the like.

The pinion walls at the oven-battery ends are of reinforced concrete, tied into the base slab and interconnected by longitudinal tie rods laid into the oven-top brickwork. The load on these tie rods during the heating-up period of the battery is controlled by load cells which ensures that the pinion walls remain in their original position. Expansion joints are provided to cope with the slight heat expansion.

The oven battery itself is built entirely of refractory brickwork. The lower part of the regenerator walls and the oven top, where temperatures occur within the critical expansion range of silica, are of high-quality fireclay material, and the upper part of the regenerators and the heating walls are of silica refractory manufactured by a special method from selected silica ganister (Figure 4—61).

Continuous expansion joints absorb the expansion of the silica material, in the longitudinal direction of the oven battery, between the battery pinion walls. Between the pusher and coke side, the brickwork is held by buckstays at the oven ends and connected by cross-tie rods in the nozzle deck and oven top. Expansion during heating is controlled by maintaining established

loads on the helical springs provided at the ends of the cross-tie rods.

There is no bonding between the refractory material and the concrete nozzle deck, thus leaving the fireclay to expand under a predetermined load. The thermal expansion of silica is about twice that of fireclay, and a specially designed sliding joint is provided between the fireclay and silica material in the regenerator walls. A special bracing system is utilized which maintains predetermined loads on the fireclay and silica refractories and transfers these loads to the buckstays.

A second sliding joint is provided between the upper silica layer above the oven chambers and the fireclay material of the oven top. An effective bracing system like that utilized for the sliding joints in the regenerator walls transfers the loads to the buckstays. With these spring-loaded bracings, the entire bench floor acts as a horizontal girder, creating a horizontal thrust against the buckstays at bench level. These loads are required to prevent the buckstays from deflecting and are essential for the bracing of the brickwork of the high ovens (Figure 4—62).

The stability of the heating wall is influenced by two primary factors: the overall width and the weight of the oven-top brickwork. As ovens increase in height, the distance between centers and the thickness of the oven top are also increased. Not only does the added weight improve the stability of the heating wall, but also the oven-top surface temperature decreases, resulting in improved operating conditions and lower top-maintenance costs.

The heating walls are of interlocked header and liner (stretcher) construction (see Figure 4—63). The headers and liners are of almost the same weight, thus facilitating manufacture and the laying of the bricks. The headers are interlocked, not only with each other, but with the liners as well. Since there is a greater tendency for vertical joints to open up as compared with horizontal joints, the former are backed up by header bricks, to avoid through joints between the oven and the heating flue.

The continuous regenerator walls from pusher to coke side do not require any expansion joints. These pillar walls are of high mechanical strength, and tightness is assured by adequate thickness and by tongue-and-groove joints staggered both vertically and horizontally. The low pressure differential between the regenerators, a special feature of the Otto oven, is an additional safety factor.

The regenerator cells are filled with checker bricks designed to offer minimum resistance to flow and maximum surface for regenerative heat exchange (Figure 4—64).

FIRMA CARL STILL

Firma Carl Still G.m.b.H. & Company KG, Recklinghausen, West Germany, is an independent engineering company owned by the Still family since its founding in 1898. The firm's original activities consisted of the design and construction of complete coke-oven plants, with facilities for coke-oven-gas purification and recovery of chemical by-products. Firma Carl Still has obtained a world-wide reputation because of its proprietary developments in these fields. To handle company

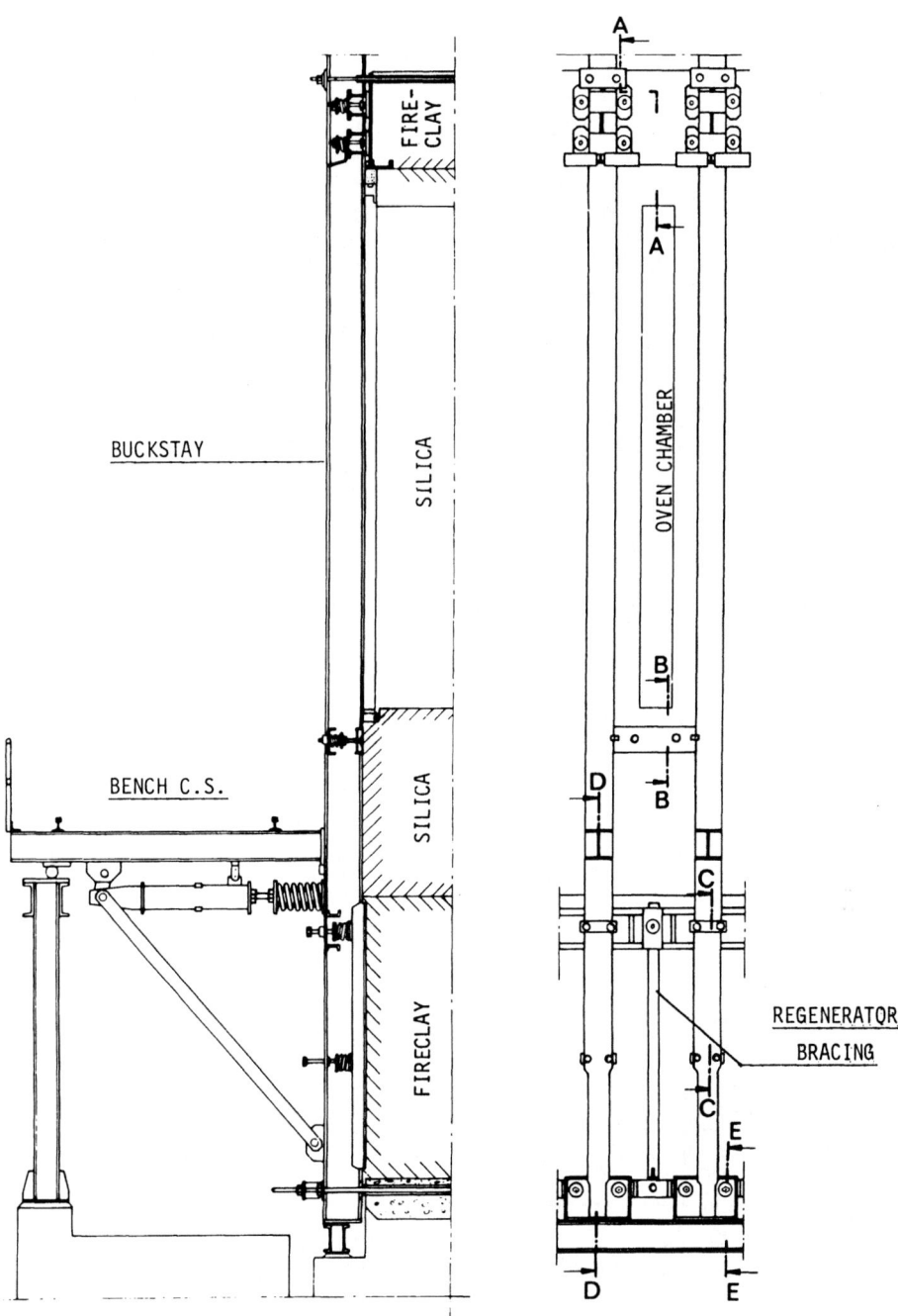

FIG. 4—62A. Bracing methods used in the construction of Otto ovens. See Figure 4—61B for enlarged details. (Courtesy Dr. C. Otto and Company.)

(Continued on next page)

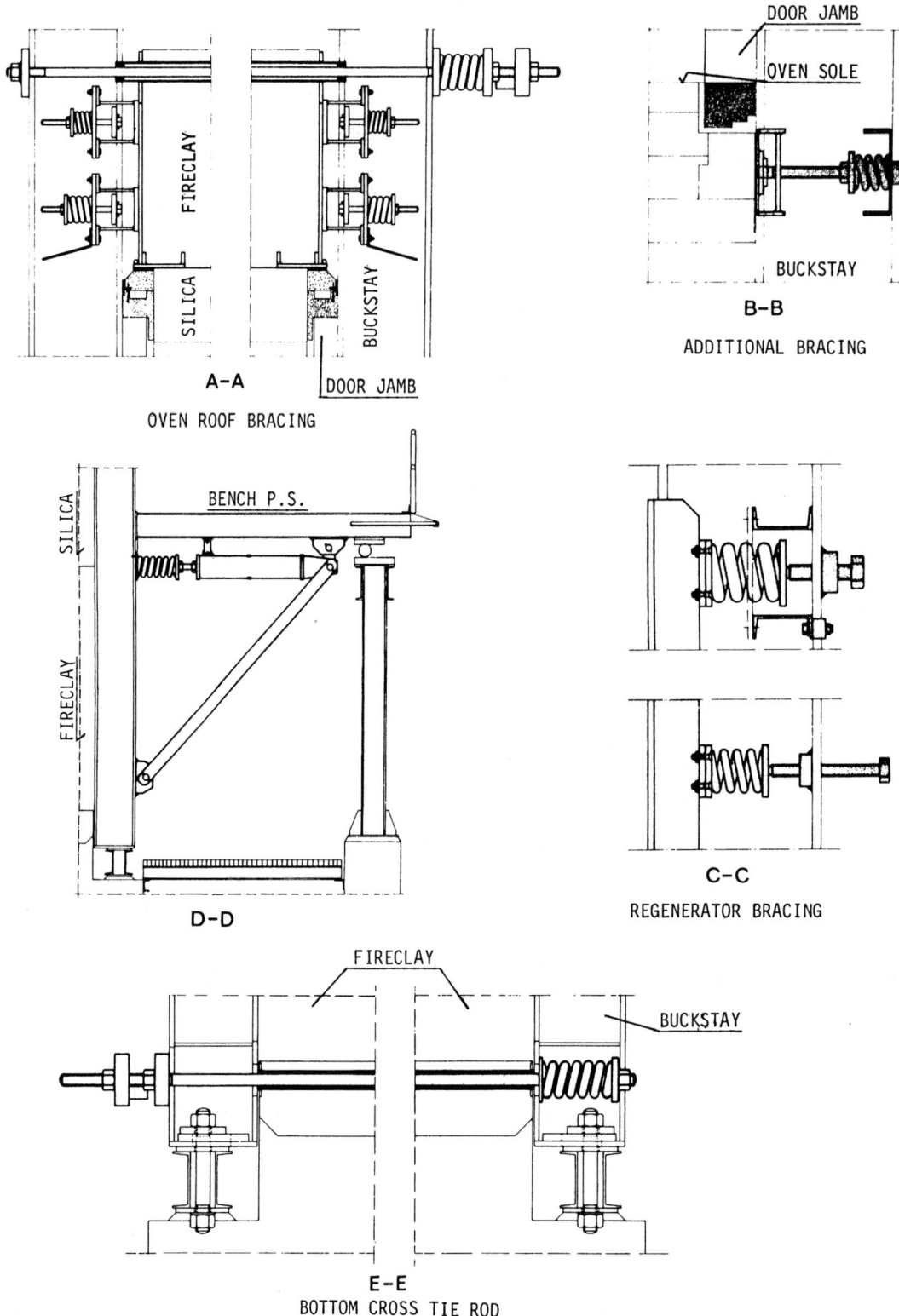

FIG. 4—62B. Enlarged details of bracing methods shown in Figure 4—62A. (Courtesy, Dr. C. Otto and Company.)

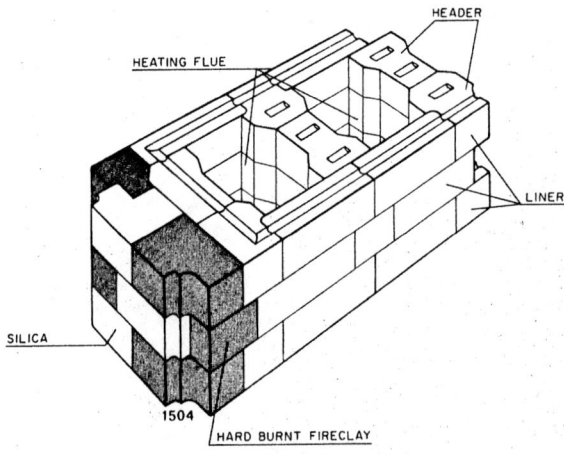

FIG. 4—63. Design of the heating walls of Otto coke ovens. (Courtesy, Dr. C. Otto and Company.)

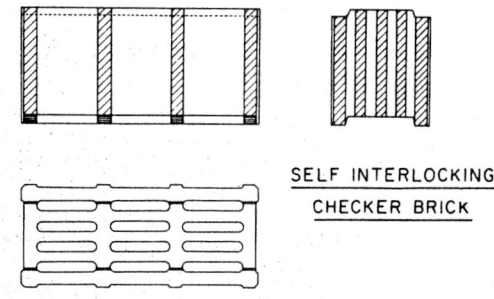

SELF INTERLOCKING
CHECKER BRICK

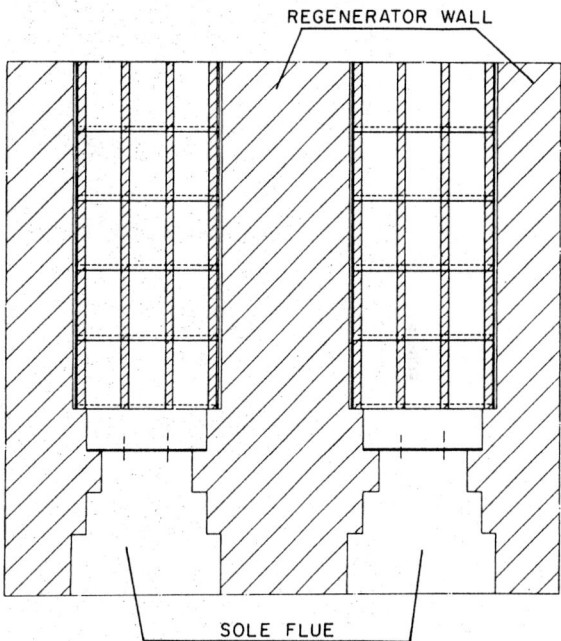

FIG. 4—64. Arrangement of the regenerators of an Otto oven, showing design of self-interlocking checker brick. (Courtesy, Dr. C. Otto and Company.)

activities in the Western Hemisphere, a subsidiary company, Carl Still Corporation (Pittsburgh, Pennsylvania, USA), was founded in 1974. Firma Carl Still is today a leading designer-builder of high-capacity coke-oven plants world-wide, with more than 3000 oven chambers higher than 5 metres (about 16½ feet) currently in operation.

GENERAL DESCRIPTION OF A FIRMA CARL STILL COKE OVEN

The basic concept in the design of a Still coke oven is multi-stage heating in the combustion flues of a "semi-divided" oven (Figure 4—65). The initial patent for multi-stage heating, granted to Dr. Carl Still in 1910, is the basis for vertical temperature control in Still tall coke ovens. This system has proven successful in the operation of oven chambers up to 7.65 m (25 ft) high, and with effective working volumes of 52.14 m³ (1842 cu ft).

The basic Still oven can be operated efficiently on rich gas (coke-oven gas), lean gas (blast-furnace gas), or a combination of both gases. It can be constructed for introduction of rich gas through headers at the side of the heating walls (side-fired), or through a manifold in the oven basement (underjet).

Multi-Stage Heating—The fundamental principle of Still multi-stage heating is controlled mixing and burning of the fuel gas and combustion air at different elevations in the vertical heating flues; in essence, a series of burners stacked vertically. The number of burners required, and their relative positions in the vertical flues, are direct functions of the height of the oven chamber (Figure 4—66).

The inherent temperature-control capability of a well-designed multi-stage heating system provides outstanding benefits in cost, product quality, and operating control. Operating experience over the years has proven that:

(1) The vertical temperature profile of the oven walls can be modified during operation, as required by conditions, from the control panel by adjusting the stack draft and, therefore, the ex-

cess air.

(2) Even in the tallest ovens, multi-stage heating produces evenly coked-out charges and uniform coke quality from a charge of either wet or pre-heated coal.

(3) Multi-stage heating eliminates hot spots and reduces roof-carbon formation.

(4) With close control of temperature, the formation of NO_x can be held to a minimum.

The high efficiency of multi-stage heating means a lower gas-flow rate and excellent fuel economy. Multi-stage heating is the basis for Still heating warranties covering both heat-input requirements per ton of coal, and temperature distribution in the heating walls.

Semi-Divided Design—The basic Still design ovens are semi-divided (Figure 4—67). The vertical heating flues (1) are connected at the top by a horizontal flue (2). Air enters at the bottom of the regenerators along the sole flue (6a, 6g) and passes upward through the checker bricks (7, 8) into the vertical flues. Rich gas from the rich-gas distribution channel (4) is led directly into the vertical heating flues. For lean-gas firing, the

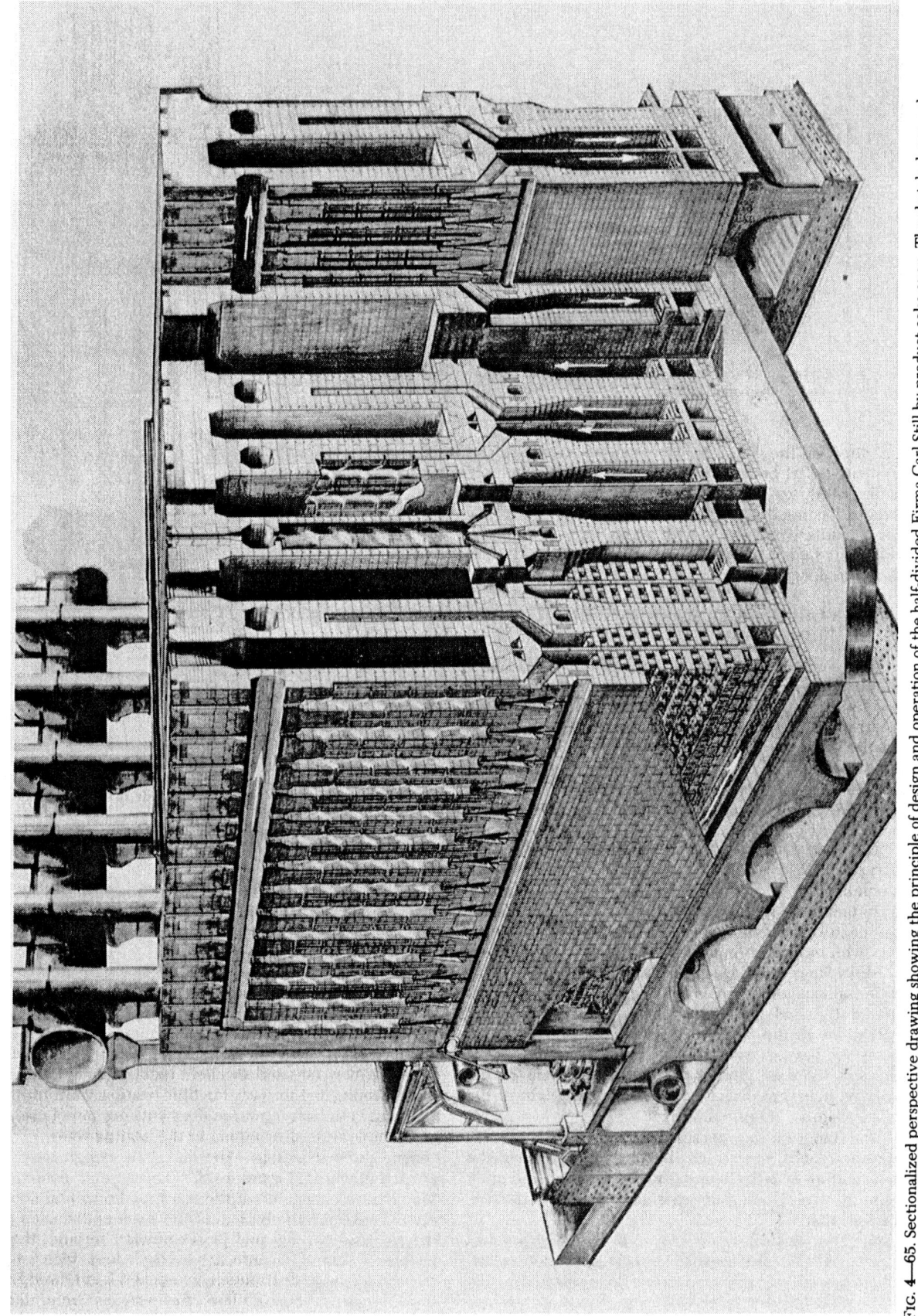

Fig. 4—65. Sectionalized perspective drawing showing the principle of design and operation of the half-divided Firma Carl Still by-product coke oven. The sketch shows only some ovens (pusher side). (Courtesy, Firma Carl Still.)

lean gas and air enter separate regenerator sections (7 and 8 respectively) and no rich gas is introduced.

A partition wall (3) running longitudinally through the battery divides the regenerators and vertical heating flues into two parts. Each part serves approximately one half of the oven chamber. Combustion of the gas and air takes place in the flues of one half of the heating wall. The combustion passes along the horizontal flue (2), then down through the other half of the wall, through the regenerators (7' and 8'), and into the waste gas flues via the regenerator sole flues (6g', 6a'). Each side is reheated alternately by reversing the flow of the underfiring gas and air from one side to the other, approximately every 20 minutes.

All heating-wall flues for the intake of lean gas and air have short connecting passageways (5a, 5g) from the regenerators and are situated side by side on each half of the battery. Fuel and waste-gas flues do not cross within the heating wall. This design minimizes pressure differentials and, therefore, leakage of fuel and air to waste gas.

The number of pusher-side flues is greater than the number of coke-side flues because of the required higher heat input per flue on the coke side, and the consequent larger size of the coke-side flues. The exact number of flues on each side depends on such design parameters as oven taper, heat input, and waste-heat balance.

Uniform distribution of air and lean gas on the combustion side, and waste gas on the exhaust side, is achieved by the design of the specific taper of the aerodynamically shaped sole flues, the precise sizing and contouring of sole-flue ports and the cellular construction of the regenerators.

The interlocked gas-tight division wall (3) in the regenerator section eliminates any possibility of fuel-gas leakage from the combustion side into the waste gases on the exhaust side.

STILL OVEN TYPES

Still ovens can be designed in various configurations for introduction of the heating gases and removal of the waste gases, to provide the overall design best meeting the customer's requirements.

Side-Fired Oven—Figure 4—65 shows schematically a cross section through a side-fired battery. One lean-gas header for each side is located beneath the alley floor, and one rich-gas header for each side is located in the alley.

The oven is semi-divided, and identical sets of waste heat-air valves and waste heat-air-lean gas valves are provided on the pusher side and the coke side. One pair of sole flues for each regenerator runs from either side to the center of the regenerator. Six waste-gas flues are provided between the regenerators and the battery base slab. When heating is on the pusher side, waste gases are carried away in the coke-side flues, and vice versa. To distribute the waste gases evenly, each waste-gas flue is fed by one third of the waste-gas valves on that side of the battery. The six waste-gas flues are connected to a common stack flue at one end of the battery, leading to the stack. The regenerator pillar walls can be designed either for combined silica-fireclay-construction (shown in Figure 4—65) or for all-silica

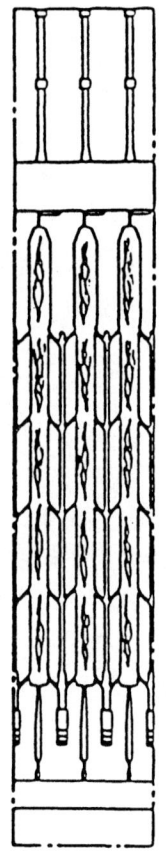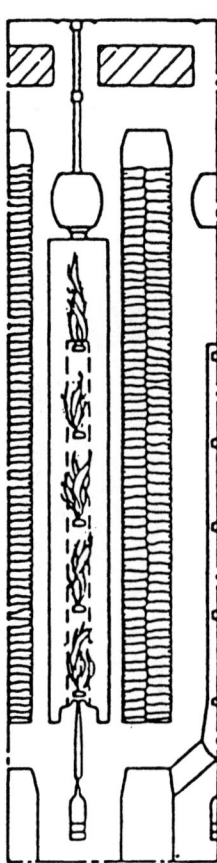

FIG. 4—66. Diagram showing how combustion is introduced at successive levels in the heating flues of a Carl Still oven to provide uniform heating from top to bottom of the heating walls. (Courtesy, Firma Carl Still.)

construction. For silica-fireclay construction, a temporary heating-up anchor is provided in the uppermost fireclay layer of each regenerator pillar wall over the whole length of the regenerators. This anchor controls expansion of the silica and fireclay during heating up.

Side-fired ovens can be designed for underfiring with rich gas only, for underfiring with lean gas or enriched lean gas only, or for underfiring with a combination of rich gas and lean or enriched lean gas. The oven depicted in Figure 4—65 can be operated on all three modes of underfiring.

The number and arrangement of waste-gas flues can be varied according to installation requirements. Instead of six flues, as shown in Figure 4—65, one flue on each side could serve the same purpose. In this case, the flues are much larger, since one flue carries all of the waste gas, instead of only one third of the total as in the Figure 4—65 arrangement.

A different concept is embodied in a recently completed 5 m (about 16½-foot) battery in North America, designed for foundry-coke production. The battery is semi-divided, of the side-fired type, for rich gas only. It uses the so-called "reversing in the flues" principle. One waste-gas flue each on pusher side and coke side is arranged so that the waste gases enter the waste-gas flues directly from the sole flues through vertical pas-

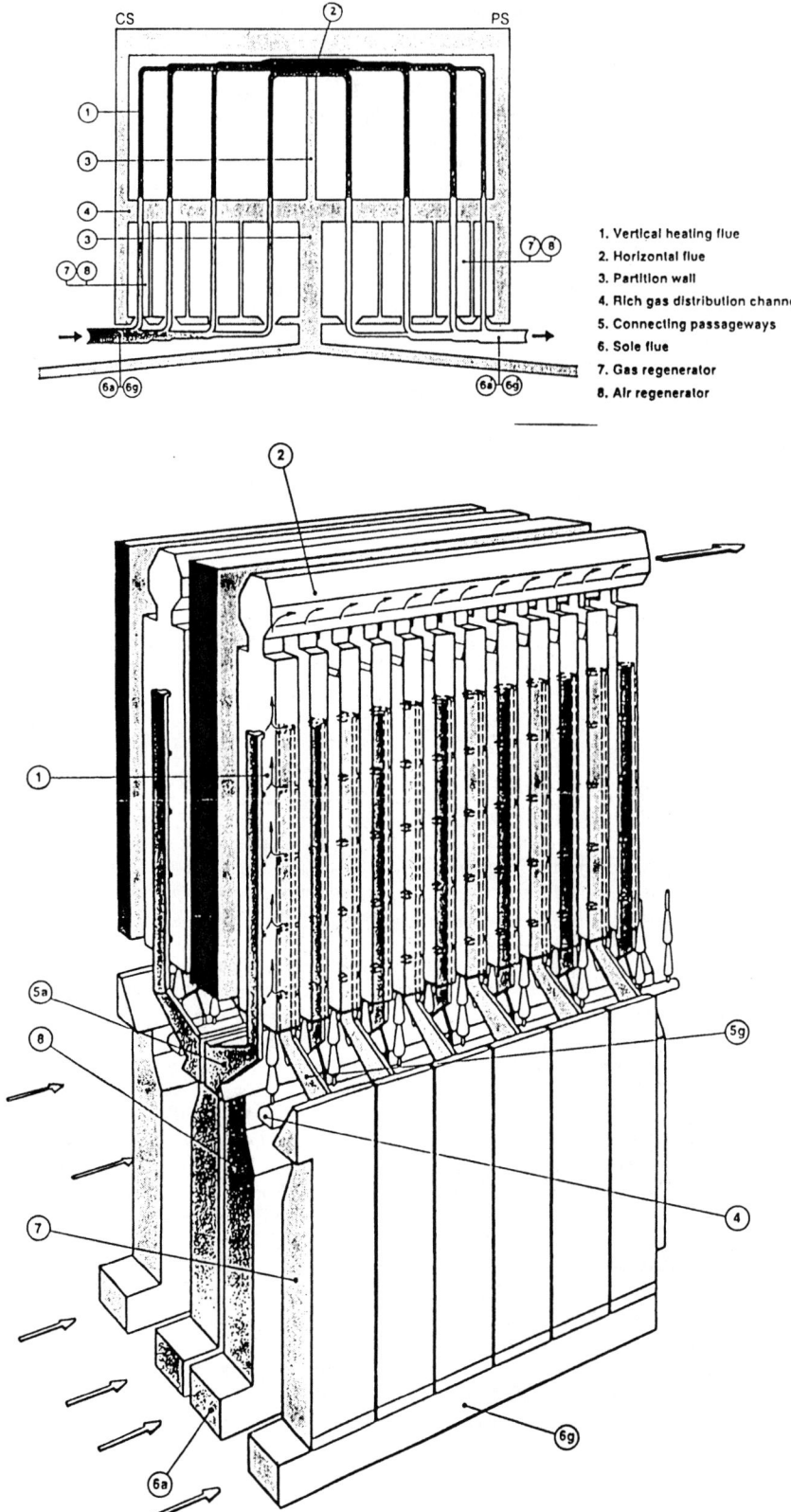

1. Vertical heating flue
2. Horizontal flue
3. Partition wall
4. Rich gas distribution channel
5. Connecting passageways
6. Sole flue
7. Gas regenerator
8. Air regenerator

Fig. 4—67. Schematic representation of the design concepts embodied in a Carl Still semi-divided coke oven. Legend: 1, Vertical heating flue; 2, Horizontal flue; 3, Partition wall; 4, Rich gas distribution channel; 5, Connecting passageways; 6, Sole flue; 7, Gas regenerator; 8, Air regenerator. (Courtesy, Firma Carl Still.)

sages. In this design, the conventional waste-gas boxes are replaced by relatively small air-inlet boxes on pusher side and on coke side. Both waste-gas flues join in a common stack flue at one end of the battery. Reversing of the heating from one side to the other is accomplished by two waste-gas reversing dampers in the two waste-gas flues outside the battery limits.

Another battery design uses the "single waste-gas flue" concept. This design also maintains the semi-divided Still oven with multi-stage heating. It requires only one waste-gas flue, which can be located on either side of the oven. Through a special regenerator bus-flue arrangement, the valves for waste gas, air, and lean gas are located only on the side where the waste-gas flue is located. This concept can be used for batteries fired with rich gas only, as well as for lean-gas-fired batteries. It uses either a single stack flue from the mid-point of the waste-gas flue, or two stack flues from the quarter points.

Underjet Ovens—Still underjet coke ovens can be designed with two waste-gas flues (either with or without reversing in the flues), or with a single waste-gas flue as described above for the side-fired ovens. With the exception of the supply of rich gas for underfiring, the underjet design has most of the features of the side-fired design.

Heating a Still Oven—A Still design oven, whether underfired with rich gas only or with both rich and lean gas (compound type), is heated with great uniformity both vertically and horizontally throughout the oven chamber.

When underfiring with lean gas, alternate sets of regenerators on the combustion side are used to pre-heat gas and air. From the regenerators, preheated air and lean gas pass through short, direct, symmetrical passages in the corbel area to their respective vertical flues within the binder bricks. Air and gas emerge through exit slots placed opposite each other in the heating flues, creating multiple combustion points at different elevations (multi-stage heating). The products of combustion pass through the horizontal flues to the exhaust side of the oven, where they are drawn by stack draft down through slots in the vertical flues and then through the regenerators into the waste-heat flues.

When a Still oven is fired with rich gas, all regenerators on the combustion side of the battery are used to preheat air. The air rises through the flues formed by the binder bricks, and emerges through the multi-stage exit slots into the heating flue. Combustion at many levels creates the long, continuous, stable flame characteristics achievable only by multi-stage heating. For a side-fired battery, rich gas is distributed through a duct commonly called a gas gun located horizontally in the corbel section and connected to a nozzle in the base of each heating flue. A Still oven can also be underfired with rich gas that is diluted with lean gas, or with a combination of rich and lean gas. Diluted rich-gas underfiring is most suitable for extended coking times. A long flame is achieved and heat distribution problems are avoided. When applying the combination underfiring of rich gas and lean gas, rich gas is distributed through the rich-gas system, and at the same time lean gas, or enriched lean gas, is fed through the

regenerators and burned as described for lean-gas heating. The underfiring gas to each wall, for rich gas as well as for lean gas, is metered by an orifice plate between the shut-off cock and the reversing cock.

The rich-gas nozzles in side-fired ovens are not sensitive to carbon blockage because (1) they are of relatively large size, and (2) they remain relatively cool because of their protected position and the liberal use of insulating material surrounding the gas gun flue. Any carbon that may form in the nozzles is eliminated during the "gas-off" part of the cycle by air drawn in through the three-way cock.

For an underjet oven, rich gas is distributed from the manifold at basement level through individual riser pipes to a nozzle in the base of each heating flue. Gas flow is controlled by a metering pin, metering orifice, or nozzle at the bottom of each riser pipe. To guard against cracking of the rich gas during its passage through the regenerator pillar walls and consequent carbon formation, the gas riser pipes are shielded from heat by insulating material and the gas remains relatively cool. Any carbon formed at the nozzle is eliminated during the "gas off" part of the cycle by the forced-air decarbonizing system.

Still ovens operate efficiently on both lean gas and rich gas. If desired, a Wobbe system can be installed that uses the Wobbe index for constant heat-input-control.

DESIGN FEATURES

The Oven Wall—Still oven walls are designed for strength, economy, and low maintenance. All brickwork in contact with the coal is high-quality silica to achieve the specified coking rates. Thickness of the heating-wall liner, except for the bottom two courses of brick, is uniform for the full height of the oven. This design results in fewer brick shapes, and consequently all bricks in the walls can be machine-fabricated, which increases their uniformity.

In Still ovens, stability and flexibility of the walls virtually eliminate wall cracking. Wall bricks are tongue-and-grooved, both vertically and horizontally. Binder bricks, which tie the wall liners together and form the vertical flues, are also tongue-and-grooved vertically and horizontally. Triple-sectioned binder construction provides a wall of great stability and flexibility (Figure 4—68).

Liberal use of insulation bricks at the oven ends helps to minimize heat losses. Oven enclosures (buckstays and flash plates) are spring-loaded to ensure even support of the refractory.

The overall design and flexibility of the heating-wall end system makes Still ovens equally suited for charging wet or preheated coal.

The Horizontal Flue—The upper end of each vertical heating flue terminates in a horizontal flue, running the length of the oven wall from pusher side to coke side. The position of the horizontal flue, in relation to the oven roof and the coal/coke line, ensures uniform heating and properly coked-out charges under all normal operating conditions.

The horizontal flue is octagonal in cross section (Figure 4—69). This configuration provides a strong, stable structure that distributes vertical loading evenly on the

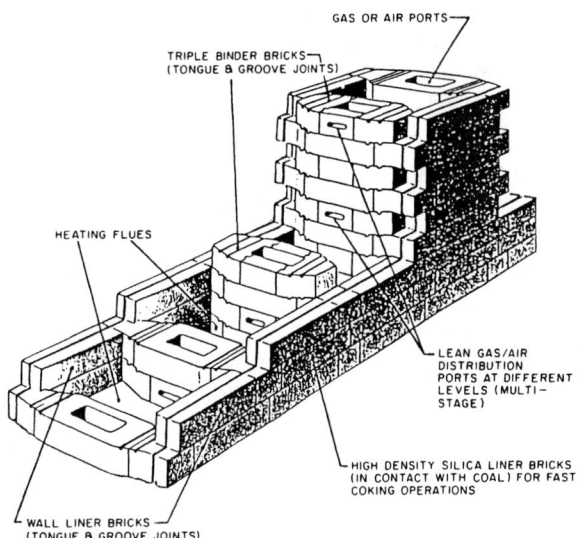

FIG. 4—68. Carl Still oven-wall construction details. (Courtesy, Firma Carl Still.)

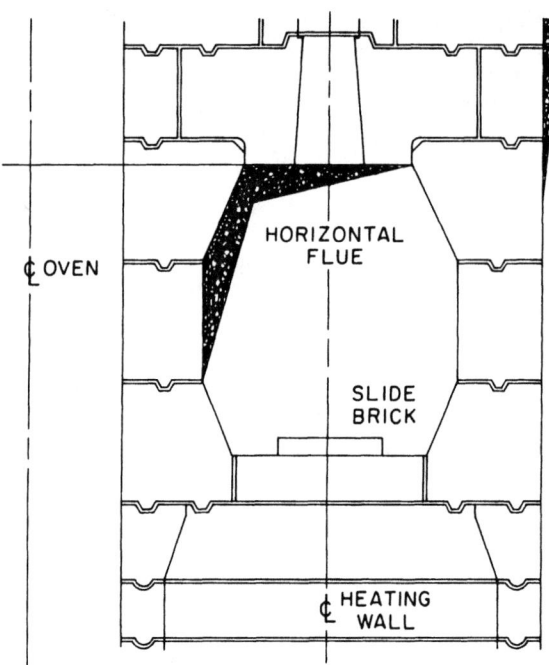

FIG. 4—69. Horizontal-flue design for a Carl Still oven. (Courtesy, Firma Carl Still.)

oven walls and easily bears the weight of modern coal charging equipment and other battery-top loads.

The openings from vertical flues into the horizontal flue are sized to distribute the flow of gases. The area of each opening can be adjusted from the oven top, during oven operation if necessary, by means of a slide brick which functions as a damper and provides additional control over the crosswall temperature profile. These adjustments are made by Still engineers during start-up and seldom need to be changed except for long-term changes in operating rates.

The Corbel Area—The corbel area below the oven chamber is of simple design and exceptionally strong. Since each vertical flue in the binder brick supplies either air or gas to the two heating flues on each side, each regenerator requires only one passage through the corbel for every other heating flue rather than two or four times as many passages required by other designs. The small number of passages permits a stronger, less complex corbel structure (Figure 4—70). Design simplicity and strength, together with properly located expansion joints, eliminate cross leakage of air and gas and prevent premature failure of the corbel.

For a side-fired battery, the rich-gas duct or gun flue is protected from excessive heat by insulation brick. This protection prevents cracking of hydrocarbons in the rich gas and formation of carbon. The gas-gun nozzles are inherently insensitive to carbon blockage (as explained previously), and carbon that may form is eliminated by air introduced during the "gas off" part of the firing cycle. Even distribution of gas to the flues is ensured by the precisely calculated aerodynamic taper of the gas-gun, which eliminates any tendency to starve the end flues.

For an underjet-fired battery, the rich-gas risers are enclosed in the regenerator pillar walls with tongue-and-grooved joints designed to prevent cross leakage of air and gas. Insulation bricks shield the gas risers from excessive heat to prevent cracking of hydrocarbons.

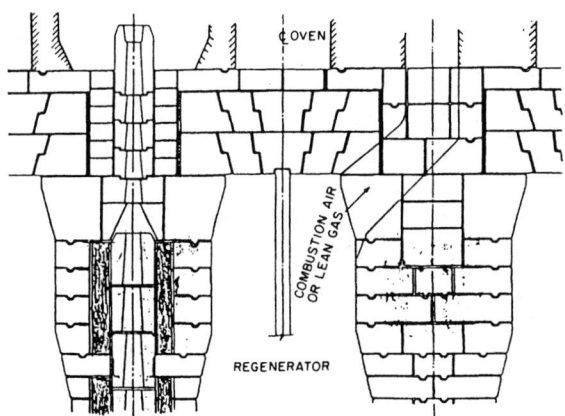

FIG. 4—70. Cross-section of the corbel area of a Still compound underjet coke oven. (Courtesy, Firma Carl Still.)

The Sole Flue—The design of the sole flue at the base of the oven assures even distribution of gases into each cell of the regenerator chambers (Figure 4—67). The taper of the sole flue and the precise size and contour of its ports distribute both the upflowing air and lean gas and the downflowing waste gas evenly throughout the length of the regenerators. As a result, all regenerator cells are fully utilized, and the total heat-exchange capacity of the regenerators is used efficiently. At the same time, the design of the system ensures that the end flues will receive their full required apportionment of lean gas and air.

Throughout the Still oven, design of the brickwork

simplifies the shapes and minimizes the number of different silica and fireclay shapes required. A high proportion of machine-made bricks is feasible, with the result that bricks are more uniform in their physical dimensions, composition and density.

The Fuel-Gas Piping—For large Still batteries and large throughput, the rich-gas and lean-gas headers on both sides of the semi-divided battery may consist of two sections each. Batteries designed for single waste-gas flues provide a lean-gas header only on one side, whereas rich gas is supplied as before.

Both lean and rich gas are supplied to the ovens by way of reversing valves. The rich-gas reversing valves are of the three-way type. During the time of gas off, forced air for decarbonizing will enter through these valves in underjet batteries, and decarbonizing air will enter side-fired ovens by natural draft. (In a very few installations, Still side-fired batteries require forced air.) Still coke-oven batteries customarily provide a quick-opening shut-off valve of the plug type between the fuel-gas header and the reversing valves.

All Still batteries can be operated with a broad range of lean-gas enrichment. Where required, Still batteries are designed for "dilution underfiring". In this mode, rich gas is diluted to some degree with lean gas in the rich-gas distribution system. Due to gas pressure required, this mode is used for extended coking times.

Where desired for underjet batteries, separate headers are provided for pusher and coke sides to supply the first and last heating flues independently with rich gas or diluted rich gas.

Reversing System—The main components of the Still hydraulic reversing system are a control unit, which is of a proprietary Still cam-operated design, and a system of single-acting hydraulic cylinders which are connected with reversing rods to the rich-gas reversing valves, to the lean-gas reversing valves, and to the air intake and the waste-gas control gates on both sides of the oven. In addition, there are a fluid-storage tank, hydraulic pumps (including spares), and a hydraulic-pressure-storage station for emergency reversal in case of a power failure. The valves and gates are opened by the cylinders, and closed by counterweights located at the opposite end of the reversing rods. Closing of the valves and gates by the counterweight makes the reversing system "fail safe" in an emergency, in which case the oil pressure in the respective cylinders is automatically released.

This system also includes a tie-in for the fuel-gas-pressure safety device. In case of low gas pressure, the oil pressure in the respective cylinders is released and counterweights close the gas-reversing valves.

When required, the Still hydraulic reversing device is provided with an automatic change-over from one mode of fuel-gas underfiring to another. When underfiring with rich gas, actuation of a selector switch will cause an automatic change-over to lean-gas heating during the reversing process (and vice versa for lean-gas to rich-gas underfiring). With this automatic feature, the system can also be designed so that the coke ovens provide the "bubble" needed for control of the fuel balance in the steelworks.

Oven Enclosures—Enclosures for Still batteries incorporate a heavy steel buckstay which is tied in across the battery at top and bottom with dual steel spring-loaded tie rods. Oven openings are enclosed with a one-piece iron block type jamb, to which the hooks for fastening the oven doors are attached. Still offers various types of oven doors as needed to meet customer specifications.

Experience has shown that for chambers over 3.5 m (11½ ft) height, additional support for the steel buckstays is desirable in order to counteract pressure due to expansion of the brickwork and pushing of the coke. Still has developed a support system which ties in the operating benches to provide additional support for the buckstay, as illustrated in Figure 4—71.

Performance Data—Main dimensions and performance data of a modern Still by-product coke-oven plant, built for metallurgical-coke production in West Ger-

Table 4—III.
Performance Data of a Firma Carl Still
Coke-Oven Battery*

	Rich-Gas Heating	Lean-Gas Heating
Throughput per day—dry coal	2646 mt (2917 t)	2655 mt (2926 t)
Tons per oven per day	51 mt (56 t)	51.0 mt (56.5 t)
Coal Characteristics		
Volatile matter (dry and ash-free) %	27.9	27.2
Moisture %	9.5	9.0
Ash %	7.02	6.57
Less than 0.5 mm (% by weight)	26.1	29.5
Less than 3.15 mm (% by weight)	68.3	69.8
Oven bulk density dry	785 kg/m³ (49 lb/cu ft)	781 kg/m³ (49 lb/cu ft)
Gross coking time	13 h	13 h
Coking rate	30.76 mm/h (1.21 in./h)	30.76 mm/h (1.21 in./h)
Heat consumption (converted to 10% coal moisture)	554 kcal/kg (996 Btu/lb)	580 kcal/kg (1043 Btu/lb)
Average flue temperature	1383°C (2521°F)	1387°C (2529°F)
Average chamber wall temp.	1170°C (2138°F)	1183°C (2164°F)
Waste-gas temperature	283°C (541°F)	244°C (471°F)
Waste-gas analysis (% by volume)		
CO_2	7.7	18.6
O_2	4.4	3.5
Gas temperature in collection main	102°C (215°F)	103°C (217°F)
Average gas space temperature	812°C (1494°F)	807°C (1484°F)
Coke Characteristics		
Moisture %		5.3
Volatile matter %		0.6
Less than 20 mm % by weight		4.1
More than 80 mm % by weight		16.5

*Source: Stahl und Eisen 94/1974, Volume 8, Pages 309 through 315 "Erste Betriebserfahrungen mit einer Grossraum-Hochleitungs-Batterie in einer Huetten-kokerei" by Bernhard Bussmann.

many, are illustrated in the following description and in Table 4—III.

There are two batteries, each with 52 ovens. The ovens are about 16 m (52 ft) long between door linings, the total height is 6.1 m (20 ft), and the average oven width is 400 mm (15.75 in.).

NIPPON STEEL CORPORATION

Nippon Steel Corporation (through Brown and Root, Inc.) offers three coke-oven designs (Basic, M-type and S-type) that permit the coke producer to select sizes up to 6.5 metres (21 ft. 4 in.) in 0.5-metre height increments and from full gas-gun heating to full underjet heating.

The M-type oven was developed from the Basic type, while the S-type was independently designed. A comparison of the principal design elements of the three types is given in Table 4—IV.

M-Type Coke Oven—As a coke oven designed for multiple-stage heating, the M-type was developed to provide fine control of heating for tall ovens (5.5-metre or 18-ft or taller). In 1968, the first 4.4-metre (18-ft) M-type coke-oven battery was commissioned in Japan and is still in operation without any major repairs performed or anticipated.

M-Type Heating—The M-type coke-oven design supplies coke-oven gas, blast-furnace gas, and combustion air by the underjet method to each individual flue in a flue wall. (The gun-fired design M-Type may also be used, depending on user requirements and fuel availability. However, this will not be described here.)

With **blast-furnace-gas heating** (Figure 4—72), blast-furnace gas and air flow to respective regenerators through underjet pipes embedded in the nozzle deck and passing through the sole flue. The blast-furnace gas is enriched with coke-oven gas to a predetermined calorific value in order to eliminate deposits on the underjet orifices and pipes, before or after enriching with coke-oven gas. The heated, enriched gas then enters the battery basement through the single blast-furnace-gas distributing pipe (2) located below the basement floor. Combustion air for heating with blast-furnace gas is also introduced through the battery basement in the air-distributing pipe (1) below the basement floor. The combustion air is drawn in through a filter media by a forced-air fan which discharges it at constant pressure into the distributing pipe.

At each flue wall, two blast-furnace-gas and two combustion-air headers come off the respective distribution pipes. One pair of blast-furnace-gas combustion-air headers supplies the even-numbered flues and the other pair supplies the odd-numbered flues in a specific heating wall. With a hairpin-flue heating design, all even-numbered flues burn and then all odd-numbered flues burn. (This grouping of two pairs of blast-furnace-gas/combustion-air headers underfires a full heating wall as well as the adjacent heating wall [For a specific oven, its two heating walls are fired from one common grouping of blast-furnace-gas/combustion-air headers as shown in Figure 4—73].) Before each blast-furnace-gas header and each combustion-air header there are one manual shutoff cock and one reversing valve (controlled by the reversing system). As gas passes through the valves and headers it is distributed to each on-

burning flue underjet pipe (5) through an orifice (8) and into a riser pipe. The riser pipe passes through the nozzle deck and through the sole flue into a regenerator half cell (11). It should be noted that the orifices can easily be removed and replaced from the basement.

Each full regenerator (10) for blast-furnace gas (also Figure 4—74) is divided into two regenerator sub-cells (11) (not shown in Figure 4—74) for a flue wall except for the end regenerators which are half cell size. Each full regenerator is separated from the adjacent ones (which are for air) by all-silica refractory support walls thereby preventing crossleakage. This is also prevented by the uniform pressures of the combustion air and blast-furnace-gas which eliminates differential pressures. Within each full regenerator, the fireclay checker brick have been designed to divide the cell into two half cells thereby allowing each blast-furnace-gas riser pipe to conduct the gases through a half cell and leave the top with a minimum pressure loss through the regenerator.

At the top of the regenerator chamber the silica bricks form a partition wall (12) to maintain the flow to the proper flue. Each regenerator subcell top has two gas-outlet ports: one to each of the identical flues of adjacent flue walls. The heated blast-furnace gas then rises through the flue duct to the base of the flue (Figure 4—73).

The blast-furnace-gas flue duct branches into three outlets per flue (14) in order to provide uniform combustion throughout the height of the flue wall. Blast-furnace gas is introduced at the base of the heating flue, at a second port at ¼ height, and at the ½ height point. The top port has been provided with a sliding brick which is adjusted only when significant production rate changes are required. This sliding brick (13) is adjusted from the top of the battery to provide the desired vertical temperature differential.

In an identical manner to the blast-furnace gas, the combustion air flows from the basement air header to the three flue-outlet ports. Briefly described are the pressurized air flows through reversing valves into the flue-wall header (4) and through regulating orifices (7). Combustion air passes through the riser pipe and into a subcell (11) of an air regenerator (10) which is adjacent to a blast-furnace-gas regenerator. The air is heated in the regenerator subcell and finally discharges at the top into one of two heating-flue ducts. The heating-flue duct has three discharge points at the same elevation as the blast-furnace gas but from the opposite wall in the heating flue (Figure 4—73).

The design of the battery heating system is such that, should the combustion-air blower fail, the reversing system would advance to cause the waste-heat (air) boxes to open allowing sufficient air to flow, by natural draft, to sustain combustion for an indefinite time. The only detriment with natural draft is the inability to regulate the air to each flue in a flue wall.

As the air and blast-furnace gas discharge into the flue, controlled combustion occurs, with the hot products of combustion rising to the top of the flue, turning the hairpin and discharging down the off-burning flue. The waste gases exit through the six available flue ports (3 air, 3 blast-furnace gas during combustion) and passes to the regenerator (and its subcells). The waste gases

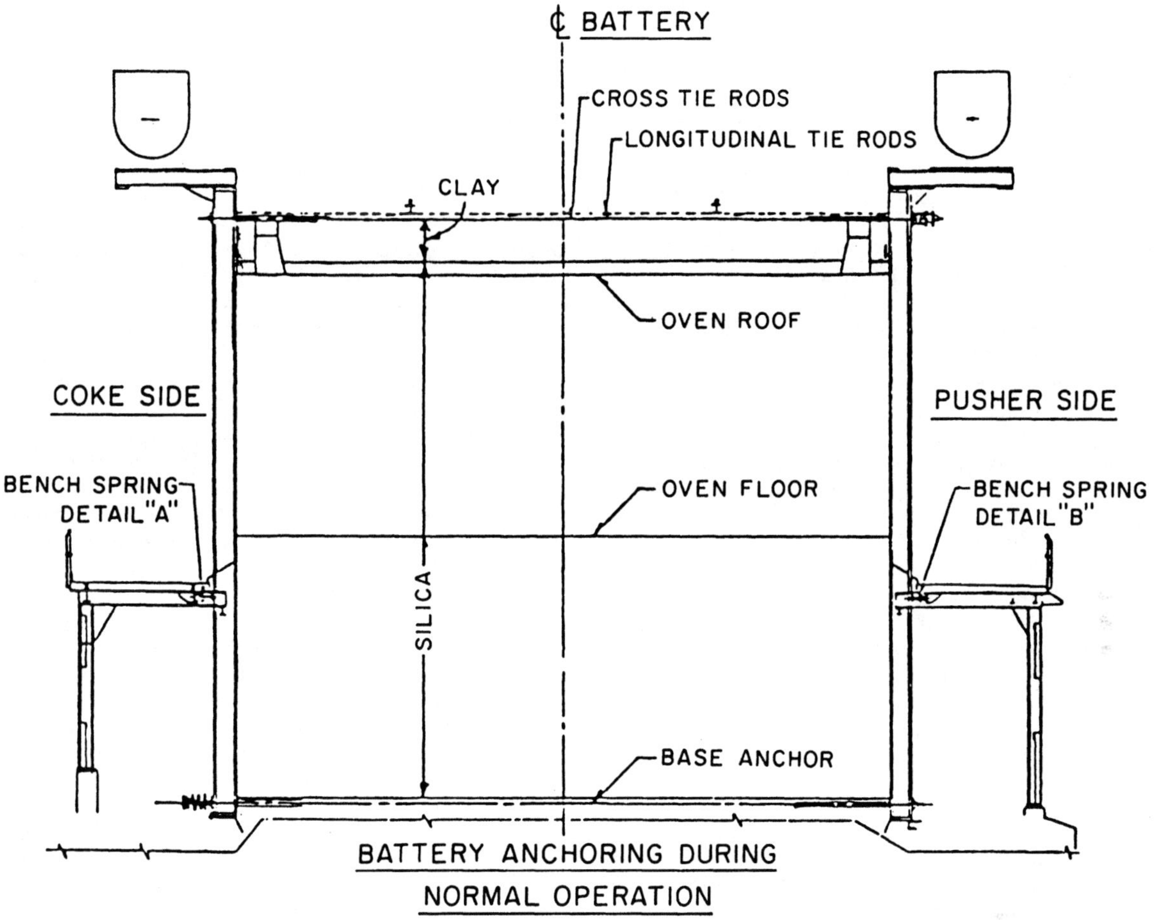

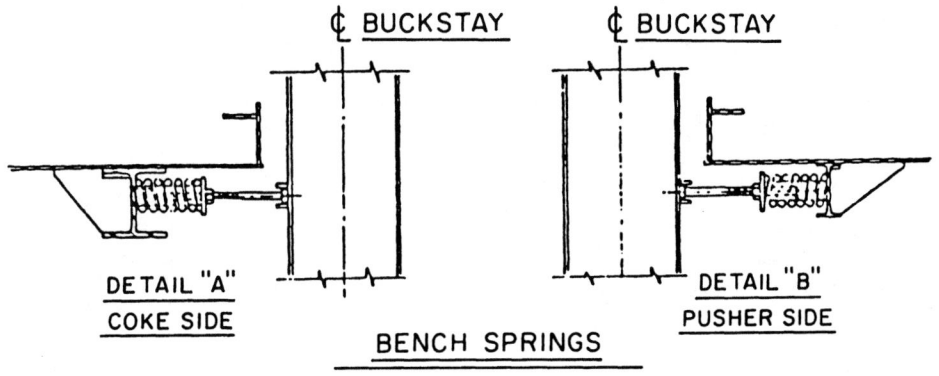

FIG. 4—71. Still oven-anchoring system. (Courtesy, Firma Carl Still.)

Table 4—IV. Comparison of Design Elements of Nippon Steel Corporation Coke Ovens

Basic Type	M-Type	S-Type
Compound heating (BFG/COG)	Compound heating	Compound heating
Hairpin-flue arrangement	Hairpin flue	Half-divided flue wall
Single-stage combustion	Multiple-stage combustion	Single-stage combustion
Number of collecting mains (all types)	1 (normal) or 2	
Number of charging holes (all types)	4 or 5	
Waste-heat canals (all types)	2 (normal)	

flood all regenerators in a bank during the off-burning cycle (Figure 4—73) and reheat the checker brick. They discharge into a pair of full-length sole flues finally exiting through single, internally divided waste-heat boxes located at each side of the battery flue wall.

With the combination of multistage heating, sliding-brick adjustment, hairpin flues, and individual flue orifices, the final control of flue-wall heating both horizontally and vertically is available to the battery operator. Crosswall temperatures and vertical-flue temperatures can be regulated by the front end control of the air and blast-furnace gas.

When heating with coke-oven gas, the gas is supplied to the battery from underjet piping, while combustion air is provided through natural draft (Figure 4—72). Clean coke-oven gas passes through a gas preheater prior to entering the battery basement. After being heated it enters the basement below the floor level in the main distributing pipe (3).

From the main distributing pipe, a pair of coke-oven-gas headers branch to each flue wall. Each pair of headers supplies gas to the even-numbered flues and the other header supplies gas to the odd-numbered flues. For each header there is one manual shutoff cock and one reversing valve. The gas flows through the on-burning (open) reversing valve and into the flue-wall header and through the individual underjet pipes (6). Each underjet (riser) pipe has an interchangeable orifice (9) to meter the gas to each flue. The gas riser duct passes through the partition wall between the adjacent regenerators (air/waste gas) directly to the base of the flue. At the base, coke-oven gas discharges through a single nozzle port. Each flue has a single nozzle situated in either a high or low position, beginning with a low nozzle at the first pusher-side flue and alternating high/low on subsequent flues across the wall to the coke side. However, the final coke-side flue has a low burner rather than a high one in order to provide better heating to the coke-side end (Figure 4—72).

Combustion air for the burning of coke-oven gas is normally introduced by natural draft through the reversing boxes on both sides. (Combustion air for coke-oven-gas burning could be supplied by the forced-air blower system, but it would require additional piping and reversing system capacity.) Air is drafted in through a single, internally divided reversing box at each side of the battery, flooding the pair of sole flues. Rising from the sole flues, the air is heated as it passes through the subcell regenerator checkers of the full bank of regenerators (coke side to pusher side). Hot air discharges at the top of each regenerator subcell into the air-riser ports as well as the blast-furnace-gas ports

described previously in the blast-furnace-gas heating section. Therefore, each flue discharges, at six ports, air which combusts the coke oven gas and conducts the heat through the hairpin and down the off-burning flue.

The waste-gas products discharge down the off flue and out the six flue ports, through the regenerators where it reheats the checker brick. Finally, the waste gas exits the regenerator bottom into a sole-flue pair, ultimately discharging into either end's waste-gas canal.

It is possible to manually change over from coke-oven-gas heating to blast-furnace-gas heating and vice-versa in a matter of minutes, depending upon the number of ovens in the battery. This reversal can also be performed automatically in less time through hydraulic-cylinder actuation and linkage. The system can also be designed to permit the underfiring of part of the battery with blast-furnace gas and the balance with coke-oven gas if required.

Basic-Type Oven—The Basic-type coke oven of Nippon Steel Corporation was the forerunner of the M-type for the standard 4-metre (13 ft. 1 in.) ovens and is still employed. The significant difference between the M-type and the Basic type is the single-stage heating for both coke-oven gas and blast-furnace gas.

Therefore, the blast-furnace gas and its forced combustion air as well as coke-oven gas enter through basement distributing pipes with two flue-wall headers for each gas per flue wall. Combustion air for coke-oven gas enters by natural draft through the sole flues. The silica-brick-walled regenerators are divided into subcells passing combustion air, or blast-furnace gas upward or waste gas downward.

However, the air and blast-furnace gas, upon reaching the top of the regenerator subcell, enter riser ducts which convey it to the base of the appropriate flue. There is no flue-wall duct nor additional ports, all combustion occurring at the base of the flue; all blast-furnace-gas burners are low profile, while coke-oven-gas burners are alternating high and low as described above. Finally, the waste-heat products turn the hairpin and discharge into the regenerators via the blast-furnace-gas and air ports in the base of the off-flue.

S-Type Oven—This type oven is "half divided" to simplify oven structure. It is available for both tall ovens and standard ovens.

Coke-oven gas is introduced, through the underjet pipe installed in the basement, from the duct passing through the pillar wall of the regenerator into the flues. Blast-furnace gas and air are introduced into the regenerator by way of the reversing box. The heating of the S-type oven with the upper horizontal flue can be performed with single-stage combustion in heating with

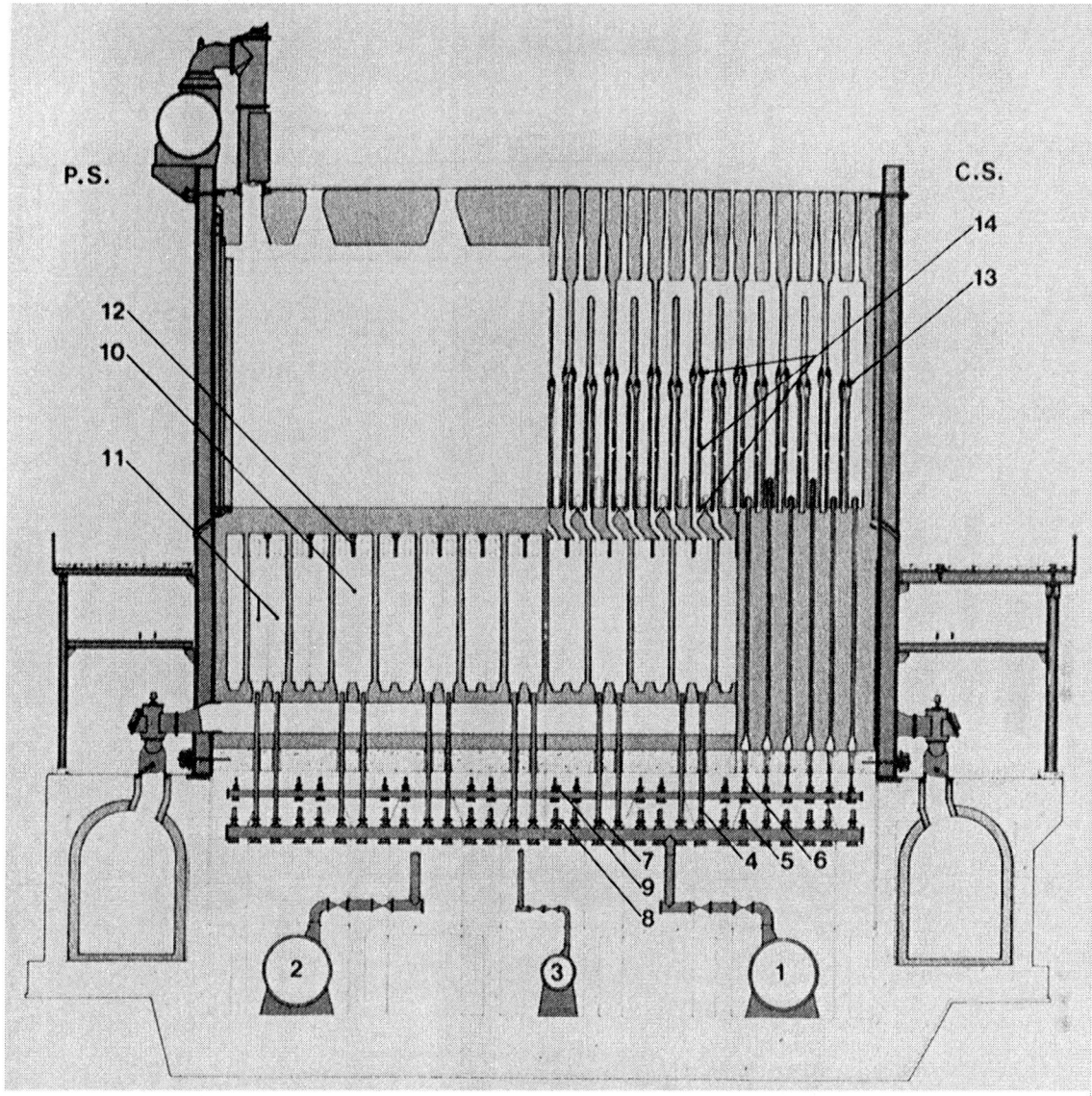

FIG. 4—72. Nippon Steel Corporation M-type coke oven (full underjet design). Legend: 1, Air-distributing pipe; 2, Blast-furnace-gas distributing pipe; 3, Coke-oven-gas distributing pipe; 4, Air underjet pipe; 5, Blast-furnace-gas underjet pipe; 6, Coke-oven-gas underjet pipe; 7, Adjusting device for air; 8, Adjusting device for blast-furnace gas; 9, Adjusting device for coke-oven gas; 10, Regenerative cell; 11, Regenerative subcell; 12, Partition brick; 13, Sliding brick; 14, Three-stage burner. (Courtesy, Nippon Steel Corporation.)

blast-furnace gas and alternate high and low burners when heating with coke-oven gas.

The sole flue, inclined up from the outer side to the oven center, is designed so that equal static pressure can be obtained at any sole-flue section. Steel plates in the duct connecting the sole flue to the regenerator regulate gas and air. The plate has openings matched to gas and air distribution rates and can be inserted from outside the oven.

Each duct connects with one regenerator cell, which in turn ties into one heating flue of a heating wall. Thus, the steel plates control the amount of gas and air being supplied to the individual heating flues.

Heating flues for the oven are divided into a group of "N" on the coke side and "N+1" or "N+2" on the pusher side where "N" can be even or odd. When the

coke-side group is in combustion, the pusher-side group carries waste gas. Gas-flow direction is reversed in the next period. In the upper section of the heating-flue group, a horizontal flue either collects or distributes waste gas. Two sliding bricks in the upper section of each flue regulate gas flow. They are adjusted from the oven top.

Combination combustion is used in the end flues to elevate their temperatures. Apart from the inner-flue burners for heating with coke-oven gas, both end flues are equipped with a separate system for coke-oven-gas burning to permit combined blast-furnace and coke-oven gas combustion in both end flues for heating with blast-furnace gas. The end flues are also equipped with devices to introduce additional forced air.

Reversing Mechanism—Mechanical reversing ma-

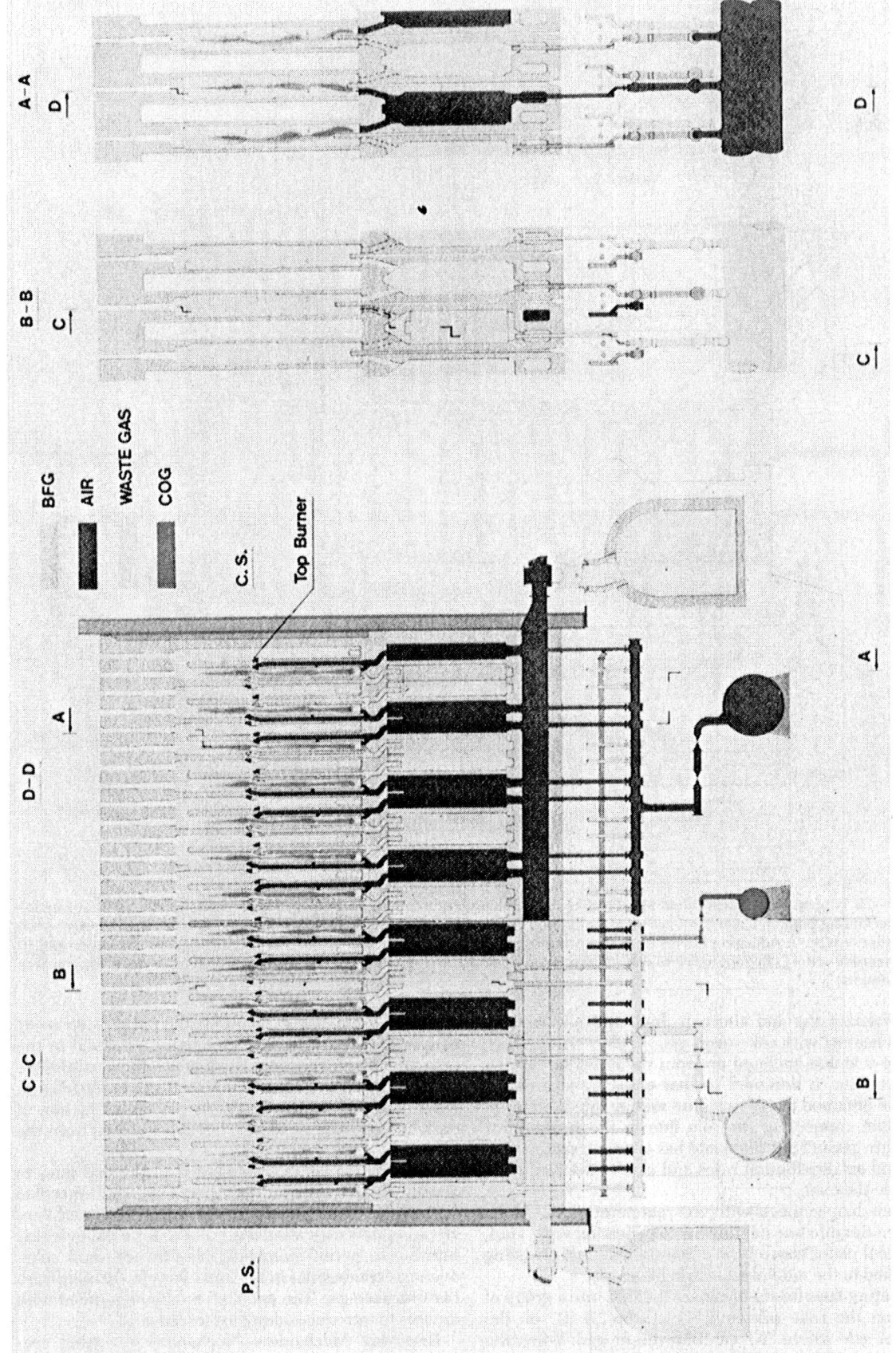

Fig. 4—73. Air, fuel and waste-gas flows in a Nippon Steel Corporation M-type coke oven fired with blast-furnace gas. (Courtesy, Nippon Steel Corporation.)

chines were employed on the standard Nippon Steel Corporation ovens. In the past twenty years they have been gradually replaced by hydraulic-cylinder systems, while all new coke-oven batteries have been equipped exclusively with hydraulic-cylinder systems.

Refractories—Nippon Steel Corporation design for M-type ovens employs approximately 800 shapes. Silica brick which are used normally weigh 1793 kg/m³ to 1841 kg/m³ (112 to 115 lb/ft³). As described above, the regenerator pillar walls are all silica down to the sole-flue level to assure uniform expansion and eliminate differential movement. Also the end-flue-jamb brick have been designed for either silica or fireclay refractories.

Collecting Main and Goosenecks—All Nippon Steel Corporation coke ovens in operation have but one collecting main on the pusher side, with an auxiliary charging main on the coke side. The auxiliary main is employed only during the charging of coal into an oven, when some of the gas is exhausted from the oven, combusted, evacuated from the battery and wet scrubbed. During normal coking (after charging), only the single collecting main handles the gas.

There are no mechanical gooseneck cleaners in operation on the Nippon Steel Corporation batteries. In place of mechanical cleaning, a successful continuous self-cleaning gooseneck, requiring only nominal inspection and maintenance, has been designed.

Waste-Heat Ducts—Because of the fully open sole flues to both waste-gas ducts, the need to mechanically balance the waste-gas flow is not necessary. Therefore, only one suction pressure-control damper is needed at the canal inlet to the stack. This hydraulically controlled damper regulates the entire battery draft.

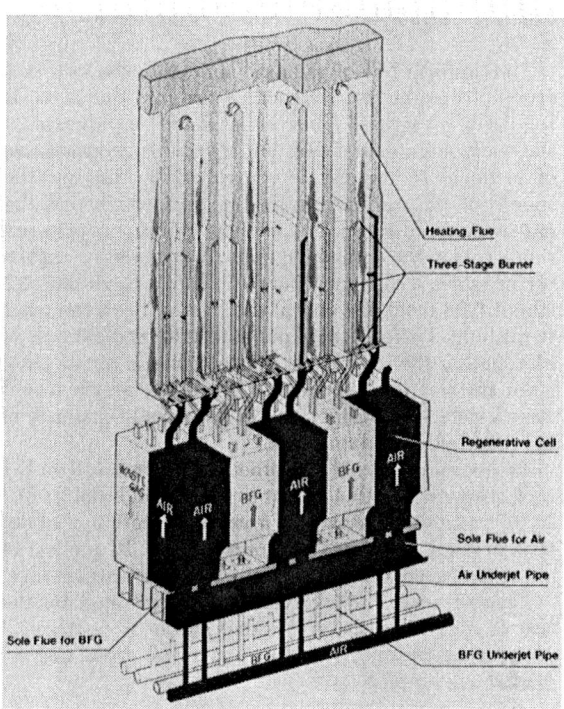

FIG. 4—74. Fuel and air flows in a Nippon Steel Corporation M-type oven when heating with blast-furnace gas. (Courtesy, Nippon Steel Corporation.)

SECTION 7

RECOVERY OF COAL CHEMICALS

COLLECTION OF VOLATILE PRODUCTS FROM OVENS

Collecting Main and Suction Main—In the collection system for the recovery of the volatile products from coal, the first operation reduces the temperature of these products, which are generally referred to as **foul gas**. This takes place in a system of gas mains through which the foul gas passes.

The foul gas passes out of the oven chamber through a refractory-lined ascension pipe and into a gooseneck which connects into the **collecting main** through a damper valve. The collecting main serves an entire battery of ovens, running parallel with the battery and extending above it on one side or on both sides depending on whether it is a single or double collecting-main operation.

The gas and vapors ordinarily leave the oven at temperatures in the range 600°-700°C (1100°-1300°F), and are shock-cooled by spraying with flushing liquor in the goosenecks and further cooled by spraying again with flushing liquor at different points along the collecting main. The cooling is effected by the evaporation of a portion of the water from the flushing liquor which removes some of the sensible heat from the gas and condenses some of the vapors, with the resultant condensation of heavy tar from the gas.

The gas and remaining vapors pass from the collecting main through one or more **cross-over mains** into the **suction main**. A pressure-regulating valve, automatically controlled, is located in each cross-over main. After the gas and vapors have passed this control valve their temperature drops to a range of 80° to 100°C (175° to 212°F), as a result of atmospheric cooling and further evaporation of the flushing liquor.

The **flushing liquor**, used for cooling in the spray system, is liquor which has been condensed in the mains, collected and recirculated, amounting to 3 to 4.5 cubic metres (800 to 1 200 gallons) per ton of coal carbonized on coke-oven batteries with a single collecting main and up to about 7.5 cubic metres (2 000 gallons) on those with two mains. The flushing liquor, which cools and condenses various vapors in the gas, provides a carrying medium for the condensable tars and other compounds formed in the operations. These liquid materials flow from the collecting main through a seal into a downcomer and are delivered through the return flushing liquor lines to a collecting unit customarily called a flushing-liquor decanter (Figures 4—75 and 4—76).

The uniform flow of gas and vapors into the system is accomplished by the charging of coal into the ovens at regularly prescribed intervals and the withdrawal of the evolved gases at a constant rate. This constant rate of removal of the gas is controlled by varying the speeds of the turbines and exhausters which pull the gas away from the ovens and by automatic pressure regulators in the cross-oven main. This pressure regulator provides a slight pressure of about one millimetre (about 0.04 inch) of water at the base of the oven prior to pushing. This control of pressure is for the purpose of eliminating the infiltration of atmospheric air or gases from the heating system into the oven, which would have a deleterious effect on the quality and quantity of the coke and coal chemicals.

As a consequence of this practice the pressure on the collecting main is about eight to twelve millimetres (0.3 to 0.5 inches) of water with a variable suction of about two to three hundred millimetres (8 to 10 inches) of water in the cross-over main after the regulating valve.

These pressure differentials are maintained by the use of either low-speed positive turbo or centrifugal type exhausters designed to remove the gases and vapors at a controlled rate.

RECOVERY OF CRUDE COAL TAR

Flushing-Liquor Decanter Tank—The flushing-liquor decanter tank (Figure 4—76) serves a two-fold purpose in the processing of the liquid condensates and recirculating liquor in the primary liquid system:

(a) It provides a settling basin in which the velocity of the tar and liquor is reduced to permit separation of the tar and liquor by the difference in specific gravity.

(b) It serves as the first settling point for carbonaceous and other finely divided material that is carried along with tar and liquor from the collecting main.

The flushing-liquor decanter is a rectangular steel tank, inclined at one end to facilitate removal of solid accumulations. The tar and flushing liquor enter the decanter and flow into a trough which is designed to minimize agitation of the mixture in the decanter. The mixture overflows the trough into the main compartment, where the velocity is reduced to permit the tar, which has a higher specific gravity than the flushing liquor, to settle to the bottom. The liquor flows over a fixed weir at the opposite end of the decanter and into the connecting lines to the flushing-liquor pumps. The tar leaves the bottom of the decanter through an adjustable seal, known as the decanter valve, which can be raised or lowered. Tar quality is controlled by adjusting this seal either upwards or downwards to regulate the retention time of the tar in the decanter. Carbonaceous deposits in the bottom of the decanter are continuously removed by scrapers dragged slowly along the bottom by two endless chains.

Normally, the tar recovered from the flushing liquor contains 2 to 5 per cent of water. When the water content of the tar is in excess of 5 per cent, further decantation or blending may be required to reduce the water content. This is usually accomplished by placing tar-receiving tanks and separating tanks in the process lineup prior to the tar-storage tanks. Tar-receiving tanks are simply intermediate storage tanks which receive the tar from the decanters prior to being pumped into the tar-storage tanks. Depending on the water content of the tar in a receiver, the tar may be pumped directly to storage, heated to lower the water content, or be pumped back into the decanter system.

Primary Cooler—The non-condensed gas and vapors leaving the collecting and suction mains at a temperature of 75°-80°C (167°-176°F) require further cooling to 35°C (95°F) to remove additional tar and a major portion of the water vapor and to reduce both volume and temperature of the gas before its admission to the exhausters. This cooling may be conducted in either direct or indirect primary coolers.

The **direct primary cooler** (Figure 4—77) consists of a tall, cylindrical scrubbing tower fitted with hurdles or baffles usually constructed of wood. The top portion is equipped with a series of spray nozzles and the lower portion contains a chamber to collect the liquor and condensate.

The gas enters the bottom of the tower and the cooling liquor is pumped into the top of the tower through the spray system to provide a downward flow of cooling liquor in counter-current flow to the gas stream. This direct contact between the gas and liquor provides for exchange of heat which is transferred from the hot gas to the cold liquor. This heat is removed from the liquor by indirect heat exchange, through tubular heat exchangers, with circulating water. As a result of this cooling, 20 to 25 per cent of the total tar recovered is condensed along with a considerable quantity of weak liquor containing ammonia. These condensates are processed either separately or in conjunction with the tar and liquor condensates from the collecting main.

There are two different indirect-cooler types:

(a) The vertical-tube type, which preferably will be installed when only contaminated cooling water is available (see Figure 4—78), and

(b) The horizontal-tube type (Figure 4—79), preferably being operated with a closed cooling-water circuit. The horizontal type provides the best temperature approach between cooling water in and gas out. High efficiency is maintained by a specific continuous flushing circuit, which simultaneously effects naphthalene removal.

Electrostatic Precipitator—The gas leaving the primary coolers still contains small amounts of tar that would cause difficulty in the operation of subsequent units in the recovery system. The method used for removal of this entrained tar is through electrostatic precipitation. The electrostatic precipitator may be placed before or after the exhausters. The preferred location is after compression of the gas (following the exhausters) to prevent infiltration of air into the precipitator. In the electrostatic precipitator (Figure 4—80) removal of tar fog from gas is achieved by passing the gas between electrodes having a high electrical potential. The discharge electrode is of small cross section, such as a wire or a series of points, in order to develop the high-intensity electrical field at its surface which is required for ionization of gas. The collecting electrode has a large cross section and serves as a collector for the suspended particles which are ionized and transferred to this electrode. In this operation, the electrostatic pre-

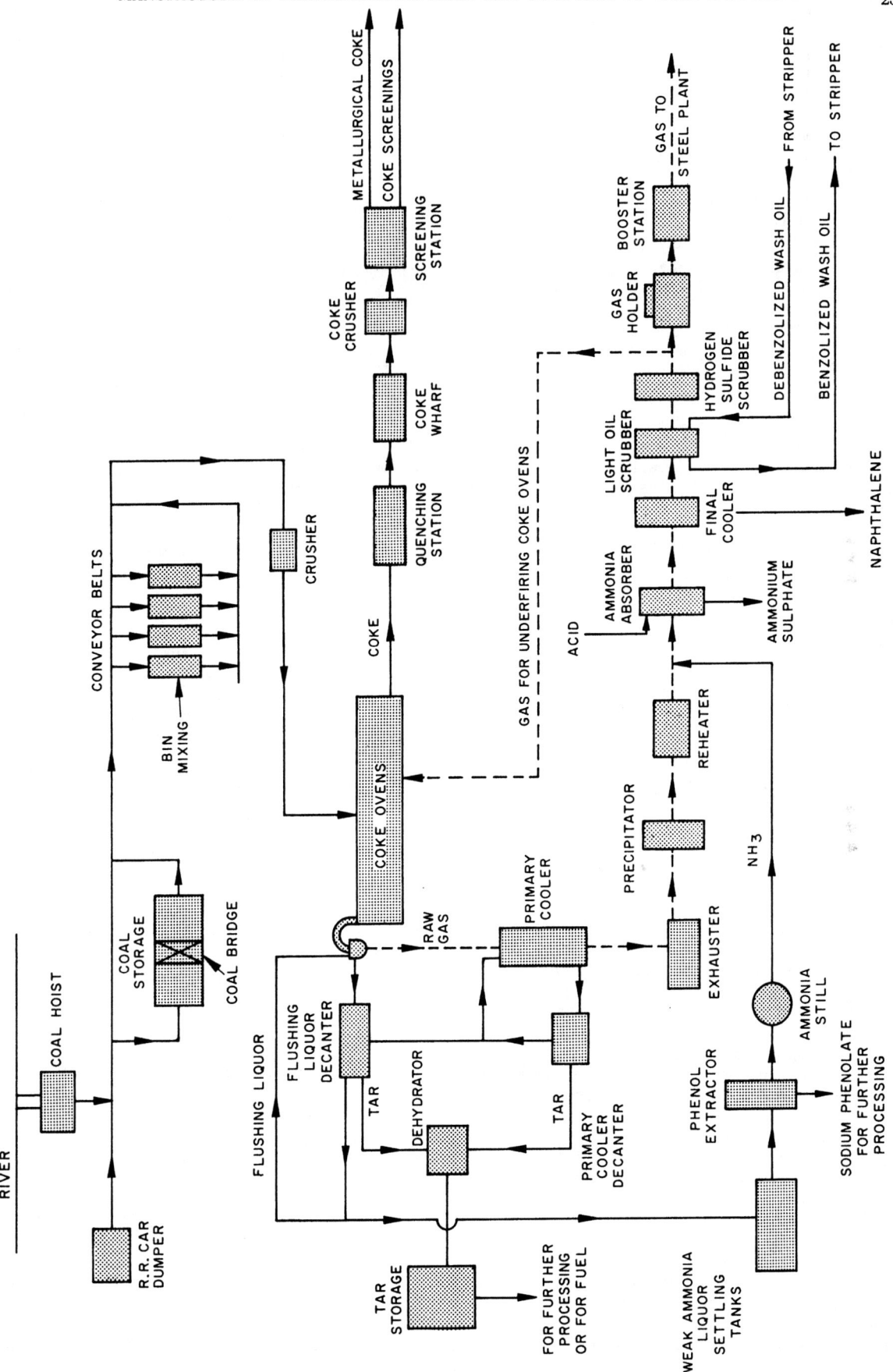

FIG. 4—75. Flow sheet showing the major steps involved in the carbonization of coal by the by-product process and the subsequent recovery of coal chemicals from the gases generated at the ovens.

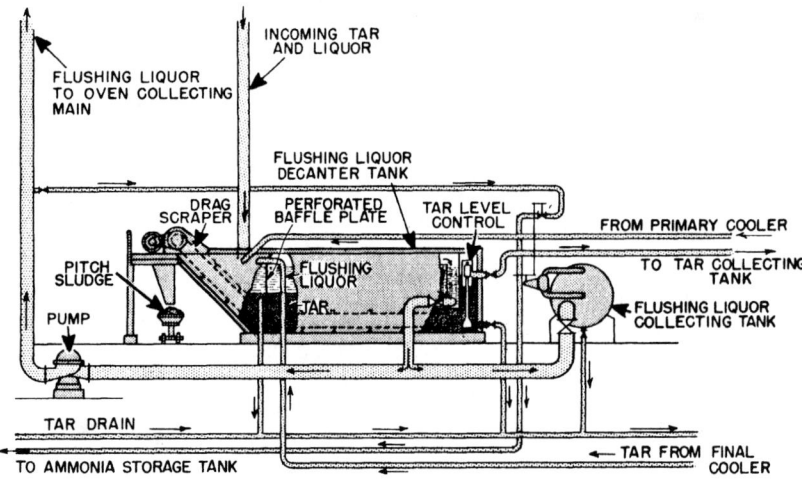

FIG. 4—76. Schematic representation of a flushing-liquor decanter tank or hot-tar drain tank.

cipitator, in addition to its function as a collecting unit for dispersoids, also serves to catalyze the formation of vapor-phase gums formed by oxides of nitrogen and unsaturated hydrocarbons.

RECOVERY OF AMMONIA

The ammonia formed during coking appears in both the coke-oven gas and the weak liquor condensed from the gas. Typically, 20 to 25 per cent of the total ammonia is found in the liquor.

The recovery of this ammonia can be accomplished by two different methods:

(a) The **semi-direct process** in which the ammonia in the weak liquor produced during carbonization is removed by distillation and alkali treatment and added to the gas stream, the gas containing all of the ammonia being then passed through an absorber containing an absorbing solution for the ammonia; or

(b) The **indirect process** in which the ammonia is removed from the gas by scrubbing with water and then removed from the water by distillation and treatment with an alkali, after which the ammonia and steam are passed through the absorber.

The ammonia present in the weak liquor is in two forms classified as "free" and "fixed". The free ammonia is that which is readily dissociated by heat, such as the ammonium carbonate, sulphide, cyanide, and so on, while the fixed ammonia is that which requires the presence of a strong alkali to effect displacement of the ammonia from the compound in which it is present, such as ammonium chloride, thiocyanate, ferrocyanide, sulphate, and so on. The operation to recover this ammonia is carried out in an ammonia still.

Conventional Ammonia Still—In the processing of the liquor, a uniform flow of liquor is fed to the top of the "free leg" of the ammonia still (Figure 4—81) and passes down the column over a series of trays. This liquor is contacted by an upward flow of steam which vaporizes the ammonia and acidic gases. The vapor leaves the top of the "free leg" at a temperature of 96° to 100°C (205° to 212°F), and passes into a dephlegmator to cool the vapor and remove excess water which is returned to the still. The vapor leaving the dephlegmator consists of ammonia which may vary between 10 to

25 per cent, depending upon the vapor temperature, with the balance consisting of water and some acidic gases and neutral oils.

The liquor leaving the base of the "free leg" passes into the "lime leg" where it is treated with "milk of lime". The calcium hydroxide reacts with the fixed ammonium salts of which ammonium chloride is the main constituent.

The liquor then flows into the "fixed leg" which consists of a series of trays to provide effective stripping of the ammonia by a counter-current flow of steam. The steam pressure required at the base of the still is of the order of 21 to 28 kPa (about 3 to 4 psi). Some ammonia stills are operated with caustic soda (NaOH) instead of lime to avoid the column fouling caused by lime salts.

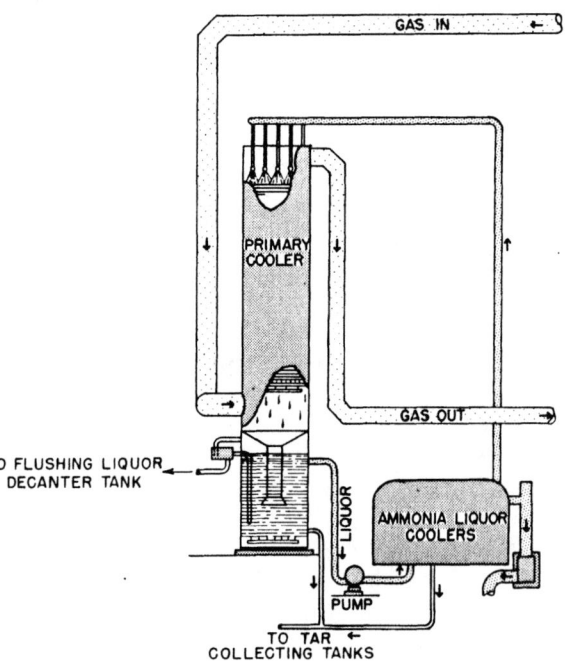

FIG. 4—77. Schematic diagram of a direct primary cooler. (Courtesy, Koppers Company, Inc.)

However, the cost of caustic soda is several times that of lime.

The vapors leaving the ammonia still are added to the gas stream so that all of the ammonia can be recovered in the ammonia absorbers.

USS CYAM System—New environmental regulations on water discharged from coke plants have prompted development of more efficient ammonia stills. The technology for meeting these standards typically consists of a highly efficient still system followed by a biological treatment plant. The USS CYAM system was developed to provide the efficient cyanide and ammonia removal required to allow a biological treatment plant to function efficiently. The combination can achieve a high degree of removal of ammonia, cyanides, sulphides and organics including phenols.

In the CYAM* system, the waste-water flow path is similar to that through a conventional ammonia still. However, ammonia from the fixed still is not introduced into the free still. This allows the bottom section of the free still to become neutral or slightly acid, greatly enhancing the volatility of acid gases in the free still.

A second feature of the process is its relative freedom from fouling by lime salts, achieved by a combination of process design and an anti-fouling additive. The system can operate a year or more without cleaning, which

---

*US Patent Nos. 4,104,131, 4,111,759 and 4,260,462.

negates the need for a spare still. The CYAM system can also use caustic soda (NaOH) and achieve high cyanide removal, if higher alkali cost is accepted.

The flow sheet (Figure 4—82) shows a typical arrangement of the CYAM system in which steam jet compressors are used to recover heat (low-pressure steam) from the effluent stream. Overhead vapor from the fixed still, typically 70 to 84 kPa (10 to 12 psig), flows into the free-still reboiler where it is partially condensed to generate ammonia-free stripping vapors for the free still. Overhead vapor from the free still, normally at about 28 to 42 kPA (4 to 6 psig), is passed through a partial condenser to achieve the desired concentration, which is typically in the range of 12 to 16 weight per cent ammonia. Remaining vapors from the free and fixed stills are then sent to the coke-oven gas upstream of ammonia-recovery equipment. Stripping-steam requirements are about 0.12 to 0.14 kilograms of steam per litre of feed (1.0 to 1.2 pounds of steam per gallon of feed). Lime requirements are about 2.2 kilograms of 90 per cent CaO per kilogram of fixed ammonia. (2.2 pounds of CaO per pound of fixed ammonia.)

Ammonia nitrification and removal of carbonaceous material can be achieved in a single totally-mixed aeration basin when CYAM effluent is treated in a biological treatment plant. Ammonia concentration can be reduced from the 25 to 100 millilitres per cubic metre (25 to 100 ppm) range normally produced by the CYAM

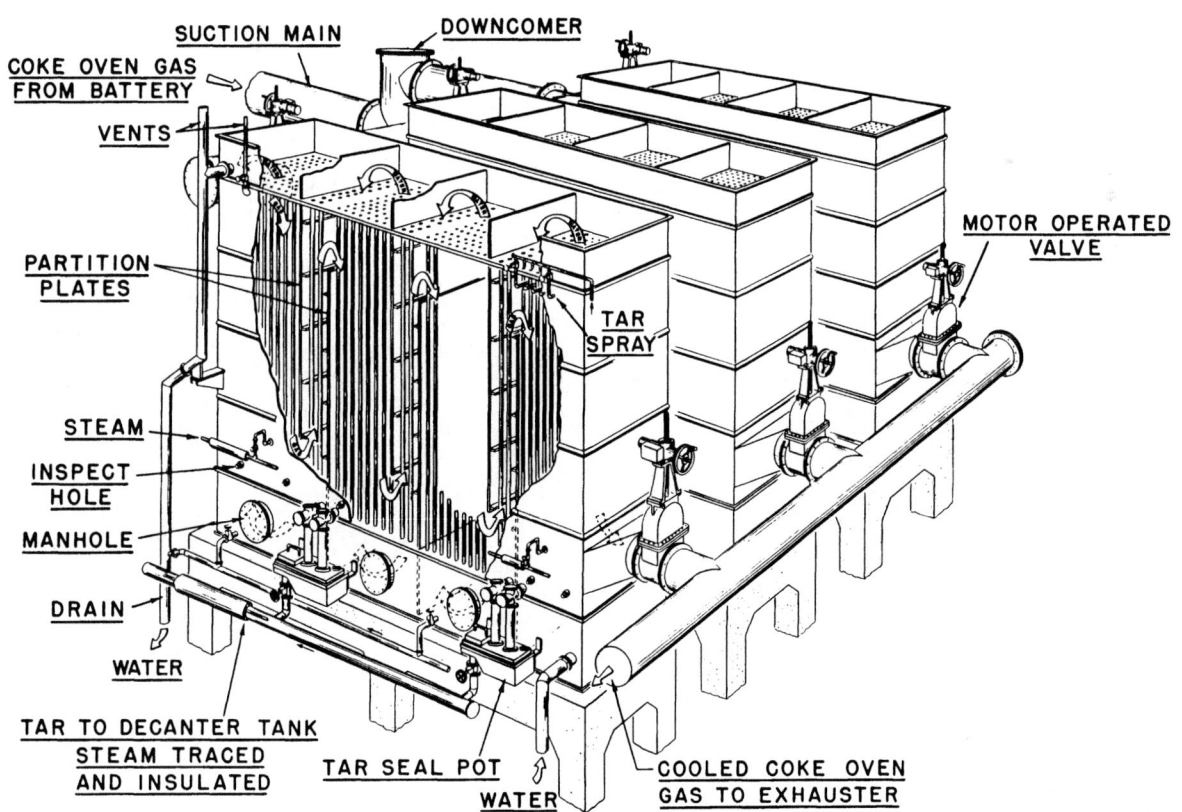

FIG. 4—78. Three indirect primary coolers of the vertical-tube type. (Courtesy, Koppers Company, Incorporated.)

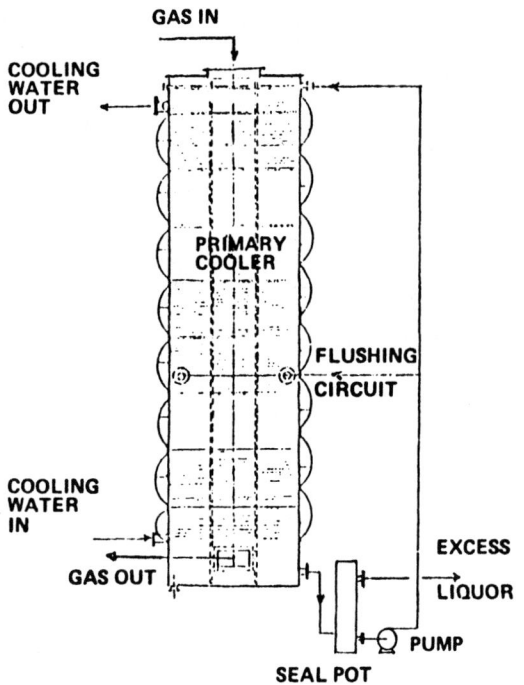

FIG. 4—79. Schematic diagram of a horizontal-tube-type indirect cooler. (Courtesy, Didier Engineering G.m.b.H.)

system to about 10 millilitres per cubic metre (10 ppm). Free cyanide concentration (CN_A) in treated water from the CYAM system is about 1 to 2 millilitres per cubic metre (1 to 2 ppm) and is further reduced to about 0.25 millilitres per cubic metre (0.25 ppm) in the biological-treatment plant. Sulphide removal is essentially complete in the CYAM system, and effluent from the biological treatment plant is normally below 0.3 millilitres per cubic metre (0.3 ppm).

RECOVERY OF AMMONIA AS AMMONIUM SULPHATE

Saturator—At coke plants built prior to about 1930, ammonia-absorbing facilities consist mainly of devices called saturators. These are large dome-shaped, cast iron, lead-lined vessels. Gas is admitted to the saturator through a distributor called a "cracker pipe". This arrangement provides for direct contact between the ammonia and dilute acid, which react according to the following to form ammonium sulphate:

$$2NH_3 + H_2SO_4 = (NH_4)_2SO_4$$

More modern processes have superceded the saturator.

Ammonia Absorber—In an Otto-type absorber, coke-oven gas enters the ammonia absorber (Figure 4—83) near the bottom and is sprayed with a dilute solution of sulphuric acid as it rises to the top of the absorber.

As the dilute sulphuric acid sprays the gas rising through the ammonia absorber, the ammonia in the gas combines with the acid to form ammonium sulphate. The resulting solution drains to a crystallizer from which it is recirculated to the absorber. A constant flow of sulphuric acid is added to the ammonia absorber to replace the acid neutralized by the ammonia in the coke-oven gas. After the solution becomes super-

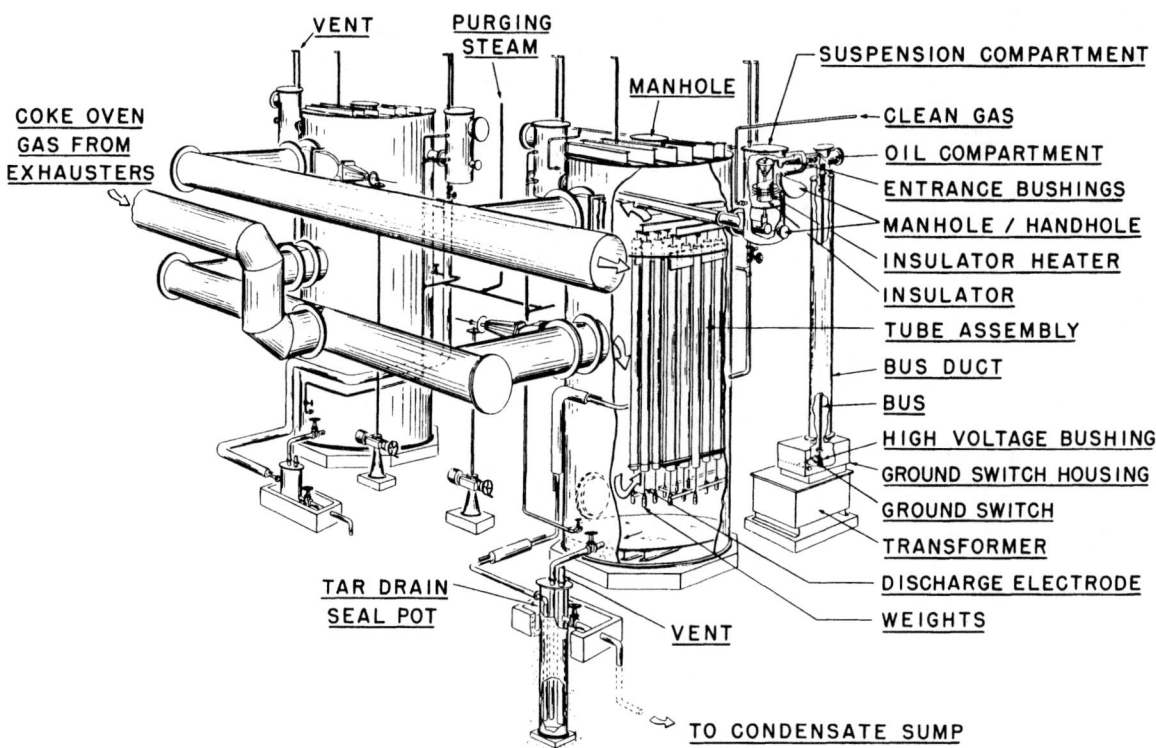

FIG. 4—80. Electrostatic precipitators of the tube type. (Courtesy, Koppers Company, Incorporated.)

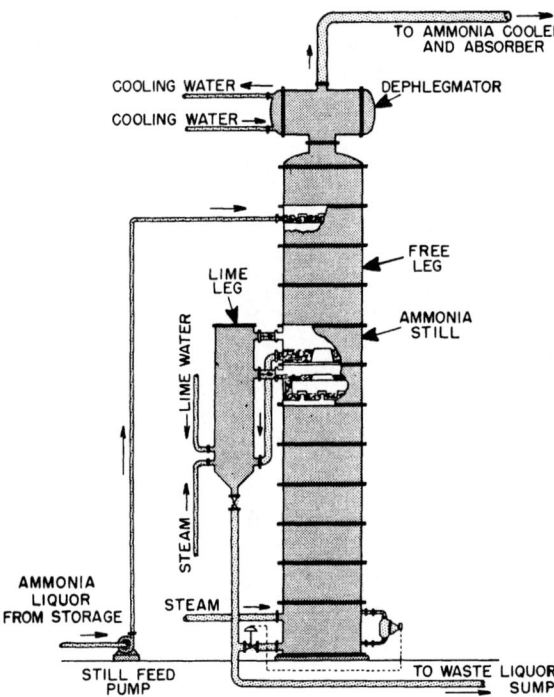

FIG. 4—81. Diagrammatic representation of the essential parts and operation of an ammonia still. (Courtesy, Koppers Company, Inc.)

saturated, crystals of ammonium sulphate are precipitated in the crystallizer and accumulate as a slurry in the bottom. A portion of this slurry is removed from the crystallizer and is pumped to the slurry tank where the salt settles, the liquid overflows and returns to the ammonia absorber. The concentrated slurry is withdrawn from the bottom of the slurry tank and is fed continuously or in batches to the centrifugal dryers. These dryers are arranged to perform the following sequence of operations automatically:

(a) rinse the dryer-basket screen with water;

(b) feed the slurry into the basket;

(c) neutralize the acid remaining in the salt with a dilute solution of aqueous ammonia;

(d) rinse the salt with water to remove excess ammonia;

(e) centrifuge the water from the salt in the basket;

(f) remove the dried salt from the basket; and

(g) discharge it onto a conveyor belt.

The liquid portion of the slurry is recovered and returned to the ammonia absorbers. The partially-dried ammonium sulphate is conveyed to heated rotary-drum dryers for final drying to a content of approximately 0.1 per cent water. All equipment is stainless steel.

Another process used for the recovery of ammonia is the Wilputte low-differential controlled-crystallization process for producing ammonium sulphate, the equipment for which is constructed of stainless steel throughout (Figure 4—84). In this process, the gas is passed

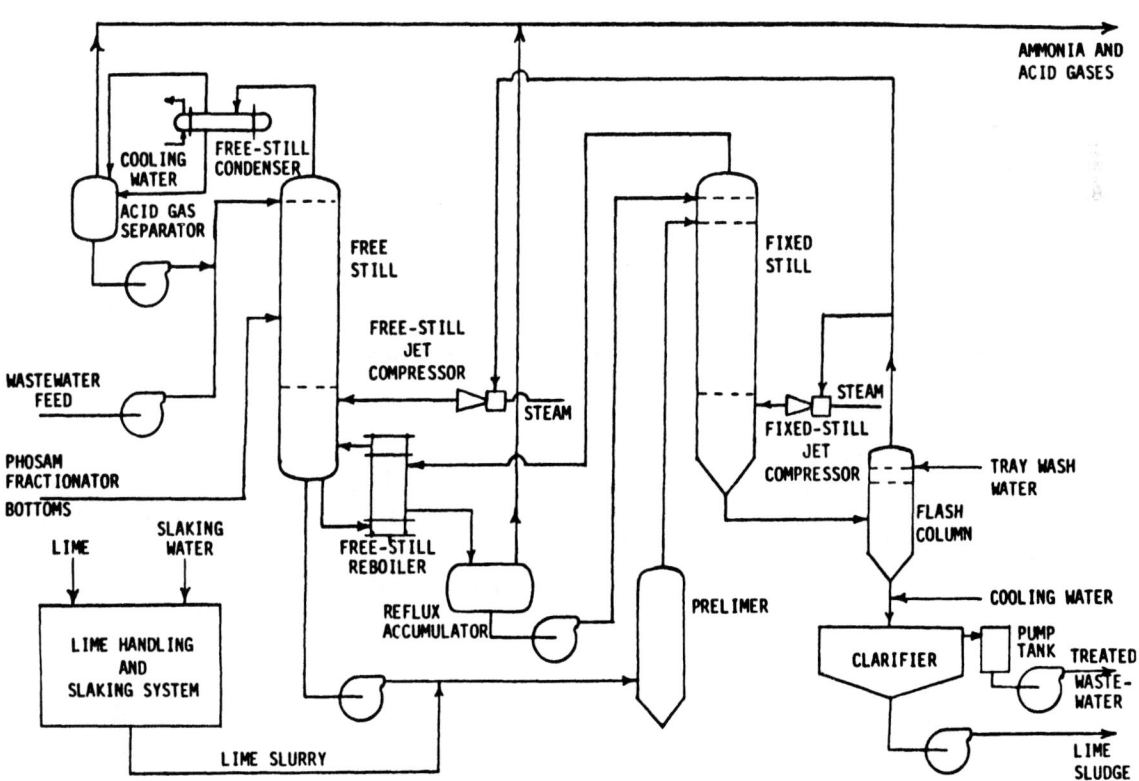

FIG. 4—82. Flow diagram for the USS CYAM system.

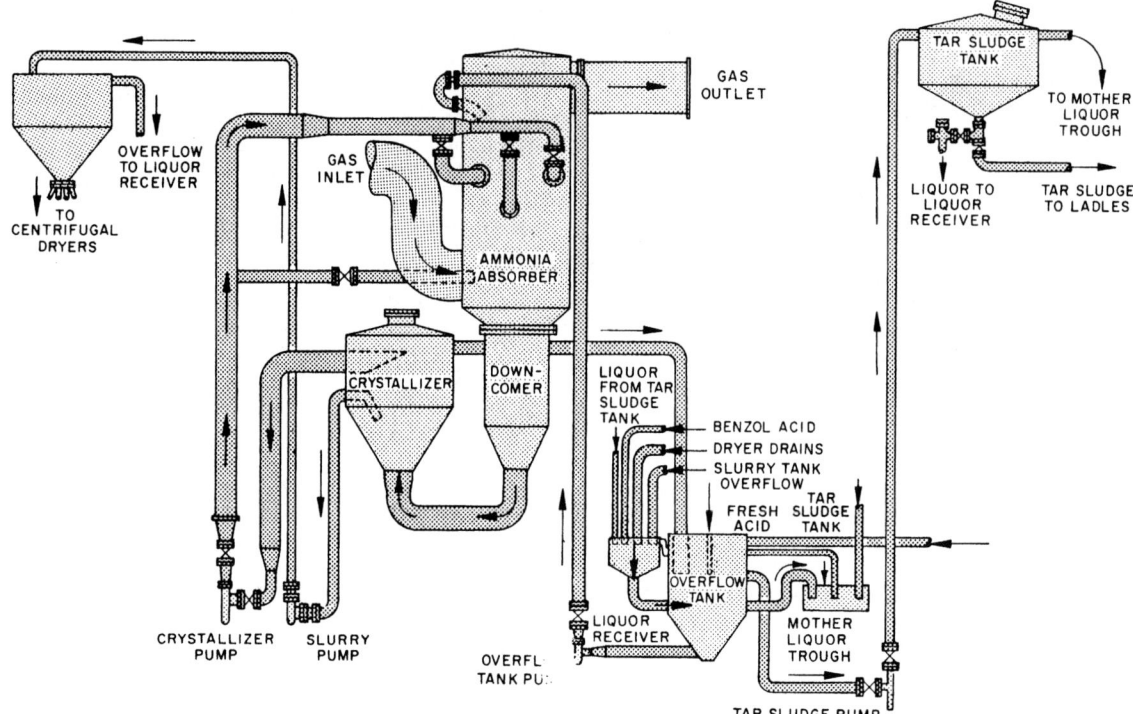

FIG. 4—83. Schematic flow sheet of the operation of an Otto spray-type absorber for the recovery of ammonia as ammonium sulphate. (Courtesy, Otto Construction Corporation).

through a spray-type absorber over which is circulated a 6 per cent solution of sulphuric acid nearly saturated with ammonium sulphate. The acid-entrainment arrestor is an integral part of the absorber. Leaving the absorber, the solution is delivered to the solution-circulating system of a crystallizer in which crystallization takes place by the combined cooling and concentrating effects of vacuum evaporation.

By the variation of circulating rate and the degree of concentration, the size range of the product can be controlled closely within narrow limits. As the crystals grow in size, they settle or gravitate to the bottom of the suspension tank from which they are delivered to a slurry feed tank and from there to a continuous centrifuge or filter, followed in either case by a dryer. The accumulation of deposits of hard salt within the equipment is minimized and "killing" for removal of such deposits is very infrequent. The product can be made with a size consist favorable to almost any type of application.

RECOVERY OF AMMONIA AS ANHYDROUS AMMONIA

The USS PHOSAM Process* is a development by United States Steel Corporation for recovery of the ammonia in coke-oven gas and gas liquor as high-quality anhydrous ammonia. It is applicable to the recovery of ammonia from any gas or vapor stream and is especially advantageous when H_2S, CO_2, or other acidic gases are present. A sharp separation of ammonia from these

*U.S. Patent Nos. 3,024,090, 3,186,795 and 3,985,863.

contaminants is obtained by using a highly selective absorbent solution.

Figure 4—85 shows the essential features of the process, as applied to the direct recovery of ammonia from coke-oven gas. Coke-oven gas, from primary coolers, exhausters, and conventional gas-cleaning equipment, is passed through a two-stage spray-type absorber. Ammonia is scrubbed from the gas by counter-current contact with ammonia-lean phosphate solution, which enters the top of the absorber. The gas leaves the absorber with 98 to 99 per cent of its ammonia removed, and is suitable as such for further processing. The absorbing solution is stable and non-volatile, and does not require periodic replacement or purification. It is highly selective for ammonia, rejecting acidic gases such as hydrogen sulphide and organic compounds that are present in the feed gas.

Absorber pressure drop is low, normally 980 to 1470 pascals (100 to 150 millimetres of water), and the operating pressure is that imposed by downstream equipment. Depending on the temperature and humidity of the inlet gas, the gas may be heated slightly or cooled in passing through the absorber, where ammonia is absorbed and water is evaporated.

It should be noted that the placement of the PHOSAM absorber just downstream from exhausters and tar precipitators is conventional, but not the only possible location. The absorption of ammonia is carried out at relatively high temperatures of 40° to 60°C (104° to 140°F) always above the naphthalene dewpoint established in the primary coolers. Therefore, there is no need for naphthalene scrubbers, H_2S scrubbers, nor

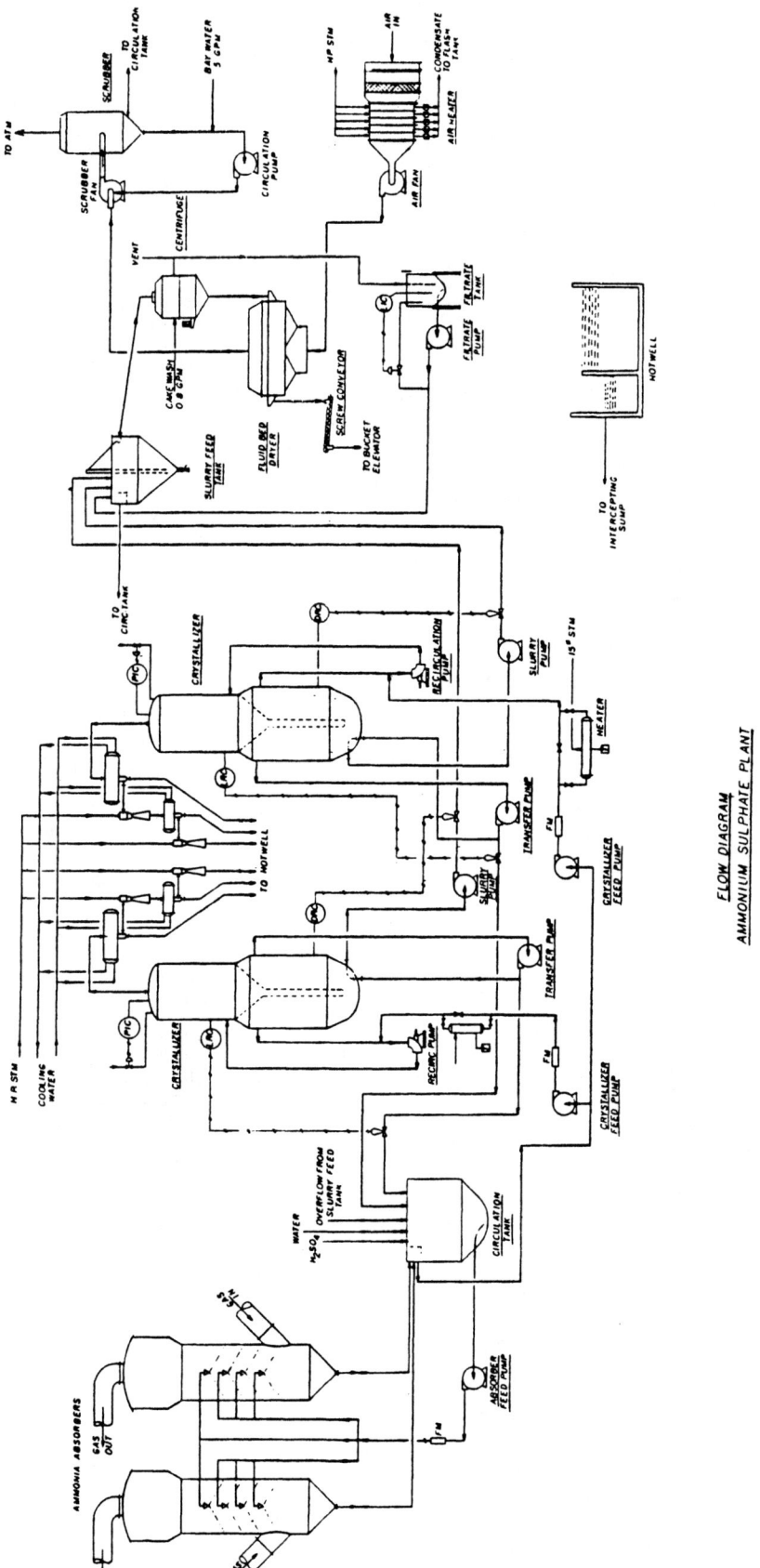

FIG. 4—84. Flow diagram of the Wilputte low-differential controlled crystallization process for producing ammonium sulphate. (Courtesy of Wilputte Corporation, a Salem Company.)

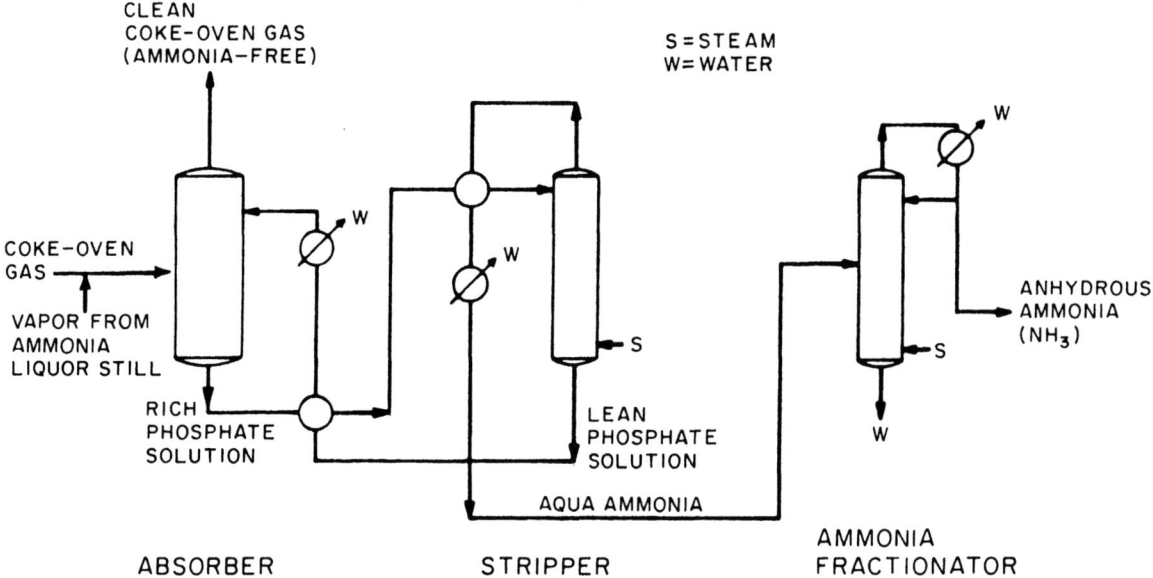

FIG. 4—85. USS PHOSAM process for producing anhydrous ammonia.

secondary coolers ahead of the PHOSAM ammonia absorber. When ammonia-liquor stills are operated, the
vapor from the still is added to the absorber feed gas.

The ammonia-rich solution from the absorber is
pumped through heat exchangers into the stripper, recovering heat from the stripper bottoms and overhead
vapor. In the stripper, the solution is counter-currently
contacted with steam at elevated pressure, stripping
out the absorbed ammonia and regenerating the lean
solution. The lean solution is cooled and returned to the
absorber. Automatic controls make it easy to operate
the plant efficiently, despite variations in the feed gas,
and to control the water content of the solution.

The overhead vapor from the stripper is condensed
to form an aqueous ammonia feed to the fractionator
where anhydrous ammonia is produced by pressure
fractionation, again by using direct steam. The overhead product is liquid anhydrous ammonia, of high purity, ready for shipment.

Both the distillation operations are conducted under
pressure for reasons of economy. The optimum pressures depend somewhat on the cooling-water temperature, but are generally in the range 1240 to 1515 kPa
(180 to 220 pounds per square inch) gage. This means,
in turn, that steam at a gage pressure of 1380 to 1725
kPa (200 to 250 pounds per square inch) is required.

Stripper and fractionator towers in the PHOSAM
Process are quite small in diameter, and, together with
the associated heat exchangers and vessels can be accommodated in an area of about 9.1 by 21.3 metres (30
by 70 feet).

When indirect processes for ammonia removal from
coke-oven gas are used (water wash or liquors associated with desulphurization processes), ammonia-rich
vapors generated in these systems can be passed
through a "hot PHOSAM absorber" to recover the ammonia. The "hot absorber" would operate in the range
of 80° to 100°C (176 to 212°F). The absorbing solution

(ammonia-lean phosphate solution) from the "hot absorber" would be processed in a conventional
PHOSAM stripper and fractionator to produce liquid
anhydrous ammonia.

RECOVERY OF PHENOL

The water, i.e., weak ammonia liquor, recovered
with the volatile products of coal carbonization contains 0.5 to 3.0 grams per cubic decimetre (about 0.031
to 0.187 pounds-mass per cubic foot or 0.004 to 0.025
pounds-mass per gallon) of phenol (sometimes called
carbolic acid) and its homologues. In order to recover
this phenol, a solvent extraction process is used.

Solvent Extraction Process—This method is based on
the principle that phenols are more soluble in benzene
or light oil than in water and that the phenols can be
extracted from benzene or light oil with caustic soda.

In the operation of a solvent extraction process for
the removal of phenols from weak ammonia liquor
(Figure 4—86) the liquor is pumped into the distributor
header located near the top of the ammonia-liquor
scrubber. This scrubber has wooden grids to ensure
good contact between the counterflow liquids.

The liquor passes downward through the scrubber
and comes in contact with a countercurrent flow of
benzene or light oil. The benzene or light oil is immiscible with and has a lower specific gravity than the
liquor; therefore, it rises to the top of the scrubber as it
absorbs the phenol from the liquor. When the ammonia
liquor reaches the base of the scrubber, the phenol
extraction is complete, and the liquor flows to a storage
tank for further processing.

The benzene or light oil reaching the top of the
scrubber is phenolized. This solvent then flows to the
caustic washer through an overflow line connecting the
ammonia-liquor scrubber and the light-oil caustic-
treatment tower.

The light-oil caustic-treatment tower is divided into

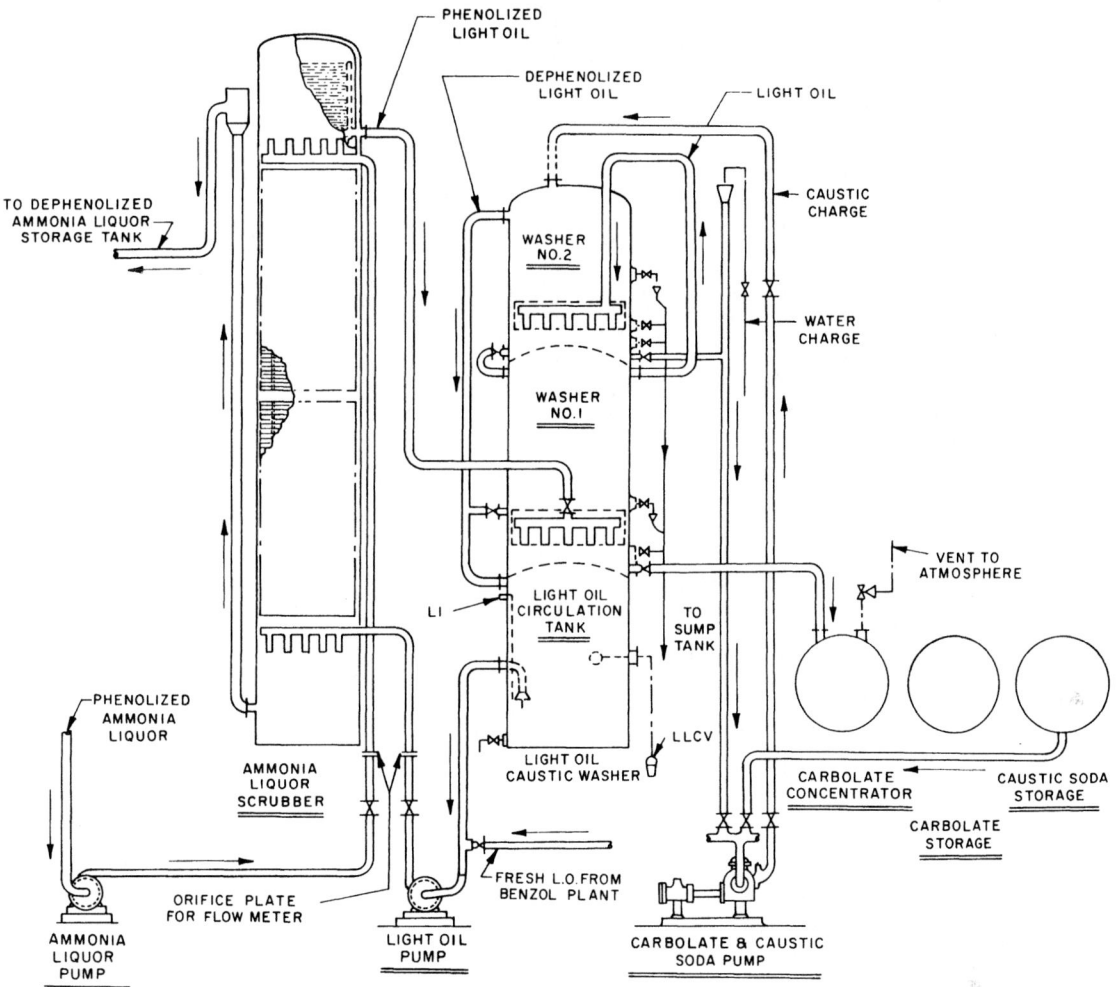

FIG. 4—86. Diagrammatic representation of the equipment for recovery of phenol by the solvent extraction process.

three compartments. The bottom compartment is the benzene or light-oil circulation tank which is the pumping chamber for the dephenolized benzene or light oil. The upper two sections are the caustic-washing compartments, packed with ceramic tile. In these two compartments, the phenolized benzene or light oil passes through the caustic (sodium hydroxide) to remove the phenols by the chemical reaction between the caustic (base) and the phenols (acid).

The phenolized benzene or light oil passes from the ammonia scrubber to the distributor header on No. 1 washer, through No. 1 washer to a distributor header on No. 2 washer, and through No. 2 washer to the overflow line, where the benzene or light oil, now dephenolized is returned to the circulation tank.

After a week or two, the caustic in No. 1 washer is saturated with phenols. When this happens, the recovery operation is shut down so that the plant can be recharged with fresh caustic-soda solution.

The sodium phenolate in No. 1 washer is drained into the carbolate concentrator. When this is complete, the partially phenolized caustic in No. 2 washer is drained to No. 1 washer, leaving No. 2 washer empty to

receive a fresh supply of caustic-soda solution. The plant is then ready to resume operation.

The sodium carbolate in the concentrator is boiled to remove entrained solvent and moisture. It is then neutralized with carbon dioxide to liberate crude phenols and phenol homologues.

RECOVERY OF COKE-OVEN LIGHT OIL

Light Oil—The gas leaving the ammonia absorbers contains light oil. This oil is a clear yellow-brown oil with a specific gravity of about 0.880. It is a mixture of all those products of coal gas with boiling points mostly ranging from 0°C to 200°C (32°F to 390°F), containing well over a hundred constituents (Table 4—V). Most of these are present in such low concentrations that their recovery is seldom practicable. Many of the constituents, such as olefin and diolefin hydrocarbons, some straight chain and cyclic paraffins, sulphur, nitrogen and oxygen compounds, are present in small quantities. The principal usable constituents are benzene (60-85 per cent), toluene (6-17 per cent), xylene (1-7 per cent), and solvent naphtha (0.5-3 per cent). Light oil constitutes approximately one per cent of the coal carbon-

Table 4—V. Fractions of Coke-Oven Light Oil, and Boiling Points of Some of Their Constituents

CONSTITUENT GROUP	FORERUNNINGS (2% of Light Oil)			CRUDE BENZENE (60% of Light Oil)			CRUDE TOLUENE (18% of Light Oil)		
AROMATICS	Traces of Benzene			Benzene	C_6H_6	80.1°C.	Toluene	$C_6H_5CH_3$	110.6°C.
PARAFFINS	n-Pentane	C_5H_{12}	36.1°C.	n-Hexene 2-Methylhexane n-Heptane	C_6H_{14} C_7H_{16}	68.8°C. 90.3°C. 98.4°C.	n-Heptane n-Octane	C_7H_{16} C_8H_{18}	98.4°C. 125.6°C.
CYCLOPARAFFINS NAPHTHENES	Cyclopentane	C_5H_{10}	51.0°C.	Cyclohexane	C_6H_{12}	80.8°C.	Methylcyclohexane Cycloheptane	$C_6H_{11}CH_3$ C_7H_{14}	100.3°C. 120.3°C.
UNSATURATED (Olefins-Diolefins and Aromatic Hydrocarbons with Unsaturated Side Chains)	Butene-1 Pentene-1 Amylenes n-Hexylene Cyclopentadiene-1,3 Butadiene-1,3	C_4H_8 C_5H_{10} C_6H_{12} C_5H_6 C_4H_6	-6.5°C. 30.1°C. 20 to 41.0°C. 64.0°C. 42.0°C. -5.0°C.	Hexene-2 Hexadiene-1,3 N-Heptylene Cyclohexene Unidentified Compounds	 C_7H_{14}	 99.0°C. 85.0°C.	Unidentified Compounds		
SULPHUR COMPOUNDS	Carbon Disulphide Hydrogen Sulphide Hydrogen Cyanide Carbonyl Sulphide Methyl Mercaptan Ethyl Mercaptan Dimethyl Sulphide	CS_2 H_2S HCN COS CH_3SH C_2H_5SH $(CH_3)_2S$	46.3°C. -59.6°C. 26.0°C. -50.2°C. 7.6°C 34.7°C. 36.2°C.	Thiophene Diethyl Sulphide	C_4H_4S $(C_2H_5)_2S$	85.0°C. 91.6°C.	Methylthiophene	C_5H_6S	112 to 115°C.
NITROGEN & OXYGEN COMPOUNDS							Pyridine	C_5H_5N	115.3°C.

CONSTITUENT GROUP	CRUDE NO. 1 SOLVENT (8% of Light Oil)			CRUDE NO. 2 SOLVENT (6% of Light Oil)			CRUDE RESIDUE (6% of Light Oil)		
AROMATICS	o-Xylene m-Xylene p-Xylene Ethyl Benzene	$C_6H_4(CH_3)_2$ $C_6H_5C_2H_5$	144.0°C. 139.1°C. 138.4°C. 136.2°C.	n-Propyl Benzene Ethyl Toluenes Mesitylene Pseudocumene Hemimellitene Cymenes Durenes	C_9H_{12} $C_6H_3(CH_3)_3$ $C_{10}H_{14}$ $C_6H_2(CH_3)_4$	158.6°C. 161.2 to 164.9°C. 164.6°C. 169.2°C. 176.2°C. 175.5 to 177.3°C. 196.0 to 198.0°C.	Wash Oil Naphthalene Solvents Pitch Residue		275 to 360°C. 218°C. Above 200°C.
PARAFFINS	n-Octane n-Nonane	C_8H_{18} C_9H_{20}	125.6°C. 150.7°C.	n-Decane	$C_{10}H_{22}$	174.0°C.			
CYCLOPARAFFINS NAPHTHENES	Cycloöctane	C_8H_{16}	150.0°C.	Cyclononame	C_9H_{18}	172.0°C.			
UNSATURATED (Olefins-Diolefins and Aromatic Hydrocarbons with Unsaturated Side Chains)	Octylene Styrene Unidentified Compounds	C_8H_{16} C_8H_8	126.0°C. 146.0°C.	Coumarone Dicyclopentadiene Indene	C_8H_6O $C_{10}H_{12}$ C_9H_8	175.0°C. 170.0°C. 182.0°C.			
SULPHUR COMPOUNDS	Thioxenes	C_6H_8S	137 to 146.0°C.	Trimethylthiophene Thiophenol Tetramethylthiophene		160 to 163.0°C. 169.5°C. 182 to 184.0°C.			
NITROGEN & OXYGEN COMPOUNDS	Picolines	C_5H_7N	131 to 143.0°C.	Cresols Dimethyl Pyridines Phenol	C_7H_8O C_7H_9N C_6H_6O	190 to 203.0°C. 143 to 164.0°C. 182.0°C.			

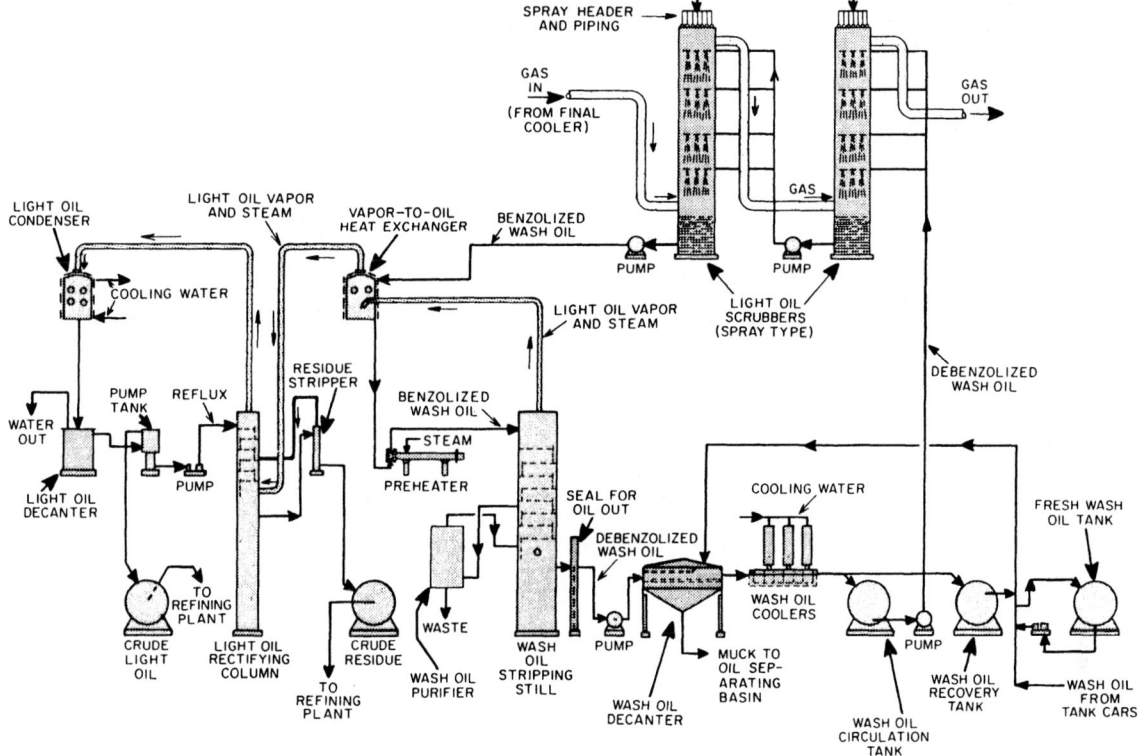

FIG. 4—87. Flow sheet of a light-oil recovery system. (Courtesy, Wilputte Corporation, a Salem Company.)

ized.

Light-Oil Recovery Processes—The removal of light oil from coal gas is generally the last step in the coal-chemical recovery process.

There are three general methods used for the recovery of light oil:

(a) Refrigeration and compression involving temperatures below–70°C (–94°F) and pressures of 10 atmospheres (7600 millimetres Hg).

(b) Adsorption by solid adsorbents involving the removal of light oil from the gas by passing it through a bed of activated carbon and recovering the light oil from the carbon by heating with direct or indirect steam.

(c) Absorption by solvents involving washing the coal gas with a petroleum wash oil, a coal-tar fraction, or other absorbent, followed by steam distillation of the enriched absorbent to recover the light oil. A process employing petroleum wash oil and another using a coal-tar fraction will be described here.

Process Using Petroleum Wash Oil—The practice of using petroleum wash oil (Figure 4—87) is the one almost universally followed in the United States due to the availability and low cost of petroleum wash oil. The efficiency of recovery varies widely with the seasons since one of the major considerations is the temperature of the coal gas and wash oil entering the absorbing process. Another consideration is the ratio of wash oil to gas. The absorption equipment should be of reasonable design as to size and contact time. The oil-and-gas ratio varies depending on the equipment design and light-oil content of the gas prior to light-oil removal.

Typical operating conditions are as follows: the temperature of gas entering the absorption process is 15° to 30°C (59° to 86°F), the temperature of wash oil entering the process is 17° to 32°C (60° to 90°F), and the wash oil circulated per metric ton of coal carbonized is about 525 to 835 cubic decimetres (150 to 200 gallons per net ton).

The boiling point of the wash oil should be well above 200°C (390°F) so as to permit an effective separation of light oil from wash oil in debenzolization. The oil should not thicken and should have a low viscosity to permit its distribution in the scrubbing towers. It should not deteriorate readily but maintain its initial properties as long as possible to keep makeup oil at a minimum. It must be especially stable with respect to the repeated heating which takes place in the recycling of the oil in the process. Its absorptive capacity should be very high and it should not react with or contaminate the coal gas. The specific gravity should be low enough to permit effective separation of wash oil and water in the processing and keep emulsification of the two to a minimum. The specific heat should be low because the oil is subjected to repeated heating and cooling as it is recycled in the process.

The petroleum wash oil normally used for this absorption process has a boiling range of 270° to 350°C (518° to 622°F). Other specifications which are general for petroleum wash oil include a specific gravity of about 0.830, a viscosity of 45 seconds Saybolt at 38°C (100°F), a pour point of 2°C (35°F), an emulsification of 95 per cent separation in 50 seconds, a flash point of 150°C (300°F), fire point of 168°C (335°F), and a low

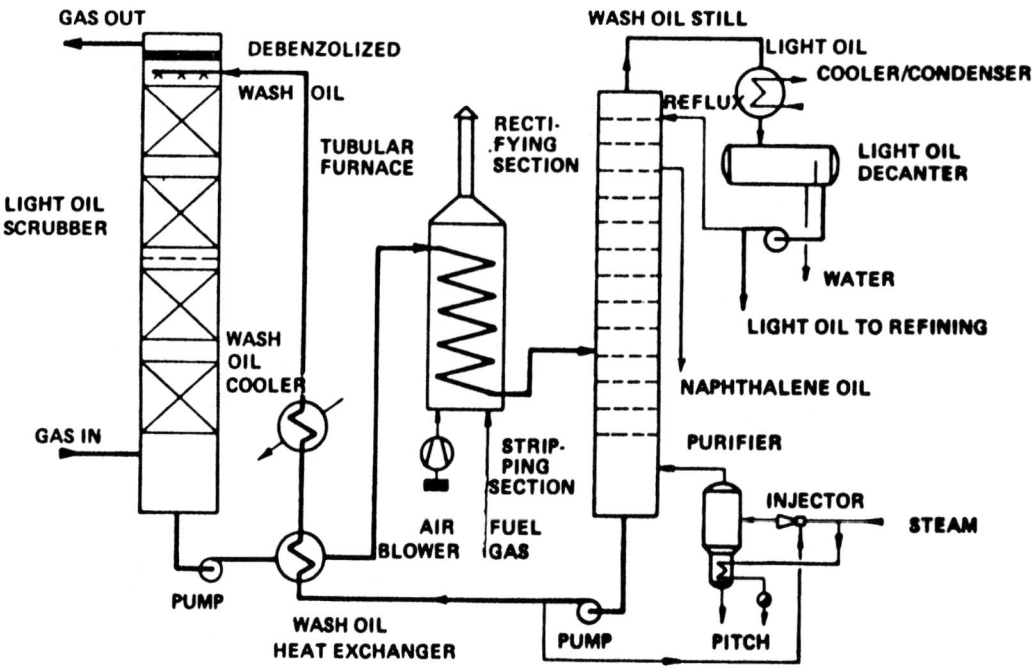

FIG. 4—88. Flow sheet of a light-oil recovery system with coal-tar wash oil. (Courtesy, Didier Engineering G.m.b.H.)

residue under 0.10 per cent when heated for a period of five days at approximately 150°C (300°F).

Process Using Coal Tar—Coal-tar wash oil having a boiling range of 230°C (446°F) through 300°C (572°F) is used for light-oil removal in a packed-type absorptive tower. The benzolized oil which has a perfect heat stability is pre-heated by the regenerated oil and finally heated to 185°C (365°F) in a tubular furnace, before being flashed into the wash-oil still for regeneration. Through oil-to-oil heat exchange and because of the use of a tubular furnace, steam consumption is minimized. There are two processing sections in the still: the top one for light-oil rectification, including top reflux and side-stream draw-off for the naphthalene-oil fraction, and the bottom one for wash-oil debenzolization. A split stream of wash oil is passed through a purifier where by direct steam injection volatile components are returned into the bottom section of the still and heavier components, i.e., pitch 70°C (158°F) S.P., are withdrawn from the purifier bottom. The activated wash oil is reused for light-oil absorption after heat exchange and cooling. A typical flow sheet (Figure 4—88) shows the operation of the process.

Final Cooler—The first step in the recovery of light oil by absorption in a liquid medium is that of cooling the gas leaving the ammonia absorbers at a temperature of 50° to 60°C (112° to 140°F) by direct contact with water in a tower scrubber called a final cooler. The facilities are so named since the gas is here given its final cooling in the coal-chemical processing. This is necessary to remove naphthalene from the gas and also cool the gas prior to its admission to the wash-oil scrubbers.

The tower consists of a tall cylindrical shell of steel approximately 3 to 4.5 metres (10 to 15 feet) in diameter and about 15 to 23 metres (50 to 75 feet) in height filled with a suitable packing material, either metallic or wooden. The gas enters near the bottom of the tower and passes up through the tower and out near the top. The cooling water enters the top of the tower through a spray system and the water passes down through the tower, coming in direct contact with the gas in a countercurrent manner. The water leaves the tower at the bottom through a sealed outlet pipe to prevent escape of any gas. The heat from the gas is transferred to the water, which in turn is cooled in an induced-draft water-cooling tower with air or in an atmospheric water-spray cooling operation. Cooling of the water depends upon air circulation and the vaporization of a part of the water in circulation, the latent heat of vaporization of the water being responsible for additional cooling. Operating practice is to cool the water, and in turn cool the gas, to as low a temperature as practicable, depending upon atmospheric temperature, since most effective absorption of light oil is obtained at low temperatures. Cooling is not carried below 15°C (60°F), since below that temperature petroleum absorbing oil becomes too viscous to flow freely.

This direct-cooling operation causes the condensation of a major portion of the naphthalene and any entrained tar and vapor-phase gums. The naphthalene is recovered in a sump operation and is either added to the tar or refined directly to provide a salable product.

In some of the more modern facilities, the lower part of the final-cooler tower is redesigned to permit the outlet water to come in direct contact with tar in order to dissolve the naphthalene as it is being removed from the gas.

This cooling operation may also be carried out using petroleum wash oil instead of water. A small bleed stream of debenzolized wash oil from the light-oil strip-

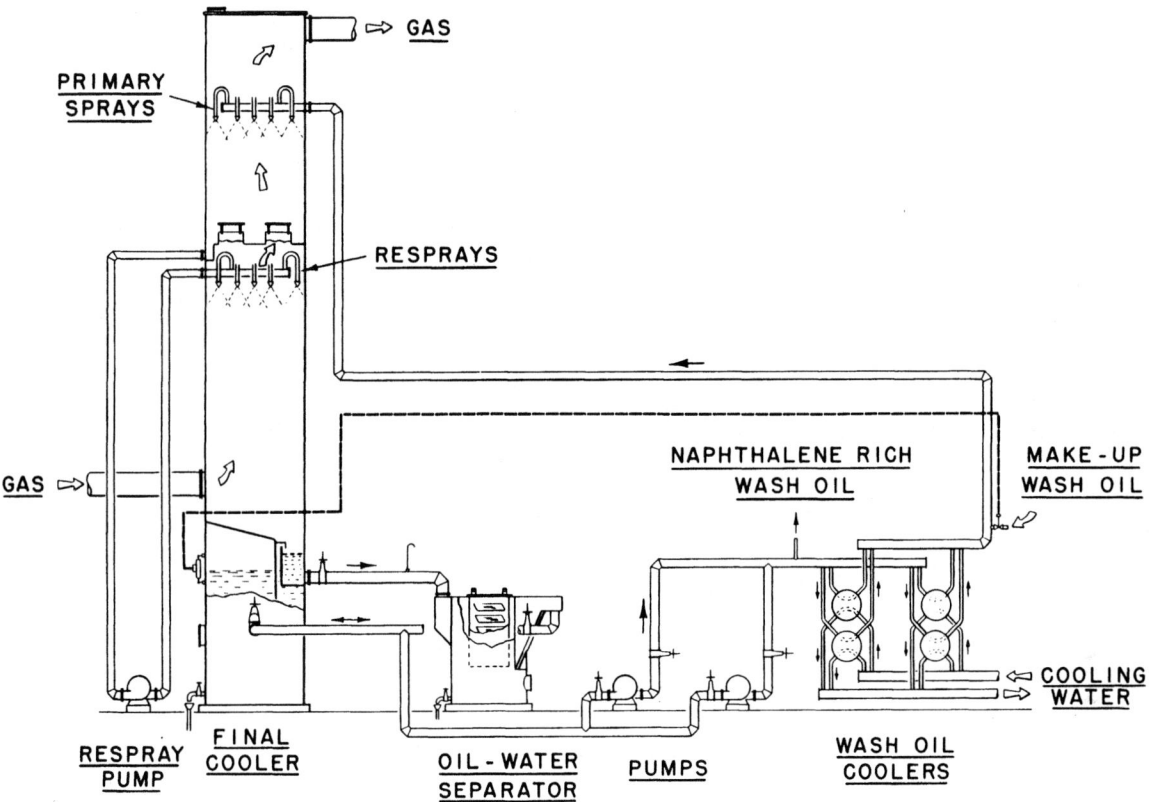

FIG. 4—89. Final cooler of the wash-oil spray type. (Courtesy, Koppers Company, Incorporated.)

per may be used for this purpose. In this case, the wash oil will absorb most of the naphthalene so that the gas leaves with less than about 460 to 685 grams per cubic metre (2 to 3 grains per standard cubic foot) of naphthalene in the gas. However, the quantity of wash oil used is small enough so that the light oils are not absorbed to any great extent in the final cooler.

A typical final cooler is divided into two spray compartments (see Figure 4—89). Each compartment is provided with its own set of recirculation pumps in order to recirculate wash oil from the top-section drains to the bottom section. The wash oil which is recirculated in the bottom section is cooled in a series of wash-oil coolers. An oil-water separator is provided at the bottom of the column to decant condensed water from the wash oil.

The naphthalene which is absorbed in the wash oil may be stripped out by feeding the wash oil into the top of the wash-oil scrubber at a point which is about 16 trays above where the benzolized wash oil is introduced. Otherwise, the naphthalene may be stripped out in a separate still.

Wash-Oil Scrubber—The second step in the recovery of light oil is its absorption in the liquid petroleum wash oil. The gas comes in direct contact with the wash oil in one or more tall scrubbing towers. The gas passes from the first tower to the last in series and the wash oil travels from the last tower to the first in reverse series. The flow of gas and wash oil is counter-current in each tower. The steel towers are approximately about 4.5 to

6.7 metres (15 to 22 feet) in diameter and about 30.5 metres (100 feet) in height.

The wash oil is introduced through a number of sprays in the top of the tower and comes into direct contact with the gas, which flows from the bottom to the top. An oil-storage tank is provided in the base of the tower to receive the oil and maintain a surge capacity for pumping the oil away. The oil passes from the gas compartment to oil storage through a sealed pipe. It is pumped from the base of one tower to the spray system in the top of the next tower in the series. From the last tower the oil is pumped at a controlled rate to the stripping stills for separation of light oil from wash oil. Wash oil prior to light-oil absorption is called debenzolized and, after absorption, benzolized. The benzolized wash oil contains 2 to 3 per cent light oil. The debenzolized wash oil is cooled in indirect coolers to a temperature several degrees higher than that of the gas entering the scrubbers, which is 15° to 25°C (60° to 75°F). This is to prevent condensation of water from the gas, which would form an emulsion with the oil, causing clogging of the free space in the packing of the tower. The rate requirement for the circulation of oil through the scrubbers is a function of the vapor-pressure distribution between the light oil dissolved in the absorbent oil and that remaining in the gas at the temperature of operation. From 90 to 95 per cent of the light-oil content of the gas is recovered in this operation. The wash oil, after being cooled, passes through a large decanting tank which acts as a settling com-

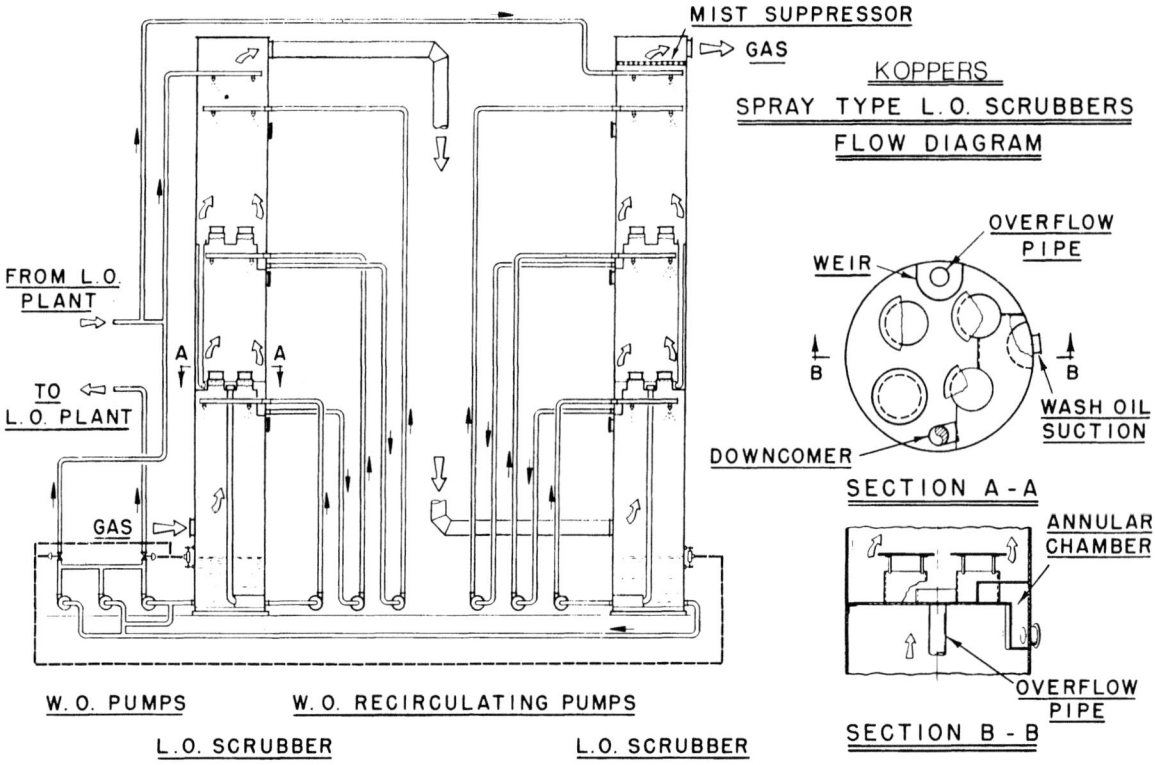

FIG. 4—90. Flow diagram of spray-type light-oil scrubbers. (Courtesy, Koppers Company, Incorporated.)

partment for the emulsified and resinous materials present in the wash oil. This material accumulates in the bottom of the tank, and the wash oil decants off at a higher level to a small receiving tank, from which it is pumped to the top of the first scrubbing tower in the series. Generally, two decanting and two receiving tanks are provided to permit cleaning the residue from the tanks periodically.

Recent designs of wash-oil scrubbers are not fitted with hurdles and packing used in the previous type. Contact between the gas and absorption oil is accomplished by the use of single conical sprays placed at three or four different elevations within the tower, as shown in Figure 4—90.

Each of the sets of sprays is provided with its own set of recirculation pumps. The wash oil is recirculated in each set to obtain adequate contact between the gas and liquid in the scrubber. The wash oil is collected and recirculated by chimney trays at the bottom of each spray chamber except at the bottom where the oil is pumped from the scrubber sump.

Debenzolization of Wash Oil—In the debenzolization step, the light oil (2 to 3 per cent) in the benzolized wash oil is separated by steam distillation. The carryover of absorbing oil into the light oil is kept to about 5 per cent and the debenzolized absorbing oil contains 0.2 per cent light oil.

In the straight steam-distillation process at atmospheric pressure, the benzolized wash oil is preheated to approximately 100°C (212°F) with a vapor-to-oil and an oil-to-oil heat exchanger. Heating is continued to 145°C (295°F) with an indirect preheater of the shell-and-tube type, with the oil flowing through the tubes, using steam as the heating medium on the shell side. The preheated oil is introduced near the top into a multi-plate bubble-cap fractionating column leaving several plates above the feed to keep entrainment of wash oil to a minimum. The benzolized wash oil flows down the column countercurrent to upward flow of live steam, which is introduced in the base of the still column. The debenzolized wash oil leaving the base of the column through a sealed outlet at a temperature of 145° to 150°C (290° to 300°F), passes through the oil-to-oil exchanger in which it is cooled to 100°C (212°F) giving up its heat to the incoming benzolized wash oil. Water separates out at this point and is drained off. The wash oil passes through a pumping tank and is pumped at 100°C (212°F) to cooling coils for cooling prior to being used again as an absorbent for the light oil in the scrubber towers.

The mixture of steam and light-oil vapors leaving the top of the column flows through the tubular vapor-to-oil heat exchanger which recovers heat and also acts as a partial condenser. Sufficient heat is imparted to the incoming benzolized oil to raise its temperature 25°C (45°F) and, at the same time, the vapors are cooled to cause a portion of the steam and high-boiling constituents of light oil to condense (the condensate of which carries along some of the wash oil which was carried over the top of the column as entrainment). The mixture of oil and water is separated in a gravity separator tank, the water flowing to the sump system and

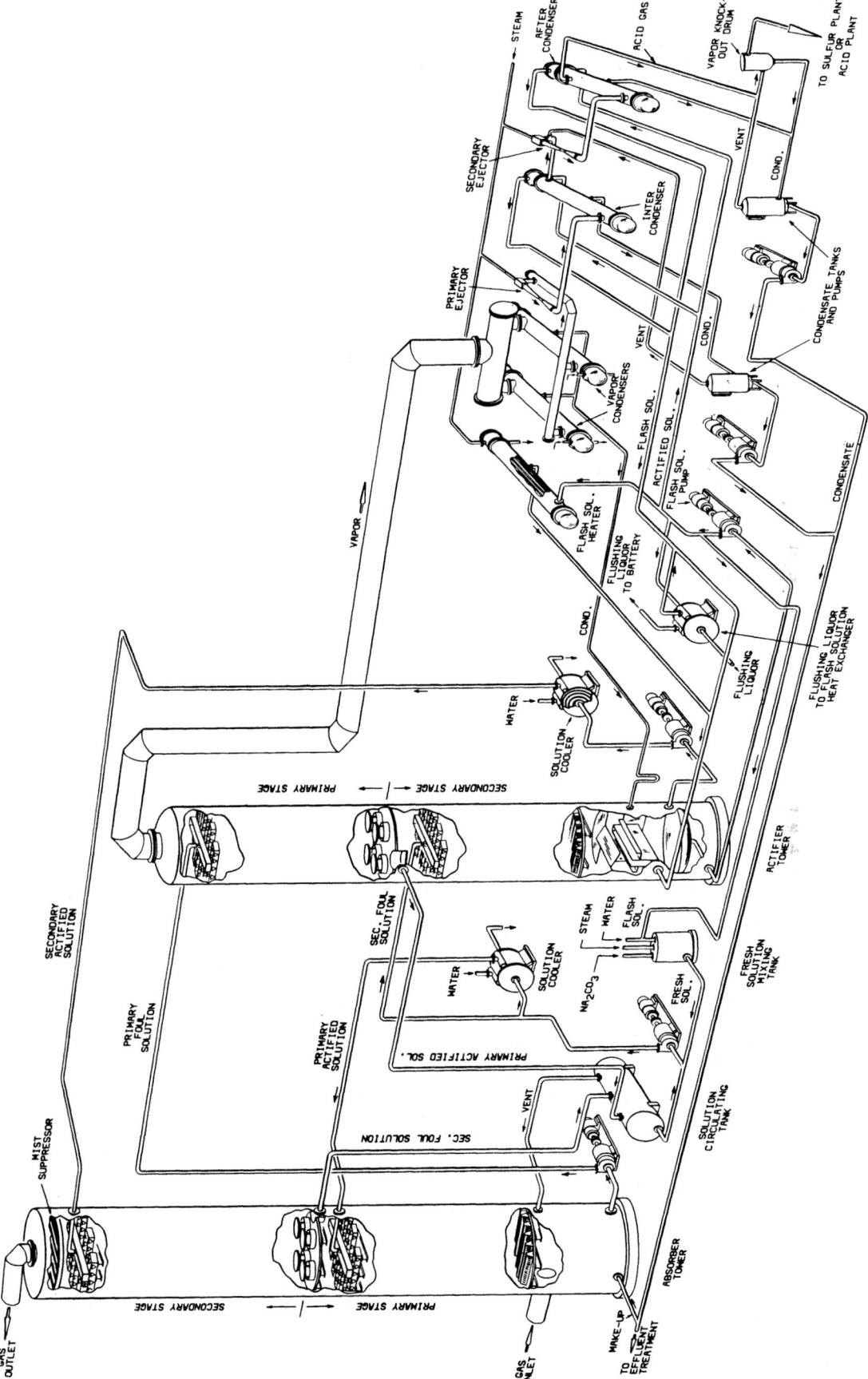

FIG. 4—91. Flow diagram of two-stage vacuum carbonate process for removal of hydrogen sulphide from coke-oven gas. (Courtesy, Koppers Company, Incorporated.)

the oil being returned to the debenzolized oil streams. The mixture of steam and light-oil vapors leaving the vapor-to-oil heat exchanger passes to a water-cooled condenser, which is of a multi-pass design, with the vapor and water flowing countercurrent to each other. The condensate flows to a gravity separator effecting a separation of the light oil and water, the light oil flowing to storage and the water to the sump system.

In some designs, an additional fractionating column is added to the debenzolization process for rectifying the light-oil vapors from the vapor-to-oil heat exchanger. In this case, the condensate of the vapor-to-oil heat exchanger, after separation of the water, is also introduced into the rectifying column. The mixture of steam and light-oil vapors enters the multi-plate bubble-cap rectifying column near the middle section. The light oil is separated into two fractions: the distillate containing forerunnings, benzene, toluene, xylene, and low-boiling solvent, and a residual fraction containing an admixture of high-boiling solvents, naphthalene and wash oil. A portion of the distillate is returned to the top of the column as reflux, the control point being the vapor temperature at the top of the column.

In the more modern debenzolization processes, the benzolized wash oil is processed at a temperature of 90° to 120°C (195° to 250°F) and this eliminates the need for an oil-to-oil heat exchanger in the process lineup.

HYDROGEN SULPHIDE REMOVAL

The first successful liquid-purification process for removal of hydrogen sulphide from coke-oven gas was the Seaboard process introduced in 1921 by Koppers Company, Incorporated. In this process, the coke-oven gas is washed with a sodium-carbonate solution in an absorber to remove the hydrogen sulphide from the gas. The solution is revivified by spraying it into an actifier where it is blown with air which transfers the hydrogen sulphide to the air.

The Seaboard process was replaced by the Koppers vacuum carbonate process (see Figure 4—91) which removes hydrogen sulphide from coke-oven gas and recovers it in the form of a concentrated acid gas stream from which the hydrogen sulphide can readily be converted to either elemental sulphur or sulphuric acid.

Although coke-oven-gas technology has been practiced for a number of years, modern sulphur-emission specifications require improvement of the old processes or development of new technology to meet regulatory requirements.

Commercial coke-oven-gas desulphurization processes can be divided into two categories: those processes which use wet oxidation to produce sulphur and those which absorb and strip H_2S for subsequent conversion into sulphur or sulphuric acid.

Wet Oxidation Processes—All wet oxidation processes utilize a reduction-oxidation catalyst to facilitate the wet oxidation of hydrogen sulphide to elemental sulphur or sulphate. All of these processes are characterized by a very efficient removal of sulphur, but have the disadvantage of producing highly contaminated waste products which require elaborate waste-treatment facilities as a part of the process. For example, all

the hydrogen cyanide that is absorbed is converted into thiocyanates and a portion of the sulphur is converted into thiosulfate requiring treatment in the waste-treatment facility. In addition, the solution is relatively sensitive to contamination and must be controlled to a uniform composition.

The major commerical wet-oxidation processes which are used in the world today, along with their reduction-oxidation catalysts, are listed in Table 4—VI.

Table 4—VI. Major Wet-Oxidation Desulphurization Processes and Their Catalysts

Process	Catalyst
Takahax Process	Naphthoquinone-sulphonic acid
Holmes-Stretford Process	Anthraquinone-disulphonic acid and sodium ammonium vanadate
Fumaks Process	Picric acid
Thylox and Giammarco Vetrocoke	Sodium thioarsenate

The Takahax Process, Figure 4—92, can be designed in four different ways, depending upon the alkaline solution used in the absorber for the end product desired. In the desulphurization process, crude coke-oven gas is counter-currently contacted with the alkaline solution in the absorber. H_2S removal of up to 90 to 99 per cent and HCN removal of 85 to 95 per cent can be obtained. Solution from the bottom of the absorber is sent to a regenerator where it is air-blown to regenerate the catalyst. In order to maintain the absorbability of the solution, part of the solution is split from the system and treated by the waste-solution-treatment process. The desulphurization process is similar in all four types, except that Types A and B use ammonia and Types C and D use sodium carbonate as the alkaline solution. The main difference in the four types of processes is in the waste-treatment systems.

(1) **Type A Takahax** oxidizes the waste solution with compressed air to form ammonium sulphate. Ammonium sulphate crystals are recovered in a conventional crystallizer.

(2) **Type B Takahax** incinerates the entire waste solution, along with the sulphur cake produced in the absorber, to produce SO_2 gas that is converted to sulphuric acid.

(3) **Type C Takahax** produces sulphur cake from the absorber and a portion of the sulphur is bled off and sent to a combustion furnace along with the waste solution. In the submerged combustion furnace, fuel is burned in a reducing atmosphere to convert the sulphur compounds to H_2S. The H_2S gas and the remaining sulphur are sent to the sulphuric-acid plant.

(4) **Type D Takahax** produces sulphur cake, and a portion of the sulphur cake and waste solution is sent to the submerged-combustion furnace as in Type C. H_2S produced in the combustion furnace is re-absorbed in the Takahax absorber, so that the entire discharge of sulphur from the process is elemental sulphur. Sulphur cake is melted in an autoclave to purify the product, producing a sulphur with a purity of 98 to 99 per cent.

All of the wet oxidation processes produce a low-

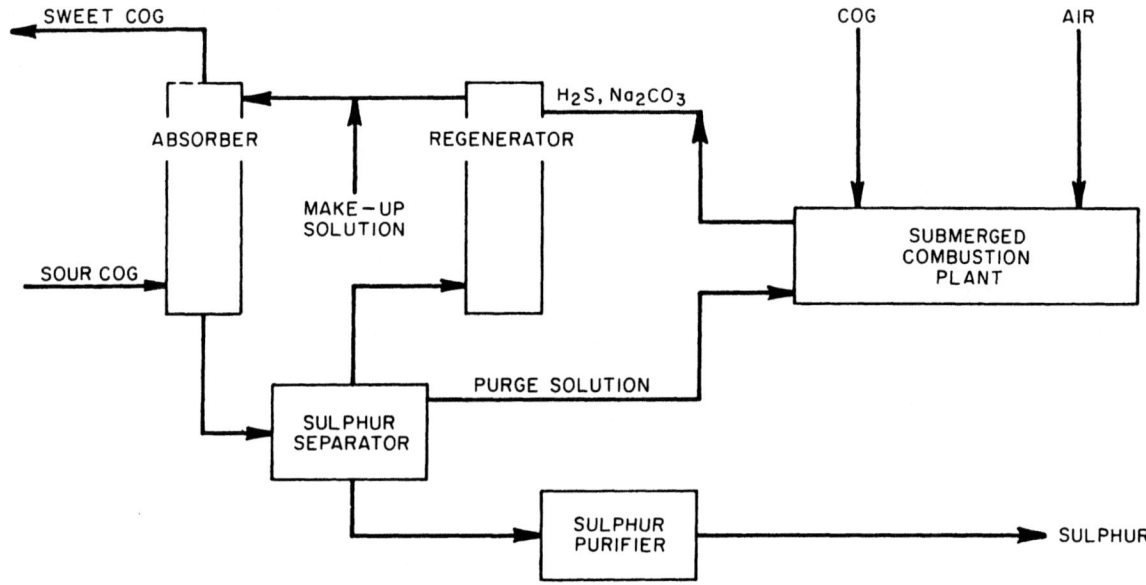

FIG. 4—92. Flow diagram of the Takahax process for desulphurization of coke-oven gas.

grade sulphur and a highly contaminated waste stream requiring elaborate waste-treatment facilities. Operation of these processes requires a more sophisticated chemical control than the absorption stripping processes.

The **Holmes-Stretford Process** is used to economically remove H_2S from coke-oven gas. In addition to the basic H_2S absorption plant, an effluent-treatment plant is included to convert the plant waste streams for recycling through the plant. This process will remove all but a trace of HCN, reduce the H_2S content to substantially less than 0.23 grams per cubic metre (10 grains per hundred cubic feet) of gas, and produce a very saleable grade of liquid sulphur.

An elementary flow diagram of the process is shown in Figure 4—93.

The process provides for scrubbing coke-oven gas with an alkaline solution of sodium ammonium vanadate which oxidizes the H_2S in the gas to free sulphur. Sulphur is collected as a froth which is washed, concentrated and purified to a final molten product.

The reduced vanadate solution is re-oxidized with anthraquinone disulphonic acid and air to its original state and recycled to the H_2S scrubber.

The basic plant utilizes the Holmes-Stretford process parts which are available under license granted by the British Gas Corporation. The complete plant consists of:

(1) Basic hydrogen sulphide removal system;
(2) Sulphur recovery and purification system; and
(3) Spent-liquor reprocessing system.

The combined systems result in a plant with economical operating costs, a saleable grade of sulphur product, and no effluent waste streams.

ABSORPTION STRIPPING PROCESSES

The absorption stripping processes are characterized by generally lower H_2S removal, but, since air is not included in the regenerating system, contaminated waste products are minimized or eliminated. The processes can be operated to produce sulphuric acid or a very high-purity elemental sulphur. There are four principal absorption stripping processes in commercial use today: the vacuum-carbonate process, the Sulfiban process, the Carl Still or Diamex process and the DESULF process.

Vacuum Carbonate Process—In this process, the gas is contacted counter-currently with a solution of sodium carbonate in an absorber tower to remove the hydrogen sulphide and other impurities such as hydrogen cyanide. The foul solution from the base of the absorber is circulated over the actifier where the hydrogen sulphide is removed by counter-current stripping with water vapor under vacuum. The actified solution is pumped from the base of the actifier through a cooler back to the absorber.

The actifier is maintained under a relatively high vacuum to reduce the quantity of heat required to generate the water vapor for actifying the solution. The mixture of hydrogen sulphide and water vapor leaving the top of the actifier is passed through a condenser where most of the water vapor is condensed. The vacuum in the actifier and condenser is maintained by a two-stage steam ejector. After the ejector system the concentrated acid gas stream is further cooled and processed in a Claus plant to produce elemental sulphur, or in a sulphuric-acid converter.

The heat requirements to generate the water vapor to strip the hydrogen sulphide from the foul solution in the actifier can be obtained almost completely from indirect heat exchange of the solution in the base of the actifier with the vapors at the discharge of the steam-jet vacuum pumps, and supplemental indirect heat exchange with the flushing liquor from the coke-oven batteries or from other waste-heat sources.

The single-stage vacuum carbonate process is eco-

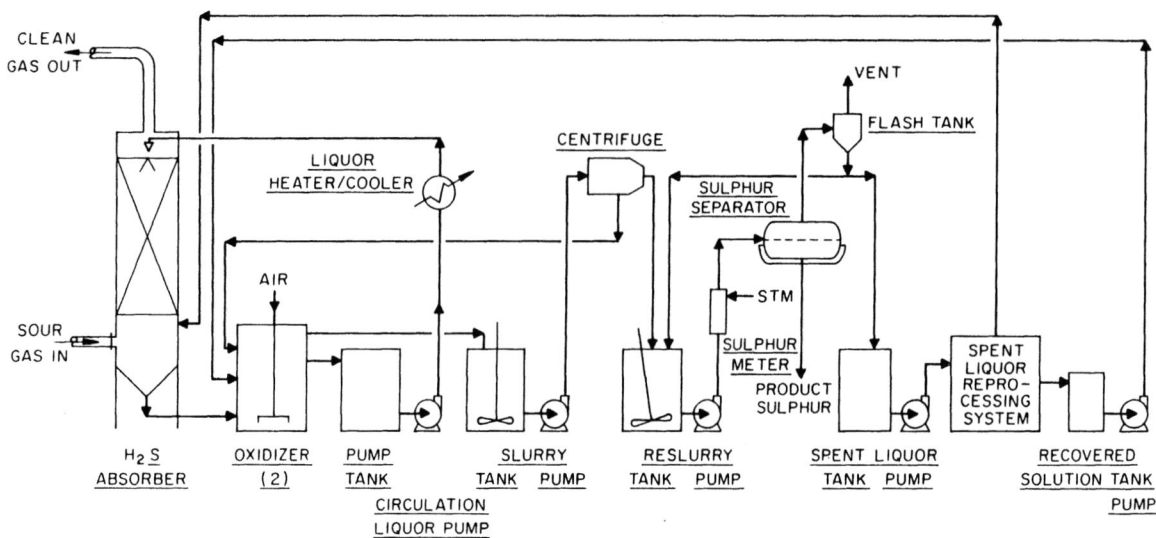

FIG. 4—93. Flow diagram of the Holmes-Stretford process for desulphurization of coke-oven gas. (Courtesy, Wilputte Corporation, a Salem Company.)

nomical for removal of approximately 90 per cent of the H_2S from typical coke-oven gas. In order to meet more stringent requirements in some locations, the Koppers two-stage vacuum carbonate process was developed to remove approximately 98 per cent of the H_2S, producing a clean gas with no more than 0.23 grams per cubic metre (10 grains of H_2S per 100 standard cubic feet).

In the two-stage process (Figure 4—94), the coke-oven gas flows upward through two sections of a packed absorber. In the lower section the gas is contacted with sodium carbonate solution from the upper stage of a two-stage vacuum actifier tower, and about 90 per cent of the H_2S is removed. The gas then flows to the upper stage of the absorber where it is contacted with a smaller stream of sodium carbonate solution from the lower stage of the actifier. This solution has been very highly actified, and removes 80 to 90 per cent of the remaining H_2S, for an overall removal of 98 per cent.

The same stripping steam is used in both sections of the two-stage actifier, resulting in high removal efficiency and low operating cost. The condensing system, vacuum system, and heat-exchange system are substantially the same as those used for the single-stage process.

The choice between the Koppers single-stage and two-stage vacuum carbonate processes depends upon the stringency of the regulatory requirements which must be met.

Sulfiban Process—The Sulfiban process was developed jointly by Bethlehem Steel Corporation and Applied Technology Corporation. The process utilizes monoethanalamine (MEA) as the absorbing solution. In the Sulfiban process, coke-oven gas is contacted counter-currently with an aqueous solution containing 13 to 18 weight per cent MEA. The process is reported to remove organic sulphur as well as H_2S, and ATC will guarantee a total sulphur content in the treated gas of

less than 0.23 grams per standard cubic metres (10 grains per 100 standard cubic feet).

Rich solution from the absorbers is pumped to the still, where H_2S and HCN are steam-stripped and the regenerating solution from the bottom of the still is cooled and recycled to the absorber. The steam for the stripping operation is produced in a steam-heated reboiler connected to the still. A small amount of insoluble compounds, mainly ferrocyanide, are formed and must be removed from the system. About 3 per cent of the circulating stream is sent to a reclaimer which is heated with high-pressure steam to totally evaporate the water and MEA. Vapors from the reclaimer are sent to the still and solids are periodically removed for disposal.

Acid gases from the top of the still, containing both H_2S and HCN, are sent to a cyanide reactor where the cyanide is catalytically converted to ammonia so that it will not interfere with the operation of the sulphur plant. This ammonia could be recovered with a small USS PHOSAM absorber. Where a USS PHOSAM plant already exists, the integration of this small absorber into the system would be inexpensive and advantageous.

Gas from the cyanide reactor passes to a plant for conversion to a high-purity sulphur.

Carl Still or Diamox Processes—Both the Carl Still and Diamox processes utilize aqueous ammonia as the absorbing solution to desulphurize coke-oven gas. The difference in the two processes is primarily in materials of construction and in some operating refinements. Either process can be synergistically combined with a USS PHOSAM process to recover both ammonia and high-grade elemental sulphur.

Coke-oven gas following the naphthalene scrubber passes through the H_2S scrubber, where it is counter-currently contacted with ammonia solution. The ammonia-absorbing solution is made up of ammonia liquor from the free leg of the coke-plant ammonia still,

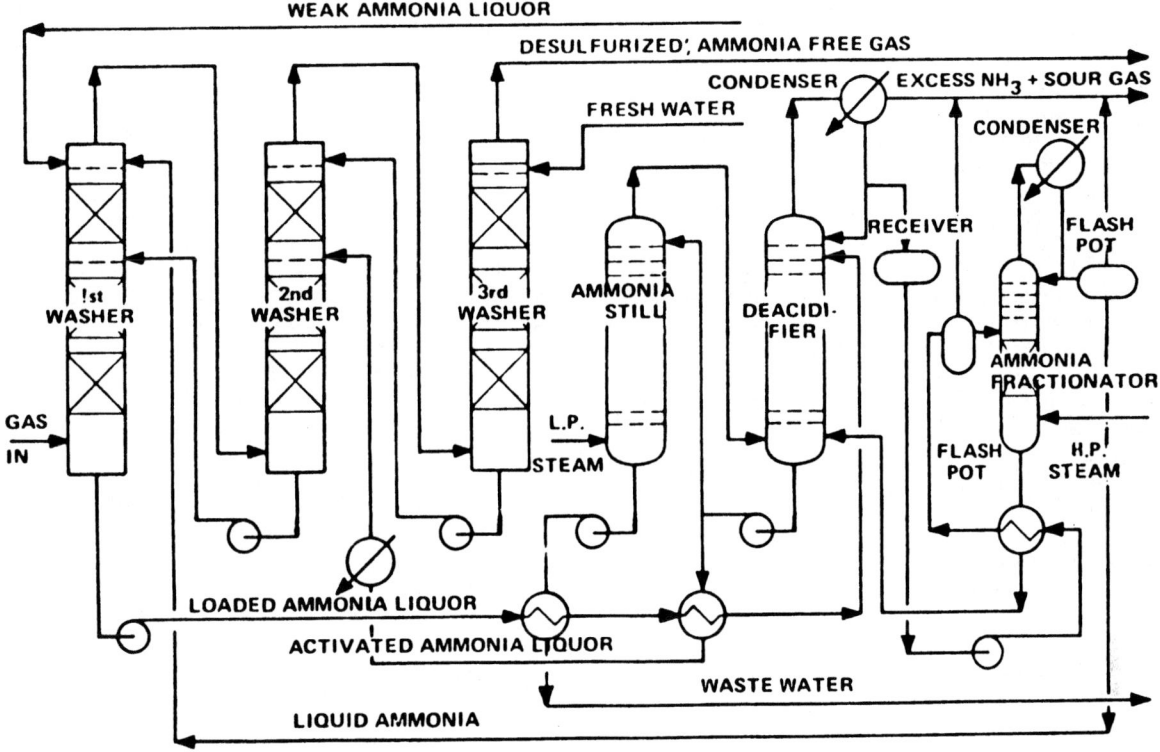

FIG. 4—94. Flow sheet of the DESULF process. (Courtesy, Didier Engineering, G.m.b.H.)

recycled solution from the deacidifier, and ammonia from the top of the PHOSAM stripper. If the H_2S absorber is operated with no recycled ammonia from the USS PHOSAM stripper, it will remove H_2S to a level of 0.92 to 1.14 grams per standard cubic metre (40 to 50 grains per 100 standard cubic feet). If about 15 per cent of the ammonia from the PHOSAM stripper is recycled to the H_2S absorber, the H_2S concentration will be reduced to about 0.23 grams per standard cubic metre (10 grains per 100 standard cubic feet). The only penalty for the improved H_2S removal is 15 per cent additional capacity in the PHOSAM stripper. Absorbing solution from the bottom of the H_2S absorber is pumped to the deacidifier, where it is contacted with steam from the coke-plant ammonia still, augmented by additional low-pressure steam.

Stripping solution from the bottom of the deacidifier is cooled and recycled to the H_2S absorber. Acid gases from the top of the deacidifier, containing H_2S, ammonia, HCN and CO_2, are sent to a small PHOSAM absorber, where the ammonia is selectively removed. Acid gases from the PHOSAM absorber pass to an HCN reactor, where the HCN is catalytically decomposed. Gases from the HCN reactor then pass to a plant for production of high-purity sulphur.

The integration of the Carl Still process with the USS PHOSAM process permits reduction of H_2S in the coke-oven gas from about 1.03 to 0.26 grams per standard cubic metre (about 45 grains to 10 grains per 100 standard cubic feet). In addition, removal of ammonia from the acid gases is advantageous to the operation of the sulphur plant. The successful combination of these

two processes offers a new technology in coke-oven-gas desulphurization.

The Diamox process is similar to the Carl Still process, except that it operates with lower ammonia concentrations so that ammonia recycle from the PHOSAM stripper and the acid gas PHOSAM absorber can be eliminated.

DESULF Process—This dominating coke-oven-gas liquid-desulphurization process in Europe, which started in the 1940's, uses ammonia water. This process, in fact, combines hydrogen sulphide removal with ammonia removal. The basic process has been updated recently to achieve lower H_2S figures in the purified gas and to adjust to the climatic conditions—higher temperatures—of other countries.

In the DESULF process (Figure 4—94) the gas is contacted counter-currently by the liquor in three absorptive towers operated in series where the liquor's ammonia content decreases in the direction of the gas flow. The first washer provides for H_2S bulk removal; in the second washer, ammonia and hydrogen sulphide are absorbed; while the third washer effects final removal of NH_3 (0.02 g/m^3n or 0.82 gr/100 SCF) and H_2S (0.2 g/m^3n or 8.2 gr/100 SCF). The absorption liquor streams are as follows: fresh water is used in the third tower and pumped from the tower bottom up to the top of the second washer, where activated ammonia liquor is also added. The liquor is pumped to the middle section of the first washer, while weak ammonia liquor and liquid ammonia enter its top section.

The loaded ammonia liquor is preheated and activated in the deacidifier which is equipped with bubble-cap trays. Activated ammonia liquor is pumped out of the bottom section and, after heat exchange and cooling, reused in the second washer. Surplus liquor is pumped from the deacidifier into the ammonia still, where the ammonia is removed by direct low-pressure steam injection. After heat exchange, the liquor is discharged into the waste-water system. The vapors from the ammonia still are directed into the bottom section of the deacidifier. Top vapors from the deacidifier partially condense, so that H_2S and the like remain as sour gas, while steam and ammonia mostly condense. A controlled proportion of the condensate is refluxed; most of the flow is pumped into the ammonia fractionator after heating and flashing off any still-entrained sour gases. The fractionator is operated at increased pressure. The concentrated ammonia liquor is fractionated by direct steam injection into lean liquor, which is returned to the deacidifier bottom, and into ammonia vapor, which is liquified in the condenser and then directed into the first washer.

This process can be easily adjusted to changed operating conditions by modifying the ratio of activated ammonia liquor to liquid ammonia.

COKE-PLANT WASTE WATER

Public Law 92-500 has mandated that waste waters must be treated to remove certain contaminants before these waters can be discharged to public waterways. The United States Environmental Protection Agency (USEPA) was given the responsibility of controlling and regulating these discharges and of developing uniform guidelines for required treatment of municipal and industrial point sources. Discharges from coke plants are one of the many areas in the steel industry that come under these Federal guidelines.

Coke-plant waste water is made up of blowdown from the flushing-liquor system, discharges from the primary and final coolers, decanted water from the light-oil decanters, discharges from the tar-recovery and the chemicals-recovery operations, and finally from other small miscellaneous sources throughout the coke plant. Because the carbonaceous and nitrogenous compounds in coke-plant waste water are amenable to biological treatment, this method has been widely used to reduce or remove these contaminants from the water before discharge to public waterways. A simplified block diagram of this process is presented in Figure 4—95. Heavy solids or grit are removed from the waste waters by use of bar screens or other similar devices, followed by oil and tar removal in settling tanks. Oil is skimmed from the top and tar is recovered from the bottom of these tanks. From the settling tanks, the waste water is normally steam-stripped to reduce ammonia and acid gases, primarily hydrogen cyanide and hydrogen sulphide, prior to biological treatment. The CYAM process, discussed earlier in this chapter, is a particularly efficient steam-stripping system. It is composed of a free-ammonia still and a fixed-ammonia still. The free ammonia and the acid gases are removed in the free still whereas the fixed ammonia is removed in the fixed still after addition of lime. The waste water stream from the stills is clarified and then flow and

concentration equalized before undergoing biological treatment.

In most instances, biological treatment or activated-sludge treatment of coke-plant waste water is an aerobic process, or one which requires oxygen. Oxygen is normally obtained from the air by mechanical mixing of the water and air or sparging with air from air blowers. Some processes that use pure oxygen gas have also been developed. The biological process is one where active bacteria, in the form of solids, are suspended in a containment vessel commonly referred to as an aeration basin. The raw waste water, sometimes diluted with relatively clean process water to reduce the waste-water toxicity, enters the aeration basin where it contacts the suspended bacteria, which absorb the contaminants from the water and biologically oxidize them ultimately to carbon dioxide and water in the case of carbonaceous compounds and to nitrate in the case of ammonia and other nitrogenous compounds. Generation of biological solids, which must be periodically removed from the system to maintain control, is a direct result of this process. Typically, waste water is held in the aeration basin about two days to complete the biological process, while the biological solids are kept in the system about 50 days. The former is referred to as the hydraulic-retention time of the biological system, and the latter is referred to as the solids-retention time or sludge age of the system. These two parameters are very important control points which need to be maintained if a quality discharge is to be realized.

Following biological treatment, the waste water is clarified and usually filtered through dual media filters or other filtering devices, or routed through large settling or polishing ponds to reduce the suspended-solids content before finally discharging into a public waterway.

Table 4—VII gives a typical composition of a combined waste-water stream from a coke plant before it has been given any type of treatment. Following the treatment steps outlined above, the waste-water composition can be expected to contain the levels of contaminants indicated in the last column of Table 4—VII. This complete treatment process yields a waste water that is harmless to the aquatic ecology and meets current guidelines for discharge to public waterways.

USES OF COKE, COKE-OVEN GAS AND COAL CHEMICALS

Metallurgical Coke—Coke is used for production of iron in blast furnaces and the coke breeze as a fuel for sintering plants and for steam generation in boiler houses.

Fuel Gas—After the recovery of coal chemicals, the gas provides fuel for heating the coke ovens, and the excess gas is used as a fuel in various steelmaking operations. When practicable, other gas of lower calorific value may profitably replace the coke-oven gas for firing coke ovens.

Ammonium Sulphate—The ammonium sulphate recovered from coke-oven gas is used for admixture with phosphate and potash constituents to provide balanced agricultural fertilizers for the various requirements, or

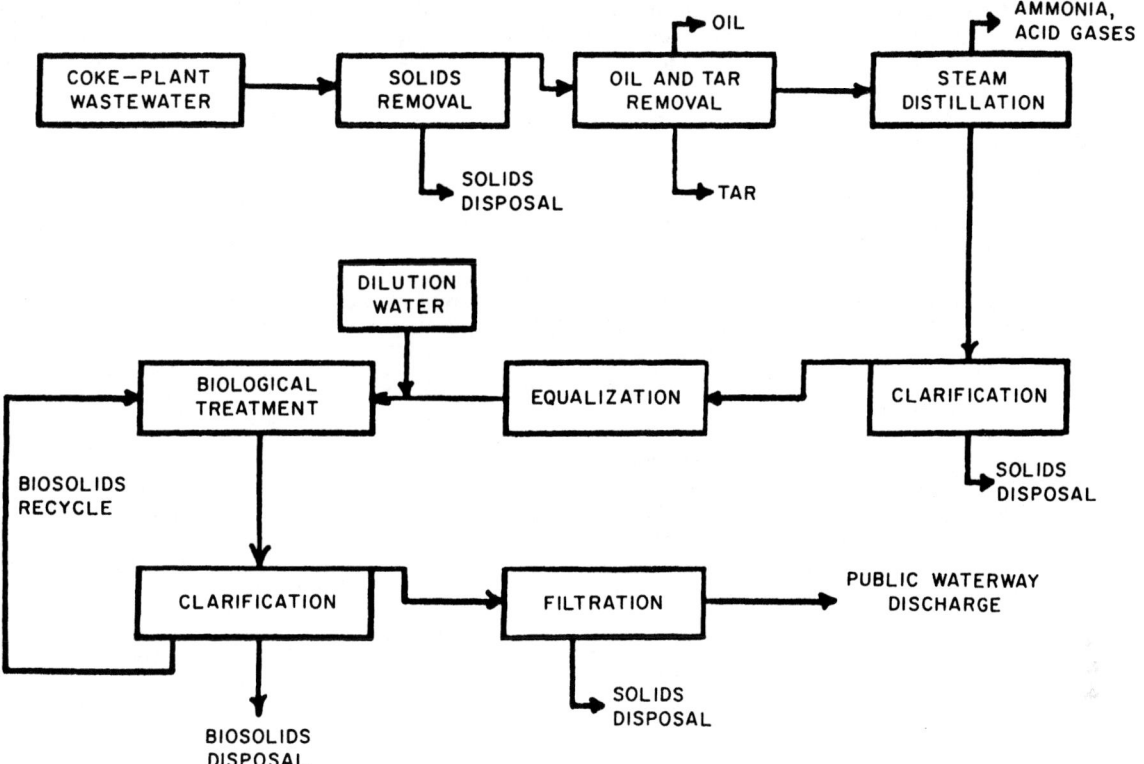

FIG. 4—95. Simplified diagram of a system for biological treatment of coke-plant waste waters.

it may be used for direct application where nitrogen is the only requirement at the time of use.

Anhydrous Ammonia—The ammonia recovered from the coke-oven gas and waste liquor as anhydrous ammonia liquid is a main source of nitrogen in making fertilizer. Ammonia is a major fertilizer in its own right, being introduced directly into soils. Special high-purity ammonia is used in chemical synthesis, in refrigeration and in steel plants for generating reducing atmospheres.

Phenol—Phenol (C_6H_5OH), sometimes called carbolic acid, is recovered from both coal tar and ammonia liquor. Its most important use is in the manufacture of resinous condensation products by reaction with formaldehyde, e.g., "Bakelite." As a chemical intermediate it is used in the preparation of synthetic tannins, dye intermediates, perfumes, plasticizers, picric acid, salicylic acid, and in the refining of lubricating oils.

Ortho Cresol—Ortho cresol ($CH_3C_6H_4OH$) is also used in the production of synthetic resins to control the plasticity of the resin. It is nitrated to produce insecticides and weed killers. It is used in various organic syntheses and in the production of artificial flavors and perfumes.

Meta-Para Cresol—Meta-para cresol ($CH_3C_6H_4OH$) is used chiefly in the production of synthetic resins and the plasticizer tricresyl phosphate. It is also used in organic synthesis and in the production of insecticides, dyestuffs, pharmaceuticals and photographic compounds.

Naphthalene—A large percentage of the naphthalene produced from coal is converted to phthalic anhydride. The principal use of the anhydride is in plasticizers, such as dioctyl phthalate and disoctyl phthalate, for use in synthetic resins. Nearly as much phthalic anhydride is used, however, in the alkyd resins, the important outlet for which is in coatings. In addition, polyester resins, dyes, agricultural chemicals, pharmaceuticals, insect repellents, beta-naphthol, surface-active agents, tanning agents, and insecticides consume large volumes of phthalic anhydride.

Table 4—VII. Typical Composition of Coke-Plant Waste Water Before and After Treatment

	Before	After
Ammonia	1500-2000 ml/m³ (ppm)	<10 ml/m³ (ppm)
Phenol	800-1200 ml/m³ (ppm)	<0.1 ml/m³ (ppm)
Thiocyanate	600-700 ml/m³ (ppm)	<1.0 ml/m³ (ppm)
Total Cyanide	200-400 ml/m³ (ppm)	<1.0 ml/m³ (ppm)
Oil/Grease (Freon Extractables)	2000-4000 ml/m³ (ppm)	<15 ml/m³ (ppm)
Total-Suspended Solids	300-1500 ml/m³ (ppm)	<25 ml/m³ (ppm)
pH	8-9	7-7.5
Temperature	54° to 77°C (130° to 170°F)	24° to 29°C (75° to 85°F)

Creosote—Creosote (from coal tar) constitutes a large part of the distillate from tar, and is a blend of different fractions to meet specifications established by the American Wood Preservers Association. Practically all of it goes into the pressure impregnation of wood, such as piling, poles and railroad ties.

Pitch—Pitch is employed as a binder in making carbon electrodes, as roofing pitch, fiber pitch, road tars, and in pipe-line enamels. Pitch is also used as an open-hearth fuel when mixed with virgin tar to maintain fluidity.

Benzene—Recovered from refining light oil, benzene is industrially the most important member of the aromatic family. The principal use is for the manufacture of styrene, which goes into polystyrene resins and important grades of synthetic rubber. Other important uses of benzene are for manufacture of phenol, nylon, synthetic detergents, aniline, and a host of other important organic chemicals, including DDT, maleic anhydride, benzene hexachloride, mono- and dichlorobenzene, and nitrobenzene. The uses for some of these benzene derivatives are evident, but others are intermediates, in turn, for additional chemicals, plastics, dyes, fungicides, and the like.

Toluene—The principal uses for toluene are in the manufacture of synthetic organic chemicals, detergents, resins, plasticizers, explosives, solvents, dye intermediates, and pharmaceuticals, and for conversion to benzene.

Toluene is converted through synthesis to phenol, benzoic acid, benzoyl chloride, benzoates, para cresol, dye intermediates, toluene sulphonates, di-isocyanates and trinitrotoluene.

Considerable quantities of toluene are also used as a diluent in various types of coatings.

Bibliography

Anon., Coke ovens: to build or not to build—the steelmakers dilemma. 33/The Magazine of Metals Producing 5, April 1967, 53-72.

Blauvelt, W.H., The by-product coke oven and its products. Trans. Am. Inst. of Mining and Met. Engrs 61, 1919, 436-453.

Childs, L., and Bridges, E. E., Recovering waste heat from coke by dry quenching. Waste heat recovery, Inst. of Fuel, Chapman and Hall (London), 1963.

Denig, F., Industrial coal carbonization—chemistry of coal utilization I. John Wiley and Sons, New York, 1945.

Dept. of the Interior, U. S. Geological Survey, Mineral Resources of the U. S.: Part II, Non-metallic Products. Washington, D.C., 1907.

Didier-Werke A. G., Private communication.

Finney, C. S., and Mitchell, John, History of the coking industry in the United States. Jour. of Metals 13, No. 4 (April), 285-291; No. 5 (May), 373-378; No. 6 (June), 425-430; No. 7 (July), 501-504; No. 8 (August), 559-561; (1961).

Foxwell, G.E., Dry coke cooling, Jour, Inst. of Fuel, Oct. 1949; 346-353.

Gayle, John, et al., Carbonizing tests in the Tuscaloosa oven: properties of wet and dry quenched cokes. U. S. Bureau of Mines RI 5479, 1959.

Helm, Edward J., New developments in American coke-oven construction and operation. Blast Furnace and Steel Plant 54, June 1966, 493-495.

Hersche, W., The recovery of heat from incandescent coke in gas and coke works. Sulzer Technical Review 39, 1957, 15-29.

Johnson, D. L., Glund's international handbook of by-product coke industry (American Ed.) Chemical Catalog Co. New York, 1932.

Koppers Company, Inc., Koppers-Becker coke ovens. The Company, Pittsburgh, Pa. 1944.

Koppers Company, Inc., Private communication.

Koppers, Heinrich, G.m.b.H., Private communication.

Kruchinin, M. S., and Nazarov, V. V., Experience of commissioning and running a coke dry cooling plant. Coke and Chemistry 3 (1968), 18-22.

Meissner, C. A., The modern by-product coke oven. Yearbook, Am. Iron and Steel Inst., 1913; 118-178.

Myhill, A. R., Dry coke cooling. Coke and Gas 20, Mar. 1958; 102-105 and 110: April 1958; 153-157 and 166.

Otto, Carl, The influence of the past on the coke oven of today. Blast Furnace and Steel Plant 28, Jan. 1940, 53-57, 96-97.

Otto, Dr. C., and Company, Private communication.

Process Plants Construction Division, Allied Chemical Corp., Private communication.

Ress, Franz Michael, Geschichte der kokereitechnik. Verlag Gluckauf, G.m.b.H., Essen, Federal Republic of Germany, 1957.

Russell, C. C., Discussion, Proc. Blast Furnace, Coke Oven and Raw Materials Committee, Am. Inst. of Mining and Met. Engrs 13, 1954, 104-105.

Sheridan, Eugene T., and DeCarlo, Joseph A., Coal carbonization in the United States, 1900-1962. U. S. Dept of the Interior, Bureau of Mines Information Circular 8251 (1965).

Still, Carl, Firma, Private communication.

Wilputte Corporation, a Salem Company, Private communication.

CHAPTER 5

Iron Ores

SECTION 1

THE NATURE AND OCCURRENCE OF IRON ORES

Of the many natural raw materials that are required for the making of iron and steel, the most important as to tonnage and value is iron ore. For each net ton of raw steel produced by the American steel industry, approximately one net ton of iron ore (including agglomerates) is consumed, according to statistics of the American Iron and Steel Institute.

The gross ton or long ton (2240 lb) formerly was the usual commercial weight unit for iron ore. The net ton or short ton (2000 lbm) was sometimes used in statistics. The metric ton of 1000 kgm (approx 2205 lbm) will be used throughout this chapter except where net tons or long tons are expressly specified.

IRON-BEARING MINERALS

A large number of minerals contain iron; however, only a few are used commercially as sources of iron. Minerals containing important amounts of iron may be grouped according to their chemical compositions into oxides, carbonates, sulphides and silicates. Table 5—I illustrates the various classes of iron minerals and indicates the mineral species commonly used as sources of iron. Oxide minerals are the most important sources of iron, followed by carbonates, sulphides and iron silicates. In the descriptions of the important iron-ore minerals or mineral groups that follow, the chemical compositions are for the pure minerals: the iron content of commercial ores or concentrates generally is lower, due to the presence of gangue and other impurities.

Magnetite—Chemical composition, Fe_3O_4, corresponding to 72.36 per cent of iron and 27.64 per cent of oxygen; color, dark gray to black; specific gravity, 5.16 to 5.18. It is strongly magnetic, sometimes possessing polarity so it will act as a natural magnet. The magnetic property of magnetite is important, for it permits exploration by magnetic methods and makes possible the magnetic separation of magnetite from gangue materials to produce a high-quality concentrate. Magnetite occurs in igneous, metamorphic, and sedimentary rocks. It has become increasingly important as a source of iron, as a consequence of the continued improvements in magnetic concentration techniques and in the expanded use of the high-grade products. At times, magnetite contains titanium in small amounts as inclusions of ilmenite. When the titanium content reaches 2 to 15 or more per cent, the magnetite is termed **titaniferous magnetite**.

Hematite—Chemical composition, Fe_2O_3, corresponding to 69.94 per cent of iron and 30.06 per cent of oxygen; color, steel gray to dull red or bright red; earthy to compact or crystalline; specific gravity, 5.26. Common varieties are termed **crystalline, specular, martite** (pseudomorphic after magnetite), **maghemite** (magnetic ferric oxide), **earthy, ocherous, and compact.**

Table 5—I. Chief Iron-Bearing Minerals

Class and Mineralogical Name	Chemical Composition of Pure Mineral	Common Designation
Oxide		
Magnetite	Fe_3O_4	Ferrous-ferric oxide
Hematite	Fe_2O_3	Ferric oxide
Ilmenite	$FeTiO_3$	Iron-titanium oxide
Limonite	$\left\{ \begin{array}{l} HFeO_2{}^\circ \\ FeO(OH)^{\circ\circ} \end{array} \right.$	$\left. \begin{array}{l} \\ \end{array} \right\}$ Hydrous iron oxides
Carbonate		
Siderite	$FeCO_3$	Iron carbonate
Silicate		
Chamosite	Various	
Stilpnomelane	and	
Greenalite	sometimes	Iron silicates
Minnesotaite	complex	
Grunerite		
Sulphide		
Pyrite (iron pyrites)	FeS_2	
Marcasite (white iron pyrites)	FeS_2	Iron sulphides
Pyrrhotite (magnetic iron pyrites)	FeS	

°Goethite
°°Lepidocrocite

253

Hematite is one of the most important iron minerals. It has a wide occurrence in many types of rocks and is of varying origins. It occurs associated with vein deposits; igneous, metamorphic, and sedimentary rocks; and as a product of the weathering of magnetite. Some low-grade deposits of disseminated crystalline hematite have been successfully treated by both gravity and flotation techniques to produce high-quality concentrates.

Hydrous Oxides—Limonite is the name commonly given to hydrous iron oxides that mineralogically are composed of various mixtures of the minerals **goethite** or **lepidocrocite.** The chemical formula for goethite is $HFeO_2$ and that for lepidocrocite is $FeO(OH)$. Goethite contains 62.85 per cent of iron, 27.01 per cent of oxygen, and 10.14 per cent of water; specific gravity ranges from 3.6 to 4.0; color, commonly yellow or brown to nearly black; compact to earthy and ocherous. In non-technical parlance, the term "limonite" is used to denote unidentified oxides with a variable moisture content due to absorbed or capillary water. It is a secondary mineral, formed commonly by weathering, and occurs in association with other iron oxides and in sedimentary rocks. Limonites are important sources of iron throughout the world.

Ilmenite—Chemical composition, $FeTiO_3$, corresponding to 36.80 per cent of iron, 31.57 per cent of titanium, 31.63 per cent of oxygen. This is commonly considered an iron titanate. Ilmenite is often associated in small amounts with magnetite. Although generally mined as a source of titanium rather than as an ore of iron, iron may be recovered as a by-product.

Siderite—Chemical composition, $FeCO_3$, corresponding to 48.20 per cent of iron, 37.99 per cent of CO_2 and 13.81 per cent of oxygen; specific gravity, 3.83 to 3.88; color, from white to greenish gray and brown. Siderite commonly contains variable amounts of calcium, magnesium or manganese. Siderite varies from dense, fine grained and compact to crystalline. The siderite ores are sometimes termed **"spathic iron ore"** or **'black-band ore."** Carbonate ores are commonly calcined before they are charged into the blast furnace. They frequently contain enough lime and magnesite to be self-fluxing.

Silicate Group—There are a large number of silicate minerals containing small amounts of iron associated with other bases but there are comparatively few silicates with iron as the principal base. They often have a rather complex chemical formula, with specific gravities higher than 2.8, and occur in various shades of green or black. Important iron-silicate minerals are **chamosite, stilpnomelane, greenalite, minnesotaite,** and **grunerite.** The iron silicates, while of limited importance in themselves as a source or iron ore, are of interest as a primary source of oxide iron ores which form through weathering or hydrothermal oxidation of the silicate minerals. They have a wide distribution in sedimentary rocks and metamorphic iron formations.

Sulphide Group—Iron occurs in a large number of sulphide minerals. The principal iron-sulphide minerals are: pyrite, pyrrhotite and marcasite. **Pyrite** (iron pyrites), chemical composition FeS_2, corresponding to 46.55 per cent of iron and 53.45 per cent of sulphur; specific gravity, 4.95-5.10; color, pale brass yellow; is the most widespread of the iron sulphides and occurs in

sedimentary, metamorphic, and igenous rocks and in veins. **Pyrrhotite** (magnetic pyrite), a sulphide of iron that varies in chemical composition from FeS to $FeS + S$, typically contains 60.4 per cent of iron and 39.6 per cent of sulphur; its color is bronze yellow to copper red, frequently tarnished. Pyrrhotite is often considered to be an indicator of nickel deposits because of its common association with the nickel sulphide, pentlandite. When nickel occurs in pyrrhotite it is generally in the form of fine inclusions of pentlandite. **Marcasite** (white iron pyrites), chemical composition FeS_2, corresponding to an iron content of 46.55 per cent and a sulphur content of 53.45 per cent, is pale brass yellow in color and is commonly associated with limestones, clays, and lignite deposits. It differs from pyrite only in its crystal structure and greater chemical instability.

Iron sulphides are sometimes mined as a source of sulphur. More commonly they are mined because of their association with other valuable metallic elements such as copper, nickel, zinc, gold, silver, etc. Iron is sometimes recovered as a by-product after the removal of the more valuable metals and the sulphur. The sulphides are of growing importance as sources of by-product iron principally from pyrite and pyrrhotite. By-product iron or iron oxide has been produced from sulphide ores by International Nickel, Cominco, Tennessee Copper Company, Falconbridge Nickel Mines Ltd. and Canadian Industries, Ltd.

GEOLOGY OF IRON-ORE DEPOSITS

Geologic Ages of Iron-Ore Deposits—Iron ores have a wide range of formation in geologic time as well as a wide geographic distribution. They are found in the oldest known rocks of the earth's crust, with an age in excess of 2.5 billion years, as well as in rock units formed in various subsequent ages; in fact, iron ores are forming today in areas where iron oxides are being precipitated in marshy areas, and where magnetite placers are being formed on certain beaches.

Many thousands of iron deposits are known throughout the world. The deposits range in size from a few tons to many hundreds of millions of tons. Many of the world's largest deposits of iron ore are located in the oldest geologic series—the Pre-Cambrian. Table 5—II illustrates the geologic age for selected iron deposits.

Genesis of Iron-Ore Deposits—Iron ores occur in a wide variety of geological environments in igneous, metamorphic or sedimentary rocks, or as weathering products of various primary iron-bearing materials. For convenience of study and comparison, iron ores may be grouped into types of similar geological occurrence, composition and structure. The following simplified classification based on genesis of the deposits and geological environment shows the chief modes of occurrence of iron ores and illustrates the varied geology of iron-ore deposits. Some of the examples listed may be approaching the end of their productive life or they may no longer be active, but they are generally well known and have been described in the geological literature.

IGNEOUS ORES

Iron ores may be formed by crystallization from liq-

Table 5—II. Geologic Age of Selected Iron-Ore Deposits

Geologic Age	Deposit	Location
CENOZOIC ERA		
QUATERNARY PERIOD		
Recent Epoch	St. Lawence Magnetite placers	Quebec, New York
TERTIARY PERIOD		
Pliocene Epoch	° Kerch Oölitic Limonite	Crimea, Russia
	El Tofo Magnetite	Chile, South America
Miocene Epoch	Honshu and Hokkaido gravel placers	Japanese Archipelago
Oligocene Epoch	Cheikh-ab-Charg Hematite	Persia
Eocene Epoch	Upper Assam Clay Ironstones	India
	Bahariya Hematite	Egypt
MESOZOIC ERA		
CRETACEOUS PERIOD	° Salzgitter Limonite and Hematite	Germany
	Bilbao Hematite	Spain
	Algerian and Moroccan Magnetite and Hematite	North Africa
JURASSIC PERIOD	° Minette Limonite and Hematite	France, Germany and Luxemburg
	Iron Springs Magnetite	Utah
TRIASSIC PERIOD	Kashmir Calcareous Iron Ore (Hematite)	India
PALEOZOIC ERA		
PERMIAN PERIOD	Damuda Sandstone (Hematite)	India
CARBONIFEROUS PERIOD	° Black Band Ironstones	British Isles
	Ohio Siderite Ores	Ohio
DEVONIAN PERIOD	° Siegerland Siderite	Germany
	Oriskany Limonite and Hematite	Virginia
SILURIAN PERIOD	° Clinton Hematites	Alabama
ORDOVICIAN PERIOD	° Wabana Oölitic Hematites	Newfoundland
CAMBRIAN PERIOD	Residual Limonites of the Appalachians	Georgia, Virginia, Alabama, Tennessee
PRE-CAMBRIAN ERA°°	° Minas Gerais and Serra dos Carajas Hematite	Brazil
	° Krivoi Rog Hematites	Ukraine, Russia
	° Bihar, Orissa and Bastar Hematites	India
	° Labrador Hematite	Quebec and Labrador
	° Lake Superior Taconites and Jaspilities, Hematites and Magnetites	Michigan, Wisconsin, Minnesota, Ontario
	° Cerro Bolivar and El Pao Hematites	Venezuela
	° Kirunavaara Magnetite	Sweden
	° Hamersley Range Hematites	Western Australia
	° Nimba Range Hematite	Liberia and Guinea
	° Sishen Hematite	South Africa

° Well-known, important deposits.
°° Represents the time span of the combined Proterozoic and Archeozoic Eras. (see Table 5—VI).

uid rock materials, either as layered-type deposits that possibly are the result of crystals of heavy iron-bearing minerals settling as they crystallize to form iron-rich concentrations, or as bodies which show intrusive relationship with their wall rocks. These ore bodies may be tabular or irregular and are composed largely of magnetite with varying amounts of hematite. Igneous ores are usually high in iron content and are often high in phosphorus or titanium content. Deposits of this type include the ores at Kiruna, Gallivare and Grangesberg, Sweden; Iron Mountain and Pea Ridge, Missouri; titaniferous magnetite deposits at Lake Sanford, New York; Talberg, Sweden; and Bushveld, Transvaal, Union of South Africa.

CONTACT ORES

Iron-ore deposits formed at or near the contact between igneous rocks and sedimentary rocks, the latter usually limestones, are commonly composed of mag-

netite and hematite with associated carbonates and pyrite. The ore deposits are commonly in the sedimentary rocks as irregular or tabular replacement bodies. Deposits of this type include: Cornwall, Pennsylvania; Iron Springs, Utah; Mount Magnitnaya, Russia; Marcona, Peru; Dungun, Malaya; Larap, Philippine Islands; and Marmora, Canada.

HYDROTHERMAL ORES

Iron-ore deposits formed by hot solutions which transported iron and replaced rocks of favorable chemical composition with iron minerals to form irregular ore bodies, commonly in limestones, are termed hydrothermal deposits. The iron often occurs as siderite ($FeCO_3$) or sometimes as oxides. Examples of hydrothermal deposits include those at Bilbao, Spain; Cumberland, England; Kenifra, French Morocco; Ouenza, Algeria; Rudabanya, Hungary; hard ores of the Marquette Range, Michigan; and Iron Monarch ores of the Middleback Range, Australia.

SEDIMENTARY ORES

Bedded Ores—Sedimentary bedded iron ores, often composed of oölites of hematite, siderite, iron silicate or, less commonly, limonite in a matrix of siderite, calcite or silicate, have a wide geographic distribution associated with other sedimentary rocks. The Clinton ores in Alabama also have fossil fragments coated or replaced with hematite, or sand grains and pebbles coated with hematite in a hematite and calcite matrix. These ores often have a fairly high phosphorus content and may be self-fluxing. Ores of this type include the Wabana ores, Newfoundland, Canada; Clinton ores, Alabama; Minette ores, France; Jurassic ironstones, England; and Kerch ores, Russia.

Siderite Ores—These ores consist of beds of siderite or siderite nodules associated with shales. They are common in the coal measures and are often termed **clay ironstones** or **black-band ironstones**. These ores commonly contain associated sulphides and often have a fairly high sulphur and phosphorus content. They were formerly of considerable importance in Great Britain and Germany.

Placer Ores—Iron oxides, when compact, are rather resistant to weathering and erosion and under favorable conditions may form placer deposits which, in a few instances, constitute iron ores. In general they are of rather minor importance as sources of iron. Magnetite sands are mined in Japan and are known in many areas. Deposits of iron-bearing rubble cemented with limonite and clay, that occur on hill slopes in tropical areas adjacent to pre-existing iron deposits, are termed **canga**. These are sometimes mined as ore. Placer iron-ore deposits include beach sands in Japan and New Zealand. Canga occurs in Brazil, India and Venezuela.

Bog Iron Ores—Bog ores occur in many swampy areas particularly in glaciated areas in Europe, Asia and North America. They occur commonly as dark-brown, cellular masses, or granular or fine particles of limonite. In years gone by, such deposits were mined rather widely to supply small iron furnaces locally. However, they have long ago ceased to be important commercially.

Metamorphosed Iron Formations—These include the metamorphosed bedded ferruginous rocks composed usually of alternating thin layers of chert or fine-grained quartz and ferric oxides. The iron is normally present in the mineral form of magnetite or hematite, along with lesser amounts of iron silicates and iron carbonates. Essentially all of the Pre-Cambrian sedimentary iron formations are of this type: these include the magnetic and nonmagnetic taconites of Minnesota; jaspilites of Michigan; itabirites of Brazil; magnetite gneisses of New York; quartz banded ores of Norway and Sweden; iron formations of Quebec and Labrador; banded hematite quartzites of Africa; hematite jaspers of India; and the banded iron formations of Manchuria, Korea and Australia. The metamorphosed types also include those in which the original form of the ores has been obscured by extensive recrystallization. In recent years, some of these iron formations have become important economically as iron ores because of their amenability to beneficiation by fine grinding and by concentration of the ore minerals by gravity and magnetic methods.

Residual Ores—Residual ores are commonly products of the surficial weathering of rocks but may include ores formed by hydrothermal oxidation and leaching. Ores of this type were formed extensively in Pre-Cambrian iron formations by leaching of silica, which commonly constituted in excess of 50 per cent of the rock. Oxidation changes iron carbonate, silicate minerals, and magnetite to hematite or limonite. Examples of residual ores include: iron **laterites** formed by the weathering of basic igneous rocks as in Cuba, Conakry (Guinea, Africa), and Mindanao in the Philippine Islands; the **brown ores** of Alabama, Georgia and Missouri; **soft hematite** and **limonite** Pre-Cambrian ores such as occur in the Lake Superior District; Estado Bolivar, Venezuela; Western Australia; Krivoi Rog, Russia; Minas Gerais, Brazil; Quebec-Labrador, Canada; Nimba, Liberia; Nimba, Guinea; and India.

DEFINITION OF THE TERM "ORE"

The term "ore" is often used loosely—as it has been up to this point—in discussions of a general nature. However, a more restrictive definition must be used in discussions of ores used by the iron and steel industry and particularly in discussing iron-ore reserves.

Iron is one of the more abundant and widely distributed elements in the earth's crust, constituting not less than 4 per cent of the total. Its supply is essentially limitless in almost all regions of the world. However, most of this iron, because of its position or concentration in the earth's crust, is not available to us, and much of it is in a form that cannot be used in current iron-making practices. That part of the total iron in the earth's crust that is available to industry, both economically and spatially, may correctly be termed "ore." However, what constitutes "ore" varies widely form place to place and time to time. For example, the principal ores now being mined and used in France contain less than 25 per cent of iron. Such material would not be considered suitable furnace feed in most other sections of the world. However, in this particular case, the spatial proximity of this ore to sources of coking coal and to an important market area for steel, as well as its adaptability to the furnace practices developed in that area, has given it a favorable price-quality relationship with respect to higher grade but more distant ores. Much the same situation existed in the Birmingham, Alabama, area, where ores containing only 37 per cent of iron were mined and used locally, although such low-grade, high-phosphorus ore would not be considered as commercial ore in other mining districts. There are many factors which enter into determining what iron-bearing material can be classed as an "ore," but basically it is a question of economics. Keeping this concept in mind, a logical definition of "iron ore" for commercial purposes is as follows: "iron-bearing material that can be economically used at a particular place and time under then current cost and market-price conditions."

With the advent of improved methods of beneficiation, concentration and agglomeration, the variety of iron-bearing materials that can now be used has been broadened and many low-grade material types which were once uneconomic, may now be considered as "ore". The beneficiation methods discussed in Section 4 are used not only to make iron-ore concentrates from taconites, but also are used to improve high-grade ores by controlling particle size and by reducing the content of gangue minerals. These methods are currently being used on a world-wide basis in almost every major iron-ore producing district. Ferruginous by-products derived from the processing of materials for their content of sulphur, titanium, copper, or other metals would not be considered as iron ore in the strict sense of the term, nor would beach sands unless they were actually being utilized, as for example, beach sands in New Zealand. Other iron-bearing material would also have to be denied classification as ore because of impurities, geographic inaccessibility, or, in some cases, simply because there is insufficient available information for a valid appraisal.

IRON-ORE RESERVES

The potential iron "ore" known to exist is called iron-ore reserve. Such reserves can best be specified by using the definition of "ore" given in the preceding paragraph. Because the iron content of different ore bodies is sometimes widely different, it is often more realistic to compare them on the basis of the iron they contain rather than on the basis of the amount of ore. For this reason, reserves may be reported as "iron ore" or as "iron-in-ore," depending upon the use to be made of the information.

Reserves are defined as those resources that can be economically mined at the time of determination. Identified resources include both reserves and other iron-bearing materials that may become profitable to mine under future economic conditions. Identified resources are those whose "location, grade, quality and quantity are known or estimated from specific geologic evidence. Identified resources include economic, marginally economic and sub-economic components", and depending upon the degree of geologic certainty, each of those "economic divisions can be further subdivided into measured, indicated and inferred."

Based on these definitions, the United States Bureau of Mines has compiled an estimate of the identified world iron-ore resources in terms of contained iron-in-ore. Their data are shown in Table 5—III. In Section 2 of this chapter, the geologic descriptions of the individual countries contain resource data in terms of ore grades and metric tons of ore.

SECTION 2

MAJOR IRON-ORE DEPOSITS SUPPLYING FURNACES IN THE UNITED STATES AND OTHER SELECTED MAJOR DEPOSITS

Following are brief descriptions of the major sources of iron ore available to the United States' iron and steel industry and the particular markets these sources serve. Also covered are the principal sources of iron ore in international commerce. Included in this category are iron-ore sources of North America (discussed in subsections A, B and C) as well as certain selected ore deposits in South America (subsection D) and West Africa (subsection E) which, because of their location or ownership, supply ore to the United States' market. Additionally, iron-ore deposits of Australia, Europe and Asia are mentioned and briefly described in subsection F. Information on other world iron-ore deposits can be obtained from the bibliography at the end of this chapter.

It should be borne in mind that the iron-ore industry is changing rapidly as new steel plants are built, as ironmaking technology advances, as transportation facilities improve, and as new ore sources of major importance are developed. The following comments refer to conditions as they existed in 1981.

The principal iron-ore deposits of North America are in the Lake Superior District of the United States and the Quebec-Labrador Region of Canada, with lesser deposits, which have been important as local sources of iron ore, in Missouri, Texas, New York, Nevada, Utah, Wyoming, California, Alabama, British Columbia, Ontario and Mexico. Each iron-ore source has its market area which may overlap market areas with other iron-ore sources. The Lake Superior ores and Ontario ores serve the Great Lakes Region furnaces in the United States and Canada; Quebec-Labrador ores are shipped to the Great Lakes Region furnaces and Eastern seaboard plants, as well as being exported to Europe and Japan; ores from Missouri, Texas and Mexico are used locally; the Utah, Wyoming and California ores serve western United States plants; and mines in British Columbia have produced co-product iron-ore and copper concentrates which largely have been exported to Japan.

The general ore reserves for North America are included in Table 5—III.

In addition to mined iron ore, by-product iron or iron oxide is currently available from several sources in both Canada and the United States.

A. DEPOSITS OF THE UNITED STATES

The areas of principal iron ore reserves and resources in the United States are shown in Figure 5—1. Some are being mined actively while others, for economic reasons, are either inactive or have never been developed. They conveniently can be grouped into five areas: Lake Superior, Northeastern, Missouri, Southeastern and Western.

LAKE SUPERIOR DISTRICT

The iron-mining region known as the Lake Superior

District is situated on the southwestern margin of a vast area of Pre-Cambrian rocks, geologically termed the Canadian shield, which covers much of Eastern Canada. The district includes several separate mining centers, known locally as ranges, some of which lie within the United States and others in Canada (Figure 5—2). The ranges in the United States include the Cuyuna, Mesabi, and Vermilion ranges in northern Minnesota, the Penokee-Gogebic in northern Wisconsin and adjacent Michigan, and the Menominee and Marquette ranges in northern Michigan. The Canadian ranges include the Atikokan, west of Lake Superior, the Michipicoten range east of Lake Superior, and the Moose Mountain, north of Lake Huron. In addition to these areas, the iron-bearing formations occur on the Gunflint Range near Port Arthur, north of Nakina, at Lake St. Joseph, the Onoman ranges and at Kaministikwia, etc., all in western Ontario. The Lake Superior District ores are accessible to the steel furnaces in the southern Great Lakes areas by water transportation after short rail hauls to lake ports.

The Lake Superior District is one of the great iron-mining regions in the world, both on the basis of past production and future potential. Iron was first discovered in 1844 near the site of the City of Negaunee, Michigan, by a government survey party with William A. Burt in charge. Some iron ore was mined in 1848 when an attempt was made to make iron in a local forge. This project was unsuccessful. Attempts were made to ship iron ore in the early 1850's, but regular shipments to the lower lakes did not start until 1856. Iron-ore explorations in the Lake Superior District were pressed with considerable vigor after the Civil War. Ore was discovered on the Vermilion Range, Minnesota, in 1865; the Menominee Range in 1872; Crystal Falls, Florence and Iron River Districts in 1880; the Gogebic Range in 1882; the Mesabi Range in 1890; the Michipicoten District in 1898; and the Cuyuna Range in 1903. Iron-formation materials have been known in the Atikokan District since 1889 and at Steep Rock Lake since 1901; however, the Steep Rock ore body was not discovered until 1938.

General Geology—The rocks of principal importance to the iron-mining industry in the Lake Superior region are of Pre-Cambrian age. For convenience of reference, these rocks are divided into Archean and Proter-

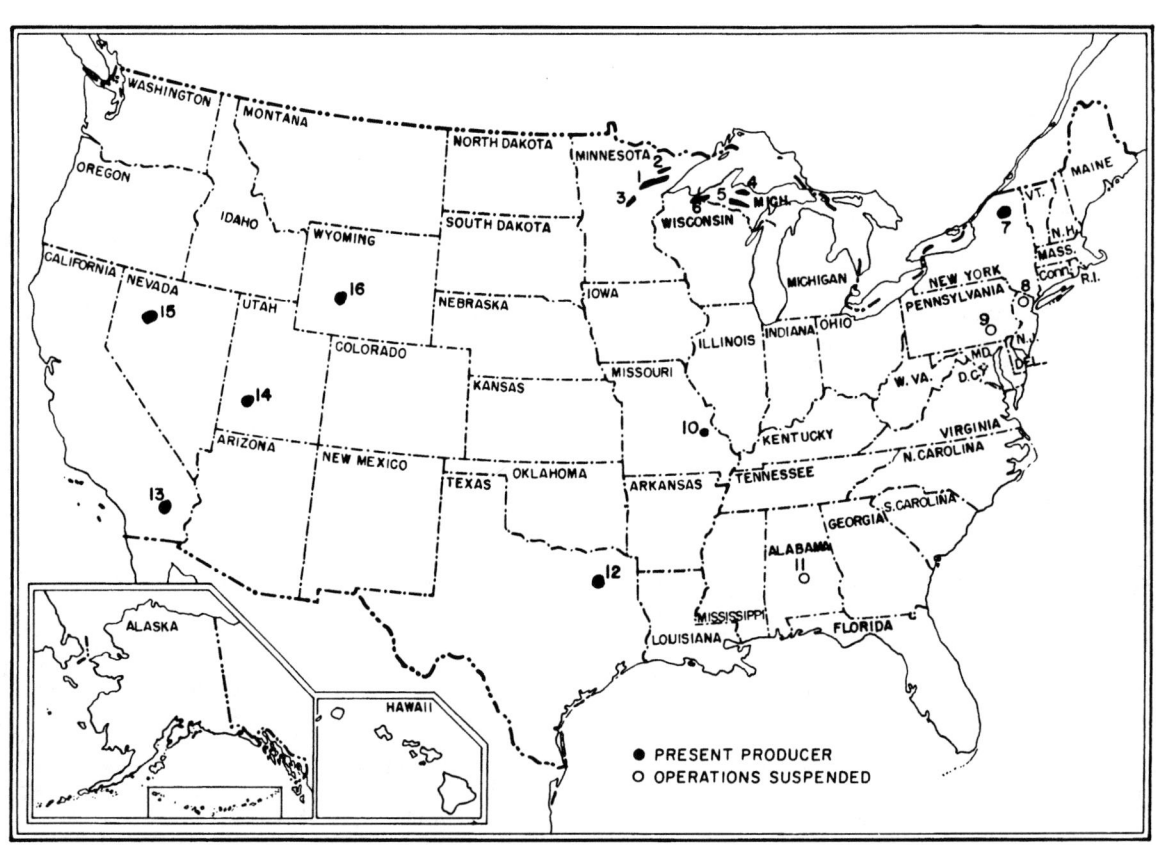

● PRESENT PRODUCER
○ OPERATIONS SUSPENDED

KEY

1. Mesabi Range	9. Pennsylvania (Cornwall
2. Vermilion Range	and Grace Mines)
3. Cuyuna Range	10. Missouri and Pea Ridge
4. Marquette Range	Mines
5. Menominee Range	11. Birmingham District
6. Penokec-Gogebic Range	
7. Adirondacks (Benson Mines,	12. Texas Brown Ore Fields
Lyon Mt., Port Henry, Lake	13. Eagle Mountain Mine
Sanford)	14. Iron Springs District
8. New Jersey (Scrub Oaks and	15. Mineral Basin
Washington Mines)	16. Atlantic City Mine

FIG. 5—1. Principal iron-ore deposits of the United States. No iron ore is being mined at sites 5, 6, 8, 9 and 11.

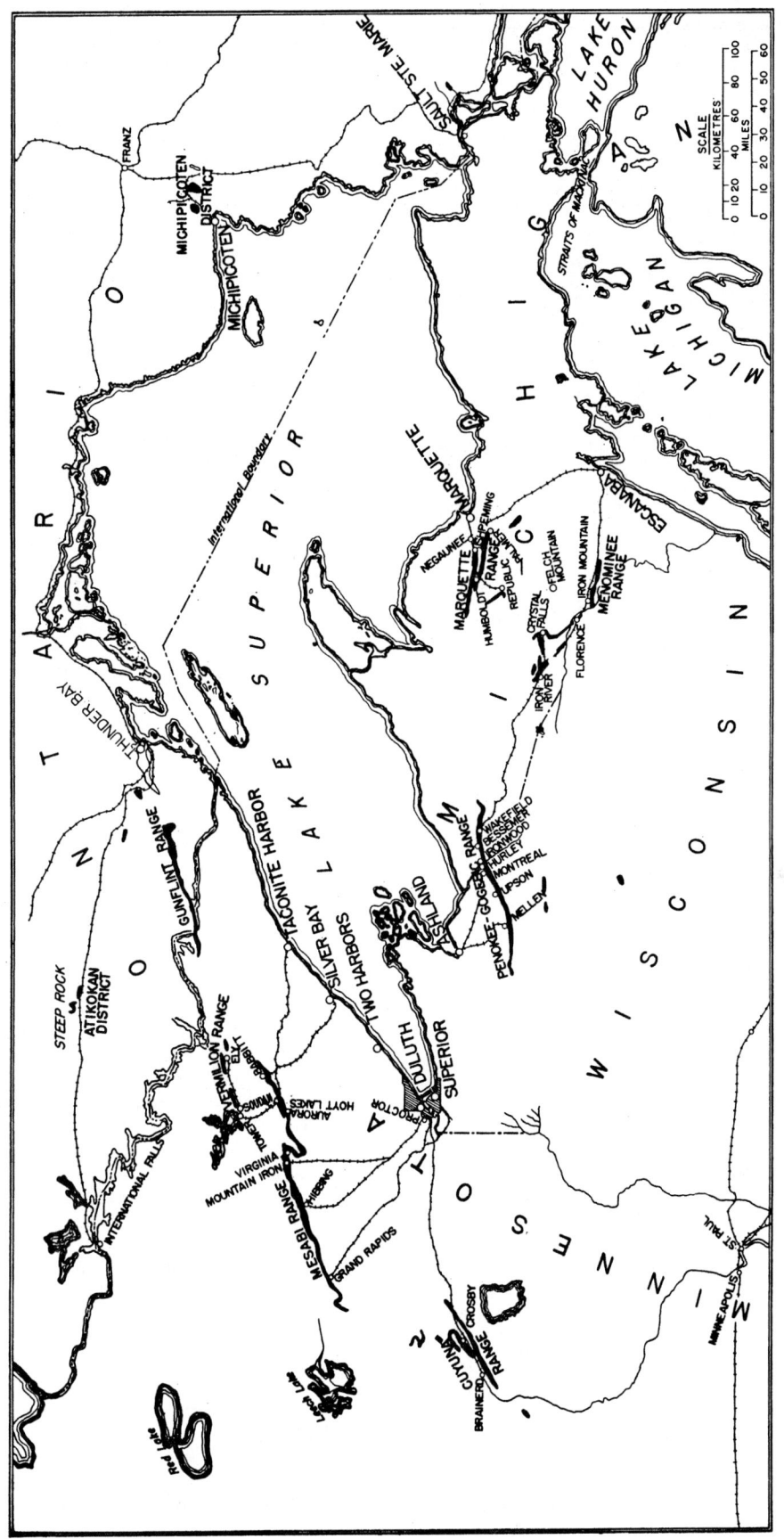

Fig. 5—2. Map of the Lake Superior District, showing location of individual iron-ore ranges in the United States and Canada, and the principal lake ports involved in the shipment of iron ore to lower lake ports.

Table 5—III. Identified World Iron-Ore Resources (million of short tons of contained iron)

	Reserves[1]	Other	Total
North America:			
Canada	12,000	17,000	29,000
United States	4,000	15,600	19,600
Other[2]	400	200	600
Total	16,400	32,800	49,200
South America:			
Brazil	18,000	11,000	29,000
Venezuela	[3]1,400	2,500	3,900
Other	1,400	16,700	18,100
Total	20,800	30,200	51,000
Europe:			
France	1,800	1,800	3,600
Sweden	2,200	800	3,000
U.S.S.R.	31,000	26,000	57,000
Other	3,600	1,400	5,000
Total	38,600	30,000	68,600
Africa:			
Liberia	700	100	800
South Africa, Republic of	1,200	1,800	3,000
Other	1,700	2,500	4,200
Total	3,600	4,400	8,000
Asia[4]			
China, People's Republic of	3,000	4,100	7,100
India	6,200	2,500	8,700
Other	2,000	1,000	3,000
Total	11,200	7,600	18,800
Oceania:			
Australia	11,800	[5]8,200	20,000
Other	200	800	1,000
Total	12,000	9,000	21,000
World Total[6]	103,000	114,000	217,000

[1]Recoverable iron.
[2]Mexico, Central America, and Puerto Rico.
[3]As of Dec. 31, 1977.
[4]Including Middle East and Far East.
[5]Minimum estimate.
[6]Rounded to nearest million.

ozoic Eons with subdivisions of Early, Middle and Late within each (Table 5—IV). These divisions are recognized both by geological criteria and by radiometric dating. Each unit has a stratigraphic succession which is worked out in more or less detail for each of the range areas. Rocks of the Archean are dated as older than 2.5 billion years, and the Proterozoic rocks range in age from 2.5 to 0.6 billion years. Rock formations in the Lake Superior District display a wide variety of lithologic types including metamorphic sedimentary rocks, quartzites, slates, dolomites and iron formations associated with volcanic extrusive and pyroclastic rocks, many of which are grouped under the general term of greenstones. Intrusive rocks in the region include granites and basic intrusives of various types.

Table 5—IV is a general summary of the geochronology and succession of the principal Pre-Cambrian stratigraphic units and geologic events associated with iron ores in the Lake Superior region. The table incorporates the principal unit names and their relative stratigraphic positions as presently used, however, the Lake Superior region is the focus of continuing geologic studies which in the future will undoubtedly result in changes being made as the results of more geologic investigations and more radiometric age datings become available. Also included in an adaptation of the major stratigraphic and geochronologic divisions as proposed by various workers. In the column summarizing the IUGS subcommission's work, the terms Early, Middle and Late have been substituted for Proterozoic I, II and III as well as for those subdivisions proposed for the Archean.

Iron-ore deposits in the Lake Superior region are related to sedimentary banded iron formations of the Archean and Proterozoic Eons in rocks of the Keewatin system and Animikee group in the United States and in the Keewatin and Timiskamian systems in Canada. The term iron formation is applied to banded sedimentary rocks which are composed of layers rich in iron minerals and layers which are largely silica either in the form of chert or finely granular quartz. The iron formations are variously termed ferruginous cherts, ferruginous carbonates, carbonate cherts, jaspers, ferruginous slates and taconite, depending upon mineralogy and texture. Mineralogically the iron formations consist of iron oxides as hematite, geothite or magnetite, iron carbonate (siderite), and the iron silicates—greenalite, minnesotaite, stilpnomelane, grunerite, fayalite, and iron amphiboles and pyroxenes. The iron formation, as originally deposited, probably consisted of layers of iron oxides or iron carbonates and iron silicates interbedded with layers predominantly chert. Upon compaction and metamorphism the original sediments were transformed into materials which are termed magnetic taconites, ferruginous cherts, jaspers, carbonate chert, etc. The magnetite taconites in Minnesota and the jaspers in Michigan are examples of metamorphosed iron formations which are currently important as sources of iron ore. The term magnetite taconite is applied to iron formations in which the principal iron mineral is magnetite associated with varying but lesser amounts of iron carbonate and iron silicate. The silica bands are either chert or granular quartz. The term jasper is applied to the iron formation in Michigan where the principal iron mineral is a flaky, crystalline hematite, steel-gray in color, associated with the finely granular quartz. The jaspers in Michigan are either red-brown or gray-white in color. In some cases, the metamorphism of the iron formation resulted in the extensive development of iron silicate minerals through the chemical combination of iron and silica, so that only a minor part of the total contained iron is in the form of magnetite. In these cases, the material is not of interest as an iron ore.

Iron ores occur within the iron formation either as soft, porous ores which formed by the oxidation and leaching of the iron formation in structurally favorable zones, or as hard ores formed by movement of iron minerals by hot solutions within the iron formation whereby iron oxides (hematite and magnetite) have replaced the silica in favorable areas to form ore bodies (see Figures 5—3, 5—4 and 5—5).

The Pre-Cambrian iron formations in Minnesota were exposed to erosion during the Cretaceous period which resulted in the local development of a conglomeratic ore along the Mesabi Range in which the pebbles were derived from the iron formation or ores and are cemented by secondary limonite.

Ores which may be of Cretaceous age occur in southeastern Minnesota in the Spring Valley District where a hard, cellular limonite occurs in small deposits on the weathered surface of Paleozoic limestones associated with Cretaceous sands and gravels. These ore materials are believed to have originated in a manner similar to bog ores. The ore bodies are from a few thousand tons to a few hundred thousand metric tons in size with the average deposit about 40 000 metric tons. Ores were mined by open-pit methods, and crushed and washed to remove associated clay. They are now unimportant commercially.

The Lake Superior region has been heavily glaciated and a mantle of glacial detritus covers essentially all of the area. The glacial deposits range in thickness from zero to in excess of about 90 metres (300 feet).

Types of Iron Ore—Iron ores of the Lake Superior region vary widely in mineralogy and chemical composition and physical characteristics. The ore types include (1) merchantable furnace ores and lump ores (commonly termed "soft ore" and "hard ore," respectively), (2) beneficiating ores (material which can be upgraded easily by gravity methods), (3) magnetite taconite which is concentrated by magnetic methods, (4) hematite-bearing jaspers which can be concentrated by flotation techniques and (5) potentially concentratable oxidized iron formations (ferruginous chert and semitaconite). The term "merchantable ore" is used to designate ore which is shipped as mined, without beneficiation. Until approximately 1950, iron ores of the Lake Superior region were restricted to the merchantable furnace ores, lump-grade ores and beneficiating ores. Since 1956, magnetite taconite concentrates have contributed substantially to the production of iron ores from Minnesota and Michigan, and, since 1974, jaspers in Michigan have contributed important tonnages to the production of non-magnetic iron-ore concentrates.

The soft, porous hematitic and limonitic natural ores, including merchantable ores and beneficiating ores, were derived by the leaching of silica from the iron formation by solutions moving through the rock. In general, the ore deposits are related to channelways which would tend to localize the flow of downward-moving ground waters, such as trough-shaped structures, either folds, dike and slate intersections or fault troughs, or fault and fracture zones. On the Mesabi Range, oxidation and leaching have extended to a maximum depth of around 245 metres (about 800 feet) although the ore bodies are commonly less than about 150 metres (around 500 feet) in depth; on the Gogebic Range, iron-ore bodies have been found to depths in excess of 1525 metres (about 5000 feet); and on the Marquette Range to depths over 915 metres (about 3000 feet).

The original iron formation has an average iron content of about 30 per cent and an average silica content of about 50 per cent. Where the leaching of silica has sufficiently advanced so that the remaining silica is reduced to less than 10 per cent, merchantable ores result. Original iron-formation horizons which contained appreciable alumina commonly form "painty ore" or "paint rock" upon oxidation and leaching. These materials are comparatively high in alumina and moisture, low in natural iron and have a poor structure which is fine-grained and sticky. The "paint rock" and "painty ores" are generally not marketable.

Beneficiating ores occur in areas where the leaching has been only partly complete thus resulting in a crude ore material that can be beneficiated by a combination of crushing, screening and gravity separation methods (such as the use of jigs, spirals, cyclones and heavy density media) to yield an iron ore.

Lake Superior hard ores were produced at the Pioneer Mine of United States Steel Corporation, located at Ely, Minnesota; the Cliffs Shaft of the Cleveland-Cliffs Iron Company, at Ishpeming, Michigan; and the Champion Mine of North Range Mining Company at Champion, Michigan, on the Marquette Range in 1966, but these mines have since been closed. These ores are compact, fine-grained, gray hematite with some magnetite at the Cliffs Shaft. The lump grade ores contain from 60.3 to 62.7 per cent iron, 4.2 to 7.6 per cent silica, and 0.079 to 0.20 per cent phosphorus—all on a dry basis. The lump ores occur within the banded siliceous iron formation. They appear to have been formed by the movement of iron by hot solutions to favorable areas where silica in the iron formation is replaced by iron oxides.

Magnetite taconite is the rock name applied to bedded sedimentary iron formations in which the principal iron mineral is magnetite. It is a hard, dense, compact, fine-grained rock, commonly containing from 40 to 55 per cent silica and 15 to 35 per cent iron in the form of magnetite. Magnetite taconite ores are concentrated by magnetic methods after fine grinding.

Jasper ore is the term applied to iron formations which are composed of steel-gray, crystalline or specular hematite, and finely granular quartz, with minor iron silicates. Jaspers are concentrated by froth flotation methods.

Oxidized iron formations represent potential iron ores in the Lake Superior Region. They are commonly composed of fine grained to earthy hematite, goethite, and chert, and range from a rather friable material to a hard, compact rock. These materials are termed oxidized taconite, nonmagnetic taconite or ferruginous cherts or, when they are somewhat friable, "semitaconite".

Ore Grades—All merchantable ores shipped from the Lake Superior District were classed either as furnace ores (for use in the blast furnace) or open-hearth ores. The furnace ores are commonly blended products from more than one mine, and sometimes represent a blend of natural ores, coarse or fine, and beneficiated products. The natural ores are shipped as either standard unscreened ore, or in their coarse and fine screened fractions.

Furnace ores are commonly shipped as specific grades on a basis of chemical composition. Each ore producer establishes grades of guaranteed composition

Table 5—IV. Generalized Geochronology and Stratigraphic Succession of the Lake Superior Region (see also next page).*

Eon	Period/System	Orogeny	Cuyuna Range	Mesabi Range	Vermilion Range	Gunflint Range	Gogebic Range
PROTEROZOIC	KEWEENAWAN	Grenville		Duluth Gabbro	Duluth Gabbro	Duluth Gabbro	Gabbros & Granites
			Volcanics	North Shore Volcanics Puckunge Sandstone	North Shore Volcanics		Sandstone Shales and Conglomerates Volcanics Sandstone
		Penokean	Gneiss (reactivated)				Granite
			(No equivalent rocks)	(No equivalent rocks)	(No equivalent rocks)	(No equivalent rocks)	(No equivalent rocks)
		Animikie Group (Minnesota)	Rabbit Lake Formation Volcanics	Virginia Formation		Rove	Tyler Slate
			Trommald Iron Formation Mahnomen Slates and Quartzites	Biwabic Iron Formation Pokegama Quartzite		Gunflint Iron Formation Kakabeka Quartzite	Ironwood Iron Formation Palms Quartzite
			Trout Lake Dolomite				Bad River Dolomite Sunday Quartzite
ARCHEAN	TIMISKAMIAN	Algoman	Granite	Granite	Granite	Granite	Granite
		Laurentian		Knife Lake Slates	Newton Lake Fm. Knife Lake Slates		
	KEEWATIN		Greenstone	Greenstone	Upper Ely Greenstone Soudan Iron Fm. Lower Ely Greenstone Gneiss		Greenstone

*Adapted from Marsden (1968), Schmidt (1973) Sims (1976), Anderson (1970), Dutton (1970), Morey (1973), Cannon and Gair (1970).

Table 5—IV (Continued).

Marquette Range	Menominee Range Iron and Dickinson Counties	Leith, Lund & Leith (1935)	USGS (James, 1971)	CANADA (Stockwell, 1972)	Subcommission IUGS (Sims, 1979)
		Post Keweenawan	PHANEROZOIC	PHANEROZOIC	PHANEROZOIC
	Diabase Dikes	Killarney Granite	800*	Hadrynian	LATE
				955* (Grenvillian)	900*
		KEWEENAWAN	PRECAMBRIAN (Y)	Helikian	MIDDLE
Granite & Basic Intrusives	(Metamorphism)		1600*	1735* (Hudsonian)	1600*
Marquette Range Supergroup — Michigamme Formation: (No equivalent rocks) Upper Slate mbr. Bijiki Iron fm. mbr. Lower Slate mbr. Clarksburg Volcanics Greenwood Iron fm. mbr. Goodrich Quartzite Negaunee Iron Formation Siamo Slate Ajibik Quartzite Wewe Slate Kona Dolomite Mesnard Quartzite Enchantment Lake Fm	Marquette Range Supergroup — Paint River Group: Fortune Lake Slate Stambaugh Formation Hiawatha Graywacke Riverton Iron Fm Dunn Creek Slate; Baraga Group: Badwater Greenstone Michigamme Slate Fence River Formation Hemlock Formation Goodrich Quartzite; Menominee Group: Vulcan Iron Formation Felch Formation; Chocolay Group: Randville Dolomite Sturgeon Quartzite Fern Creek Formation	ALGONKIAN TYPE — HURONIAN	PROTEROZOIC — PRECAMBRIAN (X)	PROTEROZOIC — Aphebian	PROTEROZOIC — EARLY
Granite	Granite	Algoman gr.	2500*	2480* (Kenoran)	2500*
	Seven Mile Amphibolite Solberg Schist East Branch Arkose	KNIFE LAKE	PRECAMBRIAN (W)	ARCHEAN	LATE
		Laurentian			2900*
		ARCHEAN TYPE — KEEWATIN		ARCHEAN	ARCHEAN — MIDDLE
					3500*

*Radiometric dates in millions of years

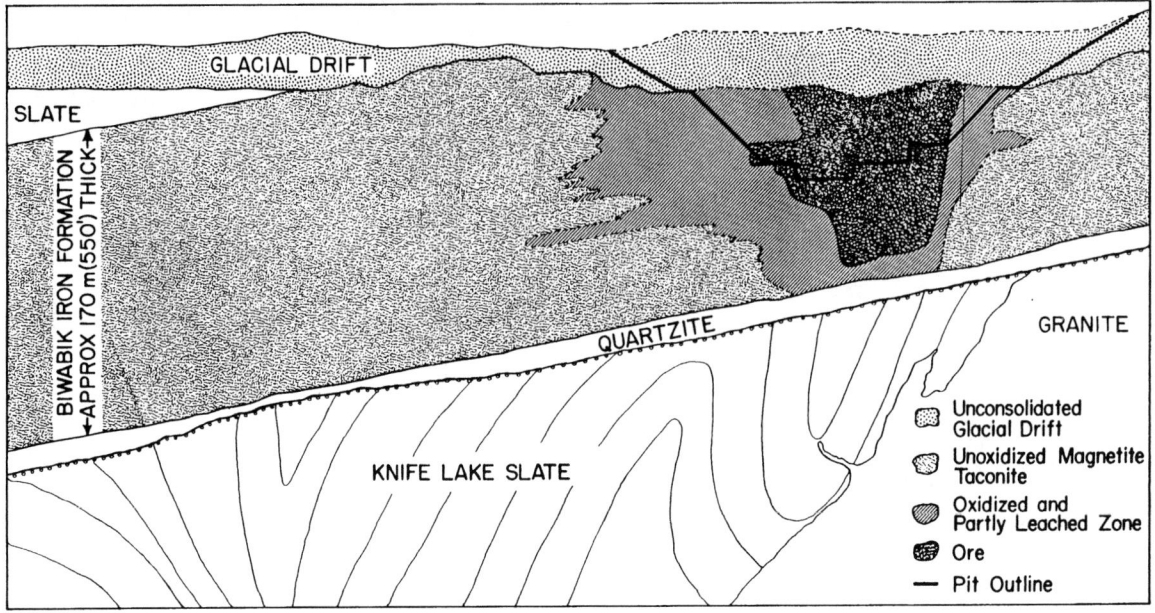

FIG. 5—3. Typical cross-section of an ore body on the Mesabi Range.

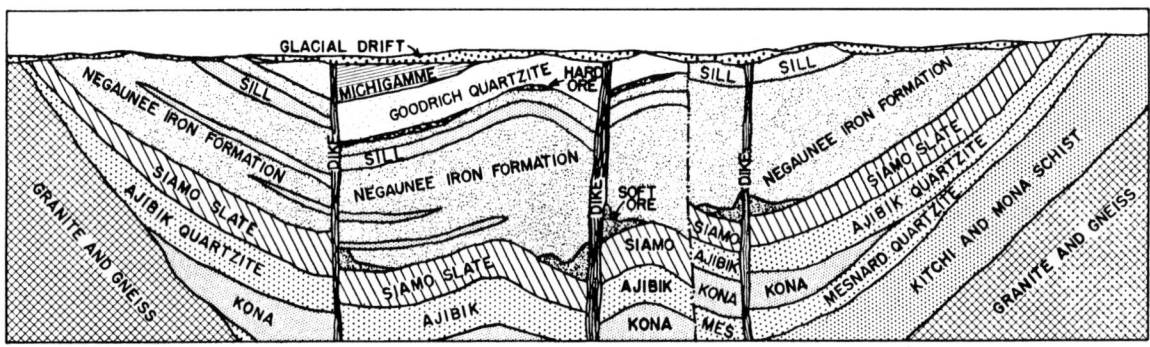

FIG. 5—4. Diagrammatic cross-section of the Ishpeming area of the Marquette Range in Michigan, looking west.

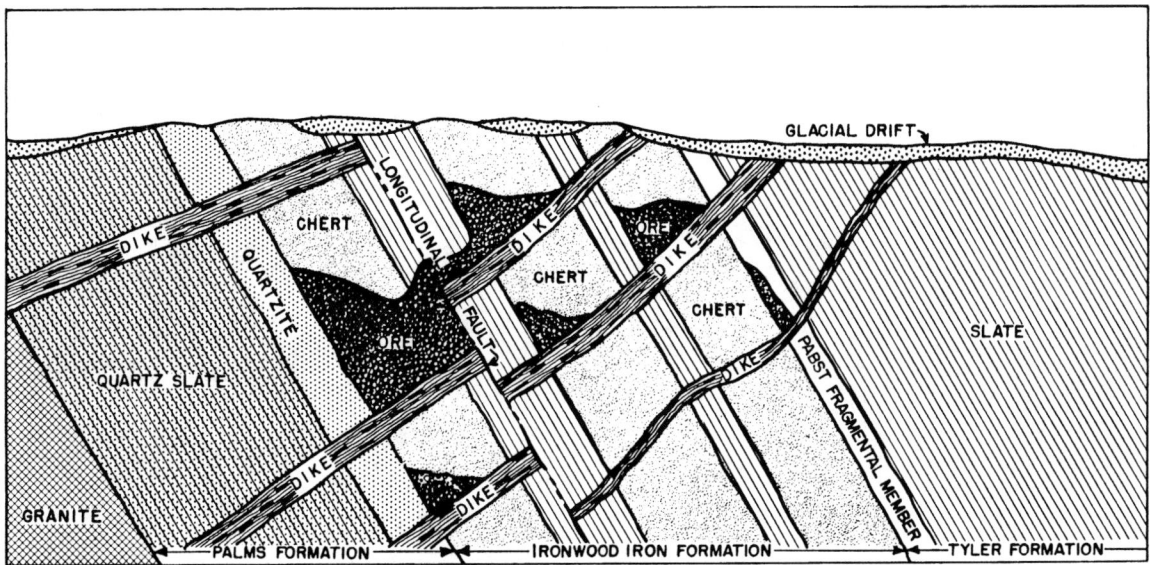

FIG. 5—5. Typical cross-section of the geologic structure of the Penokee-Gogebic Range, showing how ore deposits are located in trough-like structures formed by intersection of dikes with the iron formation.

that are referred to by grade names, such as Tracy, Pearson, Beaver, and Oliver Coarse. Historically, the furnace ores have been classed as Bessemer (phosphorus not over 0.045 per cent). Low-Phosphorus Non-Bessemer (phosphorus over 0.045 per cent but less than 0.180 per cent), High-Phosphorus Non-Bessemer (phosphorus over 0.180 per cent), Manganiferous (manganese over 2 per cent) and Siliceous (silica 18 per cent or over). The distinction between Bessemer and Non-Bessemer grades is unimportant today, as no ore is sold for making iron (hot metal) for use in the Bessemer process.

Open-hearth ores include lump ores and agglomerates in the form of pellets and nodules with suitable chemical and physical characteristics.

The Mesabi Range—The Mesabi Range, located about 105 kilometres (65 miles) north of Duluth in Minnesota, consists of a belt of iron formation about 193 kilometres (120 miles) long extending from the town of Grand Rapids near the west end to the town of Babbitt on the east end. Iron-bearing rocks of the iron formation have been known in this area since about 1866, but commercial iron ore was not discovered until 1890. The first shipment of ore was made in 1892.

The iron-ore deposits of the Mesabi Range occur in the Animikie Biwabik iron formation which is overlain by a thick slate series and underlain by quartzite, as shown on the diagrammatic section, Figure 5—3. The iron formation ranges from about 120 to 230 metres (approximately 400 to 750 feet) in thickness and is commonly called "taconite". Where it is magnetic, it is termed "magnetite taconite", and where it has been weathered and the magnetite, iron carbonate and iron silicates are largely oxidized to limonite and hematite, it is commonly termed "oxidized iron formation" or "oxidized taconite". The structure of the rocks on the Mesabi Range is comparatively simple. Rocks generally dip gently to the south, commonly at an angle of less than 10 degrees. In local areas the iron formation dips more steeply or is cut by faults.

The iron-ore materials of the Mesabi Range include: (1) merchantable furnace ores, (2) beneficiating ores, (3) magnetite taconites, and (4) as a potential ore, favorable portions of the oxidized iron formation. From 1892 through 1980 there were a total of 3 335 124 000 metric tons of iron ores and concentrates shipped from the Mesabi Range. Of that total, 51.1 per cent was direct shipping or merchantable ore, 22.3 per cent was beneficiation ore and 20.6 per cent was taconite (magnetic) concentrates. In 1980, of the total shipments of 45 852 916 metric tons, less than 0.1 per cent was merchantable furnace ore, 5.7 per cent was beneficiating ore and 94.6 per cent was taconite concentrates.

The merchantable, natural ores of the Mesabi Range are generally soft, porous, blue, red, yellow or brown containing various mixtures of hematite and goethite. These ores were largely produced from the portion of the Mesabi Range between Hibbing and Aurora. The natural ores were crushed, screened and graded to give a variety of structure and chemical compositions to meet customer requirements. Most of the natural ores still being mined are beneficiated by washing or gravity methods.

Magnetite taconites themselves can be considered as

Table 5—V. Compositions of Pellets Produced from Concentrates Originating with Magnetic Taconites (1980)*

	Fe	P	SiO₂	Mn	Al₂O₃	CaO	MgO	S
Minntac Pellets	63.29	.010	5.84	.16	—	—	—	—
Reserve Pellets	61.71	.040	6.90	.20	.45	.39	.54	.002
Erie Pellets	61.71	.015	6.00	.20	.37	.39	.54	.004
Eveleth Pellets	64.43	.014	5.40	.09	.40	.31	.38	.002
Hibbing Taconite Pellets	64.69	.003	4.80	.07	.16	.18	.23	.001
Itasca Pellets (Butler)	64.44	.012	4.95	.08	.18	.16	.32	.001
Minorca Pellets	64.77	.013	4.62	.09	.15	.13	.21	.004
National Pellets	64.20	.009	5.02	.08	.17	.27	.42	.002

*Dry Basis

ore when they contain sufficient magnetite in grain sizes which allows their commercial concentration by magnetic methods following fine grinding. The fine-sized magnetic concentrates must be agglomerated to produce a suitable blast-furnace feed (see "Agglomeration Processs" in Section 4 of this chapter for a description of pelletizing, sintering and nodulizing).

There are two types of magnetite taconite: (1) magnetite associated with minnesotaite, stilpnomelane and cherty quartz with minor greenalite and carbonates, and (2) magnetite associated with iron amphiboles (grunerite, cummingtonite, actinolite), pyroxenes, garnet, fayalite and finely granular quartz. The first type occurs westward from near Aurora, Minnesota; the second type from Aurora eastward, in an area where the iron formation has been metamorphosed by the intrusion, on the south, of the Duluth Gabbro.

Magnetite taconites are concentrated and agglomerated at several locations in Minnesota. The taconite pellets produced at some of these locations in 1980 had the compositions shown in Table 5—V.

Research and pilot-plant work is in progress in an effort to develop a commercial process for the concentration of another type of low-grade ore, the somewhat coarser-grained portions of the oxidized Biwabik iron formation. This is termed "oxidized cherty iron formation" or "oxidized cherty taconite", "semi-taconite" when it is somewhat friable and does not respond to gravity concentration. Work is being directed toward the conversion of hematite and limonite to magnetite, which can then be concentrated by the usual magnetic methods to yield a high-grade iron concentrate. Flotation methods of concentration are also being investigated.

Vermilion Range—The 35-kilometre (22-mile) long Vermilion Range is situated in northeastern Minnesota about 120 kilometres (75 miles) north of Duluth, 90 kilometres (55 miles) from Lake Superior and 24 kilometres (15 miles) north of and parallel to the eastern end of the Mesabi Range. The inactive mining districts of Tower and Soudan are located on the western end of the range and Ely is on the eastern end. Iron ore was mined from 1884 through 1966. Mining was by underground methods. The Soudan mine on the west produced a hard, compact open-hearth lump ore and several mines near Ely produced both furnace ore and open-hearth lump ore.

The iron-bearing materials of the Vermilion Range are enclosed in volcanic rocks of middle Archean (Keewatin) Age and are intruded by basic and felsitic dikes whereas, by contrast, the iron-making materials of the Mesabi Range 25 kilometres to the south are of middle Pre-Cambrian (Animikie) Age.

Rocks associated with the iron ores above the 460-metre level, and to a lesser extent below, are strongly oxidized and leached. The volcanic rocks, dikes and slaty iron formation become a fine-grained, soft, red-brown, clay-like material termed "paint rock". The ore is composed of angular fragments of hard, steel-gray hematite which in the upper part of the deposit is in a soft, red, hematite and limonite matrix but, in depth, becomes secondary hard, gray hematite. Thus, in some areas, the ore is hard, compact hematite. Locally, there was secondary pyrite in the Ely deposit requiring care in mining to keep from obtaining objectionable amounts of sulphur. In fact, some material was unusable because of its high pyrite content.

Iron ore at one of the mines near Ely was washed and screened to give three shipping products; lump ore, and coarse and fine blast-furnace ores. The lump ores contained about 62.7 per cent of iron with 4.25 to 5.7 per cent of silica, and the furnace ores contained from 59 to 60 per cent of iron and 5 to 7 per cent of silica, all on a dry basis.

The Cuyuna Range—This range is located in Minnesota, 160 kilometres (100 miles) southwest of Duluth, and has a maximum length of about 110 kilometres (68 miles). Mining activity has been largely confined to an area about 16 kilometres (10 miles) long and 4.8 kilometres (3 miles) wide near Crosby and Ironton. Iron ores are found in an iron formation which is overlain by slate and underlain by a series of quartzites and slates. The iron formation, which consists of interbedded iron-rich layers and cherty layers, ranges from about 14 to 152 metres (45 to 500 feet) in thickness, and is probably the equivalent of the iron formation on the Mesabi Range to the north. The iron formation of the Cuyuna Range differs from that of other ranges in the Lake Superior District because of its higher manganese content. The original unoxidized iron formation contains from 18 to 35 per cent iron and from about 1 to 8 per cent manganese. The Cuyuna iron ores and manganiferous iron ores were formed by the leaching of silica and the development of iron and manganese oxides as porous red-brown to red and black ore. The ores appear to be related to the present erosion surface and rarely are over 152 metres (500 feet) deep.

A small amount of merchantable and manganiferous iron ores are produced from the Cuyuna Range, with about 60 per cent of the material currently shipped requiring beneficiation by gravity concentration to meet grade requirements.

The ores produced from the Cuyuna Range contain from 37 to 44 per cent of iron, from 4½ to 13½ per cent of manganese, and from 11 to 29 per cent of silica on a dry basis. In 1980, shipments were about 106 000 tons of washed ore and heavy media concentrates. The Cuyuna ores, although of marginal character, are of interest because of their relatively high manganese content.

Marquette Range—The east trending Marquette Range is located in the northern part of the Upper Peninsula of Michigan with its eastern end 16 kilometres (10 miles) west of the Lake Superior port of Marquette. The range is approximately 49 kilometres (30 miles) long and 9.5 kilometres (6 miles) wide and includes the towns of Negaunee, Ishpeming, Palmer, Humboldt, Republic and Michigamme (see Figure 5—2). Mining of natural ores was centered in the vicinity of Negaunee, whereas the mining of concentrating materials is in the vicinities of Republic, Ishpeming, Palmer and Tilden. The first iron ore discovered in the Lake Superior District was found near Negaunee in 1844. The major portion of the direct-shipping ores were mined by underground methods whereas the lower-grade taconite ores are mined by open pit. The ores of the Marquette Range included both hard, lump, open-hearth, and soft furnace ores and taconites. Natural ores are no longer produced and all current production is from magnetic and non-magnetic taconites. The Cleveland-Cliffs' Mather Mine, the last underground mine on the range, was closed permanently in 1979.

The Marquette Range is a large structurally complex synclinal basin comprised of metasedimentary and metavolcanic rocks which are collectively referred to as the Marquette Range Supergroup. These rocks have been tightly folded into a westward plunging syncline which opens toward the west. The Negaunee Iron Formation, the unit of major economic importance, attains thicknesses of 760 metres (2500 feet) on the eastern end of the syncline near the city of Negaunee, but thins rapidly toward the west. The entire sequence has been faulted and cut by younger intrusives as shown in Figure 5—4. These rocks of the Marquette Range Supergroup are bracketed by the radiometric dates of 1.9 to 2.5 billion years and are usually correlated with the Animikie series of the Mesabi Range on the north shore. These rocks are divided into four groups of which two contain iron formations which have been the source of commercial iron ores. The principal source of both merchantable and concentrating ores has been the Negaunee iron formation, but there has been a lesser production from the stratigraphically higher Bijiki iron formation (Table 5—IV).

The open-hearth lump ores were commonly mined from the upper part of the iron formation in areas where the silica of the iron formation had been replaced by iron possibly as a result of the action of hydrothermal solutions. The lump ores were commonly hard, blue hematite with some magnetite and ranged in composition from about 60 to 62.6 per cent iron and 5.7 to 7.6 per cent silica.

The soft, blast-furnace ores commonly occur in trough-like structures formed by folding and faulting. The complicated structure of the Range gives a variety of structural situations favorable for ore development. The soft ores are usually porous, from fine-grained to lumpy, and are blue, red, brown or yellow in color. They range in composition from about 56 to 61 per cent iron and from 6 to 11.7 per cent silica on a dry basis, corresponding to about 50.5 to 54 per cent iron on a natural ore basis. In recent years most of this natural ore was used as pellet plant feed.

The metamorphosed Negaunee iron formation contains both specular hematite and magnetite facies and these are the present source of the iron minerals for the production of concentrates and pellets. Much of this iron formation requires fine grinding (up to 90 per cent minus 0.019 mm or minus 500 mesh) for the liberation of the ore minerals. Concentration is either by selective flotation or magnetic methods or a combination of both. The concentrates are agglomerated into pellets ranging from 63.20 to 63.82 per cent iron and 4.72 to 6.25 per cent silica. In 1980 there were three operating mining and pelletizing operations which produced 13 852 431 metric tons of pellets.

Menominee Range—The Menominee Range is located along the southern boundary of the Upper Peninsula of Michigan in the vicinity of Iron Mountain and in the adjacent part of northern Wisconsin. The major production has been from underground mines.

Iron ores of the Iron River, Crystal Falls and Florence Districts, near towns of these same names, are associated with the Riverton iron formation of the Paint River Group; whereas the ores of the Amasa area, near Amasa, Michigan, Felch Mountain, east of Randville, Michigan, and the Menominee area, extending from Iron Mountain to Waucedah, are in the Vulcan iron formation of the Menominee Group (Table 5—IV). Although merchantable ores are no longer produced from these iron formations, important tonnages were produced in the past, particularly from the Menominee area. Concentrates are produced at a mine in the Felch Mountain area, where the iron formation consists of steel-gray hematite and magnetite in a granular quartz. The material is relatively coarse-grained so that liberation of the iron minerals is obtained by grinding to about minus 0.208 millimetre (65 mesh) which allows the coarser fraction to be concentrated magnetically and in Humphrey spirals and the finer fraction by flotation. The concentrates contain about 62.58 per cent iron, 7.25 per cent silica, and 1.13 per cent lime.

The Vulcan iron formation in the Menominee District is well banded with ferruginous layers and cherty layers. The iron formation members commonly contain from 30 to 40 per cent iron as fine, compact grains of hematite associated with finely granular quartz. Research is in progress by public and industrial research groups in an effort to develop a commercial process for the beneficiation of this material by means of a reduction roast to produce artificial magnetite which can be separated magnetically.

Penokee-Gogebic Range—This range is located in northern Wisconsin and in the western end of the Upper Peninsula of Michigan. It is about 130 kilometres (80 miles) in length and strikes in an east-west direction with the towns of Mellen and Hurley, Wisconsin; and Ironwood, Bessemer and Wakefield, Michigan, situated along the Range. The district was discovered in 1882.

The iron ore occurs in the Ironwood iron formation of Animikie age (Table 5—IV). It consists of banded ferruginous cherts and carbonate cherts with some slaty iron formation. The iron formation dips to the north at angles of about 50 to 70 degrees. Iron mining has been confined to an area about 24 kilometres (15 miles) in length, extending from Montreal, Wisconsin, to Wakefield, Michigan, where the ores are, for the most part, soft and rather fine-grained red hematite with some limonite. Locally, there are some hard blue ores and in restricted areas manganiferous iron ore. The iron ores occur in trough-like structures formed by the intersection of basic dikes which cut the iron formation at high angles to the bedding (see Figure 5—5). The trough structures commonly pitch to the east with the ore extending for long distances along the pitch. Iron ore is known to occur to vertical depths of 1 525 metres (about 5 000 feet).

West of the productive section of the range between Montreal and Upson, Wisconsin, although the iron formation is commonly oxidized, there are no significant deposits of merchantable ore. Between Upson and Bad River, near Mellen, Wisconsin, the iron formation largely consists of magnetite taconite similar to that of the east central part of the Mesabi Range, with comparable concentrating characteristics. In this section, which has a length of about 22.5 kilometres (14 miles), the iron formation commonly contains magnetite with minor iron silicates—minnesotaite and stilpnomelane—associated with fine-grained cherty quartz. West of Bad River the iron formation has been more strongly metamorphosed as evidenced by the occurrence of larger amounts of iron silicates as amphiboles and pyroxenes.

The underground mining of iron ore on the Penokee-Gogebic Range ceased with the closing of the Peterson Mine in February, 1966, although small shipments of stockpiled ore were made in 1966 and 1967. The future possibilities of the Range appear to rest largely in the magnetite taconites. The iron formation in the oxidized portion of the Range is very fine-grained and does not respond well to concentration.

A number of relatively small taconite occurrences located south of the Penokee Range in Wisconsin have been investigated by iron-ore producers from time to time. A small mine was developed near Black River Falls, Wisconsin.

NORTHEASTERN UNITED STATES

In the northeastern United States, hundreds of iron deposits of diverse types have been worked at various times since the Colonial period. Operations were gradually restricted to include only the larger and more important deposits, and these recently have been phased out. In New York State, the Benson Mines at Star Lake (Jones & Laughlin Corporation) were closed in 1977 and the Port Henry Mine at Minesville (Republic Steel Corporation) ceased mining in 1971. In Pennsylvania, the Cornwall and the Grace Mine (Bethlehem Cornwall Corporation) at Cornwall and Morgantown respectively, were closed in 1977. The last active mine in New Jersey (the Scrub Oaks) operated by the Alan Wood Steel Company was closed in 1965. A famous old mine (The Chateaugay at Lyon Mountain, New York)

**Table 5—VI. Typical Thickness and Composition of Ferruginous Sandstone Beds
of the Red Mountain Formation**

Seam		Thickness Metres (Feet)	Composition (Per Cent)			
			Iron	Silica+ Alumina	Lime	Phosphorus
Upper Ferruginous Sandstone		10.7-12.2 m (35'-40')	23.2	54.9	4.3	0.08
Ida		0.6-1.8 m (2'-6')	31.0	43.7	4.1	0.16
Big Seam	Upper Bench	1.8-2.9 m (6'-9.5')	36.7	18.1	14.9	0.32
	Lower Bench	1.5-2.7 m (5'-9')	30.6	29.8	13.4	0.26
Irondale		1.5-2.7 m (5'-9')	30.6	21.6	15.7	0.22

was closed on June 30, 1967. The MacIntyre Mine at Tahawus, in the Lake Sanford area of New York (Titanium Division of National Lead Company) is currently operating. Most of the iron ore shipped from the northeastern United States mines was in the form of concentrates, pellets, or sinter (see Section 4). In general, these materials contain over 62 per cent iron.

Iron-ore concentrates are produced from the titaniferous-magnetite deposit at the MacIntyre Mine, at Tahawus, New York. This deposit contains ores associated with basic igneous rocks and consists of large irregular bodies of magnetite and ilmenite. The ilmenite occurs in relatively coarse grains so it can be concentrated by gravity and magnetic methods. This deposit is mined principally as a source of titania with magnetite concentrates obtained as a by-product.

MISSOURI DEPOSITS

The iron-ore deposits of Missouri occur in the southeastern part of the state, near the towns of Bourbon and Iron Mountain, about 80 to 120 kilometres (50 to 75 miles) southwest of St. Louis. Iron ore has been produced from a number of mines in this area since 1815. As of the end of 1980, the sole remaining operation was the underground mine at Pea Ridge in Washington County.

At Pea Ridge the ore occurs in a steeply dipping arcuate-shaped ore body emplaced in Pre-Cambrian porphyries. The ore is composed principally of magnetite with some specular hematite with minor quartz, apatite and pyrite. The hematite is reported to be along the upper portion of the ore lenses and along the footwall. The crude ore grades are 56 per cent iron with 0.4 to 1.0 per cent phosphorus. The ore deposit has been interpreted to have been the result of injection of a magmatic differentiate with associated late stage hydrothermal activity. The crude ore is concentrated magnetically after grinding from 70 to 80 per cent minus 0.043 millimetre (minus 325 mesh) and the product is used to make pellets (64 per cent Fe). Mining is by underground methods, beginning about 520 metres (about 1 700 feet) below the surface. The plant

capacity is reported to be 2.0 million tons of pellets annually.

SOUTHEASTERN UNITED STATES

Hematite Ores—Red hematite ores of Alabama, along with smaller reserves of limonite or brown ores of the southeastern states, constitute the largest known potential source of domestic ore outside the Lake Superior district. Largest of the known red-ore deposits occur in the Birmingham district adjacent to abundant fluxstone deposits in proximity to the principal Alabama coal fields. This favorable grouping of major raw materials was largely responsible for the development of Birmingham as a steel-manufacturing center.

The red hematite ores, termed "red ores" in Alabama, occur in bedded deposits of sedimentary origin, within the Clinton formation of Silurian Age. This formation extends throughout the entire length of the Appalachian Valley from Alabama to Canada, but only in the vicinity of Birmingham were the ore beds of the commercial thickness and quality required for exploitation.

In the Birmingham District, Clinton beds occupy the upper half of the Red Mountain formation which consists of shale and sandstone beds about 76 metres (250 feet) thick.

All stratigraphic seams within the formation are ferruginous to varying degrees, but those which contained sufficient concentration of iron to be of some interest as sources of iron ore were found only in four horizons, as listed in Table 5—VI .

The Ida seam is separated stratigraphically from the overlying Upper Ferruginous Sandstone by 5.5 m (18 ft) of shale and from the underlying Big Seam by 7.5 to 9 metres (25 to 30 feet) of slightly ferruginous sandstone and shale.

In the vicinity of Birmingham, the ore beds which were of commercial interest varied in thickness from 1.8 to 2.9 metres (6 to 9½ feet) throughout a strike length of about 19.3 kilometres (12 miles). The beds at the surface dip 15 to 20 degrees toward the southeast, but they flatten to low angles about a mile down dip

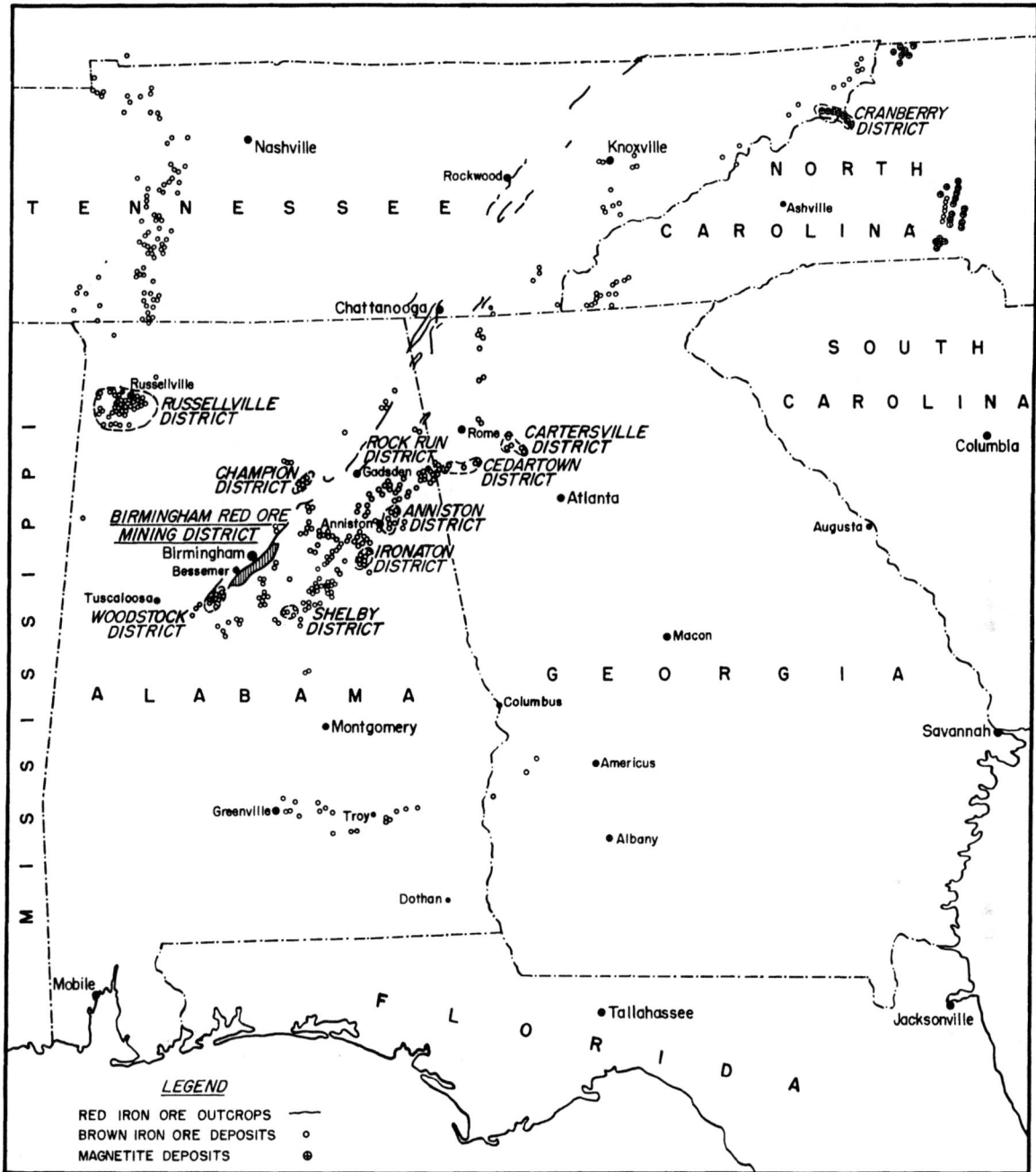

FIG. 5—6. Distribution of iron ores in the southeastern United States.

from the outcrop. Mining was extended nearly 5 kilometres (3 miles) down dip to depths of 550 metres (about 1 800 feet).

Red ore was mined at Birmingham almost continuously from about 1865 to the mine closure in 1961. Between 1900 and the end of World War II, annual production averaged about 4 million metric tons. By selective mining of only the upper bench of Big Seam, an average iron content of 37 per cent was maintained. Large resources of ferruginous material remain; however, they are not economically treatable with current beneficiation techniques.

Extending northeast from Birmingham, the red-ore beds outcrop intermittenly across northeastern Alabama, northeastern Georgia, across eastern Tennessee and into western Virginia. The ores have been mined on a small scale in some localities, but the beds are usually too thin or the ore too low in iron content to permit profitable mining under prevailing economic conditions.

Brown Ores—The occurrence of brown ores, as shown in Figure 5—6, is widespread throughout most of the southeastern states. Historically, these ores contributed about 14 per cent of the total ore consumed in the southeastern furnaces and future resources are inferred to be adequate to maintain this proportion, should the red ores again become economic.

Iron content of the brown ores varies between 40 per cent and 55 per cent, phosphorus between 0.27 per cent and 1 per cent, silica between 2 per cent and 28 per cent. The ore generally occurs as nodules, lumps and boulders in a residual clay matrix. Individual deposits are usually discontinuous and are invariably irregular and erratic in areal extent and thickness. Concentration of ore in the host material varies widely.

Brown-ore deposits were worked by small-scale open-pit or stripping operations. Clay and sand were separated from bank material in log washers, followed by hand picking or heavy-media concentration to upgrade the shipping product.

Since the end of World War II, high-grade imported ores have replaced the red ore in the southern iron and steel industry. These foreign ores contain about 60 per cent iron, are of uniform quality, and lend themselves particularly to the production of quality steel.

Reduced red-ore production from mines of the Birmingham District stimulated interest in development of methods to upgrade the iron content of these local ores. However, conventional beneficiation methods have so far failed to produce a concentrate that is economic.

WESTERN UNITED STATES

Iron-ore deposits in the western states are numerous, widely scattered, and of diverse origin. Although important as a source of supply for the western iron and steel industry, they do not begin to compare in size with the iron ranges in the Lake Superior District or with the "red ore" deposits of the Birmingham District. Most of them require a certain amount of concentration to reduce the quantity of contaminants and gangue minerals sufficiently to produce a satisfactory blast-furnace feed.

Because of the geographical location of the iron-ore deposits of the western states and their dependence on comparatively costly rail transportation, western iron-ore production is limited, for the most part, to the requirements of the local steel plants.

Only a few mines have shipped ores outside of their own captive markets, notably Kaiser Steel Corporation's Eagle Mountain Mine in southern California and several small producers in Nevada, with most such shipments going to Japan in the 1950's and early 1960's, and lesser tonnages to iron and steel plants on the West Coast and in the north central states. Such outside shipments accounted for only a few tens of thousands of tons in 1978.

Table 5—VII lists the five western integrated steel plants and their principal sources of supply for iron ore. These plants have an aggregate capacity of about 7.25 million metric tons (eight million net tons) of iron and steel products. Because of this limited market, only a few of the many western iron-ore deposits are being actively exploited, the principal ones being as follows:

Eagle Mountain, California—The Eagle Mountain

Table 5—VII. Blast Furnaces of the Western United States and Their Principal Sources of Iron Ore Supply

Plant	Location	Source of Iron Ore
Fontana	California	Eagle Mountain Mine, California
Pueblo	Colorado	Iron Springs District, Utah
Geneva	Utah	Iron Springs District, Utah; and Atlantic City Mine, Wyoming
Houston	Texas	Brown Ore from Northeastern Texas, Mexico, and Chile
Daingerfield	Texas	Brown Ore from Northeastern Texas

iron deposits are located some 290 kilometres (180 miles) east of Los Angeles, in Riverside County, California. The area's geologic column comprises a thick series of metasediments resting on Pre-Cambrian gneisses and schists that have been intruded by sill-like masses of porphyritic quartz monzonite. The metasediments are composed of three quartzite units that are separated by a schistose meta-arkose and by two ore zones. They have been folded into a large east-trending anticlinal structure that extends completely across the Eagle Mountains. Host rocks for the ore zones were originally a series of limestones and dolomites with associated quartzites. Magnetite plus pyrite comprises the primary mineralization that forms replacement lenses and stringers. Oxidation of primary ore to hematite and goethite is extensive, especially at higher elevations within the deposits. Associated skarn minerals include actinolite, tremolite, local phlogopite, and minor tourmaline. The ore deposits are in a zone with a strike length of over 10 kilometres (6 miles). Drilling has shown that the ore zone continues eastward in a down-faulted block that is covered by 1 000 metres (1 600 feet) of alluvium and Tertiary lake beds.

In late 1981, Kaiser Steel Corp. announced its intention to close the Eagle Mountain Mines.

The Eagle Mountain Mine obtained its ore from several major open-pits located along the 10-kilometre (6-mile) long ore zone. Mined ore varied widely in size from coarse to fine, and in iron content from 15 per cent to 60 per cent. Mining and milling operations were designed for an annual production of 2.36 million metric tons (2.6 million net tons) of pellets (63 per cent iron) and 1.09 million metric tons (1.2 million net tons) of concentrated ore (56 per cent iron).

Mineral Basin, Nevada—Several small deposits have been mined as open-pits in the Buena Vista area of Mineral Basin, about 32 kilometres (20 miles) southeast of Lovelock. These deposits occur as irregular magnetite bodies in metavolcanic and dioritic rocks of Jurassic age. Three types of occurrences are evident: (1) tabular bodies of high-grade, direct-shipping ore resembling veins in faulted or sheared zones, (2) crackled or brecciated zones exhibiting narrow vein-like fracture fillings, and (3) disseminated magnetite extending outward from and between the high-grade and brecciated zones. Most of the ore tonnage potential is contained in

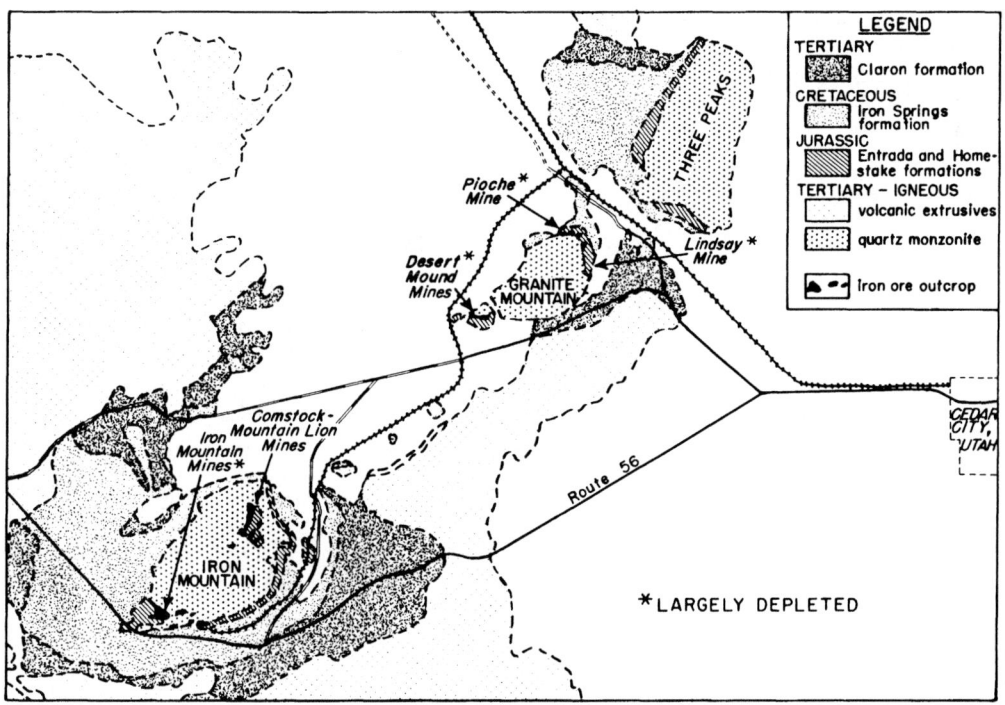

FIG. 5—7. Geologic map of the area of southwestern Utah in which the iron mines discussed in the text are located.

the low-grade disseminated zones. Vanadium occurs in the magnetite in amounts ranging from 0.30 to 0.50 per cent V_2O_5. The host rocks are extensively altered to scapolite; hornblende and chlorite are other common gangue minerals. Narrow, barren, post-mineral dikes of basic composition cut through the mineralized areas. Apatite, occurring as crystals up to 2.5 centimetres (one inch) or more in length, is commonly associated with high-grade magnetite.

Iron Springs District—This district is located in southwestern Utah near Cedar City. The iron deposits are of contact metamorphic origin and, in large part, have replaced a limestone horizon intruded by three laccoliths, now partially eroded and termed "Iron Mountain", "Desert Mound", and "Three Peaks" (see Figure 5—7). These laccoliths, consisting of quartz monzonite, domed the overlying sediments, and except for preintrusive folding and faulting, the quartz monzonite now forms the footwall of the limestone horizon with the regularity of a sill. The principal ore bodies are replacement deposits in the limestone. In places the ore extends from footwall to hanging wall and ranges in composition from 35 to 65 per cent iron. An overlying siltstone, where crackled or broken, often contains veinlets of magnetite. This lean ore averages about 22 per cent iron. In addition, there is a series of small high-grade veins of magnetite, with crystals of apatite, in the quartz monzonite. Figure 5—8 gives an idealized cross section of an Iron Springs ore body.

The gangue minerals in the Iron Springs District are unusual for a contact-metamorphic deposit. The "usual" skarn minerals, garnet, pyroxene and lime silicates, are scarce. The principal gangue minerals are mica, some apatite and locally pyrite with unreplaced

limestone and impure limestone at its base and top.

The largest remaining ore body, the Rex, contains about 90 700 000 metric tons (100 million net tons) of measured and inferred ore. The entire district is credited with a potential of 308 million metric tons (340 million net tons), of which most may require mining by underground methods.

The only active operation is the open pit Mountain Lion mine which produces ore grades of 49 to 55 per cent iron.

Sunrise Mine, Wyoming—The Sunrise Mine is in the Hartville iron district of southeastern Wyoming, 11.3 kilometres (7 miles) north of Guernsey. The ore bodies occur in Pre-Cambrian iron formation associated with dolomite, quartzite and schist. The ore is soft and hard hematite, which grades into lean banded siliceous ore and ferruginous chert, with carbonate pinite schist, pyritic graphitic schist, and dark-colored banded jasper. The ore bodies vary from a few metres to about 30 metres wide (a few feet to about 100 feet wide), and some extend more than 300 metres (1 000 feet) in length. In many respects, these ore bodies resemble the hematite ore occurrences in Archean iron formation at the Soudan Mine, Minnesota.

The Pre-Cambrian rocks in the Hartville district are overlain unconformably by horizontal or gently-dipping Mississippian limestone, which ranges up to 91 metres (300 feet) in thickness. Near the Sunrise Mine this limestone contains detrital masses of reworked hematite and secondary copper ore. Both hematite and copper ore originally cropped out at the surface, and the first mining from 1880 to 1887 was for copper.

Subsequently, the iron was mined by open pit, followed by shafts to facilitate mining by glory-hole meth-

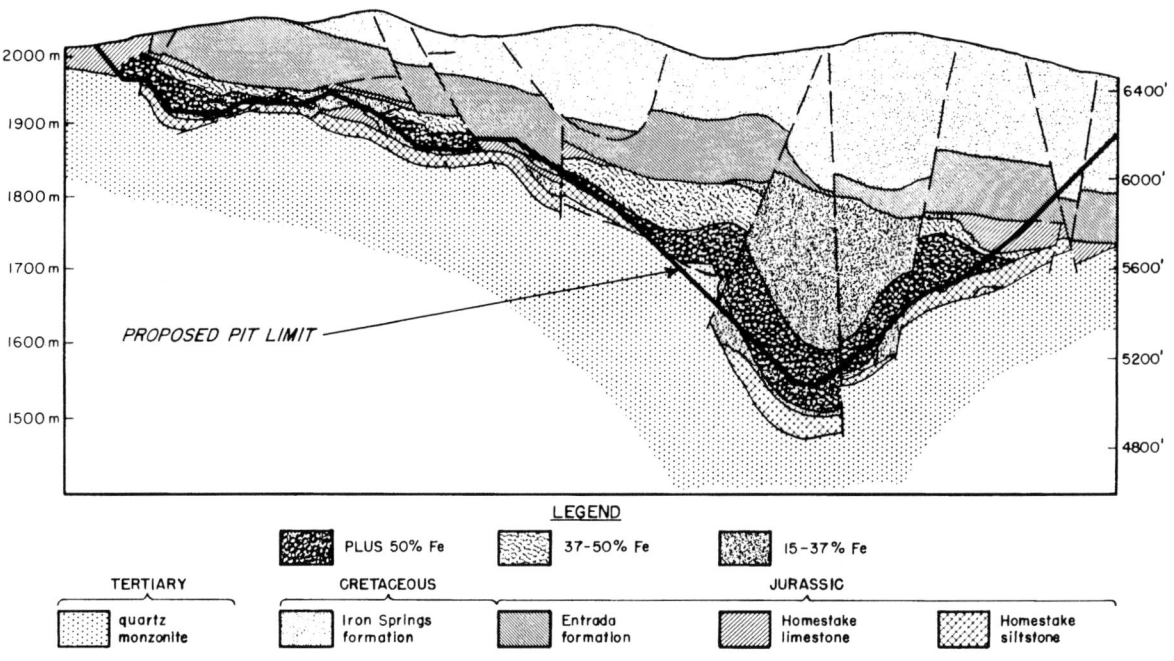

FIG. 5—8. Idealized cross-section of an Iron Springs ore body.

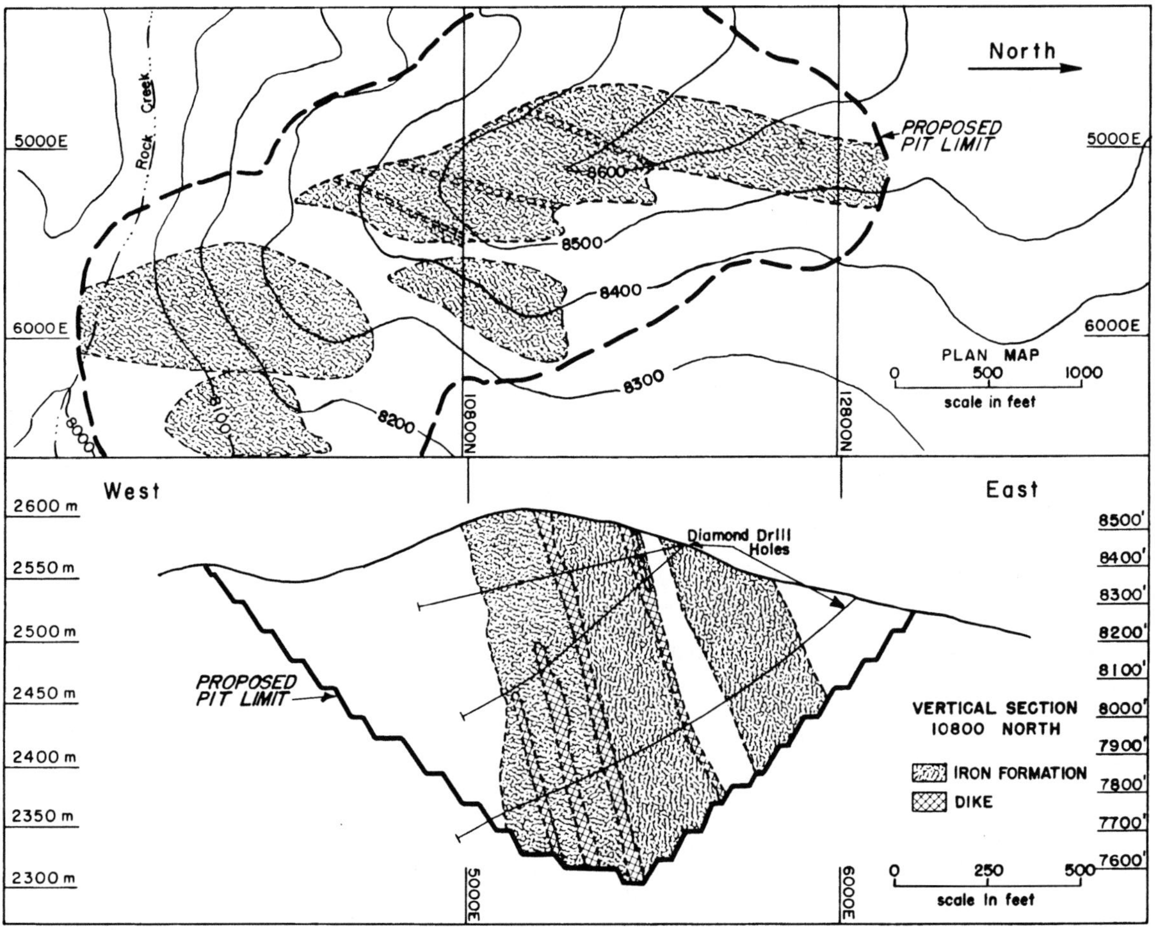

FIG. 5—9. Plan and cross-section of the Atlantic City, Wyoming iron-ore deposit.

ods. With increased depth, the underground methods of sub-level-caving and block-caving were employed. Shipments to Pueblo, Colorado, during 1978 totalled 399 000 metric tons (440 000 net tons) of beneficiated ore averaging approximately 56 per cent iron. The mine has been inactive since mid-1980.

Atlantic City Mine, Wyoming—The Atlantic City iron deposit is located about 42 kilometres (26 miles) south of Lander, near the Continental Divide.

This iron occurrence is a steeply-dipping Pre-Cambrian iron formation, averaging approximately 46 metres (150 feet) in thickness. It is exposed in a north-south direction for more than 3048 metres (10 000 feet). In the principal mining area, complex drag folding has locally thickened the ore to more than 150 metres (500 feet) in width. To the north, it is overlain, unconformably, by Cambrian quartzites, limestones and dolomites. It lenses out to the southward and is interbedded with quartz, mica, hornblende and andalusite schists. Faulting is prominent, with associated metagabbro dikes (see Figure 5—9).

This iron formation averages about 30 per cent iron and is being mined by open-pit methods. The mined material is concentrated magnetically and pelletized before shipment to Provo, Utah. Shipments in 1979 were approximately 1.6 million metric tons (1.8 million net tons) of 62 per cent iron pellets. The original reserves of crude ore were measured at more than 136 000 000 metric tons (150 000 000 net tons).

Texas Brown Ores—These ores are found in many small deposits scattered over 22 counties of northeastern Texas, largely east and southwest from Daingerfield. The Sabine River divides the deposits into what are referred to as the North and South Basins.

These ores are of sedimentary origin and occur near the tops of flat-topped, sand-covered hills characteristic of much of the landscape of eastern Texas. The North Basin deposits occur in green-sand composed of a mixture of a granular iron-silicate mineral termed glauconite, in part altered to limonite and siderite and mixed with varying quantities of quartz, sand and clay. The glauconite sands range up to 30 metres (100 feet) in thickness and average about 7.6 metres (25 feet), and include a weathered, limonite-rich zone averaging 2.7 metres (9 feet) thick. The sands of the South Basin contain abundant marine fossils with much glauconitic clay and much less quartz sand. These sands were deposited in deeper water than those of the North Basin and contain more limonite as laminated and massive ore. These shallow deposits in the aggregate are estimated to contain an indicated and potential reserve of 180 million metric tons (200 million net tons).

The ores range from 12 per cent to 34 per cent of iron. They are concentrated to an average grade of about 42 per cent of iron before shipment.

B. IRON ORES OF CANADA

Canada ranked 4th in world iron-ore production in 1979 with total shipments of 60.2 million metric tons.

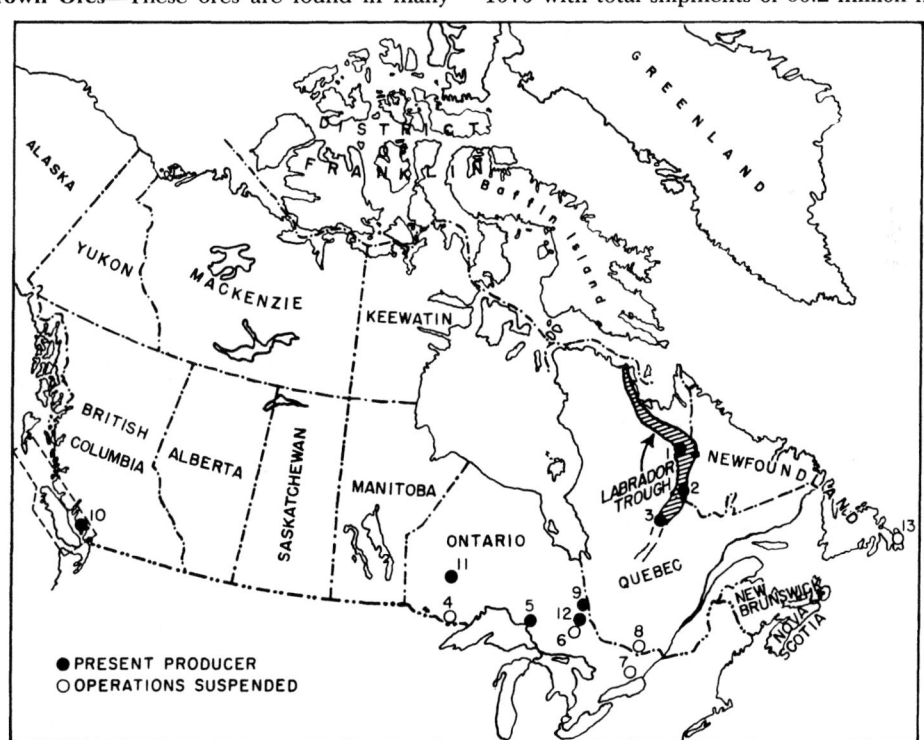

1. Knob Lake	5. Michipicoten	10. British Columbia
2. Wabush-Carol Lake- Mt. Wright-Fire Lake	6. Moose Mountain	11. Bruce Lake
3. Lac Jeannine	7. Marmora	12. Temagami Lake
4. Steep Rock Lake	8. Hilton	13. Wabana
	9. Kirkland Lake	

FIG. 5—10. Map showing location of recent sources of iron ore in Canada.

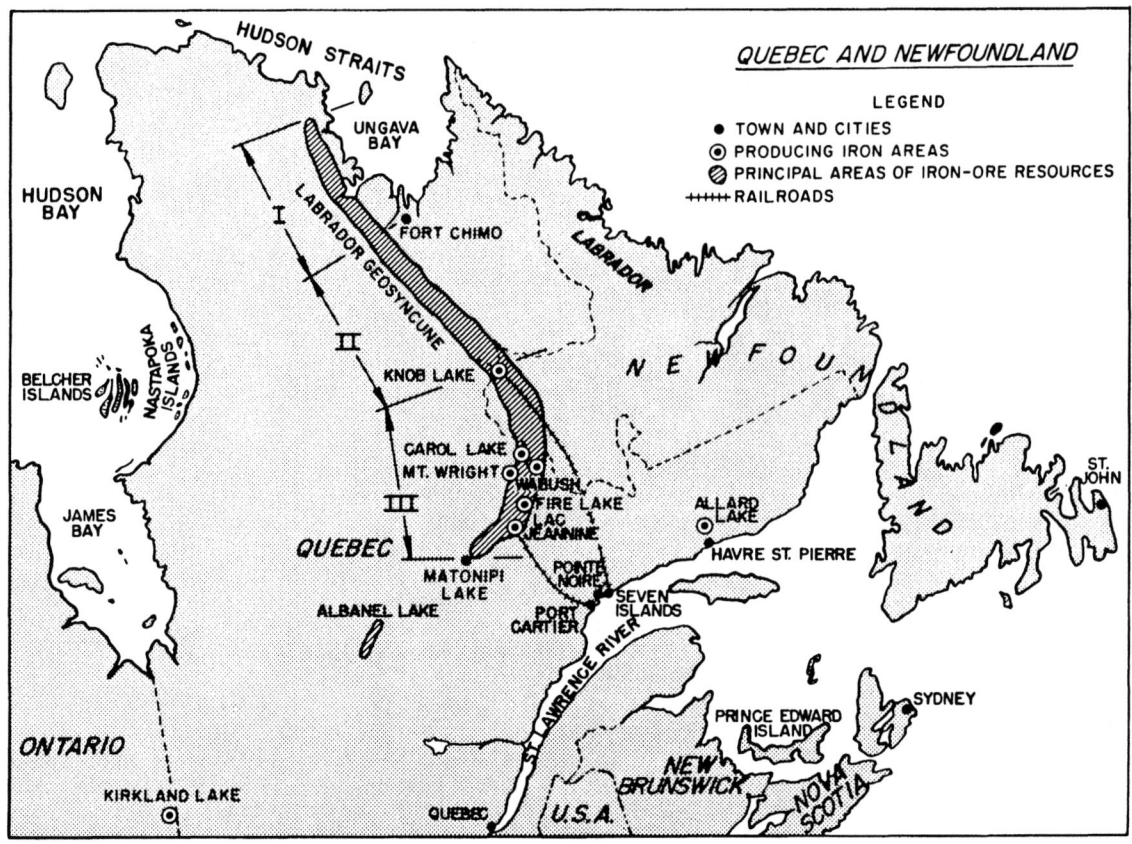

FIG. 5—11. Map showing areas of iron-ore production and principal iron-ore resources in Quebec and Newfoundland.

Of this tonnage, 36.9 million were exported. Iron Ore Company of Canada (hereafter abbreviated as IOC) shipped 3 to 4 million tons per year as direct ore. The remainder of the Canadian ore production was concentrated and some of it was pelletized.

In 1979, the Quebec-Labrador region produced 86 per cent of the total Canadian iron-ore tonnage followed by Ontario with 13 per cent. The remainder was from properties in British Colombia and minor by-product producers (see Figure 5—10). Fifty-one per cent of the ore was shipped as pellets, 40 per cent as concentrates, pellet feed or sinter feed, 6 per cent as direct shipping ore and the remaining 3 per cent as sinter or partially reduced products.

Sixty-one per cent of the ore was exported including 29 per cent (17.6 million metric tons) shipped to the United States of America. Other destinations included: the Netherlands (7.7 per cent), Japan (5 per cent), Italy (3 per cent), West Germany (2.6 per cent) and the United Kingdom (7.8 per cent) plus small lots shipped to other countries.

The Quebec-Labrador area is the leading iron-ore producer in Canada. It contains extremely large reserves of concentrating ore and will continue to be Canada's prime producer for many years. The Labrador geosycline extends from the northwest side of Ungava Bay southward to Knob Lake, Wabush Lake and then southwest to Lake Albanel, a distance of 1600 km. This arcuate shaped structure is divided into three gen-

eral zones (see Figure 5—11): (I) The northern zone extends from northwest Ungava southward to Fort Chimo. The iron formation of this area is characterized as a medium-grained hematite-magnetite quartz rock amenable to beneficiation. It occurs in long synclinal or overturned to recumbent folds with the iron-formation beds repeated across the strike. The potential tonnage in this area is very large but there has been little definitive evaluation of the many known prospects. (II) The central area of the Labrador geosyncline extends southward from the vicinity of Fort Chimo to Knob Lake. Over 47 deposits have been investigated in the central area. Estimates indicate that over 25 billion tons amenable to open-pit mining may exist. The central zone is characterized by Lake Superior type iron formations similar to the Mesabi Iron Range of northern Minnesota where rich hematite-geothite deposits occur within fine to medium grained, bedded, hematite, magnetite chert rock with iron silicates. Some of the ores are direct shipping, others are amenable to known methods of concentration such as magnetic, spiral and flotation. Deposits may contain up to 25 per cent magnetite. The central area contains the Schefferville/Knob Lake mine of IOC which produced 3.3 million metric tons of direct ore and 2.8 million metric tons of pellet feed during 1980. (III) The southern zone extends from north of Wabush Lake to Mt. Wright and Fire Lake and then southwestward to Matonipi Lake. The iron formations of this area lie within the Greenville Belt and are characterized by medium- to coarse-grained specular hematite, magnetite, quartz and minor iron silicates. They are all concentrating ores but due to their coarse grain size are all amenable to gravity concentration at around 0.6 mm or 28 mesh. Known deposits are highly folded.

In 1979, Quebec Cartier produced 15 000 000 metric tons of concentrates from Mt. Wright; IOC at Carol Lake produced 8 331 000 tons of concentrates and 10 794 000 tons of pellets; Wabush Mines at Wabush Lake produced 5 480 000 tons of pellets; and SIDBEC Normines at Fire Lake produced 100 000 tons of concentrates and 4 252 000 tons of pellets.

The Quebec Cartier mine at Lac Jeannine near Gagnon was closed in 1977: however, it was replaced by the Fire Lake mine 64 km (about 40 miles) northeast of Gagnon in a Joint Venture of U.S.S.C. and Sidbec-Normines with a design capacity of 6 million tons per year.

Ontario production of iron ore in 1979 totaled 7 881 000 metric tons from 7 mines and one by-product producer (International Nickel Company). Of the 7 listed Ontario producers during 1979, two mines, Steep Rock at Atikokan and National Steel's Moose Mountain mine at Capreol with a combined production of 1 210-000 tons were closed. Mines closed in recent years include the Hilton and Marmora in Ontario and the Craigmont Mine in British Columbia. Steep Rock has been investigating a low-grade magnetite deposit at Bending Lake, Ontario where they propose mining and concentrating onsite and conveying the concentrates 64 kilometres (40 miles) by pipeline to the existing Steep Rock pellet plant for agglomeration.

Algoma Steel Corporation, Ltd. has mined siderite from various mines on the Michipicoten Range since 1939. The original mining started in 1900 when hematite and limonitic iron ore were produced from the Helen mines. The ores of the Michipicoten are of Keewatin age and occur within a series of steeply dipping flows and pyroclastics. Major regional faults have broken the range into a series of large segments from 2.5 km to 5 km long (1½ to 3 miles). Calcining the sideritic ore at Wawa on this range raises the grade from 35 per cent to 50 per cent iron. In 1979, 1 700 000 metric tons of sinter were shipped.

In 1979 International Nickel (Inco) shipped approximately 500 000 tons of pellets from their iron-ore-recovery plant at Copper Cliff, Ontario near Sudbury.

Canada's iron-ore reserves are substantial. The Quebec-Labrador region alone contains an estimated 40 to 60 billion tons of concentrating ores. Large resources of concentrating magnetite-bearing quartzites have been reported in the Canadian Arctic on the Melville Peninsula.

C. IRON-ORE DEPOSITS OF MEXICO

There are more than 30 iron-ore occurrences of more than one million metric tons each reported in Mexico; however, the principal iron-ore occurrences are located in the Las Truchas, Pena Colorado and El Encino districts along the Pacific coast of Central Mexico and at the La Perla and Hercules localities in Northcentral Mexico (see Figure 5—12). The deposits are mostly massive, irregular and dike-like bodies in igneous and sedimentary rock and as metasomatic replacements in limestones. They are composed of hematite, magnetite, goethite with minor pyrite and chalcopyrite. Reserves are reported to be about 500 million metric tons containing 170 million metric tons of iron. In 1979 ore production was 5 178 000 metric tons of concentrates and pellets, all for domestic consumption.

Mexico has announced plans for an expanded production of steel to utilize its large resources of natural gas. It is anticipated that the domestic iron-ore resources will be exhausted about the year 2000 and Mexico is presently preparing the facilities to import ores from overseas.

D. IRON-ORE DEPOSITS OF SOUTH AMERICA

ARGENTINA

The principal iron deposits of Argentina (Figure 5—12) are located near Zapla, 1400 kilometres northwest of Buenos Aires within the Andean Cordillera and at Sierra Grande, 900 kilometres south of Buenos Aires. At Zapla the ores occur in lenticular beds intercalated with metamorphosed lower Paleozoic sediments and consist of hematite with minor goethite and magnetite. At Sierra Grande the ore occurs in the bedded lenticular deposits up to 12 metres in thickness and with magnetite as the principal ore mineral below the near-surface zones which have been altered to hematite. The structure of the deposits is such that at Zapla, mining is both open-pit and underground whereas at Sierra Grande mining will be by underground methods. Measured and indicated reserves at Zapla are 18 million metric tons of 40 per cent Fe and at Sierra Grande the indicated reserves of magnetic ore are 200 million metric tons with crude ore grades averaging 55 per cent iron, 5.95 per cent silica, 4.85 per cent alumina, 3.27 per cent lime and 1.43 per cent phosphorus.

The mining at Zapla averages about 500 000 tons per year. The mine at Sierra Grande is scheduled to produce 3.5 million tons annually to yield 2.5 million tons of pellets averaging 69.5 per cent Fe and 0.1 per cent phosphorus. The ore is concentrated magnetically at the mine and transported by pipeline 33 kilometres to a pellet plant at the port of Punta Colorada. The first shipment of pellets was made in September 1979. The ore and pellet production is principally for domestic consumption.

BOLIVIA

Bolivia is not an iron-ore producer although it does contain large iron-ore resources which have at various times been proposed for development. The principal deposits (Figure 5—12) are located in the Cerro Mutun near Puerto Juarez in Southeastern Bolivia, at a distance of about 1500 kilometres from the Atlantic ports which could be accessible by rail across Brazil, and at a distance of 2500 kilometers from the estuary of the Rio Platte which could be accessible by the Rio Parana. Small amounts of iron ore of up to 80 000 tons per year have been shipped by barge to Argentine ports.

The ore consists of bedded hematite with intercalated siliceous beds which have been enriched

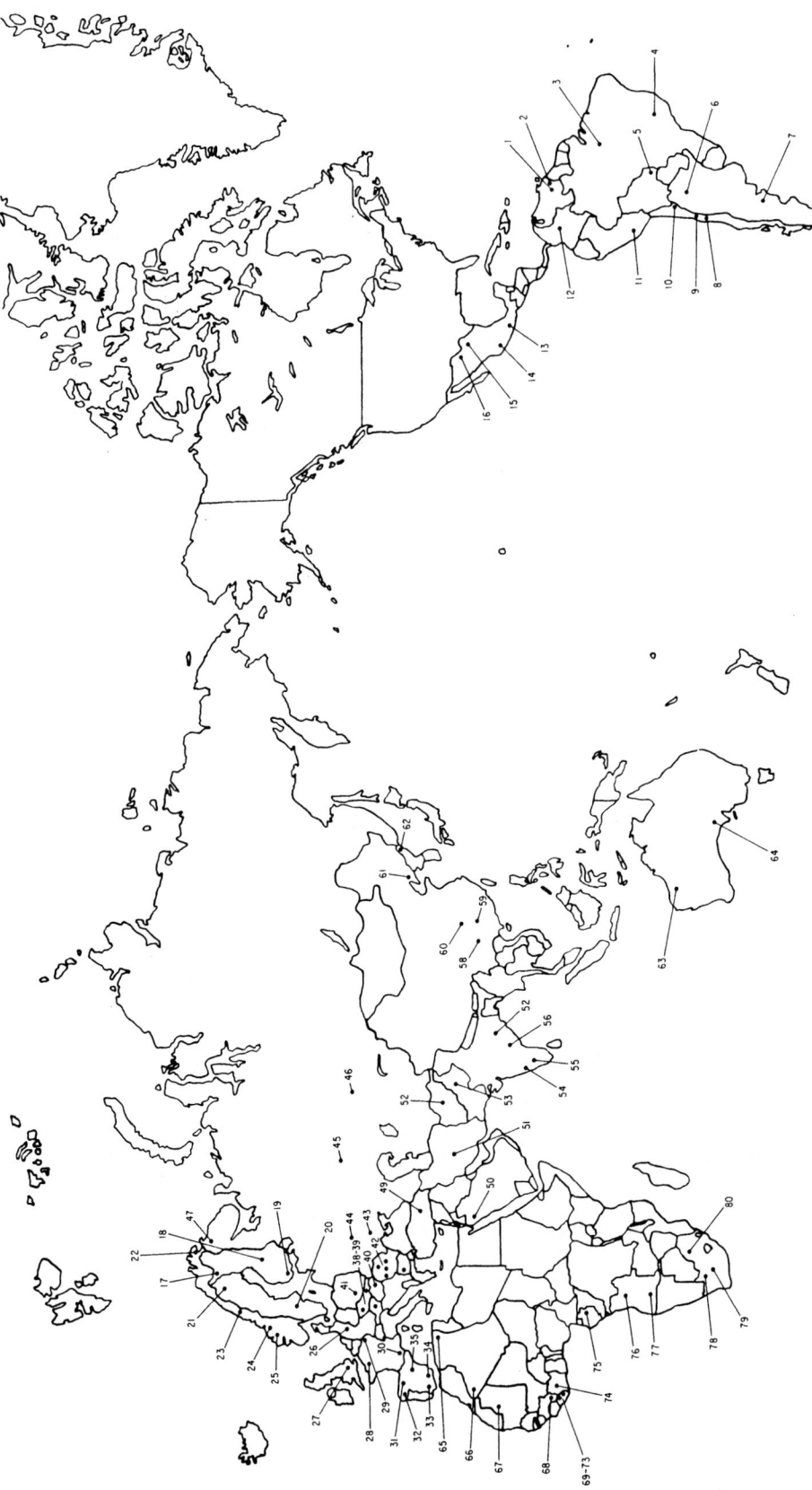

FIG. 5—12. Outline map showing locations of world's principal sources of iron ores. (See Figures 5—1, 5—2, 5—6 and 5—7 for deposits in the United States; Figures 5—10 and 5—11 for Canadian sources and Figure 5—13 for more detailed map of sources in Australia and Tasmania.)

(see next page for site identification.)

residually by weathering and the removal of silica. Ore grades are reported to vary from 45 to 55 per cent Fe, 11 to 29 per cent SiO_2 and 0.03 to 0.17 per cent P. The indicated and inferred resources are reported to range up to 40 000 million tons of residual and detrital ore.

BRAZIL

The largest ore reserves in South America are found in Brazil. Iron-ore deposits are reported at a number of different localities within the crystalline rocks of the Brazilian shield but the largest known concentrations are in the Quadrilatero Ferrifero of Minas Gerais and in the Serra dos Carajas of the state of Para in the Amazon basin (Figure 5—12). The ores occur as dense compact and powdery hematite-magnetite developed largely in situ by the weathering of Pre-Cambrian fine-grained and laminated sedimentary iron formation called itabirite. The itabirites are composed of magnetite, hematite and quartz with minor goethite, martite and gibbsite. The near surface portion of the ore bodies are frequently hydrated and covered with canga caps of variable thicknesses.

The ore reserves of Brazil probably exceed 27 billion metric tons, of which 11 billion tons are located within the Quadrilatero Ferrifero and 16 billion are in the Serra dos Carajas. The Serra dos Carajas ores contain plus 65 per cent Fe and the Quadrilatero Ferrifero ores include over 3 billion tons of plus 66 per cent Fe.

The principal producing iron mines are open-pit operations located in the Quadrilatero Ferrifero. The ore is transported by rail either to the port of Tubarao, near Vitorio, a distance of approximately 598 kilometres or to Rio de Janeiro and Sepetiba Bay, a distance of about 640 kilometres. Also, ore concentrates are transported as a slurry through a 396-kilometre long, 0.5-meter diameter pipeline from the Germano mine in the eastern Quadrilatero Ferrifero to the port of Ponta Ubu. Total mine production for Brazil in 1980 was approximately 78.8 million metric tons, of which about 80 per cent was for export in the form of direct ore, fines, concentrates and pellets. Principal producing mines were in the Quadrilatero Ferrifero at Itabira, Aguas Claras, Germano, Casa de Pedra, Fabrica and the Trindade group. Principal export markets were Western Europe, Japan, Eastern Europe and the United States. The principal domestic consumption was at the national steel plant at Volta Redonda.

The undeveloped Serra dos Carajas iron-ore deposits in Para are planned for development by open-pit mining and the construction of all of the necessary infrastructure including townsites, an 887-km long railroad and a new deepwater port at Sao Luis on the Atlantic.

CHILE

The principal iron-ore deposits of Chile are located along the western slopes of the Andes in a narrow 1000-kilometre long area extending northward from the Province of Coquimbo to the Province of Antofagasta (Figure 5—12). All of the known ore deposits except El Laco are located relatively close to the coastal waters of the Pacific. The El Laco deposit is located about 300 kilometres inland from the port of Antofagasta and on the Andean Altiplano at elevations of 4300 to 4900 metres.

Most of the ore deposits occur as segregation and replacement bodies related to the intrusive Andean granodiorites. These ores are composed of massive magnetite-hematite with the gangue minerals of amphiboles and apatite. The El Laco hematite-magnetite ores have been described as being near surface magmatic intrusions, some of which may have broken through to the surface. In the Province of Arauco there are reported occurrences of bedded, finely laminated itabirites containing magnetite, hematite and quartz containing 30 to 40 per cent Fe.

Measured and indicated reserves of ores and concentrates-in-ores are about 300 million metric tons, of which 80 to 100 million tons are located in the operating mines. Crude ore grades range from 58 per cent to 62 per cent Fe. Not included are the inferred reserves of plus 250 million metric tons of reported 65 per cent Fe, high-phosphorus ores at El Laco.

Shipments in 1980 were 8.6 million metric tons of coarse ore, fines and pellets coming principally from open-pit mining operations of the CAP mines at El Romeral, Algarrobo and the Santa Fe group. Principal exports were to Japan with lesser amounts to Europe, USA and Argentina. Approximately 12 per cent was consumed domestically.

In 1980, the pellet production at Guacolda on Huasco Bay using ore from Algarrobo was 3.3 million metric tons. Reported pellet production capacity is 3.4 million metric tons. Pellet grades are 65 per cent Fe, 0.5 per cent P and 0.3 per cent S.

COLOMBIA

The principal iron-ore deposits in Colombia (Figure 5—12) are near Paz de Rio, Boyaca, approximately 220 kilometres northeast of Bogotá. The deposits are bedded minette-type ores associated with sediments of Tertiary age. The bed thickness ranges from 1 to 10 metres but the average thickness of those being mined is about 4.5 meters. Ore grades average 44 per cent of iron, 11 per cent of silica, 1.0 per cent of phosphorus and 0.1 per cent of sulphur. Measured reserves of 73 million metric tons were reported in 1980. The ore is mined underground and is crushed and screened. Fines generated prior to 1968 and stockpiled are now being used as sinter feed. Production in 1980 was 505-000 metric tons of ore and concentrates. Metallurgical coal and limestones are found in the vicinity of Paz de Rio.

PERU

The principal iron deposits of Peru (Figure 5—12) are located along the coast about 400 kilometres south of Lima and along the contact zones of the Andean intrusive granodiorite with which they are believed to be genetically related. They occur either in tabular-shaped ore bodies, as contact-metamorphic replacements of calcareous zones within the enclosing rocks, or as disseminated magnetite in siliceous metasediments. The iron-bearing minerals of these ore deposits are primarily magnetite and hematite with occasional pyrite and traces of copper. In the near-surface zones (upper 25 metres or about 80 feet), the ores have been oxidized to hematite (martite)-limonite, with a corresponding oxidation of sulphur to sulphates.

In 1980 the measured-indicated reserves were 230 million metric tons of ore grading 50 to 55 per cent Fe. Additional reserves of 500 million metric tons of similar grades are inferred at depth. The deposits are located at an elevation of about 800 metres on the Pampa de Marcona, 15 to 20 kilometres inland from the port of San Nicolas.

The mining operation at Marcona is by open-pit methods and transportation to the coast is by a combination of truck hauls and a 15-km long, down-hill conveyor. Prior to 1963 the mine production was shipped in the form of natural ores from San Juan Bay. In 1963 a magnetic concentrator and pellet plant with an annual 1.2-million-ton-per-year capacity were put into operation in St. Nicolas Bay. That has subsequently been expanded to its present capacity by utilizing both magnetic and gravity concentration procedures.

Total export shipments during 1980 were 6 323 594 natural metric tons of which 3.1 million metric tons were sinter feed, 1.2 million tons were pellets and 1.6 million tons were pellet feed. Sixty-nine per cent of the products were shipped to Japan and South Korea, 4 percent to the USA, 2.2 per cent to Latin America and 18 per cent to Eastern Europe. Additionally, 436 498 metric tons or 7.0 per cent was shipped to domestic markets. The maximum annual rate of shipments of about 10.5 million metric tons was achieved in 1975.

VENEZUELA

The Venezuelan iron ore deposits are associated with the itabirite-type rocks of the Sierra de Imataca which underlie a series of ridges lying south of and generally parallel to the Rio Orinoco and which extend for about 500 kilometres from the Guyana frontier on the east to beyond the Rio Caura on the west. There are two types of ore bodies: the hard, massive, crystalline replacement ore bodies such as those of El Pao, and the residual, in situ deposits such as those found at Cerro Bolivar and San Isidrio. The replacement types are primarily compact ores, consisting of dense crystalline hematite and magnetite. The residual ores consist of granular hematite with surficial cavernous limonitic canga cappings.

Measured indicated reserves of 1.7 billion metric tons of plus 55 per cent Fe residual type ores are found within the Cerro Bolivar district to the west of the Rio Caroni and 100 million metric tons of compact ores of similar grades are within the El Pao district to the east of the Rio Caroni. Mining is by open-pit methods. The ore is moved by rail from El Pao and Cerro Bolivar to separate deep-water ports on the Rio Orinoco where it is passed through treatment plants to produce various sized products for domestic use or export. The export products are loaded into ocean-going vessels and transported 273 kilometres down the Rio Orinoco to the Atlantic. The domestic products are shipped to a local steel plant. Product grades are plus 60 per cent Fe.

Total mine production for 1980 was 15.9 million metric tons, of which 72 per cent was from the Cerro Bolivar and Cerro Altamira mines and 28 per cent from El Pao. Shipments for the year were nearly 14 million tons, of which 20 per cent were for domestic consumption and the remainder for exports to the United States and Europe.

E. IRON-ORE DEPOSITS OF WESTERN EUROPE

AUSTRIA

Iron-ore production in Austria is from the Erzberg mine near Leoben (Figure 5—12). The ore deposits are of the Bilbao type and average iron content is 32 per cent. Remaining reserves are estimated to be 275 million metric tons. Production in 1980 was 3.2 million metric tons of which 2.6 million was from open-pit and the remainder was underground, and which represented about 40 per cent of the domestic consumption.

FEDERAL REPUBLIC OF GERMANY

The principal iron-ore occurrences in F. R. G. occur in the northern and northeastern parts of the country in the Salzgitter, Gifhorn and Peine-Ilsede districts (Figure 5—12). The ores consist of goethite, siderite, chamosite and hematite in bedded deposits of Cretaceous Age. Ore grades range from 24 to 33 per cent Fe, 6 to 18 per cent SiO_2, 0.04 to 0.8 per cent P and 11 to 27 per cent CaO. Mining is underground and the ore is shipped to nearby consuming centers where it is upgraded. Production in 1980 was 1.9 million tons averaging 31 per cent Fe from 5 mines which supplied about 3 per cent of the domestic iron consumption.

FINLAND

The iron-ores types of Finland (Figure 5—12) include the magmatic ilmenite-magnetite deposits of Otanmaki in the south central part of the country, the magnetite metasomatic skarn deposits at Rautuvaara in the northwest and, the metamorphic bedded deposits of the type found at Jussaro in southern Finland. The Otanmaki deposit is a series of hundreds of ore lenses made up of mixtures of 27 to 31 per cent of ilmenite and 38 to 40 per cent of magnetite with minor pyrite. The magnetite contains an average of 0.62 per cent of V. The ore lenses are of varying sizes and are scattered throughout a zone between gabbros and amphibolites. The Rautuvaara deposit with an average grade of 37 per cent Fe is one of the higher grade occurrences of the skarn type of deposit. Ore minerals include magnetite, hematite, pyrrhotite along with gangue minerals of amphibole, diopside, plagioclase and olivine. The quartz-banded iron ores, containing 25 per cent Fe of the Jussaro deposit consist of several 7 to 16 m thick steeply dipping beds of magnetite in migmatites and veined gneisses. In 1980 the Otanmaki mine produced 288 000 metric tons of concentrates from 1 552 000 tons of ore and the Rautuvaara mine produced 522 000 tons of concentrates from 1 238 000 tons of ore. Reserves at the Otanmaki mine are reported to be 17 000 000 tons and at Rautuvaara to be 18 000 000 tons. Reserves at the Mustavaara deposit of the Otanmaki type, located in the north central part of the country, are reported to be 38 000 000 tons of ore containing 22 per cent of iron. The Jussaro deposit with estimated reserves of 120 million metric tons, is not currently being exploited.

FRANCE

The principal iron-ore deposits are in Lorraine in Eastern France, Anjou-Brittany and Normandy in the northwest and in smaller occurrences in the Pyrenees in the south (Figure 5—12). The ores in the Lorraine Basin are the classical minette bedded deposits of oolitic goethite of Lias Age. They occur in units of up to 30 metres thick and with lateral extensions of 20 to 30 kilometres. Overall ore grades are 31 to 32 per cent Fe, 0.5 to 0.7 per cent P and 4.5 per cent Al_2O_3. Two ore types are recognized: siliceous ores which average 17 per cent SiO_2, 0.05 per cent sulphur and 10 per cent CaO; and calcareous ores which average 8.0 per cent SiO_2, 0.02 per cent sulphur and 17 per cent CaO. Principal mining is underground. The Lorraine resources are estimated to be 9000 million tons of which 4000 million are reserves divided about equally between siliceous and calcareous types. The Normandy-Anjou-Brittany deposits composed of siderite, magnetite and hematite are in sedimentary, folded strata of Ordivician—Devonian age. The mineable beds range in thickness from 1.5 to 6.0 metres and have a mean composition of 42 per cent Fe, 13 per cent SiO_2, 0.5 per cent P, 2 per cent CaO and 8.0 per cent Al_2O_3. Reserves are reported to be nearly 2000 million tons of which 1000 million are in Normandy and 900 million in the Anjou-Brittany district. Ores of the Bilboa type with hematite and siderite occur in central and southern France.

Ore production in 1979 was 29 million tons containing 9.1 million tons of contained iron for an average grade of 31.1 per cent Fe. Of this production, 95.6 per cent came from Lorraine, 4.1 per cent from the west and less than 0.3 per cent from the south. During the same period France imported 18.6 million metric tons of high-grade ore from various foreign sources. France exported about 8.6 million tons principally to the Belgo-Luxemburg economic union and to the Saar.

NORWAY

There are several iron mines in Norway producing both for export and domestic consumption (Figure 5—12). The largest is A/S Sydvaranger located near the Port of Kirkenes on the Barents Sea in arctic Norway. The next largest producer is Rana Gruber located near Moirana at the head end of the Rana Fjord in central Norway. Several smaller mines are located at Fosdalen 85 kilometres north of Trondheim and at Rodsand in the coastal areas of west central Norway.

At Sydvaranger, banded-iron formation is a part of a Proterozoic metasedimentary sequence dated at 2500 million years. The iron formations are made up of alternating 5 to 10 mm thick bands of magnetite and quartz with minor green hornblende, grunerite, epidote, biotite and hematite. The chemical composition averages Fe 32.32 per cent (95 per cent mag. iron), SiO_2 45.1 per cent, Al_2O_3 2.23 per cent, CaO 2.80 per cent, MgO 2.72 per cent and phosphorus 0.046 per cent. In the Sydvaranger area there are a number of iron-ore deposits of which Bjornevann, the one presently being mined, has a reported reserve of 250 million metric tons. The ores are mined by open-pit methods, crushed and rail hauled 8 kilometres to a magnetic concentrator and pellet plant at Kirkenes. Since 1975 total production has been in the form of pellets. Average pellet grades are Fe 65.5 per cent, SiO_2 7.5 per cent, Al_2O_3 0.4 per cent, CaO + MgO 0.86 per cent and phosphorus 0.008 per cent. Total shipments of pellets from Kirkenes in 1979 were 2 729 000 metric tons of which 68 per cent went to West Germany, 21 per cent to the United Kingdom, 8 per cent to Belgium and France and the remainder to Poland, Iceland and domestic consumers.

The iron mines of Rana Gruber are located in the valley of Dunderlandsdalen, approximately 23 kilometres inland from the head of Rana Fjord in the north central part of Norway. A number of sedimentary magnetite-hematite deposits are associated with the Caledonian iron formation of northern Norway and are composed of bedded magnetite-hematite metasediments associated with garnet-quartz mica schists. Open-pit reserves are reported to be 200 million metric tons of ore grading Fe 32 to 34 per cent, SiO_2 30 to 32 per cent and P 0.2 per cent. Additional underground reserves are projected to be 300 million tons of similar grades. Development started in 1902 utilizing the Edison dry magnetic concentration techniques. The plant closed in 1908 after having treated 468 000 tons of ore to produce 76 000 tons of concentrates. In 1918 a new plant was started using wet magnetic concentration methods and in three years treated 739 000 tons of ore to produce 259 000 tons of concentrates. The present concentration process produces a bulk gravity concentrate which is subsequently divided into a magnetic and non-magnetic fraction. In 1979 this plant treated 3 217 000 tons of ore of 33.1 per cent Fe to produce 668 000 tons of Fe 63.4 per cent and P 0.017 per cent for sinter concentrates and 500 800 tons of Fe 64.2 per cent and P 0.019 per cent for pellet concentrates. Most of the production is used locally. There are several other smaller producers at Fosdalen and Rodsand located in the Trondheim region along the western Norway coast which supply local consumers.

SPAIN

Iron ores occur widely throughout Spain, principally in the Provinces of Santander and Vizcaya in the north, Leon in the northwest, Huelva and Sevilla in the southwest, Granada and Almeria in the south and Guadalajara and Teruel in the east (Figure 5—12). Magnitnaya ore types are found in the southwest, ferruginous sandstones and the minette ore types in the northwest and ores of the Bilbao type in the northern, southern and eastern regions. Ore grades vary with the Bilbao types containing 48 to 53 per cent of iron; minette types with 53 per cent of iron; and, magnitnaya types with 53 per cent of iron. Published data suggest reserves of 1300 million tons with about 50 per cent of iron occurring in the northwest. The reported production in 1980 was 8.9 million metric tons largely for domestic consumption, but with about 7 per cent for export, principally to Europe.

SWEDEN

The principal producing iron mines in northern Sweden are at Kiirunavaara and Luossavaara near Kiruna, Leveaniemi near Svappavaara and Malmberget near Gallivare (Figure 5—12). In central Sweden the producing mines are at Grangesberg and Strassa. The principal ore type (Kiruna) is apatite bearing magnetite averaging 60 per cent of iron and with 0.05 to 5.0 per cent of phosphorus. The largest of the ore deposits, located at Kiruna, is a steeply dipping tabular shaped body which has an over-all length of 4 kilometres, a width of 100 to 200 metres and drilled depth extensions of 1800 metres. The ore is composed largely of magnetite with some hematite and minor apatite. The hematite is the result of the oxidation of magnetite by superficial weathering. Other minor minerals include amphibole, pyroxene and calcite. The titanium content is usually less than one per cent. The Kiirunavaara mine, which started as an open-pit operation, is now completely underground. On strike with the Kiirunavaara is the geologically similar but smaller Luossavaara ore body which is being mined by open-pit methods. There is a group of 5 other smaller but similar ore bodies located to the east and north of the Luossavaara which is known as the "PerGerjer" group. Four of these ore bodies are exposed at the surface and are, or have been, in production. The ores have been described as being magmatic but recent work has led to the suggestion that they may be exhalative-sedimentary in origin.

Other ore deposits in northern Sweden of the Kiruna type are being mined at Svappavaara by open-pit and at Malmberget by underground methods. Reserves at Kiruna are reported to be 1800 million metric tons of 60 to 65 per cent of iron and 0.2 per cent of phosphorus; at Svappavaara 300 million tons of 50 per cent iron and 0.5 per cent P; Malmberget 300 million tons of 55 per cent iron and 0.7 per cent phosphorus. The LKAB iron-ore shipments in 1980 were 21 066 000 metric tons which included 19 621 000 tons for export and 1 445 000 tons for domestic use. Of that total, 4 877 000 metric tons were pellets including 3 637 000 metric tons for export. Kiruna supplied 54 per cent of the total production.

In central Sweden, ore deposits of the Kiruna type are being mined from the export fields at Grangesberg. At Strassa the ore being mined is a metamorphic quartz banded rock composed of magnetite and hematite. The ores grade off into skarns which contain amphibole, micas and diopside. The Strassa deposit is one of the largest non-apatitic ore bodies in central Sweden and has an over-all grade of 35 per cent of iron and 0.013 per cent phosphorus.

In late 1977 the Grangesberg and Strassa iron ore mines of Granges in central Sweden were sold to the newly formed SSAB Gruvor (Swedish Steel Co.) in which Granges became a partner. In 1980 these mines produced 2 240 000 metric tons of concentrates from underground mines of which 1 087 000 metric tons were exported. The 1 153 000 metric tons used for domestic consumption included 348 000 tons of pellets from the Strassa mine.

UNITED KINGDOM

The iron-ore deposits in the United Kingdom (Figure 5—12) are located in the Frodingham, Lincoln, Northampton and Banbury districts of central and southeastern UK, and in the Cumbria district of the central western part of Wales. The largest tonnages of ore occur as bedded deposits of the minette type and are composed of chamosite and goethite. The deposits of Wales and western England are of the Bilbao type. Reserves are reported to be about 2900 million tons averaging 20 to 26 per cent of iron and these are located mostly in the eastern and southeastern districts. Annual production, mainly from open-pit operations, has dropped from 12 million in 1979 to 0.9 million (avg. 23 per cent of iron) in 1980. The open-pit mine at Frodingham in the eastern district is the only mine reported to be in production. Ore from this mine is being blended with high-grade imported ores.

F. IRON-ORE DEPOSITS OF EASTERN EUROPE
BULGARIA

Bulgaria has iron-ore reserves (Figure 5—12) estimated at 250 million metric tons. This ore is predominantly low-grade. The 1980 production was estimated to be 2.1 million metric tons of crude ore at 32 per cent Fe. About sixty per cent of Bulgaria's iron-ore requirements are produced domestically from the western part of the country, principally from the Kremikowtzi area. These ores are Bilbao type replacements in lower Triassic limestones and are composed of limonite (70 per cent), siderite (18 per cent), and hematite (12 per cent). The Martinovo deposit in northwestern Bulgaria contains relatively high-grade magnetite in a Devonian limestone. The USSR is the main source of imported ores.

CZECHOSLOVAKIA

Czechoslovakia produced 2.0 million metric tons of iron ore in 1980; however, they imported 11 million metric tons per year from the USSR to maintain their steel industry. Measured, indicated, and inferred reserves total an estimated 500 million tons of 30 per cent iron. Ore is produced from underground mines in the Barrandienne Basin west-southwest of Prague and the Spiss-Heimer area in the east (Figure 5—12). Ore produced from the Rudnany mines in the Spiss-Heimer area is from ore deposits which consist principally of siderite in masses and veins in schists and eruptive rocks of Devonian to Carboniferous age. Ore grades average 36 per cent iron. The Barrandienne ores are minette types in Ordivician schists and diabases. Ore grades range from 25 to 37 per cent iron, and the principal minerals are hematite, siderite with accessory chamosite and magnetite.

HUNGARY

Hungary produced 426 000 tons of ore in 1980 from the Rudabanya mine in the northeastern part of the country (Figure 5—12). The mine is partly open-pit and partly underground. Known reserves are on the order of 10 million tons. The Rudabanya ores are of the Bilbao type in Triassic dolomites. Hungary produces about 10 to 12 per cent of its domestic requirements and imports are largely from the USSR, with some from India.

POLAND

In 1980, Poland produced 94 000 tons of 31 per cent iron ore and imported 20 133 000 tons of ore, principally from the USSR (70 per cent), Sweden, and Brazil. Domestic reserves are estimated to be 350 million tons of sideritic-goethitic ore associated with Jurassic schists and clays. The principal production was from the Czestochowa Klobuck area in southern Poland (Figure 5—12).

ROMANIA

Romania produced 2.6 million tons of ore during 1980. Current reserves are estimated at 50 million tons of primarily sideritic-goethitic ores. Known deposits are Ocna de Fier in western Romania in lower Cretaceous limestones, Ghelar and Teliuc in central Romania in Devonian dolomites, and Capus (northwest) and Leuta (central) in Tertiary bedded deposits.

USSR

The USSR dominates world iron-ore production. In 1979 it produced 242 million tons of useable ore which represented 27 per cent of the world's total output. The principal producing regions (Figure 5—12) included: the Ukraine which produced over 50 per cent of the Soviet total; the Kursk region which was second largest followed by the Urals, Kazakhstan, Siberia and the Kola peninsula. Eighty-nine per cent of the Ukraine's production came from 23 underground mines plus nine large and several small open-pit operations in Krivoi Rog. The total Soviet iron-ore reserves in categories measured, indicated and inferred (A + B + $C_1 + C_2$*) are reported to be 111 billion tons averaging 34.8 per cent iron. Of that total the reserves in the measured, indicated and inferred categories (A + B + C_1) are 60 billion tons averaging 38 per cent iron including 10 billion tons of ore averaging 55 per cent iron. The total reserves are divided among: the Ukraine 31 per cent, European Russia 24.4 per cent, the Urals 15.7 per cent, Kazakhstan 15 per cent, Siberia 7.4 per cent, northwest (Karelo-Kola) 3 per cent, Soviet Far East 2.5 per cent and all others 1 per cent. Ore types range from natural soft ores to taconites from which magnetic and gravimetric concentrates are produced. About 16 per cent of the production is exported with 90 per cent of the exports going to Poland, Czechoslovakia, Romania, Hungary, Bulgaria and the German Democratic Republic.

---

*A + B + C_1 = Measured and indicated and C_2 = Inferred. (Classification used in U.S.S.R.)

YUGOSLAVIA

The iron-ore occurrences of Yugoslavia (Figure 5—12) are located near Ljubija and Vares in the Province of Bosnia - Hercegovina in the central part of the country and near Skopje in Macedonia in the southeast. The Ljubija ores are of the Bilbao type and average 43 per cent iron. The ores near Vares are reported to be of the Lahn-Dill type and average 34 per cent iron. Ores of the minette type averaging 30 per cent iron are found near Skopje. Reserves are reported to be 1000 million metric tons ranging from 20 to 46 per cent iron and averaging 40 per cent iron. About 70 per cent of the reserves are located in the Ljubija and Vares districts. In 1980 an estimated 3.8 million tons of ore were mined and were upgraded for use in pelletizing and agglomeration plants. This production represents approximately 85 per cent of the domestic consumption.

G. IRON-ORE DEPOSITS OF MIDDLE EAST AND ASIA

AFGHANISTAN

In Afghanistan large, bedded iron deposits in excess of 1 billion tons containing 60 per cent iron are reported to be located 140 km (87 miles) WNW of Kabul near Hajigak Pass (Figure 5—12). They are reportedly

of Paleozoic Age. Other iron deposits in the country are located at Nish, Haji Alam and Pagmur although reserve data are not available.

INDIA

India produced about 37 million metric tons of iron ore during 1980 of which approximately 43 per cent was for domestic consumption and the remainder was for export. Principal producing areas have been Goa, Madya Pradesh, Bihar-Orissa and Karnakata. Shipped products included both lump and fines. Reserves are reported to be 13 500 million tons of ore including 10 500 million tons of hematite and 3000 million tons of magnetite. The Kudremukh iron-ore project in Karnakata, was completed in 1980. The design capacity is to produce 7.5 million tons per year of 67 per cent Fe concentrates from 22.6 million tons of low grade 38 to 40 per cent Fe magnetite-hematite banded iron formation. The concentrates are moved in slurry form through a 67-km long pipeline to a port at Mangalore. The largest export market is Japan followed by eastern Europe.

IRAN

Iran produced 1 100 000 metric tons (1 083 000 long tons) of iron ore in 1977 for the expanding domestic steel industry. Choghart near Bafqu in central Iran is the major producer. The ore is massive, high-grade, high-phosphorus, occurring as an intrusive mass similar to the Kiruna ores. Also within the Bafqu area are the deposits of Chador Malu. With reserves estimated at 160 million metric tons (157.5 million long tons) the Bafqu area is the most important iron-ore area of Iran. Other reserves such as those of the Arak area are small and low-grade with many deleterious elements. Total reserves of iron ore in Iran are in the order of 250 000 000 metric tons.

NORTH KOREA

North Korea produced 7.4 million metric tons (7.3 million long tons) of iron ore during 1979, of which a large portion came from the Musan mine in the northeastern part of the country (Figure 5—12). The ore is described as Lake Superior type banded magnetite schists containing 40 per cent iron from which 59 to 60 per cent iron ore concentrates are made. Estimated reserves at Musan are 1.3 billion tons. Other producing iron-ore mines are found in south Hamgyong Province along the eastern coast, in the provinces of North and South Hwanghae and in North Pyongang, along the western part of the country. Ores include both magnitnaya and residual types.

PAKISTAN

The estimated iron-ore resources of Pakistan are 500 million metric tons of which the largest deposits are the Chichali minette-type ores containing an indicated 335 million tons of 31 to 34 per cent iron ore. These deposits, also known as the Kalabagh iron ores, are located in northern Pakistan along the Indus River.

PEOPLE'S REPUBLIC OF CHINA

Asian iron-ore production was dominated by the People's Republic of China with a reported 1979 output of 76 million metric tons (75 million gross tons) of ore thus making it the sixth largest producer in the world preceded by the U.S.S.R., Australia, U.S.A., Brazil and Canada. Reserves are reported to be 38 000 million metric tons of which the major portion is in the 30 to 35 per cent Fe range. Approximately 25 per cent of that total tonnage is in Pre-Cambrian banded metamorphic ores of magnetite-hematite containing 33 per cent iron and 50 per cent silica. The ores of this type are found at Anshan and in the northern and northeastern provinces (Figure 5—12). Approximately 40 per cent of total reserves are in bedded oolitic hematite-limonite ores which average 40 per cent iron. The major deposits of these ores are the Hsuanlung (Sinian) Ninghsiang (Devonian) and Chichiang (Jurassic). Metasomatic skarn deposits containing high-grade magnetite with associates sulphides of copper and cobalt are reported at Tayeh in central China. These five major deposits of three ore types represent almost 70 per cent of China's total iron-ore reserves.

The 1979 domestic consumption of iron ore was augmented by 6 million metric tons of ore imported mostly from Australia with small tonnages from Brazil.

PHILIPPINES

There are numerous small iron-ore occurrences reported throughout the Philippines (Figure 5—12). They include: primary replacements and pod-like deposits of magnitnaya type ore, concentrations in beach and dune deposits of iron minerals eroded from andesites and volcanic rocks, and, residual lateritic accumulations developed in situ on basic and ultrabasic rocks under the influence of intense tropical weathering. The magnitnaya types contain 37 to 60 per cent Fe in magnetite-hematite with small percentages of iron and copper sulphides; the beach-dune deposits contain magnetite and titano-magnetite which averages 55 to 65 per cent Fe and 4.5 to 7 per cent titania; and, the lateritic deposits contain 30 to 50 per cent iron in hematite-goethite with minor chromium, nickel and cobalt. There has been a small production from the primary type ore deposits and from the sand deposits although there was no significant recorded production during 1977-1980. It has been reported that the Santa Ines deposit in Rizal, Luzon with reserves of 30 million metric tons is to be developed to supply ore to a local steel plant. The largest potential iron resource is the residual laterites on Nonoc and Mindanao islands which contain hundreds of millions of tons of 35 to 48 per cent iron which are not economic at the present time.

SAUDI ARABIA

Iron ores are reported at several localities in Saudi Arabia (Figure 5—12). At Wadi Sawawin in the remote northwest region of the country there are indicated reserves of 350 to 400 million metric tons of 30 to 35 per cent Fe in bedded magnetite-jaspillite Pre-Cambrian beds. At Fatima between Mecca and Jidah there is an oolitic hematite deposit containing 40 million metric tons of 45 per cent Fe which reportedly can be beneficiated to 58 per cent Fe.

TURKEY

The principal iron-ore producing mine is located at Divrigi in Sivas-Malataya province in eastern Turkey. The ore occurrences are described as contact metasomatic deposits of magnetite in limestone and of irregular size and distribution. Ore grades range from 52 to 55 per cent Fe and production in 1980 was 3.3 million metric tons (3.2 million long tons). Inferred reserves for Turkey are reported at about 800 million metric tons (787 million long tons) in a number of widely scattered deposits and of varying geologic types.

OTHER ASIA

Japan, Malaysia and Thailand produced small amounts of iron ore during 1979.

The iron-ore potential of Indonesia consists primarily of laterite deposits and iron sands with estimated total resources of 500 000 000 metric tons (492 million long tons).

Estimates of iron resources in Laos range up to 1000 million metric tons of 60 per cent iron in contact metamorphic type deposits located in Xien Khoung province.

H. IRON-ORE DEPOSITS OF AFRICA

The largest deposits of iron in Africa are associated with Pre-Cambrian formations and are mostly of the Lake Superior type. Countries bordering the Atlantic have been the biggest African producers and significant additional deposits are currently waiting to be developed. Summary descriptions of the more important producing and proposed iron-mining operations are included below (see Figure 5—12).

ALGERIA

Ouenza—The Ouenza iron-ore mine which is situated in Algeria near the Tunisian border some 110 kilometres from the coast has been producing since 1921. The deposits are operated by Sonarem (Ste. Nationale Algerienne De Recherches Minieres, Algiers), which is owned by the Algerian Government. Reserves are estimated to be on the order of 70 million metric tons of ore grading 54 per cent Fe, 3.3 per cent SiO_2, 0.6 per cent Al_2O_3 and 7 per cent ignition loss. The ore is marketed through the port of Annaba to Germany, Italy and Eastern Europe at the rate of 3 to 5 million metric tons per year. About 600 000 tons per year are consumed domestically.

Gara Djebilet (Proposed)—The Algerian Government's Sonarem is currently promoting the exploitation of the Gara Djebilet iron-ore deposit in southwest Algeria about 1100 kilometres from the Mediterranean seaport of Oran. Reserves are estimated to be 2 000 million metric tons of ore grading 57 per cent Fe, 5 per cent SiO_2, 5 per cent Al_2O_3, 0.8 per cent P and 1.0 per cent S. The location of this deposit together with the high phosphorus and sulphur contents of the ore have been major obstacles in bringing this project on stream.

ANGOLA

Cassinga—The Cassinga iron-ore district is located in the southern part of Angola about 630 kilometres (391 miles) inland from the port of Mocamedes. The area contains three types of deposits: detrital high-grade hematite, enriched hematite developed in situ on itabirites. The high-grade lump and fine ore contain 62 per cent Fe, 6 per cent SiO_2, 2 per cent Al_2O_3 and 0.06 per cent P. The itabirites average 40 per cent Fe. High-grade ore reserves are indicated to be 120 million metric tons in fifteen different areas individually ranging from 0.5 million to 20 million metric tons. The itabirite resources are reported to be plus 1 000 million metric tons. Production of more than 5 million metric tons of ore was reached during the early 1970's.

Cassala (Proposed)—The Cassala deposit in north central Angola was explored and found to contain 54 million metric tons of itabirite having an average grade of 34 per cent Fe. Wet magnetic separation tests produced a 66.5 per cent Fe concentrate from it. The nearby Mount Quitungo, however, can contribute additional ore bringing the total reserves to over 150 million metric tons. The proposed port site for this ore would be Luanda, but development of the project has been delayed.

EGYPT

The principal iron-ore occurrences in Egypt are located within and near the Bahariya oasis in the Western Desert. Total reserves are reported to be 250 million metric tons, of which 129 million tons of 54 per cent iron ore are in the El Gedida deposit. The ores occur in bedded sedimentary deposits of Eocene age with thicknesses of up to 12 metres. In 1980 the production from the El Gedida mine was 1.5 million metric tons of ore which was rail shipped about 295 kilometres to the steel plant located at Helwan near Cairo.

GABON

Mekambo—The Societe des Mines de Fer (SOMIFER), which is jointly owned by the Government of Gabon (60 per cent) and Bethlehem Steel (20 per cent) together with French, German, Belgian, Dutch, and Japanese interests (20 per cent), has been engaged in evaluating the iron ores of the Mekambo district of northeastern Gabon since 1959. The estimated total ore reserves are 870 million metric tons, of which 570 million tons are located at Belinga, 200 million tons at Boka-Boka, and 100 million tons at Batoula. The grades of the deposits range from 63 to 66 per cent iron, 1.2 to 3 per cent silica, 2 to 3 per cent alumina, and 0.05 to 0.16 per cent phosphorus. The ore types consist of high-phosphorus hematite-goethite ores near the surface and overlying low-phosphorus blue hematitic ores at depth. Estimates of low-phosphorus ore in the Babiel zone of the Belinga deposits indicate there are 280 million metric tons at 64.9 per cent Fe and 0.064 per cent P. The Trans Gabon Railway is being constructed to connect a new deep-water port at Santa Clara near Libreville with Franceville in southeastern Gabon. A branch of that railway is planned to connect with the iron district in the northeast. The rail distance from Santa Clara to Belinga is 560 kilometres. The port is planned to accommodate ships of up to 250 000 dwt.

GUINEA

Guinea Nimba (Proposed)—Guinea Nimba, located in the southeastern part of Guinea, is a northern prolongation of the Nimba range in Liberia where the LAMCO iron-mining operations are located. The Guinean Government has been interested in the development of the deposit for some time through Mifergui (a consortium of European, American and Japanese interests) and exploration studies have indicated over 600 million metric tons of crude ore grading about 65 per cent Fe and 4 per cent combined $SiO_2 + Al_2O_3$. Sinter fines and pellet-feed are being considered as the marketable products. The ore would be transported from the mine over a new 18-kilometre railway to connect with the existing LAMCO railroad in Liberia and thence to the port at Buchanan.

IVORY COAST

Mont Klahoyo (Proposed)—The Mont Klahoyo iron deposit is located about 600 kilometres northwest of Abidjan in the Ivory Coast. Estimated reserves are of 670 million metric tons grading 36 per cent Fe. The ore zone is a quartz-magnetite iron formation with an average thickness of 130 metres and extending over a distance of approximately 5.5 kilometres. It has been determined that the reserves are amenable to beneficiation to produce about 300 million metric tons of concentrates with 69 per cent Fe and 2 per cent SiO_2. It is proposed that the crude ore be concentrated at the mine site and then transported by pipeline to the proposed port site of San Pedro. Pelletizing would be carried out at the port. Ownership is shared by European, American and Japanese interests together with the Government of the Ivory Coast.

LIBERIA

Bong Range—The Bong Range is located in central Liberia about 80 kilometres northeast of Monrovia. The iron mine is operated by the Bong Mining Company, which is owned by the Liberian Government (50 per cent), Exploration and Bergbau (35 per cent), Finsider (12.5 per cent) and the Liberian Public (2.5 per cent). The Bong Range is a chain of hills 35 kilometres long that rises nearly 305 metres above the surrounding plain. The principal iron formation (itabirite) has a strike length of 14 kilometres in an east-west direction. The orebody, located on the western end of the range, is mined by open-pit methods, and has estimated remaining reserves of 70 million metric tons of itabirite averaging 38.7 per cent Fe, 40.8 per cent SiO_2, 1.2 per cent Al_2O_3 and 0.06 per cent P. In addition, there are another 100 to 150 million tons of probable ore of similar grades located further along the range. Mining commenced in 1965 and has been expanded by degrees since that date. Shipments in 1980 were 6 488 888 metric tons of concentrates and pellets grading about 65 per cent Fe, 7 per cent SiO_2, 0.25 per cent Al_2O_3 and 0.02 per cent P. Both concentration and pelletization are carried out in the vicinity of the mine prior to the transportation of the products to the port of Monrovia. A second pellet plant was brought into operation in 1977 bringing productive capacity for pellets to 4 800 000 metric tons per year. Overall productive capacity for both pellets and concentrates is currently 7 500 000 metric tons per year. The products are sold principally to Germany, Italy and Holland. It has been reported that the mine will be exhausted during 1985-1990.

Mt. Nimba—The Mt. Nimba iron-ore district is located in the north central part of Liberia adjacent to the international borders with Guinea and the Ivory Coast. It is about 290 kilometres northeast of Monrovia. The mine is operated by LAMCO JV Operating Company, which is jointly owned by Bethlehem Steel (25 per cent), the Liberian Government (37.5 per cent), the Swedish LAMCO Syndicate (22.5 per cent) and the International African American Corporation (15 per cent). Five orebodies contained pre-mining reserves of 200 million metric tons of residual fine-grained Lake Superior type hematite-limonite ore containing more than 60 per cent iron. Mining, which began in 1963, is done by open-pit methods. The ore is crushed prior to being railed 274 kilometres to the port of Buchanan near the mouth of the St. John's River. The somewhat siliceous "blue" hematite ores (1 per cent loss on ignition) of the main orebody are mixed on a 60/40 basis with slightly aluminous limonite "brown" ores (4 per cent loss on ignition) from a secondary orebody. The blend produces three products after washing and screening in Buchanan. These are: washed-lump ore (+6 mm), washed sinter-fines (–6 mm + 100 mesh) and pellet-feed (–100 mesh). The grade of the pellet-feed is improved by the addition of magnetite concentrates from Mt. Tokadeh, one of LAMCO's concession areas near Mt. Nimba. Lamco's pelletizing capacity is 2.2 million metric tons per year. The pellets average approximately 63 per cent Fe, 6 per cent SiO_2, 2 per cent Al_2O_3 and 0.06 per cent P. The washed-lump and fines average about 65 per cent Fe, 4 per cent SiO_2, 1 per cent Al_2O_3 and 0.06 per cent P. While annual shipments during peak years were in the neighborhood of 12 million metric tons, they have been running at less than 10 million metric tons more recently. In 1980 shipments were 9 326 000 metric tons and were destined principally to Europe and the United States. Reserves in the main orebody are expected to be depleted by 1987 and LAMCO has been studying the feasibility of beneficiating low-grade ores from its concession areas in order to continue mining after that date.

Mano River—The Mano River iron-ore mine is located in Liberia near the Sierra Leone border about 145 kilometres north of the port of Monrovia. It is operated by National Iron Ore Company, Limited, which is owned jointly by the Liberian Government (50 per cent), Liberian Enterprises, Ltd. (35 per cent) and Liberian Mining Company (15 per cent). The Mano River ore deposits are situated on the upper slopes of hills varying from 366 metres to 564 metres above sea level. The rock type varies from hard dense rich-looking canga ore of more than 60 per cent Fe to lean limonite canga averaging over 40 per cent Fe. A

banded iron formation, which has been enriched to ore grade by weathering is also found within the ore zones. The company produces a single washed fines product grading at 58.5 per cent Fe, 4.5 per cent SiO_2, 4.5 per cent Al_2O_3 and 0.04 per cent P. Reserves are estimated to be 40 million metric tons and shipments totalled 2 416 900 tons in 1979. The product is largely sold in Europe with minor sales in the United States. The project has been in operation since 1961 and the current production capacity is 5 million metric tons.

Bie Mountains (Proposed)—The Bie Mountains iron-ore deposits which lie about 95 kilometres NNW of Monrovia near the rail line to National Iron Ore Company's (NIOC's) iron-ore mine at Mano River. The ore, which is derived from a banded iron formation, occurs in weathered, semi-weathered and non-weathered zones. The weathered and semi-weathered ore, which grades 35 to 45 per cent Fe, can be concentrated to about 64 per cent Fe by magnetic and heavy-media separation methods. The non-weathered ore, which has a grade of between 30 and 35 per cent Fe yields about a 67 to 68 per cent Fe concentrate by magnetic separation. Reserves of crude ore mineable from the three zones are estimated to be 365 million metric tons. The construction of a 15-kilometre spur line linking with the existing (NIOC) railroad to the port of Monrovia or the installation of a slurry pipe line over a distance of about 60 kilometres to Williams Port are two alternatives for ore handling under consideration.

Tokadeh (Proposed)—The LAMCO JV Operating Company has been studying the expansion of mining the Tokadeh deposit when the company's Mt. Nimba reserves are exhausted in 1987. Tokadeh is located near LAMCO's existing rail line about 19 kilometres southwest of Nimba. Reserves at Tokadeh are estimated to be 100 million metric tons of ore grading 42 to 43 per cent Fe. An additional 230 million metric tons of similar grade ore are contained in the nearby hills (Mts. Beeton, Gangra and Yulliton). Production on a limited scale began in 1973 and expansion has been studied.

Wologisi (Proposed)—The Wologisi deposits are situated near Mt. Wutivi in the Wologisi range in the extreme northwestern part of Liberia. A project proposed to develop these deposits involves the beneficiation of low-grade magnetite-bearing itabirite containing 35 to 40 per cent Fe to produce 4 million metric tons of sinter feed grading at 66.5 per cent Fe, 4 per cent SiO_2, 0.6 per cent Al_2O_3 and 0.02 per cent P. A 250-kilometre railroad or slurry pipe line would have to be constructed to the prospective port of Williams Port. Reserves are estimated to be 600 to 1 000 million metric tons of low grade (35 to 40 per cent Fe) ore.

MAURITANIA

Kedia d'Idjil—The Mauritanian Government's Societe Nationale Industrielle et Miniere (SNIM) owns and operates the Tazadit, Rouessa and F'Derik mines in the Kedia d'Idjil area immediately east of Fort Gouraud in northwestern Mauritania. Production commenced in 1963 by a French company, Miferma, which was owned prior to nationalization in 1974, by French financial companies and steel works (60 per cent) and by British, Italian and German steel works (40 per cent). The ore is predominately hematite with minor limonite. Structurally the ore varies from massive to platy and granular. The ore grade is 65 per cent Fe, 4.3 per cent SiO_2, 1.2 per cent Al_2O_3, and 0.03 per cent P. Reserves of this type of ore are estimated to be 80 million metric tons. The ore is mined by open-pit methods and transported by rail 650 kilometres to the loading port of Nouadhibou. In 1979, 9 271 000 metric tons of ore were shipped to buyers in several European countries and Japan.

Guelb (Proposed)—In view of the diminishing reserves in the Kedia d'Idjil area, the Societe Nationale Industrielle et Miniere has undertaken the investigation of the nearby Guelb deposits. These deposits consist of 25 orebodies scattered within a 104-kilometre radius of the present Fort Gouraud mining area. The ore is a low-grade Pre-Cambrian banded magnetite quartzite, which averages 37 per cent Fe, 40.4 to 46.7 per cent SiO_2, 0.5 to 2 per cent Al_2O_3 and 0.03 to 0.07 per cent P. The project involves the building of a spur line from the existing railway 37 kilometres to El Rhein and an additional 26 kilometres to Oum Arwagen, the first deposits to be mined. The ore will be mined by open-pit methods and concentrated in two stages, one at El Rhein and the other at the port of Nouadhibou. Dry magnetic separation is to be carried out at the former location, with wet gravity separation of one of the resultant fractions being performed at the latter. Current plans are to market the ore as magnetite sinter feed (65.8 per cent Fe), oxidized ore sinter feed (64.5 per cent Fe) and magnetite pellet feed (63.5 per cent Fe) with each constituting about one-third of the total volume. Reserves are 530 million metric tons of ore grading at 37 per cent Fe in the eastern Guelb deposits and an additional 980 million metric tons in the western Guelb deposits.

SOUTH AFRICA

Sishen—Sishen is the largest iron-ore producing mine in South Africa. Its ore is located in two Pre-Cambrian geological units, which dip about 10° to the west in the Sishen-Postmasburg region of the northwestern Cape Province. The mine has available reserves of nearly 1300 million tons of hematite with a Fe content between 65 per cent and 69.9 per cent for about 85 per cent of the deposit. A further 11 per cent contains 63 to 65.9 per cent Fe and the remainder has from 60 to 62.9 per cent. The state-owned South African Iron and Steel Industrial Corporation, Ltd. (ISCOR), which owns and operates the mine, has additional reserves of high grade hematite ore in the Sishen-Postmasburg region totaling about 3 000 million metric tons. About a third of ISCOR's reserves are recoverable by open-pit mining techniques. Mining began at Sishen as early as 1953 but it was not until the completion of the port at Saldanha Bay in 1976 that the mine began to contribute significantly to the world market. Sishen currently has a crushing capacity of 20 million tons per year. Crude ore is beneficiated at the mine by a heavy-media process to recover lump ore and sinter fines by floating off the lighter shale and banded ironstone. The prod-

ucts are transported 850 kilometres over South African Railway to Saldanha Bay near Capetown. The rail capacity is only 18 million metric tons per year but will be raised to 28 million metric tons. The port capacity is 28 million metric tons annually and is designed to eventually accommodate ships of up to 350 000 dwt. Sishen shipped 19 532 300 metric tons in 1980. Japan was the principal recipient of the export ore with lesser amounts going to France, Germany, Italy and the U. K. ISCOR consumed 7 040 700 tons and 196 300 tons were sold locally.

Beeshoek—Associated Manganese Mines of South Africa, Ltd. (ASSOMAN) operates an open-pit iron-ore mine at Beeshoek, which is located about 10 kilometres northwest of Postmasburg in Cape Province, South Africa. Run-of-mine grade averages 64 per cent Fe, 5 per cent SiO_2, 2 per cent Al_2O_3, and 0.08 per cent P; and reserves are estimated to be large. Shipments totalled 1 273 000 metric tons in 1980. The principal recipients of the ore are the United States and Japan. The port is Saldanha Bay.

Thabazimbi—The Thabazimbi iron-ore deposits are located in the northern Transvaal of South Africa about 483 kilometres northeast of Sishen. They are of the Lake Superior type of deposit, occurring in the banded ironstone and ferruginous chert forming the top of the Dolomite Series of the Transvaal System. The orebodies have a strike length of 3 kilometres and an average width of 27.5 metres. The mine is owned and operated by the South African Iron and Steel Corporation, Ltd. (ISCOR) and the product is destined principally for domestic consumption at the company's Pretoria steelworks. At this property, operated by both the underground and open-pit methods, the crude ore is beneficiated by the heavy-media process to produce lump ore and sinter fines. Specular hematite and geothitic hematite are mined and estimated reserves are 100 million metric tons of crude ore grading at 58 per cent Fe and 13 per cent SiO_2. The product grade is 63 to 64 per cent Fe and 6 to 7.5 per cent SiO_2. Shipments in 1980 were 2 128 700 metric tons of which 2 110 400 tons were taken by ISCOR and 18 300 tons were sold locally.

I. IRON-ORE DEPOSITS OF OCEANIA

AUSTRALIA

Hamersley Holdings—Hamersley Holdings has operating iron ore mines at Mt. Tom Price and Paraburdoo in the Hamersley Iron Province of Western Australia (Figure 5—13). The orebodies at both localities were developed in-situ by the secondary enrichment of the Pre-Cambrian Brockman iron formation. The Mt. Tom Price has an overall length of 6.4 kilometres and a width of up to 1.6 kilometres (average 0.6 kilometres). Reserves were calculated in 1980 to be 400 million metric tons of high grade (64 per cent Fe) and 250 million metric tons of beneficiating low-grade (58.5 per cent Fe) ore. Mining began in 1966 by open-pit methods. The ore is crushed and screened into several sized products prior to its rail shipment of 278 kilometres to

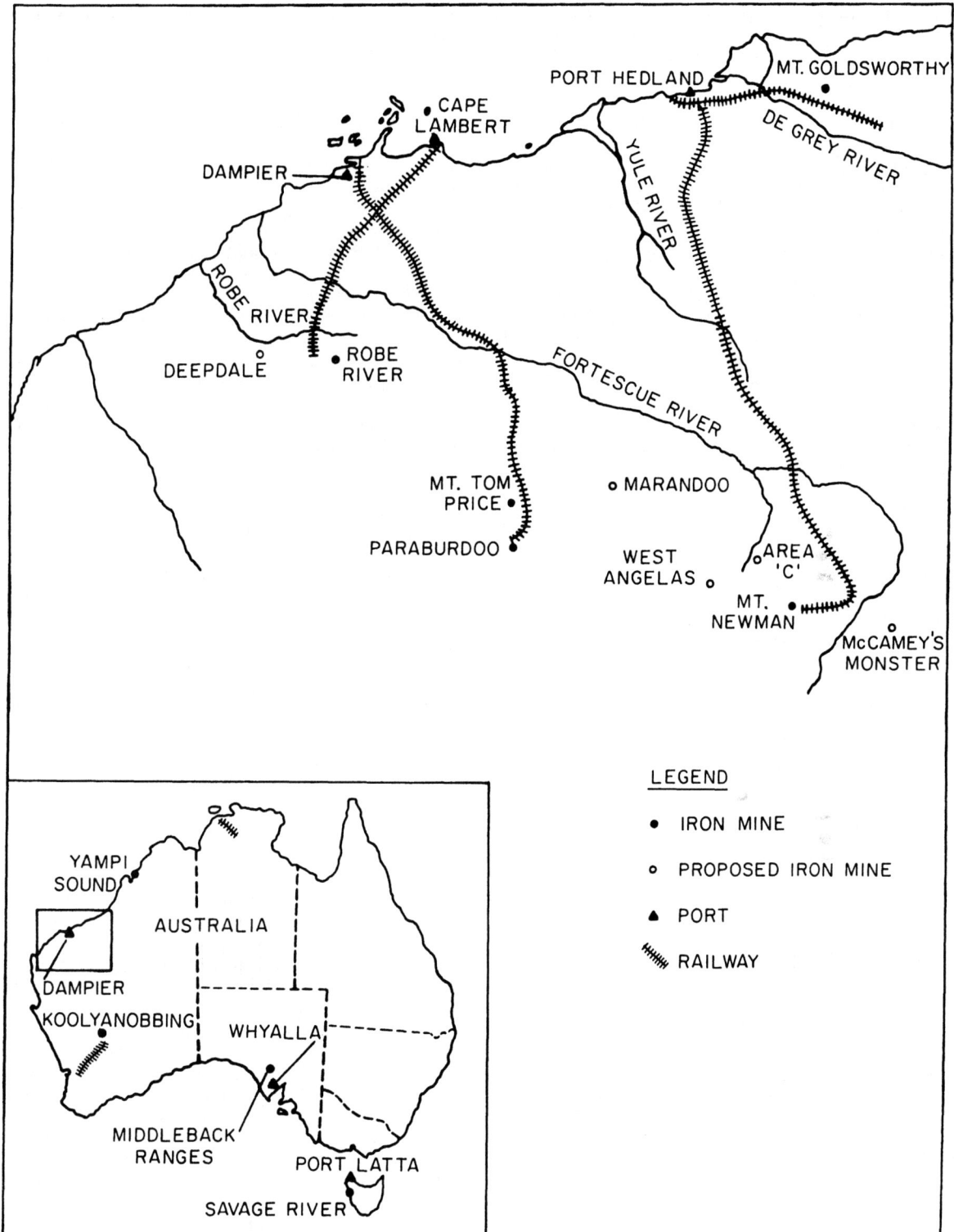

FIG. 5—13. Map showing location of principal present and proposed sources of iron ore in Australia.

the port at Dampier. Total production to date of saleable ore has been about 260 million metric tons. The Paraburdoo ore deposits, which are located 98 kilometres south of Mt. Tom Price, consist of a number of orebodies scattered throughout a 24-kilometre long area along the Paraburdoo Ridge. Production is by open-pit methods and the ore is rail-hauled 386 kilometres via Mt. Tom Price to the port of Dampier. Production started in 1973 and to date 103 million metric tons of saleable ore have been produced. A pellet plant with an annual capacity of 2.0 million metric tons was completed in 1968 at Dampier. In 1979 a heavy-media concentrator with an annual capacity of 6.0 million metric tons was commissioned at Mt. Tom Price. It will treat the low grade 58.5 per cent Fe ore to produce a saleable product in the 62 to 64 per cent Fe range. Total ore production in 1980 was 39.5 million metric tons of all products. About 65 per cent of the production was destined for Japan, 20 per cent for other Asian countries and 16 per cent for Europe. Hamersley Holdings is owned by Conzinc Rio-Tinto of Australia (82.32 per cent), Australian public (11.48 per cent) and Japanese steel mills and trading firms (6.2 per cent).

Koolyanobbing—Broken Hill Proprietary Co., Ltd., (BHP) produces iron ore at Koolyanobbing in Western Australia for internal consumption by its steelworks at Kwinana near Perth. The ore consists of hematite, geothite and magnetite and averages approximately 61.5 per cent Fe, 3.5 per cent SiO_2 and 0.14 per cent P. Production in 1979 was 1 386 000 wet metric tons. Reserves are estimated to be 34 000 000 metric tons.

Middleback Range, South Australia—Deposits of high-grade (more than 60 per cent Fe) hematite ore, manganiferous in part and associated with Pre-Cambrian banded hematite-quartz jaspillites have been mined by Broken Hill Proprietary Co. Limited since 1915 to supply its Whyalla, Newcastle and Port Kembla steel plants. Mining at the Iron Prince/Iron Baron and Iron Monarch mines is by open-pit methods, using both truck and rail haulage. Most ore is crushed to minus 102 millimetres prior to shipment; however, a small amount of low-grade ore is concentrated in a heavy-media plant. The ore and concentrates are shipped 53 kilometres to a stockpiling and shiploading facility at Whyalla. A two-million-metric-ton per year pellet plant at Whyalla began production in 1968. In 1979 total shipments of iron ore were 2 685 000 wet metric tons most of which was consumed domestically, 89 000 tons of ore were exported to South Korea and 18 000 tons of pellets were exported to Thailand. Total reserves of the Middleback Range are estimated to be 90 million metric tons.

Mt. Goldsworthy—Steeply-dipping lenses of massive hematite occurring within banded iron formation of early Pre-Cambrian age have been mined at Mt. Goldsworthy and associated deposits in the Nimingarra Range of the Port Hedland district in Western Australia since 1966. Ore reserves in the Mt. Goldsworthy vicinity (i.e., the A and B areas) are expected to be exhausted in 1983. The ore, which is mined by conven-

tional open-pit methods, is crushed and screened into coarse and fine products and shipped 113 kilometres by rail to a shiploading facility at Port Hedland. The mine is managed by Goldsworthy Mining Ltd., which is jointly owned by Utah Development Co. (33⅓ per cent), Consolidated Gold Fields Australia Ltd. (46⅔ per cent) and Mt. Isa Mines (20 per cent). In 1980 the company shipped 5.4 million metric tons (natural) of iron ore. Prospects for the company's continued mining of iron ore after 1983 depend on the successful development of the company's Area C orebody.

Area C (proposed)—Goldsworthy Mining Ltd. intends to develop its Area C orebody in the Opthalmia Range 353 kilometres south of Port Hedland. The ore in Area C is developed in the Marra Mamba formation. These ores have a lower phosphorus content than many of the other Hamersley Range ores but have high loss on ignition (LOI) which tends to restrict their use as lump ore in blast furnaces. Reserves are estimated to be 650 million metric tons of iron ore grading approximately 62 per cent Fe, 2.7 per cent SiO_2, 1.5 per cent Al_2O_3, 0.06 per cent P and LOI 5.9 per cent. The Goldsworthy group plans to construct a single-track railroad from the mine site to the existing terminal on Finucane Island, Port Hedland.

Mt. Newman—The Mt. Newman deposits are included in a mineral lease covering 777 square kilometres (300 square miles) in the Hamersley Iron Province of Western Australia. The main orebody, consisting of hard hematite with minor magnetite and martite (64 per cent Fe) is in an ore zone 5.5 kilometres long and 120 metres wide, along Mt. Whaleback. While this orebody occurs in the Brockman formation, the mineral lease also includes other orebodies within the stratigraphically lower Marra Mamba formation consisting of limonitic hematite (61.8 per cent Fe). Reserves of Mt. Whaleback are on the order of 1 300 million metric tons of high-grade ore but there are an estimated 5 000 to 6 000 million metric tons of additional material of ore potential at various grades within the lease area. Open-pit mining was commenced at Mt. Whaleback in 1969. The ore is crushed and screened prior to being railed 426 kilometres to Port Hedland.

In 1979 the company shipped 31.5 million wet metric tons of ore comprised of about equal amounts of high grade fines (62 per cent Fe) and lumpy ore (64 per cent Fe). The principal recipients of this ore were: The Far East (67 per cent), Australia (22 per cent) and Europe (11 per cent). Mt. Newman Mining Co., Pty., a BHP subsidiary, manages the Mt. Newman Joint Venture made up of Amax Iron Ore Corporation (25 per cent), Pilbara Iron Ltd. (30 per cent), Dampier Mining Co. Ltd. (30 per cent), Seltrust Iron Ore Ltd. (5 per cent) and Mitsui-C, Itoh Iron Pty. Ltd. (10 per cent).

Mt. Newman has been stockpiling low-grade material encountered during mining and by 1978 had accumulated a 30-million-ton pile of fines averaging 55.8 per cent Fe. In 1979, they opened a new heavy-media concentrator to upgrade the fines at an annual rate of 5.0 million tons of concentrate from 7.0 million metric tons of feed. In mid-1980 Mt. Newman initiated mining from the Marra Mamba formation at the rate of 1.5 million metric tons to augment the production from

Mt. Whaleback. Mt. Newman plans to increase their annual production capacity to 70 million metric tons by (1) upgrading of the low-grade fines at Mt. Whaleback, (2) expansion of the mining of the Marra Mamba formation, and (3) by the initiation of mining on the outlying ore deposits on the Brockman formation.

Robe River—The Robe River ore deposits at Pannawonica are 6 to 26 metre-thick cappings of goethitic, pisolitic iron ore on 40 to 60 metre high mesas, scattered along the broad valley of the Robe River. The ores are considered to have been deposited during Cretaceous times as a thick blanket of limonitic silt and mud in the broad estuary of the ancestral Robe River. The area was subsequently uplifted and eroded into the present day isolated mesas. The largest of the orebodies in this group contained about 50 million tons containing 57 per cent Fe. The ore is mined by open-pit methods and rail hauled 170 kilometres to Port Lambert. Because the ore deposits are scattered, the central crushing station is located at the port site rather than at the mine. In 1980 the Robe River project produced 14.8 million tons of ore consisting of 13.7 million tons of sinter fines and 1.1 million tons of pellets. Sintering and pelletizing the crude ore upgrades it to 63 per cent Fe by driving off the combined water. The Pannawonica reserves are estimated to be 150 million metric tons of 57 per cent Fe. Through the terms of the "Dampier" agreement with BHP, the Robe River group will have access to an additional 150 million tons of equivalent ore from the nearby Deepdale deposit of BHP. Additionally they have reserves of 1000 million tons of about 62 per cent Fe ore in the Marra Mamba formation at West Anglea. Cliffs Robe River Iron Associates is jointly owned by Cliffs Western Australian Mining Co. Pty. Ltd. (30 per cent), Mitsui Iron Ore Development Pty. Ltd. (30 per cent), Robe River Ltd. (17.15 per cent), Redesdale Pty. Ltd. (17.85 per cent) and Cape Lambert Iron Associates (5 per cent).

Yampi Sound—Hematite deposits occurring as lenses in Pre-Cambrian metamorphosed sediments have been mined since 1950 on Cockatoo and Koolan Islands in Yampi Sound, Western Australia. The ore body on Cockatoo Island is 2100 metres long, 5 to 40 metres thick, and is a part of the southern limb of an overturned southeast trending syncline. It crops out as a seacliff along the south shore of the island and dips southward under the sea. The ore body is known to extend to depths of 210 metres below sea level. Open-pit mining has lowered the bench levels nearly to sea level and further extraction of ore will require the construction of a retaining sea wall to permit the continued mining below sea level. The main Koolan ore body is 2000 metres long and averages 27 metres in thickness. It is also folded into an overturned syncline and anticline of which the overturned synclinal southern limb is the main ore body and which continues down dip with little change to 190 metres below sea level. Reserves at Koolan are estimated to be about 30 million metric tons of high-grade (66 to 67 per cent iron) ore. Shipments from Yampi Sound in 1979 were 3.4 million metric tons of which 2.3 million were from Koolan and 1.1 million were from Cockatoo. Sixty-four per cent of the ore was shipped to Japan, 21 per cent to the Peoples Republic of China, 7 per cent to South Korea, 1 per cent to Saudi Arabia and 6.5 per cent was consumed domestically. The mine operator is the Dampier Mining Company, a subsidiary of Broken Hill Proprietary Company, Ltd.

Deepdale (Proposed)—Broken Hill Proprietary Company, Ltd. (BHP) plans to develop Deepdale deposits by constructing a railway spur to link their proposed mining operation to the Robe River J.V. railroad at Pannawonica. BHP has entered into an arrangement (Dampier Agreement) wherein they will acquire a 50 per cent equity in the Robe River J.V. railroad and port properties at Cape Lambert in return for providing the joint ventures with 150 million tons of equivalent ore reserves. The estimated ore reserves at Deepdale are 1500 million metric tons of goethite ore grading 57 per cent Fe, 5.7 per cent SiO_2, 2.7 per cent Al_2O_3, 0.05 per cent phosphorus and 9.0 per cent LOI. Additional construction of about 24 kilometres of track will be needed to link the Deepdale deposits with the Robe River J.V. rail, thus making the total rail haul of about 192 kilometres.

Marandoo (proposed)—The Hancock-Wright partnership of Australia and CRA Limited are planning the development of the Marandoo iron-ore deposit which is located 259 kilometres south of its proposed shiploading terminal at Dixon Island in Western Australia. The ore occurs within the Marra Mamba iron formation and reserves are estimated to be 463 million metric tons averaging 62.5 per cent Fe, 2.8 per cent SiO_2, 1.9 per cent Al_2O_3, 0.06 per cent P and 3.94 per cent LOI.

Yandicoogina (Proposed)—CSR of Australia is proposing the development of the Yandicoogina deposits located about 80 kilometres northeast of the Mt. Newman mine and about 25 kilometres away from the Mt. Newman railroad. Reserves are estimated to be about 3 billion tons of pisolitic ores of the Robe River type. Grades are reported to be 58.5 per cent Fe, 4.2 per cent SiO_2, 1.5 per cent Al_2O_3, 0.05 per cent phosphorus and 9.9 per cent LOI. The length of the rail haul to Port Hedland is 237 kilometres.

TASMANIA

Savage River—There are several low-grade ore bodies containing magnetite and pyrite along with minor chalcopyrite, which occur as steeply dipping lenses in an amphibolite sill intruded into an upper Pre-Cambrian metamorphosed basic volcanic succession of rocks of the Savage River area of Western Tasmania. The central and largest of the deposits has an average surface width of 183 metres and length of 1676 metres. The deposit tapers to a width of 122 metres at the ultimate pit-bottom elevation. In 1979 the remaining crude ore reserves were calculated to be 47.6 million metric tons containing about 22 million tons of recoverable concentrates. Reserves of contained concentrates in the northern extension are 25 million tons and in the southern extension, 2.5 million metric tons. The ore is mined by open-pit methods. It is crushed and magnetically concentrated prior to being piped in slurry form 85 kilometres to Port Latta. About 5.0 million metric tons of crude ore must be mined for every 2.25 million metric tons of concentrate. Blending is carried out to maintain a uniformly low nickel content, and the concentrates are pelletized at the port. The pellets analyze at approximately 67.5 per cent Fe, 1.30 per cent SiO_2, 0.35 per cent Al_2O_3, 0.005 per cent S, 0.015 per cent P, 0.008 per cent Cu, 0.38 per cent Ti, 0.047 per cent Ni, 0.46 per cent V, 0.07 per cent Ca and 0.9 per cent Mg. The capacity of the pellet plant is 2.5 million metric tons per year, and in 1980 shipments amounted to 2.2

million metric tons. The Port Latta shiploading berth, which is more than a mile off shore, can accommodate ships of up to 100 000 tons deadweight. Operations started in 1967 and the first shipment went out in April, 1968 under a 20-year, 40.6 million metric ton pellet contract. Savage River Mine's Joint Venture is owned by Northwest Iron Co. Ltd. (50 per cent) and Dahlia Mining Co. Ltd. (50 per cent).

SECTION 3

DISCOVERY AND MINING OF IRON ORES

Introduction—The first part of this section will be confined principally to a brief discussion of modern geophysical methods for iron-ore exploration. The second part will discuss briefly the current methods used for mining iron ore.

DISCOVERY METHODS

Present geophysical techniques and instrumentation: sampling methods; drilling procedures; and some methods of geological investigation will be discussed here, but only as they apply to the search for iron ore.

Geophysics, as applied to iron-ore explorations, is primarily a reconnaissance tool that provides information that must subsequently be complemented by geological mapping, petrographic studies, drilling and the evaluation of ore analyses and treatment tests.

The geophysical techniques used in the search for iron ores, as in most geophysical mapping, are based upon the presence of measureable contrasts of physical properties between the ore minerals and the surrounding rocks. The physical properties used principally are magnetism (both remanent and induced) and density. Electrical methods (including polarization and electromagnetism) and seismic studies can sometimes be used in conjunction with magnetics or gravity surveys to obtain better definition of the ore bodies.

Iron ores in the Lake Superior region were first discovered in outcrops in the vicinity of Negaunee, Michigan and the recognition of iron ore and iron formation in outcrops first served to focus attention on the various iron ranges of the region. Prior to 1950, buried deposits were located by ground surveys to detect variations of the earth's magnetic field, using the simple dip needle or the Hotchkiss "Super-Dip," both of which were effective in detecting iron-bearing deposits that contained magnetite. The Cuyuna Range, where no outcrops occurred because the ore bodies were buried under glacial drift, is one example of a series of deposits that were located solely by dip-needle surveys.

Magnetometers—The greater sensitivity and convenience of operation of modern magnetometers since 1950 have all but retired the dip needle and Super-Dip from practical employment in exploration for iron ore. Magnetometers for this purpose have passed through several successive stages of development, the principal forms being known, in the order of their conception, as balance-type, torsion-type and flux-gate magnetome-

ters, followed in recent years by magnetometers that were conceived and developed in the field of atomic physics: these latter instruments include the rubidium-vapor, proton-precession and optical absorption magnetometers.

All of the foregoing magnetometers are used to determine the strength of the earth's magnetic field or its vertical component at a given location. The earth's field is very weak, ranging from about 0.7 oersted at magnetic poles to about 0.25 oersted at some points on the magnetic equator. In geomagnetic studies, field strength is measured in a much smaller unit than the oersted; the gamma, which is equal to 0.00001 or 10^{-5} oersted. The shape of the earth's magnetic field is not uniform, but shows large-scale regional irregularities due to variations in the shape and composition of the earth's crust and upper mantle. Variations on a smaller scale result from magnetic disturbances caused by concentrations of magnetic material near the surface and it is these local variations that are sought in searching for ore.

Magnetic Surveying—The airborne magnetometer is the primary geological tool used in the search for iron ores and iron-bearing materials in large areas. It was first developed during World War II as a metal-detecting device to locate submarines at sea and was later adapted to mineral and petroleum exploration. Airborne surveys with the magnetometer led to the discovery in 1947 of the Pea Ridge iron-ore deposits of Missouri and in 1950 to the mineralized areas at Lyon, Nevada. It has since been used on a world-wide basis in the search for petroleum and minerals. The method of conducting an airborne magnetic survey is to install a flux-gate or proton precession magnetometer in an aircraft which traverses the target area at a fixed altitude and along predetermined flight lines. The magnetometer measures the magnitude of the earth's magnetic field and these data are recorded electronically along with the position of the aircraft and its altitude. Improvements in the quality of the surveys in recent years have been brought about by refinements in the equipment, i.e., greater sensitivity and simplicity, multiple channel data recording, miniaturization of instruments and a more accurate positioning capability. The presentation and recording of the data in digital form permits the use of computers to carry out the necessary data reduction and plotting requirements needed for analyses and interpretation. Data from these records are plotted as a contour map, with lines connecting points of equal magnetic intensity on the map. The patterns formed by these lines indicate areas where magnetic anomalies (major local distortions of the earth's magnetic field) occur. The areas indicated by anomalies on the magnetic map may then be investigated in greater detail by geological surveys and by gravity measurements, electromagnetic studies or other geophysical techniques.

The detailed magnetic study of anomalous areas may involve using a magnetometer in a helicopter, or ground surveys employing hand-held or other portable magnetometers. A new electromagnetic prospecting technique known as AFMAG (audio frequency magnetics) has been used in areas where magnetic anomalies have been detected to attempt to differentiate between buried deposits of volcanic glass or low-grade iron-bearing intrusives and deposits with high remanent magnetization that represent potential ore bodies. The rubidium-vapor magnetometer, likewise, may make possible the rejection of non-economic deposits by differentiating between magnetic deposits, high in magnetic susceptibility and electrical conductivity, and buried volcanic glass and low-grade nonconducting iron-bearing intrusives of low susceptibility which, however, are capable of producing attractive magnetic anomalies.

Sampling and Drilling—In the early period of iron-ore discovery, most of the exploration of potential ore bodies was done by test pits and shafts. In recent years, correlation and evaluation of the detailed data from magnetometer or other surveys is usually followed by a carefully worked-out drilling program to provide samples that, through geological and mineralogical studies, establish the kind, quality and extent of ore that may be present, and the nature and quantity of the overburden or rock formations associated with the ore.

Considerable attention is being given to the improvement of core-drilling methods to provide better samples. The most complete and undisturbed drill sample possible at a reasonable cost is the ultimate goal of these studies. Diamond drills are employed especially in hard formations. The use of drilling muds with diamond drills has been adopted where samples of the highest quality from alternately hard and soft banded material are desired. Rotary down-hole drills and reverse circulation drills of several types can provide a rapid rate of penetration with satisfactory sample recovery in some sampling applications. Wire-line drilling is employed in about half of the core-drilling operations in the United States.

Studies are continuing of the statistical evaluation of the results of exploration drilling to provide guides for planning drilling programs, especially with regard to the most economical spacing of holes and the most desirable degree of core recovery that would provide adequate sampling at the lowest cost.

MINING OF IRON ORES

Planning and Development—The general occurrence, size and shape of an iron-ore deposit is determined during the exploration phase, which is discussed above. Knowledge of the deposit is determined in more

detail through development work. It is often necessary during the development of a mine to determine, in considerable detail, the position and nature of geological structures which affect ore distribution and availability. Currently most deposits being exploited in the United States are of low-grade materials called taconites containing about 30 per cent iron, all of which must be beneficiated or concentrated to produce acceptable shipping products. In such ore deposits, the results of laboratory concentration tests of drill samples, supplemented by pilot-plant tests and by experience with commercial-plant results, are used to determine the economics of ore treatment.

After sufficient detailed information is obtained, various combinations of operating plans are studied using maps and sections prepared for this purpose. These show the size and shape of the ore body, ore compositions and laboratory-test results. From these graphic representations, the quantities of ores and waste materials are determined by the application of volume-weight factors. Computers are commonly used in the preparation of tonnage estimates and in the preparation of detailed mining plans. Through the use of these systems, comparative evaluations of various mining methods and plans may be made to determine the most favorable plan for each particular deposit and to schedule the mining of the deposit.

Open-Pit Mining—Inasmuch as open-pit mining provides the lowest-cost operation, it is employed wherever the ratio of overburden, either consolidated or unconsolidated, to ore does not exceed an economical limit. Nearly all the large iron-ore mines in the world, with the exception of some in Europe, are worked by open-pit methods. Figures 5—14, 5—15 and 5—16 show examples of open-pit mining on the Mesabi Range.

The depth to which open-pit mining can be carried depends upon the grade of the ore, the nature of the overburden and the stripping ratio. The stripping ratio is the amount of overburden and waste that has to be handled for each unit of ore mined. The economic stripping ratio varies widely from mine to mine and district to district, depending upon a number of factors. In the case of direct-shipping ores, it may be as high as 6 or 7 to 1; whereas, in the case of taconite, a stripping ratio of less than ½ to 1 may be the economic limit.

Overburden (stripping) may consist of unconsolidated material, rock, or lean ore material. In open-pit mining, removal of overburden may continue through a large part of the life of a mine as the pit walls are cut back to permit deepening of the mine to recover ore in the bottom. Unconsolidated materials are excavated by power shovels, draglines or power scrapers, depending on local conditions. Other materials are generally excavated with power shovels.

Drilling and blasting is done to break consolidated materials into sizes capable of being handled by mining equipment and beneficiation facilities, and is also done to loosen ore banks ahead of power shovels to increase the efficiency of loading.

Taconite ore is loaded by power shovels equipped with buckets ranging in capacity up to 11.5 m³ (15 cubic yards). The ore is transported out of the pit by railroad cars, trucks, belt conveyors or combinations of

FIG. 5—14. Electric power shovel on caterpillar tracks, making a "sinking cut" and loading ore into steel dump cars for haulage from an open-pit mine to a beneficiation plant. Diesel-electric locomotives handle the dump cars. Cable carries power to shovel.

FIG. 5—15. Electric power shovel in an open-pit mine loading iron bearing material into diesel-powered truck for transfer to a processing plant.

FIG. 5—16. Electric power shovel loading material from an open-pit iron-ore mine for transfer to an ore-processing plant.

these, to a crushing plant for size reduction and then to a beneficiation plant for fine grinding and concentration by magnetic, gravity or flotation techniques. The concentrates are usually agglomerated into 6.4 to 9.5-mm (¼- to ⅜-in.) balls and fired into hard durable pellets which will withstand handling, shipping and furnace charging.

The mining of taconite poses special problems because of its extreme hardness, which necessitates considerable additional drilling and blasting and more specialized and rugged equipment, as compared with the techniques and equipment used in mining most oxidized ores. Also, the relatively low iron content of taconite makes it necessary to handle two to four times as much mined material to obtain the same quantity of iron-in-ore as from higher grade ore deposits.

Water causes a variety of problems in iron-ore-mining operations. Except in rare instances, such as in hilltop mining or mining under desert conditions, water must be collected in sumps, wells or underground workings and pumped out of the mine. Such drainage water is often utilized directly to make up for water losses in concentration operations.

Underground Mining—When the stripping ratio becomes too high for economical open-pit mining, underground mining methods may be employed. In most cases, access to underground mines is obtained through inclined or vertical shafts sunk adjacent to the deposit but far enough away to avoid the effects of surface subsidence resulting from mining operations. Some deposits are so situated that adits can be driven directly into them from hillsides or from open-pit banks.

Underground mining requires a larger capital investment per ton of annual capacity than open-pit mining because it depends upon costly shafts or tunnels, underground haulage and development workings, and elaborate pumping facilities. Moreover, the production of iron ore per man per day in an underground mine is only a fraction of that in open-pit operations, whereas the cost of supplies, maintenance, hoisting, and pumping, are all higher.

In underground mining, several methods of ore extraction may be used. Among the most common, in order of increasing cost, are: block caving, sub-level stoping, sub-level caving, top slicing, and modifications or combinations of these. All of these methods involve: drilling; blasting; transportation within the mine by rail tramming, trackless shuttle cars, scrapers, or conveyor belts; and hoisting or hauling to the surface. On the surface, the ore may be crushed, sized or concentrated prior to shipment.

In general, the higher costs of underground mining

has limited its use to ores requiring simple crushing or sizing, special ores such as open-hearth lump ore, or low-grade ores located so near a consuming market that transportation costs to the point of use were not significant. There is one operating underground mine in the United States, located at Pea Ridge, Missouri, where a magnetite ore body is being mined at depths of plus 600 metres (plus 2000 feet). The ore is crushed, concentrated and used as pellet feed.

Grading of Iron Ores—Furnace practices, as described elsewhere in this volume, require the production of natural ores and/or concentrates to meet the physical and chemical specifications required in iron-ore steelmaking processes. Prior to the shift in recent years to the large-scale production of iron-ore concentrates and pellets, the natural ores were "graded" by the producers and shippers to meet furnace demands for particular and uniform chemical composition and structure. Iron-ore merchant companies could satisfy grade requirements by purchase or exchange of ores, which were then mixed together in the correct proportions. Recognition of the importance of uniformity led to the use of elaborate ore-blending facilities at some producing and consuming points, involving systematic layering in stockpiles and recovery for shipment or consumption by cross-cutting the layers (see Figure 5—17).

In recent years in the United States, the annual production of natural ores has decreased significantly and there has been a reciprocal increase in the production of iron-ore concentrates and pellets. On a world-wide basis, many iron-ore districts continue to produce natural ores which are usually mined to supply a uniform quality of ore. Some producers ship natural ores along with sized products, concentrates and pellets while others ship only concentrates and/or pellets. In general, it has been demonstrated that the added cost of utilizing a uniform high-grade iron-ore product as furnace feed can result in significant economies in the cost of the resulting hot metal.

Of particular significance in grading of ores, aside from the iron content, is the content of silica, phosphorus, manganese and alumina. A high lime content makes some ores self-fluxing. Sulphur, copper, nickel, titanium and other deleterious constituents may require close control in some producing areas. While ores are generally priced on the basis of natural iron content (that is, the amount of iron in the ore before the free moisture is removed), penalties or premiums may be applied for varying chemical and structural quality. Practices vary considerably in the various world iron-ore markets.

SECTION 4

BENEFICIATION OF IRON ORES

Introduction—The term "beneficiation" in regard to iron ores encompasses all of the methods used to process ore to improve its chemical or physical characteristics in ways that will make it a more desirable feed for the ironmaking furnace. Such methods include crushing, screening, blending, concentrating, and agglomerating.

Because of the differences in structure and mineral content of ores from different deposits, beneficiation methods vary considerably.

Consequently, the following brief and generalized descriptions are intended only to describe how some types of ore are beneficiated, and are not to be interpreted as suitable for all ores.

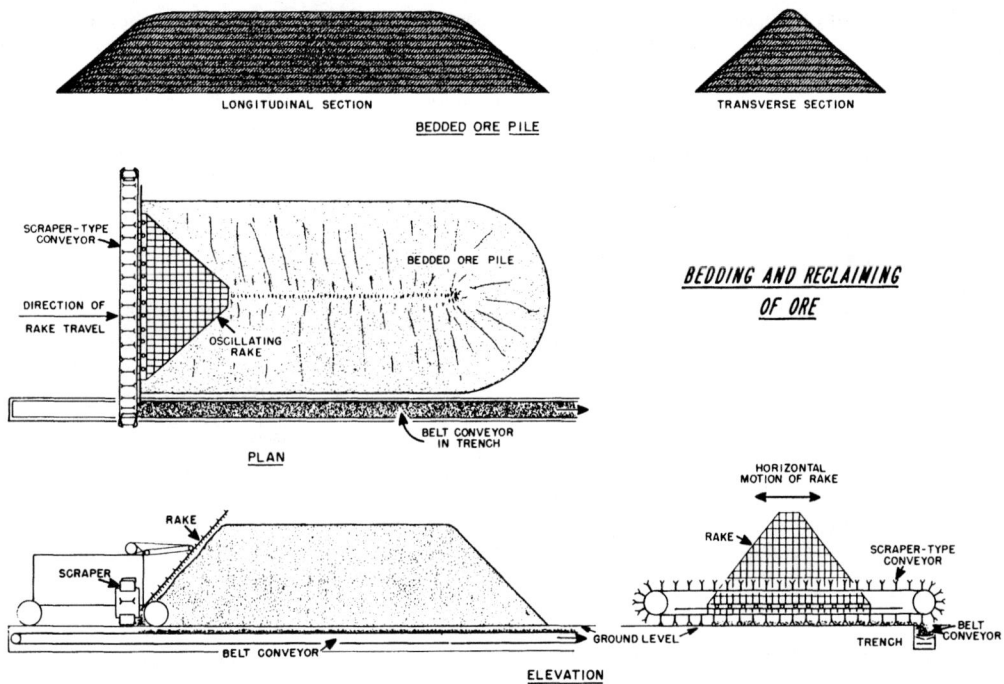

LONGITUDINAL SECTION

BEDDED ORE PILE

TRANSVERSE SECTION

SCRAPER-TYPE CONVEYOR

BEDDED ORE PILE

DIRECTION OF RAKE TRAVEL

OSCILLATING RAKE

BEDDING AND RECLAIMING OF ORE

BELT CONVEYOR IN TRENCH

PLAN

HORIZONTAL MOTION OF RAKE

RAKE

SCRAPER

RAKE

SCRAPER-TYPE CONVEYOR

BELT CONVEYOR

BELT CONVEYOR

GROUND LEVEL

TRENCH

ELEVATION

FIG. 5—17. Schematic representation of the principle of operation of a stacking and reclaiming system.

CRUSHING, SCREENING AND BLENDING

Crushing and Screening—Iron ore of merchantable grade must be properly sized prior to charging to the blast furnace. Present blast-furnace technology commonly requires crushing the ore finer than 50 mm and that fine material less than 6 mm be removed. The specific size selected is based on the characteristics of the ore and is specified so as to maintain high stack permeability and also allow sufficient time for reduction of the coarser material. Consequently, crushing and screening are an integral part of ore-producing facilities.

The fines produced in crushing require agglomeration, usually by sintering, prior to charging. The sinter produced is crushed and screened to meet size specifications compatible with the other charge components.

Most of the iron ore now mined in North America is concentrated to increase its iron content. The concentration processes range from relatively simple crushing, screening, and washing operations to sophisticated, large-scale operations involving crushing, fine grinding, magnetic separation, and flotation. Usually the large modern plants are integrated with a pellet plant.

Blending—The mining program at individual mines is developed to produce a uniform product. When it is desirable to blend ores of different compositions or size consists (as when material from different sources are to be combined), mixing may be accomplished during the numerous handling operations involved in transportation of the ore to its point of use, or special blending facilities may be employed.

Iron ores of different characteristics and compositions can be blended to a more uniform composition by

a method known as stacking and reclaiming. In one method the ore is piled by a machine called a stacker. Stacking results in "layering" of the ores (Figure 5—17). Each successive layer represents an ore that may differ in size consist or chemical composition from adjacent layers. The elongated pile is built up to a height limited by the stacking capability of the machine. The ore may be reclaimed for use by bucket-wheel excavators, front-end loaders, or a scraper-cross conveyor (Figure 5—17). Removal of ore from the face of the pile results in a stream of material that is a mixture of ore from all the layers.

CONCENTRATING PROCESSES

The earliest methods for improving the quality of iron ores consisted chiefly of processes called by the general name of "washing", by which a large proportion of the sand, clay, and rock could be removed from the crushed and screened ore. The iron (oxide) content of the washed product was considerably higher than that of the crude ore. As ore-grade requirements became more stringent, other more complex methods were adopted to upgrade ores that are not amenable to simple washing operations. The specific methods depend on the properties of the ore and include among others: heavy-media separation, spirals, flotation, and magnetic and electrostatic concentration.

The past two decades have seen the development of major mining and concentrating facilities to treat low-grade magnetite-taconite, specular hematite iron formation and oxidized taconite. These source materials are all low-grade iron formations that contain less than 35 per cent iron. They all require relatively fine grinding to produce high-grade concentrates that generally contain less than 6 per cent silica. These final concentrates are either pelletized or sintered prior to charging to the blast furnace. Their development as commercially successful operations was based on years of research on grinding, magnetic separation, flotation, and agglomeration supported by both government and commercial laboratories.

In the instance of magnetite ores, including the magnetite taconites of Minnesota and Wyoming, the iron-oxide (magnetite) in the finely ground crude ore is successfully separated from the non-magnetic gangue material by magnetic means. Flotation and selective flocculation are effective in the separation of nonmagnetic hematite from silica in certain of the Michigan jaspers. Much of the semi-taconite and other nonmagnetic iron-formation materials respond favorably to roasting in a reducing atmosphere to convert hematite to magnetite. The converted material can then be ground and the magnetite separated from the siliceous gangue by magnetic methods to produce a high-grade concentrate. Concentrates obtained by the above methods contain 62 to 67 per cent of iron, with about 5 to 9 per cent of silica and, when agglomerated, make a desirable blast-furnace feed.

The investment required to produce concentrates from low-grade ore is high. Nonetheless, in the United States, premium-quality blast-furnace feed is being produced from the magnetic taconites of the eastern Mesabi Range, plants in Michigan are operating on jasper, and experimentation with the nonmagnetic taconites and semi-taconites of the western Mesabi Range and with Michigan iron formation is continuing.

Steelmaking facilities in Alabama utilize high-grade imported ores because the technical difficulties in concentrating the low-grade local ores have not been solved.

The intermountain steelmaking centers in Colorado and Utah, that are relatively inaccessible to foreign ores, beneficiate local ores by washing and coarse magnetic separation. A major concentration plant in Wyoming is beneficiating a magnetite taconite which is pelletized and shipped to blast furnaces in Utah.

The necessity for beneficiation does not apply only to North American ores, for the problem is world-wide and will become of increasing importance as high-grade deposits become depleted. Space does not permit a review of the matter on a world-wide basis nor is it possible to discuss, except in a general way, the various processes and equipment employed in beneficiation of iron ores; further details on these and related matters will be found in the sources listed in the bibliography at the end of this chapter.

All concentration processes produce tailings that must be stored indefinitely in an environmentally acceptable manner. Usually, fine tailings are pumped in slurry form to an impoundment area and the water is recycled to the plant. This practice assures an adequate supply of process water and eliminates the pollution of surface waters.

Comminution—Size reduction is usually the first step in an iron-ore beneficiation process. Its purposes are to liberate iron minerals from waste prior to a concentration step and/or to generate a particle-size distribution suitable for further processing. If the ore is to be concentrated it must be broken to a size at which the iron oxides may be separated from gangue material. The amount of size reduction is determined by the grain size and degree of intergrowth of the iron oxides and the gangue minerals. Wash ores, which contain rather coarse iron-oxide particles loosely cemented together by fine silica and clay, require crushing to about 25 mm (1 inch) prior to concentration. On the other hand magnetic taconite contains finely disseminated magnetite and silica particles and generally requires crushing and grinding to about 90 per cent minus 53 μm (270 mesh) before a final concentrate can be produced.

Size-reduction processes usually involve several stages in series and may include concentration equipment between stages to reject waste particles from the circuit as soon as they are liberated. Run-of-mine taconite and jasper ores are broken to minus 305 to 457 mm (12 to 18 inches) by jaw or gyratory primary crushers. The primary crusher product is suitable feed for primary autogenous grinding mills, as these utilize a charge of ore chunks as grinding media. If grinding is to be accomplished by rod mills and ball mills, the primary crusher product must be further crushed to minus 18 mm (¾ inch). This requires two or three additional stages of cone crushers and sizing screens. A typical three-stage taconite crushing flow sheet is shown in Figure 5—18.

As the name implies, rod mills and ball mills contain a charge of steel rods or balls to accomplish the grinding. Usually the ore is ground as a water slurry containing some 68 to 80 per cent solids by weight. Final product size usually is controlled by hydrocyclone classifiers, sometimes supplemented by screens.

Primary autogenous mills may operate either wet or dry and accept ore having a top size of 305 to 457 mm (12 to 18 inches), which is produced from only one stage of primary crushing. Successful operation of a fully autogenous primary mill requires that there be a proper distribution of small and large ore particles in the mill. If the breakage characteristics of the ore are not compatible with this requirement a small charge of 125 to 150 mm (5 to 6 inch) steel balls may be added, in which case the operation is termed semi-autogenous grinding.

Pebble mills are used for fine grinding. They are similar to ball mills except that they are charged with 25 to 100 mm (1 to 4 inch) pebbles rather than steel balls. The pebbles are ore particles obtained from the crushing plant or from the primary grinding mill.

The choice between autogenous or rod mill-ball mill grinding is decided on the basis of physical and economic factors. Autogenous grinding is not feasible or may be difficult to control when the ore does not have the proper breakage characteristics. The obvious advantage of not requiring fine crushing for autogenous grinding may be outweighed by a higher overall grinding energy requirement.

Washing—Washing is a form of concentration that utilizes the differences in specific gravity and particle size of valuable iron-bearing minerals and gangue to separate the two. As applied to iron ores, washing removes fine clay and sand by suspending these unwanted materials in a flowing stream of water that

AS MINED ORE

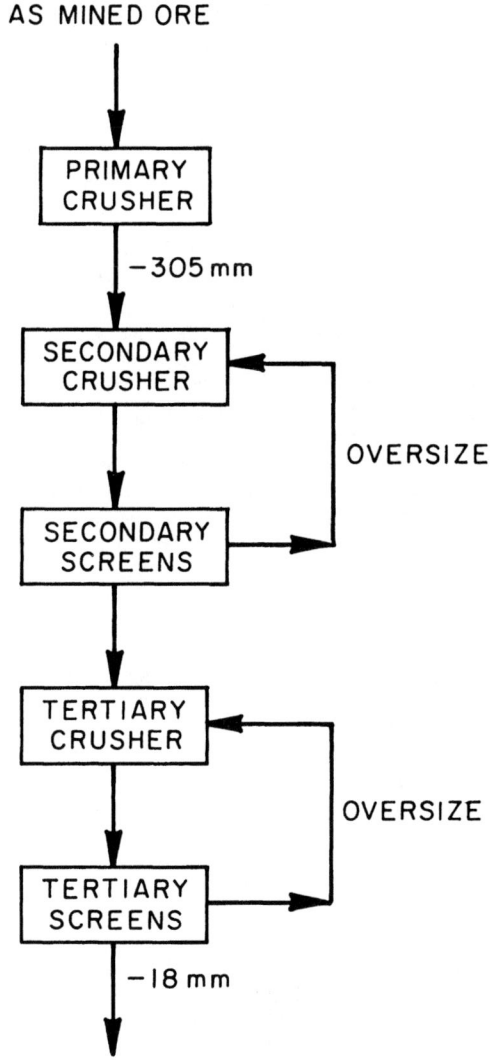

FIG. 5—18. Flow sheet of an iron-ore crushing circuit.

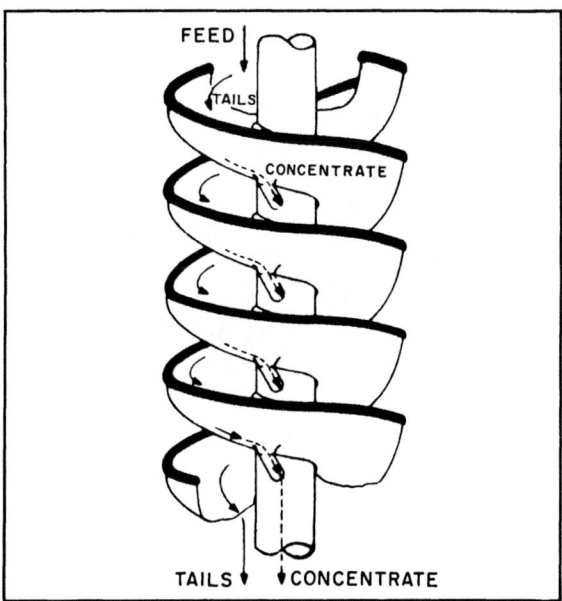

FIG. 5—19. Schematic of a spiral concentrator.

carries them out of the washer, while the heavier iron minerals sink to the bottom of the separating vessel and are removed as a concentrate. Log washers have been the most common machines used for washing iron ores, often operating in conjunction with rake classifiers or spiral classifiers that perform a supplementary concentrating operation for removal of fine-sized ore from the overflow of the log washer. Some washing plants employ spiral classifiers in one or two stages without a log washer on ores containing a minimum amount of sticky clay gangue. Hindered-settling classifiers of various types also have been used advantageously for washing iron ores of finer sizes than are normal feed for log washers and rake or spiral classifiers. Vertical-current washers and turbo washers have also been employed.

Jigging—Jigging is a more complex form of beneficiation than simple washing and involves the stratification of ore particles and gangue by subjecting the ore to alternating upward and downward pulsations of water. The gangue overflows the jig, while the iron oxide par-

ticles are removed as an underflow product, either periodically or continuously.

Heavy-Media Separation—Heavy-media separation processes operate on the "sink-and-float" principle. When a mixture of particles of two minerals of different specific gravities is placed in a fluid having a specific gravity intermediate between the specific gravities of the two minerals, the less dense mineral will float and the more dense will sink. For example, quartz (sp gr = 2.65) will float in a medium having a specific gravity of 3.00, while hematite (sp gr = 5.00) will sink.

Heavy-media separation plants for iron-ore concentration employ suspensions of magnetic material as the medium. The favored medium for concentrating coarse ore is ferrosilicon containing 15 per cent silicon and 85 per cent iron. Water suspensions containing 64 to 85 per cent by weight of finely ground ferrosilicon have specific gravities ranging from 2.20 to 3.60. The separation vessels are designed to provide continuous removal of sink products at one point and of float product (waste) at another. The medium leaving the separator must be cleaned, its specific gravity adjusted, and the suspended particles redispersed before it is returned to the separatory vessel. Medium is removed from the sink and float products by screens and water sprays and is recovered from the wash water by wet-type magnetic separators. Hydrocyclones are utilized as separatory vessels for finer ore [6 × 25 mm (¼ × 1 inch)] and usually use a slurry of water and magnetite as medium.

Spirals—A spiral concentrator is a curved-bottom trough, wound around a vertical axis in the form of a helix (see Figure 5—19). When fed at the top with a slurry of iron ore and gangue, the less dense gangue, being more readily suspended by the water, attains greater tangential velocity than the iron minerals, and migrates toward the outer rim of the spiral trough. Wash water is added along the inside rim to further

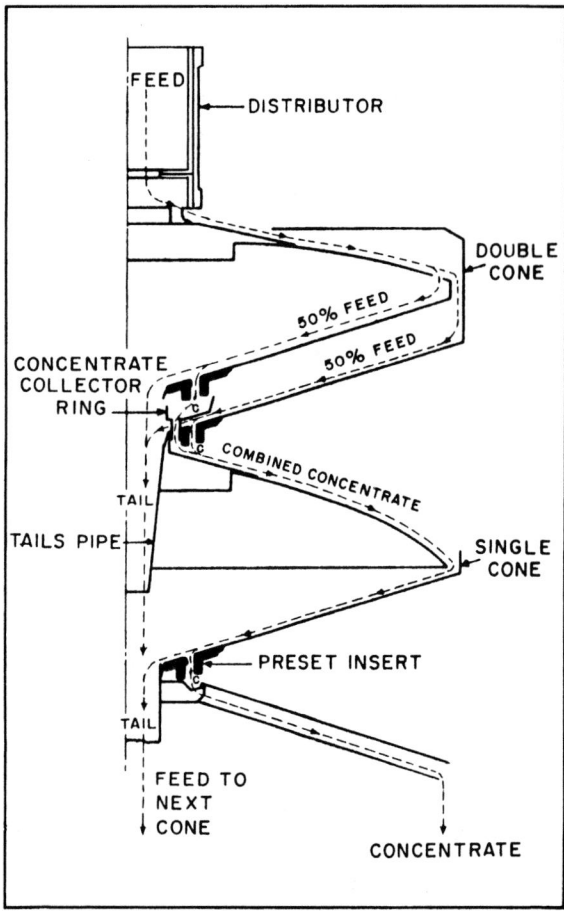

FIG. 5—20. Schematic of a section of a Reichert cone concentrator.

wash away the lighter gangue. After a few turns, a band of iron mineral forms along the inner rim, and the gangue forms bands toward the outer rim. Ports are spaced along the inner rim to collect and remove the iron minerals. The gangue remains in the spiral and discharges at the bottom. Recovery of particles smaller than 100 μm (150 mesh) is relatively poor, so that the efficient application of spirals is limited to those feeds that do not contain a large amount of minus 100 μm iron oxides.

Reichert Cone—The Reichert cone (Figure 5—20) is a flowing-film concentrator. The denser particles concentrate at the bottom of a flowing film of slurry having a solids content of about 60 per cent by weight. The separation mechanism is a combination of hindered settling of the dense particles and interstitial trickling of the fine particles. The separation element in the Reichert unit is an inward sloping 1.9-m (6.25-ft) diameter cone. Feed pulp is evenly distributed around the periphery of the cone. As the pulp flows by gravity toward the center, the fine and the heavy particles concentrate on the bottom and are removed through an annular slot near the apex of the cone. The tailing flows over the slot and is collected at the apex or center of the cone. Because the efficiency of this separation

process is relatively low, it is repeated several times within a single stacked arrangement of cones to increase the recovery. Generally, the highest-grade concentrate is the first removed.

The characteristics of the cone are capacity of up to 100 metric tons per hour and good recovery of fine iron oxide. The cone requires a well-deslimed feed and works best on specular hematite ore.

Flotation—Flotation processes depend on the fact that certain reagents added to water suspensions of finely ground iron ore selectively cause either iron-oxide minerals or gangue particles to exhibit an affinity for air. The minerals having this affinity attach to air bubbles passing through the suspension and are removed from the suspension as a froth product.

The reagents added to induce the preferential affinity for air are commonly called collectors or promoters; substances added to cause stable bubble or froth formation are known as frothers; other substances added for control purposes such as pH adjustment, or to cause better dispersion or flocculation are known as modifiers, dispersants, and depressants.

Flotation collectors are of two general types; anionic and cationic. Anionic collectors ionize in solution such that the active species (that which attaches to the positively charged mineral surface) is negatively charged. Conversely, the active ionic species in cationic flotation collectors is positively charged.

The main application of anionic flotation is to float iron-bearing minerals away from gangue material. This requires the addition of fatty acids or petroleum sulphonates as collectors. Fuel oil is often added along with the collectors to promote recovery of iron oxide grains finer than about 10 micrometres.

Cationic flotation is used to float siliceous gangue away from finely ground crude ore and to remove small amounts of gangue material from some magnetic concentrates. Cationic collectors are primary aliphatic amines or diamines, beta-amine, or ether amines, generally in the acetate form. An essential requirement for cationic iron-ore flotation is that the slime content of the ore (the very fine siliceous material less than 10 micrometres) be substantially reduced. This is accomplished by adding starch products to selectively flocculate iron minerals so that they will not be lost during desliming, and to depress the iron minerals during flotation.

Magnetic Separation—Magnetic separation of magnetite-bearing iron ore and taconite is usually accomplished by low-intensity, drum-type separators having a field strength less than 2000 gauss. Low-intensity magnetic separation is a simple and efficient means for making a clean separation between magnetite particles and gangue minerals. High-intensity magnetic separators have field strengths greater than 5000 gauss and are employed at several locations for concentrating slightly magnetic hematite ores and concentrates.

In most cases low-intensity magnetic separators treat an ore slurry that has been ground to minus 6 mm (¼ inch) or finer. Typically the machines are drum-type such as the one depicted in Figure 5—21. In this machine, magnetite particles are attracted to and held against the surface of the rotating drum until they are carried out of the magnetic field; the magnetite then

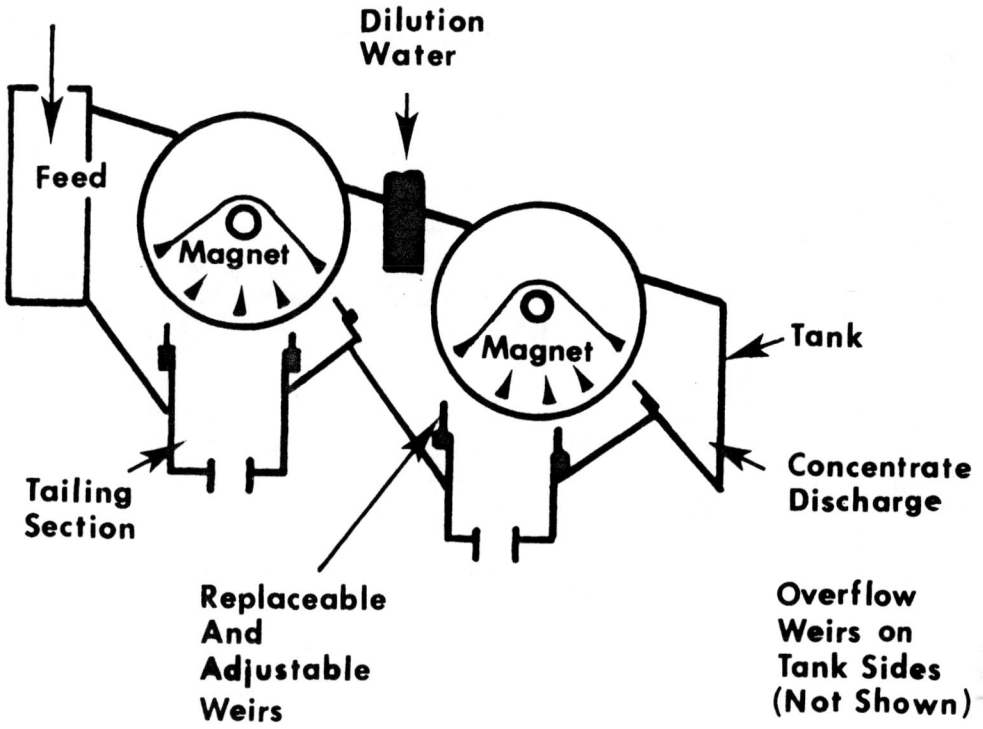

FIG. 5—21. Schematic of two-drum wet magnetic separator.

transfers from the drum surface to an appropriate concentrate receiver. Multiple stages of magnetic separation are sometimes required to eliminate fine gangue particles that become entrapped in the flocs that form when slurry containing fine magnetite particles enters the field of a magnetic separator. Often this is accomplished by using machines having two to four drums in series. Most taconite processes incorporate magnetic separation after each stage of grinding in order to reject nonmagnetic waste at as coarse a size as possible, resulting in reduced grinding costs.

The earliest drum separators used electromagnets. Today powerful permanent magnets have replaced the electromagnets, with substantial energy savings. The trend is toward larger drums of 1220-mm (48-inch) diameter compared with the more common 914-mm (36-inch) diameter.

Dry low-intensity separation systems have not been as widely applied as wet systems because the presence of water is highly advantageous as an agent for flushing fine nonmagnetic particles out of the magnetic field. In certain arid areas both grinding and magnetic separation are performed dry.

In wet high-intensity magnetic separation (WHIMS), electromagnets produce a very high-strength magnetic field that is applied to a matrix consisting of steel balls, spaced grooved plates, steel wool, or pieces of expanded metal (see Figure 5—22). The matrix is contained in a metal box and must have sufficient free area so that the ore slurry to be treated can flow through it without plugging. The sharp edges of the matrix concentrate the magnetic field to produce a high magnetic gradient. This combination results in a high attractive force that attracts and holds hematite particles. Re-

moving the matrix (which has a relatively low remanance) from the field and flushing it with water releases the hematite.

WHIMS applications include recovery of iron oxides from natural ore fines, upgrading of spiral concentrates for direct-reduction feed, and recovery of hematite from tailings.

Electrostatic Separation—In this process, high voltage (about 30 000 volts) is applied to impose a surface charge on dry iron oxides and gangue passing over an electrically grounded drum called a rotor. The iron oxides, which are electrical conductors, bleed their charge to the grounded rotor and fall into a pocket at the bottom of the rotor. Because the nonconductive gangue minerals cannot discharge they adhere to the rotor (commonly called pinning). The pinned minerals are carried by the rotor to the discharge point where they are removed by a stationary brush. This process is currently being applied at a large iron-ore beneficiation plant in Canada.

CONCENTRATION-PLANT FLOW SHEETS

The type of beneficiation process depends on the ore and the ultimate destination of the iron concentrate, i.e., direct-reduction feed requires a much lower silica content necessitating a more elaborate beneficiation flow sheet than does blast-furnace feed.

The flow sheet for United States Steel's Minntac taconite concentrator is shown in Figure 5—23. Run-of-mine ore is crushed in three stages to 90 per cent minus 18 mm (¾ inch) and is then ground in a rod mill to produce a 90 per cent minus 3 mm (⅛ inch) feed to the first stage (cobber) of magnetic separation. The

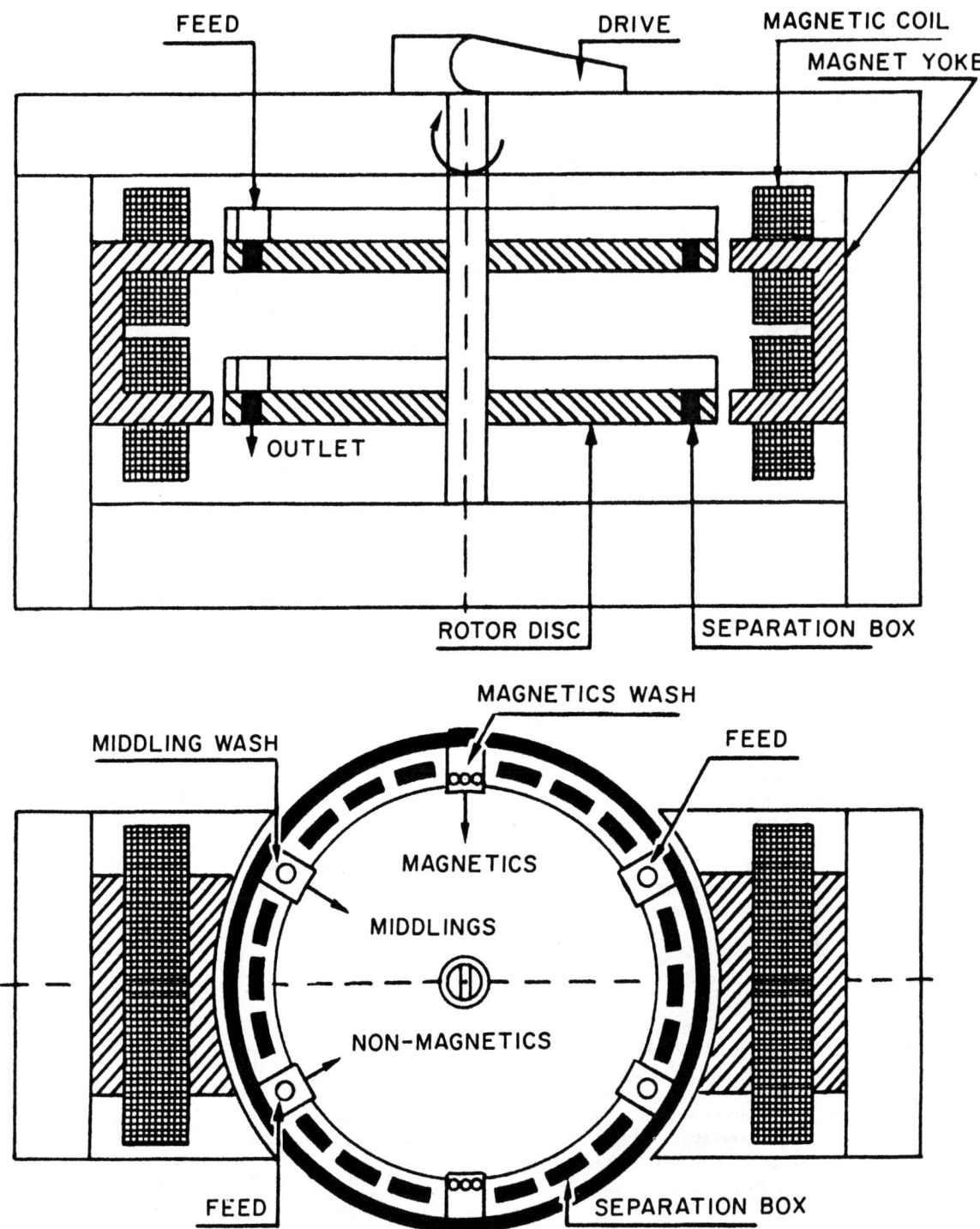

FIG. 5—22. Schematic diagram of a wet high-intensity magnetic separator.

magnetic concentrate is reground in the primary ball mill to minus 420 micrometres (35 mesh) and sent to the second stage (rougher) of magnetic separation. The second stage concentrate is reground to 70 per cent minus 53 micrometres (270 mesh) and sent to hydroseparators where some fine silica is removed in the overflow. The hydroseparator underflow goes to the finisher stage of magnetic separation. Finisher magnetic concentrate is screened, with the oversize returned to the secondary ball mill and the undersize at about 90 per cent minus 53 micrometres (270 mesh) goes to a cleaner stage of magnetic separation. The

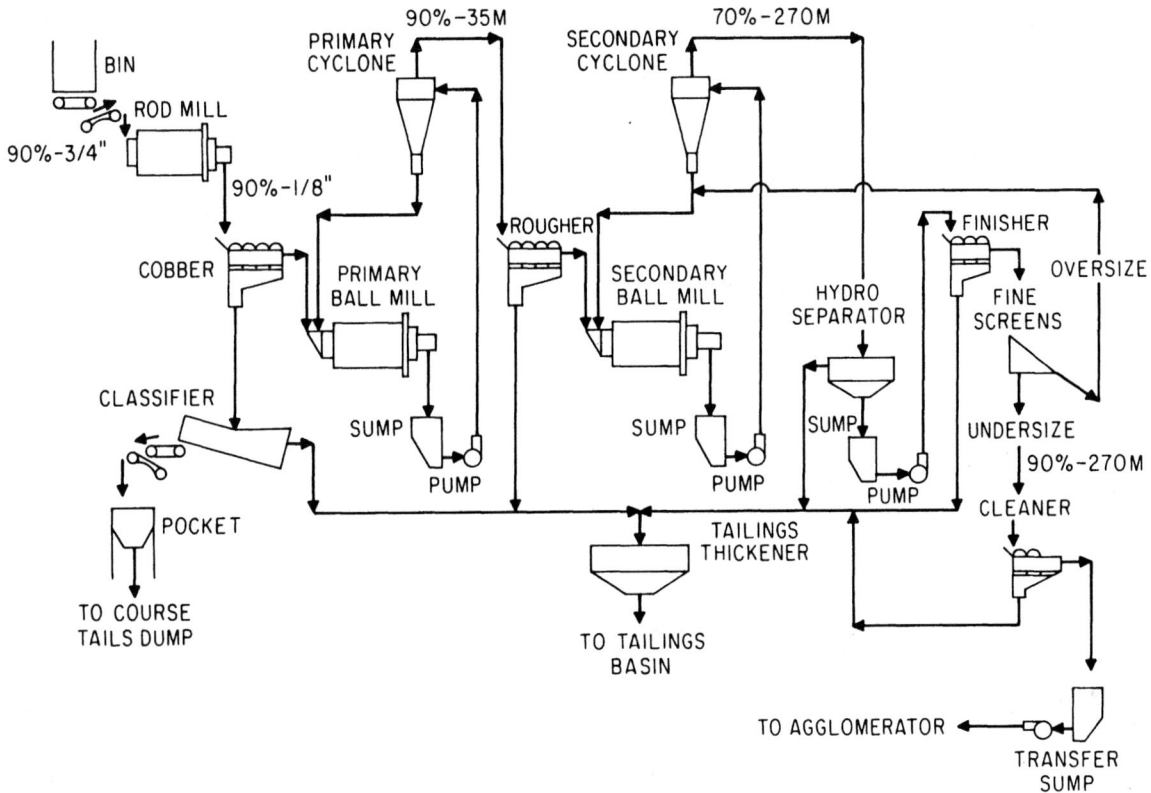

FIG. 5—23. Flow sheet for United States Steel's Minntac taconite concentrator.

FIG. 5—24. Grinding mills and magnetic separation of Minntac.

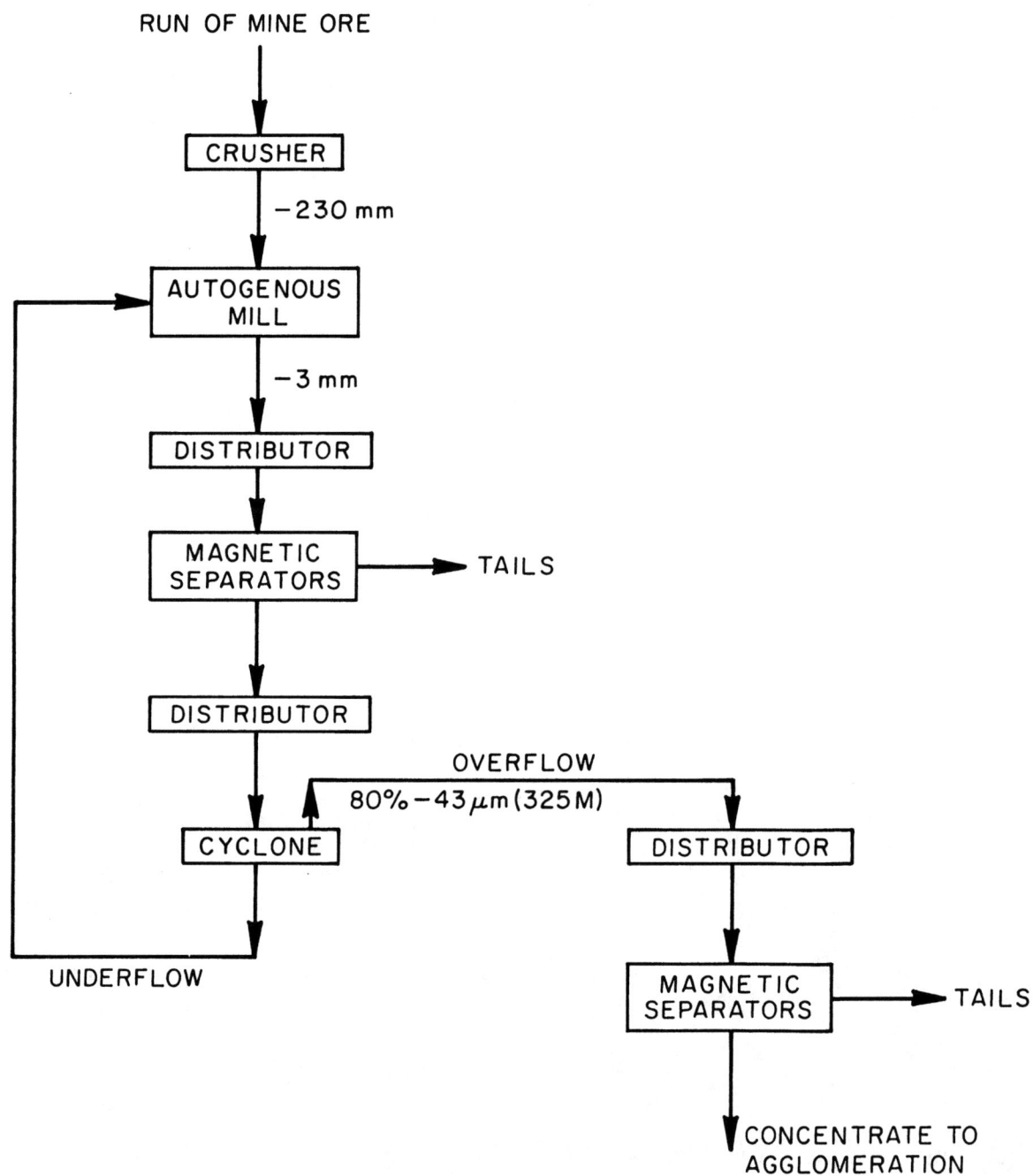

FIG. 5—25. Simplified Hibbing taconite flow sheet.

final concentrate is pumped to the agglomerator for pelletizing. Minntac grinding mills and several stages of magnetic separation are pictured in Figure 5—24.

Another method of processing taconite is shown in Figure 5—25, which is a simplified flow sheet for the Hibbing Taconite Company. Run-of-mine ore is crushed to minus 230 mm (9 inch) in a single stage and is ground in autogenous mills to minus 3 mm (⅛ inch). The autogenous mill discharge is sent to the first stage of magnetic separation. The magnetic concentrate is classified in hydrocyclones to produce an overflow con-

taining about 80 per cent minus 43 micrometres (325 mesh), which goes to the second stage of magnetic separators. The coarse cyclone underflow returns to the autogenous mill. The second stage magnetic concentrate is the final product and goes to the agglomerator.

The flow sheet of Quebec Cartier Mining Company's Mount Wright concentrator is shown in Figure 5—26. The ore contains specular hematite and quartz and is ground to minus 833 μm (20 mesh) to liberate the hematite so that it may be concentrated by spirals. Run-of-mine ore is crushed to minus 305 mm (12 inches) and is

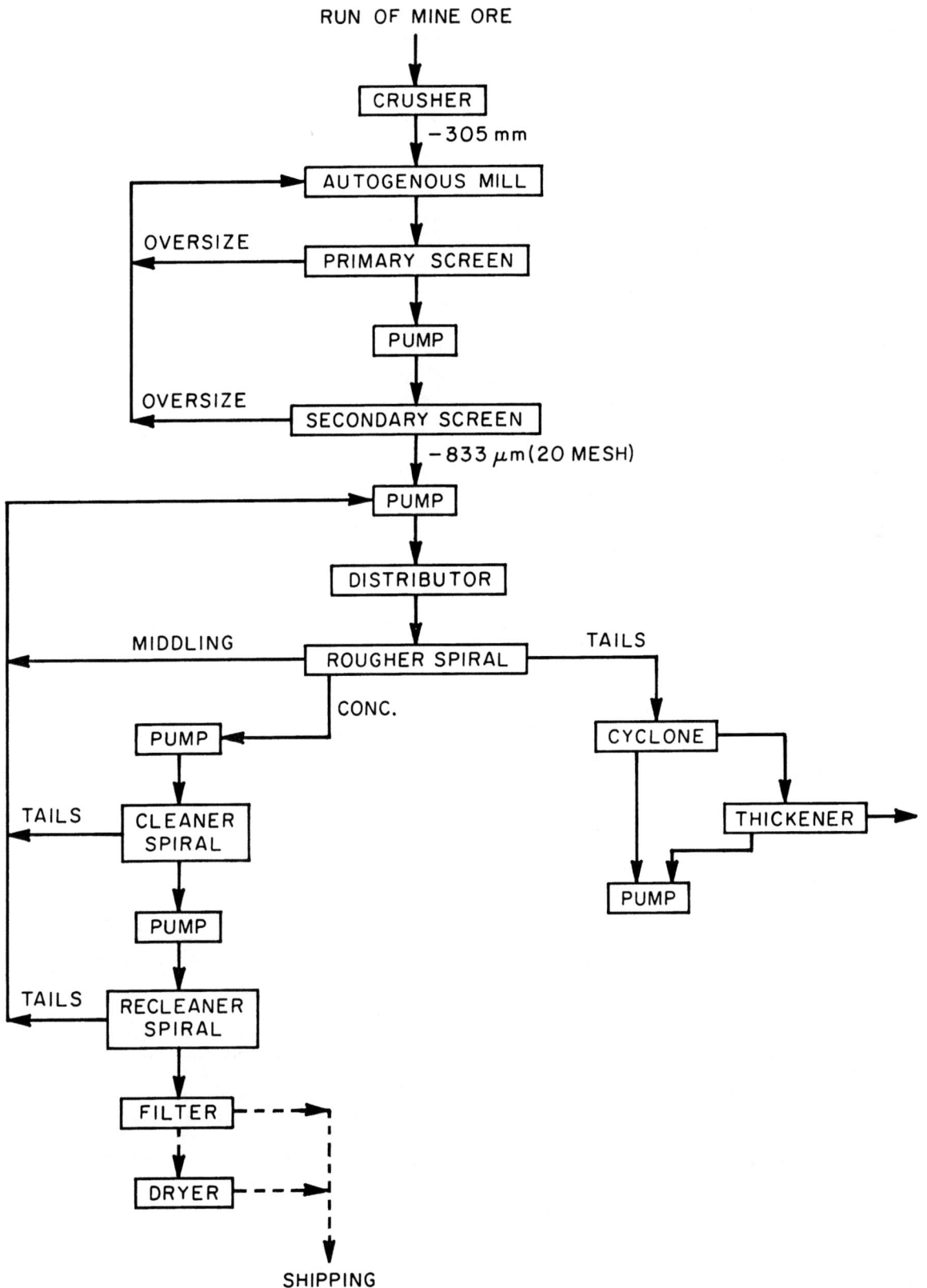

FIG. 5—26. Flow sheet for spiral concentration.

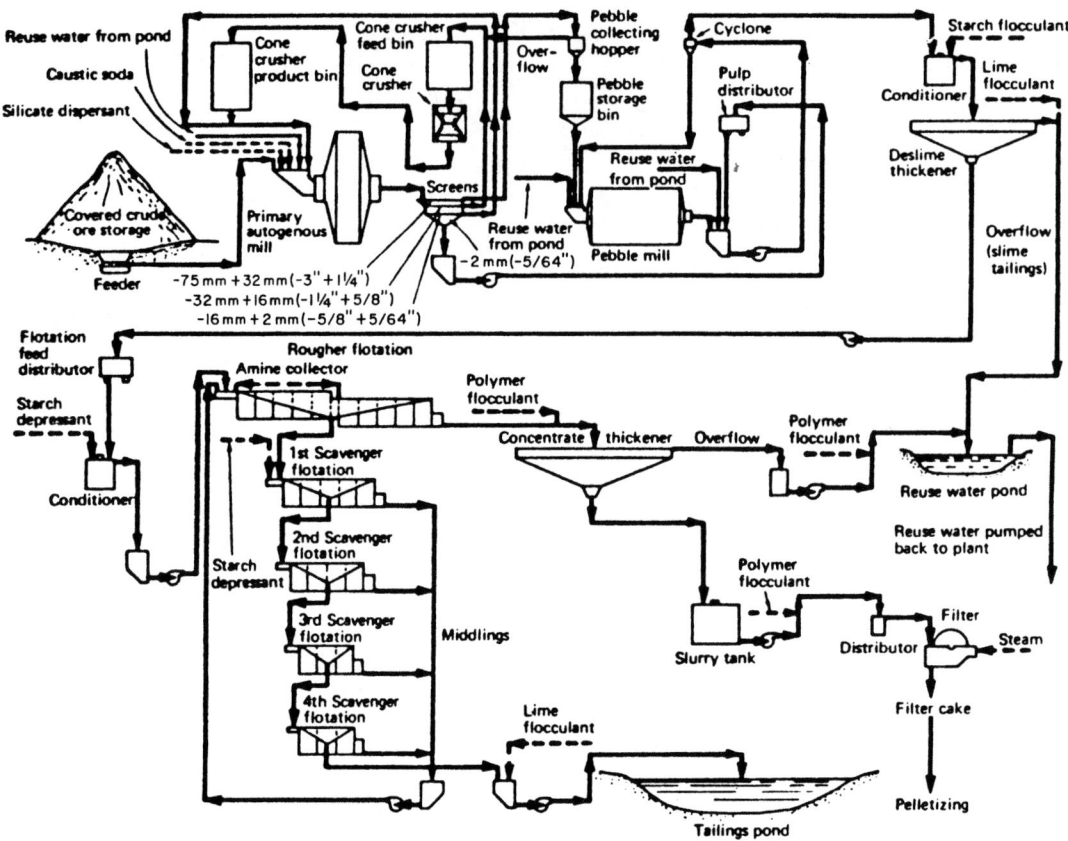

FIG. 5—27. Concentrator flow sheet of the Tilden iron-ore mine. (Courtesy, Engineering/Mining Journal.)

ground in an autogenous mill. The grinding mill product is screened in two stages to remove the plus 833 μm (20 mesh) oversize, which is returned to the mill. The screen undersize is concentrated by three stages of spirals to make a concentrate which is then dewatered on horizontal vacuum filters. The concentrate enters world markets as a minus 833 μm (20 mesh) product and is used mainly as sinter feed.

A flow sheet for the Tilden concentrator operated by the Cleveland Cliffs Iron Company is shown in Figure 5—27. The crude ore is a jasper containing fine, disseminated hematite. Run-of-mine ore is crushed and then is ground in primary autogenous mills. The mill discharge is screened to remove minus 2 mm (8 mesh) particles, which are reground to 80 per cent minus 74 micrometres (200 mesh) in pebble mills. Part of the screen oversize is used as grinding media in the pebble mills and the remainder returns to the primary mill. The 75 mm (3 inch) by 32 mm (1¼ inch) fraction of the screen oversize is crushed before it is returned to the grinding mill in order to avoid an excessive and undesirable accumulation of this size fraction in the primary-mill circuit. Reagents are added to the cyclone overflow from the pebble-mill circuits to selectively flocculate the hematite. Removal of siliceous slime is then accomplished by overflowing it at the desliming thickeners. The deslimed underflow goes to flotation, where the remaining silica is floated leaving the iron concentrate behind. The selective flocculation of iron oxides is accomplished by starch addition and the collector for silica flotation is a cationic amine.

AGGLOMERATION PROCESSES

The blast furnace is a counter-current gas-solid reactor in which the solid charge materials move downward while the hot reducing gases flow upward. The best possible contact between the solids and the reducing gas is obtained with a permeable burden, which permits not only a high rate of gas flow but also a uniform gas flow, with a minimum of channeling of the gas. The primary purpose of agglomeration is to improve burden permeability and gas-solid contact, and thereby reduce blast-furnace coke rates and increase the rate of reduction. A secondary consideration is the lessening of the amount of fine material blown out of the blast furnace into the gas-recovery system. Furthermore, in steelmaking furnaces, agglomerated materials, when they have the proper chemical composition, can substitute for lump ores used as charge ores.

A good agglomerate for blast-furnace use should contain 60 per cent or more of iron, a minimum of undesirable constituents, a minimum of material less than 6 mm (¼ inch) in size, and a minimum of material larger than 25 mm (1 inch). The agglomerate should be strong enough to withstand degradation during stockpiling, handling, and transportation to the furnace so as to arrive at the furnace skip containing a minimum of 85

FIG. 5—28. Four types of iron-ore agglomerates. The briquettes at the lower right were produced by hot-ore briquetting.

to 90 per cent of plus 6 mm (¼-inch) material. In addition, the agglomerate must be able to withstand the high temperature and the degradation forces within the furnace without slumping or decrepitating. The agglomerate should also be reasonably reducible so that it can reduce at a satisfactorily high rate in the blast furnace. There is less definite knowledge about the following properties of agglomerates: preferred shape; most suitable size within the 6 mm (¼ inch) to 25 mm (1 inch) range; minimum strength required, and most desirable mineralogical structure.

Four types of agglomerating processes have been developed: sintering, pelletizing, briquetting, and nodulizing. Their individual products are known as sinter, pellets, briquettes, and nodules (Figure 5—28). Only the sintering and pelletizing processes are of major importance as neither briquetting nor nodulizing has gained any substantial degree of commercial acceptance. Careful evaluation should be made of the processes, material to be agglomerated, and the product desired before arriving at a final decision on a commercial installation. Quite often the origin of the material to be agglomerated together with material handling and transportation considerations will dictate which process is chosen. Fine concentrates such as those made from magnetite taconite are not readily

shipped because of dusting and freezing problems but are readily made into pellets that are easy to handle and transport with minimal degradation. Consequently, if there is a considerable distance between the mine and the blast furnace it is preferable to locate pellet plants near the mine site. Materials that do not have the particle-size distribution and characteristics required for pelletizing may be agglomerated by sintering. Typical sinter-feed materials include fines generated during ore transport, flue dust, mill scale, and fine concentrates that are too coarse for pelletizing. Sinter plants tend to be located near the blast furnaces because sinter degrades badly during shipment and because the steelmaking facilities are the point of origin of many of the materials that must be agglomerated.

Energy cost and the uncertain availability of fuels are important factors in all processes and have provided the incentive for development work to reduce fuel consumption and to utilize substitute fuels. Better utilization of hot gases and heat recuperation have recently resulted in lower fuel costs and conversion of oil- and gas-fired pelletizing operations to coal firing has resulted in a more reliable fuel source.

Sintering—Sintering has been referred to as the art of burning a fuel mixed with ore under controlled conditions. The flexibility of the process permits conversion

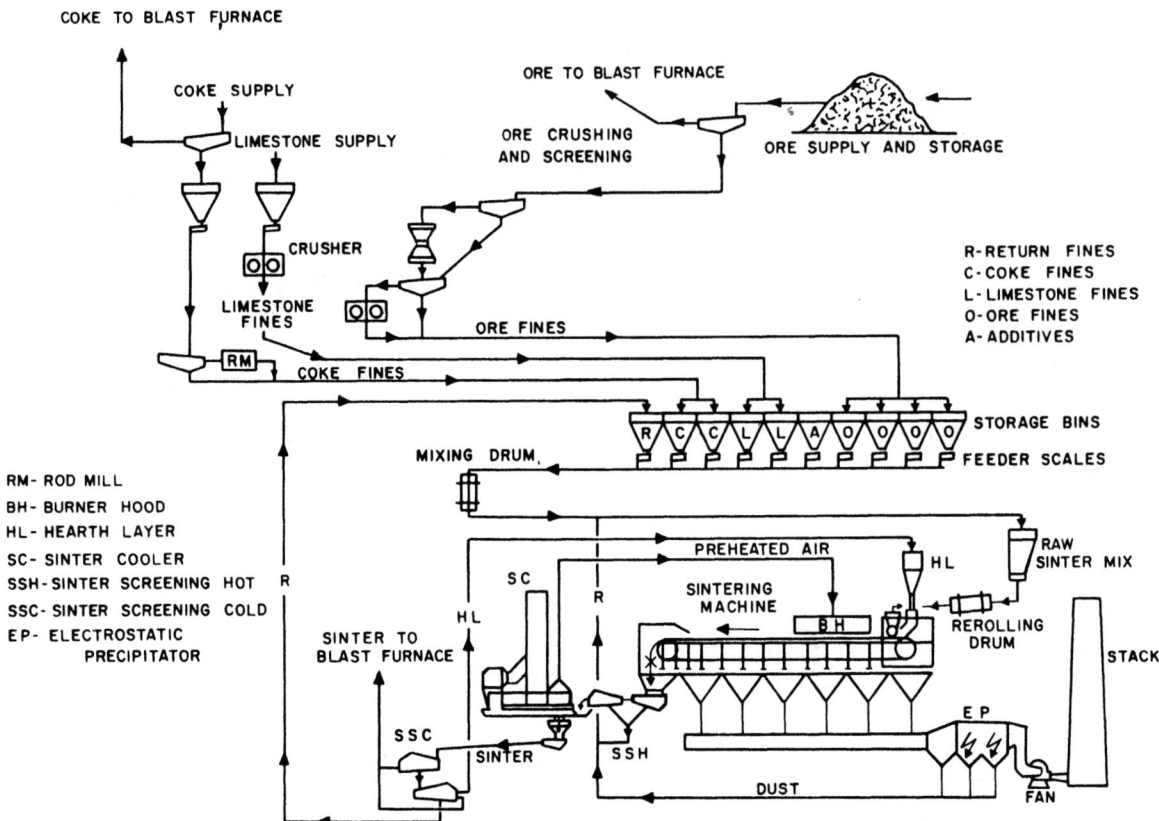

FIG. 5—29. Schematic flow diagram of continuous iron-ore sintering process (adapted from Metallgessellschaft A.G.).

of a variety of materials, including naturally fine ores and ore fines from screening operations, flue dust, ore concentrates, and other iron-bearing materials of small particle size into a clinker-like aggregate that is well suited for use in the blast furnace.

The continuous sintering process shown schematically in Figure 5—29 is carried out on a traveling grate that conveys a bed of ore fines or other finely divided iron-bearing material, intimately mixed with approximately 5 per cent of a finely divided fuel such as coke breeze or anthracite. Near the head or feed end of the grate, the bed is ignited on the surface by gas burners (Figure 5—30) and, as the mixture moves along on the traveling grate, air is pulled down through the mixture to burn the fuel by downdraft combustion. As the grates (or pellets) move continuously over the windboxes toward the discharge end of the strand, the combustion front in the bed moves progressively downward. This creates sufficient heat and temperature [about 1300°C to 1480°C (2370°F to 2700°F)] to sinter the fine ore particles together into porous clinkers. That location along the traveling grate where the combustion front touches the bottom of the bed is called the burn-through point.

Although simple in principle, sintering plants require that a number of important factors in their design and operation be observed to attain optimum performance. Intimate mixing of the feed materials is one of the most important, and balling-drum or disc-pelletizer mixers

(Figures 5—31 and 5—32) are employed to achieve this end. These mixers are operated to produce small rice-size nodules that significantly improve the permeability of the sinter bed. Improved permeability, in turn, results in more rapid and uniform sintering. Desirable mixer retention times vary from about one minute for sticky hematite ores to four minutes for more-difficult-to-ball ores.

In transferring the prepared mix from the mixer to the grate of the sintering machine it is essential to feed the material carefully to provide a uniform, homogeneous bed and to prevent compacting of the bed. Chutes must be designed to avoid a direct drop of feed onto the grate, because such a drop does tend to compact the feed. Design of surge bins and feeders for distributing the prepared mix into these bins is equally important because, if the prepared mix is compacted or segregated during handling and loading onto the grate, all of the advantages gained through good feed preparation may be lost.

Proper ignition of the sinter bed is also important. Poor ignition results in spotty burning and may leave unsintered material over the surface of the bed. Conversely, too intense an ignition flame can result in slagging over the bed and reduced sintering rates. The radiant-hood ignition furnace provides good ignition. Replacing part of the solid fuel with gaseous fuel results in sinter having a slightly improved strength and reducibility without affecting sinter-production rate. This

Fig. 5—30. Sintering machine in operation. The ignition furnace in the center background ignites the fuel in the surface layer of the sinter mix. As the bed of sinter progresses into the foreground, air is pulled down through the bed to cause the burning zone to move downward through the bed by igniting fuel in deeper and deeper zones in the mix. By the time the discharge end of the machine is reached, sintering will have taken place throughout the depth of the bed.

practice is termed "mixed firing". Where a shortage of solid fuel exists, and gas is available, use of increased amounts of gaseous fuel should be desirable. Plants using increased ignition (extended firing) have approximately 25 per cent of the length of the sinter bed covered by a gas-fired ignition-type hood. The temperature in this hood ranges from about 1150°C (2100°F) in the first section where ignition begins to approximately 800°C (1500°F) at the exit end of the hood. Depending upon the characteristics of the ore materials and the sintering conditions, daily average production rate of 22.4 to 42.9 metric tons/m²/day (2.3 to 4.4 net tons per square foot per day) of grate area are expected, and individual daily rates in excess of 48.8 metric tons/m²/day (5 net tons per square foot per day) have been attained.

Cooling of the sinter below 150°C (300°F) so that it can be handled on conveyor belts is an important part of the operation. Sinter coolers, such as the rotary-type (Figure 5—33) and shaft-type are usually used in conjunction with a water quench. The exhaust air from these coolers is normally at too low a temperature to permit the economical recovery of heat. The most recent developments in sinter cooling have been directed towards on-strand cooling. This could improve heat recuperation, sinter quality, and dust collection.

The use of sinter in the blast furnace has resulted in significant improvements in furnace performance as discussed in Chapter 15. Additional improvements have also been obtained (1) by incorporating the blast-furnace flux into the sinter rather than charging it separately to the top of the furnace, as was formerly done, and (2) by use of sized sinter. The available data on the use of fluxed sinter, sometimes called self-fluxing sinter, indicate that for each net ton of limestone removed from the blast-furnace burden and charged into the sinter plant to make a fluxed sinter, approximately 182 kg (400 pounds) of metallurgical coke are saved. The coke saving results primarily from calcining of limestone on the sintering grate rather than in the blast furnace. Limestone in the form of "fluxing fines" for the production of sinter is made by crushing and screening methods that result in a product meeting size specifications, as described in Chapter 6.

Use of sized sinter is desirable because iron-production rates in the blast furnace are further increased. Plant tests have demonstrated significant increases in iron-production rate as a result of screening out small-sized material in sinter before it is charged to the furnace. Other tests have shown that sized sinter, which contains 85 to 90 per cent of 25 mm by 6 mm (1 inch by ¼ inch) material as compared with 60 per cent

FIG. 5—31. Two disc-type pelletizing machines. Ore fines are fed to the discs by the belt conveyors (behind the operator). As the discs rotate, there is a "balling" action that causes the fines to agglomerate into pellet-like masses that are discharged from the discs over their lip at the bottom onto the belt conveyor shown at floor level that carries the pellets to the bins of the sinter machine.

FIG. 5—32. Balling drum mixer.

FIG. 5—33. Induced-draft rotary or circular sinter coolers, shown in process of construction.

in standard sinter has a much higher permeability than standard sinter and performs as well as pellets of comparable size. It also appears that crushing to minus 25 mm (1 inch) size at the sinter plant yields a more stable sinter because the smaller size fractions are more resistant to degradation.

Pelletizing—Pelletizing differs from sintering in that a "green" unbaked pellet or ball is formed and then hardened by heating. Experimental work, started many years ago by E. W. Davis and his associates at the University of Minnesota on the concentration and agglomeration of low-grade iron ores, showed that it was possible to ball or pelletize fine magnetite concentrate in a balling drum and that if the balls were fired at sufficiently high temperature (usually below the point of incipient fusion) a hard, indurated pellet (Figure 5—28) well adapted for use in the blast furnace, could be made. Consequently, despite the unquestioned benefits of sinter on blast-furnace performance, intense interest in the pelletizing process has developed because of the outstanding performance achieved by steel producers in extended operations with pellets as the principal iron-bearing material in the blast-furnace burden.

In general, the pelletizing process is desirable for agglomeration of finely divided concentrates because they are normally of such fine size that they will form into a green ball with little difficulty. Concentrates and high-grade ores that are not suitable in size for pelletizing are in some cases ground to the required size when pellets are desired as the final product.

The balling drum and the disc pelletizer are the most widely used devices for forming "green balls". Compared with the balling drum, the disc has the advantages of lighter weight and greater possibility for adjustment. Its inherent design averages out the effect of instantaneous fluctuations in the feed, whereas the drum cannot. Also, the classifying action of the disc promotes discharge of balls of more uniform size, which simplifies screening of the product. However, the capacity of the discs is low and discs generally require closer control than drums. Best control of ball size is achieved when the balling device is in closed circuit with a screen to remove and recycle the undersize material. Binders, such as bentonite, clay or hydrated lime, are generally used to raise the wet strength of green balls to more acceptable levels for handling. Bentonite consumption at the rate of 6.3 to 10 kg (14 to 22 pounds) per ton of feed is a significant cost element and considerable effort has been directed to reducing its usage and to development of cheaper substitutes. The ballability and strength of green balls are influenced by the additives and by the moisture content and particle-size distribution of the concentrates. Optimum moisture content for good balling is usually in the 9 to 12 per cent range. It appears that balling characteristics are relatively independent of the chemical composition of a concentrate, but are strongly affected by its physical properties. For example, specular hematites are more difficult to ball than magnetite concentrates because of the "plate-like" structure of the specular hematite particles. In any case, satisfactory pellet formation is usually achieved by regrinding to about 80 to 90 per cent minus 43 μm (325 mesh). Normally, any material considered for pelletizing should contain at least 70 per cent minus 43 μm (325 mesh) and have a specific surface area (Blaine) greater than 1200 cm^2/gram for proper balling characteristics.

Both the drop and compressive strengths of green pellets are important but because dried pellets are not required to withstand much handling, their compressive strength is considered most important. The strength of fired pellets is important in minimizing degradation by breakage and abrasion during handling and shipping, and in the blast furnace. Strong bonding in pellets is believed to be due to grain growth from the accompanying oxidation of magnetite to hematite, or recrystallization of hematite. Although slag bonding may promote more rapid strengthening at slightly lower firing temperatures, pellet strength is normally decreased, especially resistance to thermal shock. Fired pellet strength is most commonly determined by compression and tumble tests. Compressive strengths of individual pellets depend upon the mineralogical composition and physical properties of the concentrate, the additives used, the balling method, pellet size, firing technique and temperature, and testing procedure. The compressive strengths of commercially acceptable pellets are usually in the range of 150 to 400 kg (300 to 900 pounds) for pellets in the size range of minus 13 mm plus 9 mm (minus ½ inch plus ⅜ inch). In the tumbler test 11.4 kg (25 pounds) of plus 6 mm (¼ inch) pellets are tumbled for 200 revolutions at 25 rpm in a drum tumbler (ASTM E279-65T) and then screened. A satisfactory commercial pellet should contain not more than about 6 per cent of minus 0.6 mm (28 mesh) fines, and 90 per cent or more of plus 6 mm (¼ inch) size, after tumbler testing. A minimum of broken pellets between 6 mm and 0.6 mm (¼ inch and 28 mesh) in size is also desirable. Other important properties of fired pellets to be used for blast-furnace feed are reducibility, porosity, and bulk density. With some concentrates these can be varied within certain limits.

The flow sheet of a pelletizing process is similar in many respects to the sintering process, particularly in the materials handing area. Usually the associated mining, concentrating, and grinding installations are operated as a feed preparation section of the pellet plant. A typical pelletizing plant is shown in Figure 5—34. The three most important pelletizing systems are the traveling-grate, the grate-kiln, and the shaft furnace. Each system has been used commercially to make acceptable quality pellets and thus, capital and operating factors are usually involved in choosing one or the other. Fuel requirements for pelletizing by these systems vary from about 500 000 to 1 000 000 kJ per metric ton (Btu/long ton) of pellets depending on the feed material. Oxidation of magnetite to hematite during pelletizing will provide a significant proportion [about 300 000 kJ per metric ton (Btu per long ton)] of the heat requirement in all of the systems. Pelletizing processes are being improved constantly and further details on their technology and development may be found in the references at the end of the chapter. The production of self-fluxing and pre-reduced pellets are examples of innovations that are currently becoming accepted on a commercial scale. A brief description of important differences in the major pelletizing systems

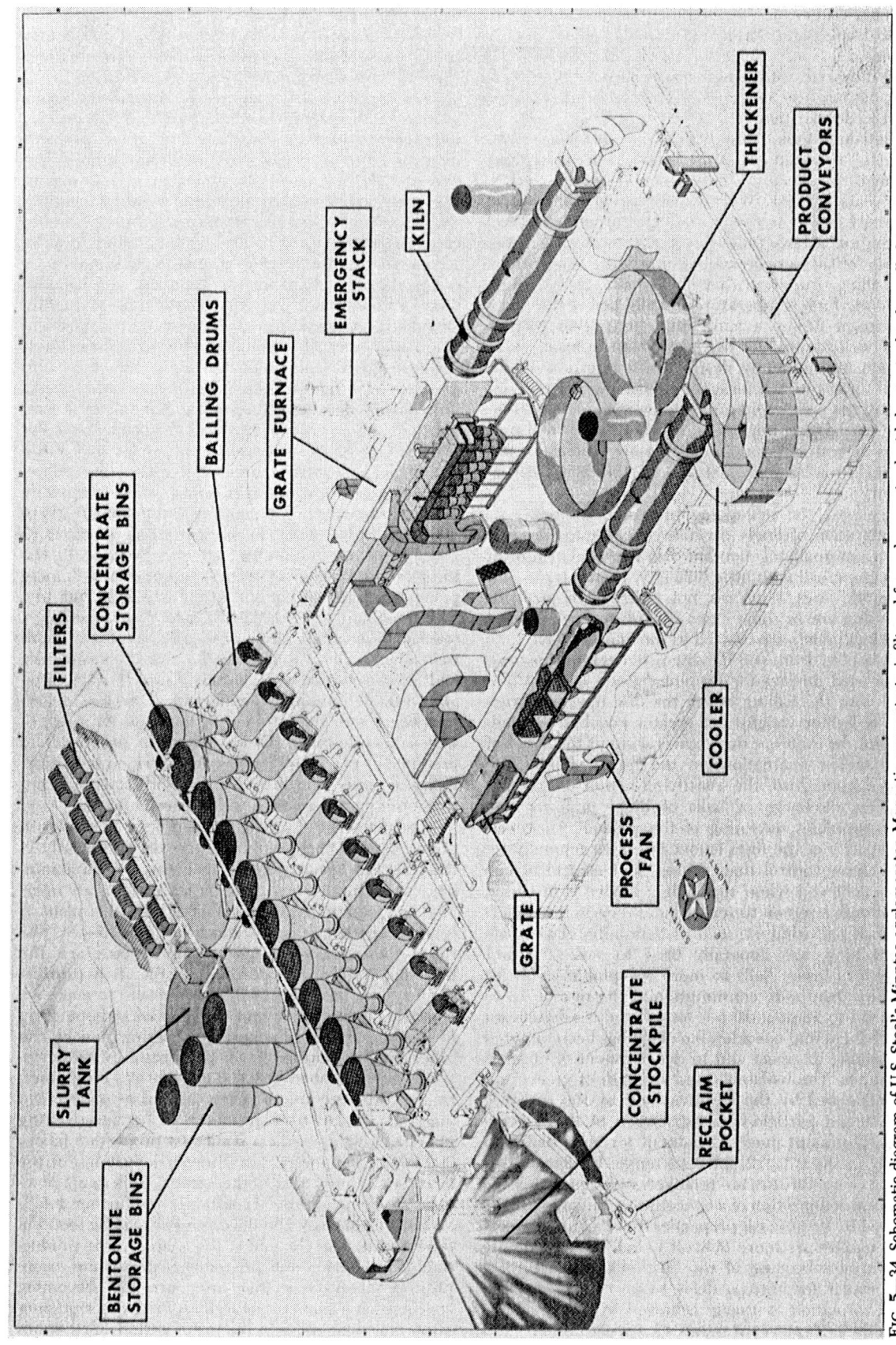

FIG. 5—34. Schematic diagram of U.S. Steel's Minntac agglomerator. Magnetic concentrate is filtered from slurry, mixed with bentonite binder, rolled into "green balls" in the balling drums and fired to the grate-kiln furnaces to make taconite pellets.

are pointed out in the discussion that follows.

Traveling-Grate—The traveling-grate system for producing pellets, illustrated in Figure 5—35, is essentially a modification of the sintering process. The green balls are fed onto the grate continuously to give a bed depth of about 300 to 400 mm (12 to 16 inches) and are dried in the first few windboxes by updraft air recuperated from the firing zone, followed by downdraft drying using recuperated air from the cooler. This arrangement of hot air flows limits pellet damage resulting from condensation of moisture in the bed. Following drying, the pellets are preheated by downdraft air from the cooling zone. Firing is done downdraft in the combustion zone by burning fuel oil or natural gas with hot air from the cooling zone. The cooling zone follows the combustion zone and uses updraft fresh air.

Fuel consumption in the traveling-grate system is about 350 000 to 600 000 kJ per metric ton (Btu per long ton) of pellets produced from magnetite and up to 1 000 000 kJ per metric ton (Btu per long ton) when pelletizing hematite. The system offers good temperature control in the firing zone. Pellet consistency throughout the bed may be achieved by recirculating some fired pellets to form hearth and side layers on the grate. The largest grate machines are 4 metres (13 feet) wide and are capable of producing more than 3 million metric tons of pellets per year. Circular grate machines have also been designed and one such machine is in operation.

Grate-Kiln System—The grate-kiln system depicted in Figure 5—36, consists of a straight grate for drying and preheating the pellets to about 1040°C (1900°F), a rotary kiln for heating to the final induration temperature of 1315°C (2400°F), and a horizontal rotary hearth for cooling and heat recuperation. Heat for firing is supplied by a central oil, gas, or coal burner at the discharge end of the kiln. Hot gases produced in the kiln are used for downdraft drying and preheating of the pellets. Hot air from the cooling zone is used for combustion.

The grate-kiln system offers good temperature control in all stages and produces a relatively consistent product. Fuel consumption is 400 000 to 600 000 kJ per metric ton (Btu per long ton) of pellets produced from magnetite, and up to 1 000 000 kJ per metric ton (Btu per long ton) of pellets produced from hematite. The largest systems have 7.6 metre (25 foot) diameter kilns and are capable of producing about 4 000 000 metric tons per year of pellets per line.

Vertical-Shaft Furnaces—Vertical-shaft furnaces are not as common as the traveling-grate or grate-kiln systems. There are several variations in shaft-furnace design but the most common is the Erie type, shown in Figure 5—37.

Green balls are charged at the top and descend through the furnace at a rate of 25 to 38 mm (1 to 1½ inches) per minute countercurrent to the flow of hot gases. About 25 per cent of the total air enters the furnace through the hot-gas inlet at temperatures from 1280°C (2340°F) to 1300°C (2375°F). Pellets in this zone of the furnace reach temperatures of 1315°C (2400°F) or higher because exothermic heat is released when the magnetite oxidizes to hematite, increasing the temperature. The remaining 75 per cent of the furnace air

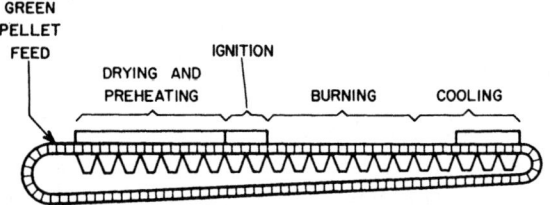

FIG. 5—35. Schematic diagram of the traveling-grate system for producing pellets.

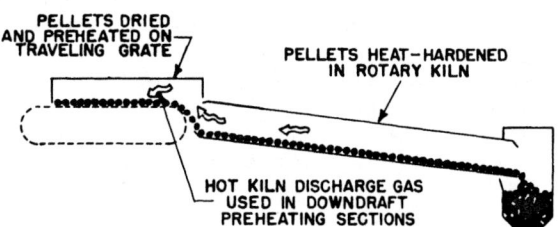

FIG. 5—36. Schematic diagram of the grate-kiln system for producing pellets.

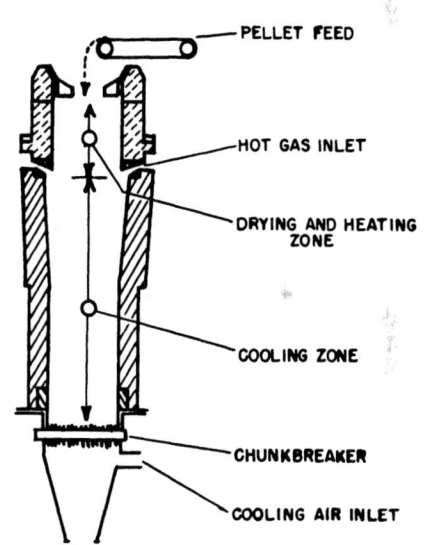

FIG. 5—37. Schematic diagram of the shaft-furnace system for producing pellets.

enters via the cooling-air inlet. Pellets discharge at about 370°C (700°F), and top-gas temperature is about 200°C (400°F).

Typical furnace capacities are 1000 to 2000 metric tons per day. Shaft furnaces are more energy efficient than the traveling-grate or grate-kiln systems. The shaft furnace is well suited for pelletizing magnetite, but not hematitic or limonitic materials.

Disadvantages of shaft furnaces are low unit-productivity and difficulty in maintaining uniform temperature in the combustion zone. Hot spots may occur which cause pellets to fuse together into large masses, producing discharge problems.

Nodulizing—In this agglomerating process, fine iron-bearing materials moving through a rotary kiln are formed into nodules or lumps (Figure 5—28) by the rolling of the charge heated to incipient fusion temperatures. This process has been used at various places throughout the world to agglomerate fine iron-bearing materials but is now of very little importance due to the developments in sintering and pelletizing. Magnetite concentrates from the Cornwall, Pennsylvania, mine were agglomerated by nodulizing in a bituminous coal-fired rotary kiln for 13 years in a plant that was shut down in 1920 because of the adoption of the cleaner sintering process. Nodulizing has the advantage that feed moisture and particle size are not critical as they are in pelletizing. In addition to the disadvantage of high fuel consumption, which is reported to vary from about 2 to 4 million kJ per metric ton (Btu per long ton) of product, the formation of rings and large balls has been a serious problem at all installations at some time in their operation and seems to be the most frequent cause of shutdown. Nodules do not give the best blast-furnace performance largely because of their nonuniform size and inferior reducibility.

Briquetting—Briquetting is an old art that has been used to agglomerate or form small or large lumps of regular shape from a wide variety of materials, including wood, coal, lignite, chars, cokes, ores, and flue dust. Various designs of punch and roll presses are or have been used. However, briquetting of cold or unheated material has not been successful as a means for producing satisfactory agglomerates for the blast furnace. Briquettes are usually formed with the use of a binder and do not possess the strength resulting from the high-temperature heating that is developed in pellets and sinter. A process known as the high-iron-briquetting (HIB) process is being applied to Venezuelan ore fines: this involves pre-reduction of the ore by fluidized-bed methods to remove about 75 per cent of the oxygen from the iron oxides, followed by briquetting of the reduced material for stockpiling and shipment (see Chapter 14 on Direct-Reduction Processes).

SECTION 5

TRANSPORTATION OF IRON ORES

Historically, since the development of high-capacity iron- and steelmaking complexes, the avenues of commerce for steelmaking raw materials such as iron ore, concentrates, and pellets have been by rail and water. The great Lake Superior Iron Ore District has occupied a very strategic location with respect to water routes along the Great Lakes to the steel centers on Lake Michigan and Lake Erie and via water to rail transfer to facilities in the Pittsburgh area. As much as 96 million tons in one year have moved by rail to Great Lakes ports in Minnesota, Wisconsin, Michigan and Ontario, and thence by vessel to lower lake ports. There are now over 200 American and Canadian Great Lakes ore vessels in service or available for service ranging in capacity per trip from 11 000 long tons to 60 000 long tons and averaging 24 300 long tons. Vessels of this type are shown in Figures 5—38, 5—39, 5—40 and 5—41.

Discovery and development of large deposits of high-grade ore close to deep-water ports outside the United States has led to the utilization of ocean-going vessels for transporting iron ores to steel centers on the coast or via rail transport to inland steel centers. On the Great Lakes, ore carriers with deadweight capacities of more than 60 000 long tons have been constructed to realize the economy of size. These vessels, 305 metres (1000 feet) in length and 32 metres (105 feet) in beam, are designed to take maximum advantage of the Poe lock in the St. Marys River at Sault Ste. Marie, Michigan and are restricted by their size to service on the four western Great Lakes. The maximum size vessel capable of transiting the Welland Canal to Lake Ontario and the St. Lawrence Seaway is 223 metres (730 feet) in length and 23 metres (76 feet) beam.

Transfer of iron ores from railroad cars to vessels has been accomplished in the Great Lakes ports by unloading into dock pockets from which ore is discharged by gravity through chutes lowered into the holds of vessels (Figure 5—38). In the more modern Great Lakes vessel loading installations constructed to accommodate larger vessels, ore is unloaded from railroad cars into storage bins or stockpiles from which it is recovered by shovels or ore-reclaiming equipment and transferred by conveyor belts for discharge into the vessel's holds (Figure 5—39). At receiving ports, the ore has historically been removed from the vessels' holds by unloading rigs or Hulett unloaders and transferred to stockpiles or to railroad cars for shipment inland (Figure 5—40). Recent trends have replaced shore-based unloading equipment through the use of self-unloading conveyor systems located onboard the vessel that transfer the ore to stockpiles or hoppers shoreside (Figure 5—41). The ore is then conveyed for utilization within the steel mill or reshipment by rail to steel mills inland.

The Great Lakes shipping season has successfully been extended into the winter months when traditionally the vessels were laid up due to ice conditions. Transportation of iron ores in modern integrated ore trains has also been accomplished.

Under consideration for the future are pipelines for transportation of fine iron-ore concentrates from mine and concentrator to agglomerating facilities located at shipping ports or at consuming centers.

FIG. 5—38. Loading pellets by gravity chutes into the Great Lakes ore vessel "Roger Blough".

FIG. 5—39. 305-metre (1000 ft) Great Lakes vessel "Edwin H. Gott"
being loaded with pellets at a modern conveyor loading dock.

FIG. 5—40. Unloading ore using Hulett unloaders from a typical Great Lakes vessel "Robert C. Stanley" on southern Lake Erie.

FIG. 5—41. Self-unloading vessel "Arthur M. Anderson" unloading pellets into a shore bin for transfer by conveyor to stockpiles.

SELECTED BIBLIOGRAPHY

WORLD IRON-ORE DEPOSITS
General
Blondel, F. and L. Marvier, eds. Symposium sur les gisements de fer du monde; XIX International Geological Congress, pp. 267-498, 1952.

XI International Geological Congress: The iron ore resources of the world, v. II, pp. 779-826, 1910.

United Nations Department of Economic Affairs: Survey of World Iron Ore Resources, pp. 302-380, 1970.

United Nations Department of Economic Affairs: Survey of world iron ore resources, pp. 290-223, 1955.

United States Bureau of Mines: Materials survey-iron ore: pp. V-76 to V-88, 1956.

IRON-ORE DEPOSITS OF THE UNITED STATES

LAKE SUPERIOR DISTRICT
American Iron Ore Association. Iron Ore (various annual editions).

Anderson, Gerald J., "The Marquette District, Michigan" in John D. Ridge, Editor, Ore Deposits of the United States, 1933/1967, (Graton-Sales Volume), Vol. I, pp. 507-517, American Institute of Mining, Metallurgical and Petroleum Engineers (1968).

Bayley, R. W. and H. L. James, "Precambrian Iron Formation of the United States" **Economic Geology**, Vol. 68, No. 7, pp. 915-923 (1973).

Dutton, Carl E., Paul W. Zimmer, "Iron Ore Deposits of the Menominee District, Michigan" in John D. Ridge, Editor, Ore Deposits of the United States, 1933/1967, (Graton-Sales Volume), Vol. I, pp. 538-549, American Institute of Mining, Metallurgical and Petroleum Engineers (1968).

Dutton, C. E., Pre-Cambrian Geology of Parts of Dickinson and Iron Counties, Michigan. Michigan Basin Geological Society, 1958.

Gair, J. E., Iron Deposits of Michigan (United States of America) in: Genesis of Precambrian Iron and Manganese Deposits, Proceeding of the Kiev Symposium, 20-25 August 1970, UNESCO.

Goldrich, S. S. et al., The Pre-Cambrian Geology and Geochronology of Minnesota. Minnesota Geological Survey Bulletin 41, 1961.

Gruner, J. W., Mineralogy and Geology of the Mesabi Range. Iron Range Resources and Rehabilitation Commission, 1946.

James, H., "Note 40—Subdivision of the Precambrian: an interim scheme to be used by the U. S. Geological Survey", Am. Assoc. of Petroleum Geologists Bulletin, Vol. 56, pp. 1128-1133, (1972).

James, H. L. et al., Geology of Central Dickinson County. U. S. Geological Survey Professional paper 310, 1961.

James, H. L., Sedimentary Phases of Iron Formation. Economic Geology, v. 29, 1954.

Lake Superior Iron Ore Association. Lake Superior Iron Ores, First and Second Editions.

Leith, C. K., Lund, R. J., and Leith, A., Pre-Cambrian Rocks of the Lake Superior Region. United States Geological Survey Professional Paper 184, 1935.

Marsden, Ralph W., "Geology of the Iron Ores of the Lake Superior Region in the United States" in John D. Ridge, Editor, Ore Deposits of the United States, 1933/1967, (Graton-Sales Volume), Vol. I, pp. 490-517, American Inst. of Mining, Metallurgical and Petroleum Engrs. (1968).

Marsden, Ralph W., Emanuelson, J. W., Owens, J. S., Walker, N. E. & Wernor, R. F. "The Mesabi Iron Range, Minnesota" in John D. Ridge, Editor, Ore Deposits of the United States, 1933/1967, (Graton-Sales Volume), Vol. I, pp. 518-537, American Institute of Mining, Metallurgical and Petroleum Engineers (1968).

Morey, G. B., Mesabi, Gunflint and Cuyuna Ranges, Minnesota

(United States of America) in: Genesis of Precambrian Iron and Manganese Deposits, Proceeding of the Kiev Symposium, 20-25 August 1970, UNESCO.

Schmidt, R. G., "Geology of the Precambrian W (Lower Precambrian) Rocks in Western Gegebic County, Michigan" U. S. Geological Survey Bulletin 1407 (1976).

Schmidt, Robert Gordon, "Geology and Ore Deposits of the Cuyuna North Range, Minnesota", U. S. Geological Survey Prof. paper 407 (1963).

Schwartz, G. M., Goldich, S. S., and Marsden, R. W., Pre-Cambrian of Northeastern Minnesota. Geological Society of America Guidebook, 1956.

Sims, P. K., "Precambrian Tectonics and Mineral Deposits-Lake Superior Region" Presidential Address, Economic Geology, Vol. 71, pp. 1092-1127 (1976).

Tyler, S. A., Development of Lake Superior Soft Ores. Geological Society of America Bulletin, v. 60, 1949.

Van Hise, C. R. and Leith, C. K., Geology of the Lake Superior Region. U. S. Geological Survey Monograph 42, 1911.

Wade, H. H. and Alm, M. R., Mining Directory. University of Minnesota Bulletin, 1961.

White, D. A., Stratigraphy and Structure of the Mesabi Range. Minnesota Geological Survey Bulletin 38, 1954.

NORTHEASTERN DISTRICT
Bayley, William S., Iron Mines and Mining in New Jersey. Geological Survey of New Jersey, v. 7, 1910.

Callahan, William H., "Geology of the Friedensville Zinc Mine, Lehigh County, Pennsylvania" in John D. Ridge, Editor, Ore Deposits of the United States, 1933-1967, (Graton-Sales Volume), Vol. I; "Iron", pp. 105-106, American Institute of Mining, Metallurgical and Petroleum Engineers (1968).

Crump, Robert M., Edward L. Beutner, "The Benson Mines Iron Ore Deposit, Saint Lawrence County, New York" in John D. Ridge, Editor, Ore Deposits of the United States, 1933-1967, (Graton-Sales Volume), Vol. I, pp. 49-71, American Institute of Mining, Metallurgical and Petroleum Engineers (1968).

Gray, Carlyle and Lapham, Davis M., Guide to the Geology of Cornwall, Pennsylvania. Pennsylvania Topographic and Geologic Survey Bulletin G35, 1961.

Gross, Stanford O., "Titaniferous Ores of the Sanford Lake District, New York" in John D. Ridge, Editor, Ore Deposits of the United States, 1933-1967, (Graton-Sales Volume), Vol. I, pp. 140-153, American Institute of Mining, Metallurgical and Petroleum Engineers (1968).

Lapham, Davis M., "Triassic Magnetite and Diabase at Cornwall Pennsylvania" in John D. Ridge, Editor, Ore Deposits of the United States, 1933-1967, (Graton-Sales Volume), Vol. I, pp. 73-94, (1968).

Leonard, Benjamin F., Magnetite Deposits of the Saint Lawrence County District, New York. United States Geological Survey Open File Report, 1951.

Newland, D. H. and Kemp, J. F., Geology of the Adirondack Magnetic Iron Ores, with a Report on the Mineville-Port Henry Mine Group. N. Y. State Museum Bull. 119, 1908.

Postel, A. Williams, Geology of Clinton County Magnetite District, New York. United States Geological Survey Professional Paper 237, 1952.

Sims, P. K., Geology of the Dover Magnetite District, Morris County, New Jersey. United States Geological Survey Bulletin 982-G, pp. 245-305, 1953 (1954).

Sims, Samuel J., "The Grace Mine Magnetite Deposit, Berks County Pennsylvania" in John D. Ridge, Editor, Ore Deposits of the United States, 1933-1967, (Graton-Sales Volume), Vol. I, pp. 108-124, American Institute of Mining, Metallurgical and Petroleum Engineers (1968).

Spencer, A. C., Magnetite Deposits of the Cornwall Type in Pennsylvania, U. S. Geological Survey Bulletin 359, 1908.

Stephenson, R. C., Titaniferous Magnetite Deposits of the Lake Sanford Area, New York State Museum Bulletin 340, 1945.

MISSOURI DISTRICT

Ballinger, H. J., and Pesonen, P. E., Investigation of Southeast Missouri Secondary Limonite Deposits—Wayne, Butler, and Ripley Counties, Missouri. United States Bureau of Mines Report of Investigations 4314, 1948.

Brown, J. S., Iron Ores of Missouri. Presented at American Institute of Mining, Metallurgical, and Petroleum Engineers Meeting, St. Louis, 1958.

Christiansen, Carl R., Iron Ore Mining in Missouri. Skillings' Mining Review no. 5, v. 51, February 3, 1962.

Crane, G. W., The Iron Ores of Missouri. Missouri Bureau of Geology and Mines, v. X, second series, 1912.

Emery, John A., "Geology of the Pea Ridge Iron Ore Body" in John D. Ridge, Editor, Ore Deposits of the United States, 1933-1967, (Graton-Sales Volume), Vol. I, pp. 359-369, American Institute of Mining, Metallurgical and Petroleum Engineers (1968).

Frommer, D. W. and Fine, M. M., Recovering Iron Concentrates from the Pea Ridge Deposit, Central Missouri. United States Bureau of Mines Report of Investigations 5550, 1960.

Hayes, W. C., Geology and Exploration of Missouri Iron Deposits. Missouri Division of Geological Survey and Water Resources Miscellaneous Publication, 1961.

Jones, E. A., Meramec Mining Company's Iron Mine at Pea Ridge, Missouri. Twenty-Third Annual Mining Symposium, University of Minnesota, and the Annual Meeting of Minnesota Section, American Institute of Mining, Metallurgical, and Petroleum Engineers, 1962.

Kline, H. D., Methods and Costs of Producing Brown Iron Ore at Two Small Southern Missouri Mines. United States Bureau of Mines Information Circular 7983, 1960.

Murphy, John E., Ernest L. Ohle, "The Iron Mountain Mine, Iron Mountain, Missouri" in John D. Ridge, Editor, Ore Deposits of the United States, 1933-1967, (Graton-Sales Volume), Vol. I, pp. 288-302, American Institute of Mining, Metallurgical and Petroleum Engineers (1968).

New Mineral Developments in S.E. Missouri. St. Louis Commerce, December, 1960.

Pettit, R. F., Jr., Calhoun, W. A. and Reynolds, B. M., Mining and Milling Methods and Costs, Ozark Ore Company. Iron Mountain Iron-Ore Mine, St. Francois County, Missouri. U. S. Bureau of Mines Information Circular 7807, 1957.

Snyder, Frank G., "Geology and Minerals Deposits, Midcontinent United States" in John D. Ridge, Editor, Ore Deposits of the United States, 1933-1967, (Graton-Sales Volume), Vol. I, pp. 264-268, American Institute of Mining, Metallurgical and Petroleum Engineers (1968).

SOUTHEASTERN DISTRICT

Brown, Andrew, North Alabama Brown Iron Ores. United States Bureau of Mines Report of Investigations 4229, 1948.

Chapman, H. H., The Iron and Steel Industries of the South. University of Alabama Press, 1953.

Pallister, H. D., Brown Iron Ore in South Alabama. Alabama Academy of Science Journal, v. 26, 1954.

Simpson, Thomas A., Tunstall R. Gray, "The Birmingham Red-Ore District, Alabama" in John D. Ridge, Editor, Ore Deposits of the United States, 1933-1967, (Graton-Sales Volume), Vol. I, pp. 187-206, American Institute of Mining, Metallurgical and Petroleum Engineers (1968).

Thoenen, J. R. and Clemmons, B. H., The Future of Birmingham Red Iron Ore. United States Bureau of Mines Report of Investigations 4988, 1953.

U. S. Bureau of Mines. Materials Survey—Iron Ore. 1956.

WESTERN DISTRICT
General

American Institute of Mining and Metallurgical Engineers. Ore Deposits of the Western United States. (Lindgren Volume) 1933.

Carr, M. S., and Dutton, C. E., Iron Ore Resources of the United States including Alaska and Puerto Rico. United States Geological Survey Bulletin 1082-C, pp. 61-134, 1959.

Luttrell, G. W., Bibliography of Iron Ore Resources of the World (to January 1955). United States Geological Survey Bulletin 1019-D, pp. 187-371, 1957.

Open Pit Mine Production and Stripping Tonnages, 1958-1961; Underground Mine Production from Individual Mines, 1956-1961. Mining World, v. 24 no. 5, pp. 95, 97-98, 1962.

Park, Charles F. Jr., "Mineral Deposits of the Pacific Coastal Region" in John D. Ridge, Editor, Ore Deposits of the United States, 1933-1967, (Graton-Sales Volume), Vol. II, pp. 1524-1525, American Institute of Mining, Metallurgical and Petroleum Engineers (1968).

Singewald, J. T., Jr., The Titaniferous Iron Ores in the United States; Their Composition and Economic Value. United States Bureau of Mines Bulletin 64.

United States Bureau of Mines. Minerals Yearbooks.

United States Geological Survey. Geological Survey Estimates, United States Iron Ore Resources. Information Service Release 2905, 1957.

Arizona

Burchard, E. F., Iron Ore on Canyon Creek, Fort Apache Indian Reservation, Arizona. United States Geological Survey Bulletin 821-C, pp 51-78, 1931.

Farnham, L. L., and Havens, R., Pikes Peak Iron Deposits, Maricopa County, Arizona. United States Bureau of Mines Report of Investigations 5319, 1957.

Joseph, P. E., Iron. Arizona University Bureau of Mines Bulletin 43, 1916.

Kerns, W. H., Kelly, F. J., and Mullen, D. H., The Mineral Industry of Arizona. United States Bureau of Mines Minerals Yearbook, 1960, v. 3, p. 107, 1961.

Stewart, L. A., Apache Iron Deposit, Navajo County, Arizona. U. S. Bureau of Mines Rept. of Investigations 4093, 1947.

California

California Division of Mines. Iron Resources of California. California Division of Mines Bulletin 129, 1948.

Carlisle, Donald, Davis, D. L., Kildale, M. B., and Stewart, R. M., Base Metal and Iron Deposits of South California. California Division of Mines Bulletin 170, chap. 8, pt. 5, pp. 41-50, 1954 (1955).

Dubois, Robert L., Richard W. Brummett, "Geology of the Eagle Mountain Mine Area" in John D. Ridge, Editor, Ore Deposits of the United States, 1933-1967, (Graton-Sales Volume), Vol. II, pp. 1592-1605, American Institute of Mining, Metallurgical and Petroleum Engineers (1968).

Harder, E. C., Iron Ore Deposits of the Eagle Mountains, California. United States Geological Survey Bulletin 503, 1912.

Hughes, M. J., Pit Operations at the Eagle Mountain Iron Ore Mine. Skillings' Mining Review, v. 49, no. 50, pp. 1, 4-7, 16, 1960.

Mining Methods at the Vulcan Iron Mine, San Bernardino County, California. United States Bureau of Mines Information Circular 7437, 1948.

Severy, C. L., Exploration of the Minarets Iron Deposit, Madera County, California. United States Bureau of Mines Report of Investigations 3985, 1946.

Shattuck, J. R. and Ricker, Spangler, Shasta and California Iron Ore Deposits, Shasta County, California. United States Bureau of Mines Report of Investigations 4272, 1948.

Trengove, R. R., Methods and Operations at the Kaiser Steel Corporation, Eagle Mountain Iron Mine, Riverside County, California. United States Bureau of Mines Information Circular 7735, 1956.

Wiebelt, F. J. and Ricker, Spangler, Iron Mountain Deposits, San Bernardino County, California. United States Bureau of Mines Report of Investigations 4236, 1948.

Montana

DeMunck, V. C., Iron Deposits in Montana. Montana Bureau of Mines and Geology Information Circular 13, 1956.

Holmes, W. T., II, Holbrook, W. F., and Banning, L. H., Beneficiating and Smelting Carter Creek, Montana, Iron Ore. United States Bureau of Mines Report of Investigations 5922, 1962.

Nevada

Geehan, R. W., Investigation of the Dayton Iron Deposit, Lyon and Storey Counties, Nevada. United States Bureau of Mines Report of Investigations 4561, 1949.

Karl, V. E., Modarelli Iron Deposit, Eureka County, Nevada. U. S. Bureau of Mines Rept. of Investigations 4005, 1947.

Mason Valley News, 1959-1962. Many articles on the Minnesota Mine, Yerington, Nevada.

Reeves, R. G., and Kral, V. E., Geology and Iron Ore Deposits of the Buena Vista Hills, Churchill and Pershing Counties, Nevada. Nevada Bureau of Mines Bull. 53-A, pp. 1-32, 1955.

Reeves, R. G., Shawe, F. R., and Kral, V. E., Iron Ore Deposits of West-Central Nevada. Nevada Bureau of Mines Bulletin 53-B, pp. 33-78, 1958.

New Mexico

Hernon, Robert M., William R. Jones, "Ore Deposits of the Central Mining District, Grant County, New Mexico" in John D. Ridge, Editor, Ore Deposits of the United States, 1933-1967, (Graton-Sales Volume), Vol. II, pp. 1226-1228, American Institute of Mining, Metallurgical and Petroleum Engineers (1968)

Kelley, V. C., Geology and Economics of New Mexico Iron Ore Deposits. New Mexico University Publications in Geology no. 2, 1949.

Kelley, V. C., Oölitic Iron Deposits of New Mexico. American Association of Petroleum Geologists Bulletin, v. 35, no. 10, pp. 2199-2228, 1951.

Paige, Sidney, The Hanover Iron Ore Deposits, New Mexico. U. S. Geological Survey Bulletin 380, pp. 199-214, 1909.

Sheridan, M. J., Lincoln County Iron Deposits, New Mexico. U. S. Bureau of Mines Rept. of Investigations 3988, 1947.

Soule, J. H., Capitan Iron Deposits, Lincoln County, New Mexico. U. S. Bureau of Mines Rept. of Investigations 4022, 1947.

Oregon and Washington

Reichert, W. H., compiler. Bibliography and Index of the Geology and Mineral Resources of Washington, 1937-1956. Washington Division of Mines and Geology Bulletin 46, 1960.

Steere, M. L., compiler. Bibliography of the Geology and Mineral Resources of Oregon, Second Supplement. Oregon Department of Geology and Mineral Industries Bulletin 44, 1953.

Twenhofel, W. H., Origin of the Black Sands of the Coast of Southwest Oregon. Oregon Department of Geology and Mineral Industries Bulletin 24, 1943.

Zapffe, Carl, A Review of Iron-Bearing Deposits in Washington, Oregon, and Idaho. Raw Materials Survey Resource Report 5, 1949.

Texas

Brown, W. F., Sampling East Texas Iron Ores. United States Bureau of Mines Report of Investigations 5488, 1959.

Eckel, E. B., The Brown Iron Ores of Eastern Texas. United States Geological Survey Bulletin 902, 1938.

Fine, M. M., Frommer, D. W. and Dressel, W. M., Preliminary Mineral-Dressing Investigation of East Texas Brown Iron Ores. United States Bureau of Mines Report of Investigations 5252, 1956.

Kenworthy, H., and Starliper, A. G., Electric Furnace Smelting of East Texas Iron Ores—A Progress Report. United States Bureau of Mines Report of Investigations 5427, 1958.

Utah

Allsman, P. T., Investigation of Iron Ore Reserves of Iron County, Utah. U. S. Bureau of Mines Rept. of Investigations 4388 (supplement to Report of Investigations 4076), 1948.

Cook, K. L., Magnetic Surveys in the Iron Springs District, Iron County, Utah. United States Bureau of Mines Report of Investigations 4586, 1950.

Leith, C. K., and Harder, E. C., The Iron Ores of the Iron Springs District, Southern Utah. United States Geological Survey Bulletin 338, 1908.

Mackin, J. Hoover, "Iron Ore Deposits of the Iron Springs District, Southwestern Utah" in John D. Ridge, Editor, Ore Deposits of the United States, 1933-1967, (Graton-Sales Volume), Vol. II, pp. 992-1019, American Institute of Mining, Metallurgical and Petroleum Engineers (1968).

Mackin, J. H., Some Structural Features of the Intrusions in the Iron Springs District—Guidebook to the Iron Springs District, Utah. Published by the Utah Geological Society, in cooperation with the United States Geological Survey, 1947.

Young, W. E., Iron Deposits, Iron County, Utah. United States Bureau of Mines Report of Investigations 4076, 1947.

Wyoming

Ball, S. H., The Hartville Iron Ore Range, Wyoming. United States Geological Survey Bulletin 315 pp. 190-205, 1907.

Bayley, Richard W., "Ore Deposits of the Atlantic City District, Fremont County, Wyoming" in John D. Ridge, Editor, Ore Deposits of the United States, 1933-1967, (Graton-Sales Volume), Vol. I, pp. 589-604, American Institute of Mining, Metallurgical and Petroleum Engineers (1968).

Bayley, R. W., Iron Deposits Near Atlantic City and South Pass, Fremont County, Wyoming. United States Geological Survey Open File Report, 1958.

Diemer, R. A., Titaniferous Magnetite Deposits of the Laramie Range, Wyoming. Wyoming Geol. Survey Bulletin 31, 1941.

Frey, Eugene, Hartville Iron District, Platte County, Wyoming. United States Bureau of Mines Report of Investigations 4086, 1947.

Hagner, Arthur F., "The Titaniferous Magnetite Deposit at Iron Mountain, Wyoming" in John D. Ridge, Editor, Ore Deposits of the United States, 1933-1967, (Graton-Sales Volume), Vol. I, pp. 665-680, American Institute of Mining, Metallurgical and Petroleum Engineers (1968).

King, R. H., Iron Deposit near Atlantic City, Wyoming. Wyoming Geological Survey Open File Report, 1968.

Lovering, T. S., The Rawlins, Shirley, and Seminoe Iron Ore Deposits, Carbon County, Wyoming. United States Geological Survey Bulletin 811-D, pp. 203-235, 1929.

Wideman, F. L., Block-Caving Methods at the Sunrise Mine, Platte County, Wyoming. United States Bureau of Mines Information Circular 7759, 1956.

IRON ORE DEPOSITS OF CANADA

Brantley, F. E., Iron ore: preprint from 1965 Minerals Yearbook, U. S. Bureau of Mines, pp. 17 and 22, 1967.

Coughlan, W. K., Geology of the Wabana Deposit: Canadian Mining and Metall. Bull., v. 59, no. 646, pp. 171-175, 1966.

Dahlstrom, C. D. A., Snake River Iron Deposit: Western Miner and Oil Rev. May 1963, v. 36, no. 5, pp. 24-25.

Dubuc, F., Geology of the Adams Mine: Canadian Mining and Metall. Bull., v. 59, no. 646, pp. 176-181, 1966.

Eastwood, G. E. P., Replacement Magnetite on Vancouver Island, British Columbia: Econ. Geol., v. 60, no. 1, pp. 124-148, 1965.

Engineering and Mining Journal. Quebec Cartier: v. 165, no. 9, pp. 75-93, 1964.

Gauvin, C. J. and V. B. Schneider, Canadian Iron Ore Industry 1965: Mineral Resources Division Department of Energy, Mines and Resources, Ottawa, Canada, 1967.

Goodwin, A. M., "Archean Volcanogenic Iron-Formation of the Canadian Shield" in: Genesis of Precambrian Iron and Manganese Deposits. Proceedings of the Kiev Symposium, 20-25 August 1970, UNESCO.

Goodwin, A. M., Structure, stratigraphy, and origin of iron formations, Michipicoten area, Algoma District, Ontario, Canada; Geol. Soc. Am. Bull., v. 73, no. 5, pp. 561-585, 1962.

Gross, G. A., (1965) "Geology of Iron Deposits in Canada". Volume I: General Geology and Evaluation of Iron Deposits, 1967. Volume II: Iron Deposits in the Appalachian and Greenville Regions of Canada, 1968. Volume III: Iron Ranges of the Labrador Geosyncline. Geol. Surv. Canada, Econ. Geol. Rept. No. 22.

Gross, G. A., Metamorphism of iron formation and its bearing on their beneficiation: Can. Mining & Metall. Bull., v. 53, no. 575, p. 193, Mar. 1960.

Gross, G. A., Iron formations and the Labrador Geosyncline: Geol. Surv. of Canada, paper 60-30, 1961.

Gross, G. A., Iron deposits near Ungava Bay, Quebec: Geol., Surv. of Canada, Bull. 82-1962.

Gross, G. A., Geology of iron deposits in Canada, v. 1: General Geology and Evaluation of Iron Deposits: Geol. Surv. of Canada, Econ. Geol. No. 22, 1965.

Gross, G. A., The origin of high grade iron deposits on Baffin Island: Canadian Mining Jour., v. 87, no. 4, pp. 111-114, 1966.

Harrison, J. M., The Quebec-Labrador iron belt, Quebec and Newfoundland: Geol. Surv. of Canada, Paper 52-20, 1952.

Jackson, G. D., Geology and Mineral Possibilities of the Mary River region, northern Baffin Island: Canadian Mining Jour., v. 87, no. 6, p. 57-61, 1966.

Jeffery, W. G., Iron deposits in B.C.: Western Miner, v. 34, no. 11, pp. 28-33, 1961.

Jeffrey, W. G. and A. Sutherland Brown, Geology and Mineral Deposits of the Major Islands on the British Columbia Coast: Canadian Mining Journal, v. 85, no. 7, pp. 51-55, 1964.

Joliffe, A. W., Stratigraphy of the Steeprock Group, Steep Rock Lake, Ontario: in Precambrian Symposium. The relationship of mineralization to Precambrian stratigraphy in certain mining areas of Ontario and Quebec: Geol. Assoc. Canada Spec. Paper 3, pp. 75-98, 1966.

Markland, G. D., Geology of the Moose Mountain mine and its application to mining and milling: Canadian Mining and Metall. Bull., v. 59, no. 646, pp. 159-170, 1966.

Metal Bulletin. Canada-Steep Rock: Iron ore Special Issue, pp. 60-62, March 1965.

Metal Bulletin. Canada-Quebec Cartier: Iron Ore Special Issue, pp. 62-65, March 1965.

Metal Bulletin. Iron Ore of Canada: Iron Ore Special Issue, pp. 67-70, March 1965.

Murphy, Daniel L., Iron ore deposits in the Mt. Wright—Lake Carheil area, Quebec: A.I.M.E. Trans., v. 223, pp. 285-290, 1962; The iron-formation of Mt. Wright-Lake Carheil: Min. Eng. v. 14, no. 9, pp. 68-70, 1962.

Peterson, E. C. and C. T. Collins, Iron Ore Minerals Yearbook 1978-1979, U.S.B.M., Volume I.

Pettijohn, F. J. (1972), "The Archean of the Canadian Shield, A Resume", Geol. Soc. of America, Memoir 135, pp. 131-136.

Rose, E. R., Iron deposits of eastern Ontario and adjoining Quebec; Geol. Surv. of Canada, Bull. 45, 1958.

Schneider, V. B., Iron Ore: Canadian Mining Journal, Annual Canadian Mineral Industry Rev., pp. 128-135, Feb., 1967.

Selleck, D. J. and W. A. Campbell, Exploration and development of the Carol ores: in Mining Symposium, 26th Ann., Duluth 1965 (AIME, Minnesota Sec., 38th Ann. Mtg.): Minneapolis, Univ. Minnesota Center Continuation Study, pp. 15-22, 1965.

Shklanka, R., Steeprock Lake iron area, District of Rainy River: Ontario Dept. of Mines Prelim. Geol. Map, p. 348, 1966.

Simony, P. S., Geology of northwestern Timagami area, District of Nipissing: Ontario Dept. Mines Geol. Rpt. 28, 1964.

Stockwell, C. H., et al., (1972) "Geology of the Canadian Shield", Chapter IV, Geology and Economic Minerals of Canada, Ed. by R. J. W. Douglas, Geol. Survey of Canada, Dept. of Mines and Econ. Minerals of Canada.

Stubbins, John B. (and Roger A. Blais and I. Stephan Zajac), Origin of the soft iron ores of the Knob Lake range: Cana-

dian Mining and Metall. Bull., v. 54, no. 585, pp. 43-58 (Canadian Inst. Min. & Met., Tr., v. 64, pp. 27-52), 1961.

Swensen, W. T., Geology of the Nakina iron property, Ontario: A.I.M.E. Trans., v. 217 pp. 451-457, 1961.

Underhill, Douglas H. and C. B. Cox (1981), "Borealis: The Geology and Feasibility of Its Magnetite Deposits in the Canadian Arctic" Metal Bulletin's Second International Iron Ore Symposium Frankfurt, West Germany.

IRON-ORE DEPOSITS OF MEXICO

White, Lane, "Mining in Mexico", Eng. and Mining Jour., Nov. 1980, pp. 64-65, 84-86, 89-104.

IRON-ORE DEPOSITS OF SOUTH AMERICA
Argentina

Skillings, David N. Jr., HIPASAM New Iron Ore Pellet Producer in Argentina, Skillings Mining Review, 18 July 1981, p.8. Panarama Minero, "Prominente Hallazgo en Materia de Hicerro", 1976 Martino, Orlando, "The Mineral Industry of Argentina", U.S.B.M. Minerals Yearbook (1978-79), Vol. III.

Bolivia

Martino, Orlando, "The Mineral Industry of Bolivia, U.S.B.M. Minerals Yearbook, 1980, Volume III.

Mining Journal, 2 February 1973, p. 90, Skillings Mining Review: 1 April 1978, p. 16, Metal Bulletin, 21 September 1979, p. 40.

Brazil

Dorr, J. van, 2nd, "Iron Formation and Associated Manganese in Brazil" in: Genesis of Precambrian Iron and Manganese Deposits, Proceedings of Kiev Symposium 20-25 August 1970, UNESCO

Dorr, J.V.N., 2d. Geology and ore deposits of the Itabira district, Minas Gerais, Brazil: U.S.G.S. Prof. Paper 341-C, 1963.

Dorr, John Van N., 2nd Supergene iron ores of Minas Gerais, Brazil: Econ. Geol. v 59, no. 7, p. 1203-1240, 1964.

Dorr, John Van N., 2d. Nature and origin of the high grade hematite ores of Minas Gerais, Brazil: Econ. Geol. v. 60, no. 1, pp. 1-46, 1965.

Gair, J. E. Geology and ore deposits of the Nora Lima and Rio Acima quadrangles, Minas Gerais, Brazil: U.S.G.S. Prof. Paper 341-A 1962.

Guild, P. W. Geology and mineral resources of the Congonhas district, Minas Gerais, Brazil: U.S.G.S. Prof. Paper 290, 1957.

Johnson, R. F. Geology and ore deposits of the Cachoeira do Campo, Dom Bosco, and Ouro Branco quadrangles, Minas Gerais, Brazil: U.S.G.S. Prof. Paper 341-B, 1962.

Mining Magazine, "Brazilian Iron Ore", March 1977, pp. 152-165.

Moore, S. L., "Geology and Ore Deposits of the Antonio dos Santos, Gongo Soco and Conceigao do Rio Acima Quadrangles, Minas Gerais, Brazil", U.S.G.S. Prof. Paper 341-I, 11-150, 1969.

Papacek, H. G. and Heep, Hans, "The Fabrica Mine Expansion Project", A.I.M.E. Annual Meeting, New Orleans, Louisiana, February 18-22, 1979.

Pomerene, J. B., Geology and ore deposits of the Belo Horizonte, Ibirite, and Macacos quadrangles, Minas Gerais, Brazil: U.S.G.S. Prof. Paper 341-D, 1964.

Reeves, R. G., Geology and mineral resources of the Monlevade and Rio Piracicaba quadrangles, Minas Gerais, Brazil: U.S.G.S. Prof. Paper 341-D, 1964.

Simmons, G. C., "Geology and Iron Deposits of the Western Serra do Curral, Minas Gerais, Brazil", U.S.G.S. Prof. Paper 341-G, G1-G57, 1968

Simmons, G. C., "Geology and Mineral Resources of the Barao de Cocais Area, Minas Gerais, Brazil", U.S.G.S. Prof. Paper 341-H, H1-H46, 1968.

Skillings Mining Review, 7 March 1981, p. 24

Skillings, David N. Jr. "C.V.R.D.'s Terminal and Iron Ore Pellet Plants at Tubarao" Skillings Mining Review, 1 Oct. 1977, pp. 12-15

Skillings, David N. Jr., "C.V.R.D.'s Iron Ore Mining and Plant Operations at Itabira" Skillings Mining Review, 24 September 1977, pp. 10-16

Skillings, David N. Jr., "C.V.R.D.'s Iron Ore Mining and Processing Operations at Itabira", Skillings Mining Review, 30 November 1974, pp. 6-12.

Skillings, David N. Jr., "MBR Second Largest Producer of Iron Ore in Brazil", Skillings Mining Review, 3 June 1978, pp. 12-17.

Skillings, David N. Jr., "Ferteco Operating New Production Facilities at Fabrica" Skillings Mining Review, 12 November 1977, pp. 10-17.

Skillings, David N. Jr., "SAMITRI" Skillings Mining Review, 21 April 1973, pp. 11-17.

Skillings, David N. Jr., "Samarco Operating Major New Iron Ore Development in Brazil" Skillings Mining Review, 7 January 1978, pp. 20-28.

Skillings, David N. Jr., Wm. H. Muller, S.A.'s Iron Ore Operations", Skillings Mining Review, 1 March 1975, pp. 3.

Skillings, David N. Jr., "C.V.R.D. Carajas Iron Ore Project in Amazon Region", Skillings, Mining Review, 12 June 1981, pp. 12-24.

Skillings, David N., Jr. Cia Valedo Rio Doce; Skillings Mining Review, v. 55, no. 37, pp. 1-7, Sept. 10, 1966.

Tolbert, G. E., J. W. Tremaine, G. C. Melcher and C. B. Gomes; Geology and Iron Ore Deposits of Serra dos Curajas, Para (Brazil) in: Genesis of Precambrian Iron and Manganese Deposits, Proceedings of Kiev Symposium 20-25 August 1970, UNESCO.

Wallace, R. M., Geology and mineral resources of the Pico de Itabirito district, Minas Gerais, Brazil: U.S.G.S. Prof. Paper 341-F, A65.

Chile

Fuller, Carlos Ruiz, Hierro: in Geologia y Yacimientos Mataliferos de Chile. Instituto de Investigaciones Geologicas, Santiago 1965.

Metal Bulletin. Chile-Bethlehem: Iron Ore Special Issue, pp. 70-71, March 1965.

Metal Bulletin, Chile-Algarrobo: Iron Ore Special Issue, pp. 71-73, March 1965.

Metal Bulletin. Chile-Santa Fe: Iron Ore Special Issue, pp. 73-75, March 1965.

Velasco, Pablo, "The Mineral Industry of Chile", U.S.B.M. Minerals Yearbook, 1980, Volume III.

Skillings Mining Review, 7 March 1981, p. 14.

Skillings, David N. Jr., "C.A.P.'s El Romeral Marking 25th Year of Iron Ore Production", Skillings Mining Review, 11 August 1979, pp. 1024.

Colombia

Anon., "Colombia unveils plans to step up development," Eng. and Mining Jour., Nov. 1980, p. 55.

Cardona, C. Diego, "Tendencias En Ferromineria Latinoamericana" Seminaris Latinamericano de Ferromineria Organizado Por el Instituto Latinamericano del Fierro y el Acero ILAFA, 30 October—1° Noviembre de 1974, Lima, Peru, pp. F/3-F/11.

Peru

Engineering and Mining Journal. Marcona—Four Months from Plan to Production. v. 155, pp. 84-88, 1954.

Fomento Digest, May 1981, p. 3, Goa, India.

Herkenhoff, E. Marcona Iron Ore From Peru: Twenty-Third Annual Mining Symposium, University of Minnesota and the Annual Meeting of Minnesota Section, American Institute of Mining, Metallurgical and Petroleum Engineers, pp. 143-149, 1962.

Metal Bulletin. Peru-Marcona: Iron Ore Special Issue, pp. 89-91, March 1965.

Peru Instituto Nacional de Investigacion y Fomento Mineros, Division de Geologia y Minas. El Fierro en el Peru, Lima, 1952.

Skillings Mining Review, Mar. 14, 1981, p. 18.

Skillings, David N., Jr. Marcona Mining Co.: Skillings' Mining Review, v. 56, no. 19, pp. 1-8, May 13, 1967.

Venezuela

Kalliokoski, J. The metamorphosed iron ore of El Pao, Venezuela: Econ. Geol. v 60, no. 1, pp. 100-116, 1965.

Metal Bulletin, Venezuela-Orinoco: Iron Ore Special Issue, p. 112, March 1965.

Metal Bulletin. Venezuela-Iron Mines: Iron Ore Special Issue, pp. 114-115, March 1965.

Mining Magazine, "Iron Ore Mining in Venezuela, July 1981, pp. 18-25 Tovar, N. Dr. Omar A., "C.U.G. Ferrominera Orinoco C.A. Description of It's Mining Operations, It's past, Present and Future".

Ruckmick, J. C. (and S. E. Luchsinger). Geologia de Cerro Bolivar: Venezuela, Direccion de Geol., Bol. de Geol., Pub. Especial no. 3, t 3, pp. 972-984, map, 1960.

Ruckmick, John C., The iron ores of Cerro Bolivar, Venezuela: Econ. Geol v. 58, no. 2, pp. 218-236, illus. (incl. g. sk. map) 1963.

Skillings, David N. Jr., "Ferrominera's Cerro Bolivar and Puerto Ordaz Operations", Skillings Mining Review, 17 November 1979, pp. 10-20 and 24 November 1979, pp. 12-21.

IRON-ORE DEPOSITS OF AFRICA
Algeria

Metals Week, 8 March 1976, p. 6.

New York Times, 12 July 1978.

Angola

Metal Bulletin, Ferrous raw materials, 28 Nov. 1975, p. 44.

Mining Magazine, Monte Quintungo iron ore deposit, August 1975, pp. 143-144.

The Tex Report, Iron ore manual, Cassinga iron ore, 1976.

Egypt

Basta, Emile Z., and Amer, Hamza, I., Geological and petrographic studies on El Gedida area, Bahariya Oasis, U.A.R., Bulletin of the Faculty of Science, No. 43, Cairo U., 1969.

Basta, Emile Z., and Amer, Hamza, I., El-Gedida ores and their origin, Bahariya Oasis, Western Desert, U.A.R., Economic Geology, Vol. 64, pp. 424-444, 1969.

El-Hinnawi, Essam E., Contributions to the study of Egyptian (U.A.R.) iron ores Economic Geology, Vol. 60, pp.1497-1509, 1965.

Nakhla, F.M., The iron ore deposits of El Bahariya Oasis, Egypt, Economic Geology, Vol. 56, pp. 1103-1111, 1961.

Said, Rushdi, The geology of Egypt. Elsevier Publishing Company, Amsterdam-New York 1962.

U.S.B.M., Minerals Yearbook, Vol. III, Area Reports, International, 1980.

Zaatout, M.A. and Abdou, H.F., A review of Bahariya iron deposits, Annals of the Geological Survey of Egypt, Vol. V, pp.71-86, 1975.

Gabon

Engineering & Mining Journal, Gabon's new rail link to boost Mn and Fe output August 1975, p. 27.

Metal Bulletin, Gabon Port project, 10 October 1978, p. 42.

Mining Magazine, Iron ore deposits at Belinga, November 1978, pp. 543-545.

The Tex Report, Iron ore manual, Mekambo Iron Mine, 1976, pp. 260-261.

Mining Magazine, Guinea's Mifergui iron ore project, October 1978, p. 145.

Skillings Mining Review, Lamco negotiating for Guinean ore shipments, 23 July 1977.

The Tex Report, Iron ore manual, Nimba iron mine development project, 1976, pp. 259-260.

World Mining, Guinea-Nimba plans to develop, August 1975, pp. 56-59.

Ivory Coast

Japan Commerce Daily, New overseas iron ore projects, Vol. 17, 3 February 1977, pp. 4-8.

Schmidt, R.C. and Kennedy, B.E., The geology of the Mont Klahoyo iron ore deposit, SME, AIME, Annual Meeting, February 1979, pp. 1-25.

The Tex Report, Iron ore manual, Mont Klahoyo project, 1976, pp. 243-246.

Liberia

Angel, Tryggve, The Lamco Iron Ore development in Liberia: Mining Congress Journal, pp. 69-72, January 1967.

Berge, J.W., K. Johansson and J. Jack, Geology and origin of the hematite ores of the Nimba Range, Liberia, Economic Geology, Vol. 72, 1977, pp. 582-607.

Danielsson, Christer and Sven Ivarsson. Resume and influence of Northern and Western African iron ore developments on world markets: Skillings Mining Review, v. 55, no. 13, pp. 1-6 and 16. Mar. 26, 1966 and v. 55, no. 14 pp. 1-9 and 17, April 2, 1966.

Danielsson, Christer, Lamco ore project swings into full production: Eng. and Mining Jour., V, 165, no. 5, pp. 91-97, May 1964.

Jacobs, W., H. G. Papacek and D. Brennecke, Inverse flotation of iron ores, Skillings Mining Review, 5 August 1978, pp. 10-24.

The Japan Commerce Daily, Putu iron ore project, Vol. 17, No. 129, p. 1, 15 July 1977.

Metal Bulletin, Bomi Hills of Liberia: pp. i-iv, July 31, 1964.

Metal Bulletin. Liberia-Lamco: Iron Ore Special Issue, pp. 78-79, March 1965.

Metal Bulletin. Mano River of Liberia: pp. 19-20, May 13, 1966.

New African, Iron ore waiting for those new mines, August 1977, pp. 803-804.

Skillings Mining Review, Iron ore shipments of companies-1977, 8 July 1978, pp. 8-9.

The Tex Report, Iron ore manual-1976, Bie Mountain iron ore deposit, pp. 263-264.

The Tex Report, Iron ore manual-1976, Wologisi iron mine development project (Liberia), pp. 246-247.

Mauritania

Audibert, J., P. Caruel and A. Chaubersky. Development of the Kedru d'Idjil Orebodies: Miferma (S.A. des Mines de Fer de Mauritanie), Islamic Republic of Mauritania: The Institution of Mining and Metallurgy, Symposium on open cast mining, quarrying and alluvial mining, Paper 20, London, 1964.

Canadian Mining Journal. Miferma—The iron mountain of Mauritania: Canadian Mining Journal, v. 87, no. 6, pp. 87-90, June 1966.

Europe (France) Outremer, Miferma: Progression de la production de la minerai de fer: 43me ann., no. 431, pp. 27-30, Dec. 1965.

The Japan Commerce Daily, Guelb iron ore project in Mauritania, 4 November 1975, pp. 3-5.

The Japan Commerce Daily, New overseas iron ore project, 31 January 1977, pp. 4-5.

Mining Journal, Mauritania's Guelb iron ore project, 7 October 1977, p. 287.

Skillings Mining Review, Iron ore shipments of companies, 8 July 1978, pp. 8-9.

The Tex Report, Iron ore manual, 1976, Guelb iron mine development project (Mauritania), pp. 258-259.

South Africa

Beukes, N.J., Precambrian iron formations of South Africa, Economic Geology, Vol. 68, 1973, pp. 960-1004.

The Point-Supplement, Sishen-Saldanha, March 1976, Vol. 5, No. 13.

Skillings Mining Review, The Sishen-Saldanha project in South Africa, 14 April 1973, Vol. 62, No. 15, 4 pp.

Skillings Mining Review, 1977 Iron ore shipments of companies, 8 July 1978, pp. 8-9.

The Tex Report, Iron ore manual-1976, Palabora iron ore (South Africa), pp. 180-182.

The Tex Report, Iron ore manual-1976. Saldanha Bay development project by Iscor, pp. 239-241.

The Tex Report, Assoman-South Africa, Vol. 9, No. 2003, pp. 11-12.

IRON-ORE DEPOSITS OF ASIA
Peoples Republic of China

Ikonnikov, A. B., Mineral Resources of China. The Geological Society of America, Microfilm Publ. 2, 1975.

United Nations, Survey of world iron-ore resources, 1970.

U. S. Bureau of Mines, Minerals Yearbook 1980, vol. III.

India

Mining Journal Review, 1981.

Skilling's Mining Review 10, 1981.

United States Bureau of Mines Yearbook 1980, vol. III.

North Korea

United Nations, Survey of world iron-ore resources, 1970.

U. S. Bureau of Mines, Minerals Yearbook, 1980, vol. III.

Pakistan

United Nations, Survey of world iron-ore resources, 1975.

U. S. Bureau of Mines, Minerals Yearbook, 1980, vol. III.

Philippines

Leonardo, Antonio R., Geology of Santa Ines iron deposits, Antipolo, Rizal, Philippines, Proceedings of the Second Geological Convention and First Symposium on the Geology of the Mineral Resources of the Philippines and Neighboring Countries, Geological Society of the Philippines, 1967, pp. 121-136.

Wright, W.S., The Surigao laterite deposits, The Philippine Geologist, 1958, Vol. 12, pp. 47-59.

Wright, W.S., Quicho, R.B., Santos-Yuigo, L., Salazar, A. and Manigque, M.D., 1958, Iron-Nickel-Cobalt Resources of Nonoc, Awasan, and Southern Dinagat Islands in Parcel II of the Surigao Mineral Reservation, Surigao, Mindanao, Publication 17, Special Projects Series, Philippine Bur. Mines, Manila.

IRON-ORE DEPOSITS OF AUSTRALIA

Australian Government Publishing Service, The Pilbara study. Canberra, 1974.

Brandt, R. T., Iron ore deposits of Mount Goldsworthy area, Port Hedland district, western Australia: Australasian Inst. Min. & Met., Proc. no. 211, pp. 157-180, Sept. 1964.

Brandt, R. T. The genesis of the Mount Goldsworthy iron ore deposits of Northwest Australia: Econ. Geol., v. 61, no. 6, pp. 999-1009, 1966.

Campana, B., F. E. Hughes, W. G. Burns, I. G. Whitcher and E. Muceniekas., Discovery of Hamersley iron deposits: Australasian Inst. Min. & Met. Proc. no. 210, pp. 1-30, June 1964.

Canavan, F., Iron ore deposits of Australia: in Eighth Commonwealth Mining and Metallurgical Congress Publications, v. 1: Geology of Australian Ore Deposits, 2nd ed., pp. 13-23, 1965.

Connolly, R. R., Iron ores of Western Australia: Western Australia, Geol. Surv., Min. Res. of Western Australia Bull. 7, 1959.

Dunn, J. A. Australian iron ore developments: Mining Mag., v. 113, no. 3, pp. 180-181, 183-185, Sept. 1965.

Harms, J. E. and B. D. Morgan, Pisolitic limonite deposits in northwest Australia: Australasian Inst. Min. & Met., Proc. no. 212, pp. 91-124, Dec. 1964.

Hughes, T. D., Iron ore deposits of Savage River; in Eighth Commonwealth Mining and Metallurgical Congress Publications, v. 1; Geology of Australian ore deposits, 2nd, ed. pp. 525-526, 1965.

Japan Commerce Daily, Marandoo ore project in Western Australia, pp. 8-9, January 27, 1977.

Hamersley Iron Pty. Ltd., Iron ore data sheet, pp. 1-12, August

1975.

MacLeod, W. N. Banded iron formations of Western Australia: in Eighth Commonwealth Mining and Metallurgical Congress Publications, v. 1; Geology of Australian ore deposits, 2nd. ed., pp. 113-117, 1965.

MacLeod, W. N., Iron ore deposits of the Hamersley iron province: in Eighth Commonwealth Mining and Metallurgical Congress Publications, v. 1: Geology of Australian Ore deposits, 2nd ed., pp. 118-125, 1965.

Matheson, R. S., P. B. Andrews, R. T. Brandt and W. K. Liddicoat. Iron ore deposits of the Port Hedland district: in Eighth Commonwealth Mining and Metallurgical Congress Publications, v. 1: Geology of Australian ore deposits, 2nd ed., pp. 132-137, 1965.

Metal Bulletin, BHP seeks outlets, April 22, 1977.

Metal Bulletin, Focus on Mt. Newman, p. 42, May 10, 1977.

Metal Bulletin. Australia-Hamersley: in Iron Ore Special Issue, pp. 47-49, March 1965.

Metal Bulletin. Australia-Mount Goldsworthy: Iron Ore Special Issue, pp. 51-55, March 1965.

Mineral Commodity Profiles, MCP-13, Iron ore, pp. 1-27, May 1978.

Mining Magazine, Expansion at Mt. Newman, May, 1976.

Owen, H. B. and Sylvia Whitehead, Iron ore deposits of Iron Knob and the Middleback Ranges: in Eighth Commonwealth Mining and Metallurgical Congress Publications, v. 1: Geology of Australian ore deposits, 2nd ed., pp. 301-308, 1965.

Reid, Ian W., Iron ore deposits of Yampi Sound: In Eight Commonwealth Mining and Metallurgical Congress Publications, v. 1: Geology of Australian ore deposits, 2nd ed., pp. 126-131, 1965.

Skillings, D. N., Savage River mines. Skillings' Mining Review, pp. 1-15, Dec. 6, 1969.

Skillings' Mining Review, The Robe River iron ore project, pp. 10-15, Oct. 30, 1971.

Skillings' Mining Review, 1977 Iron ore shipments of companies, pp. 8-9, 1978.

Skillings' Mining Review, Mount Goldsworthy iron ore project: v. 55, no. 32, pp. 1-8 and 16, August 6, 1966.

Skillings' Mining Review, Hamersley iron ore project in Western Australia: v. 55, no. 50, pp. 1-7 and 24, Dec. 10, 1966.

Steel times International, West Australia feeds the world's steel mills, pp. 53-55, Sept. 1978.

Tex Report Co., Ltd., Iron ore manual, 1977. New projects, pp. 253-262: Iron ore of Australia, p. 93.

Tex Report Co., Ltd., The Tex report, West Angelas deposit. Vol. 9, No. 1972, pp. 3-4, March 25, 1977.

IRON-ORE DEPOSITS OF EUROPE
General

Blondel, F. and L. Marvier, editors, Gismets de fer du monde, Vol. II, XIX Geological Congress, Algiers, 1952.

Bowie, S.H.U., Kvalheim, A. and H. W. Haslam, Mineral Deposits of Europe, Vol. 1, Northwest Europe, The Institution of Mining and Metallurgy and The Mineralogical Society, London, 1978.

James, Harold and Paul K. Sims, editors, Precambrian iron formations of the world, (contains 16 titles), Economic Geology, Nov. 1973, Vol. 68, No. 7, pp. 913-1179.

Marelle, Andre, Iron ore deposits of Europe, in World Survey of World Iron Ore Resources, UN, 1970, pp. 270-301.

Metal Bulletin, Iron Ore, 1969, Special Issue.

WESTERN EUROPE
Austria

Mining Annual Review, Mining Journal, 1981, p. 545.

Skillings Mining Review, 6 March 1981, p. 6.

Federal Republic of Germany

U.S.B.M. Minerals Yearbook, 1980, Vol. III, P. 403.

Skillings Mining Review, 21 February 1981, p. 6.

Finland

Isokangas, Pauli, Finland, in Mineral Deposits of Europe, vol. 1, Northwest Europe.

Skillings Mining Review, 21 March 1981, p. 18.

France

Fomento Digest, Changes in iron ore mining-France, January 1981, p. 5.

Metal Bulletin, 11 November 1980, p. 42.

Norway

Bugge, J.A.W., Norway in Mineral Deposits of Europe, Vol. I, Northwest Europe, The Institution of Mining and Metallurgy, The Mineralogical Society, London, 1978, pp. 199-249.

Mining Magazine, Rana Gruber: new pits, new plant, November 1980, pp. 386-399.

Skillings Mining Review, 4 November 1980.

Spain

Skillings, David N. Jr., C.A.M. iron ore operations at Marquesado mine and Almira port, Skillings Mining Review, 18 April 1981, pp. 16-19.

Skillings Mining Review, 1 November 1980, p. 25.

Sweden

Frietsch, Rudyard, On the magmatic origin of iron ores of the Kiruna type, Economic Geology, Vol. 73, pp. 478-485, 1978.

Grip, E., Sweden in Bowie, Mineral Deposits of Europe, Vol. 1, 1978, pp. 93-184, Northwest Europe.

Parak, Tibor, Kiruna iron ores are not intrusive-magmatic ores of the Kiruna type, Economic Geology, Vol. 70, 1975, pp. 1242-1258.

Skillings Mining Review, 12 February 1981, p. 12.

Skillings Mining Review, Exploration of Northern Sweden iron ore deposits 17 September 1977, pp. 14-16.

United Kingdom

Kingsley Dunham, K. E. Beer, R. A., Ellis, M. J. Gallagher, M.J.C. Nutt and B.C. Webb, United Kingdom, in Mineral Deposits of Europe, Vol. 1, Northwest Europe.

Mining Annual Review, Mining Journal, June 1981, p. 533.

EASTERN EUROPE
Bulgaria

U.S.B.M. Minerals Yearbook 1978-79, Vol. III, Area Reports International, p. 9.

Czechoslovakia

U.S.B.M. Minerals Yearbook 1978-79, vol. III, Area Reports International, p. 11.

Hungary

U.S.B.M. Minerals Yearbook 1978-79, Vol. III, Area Reports International, p. 12.

World Mining, Hungary, Catalog, Survey and Directory, 1979.

Poland

U.S.B.M. Minerals Yearbook 1978-79, Vol. III, Area Reports International, p. 13.

Romania

U.S.B.M. Minerals Yearbook 1978-79, Vol. III, Area Reports International, p. 11.

USSR

Alexandrov, Eugene A., The Precambrian iron formations of the Soviet Union, Economic Geology, November 1973, vol. 68, No. 7, pp. 1035-1062.

Kish, George, Ian Matley, Betty Bellaire, Economic Atlas of the Soviet Union, 1961, Univ. of Michigan Press, Ann Arbor.

U.S.B.M. Minerals Yearbook 1978-79, vol. III, pp. 30-31.

Yugoslavia

Metal Bulletin, 7 April 1981, p. 35.

Mining Journal, Mining Annual Review - 1979, p. 595.

U.S.B.M. Minerals Yearbook 1978-79, pp. 1055-1056.

World Mining, Catalog and Directory, No. 1979, p. 224.

DISCOVERY AND MINING OF IRON ORES

Cummings, J. D., Diamond Drill Handbook J. K. Smit & Sons, 1956.

McManus, C. E., Stubbins, J., New Development in Open Pit Mining, Canadian Area. Twenty-Third Annual Mining Symposium, University of Minnesota, and the Annual Meeting of Minnesota Section, American Institute of Mining, Metallurgical and Petroleum Engineers, pp. 115-120, 1962.

Moberg, N. A., New Developments in Exploration and Investigation of Iron Ore Properties. Twenty-Third Annual Mining Symposium, University of Minnesota, 1962.

Parks, R. D., Examination and Valuation of Mineral Property. Addison-Wesley Press, Inc., 1949.

Pearson, P. D., New Developments in Underground Mining Throughout the Lake Superior Iron Ranges. Twenty-Third Annual Mining Symposium, University of Minnestoa, and the Annual Meeting of Minnesota Section, American Institute of Mining, Metallurgical, and Petroleum Engineers, pp. 101-107, 1962.

Peele, R., and Church, J. A., Mining Engineers' Handbook. John Wiley & Sons, Inc., 1941.

Woodle, M. G., and Bertie, R., New Developments in Open Pit Mining, Minnesota and Michigan Area. Twenty-Third Annual Mining Symposium, University of Minnesota, and the Annual Meeting of Minnesota Section, American Institute of Mining, Metallurgical, and Petroleum Engineers, pp. 109-114, 1962.

Young, G. J., Elements of Mining. McGraw-Hill Book Co., 1946.

BENEFICIATION OF IRON ORES
General

Anon., Grinding: the search for increased efficiencies. Engineering and Mining Journal, Vol. 176, No. 5, June 1975, pp. 109-110a.

Atkins, A. R., A. L. Hinde, P. J. D. Lloyd and J. G. Mackay, The control of milling circuits. Jour. South African Mining and Met., Vol. 74, No. 11, June 1974, pp. 388-395.

Bartnik, J. A., W. H. Zabel and D. M. Hopstock, On the production of iron ore superconcentrates by high-intensity wet magnetic separation. Intl. Jour. of Mineral Processing, Vol. 2, No. 2, June 1975, pp. 117-126.

Bassarear, J. H., Crushing and grinding. Mining Engineering, Vol. 28, No. 2, February 1976, pp. 69-71.

Berkolm, R. W., How Flotation Makes High-Grade Specular Hematite Concentrate at Humboldt. Mining World, pp. 30-33, November 1961.

Burkhardt, E. S., Efficient crushing improves reduction costs. Allis-Chalmers Engrg. Rev., Vol. 40, No. 2, 1975, pp. 25-27.

Cavanaugh, W. J., Snyder process—a breakthrough in communition. Mining Congress Journal, Vol. 58, No. 12, December 1972, pp. 30-36.

Cohlmeyer, S. H., Henderson, A. S., and Morgan, R. C., The Story of Atlantic City. Twenty-Third Annual Mining Symposium, University of Minnesota, and the Annual Meeting of Minnesota Section, American Institute of Mining, Metallurgical, and Petroleum Engineers, pp. 133-138, 1962.

DeVaney, F. D., Iron Ore Beneficiation. Engineering and Mining Journal, pp. 141-143, February 1956.

Hawker, T. G., Magnetic and electrostatic techniques in mineral dressing. Mine and Quarry, Vol. 3, No. 10, October 1974, pp. 51-57.

Kelland, D. R., and E. Maxwell, Pilot investigation of high gradient magnetic separation of oxidized taconites. Preprint 75-B-119, AIME Annual Meeting, 1975.

Lawver, J. E. and D. M. Hopstock, Wet magnetic separation of weakly magnetic minerals. Minerals Science and Engineering, Vol. 6, No. 3, July 1974, pp. 154-172.

Lynch, A. J., Mineral crushing and grinding circuits. Elsevier Scientific Publishing Co., Amsterdam, The Netherlands, 1977.

McAneny, C. C., Special Report—Direct Reduction of Iron Ore. Engineering and Mineral Jour., pp. 84-99. Dec. 1960.

Ramsey, R. H., Teamwork on Taconite. Engineering and Mining Journal, pp. 82-93, March 1955.

Reserve's New Taconite Project. Engineering and Mining Journal, pp. 75-102, December 1956.

Mular, A. L. and R. D. Bhappu, Mineral processing plant design. Society of Mining Engineers of AIME, New York, N. Y., 1978.

Pickands-Mather, The Hibbing Taconite Co. story. SME-AIME Preprint 79-119, AIME New Orleans Meeting, 1979.

Reserve's Taconite Concentrator, Engineering and Mining Journal, pp. 228-229, June 1959.

Roe, L. A., Iron Ore Beneficiation, Minerals Publishing Co., 1957.

Runnels, D., Crushing, grinding and concentration—the Mt. Wright story. CIM Bulletin April 1977, pp. 97-102.

Scott, D., Beneficiation of Lake Superior Iron Ores, A Review and Appraisal of Recent Trends. Twenty-Third Annual Mining Symposium, Universtiy of Minnesota, and the Annual Meeting of Minnesota Section, American Institute of Mining, Metallurgical, and Petroleum Engineers, pp. 121-131, 1962.

Smith, R. R., Iron Ore Flotation in Michigan. Skillings' Mining Review, January 28, 1961.

Urich, D. M., Pelletizing Humboldt's Iron Concentrate by Grate-Kiln Process. Mining World, pp. 16-21, October 1961.

Veith, D. L., Superconcentration of commercial magnetic taconite concentrate by cationic flotation. United States Bureau of Mines RI 7852, 1974.

Voges, H. C., the use of heavy-medium separation in the processing of iron ores. Jour. South African Inst. Mining and Met., Vol 75, No. 11, June 1975, pp. 303-306.

Wade, H. H., and Schulz, N. F., Magnetic Roasting of Iron Ores in a Traveling Grate Roaster. Mining Engineering, pp. 1161-1165, November, 1960.

White, L., Swedish symposium offers iron ore industry an overview of ore dressing developments. Engineering and Mining Journal, April 1978, pp. 71-77.

World's Largest Taconite Plant. Engineering and Mining Journal, pp. 235-237, June 1959.

CONCENTRATION

Barthelemy, R. E., How High Tension Electrostatic Separation Recovers Iron Ore. Engineering and Mining Journal, v. 159, pp. 87-91, December 1958.

Bernstrom, B., Grinding Iron Ore in a Wet Autogenous Mill. American Institute of Mining, Metallurgical, and Petroleum Engineers Transactions, v. 223, pp. 304-311, September 1962.

Bogdanov, O. S., Shapiro, R. B., and Danilova, E. V., New Trends in the Beneficiation of Ferrous and Nonferrous Metal Ores. Contemporary Problems of Metallurgy, A. M. Samarin, Editor, Consultants Bureau, pp. 29-42, 1960.

Everard, F. and Janes, R., Dry Autogenous Grinding and Dry Magnetic Separation of Iron Ores. American Institute of Mining, Metallurgical, and Petroleum Engineers Transactions, v. 223, pp. 88-96, March 1962.

Frommer, D. W. and Fine, M. M., Recovering Iron Concentrates from the Pea Ridge Deposit, Central Missouri, United States Bureau of Mines Report of Investigations 5550, 1960.

Gaudin, A. M., Principles of Mineral Dressing. McGraw-Hill Book Co., 1939.

Holliday, R. W., Iron. United States Bureau of Mines Bulletin 556, Mineral Facts and Problems, pp. 371-398, 1956.

International Mineral Dressing Congress Proceedings, 1960. Institute of Mining and Metallurgy, pp. 675-754.

Lee, Oscar, Taconite Beneficiation Comes of Age at Reserve's Babbitt Plant, Mining Engineering, v. 6, pp. 484-488, May 1954.

Linney, R. J., Economy Through Design. American Institute of Mining, Metallurgical, and Petroleum Engineers Transactions, v. 214, pp. 909-914, 1959.

Palasvirta, O. E., and Andreachi, J. R., High-Intensity Magnetic Separation of Iron Ores. Twenty-Third Annual Mining Symposium, Universtiy of Minnesota, pp. 37-41, 1962.

Reno, H. T., Iron. United States Bureau of Mines Bulletin 585, Mineral Facts and Problems, pp. 403-421, 1960.

Roe, L. A., Iron Ore Beneficiation. Mineral Publishing Company, 1957.

Scott, D. W., and Wesner, A., Properties of Nonmagnetic Taconites Affecting Concentration. American Institute of Mining, Metallurgical, and Petroleum Engineers Transactions, v. 199, pp. 635-641, 1954.

Shale, S. J., Cornwall Keeps Its Methods Up-to-Date. Mining Engineering, v 7, pp. 670-675, July 1953.

Taggart, A. F., Handbook of Mineral Dressing. Section 2, Article 28, Iron Minerals. John Wiley and Sons, pp. 134-151, 1945.

Volin, M. E., Beebe, R. R., and Hockings, W. A., Problems of Beneficiating the Oxidized Iron Formations. The Mines Magazine, v. 51, pp. 16-20, October 1961.

Wade, H. H., Oxidized Taconite—Its Possible Utilization by Roasting and Magnetic Concentration. 23rd Annual Mining Symposium, University of Minnesota, pp. 67-72, 1962.

Watkins Cyclopedia of the Steel Industry. Steel Publications, pp. 61-72, 1963.

Webb, W. R. and Fleck, R. G., Beneficiation of Adirondack Magnetite, Blast Furnace, Coke Oven and Raw Materials Committee, American Institute of Mining, Metallurgical and Petroleum Engineers Proceedings, v. 9, pp. 220-230, 1950.

AGGLOMERATION
General

Bagnall, E. J. and C. W. Brock, Energy considerations for the preparation and processing of iron ores. Seaisi Quarterly, 4-2 (665/75, April 1975, pp. 14-22.

Bennett, R. L. and R. D. Lopez, Agglomeration of iron ore concentrates. Chemical Engineering Progress Symposium Series, Vol. 59, No. 43, 1963, pp. 40-52.

Cummins, A. B., and I. A. Given (eds.), SME Mining engineering handbook, SME-AIME, New York, N. Y., 1973.

English, A., Iron ore agglomeration—sintering and pelletizing. Iron and Steel Engineer, Vol. 38, No. 3, March 1961, pp. 113-119.

Leconte, P., R. Vidal, A. Poos and A. Decker, Softening of sinters, pellets and iron ores. CNRM, Vol 21, December 1969, pp. 21-28.

Sastry, J. V. S. (ed.). Agglomeration 77. AIME, New York, N. Y., 1977.

Sisselman, R., Iron ore in U.S.—a profile of major mining, processing facilities. Mining Engineering, Vol. 25, No. 7, September 1973, pp. 45-64.

Sintering

Brandes, G. and Rausch, H., Mixing and Conditioning Sinter-Plant Feed. Blast Furnace, Coke Oven, and Raw Materials Committee, American Institute of Mining, Metallurgical and Petroleum Engineers Proceedings, v. 18, pp. 232-249, 1959.

Chang, M. C., Pyrolysis-agglomeration. Mining Engineering, Vol. 18, No. 2, February 1966, pp. 103-105.

Davies, W. E., Some Practical Applications of Fundamental Sinter Research. Canadian Mining and Metallurgical Bulletin, v 53, pp. 173-185, March 1960.

English, Alan, Iron Ore Agglomeration—Sintering and Pelletizing. Iron and Steel Engineer, v 38, pp. 113-119, March 1961.

Knepper, W. A., Editor. Agglomeration; based on an International Symposium held in Philadelphia, Pa., April 12-14, 1961. Interscience Publishers, 1962.

Meredith, J. W. and Frankau, A. M., Control and Proportioning of Sinter Plant Raw Materials. Blast Furnace, Coke Oven, and Raw Materials Committee, American Institute of Mining, Metallurgical, and Petroleum Engineers Proceedings, v. 18, pp. 60-67, 1959.

Nyquist, Orvar, Studies of the Effect of Gangue in the Sintering of Rich Magnetite and Hematite Concentrates. Jernkontorets Annaler, v. 146, No. 2, pp. 81-145, 1962.

Porteus, J. H., A Review of Iron Ore Sintering. Iron and Steel Engineer, v. 38, pp. 144-155, May 1961.

Pritykin, D. P., Experience in Using Equipment at Sintering Plants. Metallurgist, no. 10, pp. 429-435, October 1960.

Robinson, A. W., Swedish Sintering Practice—The Holmberg System. Blast Furnace, Coke Oven and Raw Materials Committee, American Institute of Mining, Metallurgical, and Petroleum Engineers Proceedings, v. 9, pp. 246-258, 1950.

Rowen, H. E., Development of the Dwight-Lloyd Sintering Process. Journal of Metals, v. 8, pp. 828-831, July 1956.

Takahaski, Y., et al. New Sintering Process (Semi-Pellet) for Fine Iron Ores. American Institute of Mining, Metallurgical, and Petroleum Engineers Trans., v. 220, pp, 499-505, 1961.

Walter, A. R., The Use of Anthracite Coal in Sintering Iron Ore at the Bethlehem Steel Company Concentrator Plant, Lebanon, Pa. Sixth Annual Anthracite Conference of Lehigh University, Bethlehem, Pa., May 6-7, 1948, Transactions, pp. 139-158.

Wendeborn, H., The Importance of the Sintering Process in the Production of Pig Iron. MetallGesellschaft, Review of the Activities, no. 1, pp. 2-12, 1959.

Pelletizing

Banks, G. N., Campbell, R. A. and Viens, G. E., Iron Ore Pelletizing—A Literature Survey. Department of Mines and Technical Surveys, Mines Branch, Extraction Metallurgy Division, Ottawa, Canada, Internal Report EMA 62-7, April 10, 1962. Presented before the Annual Meeting of the Canadian Insitute of Mining and Metallurgy, Ottawa, April 1962.

Barrett, E. P., Shaft Furnace Reduction by the Glomerule Method. United States Bureau of Mines, Report of Investigations 3229, pp. 47-49, 1934.

Berkhahn, R. W. and Urich, D. M., Grate Kiln Pelletizing Process at Humboldt. Twenty-Third Annual Mining Symposium, University of Minnesota, pp. 25-32, 1962.

Bunge, F. H. and Wakeman, J. S., Pelletizing Butler, Groveland, and Carol Lake Concentrates. Twenty-Third Annual Mining Symposium, University of Minnesota, Minneapolis, pp. 49-59, 1962.

Davis, E. W. and Wade, H. H., Agglomeration of Iron Ore by the Pelletizing Process. University of Minnesota Mines Experiment Station Circular no. 6, 1951.

Dugge, R. D., Modelling of heat exchange phenomena during the indurating of magnetite pellets on a travelling grate. Preprint No. 75-B-51, AIME Annual Meeting, 1975.

Firth, C. V., Agglomeration of Fine Iron Ore. American Institute of Mining, Metallurgical, and Petroleum Engineers Blast Furnace and Raw Materials Committee Proceedings, v. 4, pp. 46-65, 1944.

Greenwalt, A. B. and G. F. Cofield, The circular grate system for iron ore pelletizing. Mining Congress Journal, Vol. 54, No. 8, August 1968, pp. 39-45.

Haley, K. M. and Apuli, W. E., Pelletizing on a Horizontal Grate Machine. Agglomeration, edited by W. A. Knepper. Interscience Publishers, pp. 931-957, 1962.

Jewett, R. P., Wood, C. E., and Hansen, J. P., Effect of Particle Size Upon the Green Strength of Iron Oxide Pellets. United States Bureau of Mines Report of Investigations 5762, 1961.

Kaiser, C. F., T. J. Roberts and I. A. Thompson, Operating experience with the production of fluxed pellets. Trans. SME-AIME, Vol 252, No. 4, December 1972, pp. 439-443.

Kolesanov, F. F. and Gavin, E. G., Production of fluxed pellets from sulfur-containing magnetite concentrates of Magnitogorsk and Sokolov-Sarbai ores. Stal in English, No. 4, pp. 247-250, April 1962.

Madigan, P. C., Pelletizing—a review. AMDEL Bulletin, Vol. 9, April 1970, pp. 1-50.

Merklin, K. E. and DeVaney, F. D., The Coarse Specularite—Fine Magnetite Pelletizing Process. Agglomeration, edited by W. A. Knepper. Interscience Publishers, pp. 965-975, 1962.

Merklin, K. E. and DeVaney, F. D., Production of Self-Fluxing Pellets in the Laboratory and Pilot Plant. American Institute of Mining, Metallurgical, and Petroleum Engineers Transactions, v 217, pp. 46-51, 1960. Also United States Patents 2,816,016; 2,831,210; 2,990,268.

Nicol, S. K. and Z. P. Adamiak, Role of bentonite in wet pelletizing processes. Trans. Inst. of Mining and Metallurgy, Vol. 796, No. 82, March 1973, pp. 626-633.

Rovenskii, I. I. and Berezhnoi, N. N., Travelling-Grate Machine for Firing Fluxed Iron-Ore Pellets. Stal in English, No. 3, pp. 174-177, March 1962.

Sastry, J. V. S. and D. W. Fuersteneau, Ballability index to quantify agglomerate growth by green pelletization. Trans. SME-AIME, Vol. 252, September 1976, pp. 254-258.

Urich, D. M., Pyrolysis and agglomeration trends dominated by pelletizing. Mining Engineering, Vol. 22, No. 2, February 1970, pp. 108-110.

Westwater, J. S., Pelletizing by the Grate-Kiln Method at Humboldt. Blast Furnace and Steel Plant v. 49, pp. 513-518, 530, June 1961.

Briquetting

Byrns, H. A., Briquetting Fine Ores at Woodward, Alabama. Blast Furnace, Coke Oven, and Raw Materials Committee, American Institute of Mining, Metallurgical, and Petroleum Engineers Proceedings, v. 8, pp. 158-164; Discussion, pp. 165-170, 1949.

Fournier, E., Note on the Agglomeration of Fine Ores from the Iron Mines at Rouen. Revue de I'Industrie Minerals, No. 235, pp. 435-438, 1930.

Franke, G., Present State of the Briquetting and Agglomeration of Iron Ores in Germany. Revue de Metallurgie, v. 7, pp. 953-954, 1910.

Hansell, N. V., The Briquetting of Iron Ores. American Institute of Mining, Metallurgical, and Petroleum Engineers Transactions. v. 43, pp. 394-411, 1912.

Lueck, B. F., Olson H. S. and Wiley, A. J., Modified Spent Sulphite Liquor Products as Binders and Adhesives for Briquets and Other Products. International Briquetting Association, Proceedings of the Seventh Biennial Conference, Jackson, Wyoming, August 28-30, 1961, pp. 14-25.

Moore, J. E. and Marlin, D. H., Hot Briquetting of Partially Reduced Iron Oxide Ores and Dusts. Agglomeration, edited by W. A. Knepper, Interscience Publishers. pp. 743-781, 1962.

Onoprienko, V. P., Lebedev, A. E. and Furman, D. M., Production of Fluxed Briquettes for the Iron and Steel Industry, Stal in English, No. 2, pp. 77-80, February, 1961.

Stillman. A. L., Flue Dust Briquetting by the Corrosion Process. Iron Age, v. 110, pp. 1571-1572, 1922.

Thompson, R. G., Franklin, R. L., Guseman, J. R. and Rohaus, D. E., United States Steel Hot Ore Briquetting Process. Blast Furnace, Coke Oven, and Raw Materials Committee. American Institute of Mining, Metallurgical, and Petroleum Engineers Proceedings, v. 20, pp. 316-323, 1961. See also United States patent 2,336,618.

Wolf, W. and Wysocki, H., Operating Data on the Briquetting of (Blast Furnace) Flue Dust. Stahl und Eisen, v. 81, No. 9, pp. 559-561, 1961. (Brutcher Translation No. 5158)

Nodulizing, Extrusion, and Fluid-bed

Bennett, R. L., Hagen, R. E. and Mielke, M. V., Nodulizing Iron Ores and Concentrates at Extaca. Mining Engineering, v. 6, pp. 32-38, January 1954.

Brownstead, E. F., Manufacture of Nodules at Ironton, Ohio, Using Mesabi Fines and Blast-Furnace Flue Dust. American Institute of Mining, Metallurgical, and Petroleum Engineers Proceedings, v. 8, pp. 139-142; Disc., pp. 142-145, 1949.

Cavanagh, P. E., Pelletizing of Iron-bearing Fines by Extrusion. Blast Furnace, Coke Oven, and Raw Materials Committee, American Institute of Mining, Metallurgical, and Petroleum Engineers Proceedings, v. 9, pp. 54-72, 1950.

Langston, G. B. and Stephens, F. M., Jr., Direct Reduction of Fine Iron Concentrates in Self-Agglomerating Fluidized Bed. Blast Furnace, Coke Oven, and Raw Materials Committee, American Institute of Mining, Metallurgical and Petroleum Engineers Proceedings, v. 19, pp. 205-214, 1960.

Ludwig, Carl., Agglomerating Ores by Vacuum Extrusion. Presented before the Annual Meeting, American Institute of Mining, Metallurgical, and Petroleum Engineers, February 18-21, 1952.

Oesterle, A. A., Manufacture of Nodules from Fine Ore and Limestone at Buffalo, N. Y. Blast Furnace, Coke Oven, and Raw Materials Committee, American Institute of Mining, Metallurgical and Petroleum Engineers Proceedings, v. 8, pp. 132-136; Discussion, pp. 136-138, 1949.

TRANSPORTATION OF IRON ORES

Benford, H., Thornton, K. C. and Williams, E. B., Current Trends in the Design of Iron-Ore Ships, Society of Naval Architects and Marine Engineers Transactions, v. 70, 1962.

Greenwood's Guide to Great Lakes Shipping 1982, Freshwater Press, Inc., Cleveland, Ohio.

Hussey, C. R., Transportation, Handling and Storage of Iron Ores from Mines to Docks in the Lake Superior Region. Twenty-Second Annual Mining Symposium, University of Minnesota, and the Annual Meeting of Minnesota Section, American Institute of Mining, Metallurgical, and Petroleum Engineers, pp. B38-41, 1961.

Meissner, J. F., World Development and Movement of Iron Ore. Society of Naval Architects and Marine Engineers Transactions, v. 70, 1962.

Power, R. E., Economic Factors in the Location of Agglomerating Facilities. Twenty-Second Annual Mining Symposium, University of Minnesota, and the Annual Meeting of Minnesota Section, American Institute of Mining, Metallurgical, and Petroleum Engineers, pp. B77-84, 1961.

Turner, J. R., From Dock to Steel Plants—Lower Lakes Ports. Twenty-Second Annual Mining Symposium, University of Minnesota and the Annual Meeting of Minnesota Section, American Institute of Mining, Metallurgical, and Petroleum Engineers, pp. B50-53, 1961.

Vines, F. D., Transportation, Handling and Storage of Iron Ores. Twenty-Second Annual Mining Symposium, University of Minnesota, and the Annual Meeting of Minnesota Section, American Institute of Mining, Metallurgical, and Petroleum Engineers, pp. B42-49, 1961.

Wilbur, J. S., Lower Lake Railroads and Iron Ore Industry. Twenty-Second Annual Mining Symposium, University of Minnesota, and the Annual Meeting of Minnesota Section, American Institute of Mining, Metallurgical, and Petroleum Engineers, pp. B54-66, 1961.

ENVIRONMENTAL CONSIDERATIONS

Anon., Pollution problems. Revue de Metallurgie, Vol. 71, no. 3, March 1974, pp. 293-299.

Aplin, C. L. and G. O. Argall, Tailing disposal today. Miller Freeman Publications, San Francisco, Cal. 1973.

Copper, P. S., Dust control at Steep Rock iron ore pellet plant. Skilling' Mining Review, vol. 65, no. 41, Oct. 1976, pp. 6-9.

Jones, J. D. and R. G. Jenkins, Design and construction of tailings ponds and reclamation facilities—case histories. CIM Bulletin, Vol. 71, No. 798, October 1978, pp. 55-59.

Waugh, J. H. and R. J. Triscori, Consider all the options for a sinter plant air pollution control system. Iron and Steel Engineer, Vol. 54, No. 3, March 1977, pp. 36-40.

CHAPTER 6

Fluxes In Iron- And Steelmaking

Function of Fluxes—Any metallurgical operation in which metal is separated by fusion from the impurities with which it may be chemically combined or physically mixed (as in ores) is called **smelting**. Since in iron smelting both these conditions with respect to impurities are always present, the production of crude iron involves two processes: (1) the reduction of the metal from its compounds and (2) its separation from the mechanical mixture. Many of the impurities associated with iron ores are of a highly refractory nature, that is, they are difficult to melt. If they should remain unfused, they would retard the smelting operation and interfere with the separation of metal and gangue. *To render such substances more easily fusible is the primary function of a flux.* Some elements, being reduced almost simultaneously with the iron, dissolve in it, or even combine chemically with it. Some other compounds, already combined with the metal in the raw materials, cannot separate from it unless a substance be present with which they can combine in preference to the metal under the prevailing conditions. *To furnish a substance with which these elements or compounds may combine in preference to the metal is the secondary function of a flux.*

Chemistry of Fluxes—Selection of the proper flux for a given process is chiefly a chemical problem requiring knowledge of the composition and properties of all materials entering the process. With this knowledge, selection will be governed by well-established physical and chemical laws which apply at smelting temperatures, as discussed in Chapter 13. These laws are not unlike those which govern chemical reactions at ordinary temperatures. Of most importance are the laws concerning the formation of salts from the interaction of acids and bases. Practically all of the slag-forming compounds that enter into a smelting or refining process may be classed as either "acids" or "bases" by virtue of the fact that they will react with each other to form compounds which are analogous to the salts formed in reactions taking place in water solutions. In like manner, stronger or more active "acids" and "bases" will replace weaker or less active ones in slag compounds. The substances which are conveniently considered to be the most active "bases" at the high temperatures encountered in smelting and refining are those which are compounds of the elements forming basic compounds in ordinary chemical reactions in water solutions, such as calcium, magnesium, sodium, etc.; while the most active "acid" impurities are compounds of silicon and phosphorus. These latter elements normally form acids in water solutions. In addition to these

are compounds which are analogous to the amphoteric compounds. These are capable of acting as acids or bases depending on the conditions imposed. Since one of the functions of a flux is to react chemically with unwanted impurities to form a fusible slag, it will naturally follow that to remove "basic" impurities, an "acid" flux will be required and to remove "acid" components, a "base" will be used as the flux. It is fortunate that, generally speaking, the slag compounds formed by reactions between acid and basic materials have lower melting points than the reacting compounds, so that the primary function of a flux, rendering impurities more fusible, is simultaneously fulfilled. However, in some instances, a "neutral" material may also be used to lower the slag melting temperature and to improve slag fluidity. In most ores the impurities will be both acid and basic, with the acid materials usually predominating. In certain Southern ores, however, the acids and bases are so well balanced, or can be made so by mixing, that no additional flux is required.

Acid Fluxes—Silica (SiO_2) is the only substance that is used as a strictly acid flux. For this purpose, it is available as sand, gravel, and quartz in large quantities and in a sufficiently pure state. In acid-steelmaking processes, silica is seldom added as a flux, for the silica sand used on the banks to protect the lining of the furnace supplies what may be required. On occasion, silica may be used in basic processes where excess lime has been charged, or where the raw materials are too low in silicon to produce a sufficient quantity of slag. Acid-Bessemer slags formerly were charged in the blast furnace for their fluxing or scouring effect on accumulations of lime on the furnace wall, as well as to recover the iron and manganese they contained. Gravel often is charged for the same purpose. Olivines, which are naturally occurring magnesium-silicate minerals, are sometimes used to enhance the removal of alkalies from the blast furnace. Siliceous iron-bearing materials are charged in making blast-furnace grades of ferrosilicon, but the function of silica in this case is to provide a source of silicon rather than to act as a flux.

Basic Fluxes—The chief natural basic fluxes are limestone, composed principally of calcium carbonate ($CaCO_3$), and dolomite, composed principally of calcium-magnesium carbonate ($Ca,Mg)CO_3$. Either dolomite or limestone may be used as a blast-furnace flux, the proportions of each depending on the other constituents of the slag and the amount of sulphur that the slag must remove. Where large amounts of sulphur are to be removed, limestone is preferred. The blast-furnace fluxes are either charged into the furnace top as raw stone or, if the iron-bearing materials are to be sintered, the fluxes may be crushed and added in proper proportions to the sinter mix in a form called **fluxing fines**. In the latter instance, the flux combines with some of the impurities before charging into the blast furnace, thereby lessening the amount of raw stone required in the process.

Availability and cost are important factors when choosing between limestone and dolomite for blast-furnace use. Among other factors affecting the choice is the fact that a slag high in magnesia content is not desirable as a raw material for cement manufacture, but for ballast or concrete aggregate it produces a much harder product which is less subject to degradation by abrasion. Representative compositions of limestone and dolomite are given in Table 6—I, together with typical compositions for calcined or "burnt" limestone and dolomite.

In the basic oxygen steelmaking process (BOP), lime normally is added as calcined or "burnt" lime and as burnt dolomite. The calcium oxide in either the burnt lime or the burnt dolomite fluxes the silica formed upon oxidation of silicon in the hot metal. A common practice is to use enough burnt dolomite to form a slag containing about 6 per cent magnesium oxide. The presence of this magnesium oxide in the slag decreases greatly the tendency for the slag to attack the magnesium oxide refractory in the furnace lining.

The various fluxes used in the BOP are stored in separate bins which feed weighed amounts of these materials to a discharge hopper located just above the mouth of the furnace. The flux materials, which are sized (nominally 6 × 38 mm or ¼ × 1½ inches) to minimize the amount of material carried away by the gases leaving the furnace mouth, are then fed by gravity into the furnace at specific times during the oxygen blow.

Because flux materials serve the same purpose in the Q-BOP as in the BOP, the same reasoning is used in specifying the composition of these materials in both processes. For instance, burnt lime is used to supply calcium oxide to neutralize the effects of acid-type slag constituents such as silica, and to enhance the occurrence of desirable metallurgical reactions. Therefore, the calcium oxide (CaO) content of burnt lime should be as high as possible and the minimum CaO content of burnt lime is usually specified to be 95 per cent or

greater. Similarly, fluorspar should be high in calcium flouride content, typically in the range of 50 to 75 per cent in naturally occurring fluorspar rock. Because the other components found in fluorspar are not deleterious to steelmaking operations, considering the relatively small amounts of fluorspar used, the selection of a particular source of fluorspar becomes one of economics. Because the flux materials used in the Q-BOP are pneumatically conveyed by process oxygen into the steel bath through tuyeres mounted in the bottom of the furnace, they can be significantly finer in size than those used in the BOP. The correspondingly high surface area of the finely divided particles promotes the rate and extent of desirable slag-metal reactions; thus, the Q-BOP enjoys several metallurgical advantages over the BOP. (See Chapter 16 for details of the various pneumatic steelmaking processes.) Table 6-II shows a typical size specification for bottom-injected Q-BOP fluxes. Although different sized materials can be used in the Q-BOP, the specified size has adequate surface area to accomplish the desired metallurgical reactions, without incurring the handling problems and excessive grinding costs associated with production of a finer sized material.

Table 6—II. Typical Size Distribution of Q-BOP Flux Materials

Particle Size		Amount of Particles Smaller than Indicated Size, %
(US Series Mesh Size)	(Micrometres)	
18	1000	100
140	105	95 to 85
325	44	40 to 60
—	10	35 maximum

In the basic open-hearth process, most of the lime is charged into the furnace as limestone. If lime is needed when the heat of steel is being refined, it is usually added as burnt lime. In the basic open hearth, burnt dolomite is used to repair the furnace bottom.

The effectiveness of a natural basic flux is reduced by the chemically-acid impurities it contains, since these also must be neutralized by some of the chemically-basic compounds in the flux. The effectiveness of a basic flux is expressed in terms of "available base," by which is meant the amount of basic substance that remains in the raw flux after its own acid contaminants have been satisfied.

Alumina—Although alumina is seldom employed as a flux, it deserves mention at this point because it is present in a large number of raw materials as an impurity and is therefore present in slag. In slags it may function as an acid or as a base, depending on the conditions imposed. For instance, in highly siliceous slags, it may form aluminum silicates; while in the presence of an excess of a strong base such as lime, it may form calcium aluminates.

Neutral Fluxes—For the purpose of making slags more fusible, neutral substances having low fusion points may be added. For this purpose, fluorspar is the most commonly used substance. A typical composition of fluorspar, in which calcium fluoride is the active ingredient, is given in Table 6—III.

Table 6—I. Representative Compositions of Limestone, Burnt Lime, Dolomite, and Burnt Dolomite (Natural State Determinations).

Ingredient	Limestone	Burnt Lime	Washed Dolomite	Burnt Dolomite
$CaCO_3$	95.50	—	53.60	—
$MgCO_3$	1.63	—	44.00	—
Fe_2O_3	0.23	0.40	0.08	0.11
Al_2O_3	0.07	0.10	0.06	0.15
SiO_2	0.59	1.05	0.25	0.42
Sulphur	0.060	0.030	0.008	0.007
Phosphorus	0.002	0.008	0.002	0.005
H_2O	2.00	—	2.00	—
CaO	—	95.70	—	57.50
MgO	—	1.30	—	40.50
Loss of Ignition	—	1.50	—	1.31

Table 6—III. Typical Composition of Fluorspar (Dry Basis)

Ingredient	Per Cent
CaF_2	75.0
SiO_2	2 to 4
Al_2O_3, Fe_2O_3	1.00
S	1.00
$CaCO_3$	Remainder

Sources of Fluxing Materials—As mentioned earlier, sources of silica for fluxing purposes are sand, gravel, quartz, and, in some cases, such material as used brick from silica-brick lined vessels or ladles. Highly basic slags from steelmaking operations are frequently crushed and sized for charging to the blast-furnace to utilize the fluxing compounds CaO, MgO, and SiO_2 in the slag and to recover the iron and manganese units. Siliceous ores may be used in special instances. Limestone of fluxing quality is distributed widely and underlies most of the area drained by the Mississippi and Ohio Rivers. However, the larger portion of the limestone for the Pittsburgh steelmaking district as well as that for the Lakes district is obtained from the northeastern section of the lower peninsula of Michigan and from northern Ohio and southern Illinois. Dolomite is obtained from the same regions and from quarries in the eastern United States, the latter for use in the Eastern district. Fluorspar for the steel industry comes primarily from overseas.

The quantities of various fluxing materials used by the iron and steel industry in the United States in 1981 are given in Table 6—IV.

MINING AND QUARRYING LIMESTONE AND DOLOMITE

Limestone and dolomite are taken from the earth's surface by one of two methods: underground mining and open-pit quarrying. Underground mines are now being replaced by more efficient, less costly quarrying operations where the size of machinery is not limited to the space restrictions of most underground operations. Because of its wide occurrence, limestone is probably the least expensive of all raw materials used in industry. To economically process limestone, then, a successful quarry must meet three requirements: (1) it must produce a quality stone high in calcium content (high in calcium and magnesium in the case of dolomite) with few impurities; (2) it must contain a sufficient volume of this stone either on or near the surface of the earth to keep the costly stripping of overburden to a minimum; and (3) it must be near navigable waters, highways, or rail facilities for convenient transportation.

Exhaustive steps are constantly being taken by geologists to discover new limestone deposits that meet these requirements. Once operations begin at a new quarry, a continuing program of quality analysis is initiated on stone taken from test holes and from processed stone to establish conformity with the exacting requirements of the steel manufacturer. To gain the proper chemical balance to meet specifications, stone taken from several points in the quarry may be blended on the basis of results of analyses by the chemical laboratory. As stated earlier in this chapter, stone used in iron- and steelmaking processes must have a high calcium carbonate or a high calcium-magnesium carbonate content, and a low content of silica and sulphur.

Table 6—IV. Consumption of Fluxes in 1981.*
(Thousands of Tons)

Use	Fluorspar	Limestone	Lime	Other Fluxes**	Total
In agglomerated products					
Metric tons	—	5776	—	—	5776
Net tons	—	6368	—	—	6368
In blast furnaces					
Metric tons	—	5398	—	—	5398
Net tons	—	5951	—	—	5951
In steelmaking furnaces					
Basic-oxygen processes					
Metric tons	277	292	5489	384	6442
Net tons	305	322	6052	423	7102
Open-hearth processes					
Metric tons	39	443	129	130	740
Net tons	43	488	142	143	816
Electric-furnace processes					
Metric tons	73	215	1039	155	1482
Net tons	80	237	1146	171	1634
Total					
Metric tons	389	12123	6657	669	19838
Net tons	428	13336	7340	737	21871

*Quantities in metric tons calculated from values in net tons taken from Annual Statistical Report, American Iron and Steel Institute (1981).
**Chiefly silica.

Continuous quality-control programs, then, are necessary from quarry to loading dock.

Quarrying operations for limestone and dolomite differ only in those areas where the greater hardness of dolomite requires modifications in standard excavating and processing methods. In open-pit quarrying, the stone is blasted loose from the deposit by charges of an ammonium-nitrate blasting agent or dynamite in a series of cuts made in the quarry face. In a typical blasting of stone by the ammonium-nitrate method, holes up to about 230 millimeters (9 inches) in diameter are drilled at predetermined distances and depths. A charge of dynamite of specified strength is first lowered to the bottom of the hole, attached to a detonating cord. Depending upon the force of the charge required, alternate amounts of ammonium nitrate and primers (blasting-cap or detonating-cord sensitive charges) are loaded into the hole until the required explosive charge is placed. The hole is then stemmed by filling the remaining space with drill cuttings, and the detonating-cord ends from each of the blasting holes are gathered and taken to a point a safe distance from the blast area. After the blast area is cleared of personnel and equipment, the holes are fired.

Secondary blasting is performed on those pieces of stone that are not shattered by primary blasting. Smaller holes machine-drilled in these pieces are generally charged with dynamite alone.

In most modern quarries, stone is dug from the shattered piles along the quarry face by electrically powered shovels of various capacities. The stone is shovel-loaded into large trucks or side-dumping railroad cars for the trip to the primary crusher. The recent trend toward substituting truck haulage for railroad haulage stems principally from the inherent economy and flexibility of truck haulage over short distances or within confined areas of irregular grades.

PREPARATION OF FLUXES FOR USE

The fluxes previously mentioned usually require only drying and sizing before use. Those materials that may be added to a slag after it has melted, such as fluorspar and burnt lime, are desired small in size so that they will rapidly react to produce the desired results. Limestone and dolomite for sintering or for use directly in the blast furnace and limestone for use in the basic open-hearth require careful sizing because the rate of calcination of these materials is controlled primarily by the surface exposed. The chemical reactions of calcination are:

A. For pure calcium carbonate:
$$CaCO_3 \leftrightarrows CaO + CO_2$$

B. For pure magnesium carbonate:
$$MgCO_3 \leftrightarrows MgO + CO_2$$

Both decomposition reactions are endothermic and are affected by temperature and pressure. Reaction A comes to equilibrium under a pressure of one atmosphere of carbon dioxide at a temperature of about 885°C (1625°F), and Reaction B does the same at about 400°C (750°F), so that calcination will begin at temperatures much below steelmaking temperatures. For a given limestone at a given calcining temperature the surface area (screen size and shape) determines the time for complete calcination. (Refer to Table 6—V.)

Table 6—V. Times for Complete Calcination of an Ordinary Limestone at 970°C (1780°F)

Screen Size of Stone		Calcining Time (Hr.)
mm	In.	
25	1	1 to 1½
50	2	2 to 3
75	3	3 to 4½
100	4	4 to 6
125	5	5 to 7½
150	6	6 to 9

Thus, in a blast furnace, for example, it is possible through sizing of the stone to have its decomposition take place at different elevations in the stack. In general, the preferred size of limestone and dolomite for sintering is minus 3.2 millimeters (⅛ inch), that for blast furnaces is about 12 to 40 millimeters (½ to 1½ inches), and that for open-hearth use is about 100 to 200 millimeters (4 to 8 inches).

Burnt lime is produced by the calcination of limestone. This calcination can be carried out in any one of several types of kilns: long rotary kilns, short rotary kilns preceded by a preheater, or vertical shaft kilns. In the United States, most of the lime used in steelmaking is produced in long rotary kilns; such kilns may be well over 100 metres (several hundred feet) long.

CRUSHING AND SCREENING OF LIMESTONE AND DOLOMITE

The first essential step to arrive at a marketable limestone product takes place in the crusher house. In the larger plants, the primary crusher is usually a gyratory crusher resembling a gigantic mortar and pestle. The mortar-like pit is lined with manganese steel, and the stone is crushed against the walls of the pit by the off-center gyrations of the pestle (mantle) that is usually made of chromium-molybdenum alloy steel. The large pieces of limestone or dolomitic limestone are reduced in size quickly by the action of the crusher and fall from a restricted opening at the base of the crusher onto a continuous belt conveyor. Throughput is rapid; a typical primary crusher can handle the crushing of over 1600 metric tons (1600 gross tons) of stone per hour.

If the primary crusher is located within the quarry area (Figure 6—1), the crushed stone is generally lifted on the belt conveyor to a storage pile for shipment by truck or train to the screen house. Where the crusher is located at the plant site (Figure 6—2), the stone first travels over a scalping screen from a continuous belt conveyor. The scalper screens out all pieces of stone over a specified size, and the over-size stone is shunted to a secondary crusher. After it is re-crushed, this stone is returned to the conveyor carrying stone that passed through the scalping screen. All of the stone is then lifted to the top of the screen house or mill by belt conveyor to begin its descent through a series of crushers and screens equipped with water sprays to ensure cleanliness of the product. Stone intended for a particular product is subject to a specific series of crushing and screening operations in the mill's system. As the stone reaches the predetermined size for the prod-

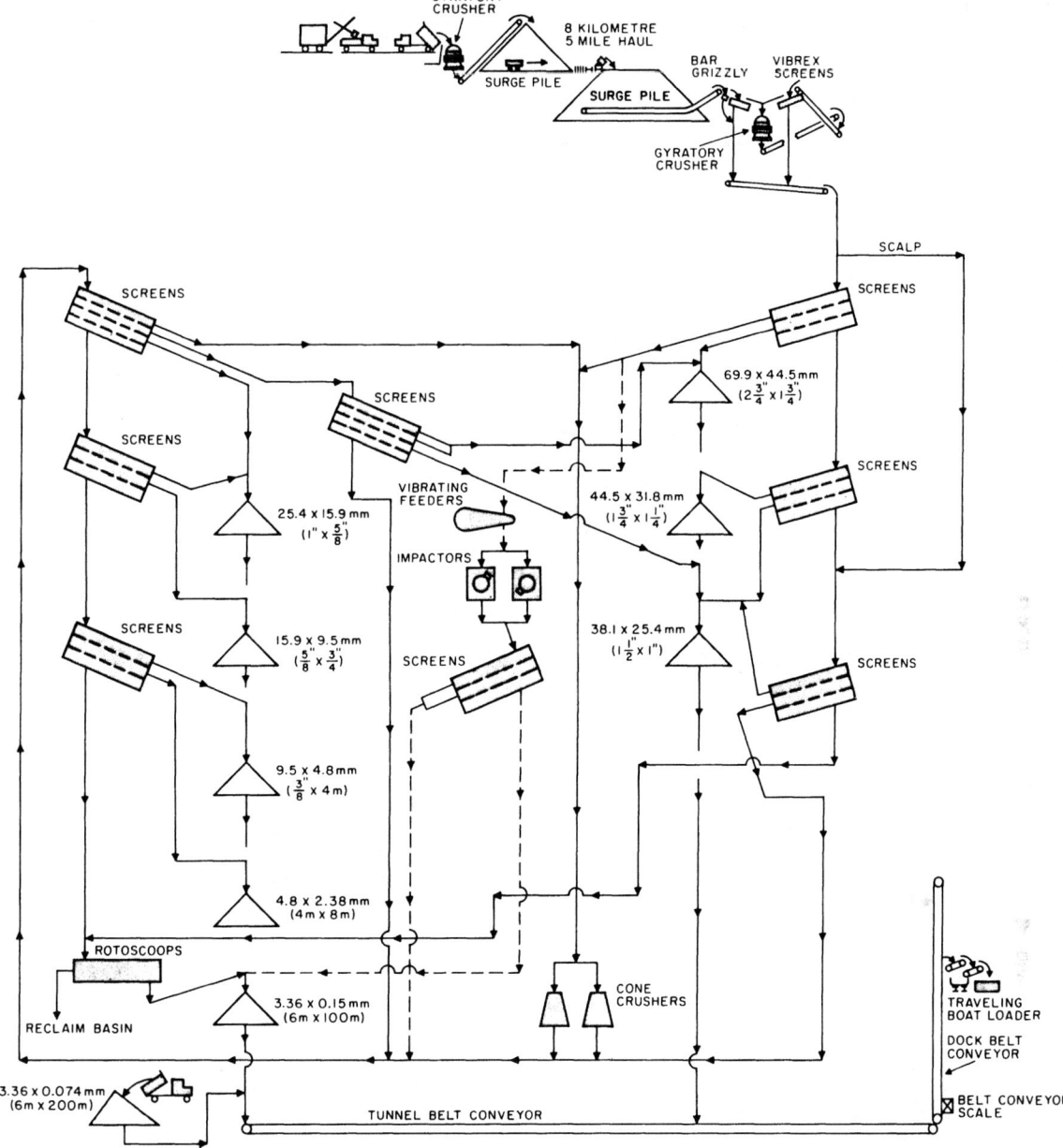

FIG. 6—1. Schematic flow diagram for an American steel plant. Primary crushing is performed at the quarry site.

uct group, it is transported by conveyor to storage piles on the shipping dock.

Thus, the ultimate size of the stone product required is determined by the amount of crushing and screening undertaken in the processing. Large open-hearth flux stone is sized and separated out early in the process; however, fluxing fines, the finely ground stone of exacting size specifications used by the steel industry in the production of sinter, are produced by refinements in crushing and screening methods. Constant samplings of stone in process are taken and tested in the chemical laboratories to ensure maintenance of the specified quality of the stone as it moves from quarry face to shipping dock.

TRANSPORTATION OF LIMESTONE AND DOLOMITE

Limestone is transported from quarry to steel plant by rail, truck, or lake carrier, depending upon the section of the country in which the consuming plant is located. Major carriers of limestone from the large quarries located on the Great Lakes are self-unloader vessels, although some tonnage continues to be transported in the conventional bulk-cargo vessels. As the name implies, the self-unloader (Figure 6—3 and 6—4) is a self-contained carrier with its own unloading equipment, negating the necessity of unloading towers, crane bridges, or other unloading machinery at shore facilities. The self-unloader is a highly versatile cargo

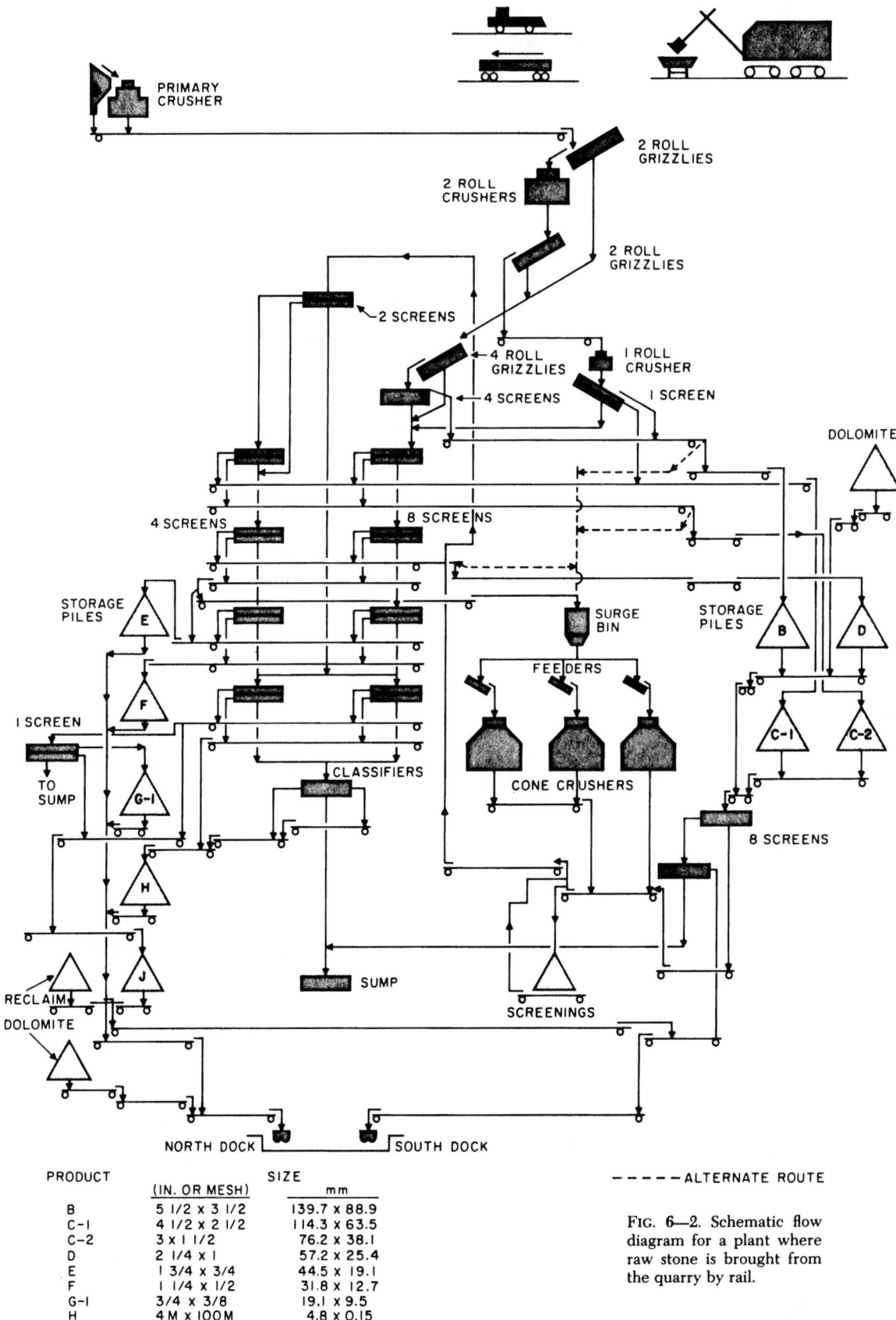

PRIMARY CRUSHER

2 ROLL GRIZZLIES

2 ROLL CRUSHERS

2 ROLL GRIZZLIES

2 SCREENS

4 ROLL GRIZZLIES

1 ROLL CRUSHER

4 SCREENS

1 SCREEN

DOLOMITE

4 SCREENS

8 SCREENS

SURGE BIN

STORAGE PILES

STORAGE PILES

B

D

E

FEEDERS

F

C-1

C-2

1 SCREEN

TO SUMP

G-1

CLASSIFIERS

CONE CRUSHERS

8 SCREENS

H

J

SUMP

RECLAIM DOLOMITE

SCREENINGS

DOLOMITE

NORTH DOCK   SOUTH DOCK

PRODUCT	SIZE	
	(IN. OR MESH)	mm
B	5 1/2 x 3 1/2	139.7 x 88.9
C-1	4 1/2 x 2 1/2	114.3 x 63.5
C-2	3 x 1 1/2	76.2 x 38.1
D	2 1/4 x 1	57.2 x 25.4
E	1 3/4 x 3/4	44.5 x 19.1
F	1 1/4 x 1/2	31.8 x 12.7
G-1	3/4 x 3/8	19.1 x 9.5
H	4M x 100M	4.8 x 0.15
J	3/4 x 6M	19.1 x 3.36

– – – – ALTERNATE ROUTE

FIG. 6—2. Schematic flow diagram for a plant where raw stone is brought from the quarry by rail.

FIG. 6—3. A self-unloading limestone carrier with its boom swung overside, discharging its cargo onto storage piles at the receiving dock.

vessel that can accurately discharge from 10 000 to 25 000 metric tons (10 000 to 25 000 gross tons) of limestone at points over 60 metres (as far as 200 feet) inland from the dock, depending upon the size of the vessel and the length of the vessel's mechanized boom. Many of the limestone carriers now in service are bulk-cargo vessels that have been converted to self-unloaders by the installation of hopper bins, stone conveyors, bucket elevators and movable conveyor booms.

Below deck of the self-unloader is a series of hoppers. The hoppers are constructed in pairs, and each empties its cargo through a gate at the bottom of the hopper that is either mechanically or manually operated. Two belt conveyors run the length of the cargo hold, directly beneath the side-by-side hoppers. These parallel hold conveyors run from the stern of the cargo hold, carrying forward the limestone dropped from the various hopper gates and discharging the stone into chutes at the forward end of the vessel. The chutes, in turn, drop the stone into a large bucket elevator that raises the limestone above deck to the boom conveyor. The stone then travels the length of the boom, to be discharged onto the dock. The boom is designed not only to swing out a full 110 degrees from the longitudinal centerline of the vessel's deck to the dock area, but also to have its discharge end raised or lowered in any position.

Hopper gates are designed to give vessel personnel complete control over the flow of limestone. This control is particularly important in keeping the vessel on an even keel during the unloading operations.

When transported by rail, limestone and dolomite are carried either in gondola or hopper-bottom open cars, usually of 50-, 70- or 90-metric ton (50-, 70- or 90-gross ton) capacity, according to the requirements of the plants where the stone is to be delivered.

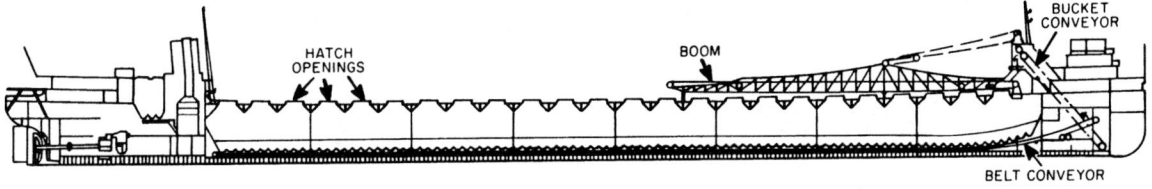

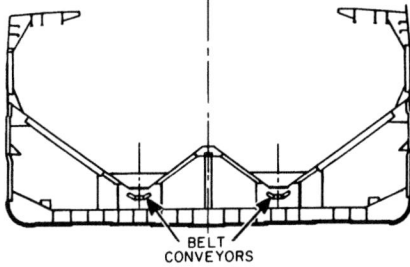

FIG. 6—4. Diagrammatic sections of a self-unloading limestone carrier with a capacity of 25 039 metric tons (25 550 gross tons). This particular vessel is slightly over 234 metres (768 feet) long, with a 22-metre (72-foot) beam. A geared steam turbine supplies 7000 shaft horsepower to propel the vessel at a service speed, loaded, of 14.54 knots per hour.

CHAPTER 7

Slags for Nonmetallurgical Uses

METALLURGICAL CHARACTERISTICS OF SLAGS

Slag Defined—Slag is the name applied to the fusible material formed by the chemical reaction of a flux with the gangue of an ore, with the ash from a fuel, or with the impurities oxidized during the refining of a metal. As indicated in Chapter 6, the compounds in the slag are formed from the neutralization of basic and acid oxides somewhat analogous to the formation of salts in water solutions during chemical reactions at ordinary temperatures. The subject of slags in general is a very large one, and the discussion in this section is intended only to introduce certain of its more important aspects. A more detailed discussion of the chemical and physical properties of slags will be found in Chapter 13 on "Physical Chemistry of Iron- and Steelmaking", in Chapter 15 on "Manufacture of Pig Iron", and in Chapters 16, 17 and 18 on the various steelmaking processes.

Metallurgical Functions of Slags—On account of their fusibility, chemical activity, dissolving power and low density, slags furnish the means by which impurities are separated from the metal and removed from the furnace in both iron- and steelmaking processes. Slag also serves as a heat insulator for liquid iron and steel, especially when the metal is being held in ladles.

In most iron- and steelmaking processes, the slag is in intimate contact with the liquid metal and chemical reactions readily occur between them. For this reason, the chemical composition of the liquid metal is controlled in part by the chemical composition of the slag. In the manufacture of pig iron in the blast furnace, the composition of the slag and the temperature of the slag and metal in the furnace hearth are the major factors controlling the sulphur and silicon content of the hot metal. In the large-scale steelmaking processes such as the basic oxygen process, the open-hearth or electric-furnace process, the slag is the primary means by which phosphorus and sulphur, impurities in the charge, are removed. For this reason, it is essential for the metallurgist to have knowledge of the properties of slags and their chemical behavior so that proper adjustments can be made to control their fusibility and fluidity.

No attempt has been made in this brief statement to do more than to call attention to the importance of slag in blast-furnace and steelmaking operations; the subject is covered in detail in the chapters referred to above.

Secondary Metallurgical Uses of Slags—In addition to their metallurgical functions in different processes, steelmaking slags, with their high iron and manganese contents, are frequently used in the blast-furnace burden. In this way, most of the useful metallics in the slag are recovered that would otherwise be wasted. The iron is almost completely recovered and about 70 per cent of the manganese is also reduced into the hot metal. Furthermore, the steelmaking slags contain considerably more lime than that required to flux the acid oxides present and this lime is useful in fluxing the acid impurities of the iron-bearing burden and the coke ash. Because the lime is already calcined, very little additional fuel is required and this usually accounts for an overall cost savings. The amount of steelmaking slag that can be recycled back into the blast furnace is limited, however, by the phosphorus content of the slag because the phosphorus also is reduced into the pig iron and, if too much is used, the phosphorus content of the hot metal could exceed the specification for that element.

Special methods or treatments have been developed or are in the course of development for the recovery of valuable alloys from electric-furnace slags. The value of these processes, as in those mentioned above, is largely one of economics; that is, they are feasible only when the value of the material recovered amounts to more than the cost of its recovery.

CHEMICAL AND PHYSICAL PROPERTIES OF SOLID SLAGS

Each of the primary metallurgical processes in the production of iron and steel also results in the production of a slag. While these slags all have common metallurgical functions, they vary widely in chemical and physical properties with resultant variations in value and usefulness in nonmetallurgical applications. Major slag products from the steel industry can be classified into two categories: iron-blast-furnace slags and steelmaking slags with respect to their origin and uses outside the furnaces in which they are formed. These slags, their properties and uses are described below under the individual category name.

BLAST-FURNACE SLAG

Blast-furnace slag has been used in all types of construction in the United States since about 1920. Blast-furnace slag is defined by the American Society for Testing and Materials (Designations C 125 and D 8) as "the nonmetallic product consisting essentially of silicates and aluminosilicates of calcium and other bases, that is developed in a molten condition simultaneously with iron in a blast furnace".

In the production of iron, the blast furnace is charged with iron-bearing materials (iron ore, sinter, pellets, etc.), flux (limestone and/or dolomite) and fuel (coke). Two products obtained from the furnace are molten iron and slag.

The principal constituents of blast-furnace slag are the oxides silica and alumina originating chiefly with the iron-bearing material, and lime and magnesia originating with the flux. These comprise 95 per cent or more of the total. Minor elements include manganese, iron and sulphur compounds, and trace quantities of several others. Compositions of most blast-furnace slags fall within the ranges shown in Table 7—I. The major oxides do not occur in free form in the slag; instead they are combined to form various silicate and aluminosilicate minerals. The chemical composition of slag from a given source (and, therefore, its mineralogic composition) varies within relatively narrow limits since the raw materials charged into the furnace are carefully selected and blended.

Table 7—I. Composition Ranges of Blast-Furnace Slags (Per Cent).

Silica (SiO$_2$)	32 to 42
Alumina (Al$_2$O$_3$)	7 to 16
Lime (CaO)	32 to 45
Magnesia (MgO)	5 to 15
Sulphur (S)*	1 to 2
Iron Oxide (Fe$_2$O$_3$)	0.1 to 1.5
Manganous Oxide (MnO)	0.2 to 1.0

*Principally in the form of calcium sulphide.

Slag comes from the furnace as a liquid at temperatures about 1480°C (2700°F), resembling a molten lava. Dependent upon the manner in which the molten slag is cooled and solidified, three distinct types of blast-furnace slag can be produced: air-cooled, expanded, or granulated (see Figure 7—I).

Air-Cooled Slag—To produce air-cooled slag, the molten blast-furnace slag is permitted to run into a pit adjacent to the furnace or is transported in large ladles and poured into a pit at some distance away from the furnace. Solidification takes place under the prevailing atmospheric conditions, after which cooling may be accelerated by spraying water on the solidified mass. After a pit has been filled with slag that has cooled sufficiently to be handled, the slag is dug, crushed, and screened to desired aggregate sizes.

The solidified slag characteristically has a vesicular structure with many nonconnected cells. Air-cooled slag crushes to angular, roughly cubical pieces with a minimum of flat or elongated fragments. The texture ranges from rough, pitted surfaces through all degrees of roughness to almost smooth, conchoidal fractures for denser pieces.

The bulk specific gravity (dry basis) of air-cooled slag used as coarse aggregate generally falls within the range of 2.0 to 2.5. Since large vesicules cannot exist in small particles, the smaller sizes have higher specific gravities.

The unit weight varies with: (a) size and grading of the slag, (b) method of measuring and (c) bulk specific gravity of the slag. Typical unit weight (compacted) of crushed and screened air-cooled slag, graded as ordinarily used in concrete, is usually in the range of about 1120 to 1360 kilograms per cubic metre (70 to 85 pounds per cubic foot). Air-cooled slag has a lower unit weight than most natural aggregates.

Air-cooled slag is highly resistant to the action of weathering. It will withstand a large number of cycles of freezing and thawing or wetting and drying with little or no adverse effect. High temperatures have very little effect on slag. An outstanding characteristic of air-cooled slag is its toughness and resistance to polishing under traffic. Air-cooled slag fines are nonplastic.

Difficulty may be encountered with air-cooled blast-furnace slag when unusual furnace conditions exist and the slag contains an excess of lime. The condition is called "falling" slag because after the slag has cooled down to ordinary temperatures it will disintegrate into a fine powder. After considerable research, it was determined that the cause of the falling slag is the presence of dicalcium silicate, 2CaO·SiO$_2$. This compound has three different crystalline forms: the alpha (α) form is stable above 1415°C (2579°F), the beta (β) form between 1415°C and 675°C (2579°F and 1247°F) and the gamma (γ) form below 675°C (1247°F). On cooling below 675°C (1247°F), the phase change is accompanied by an expansion of about 10 per cent and this produces the disintegration of the slag. From the experimental work of T. W. Parker and J. F. Ryder, the following equations were derived: if the slag composition does not satisfy either of these equations, the potential for a falling slag exists.

$$CaO + 0.8\ MgO \leqq 1.2\ SiO_2 + 0.4\ Al_2O_3 + 1.75\ S$$
$$CaO \leqq 0.9\ SiO_2 + 0.6\ Al_2O_3 + 1.75\ S$$

Expanded Slag—Treatment of the molten blast-furnace slag with controlled quantities of water accelerates the solidification and increases the cellular or vesicular nature of the slag, producing a lightweight product. Either machine or pit processes may be used to mix the water and molten slag. The solidified expanded slag is crushed and screened for use as a lightweight aggregate.

Expanded slag is either angular and roughly cubical in shape or spherical (pelletized) with a minimum of flat or elongated fragments. Due to the action of water and steam during the expanding process, the cellular structure of the aggregate is more pronounced than that of the air-cooled type of slag.

The unit weight of expanded slag (loose) usually ranges from about 800 to 1040 kilograms per cubic

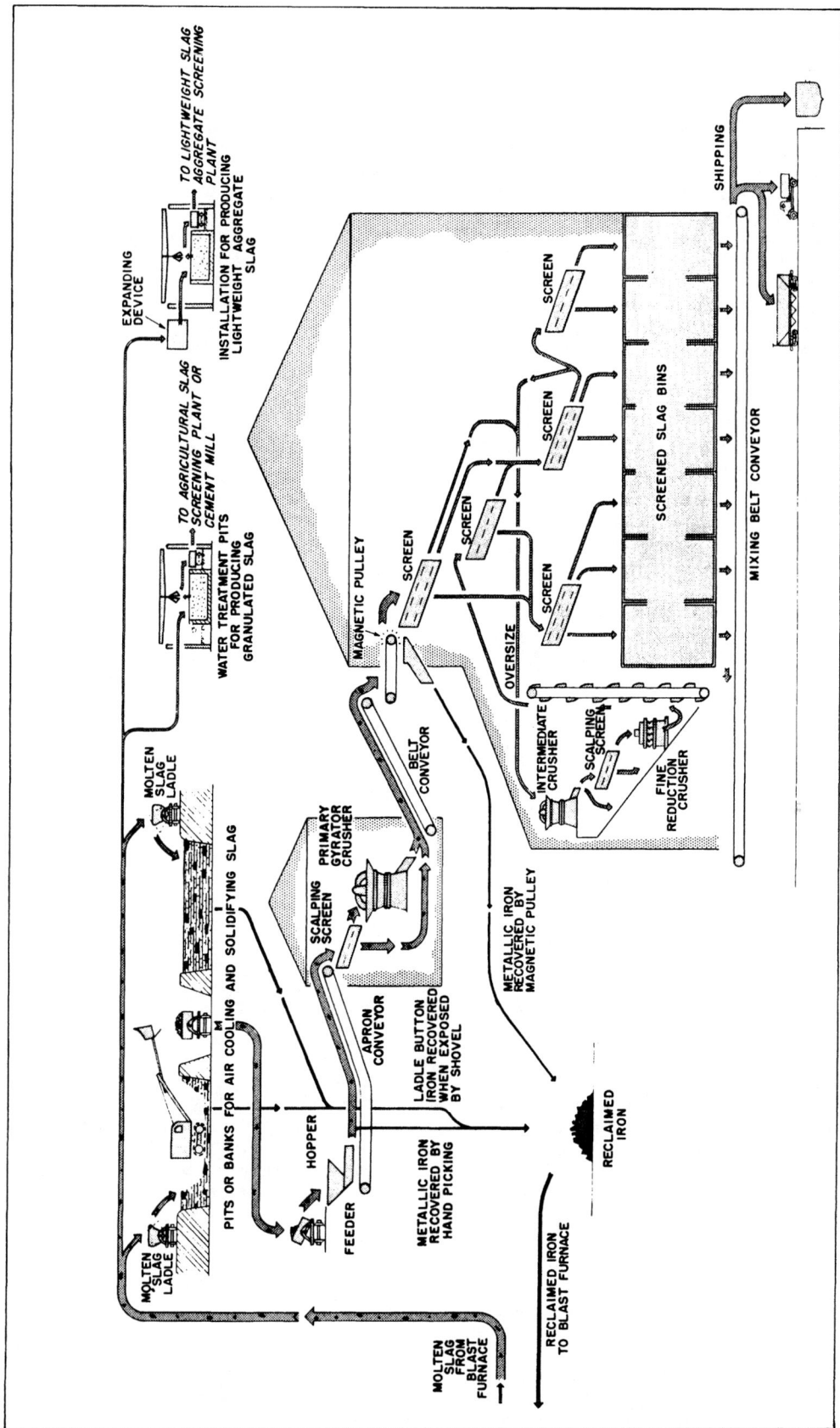

Fig. 7—I. Schematic representation of the steps involved in processing blast-furnace slag to convert it to useful forms. The purpose of the multiple screens preceding the screened-slag bins is to sort the crushed slag into appropriate sizes.

metre (50 to 65 pounds per cubic foot) for the fine aggregate, and from about 560 to 800 kilograms per cubic metre (35 to 50 pounds per cubic foot) for the coarse aggregate.

Expanded slag has the same durability characteristics as air-cooled slag.

Granulated Slag—Molten blast-furnace slag is quenched quickly in water to form a product called "granulated slag". This process is the most rapid of the cooling processes and little or no crystallization occurs. Dependent upon chemical composition of the slag, its temperature at the time of quenching in water, and the method of production utilized, the physical structure of the granulated grains may vary from a friable popcorn-like structure to grains resembling dense glass.

The granulated slag may be crushed and screened, or pulverized, for various applications.

Granulated slag has excellent hydraulic properties so that when it is compacted in the presence of moisture it will set up similar to cement.

SPECIFIC USES OF BLAST-FURNACE SLAG

Blast-furnace slag has been very successfully put to many commercial uses, many of which are summarized in Table 7—II. The steps in processing the slag to convert it to useful forms have been shown schematically in Figure 7—I.

Blast-furnace slag output in 1982, as reported by the United States Bureau of Mines, totaled 13.4 million metric tons (14.8 million short tons). The value of the 1982 output was 64.9 million dollars. In the processing of slag to produce screened slag products (Figure 7—I), slag that has passed through the primary crusher is passed over a magnetic pulley to separate particles of iron and iron-bearing materials from the crushed slag; the iron-bearing material, with the iron particles, is referred to as **iron slag** and the mixture contains an average of 60 per cent iron.

Uses of Air-Cooled Slag—Air-cooled slag has been used extensively as an aggregate in all types of concreting operations. Many miles of concrete pavements, bridges (both highway and railroad), concrete buildings (both industrial and office) and various concrete products (including prestressed sections) are constructed with slag concrete. Since air-cooled slag is light in weight, it often is the preferred material where high strength, durability, light weight and economy are important factors.

Slag concrete compares favorably in strength—both compressive and flexural—with that made with other types of high-quality aggregates. Under normal conditions, 28-day strengths with slag aggregate of 27 580 kPa to 41 370 kPa (4000 to 6000 psi) in compression and 4135 kPa to 5170 kPa (600 to 750 psi) in flexure, are readily obtained with a cement factor of about 280 to 390 kilograms per cubic metre (5 to 7 sacks of 94 pounds each per cubic yard).

Because of the rough surface and angular shape of slag particles, excellent bond between the aggregate and the mortar is provided. Further, compatibility between the slag aggregate and cement mortar with respect to expansion coefficient assures continued presence of this desirable bond relationship irrespective of temperature variations to which concrete is normally

Table 7—II. Principal Uses of Blast-Furnace Slag

Air-Cooled Slag

Railroad ballast
Aggregate in concrete and concrete products
Aggregate in bituminous-paving applications
Aggregate in bases and subbases
Aggregate in surface courses, shoulders and berms
Slope protection
Anti-skid material
Manufacture of portland cement
Porous backfill and underdrains
Embankments and fills
Sewage trickle-filter media
Roofing granules
Mineral wool
Glass sand

Expanded Slag

Aggregate in lightweight concrete and
   lightweight-concrete products
Aggregate in lightweight fill and embankment
Manufacture of portland cement
Thermal-insulation fill

Granulated Slag

Aggregate in lightweight concrete and
   lightweight-concrete products
Aggregate in bases and subbases
Aggregate in fill and embankment
Manufacture of portland cement
Manufacture of slag cement
Manufacture of portland blast-furnace-slag cement
Soil conditioner

subjected during field service. Chemical similarity between slag and cement and the surface cementitious properties of noncrystalline slag phases may also result in the establishment of chemical bonds. Many and varied structures made of air-cooled-slag concrete have given excellent performance through long years of service and are the best indicators of the high durability of air-cooled-slag concrete. Air-cooled-slag concrete, on the basis of field experiences and many laboratory tests to determine its fire resistance, has demonstrated its superiority to concrete with natural aggregates. Air-cooled-slag concrete will weigh approximately 160 kilograms less per cubic metre (10 pounds less per cubic foot or 250 pounds less per cubic yard) than concrete made with other types of aggregates for comparable conditions of quality.

The characteristic rough surface and angular shape of slag particles, together with the fact that they are hydrophobic in nature, makes them exceptionally suitable for bituminous construction of all types. These characteristics are favorable to an excellent bond between the bitumen and aggregate, and also permit the application of a sufficiently heavy film of bitumen on the aggregate surface to assure durability and long life in the pavement. Slag sand is particularly valuable for use as fine aggregate in bituminous concrete of both the hot- and cold-mix types.

Because of the inherent characteristics of air-cooled

slags, bituminous-concrete surfaces constructed with them exhibit good skid resistance under wet or dry conditions throughout the entire service life of the pavement.

The physical properties of slag which make it highly desirable for use in bituminous mixes also favor its use in base courses. The particle shape provides extremely high stability for any type of base course: macadam, dense-graded aggregate, bituminous stabilized (black) base, or soil-aggregate bases.

High-quality air-cooled-slag aggregate bases can accommodate very heavy, channelized wheel loads with high wheel loading, and actual field experience has demonstrated their excellence. They are particularly effective in climates and locations where frost action is a potential problem.

As railroad ballast, slag has been used successfully by the leading railroads in the United States for many years. Many of the most heavily traveled tracks in the country are ballasted with slag. More than 1.4 million metric tons (1.5 million short tons) were so used in 1982.

Slag also has proved of value as a trickle-filter medium for sewage disposal. This is especially true in the northern states where the filter medium is subjected to severe freezing and thawing as well as wetting and drying. Slag has withstood the extreme exposure in sewage-disposal filter beds to a remarkable degree with practically no breakdown, even after as much as thirty years of service.

Slag has been in great demand for many years for use as a covering for built-up bituminous roofs and as granules for covering asphalt shingles.

A large percentage of mineral wool produced in the United States is made with air-cooled slag. Slag is remelted in a small cupola furnace and the molten product is blown into wool by a jet of high-pressure steam or compressed air or spun into filaments by the action of a high-speed rotating disc.

Air-cooled slag is used as an ingredient in the manufacture of transparent aluminum glass, black opaque glass, and amber glass, particularly for bottles. An important requirement of slag for glassmaking is a low free-iron content.

Uses of Expanded Slag—Expanded slag, both fine and coarse, is used as aggregate in the manufacture of lightweight concrete for structural purposes and floor fills, concrete products, and masonry units.

The major application of expanded slag is in lightweight-concrete masonry units. Expanded-slag concrete masonry units possess light weight, durability, strength, low shrinkage and other desirable properties for high-quality building construction. The thermal-insulating properties of slag-concrete masonry units are exceptionally good, and these units exhibit greater fire resistance than those made with other types of aggregate.

Structural lightweight concrete made with expanded slag exhibits high strength, low unit weight, durability, low heat transmission, good bond, low creep, and is free from alkali reactivity. Expanded-slag concrete is excellent for precast and prestressed shapes because of its light weight and high early compressive and flexural strengths.

Expanded slag is used for embankments of all types since it compacts easily and has inherent cementing action along with relatively high permeability. It is particularly valuable as a base, embankment or fill over weak subgrades, minimizing the deadweight load and providing some cementitious structural strength.

Expanded blast-furnace-slag use in cement manufacture has increased significantly in recent years. Any of the expanded slags can be used as a raw material in the manufacture of portland cement. The more glassy materials have marked cementitious properties after grinding to cement sizes. These are being used both in the manufacture of portland blast-furnace-slag cements and as a separately ground and marketed cementitious material to replace part of the portland cement at the mixer.

Uses of Granulated Slag—At the present, granulated slag is used mainly in the manufacture of cement, construction of highway bases and fill, as aggregate for concrete products, and for soil conditioning.

Properly compacted, granulated slag provides an excellent base for pavements, runways, parking areas, and so on. Because of the cementitious properties of granulated slag, the bearing value increases with age like a cement-treated or pozzolanic base.

Processed granulated slag is used in the manufacture of concrete masonry units. The resultant products, excepting texture and weight, possess properties similar to those manufactured with expanded slag. For use in concrete-masonry units, granulated slag is usually blended with other aggregates—frequently with air-cooled slag.

Granulated blast-furnace slag as a soil conditioner neutralizes soil acids through the effective liming action of its silicates and aluminosilicates of lime and magnesia. It also contains many minor elements such as manganese, titanium, sulphur and boron which stimulate plant growth.

One of the earliest uses for granulated slag was in the manufacture of cement. Large quantities have been used in the manufacture of portland as well as other types of cement. For portland cement, it is used as a part of the raw material that is burned in the usual way to form clinker which, in turn, is ground into cement.

Portland blast-furnace-slag cement (ASTM Designation C 595) is made by intergrinding portland-cement clinker with granulated slag, or separately grinding granulated slag and blending it with portland cement. Concrete made from portland blast-furnace-slag cement is very workable, has a low slump loss in hot-weather construction, has a low heat of hydration and is moderately resistant to sulphate attack.

Slag cement (ASTM Designation C 595) is a product wherein the principal ingredient is finely ground granulated slag. It is used in combination with portland cement in making concrete and with lime in making masonry mortar.

Further, there is a slag cement which is composed entirely of finely ground granulated slag. This cement is used in combination with conventional portland cement.

Granulated slag, which is relatively light in weight and has good permeability, makes an ideal backfill for bridge abutments. It is extensively used as the fill within foundation walls over which floor slabs are subsequently constructed.

STEELMAKING SLAGS

The composition of steelmaking slags will vary depending upon the type of steelmaking process in which they originate. There are three basic steelmaking processes: basic oxygen, electric-arc and open-hearth. Even greater variation can be introduced by the composition of the iron-bearing feed and the batch nature of the steelmaking processes. The major chemical components of steelmaking slags are calcium silicates, lime-iron compounds and lesser amounts of free lime and magnesia. The free lime content is highly variable, even for a given process.

Some steelmaking slags are being recycled to blast furnaces for recovery of the iron units they contain and for their fluxing properties. Chemical composition, specifically phosphorus and zinc content, is critical in establishing the amount that can be recycled. An additional benefit is a reduction in energy units compared to a conventional flux charge because the lime and magnesia in steelmaking slags are in precalcined form.

The majority of steelmaking slags are air cooled. The United States Bureau of Mines reported use of 4.3 million metric tons (4.7 million short tons), valued at 14.6 million dollars, in 1982 for base and fill materials, ballast, bituminous paving, and soil conditioning.

Volume changes of detrimental proportions have been observed in some field applications with some steelmaking slags. This expansion is attributed to the hydration of free lime or magnesia, or the dicalcium silicate phase change, or a combination of these actions. Steelmaking slags are frequently weathered in stockpiles to allow the bulk of any volume change to take place prior to use in construction.

The high bulk density in excess of 2.4 metric tons per cubic metre (2 short tons per cubic yard) and high specific gravity (in excess of 3.0) of steelmaking slags are due to the presence of FeO and MnO.

In general, the uses of steelmaking slags are limited to nonconfined applications due to the volume-change potential of some of the slags. Railroad ballast is one of these applications. The high weight and interlocking properties are of benefit for this use. Like air-cooled blast-furnace slag, steelmaking slags exhibit excellent skid-resistant properties in surface-course bituminous-concrete mixes. Acceptance is growing for these applications.

Soil Conditioning—Basic-steelmaking slag has also found an important use as a soil conditioner. The first basic slag used as soil conditioner was from the Thomas-Gilchrist or basic Bessemer process of steelmaking. The possibilities of this slag as a soil conditioner and stimulant to plant growth were investigated first in England in 1884. Its use was adopted in Scotland in 1890, and the practice spread rapidly throughout Europe in succeeding years.

The chief food elements that plant life or vegetation obtains from the soil are potassium, phosphorus and nitrogen, but other chemical elements such as iron, sulphur and manganese are essential. For plants to thrive, soil conditions must be favorable, particularly with respect to humus content and basicity. Thus, if the soil is in clods, or if it is too acid or basic, unsatisfactory crop results are obtained irrespective of the amount of fertilizer applied. An important function of slags is to correct such adverse conditions.

The chemical bases in slags consist mainly of lime, with some magnesia and smaller amounts of other basic compounds. The lime in slag, however, is in loose chemical combination with silica, iron and manganese, so that it does not "burn" like ordinary agricultural lime nor revert to carbonate, but remains in a stable, almost neutral form, available as needed over long periods of time until exhausted.

While slags used for agricultural purposes cannot be called fertilizers, they do contain some fertilizing elements and, if of a suitable grade and properly prepared and applied to the soil, do promote conditions essential to profitable farming. Hence, they rightly are called soil conditioners.

Bibliography

Am. Institute of Mining and Metallurgical Engineers, Iron and Steel Div., Committee on Physical Chemistry of Steelmaking, Basic open hearth steelmaking; 2nd ed. (Seeley W. Mudd series). N.Y., The Institute, 1951.

Barton, W. R., Construction materials: aggregates—slag. Industrial minerals and rocks, published by AIME (1975), 109-127.

Bishop, H. L.; King, T. B.; and Grant, J. J., The role of slag composition in open hearth desulphurization and oxidation. Yearbook of the Am. Iron and Steel Institute, 1955—249-266.

Drake, H. J., and Shelton, J. E., Disposal of iron and steel slag. 4th Mineral Waste Utilization Symposium Proceedings (1974), 303-308. (Symposium co-sponsored by U. S. Bureau of Mines and IIT Research Inst.)

Emery, J. J., Slags. 5th Mineral Waste Utilization Symposium Proceedings (1976), 291-300. (Symposium co-sponsored by U. S. Bureau of Mines and IIT Research Inst.)

Hodges, P. C., Production and preparation of blast-furnace flux. Am. Institute of Mining and Metallurgical Engineers, Iron and Steel Div., Trans. **120**, 121-133 (1936).

Lambing, L. A., Discussion: Influence of various limestones. Am. Institute of Mining and Metallurgical Engineers, Open Hearth Proc. **21**, 84-85 (1938).

Lightner, M. W., Current concepts of open hearth slag control. Am. Institute of Mining and Metallurgical Engineers, Open Hearth Proc. **40**, 304-314 (1957).

National Slag Association, Processed blast-furnace slag: the all-purpose construction aggregate. NSA Bulletin 178-1 (1978).

Osborn, E. F.; De Vries, R. C.; Gee, K. H.; and Kramer, H. M., Optimum composition of blast-furnace slag as deduced from liquidus data for the quaternary system $CaO—MgO—Al_2O_3—SiO_2$. Am. Institute of Mining and Metallurgical Engineers, Trans. **200**, 33-45 (1954).

Turkdogan, E. T. and Pearson, J., Activities of constituents of iron and steelmaking slags. Part I—Iron Oxide, J. Iron and Steel Institute **173**, 217-223 (1953); Part II—Manganous Oxide, J. Iron and Steel Institute **173**, 393-398 (1953); Part III—Phosphorus Pentoxide, J. Iron and Steel Inst. **173**, 398-401, (1953).

Turkdogan, E. T. and Pearson, J., Reaction equilibria between metal and slag in acid and basic open hearth steelmaking. J. Iron and Steel Institute **176**, 59-63 (1954).

U. S. Bureau of Mines, Iron blast-furnace slag: Production, processing, properties and uses, by G. W. Josephson, F. Sillers, Jr., and D. G. Runner (Bulletin 479). Wash., Govt. Printing Office, 1949.

U. S. Department of Interior, Bureau of Mines, Minerals Yearbook, Volume I, Metals and Minerals, 1982.

CHAPTER 8

Scrap for Steelmaking

Scrap consists of the by-products of steel fabrication, and worn out, broken or discarded items containing iron or steel. Scrap in the form of iron or steel is one of the two principal sources of metallics in steelmaking; the other principal source is iron from the blast furnaces, either molten as it comes from the blast furnace ("hot metal") or in solid pig form. Scrap is of great practical value. Every ton of scrap consumed in steelmaking is estimated to displace and conserve for future use 3½ to 4 tons of other natural resources including iron ore, coal and limestone. On the average, the steel industry consumes about one-fifth more pig iron than scrap. According to published figures of the American Iron and Steel Institute, the steel industry consumes between 50 and 56 million tons of iron and steel scrap in producing 100 million tons of raw steel.

The various steelmaking processes differ widely in their abilities to consume scrap. The basic oxygen steelmaking processes use 10 to 30 per cent scrap while the open-hearth processes utilize 25 to 100 per cent. The electric furnace usually is charged almost entirely with cold scrap. Table 8—I summarizes the consumption of scrap by several steelmaking processes and the blast furnace during 1981 in the United States.

Types and Sources of Scrap—Scrap iron and steel may be classified as originating from two sources: home scrap produced as unsalable products resulting from steelmaking and finishing operations, and purchased scrap.

Home Scrap (also called "revert scrap") includes such items as pit scrap; ingots too short to roll; rejected ingots; crop ends from slabs, blooms and billets; shear cuttings from trimming flat-rolled products to specified size; products irrecoverably damaged in handling or finishing; ends cut from structural shapes, rails, bars, pipe or tubing to bring them to standard or exact or-

dered length; turnings from machining operations, broken molds, obsolete machinery, dismantled buildings, steel shot recovered from slag, and so on. Bloom and slab crops constitute the largest single item of home scrap.

In general, about 35 million tons of home scrap would result from the manufacture of 100 million tons of raw steel and the processing of this steel into finished mill products.

Purchased Scrap is used to supplement home scrap to provide the total iron and steel scrap used to produce steel. Purchased scrap is divided into two general classifications: (1) dormant scrap and (2) prompt industrial scrap.

Dormant scrap comprises obsolete, worn out or broken products of consuming industries. Typical examples of dormant scrap are: discarded steel furniture, washing machines, stoves, tin cans and other outdated consumer goods; beams, angles, channels, girders, railings, grilles, pipe, etc., arising from the demolition of buildings; useless farm machinery; obsolete, broken or damaged industrial machinery; old ships; railroad rails and rolling stock that have outlived their usefulness; wrecked automobiles, and so on. This type of scrap, because of its miscellaneous nature, requires careful sorting and classification to prevent the contamination of steel in the furnace with unwanted chemical elements from alloys that may be present in some of the scrap. It should also be of such physical size as to facilitate handling and loading into charging boxes. The need for proper classification and preparation of dormant scrap is emphasized by the existence of over 70 different specifications covering various grades of scrap for use in blast furnaces, acid and basic open-hearth furnaces, electric furnaces, the basic oxygen steelmaking process, gray-iron foundries and elsewhere. In addition, the Association of American Railroads has forty-five specifications applying to scrap of railroad origin. These all have been prepared to facilitate proper classification of scrap for different uses.

Prompt industrial scrap is generated by steel consumers in making their products. It may consist of the unwanted portions of plate or sheet that has been cut or sheared to the desired final size and shape, trimmings resulting from stamping and pressing operations, machine turnings, rejected products scrapped during manufacture, short ends, flash from forgings, and other types of scrap. Prompt industrial scrap can usually be identified easily as to source and composition, provided that proper segregation plans are in effect in the consumer's plant, the scrap dealer's yard,

Table 8—I. Consumption of Scrap by Steelmaking Processes in the United States During 1981*

| Process | Scrap | |
	Thousands of Metric Tons	Thousands of Net Tons
Basic oxygen	21 006	23 160
Open hearth	6 760	7 453
Electric	31 217	34 418
Blast furnace	3 779	4 167

*From: Annual Statistical Report (1981), published by the American Iron and Steel Institute.

and in the steel plant.

According to the 1981 Annual Statistical Report of the American Iron and Steel Institute, the domestic steel industry in that year produced 35 331 000 metric tons (38 954 000 net tons) of scrap as a result of its own operations, and received 33 309 000 metric tons (36 724 000 net tons) from outside sources. In the same year, it consumed 63 912 000 metric tons (70 465 000 net tons) of scrap in producing 109 959 000 metric tons (120 828 000 net tons) of raw steel.

Physical Preparation of Scrap—Scrap is classified according to its physical size and chemical composition. Pieces too large to be accommodated by charging-machine boxes must be cut into satisfactorily smaller sizes; shears, flame-cutting, impact devices and other means may be used, depending upon the type of scrap being handled. Sometimes, very large pieces of scrap that cannot pass through the furnace doors may be charged into an open-hearth furnace by overhead crane when the furnace roof is off during rebuilding. Sheet shearings, punchings and similar types of relatively thin and usually small pieces of scrap may be compressed into block-like bundles in specially-designed hydraulic baling presses; since about half of the steel rolled in the United States is in the form of relatively thin flat-rolled products, large quantities of scrap require baling.

The proper physical preparation of scrap is to increase the amount of scrap that can be loaded at one time into a charging box for placing the scrap in the steel-producing furnace. The denser the load in the charging boxes, the fewer the number of boxes that need to be loaded, transported and their contents dumped into the furnace, and the less the time consumed in charging the furnace. Delays in charging can result in a corresponding loss of steel production.

The prime grade of purchased scrap for production of basic oxygen process and open-hearth steel must be at least about 3 millimeters (⅛ inch) thick, no more than 460 millimeters (18 inches) wide, nor more than about 1.5 metres (5 feet) long. Electric furnaces require purchased scrap of smaller dimensions ranging in size from one metre (3 feet) down to punchings; further details relating to scrap for electric-furnace melting are given in Chapter 18. Short lengths of turnings are preferred for blast-furnaces.

The iron and steel industry's receipts and consumption of iron and steel scrap by grades in 1981 are listed in Table 8—II. The Yearbook of the Institute of Scrap Iron and Steel gives definitions of the various grades.

Chemical Composition of Scrap—Certain chemical elements are desirable constituents of scrap for steelmaking, especially when used in electric furnaces. In general, however, scrap for all steelmaking processes should be free from unknown and unwanted elements referred to as "tramp alloys."

The segregation of home scrap according to its chemical composition is relatively simple. Purchased scrap, especially dormant scrap, presents some problems since a large percentage of it is of unknown origin and composition. While it would be impractical to

chemically analyze each individual piece of the huge amounts of dormant scrap consumed every year, the chemical analysis of selected samples of individual lots sometimes is employed by steel plants in the classification of scrap. Spectrographic analysis sometimes is employed instead of chemical analysis because it is more rapid; however, both are relatively time consuming and expensive and both require careful selection and preparation of samples. Some less costly but less accurate tests are commonly used; these include magnetic tests, spark tests, spot tests and pellet tests.

When the chemical composition of scrap is known, the scrap can prove to be a valuable source of alloying elements needed in the production of alloy steels. Full advantage is taken of this source in the production of alloy steels in the electric furnace. In the basic oxygen furnace and the open-hearth furnace, however, the preponderance of production consists of carbon and low-alloy steels. In general, unidentified alloying elements in scrap can be a source of trouble.

Tin, copper, nickel and other elements present in scrap will alloy readily with steel and, in many instances, render it unfit for its intended use. Relatively small amounts of these metals can contaminate an entire heat of steel. Tin and copper in certain ranges of composition cause brittleness and bad surface conditions in steel. Nickel and tin not only contaminate heats into which they may be unintentionally introduced, but may deposit a residue in the furnace that is absorbed by successive heats with resultant contamination. The foregoing examples represent some of the difficulties caused by only a few of the chemical elements that may enter steel from poorly prepared or carelessly classified scrap.

Table 8—II. Consumption of Iron and Steel Scrap in the United States (1981) by Grade*

Grade	Thousands of Metric Tons	Thousands of Net Tons
Carbon steel		
Low phosphorus plate and punchings	395	436
Cut structurals and plate	1 230	1 356
No. 1 heavy melting steel	21 726	23 954
No. 2 heavy melting steel	3 344	3 687
No. 1 and electric-furnace bundles	7 824	8 626
No. 2 and other bundles	1 895	2 089
Turnings and borings	1 537	1 695
Slag scrap (iron content)	4 041	4 455
Shredded or fragmentized	3 340	3 682
All other carbon steel scrap	11 669	12 865
Stainless steel	858	946
Alloy steel (excluding stainless)	1 619	1 785
Iron scrap	2 859	3 152
Other grades	1 575	1 737
Total Scrap	63 912	70 465

*From: Annual Statistical Report (1981), published by the American Iron and Steel Institute.

CHAPTER 9

Addition Agents Used In Steelmaking

Definitions—Steelmaking involves the deliberate addition of various chemical elements to the molten metal to effect several desirable ends. These ends may include deoxidation of the molten metal to the desired degree, control of grain size, control of non-metallic inclusion shape, improvement of the mechanical and physical properties and corrosion resistance of the steel, increase of the response of the steel to subsequent heat treatment, or attainment of other specific effects that are discussed elsewhere in this book. Originally, the chemical element to be incorporated into the steel was added to the bath in the form of an alloy that consisted principally of iron but was rich in the desired element. Such alloys, because of their high iron content, became known as **ferroalloys,** and most of the available types were produced in the iron blast furnace. Eventually, the production of alloys for steelmaking purposes began to be carried out in electric-reduction and other types of furnaces as well, and a number of alloys now produced contain relatively little iron. For this reason, the term **addition agent** is preferred to describe any of the materials added to molten steel for altering its composition or properties; under this definition, the ferroalloys form a special class of addition agents.

The more common addition agents definitely in the ferroalloy class include alloys of iron with aluminum, boron, chromium, columbium*, manganese, molybdenum, nickel, nitrogen, phosphorus, selenium, silicon, tantalum, titanium, tungsten, vanadium, and zirconium. Some of these chemical elements and others are available in addition agents that are not ferroalloys, as well as in almost pure form; these include relatively pure metals such as aluminum, calcium, cobalt, copper, manganese and nickel; oxides of molybdenum, nickel and tungsten; carbon, nitrogen and sulphur in various forms; and alloys consisting principally of combinations of two or more of the foregoing elements. The more important of the addition agents (including ferroalloys and rare-earth alloys) will be discussed individually later in this chapter.

Use of Addition Agents—Addition agents may be added with the charge in the steelmaking furnace, or in the molten bath near the end of the finishing period, or in the ladle or in the molds. Timing of the alloy

additions is dependent on the effect of the addition on the temperature of the molten metal, ease with which specific addition agents go into solution, susceptibility of a particular addition agent to oxidation, and formation of reaction products.

Economy in manufacture of alloy steels requires consideration of the relative affinity of the alloying elements for oxygen as compared with the affinity of iron for oxygen. For example, copper, molybdenum, or nickel may be added with the charge or during the working of the heat and are wholly recovered. Easily oxidized materials such as aluminum, chromium, manganese, boron, titanium, vanadium, and zirconium are normally added in the ladle in order to minimize oxidation losses.

To offset the chilling tendency of large additions and to minimize or eliminate the necessity for preheating, some addition agents such as ferromanganese and ferrochromium can be obtained mixed with chemical reagents to provide exothermic reactions that permit these agents to be added to the ladle without undue chilling of the steel.

The agents to be added to the bath should be lump size (say about 125 mm or 5 inches) in order to penetrate the slag easily. For ladle additions, the alloy should have a maximum size of approximately 50 mm or 2 inches to assure rapid solution.

Storage Facilities for Addition Agents—From the standpoint of material handling and of conservation and identification, it is advisable to store addition agents in properly designed bins in which they are protected from the weather. The location and design of bins should make the contents quickly available and with a relatively low handling cost. All bins should provide identification of the contents since confusion may be costly due to failure to meet specified chemical composition of the finished steel. Certain addition agents are more easily broken on handling than others, and caution should always be exercised to avoid production of fines.

COMPOSITIONS AND SOME USES OF VARIOUS ADDITION AGENTS

Ferromanganese is the most common of the ferroalloys used in steelmaking. **Standard ferromanganese** is the form commonly used to introduce the element into molten steel. It contains 74 to 82 per cent

---
*Known also as niobium.

manganese, and contains not more than 1.25 per cent silicon, 0.35 per cent phosphorus, 7.50 per cent carbon, and 0.05 per cent sulphur. **Low-phosphorus ferromanganese** suitable for addition to acid open-hearth steel should not contain over 0.10 per cent phosphorus. **Low-carbon ferromanganese** is used when it is important to limit the amount of carbon entering the steel from the ferromanganese addition. This low-carbon product is available in several grades containing increasing amounts of carbon, e.g., 0.10 per cent, 0.20 per cent, 0.30 per cent, and 0.75 per cent (all maxima), and the lower the carbon content the higher the price per pound. All of the foregoing grades of low-carbon ferromanganese contain 80 to 90 per cent manganese. **Electrolytic manganese metal** (about 99.9 Mn) and "Massive Manganese"* (92 to 94 per cent Mn) are also being used as a source of low-carbon manganese. **Medium-carbon ferromanganese** contains a maximum of 1.5 per cent carbon and 80 to 85 per cent manganese. If low-carbon and medium-carbon ferromanganese are used as bath additions (to the furnace) such additions should be made after the bath has been deoxidized. Exothermic mixtures of ferromanganese with suitable chemical reagents permit large manganese additions to be made to the ladle without undue chilling effect on the molten steel.

Silicomanganese is used by some open-hearth furnace operators as a furnace addition to **block the heat** (retard the oxidizing reactions taking place in the furnace toward the end of the finishing period) because of the shorter holding time required from time of addition to tap, as compared with the use of ferrosilicon followed by ferromanganese. Silicomanganese contains up to 75 per cent manganese, 18 to 20 per cent silicon, and a usual maximum of 1.5 per cent carbon.

Ferromanganese-silicon, used in the making of stainless steels in electric-arc furnaces, contains about 65 per cent of manganese and about 30 per cent of silicon: the carbon content is 0.08 per cent maximum.

Blast-Furnace Ferrosilicon—The low-silicon grades of ferrosilicon, which usually start at 10 per cent silicon and ordinarily do not exceed 17 per cent silicon, generally are blast-furnace products, and contain 1.50 per cent maximum carbon.

Electric-furnace ferrosilicon, made by an electric-furnace process, is graded according to silicon content. The principal grades nominally contain, respectively, 25 per cent, 50 per cent, 65 per cent, 75 per cent, 85 per cent, and 90 to 95 per cent silicon. The grades containing the most silicon can be obtained with special low aluminum content. The 50 per cent silicon grade is by far the most widely used, and is employed both as a blocking addition in the furnace and as a ladle addition. The 10 to 15 per cent and 25 per cent silicon grades are sometimes used in the open-hearth furnace for blocking the heat, and may also be used as deoxidizers added prior to addition of other more expensive alloys. Boron-bearing and calcium-bearing types of electric-furnace ferrosilicon are available, as well as high-purity grades, in certain silicon-content ranges. When ferrosilicon containing more than about

---

*"Massive Manganese" is a trademark of Diamond Shamrock Corporation.

60 per cent silicon is added to liquid steel, the liquid steel temperature increases because of the large amount of heat produced by the heat of solution of the silicon in the metal.

Ferrochromium, containing 65 to 72 per cent chromium and a maximum of 2 per cent silicon, is classified into grades by carbon content. The respective grades contain 0.015, 0.025, 0.05 and 0.75 per cent carbon, along with the most commonly used and cheapest grade containing 50 to 69 per cent chromium and 4.50 to 8.50 per cent carbon. Ferrochromium containing the higher amounts of carbon is used as a furnace addition. As in the case of ferromanganese, the alloys of lower carbon content are the more expensive. Several silicon-bearing ferrochromium alloys containing from 58 to 65 per cent chromium, 10 to 14 per cent silicon, and 4 to 6 per cent carbon are occasionally used as blocking additions to the furnace in the open-hearth process. Ferrochromesilicon alloys, sometimes called chrome-silicides, that contain a maximum of 0.05 per cent carbon are available in two grades: one contains 36 per cent chromium and 40 per cent silicon; the other, 40 per cent chromium and 43 per cent silicon. Briquettes of ferrochromium in which are incorporated chemical reagents to provide exothermic reactions permit chromium to be added to the ladle in this form without an undue chilling effect.

Ferrovanadium is available in commercial grades containing 50 to 80 per cent vanadium and usually is added to killed steel in the ladle, but two proprietary products containing, respectively 84.5 and 42.5 per cent vanadium account for a large portion of the domestic vanadium-addition agents used.

Ferromolybdenum, containing 55 to 75 per cent molybdenum, is used where the higher molybdenum contents are desired. For lower molybdenum contents, molybdenum oxide or calcium molybdate may be used. All are furnace additions.

Ferrotitanium available in the following grades generally is used as a ladle addition: a 0.15 per cent carbon grade containing 40 per cent titanium, a 0.20 per cent maximum carbon grade containing 70 per cent titanium, and a 5.0 per cent maximum carbon grade containing 15 to 25 per cent titanium. Titanium scrap containing 90 per cent titanium, 4 per cent vanadium and 6 per cent aluminum is frequently used.

Zirconium is obtained from one alloy containing 12 to 15 per cent zirconium, 39 to 43 per cent silicon, and a maximum of 0.20 per cent carbon; or another containing 35 to 40 per cent zirconium, 37 to 52 per cent silicon, and a maximum of 0.50 per cent carbon. These zirconium alloys generally are added in the ladle.

Ferrophosphorus in two grades containing, respectively, 17 to 19 per cent and 23 to 26 per cent phosphorus, usually is added in the ladle.

Nickel, obtained in the form of sheared electrolytic cathodes, ingots produced from remelted cathodes, and nickel briquettes, all containing a minimum of 99 per cent nickel, are used as furnace additions. Little or no nickel is lost through oxidation when it is added to the bath, so nickel steels can be made by charging nickel-steel scrap and adding metallic nickel after the charge has been melted completely. Nickel oxide, nickel powders, sinter and ferronickel (the latter containing

15 to 50 per cent nickel, balance iron) are also used as charge material. Nickel oxide is sometimes used in conjunction with reducing slags in the electric furnace as a source of metallic nickel.

Copper usually is added to the bath in the form of virgin copper pigs or as scrap copper. It also can be recovered, with little or no loss, from copper-bearing steel scrap in the initial furnace charge. Some steel producers add copper pellets to individual molten ingots for alloying purposes when heat lots of copper steels are not required.

Lead is added to molten ingots in the form of lead shot by many steel producers to provide steel with free-machining properties.

Aluminum usually is added in the form of secondary aluminum having an aluminum content of from 85 per cent upward. It generally is used in the form of shot for addition to the ingot mold, or as shot or bars for addition to the ladle. In special cases, as when used as mold additions for the deoxidation of small ingots, the metallic aluminum content of the addition may be as high as 97.5 per cent. Limitations are placed on the copper content of aluminum used in steelmaking operations: it generally is not to exceed 4.5 per cent.

Cobalt in metallic form (97 to 99 per cent pure) is usually added to the furnace in the form of shot or rondelles in the manufacture of high-speed steels, permanent-magnet steels and other special steels in the electric furnace.

Ferrocolumbium, containing 55 to 70 per cent of columbium and up to 3 per cent silicon with a maximum of 0.30 per cent carbon, is usually added through the reducing slag in electric furnaces in the production of austenitic stainless steels of the chromium-nickel type and alloys for use at high temperatures. Small amounts of columbium added to steel in the ladle or ingot mold promote a fine-grained structure in carbon and low-alloy steels and increase the hardness and tensile strength.

Rare-Earth Metals—The rare earth metals (REM) used in steelmaking are primarily cerium and lanthanum. A commonly used form of REM is **misch metal** which contains about 96 per cent rare-earth elements with roughly 50 per cent cerium, 25 per cent lanthanum, and smaller amounts of other rare earth metals such as neodymium and praeseodymium. The most effective production method for adding REM to liquid steel is the "plunging method." In this method, a large steel bloom is prepared as shown in Figure 9—1 and a predetermined amount of REM is placed in a container which is fastened to the lower end of the bloom assembly. A heat of steel is tapped into a ladle, alloyed, and deoxidized. Then, the bloom with the REM container is lowered into the ladle of steel and the REM dissolves in the liquid steel without being exposed to air or slag. The use of REM promotes the formation of globular sulphides which results in more nearly isotropic properties in flat-rolled products.

Ferroselenium, used for the addition of selenium to stainless steels to improve their machinability, is added to the ladle during the tapping of electric-furnace heats. It contains 50 to 60 per cent of selenium.

Tantalum for steelmaking purposes is available as ferrotantalum-columbium, which has a content of approximately 20 per cent tantalum, 40 per cent columbium and a maximum of 0.30 per cent carbon. It can be used as a replacement for ferrocolumbium in some cases.

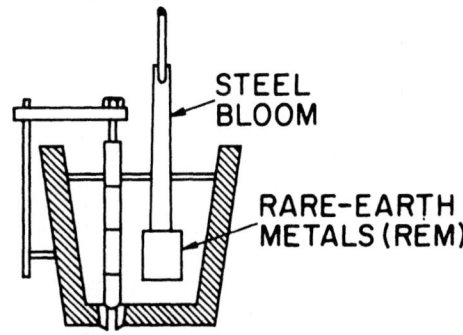

FIG. 9—1. Plunging method for adding rare earth metals (REM) into a ladle of molten steel.

Tungsten, used principally in the manufacture of high-speed tool steels in electric furnaces, is available from ferrotungsten, the standard grade of which contains 70 to 80 per cent of tungsten and a maximum of 0.60 per cent of carbon. Tungsten is also obtained from oxide forms such as scheelite and calcium tungstate.

Sulphur as required by specification is added to the ladle in the form of such addition agents as flowers of sulphur, stick sulphur, iron sulphide, or less often as manganese sulphide, sodium sulphide, etc.

Carbon—For raising the carbon content of the steel during tapping, additions may be made to the ladle in the form of coke, anthracite coal, graphite, or petroleum derivatives. It is desirable that such additions be low in sulphur and high in carbon. In some instances, a low hydrogen content is also desirable.

Boron—Boron addition agents often contain one or more deoxidizers such as manganese, silicon, aluminum, titanium and zirconium to protect the boron from reaction with oxygen and nitrogen from the air when it is added to liquid steel. Boron is used to increase the hardenability of carbon and low-alloy steels.

The foregoing brief descriptions of addition agents summarize only the principal sources of the various elements for the manufacture of steel, with some reference to the manner in which most of them are used. Specific reasons and the manner for the use of individual addition agents, or combinations of them, are discussed in other chapters describing the several steelmaking processes and the properties and heat treatment of constructional (formerly AISI) alloy steels, stainless and heat-resistant steels, and tool steels.

There are American Society for Testing and Materials specifications covering a number of addition agents. Reference should be made to the standards for details on these and other materials used in steelmaking.

LIQUID STEEL TEMPERATURE CHANGE CAUSED BY COMMON ADDITION AGENTS

When addition agents are added to liquid steel, they are (1) heated from their initial ambient temperature to the liquid steel temperature, (2) melted, and (3)

Table 9—I. Chill Factors for Common Addition Agents.

Addition Agent	Chemical Composition, Weight Per Cent	Chill Factors*	
		°C change for one kilogram addition to one metric ton of steel	°F change for one pound addition to one net ton of steel
Aluminum, Grade No. 1	95Al-2Cu-1Fe-1Mn-1Si	−0.16	−0.14
Ferroboron Grade B	78Fe-18B-1Mn-3Si	+4.02	+3.62
Carbon (Graphite)	100C	+5.96	+5.36
Low-Carbon Ferrochromium	70Cr-29Fe-1Si	+1.80	+1.62
High-Carbon Ferrochromium	68Cr-25Fe-5C-1.5Si-0.5Mn	+2.30	+2.07
Blocking Chrome	55Cr-25Fe-10Si-5C-5Mn	+1.77	+1.59
Charge Chrome	70Cr-23Fe-5C-2Si	+2.27	+2.04
Cobalt	100Co	+1.60	+1.44
Medium-Carbon Ferromanganese	81Mn-16Fe-1.5C-1.5Si	+2.01	+1.81
High-Carbon Ferromanganese	76Mn-16Fe-7C-1Si	+2.26	+2.03
Electrolytic Manganese	100Mn	+1.98	+1.78
Nickel	100Ni	+1.39	+1.25
Silicomanganese	67Mn-17Si-14Fe-2C	+1.58	+1.42
Ferrophosphorus	70.5Fe-24P-3Mn-3Si	+3.53	+3.18
16% Ferrosilicon	82Fe-16Si-.6C-1Mn	+1.73	+1.56
50% Ferrosilicon	50Si-48Fe-2Al	+0.84	+0.76
75% Ferrosilicon	75Si-23Fe-2Al	−0.57	−0.52
Sulphur	100S	−1.51	−1.36

*A positive chill factor indicates that the liquid-steel temperature decreases when the addition is made.

dissolved into the liquid steel. For most addition agents, this series of events absorbs heat from the liquid steel, causing its temperature to decrease: however, the addition of some agents causes a temperature increase. The temperature change can be expressed in terms of a **chill factor** which, in SI units, is the temperature change in degrees Celsius when one kilogram of addition agent is added to one metric ton of liquid steel and, in fps units, is the temperature change in degrees Fahrenheit caused by the addition of one pound of addition agent to one net ton of liquid steel. Values of

the chill factors for the common addition agents, calculated theoretically, are given in Table 9—I. As stated in Table 9—I, a positive chill factor indicates that the liquid-steel temperature decreases when a particular agent is added: conversely, a negative chill factor indicates that the liquid-steel temperature increases when the addition is made.

Use of the chill factors to calculate the total temperature decrease caused by a series of ladle additions when using SI units is shown in Table 9—II. Table 9—III shows the same calculation using fps units.

Table 9—II. Total Chilling Effect for a Series of Ladle Additions to 180 Metric Tons of Liquid Steel.

Addition Agent	Amount of Addition		Chill Factor, °C/kg/ metric ton	Temperature Decrease, °C
	kilograms	kilograms per metric ton of steel		
High-carbon ferromanganese	2025	11.25	+2.26	25.4
50% ferrosilicon	810	4.50	+0.84	3.8
High-carbon ferrochromium	315	1.75	+2.30	4.0
Total				33.2°C

Table 9—III. Total Chilling Effect for a Series of Ladle Additions to 200 Net Tons of Liquid Steel.

Addition Agent	Amount of Addition		Chill Factor, °F/lb/net ton	Temperature Decrease, °F
	pounds	pounds per net ton of steel		
High-carbon ferromanganese	4500	22.5	+2.03	45.7
50% ferrosilicon	1800	9.0	+0.76	6.8
High-carbon ferrochromium	700	3.5	+2.07	7.2
Total				59.7°F

CHAPTER 10

Water Requirements For Steelmaking

SECTION 1

GENERAL USES AND PROPERTIES OF WATER

Water is such a common substance and has been so generally abundant that its importance to the iron and steel industry is seldom emphasized in the discussion of metallurgical problems. Because the early large steel mills were built adjacent to ample sources of fresh water, the availability of water was taken for granted for many years. Future use of water in the steel industry should diminish because of greater use of continuous processes and, therefore, fewer operations involving heating and cooling, but the importance of water to the steel industry can only increase because of shrinking supplies and environmental regulations governing the quality of water discharged from plants.

General Uses of Water—The quantity of water associated with Earth is constant and plentiful, but the unevenness of its distribution creates problems in its use. Precipitation is nature's most important method of distributing water. General uses of this water in the United States in 1980 are shown in Table 10—I[1].

Uses of water are classified in two ways: **withdrawal** or **nonwithdrawal** and **consumptive** or **nonconsumptive**. Withdrawal uses involve the removal or diversion of waters from their sources and include all those shown in Table 10—I. Nonwithdrawal uses include navigation, recreation, and conservation of wild life. A consumptive use, in contrast to nonconsumptive, is one wherein the water does not return to its source or to other bodies of water but instead is evaporated, incorporated into a product, or lost by transpiration. Cooling with once-through water in turbine condensers of the electric-power industry and in heat exchangers of other industries has usually been considered to be a nonconsumptive use. However, there is some slight consumption in these latter uses because of evaporation at the higher temperatures which result in the receiving stream. Other withdrawal uses are consumptive and of these, irrigation accounts for the greatest losses because of evaporation and transpiration; in fact, where water is distributed by unlined irrigation canals, there is a 25 per cent loss even before use[2].

Properties of Water—Water and mercury are the only two common mineral substances that are liquids at ordinary temperatures. Water is denser as a liquid than

Table 10—I. Distribution of Water Use in the United States (billions of gallons per day).

Use	Withdrawn	Consumed	Returned
Irrigation	151	83.0	68.0
Public Utilities	34	7.1	26.9
Rural Domestic	5.6	4.0	1.6
Industrial and Miscellaneous	45	5.0	40.0
Steam Electric Utilities	220	3.3	216.7
Totals	455.6	102.4	353.2

as a solid. Its high specific heat and high latent heat of vaporization enable it to transfer heat energy very effectively.

The high kinetic energy that can be imparted to water makes it useful in driving turbines for electric-power generation, in hydraulic placer mining, and in high-pressure jets for cleaning and descaling, to name only a few uses of this property of water.

The ability of water to dissolve, entrain, and suspend many substances makes it a convenient means for transporting, mixing, grinding, separating, and cleaning many materials used by industry.

The solvent properties of water provide a basis for classifying natural waters into freshwater, seawater, and brackish water.

Freshwater includes all inland surface and ground waters except those with a high enough content of dissolved solids to make them undrinkable or unpalatable, namely over about 1500 parts per million (ppm). Freshwaters commonly are differentiated into **soft water** (low in calcium and magnesium contents) and **hard water** (high in calcium and magnesium).

Seawater includes all ocean waters and has a dissolved solids content of about 3.5 per cent (35 000 ppm), mostly sodium chloride.

Brackish waters include (1) coastal waters that are contaminated with seawater, and (2) inland-lake or well waters that have enough naturally-occurring dissolved solids to make then undrinkable.

Saline water refers to all brackish waters as well as seawater.

Brine is water that contains more than 3.5 per cent of dissolved salt as from some oil and gas wells.

SECTION 2

TREATMENT OF WATER

Some of the properties of water that make it useful for so many purposes can also be responsible for the creation of problems in its use. The ability of water to entrain, dissolve, and suspend solids, gases, and other liquids affects the purity of raw water and limits its use in many instances until suitable treatment can remove or neutralize undesirable constituents. Suspended or undissolved contaminants such as clay, sand, metal oxides, and organic materials can clog water systems, impede heat transfer, and contribute to corrosion. Calcium, magnesium, manganese, barium, sulphate, bicarbonate, carbonate, and sulphide ions are some of the constituents that may cause mineral deposits in the form of scaling in water systems. Iron and other dissolved metals may contribute to deposits also. Chlorides, mine acid (H_2SO_4), sulphate salts, hydrogen sulphide, and carbon dioxide cause or accelerate corrosion; seawater, brackish water, and brine have one or more of these constituents. Tannins derived from plant life cause damaging carbon deposits in boilers and in-terfere in manufacturing processes. Among naturally-occurring microorganisms, some bacteria cause corrosion while others obstruct water flow with their slime-like colonies, algae can clog water systems with heavy growth, and fungi can rot wood components of cooling towers. Asiatic clams have become a new problem on most inland waterways since about 1970. These latter mollusks attach themselves and grow in service-water or once-through cooling-water systems and obstruct flow.

The various undesirable properties of water may be combated by a choice of unit operations, materials of construction, operating modes, and maintenance procedures. Some of these are listed in Table 10—II. Since there are usually several possible alternatives or combinations of measures which may be applied, and because two or more problems can co-exist, selection of the proper methods for water use and treatment has become a highly specialized activity in many industrial fields.

SECTION 3

USES OF WATER IN THE IRON AND STEEL INDUSTRY

Because of the great variety of manufacturing processes, plant layouts, and availability of water sources, the water requirements of large integrated steel plants vary widely. Water usages have been estimated to range from 1500 to 45 000 gallons per ton of steel produced[3]. About two thirds of the total volume applied is used for indirect cooling purposes and does not come into direct contact with raw materials, products or flue gases.

In the following paragraphs, some specific uses of water in certain departmental areas of the iron and steel industry are discussed.

Raw-Materials Preparation—The principal solid natural raw materials of the iron and steel industry are coal (most of which is converted to coke and coal chemicals), iron-bearing materials (iron ores and agglomerates), and fluxes (principally limestone and dolomite).

Considerable quantities of water are used in coal mining, principally for dust control at the working face, and in coal-cleaning (washing) plants where solid foreign matter is removed from the as-mined coal (see Chapter 3.) Much of the water used in coal cleaning is collected in settling basins where it is clarified and returned to the cleaning cycle to minimize the amount of make-up water used in the process.

Large amounts of water are used in the beneficiation of iron ores, for cleaning (to remove clay and sand, for example) and in concentrating processes (see Chapter 5). In many operations, water of an inferior grade may be usable; some of the flotation processes, however, require water free from certain impurities and of generally higher quality. Lack of abundant water at many mining and beneficiation sites makes it necessary to recycle as much water as possible after some form of treatment that will render effluent water suitable for repeated use. Clarification is the principal treatment given to water used in the mining industry.

Coke and Coal-Chemical Plants—The chief use of water in the manufacture of coke is to quench the hot coke after it is pushed from the ovens. Large quantities are also used for cooling the by-product coke-oven gas.

The various chemical processes employed in the recovery of coal chemicals require considerable amounts of water of suitable purity: for the generation of process steam for heating stills, etc.; for cooling purposes in heat exchangers and condensers; for washing certain crystalline products; and other uses mentioned in Chapter 4.

Sintering Plant—The principal uses of water in a sintering plant are as additions for controlling the mois-

Table 10—II. Methods for Combating Undesirable Components in Water.

Components	Corrective Methods
1. Suspended solids (clay, sand, metal oxides, organic substances)	Strainers, filters, centrifuges, magnetic separators, settling basins, clarifiers, air-flotation units. Dispersants, surfactants, flocculants. Backflushing, agitation with air or steam, alkaline cleaning (to remove organic substances), and mechanical cleaning.
2. Scale-forming substances (calcium, magnesium, bicarbonate and other ions)	Softening, distillation, seeding, and precipitation. Addition of acids, phosphates, phosphorates, other dispersants, and scale inhibitors. Increasing velocity of flow. Cleaning with acid, chelant, and high-pressure water.
3. Corrosive substances (chloride ions, sulphate ions, carbon dioxide, and so on)	In-plant, pilot-plant, and/or laboratory studies to choose materials of construction for process equipment. Distillation, deaeration, and degassing. Addition of corrosion inhibitors and alkali. Freshwater purge, nitrogen blanketing, and continuation of flow during shutdowns. Chemical cleaning.
4. Microorganisms (bacteria, slime, fungi)	Clarification, chlorination, and use of other biocides. Minimization of stagnant areas. Impregnation of wood with penta chlorophenates or arsenic compounds.
5. Tannins	Use of clarifiers or activated carbon.
6. Asiatic clams	Heating to 49°C (120°F), prolonged chlorination, and use of other biocides.

ture content of the mix, for dust control, and for cooling sinter. Another important use is the washing of flue gas.

Blast Furnace—The blast furnace requires very large amounts of water for its efficient operation, notably for cooling various parts of the furnace and its auxiliaries. For example, cooling water circulates constantly through the tuyeres, hearth staves, bosh and inwall cooling plates, cinder notch, and stove valves. Considerable quantities of water are used to remove flue dust from the gases leaving the top of the furnace and for final cooling of these gases. Additional water is used for slag granulation and other purposes.

Steelmaking Furnaces—The fuel burners of an open-hearth furnace are water-cooled, as are the valves that regulate the flow of air and waste gases into and out of the furnace system. In addition, the skewback channels and doors are cooled with circulating water. The roof lances used in some furnaces for injecting oxygen onto the bath are also water-cooled.

Electric-arc furnaces are equipped with water-cooled doors and side-wall panels. Water cooling is applied also to the roof ring, the electrodes, electrode rings, and electrode clamps (see Chapter 18).

Basic oxygen steelmaking requires the use of a water-cooled lance to inject oxygen into the bath. The gases emitted by the vessel are collected in a water-cooled hood and passed through water sprays for additional cooling prior to cleaning. Cleaning to remove entrained solids is accomplished in a dry process, or in a wet process that requires additional large quantities of water.

Rolling Mills—The reheating furnaces employed to heat steel for rolling employ water quite extensively for cooling furnace doors, skid pipes, skewbacks, and so on. Water also is used extensively as a coolant for rolling-mill rolls to maintain their contours, minimize fire checking, and lengthen their service life. On hot-rolling mills, water in the form of high-pressure jets is used to remove scale from the hot steel before rolling and to keep the surface clean between certain passes. Hot-strip mills also use cooling sprays over the runout table to cool the strip to the proper temperature for coiling.

The scale removed from hot steel by the high-pressure jets referred to above falls into a flume or sluice beneath the mill, where a running stream of water carries the scale to a collection point.

The roll-neck bearings of some mills are made of materials that permit the use of water as a lubricant.

Pickling—The removal of scale from the surface of hot-rolled steel generally is accomplished by immersing the steel in a water solution of sulphuric, hydrochloric, nitric, or other acid. The large scope of most pickling operations involves the use of large quantities of water, not only for making up the pickling solutions, but also for rinsing.

Electrolytic Tinning—Part of the sequence of operations in electrolytic tinning involves the electrolytic cleaning of the steel strip in an aqueous alkaline bath, followed by acid pickling, rinsing, and scrubbing—all prior to the actual plating of tin on the steel surface. After plating, the strip is rinsed to remove the drag-out of plating solution and dried. After this, a melting operation to "flow" the tin is followed by a quenching operation. The various baths for cleaning, pickling, rinsing, and quenching require considerable amounts of high-quality water.

Steam Generation—Large amounts of water of controlled purity are required for boiler water for steam generation. The steam may be used to drive turbines for electric-power generation or to drive turboblowers, in both of which cases a great deal of water must be circulated through the condensers serving the turbines. Some steam is used as process steam for various heating, drying, and moisture-content-control operations, some of which consume large quantities of steam that add considerably to plant water requirements.

Miscellaneous Uses—Hydraulic mechanisms that operate valves, open and close furnace doors, and actuate materials-handling devices require considerable amounts of water, as do quenching tanks employed in heat-treating operations.

Ample supplies of potable water, along with water for sanitary purposes, fire protection, and landscape maintenance, must also be provided.

SECTION 4

TREATMENT OF EFFLUENT WATER

In the iron and steelmaking industry, as in any major industry, the large volumes of process water that come into direct contact with the raw materials, products, and flue gases must be treated for removal of contaminants prior to discharge or re-use of the water. As discussed previously, the main operations in an integrated steel plant that require wastewater treatment include cokemaking; ironmaking; steelmaking; hot and cold rolling; and finishing operations such as pickling, electrolytic tinning, and other coating processes. As established by the Federal Clean Water Act, any discharge to public waterways is regulated by a government permit under the National Pollutant Discharge Elimination System. The terms of each individual permit place stringent limitations on the quantities of various materials discharged from each outfall.

For the iron and steel industry, the parameters of most significance, and which are generally regulated by the terms of the discharge permits, are suspended solids, oil and grease, phenol, cyanide, ammonia, and trace amounts of heavy metals such as lead, zinc, chromium, and nickel. In addition, several organic compounds that are on the "priority pollutant" list compiled by the United States Environmental Protection Agency are regulated for cokemaking and cold rolling operations. The following discussion will describe the conventional wastewater treatment technologies employed in the iron and steel industry for effective removal of the above types of contaminants.

Control of Suspended Solids—Removal of suspended solids is necessary in the wastewaters from practically all of the production steps in the iron and steel industry, from cokemaking to product finishing. Solid particulates become suspended in process water streams during cleaning and cooling of flue gases, descaling, roll and product cooling and flume flushing in rolling mills, and product rinsing in finishing operations. The two general methods for removing suspended solids are sedimentation and filtration. Sedimentation, which is settling by gravity, can be accomplished in a large lagoon or in a clarifier specifically designed for a given application. Clarifiers are generally circular but may also be constructed of rectangular shape. An advantage of clarifiers over lagoons is that clarifiers occupy much less ground space. Also, the sludge that accumulates in lagoons must be periodically dredged out to maintain an effective volume for settling, whereas clarifiers are designed for organized continuous removal of the collected sludge from the bottom of the unit. The underflow sludge is further dewatered in one of several types of sludge filters or in centrifuges to reduce the volume for ease in handling and economy of disposal.

Coagulant aids, such as alum, ferrous sulfate, ferrous chloride, and commercial organic polyelectrolytes, are often added to the wastewater prior to clarification to promote flocculation of the solid particles, which increases their effective size to improve the settling rate.

Multi-media filtration is another method for removal of fine suspended particulates that is commonly applied to steel industry wastewaters. The water is passed through layers of filter media contained in a vessel. The system usually is comprised of a number of individual filtration units in parallel. It is desirable to design a filter system with the highest feasible flow rate through the filter media to minimize the required size and cost. This is achieved through the use of a dual filter media. In a typical system the water first passes through a relatively coarse layer of a media such as anthracite coal, and then through a layer of fine sand. Most of the particulate is removed by the coarse layer while the fine layer does the final polishing. Periodically the collected particulate must be removed from the filter media by backwashing. In this operation the influent flow of wastewater is shut off and a stream of water is passed through the filter in the opposite direction to flush out the collected solids. By having a number of filter units installed in parallel, one unit can be put through the backwash cycle without interrupting the continuous treatment of the wastewater stream. The backwash stream is usually processed through a thickener and sludge filter to remove and dewater the high concentration of suspended solids. Multi-media filters can produce a high degree of clarity in effluent streams and may be preferred over clarifiers where space is limited or where a clarifier cannot attain the required effluent solids concentration.

The quantity of suspended solids discharged with a wastewater stream can also be greatly reduced by recirculating the water back to the process to reduce the volume discharged. However, the degree of recirculation that is feasible is limited by the buildup in concentration of dissolved salts in the system, which eventually leads to plugging in equipment and piping. Therefore, a certain portion of the circulating water volume must be "blown down" to control the concentration of dissolved salts at a tolerable level.

Control of Oil—Oil, and sometimes grease, are commonly found in wastewaters from continuous casting, hot and cold rolling, pickling, and electroplating and other coating operations. The oils originate from machinery and product lubricants and coolants; hydraulic systems; and preservative coatings applied during certain phases of the production operations. If the oils are insoluble in water, they can be controlled by gravity separation and skimming. Gravity oil separators are usually rectangular chambers, in which the velocity of the water stream is slowed down sufficiently to allow time for the oil to float to the surface, from where it is removed by one of several available types of skimming devices. Insoluble oils can also be removed along with suspended solids in the multi-media filters previously described. If the oils are emulsified or water-soluble, such as found in waste cold rolling solutions or rinse waters, they can be treated by acid to break the emulsion, and then be processed by gravity sedimentation and skimming, or by air flotation. The latter is a process that is used to separate floatable materials that have a specific gravity very close to that of water and which therefore cannot be effectively separated by gravity alone. In a flotation system, gas bubbles, usually air, are released in the wastewater and attach to the oil and fine solid particles, causing them to float more rapidly to the surface from where they are skimmed off as a froth. Flotation has had only a limited application in treatment of waste water from the steel industry.

Control of Heavy Metals—Limitations on the discharge of specified heavy metals have been established by the United States Environmental Protection Agency for steel industry process waters from blast furnaces; steelmaking furnaces; and pickling, cold rolling, electroplating, and hot coating operations. The conventional method used for removal of these trace metals is alkaline precipitation followed by clarification or filtration. The solubility of heavy metals in water is a function of pH. Generally, metals become less soluble the higher the pH. Thus, to remove dissolved metals, a wastewater stream is treated with an alkaline material in a mixing tank with a pH controller. Lime, being the least expensive reagent, is most commonly used, although caustic soda or other alkali may also be employed for this purpose. After the pH is raised to a level where the dissolved metals will precipitate as hydroxides, the water passes to either a clarifier or a filter for removal of the precipitated solids. The addition of a coagulant aid might be required if the precipitate is extremely fine. If chromium is present in the hexavalent form, it must first be chemically reduced to the trivalent form before it will precipitate. This is usually done by lowering the pH to 2-3, and adding a reducing agent to convert the chromium to the trivalent form.

Coke Plant Waste Treatment—Biological oxidation is the most commonly applied technology for final treatment of coke plant wastewaters. These waters pick up significant levels of phenol, cyanide, and ammonia, plus lesser concentrations of other organic compounds, primarily as a result of condensation from coke oven gases, which contain these substances. A conventional system for treatment of coke plant waste would include ammonia distillation followed by biological oxidation. The ammonia still system may have a "free leg," where dissolved gaseous ammonia is removed by steam distillation, and a "fixed leg," where ionized ammonia is converted to free ammonia by alkali addition, and then is removed in the free leg. While this distilla-

tion process will remove a large percentage of the ammonia and cyanide from the water, enough of these contaminants will remain in the ammonia still waste, in addition to phenol and other organics, to require further treatment, such as biological oxidation, prior to disposal.

Because biological oxidation is highly sensitive to fluctuations in contaminant loadings, the effluent from the ammonia still is first passed through an equalization basin to level out the concentrations, temperature, and flow volume. A conventional single-stage biological oxidation system for treatment of coke plant waste consists of an aeration basin containing activated sludge, followed by a clarifier. The activated sludge principle is similar to that applied in sewage treatment plants. In the aeration basin a mass of microorganisms in the form of suspended solids called an "activated sludge" is supplied with oxygen, which enables it to destroy the biologically degradable contaminants in the wastewater. Populations of microorganisms can be developed that can effectively degrade phenol and other organics, cyanide, and ammonia in coke plant waste. The required oxygen is supplied either by mechanical surface aerators or by diffusion of air bubbles through the basin. The treated water overflows the basin to a clarifier, where the activated sludge is settled out to be recycled back to the aeration basin. The overflow water from the clarifier is discharged. Biological oxidation systems can be designed to treat coke plant wastewater to attain the stringent discharge limits for phenol, cyanide, and ammonia that have been established by environmental control agencies.

An alternative to treatment and discharge of coke plant wastewater, where permissible, is to dispose of all or part of the water by evaporation in quenching hot coke.

Terminal Treatment—A common practice in wastewater treatment within the steel industry is to combine wastes from several different types of operations for treatment in a so-called terminal treatment plant. This practice has been particularly successful in the handling of wastes from the various finishing operations. These wastes typically might contain suspended solids, free and emulsified oils from cold rolling, acids from pickling rinsewaters, and heavy metals from picking and coating processes. In a typical system, acid streams are mixed with the emulsified-oil streams to break the emulsions and the combined wastes are then passed through a gravity oil separator, neutralized with lime to remove acids and precipitate heavy metals, and treated for removal of solids and any remaining oils in a clarifier or filter. This general type of terminal treatment system can be applied for environmental control of a number of different steel industry processes more economically than the alternative of providing a separate treatment system for each process.

SECTION 5

WATER ECONOMY BY REUSE AND RECYCLE

It has been estimated in the past that about 98 per cent of circulated water was returned to its source at

most steel plants, as illustrated in Figure 10—1. However, every large steel plant practices to some extent

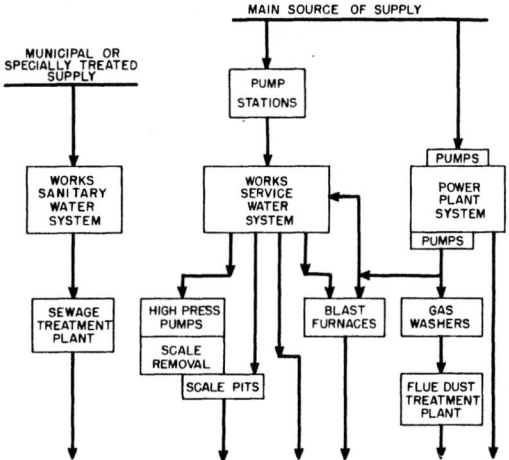

FIG. 10—1. Simplified flow diagram illustrating the use of water at an integrated steel plant with an abundant supply of high-quality water. (After Nebolsine 4).

the repeated use of the same water to limit the amount of new "make-up" water that must be supplied. A typical example of such economy is the use of cooling water discharged from the turboblower and turbogenerator condensers for non-contact cooling at blast furnaces and to pump the discharged blast-furnace-cooling water to the furnace-gas washers. Such series use of water is defined as reuse.

In the western part of the country, water supplies are not as ample as in the East and even greater economies have had to be practiced. At one Far Western plant, the main source of water consists of wells of limited capacity and must be supplemented by purchases from a local water company. Figure 10—2 shows the layout of the principal parts of the system at this plant. Water is stored in a reservoir from which it is supplied to a water-treatment plant where the solids content of the water is reduced. The treated water then goes to an

open industrial-water reservoir and a covered domestic-water reservoir. Water from the domestic reservoir is pumped to supply powerhouse and sanitary requirements, fire protection, lawn and shrubbery requirements, and some minor industrial uses. The industrial-water system consists of a number of individual systems in series, with the water discharged from one system becoming the supply of the next succeeding process system in a cascading design mode. Thus, the first system supplies water to meet the highest quality requirements and the last system supplies water of relatively poor quality.

This cascade plan constitutes reuse of water. However, the greatest economy at the plant is achieved by recirculating all of the water for each process within its own system except for the small percentage blown down to the next process in series. Such circulation and repeated use within a process is called recycle. The combination of reuse and recycle at this Far Western plant results in a water requirement of only 1100 gallons per net ton of raw steel produced.

Whereas the plant represented in Figure 10—2 was built to have virtually complete recycle, older plants are progressively converting to greater recycle to meet environmental restrictions imposed by governmental regulatory agencies at the federal, state, and local levels. Conversions to meet the mandated requirements are most costly to implement for the older plants with ample sources of water because these plants were not built originally with any water-recycle systems.

References

1. Statistical Abstract of the United States. U.S. Department of Commerce, 1982-83.
2. Vogely, W. A. and Mack, H. B., A 30-year review of water supply and demand. Proceedings of the first national conference on complete water reuse, pages 12-27, 1973.
3. Nebolsine, Ross, Steel plant waste water treatment and reuse. Iron and Steel Engineer, March, 1967.
4. Nebolsine, Ross. Water supply for steel plants. Iron and Steel Engineer, April, 1954.

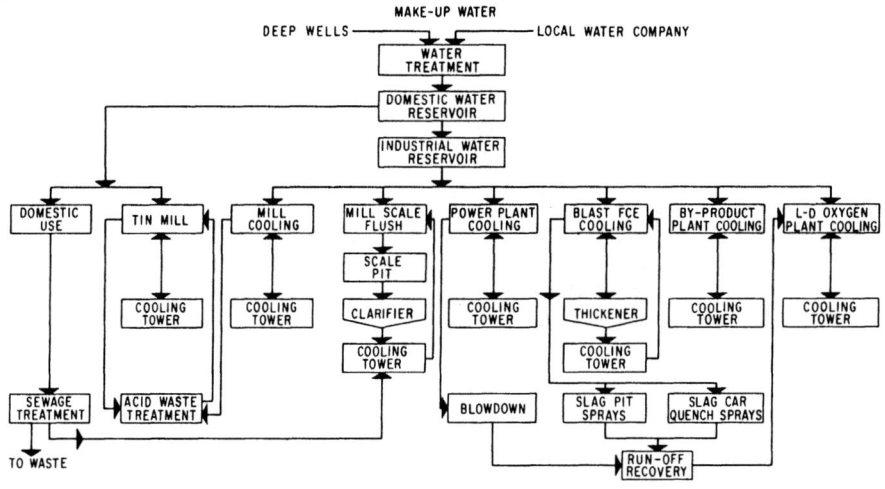

FIG. 10—2. Flow diagram of the water-supply system at a Far Western steel plant where conservation practices limit consumption of water to as little as 1100 gallons per net ton. (Courtesy, Kaiser Engineers.)

CHAPTER 11

Tonnage Oxygen For Iron-
And Steelmaking

The iron and steel industry has become one of the largest industrial users of gaseous oxygen. It consumed a total of about 5761 million Nm³* (214 446 million scf*) in gaseous form in 1980. About 10 per cent of the total oxygen used was produced by iron and steel companies in their own or leased facilities. In iron and steel plants in 1980, total consumption of oxygen was 56.7 Nm³ of gas for each metric ton of raw steel produced (1918 scf per net ton). This figure is based on the consumption of oxygen for all iron- and steel-plant uses. As shown in Table 11—I, roughly 24 per cent of the oxygen consumed by the industry in 1980 was employed elsewhere than in steelmaking furnaces. Thus, in that year, the average amount of oxygen used in the production of raw steel (by the basic-oxygen, open-hearth and electric-furnace steelmaking processes) was 43.1 Nm³ per metric ton (1458 scf per net ton).

Oxygen in the Blast Furnace—Increasing the oxygen content of the blast has proved beneficial to the operation of blast furnaces in the production of both basic pig iron and ferromanganese (see Chapter 15). However, the use of oxygen enrichment of the blast is not widespread, except in furnaces producing ferromanganese. Greater benefits are derived by oxygen enrichment in the production of ferromanganese than in the production of basic pig iron because larger quantities of high-temperature energy are required in blast furnaces making ferromanganese. For example, enrichment of the blast from the normal 21 per cent oxygen content of natural air to as high as 30 per cent has considerably increased production rates and decreased the consumption of coke per ton of ferromanganese produced. The use of simpler and more economical means of increasing productivity of furnaces producing basic pig iron, such as higher hot-blast temperatures and burden preparation, has limited the adoption of oxygen enrichment of the blast for such furnaces.

The Use of Oxygen in Steelmaking—Gaseous oxygen is used as the sole refining agent in the oxygen steelmaking processes described in Chapter 16. Since the introduction of the Basic Oxygen Process (BOP) in the 1950's, the technique of using oxygen to make steel has undergone several changes. Oxygen is now blown from the top, bottom, and even the sides of the furnace in some steelmaking processes. Furthermore, oxygen steelmaking now accounts for about 60 per cent of domestic steel production. However, these changes have had little effect on oxygen production technology. Tonnage oxygen is still produced in a cryogenic air separation plant, designed essentially the same as it was in the 1950's.

All the basic oxygen steelmaking processes presently operating require an intermittent supply of high-pressure (about 1380 Kpa or 200 psia) oxygen. To meet these requirements, the low-pressure gaseous oxygen plant is provided with a product compressor, a booster compressor, and sufficient gaseous oxygen storage to hold the output of the oxygen plant during nonblowing periods. In addition, a one-week supply of liquid oxygen is provided to assure the availability of oxygen dur-

Table 11—I. Consumption of Oxygen by the Iron and Steel Industry in 1980*

	Millions of m³	Millions of ft³
Purchased[a]	5,165	192,264
Purchased[b]	596	22,182
Total	5,761	214,446
Consumption by Uses		
Conditioning	354	13,161
Scrap preparation	44	1,635
Other burning and welding	115	4,273
Blast furnaces	618	23,007
Steelmaking		
Open hearth	554	20,626
Basic oxygen	3,516	130,906
Electric	309	11,485
Total steelmaking	4,379	163,017
Maintenance and construction	23	866
All Other	228	8,487
Total	5,761	214,446

*Consumption in millions of cubic feet from "Annual Statistical Report—1980 of the American Iron and Steel Institute. Oxygen consumed in liquid form was reduced to its gaseous equivalent in these statistics, using 60° F and 30 in. Hg as standard conditions. Consumption in millions of cubic metres was calculated, using 0° C and 760 mm as standard conditions.

[a]Purchased from vendors with facilities in or adjacent to plant or facilities located away from plant.

[b]Produced in companies' own facilities or leased facilities.

---

*Nm³ = normal cubic metres measured at 0° C, 760 mm Hg pressure.
scf = standard cubic feet measured at 60° F, 30 in. Hg pressure.

ing plant outages. These auxiliary requirements add considerably to the capital and operating cost of the oxygen plant. For example, it takes about 330 kWh of electric power to produce one metric ton of low-pressure gaseous oxygen (300 kWh/net ton). After adding oxygen compression and storage requirements, the power required increases to about 450 to 500 kWh per metric ton of oxygen (400 to 450 kWh/net ton).

Although most raw steel is produced by the basic oxygen process, open hearths still produce about eight per cent of the domestic steel. In the open-hearth furnace, gaseous oxygen can be introduced through the fuel burner when firing from either end of the furnace or through roof burners to intensify combustion and increase heat input to the furnace during the period when scrap and other solid raw materials are being charged and melted. In the refining phase of an open-hearth heat, water-cooled lances extending into the furnace through the roof are lowered to inject oxygen directly onto the metal. Such use of oxygen increases the rate of scrap melting, desiliconization and decarburization, and materially reduces heat time.

In the electric-arc furnace, oxygen is introduced to the bath through water-cooled or consumable lances. The oxygen reacts with carbon in the bath, causing a stirring action and providing some of the heat required for meltdown. It also provides means for rapid cutting of scrap. The net effect of oxygen lancing is to reduce the meltdown time, to reduce the consumption of electric power, and to reduce the carbon level in the steel when necessary. Many electric furnaces consume as much as 30 nM^3 of oxygen per metric ton of liquid steel (1000 scf/net ton) and some consume even more. Oxygen is also used in oxy-fuel burners (usually three per furnace) to reduce meltdown time, reduce electric power consumption, and to eliminate cold spots in the furnace.

Because most electric-furnace shops are relatively small (less than one million metric tons per year) and do not consume as much oxygen per ton as the basic oxygen processes, the oxygen demand by electric furnaces is not large enough to support an on-site oxygen plant. Generally, oxygen is supplied as liquid, or is piped in from an off-site plant that supplies several other customers.

Purity of Oxygen for Steel-Plant Use—The oxygen used by the steel industry is usually produced as 99.5 per cent pure oxygen and is used in large quantities in the oxygen steelmaking processes, in open-hearth and electric steelmaking furnaces, and to increase the oxygen content of air for the blast furnace and, in much smaller quantities, for welding and cutting.

However, some users do not need high-purity oxygen. Blast furnaces, for example, do not require high-purity oxygen; neither do alternative iron/steelmaking processes such as KR (Kohl-Reduction), KS (Klockner Stahlforschung), and Inred. These oxygen users could reduce operating costs by specifying 90 to 95 per cent oxygen, which requires about seven to ten per cent less power for production than high-purity oxygen.

Production of Oxygen—Gaseous oxygen of the desired purity is produced from atmospheric air by fractional distillation processes carried out at very low temperatures and elevated pressures; for instance, at

temperatures and pressures of the order of –185° C (–300° F) at 483 kPa (70 lb per sq in.) gage pressure are employed at most plants in the United States although there are oxygen plants here and abroad that operate at much higher pressures. The several commercially available processes are essentially the same, although many minor differences in arrangement are encountered, depending upon local conditions, special designs of the individual builders of process equipment, and the need for such additional products as liquid oxygen, nitrogen, or argon.

The processes start by compressing the air to an elevated pressure, followed by progressively cooling it to saturation temperature in steps in a series of highly efficient heat exchangers. Condensation and freezing out of moisture, carbon dioxide, and hydrocarbons take place as the temperature is lowered, after which hydrocarbons still remaining are removed in adsorbent traps. The cold, purified air is finally separated into its components in fractionating (distillation) columns. The requirements for heat removal by refrigeration at the low temperature level are met by expansion of a portion of the cold compressed air in an expansion turbine.

A typical flowsheet for one type of large plant for the production of low-pressure gaseous oxygen of 99.5 per cent purity, at about 69 kPa (10 lb per sq in.) gage pressure, is shown in Figure 11—1. In this plant, the flow of filtered air from the compressors enters the main heat exchangers. In the heat exchangers, the incoming air is cooled against the outgoing products; and moisture, carbon dioxide and most of the hydrocarbons are deposited as solids in the exchanger passages. The flow in the exchangers is reversed periodically between air and waste nitrogen to re-evaporate and flush out the accumulated moisture, carbon dioxide and deposited hydrocarbons. For the sake of simplicity, the reversing mechanism is not illustrated in Figure 11—1.

The streams of cooled, compressed air from the reversing heat exchangers are run through a hydrocarbon-adsorbent trap to remove hydrocarbons not already deposited. The major portion (about 85 per cent) of the air, at saturation temperature and pressure (–172.2° C, 496 kPa; equivalent to –278° F, 72 lb per sq in. gage), then is delivered to the bottom of the lower fractionating column (Figure 11—2). A side stream (totaling about 15 per cent of all the air) is taken from the cold end of the reversing heat exchangers and passed back through a separate passage in the exchangers and next sent through the expansion turbine. In the expansion turbine, the air performs work and thereby removes heat from the low-temperature system to provide the refrigeration requirements of the cycle. The expanded air enters the middle of the upper fractionating column (Figure 11—2), together with crude liquid oxygen (approximately 38 mole per cent O_2) from the bottom of the lower column. Reflux liquid for the top of the upper column is provided from the condenser located above the lower column. In the condenser, gaseous nitrogen from the lower column is condensed by the lower temperature of the liquid oxygen in the reboiler. Although nitrogen normally boils at a lower temperature than oxygen, the boiling point of nitrogen can be raised above that of oxygen by operating the lower column at a higher pressure; this fixes the

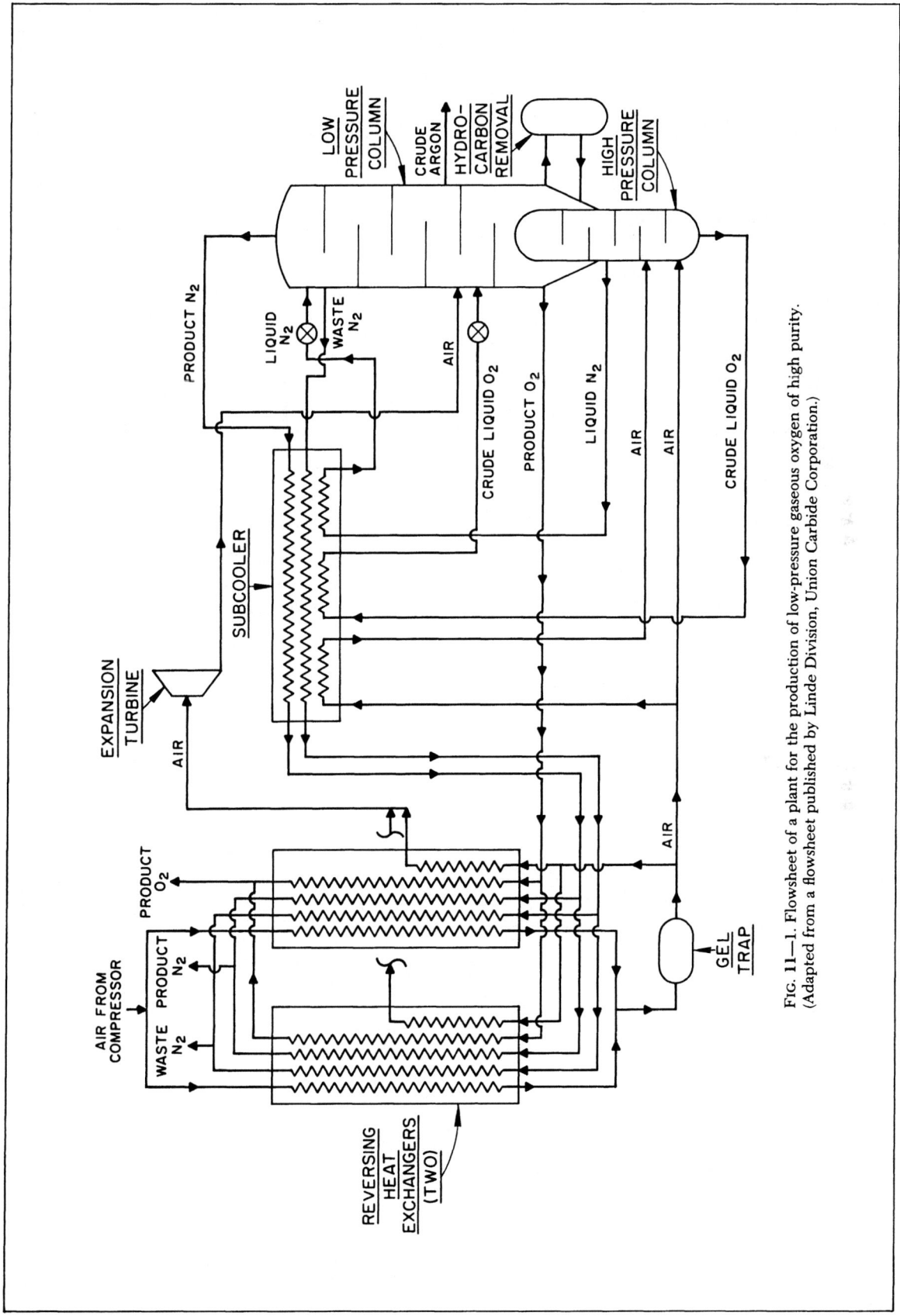

FIG. 11—1. Flowsheet of a plant for the production of low-pressure gaseous oxygen of high purity. (Adapted from a flowsheet published by Linde Division, Union Carbide Corporation.)

minimum allowable compressor discharge pressure. The heat given off by the condensing nitrogen boils the liquid oxygen (of 99.5 per cent purity) to produce vapors that rise through the trays of the upper column. The liquid flowing from tray to tray down through the upper column becomes progressively richer in oxygen (and the upward flowing gaseous product richer in nitrogen) as the oxygen from the countercurrent gas stream is stripped out by the liquid. The result is a high-purity, gaseous nitrogen product at the top of the upper column, and a high-purity gaseous oxygen product at the bottom of the upper column. These two gaseous products are brought out through non-reversing passes of the reversing heat exchangers where they remove heat from the incoming air. A waste-nitrogen product also comes off the upper column and portions are led separately through the reversing heat exchangers, thus flushing out impurities deposited in the previous cycle when incoming air passed through these units. The liquid high-purity oxygen at the bottom of the upper column is passed through another hydrocarbon removal trap (Figure 11—1). This trap removes the last traces of acetylene and other hydrocarbons that possibly were not deposited in the regenerators or the reversing heat exchangers, or adsorbed by preceding traps.

Although not shown in Figures 11—1 or 11—2, a portion of the oxygen product can be withdrawn as a liquid directly from the bottom of the upper column, and a portion of the nitrogen product can be withdrawn as a liquid from the top of the lower column. This creates an additional refrigeration load, which must be provided for in the plant design.

Plants are built for the production of both high- and low-purity oxygen with capacities up to 2300 metric tons (2500 net tons) per day, and still larger plants have been projected.

Distribution Methods—When an oxygen-generating facility is installed adjacent to an iron or steel works to supply the oxygen requirements of that works, it is generally referred to as an "on-site" oxygen plant. The low-pressure gaseous oxygen from an on-site oxygen plant is compressed to the proper pressure and distrib-

uted through pipes to the consuming points. Liquid oxygen from on-site storage or delivered to the consuming plant from off-site storage is first gasified and then distributed, at the proper pressure, through the same system of plant piping.

Although most on-site oxygen plants are owned and operated by an oxygen-generating company, a number are owned and operated by the steelmaking companies.

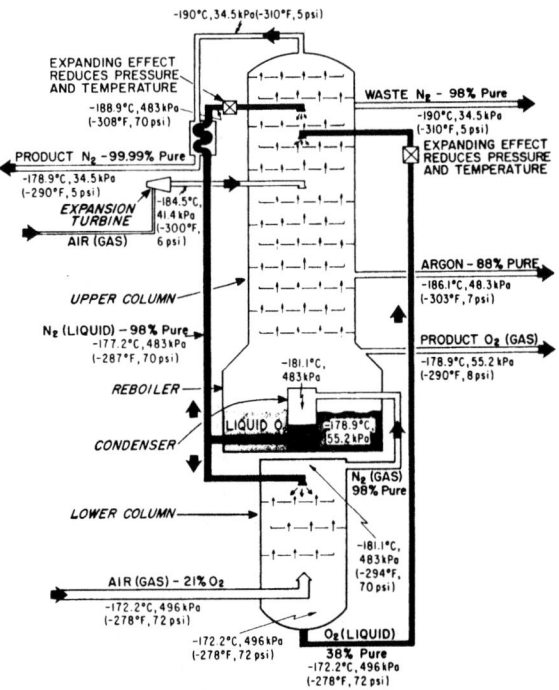

FIG 11—2. Schematic diagram showing operation of fractionating (distillation) columns in a gaseous-oxygen generating plant. The hydrocarbon trap that removes the last traces of hydrocarbons from the liquid oxygen in the reboiler, shown in Figure 11—1, has been omitted from this diagram for the sake of simplification.

CHAPTER 12

Lubrication and Lubricants

SECTION 1

INTRODUCTION

Importance of Lubrication—In spite of its size, ruggedness and complexity, most of the equipment in a steel plant can be classed as precision machinery. Nevertheless, it is sometimes called upon to operate under conditions that are less than ideal. Operating conditions that must be met may involve non-stop operation over extended periods of time; frequent starting, stopping and reversal; extreme pressure; high speeds; shock loads; and, in some instances, exposure to dirty or wet environments. Depending upon the purpose for which it is used, a machine may be required to meet any one or several of these conditions.

Friction, defined for the present purpose as the resistance to relative motion of two solid surfaces in contact, can waste power and cause wear and overheating of moving parts of machines. Frictional effects can be greatly reduced by proper lubrication which, in essence, consists of interposing a fluid film between adjacent solid surfaces that move in relation to each other. Frictional effects can be aggravated by adverse operating conditions unless proper attention is given to the original design and subsequent maintenance of machinery. Well-planned and rigidly-adhered-to lubrication programs and methods for selecting and applying lubricants can assure positive and adequate lubrication. Such programs and methods extend the life of machine elements, reduce maintenance, increase the availability and efficiency of equipment, and help to maintain the continuity of production that is essential to the economical operation of a modern steel plant.

SECTION 2

THEORY OF LUBRICATION

Kinds of Friction—The term "friction" as generally used refers to **solid friction** or the resistance to relative motion between solid surfaces in contact. Solid friction may be the result of a sliding motion of one plane surface (flat or curved) over another, when it is called **sliding friction;** or it may result from the action of one body (e.g., a ball or cylinder) rolling over a plane surface, when it is termed **rolling friction.** Rolling friction generally is considerably less than sliding friction. **Pivotal friction** is a special case of sliding friction. If friction is between two unlubricated surfaces, it is called **dry friction.**

Simplified mechanical arrangements that can result in the foregoing kinds of friction are illustrated in Figure 12—1. The frictional effect in each case is related to the magnitude of the load, to the relative roughness of the surfaces involved, and to the speed of motion of the surfaces in relation to each other. The mathematical relationships between the several factors cannot be considered here.

Internal Friction is defined as the resistance of the components (e.g., molecules) of a body (solid or fluid) to motion in relation to each other. Internal friction in solids is exemplified by the dying away of the vibrations of a bell after it has been struck; the energy of the initial blow is dissipated by being transformed into heat in overcoming the resistance of the mass of the bell metal to being set into vibratory motion. (Each metal has its own characteristic ability to absorb vibrations; this characteristic is called its **damping capacity.**) In fluids, internal friction is manifested in their resistance to flow or shear; a measure of the internal friction of fluids is their **viscosity** which, as discussed later, is an important property of lubricants.

Friction Reduction by Lubricants—If two steel blocks with smoothly ground flat surfaces are placed one on top of the other (Figure 12—2), a certain minimum horizontal force will be required to cause the top block to slide over the bottom one; this force is proportional to the pressure between the blocks and the relative roughness of the contacting surfaces. If a lubricant such as oil is placed between the blocks, a much smaller force will be required to make the top block slide, and the lubricant is said to have reduced the friction between the two blocks. The action of a lubricant in reducing sliding friction is explained by Figure 12—2, in which it is shown that, at sufficiently high magnification, even a ground or polished surface is not a plane,

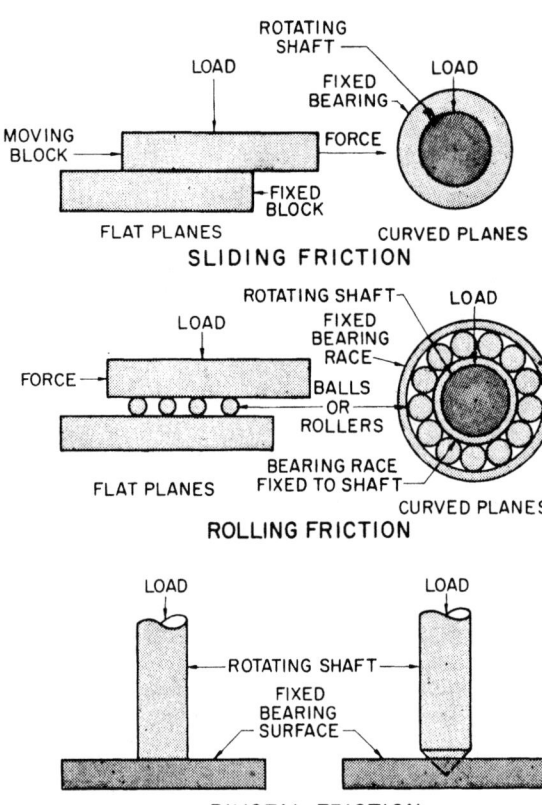

FIG. 12—1. Schematic arrangements of machine elements that result in sliding friction (top), rolling friction (middle), and pivotal friction (bottom).

but consists of a series of minute peaks and valleys (technically referred to as **asperities**). If two such surfaces are in direct contact, the peaks and valleys can interlock and prevent the surfaces from moving freely in relation to each other. If a film of lubricant of sufficient thickness is present between the surfaces, physically separating them, interlocking cannot occur and sliding of one surface in relation to the other can occur with relative ease. The effect of the lubricant has been to substitute the internal friction of the lubricant for the dry sliding friction between the metal surfaces. The resistance to slippage of molecules of a fluid over each other is far less than that of two solid surfaces. Such slippage of fluid molecules must take place within the film because the film "wets" or adheres to each of the metal surfaces and motion of the blocks must involve displacements within the body of the film.

While the foregoing example relates to sliding friction between two flat plane surfaces, the same principles apply to sliding friction on curved (cylindrical) surfaces, as exemplified by a shaft turning in a journal bearing as discussed in the next paragraph (see Section 3 for definition of a journal bearing). The mechanism of effects produced by rolling friction is more complex, but still involves relationships between load, speed, surface conditions and properties of lubricants.

Figure 12—3 shows schematically how a film of oil forms between a bearing and the journal that it supports and guides. It is assumed that the load on the

shaft is directed vertically downward. With the shaft at rest, the bottom of the journal is in metal-to-metal contact with the bearing. As the shaft begins to rotate, the journal tends to climb up the side of the bearing because of friction due to the metal-to-metal contact. As the shaft picks up speed, oil is drawn into the space between the bearing and the journal because it tends to adhere to both of them. This pumping action increases as the shaft speed continues to increase, and pressure develops in the wedge-shaped oil film and increases until the bearing and journal become physically separated with an oil film between them. When the shaft attains full speed, the pressure in the oil wedge tends to force the journal into an eccentric position with respect to the axis of the bearing. The final position of the journal at full speed depends upon the speed of rotation of the shaft, the viscosity of the oil, and the magnitude of the load. The foregoing mechanism of lubrication is referred to as the **hydrodynamic theory of lubrication.**

The thickness of the film of lubricant that can be maintained between the moving surfaces is an important consideration. When the film is thick enough to separate the surfaces completely, the condition is known as **thick-film lubrication.** When the film thickness permits partial metal-to-metal contact, the condition is called **thin-film** or **boundary lubrication.** For the expected speed and load for a given bearing, the lubri-

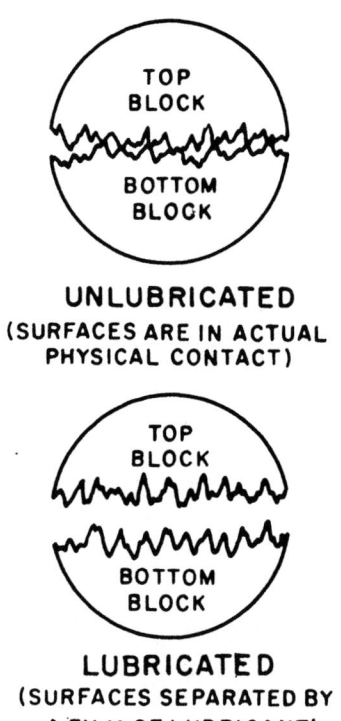

FIG. 12—2. Schematic cross-sections of supposedly smooth metal surfaces which in reality consist of a series of peaks and valleys when viewed at high magnification. Top sketch shows how peaks and valleys can interlock and resist relative motion of the two blocks. Bottom sketch illustrates how a film of lubricant physically separates the surfaces to prevent metal-to-metal contact.

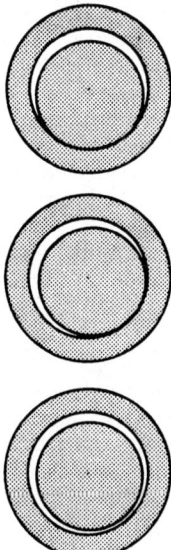

FIG. 12—3. Schematic representation of the formation of an oil wedge between a journal and its bearing due to rotation of the shaft. Top sketch shows the shaft at rest, with the journal in contact with the bearing. As the shaft begins to rotate (clockwise, in this sketch), the journal starts to "climb" the bearing wall due to dry friction (middle sketch). The bottom sketch illustrates the action of the oil wedge in eliminating metal-to-metal contact between the journal and bearing. (Clearances are exaggerated for clarity.)

cant selected should have viscosity sufficient to maintain the thick-film condition.

Other Functions of Lubricants—The primary function of lubricants—that of reducing friction between moving parts of machines—already has been discussed. Other functions of lubricants are to reduce wear, carry away heat, keep dirt away from bearing surfaces, prevent rusting, and transmit power.

Figure 12—2, referred to earlier, demonstrates how lubricants can reduce wear. If they were not separated by a lubricating film, some of the small peaks on adjacent surfaces would be broken off as one block moved over the other. Eventually, this would result in visible wear of the surfaces due to actual removal of material from them.

A certain amount of heat is always developed by the rubbing contact of moving parts in a machine, even though they are well lubricated. Some lubricating systems are designed so that the lubricant (usually oil) is circulated. As the oil performs its lubricating function, it can absorb heat generated by the moving parts. The heated oil is continuously removed from the lubricating site and replaced by cool oil. At some point in the system, the heated lubricant is cooled before it is returned to the sites requiring lubrication, as discussed in Section 6 of this chapter under "Circulating Oil Systems."

Special-Purpose Lubricants—Although this chapter is intended to consider lubricants from the principal standpoint of machine lubrication, it should be mentioned that lubrication of steel undergoing processing is required in the manufacture of many products.

For example, special types of lubricants are applied to strip steel being rolled on cold-reduction mills, where they serve the double purpose of lubricating and cooling (see Chapter 34). Lubricants also are required for either lubrication, cooling, or both, during cold drawing of steel tubes, bars and wire; hot and cold extrusion; machining; pipe threading; and many other steel-plant operations.

Lubricants of various types, with and without additions of inhibitors or other substances intended to give them special properties, are used as surface coatings to prevent rusting and damage to mill products during processing, finishing, plant storage, and transportation. They are similarly used to protect spare machinery and parts in storage and transit. These are normally referred to as process lubricants.

Although most of the hydraulic systems are designed to operate on oils of various types as the medium for transmitting power, other systems employ water and fire-resistant fluids. The fire-resistant fluids are classified into four categories—phosphate esters, oil-synthetic blends, water glycols, and water-in-oil emulsions.

SECTION 3

MACHINE ELEMENTS REQUIRING LUBRICATION

Lubrication is necessary at all points where one surface slides upon or rubs against another. This occurs in **bearings** which support sliding members or rotating shafts, in **gears** which have meshing teeth, and between **pistons** and the cylinders in which they operate. Every machine, however large and complex, has these three fundamental components that involve motion and require lubrication—bearings, gears and pistons.

There are two main types of bearings: plain, and anti-friction.

Plain bearings (Figure 12—4) are classified according to function as journal, guide, and thrust bearings.

Journal bearings support or operate against a rotating shaft. The portion of the shaft that contacts the bearing surface is known as the journal. As shown in Figure 12—4, journal bearings may be of the **solid, split, half** or **multipart** design.

Solid bearings, sometimes referred to as **sleeves** or **bushings,** are one-piece bearings.

Split bearings are divided lengthwise into two parts and can be replaced on a shaft more easily than a solid bearing because it is not necessary to slide them onto the shaft from one end.

Half bearings contact only one half of a journal, leaving the other half exposed. They are used when a load is applied downward vertically on to a shaft, as in some cranes wherein the entire weight of the crane and its load presses the bearings down against the journals of

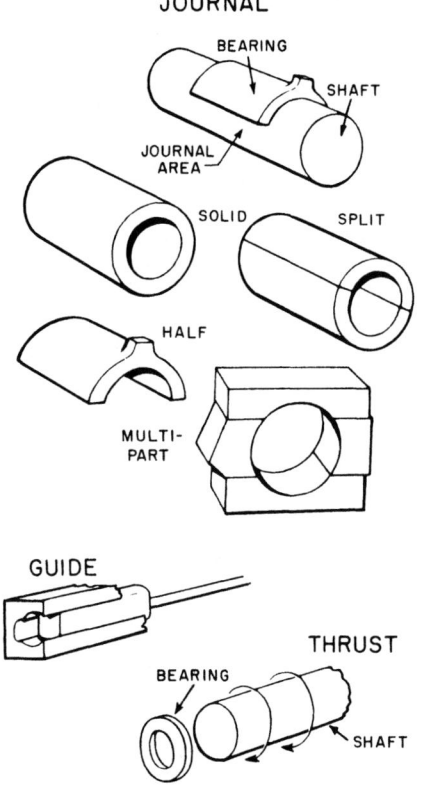

FIG. 12—4. Diagrammatic representations of various types of plain bearings. In the case of the thrust bearing, the sketch shows the shaft and bearing separated: actually, an axial load on the rotating shaft would force its end against one flat surface of the bearing while the other flat surface of the bearing would be supported against a fixed part of the machine.

the axles of the track wheels.

Multi-part bearings usually consist of four separate parts or quarters: These bearings are used on crankshafts of machines (e.g., air compressors); the

four-piece construction permits these bearings to be adjusted to restore a proper fit with the crankpin to compensate for wear.

Guide bearings guide and hold in proper position a reciprocating part of a machine such as a slide that has a back-and-forth motion.

Thrust bearings prevent a shaft from moving endwise. Plain bearings usually are made from bronze, babbitt or some other material that is softer than the steel shaft they support, so that wear will take place on the bearing in preference to the steel part that rubs against it.

ANTI-FRICTION BEARINGS

Anti-friction bearings are ones in which a series of rollers or balls is interposed between the moving parts. The rollers or balls are usually, but not necessarily, mounted in a cage or separator and enclosed between rings known as races. Whether rollers or balls are employed in their design determines whether the two principal types of anti-friction bearings are called **roller bearings** or **ball bearings**.

In the usual anti-friction bearing installation, the outer race (see Figure 12—5) is held stationary in the machine, while the inner race is fixed tightly on, and rotates with, the shaft. When the shaft is rotating, the rollers or balls are rolling in contact with both the inner and the outer races. This rolling action takes place with very little friction, compared to the sliding friction of a journal bearing.

The **straight roller bearing** (Figure 12—5) has the axis of each roller parallel to the axis of the bearings, and is intended to support only radial loads. The **tapered roller bearing** both supports the shaft and also prevents the shaft from moving endwise; it thus can support both radial and thrust loads.

Ball bearings are intended for radial loading. **Ball thrust bearings** (Figure 12—5) are designed to prevent a shaft from moving endwise.

Needle bearings differ from the others in that they have no inner race and no separator or cage. The small

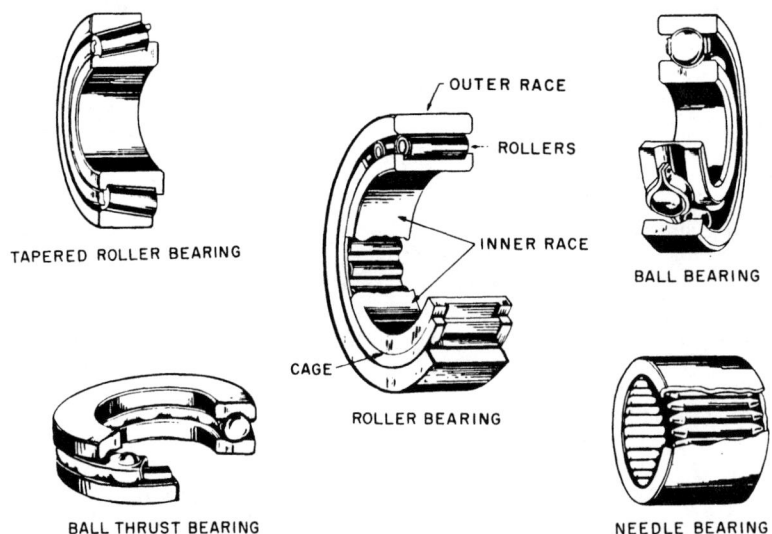

FIG. 12—5. Cut-away sketches illustrating several types of anti-friction bearings.

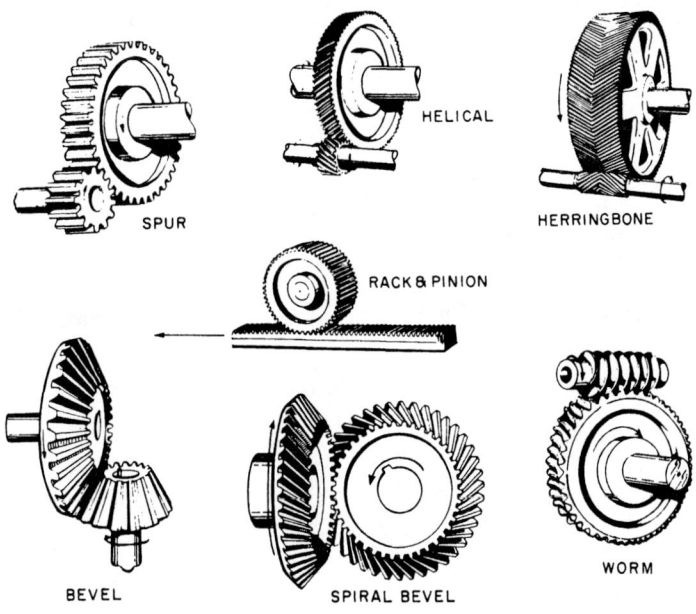

FIG. 12—6. Schematic representations of gear types.

rollers or needles are just slightly separated by the lubricant. The name "needle bearing" derives from the fact that the length of each roller is so much greater than its diameter.

The load-carrying parts of anti-friction bearings (balls, rollers, races) are made of through-hardened or case-hardened steel.

GEARS

Gears serve to transmit motion from one shaft to another; by selecting gears of the proper pitch diameter, the driven shaft may have the same or a different speed of rotation than the drive shaft. Figure 12—6 illustrates pairs of gears of the several basic types.

Spur gears have teeth parallel to the shaft or axle. A small gear that meshes with a larger gear is usually called a **pinion gear.**

Helical gears have their teeth cut at an angle on the face of the gear. When helical gears are in mesh, more teeth are in contact at one time than is the case with spur gears, and smoother operation results.

Herringbone gears have the appearance of two helical gears placed side by side, with the teeth of one gear cut in a direction opposite to the teeth of the other. These gears provide smooth operation and prevent the end thrust on a shaft that would be present if only a helical gear were used.

Bevel gears transmit motion between shafts that are at an angle to each other. **Straight bevel gears** have straight teeth on a slanted or beveled working surface. **Spiral bevel gears** have spiral teeth and give smoother operation than straight bevel gears.

In a **worm gear set** (Figure 12—6), the small element called the **worm** drives the large element known as the **worm wheel** or **worm gear.** Such sets transmit power from one shaft to another at right angles to it.

The **rack and pinion** combination is used to translate rotary motion of the pinion into reciprocating linear motion. The long rack (Figure 12—6) moves back and forth as the pinion is rotated in one way or the other.

A **screw and nut combination** may be considered a

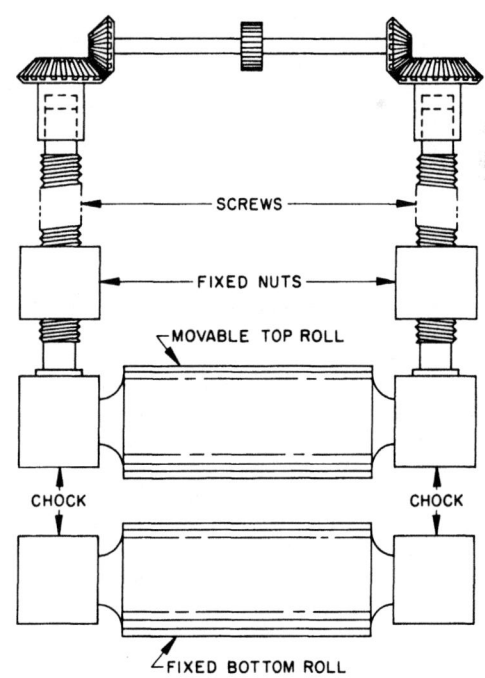

FIG. 12—7. Diagrammatic representation of the principle of operation of the screwdown of a rolling mill. Reduction gearing turns a shaft that causes the two screws to rotate, resulting in a simultaneous up or down motion of both screws as they turn in the fixed nuts.

type of gear, since the threads in the nut translate rotary motion into linear motion, as shown schematically in Figure 12—7, which illustrates how such a combination is used to adjust the position of the rolls in a rolling mill. In this case, the screw is lowered or raised by rotating it in a stationary nut. Two other types of motion are possible with the screw and nut combination: (1) if the screw is prevented from moving lengthwise while it is rotated, the nut will move along the screw provided the nut is prevented from turning with the screw, and (2) if the nut is rotated while held in a fixed position, the screw will move back or forth if it is prevented from rotating with the nut.

PISTONS

The third fundamental moving part found in machinery is the piston that operates in a cylinder. There are many types of hydraulically operated equipment in a steel plant, employing various fluids for transmitting pressure to actuate pistons in cylinders. The motion of the piston may be used to exert powerful forces for moving heavy masses; for opening and closing valves, furnace doors, etc.; for balancing the top rolls and for operating the manipulators on some rolling mills; and so on. In addition, pistons operate in the cylinders of internal-combustion engines, gas and air compressors, and pumps of various sorts.

SECTION 4

TYPES OF LUBRICANTS

Most of the lubricants used in the steel industry are derived from crude petroleum, the origins and characteristics of which are discussed in Chapter 3 under "Liquid Fuels and Their Utilization." The crude petroleum is subjected to refining which separates it into various components or fractions by the use of heat and pressure. One hundred cubic metres of crude oil will yield the following approximate amounts of products in the refining process:

Product	Cubic Metres
Gasoline	44
Fuel oils	36
Kerosene	6
Lubricating oils	3
Miscellaneous (coke, asphalt, wax, etc.)	8
Loss	3

The proportions of the products will vary with the type of crude oil being refined and the refining methods used.

LUBRICATING OILS

Lubricating oils are recovered only from paraffin-base, naphthene-base or mixed-base types of crude petroleum; these types of petroleum are defined in Chapter 3. In general, lubricating oils refined from paraffin-base crudes have the best stability and resistance to oxidation at ordinary temperatures; those refined from naphthene-base crudes are the next most stable, followed by those refined from mixed-base and aromatic-base crudes. Stability is dependent also upon the type of refining operation used and the additives or inhibitors that may be incorporated with the refined oil during subsequent compounding.

Blending—It is not practical to operate the distilling equipment of a refinery to produce directly lubricating oils having precisely the desired viscosity characteristics. Instead, the lubricating-oil fractions of different viscosity from several levels of the fractionating (distilling) tower are collected separately and later **blended** in various proportions to obtain a product of the desired viscosity.

Compounding—Compounding consists of incorporating additives with lubricating oils to impart characteristics not obtainable by straight refining operations.

Stabilizers or oxidation inhibitors are added to increase resistance of an oil to oxidation, and serve to retard formation of sludge, gum, varnish and other objectionable products of oxidation.

Inhibitors may be added to lubricating oils to prevent corrosion of metal parts by water in the lubricating system. Preferential wetting of a metal surface by such inhibitors keeps water from contacting the surfaces.

Detergent-type lubricating oils have dispersion agents added to them to keep in suspension in the oil the products of oxidation, dirt and dust that accumulate in an engine or lubricating system during operation. Since the contaminants are kept in suspension, they cannot settle out and thus are kept from forming harmful deposits.

Various agents are added to lubricating oils after refining to increase their film strength and assist them in maintaining a lubricating film between the rubbing surfaces of exceptionally heavily loaded machine elements to minimize galling and wear. Oils with these additives are referred to as **extreme-pressure oils.**

Wetting agents are added to lubricating oils to increase their spreading or wetting ability; addition of wetting agents to rolling oils used to lubricate the surfaces of steel during cold reduction is an example.

Anti-foaming agents may be added to lubricants in circulating systems to prevent foaming of oil in reservoirs.

Other agents are used to improve the performance of lubricants at low temperatures, or impart some other beneficial characteristic for a specific purpose.

Antiwear agents are added to lubricating oils to meet the demand for fluids that will serve under the most severe conditions in fluid-power systems that utilize high-performance pumps. They are especially recommended for hydraulic systems that operate under high pressure, shock loads, high temperatures, and/or high rpm.

LUBRICATING GREASES

Lubricating greases are made by combining lubricating oils with specially made metal-base soaps to form chemical gels. Water forms the bond between the soap and oil phases of greases, except in a few cases. The soaps are made by the saponification, under heat and pressure, of animal or vegetable fats or fatty acids, usually with an alkaline compound of one of the following metals: aluminum, barium, calcium, lithium, sodium or strontium. Mixed-base soaps (e.g., containing both calcium and barium soaps) also are used. An exception to the greases containing metal soaps is a series of greases consisting of mechanical gels produced by combining lubricating oils with bentonite (clay). A few greases are made using synthetic oils instead of petroleum-base oils.

Greases vary in consistency over a wide range; some have the appearance and characteristics of a very fluid oil while, at the other extreme, are the very hard greases known as **block grease**. The type and content of soap in grease is selected to give the desired consistency to the product. The lubricating properties of a grease reflect the type and quality of the lubricating oil used in its manufacture. However, the over-all lubricating properties of a grease must be ascribed to the grease as a composite, and not to either the soap phase or the oil phase by itself.

Fluid greases to be used in lubricating surfaces to resist galling and wear under heavy working loads are made by adding compounds of lead, sulphur or other elements: these are referred to as **extreme-pressure greases.**

SECTION 5

PHYSICAL AND CHEMICAL PROPERTIES

The ability of a lubricant to perform satisfactorily under a given set of conditions is related to its physical and chemical characteristics. Numerous laboratory tests have been developed for measuring chemical and physical properties in ways that permit expressing the degree to which a lubricant possesses a given property by a numerical value which enables different lubricants to be evaluated on a comparative basis.

It usually is necessary to determine several of the properties of a lubricant to establish whether or not it will meet requirements. The American Society for Testing and Materials has adopted over 60 procedures for conducting standardized tests to determine specific properties of lubricants. In addition, other methods have been developed by producers and users of lubricants and others involved in the study of lubrication problems. Space does not permit description or even listing of all of these methods here; reference may be made to the sources of information listed at the end of this chapter.

Following are some of the more important properties of lubricating oils and greases.

IMPORTANT PROPERTIES OF LUBRICATING OILS

Viscosity—Viscosity, defined as resistance of an oil to flow or shear, is measured by several methods namely Kinematic, Engler, Redwood No. 1 and No. 2, and Saybolt. In all cases, it is a time measurement of flow through a calibrated orifice at a standard temperature.

In the past the American Society for Testing and Materials (ASTM) used a standard viscosity scale of Saybolt Universal Seconds (SUS) at 37.8°C (100°F). However, a new viscosity system has been developed—International Standards Organization (ISO) grade numbers. The new standard is based on kinematic viscosities in mm²/s (centistokes, cSt) at 40°C (104°F).

This system is applicable to fluids ranging in kinematic viscosity from 2 to 1500 mm²/sec (cSt) as measured at a reference temperature of 40°C. Expressed in approximate equivalents the range would be 32 to 7000 SUS. In the category of petroleum-based fluids, this covers the range from kerosene to heavy cylinder oils.

Table 12—1 lists the 18 ISO viscosity grades and equivalent kinematic (in mm²/s or centistrokes at 40°C) and Saybolt (in SUS at 40°C or 104°F) viscosities.

The ability of a lubricating oil to form and maintain a fluid film depends primarily on its viscosity. In general, heavy (high-viscosity) oils are used on parts moving at slow speeds under high pressure because it better resists being squeezed out from between the rubbing surfaces. Light (low viscosity) oils are used when higher speeds and lower pressures are encountered, because they do not impose as much drag on high-speed parts and the high speed permits a good oil wedge to form (Figure 12—3) even though the viscosity of the oil is low.

Low-temperature conditions usually require light oils and high-temperature conditions require heavy

Table 12—I. Viscosity Conversion Chart

ISO Viscosity Grade	Kinematic Viscosity, mm²/s (Centistokes) at 40°C (104°F)	Saybolt Viscosity SUS at 104°F (40°C) (approx.)
2	1.98–2.42	32
3	2.88–3.52	36
5	4.14–5.06	40
7	6.12–7.48	50
10	9.00–11.0	60
15	13.5–16.5	75
22	19.8–24.2	105
32	28.8–35.2	150
46	41.4–50.6	215
68	61.2–74.8	315
100	90.0–110	465
150	135–165	700
220	198–242	1000
320	288–352	1500
460	414–506	2150
680	612–748	3150
1000	900–1100	4650
1500	1350–1650	7000

oils, because viscosity decreases with increasing temperature and increases with decreasing temperature. **Viscosity index** is an empirical number indicating the rate of change in viscosity of an oil with change in temperature within a given range. A low viscosity index signifies a relatively large change of viscosity with temperature while a high viscosity index signifies the opposite.

Viscosity of a lubricating oil may change with use due to oxidation, entrainment of solid particles, or dilution by contaminants. Suitable tests at intervals dictated by service requirements will determine when the viscosity has changed sufficiently to impair the lubricating properties of an oil.

Pour Point—Pour point is another property of an oil that is important if the oil is to be used to lubricate machinery that operates in unheated buildings or out-of-doors during cold weather. It is the lowest temperature at which a given oil will pour or flow under prescribed conditions when it is chilled without disturbance at a fixed rate. Some oils with high pour points will cease to flow at 4°C while others with low pour points will still flow freely at temperatures below –18°C.

Oxidation Resistance—The constituents of lubricating oils, which are a complex mixture of hydrocarbons, will combine more or less slowly with oxygen in the air, especially if subjected to heat. The slow combination of hydrocarbons with oxygen is called **oxidation,** and the new substances formed, called **products of oxidation,** usually have no lubricating value in themselves and even may affect seriously the performance of the remaining unoxidized lubricant.

Service conditions involving churning and splashing of an oil and its exposure to temperatures much above room temperature will accelerate oxidation. Lubricating oils that must remain in service for long periods of time without change, as in turbines and large circulating systems, must have high resistance to oxidation and require extra refining techniques. Where an oil remains in service only a short time, or new oil frequently is added as makeup, lower resistance to oxidation may be acceptable.

Flash Point—Flash point is the temperature at which a flash (momentary ignition) occurs at any point on the surface of an oil when a flame is passed over it, but the oil does not continue to burn.

Fire Point—The fire point is the temperature at which ignition occurs at any point on the surface of an oil when a flame is passed over it and the oil continues to burn. This temperature is about 28°C above the flash point.

Demulsibility—When water enters a lubricating system and is churned up with the oil, an emulsion is formed that has poor lubricating properties. If the oil is clean, it will separate from the emulsion if the latter is allowed a rest period; this characteristic is known as the demulsibility of the oil. Demulsibility is good if the oil separates quickly and poor if it separates slowly. The presence of products of oxidation may retard or even prevent the separation process.

Neutralization Number—High acidity makes oil detrimental to bearing surfaces, and may be indicative of undesirable chemical changes that are taking place in a lubricant. The neutralization number is the number of milligrams of potassium hydroxide necessary to neutralize the acid content of one gram of oil under standard test conditions.

IMPORTANT PROPERTIES OF LUBRICATING GREASES

Grease Hardness—Grease hardness, sometimes referred to as **consistency** or **penetration,** is a measurable characteristic affecting selection of a suitable grease for a particular lubrication requirement. The National Lubricating Grease Institute (NLGI) has established a series of numbers to rate the various grades of grease for consistency (see Table 12—II). The penetration

Table 12—II. National Lubrication Grease Institute Grease Classification

NLGI Number	ASTM Worked Penetration Units
0	355 – 385
1	310 – 340
2	265 – 295
3	220 – 250
4	175 – 205
5	130 – 160
6	85 – 115

units referred to in the table are obtained from a penetration test. **Penetration** is the consistency or hardness of a grease measured by the distance a standard cone will penetrate the grease by a free fall at a standard temperature. It is a measure of the plasticity of a grease. The grease is brought to the proper temperature in a container that is placed directly beneath the cone tip. Release of a supporting plunger allows the cone to drop into the grease and the depth to which it penetrates is measured in tenths of millimeters (penetration units). The grease is "worked" mechanically in a prescribed way before the test is performed.

Grease designated as No. 0 is soft, and the hardness increases through No. 1 and No. 2, and so on, up to No. 6 which has the consistency of a cake of soap.

Pumpability—Pumpability or grease mobility is an important property of grease that represents ease of pumping, as in centralized grease systems where a pump must force grease through lines and small passages in valves as it is distributed to machine bearings. Pumpability is related to the kinds and proportions of oils and soaps in grease, and is affected by temperature.

Water Resistance—Some lubricants must perform in the presence of water. Greases containing calcium-base and lithium-base soaps will not dissolve in water, but those containing sodium-base soaps will; sodium-base soap greases therefore are not used where there is a possibility of contact with water. Finely powdered graphite is added to some greases to improve their performance on plain bearings in the presence of water.

Dropping Point—Dropping point of a grease is the temperature at which it passes from a semi-solid to a liquid state. This characteristic is dependent upon the type of soap used in a grease. Greases containing calcium-base soaps melt in the 71°C to 99°C range.

Greases containing sodium-base soaps melt in the 135° to 177°C range. Greases with lithium-base soaps are generally used where temperatures are expected to be relatively high.

Stability—Some greases, when placed in a bearing, will retain their original hardness; that is, they are stable and will stay in antifriction bearings very well and give good lubrication for a long time. Other greases lose their hardness after being worked in a bearing for some time, become quite fluid, and may even run out if the bearing is not well sealed. Obviously, grease that retains its hardness and stays in the bearing is the more desirable; such a grease is considered to have good stability.

SECTION 6

LUBRICATION METHODS

Introductory—Because of the many differences in the properties of lubricants and in the machines to be lubricated, considerable knowledge and study are necessary for selection of the proper lubricant for a particular machine. In the larger plants, a lubrication engineer makes the selection, basing decisions on the specifications of the machine manufacturers, recommendations of lubricant manufacturers, and knowledge of the particular conditions under which a machine will operate. A lubrication chart then is prepared for each of the different items of plant equipment. Such a chart lists the different parts of a machine to be lubricated, the method of applying the lubricant, and the lubricant to be used. Sometimes the frequency of lubricant application also is listed, but in other instances this is not specified because it is dependent upon whether a machine operates continuously or intermittently and on the amount of leakage of lubricant that is characteristic of a given machine.

OIL APPLICATION METHODS

Various methods are used for applying oil to steel-plant equipment.

These methods can be classified as:
1. Once-through oiling.
2. Oil reservoirs.
3. Circulating-oil systems.
4. Other oiling methods.

ONCE-THROUGH OILING

In **once-through oiling**, the oil passes through the bearing only once and is lost for further use. Once-through methods include hand oiling, drop-feed oiling, wick-feed oiling and bottle oiling, which are illustrated schematically in Figure 12—8.

Hand oiling is the direct application of oil to a moving machine part from a hand oil can and is much used on older equipment. It is also used on newer equipment with small bearings involving little movement. The method has a disadvantage in that an excess of oil is applied which soon runs off, leaving the bearing to operate with insufficient oil until the next oiling.

When a more uniform supply of oil is required, a **drop-feed** oiler may be used. It consists of a shut-off lever, feed adjustment, oil chamber, needle valve and sight glass. The shut-off lever is adjusted to regulate the rate of feed of oil.

The **wick feed** oiler consists of an oil reservoir and a wool wick. The wick draws oil from the reservoir and feeds it into an opening in the bearing. The amount of oil delivered can be regulated by changing the size of the wick.

The **bottle** oiler has an inverted glass bottle mounted above the bearing and fitted with a sliding pin that rests on the journal. When the journal rotates, it vibrates the pin to encourage flow of oil from the bottle to the bearing through the space between the pin and its sleeve.

OIL-RESERVOIR METHODS

The reservoir methods use the same oil over and over again. All of these methods make use of a supply of oil held in the base of a gear casing or bearing, as illustrated schematically in Figure 12—9.

In the **ring oiling** method, a metallic ring, larger in diameter than the journal, rides on the journal and turns as the journal rotates. The ring, dipping into the oil, carries it to the top of the journal where it flows along and around the journal, providing lubrication before returing to the reservoir.

Chain oiling is similar to ring oiling except that a small-linked chain is used instead of a ring. The chain will carry a larger volume of oil than a ring.

An oil **collar** may be used to carry oil from the reservoir to a journal turning at a speed so high that rings and chains would slip. The collar, fastened to the journal, dips into the oil as the journal rotates, carrying the oil to an overhead scraper that removes and distributes it along the journal.

A group of bearings and gears enclosed in a single oil-tight casing usually employs **splash oiling** for lubrication. In this method, some moving part is in direct contact with the oil in the bottom of the casing. As the moving part contacts and then passes through the oil, it splashes and carries the oil up to the other parts within the casing. Splash oiling is one of the most reliable methods for lubrication. One kind of splash oiling is that in which the connecting-rod bearing and crankpin are completely immersed in oil when in their lowest position. The oil is splashed by the crankshaft and counterweight. Another kind of splash oiling is that in which a gear or worm dips into the oil and carries it to the moving part with which it is in contact; this method is often called **dip oiling**.

CIRCULATING-OIL SYSTEMS

Many large machines and certain smaller ones are equipped with circulating oil systems that make use of pumps and piping to deliver oil under pressure and

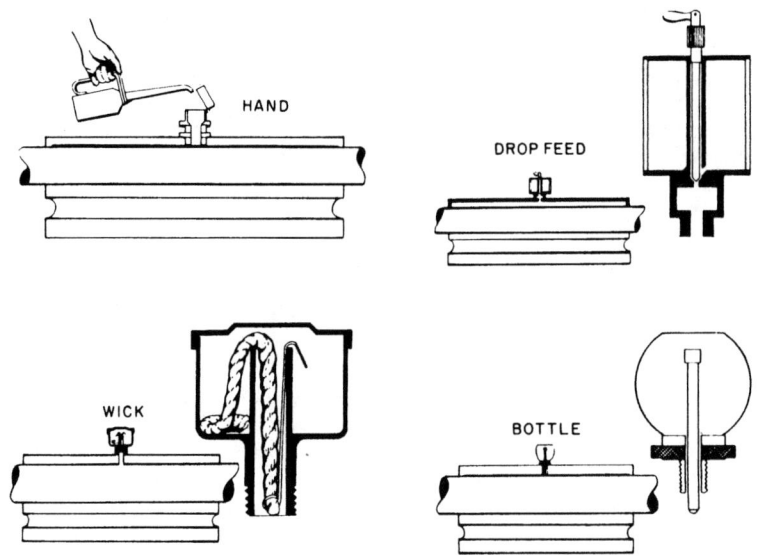

FIG. 12—8. Schematic representation of once-through oiling methods.

often in large quantities to the moving parts. Large motors and many parts of rolling mills where lubrication is of extreme importance usually are equipped with circulating systems. There are two general types of circulating systems, gravity type and pressure type.

In **gravity-type circulating systems,** the lubricant flows by gravity from overhead tanks to those parts of a machine requiring lubricant, then drains into a sump at a lower level than the machine from which it is pumped back to the overhead tanks. The oil may be strained and cooled in its passage through the system.

Figure 12—10 illustrates the principle of **pressure-** **type circulating systems,** such as used to lubricate heavy rolling-mill equipment. In this illustration, oil flows over the gear teeth and through the bearings and then falls into a settling tank located in the basement. As the oil flows slowly across this tank, much of the dirt and water settle to the bottom. The oil is then withdrawn from near the surface and pumped through a filter for additional cleaning. The pressure tank is partly filled with air; its function is to prevent sudden changes in pressure in the system. From the filter, the oil passes through an oil cooler and is delivered under pressure to the point at which it starts its free flow to

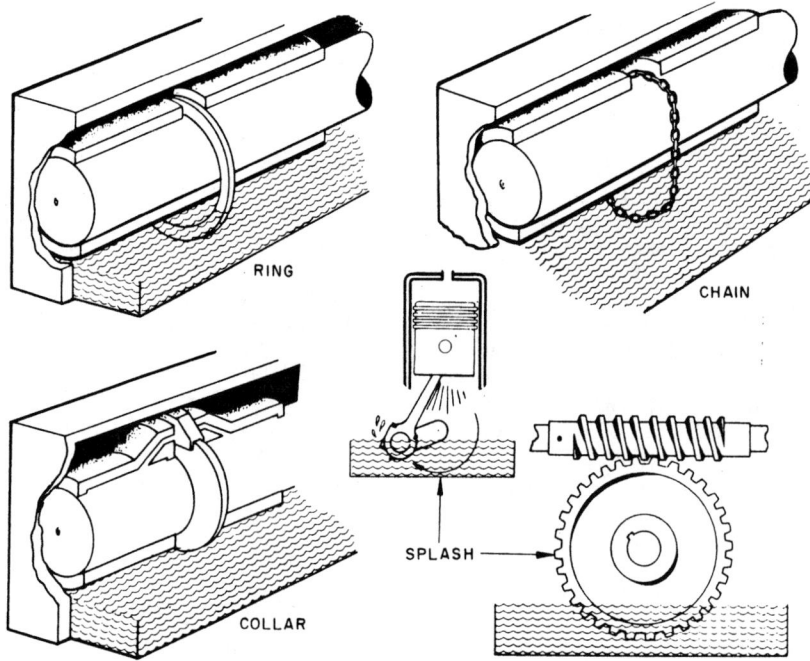

FIG. 12—9. Schematic representation of various ways in which oil from a reservoir can be transferred to bearing surfaces.

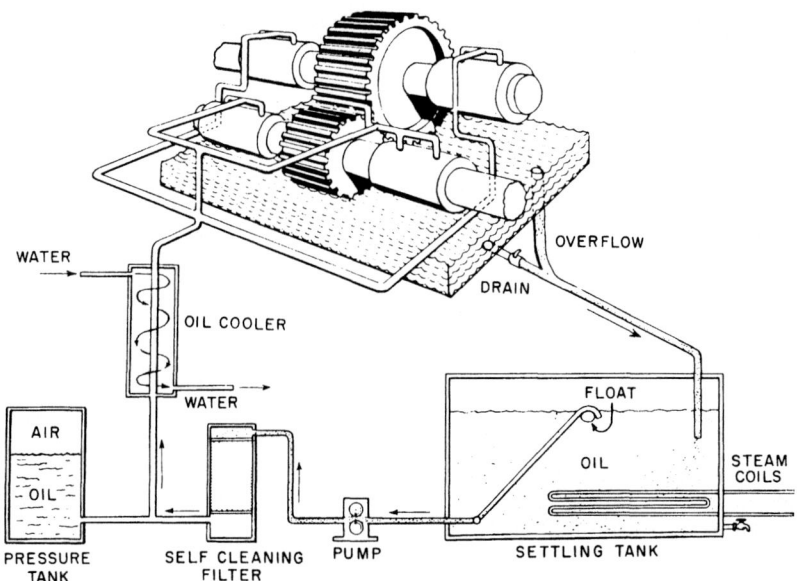

FIG. 12—10. Schematic arrangement of the principal components of a pressure-type oil-circulating system.

perform its lubricating function. The fact that the lubricant is delivered under pressure distinguishes the pressure-type systems from the gravity systems.

OTHER OILING METHODS

The **oil mist** procedure for lubrication uses compressed air to atomize oil from a reservoir and deliver it as a mist through pipes to the bearings and gears. The oil-saturated atmosphere provides lubrication, and the air passing through the system helps to carry away heat as well as prevent entrance of dirt.

Mechanical force-feed lubrication is used for lubricating large pistons and cylinders (e.g., in a large air compressor), where the lubricant must be supplied in small but accurately controlled amounts, against the pressure in the cylinder, to provide adequate lubrication without flooding. Force-feed lubricators consist of an oil reservoir and several individual pumping units that deliver controlled amounts of oil through pipes to points along the sides of the cylinder. The pumping units are operated from a common drive shaft which may be driven by an electric motor or by some moving part of the machine being lubricated. Each pumping unit can be adjusted to deliver exactly the quantity of oil desired. This type of lubricator, originally designed to deliver oil to cylinders as just described, is often used to supply fluid lubricants to bearings and open gears where delivery of small, controlled amounts of lubricant is desired.

Wool waste and special **felt pads,** kept saturated with oil, are used to lubricate track-wheel journals on cranes and railroad cars that are equipped with half bearings.

GREASE APPLICATION METHODS

Greases are semi-solids and pressure must be applied to most of them to cause them to move or flow, while oils flow of their own accord.

Under certain conditions, grease is the preferred lubricant. Some of these conditions are:

(1) Where a machine is so designed that there is no way to retain oil for the parts being lubricated. Examples of machine elements where this is the case are open gears and many open guide bearings.

(2) When the lubricant must act also as a seal to prevent the entrance of dirt into a bearing. Grease will maintain a seal at the ends of a bearing where oil would quickly run out.

(3) Where a lubricant is seldom added, as in electric-motor bearings.

(4) Where speeds are low and pressures are high, as in some types of roll-neck bearings.

There are basically three methods of applying grease: hand greasing, hand operated mechanical greasing devices, and centralized grease systems.

Hand greasing is frequently used during the assembly of a machine. Grease is spread by hand over bearings and gears to protect them from rusting and to ensure the availability of lubricant when the machine is started for the first time. Ball and roller bearings must have grease worked by the fingers into the spaces between the balls or rollers as well as receiving an over-all coating. In another hand method, melted grease is poured from a suitable container onto gears or open guide bearings.

Hand-operated mechanical devices are used to deliver grease to one point at a time; these include grease cups and fittings through which grease is applied.

The **screw-drive grease cup** has a small reservoir for holding grease and a plate that screws down into the reservoir to exert pressure on the grease and force it into a bearing. It is filled by unscrewing the plate and filling the reservoir with grease.

The **compression-type grease cup** has a spring-loaded plate that maintains pressure on the grease and forces it slowly into the bearing.

Grease fittings, which make possible the use of pressure grease guns, provide a normally closed opening into the housing of a bearing. They have replaced grease cups on most steel-plate equipment because they shorten the time required for greasing. The inner construction of all grease fittings incorporates a spring and a steel ball. When grease is forced into the fitting, the spring is compressed, allowing the grease to enter the bearing housing. When the pressure is released, the spring regains its original no-load position, forcing the steel ball against the opening through which the grease entered and effectively sealing the opening against escape of grease or entrance of dirt.

Hand-operated **grease guns,** hand-operated **bucket pumps,** or **barrel pumps** operated by air or electric motors are used to force grease into the fittings.

Spray guns similar to those used for spraying paint use compressed air to spray gear teeth with a film of grease.

Centralized grease systems provide the best method for supplying grease to a number of bearings. The number of bearings serviced by such a system can be quite large. A system consists of a centrally located grease reservoir with a pump and permanently installed piping that carries the grease under pressure to measuring and distributing valves that feed the grease to the bearings in measured amounts at predetermined intervals. Feeding of grease takes place only when the pump is operated.

The pump may be operated by hand, by compressed air, or by an electric motor. If the pump is air- or motor-operated, the system may be equipped with a timer that causes the pump to function at predetermined intervals.

Some of the advantages of centralized grease lubrication are:

(1) All bearings are assured of lubrication and each receives the proper amount of lubricant.

(2) Lubricant can be applied more frequently than would be practical with hand-operated mechanical greasing devices.

(3) Down-time of equipment is reduced since it can be lubricated while in operation.

(4) Much less time is required than for other lubricating methods.

Centralized grease systems utilize either of two general principles for supplying grease: in one, the valves that measure and distribute the grease to the bearings are located in series in the main supply lines; in the other, the valves are fed in parallel from the main supply lines. In systems with the valves in series, the flow of grease through the main supply line depends upon the successive action of each valve in the line, while in systems with the valves in parallel, the flow through the main supply lines is independent of the performance of any valve. The failure of lubricant flow through any valve in the parallel system is detected by an indicator on each valve. In systems with the valves in series, a gauge or alarm at the supply pump indicates, by an excessive pressure build-up, the failure of flow in the system.

Due to pressure drop in the supply lines and practical limitations as to pressure developed by the pump, the length and size of the supply lines limit, in some cases, the number of lubrication points that should be included in a centralized grease system. In a modern hot-strip mill, for example, it would be completely impractical to lubricate the entire rolling-mill complex by a single system; instead, several grease systems will be located strategically to supply different parts of the complex. In this way, the number of points to be fed by each system can be limited and the distances between the pump and the points can be kept within practical limits.

Bibliography

American Society of Lubrication Engineers, Handbook of Lubrication, Volume I and Volume II. The Society, Park Ridge, Illinois.

American Society for Testing and Materials, ASTM Standards. The Society, Philadelphia, Pa. (latest edition.)

Anon., Solid lubricants: specials ready for standard roles. Steel **156,** No. 20, May 17, 1965, 44-48.

Azzam, Hani T., Evaluation of solid lubricants, oils and greases. Iron and Steel Engineer **45,** No. 4, April 1968, 115-122: discussion, 122.

Barwell, F. T., Friction and its measurement. Metallurgical Reviews **4,** No. 14, 1959, 141-177.

Bisson, Edmond E., et al., Friction, wear and surface damage of metals as affected by solid surface films. National Advisory Committee for Aeronautics Report 1254, 1956.

Cichelli, A. E., Mill gearing as viewed by a lubrication engineer. Iron and Steel Engineer **35,** No. 9, September 1958, 91-102.

Cichelli, A. E., Plant lubrication, an engineering approach. Iron and Steel Engineer **42,** No. 9, September 1965, 95-106.

Devine, M. J., et al., Improving frictional behavior with solid film lubricants. Metals Engineering Quarterly **7,** No. 2, May 1967, 33-43.

Gesdorf, E. J., Development and application of spray lubrication. Iron and Steel Engineer **35,** No. 5, May 1958, 115-126.

Jones, H., and G. D. Jordan, Lubrication in iron and steelworks. Journal of the Iron and Steel Institute **185** (Part 3), March 1957, 389-408: discussion; Journal of the Iron and Steel Institute **187** (Part 2), October 1957, 128-143.

Jost, H. Peter, Modern British and European steelworks lubrication developments. Iron and Steel Engineer **43,** No. 5, May 1966, 88-108.

Ling, F. F., Mechanics of sliding surfaces. Metals Engineering Quarterly **7,** No. 2, May 1967, 1-3.

McCandless, O. G., Some concepts of mill bearing lubrication employing oil, plastic grease and oil mist. Iron and Steel Engineer **38,** No. 9, September 1961, 171-176.

Murphy, T. M., Application of oil mist lubrication in the steel industry. Iron and Steel Engineer **33,** No. 12, Dec. 1956, 77-81.

Pope, Charles L., The mechanics of lubrication. Iron and Steel Engineer **33,** No. 9, September 1956, 133-135.

Rabinowicz, Ernest, Friction and wear of metals. Metals Engineering Quarterly **7,** No. 2, May 1967, 4-8.

Ritchie, G. O., and J. A. Robertson, Special lubrication requirements of iron and steel works. Institution of Mechanical Engineers, 3rd annual meeting. Lubrication and Wear Group, Paper 18, October 1964.

Society of Automotive Engineers, Inc., SAE Handbook. The Society, New York, N.Y. (Latest edition.)

United States Steel Corporation, The Lubrication Engineers Manual. United States Steel, Pittsburgh, Pa. 1971.

CHAPTER 13

Physical Chemistry of Iron- And Steelmaking

Introduction

In writing this chapter, an attempt has been made to limit the discussion to an average level suitable for the students of metallurgy pursuing graduate or post-graduate education as well as for those with some scientific background engaged in the iron and steel industry. It is assumed that the reader has some basic knowledge of chemistry, physics and mathematics, so that the chapter can be devoted solely to the discussion of the chemistry of the processes.

The chapter consists of six parts. The first part describes briefly the fundamentals of thermodynamic concepts and physical chemistry. The second part gives many of the properties of metallurgical systems. The systems are divided into metals and alloys, slags and refractories, and gases. Among the properties listed for these systems are phase equilibria, thermodynamics, diffusivities, surface tension and viscosity. Part III of the chapter is concerned with ironmaking reactions occurring in the blast-furnace and direct-reduction processes. Part IV deals with steelmaking reactions with particular emphasis on the newer steelmaking processes such as BOP and Q-BOP. The deoxidation and ladle reactions are discussed in Part V; included are discussions on oxygen sensors, deoxidation equilibria, ladle desulphurization and degassing. In Parts III, IV and V the emphasis is on the chemical reactions and not the processes. The processes are discussed in detail in other parts of this book. Part VI discusses nonmetallic inclusions in steel.

PART I

Thermochemistry and Fundamentals

THERMODYNAMIC CONCEPTS

Ideal and Nonideal Gases—As a limiting case, a hypothetical gas which obeys the simple gas laws —**Boyle's law** (1662) and **Gay-Lussac's law** (1802)—is called an ideal gas or a perfect gas satisfying the relation

$$PV = nRT \qquad (1)$$

where P, V and T are pressure, volume and absolute temperature respectively; n is any arbitrary number of moles and R is the molar gas constant.

Based on **Avogadro's law** (1811), one mole of an ideal gas (containing $N = 6.0247 \times 10^{23}$ molecules) occupies a volume of 22.414 litres at 1 atm pressure and $273.16°K$ ($\equiv 0°C$). Using these units, $R = 0.08205$ litre-atm per deg per mole. In terms of SI units, P is in newton·metre^{-2} ($N \cdot m^{-2} \equiv$ Pa, pascal, and 1 atm $= 1.01325 \times 10^5 \ N \cdot m^{-2} \equiv 1.01325$ bar) and V in metre^3, then $R = 8.314 \ J \cdot deg^{-1} \cdot mole^{-1} \equiv 1.987$ cal deg^{-1} mole^{-1}.

In a gas mixture containing $n_1, n_2, n_3 \ldots$ numbers of moles of gases 1, 2, 3 . . . occupying a volume of V at a total pressure of P, the partial pressures of the constituent gaseous species will be

$$P_1 = P \frac{n_1}{n_1 + n_2 + n_3 + \ldots} ; P_2 = P \frac{n_2}{n_1 + n_2 + n_3 + \ldots} \qquad (2)$$

From **Dalton's law** (1801),

$$P = p_1 + p_2 + p_3 + \ . \ . \qquad (3)$$

and using the ideal gas equation,

$$p_1 V = n_1 RT, \ p_2 V = n_2 RT, \ldots \qquad (4)$$

Generally speaking, deviation from the ideal gas equation becomes noticeable with easily liquefiable gases and at low temperatures and high pressures. The behavior of gases becomes more ideal with decreasing pressure and increasing temperature. The nonideality of gases, the extent of which depends on the nature of the gas, temperature and pressure, is attributed to two major causes: (1) van der Waals' forces and (2) chemical interaction between the different species of gas molecules or atoms.

The condition for the liquefaction of a gas was first discovered by **Andrews** (1869) based on a study of the pressure-volume-temperature relationship of carbon dioxide; this is shown in Figure 13—1, where pressure-volume isotherms are shown. At temperatures below the critical temperature, T_c, each isotherm consists of three distinct parts. For example, the isotherm abcd represents the relation between pressure and volume for the vapor ab, the co-existing vapor and liquid bc and the liquid cd respectively. Since liquid carbon dioxide is relatively incompressible, the cd part of the isotherm is nearly vertical. The volumes V_1 and V_2 are the specific volumes of the saturated vapor and the liquid under their own vapor pressures respectively. With increasing temperature the difference between the specific volumes of the liquid and vapor becomes progressively smaller until at T_c the liquid and vapor can no longer be distinguished from one another. The temperature above which a vapor cannot be liquefied is called the critical temperature, T_c, the highest vapor pressure exerted by the liquid is the critical pressure, P_c, and the critical volume is the volume of one mole of a liquid (or gas) at the critical temperature and pressure, V_c. On the basis of the critical phenomena, Andrews defined vapor and gas as gaseous species below and above their critical temperatures respectively; this is indicated by the vertically shaded area in Figure 13—1.

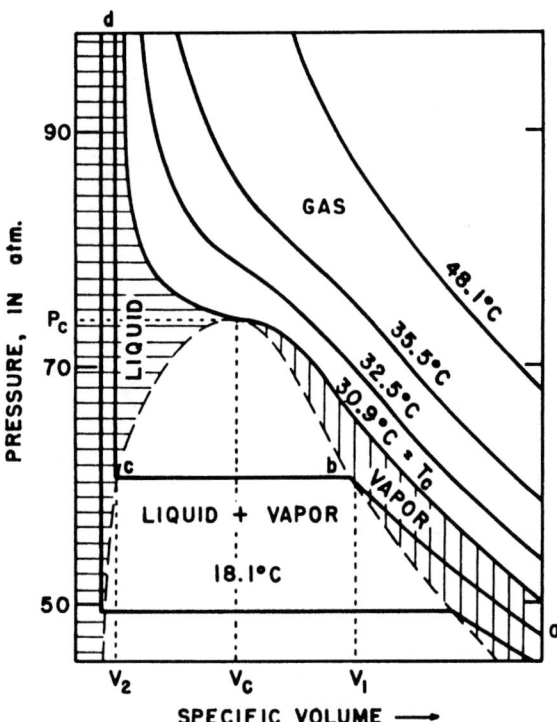

FIG. 13—1. Andrews' (1869) isotherms for carbon dioxide.

At temperatures just above the critical temperature, T_c, isotherms contain a kink, indicating the nonideal behavior of the gas. The pressure-volume-temperature relationship of one mole of gas and its transformation to the liquid state is most simply (though only approximately) formulated by the van der Waals' equation of state (1880), thus

$$\left(P + \frac{a}{V^2} \right) (V - b) = RT \qquad (5)$$

where a and b are constants for a given gas. At the critical temperature, i.e. the temperature above which a vapor cannot be liquefied, the derivatives $(\partial P / \partial V)_T$ and $(\partial^2 P / \partial V^2)_T$ must be zero and, therefore, the following expressions may be derived from equation (5) for the critical isotherm

$$P_c = \frac{a}{27b^2} , \; V_c = 3b \text{ and } T_c = \frac{8a}{27Rb} \qquad (6)$$

There are other equations of state for gases, but they will not be given here. The values of a and b can be found in reference books; the critical constants of some common gases are given in Table 13—I. It should be noted that the value of the critical ratio $P_c V_c / RT_c$, i.e. compressibility factor at the critical point, is about 0.29; this is often regarded as a universal constant for some simple gases and vapors, classified as spherical and nonpolar gases, e.g. monatomic gases and some diatomic gases.

Another cause of nonideality is the chemical interaction between the different species of gas molecules or atoms; this is discussed later.

Heat Capacity—The heat capacity of a substance is defined as the quantity of heat required to raise the temperature by one degree. The heat capacity of 1 g of a substance is called the specific heat and that of one gram-molecule (abbreviated as mole) is called the molar heat capacity.

For an ideal gas the difference between the molar heat capacities at constant pressure, C_p, and constant volume, C_v, is equal to the molar gas constant

$$C_p - C_v = R \qquad (7)$$

At low temperatures and high pressures this difference in molar heat capacities is somewhat higher than R because of departure from the ideality.

The classical law derived by **Dulong** and **Petit** (1801) from experimental observations states that "all solid elements have the same heat capacity per g-atom". Although the atomic heat capacities of most solid elements at room temperature are about 6.2 ± 0.4 cal per deg per mole, there are a number of exceptions to the rule: for example, at 25°C, atomic heat capacities of

Table 13—I. Critical constants of some gases.[1]

Gas	T_c (°K)	V_c (cm³/mole)	P_c (atm)	$P_c V_c / RT_c$
Helium	5.3	57.8	2.26	0.300
Neon	44.5	41.7	25.9	0.296
Argon	150.7	75.2	48.0	0.291
Hydrogen	33.3	64.5	12.8	0.304
Nitrogen	126.1	90.0	33.5	0.292
Oxygen	154.4	74.4	49.7	0.292
Carbon dioxide	304.3	95.6	73.0	0.279
Carbon monoxide	134.2	90.0	35.0	0.286
Methane	190.7	98.7	45.8	0.288
Water vapor	647.0	45.0	217.72	0.185

[1] Number refers to reference source in bibliography at end of chapter.

beryllium, boron, carbon (diamond) and silicon are 3.93, 2.65, 1.45 and 4.73 cal per deg per mole.

Another generalization is that known as **Kopps law** (1865) which states that "at ordinary temperatures the molar heat capacity of a solid compound is approximately equal to the sum of the atomic heat capacities of its constituents". Since little or no information is available on C_p for alloys, the approximation made in Kopp's law may be extended to alloys. That is, the heat capacity is a linear function of atom fraction.

The first detailed theoretical derivation of heat capacity was made by **Einstein** (1907); this theoretical treatment was later modified by **Debye** (1912), who showed that the following expression represents the molar heat capacity of solids satisfactorily at very low temperatures, thus

$$C_v = \frac{12}{5} \ \pi^4 R \ \left(\frac{T}{\Theta}\right)^3 \tag{8}$$

i. e.

$$C_v = 464.5 \ \left(\frac{T}{\Theta}\right)^3 \tag{9}$$

where Θ is constant for a given solid and has the dimensions of temperature and is called the characteristic temperature of the substance. Therefore, if C_v is plotted against T/Θ, the same curve should apply to all solid elements. This generalization applies essentially to isotropic solids (cubic systems). In the derivation of the Debye equation it was postulated that C_v for the solid was that contributed only by the vibrations of the atoms in a lattice assumed to be a continuous elastic medium. In this approximation, the electronic contribution to C_v and the effects of crystal structure and structural defects on C_v were not taken into consideration. From the free electron theory of metals, it is derived that the electronic contribution to heat capacity is directly proportional to the absolute temperature at low temperatures. Therefore, for solids which exhibit no anomalies, e. g. Curie point, lattice defects, etc.,

$$C_v = 464.5 \ \left(\frac{T}{\Theta}\right)^3 + \gamma T \tag{10}$$

where, depending on the material, γ has a value within the range 1×10^{-4} to about 50×10^{-4} cal per deg^2 per mole. With increasing temperature the electronic contribution to C_v becomes less pronounced. In the vicinity of room temperature, the Debye equation reduces to $C_v \rightarrow 3R$, i.e. approaching the value of Dulong and Petit.

Normally, the heat capacity is measured at constant pressure and in order to use equation (10), C_v is calculated from C_p by using the thermodynamic relation

$$C_p - C_v = \frac{\alpha^2}{\beta} VT \tag{11}$$

where α is the coefficient of thermal expansion and β is the coefficient of compressibility.

At temperatures much above 25°C, lattice and electronic contributions to heat capacity at constant pressure, C_p, are represented by an empirical equation

$$C_p = a + bT - cT^{-2} \tag{12}$$

where a, b and c are constants for a given substance

and they are evaluated from the experimental data on C_p over a wide temperature range.

Enthalpy (Heat Content)—The energy of a system, E, (frequently called internal or intrinsic energy) includes all forms of energy other than the kinetic energy. Any exchange of energy between a system and its surroundings, resulting from a change of state, is mainifested as heat and work.

Let us consider, for example, a system expanding against a constant external pressure, P, resulting in an increase of volume of ΔV; the work done by the system becomes $w = P\Delta V$. Since this is the work done by the system against the surroundings, the system absorbs a quantity of heat, q; the increase, ΔE, of the energy of the system in passing from state A to state B is,

$$\Delta E = E_B - E_A = q - P\Delta V = q - P(V_B - V_A) \tag{13}$$

$$(E_B + PV_B) - (E_A + PV_A) = q \tag{14}$$

For an infinitesimal change at constant pressure, we write

$$dE = \delta q + PdV \tag{15}$$

The quantity $E + PV$ is represented by a single symbol H, thus

$$H = E + PV \tag{16}$$

and

$$\Delta H = (E_B + PV_B) - (E_A + PV_A) \tag{17}$$

The H function is known as heat content or enthalpy. Most calorimetric measurements are made at constant pressure; hence no work other than that due to volume change is involved, and $\Delta H = q$ is called the heat content. However, the enthalpy is a more general term, and includes processes involving pressure changes.

The enthalpy is an extensive property of the system and, as in the case of energy, only the change in heat content can be measured. It is therefore, essential that a standard reference state is chosen for each element so that any change in the heat content of the element may be referred to its standard state, and this change is denoted by $\Delta H°$. The natural state of elements is by convention taken to be the reference state; on this definition, the elements in their standard states have zero heat contents. The heat of formation of a compound is then the heat absorbed or evolved in the formation of one mole of compound from its constituent elements in their standard states.

The change of enthalpy accompanying a reaction is given by the difference between the enthalpies of the products and those of the reactants; thus for an isothermal reaction

$$A + B = C + D$$

the enthalpy change is given by

$$\Delta H° = \Delta H°_C + \Delta H°_D - (\Delta H°_A + H°_B) \tag{18}$$

There are two fundamental thermochemical laws which express the first law specifically in terms of enthalpy. The first principle derived by **Lavoisier** and

Laplace (1780) states that "the quantity of heat required to decompose a compound into its elements is equal to the heat evolved when that compound is formed from its elements"; i.e. the heat of decomposition of a compound is numerically equal to its heat of formation, but of opposite sign. The second principle is that discovered by **Hess** (1840); it states that "the heat of reaction depends only on the initial and final states, and not on the intermediate states through which the system may pass."

The variation of energy, at constant volume, and of enthalpy, at constant pressure, with temperature gives the heat capacity of the system, thus

$$C_V = \left(\frac{\partial E}{\partial T} \right)_V \qquad (19)$$

$$C_p = \left(\frac{\partial H}{\partial T} \right)_P$$

From the temperature dependence of heat capacity at constant pressure, the enthalpy change is obtained by integrating equation (19),

$$H^\circ_{T_2} - H^\circ_{T_1} = \int_{T_1}^{T_2} C_p dT \qquad (20)$$

Usually the temperature dependence of C_p is formulated as in equation (12) for elevated temperatures, and the enthalpy increase is that relative to room temperature; e. g., $T_1 = 298.16°K$, thus

$$H^\circ_T - H^\circ_{298} = a(T-298.16) + \frac{b}{2}(T^2-8.89\times10^4) + c(T^{-1}-33.54\times10^{-4}) \qquad (21)$$

where the constants a, b and c are determined experimentally for each substance.

It follows from the above considerations that the temperature dependence of the enthalpy change accompanying a reaction is given by

$$\Delta H^\circ_T = \Delta H^\circ_{298} + \int_{298}^{T_2} (\Delta C_p) dt \qquad (22)$$

where ΔC_p is the difference between the heat capacities of reaction products and reactants. This relationship was first developed by **Kirchhoff** (1858). Although the heat content of a substance varies considerably with temperature, the influence of temperature on the heat of reaction is quite small, because ΔC_p is small for many reactions.

Let us consider for example the enthalpy changes accompanying the following two reactions under isothermal conditions.

$$Fe_3O_4 + 4CO = 3Fe + 4CO_2 \qquad (a)$$

$$Fe_3O_4 + 4H_2 = 3Fe + 4H_2O \qquad (b)$$

Using the compiled data given in Table 13—III, the heat contents of the reactants are calculated for 900°, 1400° and 1600°C; these are given in Table 13—II. It is seen that the reduction of magnetite to iron by carbon monoxide is an exothermic reaction, while that by hydrogen is endothermic below 1600°C. In both cases, the enthalpy change accompanying the reduction does not

Table 13-II. Enthalpy changes accompanying isothermal reduction of magnetite to iron by carbon monoxide or hydrogen at different temperatures.

Temp. °C	ΔH°_T, kcal mole				Enthalpy change for reaction, kcal	
	Fe_3O_4	CO	CO_2	H_2O	a	b
900	−261.2	−27.01	−94.40	−59.50	−8.36	23.20
1400	−259.6	−27.85	−94.65	−60.00	−7.60	19.60
1600	−237.6	−28.21	−94.75	−60.16	−18.56	−3.04

change much with temperature within the range 900° to 1400°C where there is no phase transformation. On the other hand, when liquid magnetite is reduced to liquid iron at 1600°C, the enthalpy changes become noticeably different from those for the solid-state reduction; in fact reduction by hydrogen becomes slightly exothermic.

As another example of calculating the heat balance, let us consider the reduction of liquid magnetite at 1600°C by carbon (assumed to be graphite for simplicity) introduced into the melt at 25°C producing liquid iron and carbon monoxide both at 1600°C.

$$4C(25°C) = 4C (1600°C) \quad (+31.04 \text{ kcal}) \qquad (c)$$

$$Fe_3O_4(1600°C) + 4C(1600°C) = 3Fe(1600°C) + 4CO(1600°C) \ (+124.76 \text{ kcal})$$

The net endothermic heat of reaction is $\Delta H = 155.80$ kcal. The heat required to maintain the system at 1600°C may be obtained by burning some of the carbon monoxide; if all the carbon monoxide is burnt by injecting oxygen at 25°C,

$$4CO(1600°C) + 2O_2(25°C) = 4CO_2(1600°C) \ (-240.14 \text{ kcal}) \qquad (d)$$

The heat of reaction is –240.14 kcal. From the sum of (c) and (d) the net heat of reaction is –84.34 kcal. On the other hand, the heat required to raise the temperature of one mole of magnetite from 25°C to 1600°C and to melt is 110.15 kcal. Therefore, the net heat balance for reactants magnetite, graphite and oxygen at 25°C producing liquid iron and carbon dioxide at 1600°C is

$$Fe_3O_4(25°C) + 4C(25°C) + 2O_2(25°C) = 3Fe(1600°C) + 4CO_2(1600°C) \quad \Delta H = 25.81 \text{ kcal} \qquad (e)$$

The mechanics of such a reduction is complex and perhaps not feasible in practice; this example is intended for heat balance calculations only.

Entropy—The concept of entropy is the consequence of the second law of thermodynamics. The law of dissipation of energy or degradation of available energy constitutes the basis of the second law of thermodynamics. This law states that "all natural processes occurring without external interference are spontaneous (irreversible processes)". For example, diffusion of one gas into another, passage of electricity from a high to a low potential, conduction of heat from a hot to a cold part of the system are all irreversible processes. That is, the processes cannot be reversed without some

Table 13—III. Thermodynamic data on some elements and compounds encountered in ironmaking and steelmaking processes

Units: ΔH°_{298}, cal per mole; S°_{298}, cal per deg per mole; C_p, cal per deg per mole; heats of transformation ΔH_t and fusion ΔH_f, cal per mole.

Notations: t.p., transformation point; m.p., melting point; b.p., boiling point; "*" indicate nonstoichiometric phase; underlined m.p. indicates incongruent m.p.; values in ( ) are estimated; dec., decomposes at 1 atm; sub., sublimes. Asterisk (*) refers to "Remarks" column.

$C_p = a + bT - cT^2$

Substance	$-\Delta H^{\circ}_{298}$	S°_{298}	a	$b \times 10^3$	$c \times 10^{-5}$	Temp. Range °C	t.p. °C	m.p. °C	b.p. °C	ΔH_t	ΔH_f	Remarks
Al	0	6.77	4.94	2.96	—	25–659		659	2467		2 570	
			7.00	—	—	659–2400						
AlO	−10 000	52.16										
Al₂O₃	399 600	12.2	8.22	0.44	0.87	25–1700	(1000)	2030	dec.		(26 000)	
Al₂S₃	172 900		27.49	2.82	8.38	25–1500		1100	dec.			
AlN	76 470	5.0	5.47	7.80	—	25–600		dec.	dec.			
Al₄C₃	51 500	(21.3)	24.08	31.60	—	25–320						
Al₂SiO₅(1)	1 350*	22.3	46.24	—	12.53	25–1300						*(1) Andalusite ⎱ Heats of
(2)	1 900*	20.0	45.52	2.34	16.00	25–1400						*(2) Kyanite ⎰ formation
(3)	600*	23.0	40.09	5.86	10.13	25–1300						*(3) Sillimanite ⎱ from oxides,
Al₆Si₂O₁₃			59.65	67.00	—	25–300		1810				Mullite. ⎰ Al₂O₃ + SiO₂.
B	0	1.40	4.13	1.66	1.76	25–2050		2050	(3900)		5 300	
			7.50	—	—	2027–2700						
B₂O₃*	306 100	12.9	8.73	25.40	1.31	25–450		450	(2300)		5 500	*Crystalline.
			30.50	—	—	450–1700						
B₂O₃*	305 500	18.6	2.28	42.10	—	25–450		450	(2300)			*Amorphous (glass).
			30.50	—	—	450–1700						
BN	60 700	3.67	1.82	3.62	—	25–900						
"B₄C"	12 200	6.47	22.99	5.40	10.72	25–1450						
Ba	0	16.0	5.36	3.16	—	25–370	370	710	1637	150	1 830	
			2.60	6.86	—	370–710						
			7.50	—	—	710–1600						
"BaO"	133 500	16.8	11.79	1.88	0.88	25–1700		1925	(2750)			
BaS	106 000	18.7						2200				
Ba₃N₂	87 000	36.4						dec.				
BaSiO₃	38 000*	26.8						1605				*from its oxides.
Ba₂SiO₄	64 500*	43.5	29.03	2.04	4.58	25–1700		1760			13 800	*from its oxides.
BaTiO₃		25.8	43.00	1.60	6.96	25–1700	5;120	1705		16;47		
Ba₂TiO₄		47.0										
Be	0	2.28	4.58	2.12	1.14	25–1283	1254	1286	(2400)		3 500	
			7.50	—	—	1283–2400						
BeO	143 100	3.37	8.45	4.00	3.17	25–900		2530	4120		17 000	
BeS	55 900	8.4										
Be₃N₂	140 600		8.16	30.80	—	25–500						
Be₂SiO₄	12 000*	15.4	22.84	—	—	25		1560				*from its oxides.

(Continued on next page)

Table 13—III. Thermodynamic data on some elements and compounds encountered in ironmaking and steelmaking processes—Continued

Substance	$-\Delta H°_{298}$	$S°_{298}$	$C_p = a + bT - cT^{-2}$ a	$b \times 10^3$	$c \times 10^{-5}$	Temp. Range °C	t.p. °C	m.p. °C	b.p. °C	ΔH_t	ΔH_f	Remarks
C (1)	0	1.36	4.03	1.14	2.04	25—2200		sub.	(4350)*		(33 000)*	(1) Graphite; *Sublimation point.
C (2)	−454	0.58	2.27	3.06	1.54	25—900						(2) Diamond.
CO	26 420	47.3	6.79	0.98	0.11	25—2200	−212	−205	−192			
CO_2	94 050	51.1	10.57	2.10	2.06	25—2200		sub.	−79			
CS	−55 000	50.3	7.39	1.02	0.51	25—1700						
CS_2	−21 000	56.8	12.45	1.60	1.80	25—1500		−112	46		1 050	
COS	33 900	55.3	11.33	2.18	1.83	25—1500		−139	−50		1 130	
CH_4	17 890	44.5	5.65	11.44	0.46	25—1200	−253	−183	−162			
Ca	0	9.95	5.25 2.68 7.40	3.44 6.80	— — —	25—440 440—850 850—1500	440	850	1492	270	2 070	
CaO	151 500	9.5	11.67	1.08	1.56	25—1700		2600	(3500)		19 000	
CaS	110 000	13.5	(10.20)	(3.80)	—	25—700						
Ca_3N_2	105 000		20.44	22.00	—	25—500		1195				
CaC_2	14 100	16.8	16.40 15.40	2.84 2.00	2.07 —	25—447 447—1200	447	2300		1330		
CaF_2	292 000	16.5	14.30 25.81 23.90	7.28 2.50	0.47 —	25—1151 1151—1418	1151	1418	2500	1140	7 100	
$CaAl_2$	54 000							1079				
CaSi	36 000							1240				
$CaSi_2$	36 000							1020				
Ca_2Si	50 000							<u>910</u>				
$Ca_3Al_2O_6$	1 600*	49.1	62.28	4.58	12.01	25—1500						*from its oxides.
$CaAl_2O_4$	3 700*	27.3	36.00	5.96	7.96	25—1500						*from its oxides.
$CaAl_4O_7$		42.5	66.09	5.48	17.80	25—1500						
$Ca_3B_2O_6$	60 000*	43.9	56.44 94.00	10.42	13.02	25—1487 1487—1700		1487			35 490	*from its oxides.
$Ca_2B_2O_5$	45 800*	34.7	43.75 52.29 68.20	11.50 2.40	10.69 —	25—531 531—1312 1312—1700	531	1312		1100	24 090	*from its oxides.
CaB_2O_4	29 400*	25.1	31.02 61.70	9.76 —	8.07 —	25—1152 1152—1700		1152			17 670	*from its oxides.
CaB_4O_7	42 900*	32.2	51.34 106.30	19.16 —	17.16 —	25—987 987—1700		987			27 060	*from its oxides.
$CaCO_3$ (1)	288 400	21.2	20.13	10.24	3.34	25—300	50	dec.		45		(1) Aragonite.
(2)		22.2	24.98	5.24	6.20	25—900	50	dec.		45		(2) Calcite.
$Ca(OH)_2$	237 500	19.9	19.07	10.80	—	25—400		dec.				

(Continued on next page)

Table 13—III. Thermodynamic data on some elements and compounds encountered in ironmaking and steelmaking processess—Continued

Substance	$-\Delta H°_{298}$	$S°_{298}$	$C_p = a + bT - cT^{-2}$			Temp. Range °C	t.p. °C	m.p. °C	b.p. °C	ΔH_t	ΔH_f	Remarks
			a	$b \times 10^3$	$c \times 10^{-5}$							
$Ca_2Fe_2O_5$	9 880	45.1	59.24	—	11.68	25—1435		1435			36 110	*from its oxides.
			74.20	—	—	1435—1700						
$CaFe_2O_4$	14 580	34.7	39.42	4.76	3.66	25—1237		1237			25 870	*from its oxides.
			54.90	—	—	1237—1700						
$Ca_2P_2O_9$	172 910*	57.6	48.24	39.68	5.00	25—1100	1100			3700		*from its oxides.
$Ca_3P_2O_8$	162 680*		79.00	—	—	1100—1300						*from its oxides.
$Ca_2P_2O_7$		45.2	53.03	14.76	11.16	25—1140	1140	1353		1600	24 100	
			76.15	—	—	1140—1353						
			96.80	—	—	1353—1430						
$CaSO_4$	342 400	25.5	16.78	23.60	—	25—1100	1193	1465	dec.		(6 700)	
Ca_3SiO_5	27 000*	40.3	49.85	8.62	10.15	25—1500						*from its oxides.
Ca_2SiO_4	30 200*	30.5	34.87	9.74	6.26	25—697	697;1437	2130		440;3390		*from its oxides.
			32.16	11.02	—	697—1437						
			49.00	—	—	1437—1700						
$CaSiO_3$	21 500*	19.6	26.64	3.60	6.52	25—1180	1190	1540		(1300)	(13 400)	*from its oxides.
			30.47	1.36	6.69					550		
$CaTiO_3$		22.4	32.03	—	—	1257—1700	1257					
Ce	0	16.6	5.70	3.98	—	25—730	730	804	2927	300	2 200	
			8.20	—	—	730—804						
			8.00	—	—	804—2700						
Ce_2O_3	429 310	36.0	15.1	—	—	25—100		>2600				
CeO_2	260 200	17.7						2050				
Ce_3S_4	483 180											
Ce_2S_3	346 760	43.1						1890				
CeS	133 420	18.7						2450				
CeN	78 000											
Cl_2	0	53.3	8.85	0.16	0.68	25—3000				1530	4 880	
Cl	-29 000	39.5	5.53	-0.16	0.23	25—5000		-101	-34			
Co	0	7.2	4.74	4.00	—	25—437	437;1120	1495	2877	105;0	4 100	
			2.16	7.02	—	437—1120						
			17.49	-4.92	—	1120—1495						
			9.00	—	-0.40	1495—2900						
CoO	57 100	12.7	11.54	2.04	5.72	25—1700		1805				
Co_3O_4	216 300	24.5	30.84	17.08		25—700		dec.				
Co_9S_8	197 000	(110.0)						835	dec.			
$CoS_{0.89}$	20 400	13.6						834	dec.			
Co_3S_4	75 000							625	dec.			
CoS_2	33 500							dec.	dec.			

(Continued on next page)

Table 13—III. Thermodynamic data on some elements and compounds encountered in ironmaking and steelmaking processes—Continued

Substance	−ΔH°298	S°298	$C_p = a + bT - cT^2$			Temp. Range °C	t.p. °C	m.p. °C	b.p. °C	ΔHt	ΔHf	Remarks
			a	b × 10³	c × 10⁻⁵							
Cr	0	5.7	4.16	3.62	−0.30	25—1903	1835	1903	2665		5 000	
			9.40	—	—	1903—2700						
"Cr2O3"	270 000	19.4	28.53	2.20	3.74	25—1500		(2400)	dec.			
CrO2	139 400	17.2							dec.			
CrO3	138 500	18.0						185				
"Cr2N"	27 300	8.0	11.01	16.40	—	25—500		1520				
CrN	29 400		12.20	—	—	25—500		1780				
"Cr4C"	16 400	25.3	29.35	7.40	5.02	25—1400		1890				
"Cr7C3"	42 500	48.0	56.96	14.54	10.12	25—1200						
"Cr3C2"	21 000	20.4	30.03	5.58	7.40	25—1300						
Cu	0	7.97	5.41	1.50	—	25—1084		1084	2547		3 120	
			7.50	—	—	1084—2500						
Cu2O	40 000	22.4	14.90	5.70	—	25—900	(56)	1230	dec.			
CuO	37 100	10.2	9.27	4.80	—	25—980		(1110)	dec.		(13 400)	
Cu2S	19 600	28.5	19.50	—	—	25—103	103;350	1130	dec.	920;200	2 600	
			23.25	—	—	103—350						
			20.32	—	—	350—1130						
CuS	12 100	15.9	(10.60)	(2.64)	—	25—1000		dec.				
Fe	0	6.49	3.04	7.58	−0.60	25—769	760*;910; 1392	1537	3070	326;215; 165	3 670	*Curie point.
			11.13	—	—	769—911						
			5.80	1.98	—	911—1392						
			6.74	1.60	—	1392—1537						
			9.77	0.40	—	1537—2700						
"FeO"*	63 800	13.7	11.66	2.00	0.67	25—1377		1377	dec.		7 490	*Fe0.947O, wustite in equilibrium with iron.
			16.30	—	—	1377—1700						
Fe3O4	267 800	35.0	21.88	48.20	—	25—627	627	1597	dec.	0	33 000	
			48.00	—	—	627—1597						
Fe2O3	196 800	20.9	23.49	18.60	3.55	25—677	677;777		dec.	160;0		
			36.00	—	—	677—777						
			31.71	1.76	—	777—1500						
"FeS"	22 800	14.1	5.19	26.40	—	25—138	138;325	1195	dec.	570;120	7 730	
			17.40	—	—	138—325						
			12.20	2.38	—	325—1195						
			17.00	—	—	1195—1700						
"FeS2"	42 400	12.7	17.88	1.32	3.05	25—700						

(Continued on next page)

Table 13—III. Thermodynamic data on some elements and compounds encountered in ironmaking and steelmaking processes—Continued

Substance	$-\Delta H^{\circ}_{298}$	S°_{298}	$C_p = a + bT - cT^{-2}$			Temp. Range °C	t.p. °C	m.p. °C	b.p. °C	ΔH_t	ΔH_f	Remarks
			a	$b \times 10^3$	$c \times 10^{-5}$							
Fe₈N	2 700		(26.84)	(8.16)	—	25—700						
Fe₄N	1 100	37.4	(14.91)	(6.09)	—	25—700						
Fe₂N	900		19.64	20.00	—	25—190						
Fe₃C	−5 980	24.2	25.62	3.00	—	190—1200	190			180		
FeAl₃	26 800											
FeAl₂	19 500											
FeAl	12 200											
Fe₃P	39 000											
FeSi	19 200	12.0										
FeTi	9 700											
Fe(CO)₅	54 000	80.7										
Fe₂SiO₄	346 000	34.7	36.51	9.36	6.70	25—1217		1217	dec.		22 030	
			57.50	—	—	1217—1700						
FeTiO₃		25.3	27.87	4.36	4.79	25—1367		1367	dec.		21 670	
H₂	0	31.21	6.52	0.78	−0.12	25—2700		−259	−253			
H₂O (1)	68 320	16.75	18.04	—	—	25—100		0	100		1 436	(1) Liquid.
(2)	57 800	45.13	7.30	2.46	—	25—2500						(2) Gas.
H₂S	4 800	49.1	7.81	2.96	0.46	25—2000	−170; −147	−86	−60	370;110	586	
Mg	0	7.8	4.97	3.04	−0.04	25—650		650	1105		2 140	
			7.80	—	—	650—1100						
MgO	143 700	6.5	10.18	1.74	1.48	25—1800						
MgS	83 000	10.2	8.82	0.08	0.61	25—1700						
MgC₂	−21 000	14.0						dec				
MgAl₂O₄	262 000	19.2	36.80	6.40	9.78	25—1500		2135				
MgCO₃		15.7	18.62	13.80	4.16	25—500		dec.				
Mg₃P₂O₈	110 940*											*from its oxides.
Mg₂SiO₄	15 100*	22.7	35.81	6.54	8.52	25—1500		1890				*from its oxides.
MgSiO₃	8 700*	16.2	24.55	4.74	6.28	25—1300+		1560				*from its oxides. +dinoenstatite.
Mg₂TiO₄		24.8	35.96	8.54	6.89	25—1500		1830				
MgTiO₃		17.8	28.29	3.28	6.53	25—1500		1840				
MgTi₂O₅		30.4	40.68	9.20	7.35	25—1700						

(Continued on next page)

Table 13—III. Thermodynamic data on some elements and compounds encountered in ironmaking and steelmaking processes—Continued

Substance	$-\Delta H^\circ_{298}$	S°_{298}	$C_p = a + bT - cT^{-2}$			Temp. Range °C	t.p. °C	m.p. °C	b.p. °C	ΔH_t	ΔH_f	Remarks
			a	$b \times 10^3$	$c \times 10^{-5}$							
Mn	0	7.6	5.70	3.38	0.37	25—727	718;1100;1137	1244	2010	535;545;430	3 500	
			8.33	0.66	—	727—1101						
			10.70	—	—	1101—1137						
			11.30	—	—	1137—1244						
			11.00	—	—	1244—2000						
"MnO"	92 000	14.3	11.11	1.94	0.88	25—1500		1875	dec.			
"Mn₃O₄"	331 400	35.5	34.64	10.82	2.20	25—1172	−56;1172	1565	dec.	4970		
			50.20	—	—	1172—1560						
"Mn₂O₃"	228 700	26.4	24.73	8.38	3.23	25—1100		dec.				
"MnO₂"	124 300	12.7	16.60	2.44	3.88	25—550	250	dec.				
MnS	49 000	18.7	11.40	1.80	—	25—1530		1530		0	6 240	
			16.00	—	—	1530—1700						
"Mn₄N"	31 200		21.15	30.50	—	25—500						
"Mn₅N₂"	55 200		30.55	38.40	—	25—500						
Mn₃P₂O₈	105 230*							1119				*from its oxides.
Mn₂SiO₄	11 800*	39.0						1340				*from its oxides.
MnSiO₃	5 900*	21.3	26.42	3.88	6.16	25—1200		1270				*from its oxides.
Mo	0	6.83	5.18	1.66	—	25—2617		2617	5550		6 650	
			10.00	—	—	2617—2700						
"MoO₂"	139 500	11.1	20.73	5.18	4.18	25—795		dec.				
"MoO₃"	178 200	18.6	32.00	—	—	795—1100		795	1100		12 500	
Mo₂S₃	92 500	28.5										
MoS₂	60 400	15.2										
"Mo₂N"	16 600	(21.0)	11.19	13.80	—	25—500		2690				
"Mo₃C"	−4 200	(19.8)										
N₂	0	45.77	6.83	0.90	0.12	25—2700	−237.5	−210	−196		172	
NH₃	11 000	45.97	7.11	6.00	0.37	25—1500		−78	−33.5		1 352	
Na	0	12.3	4.02	9.04	—	25—98		98	905		622	
			6.83	1.08	—	98—905						
Na₂O	100 700	17.0	15.70	5.40	—	25—920		920	dec.			
Na₂S	92 400	23.5	19.81	1.64	—	25—700		950				
NaAlO₂	20 900*	16.9	19.18	7.14	3.36	25—467					(1 600)	*from its oxides.
			20.21	4.24	—	467—1400						
Na₂CO₃	271 600	32.5	27.13	15.62	4.78	25—858	320;480	858	dec.	310	7 100	
			45.00	—	—	858—1200						
Na₂SiO₃	55 500*	27.2	31.14	9.60	6.47	25—1088		1088			12 470	*from its oxides.
			42.80	—	—	1088—1700						
Na₂Si₂O₅	60 500*	39.4	44.38	16.86	10.67	25—874	678	874		1700	8 500	*from its oxides.
			62.35	—	—	874—1700						

Table 13—III. Thermodynamic data on some elements and compounds encountered in ironmaking and steelmaking processes—Continued

Substance	$-\Delta H^\circ_{298}$	S°_{298}	$C_p = a + bT - cT^{-2}$			Temp. Range °C	t.p. °C	m.p. °C	b.p. °C	ΔH_t	ΔH_f	Remarks
			a	$b \times 10^3$	$c \times 10^{-5}$							
Ni	0	7.12	4.06	7.04	—	25—360	360	1452	2910	0	4 210	
			6.00	1.80	—	360—1452						
			9.20	—	—	1452—2900						
"NiO"	57 500	9.1	−4.99	37.58	−3.89	25—252	252;292	1984		0;0		
			13.88	—	—	252—292						
			11.18	2.02	—	292—1700						
"NiS"	22 200	16.1	9.25	6.40	—	25—300	396	>800		630		
O_2	0	49.02	7.16	1.00	0.40	25—2700	−250;−229	−219	−183	224;178	106	
P*	0	9.80	5.50	—	—	25—44		44	280		150	*white.
			5.88	—	—	44—130						
P*	4 400	5.46	4.74	3.90	—	25—500		sub.	590			*red.
P_2*	33 600	52.1	8.31	0.46	0.72	25—1700			605			*gas.
P_2O_5	356 600	27.3	8.38	54.00	—	25—358		580			(5 800)	
S	0	7.62	3.58	6.24	—	25—95	95	119	444	85	335	
			6.20	—	—	95—119						
			8.73	—	—	419—444						
S_2	−31 000	54.4	8.72	0.16	0.90	25—2700						
SO	−200	53.1	8.26	0.32	1.00	25—2700						
SO_2	70 950	59.25	11.04	1.88	1.84	25—1700		−76	−10		1 770	
Si	0	4.5	5.70	0.70	1.04	25—1412		1412	2600		12 100	
			6.10	—	—	1412—1700						
SiO	23 200	50.5	7.70	0.74	0.70	25—1700						
SiO_2*	217 000	10.0	11.22	8.20	2.70	25—575	575	(1610)		290		*alpha-quartz.
			14.41	1.94	—	575—1700						
SiO_2*	216 100	10.2	4.28	21.06	—	25—250	250	1713	dec.	200	3 600	*beta-cristobalite.
			14.40	2.04	—	250—1713						
SiS	49 000	53.4										
SiS_2								1090	(1130)			
Si_3N_4	176 300	23.0	16.83	23.60	—	25—600		dec.				
SiC	15 000	3.95	9.97	1.82	3.64	25—1700		>2700				

(Continued on next page)

Table 13—III. Thermodynamic data on some elements and compounds encountered in ironmaking and steelmaking processes—Continued

Substance	$-\Delta H°_{298}$	$S°_{298}$	$C_p = a + bT - cT^{-2}$			Temp. Range °C	t.p. °C	m.p. °C	b.p. °C	ΔH_t	ΔH_f	Remarks
			a	$b\times10^3$	$c\times10^{-5}$							
Th	0	12.8	5.17	4.56	—	25—1400	1400	1695	4227	670	4 500	
			11.00	—	—	1400—1695						
			11.00	—	—	1695—2700						
ThO$_2$	293 200	15.6	15.84	2.88	1.60	25—1700		(3000)			4 460	
Ti	0	7.3	5.25	2.52	—	25—882	882	1667	3285	950		
			7.50	—	—	882—1667						
			8.00	—	—	1667—2700						
"TiO"	123 900	8.3	10.57	3.60	1.86	25—991	991	1760		820		
			11.85	3.00	—	991—1760						
"TiO$_2$"	225 500	12.0	17.97	0.28	4.35	25—1500		1870	dec.			
"TiN"	80 400	7.2	11.91	0.94	2.96	25—1700		2950	dec.			
"TiC"	43 900	5.8	11.83	0.80	3.58	25—1700		3150				
V	0	7.0	4.90	2.58	-0.20	25—1917		1917	3350		5 050	
			9.50	—	—	1917—2700						
VO	102 000	9.3	11.32	3.22	1.26	25—1700		1700				
V$_2$O$_3$	294 000	23.5	29.35	4.76	5.42	25—1500		>2000				
VO$_2$	171 000	12.3	14.96	—	—	25—72	72	1360		1 025	13 600	
			17.85	1.70	3.95	72—1545						
			25.50	—	—	1545—2700						
V$_2$O$_5$	371 800	31.3	46.54	-3.90	13.22	25—670		670	dec.		15 560	
			45.60	—	—	—						
"VN"	60 000	8.9	10.94	2.10	2.21	25—1500		(2050)				
"VC"	(12 500)	6.8	9.18	3.30	1.95	25—1400		(2850)				
W	0	8.0	5.74	0.76	—	25—2700	720	3380	(5400)			
"WO$_3$"	200 000	19.9	(17.75)	(5.87)	—	25—1473		1473	(1850)			
WC	9 100	10.0	(7.98)	(2.17)	—	25—2700		dec.				
Zn	0	9.9	5.35	2.40	—	25—420		420	907		1 765	
			7.50	—	—	420—907						
ZnO	83 200	10.4	11.71	1.22	2.18	25—1700		1975				
ZnS	48 200	13.8	12.16	1.24	1.36	25—900	1020	dec.		(3 200)		
Zr	0	9.3	6.50	1.42	0.82	25—862	862	1857	(4400)	915	4 900	
			7.90	—	—	862—1857						
			8.00	—	—	1857—2700						
"ZrO$_2$"	259 500	12.1	16.64	1.80	3.36	25—1205	1205	2700	(4300)	1 420		
ZrN	87 300	9.3	11.10	1.68	1.72	25—1400		2950				
"ZrC"	44 100	8.5						3500				
ZrSiO$_4$		20.2	31.48	3.92	8.08	25—1500		2430				

References to the thermodynamic data: $\Delta H°_{298}$ from O. Kubaschewski, E. Ll. Evans and C. B. Alcock: "Metallurgical Thermochemistry," Pergamon Press, London, 1967. $S°_{298}$ from K. K. Kelley and E. G. King: U. S. Bur. Mines Bull. No. 592, 1961. C_p from K. K. Kelley: U.S. Bur. Mines Bull. No. 584, 1960.

change in the system brought about by external interference. For a process to be reversible, all sources of dissipation of energy must be eliminated. Let us consider an imaginary process carried out infinitesimally slowly, so that the system is always in equilibrium with its surroundings; if at any time the process is reversed by an infinitesimal change in the surroundings, the process is said to be reversible. This is, of course an imaginary limiting case for the real processes.

The thermodynamic efficiency was derived by **Carnot** (1824) by considering a reversible cyclic process occurring between temperatures T_1 and T_2 at four reversible steps, of which two are isothermal at T_1 and T_2 and two are adiabatic from T_1 to T_2 and from T_2 to T_1. When the cycle is completed, the system returns to the initial state. The application of the first law to such a reversible cyclic process gives

$$\frac{w}{q_2} = \frac{q_2 + q_1}{q_2} = \frac{T_2 - T_1}{T_2} \qquad (23)$$

which represents the efficiency of any reversible cyclic process operating between the temperatures T_2 and T_1. At the higher temperature T_2 the heat absorbed is q_2; $-q_1$ is the heat evolved at the lower temperature T_1. According to the first law, the difference between the heat absorbed and that evolved corresponds to the amount of cyclic work done, i. e. $w = q_2 + q_1$. Therefore, the efficiency of a reversible process is determined by the temperatures of source and sink; the maximum efficiency of unity can be obtained only when T_1 is at absolute zero.

By rearranging equation (23), the following is obtained:

$$\frac{q_1}{T_1} + \frac{q_2}{T_2} = 0 \qquad (24)$$

Any reversible cycle may be regarded as being made up of a number of cycles (Carnot cycles) and equation (24) may be written in its general form as

$$\sum \frac{q}{T} = 0$$

or using the calculus notation

$$\oint (\delta q/T) = 0$$

where $\oint$ refers to integration over a reversible cycle, δq being the infinitesimal heat change. For an ideal gas $\delta q/T$ is an exact differential, and its integral between states A and B is a function of these states only, and independent of the path followed in passing from state A to B. The thermodynamic quantity, entropy S, is defined such that

$$dS = \frac{\delta q}{T} \qquad (25)$$

The entropy is an extensive property, and therefore the entropy change depends solely upon the initial and final states.

In a process occurring at constant pressure,
$$\delta q = C_p \, dT,$$

therefore,

$$dS = \frac{C_p}{T} \, dT = C_p \, d \ln T \qquad (26)$$

The increase in entropy in heating a substance from absolute zero to $T°K$ is obtained by integrating equation (26)

$$S_T - S_0 = \int_0^T \frac{C_p}{T} \, dT \qquad (27)$$

The heat theorem put forward by **Nernst** (1906) constitutes the third law of thermodynamics, which in simplest form states that "for all reactions involving substances in the condensed state, entropy change ΔS is zero at the absolute zero". Stating it in another form, "the entropy of any homogeneous and ordered crystalline substance which is in complete internal equilibrium is zero at the absolute zero." That is, S_0 in equation (27) is zero at $0°K$. In imperfect crystals, glasses, solid solutions and metastable phases, entropy is not zero at $0°K$. For perfect crystals entropy is an absolute thermodynamic quantity; for convenience, entropy values are tabulated for 25°C (298.16°K); the standard entropy is usually denoted by $S°_{298}$. The entropy change accompanying a reaction at any temperature is given by

$$\Delta S°_T = \Delta S°_{298} + \int_{298}^T \frac{\Delta C_p}{T} \, dT \qquad (28)$$

When a phase change occurs, as in a transformation (t) from one crystalline state to another, in a fusion (f) or in a vaporization (v), the entropy increase is given by:

$$\Delta S_t = \frac{\Delta H_t}{T_t}; \Delta S_f = \frac{\Delta H_f}{T_f}; \Delta S_v = \frac{\Delta H_v}{T_v} \qquad (29)$$

where ΔH_t, ΔH_f and ΔH_v are the latent heats of transformation, fusion and vaporization. It has been found that for many substances the ratio $\Delta H_v/T_v = 22$ cal per deg mole and $\Delta H_f/T_f = 2$ cal per deg mole; these are known as the **Trouton** and **Richards'** rule respectively.

Free Energy—A combined form of the first and second laws may be obtained from equations (15) and (25): thus, for a reversible process against pressure only,

$$dE = TdS - PdV \qquad (30)$$

In this differential equation, E is a function of the entropy S and of the volume V. However, experimentally, it is easier to control the pressure and temperature of the system and, therefore, a new function F was defined by **Gibbs**, thus

$$F = E + PV - TS \qquad (31)$$

and, since $E + PV = H$, for constant pressure

$$F = H - TS$$

where F is known as the chemical potential or the (Gibbs) free energy of the system. When a system changes isothermally from state A to state B, the change in the free energy is

$$F_B - F_A = \Delta F = \Delta H - T\Delta S \qquad (32)$$

where ΔF refers to the change in F accompanying the reaction.

During any process which proceeds spontaneously in such a manner that the pressure and temperature of the reactants and products are the same, the free energy of the system decreases. Another criterion is that the free energy change of a system undergoing a reversible process at constant temperature and pressure and doing work only against pressure is zero. Therefore, for any system at constant temperature, pressure and mass, which does no work other than against pressure,

$$\Delta F \leq 0 \qquad (33)$$

Although any process for a system of constant temperature, pressure and mass for which $\Delta F \leq 0$ is thermodynamically possible, the reaction may not proceed at a perceptible rate if the activation energy required to overcome the resistance to reaction is too high. However, if $\Delta F > 0$ the reaction will not take place.

As already stated, the free energy change accompanying a reaction is given by:

$$\Delta F^\circ = \Delta H^\circ - T\Delta S^\circ \qquad (34)$$

where the superscript ($^\circ$) refers to the change in the thermodynamic property with respect to the standard state. Usually, the pure solid, liquid or gas in its stable form at atmospheric pressure and 298.16°K is chosen as the standard state.

It is customary to tabulate the heats of formation and entropies of substances for room temperature (298.16°K) and if the high-temperature heat capacities are known, the variation of the standard free energy change with temperature may be obtained by combining equations (22), (28), and (34), thus

$$\Delta F^\circ_T = \Delta H^\circ_{298} + \int_{298}^{T} \Delta C_p dT -$$

$$T\Delta S^\circ_{298} - T \int_{298}^{T} \frac{\Delta C_p}{T} dT \qquad (35)$$

Solutions—A solution is a homogeneous gas, liquid or solid mixture, any portion of which has the same state properties. The composition of a gas solution is usually given in terms of partial pressures of species in equilibrium with one another under given conditions. In metallic liquid and solid solutions, the concentrations of the constituent elements, usually reported in weight per cent, are converted to atomic concentrations when considering the thermodynamic properties of the solutions. The atom fraction, N_i, is given by the ratio:

$$N_i = \frac{n_i}{\Sigma n}$$

where n_i is the number of gram-atoms of element i per unit mass of the substance and Σ is the total number of atoms. Since the composition is usually given in weight per cent, n_i per 100 g of the substance is

$$n_i = \frac{\%i}{M_i}$$

where M_i is the atomic weight. For complex systems, e.g. solid and liquid solutions of oxides and liquid slags, the composition is often given in terms of concentrations of oxides.

(a). Partial molar quantities:

Contribution made by each constituent of a solution to its extensive property, e. g. volume, free energy, enthalpy, entropy, is not additive. If a solution is made of n_1, n_2, n_3 ... numbers of moles of constituents 1, 2, and 3 ... and G' is any total extensive property of the solution, the partial molar quantity is given by the partial derivative,

$$\overline{G}_1 = \left(\frac{\partial G'}{\partial n_1} \right)_{P,T,n_2 n_3 \ldots} \qquad (36)$$

That is, if an infinitesimal amount, dn_1, of component 1 is added to a large amount of solution so that the composition of the solution does not change, at constant temperature and pressure the ratio $(\partial G'/\partial n_1)_{P,T,n_2,n_3}$ defines the partial molar quantity of component 1. Similar equations can be written for other components, and from the solution of the sum of these differential equations, the following basic equations are obtained.

$$G' = n_1\overline{G}_1 + n_2\overline{G}_2 + n_3\overline{G}_3 + \ldots \qquad (37)$$

and since the molar quantity is

$$G = \frac{G'}{n_1 + n_2 + n_3 + \ldots}$$

the following is the relation for one mole of solution

$$G = N_1\overline{G}_1 + N_2\overline{G}_2 + N_3\overline{G}_3 + \ldots \qquad (38)$$

where N is the atom fraction, i.e. $N_1 + N_2 + N_3 + \ldots = 1$.

Another basic equation is

$$N_1 d\overline{G}_1 + N_2 d\overline{G}_2 + N_3 d\overline{G}_3 + \ldots = 0 \qquad (39)$$

The special form of this equation for the partial molar free energies is known as the Gibbs-Duhem equation, and for binary solutions,

$$N_1 d\overline{F}_1 + N_2 d\overline{F}_2 = 0 \qquad (40)$$

From equations (38) and (39) the following is derived for a binary system

$$\overline{G}_1 = G + (1 - N_1) \frac{dG}{dN_1} \qquad (41)$$

and similarly

$$\overline{G}_2 = G + (1 - N_2) \frac{dG}{dN_2} \qquad (42)$$

These equations show the method of evaluating the partial molar quantities $\overline{G}_1$ and $\overline{G}_2$ graphically from the plot of the molar quantity G against composition. Figure 13—2 shows the variation of G with composition; to determine $\overline{G}_1$ and $\overline{G}_2$ at a composition indicated by point A, a line is drawn tangentially to the curve at A and the intercepts at $N_1 = 0$ and $N_1 = 1$ give the values of $\overline{G}_1$ and $\overline{G}_2$ respectively.

(b). Chemical potential and activity:

Another important derivation from the Gibbs free energy equation is that obtained by combining equation (30) with the derivative of equation (31), thus

$$dE = TdS - PdV$$
$$dF = dE + PdV + VdP - TdS - SdT \qquad (43)$$
$$dF = VdP - SdT$$

For constant temperature $dF = VdP$, and inserting $V = RT/P$ for one mole of an ideal gas, the following is obtained

$$dF = RT(dP/P)$$
$$= RT \, d \ln P \qquad (44)$$

Similarly, for each component of the system, the partial molar free energy equation is:

$$d\overline{F}_i = RT \, d \ln p_i$$

On integration

$$\overline{F}_i = F_i° + RT \ln p_i \qquad (45)$$

where p_i is the partial pressure of component i in an ideal solution and $F_i°$ is the integration constant which is the free energy of the ith component at its standard state. The partial molar free energy $\overline{F}_i$ is also called the chemical potential of the ith component.

In real gases, particularly at low temperatures and high pressures, the ideal gas law is not obeyed. In order to maintain the generality of equation (45) the concept of fugacity was introduced by **Lewis** (1901). The fugacity (f_i) for a pure species i is defined in terms of chemical potential by the following isothermal relation,

$$RT \ln (f_i) = F_i + I \qquad (46)$$

where I is the integration constant so chosen that the fugacity approaches the pressure as the pressure approaches zero. In an ideal gas mixture fugacities of gas species are numerically equal to their respective partial pressures. The effect of pressure on the deviation from the ideal gas law is illustrated by the following example. For CO_2 at 100°C at P = 50 atm pressure, $f = 44.2$ atm and at P = 100 atm, $f = 79$ atm. At atmospheric pressures and moderate to high temperatures, the fugacity and the partial pressures are almost equal and, therefore, in the subsequent considerations the partial pressures are used with the assumption that the ideal gas law is obeyed.

Now, let us consider a solution in equilibrium with its vapor; the vapor pressure of the ith component in the solution is equal to the partial pressure p_i of the ith component in the gas phase. If the partial pressure of the pure ith component is $p_i°$, then the ratio $p_i/p_i° = a_i$ was defined by Lewis as the activity of the ith component in the solution. Although in many cases the pure component is chosen as the standard state, any other state can be a reference state; the choice of standard states and conversion from one to another is discussed later in this section. The activity of the component at its standard state is of course unity.

From the above definition of activity, the chemical potential derived for gases can be applied to liquid or solid solutions by using activity a_i instead of the partial pressure p_i (or fugacity f_i), thus at constant temperature and pressure and at complete internal equilibrium,

$$\overline{F}_i = F_i° + RT \ln a_i$$
$$\text{or} \qquad \Delta \overline{F}_i = \overline{F}_i - F_i° = RT \ln a_i \qquad (47)$$

The value determined experimentally is the difference $\overline{F}_i - F_i° = \Delta \overline{F}_i$, i.e. the difference between the free energy value for a given solution and that in a standard state. This difference $\Delta \overline{F}_i$ is called the relative partial molar free energy; similarly $\overline{L}_i$ and $\overline{S}_i$ are the relative partial molar heat and entropy of component i. For convenience, these terms are often referred to as the free energy, heat and entropy of solution.

The effect of temperature on the activity can be computed from the relative partial molar free energy of the ith component in a solution,

$$\Delta \overline{F}_i = \overline{L}_i - T\overline{S}_i \qquad (48)$$

Therefore,

$$\ln a_i = \frac{\overline{L}_i}{RT} - \frac{\overline{S}_i}{R} \qquad (49)$$

and for $\overline{L}$ in calories and $\overline{S}$ in entropy units (e.u.)

$$\log a_i = \frac{\overline{L}_i}{4.575T} - \frac{\overline{S}_i}{4.575} \qquad (50)$$

In most metallurgical solutions, i.e. liquid or solid metallic solutions, glasses and molten slags, the partial molar heat capacity of the solute is negligibly small. That is, $\overline{L}_i$ and $\overline{S}_i$ do not change perceptibly over a temperature range of most experimental interest. Therefore, for a given composition, $\log a_i$ is a linear function of the reciprocal of absolute temperature and the slope of the line multiplied by 4.575 gives the relative partial molar heat of solution of the ith component; the intercept multiplied by 4.575 gives the relative partial molar entropy of solution. Since $\overline{L}$ and $\overline{S}$ are relative partial molar quantities, they vary with composition as evident from the schematic plot given in Figure 13—2.

It follows from equation (38) that the molar free energy of mixing, ΔF^M, enthalpy of mixing, ΔH^M, and entropy of mixing, ΔS^M, are given by relations

$$\Delta F^M = N_1 \Delta \overline{F}_1 + N_2 \Delta \overline{F}_2 + N_3 \Delta \overline{F}_3 + \ldots \qquad (51)$$

$$\Delta H^M = N_1 \overline{L}_1 + N_2 \overline{L}_2 + N_3 \overline{L}_3 + \ldots \qquad (52)$$

$$\Delta S^M = N_1 \overline{S}_1 + N_2 \overline{S}_2 + N_3 \overline{S}_3 + \ldots \qquad (53)$$

(c). Ideal solutions—Raoult's law:

The solutions are said to be ideal, if the activity is directly proportional to the mole (or atom) fraction. This is known as Raoult's law (1887) defined by the relation

$$a_i = N_i \qquad (54)$$

A thermodynamic consequence of Raoult's law is that the enthalpy, or heat, of mixing for an ideal solu-

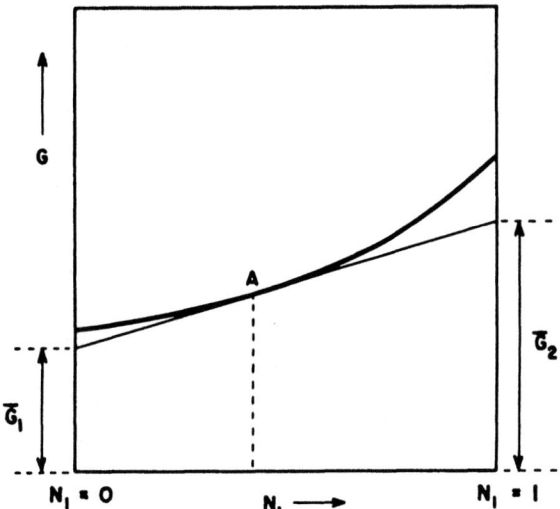

FIG. 13—2. Graphical method of evaluating the partial molar quantities from the molar quantity of a binary system.

tion, $\triangle H^{M,id}$, is zero. From equations (48), (51), (52) and (53) with $a_i = N_i$ and $\triangle H^{M,id} = 0$, the following relation is found for the entropy of formation of an ideal solution

$$\triangle S^{M,id} = -R(N_1 \ln N_1 + N_2 \ln N_2 + N_3 \ln N_3 + \ldots)$$

(55)

(d). Actual solutions—activity coefficient:

Practically all metallic solutions exhibit nonideal behavior. Depending on the nature of the elements constituting a solution, the activity vs composition relationship deviates from Raoult's law to varying degrees. Figures 13—3, 13—4 and 13—5 are examples showing the variation of activities with composition in iron-nickel, iron-silicon and iron-copper melts at 1600°C. Raoult's law for the ideal solution is indicated by the dotted lines.

The concept of activity coefficient was conceived by Lewis (1923). The activity coefficient of solute i is defined by the ratio of the activity to the mole fraction,

$$\gamma_i = \frac{a_i}{N_i}$$

(56)

If the activity is relative to the pure component i, it follows from Raoult's law that as $N_i \to 1$, $\gamma_i \to 1$.

(e). Dilute solutions—Henry's law:

The variation of activity with composition in infinitely dilute solutions is generalized by Henry's law which states that at infinitely dilute solutions, i.e. solute $N_i \to 0$, the activity is directly proportional to concentration. Provided that the solute does not dissociate at infinite dilution (valid assumption for metallic solutions), Henry's law is anticipated from calculus. Thus, $a_i \to 0$ when $N_i \to 0$; therefore the derivative $da_i/dN_i \to a_i/N_i =$ finite, i.e. at infinite dilution

$$a_i = \gamma^\circ N_i$$

(57)

where γ°_i is the activity coefficient of the ith species at infinite dilution.

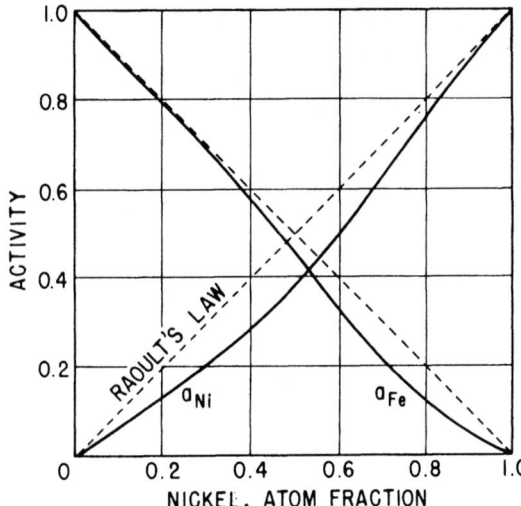

FIG. 13—3. Activities in iron-nickel binary melts at 1600°C.

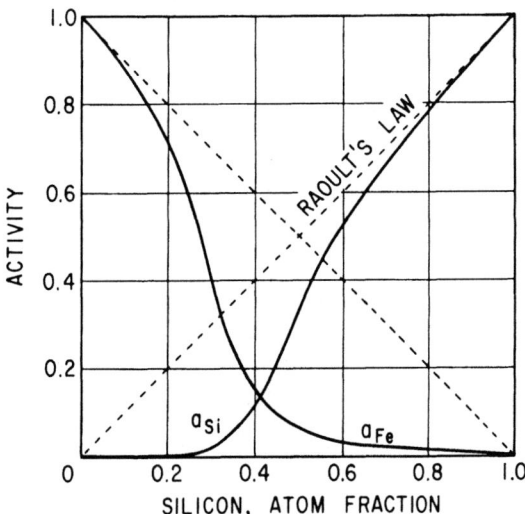

FIG. 13—4. Activities in iron-silicon binary melts at 1600°C.

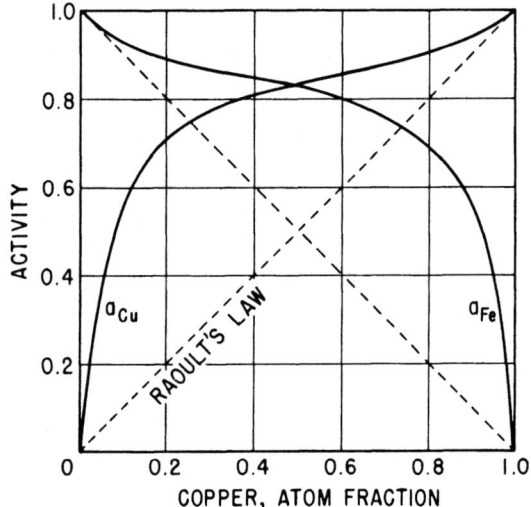

FIG. 13—5. Activities in iron-copper binary melts at 1550°C.

It should be remembered that for correct application of Henry's law the appropriate species of the solute must be used. For example, oxygen dissolves in water as molecular oxygen; therefore according to Henry's law $N_{O_2} \propto p_{O_2}$. On the other hand, oxygen dissolves in liquid or solid metals as atomic (or ionic) oxygen; therefore at low concentration Henry's law states that $N_O \propto (p_{O_2})^{1/2}$. The special form of Henry's law is also known as **Sievert's law** established from his work on the solubility of gases in metals.

It follows from Gibbs-Duhem integration (40) that if for the solute i, $a_i/N_i = \gamma_i$ (finite) at $N_i = 0$, then for the solvent 1, $\gamma_1 \rightarrow 1$ as $N_1 \rightarrow 1$. That is, the validity of Raoult's law for the solvent is predicted from Henry's law for the solute at infinite dilution.

Since the solubility of gases O_2, N_2, H_2, etc. in solid and liquid iron is small, e.g. < 1 atom per cent, Henry's law may be extended to small finite concentrations of the solute. This is illustrated, for example, by the work of Floridis and Chipman[2] (Figure 13—6) showing that

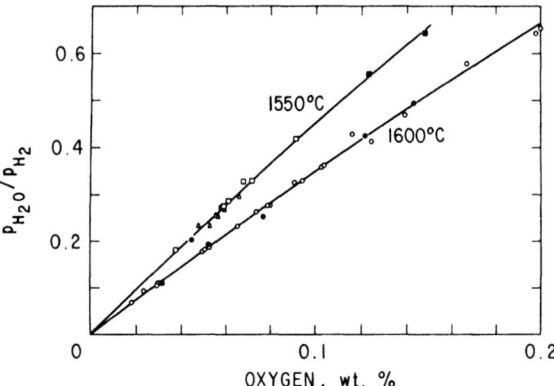

FIG. 13—6. Variation of oxygen content with the ratio p_{H_2O}/p_{H_2} in equilibrium with liquid iron. (Floridis and Chipman[2].)

up to about 0.1 weight per cent O, the oxygen content of purified liquid iron is directly proportional to the oxygen activity as determined by the ratio p_{H_2O}/p_{H_2} in the gas phase in equilibrium with the metal. However, it should be emphasized that Henry's law is a limiting law for $N_i = 0$; its application to finite solute concentrations involves approximation. The departure from Henry's law even at small solute concentrations is well demonstrated by the results of Smith[3] on the activity of carbon in austenite (Figure 13—7).

(f). Gibbs-Duhem integration:

In terms of activities, the Gibbs-Duhem equation (40) is given by

$$N_1 d \ln a_1 + N_2 d \ln a_2 + N_3 d \ln a_3 + \ldots = 0 \quad (58)$$

A more convenient form of this equation is in terms of the activity coefficient, thus

$$N_1 d \ln \gamma_1 + N_2 d \ln \gamma_2 + N_3 d \ln \gamma_3 + \ldots = 0 \quad (59)$$

For a binary system integration gives

$$\ln \gamma_1 = - \int_{N_1 = 1}^{N_1 = N_1} \frac{N_2}{N_1} d \ln \gamma_2 \quad (60)$$

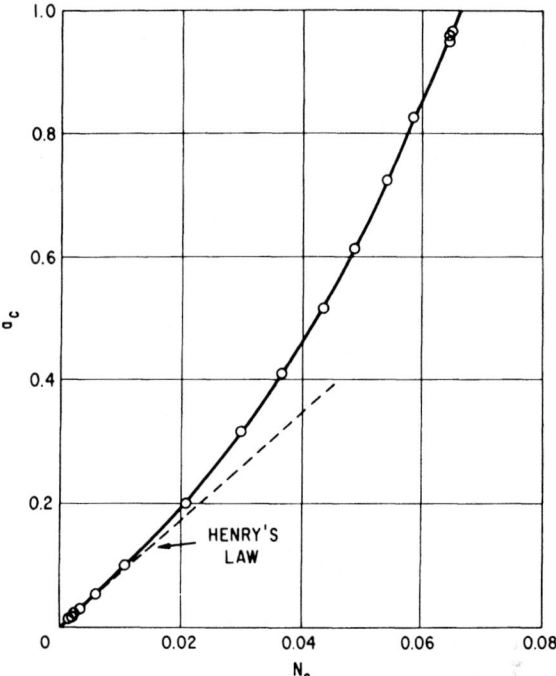

FIG. 13—7. Activity of carbon in austenite (relative to graphite) at 1000°C, demonstrating deviation from Henry's law. (Smith[3].)

If the activity of solute 2 is known, the activity of the solvent can be derived by graphical integration of the curve N_2/N_1 vs $\ln \gamma_2$. Since $\gamma_2 = \gamma_2°$ at $N_1 = 1$, the graphical integration is much facilitated; however, integration difficulty does arise near the other limit of $N_1 \rightarrow 0$.

When the activity of one of the components of a ternary system is known, the activities of the other two components may be calculated again by the application of the Gibbs-Duhem equation. Different methods of computation are discussed by Darken[4], Wagner[5] and by Schuhmann[6]. Later, Gokcen[7] demonstrated the application of these methods of calculation to multi-component systems.

(g). Conversion from one standard state to another:

For evaluating activities any particular composition can be chosen as the standard state. Using equation (47) the chemical potential of a component i in a solution for two different standard states (1) and (2) is:

Standard state (1) $\overline{F}_i = F^{\circ(1)} + RT \ln (a_i)_1$
Standard state (2) $\overline{F}_i = F^{\circ(2)} + RT \ln (a_i)_2$

From the difference, the change of standard free energy is

$$\Delta F° = F^{\circ(1)} - F^{\circ(2)} = RT \ln \frac{(a_i)_2}{(a_i)_1} \quad (61)$$

This is the free energy change accompanying the isothermal transfer of solute from state (1) to state (2).

In steelmaking processes we are invariably concerned with reactions between solutes at low concentrations in steel. Therefore, it is convenient to choose the nonphysical standard state such that $a_i \rightarrow N_i$ at infinite dilution. For pure component as the standard state (1), $\gamma_i = 1$ when $N_i = 1$; for the nonphysical standard state (2) hypothetical extension of Henry's law gives $\gamma_i = \gamma_i^\circ$ at $N_1 = 1$. Therefore, the change in standard state is represented by the expression

$$M_i \text{ (pure l)} = M_i \text{ (nonphysical state)}, \quad \Delta F = RT \ln \gamma_i^\circ \quad (62)$$

The atom fraction of the solute at 1 per cent concentration is

$$N_i = \frac{1/M_i}{1/M_i + 99/M_s}$$

where M_i and M_s are the atomic weights of the solute and the solvent. Since $1/M_i$ is much smaller than $99/M_s$, the former can be omitted, and assuming that iron is the solvent ($M_s = 55.85$)

$$N_i = \frac{0.5585}{M_i}$$

at 1 per cent of solute i in iron.

If Henry's law is obeyed at concentrations up to 1 per cent, the activity coefficient γ_i° for the infinitely dilute solution can be used, i.e. for the standard state (2) $a_i \cong 0.5585\, \gamma_i^\circ/M_i$ at 1 per cent. Inserting this value in equation (61) the standard free energy of solution of pure component i in iron with reference to the hypothetical 1 per cent solution becomes:

$$M_i \text{ (pure)} = \underline{M_i}\text{(1 wt. \%)}, \quad \Delta F_i^\circ \cong RT \ln \frac{0.5585\,\gamma_i^\circ}{M_i} \quad (63)$$

If ΔF_i° is in terms of cal per mole, and converting (ln) to (log) scale,

$$\Delta F_i^\circ = 4.575\, T \log \frac{0.5585\,\gamma_i^\circ}{M_i} \quad (64)$$

Since Henry's law is valid at infinite dilution only, appropriate correction must be made for nonideal behavior when using equation (64) for finite solute contents. It is convenient to introduce another activity coefficient defined by the ratio

$$f_i = \frac{\gamma_i}{\gamma_i^\circ} \quad (65)$$

such that $f_i \to 1$ when per cent $i \to 0$.

Now, for an example, let us evaluate the free energy change accompanying the solution of pure liquid silicon in liquid iron (1 wt per cent Si)

$$Si \text{ (l)} = \underline{Si} \text{ (1 wt per cent)} \quad (66)$$

where underscore denotes silicon in solution in iron. From the data compiled in Part II, $\gamma_{si}^\circ = 0.0046$ at 1600°C and inserting $M_{Si} = 28.09$, equation (64) gives

$$\Delta F_{Si} = 4.575 \times 1873 \log \frac{0.5585 \times 0.0046}{28.09}$$
$$= -34\,600 \text{ cal/mole Si} \quad (67)$$

In order to calculate the standard free energy of solution for other temperatures, the partial molar heat of solution of liquid silicon in liquid iron must be known. From the data cited in Part II, the enthalpy of solution of liquid silicon in liquid iron at infinite dilution is $\overline{L}_{si} = -35,000$ cal/mole Si. Making the approximation that L_{Si} at 1 wt per cent Si $\simeq L_{Si}^\circ$, the entropy of solution is

derived from

$$-34\,600 = -35\,000 - 1873 \times S_{Si} \quad (68)$$

$$S_{Si} = -0.32 \text{ cal/deg mole Si} \quad (69)$$

Since $\overline{L}_{si}$ and $\overline{S}_{Si}$ can be assumed to be independent of temperature, the variation of ΔF_{si}° with temperature is given by a linear equation

$$\Delta F_{si}^\circ = -35\,000 + 0.32T \quad (70)$$

Equilibrium Constant—Let us consider the following equilibrium occurring at constant temperature and pressure

$$xX + yY + \ldots = uU + vV + \ldots \quad (71)$$

where $x, y, u, v, \ldots$ are the numbers of gram-atoms of $X, Y, U, V, \ldots$ In terms of the partial molar free energies, the change in the free energy accompanying reaction is

$$\Delta F = (u\overline{F}_U + v\overline{F}_V + \ldots) - (x\overline{F}_X + y\overline{Y}_Y + \ldots) \quad (72)$$

Similarly, when the reactants and products are in their standard states, the free-energy change for the reference state is:

$$\Delta F^\circ = (uF_U^\circ + vF_v^\circ + \ldots) - (xF_x^\circ + yF_Y^\circ + \ldots) \quad (73)$$

The difference $\Delta F^\circ - \Delta F$ becomes,

$$\Delta F^\circ - \Delta F = -RT\,[\,(u \ln a_U + v \ln a_V + \ldots) - (x \ln a_X + y \ln a_Y + \ldots)\,] \quad (74)$$
$$= -RT \ln \frac{(a_U)^u (a_V)^v \ldots}{(a_X)^x (a_Y)^y \ldots}$$

In terms of activities of reactants and products, the equilibrium constant K is defined by the ratio,

$$K = \frac{(a_U)^u (a_V)^v \ldots}{(a_X)^x (a_Y)^y \ldots} \quad (75)$$

When discussing the Gibbs free energy concept, it was shown that $\Delta F = 0$ when the system is in a state of absolute rest, i.e., state of equilibrium. Therefore, inserting $\Delta F = 0$ in equation (74), the standard free-energy change accompanying reaction (71) is related to the equilibrium constant by

$$\Delta F^\circ = -RT \ln K = -2.303\, RT \log K \quad (76)$$

If ΔF° is in calories,

$$\Delta F^\circ = -4.575\, T \log K \quad (77)$$

It follows from equation (33) that

$$\frac{d(\Delta F^\circ/T)}{dT} = -\frac{\Delta H^\circ}{T^2}$$

or
$$\quad (78)$$
$$\frac{d(\Delta F^\circ/T)}{d(1/T)} = \Delta H^\circ$$

Combining this with equation (76) gives

or

$$\frac{d \ln K}{dT} = \frac{\Delta H°}{RT^2}$$

$$\frac{d \ln K}{d(1/T)} = -\frac{\Delta H°}{R}$$

(79)

This is commonly known as the **van't Hoff** equation.

It is evident from the relation in equation (79) that if the temperature of the system (at equilibrium) is increased, the reaction will proceed in the direction causing heat absorption. The pressure also has a similar effect on the direction of reaction. For example, consider the dissociation of carbon dioxide:

$$2CO_2 = 2CO + O_2$$

This is an endothermic reaction, therefore an increase in temperature will favor dissociation of CO_2.

Since the dissociation is accompanied by volume expansion at constant temperature and pressure, increasing pressure will reverse the reaction from right to left, i.e., dissociation of carbon dioxide becomes less with increasing pressure.

These effects on the state of equilibrium are summarized by the **Le Chatelier** principle (1885) which may be stated as follows: "If a system in equilibrium is subjected to a constant addition of heat, increase of volume, pressure, etc. which alter the equilibrium, the direction of the reaction taking place is such as to oppose the constraint, i.e., partially to nullify its effect."

If Raoult's law is obeyed, the activity coefficient is unity, i.e. $a_i = N_i$ and, therefore, for ideal solutions K is represented in terms of concentrations:

$$K = \frac{(N_U)^u (N_V)^v \dots}{(N_X)^x (N_Y)^y \dots}$$

(80)

It is this form of the equilibrium constant which was first formulated by **Guldberg** and **Waage** (1864-67) based on the concept of mass action law.

It follows from the foregoing definition of the activity that in reactions involving gases and vapors, the equilibrium constant is given in terms of the partial pressures of the reacting gases.

Phase Rule and Phase Equilibrium Diagrams—The number of variable factors which define the state of a system is called the degrees of freedom. From thermodynamic considerations Gibbs showed that provided the equilibrium between any number of phases is not influenced by external forces, e.g. gravity, electrical, magnetic, etc., but only by temperature, pressure and composition, the number of degrees of freedom (F) of a system at equilibrium is related to the number of components (C) and of phases (P) by equation

$$F = C - P + 2$$

(81)

This is known as the **phase rule**.

The application of the phase rule to heterogeneous equilibria is illustrated below by considering the solid-liquid-vapor equilibria for a system of one component. For a single phase such as water, C = 1 and P = 1 and, therefore, the number of degrees of freedom F = 2; this is a bivariant equilibrium where both temperature

and pressure can be set or changed arbitrarily. For water-water vapor, water-ice or ice-water vapor equilibrium C = 1, P = 2 and F = 1; i.e. in this univariant equilibrium either temperature or pressure is the independent variable. When the three phases, ice, water and water vapor are in equilibrium, F = 0; this is an invariant system. Using the information obtained from the phase rule, a pressure/temperature phase diagram for a single component system is drawn schematically in Figure 13—8. The curves separating two single-phase regions correspond to two-phase univariant equilibria. At the triple point O (invariant) three

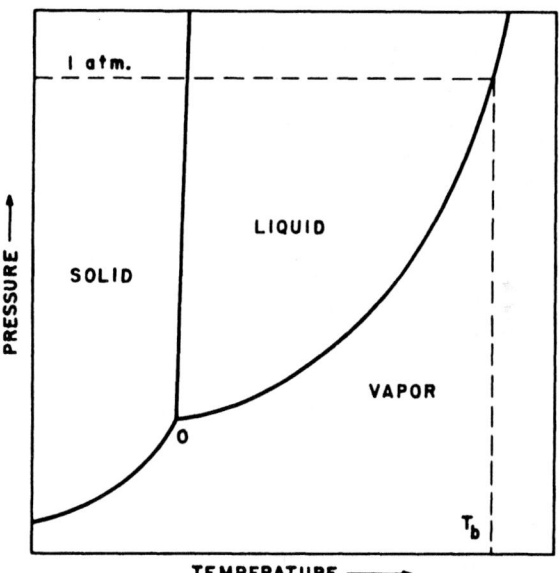

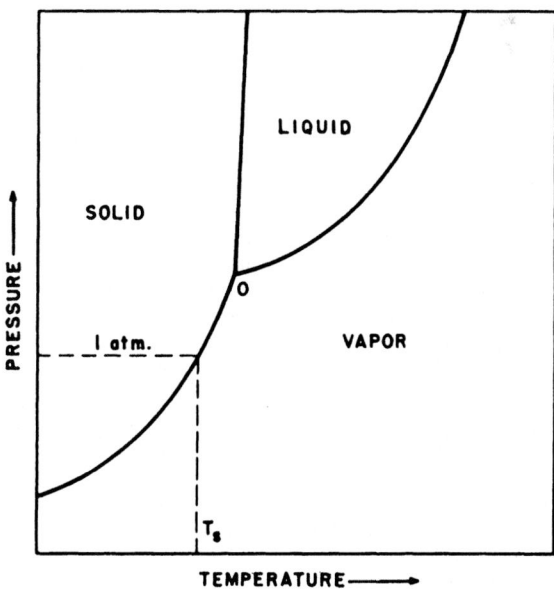

FIG. 13—8. Pressure-temperature phase equilibrium diagram for a single component system which (a) has a normal boiling temperature and (b) sublimes at atmospheric pressure.

univariant curves intersect. As indicated in Figure
13—8, if the pressure for the invariant system is greater
than 1 atm, the substance does not melt at atmospheric
pressure, but rather sublimes. For example, at atmospheric pressure solid carbon dioxide sublimes at
— 78.5°C; if pressure is increased to 5.11 atm, it melts
at — 56.4°C.

Again it follows from the phase rule that at constant
pressure in a single component system the transformation of a substance from one crystalline formation to
another, melting or boiling occur at particular temperatures characteristic of the substance, and during the
process of transformation the temperature of the system remains constant.

Metallurgical systems are generally condensed systems (solid and liquid systems), therefore the effect of
external pressure on the phase equilibria is not perceptible, at least at ordinary pressures. Since all the phase
equilibrium diagrams are for a constant pressure of 1
atm, the only remaining independent variables of the
system are temperature and composition, consequently
the phase rule takes the following form for constant
pressure:

$$F = C - P + 1 \qquad (82)$$

The derivation of the temperature/composition
phase equilibrium diagram from the free energy of the
system will not be discussed here; adequate information on this subject is available in textbooks. For the
present purpose it is sufficient to give a few schematic
examples of typical binary phase equilibrium diagrams.

Binary Systems—When the components A and B of a
binary system are mutually soluble in the liquid and
solid state, the simplest binary phase diagram is as
shown in Figure 13—9. In this system, the freezing of
a melt of composition X occurs within the temperature
range T_1 and T_2, and the composition of the liquid and
solid solutions in equilibrium with one another change
along the L_1L_2 liquidus and S_1S_2 solidus curves respectively. At an intermediate temperature T', the ratio
L'X'/X'S' gives the amount of solid solution S' relative
to residual liquid.

When there is a greater positive departure from ideality for the solid solution than for the liquid, the
liquidus and solidus curves pass through a minimum
and at a lower temperature the solid solution separates
into two phases, as shown in Figure 13—10. With further increase in positive departure from ideality in the
solid solution, a eutectic type of a diagram is obtained;
this is shown in Figure 13—11 where, at the eutectic
temperature, there are three phases, α, β and liquid.
Hence, the degree of freedom is zero; this is an invariant system and, therefore, the eutectic composition
and temperature are constant and the solidus line is
drawn parallel to the composition abscissa.

Figure 13—12 is a typical example of a peritectic
reaction with partial solid solutions. In this system, the
peritectic reaction occurs at the peritectic temperature
between the solid solution α of composition C and liquid P forming a solid solution β of composition D.
Within the field CDC'D' there are two phases α and β,
the compositions of which change with temperature
along the solubility curves CC' and DD' respectively.

In a number of binary systems the components are

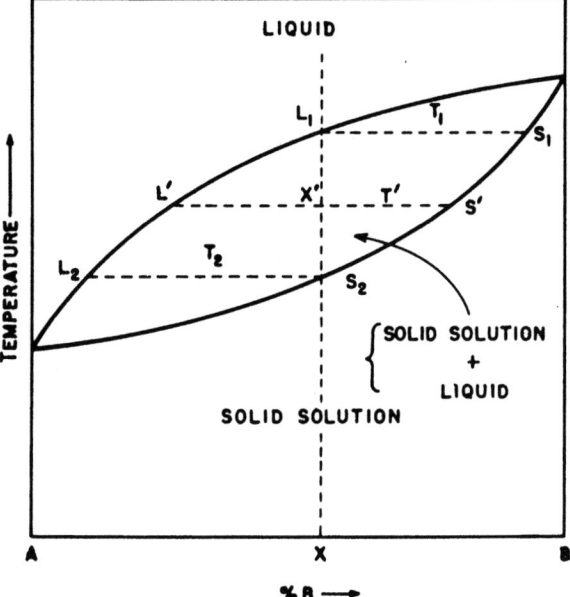

FIG. 13—9. Complete solid solution in a binary system.

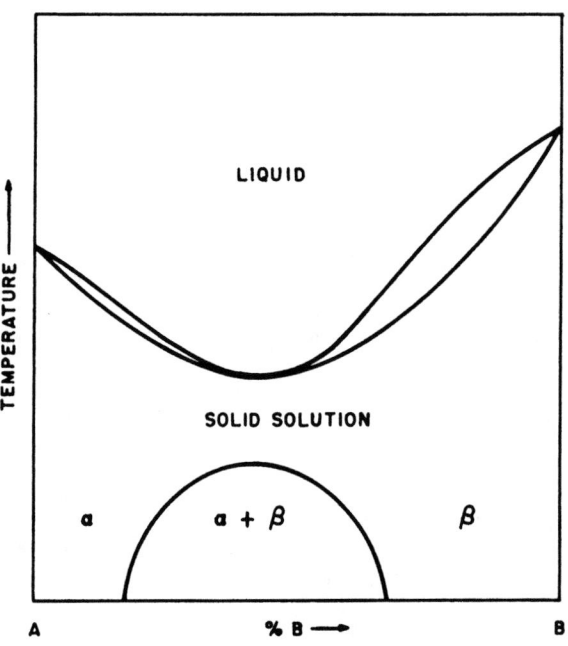

FIG. 13—10. Minimum on liquidus and solidus curves and partial miscibility in the solid state.

not completely miscible in the liquid state; a typical
example is given in Figure 13—13. With increasing
temperature the mutual solubilities of liquid (1), rich in
A, and liquid (2), rich in B, increase and above a particular critical temperature the two liquids are completely miscible. However, there are many systems
where complete miscibility can never be reached at
atmospheric pressures, e.g. Fe-Pb, Ag-S. The
activity/composition relation for a system with miscibility gap is as shown in Figure 13—14. Since within

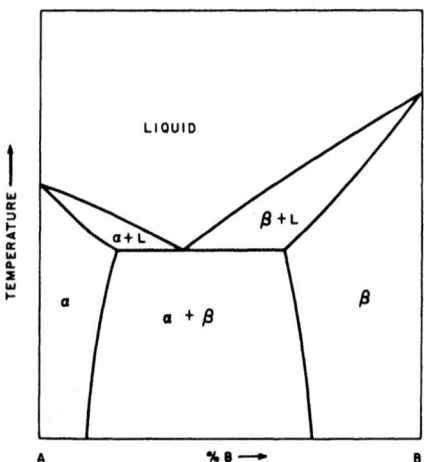

FIG. 13—11. Binary eutectic systems with partial solid solubilities.

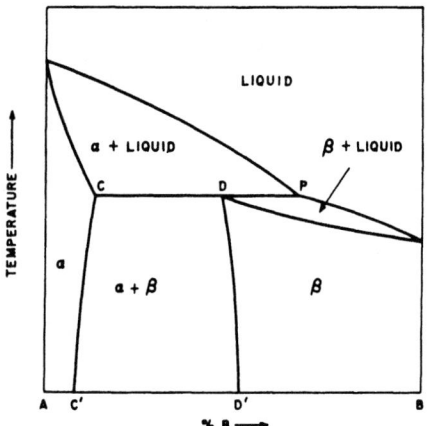

FIG. 13—12. Peritectic reaction in a binary system with partial solubilities.

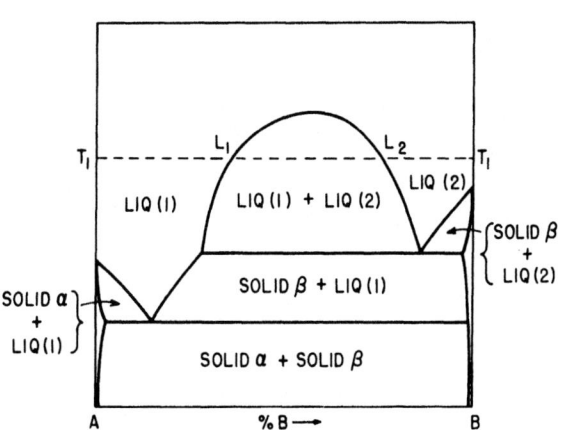

FIG. 13—13. Liquid miscibility gap in a binary system.

the composition range L_1-L_2 the liquids L_1 and L_2 are in equilibrium, the activity of A (and B) in both phases are equal as indicated by the horizontal line L_1L_2 in Figure 13—14.

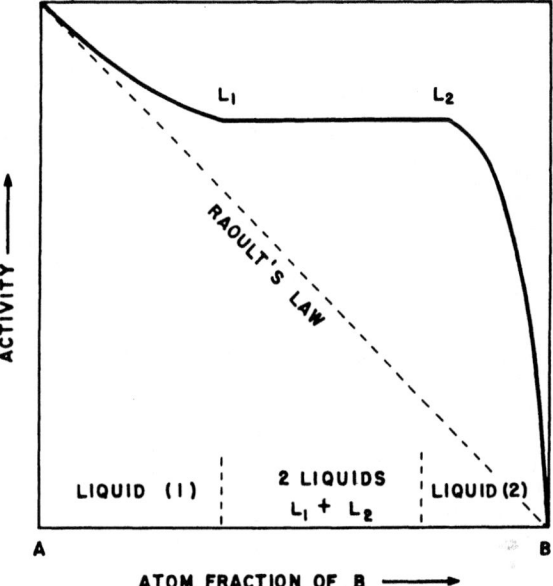

FIG. 13—14. Activity of component A at temperature T_1 of the system given in Fig. 13—13; pure liquid A is the standard state.

Figure 13—15 shows the formation of a compound X in a binary system A-B. This can be visualized as two binary eutectic systems A-X and X-B with limited solid solubilities at the terminal ends of the system. When a chemical compound does not have a fixed composition, it is said to be a nonstoichiometric compound; this is shown in Figure 13—16 where the solid solution β phase is a nonstoichiometric compound.

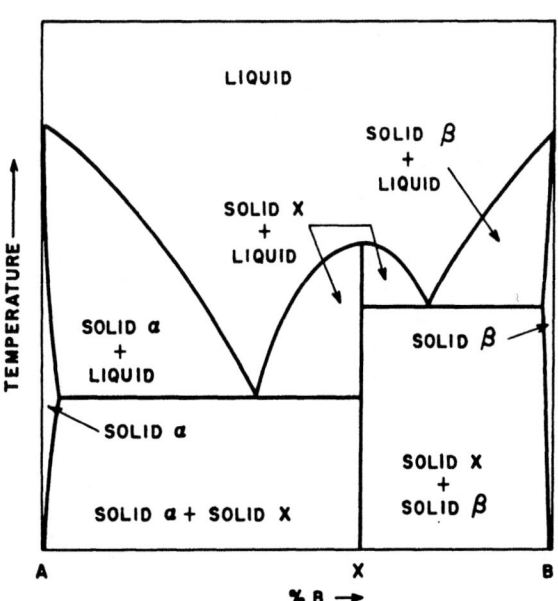

FIG. 13—15. Binary eutectic system with a compound.

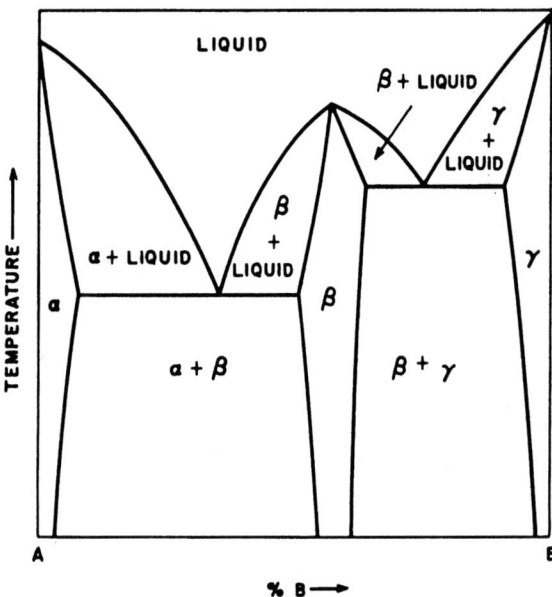

FIG. 13—16. Solid solution in a eutectic system with non-stoichiometric compound β phase.

THERMOCHEMICAL DATA

There are several compilations of thermochemical data which have periodically been updated. To facilitate thermodynamic calculations, selected values of ΔH°_{298}, ΔS°_{298} and high-temperature heat capacities are given in Table 13—III for substances frequently

encountered in the metallurgical processes. The enthalpies of formation of compounds from elements in their standard states at 298.16°K are reproduced from the compilation of Kubaschewski, Evans and Alcock[8], the entropies at 298.16°K from the compilation of Kelley and King[9] and the high-temperature heat capacities from Kelley[10].

Simplified Free-Energy Equation—The exact method of handling thermochemical data for determining the free-energy equation for a reaction is given in other publications. However the lack of accuracy of the thermochemical data for many substances of interest in high-temperature reactions hardly warrants exact thermodynamic calculations. Little is known about the high-temperature heat capacities of solid or liquid metallic solutions and slags. Furthermore, for many reactions ΔC_p is small and often can be taken to be zero, therefore error involved in approximating ΔF° as a linear function of temperature is small. For these reasons the change in the standard free energy accompanying a reaction is often given in a simplified form, thus

$$\Delta F^{\circ} = \Delta H^{\circ} - T\Delta S^{\circ} \qquad (83)$$

where ΔH° and ΔS° are the average values of enthalpy and entropy changes accompanying a reaction over a moderate temperature range for which a linear equation (83) is a reasonable approximation.

The standard free energy changes accompanying simple metallurgical reactions are given in Table 13—IV as linear functions of temperature, thus

$$\Delta F^{\circ} = A + BT \qquad (84)$$

Table 13—IV   Standard Free Energies of Some Reactions Encountered in Ferrous Metallurgical Processes*.
Notations: (g) gas, (l) liquid, (s) solid, " " nonstoichiometric compounds.

Reaction	$\Delta F^{\circ}_T = A + BT$ cal		$\pm$ kcal	Temp. Range °C
	$-A$	B		
$2Al(s) + \frac{3}{2} O_2(g) = Al_2O_3(s)$	399 500	74.71	2	25–659
$2Al(l) + \frac{3}{2} O_2(g) = Al_2O_3(s)$	402 300	77.83	2	659–1700
$Al(s) + \frac{1}{2} N_2(g) = AlN(s)$	77 000	22.25	8	25–659
$Al(l) + \frac{1}{2} N_2(g) = AlN(s)$	79 270	24.68	8	659–1700
$4Al(s) + 3C(s) = Al_4C_3(s)$	51 600	10.0	2	25–659
$4Al(l) + 3C(s) = Al_4C_3(s)$	63 700	23.0	2	659–1700
$2B(s) + \frac{3}{2} O_2(g) = B_2O_3(s)$	305 120	63.01	1	25–450
$2B(s) + \frac{3}{2} O_2(g) = B_2O_3(l)$	295 630	50.41	1	450–1700
$B(s) + \frac{1}{2} N_2(g) = BN(s)$	60 600	21.40	0.5	25–900
$4B(s) + C(s) = B_4C(s)$	13 580	1.69	1	25–900
$Ba(s) + \frac{1}{2} O_2(g) = $ "BaO"(s)	135 800	23.2	4	25–704
$Ba(l) + \frac{1}{2} O_2(g) = $ "BaO"(s)	139 000	26.6	5	704–1638
$3Ba(s) + N_2(g) = Ba_3N_2(s)$	87 000	57.4	9	25–704
$BaO(s) + SiO_2(s) = BaSiO_3(s)$	26 800	0.1	3	25–1300
$Be(s) + \frac{1}{2} O_2(g) = BeO(s)$	142 900	23.13	5	25–1283
$Be(l) + \frac{1}{2} O_2(g) = BeO(s)$	142 360	23.36	5	1283–1700
$3Be(s) + N_2(g) = Be_3N_2(s)$	134 700	40.6	12	25–700
$C(s) + 2H_2(g) = CH_4(g)$	21 550	26.16	1	25–2000
$C(s) + \frac{1}{2} O_2(g) = CO(g)$	26 700	−20.95	1	25–2000

(Continued on next page)

Table 13—IV (Continued). Standard Free Energies of Some Reactions Encountered in Ferrous Metallurgical Processes*—Notations: (g) gas, (l) liquid, (s) solid, " " nonstoichiometric compounds.

Reaction	$\Delta F°_T = A + BT$ cal		± kcal	Temp. Range °C
	−A	B		
$C(s) + O_2(g) = CO_2(g)$	94 200	−0.20	1	25–2000
$C(s) + \frac{1}{2} S_2(g) = CS(g)$	-59 000	−22.75	7	1600–1800
$C(s) + S_2(g) = CS_2(g)$	3 100	−1.73	1	25–1300
$CO(g) + \frac{1}{2} S_2(g) = COS(g)$	22 860	18.7	3	25–1200
graphite = diamond	310	1.13	0.2	25–1200
$Ca(s) + \frac{1}{2} O_2(g) = CaO(s)$	151 330	23.66	1.5	25–850
$Ca(l) + \frac{1}{2} O_2(g) = CaO(s)$	153 550	25.64	1.5	850–1487
$Ca(g) + \frac{1}{2} O_2(g) = CaO(s)$	190 100	46.62	2.5	1487–1700
$Ca(s) + \frac{1}{2} S_2(g) = CaS(s)$	129 490	22.86	1	25–850
$Ca(l) + \frac{1}{2} S_2(g) = CaS(s)$	131 780	24.94	1	850–1487
$Ca(g) + \frac{1}{2} S_2(g) = CaS(s)$	168 360	45.72	2	1487–1700
$3Ca(s) + N_2(g) = Ca_3N_2(s)$	105 000	50.0	10	25–850
$Ca(s) \alpha + 2C(s) = CaC_2(s)$	13 600	−5.9	3	25–400
$Ca(s)\beta + 2C(s) = CaC_2(s)$	11 620	−8.64	3	400–850
$Ca(l) + 2C(s) = CaC_2(s)$	13 700	−6.80	3	850–1487
$Ca(g) + 2C(s) = CaC_2(s)$	51 210	12.3	5	1487–1900
$Ca(s) + Si(s) = CaSi(s)$	36 000	−0.5	4	25–850
$Ca(l) + Si(s) = CaSi(s)$	25 750	−6.6	5	850–1444
$2Ca(l) + Si(s) = Ca_2Si(s)$	42 600	−4.6	5	850–1444
$3CaO(s) + Al_2O_3(s) = Ca_3Al_2O_6(s)$	3 900	−6.30	2	25–1550
$12CaO(s) + 7Al_2O_3(s) = Ca_{12}Al_{14}O_{33}(s)$	17 460	−49.60	2	25–1500
$CaO(s) + Al_2O_3(s) = CaAl_2O_4(s)$	4 570	−4.10	2	25–1600
$CaO(s) + CO_2(g) = CaCO_3(s)$	40 250	34.40	1	25–880
$2CaO(s) + Fe_2O_3(s) = Ca_2Fe_2O_5(s)$	9 200	−2.33	1.5	600–1435
$2CaO(s) + Fe_2O_3(s) = Ca_2Fe_2O_5(l)$	−7 560	−12.13	1.5	1435–1600
$4CaO(s) + P_2(g) + \frac{5}{2} O_2(g) = Ca_4P_2O_9(s)$	563 580	144.0	3.0	1300–1600
$3CaO(s) + P_2(g) + \frac{5}{2} O_2(g) = Ca_3P_2O_8(s)$	553 350	144.0	3.0	1300–1600
$2CaO(s) + SiO_2(s) = Ca_2SiO_4(s)$	30 200	−1.2	2.5	25–1400
$CaO(s) + SiO_2(s) = CaSiO_3(s)\alpha$	21 300	0.12	1	25–1210
$CaO(s) + SiO_2(s) = CaSiO_3(s)\beta$	19 900	−0.82	2	1210–1543
$2Ce(s) + \frac{3}{2} O_2(g) = Ce_2O_3(s)$	434 870	78.4	25	25–804
$2Ce(l) + \frac{3}{2} O_2(g) = Ce_2O_3(s)$	437 850	87.1	25	804–1700
$Ce(s) + O_2(g) = CeO_2(s)$	244 060	48.9	12	25–804
$Ce(l) + O_2(g) = CeO_2(s)$	247 500	52.0	12	804–1700
$Co(s) + \frac{1}{2} O_2(g) = CoO(s)$	55 900	16.9	2	25–1495
$3CoO(s) = \frac{1}{2} O_2(g) = Co_3O_4(s)$	43 800	35.4	3	25–1000
$9Co(s) + 4S_2(g) = Co_9S_8(s)$	316 960	159.24	2	25–778
$2Co(s) + C(s) = Co_2C(s)$	−3 950	−2.08	5	25–900
$2Cr(s) + \frac{3}{2} O_2(g) = Cr_2O_3(s)\beta$	267 750	62.10	0.5	25–1898
$2Cr(s) + \frac{1}{2} N_2(g) = Cr_2N(s)$	24 000	11.65	5	25–1898
$Cr(s) + \frac{1}{2} N_2(g) = CrN(s)$	25 500	16.70	7.5	25–1898
$23Cr(s) + 6C(s) = Cr_{23}C_6(s)$	98 280	−9.24	10	25–1400
$7Cr(s) + 3C(s) = Cr_7C_3(s)$	41 800	−6.1	10	25–1200
$3Cr(s) + 2C(s) = Cr_3C_2(s)$	20 800	−4.0	10	25–1700
$2Cu(s) + \frac{1}{2} O_2(g) = Cu_2O(s)$	39 330	16.40	1	25–1084
$2Cu(l) + \frac{1}{2} O_2(g) = Cu_2O(s)$	45 570	21.00	1.5	1084–1230
$\frac{1}{2}Cu_2O(s) + \frac{1}{4}O_2(g) = CuO(s)$	18 250	13.25	0.5	25–1000
$Fe(s) + \frac{1}{2} O_2(g) = $ "FeO"(s)	62 050	14.95	1	25–1371
$Fe(l) + \frac{1}{2} O_2(g) = $ "FeO"(l)	55 620	10.83	1	1537–1700
$3Fe(s) + 2O_2(g) = Fe_3O_4(s)$	260 770	74.75	2	25–560
$3FeO(s) + \frac{1}{2} O_2(g) = Fe_3O_4(s)$	74 620	29.90	2	560–1371
$\frac{2}{3} Fe_3O_4(s) + \frac{1}{6} O_2(g) = Fe_2O_3(s)$	19 870	11.20	2	25–1400
$Fe(s) + \frac{1}{2} S_2(g) = FeS(s)\alpha$	37 160	15.59	1	25–140
$Fe(s) + \frac{1}{2} S_2(g) = FeS(s)\beta$	35 910	12.56	1	140–906
$FeS(\beta) + \frac{1}{2} S_2(g) = FeS_2$	43 350	45.00	1.5	300–800

(Continued on next page)

Table 13—IV (Continued). Standard Free Energies of Some Reactions Encountered in Ferrous Metallurgical Processes*—Notations: (g) gas, (l) liquid, (s) solid, " " nonstoichiometric compounds.

Reaction	$\Delta F°_T = A + BT$ cal		± kcal	Temp. Range °C
	−A	B		
$4Fe(s) + \frac{1}{2} N_2(g) = Fe_4N(s)$	1 130	9.70	1	25–600
$3Fe(s) + \frac{1}{2} P_2(g) = Fe_3P(s)$	51 000	11.3	8	25–1170
$3Fe(s) + C(s) = Fe_3C(s)$	−6 200	−5.53	1	25–190
$3Fe(s) + C(s) = Fe_3C(s)$	−6 380	−5.92	1	190–840
$3Fe(s) + C(s) = Fe_3C(s)$	−2 475	−2.43	1	840–1537
$"FeO"(s) + Al_2O_3(s) = FeAl_2O_4(s)$	11 800	5.43	4	25–1371
$"FeO"(s) + Cr_2O_3(s) = FeCr_2O_4(s)$	2 600	−3.37	2	35–1371
$2"FeO"(s) + SiO_2(s) = Fe_2SiO_4(s)$	7 950	3.65	2	25–1217
$2"FeO"(s) + SiO_2(s) = Fe_2SiO_4(l)$	−14 880	−11.49	2	1217–1371
$2"FeO"(l) + SiO_2(s) = Fe_2SiO_4(l)$	−3 450	−4.58	2	1371–1700
$"FeO"(s) + TiO_2(s) = FeTiO_3(s)$	1 410	−2.54	1	900–1371
$H_2(g) + \frac{1}{2} O_2(g) = H_2O(g)$	58 900	13.10	0.5	25–1700
$H_2(g) + \frac{1}{2} S_2(g) = H_2S(g)$	21 580	11.80	0.5	25–1500
$\frac{3}{2} H_2(g) + \frac{1}{2} N_2(g) = NH_3(g)$	12 050	26.70	2	25–700
$2La(s) + \frac{3}{2} O_2(g) = La_2O_3(s)$	444 750	66.8	4	25–700
$La(s) + \frac{1}{2} N_2(g) = LaN(s)$	72 100	25.0	9	25–700
$Mg(s) + \frac{1}{2} O_2(g) = MgO(s)$	143 740	25.17	1.5	25–650
$Mg(l) + \frac{1}{2} O_2(g) = MgO(s)$	146 570	27.36	1.5	650–1120
$Mg(g) + \frac{1}{2} O_2(g) = MgO(s)$	176 800	49.10	3	1120–1700
$Mg(s) + \frac{1}{2} S_2(g) = MgS(s)$	99 650	22.80	5	25–650
$Mg(l) + \frac{1}{2} S_2(g) = MgS(s)$	101 800	25.65	5	650–1120
$Mg(g) + \frac{1}{2} S_2(g) = MgS(s)$	134 350	48.75	5	1120–1700
$3Mg(s) + N_2(g) = Mg_3N_2(s)$	109 600	47.41	3	25–650
$3Mg(l) + N_2(g) = Mg_3N_2(s)$	115 970	54.30	3	650–1120
$3MgO(s) + CO_2(g) = MgCO_3(s)$	28 100	40.6	3	25–700
$3MgO(s) + P_2(g) + \frac{5}{2} O_2(g) = Mg_3P_2O_8(s)$	501 610	144.0	3	1000–1250
$2MgO(s) + SiO_2(s) = Mg_2SiO_4(s)$	15 120	0.0	2	25–1400
$MgO(s) + SiO_2(s) = MgSiO_3(s)$	8 900	1.1	1	25–1300
$Mn(s) + \frac{1}{2} O_2(g) = MnO(s)$	91 950	17.40	3	25–1244
$Mn(l) + \frac{1}{2} O_2(g) = MnO(s)$	95 400	19.70	3	1244–1700
$Mn(s) + \frac{1}{2} S_2(g) = MnS(s)$	65 000	16.21	1.5	25–1244
$Mn(l) + \frac{1}{2} S_2(g) = MnS(s)$	69 010	18.86	2	1244–1530
$Mn(l) + \frac{1}{2} S_2(g) = MnS(l)$	62 770	15.40	2	1530–1700
$3Mn(s) + C(s) = Mn_3C(s)$	3 330	−0.26	3	25–740
$MnO(s) + SiO_2(s) = MnSiO_3(s)$	5 920	3.0	4	25–1300
$Mo(s) + O_2(g) = MoO_2(s)$	139 230	40.70	2	25–1000
$MoO_2(s) + \frac{1}{2} O_2(g) = MoO_3(s)$	38 700	19.50	3	25–1000
$2Mo(s) + \frac{3}{2} S_2(g) = Mo_2S_3(s)$	128 550	54.62	3	850–1200
$2Mo(s) + \frac{1}{2} N_2(g) = Mo_2N(s)$	15 250	13.25	3	25–1000
$2Mo(s) + C(s) = Mo_2C(s)$	6 700	0.0	8	25–1000
$Ni(s) + \frac{1}{2} O_2(g) = NiO(s)$	55 940	20.04	2	25–1452
$Ni(l) + \frac{1}{2} O_2(g) = NiO(s)$	60 140	22.47	3	1452–1900
$Ni(s) + \frac{1}{2} S_2(g) = NiS(s)$	39 980	17.20	3	300–580
$3Ni(s) + C(s) = Ni_3C(s)$	−8 110	−1.70	3	25–700
$\frac{1}{2}S_2(g) = S(g)$	−51 000	−13.9	0.5	25–1700
$\frac{1}{2}S_2(g) + \frac{1}{2} O_2(g) = SO(g)$	15 400	−1.4	3	25–1700
$\frac{1}{2}S_2(g) + O_2(g) = SO_2(g)$	86 620	17.31	0.5	25–1700
$\frac{1}{2}S_2(g) + \frac{3}{2} O_2(g) = SO_3(g)$	109 220	38.67	1.5	25–1500
$Si(s) + \frac{1}{2} O_2(g) = SiO(g)$	22 600	−19.71	3	25–1410
$Si(l) + \frac{1}{2} O_2(g) = SiO(g)$	36 150	−11.51	3	1410–1700
$Si(s) + O_2(g) = SiO_2(s)\beta$ cristobalite	215 600	41.50	3	400–1410
$Si(l) + O_2(g) = SiO_2(s)\beta$ cristobalite	227 700	48.70	3	1410–1700
$3Si(s) + 2N_2(g) = Si_3N_4(s)$	172 700	75.2	2	25–1410
$3Si(l) + 2N_2(g) = Si_3N_4(s)$	209 000	96.8	2	1410–1700

(Continued on next page)

Table 13—IV (Continued). Standard Free Energies of Some Reactions Encountered in Ferrous Metallurgical Processes*—Notations: (g) gas, (l) liquid, (s) solid, " " nonstoichiometric compounds.

Reaction	$\Delta F°_T = A + BT$ cal		± kcal	Temp. Range °C
	−A	B		
$Si(s) + C(s) = SiC(s)\beta$	13 000	0.73	2	25–1410
$Si(l) + C(s) = SiC(s)\beta$	25 100	7.91	2	1410–1700
$Sn(l) + O_2(g) = SnO_2(s)$	140 180	51.52	1	500–700
$Th(s) + O_2(g) = ThO_2(s)$	294 400	43.38	5	25–1500
$3Th(s) + 2N_2(g) = Th_3N_4(s)$	310 400	89.7	20	25–1700
$Th(s) + 2C(s) = ThC_2$	45 000	2.6	10	25–2000
$Ti(s) + ½ O_2(g) =$ "TiO"	122 300	21.3	4	300–1700
2"TiO"(s) + ½ $O_2(g) = Ti_2O_3$	114 150	19.1	4	25–1700
$\frac{3}{2} Ti_2O_3(s) + \frac{1}{4} O_2(g) = Ti_3O_5(s)$	44 250	9.85	1.5	400–1700
$\frac{1}{3}Ti_3O_5(s) + \frac{1}{6} O_2(g) =$ "TiO_2"(s)	24 330	7.67	1	25–1850
$Ti(s)\alpha + ½ N_2(g) =$ "TiN"(s)	80 250	22.2	2	25–882
$Ti(s)\beta + ½ N_2(g) =$ "TiN"(s)	80 850	22.77	2	882–1200
$Ti(s)\alpha + C(s) =$ "TiC"(s)	43 750	2.41	3	25–882
$Ti(s)\beta + C(s) =$ "TiC"(s)	44 600	3.16	3	882–1700
$V(s) + ½ O_2(g) =$ "VO"(s)	102 950	17.95	3	600–1500
2"VO"(s) + ½ $O_2(g) = V_2O_3(s)$	87 650	20.27	3	550–1112
$V(s) + C(s) = VC(s)$	24 100	1.5		900–1100
$W(s) + C(s) = WC(s)$	9 000	0.4	3	25–1700
"Zr"(s)$\alpha + O_2(g) =$ "ZrO_2"(s)	259 400	45.6	2	25–870
"Zr"(s)$\beta + O_2(g) =$ "ZrO_2"(s)	255 020	42.15	2	1205–1865
"Zr"(s)$\alpha + ½ N_2(g) =$ "ZrN"(s)	87 000	22.30	1	25–862
"Zr"(s)$\beta + ½ N_2(g) =$ "ZrN"(s)	87 930	23.11	1	862–1200
$Zr(s) + C(s) =$ "ZrC"(s)	44 100	2.2	3	25–1900

Most of the data in Table 13—IV are taken from the compilation of Kubaschewski et al. and some from Elliott et al.[11,12]. The equilibrium constant when using this simple form of the free-energy equation is given by

$$\ln K = -\frac{A}{RT} - \frac{B}{R} \tag{85}$$

Free Energies of Solution in Liquid Iron—Reactions occurring in iron- and steelmaking processes involve solutes at low concentrations in steel. For this reason it is convenient to choose a standard state for the solute, i, in iron such that the activity of i, a_i, is proportional to wt. % i at infinite dilution. The standard free energy of solution of pure component i in iron, with reference to the hypothetical 1 per cent solution is given by equation (63).

Selected $\Delta F°_i$ values given in Table 13—V are taken from those tabulated by Elliott et al.[12] with the exception of the values for Al and Si which are taken from the work of Fruehan[13,14].

Vapor Pressure of Elements—The vapor pressures of elements compiled by Honig and Kramer[15] are given graphically in Figures 13—17a, 13—17b, and 13—17c. The vapor pressure is in units of torr (mm Hg), atmosphere and newton/m² and the temperature in °C and °K. The circled point shown on most curves is the melting point of the element. The letters s (solid) or l (liquid) have been appended to the chemical symbol if the melting point falls outside the range of the graph.

Table 13—V. Standard Free Energy of Solution of Selected Elements in Liquid Iron (1 wt % solution, hypothetical).*

Element	$\Delta F°_i$, cal/g-atom
Al(l)	−15 000 − 6.67T
C(gr)	5 400 − 10.10T
Cr(s)	4 600 − 11.20T
½ H_2(g)	8 720 + 7.28T
Mn(l)	976 − 9.11T
½ N_2(g)	860 + 5.71T
½ O_2(g)	−28 000 − 0.69T
½ P_2(g)	−29 200 − 4.6T
Si(l)	−31 430 − 4.12T
½ S_2(g)	32 280 + 5.60T

THEORY OF DIFFUSION

Some knowledge of diffusion phenomena is prerequisite to the understanding of the kinetics of reactions, phase transformations, the rates of solidification, heat transfer, etc. In this section, only some of the basic principles of the diffusion theory are mentioned.

Diffusion is mass transfer leading to equalizing of chemical potentials. In discussing the basic diffusion theories, the concentration rather than chemical potential or activity of the diffusing component is used, on the assumption that the solution is ideal.

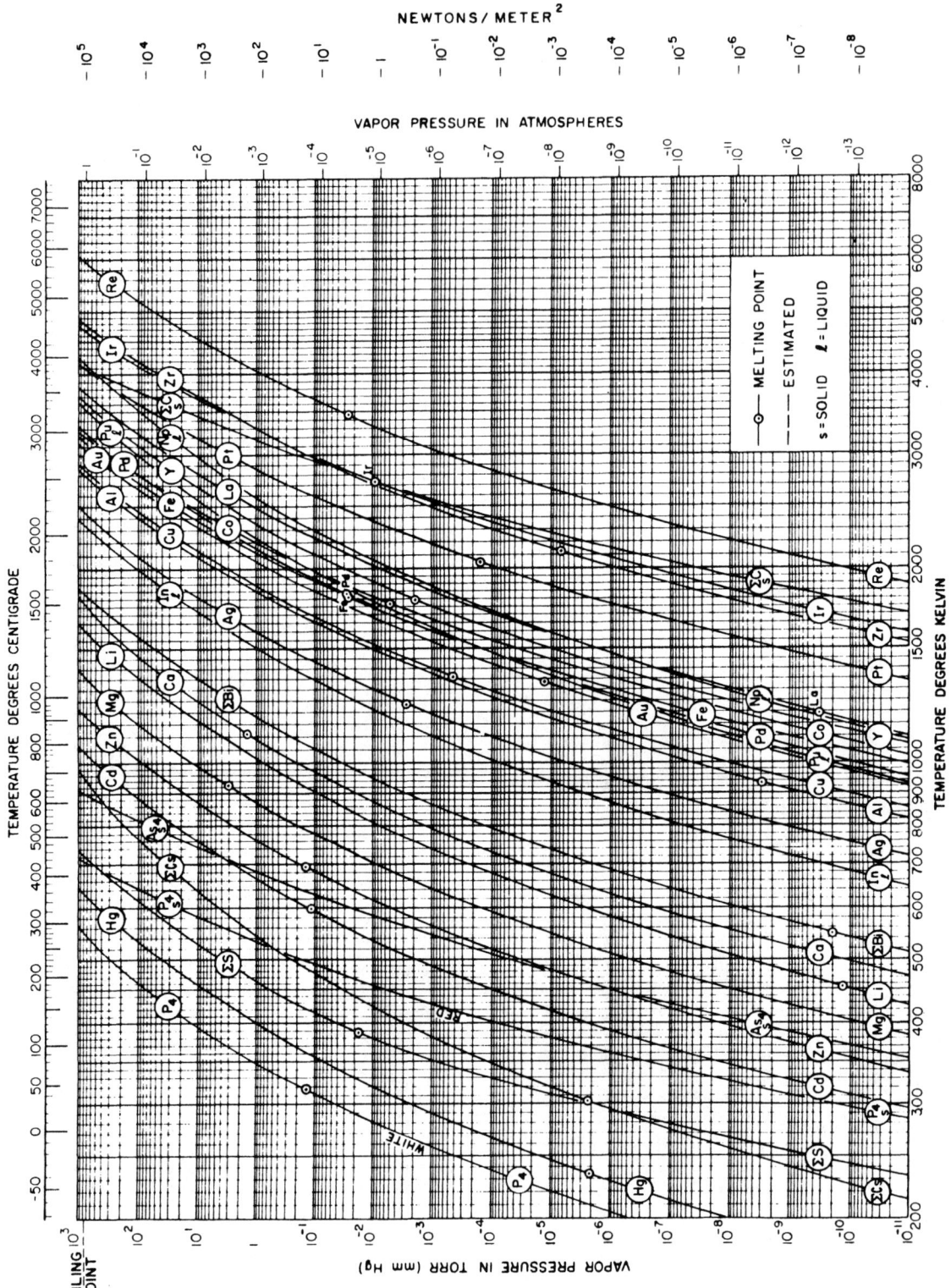

Fig. 13—17a. Vapor pressure data for the solid and liquid elements. (Honig and Kramer[15]).

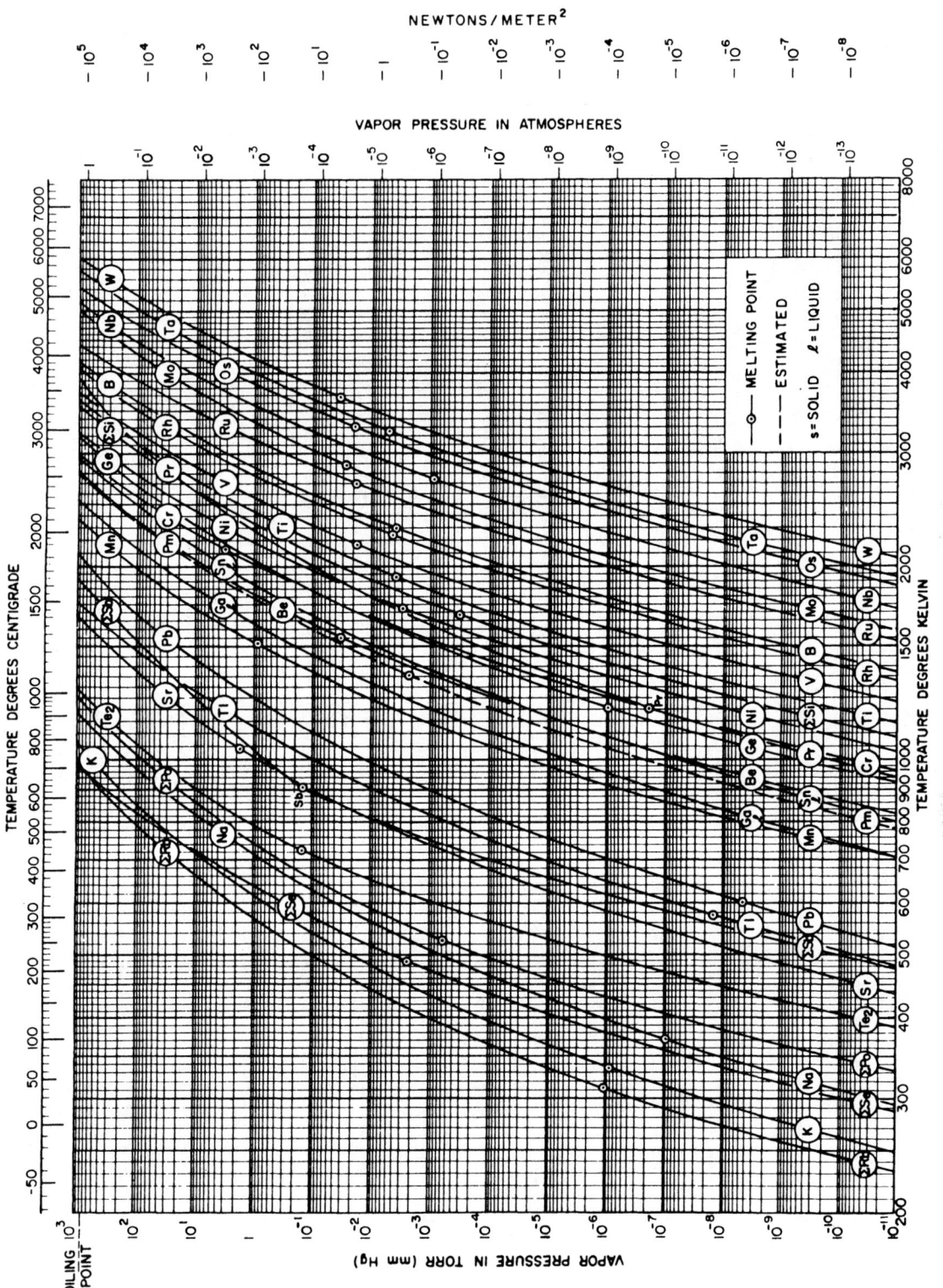

Fig. 13—17b. Vapor pressure data for the solid and liquid elements (Honig and Kramer[15]).

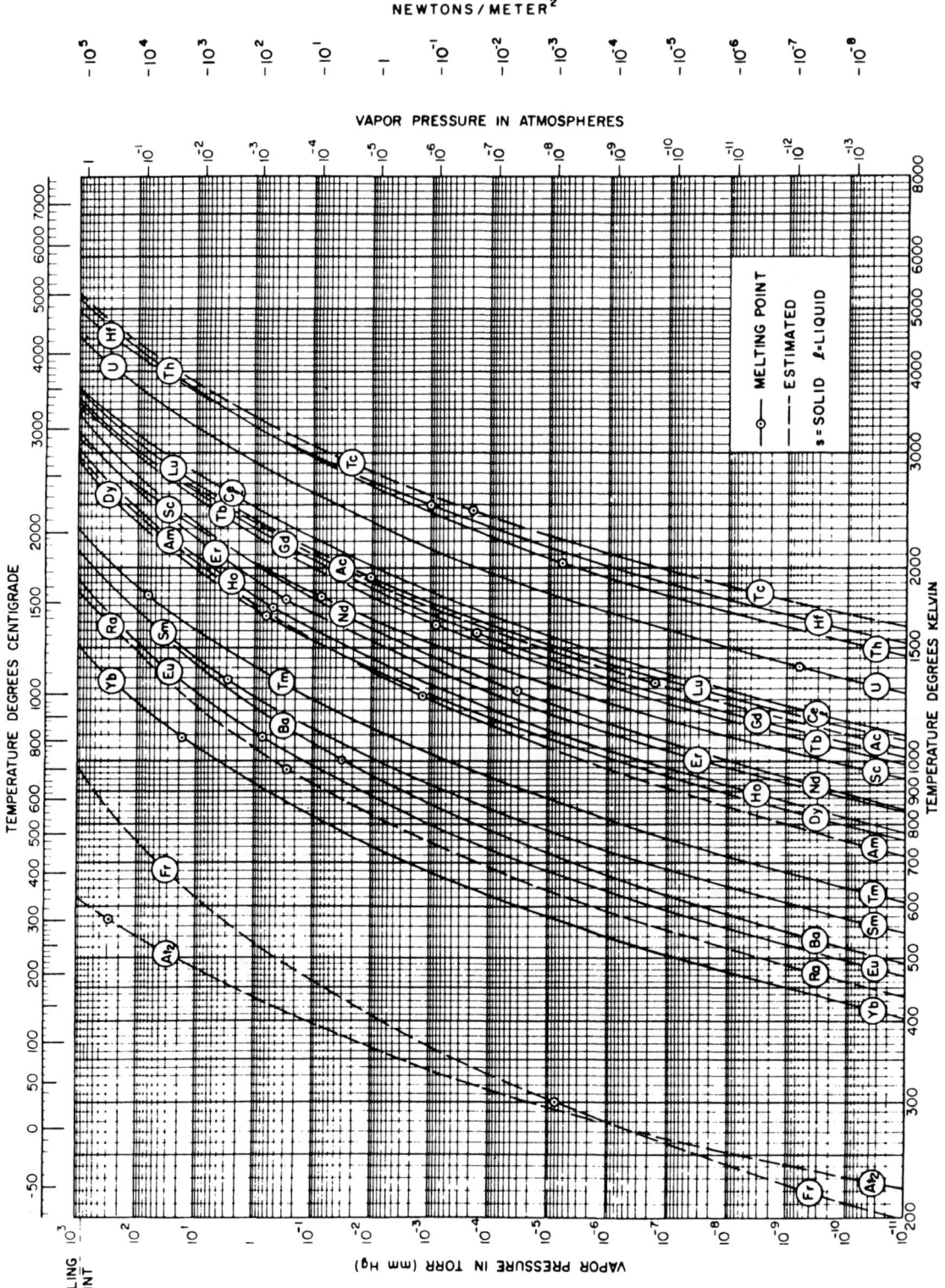

Fig. 13—17c. Vapor pressure data for the solid and liquid elements. (Honig and Kramer[15]).

FICK'S FIRST LAW—STEADY-STATE DIFFUSION

The diffusive flux J is defined as the amount of a diffusing species crossing a surface of unit area, perpendicular to the direction of flow, in unit time; this flux is equal to the product of the diffusivity, D, and the concentration gradient $\partial C/\partial x$, thus

$$J = -D\frac{\partial C}{\partial x} \quad (86)$$

where

C = amount of substance per unit volume,
x = distance in the direction of diffusion,
$\partial C/\partial x$ = concentration gradient,
D = coefficient of diffusion or diffusivity.

The diffusivity has the dimensions of $(length)^2/time$; using cgs units, D is given in cm^2/sec. The coefficient of diffusion is a function of state, i.e. varies with temperature, pressure and composition only.

Equation (86) states **Fick's first law,** and it is directly applicable for steady state diffusion with constant diffusivity. When a steady state is reached, flux J at any point along the diffusion path is constant and independent of time or distance. This is illustrated by an example in Figure 13—18, showing the diffusion of a gas through, for example, a metal diaphragm fixed inside a tube, the walls of which are impermeable to the gas. If the partial pressures p_1 and p_0 are kept constant on either side of the membrane, and $p_1 > p_0$, i.e. in terms of gas concentrations $C_1 > C_0$, the concentration gradient through the diaphragm is constant, i.e. $\partial C/\partial x = \Delta C/\Delta x = (C_0 - C_1)/\Delta x$. Since the diffusivity is assumed to remain constant, for a given temperature the flux is given by the first law:

$$J = -D\frac{C_0 - C_1}{\Delta x} \quad (87)$$

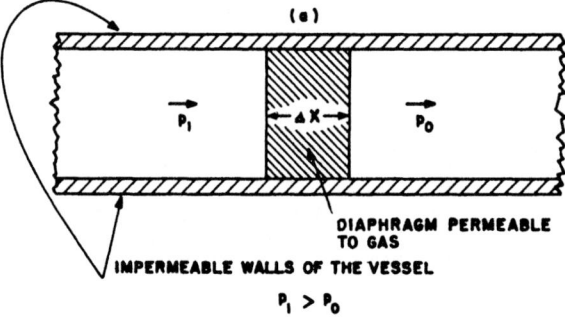

(a)

$P_1 > P_0$

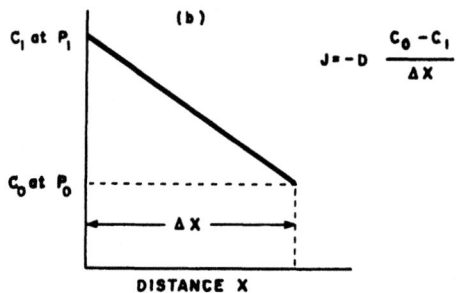

(b)

DISTANCE X

FIG. 13—18. Application of Fick's first law to a steady state diffusion through a diaphragm for constant diffusivity.

FICK'S SECOND LAW—NONSTEADY STATE DIFFUSION

In nonsteady state diffusion, the flux changes with the diffusion distance x, and time t; from the first law, this change $\partial J/\partial x$ is given by

$$\frac{\partial J}{\partial x} = -\frac{\partial}{\partial x}\left(D\frac{\partial C}{\partial x}\right) \quad (88)$$

The difference in flux is equal to $-\partial C/\partial t$, negative rate of concentration change, therefore,

$$\frac{\partial C}{\partial t} = \frac{\partial}{\partial x}\left(D\frac{\partial C}{\partial x}\right) \quad (89)$$

If the diffusivity D is independent of concentration (or substantially so under conditions of the experiment), **Fick's second law** may be written

$$\frac{\partial C}{\partial t} = D\frac{\partial^2 C}{\partial x^2} \quad (90)$$

The solution of equation (90) depends on the geometry and on the boundary conditions of the medium into which a substance is diffusing. For the present purpose, considerations are limited to a few special cases which are most frequently encountered in the study of the kinetics of metallurgical reactions to be discussed in the subsequent parts of this chapter.

(1) **Unidirectional diffusion in a semi-infinite medium.** Figure 13—19a shows a cross section of a slab with impermeable side walls. If the length of the slab is large, compared with the distance over which change in composition has occurred due to diffusion, the medium is said to be semi-infinite. At the beginning of diffusion the concentration at the surface is instantaneously brought to C_S and maintained constant throughout the diffusion time. These boundary conditions are usually abbreviated as:

$C = C_0$ at $t = 0$ and $0 < x < \infty$
$C = C_S$ at $x = 0$ and $0 < t < \infty$

where

C_0 = initial uniform concentration
C_S = constant surface concentration
x = distance from surface
t = time of diffusion

From the mathematical formula derived from these boundary conditions for constant D, the dimensionless variables have been computed and the values of $(C_x - C_0)/(C_s - C_0)$ for any value of $x/\sqrt{Dt}$ can be obtained from the appropriate tables, e.g. as given in Table 13—VI. Since these variables are dimensionless, they can be used for any diffusion problem satisfying the boundary conditions given above. For example, if C_0, C_s and D are known, the concentration of the diffusate, C_x, at distance x can be calculated for any time of diffusion using the values in Table 13—VI.

(2) **Diffusion into a slab or cylinder.** In this case, the undirectional diffusion is considered to take place through the opposite sides of a plate of finite or infinite length as shown in Figure 13—19b. For objects of finite dimensions comparable with the diffusion distance, sides of the sample must be coated with an imperme-

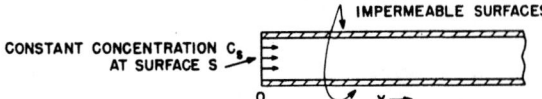

(a) UNIDIRECTIONAL DIFFUSION IN A SEMI-INFINITE MEDIUM.

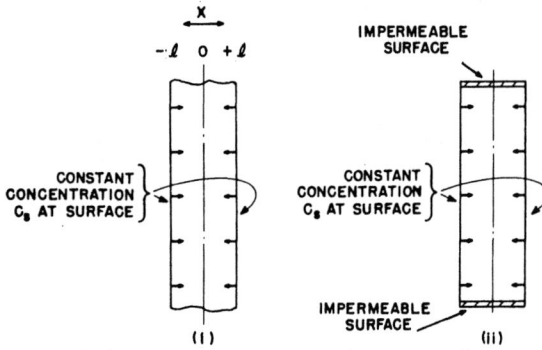

(b) UNIDIRECTIONAL DIFFUSION INTO PLATES (THICKNESS 2l) OR INTO CYLINDERS (DIAMETER 2 l) OF (i) INFINITE LENGTH OR (ii) FINITE LENGTH.

FIG. 13—19. Some of the boundary conditions for unidirectional diffusion.

Table 13—VI. Concentration Ratios $\dfrac{C_x-C_0}{C_s-C_0}$ for Certain Values of $x/\sqrt{Dt}$ in Undirectional Diffusion into a Semi-Infinite Medium for Constant D. (*).

The boundary conditions are:

$\quad$ C = C_0 at t = 0 and $0 < x < \infty$

$\quad$ C = C_s at x = 0 and $0 < t < \infty$

C_x = concentration at distance x

C_0 = initial uniform composition

C_s = surface concentration maintained constant

$\dfrac{x}{\sqrt{Dt}}$	$\dfrac{C_x-C_0}{C_s-C_0}$	$\dfrac{x}{\sqrt{Dt}}$	$\dfrac{C_x-C_0}{C_s-C_0}$	$\dfrac{x}{\sqrt{Dt}}$	$\dfrac{C_x-C_0}{C_s-C_0}$
0	1.0000	0.9538	0.5000	2.8	0.0477
0.1	0.9436	1.0	0.4795	3.0	0.0399
0.2	0.8875	1.2	0.3961	3.2	0.0236
0.3	0.8320	1.4	0.3222	3.4	0.0162
0.4	0.7773	1.6	0.2579	3.6	0.0109
0.5	0.7237	1.8	0.2031	3.8	0.0072
0.6	0.6714	2.0	0.1573	4.0	0.0047
0.7	0.6206	2.2	0.1198	4.4	0.0019
0.8	0.5716	2.4	0.0897	4.8	0.0007
0.9	0.5245	2.6	0.0660	5.2	0.0002

*This table is taken from Darken and Gurry[49]

ble substance to insure undirectional diffusion. For the boundary conditions, C_0 = initial uniform concentration, C_s = constant surface concentration and constant D, the solution of Fick's second law are given in Table 13—VII, in terms of dimensionless variables. If the diffusivity and the thickness of the plate or the diameter of the cylinder are known, the fractional saturation or desaturation of the diffusing substance can be calculated from the dimensionless variables in Table 13—VII. The fractional saturation (or desaturation) is defined by

$$F = \frac{C_m - C_0}{C_s - C_0} \qquad (91)$$

where C_m is the mean concentration of the diffusate. In Table 13—VIII, the values of $(C_x - C_0)/(C_s - C_0)$ are given for certain values of F and x/l. From these values, the concentration at any distance x and time t can be calculated, if F, l, C_s, C_m and C_0 are known.

Table 13—VII. Fractional Saturation $F=\dfrac{C_m-C_0}{C_s-C_0}$ for an Infinite Slab of Thickness 2(l) or an Infinite Cylinder of Diameter 2(l) for Certain Values of Dt/(l)^2, for Constant D. (*)

C_m = mean concentration

C_0 = initial uniform concentration

C_s = constant surface concentration

$\dfrac{Dt}{(l)^2}$	$F=(C_m-C_0)/(C_s-C_0)$		$\dfrac{Dt}{(l)^2}$	$F=(C_m-C_0)/(C_s-C_0)$	
	Slab	Cylinder		Slab	Cylinder
0.02	0.161	0.302	0.40	0.702	0.9316
0.04	0.227	0.412	0.50	0.764	0.9616
0.06	0.275	0.488	0.60	0.816	0.9785
0.08	0.320	0.550	0.70	0.856	0.9879
0.10	0.357	0.606	0.80	0.887	0.9932
0.15	0.438	0.708	0.90	0.912	0.9960
0.20	0.503	0.781	1.00	0.931	0.9979
0.25	0.560	0.832	1.50	0.980	0.9999
0.30	0.612	0.878	2.00	0.994	

*This table is taken from Darken and Gurry[49]

Table 13—VIII. Concentration Ratio $(C_x-C_0)/(C_s-C_0)$ for Certain Values of x/(l) and $F = (C_m-C_0)/(C_s-C_0)$ in Unidirectional Diffusion into Slab of Thickness 2(l) and Cylinder of Diameter 2(l), for Constant D.(*)

C_x = concentration at distance x

C_0 = initial uniform concentration

C_s = constant surface concentration

x = distance from the center line of slab or cylinder

F	Value of x/l							
	0.95	0.9	0.8	0.7	0.6	0.5	0.3	0.0

Values of $(C_x-C_0)/(C_s-C_0)$

For Slab of Thickness 2(l)

F	0.95	0.9	0.8	0.7	0.6	0.5	0.3	0.0
0.1	0.703	0.435	0.100	0.020	0.004	0.000	0.000	0.000
0.2	0.842	0.688	0.420	0.223	0.105	0.045	0.007	0.000
0.3	0.890	0.787	0.590	0.420	0.282	0.180	0.063	0.015
0.4	0.918	0.842	0.690	0.550	0.422	0.320	0.177	0.090
0.5	0.935	0.874	0.754	0.640	0.536	0.443	0.305	0.223
0.6	0.950	0.902	0.807	0.718	0.633	0.558	0.447	0.376
0.7	0.963	0.928	0.856	0.788	0.724	0.667	0.582	0.530
0.8	0.976	0.951	0.905	0.860	0.817	0.780	0.720	0.687
0.9	0.987	0.975	0.951	0.929	0.908	0.890	0.861	0.843

For Cylinder of Diameter 2(l)

F	0.95	0.9	0.8	0.7	0.6	0.5	0.3	0.0
0.1	0.435	0.097	0.006	0.001	0.000	0.000	0.000	0.000
0.2	0.726	0.475	0.112	0.025	0.010	0.002	0.000	0.000
0.3	0.823	0.653	0.363	0.160	0.059	0.016	0.001	0.000
0.4	0.875	0.755	0.525	0.330	0.185	0.095	0.023	0.003
0.5	0.908	0.821	0.647	0.483	0.342	0.230	0.095	0.035
0.6	0.931	0.864	0.732	0.605	0.483	0.378	0.228	0.140
0.7	0.950	0.901	0.805	0.712	0.619	0.535	0.405	0.320
0.8	0.968	0.937	0.873	0.811	0.750	0.695	0.605	0.541
0.9	0.984	0.968	0.937	0.904	0.874	0.847	0.800	0.767

*This table is taken from Gurry[24a]

PART II

Properties of Metallurgical Systems

THERMODYNAMIC PROPERTIES OF SOLID AND LIQUID IRON ALLOYS

Metallic Solutions—The metallic solutions considered in this section are confined to solid and liquid iron-base alloys normally encountered in steelmaking. First, the composition dependence of the activity coefficient, the enthalpy and the entropy of binary metallic solutions, and the activities of ternary metallic solutions will be discussed. This is then followed by the discussion of the thermodynamic properties of selected binary and ternary iron-base alloys, and selected phase-equilibrium diagrams. Finally the diffusion coefficient, coefficient of viscosity, thermal conductivity, surface tension and density of selected iron-base systems are given.

The activity vs composition relations derived from the experimental measurements must satisfy two important limiting laws for infinitely dilute solutions: Raoult's law and Henry's law. The departure from ideal behavior even in relatively dilute solutions is well demonstrated by the data compiled in Figure 13—20 where $\log (\gamma_x/\gamma_x^\circ)$ is plotted against the atom fraction, N_x, of the solute X (top scale) and $(1-N_x)^2$ (bottom scale) in amalgams.

The ratio γ_2/γ_x° for the solute 2, often symbolized by f_2—equation (65)—is a measure of departure from Henry's law in dilute solutions. Wagner [16] showed that it is a good approximation to take $\log \gamma_2$ (or $\log f_2$) as a linear function of N_2 in dilute solutions.

$$\ln \frac{\gamma_2}{\gamma_2^\circ} = \epsilon_2^{(2)} N_2 \qquad (92)$$

or

$$\log f_2 = e_2^{(2)} (\%i) \qquad (93)$$

where per cent i is the concentration of solute 2 in wt. per cent. The coefficient $\epsilon_2^{(2)}$ or $e_2^{(2)}$ is known as the "interaction parameter" and has been widely used. The relationship between ϵ and e for the limiting case $N_2 \to 0$ is given by [17]

$$\epsilon_2^{(2)} = 230 \frac{M_2}{M_1} e_2^{(2)} + \frac{M_1 - M_2}{M_1} \qquad (94)$$

where M is the atomic weight of the component indicated by the subscript. At higher solute contents equation (92) does not hold.

Darken [18] introduced a quadratic formalism describing the composition dependence of the activity coefficient in the terminal regions of binary metallic solutions. In the terminal region, which may extend to as much as 50 atom per cent, the logarithm of the activity coefficient of component 2, $\log \gamma_2$, is a linear function of $(1 - N_2)^2$. Similarly, for the other component, $\log \gamma_1$ is a linear function of $(1 - N_1)^2$. The examples given in Figure 13—21 demonstrate the general applicability of the quadratic formalism as stated in the following equations for the terminal regions:

For region (I), $0 < N_2 < N'_2$,

$$\log \frac{\gamma_2}{\gamma_2^\circ} = \alpha_{12}(N_1^2 - 1) = \alpha_{12}(-2N_2 + N_2^2)$$

$$\log \gamma_1 = \alpha_{12}(1 - N_1)^2$$

For ternary solutions the most convenient method of describing the activity coefficient is by means of the interaction coefficient. For dilute solutions of N_i and N_j in solvent N_1,

$$\ln \frac{\gamma_i}{\gamma_i^\circ} \simeq \epsilon_i^{(i)}N_i + \epsilon_i^{(j)}N_j \qquad (95)$$

or

$$\log f_i \simeq e_i^{(i)}(\%i) + e_i^{(j)}(\%j) \qquad (96)$$

The relationship between ϵ and e is for dilute solutions, i.e. $N_i \to N_j \to 0$, is given by

$$\epsilon_i^{(i)} = 230 \frac{M_i}{M_1} e_i^{(i)} + \frac{M_1 - M_i}{M_1}$$

$$\epsilon_i^{(j)} = 230 \frac{M_j}{M_1} e_i^{(j)} + \frac{M_1 - M_j}{M_1} \qquad (97)$$

The values of e compiled from the available data for ternary systems Fe-X-H, Fe-X-C, Fe-X-N, Fe-X-S and Fe-X-O at 1600°C are given in Table 13—IX. The references cited are recent publications where references to other work on the same subject can be found. It should be emphasized once again that equations (95) and (96) are intended for dilute solutions only. The greater the magnitude of e, the smaller is the concentration range of j for the validity of this linear relation. The maximum values of per cent j for the applicability of equations (95) and (96) are given in Table 13—IX for all the systems considered.

Like other thermodynamic parameters, ϵ is a function of temperature, and as shown by Lupis and Elliott [17], the temperature dependence of ϵ is represented by

$$\epsilon_i^{(j)} = \frac{\eta_i^{(j)}}{RT} - \frac{\sigma_i^{(j)}}{R} \qquad (98)$$

where η and σ are enthalpy- and entropy-like terms. However the data available at present are not adequate for reasonable assessment of the values of η and σ.

Since there is no better substitute, an extended form of equation (95) may be used (in the absence of direct measurements) to estimate f_i in dilute solutions of multicomponent systems.

$$\log f_i \simeq e_i^{(i)} (\%i) + e_i^{(j)} (\%j) + e_i^{(k)} (\%k) + \ldots (99)$$

At higher solute concentrations the values of $f_i^{(X)}$ are read from smooth curves in the nonlinear regions of $\log f_i^{(X)}$ vs per cent X plots and are used to obtain the sum

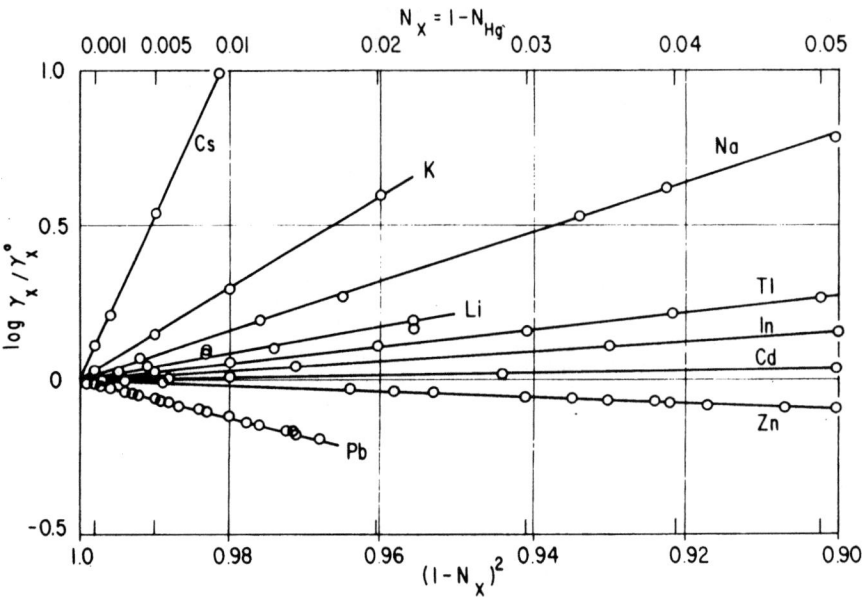

FIG. 13—20. Activity coefficient of solute X in dilute liquid amalgams at 25°C.

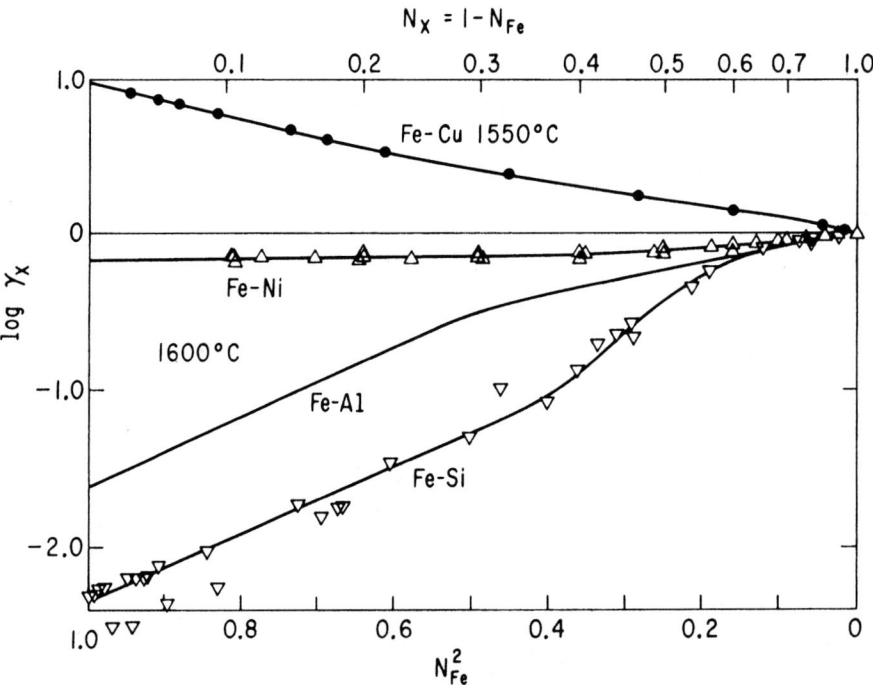

FIG. 13—21. Activity coefficient of X for various liquid Fe-X alloys at 1600°C (except for Fe-Cu at 1550°C). (Darken[18]).

$$\log f_i \simeq \log f_i^{(i)} + \log f_i^{(j)} + \log f_i^{(k)} + \ldots \quad (99a)$$

Ban-ya and Chipman[19] investigated the sulphur activity in multicomponent liquid iron alloys containing 1 to 4 per cent S, 0 to 4 per cent Si, 0 to 11 per cent Cr, 0 to 20 per cent Ni, 0 to 14 per cent Co, 0 to 14 per cent Mo and 0 to 14 per cent W. Using their values of $\log f_S^{(X)}$, $\log f_S$ is computed from equation (99a) for the

multicomponent alloys, and these are compared in Figure 13—22 with the measured values.

Thermochemistry of Iron—On heating, iron undergoes two polymorphic phase transformations. Below 911°C, it crystallizes as body-centered cubic (bcc or α), between 911° C and 1392°C as face-centered cubic (fcc or γ) and between 1392°C and the melting temperature, 1536°C, as body-centered cubic (bcc or δ). As seen

Table 13—IX Interaction parameter $e_i^{(j)} = \partial \log f_i / \partial(\%j)$ for ternary alloys Fe-X-H, Fe-X-C, Fe-X-N, Fe-X-S and Fe-X-O at 1550 to 1600°C; values in parentheses are derived from the corresponding values of $\epsilon_j^{(i)}$; $< \% j$ indicates compositions below which $\log f_i \propto \% j$ is a reasonable approximation. (From Turkdogan[20])

Solute j	$e_H^{(j)}$	$< \% j$	$e_C^{(j)}$	$< \% j$	$e_N^{(j)}$	$< \% j$	$e_S^{(j)}$	$< \% j$	$e_O^{(j)}$	$< \% j$
Al	0.013	2	0.064	2	0.002	0.5	0.035	1	−3.90	0.2
B	0.050	1	—	—	—	—	0.134	0.5	−2.6	0.05
C	0.060	1	0.22	1	0.25	0.5	0.114	0.5	−0.13	1
Co	0.002	14	0.062	10	0.011	12	0.003	10	0.007	5
Cr	−0.002	2	−0.024	25	−0.045	7	−0.011	5	−0.037	20
Cu	0.0005	12	0.018	10	0.009	10	−0.008	8	−0.016	15
H	0	—	(0.72)	—	—	—	(0.26)	—	—	—
Mn	−0.001	11	−0.007	10	−0.02	6	−0.026	3	0	—
N	—	—	(0.11)	—	0	—	(0.03)	—	(0.057)	—
Nb	−0.002	2	−0.06	2	−0.061	10	−0.013	5	−0.14	3
Ni	0	—	0.012	5	0.010	10	0	—	0.006	20
O	—	—	(−0.097)	—	0.05	—	(−0.18)	—	−0.20	—
P	0.011	0.5	—	—	0.051	—	0.029	1	0.07	0.5
S	0.008	0.1	0.057	2	0.013	—	−0.028	1	−0.091	—
Si	0.027	1	0.113	2	0.047	3	0.063	0.5	−0.14	1
Ti	0.08	0.5	—	—	−0.53	0.2	−0.072	1	−1.150	0.3
V	—	—	−0.038	20	−0.093	2	0.016	5	−0.14	5.0
W	—	—	−0.033	20	−0.002	15	0.001	10	0.008	5
Zr	—	—	—	—	−0.63	0.1	−0.053	2	—	—

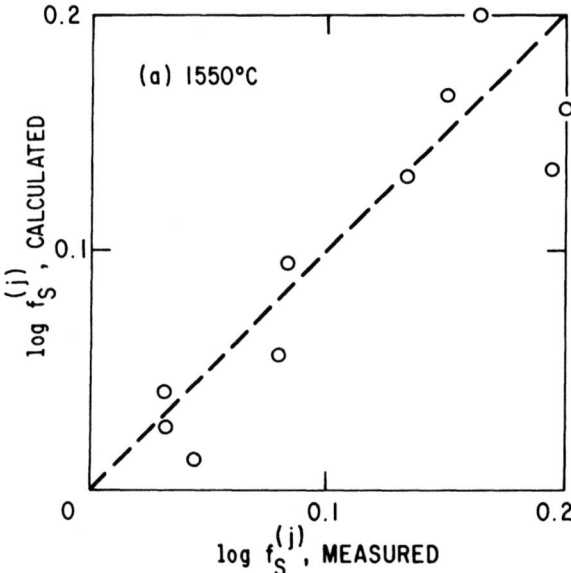

FIG. 13—22. Comparison of calculated and measured values of $\log f_S^{(j)}$ for multicomponent liquid iron alloys using data of Ban-ya and Chipman[19].

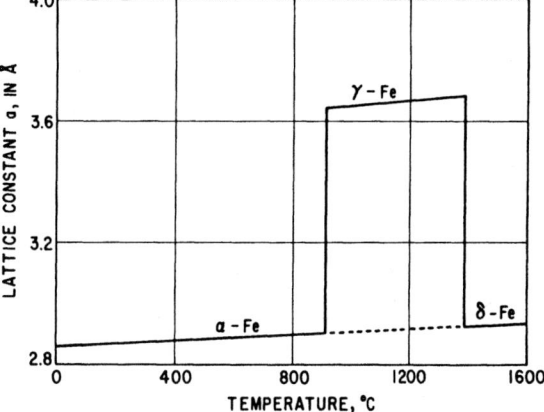

FIG. 13—23. Variation of lattice constant of pure iron with temperature.

from the x-ray data in Figure 13—23, when fcc structure (γ-iron) is formed, the lattice constant increases. In metallurgical language, α-Fe is called the ferrite phase and γ-Fe, the austenite phase. The ferrite phase goes through a magnetic transformation at 769°C, above which iron is not magnetic, however, this is not a true polymorphic transformation. Since the lattice constant of austenite is larger than that of ferrite, the elements forming interstitial solid solutions in iron, i.e. within the iron space lattice, have larger solubilities in the austenite phase, e.g., H, B, C, N, and O. In the substitutional

solid solutions, solute atoms replace the solvent atoms in the space lattice of the latter. Depending on the nature of the alloying elements in iron, the transformation temperatures vary over a wide range. The elements with fcc structure dissolved in iron favor the fcc structure and, therefore, they extend the temperature stability range of the austenite phase, e.g., Ni, Co. Similarly, a solute element with bcc structure extends the stability range of the ferrite phase, e.g., Si, Cr, V. The existence of these two crystalline forms of iron is largely responsible for the versatility of iron alloys in practical applications. The iron can be rendered malleable, ductile, tough or hard, by alloying with suitable elements and applying appropriate heat and mechanical treatments.

The thermodynamic functions for solid and liquid iron are given in Tables 13—X and 13—XI, reproduced from the compilation by Orr and Chipman[21]. The value

Table 13—X Thermodynamic Functions for Solid Iron[†] (Orr and Chipman[21]).

T,°K	Fe$_{(\alpha,\delta)}$				Fe$_{(\gamma)}$			
	C_p cal per deg g-atom	$H°_T - H°_{298}$ cal per g-atom	$S°_T - S°_{298}$ cal per deg g-atom	$\dfrac{-(F°_T - H°_{298})}{T}$ cal per deg g-atom	C_p cal per deg g-atom	$H°_T - H°_{a,298}$ cal per g-atom	$S°_T - S°_{a,298}$ cal per deg g-atom	$\dfrac{-(F°_T - H°_{a,298})}{T}$ cal per deg g-atom
298.15	5.97	0	0.0000	6.5200	—	—	—	—
300	5.98	11	0.0369	6.5202	—	—	—	—
400	6.54	637	1.8340	6.7615	—	—	—	—
500	7.10	1319	3.3535	7.2355	6.73*	3174*	5.3238*	5.4958*
600	7.66	2057	4.6975	7.7892	6.93*	3857*	6.5685*	6.6602*
700	8.27	2852	5.9222	8.3675	7.13*	4560*	7.6518*	7.6575*
800	9.07	3717	7.0754	8.9494	7.33*	5283*	8.6169*	8.5332*
850	9.61	4183	7.6408	9.2394	7.43*	5652*	9.0643*	8.9349*
900	10.30	4680	8.2088	9.5285	7.53*	6026*	9.4918*	9.3162*
950	11.29	5218	8.7905	9.8174	7.63*	6405*	9.9016*	9.6795*
1000	13.01	5820	9.4073	10.1073	7.73*	6789*	10.2955*	10.0265*
1020	14.34	6092	9.6769	10.2241	7.77*	6944*	10.4490*	10.1612*
1030	15.55	6241	9.8220	10.2827	7.79*	7022*	10.5249*	10.2276*
1042(T$_c$)	20.00	6448	10.0216	10.3535	7.81*	7115*	10.6153*	10.3067*
1050	13.03	6563	10.1318	10.4011	7.83*	7178*	10.6751*	10.3589*
1060	12.31	6690	10.2525	10.4607	7.85*	7256*	10.7495*	10.4238*
1080	11.58	6928	10.4742	10.5796	7.89*	7414*	10.8965*	10.5519*
1100	11.09	7154	10.6820	10.6982	7.93*	7572*	11.0416*	10.6780*
1184(T$_{\alpha-\gamma}$)	9.90	8030	11.449	11.1876	8.10	8245	11.6313	11.1876
1200	9.75*	8187*	11.5816*	11.2790*	8.13	8375	11.7402	11.2810
1300	9.26*	9132*	12.3382*	11.8335*	8.33	9198	12.3988	11.8434
1400	9.21*	10052	13.0199*	12.3599*	8.53	10041	13.0235	12.3714
1500	9.43*	10983*	13.6625*	12.8602*	8.73	10904	13.6188	12.8695
1600	9.67*	11938*	14.2787*	13.3372*	8.93	11787	14.1886	13.3417
1665(T$_{\gamma-\delta}$)	9.83	12572	14.6669	13.6363	9.06	12372	14.5468	13.6363
1700	9.91	12917	14.8722	13.7937	9.13*	12690*	14.7360*	13.7913*
1800	10.15	13920	15.4454	14.2318	9.33*	13613*	15.2635*	14.2207*
1809(T$_m$)	10.17	14012	15.4961	14.2704	9.35*	13697*	15.3101*	14.2585*
1900	10.39*	14947*	16.0006*	14.6536*	9.53*	14556*	15.7733*	14.6323*
2000	10.63*	15998*	16.5397*	15.0655*	9.73*	15519*	16.2673*	15.0278*

[†] $H°_{298}$ and $S°_{298}$ refer to Fe$_{(\alpha)}$ in all cases. Functions for metastable phases are indicated by (*). $H°_{(\alpha),298} - H°_{(\alpha),0} = 1073$ cal per g-atom and $S°_{(\alpha),298} = 6.52$ cal per deg g-atom.

of $H°_{298}$ pertains to α-iron in all cases; and hence when using the table $\Delta H°_{298}$ is taken as zero for all transformation of pure iron.

Iron-Carbon Alloys—Since carbon is one of the most important ingredients of steel, the study of the iron-carbon phase equilibrium diagram has received much attention during the past several decades. The phase diagram is given in Figure 13—24. There are three invariants in this system; peritectic at 1499°C, eutectic at 1152°C and eutectoid at 738°C. The phase boundaries shown by broken lines are for the metastable equilibrium of cementite, Fe$_3$C, with austenite. It was shown by Wells[22] that if sufficient time is allowed, iron-carbon alloys containing austenite and cementite decompose to austenite and graphite. The solubilities of graphite and of cementite in α-iron below the eutectoid temperature are given by the following equations using the data of Swartz[23]

$$\log \%C \text{ (graphite)} = -\frac{5250}{T} + 3.53 \qquad (100)$$

$$\log \%C \text{ (cementite)} = -\frac{3200}{T} + 1.50 \qquad (101)$$

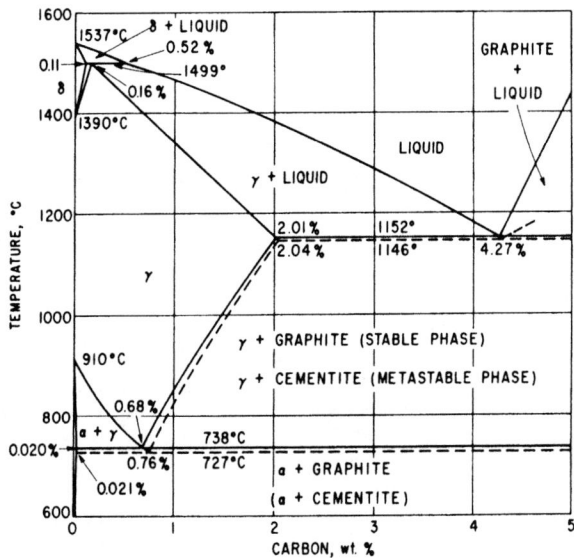

FIG. 13—24. Iron-carbon phase equilibrium diagram. Dashed lines represent phase boundary for metastable equilibrium with cementite.

Table 13—XI. Thermodynamic Functions for Liquid Iron†. (Orr and Chipman[21]).

T, °K	C_p cal per deg g – atom	$H^0_T - H^0_{a.298}$ cal per g – atom	$S^0_T - S^0_{a.298}$ cal per deg g – atom	$-(F^0_T - H^0_{a.298})/T$ cal per deg g – atom
1200	11.00*	10613*	12.8053*	10.4811*
1300	11.00*	11713*	13.6858*	11.1958*
1400	11.00*	12813*	14.5010*	11.8989*
1500	11.00*	13913*	15.2599*	12.5046*
1600	11.00*	15013*	15.9698	13.1067*
1700	11.00*	16113*	16.6367*	13.6785*
1800	11.00*	17213*	17.2654*	14.2226*
1809(T$_m$)	11.00	17312	17.3203	14.2704
1900	11.00	18313	17.8602	14.7418
2000	11.00	19413	18.4244	15.2379
2100	11.00	20513	18.9611	15.7130
2200	11.00	21613	19.4728	16.1687
2300	11.00	22713	19.9618	16.6066
2400	11.00	23813	20.4299	17.0278
2500	11.00	24913	20.8790	17.4338
2600	11.00	26013	21.3104	17.8254
2700	11.00	27113	21.7256	18.2037
2800	11.00	28213	22.1256	18.5695
2900	11.00	29313	22.5116	18.9237
3000	11.00	30413	22.8845	19.2668
3100	11.00	31513	23.2452	19.5997
3200	11.00	32613	23.5944	19.9228

(†) H^0_{298} and S^0_{298} refer to Fe$_{(a)}$. Functions for metastable liquid are indicated by (*).

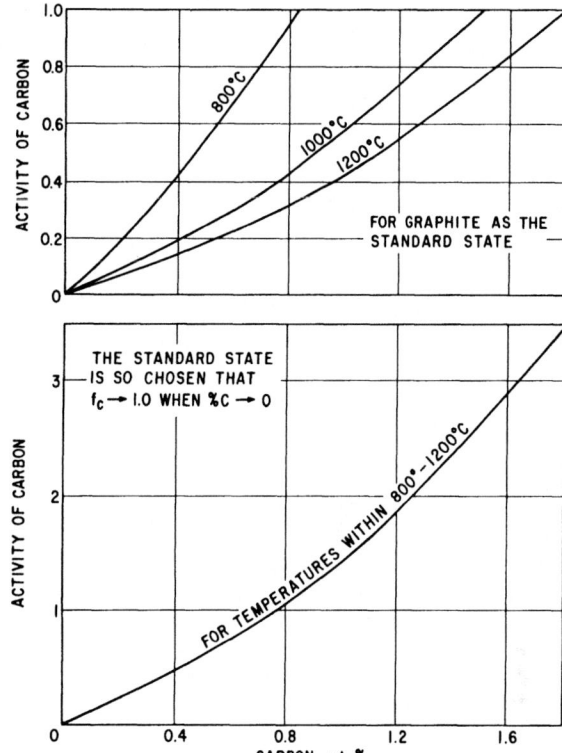

FIG. 13—25. Activity of carbon in austenite for two different standard states.

The heat of solution of graphite in α iron is 24 kcal/g-atom C over the temperature range 400 to 740°C.

The activity of carbon in austenite has been determined by numerous investigators [24, 25]. The data are summarized in a condensed form in Figure 13—25 where the carbon activity is given for two different standard states. When the standard state is chosen so that in infinitely dilute solutions the activity coefficient, f_C, approaches unity, the temperature effect on the activity vs composition relation becomes almost negligible at least within the range 800°-1200°C. The composition dependence of the activity of carbon (relative to graphite) may be approximated by the following equation

$$a_C = (\%C)\, \gamma^°_C\, e^{0.385\,(\%C)} \qquad (102)$$

where $\gamma^°_C$ is a temperature-dependent activity coefficient of carbon (relative to graphite) when $\%C \to 0$, represented by

$$\log \gamma^°_C = \frac{2336}{T} - 2.288 \qquad (103)$$

The heat of solution of graphite in γ-iron for dilute solutions of carbon is 10.7 kcal/g-atom C.

The activities of carbon in the austenite phase of Fe-Si and Fe-Mn alloys at 1000°C are given in Figures 13—26 and 13—27, taken from the work of Smith[26,27]. While manganese lowers the activity coefficient of carbon in austenite, silicon raises it.

The most reliable data so far available on the thermodynamics of iron-carbon binary melts (up to about 1 per cent C) are those presented by Richardson and Dennis[28]; their results on carbon activity are given in Figure 13—28. Of the several activity measurements made at high carbon contents, those of Rist and Chipman[29] are more reliable. The composition dependence of the activity coefficient of carbon in liquid iron is well represented by the quadratic formalism, thus

$$\log \frac{\gamma_C}{\gamma^°_C} = -2.38\,(N^2_{Fe} - 1) \qquad (104)$$

where $\gamma^°_C$ (relative to graphite is the value of γ_C at $N_C = 0$ and according to Chipman's compilation[30],

$$\log \gamma^°_C = \frac{1180}{T} - 0.87 \qquad (105)$$

The heat of solution of graphite in liquid iron at infinitely dilute solution of carbon is 5400 cal/g-atom C. With increasing carbon content, the heat of solution increases reaching a value of about 9000 cal/g-atom C at graphite saturation. The variation of f_C ($=\gamma_C/\gamma^°_C$) with carbon content of liquid iron is shown in Figure 13—29.

The solubility of graphite in liquid iron is well established and the experimental data, reviewed by Neumann and Schenck[31], are summarized by the equation:

$$[\%C] = 1.30 + 2.57 \times 10^{-3}\, t\,(°C) \qquad (106)$$

If atom fraction (N_C) is used, the same set of data can be represented by the following equation in terms of log N_C and the reciprocal of the absolute temperature:

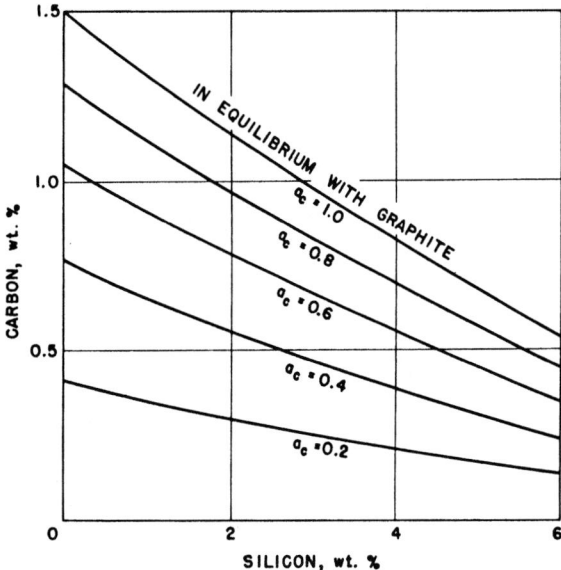

FIG. 13—26. Isoactivity curves for carbon in austenitic Fe-Si-C alloys at 1000°C. (Smith[26]).

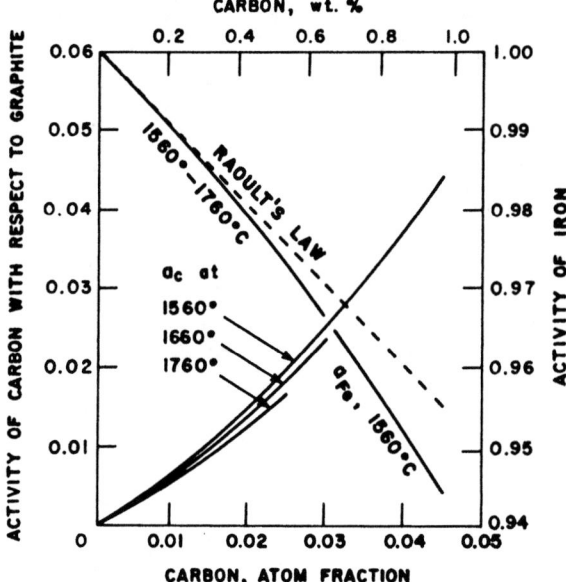

FIG. 13—28. Activity of carbon and iron in Fe-C melts. (Richardson and Dennis[28]).

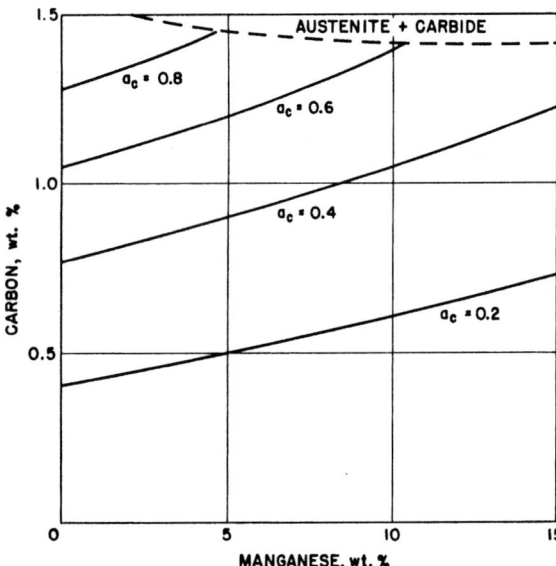

FIG. 13—27. Isoactivity curves for carbon in austenitic Fe-Mn-C alloys at 1000°C. (Smith[26]).

$$\log N_C = -\frac{560}{T} - 0.375 \qquad (107)$$

The effect of silicon, phosphorus, sulphur, manganese, cobalt and nickel on the solubility of graphite in molten iron was determined by Turkdogan et al.[32,33] and graphite solubility in iron-silicon and iron-manganese melts by Chipman et al.[34,35]. Similar measurements with iron-chromium melts were made by Griffing and co-workers[36]. The experimental data are given graphically in Figure 13—30 for 1500°C; the solubility at other temperatures can be estimated from Figure 13—30 using the temperature coefficient given in equation (106) for pure iron. In the iron-sulphur-

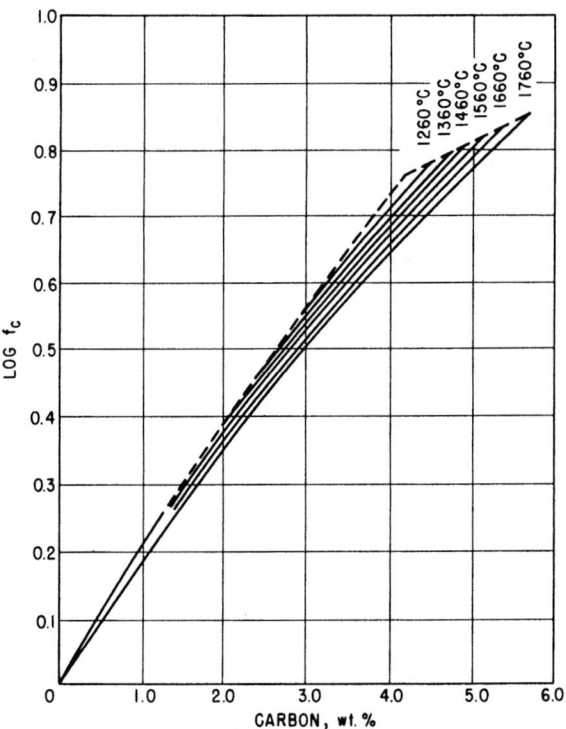

FIG. 13—29. Activity coefficient of carbon in liquid iron for the standard state $f_C \to 1.0$ when $\%C \to 0$. (Rist and Chipman[29]).

carbon system there is a large miscibility gap. For example, at 1500°C the melt separates into two liquids containing:

phase (I) 1.8 per cent S and 4.24 per cent C;

phase (II) 0.90 per cent C and 26.5 per cent S.

The effect of various elements on the activity coeffi-

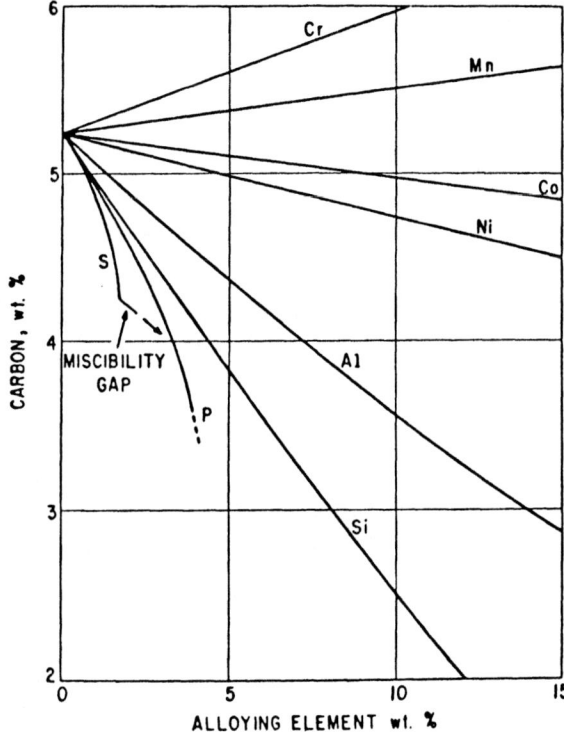

FIG. 13—30. Solubility of graphite in alloyed iron melts at 1500°C.

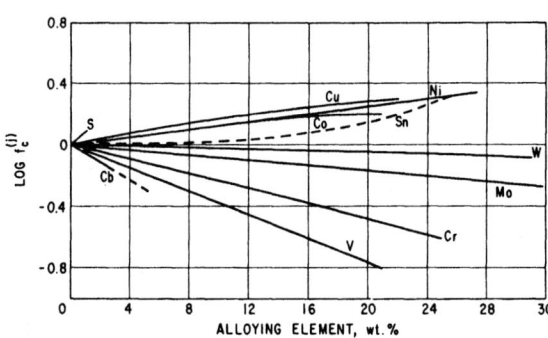

FIG. 13—31. Effect of alloying elements on activity coefficient of carbon for dilute solutions of carbon in iron at 1560°C. (Fuwa and Chipman[37]).

cient $f_C^{(j)}$ at dilute solutions of carbon in liquid iron alloys is shown in Figure 13—31, reproduced from a paper by Fuwa and Chipman[37].

Iron-Hydrogen Alloys—Initial studies on the solubility of gases in metal is attributed to Sieverts who found experimentally that the amount dissolved in the metal is directly proportional to the square root of the partial pressure of the gas. This is known as the **Sieverts' law** and it states that when a di-atomic gas reacts with a metal, it dissolves in the atomic form. Thus, in the case of hydrogen:

$$\frac{1}{2} H_2 \text{ (gas)} = \underline{H} \text{ (dissol. in metal)} \qquad (108)$$

whereby the hydrogen loses its gaseous character.

Assuming ideal solutions, the equilibrium constant of reaction (108) is:

$$K = \frac{[H]}{(p_{H_2})^{1/2}} \qquad (109)$$

The solubility of hydrogen in iron in equilibrium with $p_{H_2} = 1.0$ atm is given in Figure 13—32 based on the data compiled by Geller and Sun[38]. Representing the concentration of hydrogen in iron in terms of cm^3 of H_2 at STP/(100 g Fe × $atm^{1/2}$), the equilibrium constants for $\alpha(\delta)$, γ and liquid iron are given by the following equations:

$$\log K_{\alpha, \delta} = -\frac{1418}{T} + 1.677$$

$$\log K_{\gamma} = -\frac{1182}{T} + 1.677 \qquad (110)$$

$$\log K_l = -\frac{1637}{T} + 2.362$$

If concentration of hydrogen in iron is required in

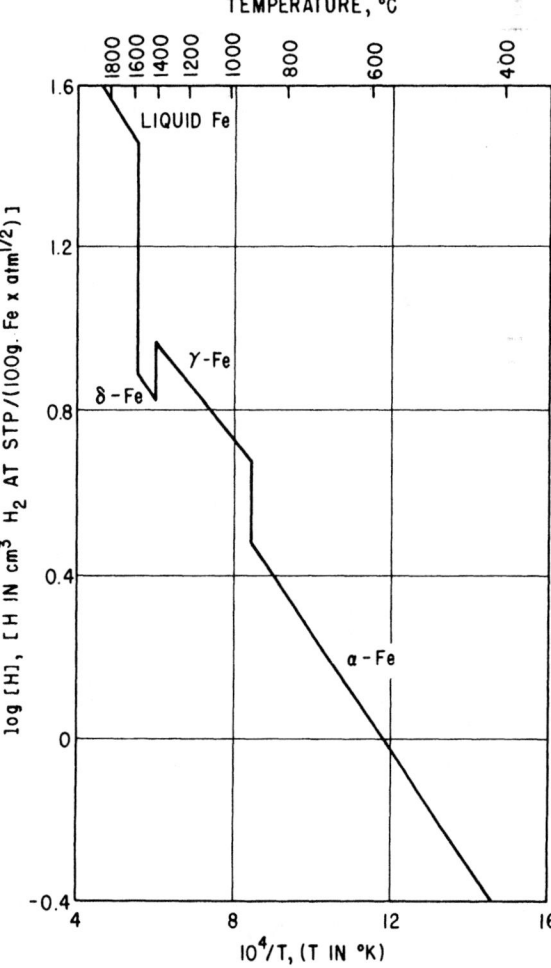

FIG. 13—32. Solubility of hydrogen in iron at $p_{H_2} = 1.0$ atm. (Geller and Sun[38]).

terms of parts per million (ppm) by weight, the number 0.0494 should be subtracted from the values of log K obtained from the above equations.

The effect of alloying elements on the solubility of hydrogen at 1 atm in liquid iron is shown in Figure 13—33 using the data of Weinstein and Elliott[39]. The variation of $\log f_H^{(j)}$ with the concentration of the alloying element X is shown in Figure 13—34.

Iron-Nitrogen Alloys—Since the initial work of Sieverts, several investigators have measured the solubility of nitrogen in α, γ, δ and liquid iron. Using the data considered to be most reliable[39-43], the solubility of nitrogen in iron at one atmosphere pressure is given in Figure 13—35 from which the following values of K = $[\%N]/pN_2^{1/2}$ are derived as functions of temperature:

$$\log K_{\alpha, \delta} = -\frac{1570}{T} - 1.02$$

$$\log K_\gamma = \frac{450}{T} - 1.95 \qquad (111)$$

$$\log K_l = -\frac{188}{T} - 1.24$$

The solubility of nitrogen in austenitic iron-silicon and iron-chromium alloys has been measured by Turkdogan and Ignatowicz [44,45] and in iron-nickel alloys by Wriedt and Gonzalez[46]. The nitrogen solubilities in these alloys are given in Figure 13—36. The effect of alloying elements on the activity coefficient of nitrogen dissolved in liquid iron at 1600°C is shown in Figure 13—37.

The phase equilibrium diagram of the iron-nitrogen binary system is shown in Figure 13—38; for the details of work on this subject, reference should be made to the papers reviewed by Hansen and Anderko[47]. It should be realized that this phase diagram is not for the iron-nitrogen system at one atmosphere of nitrogen; the phases given are in equilibrium with very high nitrogen fugacities attainable by equilibrium dissociation, for example, of ammonia

$$NH_3 = \underline{N} + \frac{3}{2} H_2 \qquad (112)$$

The 'Fe$_4$N' (γ'-phase) is a nitrogen-deficient phase. According to the work of Wriedt[46], at 500°C the composition range is from 5.77 per cent N (in equilibrium with α-iron) to 5.88 per cent N (in equilibrium with ϵ-iron nitride); the stoichiometric composition is 5.900 per cent N. The temperature dependence of the solubility of 'Fe$_4$N' in α-iron is represented by[40]

$$\log [\%N]_{Fe_4N} = -\frac{1814}{T} + 1.09 \qquad (113)$$

At temperatures below about 300°C, an intermediate nitride, 'Fe$_8$N', is formed as a metastable phase prior to the precipitation of 'Fe$_4$N' from the super-saturated α-iron; the solubility of 'Fe$_8$N' in α-iron is given by[40]

$$\log [\%N]_{Fe_8N} = -\frac{2160}{T} + 2.51 \qquad (114)$$

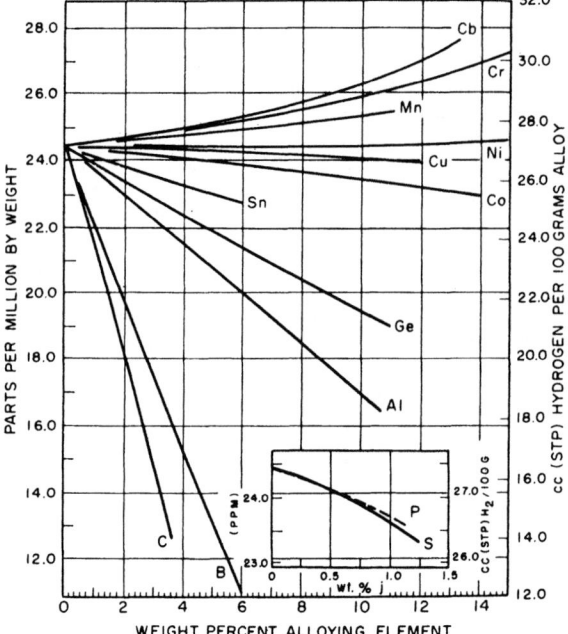

FIG. 13—33. Solubility of hydrogen at 1 atm pressure in binary iron alloys at 1592°C. (Weinstein and Elliott[39]).

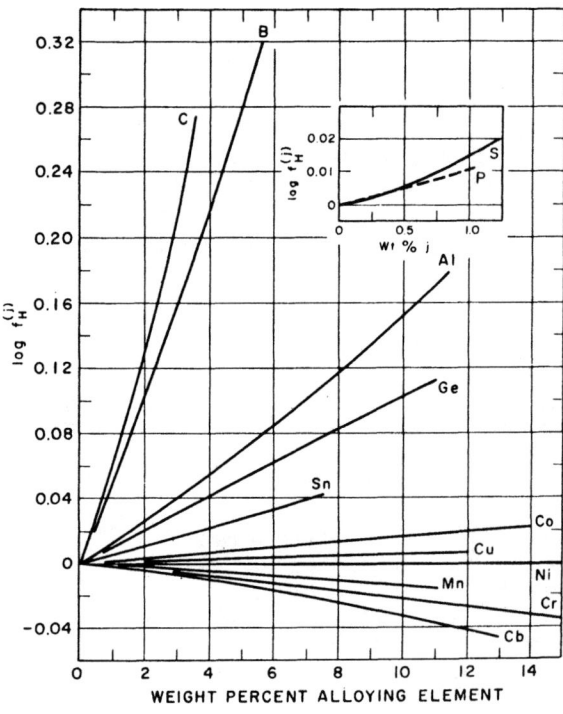

FIG. 13—34. Effect of alloying elements on the activity coefficient of hydrogen in binary iron alloys at 1592°C. (Weinstein and Elliott[39]).

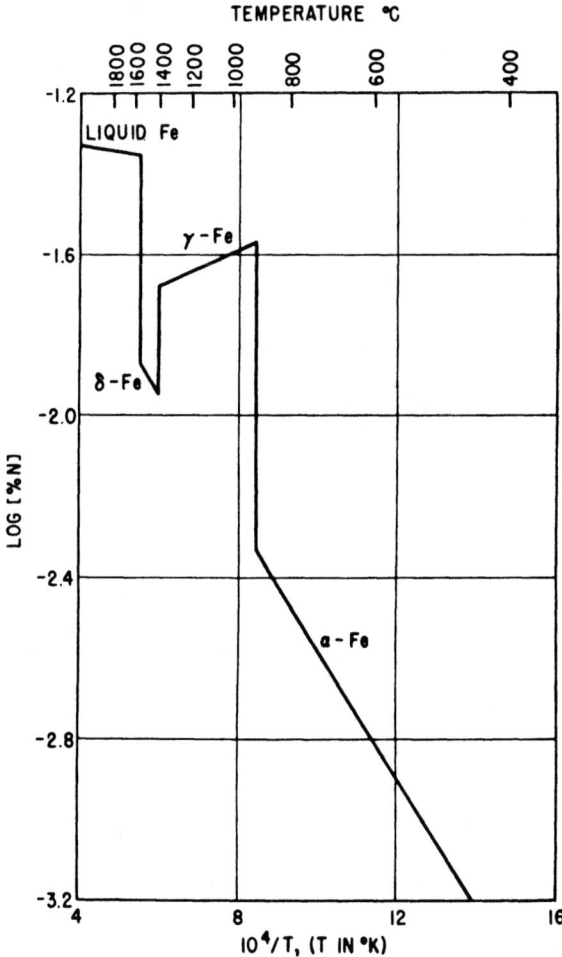

FIG. 13—35. Solubility of nitrogen in iron at $p_{N_2} = 1.0$ atm.

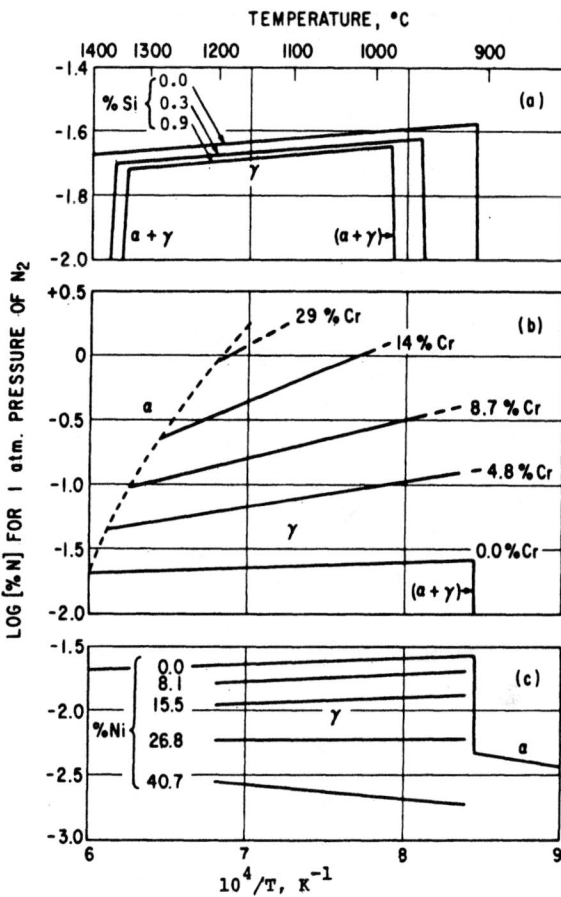

FIG. 13—36. Solubility of nitrogen in austenitic alloys at 1 atm N_2. (a), (b): Turkdogan and Ignatowicz[44,45]. (c): Wriedt and Gonzalez[46].

Iron-Oxygen Alloys—The thermodynamics of the iron-oxygen system was studied in detail by Darken and Gurry[49]. Figure 13—39 gives the temperature-composition phase diagram for a total pressure of one atmosphere. There are two characteristic features of this system. Oxygen is soluble in iron to a limited extent only; at the eutectic temperature 1527°C, the maximum solubility is 0.16 per cent O above which a liquid oxide phase, containing 22.6 per cent O, is formed. The second characteristic feature is the formation of wustite, which has a variable composition and is not stable below 560°C. The stoichiometric ferrous oxide does not exist and the wustite in equilibrium with iron has the composition corresponding to about $Fe_{0.95}O$ within the temperature range 800°–1371°C (melting temperature). That is, wustite is deficient in iron cations and the electroneutrality is maintained by the presence of some trivalent iron cations together with the divalent iron cations in wustite. Within the wustite phase, the ration Fe^{3+}/Fe^{2+} increases with increasing partial pressure of oxygen. Although magnetite in equilibrium with wustite has the stoichiometric composition Fe_3O_4, in the presence of hematite the magnetite phase becomes deficient in iron with increasing tem-

perature. The remainder of the diagram in Figure 13—39 is self-explanatory.

Figure 13—40 gives the relation between log p_{O_2} and $1/T$ for several three-phase equilibria, i.e. phase boundaries, and isoactivity curves of iron in solid and liquid oxides.

The solubility of oxygen in high-purity zone-refined iron was determined by Turkdogan et al.[50, 51]; this was later confirmed by the work of Kusano et al.[52]. Iron side of the Fe-O phase diagram is shown in Figure 13—41. At about 900°C iron (α or γ) in equilibrium with wustite contains about 2 ppm O. At the peritectic invariant (1390°C) γ-iron containing 28 ppm O is in equilibrium with δ-iron and liquid iron oxide. In δ-iron equilibrated with liquid iron oxide the oxygen content increases from 54 ppm at 1390°C to 82 ppm at the eutectic temperature 1527°C. To facilitate other thermodynamic computations, the following free energy equations are derived for the solution of oxygen in solid iron, i.e. for reaction $\frac{1}{2} O_2$ (g) = $\underline{O}$ (1 wt. %):

$$\text{bcc iron} \quad \Delta F^\circ = -37\ 190 + 10.20\ T$$

$$\text{fcc iron} \quad \Delta F^\circ = -41\ 860 + 14.46\ T \tag{115}$$

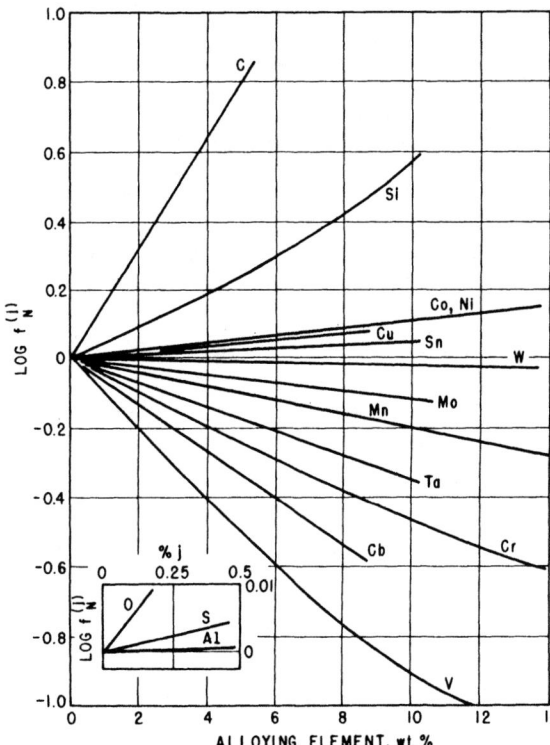

FIG. 13—37. Effect of alloying elements on the activity coefficient of nitrogen in iron at 1600°C. (Pehlke and Elliott[40] and Schenck et al[43]).

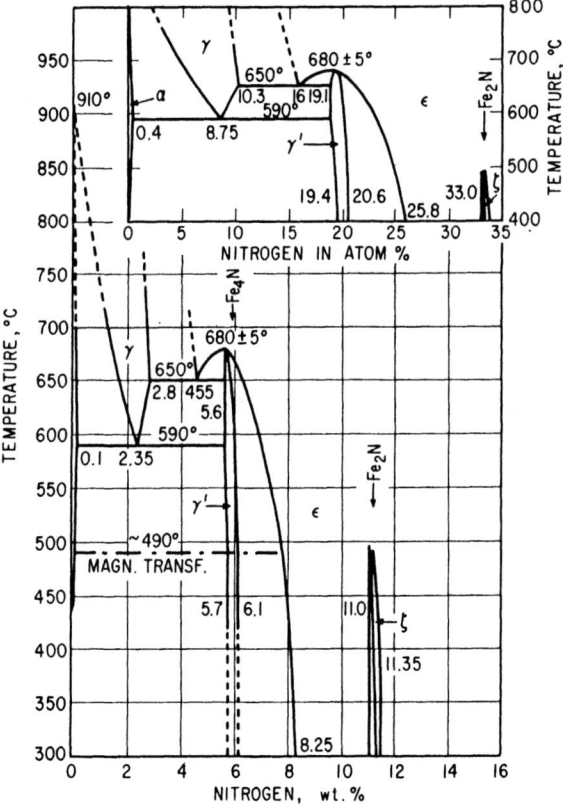

FIG. 13—38. Iron-nitrogen phase equilibrium diagram. (Hansen and Anderko[47]).

The solubility of oxygen in molten iron in equilibrium with almost pure iron oxide was determined by Chipman and co-workers[53, 54]; their results are given in Figure 13—42 and variation of solubility with temperature is represented by the equation:

$$\log [\% \; O] = -\frac{6320}{T} + 2.734 \qquad (116)$$

This also represents the equilibrium constant for the reaction

$$\text{'FeO'} = \underline{O} + Fe \qquad (117)$$

where the standard state for 'FeO' is pure iron oxide in equilibrium with liquid iron and for $\underline{O}$, $f_O \to 1.0$ when the oxygen content approaches zero.

The solution of oxygen in iron obeys Sieverts' law up to about 0.1 per cent O and the following free energy of solution is derived from the work of Dastur and Chipman[55] and Floridis and Chipman[2]:

$$\tfrac{1}{2} \; O_2 \; (g) = \underline{O} \; (1 \; wt. \; \%)$$
$$\Delta F^\circ = -28\,000 - 0.69T \qquad (118)$$

for the standard states, p_{O_2} in atm and $f_O \to 1.0$ when the oxygen content approaches zero.

The solubility of oxygen in molten iron is affected by the alloying elements in two ways. First, the solubility is limited by the formation of an oxide phase, and secondly, the activity coefficient of oxygen is raised or

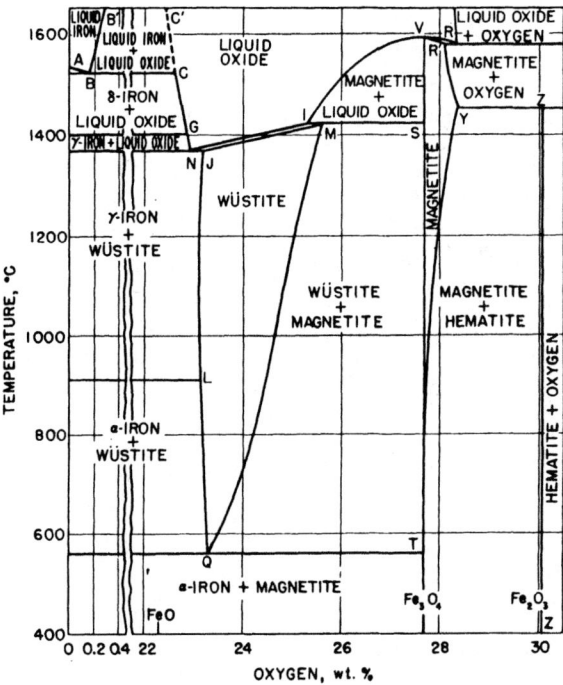

FIG. 13—39. Iron-oxygen system. (Darken and Gurry[49]).

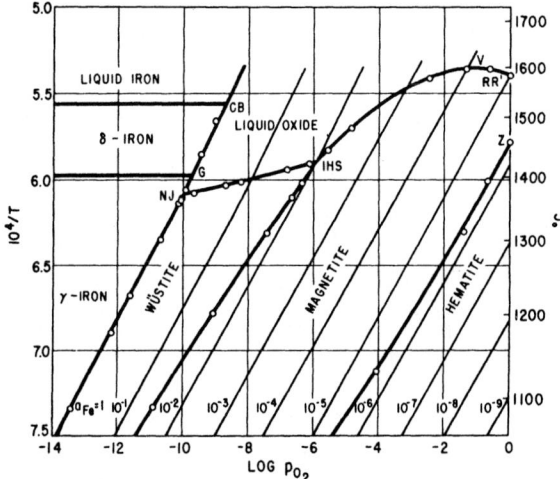

FIG. 13—40. Relation between log p_{O_2} and $1/T$ for the several three-phase equilibria, and the calculated isoactivity curves of iron in solid and liquid oxide throughout the range. (Darken and Gurry[49]).

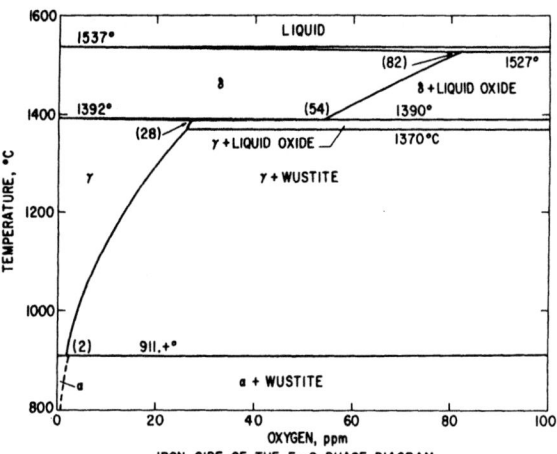

FIG. 13—41. Iron side of the Fe-O phase diagram. (Swisher and Turkdogan[51]).

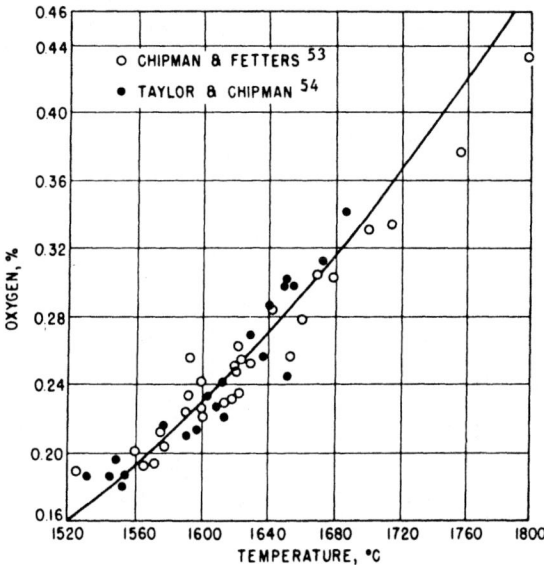

FIG. 13—42. Solubility of oxygen in liquid iron in equilibrium with almost pure molten iron oxide.

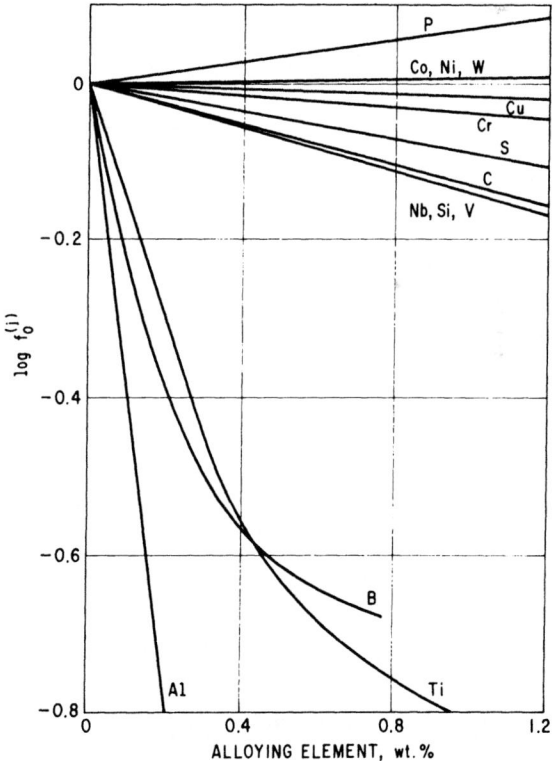

FIG. 13—43. Effect of alloying elements on the activity coefficient of oxygen in iron at 1600°C.

lowered depending on the extent of interaction between oxygen and elements dissolved in iron. As seen from the data compiled in Figure 13—43, most of the elements lower the activity coefficient of oxygen. Generally speaking, the lower the oxygen potential of the oxide of an element j, the lower is the value of $f_0^{(j)}$. Using equation (99) one may estimate the activity coefficient of oxygen in iron containing several alloying elements.

The solubility of oxides, e.g. MnO, SiO_2, Al_2O_3, etc., in liquid iron is given in Part V when discussing the deoxidation of steel.

Iron-Sulphur Alloys—The iron-sulphur phase equilibrium diagram reviewed by Hansen and Anderko is given in Figure 13—44. The solubility of sulphur in solid iron was determined by Rosenquist and Dunicz[56] and by Turkdogan et al.[57]; the iron side of the Fe-S system thus evaluated is shown in Figure 13—45.

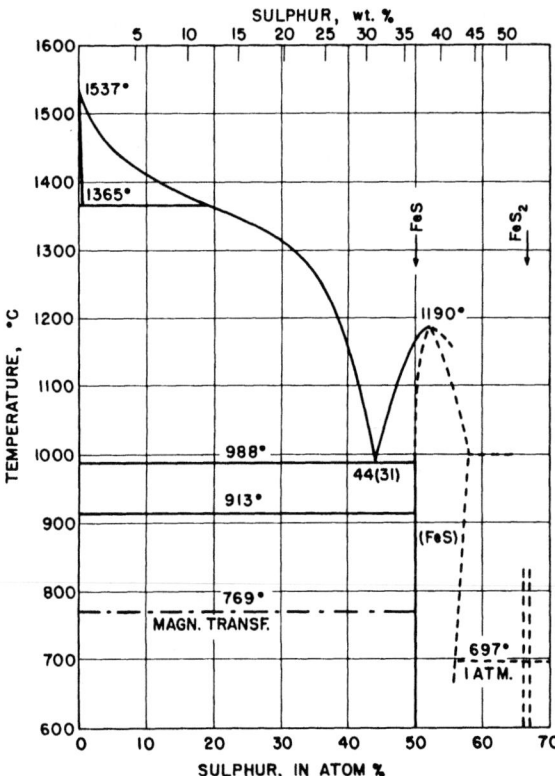

FIG. 13—44. Iron-sulphur phase-equilibrium diagram. (From Hansen and Anderko[47]).

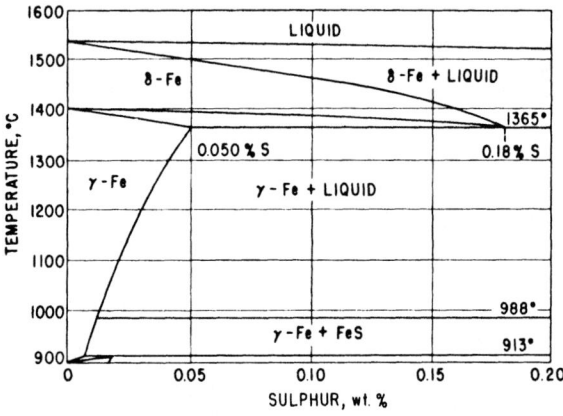

FIG. 13—45. Iron side of the Fe-S phase diagram. (Rosenquist and Dunicz[56] and Turkdogan et al[57]).

There is a peritectic invariant at 1365°C where the equilibrium phases are: γ-iron with 0.050 per cent S, δ-iron with 0.18 per cent S and liquid containing 12 per cent S. At the eutectic invariant 988°C γ-iron in equilibrium with the liquid phase contains 0.012 per cent S. The free energy of solution of sulphur in γ- and liquid iron is given by

$$\frac{1}{2} S_2 (g) = \underline{S} \ (1 \ wt. \ \%)$$

$$\gamma\text{-iron} \ \Delta F^\circ = -11 \ 710 + 2.48 T \qquad (119)$$

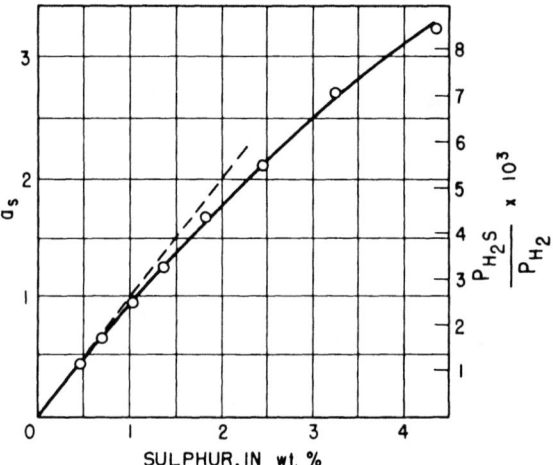

FIG. 13—46. Variation of sulphur content of iron at 1600°C with p_{H_2S}/p_{H_2} (or activity). (Sherman, Elvander and Chipman[58]).

$$\text{liquid-iron} \ \Delta F^\circ = -32 \ 280 + 5.60 \ T \qquad (120)$$

The ΔF° value for liquid iron is from Ban-ya and Chipman[19].

The variation of sulphur content of liquid iron at 1600°C with the corresponding equilibrium ratio p_{H_2S}/p_{H_2} (or sulphur activity[58]) is shown in Figure 13—46. The standard state is so chosen that the activity coefficient $f_s \to 1$ when $\%S \to 0$. Several workers investigated the effect of alloying elements on the activity coefficient of sulphur in liquid iron. The values of $f_s^{(j)}$ redetermined and compiled by Ban-ya and Chipman[18] are given in Figure 13—47.

Hot-shortness of steel is an old metallurgical problem that has been a subject of research for several decades. It has long been recognized that the presence of oxygen and, particularly, of sulphur in steel is the primary cause of hot-shortness which usually occurs within the temperature range 900° to 1100°C. The hot-short range, of course, varies with the type of steel, particularly with the %Mn/%S ratio of the steel. From the phase diagram (Figure 13—45) it is predicted that pure iron-sulphur alloys containing less than 100 ppm S should not be subject to hot-shortness during the entire hot-working operation because no liquid sulphide can form. On the other hand, an iron containing 200 ppm S will produce a small quantity of a liquid phase as its temperature falls below 1100°C, and the presence of this liquid phase along the grain boundaries will no doubt be a cause of the weakness of the metal. In the presence of oxygen, a liquid oxy-sulphide forms in γ-iron at lower sulphur contents and lower temperature. As shown in Figure 13—48, the sulphur content of iron in equilibrium with wustite and liquid oxy-sulphide reaches a maximum of 143 ppm at 1200°C; this is about one half of that in purified iron. At the ternary eutectic invariant (915°C), the sulphur in solution in equilibrium with wustite and liquid oxy-sulphide is 70 ppm.

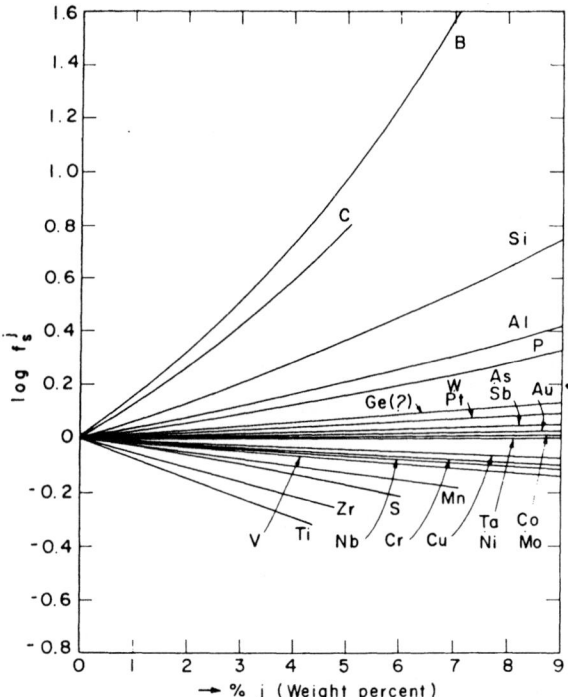

FIG. 13—47. Effect of alloying elements on the activity coefficient of sulphur in iron at 1550°C. (Ban-ya and Chipman[19]).

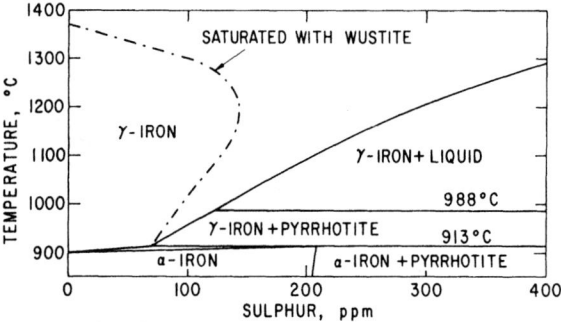

FIG. 13—48. Sulphur content of γ-Fe saturated with wustite is superimposed on binary Fe-S phase diagram. (Turkdogan and Kor[59]).

The manganese in steel is a well-known remedy to hot-shortness by virtue of forming solid oxides and sulphides. The solubility of MnS in γ-iron, determined by Turkdogan et al.[57] is represented by

$$MnS (s) = \underline{Mn} + \underline{S}$$

$$\log [\%Mn] [\%S] = -\frac{9020}{T} + 2.93 \quad (121)$$

The univariant equilibria in the Fe-Mn-S-O system in the presence of γ-iron and Mn(Fe)O phases, derived by Turkdogan and Kor[59] are shown in Figure 13—49. At manganese contents in the region above the univariant j, no liquid phase is present; below this curve liquid oxy-sulphide forms. That is, as long as the steel contains Mn(Fe)O and Mn(Fe)S in equilibrium with iron, a liq-

uid oxy-sulphide may form somewhere between 900° and 1225°C, depending on the concentration of manganese in solution in steel (in the absence of other alloying elements), e.g. for 10 ppm Mn, 900°C and for ~ 10 per cent Mn, ~ 1220°C. In the absence of oxygen, however, a very small amount of manganese (e.g. 20 ppm Mn at 1000°C and 400 ppm at ~ 1300°C) is sufficient to suppress the formation of the liquid sulphide.

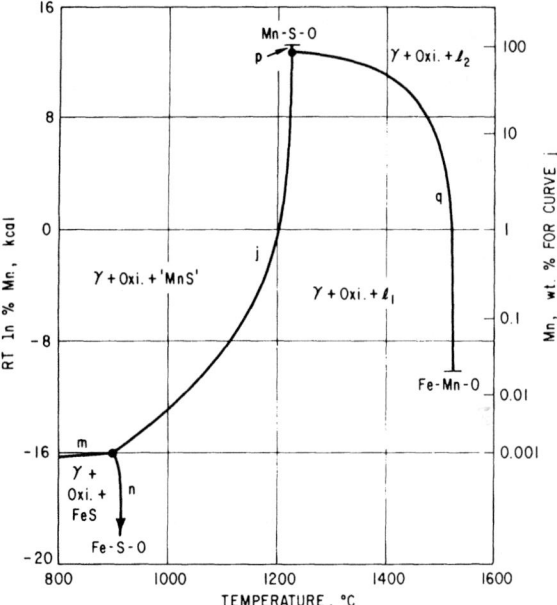

FIG. 13—49. Univariant equilibria in Fe-Mn-S-O system in the presence of γ-Fe and Mn(Fe)O phases. (Turkdogan and Kor[59]).
  j) γ-Fe, Oxi., 'MnS', l_1
  m) γ-Fe, Oxi., 'MnS', 'FeS'
  n) γ-Fe, Oxi., 'FeS', l_1
  p) γ-Fe, Oxi., 'MnS', l_2
  q) γ-Fe, Oxi., l_1, l_2

Elements of Low Solubility in Liquid Iron—A few elements of low solubility in liquid iron play some role in the steelmaking technology; a brief comment on the chemistry of such elements in liquid iron is considered desirable.

(a) Lead:

At steelmaking temperatures the vapor pressure of lead is about 0.5 atm and the solubility[60] in liquid iron is about 0.24 per cent Pb at 1500°C increasing to about 0.4 per cent Pb at 1700°C. The free-machining leaded steels contain 0.15 to 0.35 per cent Pb. Evidently lead added to such steels is in solution in the metal prior to casting and precipitates as small lead spheroids during the early stages of freezing. The break out of the furnace lining, particularly in the open-hearth furnaces, is often blamed on the presence of lead in the melt. Small amounts of metal trapped in the crevices of the lining are likely to get oxidized subsequent to tapping. The lead oxide together with iron oxide readily will flux the furnace lining, hence widen the cracks in the lining, ultimately leading to its failure. The solubility of lead in liquid iron is sufficiently high that in normal steelmaking practice there should be no accumulation of lead at the bottom of the melt, except perhaps in the early stages of melting of the lead-containing scrap.

(b) Calcium:

The boiling point of calcium is 1492°C and its solubility in liquid iron is very low. Sponseller and Flinn[61] measured the solubility of calcium in liquid iron at 1607°C for which the calcium vapor pressure is 2 atm. Under these conditions at 1607°C, 0.032 per cent Ca is in solution. They also investigated the effect of some alloying elements on the solubility; as seen from the data in Figure 13—50, C, Si, Ni and Al increase markedly the calcium solubility in liquid iron. In melts saturated with CaC_2 the solubility of calcium, of course, decreases with increasing carbon content.

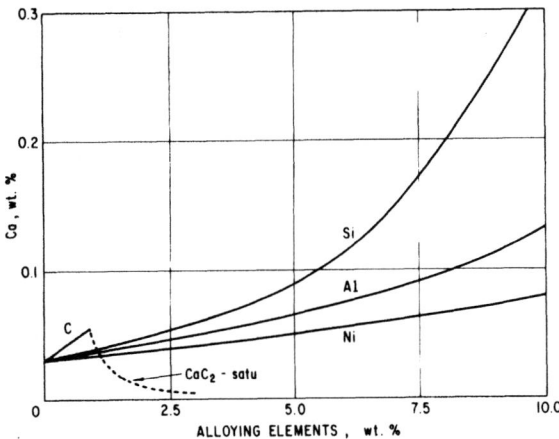

FIG. 13—50. Effect of alloying elements on the solubility of calcium in liquid iron at 1607°C and 2 atm pressure of calcium vapor. (Sponseller and Flinn[61]).

(c) Magnesium:

The solubility of magnesium in iron-carbon alloys determined by Trojan and Flinn[62], is shown in Figure 13—51 as functions of temperature and carbon content. These investigators also observed that the solubility of graphite in liquid iron increases slightly with the addition of magnesium. For example, at 1430°C the graphite solubility in iron increases from 5.03 per cent C to 5.35 per cent C when 3 per cent Mg is added.

TRANSPORT PROPERTIES OF SOLID AND LIQUID IRON ALLOYS

Unlike gases, the transport properties of solids and liquids cannot be calculated on the basis of a rigorous kinetic theory for condensed phases. Such properties have to be determined experimentally The transport properties of solid and liquid iron alloys pertinent to the study of steelmaking problems are given in tabular or graphical forms in this section.

(a) **Coefficient of Diffusion** —The temperature dependence of the diffusivity is usually given by an expression in the form

$$D = D_0 e^{-\frac{E}{RT}}$$

where E is the activation energy for the diffusion process. In Tables 13—XII and 13—XIII the values of D_0 and E are given for solid and liquid iron-base alloys

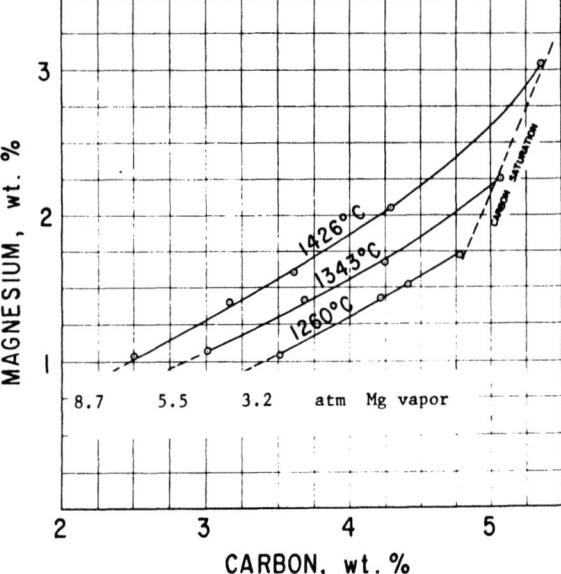

FIG. 13—51. Solubility of magnesium in liquid iron-carbon alloys at indicated temperatures and pressures of magnesium vapor. (Trojan and Flinn[62]).

(mostly for dilute solutions of the diffusate); in some cases data are available only for values of D at certain temperatures.

The general trend in the temperature dependence of diffusivity is shown in Figure 13—52. The interstitial elements, e.g. O, N, C, B, H, have diffusivities much greater than the substitutional elements. Because of larger interatomic spacing the diffusivities of interstitials in bcc-iron are greater and the heats of activations (~ 20 kcal) are smaller than those in the fcc-iron with E ~ 40 kcal. With the substitutional elements, the heat of activation is within 50 to 60 kcal for bcc-iron and within 60 to 70 kcal for fcc-iron.

In liquid iron alloys diffusivities of elements are within 10^{-5} to 10^{-4} cm²/sec with E within 5 to 10 kcal.

(b) **Coefficient of viscosity**—The coefficients of viscosity of iron, cobalt, nickel and copper measured by Cavalier[63] are given in Figure 13—53. In these experiments Cavalier was able to supercool metals 140°-170°C below their melting points and, as seen from the data in Figure 13—53, there are no discontinuities in the viscosity coefficient vs temperature lines extending below the melting temperatures of the metals. Although the viscosity data on copper obtained by Barfield and Kitchener[64] are in complete agreement with those of Cavalier, the data on molten iron by the former authors (Figure 13—54) are higher than Cavalier's values by about 1 cp. The viscosities of Fe-1.16 per cent S and Fe-C alloys determined by Barfield and Kitchener are given in Figures 13—55 and 13—56. In the iron-carbon melts the coefficient of viscosity is essentially independent of composition within the range 0.8 to 2.5 per cent C; above 2.5 per cent C, the viscosity decreases continuously with increasing carbon content. Similar observations were made with measuring the flow of iron-carbon alloys into molds[65].

Table 13—XII Diffusion coefficients of various elements in solid iron and iron-base alloys (Turkdogan[20]).

Diffusing element	Concentration wt.%	Medium	Temp. range °C	D cm²/sec	D₀ cm²/sec	E kcal/mole
Al		γ-Fe	950-1100		30	56.0
B	<0.04	γ-Fe	950-1300		2×10^{-3}	21.0
C	Dilute sol.	α-Fe	α-range		0.34	24.1
C	0.1	γ-Fe	γ-range		0.78	37.9
Cr	Dilute sol.	α-Fe	800-900		8.52	59.9
Cr	Dilute sol.	γ-Fe	900-1400		10.8	69.7
Fe	Pure Fe	α-Fe	α-range		1.9×10^{-2}	47.4
Fe	Pure Fe	γ-Fe	γ-range		1.3	67.9
H	Dilute sol.	α-Fe	15-900		8.85×10^{-4}	3.05
H	Dilute sol.	γ-Fe	900-1400		1.1×10^{-2}	9.95
Mn	<20	γ-Fe	950-1450		0.486+0.011 × % Mn	66.0
N	Dilute sol.	α-Fe	α-range & δ-range		7.8×10^{-3}	18.9
N	Dilute sol.	γ-Fe	950-1350		0.91	40.26
Ni	<20	γ-Fe	1050-1450		0.344+0.012 × % Ni	67.5
O	Dilute sol.	δ-Fe	1450	4.1×10^{-5}		
O	Dilute sol.	α-Fe	α-range		(0.24)	(23.0)
O	Dilute sol.	γ-Fe	γ-range		2.14	40.2
P	Dilute sol.	α-Fe	850-875 1410-1458		2.9	55.0
P	Dilute sol.	γ-Fe	950-1300		6.3×10^{-2}	46.2
S	Dilute sol.	α-Fe	750-890 1400-1451		1.5	48.6
S	Dilute sol.	γ-Fe	1200-1350		2.42	53.4
Si	4.5-7.1	α-Fe	1095-1347		0.44	48.0
Si	<1.9	γ-Fe	1206	4×10^{-10}		
Ti	0.8-2.5	α-Fe	1075-1225		3.15	59.2
Ti	<0.6	γ-Fe	1075-1225		0.15	60.0
V	2.0	α-Fe	1000-1450		3.92	57.6
V	Dilute sol.	γ-Fe	1100-1350		0.28	69.3
V	0.1	γ-Fe	980-1240		1.1	53.5
V	0.3	γ-Fe	980-1240		0.59	53.1
W	<10	α-Fe	1400	3×10^{-9}		
W	<3	γ-Fe	1280	3.7×10^{-10}		

Table 13—XIII Diffusion coefficient of various elements in liquid iron and iron-base alloys. (Turkdogan[20]).

Diffusing element	Concentration wt.%	Medium	Temp. range °C	D cm²/sec	D₀ cm²/sec	E kcal/mole
C	0.03	Fe	1550	7.9×10^{-5}		
C	2.1	Fe	1550	7.8×10^{-5}		
C	3.5	Fe	1550	6.7×10^{-5}		
Co	Dilute sol.	Fe	1568	4.7×10^{-5}		
Co	Dilute sol.	Fe	1638	5.3×10^{-5}		
Fe		Fe-4.6%C	1240-1360		4.3×10^{-3}	12.2
Fe		Fe-2.5%C	1340-1400		1.0×10^{-2}	15.7
H	Dilute sol.	Fe	1565-1679		3.2×10^{-3}	3.3
Mn	2.5	gr. satu. Fe	1300-1600		1.93×10^{-4}	5.8
N	Dilute sol.	Fe	1600	1.1×10^{-4}		
N	Dilute sol.	Fe-0.15%C	1600	5.6×10^{-5}		
O	Dilute sol.	Fe	1610	1.2×10^{-4}		
O	Dilute sol.	Fe	1560,1660	2.5×10^{-5}		
P	Dilute sol.	Fe	1550	4.7×10^{-5}		
S	< 0.64	gr. satu. Fe	1390-1560		2.8×10^{-4}	7.5
S	~ 1	Fe	1560-1670		4.9×10^{-4}	8.6
S	~ 1	Fe	1560-1660	1.7×10^{-4}		
Si	< 2.5	Fe	1480	2.4×10^{-5}		
Si	< 1.3	Fe	1540	3.8×10^{-5}		
Si	1.5	gr. satu. Fe	1400-1600		2.4×10^{-4}	8.2
Si	1-6	gr. satu. Fe	1600	5.4×10^{-5}		

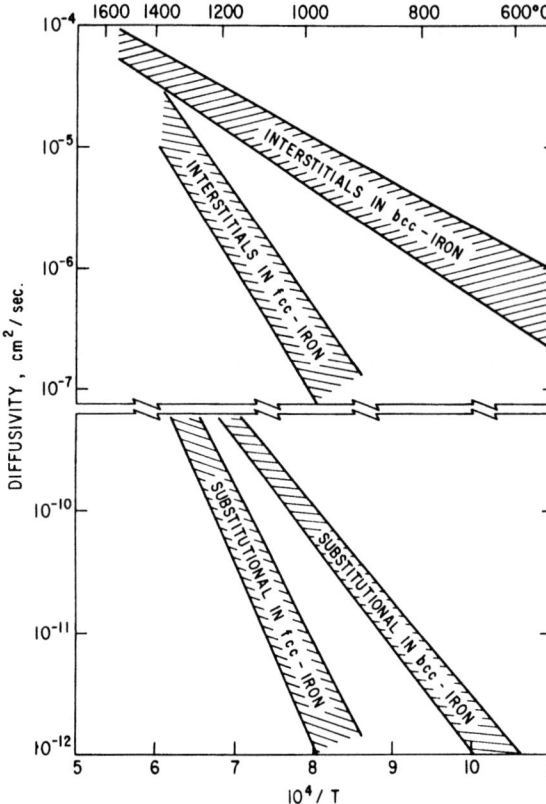

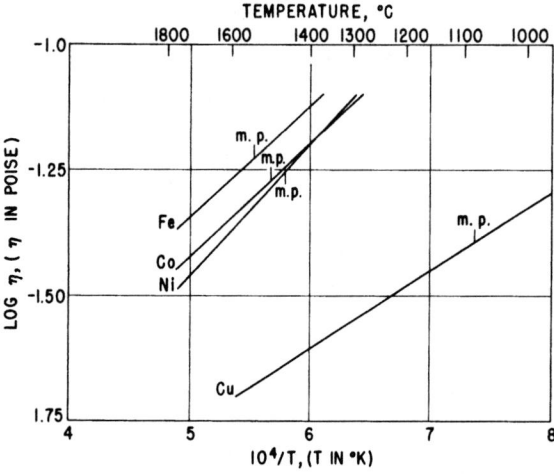

FIG. 13—52. Range of diffusivities of interstitial and substitutional elements in bcc and fcc iron.

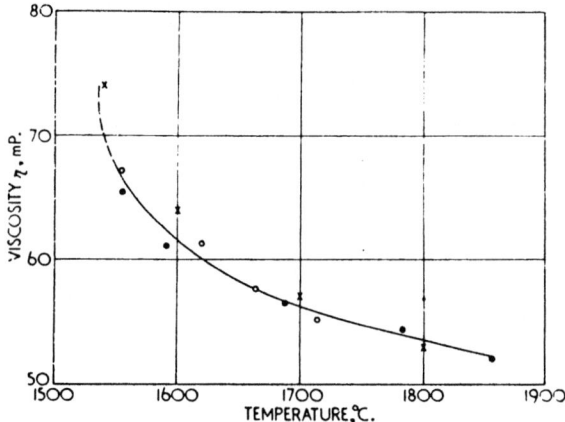

FIG. 13—54. Viscosity coefficient of liquid iron. (Barfield and Kitchener[64]).

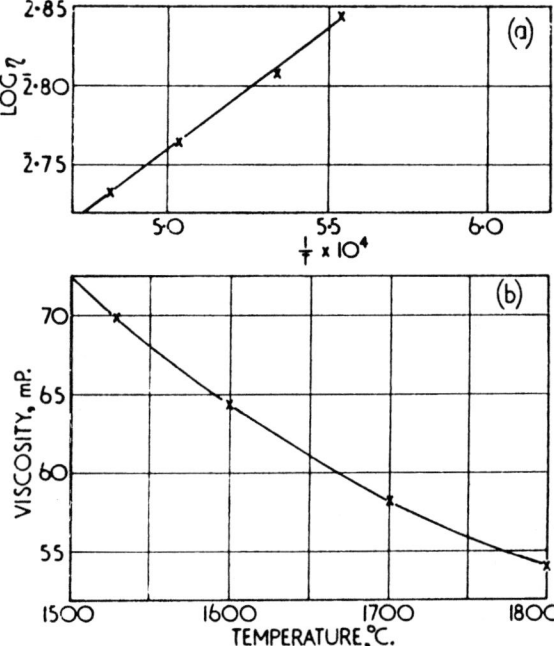

FIG. 13—55. Viscosity coefficient of liquid iron containing 1.16 per cent S. (Barfield and Kitchener[64]).

FIG. 13—53. Temperature dependence of viscosity coefficient of liquid iron, cobalt, nickel and copper. (Cavalier[63]).

(c) Thermal conductivity—The temperature dependence of the thermal conductivity of purified iron (Armco iron) measured by Powell et al.[66] is shown in Figure 13—57. The lower bands are for carbon steels and high-alloy steels[67,68]. The effect of alloying elements on the thermal conductivity becomes less at elevated temperatures.

(d) Surface tension—The surface tension of liquid iron has been measured by a number of investigators. The most reliable value perhaps is that obtained by Kozakevitch and Urbain[69] for liquid iron, cobalt, nickel and copper at 1550°C: Fe=1835 dynes/cm, Co=1936 dynes/cm, Ni=1924 dynes/cm and Cu=1300 dynes/cm. The effect of various elements on the surface tension of iron, determined by Kozakevitch and Urbain, is given in Figures 13—58, 13—59 and 13—60. The surface tension of the ternary Fe-C-S alloys at 1450°C is given in the top diagram of Figure 13—61. Carbon has no effect on the surface tension of iron (Figure 13—60); however, since carbon raises the activity of sulphur in iron, in the presence of carbon the sulphur has a greater effect on the surface tension of iron[70], as evidenced from the lower diagram in Figure 13—61.

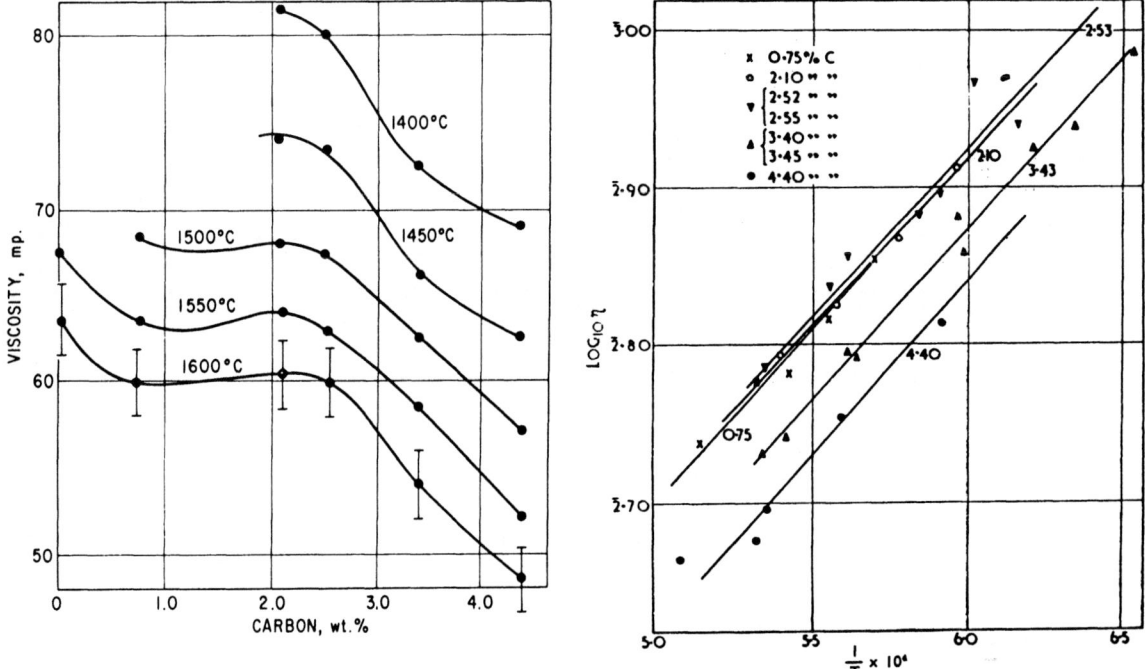

Fig 13—56. Composition and temperature dependence of the viscosity coefficient of iron-carbon melts. (Barfield and Kitchener[64]).

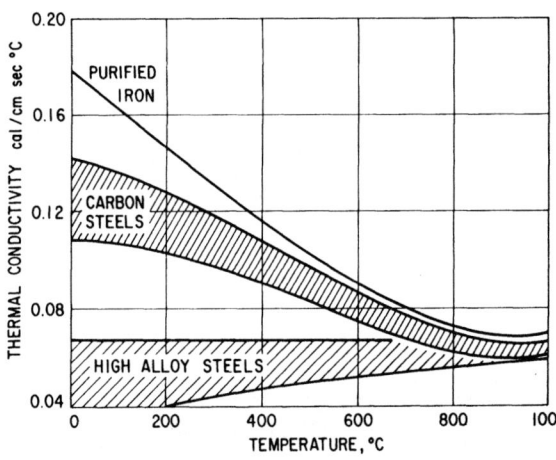

FIG. 13—57. Temperature dependence of thermal conductivity of purified iron[66] and iron alloys[67,68].

(e) **Density**—The temperature dependence of the densities of liquid iron, nickel, cobalt, copper, chromium, manganese, vanadium and titanium are given by the following equations as a linear function of temperature in °C in g/cm^3:

Iron:	$8.30 - 8.36 \times 10^{-4}T$
Nickel:	$9.60 - 12.00 \times 10^{-4}T$
Cobalt:	$9.57 - 10.17 \times 10^{-4}T$
Copper:	$9.11 - 9.44 \times 10^{-4}T$
Chromium:	$7.83 - 7.23 \times 10^{-4}T$
Manganese:	$7.17 - 9.30 \times 10^{-4}T$
Vanadium:	$6.06 - 3.20 \times 10^{-4}T$
Titanium:	$4.58 - 2.26 \times 10^{-4}T$

(122)

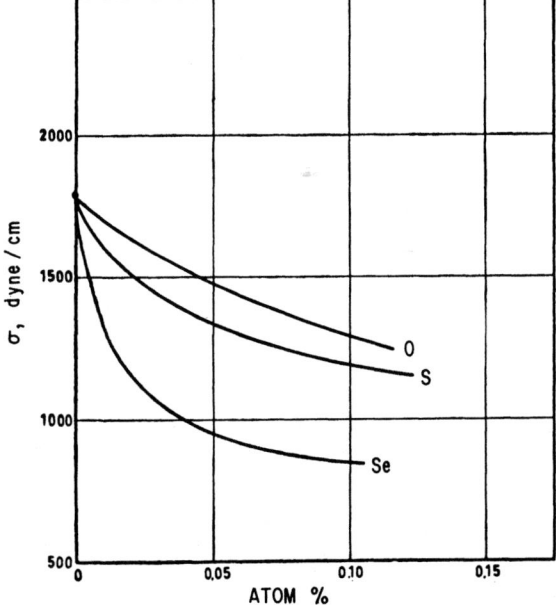

FIG. 13—58. Effect of oxygen, sulphur and selenium on the surface tension of liquid iron at 1550°C. (Kozakevitch and Urbain[69]).

The specific volume and density of liquid iron-carbon alloys are given in Figure 13—62 for various temperatures. It should be noted that the density of the liquid in equilibrium with austenite does not change much over the entire liquidus range.

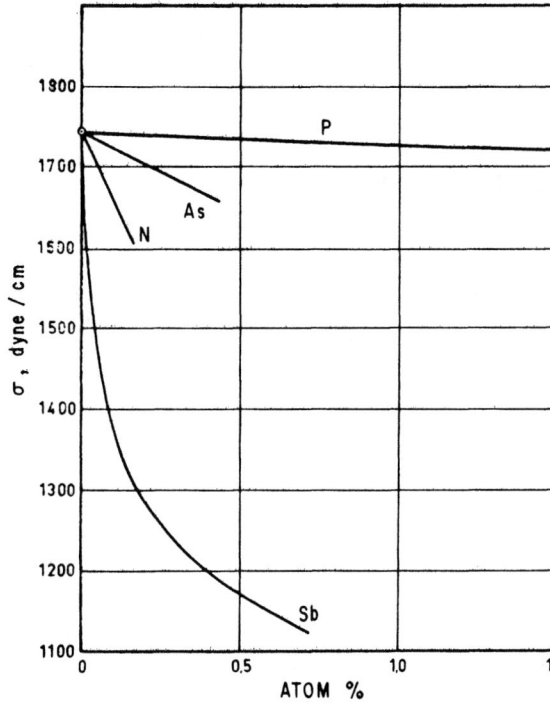

FIG. 13—59. Effect of phosphorus, arsenic, nitrogen and antimony on the surface tension of liquid iron at 1550°C. (Kozakevitch and Urbain[69]).

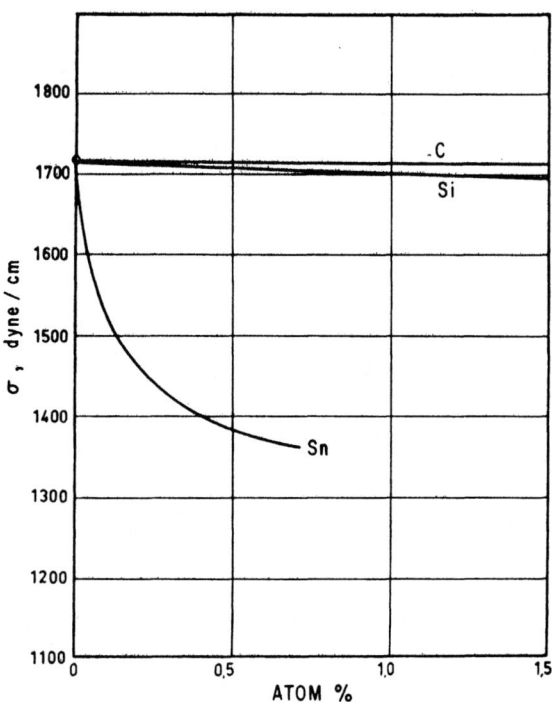

FIG. 13—60. Effect of carbon, silicon and tin on the surface tension of liquid iron at 1550°C (Kozakevitch [70]).

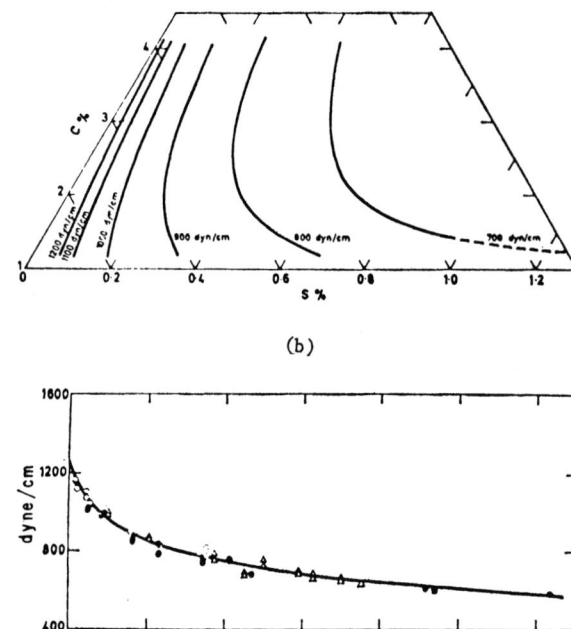

(b)

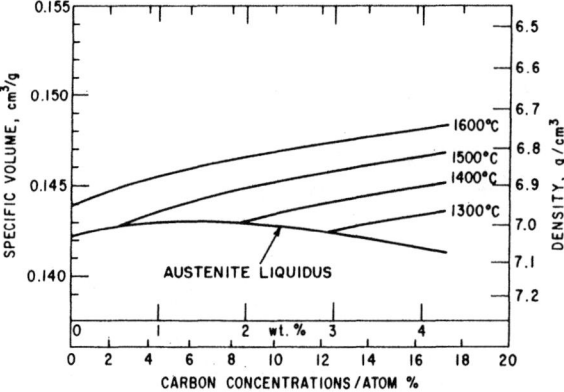

FIG. 13—61. Surface tension of liquid Fe-C-S alloys at 1450°C (Kozakevitch[70]). (a) Lines of constant surface tension. (b) Surface tension as a function of sulphur activity for carbon percentages: 0, 1.25; 0, 2.2; ∆, 4.0.

FIG. 13—62. Density of iron-carbon alloys. (Lucas[71]).

STEELMAKING REFRACTORIES AND SLAGS

The understanding of the chemistry of steelmaking processes and the study of the methods of control require adequate knowledge of the basic physicochemical properties of slags and also of the oxide systems constituting the refractory materials used in the steel industry. With this in view, data are compiled in this section on the phase-equilibrium diagrams and thermodynamic activities in binary, ternary and a selected few multicomponent oxide systems and slags.

Phase Equilibrium Diagrams—Binary and ternary phase diagrams to be discussed here are limited to the oxide systems relevant to the iron- and steelmaking processes. Whenever possible, the references are made to those authors who made the major contribution to the construction of the phase diagrams, and for the sake of brevity, no reference will be made to those authors who made minor, though valuable, contributions to the present state of our knowledge on these phase diagrams. For convenience, the interoxide phases (compounds) with stoichiometric compositions are formulated in terms of the oxides of the constituent elements. Although, for example, di-calcium silicate should be formulated as Ca_2SiO_4, in connection with oxide phase diagrams it is customary to formulate it as $2CaO \cdot SiO_2$.

(a) **Binary systems**—In phase diagrams where "FeO" (wustite) is one of the oxide components, the system is for that in equilibrium with solid or liquid iron, unless stated otherwise.

The simplest oxide system is that between wustite and MnO where there is a complete series of liquid and solid solutions (Figure 13—63).

In considering the silicate binary systems, mention should be made of the polymorphic forms of silica. The following are the major phase transformations of silica: melting temperature, 1713°C; cristobalite → tridymite transformation, 1470°C; and tridymite → quartz transformation, 867°C. However, tridymite is not a true polymorphic form of pure silica. It has been demonstrated[73-75] that tridymite is formed within the temperature range 867°-1470°C only when some metal ions dissolve in silica. It is for this reason that in all the metal oxide-silica systems tridymite appears as a stable phase over the indicated temperature range. In most of the silicate systems, except those with alkali oxides and alumina, there is a liquid miscibility gap as seen from the phase diagrams in Figures 13—64 to 13—67. Another characteristic feature is that there is very little or negligible solid-solution formation in most of the silicate systems.

In the CaO-SiO$_2$ system (Figure 13—64) there are two phases with congruent melting points, α-CaO·SiO$_2$ and α-2CaO·SiO$_2$ and two with incongruent melting points, 3CaO·2SiO$_2$ and 3CaO·SiO$_2$, the latter being stable only within the temperature range 1250°-2070°C. One characteristic feature of the di-calcium silicate is that on conversion from β to γ crystalline modification at 675°C, there is about 12 per cent volume expansion causing the solid mass to become powdery. In electric furnace practice, the reducing slag used has a composition similar to that of di-calcium silicate, and because of the transformation from β to γ form, on cooling the slag becomes powdery; in industry this type of slag is often referred to as the "white falling slag". The formation of γ-2CaO·SiO$_2$ in blast-furnace slags is highly undesirable if the slag is to be used as an aggregate for constructional purposes.

The phase diagram for the 'FeO'-SiO$_2$ system in Figure 13—65 is that in equilibrium with solid or liquid iron; this is not a true binary system. The amount of ferric oxide in the melt along the liquidus curve, in equilibrium with iron, decreases from about 12 per cent at the 'FeO' corner to about 1 per cent at silica saturation.

No further work has been reported on the MgO-SiO$_2$ diagram (Figure 13—66) since 1927[81,82]. Below 40 per cent SiO$_2$ the eutectic melting occurs at 1850°C indi-

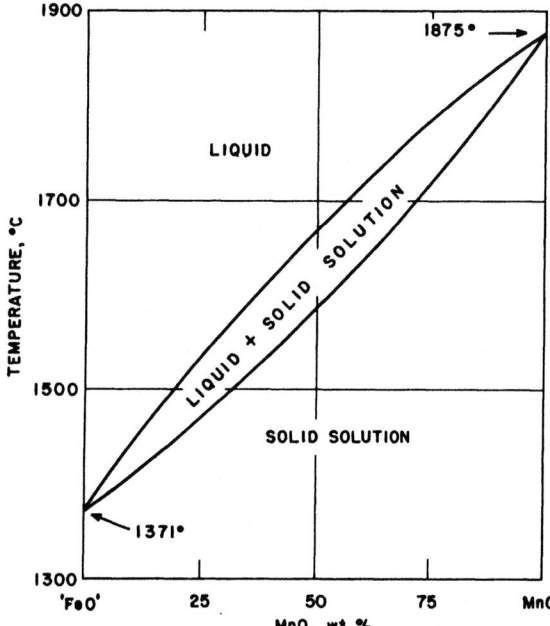

FIG. 13—63. Phase equilibrium diagram for 'FeO'-MnO system. (Fischer and Fleischer[72]).

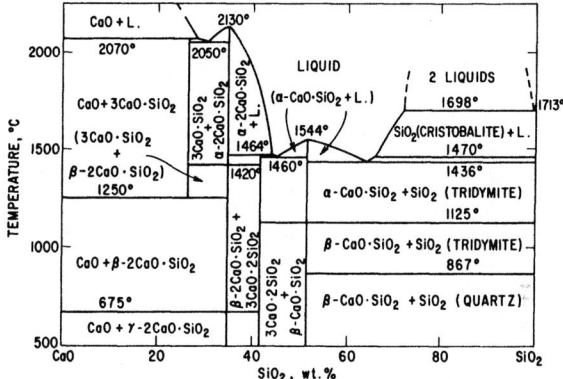

FIG. 13—64. Phase-equilibrium diagram for CaO-SiO$_2$ system. (Rankin and Wright[76] with certain modifications[67,68]).

cating that magnesia, even in the presence of appreciable amounts of silica, is a good refractory material. The refractory oxides and some silicates are often called by their mineralogical names: MgO (magnesia) is referred to as periclase; 2MgO·SiO$_2$ as forsterite; MgO·SiO$_2$ as protoenstatite.

Figure 13—67 is the phase diagram for the MnO-SiO$_2$ system re-examined by Glasser[85], making some modifications on the original diagram by Herty[83] and White et al.[84].

The diagram for the Al$_2$O$_3$-SiO$_2$ system in Figure 13—68a essentially is that given by Bowen and Greig[86] with a slight modification[87]. More recent studies suggest that mullite is a nonstoichiometric phase with a narrow solid solubility range, and at present there is divergence of opinion on the question of dissociation of mullite at its melting temperature[88,89]. Although not certain, Figure 13—68b is the suggested modification of this phase diagram.

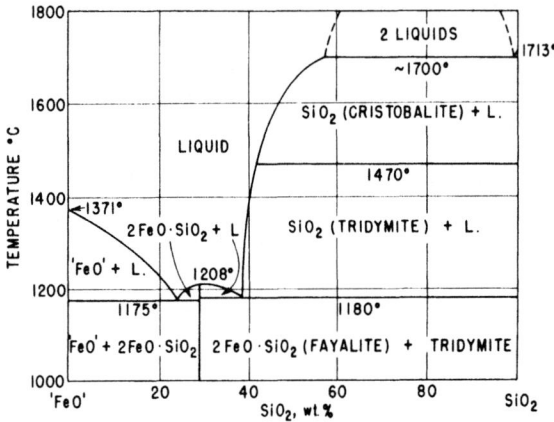

FIG. 13—65. Phase-equilibrium diagram for 'FeO'-SiO₂ system. (Original diagram of Bowen and Shairer[79] was verified by Schuhmann and Ensio[80]).

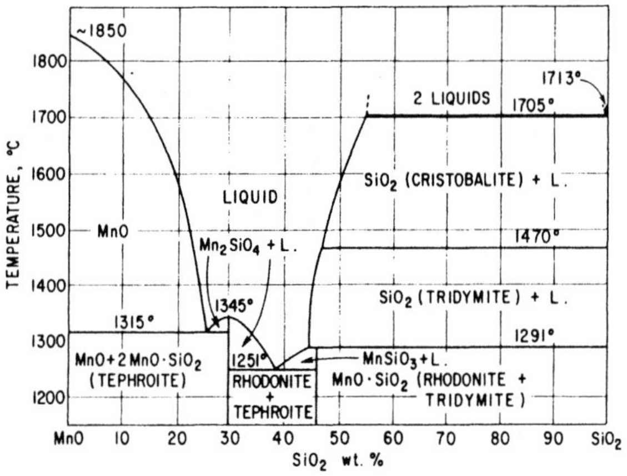

FIG. 13—67. Phase-equilibrium diagram for MnO-SiO₂ system. (Original diagram by Herty [83] and White et al.[84]; the above modification by Glasser[85]).

There are three types of aluminum silicates in considerable quantities in nature: andalusite, sillimanite and kyanite of composition $Al_2O_3 \cdot SiO_2$, but of different crystalline forms. These minerals are employed as refractories; upon heating they dissociate to mullite and silica.

Figure 13—69 is the phase diagram for the CaO-Al_2O_3 system which, first obtained by Rankin and Wright[76], has been modified by incorporating the results of subsequent investigators [90-92]. Another phase $CaO \cdot 6Al_2O_3$ melting incongruently at 1850°C has been claimed to exist in this binary system by Filemenko and Lavrov[93]; however, this finding has not so far been confirmed by any other work and, therefore, it is not included in Figure 13—69.

In the CaO-MgO system (Figure 13—70) there is small solid solubility at the terminal regions at the eutectic temperatures ∼ 2300°C. The dolomite, $MgCa(CO_3)_2$, is an important raw material for use in steelmaking furnaces as a basic refractory material. The

dolomite is calcined at temperatures of the order of 1700°C to yield a material less subject to deterioration by moisture during storage. In the manufacture of dolomite refractory bricks, iron oxide, silica and alumina are added to stabilize lime. A great deal of work has been done on the stabilization of dolomite bricks; this is discussed in detail, for example, by Chesters[95] in his book on refractories for steelplants.

The phase diagram in Figure 13—71 is for the CaO-Fe_2O_3 system at one atmosphere of oxygen; the curve within the ternary diagram gives the composition of the liquid along the hematite and magnetite liquidus curve, given as a pseudo-binary section in Figure 13—71.

The phase diagram for the CaO-P_2O_5 system determined by Tromel and co-workers[97,98] is given in Figure 13—72. In steelmaking involving the charging of high-phosphorus hot metal, the formation of tetra- or tricalcium phosphate in lime-saturated slag plays an important role in the dephosphorization of steel.

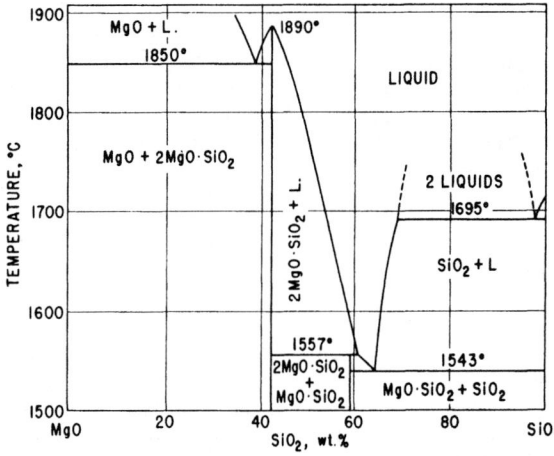

FIG. 13—66. Phase-equilibrium diagram for MgO-SiO₂ system. (Originally by Bowen and Anderson[81], modified by Greig[82]).

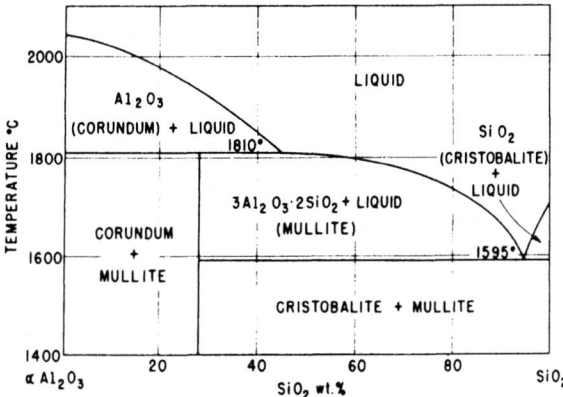

FIG. 13—68a. Phase-equilibrium diagram for Al_2O_3-SiO_2 system. (Bowen and Greig[86]).

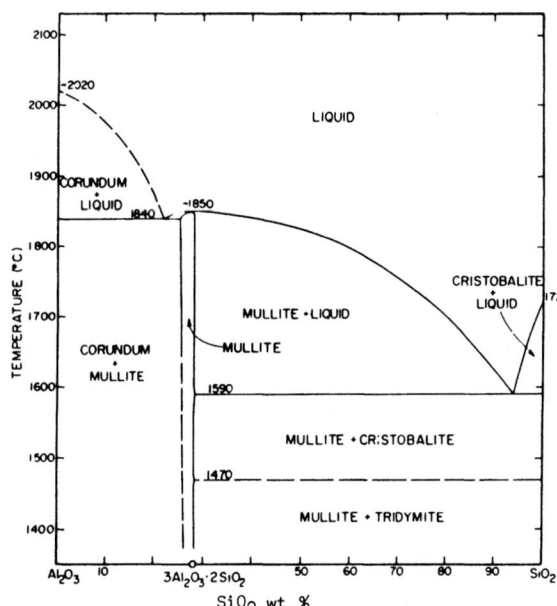

FIG. 13—68b. Phase diagram for the system Al_2O_3-SiO_2 modified on the basis of more recent data[77,78].

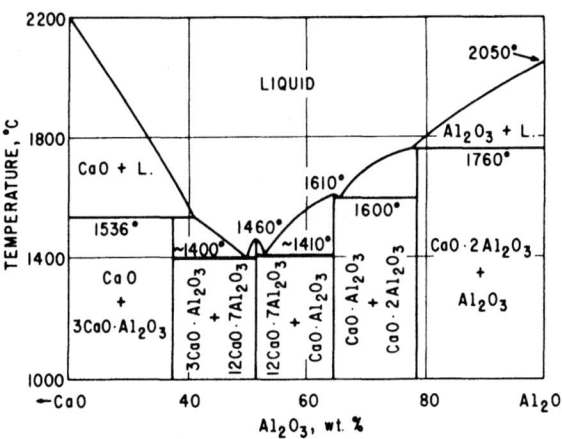

FIG. 13—69. Phase diagram for CaO-Al_2O_3 system; original diagram of Rankin and Wright[76] modified using later data[90-92].

(b) **Ternary systems**—Detailed discussion of phase equilibria in ternary oxide systems constituting the refractories for steelplants is much beyond the scope of this chapter. In order to demonstrate the complexity of the phase diagrams of ternary oxide systems, let us consider the CaO-MgO-SiO_2 system. The last revision of the data on this system was made in 1960 by Osborn and Muan[96] from whose paper Figure 13—73 is reproduced, where the liquidus isotherms are projected on the plane of the diagram. There are four ternary interoxide phases:

Mineralogical name	Formula
Diopside	$CaO \cdot MgO \cdot 2SiO_2$
Akermanite	$2CaO \cdot MgO \cdot 2SiO_2$
Monticellite	$CaO \cdot MgO \cdot SiO_2$
Merwinite	$3CaO \cdot MgO \cdot 2SiO_2$

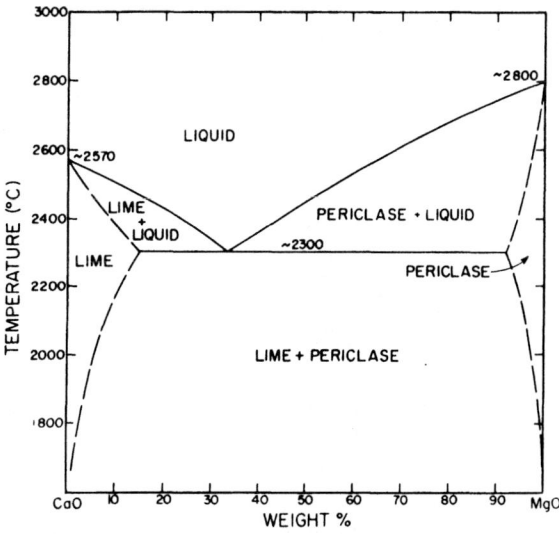

FIG. 13—70. Phase diagram for CaO-MgO system. (Doman et al.[94]).

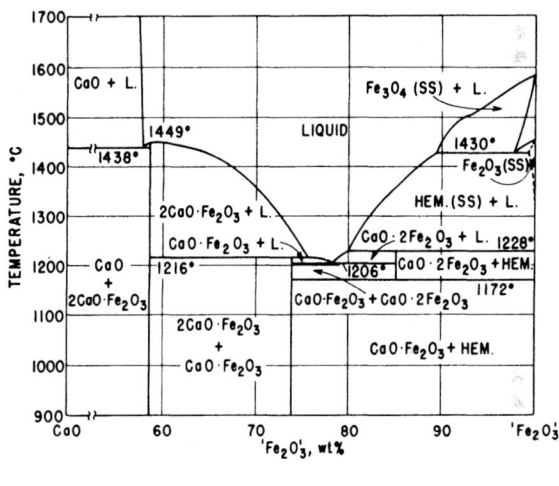

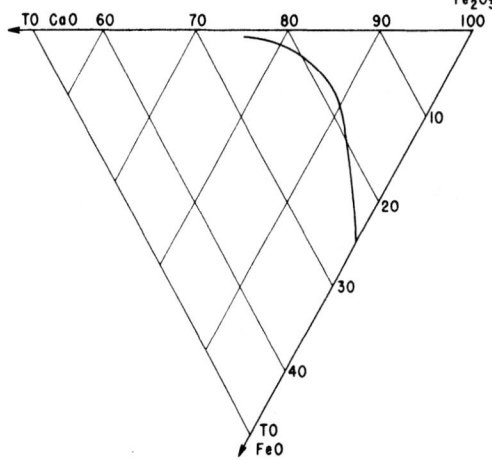

FIG. 13—71. (Above) Phase-equilibrium diagram for CaO-'Fe_2O_3' system at 1 atm O_2. (Below) Variation of the composition of the melt along the hematite and magnetite liquidus curves. (Swayze[100] and Phillips and Muan[99]).

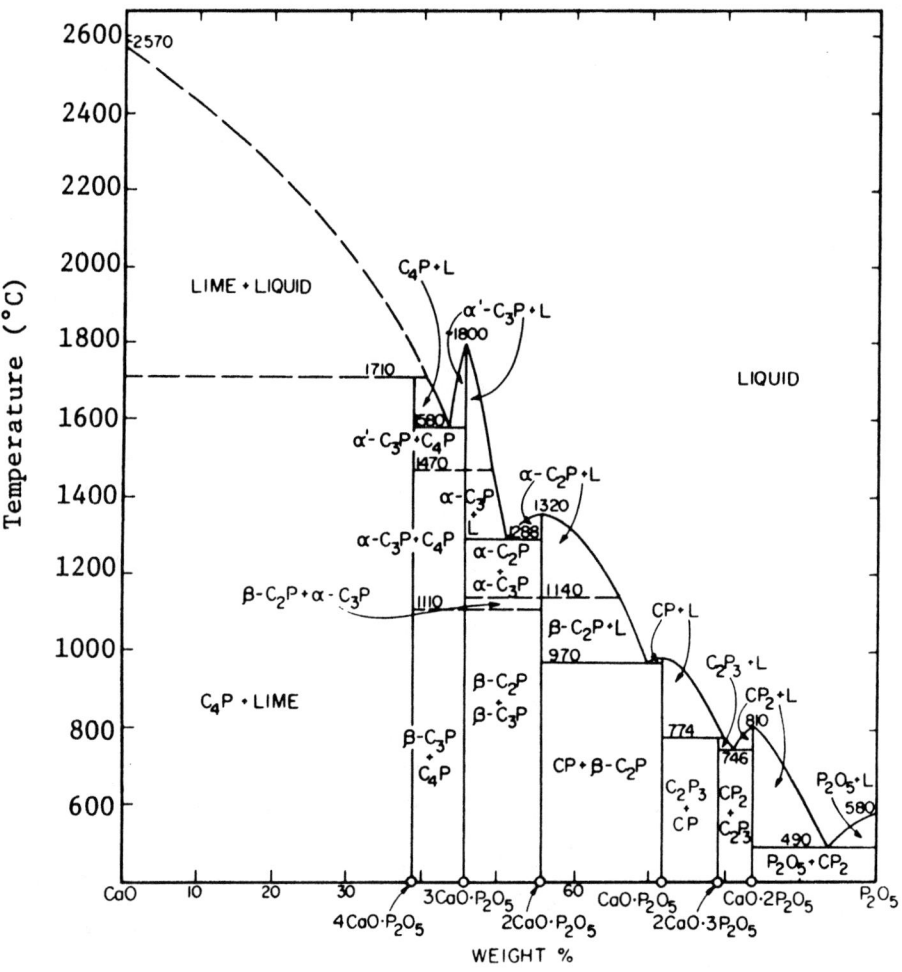

FIG. 13—72. Phase diagram for the system CaO-P$_2$O$_5$, as summarized by Tromel et al.[97,98]. Abbreviations used have the following meanings, C$_4$P = Ca$_4$P$_2$O$_9$, C$_3$P = Ca$_3$P$_2$O$_8$, C$_2$P = Ca$_2$P$_2$O$_7$, CP = CaP$_2$O$_6$, C$_2$P$_3$ = Ca$_2$P$_6$O$_{17}$, CP$_2$ = CaP$_4$O$_{11}$, L = liquid.

There are eleven ternary peritectic invariants and five ternary eutectic invariants. There are a number of oxide solid solutions in certain parts of the system. At high temperatures there is a complete series of pyroxene solid solutions between MgO·SiO$_2$ and CaO·MgO·2SiO$_2$, a limited solid solution in α-CaO·SiO$_2$. One of these solid solution regions is shown in Figure 13—74 for the section between forsterite and monticellite. A typical dolomite refractory has a composition close to a point on the 3CaO·SiO$_2$-MgO join, consisting of tricalcium silicate and periclase with a small amount of either lime or di-calcium silicate. As pointed out earlier, the presence of free lime and/or di-calcium silicate is detrimental to the dolomite bricks: lime hydrates and di-calcium silicate dusts. Temporary resistance of dolomite bricks to hydration is achieved by applying tar coating.

In relation to the melting temperatures of blast furnace type slags, reference may be made to the work of Osborn et al.[102] who measured the liquidus isotherms in parts of the quaternary system CaO-MgO-SiO$_2$-Al$_2$O$_3$. These isotherms are given in Figure 13—75 for part of the pseudo-ternary system containing 10 per cent Al$_2$O$_3$; the average compositions of the blast-furnace slags are those located within the areas A, B and C.

Muan and Osborn[103] made a comprehensive review of the available data on the phase equilibria among oxides of interest to steelmaking and refractories. The liquidus isotherms of selected ternary oxide systems given in Figures 13—74 to 13—79 are reproduced from the compilation of Muan and Osborn of particular interest to steelmaking and refractories. Poor resistance of alumina-silica refractories to slag attack is evident from relatively low eutectic temperatures obtained by the addition of CaO, MgO, FeO (and MnO) to the Al$_2$O$_3$-SiO$_2$ system. The phase diagrams for the MgO-Cr$_2$O$_3$-SiO$_2$ and MgO-Cr$_2$O$_3$-Al$_2$O$_3$ systems in Figures 13—80 and 13—81 are most appropriate for the chrome-magnesite refractories. The melting temperature of MgO-Cr$_2$O$_3$ and MgO-Al$_2$O$_3$ spinels is not lowered by modest additions of SiO$_2$, CaO and FeO. The

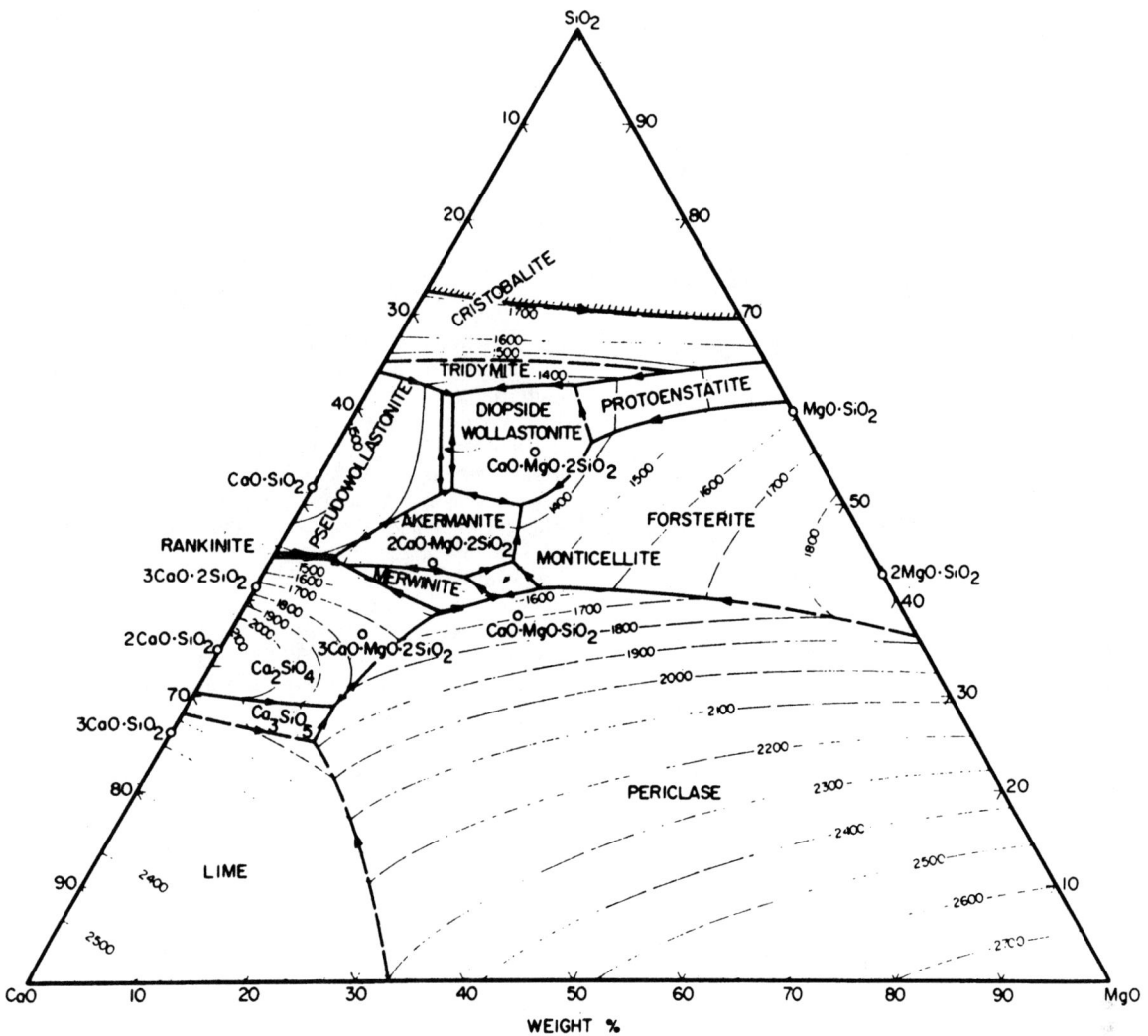

Name	Composition
Periclase	MgO
Cristobalite } Tridymite }	SiO_2
Lime	CaO
Diopside	$CaO \cdot MgO \cdot 2SiO_2$
Pyroxene	Solid solution along section $MgO \cdot SiO_2$-diopside
Akermanite	$2CaO \cdot MgO \cdot 2SiO_2$
Monticellite (Mo)	$CaO \cdot MgO \cdot SiO_2$
Merwinite (Merw)	$3CaO \cdot MgO \cdot 2SiO_2$
Forsterite	$2MgO \cdot SiO_2$

FIG. 13—73. Liquidus isotherms for the CaO-MgO-SiO_2 system. (Compiled by Osborn and Muan[96]).

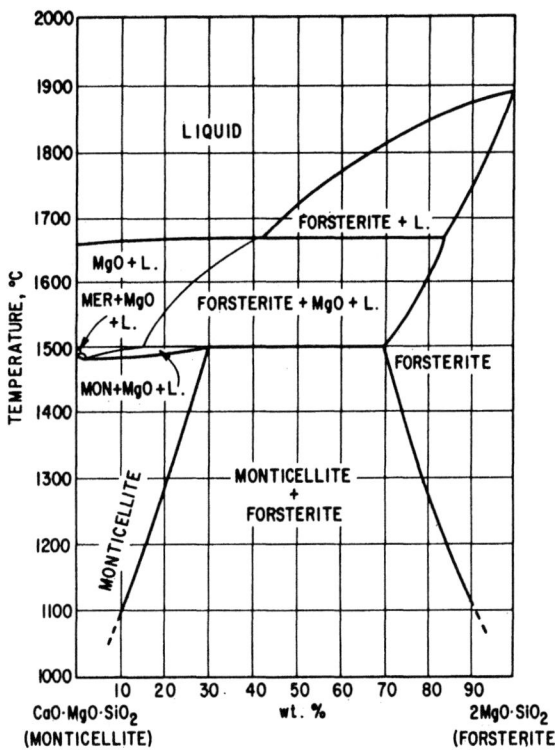

FIG. 13—74. Phase equilibrium in the section monticellite-forsterite of the CaO-MgO-SiO₂ system. (From Ricker and Osborn[101]).

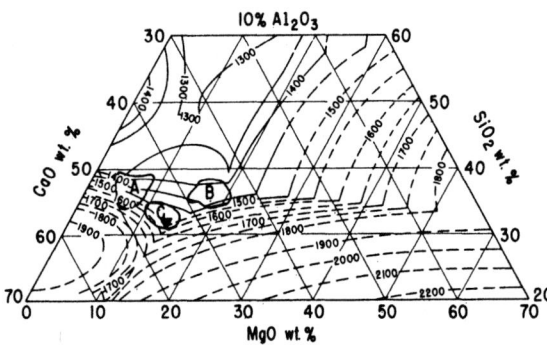

FIG. 13—75. Diagram of part of 10 per cent Al₂O₃ plane showing location of representative blast furnace slags recalculated to 100 per cent CaO + MgO + Al₂O₃ + SiO₂. Each area delimited by a heavy line (A, B, and C) includes daily average slag analyses from a particular furnace where Al₂O₃ was 10±1 per cent. (Osborn et al.[102]).

failure of chrome refractories in the steelplants is often associated with "bursting" expansion associated with absorption of iron-oxide-rich slag by the spinel grains. Increasing the Al₂O₃ and SiO₂ and decreasing Cr₂O₃ contents of the chrome-magnesite bricks is claimed to minimize the tendency to spalling.

There is a large liquid miscibility gap in the CaO-FeO-P₂O₅ system which was first investigated by Oelsen and Maetz[104]. The phase diagram in Figure 13—82 is that compiled by Tromel and co-workers[97,98].

The composition of Thomas slags containing about 25 per cent P_2O_5 and about 50 per cent CaO lies slightly below the miscibility gap univariant close to lime saturation. As shown by Olette et al.[105,106], the composition range of the liquid miscibility gap in the CaO-iron oxide-P_2O_5 system is not affected by oxygen pressure. In fact, Schwerdtfeger and Turkdogan[107] also showed that the composition range of the miscibility gap in the system equilibrated with air (Figure 13—83) is similar to that in equilibrium with iron (Figure 13—82).

THERMODYNAMIC ACTIVITIES IN MOLTEN SLAGS AND GLASSES

The systems consisting of silicates, silico-phosphates, aluminates, borates, ferrites, etc. are generally called slags and/or glasses; their structure in the crystalline or molten state has some similar characteristics. Experiments on electrical conductivity and electrolysis first carried out about 50 years ago indicated that natural silicates and slags consist of simple and complex ions. The theory of the structure of glasses in general was first studied by Zachariasen[108] and by Warren[109] using data on x-ray diffraction measurements. The structural studies were later extended to molten slags, for example, by Warren et al[110]. Since the 1930's, many papers have been published on the structure of solid and molten glasses and slags. Although these investigations revealed several important facts about the structures of these systems, no fundamental changes were made to the original concept developed by Zachariasen and Warren et al.

The basic concept is that in glasses and molten silicates the structure consists of tetrahedrally bonded Si-O networks or rings with cations randomly distributed in the interstices. In fused or vitreous silica, each silicon atom is bonded to four oxygen atoms and each oxygen atom to two silicon atoms, and the network is continuous in three dimensions. Although in the crystalline state the tetrahedral arrangement of the silicon and oxygen atoms is symmetrical, in vitreous silica the lattice is distorted; in the molten state, there is a further decrease in the symmetry of the network, and in addition, there is a general weakening of the bond strength.

In representing the composition of systems consisting of oxides, sulphides, silicates, etc., it is not necessary to know anything about the composition of the complex ions. Depending on the circumstances, the composition may be given in terms of oxides, FeO, Fe₂O₃, CaO, SiO₂...etc., in terms of atoms, Fe, Ca, Si...etc., or in terms of some assumed ions, Fe^{2+}, Fe^{3+}, Ca^{2+}, SiO^{4-}_4, PO^{3-}_4...etc. For convenience, the oxides are often used to represent the composition of the system. When considering the thermodynamics of the oxide system, the composition in terms of weight per cent of the oxides is converted to the mole fractions of the oxides, N, i.e.

$$N_i = \frac{\text{wt. \%}i}{M_i \, \Sigma n} \qquad (123)$$

where M_i is the molecular weight of an oxide i and Σn is the total number of moles of oxides in 100 g of the oxide mixture.

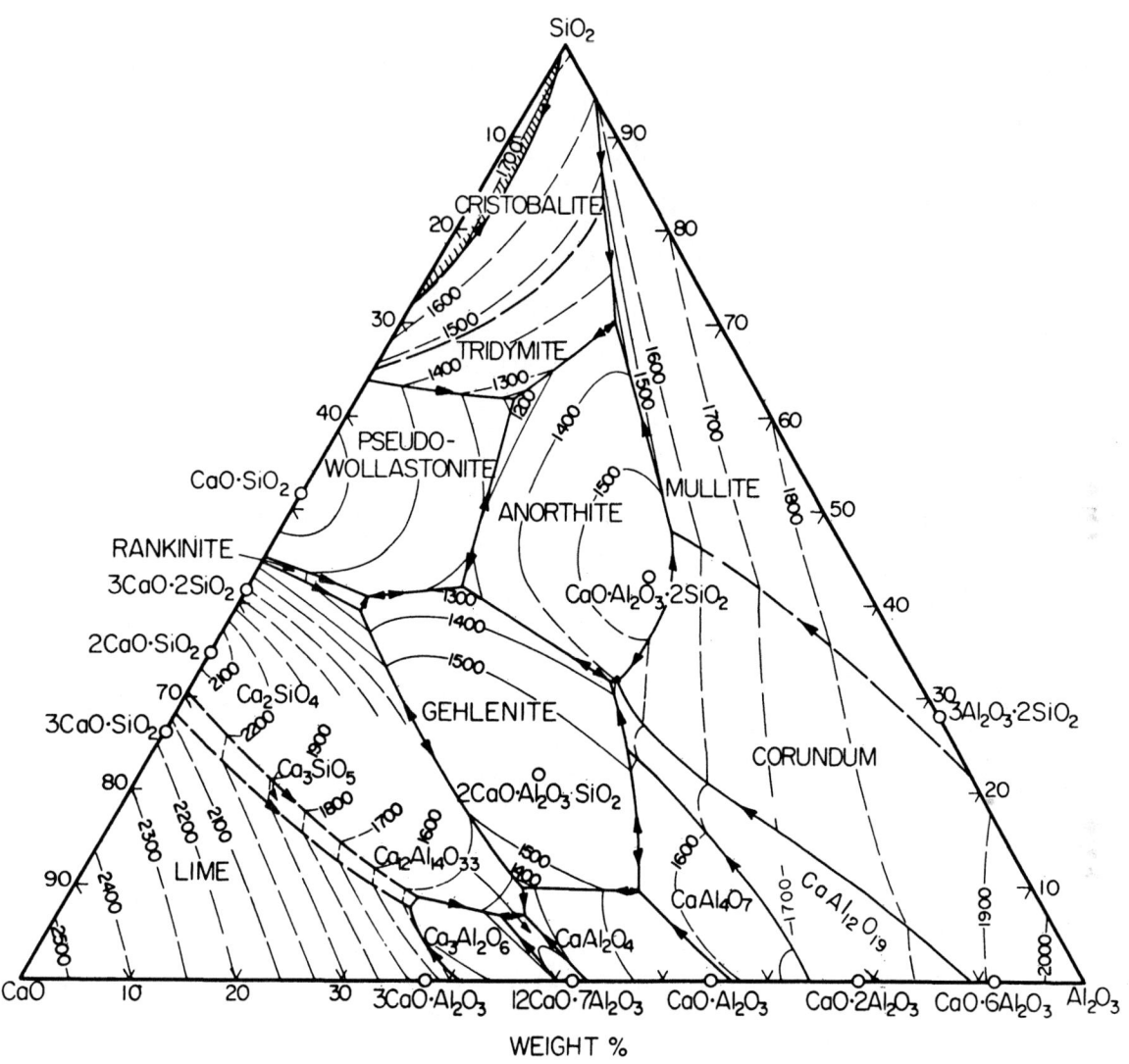

FIG. 13—76. Liquidus isotherms for the $CaO-Al_2O_3-SiO_2$ system. (Compiled by Muan and Osborn[103]).

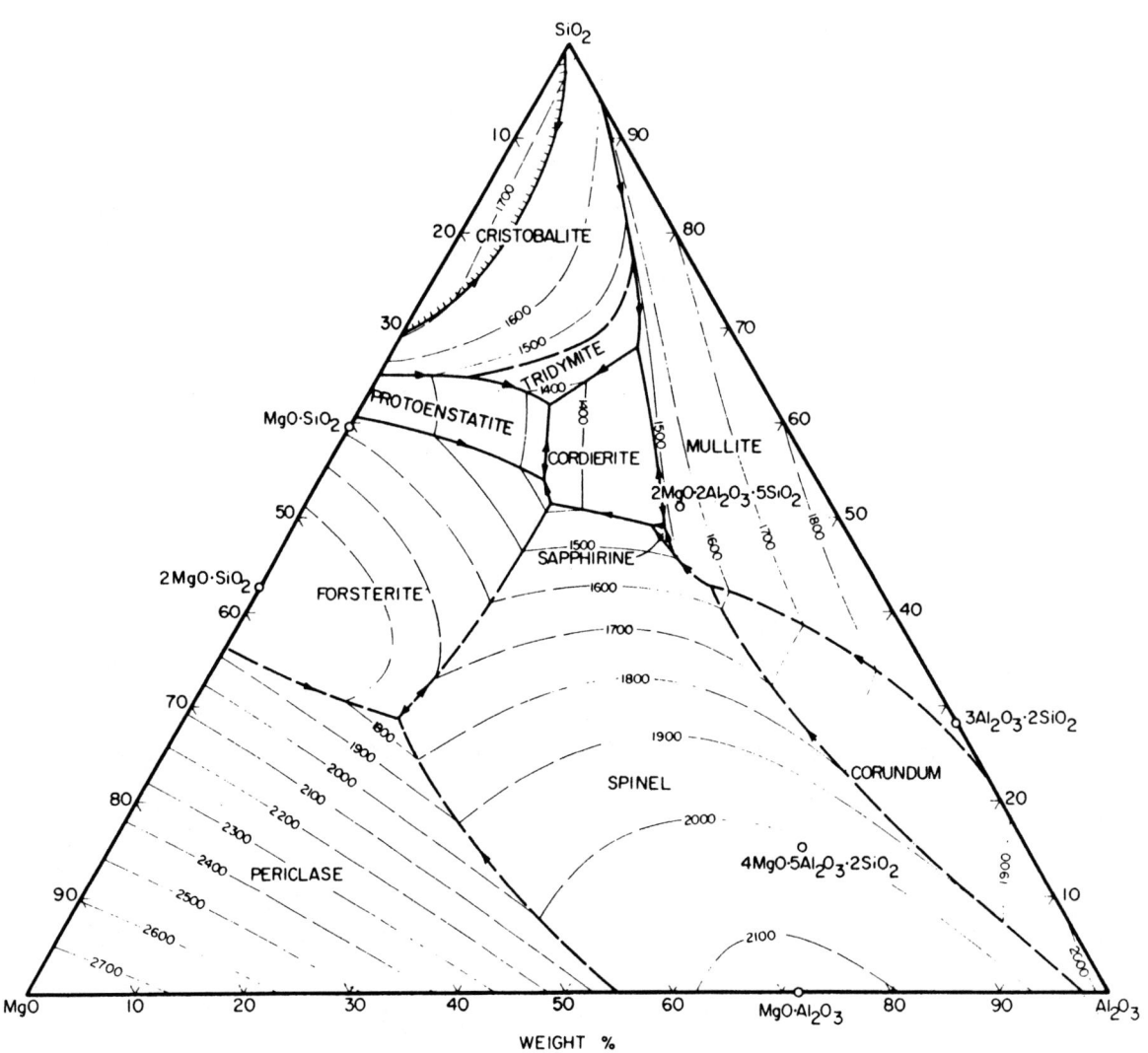

FIG. 13—77. Liquidus isotherms for the MgO-Al₂O₃-SiO₂ system. (Compiled by Muan and Osborn[103]).

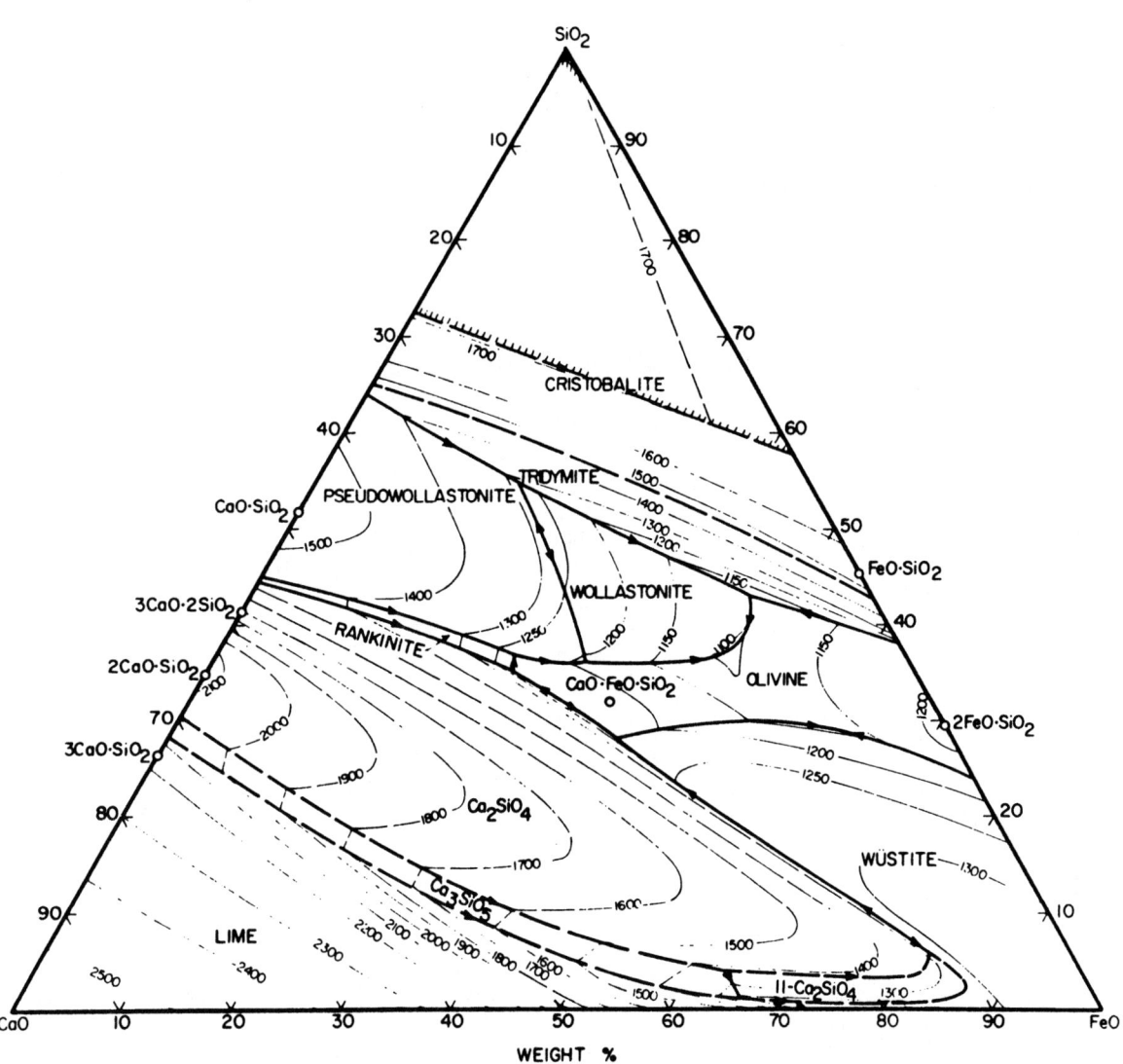

FIG. 13—78. Liquidus isotherms for the CaO-FeO-SiO₂ system. (Compiled by Muan and Osborn[102, 103]).

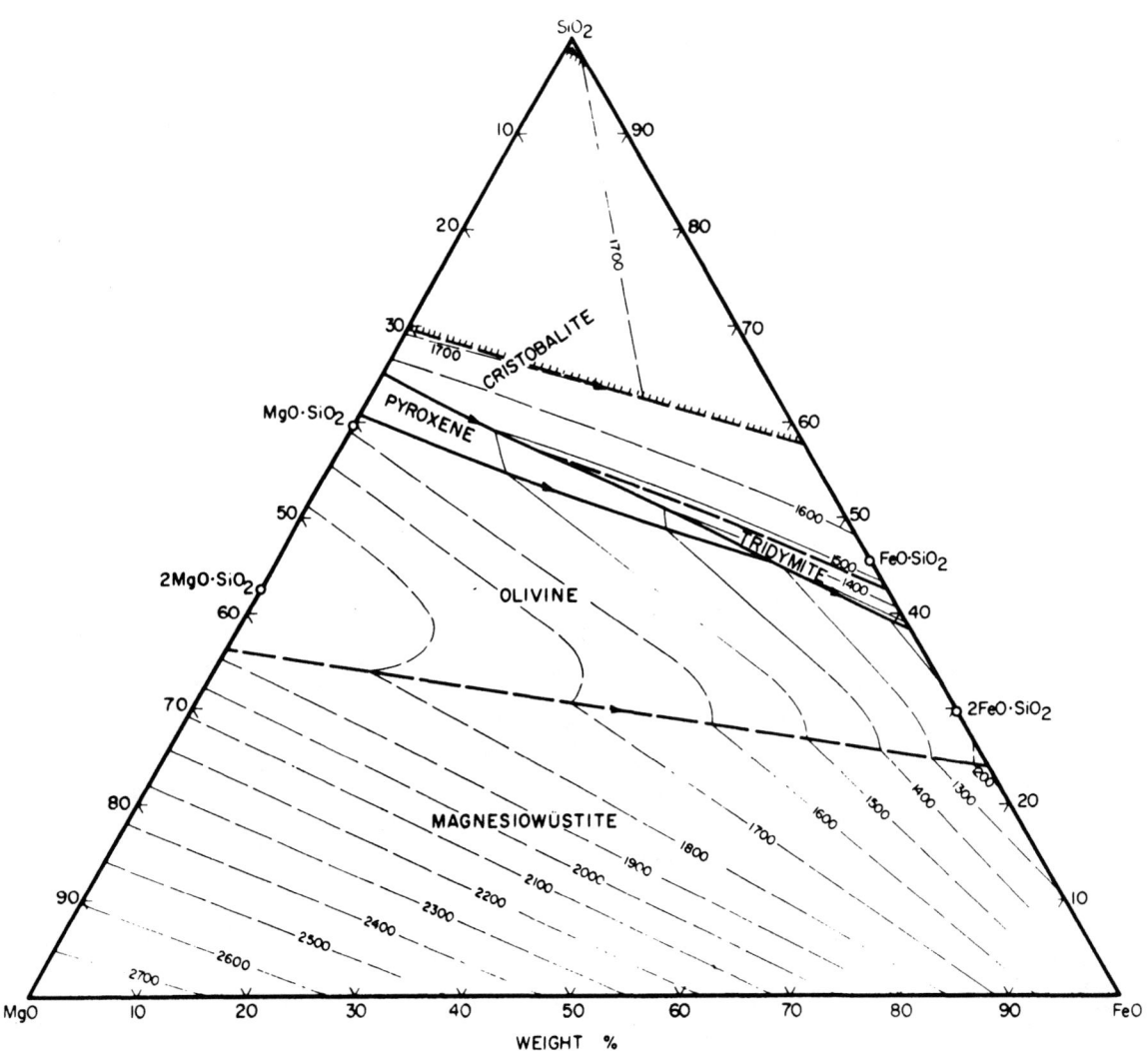

FIG. 13—79. Liquidus isotherms for the MgO-FeO-SiO₂ system. (Compiled by Muan and Osborn[103]).

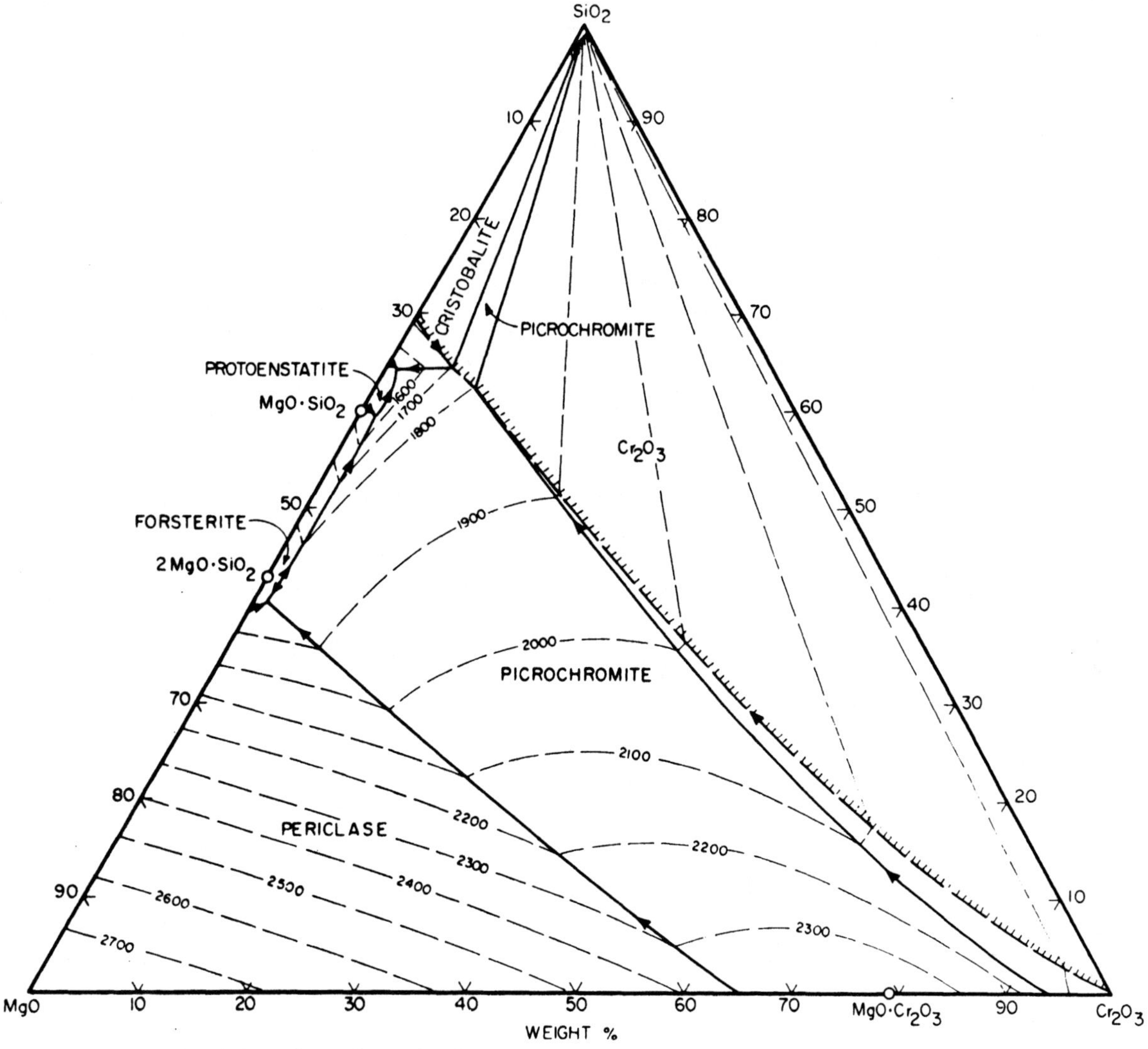

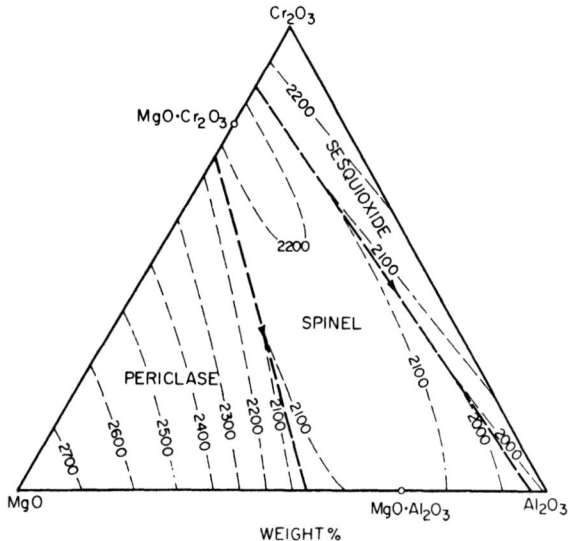

FIG. 13—80. Liquidus isotherms for the MgO-Cr₂O₃-SiO₂ system. (Compiled by Muan and Osborn[103]).

FIG. 13—81. Liquidus isotherms for the MgO-Cr₂O₃-Al₂O₃ system. (Compiled by Muan and Osborn[103]).

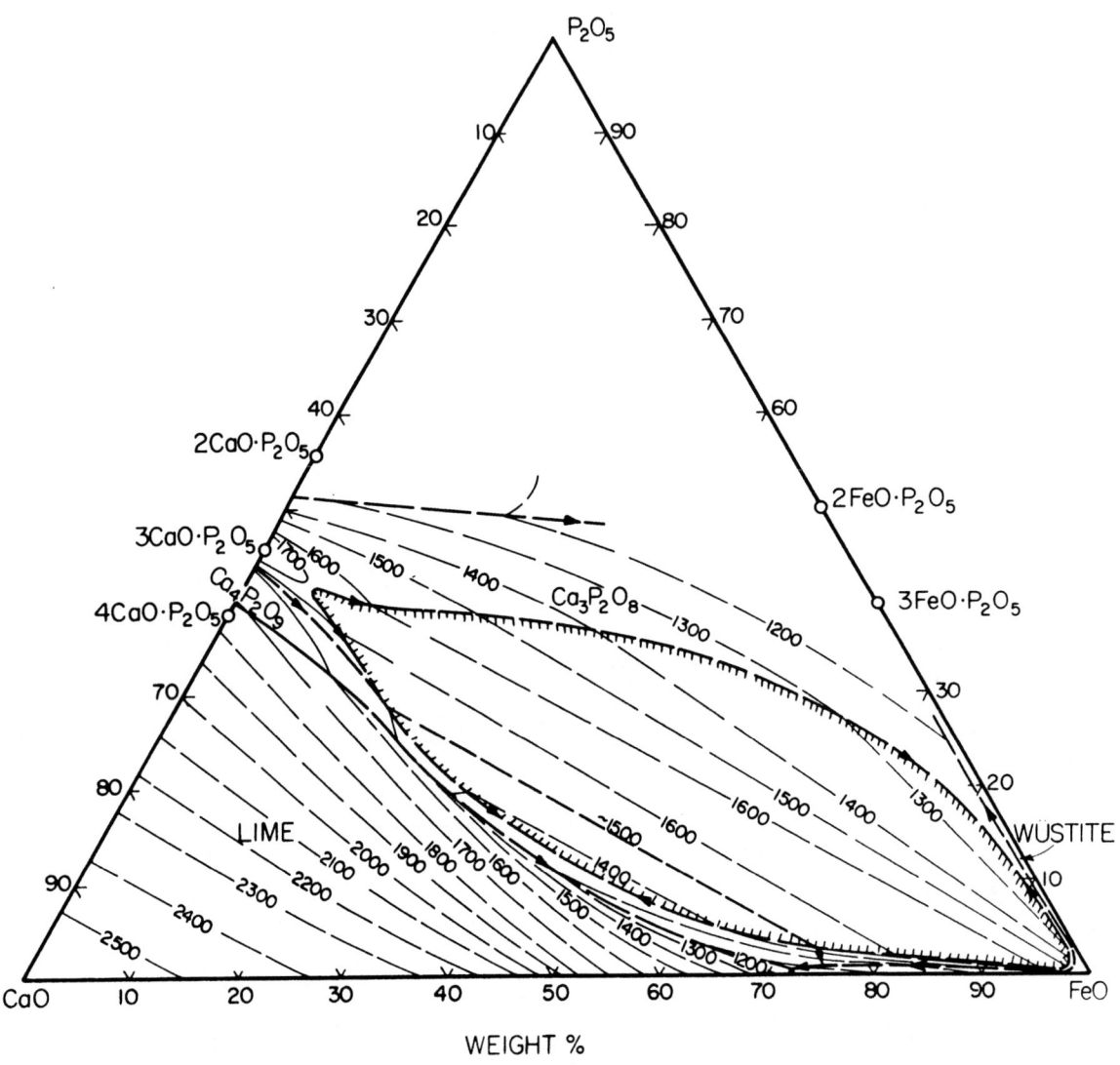

FIG. 13—82. Liquidus isotherms and the projection of the liquid miscibility gap for the
CaO-FeO-P₂O₅ system in equilibrium with liquid iron. (Compiled by Tromel et al.[97, 98]).

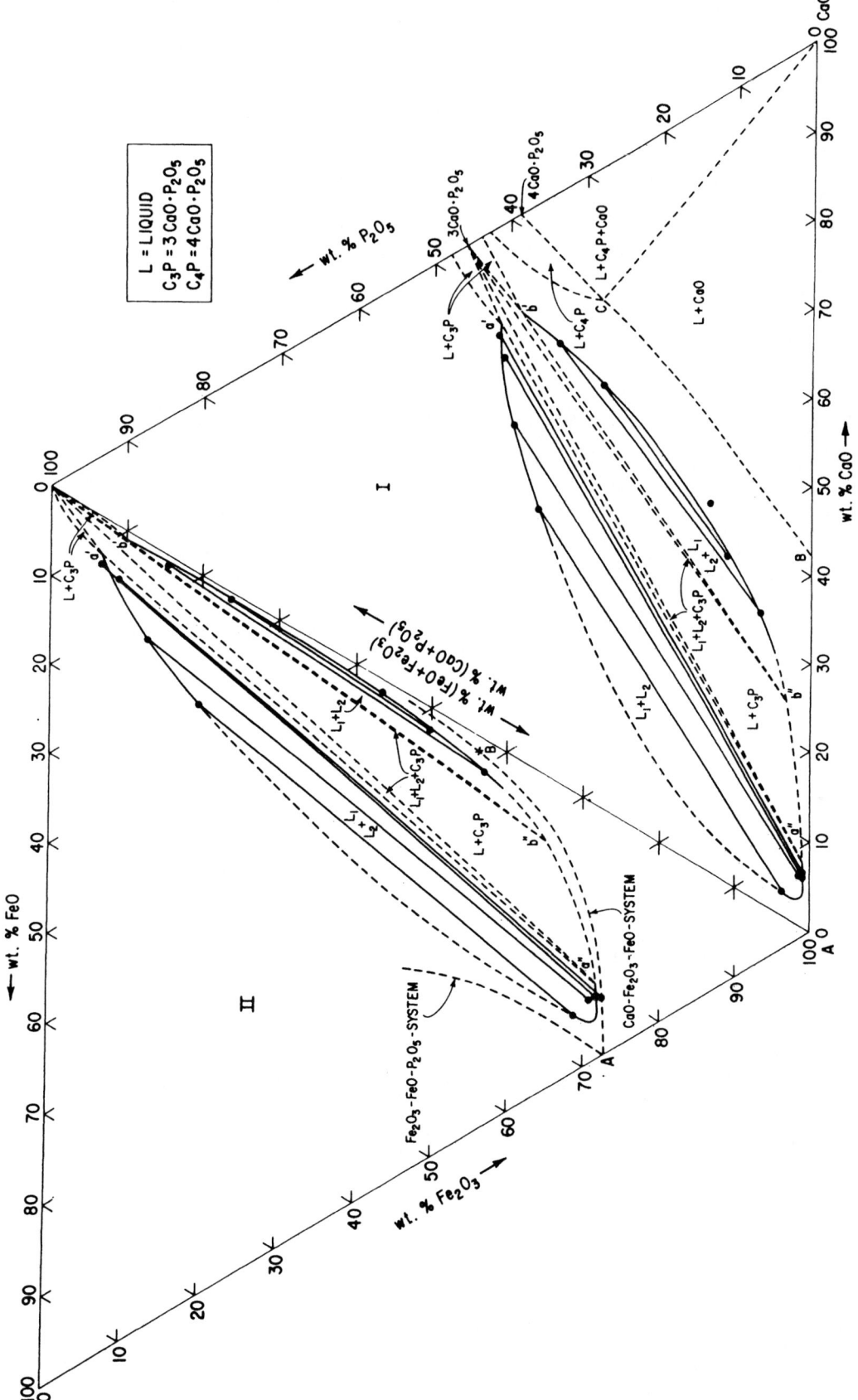

Fig. 13—83. Phase diagram for the system CaO–Fe$_2$O$_3$–FeO–P$_2$O$_5$ at p$_{O_2}$ = 0.20 atm and 1625°C. (Schwerdtfeger and Turkdogan[107]).

With the exception of iron, each of all the other elements constituting the industrial slags possesses primarily a single type of valency, e.g. manganese has a valency of two, phosphorus three, silicon four. Although oxides do not exist in molten slags in the stoichiometric form as normally formulated, e.g. CaO, SiO_2, P_2O_5, etc., it is convenient to employ the activity of an oxide in the slag relative to the pure liquid or solid oxide. Several attempts have been made in the past to rationalize the mode of composition dependence of the activities of oxides in simple and complex slags with little success. The activities determined experimentally or estimated from the phase diagrams are usually given in a graphical form for use in calculating reaction equilibria in slag-metal and slag-gas systems.

(a) **Binary Systems**—The 'FeO'-MnO solution is ideal both in the solid and liquid state[111,112], as would be expected from the phase equilibrium diagram given in Figure 13—63.

The activities in 'FeO'-CaO melts in equilibrium with liquid iron at 1550°C are given in Figure 13—84. It should be noted that this is not a true binary system, but consists of FeO, Fe_2O_3 and CaO in equilibrium with liquid iron.

The activities in binary silicates 'FeO'-SiO_2 (in equilibrium with iron), MnO-SiO_2 and CaO-SiO_2 are given in Figures 13—85, 13—86, and 13—87. The FeO activity is relative to molten iron oxide in equilibrium with liquid iron and those of SiO_2, MnO and CaO relative to the respective pure solid oxides; therefore, in melts

saturated with SiO_2 or CaO, a_{SiO_2} or a_{CaO} is unity. For a given silica content, the activities in the CaO-SiO_2 system are lower than those in the 'FeO'-SiO_2 system, the MnO-SiO_2 system being in an intermediate range. This is consistent with the differences in the free energies of formation of the respective silicates. Stating it in another manner, the interaction between calcium cations and silicate anions is much stronger than that between iron cations and silicate anions.

The activities in the CaO-Al_2O_3 system at 1500°C are given in Figure 13—88.

(b) **Ternary and Multicomponent Systems**—Larson and Chipman[119] determined the oxygen activity in CaO-FeO-Fe_2O_3 melts and computed the activities of the oxides by the use of ternary Gibbs-Duhem equations. Similar work was done by Schuhmann et al.[114] and Turkdogan and Bills[120] for the SiO_2-FeO-Fe_2O_3 system. The activities of oxides in these two ternary systems at 1550°C are given in Figures 13—89 and 13—90.

Chipman and co-workers[121] measured the oxygen content of liquid iron equilibrated with complex silicate and phosphate slags, thus evaluating the activity of ferrous oxide in steelmaking slags. The activity curves for FeO given in Figure 13—91 were computed by Elliott[122] for the CaO-FeO-SiO_2 system using the above-mentioned data for slags low in MnO, MgO and P_2O_5.

Figure 13—92 and 13—93 show the variation of the activity of manganous oxide (relative to the pure solid

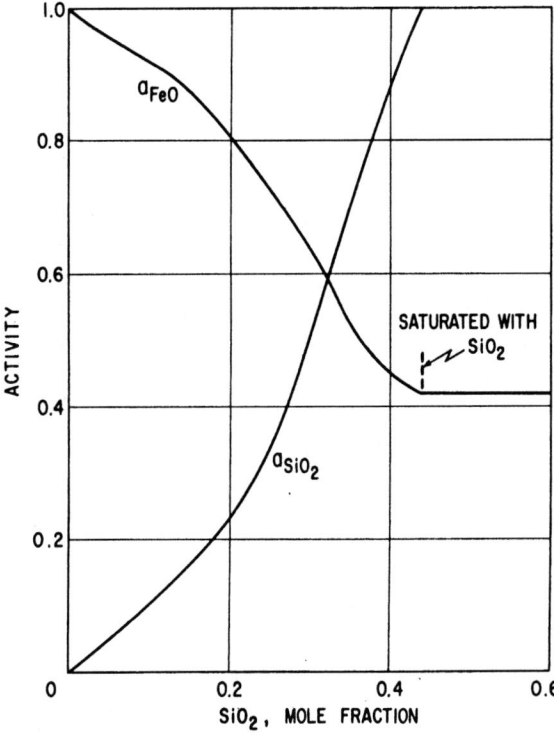
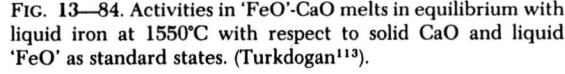

FIG. 13—84. Activities in 'FeO'-CaO melts in equilibrium with liquid iron at 1550°C with respect to solid CaO and liquid 'FeO' as standard states. (Turkdogan[113]).

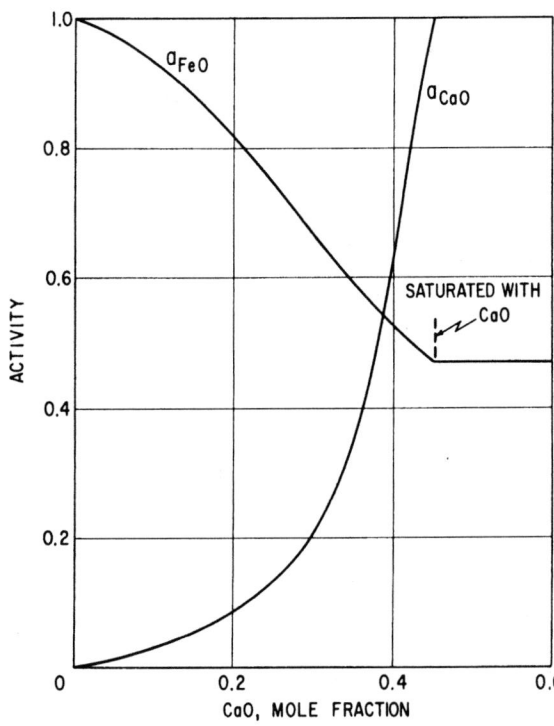

FIG. 13—85. Activities in 'FeO'-SiO_2 melts in equilibrium with liquid iron at 1550°C with reference to wustite and solid silica as standard states. (By extrapolating[115] the data of Schuhmann and Ensio[114]).

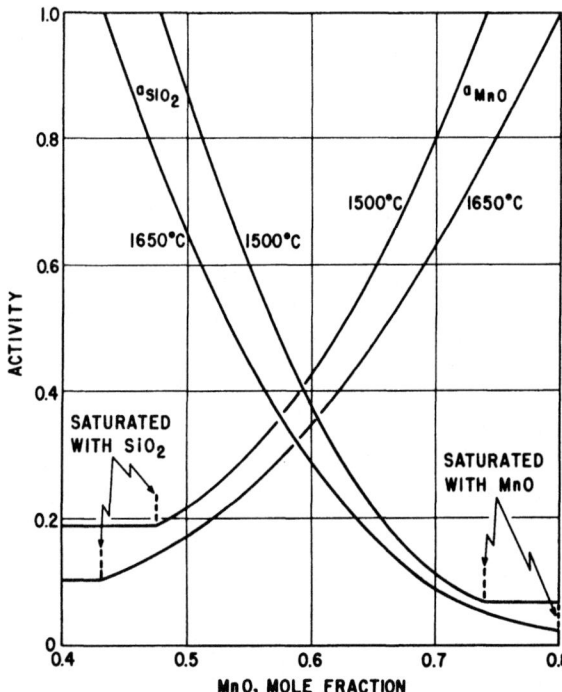

FIG. 13—86. Activities in MnO-SiO₂ melts with reference to solid MnO and SiO₂ as standard states. (Abraham et al.[116]).

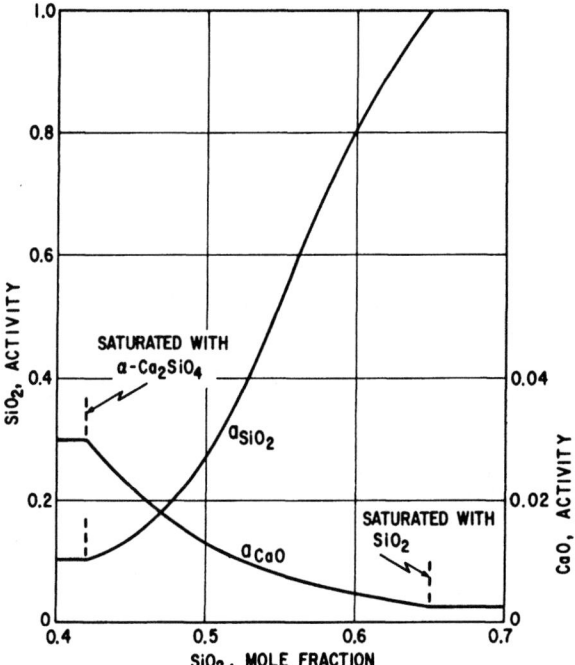

FIG. 13—87. Activities in CaO-SiO₂ melts at 1550°C with respect to solid CaO and SiO₂ as standard states. (Kay and Taylor[117]).

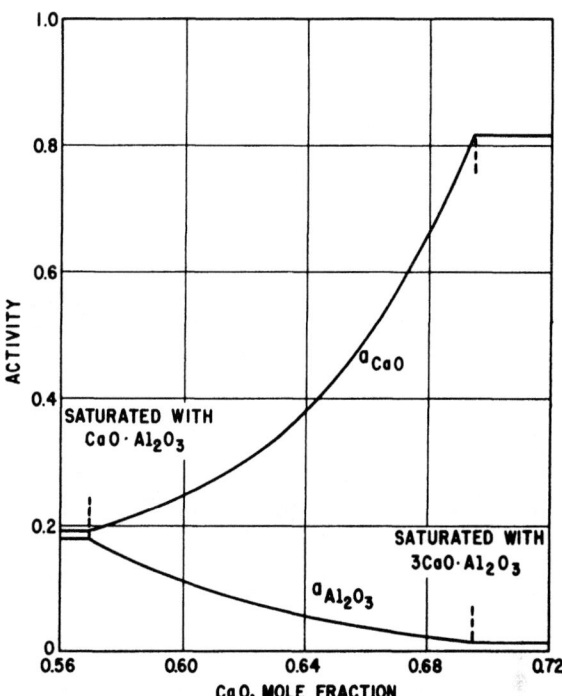

FIG. 13—88. Activities in CaO-Al₂O₃ melts at 1500°C with respect to solid CaO and Al₂O₃ as standard states. (Sharma and Richardson[118]).

MnO) with composition of CaO-MnO-SiO₂ system at 1500° and 1650°C determined by Abraham et al.[116]. The activities of manganous oxide in the quaternary system CaO-MnO-Al₂O₃-SiO₂ are given in Figure 13—94 for certain concentration ratios of CaO : SiO₂ : Al₂O₃ at 1650°C.

The activity of silica in CaO-Al₂O₃-SiO₂ melts at 1550°C determined by Kay and Taylor[117] is given in Figure 13—95.

(c) **Activities of sulphides and sulphates in oxide melts**—The equilibrium of an oxide melt with sulphur-bearing gases may be expressed by the equation:

$$\tfrac{1}{2} S_2 \text{ (gas)} + O^{2-} \text{ (melt)} = S^{2-}\text{(melt)} + \tfrac{1}{2} O_2 \text{ (gas)} \qquad (124)$$

for which the equilibrium relation is given by

$$k' = \frac{N_S}{N_O} \left(\frac{p_{O_2}}{p_{S_2}} \right)^{\tfrac{1}{2}} \qquad (125)$$

where N_S and N_O are atom fractions of sulphur and oxygen in the melt and p is the partial pressure of the gas species indicated by the subscript. Since the activity coefficients of sulphur and oxygen are omitted in equation (125), the equilibrium index k' varies with composition of the melt. In fact, it was shown by Dewing and Richardson[123] and Turkdogan and Darken[124] that in iron oxide-iron sulphide melts, k' decreases with increasing concentration of trivalent iron. While Dewing and Richardson equilibrated melts containing 20 to 50 mole per cent sulphur, in the work of Turkdogan

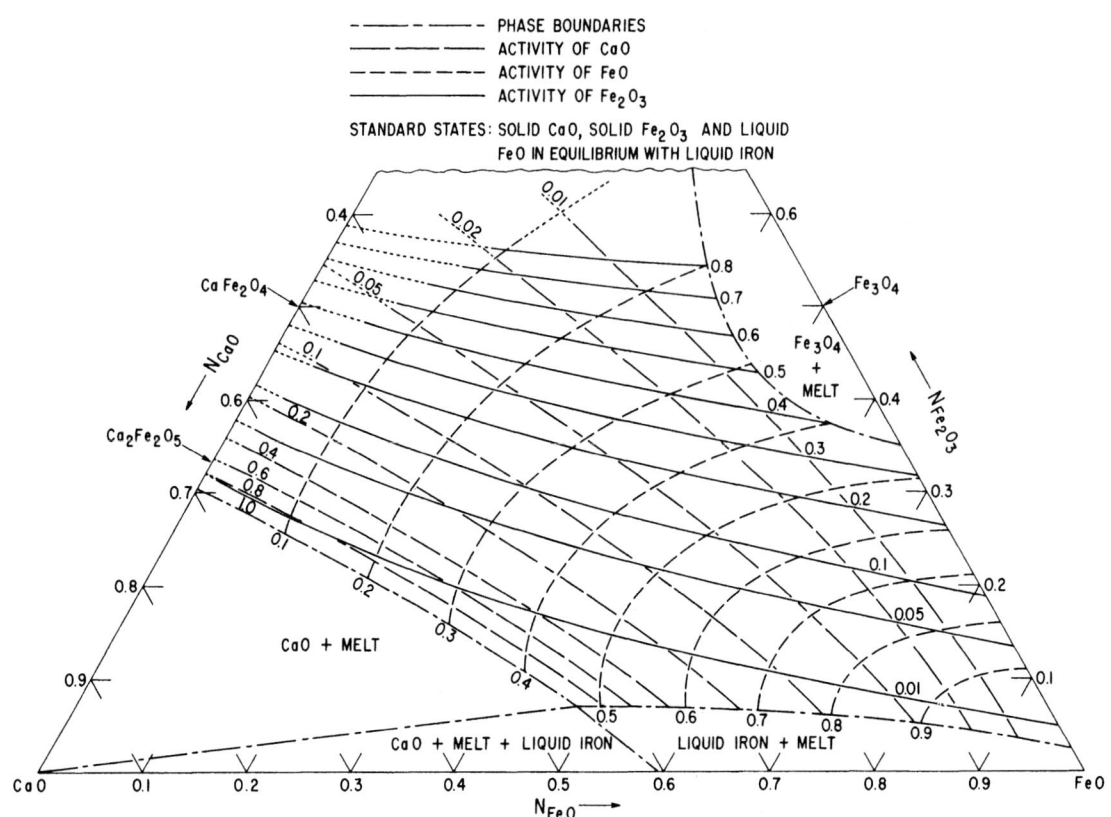

FIG. 13—89. Isoactivity curves in CaO-FeO-Fe₂O₃ melts at 1550°C. (Turkdogan[115]).

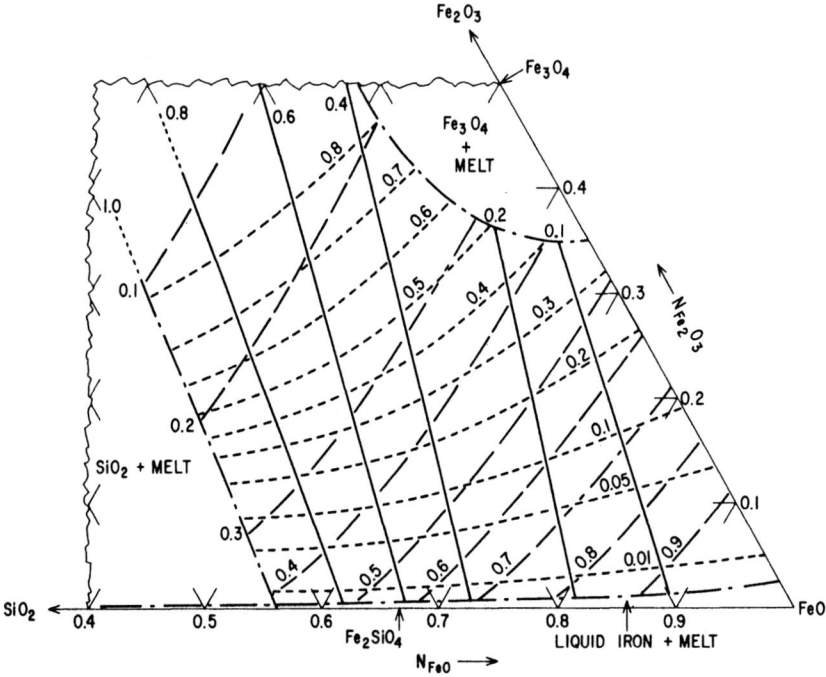

FIG. 13—90. Isoactivity curves in SiO₂-FeO-Fe₂O₃ melts at 1550°C. (Turkdogan[115]).

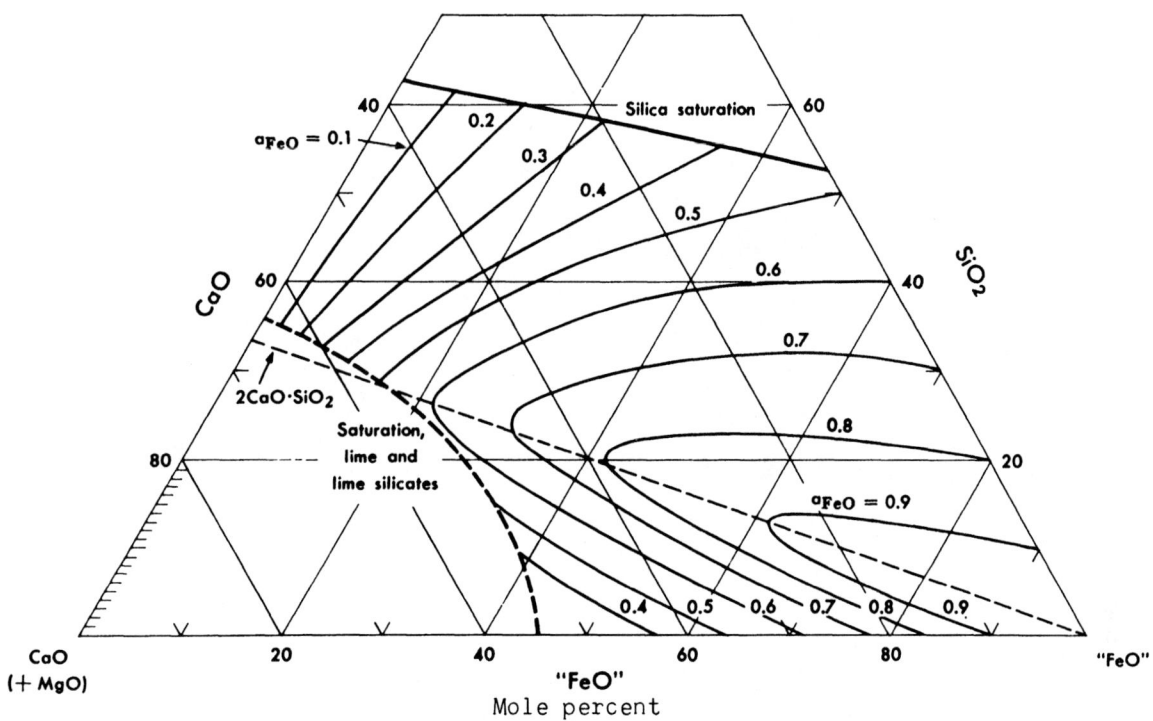

FIG. 13—91. Activity of ferrous oxide in CaO-'FeO'-SiO₂ system at 1600°C, relative to pure liquid 'FeO' in equilibrium with liquid iron. (Elliott[122]).

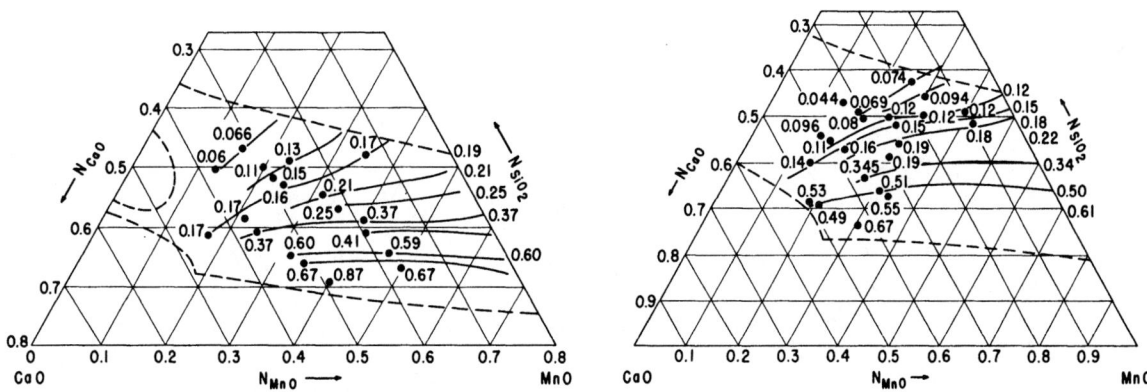

FIG. 13—92. Activities of MnO in MnO-CaO-SiO₂ melts at 1500°C with respect to solid MnO. (Abraham et al.[116]).

FIG. 13—93. Activities of MnO in MnO-CaO-SiO₂ melts at 1650°C with respect to solid MnO. (Abraham et al.[116]).

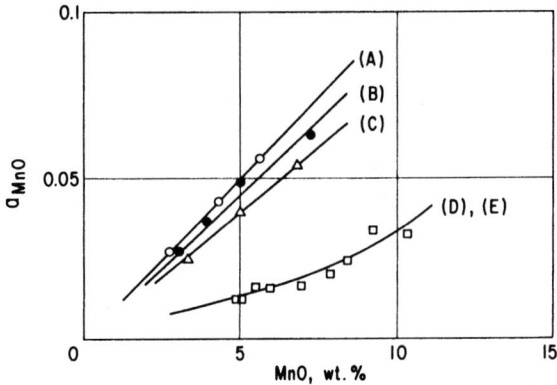

FIG. 13—94. Activities of manganous oxide at 1650°C in MnO + CaO + Al$_2$O$_3$ + SiO$_2$ melts for the following weight ratios of CaO : SiO$_2$: Al$_2$O$_3$: (A) 52.0 : 37.2 : 10.8 ; (B) 47.2 : 38.2 : 14.6; (C) 36.7 : 33.3 : 30.0 ; (D) 30.0 : 63.3 : 6.7; (E) 16.7 : 66.7 : 16.6. (Abraham et al.[116]).

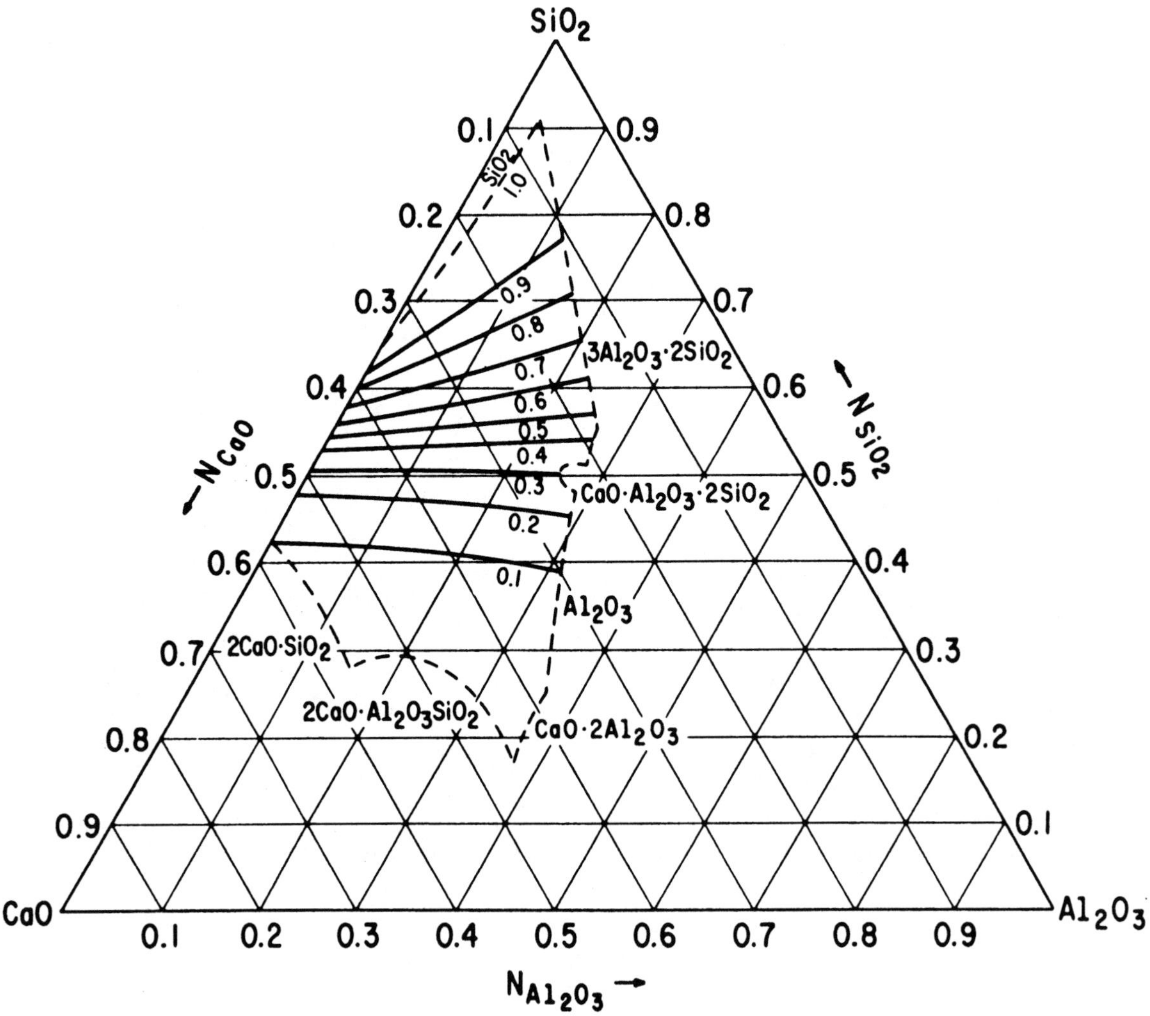

FIG. 13—95. Isoactivity curves for silica in CaO-Al$_2$O$_5$-SiO$_2$ melts at 1550°C, relative to solid SiO$_2$. (Kay and Taylor[117]).

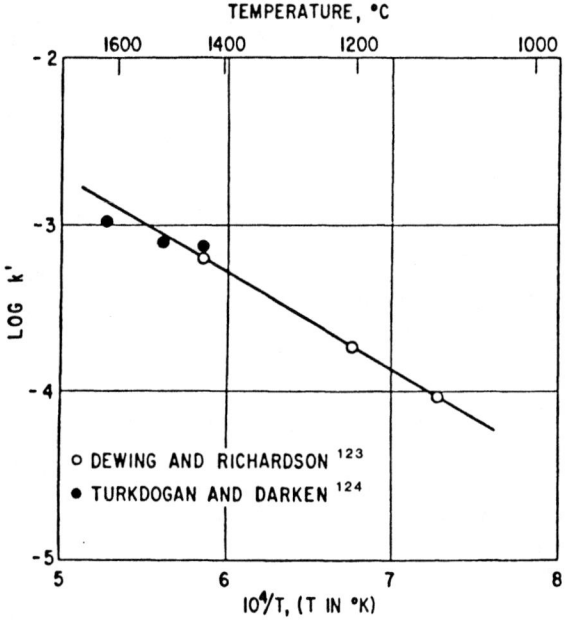

FIG. 13—96. Variation of $\quad k' = \dfrac{N_S}{N_O} \left(\dfrac{P_{O_2}}{P_{S_2}} \right)^{1/2}$

with temperature for Fe-S-O melts containing $N_{Fe^{3+}} = 0.125$.

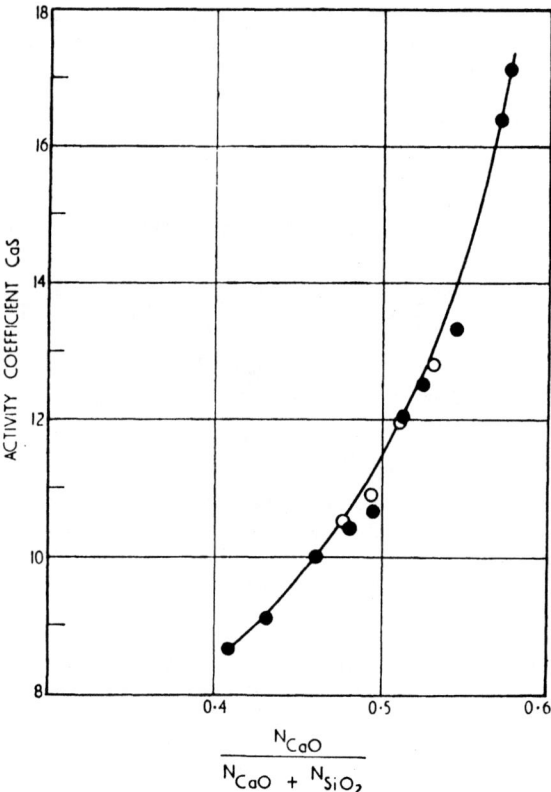

FIG. 13—98. The activity coefficient of CaS at saturation (relative to solid) shown as a function of composition of calcium silicate melts at 1500° and 1550°C. (Sharma and Richardson[126]).

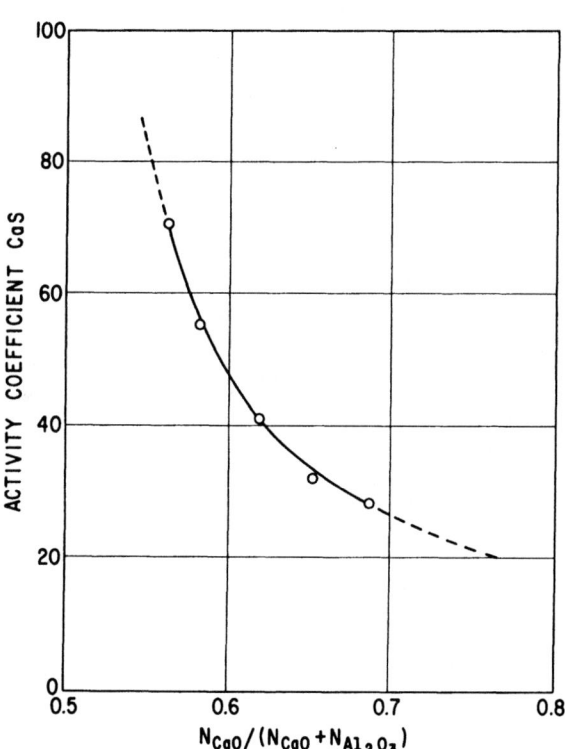

FIG. 13—97. Activity coefficient of CaS (with respect to solid CaS) in CaO-Al$_2$O$_3$ melts saturated with CaS at 1500°C. (Sharma and Richardson[118]).

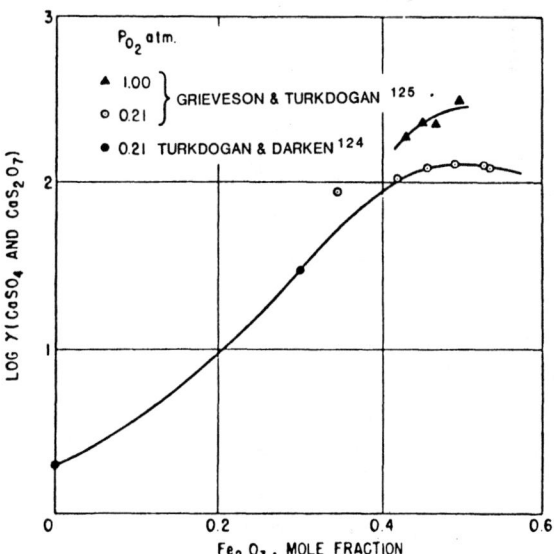

FIG. 13—99. Variation of log γ (CaSO$_4$ and CaS$_2$O$_7$) with composition of CaO-FeO-Fe$_2$O$_3$ melts at 1500°C and p_{O_2} = 0.2 and 1.0 atm. Standard states: liquid CaSO$_4$; liquid CaS$_2$O$_7$.

and Darken sulphur in the iron oxide was below 4 mole per cent, but as seen from Figure 13—96, large change in the sulphur content has no effect on the value of k' for a given temperature. Therefore, it may be stated that the activity coefficient ratio γ_S/γ_O or $\gamma_{FeS}/\gamma_{FeO}$ is essentially independent of the sulphur content of the melt.

For calcium-silicate, calcium-aluminate, calcium-ferrite, etc. melts, the sulphide and sulphate equilibrium may be expressed equally well by the equations

$$\tfrac{1}{2}\,S_2\,(gas) + CaO\,(melt) \\ = CaS\,(melt) + \tfrac{1}{2}\,O_2\,(gas) \tag{126}$$

$$\tfrac{1}{2}\,S_2\,(gas) + 3/2\,O_2\,(gas) + \\ CaO\,(melt) = CaSO_4\,(melt) \tag{127}$$

$$S_2\,(gas) + 3\,O_2\,(gas) + \\ CaO\,(melt) = CaS_2O_7\,(melt) \tag{128}$$

From the equilibrium measurements and known equilibrium constants the values of γ_{CaS} and γ_{CaSO_4} ($= \gamma_{CaS_2O_7}$) have been determined for some melts; the available data are summarized in Figures 13—97, 13—98 and 13—99 for calcium aluminate, silicate and ferrite melts respectively. On similar considerations, the activity coefficient of MnS in MnO-SiO$_2$ melts has been evaluated and the results obtained are given in Figure 13—100. As would be anticipated from the high values of $\gamma_{CaS_2O_7}$ or γ_{CaSO_4} (Figure 13—99), there is a large liquid miscibility gap in the CaSO$_4$-CaO-Fe$_2$O$_3$ system[125].

ture dependence of the self-diffusivities of O, Ca, S, Al and Si in calcium alumino-silicate melts is shown in Figure 13—101.

A few selected diffusivities (mostly self-diffusivities) are given in Table 13—XV for oxides, sulphides and nitrides. Because of the extensive nonstoichiometry in wustite and pyrrhotite, the diffusivity of iron in these phases varies with composition; this is reflected mainly on the value of D_0 as indicated in Table 13—XIV. The characteristic structural features of wustite and pyrrhotite is that of vacancies in the iron sublattice; hence iron diffuses in these phases via the cation vacant sites. As seen from the plots in Figures 13—102 and 13—103, the self-diffusivity of iron is directly proportional to the fraction of iron deficiency, i. e. fraction of cation vacancy.

(b) Coefficient of Viscosity—For the present purpose, the compilation of the data on viscosities is limited to slags which are similar to those used in iron- and steelmaking. The coefficient of viscosity of lime-alumina-silica melts has been measured by Machin and Yee[151] and Kozakevitch[152]. Figures 13—104 and 13—105 are examples showing the change of viscosity coefficient with composition and temperature. For a given alumina concentration, the viscosity of the melt increases with increasing silica content, and for a given composition, the melt becomes more viscous with de-

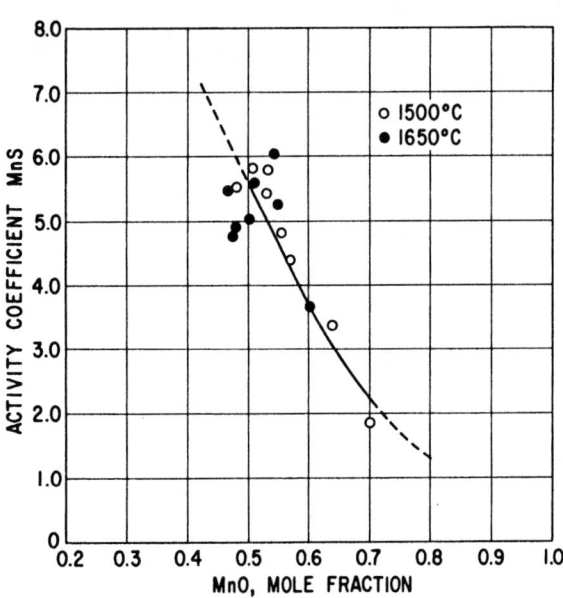

FIG. 13—100. Activity coefficients of MnS (relative to liquid MnS) in MnO-SiO$_2$ melts. (Abraham et al.[116]).

TRANSPORT PROPERTIES OF SLAGS

(a) Coefficient of Diffusion—The diffusivities in liquid slags compiled by Elliott et al.[12] are reproduced in Table 13—XIV with few additional data. The tempera-

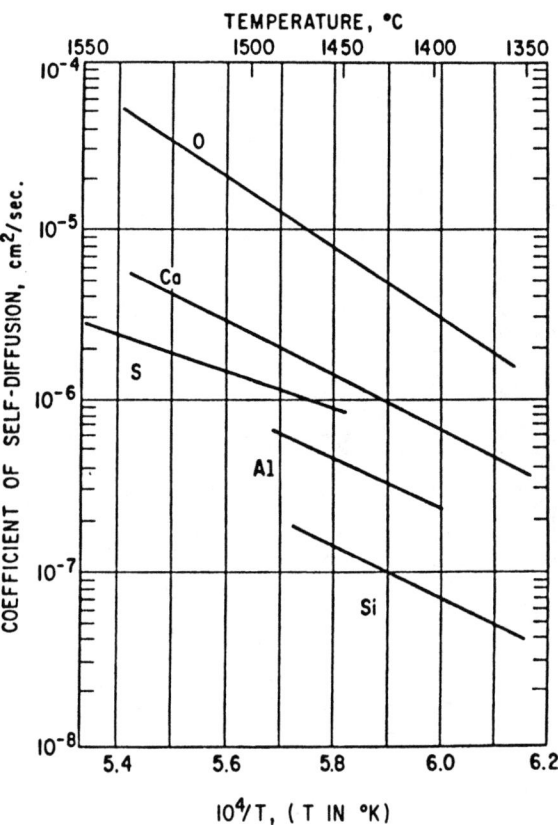

FIG. 13—101. Self-diffusivities in calcium aluminosilicate melts.

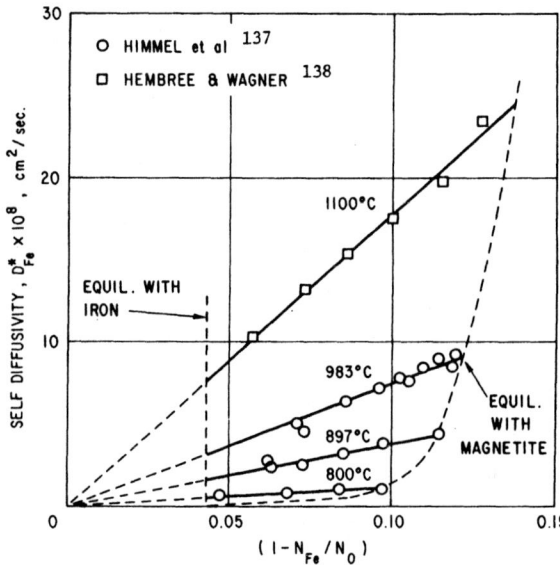

FIG. 13—102. Self-diffusivity of iron as a function of $(1 - N_{Fe}/N_0)$, fraction of iron deficiency.

creasing temperature.

The viscosity measurements have been made by Machin and co-workers[153,154] over wide temperature and composition ranges in the CaO-MgO-Al$_2$O$_3$-SiO$_2$ melts. The data relevant to blast-furnace slags are given in Figures 13—106 and 13—107.

Kozakevitch[155] found that the introduction of sulphur, up to at least 2 per cent, into the blast-furnace type of slags had no effect on viscosity. However, the addition of calcium fluoride was found to have a pronounced lowering effect on the viscosity of calcium alumino-silicate melts as shown in Figure 13—108. Bills[156] observed similar effect of calcium fluoride on the viscosity of alumina silicates.

(c) Surface Tension—The surface tensions of some binary silicate melts at 1570°C measured by King[157] are given in Figure 13—109. The results obtained by Kozakevitch[158], given in Figure 13—110, show the varying effects of different oxide additions on the surface tension of molten iron oxide at 1400°C. Boni and Derge[159] measured the surface tensions of alkaline-earth alumino-silicates; some of their results are given in Figure 13—111. They also observed that increasing the sulphur content decreases appreciably the surface tension of the melts, as shown in Figure 13—112. Simi-

Table 13—XIV. Diffusivities in liquid slags.

Diffusing Element	Composition of Slag	Temperature °C	D, cm²/sec	D₀, cm²/sec	E, Kcal/mole	Reference
Al²⁶ (self-diffusion)	43% CaO-10% Al₂O₃-47% SiO₂	1440-1520		4.3×10⁴	85	127
Al²⁶ (self-diffusion)	39% CaO-20% Al₂O₃-41% SiO₂	1398-1487		5.4	60	127
Ca	30% CaO-69.5% P₂O₅	980	3.0×10⁻⁶			128
Ca	30% CaO-69.5% P₂O₅	1000	4.7×10⁻⁶			128
Ca	39.4% CaO-21.2% Al₂O₃-38.8% SiO₂	1300	4.4×10⁻⁷		56.6	128
Ca	39.4% CaO-21.2% Al₂O₃-38.8% SiO₂	1400	1.4×10⁻⁶		56.6	128
Ca	39.4% CaO-21.2% Al₂O₃-38.8% SiO₂	1450	2.4×10⁻⁶		56.6	128
Ca	39.4% CaO-21.2% Al₂O₃-38 8% SiO₂	1500	3.8×10⁻⁶		56.6	128
Ca	37.5% CaO-10% Al₂O₃-52.5% SiO₂	1380-1560		0.045	29	129
Ca	32.5% CaO-10% Al₂O₃-57.5% SiO₂	1450-1560		0.0039	21	129
Ca	27.5% CaO-10% Al₂O₃-62.5% SiO₂	1360-1560		0.0101	28	129
Ca⁴⁵ (self-diffusion)	40% CaO-20% Al₂O₃-40% SiO₂	1350	3×10⁻⁷		70	130
Ca⁴⁵ (self-diffusion)	40% CaO-20% Al₂O₃-40% SiO₂	1400	6.8×10⁻⁷		70	130
Ca⁴⁵ (self-diffusion)	40% CaO-20% Al₂O₃-40% SiO₂	1450	1.3×10⁻⁶		70	130
Ca⁴⁵ (self-diffusion)	39% CaO-21% Al₂O₃-40% SiO₂	1350	3.5×10⁻⁷		70	130
Ca⁴⁵ (self-diffusion)	39% CaO-21% Al₂O₃-40% SiO₂	1500	2.1×10⁻⁶		70	130
Ca⁴⁵ (self-diffusion)	39% CaO-21% Al₂O₃-40% SiO₂	1540	3.4×10⁻⁶		70	130
Ca⁴⁵ (self-diffusion)	43% CaO-18% Al₂O₃-39% SiO₂	1350-1540			30	131
Fe	Iron oxide, Fe³⁺/ΣFe=0.4	1550	5.0×10⁻⁵			132
Fe	Iron oxide, Fe³⁺/ΣFe=0.12	1430-1550		6.60×10⁻³	10.7	133
Fe	Iron oxide, Fe³⁺/ΣFe=0.33	1430-1550		1.61×10⁻²	16 6	133
Na²² (self-diffusion)	24% Na₂O-76% SiO₂	850-1500		4.67×10⁻³	16.1	134
Na²² (self-diffusion)	34% Na₂O-66% SiO₂	850-1500		2.29×10⁻³	12.4	134
O¹⁷ (self-diffusion)	40% CaO-20% Al₂O₃-40% SiO₂	1372-1535		1.2×10⁶	87.9	135
O¹⁸ (self-diffusion)	40% CaO-20% Al₂O₃-40% SiO₂	1372-1535		0.93×10⁶	87.9	135
P	39.4% CaO-21.2% Al₂O₃-38.8% SiO₂	1300	10×10⁻⁷		46.6	128
P	39.4% CaO-21.2% Al₂O₃-38.8% SiO₂	1400	2×10⁻⁶		46.6	128
P	39.4% CaO-21.2% Al₂O₃-38.8% SiO₂	1450	4.5×10⁻⁶		46.6	128
P	39.4% CaO-21.2% Al₂O₃-38.8% SiO₂	1500	4.4×10⁻⁶		46.6	128
P	30% CaO-69.5% P₂O₅	980	2.7×10⁻⁶			128
P	30% CaO-69.5% P₂O₅	1000	2.8×10⁻⁶			128
S-1.5% S	50.3% CaO-10.4% Al₂O₃-39.3% SiO₂	1445-1580		1.4	49	136
S-1.5% S	42.5% CaO-9.6% Al₂O₃-47.9% SiO₂	1440	0.8×10⁻⁶			136
Si³¹ (self-diffusion)	39% CaO-21% Al₂O₃-40% SiO₂	1365	4.7×10⁻⁸		70	130
Si³¹ (self diffusion)	39% CaO-21% Al₂O₃-40% SiO₂	1430	1.05×10⁻⁷		70	138

Table 13—XV. Diffusivities in solid oxides, sulphides and nitrides.

Diffusate	Medium	Type of Diffusion	Temperature °C	D cm²/sec	D_0 cm²/sec	E kcal/mole	Reference
Fe^{55}	Wustite	Self	800-1100		$5.05 \times 10^{-2}(1-N_{Fe}/N_0)$	27.8	137,138
Fe^{55}	Magnetite	Self	900-1200		3.49×10^{-4}	33.6	139
Fe^{55}	Magnetite	Self	750-1000		5.2	55.0	137
Fe^{55}	Hematite	Self	1000-1217		4.0×10^{5}	112.0	137
Fe^{55}	$CaFe_2O_4$	Self	890-1140		3.2	72.0	140
Fe^{55}	$ZnFe_2O_4$	Self	750-1300		8.5×10^{2}	82.0	141
O^{18}	Al_2O_3 single crys.	Self	1500-1800		1.9×10^{3}	152.0	142
O^{18}	Al_2O_3 poly. crys.	Self	1400-1800		2.0	110.0	142
O^{18}	Hematite	Self	900-1250		2.04	77.9	143
O^{18}	MgO single crys.	Self	1250-1800		2.5×10^{-6}	62.4	144
O^{18}	SiO_2	Self	1000-1200		1.51×10^{-2}	71.2	145
O	ZrO_2	Interdiffus.	875-1050		1.36×10^{-4}	28.4	146
Fe	Pyrrhotite	Self	670-900		$2.97 \times (1-N_{Fe}/N_S)$	29.7	147
N	"Cr_2N"	Interdiffus.	1200	4.2×10^{-8}			148
N	"Fe_4N"	Self	504	3.2×10^{-12}			149
N	"Fe_4N"	Self	554	7.9×10^{-12}			149
N	Iron nitride ϵ	Interdiffus.	580	5.5×10^{-10}			150
N	ϵ	Interdiffus.	660	2.3×10^{-9}			150
N	ϵ	Interdiffus.	730	5.5×10^{-9}			150

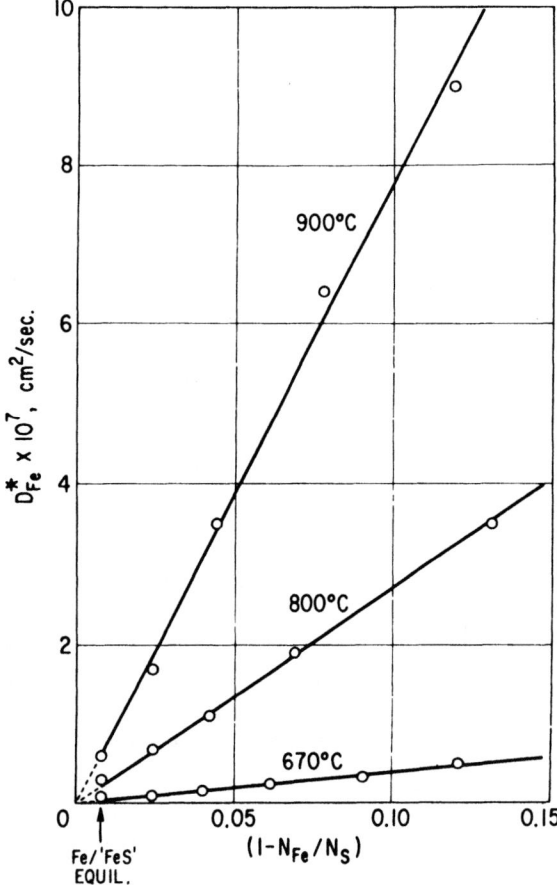

FIG. 13—103. Computed D_{Fe}^* as a function of $(1 - N_{Fe}/N_0)$, fraction of iron deficiency. (Turkdogan[147]).

larly, fluorides lower the surface tension of liquid silicates[160] as shown in Figure 13—113 for calcium silicates containing fluorides.

The variation of the surface tension of silicate melts with temperature depends on the nature of the cations present. For example, while in silicates of divalent elements the temperature coefficient of surface tension, $\partial\sigma/\partial T$, is positive, in alkali silicates $\partial\sigma/\partial T$ becomes negative. The variation of $\partial\sigma/\partial T$ with composition of some binary silicate melts is shown in Figure 13—114.

(d) Density—The densities of solid oxides and refractory materials are given in many engineering handbooks; they will not be given here.

The density of liquid iron silicates has been measured by several investigators. The composition dependence of the density of molten iron silicate in equilibrium with solid iron at 1410°C is shown in Figure 13—115. The dotted curve is calculated density for a mechanical mixture of iron oxide and silica; for most practical purposes the additivity rule would give a reasonable estimate of the density of complex slags. The density of $CaO\text{-}FeO\text{-}SiO_2$ melts in equilibrium with solid iron at about 1400°C determined by several investigators is given in Figure 13—116.

Barrett and Thomas[167] measured the densities of some calcium alumino-silicate melts over the temperature range 1350°-1650°C; these are given in Table 13—XVI. Since the data are limited, no attempt is made to draw a series of isodensity curves within a ternary composition diagram. However, they may be useful in estimating the densities of blast-furnace type of slags.

TRANSPORT PROPERTIES OF GASES

Many metallurgical reactions are controlled by transport in the gas phase. As discussed in detail in textbooks[168,169], the transport properties of gases, e. g. mass diffusivity, thermal diffusivity, thermal conductivity, and viscosity, can be predicted successfully on the basis of the rigorous kinetic theory of gases devel-

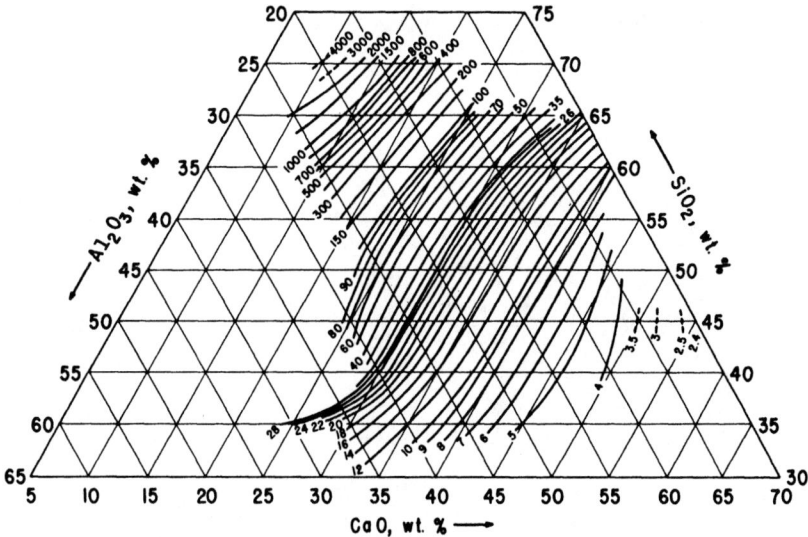

FIG. 13—104. Isoviscosity coefficients (isokoms) in poise of
CaO-Al₂O₃-SiO₂ melts at 1500°C. (Machin and Yee[151]).

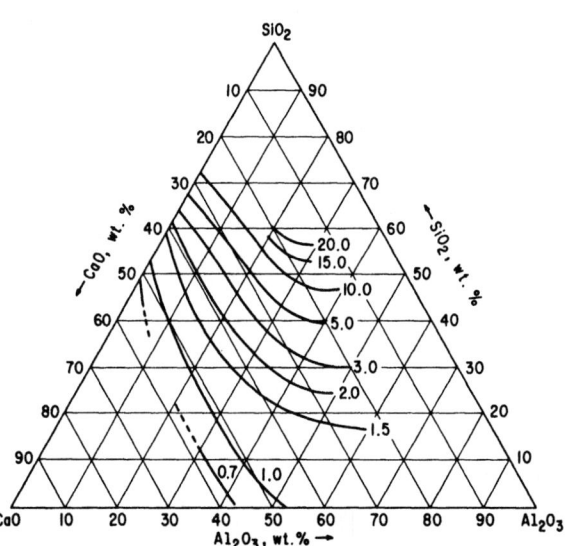

FIG. 13—105. Isoviscosity coefficients (isokoms) in poise of
CaO-Al₂O₃-SiO₂ melts at 1800°C. (Kozakevitch[152]).

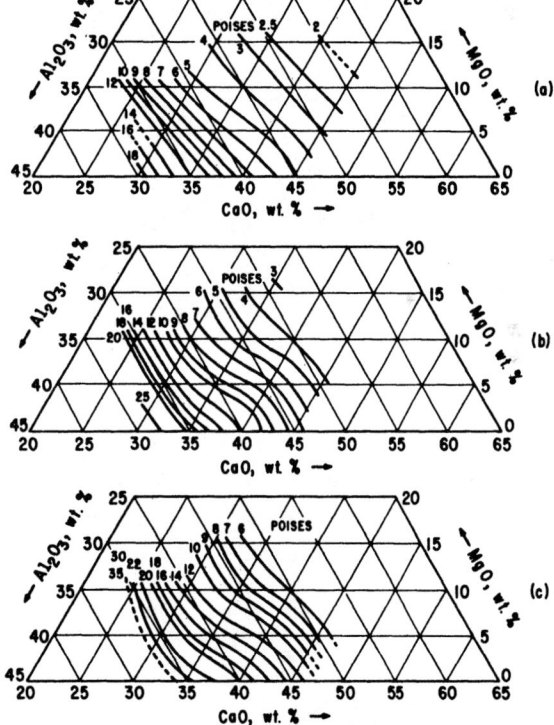

FIG. 13—106. Isokoms in CaO-MgO-Al₂O₃-35% SiO₂ melts at
(a) 1500°C, (b) 1450°C and (c) 1400°C. (Machin et al.[154]).

oped by Enskog and Chapman. Discussion of the kinetic theory of gases and derivation of the transport properties is outside the scope of this chapter. For the present purpose it is adequate to define the basic parameters and examples of their values. For computing the values of the parameters the reader is referred to Hischfelder et al.[169].

The diffusive mass flux, J, is defined as the amount of a diffusing species crossing a surface of unit area, perpendicular to the direction of flow, in unit time; the ratio of this flux to the concentration gradient is the coefficient of diffusion or diffusivity, $D = -J/(\partial c/\partial x)$. The diffusivity has the dimensions of length^2/time; using cgs units, D is given in cm^2/s.

The coefficient of viscosity, η, is defined as the shearing force per unit area required to maintain a unit difference of velocity between the adjacent layers a unit distance apart. When cgs units are used, the viscosity coefficient has the dimensions of dynes · s/cm² and this is called "poise" (millipoise, mp = 10^{-3} poise and centipoise, cp = 10^{-2} poise). In SI units, 1 poise = 0.1 N·s·m⁻².

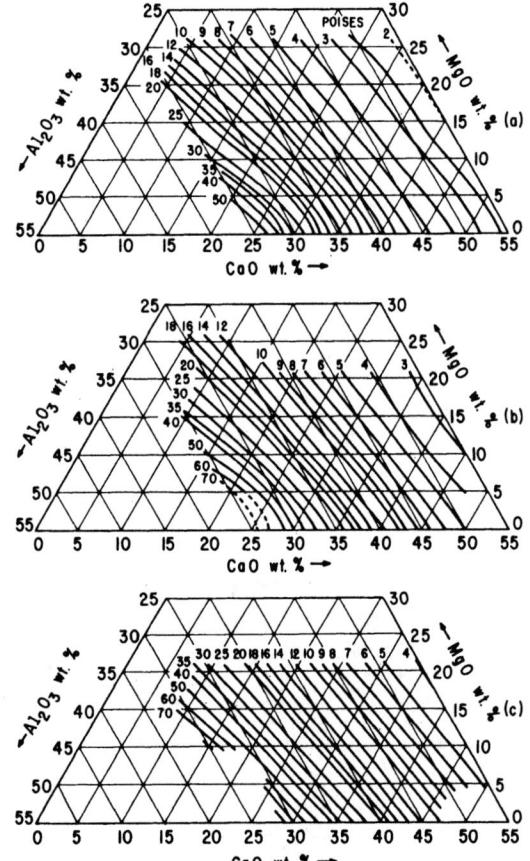

FIG. 13—107. Isokoms in CaO-MgO-Al₂O₃-45% SiO₂ melts at (a) 1500°C, (b) 1450°C and (c) 1400°C. (Machin et al [154]).

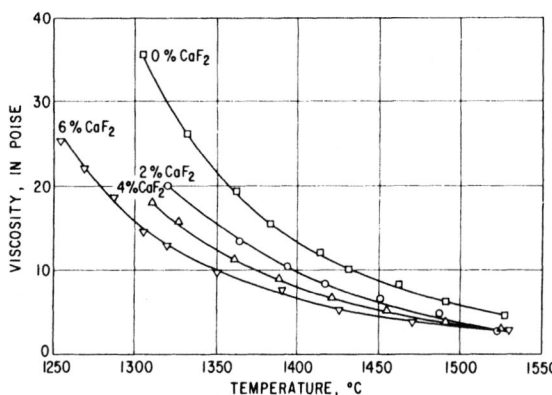

FIG. 13—108. Effect of calcium fluoride on the viscosity coefficient of aluminosilicate melts containing 44 per cent SiO₂, 12 per cent Al₂O₃, 41 per cent CaO and 3 per cent MgO. (Kozakevitch [155]).

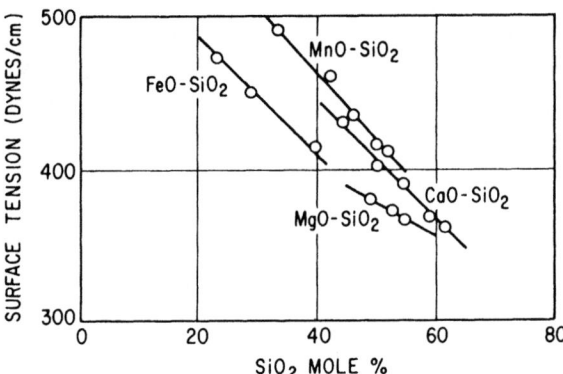

FIG. 13—109. Surface tension (in air) of binary silicates of some divalent elements at 1570°C. (King [157]).

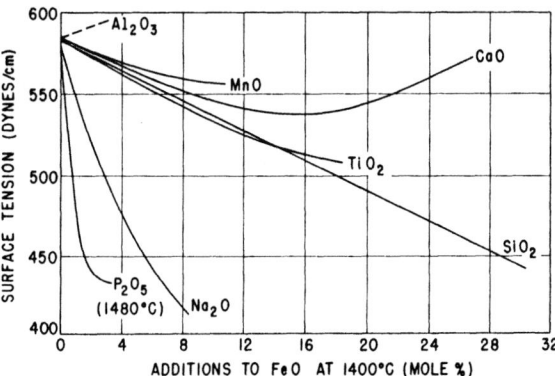

FIG. 13—110. Surface tension (in argon) of binary melts with iron oxide at 1400°C. (Kozakevitch [158]).

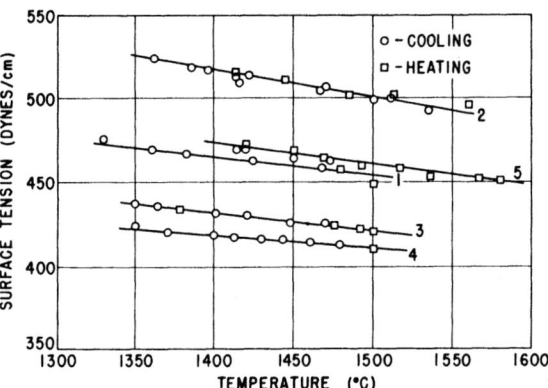

FIG. 13—111. Surface tension (in argon) plotted vs temperature for SiO₂-Al₂O₃-alkaline earth oxide slags. Numbers keyed to curves refer to slags of the following compositions. (Boni and Derge [159]).

| | Slag composition, wt per cent | | | | |
	MgO	CaO	BaO	SiO₂	Al₂O₃
1	—	31.3	—	52.4	16.3
2	—	44.8	—	40.3	14.9
3	—	—	54.0	36.2	9.6
4	—	—	69.2	22.4	8.4
5	23.6	—	—	60.0	16.4

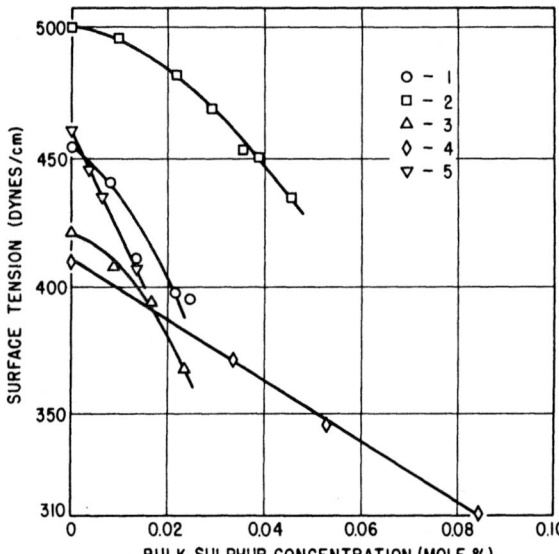

FIG. 13—112. Surface tension (in argon) plotted vs sulphur concentration for five slags at 1500°C. Numbers keyed to the curves refer to slags, the compositions of which are given in Fig. 13—111. (Boni and Derge[159]).

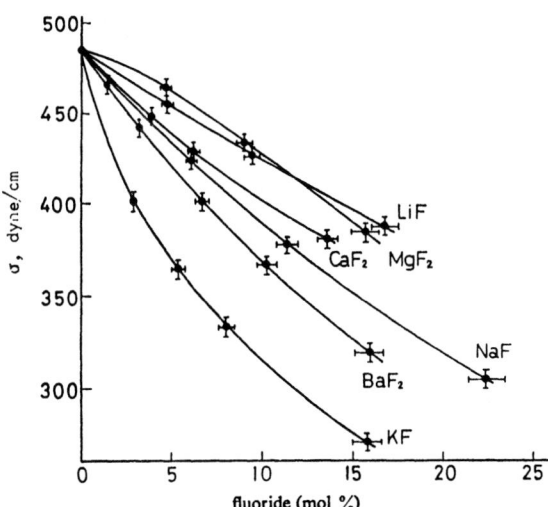

FIG. 13—113. Effect of fluoride content on the surface tension of molten CaO-SiO₂-fluoride systems (with $N_{CaO}/N_{SiO_2} = 1$) at 1550°C. (Ejima and Shimoji[160]).

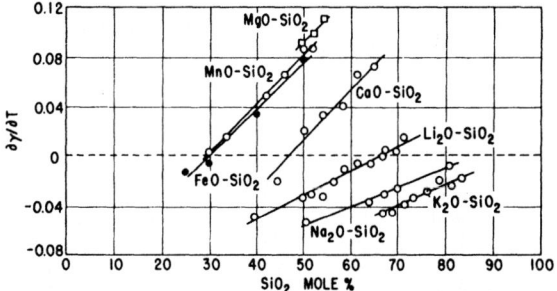

FIG. 13—114. Variation of temperature coefficient of surface tension, $\partial\sigma/\partial T$, with composition of binary silicate melts. (King[161]).

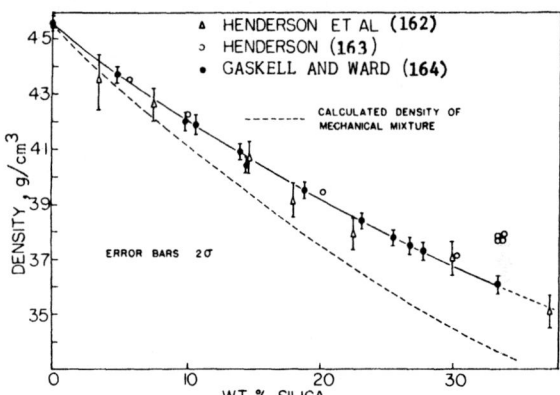

FIG. 13—115. Density-composition relationship of iron silicates in contact with solid iron at 1410°C.

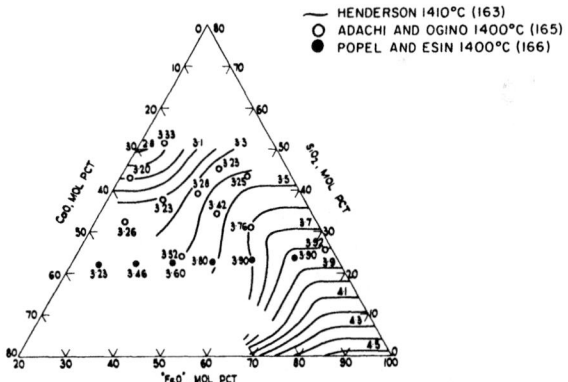

FIG. 13—116. Density of lime-iron oxide-silica melts in contact with solid iron, as a function of composition. (Henderson[163]).

The ratio of the heat flux to the temperature gradient gives, by definition, the thermal conductivity of the medium $\kappa = -J/(\partial T/\partial x)$. The thermal conductivity of gases is derived directly from the coefficient of viscosity as given below for single component and multicomponent gas mixtures; in cgs units κ is given in joule/cm · s · °C (1 cal = 4.184 J).

Monatomic gases:

$$\kappa = \frac{15}{4}\ \frac{R}{M}\ \eta \qquad (129)$$

Polyatomic gases:

$$\kappa = \left(C_p + \frac{9R}{4M} \right) \eta \qquad (130)$$

where C_p is the specific heat content, M is the molecular weight and R is the gas constant.

In the formulation of many heat transfer problems it has been found convenient to introduce a parameter called thermal diffusivity, α, defined by the ratio

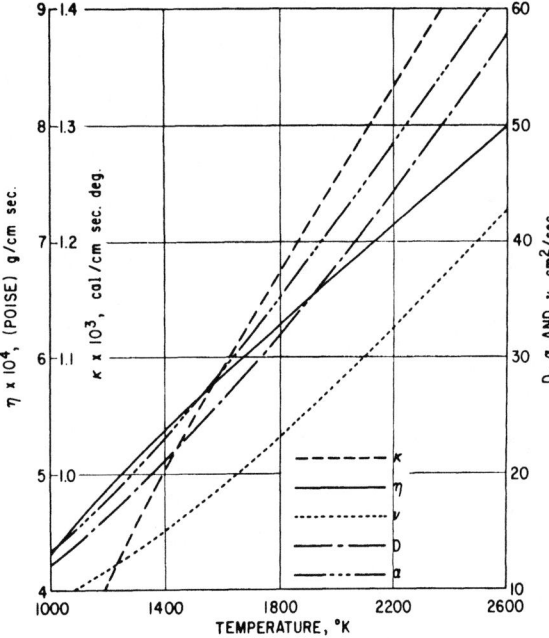

FIG. 13—117. Transport properties of helium.

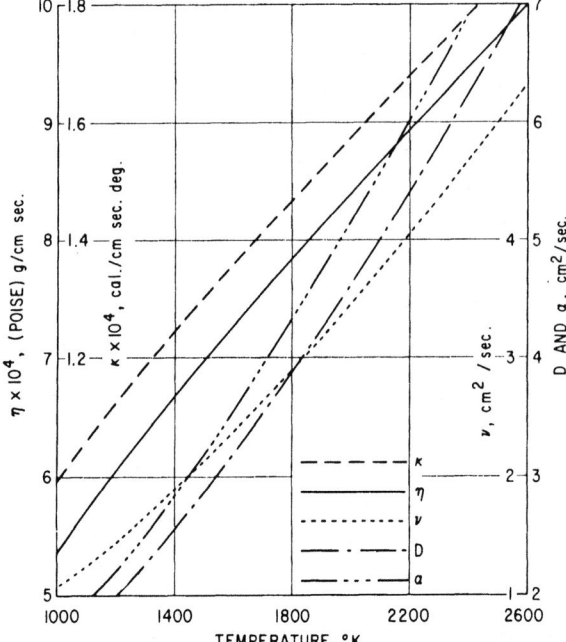

FIG. 13—118. Transport properties of argon.

$$\alpha = \frac{\kappa}{\rho C_p} \qquad (131)$$

where ρ is the density. Since Newton's law of thermal conduction is analogous to Fick's law of mass diffusion, the coefficients of thermal and mass diffusion have the same dimensions of length^2/time.

The transport properties calculated for helium and argon are given as examples in Figures 13—117 and 13—118 for temperatures from 1000° to 2600°K. The argon-metal vapor (Mn, Cr, Co, Ni and Fe) interdif-

fusivities measured by Grieveson and Turkdogan[170] are compared in Figure 13—119 with those calculated as indicated by Hirschfelder et al.[169].

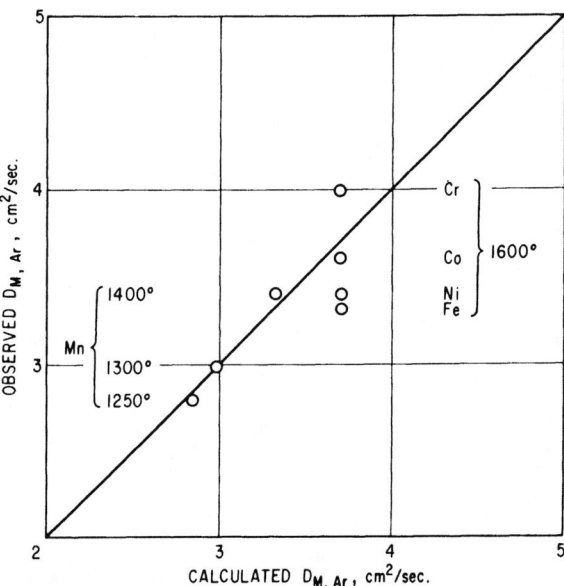

FIG. 13—119. Comparison of estimated and experimental interdiffusivities of metal vapor-gas binary mixtures at 1 atm pressure. (Grieveson and Turkdogan[170]).

Table 13—XVI. Densities of CaO-Al$_2$O$_3$-SiO$_2$ melts for the temperature range 1350° - 1650°C. (Barrett and Thomas[167])

Composition, Weight Per Cent			Density at t°C (g/cm³)
CaO	Al$_2$O$_3$	SiO$_2$	
39	19	42	$3.771 - 6.5 \times 10^{-4}t$
35	10	55	$3.215 - 4.2 \times 10^{-4}t$
34	30	36	$3.535 - 5.0 \times 10^{-4}t$
30	25	45	$3.705 - 7.9 \times 10^{-4}t$
30	10	60	$3.165 - 4.8 \times 10^{-4}t$
29	40	31	$3.280 - 3.5 \times 10^{-4}t$
25	20	55	$2.955 - 3.1 \times 10^{-4}t$
25	10	65	$3.790 - 9.8 \times 10^{-4}t$
23	15	62	$3.332 - 6.6 \times 10^{-4}t$
15	20	65	$3.150 \times 1.9 \times 10^{-4}t$

PART III

Ironmaking Reactions

Vast advances have been made in blast-furnace technology during the past two decades through plant trials and plant developments assisted by research to provide better understanding of the physical and chemical workings of the blast furnace. These advances have contributed to the current development of the large-capacity blast furnaces with high production capabilities at low coke rates. E. T. Turkdogan[171] has discussed many of these advances in the 1978 Howe Memorial Lecture and some of the material presented in this chapter is taken from that lecture. Similarly, considerable research and development work has been carried out on direct-reduction processes. It is beyond the scope of this chapter to describe all of the direct-reduction processes and only the basic general reactions will be discussed. In discussing the reactions in the blast furnace and direct-reduction processes it is helpful to make use of free energy versus temperature diagrams for metal oxides. The use of these diagrams will be described before discussing the ironmaking reactions.

Free Energy—Temperature Diagram For Metal Oxides—In order to assess the feasibility of a reduction or an oxidation reaction to occur, consideration must first be given to the free-energy changes accompanying these reactions at various temperatures and pressures. The reactions to be considered may be written in a general form as:

$$\frac{2x}{y} M + O_2 = \frac{2}{y} M_xO_y \qquad (132)$$

where M and M_xO_y is metal and metal oxide, respectively. If M and M_xO_y are at unit activities, i. e. pure metal and pure metal oxide, the equilibrium constant for reaction (132) is inversely proportional to the equilibrium partial pressure of oxygen

$$K = \frac{1}{p_{O_2}} \qquad (133)$$

where p_{O_2} is in atmospheres and the standard state for oxygen is 1 atmosphere at the temperature under consideration. The standard free-energy change accompanying oxidation is given by:

$$\Delta F^\circ = -RT \ln K = RT \ln p_{O_2}$$
$$= 2.303 \ RT \log p_{O_2}. \qquad (134)$$

where ΔF° = the standard free-energy change in calories per mole O_2
  T = temperature in K
  R = the gas constant (1.987 cal per mole per deg)

In terms of standard heat and entropy changes, the standard free-energy change is expressed as:

$$\Delta F^\circ = \Delta H^\circ - T \Delta S^\circ \qquad (135)$$

As discussed in Part I, ΔH° and ΔS° do not vary with temperature to a noticeable extent and, therefore, ΔF° varies linearly with temperature. However, at the melting and boiling points of the metal or the metal oxide, there is an abrupt change in the slopes of the free-energy lines. Using such a diagram for oxides, one can predict what oxides can be reduced by which elements from relative positions of the free-energy lines.

In Figure 13—120, the standard free-energies of formation of oxides per mole of oxygen are plotted against temperature in °C. Since the standard free energy of formation of an oxide is directly related to its equilibrium oxygen partial pressure, Figure 13—120 is often referred to as the oxygen potential diagram.

It is to be noted that all the free-energy lines in Figure 13—120 except those for C-CO, C-CO$_2$ and Si-SiO have positive slopes; this is due to the decrease in the entropy of the system when an element in a condensed phase (solid or liquid) reacts with oxygen (gaseous) to produce a condensed phase (solid or liquid oxide). In the reaction C (solid) + O$_2$ (gas) → CO$_2$ (gas) there is no volume change at a given temperature and pressure, therefore, ΔS° is almost zero. However, in reactions 2C (solid) + O$_2$ (gas) → 2CO (gas) and 2Si (solid or liquid) + O$_2$ (gas) → 2SiO (gas) there is an increase in volume for a given temperature and pressure, i. e. entropy change ΔS° is positive and, as a result, the free-energy line has a negative slope.

As discussed in Part II of this chapter, a number of elements form non-stoichiometric compounds, e. g. 'FeO', 'Fe$_3$O$_4$', 'TiO', 'VO'... etc. For the sake of clarity, these are not indicated in Figure 13—120. However, the free-energy lines drawn are for those oxides which are in equilibrium with the respective elements or other oxides. For example, in the system Fe-FeO, the latter has the composition Fe$_{0.95}$O and in the system FeO-Fe$_3$O$_4$, the composition of ferrous oxide is Fe$_{0.87}$O, at 1373°C.

The tendency of an oxide to form or to decompose at a given temperature and pressure is predicted from the relative positions of the free-energy lines. for example, at an oxygen potential

$$\Delta F^\circ = 2.303 \ RT \log p_{O_2} = -180 \text{ kcal at } 1200°C,$$

the elements Ti, Al, Mg, Ca, are oxidized, but the elements Si, V, Mn, etc. are not oxidized. Similarly, at $\Delta F^\circ = -180$ kcal and 1200°C, the oxides of Si, V, Mn, etc. will be reduced, but the oxides of Al, Mg, Ca, etc. will not be reduced. Above 700°C the free-energy line for Cu$_2$O is below that of Fe$_2$O$_3$ consequently, at temperatures above 700°C, Cu can reduce Fe$_2$O$_3$ to Fe$_3$O$_4$.

The value of carbon as a reducing agent is clearly revealed by the oxygen potential diagram. It is self evident that carbon can reduce most of the oxides. The oxides for which the free-energy lines are above that of CO can be reduced by carbon, and as the affinity of the metals for oxygen increases, i. e. ΔF° decreases, the temperature of reduction of the oxide by carbon increases. It will be observed in Figure 13—120 that,

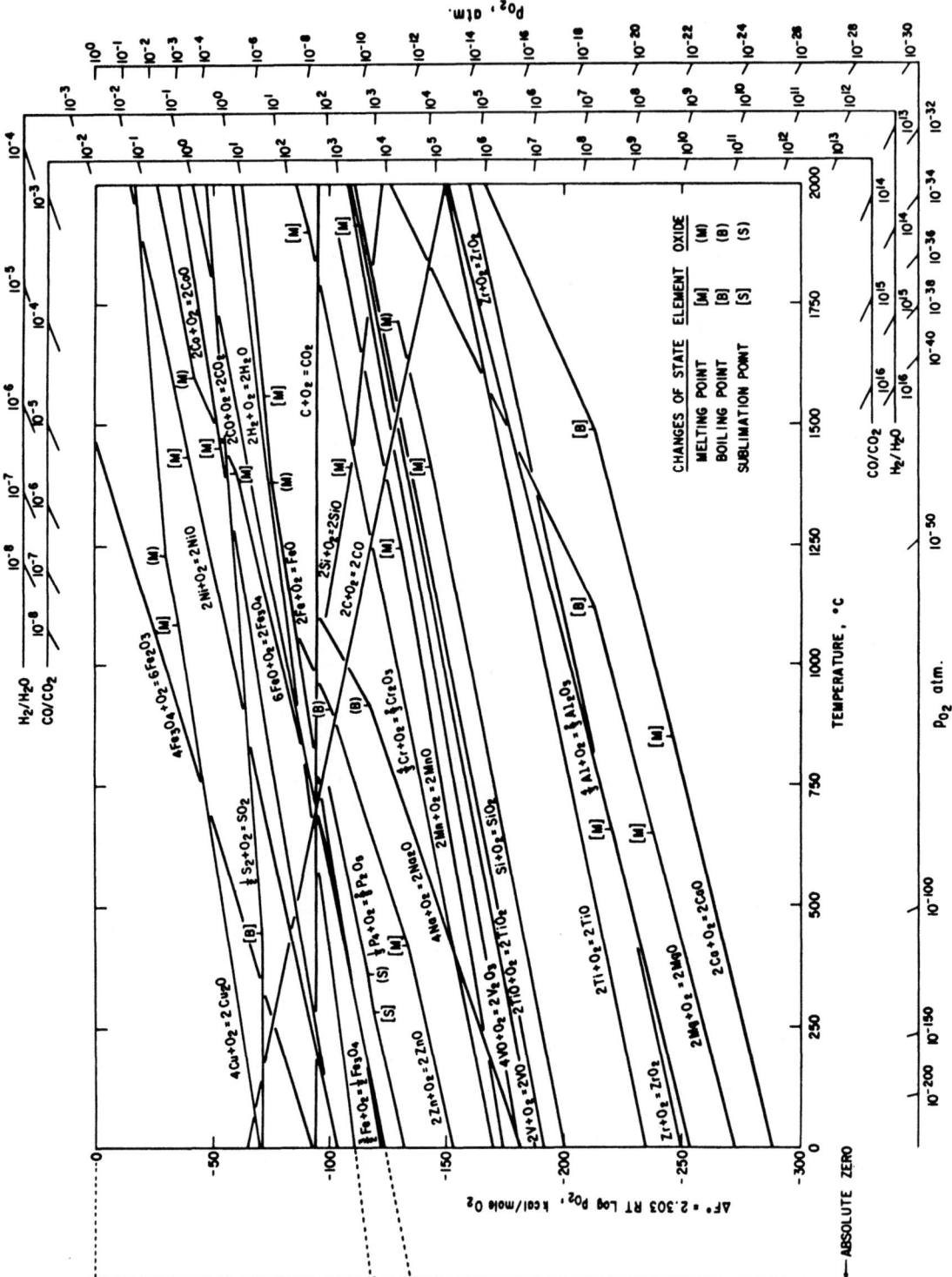

Fig. 13—120. The free energies of formation of oxides for the standard states: pure elements, pure oxides and gases at 1 atm.

although Mg has high affinity for oxygen, above the boiling point of Mg (1120°C) the slope of the $\Delta F°$ increases rapidly with increasing temperature and the Mg-MgO line cuts the $\Delta F°$ line for CO at about 1780°C; therefore, above this temperature carbon should reduce MgO to yield Mg vapor.

For quick conversion of oxygen potentials to p_{O_2} or to ratios H_2/H_2O and CO/CO_2, appropriate scales are included in Figure 13—120.

Scale for p_{O_2}—It follows from equation (134) that, for a given partial pressure of oxygen, $\Delta F°$ varies directly with temperature; at its standard state $p_{O_2} = 1$ atm, $\Delta F°$ is zero at all temperatures and at the absolute zero $\Delta F° = 0$ for all oxygen pressures. Therefore, lines drawn from the point "O", on the ordinate for the absolute zero temperature, through the points marked on the right hand side of the diagram give the isobars. For example, at 1200°C, the oxygen potential corresponding to the equilibrium of γ-Fe with wustite is −82 kcal; by drawing the line passing through this point and the point "O", the oxygen partial pressure of about 7×10^{-13} atm is read off the p_{O_2} scale.

Scale for H_2/H_2O Ratio—Since the free energy of formation of water vapor from H_2 and O_2 is known, the oxygen potentials for the H_2-H_2O-O_2 equilibrium can be calculated for any ratio H_2/H_2O. The lines for iso-gas ratios converge at point "H" on the ordinate at absolute zero temperature. Referring to the above example on the equilibrium of γ-Fe and wustite, the line drawn through $\Delta F° = -82$ at 1200°C and the point "H" intersects the H_2/H_2O scale at about 1.4/1 which is the gas ratio in equilibrium with Fe-FeO mixture.

Scale for CO/CO_2 Ratio—A grid for the iso-CO-CO_2 ratios is constructed in a similar way and these lines converge at point "C". By drawing a line through point "C" and a point on the free-energy line for an oxide, its equilibrium CO/CO_2 ratio can be read off directly from the scale for the CO/CO_2 ratio.

The above graphical relations can readily be proved by considering, for example, the reaction:

$$2H_2 + O_2 = 2H_2O \qquad (136)$$

for which the equilibrium oxygen partial pressure is given by:

$$p_{O_2} = \frac{1}{K}\left(\frac{p_{H_2O}}{p_{H_2}}\right)^2 \qquad (137)$$

where K is the equilibrium constant. In terms of oxygen potential

$$2.303\, RT \log p_{O_2} = -2.303\, RT\left(\log K + 2\log\frac{p_{H_2}}{p_{H_2O}}\right) \qquad (138)$$

$$= \Delta F° - 4.606\, RT \log\frac{p_{H_2}}{p_{H_2O}}$$

Assuming that $\Delta F°$, the standard free-energy change accompanying reaction, is a linear function of temperature then for a given ratio H_2/H_2O, the oxygen potential varies linearly with temperature. Since $\Delta F° = \Delta H° - T\Delta S°$, the oxygen potential line for any ratio H_2/H_2O intersects the ordinate at point "H" at T = 0°K, corresponding to the value of $\Delta H°$ at zero absolute temperature. Errors involved in the assumption of the linearity of $\Delta F°$ vs T plot for reaction (136) are much less than the uncertainty of the free-energy data on other oxides given in Figure 13—120.

By using the above grids, one can estimate quickly whether or not a metal will be oxidized at a given ratio of H_2/H_2O or CO/CO_2. However, for more accurate calculations, the free-energy equations should be used.

As already indicated, the data in Figure 13—120 are for pure elements and their oxides. For metals and oxides in solutions, due account must be taken of their activities in metallic and oxide solutions. Using the oxygen potential diagram as the basic tool, it is possible to generalize some aspects of the reduction and oxidation reactions occurring in smelting and refining processes.

REACTIONS IN THE BLAST-FURNACE STACK

In the blast-furnace stack there is countercurrent flow of gas and solids and the heat transfer from ascending gas to descending solids is accompanied by oxygen transfer from solids to gas. The enthalpy and temperature profiles in the stack shown schematically in Figure 13—121 are based on the concept of heat

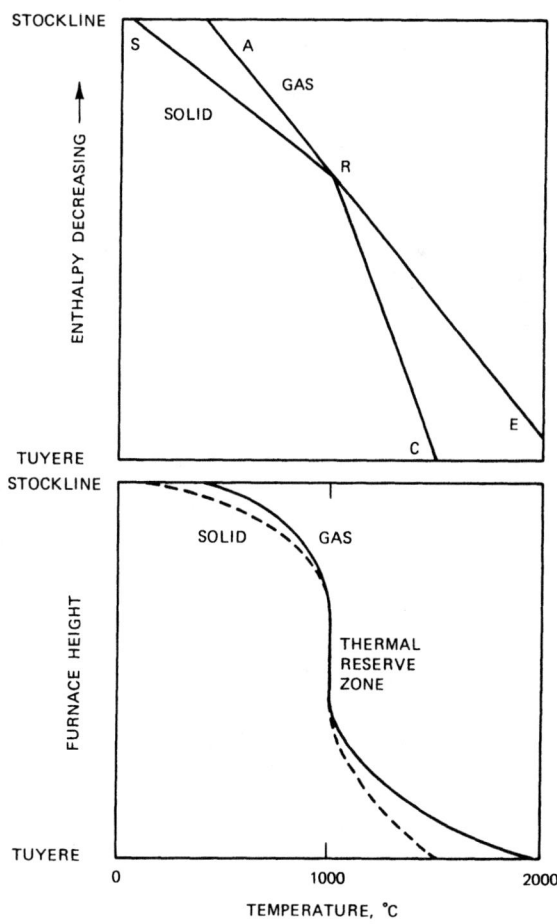

FIG. 13—121. Simplified sketch of Reichardt's diagram.

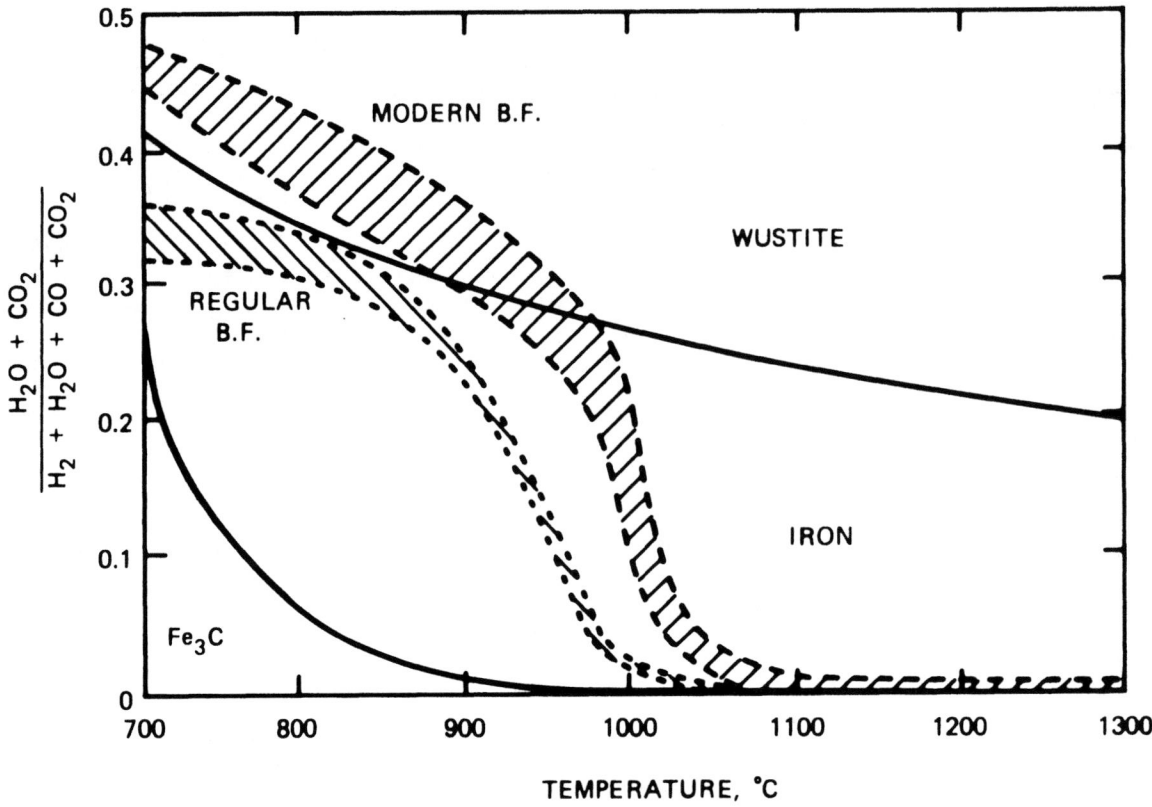

FIG. 13—122. Variations in temperature and gas-composition profiles with blast-furnace practice. (Turkdogan[171]).

balance conceived by Reichardt[172] from the consideration of the first and second laws of thermodynamics. As
a result of heat transfer, the enthalpy of the gas decreases with decreasing temperature of the ascending
gas, as represented by the line ERA. The increase in
the enthalpy of the solids with increasing temperature
during descent in the stack is represented by the line
SR. In the lower part of the furnace, the overall heat
capacity of the burden increases because of fusion, the
onset of the strongly endothermic Boudouard reaction,
and greater heat losses; hence, the slope of the line RC
is greater than that of SR. The net result is that, over
some distance in the stack, the temperature difference
between the ascending gas and descending burden
reaches a minimum, known as the "thermal pinch
point."

The above discussion is an over simplification; the
actual gas composition and temperatures in the stack
depend on blast-furnace operation. The lower curve in
Figure 13—122 is for a regular blast furnace using an
acid sinter or pellet and lump ore with a high coke rate.
The upper curve is for a basic sinter, oxygen-enriched
air blast and a low coke rate. In the older type blast-
furnace operations the gas is always reducing to
wustite, whereas in modern operations the gas is oxidizing to iron.

Reference should be made to the work of Rist and
coworkers[173, 174] who did extensive work on the
graphical representation of the heat and mass balances

in the blast furnace which are used for controlling
blast-furnace operations.

The two basic reactions in the blast-furnace stack are
the oxidation of coke by CO_2 to produce CO and heat
and the reduction of iron oxides by CO. These two
reactions have been studied extensively in the past few
years. It is beyond the scope of this chapter to discuss
these reactions in detail. For details on the reduction of
iron oxide the reader is referred to the book by von
Bogdandy and Engell[175] and the review paper by
Turkdogan[176] and for the oxidation of coke to the work
of Turkdogan and Vinters.[177] The oxidation of carbon
can be represented by the following reaction.

$$2CO + O_2 = 2CO_2 \qquad (139)$$

From the data compiled in Part II of this chapter, the
following is obtained for the standard free-energy
change accompanying reaction (139)

$$\Delta F° = -135\,000 + 41.42\,T \text{ cal per mole } O_2 \quad (140)$$

and the corresponding relation for the equilibrium constant is

$$\log K = \frac{29\,508}{T} - 9.054 \qquad (141)$$

where

$$K = \frac{p^2_{CO_2}}{p^2_{CO}\, p_{O_2}} \qquad (142)$$

For dense sintered hematite pellets the reduction of iron oxide is slow, resulting in layers of iron, wustite, magnetite and hematite as shown in Figure 13—123a. The formation of these product layers is the result of slow diffusion of gas in the product layer. [178, 179] For lump ores or pellets with high porosity there is no product layer formation but rather internal reduction as shown in Figure 13—123b. For a given form of iron oxide, the greater the driving force for reduction, the greater the tendency to form product layers. For example, reduction by H_2 is more likely to produce layers than reduction by CO.

Formation of product layers during the early stages of reduction in the stack may have an adverse effect on the reducibility of the ore at higher temperatures. It has been shown that the initial reduction temperature is important and a subsequent rise in temperature does not readily coarsen the pore structure.

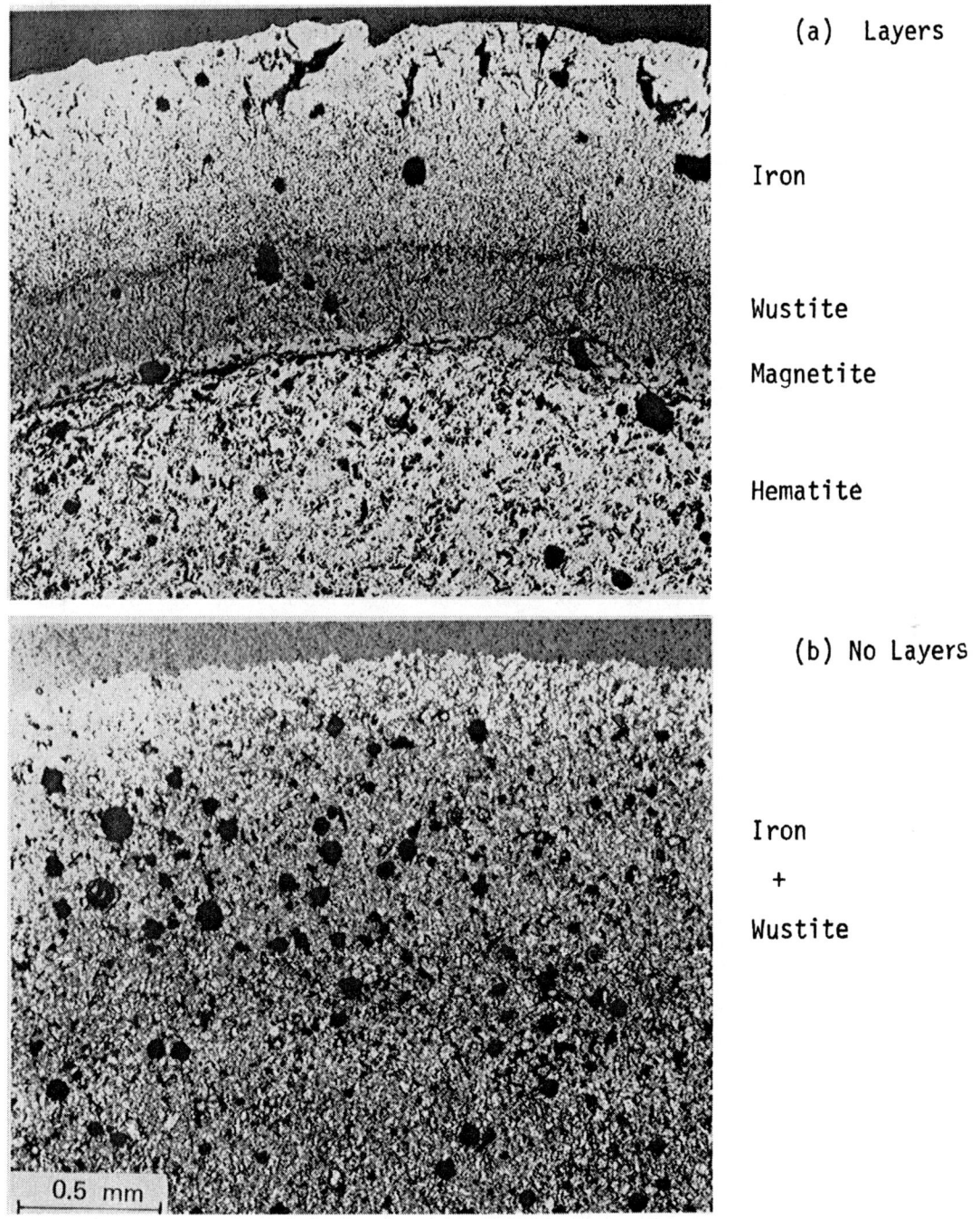

(a) Layers

Iron

Wustite

Magnetite

Hematite

(b) No Layers

Iron

+

Wustite

0.5 mm

FIG. 13—123. Polished sections of sintered hematite pellets reduced 30 per cent in hydrogen at 900°C. (a) Dense Fe_2O_3 pellet—reduction layers. (b) Porous ore pellet—no reduction layers. (Turkdogan[176]).

With the reduced coke rates in new blast furnaces more residual iron oxide is carried to the upper part of the bosh. Therefore, the high-temperature reducibility of the oxide becomes important. If an acid pellet or sinter is used, a liquid iron silicate may form at high temperatures which partially fills the pores and lowers the rate of reduction. Therefore, in a blast furnace with a low coke rate, the basicity of the sinter or pellet should be increased to insure easy reducibility of the oxide.

REACTIONS IN THE BLAST-FURNACE BOSH AND HEARTH

With respect to metal-slag systems at constant activities of carbon and carbon monoxide, two types of overall reactions may be considered: primary reactions and coupled reactions. For blast-furnace-type metal-slag systems, the following primary reactions may be considered:

$$\underline{Mn} + CO = MnO + \underline{C} \qquad (143)$$

$$\underline{Si} + 2CO = SiO_2 + 2\underline{C} \qquad (144)$$

$$\underline{S} + CaO + \underline{C} = CaS + CO \qquad (145)$$

The sulphide and silicate ions in the slag will react with carbon to produce gaseous species SiO and SiS; therefore another overall primary reaction is:

$$CaS + 2SiO_2 + 2\underline{C} = CaO + SiO + SiS + 2CO \qquad (146)$$

This and other primary reactions involving volatile reaction products may impede the ultimate approach to equilibrium because of continuous loss of materials from the system.

The reactions that do not involve the invariant components of the system, in this case C and CO, are considered coupled reactions between slag and metal phases. In terms of overall reactions, there are three coupled reactions of particular importance.

$$2MnO + \underline{Si} = SiO_2 + 2\underline{Mn} \qquad (147)$$

$$\underline{S} + \underline{Mn} + CaO = CaS + MnO \qquad (148)$$

$$\underline{S} + \frac{1}{2}\underline{Si} + CaO = CaS + \frac{1}{2}SiO_2 \qquad (149)$$

It should be noted that these overall coupled reactions are the sum of transient electrochemical reactions. For example, the transfer of sulphur from metal to slag, requiring electrons, will be accompanied by anodic oxidation of alloying elements as well as iron, leading to transient electrochemical reactions such as

$$\underline{Mn} + \underline{S} \rightarrow Mn^{2+} + S^{2-} \qquad (150)$$

$$2\underline{Si} + \underline{S} + 2(\equiv Si - O^-) \rightarrow 2(\equiv Si - O - Si \lessgtr) + S^{2-} \qquad (151)$$

Although the gas-slag-metal system may be out of equilibrium with respect to one or all of the primary reactions, the system may be at a state of partial equilibrium with respect to at least one of the coupled reactions represented by equations (147) to (149).

As an example, let us reanalyze the data obtained by Filer and Darken[180] on remelting of blast-furnace metal and slag samples in a graphite crucible at 1 atm CO. It is seen from their results in Figure 13—124 that at 1500°C the reduction of manganese and silicon from slag to metal is sluggish. It took about 18 to 20 hours for the primary manganese reaction (143) to approach equilibrium; for the primary silicon reaction (144), the time of equilibration was about 28 hours.

For a given slag basicity and temperature, the equilibrium relation k_{MnSi} for the coupled reaction (147) is given by the equation

$$K_{MnSi} = \left(\frac{[\%Mn]}{(\%MnO)}\right)^2 \frac{(\%SiO_2)}{[\%Si]} \qquad (152)$$

As is indicated by the data in Figure 13—124, the values of k_{MnSi} do not change much after about 5 hours reaction time, indicating partial equilibrium in the system relative to reaction (147).

In these experiments of Filer and Darken, the sulphur distribution ratio (%S)/[%S] was about 160 and did not change much in reaction times from 2 to 66 hours. Because the primary sulphur reaction (145) approached equilibrium in a relatively short time in their experiments at 1500°C, the coupled reactions (148) and (149) lose their significance in this particular case. In fact, as would be anticipated from the data presented in Figure 13—124,

$$k_{Mns} = \frac{(\%S)}{[\%S]} \frac{(\%MnO)}{[\%Mn]} \frac{1}{(\%CaO)} \qquad (153)$$

reaches the equilibrium value only when the primary manganese reaction is close to equilibrium. However, as shown later, the results of other experimental studies substantiate the view that prior to the establishment of equilibrium for the primary sulphur reaction, a partial equilibrium may be established between metal and slag with respect to the manganese-sulphur and silicon-sulphur reactions.

With the realization of the importance of coupled reactions in metallurgical systems, Oelsen and co-workers[181-183] made an exhaustive number of experiments on blast-furnace-type slag-metal reactions. Unfortunately, in their experiments of only 30 min duration, the initial compositions of metal and slag departed considerably from the equilibrium values for both the primary and coupled reactions. Difficulties encountered in the interpretation of such data are demonstrated in Figure 13—125 by using the results of one series of experiments reported by Oelsen et al.[183] In this series of experiments at 1550°C and unit activities of carbon and carbon monoxide, the metal initially contained 0.13 per cent, 0.94 per cent or 1.9 per cent S and the initial manganese distribution ratio $\{[\%Mn]/(\%MnO)\}_i$ was varied from about 0.7 to 3.0. Systematic variations in the distribution ratios [%Mn]/(%MnO) and (%S)/[%S], and the corresponding values of K_{Mns}, at 30 min reaction time, with the initial melt compositions are shown in Figure 13—125; the date are for slags of basicity B = 1.2.

The dotted line in Figure 13—125A with a slope of one delineates the regions of MnO reduction and Mn oxidation. For their slag basicity of 1.2, the equilibrium

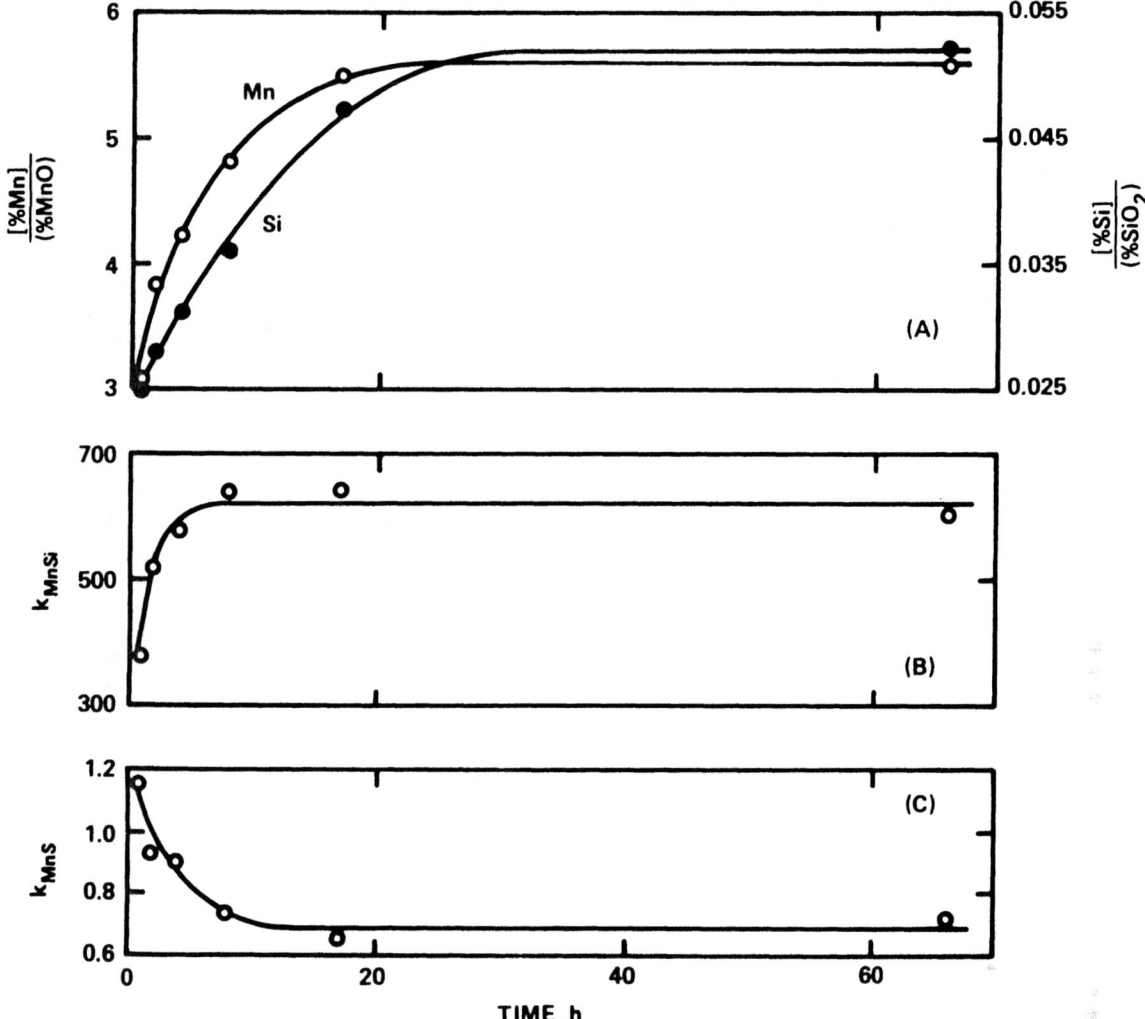

FIG. 13—124. Approach to equilibrium for primary manganese and silicon reactions and corresponding changes in k_{MnSi} and k_{MnS} with reaction time at 1500°C at unit activities of carbon and CO; data are from Filer and Darken.[180]

distribution ratios for the primary manganese and sulphur reactions are $\{[\%Mn]/(\%MnO)\}_e = 5$ and $\{(\%S)/[\%S]\}_e = 82$. Although the initial manganese ratio is below the equilibrium value, there is manganese oxidation during the 30 min reaction time, the extent of oxidation being greater with increasing initial sulphur concentration in the metal. This behavior is attributed to the transient electrochemical reaction (150). A large influx of sulphur from metal to slag in a short reaction time forces the manganese and silicon reactions away from equilibrium such that even for the coupled reactions the values of k_{MnSi} and k_{MnS} vary systematically with the initial melt composition. As is seen from the data in Figures 13—125(B) and (D), the values of k_{MnSi} and k_{MnS} can be higher or lower than the equilibrium values. For example, for a basicity of 1.2, the values of k_{MnSi} are in the range 10 to 300, depending on the initial melt composition. However, an average value of about 150 for k_{MnSi} agrees well with that derived from other data.

For small variations in the concentrations of Mn and Si in graphite-saturated iron, their activity coefficients, as affected by carbon, will remain essentially constant. Therefore, for a given temperature and slag basicity the equilibrium relation k_{MnSi} may be represented by

$$k_{MnSi} = \left(\frac{[\%Mn]}{(\%MnO)}\right)^2 \frac{(\%SiO_2)}{[\%Si]} \qquad (154)$$

Because the activity coefficients of MnO and SiO_2 are simple functions of the slag basicity B, the equilibrium relation k_{MnSi} should also depend on the slag basicity. The results of two different sets of experiments[180,184] are shown in Figure 13—126 for liquid iron equilibrated with blast-furnace-type slags in a graphite crucible at 1 atm pressure of CO. Because the temperature coefficients of the equilibrium constants for reactions (143) and (144) (for 2 moles of MnO) are similar, a small temperature effect of k_{MnSi} is masked by the scatter in the experimental data. The average line indicating the equilibrium relation for the manganese-silicon reaction

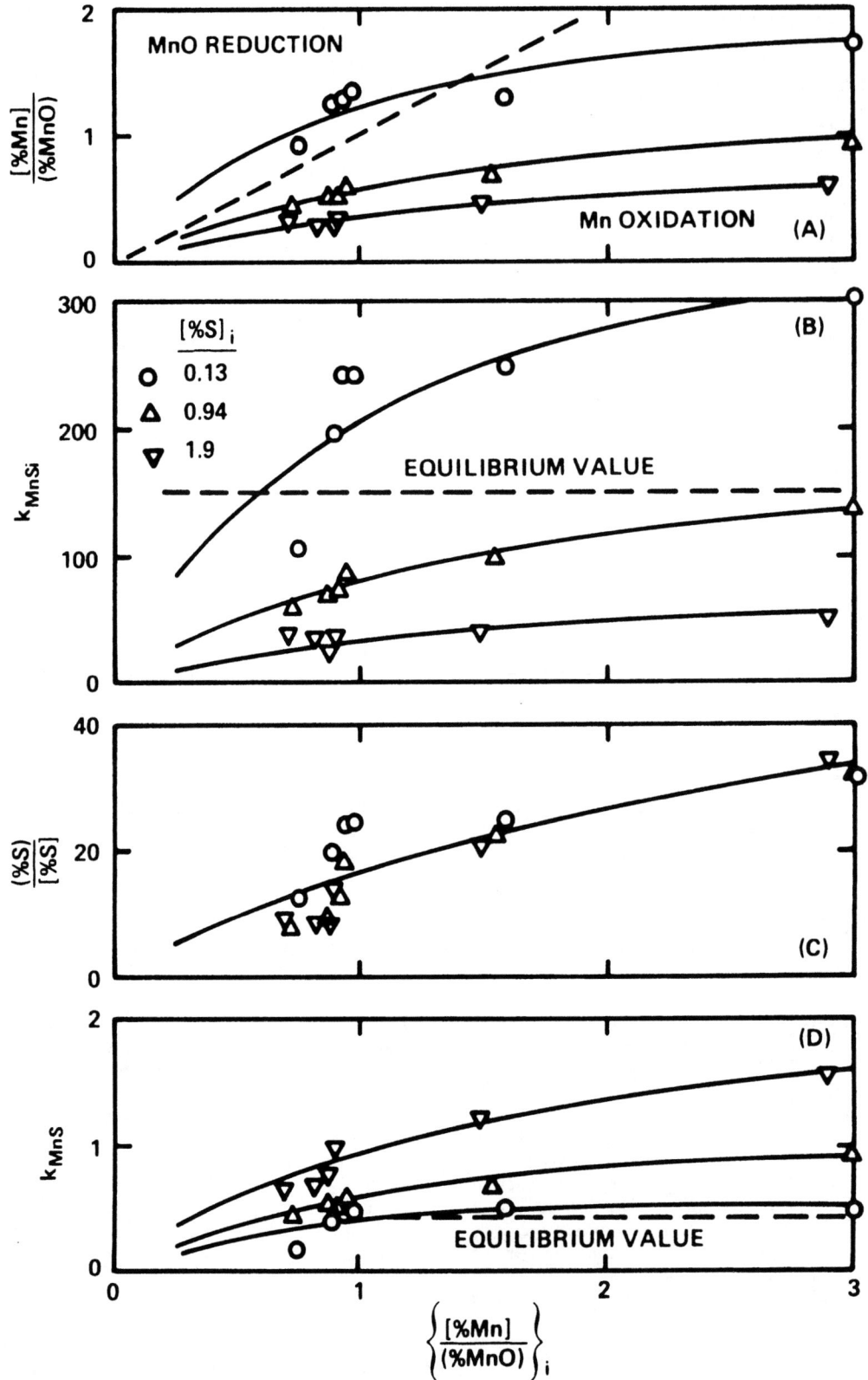

FIG. 13—125. Variations of [Mn]/(MnO), [Si]/(SiO₂), k_{MnSi} and k_{MnS} with initial melt compositions in experiments of Oelsen, et al.[181-183] at 1550°C for 30 min.

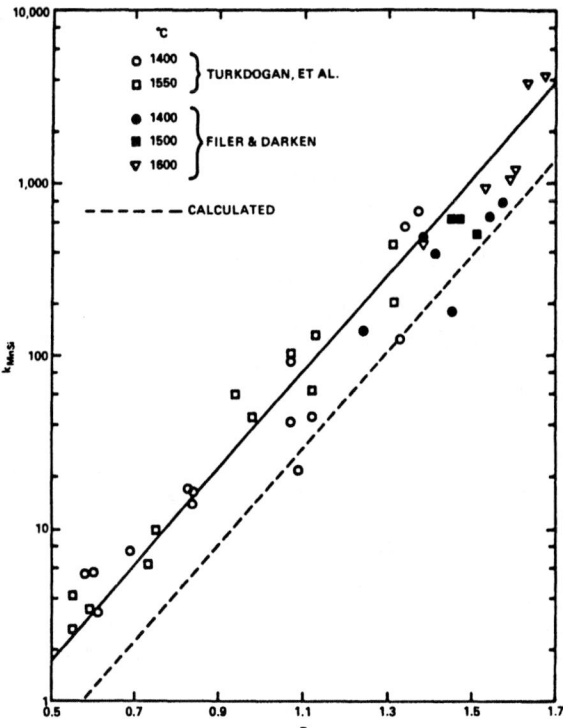

FIG. 13—126. Variation of the equilibrium relation k_{MnSi} for the manganese-silicon coupled reaction with slag basicity B for unit activity of carbon. After Turkdogan, et al.[184]

at unit activity of carbon is represented by the equation

$$\log k_{MnSi} = \log \left(\frac{[\%Mn]}{(\%MnO)} \right)^2 \frac{(\%SiO_2)}{[\%Si]} = 2.8B - 1.16. \quad (155)$$

For melts not saturated with graphite, equation (155) should be adjusted for the decrease in γ'_{Si} with decreasing concentration of carbon.

The dotted line in Figure 13—126 is calculated from the free-energy and activity data. The difference of the calculated line from the average experimental values by a factor of 2.7 is attributed to the accumulated uncertainties in the free-energy data and also to the approximations made in the calculations. An error in the calculated data may arise from the use of the free-energy equations for the solution of silicon in pure liquid iron for reactions with graphite-saturated liquid iron.

The equilibrium constant for reaction (143) is represented by the equation

$$K_{Mn} = \frac{[\%Mn] \, \gamma'_{Mn}}{a_{MnO}} \frac{p_{CO}}{a_c} \quad (156)$$

where γ'_{Mn} is the activity coefficient of manganese as affected by carbon, defined such that $\gamma'_{Mn} \to 1$ when $\%C \to 0$; for graphite-saturated iron $\gamma'_{Mn} = 0.8$. In most blast-furnace slags, the total number of moles of the constituent oxides per 100 g of slag is about 1.65. Therefore, the activity of manganese oxide, a_{MnO}, in the slag may be represented in terms of the weight per cent MnO and its activity coefficient γ_{MnO}:

$$a_{MnO} = \frac{(\%MnO) \, \gamma_{MnO}.}{70.94 \times 1.65} \quad (157)$$

For unit activity of carbon in graphite-saturated melts ($a_c = 1$), equations (156) and (157) give

$$\frac{[\%Mn]}{(\%MnO)} \, p_{CO} = 1.07 \times 10^{-2} \, K_{Mn} \, \gamma_{MnO} \quad (158)$$

where p_{CO} is in bar.

The free-energy data give for the temperature dependence of the equilibrium constant

$$\log K_{Mn} = -\frac{15\,183}{T} + 10.874. \quad (159)$$

From equations (158) and (159) and known values of γ_{MnO} (Figure 13—94) calculations are made of the equilibrium distribution of manganese between graphite-saturated liquid iron and slag for any given partial pressure of CO and the basicity B. This equilibrium relation is shown in Figure 13—127 for blast-furnace-type slags, which usually contain 10 to 20 per cent Al_2O_3. These calculated data agree well with values determined experimentally over limited ranges of slag compositions.

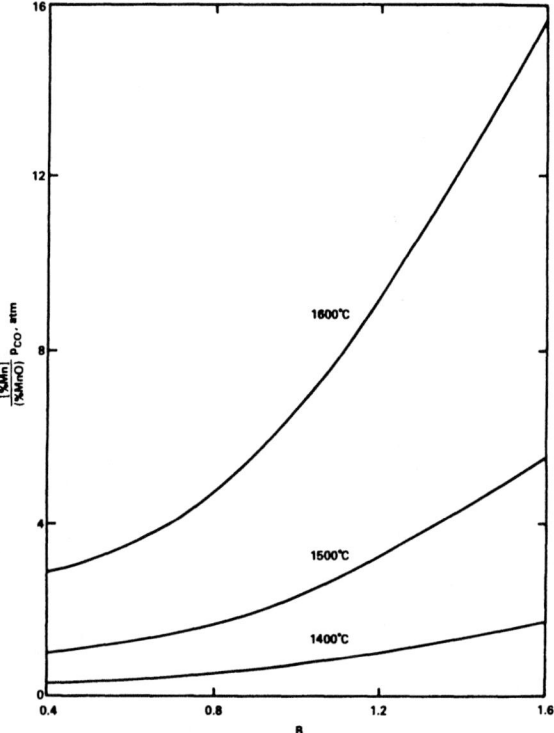

FIG. 13—127. Calculated equilibrium metal/slag manganese distribution for graphite-saturated blast-furnace-type slags containing 10 per cent to 20 per cent Al_2O_3.

The equilibrium constant for the silicon-sulphur coupled reaction (149)

$$K_{SiS} = \frac{(a_s)}{[a_s]} \left[\frac{(a_{SiO2})}{[a_{Si}]} \right]^{1/2} \frac{1}{a_{CaO}} \quad (160)$$

may be converted to a more practical form, thus

$$K'_{sis} = \frac{(\%S)}{[\%S]} \left[\frac{(\%SiO_2)}{[\%Si]} \right]^{1/2} \frac{1}{(\%CaO)} \left(\frac{\gamma_{CaS}}{\gamma_{CaO}} \right) (\gamma_{SiO_2})^{1/2} \tag{161}$$

Because of small variations in the concentration of S and Si in the metal, their activity coefficients will remain essentially constant and therefore need not be included in the above equation. Because γ_{SiO_2} and the ratio $\gamma_{CaO}/\gamma_{CaS}$ vary primarily with the slag basicity, the equilibrium relation k_{sis} for a given temperature

$$k_{sis} = \frac{(\%S)}{[\%S]} \sqrt{\frac{(\%SiO_2)}{[\%Si]}} \frac{1}{(\%CaO)} \tag{162}$$

will depend essentially on slag basicity only.

For the manganese-sulphur coupled reaction (148) a similar expression may be written

$$k_{MnS} = \frac{(\%S)}{[\%S]} \frac{(\%MnO)}{[\%Mn]} \frac{1}{(\%CaO)} \tag{163}$$

In a recent experimental study of blast-furnace reactions, together with other experimental data, Turkdogan et al.[184] derived the following relations for the temperature range 1400° to 1600°C and for basicity, B, in the range 0.5 to 1.7,

$$\log k_{sis} = \frac{6327}{T} - 4.43 + 1.4B, \tag{164}$$

$$\log k_{MnS} = \frac{6327}{T} - 3.85. \tag{165}$$

$$B = \frac{\%CaO + \%MgO}{\%SiO_2} \tag{166}$$

The effect of slag basicity on the equilibrium distribution of silicon and sulphur is shown in Figures 13—128 and 13—129.

Observation From Plant Data—Prior to the experimental studies of the past decade, it was generally believed that as the reduced iron begins to melt in the bosh region, the metal droplets pick up silicon from the slag reduced by the carbon. The research of the past ten years brought about new thoughts on the role of silicon and sulphur-bearing volatile species on the reactions in the bosh and hearth. Tsuchiya et al.[185] and Turkdogan and coworkers[184] have demonstrated that the silicon and sulphur transfer to the metal is via the formation of silicon monoxide and silicon monosulphide from the coke ash in the high-temperature region of the combustion zone. They also demonstrated experimentally that as the metal droplets pass through the slag layer, some of the silicon picked up earlier is oxidized by iron oxide and manganese oxide in the slag, accompanied by transfer of sulphur from metal to slag. Deductions made from these experimental findings are consistent with the observations made from the quenched experimental blast furnace.[186] That is, the silicon and sulphur contents of the metal droplets reach maxima near the tuyere level and decrease in the slag layer.

Because of fluctuations in the operating conditions in the furnace there will be variations in the extent of reaction of metal droplets in the slag layer. Consequently, the metal droplets of varying compositions collecting in the stagnant hearth will bring about the composition and temperature stratification that is common to all blast furnaces.

It is not uncommon to experience variations in the compositions of slag and hot metal in consecutive tappings in any given day. Despite these tap-to-tap variations, the daily averages do not change much. To evaluate the state of reactions in various blast furnaces,

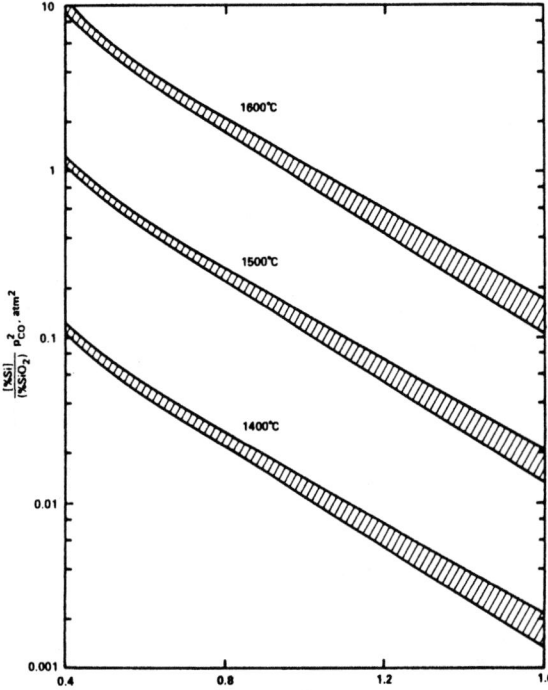

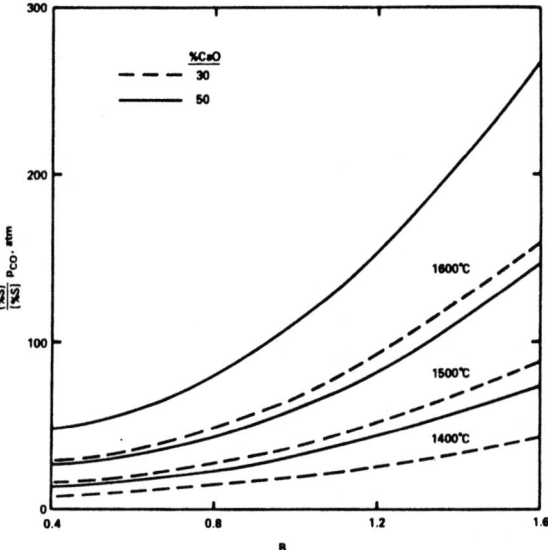

FIG. 13—128. Equilibrium metal/slag silicon distribution for graphite-saturated blast-furnace-type slags. After Turkdogan, et al.[184]

FIG. 13—129. Equilibrium slag/metal sulphur distribution for graphite-saturated blast-furnace-type slags. After Turkdogan, et al.[184]

it is adequate to use the daily average compositions of metal and slag samples as normally recorded in the plants.

In a paper by Okabe et al.[186] the daily average data are given for tap slag and metal samples taken from five blast furnaces at Chiba Works of Kawasaki Steel. These plant data, together with the daily average data for a period of a month from United States Steel's Homestead Works No. 3, Duquesne Works No. 4 and South Works No. 10 blast furnaces, were used previously to evaluate the state of slag-metal reactions in blast furnaces.[171] Deductions made from this earlier study are updated and modified here in the light of subsequent studies of slag-metal equlibria.[185]

The data analyzed are typical for most blast-furnace operations. The melt temperatures at tap are in the range 1450° to 1550°C, mostly in the range 1475° to 1525°C. The average slag compositions in weight per cent are in the range: 38-44 per cent CaO, 8-10 per cent MgO, 34-38 per cent SiO_2, 10-12 per cent Al_2O_3, 0.5-1.0 per cent MnO, 1-2 per cent S, 0.1-0.6 per cent K_2O, <0.2 per cent FeO, and minor amounts of other oxides. The sum of the concentrations of four major constituent oxides CaO, MgO, SiO_2, and Al_2O_3 is in the range 95 to 97 per cent. The basicity of these slags, defined by the ratio (%CaO + %MgO)/%SiO_2, is in the range 1.25 to 1.55. The hot-metal compositions from these blast furnaces are in the range: 0.4-0.8 per cent Mn, 0.5-1.5 per cent Si, 0.02-0.05 per cent S and other usual impurities, the amount of carbon in graphite-saturated metal depending on temperature and the concentration of the other solutes.

As is seen from the plot of the plant data in Figure 13—130, there are marked variations in the observed

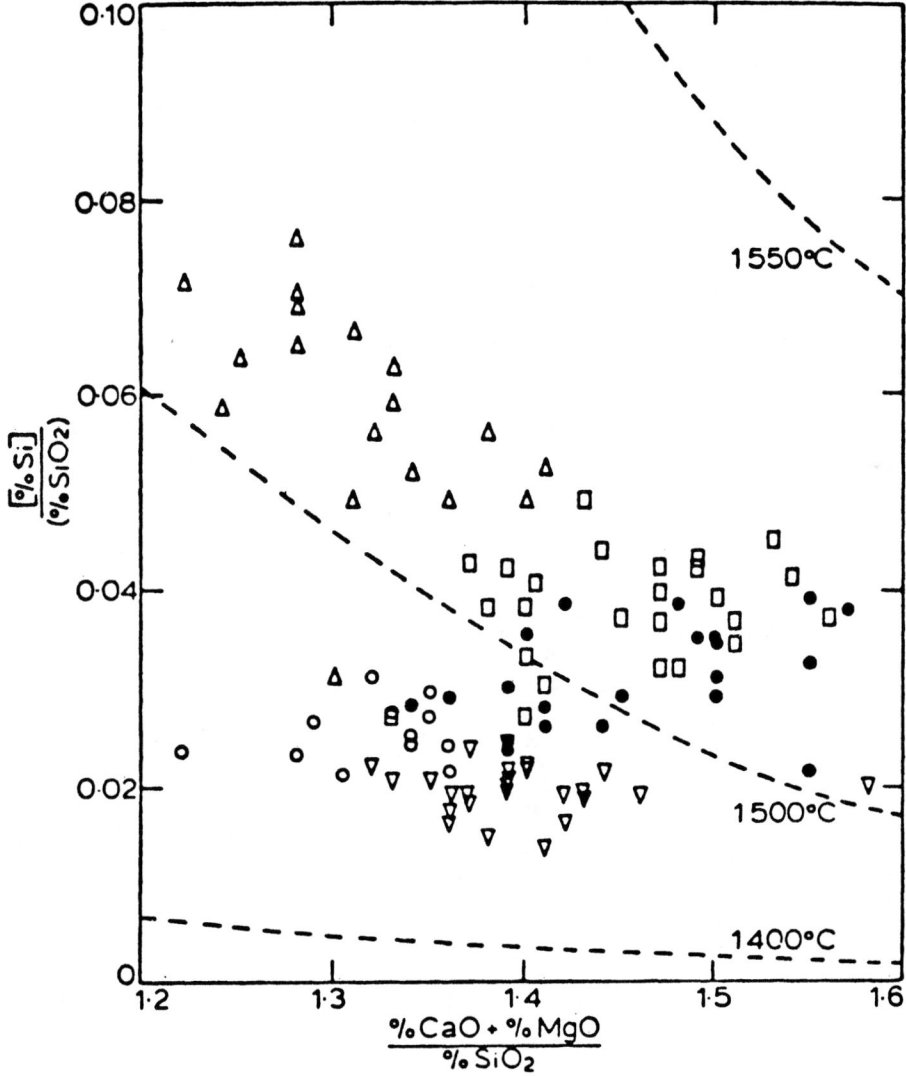

FIG. 13—130. Daily average blast furnace plant data showing trends in the variation of the silicon distribution ratio between metal and slag with slag basicity, compared with the equilibrium relation (– – –) for graphite-saturated melts at 1 bar CO, reproduced from Fig. 13-128: (△) Chiba BF1, 1500°-1525°C; (△) Chiba BF2, 1500°-1525°C; (○) Homestead BF3, 1450°-1500°C; (●) Duquesne BF4, 1475°-1550°C; (□) South BF10, 1450°-1500°C. (Turkdogan[235]).

effect of slag basicity on the silicon-distribution ratio between metal and slag. For Chiba No. 1 and South No. 10 blast furnaces, the decrease in the ratio [%Si]/(%SiO₂) with an increase in slag basicity follows the trend for the three-phase equilibrium

$$(SiO_2) + 2\underline{C} = \underline{Si} + 2CO$$

The data points are scattered about the equilibrium curve for 1 atm CO and 1510°C, which is the average tap temperature for these furnaces. Tap temperatures for Homestead blast furnace No. 3 are lower (1450° to 1500°C), and the data points are scattered about the 1450°C-equilibrium values. On the other hand, the silicon-distribution ratios for Chiba No. 2 and most of Duquesne No. 4 furnace are well below the equilib-

rium values for their tap temperatures.

The effect of slag basicity on the distribution of manganese between metal and slag is shown in Figure 13—131. The plant data segregate into three groups: Chiba No. 1-South No. 10, Homestead No. 3-Duquesne No. 4, and Chiba No. 2 furnaces. In the first and second groups the ratio [%Mn]/(%MnO) increases with increasing slag basicity; little or no basicity effect is seen for Chiba No. 2. In most cases, the manganese-distribution ratios are below the equilibrium values for the three-phase reaction

$$(MnO) + \underline{C} = \underline{Mn} + CO$$

However, at high basicities the data points for South No. 10 furnace are close to the equilibrium relation.

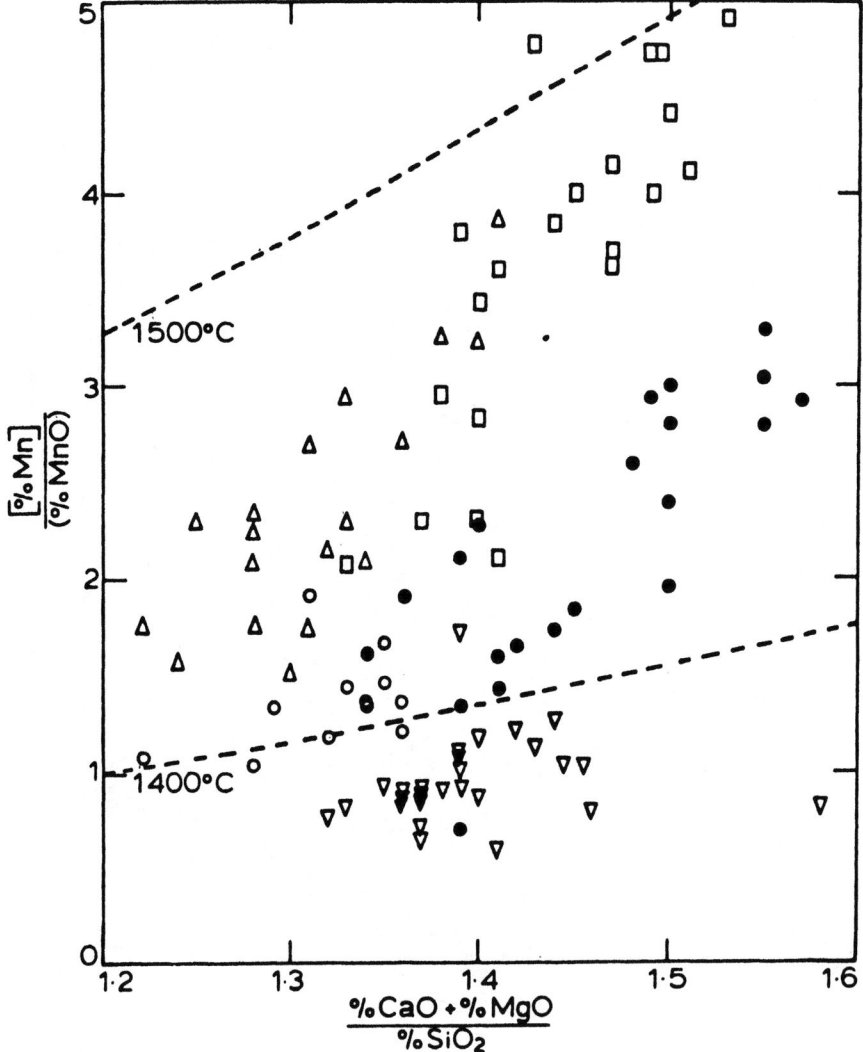

FIG. 13—131. Effect of basicity on manganese distribution ratio in various blast furnaces; symbols same as in Figure 13—130. (Turkdogan[235]).

The distribution of sulphur between slag and metal is seen in Figure 13—132 to increase with increasing slag basicity. The data points for furnaces Nos. 1, 3, 4 and 10 are within the same scatter band; the sulphur-distribution ratios for No. 2 are much lower than for other furnaces. In all cases, the ratios (%S/[%S] are well below the equilibrium values for the three-phase reaction

$$(CaO) + \underline{S} + \underline{C} = (CaS) + CO$$

for the tap temperatures and 1 atm CO. In laboratory experiments of Turkdogan, et al.[184] simulating, at least approximately, the descent of metal droplets in the slag layer, there was rapid approach to equilibrium for reaction (145). Such a fast reaction does not occur in the slag layer of the blast furnace.

Observations made from plant data for three-phase reactions may be summarized as follows:

(1) Ratio [%Si]/(%SiO$_2$) decreases with increasing slag basicity and decreasing temperature. Reaction (144) appears to approach equilibrium in some blast furnaces; in No. 2 the silicon distribution ratios are below the equilibrium values.

(2) Ratio [%Mn]/(%MnO) increases with increasing slag basicity and temperature; the manganese-distribution ratios in all cases are below the equilibrium values.

(3) Ratio (%S)/[%S] increases with increasing slag basicity and temperature; the sulphur-distribution ratios in all cases are below the equilibrium values.

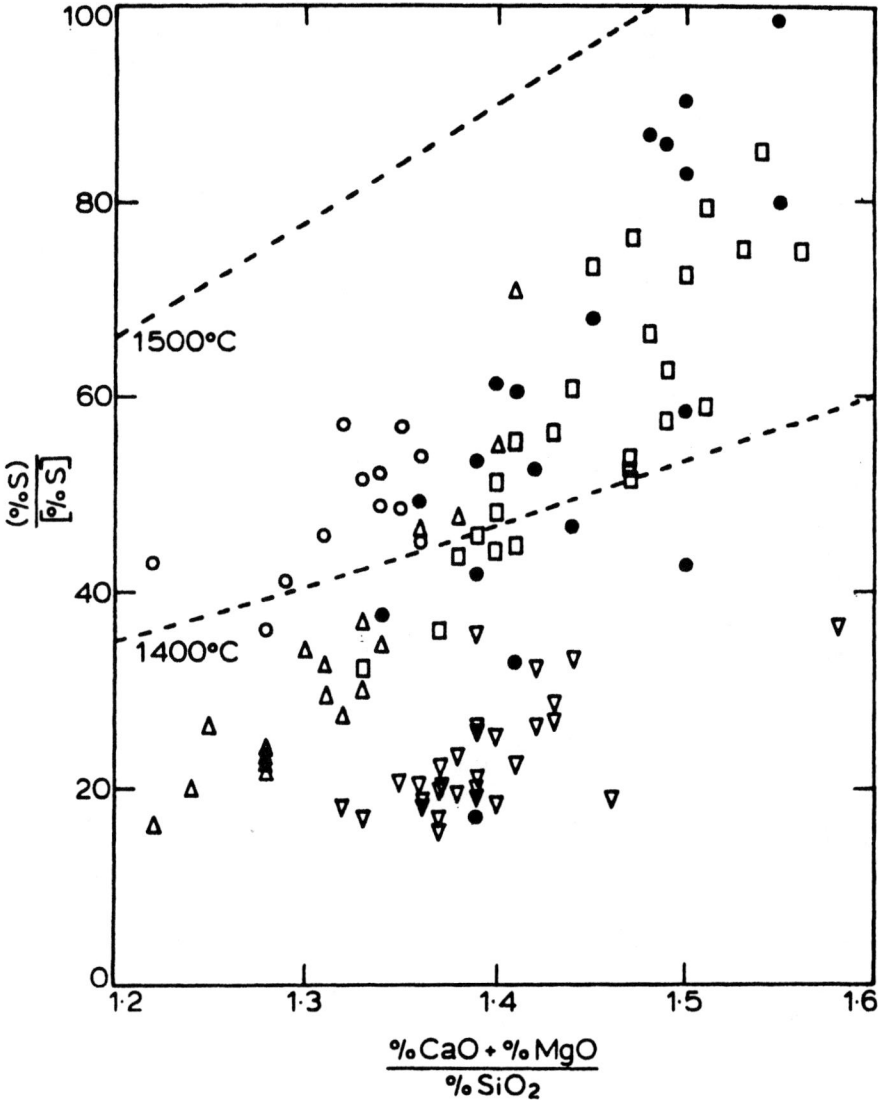

FIG. 13—132. Effect of basicity on sulphur distribution ratio in various blast furnaces; symbols same as in Figure 13—130. (Turkdogan[235]).

If there is approach to equilibrium for the manganese-sulphur coupled reaction,

$$\underline{S} + \underline{Mn} + (CaO) = (CaS) + (MnO)$$

for small changes in the concentration of CaO in slags, the sulphur-distribution ratio $(\%S)/[\%S]$ should be directly proportional to the manganese-distribution ratio. The plant data plotted in Figure 13—133 are compared with the equilibrium lines (dotted) calculated from equation (165) for slags containing 40 per cent CaO. The data for Chiba No. 1 and South No. 10 furnaces are scattered about the equilibrium line for 1600°C which is about 100°C higher than the average tap temperatures. Observed relation for other blast furnaces is scattered within the range bounded by the equilibrium lines for temperatures of 1380° to 1480°C. Despite departures from slag-metal reaction equilibria, there are, nevertheless, certain regularities in the variation of $(\%S)/[\%S]$ with the ratio $[\%Mn]/(\%MnO)$. It is difficult, however, to account for observed variations in departure from equilibrium for reaction (148) in different furnaces.

Product of the manganese and silicon-distribution ratios is seen in Figure 13—134 to increase with an increase in slag basicity, but is below the equilibrium line (dotted) reproduced from Figure 13—126. As noted previously, over the range of measurements (1400° to 1600°C), the temperature has a small effect on the equilibrium value of k_{MnSi}; the temperature effect being masked by the scatter in the experimental equilibrium values of k_{MnSi} (Figure 13—126). Of the data from five blast furnaces, the values of k_{MnSi} for South No. 10 furnace are closest to the equilibrium line. On the whole, the plant data may indicate that as the metal droplets pass through the slag layer, the direction of the reaction might be

$$2(MnO) + \underline{Si} \rightarrow 2\underline{Mn} + (SiO_2)$$

Such an intuitive deduction made from the observation of the plant data in Figure 13—134 with respect to the equilibrium relation may not be correct.

Most of the plant data analyzed indicate closer approach to equilibrium for the three-phase reaction (145). Presumably, approach to silicon equilibrium was

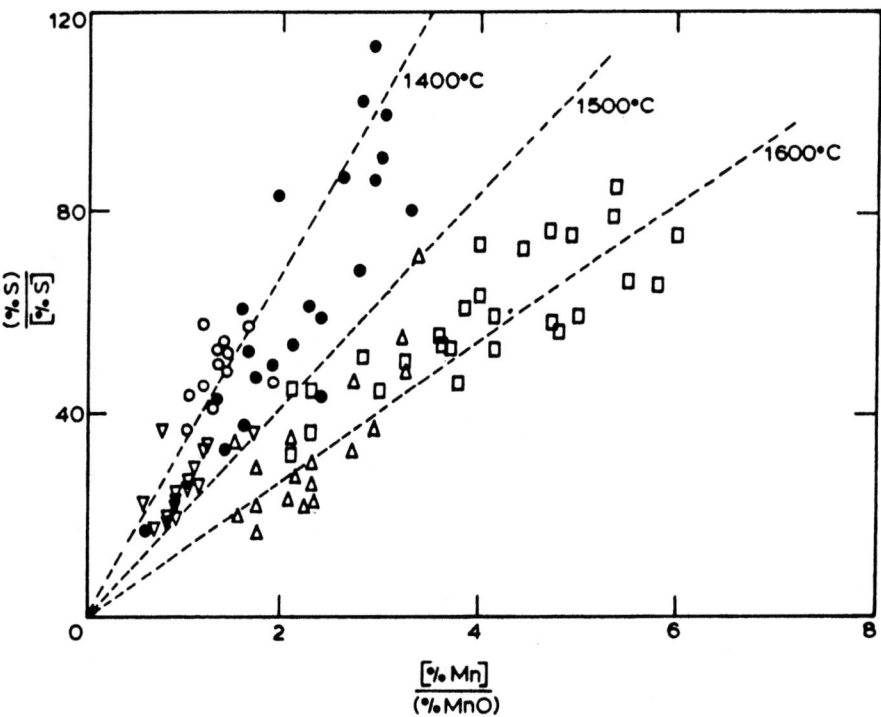

FIG. 13—133. Sulphur and manganese distribution ratios are compared with equilibrium relations for indicated temperatures and slag containing 40 per cent CaO. (Turkdogan[235]).

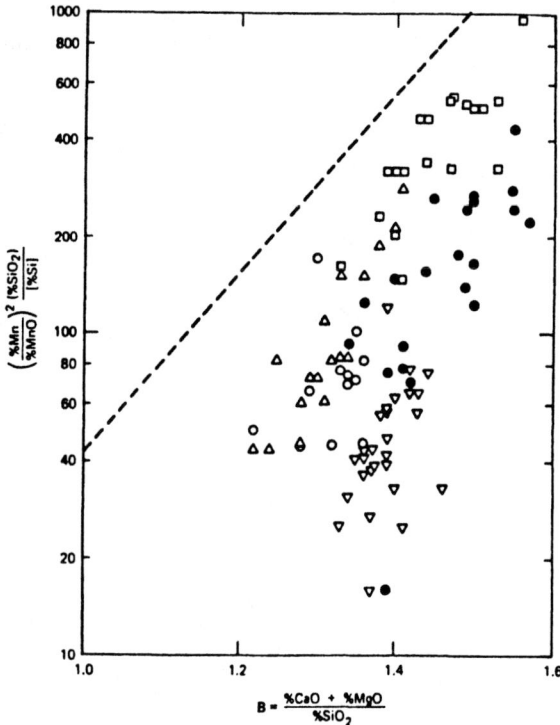

FIG. 13—134. The state of manganese-silicon reaction in various blast furnaces is compared with the equilibrium relation (– – – –).

by oxidation of silicon in iron droplets during their descent in the slag layer. Such an oxidation reaction is likely to occur more readily in the uper part of the slag layer where the concentration of iron oxide would be much higher. There is also the reduction of iron oxide by carbon (coke) in the upper part of the slag layer. Ultimate approach of the silicon reaction (144) to equilibrium precludes further oxidation of silicon by MnO in the slag, despite the chemical driving force that exists for reaction (147) in the slag layer. Observed departure from equilibrium for reaction (147) may be attributed to competition between FeO and MnO in the slag for oxidation of silicon. It is a generally accepted view that with a blast-furnace burden of low basicity, the residual iron oxide in the melting zone of the bosh is higher than with a burden of high basicity. In slags of low basicity, the higher concentration of iron oxide in the upper part of the slag layer is in competition with MnO for reaction with both silicon and carbon. Consequently, departure from equilibrium for reactions (143) and (147) is expected to be greater in slags of low basicity. The trends seen from the plant data in Figures 13—131 and 13—134 substantiate the deductions made from the foregoing argument.

Ferromanganese-Blast-Furnace Slags—Compositions of metal and slag from the ferromanganese blast furnace are within the following ranges in weight per cent: 32-46 CaO, 5-25 MnO, 12-14 Al_2O_3, 26-29 SiO_2, ~ 1.5 BaO, ~ 2 MgO, ~ 2 S, <0.3FeO. Metal composition in weight per cent: ~ 7 C, 0.4-3.2 Si, ~ 0.03 S, ~ 0.03 P, 76-80 Mn, 11-15 Fe. Using daily average compositions of slag and metal samples from a ferromanganese blast furnace, Turkdogan[187] found that the

manganese- and silicon-distribution ratios varied with slag basicity in a consistent manner. This is shown in Figure 13—135. The values of k_{MnSi} derived from these data increase with increasing slag basicity. For the entire basicity range, the values of k_{MnSi} from plant data are about one-third the equilibrium values. Observed departure from equilibrium for reaction (147) in the ferromanganese blast furnace is similar to that shown in Figure 13—134 for the iron blast furnace.

Alkali Recycle in the Blast Furnace—Problems in the operation of blast furnaces arising from the accumulation of alkalies in the stack as carbonates and cyanides have been known for over 150 years. Reactions responsible for the recycle of alkalies and their accumulation in the stack were outlined in earlier studies made by Kinney and Guernsey.[188] Thermodynamics of reactions describing alkali recycle and problems associated with alkalies in blast furnaces and methods of remedy are well documented in numerous publications, some of which are cited in the references.[189-192] The present comments are confined to the role of blast-furnace slag in the control of alkali buildup in the furnace.

The alkalies are present in the burden gangue and coke ash as complex silicates. Depending on the raw materials, the alkali input to the furnace varies over a wide range from 2 to 12 kg per metric ton of hot metal.

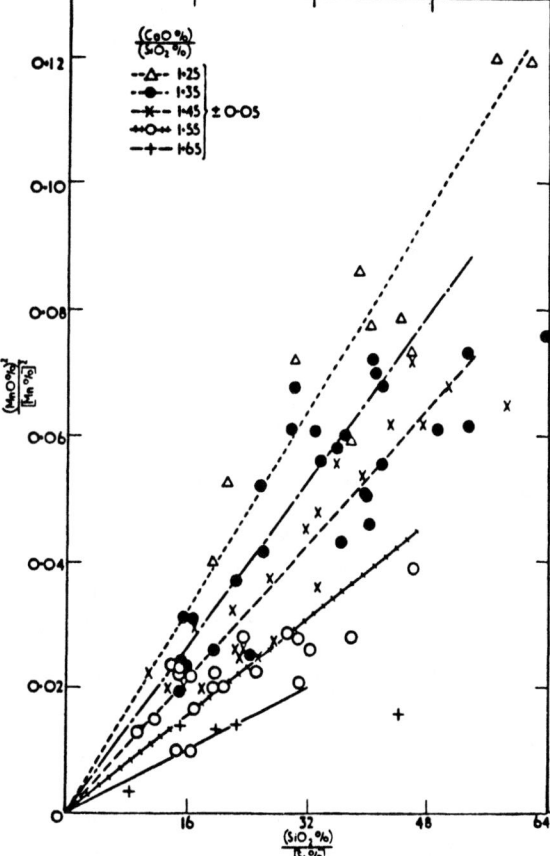

FIG. 13—135. Interrelation between manganese and silicon distribution ratios in the ferromanganese blast furnace. After Turkdogan.[187]

Because the silicates and salts of potassium are less stable than those of the sodium, under reducing conditions, the potassium is the main cause of the alkali problem. The potassium vapor is generated by the reduction of silicates in the ore and coke ash at elevated temperatures in the lower part of the bosh and the combustion zone, thus

$$(K_2SiO_3) + CO = 2K(g) + CO_2 + (SiO_2) \qquad (167)$$

The alkali vapors carried up with the ascending furnace gas react mostly with the slag and can therefore be removed from the furnace; the remainder are converted to alkali carbonates and cyanides and deposited on the burden in the stack. The following reactions are of particular importance:

$$2K(g) + 2CO_2 = K_2CO_3(s, l) + CO \qquad (168)$$

$$K(g) + 2CO + \tfrac{1}{2}N_2 = KCN(l) + CO_2 \qquad (169)$$

$$KCN(l) = \tfrac{1}{2}(KCN)_2(g) \qquad (170)$$

$$2K(g) + CO_2 = K_2O(s) + CO \qquad (171)$$

The equilibrium vapor pressures of potassium calculated for these reactions are plotted in Figure 13—136 as functions of temperature and for the corresponding p_{co}/p_{co_2} ratios$_2$ and total gas pressures normally encountered in high-top-pressure blast furnaces. The curve for potassium cyanide is for the sum of vapor species K and $(KCN)_2$; above 900°C the vapor pressure of liquid KCN exceeds the equilibrium vapor pressure of K for reaction (169).

In the lower part of the bosh where molten slag begins to form, most of the potassium vapor will dissolve in the slag by a reaction of the type

$$2K(g) + CO_2 = 2(K^+) + (O^{2-}) + CO. \qquad (172)$$

Because the activity coefficient of K_2O in the slag decreases with decreasing basicity, the potassium capacity of the slag increases as the slag basicity decreases. That is, in slags of lower basicity, the potassium solubility is higher and the potassium vapor pressure is lower as dictated by the state of equilibrium for reaction (172). Depending on the operating conditions and alkali input into the furnace, the total alkali content of the blast-furnace slag varies with basicity within the shaded area shown in Figure 13—137.

One method of minimizing the ascent of alkali vapors to the stack, hence reducing the alkali recycle, is by operating the furnace with a slag of low basicity, particularly when the alkali input is high. The slag of low basicity, however, will have an adverse effect on the composition of the hot metal produced; it will result in low manganese, high silicon and high sulphur in the metal. For a given preferred basicity, the slag mass per ton of hot metal may have to be increased when there is an increase in the alkali input to the furnace. An increase in amount of slag will increase the coke rate, hence will increase the alkali input. Obviously, with a burden of high alkali input, a compromise has to be made in adjusting slag basicity and slag mass so that

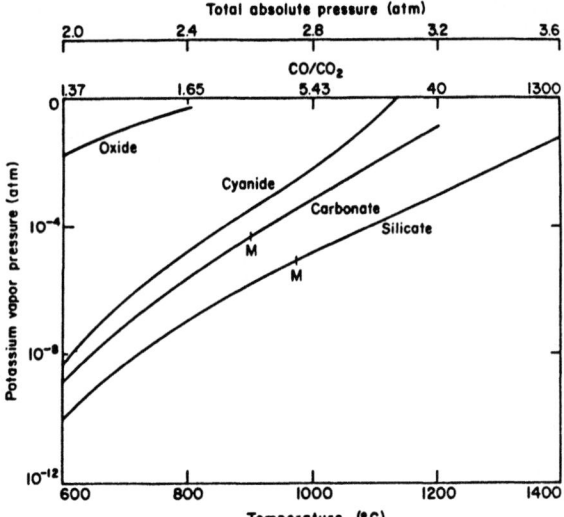

FIG. 13—136. Equilibrium vapor pressure of potassium for reactions (167)-(171) at indicated temperatures and corresponding CO/CO_2 ratios and total pressures in the blast-furnace bosh and hearth.

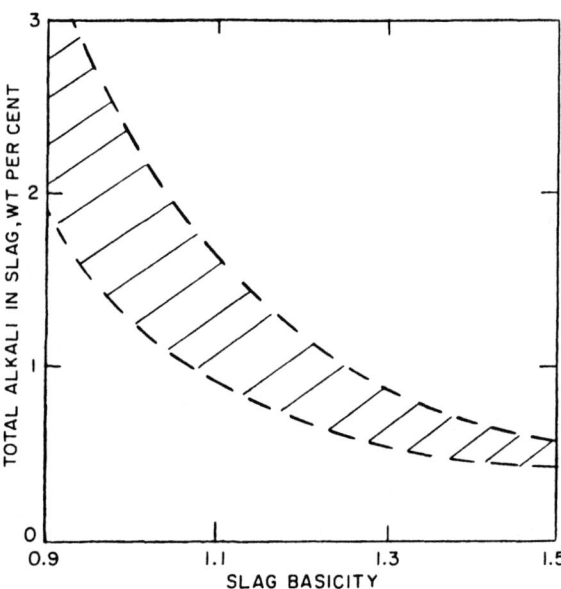

FIG. 13—137. Shaded area shows variations in the alkali content of slag with basicity from one blast furnace to another.

the alkali recycle would be low.

Even with a low alkali input some alkali buildup will occur. To prevent excessive accumulation of alkalies, which lead to scaffolding, gas channeling, furnace upsets and so on, periodic additions of gravel (silica) or olivine (magnesium silicate) are made to the blast-furnace charge. The silica reacts with alkali carbonates and cyanides, thus facilitates the discharge of the accumulated alkalies from the furnace by the slag. Practical experience also shows that periodic addition of calcium chloride to the burden also facilitates the removal of alkalies by the slag.

Other Reactions in Blast-Furnace Slags—Experience has shown that the iron ore containing 0.2 to 0.3 per cent TiO_2 often leads to operating difficulties in the blast furnace because of the precipitation of titanium carbonitrides in the furnace hearth. Strangely enough, some of the blast furnaces in China operate, apparently satisfactorily, with as much as 10 per cent TiO_2 in the slag. Because of lack of information made available on their operating conditions, the subject cannot be discussed here in a meaningful manner.

Little interest has been shown on the study of the equilibrium distribution of titanium between graphite-saturated iron and blast-furnace-type slags. However, Delve et al.[193] have studied by laboratory experiments the titanium-silicon reaction in graphite-saturated melts using blast-furnace-type slags. The equilibrium relation k_{TiSi} for reaction

$$(TiO_2) + \underline{Si} = (SiO_2) + \underline{Ti}$$

$$k_{TiSi} = \frac{[\%Ti]}{(\%TiO_2)} \frac{(\%SiO_2)}{[\%Si]}$$

(173)

increases with an increase in slag basicity. Their experimental results, for temperatures of 1500° and 1600°C and basicities $\%CaO/\%SiO_2$ in the range 1 to 2, may be summarized by the following equation

$$\log k_{TiSi} = 0.46 \left(\frac{\%CaO}{\%SiO_2} \right) + 0.39 \quad (174)$$

From the blast-furnace data compiled by Delve et al. and by Hess et al.[194] it is found that the values of k_{TiSi} are two or three times lower than those given by equation (174), which presumably represents the equilibrium relation. Observed departure from equilibrium for reaction (173) is indeed very similar to the state of silicon-manganese reaction (147) in iron and ferromanganese blast furnaces already discussed.

Many studies were made of the solubility of nitrogen and carbon in graphite-saturated silicate and aluminate melts, with as many conflicting views on the nature of the dissolution reactions. Much of the controversy was subsequently resolved by Schwerdtfeger and coworkers[195-197] through carefully conducted experiments by equilibrating graphite-saturated melts with known mixtures of nitrogen and carbon monoxide. They could interpret the solubility data in terms of three major ionic species in solution: nitride (N^{3-}), carbide (C^{2-}) and cyanide (CN^-) ions.

Schwerdtfeger et al.[197,198] also analysed some samples of blast-furnace slags for nitrogen and carbon. The total nitrogen content was found to be about 0.03 per cent. This solubility corresponds to an equilibrium gas composition of 5 per cent N_2 plus 95 per cent CO at 1450°C, which is much lower in N_2 and higher in CO than that in the blast-furnace bosh, indicating that the blast-furnace slag is far from equilibrium with regard to nitrogen dissolution. They also found that these slags contained about 0.001 per cent CN which is also 20 to 30 times below the equilibrium solubility for the prevailing gas compositions in the furnace.

Upon solidification, blast-furnace slags acquire a porous texture, resulting from gas evolution. One source of gas evolution is by the reaction in the slag

$$3(Fe^{2+}) + 2(N^{3-}) = N_2(g) + 3Fe. \quad (175)$$

Thermodynamic calculations have shown[198] that even at low concentrations of Fe^{2+} and N^{3-} in blast-furnace slags, an equilibrium pressure of several thousand atmospheres exists for reaction (175). Observed apparent low solubility of nitrogen in blast-furnace slags in comparison to the equilibrium value for calcium aluminosilicate melts may well be attributed to the presence of small amounts of iron oxide in the slag for which much higher nitrogen fugacity is needed for nitrogen dissolution.

DIRECT REDUCTION OF IRON OXIDES

During the past twenty-five years, extensive work has been carried out to develop alternatives to the blast furnace for reducing iron oxides. Most of these processes are based on direct reduction with gaseous or solid reductants. There have been over forty different processes for which extensive research, pilot plant or production plant work have been carried out. Some of these processes are described in other chapters of this book.

In this chapter only the basic aspects of the gaseous reduction of iron oxides will be discussed. For a more detailed discussion, the reader is referred to the book by von Bogdandy and Engell[175] and the review paper by Turkdogan[178].

Swelling and Pore Structure—At room temperature, the specific volumes of hematite, magnetite, wustite, and iron are, respectively, 0.190, 0.193, 0.175 and 0.127 cm^3/g. Therefore, volume expansion in the early stages of reduction followed by some shrinkage in the final stages of reduction is anticipated. However, because of the increase in pore volume and limited sintering of the reduced oxide, the extent of swelling or shrinkage varies from one type of iron-oxide pellet to another. This is manifested by variation in the experimental observations of different investigators.[199-202] Swelling is much greater for pellets reduced in CO-CO_2 than in H_2; the presence of silica generally reduces swelling, while the presence of alkali increases swelling.[203] Several investigators have attempted[204-206] to explain swelling but at the present time it is not completely understood.

The reduction of iron oxides is greatly affected by the pore structure of the oxide and of the reaction product. The structure of the pores of the reduced iron is affected by the type of oxide and the reducing conditions. For example, the pore structure obtained by hydrogen reduction is finer than that by CO reduction as shown in Figure 13—138. Also, as shown in Figure 13—139, the pore structure becomes increasingly coarser with increasing reduction temperature from 600° to 1200°C.

In addition to microscopic examination, the characteristics of the pores should be qualified by measurements of pore volume, pore surface area and effective gas diffusivity. The effect of reduction temperature and gas composition on pore surface area is shown in Figure 13—140. Again it can be seen that the pore structure is coarser for CO reduction and with increasing temperature. Also as seen in Figure 13—141, for a given type of

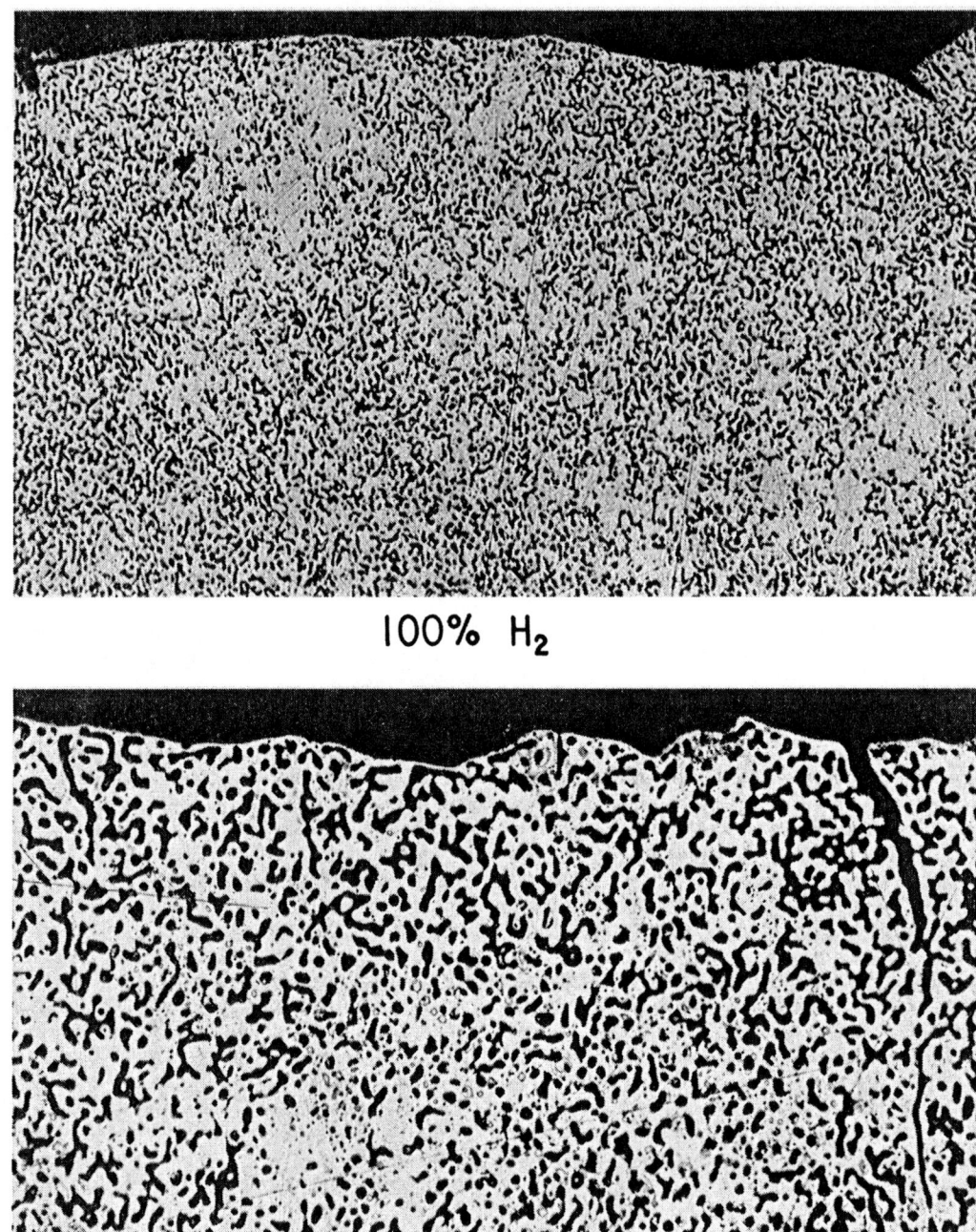

FIG. 13—138. Polished sections showing network of pores in the iron formed by H_2 or CO reduction of dense wustite at 1200°C. (Turkdogan and Vinters[207]).

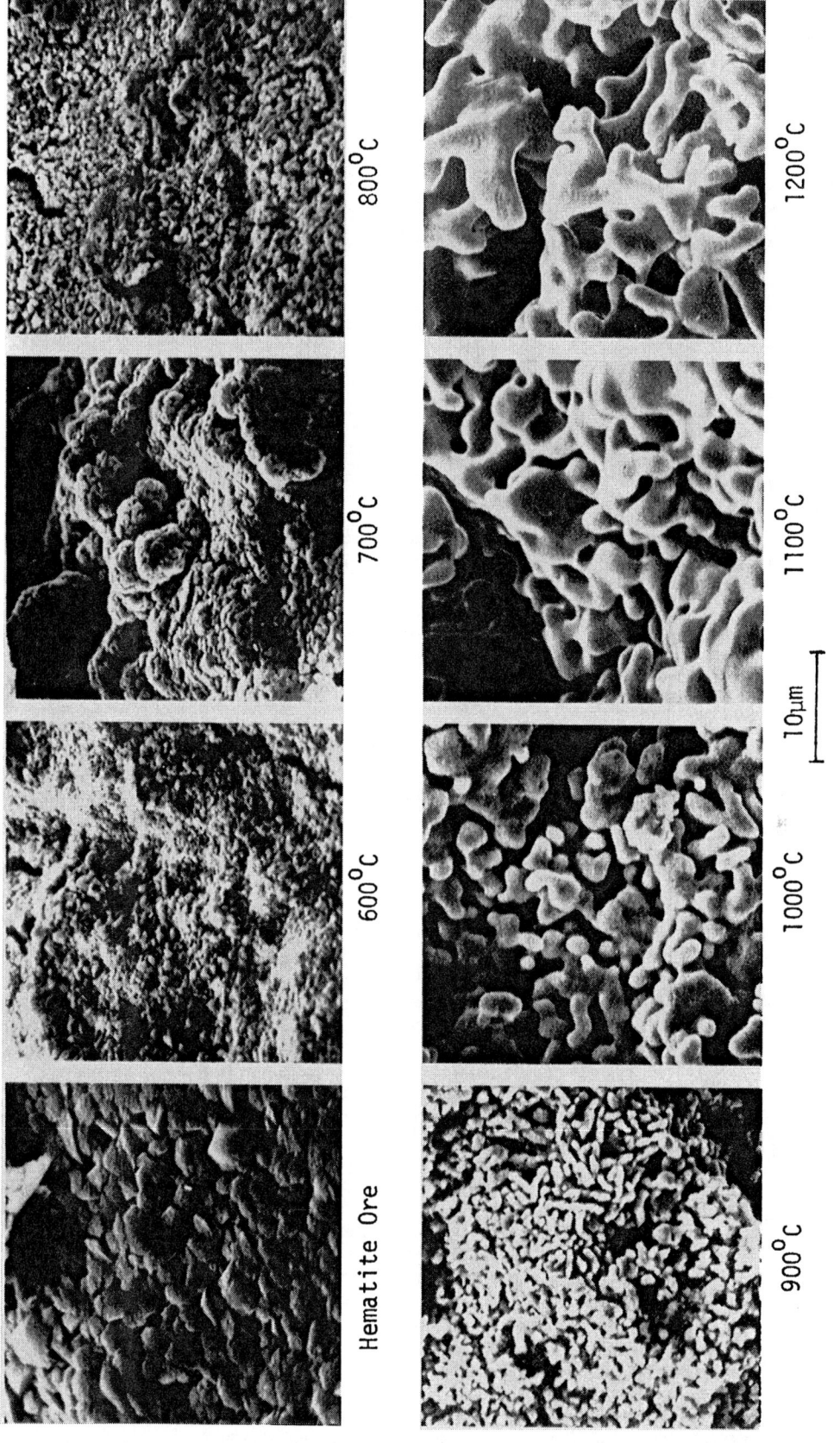

Fɪɢ. 13—139. Fracture surface of lump hematite ore and porous iron reduced in hydrogen at indicated temperatures as viewed in the scanning electron microscope. (Turkdogan and Vinters[179]).

iron oxide, such as natural hematite ore or sintered hematite pellets, the pore surface area of iron decreases with decreasing initial pore surface area of the oxide.

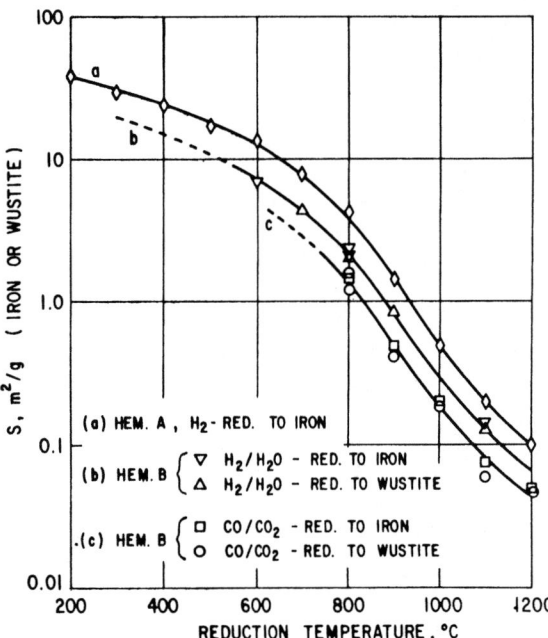

FIG. 13—140. Internal pore surface area of iron and wustite formed by reduction of hematite ores A and B as a function of reduction temperature. (Turkdogan and Vinters[207]).

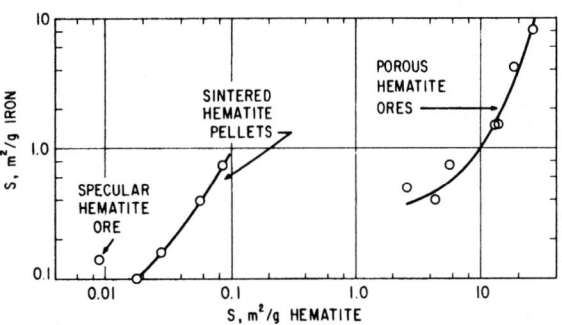

FIG. 13—141. Relations between the internal pore surface area of H_2-reduced iron at 800°C and the pore surface area of the corresponding hematite ore or sintered pellets. (Turkdogan and Vinters[207]).

The average effective diffusivity, D_e, for binary gas mixtures is given approximately by

$$\frac{1}{D_e} = \frac{1}{(D_{12})_e} + \frac{1}{(\overline{D}_K)_e} \qquad (176)$$

where $(D_{12})_e$ is the effective molecular diffusivity and $(\overline{D}_K)_e$ is the effective average Knudsen diffusivity for the average molecular weight of the gas mixture. Previous studies[179] suggest that the pore structure of reduced iron approximately conforms to the limiting ideal structure. That is, the porous material appears to have pores of uniform size that are all interconnected and intersect each other with an angle of 45 degrees,

for which the effective diffusivities are given by

$$(D_{12})_e = \frac{\epsilon}{2} D_{12} \qquad (177)$$

$$(\overline{D}_K)_e = \frac{\epsilon}{2} \overline{D}_K \qquad (178)$$

For a given porous medium, the effective diffusivity varies with temperature and pressure and is different for different binary gas pairs, that is for different values of D_{12}. However, the diffusivity ratio D_e/D_{12} is a property of a given porous material and characterizes the pore structure. Typical values of D_e/D_{12} determined from diffusivity measurements for iron formed by reduction of hematite ore in hydrogen are given in Table 13—XVII.

Table 13—XVII. Typical Values of D_e/D_{12}.
(Turkdogan et al.[179])

Reduction temperature, °C	D_e/D_{12}
600	0.020
800	0.075
1000	0.205

Decrease in the value of D_e/D_{12} with decreasing reduction temperature is consistent with pore structures observed under the microscope, and with the values of pore surface area. That is, the pore structure becomes finer with decreasing reduction temperature.

Modes and Rate of Reduction of Iron Oxides—As discussed previously, the formation of product layers in reduction of iron oxide (Figure 13—123) plays an important role in the rate and mode of reduction. When there is no layer formation, Figure 13—123b, there is sufficient gas diffusion in the wustite layer and there is internal reduction of the pellet. The relative depth of internal reduction increases with decreasing temperature, increasing porosity, and the decreasing particle size. For example, for a 1-mm diameter porous hematite, the reduction is almost completely internal and uniform.

As discussed by Turkdogan and Vinters,[179] the effect of particle size on the time to achieve a given percentage of reduction depends on the mode of reduction and hence on the type of rate-controlling process. For the modes of reduction considered, there are three limiting rate processes pertinent to the gaseous reduction of porous metal oxides: uniform internal reduction, limiting mixed control, and diffusion in the porous metal layer. As shown in the following sections, if the reduction were controlled solely by one of these, then the time, t, of reduction would be related to the particle (spheroidal) diameter, d, in one of the following ways:

Uniform internal reduction: t is independent of d

Limiting mixed control: $t \propto d$

Diffusion in porous iron: $t \propto d^2$

The t-d relations for these individual limiting rate processes are shown schematically by dotted curves in Figure 13—142. The curve drawn corresponds to the experimental observation. The rate-controlling process becomes relatively simple only when (1) the particle size is small, hence there is uniform internal reduction, or (2) the particle size is large, so that the ultimate rate control by gas diffusion in the pores of the iron layer predominates. With medium-sized particles there is mixed control and the interpretation of the rate data becomes complex. It should also be realized that depending on temperature, gas composition, particle size, and type of oxide, there may be transition from one limiting rate-controlling process to another as the reduction progresses.

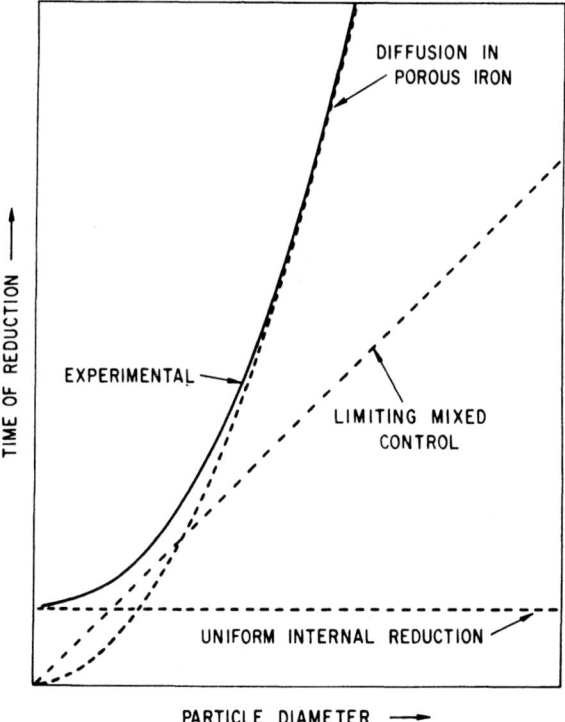

FIG. 13—142. Schematic representation of time of reduction (for a given per cent oxygen removal) as a function of particle size. (Turkdogan and Vinters[178]).

Rate of Reduction of Porous Ore Granules—Turkdogan and Vinters[207] investigated the rate of reduction of Venezuelan ore granules in detail. Typical examples of their rate data are shown in Figure 13—143, where F is the fraction of oxygen removed. At temperatures above 900°C and below 500°C, log $(1-F)$ is a linear function of time up to about 96 to 98 per cent oxygen removal. For intermediate temperatures the plots are essentially linear up to about 80 per cent oxygen removal, beyond which reduction becomes sluggish, as shown in Figure 13—143. The data in Figure 13—144 are for reduction in a 90 per cent CO—10 per cent CO_2 mixture at various temperatures. Unlike the case for H_2 reduction, there was no unusual temperature effect on reduction in CO at 800°C and above.

In the early stages of reduction of porous hematite granules, there is rapid conversion to wustite followed by internal reduction of wustite to iron. In the limiting case of almost complete gas diffusion in the pores of the oxide, the internal reduction predominates and the rate is controlled primarily by a gas-solid reaction on the pore walls. The pore walls of wustite are assumed to

be covered with a layer of iron a few atoms thick. The reduction rate is presumed to be controlled by the reaction of H_2 (or CO) with oxygen on the surface of this very thin iron layer, through which oxygen diffusion is rapid.

For uniform internal reduction, the specific rate is represented by

$$\frac{dw}{dt} = - wS\phi_i'(1 - \Theta)\,[p_i - (p_i)_e]\qquad(179)$$

where w is the amount of oxygen in the sample at time t, S is the pore surface area of wustite per unit mass of oxygen, p_i is the partial pressure of H_2 or CO in the gas, and $(p_i)_e$ is the corresponding equilibrium value for coexistence with iron and wustite phases. The integrated form of the rate equation is

$$\ln(1-F) = - S\phi_i'(1 - \Theta)\,[p_i - (p_i)_e]\,t\qquad(180)$$

Because of rapid gas diffusion in small porous particles, $(p_i)_e \to 0$ for reduction in H_2 or CO with no H_2O or CO_2 in the gas stream.

The rate should be independent of particle size as shown in Figure 13—145 and increase with internal surface area as shown in Figure 13—146.

When a gas mixture of $CO\text{-}CO_2\text{-}H_2\text{-}H_2O$ is used for reduction, as would be the case in several direct-reduction processes, two parallel reducing reactions are occurring. An equation similar to (180) is used except that there will be two terms; one for H_2 and one for CO.

Rate of Reduction of Lump Ore or Sintered Oxide Pellets—The rate of reduction of lump ore or sintered pellets in a stream of reducing gas or in a packed bed is complex, because there are several reaction processes in series which influence the overall rate of the reduction: for example, heat and mass transfer through the gas-film boundary layer, gas-solid reactions, and gas diffusion in porous product layers. Through mathematical analyses, facilitated by computer calculations, numerous equations have been derived to describe the rate of reduction of large oxide particles for various modes of reduction.[175,208-211]

In most experiments with single pellets or ore particles, heat transfer is relatively fast, and with sufficiently high-velocity gas streams, the gas-film mass-transfer resistance is small enough to be neglected. Therefore, there are primarily two major reaction steps in series which influence the rate of reduction: gas-oxide reactions, and gas diffusion in porous oxide and porous product layers. The relative effects of these rate processes depend on the particle size, gas composition, temperature, and mode of reduction, and they change with the progress of reduction. For the majority of the reaction time the rate is controlled by gas diffusion in the iron layer. Therefore, in this section only this rate limiting case is considered.

Gas Diffusion in the Porous Iron Layer—In the reduction of spheroidal oxide particles or pellets with a porous outer iron shell surrounding the unreacted oxide core, the following limiting rate equation is obtained for pore diffusion control. For the spherical ge-

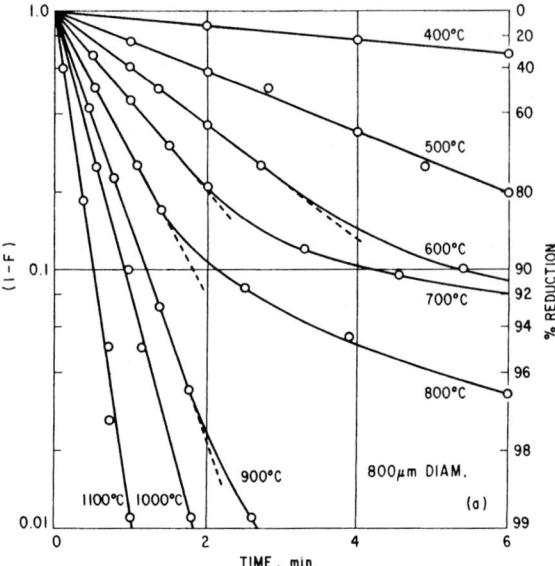

FIG. 13—143. Reduction of hematite (B) granules in H_2 at atmospheric pressure and indicated temperatures. (Turkdogan and Vinters[207]).

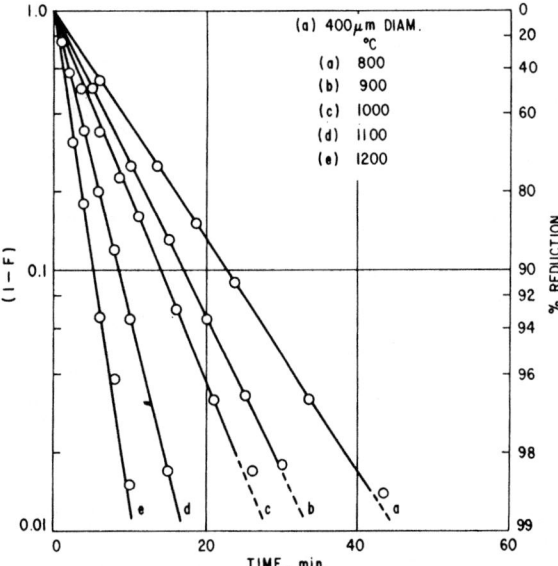

FIG. 13—144. Reduction of hematite (B) granules in a 90 per cent CO + 10 per cent CO_2 mixture at atmospheric pressure and indicated temperatures. (Turkdogan and Vinters[207]).

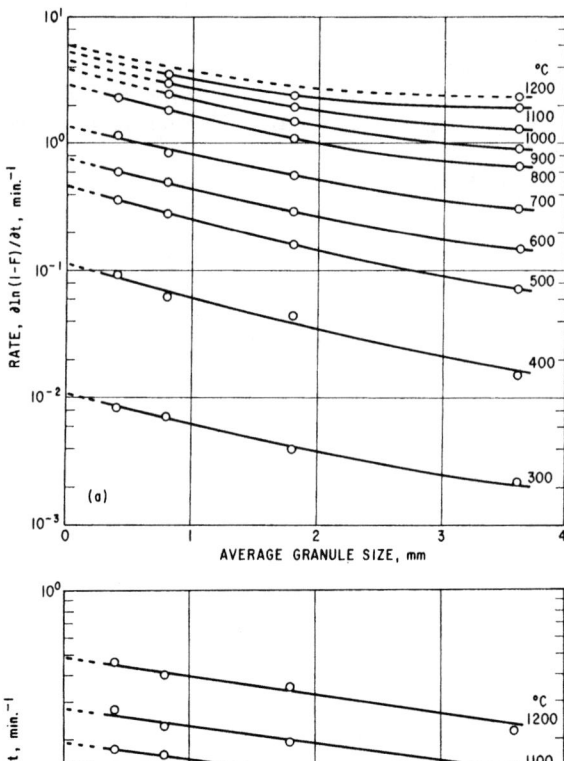

FIG. 13—145. Effect of granule size on the rate of internal reduction of hematite (B) granules in (a) 100 per cent H_2 and (b) 90 per cent CO + 10 per cent CO_2 at atmospheric pressure. (Turkdogan and Vinters[207]).

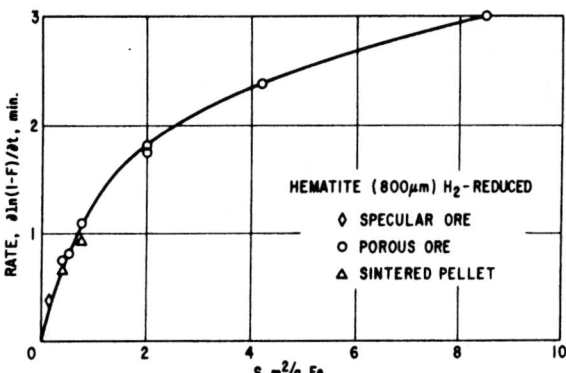

FIG. 13—146. Rate of reduction, at 800°C and atmospheric pressure of hydrogen, of different types of hematite granules (800 μm) as a function of pore surface area of iron (or wustite) formed. (Turkdogan and Vinters[207]).

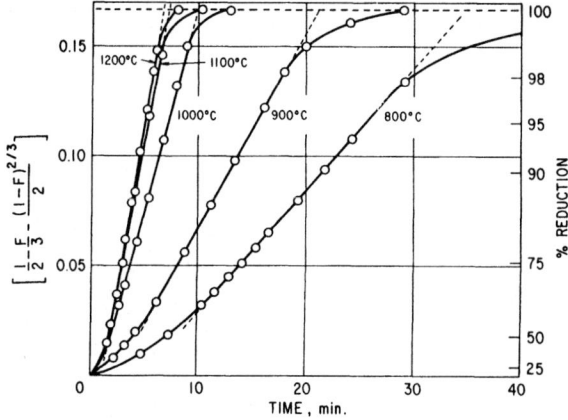

FIG. 13—147. Diffusion plot of reduction data for 15-mm diameter hematite ore at 1 atm H_2. (Turkdogan and Vinters[178]).

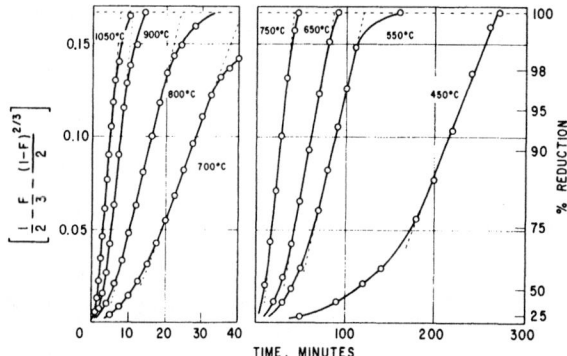

FIG. 13—148. Data for sintered synthetic hematite pellets (8 to 11 mm diameter) at 1 atm H_2. (Turkdogan and Vinters[178]).

ometry, the equimolar diffusion at constant total pressure in the reduced iron shell at any time t is given by

$$J = \frac{4\pi D_e}{RT} \cdot \frac{p_o - p_i}{\frac{1}{r'} - \frac{1}{r}} \qquad (181)$$

where r' is the radial distance of the metal/metal oxide interface from the center of the sphere, p_o is the hydrogen partial pressure on the surface of the sphere (the same as that in the gas stream for fast flow rates), p_i is that for the metal/metal oxide equilibrium. From the spherical geometry, it can be shown that

$$\left[\frac{1}{2} - \frac{F}{3} - \frac{(1-F)^{2/3}}{2} \right] = \frac{D_e}{\rho r^2} \cdot \frac{\Delta p_i}{RT} + C \qquad (182)$$

where r is the initial radius of the spheroidal pellet, assumed to undergo little or no dimensional change during reduction, and C is a constant (a negative number) and takes account of all early time departures from the assumed boundary conditions. The value of C depends on particle size, gas composition, and temperature. Only in the limiting case of very fast interfacial reactions does C approach zero.

Typical experimental data for the hydrogen reduction of spheroids of natural hematite are plotted in Figures 13—147 and 13—148 in accordance with equation (182). The middle parts of the elongated S-shape curves from 50 or 60 per cent to 95 or 99 per cent reduction are well represented by straight lines. As discussed in detail by Turkdogan,[171] deviation from a parabolic relationship, as given by equation (182), in the early and final stages of reduction is accountable and the slopes of the lines can be related to the diffusivity in the iron layer.

It can be concluded that the mode and rate of reduction of lump ores or sintered pellets in hydrogen-carbon monoxide mixtures depend on the pore structure of the iron oxide and reduced iron. The gas diffusion in the pores of the iron layer plays an important role in controlling the rate of reduction of oxide pellets. The higher the reduction temperature the coarser is the pore structure. This change in pore structure has decisive influence on the rate of reduction. The rate of reduction of wustite in hydrogen is about an order of magnitude greater than in carbon monoxide.

PART IV

Steelmaking Reactions

Primarily, five types of processes are currently employed in the production of low-alloy steels: (1) oxygen top blowing, known as LD, BOF, or BOP; (2) oxygen and lime bottom blowing, known as OBM or Q-BOP; (3) top and bottom mixed blowing; (4) open-hearth furnace; and (5) electric-arc furnace. The AOD (argon-oxygen decarburization) is a bottom-blowing process used primarily for stainless steelmaking. Even in open-hearth and electric-furnace steelmaking much of the refining is done with oxygen blowing. Although the steelmaking methods and practices differ to varying degrees in these processes, the same fundamental principles of slag-metal equilibria apply to them all.

In open-hearth and oxygen-steelmaking processes the metallic charge to the furnace consists of hot metal from the blast furnace and steel scrap. In most electric-furnace steelmaking, the steel scrap is used entirely as the metallic charge material. In all cases, the impurities in the metal, such as carbon, silicon, manganese, and phosphorus, as well as some iron are oxidized with oxygen blowing. A molten slag is formed by reaction of the oxides of iron, silicon, manganese, and phosphorus with the calcium oxide (burnt lime) that is added to the furnace together with the metallic charges. Although the oxidizing conditions in the furnace do not favor efficient desulphurization of the steel, about one-third or half of the sulphur in the metal is transferred to the slag during the course of oxidation of other impurities.

Because the oxidation of impurities is exothermic, the bath temperature increases with the progress of metal refining and brings about the melting of steel scrap in the charge and the formation of molten slag. Another important feature of the self sufficiency of the process is the evolution of carbon monoxide during decarburization of the metal, resulting in vigorous mixing of slag and metal phases, hence increasing the rate of refining and melting of scrap.

To facilitate a quantitative presentation of the compiled data on slag-metal equilibria, each reaction will be considered separately, although during steelmaking most of these reactions occur together. First, we shall discuss some specific features of the composition of steelmaking slags, as derived from laboratory equilibrium measurements.

Composition of Basic Slags—In low-phosphorus practices, the steelmaking slag consists primarily of CaO, MgO, SiO_2, and 'FeO', where '' indicate total iron as oxides in the slag converted to the stoichiometric composition FeO. The total of these oxides in the slag is about 87 to 90 per cent, the remainder being essentially MnO, P_2O_5, Al_2O_3, and S. Basic slags used in most steelmaking processes are saturated with CaO and often also with MgO, so that there would be minimum erosion of the furnace refractory lining, which is made of either calcined dolomite $(CaO \cdot MgO)$ or magnesia (MgO).

The phase equilibria and the liquid range in the $CaO-SiO_2-'FeO'$ system at 1600°C are shown in Figure 13—149. The composition of steelmaking slags at the

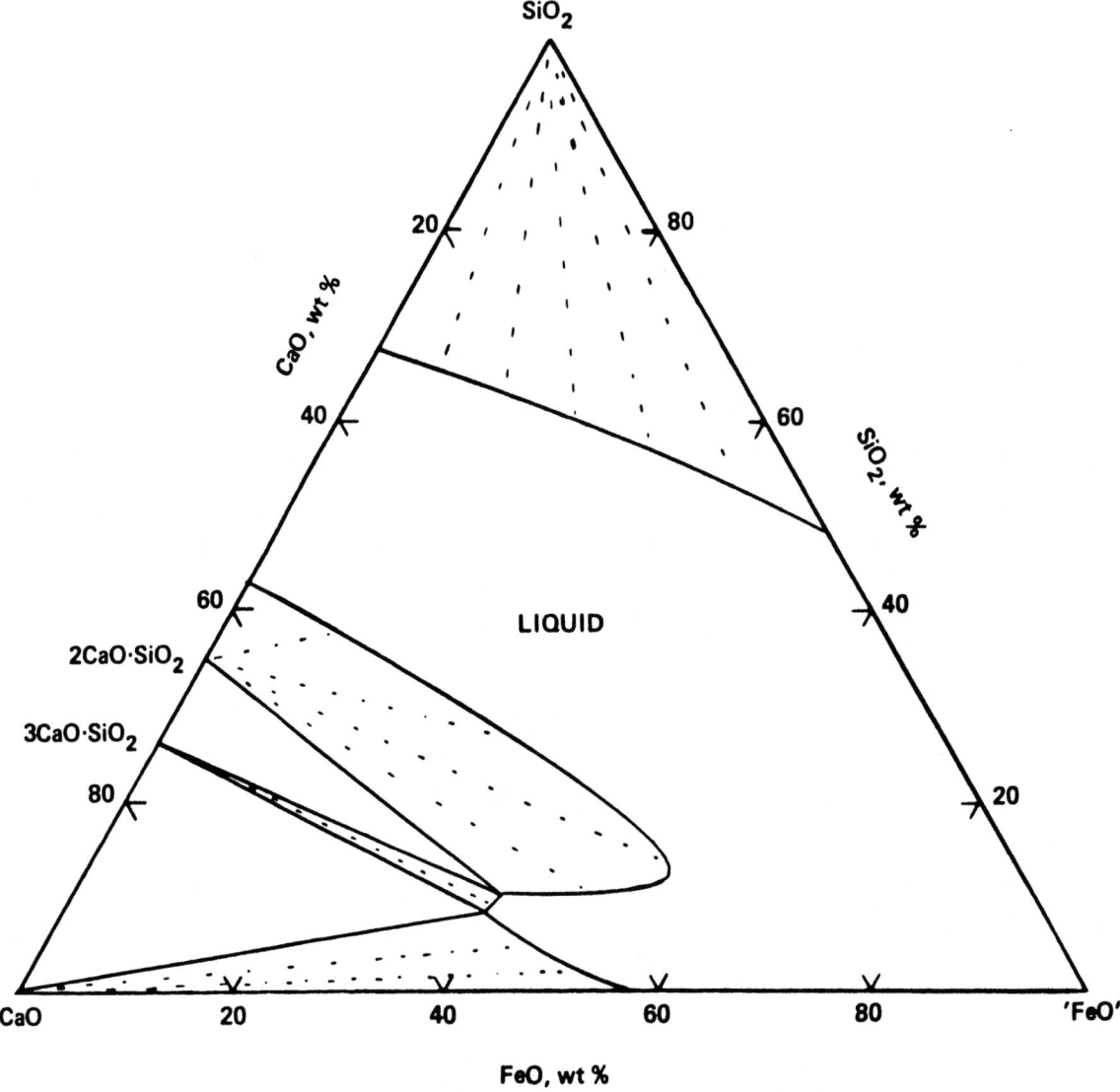

FIG. 13—149. Phase equilibria in $CaO-SiO_2-'FeO'$ ternary system at 1600°C.

time of tapping is in the region of dicalcium silicate saturation, which limits the solubility of CaO in the slag. The plant and laboratory investigations on the rate of lime dissolution in slags have been reviewed by Nilles et al.[212] for the period up to 1966. References to subsequent studies are given in a paper by Natalie and Evans,[213] together with their experimental work on the rate of dissolution of lime in stirred CaO-SiO$_2$-'FeO' slags. Consensus is that the rate of dissolution in slags is in accord with the relative reactivity of burnt lime as determined by the ASTM water-slaking test. That is, the soft burnt lime with large porosity dissolves faster in molten slags, partly because the soft burnt lime tends to crumble more readily in molten slags.

The solubility of MgO in CaO-MgO-SiO$_2$-'FeO' slags, studied to a limited extent initially by Fetters and Chipman,[214] was investigated in more detail subsequently by Tromel et al.[215] The effect of MgO on the liquidus isotherms of calcium silicates and calcium oxide at 1600°C is shown in Figure 13—150. If a slag saturation is reached at 35 per cent CaO in the absence of MgO, with the addition of 8 per cent MgO, the solubility of lime increases to 40 per cent CaO. The dot-dash curve delineates the region of double saturation of the slag with solid calcium (magnesium) silicates and solid magnesio-wustite. The curves for iso-MgO concentrations in Figure 13—151 delineate the ranges of compositions of slags saturated with magnesio-wustite or olivine (calcium, magnesium, iron silicates) at 1600°C. To facilitate the use of these data, the solubility of MgO at magnesio-wustite saturation is shown in Figure 13—152 as a function of slag basicity %CaO/%SiO$_2$

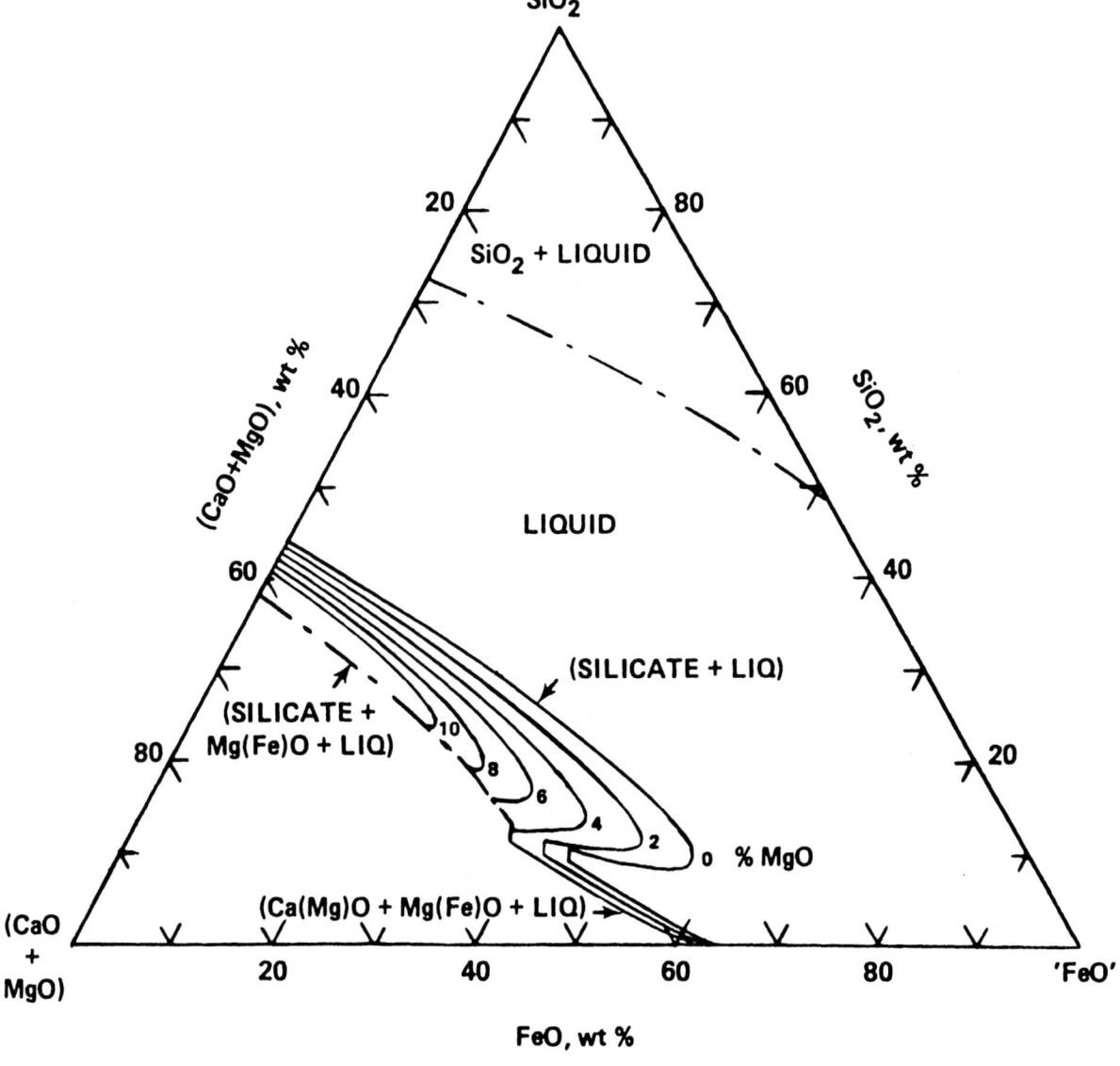

Fig. 13—150. Projection of liquidus isotherms (1600°C) of dicalcium silicate in CaO-MgO-SiO$_2$-'FeO' quaternary system on the composition diagram of pseudoternary system (CaO+MgO)-SiO$_2$-'FeO'. After Tromel, et al.[215]

for various concentrations of iron oxide in the range 10 to 60 per cent. The dotted curves delineate double saturation of the slag with two solid phases. Leonard and Herron[216] found that the addition of CaF_2 up to 10 per cent has no perceptible effect on the solubility of MgO in basic slags.

The phase boundaries in the composition diagrams in Figures 13—149 through 13—152 are for slags in equilibrium with liquid iron at 1600°C. At higher oxygen activities, that is, higher Fe^{3+}/Fe^{2+} ratios, the solubility isotherms shift to higher concentrations of CaO. Perhaps it is for this reason that a faster lime dissolution in the slag (rapid slag formation) is achieved by oxidizing the slag with a high-oxygen-lance practice in BOF steelmaking. Faster rise in slag temperature in the high-oxygen-lance practice further aids the rate of

dissolution of lime.

For low phosphorus in the charge, i.e., less than 0.2 per cent P in the hot metal, the compositions of slags at the time of furnace tapping are in the following ranges: 45 to 65 per cent CaO, 2 to 8 per cent MgO, 3 to 8 percent MnO, 4 to 30 per cent 'FeO', 10 to 25 per cent SiO_2, 1 to 5 per cent P_2O_5, 1 to 2 percent Al_2O_3, 1 to 2 per cent CaF_2, 0.1 to 0.3 per cent S and minor amounts of other oxides. In BOF (or BOP) steelmaking, the slags at tap contain 45 to 55 per cent CaO and 10 to 30 per cent 'FeO'; Q-BOP slags usually contain 55 to 65 per cent CaO and 4 to 22 per cent 'FeO', depending on the carbon content of the steel at tap.

The amount of burnt lime charged to the furnace in steelmaking is often in excess of that which is soluble in the slag, even at the end of refining. Because of the

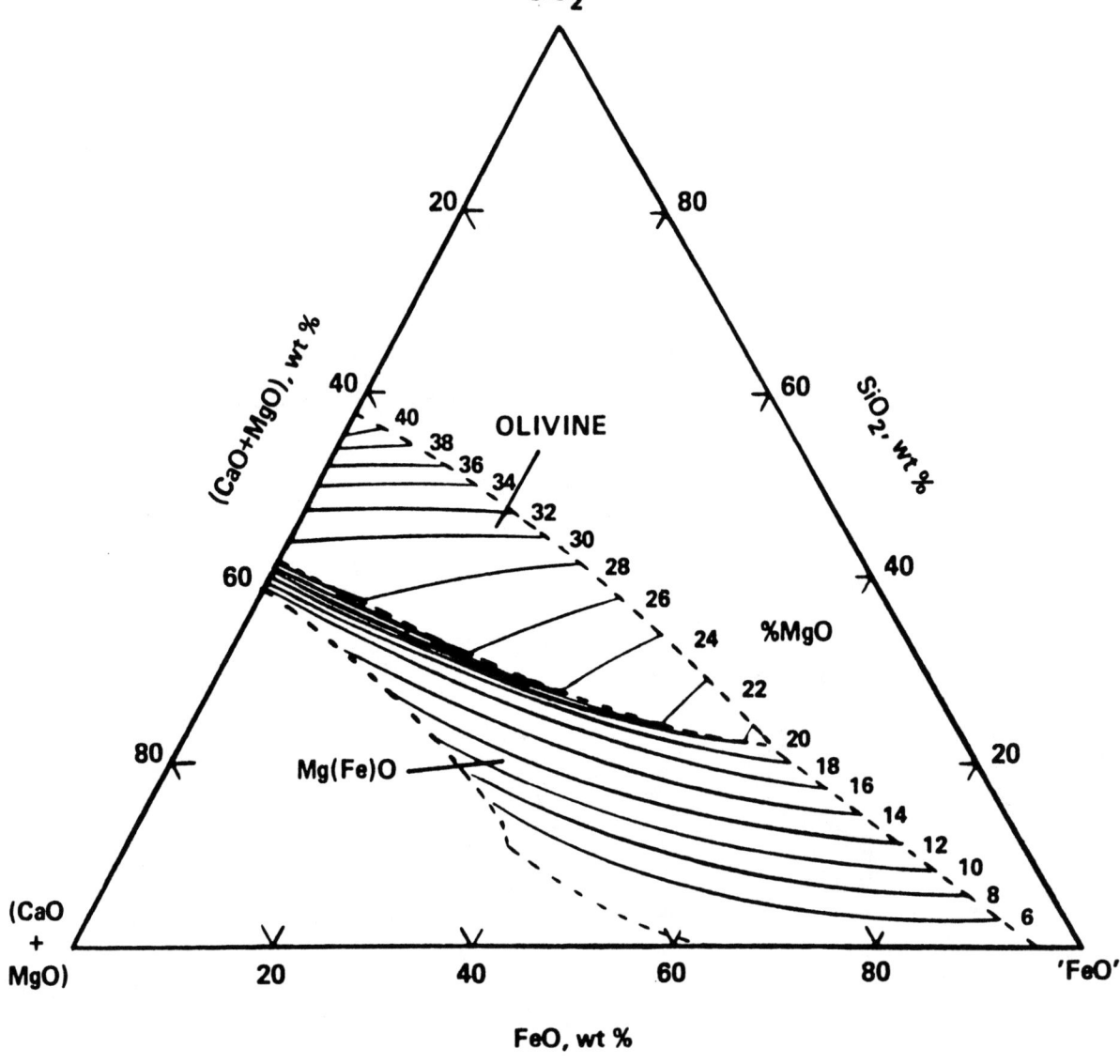

FIG. 13—151. Iso-MgO concentrations for CaO-MgO-SiO$_2$-'FeO' slags saturated with olivine and magnesio-wustite in equilibrium with liquid iron at 1600°C. After Tromel, et al.[215]

presence of undissolved lime, the slag is invariably heterogeneous. In an attempt to sample the liquid part of the slag, the slag is chilled on a steel rod immersed in the bath. However, tests have shown that the slag samples contain metallic iron shots and finely dispersed undissolved lime particles for which due corrections must be made in reporting the slag analysis. That is, the total iron oxide reported as 'FeO' is that for iron in solution as oxides in the slag. Also, the slag should be analyzed for free lime to be deducted from the total lime to give per cent CaO in solution; the concentration of free lime is usually in the range 2 to 6 per cent CaO. The slag composition often reported gives only the total lime concentration, including that part associated with CaF_2 in the slag.

In some European steelmaking practices with a high-phosphorus charge, i.e., ~ 1 per cent P in the hot metal, the slag may contain 20 per cent P_2O_5 or more, but much less SiO_2. With a high phosphorus charge usually a two-slag practice is used, known as LD/AC process. In the first stage of oxygen top blowing all the silicon, most of the phosphorus and about 70 per cent of the carbon are oxidized forming a phosphorus-rich foaming slag. After the removal of this slag, the oxygen blow is resumed with the injection of lime powder. The slag at tap has a composition similar to the single-slag practice for low-phosphorus charge. Incidentally, the phosphorus-rich slag formed in the first stage of the LD/AC process has a high citric-acid solubility, therefore is a saleable by-product as a farm-land fertilizer.

Basic open-hearth steelmaking is also a double slag process. The early slag that forms during the melt down of the scrap, followed with the hot-metal charge, is essentially an iron-manganese silicate containing about 10 per cent CaO, 5 per cent MgO and 2 to 3 per cent P_2O_5. This slag readily foams during decarburization of the melt by the iron-oxide-rich slag. Following the removal of this early slag, known as slag flush, the required amount of limestone is charged and oxygen is blown through the roof lances until the melt is decarburized to the desired level, accompanied with adequate dephosphorization and desulphurization of the melt.

Basic slags are saturated with dicalcium (magnesium) silicate and, for the low-phosphorus practice, the sum of the concentrations of CaO, 'FeO' and SiO_2 is in the

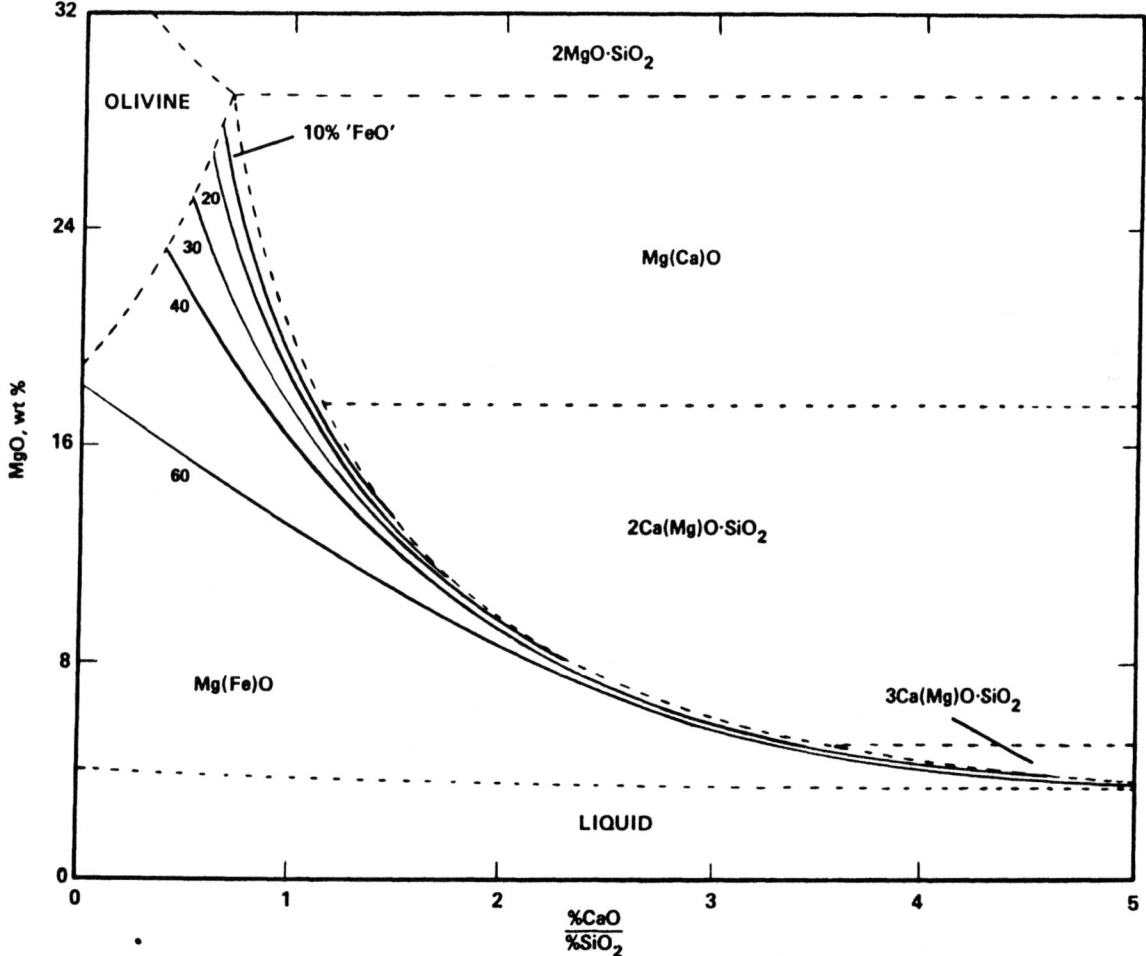

FIG. 13—152. Solubility of MgO, as magnesio-wustite, in CaO-MgO-SiO_2-'FeO' system at 1600°C as a function of slag basicity and FeO concentration, derived from the data in Figure 13—150.

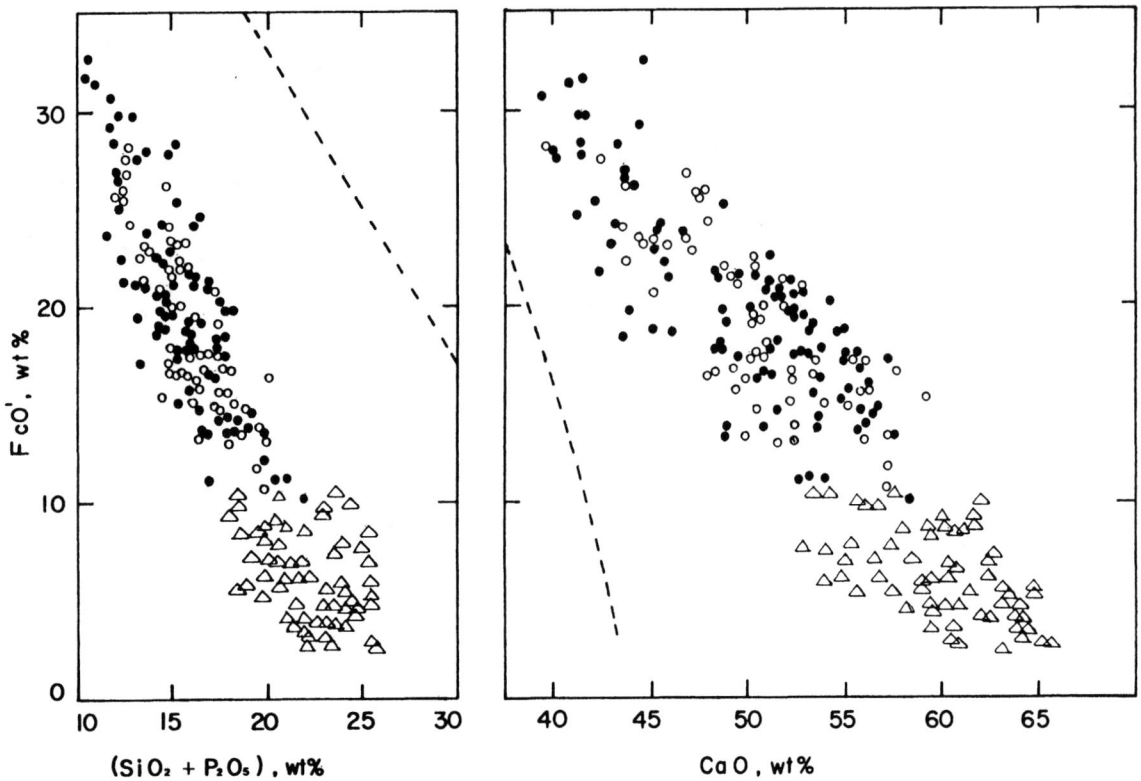

FIG. 13—153. Compositions of BOP (o, •) and Q-BOP (△) slags; (o) ≤ 1600°C, (•) > 1600°C.

range 87 to 90 per cent. In this pseudo-ternary system, the composition of the melt is that given by the univariant equilibrium for dicalcium silicate saturation at a constant temperature. This is shown in Figure 13—153, using United States Steel plant data for Gary BOP and Gary Q-BOP slags. For the sake of clarity, the data for Q-BOP slags used in the plot are only for melts containing more than 0.2 per cent C at tap; the data points for lower carbons, i.e., higher iron-oxide contents, are within the scatter band of the BOP slags. These slags contain 4 to 8 per cent MgO. The dotted line in each diagram is for dicalcium (magnesium) silicate saturation in CaO-6% MgO-'FeO'-SiO$_2$ melts in equilibrium with liquid iron at 1600°C, read off the isotherm in Figure 13—150 with readjustment of the composition to a total of 87 per cent.

The temperature effect on the compositions of saturated slags is masked by scatter in the data. One of the reasons for the scatter is that the reported analysis gives the total concentrations of CaO and SiO$_2$ which include those due to entrapped particles of lime and dicalcium silicate. It should be noted that the difference %CaO total — %CaO dissolved is much greater than the difference %SiO$_2$ total — %SiO$_2$ dissolved. Despite these minor uncertainties about the reported compositions of steelmaking slags, there are certain characteristic features to be noted. An important feature is that there is a pronounced shift in the position of the univariant equilibrium in the composition diagram for saturated steelmaking slags to higher concentrations of CaO and to the corresponding lower concen-

trations of SiO$_2$ as compared to the quaternary system CaO-6% MgO-'FeO'-SiO$_2$ in equilibrium with liquid iron. This change in the melt composition to higher concentrations of lime is due to the presence of P$_2$O$_5$, MnO, Fe$_2$O$_3$, CaF$_2$ and small amounts of other oxides which shift the saturation isotherm to higher concentrations of CaO. Another rationale is that the slag basicity changes from low to high levels as the iron oxide content of the saturated slag increases, e.g., the V ratio is 2.0 to 2.5 for slags containing 4 to 6 per cent 'FeO' and increases to 3.4 to 3.6 for slags containing 28 to 38 per cent 'FeO'. That is, in saturated slags the basicity is fixed by the concentration of iron oxide.

Slag Formation in BOF Steelmaking—Single-slag practice is used in BOF (or BOP) steelmaking. The charge to the furnace consists primarily of burnt lime, steel scrap and hot metal, and oxygen is blown through a lance at a rate of about 2 nm^3 per minute per metric ton for a period of 16 to 25 min, depending on melt size and the carbon content of the steel at tap. Normal lance height is about 2 m above the melt surface.

As an example, the plant data of van Hoorn et al.[217] are given in Figure 13—154, showing changes in compositions of metal and slag with the volume of oxygen blow for a 300 metric ton heat. In the early stages of the oxygen blow the silicon is oxidized, forming a slag of relatively low basicity. The hump on the manganese curve is characteristic of all pneumatic steelmaking processes, including the Bessemer process.

Changes in slag composition during the blow are shown in Figure 13—155a for two different practices.

In this plot the slag composition is recalculated to give 100 per cent for the sum CaO + MgO + 'FeO' + SiO$_2$. The curve I is that reported by van Hoorn et al. for the BOF practice at Hoogovens Ijmuiden B.V. This is an underoxidized-slag practice which leads to low metal dispersion in the slag and minimum slopping of the bath. This practice is considered good for fast rate of decarburization and suitable only for low sulphur and low phosphorus in the hot metal. The overoxidized-slag practice of Mannesmann is represented by curve II as reported by Bardenheuer et al.[218] In this practice a liquid slag of high basicity is obtained more readily,

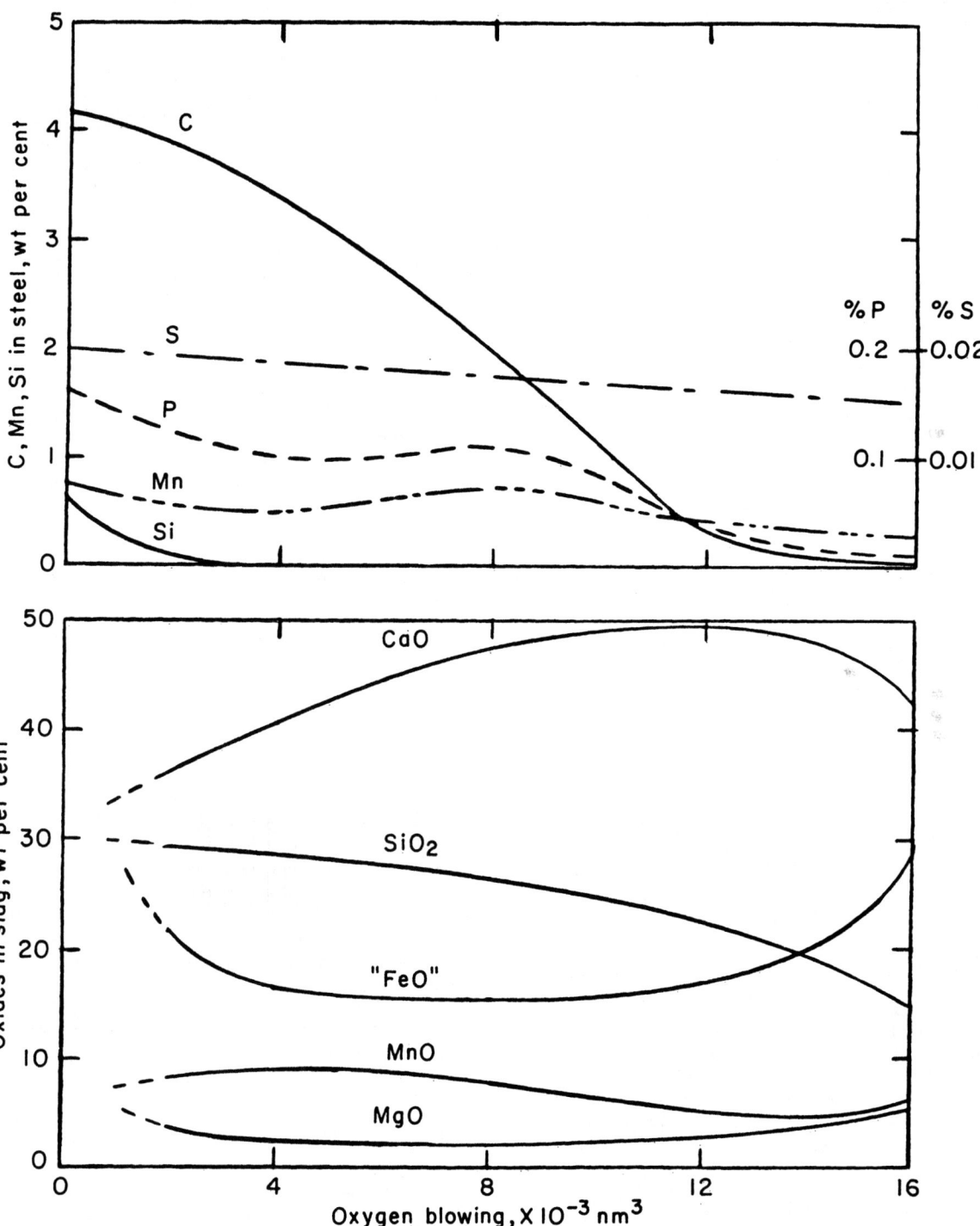

FIG. 13—154. Changes in metal and slag compositions during steelmaking in BOF for about a 300-metric-ton melt, using data of van Hoorn, et al.[217]

resulting in low magnesia pickup by the slag and faster rate of removal of sulphur and phosphorus. With this slag practice, however, there is a greater tendency to slopping. Two other practices are shown in Figure 13—155b. According to Nilles[219] a path AA for slag formation gives best refining conditions, particularly for decarburization. The path BB was considered by Baker[220] better for phosphorus and sulphur removal.

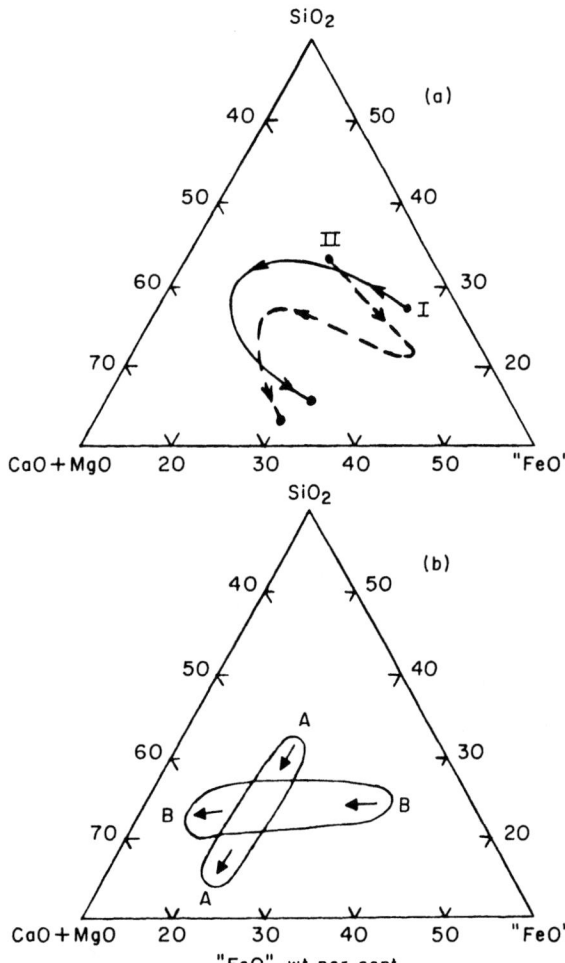

FIG. 13—155. Change in composition of slag during oxygen blowing for various BOF practices; (a): I (———) van Hoorn, et al.[217] and II (– – –) Bardenheurer, et al;[218] (b): AA Nilles. et al.[219] and BB Baker.[220]

As is seen from the equilibrium data in Figure 13—152, for double saturation of slags with magnesio-wustite and dicalcium (magnesium) silicate, the solubility of MgO decreases with an increase in slag basicity, e.g., 8 per cent MgO for V~2 and 4 per cent MgO for V~3.5. To minimise erosion of magnesia lining of the furnace by slag attack it is customary to add some magnesia as doloma to the charge together with burnt lime. The burnt lime used in some practices is that obtained by calcination of dolomitic limestone. In the early stages of the blow the slag basicity is low, therefore the solubility of MgO is high. As the basicity increases with the progress of the oxygen blow, the solubility of MgO

decreases, resulting in rejection of MgO from molten slag. However, an increase in slag basicity during refining is accompanied by an increase in slag mass, which counteracts MgO rejection; in fact, the total amount of MgO taken up by the slag increases during the later stages of the blow.

As indicated by numerous plant observations,[221-224] the addition of dolomitic burnt lime, to yield 4 to 6 per cent MgO in the slag, accelerates the rate of assimilation of lime in the slag. It was pointed out earlier that the rate of dissolution of burnt lime is controlled primarily by mass transfer in molten slag; therefore a decrease in slag viscosity caused by small additions of MgO may increase the rate of mass transfer through the diffusion boundary layer, resulting in a faster rate of lime assimilation. In addition, an increase in the MgO content of the slag moves the liquidus isotherm of the dicalcium silicate to higher concentrations of lime, thus increases the chemical driving force ΔC_{CaO} and consequently, increases the rate of dissolution of lime. Other additives such as Fe_2O_3, MnO, TiO_2 and CaF_2 will enhance the rate of lime dissolution for reasons similar to those stated above for the MgO addition.

Despite what is known about the solubility of lime in slags, there is still a tendency in melt shops to charge lime into the furnace well in excess of the solubility limit in the slag. The presence of lumps of undissolved lime interferes with the bath movement in the converter and presumably leads to less effective slag and metal mixing. The formation of dicalcium silicate coating on undissolved lime particles depletes some silica from the molten part of the slag, hence lowers the mass of molten slag, resulting in less efficient sulphur and phosphorus removal. Since almost all the silicon and phosphorus in the metallic charge (hot metal and scrap) end up in the slag during refining, the amount of burnt lime charged to the converter should be no more than that calculated from the silicon and phosphorus material balance and the lime/silica ratio that is attainable for a given level of iron oxide in the tap slag, as dictated by the univariant equilibrium in Figure 13—153.

Slag Formation in Q-BOP Steelmaking—In Q-BOP steelmaking, the oxygen and lime powder are blown through a series of tuyeres located at the bottom of the converter. In this single-slag process, for hot-metal charges containing less than about 0.2 per cent P, the slag formation begins in the melt near the oxygen + lime entrance where the oxides of iron, manganese, silicon and phosphorus react rapidly with lime powder. As the slag particles traverse the bath rapidly with CO bubbles, they react with carbon in the bath so that the concentrations of iron oxides and manganese oxide in the overlaying slag are lower than those in the slag particles at the time of their formation near the tuyere zone.

Because of the secondary reaction with carbon in the bath, the concentrations of iron oxides and manganese oxide in Q-BOP slags are lower than those in BOF (or BOP) slags. As would be expected from the difference in the direction of oxygen blow in these top and bottom blowing processes, the state of oxidation of iron in BOF slags is higher than that in Q-BOP slags. For the same reason the slag temperature in BOF is higher than the metal temperature, the reverse being the case for the

Q-BOP. Also, in the BOF there is extensive dispersion of metal droplets in the slag, whilst in the Q-BOP there is extensive dispersion of lime-rich slag particles in the metal. Aside from these physical differences, the composition of slags for top and bottom blowing processes obey the same univarient equilibrium relation for saturated slags as shown in Figure 13—153.

Slag Basicity—The oxides, e.g. SiO_2, P_2O_5, Al_2O_3, which form anion complexes in molten slags (network formers) are said to be acidic oxides. The oxides, e.g. CaO, MgO, MnO, FeO, which break down the anion complexes in the melt are known as the network modifiers and are said to be basic oxides. The ratio of the concentrations of basic oxides to those of the acidic oxides is called the "basicity" of the slag. There are numerous ways of representing the basicity of the industrial slags. Since this is an arbitrary way of expressing the general chemical behavior of the slag, there is no particular rule dictating the manner in which the basicity should be represented.

In simple slags where lime and silica are the major constituent oxides the basicity is usually defined by the concentration ratio %CaO/%SiO$_2$, which is often called the "V ratio". However, most of the industrial slags contain fair proportions of magnesia and phosphorus pentoxide; in such cases the basicity may be defined as follows: It is assumed that the concentrations of CaO and MgO are equivalent on molar basis; converting it to concentrations in weight per cent, CaO equivalence of MgO becomes 1.4 x % MgO = % CaO. On a molar basis ½ P_2O_5 is assumed to be equivalent to SiO_2; therefore in terms of weight per cent, the SiO_2 equivalence of P_2O_5 is 0.84 x % P_2O_5 = %SiO$_2$. The basicity is then given by

$$\text{Slag basicity} = \frac{\%\text{CaO} + 1.4\,(\%\text{MgO})}{\%\text{SiO}_2 + 0.84\,(\%\text{P}_2\text{O}_5)} \qquad (183)$$

Oxidation of Iron—Taylor and Chipman[225] measured the oxygen content of liquid iron that was equilibrated with the CaO(MgO)-'FeO'-SiO$_2$ slags, and derived the activity of FeO from the ratio of the oxygen concentrations in liquid iron

$$a_{FeO} = \frac{[\%\text{O}] \text{ in equilibrium with slag}}{[\%\text{O}] \text{ in equilibrium with pure liquid 'FeO'}} \quad (184)$$

From the results of the earlier[214, 225] and later[226] studies, the solubility of oxygen in liquid iron in equilibrium with liquid iron oxide at temperatures of 1530° to 1960°C is represented by the equation

$$\text{Fe}_x\text{O(l)} = x\text{Fe(l)} + \underline{\text{O}}$$

$$\log [\%\text{O}] = -\frac{6329}{T} + 2.734 \qquad (185)$$

The value of x in the formula Fe$_x$O for iron-saturated pure liquid iron oxide is 0.98 at the Fe-Fe$_x$O eutectic temperature 1527°C and about 1.00 at 2000°C. The isoactivity curves for 'FeO' are shown in Figure 13—156 for the CaO(MgO)-'FeO'-SiO$_2$ melts in equilibrium with liquid iron at 1600°C. The concentration and activity of 'FeO' for slags saturated with CaO and MgO, derived from the data in Figures 13—150 and 13—156,

are shown in Figure 13—157 as a function of the slag basicity CaO/SiO$_2$.

Oxidation of Carbon— The equilibrium constant for the carbon-oxygen reaction in liquid iron is represented by[227]

$$\text{CO(g)} = \underline{\text{C}} + \underline{\text{O}}$$

$$K = \frac{[a_C][a_O]}{p_{CO}}, \quad \log K = -\frac{1168}{T} - 2.07 \qquad (186)$$

where p_{CO} is in bar and the activities a_C and a_O are taken equivalent to [%C] and [%O], respectively, at low solute concentrations. The activity coefficient of carbon in iron increases and that of oxygen decreases with increasing concentration of carbon; the net result is that, for a given p_{CO}, the product [%C] [%O] decreases only slightly with increasing concentration of carbon. For concentrations above 0.03 per cent C, the reaction product is essentially CO with a negligible amount of CO_2. For steelmaking temperatures, around 1600°C, and 1 atm CO pressure, the equilibrium product [%C][%O] is about 0.002 for concentrations below 0.5 per cent C.

Noting from equations (185) that the solubility of liquid iron oxide in iron at 1600°C gives 0.226 per cent O in the metal, it is found from equation (186) and Figure 13—157 that for concentrations in the range 0.03 to 0.3 per cent C, $p_{CO} = 1$ atm, and for slags saturated with CaO and MgO, the slag-metal equilibrium with respect to the oxidation of carbon at 1600°C is simplified to

$$(\%\,'\text{FeO'})\,[\%\text{C}] = 0.33 \pm 0.02 \qquad (187)$$

In all steelmaking processes the state of oxidation of iron in the slag, as measured by the ratio Fe^{3+}/Fe^{2+}, is much higher than that for the slag-metal equilibrium. Such a departure from equilibrium is as would be expected for a dynamic process where the impurities in the metal are transferred to the slag via an oxidation step. The concentration of iron oxide in the slag for any given carbon level is much higher than that for the C-FeO equilibrium as given by equation (187) for 1 atm pressure of CO. Also, it has long been known that in most steelmaking processes the product [%C][%O] in the steel at tap is 1.5 to 3.0 times higher than the equilibrium value of 0.002 at steelmaking temperatures and atmospheric pressure of CO.

Increase in the oxygen content (≅ oxygen activity measured by the emf-oxygen sensor) of steel with decreasing concentration of carbon is shown in Figure 13—158a for BOP and Q-BOP steelmaking. The equilibrium relation for atmospheric pressure of CO is shown by the dotted line. Because of the ferrostatic head, the total pressure at the bottom of the converter is about 2 atm; an average pressure of 1.5 atm for CO bubbles in the melt may be assumed. With the exception of low-carbon melts in the Q-BOP, the oxygen contents of steel at all carbon levels are close to equilibrium values for an average CO pressure of 1.5 atm. Below about 0.06 per cent C the oxygen content of steel in Q-BOP steelmaking is invariably below that for BOP steelmaking.

We should also consider the relation between the concentrations of oxygen and manganese. As is seen from the plot in Figure 13—158b, the data points for BOP and Q-BOP are scattered about the equilibrium line for the reaction

$$\underline{Mn} + \underline{O} = (MnO) \qquad (188)$$

The same Mn-O equilibrium relation seems to hold at turndown for both top and bottom oxygen blowing. That is, for a given oxygen content of steel at turndown, the concentration of manganese is similar in both processes, yet the corresponding concentration of carbon in BOP for 1.5 atm CO pressure is two to three times higher than that in Q-BOP.

The difference in oxygen levels for top and bottom blowing at low carbon contents may be due in part to the continuation of carbon-oxygen reaction in Q-BOP when the nitrogen is blown through the tuyeres for 30 to 45 seconds during the rotation of the converter to the sampling or tapping position. It is doubtful, how-

ever, that much oxygen can be removed from the melt by such a mechanism because of mass transfer restrictions at very low solute concentrations and short duration of the inert gas flush. Also, a small volume of natural gas blown through the annular space around the oxygen tuyere as a coolant in the Q-BOP will not dilute CO sufficiently to bring about the observed decarburization at low oxygen contents in steel. It must therefore be conceded that the partial pressure of CO in the submerged gas stream is lowered much below the average bubble pressure of 1.5 atm by the presence of CO_2 in excess of that for the equilibrium

$$CO_2 + \underline{C} = 2CO. \qquad (189)$$

At low carbon levels, the rate of decarburization is controlled by the rate of mass transfer of carbon to the gas bubbles. Also, the rate of oxidation of iron and man-

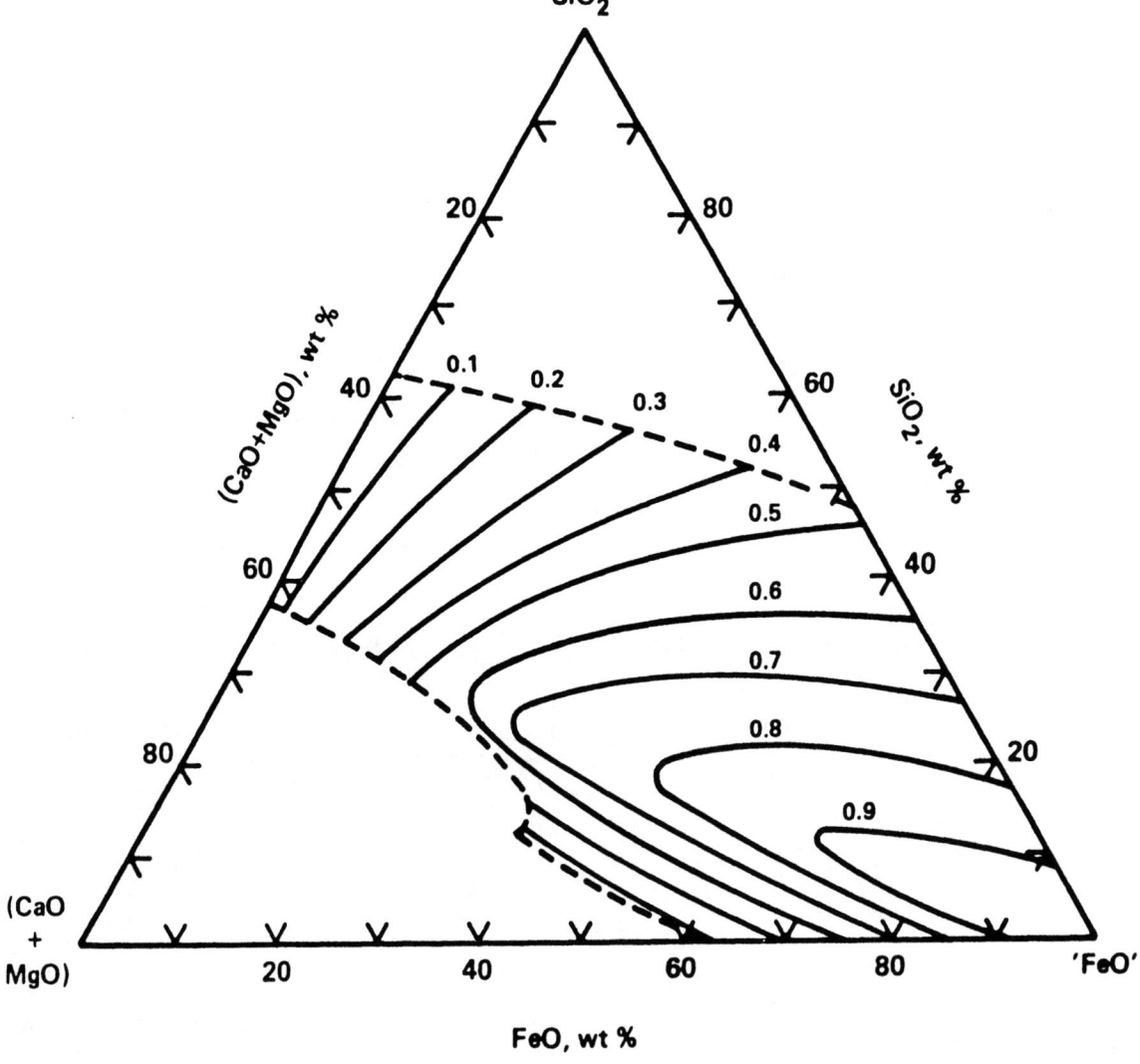

FIG. 13—156. Iso-activity curves for 'FeO' in CaO(MgO)-'FeO'-SiO₂ melts in equilibrium with liquid iron at 1600°C. After Taylor and Chipman.[225]

ganese will be limited by some kinetic effect. When the rate of oxygen bottom blowing exceeds the rate of consumption of oxygen in the bath, then the submerged gas stream will contain CO_2 in excess of the equilibrium value, thus diluting CO and allowing the decarburization to proceed as the gas bubbles traverse the

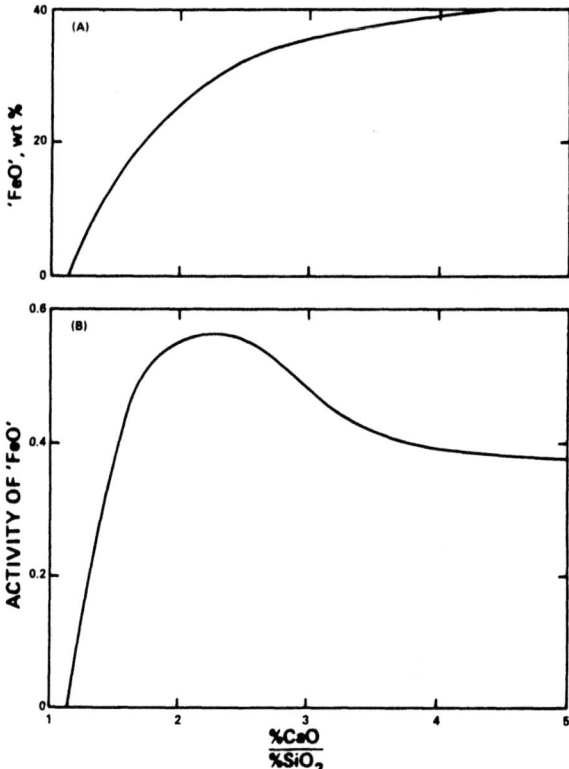

FIG. 13—157. MgO and $2Ca(Mg)O \cdot SiO_2$-saturated CaO-MgO-'FeO' SiO_2 melts in equilibrium with liquid iron at 1600°C: (A) concentration of 'FeO' and (B) activity of 'FeO'.

bath. As the concentration of carbon decreases, there will be more CO dilution by CO_2, and the driving force for decarburization will be maintained with some increase in the oxygen content of the steel, controlled by the manganese-oxygen reaction. In the top-blowing process, the gaseous oxygen encounters the slag layer directly; therefore, the oxygen that is not consumed by carbon and manganese in the steel is taken up by the slag via the reaction $Fe^{2+} \rightarrow Fe^{3+}$, instead of oxidation of CO to CO_2 as envisaged for bottom blowing. That is, in top blowing the gas bubbles in the bath contain essentially all CO at an average pressure of 1.5 atm, while in bottom blowing there is much higher concentration of CO_2 in the gas bubbles hence the partial pressure of CO < 1.5 atm.

The carbon-oxygen equilibrium relations in Figure 13—159 are for 100 per cent $(CO + CO_2)$ and for 90 per cent Ar + 10 per cent $(CO + CO_2)$ at a total pressure of 1.5 atm; the dotted lines are for manganese-oxygen equilibrium at $a_{MnO} = 0.1$ read off from the plot in Figure 13—158b. It should be noted that the equilibrium ratio p_{CO_2}/p_{CO} in the gas bubble increases with decreasing concentration of carbon in the steel, e.g., for 1600°C and 100 per cent $(CO + CO_2)$ at 1.5 atm pressure, the ratio p_{CO_2}/p_{CO} is 0.029 at 0.1 per cent and 0.237 at 0.01 per cent C. For this reason the calculated equilibrium curves depart slightly from linearity in this log-log plot. The carbon-oxygen equilibrium curves terminate at the oxygen solubility limit for liquid iron saturated with iron oxide. It is self evident that in the BOF steelmaking the metal cannot be decarburized below 0.02 per cent C. In fact, because of loss of iron due to oxidation at low carbon levels, in the BOF the steel is decarburized to levels no lower than 0.03 per cent C.

For certain applications, e.g., sheet steel for motor lamination, the steel should contain less than 0.005 per cent C. The decarburization of steel to such low levels of carbon, even with oxygen bottom blowing, will be at the expense of excessive oxidation of iron and result in high

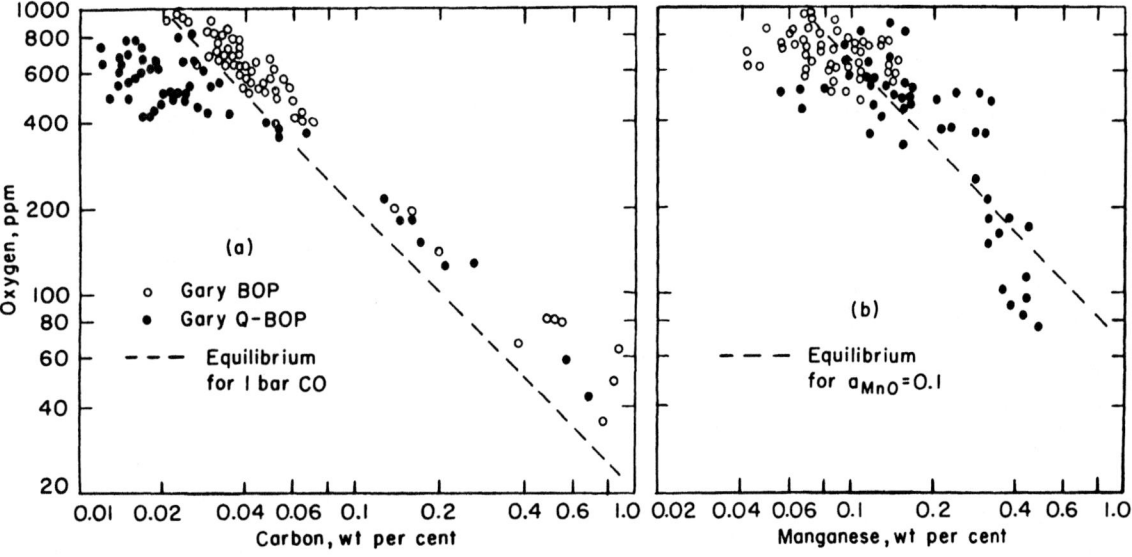

FIG. 13—158. Oxygen content of steel ($\simeq$ oxygen activity measured by the emf − oxygen sensor) in BOP and Q-BOP furnaces as a function of (a) the carbon content of steel and (b) the manganese content of steel at first turndown.

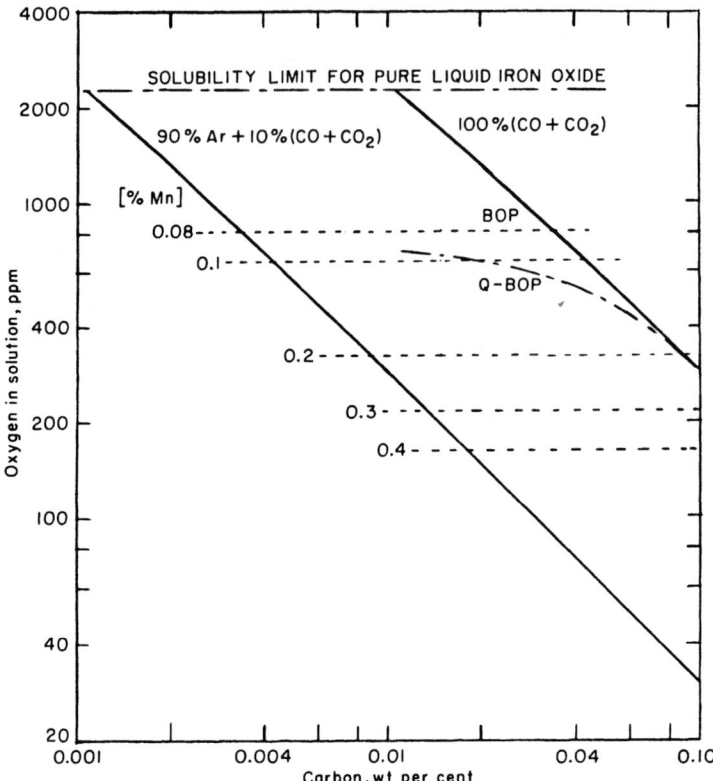

Fig. 13—159. Calculated data for 1600°C showing the carbon-oxygen relations for 100 per cent (CO + CO₂) and 90 per cent Ar + 10 per cent (CO + CO₂) at a total pressure of 1.5 bar; dotted horizontal lines are for manganese-oxygen equilibrium at activity of $a_{MnO} = 0.1$ for indicated concentrations of manganese in the steel.

melt temperatures. This problem can be circumvented by bottom blowing with an argon-oxygen mixture to lower the partial pressure of CO in the submerged gas stream, hence achieve decarburization at low carbon levels without excessive oxidation of iron and residual manganese in the steel. The carbon-oxygen equilibrium curve drawn in Figure 13—159 for a 90 per cent Ar + 10 per cent CO mixture at a total pressure of 1.5 bar shows that the equilibrium oxygen content would be 700 ppm at about 0.004 per cent C in the steel.

Oxidation of Silicon—Depending on the steelmaking practice and the type of hot metal used, the silicon content of the charge varies from about 0.5 per cent to 1.5 per cent. The oxidation of silicon to silica is an exothermic reaction and provides the heat necessary to raise the temperature of the bath during oxygen blowing. From the heats of solution of liquid silicon and gaseous oxygen in liquid iron and the heat of formation of silica given in Parts I and II, the heat generation upon oxidation of silicon is readily computed.

At 1600°C	ΔH kcal
$Si(l) = \underline{Si}$	−35
$O_2(g) = 2\underline{O}$	−56
$Si(l) + O_2(g) = \overline{Si}O_2(s)$	−226
$\underline{Si} + 2\underline{O} = SiO_2(s)$	−135

The silica together with iron oxide flux the lime added with the charge forming the slag. The lime and scrap-metal additions are adjusted according to the silicon content of the charge such that a slag with a desired basicity ratio of 2.5 to 3.5 and an optimum bath temperature are achieved.

Silicon is readily oxidized in the early stages of blowing, and at tap the silicon content of the steel is below 0.005 per cent as would be expected from consideration of the reaction equilibrium. For the silicon-oxygen reaction and indicated standard states,

$$\underline{Si} \text{ (l wt. %)} + 2 \underline{O} \text{ (l wt. %O)} = SiO_2(s) \quad (190)$$

the equilibrium constant is 1.62×10^4 at 1650°C which is the upper tap temperature.

Oxidation of Manganese—The oxidation of manganese in iron by the slag is an iron-manganese exchange reaction between slag and metal,

$$\underline{Mn} [1\%] + (FeO) (l) = (MnO) (s) + Fe (l). \quad (191)$$

For the indicated standard states, the equilibrium constant and its temperature dependence are

$$K = \frac{(x_{MnO})}{(x_{FeO})} \frac{1}{[\%Mn]} \frac{\gamma_{MnO}}{\gamma_{FeO}}, \log K = \frac{7784}{T} - 3.590. \quad (192)$$

In an earlier study of manganese reaction[228], it was found that in a plot of k_{Mn} vs B', the experimental data of Winkler and Chipman[229] agree well with the plant data from various basic open-hearth furnaces. For the basicity B' from 1.0 to 4.5 and steelmaking temperatures, ~ 1600°C, the following relation was found:

$$B' \, k_{Mn} = 7 \pm 0.5 \qquad (193)$$

where the basicity B' is as given in equation (183).

As discussed previously (Figure 13—153), the concentration of iron oxide in Q-BOP slags being lower than in BOP slags, it follows from the equilibrium relation in equation (192) that for slags of similar basicity the distribution ratio [%Mn]/(%Mn) for Q-BOP steelmaking would be higher than for BOP steelmaking, as seen from the plot in Figure 13—160.

The difference in the levels of manganese concentrations in these processes is further demonstrated in Figure 13—161 by the plant data from Kawasaki Steel.[230]

Sulphur Reaction—The sulphur reaction in steelmaking can be related to the slag metal equilibrium for the reaction.

$$\underline{S} \text{ (metal)} + O^{2-} \text{ (slag)} = \underline{O} \text{ (metal)} + S^{2-} \text{ (slag)} \quad (194)$$

For given temperature and slag composition, the equilibrium relation is represented by

$$K_S = (N_S) \frac{[a_O]}{[a_S]} \qquad (195)$$

where $[a_O]$ and $[a_S]$ are the activities of oxygen and sulphur in iron. The variation of K_S with temperature and slag composition derived by Turkdogan[231] using several slag-metal equilibrium data of Chipman et al., is shown in Figure 13—162.

In practice, it is convenient to use the iron oxide content of the slag instead of [%O], and the sulphur reaction can equally well be represented by

$$\underline{S} + Fe + O^{2-} = S^{2-} + FeO \qquad (196)$$

and the composition-dependent equilibrium ratio becomes

$$K_F = \frac{(\%S)}{[\%S]} (\%FeO)_t \qquad (197)$$

where $(\%FeO)_t$ is the total iron oxide content of the slag (= $0.9 \times \%Fe_2O_3 + \%FeO$). Using these practical units, in Figure 13—163 K_F is plotted against the basicity ratio B. The scatter is as would be expected from the steelmaking data; however, the results indicate that under BOH steelmaking conditions the sulphur reaction approaches close to equilibrium.

For BOP and Q-BOP steelmaking processes it has been shown that the sulphur reaction is close to slag-metal equilibrium. In Figure 13—164, K_F for plant data are plotted versus $\%SiO_2 + \%P_2O_5$ and the data are scattered within a band shown by the dotted curves for slag-metal equilibrium.

Since the sulphur reaction is close to equilibrium and the iron-oxide content can not be varied for a given steel, there is little the steelmaker can do to insure low sulphur in the steel. A possible solution of the sulphur problem in steelmaking is to keep the sulphur input into the furnace at a low level. The hot metal, steel

scrap and sometimes burnt lime are the principal sources of sulphur. The burnt lime containing 0.03 per cent S (maximum) is usually considered satisfactory. The sulphur content of the hot metal varies over a wide range, depending on the type of ore and the blast-furnace practice employed, normally within the range 0.03 per cent to 0.05 per cent.

The reducing conditions in the hot metal are most favorable for ladle desulphurization by lime, calcium carbide or soda ash. The overall reaction occurring in the lime treatment of hot metal is represented by

$$\underline{S} + CaO(s) + \underline{C} = CaS(s) + CO(g) \qquad (198)$$

At $p_{CO} = 1$ atm and 1500°C the equilibrium sulphur content of graphite-saturated iron is about 6 ppm in the presence of CaO and CaS. In the presence of silicon in the metal desulphurization occurs by the following reaction

$$\underline{S} + 2CaO(s) + 1/2 \, \underline{Si} = $$
$$CaS(s) + 1/2 \, Ca_2SiO_4 \qquad (199)$$

At 1 per cent Si in the hot metal the equilibrium sulphur content for this reaction is about 0.1 ppm.

For effective desulphurization of hot metal, the ladle additions should be well mixed with the metal. In order to benefit from the lime or soda treatment, the hot metal must be skimmed of kish and slag immediately before charging into the BOP vessel.

Another possible method of reducing the sulphur content of the steel is by secondary steelmaking ladle treatments which are discussed later.

Oxidation of Phosphorus—The phosphorus reaction between metal and slag may be represented by

$$2\underline{P} + 5FeO = P_2O_5 + 5Fe \qquad (200)$$

$$k_p = \frac{(\%P_2O_5)}{[\%P]^2 \, (\%\Sigma FeO)^5} \qquad (201)$$

This reaction equilibrium is well represented by the following empirical relation, which is based on the work of Balajiva and co-workers[233,234]

$$\log k_p = 10.78 \log(\%CaO) - 0.00894T - 6.245 \quad (202)$$

where T is temperature, °C. This relation derived from scattered experimental data lacks generalization. Also the value of k_p is much too sensitive to the concentration of iron oxide.

In a recent study a new concept was introduced by Turkdogan[235] to describe the effect of slag composition on the isothermal equilibrium distribution of phosphorus between slag and the metal. It was shown through a thermodyamic argument that for isothermal conditions the equilibrium distribution of phosphorus between slag and metal, i.e., $(\%P)/[\%P]$ = $(\%P_2O_5)/[\%P]$, is a function of the concentrations of silica and iron oxide, the relation being independent of basicity for slags saturated with dicalcium (magnesium) silicate. It was shown that there is close agreement between the results of independent laboratory experiments demonstrating that for a given temperature the equilibrium relation

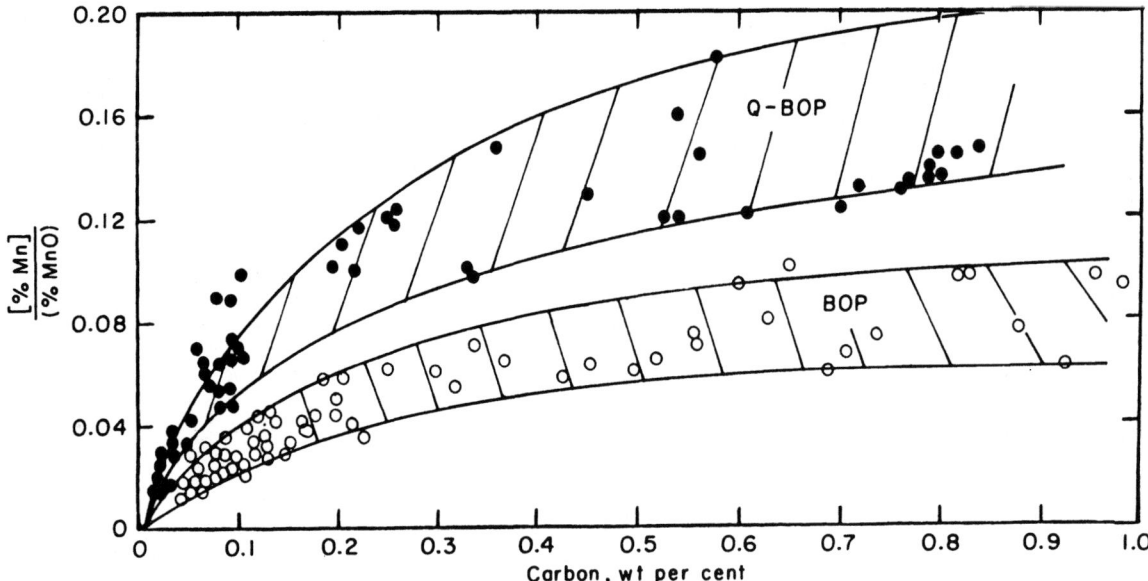

FIG. 13—160. Variation of the manganese-distribution ratio with the concentration of carbon in steel at tap in BOP and Q-BOP steelmaking using data from United States Steel's Gary plant.

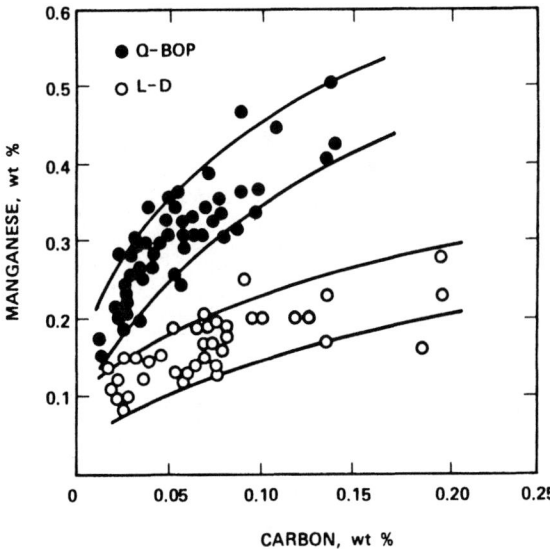

FIG. 13—161. Relation between manganese and carbon contents of steel at turndown in Kawasaki Q-BOP and L-D steelmaking.[230]

$$k_{PS} = \frac{(\%P_2O_5)}{[\%P]} (1 + \%SiO_2) \qquad (203)$$

is a single function of the iron oxide content of the slag that is saturated with dicalcium silicate and magnesia.

Using the plant data for BOF and Q-BOP steelmaking at temperatures of 1590-1610°C, the variation of log k_P with the concentration of dissolved lime is shown in Figure 13—165. The values of k_P for the plant data lie well above the broken curve representing the laboratory data. Evidently, equation (202) de-

rived from the laboratory experimental data cannot be extended to the compositions of industrial steelmaking slags.

The same BOP data are plotted in Figure 13—166a in accord with the relation (203). The equilibrium relation for 1600° is shown by the dotted curve. In accord with the equilibrium data, the k_{PS} is a weak function of the iron-oxide content of the slag. The values of k_{PS} are scattered within the range 800 to 3200, the average value of k_{PS} being about 2000 which is about one-quarter of the equilibrium value for an average melt temperature of 1600°C.

In view of the relation in Figure 13—153 for univariant equilibrium in saturated steelmaking slags, the phosphorus distribution ratio $(\%P_2O_5)/[\%P]$ alone should be a function of the iron-oxide content of the slag. This is shown in Figure 13—166b for the BOP and the Q-BOP data. The dotted line for the average melt temperature of 1600°C is derived from the equilibrium relation. The data points from the Q-BOP are scattered about the equilibrium curve for the phosphorus distribution ratio which becomes independent of the iron-oxide content above about 16 per cent 'FeO'. Most of the data points from the BOP are again below the equilibrium curve. That is, in the Q-BOP the steel is dephosphorized to levels (close to equilibrium) which are lower than those achieved in the BOP, even with Q-BOP slags containing lower concentrations of iron oxide as compared to the BOP slags.

Effect of Carbon on Distribution Ratios—It has already been established from the equilibrium and plant data for basic slags that the basicity attained in the molten portion of the saturated slag depends on the concentration of iron oxide, which in turn depends on the aimed concentration of carbon in steel at tap. This interdependence of basicity and iron-oxide content of the slag, hence the aimed carbon content of the refined steel, has a decisive influence on the weight of the slag

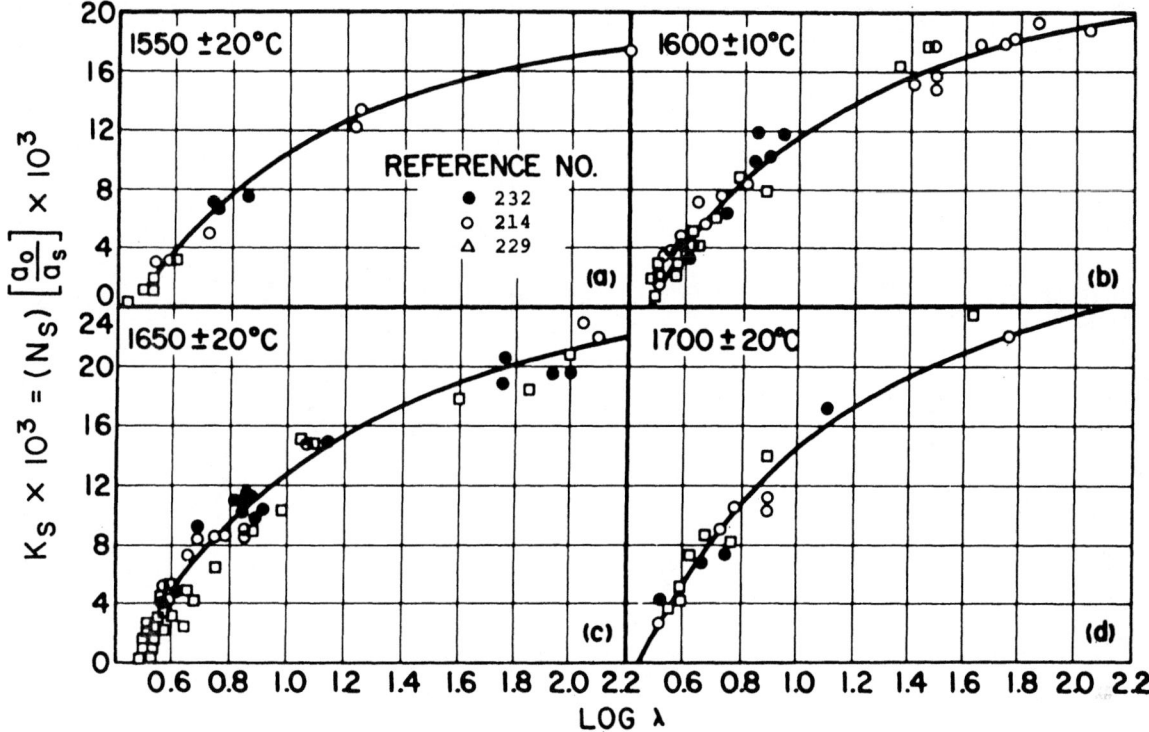

FIG. 13—162. K_S related to slag composition for mean temperatures of (a) 1550 ± 20°C; (b) 1600 ± 10°C; (c) 1650 ± 10°C; (d) 1700 ± 20°C. (Turkdogan[230]).

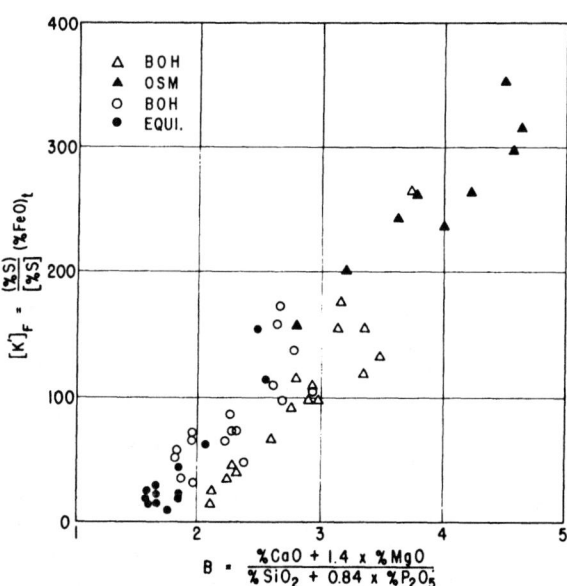

FIG. 13—163. The sulphur distribution ratio times iron oxide as a function of slag basicity.

that can be sustained molten in the converter. This is illustrated by the following example for a fixed amount of silicon in the hot-metal and scrap charge.

Noting that the composition of the saturated slag is in fact fixed by the concentration of carbon in steel at tap, the following average values for high and low-carbon

steels, with the sum %CaO + %SiO₂ + %'FeO' being 87 per cent, will be taken.

High-carbon melt: 8%'FeO' and %CaO/%SiO₂ = 2.5

Low-carbon melt: 25%'FeO' and %CaO/%SiO₂ = 3.5

The total weight of molten slag w_t is given by the sum

$$w_t = w_{SiO_2} + w_{CaO} + \left(\frac{\%'FeO'}{100}\right)w_t + w' \qquad (204)$$

where w_{SiO_2} is the weight of silica from oxidation of silicon in the charge, w_{CaO} the amount of burnt lime charged to the furnace and w' the sum of the amounts of other oxides assumed to be the same for both high- and low-carbon heats. Assuming that the lime added dissolves completely in molten slag, i.e., charged V ratio equal actual V ratio in molten slag, the following relations are obtained from the material balance.

High-carbon melt: $0.92w_t^h = 3.5w_{SiO_2} + w'$
$$\qquad (205)$$
Low-carbon melt: $0.75w_t^l = 4.5w_{SiO_2} + w'$

where superscripts h and l indicate the total slag weight for the high and low-carbon heats, respectively. Noting from equation (205) that

$$w_t^h = \frac{0.75w_t^l - w_{SiO_2}}{0.92}$$

for a fixed amount of phosphorus in the charge for high- and low-carbon heats, the following relation is obtained

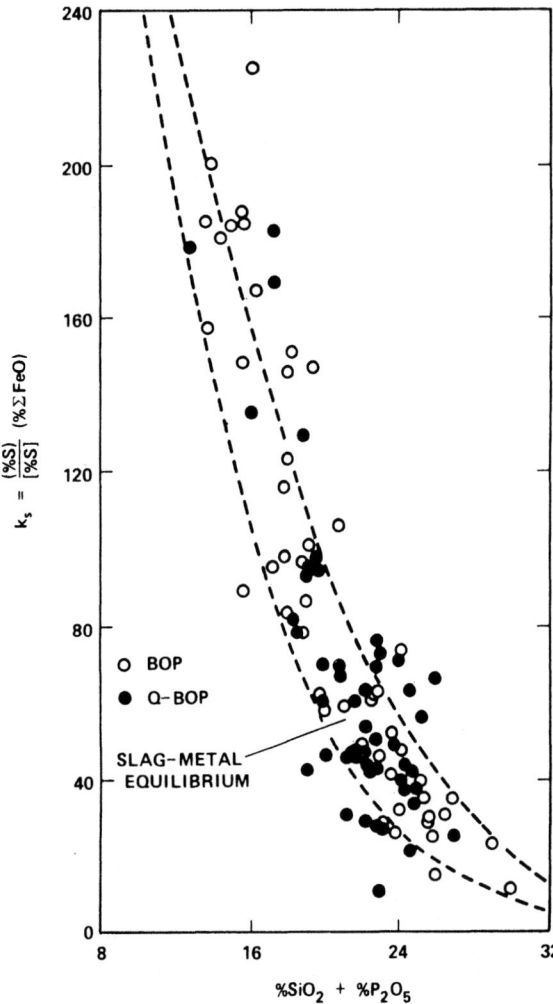

FIG. 13—164. Values of K for sulphur reaction from BOP and Q-BOP plant data compared with the relation for the slag-metal equilibrium at 1600°C.

for the ratio of $\%P_2O_5$ in the slag for low- and high-carbon heats

$$\frac{(\%P_2O_5)_l}{(\%P_2O_5)_h} = \frac{0.75w_t^l - w_{SiO_2}}{0.92w_t^l} =$$

$$0.82 - 1.09\,\frac{\%(SiO_2)_l}{100} \qquad (206)$$

where the subscripts l and h refer to slag composition for low- and high-carbon heats. Inserting $(\%SiO_2)_l = 13.8$ gives for the ratio of the total slag weight

$$\frac{w_t^h}{w_t^l} = \frac{(\%P_2O_5)_l}{(\%P_2O_5)_h} = 0.67. \qquad (207)$$

From the relation in Figure 13—166b can be taken the ratio

$$\frac{\left[(\%P_2O_5)/[\%P]\right]_l}{\left[(\%P_2O_5)/[\%P]\right]_h} \simeq 3 \qquad (208)$$

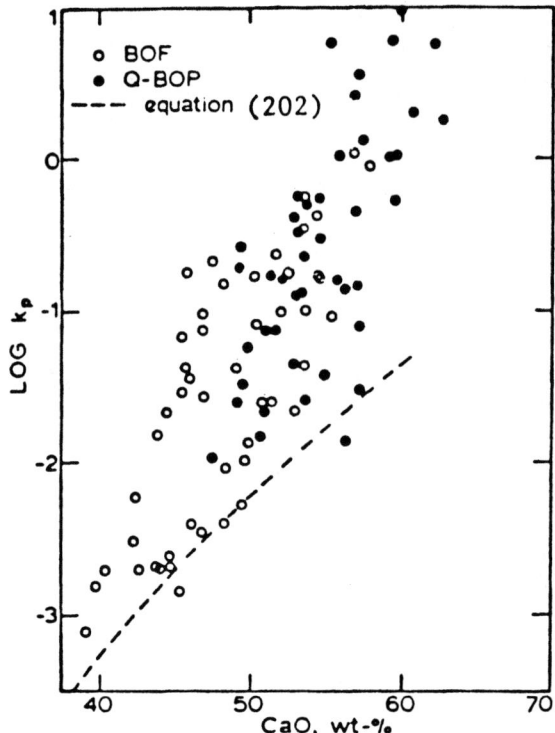

FIG. 13—165. Variation of k_p for the phosphorus reaction with the concentration of dissolved CaO in BOF and Q-BOP steelmaking for tap temperatures of 1590°-1610°C. (Turkdogan[235]).

Combining this with equation (207) gives for the ratio of the residual phosphorus in high- and low-carbon heats

$$\frac{[\%P]_h}{[\%P]_l} \simeq 4.5 \qquad (209)$$

It is seen from this illustration that for a given total silicon and phosphorus input with the charge, the residual phosphorus at tap in high-carbon heats will be 4 to 5 times higher than that in the low carbon heats, a prediction that is in general accord with practical observations. For the reasons given, the slag weight cannot be increased at will by simply increasing the amount of burnt lime in the charge. Because of the constraint on the composition of the saturated slag, imposed by the concentration of carbon in the steel, the slag weight can be increased only by increasing the amount of silicon in the charge together with a corresponding increase in the amount of lime charge to maintain saturation. Even if the slag weight for high-carbon melts were kept at the same level as that for low-carbon melts, the ratio $[\%P]_h/[\%P]_l$ would still be high, about 3. To insure low residual phosphorus in high-carbon steels at tap, the total phosphorus in the charge should be decreased. Alternatively, the steel should be blown to low carbon levels, then recarburized in the ladle, which is in fact a common practice in steelmaking.

Similar calculations made for the sulphur distribution ratio using the equilibrium relation in Figure 13—164 gives

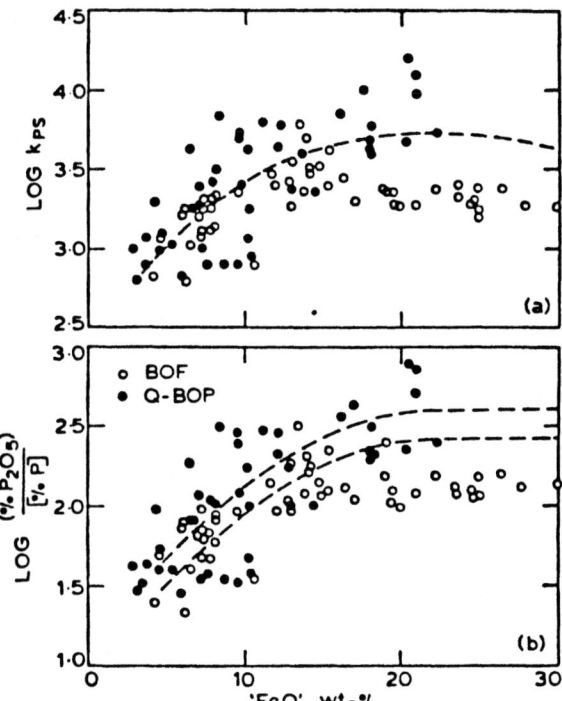

FIG. 13—166. Plots showing the state of the phosphorus reaction in BOF and Q-BOP at tap at temperatures of 1590°-1610°C relative to the equilibrium relation (---). (Turkdogan[235]).

$$\frac{[\%S]_h}{[\%S]_l} \simeq 1.1 \qquad (210)$$

In fact, within the scatter in the plant data, not much difference is seen in the concentrations of residual sulphur in high- and low-carbon heats. The equilibrium relation for the manganese reaction, equation (193), and the above method of calculation gives

$$\frac{[\%Mn]_h}{[\%Mn]_l} \simeq 3.3 \qquad (211)$$

which is again in general accord with the trends seen in high-carbon and low-carbon steelmaking.

The conclusion to be drawn from the foregoing analysis is that, with saturated basic slags, the role of slag in the refining of steel, hence the residual concentration of impurities in the steel, is dictated to a large extent by the aimed level of carbon in the steel at the time of furnace tapping. The higher the concentration of carbon in the steel, the higher would be the residual concentrations of phosphorus and manganese, the effect on the residual sulphur being small. Another general observation is that, despite the compositional complexity of basic steelmaking slags, the distributions of phosphorus, manganese and sulphur between slag and metal achieved during steelmaking are relatively simple functions of slag composition and not much affected by temperature within the range 1560° to 1630°C usually encountered in steelmaking.

REACTIONS IN AOD STEELMAKING

In the AOD process the steel is first melted in an electric furnace and has a relatively high carbon content, 0.4 to 1 per cent. The melt is then put into an AOD vessel where an argon-oxygen gas mixture is blown through tuyeres located near the bottom of the vessel. The blow is usually done in several stages. Typically, for a UNS S40900 (AISI Type 409) grade steel containing 11 per cent chromium an O_2/Ar ratio of $\frac{3}{1}$ is used down to 0.2 per cent carbon and then a $\frac{1}{3}$ ratio to 0.03 per cent carbon. For an 18 per cent chromium steel an $O_2/Ar = \frac{3}{1}$ is used to 0.6 per cent carbon then a $\frac{1}{1}$ to 0.4 per cent carbon and finally a $\frac{1}{3}$ to 0.03 per cent carbon. Nitrogen can be used in place of argon during the early stages of the blow.

In the AOD stainless steelmaking process, the carbon is oxidized in preference to chromium because the CO formed is diluted with argon. The overall reaction describing the process can be written as

$$Cr_2O_3 \text{ (s)} + 3\underline{C} \text{ (in Fe)} \rightarrow 2\underline{Cr} \text{ (in Fe)} + 3CO \text{ (g)} \quad (212)$$

The temperature dependence of the equilibrium constant (K) for the reaction is given by

$$\log K = -\frac{40\,970}{T} + 27.31 \qquad (213)$$

Laboratory investigations and studies of plant data indicate that reaction (212) does not reach equilibrium; chromium oxidation occurs at carbon levels higher than those expected with argon dilution of CO in the bath.

Studies of Choulet, et al.[236] indicated that for argon to be effective in lowering chromium oxidation, the gas had to be injected deep into the bath. They also showed that the carbon content at which significant chromium oxidation occurred was higher than predicted by the equilibrium for reaction (212). Fulton and Ramachandran[237] obtained similar results and showed that the carbon content at which chromium oxidation occurred depended on blowing conditions. Nelson[238], in his AOD patent, also emphasized that the effectiveness of argon in protecting the chromium from oxidation depended on the operating practice employed. These findings indicate that the decarburization of stainless steel in the AOD process is not determined by the equilibrium for reaction (212) alone, and that the actual reaction sequence and kinetics must be considered in developing a reaction model for the process.

In laboratory experiments, Fruehan[239] has shown that when O_2-Ar gas mixtures are injected into a shallow melt, less than 10 cm deep, nearly all of the oxygen is consumed in the oxidation of chromium. It was therefore concluded that in the tuyere zone of the AOD vessel oxygen is consumed by chromium oxidation and as the Cr_2O_3 particles rise they oxidize the carbon. Further laboratory tests indicated that the rate of reduction of Cr_2O_3 by carbon dissolved in iron is controlled by liquid-phase mass transfer[240].

In the model developed it is assumed that most of the oxygen blown is consumed in the oxidation of chromium in the vicinity of the tuyere zone. As the chromic oxide particles move away from the tuyere zone with the argon bubbles, reaction (212) occurs. The rate of

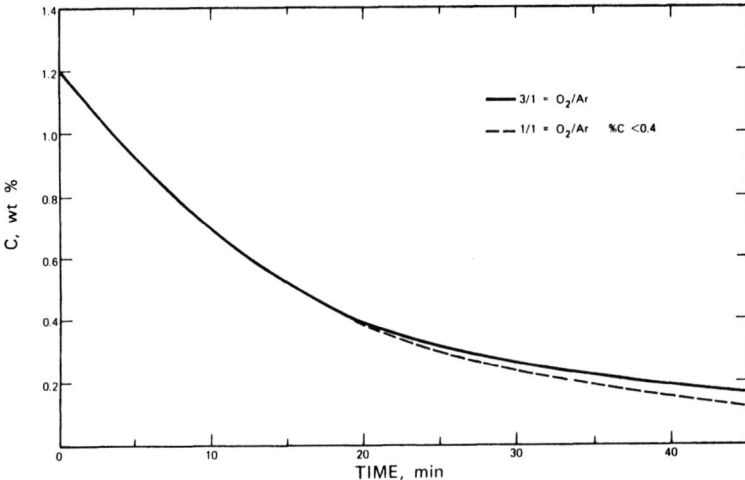

FIG. 13—167. Calculated rate of decarburization for an 18-8 type stainless steel (Fruehan[240]).

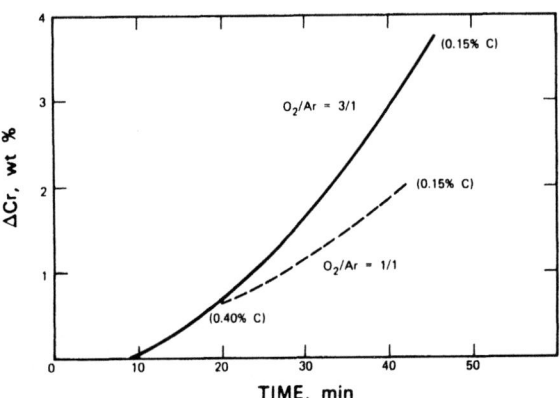

FIG. 13—168. Calculated rate of Cr oxidation for an 18-8 type stainless steel. (Fruehan[240]).

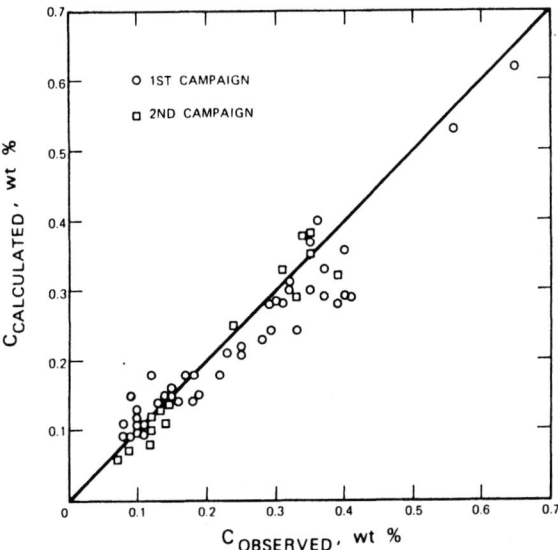

FIG. 13—169. Calculated and observed carbon contents when processing a UNS S40900 (AISI Type 409) grade stainless steel, $O_2/Ar = 3/1$. (Fruehan[240]).

this reaction is assumed to be controlled by the mass transfer of carbon from the bulk-metal phase to the melt-bubble interface. By introducing an empirical rate parameter, α, which is proportional to the average liquid-phase mass-transfer coefficient for carbon transport to the Ar-CO bubbles passing through the bath, the average rate of oxidation of carbon can be represented by

$$\frac{d\%C}{dt} = -\alpha[\%C - (\%C)_e] \qquad (214)$$

where

%C = time-dependent average carbon content of the bath.

$(\%C)_e$ = the average carbon content at the bubble surface in local equilibrium with the average chromium content and the average CO partial pressure in the gas bubbles.

The value of $(\%C)_e$ depends on the O_2/Ar blowing rate, K, the chromium content and the rate of decarburization. For details the reader is referred to the original work. Equation (214) turns out to be a nonlinear differential equation which can be solved numerically. The rate of chromium oxidation is then easily calculated by

assuming that the oxygen not used for decarburization oxidizes chromium.

The value of α is determined from operating data. However, once α has been determined for a given furnace its value can be computed for other operating conditions, because it is independent of temperature, O_2 to Ar ratio, and Cr content. It is approximately proportional to total gas-flow rate and the bath depth. A typical value for α is about 0.1 (minutes)$^{-1}$ for the United States Steel 100-net-ton AOD at South Works.

Equation (214) was solved for various operating conditions for producing 18-8 type and UNS S40900 (AISI Type 409) stainless steels in the 100-net-ton AOD at United States Steel's South Works. A typical decarburization curve is shown in Figure 13—167 for an 18-8 steel and chromium oxidation in Figure 13—168. The predicted results were in good agreement with operating data as demonstrated in Figure 13—169 where the calculated carbon is plotted versus observed carbon.

PART V

Ladle Treatment of Steel

With the stress for cleaner steels, lower sulphur contents and closer control of alloying elements the use of post-steelmaking treatments has grown in the last twenty years. Still by far the most common treatment is deoxidation. With the advent of the oxygen sensor and a greater knowledge of the equilibria and kinetics of deoxidation reactions, the ladle treatment processes have been improved. Degassing processes have also been developed in the last twenty years which have helped make cleaner steels. These processes have included vacuum-type processes as well as inert-gas-flushing (argon) practices. The Q-BOP vessel offers a unique opportunity for inert-gas flushing. In the past there has been an increased use of ladle refining in general. An example of a relatively new post-treatment of steel is desulphurization by injection of a desulphurization agent such as Mag-Spar, a mixture of magnesium and CaF_2, with argon as in the Thyssen Niederrhein (TN) Process.

In this part of the chapter the basic chemical reactions involved in these treatments are discussed. As in the other steelmaking operations discussed in this chapter, the operating details are not given, but rather the emphasis is on the chemistry of the reactions involved. Since deoxidation is still the most important of these processes much of this part of the chapter is devoted to deoxidation and related subjects. In the second section, degassing is discussed and in the third other ladle refining processes are discussed.

DEOXIDATION OF STEEL

Deoxidation is the last stage of steelmaking. In all types of steelmaking practices, the steel bath at the time of tapping contains 400 to 800 ppm O. Deoxidation is carried out during tapping by adding into the tap-ladle appropriate amounts of ferromanganese, ferrosilicon and/or aluminum or other special deoxidizers. If at the end of the blow the carbon content of the steel is below specification, the metal is also recarburized in the ladle. However, large additions in the ladle are undesirable, because of the adverse effect on the temperature of the metal.

Perhaps the foremost importance is the choice of a particular type of deoxidation which will ensure the production of a desired type of ingot. Eight typical conditions of commercial ingots, cast in identical bottle-top molds, in relation to the degree of suppression of gas evolution are shown schematically in Figure 13—170. The dotted line indicates the height to which the steel originally was poured in each ingot mold. Depending on the carbon and particularly on the oxygen content of the steel, the ingot structures range from that of a fully killed or dead-killed ingot (No. 1) to that of a violently rimmed ingot (No. 8). Included in the series are the four main types of ingots that are produced commercially as indicated in Figure 13—170, i. e. killed steel (No. 1), semikilled steel (No. 2), capped steel (No. 5), and rimmed steel (No 7).

Rimmed steel usually is tapped without having made additions of deoxidizers to the steel in the furnace, and only small additions to the molten steel in the ladle, in order to have sufficient oxygen present to give the desired gas evolution by reacting in the mold with carbon. The exact procedures followed depend upon whether the steel has a carbon content in the higher ranges, e. g . 0.12 per cent to 0.15 per cent, or in the lower ranges, e. g. below 0.1 per cent. When the metal in the ingot mold begins to solidify, there is a brisk evolution of carbon monoxide, resulting in an outer ingot skin of relatively clean metal low in carbon and other solutes. Such ingots are best suited for the manufacture of steel sheets.

Capped-steel practice is a variation of rimmed-steel practice. The rimming action is allowed to begin normally, but is then terminated at the end of a minute or more by sealing the mold with a cast-iron cap. In steels

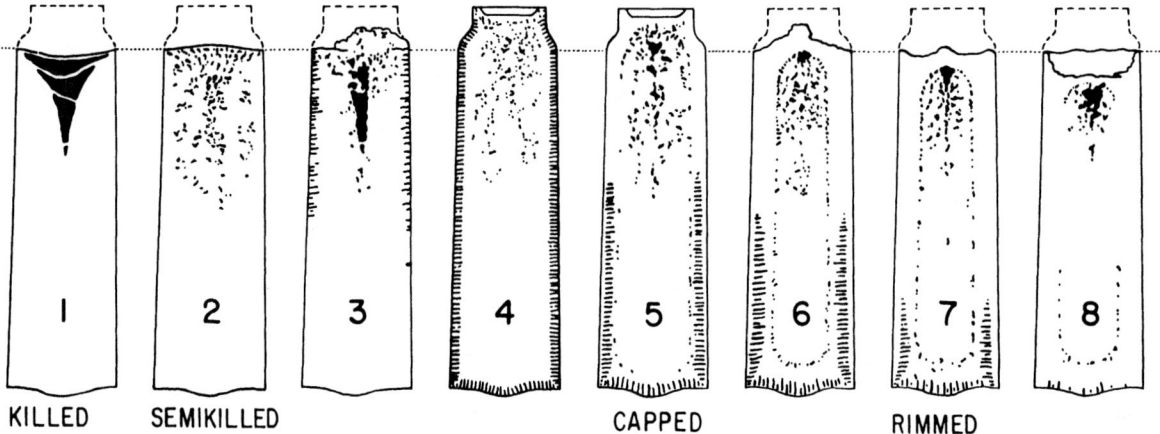

KILLED   SEMIKILLED   CAPPED   RIMMED

FIG. 13—170. Series of typical ingot structures.

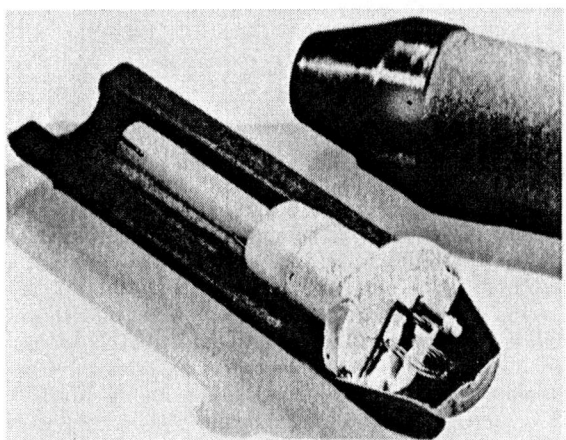

FIG. 13—171. Prototype disposable United States Steel oxygen sensor.

with a carbon content greater than 0.15 per cent, the capped ingot practice is usually applied to steel for sheet, strip, wire and bars.

Semikilled steel is deoxidized less than killed steel, and there is enough oxygen present in the molten steel to react with carbon forming sufficient carbon monoxide to counterbalance the solidification shrinkage. This steel generally has a carbon content within the range of 0.15 per cent to 0.30 per cent and finds wide application in structural shapes.

Killed steel is deoxidized to such an extent that there is no gas evolution during solidification. Aluminum is used for deoxidation, together with ferroalloys of manganese and silicon; in certain cases calcium silicide or other special strong deoxidizers are used. In order to minimize piping, almost all killed steels are cast in hot-topped big-end-up molds. Killed steel generally is used when a homogeneous structure is required in the finished steel. Alloy steels, forging steels and steels for carburizing are of this type, when the essential quality is soundness. In producing certain extra-deep-drawing steels, a low-carbon (< 0.1 per cent C) steel is killed, usually with a substantial amount of aluminum that is added in the ladle, in the mold, or both.

Although the deoxidation of steel by aluminum suppresses the formation of carbon monoxide during solidification, and hence suppresses blowholes, there are many steel-processing operations where aluminum killing of steel is undesirable. For example, it is widely recognized that certain alloy steels to be cast as large ingots should not be subject to aluminum killing, because of the piping and of deleterious effects of alumina inclusions on the subsequent processing of ingots for certain applications, e. g. generator-rotor shafts. It has long been recognized from the early days of the continuous-casting operation nearly three decades ago that casting difficulties and poor surface conditions are often experienced with aluminum-killed steels. It is for these reasons that other forms of deoxidation are often preferred in a number of steel-processing operations, e. g. silico-manganese deoxidation and/or vacuum-carbon deoxidation.

Oxygen Sensors—Prior to the development of oxygen sensors, the control of deoxidation was based on the total oxygen content of steel determined by analysis of samples taken with bomb or lolly-pop samplers or else more commonly simply based on the carbon content of the steel at tap. The analysis of cast steel for total oxygen is an important postmortem for guidance to steel cleanliness but is obviously too late in many cases. It became clear that what was needed was a device which could measure the dissolved oxygen content of steel rapidly so that the correct amount of deoxidant could be added. Research into solid oxide electrolyte galvanic cells led to the development of oxygen sensors in the late sixties. A solid oxide electrolyte is a semiconductor which conducts oxygen ions at high temperatures.

An oxygen sensor was developed by United States Steel Corporation which has proven to be a highly accurate and reliable instrument. A picture of one version of the U. S. Steel oxygen sensor is shown in Figure 13—171. The basic galvanic cell development is described by Fruehan and co-workers[241] while the development of the commercial sensor for steelmaking operations is discussed by Russell et al.[242]

The galvanic cell consists of a silica tube, with an internal diameter of about 3 mm, into which a small ZrO_2 (4 per cent CaO) disk is sealed into one end. A mixture of Cr-Cr_2O_3 is the reference electrode with Mo wire as electrical contact. There is also a Mo rod for steel contact and a thermocouple included in the sensor. The sensor is a disposable immersion device used in a similar manner as an immersion thermocouple.

The galvanic cell used is represented by

$$Cr\text{-}Cr_2O_3 \ (s) \mid ZrO_2 \ (CaO) \mid \underline{O} \ (in \ Fe \ alloy) \quad (215)$$

The activity of oxygen (a_O) in liquid iron alloys is related to the electromotive force (E) by

$$E \ = \ \frac{RT}{nF} \ \ln \ \frac{a_O}{K(p_{O_2})^{1/2}} \quad (216)$$

where p_{O_2} is the reference oxygen pressure in equilibrium with Cr-Cr_2O_3, n = 2, F is the Faraday constant (23 061 cal per volt equivalent) and K is the equilibrium constant for the reaction

$$\tfrac{1}{2} \ O_2 \ (g, \ atm) = \underline{O}(1 \ wt \ \% \ in \ Fe) \quad (217)$$

The temperature dependence of the equilibrium constant K for reaction (217) and $p_{O_2}^{1/2}$ for the Cr-Cr_2O_3 equilibrium is given by

$$\log K \ = \ \frac{6120}{T} \ + \ 0.15 \quad (218)$$

$$\log (p_{O_2}^{1/2}, \ atm^{1/2}) = - \ \frac{19 \ 700}{T} \ + \ 4.47 \quad (219)$$

Therefore, the activity of oxygen, which is taken equal to the concentration of oxygen in weight per cent in liquid iron is given by

$$\log a_O = 4.62 - \frac{13 \ 580 - 10.08 \ E}{T} \quad (220)$$

where E is expressed in millivolts and T is in degrees

Kelvin. The activity coefficient of oxygen in iron alloys is estimated from the equation,

$$\log f_0 = \sum_j e_0^{(j)} (\%j) \qquad (221)$$

where $(\%j)$ are the concentrations of the alloying elements and $e_0^{(j)}$ are the interaction coefficients given in Table 13—IX. However, it should be remembered that in low-carbon steelmaking the concentrations of manganese, silicon, carbon and other elements are small enough that the above correction becomes insignificant for all practical purposes.

The important feature of the probe is that within about 4 seconds of immersion in the melt, the electromotive force quickly achieves a steady value ($\pm$ 2 mv) and remains steady for as long as 30 seconds or more. Typical emf traces obtained from oxygen sensor readings taken in a BOP furnace are shown in Figure 13—172. The readings were taken at various stages of

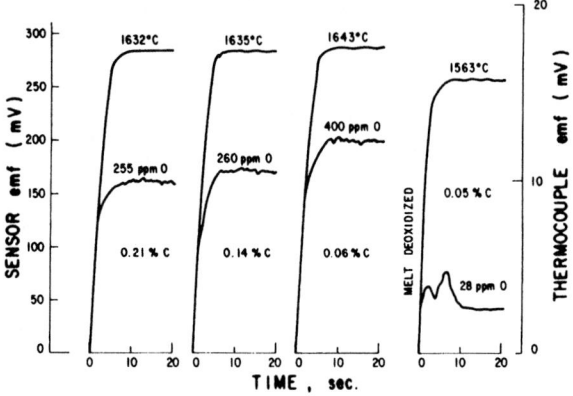

FIG. 13—172. Oxygen sensor traces for BOP furnace heat.

the blow along with samples of the melt. The final reading was after deoxidation with 1 kilogram of aluminum per metric ton (2 pounds of aluminum per net ton) of metal. The oxygen sensor has been extensively tested and is commercially produced for use in many steelmaking facilities. The oxygen sensor has been used to improve deoxidation practices for many types of steel produced in the BOH, BOP and Q-BOP. An example of its use is given by Dukelow and co-workers[243] in their work on rimmed, capped and semikilled steels.

Oxygen sensors have been developed also by other workers They are basically the same as the one described above. The most successful ones also use solid reference electrodes in preference to gas as does the one developed by United States Steel. The only difference is that in some Mo-MoO$_2$ is used as the reference electrode in place of the Cr-Cr$_2$O$_3$. Oxygen sensors are discussed further later in this chapter under "Observations in Practice."

SOLUBILITY OF DEOXIDATION PRODUCTS

Selected data for the deoxidation equilibria in Fe-X-O melts at 1600°C are presented in Figure 13—173, reproduced from a recent compilation[224].

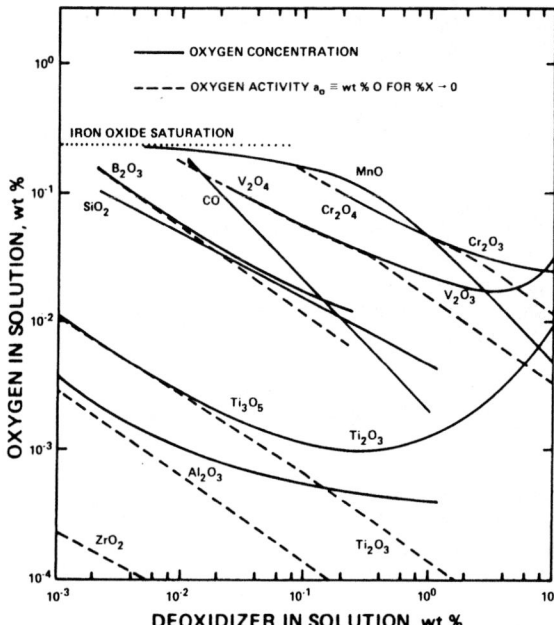

FIG. 13—173. Deoxidation equilibria in liquid iron alloys at 1600°C. (Turkdogan[244]).

The solubility of the deoxidation products in liquid iron alloy is represented in a general form by

$$M_xO_y = x\underline{M} + y\underline{O} \qquad (222)$$

for which the equilibrium constant is

$$K = \frac{[a_M]^x [a_O]^y}{a_{M_xO_y}} \qquad (223)$$

The composition of the oxide phase in equilibrium with the melt depends on temperature and composition of the alloy. For example, at the univariant point at 1600°C iron containing about 3 per cent Cr is in equilibrium with co-existing oxides chromite, FeCr$_2$O$_4$, and chromic oxide, Cr$_2$O$_3$. Below 3 per cent Cr, chromite is the equilibrium oxide phase and above 3 per cent Cr, chromic oxide is formed as the equilibrium phase. The points marked on the curves give compositions corresponding to the three-phase univariants at which two oxide phases are in equilibrium with the melt.

With the exception of Fe(Mn)O, all other solid deoxidation products have essentially stoichiometric compositions; in such cases the activity of the oxide is unity by definition.

Since, in general, we are concerned with dilute solutions of added elements in steel, the solute activities are chosen such that at infinitely dilute solutions $a_M \equiv$ wt %M and $a_O \equiv$ wt %O in the metal. Now, inserting the activity coefficients $f_M = a_M/\%M$ and $f_O = a_O/\%O$ and taking $a_{oxide} = 1$, equation (223) simplifies to

$$K = [\%M]^x[\%O]^y f_M^x f_O^y \qquad (224)$$

where $f_M = f_O \to 1$ when %M $\to$ 0.

The effect of alloying elements on the activity coefficient of oxygen in iron is shown in Figure 13—43. With the exception of B and Ti, log f_O^M is essentially a linear function of wt %M (added element) over the indicated composition range.

As is seen from the data in Figure 13—173 for deoxidation equilibria, in several systems the solubility of oxygen initially decreases with increasing content of the deoxidizer; further increase in the concentration of the deoxidizer brings about an increase in the oxygen solubility. The minimum on the oxygen solubility in equilibrium with the deoxidation product is caused by changes in the activity coefficients f_M and f_O with %M. In the system under consideration, as the concentration of the alloying element increases, so does its activity; however, the alloying element decreases the activity coefficient of oxygen; net result is that a minimum may occur in the oxygen solubility. The general trend is that the minimum oxygen content decreases as the stability of the deoxidation product increases; also, the alloy content at which the minimum occurs decreases with increasing stability of the oxide. The temperature dependence of the equilibrium constant is given in Table 13—XVIII for several deoxidizers.

Complex Oxides—It has long been recognized that the deoxidation is more extensive when the reaction product is dissolved in another oxide that is intermixed with the metal in the ladle treatment of the steel. This is achieved by using either the alloys of the deoxidants or a lime addition with the deoxidant. The deoxidation reaction equilibria considered below are for three cases of practical importance.

1. *Si/Mn deoxidation:* The use of silico-manganese in ladle deoxidation goes back to the days of the Bessemer steelmaking. The equilibrium relations between the reaction product molten manganese silicates and the residual manganese, silicon and oxygen in solution in the metal are well known. The equilibrium data for simultaneous deoxidation of steel with silicon and manganese at 1600°C are given in Figure 174a.[245] The critical silicon and manganese contents of iron in equilibrium with silica-saturated manganese silicate melts are given in Figure 13—174b for several temperatures. For alloy compositions above a given isotherm, the manganese does not participate in the reaction; instead silica is formed as the deoxidation product. In the region below the isotherm, the reaction product is molten manganese silicate (with little iron oxide), the composition of which is determined by the ratio [%Si]/[%Mn]^2 in the metal as deduced from the equilibrium constant for the reaction.

Table 13—XVIII. Deoxidation solubility products in liquid iron.

Equilibrium constant K (*)	Composition range	K at 1600°C	log K
$[a_{Al}]^2[a_O]^4$	< 1 ppm Al	1.1×10^{-15}	$-\dfrac{71\,600}{T} + 23.28$
$[a_{Al}]^2[a_O]^3$	> 1 ppm Al	3.2×10^{-14}	$-\dfrac{67\,260}{T} + 22.42$
$[a_B]^2[a_O]^3$	—	1.3×10^{-8}	
$[a_C][a_O]/p_{CO}$	> 0.02% C	2.0×10^{-3}	$-\dfrac{1\,168}{T} - 2.07$
$[a_{Cr}]^2[a_O]^4$	< 3% Cr	4.0×10^{-6}	$-\dfrac{50\,700}{T} + 21.70$
$[a_{Cr}]^2[a_O]^3$	> 3% Cr	1.1×10^{-4}	$-\dfrac{40\,740}{T} + 17.78$
$[a_{Mn}][a_O]$	> 1% Mn	5.1×10^{-2}	$-\dfrac{14\,450}{T} + 6.43$
$[a_{Si}][a_O]^2$	> 20 ppm Si	2.2×10^{-5}	$-\dfrac{30\,410}{T} + 11.59$
$[a_{Ti}]^3[a_O]^5$	0.01-0.25% Ti	7.9×10^{-17}	
$[a_V]^2[a_O]^4$	< 0.1% V	8.3×10^{-8}	$-\dfrac{48\,060}{T} + 18.61$
$[a_V]^2[a_O]^3$	> 0.3% V	3.5×10^{-6}	$-\dfrac{43\,200}{T} + 17.52$
$[a_{Re}]^2[a_O]^3$		$\sim 10^{-20}$	

(*) Activities are chosen such that $a_M \equiv$ %M and $a_O \equiv$ %O when %M $\to$ 0.

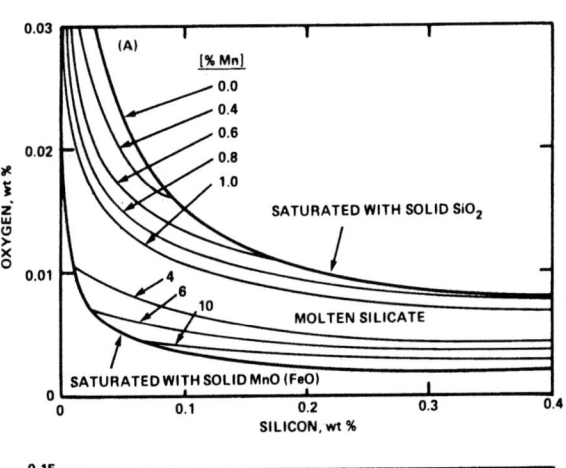

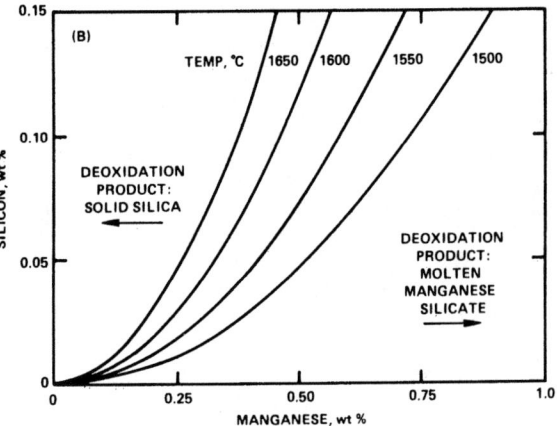

FIG. 13—174. Equilibrium relations for deoxidation of steel with silicon and manganese at 1600°C. (Turkdogan[245]).

$$\underline{Si} + 2MnO = 2\,\underline{Mn} + SiO_2,$$

$$K = \frac{[\%Mn]^2}{[\%Si]} \frac{(a_{SiO_2})}{(a_{MnO})^2} = 118 \text{ at } 1600°C, \qquad (225)$$

and the activities of the oxides in the MnO-SiO₂ melts.

2. *Al/Si/Mn deoxidation:* The Si/Mn deoxidation is used primarily in the production of rimmed and capped steels. For most grades of semikilled steels, the residual dissolved oxygen in the range 30 to 60 ppm is obtained by simultaneous deoxidation with Al, Si and Mn. The equilibrium relations for this complex deoxidation have been computed by Fujisawa and Sakao[246] from the free-energy data for the oxides and the activities of the constituent oxides in the MnO-SiO₂-Al₂O₃ system. The calculated equilibrium relations for 1550° and 1650°C are given in Figure 13—175 for steels containing [%Mn] + [%Si] = 1. Depending on the concentration of deoxidants in solution, the deoxidation product is (i) liquid aluminosilicate, (ii) galaxite (MnO·Al₂O₃, mp ∿ 1850°C), (iii) corundum (Al₂O₃, mp 2054°C) or (iv) mullite (3Al₂O₃·2SiO₂, mp ∿ 1850°C). The variation of oxygen activity with the concentration of aluminum and the ratio [%Mn]/[%Si] in liquid steel at 1550°C is shown in Figure 13—176. For a semikilled steel to contain, for example, 1 per cent Mn, 0.025 per cent Si and 40 ppm O in solution, and a liquid aluminosilicate deoxidation product, the residual aluminum in solution would be negligibly small, below 2 ppm Al.

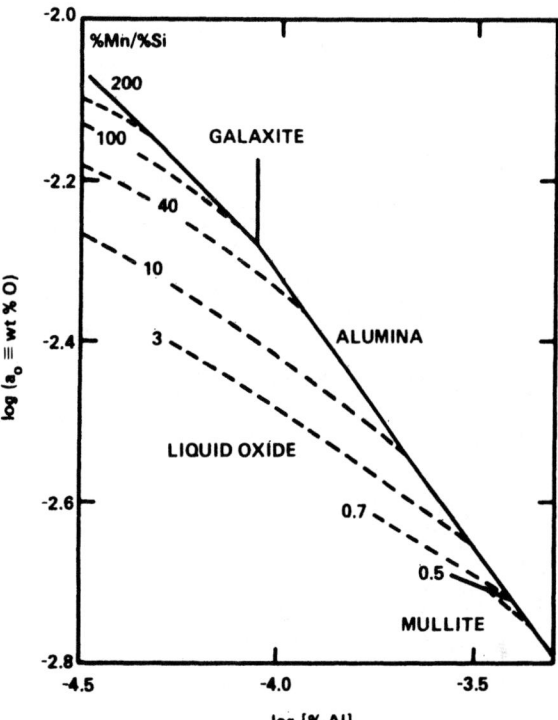

FIG. 13—176. Equilibrium oxygen activity as a function of %Al and %Mn/%Si ratio in liquid steel at 1550°C. After Fujisawa and Sakao.[246]

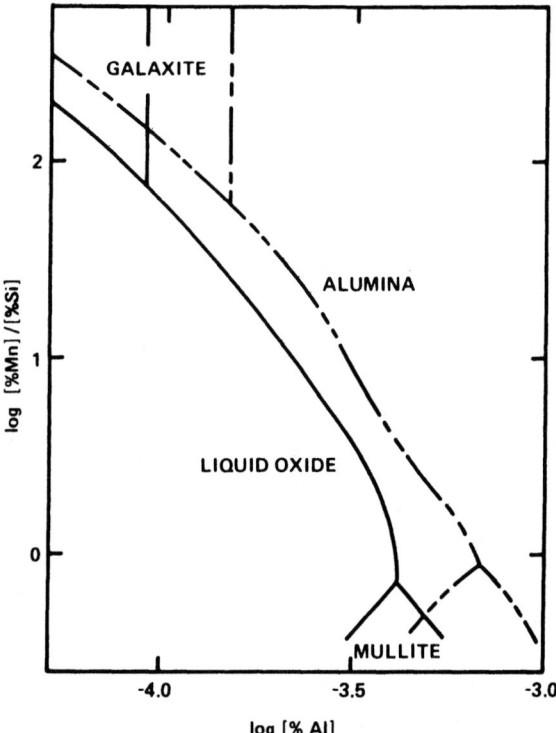

FIG. 13—175. Equilibrium relations for deoxidation of liquid steel with Al-Si-Mn for %Si + %Mn = 1. After Fujisawa and Sakao.[246]

3. *Aluminum-lime deoxidation:* Complete deoxidation of steel with aluminum to produce killed steel is much facilitated by ladle addition of burnt lime because of the formation of molten calcium aluminate as the reaction product,

$$CaO + 2\underline{Al} + 3\underline{O} = CaO·Al_2O_3. \qquad (226)$$

The calculated equilibrium relations between the oxygen activity in the deoxidized metal and composition of the molten calcium aluminate are shown in Figure 13—177 for 1600°C and concentrations of dissolved Al in the range 0.001 to 0.05%. To facilitate the fluxing of the deoxidation product alumina with lime, some fluorspar should be added (10 to 15 per cent) together with lime.

The deoxidation with a mixture of aluminum and burnt lime is particularly helpful in the production of semikilled or killed steel low in residual (dissolved) silicon and aluminum. For example, the steel deoxidized with aluminum alone to a level of 4 ppm oxygen activity will contain 0.02 per cent Al as residual aluminum in solution. On the other hand, with aluminum + lime deoxidation, forming lime-saturated aluminate liquid reaction product, the same level of deoxidation will be achieved with only 0.001 per cent Al left in solution. For 100 per cent efficient usage of the additives in deoxidation, the recommended mixture is 3 parts CaO + 1 part Al by weight to give lime-saturated calcium aluminate liquid reaction product.

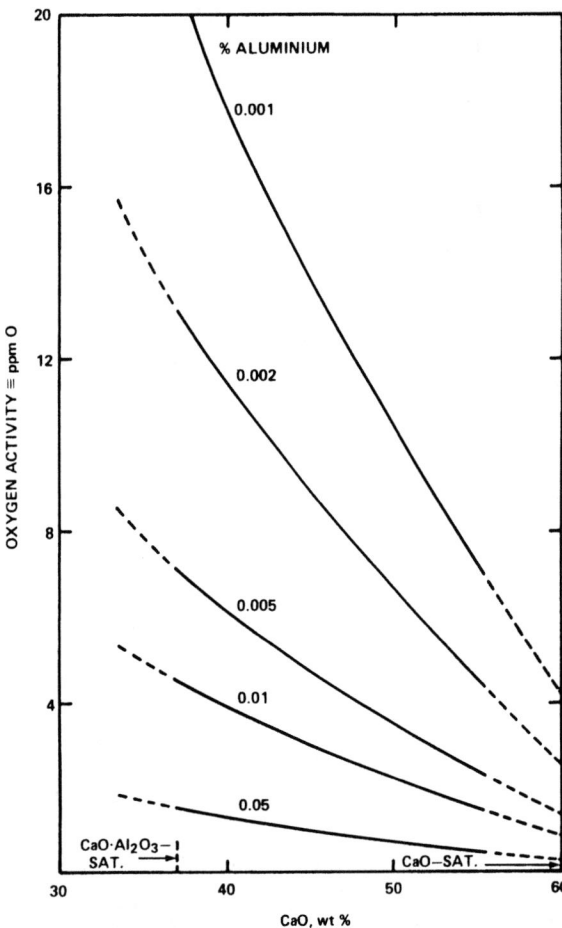

FIG. 13—177. Deoxidation equilibrium with aluminum and calcium aluminate at 1600°C. (Turkdogan[244]).

DESULPHURIZATION

Solubility Product of Simple Sulphides—Before discussing reaction equilibria in ladle desulphurization of steel, the solubility of a few selected sulphides in oxygen-free liquid iron as limiting cases will be considered.

The solubilities of LaS, CeS, Zr_3S_4 and TiS in liquid iron alloys were measured by Ejima et al.[247] by equilibrating the melt with the respective sulphide that formed the inner lining of the melt container. Measured concentrations of sulphur and the alloying elements in solution saturated with pure solid sulphides at 1600°C are shown in Figure 13—178; the line for pure liquid MnS is calculated from the available free-energy data. The dotted lines indicate the limiting theoretical slopes for low solute concentrations when the activity coefficients f_X and f_S approach unity.

The solubility products for MgS and CaS are much lower than those of the rare-earth sulphides. The solubilities at 1600°C estimated from the free-energy data are 37 ppm S for 1 bar Mg vapor and less than 0.001 ppm S for 1 bar Ca vapor. Because of their high vapor pressures, these alkaline-earth elements are not suitable for effective ladle desulphurization of steel.

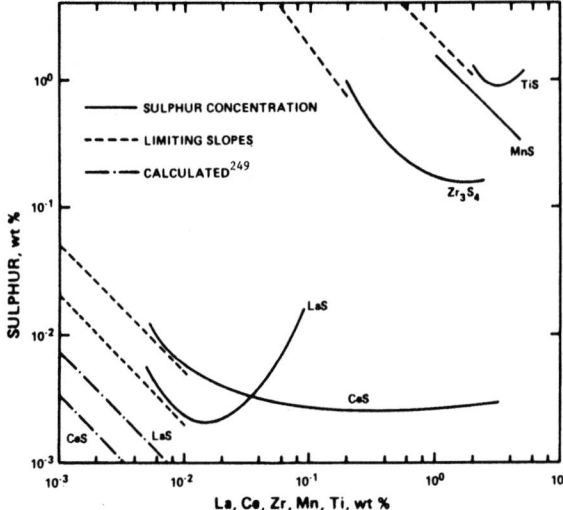

FIG. 13—178. Desulphurization equilibria in oxygen-free liquid iron alloys at 1600°C. (Turkdogan[244]).

Oxide/Sulphide Equilibrium—Although there are many variations of the methods of ladle desulphurization of steel, a feature common to them all is the necessity for effective deoxidation of the steel. All desulphurizing elements react also with oxygen in the metal; therefore, the overall reaction to be considered is that involving an oxide and a sulphide phase,

$$\underline{S} + MO = \underline{O} + MS. \qquad (227)$$

For the limiting case of unit activities of MO and MS, the equilibrium constant is given by the ratio of the activities of oxygen and sulphur in liquid steel,

$$K = \frac{[a_O]}{[a_S]} \qquad (228)$$

Examples are given in Figure 13—179 for alkaline-earth sulphide/oxide and cerium oxysulphide/oxide equilibria. For a given level of deoxidation, the higher the equilibrium ratio a_O/a_S, the lower will be the residual sulphur in the steel. Therefore, good desulphurization is expected with barium oxide, whereas no desulphurization can be achieved with MgO.

In rare-earth-treated steels containing oxides Re_2O_3, the sulphide precipitates as an oxysulphide Re_2O_2S. Fruehan[248] determined the oxygen and sulphur activities of coexisting phases Ce_2O_3 and Ce_2O_2S by equilibrating them with CO-CO_2-SO_2 gas mixtures at temperatures of 900° to 1400°C, using the available free-energy data for cerium sulphides, to calculate the phase-predominance diagram shown in Figure 13—180 for 1627°C. The activities are defined such that at very low concentrations of cerium, $a_O \rightarrow \%O$ and $a_S \rightarrow \%S$. In rare-earth-treated steels, the metal is hardly ever free of Ce_2O_3, therefore Ce_2O_2S is expected to form. If all the particles of Ce_2O_3 were coated with a layer of Ce_2O_2S, at sufficiently low oxygen activities some cerium sulphide might be found on the outer layer of the inclusions.

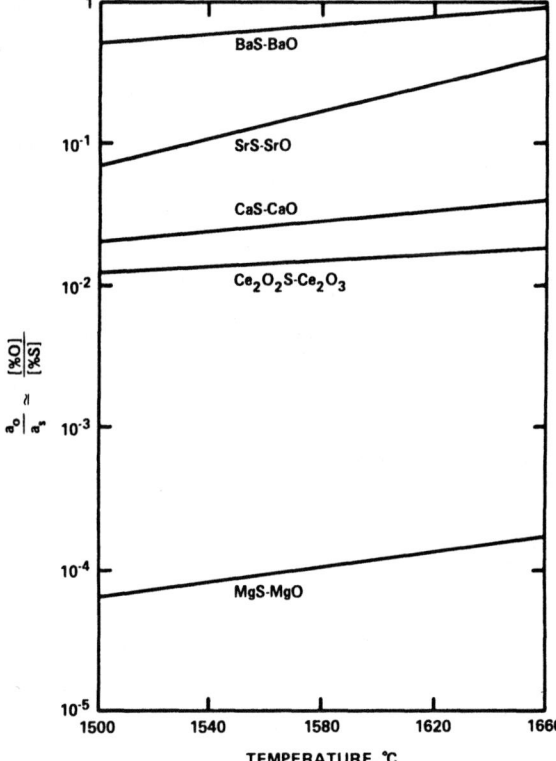

FIG. 13—179. Oxygen/sulphur activity ratio in liquid iron for the indicated sulphide-oxide equilibrium at 1600°C. (Turkdogan[244]).

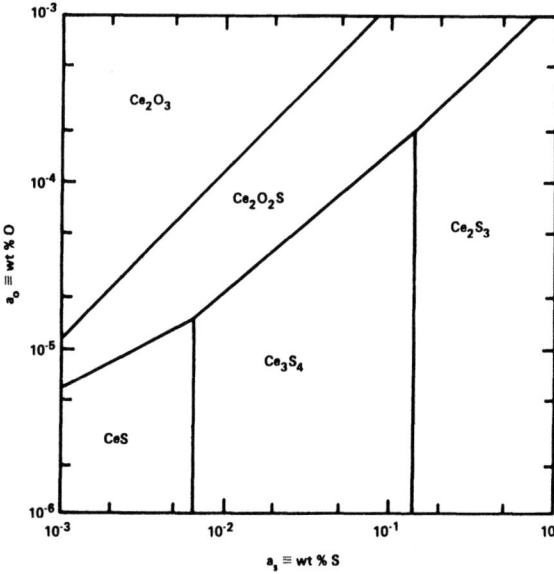

FIG. 13—180. Phase equilibria for the Fe-Ce-O-S system at 1627°C. After Fruehan.[248]

Slag-Metal Equilibrium—The lime and aluminum are the two essential ingredients of ladle additions for effective deoxidation and desulphurization of the steel. From the point of view of reaction equilibria, the extent of deoxidation and desulphurization is determined by temperature, slag composition, and concentration of residual aluminum in solution in steel. Because of practical importance, only the reaction equilibria involving calcium-aluminate-based slag will be considered.

The overall reaction is

$$CaO + \frac{2}{3}\underline{Al} + \underline{S} = CaS + \frac{1}{3}Al_2O_3. \qquad (229)$$

Despite the practical importance, no direct measurements have been made of this slag-metal reaction equilibrium. However, the equilibrium relations for reaction (229) can be derived with some confidence from the available equilibrium data for the appropriate gas-slag and gas-metal reactions and the free-energy data. The calculations are based partly on the sulphide capacity data for $CaO\text{-}SiO_2$, $CaO\text{-}Al_2O_3$ and $CaO\text{-}Al_2O_3\text{-}SiO_2$ melts (Figure 13—181), compiled by Abraham and Richardson;[249] the curves in Figure 13—181 are terminated at compositions of melts (given in mole fractions x_{CaO}) saturated with a solid oxide.

From the known free energies of solution of gaseous oxygen and sulphur in liquid iron, the equilibrium constant is evaluated for the following exchange reaction:

$$\frac{1}{2}O_2 + \underline{S}[1\ wt\%] = \frac{1}{2}S_2 + \underline{O}[1\ wt\%] \qquad (230)$$

$$K = \frac{[a_O]}{[a_S]}\left(\frac{p_{S_2}}{p_{O_2}}\right)^{1/2}, \ \log K = \frac{935}{T} + 1.375 \qquad (231)$$

The isothermal equilibrium relation for the slag-metal reaction

$$\underline{S} + O^{2-} = S^{2-} + \underline{O}, \qquad (232)$$

is represented by the following equation for a given slag composition:

$$k_S = KC_S = (\%S)\frac{[a_O]}{[a_S]} \qquad (233)$$

where the activities are defined such that at infinite dilution of alloying elements and low concentrations of oxygen and sulphur, $a_O \to$ wt. %O and $a_S \to$ wt. %S, and C_S is the sulphide capacity of the slag.

$$k_S = \frac{(\%S)}{[\%S]}[a_O]. \qquad (234)$$

From C_S in Figure 13—181 for calcium aluminates and equations (231) and (234), the equilibrium relation k_S is calculated as a function of slag composition for 1500° and 1650°C. The results are given in Figure 13—182; the dotted curve represents the experimental data of Schurmann et al.[250] In their experiments, the oxygen activity, controlled by aluminum addition, was measured by an oxygen-emf cell with a thoria electrolyte. The values of k_S given by Schurmann et al. are about three times greater than those computed from the sulphide capacity data.

The equilibrium relation shown in Figure 13—183 is obtained for the sulphur distribution ratio as a function

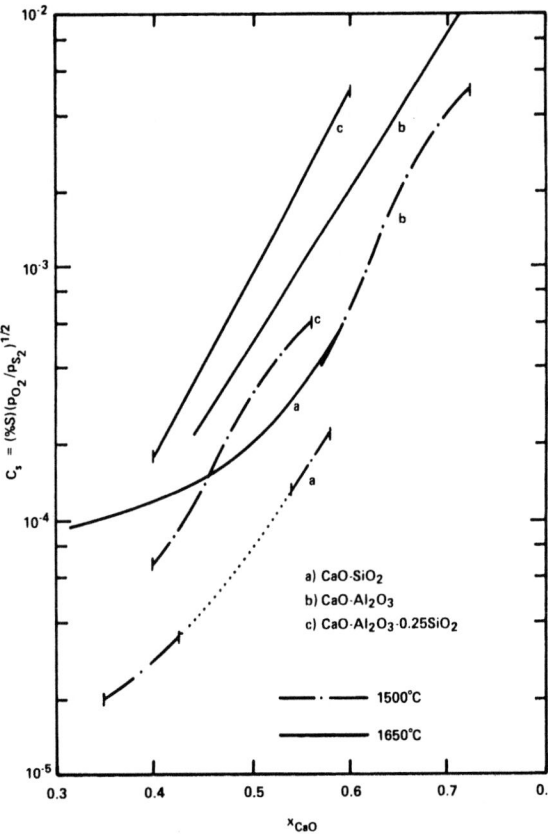

FIG. 13—181. Sulphide capacity of CaO-SiO₂, CaO-Al₂O₃ and CaO-Al₂O₃-0.25 SiO₂ melts from the data of Abraham and Richardson.[249]

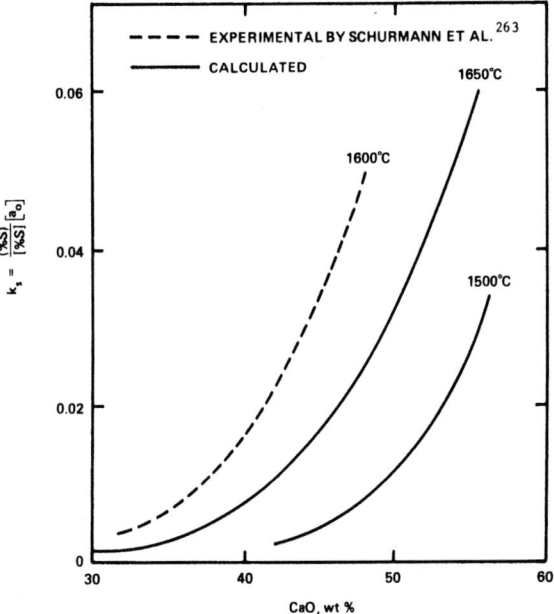

FIG. 13—182. Sulphur distribution equilibrium between liquid iron and calcium aluminate at indicated temperatures and compositions.

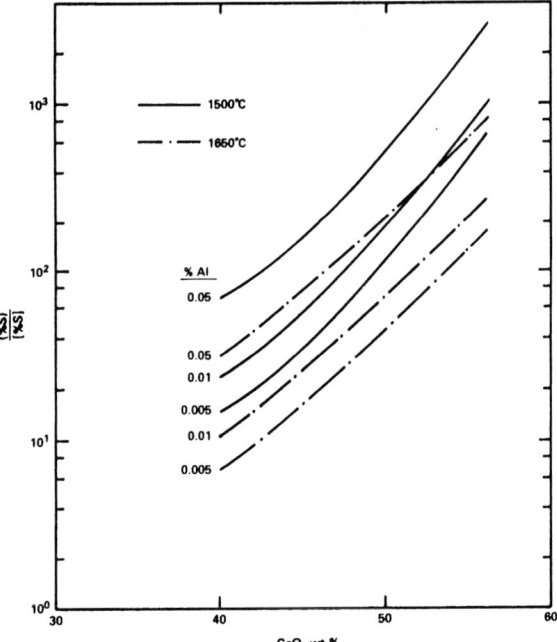

FIG. 13—183. Variation of the equilibrium sulphur distribution ratio with temperature and concentration of aluminum in solution in steel and calcium oxide in calcium aluminate melts. (Turkdogan[244]).

of the concentration of dissolved aluminum in liquid steel and calcium oxide in calcium-aluminate melts. The concentration of dissolved sulphur in steel in equilibrium with CaS-saturated calcium aluminate is given in Figure 13—184. The aluminate slag formed in the ladle practice is usually saturated with lime; therefore, the concentration of aluminum and the melt temperature are the two independent variables that affect the extent of ladle desulphurization. By appropriate interpolation and extrapolation of the equilibrium data cited, calculations are made of the relation between [%S] and [%Al] at temperatures of 1550°, 1600°, and 1650°C for the aluminate containing 56 per cent CaO which is close to lime saturation; the results are given in Figure 13—185. As discussed later, low levels of sulphur obtained, less than 0.002 per cent in ladle desulphurization with aluminum and lime, are consistent with the calculated relations shown in Figure 13—185.

HYDROGEN PICKUP

The moisture in the furnace atmosphere and in the scrap, lime and fluxes charged to the furnace is the primary source of hydrogen pickup in steelmaking. Although most of the hydrogen is removed during the "carbon boil", there is hydrogen pickup by the steel during the ladle treatment for deoxidation, recarburization, desulphurization and alloying additions, because the added materials are not moisture free. The extent of hydrogen pickup from water vapor depends on the activity of oxygen in the steel as indicated by the reaction

$$H_2O(g) = 2\underline{H} + \underline{O} \tag{235}$$

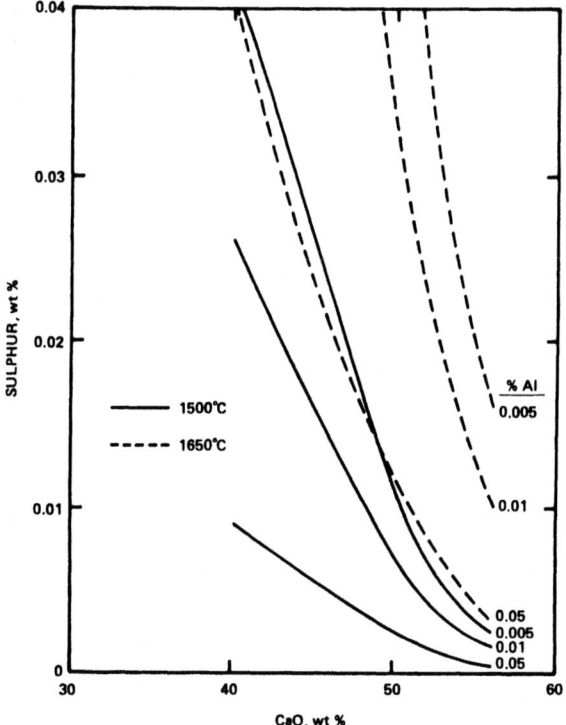

FIG. 13—184. Sulphur content of liquid steel in equilibrium with CaS-saturated calcium aluminate at indicated temperatures and compositions. (Turkdogan[244]).

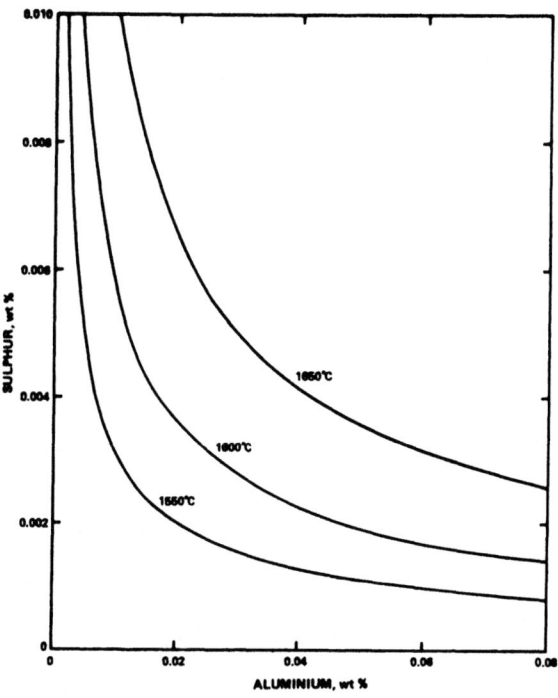

FIG. 13—185. Calculated relation between concentrations of dissolved sulphur and aluminum in liquid iron in equilibrium with calcium aluminate melt containing 56 per cent CaO. (Turkdogan[244]).

for which the following equilibrium constant is derived from the available thermodynamic data

$$\log \frac{[H]^2[O]}{p_{H_2O}} = - \frac{10\ 610}{T} + 11.913 \qquad (236)$$

where p_{H_2O} is in bar and the concentrations ($\equiv$ activity) [H] and [O] are in ppm. The equilibrium concentrations of hydrogen and oxygen computed for various partial pressures of water vapor at 1600°C are given in Figure 13—186. It is seen that the deoxidized steel is more susceptible to hydrogen pickup.

A simple material balance indicates that even a small amount of moisture in the ladle additives could be responsible for hydrogen pickup by the steel. The importance of ladle additions being free of moisture cannot be overemphasized.

RATE PHENOMENA

Various aspects of rate phenomena have been studied in relation to improved methods of ladle treatment of steel: dissolution of metallic additives, separation of reaction products from liquid steel; minimization of the reoxidation of steel during casting, with the ultimate objective of producing cleaner steels; and the formation of a type of residual inclusions that are not detrimental to the mechanical properties of the product. These aspects of the rate phenomena are discussed briefly in the following sections.

Mechanism of Dissolution of Additives—The adjustment of steel composition by the late ladle additions, subsequent to tapping, can be a problem because of the buoyancy effect and retarded melting of the additives. In an experimental and mathematical study, Guthrie et al.[251] showed that severe thermal and fluidynamic contacting resistances are encountered when buoyant objects are projected into liquid steel. The thermal effect arises from the formation of a solid

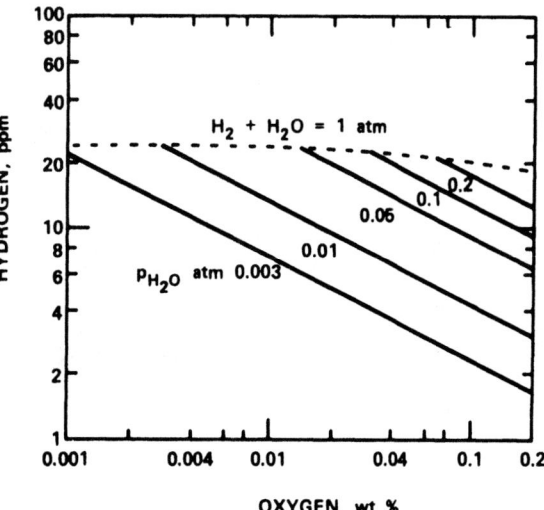

FIG. 13—186. Concentrations of hydrogen and oxygen in liquid iron at 1600°C in equilibrium with indicated compositions of H_2-H_2O mixtures.

shell of steel around the metal particle added to a bath of molten steel because of a rapid withdrawal of heat by the immersed object from the adjacent liquid. The depth of penetration and residence time of an object in the steel bath depend on the entry velocity and size and mass of the object. These variables, as well as the thermal properties of the added metal, determine the extent of melting that will take place before resurfacing occurs. The predictions made by Guthrie et al. from their mathematical analysis are summarized in Figure 13—187 for aluminum spheres projected into a quiescent bath of liquid steel at indicated entry velocities of 5 to 100 m s^{-1}.

The dotted lines show the time of residence as a function of the diameter of the aluminum sphere for indicated entry velocities. The bottom curve (full line) is for the start of melting of the aluminum inside the solid steel shell, the middle curve is for complete melting of the aluminum, and the top curve is for the start of melting of the steel shell. For example, a 2-cm-diameter aluminum sphere projected with an entry velocity of 50 ms^{-1} will remain in the steel bath for 0.8 s, which is the time required for complete melting of the aluminum inside the solid steel shell. For a 1-cm-diameter sphere with an entrance velocity of 100 m s^{-1}, the aluminum will be completely molten and the steel shell partly melted in a residence time of 0.7 s.

These findings of Guthrie et al. indicate that light

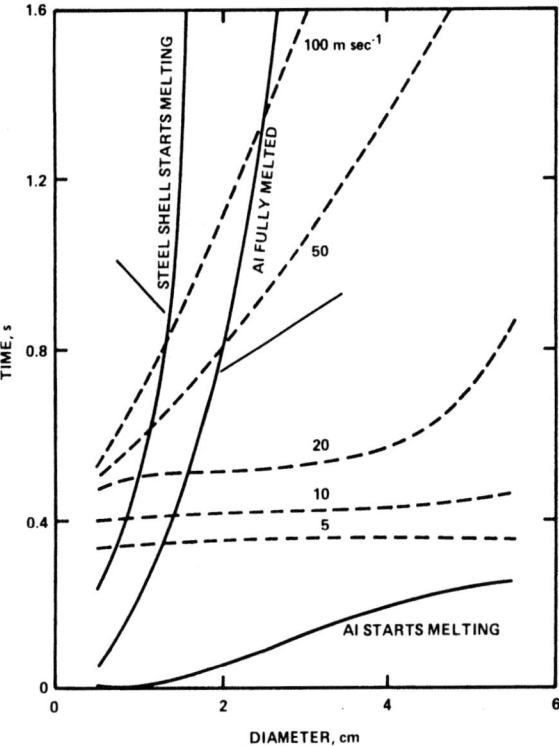

FIG. 13—187. Calculated residence time for indicated entry velocities as a function of the diameter of added aluminum sphere (- - - -); start and completion of the melting of aluminum and start of the melting of the steel shell (———). After Guthrie et al.[251]

metals, such as aluminum, added to a quiescent liquid steel bath will generally refloat before complete dissolution in the bath, even when projected with high entry velocities, as with bullet shooting methods.[252] In practice, the ladle additions are made into the tapping stream; in the late ladle additions argon is injected deep into the melt to facilitate dissolution of additives before resurfacing. Nowadays, moderate ladle additions are made by the 'wire' feed technique. It should be emphasized that any late ladle additions, accompanied with argon stirring, should be done under a cover of a nonoxidizing slag, such as calcium aluminate, with a lid on the ladle to curtail air oxidation of the bath.

NUCLEATION, GROWTH, AND FLOTATION

The extent of supersaturation of liquid steel with oxygen and deoxidant needed for the spontaneous nucleation of the deoxidation product may be estimated from the theory of homogeneous nucleation. Turpin and Elliott[253] gave several examples of estimating the supersaturation needed for homogeneous nucleation of various types of oxides in liquid steel. For example, for liquid iron-alumina or iron-zirconia systems, the interfacial energy is in the range 1500-2000 mJ m^{-2}, for which the supersaturation ratio R_S is of the order of 10^5 to 10^{10}. In fact, experimental observations of von Bogdandy et al.[254] have shown that the spontaneous nucleation of these oxides in liquid iron occurs only when R_S is more than about 10^8. Yet, in practice, spontaneous nucleation occurs readily even though R_S in the melt is 50 or less for homogeneous distribution of the added deoxidant. The deoxidizers added to steel always contain some oxide particles which will serve as nucleation sites; hence little or no supersaturation is needed for the initiation of the deoxidation reaction.

The rate of the deoxidation reaction is probably controlled by solute diffusion to the oxide particles dispersed in the melt; examples of such calculations are given elsewhere.[255] The inclusion counts in steel samples give values in the range 10^5 to 10^7 particles cm^{-3}, for which the time of completion of deoxidation is of the order of a few minutes to reach the equilibrium levels of 10-20 ppm O. If the equilibrium oxygen concentration is below 0.1 ppm, as would be the case when the steel is deoxidized with rare-earth elements, approach to equilibrium may be slow because of the small driving force for diffusion.[256] However, for most ladle practices, the diffusion-controlled rate of deoxidation is expected to be relatively fast.

The slow step in deoxidation is the flotation of inclusions that are less than 20 μm diameter. Rapid inclusion growth and flotation is enhanced by stirring of the melt. For example, Miyashita et al.[257] have shown that while the rate of flotation of inclusions in unstirred melts is in accord with Stoke's law, the rate of rise is much faster in stirred melts. Also, experiments have shown that in stirred melts the inclusions are trapped on surfaces of the ladle lining.[258, 259] Various methods of stirring have been applied in steel plants: electromagnetic, mechanical, and argon bubbling. Reference may be made to review papers[260] on the practical merits of these processes. Reference may be made also to Szekely[261] for recent studies of better understanding of the fluidynamics of mixing in stirred melts through mathematical modeling and computer-aided calculations. For the present it will suffice to state that the experimental and plant studies have shown that the rate of flotation of deoxidation products in stirred melts can be represented by an equation of the form

$$C_t = C_o \exp (- kt) \qquad (237)$$

where C_0 is the initial concentration of inclusions, C_t the concentration at time t, and k the apparent flotation-rate constant for a given type and intensity of stirring.

OBSERVATIONS IN PRACTICE

A few examples are given here of some technical aspects of ladle refining of steel, with emphasis on deoxidation and desulphurization, and implementation of research findings for further developments in ladle practices.

Oxygen Sensors—The basic features of commercial oxygen sensors are essentially the same: lime-stabilized zirconia as the electrolyte and $Cr\text{-}Cr_2O_3$ or $Mo\text{-}MoO_2$ as the reference oxygen electrode. However, there are variations in the quality of performance of oxygen sensors, partly because of variations in sensor design and properties of the electrolytes used. It is well to remember that for accurate measurement of oxygen activity with the emf cell, the solid oxide electrolyte should (i) be an ionic conductor with negligible electronic conductivity, (ii) have a negligibly small permeability to oxygen ions, and be free of microcracks and pores, for negligible diffusion of gaseous oxygen, to avoid polarization at the electrolyte-electrode interface; and (iii) not react with liquid steel or the reference electrode. As shown by Janke and Richter,[262] these effects become more noticeable at low oxygen activities in the steel, particularly for the $ZrO_2(MgO)$ electrolyte. They showed that by making an appropriate correction for the partial electronic conductivity, oxygen activities at low levels can be reliably measured with the stabilized zirconia electrolytes. The yttria-doped thoria electrolytes, which have negligibly small electronic conductivity, were shown to give more reliable results.

Many plant tests have been made to evaluate the applicability of the oxygen sensor in steelmaking and develop improved deoxidation practices. It is generally agreed that the adjustment of silicon and aluminum additions on the basis of the measurement of dissolved oxygen in the furnace by the oxygen probe leads to a better control of the concentrations of dissolved silicon and aluminum at desired levels for rimmed, semikilled and killed steels. In steels containing more than 0.1 per cent C, the oxygen content of the steel does not vary much from one heat to the next, perhaps 150 to 300 ppm O, and therefore the ladle additions can be judged from the carbon content at tap (>0.2 per cent C) with good reproducibility of results without using the oxygen sensor.

However, there are large variations in the oxygen content of low-carbon steels (<0.1 per cent C), as much as 500 to 900 ppm O for <0.05 per cent C; therefore, the use of the oxygen sensor in low-carbon steels was found to be of considerable help in the control of deoxidation and concentrations of residual Mn, Si and Al. After partial or complete deoxidation in the tap ladle, the oxygen activity in the melt should be measured again to determine the amount of mold additions to control rimming action, or to adjust the concentration of aluminum in the steel, as in the aluminum trim prior to the desulphurization treatment. However, if the residual aluminum in the tap ladle is more than about 0.02 per cent, instead of measuring the oxygen activity by the sensor, the steel sample should be analyzed for aluminum for better control of the aluminum trim.

In addition to the use of oxygen sensors, a good furnace-tapping practice is essential for better control of deoxidation and the residual concentrations of Al, Si and Mn in the tapped steel. Such a practice would result in little or no slag carryover into the ladle and minimum flare of the metal stream to suppress air oxidation of the ladle additives. These objectives can now be accomplished by using, for example, the sliding-gate-type tap holes in open-hearth and electric-arc furnaces, or a floating refractory ball or cube for the BOF tap hole. Also, in some practices, the freeboard in the covered ladle is flushed with argon to minimize air oxidation of the metal stream during tapping.

ARGON STIRRING

Argon stirring of the melt is now a common practice in ladle refining of steel. The manner of argon injection and the flow rates used depend on the objectives to be accomplished. In the North American steel plants, various terms are used to identify the specific objectives of argon injection.

Argon Rinse—To homogenize the melt temperature and composition and also assist the flotation of deoxidation products, the argon is blown through the melt at a rate of 0.08-0.13 nm^3 min^{-1} for 3 to 5 min.

Argon Trim—The gas is blown at a rate of 0.30-0.45 nm^3 min^{-1} to facilitate the dissolution of ladle additions.

Argon Stir—The argon is blown at a rate of 0.3-0.5 nm^3 min^{-1} to achieve slag-metal mixing in ladle desulphurization of steel.

Powder Injection—Powdered desulphurizing materials such as magnesium-fluorspar (Mag-Spar), calcium silicide (Cal-Sil) or prefused calcium aluminate, are injected with a single or a double-port lance deep into the steel bath in the ladle with argon flowing at rates in the range 0.9-1.8 nm^3 min^{-1}.

Following the development work done at DOFASCO,[263] the porous refractory plug installed at the bottom of the ladle is now being used extensively for argon rinse, trim and stir. There should always be sufficient slag cover on the melt to minimize air oxidation of the melt during argon injection. In fact, experience shows that prolonged argon stirring results in loss of dissolved aluminum and decrease in steel cleanliness.

The efficiency of coalescence of inclusions accompanying their collision depends on the intensity of stirring, as suggested by equation (237) and also on the surface properties of the particles. For example, the force of adhesion of alumina particles, with a contact angle of 140° in liquid iron, is about twice that for silica particles for which the contact angle is about 115°. Therefore, the alumina inclusions will cluster and float out of the melt more readily than the silica or molten oxide inclusions, as substantiated experimentally by Torssell and Olette.[264]

MOLD ADDITIONS

For certain grades of steel the mold additions are made, both in continuous casting and ingot casting, to control rimming action, deoxidation or morphology of inclusions that are formed during solidification. One example of mold addition to be mentioned here is for ingot casting of rim-stabilized steel.

The rim-stabilized steel consists of a rimmed outer zone and an aluminum-killed core having an aluminum content ranging from about 0.025 to 0.070 percent. After being teemed into an ingot mold, the steel is allowed to rim for 2 to 3 min, then aluminum is added during the backpouring step of teeming to produce the composite ingot. In most practices solid aluminum is used in the form of granules, shot, heavy wire or rod. As discussed earlier, this manner of mold addition is inherently inefficient because of the buoyancy effect and sluggish melting of the additives.

In the process developed at United States Steel Corporation (MA-RK and MA-RS process)[265] molten aluminum is added into the rimming ingot during the teeming operation. MA-RK is an acronym for molten aluminum-rimmed killed steel and MA-RS for molten aluminum-rimmed semikilled steel. In both processes, after 2 to 3 min rimming action, the ingot is topped off with additional steel while at the same time molten aluminum is injected into the steel stream. For severe deep-drawing applications, the interior of the ingot is killed, with residual aluminum in the range of 0.025-0.050 per cent. For less severe deep-drawing applications, only enough molten aluminum is added to produce a semikilled core that forms a domed top on freezing.

Although rim-stabilized steels are inherently dirty as regards to inclusion counts, the use of molten aluminum for mold addition produces a more uniformly distributed, galaxy type of less refractory inclusions that are less harmful in drawing, forming, and enameling applications.

LADLE DESULPHURIZATION

In addition to cleanliness, as manifested by low total oxygen and sulphur contents, it is necessary to control the composition, shape and size distribution of inclusions to achieve the desired properties in the high-strength, low-alloy steels (HSLA). The HSLA steels should have good cold bending, impact and through-thickness properties, good weldability, and resistance to lamellar tearing for heavy-plate welded constructions, automotive applications and line-pipe applications in arctic weather conditions. The ladle desulphurization and treatment of liquid steel with calcium or rare-earth alloys is necessary for HSLA steels, as well as controlled rolling to plates and sheets.

Knowledge acquired through research in the mid-

1950's to mid-1960's has been of much help in bringing about the developments of today in the ladle refining of steel. The basic requirements for the process are (i) extensive deoxidation with aluminum, (ii) mixing of liquid steel with desulphurizing additives, (iii) good mixing of the ladle slag (calcium aluminate based) with the steel, and (iv) minimum reoxidation of the bath. Two ladle refining processes are mentioned here as examples, with emphasis on the TN and other related processes.

In the ASEA/SKF process, after the vacuum-carbon deoxidation, the steel is killed with aluminum and desulphurized mainly by mischmetal (REM) while the bath is inductively stirred.[266] In the TN (Thyssen-Niederrhein)-process, the Mag-Spar or Cal-Sil is injected with argon into the Al-killed steel under a cover of a non-oxidizing slag (essentially calcium aluminate), followed by a short argon rinse.[267] The rate of desulphurization in the TN process is about 4 times faster than in the ASEA/SKF process, because the volume power density of stirring is greater in the TN process. For practical details on these processes, reference may be made to the papers cited.[268, 269] The present discussion is confined to certain fundamental aspects of the workings of the process.

The overall reaction (229) is the key to ladle desulphurization with a calcium aluminate-rich slag intermixed with the steel by stirring, preferably with argon injection. The extent of desulphurization that can be achieved is readily calculated from the equilibrium data presented in Figures 13—183 to 13—185 as limiting cases. For example, for the aluminate slag containing 56 per cent CaO and saturated with CaS (= 2 per cent S at 1600°C), the equilibrium dissolved sulphur in the steel at 1600°C is about 0.002 per cent for the concentration of dissolved aluminum in the range 0.04-0.05 per cent. The minimum quantity of slag required is that for CaS saturation. For the case considered, the minimum quantity is 5 kg slag per metric ton of steel for $\%\Delta S$ = [%S] initial—[%S] equilibrium = 0.01%. In practice about 15 to 25 kg slag per metric ton of steel is used, so that the slag would not be saturated with calcium sulphide, which in turn allows a faster rate of reaction and more extensive sulphur removal; also, a heavy slag cover would retard the reoxidation of the bath.

With lime-saturated aluminate slag and an uptake of 1.0 to 1.5 per cent S and steel containing 0.04 to 0.05 per cent residual aluminum, it should be possible to desulphurize steel to less than 10 ppm S at temperatures of 1550° to 1570°C. However, such a product can be made only by stepwise removal of sulphur at various stages of steelmaking, such as (i) desulphurization of hot metal, (ii) desulphurization during steelmaking, and (iii) some sulphur removal in the tap ladle so that the steel would contain 0.008 to 0.01 per cent S before the final stage of the treatment as in the TN process or modifications thereof.

Numerous claims have been made on the importance of injecting magnesium, as Mag-Spar, or calcium, as Cal-Sil, for more efficient ladle desulphurization and control of inclusion shape with residual magnesium or calcium in the metal. The validity of these popular views, however, may be questionable. In fact, our experience at United States Steel's Texas Works has shown that with lime-saturated calcium aluminate plus fluorspar top slag, the argon injection alone is adequate to achieve 0.001 per cent S in the steel.

First, let the solubility of magnesium and calcium in oxygen-free liquid iron be estimated. According to the measurements of Trojan and Flinn,[270] the solubility of magnesium decreases with decreasing carbon content of the iron. The extrapolation of their data to 1600°C gives for the solubility at 20 bar Mg vapor 1 per cent Mg at 1.6 per cent C and 1.75 per cent Mg at 3.00 per cent C; linear extrapolation to zero carbon content gives 0.15 per cent Mg at 20 bar. From these extrapolated data, the solubility is estimated to be about 75 ppm Mg at 1 bar Mg vapor and 1600°C. Sponseller and Flinn[271] equilibrated liquid iron and liquid calcium under pressure at various temperatures. They give 320 ppm Ca for the solubility at 1607°C for which the calcium vapor pressure is 2 bar. Therefore, the solubility is estimated to be about 160 ppm Ca at 1 bar Ca vapor and 1600°C.

As magnesium or calcium dissolves in the steel, it reacts with the slag and the refractory lining of the ladle. Because of intimate mixing of slag and metal, the following reaction in estimating the solubility of, for example, calcium in liquid steel should be considered:

$$CaO(slag) = \underline{Ca} + \underline{O} \qquad (238)$$

For CaO-saturated slag, the solubility product at 1600°C is about

$$[ppm, Ca] [ppm O] = 0.0054 \qquad (239)$$

For steel containing 0.04 per cent Al in equilibrium with CaO-saturated calcium aluminate, the oxygen activity is equivalent to about 0.3 ppm O, for which the equilibrium concentration of calcium in solution is about 0.02 ppm Ca, an insignificant amount. For the limiting equilibrium conditions considered, calcium cannot be considered as having a direct effect on the sulphide morphology during the solidification of steel. A similar statement would apply to magnesium.

Although hardly any calcium is expected to remain in solution, even in aluminum-killed steel, the product of the ladle-treated steel always contains some calcium. The residual calcium in liquid steel, prior to casting, is believed to be in the form of dispersed particles of CaO, CaS, and liquid calcium aluminate. In the ladle desulphurization with argon injection of Cal-Sil, the product contains usually 40 to 60 ppm Ca; with argon injection alone, the product contains 3 to 8 ppm Ca. In both practices there is extensive slag-metal mixing. When the gas injection is stopped, relatively large, dispersed-slag particles readily float out of the melt, hence the calcium content of the product is low. On the other hand, finely dispersed particles of calcium sulphide, oxide and aluminate that are formed with the injection of Cal-Sil do not float out quickly, resulting in higher concentrations of total calcium in the product. Finely dispersed particles of calcium oxide and aluminate are believed to play an important role in suppressing the formation of manganese sulphide stringers during solidification of the steel.

The calcium oxide formed by the oxidation of calcium vapor will react with sulphur in the melt to an extent determined by the state of equilibrium for reaction (229), similar to what would be achieved by intermixing the top slag with the metal by stirring with argon alone.

However, some beneficial effects are realized from the injection of magnesium or calcium compounds: (i) the metal vapor generated increases the volume power density of stirring, and (ii) the metal vapor escaping from the melt counteracts the air oxidation of the bath. Because the vapor pressure of magnesium (bp 1105°C) is much higher than that of calcium (bp 1491°C), the stirring of the bath is more violent when Mag-Spar is injected instead of Cal-Sil; therefore, a faster rate of desulphurization is expected with the injection of Mag-Spar. In fact, as reported by Bruder et al.,[269] the rate of desulphurization per mole of Mg injected is about twice that for per mole of Ca injected. Similar observations have been made in the operation of the TN process at Texas Works of U.S. Steel Corporation.

For more efficient usage of additives, prevention of nitrogen pickup, improvement of steel cleanliness and better control of the process, air aspiration into the freeboard of the ladle should be avoided in the ladle desulphurization of steel. With the elimination of air oxidation of the bath, the Cal-Sil injection would not be necessary. Such improvements have been made in some plants by various modifications of the process. For example, the ladle top is contained completely under a cover that is flushed with argon. In another single ladle practice, the ladle is placed in a tank that is maintained at a low pressure during argon stirring of the bath for desulphurization with the top slag. Then, there is the double-ladle practice where the desulphurization is incorporated with the ladle-to-ladle degassing. The added advantage of desulphurization under a reduced pressure is the removal of some hydrogen and nitrogen from the steel. Also, under a reduced top pressure, rigorous slag-metal mixing is achieved even with a slow rate of argon injection through the porous plug.

REACTIONS IN DEGASSING PROCESSES

Within the last twenty years there has been a large increase in the use of degassing processes. This has been in part due to the growth of continuous casting for which in many cases conventional deoxidation practices cannot be used. Degassing can be used to remove oxygen (via CO), hydrogen and nitrogen from steel. Degassing processes can be divided into two basic types; vacuum degassing and inert-gas flushing processes.

Vacuum Degassing—The chemical reactions occurring during vacuum treatment of steel are relatively simple

$$\underline{C} + \underline{O} = CO \ (g) \qquad (240)$$

$$\underline{N} = \tfrac{1}{2} \ N_2 \ (g) \qquad (241)$$

$$\underline{H} = \tfrac{1}{2} \ H_2 \ (g) \qquad (242)$$

The reactions proceed because the gases are formed at very low pressures (vacuum).

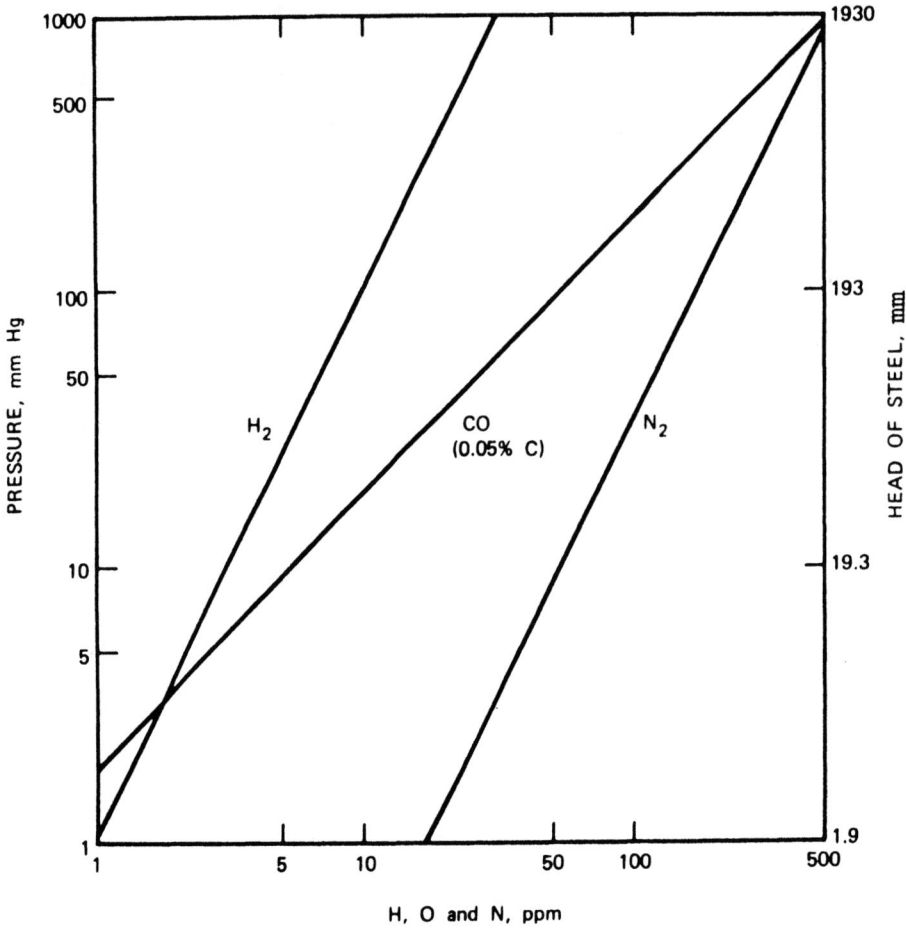

FIG. 13—188. Concentrations of hydrogen, oxygen (for 0.05% C) and nitrogen in steel at 1600°C in equilibrium with H_2, CO and N_2 pressures and the corresponding head of steel in cm.

The theory of homogeneous nucleation of bubbles has been substantiated experimentally with aqueous solutions of low surface tension. The application of this theory to liquid metals of high surface tension indicates that an excess pressure of the order of 10^4 atm is needed for bubble nucleation. Such a prediction is not consistent with experimental observations. In levitated liquid iron-carbon-oxygen alloys, not in contact with a refractory surface and presumably free from inclusions, the burst of gas evolution has been observed at carbon and oxygen contents which correspond to the CO-supersaturation of the melt at a pressure of 20 to 100 atm.[272] If this is a manifestation of homogeneous bubble nucleation, the supersaturation required is some orders of magnitude less than that predicted by the theory.

In practice, the bubble nucleation occurs primarily in crevices of the refractory lining, provided the solute content is sufficiently high to sustain the required excess pressure in the bubble under a given head of liquid metal. This is demonstrated in Figure 13—188 showing the concentrations of hydrogen, oxygen (for 0.05 per cent C) and nitrogen in steel at 1600°C in equilibrium with indicated H_2, CO and N_2 pressures and the corre-

sponding head of steel above which there is vacuum (or a very low gas pressure).

The higher the head of liquid, the higher will be the solute concentration for the bubble to grow. For example, for bubble growth under a 5-cm head of steel, there should be more than 5 ppm H or more than 100 ppm N. Partially deoxidized steel containing, for example, 0.05 per cent C will have about 120 ppm oxygen in solution and, consequently, CO bubbles will grow even under a much higher head of steel. Once the growth of gas bubble has been established, other solutes such as hydrogen and nitrogen will be flushed out of the metal. Degassing by purging of the metal with an inert gas such as argon is frequently used in the metals industry. Combination of argon purging and vacuum treatment as in the R-H (Ruhrstahl-Heraeus) process further enhances the efficiency of degassing.

Due to kinetic factors, the equilibrium is often not achieved in these processes. The rates of reactions are usually limited by mass transfer in the liquid or by a slow chemical reaction as in the case of nitrogen. These two limiting factors can be overcome when degassing is accomplished by exposing the molten metal in a vacuum chamber over as great a surface area as possible.

The commercial processes used in degassing of steel have been compiled in a comprehensive review paper by Flux.[273]

In the present commercial stream-degassing processes, a stream of steel enters a vacuum chamber, breaks into a spray of droplets and falls into a collecting vessel at the base of the chamber. Degassing takes place from the droplets in a contact time of less than a second, and from the liquid pool in the collecting vessel. In present practices the chamber pressure is maintained at about 1 mm Hg. However, as demonstrated by Olsson and Turkdogan[274], effective vacuum-carbon deoxidation can be achieved even at much higher chamber pressures, provided there is good stream break up to give droplet sizes of the order of 100 μm diameter. Such a stream break up was achieved by injecting inert gas bubbles into the liquid stream just before passage through a nozzle to a chamber at a lower pressure. The oxygen removal from 0.1 per cent carbon steel was shown to be essentially independent of chamber pressure up to about 200 mm Hg and was controlled by diffusion in the metal drops.

In vacuum degassing of metals, there will be some loss of certain alloying elements such as manganese and chromium which have relatively high vapor pressures. The rate of loss of such elements will be less than that calculated for free vaporization because of the diffusional effects. It should also be kept in mind that the reduction of some refractory oxides and inclusions such as MgO and SiO_2 by carbon takes place in steel during vacuum treatment. In fact, Philbrick[275] has observed that with stream degassing, the carbon loss is about twice that accounted for by the decrease in the oxygen content.

In some steelmaking practices, vacuum treatment is applied to fully killed steel to improve steel cleanliness. Effective stirring achieved during the repeated recycle of metal from the ladle to the vacuum chamber, as in the D-H and R-H processes, apparently aids the coalescence of small inclusions and facilitates their separation from the melt prior to casting. In fact, it should be noted that when a killed steel is treated there is no carbon-oxygen degassing occurring because there is so little oxygen in solution, but separation of the inclusions is facilitated by agitation of the bath during the vacuum treatment. However, there is some hydrogen and nitrogen removal.

PART VI

Inclusions in Steel

Processing of steel to various products by rolling, forging, drawing, forming or welding, and the resulting mechanical properties and performance in service such as resistance to corrosion, cracking and fatigue failure, are much affected by the presence of nonmetallic inclusions in the steel. The inclusions in steel are of major concern in obtaining the desired surface finishing qualities for deep drawing, coating, enameling, painting, and so on. Also, for special applications as in steel tire cords and thin-walled stainless-steel tubes for nuclear reactors, there are severe restrictions on inclusion counts and size. Reference may be made to Kiessling[276] and Pickering[277] for in-depth review of work done on nonmetallic inclusions in steel. However, a brief discussion here of certain aspects of inclusions in steel is considered appropriate in relation to steel cleanliness.

Deformability—Although demand for ultra-clean steel will continue to increase, for many large-tonnage applications the emphasis is not so much on the total cleanliness, as on rendering inclusions less detrimental to the processing of the steel. The deformability of the inclusion is characterized by the deformability index, v,

$$v = \frac{\text{true elongation of the inclusion}}{\text{true elongation of the steel}} \quad (243)$$

Above a certain critical temperature, the index v rises rapidly from almost zero to 1 or higher. During hot rolling at temperatures where $v \geq 1$, the inclusions are plastically deformed to long stringers. On cold rolling, thin glassy stringers are broken up to small inclusions which are less detrimental to the properties of the steel.

The deformability-transition temperature depends on the composition of the inclusions. For glassy silicate inclusions the transition temperature increases from 700° to 1200°C with increasing concentration of silica. The presence of calcium in silicate inclusions decreases the transition temperature and increases the deformability index. Also, the formation of undesirable alumina clusters is prevented by treating the aluminum-killed steel with calcium or a calcium aluminate slag. The crystalline inclusions such as corundum (Al_2O_3), galaxite ($MnO \cdot Al_2O_3$), mullite ($3Al_2O_3 \cdot 2SiO_2$), rhodonite ($MnO \cdot SiO_2$) and tridymite (SiO_2) are non-deformable, therefore responsible for crack initiation and propagation during hot working. As shown, for example, by Ekerot[278] inclusion of spessartite composition ($3MnO \cdot Al_2O_3 \cdot 3SiO_2$) has a high deformability index and a low transition temperature in the range 700° to 800°C. Therefore, inclusions of spessartite composition are preferred in steels deoxidized with Al and Si-Mn additions in the ladle.

The manganese-sulphide stringers formed during hot rolling do not shatter to fine inclusions upon cold rolling. The interdentritic precipitation of MnS particles, which upon rolling become thin sheets of sulphide inclusions, is detrimental to the through-thickness properties of the steel. The control of sulphide precipitation during solidification is discussed later.

Oxide Inclusions—In an oversimplified classification, the nonmetallic inclusions are said to be of **indigenous** and of **exogenous** origin. With the advent of continuous casting and mold addition practices, a simple classification is no longer adequate to describe the origin of inclusions. With argon rinsing, the oxide inclusions remaining dispersed in the metal at the time of teeming are less than about 5 μm diameter. Following the argon rinse, the total concentration of oxygen in aluminium-killed steel should be about 20 ppm or less for a good ladle practice, corresponding to about 0.01 volume per cent exogenous alumina inclusions. The semikilled steel with 40 ppm dissolved oxygen will contain about 0.044 volume per cent indigenous silicate inclusions, of sizes less than 2 μm, formed during solidification. For both limiting cases, the steel is considered clean. In continuous casting and ingot casting practices, much lower cleanliness rating is not unusual, for example, 0.1 to 0.3 volume per cent inclusions of sizes in the range 10 to 50 μm diameter. The large inclusion counts and sizes frequently observed in practice are due primarily to reoxidation of the metal stream during teeming and/or poor practice of mold additions.

The steel reoxidation is markedly curtailed by employing submerged pouring tubes with argon shroud for the metal stream from the ladle to the tundish and from the tundish to the mold. However, some reoxidation still occurs because of air aspiration through the leaky joints of the pouring tube to the ladle or the tundish. The flux additions are made to the tundish and the mold to assimilate alumina or other reoxidation products. The tundish is designed to give an optimum liquid-flow pattern that will minimize the entrainment of scum into the metal stream. Further improvement is made by flowing argon through a porous well around the tundish nozzle opening to suppress vortexing of the stream, hence avoid scum entrainment. A more efficient method of eliminating reoxidation and scum formation is by complete enclosure of the tundish and continuously flushing with argon. There is also the reoxidation and scum formation resulting from interaction of liquid steel with the refractory lining of the tundish. For the casting of aluminum-killed low-sulphur steel the tundish should be lined with magnesia or other basic refractory materials. The foregoing are just a few examples of measures taken in practice to curtail reoxidation of the steel during casting.

The macrosegregation that occurs during solidification of steel is responsible for clustering of inclusions in certain regions of the casting, such as the A and V segregation pattern in ingot casting and center-plane segregation in continuous casting of slabs. This aspect of inclusions will not be discussed here.

Blowholes—The blowhole is another form of inclusion that is formed during solidification of rimmed or semikilled steel. To control the gas evolution in capped-steel casting practice, or to suppress the blowhole formation in continuous casting of semikilled steel, the steel composition and the required level of deoxidation can be formulated by considering the reactions that occur within the impurity-enriched interdendritic regions of the solidifying steel. This subject has been discussed by Turkdogan[279] on previous occasions, and studied further by others;[280, 281] therefore, a brief statement here of the basic concept will suffice.

The solidification inclusions are the products of reac-

tions occurring during solidification of steels. The formation of solidification inclusions is a consequence of microsegregation that accompanies dendritic freezing of alloys. Advances made in the understanding of the morphology of dendritic solidification are well documented in textbooks by, for example, Chalmers[282] and by Flemings.[283]

Microsegregation resulting from solute rejection during dendritic freezing in binary alloys has been formulated by Scheil[284] for two limiting cases. When there is complete diffusion in the solid (dendrite arms) and in the liquid phase between the dendrite arms, the concentration of solute in the liquid is given by

$$C_l = \frac{C_o}{1 + g(k-1)} \qquad (244)$$

where
C$_o$ = initial solute concentration,
C$_l$ = solute concentration in enriched liquid,
k = solid/liquid solute distribution ratio,
g = volume fraction of liquid solidified.
This limiting equation will apply to carbon, oxygen, hydrogen and nitrogen in iron, because of their high diffusivities.

When there is negligible diffusion in the solid and complete diffusion in the liquid phase, the solute enrichment in the liquid is given by

$$C_l = C_o(1-g)^{k-1} \qquad (245)$$

This equation is more appropriate for substitutional elements in iron, e.g. manganese, silicon, phosphorus, because of their low diffusivities in solid alloys.

For dilute solutions, the solute distribution ratio k is about 2/3 for the Fe-Mn and Fe-Si systems, 0.2 for the Fe-C system, and essentially zero for the Fe-O system. For the purpose of demonstrating the consequences of microsegregation in the solidification of steel, let the simplifying assumption be made that the solute distribution ratios for low-alloy steels are similar to those in the iron-base binary alloys, and not affected by the other solutes. Calculated solute content in the interdendritic liquid is shown in Figure 13—189 as a

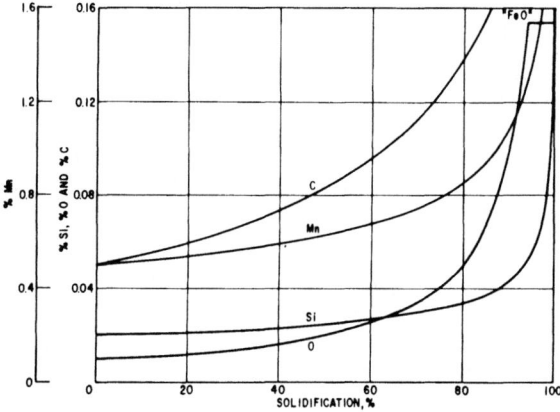

FIG. 13—189. Solute enrichment in solidifying liquid steel if no reaction takes place between the solutes in iron.

function of per cent solidification in a given volume of the liquid.

In the case considered, oxygen in solution is the only element which may react with the other solutes when their concentrations become sufficiently high in the enriched liquid. On the basis of this reasoning, Turkdogan[279] has pointed out that in calculating the solute enrichment in the entrapped freezing liquid, due account must be taken of the deoxidation reactions. Because of the complexity of the problem, certain simplifying assumptions have to be made in computing the solute enrichment controlled by solute distribution between solid and liquid phases coupled with reactions in the enriched liquid. The primary assumption is that there are sufficient nuclei present in the system, and therefore, little or no supersaturation is needed for the formation of the reaction products.

An example of the calculations is given in Figure 13—190 showing the oxygen content of entrapped liquid in the solid-plus-liquid mushy zone of the ingot, controlled by the Si-Mn deoxidation reaction, as a function of the local percentage of solidification. The dotted curve gives the oxygen enrichment in the absence of deoxidation reaction.

When the carbon content of the interdendritic liquid becomes sufficiently high, it will react with oxygen forming carbon monoxide. Similarly, at sufficiently high concentrations of hydrogen and nitrogen in the enriched liquid, there will be gas evolution. This gas formation results in the displacement of liquid to neighboring interdendritic cells, hence the formation of blowholes. The equilibrium oxygen content of steel for the carbon-oxygen reaction at 1 atm CO at various stages of solidification for several initial carbon concentrations are superimposed in Figure 13—191 on the curves reproduced from Figure 13—190. If the concentration of silicon, manganese or other deoxidizers is sufficiently high, the oxygen in solution in the enriched liquid will be maintained at low levels during solidification, such that the carbon-oxygen reaction will not take place; thus the absence of blowholes (if the hydrogen and nitrogen contents are sufficiently low). As shown in

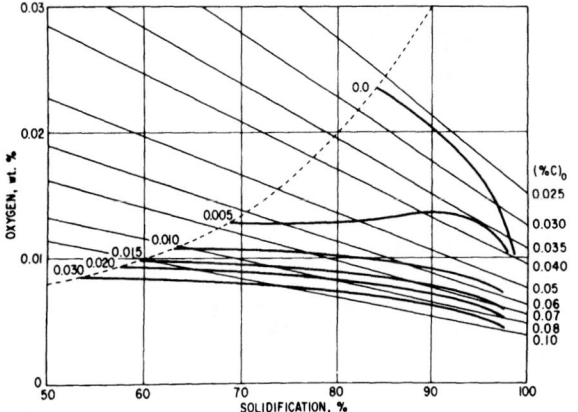

FIG. 13—191. Oxygen content of enriched liquid controlled by carbon in solution for 1 atm CO compared to that controlled by Si-Mn deoxidation.[279]

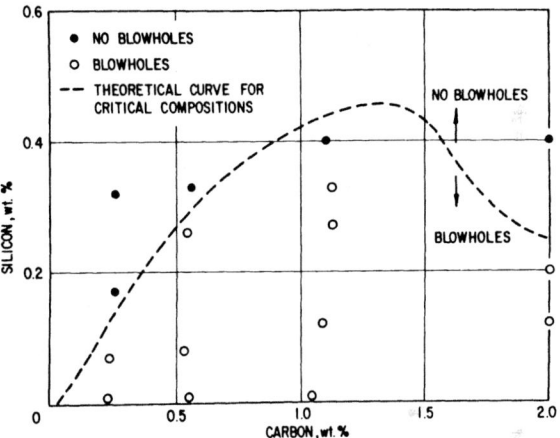

FIG. 13—192. Estimated critical Si and C contents of 0.5 per cent Mn-steel delineating region of no blowhole formation.[279]

Figure 13—191, for 1 atm pressure of CO, a particular carbon line tangential to the Si/Mn-deoxidation curve gives the critical carbon content of the steel below which blowholes will not form.

In a composition plot in Figure 13—192, the sections of laboratory ingots which showed blowholes are indicated by open circles and the castings free of blowholes by filled circles. The critical curve derived from a theoretical analysis, briefly outlined here, is well supported by the experimental observations.[285, 286] Furthermore, practical experience in continuous casting of low-carbon and low-silicon steels, deoxidized with silicon, manganese and some aluminum, has substantiated the applicability of the critical Si-C curve (for a given manganese level) in the production of castings free of blowholes.

In estimating the critical composition of the semikilled steel that will solidify without gas evolution, due account should be taken of hydrogen and nitrogen present in the steel. With $k = 0.38$ for the δ/l nitrogen

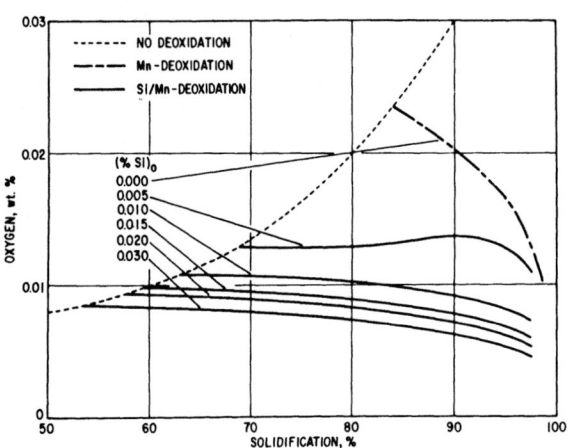

FIG. 13—190. Oxygen content of enriched liquid controlled by Si-Mn deoxidation during solidification. After Turkdogan.[279]

distribution ratio and k = 0.30 for the hydrogen distribution ratio, the nitrogen and hydrogen enrichment in the freezing liquid is calculated from equation (244) for the initial concentrations of 20 and 50 ppm N, and 2 and 5 ppm H. Calculated results are given in Figure 13—193 in terms of the pressures of N_2 and H_2 that are in equilibrium with the corresponding solutes in the enriched liquid at the freezing temperature of 1525°C. Since the total pressure, $p_{N_2} + p_{H_2}$, is below atmospheric, no gas-bubble formation is expected in killed steels. However, in the case of semikilled steels, CO may form during freezing, therefore, all three gaseous species should be taken to be present in the gas pocket so that the total pressure is equal to the sum $p_{co} + p_{N_2} + p_{H_2}$

The critical concentrations of Si and C thus computed for various initial concentrations of nitrogen and hydrogen are given in Figure 13—194. It is seen that an increase in hydrogen and nitrogen contents has to be accompanied by an appropriate increase in silicon or a decrease in carbon content of semikilled steel so that blowholes may not form. It is also seen that an increase in hydrogen content, for example, from 2 to 5 ppm promotes the blowhole formation more readily than an increase in nitrogen content from 20 to 50 ppm.

This conceptual analysis of what happens during solidification of steel should be helpful in adjusting the composition of semikilled steels for controlled gas evolution, hence the blowhole formation that is confined to the upper part of the ingot to avoid piping due to the solidification shrinkage. For example, in semikilled steels with silicon and carbon contents slightly below the critical curve (Figure 13—194) for 1 atm total pressure, there will be blowholes in the upper part of the ingot, but not in the lower part, because of the ferrostatic pressure of the liquid metal exerted on the lower section of the solidifying ingot. On the other hand, for continuous casting of the semikilled steels, the silicon and carbon contents should be above the critical curve for atmospheric pressure so that the

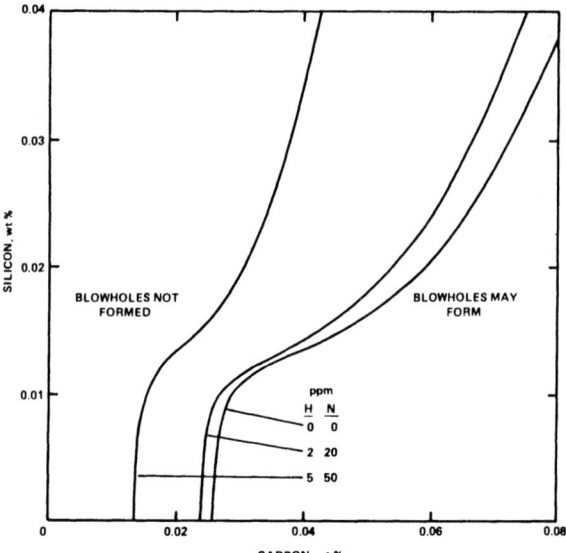

FIG. 13—194. Effects of hydrogen and nitrogen in liquid steel on critical %Si vs %C for the formation of blowholes for $p_{H_2} + p_{N_2} + p_{co} = 1$ atm.[279]

blowholes do not form close to the surface of the solidified skin where the total pressure is only slightly above atmospheric during the early stages of freezing in the mold. When the steel composition and the level of deoxidation is such that there are no blowholes near the surface of the continuously cast slab, no gas evolution and blowholes are expected to occur in the interior of the casting because of the ferrostatic pressure, except for those due to lack of liquid flow into the shrinkage cavities.

When there are strong convection currents in the liquid pool of the solidifying ingot, the extent of the solid-plus-liquid mushy zone will be reduced, and in the limiting case the solid and liquid phases will meet at a plane front interface parallel to the mold wall. It is for this reason that the violent CO evolution which occurs during freezing of rimming steel is responsible for nondendritic freezing of the rim with a plane solidification front. Under these conditions, the solute enrichment in the liquid will be confined to the diffusion layer adjacent to the solidifying front. The deoxidation reactions occurring in this type of freezing have been studied by Nilles.[287] The critical oxygen, carbon and manganese contents of silicon-free rimming steels solidifying nondendritically are compared in Figure 13—195 with those computed for dendritic freezing with no blowhole formation.

In the production of rim-stabilized steel ingots, aluminum (about 1 kg/metric ton of steel) is added to the liquid core of the solidifying ingot after the solidification of 35 to 40-mm-thick rim. This controlled deoxidation stops the rimming action to yield a rimmed ingot essentially free of blowholes in the interior, except in the upper part of the ingot where some gas evolution is allowed to avoid piping.

Oxidation During Casting—As the deoxidized steel is teemed from a ladle to a mold, or to a tundish then to a mold, some oxygen pickup by steel exposed to air is

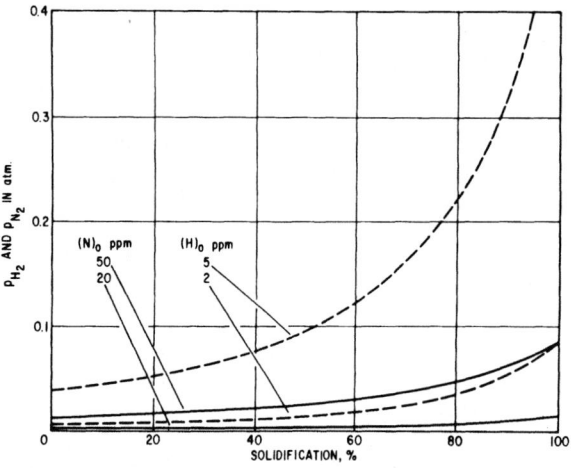

FIG. 13—193. Equilibrium partial pressures of H_2 and N_2 corresponding to their concentrations in enriched liquid at indicated stages of solidification.[279]

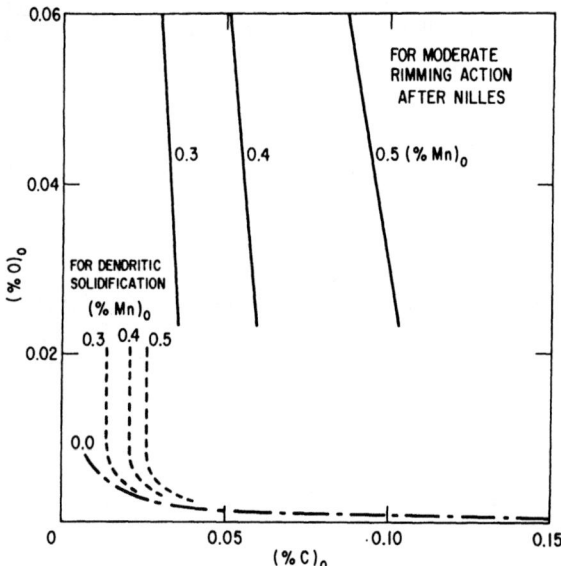

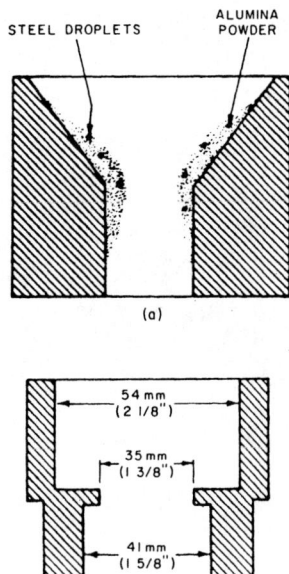

FIG. 13—195. Initial concentrations of oxygen and carbon in steel containing 0.3 to 0.5 per cent Mn for moderate rimming action compared with those for the formation of interdendritic blowholes.[256]

FIG. 13—196. (a) Alumina buildup in a regular tundish nozzle; (b) annular nozzle design. After Golas & Singh.[288]

imminent. Although the steelmakers have been aware of the oxidation of liquid streams during casting for many years, the solutions to the problem were not sought until the advent of continuous casting. Even today, the continuous casting of aluminum-killed steels is not trouble free.

The extent of oxygen pickup in open pouring of steel is difficult to estimate as it depends much on the flow conditions and the type of teeming practices. In addition to the oxygen pickup by the liquid-steel stream, there is oxygen transfer to the liquid steel in the mold by entrained air bubbles.

In many continuous-casting practices, the liquid steel is poured from the tundish into the mold through a refractory tube immersed in the liquid pool in the mold. Also, the liquid pool is covered with a low-melting flux to suppress oxygen pickup from the air. In addition, the mold flux assimilates alumina clusters in the liquid pool and also acts as a mold lubricant, thus improving the surface quality of the slab.

Pouring killed steel through a tundish nozzle or immersed tube is often troublesome. Even when the necessary precautions are taken to suppress oxidation, there is a buildup of alumina, zirconia or other refractory oxides on the inside walls of the nozzle or immersed tube leading to subsequent blockage. There are two sources of alumina: those which did not float out of the steel in the ladle or the tundish, and those formed by oxidation of the residual aluminum in the steel with air that is aspirated into the liquid stream through the pores of the refractory nozzle or through the leaky joint of the nozzle to the tundish. Whatever is the source of alumina, they will deposit on the inside walls of the nozzle because of the stagnant layer at the interface between the nozzle wall and the liquid-metal stream. The cross section of a typical tundish nozzle

shown schematically in Figure 13—196a illustrates the alumina buildup which ultimately leads to nozzle blockage.

The deposition of alumina, or other refractory inclusions, may be avoided by preventing contact between the liquid-metal stream and the nozzle wall at the constricted region. This can be achieved in a nozzle of the type shown in Figure 13—196b. In the design of the so-called annular nozzle, developed by Golas and Singh,[288] the use was made of the concept of *vena contracta* that describes the convergence of the liquid stream as it passes through an orifice. The plant trials made at the continuous-casting facilities of U. S. Steel have shown that the alumina deposition could be alleviated and consequently a uniform casting speed could be maintained, by using the annular nozzle for the tundish.

Another consequence of air oxidation of aluminum-killed steels is the contamination of the castings with alumina clusters. A scanning electron micrograph of a colony of alumina clusters shown in Figure 13—197 is a typical example of clusters observed by many investigators. The dendritic nature of alumina inclusions is clearly evident, and furthermore, the dendrites constituting a cluster seem to have grown from a single nucleus. The occurrence of alumina clusters in steel castings has been eliminated to a large extent by using mold fluxes which are capable of absorbing the alumina scum that accumulates in the liquid pool in the mold.

Sulphide Inclusions—In steels that are not treated with oxides or alloys of alkaline-earth or rare-earth elements, sulphide and oxysulphide inclusions are formed during solidification in a manner similar to that discussed above for the formation of interdendritic oxide and silicate inclusions and blowholes. The metallographic examinations made by Turkdogan and

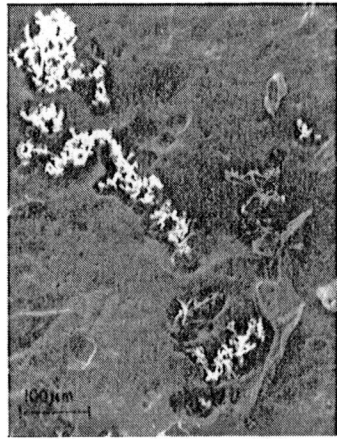

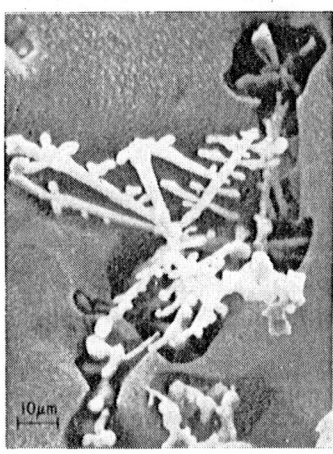

FIG. 13—197. Alumina clusters in continuously cast steel.

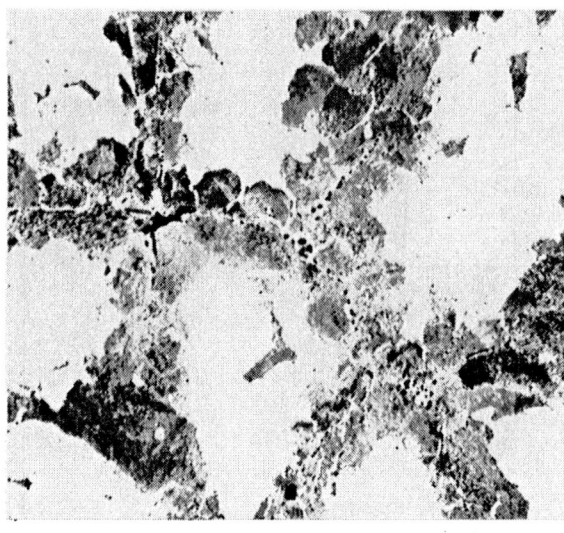

X 500

FIG. 13—198. Manganese sulphide inclusions in the interdentritic regions in the as-cast condition (0.25% C, 1.5% Mn and 0.05% S). (Turkdogan and Grange[286]).

Grange[286] of as-cast (Figure 13—198) and annealed samples of steels substantiated the view that the manganese-sulphide inclusions in steel are indeed found only in the interdendritic regions which ultimately become the grain boundaries in the steel. The sulphide inclusions accumulated at the planes of the interdendritic network form sheets and stringers of sulphide inclusions upon hot rolling, a feature that is responsible for poor transverse mechanical properties.

When the sulphur content of the steel is less than about 0.01 per cent, there is not enough sulphur in the system for the formation of a continuous network of interdendritic sulphide inclusions. However, the macrosegregation that is inherent to the solidification process, both in continuous and ingot casting, is responsible for the accumulation of impurity-enriched liquid in certain regions of the casting, resulting in localized precipitation of relatively large manganese (iron) sulphides, even when the steel contains as little as 0.002 per cent S.

The control of the so-called shape and morphology of sulphide inclusions is in fact the suppression of micro- and macrosegregation of dissolved sulphur in the en-

riched liquid, hence prevention of the formation of manganese and iron sulphides during solidification. This is achieved by having in the liquid steel dissolved elements such as rare earths or dispersed particles of oxide mixtures such as calcium aluminates or rare-earth oxides which have strong affinity and high capacity for sulphur.

In line-pipe-quality steels containing 1.2 to 1.5 per cent Mn, no MnS inclusions are observed when the steel is desulphurized to levels below 0.002 per cent S. In such steels the inclusions consist of dispersed particles ($< 5 \mu m$ diameter) of calcium aluminate which absorb dissolved sulphur during cooling and solidification of steel. The X-ray scan spectra of inclusions shown in Figure 13—199 are typical examples for continuously cast line-pipe steels desulphurized to 0.001 per cent S with Cal-Sil injection under a calcium-aluminate slag. There are variations in composition from one particle to another, or sometimes within the same particle. The observed variation in composition is a manifestation of phase separation that occurs during solidification of molten CaO-Al_2O_3-CaS inclusions. When similar steels were treated with misch metal in the ladle, subsequent to desulphurization with calcium aluminate, the inclusions consisted of rare-earth aluminates and oxysulphides, some of which contained calcium, as is seen from the X-ray scan spectra in Figure 13—200. None of the inclusions in these low-sulphur steels contained any detectable amount of manganese or other alloying elements that are present in the line-pipe-quality steels.

The total concentration of calcium in the ladle-treated steel may be as high as 0.006 per cent Ca as calcium aluminate. This amount of dispersed calcium aluminate is not sufficient to suppress the formation of

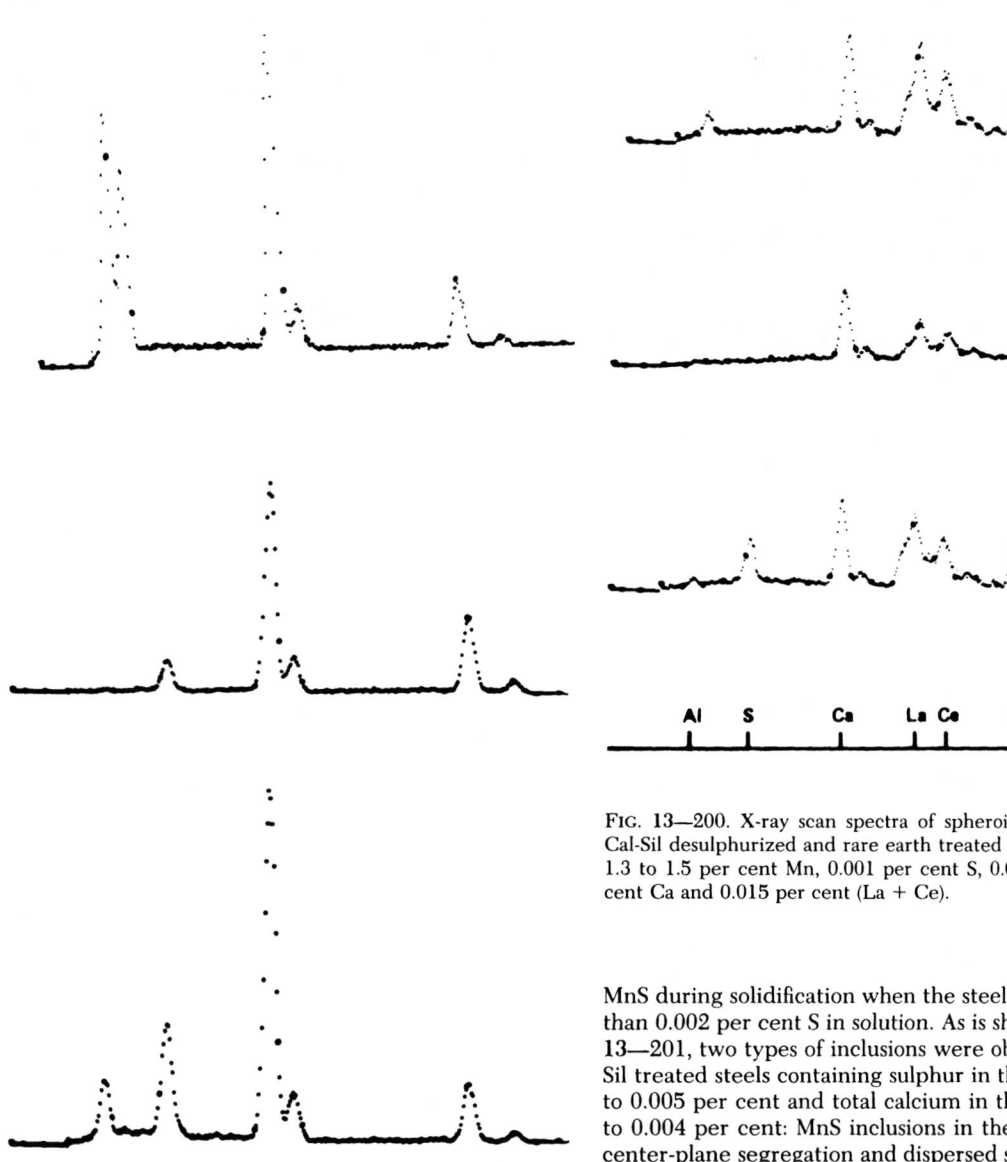

FIG. 13—200. X-ray scan spectra of spheroidal inclusions in Cal-Sil desulphurized and rare earth treated steels containing 1.3 to 1.5 per cent Mn, 0.001 per cent S, 0.002 to 0.004 per cent Ca and 0.015 per cent (La + Ce).

FIG. 13—199. X-ray scan spectra of spheroidal inclusions in Cal-Sil desulphurized steels containing 1.3-1.5 per cent Mn, 0.001 per cent S and 0.004-0.006 per cent Ca.

MnS during solidification when the steel contains more than 0.002 per cent S in solution. As is shown in Figure 13—201, two types of inclusions were observed in Cal-Sil treated steels containing sulphur in the range 0.002 to 0.005 per cent and total calcium in the range 0.001 to 0.004 per cent: MnS inclusions in the region of the center-plane segregation and dispersed spheroidal particles of $CaO\text{-}Al_2O_3\text{-}CaS$ inclusions. Such steels should be treated with small additions of misch metal in the ladle just before casting, to prevent MnS formation during solidification.

Upon addition of misch metal to an aluminum-killed steel, almost all the residual oxygen in solution will be converted to rare-earth oxides, Re_2O_3, or more likely to rare-earth aluminates, which remain in suspension in liquid steel and react with dissolved sulphur to form oxysulphides, Re_2O_2S, during cooling and solidification of steel. It is recommended, as a rule of thumb, that the amount of misch metal to be added be about 6 times the sum of the total concentrations of oxygen and sulphur in the steel at the end of the ladle desulphurization. For steel containing, for example, 30 ppm total O and 60 ppm total S, it is sufficient to add 0.5 kg misch metal per metric ton of steel. With steel cleanliness in mind, it is suggested that the rare-earth treatment be confined to aluminum-killed steels containing less than 0.006 per cent S. Below 0.002 per cent S in the steel, the rare-earth treatment should not be necessary.

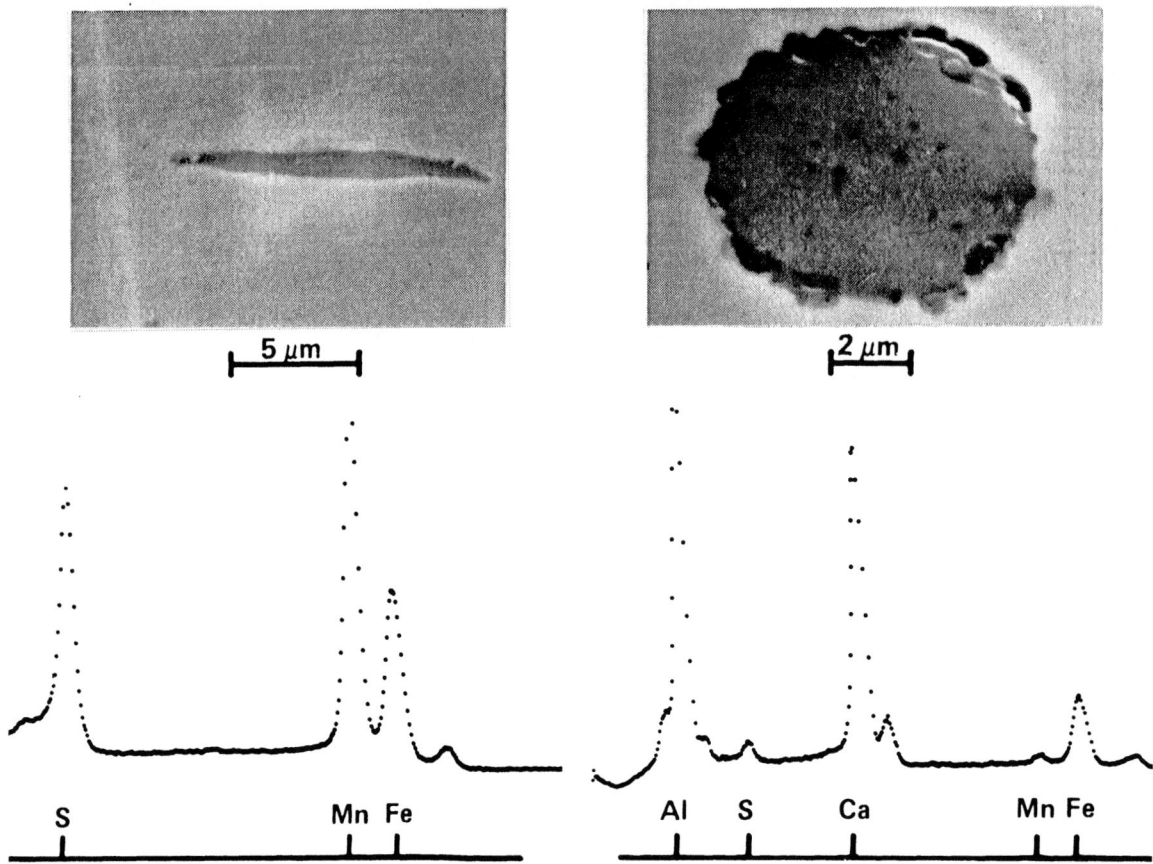

FIG. 13—201. Typical manganese sulphide and calcium aluminate (sulphide) inclusions in Cal-Sil desulphurized steels containing 1.3 to 5 per cent Mn, 0.002 to 0.005 per cent S and 0.001 to 0.003 per cent Ca. (Turkdogan[244]).

References

1. "Handbook of Chemistry and Physics," Chemical Rubber Publishing Co., Cleveland, 50th Edition, 1969-1970.
2. T. P. Floridis and J. Chipman, Trans. Met. Soc. AIME, 1958, vol. 212, pp. 549-53.
3. R. P. Smith, J. Am. Chem. Soc., 1946, vol. 68, pp. 1163-75.
4. L. S. Darken, J. Am. Chem. Soc., 1950, vol. 72, pp. 2909-14.
5. C. Wagner, "Thermodynamics of Alloys," Addison-Wesley Press, 1952.
6. R. Schuhmann, Acta Met., 1955, vol. 3, pp. 219-26.
7. N. A. Gokcen, J. Phys. Chem., 1960, vol. 64, pp. 401-06.
8. O. Kubaschewski, E. L. Evans and C. B. Alcock, "Metallurgical Thermochemistry," Pergamon Press, 1967.
9. K. K. Kelley and E. G. King, U. S. Bur. Mines Bull. No. 592, 1961.
10. K. K. Kelley, U. S. Bur. Mines Bull. No. 584, 1960.
11. J. F. Elliott and M. Gleiser, "Thermochemistry for Steelmaking," Addison-Wesley Pub. Co., Inc., 1960.
12. J. F. Elliott, M. Gleiser and V. Ramakrishna, "Thermochemistry for Steelmaking," vol. II, Addison-Wesley Pub. Co., Inc., 1963.
13. R. J. Fruehan, Met. Trans., 1970, vol. 1, pp. 865-70.
14. R. J. Fruehan, Met. Trans., 1970, vol. 1, pp. 3403-10.
15. R. E. Honig and D. A. Kramer, "Physicochemical in

Metals Research," vol. IV, Part I, pp. 512-04, Interscience Pub. 1970.
16. C. Wagner, "Thermodynamics of Alloys," Addison-Wesley Press, Cambridge, Mass., 1952.
17. C. H. P. Lupis and J. F. Elliott, Trans. Met. Soc. AIME, 1965, vol. 233, pp. 257-58.
18. L. S. Darken, Trans. Met. Soc. AIME, 1967, vol. 239, pp. 80-89.
19. S. Ban-ya and J. Chipman, Trans. Met. Soc. AIME, 1969, vol. 245, pp. 133-43 and pp. 391-96.
20. E. T. Turkdogan, in "BOF Steelmaking" (R. D. Pehlke, et al), pp. 1-190. Iron and Steel Soc. AIME, New York, 1975.
21. R. L. Orr and J. Chipman, Trans. Met. Soc. AIME, 1967, vol. 239, pp. 630-33.
22. C. Wells, Trans. ASM, 1938, vol. 26, pp. 289-344; author's closure, pp. 351-57.
23. J. C. Swartz, Trans. Met. Soc., 1969, vol. 245, pp. 1083-91.
24. H. Dünwald and C. Wagner, Z. anorg. allgem. Chem., 1931, vol. 199, pp. 321-46.
24a. R. W. Gurry, Trans. AIME, 1942, vol. 150, pp. 172-182.
25. J. Chipman, Trans. Met. Soc. AIME, 1967, vol. 239, pp. 2-7.
26. R. P. Smith, Trans. Met. Soc. AIME, 1960, vol. 218, pp.

62-64.

27. R. P. Smith, J. Am. Chem. Soc., 1948, vol. 70, pp. 2724-29.
28. F. D. Richardson and W. E. Dennis, Trans. Faraday Soc., 1953, vol. 49, pp. 171-80.
29. A. Rist and J. Chipman, Rev. Met., 1956, vol. 53, pp. 796-807.
30. J. Chipman, Met. Trans., 1970, vol. 1, pp 2163-68.
31. F. von Neumann and H. Schenck, Arch. F. D. Eisenhuttenw., 1959, vol. 30, pp. 477-83.
32. E. T. Turkdogan and L. E. Leake, J. Iron and Steel Inst., 1955, vol. 179, pp. 39-43.
33. E. T. Turkdogan and R. A. Hancock, J. Iron and Steel Inst., 1955, vol. 179, pp. 155-59; 1956, vol. 183, pp. 69-72.
34. J. Chipman, R. M. Alfred, L. W. Gott, R. B. Small, D. M. Wilson, C. N. Thomson, D. L. Guernsey and J. C. Fulton, Trans. ASM, 1952, vol. 44, pp. 1215-30.
35. J. Chipman and T. P. Floridis, Acta Met., 1955, vol. 3, pp. 456-59.
36. N. R. Griffing, W. D. Forgeng and G. W. Healy, Trans. Met. Soc. AIME, 1962, vol. 224, pp. 148-59.
37. T. Fuwa and J. Chipman, Trans. Met. Soc. AIME, 1959, vol. 215, pp. 708-16.
38. W. Geller and T. N. Sun, Arch. F. D. Eisenhuttenw., 1950, vol. 21, pp. 423-30.
39. M. Weinstein and J. F. Elliott, Trans. Met. Soc. AIME, 1963, vol. 227, pp. 382-93.
40. R. D. Pehlke and J. F. Elliott, Trans. Met. Soc. AIME, 1960, vol. 218, pp. 1088-1101.
41. J. D. Fast and M. B. Verrijp, J. Iron and Steel Inst., 1955, vol. 180, pp. 337-43.
42. N. S. Corney and E. T. Turkdogan, J. Iron and Steel Inst., 1955, vol. 180, pp. 344-48.
43. H. Schenck, M. Frohberg and H. Graf, Arch. F. D. Eisenhuttenw., 1958, vol. 29, pp. 673-76; 1959, vol. 30, pp. 533-37.
44. E. T. Turkdogan and S. Ignatowicz, J. Iron and Steel Inst., 1957, vol. 185, pp. 200-06.
45. E. T. Turkdogan and S. Ignatowicz, J. Iron and Steel Inst., 1958, vol. 188, pp. 242-47; 1961, vol. 199, pp 287-96.
46. H. A. Wriedt and O. D. Gonzalez, Trans. Met. Soc. AIME, 1961, vol. 221, pp. 532-35.
47. M. Hansen and K. Anderko, "Constitution of Binary Alloys," 2nd ed., McGraw-Hill, New York, 1958.
48. H. A. Wriedt, Trans. Met Soc. AIME, 1969, vol. 245, pp. 43-46.
49. L. S. Darken and R. W. Gurry, J. Am. Chem. Soc., 1945, vol. 67, pp. 1398-1412; 1946, vol. 68, pp. 798-816.
50. M. T. Hepworth, R. P. Smith and E. T. Turkdogan, Trans. Met. Soc. AIME, 1966, vol. 236, pp. 1278-83.
51. J. H. Swisher and E. T. Turkdogan, Trans. Met. Soc. AIME, 1967, vol. 239, pp. 426-31.
52. A. Kusano, K. Ito and K. Sano, Iron and Steel Institute of Japan, 1970, vol. 10, pp. 78-82.
53. J. Chipman and K. L. Fetters, Trans. ASM, 1941, vol. 29, pp. 953-67.
54. C. R. Taylor and J. Chipman, Trans. AIME, 1943, vol. 154, pp. 228-45.
55. M. N. Dastur and J. Chipman, Trans. AIME, 1949, vol. 185, pp. 441-45.
56. T. Rosenqvist and B. L. Dunicz, Trans. AIME, 1952, vol. 194, pp. 604-08; J. Iron and Steel Inst., 1954, vol. 176, pp. 37-57.
57. E. T. Turkdogan, S. Ignatowicz and J. Pearson, J. Iron and Steel Inst., 1955, vol. 180, pp. 349-54.
58. C. W. Sherman, H. I. Elvander and J. Chipman, Trans. AIME, 1950, vol. 188, pp. 334-40.
59. E. T. Turkdogan and G. J. W. Kor, Met. Trans. of AIME, 1971, vol. 2, pp. 1561-70.
60. A. E. Lord and N. A. Parlee, Trans. Met. Soc. AIME, 1960, vol. 218, pp. 644-46.
61. D. L. Sponseller and R. A. Flinn, Trans Met. Soc., AIME, 1964, vol. 230, pp. 876-88.
62. P. K. Trojan and R. A. Flinn, Trans. ASM, 1961, vol. 54, pp. 549-66.
63. G. Cavalier, Proc. Nat. Phys. Lab., No. 9, 1958, vol. 2, 4D.
64. R. N. Barfield and J. A. Kitchener, J. Iron and Steel Inst., 1955, vol. 180, pp. 324-29.
65. Third Report of the Steel Castings Research Committee, Iron and Steel Inst., Spec. Rept., 1938, No. 23, pp. 45-60.
66. R. W. Powell, M. J. Hickman, R. P. Tye, M. J. Woodman, Proc. Intl. Res. on Thermodynamic and Transport Properties, ASME, (1962), pp. 466-73.
67. Aerospace Structural Methods Handbook, 1969 Pub. Air Force Materials Lab., Wright-Patterson Air Force Base, Ohio.
68. Metals Handbook, 8th edition, ASM, 1961.
69. P. Kozakevitch and G. Urbain, J. Iron and Steel Inst., 1957, vol. 186, pp. 167-73.
70. P. Kozakevitch, Surface Phenomena of Metals, SCI Monograph 28, London 1968, pp. 223-45.
71. L. D. Lucas, Compt. Rend., 1959, vol. 248, pp. 2336-38.
72. W. A. Fischer and J. Fleischer, Arch. F. D. Eisenhuttenw., 1961, vol. 32, pp. 1-10.
73. O. W. Florke, Naturwissen, 1956, vol. 43, pp. 419-20.
74. M. L. Keith and O. F. Tuttle, Am. J. Sci., 1952, "Bowen Volume," Part 1, pp. 203-52.
75. S. B. Holmquist, J. Am. Ceram. Soc., 1961, vol. 44, pp. 82-86.
76. G. A. Rankin and P. E. Wright, Am. J. Sci., 1915, vol. 39, Ser. 4, pp. 1-79.
77. A. Muan and E. F. Osborn, Year Book of Am. Iron and Steel Inst., 1951, pp. 325-59.
78. J. H. Welch and W. Gott, J. Am. Ceram. Soc., 1959, vol. 42, pp. 11-15.
79. N. L. Bowen and J. F. Schairer, Am. J. Sci., 1932, vol. 24, Ser. 5, pp. 177-213.
80a. R. Schuhmann and P. J. Ensio, Trans. AIME, 1951, vol. 191, pp. 401-11.
80b. E. J. Michal and R. Schuhmann, Trans. AIME, 1952, vol. 194, pp. 723-28.
81. N. L. Bowen and O. Anderson, Am. J. Sci., 1914, vol. 37, Ser. 4, pp. 487-500.
82. J. W. Greig, Am. J. Sci., 1927, vol. 13, Ser. 5, pp. 133-54.
83. C. H. Herty, Metals and Alloys, 1930, vol. 1, pp. 883-89.
84. J. White, D. Howat and R. Hay, J. Royal Tech. Coll. (Glasgow), 1934, vol. 3, pp. 231-40.
85. F. P. Glasser. Am. J. Sci., 1958, vol. 256, Ser. 5, pp. 398-412.
86. N. L. Bowen and J. W. Grieg, J. Am. Ceram. Soc., 1924, vol. 7, pp. 238-54.
87. J. F. Schairer, J. Am. Ceram. Soc., 1942, vol. 25, pp. 241-74.
88. S. Aramaki and R. Roy, Nature, 1959, vol. 184, pp. 631-32.
89. J. H. Welch, Nature, 1960, vol. 186. pp. 545-46.
90. W. Bussem and A. Eithel, Z. Krist., 1936, vol. 95, pp. 175-88.
91. B. Tavasci, Tomind Ztg., 1937, vol. 61, pp. 717-19, 729-31.
92. J. R. Goldsmith, J. Geol., 1948, vol. 56, pp. 80-81.
93. N. E. Filomenko and I. V. Lavrov, J. Appl. Chim., (USSR) 1950, vol. 23, pp. 1040-46.
94. R. C. Doman, J. B. Barr, R. N. McNally and A. M. Alper, J. Am. Ceram. Soc., 1963, vol. 46, pp. 313-16.
95. J. H. Chesters, "Steelplant Refractories," The United Steel Companies Ltd., Sheffield, 1957.
96. E. F. Osborn and A. Muan, "Phase Equilibrium Diagram," Am. Ceram. Soc., Columbus, Ohio, 1960.

97. G. Tromel and H. W. Fritze, Arch. F. D. Eisenhuttenw., 1959, vol. 30, pp. 461-72.
98. G. Tromel, W. Fix and H. W. Fritze, Arch. F. D. Eisenhuttenw., 1961, vol. 32, pp. 353-59.
99. B. Phillips and A. Muan, J. Am. Ceram. Soc., 1958, vol. 41, pp. 445-54.
100. M. A. Swayze, Am. J. Sci., 1946, vol. 244, pp. 1-30.
101. R. W. Ricker and E. F. Osborn, J. Am. Ceram. Soc, 1954, vol. 37, pp. 133-39.
102. E. F. Osborn, R. C. DeVries, K. H. Gee and H. M. Kramer, Trans. AIME, 1954, vol. 200, pp. 33-45.
103. A. Muan and E. F. Osborn, "Phase Equilibria Among Oxides in Steelmaking," Addison-Wesley Pub. Co. Inc., 1965.
104. W. Oelsen and H. Maetz, Mitt. Kaiser-Wilhelm Inst. f. Eisenforsch., 1941, vol. 23, pp. 195-245.
105. M. Olette and G. Vancon, Circ. Inform. Tech., no. 7-8, pp. 1739-54.
106. F. J. Drewes and M. Olette, Arch. F. D. Eisenhuttenw., 1967, vol. 38, pp. 163-75.
107. K. Schwerdtfeger and E. T. Turkdogan, Trans. Met. Soc. AIME, 1967, vol. 239, pp. 589-90.
108. W. H. Zachariasen, J. Am. Chem. Soc., 1932, vol. 54, pp. 3841-51; Ind. Chem. Phys., 1935, vol. 1, pp. 162-63.
109. B. E. Warren, Z. Krist., 1933, vol. 86, pp. 349-58; Phys. Rev., 1934, vol. 45, pp. 657-61; J. Am. Ceram. Soc., 1934, vol. 17, pp. 249-54; 1935, vol. 18, pp. 269-76.
110. a. B. E. Warren, J. Appl. Phys., 1937, vol. 8, pp. 645-55.
     b. B. E. Warren and J. Biscoe, J. Am. Ceram. Soc., 1938, vol. 21, pp. 259-65 and 287-93.
     c. B. E. Warren and A. G. Pincus, J. Am. Ceram. Soc. , 1940, vol. 23, pp. 301-04.
     d. J. Biscoe, A. G. Pincus and B. E. Warren, J. Am. Ceram. Soc. 1941, vol. 24, pp. 116-19.
111. P. K. Foster and A. J. E. Welch, Trans. Faraday Soc., 1956, vol. 52, pp. 1636-42.
112. J. Chipman, J. B. Gero and T. B. Winkler, Trans. AIME, 1950, vol. 188, pp. 341-45.
113. E. T. Turkdogan, Trans. Met. Soc. AIME, 1961, vol. 221, pp. 1090-95.
114. R. Schuhmann and P. J. Ensio, Trans. AIME, 1951, vol. 191, pp. 401-11.
115. E. T. Turkdogan, Trans. Met. Soc. AIME, 1962, vol. 224, pp. 294-98.
116. K. P. Abraham, M. W. Davies and F. D. Richardson, J. Iron and Steel Inst., 1960, vol. 196, pp. 82-89.
117. D. A. R. Kay and J. Taylor, Trans. Faraday Soc., 1960, vol. 56, pp. 1372-86.
118. R. A. Sharma and F. D. Richardson, J. Iron and Steel Inst., 1961, vol. 198, pp. 386-90.
119. H. Larson and J. Chipman, Trans. AIME, 1953, vol. 197, pp. 1089-96; 1954, vol. 200, pp. 759-62.
120. E. T. Turkdogan and P. M. Bills, J. Iron and Steel Inst., 1957, vol. 186, pp. 329-39.
121. T. B. Winkler and J. Chipman, Trans. AIME, 1946, vol. 167, pp. 111-33.
122. J. F. Elliott, Trans. AIME, 1955, vol. 203, pp. 485-88.
123. E. W. Dewing and F. D. Richardson, Trans. Faraday Soc., 1959, vol. 55, pp. 611-16.
124. E. T. Turkdogan and L. S. Darken, Trans. Met. Soc. AIME, 1962, vol. 221, pp. 464-74.
125. P. Grieveson and E. T. Turkdogan, Trans. Met. Soc. AIME, 1962, vol. 224, pp. 1086-95.
126. R. A. Sharma and F. D. Richardson, J. Iron and Steel Inst., 1962, vol. 200, pp. 373-79.
127. J. Henderson, L. Yang and G. Derge, Trans. Met. Soc. AIME, 1961, vol. 221, pp. 56-60.
128. O. A. Esin, E. S. Vorontsov and S. K. Chukmarev, Zhur Fiz. Khim, 1957, vol. 31, pp. 2322-27.
129. N. McCallum and L. R. Barrett, Trans. British Ceram. Soc., 1952, vol. 51, pp. 523-48.
130. H. Towers, M. Parsi and J. Chipman, Trans. AIME, 1953, vol. 197, pp. 1455-58; 1957, vol. 209, pp. 769-73.
131. K. Niwa, Nippon Kinzoku Gakkai-Si, 1957, vol. 21, pp. 304-08.
132. P. Grieveson and E. T. Turkdogan, Trans. Met. Soc. AIME, 1964, vol. 230, pp. 1609-14.
133. K. Mori and K. Suzuki, Trans. Iron and Steel Inst. Japan, 1969, vol. 9, pp. 409-12.
134. Y. P. Gupta and T. B. King, Trans. Met. Soc. AIME, 1967, vol. 239, pp. 1701-07.
135. T. B. King and P. J. Koros, Trans. Met. Soc. AIME, 1962, vol. 224, pp. 299-306.
136. T. Saito and Y. Kawai, Sc. Repts. Tohuku Univ., 1953, Series A5, pp. 460-68.
137. L. Himmel, R. F. Mehl and C. E. Birchenall, Trans. Met. Soc. AIME, 1953, vol. 197, pp. 827-43.
138. P. Hembree and J. B. Wagner, Trans. Met. Soc. AIME, 1969, vol. 245, pp. 1547-52.
139. J. O. Edstrom, Jernkont. Ann., 1957, vol. 141, pp. 809-36; 1958, vol. 142, pp. 401-28.
140. J. F. Hedvall, C. Brisi and R. Lindner, Arkiv Kemi, 1952, vol. 4, pp. 377-80.
141. R. Lindner, Arkiv Kemi, 1952, vol. 4, pp. 381-84.
142. Y. Oishi and W. D. Kingery, J. Chem. Phys., 1960, vol. 33, pp. 480-86.
143. W. C. Hagel, Trans. Met. Soc. AIME, 1966, vol. 236, pp. 179-84.
144. Y. Oishi and W. D. Kingery, J. Chem. Phys., 1960, vol. 33, pp. 905-06.
145. E. W. Sucov, J. Am. Chem. Soc., 1963, vol. 46, pp. 14-20.
146. C. J. Rosa and W. C. Hagel, Trans. Met. Soc. AIME, 1968, vol. 242, pp. 1293-98.
147. E. T. Turkdogan, Trans. Met. Soc. AIME, 1968, vol. 242, pp. 1665-72.
148. K. Schwerdtfeger, Trans. Met. Soc. AIME, 1967, vol. 239, pp. 1432-38.
149. K. Schwerdtfeger, P. Grieveson and E. T. Turkdogan, Trans. Met. Soc. AIME, 1969, vol, 245, pp. 2461-66.
150. B. Prenosil, Kovove Materialy, 1965, vol. 3, pp. 69-87.
151. J. S. Machin and T. B. Yee, J. Am. Ceram. Soc., 1948, vol. 31, pp. 200-04.
152. P. Kozakevitch, "Phys. Chem. Process Metallurgy," pp. 97-116, Interscience Pub., New York, 1961.
153. J. S. Machin and D. L. Hanna, J. Am. Ceram. Soc., 1945, vol. 28, pp. 310-16.
154. J. S. Machin, T. B. Yee and D. L. Hanna, J. Am. Ceram. Soc., 1952, vol. 35, pp. 322-25.
155. P. Kozakevitch, Rev. Met., 1954, vol. 51, pp. 569-87.
156. P. M. Bills, J. Iron and Steel Inst., 1963, vol. 201, pp. 133-40.
157. T. B. King, J. Soc. Glass Tech., 1951, vol. 35, pp. 241-59T.
158. P. Kozakevitch, Rev. Met., 1949, vol. 46, pp. 505-16; 572-82.
159. R. E. Boni and G. Derge, Trans. Met. Soc. AIME, 1956, vol. 206, pp. 59-64.
160. A. Ejima and M. Shimoji, Trans. Faraday Soc., 1970, vol. 66, pp. 99-106.
161. T. B. King, "Physical Chemistry of Melts," Inst. Min. Met., London, 1953, pp. 35-41.
162. J. Henderson, R. G. Hudson, R. G. Ward and G. Derge, Trans. Met. Soc. AIME, 1961, vol. 221, pp. 807-11.
163. J. Henderson, Trans. Met. Soc. AIME, 1964, vol. 230, pp. 501-04.
164. D. R. Gaskell and R. G. Ward, Trans. Met. Soc. AIME, 1967, vol. 239, pp. 249-52.
165. A. Adachi and K. Ogino, Tech. Repts. Osaka Univ., 1962, vol. 12, pp. 147-52.
166. S. I. Popel and O. A. Esin, Zh. Prikl. Khim, 1956, vol. 29, pp. 651-55.
167. L. R. Barrett and A. G. Thomas, Trans. Soc. Glass Tech.,

1959, vol. 43, pp. 179-90.

168. S. Chapman and T. G. Cowling, "The Mathematical Theory of Nonuniform Gases," Cambridge University Press, 1939.

169. J. O. Hirschfelder, C. F. Curtiss and R B. Bird, "Molecular Theory of Gases and Liquids," John Wiley and Sons, Inc., 1954.

170. P. Grieveson and E. T. Turkdogan, J. Phys. Chem., 1964, vol. 68, pp. 1547-51.

171. E. T. Turkdogan, Met. Trans. B, 1978, vol. 9B, pp. 163-79.

172. P. Reichardt, Arch. F. D. Eisenhuttenw., 1927/28, vol. 1, pp. 77-101.

173. A. Rist and N. Meysson, Rev. Met., 1965, vol. 62, pp. 995-1039; Engl. Trans. BISI 4437.

174. N. Meysson, A. Maaref and A. Rist, Rev. Met., 1965, vol. 62, pp. 1161-1180; Engl. Trans. BISI 4786.

175. L. von Bogandy and H. J. Engell, "The Reaction of Iron Oxides," Springer-Verlag, 1971.

176. E. T. Turkdogan, Ironmaking Proceedings, AIME, 1972, vol. 31, pp. 438-58.

177. E. T. Turkdogan and J. V. Vinters, Carbon, 1970, vol. 8, pp. 39-53.

178. E. T. Turkdogan and J. V. Vinters, Met. Trans., 1971, vol. 2, pp. 3175-88.

179. E. T. Turkdogan, R. G. Olsson and J. V. Vinters, Met. Trans., 1971, vol. 2, pp. 3189-96.

180. E. W. Filer and L. S. Darken, Trans. AIME, 1952, 194, 253.

181. W. Oelsen, Arch. F. D. Eisenhuttenw., 1964, 35, 1, 699, 713.

182. W. Oelsen and H. G. Schubert, Arch. F. D. Eisenhuttenw., 1964, 35, 1039, 1115.

183. W. Oelsen, H. G. Schubert, and O. Oelsen, Arch. F. D. Eisenhuttenw., 1965, 36, 779.

184. E. T. Turkdogan, G. J. W. Kor, and R. J. Fruehan, Ironmaking & Steelmaking 7, 268 (1980).

185. N. Tsuchiya, M. Tokuda, and M. Ohtani, Met. Trans. B7, 315 (1976).

186. K. Okabe, et al., Kawasaki Steel Tech. Res. Lab. Report, May 4, pp. 1-43 (1974); English Transl. BISI 13657; Nov. 27, pp. 1-11 (1974); English Transl. BISI 13658.

187. E. T. Turkdogan, J. Iron Steel Inst. 182, 74 (1956).

188. S. P. Kinney and E. W. Guernsey, Blast Furnace Steel Plant, p. 395 (1925).

189. "Alkalis in Blast Furnaces" (N. Standish and W.-K. Lu, eds.), McMaster University, Hamilton, Ontario, 1973.

190. K. P. Abraham and L. J. Staffansson, Scand. J. Metall. 4, 193 (1975).

191. J. Davies, J. T. Moon, and F. B. Traice, Ironmaking & Steelmaking 5, 151 (1978).

192. M. Scheider and J. M. Steiler, Rev. Met. 76, 621 (1979).

193. F. D. Delve, H. W. Meyer, and H. N. Lander, in "Phys. Chem. Process Metallurgy" (G. R. St. Pierre, ed.), p. 1111. Interscience, New York, 1961.

194. W. Hess, S. Mayer, and P. Schulz, Stahl Eisen 93, 107 (1973).

195. K. Schwerdtfeger and H. G. Schubert, Arch. F. D. Eisenhuttenw. 45, 437, 649, 905 (1974).

196. K. Schwerdtfeger and H. G. Schubert, Met. Trans. B8, 535 (1977).

197. K. Schwerdtfeger, W. Fix, and H. G. Schubert, Ironmaking & Steelmaking 5, 67 (1978).

198. W. Fix, A. Moradoghli-Haftwani, and K. Schwerdtfeger, Arch. F. D. Eisenhuttenw. 46, 363 (1975).

199. J. O. Edstrom, J. Iron Steel Inst., 1953, vol. 175, pp. 289-304.

200. H. Schenck, A. Majidic, and U. Putzier, Arch. F. D. Eisenhuttenw., 1967, vol. 38, pp. 669-75.

201. T. Fuwa and S. Ban-ya, Trans. Iron Steel Inst., Japan, 1969, vol. 9, pp. 137-47.

202. O. Burghardt, H. Kostmann, and B. Grover, Stahl Eisen, 1970, vol. 90, pp. 661-66.

203. W. Wenzel, H. W. Gudenau, and M. Ponthenkandath, Aufbereitungstechnik, 1970, vol. 11, pp. 154-57; pp. 492-94.

204. L. Granse, Proc. Intern. Conf. Science and Technology of Iron and Steel, Trans. Iron Steel Inst., Japan, 1971, vol. 11, pp. 45-51.

205. J. T. Moon and R. D. Walker, Ironmaking & Steelmaking, 1975, vol. 2, pp. 30-35.

206. H. vom Ende, K. Grebe, and S. Thomalla, Stahl Eisen, 1970, vol. 90, pp. 667-76.

207. E. T. Turkdogan and J. V. Vinters, Met. Trans., 1972, vol. 3, pp. 1561-74.

208. B. B. Seth and H. U. Ross, Trans. Met. Soc. AIME, 1965, vol. 233, pp. 180-85.

209. R. H. Spitzer, F. S. Manning, and W. O. Philbrook, Trans. Met. Soc. AIME, 1966, vol. 236, pp. 726-42.

210. J. Szekely and J. W. Evans, Met. Trans., 1971, vol. 2, pp. 1691-98 and 1699-1710.

211. R. H. Tien and E. T. Turkdogan, Met. Trans., 1972, vol. 3, pp. 2039-48.

212. P. Nilles, E. Denis, P. Dauby, and N. Bach, J. Metals 19 (1), 18 (1967).

213. C. A. Natalie and J. W. Evans, Ironmaking Steelmaking 6, 101 (1979).

214. K. L. Fetters and J. Chipman, Trans. AIME 145, 95 (1941).

215. G. Tromel, K. Koch, W. Fix, and N. Grobkurth, Arch. F. D. Eisenhuttenw, 40, 969 (1969); Engl. Trans. BISI No. 8306.

216. R. J. Leonard and R. H. Herron, Proc. NOH-BOS Conference, AIME 60, 127 (1977).

217. A. I. van Hoorn, J. T. van Konynenburg, and P. J. Kreyger, in "Role of Slag in Basic Oxygen Steelmaking Processes" (W.-K. Lu ed.), p. 2-1. McMaster University Press, Hamilton, Ontario, 1976.

218. F. Bardenheurer, H. vom Ende, and K. G. Speith, Arch. F. D. Eisenhuttenw. 39, 571 (1968)

219. P. Nilles, E. Denis, F. Merken, and P. Dauby, CRM Metall. Rep. 27, 3(1971).

220. R. Baker, Brit. Steel Corp. Report, code CAPL/SM/A/31/74.

221. N. I. Yaroshenko, R. V. Starov, E. B. Tret'Yakov, V. K. Didkovaskii, and I. D. Podoprigora, Steel USSR 1, 690 (1971); 3, 289 (1973).

222. H. Gaye, P. V. Riboud, R. D. Schmidt-Whitley, P. Papier, and D. Mansuy, "Advantages of Using Charge Dolomite for Converter Lining Life", IRSID Rep. RP. FAR. 34, January 1976.

223. R. J. Leonard and R. H. Herron, Steelmaking Conference AIME 60, 127 (1977).

224. D. W. Coato and J. G. Selmeczi, Electric Furnace Conference, AIME 37, 258 (1979).

225. C. R. Taylor and J. Chipman, Trans. AIME 154, 228 (1943).

226. P. A. Distin, S. G. Whiteway, and C. R. Masson, Can. Metall. Quarterly 10, 13 (1971)

227. T. Fuwa and J. Chipman, Trans. Met. Soc. AIME 281, 887 (1960).

228. E. T. Turkdogan, in "The Making, Shaping and Treating of Steel," 9th edition (H. E. McGannon, ed.), p. 378. U. S. Steel Corporation, Pittsburgh, 1970.

229. T. B. Winkler and J. Chipman, Trans. AIME 167, 134 (1946).

230. M. Kawana, "Pure Oxygen Bottom Blown Converter Steelmaking Process," TEKKO KAI, Japan, January 1978.

231. E. T. Turkdogan, J. Iron and Steel Inst., 1955, vol. 179, pp. 147-54.

232. N. J. Grant and J. Chipman, Trans. AIME, 1946, vol. 167,

pp. 134-39.

233. K. Balajiva, A. G. Quarrell and P. Vajragupta, J. Iron and Steel Inst., 1946, vol. 153, pp. 115-50.

234. K. Balajiva and P. Vajragupta, J. Iron and Steel Inst., 1947, vol. 155, pp. 562-67.

235. E. T. Turkdogan, "Physicochemical Properties of Molten Slags and Glasses", The Metals Society, London, 1983.

236. R. J. Choulet, F. S. Death and R. N. Dakken, Argon-Oxygen Refining of Stainless Steel, Canadian Metallurgical Quarterly, 1972, vol. 10, pp. 129-36.

237. J. C. Fulton and S. Ramachandran (Allegheny Ludlum, Inc.,) Decarburization of Stainless Steels, presented at the Annual AIME Meeting, Chicago, 1973.

238. E. C. Nelson, Decarburization Process for High-Chromium Steel, U.S. Patent No. 3,046,017.

239. R. J. Fruehan, Met. Trans. B., 1975, vol. 6B, pp. 573-8: and Met. Trans. B., 1977, vol. 8B, pp. 429-33.

240. R. J. Fruehan, Ironmaking and Steelmaking, 1976, vol. 3, pp. 153-8.

241. R. J. Fruehan, L. J. Martonik and E. T. Turkdogan, Trans. Met. Soc. AIME, 1969, vol. 245, pp. 1501-09.

242. C. K. Russell, R. J. Fruehan and R. S. Rittiger, J. of Metals AIME, Nov. 1971, pp. 44-47.

243. D. A. Dukelow, J. L. Steltzer, and G. F. Koons, J. of Metals AIME, Dec. 1971, pp. 22-25.

244. E. T. Turkdogan, Arch Eisenhüttenwes., 1983, 54, 1, 45.

245. E. T. Turkdogan, "Physical Chemistry of High Temperature Technology". Academic Press, New York, 1980.

246. T. Fujisawa and H. Sakao, Tetsu-to-Hagane 63, 1494 (1977).

247. A. Ejima, K. Suzuki, and N. Harada, Tetsu-to-Hagane 61, 2784 (1975); Trans. Iron Steel Inst. Japan 17, 349 (1977).

248. R. J. Fruehan, Met. Trans. 10B, 143 (1979).

249. K. P. Abraham and F. D. Richardson, J. Iron Steel Inst. 196, 313 (1960).

250. E. Schurmann, H. Bruder, K. Nernberg, and H. Rechter, Arch. F. D. Eisenhuttenw. 50, 139 (1979).

251. R. I. L. Guthrie, L. Gourtsoyannis, and H. Henein, Can. Metall. Quarterly 15, 145 (1976).

252. T. Tanoue, Y. Uneda, H. Ichikawa, and T. Aoki, Sumitomo Metals 25 (2), 146 (1973).

253. M. L. Turpin and J. F. Elliott, J. Iron Steel Inst. 204, 217 (1966).

254. L. von Bogdandy, W. Meyer, and I. N. Stranski, Arch. F. D. Eisenhuttenw. 32, 451 (1961).

255. E. T. Turkdogan, J. Iron Steel Inst. 204, 914 (1966).

256. E. T. Turkdogan, in "Sulfide Inclusions in Steel" (J. J. deBarbadillo and E. Snape, eds.), p. 1. ASM, Metals Park, Ohio, 1975.

257. Y. Miyashita, K. Nishikawa, T. Kawawa, and M. Ohkubo, Jpn-USSR Jt. Symp. Phys. Chem. Metall. Processes, 2nd p. 101. Iron Steel Inst. Japan, Tokyo, 1969.

258. N. Lindskog and H. Sandberg, Scand. J. Metall. 2, 71 (1973).

259. T. A. Engh and N. Lindskog, Scand. J. Metall. 4, 49 (1975).

260. "Secondary Steelmaking", Met. Soc. London, 1978.

261. J. Szekely, "Fluid Flow Phenomena in Metals Processing". Academic Press, New York, 1979.

262. D. Janke and H. Richter, Arch. F. D. Eisenhuttenw 50, 93 (1979).

263. B. A. Strathdee and D. McFarlane, Proc. Nat. O.H. & B.O.F. Conference, AIME, 52, 163 (1969).

264. K. Torssell and M. Olette, Rev. Metall. 66, 813 (1969).

265. J. W. Bales and M. A. Orehoski, "Method of Making Rim-Stabilized Ingots". U. S. Patent 3,754,591, Aug. 28, 1973; U. S. Patent 3,865,643, Feb. 11, 1975.

266. B. Tivelius, T. Sohlgren and C. Wretlind, in "Secondary Steelmaking", p. 54. Metals Society, London, 1978.

267. E. Forster, W. Klapdar, H. Richter, H. W. Rommerswinkel, E. Spetzler, and J. Wendorff, Stahl u. Eisen 94, 474 (1974).

268. B. Tivelius and T. Sohlgren, Steelmaking Proceedings, AIME, 61, 154 (1978).

269. R. Bruder, H. P. Haastert, H. Richter, and E. Schulz, in "McMaster Symposium on Iron and Steelmaking, No. 7" (J. S. Kirkaldy, ed.), p. 2-1. McMaster University Press, Hamilton, 1979.

270. P. K. Trojan and R. A. Flinn, Trans. ASM 54, 549 (1961).

271. D. L. Sponseller and R. A. Flinn, Trans. Met. Soc. AIME 230, 876 (1964).

272. P. A. Distin, G. D. Hallett and F. D. Richardson, J. Iron Steel Inst., 1968, vol. 206, pp. 821-33.

273. J. H. Flux, "Vacuum Degassing of Steel," I, The Iron and Steel Institute, Special Report 92, 1965, pp. 1-23.

274. R. G. Olsson and E. T. Turkdogan, J. Iron Steel Inst., 1973, vol. 211, pp. 1-8.

275. H. S. Philbrick, "Vacuum Metallurgy Conference 1963," p. 314 (1964), Boston, Amer. Vac. Soc.

276. R. Kiessling and N. Lange, "Non-Metallic Inclusions in Steel", Parts I to IV. Metals Society, London, 1978.

277. F. B. Pickering (ed.), "Inclusions". Institution of Metallurgists, London, 1979.

278. S. Ekerot, Scand. J. Metall. 3, 21, 151 (1974).

279. E. T. Turkdogan, Trans. Met. Soc. AIME 233, 2100 (1965).

280. D. Burns and J. Beech, Ironmaking & Steelmaking 1, 239 (1974).

281. A. Palmaers, J. Defays, and L. Philippe, C. R. M. No. 55, 15 (1979).

282. B. Chalmers, "Principles of Solidification," John Wiley New York, 1966.

283. M. C. Flemings, "Solidification Processing," McGraw-Hill, New York, 1974.

284. E. Scheil, Z. Metallk., 34, 70 (1942).

285. E. T. Turkdogan, J. Metals. AIME, 19, 38 (1967).

286. E. T. Turkdogan and R. A. Grange, J. Iron Steel Inst., 208, 482 (1970).

287. P. Nilles, J. Iron Steel Inst., 202, 601 (1964).

288. E. A. Golas and S. N. Singh, "Method of Single Piece Annular Nozzle to Prevent Alumina Buildup During Continuous Casting of Al-Killed Steels," U. S. Patent 4,117,959, October 3, 1978.

CHAPTER 14

Direct Reduction and Smelting Processes

SECTION 1

INTRODUCTION

During the past century, many efforts were made to develop processes for producing iron for steelmaking that could serve as alternatives to the conventional blast furnace. Many of these projects were stimulated by a desire or necessity to employ lower-grade ores and available fuels that are unsuitable for the blast furnace. Today, processes that produce iron by reduction of iron ore below the melting point of the iron produced are generally classified as **direct-reduction processes,** and the products referred to as **direct-reduced iron (DRI).** The processes that produce a molten product (similar to blast-furnace hot metal) directly from ore are generally classified as **direct-smelting processes.** In some of the more ambitious projects, the objective is to produce liquid steel directly from ore and these processes are generally classified as **direct-steelmaking processes.** These broad categories are clearly distinguished by the characteristics of their respective products, although all of these products may be further treated to produce special grades of steel in the same refining or steelmaking process.

While the blast furnace is expected to remain the world's chief source of iron units for steelmaking, as long as adequate supplies of suitable coking coals remain available at competitive cost, DRI contributes three to four per cent of the world's total ironmaking capacity with about 20 million annual metric tons (22 million annual net tons) of DRI capacity having been installed in the free world as of 1983.

The major part of DRI production is used as a substitute for scrap in the electric-arc steelmaking furnace (EAF). DRI derived from virgin iron units is a relatively pure material which dilutes contaminants in the scrap and improves the steel quality. The availability of low-cost scrap and the high cost of energy restrict the use of DRI in most highly industrialized countries. Direct processes are especially favored in those locations that are endowed with abundant reserves of inexpensive natural gas, non-coking coals, and/or hydroelectric power, and which have access to suitable iron ores or agglomerates.

This chapter presents a review of the alternative processes that have achieved some measure of pilot or commercial success, with particular emphasis on those now in commercial operation.

SECTION 2

HISTORICAL DEVELOPMENT AND BACKGROUND

Sponge iron provided the main source of iron and steel for many centuries before the blast furnace was developed around 1300 A.D. As described in Chapter 1, sponge iron was produced in relatively shallow hearths or in shaft-furnaces, both of which used charcoal as fuel. The product of these early smelting processes was a spongy mass of coalesced granules of nearly pure iron intermixed with considerable slag. Usable ar-

ticles of wrought iron were produced by hammering the spongy mass, while still hot from the smelting operation, to expel most of the slag and compact the mass. By repeated heating and hammering, the iron was further freed of slag and forged into the desired shape.

All of the methods whereby low-carbon wrought iron can be produced directly from the ore are referred to as **direct processes.** After the development of the blast

507

furnace, which made large quantities of iron having a high carbon content available, low-carbon wrought iron was produced by refining this high-carbon material. Because two or more steps were involved, the processes came to be known as **indirect processes.**

Direct methods are still in use, and, indeed, have never been wholly abandoned even by the most advanced nations. The ease with which iron ores are reduced makes the direct processes appear enticingly simple and logical, primarily because the reduction takes place at relatively low temperatures.

In modern times, sponge iron itself has found increasing use in various industrial processes, other than in the manufacture of wrought iron. The iron produced in sponge form has a very high surface area and is used extensively in the chemical industry as a strong reducing agent. It is chemically much more active than steel or iron in the form of millings, borings, turnings, or wire. Sponge iron may be produced as a granular material or as a sintered mass, depending upon the methods of manufacture. In the purified granular form, in which it is commonly known as **powdered iron,** it is used in the manufacture of many useful articles by the techniques of powder metallurgy where (1) iron powders are first compacted by pressure alone into the approximate shape of the finished article; (2) the compact is then "sintered" at a temperature ranging between approximately 950 and 1095°C (1800 and 2000°F) in furnaces provided with a protective atmosphere to prevent oxidation; and (3) the sintered articles are then pressed or machined to their final shape.

Iron powders are produced, not only by direct-reduction of iron ores or oxides using solid carbonaceous reducing agents and gaseous reducing agents such as carbon monoxide and hydrogen, but also by electrolytic processes, and by thermal decomposition of iron carbonyl which has the chemical formula $Fe(CO)_5$.

In its modern usage, sponge iron is referred to as direct-reduced iron (DRI). Today, the major portion of DRI production is melted in electric arc furnaces for steelmaking. In addition, minor amounts may be charged to the ironmaking blast furnace. The attempts to develop large-scale direct processes have embraced practically every known type of apparatus suitable for the purpose, including pot furnaces, reverberatory furnaces, regenerative furnaces, shaft-furnaces, rotary and stationary kilns, retort furnaces, rotary-hearth furnaces, electric furnaces, various combination furnaces, fluidized-bed reactors, and plasma reactors. Many different kinds of reducing agents, such as coal, coke, graphite, char, distillation residues, fuel oil, tar, producer gas, coal gas, water gas, and hydrogen, have also been tried. To facilitate classification and discussion of the different processes, broad categories are shown on Table 14—I. The direct reduction (DR) processes which employ reducing gases generated externally from the reduction furnace are described in Section 3. The reformed natural gas processes exemplify this category. The DR processes in which the reducing gas is generated from hydrocarbons in the reduction furnaces are described in Section 4. In this category, the furnace design facilitates the high heat transfer rate required for the endothermic gasification of carbon by iron oxide. Some of the direct smelting processes that were

developed are also listed in Table 14—I and are reviewed in Section 5. Section 6 describes plasma processes whereby electric plasma supplants oxygen and fuel to heat gases and solids.

No effort is made here to evaluate or compare the different processes on either an economical or technical basis because in many cases factors associated with location, including cost and availability of fuels and iron-bearing materials, availability of trained manpower, and size and proximity of markets, may be overriding. A detailed treatment including quantitative discussions of these important considerations is outside the scope of this book. Further details on the processes described in these sections and on other direct processes are given in the reference material listed at the end of this chapter.

DEFINITIONS AND TERMS

Chemical Reactions—The reduction of iron ore in any direct process is accomplished by the same chemical reactions that occur in the blast furnace (see Chapter 15). The typical reduction reactions occur as follows:

$$3Fe_2O_3 + H_2 = 2Fe_3O_4 + H_2O \qquad (1)$$
$$3Fe_2O_3 + CO = 2Fe_3O_4 + CO_2 \qquad (2)$$

$$Fe_3O_4 + H_2 = 3FeO + H_2O \qquad (3)$$
$$Fe_3O_4 + CO = 3FeO + CO_2 \qquad (4)$$

$$FeO + H_2 = Fe + H_2O \qquad (5)$$
$$FeO + CO = Fe + CO_2 \qquad (6)$$

$$3Fe + CO + H_2 = Fe_3C + H_2O \qquad (7)$$
$$3Fe + 2CO = Fe_3C + CO_2 \qquad (8)$$

$$CO_2 + C = 2CO \text{ (Boudouard Reaction)} \qquad (9)$$
$$H_2O + C = CO + H_2 \qquad (10)$$

$$FeO + C = Fe + CO \qquad (11)$$

$$3Fe + C = Fe_3C \qquad (12)$$

In cases where the reduction is carried out below about 1000°C (1830°F), the reducing agents for the iron oxide comprise the gases CO and H_2 of reactions (1) to (6), and the DRI produced will be porous and have about the same size and shape as the original iron-ore particle or agglomerate. The metallic iron also absorbs carbon according to reactions (7) and (8). Typically, DRI from the gas-based processes contains 1 to 2.5 per cent carbon as cementite.

Above about 1000°C (1830°F), the gaseous reduction of solid iron oxides continues. However, carbon now reacts with CO_2 and H_2O according to reactions (9) and (10). This forms CO and H_2, thus renewing the reducing potential of the gas. The resultant net reaction (11) is an important mechanism for processes that produce DRI directly from solid coal without prior gasification of its fixed carbon.

At about 1200°C (2190°F), considered to represent close to the upper limit for the direct-reduction process, a pasty, porous mass forms. Above this temperature, the metallic iron formed will absorb any carbon present with resultant fusing or melting of the solid

Table 14—I. Classification of Processes

Direct-Reduction Processes—Reducing Gas Generated Externally from Reduction Furnace (Section 3)	**Direct-Reduction Processes—Reducing Gas Generated from Hydrocarbons in Reduction Furnace (Section 4)**

Shaft-Furnace Processes, Moving Bed
  Wiberg-Soderfors
  Midrex
  HYL III
  Armco
  NSC
  Purofer
Shaft-Furnace Processes, Static Bed
  HYL I and II
Fluidized-Bed Processes
  FIOR
  HIB
	Kiln Processes
	  Krupp-Renn
	  Krupp-CODIR
	  SL/RN
	  ACCAR
	  DRC
	  LS-RIOR
	Rotary-Hearth Processes
	  INMETCO
	  Salem
	Retort Processes
	  Hoganas
	  Kinglor-Metor
	Shaft-Furnace Process, Moving Bed
	  Midrex Electrothermal

Direct Smelting Processes (Section 5)	**Plasma Processes (Section 6)**

Electric-Furnace Smelting Processes
  Pig Iron Electric Furnace
  DLM
Oxy-Fuel Smelting Systems
  INRED
  KR
  Kawasaki
  CGS
	Nontransferred Arc
	  Plasmasmelt
	  Plasmared
	Transferred Arc
	  ELRED
	  EPP
	  SSP
	  The Toronto System
	  Falling Film Plasma Reactor

even though the melting point of pure iron is 1530°C (2785°F). Thus, the direct-smelting processes all operate with product temperatures higher than 1300°C (2370°F) because at this temperature, carbon is absorbed rapidly and a liquid hot metal forms that can be handled effectively during subsequent processing.

The direct-steelmaking processes all operate with product temperatures higher than 1530°C (2785°F) to produce steel that will remain fluid as needed during subsequent treatment.

Effectiveness of Reduction Reactions for DRI—The several terms associated with measuring the effectiveness of the reduction reactions for DRI are defined as follows:

1. **Per cent Total Iron** (Fe_T) refers to the total amount of iron in a sample. All forms of iron (metallic iron and iron chemically combined with other elements such as oxygen) are included.

$$Fe_T = \frac{\text{weight of iron}}{\text{total weight of sample}} \times 100\%$$

2. **Per cent Metallic Iron** ($Fe°$) refers to iron chemically uncombined in a sample and combined as cementite (Fe_3C).

$$Fe° = \frac{\text{weight of metallic iron}}{\text{total weight of sample}} \times 100\%$$

3. **Per cent Metallization** refers to that portion of the total iron present as metallic iron. Per cent metallization is not synonymous with either per cent reduction or per cent metallic iron.

$$\text{Per cent Metallization} = \frac{\text{total weight of metallic iron}}{\text{total weight of iron}} \times 100\%$$

$$= \frac{Fe°}{Fe_T} \times 100\%$$

4. **Per cent Reduction** is based on the premise that before reduction the iron was present as hematite (Fe_2O_3), its highest state of oxidation. Per cent reduction refers to the oxygen (combined as iron oxide) that has been removed in producing the DRI.

Per cent Reduction =

$$\frac{\left(\frac{\%\ Oxygen}{\%\ Fe_T}\right)_{Fe_2O_3} - \left(\frac{\%\ Oxygen}{\%\ Fe_T}\right)_{DRI}}{\left(\frac{\%\ Oxygen}{\%\ Fe_T}\right)_{Fe_2O_3}} \times 100\%$$

Per cent Reduction =

$$\left[1 - 2.327 \times \left(\frac{\%\ Oxygen}{\%\ Fe_T}\right)_{DRI}\right] \times 100\%$$

As defined above, the per cent reduction gives the state of the iron contained in the DRI. However, confusion can arise because some investigators define per cent reduction in reference to the initial oxide state of the ore which may be magnetite (Fe_3O_4). In that case, per cent reduction defines the amount of reduction performed in the DRI process.

5. **Per cent Gangue** refers to the non-iron compounds (SiO_2, Al_2O_3, CaO, MgO, etc.) in the DRI.

$$Per\ cent\ Gangue = 100\% - \%FeO - \%Fe° - \%C$$

6. **Per cent Oxygen** refers to the oxygen combined as iron oxides. Oxygen combined with the gangue constituents is excluded.

$$Per\ cent\ Oxygen = \%\ FeO + \%Fe° - \%Fe_T$$

7. **Per cent Wustite** refers to the amount of wustite (FeO) present in the sample.

$$Per\ cent\ FeO = \frac{total\ weight\ of\ FeO}{total\ weight\ of\ sample} \times 100\%$$

The following analysis of a DRI material illustrates the above definitions:

Material	Composition, Per Cent
Fe_T	91.8
FeO	8.3
Fe°	85.4
SiO_2	3.1
Al_2O_3	0.4
MgO	0.4
CaO	0.5
C	1.6
S	<0.005

Based on this analysis, the following terms have been calculated:

$$Per\ cent\ Metallization = \frac{85.4}{91.8} \times 100 = 93\%$$

$$Per\ cent\ Oxygen = 8.3 + 85.4 - 91.8 = 1.9\%$$

$$\begin{matrix}Per\ cent\\ Reduction\end{matrix} = \left[1 - 2.327 \times \frac{1.9}{91.8}\right] \times 100 = 95.2\%$$

$$Per\ cent\ Gangue = 100 - 8.3 - 85.4 - 1.6 = 4.7\%$$

DRI QUALITY

As mentioned previously, the EAF process consumes the major portion of the world production of DRI. Since the technologies of both the DR and EAF processes are continuing to evolve, the industry has not formalized uniform quality specifications. Nevertheless, various relevant factors have been published.

The portions of iron oxide and gangue materials in DRI above certain minimums increase the power requirements of melting in the EAF compared to an equivalent quantity of scrap. A portion of the iron oxide (FeO) reacts with carbon in the EAF to produce metallic iron and carbon monoxide according to reaction (11), which is endothermic. Furthermore, the gangue and the associated flux require energy for melting. However a metallization that is too high decreases both the fuel efficiency and the productivity of the DR process, and some of the gangue in DRI substitutes for slag building agents normally present in the EAF charge.

The metallization of DRI normally ranges between 90 and 95 per cent (average 91 to 93 per cent) depending on the process and on the reducibility of the original iron oxide. This probably represents optimization in the various plants. Based on the level of metallization, the carbon content of the DRI is controlled in some DR processes to be between 1 and 2.5 per cent to facilitate reduction of FeO by carbon during melting. This not only increases iron recovery, but also, through the generation of gaseous carbon monoxide, promotes the foamy slag practice of the EAF. However, alternative EAF practices can achieve the same effects without DRI or with DRI of lower carbon content. Of the possible carbon forms associated with DRI, the combined carbon cementite (Fe_3C) is more desirable than loose carbon fines or soot which may not be useful to the process. Any iron oxide from DRI that is not reduced in the EAF enters the slag.

The gangue in DRI that substitutes for the normal amount of slag-building agents does not penalize the EAF operation. This amount of gangue is usually between 2 and 4 per cent. However, the actual amount depends on the proportion of silica (SiO_2) in the gangue which must be fluxed and on the percentage of iron units in the charge which are derived from DRI. DRI with a low gangue content is derived from premium sources of iron oxide.

Another possible impact on the operation of the EAF pertains to fines in the DRI. Fines affect iron recovery, increase the dust loading, and contaminate electrical parts. Manufacturers and users of DRI vary in their specification on allowable fines. However, the size specification is usually in the range of no more than five per cent minus 5-mm or five per cent minus 3-mm fines.

For DRI which satisfies a size specification, screening and blending assure that fines will be distributed uniformly throughout the DRI charge to the EAF. DRI containing larger amounts of fines is usually compacted in briquetting molds. The total amount of fines produced in a DRI plant depends on the properties of the original oxide feed, on the process employed, and on the amount and severity of handling.

While most quality requirements of DRI relate to an optimization of productivity and energy consumption, the absence of contaminating residuals such as copper and tin is an important advantage for upgrading steel. Furthermore, phosphorus, manganese and vanadium contained in the ore remain in the gangue in the gas-based processes and are usually not a factor in DRI-EAF steelmaking operations. However, phosphorus reduction may be affected by the EAF practice that is employed. The sulphur content of DRI is also relatively low, depending on the amount of sulphur in the fuels and reductants contacting the iron. Sulphur contents of DRI vary from less than 0.005 per cent in the DR processes employing sulphur-free gas to about 0.02 per cent in DR processes employing sulphur-bearing coal and limestone together with iron oxide in the charge mix. In the latter case, coals containing more than about 1.5 per cent sulphur probably necessitate desulphurization during steelmaking.

SPECIAL PRECAUTIONS

DRI, whether in the form of particles, pellets, or molded briquettes, tends to revert to the oxide state when exposed to natural environments. The large pore surface area of DRI makes it susceptible to spontaneous reoxidation. Spontaneous reoxidation is undesirable because of the possible temperature rise during storing or shipment. Even if ignition can be prevented, DRI will gradually lose its high degree of metallization because of reoxidation due to weathering. When the oxygen in the gas in contact with the DRI becomes depleted, the DRI also reacts with moisture to produce hydrogen gas. This could produce a flammability or explosion hazard especially in confined spaces. The reoxidation and hydrogen evolution rates increase with the amount of moisture. Furthermore, the high electrical conductivity of seawater severely aggravates the corrosion problem. Because of the potential hazards, the loss in value, and the possible harmful effects during subsequent melting, producers and users usually protect DRI from contact with water. DRI produced in geographic areas with wet climates is usually consumed on site.

The stability and behavior characteristics of DRI depend on a number of variables affecting the pore surface area. These include the mineralogy of the oxide ore or agglomerate, reduction temperature, time at the reduction temperature, gas composition, age, and previous history of exposure and handling. As one example, the DRI stability increases with reduction temperature because the pore surface area decreases as the temperature is raised. Other methods of improving the stability involve making the iron in the pore surfaces unavailable for reoxidation or decreasing mechanically the pore surface area.

To prevent the deterioration of DRI, the industry has studied the techniques used for the protection of structural metals from corrosion. Impervious coatings have been developed utilizing oxide passivation, chemical solutions, oily organic rust inhibitors, paints, etc. Midrex has patented the CHEMAIRE® process, a sodium silicate-air passivation treatment.

While some coatings potentially contaminate steel, passivation produces an iron oxide coating which is consistent with the steelmaking process. Passivation involves treatment with a mildly oxidizing gas at temperatures below 540°C (1000°F) for the formation of a tightly bonded magnetite layer, which is essential for the protective process to work. Passivation to increase stability from spontaneous combustion in a dry atmosphere can be achieved by oxidation without measurable losses in the state of metallization. It has been observed that given sufficient time, the internal surface area of porous DRI pellets or lumps also acquires the oxide coating which tends to increase the stability of fines generated in subsequent handling of DRI. Some autogenous passivation occurs during aging of DRI in a protected ambient atmosphere. Passivation is usually performed at temperatures between 90 to 200°C (200 to 400°F) with various gases, including combustion gases and reducing gas, containing less than about two per cent oxygen.

At the current state of the art, passivation or economical coatings acceptable in the subsequent melting process are not completely reliable. Incidents of heating of DRI in the holds of ships have been reported. The problem may be intensified by contamination with seawater or by a lack of uniformity in the DRI as produced or treated. Therefore, to avert possible catastrophes, an emergency supply of inert gas should be available for ship holds or enclosed storage bins. Such storage areas could be flooded with the inert gas should the temperature rise above a safe minimum or should an explosion hazard arise. The inert medium should probably be nitrogen rather than flue or exhaust gas, which contains carbon dioxide. It is possible that carbon dioxide can be a source of oxygen for the oxidation of DRI; however, the generation of poisonous carbon monoxide is the more immediate concern.

While coating procedures attempt to protect the iron in the pores from reoxidation, hot-molded briquetting mechanically decreases the pore surface area of DRI. In the HIB and FIOR processes, the –2-mm product is briquetted while the material is hot and malleable. The individual briquettes attain densities of 5200 to 6000 kg/m³ (320 to 370 lb/ft³). Thus, this DRI product is capable of minimal exposure to moisture. It has been observed that hot-molded briquettes: (1) can be stored in the open prior to loading; (2) can tolerate a fine spray of fresh water to control dust accumulation; and (3) can be unloaded under less restricted weather conditions. Still, hot-molded briquettes require adequate surface ventilation and should not be loaded if the briquette temperature exceeds safety standards. Ship holds or bilges that are used to store the briquettes should be dry and weatherproof. Also, because the briquettes may deplete oxygen in the storage compartments, caution should be exercised when entering. In 1984, only hot briquetted material is designated safe for ocean shipping without provision for hold inerting. New technology is extending the hot-molded DRI concept to processes now producing DRI in the form of larger coarse particles.

DIRECT REDUCTION PROCESSES—REDUCING GAS GENERATED EXTERNALLY FROM REDUCTION FURNACE

In this section, a process description and a simplified process flowsheet are given for each DR process to illustrate the types of equipment used and to describe the flow of materials through the plant. The discussion does not mention all the variations of the flowsheet which may exist or the current status of particular plants.

In the majority of the DR processes described in this section, natural gas is reformed in a catalyst bed with steam or gaseous reduction products from the reduction reactor. Partial oxidation processes which gasify liquid hydrocarbons, heavy residuals and coal are also discussed. The reformer and partial oxidation gasifier are interchangeable for several of the DR processes. A brief summary of the gas generation schemes facilitates the DR process descriptions.

Reformed Natural Gas—Reducing gas, which is rich in carbon monoxide and hydrogen, is produced by reforming naural gas feedstock, which is primarily methane, in catalyst-filled reformer tubes. The carbon monoxide and hydrogen are generated in the methane-steam reforming processes according to the reaction:

$$CH_4 + H_2O = CO + 3H_2 \qquad (14)$$

High steam/carbon ratios were once required to prevent carbon formation in the reformer tubes, and catalyst deterioration. The resultant reducing gas then had to be cooled to condense excess water vapor, and then reheated in a second step before the reducing gas could be used to reduce iron oxide. However, modern catalysts now permit operation with steam/carbon ratios approaching the stoichiometric ratio, thus producing a mixed gas of carbon monoxide and hydrogen of 95 per cent purity (wet basis). As produced, the gas contains hydrogen and carbon monoxide in a ratio of about 3.3 to 1.0. Thus, energy saving is achieved by lower steam usage and elimination of reheating of the reducing gas. This near-stoichiometric operation is sometimes referred to as one-step reforming.

In the natural gas-based reducing processes that recycle part of the reducer-reactor off-gas through the reformer, (for example, the Midrex process) carbon monoxide and hydrogen are generated according to the following reactions:

$$CH_4 + CO_2 = 2CO + 2H_2 \qquad (15)$$

$$CH_4 + H_2O = CO + 3H_2 \qquad (16)$$

Carbon dioxide and water vapor are also present in the reformer off-gas as products of these reduction reactions. The reforming of natural gas with reducer-reactor off-gas produces a gas containing hydrogen and carbon monoxide in a ratio of about 1.5—1.6 to 1.0. This ratio can be varied by controlling the amount of water vapor in the feedstock.

The conversion efficiency of reformers is very high as measured by the approach to equilibrium of the gaseous products. In that regard, the methane breakthrough is roughly 0.5 to 1.5 per cent for reactions (14) to (16) at conditions mentioned above. Methane break-

through increases with pressure but decreases with increasing temperature and increasing steam/carbon ratio.

The efficiency of conversion in a reformer is contingent on the use of a sulphur-free feedstock to prevent poisoning and rapid deactivation of the catalyst. For that reason, reformers employ sulphur guards such as activated carbon absorbent. In the Midrex reformer, the reducer-reactor off-gas used in reforming may acquire sulphur from the iron oxide feed. This sulphur may be removed by first using the gas to cool the DRI product which absorbs the trace amount of sulphur involved. Because of high purity fuel and additional precautions for sulphur, DR processes employing reformers produce a very low sulphur product.

Regarding the reformer energy balance, reactions (14) to (16) are endothermic and take place at elevated temperature, 850 to 1000°C (1550 to 1850°F). The required energy is supplied by recuperation from the flue gas in a convection section of the reformer and by radiant burners which heat the outside of the catalyst-filled tubes. For one-step steam reforming, the methane is proportioned, about 40 per cent for fuel and 60 per cent for process gas. However, the DR reducer-reactor off-gas which contains combustible carbon monoxide and hydrogen gases may substitute for a portion of the total amount of the fuel required.

Partial Oxidation for Gasification of Hydrocarbons—The partial oxidation of fuels, especially coal or oil, is effected in gasifiers by reaction with a gas containing a high percentage of oxygen:

$$C_nH_m + \frac{n}{2} O_2 = n\ CO + \frac{m}{2} H_2 \qquad (17)$$

Compared to the reformer reactions, partial oxidation produces less gas per unit of hydrocarbon. For example, a unit of methane gives four units of gas in reaction (14) compared to three units of gas in reaction (17). Nevertheless, while additional fuel is needed to sustain the reformer reactions (14) to (16), reaction (17) of the gasifier is exothermic. Furthermore, some excess oxygen is required to retard soot formation. The resulting formation of carbon dioxide and steam provides even more energy to the system.

The high energy level of partial combustion processes allows the use of steam or recycled off-gas from the reducer-reactor to replace a part of the gaseous oxygen, to increase gas production, and to moderate the temperature of the gas produced. The gas may be enriched by (1) passing it through a bed of hot coke or by (2) mixing it with processed rich gas from the reducer-reactor. In the latter case, installed equipment is used to condense moisture and remove CO_2 from the reducer-reactor off-gas. Because of the smaller amount of hydrogen derived from steam, partial oxidation processes produce a gas with a lower hydrogen to carbon monoxide ratio than steam reforming of natural gas. Furthermore, the process employs fuels of higher molecular weight containing more carbon than methane. Example ratios of hydrogen to carbon monoxide in partial-oxidation gasifier gas vary from about 0.4 to 1.2 in fuels ranging from bituminous coal to light oil.

The various gasifiers employ alternative concepts comprising entrained beds, fluidized beds, packed beds, partial combustion burners and electric plasma. Various equipment members are protected by a water cooling system which is incorporated into the steam plant to recover the energy. The units may produce a liquid slag or a dry ash dust which must be separated from the gas in cyclones. Limestone may be mixed with sulphur-bearing fuel to effect desulphurization in the gasifier. In another method of desulphurization, the product gas may be passed through a bed of dolomite. It is desirable to perform the gas cleaning steps while the gasifier gas is hot. This saves fuel associated with reheating the gas and precludes carbon deposition which may occur at intermediate temperatures in the gas reheater. In an alternative to gas desulphurization, the DRI absorbs the sulphur in the reducer-reactor, and desulphurization is performed in the subsequent steelmaking operation.

SHAFT-FURNACE PROCESSES, MOVING BED

Wiberg-Soderfors Process—The Wiberg-Soderfors process (Figure 14—1) was the earliest successful shaft-furnace process. The largest existing commercial installation has an annual capacity of 18 000 metric tons (20 000 net tons). This plant went into production in 1952 in Sweden where several other plants have since been constructed.

The largest Wiberg furnaces are 24.4-metre (80-foot) tall structures of firebrick surrounded by insulation and enclosed in a steel-plate shell. The inside diameter of the bottom is nearly 2.8 metres (9 feet) and the inside diameter of the top is about 1.1 metre (3.7 feet).

The reducing gases (about 20 to 30 per cent hydrogen and 70 to 80 per cent carbon monoxide) are produced at about 1035°C (1900°F) in an electrically heated coke or charcoal carburettor and passed through a bed of limestone or dolomite to remove sulphur before entering the reduction zone (lower section) of the shaft. The hot gases, at approximately 1010°C (1850°F) reduce the descending iron oxide (FeO).

Approximately three-fourths of the partly spent gases are removed as they leave the FeO-to-Fe reduction zone and are recycled to the carburettor for regeneration. The remaining one-fourth of the gases enter the pre-reduction zone (the middle section of the shaft) where hematite and magnetite are reduced to FeO at about 870°C (1600°F). The off gases from the middle zone are burned in the preheating zone (the upper section of the shaft) and heat the ore charge to about 1010°C (1850°F). The final waste gases are discharged at about 38°C (100°F).

The feed to the shaft-furnace is lump ore, sinter, or pellets, all preferably less than 64 mm (2½ inches) in diameter. The product (about 90 per cent reduced) is first cooled to between 95 and 150°C (200 and 300°F) in a water-jacketed cooling chamber at the bottom of the shaft and is then discharged into steel cars that trans-

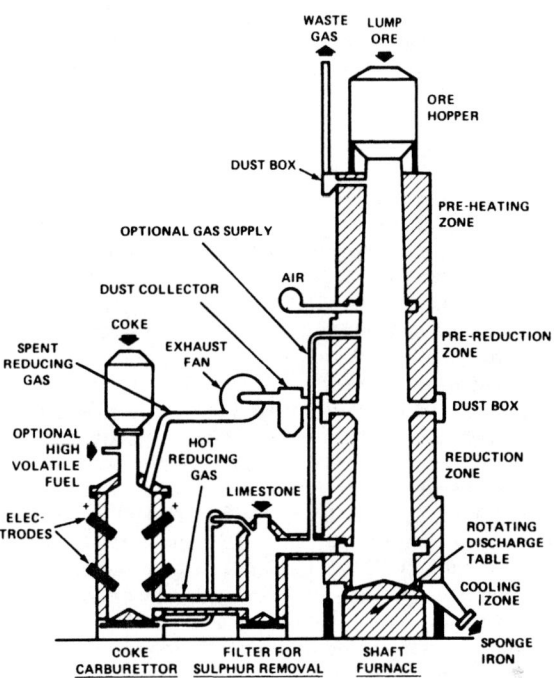

FIG. 14—1. Schematic cross-section showing the principle of operation of the Wiberg-Soderfors process.

port it to the steelmaking furnaces.

The heat requirements for the process are 10.9 to 13.1 million kilojoules per metric ton (9.4 to 11.3 million Btu per net ton) of DRI produced. These energy consumption figures do not include electric energy associated with motors, compressors, blowers, and so on.

Midrex Process—The Midrex process was developed by Surface Combustion Division of Midland-Ross Corporation in the mid-1960's. The Midrex Division became a subsidiary of Korf Industries in 1974. Midrex was subsequently acquired by Kobe Steel, Ltd. in 1983.

The first commercial Midrex plant was installed near Portland, Oregon and started production in 1969. The plant included two shaft reduction furnaces of 3.4 metre (11.2 foot) inside diameter and had a total capacity of 300 000 metric tons (330 000 net tons) per year. The average energy consumption of this early plant was about 15 million kilojoules per metric ton (12.9 million Btu per net ton) of DRI. Many difficult engineering and operating problems were solved during the first several years of operation of this plant that contributed significantly to the design, construction and operation of larger Midrex plants throughout the world during the 1970's.

The Midrex DR plants comprise the 4.88 metre (16 foot) inside diameter Midrex Series 400 and the 5.5 metre (18 foot) inside diameter Series 600-shaft furnace modules. The number of the series originally designated the DRI capacity in thousands of metric tons per year. However, the Series 600 modules may be installed at reduced capacity with the possibility of uprating as more reformer capacity is added at a later date.

By 1983, more than twenty Midrex modules were installed having a total capacity of about 9 million metric tons per year (9.9 million net tons per year); however, not all of this capacity is operating. Future construction is expected in the Middle East, Oceania, and the Soviet Union where supplies of natural gas are available.

The Midrex DR flowsheet is shown on Figure 14—2. The main components of the process are the DR shaft furnace, the gas reformer, and the cooling-gas system. Solid and gas flows are monitored so that the process variables can be controlled within operating limits. The temperature and composition of each gas stream to the shaft furnace are controlled within specification limits to maintain optimum bed temperature for reduction, degree of metallization, carbonization level (Fe$_3$C content), and to ensure the most efficient utilization of the reducing gas.

The DR furnace is a steel vessel with an internal refractory lining in the reducing zone. The charge solids flow continuously into the top of the furnace though seal legs. The reduction furnace is designed for uniform mass movement of the burden by gravity feed, through the preheat, reduction, and cooling zones of the furnace. The cooled DRI is continuously discharged through seal legs at the bottom of the furnace. The use of seal legs for feeding and discharging solids eliminates the need for complex lock hoppers. Inert gas is injected into the seal legs to prevent escape of process gases. On discharge from the shaft, the DRI is screened for removal of fines. Special precautions are undertaken to minimize any danger of spontaneous ignition of the pyrophoric DRI product during extended storage or shipment. Either the patented Midrex CHEMAIRE process or a hot briquetted iron process may be employed to protect the DRI. The reduced fines are finally briquetted to make them a usable DRI product.

Reducing process gas, about 95 per cent combined hydrogen plus carbon monoxide, enters the reducing furnace through a bustle pipe and ports located at the bottom of the reduction zone. The reducing gas temperature ranges between 760 and 927°C (1400 and 1700°F). The reducing gas flows countercurrent to the descending solids. Iron oxide reduction takes place according to reactions (1) to (6).

The partially spent reducing top gas, containing about 70 per cent carbon monoxide plus hydrogen, flows from an outlet pipe located near the top of the DR furnace into the top-gas scrubber where it is cooled and scrubbed to remove dust particles. The largest portion of the top gas is recompressed, enriched with natural gas, preheated to about 400°C (750°F), and piped into the reformer tubes. In the catalyst tubes, the gas mixture is reformed to carbon monoxide and hydrogen according to reactions (15) and (16). The hot reformed gas (over 900°C or 1650°F) which has been restored to about 95 per cent carbon monoxide plus hydrogen is then recycled to the DR furnace.

The excess top gas provides fuel for the burners in the reformer. Hot flue gas from the reformer is used in the heat recuperators to reheat combustion air for the reformer burners and also to preheat the process gas before reforming. The addition of heat recuperators to these gas streams has enhanced process efficiency,

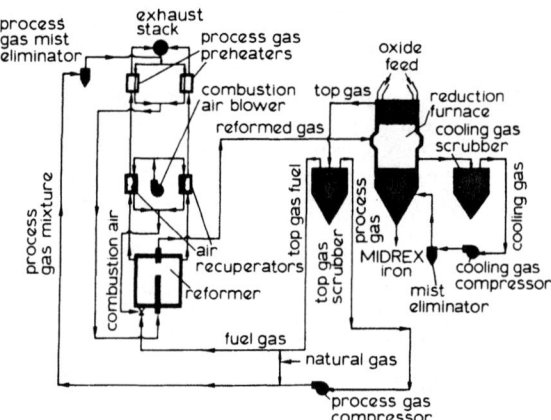

Fig. 14—2. Midrex standard DR process flowsheet.

helping to decrease annual fuel usage to a reported low figure of 11.4 to 11.6 million kilojoules per metric ton of DRI (9.8 to 10.0 million Btu per net ton).

Cooling gases flow countercurrent to the burden in the cooling zone of the shaft furnace. The gas then leaves at the top of the cooling zone and flows through the cooling-gas scrubber. The cleaned and cooled gas is compressed, passed through a demister, and is recycled to the cooling zone.

An alternative flowsheet uses cold shaft furnace top gas for cooling prior to its introduction into the reformer. Thus the DRI absorbs sulphur in the top gas that came from the raw materials. This helps to prevent sulphur poisoning of the catalyst.

HYL III Process—The HYL III DR process of HYLSA, S.A. of Monterrey, Mexico evolved from the original HYL process. Retained features of the original process comprise the catalytic reformer, the gas reheater, and the off-gas handling system which condenses water and removes particulates in a scheme that recycles the reducer-reactor off-gas. In the HYL III process, a single shaft furnace with a moving bed is used in place of the four original fixed bed reactors.

A retrofit of the second HYL plant of HYLSA in Monterrey, Mexico started production in 1980. This first HYL III plant (designated 2M5) is rated at 275 000 metric tons (300 000 net tons) per year of DRI. A second plant of 500 000 metric tons (550 000 net tons) per year capacity (3M5) started production in 1983 at the same works. While the first two plants represent a conversion of an original HYL facility, a new construction of four 500 000 metric ton (550 000 net ton) per year modules is underway at the SICARTSA plant in Las Truchas, Mexico. Startup is scheduled for 1984. Further conversions and new constructions are expected.

The HYL III plants operate with iron ore, iron oxide pellets or mixtures of the two. Natural gas consumption equivalent to about 10.7 million kilojoules per metric ton (9.2 million Btu per net ton) of DRI is quoted for a greenfield construction. The power consumption of gas compression is 90 kWh per metric ton (99 kWh per net ton) of DRI.

The HYL III process flowsheet is shown in Figure 14—3. The main equipment comprises a DR shaft furnace, a gas reformer, and a gas reheater. The principles

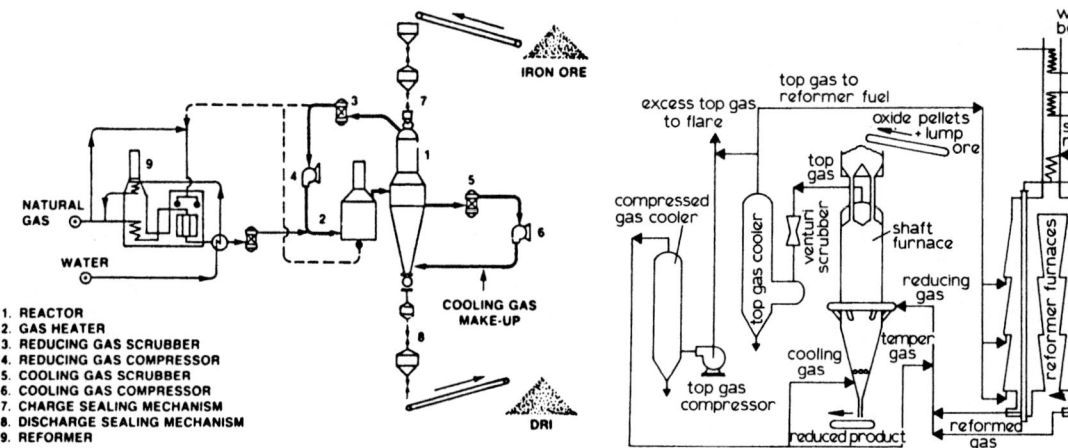

DIRECT REDUCTION AND SMELTING PROCESSES

1. REACTOR
2. GAS HEATER
3. REDUCING GAS SCRUBBER
4. REDUCING GAS COMPRESSOR
5. COOLING GAS SCRUBBER
6. COOLING GAS COMPRESSOR
7. CHARGE SEALING MECHANISM
8. DISCHARGE SEALING MECHANISM
9. REFORMER

FIG. 14—3. HYL III DR process diagram.

FIG. 14—4. Armco DR process flowsheet.

of operation of the shaft furnace are similar to the Midrex shaft furnace described previously. Continuously descending iron-bearing material is reduced in an upper zone by the countercurrent flow of gas which is rich in carbon monoxide and hydrogen. Reduction is accomplished by reactions (1) to (6). The reducing gas is introduced through a distribution system about the circumference of the shaft at an intermediate height. A proper selection of iron oxide feed stock permits operation at 950°C (1750°F). The addition of 5 per cent nonsticking ore alters the sticking tendency of iron oxide pellets and improves performance by promoting uniform descent of the burden. This increases productivity and decreases fuel consumption.

After reduction, the hot DRI continues to descend through a constant pressure zone which separates the upper reducing zone from the lower cooling zone. In the cooling zone, the DRI is cooled to below 50°C (120°F) by an independent gas stream. The cooling gas is withdrawn at the top of the cooling zone. After cooling, cleaning and compressing, this gas is recirculated to the bottom of the shaft furnace. The composition and temperature of gas flow to the shaft furnace are carefully controlled to permit independent control of the metallization and carbon content of the DRI. It is claimed that a high reduction temperature and the formation of an iron carbide shell protect the DRI from spontaneous reoxidation.

The HYL III shaft furnace operates at about 5 kg/cm² (5 atm). For this reason, the design incorporates special pressure lock systems for charging iron oxide feed materials at the top and for discharging cold DRI at the bottom. Possible advantages of high pressure operation are enhanced reduction kinetics, higher gas throughput, and condensation at elevated pressure which lowers the moisture content of the recirculated top gas.

Insofar as reducing gas is concerned, the HYL III process employs catalytic steam reforming of natural gas. As in the original HYL process, excess steam is used. Therefore, the reformed gas is cooled to condense water which increases the carbon monoxide plus hydrogen content to a high percentage.

The sensible energy of the reformed gas is recovered

during cooling by heat exchangers connected to the steam system. The usual heat recovery systems in the flue gas stacks of the reformer and gas reheater are also used. The cold reformed gas is mixed with compressed top gas from the shaft furnace. This top gas had previously been processed to remove a substantial part of its moisture and particulates. The mixed reducing gas is then reheated and introduced into the shaft furnace. Along with natural gas, excess shaft-furnace off-gas (over that amount recirculated to the process) is used as fuel in the reformer and gas heater.

HYLSA may install a system to remove carbon dioxide from the recycled shaft furnace off-gas in the future. This will provide further enrichment of the recycle gas with natural gas. This new procedure would extend the operability of the plant to fuels containing higher proportions of carbon.

Armco Process—In this process, fired iron-oxide pellets and lump ore are reduced in a continuous verticalshaft-furnace by reformed natural gas. A pilot plant was operated in Kansas City, Missouri between 1964 and 1966. A commercial plant based on this development was completed by Armco at their Houston, Texas, Works and started up in 1972. The reducing gas for this commercial plant is produced by catalytic reforming of natural gas with steam. The commercial flow sheet, which is shown schematically in Figure 14—4, is based on reforming with a very low steam-to-hydrocarbon ratio so that the hot reformed gas can be used directly, after tempering with recycled off gas, without quenching and reheating to condense excess steam. This plant, which was originally designed to produce 1000 metric tons per day (1100 net tons per day) of DRI, was modified in 1977 to provide improved cooling-zone performance. These modifications, which included decreasing the shaft diameter from 5 metres (16.5 feet) to 4.2 metres (13.7 feet) and introducing part of the hot reducing gas into the center of the shaft, provided a more uniform and stable product. The shaft is 24 metres (78 feet) high. The productivity did not decrease in proportion to the decrease in shaft crosssectional area in the modified shaft because introduction of the hot reducing gas in the interior of the shaft enabled the desired per cent reduction of DRI to be obtained at close to the design projection. Energy consumption for the Armco process is about 11.9 million kilojoules per metric ton (10.2 million Btu per net ton) of DRI.

As shown in Figure 14—4, the Armco DR process employs a shaft furnace as the reduction reactor with a steam-natural gas reformer to supply the reducing gas to the reactor. In operation, the charge is conveyed to a surge hopper mounted on top of the shaft furnace. The feed is introduced through seal legs which connect the hopper with the top of the shaft furnace. Inside the furnace, the charge descends by gravity through the reducing and cooling zones. Residence time is controlled by the rate of product removal from the bottom of the shaft furnace. A seal leg is also used for product discharge. The feed and discharge seal legs are purged with a low flow of furnace gas controlled by a steam ejector to achieve atmospheric pressure.

Reducing gas is produced in a methane-steam reformer. The reformer tubes are charged with a catalyst that permits operation at a steam/carbon ratio of 1.4:1, which is just above the carbon deposition range. The reformed gas at a temperature of 870 to 955°C (1600 to 1750°F), has a hydrogen plus carbon monoxide content of about 88.5 per cent with about 8.4 per cent water vapor. The reducing gas is tempered by the introduction of cooled, compressed top gas. The reducing gas enters a bustle pipe from which it is injected into the reactor through symmetrically spaced ports. The gas then flows countercurrent to the burden and reduction is accomplished by reactions (1) to (6).

Top gas is drawn from the reactor and is cleaned and cooled to remove dust and most of the water formed during the reducing reaction. About 40 per cent of the top gas is compressed; some of the steam is injected into the cooling zone of the reactor and the remainder used to temper the reformed gas. The remaining 60 per cent of the top gas is used as the reformer fuel gas. Waste heat recovery coils located in the stack gas flues generate process steam and preheat the methane-steam feed to the reformer tubes.

As with other processes, DRI from an Armco plant must be used reasonably soon after production, and protected from water in any interim storage, otherwise a passivation operation is necessary to prevent degradation of the product by reoxidation.

NSC Process—The Nippon Steel Corporation (NSC) began fundamental research of their shaft-furnace DR process on a small scale in 1970 in conjunction with the Iron and Steel Institute of Japan. At the same time, the Hirohata Works and Texaco, Incorporated, were jointly developing a fuel-oil partial-oxidation gasifier for producing reducing gas for blast furnace injection. These efforts formed the basis for the design and construction of a shaft-furnace pilot plant that started operation in 1971. Reducing gas was generated by the partial oxidation of fuel oil.

A demonstration plant with a design capacity of 150 000 metric tons (165 000 net tons) per year of DRI was built by NSC at their Hirohata Works, Figure 14—5. the plant was operated in 1977. The shaft furnace, which exhibits lines similar to a blast furnace, has a 2.1 metre (6.9 foot) inner diameter at the lower gas injection point, and is 9 metres (29.5 feet) high. The shaft

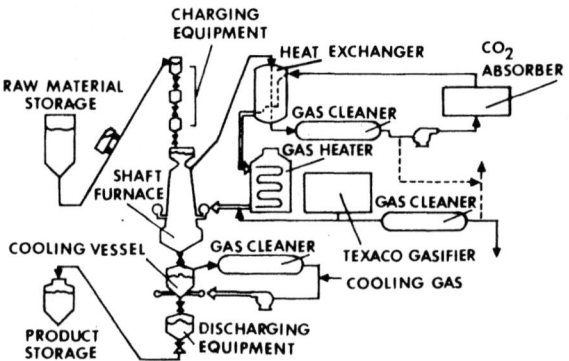

Fig. 14—5. NSC DR process flowsheet.

furnace is pressurized up to 5 bars (5 atmospheres) to gain advantages in the recycle system and blowing rate of gas as discussed previously for HYL III. The shaft furnace is equipped with a lock hopper charging system which is pressurized with seal gas to prevent leakage of reducing gas. The burden in the shaft furnace is supported at the bottom by a table. Hot DRI is removed periodically by the controlled action of scrapers. The DRI collects in an underlying conical section below the table and subsequently flows into a separate but connected cooling vessel. A flow of recirculated gas cools the DRI to below 100°C (212°F); carburization is effected in the same unit. NSC proposes briquetting to protect their DRI product from reoxidation.

A Texaco partial oxidation gasifier is employed in the process. It is designed for the production of 16 700 Nm³ per hour (10 400 scfm) of gas from 5 metric tons (5.5 net tons) per hour of light oil. The generated gas temperature exceeds 1100°C (2000°F temperature) and contains about 87 per cent hydrogen plus carbon monoxide in a ratio of about 10.5 to 1. Recycled shaft-furnace off-gas tempers the generated gas temperature to between 800 and 1000°C (1475 and 1830°F). The mixed gas is then injected into the shaft furnace. Provisions are made to remove carbon soot from the gasifier product. However, a certain amount of soot retards sticking and promotes burden movement in the shaft furnace. The process operates with lump ores and oxide pellets which are reduced by reactions (1) to (6) as they descend by gravity through the shaft furnace countercurrent to the upward flow of gas. Most of the lean gas from the top of the shaft is cooled in a recuperative heat exchanger, cleaned, recompressed, and stripped of CO_2. This gas is then reheated in the recuperative heater followed by a fixed heater. The outlet temperature of the fired heater is controlled to yield the desired temperature of the reducing-gas entering the shaft furnace. The flow of process fuel to the gasifier is equivalent to 9.5 million kilojoules per metric ton (8.2 million Btu per net ton) of DRI. Additional fuel and power is required for heating and compressing gas and for operation of the unit for CO_2 removal.

The shaft furnace is equipped with a probe for sampling the solids and gas at various elevations. A computer assists in the precise control of operating variables for the plant. The operation confirms the results of single particle reduction studies which indicate that reduction kinetics level off after the pressure is increased by a few atmospheres with no further benefits as the pressure is raised higher.

An NSC plant based on natural gas has been announced for startup in Malaysia in 1985.

Purofer Process—The Purofer process of Thyssen Purofer GmbH, a subsidiary of Niederrhein-Thyssen, AG, West Germany, was developed by August Thyssen-Hütte at Oberhausen in the 1960's.

A demonstration plant with a design capacity of 500 metric tons per day (550 net tons per day) of DRI was installed at Oberhausen and started operations in 1970. Data from this plant were used for the engineering design of two commercial plants having a design capacity of about 1100 metric tons (1200 net tons) per day of DRI in Iran and Brazil. The reduction shafts in these plants are reported to have a height of 13 metres (43 feet) and a working volume of 180 cubic metres (6360 cubic feet). The plant at Ahwaz, Iran is based on natural-gas reforming and the plant at Santa Cruz, Brazil is based on partial oxidation of heavy residual fuel oil with an oxygen-steam mixture using the Texaco partial-oxidation process. Both commercial plants started operations in 1977. Energy consumption is projected to be about 12.8 million kilojoules per metric ton (11 million Btu per net ton) of DRI for the plant in Iran and 14.7 million kilojoules per metric ton (12.6 million Btu per net ton) of DRI for the plant in Brazil.

The Purofer DR process utilizes a shaft furnace as the reduction reactor. However, instead of using a catalyst-filled, tube-type gas reformer, the Purofer process uses a regenerative-stove-type system. Use of fluid fuels other than natural gas is thus claimed as a real possibility for the process with this type of reformer.

Reducing gas is produced by partial oxidation of methane to carbon monoxide and hydrogen using scrubbed and cooled reduction-furnace top gas, with carbon dioxide used as the source of oxygen according to reaction (15). Alternatively, a carbon monoxide and hydrogen-rich gas can be produced by methane-steam reforming, according to reaction (14). Operating temperatures in the reformer reaction zone range between 1250 and 1400°C (2280 and 2550°F), under which conditions there is little or no carbon formation and very little carbon dioxide and water vapor are formed.

As shown in Figure 14—6, generation of the reducing gas is carried out in a horizontal chamber filled with checker-work refractory impregnated with a nickel catalyst. Two such chambers, each with an in-line heat exchanger, are employed, one being in the reforming cycle while the other is being heated. The hot gas leaves the reforming chamber at 1250-1400°C (2280-2550°F) and passes through the heat-exchanger section wherein the temperature is lowered to 925 ± 50°C (1700 ± 90°F). Heat is lost to the checker-work refractory in the heat exchanger. While one line is operating as the reformer, the reforming chamber of the other is being heated by firing with reducer off-gas or some other fluid fuel. Combustion air passes over the checker-work refractory in the heat-exchanger section, which serves the two-fold purpose of cooling the refractory while preheating the air. The cycle time of the reformers is about 30-60 min. An additional heat exchanger, located between the parallel reformers, further dampens any temperature fluctuations to ± 10°C (18°F).

Advantages claimed for the Purofer reforming system are its immunity to poisoning of the nickel catalyst by sulphur and to the build-up of carbon, in the form of

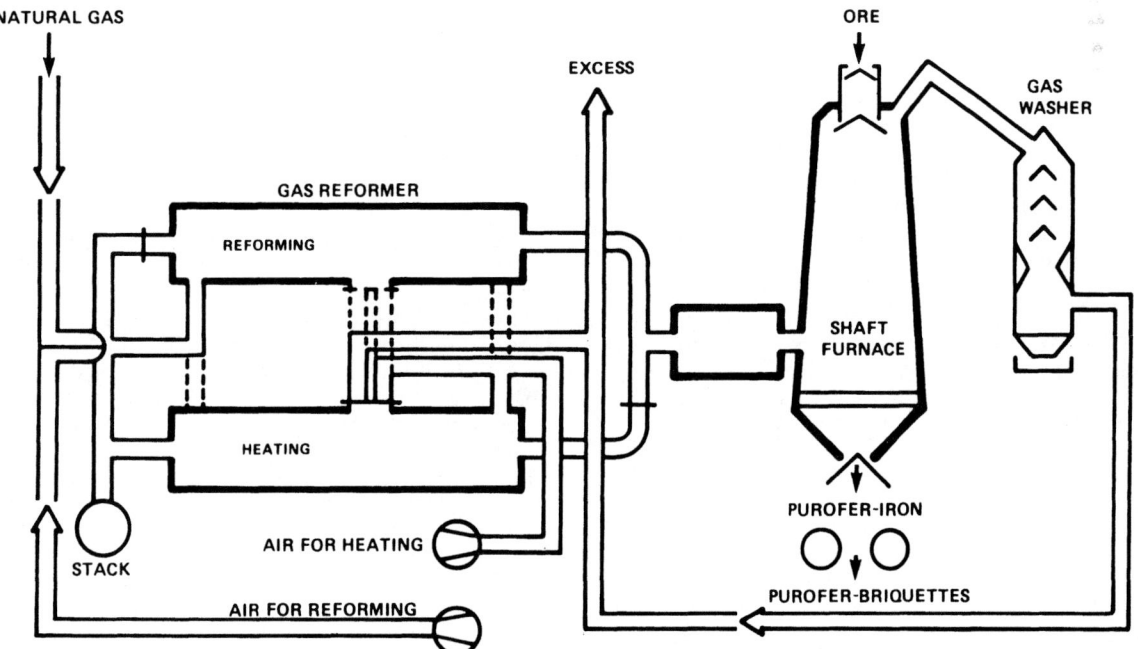

FIG. 14—6. Schematic flowsheet of Purofer process.

soot, on the catalyst because any build-up of these elements during the reforming cycle is burned off during the firing cycle. This would increase the potential range of fuels that might be used in the system.

The charge enters the Purofer shaft furnace through a double-bell seal system. The charge descends through the shaft, where it is heated and the iron oxide is reduced according to reactions (1) to (6) by the ascending stream of reducing gas. Similar to the NSC design, the Purofer process eliminates the integral cooling section prevalent in other processes. Furthermore, scraper bars move hot material from the bottom of the shaft. Purofer has recommended the hot briquetting of all the DRI to avoid the necessity of passivating to prevent reoxidation. Cooling would then follow the briquetting operation.

Shaft-furnace top gas is cleaned and cooled to remove the dust and condense most of the water formed in the reducing reaction. For methane-top gas reforming, about 70 per cent of the top gas is compressed and blended with the natural gas feed. It is then fed to the reforming chamber. The remaining top gas is used to fire the reforming chamber that is on a reheat cycle. Some natural gas is also required as supplemental fuel in the reheat cycle. For methane-steam reforming, all the top gas is used to fire the reforming chamber that is on a reheat cycle. Natural gas is used to make up any fuel shortage.

While the plant in Iran follows the schematic flowsheet of Figure 14—6, the modified plant in Brazil allows the use of heavy residual oil, and a Texaco partial oxidation gasifier is employed for that purpose. The concepts are similar to the NSC process (Figure 14—5). However, rather than keeping the gas hot, the freshly generated gas is scrubbed to remove soot and cooled. An acid gas removal system strips sulphide compounds and carbon dioxide from the gas. This gas is then mixed with a portion of the shaft furnace off-gas which has been cooled to remove water and also stripped of carbon dioxide. The mixed gas is heated and introduced into the shaft furnace. The inefficiency of gas cooling and reheating is a compromise that allows the use of the residual heavy oil in the Purofer process.

In the Brazilian plant, the hot DRI product is transported directly to the electric arc melting furnaces in sealed containers which improves the energy efficiency of the plant. Hot-molded briquetting is used to protect the stockpiled inventory of DRI.

SHAFT-FURNACE PROCESSES, STATIC BED

HYL Process—The HYL process was developed by Hojalata y Lamina S. A. (HYLSA) of Monterrey, Mexico. In the HYL process, lump ore and fired pellets are reduced in fixed-bed retorts by reformed natural gas.

The first commercial HYL plant was installed at Monterrey and started production late in 1957. This plant has a capacity of 200 metric tons per day (220 net tons per day) of DRI, and the reactors are about 2.5 metres (8 feet) in diameter and hold about 15 metric tons (16.5 net tons) of ore in a 1.5-metre (5-foot) deep bed. By 1980, fifteen HYL plants with a total capacity of about 10 million metric tons (11 million net tons) per year of DRI were scheduled to be in operation worldwide. The reactors in the most recent plants are 5.4

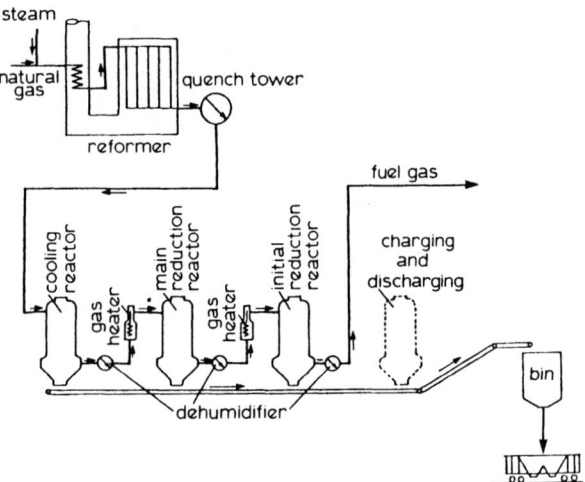

FIG. 14—7. HYL DR process flowsheet.

metres (17.5 feet) in diameter and 15 metres (50 feet) high. Design capacity is about 1900 metric tons per day (2100 net tons per day) of DRI having an average reduction of about 90 per cent. The energy consumption in the most recent plants is 14.9 million kilojoules per metric ton (12.8 million Btu per net ton) of 90 per cent reduced DRI. The more recent plants of the HYL II design (1) use high-temperature alloy tubes in the reducing gas-reheating furnaces, which permits heating the gas to higher temperatures, and (2) reduce the number of heating furnaces from the original four units to two units. This eliminates thermal cycling of the earlier designs.

In the HYL II process, reducing gas (rich in carbon monoxide and hydrogen) is generated, typically by nickel-based catalytic reforming of natural gas, which is mixed with steam before entering the reformer, Figure 14—7. Commercial HYL operations use excess steam over stoichiometric requirements, as shown in reaction (14), to prevent carbon formation and to promote long catalyst life.

HYL's practice of using the cold reducing gas for both product cooling and carburization negates any advantage of using catalysts that permit near-stoichiometric steam use.

Much of the water vapor in the reformed gas must be removed by quenching to achieve a hydrogen-rich reducing gas. Reformer heat is supplied by the combustion of reduction-process tail gas. Process steam for the reformer is produced in a waste heat boiler, using the heat in the reformer flue gas. With added boiler capacity, additional steam can be generated to feed steam turbine driven equipment, thus lowering the electrical-energy requirements.

The reducing section consists of a set of four reactors, three of which are in operation, while the fourth is engaged in discharging and charging operations. The HYL process is a cyclical batch operation, and the three on-line reactors operate in series. The reduction of the charge is performed in two stages, an initial reduction stage and a main reduction stage. Cooling and carburization (Fe_3C), and the final adjustment of metallization are performed in the third stage. Each stage of operation takes about three hours. An intricate system of valves permits the reactors to be connected in any desired order so that any one reactor can be connected in its correct process stage.

The flow of reducing gas is countercurrent to the iron oxide charge stages. The quenched fresh gas from the reformer is used first in the reactor, in the cooling stage, that will be discharged as its next operation. The gas then flows through the reactor that is in the main reduction stage, and finally through the reactor that has most recently been charged for the initial reduction stage. Reduction is accomplished according to reactions (1) to (6). Because water is formed during reduction, the reducing gas is quenched when it leaves each reactor to condense the water and enhance the reduction potential of the gas. Before entering the reactors in the reduction stage, the quenched process gas is heated to about 815°C (1500°F) in an indirectly heated, gas-fired furnace. The gas is further heated to about 1050°C (1925°F) by the combustion of residual unreformed hydrocarbons with the controlled injection of air at the entrance to the reactor. The process tail gas is used to provide fuel for the reformer and for the gas-heating furnaces.

The reducing temperature in the HYL process is above 980°C (1800°F). The advantages of using this high reducing temperature are improved reduction efficiency for a hydrogen-rich reducing gas and a more stable product with a reduced pyrophoric tendency.

As the product in the cooling stage passes through the temperature zone of about 550°C, (1020°F), carbon is deposited on the reduced product as Fe_3C (cementite), see reactions (7) and (8). The advantage cited for this cementite shell is protection against reoxidation. HYL practices carbon-level control between 1.0 and 2.5 per cent by adjusting the time that the product remains at this carburization temperature. Lower carbon content is attained by controlling cooling-gas composition.

FLUIDIZED-BED PROCESSES

FIOR Process—The FIOR (Fluid Iron Ore Reduction) process is a continuous direct-reduction process developed by Esso Research and Engineering Company (ERE), a subsidiary of the Standard Oil Company of New Jersey (renamed Exxon Corporation in 1973). Development was started in 1955 in cooperation with Arthur D. Little, Inc. and in the early stages it was called the Esso-Little process. After 1960, ERE continued development on their own and pilot plants were subsequently built at Baton Rouge, Louisiana (1961) and Dartmouth, Nova Scotia (1965). The FIOR process is now under the direction of the Davy McKee Corporation.

The FIOR flowsheet is shown in Figure 14—8. The process reduces iron ore fines in a series of four fluid bed reactors. In addition, the process uses a reformer to produce fresh reducing gas and a briquetting section in which reduced iron ore fines are compacted. The fluid bed reactors are built in a tower structure thus enabling gravity to enhance the flow of solids between stages.

Ore that has been properly sized and dried is raised

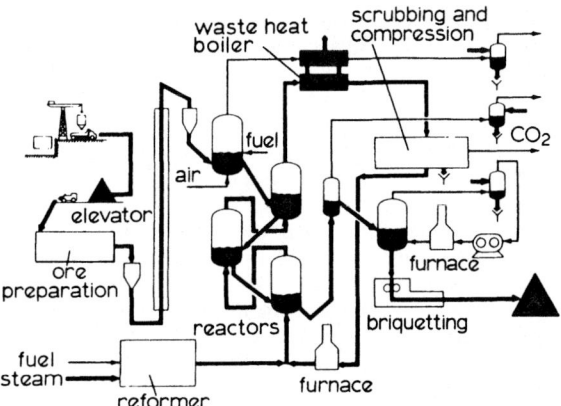

FIG. 14—8. FIOR fluid-bed DR process flowsheet.

to the top of the structure by either a pneumatic lift or a skip hoist. Since the FIOR process operates at elevated pressures (reportedly 10 atmospheres), pneumatically lifted ore could be fed directly into the first reactor, but ore raised by skip hoist would require the use of an intermediate surge-hopper system in which the pressure can be increased to operating pressure.

The ore is fed continuously into the preheating reactor, where residual water is driven off and the ore is heated to about 800°C (1470°F). The combustion products are used to fluidize and heat the ore. As the ore level builds up in the preheater, it overflows into the first of the three reduction reactors. In the first, the operating temperature is about 700°C (1290°F) and about 10 per cent reduction takes place before the ore overflows successively into the second and third reactors. The 91 to 93 per cent metallized ore fines leave the third-stage reactor at a temperature of about 750°C (1380°F).

Compressed reducing gas enters the bottom of the lowest reactor and flows countercurrent to the descending ore. Thus, the richest reducing gas, a blend of about 75 per cent recycled reactor top gas and 25 per cent fresh gas from the reformer, is mostly hydrogen with some carbon monoxide, carbon dioxide, and water vapor. Fresh reducing gas is produced by methane-steam reforming in a catalyst-tube reformer. The reformer conditions are controlled to produce a gas of 90 per cent hydrogen. The entire reducing-gas stream is heated to operating temperature in a furnace. Reactor top gas is burned as the fuel in the indirect-heat furnace.

Reactor top gas leaves the first-stage reduction reactor and is scrubbed and cooled to remove dust and most of the water produced in the reduction reaction. Some of the top-gas stream is bled off for use as process fuel gas, and in this way, a gradual build-up of carbon dioxide in the recycling gas stream is avoided. The recycled top-gas stream is compressed and blended with fresh gas from the reformer before the combined stream enters the reducing-gas furnace.

Direct-reduced iron fines are discharged from the final reactor into the briquetting feed bin from which they are fed, still hot, to the briquetting press. The pillow-shaped briquettes are formed by compacting the hot DRI fines between the two rolls of the press. The briquettes are discharged from the press in a continuous sheet, joined by a thin web of compacted DRI fines. The sheet is broken into individual briquettes in a trommel screen. The DRI briquettes are then cooled in a rotary cooler. A thin film of iron oxide forms on the surface of the briquettes in the cooler, rendering them inert.

HIB Process—The HIB (for high iron briquette) process is a continuous direct-reduction process developed by United States Steel Corporation. Development was started in 1952, and, in the pilot-plant stages at South works in Chicago, it was called the Nu-Iron process. A schematic flowsheet of the HIB process is shown in Figure 14—9.

In the process, minus-2 mm (minus-10 mesh) iron ore is reduced by reformed natural gas in a two-stage fluidized-bed reducer at about 700°C (1300°F) and 360 kPa (52 psi) pressure. Reducing gas for the process could also be manufactured from naphtha, fuel oil, pulverized coal, or other fuels by several commercially proven methods. Dry iron ore is preheated to about 810°C (1500°F) in fluidized bed preheaters consisting of two stages and fed into the first reduction stage where Fe_2O_3 is reduced to FeO by partially spent fluidizing gas from the final reduction stage. Solids from the first reduction stage flow continuously into the final reduction stage where FeO is reduced to Fe by fresh reduc-

ing gas that enters at about 750°C (1380°F). The product is briquetted in continuous roll presses while still hot and is then cooled by inert gas in a continuous shaft cooler. The off gas from the first reduction stage is scrubbed and used as a fuel in the plant. This two-stage iron reduction flowsheet is well suited to producing 75 per cent reduced high-iron briquettes as a prereduced blast-furnace burden. Modifications of this flowsheet, including a third reduction stage in series with the other reduction stages, would be made to produce a more highly reduced product (about 90 per cent reduced) for steelmaking. The flowsheet could also be modified to include recycling a portion of the spent off gas from the first reduction stage.

A plant based on the HIB process designed to produce about 900 000 metric tons (one million net tons) per year of 75 per cent reduced briquettes was constructed at Puerto Ordaz, Venezuela and started operations in 1971. The plant comprises three identical trains with the t 'o-stage reduction vessels being 6.7 metres (22 feet) ii. inside diameter and 30 metres (100 feet) high. The annual production of this plant was about 300 000 metric tons (330 000 net tons) per year as of 1977. The energy consumption for the process is projected to be 19.9 million kilojoules per metric ton (17.1 million Btu per net ton) of 75 per cent reduced HIB and 21.4 million kilojoules per metric ton (18.4 million Btu per net ton) of 93 per cent reduced HIB.

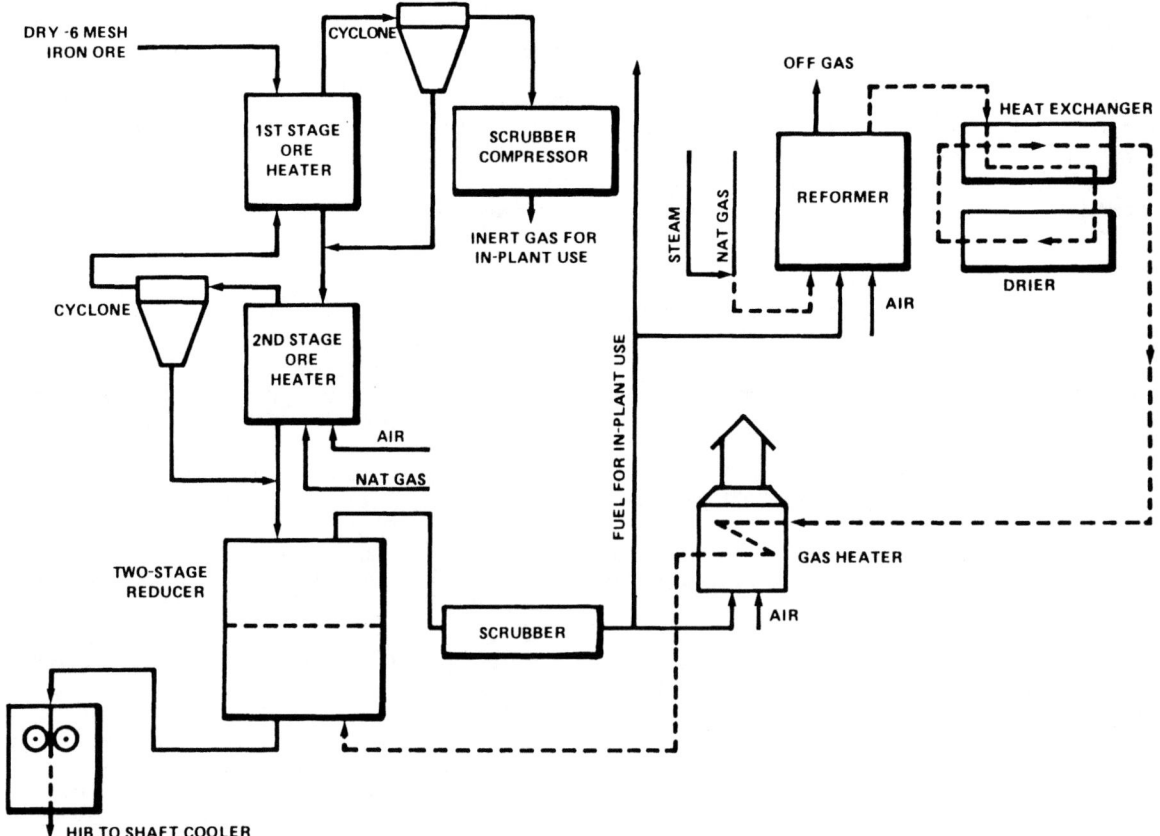

FIG. 14—9. Schematic flowsheet of HIB process.

DIRECT REDUCTION PROCESSES—REDUCING GAS GENERATED FROM HYDROCARBONS IN THE REDUCTION FURNACE

KILN PROCESSES

The rotary kiln is a versatile furnace that finds widespread use in the production of DRI. The furnace is designed to use hydrocaron fuels in the iron oxide reduction chamber without prior gasification. Although oil and natural gas are used, the primary fuel source is coal which is the least expensive form of energy in many locations.

The rotary kiln furnace is a revolving horizontal cylinder comprising a shell with an internal refractory lining. Seals at each end join the rotating cylinder to the stationary equipment for adding materials and discharging product from the furnace. The furnace is tilted at an angle of 3 to 4 per cent from the horizontal toward the discharge end. The burden travels through the rotary kiln by rotation of the kiln and gravity. Rotary kilns for DRI vary in shell diameter from 3.5 to 5.0 metres (11.5 to 19.7 feet) and in length from 50 to 125 metres (164 to 410 feet).

In rotary-kiln DR processes, coal, flux (if required) and iron oxide are metered into the high end of the inclined kiln. The burden first passes through a preheating zone where coal devolatilization occurs, flux is calcined, and the charge is heated to the operating temperature for reduction. In the reduction zone, iron oxide is reduced by carbon monoxide as represented by reactions (2), (4) and (6). Reduction by carbon monoxide is the predominant reaction because, at the elevated bed temperature, part of the carbon dioxide reacts with the carbon in the coal by reaction (9), the Boudouard reaction. Chemical reactions, other than those cited, occur within the kiln bed but these are beyond the scope of this discussion.

A portion of the process heat is usually provided by a burner located at the solids discharge end of the rotary kiln. In an all-coal DR process, pulverized coal may be supplied to the burner. Fuel oil and natural gas, however, are viable alternatives. The burner operates with a deficiency of air to maintain a reducing atmosphere in the kiln. Additional process heat is supplied by combustion of the volatile matter of the coal and of the carbon monoxide emerging from the kiln bed. Combustion air is supplied through ports spaced along the length of the rotary kiln. The air flow is controlled to maintain a uniform temperature profile in the reduction zone and a neutral or slightly reducing atmosphere above the bed. The kiln gas flows countercurrent to the flow of solids.

The combustion of gases transfers energy to the bed and sustains the preheating, calcination, and reduction of the solids. Radiation from the gases and refractory walls is the most important mechanism for heat transfer to the bed. However, convection from the gases and from superheated refractories as they rotate under the bed augment the heat transfer. Because radiant heat transfer predominates, the heat transfer is more efficient in the product end of the kiln where the temperature is higher.

Preheating of solids at the feed end accounts for a substantial amount of kiln length because of lower temperature and lower efficiency. In that regard, the productivity of a rotary kiln is increased by preheating the iron oxide feed or by underbed injection of air for combustion of coal charged with the iron oxide feed or fluid hydrocarbons that are injected simultaneously. Rotary kilns that employ the concept of underbed injection of fuel utilize a valved manifold system which rotates with the shell. Other methods of increasing energy utilization comprise coal feeding mechanisms which (1) inject part of the coal axially upstream at the product end or (2) admit part of the coal through devices installed on the shell. Both of these methods ensure that most of the volatiles are released in a section of the kiln which is hot enough to sustain ignition.

The main components in the flowsheets of these rotary-kiln systems are similar, consisting of a solids-feed system, the rotary kiln, a product cooler, screens, magnetic separators, and off-gas cleaners. The basic technology for rotary-kiln reduction emphasizes the importance of correct selection of raw materials.

The iron oxide feed (lump ore or pellets) should fulfil certain requirements regarding chemical composition, size distribution, and behavior under reducing conditions in the rotary kiln. The feed should have a high iron content (preferably close to 67 per cent for hematite ores), and correspondingly the gangue content should be low, so that costs for further processing in the electric furnace are kept as low as possible. Sulphur and phosphorus contents should also be low. The minimum size of feed ore should be controlled also, preferably at about plus 5mm. Besides being elutriated from the kiln, fine ore contributes to the operating problem of accretion build-up (ringing) in the kiln. Reduced fines in the product also reoxidize more rapidly. Reducibility of the ore, a measure of the time required to achieve a desired degree of metallization under a standard set of conditions, exerts a strong influence on the throughput capacity of the kiln. The behavior of the ore under reducing conditions is important, especially with regard to swelling, which is experienced in some feeds, and the subsequent decrepitation of such materials in kiln travel.

In the selection of coals, important factors are reactivity, volatile-matter content, sulphur content, ash content, and ash-softening temperature. Coal reactivity is indicative of the coal's reduction potential. With increased reactivity, the throughput rate of the rotary kiln can be expected to increase within certain limits, as influenced by the complexity of the multiple reactions that take place. In a simplified explanation, the process seeks a bed temperature where heat transfer to the bed balances the kinetics of reaction (9). A higher coal reactivity allows the bed to operate at lower temperature which enhances heat transfer. In coal selection for DR processes, consideration should be given to the fact that the volatile-matter content generally in-

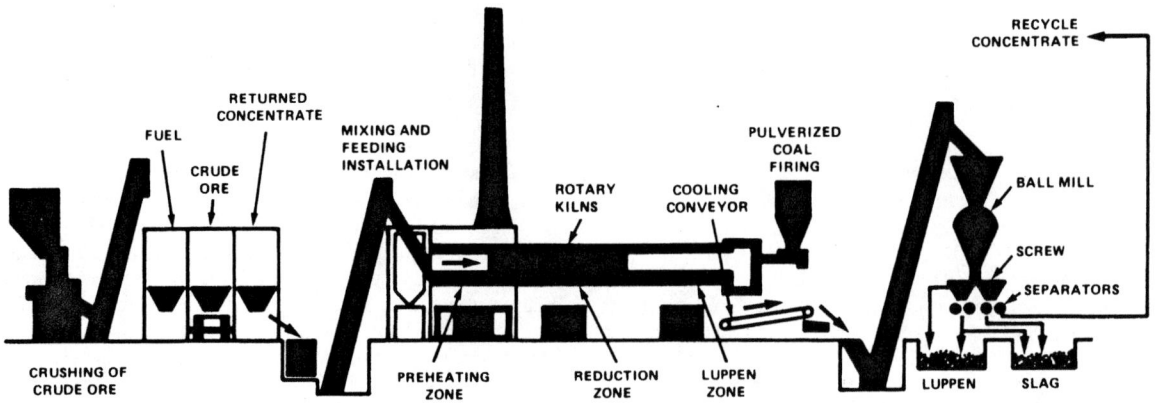

FIG. 14—10. Schematic representation of the Krupp-Renn process.

creases with reactivity. However, this concept has its limitations, because coals with a high volatile-matter content will generate more gas than can be used in the process for reduction and fuel. Therefore, for overall heat economy, the recovery of the sensible and chemical heat contained in the waste gas would have to be considered. In general, a low-sulphur-content coal is preferred as it prevents sulphur pick-up by the DRI. Dolomitic limestone is used as a scavenger for sulphur, but has the adverse effects of increasing the heat load for calcining the flux and of decreasing the throughput rate by occupying kiln space that would otherwise be used by the process reactants. Since coal ash is also a non-reactant material that takes up kiln space, a coal with low ash content is desirable. A high ash-softening temperature is also desirable to prevent, or at least minimize, the build-up of accretions in the kiln.

The solids discharged from the rotary kiln are cooled, then screened and separated magnetically. DRI fines are briquetted and used together with the normal-sized DRI in steelmaking. A carbon char is separated and recycled to the rotary kiln to increase fuel efficiency. The tailings in the product comprise the ash and the lime which contains the sulfur.

The hot waste gas from the kiln contains dust and volatile matter and must be cleaned. The gas contains considerable sensible and chemical energy which can be recovered for credit. However, several first-generation coal-based DR plants do not recuperate heat. In practice, the heavy fines are usually settled out in a gravity separation chamber that can also serve as an afterburner. The waste gas is then cooled and cleaned before being released to the atmosphere. Dust from the settling chamber is transported to a waste-disposal area.

Krupp-Renn Process—The Krupp-Renn process was developed in the 1930's to treat high-silica ore with a basicity ratio as low as 0.2 to 0.3, without the addition of limestone. In this process, a mixture of minus 64-mm (2.5-in.) ore and a fine-grained carbonaceous reducing agent (e.g., coke breeze or bituminous-coal fines) is fed continuously into a rotary-kiln (Figure 14—10). The maximum temperature of the solids in the kiln is about 1230 to 1260°C (2250 to 2300°F), which is sufficient to covert the gangue in the ore to a very viscous high-

silica slag, and also to effect coalescence of the sponge iron obtained from reduction of the ore. The reduced iron welds into nodules called "luppen" which become embedded in the pasty slag. This product is discharged from the kiln. After cooling, it is crushed and the luppen are magnetically separated from the slag. Recovery of iron in the luppen varies between 94 and 97.5 per cent.

High-titania ore can also be used in this process, and iron can be separated from titanium since the latter is not reduced. Almost any solid carbonaceous fuel can be used as a reducing agent.

Since a large part of the sulphur contained in the reducing agent goes into the luppen, the sulphur content of the metal is usually too high for economical conversion of the luppen into steel by any of the conventional steelmaking practices. In at least one instance, however, magnetic ores containing 55 to 60 per cent iron are reduced to luppen and are successfully refined into steel. In some places, the process is used to concentrate low-grade iron ores containing up to 30 per cent silica; the luppen being fed into blast furnaces.

It has been stated that an annual production of 450 000 metric tons (500 000 net tons) of iron has been achieved using the Krupp-Renn process.

Krupp-CODIR Process—The Krupp-CODIR process of Krupp Industries, West Germany, stems from the original Krupp-Renn process. The process operates at a lower temperature than the Krupp-Renn thus producing a standard DRI product. Furthermore, limestone or dolomite in the furnace charge absorbs a substantial part of the sulphur introduced with fuel.

A Krupp-CODIR plant designed for a capacity of 150 000 metric tons (165 000 net tons) started operation in 1973 at the Dunswart Iron and Steel Works, Ltd. at Benoni, Republic of South Africa. The reduction kiln in this plant is about 4.0 metres (13 feet) inside diameter and 74 metres (243 feet) long. Ultimately, Krupp expects to be able to produce 200 000 to 250 000 metric tons (220 000 to 275 000 net tons) of DRI from a single module plant that is 5.5 metres (18 feet) inside diameter and 95 metres (312 feet) long. Larger capacities would be obtained by using more than one module in parallel. The energy consumption is about 15.9 million kilojoules per metric ton (13.7 million Btu

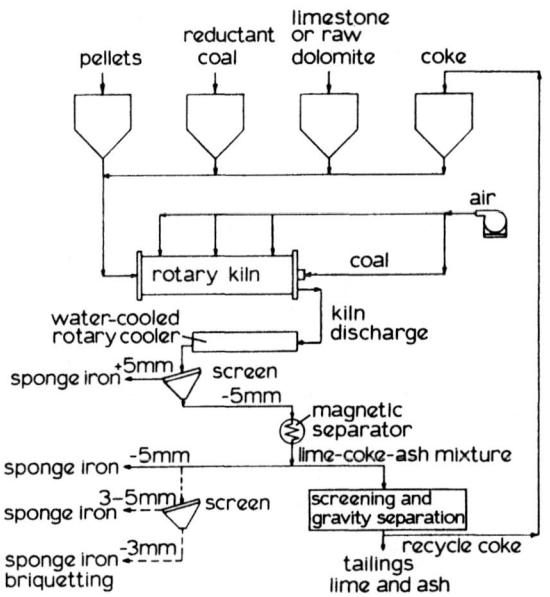

FIG. 14—11. Krupp-CODIR DR process flowsheet.

per net ton) of DRI when low-volatile anthracite is used for the reduction coal. As mentioned previously, the gross energy requirement increases when higher volatile matter coals are used.

In the flowsheet of the process (Figure 14—11) lump ore or oxide pellets, solid reductant, dolomite or limestone as flux if needed, and recycled char are metered from bins into the rotary kiln. The feed size of the solids is closely controlled to expedite separation later in the process. Typically, the preheating zone extends from 25 per cent to 40 per cent along the CODIR process kiln.

Primary heat is supplied to the kiln by the combustion of finely pulverized coal injected at the solids discharge end of the kiln. Secondary heat is supplied by metering air into the kiln gas space through tubes spaced along the entire length of the kiln. The secondary air is introduced axially (along the kiln's centerline) in the direction of the kiln gas flow and serves to burn combustibles released from the bed. In this way, a uniform charge-temperature profile between 950 and 1050°C (1740 and 1925°F) is achieved in the reduction zone of the kiln.

The DRI, char, coal ash, and spent flux are discharged via an enclosed chute from the rotary-kiln burner hood into a sealed rotary cooler. Cooling is accomplished by spraying a controlled amount of water directly onto the hot solids and by spraying additional water on the outside of the cooler shell. The cooled solids are discharged over a 5-mm screen. Because of close control of the feed size, the plus 5 mm fraction is all product. The minus 5 mm fraction is processed through further screening at 3 mm and magnetic separation to separate the finer DRI from recycled char, spent flux, and coal ash. Minus 3 mm DRI may be cold briquetted for transportation, and 5–3 mm DRI is combined with the plus 5 mm fraction. Char is separated by gravity for return to the kiln feed, and the ash and

spent flux are separated for disposal.

SL/RN Process—A forerunner to the SL/RN process, the R-N process (for Republic Steel Company and National Lead Corporation) was developed originally in Norway, primarily to recover TiO$_2$ from titanium-bearing ore for the production of paint pigments. However, further development showed that other iron-bearing ores could also be treated successfully to produce iron. Subsequently, a pilot plant was built in the United States and in 1964 Lurgi Chemie acquired the R-N patents and world rights and developed the technology further in cooperation with the Steel Company of Canada, Ltd. (Stelco) to form the SL/RN process.

The SL/RN process follows typical rotary-kiln operations described earlier. The largest commercial SL/RN plant, designed to produce 360 000 metric tons (400 000 net tons) of DRI per year, was installed by Stelco at Griffith Mine in Ontario, Canada and began operating in 1975. The reduction kiln in this plant is 6 metres (19.7 feet) inside diameter and 125 metres (410 feet) long. The energy consumption at this plant is about 22 million kilojoules per metric ton (19 million Btu per net ton) of product when the process is operated with high-volatile sub-bituminous coal. This relatively high consumption occurs because most of the volatile matter in the reduction coal leaves the kiln and is not recovered.

Other commercial installations based on the SL/RN process include one installed in 1970 by New Zealand Steel Limited at Glenbrook, Auckland, New Zealand for recovery of iron from native iron sands and a plant installed at the Fukuyama Works of Nippon Steel in Japan in 1974. The New Zealand Steel Limited plant was subsequently modified to include a multiple-hearth furnace for preheating the iron-sands feed and charring the reduction coal. The Fukuyama plant was designed to process waste oxides generated at the Fukuyama Works. Other plants have been constructed. The more recent plants are operating or under construction in Peru, India, and South Africa.

The SL/RN process flowsheet is shown in Figure 14—12. The charge consists of lump ore or pellets, coal,

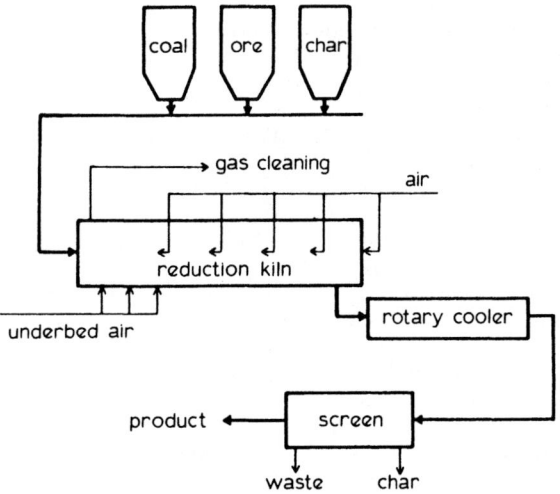

FIG. 14—12. SL/RN DR process flowsheet.

recycled char, and flux if sulphur needs to be scavenged from the coal. In the kiln preheat zone, the charge is heated to about 980°C (1800°F) by counter-flowing hot freeboard gases. For high kiln efficiency, the preheat zone is made as short as possible, usually 40 to 50 per cent of the kiln length. Reduction begins when the charge reaches a temperature in excess of 900°C (1650°F) when the carbon gasification Boudouard reaction, reaction (9), starts generating carbon monoxide. To maintain a uniform reduction-zone temperature by burning combustibles released from the bed, air is blown by shell-mounted fans into the freeboard gas streams, through burner tubes spaced uniformly along the length of the kiln. Air is introduced axially into the kiln, and additional combustion air is blown into the kiln through a central air port at the discharge end of the kiln. Auxiliary fuels (natural gas or fuel oil) are burnt at the discharge end for plant start-up, but injection of coal at the discharge end is usually not practiced in currently operating first-generation SL/RN plants. In the SL/RN practice all the coal feed is charged at the feed end, with the result that a uniform temperature distribution is established along the reduction zone, and combustion takes place at the bed surface.

The solids are discharged from the rotary kiln via a transfer chute into a sealed rotary cooler. Water sprays on the cooler shell reduce the temperature of the solids to about 95°C (200°F) in a non-oxidizing atmosphere. Internal lifters aid heat transfer in the cooler. Cooler discharge materials are continuously separated into DRI, DRI fines, and non-magnetics, by a system of screens and magnetic separation. Char is separated from the waste by gravity separation.

The SL/RN process kilns are now equipped with nozzles for underbed injection of about 25 per cent of the process air in the preheating zone of the kiln, as shown in Figure 14—12. The air is available for the combustion of the volatile matter in the coal within the bed in the preheat zone. As a result, the length of the preheat zone of the kiln is reduced because of the improved heat transfer and fuel utilization. More of the kiln length can therefore be used as a reduction zone, thus increasing capacity; accretion build-up frequency is decreased, significantly increasing the on-stream time, and the total fuel requirement is reduced. Underbed injection of a liquid fuel is also a viable option.

ACCAR Process—The Allis-Chalmers Controlled Atmosphere Reactor (ACCAR) produces highly metallized DRI in a ported rotary kiln. Liquid, solid, and gaseous fuels, singly or in combination, are used directly in the kiln without an external reformer or gasifying plant. The ACCAR kiln is equipped with an intricate port system and with valving arranged radially around the circumference of the kiln and spaced uniformly along its length, for liquid and/or gaseous fuel injection under the bed and for air injection above it. The flow rates through the ports are closely controlled to maintain optimum temperature profile and gas composition along the length of the kiln. Versatility in the use of fuel is claimed as an advantage for this process as it permits use of the most economical fuels available.

The original ACCAR development work started in the late 1960's and was based on hydrocarbon gases and liquids. Work was started in a 0.6-metre (2-foot) diameter by 7-metre (23-foot) long pilot plant at Milwaukee, Wisconsin and continued in a 2.5-metre (8-foot) diameter by 45-metre (148-foot) long demonstration plant at Niagara Falls, Ontario, Canada.

The first commercial ACCAR plant for operation with natural gas and fuel oil was established through modification of an existing SL/RN plant in Sudbury, Ontario. The plant started operating in 1976. The reduction kiln is 5 metres (16.4 feet) inside diameter and 50 metres (164 feet) long and is claimed to be capable of producing 233 000 metric tons (257 000 net tons) of DRI per year with a total energy consumption of 15.7 million kilojoules per metric ton (13.5 million Btu per net ton of DRI.

A second ACCAR plant of 150 000 metric ton (165 000 net ton) per year capacity started operations in 1983 in India. The plant operates with coal and oil. The oil can be injected under the bed for two-thirds of the kiln length. The kiln design was modified to permit the addition of some coal at the discharge end.

With 80 per cent of the hydrocarbon fuel replaced by coal, ACCAR has projected that a reduction kiln 6 metres (19.5 feet) inside diameter by 122 metres (400 feet) long will be capable of producing 600 000 metric tons (660 000 net tons) per year of DRI . The total energy consumption is predicted to be about 18.6 million kilojoules per metric ton (16 million Btu per net ton) of DRI.

Because gaseous and liquid fuels are in short supply, operation of the ACCAR process with coal is of primary interest. Coal has been used successfully to supply from 80 to 90 per cent of the fuel for the ACCAR kiln, with the remaining fuel requirements being supplied by liquid and/or gaseous fuels. As shown in Figure 14—13, coal and lump ore and/or oxide pellets are metered into the feed end of the rotary kiln. The solids are heated to reduction temperature by the counter-current flow of hot gas. Volatile matter is released from the coal during heat-up and is carried out in the kiln exhaust gas. As the coal and iron oxide travel through the kiln, reduction is accomplished by the carbon and carbon monoxide reduction mechanisms discussed previously. The coal feed is controlled so that it is essentially all consumed as the burden enters the final stages of reduction. Combustibles released from the bed are burned in the kiln freeboard with air introduced through the port in the kiln shell.

The final degree of reduction is achieved by introducing liquid and/or gaseous fuel through the kiln-shell ports near the product end of the kiln as they pass under the solids bed. In passing through the bed, this fuel is cracked to form hydrogen and carbon monoxide to complete the iron oxide reduction and to provide a protective atmosphere for the highly metallized product. This method of fuel injection permits operation without the excess of coal required to maintain a reducing atmosphere in the bed in other coal-based DR processes. Thus char recycling is eliminated.

Solids are discharged from the rotary kiln into a rotary cooler, where cooling is accomplished with external water sprays. The DRI is separated from the coal

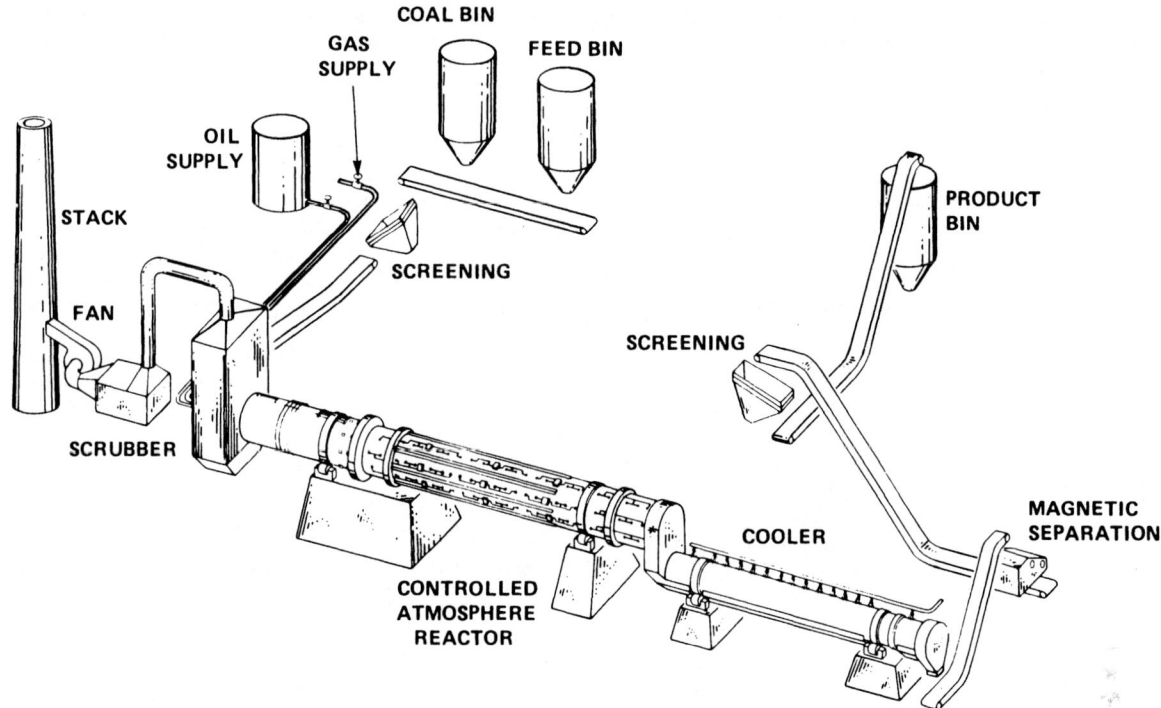

FIG. 14—13. Schematic flowsheet of ACCAR process.

ash by magnetic separation, and is then screened to achieve coarse and fine product separation.

If waste heat recovery is not practiced, the kiln off-gas is cleaned by first burning any combustibles, then the heavier solid particles are removed in a dry dust collector, and the fine particles are removed in a wet scrubber, which also cools the gas before it is released to the atmosphere.

DRC Process—The DRC process of the Davy/Direct Reduction Corporation (DRC), a Division of Davy McKee, stems from the Hockin Process of Western Titanium, Ltd. of Australia. Formerly Azion, DRC operated a pilot in Rockwood, Tennessee as a member of the Amcon group of Consolidated Gold Fields. The first commercial furnace was constructed for Scaw Metals in Germiston, South Africa.

Operations at Scaw Metals started in July of 1983. The kiln size is 4.5 metres (14.75 feet) shell diameter by 60 metres (197 feet) long. The rated annual capacity is 75 000 metric tons (83 000 net tons). Reportedly, the plant exceeds this capacity. The coal consumption is 0.72 ton per ton of DRI based on 55 per cent fixed carbon. Davy/DRC anticipates construction of larger kilns in the future. They also stress high operability and control of their plant (including sulphur in the product) through complex considerations regarding raw materials and proprietary design and operating features for the rotary kiln and ancillaries.

As shown in the process flowsheet (Figure 14—14), ore, coal, recycled char, and flux, if required, are continuously metered into the rotary kiln. Passage of the burden through a preheat zone and a reducing zone in the kiln follows typical rotary-kiln operations. Some minus 9.5 mm coal (about 12.5 per cent of the total coal)

is blown by low-pressure air into the discharge end of the kiln. Process heat is supplied by burning combustibles in the kiln freeboard. Combustion air is blown into the kiln by shell-mounted fans via tubes spaced along the length of the kiln. The bed and gas temperature profiles are controlled by adjustment of the air input through the tubes.

In the reduction zone, iron oxide is reduced by carbon from the coal, with reactions (2), (4), (6) and (9) controlling the complex gas-solids reactions. The maximum kiln bed temperature in the reduction zone is about 1060°C (1940°F) with a maximum kiln gas temperature of about 1160°C (2120°F). Emphasis is placed

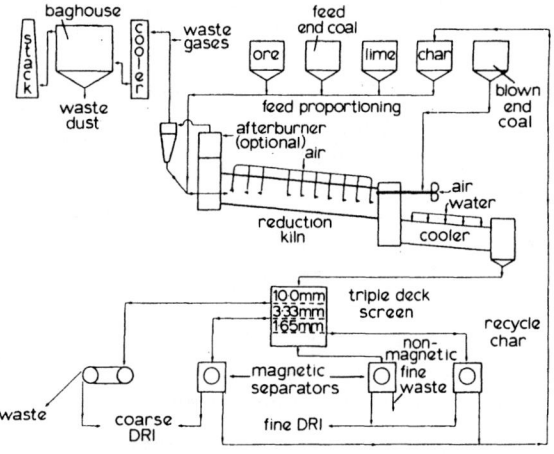

FIG. 14—14. DRC DR process flowsheet.

on maintaining a high ratio of carbon monoxide to carbon dioxide in the kiln bed to achieve a high degree of reduction. Control of the rate of heat transfer to the bed and control of the bed temperature are also critical for steady operation of the kiln, to achieve stable process chemistry and favorable reaction kinetics. The hot waste gases leave the kiln at about 800°C (1470°F).

Solids are discharged from the kiln via a sealed transfer chute into a sealed rotary cooler. Cooling is achieved by spraying water on the outside shell of the cooler. The cooled material is processed through screening and magnetic-separation circuits. Coarse and fine DRI and char that is recycled are thus separated from the fine waste.

LS-RIOR Process—The LS-RIOR direct-reduction process has been developed on a pilot scale by the Lummus Company and Sumitomo Heavy Industries, Ltd. The process controls sulphur pickup by the DRI and thus can utilize high sulphur fuels. As shown in Figure 14—15, the process preblends finely ground iron ore or concentrates in a pelletizer with petroleum residuum that serves as both a binder and a source of internal reductant for the iron oxide. The green pellets,

which are coked and preheated on a travelling grate, are discharged directly into a rotary kiln. Petroleum coke and either dolomite or limestone flux are also fed into the rotary-kiln. Petroleum coke serves as an additional reductant source and as a fuel source. After calcination, the flux materials combine with sulphur from the fuel/reductant. In order to supply process heat, air flow is controlled to partially combust volatiles and excess reducing gases are released from the kiln bed. Kiln exhaust gas is burned in an external incinerator to supply heat to the coking/preheat grate. Further heat recuperation can be accomplished by steam generation in a waste-heat boiler.

Reduced pellets are cooled and separated according to typical rotary-kiln operations. Magnetic material is recycled to the feed-grinding system; recycle coke is returned to the rotary-kiln feed; and spent flux and ash are sent to solid-waste disposal.

ROTARY HEARTH PROCESSES

INMETCO Process—The INMETCO process is currently operated by the International Metal Reclama-

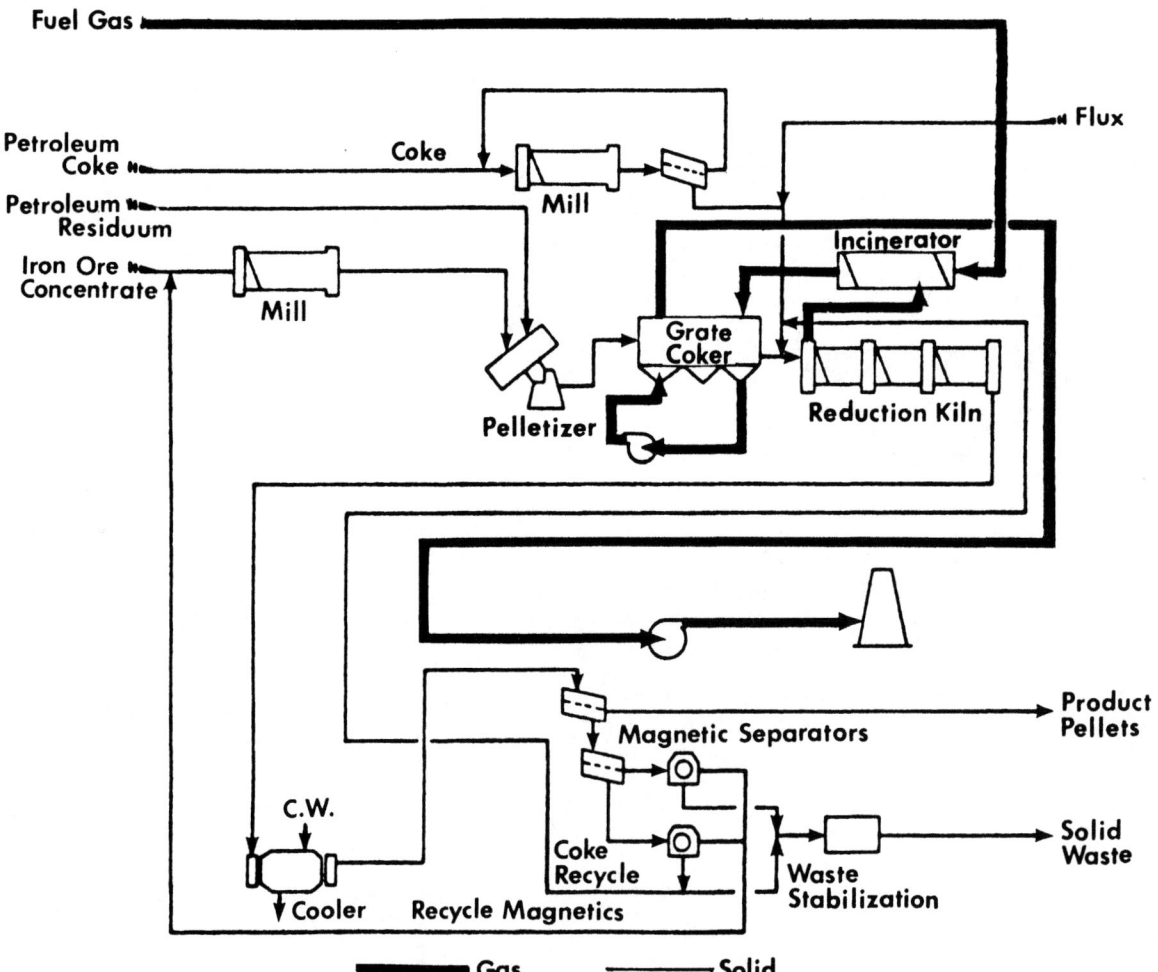

FIG. 14—15. Schematic flowsheet of LS-RIOR process.

tion Co., Inc., in Ellwood City, Penns; lvania for the recovery of waste dusts. The process, developed by INCO, uses pellets of iron oxide and coke or coal mixture with modest green strength. The pellets are fed into a rotary-hearth furnace which is separated by curtains into oxidizing, reducing, and neutral discharge zones. Burners above the bed are fired with oil or natural gas (coal burners have been tested and can be used as well). The shallow bed depth of one to three pellet diameters results in high heat transfer rates. Residence time for the pellets in the bed is approximately 15 to 30 minutes.

As of 1980, the process operated at 80 000 metric tons (88 000 net tons) of feed per year in the recovery of metals from waste dusts. An operating rate up to 250 000 metric tons (275 000 net tons) of feed per year per single unit has been proposed.

Salem Direct-Reduction Process—To avoid the accretion problem common to rotary-kiln direct reduction processes, the Salem Furnace Co. proposes using a rotary-hearth furnace as the reduction reactor for the Salem DR process (Figure 14—16). The rotary-hearth is surmounted by a stationary roof, which is fitted with a series of water-cooled rabble arms fixed across one radius. Lump ore, or pellets, and coal, are fed through a chute at the periphery of the hearth. As the hearth rotates several times, the burden is moved by the rabble arms toward a central-discharge heat soaking pit. In this manner, the material that is turned and displaced by a rabble makes one complete revolution as a quiescent bed before it reaches the next rabble. As an alternative, some or all of the coal could be introduced onto the rotary-hearth through a chute (or chutes) located at an intermediate point of the solids travel.

Volatile matter in the coal and combustibles released from the bed are burned in the freeboard between the roof and the hearth layer to provide the heat for the charge and the endothermic-reduction reactions. Combustion air, preheated by the process exhaust gases, is admitted to the freeboard through ports in the roof and side walls. The reduced product is discharged through the centrally located soaking pit into a rotary cooler, cooled by external water sprays. The DRI is separated from the coal ash and residual char by screening and magnetic separation.

RETORT PROCESSES

Hoganas Process—The E. Sieurin, or Hoganas, process was developed at Hoganas, Sweden in 1910 and is

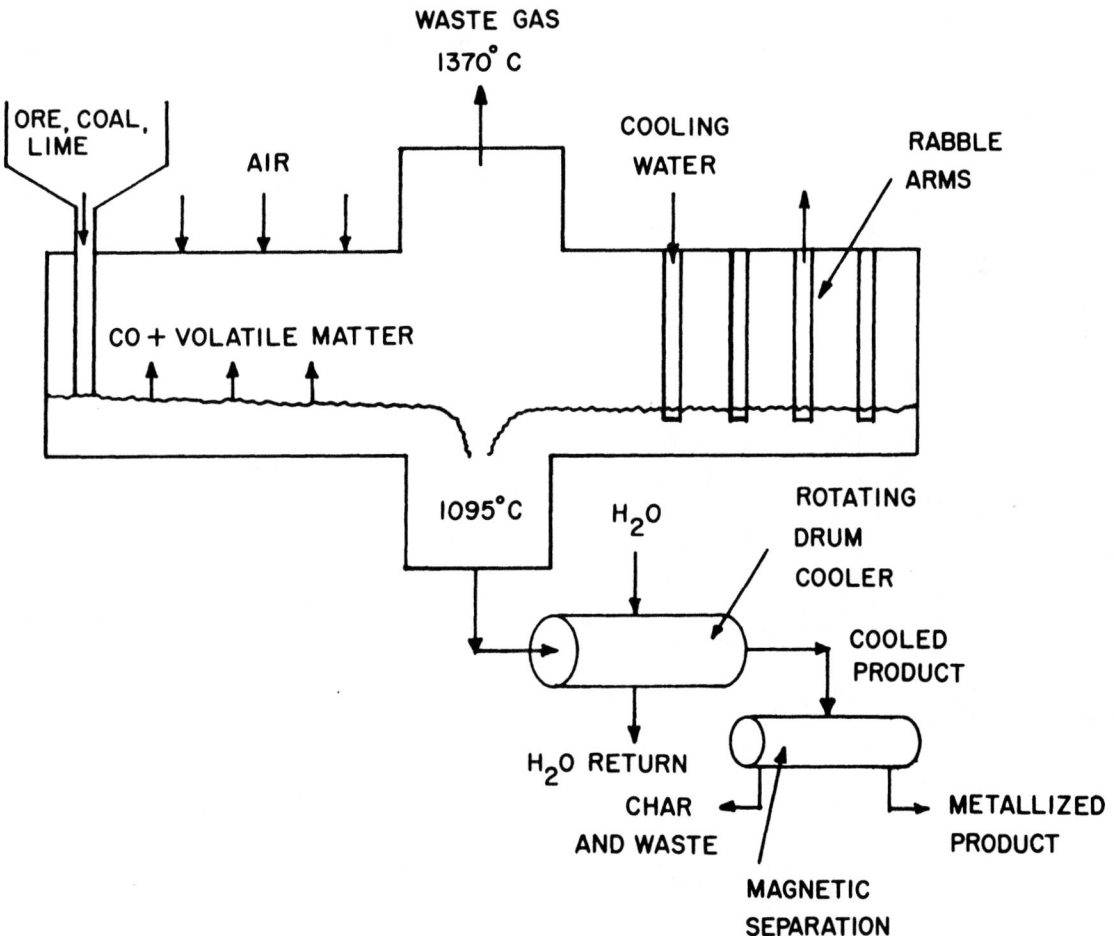

FIG. 14.—16. Salem rotary-hearth direct reduction process.

still in commercial use.

Alternate layers of fine-grained high-grade iron ore, dry coke breeze, and limestone are charged into cylindrical ceramic containers called **saggers.** The saggers are then heated to a maximum temperature of 1260°C (2300°F) in a furnace of the tunnel-kiln type used for burning brick. The furnace is heated by burning producer gas and the carbon monoxide evolved by the reduction of the ore. The containers are cooled in the furnace, removed, and the reduced iron is separated and cleaned. Total retention time of a container in the tunnel kiln is about 80 hours. Most of the sponge iron produced is sold as iron powder.

The largest Hoganas plant in operation has two tunnel kilns and is capable of producing 38 000 metric tons (41 000 net tons) annually.

Kinglor-Metor Process—The Kinglor-Metor process is based on the concept of producing iron continuously by heating a mixture of ore and coal in an externally-fired rectangular shaft or retort. Earlier attempts to implement this concept failed because the reduction reactions are highly endothermic and the production was severely limited by the slow rate of heat flow into the charge through the retort walls which were made of firebrick. Kinglor-Metor overcomes these limitations by constructing the walls of the retorts with highly conductive silicon carbide and by burning some of the carbon monoxide generated during reduction with air in a preheating zone in the upper part of the retort. Figure 14—17 shows a schematic flowsheet of the process.

A pilot plant comprising two reactors was installed at Buttrio, Italy by Danieli & Cie., SpA and started operations in 1973. The reactors are essentially vertical shafts of conical shape about 11 metres (33 feet) high with a top diameter of 0.4 metre (1.3 feet) and a bottom diameter of 0.7 metre (2.3 feet). The energy consumption is claimed to be about 16 million kilojoules per metric ton (13.8 million Btu per net ton) of DRI including recovery of 0.5 million kilojoules per metric ton (0.43 million Btu per net ton) of DRI from the reactor off-gas. The pilot-plant operations demonstrated the process to be simple to construct, easy to operate, and flexible with respect to feed and reductant requirements.

A commercial plant capable of producing 40 000 metric tons (44 000 net tons) per year has been installed by Ferriere Arvedi & Cie., SpA in Cremona, Italy. The plant consists of two identical 20 000 metric tons (22 000 net tons) per year modules. Each module contains six vertical retorts 13 metres (43 feet) high, 12.5 metres (41 feet) long, and 3 metres (8.8 feet) wide. At this plant, ore and coal are fed continuously into a silicon-carbide reactor that is heated to about 1100°C (2010°F) with natural gas radiant burners. Solid fuel requirements of about 8.5 kilojoules per metric ton (7.4 million Btu per net ton) of DRI and gaseous-fuel requirements of about 7.9 kilojoules per metric ton (6.8 million Btu per net ton) are claimed for the process. A plant has also been installed in Burma.

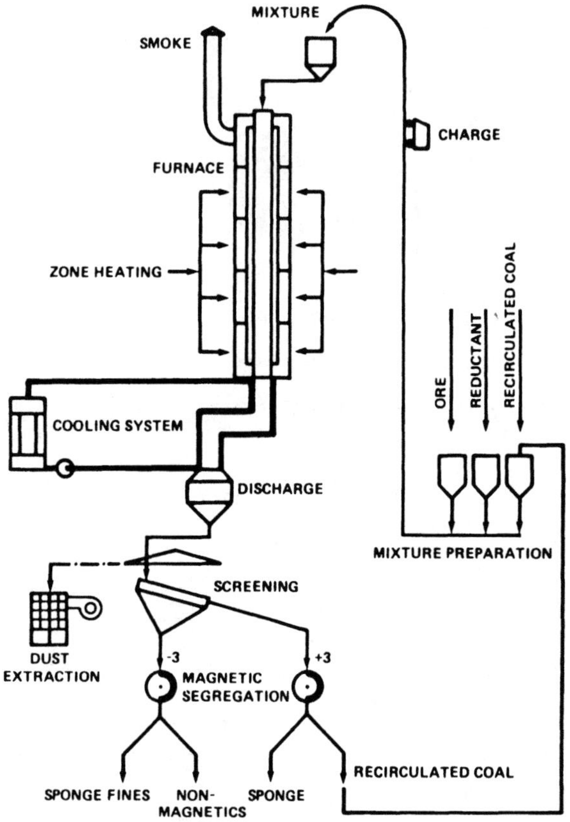

SHAFT FURNACE PROCESS, MOVING BED

Midrex EDR Process—In the Midrex EDR (Electrothermal Direct Reduction) process, the energy for the reduction of iron oxide with coal is supplied by electrical-resistance heating of the charge between electrodes in the walls of a shaft-furnace. A medium-calorific-value fuel gas is also produced for export. The gas has a heating value of about 7900 kilojoules per normal cubic metre (200 Btu per standard cubic foot). A pilot plant built in 1977, based on the Midrex EDR process, has produced 6 metric tons (6.6 net tons) of DRI per day.

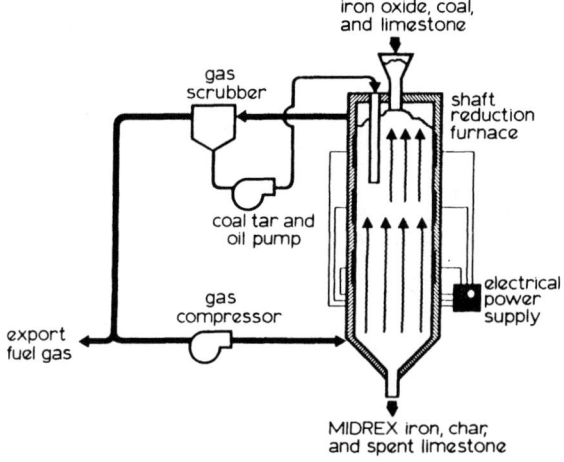

FIG. 14—18. MIDREX EDR process flowsheet.

FIG. 14.—17. Schematic flowsheet of the Kinglor-Metor process.

The Midrex EDR process is shown in Figure 14—18. The shaft-furnace is rectangular in cross section and has three distinct zones, for preheating, reduction, and cooling. Iron oxide, coal, and limestone are fed into the top of the furnace. As the burden flows down the pre-heat zone, it is heated by upflowing gas from the lower zones. The gas contains sufficient heat and reductants to partially reduce the iron oxide. The limestone is calcined into lime. The coal is devolatized and the volatiles leave the furnace in the top gas. As the burden enters the reduction zone, its temperature approaches 860°C (1580°F).

As the burden descends through the cooling zone, it is cooled and carburized by the upflow of cooling gas. The carbon monoxide in the cooling gas reacts with the metallic iron to form a small percentage of iron carbide. The cooled burden leaves the shaft furnace at about 65°C (150°F). The product mixture then flows through a magnetic separator to separate the DRI from the char and spent lime. The DRI product is comparable to that produced in the natural gas based Midrex process (see Section 3). Key advantages of Midrex's EDR process are uniformly high metallization (normally 92 per cent) and the balanced carbon content, which is adjustable between 1 and 1.5 per cent carbon, in the form of iron carbide.

SECTION 5

DIRECT SMELTING PROCESSES

ELECTRIC-FURNACE SMELTING PROCESSES

Pig Iron Electric Furnace Process—A low-shaft electric furnace is used in this process to reduce iron ore and produce molten pig iron. The furnace, shown schematically in Figure 14—19, consists of a melting chamber with three or more consumable electrodes extending downward through the roof, either in line or in a triangular arrangement. Openings are provided in the roof for charging the burden of agglomerated or lump ore, coke or coal, and limestone. The original furnace of the electric furnace process was operated in Oslo, Norway at the Christiania Spigerverk in the 1920's. It was known as the Tysland-Hole process.

Reduction of the ore and melting of the resultant product take place continuously in the hearth. A 10 000-kilowatt furnace can produce about 90 metric tons (100 net tons) of pig iron daily, with a power consumption of 2750 kilowatt-hours per metric-ton (2500 kilowatt-hours per net ton) of pig iron. Coke consumption, with lump ore of 55 per cent iron content, is 425 kilograms per metric ton (850 pounds per net ton) of pig iron. This low coke consumption, together with the use of a basic slag, helps to decrease the sulphur content of the pig iron.

The electric furnace process is commercially practical in countries such as Norway, Sweden, Italy, Switzerland, Yugoslavia, India and Venezuela, where low-cost electric power is available and/or where metallurgical coals are scarce and expensive. The process was used in Japan during the 1950's. The development of their ferroalloy industry was based partially on the use of the existing pig iron capacity.

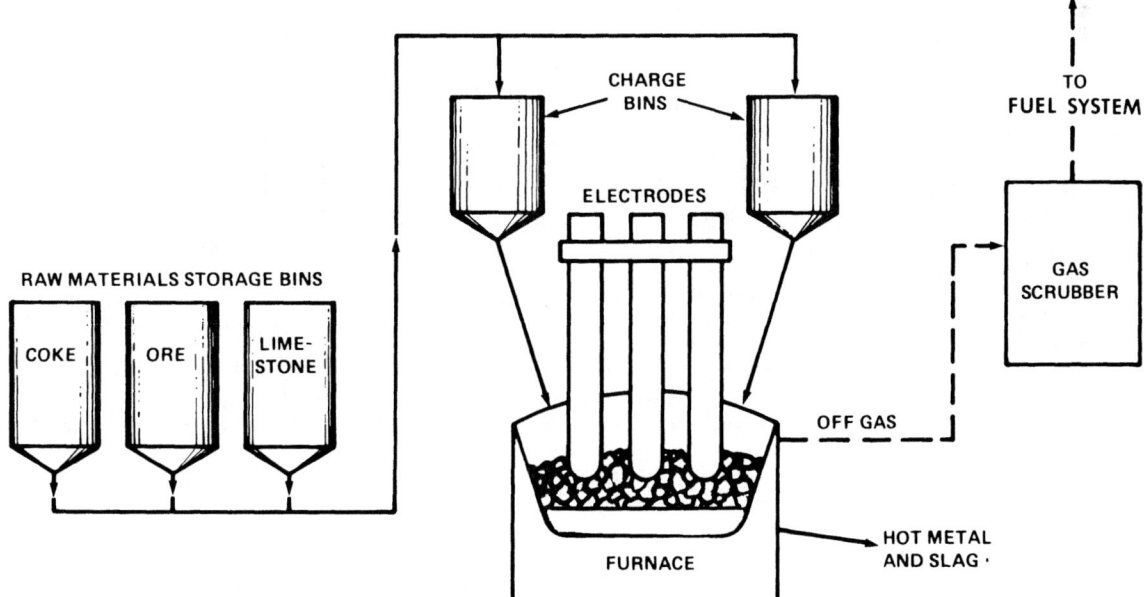

FIG. 14—19. Schematic representation of the electric furnace process for producing molten pig iron.

Modern electric furnaces for smelting pig iron employ the submerged arc concept with immersion of the electrode tips into the charge materials of the furnace. The process employs the Soderberg self-baking electrode system in which carbon electrode paste softens with mild heating and molds to the shape of a steel casing. The steel casing initially supports the electrode paste. A small portion of the power delivered to the furnace bakes the electrode giving strength and desirable electrical properties as the assembly descends into the furnace. Usually, three electrodes are arranged in a triangular configuration. The submerged arc furnace operates at lower voltages and lower power per cross-sectional area of hearth than the electric arc steelmaking furnace. Furthermore, slag composition in the submerged arc furnace is an important consideration. The objective is to optimize the metallurgical process while obtaining desirable electrical properties.

Similar to the iron-making blast furnace, ore, coke and fluxes are charged through the roof of the furnace. Other solid fuels including coal can substitute for a part of the coke. The materials form a low bed in the furnace and descend as the material melts and coke is consumed in the vicinity of the electrode tips. The reduction reactions proceed similarly to the iron-making blast furnace. However, in the absence of air blast to burn coke and provide additional quantities of carbon monoxide; and because of the short distance for gas to reach the top of the bed, a higher fraction of the iron reduction in the electric furnace involves the endothermic gasification of carbon (see Chapter 15).

Molten pig iron and slag accumulate at the bottom of the furnace and are tapped on a periodic basis. If the furnace top is closed to preclude the admission of air, the off-gas from the furnace is high in carbon monoxide and low in nitrogen. This raises its value as a fuel.

Modern submerged arc pig iron smelting furnaces have improved on the energy consumption and productivity of the original Tysland-Hole process. In the Elkem rotary-kiln electric-furnace process, preheating of the feed has reduced power consumption to 2000—2500 kilowatt-hours per metric ton (1800—2300 kilowatt-hours per net ton) and even greater savings are realized with both preheating and prereduction where power requirements of 1200—1800 kilowatt-hours per metric ton (1100—1640 kilowatt-hours per net ton) are reported. Furthermore, excess coal chars from the rotary kiln can provide carbon units to the electric furnace. Only partial prereduction is usually performed in the rotary kiln. A fuel burner may supply additional energy. Otherwise, the rotary kiln operation is similar to that discussed in Section 4. As a result of preheating and prereduction, a large electric furnace of 60 000 kVA (40 000 kW) transformer rating can be expected to produce an annual capacity exceeding 200 000 metric tons (220 000 net tons).

DLM Process—In the DLM or McDowell-Wellman process, the desired proportions of finely ground iron ore, flux, and coal are pelletized, and the pelletized feed is then dried, preheated, and partially reduced on an in-line travelling grate, equipped with a recycle hot draught, and a gas after-burner to control pollution. The reduced pellets (50 to 60 per cent reduction) are then charged hot to a submerged-arc smelting furnace.

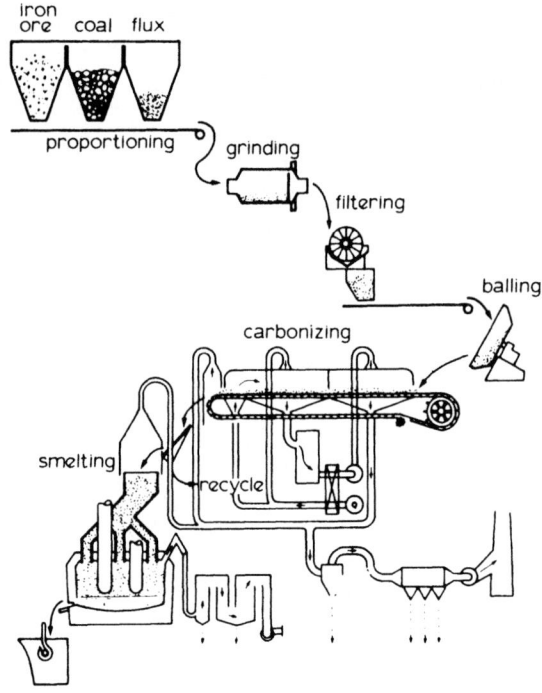

Fig. 14—20. Schematic representation of DLM iron smelting process.

The liquid iron produced in the smelting operation is the equivalent of blast-furnace iron and, therefore, then can be refined to liquid steel in a conventional BOP operation. A flowsheet of the process is shown in Figure 14—20.

In 1966, McWane Cast Iron Pipe Company signed a contract with McDowell-Wellman Engineering Company to build a 227 000 metric ton (250 000 net ton) per year plant at Mobile, Alabama. At that time the name of the process was the Dwight-Lloyd-McWane (DLM) process. The plant started operations in the spring of 1969, but was shut down in January 1970. There were reported to be some problems with the Dwight-Lloyd travelling grate operation, and the electric-furnace operation. Certain changes in practice, however, might make the process successful. It is on this basis that the process is discussed.

OXY-FUEL SMELTING SYSTEMS

INRED Process—Developed by Boliden AB, Sweden, the INRED process (Figure 14—21) reduces iron ore concentrates to molten iron in two stages in a single reactor.

Flash smelting of the concentrate by coal and oxygen is accomplished in the first stage where close to 90 per cent of the process energy is supplied. In the first stage, the ore is prereduced to FeO, the charge is superheated, and coal, supplied at the same time, is partly burned, the remainder forming coke. In the second stage, the prereduced and heated material is collected with the coke from the first stage in the lower part of

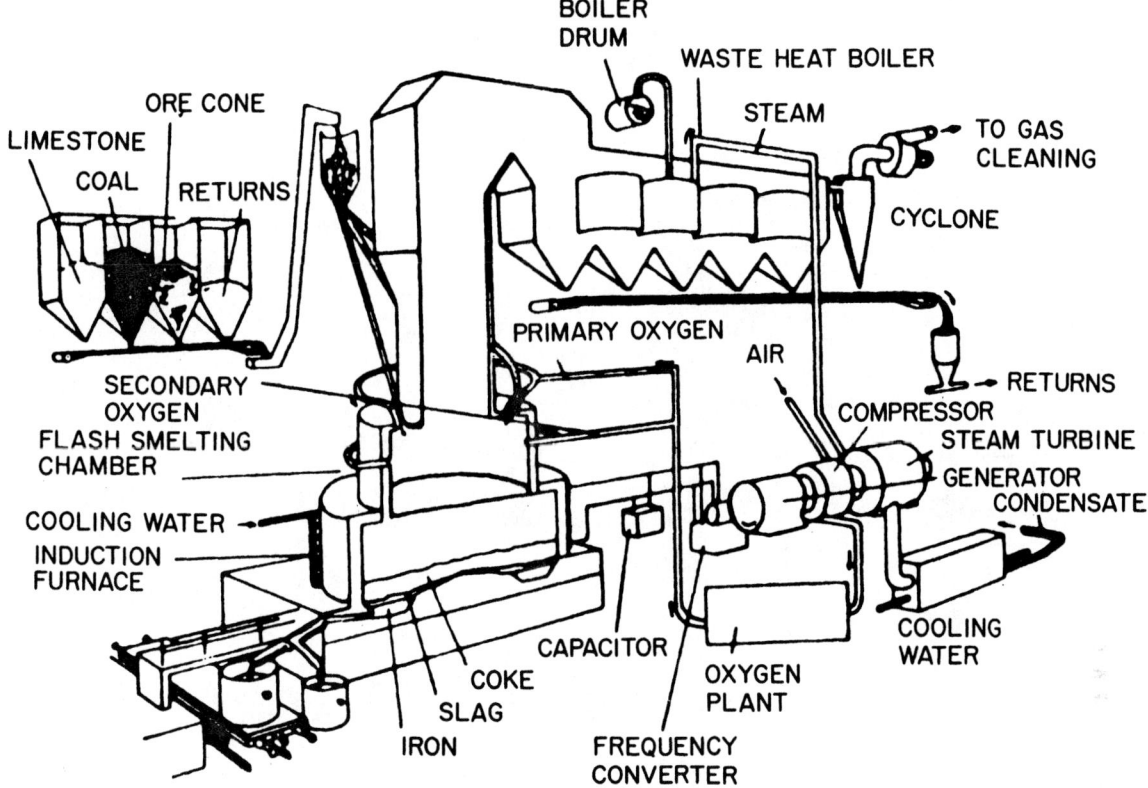

FIG. 14—21. Schematic representation of INRED process.

the reactor. Here, the remaining part of the energy is supplied by electricity. The pilot unit included induction heating; Boliden AB now uses submerged arc heating with self-baking electrodes. The flash-smelting chamber is constructed with diaphragm walls for steam generation. The central gas off-take at the top of the reactor is also similarly designed. Steam generated in the process supplies the energy for electricity generation for the arc heater and also the energy for producing the oxygen required for the process. The temperature in the flash-smelting (upper) section of the reactor is about 1900°C (3450°F), while in the lower section it is about 1450°C (2640°F).

The charge (ore, coal, and limestone) and a proportion of the oxygen are blown into a flash-smelting chamber through a circular array of nozzles on top of the chamber. The nozzles are directed so that the interaction of the jets produces a swirl in the center of the chamber. The combustion of coal is controlled in order to produce an amount of coke sufficient for final reduction and carburization of the iron in the lower section of the reactor. The iron produced is higher in sulphur and lower in carbon and silicon than blast furnace hot metal. Special provisions are required to convert the metal to steel.

KR Process—The KR process, a joint development of Korf Engineering GmbH and VOEST Alpine AG, produces hot metal directly from untreated raw coal and without the blast furnace requirement of coke. The process accomplishes this objective by conducting the

blast furnace functions of (1) preheating and gaseous reduction and (2) smelting in two separate reactors. Starting in 1981, a pilot plant of 5 to 8 metric tons (5.5 to 8.8 net tons) per hour capacity has provided information regarding raw materials, refractories, and operating and control procedures. An economic commercialization of a 300 000 metric ton (330 000 net ton) per year plant is now expected.

The process (Figure 14—22) comprises a two-stage operation in which DRI from a shaft furnace is charged without cooling into a connected melter gasifier. Partial combustion of coal with oxygen in the fluidized bed of the melter gasifier produces reducing gas for the shaft furnace. A portion of the energy from the combustion process is utilized to melt the DRI, coal ash, and flux to produce hot metal and slag which are tapped on a periodic basis. Partially-spent fuel gas from the shaft furnace is exported along with excess gas produced in the melter gasifier.

The melter gasifier operates at a pressure of about 3 to 5 bars (3 to 5 atmospheres). Along with limestone, coal is injected into the freeboard where it is rapidly heated to between 1000 and 1200°C (1830 and 2190°F). The volatile matter is driven off, and shattered fixed-carbon particles, of finer size than the original coal, fall into the lower gasification zone. In this region, a gas with a high percentage of oxygen is injected through blast furnace type tuyeres to convert carbon to carbon monoxide, which along with the preheated DRI feed, provides the energy to sustain the heating and melting

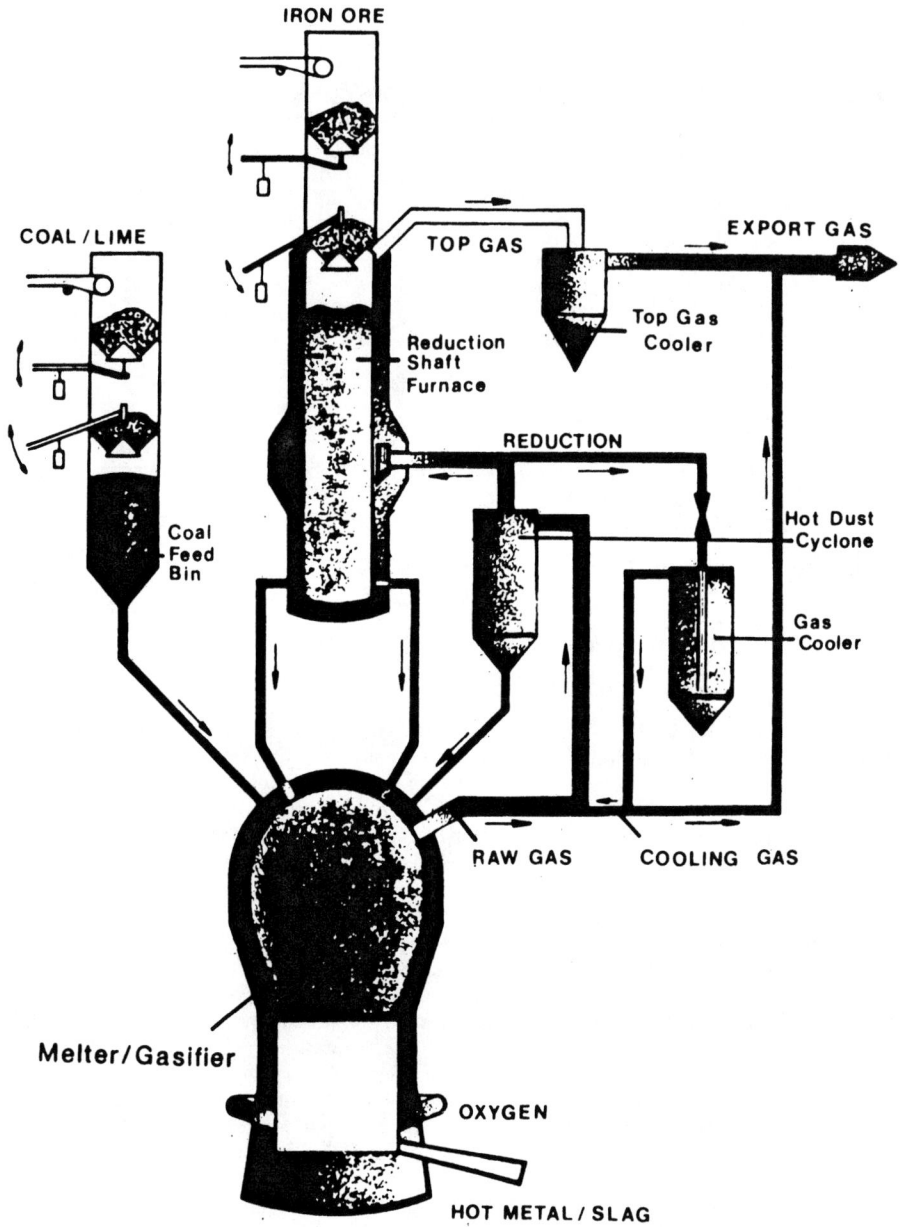

FIG. 14—22. Schematic representation of KR process.

of slag and hot metal. In this zone, the residual iron oxide (FeO) is also reduced to metal according to the net reaction (11).

The mixture of carbon monoxide from the fluidized bed and cracked volatiles from the coal contains more than 94 per cent carbon monoxide plus hydrogen. The hot raw gas is cleaned in a cyclone to recirculate entrained fines. The gas is cooled to between 800 and 900°C (1470 and 1650°F) and then introduced into the shaft furnace.

The shaft furnace operates in a manner similar to the moving bed processes such as the Midrex process described in Section 3. Lump iron ores or iron oxide pellets enter the furnace through a lock hopper system. The burden is reduced as it descends by gravity by reaction with the upward flowing gas. Screw conveyors discharge the hot DRI into downcomers leading to the melter gasifier.

The off-gas from the shaft furnace is cleaned and used for fuel gas. Depending on the coal, the melter gasifier may generate surplus gas which is added to the fuel gas. The calorific value of the gas in about 8400 kilojoules per normal cubic metre (215 Btu per standard cubic foot). One example in which a mixture of 50 per cent anthracite and 50 per cent bituminous coal was used gave a coal requirement of one ton per ton of hot metal. In one mode of operation, 42 per cent of the total energy input of the coal contributed to the iron making and the balance of 58 per cent went to the export fuel gas. As an alternative, 12 per cent of the total energy input of the coal could run an oxygen plant thus decreasing the net energy exported to 46 per cent.

As claimed, the KR process produces hot metal with carbon and silicon contents approaching blast furnace standards. However, the sulphur content is much higher because (1) by using raw coal, no sulphur is removed in the coking process and (2) nearly all the sulphur enters the slag and hot metal. In that regard, organic sulphur gasifies but is absorbed by the DRI and returned as iron sulphide to the melter gasifier system.

Kawasaki Steel Process—The Kawasaki Steel direct reduction process uses fine ore, low grade coke, pulverized coal, air and oxygen. A screw feeder charges fine ore into a preliminary reduction furnace (fluidized bed). Hot carbon monoxide produced in the melting furnace reduces the material between 60 and 70 per cent. The main smelting-reduction furnace is a shaft furnace with two levels of tuyeres. The prereduced iron ore is blown into the furnace through the upper tuyeres while hot air, oxygen, and nonmetallurgical coal are injected through the lower tuyeres. Low grade coke is loaded at the top of the furnace. The process maintains a temperature of 1450 to 1550°C (2640 to 2820°F) at the bottom of the furnace. Molten iron accumulates at the bottom of the furnace and impurities, which are drawn off periodically, float on top.

The main furnace at the experimental facility at Chiba Works measures 0.4 metres (1.3 feet) across and 2.4 metres (7.9 feet) high. Pig iron was produced at 0.25 metric tons (0.275 net tons) per hour. Kawasaki Steel estimates that a commercial plant will have a capacity of 4 metric tons/per cubic metre per day. The Kawasaki Steel process is also proposed for ferro-alloy manufacturing.

CGS Process—In the CGS (Creative Gas and Steel) process, developed by Sumitomo Metal Industries, Ltd., coal and oxygen are injected into hot metal contained in an apparatus similar to a steelmaking converter. Dissolution of carbon into the metal occurs with 2.5 to 3 per cent of carbon maintained in the metal. Carbon is simultaneously oxidized from the metal to produce reducing gas containing 20 to 30 per cent hydrogen, 70 to 80 per cent carbon monoxide, and a very low percentage of carbon dioxide. At the same time, excess energy is used to melt the DRI produced in the shaft-furnace. Hot metal and slag are removed from the furnace on a batch or continuous basis. Lime-bearing materials are used to flux the DRI gangue and coal ash which cause high sulphur removal by the slag. Gas from the gasifier is cleaned and passed through a CO-shift converter to increase the hydrogen content. The gas is then introduced into a shaft-furnace for the production of DRI. Excess gas produced is exported as fuel or chemical feedstock.

Pilot work has been performed on an 8 metric ton (8.8 net ton) per day gasifier. Melting capacity is 1.5 metric tons (1.6 net tons) of hot metal per square metre of hearth area which is projected to increase to 5 metric tons (5.5 net tons) per square metre of hearth area in large production units. System pressure is 1.5 kg/cm² per square centimetre (21 pounds per square inch).

SECTION 6

PLASMA PROCESSES

In plasma smelting for direct reduction, gases and solids are passed through an arc, much like a welding arc, and are heated. This eletric heating replaces oxygen in conventional systems that use oxy-fuel burners.

NON-TRANSFERRED ARC

In the plasma processes using a non-transferred arc technology, a plasma torch consisting of a pair of tubular, water-cooled copper electrodes discharges an electric arc which is magnetically rotated at very high speeds. The electrodes are spaced closely together and during operation, a process gas is injected through the narrow gap between the electrodes. Virtually any reducing, oxidizing, or inert gas can be heated by the electrodes. The arc current can be varied independent of gas flow rate and, thus, process temperatures can be controlled. The torches can be powered by d-c or a-c current, but generally, a d-c power supply is used.

Plasmasmelt Process—The SKF Plasmasmelt process produces molten iron from prereduced iron ore. SKF conducts pilot test work with iron ore at Hofors, Sweden. A variation of this concept, Plasmadust, is commercialized in Landskrona, Sweden. The plant is equipped with three 6-megawatt plasma torches to process 70 000 metric tons (77 000 net tons) of dust per year to recover molten iron, and zinc and lead which are volatilized. Another plant, Plasmachrome, is also expected to be built. To increase efficiency, the plants will export hot water or steam to the local community.

In the Plasmasmelt process (Figure 14—23), prereduced iron ore is reduced to molten iron in a lowshaft reactor similar in some respects to a low-shaft blast furnace. The minus 2-mm ore is prereduced to 50 per cent and injected together with coal and flux into the hearth of the reactor where it is smelted to molten iron using plasma-arc torches having an electrical consumption of about 1200 kWh per metric ton (1090 kWh per net ton) of iron. The shaft is filled with coke through a charging mechanism on top of the shaft. The hot metal produced is similar to blast-furnace hot metal, being saturated with carbon. The principal reductant in the shaft is coal, although a small amount of coke is also consumed. The hot gases from the shaft, which are essentially carbon monoxide and hydrogen, pass through the two-stage fluid bed that produces the 50 per cent prereduced iron. The off-gas from the shaft (at about 1000-1200°C, 1830-2200°F) is cleaned and cooled to about 650°C (1200°F) before being introduced into the two-stage fluid bed. The exit gas from the fluid bed reducers, containing over 30 per cent carbon monoxide and hydrogen, is used for drying and preheating the ore. The hot metal and slag are tapped intermittently, as in a blast furnace. The liquid iron produced in the smelting operation is equivalent to blast-furnace iron and, therefore, can be refined to liquid steel in conventional steelmaking operations. The process consumes approximately 200 kilograms of coal and 50 kilograms of coke per metric ton (400 pounds of coal and 100 pounds of coke per net ton).

Plasmared Process—The Plasmared process, also developed by SKF, Sweden, produces DRI in a shaft furnace with a moving bed (see Section 3). However, the process is unique in that a plasma gasifier produces the reducing gas for the reduction furnace. An old Wiberg DR facility was converted to Plasmared and started operations in 1981. The plant produces 50 000 metric tons (55 000 net tons) of DRI per year and is capable of using propane, oil, coal, and other fuels in the gasifier.

Initially, a coal/water slurry was introduced into a slagging coal gasifier along with recycled process gas through a plasma torch. Now, pregasification with oxygen is also performed. Hot gas from the gasifier is passed through a dolomite bed for sulphur removal. The hot gas is then introduced into a shaft-furnace where DRI is produced from oxide pellets or lump ore. Off-gas from the shaft-furnace is cooled and cleaned prior to carbon dioxide removal. The gas is then recycled to the gasifier.

TRANSFERRED ARC

In transferred arc plasma systems, the furnace is similar in configuration to that of an electric arc furnace. Gas, possibly containing entrained solids, is passed through the center of hollow carbon electrodes or through water-cooled tungsten electrodes, depending on whether the electrode system is consumable or nonconsumable. Either a-c or d-c power supplies may be used. The electrode is usually the cathode (negative

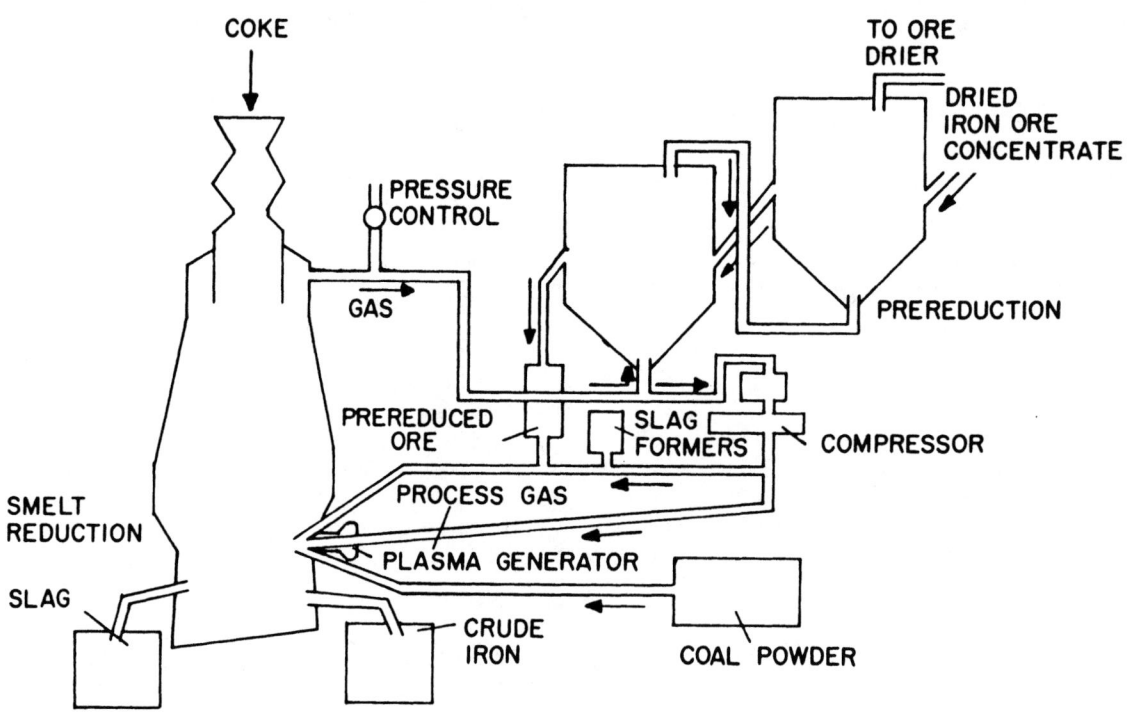

FIG. 14—23. Schematic representation of SKF Plasmasmelt process.

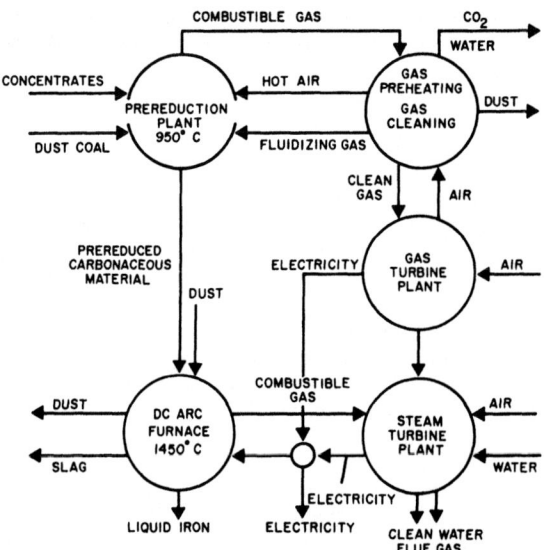

FIG. 14—24. ELRED process flowsheet.

connection in d-c systems). An anode pad underlies the slag and metal in a d-c system.

ELRED Process—The ELRED process (Asea/Stora, Sweden) comprises the prereduction of fine grained (less than 0.1 mm) iron ore concentrates with pulverized coal (0.2 to 0.3 mm) in a fast fluidized bed, followed by final reduction and smelting to liquid iron of the prereduced concentrate with associated char in the plasma beneath the electrode in a d-c arc furnace operating with an open bath. The combustible flue gases from both reduction stages are cleaned and used to generate electricity in a combined-cycle power plant to provide the power requirements for the process and a small surplus.

Figure 14—24 shows the ELRED flowsheet. Dried iron ore concentrate is fed pneumatically into venturi preheaters, where the feed is heated to about 700°C (1300°F) by contact with off-gas from the fluid-bed reactor. Combustion air and dried coal powder are injected into the main section of the fluid bed. The fluid bed operates under pressure, approximately 7 bars (7 atmospheres), and at a temperature of about 950-1000°C (1740-1830°F). The coal powder and combustion air generate the required heat to maintain the bed temperature and also generate the gas mixture of carbon monoxide and hydrogen necessary for reduction. About one-third of the fluid-bed reactor off-gas is cooled, cleaned, carbon dioxide scrubbed, reheated, compressed, and used to fluidize the lower section of the fluid bed. The bulk of the fluid-bed reactor off-gas is used to preheat the combustion air, the recycle off-gas for the fluid-bed reactor, and the boiler feed water for the steam turbine before it is combusted and expanded in the gas-turbine plant which drives the generators supplying part of the electrical power. The off-gas from the d-c arc smelter and the exhaust gases from the gas turbine provide the energy for raising the steam used to drive the steam turbine, which, in turn, drives the other generator to supply the other part of the electrical power in the combined-cycle power system. The

combined-cycle plant is an integral part of the ELRED process. Its inclusion helps to account for the relatively high thermal efficiency of the process.

Because there is a high concentration of char in the fluid bed, there is little tendency for plugging or defluidization of the bed that might otherwise be the consequence of operation at such a high temperature. This, together with the high superficial velocity of the fast fluid bed, inhibits material build-ups in the system.

The prereduced material, having about 65 per cent metallization (about 75 per cent reduced), is continually discharged from the bottom of the reactor and directly transferred to the d-c arc furnace for final reduction. A hollow carbon electrode is located in the center of the furnace roof and is connected to the negative pole of the rectifier. The positive pole is connected to a bottom electrode, which is in direct contact with the iron melt. The arc, which is submerged in foaming slag, extends vertically down toward the bath. The process is classified as a plasma process because of this feature of injection through a hollow electrode.

The prereduced material, which contains sufficient char to complete the smelting and carbonizing of the hot metal, is mixed with suitable fluxes and charged into the furnace at about 600-700°C (1100-1300°F). The material is fed through the electrode and passes through the plasma under the electrode. Melting, carbonization, and final reduction take place there. The hot metal produced contains 3.5 per cent carbon, 0.05 per cent silicon, 0.05 per cent manganese, and 0.3 per cent sulphur, when using coal containing 1.0 per cent sulphur. The slag basicity is about 1.2, as in a blast furnace; however, the FeO content of the slag is about 10 per cent. The metal produced requires external desulphurization in preparation for charging to the BOP. The use of ELRED hot metal in basic oxygen steelmaking reduces by about one half the scrap consumption that would be required with the use of blast furnace hot metal.

EPP Process—In the EPP (Expanded Precessive Plasma) process, developed by the Foster Wheeler-Tectronics, England, a d-c constricted-arc plasma gun is rotated to form a cone of very high temperature in which endothermic reactions can take place.

The system operates in the transferred-arc mode, with the plasma gun acting as the cathode and the furnace hearth acting as the anode return. The operation of the plasma gun requires a small amount of plasma gas, typically argon. The plasma gun is mounted in a hemispherical bearing with the longitudinal axis of the gun offset from the vertical by between 5 and 15°. The electrical power and gas are supplied to the gun by flexible connections, and movement of the gun assembly can be controlled by a hydraulic motor mounted over the gun. The offset angle and the speed of the precession of the gun are both determined by the process requirements, but normally the gun operates in the range 1000 to 1500 revolutions per minute at an offset of 10°. In application, the plasma-gun is mounted on the roof of a refractory-lined furnace with the offset and water-cooled section of the gun protruding into the furnace space. The plasma arc, normally 500 to 750 mm (20 to 30 inches) long, is generated between the gun electrode and the furnace hearth. The precession of the gun causes the plasma arc to generate a conic shape. A 10.8-MVA ferrochrome smelter employs these concepts at the Krugersdorp plant of Middelburg Steel and Alloys in South Africa.

In the proposal for the application of the EPP process to steelmaking, it was suggested that dried iron ore concentrate, coal, and lime, of no more than 0.5 mm (0.02 inches) should enter the furnace through a series of holes in the furnace refractory. In this proposal, it was assumed that a low-carbon, low-sulphur liquid steel could be produced in the plasma reactor in one step, using bituminous coal as the reductant. Sufficient lime was to be added to form a slag with a lime/silica ratio of 1.0. A high sulphur metal would be expected from this type of operation and would probably necessitate a desulphurization step.

SSP Process—The SSP (Sustained Shockwave Plasma) process is similar to the EPP process; however, the plasma of the SSP is processed at up to 30 000 revolutions per minute by an electromagnetic field rather than the mechanical device of the EPP process. It is claimed that electrical shockwaves are induced within the expanded, orbiting plasma thereby providing an extremely favorable thermodynamic and kinetic medium for inflight reduction with the plasma. However, experiments involving the reduction of Minnesota taconite concentrate with coal do not appear to substantiate these claims. Development so far involves a 100 kilowatt unit at the University of Minnesota.

The Toronto System—A hybrid process involving the use of consumable electrodes is currently being developed at the University of Toronto, Canada. The 100 kilowatt system uses three consumable electrodes that are hollow and composed of graphite. An arc is struck between the tips of the inclined electrodes with electricity supplied through a three-phase a-c system. The position of the electrodes is determined by the current used. A plasma-stabilizing gas, i.e., argon, can be used to extend the arc region between the electrodes. The "extended arc furnace" has been used to smelt iron ores, chromium ores, and high-grade manganese slag, and also has been used to melt metal fines.

Falling Film Plasma Reactor—In the falling film plasma reactor (Bethlehem Steel), an arc is struck between a tungsten cathode and a water-cooled copper cylinder that serves as the anode. A rotating jet of gas (argon or nitrogen) stabilizes the arc. Solid reactant particles are pneumatically injected into the stabilizing gas and are then centrifuged onto the inner walls of the anode, and form a molten film. The resultant film flows down the walls and protects the anode of the falling film reactor while also providing a long residence time for the material in the reactor. Consequently, good heat transfer, high conversion rates, and high energy utilization result, with an 80 per cent thermal efficiency claimed. The arc length of the torch varies up and down the anode column according to the voltage and other factors. A pilot plant used a d-c welding transformer to test the process.

The falling film plasma reactor has been used to produce iron and steel alloys, ferro-vanadium, molybdenum, and ferro-chromium.

Bibliography

General

Davis, C. G., McFarlin, J. F., and Pratt, H. R., Direct-reduction technology and economics, Ironmaking and Steelmaking, **9**, (3), 1982, 93-129.

von Bogdandy, L., and Engell, H. J., The reduction of iron ores, scientific basis and technology, Verlag Stahleisen GmbH, Dusseldorf, translated by Springer-Verlag, New York, 1971.

ACCAR

Albert, A. A., Jr., Application of ACCAR technology at Sudbury Metals Company and Niagara Metals Limited, CIM Conference, Vancouver, BC, 1977.

Rierson, D. W., and Albert, A. A., Jr., Development of the ACCAR process at Allis Chalmers, AIME Iron Production Proceedings, **36**, 1977, 455-467.

MacRae, D. R., Symposuim on coal-based direct reduction process, AISI, 1980.

Albert, A. A., Jr., Direct reduction of iron ores by Allis Chalmers Corporation, AISE Convention, Pittsburgh, Pa., 1983.

ARMCO

Brown, J. W., Campbell, D. L., Saxton. A. L., and Carr, J. W., Journal of Metals, **18**, 1966, 237-242.

Cruse, C. L., Kerschbaum, A. P., Marshall, W. B., and Moon, R. E., Development of Armco's direct-reduction processes, Journal of Metals, **21**, October 1969, 55-59.

Crestani, A. B., et al., paper presented at AISI 82nd general meeting, Washington, DC, May 1974.

Labee, C. J., Iron and Steel Engineer, **51**, November 1974, 73-75.

CGS

Tanoue, T., et al., Coal gasification using molten iron and its application to steelmaking, Sumitomo Metal Industries, Osaka, Japan.

DLM

Ban, T. E., Violetta, D. C., and Thompson, C. D., Travelling grate processes for the direct reduction of iron ores, Chemical Engineering Processing Symposium Series No. 43, **59**, AIChE, 1963.

Ban, T. E., and Violetta, D. C., DLM, new commercial ironmaking process, Iron and Steel Engineer, **45**, September 1968, 101.

DRC

Keran, V. P., Baker, A. C., Ridley, A. J., and Boulter, G. N., AIME Ironmaking Proceedings, **39**, 1980, 412-419.

Baker, A. C., et al., Process control in the direct reduction kiln, paper presented at AIME Ironmaking Conference, Pittsburgh, Pa., 1982.

Ridley, A. J., The integration of a coal-based direct reduction plant into a mini-mill, Metal Bulletin, Third International Mini-Mills Conference, New Orleans, La., March 1984.

ELRED

Collin, P. H., and Stickler, H., Iron and Steel Engineer, **57**, March 1980, 43-45.

EPP

Heanley, C. P., and Cowx, P. M., The smelting of ferrous ores using a plasma furnace, AIME Electric Furnace Proceedings, **40**, 1982.

Falling Film Plasma Reactor

MacRae., D. R., Plasma processing in extractive metallurgy: the falling film plasma reactor, Symposium Series No. 186, AIChE, **75**, 1979.

FIOR

Greaves, M. J., Importing and stockpiling energy through direct reduced iron, Arthur G. McKee and Co. (now Davy McKee Corp.), Cleveland, Ohio.

Cooke, H., Metal Bulletin Monthly, **10**, (114) June 1980, 29-30.

HIB

Reed, T. F., Nu-Iron, a fluidized-bed reduction process, Journal of Metals, **12**, 1969, 317-320.

Davis, W. L., Jr., Feinman, J., and Gross, J. H., Briguetacion del mineral de hierro de alto tenor dor el processo de reduccion directa en lechos fluidazos, ILAFA Memoria Tecnia del Seminario Latinamericano, La Reduccion Directa de los Minerales de Fierro y Su Application en America Latino, Ciudad de Mexico, Mexico, November 1973.

Hoganus

Eketorp, S., Hoganas sponge, iron process, Jernkontorets Annaler, **129**, 1945, 703-721, BISITS translation 275.

HYL II

Lawrence, R., Jr., The HyL direct-reduction process—past, present and future, UNESCO Economic Commission for Europe Steel Committee, Seminar on Direct Reduction of Iron Ore—Technical and Economic Aspects, Bucharest, Romania, September 1972.

Quintero, R. G., AIME Ironmaking Proceedings, **37**, 1978, 137-148.

Quintero, R. G., New HyL technology, HYLSA, S.A., Monterrey, Mexico, 1979.

HYL III

Quintero, R. G., Operational results of HYL III process, Skillings Mining Review, April 25, 1981, 12.

Carrillo, D., Trevino, A., and Price, F., Use of lump ore in HYL III direct reduction process: laboratories and plant comparison, paper presented at AIME Conference, Denver, Colorado, December 1981.

Philbrook, W. O., A close look at HYL III process in operation, Iron and Steelmaker, **9**, August 1982, 12.

Pena, J.M., Operating results of the HYL III 3M5 plant in Monterrey, Mexico, paper presented at AISE Convention, Pittsburgh, Pa., 1983.

INMETCO

Hanewald, R.H., Operating experience at INMETCO and application of the process to the production of DRI, paper presented at AISE convention, Pittsburgh, Pa., 1983.

INRED

Elvander, H. I., Edenwall, I. A., and Hallestam, S. C. J., Boliden iron process for smelting reduction of fine-grained iron oxides and concentrates, Ironmaking and Steelmaking, No. 5, 1979.

Westman, R. A., Elvander, H., and Milsson, C. G., The INRED process, description of the 60,000 tonne per year demonstration plant at MEFOS, Lulea, Sweden, paper presented at AIME Annual Meeting, Atlanta, Ga., 1983.

Kawasaki

Smith, C., The less fussy blast furnace, Financial Times, March 7, 1983.

Anon., New ironmaking/ferroalloy reduction process, Iron and Steel Engineer, May 1983.

Kinglor-Metor

Anon., Kinglor-Metor: the mini process, Metal Bulletin Monthly, December 1974, 53-57.

Ferrari, R., and Colautti, F., The Kinglor-Metor process—direct reduction using a solid reducing agent, Iron and Steel Engineer, May 1975, 57-60.

Barbi, A., The Kinglor-Metor process: commercial operation of the world's smallest DR unit, Iron and Steel International, **50**, August 1977, 229, 231-233, 236.

KR

Anon., The KR process, Korf technology for production of hot metal and reducing gas on the basis of coal, Korf Engineering GmbH, June 1983.

Braun, H. G., and Matthews, W., KR process: A new development in ironmaking, Iron and Steel Engineer, August 1984, 41-45.

Krupp CODIR

Meyer, G., Wetzel, R., and Bongers, U., The Krupp sponge-iron process—its products and applications, UNESCO Economic Commission for Europe Steel Committee, Seminar on Direct Reduction of Iron Ore—Technical and Economic Aspects, Bucharest, Romania, September 1972.

MacRae, D.R., Symposium on coal-based direct reduction process, AISI, 1980.

Anon., Sponge iron—direct route to steel, Krupp Industries and Stahlbau, 1978.

Krupp-Renn

Lehmkuhler, H., Treatment of low-grade siliceous iron ores by the Krupp-Renn process of the industrial pilot plant of Fried. Krupp AG, Stahl and Eisen, **59**, 1939, 1281-1288, Brutcher translation 1057.

Johannsen, F., The Krupp-Renn process, A Study of the Iron and Steel Industry in Latin America, United Nations Department of Economic Affairs, Bogota Meeting, **2**, 1954, 192-200.

Fastje, D., Reconstruction and initial operational results of the Krupp-Renn plant in Salzgitter-Watenstedt, Stahl and Eisen, **78**, 1958, 784-792.

Midrex

Sturgeon, J. H., The commercial production of prereduced pellets by the Midland-Ross process, AIME Ironmaking Proceedings, 1968.

Morgan, E. R., Reduccion directa continua de los oxidos de fierro dor el process Midrex, ILAFA, Memoria Tecnia del Seminario Latino Americano, La Reduccion Directa de los Minerales de Fierro y Su Application en America Latino, Ciudad de Mexico, Mexico, November 1971.

Rourke, E. T., Nickel Topics, **30**, (2), 1977, 5-6.

Rouillier, L., and Boote, A. J., Second reduction plant at the Sidbec-Dosco Contrecoeur Works, Iron and Steel Engineer, September 1978, 29-34.

Anon., Iron and Steelmaker, **6**, December 1979, 32.

Dayton, S., Engineering and Mining Journal, **180**, January 1979, 80-84.

Faucher, A., Marquis, A. H., and Dancy, T. E., Iron and Steelmaker, **6**, May 1979, 35-40.

Anon., Direct reduction plants producing steelmaking-grade DIR, Iron and Steelmaker, **10**, July 1983, 12.

Anon., Direct from Midrex, **9**, (1), fourth quarter 1983.

Midrex EDR

Goette, E. E., and Sanzenbacher, C. W., Midrex electrothermal direct reduction process, Midrex Corp., 1980.

NSC

McManus, G. J., Nippon Steel scales up in direct reduction, Iron Age Magazine, July 17, 1978.

Muraki, J., Otsuki, N., Hachisuka, K., Nishida, N., and Hara, Y., Nippon Steel direct reduction process, Nippon Steel Technical Report No. 12, December 1978.

Pig Iron Electric Furnace

Anon., Electric smelting plant at Choinez, Journal Iron and Steel Institute, **156**, 1947, 293-298.

Durrer, R., Electric smelting, Journal Iron and Steel Institute, **156**, 1947, 256-260.

Ballon, A., The electric reduction furnace, A Study of the Iron and Steel Industry in Latin America, United Nations Department of Economic Affairs, Bogota Meeting, **2**, 1954, 172-175.

Sem, M. O., Electric smelting of pig iron, A Study of the Iron and Steel Industry in Latin America, United Nations

Department of Economic Affairs, Bogota Meeting, **2**, 1954, 175-176.

Anderson, H. C., Skretting, H., and Svana, E., Alternative routes to steel, Iron and Steel Institute, London, May 1971.

Gundersen, J., and Skretting, H., A comparison of electric smelting on pilot and industrial scale, Journal of South African Institute of Mining and Metallurgy, **79**, No. 1, August 1978.

Plasmasmelt/Plasmared

Herlitz, H.. G., Johannson, B. and Santen, S. O., A new family of reduction processes based on plasma technology, Iron and Steel Engineer, March 1984, 39-44.

Purofer

Pantke, H. D., Iron ore reduction by the Purofer process, Journal of Metals, **17**, January 1965, 40-44.

von Bogdandy, L., Pantke, H. D., and Pohl, V., Journal of Metals, **18**, 1968, 519.

Pantke, H. D., and Lange, G. H., The Purofer process, UNESCO Economic Commission for Europe Steel Committee, Seminar on Direct Reduction of Iron Ore—Technical and Economic Aspects, Bucharest, Romania, September 1972.

Pantke, H. D., et al., Paper presented at Annual Meeting of Verein Deutsche Eisenhuettenleute, Dusseldorf, November 1972.

Pantke, H. D., and Lange, G. H., Purofer is setting new standards for direct reduction, AIME Ironmaking Proceedings, **36**, 1977.

Pantke, H. D., Further progress in direct reduction with Purofer, AIME Ironmaking Proceedings, **37**, 1978.

SL/RN

Stewart, A., and Work, H. K., R-N direct-reduction process, Journal of Metals, **10**, July 1958, 460-464.

Anon., Direct reduction of iron ores in a rotary kiln, Mining Congress Journal, December 1958.

Sibakin, J. G., Development of the SL direct-reduction process, Yearbook AISI, 1962, 187-228.

Meyer, K., Heitmann, G., and Janke, W., The SL/RN process for production of metallized burden, Journal of Metals, June 1966, 748-752.

McAdam, G. D., Dall, R. E. A., and Marshall, T., Direct reduction of New Zealand iron sands concentrates, New Zealand Journal of Science, **12**, December 1969.

Murdock, C. H., and Littlewood, R., Paper presented at Second Latin American Seminar on Direct Reduction, Porto Alegre, Brazil, ILAFA, May 1975.

Venkategwaran, V., and Brimacombe, J. K., Mathematical model of the SL/RN direct-reduction process, University of British Columbia, Vancouver.

Meadowcroft, T. R., and Brimacombe, J. K., Paper presented at Darken Conference, Monroeville, Pa., U. S. Steel Corp., August, 1976.

Thom, G. G., and Wilson, K., New SL/RN reduction plant at Griffith Mine, Stelco, Ltd., Hamilton, Ont., October 1976.

Bold, D. A., and Evans, N. T., Direct reduction down under: the New Zealand story, Iron and Steel International, June 1977, 146-152.

MacRae, D. R., Symposium on coal-based reduction process, AISI, 1980.

SSP

Moore, J. J., and Reid, K. J., Industrial plasma for iron and steelmaking, AIME Electric Furnace Proceedings, **40**, 1982.

Toronto System

Sommerville, I. D., McLean, A., and Alcock, C. B., Smelting and refining of ferroalloys in a plasma reactor, AIME Electric Furnace Proceedings, **41**, 1983.

Wiberg-Soderfors

Wiberg, F. M., Method of treating solid materials with gases, U. S. Pat. 1849 561, March 15, 1932.

Wiberg, M., Reduction of iron ore by carbon monoxide, hydrogen and methane, Jernkontorets Annaler, **124**, 1940, 179-212, Brutcher translation 1417.

Stahlhed, J., Production of sponge iron according to the Wiberg-Soderfors Method, A Study of the Iron and Steel Industry in Latin America, United Nations Department of Economic Affairs, Bogota Meeting, **2**, 1954, 204-209.

CHAPTER 15

The Manufacture of Pig Iron In the Blast Furnace

SECTION 1

PRODUCTION AND KINDS OF PIG IRON

As mentioned in Chapter 1, the extraction of iron from its ores dates back to prehistoric times. However, it was not until the 14th century that furnaces were developed that could both reduce the iron and melt it so that the product could be cast from the furnace in liquid form. Today's modern large blast furnaces with their very high production rates and excellent fuel effi-

ciency are the result of years of technical and engineering development of this same metallurgical process that first occurred over 500 years ago. Other processes for extracting iron from its oxides are described in Chapter 14.

The term **pig iron** is generally applied to the metallic product of the blast furnace when it contains over 90

Table 15—I. Production of Pig Iron and Ferroalloys for Selected Years in the Three Major Steel-Producing Countries and the World.[1]

	(Thousands of Tons)							
	USA[2]		Japan		U.S.S.R.		World	
Year	Metric Tons	Net Tons	Metric Tons	Net Tons	Metric Tons	Net Tons	Metric Tons	Net Tons
1981	66 728	73 570	80 030	88 236	107 377	118 387	501 088	552 468
1980	62 343	68 721	87 039	95 945	108 999	120 151	507 719	559 779
1979	78 928	87 003	83 824	92 400	108 974	118 257	528 916	583 149
1978	79 541	87 679	78 589	86 629	110 701	122 027	507 219	559 371
1977	73 780	81 328	85 885	94 672	107 367	118 352	485 781	535 842
1976	78 807	86 870	86 576	95 433	105 383	116 165	487 678	537 573
1975	72 505	79 923	86 877	95 765	102 967	113 502	470 785	518 952
1974	87 007	95 909	90 437	99 689	99 867	110 084	504 342	555 942
1973	91 814	101 208	90 006	99 215	95 932	105 747	496 026	546 775
1972	81 102	89 400	74 054	81 631	92 399	101 852	450 617	496 720
1971	74 110	81 692	72 744	80 187	89 253	98 385	424 862	468 330
1970	83 294	91 816	68 047	75 009	85 932	94 724	428 783	472 652
1965	80 611	88 858	27 502	30 316	66 198	72 971	327 078	360 542
1960	61 078	67 327	12 341	13 604	46 756	51 540	253 737	279 697
1955	70 570	77 790	4 751	5 237	33 000	36 376	191 476	211 060
1950	59 366	65 440	2 299	2 534	19 504	21 499	132 795	146 381
1945	49 138	54 165	3 115	3 434	17 053	18 798	83 284	91 805
1940	42 999	47 398	4 012	4 422	14 968	16 499	97 166	107 107
1935	21 715	23 937	2 784	3 069	12 507	13 787	71 054	78 985
1930	32 261	35 562	1 687	1 860	4 982	5 492	78 086	86 075
1925	37 290	41 105	933	1 028	1 309	1 443	76 868	84 732
1920	37 518	41 357	732	807	115	127	63 858	70 391
1915	30 396	33 506	336	392	3 685	4 062	60 658	66 864
1910	27 742	30 580	N.A.[3]	N.A.[3]	N.A.[3]	N.A.[3]	72 265	79 659
1905	23 290	25 673	N.A.	N.A.	N.A.	N.A.	59 289	65 355
1900	14 010	15 443	N.A.	N.A.	N.A.	N.A.	43 059	47 464

[1] Based on information obtained from American Iron and Steel Institute Annual Statistical Reports.
[2] See Appendix D for more detailed figures on U.S. production.
[3] N.A. means not available.

per cent iron. It is used to distinguish this product from other blast-furnace products such as ferromanganese, spiegeleisen, ferrophosphorus, ferrosilicon and silvery iron. This term arose from the old-fashioned method of casting blast-furnace iron into molds arranged in sand beds in such a manner that they could be fed from a common runner. Because the group of molds resembled a litter of suckling pigs, the individual pieces of iron were referred to as **pigs** and the runner was referred to as a **sow**.

Table 15—II. Annual Blast-Furnace Production in the United States for Selected Years, by Grade.[1]
(Thousands of Tons)

Year	Basic Metric Tons	Basic Net Tons	Bessemer Metric Tons	Bessemer Net Tons	Foundry Metric Tons	Foundry Net Tons	Malleable Metric Tons	Malleable Net Tons	All Other, Including Ferroalloys and Silvery Iron Metric Tons	All Other Net Tons	Total Metric Tons	Total Net Tons
1981	64 548	71 166	51	56	1 107	1 221	843	929	180	198	66 728	73 570
1980	60 749	66 964	35	39	647	713	881	799	187	206	62 343	68 721
1979	76 246	84 047	11	12	1 443	1 591	910	1 003	318	350	78 928	87 003
1978	76 654	84 497	268	295	1 684	1 856	427	471	508	560	79 541	87 679
1977	70 940	78 198	230	254	1 461	1 610	626	690	523	576	73 780	81 328
1976	75 200	82 894	1 043	1 150	1 351	1 489	726	800	487	537	78 807	86 870
1975	68 865	75 911	986	1 087	1 198	1 321	738	814	717	790	72 505	79 923
1974	82 729	91 193	998	1 100	1 386	1 528	1 254	1 382	640	706	87 007	95 909
1973	87 273	96 202	1 127	1 242	1 434	1 581	1 227	1 352	754	831	91 814	101 208
1972	76 168	83 961	1 212	1 336	1 794	1 977	1 082	1 193	846	933	81 102	89 400
1971	70 163	77 341	1 102	1 215	1 101	1 214	1 212	1 336	532	586	74 110	81 692
1970	78 415	86 438	1 313	1 447	1 549	1 707	1 284	1 415	734	809	83 294	91 816
1960	52 853	58 260	3 088	3 404	1 331	1 467	2 425	2 673	1 381	1 522	61 078	67 327
1950	45 250	49 880	7 340	8 091	2 545	2 805	2 886	3 181	1 345	1 483	59 366	65 440

[1]Based on information obtained from American Iron and Steel Institute Annual Statistical Reports.

Table 15—III. Chief Metallic Products of the Blast Furnace[a]

	Composition Range				
	Silicon (%)	Sulphur (%)	Phosphorus (%)	Manganese (%)	Total Carbon[b] (%)
IRON FOR STEELMAKING					
Basic Pig	1.50	0.05 max.	0.400 max.	1.01 to 2.00	3.5 to 4.40
In steps of	0.25	—		0.50	—
Acid Pig, Bessemer	1.00 to 2.25	0.045 max.	0.04 to 0.135	0.5 to 1.00	4.15 to 4.40
Acid Pig, Open-Hearth	0.70 to 1.50	0.045 max.	Under 0.05	0.5 to 2.50	4.15 to 4.40
Oxygen Steelmaking Pig	0.20 to 2.00	0.05 max.	0.400 max.[c]	0.4 to 2.50	3.5 to 4.40
MERCHANT IRON FOR FOUNDRIES					
Low Phosphorus	0.50 to 3.00	0.035 max.	0.035 max.	1.25 max.	3.0 to 4.50
Intermediate Low Phosphorus	1.00 to 3.00	0.050 max.	0.036 to 0.075	1.25 max.	3.0 to 4.50
Bessemer	1.00 to 3.00	0.050 max.	0.076 to 0.100	1.25 max.	3.0 to 4.50
Malleable	0.75 to 3.50	0.050 max.	0.101 to 0.300	0.50 to 1.25	3.0 to 4.50
Northern Foundry	3.50 max.	0.050 max.	0.301 to 0.700	0.50 to 1.25	3.0 to 4.50
Southern Foundry	3.50 max.	0.050 max.	0.700 to 0.900	0.40 to .75	3.0 to 4.50
All grades in steps of	0.25	—	—	0.25	—
FERROALLOYS					
Spiegel (3 grades)	1.0 to 4.5	0.05 max.	0.14 to 0.25	16 to 30	6.5 max.
Standard Ferromanganese (3 grades)	1.2 max.	0.05 max.	0.35 max.	74 to 82	7.5 max.
Ferrosilicon, Silvery Pig	5.00 to 17.00	0.06 max.	0.300 max.	1.00 to 2.00	1.5 max.
Ferrophosphorus	1.5 to 1.75	Under 0.05	15 to 24	0.07 to 0.50	1.10 to 2.0
FOREIGN PRACTICE					
Basic Bessemer (Gilchrist or Thomas)	0.3 to 1.00	0.20	1.9 to 2.5	0.7 to 2.5	3.50 to 4.0
Duplex Iron	1.2 to 1.75	Under 0.060	0.7 to 1.5	0.4 to 0.90	4.00 to 4.20

(a) Further information in: Steel Products Manual—Section 1—Pig Iron and Blast-Furnace Alloys, published by the American Iron and Steel Institute and ASTM Standards Part I—Ferrous Metals (Specifications), published by the American Society for Testing and Materials (latest editions).
(b) Carbon not specified.
(c) Up to 2.00 per cent phosphorus may be used by double slagging in the basic oxygen furnace.

The importance of pig iron in the 20th century can be seen from Table 15—I, which shows that the world-wide annual production of pig iron and ferroalloys increased from less than 43 million metric tons (47 million net tons) to more than 528 million metric tons (582 million net tons) between 1900 and 1979. It is interesting to note that the annual production rate of pig iron and ferroalloys in the United States reached a maximum in 1973 that was 1.5 times its 1950 production rate while the world-wide annual production rate reached a peak in 1979 that was about 4 times its 1950 rate. The rapid growth of the steel industry in foreign countries since 1950 has been accompanied by a substantial increase in the size of blast furnaces, a significant improvement in the preparation of burden materials, a better distribution of the burden materials in the furnace, a more elevated top-gas pressure, and a higher hot-blast temperature.

Kinds and Grades of Pig Iron—Most of the iron produced in blast furnaces is transported to the steelmaking shop while it is still liquid and is then used directly for the manufacture of steel. In the liquid form the iron is generally referred to as **hot metal**. Because the basic open-hearth and the basic oxygen processes are most commonly used for converting the iron into steel, most of the iron is of the basic grade.

Table 15—II shows the annual production of the different grades of pig iron in the United States for selected years between 1950 and 1981. During this period the basic grade increased from 76 per cent to 97 per cent of the total production, and the bessemer grade decreased from 12.5 per cent to practically nothing. The demand for foundry and malleable iron also decreased because many foundries began using steel scrap instead of pig iron in cupolas and electric furnaces to produce iron for castings. The production of ferroalloys and silvery iron in blast furnaces has also decreased as it has become more feasible to make these grades in electric smelting furnaces.

The general composition ranges for the different metallic products of the blast furnace are shown in Table 15—III.

OUTLINE OF THE BLAST-FURNACE PROCESS

Blast Furnace Proper—The blast furnace is a tall shaft-type furnace with a vertical stack superimposed over a crucible-like hearth. Iron-bearing materials (iron ore, sinter, pellets, mill scale, steelmaking slag, scrap, etc.), coke and flux (limestone and dolomite) are charged into the top of the shaft. A blast of heated air and also, in most instances, a gaseous, liquid or powdered fuel are introduced through openings at the bottom of the shaft just above the hearth crucible. The heated air burns the injected fuel and most of the coke charged in from the top to produce the heat required by the process and to provide reducing gas that removes oxygen from the ore. The reduced iron melts and runs down to the bottom of the hearth. The flux combines with the impurities in the ore to produce a slag which also melts and accumulates on top of the liquid iron in the hearth. From time to time, the iron and slag are drained out of the furnace through tapping holes. To produce a metric ton (1000 kilograms) of pig iron requires about 1.7 metric tons of ore or other iron-bearing material, 450 to 650 kilograms of coke and other fuel, about 250 kilograms of limestone or dolomite and 1.6 to 2.0 metric tons of air. Stated in fps units, to produce a net ton (2000 pounds) of pig iron requires about 1.7 net tons of ore or other iron-bearing material, 900 to 1300 pounds of coke and other fuel, about 500 pounds of limestone or dolomite and 1.6 to 2.0 net tons of air. Very often, the limestone and dolomite are precalcined by mixing them with the iron-bearing material to be sintered or pelletized to improve the efficiency of the blast-furnace process. For every metric ton of iron, the furnace also produces 200 to 400 kilograms of slag, 25 to 50 kilograms of flue dust and 2.0 to 3.0 metric tons of blast-furnace gas: in fps units, for every net ton of iron produced, the furnace also produces 400 to 800 pounds of slag, 50 to 100 pounds of flue dust and 2.0 to 3.0 net tons of blast-furnace gas.

Figure 15—1 is a schematic representation of a typical American blast-furnace plant showing the receiving and storage of raw materials, the weighing of the burden and the charging of the blast furnace. It shows the hot metal from the furnace going to a steelmaking shop or a pig-casting machine. The slag goes to a water-spray granulator, a dry slag pit or a slag dump. The gas from the top of the furnace goes through the gas-cleaning system, and then a portion goes to fire the hot-blast stoves with the balance being used in other parts of the plant. The dust is removed from the gas in the cleaning system and goes to the sinter plant to be agglomerated for recycling back into the blast furnace. Also shown are the boiler house that generates the power for the operation of the furnace and the turbo blowers that compress the blast air which goes first to the stoves to be heated and then to the blast furnace through the tuyeres.

DESCRIPTION OF THE CHARGE MATERIALS

Iron-Bearing Materials—The function of the iron-bearing materials is to supply the element iron which is 93 to 94 per cent of the pig iron produced. The major iron-bearing materials are ore, sinter and pellets. Most of the ore is the ferric oxide known as hematite (Fe_2O_3) or the hydrated ferric oxides known as limonite or goethite ($Fe_2O_3 \cdot XH_2O$). However, in some instances, the ores contain magnetite (Fe_3O_4) or siderite ($FeCO_3$). Chemically pure ferric oxide contains 70 per cent iron, but most ores contain only 50 to 65 per cent iron because they contain 2 to 10 per cent gangue (which consists mostly of alumina and silica) and chemically combined water. Most of the iron bearing materials are

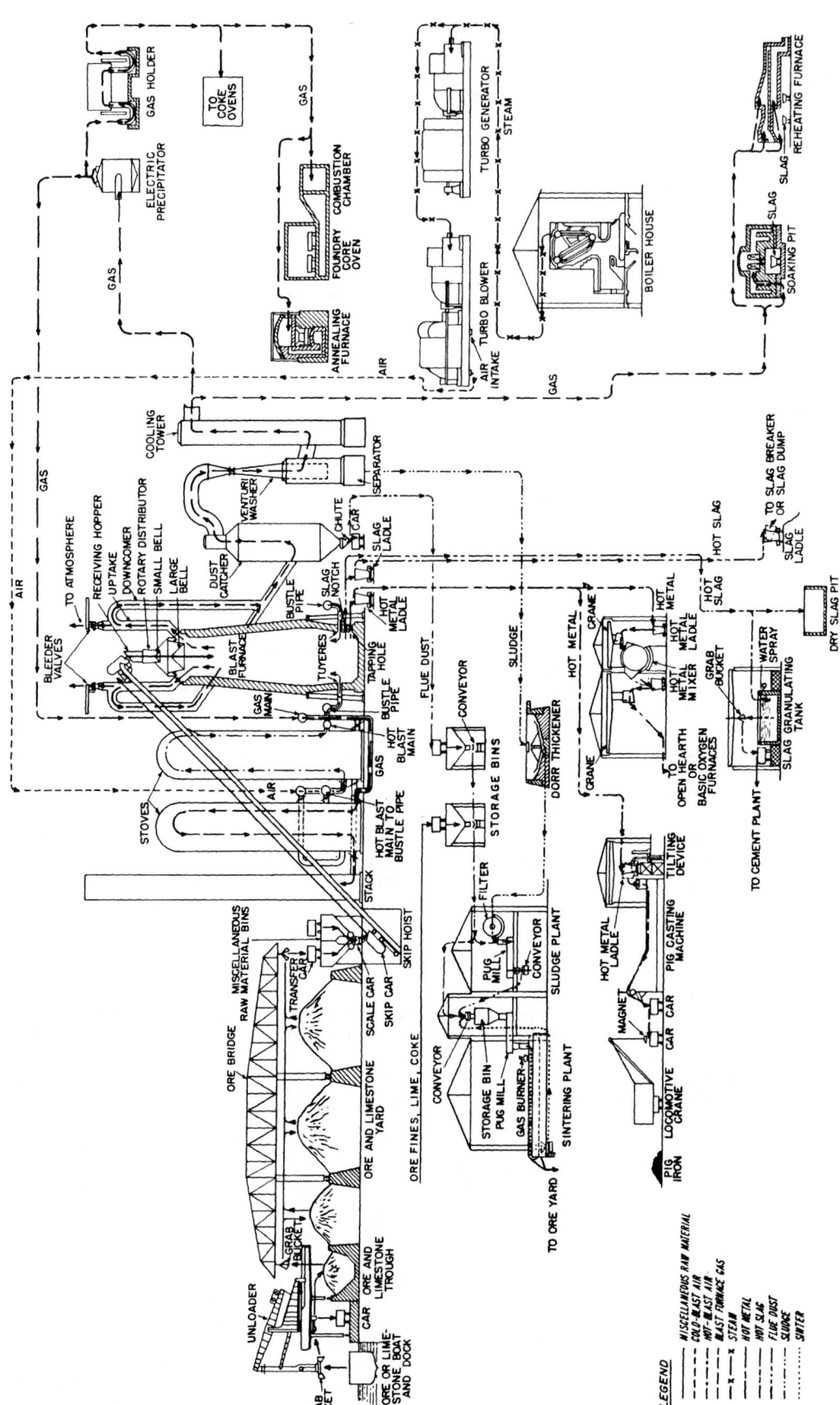

FIG. 15—1. Flow diagram depicting the principal units and auxiliaries in a modern blast-furnace plant, and showing the steps in the manufacture of pig iron from receipt of raw materials to disposal of pig iron and slag, as well as the methods for utilizing the furnace gases.

screened to remove fines, to permit the achievement of higher wind and production rates, and to permit smooth burden movement. The portion of the ore that is too fine to be charged directly is usually agglomerated in a sintering plant. In the sintering process, which is described in Chapter 5, a fine carbonaceous fuel is added to the fine ore and the mixture is placed on a grate and ignited. Air drawn through the mixture burns the fuel at a temperature high enough to frit the small particles together into a cake so that they can be charged into the blast furnace satisfactorily. For best results, pulverized flux is added to the sinter mix to combine with the gangue of the ore in the sintering process. Sinter usually contains 52 to 60 per cent iron.

Pellets are agglomerates made from very finely divided (minus 0.074-mm or minus 200-mesh) iron-ore concentrates to which a small quantity of fuel and a binder have been added. The mixture is then balled to form "green" pellets slightly larger than 6 mm (0.24 inch) but smaller than 20 mm (0.80 inch) in diameter. The fuel may be provided by "coating" the green pellets (made originally in a balling drum) with fine particles of a suitable carbonaceous material. In either case, the pellets are then hardened by firing in a shaft-type furnace or kiln or on a traveling grate. Pellets usually contain from 60 to 67 per cent of iron. Chapter 5 describes the methods used for pelletizing.

The minor iron-bearing materials are roll scale, open-hearth-furnace or basic-oxygen-furnace slag, and scrap.

Roll scale consists of oxides that form on the surface of steel during heating for rolling, and is usually a source of relatively pure iron oxide except from mills where it gets contaminated with hot-top brick or other refractories.

Basic open-hearth and oxygen furnace slags contain about 25 per cent iron and an excess of bases over acids. Consequently, they can replace a certain quantity of basic fluxes in the blast-furnace burden. Using the slag increases the total quantity of slag produced per ton of hot metal. Basic open-hearth slag also contains sufficient manganese to make it a useful source of this element. Its use is limited by the specification for the maximum phosphorus content of the iron pro-

duced. It was not used in blast furnaces producing "blowing iron" for the acid-Bessemer process, because the manganese it contained would adversely affect the properties of the slag in the converter; however, it may be desirable for increasing manganese in otherwise excessively low manganese hot metal to prevent slopping in the basic oxygen furnace.

Scrap recovered from steelmaking slag and blast-furnace runner scrap are charged for recovery of the iron units they contain. Also, small scrap that is not amenable to use in steelmaking is charged into the blast furnace.

Coke—The main function of coke is to produce the heat required for smelting and also the chemical reagents carbon and carbon monoxide for reducing the iron ore. In addition, it supplies the carbon that dissolves in the hot metal (about 40 to 45 kilograms per metric ton of iron, equivalent to 80 to 90 pounds per net ton), and because the coke retains its strength at temperatures above the melting temperature of pig iron and slag, it provides the structural support that keeps the unmelted burden materials from falling into the hearth.

Because of chemical equilibrium limitations, all of the carbon monoxide produced in the blast furnace can not be consumed in the reduction of the burden. Consequently, the gas issuing from the top of the furnace contains sufficient carbon monoxide to have a calorific value of 3 to 4 million joules per normal cubic metre (76 to 101 Btu per standard cubic foot). This gas is used to preheat the blast air and to generate power for running the blowers: thus, much of the energy is returned to the blast-furnace operation. The excess gas is used in other portions of the plant.

Fluxes—The major function of the fluxes, limestone and/or dolomite, is to combine with the ash in the coke and the gangue in the ores to make a fluid slag that can be drained readily from the furnace hearth. The ratio of basic oxides to acid oxides must be controlled carefully to preserve the sulphur-holding power of the slag as well as the fluidity. In instances where the acids in the coke ash and ore gangue are not sufficient to make enough slag volume, silica gravel or quartzite is added with the charge.

SECTION 3

CONSTRUCTION OF THE FURNACE PROPER

TERMINOLOGY

As an introduction to a discussion of blast-furnace construction, it is desirable to establish an understanding of the terms applied to the important dimensions of the furnace. These are shown graphically in Figure 15—2. However, the following terms may require additional explanation:

Hearth Diameter: The diameter of the circle determined by the inner noses of the tuyere coolers.

Hearth Line: The horizontal line at the intersection of a vertical line through the nose of the tuyere cooler

and sloping line of the bosh. With ceramic-lined boshes, the slope of the bosh is determined by the line through the noses of the bosh plates.

Height of Hearth: The vertical distance between the hearth line and the centerline of the tapping hole. The latter is determined by the center of the tapping-hole opening in the hearth jacket.

Bosh Angle: The acute angle formed by a horizontal line and the slope of the bosh.

Bosh Line: The horizontal line at the intersection of the slope of the bosh and the vertical section of the lower stack or, if there is no vertical section, at the intersec-

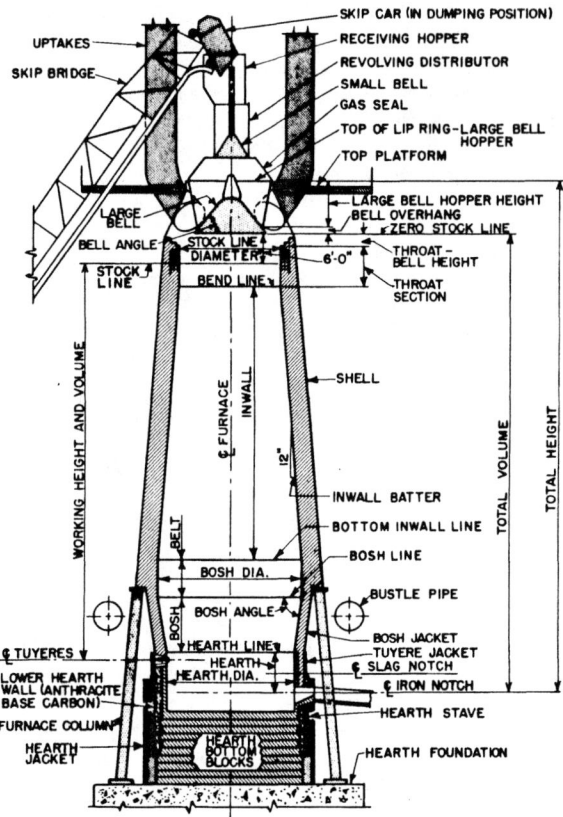

FIG. 15—2. Identification of the principal dimensions of a blast furnace.

tion of the slope of the bosh and the batter of the inwall.

Bosh Diameter: The diameter of the inside face of the lining at the bosh line. (This is also the diameter of the straight section above the bosh).

Height of Bosh: The vertical distance between the hearth and bosh lines.

Inwall Batter: The negative slope of inwall expressed numerically as the base of a right triangle whose altitude is 12 inches and whose hypotenuse is the slope of the inwall.

Bottom Inwall Line: The horizontal line through the intersection of the vertical line of the straight section and the inwall batter. (In furnaces without a straight section above the bosh, the bottom inwall line coincides with the bosh line.)

Bend Line: The horizontal line at the upper termination of the inwall batter.

Height of Inwall: The vertical distance between the bottom inwall line and the bend line.

Zero Stockline: The horizontal line at the bottom of the large bell when closed. Accordingly, a 1.8-meter (6-foot) stockline, for example, is the horizontal line 1.8 metres (6 feet) below the large bell when closed.

Stockline Diameter: The diameter from face to face of the brickwork, or imbedded armor where used, at a plane 1.8 metres (6 feet) below zero stockline.

Height of Straight Section above Bosh: The vertical distance between the bosh line and the bottom

inwall line.

Height of Throat Section: The vertical distance from the bend line to the top of the armor or lining.

Throat-Bell Height: The vertical distance between the upper termination of the throat section and the bottom of the large bell when closed.

Height Between Bottom of Large Bell and Top of Hopper: The vertical distance between the bottom of the large bell closed and the intersection of the hopper or the hopper extension with the gas seal.

Height of Large Bell Hopper: The vertical distance between the inner large-bell seat and the intersection of the hopper or the hopper extension with the gas seal.

Bell Overhang: The vertical distance between the bottom of the large bell closed and the inner bell seat.

Annular Space: The difference between the stockline radius and the large-bell radius.

Total Height of Furnace: The vertical distance between the centerline of the tapping hole and the intersection of the large-bell hopper or hopper extension with the gas seal.

Working Height of Furnace: The vertical distance between the centerline of the tuyeres and the line 1.8 metres (6 feet) below the large bell closed.

Volume Below Tuyeres: The cubical content between horizontal planes through the centerline of the tap hole and the centerline of the tuyeres.

Working Volume: The cubical content between a horizontal plane 1.8 metres (6 feet) below the large bell closed and a plane through the centerline of the tuyeres.

Volume Above 1.8 metre (6-Foot) Stock Level: The cubical content between a horizontal plane 1.8 metres (6 feet) below the large bell closed and the bottom of the large bell closed.

Total Volume of Furnace: The cubical content between a horizontal plane at the centerline of the tapping hole and the bottom of the large bell closed.

PARTS OF A BLAST-FURNACE PLANT

A schematic cross-section of a typical American blast-furnace plant is shown in Figure 15—3. Items A through C are the equipment for assembling the raw materials and placing them in the **stockhouse bins** from which they are proportioned out to the blast furnace. The charge materials are weighed either by a **scale car** or **weigh hoppers** fed by conveyor belts. The charge is then conveyed to the top of the furnace by **skip cars** or, in a few new installations, by conveyer belts. At the top of the furnace is a **receiving hopper** which feeds a **rotating distributor** to help get uniform distribution of materials in the furnace stack. There are normally two conically shaped **bells** that provide a gas-tight **lock hopper** to prevent gas from escaping while the furnace is being charged. The upper portion of the furnace (J-8) where the burden is preheated is called the **stack,** the inverted conical section where melting starts (J-9) is called the **bosh,** and the lower portion where the metal and slag collect (J-12) is called the **hearth.** The metal is removed from the **iron notch** into **transfer ladles** and the slag is either drawn off into **slag pots,** drawn off into a **dry pit** where it is allowed to solidify, or granulated with a stream of water and flushed into a **wet pit**. The gas leaves the top of the furnace through the **uptakes** (J-2), is conducted through a **dust catcher** (0), and a **cleaning system** (R-1, 2 and 3). Clean gas is burned in the **stoves** (U) to preheat the incoming blast air which is conducted by the **hot-blast line** (Q) to the **bustle pipe** (J-13) circling the furnace. The air enters the furnace through water-cooled openings called **tuyeres** (J-10) located at the top of the hearth.

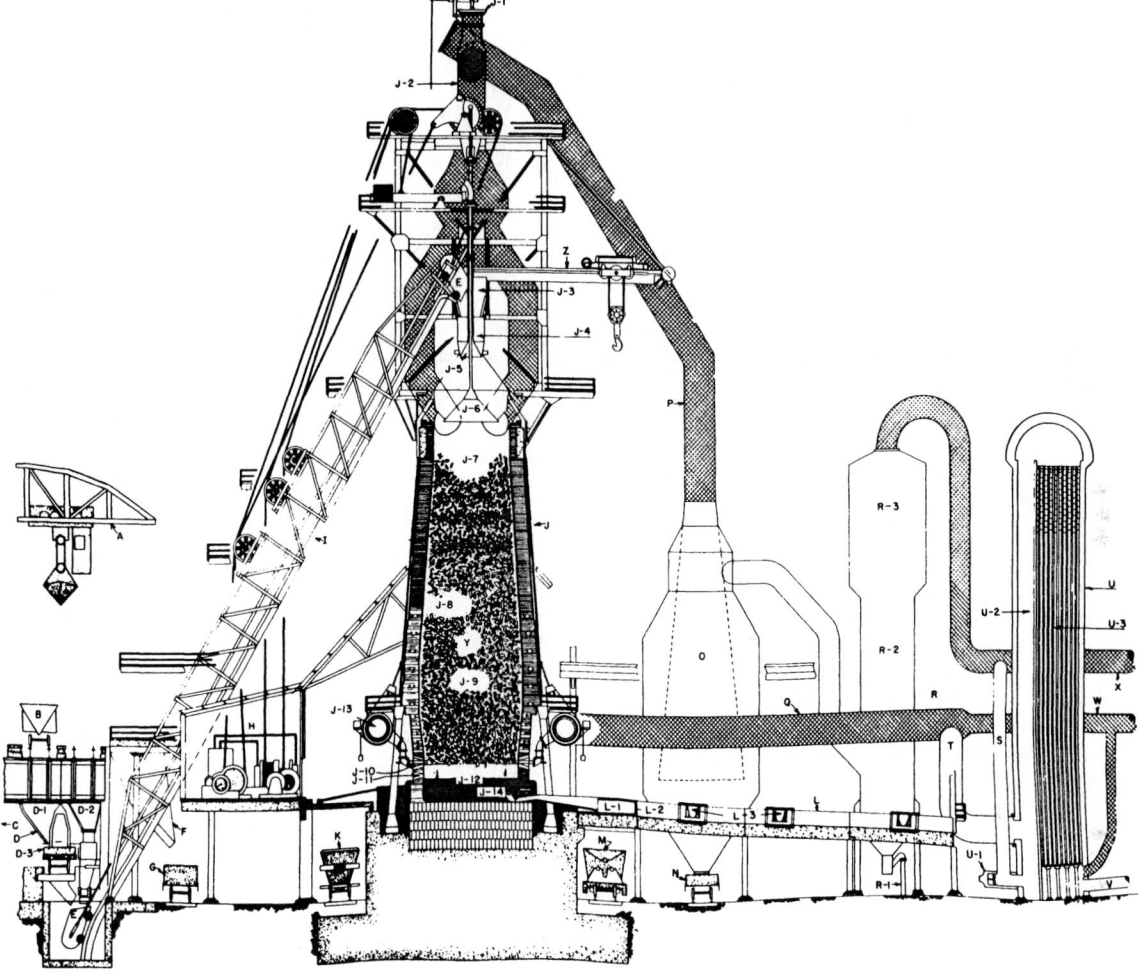

Fig. **15**—3. Idealized cross-section of a typical modern blast-furnace. Details may vary from plant to plant.

Legend

A. Ore bridge
B. Ore transfer car
C. Ore storage yard
D. Stockhouse
    D-1 Ore and limestone bins
    D-2 Coke bin
    D-3 Scale car
E. Skip
F. Coke dust recovery chute
G. Freight car
H. Skip and bell hoist
I. Skip bridge
J. Blast furnace
    J-1 Bleeder valve
    J-2 Gas uptake
    J-3 Receiving hopper
    J-4 Distributor

J-5 Small bell
J-6 Large bell
J-7 Stock line
J-8 Stack
J-9 Bosh
J-10 Tuyeres
J-11 Slag notch
J-12 Hearth
J-13 Bustle pipe
J-14 Iron notch
K. Slag ladle
L. Cast house
    L-1 Iron trough
    L-2 Slag skimmer
    L-3 Iron runner
M. Hot-metal ladle
N. Flue dust car
O. Dust catcher

P. Downcomer
Q. Hot blast line to furnace
R. Gas washer
    R-1 Sludge line to thickener
    R-2 Spray washer
    R-3 Electrical precipitator
S. Gas offtake to stove burner
T. Hot blast connection from stove
U. Stove
    U-1 Gas burner
    U-2 Combustion chamber
    U-3 Checker chamber
V. Exhaust gas line to stack
W. Cold blast line from blower
X. Surplus gas line
Y. Stock—Iron ore, coke, limestone
Z. Jib boom crane

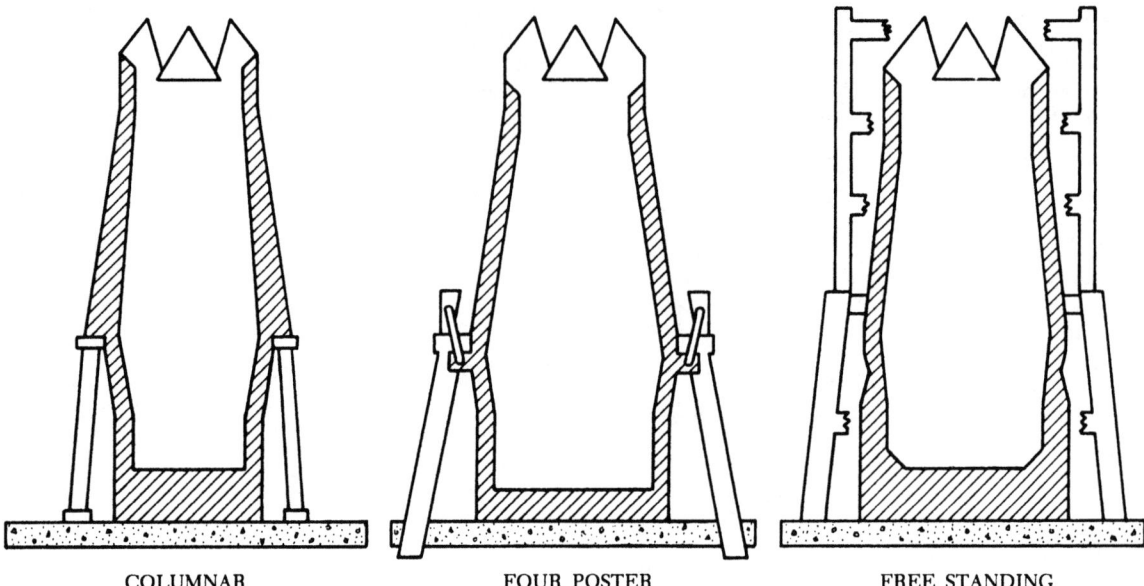

| COLUMNAR | FOUR POSTER | FREE STANDING |

FIG. 15—4. Blast-furnace support structures. (After Anderson and Tanner: see bibliography.)

FOUNDATION

The weight of a large modern blast furnace is in excess of 10 000 metric tons (11 000 net tons) and at any one time it might contain as much as 5000 metric tons (5500 net tons) additional weight in product and burden material. Consequently, the foundation must be capable of supporting this type of load. H-beam piles usually are driven down into bedrock and on top of these is placed a reinforced-concrete pad approximately 3 to 4 metres (about 9 to 13 feet) thick.

SUPPORT STRUCTURE

For most of the modern blast furnaces in the United States, the hearth rests directly on the foundation and the bosh is supported by the bosh shell. The stack of the furnace is supported by a **mantle ring** which is independently supported by columns (usually 8 to 10) that rest on the foundation. The columnar supported blast furnace is shown schematically in Figure 15—4. The stack of the furnace is completely encased in a steel shell made from 25- to 50-mm (1- to 2-inch) thick plates butt welded together. In some of the older furnaces, the steel plates are overlapped in "shingle" fashion and riveted together. With the columnar supported blast furnace, all of the top charging gear, the top-gas uptakes and the top of the skip incline are supported from the steel shell of the stack.

Recently, a few new blast furnaces have been built in the United States with a modification of this general design. This is called the 4-poster construction and is also shown in Figure 15—4. In this type of construction, the hearth and bosh are supported directly from the foundation, but the mantle which supports the stack is suspended by tension members from a heavy box girder that encircles the furnace. The box girder is supported from the foundation by four tubular steel posts filled with concrete. The design is such that the box girder will remain in position even if one of the posts is

completely destroyed.

Many foreign blast furnaces use a free-standing type of construction which is also shown in Figure 15—4. In this design, the hearth rests on the foundation and the hearth shell supports the bosh shell which must be a continuously welded structure, although it will be equipped with openings for various types of coolers. The steel shell of the stack is then supported mainly by the steel bosh shell. A square frame steel structure is built around the furnace, and this supports some of the weight of the stack and of the weight of the top charging gear and uptakes. It also supports the top of the skip incline or charging conveyor-belt gallery, whichever is used.

More information about blast-furnace support structures can be obtained from the Anderson and Tanner article listed in the bibliography at the end of this chapter.

HEARTH

In a large number of blast furnaces, the construction of the hearth is similar to that shown in Figure 15—5A. First, a leveling course of brick is laid on the concrete foundation pad so that the bottom of the hearth will be perfectly flat, and the ceramic block can be fitted together so tightly that an absolute minimum of mortar will be required to seal the joints between them. The number of rows of bottom block may vary, but in most blast furnaces there are between 8 and 10 rows. Generally, high-duty fireclay brick is used; although, in some hearths, there are a few rows of Cone 23 brick or dense 60 per cent alumina brick immediately below the top course of bottom block.

The bottom block are enclosed in a ring wall made of brick shapes, and this ring wall extends vertically all the way up to the hearth line so that it forms the crucible where the liquid hot metal is collected. The ring wall is made of either ceramic or carbon brick and is usually 575 to 900 mm (22½ to 36 inches) thick. The

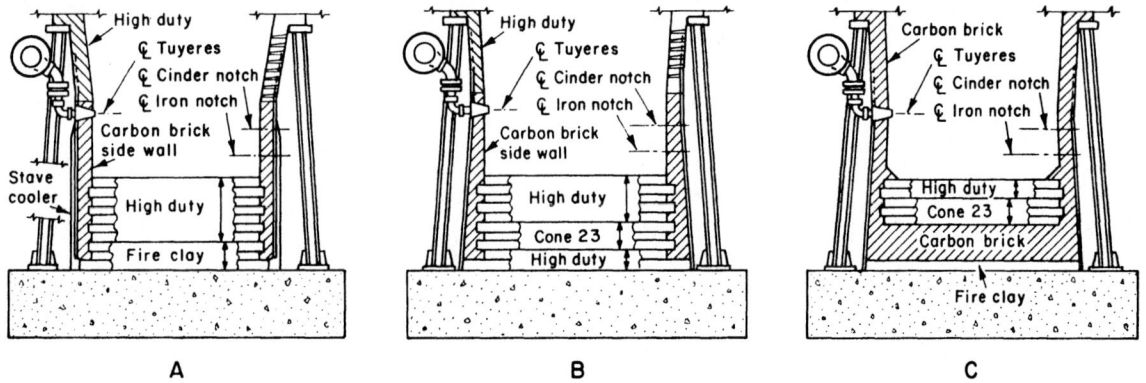

FIG. 15—5. Types of blast-furnace hearth construction. (After Saunders: see bibliography.)

most common construction is with a ring wall of anthracite-carbon brick.

The lower hearth wall and upper hearth wall, up to the cinder notch, are usually cooled by the use of cast-iron segments with internal hair-pin cooling pipes called **hearth staves**. A mortar, generally carbonaceous in make up, is used to assure good contact between the hearth wall and the staves with the elimination of any air gaps. In the vicinity of the iron notch and the cinder notch, the staves are generally designed with a removable section as a means for making replacements if a break-out or a burn-through should cut the stave in that location.

The entire hearth is encased in a steel **hearth jacket**, 25 to 50 mm (1 to 2 inches) thick. A packing material is used to fill the space between the hearth staves and the hearth jacket. In many blast furnaces, the carbon-brick ring wall extends all the way to the hearth jacket, and carbonaceous mortar is used to make good thermal contact between the carbon brick and the steel hearth jacket as shown in Figure 15—5B. Instead of using cooling staves inside the jacket, water sprays are used on the outside of the jacket to provide the necessary cooling. There is generally a slope to the hearth-jacket wall for better adherence of the water to the steel shell. Another modification of this cooling concept is the use of channels welded to the outside of the hearth jacket with a continuous flow of cooling water passing through the channels.

In recent years, another type of cooling has been introduced which originated in Russia and has gained wide-spread acceptance in foreign countries, particularly in Japan. In this system, the cooling staves are relatively short and instead of using hair-pin-shaped cooling pipes where the water enters the top of one leg and is discharged from the top of the other, the cooling pipes are straight and vertical so that the cooling medium enters near the bottom of the stave and is discharged from the top. With this system, numerous holes are required in the hearth jacket for the entrance and exit piping. The staves, which are only about 1 meter (3 feet) in length, are arranged in rows one above the other and in some furnaces these rows extend all the way up to the middle of the stack. The water leaving the top of the staves in the lower rows enters the bottom of the staves directly above. In this way, the evaporation of the water and the generation

of steam in the cooling pipes provides the energy to make the water circulate from the bottom to the top of the cooling system. A more thorough explanation of this evaporative cooling system can be found in the Berczynski article listed in the bibliography at the end of this chapter.

In several blast furnaces, carbon refractories have been used throughout the entire bottom portion of the hearth, as shown in Figure 15—5C. In such instances, large carbon beams, 3 to 6 metres (10 to 20 feet) in length, are keyed into position to keep them from floating. One leveling course of ceramic brick is generally used below the carbon and several courses of high-duty or super-duty ceramic brick are used above it. When the hearth diameter is small, the thermal conductivity of the carbon is adequate to remove sufficient heat from the center of the hearth to the hearth-jacket cooling system. However, in larger blast furnaces, with hearth diameters in excess of 8 metres (26½ feet), additional hearth cooling is provided by passing air or water through stainless-steel pipes imbedded in the carbon. A complete review of modern hearth construction is in the Saunders article listed in the bibliography at the end of this chapter.

About 1 metre (3 feet) above the top course of bottom blocks is the **iron notch**, which is the opening for removing the liquid iron from the blast furnace. Figure 15—6 shows the details of construction for a typical iron notch. The opening on the inside of the furnace is larger than that on the outside so that the tapping hole can be drilled horizontally or at different angles. Although most of the older furnaces had only one iron notch, modern large blast furnaces have two or three and a few very large ones have four.

Further up the hearth wall about 1 or 1.5 metres (3 to 5 feet) above the iron notch level is the **cinder notch** or **slag notch**. This originally was used to withdraw slag from the blast furnace between iron casts. Formerly, before burdens were beneficiated, slag volumes were quite large and it was a great advantage to remove slag, which is lighter than iron and floats on top of it, before casting to decrease the undesirable high liquid level in the hearth and to avoid having to remove a large volume of slag through the tapping hole. In those days, fireclay refractories were used in the tapping hole and the slag eroded it quite rapidly so that it was difficult to plug the tapping hole after casting if too much slag was

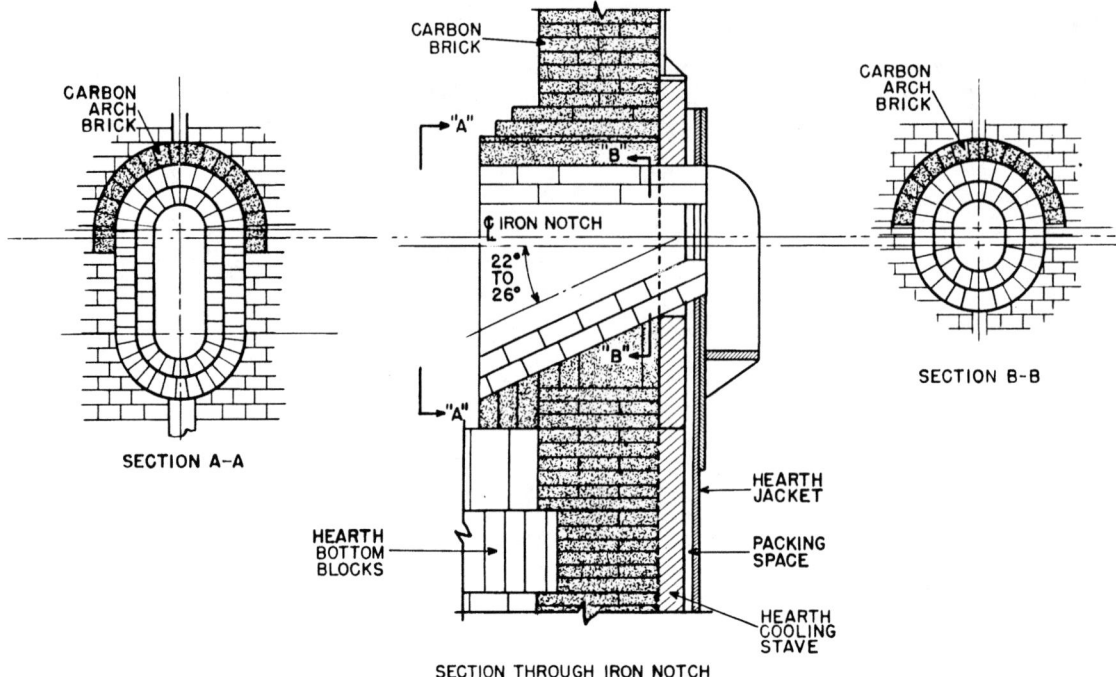

FIG. 15—6. Schematic section through iron notch of a blast furnace. Refractories other than carbon arch brick are fireclay refractories. The iron notch may also be constructed of all carbon brick or block.

removed through it. Today, most blast furnaces have carbonaceous clays for the tapping hole which are eroded much less by slag. Consequently, in many modern furnaces, the cinder notch is seldom used. Because liquid slag does not dissolve copper as liquid iron does, the slag is withdrawn through a water-cooled copper member called a **monkey**. The slag-notch assembly, shown in Figure 15—7, consists of three water-cooled members, the **monkey cooler,** the **intermediate cooler** and the **monkey.**

When a blast furnace has been in operation for an extended period of time, the portion near the center of the hearth that cannot be cooled adequately gradually erodes away, leaving a dish-shaped cavity in the hearth blocks that may be as much as 2 to 3 metres (6½ to 10 feet) deep. This cavity fills with liquid metal which remains in the furnace hearth during tapping because its elevation is well below that of the iron notch. This metal is known as a **salamander** (or **bear**).

The upper portion of the hearth, usually above the hearth-cooling staves but below the hearth line, is known as the **tuyere breast**. It is enclosed in a steel shell known as the **tuyere jacket** or **breast plate**. This jacket contains reinforced circular openings where the **tuyere coolers** are inserted.

When the hearth walls are made of ceramic brick, there are usually one or two rows of horizontal **cooling plates** inserted in the brickwork below the tuyere level and one above. When the hearth walls are made of carbon, the cooling is generally external with jacketed cooling channels on the outside of the shell. With the evaporative type of cooling system, internal staves are used in the tuyere breast between the tuyere coolers.

Figure 15—8 shows the arrangement of the tuyere-

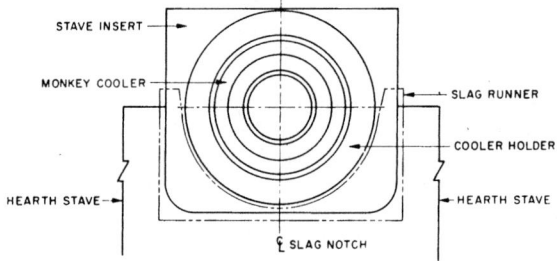

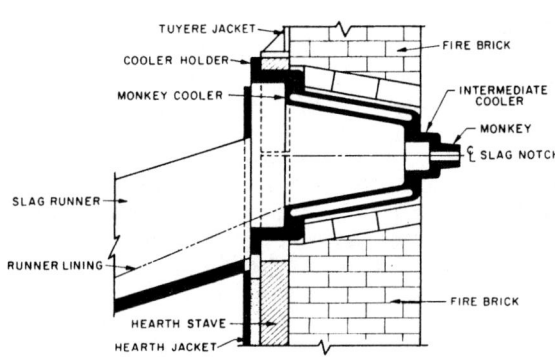

FIG. 15—7. (Below) Section through slag notch of a blast furnace. (Above) Developed view of slag notch.

cooler holder which fits in the opening of the steel shell, the tuyere cooler that fits into the holder and the tuyere that fits inside the cooler. The surfaces where

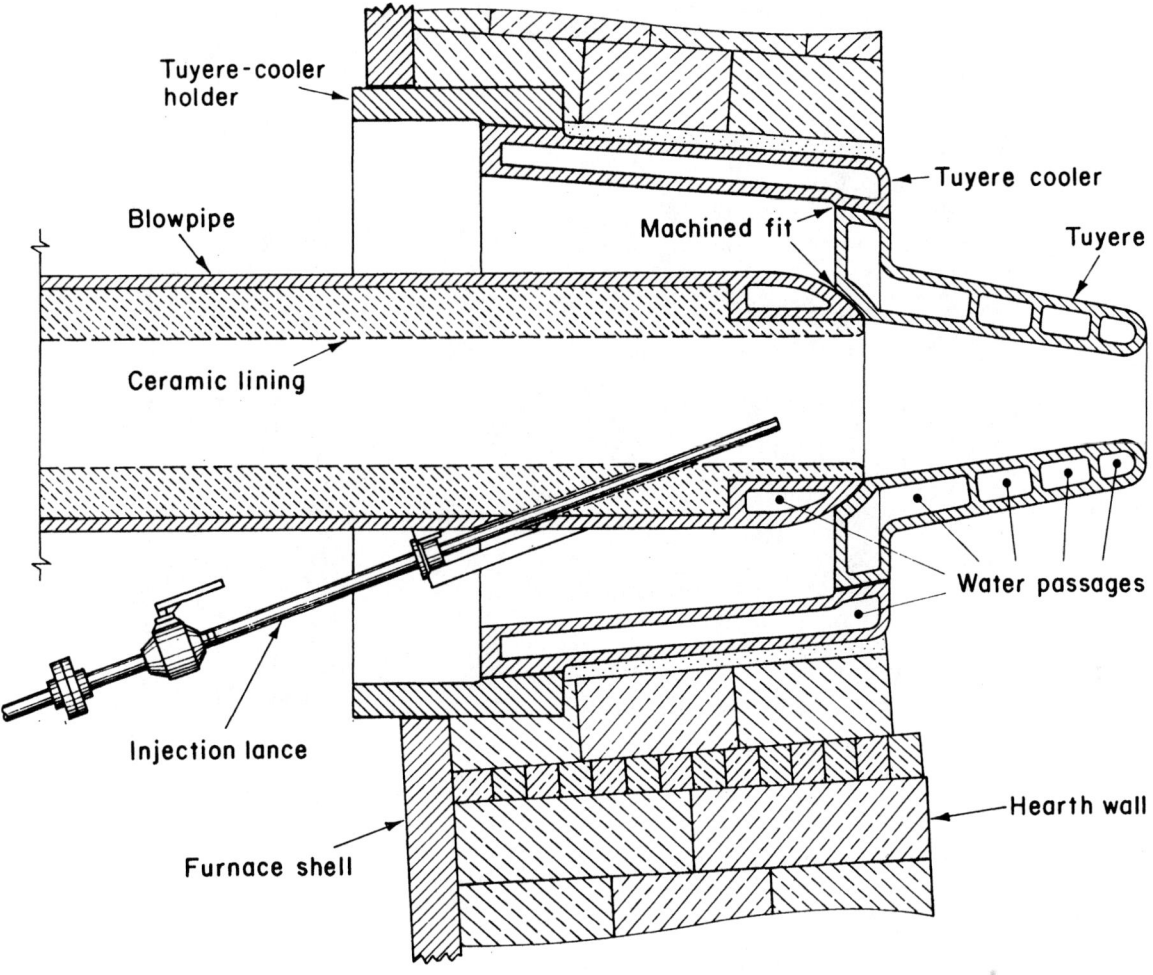

FIG. 15—8. Tuyere and blowpipe assembly.

the tuyere and the cooler contact each other are ma-chined to give an air-tight fit. The **blowpipe** which con-nects the hot-blast system to the tuyere fits into a ma-chined spherical seat at the base of the tuyere. The tuyere cooler and the tuyere are water-cooled and on many modern blast furnaces utilizing hot-blast temper-atures near 1100°C (2012°F), there are baffles in the tuyere water passages to keep the water velocity above 20 metres per second (about 3940 fpm) to improve the rate of heat transfer. In the example shown in Figure 15—8, the nose of the blowpipe is also water-cooled, although in most of the older furnaces this is not the case. The **fuel injection lance** enters through the wall of the blowpipe and usually discharges the fuel slightly off the centerline and about 50 mm (2 inches) back from the nose of the blowpipe.

The blowpipe is held tightly against the tuyere by tension in the **bridle rod** which connects the tuyere stock to the hearth jacket as shown in Figure 15—9. The **bridle spring** on the end of the bridle rod allows limited motion as the blowpipe expands and contracts with changes in hot-blast temperature. The blowpipe itself is an alloy-steel tube lined with refractory mate-rial to prevent the metal from becoming too hot. In

former days, when hot-blast temperature did not ex-ceed 650°C (about 1200°F), unlined cast-iron blowpipes were used, but these are not suitable for the high hot-blast temperatures used today.

At the back of the tuyere stock on the centerline of the blowpipe and tuyere is a small opening through which a rod can be inserted for cleaning material out of the blowpipe. The opening is closed by a cap that can be opened when necessary but is gas-tight when closed. In this cap, called a **tuyere cap** or **wicket** there is a glass-covered **peep sight** that permits the operator to inspect the interior of the furnace directly in front of the tu-yere. The upper part of the stock is connected by a swivel joint to the refractory-lined nozzle of the goose-neck to which it is clamped by lugs and keys that fit into seats of hanging bars. Each gooseneck in turn is connected by flanges and bolts to a neck extending radially from the inside diameter of the **bustle pipe**. The bustle pipe is a large, circular, refractory-lined and insulated pipe that encircles the furnace at above man-tle level and distributes the heated blast from the hot-blast main to each tuyere connection. The general ar-rangement of the bustle pipe, tuyere stock and blowpipe can be seen in Figures 15—9, and 15—10.

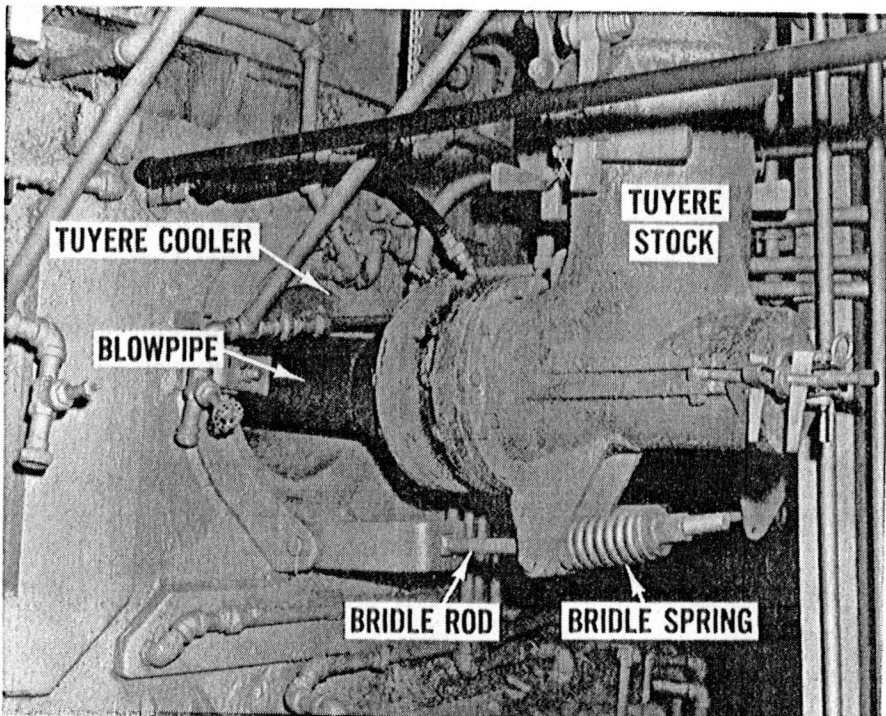

FIG. 15—9. Arrangement of tuyere stock, bridle and blowpipe.

BOSH

The bosh of the blast furnace is the region just above the tuyere breast and can be described as the frustum of an inverted cone that connects the tuyere breast to the level of maximum diameter of the stack. In the blast furnaces of 100 or more years ago, the bosh angle was very shallow so that the bosh supported much of the weight of the burden material and had only enough slope to allow the melted slag and iron to flow down into the hearth. On modern blast furnaces, however, the bosh angle is much steeper and is usually about 80 degrees. Many of the blast furnaces in the United States have bosh angles between 79 and 82 degrees, but the most recently constructed ones tend to have smaller bosh angles. The article by M. Shimizu listed in the bibliography at the end of this chapter states that the ideal bosh angle for a 12 000 metric-ton-per-day (13-200 net-ton-per-day) blast furnace is 76 degrees. This angle is reported to give smoother and more uniform movement of the burden material.

In many of the blast furnaces in the United States, the bosh is made up of a series of steel rings, called **bosh bands,** that are lined with ceramic brick. Each successive band is larger in diameter than the one beneath it so that the inner surface of the lining conforms to the desired bosh angle. This type of construction is shown in Figure 15—10. The lining is about 650 to 700 mm (26 to 28 inches) thick and is cooled by wedge-shaped water-cooled copper plates fitted through the steel bosh bands. The spacing of these plates is customarily on 450 to 600 mm (18 to 24 inch) centers vertically and 1 to 1.2 metres (3¼ to 4 feet) centers horizontally as measured at the periphery of the bands. After the fur-

nace has been in operation for a few years, portions of the ceramic brick are eroded away by the action of slag and iron oxide, but the cooling effect of the bosh plates causes slag-like material to freeze and adhere to the remaining brick. This action tends to preserve the bosh lines to a certain extent.

In many of the newer blast furnaces, particularly those using elevated top pressure, the bosh is enclosed in a solid steel shell. When a ceramic lining is used in a steel bosh shell, the copper cooling plates are fitted into

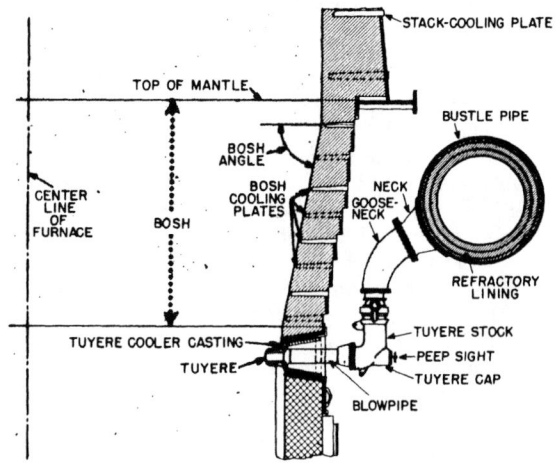

FIG. 15—10. Section through the bosh and upper part of the hearth wall of a large blast furnace, showing principle of construction as well as details of connection between the tuyere and bustle pipe.

the lining through openings in the shell and are held in place by cast-iron boxes fastened to the shell. However, in most of the newer furnaces, the solid-steel bosh shell is lined with carbon brick shapes. The lining is about 425 to 550 mm (15 to 22½ inches) thick and a carbonaceous material of high thermal conductivity is placed between the brick and the steel bosh shell. With this type of lining, there are no openings in the bosh shell, and cooling is accomplished with a cascade of water sprayed onto the outside of the shell or with water flowing through channels welded to the outside surface of the shell. Figure 15—11 shows the arrangement of a blast furnace with a carbon lining in the hearth and bosh. The lower hearth wall is cooled by a stave cooler inside the hearth shell and the upper hearth wall and bosh are cooled by channel coolers on the outside of the shell.

In a few American blast furnaces and in a large number of foreign ones, a carbon bosh lining is used, but the cooling is accomplished by specially shaped stave coolers placed inside the shell in such a way that they cover the entire inner surface of the bosh shell. In these instances, it is necessary to have small openings through the shell for the inlet and outlet water piping; however, suitable closures can be made around the piping to prevent gas leakage at very high bosh-gas pressures. The staves can be of the circulating cold-water type or the Russian evaporative type described in connection with the cooling of the hearth walls. In either case, the inside surface of the stave is protected by a refractory lining of ceramic or carbon brick. The general arrangement of stave coolers inside the bosh is shown in Figure 15—12.

It is believed that after several months of operation the original lining of the inner surface of the bosh staves is almost completely replaced by a slag-like material that freezes on the staves.

STACK

In most blast furnaces, the widest dimension of the inside of the lining is at the top of the bosh, and there is usually a vertical straight section just above the bosh with the same diameter as the top of the bosh. This section is referred to as the **belt** or **belly,** and above the belt the lining of the stack tapers inward until, at the bend line, the diameter is the same as that at the stockline. In some blast furnaces, the sloping angle of the bosh continues up into the stack for about 1 meter (3¼ feet) so that the widest dimension of the inside of the lining is about 1 metre (3¼ feet) above the top of the bosh. In such cases, there is usually no vertical straight section.

The refractory lining of the stack is supported at the bottom by a steel plate, called the **mantle plate.** In many cases, there are horizontal stave coolers at the top of the bosh in back of the brickwork, to prevent overheating of the mantle plate and the lining in that region. Figure 15—13 shows the lines of one of United States Steel Corporation's most modern blast furnaces, Fairfield No. 8. The stack lining is thickest right at the mantle. Above the straight section the brick lining, which is 914 mm (3 feet) thick, is corbeled inward to produce the inwall batter. The brick are supported in part by horizontal water-cooler copper plates placed in

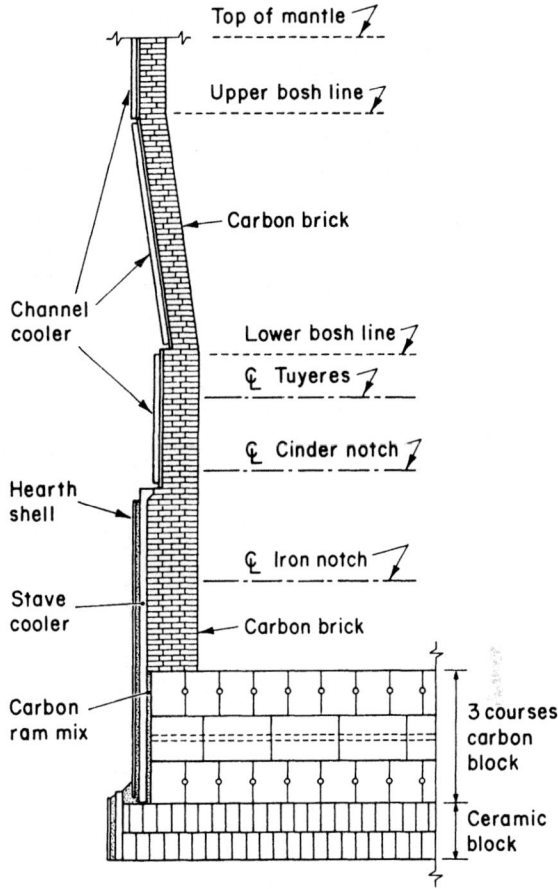

FIG. 15—11. Simplified section of a blast furnace with a carbon lining in the hearth and bosh.

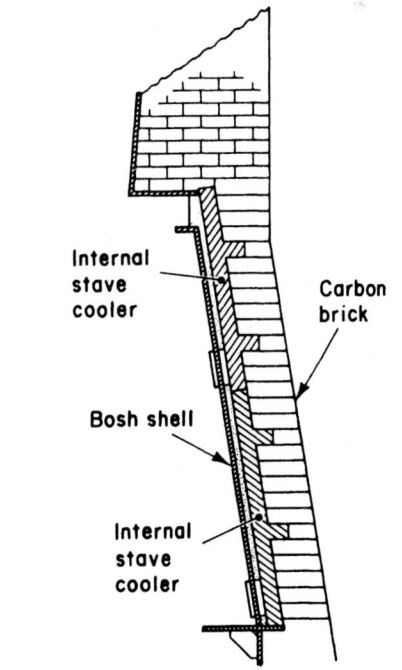

FIG. 15—12. Schematic section of carbon bosh with internal stave coolers.

Dimension	Designation	Measurement	
Working volume		2195.1 m³	77 520 ft³
Volume below tuyeres		341.6 m³	12 064 ft³
Total volume		2536.7 m³	89 584 ft³
Hearth area	a	74.7 m²	804 ft²
Hearth dia.	b	9754 mm	32'- 0"
Straight section dia.	c	11 735 mm	38'- 6"
Stockline dia.	d	8306 mm	27'- 3"
Throat dia.	e	3135 mm	10'- 3 7/16"
Overhall ht. (Centerline iron notch to top ring)	f	35 281 mm	115'- 9"
Working ht. (Centerline tuyeres to stockline)	g	25 908 mm	85'- 0
Top ring thickness	h	203 mm	0'- 8"
Top cone ht.	i	3396 mm	11'- 1 11/16"
Top ring to stockline	j	4801 mm	15'- 9"
Protective brick ht.	k	3810 mm	12'- 6"
Stack ht.	l	17 983 mm	59'- 0
Straight section ht.	m	3048 mm	10'- 0
Bosh ht.	n	4267 mm	14'- 0
Crucible ht.	o	6401 mm	21'- 0
Centerline notch to hearth floor	p	1219 mm	4'- 0
Bottom block depth	q	2743 mm	9'- 0
Lumnite thickness	r	229 mm	0'- 9"
Bottom block dia.	s	10 973 mm	36'- 0
Steel shell ID	t	11 125 mm	36'- 6"
Stack lining brick	u	914 mm	3'- 0
Castable packing	v	76 mm	0'- 3"
Bosh lining	w	457 mm	1'- 6"
Hearth wall		686 mm	2'- 3"
Bosh angle (theta)	θ	1.3449 rad	76°55'-51"
Stack angle (phi)	φ	1.4760 rad	84°34'-12"
Top cone angle (psi)	ψ	0.7850 rad	45°

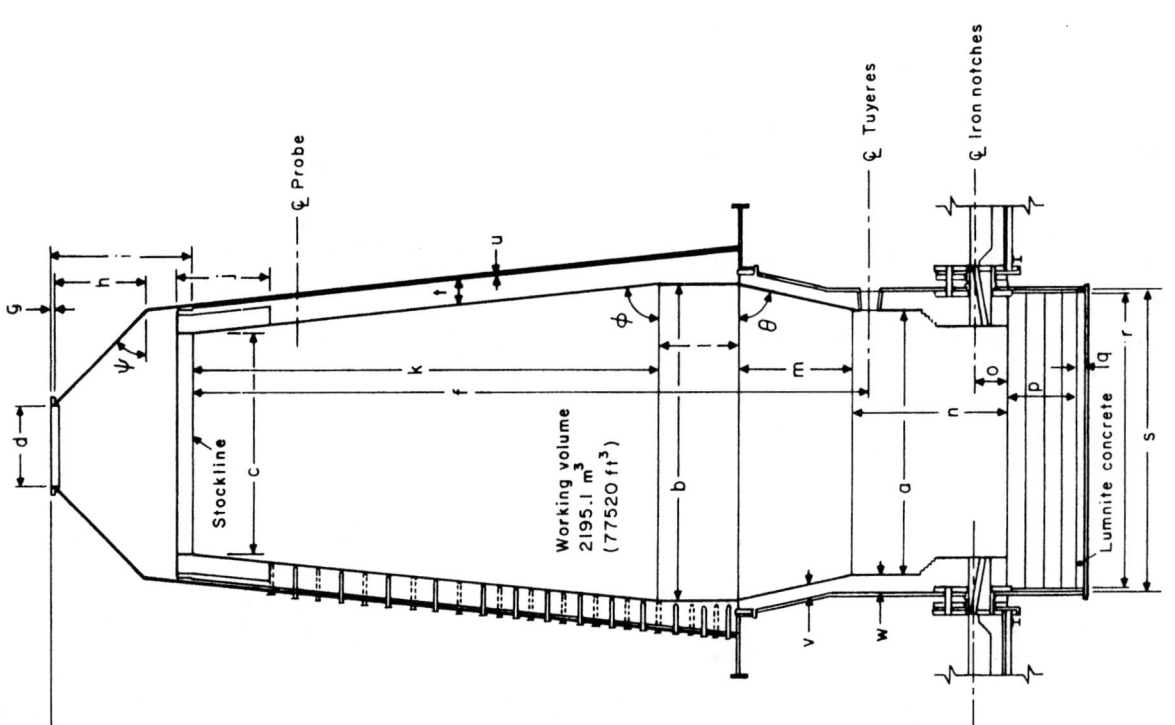

FIG. 15—13. Lines of United States Steel Corporation's Fairfield No. 8 blast furnace.   The furnace has 26 tuyeres, 2 iron notches and 1 cinder notch.

rows around the stack. The first three rows are spaced vertically 460 mm (1½ feet) apart, measuring from centerline to centerline of cooling plates. For the next 12 rows, the spacing is 610 mm (2 feet) and for the top 9 rows the spacing is 914 mm (3 feet). All in all, there are 719 stack plates. The cooling-water requirement for these is 20 060 litres per minute (5300 gpm). The entire cooling-water requirement for this furnace is shown in more detail in Table 15—IV.

After a blast furnace has been in operation for several years, the inner face of the stack lining is usually worn back significantly, particularly just above the bosh, and it is not uncommon to have to replace the stack lining before the hearth lining is completely worn out. In some instances, the working lines of the inwall have been restored to the original dimensions without completely replacing all of the old brickwork by removing the deteriorated portion of the brick and then installing a special high-purity calcium-aluminate (castable) cement over the remaining brick by using a hydraulic-gun placement technique called "guniting." A "gunite" repair of this kind may extend the life of the stack lining one or a little more than one year.

The stockline, which is located near the top of the stack, is generally protected from the impact of the charged material by steel armor. The armor is often in the form of L-shaped or Z-shaped sections, either cast or fabricated, and is placed between the rows of brick with a vertical section covering the face of the brick. In some blast furnaces, the armor is in the form of a steel cylinder made up of curved sections and is hung by chains from inside the top dome of the furnace in such a way that it is directly in front of the brick. The charged material falling from the large bell strikes the inside surface of the armor instead of the surface of the brick. Figure 15—14 shows the arrangement of imbedded and hanging stockline armor. In some of the very modern furnaces, the stockline armor is attached to push rods so that it can be moved to various positions while the burden is dumped into the furnaces. In this way, it not only protects the stockline refractories but also helps to position the charged materials within the furnace.

TOP

In early times, the tops of blast furnaces were open, and the gas from the furnace was allowed to escape into the air and burn. This made the charging of the

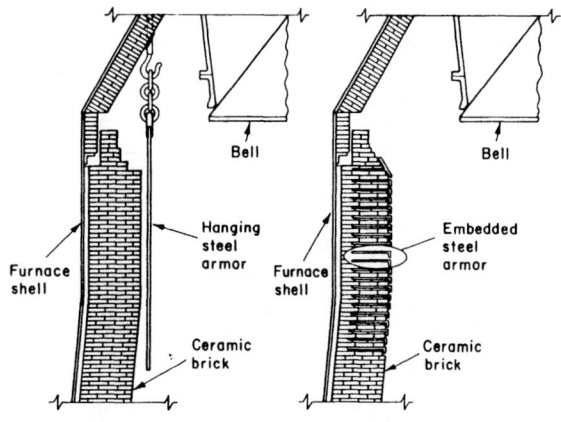

a) HANGING STEEL ARMOR   b) EMBEDDED STEEL ARMOR
   (schematic arrangement)   (schematic arrangement)

FIG. 15—14. Schematic arrangement of (above) hanging steel armor and (below) imbedded steel armor.

furnace somewhat hazardous, and it was also very wasteful of energy. The first attempts to recover the heat from the off gas were made in 1829, when tubes were placed around the inwalls of the furnace and the blast air was passed through these tubes to absorb some of the heat before it was blown into the furnace through the tuyeres. In 1845, a plan was devised for using a separate stove to transfer the heat from the top gas to the air blast. This was accomplished by using the suction created by a chimney attached to the stove to draw the hot, burning gas through the stove. At first, the blast-furnace gas was drawn out of the furnace through openings just below the stockline so that the top could be left open for charging. However, in 1850, it was found that the top could be enclosed so that the gas could be drawn off above the stock level, and a bell and hopper arrangement could be used for charging the furnace. While the furnace was not being charged, the bell kept the bottom of the hopper closed so that gas could not escape. The charge materials were placed in the hopper while the bell was closed and then it was opened only long enough for these materials to be dumped into the furnace.

With only a single bell and hopper, large quantities of gas escaped every time the bell was opened; consequently, it was not long before it was realized that by using a second bell and hopper above the first, a gastight space could be provided between the two bells to prevent gas from escaping when the lower bell was opened. The upper bell and hopper did not have to be as large as the lower one because several loads could be deposited through it onto the lower bell and then it could be closed before the lower bell was opened. The operation of this type of closure is shown schematically in Figure 15—15. The raw material is taken to the furnace top by a skip hoist or a conveyor belt and dropped into the upper hopper. With the large bell closed, the small bell is lowered and the charge material is dropped into the large-bell hopper. This procedure is repeated several times and then, with the small bell held closed, the large bell is lowered and the material is

Table 15—IV. Cooling-Water Requirements for Fairfield No. 8 Blast Furnace.

Section of Furnace	Litres per Minute	Gallons per Minute
Top gear	2 044	540
Stack plates	20 060	5 300
Bosh channels	12 188	3 220
Tuyere jacket	5 905	1 560
Tuyeres and tuyere coolers	21 347	5 640
Hearth and subhearth	18 168	4 800
Stove valves	5 375	1 420
Total	85 087	22 480

dumped into the blast furnace without allowing any of the gas to escape.

The bells are made of steel, and the slope of the sides is usually between 50 and 55 degrees. The bells are supported by **bell rods** which are attached to counterbalances through a lever arrangement that restricts their motion to a vertical direction only. The small-bell rod is hollow and the large-bell rod passes through it. Packing material is used between the two bell rods to prevent the escape of gas. In most blast furnaces, the small-bell hopper is rotated through a preprogrammed cycle after each skipful so that the charge is not always deposited in the same location but is distributed more evenly around the circumference of the bell.

The bells are generally hard-surfaced in the area where they are subjected to the most severe wear from the impact of the charge materials. Hard surfacing is also applied to the seating surfaces of the bells and hoppers. This is usually up to 9.5 mm (about ⅜ inch) thick and is deposited with a high-alloy welding rod.

With the use of very high top pressure, particularly in excess of 100 kPa (1.02 kg/cm² or 14.5 psig), it is extremely difficult to maintain a gas-tight seal with the conventional bell-and-hopper arrangement. Consequently, in many modern blast furnaces, the charge is placed on the bells through seal valves in such a manner that the charge material does not come in contact with the valve seat. One such top closure is shown in Figure 15—16. Here the charge materials, from either a skip hoist or a conveyor belt, are discharged onto a rotating chute that distributes them in fairly equal proportions into four separate funnels positioned directly above the four quadrants of the small bell. The four openings in the small-bell hopper through which the

charge material enters the hopper are equipped with seal valves which remain in the open position while material is discharged through the chute. In the open position, the seats of these valves are pulled back out of the path of the charge materials. After the small-bell hopper has received a charge, the seal valves are closed, the small-bell hopper is pressurized and the small bell is opened to dump the material into the large-bell hopper. The small bell is then closed before the small-bell hopper is depressurized for its next load. In this way, the large-bell hopper is always pressurized, so the large bell is not required to make a gas-tight seal.

A more recent innovation in top-charging devices is the Paul Wurth top, in which the charge materials are deposited into hoppers located at the top of the furnace. These can be depressurized for loading either from a conveyor belt or from a skip hoist and repressurized for discharging the material into the furnace. There are at least two hoppers, so that while one is being loaded the other can be discharging into the furnace. A diagram of this top is shown in Figure 15—17. As the charge material enters the furnace, it is directed by a rotating chute to various locations on the top of the stockline. The angle of the chute and the speed of rotation can be regulated to distribute the charge in almost any desired pattern. This type of charging device is described in the Legille and Peters article listed in the bibliography at the end of this chapter.

HOISTING APPLIANCES

Very early blast furnaces were usually located near the side of a hill, and a bridge or trestle spanned the gap between the hill and the top of the furnace so that

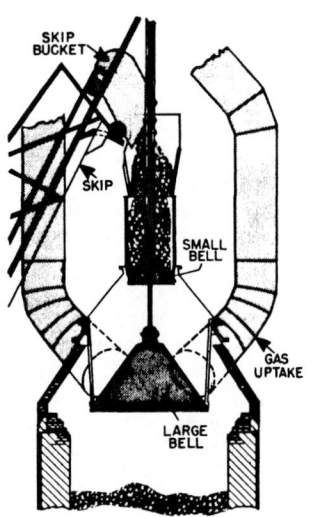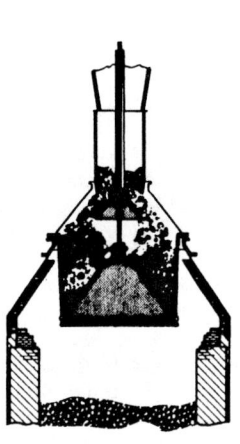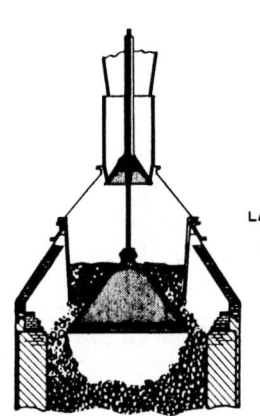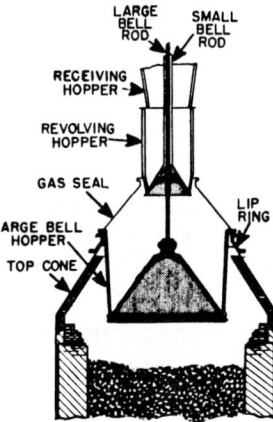

SMALL BELL AND LARGE BELL BOTH CLOSED, SKIP BUCKET TIPPED TO DUMP CHARGE IN HOPPER ABOVE SMALL BELL. GAS FLOWING FROM TOP OF FURNACE THROUGH UPTAKES LOCATED IN DOME (TOP CONE).

LARGE. BELL REMAINS CLOSED WHILE SMALL BELL OPENS TO ADMIT CHARGE TO LARGE BELL HOPPER.

SMALL BELL CLOSED TO PREVENT ESCAPE OF GAS TO ATMOSPHERE AND LARGE BELL OPEN TO ADMIT CHARGE TO THE FURNACE.

BOTH BELLS CLOSED, READY TO REPEAT CHARGING CYCLE. NOTE THAT ROD SUPPORTING LARGE BELL PASSES THROUGH HOLLOW ROD SUPPORTING SMALL BELL, PERMITTING INDEPENDENT OPERATION OF BELLS.

FIG. 15—15. Diagram of progressive steps by which the double bell and hopper permits charging materials into a blast furnace without escape of furnace gases.

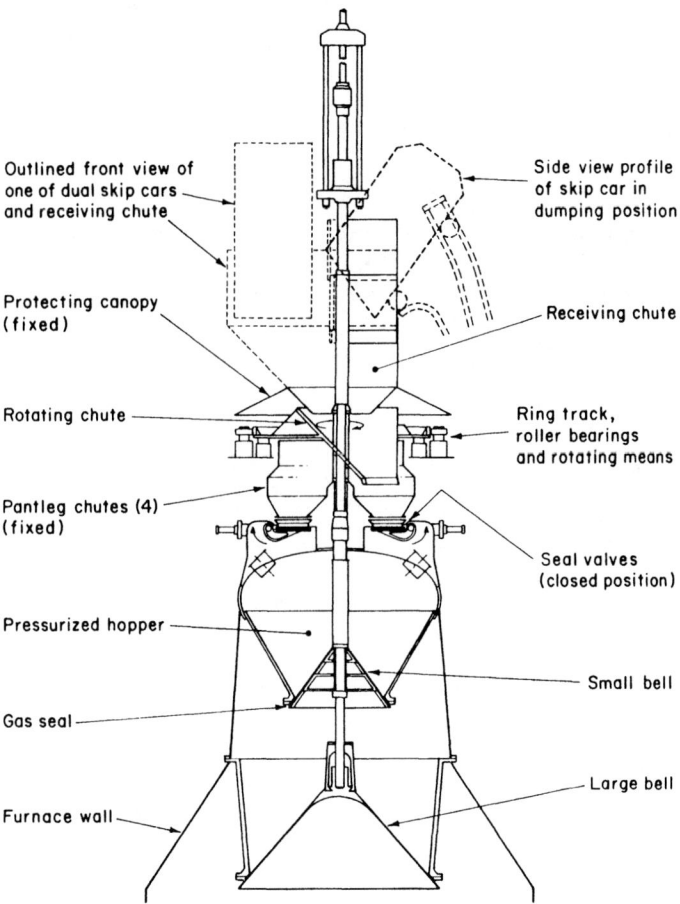

FIG. 15—16. Diagram of blast-furnace top with a rotating chute to distribute charge material in uniform proportions into the four quadrants of the small-bell hopper, and with seal valves that are closed to pressurize the small-bell hopper before it is discharged onto the large bell.

loads of charge materials could be taken up in wheelbarrows and dumped into the charging hole. Later, blast furnaces were charged with a bucket hoist. The bucket was loaded in the stockhouse and hoisted to the furnace top on a carriage traveling on inclined rails. After the bucket was positioned on the furnace top, the bottom of the bucket, which was in the form of a small bell, was lowered to discharge the material into the furnace.

When the bucket was not on the furnace top, a counterweighted gas seal (nicknamed the McKinley hat) prevented gas from escaping. This type of apparatus was known as a **Neeland top.**

Most of the blast furnaces in the United States today are charged by a skip hoist which consists of two skip cars (or skip tubs) running on parallel inclined tracks. The skip cars counterbalance one another so that while one is being loaded in the stockhouse the other is being discharged into the furnace. Figures 15—1, 15—2 and 15—3 show the general arrangement of the skips and the method by which they operate between the stockhouse and the furnace top.

At most of the modern blast furnaces in Japan and at many of the European blast furnaces, a conveyor belt is used to transport the charge materials from the stockhouse to the furnace top. This type of charging is also used at several of the most modern blast furnaces in the United States.

STOCK DISTRIBUTION

One of the significant functions of the blast-furnace charging system is the proper distribution of the stock to assure good contact between the rising gas and the charge materials. With the earliest development of two-skip charging, the charge material formed into two peaks diametrically opposite one another on the small bell. When the material was dumped onto the large bell and then into the furnace, the peaks persisted. The normal segregation of the coarse particles from the fines caused the fines to accumulate directly below the peaks while the coarse particles rolled to the other parts of the stockline. This uniform segregation pattern caused the gas to flow preferentially through the coarser material causing channeling so that the fines did not get sufficient contact with the gas. To counteract this effect, the McKee distributor was developed; this is a rotating small-bell hopper that can be programmed to rotate different amounts for each charge so that the peaks do not always occur in the same loca-

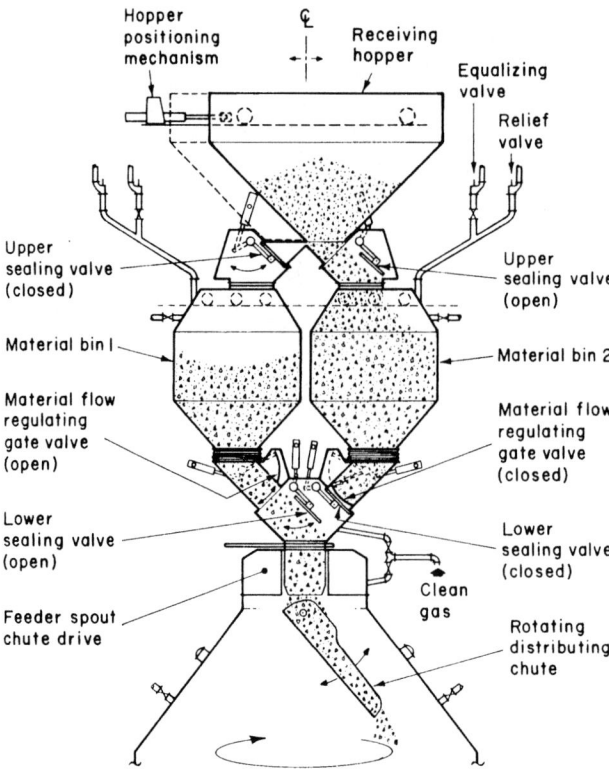

FIG. 15—17. Blast-furnace top with pressurized hoppers and a distributing chute instead of bells. (After Legille and Peters: see bibliography.)

tion.

The McKee distributor which corrected the circumferential segregation pattern did not, however, eliminate the radial segregation which was caused by discharging the material from the large bell close to the periphery of the stockline. Because of the tendency of the coarser material to roll more than the fine material, the coarser material accumulated toward the center and the finer material concentrated near the periphery. To overcome severe segregation of this kind, many model studies were made, and the sequence of charging materials on the large bell was varied to obtain the best pattern that could be obtained with this simple equipment. In the new, very large blast furnaces, it is too difficult to obtain a suitable distribution with just arranging the order of placing different materials on the large bell. Consequently, most large furnaces employ moveable stockline armor, or rotating chutes as in the Paul Wurth top, to accomplish proper control of the positioning of burden material.

TOP OPENINGS

In addition to the opening in the blast-furnace top through which the burden materials are charged, there are generally four openings for large, vertical pipes called **uptakes** through which the top gas is withdrawn. The large diameter and the vertical height of these uptakes prevents excessive amounts of charge material from blowing over into the gas-cleaning system. The uptakes are lined with 115 to 150 mm (about 4½ to 6 inches) of refractory brick or with 75 to 100 mm (about 3 to 4 inches) of monolithic castable cement applied over wire mesh that is anchored to the inner surface of the metal of the uptakes. The tops of the uptakes join together in pairs, and each pair is connected to a large duct called a **downcomer** that descends into the dustcatcher. The ducts connecting the uptakes to the downcomer are called **offtakes**. The total cross-sectional area of the downcomer is usually slightly less than that of the four uptakes.

Each of the pairs of uptakes that are joined together into a single duct is equipped at its uppermost point with a valve called a **bleeder valve.** Some of the offtakes are also equipped with bleeder valves. There are two general types of bleeder valves: those that open out, and those that open in. The open-in valves are operated by air or hydraulic cylinders and their movement is actuated by pressure impulse lines that cause the valves to open to relieve the pressure when it becomes too high. Manual switches are also provided to open these valves when the furnace has to be shut down and it is necessary to create a draft through the furnace stack. The open-out bleeders are usually counterweighted so that they will open without the need of another source of power when the pressure in the furnace significantly exceeds the planned operating pressure. This is to prevent serious damage to the furnace top in case of a bad slip or explosion.

Openings in the top are also required for the **stockline indicators,** which are usually steel rods passing through a small **try hole.** The ends of the rods are weighted to overcome the friction of the packing in the try hole and they are equipped with a broad base so that they will rest on top of the stock. The motion of the rod is transmitted to a recorder in the instrument room and shows the rate of stock descent. There are usually two stockline indicators, positioned 180 degrees apart. Many new stockline measuring devices which make use of nuclear gauges, sonar, infra-red and visible light have been used experimentally, but the development of such instruments has been very difficult because of the adverse conditions in which they must operate.

Several blast furnaces are also equipped with **temperature-sensing probes** that traverse one or two radii of the stockline and show if there are regions of severe gas channeling. Some of these probes operate just above the stockline and some are positioned about 1.5 to 2 metres (about 5 to 7 feet) below the stockline.

SECTION 4

CONSTRUCTION OF BLAST FURNACE AUXILIARIES

This section will discuss the design and operation of the stoves used to preheat air for the blast furnace, off-gas cleaning equipment, equipment for transporting hot metal, methods for slag disposal, receipt and handling of raw materials, and the instrumentation useful in control of the blast-furnace process.

STOVES

Before the blast air is delivered to the blast-furnace tuyeres, it is preheated by passing it through regenerative stoves that are heated primarily by combustion of the blast-furnace off gas. In this way, some of the energy of the off gas that would otherwise have been lost from the process is returned to the blast furnace in the form of sensible heat. This additional thermal energy returned to the blast furnace as heat decreases the amount of fuel that has to be burned for each unit of hot metal and thus improves the efficiency of the process. An additional benefit resulting from the lower fuel requirement is an increase in the hot-metal production rate.

In many modern blast furnaces, the off gas is enriched by the addition of a fuel of much higher calorific value, such as natural gas or coke-oven gas, to obtain even higher hot-blast temperatures, often in excess of 1100°C (2012°F). This decreases the fuel requirement and increases the hot-metal production rate to an even greater extent than is possible when burning off gas alone to heat the stoves.

Most blast furnaces are equipped with three hot-blast stoves, although in a few instances there are four. The stoves are tall, cylindrical steel structures lined with insulation and almost completely filled with checker brick where heat is stored and then transferred to the blast air. Each stove is about as large in diameter as the blast furnace, and the height of the column of checkers is about 1½ times as tall as the working height of the blast furnace. At the newer furnaces, the relation of the stove size to the furnace size is even larger. For example, one typical American blast furnace, which has a hearth diameter of 9.75 meters (32 feet) and a working height of 25.9 metres (85 feet), is equipped with three stoves. Each stove has an inside diameter of 10.36 metres (34 feet) and a checker height of 40 metres (131¼ feet).

Figure 15—18 shows the construction of a conventional two-pass hot-blast stove. In this type of stove, the oval-shaped combustion chamber occupies about 10 per cent of the total cross-sectional area of the stove and extends from the bottom of the stove to within about 4 metres (13 feet) of the top of the stove dome. The combustion chamber is separated by a sturdy brick breast wall from the balance of the stove which is filled with checker brick resting on a steel grid supported by steel columns.

Just inside the steel shell is an insulating lining and this is usually very thick on the side near the combustion chamber. The combustion chamber is completely surrounded by a brick well wall which is lined with super-duty brick containing 50 to 60 per cent of alumina. For very high hot-blast temperatures in excess of 1200°C (2192°F), the entire combustion chamber and the dome are lined with this type of brick: also, the top 8 to 10 metres (26.3 to 32.8 feet) of checkers are superduty brick. However, in many hot-blast stoves where the temperatures are only about 1000°C (1832°F), the upper portions of the combustion chamber, the dome lining and the upper checkers are made of semi-silica refractory.

In erecting the dome lining, arch brick are used and a space is provided between the brick and the dome to allow for expansion of the ring wall from which it is supported. In some stoves, there is an offset in the steelwork at the top of the ring wall so that the dome brick can be supported independently.

Many older hot-blast stoves that were built before gas cleaning was very efficient are equipped with basket-weave checkers with very large flue openings, as large as 140 mm by 140 mm (5½ inches by 5½ inches). However, with better gas-cleaning facilities, it became possible to use checkers with smaller flue openings without as much danger from plugging the flues with dirt. With smaller flues, heat-transfer rates could be increased because the ratio of heating surface to checker weight increased and more checker weight could be installed in the available space. However, with the smaller flue openings, it became very important to lay up the checkers so that the flues matched perfectly. Misaligned flues would increase the pressure drop through the stoves significantly and would prevent effective use of all the heat-storage capacity. Several of

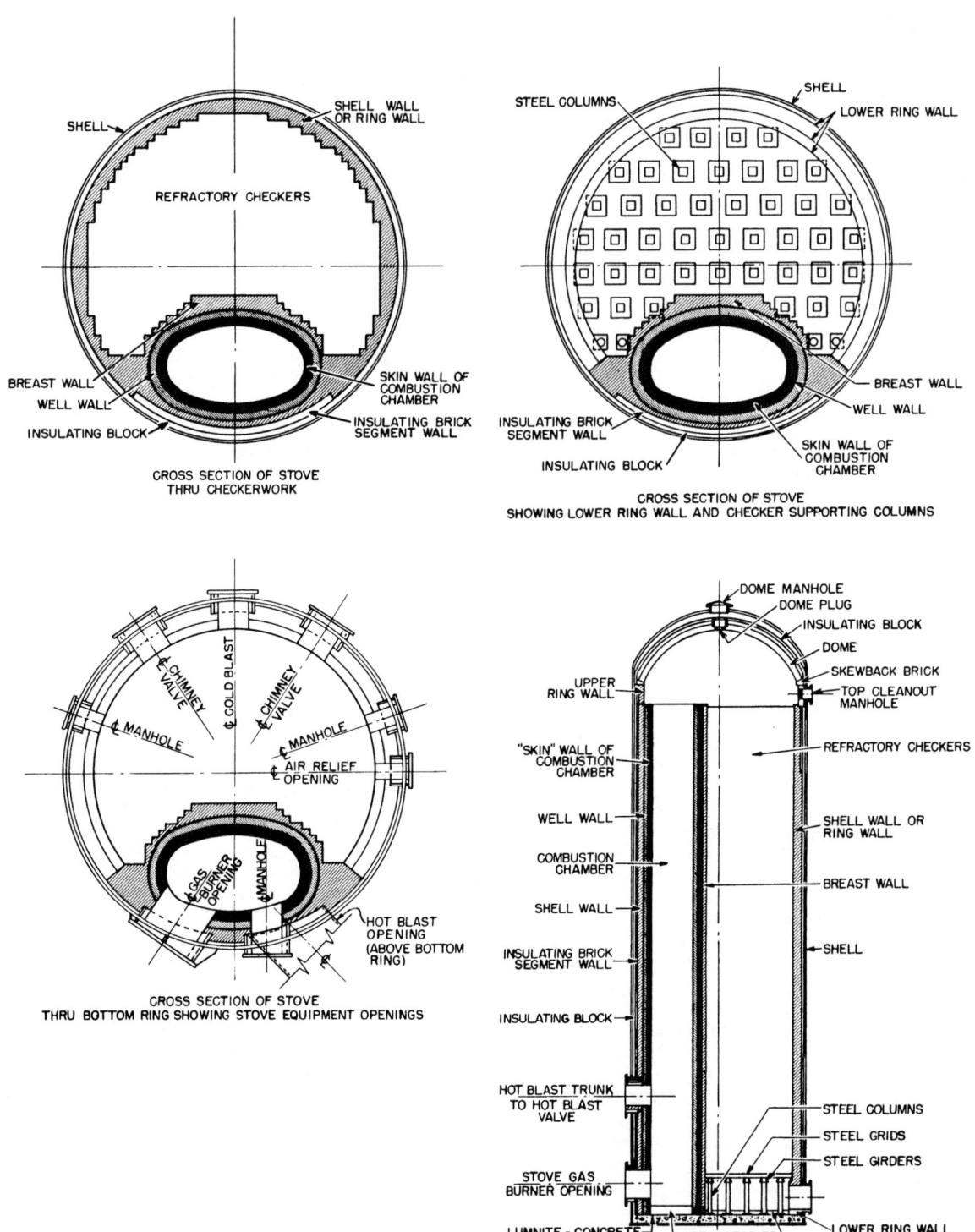

FIG. 15—18. Schematic diagrams (not to scale) showing principles of design and materials of construction for hot-blast stoves of the side-combustion type for blast furnaces. (Left) Vertical section through stove. (Right) Series of schematic sections at different elevations of stove. Top section shows lower ring wall and checker-supporting columns: middle section is through bottom ring, showing openings for stove equipment; bottom section is through the checkerwork (not shown). The spaces marked "refractory checkers" are filled with checkerwork of one of the designs shown in Figure 15—19, to provide continuous vertical flues extending from top to bottom of the checkers.

the types of checkers used presently are shown in Figure 15—19. For most of these checkers except the basket-weave type, the ratio of surface area to weight is 0.020 to 0.023 square metre per kilogram (0.10 to 0.11 square foot per pound) and the heat capacity to surface area ratio is 23.8 to 25.9 kilojoules per square metre or 5.7 to 6.2 kilogram-calories per square metre (2.1 to 2.3 Btu per square foot).

The burner for the blast-furnace stove is located near the bottom of the combustion chamber, as shown in Figure 15—18. On the majority of hot-blast stoves in the United States, the burners are external to the combustion chamber. There is a burner shut-off valve between the burner and the stove that is closed to isolate the burner when the stove is on blast but open when the stove is being fired. The gas and combustion air are partially mixed in the metallic burner but, because of their high velocity through the burner, actual ignition probably does not occur until inside the stove. The mixture of gas and air impinges on the **target wall** directly opposite the burner port and then makes a 90-degree turn. Combustion continues while the gas ascends up the combustion chamber as shown in Figure 15—20a. When a stove is to be heated from the cold condition, an ignitor must be used to start combustion but, during normal operation, the residual heat in the target wall is sufficient to cause ignition.

In several modern hot-blast stoves, ceramic burners are used. These burners, with their mixing chamber, are installed inside the combustion chamber and the firing is upward in a vertical direction instead of a horizontal direction as with the conventional metallic burner. A diagram of a ceramic burner is shown in Figure 15—20b. With this type of burner, shut-off valves are required in both the gas main and the combustion-air duct: these valves must be capable of withstanding the force of the blast pressure. The benefits of ceramic burners and the special design features required for their successful use are described in the article by Schick and Palz listed in the bibliography at the end of this chapter.

The port through which the hot-blast air exits from the stove is located on the side of the combustion chamber about 4 to 7 metres (about 13 to 23 feet) above the burner. Between the stove and the hot-blast main is a water-cooled **hot blast valve** that prevents the high-pressure air in the main from entering the stove when the stove is being heated. The hot-blast valve is usually located a short distance away from the stove to reduce the amount of radiation it receives from the combustion gases. In many installations, the cold mixer air that is used for controlling the temperature of the hot blast is mixed with the hot air from the stove on the stove side of the valve to prevent the valve from being exposed to air at the maximum temperature obtained in the stove dome. Most of the hot-blast valves are of

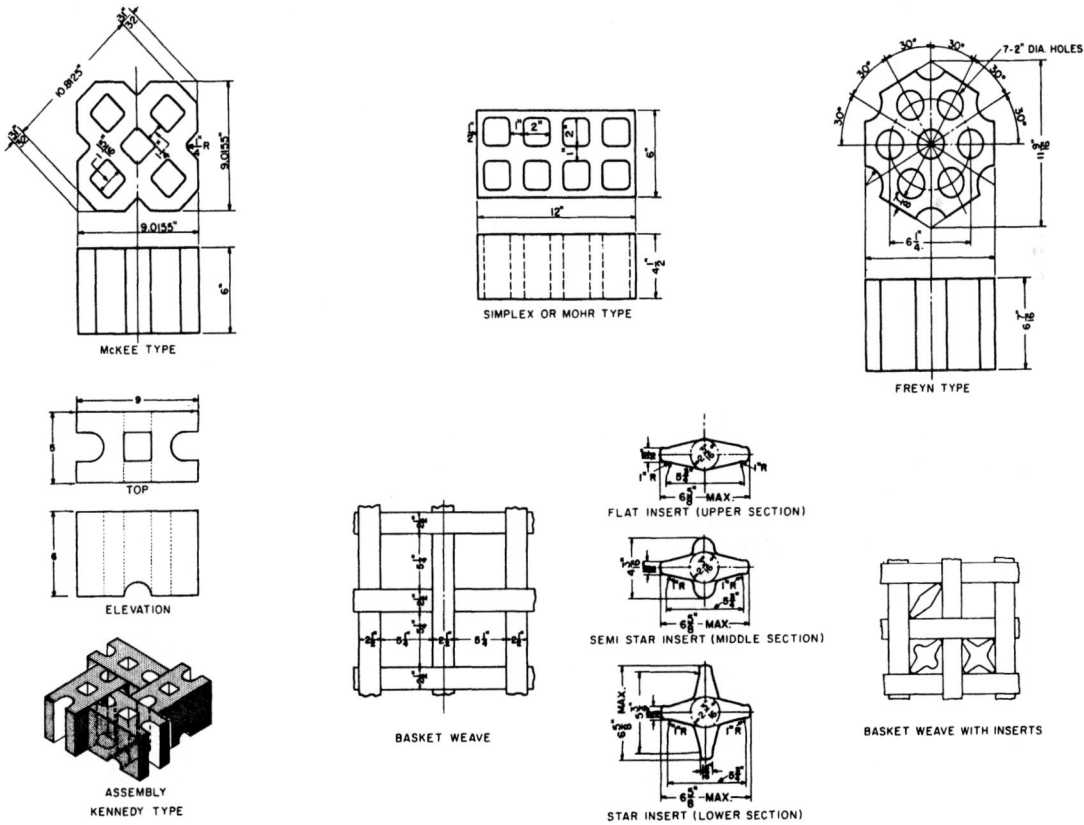

FIG. 15—19. Five different designs of checker-building shapes employed in the construction of modern blast-furnace stoves. The basket-weave design with inserts is no longer popular.

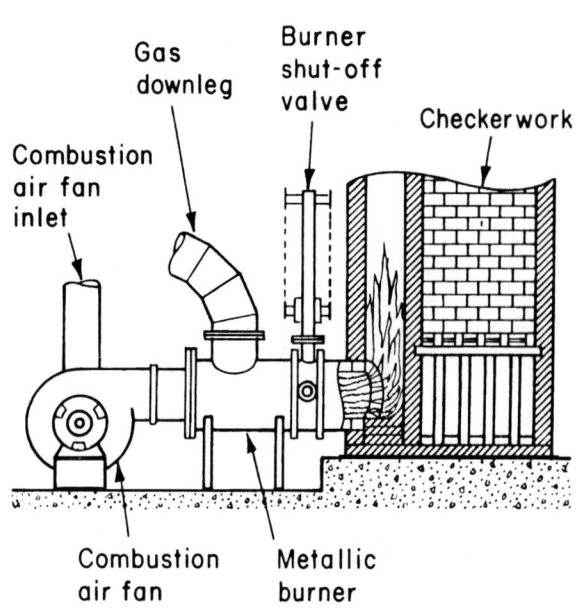

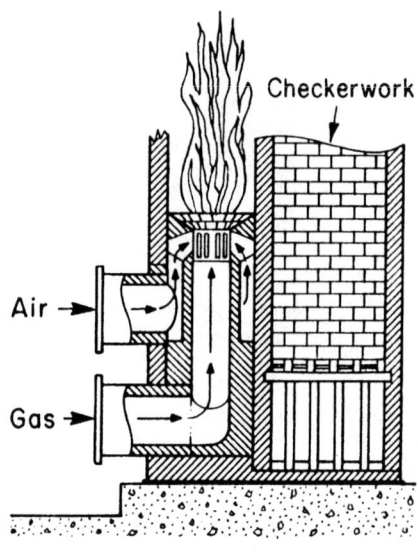

a) TYPICAL METALLIC BURNER   **b) CERAMIC BURNER**

FIG. 15—20. Hot-blast-stove burners. (After Schick and Palz: see bibliography.)

the gate type shown in Figure 15—21a or of the mushroom type shown in Figure 15—21b, and are from 1.2 to 2.0 meters (about 4 to 6½ feet) in diameter.

There is an opening in the dome of the hot-blast stove through which a thermocouple or a radiation-type termperature detector can be inserted: some such device is necessary for controlling the dome temperature and for following the operation of the stove.

In the plenum chamber below the grid that supports the checkers are the openings to the chimney and to the cold-blast main. In most modern stoves, there are two chimney valves, ranging in size from 1.5 to 2.0 metres (about 5 to 6½ feet) in diameter, which are opened when the stove is being heated so that the products of combustion will be drawn out to the stove stack. When the stove is on blast (heating the blast air), the chimney valves must be closed. Figure 15—22a shows a diagram of a typical chimney valve. The seats of the valve are arranged so that when the stove is on blast, the pressure in the stove holds the seats together to prevent leakage. When the stove is to be taken off blast and put on heating, there is a **blow-off valve** that must be opened to relieve the pressure. Because of the need to depressurize the stove rapidly, the air must exit at a very high velocity. Consequently, the blow-off valves must be equipped with mufflers to keep the noise level within tolerable limits.

The cold-blast valve is the type that is held closed by the pressure in the cold-blast main. A diagram of a cold-blast valve is shown in Figure 15—22b. Before this valve can be opened, the small ports in the valve disc must be opened to pressurize the stove and equalize the pressure on each side of the valve.

At several very modern blast furnaces, particularly in Europe and Japan, the stoves are equipped with combustion chambers completely external to the stove shell. The advantage of this design is that the entire stove shell can be filled with checkers. Furthermore, the thermal pattern in the stove is much more symmetrical and there are far less stresses that tend to distort and rupture the brickwork. However, there have been many stress-induced problems that have caused rupturing in the steelwork of the junction section between the combustion chamber and the stove. As a result, frequent repairs to the steelwork are required in this location.

Between the hot-blast stoves and the blast-furnace blower is the cold-blast main. It is unlined because the temperature of the cold blast is usually between 150° and 250°C (302° and 482°F), the temperature resulting from the heat of compression at the blower. At the stove end of the main are the cold-blast valves for the stoves and the **mixer line** equipped with a butterfly valve that is controlled by a thermocouple in the hot-blast main and proportions the amount of air delivered to the stove and the amount by-passing it. When a heated stove first goes on blast, the temperature of the heated air is much higher than the desired hot-blast temperature, so a significant portion of the air must by-pass the stove. As heat is removed from the stove and the temperature decreases, the mixer-line butterfly valve must gradually close and force more of the air through the stove. In some automatic stove-changing systems, the position of the valve is used as the signal that initiates a stove change.

The cold-blast main is also equipped with a **snort**

valve, usually located near the blast furnace, that is opened when it is necessary to decrease the blast pressure rapidly. This discharges the cold-blast air to the atmosphere and keeps a positive pressure on the cold-blast line so that gas from the furnace cannot travel back to the blower. Because of the rapid discharge of air when the snort valve is opened, it also must be equipped with a muffler.

For generating the blast air, most blast furnaces are equipped with centrifugal turboblowers provided with three or four stages. For some of the very large blast furnaces, two blowers will operate in parallel. However, with very large blast furnaces an axial blower can be used more efficiently. There are still several blast furnaces in the United States where reciprocating gas engines are used and some with reciprocating steam engines. At plants where the blast is enriched with oxygen, the oxygen can be added at atmospheric pressure to the inlet of the turboblower or it can be added under pressure in the cold-blast main. At plants with reciprocating steam engines, oxygen must be added in the hot-blast main, because there is oil and grease in the cold-blast main that could ignite. Moisture is added in the cold-blast main when it is required for blast moisture control.

BLAST-FURNACE GAS CLEANING

As the blast-furnace gas leaves the top of the furnace, it contains dust particles varying in size from about 6 mm (about ¼ inch) to only a few microns (1 micron equals 0.001 mm or about 0.00004 inch). In instances where the burden "slips" in the furnace, there are also some relatively large lumps of coke (up to about 50 mm or 2 inches in size) and burden material blown out with the gas. The dust that is carried out of the top of the blast furnace in the gas stream is generally referred to as **flue dust** and is made up of fine particles of coke and burden material and also extremely fine particles of chemical compounds that are formed from reactions within the blast furnace and are condensed from the vapor phase. Before the blast-furnace gas can be burned in either the hot-blast stoves or the boiler house, it must be cleaned to remove most of the flue dust and prevent plugging and damaging of the checkers or burners and to keep the dust from being discharged into the atmosphere with the products of combustion.

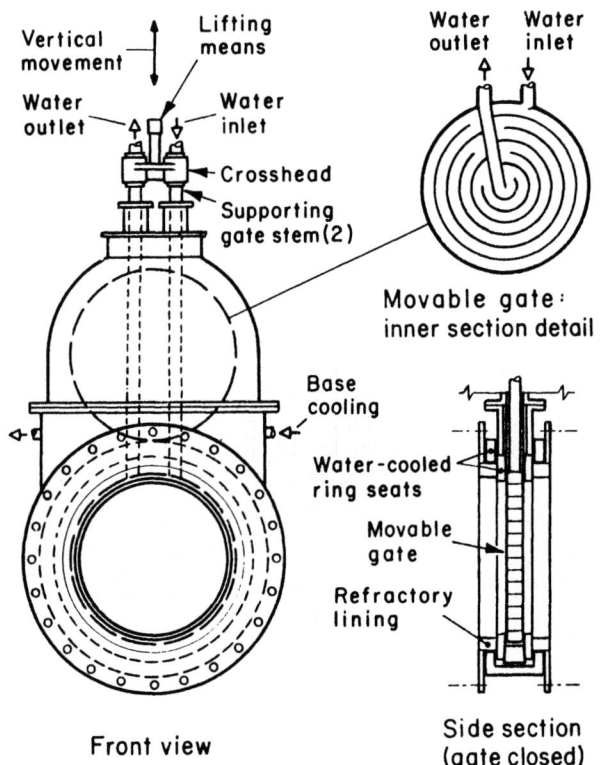

a) HOT BLAST GATE VALVE

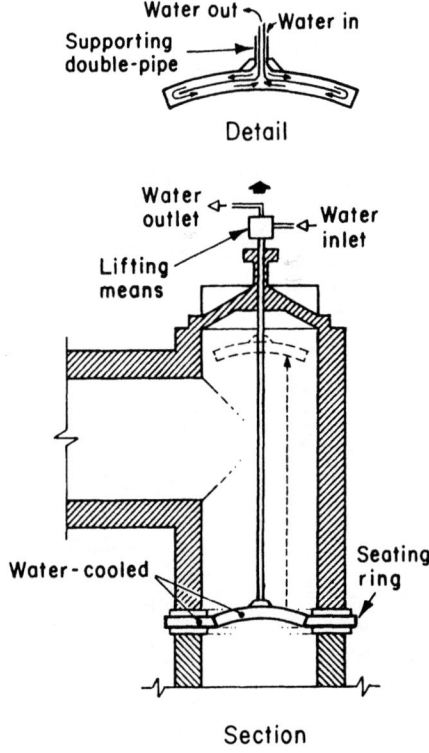

b) HOT BLAST MUSHROOM VALVE

FIG. 15—21. Hot-blast valves. (Courtesy, AIME.)

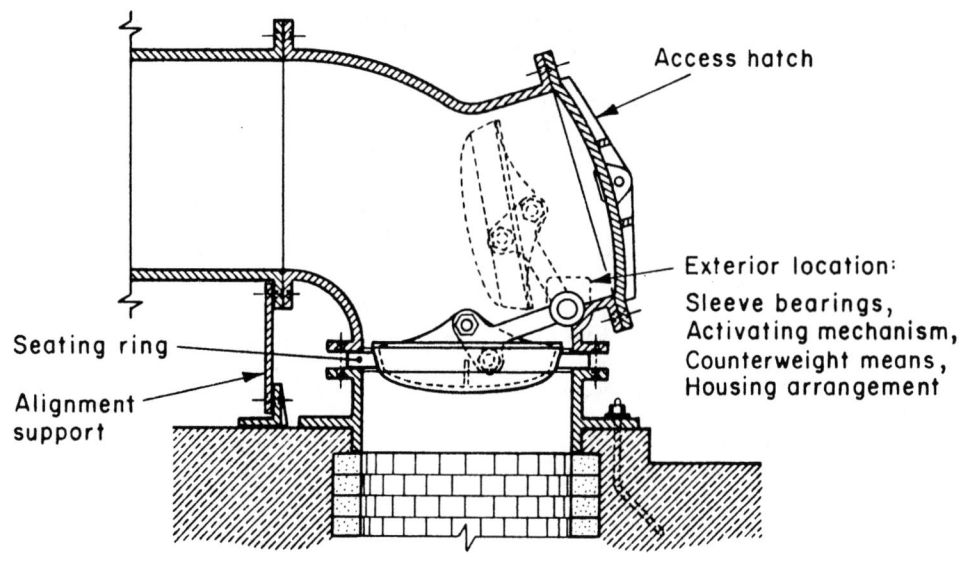

a) CHIMNEY VALVE

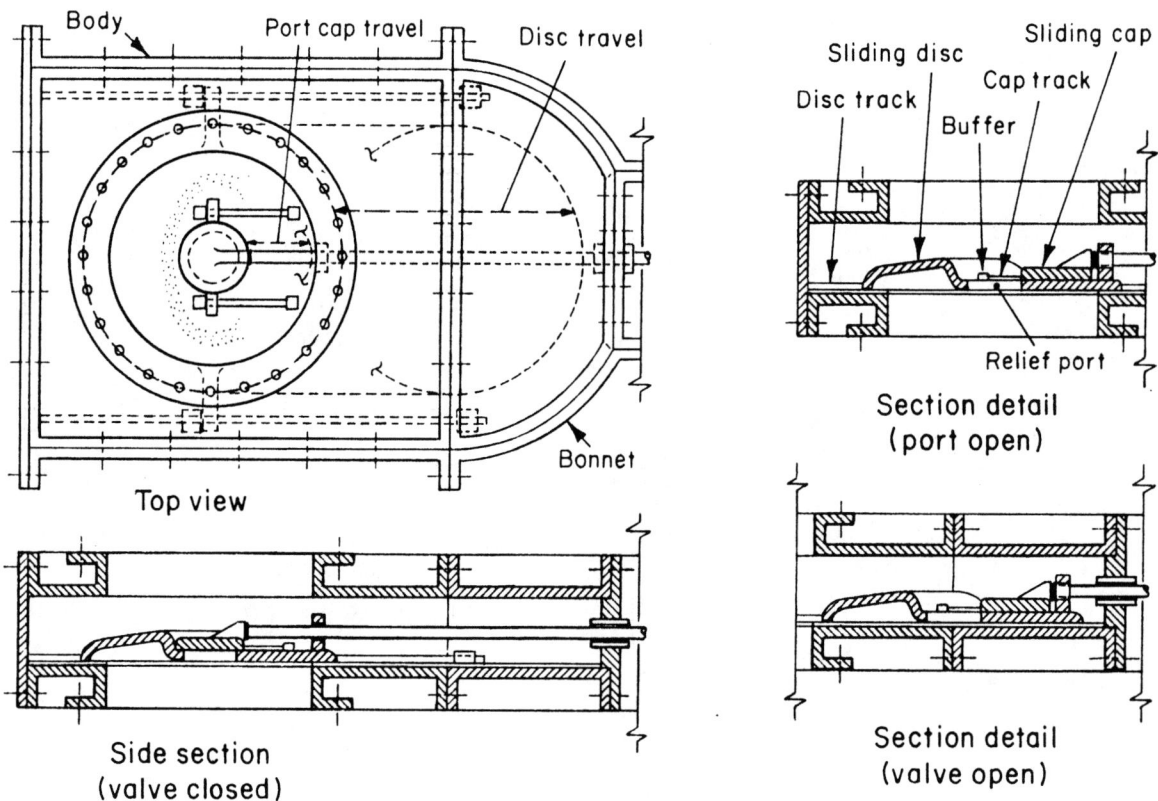

b) COLD BLAST VALVE

FIG. 15—22. Blast-furnace-stove valves. (Courtesy, AIME.)

For the first step in the cleaning process, the gas is usually always passed through a dry **dustcatcher**, where almost all of the particles coarser than 0.8 mm (0.03 inch) are removed. The dustcatcher is a large, cylindrical structure about 10 to 12 meters (about 33 to 39 feet) in diameter and 20 to 30 metres (about 66 to 98 feet) high. It is usually lined with brick or gunite to insulate it and prevent condensation of moisture in the gas so that the dust will remain relatively dry and will not ball up but will flow freely into the conically shaped section at the bottom of the dustcatcher.

The gas is conducted to the dustcatchers by a single downcomer and enters through the top by a vertical pipe that carries the gas downward inside the dustcatcher (see Figure 15—23). This pipe flares outward at its lower extremity like an inverted funnel, so that as the gas passes downward its velocity (and thus its dust-carrying potential) decreases, and most of the dust coarser than 0.8 mm (0.03 inch) drops out of the gas stream and is deposited in the cone at the bottom of the dustcatcher. Because the bottom of the dustcatcher is closed, and the gas outlet is near the top, the direction of travel of the gas must reverse 180 degrees. This sudden reversal in the direction of flow causes more of the dust to drop out of the gas stream. In a dry dustcatcher of this type, the efficiency of dust removal depends on the size of the dust particles and the ratio of the gas velocity to the size of the dustcatcher.

The dust that accumulates in the dustcatcher is generally removed through a **pug mill** attached to the bottom cone below a shut-off valve. Water is added to the dust as it passes through the pug mill to moisten it and prevent it from blowing into the atmosphere as it is discharged. On some installations, the dust is discharged directly into trucks or railroad cars as it leaves the pug mill and in others it is dumped onto the ground and removed with excavating equipment.

Each blast furnace has a set schedule for emptying the dustcatcher that depends on the type of burden being used and the amount of dust that is collected. This operation, which is called **dumping** the dustcatcher, is performed by opening the valve or valves at the bottom of the cone and allowing the dust to discharge through the pug mill. Care is taken to assure that excessive amounts of gas are not discharged with the dust.

After the gas has passed through the dry dustcatcher, it generally goes to a **wet-cleaning system** where the very fine particles of dust are literally scrubbed out of the gas with water. Formerly, blast furnaces had two stages of wet cleaning with the primary stage removing 90 to 95 per cent of the dust in the gas as it came from the dustcatcher and the secondary stage removing 90 to 95 per cent of the dust in the gas as it came from the primary stage. However, in most modern blast-furnace plants, high-energy high-efficiency scrubbers are used that clean the gas to a dust content of less than 12 milligrams per cubic metre (0.005 grain per standard cubic foot), in one operation.

For the primary wet-cleaning operation, stationary-spray towers, revolving-spray towers, Feld washers, baffled-spray towers and venturi washers were used. In the **stationary-spray tower**, the gas entered near the bottom and passed upward through three or four banks

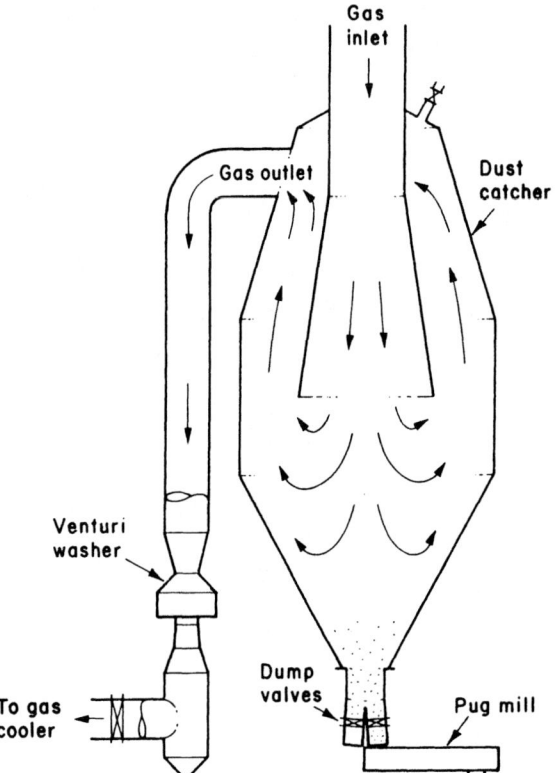

FIG. 15—23. Dustcatcher and venturi washer.

of ceramic tile to break up the gas flow. Water was sprayed over the entire cross-section of the washer above each bank of tile and a centrifugal-type water separator was located above the topmost bank to remove entrained water from the gas. In the **Feld washer**, the screen of water through which the gas passed was formed by rotating cones partially immersed in water trays instead of by water sprays on tile.

In the **venturi washer**, the gas from the dust catcher enters from the top (Figure 15—23) and passes downward through the venturi throat (see Figure 15—24). As the gas passes through the narrow throat of the venturi, it is sprayed with water. There are two sets of water sprays, one operating at low pressure and entering the unit at right angles to the gas flow, and the other operating at high pressure and directed upward at an angle of 110° to the gas flow. The washer is lined with a suitable abrasion-resistant material to withstand the erosive effect of dust-laden gas. The venturi-type washer can clean gas to an average dust content of 120 milligrams per cubic meter (0.05 grains per cubic foot).

The gas passes from the venturi washer into a water separator directly beneath it, and thence to a cooling tower where its temperature is lowered by passing through water sprays. The cooled gas then passes through a moisture eliminator before going to the secondary cleaning system.

Secondary wet cleaners have consisted of wet-type Cottrell precipitators, high-speed disintegrators and Theissen disintegrators. Formerly, Theissen disintegrators were used almost exclusively, but on later installa-

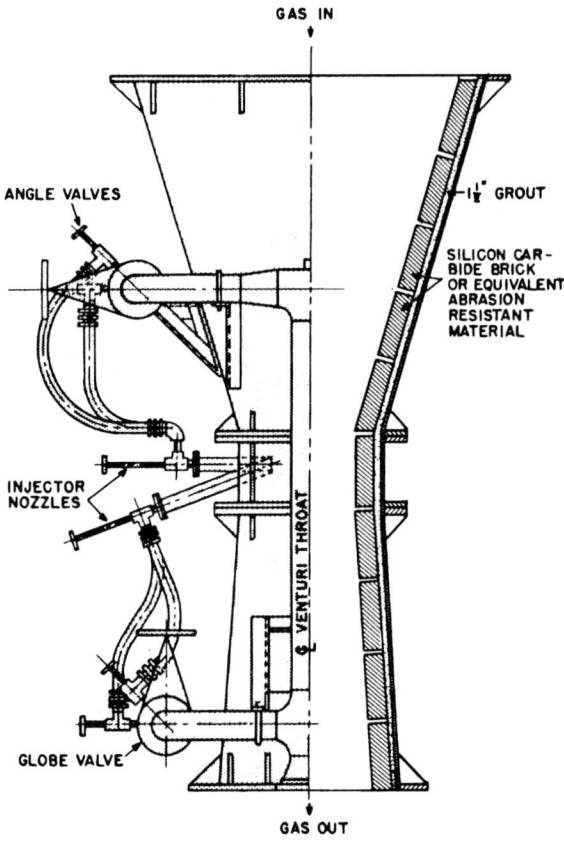

GAS IN

ANGLE VALVES

1¼" GROUT

SILICON CAR-
BIDE BRICK
OR EQUIVALENT
ABRASION
RESISTANT
MATERIAL

INJECTOR
NOZZLES

VENTURI THROAT

GLOBE VALVE

GAS OUT

FIG. 15—24. Schematic arrangement of a venturi-type unit for washing blast-furnace gas, right side cut away to show lining.

tions they have been replaced by high-speed disintegrators or by the wet-type Cottrell precipitators, with preference being given to the latter type.

In the **Cottrell wet method** of cleaning, the primary-cleaned gas is forced to pass through narrow channels or ducts across which an electrostatic field is maintained. The dust particles are "precipitated" or separated from the gas through the action of electrostatic charges. The molecules of gas are ionized and, in turn, induce electrostatic charges upon the surfaces of the small dust particles. Under these conditions, the dust particles are attracted toward the electrode of opposite polarity. There is also an "electrical wind" or corona effect. Current is supplied from an alternating-current source, stepped up in voltage by a transformer, and made unidirectional by a mechanical rectifier or by vacuum tubes. In most precipitators, cleaning cool gases, the unit collecting electrode is a vertical tube, 200 to 300 mm (8 to 12 inches) in diameter or it may be a pair of parallel steel plates through which the gas is forced upward, or sideways. In the former type, the discharge electrode is a wire suspended coincident with the long axis of the tube and, in the latter, it consists of multiple wires midway between the plates. A thin film of water flows over the inside edge of each tube or plate which washes it free of dust that has been deposited thereon. The precipitator usually is divided into two units with valving so arranged that one unit

may be shut down for inspection or repairs while the other unit is operating. In some installations, the precipitator is mounted directly above the primary washer. The rotary disintegrator has the advantage of lower first cost but has higher power cost than the wet-type Cottrell precipitator. The cleanliness of the gas may be slightly in favor of the precipitator.

Experiments conducted in 1958 showed that as the pressure drop through an orifice or venturi-type washer is increased, the cleanliness of the gas improves. Figure 15—25 is a graph from a technical article by Campbell and Fullerton (see bibliography at end of chapter) that indicates that when the pressure drop is increased to 25 kPa (about 200 mm of mercury or about 3.9 pounds per square inch) the gas cleanliness will improve to 10 milligrams per cubic metre (about 0.004 grains per cubic foot), which is as clean or cleaner than the cleanliness that was attained formerly with two stages of wet cleaning. With the increase in operating pressure of most modern blast furnaces, high-energy scrubbers have replaced most of the old two-stage systems. In many installations, the throat of the venturi is equipped with a variable opening that is adjusted automatically to maintain the pressure drop uniform even if there is a significant change in the wind rate. The venturi washer is generally followed by a **gas-cooling tower** where water sprays are used to lower the gas temperature and decrease the water-vapor content. The cooler is often followed by a **demisting chamber** to remove entrained water.

At the Fairfield No. 8 blast furnace, which operates with an elevated top pressure as high as 207 kPa (about 2.1 kilograms per square centimeter or about 30 psig), a relatively new type of two-stage venturi gas washer and cooler is used that automatically controls the top pressure. With this washer, the dirty gas enters at the top and passes through a relatively low-pressure-drop venturi that is flooded with water. The partially cleaned gas from this first stage venturi has sufficient pressure so that it can be used for pressurization in the lock hoppers on top of the furnace. The diagram of this washer, shown in Figure 15—26 depicts the manner in which it operates. After the first-stage venturi, the primary-cleaned gas makes two 180-degree directional changes and then passes vertically downward through three high-pressure-drop venturi throats. The pressure drop through these three units is controlled by a conical plug valve actuated by pressure impulse lines from the blast-furnace top. The secondary-cleaned gas impinges on rotary spin vanes and again reverses its direction of flow 180 degrees. This final reversal of flow and the centrifugal force of the spin vanes remove the remaining entrained water.

The water used in the wet gas-cleaning systems contains between 30 and 50 per cent of the total dust removed from the gas as it leaves the gas washer. Consequently, the suspended particles of dust must be removed from the water before it can be reused or discharged from the system. In a great many older plants, settling basins were used, where the dirty water entered at one end and the clear water flowed out over a weir at the other end. These units were usually in pairs, so that when one was full it could be drained for removal of the solids while the other one was in use. In

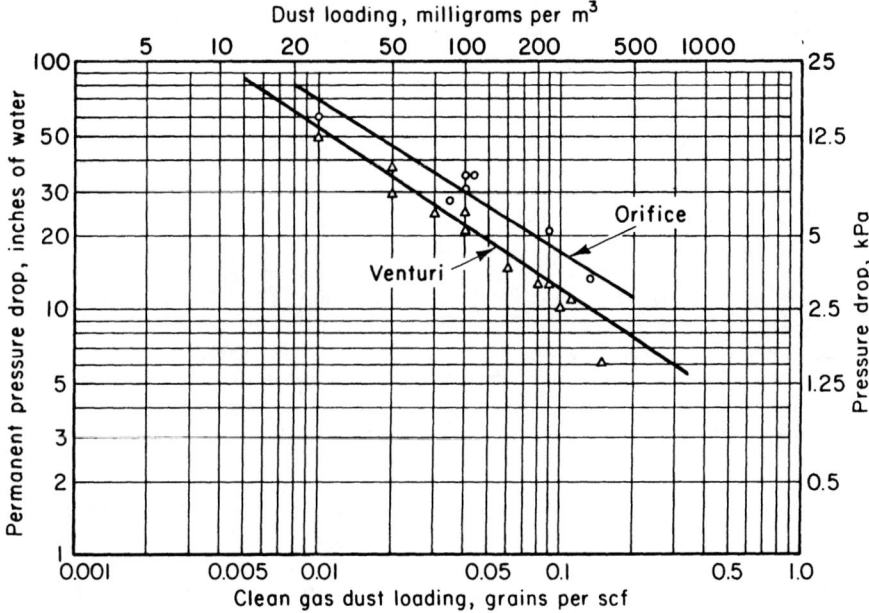

FIG. 15—25. Effect of pressure drop on scrubber performance. (Courtesy, AIME.)

most plants today, **thickeners** are used to remove the solids from the water because they require less space than the settling basins formerly used.

A typical thickener consists of a circular reinforced-concrete tank which may contain one or several compartments in each of which one or several arms revolve. The arms are driven from a central vertical shaft. Water enters at the center of the thickener and leaves over a continuous weir which follows the circumference of the compartment. Each arm carries a series of rakes or vanes set at such an angle that the solids which settle out are gently pushed toward the center of the compartment. The thickened solids, containing about 60 per cent water, are pumped out and delivered either pneumatically or by a pump to a filter for further processing. The thickener may be built with either an open or a closed top. Two types of filters are in general use, the cylindrical or drum type and the disc type. Both operate upon the underflow from the thickener, discharged into a basin into which an arc of the slowly rotating filter dips. The drum or disc consists of a framework over which canvas is stretched, and a partial vacuum is applied on the inside of that part which dips in the basin, thereby drawing liquid through the canvas while the solids are retained by the canvas. When the drum or disc reaches another point in its rotation, a slight air pressure is applied on the inside of the canvas thereby bulging it and causing some of the filter cake to crack and drop off into a chute, the balance being scraped off by a scraper. The filtrate returns to the thickener and the filter cake, containing about 25 per cent of moisture, is delivered by railroad car to the sintering plant or, if the sintering plant is located adjacent to the furnace, the filters are designed as an integral part of the sintering plant itself.

BLAST-FURNACE CAST HOUSE

The operation of a blast furnace is a continuous proc-

ess, and the furnace continues to produce liquid iron and slag as long as it is in operation. The iron and slag accumulate in the hearth, but because there is a limit to the amount that can be tolerated before it interferes with the furnace operation, they must be removed

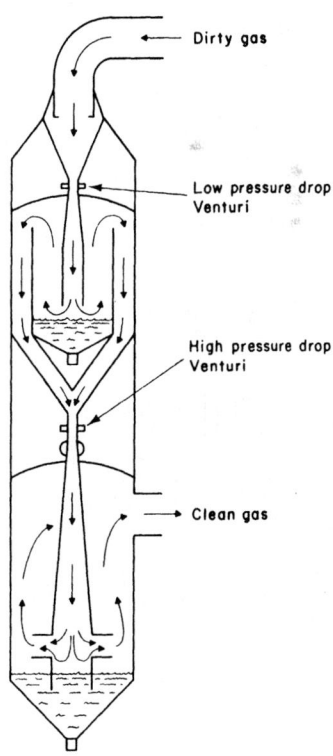

FIG. 15—26. Schematic vertical cross-section of an automatic two-stage venturi gas washer that controls blast-furnace top pressure.

from the furnace at regular intervals. The iron notch which is used for tapping the hot metal from the furnace is located slightly above the floor of the hearth. Most modern furnaces are equipped with more than one iron notch. There is also a cinder notch for removing slag from the furnace, and it is located in a plane between 1 and 2 metres (3.3 to 6.6 feet) above the iron notch.

Formerly, before the blast-furnace burdens were beneficiated as well as they are currently, the weight of slag produced was more than half the weight of the hot metal. The lower density of the slag caused it to fill up the space in the hearth above the metal, and it would interfere with the penetration of the blast air and the combustion process at the tuyeres long before the accumulation of iron had reached the desired amount for casting. Consequently, it was necessary to remove the excess slag through the cinder notch (generally referred to as the monkey) once or twice between casts. However, in recent years, since better-prepared burdens have been used and slag volumes have decreased to 200 to 300 kilograms per metric ton (400 to 600 pounds per net ton), the monkey is seldom used and the slag is removed through the iron notch.

Because it may be required at some time during a campaign, even if only for emergency, every blast furnace is equipped with a slag notch or cinder notch, and provision must be made in the casthouse for runners to conduct the slag to cinder pots or a slag pit. With the location of the cinder notch well above the level of iron in the hearth, it can usually be opened by a pricker bar if it is to be opened on a regular basis between each cast. After the slag has been flushed out of the furnace, the cinder notch is closed again by inserting a water-cooled **bott** that fits into the opening of the monkey cooler and freezes enough slag there to seal the opening. Figure 15—27 shows the mechanism that guides and supports the bott above the cinder runner.

Until 1962, all of the blast furnaces in the United States were equipped with only one iron notch; however, all of the new large blast furnaces built since then have been provided with two or more iron notches. When the furnace is in operation, the iron notch is completely filled with a refractory material called **tap-hole clay**. To cast the hot metal from the furnace, a tapping hole is drilled through this material, and after the cast has been completed, the hole is plugged again with fresh clay that is extruded into the hole from a **mud gun**. The mud gun consists of a hollow, cylindrical barrel and a plunger that pushes the clay out through a

FIG. 15—27. Mechanical slag bott withdrawn from slag notch, with stream of molten slag running down slag trough.

FIG. 15—28. General view inside the cast house of a blast-furnace plant. The mud gun is shown in position for closing the tap hole. Suspended diagonally near the center of the illustration is the drill used to drill through the clay in the tap hole when the furnace is to be cast. The bott-operating mechanism for closing the slag notch is at the extreme left. The bustle pipe and connections to several of the tuyeres are also visible.

nozzle into the tap hole. The plunger is operated either electrically or hydraulically (and in a few instances by steam). Formerly, only water-based bloating clays were used in the tapping hole, but at modern large blast furnaces where both hot metal and slag are removed through the tapping hole, this type of clay erodes too rapidly and the hole enlarges so much that it is difficult to control the cast. Consequently, at most large blast furnaces, this type of clay has been replaced by carbonaceous material to which a plasticizing agent has been added. Figure 15—28 is a photograph taken in a casthouse facing the blast furnace and shows the mud gun inserted into the tapping hole in the position used for stopping (plugging) the hole. To the left side of the mud gun is the tapping hole drill in its retracted position. For tapping, the mud gun swings away to the right and the drill swings into position so that the hole is drilled at the desired angle.

As the hot metal leaves the tapping hole it is discharged into the **trough** which is a long, narrow basin, 1 to 1.5 metres (about 3 to 5 feet) wide and 8 to 12 metres (about 26½ to 39½ feet) long. The trough generally has a slightly sloping bottom inclining 1.5 to 2 mm per metre (0.018 to 0.024 inches per foot) away from the furnace. At the far end of the trough there is a **dam** to hold back the hot metal until the depth of metal in the trough is sufficient (about 300 mm or 12 inches) to contact the bottom of a refractory **skimmer block.** The skimmer holds back the slag and diverts it into the **slag runners.** The hot metal flows over the dam and down the **iron runner** where, by a series of gates, it is directed in sequence to the train of ladles positioned under stationary spouts along the runner. At several large blast furnaces, a tilting spout is used, positioned between two hot-metal tracks. The spout is first tilted to fill the ladle on one track and then to fill the one on the other track. While the second ladle is being filled, the first one can be replaced with an empty one so that the cast can be continued uninterrupted while several ladles are filled.

At some of the very large blast furnaces that must be cast very frequently (12 to 15 times per day), hot metal is held in the trough and not drained. Because of improved trough refractories, this is becoming common

FIG. 15—29. A submarine- or torpedo-type hot-metal transfer car.

practice even on smaller furnaces. At several modern blast furnaces, the troughs are removable. Therefore, when the trough is in need of repairs it can be removed and another one put in its place so that there is no delay while it is being relined, a job that requires more than thirty-six hours.

Hot-Metal Transportation—The hot metal is transported from the blast furnace to the steelmaking shop in refractory-lined ladles that have a course of insulating material between the lining and the steel shell. There are two types of ladles: the **open-top type** and the **Pugh thermos-bottle type.** The open-top ladles are shaped like large flower pots equipped with lifting trunnions and special support trunnions that rest in accommodating brackets on the railroad car. The **Kling-type** ladles are also open-top, but they are somewhat spherical in shape so that the opening is slightly reduced in cross-sectional area and less heat is lost through radiation. These ladles have a capacity of about 100 to 110 metric tons (110 to 121 net tons) of hot metal. The Pugh-type ladle (shown in Figure 15—29) is shaped a little like a submarine and for that reason these ladles are often referred to as **submarine ladles** or **subs:** they are also sometimes called **torpedo** ladles because they have somewhat the shape of a torpedo.

These ladles have a relatively small opening in relation to their capacity which is usually 200 to 250 metric tons (220 to 275 net tons) although some have been built much larger. The submarine or torpedo ladle differs from the open-top type in that it is not removed from the railroad car for pouring. The ladle itself is an integral part of the railroad car and, to empty it, it is rotated about its longitudinal axis by a motor built on the railroad car. Because of its large capacity in relation to its small opening, the torpedo or submarine ladle loses very little heat through radiation from the opening and, in many instances, it may be used to hold hot metal for the steelmaking process, and the hot-metal mixer is eliminated.

In merchant-iron plants where the hot metal is not converted into steel, it is taken to a **pig machine** and cast into small molds that are moved consecutively and continuously under a pouring spout. The molds are attached to an endless chain. In some designs, the chains are attached to wheels that ride on the track, while in others the chain forms a track that rides on stationary rollers. The chain passes over a head and a tail sprocket wheel so that the molds are upside down on the return side. The molten iron is poured into the molds near the tail sprocket, and it solidifies and is cooled by water as the chain carries it to the head sprocket. As it passes over the head sprocket, the iron falls from the mold into a railroad car. On the return travel, the molds are sprayed with a lime wash to prevent the iron from sticking to them. Many years ago, most blast-furnace plants were equipped with pig machines that were used for disposing of hot metal when it was not required by the steelmaking shop. However, the cost of maintaining such expensive equipment just for stand-by purposes was not economical; so that in most plants at present such hot metal is poured in relatively thin layers into dry pits. The solidified iron is then broken into pieces that can be used as cold iron in the steelmaking shop. Operation of a pig machine is illustrated in Figure 15—30.

Slag Disposal—Blast-furnace slag is handled in many different ways, depending upon the amount of space available around the furnace and on the type of product that is to be made from the slag. At most of the older blast furnaces, the slag was run into **cinder pots** mounted on railroad cars and transported outside the plant to a slag dump. The cinder pots (also called **slag pots**) are unlined, thimble-shaped, cast-iron or steel castings mounted on railroad cars. The pot fits into a bracket with trunnion mountings set on a rack and pinion so that it can be rotated and dumped by a compressed-air-operated cylinder mounted on the railroad car. The slag from the pots can be dumped over a bank into pits and later excavated, crushed and screened to be used as ballast or aggregate, or it can be poured into the path of a high-pressure stream of water that expands it into a popcorn-like material to be used as lightweight aggregate. Figure 15—31 shows a train of slag pots positioned under the spouts of the slag runners of the blast-furnace casthouse.

At many modern blast furnaces, the slag is run into pits adjacent to the casthouse. There are usually two pits so that one can be excavated while the other is being filled. It usually requires two or three days to fill a pit, and after each run of slag water sprays are used to assure that the slag will solidify rapidly enough to be ready for excavating on schedule. Figure 15—32 shows a hard-slag pit adjacent to a blast furnace. At several blast furnaces, the slag runner directs the slag in front of high-pressure water sprays that granulate it into small pieces and then wash it into a wet pit. The rapid cooling of the slag gives it a glassy structure, and when it is finely ground, it has hydraulic-bonding properties and can be used in cement. Similarly, a few furnaces have been equipped with machines to form pelletized, lightweight slag granules. In these installations, the molten slag is diverted from the slag runner to water-cooled, rotating drums which throw the slag through the air onto a pile a short distance from the machines. The slag is rapidly cooled by the water used as external coolant for the drums and by the air while it is in flight. The cooled slag is similar in properties to the granulated slag except the particles are more spherical in shape. More information about the specific uses of slag is presented in Chapter 7.

Receipt of Blast-Furnace Raw Materials—The iron-bearing materials—ore and pellets—generally come from a considerable distance and must be transported either by rail or water or a combination of both. Coke is often received by rail and is consumed as rapidly as it is received. However, in the case of ore and pellets, weather conditions are often such as to prevent mining and transportation during the colder months, so during

FIG. 15—30. Pig-casting machine in operation, showing molten iron from transfer ladle being divided into two streams by the split runner to simultaneously fill molds in each of the two strands.

FIG. 15—31. Molten slag pouring from runner in cast house into slag ladle.

FIG 15—32. Slag-cooling pit adjacent to a blast furnace.

the warm months it is necessary to store approximately one-half year's supply adjacent to the furnaces. Because many of the furnaces in this country receive their iron-bearing material from the Lake Superior District, as discussed at length in Chapter 5, the transportation of ore from there to the lower lake plants is described very briefly here as being typical of water transporation of iron-bearing materials. At the head of Lake Superior, ore is loaded by gravity, from overhead bins arranged on long piers extending into the harbor, into specially constructed ore boats. Upon reaching their destination, the vessels have their ore cargo unloaded by **Hulett unloaders** or by special **unloading rigs** if the plant is large enough to justify this special equipment, or, if the plant is smaller, by an **ore bridge** which extends over the vessel from the dock and permits the ore-bridge bucket to unload the vessel.

The Hulett unloader (see Chapter 5) consists of a carriage mounted on wheels which permit it to move either parallel to or at right angles to the dock. On the harbor side, a grab bucket is attached to a counter-weighted, pivoted arm. The operator rides in a compartment near the bucket and controls its movement from this location. The bucket itself can be rotated in any direction. The bucket moves down through the open hatch of the vessel, grabs its load, and then moves up and back to deposit its load into a hopper. A rotating feeder under the hopper places the ore into a small car that runs either to a hopper over railroad cars or to an ore trench in the ore yard. The ore bridge picks the ore from the ore trench and distributes it onto the ore pile.

The unloading rigs (see Chapter 5) perform the same function as the Huletts but differ somewhat in construction. Instead of the bucket being carried by a rigid arm, it is hung from a trolley resembling a crane trolley; consequently, it cannot be rotated and it cannot clean out the hold of the vessel as completely as the Hulett, and the additional use of a small bulldozer and hand labor is necessary to complete the job.

In recent years, self-unloading ore boats have also been placed into service. The hold of these boats is compartmented into bins that are equipped with feeders that can discharge the contents of the bins onto a conveyor belt that runs along the bottom of the ship's hold. This conveyor discharges onto a series of inclined conveyors that raise the cargo out of the hold. A long conveyor mounted on the boom of the ship's crane can be swung out from the side of the ship in any direction and discharge the cargo directly into the ore trench.

Car Dumper—For those plants located some distance from the point of vessel unloading, or where ore and pellets are received directly from the mines by railroad, a car dumper is the usual means to unload them. The material, arriving at the plant in trainload lots, is switched to a siding ahead of the car dumper and the cars are unloaded one by one in rapid succession.

In one type of dumper, illustrated in Figure 15—33, a car is pulled up an incline to the platform of the dumper by a steel cable of a "mule" or "barney." There it is lifted bodily and turned over so as to empty its contents into a large transfer car which, in turn, discharges into an unloading trough at the designated location in the stockyard for that particular grade. The

dumper then resumes its former position and the empty car is pushed off the platform, by the next car of ore or pellets, to an incline, down which the empty cars move to a siding. The transfer cars are usually designed to hold two railroad cars of ore or pellets.

Another type of car dumper shown in Figure 15—34 is the rotary dumper in which the car is held in position on the track with clamps and the entire supporting structure rotates the car about its longitudinal axis. Many of the modern dumpers discharge the materials into bins from which they are fed onto conveyor belts and taken either to storage space or directly to the highline bins.

Ore Yard and Ore Bridges—The ore yard (Figure 15—35) is a large space, sometimes with concrete bottom and sides, which parallels the furnace-bin structure and serves as storage for approximately a half-year's supply of burden material.

Its width is determined by the practical maximum span of the **ore bridge,** which for a bridge supported at each end amounts to about 107 metres (about 350 feet).

FIG. 15—33. Car dumper discharging iron ore from a railroad car into a transfer car.

Other types of bridges are supported by two legs which run on tracks that run longitudinally through the ore yard (dividing the yard into three separate piles); such bridges have cantilevers on the outer sides of the two supporting legs. This type gives a longer trolley travel but does not necessarily store more ore per running foot because of the space occupied by the supporting structures. A trolley running on the bridge proper carries a grab bucket of about 15 metric tons (16.5 net tons) capacity which lifts the ore from the dumping trough and distributes it on the pile, or removes the ore from the pile and deposits it either directly into the bins or into a bin car or transfer car (Figure 15—36) which in turn distributes the ore. Some bridges are equipped with a bin mounted on the bridge into which the bucket dumps, permitting the bridge to keep working while the bin car is dumping its load. In order to obtain the maximum blending, the ore is stocked in horizontal layers and removed in vertical slices. For this reason, the use of ore direct from hopper cars unloaded from the trestle into the bins is undersirable. The details of the ore-handling system will vary considerably but the general scheme is essentially as stated.

Trestle and Stockhouse—The function of the trestle and stockhouse is to deliver the correct quantities of iron-bearing materials, flux and coke to the furnace as expeditiously as possible to keep the blast furnace at top operating performance.

The constructional details of the **trestle** differ considerably from plant to plant. A trestle of modern design consists of a reinforced-concrete wall on the stock yard side and steel columns on the furnace side, between which is a crosswork of transverse and longitudinal girders. On the top, these girders support three or four railroad tracks, and on the bottom, they support the bins. The track nearest the stock yard carries a side-dump bin car, the next track carries railroad cars containing such materials as limestone, dolomite, scale, scrap, etc., which are unloaded by manual operation of the hopper gates of the cars, and the next one or two tracks carry coke cars. When there are more than six furnaces in a line, it is desirable to have two coke tracks so that the coke cars not fully unloaded may be by-passed. Where the coke bins are filled by a belt conveyor (as described later), one coke track for emergency use is sufficient. This construction results in two

Fig. 15—34. Rotary car dumper with car in inverted position. Entire supporting structure rotates clockwise to right car, which is then unclamped and moved away to be replaced by the next loaded car.

FIG. 15—35. View of an ore yard serving the blast furnaces in the background.

lines of bins in the stockhouse, one line containing ore, flux and miscellaneous materials and the other containing coke.

At most blast furnaces, a **scale car**, equipped with scales to weigh each material drawn, runs underneath the first line of bins. The car may have a capacity of somewhat over 35 metric tons (up to 40 net tons) and carries two pockets. The ore and miscellaneous material bins are equipped with gates on the bottom which may be operated either mechanically or by hand (Figure 15—37). The car delivers its load by discharging into chutes leading to the skips. (See Figure 15—38 for an illustration of a typical scale car stockhouse.) Underneath the row of coke bins, conveyor belts, one for each side, fed by vibrating feeders at each coke bin, run toward the skip pit. Each conveyor discharges the coke onto a vibrating screen where the coke fines are screened out before the coke is discharged into a weigh hopper. Here the coke is automatically weighed and an electrical signal from the load cells under the hopper stops the belt and feeder at the designated weight. At some older furnaces, the coke is measured by volume and the signal to stop the feeder and belt are given by electrodes suspended in the hopper when they are contacted by the rising level of the coke. From the weigh hoppers, the coke is discharged directly into the skips. Considerable time is saved by this method, as the operator can be drawing ore while the coke loading is being taken care of automatically. The undersized coke from the vibrating screen is removed by a small bucket hoist, or by conveyor belt, to a storage bin where it is rescreened into coarse and fine. The coarse may be

recharged separately into the furnace or may be sold as domestic coke. The fines are used to fire boilers, as fuel in the sintering processes, and to make up bottoms of soaking pits.

Automatic Stockhouse Operation—As has already been described, the charging of coke has been made an automatic operation on many furnaces. Until recently, however, the weighing and assembly of iron-bearing

FIG. 15—36. Transfer or bin car of side-discharge type, in position to discharge load of ore into stockhouse bins. The ore must pass through the gratings shown, which effectively prevent oversize material from entering the bins.

FIG. 15—37. Scale car in stockhouse, showing overhead ore bin gates from which car is filled, and chutes that carry coke from bins directly to skips.

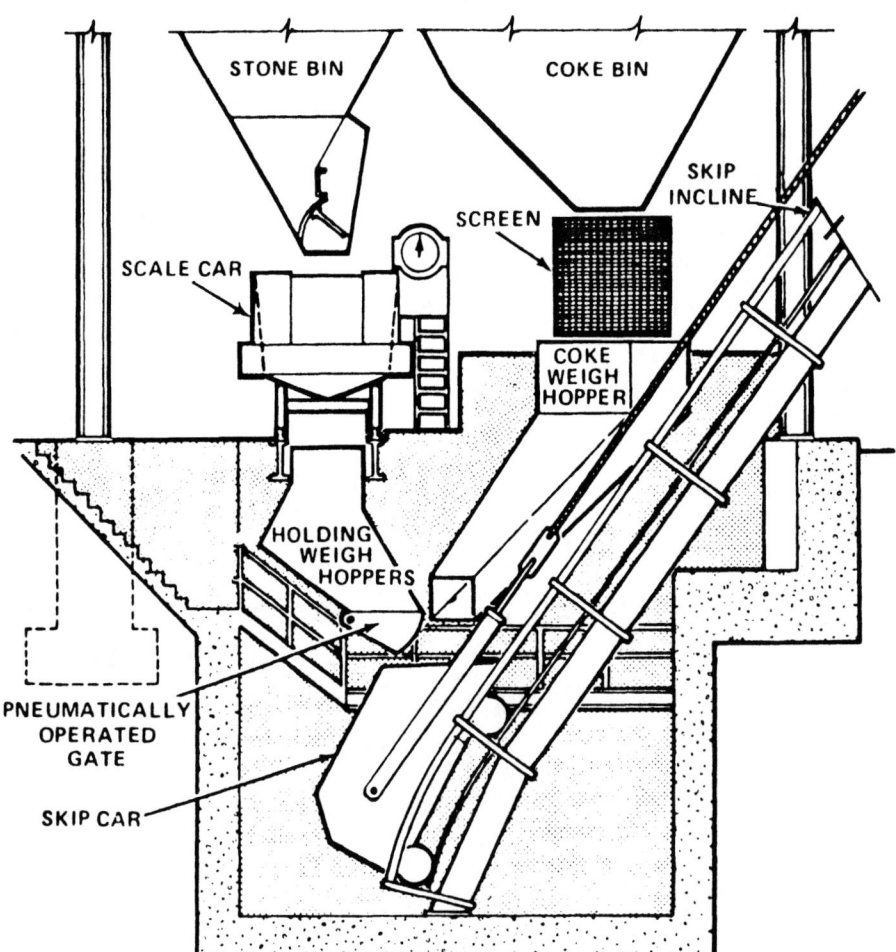

FIG. 15—38. Schematic of a typical scale car stockhouse.

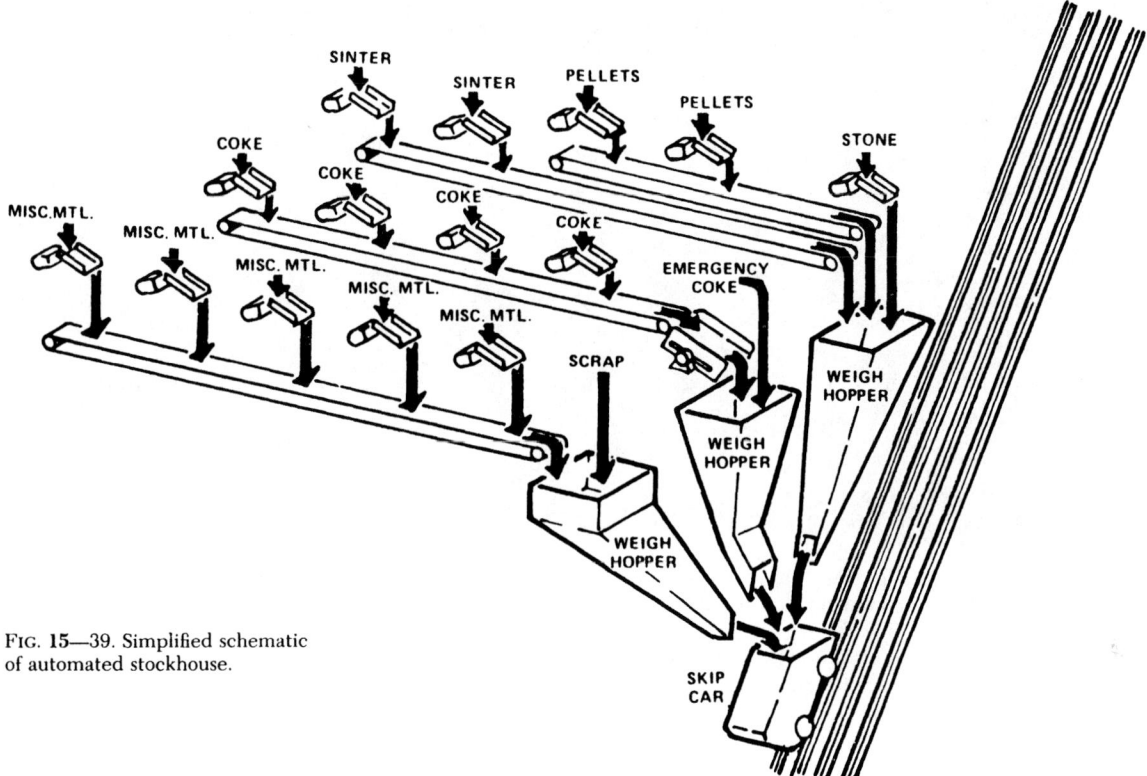

FIG. 15—39. Simplified schematic
of automated stockhouse.

and fluxing materials and their delivery to the skip pit
were functions performed by the scale car—the operation of which was independently and manually controlled. Automatically controlled systems (see Figure
15—39) for assembling, weighing and transporting
iron-bearing materials and fluxing material as well as
coke have been designed to serve new furnaces or have
been adapted to existing stockhouses, thereby eliminating the necessity for a scale car. The sequencing of
the drawing of the burden materials is controlled by
computer, and the materials are transported to the
weigh hoppers by conveyor belts. Another innovative
refinement used in computer-controlled stockhouses is
the **coke-moisture gauge** that determines the moisture
content of the coke and adjusts the ordered weight so
that the proper "dry" weight of coke is obtained. The
computer also compares the actual dry weights with
the ordered dry weights and makes a compensating
adjustment if they do not agree.

Instrumentation and Control—The blast-furnace operator is aided and guided in establishing and maintaining the best possible balance between the many variable factors affecting furnace operation by numerous
automatic devices which indicate or record conditions
at various points in the furnace system and may operate control mechanisms that automatically regulate important variables

Instruments considered essential to proper operation
of a modern blast-furnace plant include:

Blast-pressure recorder and indicator.
Blast-temperature recorder and indicator.
Stockline recorder and visual indicator.
Top-temperature recorder.

Top-pressure recorder and indicator.
Blast-volume recorder and indicator.
Stove-stack temperature recorder.
Stove-dome temperature recorder.
Automatic hot-blast temperature controller.
Automatic combustion controls for stove burners
with dome and stack temperature correctives.
Sequence recorder of large-bell movement and re-
volving distributor operation.
Annunciator for alarms warning of low gas pressure,
high stove-dome temperature, high stack temper-
ature, low blast pressure, and so on.

In addition, some plants employ multipoint inwall-
temperature recorders, top-gas analyzers, horizontal
in-and above burden temperature probes, and other
instruments for measuring, detecting and/or recording
conditions of interest to the operators of a particular
plant or furnace.

In many of the modern blast-furnace plants, the
reading of these instruments is done automatically, and
the information is fed to computers that calculate factors about the permeability of the burden, the ratio of
indirect reduction, direct reduction and other operation variations that are particularly useful in maintaining the uniformity of hot-metal composition and production rate.

Blowers—The air blast for blowing the blast furnace
can be made by a number of different types of blowers;
however, most of the blast furnaces in the United
States are currently serviced by steam-turbine-driven
centrifugal blowers called **turbo blowers.** A typical

Fig. 15—40. View of a steam-turbine-driven axial-flow blast-furnace blower, with upper half of casing removed. Vaned rotor of air compressor is at left, steam-turbine rotor at right. (Courtesy, Ingersoll-Rand Company.)

turbo blower is illustrated in Figure 15—40. There are still, however, a few blast furnaces where steam-engine-driven reciprocating compressors are employed. With the introduction in recent years of very large blast furnaces and the use of very high blast pressures, several installations of steam-turbine-driven axial-flow blowers have been made. The axial-flow blower has an advantage of greater efficiency than the centrifugal blower. The average efficiency of the centrifugal blower is about 75 per cent, whereas that for the axial blower is about 85 per cent. However, the axial blower attains its greatest efficiency over a much narrower range than that for a centrifugal blower. Consequently, in instances where an axial blower is used, the range for operating the blast furnace in regard to blast pressure and wind rate is very restricted because it must be kept in a range compatible with the characteristics of the blower.

The blast air is conducted to the blast furnace by the **cold-blast line** that leads to the inlet of the hot-blast stoves. The line connecting the stoves to the bustle pipe around the furnace is called the **hot-blast main.** The cold-blast line is provided with a check valve that closes automatically if the pressure from the blower falls below a preset value to prevent gas from coming back from the furnace into the blower. It is also provided with a **snort valve,** usually near the furnace, that releases the cold-blast air to the atmosphere. This valve is used by the blast-furnace operator to decrease the pressure at the furnace without interfering with the operation of the blower.

The volume of the air blast generally is measured at the inlet of the turbo blower. (When reciprocating compressors were used, the air volume was usually calculated from the speed of the compressor and the volume of the compressor cylinders.) When oxygen is added to the air blast, it is introduced into the cold-blast line after the blower if there is sufficient pressure in the oxygen line. Where the oxygen pressure is low, it is introduced into the blower intake. Instruments keep the ratio of oxygen to the volume of the intake air at the desired proportion. (With steam-engine-driven reciprocating compressors, there is so much oil and grease in the cold-blast line that oxygen cannot be added there and must be introduced instead directly into the hot-blast main.)

In an attempt to improve the efficiency of blowers, several experimental gas-turbine-driven blowers have been tried; however, there are none of these operating at present in the United States.

SECTION 5

CHEMISTRY OF THE BLAST-FURNACE PROCESS

PRODUCTION OF HEAT AND REDUCTION OF IRON

When the burden materials and coke that are charged into the top of the blast furnace descend through the stack, they are preheated by the hot gases ascending from the hearth. As a result of this preheat, the coke burns with great intensity when it reaches the lower portion of the furnace adjacent to the tuyeres and comes in contact with the hot-blast air. However, because of the very high temperature (1650°C or about 3000°F) and the large quantity of carbon (C) present as coke, the carbon dioxide (CO_2) formed is not stable and immediately reacts with additional carbon to form carbon monoxide (CO). Consequently, the combustion of carbon (coke) in the blast furnace can be expressed by the chemical equation:

$$C + \tfrac{1}{2}O_2 = CO \quad \Delta H = -110\,458 \text{ kJ/kmol*} \quad (1)$$

In modern blast-furnace operation, between 250 and 400 kilograms of carbon react in this manner for every metric ton of hot metal produced: this is equivalent to 500 to 800 pounds of carbon per net ton of hot metal. This reaction is the main source of heat for the smelting operation and also produces a reducing gas (CO) that ascends into the furnace stack where it preheats and reduces most of the iron oxide in the burden as it descends to the hearth.

Any moisture (H_2O) in the blast air also reacts with some of the carbon from the coke in the combustion zone. This reaction does not produce heat as combustion does but, rather, consumes heat. However, for every unit of carbon, this reaction produces more reducing gas than that produced when carbon is burned in air. (When carbon burns in air, it produces only one unit of CO, but when it reacts with H_2O, it produces one unit of CO and one unit of H_2). Consequently, in certain instances, where the inherent reduction rate of the burden materials is lower than normal and where a relatively high hot-blast temperature is available—between 1000°C and 1100°C (1832°F and 2012°F)—it has been advantageous to keep the moisture content of the blast at a uniformly high level by moisture additions to increase the reducing power of the blast-furnace gas. This chemical reaction is expressed by the following equation:

$$C + H_2O = CO + H_2$$
$$\Delta H = +131\,378 \text{ kJ/kmol} \quad (2)$$

The ascending gases start to reduce the iron oxide of the burden in the upper portion of the blast furnace where the temperature is below 925°C (1700°F). At this temperature, chemical equilibrium prevents all of the CO and H_2 from being used for reduction. Consequently, the molecular ratio of CO or H_2 to iron oxide must be approximately three times the amount shown by the following stoichiometric equations:

$$\tfrac{1}{2}\,Fe_2O_3 + 3/2CO = Fe + 3/2CO_2 \quad (3)$$
$$\Delta H = -12\,866 \text{ kJ/kmol}$$

$$\tfrac{1}{3}\,Fe_3O_4 + 4/3CO = Fe + 4/3CO_2 \quad (4)$$
$$\Delta H = -3\,904 \text{ kJ/kmol}$$

$$FeO + CO = Fe + CO_2 \quad (5)$$
$$\Delta H = -16\,108 \text{ kJ/kmol}$$

$$\tfrac{1}{2}\,Fe_2O_3 + 3/2H_2 = Fe + \quad (6)$$
$$3/2H_2O \quad \Delta H = +48\,953 \text{ kJ/kmol}$$

$$\tfrac{1}{3}\,Fe_3O_4 + 4/3H_2 = Fe + 4/3H_2O \quad (7)$$
$$\Delta H = +51\,041 \text{ kJ/kmol}$$

$$FeO + H_2 = Fe + H_2O \quad (8)$$
$$\Delta H = +25\,104 \text{ kJ/kmol}$$

In the past, this type of reduction was called "indirect reduction" in contrast to the type occurring at

---

*To convert kJ/kmol to cal/g, multiply by 0.2390: to convert kJ/kmol to Btu/lb mol, multiply by 0.4299.

higher temperatures that was called "direct reduction." However, this nomenclature has become confusing because these same chemical reactions are called "direct reduction" in describing the Wiberg, the HIB, the FIOR and similar processes (see Chapter 14). For this reason, these terms are not used as generally as they were in the past.

The portion of iron oxide that is not reduced in the upper part of the furnace where the temperature is relatively low must be reduced in the lower part where the temperature is very high. Since CO_2 and H_2O are not stable at these temperatures in the presence of large quantities of coke, they react with C almost as rapidly as they form. Consequently, the over-all reduction reaction in this part of the furnace can be represented by equation (10) no matter whether H_2 or CO does the reduction. As can be seen, equation (10) is obtained by algebraically adding either equation (5) and (9) or equations (8) and (2).

$$FeO + CO = Fe + CO_2 \quad \Delta H = -16\,108 \text{ kJ/kmol} \quad (5)$$
$$\underline{CO_2 + C = 2CO \quad \Delta H = +172\,590 \text{ kJ/kmol} \quad (9)}$$
$$FeO + C = Fe + CO \quad \Delta H = +156\,482 \text{ kJ/kmol} \quad (10)$$

$$FeO + H_2 = Fe + H_2O \quad \Delta H = +25\,104 \text{ kJ/kmol} \quad (8)$$
$$\underline{H_2O + C = CO + H_2 \quad \Delta H = +131\,378 \text{ kJ/kmol} \quad (2)}$$
$$FeO + C = Fe + CO \quad \Delta H = +156\,482 \text{ kJ/kmol} \quad (10)$$

Reaction (10) absorbs a large amount of heat; consequently, the larger the amount of reduction occurring in this way, the larger the quantity of heat that must be supplied to the furnace. Notice also that reaction (10) produces CO, which is the gas used in reactions (3), (4) and (5) in the blast-furnace stack. In most instances the most efficient operation is obtained when about one-third of the reduction is done according to equation (10) and the balance according to reactions (3) to (8).

The heat for the process is not produced entirely by the combustion of coke, since at most blast furnaces about 40 per cent is supplied by the sensible heat of the hot-blast air. A portion of the fuel can be economically injected through the tuyeres as natural gas, tar, fuel oil or pulverized coal. In such instances, the carbon in the fuel burns to CO but because of the large amount of coke present the hydrogen remains as H_2 and is not oxidized until it reduces iron oxide somewhere above the tuyeres.

REDUCTION OF MANGANESE, PHOSPHORUS AND SILICON

At the temperature in the upper part of the blast-furnace stack some of the higher oxides of manganese are reduced by CO according to the following chemical equations:

$$MnO_2 + CO = MnO + CO_2 \quad (11)$$
$$\Delta H = -147\,904 \text{ kJ/kmol}$$

$$Mn_3O_4 + CO - = 3MnO + CO_2 \quad (12)$$
$$\Delta H = -51\,254 \text{ kJ/kmol}$$

However, the lower oxide of manganese (MnO) cannot be further reduced by CO or hydrogen at any of the temperatures that are encountered in the stack. Consequently, the only chemical equation that can be used to express the final reduction of this element is as follows:

$$MnO + C = Mn + CO \qquad (13)$$
$$\Delta H = +274\,470 \text{ kJ/kmol}$$

This reaction takes place only at temperatures above 1500°C (2732°F) and absorbs large quantities of heat. At higher temperatures the percentage of Mn that can be reduced in a blast furnace increases, but in most basic practices this amounts to about 65 to 75 per cent of the manganese charged. The manganese that is reduced dissolves in the hot metal while the unreduced portion remains as part of the slag.

The reduction of SiO_2 takes place only at very high temperatures according to the following chemical equation:

$$SiO_2 + 2C = Si + 2CO \qquad (14)$$
$$\Delta H = +658\,562 \text{ kJ/kmol}$$

The rate of this reaction is relatively slow but accelerates with an increase in temperature. For any particular burden and slag composition the silicon content of the hot metal is proportional to the hot-metal temperature. The percentage of silicon in the hot metal can be increased by increasing the silicon content of the charge and the coke input. The percentage of silicon (Si) in the hot metal is also affected by production rate —silicon decreases as production rate increases.

The reduction of phosphorus is most conveniently expressed by the equation:

$$P_2O_5 + 5C = 2P + 5CO \qquad (15)$$
$$\Delta H = +995\,792 \text{ kJ/kmol}$$

The final reduction of phosphorus also takes place only at very high temperatures; however, unlike Mn and Si the phosphorus is almost completely reduced. For this reason, almost all of the phosphorus in the charge will dissolve in the hot metal. The only means of controlling the phosphorus content of the hot metal is by limiting the amount in the charge.

ELIMINATION OF SULPHUR

Sulphur enters the blast furnace mainly in the coke and is released into the blast-furnace-gas stream as H_2S or a gaseous compound of carbon monoxide and sulphur (COS) when the coke is burned. As the gas ascends through the stack some of the sulphur combines with lime of the flux and some combines with the iron. The exact mechanism of the reaction by which sulphur combines with iron is not known; however, it is generally believed to be as follows:

$$FeO + COS = FeS + CO_2 \qquad (16)$$
$$\Delta H = -80\,124 \text{ kJ/kmol}$$

The sulphur that combines with the iron must be removed at the very high temperatures that exist in the hearth. This is done by reduction of the iron sulphide in the presence of a basic flux such as lime (CaO). The chemical equation for this reaction is often written as follows:

$$FeS + CaO + C = CaS + Fe + CO \qquad (17)$$
$$\Delta H = +182\,422 \text{ kJ/kmol}$$

The amount of sulphur removed depends on the temperature in the hearth, the slag volume, and the ratio of basic oxides lime (CaO) and magnesia (MgO) to acid oxides silica (SiO_2) and alumina (Al_2O_3) in the slag. For a more complete explanation of the effect of slag composition on fluidity and melting temperature, please refer to Chapter 13 on "Physical Chemistry of Iron- and Steelmaking."

REACTION OF LESS-ABUNDANT ELEMENTS

In addition to the elements that are normally considered in reporting the chemical composition of an iron-bearing material (that is, Fe, P, Mn, SiO_2, Al_2O_3, CaO, MgO and S) there are a number of the less-abundant elements that undergo chemical reactions in the blast furnace. Some of these can cause considerable operating difficulty and some can contaminate the product and make it unsuitable for certain steelmaking applications. The source of these elements is not only from natural iron ores, but also from waste materials such as scrap, steelmaking dust, grindings, and so on, that are recycled to the blast furnace. Some of the more important of these elements are arsenic, barium, chlorine, chromium, cobalt, copper, fluorine, lead, molybdenum, nickel, potassium, sodium, tin, titanium, vanadium and zinc.

Arsenic is found in a number of iron ores, particularly ores from Russia and Turkey. The behavior of arsenic is very much like that of phosphorus, in that it is almost completely reduced and dissolves in the hot metal. It increases the fluidity of the hot metal and, for that reason, it appears to increase the wear of refractories. It is not completely removed during the steel refining process and imparts brittleness to the steel.

Barium occurs as a very basic oxide in some iron and manganese ores. It is not reduced in the blast furnace but becomes part of the slag, increasing the slag basicity. It may cause difficulty in controlling the metal composition if the operator is not aware of its presence.

Chlorine occurs as alkali chlorides in several iron ores and as a contaminating compound in ores processed with sea water. In the high-temperature zone of the blast furnace these compounds are volatilized and as they rise toward the top of the furnace they condense around cooling plates and cause corrosion. They also condense in uptakes and downcomers where they form accretions that can eventually restrict the passage of the top gas.

Chromium is found in some ores and is reduced to a certain extent depending on the basicity of the slag and the operating temperature. Normally, about 50 to 60 per cent of the chromium is reduced into the hot metal.

Cobalt, copper and **nickel** occur in several different ores and particularly in Cuban ores. They are also present in iron-bearing tailings from the copper industry that are sometimes sintered and used in blast furnaces to recover the iron. All three of these elements are reduced almost completely into the hot metal and are not oxidized in the steel refining process. Consequently, in operations that produce steel that must meet stringent ductility specifications, such ores cannot be used.

Fluorine compounds are found in several western United States ores and behave somewhat like chlorine compounds.

Lead occurrence in iron ores is quite rare, although there are some ores from Spain and Portugal that contain appreciable amounts. Lead is reduced in the blast furnace; however, it has a very high vapor pressure and a large portion of it vaporizes. As the lead vapor rises through the furnace stack, it oxidizes into small particles that escape to the dustcatcher. Molten lead is not soluble in molten pig iron (hot metal) and in blast furnaces using lead-bearing ore, the liquid lead collects in a pool below the hot metal in the hearth. In several furnaces in Portugal, a special tapping hole is provided so that the lead can be drained from the hearth every few weeks.

Molybdenum and **tungsten** occur very rarely and only in such minute quantities that they can be ignored. If any compounds of these elements were present in the blast furnace, they would be at least 90 per cent reduced into the hot metal.

Potassium and **sodium** are alkalies that occur quite frequently in iron ores and also to a large extent in coke ash. They cause particular operating difficulty because they are reduced in the blast furnace at a temperature below the temperature of the hot metal but above their own vaporization temperature. As explained in an article on the effect of alkalies on blast-furnace operation by R. L. Stephenson (see bibliography at end of chapter), the alkali metals are vaporized near the hearth of the blast furnace and condense on the coke and burden materials as well as on the blast-furnace inwalls. Thus, these elements recycle and build up in the furnace, causing deterioration of the coke and forming accretions on the walls.

Tin is an element that enters the blast furnace mostly by way of recycled materials such as scrap or sintered dusts. It is almost entirely reduced and dissolves in the hot metal.

Titanium is found with iron in several important iron-bearing deposits in the Adirondack Region and in Alaska and Russia. It also occurs in iron sands in Australia and Japan. It cannot be easily separated from iron by mechanical means because it is usually in chemical combination with the iron as $FeTiO_3$. The titanium oxide is reduced to titanium carbide in the blast furnace and also to carbonitrides. These materials have a limited solubility in iron and often can be found in iron salamanders as their presence makes the iron viscous and raises its solidification temperature. If the TiO_2 content of the burden exceeds 13 kilograms per metric ton (26 pounds per net ton) of iron, the slag and the metal are usually very viscous and the iron forms skulls in the trough, the runners and the ladle. Because of this tendency for causing the metal to solidify at a higher temperature, titanium-bearing materials are sometimes added to the burden in the range of about 5 to 7 kilograms per metric ton (10 to 14 pounds per net ton) of iron, when it is determined that the bottom of the hearth is eroding too rapidly, to start solidification of the salamander and decrease the rate of hearth erosion.

Vanadium occurs and behaves in a manner somewhat similar to chromium. About 50 per cent of the vanadium in the burden is reduced and enters the hot metal.

Zinc is a very troublesome element and causes many serious operating problems. It is found in several iron ores and in copper residues that are sintered for blast-furnace use. It is also a component of steelmaking flue

Table 15-V. Blast-Furnace Energy Balance.

	Gigajoules per Metric Ton of Hot Metal	Million Btu per Net Ton of Hot Metal
Energy Input		
Blast sensible heat	1.64	1.41
Coke combustion	2.10	1.81
Injected fuel combustion	0.27	0.23
Decomposition of water	−0.26	−0.22
Total	3.75	3.23
Energy Output		
Reduction of iron oxides	1.17	1.01
Reduction of metallords	0.15	0.13
Slag sensible heat	0.48	0.41
Hot metal sensible heat	1.36	1.18
Top gas sensible heat	0.20	0.17
Vaporization of moisture from burden	0.13	0.11
Furnace heat loss	0.26	0.22
Total	3.75	3.23

dust, particularly where galvanized scrap is used. Its chemical reactions in the blast furnace are similar to those of sodium, for it is reduced and vaporized at a temperature below that of the hot metal, and it condenses before it can leave the blast furnace. Some of the ZnO goes to the dustcatcher with the flue dust but the balance recycles in the furnace or forms accretions on the walls and reacts with the refractory lining. Because of the trouble caused by this element, the zinc content of the burden must be kept below 0.4 kilogram per metric ton (0.8 pound per net ton) of hot metal.

Selenium and **tellurium**, though somewhat rare, may be present in some raw materials; in their reactions they are somewhat similar to sulphur but with an even greater tendency to remain with the metal.

BLAST-FURNACE MATERIAL AND ENERGY BALANCE

A better appreciation of the quantity of material and the amount of energy required for producing hot metal in the blast furnace can be obtained by referring to the material balance in Figure 15—41 and the energy balance in Table 15—V. These balances are for a modern blast furnace with a working volume of 2889 cubic metres (102 000 cubic feet) operating with a wind rate of 6565 normal cubic metres per minute (245 000 standard cubic feet per minute) and a hot-blast temperature of 1056°C (1933°F). The top pressure is 206.8 kPa (30 psig), and the blast air is enriched with oxygen to about 22.9 per cent oxygen. The production rate of this furnace is about 8725 metric tons of hot metal per day (9615 net tons of hot metal per day).

The net energy introduced into the process by the combustion of coke and tuyere-injected fuel to CO and H_2 and the sensible energy of the blast is 3.75 gigajoules per metric ton of hot metal (about 3.23 million Btu per net ton of hot metal). This energy is re-

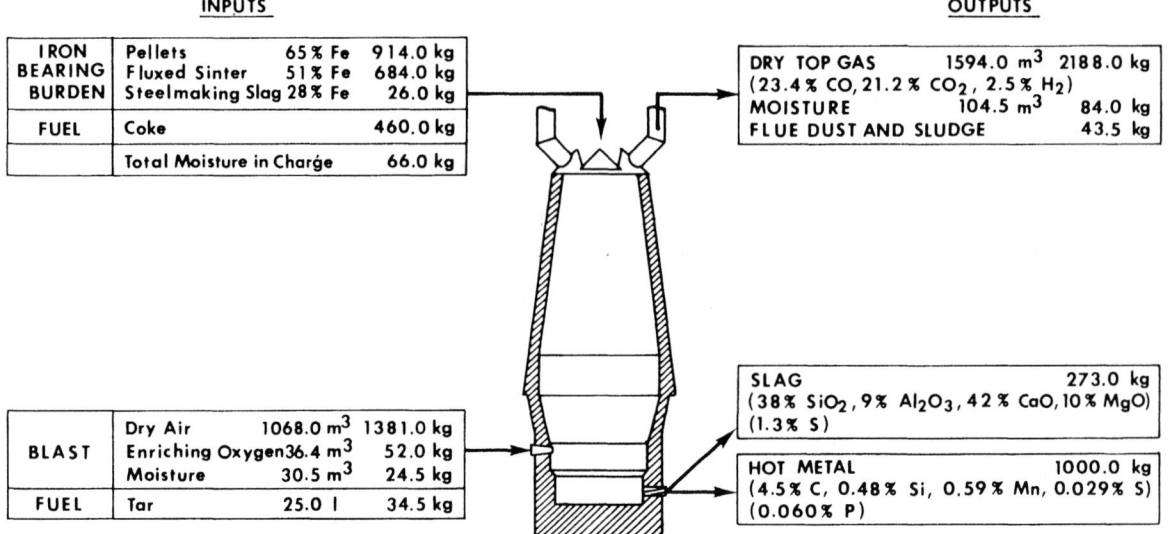

BLAST FURNACE MATERIAL BALANCE

(All Quantities in Amount per Metric Ton of Hot Metal)

INPUTS

IRON BEARING BURDEN	Pellets	65% Fe	914.0 kg
	Fluxed Sinter	51% Fe	684.0 kg
	Steelmaking Slag	28% Fe	26.0 kg
FUEL	Coke		460.0 kg
	Total Moisture in Charge		66.0 kg

BLAST	Dry Air	1068.0 m³	1381.0 kg
	Enriching Oxygen	36.4 m³	52.0 kg
	Moisture	30.5 m³	24.5 kg
FUEL	Tar	25.0 l	34.5 kg

OUTPUTS

DRY TOP GAS   1594.0 m³ 2188.0 kg
(23.4% CO, 21.2% CO_2, 2.5% H_2)
MOISTURE   104.5 m³   84.0 kg
FLUE DUST AND SLUDGE   43.5 kg

SLAG   273.0 kg
(38% SiO_2, 9% Al_2O_3, 42% CaO, 10% MgO)
(1.3% S)

HOT METAL   1000.0 kg
(4.5% C, 0.48% Si, 0.59% Mn, 0.029% S)
(0.060% P)

FIG. 15—41. Blast-furnace material balance.

quired for the reduction of iron oxides and metalloids, the sensible heat of hot metal, slag and top gas, vaporization of the burden moisture, and furnace heat loss. In the process, top gas is produced with a chemical heat content of 5.1 gigajoules per metric ton of hot metal (4.4 million Btu per net ton of hot metal), which is used for heating the blast air before it enters the tuyere, and for other uses external to the blast furnace.

SECTION 6

OPERATION OF THE FURNACE

Blowing In—The process of starting a blast furnace is called blowing in. It is made up of several steps which include drying out the lining, filling the furnace with a specially arranged high-coke blow-in charge, igniting the coke, and gradually increasing the wind rate with frequent casts to assure raising the temperature of the hearth. During this period, the ratio of burden to coke is adjusted according to a predetermined schedule until good-quality hot metal is obtained and normal operations are established.

Drying—Newly constructed furnaces must be carefully dried before the coke is ignited because the large amount of water contained in the slurry used for bricklaying and the water absorbed by the brickwork must be driven off to avoid extreme thermal shock. Furthermore, if the water from these sources is not removed from the furnace before it is put into operation, it will absorb more heat than that provided for the blow-in charge and will prevent the hearth from reaching the desired temperature. In such instances, metal and slag entering the hearth might freeze there and it would be impossible to remove them from the furnace.

Any one of several methods can be used to dry a furnace and stoves. The usual method for stoves, where

natural gas is available, is to put a gas pipe in the lower combustion chamber and start with a small flame and increase gas input for several days until a small quantity of blast-furnace gas can be used, keeping the natural gas as a pilot light. It is desirable to increase the heat slowly for at least 10 days to 2 weeks in a new stove before starting to bring the unit up to operating temperatures. Stoves which have had previous service have been warmed up in from 36 to 72 hours without apparent difficulty. Another method of drying or heating, where natural gas is not available, is a wood fire built in the bottom of the combustion chamber. The wood fire requires consistent attention until wall temperatures are sufficient to insure proper combustion of blast-furnace gas.

The best method for drying a blast furnace is the use of hot blast. It is simple and drying is under control at all times. In applying this method, the conventional hot-blast system is used except that initial blast temperature is held to about 205°C (400°F) and wind volume at a low blast level. Temperature is slowly increased for a couple of days to 425°C (800°F) and maintained at that temperature for a few additional days. The entire operation can be accomplished in a week. Some operators install elbows and pipes inside a few of the tuyeres to direct heated air down to the hearth. It is desirable during the latter part of the period to turn low-pressure steam into the hearth-cooling staves and bosh plates to assist drying. The large bell is closed while drying, and furnace bleeders are adjusted to retain as much heat as possible. No pressure reading should be observable at the blast-pressure gauge while drying.

In single-furnace plants where blast-furnace gas is not available, the method just described must be modified because an auxiliary fuel of either natural gas or oil must be used to heat the stoves. However, for the low level of heat required, this can be done satisfactorily.

Other less desirable methods of drying out are:

(1) **Dutch Ovens**—Two, three or more furnaces or ovens are constructed outside the blast furnace and the products of combustion from the ovens plus excess heated air are directed through pipes into the tapping hole and some tuyere openings. Other tuyere and cooler openings are blocked off and draft regulated by adjusting furnace bleeders. As in hot-blast drying, the large bell remains closed. The dutch ovens are fired with coke, coal or wood and require a crew to maintain the fires, haul fuel, and clean out ashes. Temperature control is more difficult in this case, but some regulation is maintained by the intensity of the oven fires and adjustment of furnace bleeders.

(2) **Gas**—The least desirable method is an open gas flame inside the furnace. The practice was to install a gas pipe through the tapping hole and ignite the gas by a small wood fire maintained at all times. Tuyere openings are equipped with shutters to regulate air input, and the top temperature of the furnace is regulated as in the two previous methods. Obvious hazard of this method precludes further discussion and it may be regarded as obsolete.

(3) **Hearth Fire** —Another method, for lack of a better name, will be called herein the **hearth-fire** method. It simply consists of a wood, coke or coal fire built in the hearth of the furnace and controlled similarly to the other methods with tuyere shutters and bleeders. Fire temperature regulation is difficult and frequent replenishment of fuel interrupts the process. If used for drying only, this method is as simple as stated, but it could be applied as the start of a controlled or slow blow-in process. Some operators claim that improved "warm-up" is achieved through this technique. After drying, as noted above, a very heavy coke blank is charged, followed by a regular blow-in burden. A low blast is then maintained, and hearth and inwall temperatures slowly raised before wind rates are increased to conventional blow-in rates. Several days may elapse before actual blow-in takes place. In theory, the scheme has merit in that there is less chance of damage to brick from thermal shock with reduced tendency for the brick to spall; however, the process is time-consuming and more expensive. Superior campaign life has been obtained through the use of conventional methods, and no evidence to date has been observed to indicate superior results from slow blow-in techniques.

Filling—At the conclusion of the drying operation, furnace bells and bleeders are opened, blow pipes taken down, some tuyeres and coolers removed, if desirable, and paraphernalia around the cast house used in the drying process is cleared away. In a relatively short time the interior of the furnace will have cooled sufficiently to permit entering the hearth to prepare for filling. In the event a hearth fire was used for drying, all ashes and refuse are cleaned out.

An inspection of the furnace-cooling system will now be made and water turned on. It is usual practice to maintain a constant watch on all cooling members starting at this time. In cases where the furnace lining is not new and the furnace is empty, the above inspection affords an excellent opportunity to find leaking cooling members. Any plates considered suspicious or showing a sign of moisture must be replaced before filling is started. Many hours of grueling and frustrating work may be avoided by attention to these details. In fact, this checkout detail applies to all mechanical, electrical and physical equipment before filling is started. Usually, a check list is prepared, and each item checked off as reports are received that the item is satisfactory.

Other preparations have preceded and some are going ahead at this point coincident to the start of the filling operation. For example, iron ladles are being heated, stockhouse bins filled according to prearranged plan, cinder ladles, if used, are being prepared for service, furnace gas lines inspected and vents opened or closed as directed. Steam purging lines are checked to make sure steam will be available when needed and the cast-house iron and slag runners prepared. Activity in the power house and blowing-engine house is going on with equal intensity. Perhaps turbines were opened for inspection and numerous preventative maintenance chores performed. Boilers are being warmed up, turboblowers warmed and turned over and over, speed-trip mechanism checked out, water turned into condensers, pumps inspected and placed in operation, wind schedules posted and reconfirmed with blast-furnace personnel, boiler-water treating plant started and switch house and sub-station activated.

Some of the final activities before starting up a blast furnace are installing and drying the lining materials for the troughs and runners, setting the skimmers and gates and making certain that the mud guns and drills are mounted correctly so that they will contact the iron notch properly. In some instances where the blast furnace is to be operated at very high pressure, the furnace is pressurized with cold-blast air prior to the start up to determine if there are any leaks that must be eliminated.

There are many different arrangements for placing the start-up charge in the furnace. In most cases, it consists of only coke and a small amount of flux below the mantle. Above the mantle there is a gradual increase in the burden-to-coke ratio. Very often, wooden cribbing (usually old railroad ties) is placed in the hearth and around the tuyeres so that there is sufficient open space for the blast air entering the furnace. In some blow-ins, blast-furnace slag and limestone are charged with the coke in the bosh region. The purpose of the slag is to provide a material that will melt readily and carry heat down into the hearth. The limestone is to combine with the coke ash and form a fusible slag. To increase the slag volume during the early period of the start up, silica gravel is sometimes added.

Percent of working volume	Metric tons of coke charged *	Sinter to coke ratio	Blast furnace slag to coke ratio	Gravel to coke ratio	Limestone to coke ratio
9	80 (88.2)	1.70	0.04	0.110	0.09
10	95 (104.7)	1.20	0.12	0.100	0.12
5	63 (69.5)	0.70	0.18	0.085	0.12
7	80 (88.2)	0.50	0.21	0.085	0.12
6	67 (73.8)	0.30	0.23	0.077	0.13
8	95 (104.7)	0.20	0.20	0.068	0.14
7	80 (88.2)	0.10	0.20	0.068	0.14
5	63 (69.5)	0.05	0.21	0.068	0.14
5	63 (69.5)	0.01	0.17	0.064	0.15
16	190 (209.4)	0	0.13	0.054	0.14
22	285 (314.1)	0	0.03	0.042	0.15

Wooden cribbing in hearth

* Net-ton equivalents in parentheses

FIG. 15—42. Start-up filling for a large blast furnace with a working volume of 2800 m³ (98 880 ft.³).

Figure 15—42 shows a start-up filling that was used in a new, large blast furnace with a working volume of 2800 cubic metres (98 880 cubic feet). In this instance, the burden was 100 per cent fluxed sinter. Consequently, as the sinter-to-coke ratio was increased toward the top of the furnace, the limestone-to-coke ratio was decreased and the gravel-to-coke ratio was increased slightly.

In calculating the relative amounts of coke, burden and flux, it is anticipated that the first few casts of iron will contain 3.5 to 4.5 per cent of silicon. Consequently, the amount of basic flux must be planned accordingly so that the slag will not be too limy. As the furnace starts to move normally and the wind rate is increased, the silicon content of the hot metal will drop and then additional flux must be provided for the silica that remains unreduced in the slag.

Lighting —Several different methods can be used to light the blast furnace; however, the one most commonly used at present is igniting the coke with hot-blast air. In this method, a relatively low blast volume is first used at a temperature ranging between 550°C and 650°C (1022°F and 1202°F) and within a matter of minutes the coke at these tuyeres will ignite. To use this method, it is necessary to have the stoves preheated. In a single-furnace plant or in a multi-furnace plant where all of the furnaces are off, this may present a problem. Frequently where blast-furnace gas is not available, natural gas is used to preheat the stoves until the gas from the blast furnace can be used.

Another method, used less frequently, is to place easily combustible material in front of the tuyeres that can be ignited with torches or red-hot bars. The natural draft through the furnace will usually provide enough air to raise the temperature of the coke near the tuyeres to the combustion temperature, and then a light flow of blast air can be started.

Before the furnace is lighted, the gas system is isolated from the furnace by a goggle valve, usually located downstream of the gas-cleaning system. The gas-cleaning system is purged from this goggle valve to the furnace with steam or some type of inert gas. During the early stages of the blow-in, the bleeders at the top of the furnace are kept open and a purging gas is kept in the gas-cleaning system up to the valve that separates the gas-cleaning system from the main plant gas system. Gas-cleaning systems may vary from furnace to furnace, and sometimes separate sections of the gas-cleaning system are isolated from one another and purged separately. In most instances, there is no isolating valve between the dustcatcher and the furnace and, consequently, the dustcatcher is purged with steam that is vented through the blast-furnace downcomer and discharged through the furnace bleeders. When the wind rate at the blast furnace has been increased sufficiently to maintain a good, steady flow of gas, the purging medium is shut off and the bleeders are closed. When a positive pressure of gas is obtained up to the isolating goggle valve, it is opened and the gas is allowed to flow into the main gas system. Significant precautions must be taken throughout this period, and the area around the furnace, the gas-cleaning system and the valves must be monitored for potential gas leaks.

When the blast furnace is started up, the wind rate is only a small fraction of what it is to be at normal operation. Consequently, it is customary to open only a few of the tuyeres so that the velocity through each tuyere will be sufficient to carry the blast well into the furnace and keep the hot gases from channeling up along the walls. This is usually accomplished by plugging up the tuyeres that are not to be used with clay balls and poking them open later as the wind rate is increased. In some instances, bushings have been placed inside the tuyeres to decrease the diameter of the opening during the early stages of the blow-in. These are then knocked out when the higher wind rates are used. At first, the wind rate is increased every few hours so that by the end of the first 24 hours it is approximately 40 to 50 per cent of full wind. After this, it is increased about 15 per cent per day until the ultimate rate is attained.

When the furnace is started, the iron notch is usually left open, and some of the gas generated by the combustion of the coke exits through the tapping hole. At first, this gas must be ignited, but as the wind rate increases it heats the coke in the hearth and ultimately discharges at a high-enough temperature to remain lit. The tapping hole generally remains open until slag begins to appear. This is indicated by a decrease in gas volume at the tapping hole and a puffing or pulsation. When this occurs, the mud gun is swung into position and the hole is plugged. Every few hours after that, the hole is reopened to remove any slag that has formed. On furnaces with more than one iron notch, a different hole is opened with each successive attempt.

Within approximately 24 to 28 hours after the start up, the first iron will be cast. This iron will be very high in silicon content because of the slow operating rate and the high ratio of silica to iron oxide in the materials in the first portion of the furnace charge. As the wind rate is increased, and the burden-to-coke ratio approaches the normal operating level, the silicon content of the metal decreases. Consequently, the ratio of flux to the other materials must be adjusted to provide sufficient basic oxide to flux the unreduced silica that reports to the slag.

Generally during the blow-in period, it is advisable to keep the ratio of burden to coke in a range that will produce high-temperature hot metal at about 1500°C (2732°F) to assure that some unforeseen problem will not cause a freeze-up. The hot-blast temperature is not pushed to the maximum output of the stoves during this period because the operators prefer to have a reserve that will enable them to introduce additional heat into the hearth in case of an emergency.

ROUTINE OPERATIONS

After a blast furnace has been blown in, it is expected to remain in operation continuously for at least 5 or 6 years with only short outages for maintenance or repair of some of the equipment that is subjected to extreme wear or for replacement of damaged cooling devices. Once the wind rate and the hot-blast temperature have reached their normal operating level, a routine casting schedule is established. This schedule depends to a great extent upon the physical quality and the chemical composition of the burden materials that are available. For example, if the burden materials are low in iron content and high in gangue content, the hot-metal production rate will be relatively slow and it may be necessary to flush slag from the furnace between casts. In this case, the furnace may be cast only 6 times a day.

After each cast, the trough may be drained so that slag and scrap can be removed from the trough and runners. Between casts, the trough and runners are cleaned and resurfaced with a protective material. Gates and dams are also reset and dried out in preparation for the next cast.

At furnaces where very good quality burden materials are used and the hot-metal production rates are quite high, the furnace may be cast as often as 13 or 14 times a day. In such instances, the trough may not be drained after each cast, but the hot metal left in the trough may be protected from losing too much heat until the next cast.

On some large blast furnaces with multiple tapping holes, the casting can be made almost continuous. In such cases, two tapping holes are used alternately while the trough at the third is rebuilt or repaired. The drill used for opening the tapping hole is of such a size that the hot metal flows out through it at a slightly slower rate than that at which it is accumulated in the hearth. As the cast proceeds, the tapping hole erodes slightly so that the casting rate then exceeds the rate at which hot metal accumulates in the hearth. When this occurs, the tapping hole is plugged up by the mud gun, and the other hole is opened.

SAMPLING THE IRON

Sampling the pig iron for chemical analysis is an important function at each cast. The samples are taken from the main runner beyond the skimmer. The first sample is taken when the first ladle is between three-quarters and completely full, and additional samples are taken at just about the time that each subsequent ladle is completely filled. The samples are rushed to the chemical laboratory for analysis for silicon, sulphur, manganese and phosphorus so that the results will be available before the hot metal is used in a steelmaking process. From observation of the sparking of the liquid metal in the runner, the silicon content can be estimated fairly accurately, and from observation of a test piece poured in the cast house, the sulphur content can also be estimated. In shops where the laboratory facilities are not able to provide rapid analyses, these methods for estimating silicon and sulphur contents may serve as a preliminary analysis of the cast for making decisions regarding disposition of the hot metal. In most plants, a sample of the slag is also analyzed at each cast. In some older plants, the composition of the slag is estimated from observation of a slag sample, solidified in a test mold, and portions of the slag samples from each cast are combined for a daily analysis by the chemical laboratory.

CHARGING THE FURNACE

The charging of the blast furnace is an extremely important part of the blast-furnace operation and must be done correctly to obtain the best operating performance. Several requirements for satisfactory charging include: keeping the furnace filled to the desired level, drawing the correct weight of each material for each skip load and keeping the charge materials in the proper sequence. At a large number of the plants in the United States, the blast furnaces are charged by a scale-car operator who draws the burden materials from the stockhouse into the scale car until the indicator on the scale shows that he has drawn the ordered amount of that particular material. He then deposits the material into the skip according to the sequence that has been established for him to follow. A signal from the stockline recorder lets him know when to activate the skip hoist for each round of charges. In these instances, the opening and closing of the small bell are automatically interconnected with the travel of the skips, and as soon as the prescribed number of skip loads has been deposited through the small bell onto the large bell, the large bell opens (with the small bell closed) and the material is charged into the furnace. At a few of the older furnaces, both the coke and the burden materials are measured out with the scale car, but in most instances, the coke is charged automatically by two weigh hoppers. The coke is fed to the weigh hoppers over screens for removal of fines, and when the load cells on which the hopper is mounted indicate that the hopper has received the correct weight, the coke feeders are shut off automatically. At the proper position in the charging sequence, the coke is automatically discharged into the skips, and the feeders are started so that a weigh-hopper full will be ready when the sequence requires another skip load of coke.

In some plants, nuclear gauges are used to determine the moisture content of the coke in the weigh hopper, and these gauges generate an electronic signal that automatically resets the shut-off point of the coke feeder so that the weight of dry coke in each skip load is kept as nearly constant as possible. At many of the more modern blast furnaces, the charging of all the materials is done automatically with belt feeders drawing the burden materials into weigh hoppers, and with the weigh hoppers discharging into the skips in accordance with the established charging sequence.

The size of the charge depends to a large extent on the size of the blast furnace, but in some cases special consideration must also be given to other factors such as the size of the large-bell hopper and the capacity of the skip hoist. The total volume of all the materials in a full charge is generally the equivalent of a layer about 1 to 2 metres (3.3 to 6.6 feet) deep on the cross-sectional area of the stockline.

The sequence of the skip loads of different materials in the charge is an important factor in determining the radial distribution in the furnace which in turn affects the contact between the hot gases rising in the furnace and the solid particles of the charge. The gas-solid contact must be optimized to obtain the required heat transfer and to permit desirable chemical reactions.

Because of variations in the quality of the burden materials and also in the physical characteristics of different blast-furnace tops, charging sequences will differ from plant to plant. One of the popular sequences for the skip loads of materials, for example, is burden-coke-limestone-burden-coke-coke. After these six skip loads are placed on the large bell in this order, the large bell is opened and the charge is dumped into the furnace. This particular filling sequence is usually designated by a set of symbols as follows: OCSOCC/. Each letter denotes a skip load of material, and the slash (/) indicates the point in the sequence when the large bell is dumped. The letter O is used to denote iron-bearing material. Originally, this letter stood for ore, but currently the letter O is used to denote sinter or pellets also. The letter S (originally used as the symbol for limestone) denotes flux such as limestone or dolomite. C is the designation for coke, and in some instances M is used to indicate miscellaneous material. At some plants, this filling is split into two sections and in such cases is called a **split filling** and its designation in symbols is OCS/OCC/.

Generally, the material charged on the large bell first is thrown close to the furnace wall. In the preceding example of a popular filling, the first material on the large bell is ore (iron-bearing material). The reason for this is that iron-bearing materials are usually the finest-size material in the charge and, by keeping them close to the walls, the hot gases are prevented from channeling up the furnace walls and causing damage to the refractory lining. It also assures better contact between the hot gases and the iron-bearing material. However, there are periods when it is desirable to place a larger percentage of coke near the walls to prevent the formation of accretions or to remove small build-ups of material. In such instances, the coke is placed on the large bell first and the sequence, designated by the symbols COS/CCO/, is called a **reverse filling.**

Another fairly popular type of filling, which is called block or layer filling, is one in which the coke portion of the charge and the iron-bearing burden portion of the charge are placed on the large bell separately and dumped into the furnace separately. The designation for this filling is OOO/CCC/. A similar sequence, but with individual skips being charged from the large bell, such as O/O/O/C/CC/, is generally used at blast furnaces equipped with movable stockline armor. The charging of individual skips in conjunction with the positioning of the movable armor provides the flexibility to charge specified materials either closer to or farther from the furnace inwall, as desired.

OPERATION OF THE STOVES

The temperature of the hot metal is a matter of great importance in controlling the uniformity of the hot-metal quality and, therefore, it is important for the blast-furnace operator to keep the temperature in a specific range. Before the blast-furnace burden materials were beneficiated as well as they are today, there were significant changes in the size and composition of the burden materials from day to day, and this had a significant effect on the extent of exothermic "indirect" reduction of the burden in the blast-furnace stack. When the quality of the material suddenly changed for the worse, the percentage that was reduced by the exothermic "indirect-reduction" reaction decreased and a larger percentage of the burden came to the hearth unreduced and had to be reduced by the endothermic "direct-reduction" reaction that occurs there. When this happened, additional heat was required in the hearth, and the only way it could be supplied quickly was to raise the hot-blast temperature until additional coke could be charged and worked down to the hearth or until the burden could be restored to its normal quality. For this reason, operators were reluctant to use the full capacity of the stoves for normal operation because they preferred to have reserve capacity for raising the hot-blast temperature during such critical periods.

Today, however, with the use of well-prepared burdens and good control of burden distribution, the furnace operation is much more uniform. Consequently, furnaces usually are operated with the maximum hot-blast temperature that the stoves can maintain or that the particular burden materials will accept without causing premature melting and poor burden movement. With higher hot-blast temperature, the blast-furnace operation is more efficient because a larger percentage of the heat consumed is furnished by the blast-furnace off gas, and less solid or liquid fuel is required.

In the operation of the hot-blast system, the ceramic checkerwork of the stoves is heated by the combustion of blast-furnace gas, and then the air from the blowers is passed through the stoves and is heated by the hot checkerwork. In the heating cycle, the stoves are fired until the temperature of the exit gases at the stack valves has reached an established maximum temperature of about 400°C to 450°C (about 750°F to 840°F). The general arrangement of a stove can be seen by referring to Figure 15—18. During the heating cycle the temperature at the dome of the stove is controlled so that it does not exceed a maximum which is determined primarily by the type of refractory material used to line the dome. If the dome temperature reaches this maximum before the stack temperature reaches its maximum, excess air is added through the burner to hold down the flame temperature and prevent the dome from being overheated while the firing is continued until the stack-gas temperature reaches its limit. However, if the dome temperature does not increase rapidly enough to reach its maximum allowable temperature by the time the stack-gas temperature reaches its maximum, the blast-furnace gas is usually enriched with a fuel of higher calorific value to obtain a faster heating rate.

After the stove has been heated, it is ready to be put on blast. This is done by first shutting off the gas and the air supply to the burner and then closing the burner shut-off valve and the chimney valves. The cold-blast valve is then opened in such a way that the air entering the stove will bring it to a pressure equal to the blast pressure without reducing the blast pressure excessively. At some very modern installations, the blower controls are switched from constant-volume

FIG. 15—43. That part of the panel board to the right of the outline of the blast-furnace contains the controls related to the automatic changing of stoves on a large modern furnace. The sequence of operations involved in automatic stove changing is described in the text.

control to constant-pressure control during a stove change. In such a system, the blower speeds up so that the stove can be filled and pressurized rapidly without causing a detectable decrease in the blast pressure. After the stove is filled, the mixer valve (which controls the amount of cold air which is by-passed around the stove to be mixed with the very hot air from the stove to produce the desired hot-blast temperature) is set at approximately the correct opening. The hot-blast valve is then opened to put the stove on blast and, once the stove is on blast, the hot-blast-temperature controller automatically adjusts the mixer valve opening to maintain the desired hot-blast temperature.

The spent stove is then taken off blast by closing first the cold-blast valve and then the hot-blast valve. The blow-off valve is then opened to depressurize the stove and, after depressurizing, the chimney valves are opened and the blow-off valve is closed. Next, the burner shut-off valve is opened, and the air supply to the burner is turned on. Finally, the gas shut-off valve is opened to obtain the desired gas-flow rate.

At some of the older blast furnaces, the stove changing is done manually and the operation requires about 20 to 25 minutes. However, at most modern furnaces, the stove valves are motorized and the valve changing is automated so that only about 3 minutes are required for a stove change. With the shorter changing time, the heating time can be increased so that higher hot-blast temperatures can be used and greater efficiency can be obtained. The automatic stove-changing cycle can be initiated either by having the stove tender push a button when the change is required or by a completely automatic electronic signal. This signal can be based on the extent of the mixer-valve opening (as, for example, when the mixer valve is 85 per cent closed), on the dome temperature, or strictly on a time cycle.

Generally, each blast furnace is equipped with three hot-blast stoves, and each stove is kept on blast about ¾ to 1½ hours. Thus, the amount of heat that is extracted from the stove while it is on blast must be put back into the stove in the heating period which is just twice the on-blast time minus twice the stove-changing time. At some furnaces, there are four stoves. With the extra stove, the firing rate does not have to be as great because the heating cycle is three times the on-blast cycle minus twice the stove-changing time. Another advantage of the extra stove is that if there is a problem with the stove equipment, the stoves can be repaired one at a time without significantly affecting the operation of the furnace. Figure 15—43 shows the control panel board for the automatic stove-changing system for a modern blast furnace equipped with four stoves. A more thorough explanation of the operation of blast-furnace stoves can be found in the article by Agarwal, Karnavas and Grina listed in the bibliography at the end of this chapter.

BLAST-FURNACE IRREGULARITIES

In spite of the many improvements in burden materials and operating procedures that have been made in recent years, the blast furnace does not always run as smoothly as the casual observer may be led to believe. Furnace upsets are not as frequent as they were in former years; however, irregularities still do occur that cause considerable concern and often require quick thinking and the use of good judgment and skill on the part of the operator to prevent serious trouble.

Slips—Slips are caused initially by **hanging** or **bridging** of the burden material in the stack of the furnace. When this occurs, the material below the "hang" continues to move downward, forming a space that is void of solid material but filled with hot gas at very high pressure. This space continues to grow until the hang finally collapses. In severe cases, the sudden downward thrust of the hanging material forces the hot gas upward with the force of an explosion. This sudden rush of gas opens the explosion bleeders and sometimes is so great that it causes severe damage to the furnace-top gear.

The hanging that precedes slipping is caused by any of a number of different conditions in which the permeability of the charge is decreased because some of the material plugs up the interstices between the particles of the charge and bonds them loosely together. When there is a high percentage of fines in the burden and the velocity of the furnace gas is relatively high, the fines will plug up the openings between the other particles and cause hanging. In some instances, slaggy material that has been melted is blown upward in droplets and when it contacts the cold particles of the charge, it resolidifies and plugs up the openings between the particles and tends to cement them together. In some cases, the carbon-monoxide decomposition reaction

$$2CO = CO_2 + C$$

will be catalyzed, and the carbon deposited as soot will plug up the openings between the particles and will hold the particles together. In other instances, where the alkali content of the burden is high, the alkali compounds will be reduced to alkali vapor that ascends with the furnace gas and condenses in the cooler portion of the charge to cause the same type of hanging condition. Another type of hanging sometimes occurs in furnaces that are being run very efficiently and are being pushed to their best production rate. Under these conditions, if there is a slightly unfavorable change in the gas distribution, the strength of the coke or the particle size of the burden, the iron oxide will not reduce to metallic iron rapidly enough, and iron oxide will melt and run down as a liquid onto the coke particles. When this occurs, the liquid iron oxide will be reduced to solid iron and considerable heat will be consumed by the reduction. Thus, the coke particles will be cemented together and the permeability of the mass will be significantly decreased so that hanging will result.

When the burden is not moving properly through the furnace, the operator must take corrective steps immediately to prevent a disastrous slip and determine the cause of the hanging so that changes can be made in the operating procedure to prevent the hang from recurring.

Scaffolding—Scaffolding is the term used when accretions or scabs build up on the furnace walls and cause a decrease in the cross-sectional area of the furnace stack. Scaffolding can occur relatively high in the furnace stack or relatively low, near the top of the bosh.

The formation of scaffolds near the top of the bosh often results because of excessive fines in the burden material and a limier than normal chemical composition of the slag. The solution of lime into the slags formed in the furnace stack increases the slag viscosity. Since the slag often carries some of the fine particles from the burden in suspension, the increase in viscosity causes this mixture of fines and slag to adhere to the upper bosh walls, and this deflects the hot gases farther into the center of the furnace. With less hot gas along the walls, the accretions tend to cool down, and they will then grow until they block a large percentage of the furnace cross-section and render it ineffective for smelting. This increases the fuel rate, decreases the hot-metal production rate and promotes hanging and slipping. Scaffolds can also be caused by alkali or zinc compounds that are reduced to metallic vapor near the bottom of the furnace and rise with the furnace gases to the cooler top portion where they are reoxidized to very fine solid particles. These very fine particles adhere to the furnace inwalls and entrap other fine material there to start the formation of a scaffold. Such scaffolds decrease the productive capacity and distort the gas flow so that poor fuel efficiency is obtained. Also, scaffolds can dislodge from the walls and descend into the hearth, causing serious furnace upsets and, consequently, poor quality metal.

Channeling—For satisfactory blast-furnace operation, the solid materials must be properly distributed both radially and circumferentially so they will be correctly preheated and chemically reduced by the gases rising from the furnace combustion zones. When the burden is charged into the furnace from a large bell, the finer material will tend to fall down directly below the rim of the bell, and the coarser particles will roll toward the walls and the center. If, for example, the percentage of fines in the burden materials increases significantly, too much of the burden will be deposited directly below the rim of the bell and very little will reach the walls. As a result, the ratio of burden to coke will be very low near the walls and, since the permeability there will increase because of the increased percentage of coke, the flow of hot gas along the walls will significantly increase. Because the heat capacity of the coke is less than that of the burden material, the gases will retain their heat and will cause an excessive rise in the wall temperature. This will increase the fuel rate, because the reducing gas will leave the furnace unused and it will also cause rapid erosion of the furnace lining. In extreme cases, the lining will wear away so rapidly that hot spots will form on the outer steel shell. When this occurs, water must be sprayed on the shell or new cooling plates must be inserted to start the solidification of materials against the shell to protect it. Also, the charging distribution must be changed to eliminate the channeling. In some instances, channeling may have been so severe that the damage to the lining cannot be repaired without taking the furnace completely out of operation for a stack reline. Chanelling is no longer a major problem because of changes in furnace top design, charging sequence, and raw material preparation.

Breakouts—Breakouts are caused by failure of the walls of the hearth, with the result that liquid iron or slag or both may flow uncontrolled out of the furnace and cause considerable damage to the furnace and surrounding auxiliaries. Slag breakouts are usually not as serious as iron breakouts, since there is not so much danger from explosions as is the case when molten iron and water come into contact. With either type of breakout, it is essential, if possible, to cast the furnace, thereby draining off as much liquid material as possible, and to take the blast off the furnace.

A slag breakout may be chilled by streams of water, and the hole where the breakout occurred may be closed by replacing the brick, pumping fireclay grout into the opening, or ramming a plastic cement or asbestos rope into it.

However, there is practically no control over an iron breakout, and the iron runs out of the hole until the furnace is dry. After the accumulated iron has been cleared away, brick, gainster, and fireclay, either used separately or in combination, may be used to close the hole. It may be necessary to renew one or several of the hearth-cooling staves or the breakout may be so severe that the operator will have to take the furnace out of blast for a complete hearth repair. The use of carbon brick in the hearth sidewall has given indications of providing a construction that is not as susceptible to this trouble as ordinary firebrick construction.

FANNING

Occasionally, during the campaign of a blast furnace, situations arise when the full productive capacity of the furnace is not required for a period of time. When this occurs, the problem can be solved by shutting the furnace down or curtailing the operation of the furnace by reducing the quantity of wind being blown. The wind rate is usually reduced until the hot-blast pressure at the tuyeres is very low. A positive pressure must be kept in the hot-blast system to assure that there is no danger of gas coming back from the blast furnace into the blower system. When the wind volume is less than 20 to 25 per cent of normal, the technique is known as **fanning**. Fanning has the advantages of keeping the gas system pressurized and furnishing a small quantity of blast-furnace gas for use as fuel, and enables a resumption of near full operation on short notice. This technique is used for emergency situations or short periods only. Prolonged use, such as 8 hours out of every 24, or on week-ends, results in a hearth build-up and frequently promotes inwall scab formation. A careful economic study is necessary before adoption of this method.

BACK DRAFTING

Occasionally it is necessary to take the furnace out of blast for short periods, often less than two hours, to perform various maintenance functions such as replacing tuyeres or repairing skip cables. In such instances, the furnace is not banked but is **back drafted**. In this operation, as soon as the wind is stopped, the bustle pipe is put under negative pressure. This is done usually by opening the chimney valve and the hot-blast valve to a stove that has already been prepared by heating it to temperature and then shutting off the gas valve. As the furnace gas is drawn back into the stove, air is admitted through the peep sights and stove burner, and the operator makes certain that the gas burns in the stove. During the operation, the bleeders at the top of the furnace also are opened to pull some of the furnace gas out through the top.

At several furnaces, a special back-draft stack is installed so that it is not necessary to draw the furnace gas back through a stove. This stack is connected to the bustle pipe or to the hot-blast main. In some instances, it is closed by a water-cooled gate valve at the level of the bustle pipe and in others it is closed by an uncooled cap valve at the top of the stack. Opening the valve allows the furnace gas to draft to the atmosphere where it burns without difficulty.

BANKING

Plans for an extended shut down or interruption to furnace operation either for a breakdown, scheduled repair or because business conditions indicate a pause in production is desirable, may influence management to bank a blast furnace. The word banking is applied because of similarity to the operation of banking a fire. The origin is lost in antiquity: however, generally it means covering a fire with ashes or fresh fuel to restrict air, reduce the combustion rate, and to preserve the fuel for future use.

Also, banking is resorted to as an emergency measure when some unforeseen event requires a furnace shut down. In this case, the blast is taken off, the blowpipes are dropped and the tuyere openings are plugged with clay to prevent air from drafting through. Thus, hearth heat is preserved and the furnace can be returned to operation with a minimum effort. If the down time exceeds four or five days duration, some difficulty can be expected in resuming operation although instances are on record that no trouble was experienced after a seven-day bank.

More common is the furnace bank carried out as a planned event. Preparation made depends upon the length of time anticipated. If for only a few days, an extra blank or two of coke may be charged without flux and the furnace taken off when the coke descends to the bosh zone. If for a slightly longer time, the ore and stone burden may be reduced 5 to 10 per cent following the coke blank possibly for ten or fifteen charges before normal charge weight is resumed.

A banking burden for a shut down for an undetermined length of time is very similar to a blow-in burden. Prior to the start of a banking burden, miscellaneous iron-bearing materials are removed from the charge and a large reduction in the amount of limestone charged is made. Extra coke also may be charged ahead of the banking burden. The purpose is to develop a hot, siliceous slag which has a tendency to clean off the lime accumulation on the bosh walls and prevent an excessively limey slag during blow-in. Limey slag is more viscous and is apt to cause some operating problems early in the blow-in period. Often during the initial warm-up period, temperatures may be very high in the bosh which will increase reduction of silica to silicon with the result that slag contains a higher proportion of lime. For this reason, effort is made to have a hot, siliceous slag at the time the furnace is banked and a like condition upon the resumption of operation. Following initial preparatory charges noted above, a

heavy coke blank is charged and subsequent charging parallels a characteristic blow-in burden. Charging continues until the coke blank reaches the upper bosh area of the furnace. At this time, the final cast is made. Effort is made to drain the hearth until a dry blow of the tapping hole is observed to insure a clean hearth for the future start up and eliminate as much as possible the need for melting cold slag early in the blow-in period. Prior to the last cast, the furnace dustcatcher is emptied. Accumulated dust has a tendency to consolidate into a rock-like mass if undisturbed for a time and could present a difficult problem after operation began again.

About the end of the cast, before the furnace is taken off, a heavy blanket of ore may be dumped in the furnace to cover the upper burden surface, thus reducing the natural drafting tendency of the furnace.

At the conclusion of the cast, the tapping hole is plugged, wind is taken off the furnace, bleeders opened, steam is turned into the dustcatcher, the furnace is isolated from the common gas system and stove valves manipulated to draft gas back through the bustle pipe, hot-blast main and out through the stove chimney. Furnace operators quickly drop the blowpipes and plug the tuyeres with clay. Many operators prefer to remove the tuyeres to avoid any chance of a stray water leak permitting water to accumulate in the furnace and also to provide an opportunity to observe the coolers for possible leaks. Clay is solidly packed into the tuyere openings and backed up with sand to eliminate any chance of air filtering in. Often this is followed by bricking up the openings as further insurance against air infiltration.

As soon as the blowpipes are down after the final cast, blowing engines are stopped, and stove-burner valves, chimney and hot-blast valves are closed to preserve heat as long as possible. As a precaution, blow-off valves are opened slightly to prevent a pressure build-up from developing in the stoves resulting from an undetected water leak or from some unsuspected source.

Within a day or two, the furnace bells are opened and steam is shut off in the dustcatcher. Daily inspection of the stock level is important. A slow stock movement is an indication that air is infiltrating and coke is being consumed. A movement of a few feet can be expected but a continual drop is undesirable and may force operators to spray the bosh with a sealing material. A thin mixture of water, clay and water glass is sometimes used since the material is cheap and does an effective job: however, a "mothballing" material can be and is used by some operators who report good results from the practice.

If the furnace is banked for an extended period, after a lapse of a month to six weeks, water flow will be reduced on the cooling members and finally, after 2 to 3 months, turned off entirely except for the hearth staves.

Very often after a bank of six or more weeks, when tuyeres are opened, all signs of fire in the tuyere area will have disappeared. Furnace operators are pleased when this condition is discovered since full benefit of the coke blank will be available to supply heat when operation begins.

BLOWING IN FROM BANK

Blowing in a banked furnace has been developed to the point where operators accept the process as a routine event. Present knowledge of the art has grown immensely in the past thirty years. In the years preceding, blowing in a banked furnace was a task approached with misgiving and the performance was usually a disappointing experience.

About the only obstacle which otherwise upsets a smooth routine operation is infrequently encountered when blowing in older furnaces, where an undetected water leak may cause extended delays and set-backs in returning to operation. Accompanying this difficulty and often resulting from the water leaks noted above are other delays from break-outs around cooling plates and coolers. In these older furnaces there may be only a little brickwork remaining in the slag-line area and the natural carbon and lime build-up is dissolved from the action of the siliceous slag intentionally developed. This condition is conducive to opening up of interstices between the bricks and is aided by temperature changes as bosh heat is increased during blow-in.

Skill, experience and good judgment on the part of operators is of inestimable value and is never more important to the success of any undertaking than in bringing in a furnace from bank.

A discussion of methods of blowing in a furnace from an extended emergency bank where no preliminary preparation has been made is omitted because of the numerous circumstances which affect the actions taken by operators. The job, at best, is a lengthy, laborious undertaking.

Blowing in from bank, when adequate preparation was made, follows somewhat the pattern of blowing in a new or relined furnace. The exception which alters the technique is that the hearth level is higher and cold slag in the hearth must be heated and liquefied to prevent chilling of the new slag which begins to form when wind is put on the furnace. Under these circumstances, operators must be sure "communication" is established between tapping hole and the areas in the hearth where liquid is being formed in order to get it out of the furnace as fast as it is made. Working with this goal in mind, heat is brought down into the hearth as the new slag forms, which transfers some heat to the bottom, and is removed before it has a chance to chill.

Several methods are used in starting: all of them have proven successful and the operators are apt to choose any one of them which he feels may fit his particular problem. But, in all cases, when wind is first put on, only a few tuyeres will be opened. These are directly above and on each side of the tapping hole or cinder notch. As the blow-in progresses, following each cast, a tuyere next to an open tuyere will be opened and wind volume increased. If the following cast is not up to expectation, that is, if the volume or temperature of the slag is lower than it is thought it should be, then opening of the next tuyere will be postponed until hearth heat is built up again.

In general, earlier preparations are similar to blowing in a relined furnace. Coolers are opened up and coke ash and refuse cleaned away. Operators make sure good clean coke is in front of the coolers before installing tuyeres. Then all but the few mentioned above are securely packed with clay or clay and brickbats to prevent any blast air from entering until the plug is removed.

The tapping hole is dug and burned back to where coke is visible and then made up similar to the method used on a new furnace. Some operators lightly plug the hole: others let the furnace gas blow out through the hole until slag appears to insure that the hot gas warms the hearth in the tapping-hole area. Another method has produced good results and is quite unique. The hole is made up and a blow-in pipe is installed to blow air or an air-oxygen mixture into the hole. Tuyeres are closed during this 4 to 8 hour operation. Then the pipe is pulled, the hole plugged and air is blown in a few tuyeres as above. Still others prefer to remove the cinder notch and build a tapping hole in its place and construct a runner to carry slag and iron produced until hearth heat is built up enough to permit normal casting.

In any method, casts are made at 2 to 3 hour intervals until operators are certain the hearth is heated sufficiently to keep the contents fluid. The tapping hole is drilled straight in early in the blow-in and the angle slowly increased as temperatures increase and the entire body of the hearth begins to return to normal operating temperature.

Burden regulation follows blow-in practice with regular increases in iron-bearing materials as the blow-in progresses.

BLOWING OUT

If business conditions deteriorate to the extent production is no longer required, the decision may be made to blow out a furnace. Starting the furnace again under conditions approaching those of starting a new furnace is generally accomplished faster and with less effort than starting from a bank: however, the cost in connection with blowing out, raking out and cleaning preparatory to starting is likely to exceed the cost of banking.

When a furnace has reached the end of its campaign (lining worn out), it is usually blown out except under most unusual circumstances.

About 12 to 16 hours before the last cast is to be made, the operation is discontinued for a short period of time to permit the installation of water sprays in the top of the furnace and thermocouples in the uptakes. The burden composition is then changed to produce a very siliceous slag to help in removing as much lime as possible from the bosh and hearth walls. The purpose of this is to prevent the formation of calcium hydroxide which would occur if lime were to come in contact with the cooling water during the later stages of the blow-out. (The formation of calcium hydroxide from lime imbedded in the lining could generate sufficient force to crack the steel hearth shell or to lift the furnace off its columns.)

After installation of the blow-out equipment, the blast is put on and charging is continued. If a gravel blow-out is the method chosen, a heavy coke blank will be charged first. The volume will be equivalent to approximately the volume of the bosh. Following the coke blank, washed and screened silica gravel, about minus 50 mm plus 25 mm (minus 2 to plus 1 inch) size is charged. The furnace is kept full early in the blow-out and then the stockline is permitted to drift down about 6 to 9 metres (20 to 30 feet) towards the end (when all the iron-bearing burden has been reduced). During blow-out, water is judiciously used to control top temperature: however, the additional charges of gravel are very effective in keeping top temperature low. A decrease in the wind rate is required as the height of the column of material in the furnace decreases. From the time that the special spray equipment is installed until the blow-out is completed requires only about 6 to 8 hours.

A coke blow-out is the same as above except that coke is used instead of gravel. Sometimes plus 20 mm minus 25 mm (plus ¾ to minus 1 inch) coke screenings are used: this size is commonly called domestic coke.

Following the last cast, the stock is watered down as described above. When cooling has progressed far enough, sluice ways frequently are built from a couple of cooler openings to an open-top railroad car and the contents of the furnace washed out with high-pressure-water jets.

DRAINING THE SALAMANDER

Subsequent to blowing out and when the furnace is to be completely relined, the salamander may be drained. The operation saves days and even weeks in relining time which otherwise might be lost in blasting out the heavy chunk of solid iron that would be formed if the metal (that accumulated in the hearth as bottom block eroded during the campaign) were permitted to solidify. Initial preparation is made before blowing out by drilling a predetermined distance into the furnace bottom below the hearth staves and installing a trough or runner for the iron. If the furnace layout is amenable, pig beds will be constructed and made ready for use immediately after the last cast is completed. In case an outside area is not available for a pig bed, then, after the last cast, ladle tracks will be removed and the pig beds arranged under the cast house. The 8- to 12-hour delay in this case is negligible since residual heat in the hearth will keep the salamander molten for several days.

When all is ready, a long oxygen lance is inserted in the drilled hole and the remaining brickwork is burned through into the pool of iron. Usually the flow of iron is slow and several hours are required to empty out an accumulation of from four hundred to six hundred tons.

SECTION 7

THE BLAST-FURNACE BURDEN

The regulation of the proportions of ore, pellets, sinter, flux, coke and miscellaneous materials charged into the blast furnace is called **burdening**. Proper burdening is essential to keep the operation of the furnace at maximum efficiency and to control the hot-metal composition.

A term often used in relation to burdening is the **burden ratio** which is the ratio of the weight of iron-bearing materials per charge to the weight of coke. However, since the widespread use of fluxed sinter and fluxed pellets, this term has become somewhat confusing because, in some instances, the weight of the iron-bearing materials also includes the weight of the fluxes.

A typical burden calculation is shown in Table 15—VI. The amount of each material in the charge is shown in the first column and the percentage of each of the components (on a natural basis) silica, alumina, lime, magnesia, phosphorus, manganese, iron and sulphur are shown in the appropriate columns. With this information, the total amount of each constituent is determined and thus the total amount of pig iron that will be produced and the composition and amount of slag that also will be produced can be calculated. In the example given in Table 15—VI, it is planned to make hot metal with a chemical composition of 0.80 per cent of silicon and 0.03 per cent of sulphur. At this silicon content, the carbon content of the iron will be about 4.50 per cent; consequently, the remaining constituents that report to the hot metal will be 94.67 per cent ($100\% - 4.50\% - 0.80\% - 0.03\% = 94.67\%$). The remaining constituents reporting to the hot metal, 100 per cent of the iron, 100 per cent of the phosphorus and 70 per cent of the manganese, total 35 825 kilograms (78 815 pounds) representing the sum of 35 650 kilograms (78 430 pounds) of iron, 31 kilograms (68 pounds) of phosphorus and 144 kilograms (317 pounds) of manganese, so that the total hot metal produced will be 37 842 kilograms or 37.84 metric tons, equivalent to 83 252 pounds or 41.7 net tons.

To calculate the amount of slag and the slag composition, all of the constituents reporting to the slag are totaled, and this is assumed to be 99 per cent of the total slag, because there are generally a few constituents that are also present but are not included in the chemical analysis. In determining the amount of silica reporting to the slag, the amount reduced to metallic silicon must be deducted from the total silica. In the example shown, the silicon in the hot metal is 303 kilogram (667 pounds); consequently, 648 kilograms (1426 pounds) of silica was reduced ($303 \times 2.14 = 648$ kilograms) and only 4136 kilograms (9099 pounds) reported to the slag (4784 kilograms−648 kilograms = 4136 kilograms (9099 pounds]. The amount of manganese oxide (MnO) is determined by multiplying the 30 per cent of unreduced manganese by 1.28 to convert it to the oxide. In this example, 79 kilograms (174 pounds) of MnO report to the slag (206 kilograms $\times 0.30 \times 1.28 = 79$ kilograms (174 pounds). The balance of the calculations are self-explanatory.

If the calculation indicates that the sulphur content of the slag is higher than that which experience shows it can absorb, then additional slag-making constituents and fluxes must be added. In the past, burden calculations were made only when there were major changes in the raw materials or in the grade of iron to be produced. On an hour-to-hour basis, small adjustments to

Table 15—VI. Burden Calculation for a Blast Furnace.
A. Calculation Using SI Units

Furnace Charge		Silica		Alumina		Lime		Magnesia		Phosphorus		Manganese		Iron		Sulphur	
Material	kg	%	kg	%	kg	%	kg	%	kg	%	kg	%	kg	%	kg	%	kg
Pellets	34 500	5.87	2 025	0.17	59	0.37	128	0.48	166	0.014	5	0.16	55	63.42	21 880	0.003	1
Sinter	23 000	5.97	1 373	1.60	368	10.20	2 346	1.93	444	0.038	9	0.27	62	55.08	12 668	0.014	3
BOP Slag	3 550	14.54	516	0.68	24	35.61	1 264	6.66	236	0.400	14	2.52	89	26.78	951	0.092	3
Dolomite	2 000	0.19	4	0.08	2	30.23	605	21.37	427	0.001	—	—	—	0.05	1	0.040	1
Coke	20 000	4.33	866	2.79	558	0.50	100	0.50	100	0.015	3	—	—	0.75	150	0.710	142
Total	83 050		4 784		1 011		4 443		1 373		31		206		35 650		150

PIG IRON PRODUCED

Theoretical		
	kg	%
C	1 703	4.50
Fe	35 650	94.21
Si	303	0.80
P	31	0.08
Mn	144	0.38
S	11	0.03
Total	37 842	100.00

SLAG PRODUCED

Theoretical		
	kg	%
SiO_2	4 136	36.6
Al_2O_3	1 011	9.0
CaO	4 443	39.3
MgO	1 373	12.2
MnO	79	0.7
S	139	1.2
Total	11 181	99.0
Total	11 294	100.00

OPERATING RATIOS

Coke Rate = 529 kg/ton (1058lb/net ton)

Slag Rate = 298 kg/ton (596lb/net ton)

Slag Basicity Ratio $\dfrac{CaO + MgO}{SiO_2 + Al_2O_3} = 1.13$

(Continued on next page)

Table 15—VI. (Continued)
B. Calculation Using fps Units.

Material	Furnace Charge lb	Silica %	Silica lb	Alumina %	Alumina lb	Lime %	Lime lb	Magnesia %	Magnesia lb	Phosphorus %	Phosphorus lb	Manganese %	Manganese lb	Iron %	Iron lb	Sulphur %	Sulphur lb
Pellets	75 900	5.87	4 455	0.17	129	0.37	281	0.48	364	0.014	11	0.16	121	63.42	48 136	0.003	2
Sinter	50 600	5.97	3 021	1.60	810	10.20	5 161	1.93	977	0.038	19	0.27	137	55.08	27 932	0.014	7
BOP Slag	7 810	14.54	1 136	0.68	53	35.61	2 781	6.66	520	0.400	31	2.52	197	26.78	2 092	0.092	7
Dolomite	4 400	0.19	8	0.08	4	30.23	1 330	21.37	940	0.001	—	—	—	0.05	2	0.040	2
Coke	44 000	4.33	1 905	2.79	1 228	0.50	220	0.50	220	0.015	7	—	—	0.75	330	0.710	312
Total	182 710		10 525		2 224		9 773		3 021		68		455		78 430		330

PIG IRON PRODUCED

	Theoretical lb	%
C	3 746	4.50
Fe	78 430	94.21
Si	667	0.80
P	68	0.08
Mn	317	0.38
S	24	0.03
Total	83 252	100.00

SLAG PRODUCED

	Theoretical lb	%
SiO_2	9 099	36.6
Al_2O_3	2 224	9.0
CaO	9 775	39.3
MgO	3 021	12.2
MnO	174	0.7
S	306	1.2
Total	24 599	99.0
Total	24 847	100.00

OPERATING RATIOS

Coke Rate = 1 058 lb/net ton

Slag Rate = 596 lb/net ton

Slag Basicity Ratio $\dfrac{CaO + MgO}{SiO_2 + Al_2O_3}$ = 1.13

the burden were made by the furnace foreman, in accordance with his experience, to compensate for changes in the chemical composition of the hot metal or the appearance of the slag. However, at present, the burden calculations are made by computer, and at some plants where the composition of the raw material changes from hour to hour, the computation is made every few hours. By entering into the computer the slag-composition range desired for good slag fluidity and the limit of sulphur desired in the slag, the computer will actually notify the operator of the burdening changes that must be made. In fact, on some blast furnaces equipped for automatic charging, these changes can be made automatically.

In computer burdening, the calculations show not only the amounts of material required to give slag of the desired composition but also the changes in fuel ratio required as different materials are added to or removed from the burden. These calculations require a knowledge of the wind rate, the hot-blast temperature, the blast moisture and the injected-fuel rate. This information can be transmitted to the computer by manual entries or by direct readings from the control instruments.

At blast furnaces of United States Steel Corporation, the fuel consumption is approximated through the use of the USS standard coke-rate formula developed by R. V. Flint and described in his paper listed in the bibliography at the end of this chapter. Table 15—VII is a listing of the coefficients used in this formula.

MODERN TECHNIQUES FOR IMPROVING BLAST-FURNACE OPERATING PERFORMANCE

During the twenty-five years between 1950 and 1975, many new techniques were adopted by the steel industry that greatly improved blast-furnace operating performance and increased the efficiency. For example, during that period the average hot-metal production rate per unit of volume for all of the blast furnaces in the United States more than doubled. During the same period, the coke rate decreased from 925 to 550 kilograms of coke per metric ton of hot metal (1850 to 1100 pounds per net ton). In the 1980's, furnaces as large as 14 metres (46 feet) in diameter produce more than 10 000 metric tons (about 11 000 net tons) of hot metal per day, at a fuel rate of less than 450 kilograms per metric ton (900 pounds per net ton).

Beneficiated Charge Materials—The physical and chemical composition of the materials used in blast furnaces has improved significantly since 1950. As described in Chapter 3, the practice of washing coal to remove ash and sulphur has greatly improved the chemical composition of the coke. Coke strength has been improved also by suitable pulverizing and blending of the coals. The Precarbon process, used at several modern American blast furnaces, is a process in which coal is pulverized and then heated to approximately 200°C (392°F) and transferred by hot gases to the coke ovens. Furnaces using this process have discovered that the ASTM stability rating of the coke can be raised 5 percentage points (from 53 to 58 ASTM stability) and at the same time the productivity of the coke ovens can be increased. The use of high-strength coke, crushed to a top size of 65 mm (about 2½ inches) and screened to

remove the minus 25 mm (about 1 inch) fines will significantly improve the production rate and furnace efficiency. The uniformly sized coke improves the permeability of the stock column and the rate of heating of the burden, and the improved coke strength prevents the formation of fines which would distort the gas flow and interfere with the gas-solid contact.

At many of the blast furnaces, the ores that are used are first crushed to minus 50 mm (about minus 2 inches) and screened to remove the minus 10 mm (about minus ⅜ inch) fines. In some instances, the ore is crushed to minus 25 mm (about 1 inch) because the ideal size for blast-furnace ore is generally conceded to be minus 25 mm plus 10 mm (minus 1 inch, plus ⅜ inch). The fines that cannot be used directly in the blast furnace are either sintered or pulverized and pelletized. The iron-bearing material from the taconite deposits and many others must be pulverized to finer than 0.1 mm (about 0.004 inch) so that the iron oxide can be satisfactorily separated from the gangue, and consequently these fines are pelletized.

The incorporation of fluxes in the agglomerates provides a means for further improvement in the blast-furnace operation because it precalcines the flux. In addition, it often improves the reduction rate of the agglomerate and helps to prevent the formation of low-melting iron silicate. This keeps the agglomerate from melting before it has been sufficiently reduced to metallic iron, and for this reason it is believed that it permits the use of higher flame temperature and improves the fuel effeciency in the blast furnace.

High Hot-Blast Temperatures—In the fifteen-year period between 1950 and 1965, the average hot-blast temperature for all the blast furnaces in the United States increased from about 550°C (about 1025°F) to about 815°C (1500°F). This improvement was accomplished through the use of better stove-firing techniques and better stove-changing equipment. Furthermore, improvements in burden materials, the use of tuyere-injected fuels and the control of blast moisture made it possible for the blast furnaces to accept the higher hot-blast temperature. The 265°C (about 475°F) increase in hot-blast temperature accounted for a decrease in the average coke rate of about 75 kilograms per metric ton of hot metal (150 pounds per net ton). For the period between 1965 and 1975, the further increase in hot-blast temperature for the United States blast furnaces was almost insignificant, and the maximum hot-blast temperature used at any furnace did not exceed 1100°C (2012°F). However, at many of the blast furnaces in Japan and several in Europe, the use of hot-blast temperatures between 1200°C and 1250°C (2200°F and 2280°F) was adopted. Some blast furnaces in the United States are now using temperatures in this range.

Fuel Injection—With the development of techniques for increasing hot-blast temperatures to the range of 1000°C to 1100°C (1832°F to 2012°F) and the need for controlling the flame temperature because of the type of burden materials in use, it was discovered that hydrocarbon fuels could be injected into the blast furnace through the tuyeres to control the flame temperature, increase the reducing power of the bosh gas and at the same time replace some of the coke. In the

Table 15—VII. Carbon Rate Formula—Summary of Variables and Coefficients.

Category	No.	Description (1)		Unit of Measure (2)		Carbon Coefficient (3)		Approximate actual practice range studied			
								Minimum (4)		Maximum (5)	
		SI	fps	SI	fps	SI	fps	SI	fps	SI	fps
Slag	1	Slag from coke ash		+1 kg per mtHM*	+1 lb per NTHM**	+0.60	+0.60	30	60	175	350
	2	Slag from uncalcined $CaCO_3$ and $MgCO_3$		+1 kg per mtHM	+1 lb per NTHM	+0.60	+0.60	0	0	500	1000
	3	Slag—all other		+1 kg per mtHM	+1 lb per NTHM	+0.15	+0.15	100	200	900	1800
Metallic-bearing materials size consist	4	−1 mm	−20-mesh	+1 kg per mtHM	+1 lb per NTHM	+0.08	+0.08	75	150	600	1200
	5	−1 mm, −10 mm	+20-mesh, −⅜ in.	+1 kg per mtHM	+1 lb per NTHM	+0.04	+0.04	200	400	900	1800
	6	+25 mm, −50 mm	+1 in., −2 in.	+1 kg per mtHM	+1 lb per NTHM	+0.03	+0.03	50	100	750	1500
	7	−50 mm, −100 mm	+2 in., −4 in.	+1 kg per mtHM	+1 lb per NTHM	+0.06	+0.06	0	0	750	1500
	8	+100 mm	+4 in.	+1 kg per mtHM	+1 lb per NTHM	+0.10	+0.10	0	0	300	600

Note: +10 mm, −25 mm (+⅜ in., −1 in.) size fraction is the reference or "Optimum" size; all other size fractions are less desirable and, thus, require a plus carbon coefficient.

Category	No.	Description		Unit of Measure		Carbon Coefficient		Minimum		Maximum	
		SI	fps	SI	fps	SI	fps	SI	fps	SI	fps
Other	9	Charged sulphur in total burden		+1 kg per mtHM	+1 lb per NTHM	+5.00	+5.00	3.5	7	17.5	35
	10	Charged free metallics		+1 kg per mtHM	+1 lb per NTHM	+0.30	+0.30	0	0	350	700
	11	Charged iron contained in iron silicates		+1 kg per mtHM	+1 lb per NTHM	+0.30	+0.30	?	?	?	?
	12	Charged combined water		+kg per mtHM	+lb per NTHM	+0.45	+0.45	0	0	115	230
Hot-Metal Analysis Variables	13	Silicon		+1 per cent	+1 per cent	+60.	+120.	0.4	0.4	3.5	3.5
	14	Manganese		+1 per cent	+1 per cent	+10.	+20.	0.1	0.1	3.0	3.0
	15	Phosphorus		+1 per cent	+1 per cent	+10.	+20.	0.1	0.1	1.8	1.8
	16	Sulphur		+1 per cent	+1 per cent	Curvilinear	Curvilinear	0.020	0.020	0.060	0.060
Operating-Practice Variables	17	Hot-blast temperature		+1°C	+1°F	−0.18	−0.20	315	600	1090	2000
	18	Wind rate		+1 Nm³/min per m³ wind vol.	+1 ft³/min per ft³ wind vol.	+50.	+100.	0.75	1.5	1.4	2.8
	19	Blast humidity		+1 kg H_2O/mtHM	+1 lb H_2O/NTHM	+0.40	+0.40	10	20	100	200
	20	Days on stack lining		+1 day	+1 day	+0.015	+0.03	15	15	2700	2700
	21	Metallic loss in runner and ladle scrap		+1 per cent	+1 per cent	+5.	+10.	0	0	3.0	3.0

*"kg per mtHM" stands for "kilogram per metric ton of hot metal."
**"lb per NTHM" stands for "pound per net ton of hot metal."

presence of large quantities of coke the hydrocarbon fuels can burn only to carbon monoxide and hydrogen; consequently, they produce less heat than the coke they replace so that they control the flame temperature, but the reducing gas they produce is more effective than that produced by combustion of coke. Many different fuels have been tried—natural gas, coke-oven gas, oil, tar and pulverized coal, even slurries of coal in oil. The equipment for injecting natural gas is the least expensive. The gas is generally fed at a pressure well above that of the air blast to a pipe encircling the furnace, and from this circle pipe individual lines carry the gas to each tuyere. Gas is introduced into the furnace through a lance entering the side of the blowpipe (Refer to Figure 15—8) and combustion takes place in the zone just inside the furnace. The rate of injection is controlled by the rate of gas flow into the circle pipe. Check valves prevent air from backing up into the gas lines and automatic shut-off valves between the circle pipe and the tuyeres close when the blast-air pressure exceeds or drops below a predetermined range. Although the use of natural gas is the least complicated of all the hydrocarbon injectants, the shortage of natural gas in most locations has made it necessary to resort to other fuels.

When oil or coal are used, they are also introduced into the air blast by a lance entering the air stream through the sides of the blowpipes. It is most desirable to have the injected fuel completely gasified and combusted before it leaves the raceway just inside the furnace. When oil is used it is fed to a circle pipe and then to each tuyere in much the same manner as gas. Steam is used to atomize the oil, which enhances combustion. Coal must be pulverized and then conveyed to the tuyeres in high-pressure air, by dense phase transport. The high cost for the equipment for coal injection has made it also very unattractive at most blast-furnace locations: consequently, the most widely used injectant is No. 6 fuel oil, formerly known as Bunker "C" fuel oil.

Oxygen Enrichment of the Blast—Although the enrichment of blast air with oxygen to increase hot-metal production rates is by no means a new idea, the use of oxygen for this purpose had little commercial interest 25 years ago, because of the high cost of oxygen. Furthermore, when the blast air is enriched with oxygen, the flame temperature increases and high flame temperatures were incompatible with the relatively low quality burden materials used then. However, with the discovery of methods for controlling flame temperature with blast moisture and fuel injection, the use of oxygen enrichment has become quite common. For every percent of oxygen above that for normal air blast (about 21 per cent), the production rate can be increased about 2 to 4 per cent. In instances where the burden materials have good reducibility, that is, they will reduce rapidly, the flame temperature can be increased significantly and the fuel efficiency can be improved. However, in very efficient furnaces, the fuel rate will increase with oxygen enrichment.

In the manufacture of ferromanganese where the high-temperature heat requirements are much greater than those for basic pig iron, oxygen enrichments as high as 30 per cent in the blast are now used commercially.

High-Pressure Operation—One of the limiting factors in attempting to increase the wind rate of a blast furnace is the lifting effect that is caused by the large volumes of gases blowing upward through the burden. This lifting effect prevents the burden from descending normally and causes a loss rather than an increase in production. To increase production rates above normal, many furnaces are equipped with septum valves in the top-gas system to increase the exit-gas pressure. This increase in pressure compresses the gases throughout the entire system and permits a larger amount of air to be blown. With this increase in the quantity of air blown per minute, there is a corresponding increase in production rate.

When the pressure of the top gas is thus increased, the pressure of the inlet air blast is increased proportionately. At many furnaces, if the top pressure were increased, it would be necessary to use a larger blower capable of delivering the increased blast volume at the higher pressure.

The furnace shell, stove shells, dustcatcher, primary washer, and gas mains must be structurally strong enough to withstand the increased pressure. The throttling valve that is used to increase top pressure is located beyond the primary gas washer where the sand-blasting effect of the gas has been reduced by removal of a large portion of the dust carried by the gas from the furnace. The exit water line from the primary washer must be equipped with a regulator so that the gas pressure within the washer will not blow the water seal. Clean gas or nitrogen is piped into the space between the large and small bells to equalize the pressure in that space with the pressure in the furnace. The equalization of pressure in this location permits the large bell to be opened and reduces erosion that is caused by top gas leaking past the seat of the large bell. The pressure between the bells is reduced to atmospheric by a by-pass valve to permit opening of the small bell. With the use of high top pressure, many maintenance problems have been encountered with the bells, hoppers, bleeders, and so on. Experiments are still being conducted with abrasion-resistant materials of construction and such new designs as a 3-bell top to improve high-pressure operation.

At some blast furnaces, top pressures in excess of 200 kPa (2 kilograms per square centimeter or 29 psig) have been used successfully. At some of these furnaces, expanding turbines have been installed in the top-gas system, to recover some of the energy of compression to decrease the cost of operating with such high pressure.

Improved Burden Distribution—As a means of obtaining better control of burden distribution in the blast-furnace stack and thereby improving the gas-solid contact and the fuel efficiency, several new developments have been used in recent years.

One such development is a device, **movable stockline armor,** that can be shifted in and out to change the diameter of the circle of armor and, in this way, change the position where each material will be deposited as it slides off the bell and rebounds from the armor. A second development is the elimination of the large bell entirely, and the use of a rotating chute that can be set at different angles and rotated at different speeds so that each charge can be placed exactly where it will give the best distribution. Although it is difficult to install these improvements on older blast furnaces, such devices are being incorporated into the designs wherever new furnaces are built.

BIBLIOGRAPHY

Agarwal, J. C., J. A. Karnavas and J. A. Grina, Firing of blast-furnace stoves. Iron and Steel Engineer, 40, No. 7,75-81 (1963).

Am. Institute of Mining, Metallurgical and Petroleum Engrs., Inc, Blast furnace—theory and practice (Julius H. Strassburger, ed.; Dwight C. Brown, Terence E. Dancy and Robert L. Stephenson, co-editors). New York, Gordon and Breach Science Publishers, 1969.

American Inst. of Mining, Metallurgical and Mining and Petroleum Engineers, Iron and Steel Div. Ironmaking Proceedings—Proceeding of various Ironmaking Conferences through 1983.

Am. Iron and Steel Institute, Annual statistical reports, N.Y., The Institute.

Anderson, M., and E. R. Tanner, Advantages and disadvantages of various blast-furnace supporting structures. TSM-AIME Ironmaking Proc. 32, 177-192 (1973).

Austin, J. B., Efficiency of the blast-furnace process. Am. Institute of Mining and Metallurgical Engineers, Iron and Steel Div. Trans. 131, 74-98; discussion, 98-101 (1938).

Ball, D. F. et al., Agglomeration of Iron Ores, American Elsevier Publishing Co., Inc., New York, 1973.

Berczynski, P. A., Cooling of blast furnace by evaporation. TSM-AIME Ironmaking Proc. 28, 14-24 (1969).

Biswas, A. K., Principles of Blast Furnace Ironmaking 1981. Published by Cootha Publishing House, Australia Library of Congress Catalog Card No. 80-65943.

Campbell, W. W., and R. W. Fullerton, High-energy wet scrubbers can satisfactorily clean blast-furnace top gas. AIME Blast Furnace, Coke Oven and Raw Materials Conference Proc. 18, 329-335 (1959).

Dobscha, H. F., Effect of sized and sintered Mesabi iron ores on blast furnace performance. Am. Institute of Mining and Metallurgical Engineers, Blast Furnace, Coke Oven and Raw Materials Committee Proc. 7, 49-57; discussion, 57-67 (1948).

Elliot, G. D., J. A. Bond and T. E. Mitchell, Ironmaking from high-sinter burdens. Iron and Steel Institute Journal 175, 241-247 (1953).

Ergun, S., Pressure drop in blast furnace and in cupola. Industrial and Engineering Chemistry 45, No. 2, 477-485 (1953).

Flint, R. V., Effect of burden materials and practices on blast furnace coke rate. Blast Furnace and Steel Plant 50, No. 1 (Jan. 1962), 47-58, 74-76.

Gumz, W., Gas producers and blast furnaces N.Y., Wiley, 1950.

Hazard, P., Fuel oil injection in a blast furnace. Iron and Steel Institute Journal 199, Part I, 127-133 (1961).

Hoffman, C. F., Manufacture of low-silicon pig iron using high blast temperatures. Am. Institute of Mining and Metallurgical Engineers, Open Hearth Conference Proc. 23, 146-150 (1940).

Holbrook, W. F. and T. L. Joseph, Relative desulfurizing powers of blast-furnace slags. Am. Institute of Mining and Metallurgical Engineers, Iron and Steel Div. Trans. 120, 99-117; discussion, 117-120 (1936).

Holbrook, W. F., C. C. Furnas and T. L. Joseph, Diffusion of sulphur, manganese, phosphorus, silicon, and carbon through molten iron. Industrial and Engineering Chemistry 24, 993-998 (1932).

Jacobs, C. B., J. F. Elliott and M. Tenenbaum. Significance of minor elements in iron-bearing raw materials for integrated

steel plants. Blast Furnace and Steel Plant, 42, No. 6, 666-680 (1954).

Johnson, H. W., Correlations of some coke properties with blast furnace operation. Am. Institute of Mining and Metallurgical Engineers, Blast Furnace and Raw Materials Committee Proc. 1, 12-45; discussion, 46-48 (1941).

Joseph T. L., Oxides in basic pig iron and in basic open-hearth steel. Am. Institute of Mining and Metallurgical Engineers, Iron and Steel Div. Trans. 125, 204-245 (1937).

Joseph T. L., Porosity, reducibility and size preparation of iron ores. Am. Institute of Mining and Metallurgical Engineers, Iron and Steel Div. Trans. 120, 72-90; discussion, 90-98 (1936).

Joseph, T. L., and K. Neustaetter. The use of carbon in the blast furnace and heat balances. Blast Furnace and Steel Plant, 35, No. 8, 944-948 (1947).

Knepper, W. A., ed., Agglomeration, Interscience Publishers, New York, 1962.

King, C. D., Seventy-five years of progress in iron and steel. N. Y., Am. Institute of Mining and Metallurgical Engineers, 1948.

Legille, E., and K. H. Peters. Operation of a blast furnace incorporating a Paul Wurth bell-less top-charging system and its application to large blast furnaces. TSM-AIME Ironmaking Proc. 32, 144-162 (1973).

Macdonald, N. D., Effect of screened sinter on furnace productivity. Am. Institute of Mining, Metallurgical and Petroleum Engineers; Blast Furnace, Coke Oven and Raw Materials Proceedings 20, 2-10; discussion, 10-15 (1961).

Martin, P. V., Effect of the solution-loss reactions on blast-furnace efficiency. Am. Institute of Mining and Metallurgical Engineers, Iron and Steel Div. Trans. 140, 31-58; discussion, 59-64 (1940).

McMaster University Symposium— Alkalis in blast furnaces. Proc. edited by N. Standish and W. K. Lu. McMasters University Press (1973).

Molerus, O., and Hufnagel, W., eds., Agglomeration 81 (preprints), NMA Nürnberger Messe—und Austellungsgesellschaft, Nürnberg, West Germany, 1981.

Negomir, J. M., and J. W. Carlson, Oxygen teams up with fuel to hike iron output. Iron Age 188, Oct. 5, 1961, 69-71.

Osborn, E. F., R. C. De Vries, K. H. Gee and H. M. Kraner, Optimum composition of blast furnace slag as deduced from liquidus data for the quaternary system CaO-MgO-Al2O3-SiO2. Am. Institute of Mining and Metallurgical Engineers, Blast Furnace, Coke Oven and Raw Materials Committee Proc. 12, 281-315; discussion, 315-317 (1953).

Peacey, J. G. and Davenport, W. G., The Iron Blast Furnace 1979. Published by Pergamon Press.

Rice, O. R., Blast furnace gas conditioning. Iron and Steel Engineer 19, 66-89 (Dec. 1942).

Saunders, L. M., The contour of blast-furnace hearths. Iron and Steel Engineer, 54, No. 5, 44-49 (1977).

Sastry, K. V. S., ed., Agglomeration 77, (two volumes) American Institute of Mining, Metallurgical, and Petroleum Engineers, Inc., New York, 1977.

Schick, F. T., and H. Palz. Ceramic burners for blast-furnace stoves. Iron and Steel Engineer, 51, No. 4, 41-44 (1974).

Shimizu, Masaharu. The determination of blast-furnace profile. ICSTIS Proc. Suppl. Trans. ISIJ, vol. 2, 172-175 (1971).

Snow, R. B., Melting temperature charts for the system CaO-MgO-Al2O3-SiO2 as related to the MgO and Al2O3 content of blast furnace slags. Am. Institute of Mining, Metallurgical and Petroleum Engineers; Blast Furnace, Coke Oven and Raw Materials Proceedings 21, 125-138 (1962).

Stapleton, J. M., Results obtained from surveys of gas at furnace tops. Am. Institute of Mining and Metallurgical Engineers, Blast Furnace and Raw Materials Committee Proc. 2, 89-118 (1942).

Stephenson, R. L., Improved productivity and fuel economy through analysis of the blast furnace process. Iron and Steel

Engineer 39, No. 8, 91-98, 1962.

Stephenson, R. L. Optimum blast-furnace hot-blast temperature. ISS-AIME Ironmaking Proc., 36, 482-489 (1977).

Stephenson, R. L., and F. C. Langenberg. Distribution of burden materials in blast-furnace model. AIME Blast Furnace, Coke Oven and Raw Materials Conference Proc. 12, 265-275 (1953).

Stephenson, R. L. Effect of alkalies on blast-furnace operation. Proc. of Symposium on Alkalis in the Blast Furnace, McMasters University. McMasters University Press (1973).

Symposium on blast furnace blowing-in practice. Am. Institute of Mining and Metallurgical Engineers, Blast Furnace, Coke Oven and Raw Materials Committee Proc. 12, 216-245 (1953).

Thompson, R. G., R. L. Franklin, J. R. Guseman and D. E. Rohaus, United States Steel hot-ore briquetting process. Am. Institute of Mining, Metallurgical and Petroleum Engineers; Blast Furnace, Coke Oven and Raw Materials Proceedings 20, 316-323; discussion, 323-328 (1961).

U. S. Bureau of Mines:

Blast-furnace stock column, by S. P. Kinney (TP 442) 1929.

Composition of materials from various elevations in an iron blast furnace, by S. P. Kinney (TP 397) 1926

Effect of sized ore on blast-furnace operation, by S. P. Kinney (TP 459 (1930).

Flow of gases through beds of broken solids, by C. C. Furnas (Bull 307) 1929.

Heat transfer from a gas stream to a bed of broken solids, by C. C. Furnas (Bull 361) 1932.

Iron blast furnace, by T. L. Joseph (IC 6779) 1934.

Iron blast-furnace reactions, by S. P. Kinney, P. H. Royster, and T. L. Joseph (TP 391) 1927.

Oxides in pig iron; their origin and action in the steel-making process, by C. H. Herty, Jr. and J. M. Gaines, Jr. (Bull 308) 1929.

Solubility of carbon in iron-manganese-silicon alloys, by C. H. Herty, Jr. and M. B. Royer (RI 3230) 1934.

Wagstaff, J. B., Further studies of the tuyere zone of the blast furnace. Am. Institute of Mining and Metallurgical Engineers, Metals Branch Trans. 197, 895-902 (1953).

White, R. H., Recent advances in iron production techniques. Presented before Minnesota Section, American Institute of Mining, Metallurgical and Petroleum Engineers, Duluth, Minn., January, 1962.

CHAPTER 16

Oxygen Steelmaking Processes

SECTION 1

INTRODUCTION

Oxygen steelmaking processes are concerned mainly with the refining of a metallic charge consisting of hot metal (molten pig iron) and scrap through the use of high-purity oxygen to rapidly produce steel of the desired carbon content and temperature. Various steelmaking fluxes are added during the refining process to reduce the sulphur and phosphorus contents of the metal bath to the desired level. The oxygen processes typically produce a 180-metric ton (200-net ton) heat of steel every 45 minutes, with only about 15 minutes of this total time being used for the actual refining. Charging of the furnace, sampling and testing for steel composition, measuring the steel temperature, tapping the steel, and pouring off the slag generally account for the balance of the time. In the oxygen processes, the oxidation of silicon, carbon, manganese, phosphorus, and iron provide the energy required to melt the scrap, form the slag, and raise the temperature of the bath to the desired temperature. The quantities of hot metal, scrap, and fluxes charged for a given heat are calculated such that at the end of blowing the prescribed oxygen, the steel bath will have the desired carbon content and temperature.

Each steelmaking process has been devised primarily to provide some means whereby controlled amounts of oxygen can be supplied to the molten metal undergoing refining. The oxygen combines with the unwanted elements (with the exception of sulphur) and, unavoidably, with some of the iron, to form oxides, which either leave the bath as gases or enter the slag. The mechanism by which sulphur is removed does not involve direct reaction with oxygen but depends instead on the conditions of the slag (primarily basicity, state of oxidation, and temperature). As the refining of the hot metal proceeds and the carbon content decreases, the melting point of the bath increases. Sufficient heat must be generated from the oxidation reactions to keep the bath molten.

The physicochemical principles governing the chemical reactions involved in steelmaking were given in some detail in Chapter 13, and will be discussed only in general terms herein.

PRINCIPLES AND TYPES OF OXYGEN PROCESSES

Oxygen steelmaking is an outgrowth of pneumatic steelmaking which was discussed in Chapter 1. In the oxygen steelmaking processes, high-purity oxygen is blown under pressure through, onto, or over a bath containing hot metal, steel scrap, and fluxes to produce steel.

There are various ways in which the oxidizing gas can be supplied. The three methods presently used in commercial processes are shown schematically in Figure 16—1. One type of converter is **top-blown**, the oxidizing gas being introduced by a pipe or lance containing special nozzles to impart a supersonic velocity to the exiting oxygen jets. This pipe or lance is water-cooled and is lowered into the vessel through the mouth opening (see Section 3).

In the **bottom-blown** processes, oxygen is introduced through a number of tuyeres in the bottom of the converter. The oxygen is blown through tuyeres consisting of two concentric pipes. The oxygen passage is through the center of the inner pipe, and a hydrocarbon coolant is injected through the annulus between the two pipes. Most bottom-blown processes use methane or propane as the hydrocarbon coolant, while other bottom-blown processes use fuel oil. The hydrocarbon coolant decomposes at liquid-steel temperatures, absorbing heat as it exits the tip of the tuyere, thereby protecting the oxygen tuyere from premature overheating and excessive burnback (see Section 4).

The third group of processes involves **combination blowing.** One general class of combination-blown proc-

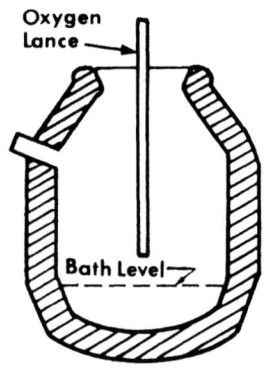

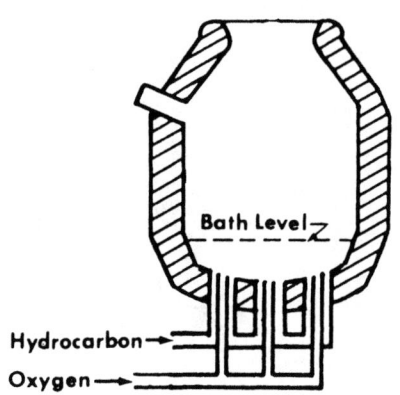

Top-Blown (BOP) Process

Oxygen of commercial purity, at high pressure and velocity, is blown downward vertically into the bath through a single water-cooled pipe or lance, indicated by arrow.

Bottom-Blown (Q-BOP) Process

Oxygen of commercial purity, at high pressure and velocity, is blown upward vertically into the bath through tuyeres surrounded by pipes carrying a hydrocarbon such as natural gas.

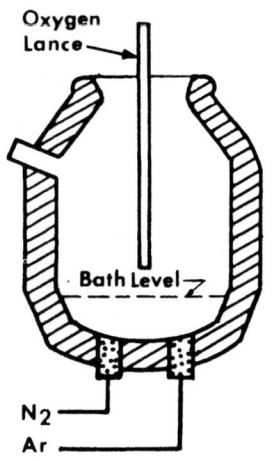

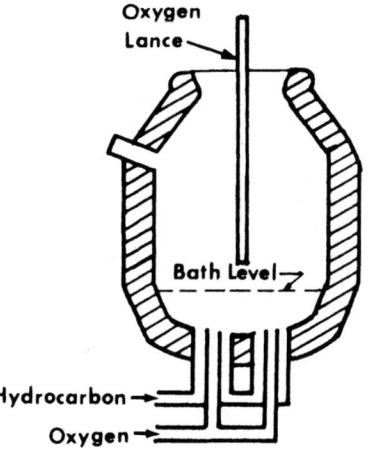

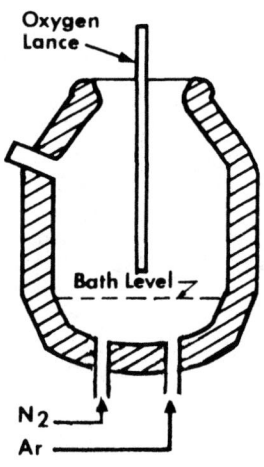

Top lance plus permeable elements in bottom

Top lance plus cooled bottom tuyeres

Top lance plus uncooled bottom tuyeres

Combination-Blown Processes

Oxygen is blown downward into the bath, and oxygen and/or other gases and/or are blown upward through permeable elements or tuyeres.

FIG. 16—1.Three principal ways in which oxygen and/or other gases are supplied to oxygen processes.

esses utilizes oxygen through a top lance, and an inert gas through tuyeres or permeable elements in the furnace bottom to stir the bath. A second class of combined blowing processes utilizes some of the oxygen through a top lance or tuyeres mounted in the top cone of the vessel, and the balance of the oxygen through Q-BOP type tuyeres mounted in the vessel bottom. These processes generally can also switch the bottom gas from oxygen to argon or nitrogen for stirring purposes (see Section 5).

In the 1960's and early 1970's, the top-blown oxygen steelmaking process was the preeminent pneumatic steelmaking process. By 1978, the bottom-blown oxygen process emerged as a new technology with specific metallurgical advantages. By this time, this process had been applied in some greenfield installations and in some conversions from the open hearth steelmaking process. However, by 1982, a multitude of combination-blown processes emerged, mainly by modification of existing top-blown or bottom-blown oxygen steelmaking processes. This came about because of the desire to approach the metallurgical advantages of bottom-blowing without abandoning the advantages of top-blowing.

The main raw materials for these processes are discussed in Section 2. The plant layout and main sequence of operation are similar for the different processes and are discussed in detail in Section 3, the section on the top-blown process. The major differences between the processes are in the design of the furnace and the equipment for introducing the oxygen and fluxes, and also the main refining steps. These differences will be pointed out in Sections 3, 4, and 5. The methods of controlling the different processes are similar and are discussed in Section 6.

SECTION 2

RAW MATERIALS

The three types of oxygen steelmaking processes use the same types of raw materials except that, in bottom-blowing and combination blowing, tuyere coolants and inert gases are added. However, the relative amounts of hot metal and scrap vary with the different processes. The fluxes are usually in lump form when fed from hoppers to the mouth of the furnace. They are in powder form when injected with oxygen through the bottom tuyeres or through the top lance.

OXYGEN

The availability of a continuous supply of low-cost tonnage oxygen was the most important factor in the development of satisfactory high-speed oxygen processes. The composition of the gaseous oxygen required is most important for the production of quality steel. Oxygen for these processes must be at least 99.5 per cent pure, and ideally 99.7 to 99.8 per cent pure. The remaining 0.2 to 0.3 per cent is made up of approximately 0.005 per cent nitrogen and approximately 0.2 per cent argon. The percentage of nitrogen is critical since very small amounts of dissolved nitrogen in the steel can cause strain-aging in certain products.

The mass flow rate of the oxygen stream is also important. In the top-blown processes, the oxygen must be blown at supersonic velocity in order to penetrate the slag and metal emulsion. High flow rates also keep the violent reactions of slag and metal from plugging up the nozzles of the lance. The oxygen flow rate also is very critical in determining the proper lance height above the bath, which has a pronounced effect on the state of oxidation of the slag, and also on the sulphur and phosphorus removal.

HOT METAL

Hot metal consists of the element iron combined with numerous other chemical elements, the most common of which are carbon, silicon, manganese, phosphorus, and sulphur. Hot metal generally contains 4.0 to 4.5 per cent carbon, 0.3 to 1.5 per cent silicon, 0.25 to 2.2 per cent or more manganese, 0.03 to 0.05 per cent sulphur (if external hot metal desulphurization is used, these values will be lower), and 0.05 to 0.20 per cent phosphorus.

The composition of hot metal depends on the blast furnace charge and the practice chosen, especially where silicon, manganese, and phosphorus are concerned. The carbon content depends on the temperature and on the concentrations of solute elements present in the hot metal.

At the time the hot metal is charged into the ladle, it cools and reaches saturation with respect to carbon. The excess carbon leaves the bath in the form of a graphite kish. Good practice requires that this graphite kish be removed, along with the blast furnace slag, which is saturated with sulphur (often ranging from 1.0 to 3.0 per cent). The removal of these undesirable components is necessary so that an accurate hot metal weight may be obtained for the process charge calculation and to obtain lower sulphur contents in the steel.

Hot metal is the primary source of iron units and fuel for the refining process for converting iron to steel. The primary source of heat inputs for the process is shown in Figure 16—2. Hot metal is usually produced in blast furnaces. The iron is cast into holding ladles for transport to the steelmaking facilities. Brick-lined torpedo (submarine-shaped) ladles are preferred because of their insulating qualities and consequent lower heat loss from the iron. From the blast furnace casthouse, the hot metal goes either to a desulphurization station or directly to the steelmaking shop. In the steelmaking shop, the hot metal is either poured directly to charging ladles to be charged into the steelmaking furnace or to large hot metal storage vessels, commonly referred to as hot metal mixers. Some steelmaking shops desulphurize the hot metal in the charging ladle prior to charging into the furnace.

Higher iron temperatures permit the use of less hot

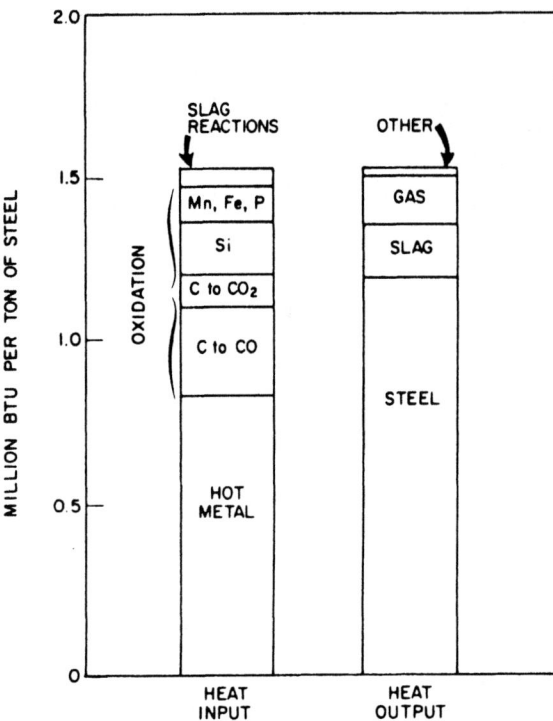

FIG. **16**—2. Chart showing the heat balance for the Basic Oxygen Process.

metal and more scrap for any particular charge. This is advantageous because scrap is generally less expensive than hot metal in North America. Blast furnace iron is normally cast at temperatures between 1480 and 1510°C (2700 and 2750°F). The temperature of the hot metal will decrease approximately 165°C (300°F) between the blast furnace cast and charging the steelmaking vessel.

The iron temperature is usually measured by pouring a given amount of the hot metal into a charging ladle, and then inserting an immersion-type thermocouple into the hot metal. Usually about three-fourths of the total amount of iron required for a charge is poured into the charging ladle before measuring the temperature. This practice permits late modifications to the scrap to hot metal ratio in the charge.

Carbon—Carbon is a very important element in hot metal because the oxidation of carbon provides considerable heat for the process. Variations in the carbon content must be known if the process is to be properly controlled. The carbon content of the hot metal is dictated primarily by the other elements in the iron, especially silicon and manganese, and will vary significantly with hot metal temperature. As the temperature of the hot metal decreases, the solubility of carbon in the hot metal decreases.

Silicon—Hot metal silicon content is probably the single most important chemical factor affecting the proportions of material charged to individual heats. The exothermic reaction of the oxidation of silicon generates significant process heat.

The amount of silicon in the hot metal has a significant effect on slag volume as most steelmaking shops operate with a slag basicity ratio ranging from 2.5 to 4.0. Small variations in silicon content can also greatly affect the quantity of scrap required to produce a balanced charge to provide a specific steel carbon content and temperature at turndown. Typical relationships between the silicon level in the hot metal and the amount of scrap in the charge is shown in Figure 16—3. The amount of scrap melted varies from shop to shop because of different types of scrap used (low-and high-carbon steel, broken molds, pit scrap, etc.) and different flux practices. Most North American shops operate with approximately 0.5 to 1.3 per cent silicon in the hot metal.

The most desirable hot metal silicon level is a function of many site-specific factors such as the required blast furnace production levels and available capacity, the presence or absence of hot metal desulphurization facilities, and the relative price of hot metal and scrap. Higher hot metal silicon levels permit a higher scrap-to-hot-metal ratio to be employed and consequently higher steel tonnage to be produced for a fixed hot metal supply. Increased scrap melting generally results in lower steelmaking costs. However, high hot metal silicon levels have an adverse impact on blast furnace productivity, flux usage, and the steelmaking furnace yield. Therefore, the total combined costs of ironmaking and steelmaking must be considered to select the optimum hot metal silicon level.

Manganese—The U.S. industry average for hot metal manganese content is 0.70 per cent, with the range varying between 0.25 to 2.15 per cent.

Increased hot metal manganese levels generally result in higher residual manganese levels in steel at turndown, resulting in lower ferromanganese ladle additions. Slag produced in oxygen steelmaking processes is suitable for charging into blast furnaces. Approximately 75 per cent of the manganese can be re-

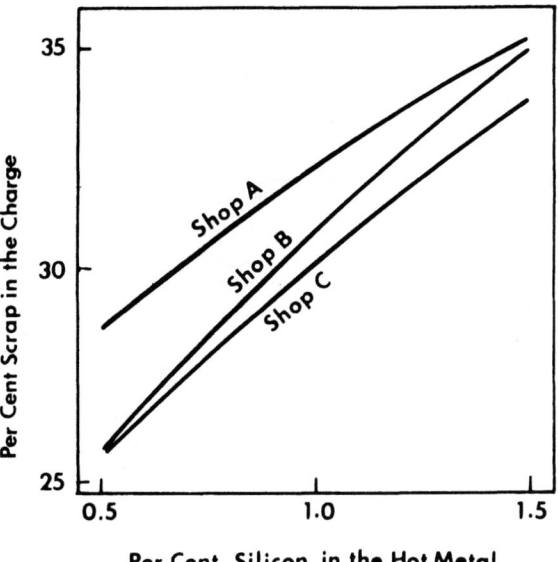

FIG. **16**—3. Scrap charge as a function of silicon content of hot metal for three different BOP shops (aim carbon = 0.05 per cent).

covered, along with some of the iron and fluxes, but the recycled phosphorus can cause problems.

The manganese level in hot metal has a strong effect on slag conditioning. High manganese levels can produce a very fluid slag, readily fluxing the burnt lime and eliminating the need for large quantities of fluorspar. However, manganese contents greater than 1.0 per cent tend to produce slopping from the vessel because of the high slag fluidity, resulting in yield losses.

Sulphur—Sulphur is one of the most troublesome elements affecting steel quality, and the hot metal sulphur content has a direct effect on steel sulphur content. The sulphur content of the hot metal generally varies inversely with silicon content and blast furnace productivity. A recent trend has been to operate the blast furnaces at higher production rates using lower silicon levels and higher sulphur levels, and then to desulphurize the hot metal in an external facility.

External desulphurization is favored by the reducing condition of the hot metal and also by the possibility of vigorously mixing the hot metal and desulphurization reagents to minimize the time required for effective desulphurization. Typical hot metal desulphurization reagents include calcium carbide, salt-coated magnesium, lime-magnesium, and lime-aluminum.

Several elements can affect the activity of the sulphur in the hot metal. Carbon and silicon increase the activity of sulphur, while manganese has a slightly negative effect. The removal of sulphur from hot metal is easier than from steel.

In general, hot metal can readily be desulphurized from sulphur levels of 0.04 to 0.06 per cent to sulphur levels of 0.015 to 0.020 per cent by external desulphurization. Levels of 0.005 per cent sulphur can be obtained by using large quantities of desulphurizing reagents. However, unless the scrap charge is very low in sulphur content, the sulphur level of the steel at tap, even with excessive lime usage, will not be much below 0.01 per cent.

Some benefits which may be realized in the steelmaking operations with the use of externally desulphurized hot metal include:
1. increased blast furnace productivity;
2. decreased blast furnace coke rate;
3. decreased blast furnace sinter rate;
4. decreased blast furnace and steelmaking slag volume;
5. decreased steelmaking flux consumption;
6. decreased steelmaking oxygen consumption;
7. increased steelmaking vessel lining life;
8. increased yield;
9. improved steel quality;
10. reduced production cost.

Phosphorus—Phosphorus is very sensitive to the iron oxide content in the slag. Although phosphorus is present in relatively small percentages in hot metal, it has a very great influence on steel quality. The removal of phosphorus requires low bath temperatures and conditions that are oxidizing, whereas the removal of sulphur is favored by reducing conditions and higher temperatures. Hot metal with a phosphorus level of 0.10 per cent or less can be reduced in oxygen steelmaking furnaces by 80 to 90 per cent with the proper flux charge for a low carbon steel. Higher phosphorus iron requires additional flux and often results in longer production time due to the need for reblows and working the phosphorus level down. Some of the adverse economic effects of high phosphorus iron are:
1. additional hot metal must be used because of the effect of additional flux on the heat balance;
2. longer heat times, reblows, and higher slag iron oxide levels result in increased refractory consumption;
3. reduced metallic yields because of a larger slag volume and higher slag iron oxide levels.

SCRAP

Scrap produced within the mill itself—sheet scrap, trimmer scrap, slab ends, bloom ends, and ingot butts—is often used and generally does not affect the final quality of the steel. Other scrap, such as scrapped auto bodies, is purchased. However, large amounts of undesirable elements can be contained in this type of scrap.

Hot metal composition and temperature are the two most important variables that determine the percentage of scrap to be charged for a heat. Typically, most pneumatic furnaces consume 20 to 35 per cent of the total metallic charge as cold scrap. Some of the newer processes using special post-combustion lances and carbon injection can melt significantly higher percentages of cold scrap.

Scrap metal quality also has a very significant effect both on the smoothness of refining a heat and the final steel quality obtained. It is important that scrap composition is considered when calculating a scrap charge for any particular heat. Residuals in the scrap, such as tin, are not oxidized in the steelmaking process, and usually are undesirable in the final product.

Scrap density is also very important. Scrap is often classified as "light" or "heavy". Most steelmaking shops use a 1:1 ratio of light to heavy scrap in their charge. It is desirable to charge the light scrap into the furnace first, then the heavy scrap, to save wear and tear on the furnace refractories.

Scrap is usually less expensive than hot metal, and therefore, increased scrap melting is generally a desired objective. The use of scrap also increases the product yield because the scrap is nearly 100 per cent iron. There are several ways to increase the scrap metal charge. The scrap can be preheated inside or outside the steelmaking vessel. Some shops increase scrap melting by charging fuels into the furnace before or during a heat. Ingots can be preheated in large soaking pits before charging the scrap into the furnace.

FLUX MATERIALS

The fluxes used in the oxygen steelmaking processes are burnt lime, burnt dolomite, and fluorspar. They are granular in form when charged through the mouth of the vessel or in powdered form when injected through bottom tuyeres or through the oxygen lance.

Burnt Lime—Burnt lime is produced by calcining high-calcium limestone. Calcining is the removal of carbon dioxide and organic materials from limestone ($CaCO_3$) to form lime (CaO). The limestone is calcined to burnt lime by three general methods: rotary kiln, shaft kiln, and rotary hearth kiln.

High calcium burnt lime consumption ranges from 100 to 200 pounds per net ton of steel produced. The wide range of lime consumption is partly due to the variability in hot metal silicon content, variations in the sulphur and phosphorus content of the hot metal, and variations in the desired composition of the product mix.

Burnt Dolomite—Most of the dolomitic lime (CaO-MgO) produced in the United States is produced by calcining dolomitic stone in rotary kilns.

The addition of dolomitic lime to the flux charge contributes to extending the furnace lining life by satisfying part of the MgO solubility of the slag so that very little MgO is dissolved from the furnace lining. High-magnesia slags are less corrosive on the magnesite or dolomite lining material. The solubility of MgO in BOP slags for different lime-silica ratios is shown in Figure 16—4. Most shops operate with a 4 to 6 per cent MgO level in the slag. If the MgO level is too high, the slag becomes thick and impairs the removal of the sulphur. A high level of MgO also adversely affects the phosphorus removal. Dolomitic lime consumption is usually 30 to 35 pounds per net ton of steel.

Fluorspar—The chief function of fluorspar is to aid in promoting rapid lime solution in the slag. Fluorspar is a slag fluidizer and lowers the melting point of the slag.

COOLANTS

The additional of coolants at the end of oxygen blowing may be necessary to lower the temperature of the heat quickly to the desired level. The most commonly used coolants are raw limestone chips and iron

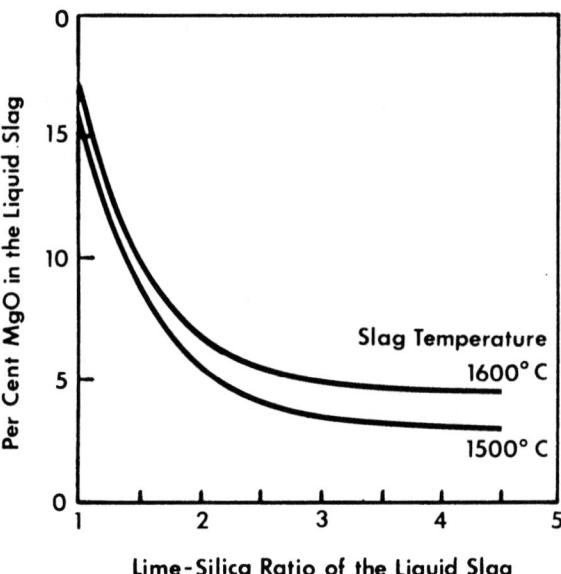

FIG. 16—4. Graph showing the solubility of MgO in BOP slags for different lime-silica ratios.

ore. Some shops use lime to produce relatively small temperature losses. Shops with dynamic control (see Section 6) add the coolants before the end of the blow. The amount of coolant is determined by the process control computer based on a temperature and carbon determination taken from the bath about three minutes before the end of the oxygen blowing.

SECTION 3

THE TOP-BLOWN OXYGEN STEELMAKING PROCESS

Equipment Design—The most common arrangement for oxygen steelmaking used in the United States today is a barrel-shaped furnace, lined with basic refractories, which is concentric in design. Some early vessels of an eccentric design were built, but are uncommon in the United States (Figure 16—5). The furnace, which is tiltable about a fixed horizontal axis, is held upright while the oxygen is injected. A safety lining next to the furnace shell plates consists of a single course of burned magnesite brick. The inner, or working, lining is of unfired tar-bonded or burned impregnated magnesite or dolomite brick. A taphole is provided to facilitate separation of slag and metal during tapping.

The dished solid bottom of the furnace is built up as follows: single or multiple courses of burned magnesite brick next to the shell and a course of tar-bonded or burned impregnated magnesite brick to form the working surface.

The oxygen-jet equipment consists of a tubular, water-cooled, copper-tipped retractable lance kept perpendicular above the center of the bath. At the top of the lance, armored rubber hoses are connected to a pressure-regulated oxygen source and to a supply of recirculated cooling water. The tip of the lance contains from three to five converging-diverging nozzles to impart a supersonic velocity to the exiting gas jets. The nozzles are generally angled from 7 to 10° to the centerline of the lance pipe and equally spaced around the lance tip, as shown in Figure 16—6. The general standards for a typical 200-net ton furnace are a 3-hole lance and an oxygen blowing rate of 15,000 to 25,000 scfm. The nozzles are generally designed to operate with an exit velocity of mach 1.7 to 2.3.

Principle of Operation—After the furnace is charged with the proper amount of scrap and molten iron, it is turned upright and the oxygen lance is lowered to a predetermined position above the surface of the bath. Pure oxygen gas issues from the jet nozzle at high velocity under a pressure that is normally held between 965 and 1240 kPa (140 and 180 psi). The action of the oxygen jet is partly chemical and partly physical. The oxygen immediately starts reacting with the silicon, generating heat and producing molten silica. Slag-forming fluxes (chiefly burnt lime, dolomitic lime, and fluorspar) are added in controlled amounts from an

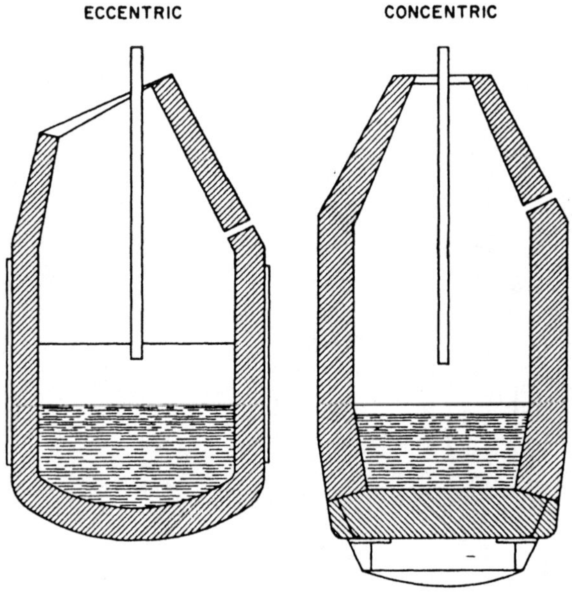

ECCENTRIC   CONCENTRIC

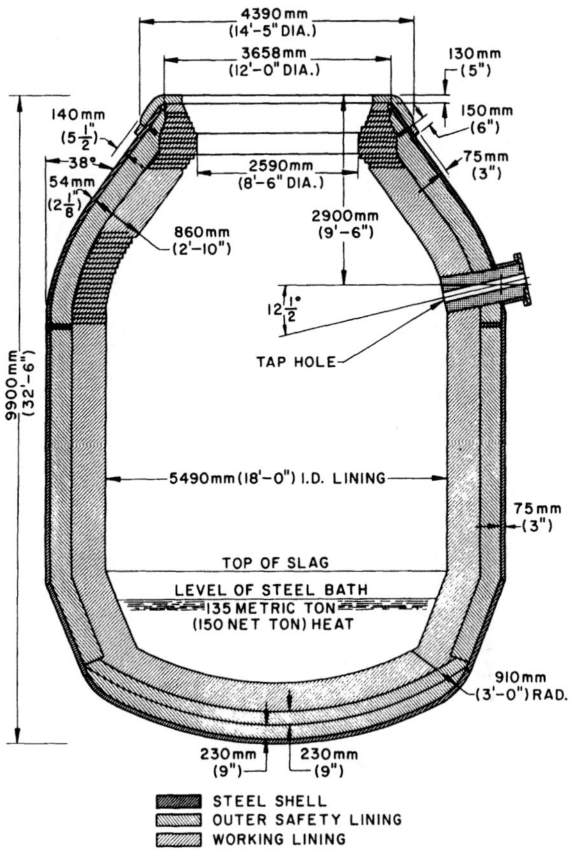

4390mm
(14'-5" DIA.)

3658mm
(12'-0" DIA.)

130mm
(5")

150mm
(6")

140mm
(5½")

75mm
(3")

38°

54mm
(2⅛")

2590mm
(8'-6" DIA.)

2900mm
(9'-6")

860mm
(2'-10")

12½°

TAP HOLE

9900mm
(32'-6")

5490mm (18'-0") I.D. LINING

75mm
(3")

TOP OF SLAG

LEVEL OF STEEL BATH
135 METRIC TON
(150 NET TON) HEAT

910mm
(3'-0") RAD.

230mm   230mm
(9")   (9")

STEEL SHELL
OUTER SAFETY LINING
WORKING LINING

Fig. 16—5. (Above) Schematic sections of types of vessels used when blowing oxygen vertically downward into the metal through a water-cooled lance. (Right) Diagrammatic section of an actual concentric type of vessel.

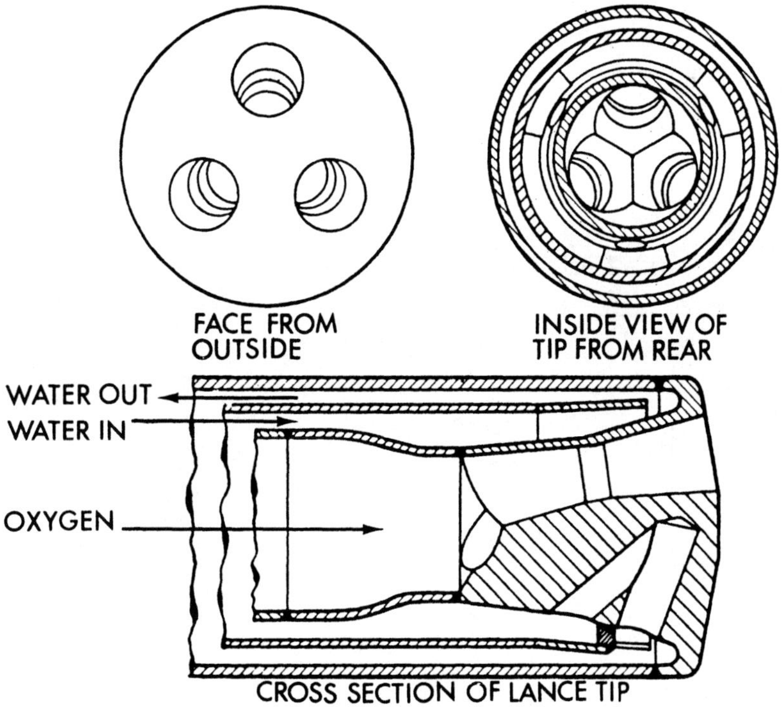

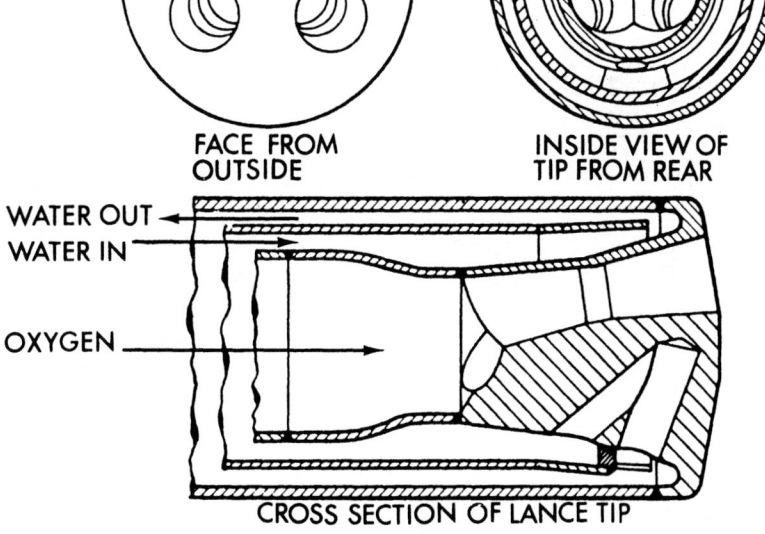

FACE FROM
OUTSIDE

INSIDE VIEW OF
TIP FROM REAR

Fig. 16—6. Three views of an oxygen lance.

WATER OUT
WATER IN

OXYGEN

CROSS SECTION OF LANCE TIP

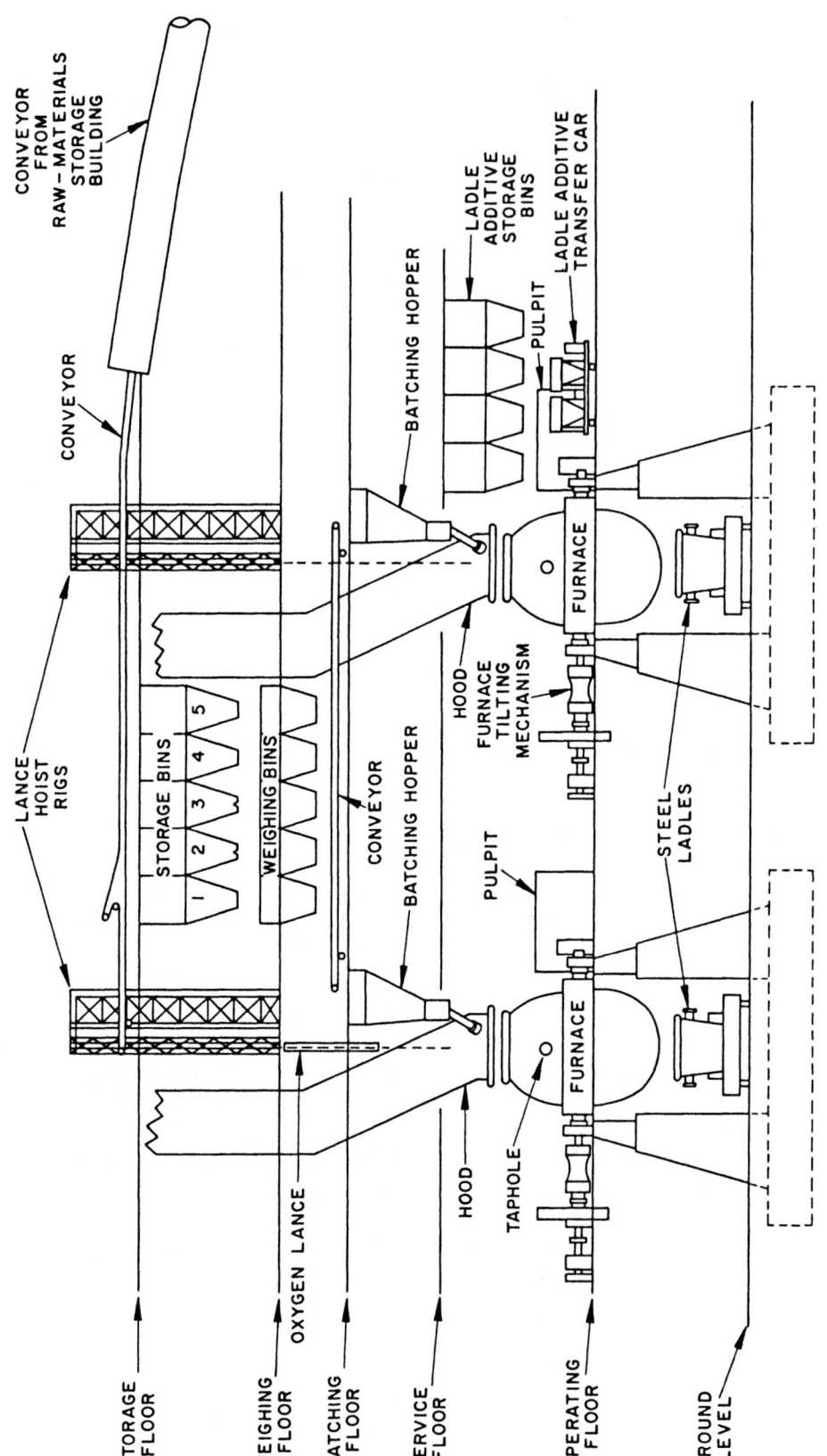

Fig. 16—7. Schematic elevation showing the principal operating units of the basic oxygen process steelmaking shop described in the text. Storage bins contain: No. 1, limestone; No. 2, fluorspar; No. 3, ore; No. 4, calcitic lime; and No. 5, dolomitic lime.

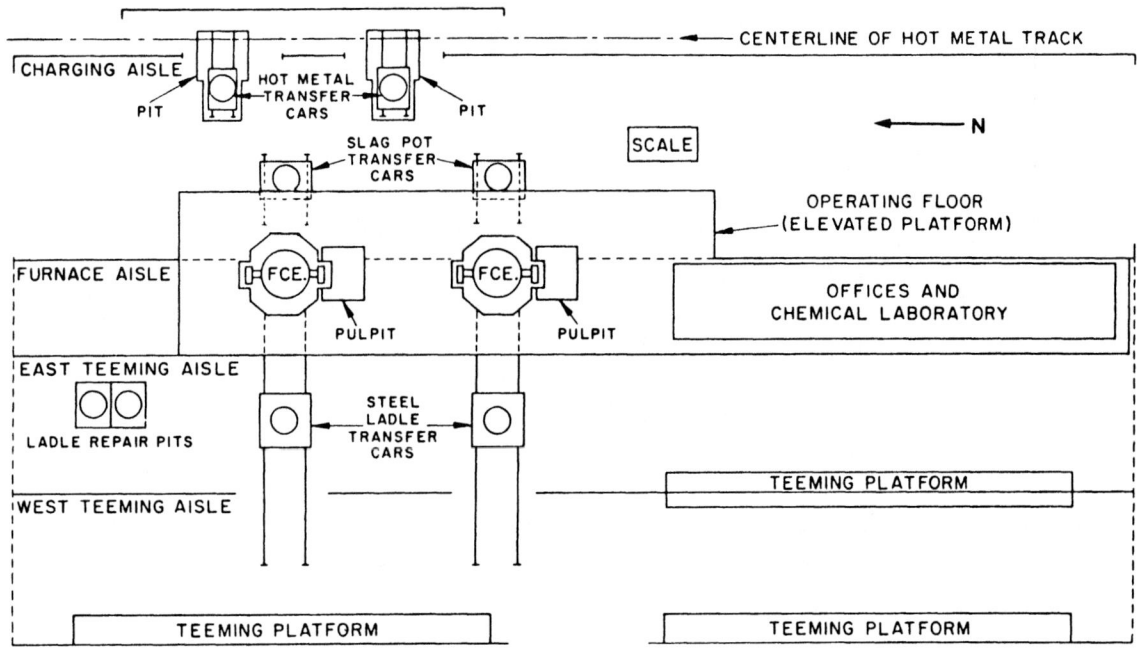

FIG. 16—8. Schematic diagram, not to scale, showing relationships of the locations of the principal operating equipment in the four parallel main areas of the BOP shop described in the text.

overhead storage system immediately after oxygen ignition. These materials, which serve to produce a slag of the desired basicity and fluidity, are added through an inclined chute built into the side of a water-cooled hood that covers the furnace (see Figure 16—7). As the blow progresses, an emulsion is formed with the slag and metal, and the other metalloids are oxidized. Carbon monoxide is evolved, which gives rise to a vigorous boiling action and accelerates the refining metallurgical reactions.

Plant Layout—Steelmaking facilities employing the basic oxygen process commonly are referred to as BOP (or BOF for basic oxygen furnace) shops. There are numerous differences among BOP shops with respect to type and location of raw-materials-handling, steel- and slag-handling and gas-cleaning equipment. For this reason, the following description based upon one BOP shop may not be typical.

The main building of the BOP shop to be described runs north-south and is composed of four parallel main areas—charging aisle, furnace aisle, east teeming aisle and west teeming aisle (Figure 16—8). Flux and ore storage facilities lie to the south and are connected to the BOP shop by a covered conveyor-belt system. Gas-cleaning and dust-collecting facilities are located north of the BOP shop. All raw materials used by the BOP furnaces—molten iron, steel scrap, and fluxes such as burnt lime, dolomitic lime, iron ore and fluorspar—enter the shop from the south. Fluxes are handled by conveyor, molten iron from the blast furnaces by railroad submarine ladle cars, and steel scrap by both rail and truck.

The scrap yard occupies the south end of the charging aisle and is large enough to maintain about a 24-hour supply. Scrap received in railroad cars is loaded into scrap boxes on transfer cars by an overhead crane equipped with a deep-field magnet. Since the exact weight of scrap to be charged into a furnace must be known, the scrap boxes are loaded on a large weighing scale, the readout of which is visible to the crane operator. The scrap boxes are moved on their cars from the scrap-loading aisle into the charging aisle where each box is raised to the charging floor by an overhead crane that charges the scrap directly into the furnace.

Molten iron is delivered to either of two hot-metal-handling stations adjacent to the charging aisle and opposite the furnaces. Each station consists of a reladling pit, operator's pulpit, and hot-metal-transfer ladles on transfer cars. After the transfer ladle has been filled from a submarine ladle, an overhead crane raises the ladle to the furnace.

Fluxing materials traveling over the covered conveyor-belt system are deposited in their respective storage bins at the storage-floor level in the furnace aisle by a tripper conveyor (Figure 16—7). From these bins, they are fed by gravity and automatically weighed as they drop into weigh hopper bins on the next lower level. The fluxes fed from the weigh-hopper bins are dispensed by another conveyor belt into surge hoppers located directly above the furnaces. This entire system can be controlled from the pulpits next to the furnaces or from local stations at the bins.

The BOP furnaces in the shop being described are of all-welded steel construction. The trunnion ring is separate from the steel shell. Each furnace is slightly more than 9900 mm (32½ feet) in height. Each can be rotated by motors operated from the pulpit adjacent to each furnace.

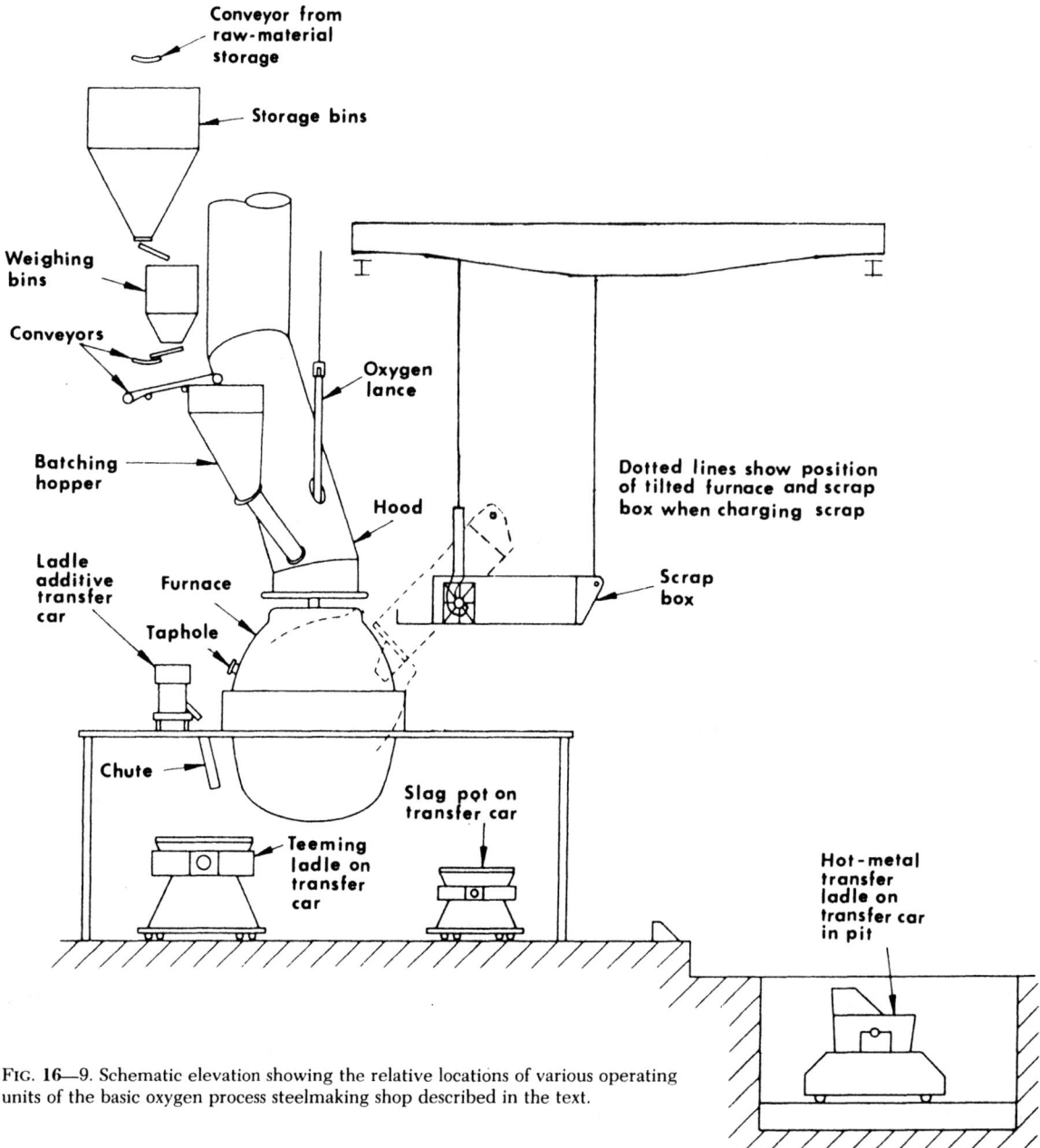

FIG. 16—9. Schematic elevation showing the relative locations of various operating units of the basic oxygen process steelmaking shop described in the text.

Beneath the furnace, teeming-ladle and slag-pot transfer cars, on tracks, run east and west (Figure 16—9). Teeming ladles are positioned by remote control to the west side and slag pots to the east side. After the ladles and slag pots have been filled, they are moved on their transfer cars from beneath the furnaces, the teeming ladle to either of the pouring aisles and the slag pot to the charging aisle. Overhead cranes then move the teeming ladle to a line of empty ingot molds in one of the pouring aisles, while the slag pot is moved to the slag dump yard.

Each furnace is equipped with two oxygen lances— one in use and one on stand-by to protect the operation against lance failure during a heat. Automatic controls prevent movement of the furnace while a lance is in operating position, or lowering of a lance when the furnace is not in the vertical position. These controls automatically remove the lance from the furnace for any of a variety of unacceptable operating conditions.

The hoods and stacks above the furnace leading to the gas-cleaning and dust-collecting facilities are water-cooled in their entirety. Of membrane-type construction, they are cooled by water flowing in 38-mm (1½-inch) steel tubes spaced on 50-mm (2-inch) centers and circular in cross-section, as shown in Figure 16—10. The cooling system of the hood is divided into four

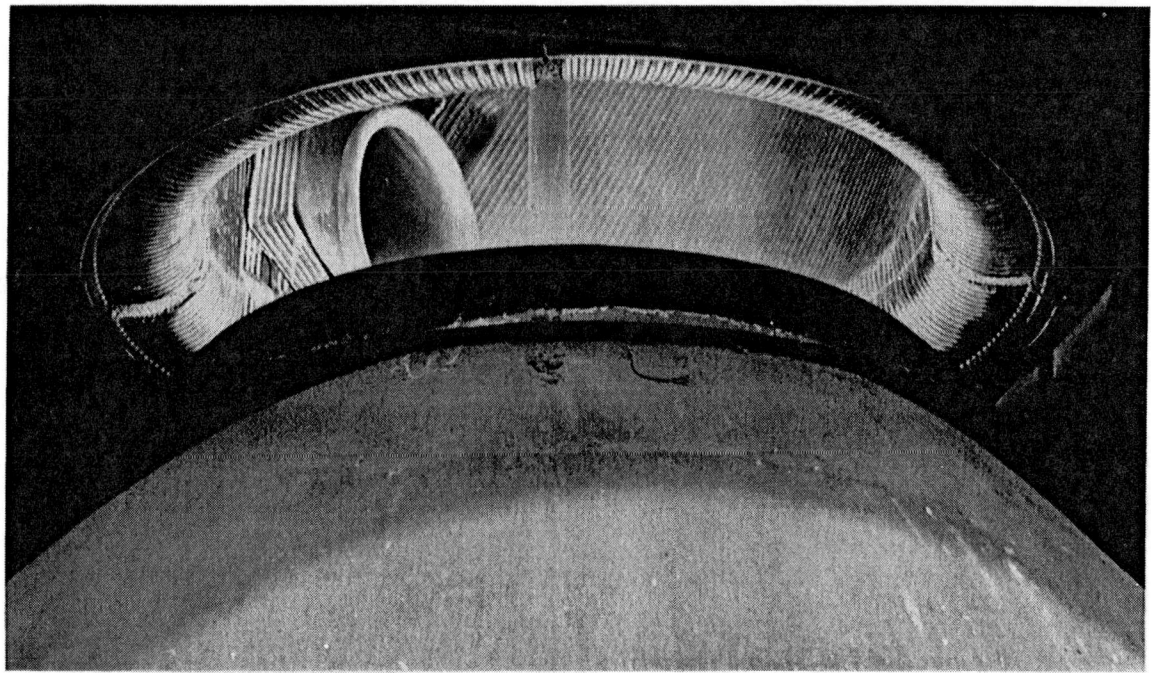

Fig. 16—10. Partial interior view of the hood of a basic oxygen process steelmaking furnace, showing piping of water-cooling system that protects the hood and stack from the hot gases emitted from the furnace during blowing with oxygen. Discharge opening of the chute for adding fluxes, etc., is visible at the left of the hood. The lance is shown lowered into the furnace. This photograph was taken during the burning-in of a new lining, which is accomplished by charging incandescent coke into the furnace and burning it with oxygen.

sections, each equipped with variable orifices to proportion water flow according to requirements.

The gas-cleaning and dust-collecting facilities usually consist of a quencher, Venturi-type washer and cooling tower in series, supplemented by a clarifier or thickening basin. The clarifier or thickening basin is 4570 mm (15 feet) high and 26 metres (85 feet) in diameter: dust entrained in the water settles out as a sludge. Cleaned water leaving the clarifier contains less than 0.085 grams per litre (5 grains per gallon) of settleable solids. Cleaned gases passing into the atmosphere contain not more than 0.11 grams per cubic metre (0.05 grains per cubic foot) of solids.

OPERATION OF THE PLANT

Sequence of Operations—Although minimum expected weights of both steel scrap and molten iron are prepared for charging into the furnace well in advance of the start of a heat, the final determination of the exact weight of each is made by a computer. To enable the computer to calculate what raw materials are required for a given heat, it must be provided with data based upon specifications for the finished steel. These are fed into the computer memory by a communications system connected with the plant's production-planning department. The specifications are held in memory until the composition and temperature of the molten iron to be charged into the furnace are secured by actual analysis and temperature measurement and transmitted to the computer.

Utilizing these data, the computer calculates the re-

quired weights of molten iron, steel scrap, and fluxes (burnt lime, dolomitic lime, and fluorspar). It also determines the amount of oxygen that will be blown during the heat. Any weight changes in the already-prepared charges of scrap and molten iron that may be called for by the computer are made very quickly. Scrap is first dumped into the furnace, which is inclined toward the charging aisle. The scrap box is raised and tilted with the charging crane (Figure 16—9) so that the load is deposited in the furnace. The overhead crane then transports the hot-metal ladle to the furnace mouth and pours its charge of molten iron on top of the scrap (Figure 16—11).

The furnace is immediately returned to an upright position, the lance is lowered into the furnace, the flow of oxygen is started, and the steelmaking process is under way (Figure 16—12). Within seconds after the oxygen is turned on, reaction with the impurities of the charge commences. At this point in the process, the prescribed weight of fluxes is added to the furnace through a flux chute in the hood (Figure 16—10). Fumes are drawn off into the hood by a fan. Under normal operating conditions, the time elapsed from the charging of scrap to the start of the oxygen blow averages less than three minutes.

When the blow is completed, as determined by the furnace operator utilizing the results of calculations by the computer, the lance is withdrawn and the furnace is tilted to a horizontal position toward the charging aisle. A temperature reading is secured with an immersion-type thermocouple, and a sample of the steel is obtained and sent to the nearby chemical labo-

FIG. 16—11. Hot-metal transfer ladle is lifted and tilted by overhead crane to pour the molten iron it contains into the mouth of a basic oxygen process steelmaking furnace.

FIG. 16—12. Basic oxygen process steelmaking furnace in operating position, with oxygen being blown into the furnace at the beginning of a heat. Gases generated by the oxidation reactions are captured by the hood, cooled and cleaned before discharge into the atmosphere.

FIG. 16—13. Molten steel pouring from the taphole of a basic oxygen process steelmaking furnace into a teeming ladle positioned beneath the furnace on a remotely controlled transfer car. Chute for addition agents is visible at the right above the ladle. Additives are weighed and fed to the chute by the weigh hopper car on the furnace operating floor (upper right).

ratory for analysis. This analysis determines the individual chemical elements in the steel, and is usually available within three to five minutes. If the steel in the furnace is too hot, the furnace is returned to the vertical position and either ore or limestone is added as a coolant through the flux chute in the hood. The vessel can also be rocked to lower the steel temperature. If the steel is too cold, the lance is again lowered into position and oxygen is blown for a short period.

Tapping Practices—The furnace then is tilted in the opposite direction toward its taphole side, and the steel is tapped into the teeming ladle (Figure 16—13). During tapping, alloying additions are added to the ladle through a chute. After the conclusion of tapping, the final step in the operating cycle is to tilt the furnace toward the charging aisle to invert it and dump slag remaining in the furnace into a slag pot. From this inverted position, the empty furnace is turned to the charging position and is ready to receive its charge for the next heat.

Tapping of an oxygen steelmaking furnace is very important in that yield should be maximized and furnace slag carryover should be minimized. To achieve these goals, the tapping operation must be highly controlled to minimize critical tonnage and to control slag carryover. Critical tonnage is defined as that quantity of steel remaining to be tapped when slag entrainment into the tap stream begins. There are several tech-

niques to accomplish these goals:

1. plug the taphole prior to tapping to prevent the potential initial contamination of 540 to 630 kg (1200 to 1400 pounds) of slag as the taphole passes through the slag layer during furnace rotation;
2. maximize furnace rotation (slag at furnace mouth) to maximize the height of steel over the taphole and monitor furnace position because it will continue to change throughout the lining campaign;
3. minimize slag volume in the furnace to permit increased rotation during the tapping;
4. utilize furnace mouth buildup to provide for increased furnace rotation;
5. use barrel taphole location rather than the conventional cone taphole location, which is most common in North America, to lessen the effect of vortexing;
6. maintain taphole diameter as constant as possible, with a smaller taphole preferred.

Chemistry of the Process—The basic oxygen process is characterized by: (1) use of gaseous oxygen as the sole refining agent; (2) a metallic charge composed largely of blast-furnace iron in a molten condition, thus greatly reducing the thermal requirements of the process; and (3) chemical reactions that proceed rapidly in a bath of comparatively low surface-to-volume ratio, thus minimizing external heat losses. This combination results in an extremely versatile autogenous process that requires

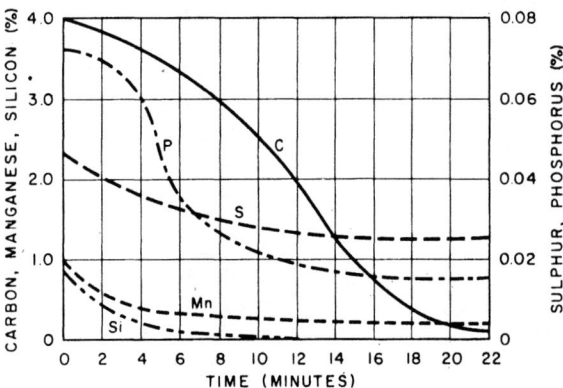

FIG. 16—14. Schematic representation of progress of refining in a top-blown basic-lined vessel.

no external fuel, yet provides a capability for melting an appreciable quantity of scrap. The use of pure oxygen for refining, however, in no way alters the fundamental chemical reactions and equilibria involved in steelmaking. The reactions of greatest importance are those pertaining to the elements carbon, silicon, manganese, phosphorus, sulphur, nitrogen, and oxygen. Changes in bath content of the five first-named elements during the blowing of a basic oxygen heat are shown schematically in Figure 16—14.

Carbon—The mechanics of carbon elimination involve oxidation of carbon for removal as carbon monoxide and carbon dioxide. Steel carbon contents as low as 0.03 per cent and as high as 1.0 per cent are made in the BOP.

Silicon—Silicon is oxidized to silica and transferred to the slag in the basic oxygen process. Silicon oxidation is important mainly due to its thermal effects. Only a trace of silicon remains in the steel at the end of the refining period.

Manganese—Under certain circumstances, the level of residual manganese is sufficient to eliminate the necessity of adding ferromanganese in the ladle. Residual manganese is closely related to the manganese level of the molten iron used in the charge.

Phosphorus—A high degree of slag fluidity and excellent slag-metal contact provide efficient removal of phosphorus in the slag. Typical phosphorus removal for the BOP is about 90 per cent, i.e., hot metal containing 0.10 per cent phosphorus can be reduced to 0.01 per cent for a low-carbon steel.

Sulphur—Factors that favor better sulphur removal

in oxygen steelmaking include high temperatures, high slag volumes, and a high base to acid ratio in the slag. Published data show an approximate average residual-sulphur content of 0.020 per cent in the steel when starting with an average hot metal sulphur content of 0.030 to 0.040 per cent. Typical sulphur removal for the BOP is about 50 per cent, but depends a great deal on starting sulphur levels in the hot metal and scrap sulphur content.

Nitrogen—Because the basic oxygen process uses a refining agent containing practically no nitrogen, and the vigorous flushing of the bath with carbon monoxide removes nitrogen, the steel produced is distinguished by its exceptionally low nitrogen content—generally about 0.004 per cent or lower.

Oxygen—Oxygen content will range from approximately 0.04 per cent at 0.10 per cent steel carbon content to about 0.09 per cent at 0.04 per cent carbon.

Residual Alloy Elements—Copper, nickel or tin are usually considered undesirable in low-carbon steels because they adversely affect ductility. The main source of these unwanted elements is purchased scrap. The generally high consumption of hot metal used in the basic oxygen process therefore results in a low residual alloy content.

Temperature Control—Temperature in the BOP process is controlled by using a thermally balanced charge of hot metal and scrap such that the aim steel temperature and composition are achieved at turndown. If a heat is cold, it is generally reblown to increase temperature. If the heat is hot, it can be cooled by rocking the furnace or by charging limestone chips or other coolants.

End-Point Control—A computer generally is used to establish when the end point of a particular heat is reached. Numerous articles in the bibliography at the end of this chapter discuss various ways in which computer control is being applied to the process.

Product Characteristics—Carbon steel made by the basic oxygen process has a wide variety of applications, including welded and seamless tubular products, wire products, tin plate, hot and cold-rolled sheets, plates, structurals, and bar products. This is especially true in applications where desired quality characteristics are directly related to chemical composition—particularly minimum sulphur, nitrogen, or residual alloy content. The basic oxygen process has been adapted to the manufacture of some of the higher alloy steels, including stainless grades, although the practice has not been adopted to any great extent.

SECTION 4

THE BOTTOM-BLOWN OXYGEN STEELMAKING OR Q-BOP PROCESS

The bottom-blown oxygen steelmaking process, also called the Q-BOP process, was developed in the late 1960's and early 1970's. Although it has many similarities to the top-blown process, there are several important differences. The primary advantage is that the Q-BOP process operates much closer to equilibrium

conditions between the metal and the slag and is therefore much lower in oxidation potential than the BOP. The manifestations of this characteristic of the Q-BOP are lower iron losses as FeO to the slag (Figure 16—15), higher manganese recoveries (Figure 16—16), less slopping, faster blow times, improved phosphorus and sul-

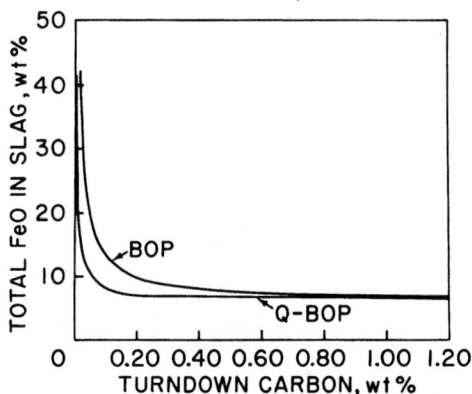

FIG. 16—15. Slag FeO content versus carbon content at first turndown for the BOP and Q-BOP. Curve for Q-BOP applies to Gary Works' 180-metric-ton (200-net-ton) Q-BOP. (After Papinchak and Weaver: see bibliography)

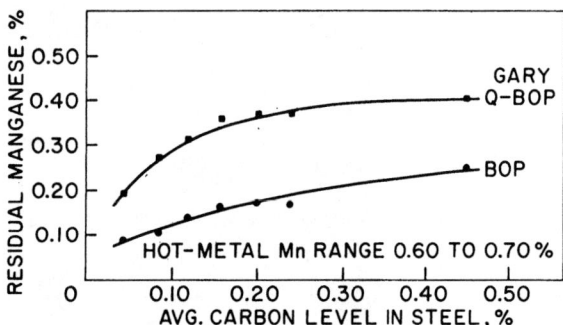

FIG. 16—16. Residual manganese versus steel carbon content for the BOP and Q-BOP. (After Papinchak and Weaver: see bibliography)

phur control, and lower dissolved oxygen and nitrogen contents. Because of these improvements, yield of the Q-BOP is 1.5 to 2.0 per cent higher than the BOP. The lower FeO generation, however, results in lower scrap melting in the Q-BOP as compared to the BOP. Process control of the Q-BOP is much easier than the BOP because of the highly consistent metallurgical behavior of the Q-BOP reactions and the absence of the variability caused by oxygen lance practices as encountered in the BOP.

PRINCIPAL FACILITIES

Plant Layout—The plant layout for a bottom-blown shop is similar to that of a top-blown shop. One major difference is that the building does not have to be as tall as that for a top-blown shop because space above the furnace for the top-blowing oxygen lance is not required in the bottom-blown shop. As a result, it is generally much easier to retrofit the bottom-blown process into an existing open-hearth shop as compared with the top-blown process.

Process Gases—The major process gas used in this process is oxygen and the amount used is 5 to 10 per cent less than that used in the top-blown process. The lime used in this process is generally powdered lime that is mixed with the oxygen and injected into the liquid iron bath with the oxygen. A hydrocarbon fluid such as natural gas or propane is used to cool the oxygen tuyeres during the oxygen blow. Upon completion of the oxygen blow, nitrogen gas is usually blown through the tuyeres while the furnace is rotated to a horizontal position for sampling. This prevents slag and metal from entering and plugging the tuyeres during the turndown.

THE FURNACE

Furnace Design—The Q-BOP furnace consists of a steel shell lined with refractory; the tuyeres are mounted in the furnace bottom, which is removable from the rest of the furnace (Chapter 2). Conventional BOP refractories are used in the barrel and cone of the Q-BOP because the service is quite similar to that in

the BOP. A bottom is generally constructed of a steel plate on which the tuyeres are mounted. The refractory is placed on the steel plate around the tuyeres. Either brick or monolithic refractory may be used.

A Q-BOP furnace bottom lining generally lasts about 500 heats and a Q-BOP furnace barrel or cone lining lasts approximately 1000 to 1500 heats. Sufficient dolomitic lime is normally added to the charge to produce a slag that contains from 4 to 6 per cent MgO. This decreases the attack of the slag on the furnace refractories and extends lining life.

Tuyere Design—Considerable experimentation has been done to permit bottom blowing of hot metal with

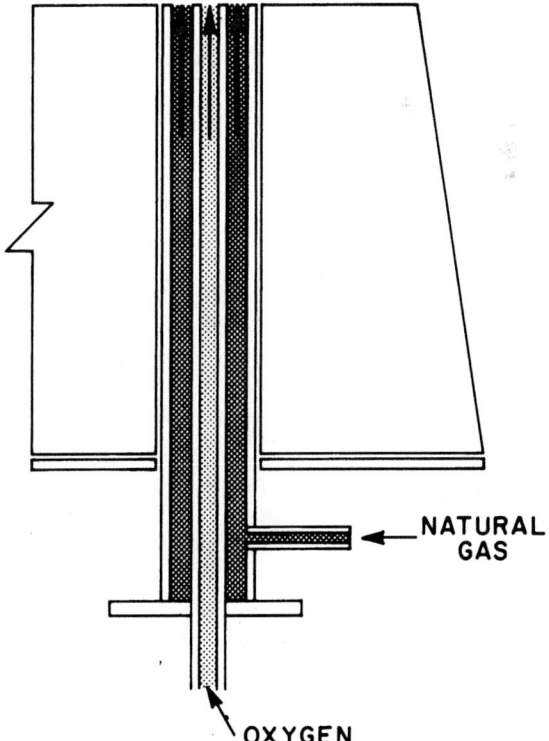

FIG. 16—17. Schematic drawing of a Q-BOP tuyere. (After Papinchak and Weaver: see bibliography)

FIG. 16—18. Charging scrap into a 180-metric-ton (200-net-ton) Q-BOP furnace.
(After Hubbard and Lankford: see bibliography)

pure oxygen. Unless special precautions are taken, the tuyeres used to conduct the oxygen through the furnace bottom are very rapidly eroded because of the large amount of heat generated when the oxygen comes into contact with the hot metal. In the Q-BOP process, a special type of tuyere is used (Figure 16—17). This tuyere consists of a central pipe surrounded by another pipe. Oxygen flows through the central pipe while a hydrocarbon fluid is blown through the annular space between the two pipes. When the hydrocarbon comes near the molten bath, it decomposes endothermically and absorbs enough heat to cool the tuyere. With this concentric tuyere, pure oxygen can be introduced into a iron bath without excessive erosion of the tuyere or the refractory surrounding the tuyere.

OPERATION OF THE PLANT

Sequence of Operations—After the steel and slag from the previous heat have been removed from the furnace, the furnace is tilted to a horizontal position and the furnace lining and tuyeres are inspected. If required, the furnace lining is repaired, generally by gunning. Occasionally, a tuyere may not be cooled sufficiently by the hydrocarbon fluid and the tuyere will burn back too rapidly, requiring that it be removed from service by plugging it with refractory. During this part of the heat, a small flow of nitrogen is maintained through the oxygen pipe and annulus of each tuyere to prevent the tuyere from being plugged by slag and metal in the furnace.

The furnace is now tilted upward and steel scrap is charged into the furnace (Figure 16—18). Because the Q-BOP is a bottom-blown process with very good mixing, it is able to melt larger pieces of scrap, including ingots and large butts, compared with the BOP.

After the scrap is charged, the flow rate of nitrogen through the tuyeres is increased and the hot metal is charged into the furnace. The furnace is then brought into a vertical position, oxygen and hydrocarbon coolant are blown through the tuyeres, and the refining of the heat is started. Within about 1 minute, powdered lime is added to the oxygen being blown into the furnace and this continues for most of the oxygen blowing period. The gas coming out of the furnace mouth is primarily carbon monoxide. This gas contains iron-oxide fume and it is passed through a gas-cleaning system to remove this fume.

After the required amount of oxygen has been blown, the gas flow through the tuyeres is changed to nitrogen. The furnace is then turned down to a horizontal position for sampling and temperature measurement and the nitrogen flow through the tuyeres is significantly reduced. The steel sample is generally analyzed for carbon, sulphur, phosphorus, and other elements. If the steel composition and temperature are within the desired ranges for the grade being made, the steel is tapped into the ladle together with deoxidizers and other ladle additions. Otherwise, the furnace is returned to the vertical position and reblown with more oxygen and possibly more lime to obtain the desired composition and temperature. The furnace is then tapped and finally turned over to dump the slag remaining in the furnace into a slag pot.

Chemistry of the Process—The reactions in the bottom-blown processes are similar to the reactions in the BOP. The major differences are that the FeO level in the slag is lower and sulphur removal is better. Oxygen gas is blown into the hot metal and the oxygen reacts with the carbon to form carbon-monoxide gas, which bubbles out of the steel. Only a small amount of the iron in the hot metal is oxidized by the oxygen gas. However, essentially all of the silicon and much of the manganese in the hot metal are oxidized by the oxygen. The silicon dioxide, manganese oxide and iron oxide formed by this oxidation combine with the powdered burnt lime (calcium oxide and dolomitic lime) that is injected into the hot metal with the oxygen. All of these oxides combine to form a liquid slag that removes much of the sulphur and phosphorus from the hot metal.

Thermal Requirements—In general, slightly less heat is produced in the bottom-blown oxygen steelmaking process than in the top-blown process because less iron and manganese are oxidized in the bottom-blown process for the same turndown carbon content, and the bottom-blown process does not oxidize as much carbon to carbon dioxide. Also, the hydrocarbon coolant extracts heat from the bath. As a consequence, the Q-BOP melts about 4 per cent less scrap compared with the BOP when using the same hot metal.

End-Point Control—End-point control of the Q-BOP is easier because the absence of an oxygen lance eliminates one process variable that can have a significant effect on the FeO content of the slag and the composition of the off gases. The reproducibility of the thermochemical reaction in the Q-BOP is much higher than in the BOP because of the closer approach to equilibrium.

Product Characteristics—The Q-BOP produces the same steel grades that are produced in the BOP. Because the Q-BOP is bottom-blown, bath carbon contents of 0.01 per cent at tap with no substantial decrease in yield are easily attained, compared to the BOP, which can only achieve bath carbon contents of less than 0.03 per cent with very high FeO contents in the slag and considerable decrease in yield. Also, the nitrogen and sulphur contents of Q-BOP steels are lower than those for the BOP, while the hydrogen content for the Q-BOP ranges from 4 to 8 ppm because of the hydrocarbon tuyere coolant. However, gas purging techniques in the furnace can be used to significantly reduce these hydrogen levels.

SECTION 5

THE COMBINATION-BLOWN STEELMAKING PROCESSES

INTRODUCTION

Many reasons for the trend towards bottom-blowing exist. Some of the advantages of bottom-blowing in comparison with top-blowing are:
1. a lower slag iron content and a better iron yield;
2. a higher phosphorus and sulphur partition between the metal and the slag;
3. a higher manganese recovery, a lower dissolved oxygen content in the steel, thus causing a higher recovery of ladle additions;
4. the possibility of producing low carbon steels without the overoxidation of the metal and slag;
5. a lower nitrogen content;
6. a quieter blow and no unmelted scrap at the end of the blow;
7. a higher process control reproducibility.

Some of the drawbacks of bottom-blowing compared to top-blowing are:
1. a lower scrap melting capability;
2. when hydrocarbon cooled tuyeres are used, there is a higher hydrogen content at turndown unless special purging practices are used;
3. a higher bottom-wear rate;
4. decreased furnace availability.

Basically, there are advantages for top-blowing or bottom-blowing, depending on the desired grade of steel to be produced and many site specific conditions. As a result, combination blowing techniques have been developed that have some of the advantages of both top and bottom blowing.

These numerous combination-blown processes differ by:
1. the mode of bottom injection (permeable elements, small tubes or pipes, and annular tuyeres);
2. the mixing intensity;
3. the nature of the bottom-blown gas;
4. the use of top charged lump lime or injected powdered lime;
5. the means employed for increasing the scrap melting rate (post combustion lance, scrap preheating, and carbon injection).

Depending on the steel grade to be produced, or on the hot metal availability, certain techniques may be preferred to others. There are three basic types of combination-blown processes in use today. All three systems use top-blown oxygen and one of three bottom systems:
1. a bottom using a multiplicity of permeable elements;
2. a bottom using a multiplicity of cooled tuyeres;
3. a bottom using a multiplicity of small uncooled tuyeres.

The first and third bottom systems can only use inert gases such as nitrogen or argon. The cooled tuyere system uses oxygen during the blow and can also switch to inert gases for stirring purposes.

TOP-BLOWN OXYGEN—
PERMEABLE ELEMENT BOTTOM

One such system that falls into this category is the lance bubbling equilibrium process, commonly called the "LBE" process, which was developed at ARBED/IRSID. In this process, oxygen is blown through an oxygen lance, while an inert gas is introduced through the bath by permeable elements located in the vessel bottom (Figure 16—19). The LBE process can:

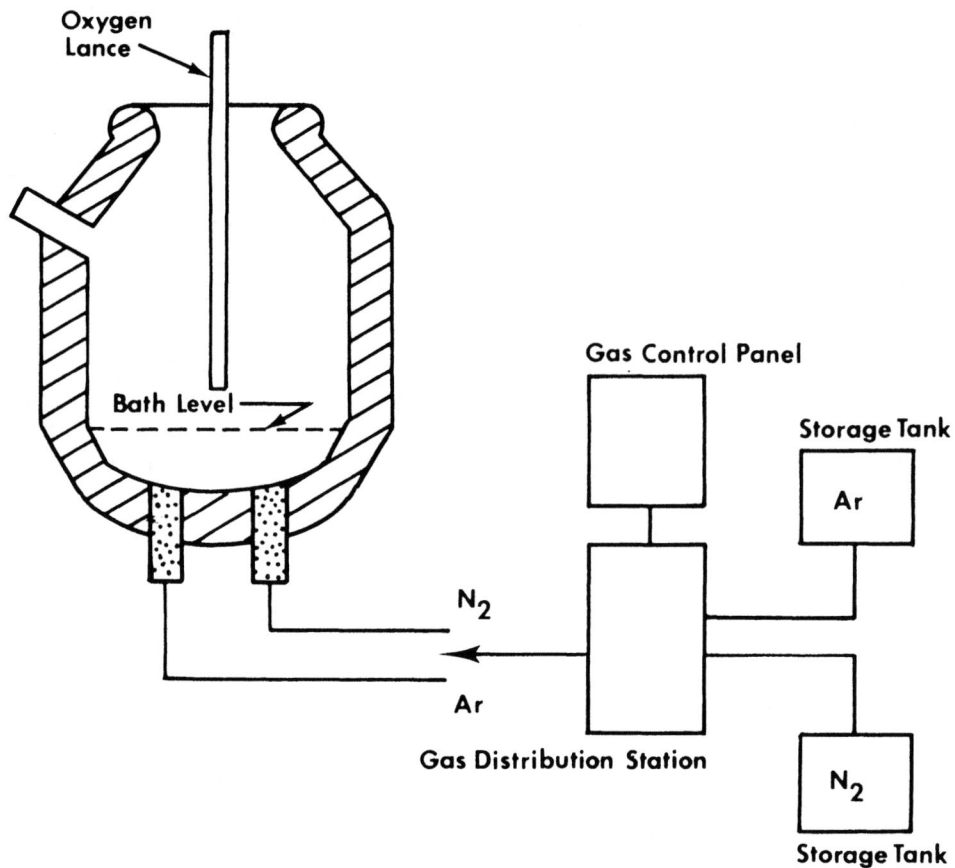

FIG. 16—19. Schematic drawing of LBE configuration showing the gas delivery system.

1. improve ingot yield (reduced slopping and lower FeO levels in the slag);
2. improve alloy recoveries;
3. shorten the heat times;
4. reduce the fluorspar consumption;
5. produce steels with lower sulphur levels;
6. easily produce low carbon steel (0.02 per cent);
7. lower the capital costs when compared to a full bottom-blown process.

Bath Agitation—Processes which employ submerged gas injection to increase bath agitation make possible increased decarburization efficiency while decreasing oxidation of the bath metallics. Improved bath agitation, when compared to top-blown processes, results in:
1. decreased iron oxidation;
2. increased manganese retention;
3. closer control of bath oxidation and lower dissolved oxygen levels;
4. reduced flux usage.

Iron Oxidation—Most processes using permeable elements are able to more economically decarburize to low carbon levels than the top-blown processes, with decreased iron and manganese oxidation at a given carbon level (Table **16**—I).

Decreased iron oxidation in the LBE results in ap-

proximately a 0.3 to 0.4 per cent increase in overall yield.

Dissolved Oxygen—Combination-blown processes utilizing permeable elements produce steel that contains less dissolved oxygen in the bath at a given carbon level than steel produced by the BOP. This low dissolved oxygen level is due to the good mixing ability of the process, which drives the reactions toward equilibrium. Less deoxidizing material must be added to the ladle and the final steel is cleaner because it contains fewer products of deoxidation compared to top blowing. A decrease in the quantity of dissolved oxygen can decrease the amounts of the manganese and aluminum that must be used as ladle additions. The decrease of dissolved oxygen aids in the recovery of both these metals.

Table 16—I. Effect of the LBE process on the slag FeO content.

Carbon %	Slag FeO content, %	
	Conventional BOP	LBE
<0.05	14.3	11.5
0.05-0.10	9.8	7.6
0.10-0.20	8.9	7.5

Table 16—II. A comparison of the BOP and LBE processes.

BOP	LBE	LBE-BOP	
1200	1400	+200	Heats on lining (average)
88.2	89.3	+1.1	Yield (%)
0.03	0.02	−0.01	Low carbon produced commercially (%)
100.1	134.7	+34.6	Burnt lime consumption (lb/ton)
78.7	23.6	−55.1	Dolomitic lime consumption (lb/ton)
70.9	72.5	+1.6	Mn efficiency (%)
40.2	42.0	+1.8	Al efficiency (%)
32.2	31.1	−1.1	Scrap consumption (without post-combustion) %
. . .	34.6	+2.4	Scrap consumption (with post-combustion) %

Phosphorus Removal—Phosphorus removal in this combined blowing process is equal to that of the BOP.

Inert Gases—The use of nitrogen during the blow, in systems with permeable elements, and as a post stirring gas, produces first turndown nitrogen contents comparable to the BOP.

Yield—Yield in the LBE process averages 0.6 to 1.0 per cent higher than in the BOP. Half of this increase in yield is due to the decreased iron and manganese oxidation during refining, and the remainder is due to the lower slag volumes and the control of the operating process. Slopping in this process is less frequent and less severe than in the BOP because of the increased stirring, which brings the reactions closer to equilibrium.

Scrap Melting—Scrap melting in combination-blown processes is less than that for the BOP. The LBE requires an average of 1.0 to 1.5 per cent more hot metal than the BOP, assuming equivalent charge materials and endpoint conditions. More heat is required because of the decrease in the slag FeO levels. Recent advances to improve the scrap melting include using a post-combustion lance to burn some of the waste carbon monoxide into carbon dioxide. With this combustion taking place inside the vessel, scrap usage can be increased 1.0 to 2.5 per cent over the BOP.

Summary of Performance—The LBE, with its top oxygen lance and permeable elements through which an inert gas is passed into the bath, compares favorably with the standard BOP, as shown in Table 16-II. The improved bath agitation results in an increased yield and decreased consumption of fluxes and alloying additions. The use of a post-combustion practice can increase scrap usage over the present BOP level.

TOP-BLOWN OXYGEN— COOLED BOTTOM TUYERES

Another system of combination blowing uses top-blown oxygen and hydrocarbon-cooled bottom tuyeres. These bottom tuyeres consist of two concentric pipes—oxygen is introduced through the center pipe, and a cooling hydrocarbon shielding gas is injected through the annulus.

Processes which use cooled bottom tuyeres are considered to have strong stirring actions. The more the stirring intensity is increased during the blow, the more the metal-slag equilibrium is approached. Various ways have been devised for controlling the slag behavior when refining low phosphorus hot metal with high stirring intensities. These techniques include powdered lime injection and pressure modulation. One such system, developed by Kawasaki Steel of Japan, is known as the K-BOP. This system consists of a top-blown oxygen lance and annular bottom tuyeres. The bottom tuyeres introduce oxygen and a hydrocarbon shielding gas into the system, and at certain points in the process, powdered lime is injected through the bottom tuyeres. In K-BOP converters, at least 20 per cent of the total oxygen must be blown through the bottom tuyeres. Although best results are obtained with a relatively large bottom gas flow rate, the flow rate must be limited because of increases in metal ejection (lance skulls and iron yield) and the increased difficulty in controlling the slag formation.

Converting a BOP to combination blowing using cooled bottom tuyeres requires installing pipes for the gas supply, tuyeres on the bottom of the vessel, and a flux injection system. Also, raw coal can be injected into the bath through either cone tuyeres mounted in the furnace bottom or through the top lance to enhance scrap melting.

TOP-BLOWN OXYGEN— UNCOOLED BOTTOM TUYERES

The last classification of combination blowing uses top-blown oxygen and uncooled bottom tuyeres. Inert gases are injected into the bath to increase the bath agitation.

One type of vessel uses a top oxygen lance and a furnace bottom, in which many small pipes are set, to introduce an inert gas such as argon into the bath. With low amounts of stirring gas, improved metallurgical results at turndown are observed only in steels with a carbon content greater than 0.06 per cent. As more argon is injected, the metallurgical relationships approach the Q-BOP. The benefits of this system are similar to the LBE process because of the large FeO difference in the slag. Argon gas currently must be limited to low amounts of stirring gas because argon costs are very high. Also, excessive cooling occurs under high argon flow, causing partial clogging of the pipes or tuyeres.

PROCESS CONTROL

Many process control strategies are available to a steelmaker. The control schemes are usually based on statistical, predictive-adaptive control from static models, and dynamic control, which uses continuous or periodic measurement of process variables such as carbon removal rate, temperature, and carbon. Differences in control strategy result from differences in (1) the final product mix, (2) decarburization reactions, (3) the plant metallurgical requirements, (4) and the time required to take corrective action for missed temperature, sulphur, or phosphorus specifications.

STATIC CHARGE CONTROL

Good control of high-purity oxygen processes requires more than the availability of an on-site process control computer programmed with a charge-control model. Any control scheme is only as good as the information with which it is supplied. To obtain good charge control in all steelmaking processes requires precise attention to a number of factors including:
1. accuracy of the inputs to the computer;
2. consistency of practices and quality of materials in use;
3. reliability of the computer system and the measuring devices;
4. convenience of the methods and devices being used to obtain the calculations;
5. close adherence to the computer recommendations and to established shop practices.

The objectives and expected benefits of good charge control include:
1. improved control of carbon, phosphorus, sulphur contents, and temperature at tap;
2. decreased time required to make a heat;
3. improved heat-size control;
4. decreased use of hot metal, coolants, and oxygen;
5. decreased use of fluxes;
6. improved yield;
7. improved steel quality.

Pinpoint control of the end-point carbon and temperature is required to prevent time-consuming reblows or cooling procedures, which add costly minutes to the heat time and reduce the potential productivity of the process. Good control of the process can decrease material costs because of decreased oxygen, hot metal, coolants, and ferro-alloy consumption.

Objective—Static control prescribes the proper combination of the charge materials—hot metal, scrap, fluxes, and oxygen—required to meet the desired end-point conditions. The static control model is thermochemical in nature, based on the chemical and thermal behavior of the metallurgical systems that comprise the process. About 90 to 95 per cent of the model is usually based on metallurgical theory, with the remainder being based on empirical relationships (tuning) that reflect the individual practices of a specific steelmaking

shop. The tuning mainly concerns the composition of the charge materials, the iron oxide content of the slag, the heat losses occuring in the vessel during the course of blowing, and the oxygen usage. Q-BOP charge models are essentially the same as BOP models, only with slight modifications that are characteristic only of the Q-BOP or combination blowing, such as natural gas injection and differences in slag chemistry. A typical static charge control model is described below.

Operation of the Model—To take maximum advantage of the charge model, the entire sequence for a heat requires three calculations—preliminary, trim, and oxygen trim. The preliminary calculation is performed well before the heat is to be made. This calculation indicates to the furnace operators the quantity of charge materials to be ordered for the heat. These materials are hot metal, scrap, fluxes, and an estimated required oxygen. This calculation is made using the aim carbon and temperature required by the heat order, and an estimate of the hot metal composition and temperature.

If special scrap items and alloy additions are to be used in a heat, the expected weights of these materials must be entered in the preliminary calculation. Special scrap items such as pit scrap, cold iron, hot crop, silicon crop, and pit iron must be entered in the preliminary calculation because of their effect on the heat balance, and therefore on the relative amounts of hot metal and scrap. Alloy additions like molybdenum and nickel oxide have a cooling effect similar to ore and therefore the heat will require more hot metal.

When the actual weights of the hot metal and scrap that will be charged are known, as well as the actual hot metal composition and temperature, a trim calculation is performed. The trim calculation exposes differences between actual and ordered hot metal and scrap weights, and between actual and assumed hot metal composition and temperature. This indicates whether the heat will be hot or cold as charged, and also indicates the necessary quantities of coolants or fuels needed to reestablish the thermal balance. The amounts of burnt lime, dolomitic lime, and spar are recalculated based on the trim input information. The trim calculation is performed as soon as the necessary information becomes available, usually near the time the furnace is charged for the heat.

The oxygen trim is performed after the heat is charged to the furnace. In addition to all the inputs required for the trim calculation, the oxygen trim requires the actual weight of fluxes that will be charged, as well as the quantity of fuel or coolant that will be added. This calculation indicates whether the heat will be hot or cold, and also indicates any adjustment needed in the quantity of oxygen required for the endpoint carbon level. This calculation also indicates if a heat should be overblown to a new, lower carbon level in order to reach the aim temperature. The oxygen required for this new aim carbon level is also indicated.

DYNAMIC CONTROL

Use of static charge control models will permit meeting the aim carbon and temperature of the steel at first turndown about 50 to 80 per cent of the time, depending on the shops' product mix.

There are several dynamic control sampling systems in use today. These systems generally are based on an initial static charge calculation, along with a dynamic charge calculation based on in-blow measurements. The off-gas analysis, sonic, and sublance systems are becoming more common and are described below in minor detail.

Sublance—A fairly common method of dynamic control involves using the sublance in conjunction with computer models. A water-cooled lance is lowered into the bath and measures the steel temperature and carbon content, and also obtains a steel sample.

Generally, the sublance is used to obtain a steel carbon and temperature measurement about 2 to 3 minutes before the end of the blow. Then an analysis is performed by the process control computer to determine the corrective actions that are necessary to achieve the desired endpoint carbon and temperature at turndown. Use of the sublance with good static and dynamic charge control models will permit first turndown hit rates well in excess of 90 per cent. However, sublance systems are expensive and can be adversely affected by large scrap charges and stiff steelmaking slags. Sublances are used extensively in Japan where low scrap charges and primarily low-carbon product mixes, with very thin slags, are common.

Off-Gas Analysis—This system takes a sample of the off-gas and produces a readout of the carbon level by continuously recomputing the carbon balance. This is achieved by monitoring the carbon monoxide and carbon dioxide contents of the exhaust gases and relating the carbon removal to the carbon level in the charge.

Off-gas analysis is a relatively inexpensive sampling process that is based on known carbon content in the bath and the carbon removed in the waste gas to calculate the instantaneous carbon content in the bath. The sample must be dust-free, and the system is rapid. However, the sample produces no direct temperature data and the hood parameters often affect the results. A mass spectrometer is often used for the gas analysis.

Sonic Analysis—Sound intensity can be used as a measure of the carbon level. This system detects a sound difference as the carbon monoxide evolution decreases at the end of the blow. Sonic indication can also be used to detect the onset of slopping. The main part of this system is a condenser microphone, which is contained in a water cooled probe and placed in a dust-free location in the hood.

Installing a sonic control system is relatively easy and costs less than an off-gas analyzer. However, like the off-gas analyzer, there is no direct temperature data. A sonic analyzer must be "tuned" to each furnace, and its performance is often affected by extraneous sounds.

Bibliography

American Institute of Mining, Mechanical and Petroleum Engineers, Sixty-fifth Steelmaking conference proceedings, Volume 65, Pittsburgh meeting, 28-31 March, 1982. The Steelmaking Division, ISS/AIME. Copyright 1982 by AIME.

American Iron and Steel Institute, The basic oxygen process. Contributions to the metallurgy of steel. Washington, D.C., The Institute, 1960.

Anon., Q-BOP, from blow to go in 90 days. Journal of Metals 24, pages 31-37, March 1972.

Anon., Basic oxygen steelmaking—a new technology emerges? Metals Society Book 197, 1979.

Behrens, K. F., J. Koenitzer and T. Kootz, The effects of lime properties on basic oxygen steelmaking. Journal of Metals 17, No. 7, pages 776-781, July 1965. Discussion by J. McNamara, 781-784.

Blum, Bernard, Control of basic oxygen steelmaking. 33/The Magazine of Metals Producing 3, No. 4, pages 75-80, April 1965.

Brotzmann, K., The bottom-blown oxygen converter—a new method of steelmaking. Technik Forschung 21, pages 718-720, 1968. BISI Translation No. 7255.

Cesselin, Ph., P. Vayssiere, J. C. Helion, B. Thome and J. Francais, Automation of basic oxygen steelmaking: application to the refining of phosphorous hot metal by the O.L.P. process (in French). Proc. International Conference on Iron and Steelmaking Vol. 1, published jointly by Centre National de Recherches Metallurgiques (CNRM), Liege, Belgium, and Verein Deutscher Eisenhuttenleute (VDEh) Dusseldorf, West Germany, pages 279-285, Summary (in English) in Journal of Metals 17, No. 7, pages 722-724, July 1965.

Hill, J., and J. M. Edge, The Q-BOP at Fairless Works, I&SM (Iron and Steelmaker) 3, pages 37-40, June 1976.

Hubbard, H. N., Jr., and W. T. Lankford, Jr., Development and operation of the Q-BOP process in the United States Steel Corporation. Iron and Steel Engineer 50, pages 37-43, October 1973.

Knueppel, H., K. Brotzmann and H. G. Fassbinder, The oxygen bottom-blowing converter, a new process of steelmaking. Stahl and Eisen 93, pages 1018-1024, 1973.

Leroy, P., The LWS process. Rev. Met. 67, pages 181-193, 1970: BISI Translation No. 9923.

Meyer, H. W., M. M. Fisher and W. F. Porter, Recent progress in basic oxygen furnace dynamic control. Proc. International Conference on Iron and Steelmaking, Vol. 1, published jointly by Centre National de Recherches Metallurgiques (CNRM), Liege, Belgium, and Verein Deutscher Eisenhuttenleute (VDEh), Dusseldorf, West Germany, pages 238-243, 1965. Summary (in English) in Journal of Metals 17, No. 7, pages 716-718, July 1965.

Meyer, H. W., W. R. Porter and J. Szekely, Slag-metal emulsions and their importance in BOF steelmaking. Journal of Metals 20, pages 35-42, July 1968.

Miltenberger, R. S., Measurements for metallurgical control of the basic oxygen process. Journal of Metals 17, No. 7, pages 761-763, July 1965, discussion by C. L. Meloy, page 763.

Nelson, F. D., W. P. Holloway, C. F. Thebo and M. E. Nickel, Progress of new basic oxygen shops—a panel discussion: I, at Inland Steel Co.; II, at Wheeling Steel Corp.; III at Bethlehem Steel Corp.; IV, at Wisconsin Steel Works, International Harvester Co. Journal of Metals 17, No. 7, pages 785-800, July 1965.

Nilles, P., and M. Boudin, Oxygen steelmaking in bottom-blown converters. Ironmaking and Steelmaking (Quarterly), No. 1, pages 22-27, 1974.

Nilles, P. E., Control of the OBM/Q-BOP process. Iron and Steel Engineer 53, pages 42-47, March 1976.

Oakey, J. D., P. B. Hunter, C. H. Marty and J. A. Kramer, Charge control at the Lorain BOP shop. Journal of Metals 26, pages 12-17, June 1974.

Papinchak, M. J., and T. M. Weaver, Current status of the Q-BOP oxygen steelmaking process. I&SM (Iron and Steelmaker) 5, pages 12-17, October 1978.

Pearson, J., The theoretical basis of oxygen steelmaking. Iron and Coal Trades Review 181, pages 1407-1413, Dec. 30, 1960.

Pehlke, R. D., et al, BOF steelmaking, AIME, 1983.

Peters, A. T., Ferrous Production Metallurgy, John Wiley and Sons, 1982.

Strassburger, J. H., Tonnage oxygen for increased iron and steel production. American Iron and Steel Institute Yearbook, pages 214-249: discussion, pages 250-259; 1948.

Trentini, B., Comments on oxygen steelmaking. Trans. Met. Soc., AIME, 242, pages 2377-2388, 1968.

Yamamoto, N., H. Hasimoto, S. Matsunaga and T. Torigoe, Computer control of basic oxygen converters. Proc. International Conference on Iron and Steelmaking, Vol. I, published jointly by Centre National de Recherches Metallurgiques (CNRM), Liege, Belgium, and Verein Deutscher Eisenhuttenleute (VDEh), Dusseldorf, West Germany, 244-250 (1965). Summary in Journal of Metals 17, No. 7, 720-721 (July 1965).

CHAPTER 17

Open-Hearth Steelmaking Process

Open-hearth furnaces, as the name implies, are open for visual inspection of the hearth after every heat of steel is tapped from the furnace. Open-hearth steelmaking furnaces have several unique features. Among these are:

(1) The ability to use any available gaseous or liquid fuel, and even pulverized coal if necessary. Changing from one type of fuel to any other without interrupting steel production is possible. Availability, convenience, and price govern selection of the type of fuel.

(2) Metallic materials for the furnace charge can range from 5 per cent cold pig iron plus 95 per cent steel scrap to as high as 70 per cent molten blast-furnace iron (hot metal).

(3) All compositions of carbon and alloys steels, except the high-chromium nickel stainless and heat-resisting steels, have been melted in open-hearth furnaces.

(4) Any lost refractory material from the hearth and banks can be replaced promptly.

(5) In contrast to production with oxygen steelmaking furnaces, open-hearth steel production diminishes only moderately when a shortage of hot metal requires a sudden change from a high hot-metal charge to a high scrap charge.

Yet, all of these favorable features cannot prevent eventual replacement of open-hearth steelmaking by a balanced combination of basic oxygen and electric-arc steelmaking furnaces.

LAYOUT OF MAIN FURNACE BUILDING

Open-hearth furnaces are arranged lengthwise parallel to the centerline of the building with space left at the ends of the line for storage or other purposes. (Figure 17—1). This arrangement facilitates charging of the furnace and handling of the finished molten steel. The part of the floor area in front of the row of furnaces is called the charging floor which, in modern shops, is elevated about 6 metres (20 feet) above the general yard level (Figure 17—2). Running the full length of the charging floor in modern open-hearth shops are three sets of tracks. Just in front of the line of furnaces is a standard-gage track onto which diesel locomotives move buggies carrying boxes loaded with scrap and other solid materials for charging into the furnace. Parallel with this standard-gage track is a very broad-gage track (about 9 metres or 30 feet) on which charging machines operate. Farthest from the furnace is the third set of tracks which is used exclusively for moving ladles of molten blast-furnace iron (hot metal) into position for pouring the metal into the furnace to complete the final charge.

Adjacent to the charging floor is a lean-to equipped with standard-gage tracks where loaded charging buggies are stored until needed at the furnaces. Beneath the floor of this lean-to are waste-heat boilers and fans that regulate the flow of combustion air to the furnaces.

On the crane runway above the charging floor are electric overhead traveling cranes whose primary function is pouring molten blast-furnace iron into the open-hearth about one hour after completion of the cold charge. The lifting capacity of the main hoist is usually about one-half of the total weight of metallic materials in the complete furnace charge.

CHARGING SIDE

On the charging floor diesel locomotives move buggies carrying boxes loaded with scrap and other solid materials to be charged into the furnace. On a very broad-gage track, the charging machines operate electrically to:

1. Pick up the loaded boxes, one at a time
2. Move them into the furnace through an open door
3. Rotate the box to dump its contents
4. Withdraw the empty box
5. Place the empty box on the buggy and
6. Disengage and move the next box into position in

FIG. 17—1. General view of the charging side of a group of basic open-hearth furnaces.

front of the charging door.

These six sequential movements for each charging box are seldom accomplished in less than one minute. Because cold charge materials compose one-third to one-half the total weight required for a single heat of open-hearth steel, completing the charging of all cold materials usually requires one to three hours. This lengthy charging procedure deters efforts to increase open-hearth furnace productivity.

Molten blast-furnace iron arrives at one end of the open-hearth shop in 181-272 metric ton (200-300 net ton) Pugh-type ladles, (Chapter 15) Thereafter the hot metal is:

(1) retained in the Pugh ladles until an individual open-hearth is ready for its hot metal charge or

(2) promptly transferred into a hot-metal mixer.

Hot-metal mixers which are horizontal, cylindrical vessels lined with super-duty fireclay and high-alumina refractory brick usually hold 907-1814 metric tons (1000-2000 net tons) of hot metal. Addition of hot metal to an open-hearth furnace occurs about one hour after completion of the cold charge. Now, the furnace proceeds with melting and refining to attain the specified tapping temperature and composition of steel.

POURING FLOOR OR PIT SIDE

Usually as long as the charging floor but at general yard level, the pouring floor, or pit side, of the furnace building extends along the tapping side of the furnaces. On this tapping side, jib-cranes of 4.5 to 6.4 metric tons (5 to 7 net tons) capacity handle such items as furnace tapping spouts. On the pouring side are facilities for relining, drying, and preheating ladles. Situated along the inside wall of the furnace building and opposite the

tapping side of the furnaces, the pouring platforms are at a height convenient to the tops of the ingot molds standing on mold cars.

MOLD YARD

Located as close as possible to the pouring platforms, the mold yard houses facilities for cleaning and replacing molds on mold cars. Before a furnace is tapped, a diesel locomotive moves the clean molds to their designated platform for pouring. After pouring is complete, a locomotive moves the ingot cars to a special purpose crane designed solely for stripping the partially or completely solidified ingots from their molds.

OPEN-HEARTH FURNACE CONSTRUCTION

Similar with respect to form and arrangements, basic and acid furnaces differ only in the materials used in constructing their hearths. Acid furnaces use high-purity silica sand, whereas basic furnaces use high-purity magnesite. Yet, from the standpoint of annual steel production, acid open hearths are no longer important.

The open-hearth furnace is both reverberatory and regenerative. It is reverberatory in that the charge is melted on a shallow refractory hearth, 305 to 457 mm (12 to 18 inches) depth of molten metal, by a flame passing over the charge so that both the charge and roof are heated by the flame. Open-hearth furnaces are regenerative in that the hot combustion products leave the furnace chamber through passages leading to checker chambers containing fire brick. These brick are arranged to give a large contact area with the hot gases, which transfer part of their heat to the brick. Figure 17—3 schematically shows the flow and reversal

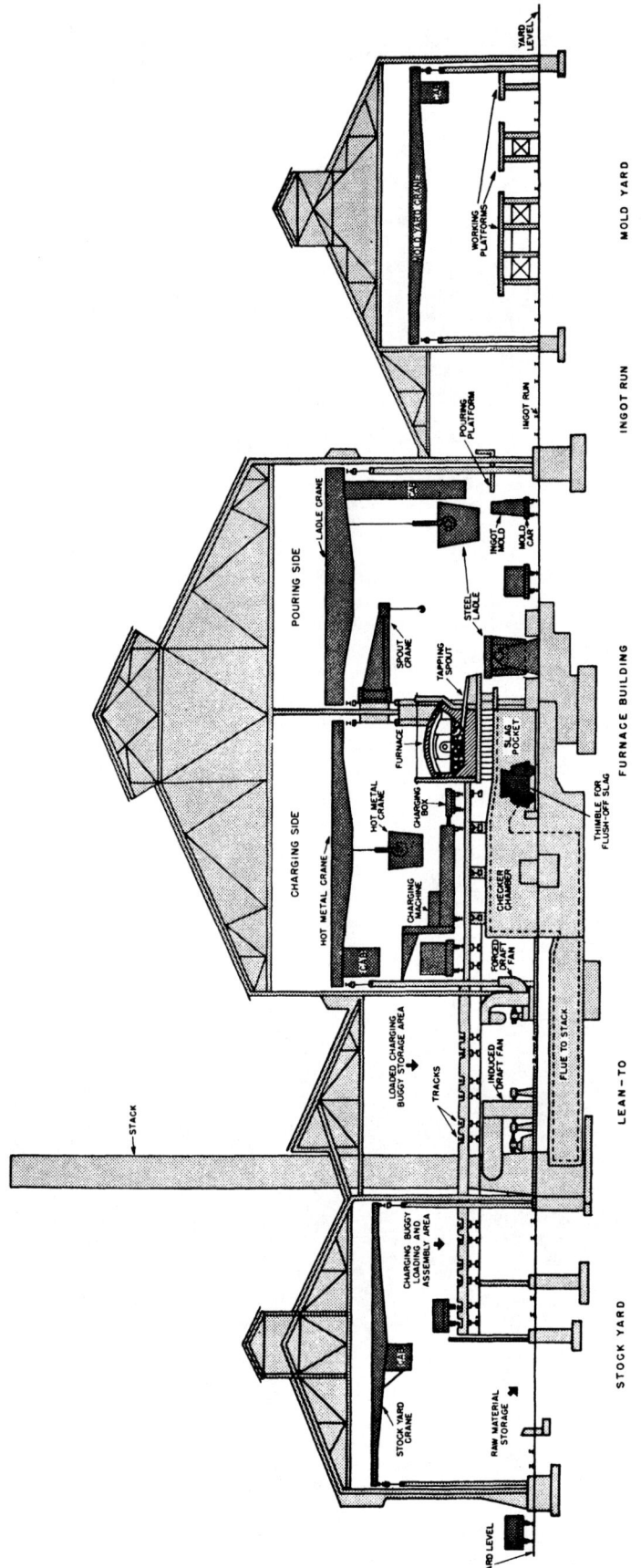

Fig. 17—2. Diagrammatic cross-section of an open-hearth plant, showing the relative locations and sizes of the various buildings.

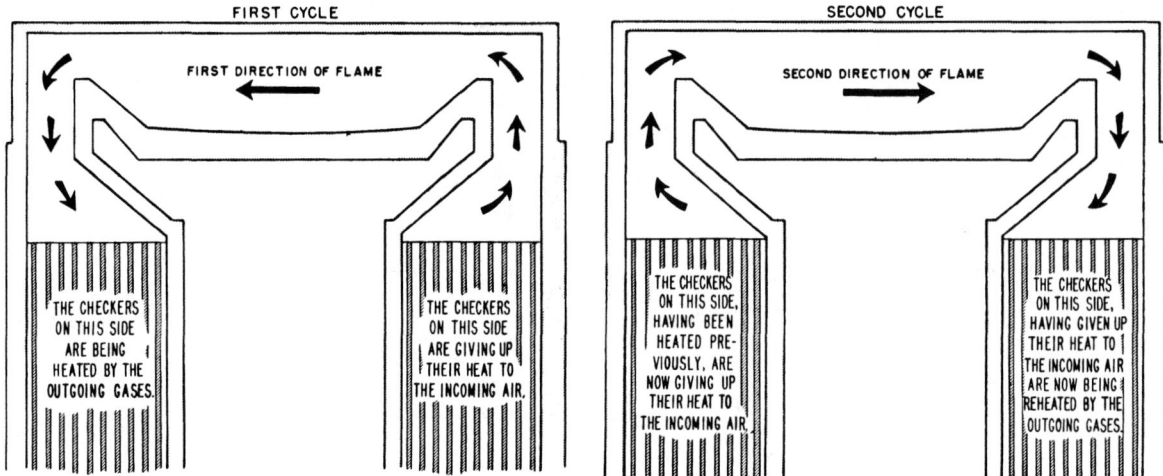

FIRST CYCLE

FIRST DIRECTION OF FLAME

THE CHECKERS ON THIS SIDE ARE BEING HEATED BY THE OUTGOING GASES.

THE CHECKERS ON THIS SIDE ARE GIVING UP THEIR HEAT TO THE INCOMING AIR.

SECOND CYCLE

SECOND DIRECTION OF FLAME

THE CHECKERS ON THIS SIDE, HAVING BEEN HEATED PRE-VIOUSLY, ARE NOW GIVING UP THEIR HEAT TO THE INCOMING AIR.

THE CHECKERS ON THIS SIDE, HAVING GIVEN UP THEIR HEAT TO THE INCOMING AIR ARE NOW BEING REHEATED BY THE OUTGOING GASES.

FIG. 17—3. These diagrams show the function of the checkers in preheating air for combustion in regenerative furnaces. Fuel is admitted to the furnace through end burners at the same end as the preheated air for each cycle.

of outgoing gases and incoming combustion air. Auto-matically, the direction of flow is reversed on a time cycle or by a maximum temperature limit for the checker brick in the outgoing regenerator.

By 1960, all refractories above charging-floor level were basic materials, except for the sprung-arch silica brick roof. The only shortcoming of silica brick for open-hearth roofs is its low melting temperature at 1520°C (3115°F). This temperature is only 93°C or 200°F above that of the molten metal and only 244°C or

500°F below the flame temperature. Under these con-ditions, service life of a silican brick roof is roughly 1000 hours. After 15 years of intensive experimen-tations, the industry changed rapidly to an all-basic roof. Major changes in roof design that led to success included use of 63.5 to 76.2 mm (2.5 to 3 inches) rise per foot of roof span, as compared with 38.1-mm (1.5-inches) rise for silica-brick roofs, plus control on the contour of the arch by longitudinal hold-down beams to prevent upward buckling of any portion of the roof,

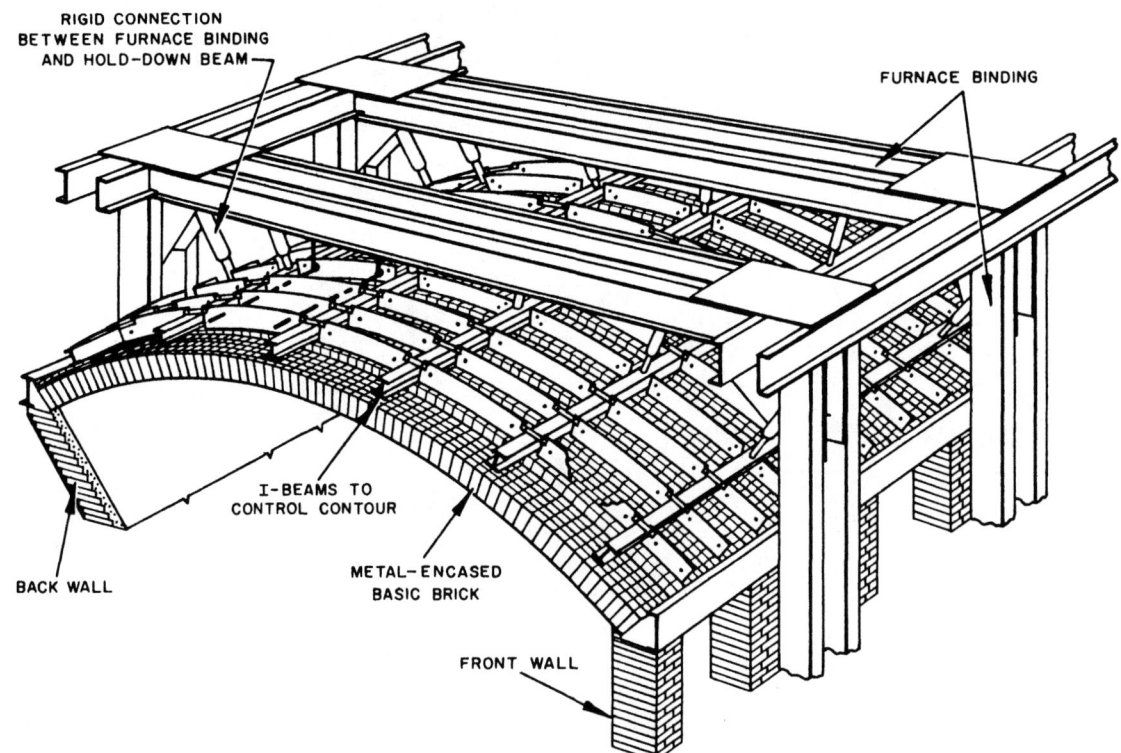

RIGID CONNECTION BETWEEN FURNACE BINDING AND HOLD-DOWN BEAM

FURNACE BINDING

I-BEAMS TO CONTROL CONTOUR

METAL-ENCASED BASIC BRICK

BACK WALL

FRONT WALL

FIG. 17—4. Simplified diagram of the method of installing a basic roof on an open-hearth furnace.

Now, service life of basic roofs is 7000 to 10 000 hours.

Success with basic roofs quickly led to development of roof-mounted, water-cooled lances for direct injection of oxygen into the molten metal. Figure 17—5 shows the design of an oxygen roof lance. In these

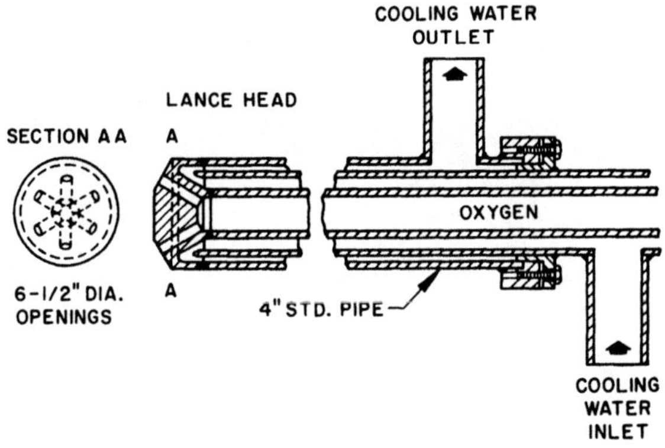

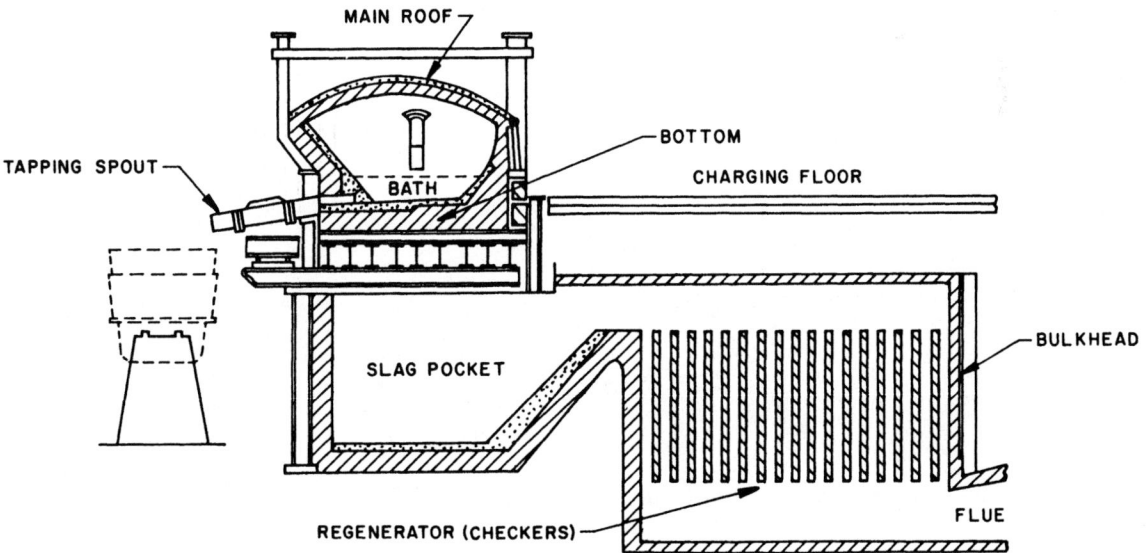

FIG. 17—5. Diagrammatic construction of oxygen roof lance.

lances, flow of oxygen to the furnace begins immediately after the addition of hot metal and is continued throughout the refining period. Using roughly 1400 cubic feet of oxygen per ton of steel doubled production and reduced by half the consumption of fuel per ton of steel. Chemical reaction of gaseous oxygen with the molten metal is highly exothermic, whereas the chemical reaction with iron ore is highly endothermic. Prior to adoption of oxygen roof lances, iron from ore

averaged about 7 per cent of the total metallics in the furnace charge.

MAINTENANCE

Unfortunately, because of high dust load and high temperature of the off-gas when leaving the hearth area, certain furnace parts require cleaning or repair before complete replacement of a basic roof is necessary. Figure 17—6 shows the furnace areas that are repaired or cleaned. (Minor shut-downs for wall repairs and cleaning of slag pockets can be accomplished in one to three days). After about one hundred to two hundred heats, front-walls and end-walls must be replaced. About every two hundred heats, accumulations in slag pockets and checkers must be removed. However, when replacing the main roof becomes necessary, the furnace is shut down for four to seven days.

HEARTH REPAIR

If hearth and banks are carefully dressed and patched after every heat, a life of over 10 years is attainable. After the tapping spout is detached from the furnace, any remaining slag drains into the cinder pit. Slag and steel held back by lime ridges on the hearth are removed by blowing with either compressed air or steam. After the ridges are removed, the hearth and banks are then dressed with dolomite or patched with magnesite if deep erosion has occurred at any spot. After patching, fuel is fired at a high rate to sinter in place the applied dolomite or magnesite. During this brief period, the tap hole is cleaned, filled with dolomite, and sealed with a plug of loam or clay. The furnace is now ready for charging the next heat.

FIG. 17—6. Vertical section across the width of an open-hearth furnace, not to scale, indicating names and relative locations of principal parts. Upper section is through taphole of furnace: bottom section is through slag pocket and regenerator.

ENVIRONMENTAL

Open-hearth shops built between 1950 and 1957 were usually equipped with venturi scrubbers or electrostatic precipitators for final cleaning of off-gas just before it enters the stack for discharge to the atmosphere. Open-hearth shops have not been built in the United States since 1957. Subsequently, air-pollution regulations required installation of cleaning devices on all open-hearth furnaces. These regulations caused premature conversion of open-hearth shops, in less than their expected life of 60 to 70 years, to electric-arc or basic oxygen steelmaking.

CHEMISTRY OF BASIC OPEN-HEARTH STEELMAKING

In simple, elementary terms, steelmaking is an oxidation process in which undesirable elements are removed from iron by reaction with oxygen. Yet there is a vast array of complicating circumstances thoroughly discussed in Chapter 13 on "The Physical Chemistry of Iron- and Steelmaking."

CHAPTER 18

Electric-Furnace Steelmaking

HISTORY AND PRESENT STATUS OF ELECTRIC-FURNACE STEELMAKING

Section 1 presents a general overview of the growth of electric-furnace steelmaking from the late 1800's to the present. The **first subsection** discusses the various methods of electric heating and the types of electric-furnaces used in steelmaking operations. The **second subsection** contains a history of the direct-arc electric-furnace, the electrode, and the field of application of electric-arc furnaces. The **third subsection** discusses the advantages and disadvantages of electric-furnace steelmaking. Then, in the **fourth subsection,** recent trends in the computerization of electric-furnace facilities and the development of D.C. electric-arc furnaces are discussed.

METHODS OF ELECTRIC HEATING AND TYPES OF ELECTRIC-FURNACES

Methods of Electric Heating—Electric current can be used for heating steel in three ways: (1) by passing electric current through an ionized gaseous medium and using the heat radiated by the generated arc; (2) by passing current through solid conductors and using the heat generated as a result of the conductors' inherent resistance to the flow of current; and (3) by bombarding the steel surface with a high intensity electron beam and using the heat generated by the conversion of energy at the relatively small area of electron impingement. Since the third practice has not been developed sufficiently for high tonnage capacities, however, it will not be discussed further.

The first practice, **arc heating,** can be applied through two methods: (a) arcs pass between electrodes supported in the furnace above the metal. In this method, known as **indirect-arc heating,** the metal is heated solely by radiation from the arcs. Or, (b) arcs pass from the electrodes to the metal. In this method, known as **direct-arc heating,** the current flows through the metal charge so that the heat developed by the electrical resistance of the metal, though relatively small in amount, is added to that radiated from the arcs. In the direct-arc heating method, two types of furnaces are used: furnaces with **non-conducting bottoms;** and furnaces with **conducting bottoms.**

In a direct-arc furnace with a non-conducting bottom, the current (alternating-current) passes from one electrode down through an arc to the metal charge, through the metal charge and up through an arc to an adjacent electrode. In a furnace with a conducting bottom, the current (direct-current) passes from an electrode down through an arc into the metal charge and then out of the furnace through an electrode forming part of the bottom in contact with the bath.

The second practice, **resistance heating,** can be applied through three methods: (a) the **indirect method** in which the steel is heated by radiation and convection from resistors through which the current is passed; (b) the **direct method** in which current is passed directly from a power source through the metal; and (c) the **induction method** in which current is induced in the steel by an induction coil connected to a power supply. Neither the indirect nor the direct method of resistance heating is practical for steel-melting operations. However, the induction method is employed successfully in special steel-melting applications. The steel charge in the induction method acts as the secondary circuit for current which is generated from a primary induction coil.

Types of Electric-Furnaces—Numerous types of furnaces using electric current as the source of heat have been developed, but relatively few have survived as practical steelmaking tools.

From among the types of furnaces listed below, only two have proven to be practical for melting steel: (1) the three-phase A.C. (alternating-current) direct-arc electric-furnace, described more fully in Sections 2, 3 and 4 of this chapter; and (2) the induction furnace, described in Section 5. Recently, D.C. (direct-current) direct-arc electric-furnaces have also been developed in smaller sizes (10 to 50 tons) for commercial use. The D.C. direct-arc furnace is described at the end of this section.

The following are general descriptions of electric-furnaces listed according to the principles of heating they employ:

Arc Furnaces

1. **Indirect-Arc Furnaces**—The metal charge is heated by an alternating-current arc passing from one electrode to another above the metal. Fur-

naces may be stationary, oscillating, or rolling.

A. Single-Phase Furnaces
  (1) Rolling furnaces with horizontal electrodes.
  (2) Furnaces for special purposes.

B. Two-Phase Furnaces
  (1) Straight arc. Not used for making steel.
  (2) Deflected arc. Not used for making steel.

C. Three-Phase Furnaces
  (1) Straight arc. Not used for making steel.
  (2) Repel-arc. Not used for making steel.

2. **Direct-Arc Furnaces**—The current passes from the electrode or electrodes through arcs to the metal charge.

A. Alternating-Current Direct-Arc Furnaces
  Current passes from one electrode down through an arc and the metal charge, then from the charge up through an arc to another electrode. Although single-, two-, and three-phase current can be used for steelmaking, furnaces employing three-phase current are used almost exclusively.

B. Direct-Current Direct-Arc Furnaces
  Current passes from one electrode through an arc and the metal charge to an electrode in the bottom of the furnace.

Resistance Furnaces

1. **Indirect Heating**—Current is passed through resistors, thus heating the furnace charge by radiation and convection. Such furnaces are used for heat treating and maintaining the temperature of molten metal, but not for melting steel.

2. **Direct Heating**—Current from low-voltage transformers is passed through the steel. This method is used to heat steel for hot working, but is not practical for melting steel.

3. **Induction Heating**—Current is induced in the steel by an oscillating magnetic field.

A. Low-Frequency Induction Furnaces—The principle of a transformer is used, with the metal charge forming the secondary circuit and a coil with an iron core forming the primary circuit.

B. Medium- and High-Frequency Induction Furnaces—Current of medium- or high-frequency is passed through a coil surrounding a crucible containing the charge.

HISTORY OF ELECTRIC-FURNACE STEELMAKING

History of the Direct-Arc Electric-Furnace—Dr. Paul Heroult developed and patented the first A.C. direct-arc electric-furnace in the late 1800's. Aside from various innovations, developments and refinements in the design of the furnace components, the basic design principle remains the same as it was originally developed and patented. The primary concept of the furnace developed by Dr. Heroult involved the use of two or more electrodes, with the electric current passing from one electrode through an arc to the charge, then flowing through the charge and passing through an arc to the other electrode or electrodes. Accordingly, this type of furnace is often referred to as the "Heroult" type. In the early 1900's, United States Steel Corporation acquired the rights to the Heroult patents and subsequently entered into the electric-furnace construction field. Thus, the name "Heroult" is still used as the trademark for the electric-furnaces designed and built by the American Bridge Division of United States Steel Corporation (Figure 18-1).

The first electric-furnace used to produce steel in this country was a Heroult direct-arc furnace that began operations in 1906 at Halcomb Steel Company in Syracuse, New York. This furnace, which had a 4-ton capacity, was rectangular and had two electrodes operating on a single-phase alternating-current.

In 1908, United States Steel Corporation acquired an exclusive American license for the Heroult furnace. On May 10, 1909, a 15-ton three-phase Heroult furnace was commissioned at United States Steel Corporation's South Works in Chicago, Illinois. It was the largest electric steelmaking furnace in the world at that time, and the first with a round shell.

When the first electric-furnaces were installed, large amounts of electric power were not available. Since the melting operation required about three times the power required by the refining period, it was thought then that the cost of this extra electric power was too high to permit economical melting of cold charges. Thus, the early installations at Halcomb and United States Steel were provided with a separate source of hot-metal to provide a molten charge. The former had a tilting, 27-metric ton (30-net ton) basic open-hearth furnace; and the latter had a Bessemer converter department. Later, as more electric power became available, cold melting became feasible. At the present time there is no appreciable tonnage of steel produced in electric-furnaces using hot-metal charges.

At first, replacing cold-charged open-hearth furnaces in non-integrated plants with basic-lined electric-arc furnaces for ingot production accounted for an appreciable increase in electric-furnace steelmaking capacity. Then, as the BOP and Q-BOP processes developed, the availability of scrap, especially of the heavy type, increased because of the reduced consumption of scrap in the BOP and Q-BOP operations. This situation further favored the use of electric-arc furnaces. Other reasons for preferring electric-arc furnaces included the increasing demand for alloy steels for new applications that previously used carbon steels, and the ability of the electric-furnace to operate on an intermittent basis. Because of its flexibility with respect to charge materials and its suitability for intermittent operation, the electric-furnace has also found a place in large integrated mills. It is often used to supplement the output of other steelmaking processes in short-term demand peaks, where the startup of large hot-metal units to meet the temporary need for additional steel would be uneconomical.

History of the Electrode—In his experimental work around 1800, the English scientist Sir Humphrey Davy created an electric-arc, using the current from a storage battery, and electrodes made from wood, charcoal, and syrup of tar, molded under about 0.7 megapascals (100 psi) of pressure.

In 1907, Dr. Heroult imported electrodes from Sweden for use in this country because the largest electrode made in the United States at that time was only 305 mm (12 inches) in diameter. Since importation was both slow and expensive, he then built a plant to produce carbon electrodes up to 610 mm (24 inches) in diameter. This plant supplied the electrodes for the large Heroult electric-arc steelmaking furnace installed at United States Steel's South Works in Chicago.

Field of Application of Electric-Arc Furnaces—At first, the electric-arc furnace was used chiefly in the production of tool steels, displacing the old crucible process. The electric-arc furnace then became the accepted means for producing high-alloy, stainless, bearing and other high-quality steels. Its field was later enlarged to include the production of low-alloy steels. Currently plain carbon steels are commonly produced in electric-furnaces.

In recent years, mini-mills or market-area mills have sprung up to service local areas with carbon steel merchant products. Mini-mills can be highly competitive because of low investment costs, low man-hours per ton of product, low scrap prices and favorable electric power rates.

The importance of electric-furnaces as steelmaking units in ingot-producing and continuous-casting plants in the United States is indicated by the annual tonnages shown in Table 18—I, which also shows the increasing trend in production of electric-furnace steels during the period of 1936 through 1982. It should be noted that these figures concern raw steel production and do not include steel made in electric-furnaces by foundries.

Electric-furnaces currently produce about thirty per cent of the total tonnage of raw steel produced in this country. These furnaces produce practically all of the stainless, constructional alloy, tool, and special alloy steels used in the chemical, automotive, aviation, machine-tool, transportation, and food processing industries. Also, electric-arc furnaces used in mini-mills play a large part in the production of carbon steel merchant products.

Almost all electric-arc furnaces used for ingot and continuous-caster steel production are basic-lined furnaces. The steel-casting industry originally found that acid-lined electric-arc furnaces were better suited to meet their needs with regard to operations control and product quality, provided a sufficient supply of selected

FIG. 18—1. A modern, three-phase Heroult electric-arc furnace. The transformer vault is in the left background. The elbow-like offtake at the right side of the roof collects fumes during operation and passes them to gas-cleaning equipment.

Table 18—I. Annual Production of Electric-Furnace Steels[1]
(Raw Steel and Steel for Castings—Thousands of Tons[2])

Year	Electric-Furnace Steel		Total Steel Production[3]		Percentage Represented by Electric-Furnace Steel
	Metric Tons	Net Tons	Metric Tons	Net Tons	
1982	21 009	23 158	67 656	74 577	31.1
1981	30 976	34 145	109 615	120 828	28.3
1980	28 274	31 166	101 457	111 835	27.9
1979	30 778	33 927	123 689	136 341	24.9
1978	29 245	32 237	124 359	137 081	23.5
1977	25 294	27 882	113 700	125 333	22.2
1976	22 328	24 612	116 121	128 000	19.2
1975	20 575	22 680	105 818	116 642	19.4
1974	26 008	28 669	132 195	145 720	19.7
1973	25 183	27 759	136 803	150 799	18.4
1972	21 519	23 721	120 874	133 241	17.8
1971	18 997	20 941	109 264	120 443	17.4
1970	18 291	20 162	119 307	131 514	15.3
1969	18 263	20 132	128 151	141 262	14.3
1968	15 253	16 814	119 260	131 462	12.8
1967	13 689	15 089	115 406	127 213	11.9
1966	13 490	14 870	121 654	134 101	11.1
1965	12 523	13 804	119 262	131 462	10.5
1964	11 501	12 678	115 281	127 076	10.0
1963	9 906	10 920	99 120	109 261	10.0
1962	8 176	9 013	89 202	98 328	9.2
1961	7 860	8 664	88 917	98 014	8.8
1960	7 601	8 379	90 067	99 282	8.4
1959	7 741	8 533	84 773	93 446	9.1
1958	6 038	6 656	77 342	85 255	7.8
1957	7 231	7 971	102 253	112 715	7.1
1956	7 839	8 641	104 522	115 216	7.5
1955	7 303	8 050	106 173	117 036	6.9
1954	4 931	5 436	80 115	88 312	6.2
1953	6 604	7 280	101 251	111 610	6.5
1952	6 167	6 798	84 522	93 168	7.3
1951	6 479	7 142	95 436	105 200	6.8
1950	5 478	6 039	87 848	96 836	6.2
1949	3 432	3 783	70 740	77 978	4.9
1948	4 588	5 057	80 413	88 640	5.7
1947	3 436	3 788	77 015	84 894	4.5
1946	2 325	2 563	60 421	66 603	3.8
1945	3 136	3 457	72 304	79 702	4.3
1944	3 845	4 238	81 322	89 642	4.7
1943	4 163	4 589	80 592	88 837	5.2
1942	3 606	3 975	78 047	86 032	4.6
1941	2 603	2 869	75 150	82 839	3.5
1940	1 542	1 700	60 766	66 983	2.5
1939	933	1 029	47 898	52 799	1.9
1938	513	566	28 805	31 752	1.8
1937	859	947	51 380	56 637	1.7
1936	785	865	48 534	53 500	1.6

[1] Figures in net tons from Annual Statistical Reports of American Iron and Steel Institute. Figures in metric tons calculated.

[2] The figures include only that portion of production of steel for castings used by foundries which were operated by companies also producing raw steel for subsequent processing into wrought-steel products.

[3] Total steel production by all processes, including electric-furnace.

scrap was available. More recently, however, more steel-casting operations have utilized basic-lined furnaces and less selected scrap.

ADVANTAGES AND DISADVANTAGES OF ELECTRIC-FURNACE STEELMAKING

Advantages of Electric-Furnaces—The increasing number of basic-lined electric-arc furnace installations shows that these furnaces provide certain advantages as steelmaking units. Among the more important considerations for selecting this method over other steel production methods is that practically all grades of steel can be produced in the basic-lined electric-arc furnace. These grades include: the plain carbon steels of rimmed, capped, semi-killed and killed types; low-alloy UNS Gxxxxx series (former AISI and SAE 13xx, 40xx, 41xx, 43xx, 45xx, 46xx, 47xx, 48xx, 50xx, 51xx, 52xx, 61xx, 81xx, 86xx, 87xx, 88xx, 92xx, 93xx and 94xx series); constructional steels; high-manganese steels (up to 14 per cent); high-silicon steels (up to 5 per cent); aluminum steels (up to 4.5 per cent); the entire range of stainless steels; super-alloy steels for high-temperature applications; and high-speed and other alloy tool steels. When producing some grades of stainless steel, however, it is more economical to supplement the electric-furnace with an AOD (argon-oxygen decarburization) vessel.

The basic-lined electric-arc furnace may be selected as the more economical steel producing method when:
1. Carbon and low-alloy steel production requirements are insufficient to justify using the combination of a blast furnace and a basic oxygen furnace to produce steel.
2. Facilities are installed in industrial areas of high steel scrap availability and low-cost electric power, but at a distance from natural sources of coke, limestone and high-grade iron ores.
3. The nature of subsequent processing is such that steel-production requirements are intermittent, or molten metal must be supplied within controlled time limits.
4. Neither hot metal nor primary mills are available, but low-cost, low-tonnage production installations such as mini-mills can be built. Mini-mills use electric-arc furnaces for producing merchant steel products because of the low investment and operating costs incurred, and the ability to turn these furnaces on or off depending on the order book. In most cases, mini-mills are competitive with foreign imports due to the lower price of scrap and electricity, and low man-hours per ton of product.

Disadvantages of Electric-Furnaces—The disadvantages of electric-furnace steelmaking include:
1. The inability to produce low residual steels from high residual scrap.
2. The inability of one electric-furnace to keep pace with a continuous-caster for sequence casting in a long series of heats.
3. Electric-furnaces are not normally suited for production of over 1 500 000 tons per year at a single location.
4. The nitrogen contents of the steels produced are usually about twice as high as those made in BOP or Q-BOP furnaces.

RECENT TRENDS IN ELECTRIC-FURNACE STEELMAKING

Trends Toward Computerization—The computerization of arc furnace meltshops is advancing rapidly. Many computer systems are installed, or are in the process of being installed, that range in complexity from performing relatively simple melting profile guidelines to complete meltshop control.

The current trend in computer system configuration is to install hierarchical computers. In these systems, up to four distinct levels of computers report to each other in a pyramid-like hierarchy. The top level computer handles management functions and the bottom level computer is the computer-furnace interface. All the computers communicate back and forth within the pyramid, forming a total computer system.

Management decision making and operational control of electric-furnace facilities are ideally suited to hierarchical computerization. Because of the large data generating capability of mainframe computers, coupled with localized micro-processors or mini-computers, information can be easily organized and summarized. These computers form the basis of a comprehensive plant management information system, with direct feedback loops to furnace operators.

Management control can key on such items as scrap, raw materials, electrodes, melting energy, refractories, fluxes and labor which contribute to over 80 per cent of conversion costs. Attention to these key performance areas is essential to achieve meaningful results.

Operational control is the process of ensuring that specific tasks are carried out efficiently. This control uses a feedback technique wherein any deviation from expected or standard performance triggers immediate corrective actions. Computerized control of the melting process and data logging functions provide statistical means by which the entire process can be optimized. Typical operational controls include:
1. Calculating the minimum cost of charge materials and providing instructions for layering the raw materials in the scrap bucket.
2. Providing on-line control of the melting process, including—scrap preheating; oxy-fuel burner control; oxygen assisted melting control; controlling the electrical energy and melting profile; thermal model; metallurgical model; calculating minimum cost of alloy additions; tapping temperature as a function of ladle additions, ladle preheating temperature and desired temperature in the ladle; automated furnace additions; monitoring electrical controls and water-cooling; process recording and data logging; recording the consumption of materials and outputs (liquid steel, slag, off-gases, etc.); recording and planning preventive maintenance based on delay input data; and controlling electrical power distribution for the entire plant.
3. Providing data logging functions, including—heat number; steel grade and chemical analysis; date; shift; furnace number; names of operators; the condition of bottom, walls and roof; scrap weights

and types; flux weights and types; alloy weights and types; off-gas temperatures and volumes; slag weights and chemical analyses; natural gas; oxygen; water temperatures; electrical energy; ladle temperature before tapping; predicted temperature of steel in the ladle versus actual temperature; tapping time; delays; and so forth.

Similarly, the continuous-casting process can be computerized and dynamic controls can be established for coordinating the melting and casting processes.

Recent Development of Direct-Current (D.C.) Electric-Arc Furnaces—The progress in power thyristor technology gave impetus to the recent development of a D.C. electric-arc furnace. This type of furnace uses one graphite regulated electrode and a bottom, air-cooled electrode. A 12-ton prototype furnace was commissioned in 1982 by M.A.N.-GHH Sterkrade and Brown-Boveri & Cie (Federal Republic of Germany).

The indicated operating characteristics of the D.C. furnace are:

1. Less electrode consumption (approximately 50 per cent of a comparable three-phase A.C. electric-arc furnace).
2. Smaller electrical utility system requirement (approximately 50 per cent of three-phase A.C.

electric-arc furnace SCMVA requirement).
3. Lower noise level. (The duration of the peak noise level is shorter than that of a three-phase A.C. furnace. Also, once the liquid phase has been reached, the D.C. furnace operates with less noise emission than a comparable three-phase A.C. furnace).
4. With no skin effect in direct-current, there is almost a uniform load across the full cross-section of the electrode.
5. There are no dead or hot zones as in a three-phase furnace.
6. The D.C. arc burns steadily at the center of the electrode tip, causing the electrode to be concavely excavated.
7. The wet heel process is preferred in order to maintain contact with the bottom electrode.
8. Bath movement is sufficient to ensure a metallurgically homogeneous melt.
9. Other consumption data such as power consumption, refractory wear, and personnel costs, are about the same as for a three-phase A.C. electric-arc furnace.

It has been forecasted that in the future, D.C. electric-arc furnaces will be available with a melting power of up to 60 MW for a 120-metric ton heat.

SECTION 2

METALLURGICAL PROCESSES IN THREE-PHASE DIRECT-ARC ELECTRIC-FURNACES

Section 2 contains a discussion of the various metallurgical processes involved in steelmaking in three-phase direct-arc electric-furnaces. The **first subsection** explains the methods used in segregating and selecting scrap. The **second subsection** describes typical stocking and charging facilities in electric-furnace shops, as well as how electric-furnaces are charged with scrap. The **third subsection** explains the primary differences between the steelmaking processes in basic-lined and acid-lined furnaces. The **fourth subsection** describes the various phases of the steelmaking process in basic-lined electric-furnaces. The **fifth subsection** then describes the phases involved in acid-lined electric-furnaces. Finally, the **sixth subsection** briefly discusses repairs to electric-furnaces after steel has been tapped.

THE SEGREGATION AND UTILIZATION OF SCRAP

Scrap Segregation—It is absolutely necessary to segregate or separate the available scrap into stock piles of identified grades for several reasons: (1) to conserve the valuable alloy content of steel scrap; (2) to economically use virgin alloys; and (3) to insure that only the elements desired in the finished steel are introduced in making the steel. When the "product mix" (grades of steel produced) varies substantially, the classification of scrap by alloy content must be much more extensive than for a specialty plant producing the same grade continuously. One plant, making various grades of steel by the basic process, including alloy and stainless steels, has found it necessary to segregate their scrap stock into 300 classifications.

The scrap stock may be revert or home scrap from rolling mills and forge shops of the same plant where the electric-furnace operates. Or it may be obtained

from scrap dealers, customers, other steel producers, or sister plants. Economically producing the wide range of steels common to modern practice requires careful selection of the scrap along with a scrap segregation and control program.

The term "segregation" as used here may be defined as the separation of a mass of mixed scrap into piles of individual compositions. Close adherence to a definite scrap segregation program is essential if the greatest benefits are to be obtained with respect to alloy conservation and melting close to the desired composition. Consistent melting practice further aids in maintaining mill schedules for the production and delivery of steel, obtaining optimum tonnage from a given unit, and maintaining steel quality at the high standards demanded by the steel industry.

Any disregard of the scrap segregation program will lead to the loss of alloys. Non-oxidizable elements such as nickel and copper may enter and remain in the bath, causing the heat to be scrapped or, at best, diverted to another order which did not originally require the use of so many valuable elements. For example, chromium can be oxidized from a heat ordered as nickel-molybdenum steel, but this is a costly and wasteful process. On the other hand, nickel cannot be oxidized from a chromium-molybdenum heat. Therefore, if nickel is present and the order restricts the nickel content to no nickel or to very low levels of nickel, the heat must be diverted or scrapped.

Methods for Insuring the Proper Segregation of Scrap—The segregation of home scrap is comparatively easy, but the introduction of outside or foreign scrap imposes a problem. Several methods are available for testing the scrap to determine if its contents meet required specifications. These methods include chemical analyses of selected samples, spectrographic analysis, and less costly, less accurate methods such as magnetic tests to separate magnetic from non-magnetic scrap, and the "spark" test. The latter test is made by holding a piece of scrap against a grinding wheel and observing the shower of sparks. A trained observer can differentiate between the various kinds of steel scrap by noting the color of the sparks, the length of the spark lines, and the characteristics of the ends of these lines.

There are also rough tests used. For example, in the "spot" test, the application of a chemical solution to a clean surface of a scrap sample permits rough estimates of the amount of nickel or other elements present. Another rough test is the "pellet" test in which elements may be detected by using a magnifying glass to view oxidized particles from "spark" tests.

Physical Requirements of Scrap—The size of the scrap and its bulk density have an important influence on both the technical and economic aspects of melting in the electric-arc furnace. Light scrap (bundles, turnings, punchings) has considerably less weight per unit volume compared to heavy scrap (ingots, ingot butts, crop ends). If too much light scrap is used for a given heat, more charging buckets have to be used to load the furnace, resulting in lost time and heat. Very light scrap is very prone to considerable oxidation during storage, leading to lower yield, greater power consumption, and unpredictable melt-in carbon in the electric-arc furnace. Light scrap also tends to weld together and stick to the furnace wall, necessitating additional measures to ensure complete melting. A first bucket charge, made up entirely of light scrap, requires that lower power be used in the early stages of melting to ensure that the electrodes do not bore through the charge and reach the hearth before a large enough pool of metal has been formed. It is possible to overheat the pool and thus damage the hearth refractories. A charge made up entirely of heavy scrap is also objectionable because it does not permit the shielding of roof and walls during the melt-down period to the same extent as a mixed charge of greater volume, and results in the decrease of refractory life. Another physical requirement, especially with heavy scrap, is that the pieces are not too long. Long pieces can create problems when charging as they can obstruct the closing of the furnace roof which results in loss of time and heat. Large pieces of scrap may cause electrode breakage as they fall or shift during meltdown of the charge. Large pieces may also form bridges which support cold scrap, the sudden collapse of which can result in metal overflowing at the furnace doors or even causing explosions.

In practice, the steelmaker makes up the furnace charge with a mix of scrap of various types to obtain the smallest number of bucket charges, the most rapid melting, the lowest power utilization, and the lowest electrode consumption consistent with the price of the scrap mix charged. The objective is to minimize the disadvantages of each type by proper loading of scrap in the bucket and charging of big pieces directly to the furnace, aiming to achieve proper placement of the scrap in the furnace.

Selection of the Scrap Charge—In an efficient scrap-segregation program, the available scrap is segregated according to physical characteristics and is kept relatively free of contaminants such as water, oil and dirt. The scrap charge should then be made as follows:

1. Grades of scrap must be selected which together contain elements necessary to make the heat ordered. The charge may contain all or part of the specified elements, but no scrap should contain elements not in the specification. This is imperative since the scrap may contain an element which cannot be oxidized by regular practice.
2. For economical operation, each element contained in the scrap should weigh as closely as possible to the number of pounds required to meet the lower range of the chemical specifications for that element in the ordered heat.
3. The total amounts of virgin alloys needed are calculated from the amounts of the various elements contained in the combination of scrap selected for a heat. This calculation is based on the weight of the heat to be made (furnace capacity and amount specified by an order) and the specification itself. In making these calculations, allowance must be made for adjustments during the making of the heat; for example, losses when melting some alloys and the possible absorption of certain alloying elements from the bottom and banks of the furnace. In regular practice, a heat made to low-alloy specifications occasionally finishes outside of the specified composition ranges when the heat is made immediately following a heat of high-alloy specifications, such as stainless grade, or high-manganese steel, etc. The usual practice to overcome this difficulty is to make a so-called "wash heat," i.e., following a heat made to high-alloy specifications, a heat is made of medium-alloy specifications containing the same elements. The wash heat absorbs any elements in the furnace bottom and banks remaining from the previous heat and thus prevents high residuals in the low-alloy heat which follows. Hence, the wash heat should be one in which the increase in certain elements is not harmful to its own composition.
4. In efficient operations, the scrap selected for a given charge is not only satisfactory for that charge but is also selected on the basis of the scrap available for efficient operation over a period of time. If only heavy or medium scrap were used for a few heats, it is probable that these particular heats could be made quickly. If, however, a large inventory of light scrap (turnings, punchings, etc.) were accumulating at the same time, excessive amounts of this light scrap would have to be charged in later heats, using more than the usual number of bucket charges. In such cases, the delays and damage to refractories would far offset any gains in the few heats made with heavy and medium scrap.
5. Direct-reduced iron (DRI) can be used to replace part or all of the steel scrap charge. In some operations, DRI is preferred to scrap because it has a known uniform composition and contains no residual elements such as chromium, copper, nickel, and tin. Also, when DRI is melted, it forms a foamy slag because it contains both carbon and

iron oxide. However the price of steel scrap in North America is usually lower than that of DRI so that the use of DRI usually cannot be justified except when it is required to obtain low concentration of residual elements. When scrap and DRI prices come closer together, it may be desirable to use about 30 per cent DRI and 70 per cent scrap in the charge. The DRI would be added after the scrap is melted to form a foamy slag and permit the use of high power after the bath is flat, thereby increasing productivity as well as facilitating the adjustment of the composition of the steel. DRI is usually added continuously through a hole in the furnace roof.

6. Probably no part of the routine of charging the furnace is more important than the loading and weighing procedures. The proper type of scrap must be selected and then weighed correctly so that the final product will be of the right composition. The bucket must be filled with scrap in a predetermined sequence. The bottom of the bucket should contain a thin layer of light scrap onto which heavy pieces of scrap are charged. The light scrap provides cushioning on the bottom of the furnace during charging. Placing large pieces low in the furnace prevents them from falling against the electrodes during melting. Coke or other carbonaceous materials should then be charged after the heavy scrap. Since fluxes are normally nonconducting, they should be placed in such a way that they will fall outside of the pitch circle of the electrodes in the furnace. Medium size scrap can be charged next. Light scrap should be charged on top to ensure a quick boredown of the electrode tips into the scrap after power on. This will reduce roof wear, allow early shift to higher voltage setting, and reduce noise emission.

Scrap Practice—When melted and refined in an electric-arc furnace, 100 tons of steel scrap usually produce about 90 tons of continuous cast steel or ingots, or a yield of about 90 per cent. The 10 per cent loss consists primarily of: (1) metal lost in the slag as iron oxide and metallic steel droplets; (2) iron vaporized from the bath and collected in the gas cleaning system; and (3) steel lost during casting or teeming. As part of a program to minimize these losses, it is desirable to accurately weigh the steel scrap used in making the steel, and to compare this weight with the weight of the steel produced.

It usually takes two or three buckets of scrap to make a heat of steel in an electric-arc furnace. In a two-bucket steel charge practice, about 60 per cent of the total scrap charge is put into the first bucket and 40 per cent in the second bucket. If the available scrap is mostly light scrap, then a three-bucket charge practice is used, with 40 per cent of the scrap in the first bucket and 30 per cent each in the second and third buckets.

STOCKING AND CHARGING FACILITIES

Most modern designs of electric-furnace plants call for a lower level stockhouse located either as an extension to the end of the furnace building, or adjacent and parallel to the furnace building. At either of these two locations, drop-bottom charging buckets, ranging in capacity from 2.8 to 113 cubic metres (100 to 4 000 cubic feet) or larger as required by furnace size, are loaded at ground level and then moved either by an overhead crane, or by transfer car and overhead crane to the furnace floor. From the furnace floor, these buckets are then charged into the furnace. Some older plants originally equipped with door-charged electric-furnaces have two-level stockhouses. Usually, the higher level of such stockhouses is used for charging cars that can be loaded on the furnace level, and the lower level is used for storage bins.

In electric-furnace practice, the different lots of alloy scrap are kept in separate bins. During normal times, between twenty and forty bins are required. When the use of virgin alloys is restricted, scrap must be segregated further and, depending on the grade of steel being made, considerably more kinds of scrap must be used. Stocking out-of-doors may be required, however, because few stockhouses are designed for this latter practice.

Since the charge should be dry before it is placed in the furnace, a covered stockyard is advantageous. For a 450 000 metric ton (500 000 net ton) plant, at least two 18-metric ton (20-net ton) cranes equipped with lifting magnets are required in the stockyard.

Tracks for both the incoming supplies and the charging cars are important factors in the plant layout. The daily supply of scrap for a 450 000 metric ton (500 000 net ton) annual capacity shop will require a minimum of thirty-two 45 metric ton (50 net ton) railroad cars, in addition to those required for limestone, ore, brick, etc. Storage tracks located nearby are also required for extra stocks of these materials to provide for possible times of irregular railroad movements.

Charging the Furnace—Electric-arc furnaces have removable roofs so scrap can be quickly and easily charged into the furnaces. Depending on the method used for opening the furnace roof for top-charging, a furnace is designated as either of two types: the swing type; or the gantry-lift type. Most modern furnaces are designed with a swing-roof, although many gantry-lift types of furnaces are still in operation.

Top-charging has the advantage of speed, since the entire charge can be quickly placed in the furnace by drop-bottom buckets. Furnaces may be filled completely with light scrap very quickly. However, scrap dropping from a considerable height results in a shock to the furnace bottom. It is therefore desirable to load the bucket with a layer of light scrap on the bottom to cushion the fall of the larger pieces of scrap.

Currently, scrap charges of 90 metric tons (100 net tons) or more are being charged into large furnaces with one bucket. However, extremely large pieces of scrap, such as large ingot butts and broken roll sections, are preferably charged by magnet onto the bottom of the furnace prior to bucket charging. Small trackless mobile machines are used in some larger shops for charging additions and for stirring the bath.

To charge the furnace, the power is turned off and the electrodes and roof are moved out of the way. The clam-shell scrap bucket is positioned above the furnace and the charge dropped in.

Alloying materials that are not easily oxidized, such as copper, nickel, and molybdenum, can be and usually are charged in the furnace prior to the melt-down.

It is desirable to melt-down with excess carbon in the bath in order that some carbon may be worked out by ore additions or oxygen injection. If the metallic charge is too low in carbon, a recarburizer in the form of coke or scrap electrodes is charged with the scrap to allow for a carbon content of the bath at melt-down that will be 0.15 to 0.25 per cent higher than the carbon content of the finished steel.

Although ore or mill scale is still used to a limited extent to lower the carbon content, the use of gaseous oxygen injected into the bath is much more common. Operators may charge the ore either with the scrap, or when the charge is partially melted, or when the charge is completely melted.

GENERAL COMPARISON OF BASIC AND ACID PROCESSES

Both the basic and the acid processes are used for making steel in electric-furnaces. However, almost all furnaces used for continuous cast and ingot steel production, and a large percentage of the foundry furnaces, are now basic-lined furnaces to use combinations of high-alloy steel scrap, lower grades of alloy scrap, and plain-carbon steel scrap to produce steels that will meet strict chemical, mechanical-property and cleanliness requirements.

The basic electric-arc furnace uses either a bottom consisting of a burned-magnesite brick subhearth with a working surface about 152 to 305 mm (6 to 12 inches) thick of high-magnesia (typically 95 to 98 per cent MgO) ramming material (Figure 18—2), or a complete brick bottom made from impregnated magnesite brick. The use of dolomite is generally confined to bottom

fettling and door banking; materials of higher magnesia content generally are used for bank repairs, being applied with refractory "guns" or "slingers." In older practices, sidewalls are lined with direct-bonded magnesite-chrome brick or burnt-magnesite brick with fused-cast or rebonded magnesite-chrome brick in the sidewall "hot spots" to provide more even sidewall wear and, consequently, longer overall refractory life. In more modern practices, special magnesite-carbon refractories or water-cooled panels are used in the walls to prolong life. (A more complete description of these practices is contained in Chapter 2 on "Refractories.")

Basic electric-arc furnace roofs are often constructed with high-alumina brick, and with high-alumina rammed or castable materials for the center section around the electrodes. Roofs are also made of direct-bonded magnesite-chrome brick or water-cooled panels to extend roof life, particularly in large, high-powered furnaces or in furnaces melting direct-reduced materials.

Basic electric-arc furnaces of the type just described are used to produce practically all of the electric-furnace steel made in ingot form and continuous-cast shapes. In addition, basic electric-arc furnaces are increasingly used in foundries for making steel castings.

Before the location for an electric-furnace plant for the production of ingots by the basic process can be selected, its capacity and probable growth must be known because an ample electric-power supply must be available. For example, if it is assumed that a plant will have an initial annual capacity of 450 000 metric tons (500 000 net tons) of ingots per year, a power supply of at least 70 000 kW must be available. This amount of power is not available in the generating sta-

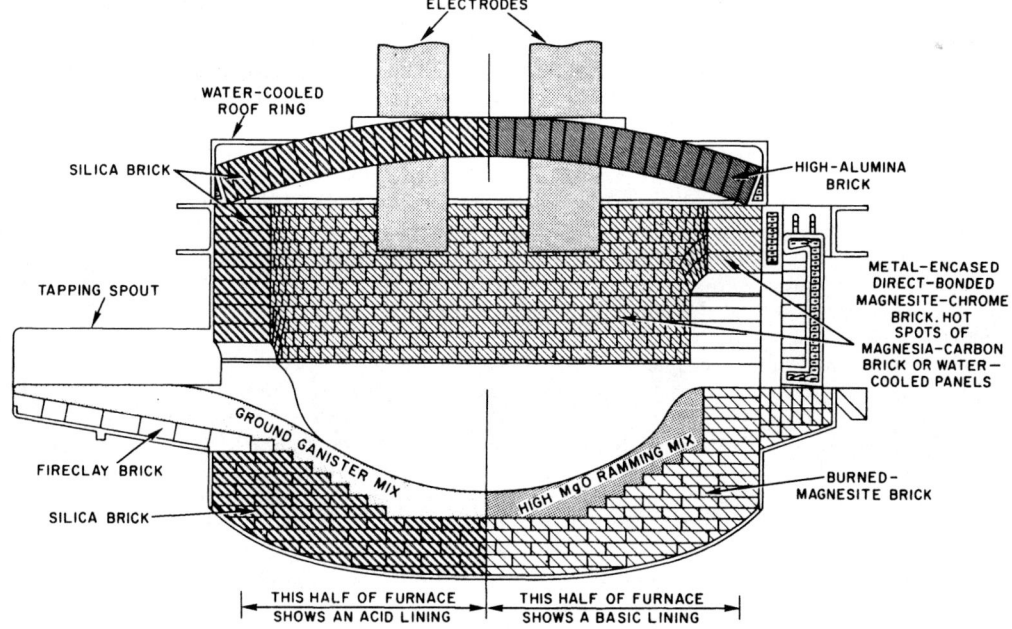

FIG. 18—2. Schematic cross-section of a Heroult electric-arc furnace with a dished-bottom shell and stadium-type sub-hearth construction, indicating typical refractories employed in (left) an acid lining, and (right) a basic lining. Although only two electrodes are shown here, furnaces of this type (which operate on three-phase current) have three electrodes. (See Figure 18—6 for other types of subhearth construction applied to furnaces with shells having dished bottoms.) (Courtesy, Harbison-Walker Refractories Company.)

tions of the steel plant, however, and due to the highly fluctuating loads, poor power factor and load factor, it is not desirable on some utility systems. For these reasons, a source of power must be found before deciding on a plant location. Also, the type of equipment selected for air pollution control is based on economic and processing factors. For example, a shop-evacuation system, although the most expensive, may be necessary if most of the heats will require multiple-slag practices and if ladle additions during tapping create excessive fumes.

The acid electric-furnace, with an acid bottom of impure silica in the form of ground ganister and silica-brick sidewalls and roof (Figure 18—2), is seldom used outside of steel foundries and forging shops. Straight-carbon, low-alloy and some high-alloy steels can be made by the acid process.

A comparison of the features of acid and basic electric-furnace steelmaking processes shows the following differences:

1. Slags in the acid process are more siliceous than those in the basic process. Since acid slags do not react with the steel bath to remove phosphorus or sulphur from the steel, the use of more expensive, carefully selected scrap and other raw materials of low phosphorus and sulphur contents is imperative. Conversely, basic slags react with and retain phosphorus and sulphur. Practically all types of scrap and raw materials can be used in the basic electric-furnace.

2. Oxidizing action occurs faster in the acid process. The time required for "working the heat" is also shorter than in the basic process, mainly because selected scrap is used in the acid furnace as compared with the wide variety of scrap used in the basic process.

3. Iron loss is lower with the single-slag acid process than with the multiple-slag basic process because less metal is trapped in the much smaller volume of slag.

The metallurgical differences between the basic and acid processes indicate that none of the advantages of one process over the other is sufficient to justify the selection of one or the other process without full consideration of all factors. Many of these factors have already been discussed, and explain the present preference for the basic process in the majority of applications. The next two subsections more fully describe the processes involved in basic- and acid-lined electric-arc furnaces.

THE BASIC ELECTRIC-FURNACE PROCESS

The process of making steel in the basic-lined electric-arc furnace can be divided into: (1) the melt-down period; (2) the oxidizing period; (3) the composition and temperature adjustment period; and (4) the tapping period. These four periods are fully discussed below.

The Melt-Down Period—When charging has been completed, the bank in front of the charging door is built up with refractory material (dolomite) to form a dam to keep the molten metal from slopping out the door (or doors, if the size of the furnace warrants more than one door). The door (or doors) is closed and the electrodes are lowered to about 25 mm (1 inch) above the scrap. The main circuit breaker is closed, an intermediate voltage is selected with proper current setting on the rheostats, and the arcs are struck under automatic control. After 1 to 3 minutes (to allow the electrodes to bore into the scrap), maximum voltage and current should be applied for the fastest possible melting of the scrap. The initial slow start is to shield the lining and roof from the heat of the arc.

The melting period in the basic electric-furnace is the most expensive period in its operation because power and electrode consumption are at the highest rate during this interval.

The electrodes melt the portion of the charge directly underneath and around them, and continue to bore through the metallic charge, forming a pool of molten metal on the hearth. From the time the electrodes bore through the scrap and form the pool of molten metal on the hearth, the charge is melted from the bottom up by radiation from the pool, by heat from the arc, and by the resistance offered to the current by the scrap. Burnt lime is usually blown on top of the bath near the end of the melting period of the first scrap charge. If blowing facilities are not available, burnt lime is usually added to the furnace before the second scrap charge is made. Heating of the charge material is continued until it is completely melted.

The Oxidizing Period—Oxidation occurs in varying degrees from the time the molten metal begins to form until the entire charge is in solution. During this period, phosphorus, silicon, manganese, carbon, and iron are oxidized. Oxygen for these, as well as for other oxidizing reactions, is obtained from:

1. Oxygen gas injected into the bath.
2. Oxygen in the furnace atmosphere.
3. Calcination of limestone. (However, because of the high energy cost for calcining limestone in the electric-furnace operation, this method is unlikely to be used as a partial replacement for burnt lime. Only small amounts of limestone are used on a flat low-carbon steel bath to minimize overheating of the furnace roof by heat radiation from a smooth bath.)
4. Oxides of alloying elements added in the furnace.
5. Ore, cinder and scale (if charged or added later).

Care must be exercised in adding the aforementioned oxidizing materials when they are contaminated with materials that could cause violent chemical reactions or explosions. The oxidizing materials added to the bath react with carbon in the bath to form carbon monoxide. If too much of the oxidizing material is added at one time, the carbon monoxide formation may be vigorous enough to eject slag and steel out of the furnace. The direct use of oxygen gas (item 1, above) is extremely important in modern practice from the standpoint of rapidly removing carbon from the bath.

As mentioned previously, enough carbon should be present in the charge so that the carbon content of the steel, when melted, is 0.15 to 0.25 per cent higher than the desired tap carbon content. This excess carbon is removed with oxygen and forms carbon monoxide gas, which bubbles out of the steel. This bubbling is called the "carbon boil," which stirs the bath and makes it more uniform in composition and temperature. This facilitates meeting steel composition and temperature specifications. The carbon boil also removes some of the hydrogen and nitrogen from the steel, which is generally desirable.

During the oxidation period, the reactions that occur in the bath of the basic electric-arc furnace are similar to those in the basic open-hearth and the basic oxygen furnace, except that the electric-furnace bath can be made hotter. Hence, there is more chance of phosphorus reversion unless the slag is strongly basic.

The Composition and Temperature Adjustment Period—Most carbon steel and low-alloy steel grades made in electric-arc furnaces are made by a single slag process. This means that the slag that forms during melting of the charge is not replaced by another slag as in the double slag process. In the single slag process, the steel is finished by adjusting the composition and temperature to the desired values, after which the steel is tapped from the furnace into a ladle. As soon as a charge is melted, a steel sample is taken and analyzed in the chemical laboratory. From this analysis, the required steel composition adjustments are determined. Several temperature measurements of the steel are made to determine the temperature adjustment required.

To increase productivity, the oxidizing period often overlaps the composition and temperature adjustment period. If further refining of the steel is required to obtain low sulphur and oxygen contents, this is accomplished by means of a ladle metallurgy treatment after the steel has been tapped from the furnace.

In the past, most high quality carbon and alloy steels were made in the electric-arc furnace by a double slag practice. In the double slag practice, as much of the original oxidizing slag was removed from the furnace as possible and a reducing slag was made by putting burnt lime, fluorspar, coke, and sand on top of the steel in the furnace. Calcium carbide formed in the slag aided in removing sulphur from the steel. However, recently developed ladle metallurgy treatments are now preferred for the production of high quality steels.

Slag control is a very important factor in electric-furnace steel production. The electric-arc furnace permits the slags to be controlled to meet almost any desired characteristic, a fact that is the real basis of the flexibility of the arc furnace.

As pointed out earlier, the function of the melt-down slag is to remove phosphorus and sulphur from the steel. The lime-silica ratio should normally be between 2.0 and 4.0. The iron-oxide content of the slag varies with the carbon in the steel at the end of the boil, and may range from 13.0 to 20.0 per cent for medium-carbon steels. A typical composition of a melt-down slag is as follows:

Constituent	Per Cent
Lime, CaO	40.9
Silica, SiO_2	13.4
Iron Oxide, FeO	14.8
Alumina, Al_2O_3	3.5
Magnesia, MgO	8.2
Manganous Oxide, MnO	12.7

Phosphorus Pentoxide, P_2O_5	0.6
Sulphur, S	0.1

The lime-silica ratio and iron oxide content can be estimated from the appearance of slag "pancakes," and a skilled operator can judge these values very closely. A "pancake" is a slag sample prepared by pouring molten slag into a small, flat iron dish possibly 100 mm in diameter and 12 mm deep (4 inches in diameter and ½ inch deep). In solidifying, the pancake acquires visible markings characteristic of its composition. Rapid chemical analyses of slags can also be made by specialized spectrographic techniques.

The CaO content of the slag is also important. For good sulphur removal, the CaO content should be at least as high as 40 per cent, although 50 per cent is preferable.

A slag control practice has been developed recently that protects the furnace refractories from the arcs and allows more power to be applied to the arcs. This is a "foamy slag" practice which involves a controlled boil. By adding carbon to the slag at a controlled rate, a moderate carbon boil is produced that results in the formation of a foamy slag layer 300 mm (12 inches) or more thick. Coke, coal, and charcoal are used for this purpose. Sometimes limestone is used instead of carbon to form the foamy slag.

The foamy slag protects the furnace refractories from heat radiated from the arcs. Normal practice requires that the power to the arcs be decreased as soon as most of the scrap has been melted and the furnace walls can be seen. However, this decrease in power slows down the heat. By using a foamy slag, this decrease in power is unnecessary and furnace productivity can be increased by the use of high power.

Another recently developed slag practice is the "slag and a half" practice. It may be considered to be almost the equivalent of a double slag practice. The slag and a half practice consists of raking off the original oxidizing slag, making up a new slag with burnt lime and fluorspar, heating this with the arcs for several minutes, and then tapping through a very large (300 by 450 mm, or 12 by 18 inches) tap-hole that permits the slag to mix with the steel during tapping. This is done to decrease the sulphur content of the steel. A 50 per cent decrease in sulphur content has been reported for this practice.

The Tapping Period—In tapping a heat, the electrodes are raised high enough to clear the bath in a tilted position, the tap-hole is opened, and the furnace is tilted by a control mechanism so that the steel is drained from the furnace into a ladle. The ladle is usually held by a teeming crane during tapping to minimize exposure of the stream to air and minimize erosion of ladle refractories. The slag may be tapped before, with, or after the steel, depending on the particular operation. A clean, round tap-hole and a clean, smooth tapping spout reduce the possibility of having a ragged, easily oxidized stream when tapping.

Often the steel is to be refined further by a ladle metallurgy treatment to remove sulphur and oxygen from the steel. The slag on the steel in the furnace contains iron oxide, which greatly inhibits the removal of sulphur and oxygen. For this reason, it is often necessary to exclude as much furnace slag as possible from the ladle during tapping. In the past, this has been done by adding material such as burnt lime to thicken the slag in the furnace just before tapping and to move the ladle away from the tapping stream at the first sign of the presence of slag in the tapping stream.

Because of vortex action that occurs at the tap-hole when the steel is tapped from the furnace, some slag always mixes with the steel that is last to leave the furnace. To get a slag-free tap, it is necessary to leave some steel in the furnace after tapping. A convenient way to avoid wasting this steel is to adopt a "wet heel" practice and save this steel for the next heat. The furnace is then repaired at the slag line and the next heat is made in the furnace. After a series of wet heel heats are made, the furnace is completely emptied to permit the furnace bottom to be repaired. When using the wet heel practice, the number of heats made between bottom repairs varies from three to twelve. The wet heel practice also has the advantage of improving arc stability during melting of the scrap.

Several methods of slag-free tapping are in current use. One method involves the use of a slide gate on the furnace tap-hole. The slide gate is closed just before any slag comes out. By weighing the steel in the ladle during tapping, better control of when to close the gate may be obtained.

Another slag-free tapping method is the eccentric bottom tapping method. A short spout is attached to the front of the furnace which contains a horizontal channel connected to the liquid steel bath. The bottom of a short vertical hole leading from the end of the horizontal channel is closed with a graphite plate while the steel is being made in the furnace. To tap the furnace, the graphite plate closure is opened. Just before any slag comes out, the furnace is quickly back-tilted and any steel remaining in the furnace forms the wet heel.

THE ACID ELECTRIC-FURNACE PROCESS

As previously mentioned, the acid electric-furnace process is employed by the foundry industry mainly for the production of iron and steel for castings. Iron making in the electric-arc furnace is basically a melting process and is not covered in this discussion. Four major variations of steelmaking in the acid process are used: (1) partial oxidation; (2) double slag; (3) complete oxidation with silicon reduction; and (4) complete oxidation (with a single slag).

The partial oxidation method is used chiefly to produce low-priced steel castings that do not require any acceptance tests other than superficial surface inspection because it is the cheapest method of making steel for such castings. The double-slag method is employed where it is desirable to have positive control to keep the FeO content of the finishing slag to a low value (about 10 per cent). Silicon in the slag can be reduced to enter the metal in the acid electric-furnace practice. This procedure is employed in European practice but is not usually followed in this country.

The great majority of American steel foundries employ the complete oxidation method. Thus, this method will be used as the basis for the ensuing discussion.

Except for the selection of scrap (which must have low phosphorus and sulphur contents), the melting of the charge in the acid electric-furnace is similar to that of the basic electric-furnace. As with the basic electric-furnace, the electrodes melt the scrap and bore their way through to the hearth of the furnace. If the pool of molten metal formed on the bottom does not cover a sufficient area to extend beyond the area affected by the electrodes, the arc will act on the sand in the hearth of the furnace. The conductivity of a nonmetal increases with temperature; consequently, the furnace hearth is a fairly good conductor of electricity if it is hot. When the arc works on the hearth or bores a hole into the hearth in this manner, it is called "pulling bottom" and is generally indicated by the appearance of white smoke accompanied by bright yellow flames around the electrode ports. When this occurs, the electrodes should be raised out of the charge, and enough clean scrap should be added and melted to form a pool extending out from under the electrodes. If enough heavy scrap is used and is packed compactly enough on the bottom of the furnace, there is very little danger of pulling bottom.

As soon as the charge is melted or nearly melted, it is time to start working the heat. A little iron ore and silica sand should be spread over the bath at this time. If a high percentage of returned foundry scrap is charged, which is often the case, very little silica sand will need to be added because the oxidation of silicon and manganese in the scrap will form almost enough slag. Some iron will also be oxidized in the melt-down, forming FeO which will contribute to the slag. Enough slag-making material should be added to each heat to cover the metal with a layer that is at least 3 mm (1/8 of an inch) thick.

Although such a thin layer of slag may cause some difficulties, it is never advisable to have more slag on the metal than is necessary at any time, because this will slow down the deoxidation of the steel. If silica sand is not added to the slag, the bath will take silica from the hearth.

Slag samples taken from the furnace at melt-down should be a glossy, black color, indicating a high iron-oxide (FeO) content, which is necessary if a boil is to be expected later. A sample of the metal should be taken from the furnace at the same time as slag samples are taken so that the carbon content can be determined. The carbon content after melt-down, as in the basic process, should be higher than the carbon desired in the finished steel after it has been killed by additions of silicon and manganese near the end of the heat. The excess carbon will be removed by the boil.

If the slag test taken from the furnace has a brown or greenish color instead of black, there is insufficient FeO present. Ore is then added shovel by shovel until a black slag results. If the charge is made up of a large percentage of returned scrap, the ore should be added before all of the scrap is melted. The FeO added in this way will be taken up by the bath, and the oxygen will react with silicon and manganese in the melting metal to form silicon dioxide (SiO_2) and manganous oxide (MnO), respectively. Because these oxides are of lower specific gravity than the molten metal, they will rise to the top of the bath and, because of their chemical properties, will unite to form a slag, a liquid mixture of these oxides.

After the bath is covered with a black or oxidizing slag and the carbon content is high enough, the temperature should be increased until the steel is hot enough to boil. The "boil" is the reaction between the carbon and oxygen dissolved in the steel, and is necessary in the manufacture of clean, high-quality steel. The boil stirs the bath, making it more uniform in composition and temperature. The decarburizing process can be accelerated by injecting oxygen gas into the bath.

Enough ore and carbon should be in the bath, either naturally or by additions, to maintain the boil for at least ten minutes. A sample of the metal is taken from the bath and the carbon level is determined by a fracture test or by some rapid analytical test. Silicon and manganese should then be added as ferroalloys for deoxidation. The heat should be tapped soon after these additions are completely melted and diffused through the bath.

The temperature at which the steel is tapped from the furnace depends largely on the size of the castings to be poured and the equipment for handling the molten metal. The steel must be hot (approximately 1620°C) if it will be poured into many small castings. However, if the steel is poured directly from a large ladle into a mold for a large casting, it may be about 55°C (100°F) colder when tapped from the furnace.

The preceding discussion refers specifically to the production of plain or carbon steel for casting. Yet alloy cast steels present additional problems. Fortunately, three of the alloys commonly used in steel foundry practice, copper, nickel and molybdenum, can be added at any time without loss due to oxidation and subsequent absorption by the slag during the steelmaking process. If these alloys are not added with the cold charge, they should be added from 15 to 30 minutes before the heat is tapped to give ample time for their solution and uniform distribution throughout the bath of molten metal.

When the steel order calls for manganese in excess of 1.25 per cent, it is difficult to maintain this manganese content in the metal under an acid slag. Lime can be added to the slag before tapping, thus decreasing its acidity and ability to absorb manganese.

Alloys such as aluminum, chromium, titanium, zirconium, vanadium and boron are added in the ladle. These alloys are commonly placed in paper sacks, which are then thrown into the ladle as the steel is being tapped so that each sack hits the metal stream.

REPAIRS TO FURNACE BOTTOM AND BANKS

Once the furnace is tapped, it is tilted back to its stationary operating position, or perhaps to a back-tilted position to provide a better view of the bottom. After the doors are opened to permit smoke and fumes to clear out, the furnace is inspected for damaged areas on the banks, bottom, roof, around the tap-hole, and on any other location, such as on furnace skulls. Since the slag erodes the basic lining, it is usually necessary to build up the "cinder line" where it has been eroded by the slag. If the refractory lining requires repairs, patching is done immediately to allow the material to be sintered into place by the heat of the furnace.

SECTION 3

MECHANICAL FEATURES OF
THREE-PHASE DIRECT-ARC
ELECTRIC-FURNACES AND AUXILIARY EQUIPMENT

Section 3 discusses the various mechanical features of three-phase direct-arc electric-furnaces and auxiliary equipment. The **first subsection** presents a general overview of the design and construction of direct-arc electric-furnaces. The **second subsection** then gives a detailed description of the mechanical features of these electric-furnaces and the auxiliary equipment used with them. The **third subsection** briefly discusses other design considerations, including the most recent technological advancements in electric-furnace auxiliary equipment. The **fourth subsection** describes the types of electrodes used in electric-furnace steelmaking. Finally, the **fifth subsection** presents a discussion of the various environmental controls used in the electric-furnace steelmaking process.

GENERAL INFORMATION ON
FURNACE DESIGN AND CONSTRUCTION

The direct-arc electric-furnace is the type most commonly used today. It is primarily a scrap-melting furnace, although iron in the form of molten blast-furnace metal (hot-metal) and direct-reduced materials (i.e., direct-reduced iron pellets, or DRI) are also used for the charge. A three-phase transformer, equipped for varying the secondary voltage, is used to supply energy to the furnace from the electric-power system over a suitable range of power levels. Cylindrical solid graphite electrodes, suspended from above the shell and extending down through ports in the furnace roof, are used to conduct the electric current inside the furnace shell.

As with all steel-mill equipment, the emphasis in design of electric-arc furnaces must be on rugged construction to provide safe, reliable, efficient operation with minimum maintenance requirements under severe operating conditions.

The trend in recent years to ultra-high-power operation has had a considerable impact on arc furnace design. The term "ultra-high-power" is a relative one and, within current technological limitations, is a function of transformer size and the ability of the furnace walls and roof to provide reasonable life. The greatest impact has been experienced in the design of the larger furnaces, wherein arc currents of the order of 70,000 to 100,000 amperes or more must be considered in lieu of the 30,000 to 60,000 amperes for which these sizes of furnaces were formerly designed. Since magnetic induction effects are a function of the square of the current, the effect of induced heating on the design of the furnace structures and the electrical components is evident.

The structural and mechanical parts of a modern electric-arc furnace consist basically of: a shell to contain the charge, with a refractory lined hearth, water-cooled wall panels and a water-cooled roof; mechanisms to position the electrodes to maintain the arc for melting and refining; mechanisms to tilt the furnace for

tapping and deslagging; and mechanisms to remove the roof for top-charging and to operate the door on the shell. Figure 18—3 shows a cutaway sketch of a Heroult electric-arc steelmaking furnace as designed and built by American Bridge. Figure 18—4 shows a plan and elevation of a similar furnace.

In the past, Heroult furnaces were designed with electro-mechanical operation of the various mechanisms. Developments in the design of hydraulic systems and components along with the formulation of compatible fire-resistant hydraulic fluids and seals improved the reliability of hydraulic systems for the operation of electric-arc furnaces. Thus, more modern designs utilize hydraulic systems for tilting, roof lifting

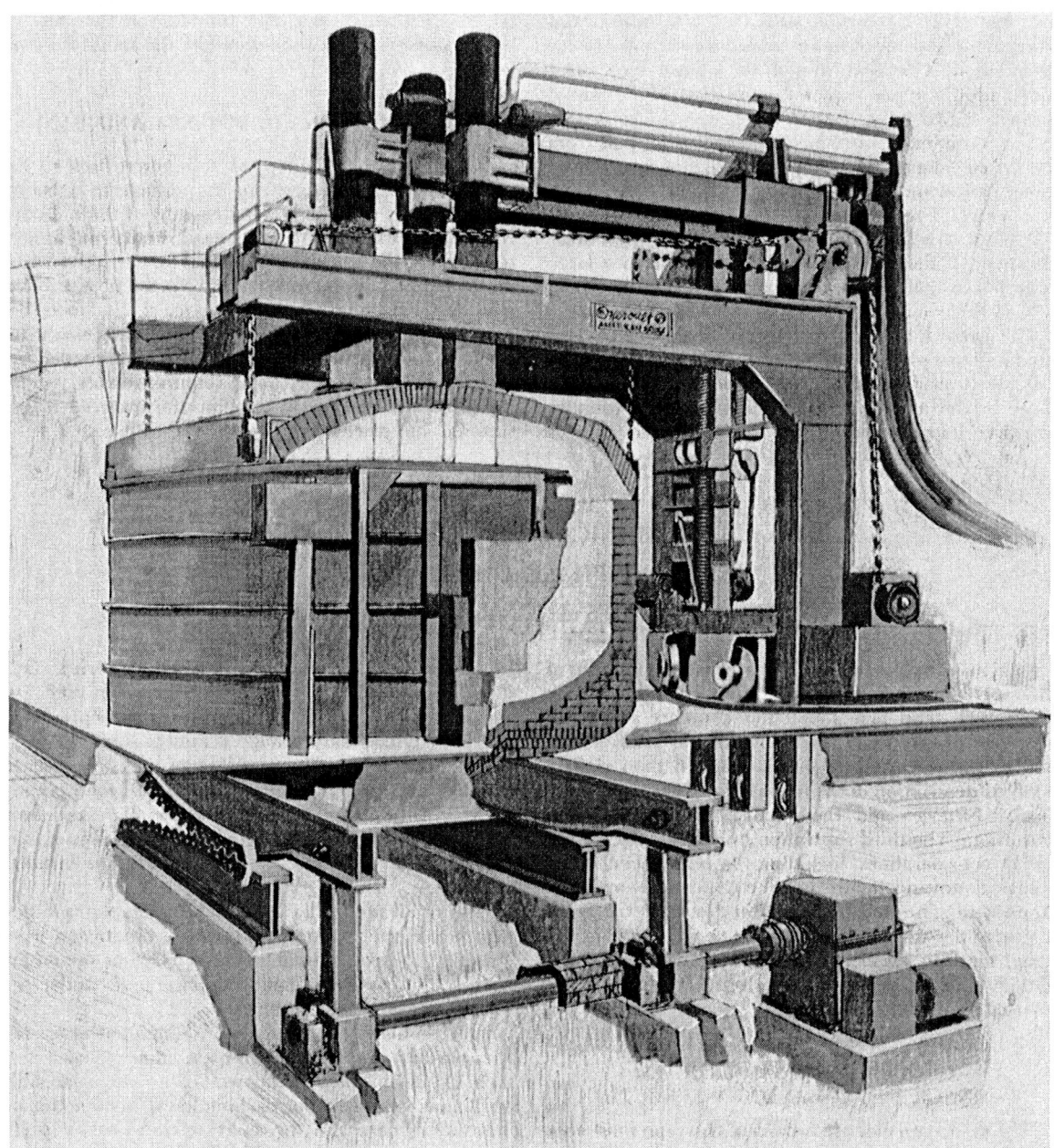

FIG. 18—3. Overall sketch of a Heroult electric-arc furnace, cut away to show sections of the bottom, sidewall and roof.

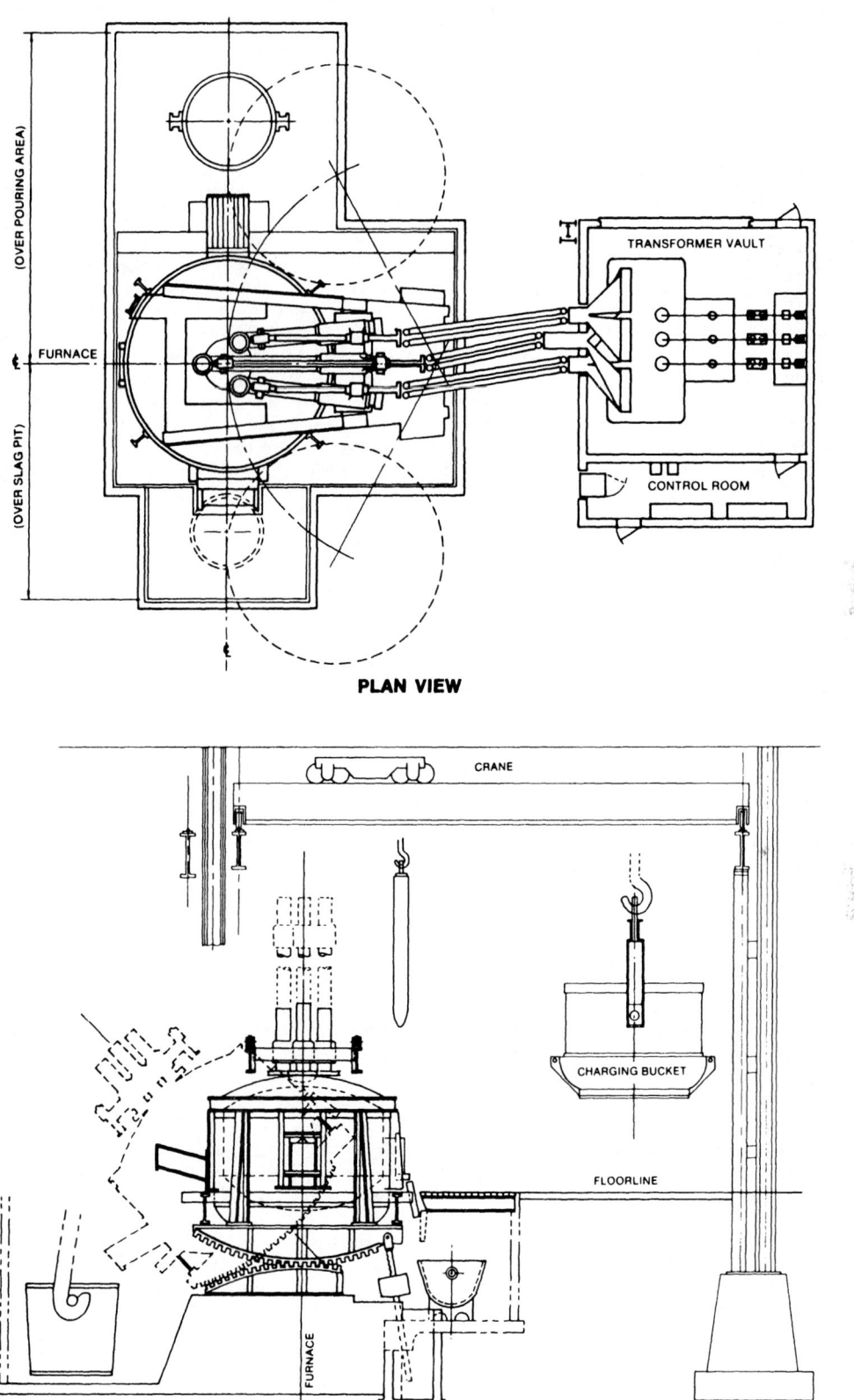

PLAN VIEW

SIDE DOOR ELEVATION

FIG. 18—4. Schematic plan and elevation of a Heroult electric-arc furnace.

and swinging, door opening, electrode unclamping, and auxiliary functions. The use of hydraulic mechanisms offers some economic advantage in equipment cost, particularly in the use of only one power unit (pump) to service all mechanisms in place of the several drive units required for the electro-mechanical design. Also, since the various mechanisms do not operate concurrently, the pumping unit can be sized to the single largest power requirement.

MECHANICAL FEATURES OF DIRECT-ARC ELECTRIC-FURNACES AND AUXILIARY EQUIPMENT

The following descriptions of the components and operating mechanisms of electric-arc furnaces pertain to the Heroult design.

Rockers—As can be seen in Figures 18—3 and 18—4, the furnace structure on larger furnaces is mounted on a toothed curved top rocker and a toothed flat bottom rail which permit forward and backward tilting of the furnace. Small furnaces can have curved top and bottom rockers. The radius of rocker curvature is selected in relation to the center of gravity of the complete mass to be tilted, i.e., furnace structure, electrodes, refractory lining, metal charge, and so on, so that there is no tendency for the furnace to overturn at any degree of tilt under normal operating conditions.

Most furnaces are designed to tilt 40 degrees forward for tapping and 10 degrees backward for deslagging. The angle of forward tilt is generally equal to or slightly greater than the angle of the hearth bank to insure complete draining of the hearth. In turn, the maximum angle of the hearth bank is dictated by the maximum angle of repose of commercially available and economically usable refractory fettling materials. The greater the bank angle, the larger the hearth capacity for a given depth.

Tilt Platform—The furnace platform is of heavy welded and bolted steel construction, and is rectangular in configuration. This platform completely surrounds the furnace shell, providing a complete safety closure with the melt shop's operating floor when the furnace is in a normal, level operating position.

Incorporated into the furnace platform are two curved rockers through which the entire furnace structure is supported on two heavy welded steel flat rails. Both the rockers and rails have machined tread plates and meshing, tooth segments, with flanges on both upper rockers. The tooth segments and rocker flanges assure permanent alignment of the rockers and rails.

Tilting Mechanisms—Furnaces may be tilted by either electro-mechanical or hydraulic mechanisms.

In the first case, a dual rack-and-pinion mechanism is attached to the rockers, and is driven by a variable-speed motor through a speed reducer. A direct-current motor is normally used for this application. The entire mechanism is designed to withstand the stalled torque of the driving motor. Electrical limit switches and mechanical stops are provided to limit the tilting motion within the design range. The rockers and tilting mechanism must be designed to accommodate thermal expansion of the furnace structure.

In the second case, the furnace is tilted by two heavy-duty double acting hydraulic cylinders. The tilt cylinders are intermediate-trunnion-mounted on the furnace foundations through trunnion blocks. The piston rods are pin-connected to the back ends of the two rockers through heavy-duty self-aligning radial bronze bushings. Fabricated steel shrouds are provided to protect the piston rods from slag and metal splash during the deslagging operation.

Shell Structure—Furnace shells are cylindrical in shape with spherically dished bottom plates. Shell construction is generally of two basic types: the integral type, with the stays and yokes solidly welded to a one-piece plate cylinder; and the "cage" type, with loose sidewall plates to minimize shell distortion and possible rupture resulting from thermal expansion and contraction, and to facilitate repairs of damage from sidewall burn-throughs.

The latter shell structure consists of a heavy welded structural-steel cage of vertical buckstays and horizontal circumferential yokes, lined with rolled-steel plates. Below the sill line, the lining plate is welded to and made an integral part of the structure. The dished and flanged bottom plate is hung from the hearth lining plate through a circumferential bolted connection. Above the hearth line, the lining is divided into sectors to span between buckstays, and is loosely fastened to the cage structure. The cage type of shell construction is mostly used for the larger size furnaces which need greater shell reinforcement, thus increasing the possibility of distortion of an integral type shell from unequal temperature changes between the shell plate and the reinforcing members. The cage type of construction also facilitates the replacement of portions of the shell plate which are accidentally damaged by overheating, which occurs more often in high-powered furnaces.

If desired, the cage-to-rocker connections can be designed so that the shell can be readily removed and replaced with another shell. This removable-shell feature has certain advantages in reducing furnace downtime for relining by replacing the shell at the end of a lining campaign with another pre-lined shell. Furnace flexibility and economy can also be extended in the production of certain grades of alloy steel sensitive to contamination by using a special shell reserved for such grades. The removable-shell concept has also been used in carbon-steel mini-mills where high furnace availability is required. Because of the weight of a fully-lined shell, the practical application of this feature is generally limited to the smaller-size furnaces.

Where limited crane capacity prevents the use of removable shells, furnaces with split interchangeable sidewall shells have been used (Figure 18—5). The split is approximately 300 mm (1 foot) above the sill level and steps up over the spout; thus, the spout stays with the fixed bottom portion of the shell. The upper shell sidewalls are removed for rebricking, and a spare upper shell which has been rebricked off the furnace is installed in its place. The split shell thereby reduces downtime for relining and increases furnace productivity.

Some split shell designs divide the shell into three parts, making the slag line section of the wall replaceable as well. The hearth section can also be made re-

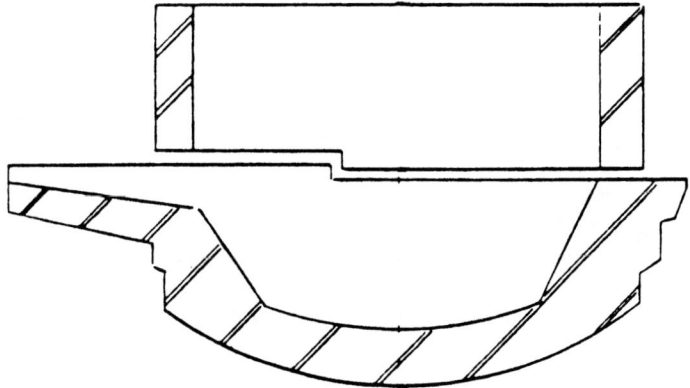

FIG. 18—5. Illustration of the removable, split shell design used in some electric steelmaking furnaces.

placeable, but is not as common because of the much greater lifted weight and the much longer normal life of the hearth lining as compared to the sidewall lining.

Another development in furnace shell design that increases productivity by increasing furnace availability is the application of water-cooling to those portions of the upper sidewall lining subject to the greatest wear, especially at the slag line. Various designs are used for these water-cooled elements, including: large hollow welded steel panels with internal baffling to direct the flow of cooling water; cast iron panels with pipes cast in place for water-cooling; and pipe-type panels with water circulation within the pipes. Panels installed at the slag line are usually made from copper. The elements are installed inside the furnace shell above the hearth line, and usually cover at least 70 per cent of the sidewall area. Panels are designed for rapid replacement on the furnace. The optimum surface area of an individual panel is between 2 to 3 square metres, (about 6½ to 10 square feet). Generally the hot face of these elements is designed to retain a thin layer of refractory material. This refractory material is maintained by periodic air gun application of refractory gunning material and/or splash from slag. Flanges and gaps between the refractory lining and the water-cooled panels are designed to prevent leaking water from entering the hearth refractories.

The provision of suitable water supplies for water-cooled linings is the major capital investment involved in moving to this technology. Large quantities of water are involved. The water must be of good quality and its temperature must be adequately controlled.

The use of water-cooled panels greatly reduces refractory brick and gunning material consumption, increasing shell lining life to 1 500 heats and above. Water-cooled panels also allow an increase of 10 to 30 per cent in the furnace volume, depending on the furnace and cooling system used. This results in larger charging volumes, a possible reduction in scrap charges, and/or less scrap leveling after charging. Because of the reduction of heat, water-cooled panels also improve working conditions around the furnace and

decrease the amount of stress and distortion on the shell.

Table 18—II relates the usual heat sizes to various shell diameters. The upper portion of each range applies to plain carbon grades of steel with minimal slag requirements, and the lower portion applies to types and grades of steel requiring larger slag volume. Although not listed, several furnaces about 10 and 12 metres or 32 and 38 feet in diameter have been built and are in operation.

The nominal molten steel capacity, i.e., hearth volume, of a furnace shell is established to a large degree by the shell diameter. The hearth usually has the general form of an inverted truncated cone and, accordingly, the incremental change in volume decreases rapidly with increasing depth. Considerable increases in capacity can be realized, however, by banking the door opening and running the metal line above the sill. However, this practice has its practical limitations from a safety standpoint.

Scrap capacity of a furnace is a function of shell volume and scrap density, and for a given shell diameter and scrap density, is roughly proportional to the height of the shell above the sill line. The standard shell heights for the various sizes of Heroult furnaces are established to provide a scrap capacity of slightly more than 60 per cent of the nominal molten-steel capacity at a scrap density of 60 pounds per cubic foot. With scrap of this average density or higher, the total metal requirement for a heat, including melting loss, can be top-charged into the furnace in a maximum of two bucket loads, consisting of the initial charge and one back charge. Charging time is thus minimized. Design of the shell height can be varied within limits for a particular installation to suit unusual conditions such as low scrap density or overhead clearance restrictions.

Shell Bottom—Figure 18—6 shows three types of subhearth construction. The shell bottom can be laid first with a layer of clay, magnesite or silica brick as required, over which a ramming mix is laid. When the stadium type of construction is used for the subhearth, successive courses of brick are set back to provide

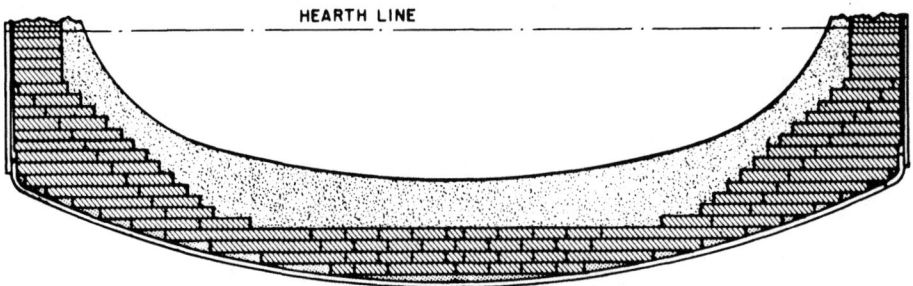

STADIUM-TYPE SUBHEARTH CONSTRUCTION

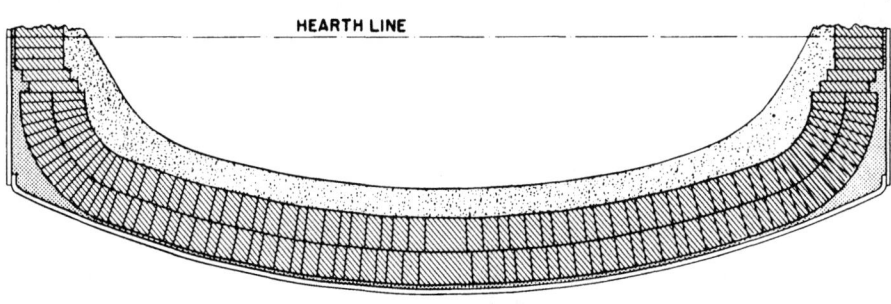

INVERTED-ARCH TYPE SUBHEARTH CONSTRUCTION

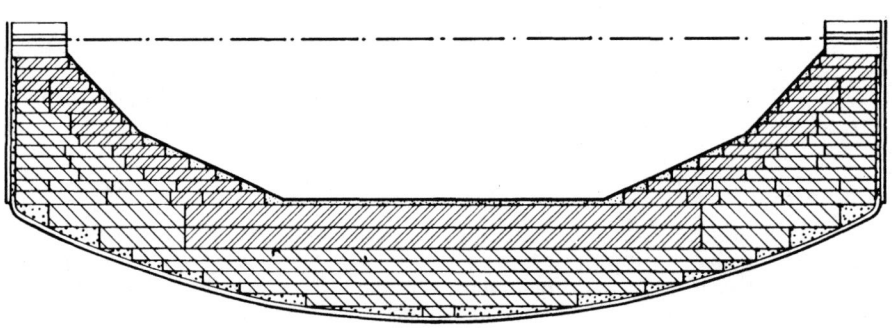

BIG-KEY, ALL-BRICK SUBHEARTH CONSTRUCTION WITH NO RAMMING MIX

FIG. 18—6. Schematic representation of vertical cross-sections of the three types of subhearth construction employed in the lining of dished-bottom electric-arc steelmaking furnaces. (Courtesy, Harbison-Walker Refractories Company.)

ledges for the material that will form the working bottom or hearth, as shown in the first diagram. In the second diagram, the inverted-arch type of subhearth construction is shown where the subhearth is constructed of shaped bricks that follow the contour of the bottom. The third diagram shows the more modern big-key, all-brick subhearth construction, where no

ramming mix is required because the hot face brick is impregnated with tar. Because larger brick is used and no ramming mix is necessary in this type of subhearth construction, there is considerably less down-time required for hearth replacement.

When magnetic-induction stirring equipment is applied to a furnace, the bottom dish must be made of

Table 18—II. Heroult Furnace Sizes, Capacities and Transformer Ratings

Inside Shell Diameter		Nominal Steel Capacity		Usual Transformer Capacity (MVA)
mm*	ft and in.	Metric Tons	Net Tons	
2135	7'0"	2.7 to 4.5	3 to 5	2 to 3
2440	8'0"	4.5 to 7.3	5 to 8	3 to 5
2745	9'0"	7.3 to 10.0	8 to 11	4 to 6
3050	10'0"	10.0 to 14.5	11 to 16	5 to 9
3355	11'0"	13.6 to 19.1	15 to 21	7.5 to 12.5
3660	12'0"	18.1 to 25.4	20 to 28	10 to 15
4115	13'6"	22.7 to 36.3	25 to 40	15 to 20
4570	15'0"	27.2 to 45.4	30 to 50	18 to 25
4875	16'0"	36.3 to 54.4	40 to 60	20 to 30
5180	17'0"	45.4 to 63.5	50 to 70	25 to 40
5485	18'0"	54.4 to 77.1	60 to 85	30 to 50
5790	19'0"	68.0 to 90.7	75 to 100	35 to 55
6095	20'0"	77.1 to 108.9	85 to 120	40 to 65
6705	22'0"	117.9 to 154.2	130 to 170	50 to 75
7315	24'0"	163.3 to 190.5	180 to 210	60 to 90
7925	26'0"	199.6 to 226.8	220 to 250	75 to 115
8535	28'0"	244.9 to 281.2	270 to 310	90 to 135
9145	30'0"	290.3 to 326.6	320 to 360	100 to 150

*To nearest 5 millimetres.

material having low permeability to permit penetration of the magnetic flux produced by a rotating field and to prevent overheating of the furnace shell by induced eddy currents. Austenitic stainless steel is used for this purpose. For the smaller-size furnaces, for which the bottom dish can be formed and shipped in one piece, UNS S30400 (AISI Type 304) steel can be used. For the larger-size furnaces requiring field welding of the plate, however, a stabilized steel such as UNS S32100 (AISI Type 321) is used to avoid precipitation phenomena at the welds and the consequent increase in permeability which would result in excessive localized induction heating in the intense magnetic field generated by the stirrer.

Shell Openings—Openings are provided in the side of the shell structure for the rear slagging and working door, and a tap hole. In past years another side door for hand-charging was a standard feature on most furnaces and had its greatest utility in making high-alloy steels, which require more refining work than other grades. It is now an optional feature and, in order to reduce heat loss, is not usually provided on furnaces used for making carbon steels. Other openings can also be provided in the shell wall for such functions as: oxygen, lime and carbon injection; oxy-fuel burners; and inspection and fettling of the walls. All shell openings are structurally reinforced and water-cooled where necessary to maintain shell integrity.

Door openings are equipped with water-cooled doors and door frames. The door fits into a recess in the front surface of the frame, which is sloped back a few degrees from the vertical plane so that the door will tend to lay against the frame. Thus, the door maintains a reasonably tight closure irrespective of the condition of the surface of the sill on which the door rests in the closed position. Loose-fitting, easily-removable guide plates attached to the shell structure are also provided to hold the door against the frame in both open and closed positions. Either an electro-mechanical winch mechanism driven by an alternating-current motor or a hydraulic mechanism is used to raise and lower the door through alloy-steel chains which are connected by pins to the door to facilitate quick door changes. Although hydraulic mechanisms may be more economical, they may drift because of fluid leakage, and do not allow as precise a control as electro-mechanical mechanisms.

The rear door is designed for a 75 to 150 mm (3 to 6 inch) thick refractory lining on the hot face. The frames may be equipped with integral water-cooled lintels to support the portion of the refractory sidewall lining above the door opening. These lintels may be flat, facilitating the installation of the sidewall lining above the door opening by making sprung arches unnecessary; or the lintels may be curved to serve as a form for the construction of a sprung arch. In either case, the lintel serves to cool and protect the refractory above it. One advantage of the curved lintel over the flat one is that a defective frame can be removed and replaced with much less chance of lining damage. The legs of the door frame usually are not extended down to the sill, but are, instead, supported on refractory pedestals 150 to 230 mm (6 to 9 inches) above the sill to both avoid damage from molten slag or metal and to facilitate removal of the frame by knocking out these supports.

Trough-shaped, cast-iron sill plates and slag aprons are usually provided at the door opening, although some operators prefer refractory lined sills and aprons.

A deep tapping spout with a replaceable end section is provided, and is bolted to the shell structure at the tap-hole opening. The spout is usually inclined upward from the horizontal to decrease the height from operating-floor elevation to the floor of the tapping area, which accommodates the ladle under the spout when the furnace is in the full forward-tilt position. When the furnace is to be used for a process which requires slag-free tapping, it is desirable to operate with a wet heel practice (approximately 20 per cent of the heat is kept in the furnace after tapping) which facilitates tapping slag-free. One of three different tapping schemes is normally employed for tapping slag-free:

1. A submerged tap-hole can be used, with a tap-hole below the metal line which permits tapping the molten metal first.

2. An eccentric bottom tapping hole can be used which provides for tapping metal first. The eccentric bottom tapping furnace has additional advantages--it only requires approximately 10 degrees forward tilt to drain the furnace; a greater use of water-cooled wall panels can be made at the front of the furnace; and the tap stream is a much tighter stream, thus reducing nitrogen pickup.

With these first two tapping schemes, however, it is desirable to have a fast backtilt speed to stop the metal flow and prevent slag carryover.

3. A slide gate can be used to control the metal flow from the furnace. The slide gate permits stopping the metal flow in about one second. No spout runner is necessary with a slide gate and the tap stream is a much tighter stream, thus reducing nitrogen pickup. The wet heel practice is not necessarily required when a slide gate is used.

Swing Platform, Side Frame, and Roof Lift Arms—The furnace superstructure, located at the side of the shell adjacent to the electrical-equipment vault, consists of a movable structure mounted on wheels running on a curved track on the tilt platform. Incorporated with this structure are: the two roof lift arms which are cantilevered out over the shell; a platform commonly called the "roof platform" interconnecting the outboard ends of these arms and providing access to the electrode holders; the "side frame" in which the electrode-positioning masts operate; and the swing platform on which the electrode-positioning, roof-lift and swing machinery are mounted.

Since parts of the superstructure are near the main power conductors that carry the large currents across the top of the furnace to the electrodes, this structure must be designed to avoid excessive heating from hysteresis, eddy currents and circulating currents induced by intense magnetic fields created by the currents flowing in the conductors. Non-magnetic materials and/or water-cooling, and electrically insulated joints are used in the design of the superstructure to prevent overheating by induction.

Roof—The furnace roof consists of a welded-steel, water-cooled circular ring forming a circumferential retainer for a dome-shaped refractory structure. The steel roof ring is essentially a hollow right triangle in cross-section, with the hypotenuse forming the interior side of the ring and serving as the skewback for the refractory dome, thus obviating the need for special skewshaped brick in the construction of the dome. Water is circulated through the interior of the hollow ring for cooling both the ring and the adjacent roof refractories. The ring is slightly larger in outside diameter than the inside diameter of the furnace shell so that the weight of the roof is carried on the steel shell structure

FIG. 18—7. A clam-shell scrap bucket is shown as it top-charges a Heroult electric-arc furnace. The furnace has a swing-type roof which has been moved aside to permit top-charging.

rather than on the furnace-sidewall refractories. The ring is equipped with lifting and stacking lugs and is reinforced to resist warping and distortion from lifting stresses. Retainers are provided on the top of the shell structure to prevent the roof from sliding when the furnace is tilted. At least two roof rings are necessary for each furnace so that, at the end of a campaign, the old roof can be replaced with a new one with a minimum of furnace downtime. Some shops carry more than one spare roof per furnace to provide for the possibility of a premature roof failure. Water-cooled roofs wherein only the center delta is refractory are being used in increasing numbers.

Roof-Removal Mechanisms—The furnace roof is removed from the shell for top-charging by two separate mechanisms: one to lift the roof off the shell; and the other to swing the roof to one side to clear the top of the shell (Figure 18-7).

The roof-lifting mechanism consists of either hydraulic cylinders or a dual drum winch driven by an alternating-current motor and equipped with alloy-steel chains running out over the furnace on sheaves (mounted on the roof-lift arms) and attached to the roof ring at four points. The maximum lift varies from 200 to 300 mm (8 to 12 inches), depending on furnace size. Electrical limit switches control the operation of the mechanism in both "raise" and "lower" directions. Electrical interlocks are provided to prevent raising of the roof unless the furnace is locked in the level position.

The roof-swinging mechanism consists of either a hydraulic cylinder or a direct-current motor driving, through a speed reducer, one of the wheels on which the superstructure is mounted. The swing motion in both directions is controlled both by electrical limit switches and mechanical stops. Electrical interlocks are provided to prevent swinging the roof unless the roof and the electrodes are in the fully raised position and the furnace is locked in the level position.

A furnace may be designed for swinging the roof forward (over the spout) or backward (over the operating floor). Forward swing is the generally preferred design. However, overhead-clearance limitations in some two-aisle installations, i.e., where the furnace is located in one building aisle and taps into an adjacent aisle, or problems of charging-crane access and crane-operator visibility in certain single-aisle arrangements, may require the back-swing design. These situations are commonly encountered when installing a modern top-charged electric-furnace in an existing open-hearth building or in an existing electric-furnace shop designed for door-charged furnaces.

Pneumatic- or hydraulic-cylinder-operated locks are provided to: (1) lock the superstructure in the normal "swung in" position; (2) lock the furnace in position for charging; and (3) provide supplemental support for the main platform to compensate for the unbalanced loading imposed by the superstructure in the "swung out" position.

Electrode-Positioning Mechanisms—The electrode-positioning mechanisms serve to raise and lower the electrode columns, or "strings" as they are commonly called, to maintain an arc at the power level selected for the particular stage of the heat, and to raise the electrodes to clear the top of the shell when the superstructure is swung for charging. The range of electrode travel, or "electrode stroke," is dependent on various factors such as hearth depth, shell height, required "bottom approach" and any special requirements for the particular installation. "Bottom approach" is the minimum distance from the tip of an electrode string to the bottom of the hearth, with the positioning mechanism in the fully lowered position, at which the major electrode slip, i.e., slipping a joint in the electrode string from about 150 mm (6 inches) above the holder to about 150 mm (6 inches) below the holder, can be made and still have sufficient stroke to raise the string clear of the shell for top-charging.

Each of the three electrode-positioning mechanisms consists of: (1) a steel mast structure to support the electrode holder, the main power conductors and the electrode-clamping mechanism; (2) a cable winch driven by an electric motor or a hydraulic cylinder to move the mast structure in the vertical direction; and (3) a mast-guiding system to maintain proper vertical alignment.

The masts are of the "inverted L" type consisting of a vertical stem, to the top end of which is affixed a horizontal arm extending out over the top of the furnace shell to support an electrode holder. The arms are attached to the top of the stems through clamp-type connections incorporating heat-resistant dielectric insulation which effects the electrical separation of the main power circuit components from the furnace structure. Similarly, in some designs for large furnaces, the upper part of the center-phase mast stem is of low permeability steel and/or is water-cooled. This design concept for insulation has the advantage of simplicity and a commensurate high degree of reliability with the insulation located away from the high-temperature area.

Each mast arm is constructed with an internal large-diameter pipe extending the full length of the arm. The mast arm is cooled by running water between the pipe and the outside arm structure. The electrode-clamping mechanism is mounted inside the pipe to protect it from heat and dirt.

All drive mechanisms actually drive the masts in the raise direction only. The force of gravity is used to move the masts in the downward direction, but the rate of acceleration and velocity of movement is controlled by the drive system. In hydraulic electrode-positioning, two types of systems are available: double-acting by trunnion mount; and single-acting by cylinder mount. In hydraulic electrode-positioning, trunnions or cylinders drive each mast and are incorporated in some designs into the mast stem structure. Control of the flow of fluid to and from the trunnions or cylinders is accomplished variously by valves of several types operated by electric solenoids and variable-speed motor-driven hydraulic pumps. The speed of response can be extremely fast with hydraulics, however; sometimes the drives must be desensitized in order to prevent mast structure oscillations. Also, high maintenance is required to prevent fluid leaks and the potential danger of mixing such fluids with the hot metal. Yet in contrast to electro-mechanical drives, hydraulic drives are intrinsically safe so that no mast-catching system is necessary.

Electro-mechanical drives are usually of the motor-driven cable-winch type, of which there are two different kinds. The first kind is a variable-speed reversing direct-current motor, driving the cable directly through a speed reducer, with the motor energized from a variable-voltage reversing-polarity D.C. power supply. Through SCR control, the motor is able to drive in both directions. The speed of response is based on the moment of inertia of the system. Also, through the use of low efficiency gear boxes, the electrode can maintain its position with low current versus the large current necessary with high efficiency gear boxes. The second type of motor-driven system consists of a constant-speed alternating-current motor that drives the drum in an upward direction only through a magnetic eddy-current coupling and speed reducer, and relies on gravity for its downward motion. The eddy current coupling is a heat source particularly when the electrode has little movement and this heat must be dissipated through either air or water. The coupling is energized from a variable-voltage D.C. power supply; and the direction, acceleration and velocity of mast travel is controlled by the amount of coupling slippage as a function of the magnitude of coupling energization.

When electro-mechanical drives are used, the electrode masts are supported by multi-part cables running over sheaves mounted on the bottom ends of the mast stems. The cables are attached at one end to the electrode-winch cable drums, and dead-ended at the other end to devices which prevent the free fall of the masts in case of winch-cable breakage. Each mast is guided in vertical travel by two nests of rollers, each bearing on all four surfaces of the square-section mast stem. One nest is located at the top of the side frame and one at the bottom. This guiding system provides a high degree of lateral and torsional rigidity with minimal friction and wear.

A motor brake is provided to hold the mast in position when the motor is deenergized. Since the motor armature, brake wheel, and couplings constitute about 90 per cent of the electrode-positioning system inertia, these components should be selected for the lowest possible inertia consistent with practical design considerations to provide fast response of the electrode-positioning system. The drive motor is conservatively rated to provide sufficient thermal capacity for the repeated rapid start-stop operation required for good control of the arc. Winches should therefore be designed for maximum electrode-travel speeds in the range of 0.03 to 0.10 metres per second (6 to 20 feet per minute), depending on the requirements of the particular installation.

The electrode-winch cables are made from high-strength, extra-flexible, steel-core wire rope. These cables are subject to wear as is normal in any hoisting operation, and should be replaced on a regular maintenance schedule, based upon operating hours. It is important that the drums and sheaves be of large enough diameter relative to the cable size to avoid excessive wear. A device is provided for each of the three masts to prevent their free fall in the event of cable breakage. This device consists of a toothed, pivoted, spring-loaded pawl, mounted on the side frame structure, and a similarly toothed rack welded to the mast over its full travel range. The winch cable is dead-ended to the pawl so that under normal operating conditions the pawl is held out of engagement with the rack by the pull of the cable. In case of cable breakage, the pawl is spring-driven into engagement with the rack, stopping the free fall of the mast. Since the pawl cannot operate instantaneously, it must be mounted on suitable shock absorbers to dissipate the energy of impact in stopping the falling mast.

Electrode Holders—The electrode holders, bolted to the outboard ends of the mast arms, serve to mechanically support the electrode columns and to transfer the electric current from the "bus tubes," which run across the top of the furnace, to the electrodes (Figure 18—8). The holders are made of cast high-conductivity copper with cast-in pipe coils for water cooling. Two large electrode contact pads, machined to suit the diameter of the electrode, are an integral part of the holder body. For a given size of electrode, the holder depth, i.e., the length of the contact pads, is designed to suit the magnitude of electrical current for the particular installation.

The electrodes are held in the holders by spring-loaded clamping mechanisms which act to push the electrodes onto the holder contact pads. The mechanisms are mounted inside the water-cooled mast arms to protect them from heat and dirt. The mechanisms are designed to provide a clamping pressure of sufficient magnitude to effect low contact resistance between the holder and the electrode, consistent with the crushing-strength limitations of the electrode material.

Short-stroke pneumatic or hydraulic cylinders are mounted on the back end of the mast arms to compress the clamping springs and release the electrode strings. These cylinders are operated by valves located at the furnace operator's control station. "Slipping" of the electrode strings is usually accomplished by supporting the string from the overhead crane, lowering the string to the desired position, and reclamping. An interlock is provided to prevent opening the clamps when power is on the electrodes. Alternatively, some operators will slip electrodes on the scrap. Particular care must be exercised, however, when slipping electrodes on scrap to prevent electrode breakage.

Bus Tubes—Heavy-wall water-cooled copper bus tubes are used to carry the electrical current across the top of the furnace to the electrode holders. These tubes are mounted on insulated non-magnetic supports, remote from the furnace structure to minimize heating and losses from magnetic induction effects. On modern furnaces these tubes are arranged in a triangular configuration to minimize impedance unbalance among the phases.

Cranes—If the plant is designed for the use of hot metal, or uses the practice of refurnacing employed by many shops that produce stainless steels, a large, four-girder, two-trolley crane is required for handling the ladles in the charging aisle. The capacity of this crane may be as large as the heat size, and it can also serve as the utility crane for the charging floor.

Cranes in the charging bay should have ample capacity for handling loaded charging buckets of whatever size is necessary. Top-charging buckets are normally sized to have approximately 85 per cent of the volumetric capacity of the furnace. With scrap having an average density of 800 to 960 kilograms per cubic metre (50 to 60 pounds per cubic foot), 50 to 60 per cent of the nominal capacity heat weight of the furnace can be accommodated in one bucket, and crane capacities must be chosen accordingly.

OTHER DESIGN CONSIDERATIONS

The structural and mechanical design of modern electric-arc furnaces should take into account the possible effects of the application of certain auxiliary equipment now or in the future, as well as possible changes in process and operating techniques in this time of rapidly changing technology. These possible changes and/or greater use may include: increased power levels; pellet- and fragmentized-scrap charging equipment; lime- and carbon-injection equipment; water-cooled or consumable-pipe oxygen lances; oxy-fuel burners; water-cooled electrodes; and higher density refractories. The greater use of such technology can have an appreciable effect on the loads imposed on the furnace structure, operating mechanisms and location of the center of gravity, as well as on the actual config-

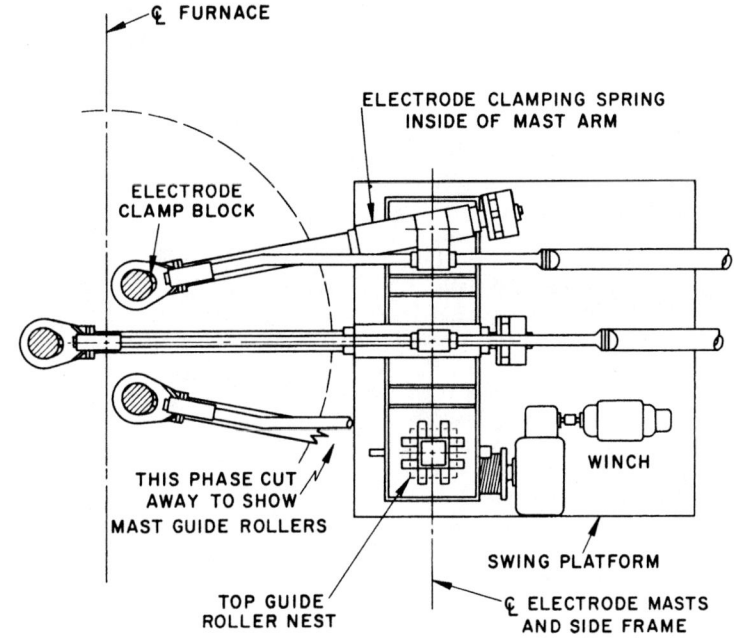

PLAN

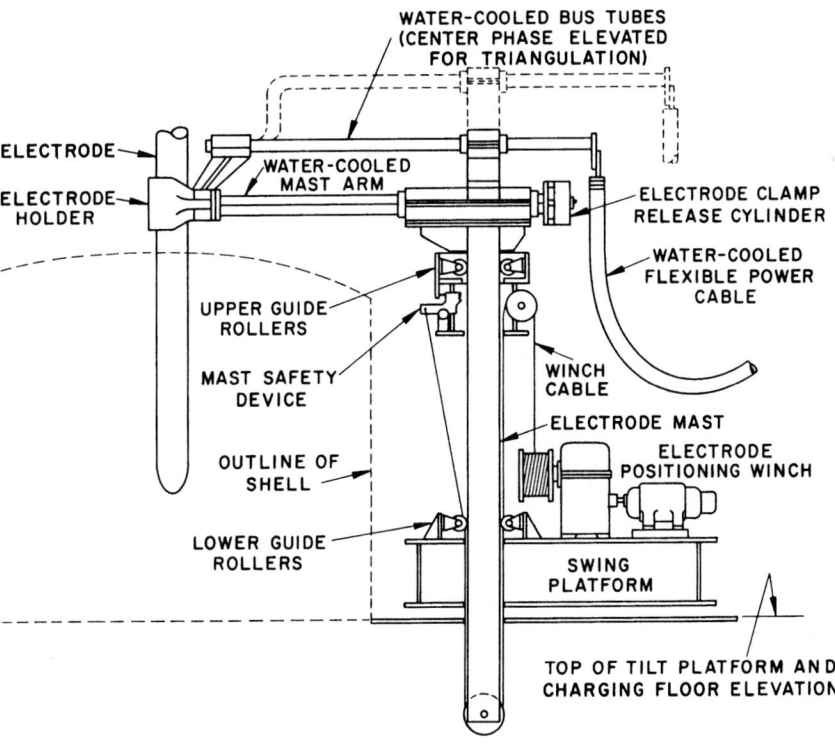

ELEVATION

FIG. 18—8. Schematic arrangement of the electrodes, their supporting masts, and the electrical power leads for an electric-arc steelmaking furnace.

uration of the furnace structure.

ELECTRODES

To the steelmaker, there are two kinds of electrodes. Graphite electrodes, made from petroleum coke at high heat to form artificial crystalline carbon, are used in electric steelmaking furnaces. Carbon (amorphous) electrodes are widely used in submerged-arc furnaces for the manufacture of ferroalloys, silicon metal, aluminum, calcium carbide, phosphorus, and so on. Both graphite and carbon electrodes are used in electric-arc furnaces because of their infusibility, chemical inertness, electrical conductivity, mechanical strength, and resistance to thermal shock.

Carbon electrodes can be divided into two types: one, made of calcined petroleum coke, is used in aluminum reduction; the other, based on calcined low-ash anthracite coal, is suitable for the other uses listed. Carbon and graphite electrodes are essentially the same chemically, but differ widely in their purity and electrical and physical properties.

In manufacture, both carbon and graphite electrodes start with mixtures of raw materials properly proportioned with a suitable bonding material, such as hot pitch or tar, and extruded or molded while still hot into "green" shapes. These are cooled, packed in powdered petroleum coke in furnaces, and baked at the desired temperatures. Carbon products are usually gas-fired in kilns to approximately 850°C (1560°F). Graphite products undergo this same gas-firing, followed by repacking in electric-resistance furnaces, where they are heated to temperatures above 2800°C (5000°F). This second treatment in the electric-resistance furnace results in the formation of graphite (crystalline carbon) and, through volatilization, removes most of the impurities. Electrodes of both types, after the foregoing processes, are turned to uniform diameter, then bored and tapped—usually on both ends.

The addition of pitch-impregnation in autoclaves to the manufacturing process for graphite electrodes has provided the higher density, higher current-carrying capacity electrodes required for ultra-high-power operation of arc furnaces. Such impregnated electrodes are termed premium grade.

Carbon and graphite electrodes vary widely in physical shape, dimensions and properties. These must be selected carefully, depending upon intended usage. A general comparison of the two types is shown in Table 18—III. Where a single value is given it may be taken as the average for a variety of sizes and shapes.

Carbon electrodes cost less than one-half as much per unit weight as graphite electrodes. However, the field of application for carbon electrodes is limited by their capacity for carrying electric current and to those applications requiring a large area of working surface. The electrical conductivity of graphite electrodes is about six times that of carbon electrodes so that a graphite electrode of less than one-half the cross sectional area of carbon will carry the same current.

Because of their greater current-carrying capacity, graphite electrodes are used almost exclusively in electric steelmaking furnaces. Figure 18—9 indicates the typical current-carrying ranges for graphite electrodes

Table 18—III. Comparison of Carbon (Amorphous) and Graphite Electrodes.

Properties	Large Carbon Electrodes	Graphite Electrodes
Specific resistivity		
ohm-cm	40×10^{-4}	5.75×10^{-4}
ohm-inches	1.6×10^{-4}	2.26×10^{-4}
Apparent Density		
grams/cubic cm	1.58	1.68
pounds/cubic foot	98.6	105
Flexural Strength		
megapascals	4.96	9.31
pounds/square inch	720	1350
Modulus of Elasticity		
megapascals	6.62×10^3	8.28×10^3
pounds/square inch	0.96×10^6	1.2×10^6
"Practical" oxidation point	370°C (700°F)	480°C (900°F)
Size ranges		
diameter		
millimetres	889-1397	31.75-711
inches	35-55	1¼-28
Length		
millimetres	1524-2794	610-2794
inches	60-110	24-110
Weight		
kilograms	2722-7030	0.9-1905
pounds	6000-15500	2-4200

of various diameters in direct-arc applications.

As an electrode is consumed from the bottom of a column in arc furnace operation, a new one is added to the top of the column by means of a threaded connecting nipple. Graphite nipples are generally used for joining both carbon and graphite electrodes. Tapered nipples and sockets are used to join graphite electrodes. The tapered design facilitates the joining of two sections, which is particularly important when this work

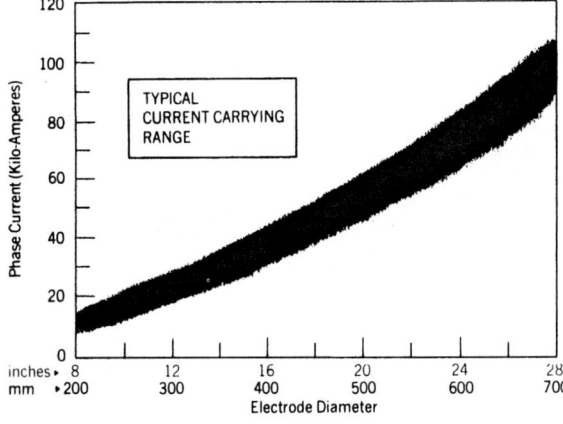

FIG. 18—9. Typical current-carrying ranges of graphite electrodes in open-arc applications for electrode diameters ranging from 200 to 700 mm (8 to 28 inches). (Courtesy, Union Carbide Corporation, Carbon Products Division.)

must be done on top of the furnace.

Pitch reservoir nipples, i.e., graphite nipples having a small amount of pitch put in a small hole drilled into the threaded body of the nipple, were developed to assure the maintenance of tight joints. When the joint becomes hot, the pitch melts and flows into the threads where it bakes in place, securing the joint. The disadvantage of using such nipples is apparent if and when it becomes necessary to loosen a joint, for example, to remove a broken or split stub.

Securing and maintaining tight electrode joints is very important to successful furnace operation. Loose joints cause overheating of the electrode due to increased electrical resistance at the area of contact which, together with the reduced mechanical strength of the electrode column, causes a large percentage of the electrode breakage that is experienced. Since electrode consumption constitutes one of the largest items of cost in electric-furnace operation, the importance of tight joints is obvious. Much work has been done on this problem, and special wrenches have been devised to attempt to assure proper tightening torque. The tightening torque recommended for 500 mm (20 inch) diameter electrodes, for example, is approximately 1355 N-m (1000 lb-ft). Because of the obvious difficulty in consistently obtaining these high tightening torques when adding electrode sections on top of a hot furnace, some electric-furnace plants have been designed with off-furnace electrode make-up facilities. In these facilities, the whole electrode string is removed from the furnace with the overhead service crane and set into a stand, and a new string of proper length is placed in the electrode holder on the furnace. A new section can then be added to the removed string, taking the time and care necessary to assure a proper joint. This string is then used to replace another string on the furnace when an electrode addition is required.

Ultra-high-power operation also requires consideration of matching the "hand" of the joint threads (left or right hand) to the electrical phase sequence so that the electro-magnetic interaction among the electrode strings tends to tighten rather than loosen the joints.

Electrode strings and sections are handled with a special steel lifting nipple that screws into the top electrode socket. The nipple has a threaded pin connecting it to the hook ring. The thread of the pin has the same pitch as the electrode socket thread. Thus, as an electrode section is turned to screw it into connection with the string, the lifting nipple lowers at the same rate, eliminating the need for lowering the section with the crane hoist and the consequent possibility of broken socket threads. The lifting nipples are removed when the electrode strings are in place on the furnace since the difference in thermal-expansion characteristics of the steel nipple and graphite electrode may cause the electrode to split at the socket. A special lifting nipple made of a material with thermal expansion characteristics similar to graphite has been developed to solve this problem.

ENVIRONMENTAL CONTROLS

Considerable effort has been expended to minimize the effect of steelmaking processes on the surrounding environment. Particular emphasis has been placed on minimizing noise levels and protecting the quality of air and water. However, before making modifications to existing plants or installing new plants, it is usually necessary to obtain construction and operating permits, to ensure that the plans completely comply with current and future governmental standards.

It is apparent that these regulations are becoming more stringent and are entailing a high level of capital spending. Nevertheless, in few cases can such expenditure be justified in narrow commercial terms. In many plants, where old equipment is involved, it is not economical to equip them with modern pollution abatement facilities. Therefore, they must either be replaced with modern facilities or be shut down and abandoned. The enormous amount of capital expenditure required to satisfactorily meet environmental regulations has undoubtedly retarded the rate of modernization and expansion of the American steel industry.

It is estimated that 10 to 15 per cent of the total capital investment for a new, fully integrated steel plant is spent on environmental controls. The electrical power requirements to operate air and water pollution control equipment in a modern integrated steel plant may exceed 100 kWh/ton. However, it is difficult to compare the cost of maintaining and operating environmental control equipment in different areas due to the lack of uniformity in regulations.

Due to the high cost of installing and operating environmental control equipment, many companies are installing electric-arc furnaces rather that the conventional coke plant, blast furnace, and basic oxygen furnace integrated facilities.

Water Controls—Legislation has established standards to define the maximum levels of pollutants which may be discharged from a source, after the most practical and economically viable technology has been applied. Though the range of permissible effluent pH values is the same for both new and existing plants (6.0 to 9.0), these standards tend to differ. For an existing plant, .0312 constitutes the maximum value of total suspended solids for any one day. This value drops to .0104 for the maximum average value of 30 consecutive daily values. For a new plant, total suspended solids must not exceed .0156 for any one day, or an average of .0052 for 30 consecutive days. New plants must also abide by flourine and zinc regulations. Flourine must be kept at or below .0126 for any one day, and .0042 for the average of 30 consecutive days. Zinc levels must not exceed .0030 for any one day, or .0010 for the average of 30 consecutive days.

The use of water-cooled linings increase the water requirements for electric-arc furnace operation and, depending upon the specific design of the water-cooling system, may require more elaborate chemical treatment of the water. In all cases, however, the water systems should be of the closed recirculating design which meets current environmental standards. Although water-cooled linings significantly increase the water requirements for an electric-arc furnace, the typical hourly value of water consumption remains small compared to that needed for a basic oxygen furnace. A commitment to the attainment of a high level of water quality should always be made, however, because loss of production arising from stoppages caused by the failure of water-cooled components can be critical, particularly when the furnace is linked to a continuous-casting machine.

Noise Controls—Regulations set by the United States Occupational Safety and Health Administration (OSHA) allow a 90 dB exposure level for eight hours per day. Revisions to noise standards are under consideration.

Considerable interest has been expressed in revising the noise regulations relating to the steel industry. To determine the actual noise levels which exist in electric-arc furnace shops, these shops have been closely studied and noise measurements have been taken under varying operating conditions. In electric-furnace operations, the primary source of noise is the arc, and the highest noise intensity occurs during the early melt-down period as a result of rapid arc fluctuations and constant arc restriking. Noise is emitted from the furnace through all the furnace openings such as electrode ports, fume ventilation ducts and furnace slag doors.

Most plants require personnel to use a form of ear protection, such as ear plugs or ear muffs, during peak periods of noise generation. These protective devices, when properly worn, do prevent aural damage. Some methods have been advocated for actually reducing noise levels that arise during plant operation.

To reduce noise levels, physical devices have been constructed in most plants. Personnel enclosures such as control pulpits are acoustically treated to suppress noise levels. These pulpits are designed for personnel who are continuously involved in furnace operation. Further reductions in noise level may be achieved by installing vertical sound barrier panels within the roof trusses located above a furnace. These panels serve to prevent deflection of noise from the furnace to areas such as the teeming aisle and roof reline area. In addition to locating barriers in the roof, another attempt to reduce noise levels has involved the installation of sound-absorbent plates on the building walls opposite the furnace charging side.

Fume Control—Almost without exception, the design of every proposed new electric-arc furnace installation includes consideration of an acceptable fume-control system. However, two primary problems have been encountered in the development of fume-control systems for electric-arc furnaces:

1. Efficient collection of furnace fumes often detrimentally affects the metallurgical conditions within the furnace. The collection of furnace fumes also requires a high level of maintenance, which often interferes with normal furnace maintenance and repair.
2. The low-cost, low maintenance methods used to efficiently clean the collected furnace gases in small non-integrated plants and foundries cannot be applied in large integrated plants.

Of the two problems, the one dealing with collection is the more difficult. Various means have been used for fume collection, including: hoods mounted directly on the furnace; hoods mounted separately over top of the furnace; offtakes that apply suction to the furnace directly through an opening in the roof or sidewall; a variation of the direct offtake, called a "snorkle," that collects fumes by stack effect through an opening in the furnace roof; and total evacuation of the building in which the furnace is housed. The hood on top of the furnace appears to offer the best functional advantage from the standpoints of efficiency of collection and minimal effects on metallurgical conditions but, from the standpoint of maintenance, appears to be the least desirable.

For the cleaning of gases emitted by the electric-arc furnace, various devices have been used, including: wet centrifugal, dynamic, orifice and venturi types; dry electrostatic precipitators; and dry centrifugal and cloth filter types. Electric-furnace gases have the characteristic of being very dry with very high electrical resistivity and, consequently, require conditioning for effective cleaning by electrical precipitation. Such gases carry a widely varying dust loading of submicrometre (submicron) mean particle size, and the cloth-type filter appears to be the most efficient and economical means of cleaning them. The status of the cloth filter for this application has been particularly enhanced by the development of synthetic and glass-fiber cloths for filter media.

Disposal of collected dust presents a third problem in fume control. Attempts to recharge the dust into the electric-furnace and to sinter the dust for charging into the blast furnace have been unsuccessful. Collected dust currently is being dumped, with some attempts to agglomerate it in pug mills to prevent objectionable dusting from the dumps. The value of the alloy contents in dust from alloy-steelmaking operations, however, indicates the desirability of developing an economical recovery process.

SECTION 4

ELECTRICAL SYSTEMS AND EQUIPMENT
OF THREE-PHASE DIRECT-ARC
ELECTRIC-FURNACES

Section 4 describes the electrical systems and equipment required in direct-arc electric-furnace installations. The **first subsection** discusses the utility system power supplies and services required for electric-furnace installations. The **second subsection** describes the electrical configurations and components of the five possible types of primary power systems for direct-arc electric-furnaces. The **third subsection** describes the arc and electrode regulators. The **fourth subsection** presents a refractory erosion index. Finally, the **fifth subsection** discusses the electrical energy requirements of these furnaces. This last subsection includes a theoretical heat chart that illustrates the power consumption and tap-to-tap time of a typical operation in a three-phase direct-arc electric-furnace.

UTILITY SYSTEMS

When planning to install electric-arc furnaces, one of the first things to check is the utility company's ability to supply the load and under what conditions and terms the company can do so. The furnace designer or consulting engineer, utility company, and owner should meet in the early planning stages to assure a mutual understanding of all requirements and restrictions. Technical features that should be reviewed include existing system service voltage, short-circuit MVA, voltage variations, frequency variations and harmonics plus permissible voltage variations, flicker, load swings, power-factor correction, demand load, period, and the economic aspects of the billing method and rates.

The electrical design of a modern arc furnace installation involves a study encompassing the complete power circuit, from the utility company's generators to the arcs in the furnace and the selection of suitable circuit components as required to put the desired power into the furnace at optimum conditions and, if required, to restrict cyclic voltage changes (flicker) to acceptable limits at some designated point in the power system and to correct power factor if desirable.

The characteristics of the utility company's primary power services generally used for arc furnace installations range from 11 000 to 138 000 volts, with short-circuit capacities from 250 MVA to 10 000 MVA and higher. The utility company's service voltages between 11 000 and 46 000 normally are applied directly to the furnace transformer. Utility company voltages higher than 46 000 have normally been stepped down before being applied to the furnace transformer. However, the future use of higher primary voltages directly on the furnace transformer, particularly for large ultra-high-powered furnaces, is probable with the continued development of furnace transformers, switchgear and cables.

To reduce the surges on the utility system, someone will occasionally suggest the use of a higher-impedance step-down transformer between the utility system and the high-voltage furnace bus. This reasoning is completely erroneous, as the furnace builder would simply select a higher no-load furnace voltage to maintain the desired arc characteristics. This, if anything, would produce slightly larger swings between no-load and short circuit.

PRIMARY POWER SYSTEMS

There are many different possible arrangements of one-line diagrams for electric-arc furnaces. Figure 18—10 shows five basic arrangements which cover the most commonly used systems. In order to simplify the one-line diagrams, no attempt has been made to show all the electrical equipment normally supplied.

Power System No. 1 (Off-Load Furnace Transformer)—This arrangement uses an off-load furnace transformer in which the primary winding normally is extended and tapped at four, six or eight positions. The primary can be connected in delta or wye through a no-load delta-wye switch. The secondary is always permanently connected in delta. The secondary will have multiple coils connected in parallel due to the high secondary current required. A current transformer is placed in one leg of each phase of the secondary coils. Extreme care must be exercised in the selection of the turns ratio of this current transformer. Normally, four, six or eight voltage-tap positions are made available to

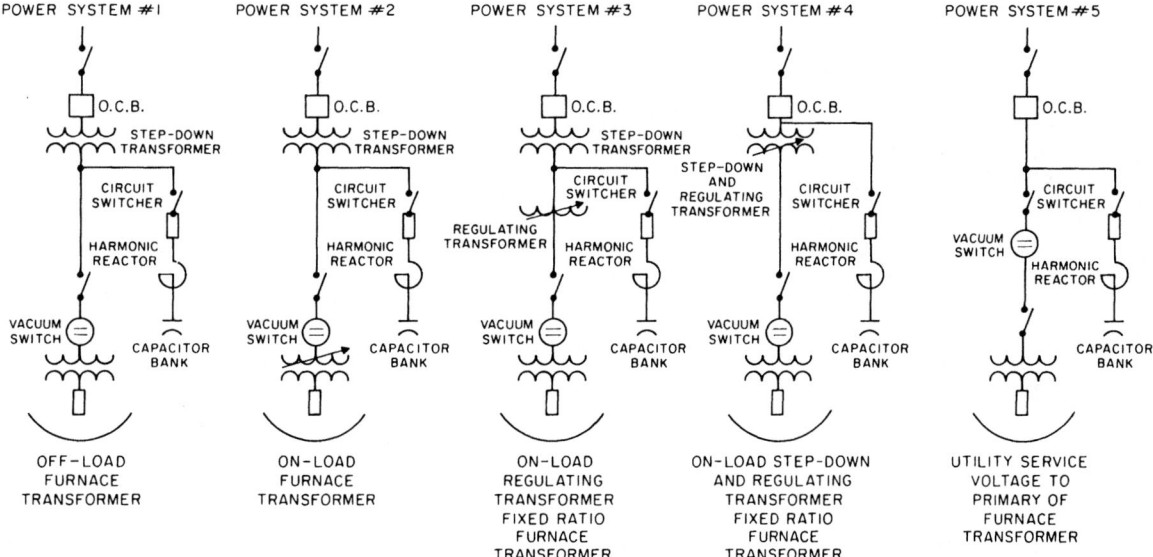

FIG. 18—10. One-line diagrams for electric-arc furnace power systems.

the furnace operator. Some of these taps are with the primary connected in delta and others with the primary connected in wye. The proper method of changing voltage-tap positions in order to minimize transient surges is to raise the electrodes until the arc is broken, open the vacuum switch, change the transformer tap and/or delta-wye position, close the vacuum switch and lower the electrodes.

The vacuum switch on the primary side of the furnace transformer is used because of its ability to open the circuit, with or without load, approximately 40 000 times before a major rebuild is necessary. To minimize transient switching surges, the furnace transformer should have surge protection at both the primary and secondary windings. Primary surge protection can consist of zinc-oxide non-linear gapless lightning arrestors with a parallel circuit consisting of capacitors in series with a resistor. Secondary surge protection can consist of capacitors for each phase plus varistors for each individual riser bar.

The step-down transformer reduces the utility system voltage to a usable furnace bus voltage. This furnace bus can be used for multiple furnaces, and several step-down transformers can be connected in parallel to supply the furnace bus. When multiple step-down transformers are connected in parallel to a single furnace bus to operate multiple furnaces, it has the advantages of having a stiffer furnace bus with a higher short circuit MVA and provides flexibility in case of a step-down transformer failure. However, it has the disadvantage of increased interaction between furnaces particularly during melt-down.

A supply-voltage oil circuit breaker is shown ahead of the step-down transformer.

The shunt capacitor bank is shown with a harmonic detuning reactor and is protected by a circuit switcher for switching under load. Large static capacitor banks require special engineering and consideration, such as:

1. Detuning reactors may be required to prevent harmonic disturbances.
2. The capacitor bank should be connected in ungrounded wye with an unbalance detection system to sound an alarm at the first level of unbalance and automatically disconnect the bank at the second level of unbalance. The unbalance is caused by a neutral shift following individual capacitor can failures. This neutral shift can cause overvoltage on the remaining capacitor cans.
3. The step-down transformer should be energized with the capacitor bank turned off. If the capacitor bank were on, the in-rush current to the step-down transformer would be magnified.
4. Any back-up oil circuit breaker should be equipped with grid resistors and rated for tripping capacitor banks in an emergency.
5. A vacuum switch or circuit switcher is the normal device used for load switching of capacitor banks. The circuit switcher should be motor-operated with preinsertion resistors.
6. A safety grounding switch should be provided to ground all capacitors. This switch should have a timer to prevent closing for five minutes after the bank has been disconnected. This will give the capacitors a chance to discharge.
7. A reclosing timer should be provided to prevent closing the bank immediately after opening, which could cause an overvoltage condition.
8. The capacitor bank should be energized approximately one minute after an arc has been sensed in all three phases, thus preventing energization with the furnace operating single-phased.
9. Capacitor cans can be selected with a rating above the rated system voltage, thus providing an added margin of safety to withstand overvoltages and harmonics.
10. Space should be provided in the capacitor bank racks for future capacitor cans to detune the bank if harmonics prove a problem.
11. Individual current-limiting fuses are preferred for each capacitor can.
12. Current transformers should be provided for relaying. They should be provided with secondary lightning arrestors.
13. Switching reactors should be provided to prevent interaction between multiple capacitor banks.
14. Static capacitor banks will increase the no-load voltage on the furnace bus.

Consideration should be given to making a transient network analysis study for all complex ultra-high-powered furnace systems. The study should analyze system voltage changes, relaying, harmonics, ferro resonance and switching surges. These studies are particularly valuable when static capacitor banks are used.

Power System No. 2 (On-Load Furnace Transformer)—This arrangement is the same as that of Power System No. 1 described above, except that the furnace transformer has an on-load tap changer in the primary winding. Vacuum tap changers can be used to extend the life of the contacts. This on-load tap changer saves time when changing tap positions as it is not necessary to break the arc and open the vacuum switch to change taps. On-load tap changing also has the advantage of implementing computer demand controllers and programmed power melting profiles.

To minimize transient voltages, it is still recommended that the arc be broken by raising the electrodes, then opening the vacuum switch, if the power in the furnace is to be turned off for such functions as tapping, back charging, and so on. Some on-load furnace transformers have been built with a full range delta primary, thus eliminating the need for an off-load delta-wye switch.

Power System No. 3 (On-Load Regulating Transformer with Fixed-Ratio Furnace Transformer)—This arrangement is similar to Power System No. 2 described above, except that a separate regulating transformer has been put in the outdoor substation to do the tap changing. This permits the furnace transformer to have a fixed ratio with the primary and secondary windings permanently connected in delta. Primary-current transformers can be used to provide greater accuracy in metering. Another advantage is that the furnace transformer is physically smaller and of a simpler design. A disadvantage is that an extra transformer tank is required, which increases the equipment and installation costs.

Note that the static capacitor bank has been shown off of the fixed-voltage bus rather than the variable-voltage bus. This was done intentionally because an on-load tap changer ahead of capacitor banks, particularly if two or more furnaces are fed from the same bus, can increase the possibility of a tuned circuit and tap changer switching surges.

Power System No. 4 (On-Load Step-Down and Regulating Transformer with Fixed-Ratio Furnace Transformer)—This arrangement is similar to Power System No. 3 described above, except that it combines the regulating transformer with the step-down transformer in one tank. The advantage of this system is that only one transformer tank is required in the substation.

Power System No. 5 (Utility Service Voltage to Primary of Furnace Transformer)—This arrangement delivers the utility-company service voltage directly to the furnace transformer which can be equipped for off-load or on-load tap changing. The highest service voltage used for this arrangement in the United States to date is 69 000. However, voltages as high as 220 000 have been used in other countries.

Flicker—The problem of flicker in power systems serving arc furnace installations must be reviewed for each arc furnace installation. Flicker is basically a sudden change in voltage which takes place on a repetitive basis every few cycles and has an adverse effect on incandescent lamps and television sets. The magnitude of flicker will vary with the characteristics of the charged material in the furnace. Pre-reduced pellets will cause only approximately one-half the flicker experienced with an all-scrap heat charge. The types of scrap, operating practices and power factor can also cause the amount of flicker to vary.

Static Var Control Systems—Static var control systems have been used to compensate for the reactive loading of electric-arc furnaces to minimize the effects of voltage swings caused by the random reactive variations of the load, and to improve the average power factor. One type of static var control uses a thyristor power controller connected in series with a reactor bank to vary the level of lagging vars drawn by the reactor bank. Fixed capacitor banks for power factor correction are used and tuned for various harmonics. As the arc furnace reactive current increases, the thyristor-reactor current is decreased, so that the leading current drawn by the fixed capacitor bank cancels the lagging current drawn by the furnaces. The measurement and control is done for each individual phase so that the effect of furnace unbalance is minimized. The net reactive current supplied by the static var control system will be substantially equal in magnitude and opposite in phase to the reactive component of load current so that the fluctuations in voltage due to the changes in load current are substantially eliminated. An additional control function uses total line current and bus voltage to hold the power factor of the system constant at the primary of the step-down transformer.

Switchgear—Based on 365 days per year operation, the switching of primary power to electric-arc furnaces can occur 15 000 to 40 000 times per year, depending on whether off-load or on-load furnace transformer tap changing is used. A vacuum switch normally is used for furnace-switching duty and should be limited to secondary overload protection. The vacuum switch offers comparatively low initial cost, long life and ease of maintenance. Consequently, the vacuum switch has been used widely in recent installations of electric-arc furnaces of all sizes. Preferably a back-up breaker with high interrupting capacity should be provided for primary fault interruption duty. This makes it possible to use the vacuum switch for furnace-switching duty only.

Vacuum switches have fast interrupting characteristics. Therefore, steep wave-front high-voltage switching surges should be investigated and proper surge protection used with the furnace transformer, as outlined above in Power System No. 1.

Standard time-over-current relays are used most commonly for fault and overload detection. Where a back-up breaker is used, the time elements of the relays are used to trip the furnace vacuum switch on overload, and the instantaneous elements to trip the back-up breakers for fault protection. A system of relaying that provides a measure of differential protection for the primary circuit and transformer has been developed and used in numerous installations, particularly where relay system coordination presents a problem. Relay coordination may present a problem, however, because normal furnace surges may appreciably exceed rated current, depending upon furnace size and other factors.

The Furnace Transformer—Transformers for electric-furnaces are similar in design to large power transformers, except that they usually are designed so that the primary winding can be connected in either delta or wye. The primary winding is also constructed with several taps to provide various secondary voltages for melting and refining. Because of the widely fluctuating load, which at times approaches short-circuit conditions, special bracing of the windings and extra insulation between turns is required. The latter is needed to withstand the high-voltage surges set up by switching operations. In normal practice, with off-load tap changing, there may be 125 switching operations per 24 hours, and the resulting surges may reach five times the normal voltage. These transformers are designed to suit the furnace. Their specifications usually are written by the furnace manufacturer.

All transformers are rated in kVA capacity. The characteristics of an alternating-current circuit may result in the voltage and current being out of phase. The ratio of the kilowatts in the circuit to kilovolt-amperes is called the power factor. The unit "kVA," an abbreviation of "kilovolt-amperes," represents the product of the impressed voltage and the current in the circuit, and is a measure of the apparent power.

$$kVA = kilovolts \times amperes$$

$$kilowatts = kilovolts \times amperes \times power factor$$

On a typical electric-arc furnace transformer, the secondary coil leads extend in the form of bars through the transformer case. A 75 000 kVA transformer would typically have eight secondary coils per phase, with each coil end brought out through the transformer tank. This would give 48 transformer riser bars.

Furnace transformers are provided with various protective devices, such as:

1. Sudden pressure relay to minimize damage on an internal fault. The relay normally will be provided with alarm and trip contacts with a lock-out device.
2. Dial-thermometer type transformer-winding hot-spot indicators for the primary and secondary windings, complete with detection elements, 15 metre (50 foot) capillary tubes and dial instruments with independent alarm and trip contacts set at 105° C (221° F) and 115° C (239° F). The capillary tube permits mounting the indicator in view of the operator in the control room.
3. Oil-flow switch with alarm contacts.
4. Transformer-oil temperature gauge with alarm contacts.
5. Transformer-oil level gauge with alarm contacts.
6. Tap changer oil-level gauge with alarm contacts.
7. Standard furnace transformer accessories, including drain valve with sampling cock, breather and filter connector, grounding pads, lifting accessories, mechanical relief alarm contacts, and nameplate. All of these accessories must be in accordance with appropriate manufacturing standards.
8. Primary and secondary surge protection.

Transformer Cooling—All arc furnace transformers, except perhaps very small ones, are oil-insulated. This oil, which is a highly refined petroleum product having a high dielectric characteristic, helps to electrically insulate the various parts of the transformer windings from one another and from the iron core and case. At the same time, the oil serves as a cooling medium to carry away the heat generated in the windings and the core by resistance loss and eddy-current and hysteresis losses. Special non-flammable synthetic fluids can be employed in place of the flammable petroleum oil. But the minimal fire hazard in a properly installed furnace transformer seldom warrants the appreciable added cost of the non-flammable fluid.

The transformer oil must in turn be cooled by some external means to maintain the temperature of the windings at a level commensurate with the thermal capabilities of the materials used to insulate the winding conductors and the leads. Furnace transformers customarily have been designed in the past for a maximum top oil temperature of 80°C (55°C rise above 25°C cooling medium temperature). This is equivalent to a maximum top oil temperature of 175°F (100°F rise above 75°F cooling medium temperature). Improvements in insulating materials, however, permits an allowable increase in maximum top oil temperature of 10°C (18°F) and, accordingly, most modern transformers are designed for a 65°C (115°F) top oil temperature rise.

Most water-cooled furnace transformers utilize externally mounted straight-tube oil-to-water heat exchangers. The water tubes are made of corrosion-resistant metal and great care is exercised in the design and construction of the heat exchangers to minimize water-to-oil leaks. The tubes may be of double wall construction with the space between the inner and outer walls open to the atmosphere at the end head so that a leak from either the oil or the water side can be detected readily. Cooling water preferably is drained

to the atmosphere, and water pressure in the heat exchanger maintained at a minimum commensurate with the required flow for proper cooling. The heads of the heat-exchanger casings are removable for cleaning of the water tubes. This type of construction, as compared to the past practice of locating cooling coils inside the main transformer tank, makes visual inspection possible and facilitates cleaning of the tubes to maintain heat-exchanger efficiency.

For economic reasons, transformers up to and including 7500 kVA capacity are generally convection oil-to-water cooled, wherein the heat exchangers are mounted high on the tank and oil flows through the heat exchangers by thermal convection (Type OW). Transformers above 7500 kVA capacity, however, are generally forced oil-to-water cooled, wherein the oil is pumped through the heat exchangers by sealed electric-motor-driven pumps (Type FOW).

Furnace transformers of all capacities can be built with forced oil-to-air cooling, wherein the oil is pumped through radiator-type heat exchangers through which cooling air is blown by fans (Type FOA). It is necessary, however, in the installation of such transformers either to duct outside air to the transformer or to locate the radiators outside. In the latter case, long oil-line runs are required. FOA units are generally employed only where suitable cooling water is unavailable or where climatic conditions make the use of water less preferable.

Tap Changers for Off-Load Tap Changing—The primary winding of the transformer generally is provided with four or more taps, when the primary is connected in delta, and four or more lower refining voltages when the primary is connected in wye. The primary-winding taps and end leads are connected to a motor-driven tap changer and delta-wye switch located in a separate oil-filled compartment mounted externally on the main transformer tank. The ratio of secondary voltages obtained with the primary connected in wye to those obtained with the primary connected in delta is $1:\sqrt{3}$. Usually the transformers are designed with winding and tap changer current capacities such that only the top several delta voltage taps are rated at full transformer kVA capacity, and the kVA capacity of the lower delta voltage taps and the wye voltage taps decreases with decreasing voltage. Rated secondary current at the lowest full-capacity tap is the maximum rated secondary current of the transformer and is the current value on which the current-capacity requirements of the secondary circuit and electrodes are based. Off-load tap changers require the primary circuit to be interrupted before taps can be changed.

On-load tap changers increase furnace arcing time because taps can be changed without interrupting the primary circuit. Also, on-load tap changers are more readily adapted for automatic power demand control and computerization of melting profiles. For on-load tap changing, the primary winding of the transformer generally is provided with seventeen taps (nine taps and eight bridging points) when connected in delta. An additional seventeen taps can be provided when the primary is reconnected in wye. However, many European transformer designs eliminate the off-load delta-wye switch and provide a full range of primary delta voltages with on-load tap changing.

The top secondary phase-to-phase no-load secondary voltage for steelmaking arc furnace transformers ranges typically from approximately 250 volts for a 3000 kVA transformer to approximately 800 volts for a 100 000 kVA unit, although these voltages are subject to some variation to suit the characteristics of the particular furnace and power system to which the transformer is applied. The secondary voltage range for a given transformer is typically about 60 per cent, i.e., the lowest voltage with the primary winding connected in wye is approximately 40 per cent of the highest voltage with the primary connected in delta. Again, this range is subject to variation depending upon the requirements of the particular installation. Furnaces used for stainless and high-alloy steel production generally require a wider voltage range to suit the extensive refining requirements than do those furnaces engaged in carbon-steel production.

Primary Reactors—To maintain a stable arc, it is necessary to have a certain minimum amount of reactance in the circuit through which power is furnished to the arc. For the smaller-size furnaces with physically shorter secondary-circuit components, the inherent reactance of the total circuit including the primary circuit and the transformer usually is not high enough to effect arc stability, particularly on the higher-voltage higher-current melting taps where the inherent transformer reactance is at a minimum. Accordingly, it is necessary to provide supplementary circuit reactance for these smaller-size furnaces. Preferably, this supplementary circuit reactance should be variable so that, on the lower-voltage lower-current refining taps where the total reactance requirements are less and the inherent transformer reactance is higher, the value of the supplementary reactance can be reduced to maintain maximum power factor. This supplementary reactance is generally provided in the form of a three-phase iron-core reactor with a tapped winding, mounted inside the main transformer tank and wired into the primary circuit in series with the primary windings. A terminal board usually is incorporated into the interconnection circuit, providing a selection of several supplementary-reactance values for each voltage tap. On some transformers, however, a second tap changer has been provided for the reactor to give the operator an easier, wider, and more flexible selection of supplementary-reactance values for each voltage tap. The reactor windings are usually tapped to provide incremental changes in supplementary-reactance value of 5 per cent (per unit).

Supplementary reactors usually are provided for furnaces operating on 60 hertz power-supply systems and utilizing transformers of 7500 kVA capacity and below. For lower-frequency power systems, the use of supplementary reactors must be extended to larger furnaces with larger transformers.

For larger furnaces the problem is reversed, in that the physically longer secondary-circuit components tend to result in a total circuit reactance considerably higher than that required for arc stability. An important consideration in the design of large furnaces is, therefore, the minimization of secondary-circuit impedance.

THE ARC AND ELECTRODE REGULATORS

The Arc—The establishment of a stable arc as differentiated from a spark or momentary discharge between two materials depends upon the emission of electrons from the negatively charged material and the flow of electrons through the separating conductive gaseous medium under the influence of the potential gradient between the two materials. In the case of air-type electric-arc furnaces, the conductive gaseous medium is ionized atmospheric air, to which is added the ionized vapors from the electrodes and charge materials. The degree of ionization of the gaseous medium is a function of the nature of the medium; namely, the elements present, temperature, pressure, and the strength of the electrostatic field. If after the arc has been struck conditions are not suitable for sustaining ionization, the arc will be extinguished and must be restruck. This accounts for the erratic behavior of an electric-arc furnace when starting to melt a cold charge, particularly if the furnace is also cold. Sometimes it is necessary to desensitize the electrode regulator during this initial melting period to avoid instability and the possible resulting damages to the electrode positioning mechanisms and electrode breakage.

Electrode Regulators—Power input in the furnace is maintained at the level selected by the furnace operator or computer by a regulator which acts on the electrode-positioning motors to maintain the length of the arc in each phase commensurate with the corresponding voltage tap and current setting. The operator's principal control of power input is by the transformer voltage tap selection. Some change in power input can be effected by current rheostat adjustment, but the practical range of such adjustment is small.

Considering that the voltage (phase-to-phase) is essentially constant on a given voltage tap and that under such constant voltage conditions arc length and arc power are functions of the current, it could be assumed that satisfactory regulation could be obtained with a simple current-sensitive device to position each electrode to maintain the proper current level for each voltage tap. In fact this is the basis on which the early regulators were designed. However, the furnace power circuit is basically ungrounded and the electrodes arc to a floating electric neutral. Accordingly, the three phases are interdependent and equal current magnitudes in the three phases do not necessarily mean equal arc lengths and balanced power input. Therefore, furnace regulators have been of the "impedance" type, with input signals of both voltage-to-neutral and phase current acting to maintain the voltage-to-current ratio, or phase impedance, in accordance with the operator's settings. Since all impedance components of the power circuit are essentially fixed, except the arc resistance which is a function of arc length, the impedance-type regulator acts to maintain the arc length of each phase in accordance with the voltage and current settings. For each phase a line-to-neutral voltage signal, taken from the secondary bus, and a secondary current signal taken from a current transformer mounted inside the transformer, are compared; and the resultant error signal (difference) produces an output voltage proportional to the magnitude of the error signal and with the correct polarity to drive the electrode motor in a direction to raise or lower the electrode until the error signal is reduced to zero.

One of the first impedance regulators used contact-type constant-potential magnetic regulators, the so-called balanced-beam type, in which a voltage-excited solenoid and a current-excited solenoid act in opposition on a pivoted contact beam to energize the electrode-positioning motor at full voltage in a direction to correct any unbalance. Next came the variable-voltage rotating regulators which made possible a proportioning-type regulator that varies the speed according to the magnitude of error between the voltage reference signal and the current feedback signal. In recent years, all-solid-state SCR (silicon-controlled rectifier) regulators have been used.

Electrode-positioning controls normally are equipped with worm-drive reducers which provide a more positive position control and prevent downward drift without the use of a large up bias voltage.

The most recent type of regulator is based on regulating resistance. The majority of the resistance in the secondary circuit is in the arc itself; therefore, a resistance regulator offers more effective regulation. For example, if the arc circuit resistance equals the resistance at setpoint conditions (0.707 power factor at the secondary of the furnace transformer), under short-circuit conditions when the circuit resistance is about 10 per cent of setpoint value, the net error signal with impedance sensing is about 40 per cent as compared to 127 per cent with resistance sensing. This difference increases rapidly as the secondary-circuit power factor at setpoint is reduced below the 0.707 level. The increased error signal with resistance sensing results in faster regulator response in raising electrodes from a cave-in condition and in repositioning the electrodes to achieve setpoint conditions.

Another recent regulator feature has been the integrating error regulator (IER). The IER introduces an inverse time-delay action prior to reversal of the electrode-drive direction. The regulator has an instant pickup error (IPE) adjustable sensitivity setting from 40 per cent to 380 per cent. Normal setting is about 110 per cent. Large arc-resistance errors greater than the IPE level overcomes the inverse time delay and results in instant switching of the output speed command signal to 100 per cent in the appropriate direction. In operation, with an arc-resistance error less than the IPE setting, output speed command to the electrode drive is zero, but the error is integrated by the IER amplifier stage. After a time delay based on the inverse time-delay curve, the output speed command is switched to the proportional characteristic, causing the electrode to be driven in a direction to correct the arc-resistance error. A 50 per cent or greater arc-resistance error produces a 100 per cent electrode speed. As the arc gap length and arc resistance approach set-point value, the electrode speed is proportionally reduced. Where the arc-resistance error becomes less than about 10 per cent, or reverses polarity, dropout occurs and the output speed command is switched off. There is a small additional electrode coasting as it comes to a stop with zero speed command signal. Low values of arc-resistance error under plus or minus 10 per cent are corrected by the electrode moving in small, incremental steps of approximately 2.54 mm (0.10 inch) after the end of each inverse time-delay period. There is no deadband, but the time delay between steps increases inversely as the error decreases.

A design feature of the IER is minimization of unnecessary electrode drive action and avoidance of operating conditions which cause motor overheating. With proper adjustment and normal melting procedures, the electrode drives are at a standstill except for brief periods of movement. While at standstill, the holding current is minimal if worm-gear reducers are used. Time between electrode position changes varies from a fraction of a second during bore-in or with an unstable arc, up to a minute or more for a stable arc on a flat bath. During bore-in, the electrode moves downward in frequent small increments, but raises only in case of a cave-in. Elimination of unnecessary oscillatory action gives improved operation under semi-stable conditions of the arc, and greatly reduces heating of the electrode drive.

Electrode-drive overheating due to operating at stall with high armature current but insufficient torque to break away is not possible with a properly adjusted integrating error regulator (IER). When IER pickup occurs, a full speed command signal is applied to the electrode to insure breakaway under worst case conditions. As the electrode moves toward the setpoint value of arc resistance, the speed command drops to zero before the proportional speed command reduces to a value at which the electrode drive could stall out.

Recent regulators have been provided with preset rheostats for each tap position and each phase. The preset rheostats provide the same function as the three conventional operator's rheostats, but with a separate set provided for each voltage tap position and each phase at a location remote from the operator's station. This not only simplifies the operator's functions but also makes it possible to develop more uniform melting profiles with different operators. More recently, microprocessors have been furnished to control the various functions in the IER regulator. This simplifies integration into a hierarchical computer system.

Operator's Controls—An operator's control panel and/or console usually contains all the instruments, rheostats and switches for control of the power input to the furnaces. This master control station usually is located near the charging side of the furnace. Controls for tilting, roof lift, roof swing, furnace locks, electrode clamps and doors are usually with or adjacent to the power-input controls. A second tilting station is normally provided near the spout for tapping the heat.

The following devices are normally included with the operator's controls:

1. A control switch and indicating lights for the primary backup oil circuit breaker.
2. A control switch and indicating lights for the primary vacuum switch.
3. A control switch and indicating lights for the transformer tap changer.
4. A control switch and indicating lights for the electrode-drive power supplies.
5. Three individual-phase electrode-control switches.

6. A three-phase master electrode-control switch.
7. Three arc-current adjusting rheostats.
8. Three arc-current indicating ammeters.
9. Three indicating lamps for electrode-to-ground voltage indication.
10. One secondary-voltage indicating meter and a transfer switch for selecting three line-to-line and three line-to-ground voltage measurements.
11. One indicating polyphase wattmeter.
12. One indicating polyphase varmeter.
13. One polyphase integrating watthour meter with heat-log reset dial.
14. One combination polyphase recording strip chart watthour and varhour meter.

In addition, an annunciator usually is included to provide an alarm and indication for the various trouble-indicating devices on the transformer and vacuum switch. Fault-finder panels have also been used. These fault-finder panels monitor, with a neon bulb, all the limit-switch and permissive-relay contacts. This fault-finding equipment both helps the operators to identify incomplete operations and directs maintenance personnel to equipment that might have malfunctioned. Today, many of the functions and devices previously used at the operator's control station (rheostats, strip charts, annunciators, fault-finders, etc.) can be incorporated into micro-processors and computer controls with CRT's.

The Secondary Circuit—An electric-arc converts large quantities of electric power into heat in a space of small volume. An electrical characteristic of the arc is that its voltage drop decreases as the current increases as shown in curve A (Figure 18—11). Hence, the arc is inherently unstable.

The condition for stability of an electric circuit is a positive volt-ampere characteristic, that is, an increase in voltage drop in the circuit simultaneous with an increase in current. Curve B (Figure 18—11) shows the curve for this part of the secondary circuit; and since this reactance is in series with the arc circuit, the two curves combine and form the total curve C. The dotted line a-a shows the lower limit of stability, and stable arc conditions are obtained by adjusting the length of the arc for some amount of current higher than the critical value indicated by the line a-a. In general, the circuit is stable with about 50 per cent reactance-volts drop in the secondary circuit. This does not result in a power loss, but instead results in a maximum power factor of about 0.866.

The secondary circuit thus consists of a fixed reactance of supply lines, cable, transformers, bus and furnace leads, and a variable resistance consisting of the fixed resistance of these same circuit elements plus the variable resistance of the arc itself. The voltage applied to the circuit may be varied by adjustment of the transformer taps. The power at any given voltage is varied by changing the resistance of the arc, which is a function of arc length. Maintenance of a desired power input is accomplished by an automatic control system that raises or lowers the electrodes to maintain the current at the value that has been selected by the furnace operator or computer.

If the kilowatt and the kilovolt-ampere inputs to the circuit are plotted for any given voltage with variations

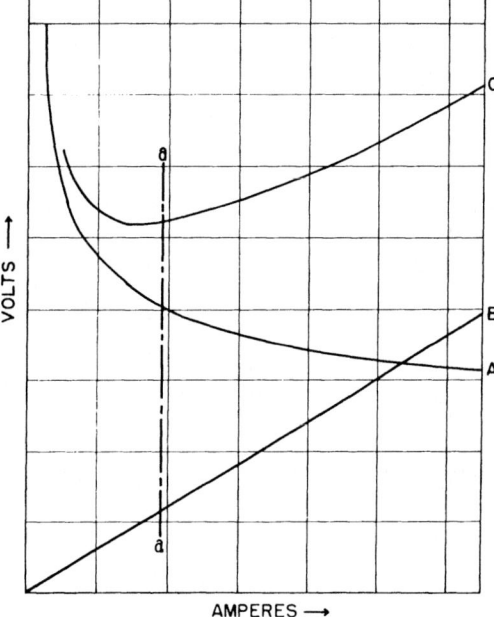

FIG. 18—11. Diagram depicting the volt-ampere characteristics of an arc with respect to its stabilizing element.

in the current input, the curves will be similar to those shown on Figure 18—12. As the current increases, the kilovolt-ampere input increases in a straight line, since it is the product of kilovolts and amperes. The kilowatt input, however, increases only until a current is reached that will result in a power factor of 0.707. Further increases in current beyond this point causes a decrease in kilowatt input with increased kilovolt-ampere input and lower power factor.

Since a part of the power input is dissipated in I^2R heat losses in the transformer bus and leads, the actual kilowatt input to the furnace itself will be less than that shown on curve A (Figure 18—12) and will follow curve B, reaching a maximum at some circuit power factor higher than 0.707, depending upon the resistance of the other circuit elements. Since the losses to these other elements of the circuit vary as the square of the current, the result is a falling efficiency curve as shown (Figure 18—12).

It is perhaps unfortunate that current provides the most convenient factor for automatic control of the circuit. Furnace operators may be misled into associating heat input with amperes rather than kilowatts, and in their attempts to get high heat input into a furnace, use current settings beyond those that give the best kilowatt input for the particular transformer-tap setting and time period in the melting profile.

The proper current for each voltage setting should be determined based on what portion of the melting or refining period the particular voltage tap is to be used. This information should be used by all melters or be incorporated in a computer melting profile.

In examining the electrical operating characteristics of the arc furnace circuit, it is necessary to define the point at which the measurements are made. Since all of the circuit elements have both resistive and reactive impedance components, but are predominantly reactive (except the arcs themselves which are considered to be almost purely resistive), under any given steady state arcing condition kW and kVA will be a minimum (and essentially equal) at the arc, with a unity power factor, and will increase with a decreasing power factor as the point of measurement is moved back through the circuit to the infinite bus of the supply power system. At the arc-length setting which effects maximum power input to the furnace on a given voltage tap, i.e., maximum power generated in the arc, the effective resistance of the arc is equal to the impedance of the circuit from the infinite bus to the arc. At the primary terminals of the furnace transformer, which is the usual point of metering for the furnace operator, the power factor under maximum arc power conditions varies from about 77 to 82 per cent, being higher for the smaller size furnaces and on the higher voltage taps. At the infinite bus under these same maximum arc power conditions, the power factor is in the range of 74 to 78 per cent.

Although under maximum arc power conditions maximum thermal energy is being put into the furnace, experience in high-power furnace operation has shown that this is not necessarily the optimum operating condition for all stages of the heat. Near the end of the melting period and during superheating and refining, when the charge is melted or nearly melted, it has been found that more heat can be put into the charge with appreciably less sidewall-refractory wear if the furnace is operated at a lower power factor, beyond the maximum arc power point, to effect short high-current low-voltage arcs that actually cavitate the molten-metal surface and drive the heat down into the bath. This same mode of operation also applies to the melting of continuously charged pellets or fragmented scrap. Alternately the recent use of foamy slags with copper slag line water-cooled panels minimizes refractory erosion and permits the use of longer arcs with higher

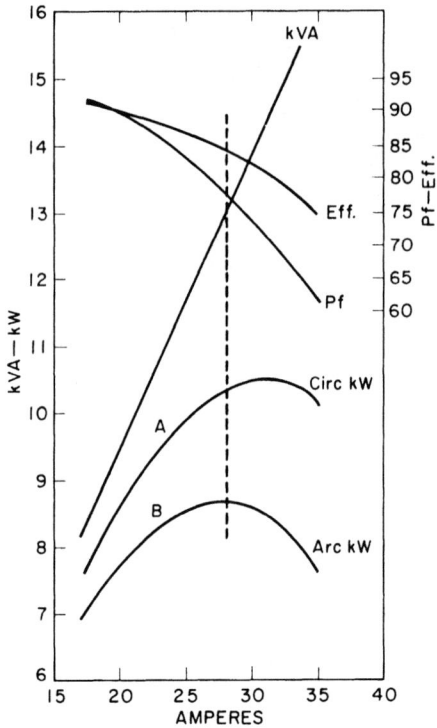

FIG. 18—12. Curves illustrating electrical characteristics of the electric-arc furnace circuit.

power factors during most of the flat bath operations.

Total Circuit Analysis—Assuming that the arcs are essentially variable linear resistances and that the impedances in the three legs of the circuit are balanced, but excluding consideration of feeder capacitances and transformer excitation currents, an electric-arc furnace circuit can, for analytical purposes, be reduced to the diagram shown in Figure 18—13.

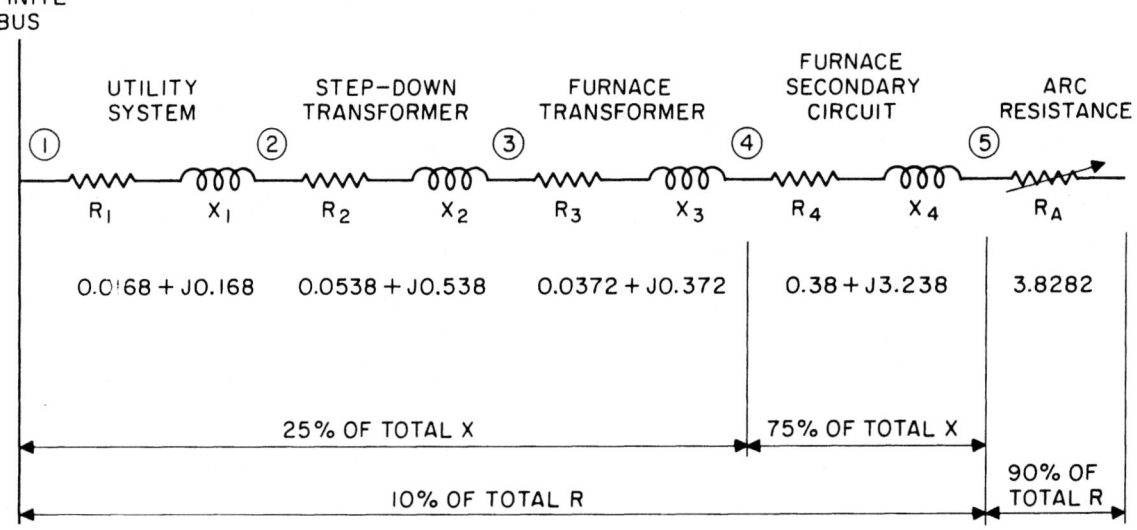

FIG. 18—13. Typical values for total circuit resistance and reactance for a 5.5 metre (18 foot) diameter electric-arc furnace. All impedance values are in milliohms and are referred to the secondary side of the furnace transformer.

The total circuit resistance and reactance must be considered when evaluating power in an electric-arc furnace.

The derivation of the mathematical formulas used in analyzing high-power alternating current circuits is complex and cannot be undertaken here: sources such as those listed at the end of this chapter should be consulted. In the following discussion, the symbol R (with suitable subscript) represents resistance and J (with suitable subscript) represents reactance. Figure 18—13 graphically shows the resistance and reactance components that make up the total circuit resistance and reactance of an electric-arc furnace circuit. These components are indicated from left to right: the infinite bus, followed by the components for the utility system $(R_1 + JX_1)$, the step-down transformer $(R_2 + JX_2)$, the furnace transformer $(R_3 + JX_3)$, the furnace secondary circuit $(R_4 + JX_4)$, and the arc resistance R_A. Also indicated are the circled numbers ① through ⑤ which will be used to relate voltage, real power in kilowatts (kW), reactive power in kilovars (kVAR), apparent power in kilovolt-amperes (kVA), and power factor (PF) as measured at particular points in the circuit.

The values for $(R_1 + JX_1)$, $(R_2 + JX_2)$ and $(R_3 + JX_3)$ can be obtained from the utility company and the transformer manufacturer. For this example: the utility system has been assumed at 2 000 short-circuit MVA; the step-down transformer has been assumed to have 8 per cent impedance at 50 MVA; and the furnace transformer has been assumed to have 5 per cent impedance with 610 volts (5.52 per cent at 580 volts) and 50 MVA. All values are in milliohms and refer to the secondary side of the furnace transformer. The resistance

has been assumed at 10 per cent of the reactance for the utility system, step-down transformer and furnace transformer. For this example, the secondary voltage has been assumed at 580 volts which will produce approximately 50 000 kVA at ③.

The assumption of balanced circuit impedances usually is reasonably valid for the primary circuitry, but relates to a subject of considerable discussion and some controversy in the design of the secondary circuit. However, studies have indicated that approximate results, close enough for practical consideration, can be achieved by the calculation of the three secondary-phase impedances under the assumption of balanced currents and the use of the average of these three-phase impedances for determining $(R_4 + JX_4)$.

In ultra-high-powered electric-arc furnaces, low secondary-circuit reactance (X_4) is always a main topic. The general inductance formula indicates that inductance decreases with a decrease in D_m (mutual geometric mean distance). It is therefore desirable to keep the distances between phases as small as practicable. For balanced power input, the electrodes, bus tubes and cable spacing are all arranged in equilateral triangles. In Figure 18—14, the electrode circle for a 5.5 metre (18 foot) furnace is shown as 1 219 mm (4 feet, 0 inches) with a calculated impedance of $0.309 + J0.606$ milliohms. The bus tubes and cables are spaced 508 mm (20 inches) between phases and have calculated impedances of $0.010 + J1.058$ milliohms (bus tubes) and $0.021 + J1.174$ milliohms (cables). This gives a 60 hertz secondary circuit impedance of $0.34 + J2.838$ milliohms. To complete the resistance and reactance of the secondary circuit under operating conditions, an allowance must be made for the furnace

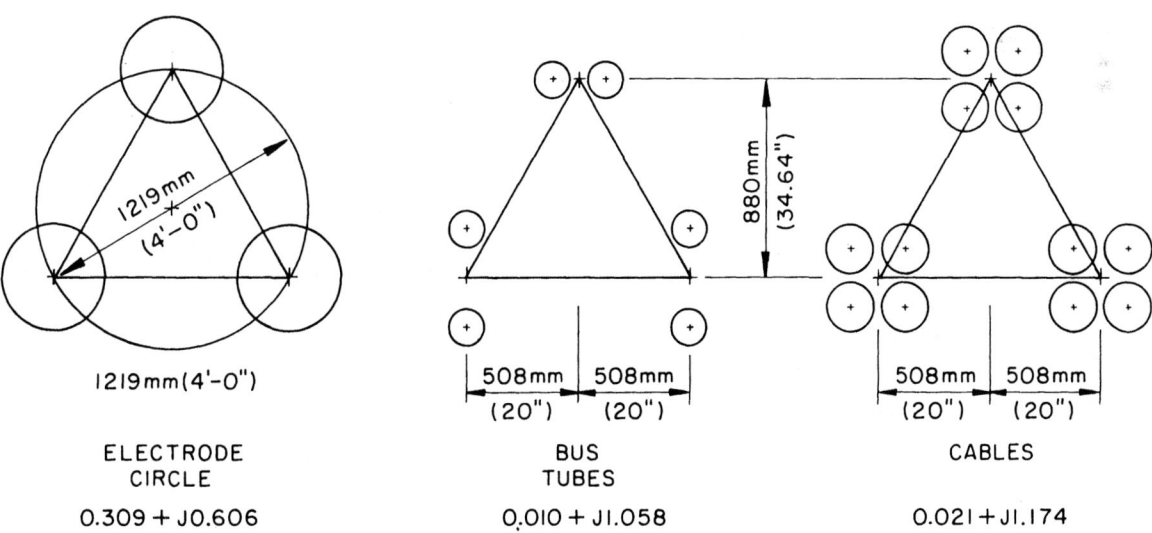

TOTAL SECONDARY CIRCUIT IMPEDANCE:

60 HERTZ ELECTRODES, BUS TUBES & CABLES	= 0.34 + J2.838
DELTA CLOSURE & HARMONICS	= 0.04 + J0.400
TOTAL $R_4 + X_4$	= 0.38 + J3.238

FIG. 18—14. Electrode circle, bus tube and cable spacing for a 5.5 metre (18 foot) diameter electric-arc furnace.

transformer delta closure and harmonics. Harmonics vary over a wide range (10 to 25 per cent) based on arc stability (early scrap melting vs. flat bath operation) and utility system parametres. For this example, the harmonics have been assumed to be approximately 10 per cent of the 60 hertz value. The total allowance for harmonics and delta closure have been assumed at 0.04 + J0.40 milliohms, which gives a total secondary circuit operating impedance of 0.38 + J3.238 milliohms for the (R_4 + X_4) component of Figure 18—13. As can be seen from Figure 18—13, the secondary circuit contains approximately 75 per cent of the total reactance, the other 25 per cent being divided between the utility system, step-down transformer and furnace transformer. For this operating position, the resistance of the arc represents approximately 90 per cent of the total circuit resistance. The majority of the remaining 10 per cent is in the electrode columns.

Assuming a power factor of 0.707 at point ① in Figure 18—13, the infinite bus, then:

$$R_A = (X_1 + X_2 + X_3 + X_4) - (R_1 + R_2 + R_3 + R_4)$$

$$R_A = 3.828 \text{ milliohms}$$

$$X_T = X_1 + X_2 + X_3 + X_4 = 4.316 \text{ milliohms}$$

$$R_T = R_1 + R_2 + R_3 + R_4 = 4.316 \text{ milliohms}$$

If $X_T = R_T$ the angle $\theta = 45°$ or a power factor of 0.707 at ①

$$Z_T = \frac{R_T}{\cos \theta ①} = 6.1047 \text{ milliohms}$$

$$I_{sec} = \frac{\frac{E_{L \cdot L}}{\sqrt{3}}}{Z_T} = \frac{\frac{580}{\sqrt{3}}}{6.1047 \times 10^{-3}} = 54\,855 \text{ A}$$

Arc Power = P_A = $3I^2R_A$ = 34 556 kW

Arc Voltage = $E_{A\,(L \cdot N)}$ = IR_A = 210 volts

$$kW① = \frac{3I^2(R_3 + R_4 + R_A)}{1000} = 38\,324$$

$$kVAR① = \frac{3I^2(X_3 + X_4)}{1000} = 32\,588$$

$$kVA① = \frac{3I^2\,[\,(R_3 + R_4 + R_A) + J(X_3 + X_4)\,]}{1000}$$
$$= 50\,306$$

Power Factor① = 0.762

Note that point ③ is the primary of the furnace transformer where the operators' instruments normally are connected.

When discussing power input in an electric-arc furnace, it is obvious that the entire power system, secondary circuit component parts, spacing, length of circuits, and harmonics must be analyzed together.

REFRACTORY EROSION INDEX

Refractory erosion can be expressed by a simplified formula, wherein refractory index:

$$R_I = \frac{P_A \times (E_A - 30)}{d^2}$$

In this formula,

 R_I = refractory index

 P_A = power in arc (one phase)

 E_A = arc voltage, line to neutral

 30 = anode and cathode drop

 d = distance in centimetres from the edge of the electrode to the refractory.

While this number cannot be related directly to the number of heats obtainable on the sidewall refractory, it does have empirical data value in evaluating how the melting profiles should be developed for a given product and charge material.

The advantage of a small electrode circle becomes obvious from this formula, because the larger the distance (d) between the edge of the electrode and the refractory, the smaller the refractory index number. The smaller refractory index number indicates less wear. A decrease in arc power and arc voltage also will reduce refractory wear. Decreasing arc voltage also decreases arc length. For a particular voltage-tap position, the refractory index can be lowered by increasing the current for that particular voltage tap. As shown in Table 18—IV, when operating on tap 2 with a power factor at the primary of the furnace transformer of 0.79, the secondary current is 50 464 Amps with an arc voltage of 216, arc length 169 mm (6.67 inches) and a refractory index, R_I, of 78 000. This has been an accepted standard practice when scrap is protecting the refractories. When the scrap is no longer protecting the refractories, as is the case with the UHP short arc concept, the current is increased which shortens the arc, lowers the power factor, and lowers the refractory index as shown in Table 18—IV under Low Refractory Index. However, with the advent of the foamy slag concept, coupled with copper slag line water-cooled panels, the arc can remain relatively long with a higher power factor and increased efficiency. The long arc foamy slag practice also reduces electrode consumption due to lower currents with less tendency to blow butts off the ends of the electrode columns.

ELECTRICAL ENERGY REQUIREMENTS

Energy requirements for actual melting are variable and depend upon the definition of when the charge is

Table 18—IV. Refractory Erosion Comparison Chart.

	High Refractory Index	Low Refractory Index
Tap Position	2	2
Rated no-load voltage	560	560
Secondary current, I (amperes)	50 464	60 910
kVA③, primary of furnace transformer, 3 phase	45 349	52 884
kW③	36 000	34 388
kVAR③	27 578	40 177
Power factor③	0.7938	0.6503
Arc power P_A, 3 phase	32 817	29 751
Arc resistance, line-neutral, R_A	4.295×10^{-3}	2.673×10^{-3}
Arc voltage, line-neutral, E_A	216	163
Arc length		
millimetres	169	120
inches	6.67	4.74
Refractory index, R_I	78 000	50 000

Table 18—V. Detailed Heat Chart for Producing 72.6 Metric Tons (80 Net Tons) of Mild Carbon Steel in a 5.5 Metre (18 Foot) Diameter Electric-Arc Furnace, Using Light-Density Scrap.*
Practice is based on the use of electrical energy for melting and finishing the heat in the electric-arc furnace with a limited use of oxygen for a carbon boil at the end of the heat. The furnace is assumed to be equipped with water-cooled slag line panels, wall panels and roof panels while being operated with the foamy slag long arc practice. (Calculations are for estimate of time requirements)

	Power Consumption (kWh)	Duration of Operation (min)
1. Fettle		10
2. Charge first bucket, 36.3 t (80 000 lb)		5
3. Bore in—Tap 4, 31 000 kVA, 0.75 PF	290	1
31 000 kVA × 0.75 PF × 0.75 AF × [1 min/(60 min/h)]		
4. Start melt—Tap 1, 55 000 kVA, 0.79 PF	4792	9
$\dfrac{36.3 \text{ t} \times 132 \text{ kWh/t} \times 60 \text{ min/h}}{55\,000 \text{ kVA} \times 0.79 \text{ PF} \times 0.80 \text{ AF}}$		
5. Melt (70% first bucket)—Tap 2, 52 000 kVA, 0.75 PF	4980	9
$\dfrac{[(0.70 \times 36.3 \text{ t} \times 396 \text{ kWh/t}-(290 \text{ kWh} + 4792 \text{ kWh})] \times (60 \text{ min/h})}{52\,000 \text{ kVA} \times 0.75 \text{ PF} \times 0.85 \text{ AF}}$		
6. Charge second bucket, 22.7 t (50 000 lb)		5
7. Bore in—Tap 4	290	1
8. Start melt—Tap 1	4792	9
9. Melt (100% second bucket)—Tap 2	3907	7
$\dfrac{[(1.00 \times 22.7 \text{ t} \times 396 \text{ kWh/t})-(290 \text{ kWh} + 4792 \text{ kWh})] \times (60 \text{ min/h})}{52\,000 \text{ kVA} \times 0.75 \text{ PF} \times 0.85 \text{ AF}}$		
10. Charge third bucket, 22.7 t (50 000 lb)		5
11. Bore in—Tap 4	290	1
12. Start melt—Tap 1	4792	9
13. Melt (100% third bucket)—Tap 2	3907	7
14. Finish melting first bucket—Tap 3, 343 000 kVA, 0.71 PF	4312	10
$\dfrac{0.30 \times 36.3 \text{ t} \times 396 \text{ kWh/t} \times (60 \text{ min/h})}{43\,000 \text{ kVA} \times 0.71 \text{ PF} \times 0.90 \text{ AF}}$		
15. Refine (flat bath)—Tap 5, 33 000 kVA, 0.71 PF	5392	15
$\dfrac{81.7 \text{ t} \times 66 \text{ kWh/t} \times (60 \text{ min/h})}{33\,000 \text{ kVA} \times 0.71 \text{ PF} \times 0.90 \text{ AF}}$		
16. Delays		10
17. Tap		7
	37 744	2 h 0 min

Power consumption for this heat equaled 520 kWh per metric ton (472 kWh/net ton).
Scrap yield for this heat was 89 per cent.

*t = metric tons; AF = arc factor (reflects effect of arc variations during different phases of melting and refining); PF = power factor at primary terminals of furnace transformer.

melted. Generally, it is assumed that the charge is melted when no further scrap remains and the metal has reached a temperature of 1565° C (2850° F).

The cold scrap charge usually is melted down at maximum permissible power input.

Following the melt-down period, the power input is reduced while the heat is worked and refined. The refining procedure depends on the composition of the steel, quality of scrap, furnace condition, and shop practice.

The rapidity of melting is a function of the amount of energy introduced in a given time.

Detailed Heat Chart—Melting profiles and power factor will change with the type of charge material, depending upon whether it consists of light-density scrap, heavy scrap and ingots, pellets or hot metal. A theoretical heat chart is shown in Table 18—V for producing merchant products with light-density scrap. This chart indicates a production rate of twelve heats per 24-hour day. This theoretical heat chart should be considered as a starting point and guide for developing a heat chart for a particular application. Every shop will develop its own method of producing a heat, depending upon product, type of charge material and slag practices.

In analyzing the heat chart in Table 18—V, all taps should be changed based on kWh input, depending upon the amount and type of scrap charged and fluxes. A thermal model, plus the rate of temperature rise in the discharge water from the water-cooled wall panels opposite the electrodes, can be useful for computer control of the practice.

A description of items in the heat chart is as follows:

Item 1: Fettling and performing any miscellaneous maintenance functions that might be required.

Item 2: Charging represents filling the hearth and shell to approximately 85 per cent water-level full, based on scrap density of 800 kg/m³ (50 lb/ft³).

Item 3: Boring-in is performed on a lower power tap with a medium power factor in order to get a hole in the scrap approximately 600 mm (2 feet) deep to shield the roof and walls from the arc. This takes about one minute and puts 290 kWh into the heat. After the hole is in the scrap, the power input can be raised to the highest level.

Item 4: Start melting is at the highest power level used with a high power factor. With this power level and power factor, it is possible to develop three large "chimneys" in the scrap with a molten pool at the bottom before exposing any refractory on the walls.

Item 5: Melting on Tap 2 is used with a medium power factor to protect the side walls from erosion. Only 70 per cent of the first bucket is melted; therefore, the walls should still be protected with scrap.

Items 6 through 13: These are basically the same as Items 2 through 5, except they are performed for the second and third buckets of scrap.

Item 14: Tap 3 is used to finish melting to a flat bath. This lower-voltage tap with a lower power factor is used to minimize refractory erosion.

Item 15: Refining is completed with a lower power level.

Item 16: Delays are a non-scheduled outage and the time will vary from heat to heat.

Item 17: Tapping the heat.

The utilization factor (UF) represents the percentage of the total heat time during which the power is on. For this theoretical heat, it is 63 per cent. A utilization factor between 60 and 75 per cent is considered to represent reasonable practice.

Carbon Steels (Merchant Products)—The heat chart for mild carbon steel in Table 18—V is based on electrical energy (with a minimum use of oxygen) for melting and finishing the heat in the electric-arc furnace and shows 520 kWh per metric ton. This electrical energy can be reduced to approximately 400 kWh per metric ton by using scrap preheating, oxy-fuel burners, and oxygen assisted melting. This practice requires between 20 to 30 NM³/metric ton of oxygen plus extra carbon units. However, production can be increased to 16 to 18 heats per day with an average tap-to-tap time of 80 to 90 minutes. There will normally be an overall reduction in energy costs with this practice.

Alloy Steels—The same productivity can be achieved for alloy steels as for carbon steels if the electric-arc furnace is used simply as a melting unit. The heat should be tapped slag free and finished in a ladle metallurgy station with arc heating and degassing if required. The arc heating station would normally require between 30 and 40 kWh/metric ton.

INDUCTION ELECTRIC-FURNACE PROCESSES

Section 5 discusses the types of induction electric-furnaces used in steel and iron making operations. The **first subsection** describes the two types of induction electric-furnaces. The **second subsection** describes three applications of induction furnaces. Finally, the **third subsection** discusses operating conditions and practices of induction furnaces.

TYPES OF INDUCTION FURNACES

The steel industry uses two types of induction electric-furnaces. These are the "channel" and "coreless" types.

Channel Induction Furnaces—These furnaces basically consist of a vessel to which one or more inductors are attached. The inductor is in principle a transformer whereby the secondary winding is formed by a loop of liquid metal confined in a closed refractory channel. In this furnace, energy is transformed from the power system at line frequency, through a power supply to the inductor and converted into heat.

One advantage of this type of furnace is that the vessel or upper case can be built in any practical size and shape to suit the application. The disadvantages include the power input limitation per inductor, and the necessity to maintain a liquid heel in the furnace at all times. The practical power input limit for a twin loop inductor is 2500 kW when applied for ferrous metal. A heel must be maintained to avoid refractory problems in the inductor which may occur from reheating after cool-down or from plugging the channel with slag during emptying, thus interrupting the loop and preventing power absorption. These disadvantages restrict the application of the inductor furnace to a receiver or holding vessel of liquid metal with limited capability for melting or alloy changes. For example, the inductor furnace is used to store and modify blast furnace iron with scrap additions and superheating before the iron is used in the BOF.

Coreless Induction Furnaces—These furnaces basically consist of a cylindrical crucible surrounded by a power coil which is supplied with energy either directly from the network (line frequency) or through a frequency converter. The magnetic field generated by the coil carries the energy to the charge. Figure 18—15 shows a typical coreless induction furnace as it is being tapped.

FIG. 18—15. One of three 50 ton coreless induction furnaces rated 16 MW, shown tapping iron into a torpedo ladle for transport to the BOF shop. (Courtesy, Brown-Boveri Cie, Mannheim, West Germany).

Furnace size, operating frequency and power density are linked together, whereby small furnaces require a higher frequency and large units generally operate at line frequency. The maximum power density for a given furnace size is dictated by the stirring action in the bath. For a given metal and frequency, the metal motion is directly proportional to the power applied to the furnace. This motion is reduced, however, by one over the square root of the frequency ratio if the ap-

plied frequency to the furnace is increased. For example, at 60 hertz the maximum practical power that can be applied to a furnace is 350 kW per metric ton crucible capacity; if the frequency were raised to 960 hertz, the power could be increased to 1400 kW per metric ton crucible capacity, resulting in the same metal motion.

Coreless induction furnaces operating in the steel industry range in size from a few pounds (as those used in laboratory or investment applications) to 70 metric tons (as those used to melt iron to feed a BOF), and have a power range from a few kW to 22 000 kW.

The coreless induction furnace is used in the steel plant for: remelting scrap; remelting alloys; and producing synthetic iron from steel scrap for use in converters to make steel. It is not commonly used as a melting unit, however, when major refining work is needed. Any attempt to blow oxygen may result in a sudden and violent reaction, since the temperature due to the stirring action is uniform throughout the bath and the surface is relatively small, thus concentrating the escape area of carbon monoxide.

APPLICATIONS OF INDUCTION ELECTRIC-FURNACES

Remelting—In the remelting process, the charge material has basically the same chemical analysis as the final melt. No major refining or metallurgical work is involved. The induction furnace is preferred for this process since reactions are more predictable than in other furnaces such as the arc furnace. However, to obtain successful results, one must still observe the metallurgical reactions which take place at elevated temperatures between the melt, the atmosphere, the slag generated or applied, and the furnace refractory.

When remelting alloy steel, certain elements which are readily oxidized or have a tendency to vaporize must be added under certain conditions. The following rules should be observed: (1) readily oxidized alloys should be added as late as possible; and (2) alloys, such as ferromolybdenum, that do not dissolve easily should be added early and during periods of great stirring action.

If it is necessary to oxidize the heat, this can be done with the addition of iron ore or mill scale, preferably with the initial charge. If conditions dictate that iron ore or mill scale be added to the melt, it should be fed in small quantities during periods of good stirring action and when the bath is at a low temperature in order to avoid a violent reaction.

Deoxidation of the heat should occur at the end of the melting process with manganese, silicon or aluminum additions.

Furnace refractories used in remelting applications can be either acid (silica), basic (MgO), or neutral (alumina). The selection of the material depends on the type of slag one expects to result, and the temperature at which the heat must be poured. For example, silica, which is a preferred induction furnace lining, cannot be used if the slag generated during the melting is highly basic or if the final temperature exceeds 1590°C (2900°F).

Premelting—Premelting alloys, such as ferrosilicon or ferromanganese for example, may offer an advantage when additions must be made to large ladles while temperature maintenance is critical. It may also prove to be advantageous to add alloys in liquid form to gain time and assure more uniform distribution. Medium-frequency induction furnaces are typically applied for this task.

Production of Synthetic Iron—In the production of synthetic iron, coreless induction furnaces are used to either supplement or substitute blast furnace output. Because of their size, only line frequency furnaces are applied. Silica is used as the lining material since the operation is generally acid and tapping temperatures are in the range of 1480 to 1540°C (2700 to 2800°F.)

CONDITIONS AND PRACTICES IN INDUCTION ELECTRIC- FURNACE OPERATION

In operation, typically 50 to 70 per cent of the crucible capacity of an induction furnace is tapped, and a heel is left in order to improve metal output.

Charge material size is limited by the furnace diameter. To avoid bridging the charge, the maximum dimension of any piece should not exceed 70 per cent of the furnace crucible diameter. Charges are typically dropped from a charge bucket into the furnace with a charge weight between 10 to 20 per cent of the furnace capacity, depending on charge density. Since the charge normally drops into a liquid heel, it must be dry. Therefore, either preheaters or dryers must be included in the charge make-up system.

Coreless induction furnaces operate with approximately 75 per cent overall efficiency. If the charge

weight is known, energy counting is usually applied for temperature control. For example, if a 10 ton charge is dropped into the furnace and is to be melted to a temperature of 1500° C, the energy input requirement shows that 380 kW hours per metric ton for a given analysis is needed. Considering a furnace efficiency (electrical and thermal) of 75 per cent, one would set the energy counter to 5067 kW hours. After the energy is consumed, the furnace will be at the proper temperature.

Alloys are added separately, either from a container or through feeder conveyors with weighing devices. Alloy recovery and mixing are successful, due to the inherent stirring action and absence of extreme temperature differences.

Lining performance can be monitored by voltage and power measurements, and by the number of capacitors switched in for power factor correction. The furnace can also be emptied and the lining measured to gain this information.

The operating conditions and practice will affect the cost of the coreless induction furnace. A furnace used constantly for melting and one which has little holding time will consume the lowest amount of energy per ton of metal produced. In practice, the furnace must be shut off, or its power reduced for tapping and slagging. These off-times should be kept to a minimum. Therefore, mechanical skimming devices should be applied and ladles should be positioned at appropriate times. The cover should be kept closed as much as possible to reduce radiation loss. Also, mechanical charging systems and automated power supply controls will result in the lowest possible labor cost.

SECTION 6

VACUUM, ATMOSPHERE, AND CONSUMABLE-ELECTRODE MELTING FURNACES

Section 6 focuses on two special types of electric-furnaces: vacuum furnaces; and electroslag remelting furnaces. The **first subsection** describes the components and processes in two kinds of vacuum furnaces. The **second subsection** then describes the components and processes in the electroslag remelting furnace.

VACUUM ELECTRIC-FURNACES

Methods have been developed for the commercial melting of certain metals and alloys in vacuum or under controlled atmospheres. Vacuum-melting techniques are employed in the case of some steels to obtain improved physical and mechanical properties which cannot be obtained by any other means. Some metals, notably titanium, can only be melted successfully under such conditions. Vacuum furnaces are heated by electrical induction, the electric-arc principle, electrical resistance, and by gas; however, only the first two have been used on any sizeable scale for melting steels. This discussion will be confined to processes employing one or the other of these two principles.

Vacuum and Atmosphere Melting in Induction Furnaces—This method employs a high-frequency coreless induction unit (see Section 5), enclosed in a container or tank that can either be evacuated or filled with an atmosphere of any desired composition and pressure. Electrical and mechanical mechanisms and controls aid in making additions to the melt, and in tilting the furnace to pour the molten metal into the ingot mold, which is also enclosed in the tank or container with the furnace (Figure 18—16). Most of the vacuum furnaces in operation in the United States have an approximate capacity of 225 to 450 kilograms (one-quarter to one-half net ton), although larger units that melt up to 60 metric tons (65 net tons) have proved practicable.

The electrical frequency employed by coreless induction furnaces for vacuum and atmosphere melting depends upon the capacity of the melting unit, as in the case of furnaces used in ordinary melting processes. Also, most of the crucibles for vacuum melting have basic linings.

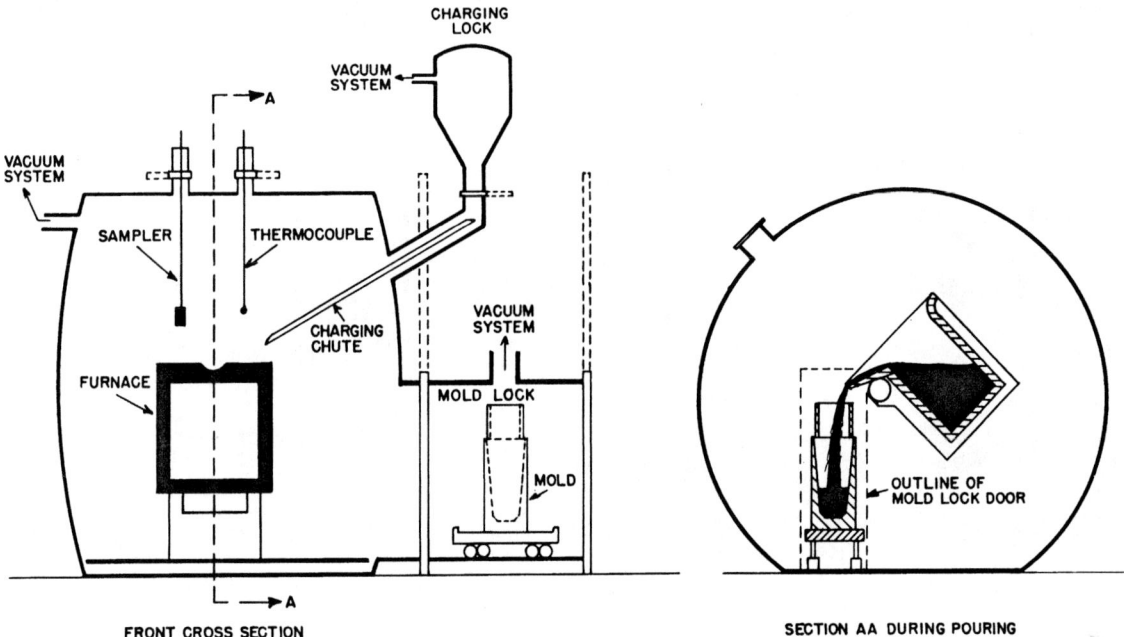

FIG. 18—16. Schematic arrangement of furnace in vacuum chamber equipped with charging and mold locks, for vacuum induction melting.

Vacuum melting has often been employed as a remelting operation for very pure materials, and as the first stage in duplex refining operations in which the product from the vacuum induction furnace is remelted in a vacuum consumable-electrode furnace. However, it is generally more useful in those applications where some refining is also done. Oxygen, hydrogen, and to a smaller extent, nitrogen, can be removed from the molten metal in vacuum melting. Carbon may also be removed when alloys with low carbon content are being produced, i.e., some stainless steels.

The low pressures, often as low as 5 micrometres (5 microns), are usually obtained by a combination of the following types of vacuum pumps: mechanical; steam; oil-diffusion; and oil ejector.

The control of pressure and composition of the gas over a melt makes it possible to deoxidize the melt with carbon or hydrogen, both of which produce gaseous deoxidation products, and thus prevent the formation of solid nonmetallic inclusions in the finished steel. Also, when melting in a vacuum, the absence of nitrogen from the atmosphere over the melt prevents the formation of nitrides and carbonitrides that appear in many steels and high-temperature alloys melted under ordinary atmospheric conditions. The exclusion of oxygen by vacuum melting prevents oxidation losses and permits very close control of the composition of alloys containing easily-oxidized components.

The volatility of certain alloying elements such as chromium, aluminum and manganese may result in high losses of these elements when they are added to steel under a high vacuum. These losses may be minimized by replacing the vacuum with an inert gas used as the atmosphere over the melt during the period when the elements are being added.

Consumable-Electrode Melting (in Vacuum)—While the consumable-electrode melting process (also called vacuum arc remelting, abbreviated VAR), has other applications, only its use in the production of special steels will be considered here. This is a refining process used to produce special-quality alloy and stainless steels (originally produced by any suitable conventional steelmaking process) by casting or forging them into an electrode that is remelted in a vacuum. Some of these special steels include bearing steels, heat-resistant alloys, ultra-high-strength missile and aircraft steels, and rotor steels.

Figures 18—17 and 18—18 respectively show a consumable-electrode furnace and the principle of its operation. The furnace in Figure 18—17 is capable of producing round ingots 813 mm (32 inches) in diameter and weighing up to 9300 kilograms (20 500 pounds). Some larger furnaces that are in operation produce ingots as large as 1525 mm (60 inches) in diameter and weighing as much as 45 360 kilograms (100 000 pounds).

A consumable-electrode furnace consists of two sections: a water-cooled tank above ground level that encloses the electrode; and a water-cooled copper mold in the lower section below ground level in which melting and solidification of the ingot take place. Direct-current is employed for melting. The oxidized scale on the surface of the electrode is removed chemically or mechanically before attaching it to the furnace components. Also, the diameter of the electrode in relation to the crucible is important with respect to product quality and melt rate. The electrode is usually connected to the negative terminals and the copper mold to the positive terminals of the electrical circuit. To achieve low pressures (sometimes as low as 10 micrometre or 10

FIG. 18—17. Worker checking alignment of the electrode in a consumable-electrode vacuum melting furnace prior to start of remelting operation.

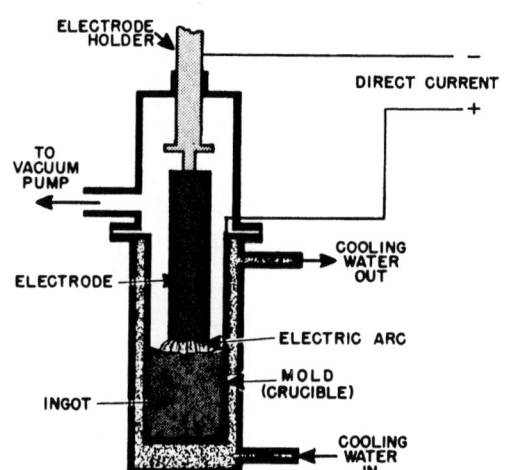

FIG. 18—18. Schematic representation of the principles of design and operation of a consumable-electrode furnace for the remelting of steels in a vacuum.

micron pressure), a combination of various kinds of vacuum pumps is used.

Steam-ejector vacuum pumps, or a combination of mechanical and oil-diffusion pumps, first evacuate the furnace to an absolute pressure of about 10 micrometres (10 microns) of mercury. Power is then turned on, and an arc is struck between the electrode and a starting block that is placed in the mold before the operation begins. Heat from the arc progressively melts the end of the electrode. The melted metal is transferred in droplet form across the arc and deposited in a shallow pool of molten metal on the top surface of the ingot being built up in the mold. Solidification of the molten metal takes place almost immediately. The rate of descent of the electrode is automatically controlled to keep the proper distance needed to maintain the arc between the end of the electrode and the top of the ingot as the end of the electrode is "consumed" or melted away.

Steel that has been remelted by this process accrues the following benefits:

1. Contents of gases (hydrogen, oxygen, and, to a lesser extent, nitrogen) are substantially reduced.
2. Cleanliness of steel is improved (fewer non-metallic inclusions).
3. Center porosity and segregation in the ingot resulting from the process are practically eliminated, because of the progressive melting and solidification of the steel and the resulting effect on crystal formation.

4. Hot workability of the metal is improved.
5. Mechanical properties of the remelted steel (ductility, impact strength, fatigue strength, creep and rupture strength) at both room temperature and at elevated temperature are improved.

ELECTROSLAG REMELTING FURNACES

The electroslag process was developed over 50 years ago in the United States for melting special alloys that were difficult to produce in conventional arc furnaces. The original concept involved remelting feedstock comprised of a folded metal strip encasing specific metal alloying powders required to produce the intended steel composition. The process attracted little attention until the feedstock was changed to a solid, premelted electrode and the ingot size was increased. Furnace size has gradually increased and ingots weighing up to 125 metric tons (about 137 net tons) are being produced. Electroslag remelting (ESR) furnaces are reportedly under construction for remelting 180 metric ton (200 net ton) ingots. The USSR uses ESR to produce the ultraservice steels that are generally melted in the United States by vacuum arc remelting (VAR). Other uses of ESR include steels for rotor forgings, rolls, offshore platforms, molds and dies, armor, nuclear containment vessels and special casting shapes (electroslag casting). Interest in ESR in the United States is growing and a number of steel companies have installed or are installing ESR facilities for producing ingots up to 1650 mm (65 inches) in diameter. In addition, several facilities for producing nickel-based superalloys, alloy steels and stainless steels are presently operating.

Electroslag remelting, like vacuum arc remelting, is a secondary refining process for electrode ingots of essentially the same composition as that which is desired in the finished product. The basic requirements of ESR furnaces are relatively simple: (1) a high amperage, low voltage electrical supply (either alternating- or direct-current); (2) an electrode feed mechanism (one or more electrodes can be melted simultaneously depending upon the circuitry used); and (3) an open-bottom, water-cooled copper mold that contains the molten slag and metal. In this furnace, the mold rests on the starting plate at the beginning of melting and moves steadily upward as melting progresses.

Round, square, rectangular, or hollow ingots can be produced by ESR furnaces, as well as various-shape castings. The profile and dimensions of the electrodes need not precisely match those of the mold. In contrast to the electrodes used in VAR furnaces, the electrodes for ESR furnaces do not need a prepared surface. They can be remelted with the surface developed during casting in cast-iron molds or refractory tubes, or during continuous casting.

The electroslag remelting process can be started by placing a quantity of prepared refining slag on the mold baseplate, and then adding molten slag (hot start), or striking an arc between the electrode and the baseplate to form a pool of molten slag (cold start). The electrode remains immersed in the slag which is electrically conductive. Electrical-resistance heating of the slag increases its temperature until droplets of molten metal form and fall from the electrode through the refining slag to form a pool in the mold. As melting proceeds, the molten pool progressively solidifies but retains a molten depth equal to roughly one-half the diameter of the mold.

In ESR, the slag performs three functions:
1. It offers resistance to the passage of an electric current and thereby generates heat to melt the electrode material.
2. It protects the molten metal from oxidation and rapid heat loss by radiation.
3. It can be varied in composition to promote desulphurization or dephosphorization of the metal droplets and to assist in flotation of inclusions.

The slag composition can be varied within fairly wide ranges, but primarily consists of calcium fluoride (CaF_2) with additions of lime (CaO) and alumina (Al_2O_3). A typical slag would contain 70 per cent CaF_2 and 30 per cent CaO, or 60 per cent CaF_2, 30 per cent CaO, and 10 per cent Al_2O_3. Although the system operates under a molten fluoride slag, no serious fume problem appears to exist; however, a collection hood should be installed above the mold as a precaution.

Bibliography

American Institute of Mining, Metallurgical and Petroleum Engineers, Electric Furnace Steel Conference Proceedings, 1943 to present (published annually).

Anonymous, The electric-arc furnace. Brussels: International Iron and Steel Institute Committee on Technology, 1981.

Babos, Luciano and Antonio Martegani, Energy conservation in dedusting plants. Iron & Steel International, February, 1983.

Bowman, B., Optimum use of electrodes in arc furnaces. Metallurgical Plant and Technology, January, 1983, 30-39.

Bowman, B., and P. J. Salomon, Present status, trends, and development of the electric-arc furnace in the iron and steel industry. Elektrowaerme Int. ed. B Vol. 39, February 1981, 34-40.

Brokmeier, Ing. K. H., Induktives Schmelzen. Brown-Boveri & Cie, Aktiengesellschaft, Mannheim, West Germany, 1983.

Caine, K. E., Jr., A review of new electric-arc furnace technologies. Iron and Steel Engineer, October, 1983, 45-47.

Elliott, J. F., The chemistry of electric-furnace. Iron and Steelmaker, January, 1975, 24-31.

Essmann, H., and D. Gruenberg, Technical and economical aspects regarding the practical applications of a D.C. electric-arc furnace. AIME Electric Furnace Steel Conference Proceedings, 1983.

Howard, E. C., Arc furnace power: Existing and future installations. Iron and Steel Engineer, October, 1983, 42-44.

Isenberg-O'Laughlin, J., Electric-furnace update '82: bearing down on melting costs. 33 Metal Producing, November, 1982, 45-58.

Isenberg-O'Laughlin, J., editor, World steel industry data handbook, Volume 5, USA. 33 Metal Producing, 1982.

Maurer, G., J. P. Motte and J. Antoine, Scrap Melting in an electric-arc furnace supplied with direct current. 40th Electric Furnace Proceedings, Vol. 40, 1982, 69-75.

Raley, et. al., Electrode evaluation by a modern arc furnace. Electric Furnace Proceedings, Vol. 40, 1982, 347-355.

Schwabe, W. E., The status of the ultra-high power electric furnace. Iron and Steelmaker, July, 1975, 11-19.

Strohmeier, B., W. Peters and M. Riess, Application of a process computer for operational and management control at ISCOR's UHP arc furnace shop. AIME Electric Furnace Steel Proceedings, 1982, 101-110.

Trageser, J. J., Power usage and electric circuit analysis for electric-arc furnaces. Conference Record IAS Annual Meeting, published by IEEE, 1978, 1261-1267.

Wheeler, F. M., and A. G. W. Lamont, Current trends in electric meltshop design. Iron and Steelmaker, January, 1979, 36-43.

Wunsche, E. R. and R. Simcoe, Electric arc furnace steelmaking with quasi-submerged arcs and foaming slags, Iron and Steel Engineer, April 1984, 35-42.

Zangs, L., Water-cooled linings for direct-arc melting furnaces. Steel Times, October, 1978, 912-916.

CHAPTER 19

Secondary Steelmaking or Ladle Metallurgy

SECTION 1

EXTERNAL STEEL-REFINING PROCESSES

Introduction—The purpose of Secondary Steelmaking (also referred to as Ladle Metallurgy) is to produce "clean" steel, steel which satisfies stringent requirements of surface, internal and microcleanliness quality and of mechanical properties. Ladle metallurgy is a secondary step of the steelmaking process often performed in a ladle after the initial refining process in a primary furnace is completed.

Increasingly stringent requirements have been demanded of a growing percentage of steels, such as alloy steels and steels used in arctic linepipe and jet aircraft engine parts. To fulfill their functions these steels must meet the more stringent requirements, which only the process of ladle metallurgy can satisfy economically. As the demand for such high quality steels increased, ladle metallurgy became a routine step in the production of steel.

Although satisfactory for making steels for most applications, conventional steel making and refining practices such as BOP, Q-BOP, open hearth and electric furnaces could not consistently achieve the high specifications the special steels had to meet. Steelmakers tried to produce these special steels by modifying and sometimes prolonging the primary furnace refining practices to save themselves the extra step ladle metallurgy required; but such methods failed to achieve the desired results and were impractical. To remain competitive and maintain production, steelmakers reluctantly accepted the secondary steel refining processes. However, when secondary steel refining practices began operating with some regularity, steelmakers realized that these practices offered advantages aside from the production of an especially high quality steel. Steelmakers found they were able to exercise a control over the many processing conditions which contributed to an even higher quality steel including:

1. Teeming temperature, especially for continuous-casting operations;
2. Deoxidation;
3. Decarburization (ease of producing steels to low carbon levels of less than 0.03 per cent);
4. Additional adjustment for chemical composition;
5. Increasing production rates by decreasing refining times in the furnace.

PURPOSE OF SECONDARY STEELMAKING PROCESSES

Secondary steelmaking processes are adopted primarily to achieve various objectives. The fulfillment of these objectives results in a steel which meets the desired stringent requirements. These objectives (also called functions or goals of secondary steelmaking) include:

1. Control of gases: degassing (decreasing the concentration of oxygen and hydrogen in steel);
2. Low sulphur contents (normally less than 0.010 per cent and to as low as 0.002 per cent);
3. Microcleanliness (removal of undesirable non-metallics, primarily oxides and sulphides);
4. Inclusion morphology (Since steelmakers cannot remove undesirable oxides completely, this step allows steelmakers to change the composition and/or shape of the undesired matter left in the steel to make it compatible with the mechanical properties of the finished steel.);
5. Mechanical properties (Charpy V-notch toughness, transverse properties, and ductility in through-thickness of plates).

PREREQUISITES FOR EFFECTIVE UTILIZATION OF SECONDARY STEELMAKING TECHNOLOGY

Although secondary steelmaking processes have the capability to extend the refining capabilities of modern steel-producing facilities, various prerequisites must be met for effective utilization of these processes.

1. Temperature and **chemical composition** of the raw steel must meet certain requirements in the primary furnace and must be maintained through tap time into the ladle or secondary vessel to produce quality steels.

2. Slag—

a. Furnace slag—The carryover of steelmaking slag into the steel ladle must be minimized during tapping to improve the effectiveness of the secondary steelmaking processes. Furnace slag in the ladle affects the deoxidation of the steel because part of the deoxidizing elements (i.e. silicon or aluminum) are used up in deoxidizing the slag. Minimizing its carryover will improve and make more predictable the ladle deoxidation. It will also improve alloying efficiency. In BOP steelmaking, carryover of furnace slag during tapping can be minimized by plugging the taphole with burlap prior to tap, utilizing maximum furnace rotation during tap, and using a suitable device—taphole slide gate, bott, or refractory floating plug—that minimizes the carryover of the slag that is entrained in the steel stream toward the end of the tapping sequence.

b. Synthetic slag—Efficient ladle desulphurization of steel and ladle refining to produce ultraclean steels are attained only when the steel is treated under a basic, nonoxidizing slag. Ladle-refining methods were also developed whereby the addition of a nonoxidizing slag to the ladle as a supplement to low-cost argon gas-stirring treatments produced a cleaner steel. Synthetic slags must meet the following general requirements:

1. low oxygen potential;
2. low melting point;
3. moderate fluidity;
4. large solubility for alumina and sulphur.

The slags having these characteristics are generally found in the CaO-SiO_2-Al_2O_3 system. Synthetic slags of this type are added to the ladle during or after tapping to provide some refining of the steel. As suitable nonoxidizing top slag coverings on the ladle, they retain the nonmetallics transported to the top slag by the argon-rinsing treatment.

3. **Accurate assessment of temperature, chemical composition, and quantity** of steel in the ladle are important. Only by knowing this can the proper weight of additions be made. Heat-size control and subsequent, precise chemical composition control are also dependent on accurate charge-control measurements and good tapping practices, as is the provisioning for further processing.

4. **The use of higher quality refractories** in the liquid-steel-handling facilities contributes to attaining the best results from secondary steelmaking processes. Many of the metallurgical functions to be performed are directly related to the state of oxidation of the steel during treatment. It is of particular relevance, therefore, that lower levels of oxygen activity are achieved when the steel is contained in ladles and tundishes lined with more thermodynamically stable, neutral to basic (high alumina to dolomite), refractories than in conventional fireclay-lined systems. The interaction between the refractory linings and the metal being treated can hinder the deoxidation, desulphurization and cleanliness treatments the steel undergoes in an otherwise well-developed ladle-metallurgy practice.

5. **Because the steel ladle is a heat sink** and its effect on steel temperature control can be significant, ladle preheating facilities and ladles equipped with slide gates are preferred to help recycle hot ladles.

PROCESSES, TREATMENTS, OBJECTIVES

Processes—Processes (also referred to as systems) of secondary steelmaking define the various ways in which the liquid steel is handled in order to achieve the objectives of secondary steel refining. Particular processes are defined partly by the equipment they use to

manipulate the steel and may involve one or more treatments to achieve the objectives of secondary steel making, such as producing a clean steel. Some processes perform certain functions better than others. (See Figure 19—1.) There can be considerable overlap in the metallurgical functions that various secondary steelmaking processes achieve. To determine which process to install, the steelmaker must reflect on the capability and flexibility of the particular process in its ability to perform the various metallurgical functions required for a given product mix. Consideration must also be given to the way in which the process affects the other operations of the shop, the tonnage throughput expected, and the relative capital and operating costs associated with the process.

Secondary steelmaking processes can be generally classified according to the metallurgical functions they perform. (These metallurgical functions have been listed above under "Purpose of Secondary Steelmaking Processes" and will be discussed below in detail under Objectives.) In this chapter, the processes are divided into three broad categories:

1. vacuum degassing;
2. vacuum degassing and refining with the capability for supplemental reheating of steel;
3. nonvacuum ladle or vessel refining without reheating.

Treatments—Treatments (also referred to as practices or methods) contrast with processes insofar as treatments are the methods by which a given process accomplishes any or a variety of its objectives. Steelmakers implement a treatment within a process to accomplish some objective such as argon bubbling (see Figure 19—2).

Objectives—The objectives of secondary steelmaking are what the processes and treatments are designed to accomplish. They are the goals steelmakers must achieve in order to produce a high quality steel.

Degassing: Liquid steel absorbs gases from the atmosphere and from the materials used in the steelmaking process, which cause embrittlement, voids, inclusions, and other undesirable phenomena in the steel after it solidifies. The major gases to be eliminated include hydrogen and oxygen. Vacuum degassing is a treatment that refers in the broadest sense to the exposure of molten steel to a low-pressure environment to remove gases (chiefly hydrogen and oxygen) from the steel.

Oxygen, as a principal refining agent, plays an important role in determining the final composition and properties of steel. Oxygen dissolved in steel greatly influences the consumption of deoxidizers and thus affects the quality of the steel. The control of oxygen in liquid steel is a prime objective in steelmaking because it enables the desired final chemical composition and solidification structures (for example, rimming steel) to be achieved easily.

The cleanliness of the steel is improved by lowering its oxygen content. If the oxygen content in molten steel is sufficiently high during vacuum degassing, the oxygen will react with some of the carbon in the steel to produce carbon monoxide (CO). The evolved carbon monoxide escapes and is removed from the system by the vacuum pumps along with other gases. When used

**Capability of Secondary Steelmaking Processes
To Perform Metallurgical Functions**

Secondary Steelmaking Processes

Metallurgical Functions	Stream Degassing	DH Degassing	RH Degassing	RH-OB Degassing	Ladle Refining Furnace, Reheat Only	Ladle Refining Furnace, Reheat and Vacuum Degas	Argon/Oxygen Decarburization, AOD	Argon Bubbling, CAS	Argon Bubbling, CAB	Lance Powder Injection	Cored Wire Injection	REM Canister
Composition Control	O	●	●	●	X	X	O	●	●	O	●	O
Temperature Control	–	●	●	●	X	X	●	O	O	–	–	–
Deoxidation (O_2)	O	●	●	●	–	X	●	O	●	●	●	O
Degassing (H_2)	X	●	●	●	–	●	O	–	–	–	–	–
Decarburization	O	●	●	X	O	●	X	–	–	–	–	–
Desulfurization	–	–	–	–	X	X	●	–	●	X	O	O
Microcleanliness	O	●	●	●	–	X	O	O	●	●	●	O
Inclusion Morphology	–	O	O	O	–	O	O	–	●	X	X	X

Legend:
- – None
- O Good
- ● Better
- X Best

FIG. 19—1. Relative efficiency of secondary steelmaking processes.

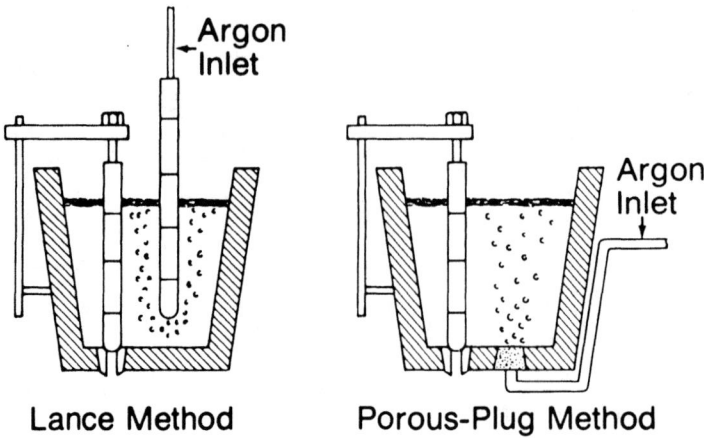

FIG. 19—2. Schematic methods for using the argon-stirring process. (Left) The lance method. (Right) The porous-plug method.

for this purpose (to eliminate oxygen by reaction with carbon), the vacuum-degassing process is often referred to as vacuum-carbon deoxidation. It is difficult and time consuming to produce steel with a carbon content below 0.03 per cent by conventional steelmaking procedures. However, if for example, undeoxidized molten steel at about 0.04 per cent carbon is exposed to a vacuum, carbon is readily removed to a level of about 0.01 per cent by reaction with oxygen in the steel.

In undeoxidized molten steel, the carbon and oxygen contents will approach equilibrium at a given temperature and pressure (See Figure 19—3) according to the following reaction:

$$C + O = CO$$

The equilibrium constant is:

$$K = p\,CO/[\%C][\%O]$$

For carbon contents below about 0.5 per cent and at steelmaking temperatures, the product of $(\%C)(\%O)$ is about 0.002 for one atmosphere of pressure of carbon monoxide. If the steel is subjected to lower and lower pressures, the equilibrium between carbon and oxygen will change and they will react in an effort to establish a new equilibrium. Carbon monoxide produced by this reaction escapes from the system as a gas, and thus, most of the oxygen is no longer available to form nonmetallic inclusions with other substances that may be later added to the steel.

Strong deoxidizers such as aluminum, titanium and silicon, when added to molten steel are effective in reducing the oxygen content so that carbon can no longer react with oxygen when the steel is vacuum degassed. However, strong deoxidizers form nonmetallic inclusions as a product of their reactions with oxygen, and these inclusions may become trapped in the steel during solidification and impair its cleanliness and mechanical properties. Figure 19—4 shows the equilibrium relationship between various deoxidizing elements and oxygen in molten steel at atmospheric pressure (1.013 kPa or 760 mm Hg or 760 Torr), and the influence of changes in pressure (vacuum treat-

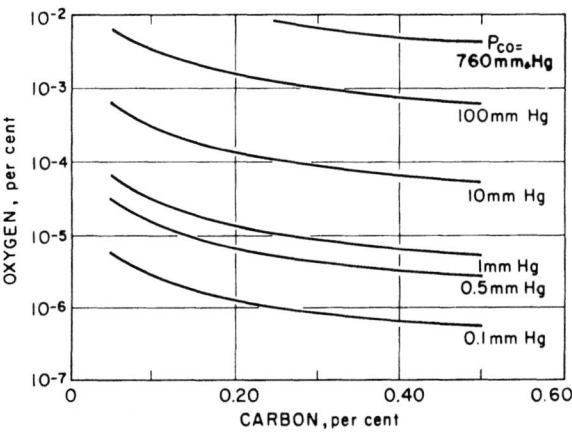

FIG. 19—3. Carbon-oxygen equilibrium relationship in liquid steel at 1600°C (2912°F) for various partial pressures of carbon monoxide in the atmosphere above the liquid steel.

ment) on the equilibrium between carbon and oxygen. At 0.1 atmosphere pressure under equilibrium conditions, oxygen will react with carbon rather than silicon. Although aluminum and titanium are stronger deoxidizers than carbon at 0.1 atmosphere, when the pressure is lowered to 0.01 atmosphere carbon becomes a stronger deoxidizer than either.

Hydrogen is a particularly troublesome gas. It is the cause of bleeding ingots, embrittlement, low ductility, and the presence of blowholes. In solid steel it can cause internal ruptures called thermal flakes. Until relatively recently, effective boiling periods in the steelmaking vessel and the drying of addition agents were necessary precautions taken during steelmaking to limit the amount of hydrogen liquid steel absorbed. Even with these precautions, after solidification the steel had to be subjected to lengthy and complicated heating and cooling cycles to promote the diffusion of hydrogen that the steel might have absorbed.

Exposing molten steel to a low pressure will lower its hydrogen content to such low levels that the steel will

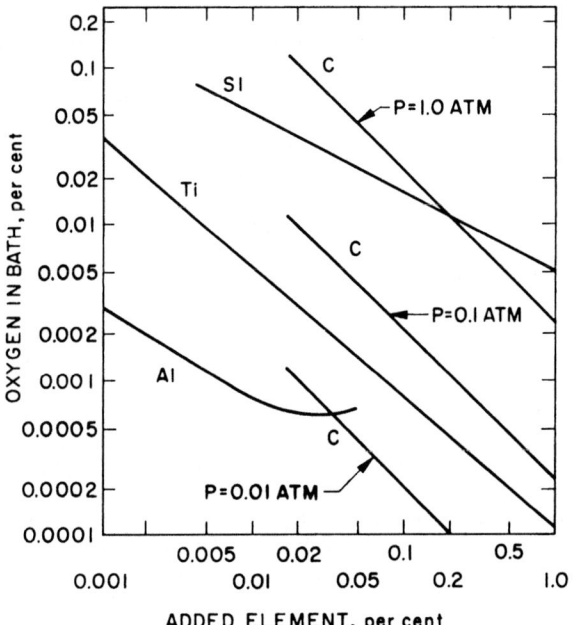

FIG. 19—4. Equilibrium relationship between total oxygen content of a steel bath and the content of various deoxidizing elements. The carbon-oxygen relationship is shown for three different partial pressures of carbon monoxide in the atmosphere above the bath.

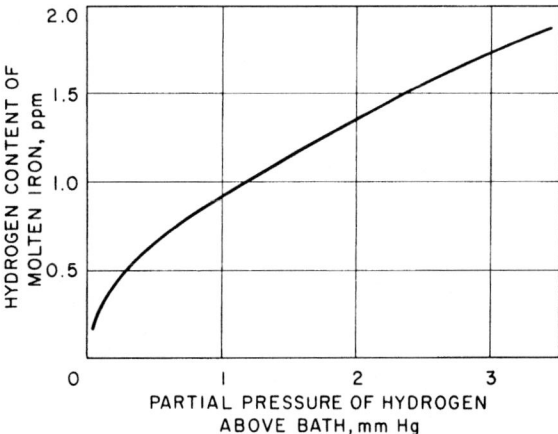

FIG. 19—5. Equilibrium solubility of hydrogen in iron at 1600°C (2912°F), as related to partial pressure of hydrogen in the atmosphere above the molten metal, according to Sievert's Law.

not develop thermal flakes after it has solidified. The amount of hydrogen that can be contained in molten steel under equilibrium conditions is proportional to the square root of the partial pressure of hydrogen in the atmosphere above the bath of molten steel (Sievert's law for gases dissolved in steel); this is expressed by the equation:

$$\% \ [H] = K \ \sqrt{p_{H2}}$$

where % [H] is the per cent of hydrogen dissolved in the molten steel, K is a constant (= 0.0027 at 1600 degrees C and 1 atmosphere of hydrogen), and p_{H2} is the partial pressure of hydrogen in the atmosphere above the molten steel (expressed in atmospheres). Thus, reducing the partial pressure of hydrogen above the molten steel by vacuum treatment enhances the removal of hydrogen from the steel.

Figure 19—5 shows the relationship between hydrogen content in liquid pure iron at 1600° C (2912° F) and the partial pressure of hydrogen above the bath.

Other conditions affecting the amount of hydrogen removal in relation to the kinetic conditions of the various vacuum-degassing processes include: surface area exposed to a vacuum, degassing pressure, steel chemical composition, extent of deoxidation with strong deoxidizers prior to vacuum degassing, and hydrogen pickup from contaminating environments such as nondried alloy additions, refractories, and slags during or after degassing. In addition, pouring the steel in the air can increase the hydrogen content of vacuum-degassed steel.

Nitrogen may have an undesirable effect on the properties of steel depending on its composition, subse-

quent treatment, and intended use. Nitrogen contents should be as low as possible in low-carbon steels intended for deep-drawing applications. Low nitrogen contents (0.004 per cent maximum) are particularly important, especially with the new sheet steels developed for processing in continuous-annealing lines.

Nitrogen, like hydrogen, also obeys Sievert's law:

$$\% \ [N] = K \ \sqrt{p_{N2}}$$

One would expect to remove nitrogen by reducing the partial pressure of nitrogen above the liquid bath. Unfortunately, this does not readily occur. Although some nitrogen is removed from molten steel by inert gas flushing or vacuum degassing, the amount is extremely slight. Instead, nitrogen tends to form stable nitrides that cannot be removed effectively by commerical methods of vacuum treatment. As a result, low nitrogen contents are attained mainly by control of primary steelmaking practices.

Decarburization: Producing steels with carbon contents below 0.03 per cent by conventional steelmaking procedures is difficult and time consuming. But if undeoxidized molten steel at about 0.04 to 0.06 per cent carbon content is exposed to a vacuum, carbon is readily removed to a level of 0.01 per cent and lower by reaction with oxygen in the steel. This is due to the fact, as explained earlier, that carbon and oxygen react at the low pressure to produce carbon monoxide that, essentially insoluble in steel, readily escapes.

Ladle Desulphurization: To make external steel-refining processes as effective as possible in removing sulphur, the following factors must be considered in addition to the type of desulphurizing agent used:

1. Oxygen in the steel—Oxygen is a deterrent in sulphur removal. To increase the efficiency and effectiveness of a desulphurizing agent, the steel should be deoxidized, preferably with aluminum, to the lowest practicable oxygen content. Deoxidizing the steel prior to the desulphurizing process is preferred, but in some processes deoxidation

is performed concurrently with the desulphurizing treatment.

2. Oxides in the furnace slags—The FeO in the furnace slag, like the oxygen in the steel, interferes with the efficiency of the desulphurizing agents. Since other oxides in the slag are unstable during the subsequent desulphurization process, the methods used should minimize or eliminate the carryover of oxidizing furnace slags in the desulphurizing process.

3. Oxides in ladle linings—In some desulphurizing processes, the oxides in regular fireclay are unstable and are reduced in contact with liquid steel, thereby decreasing both the efficiency of the desulphurization agents and the cleanliness quality of the steel.

4. Degree of mixing of reagent and steel—For the desulphurizing reaction to proceed efficiently, the desulphurizing agent and the steel must mix intimately. Conventional steel desulphurization always involves liquid slag and metal reactions.

5. Initial and final sulphur contents—As in other chemical reactions, the law of mass-action applies in the desulphurizing process. Consequently, it is increasingly difficult to attain very low sulphur levels (below 0.006 per cent) because the efficiency of the desulphurizing agent diminishes as lower sulphur levels are reached.

Inclusion Morphology: If non-metallic inclusions do remain in the steel, steelmakers may change their chemical composition and/or shape to make them as innocuous to the desired properties of the resulting steel as possible. Rare earth metals (REM) and calcium alloys are very effective in changing the morphology of the sulphide inclusions from the stringer type to the globular sulphides and this change results in a more isotropic steel. An effective production method for adding REM to molten steel is the plunging method, which will be discussed in detail under Ladle Injection in Nonvacuum Ladle Refining.

EQUIPMENT: LADLES, VESSELS, SPECIAL FURNACES

Secondary steelmaking processes relate to all operations of the post-treatment of liquid steel and occur outside the primary steelmaking furnace. They are usually performed in ladles, vessels or special furnaces. Primary furnaces, far more rugged than secondary steelmaking vessels, have the capacity to contain the violent reactions of primary steelmaking. Their linings are different in order to facilitate their capacity to produce raw, liquid steel. Whereas primary furnaces are designed to produce raw, liquid steel, secondary steelmaking vessels are not. Secondary steelmaking vessels enable secondary steelmaking processes to achieve other refinements of steel impossible with the primary furnaces.

Ladles—A ladle was a device originally used primarily to transfer liquid steel from the primary furnace to another container. Liquid steel is always tapped from a primary furnace into a tapping ladle. From there it may be poured into molds, other ladles, vessels or special furnaces, or it may undergo treatments in the tapping ladle itself.

After the molten steel in the primary furnace has been tapped into a ladle, a ladle crane moves the ladle to an area where a secondary steelmaking process can be performed, directly to the pouring platform where the steel is poured into ingot molds, continuous casting tondishes, or to a continuous casting line. The steel may be reladled (i.e., poured from one ladle into another) before teeming to insure proper mixing of alloys especially when large alloy additions are added and to reduce temperatures.

Vessels—A vessel is a container, as is a ladle or a furnace, in which liquid steel can be contained while undergoing various treatments. It is designed for secondary steel refining, but is not designed for the transfer of molten steel as is a ladle. A vessel enables the steelmaker to contain the steel while making finer adjustments to it beyond the scope of the primary process.

Special Furnaces—Some secondary processes such as the ESR (Electro Slag Remelting) or VAR (Vacuum Arc Remelting) processes occur in special furnaces. These special furnaces are not as rugged as primary furnaces, yet have the capability for furnishing heat to the secondary steel making process which uses it.

DEVELOPMENT OF MAJOR SECONDARY STEELMAKING PROCESSES

The following is an overview of the chronological development of the major secondary steelmaking processes. The ladle-to-ladle and ladle-to-mold vacuum-degassing processes, first introduced in the 1950's, were developed primarily to decrease the hydrogen content of alloy steels to prevent flake cracking in forgings and heavy-gage plate steels. The Dortmund Hoerder (DH), and Ruhrstahl-Heraeus (RH) processes developed in the early 1960's were designed with the objectives of degassing a ladle of steel without total enclosure of the ladle in a vacuum chamber. They also had the advantage of being able to vacuum-decarburize via the vacuum-carbon deoxidation reactions at reduced pressures, to low carbon contents. Significant alloying and coolant additions could be made to the steel during the course of the treatment.

In the 1970's Nippon Steel Corporation recognized that the RH vacuum-degassing equipment could be utilized as a reactor vessel in which metallurgical functions other than degassing for the removal of hydrogen from steel could be accomplished at vessel pressures considerably higher than those at which conventional degassing processes operate. Their development of RH and RH-OB treatments for low carbon steels enabled them to effectively deoxidize and decarburize these grades while also decreasing the consumption of ferroalloys, improve steel cleanliness, attain a uniformity in the temperature of the steel in the tundish for continuous casting, and attain an overall improvement in the performance of their basic-oxygen-steelmaking operations.

The ladle-refining furnace, introduced by ASEA in the early 1960's, was followed by the development of other similar processes and provided an extension to vacuum degassing. These combined processes, which have the capabilities for controlled stirring, alloying, and reheating of the steel in the ladle, possess the greatest flexiblity in secondary steelmaking available in the mid-1980's.

The argon-oxygen decarburization process (AOD) was introduced by the Linde Division of Union Carbide Corporation in the late 1960's for the production of stainless and other grades of high alloy steel. After its introduction, a number of AOD operators have found the process attractive for refining certain grades of carbon and low alloy steels for special-requirement applications. It has been claimed that the AOD process is capable of performing all the preferred metallugical functions, except for sulphide shape-control, in a single vessel operation while the quality of AOD refined steel is claimed to be equivalent to that of other secondary steelmaking processes.

Ladle-injection processes for injecting powdered metals or slags into a ladle of steel were introduced in the early 1970's. The primary purpose in the development of these secondary steelmaking processes, developed almost simultaneously in Germany (TN process developed by Thyssen-Niederrhein) and in the Scandinavian countries (Scandinavian Lancers process), was to attain rapid and efficient desulphurization of steel in the ladle. These processes inherently assist in the removal of the products of deoxidation from the steel and have been found to be favorable for the modification of the non-metallic sulphide and oxide inclusions in the steel.

The addition of aluminum by plunging of an aluminum doughnut into the ladle and the addition of cerium to steel by plunging of a rare-earth metal canister deep into the ladle are also recognized as operational practices for the ladle treatment of steel.

VACUUM-DEGASSING PROCESSES

Vacuum-degassing processes, in the broadest sense intended here, refer to the exposure of molten steel to a low-pressure environment to remove gases (chiefly hydrogen and oxygen) from the steel. The effectiveness of any vacuum degassing operation depends upon the surface area of liquid steel that is exposed to low pressure. The mechanisms of hydrogen and oxygen removal from liquid steel are related directly to surface area.

Hydrogen removal is a diffusion and partial pressure phenomenon. Oxygen removal is a function of chemical reaction of oxygen with carbon and the partial pressure of carbon monoxide.

The processes by which a degassing treatment is accomplished also achieve a host of other objectives including: composition and temperature control; decarburization; microcleanliness; and inclusion morphology. Under the vacuum-degassing treatments three processes which primarily use this treatment will be discussed: stream degassing, recirculation degassing, and vacuum degassing in the ladle.

Before discussing the various processes, some general considerations which affect the overall efficiency of the treatment are discussed including: temperature losses, vacuum pumping capacities and deoxidizer and alloy additions.

Temperature Losses—Since the use of any commercial vacuum-degassing system results in some drop in temperature of the liquid steel during the degassing operation, steel intended for degassing must be tapped at temperatures higher than would be used if the steel were not to be subjected to vacuum treatment.

Heat losses during degassing occur through radiation and through absorption of heat by ladles or vacuum vessels, and are greater with some degassing systems than with others. Radiation shields and a layer of slag covering the steel are used in some systems to minimize heat loss by radiation. Ladles and vacuum vessels are preheated to high temperatures to minimize the amount of heat they will absorb from the molten steel and to prevent formation of skulls caused by a layer of steel solidifying where it comes in contact with the relatively cool refractory linings of ladles or vessels. A few degassing systems have heating units intended to offset, at least partially, the heat losses inherent in the systems. These systems will be discussed in the section, "Vacuum Degassing with Supplemental Reheating." The amount of heat lost from any degassing system is related directly to the duration of the process, and the time required always is kept to a minimum consistent with adequate removal of gases from the steel.

Vacuum Pumping Capacities—Steam ejectors provide the pumping means for creating the vacuum in most modern degassing systems. They are installed in series to provide multistage pumping. The number of stages varies from four to six in the several existing systems. The required pumping capacity is dependent upon the grades of steel to be degassed, their level of oxidation prior to degassing, and the characteristics of the vacuum degassing system used. Steels of low carbon content and in an unoxidized or partly deoxidized condition require larger pumping capacities than do steels of higher carbon content in a deoxidized condition, when treated at the same rate.

The vacuum-induction melting process, particularly the process formerly called the Therm-I-Vac process (described later under Vacuum Degassing with Supplemental Reheating) uses six steam-ejector pumping stages in series during its reheating and refining phase, during which pressure can be as low as 0.010 Torr. Throughout this chapter, Torr will be the standard reference of pressure. Conversion factors to microns and micrometers of Hg are: 1 mm Hg = 1 Torr = 1000 microns; 0.01 mm Hg = 0.01 Torr = 10 microns.

The D-H and R-H systems (described in this section under Recirculation Degassing) normally use five-stage steam ejectors in series that can reduce pressure to as low as 0.050 Torr. In most of the other systems to be described in this section, four-stage steam ejectors in series are used with the pressures ranging from 0.3 to 1.0 Torr during degassing.

Deoxidizer and Alloy Additions—All vacuum degassing systems provide for adding deoxidizers and alloy additions to the molten steel while maintaining low pressure. However, the amount of the additions is limited for each system by (1) the cooling effect of the additions; (2) the ability of a given system to provide adequate mixing of the additions with the steel; (3) provisions of systems to replenish the heat lost during vacuum treatment.

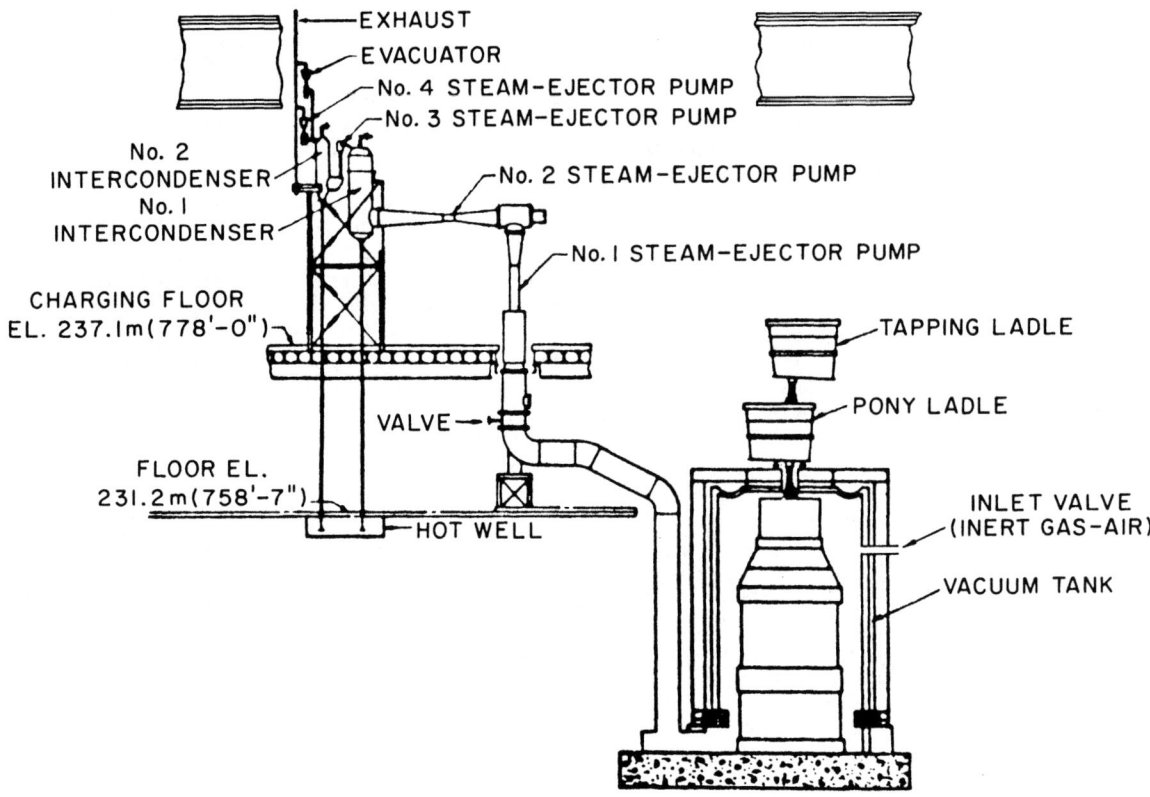

FIG. 19—6. Schematic arrangement of a vacuum-casting installation for casting large forging ingots.

STREAM DEGASSING PROCESSES

The following systems are included in the stream-degassing processes:

    a. ladle-to-mold degassing;
    b. ladle-to-ladle degassing;
    c. tap degassing.

a. Ladle-to-Mold Degassing—One variation of the ladle-to-mold or vacuum casting method is used principally for casting large ingots for forging. Figure 19—6 illustrates diagrammatically the equipment used in this method. It consists of a vacuum tank and base, a transfer or "pony" ladle, a mold, an exhaust line with a vacuum valve, and a pumping system that employs four-stage steam ejectors. To vacuum cast by this method, an ingot mold is placed on the base and fitted with a hot top, and the mold, if cold, is heated to the desired temperature by gas burners. The pony ladle, equipped with a nozzle and stopper-rod assembly, is preheated and a sleeve surrounding the nozzle on the bottom of the ladle is sealed by an aluminum diaphragm. A vacuum tank that completely encloses the mold is placed on the base, resting on a seal that makes the joint between the tank and base vacuum-tight. A preheated refractory collar is fitted into an opening in the top of the tank to direct the pouring stream from the pony ladle nozzle into the mold. The pony ladle is placed on top of the tank, resting on a seal that makes

a vacuum-tight joint between the pony ladle and tank. The steam ejectors are then started and the vacuum tank is evacuated to a pressure of about 0.10 to 0.20 Torr.

A heat of steel with about 22° C to 33° C (40° F to 60° F) superheat is tapped from the steelmaking furnace into the tapping ladle that is then transferred to a position above the pony ladle. Steel is bottom-poured from the tapping ladle into the pony ladle until the latter is about two-thirds full. The stopper rod of the pony ladle is then raised, and the stream of molten steel ruptures the aluminum diaphragm and passes through the refractory collar into the mold in the vacuum tank. As the stream of molten steel enters the evacuated space, it breaks up into tiny droplets, exposing an enormous surface to the degassing influence of the vacuum. As the steel first enters the evacuated space, the pressure inside the tank surges to about 0.8 Torr, and then declines to about 0.5 Torr. Some newer vacuum-casting facilities are equipped with a five-stage steam ejector system capable of maintaining a pressure of 0.10 Torr during the pouring of the steel. After the ingot mold is filled, a valve in the vacuum line between the tank and the pumping system is closed and pressure in the vacuum tank is restored to atmospheric pressure by admitting a mixture of nitrogen and air. Precautions must be taken to prevent gas and dust explosions when restoring the tank to atmospheric pressure because combustible gases and dust are generated during vacuum casting. Successive heats of steel may be poured by this method from several tapping ladles into the pony ladle in casting very large ingots: ingots 3400 mm (134 inches) in diameter and weighing over 320 metric tons (350 net tons) have been cast by this method.

b. Ladle-to-Ladle Degassing—Besides being used for removing hydrogen from steel, ladle-to-ladle degassing can also remove oxygen by vacuum carbon deoxidation. Figure 19—7 shows diagrammatically the components of a ladle-to-ladle degassing system. A description of its use for vacuum carbon deoxidation gives a good overall description of the operation of the system.

A preheated teeming ladle with either a nozzle and stopper rod or a sliding-gate mechanism is placed inside the tank. The opening in the tank cover is then sealed with an aluminum disc. The required amounts of ferrosilicon and aluminum are placed in the alloy-additions hopper. The steam-ejector pumping system is started and the tank is evacuated to a pressure of about 0.1 Torr. A heat of steel is superheated in the steelmaking furnace to a temperature ranging from about 45° to 72° C (80 to 130° F) above the conventional tapping temperatures (small heat sizes require the higher tapping temperatures). The heat is then tapped into a tapping ladle and is usually partly deoxidized with ferrosilicon.

The tapping ladle is placed on top of the vacuum tank, resting on a seal between the two. When the stopper rod of the tapping ladle is raised and steel flows from its nozzle, the steel melts the aluminum disc that sealed the opening in the top of the tank and flows through the refractory collar into the teeming ladle in the evacuated tank. During this operation, the remainder of the ferrosilicon and all of the aluminum are added to the steel by the alloy-additions hopper without breaking the vacuum. At the start of pouring, the pressure inside the tank surges to as high as 2.0 Torr, then declines rapidly to a stable level of about 0.8 Torr. After the degassing operation is completed, pressure in the tank is raised to atmospheric pressure by admitting nitrogen following which the tank cover is removed, the ladle of treated steel is removed, and the steel is teemed in the conventional manner. As in the ladle-to-mold degassing processes, precautions are taken to prevent gas or dust explosions when restoring the tank to atmospheric pressure because of the combustible gases and dust generated during degassing.

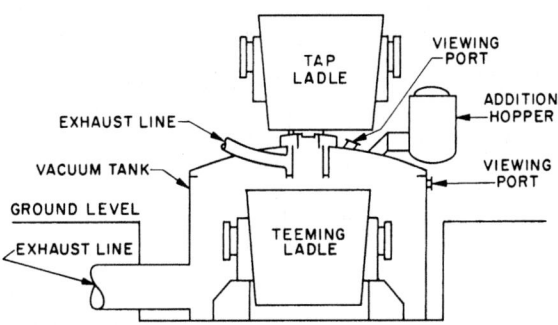

FIG. 19—7. Schematic arrangement of equipment used in the ladle-to-ladle stream degassing process.

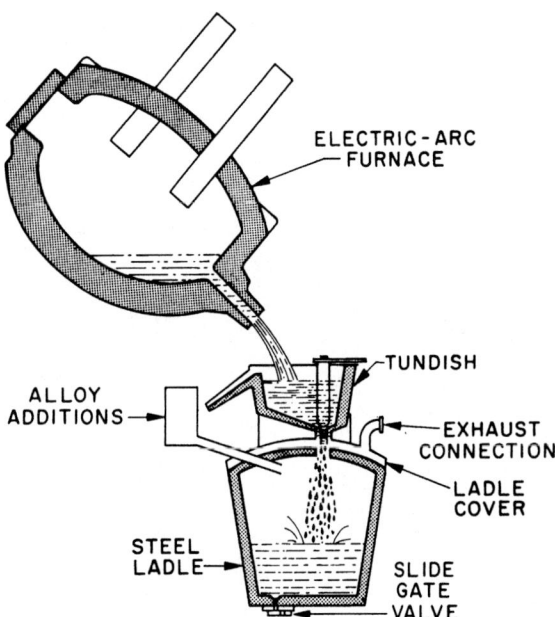

FIG. 19.—8. Diagrammatic arrangement of equipment used in the tap-degassing process.

c. Tap Degassing—Figure 19—8 is a sketch of the equipment used in tap degassing. It consists of a special ladle with a flange at the top and an "O"-ring seal plate for the nozzle opening at the bottom, a ladle cover, a stoppered tundish having an aluminum-disc seal just below its nozzle, a flexible exhaust line for the vacuum system, and a steam-ejector pumping system.

To degas a heat of steel by this method, the nozzle of the special ladle is sealed and its cover is put in place. The "O"-ring forms a seal between the ladle and cover. Part of the stopper-rod assembly, if the ladle is so-equipped, extends through the ladle cover and this region must be made air-tight by "O"-ring seals. The preheated, stoppered tundish is placed on top of the ladle cover, a suitable seal making the joint between the two air tight. The assembly is then positioned where the tapping stream from the steelmaking furnace can be collected in the tundish and the flexible exhaust line is connected to the exhaust opening in the ladle cover. The steam ejectors are started and the interior of the ladle is evacuated to between 0.2 to 0.5 Torr pressure. Molten steel, superheated about 22° C (40° F) higher than in conventional practice, is tapped from the steelmaking furnace into the tundish. When the tundish is almost full, its stopper rod is raised, the steel melts the aluminum disc and flows into the evacuated ladle. Pressure inside the ladle rises to between 7.0 to 9.0 Torr. During degassing, the required deoxidizers are added to the steel in the ladle. In some installations, the ladle inside the vacuum chamber is equipped with a porous plug through which argon is blown during the reladling process. After degassing is completed, the ladle is restored to atmospheric pressure with nitrogen and the flexible exhaust line is removed. The steel is then teemed in the conventional manner.

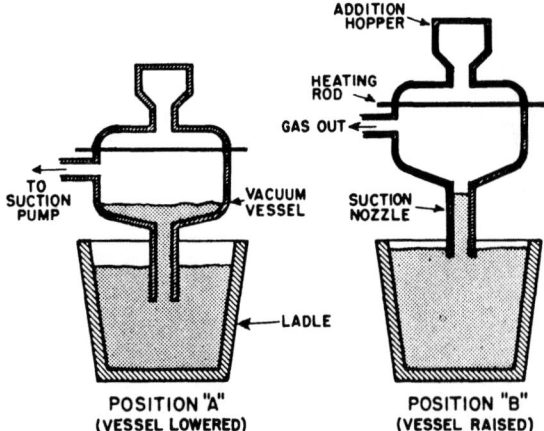

FIG. 19—9. Schematic diagram showing principle of operation of the D-H (Dortmund-Horder) process. As the evacuated vessel is lowered and then raised, atmospheric pressure causes molten steel to rise into the evacuated space and then partially run back into the ladle.

RECIRCULATION DEGASSING

Recirculation degassing is a method in which the liquid metal in a ladle is forced by atmospheric pressure into an evacuated degassing chamber where it is exposed to low pressure and then flows back into the ladle. The metal may recirculate through the chamber 40 to 50 times to achieve the desired levels of degassing. The vacuum environment is used to recirculate the steel as well as to serve as the means by which the degassing is accomplished. Three major processes are discussed:

a. Dortmund-Horder-Huttenunion Process (D-H);
b. Ruhrstahl-Heraeus Process (R-H);
c. RH-OB Process.

a. D-H Process—Figure 19—9 is a diagrammatic sketch of the equipment used in this system. It consists of a refractory-lined vacuum vessel with a "snorkel" tube, an electric-resistance graphite heating rod, an alloy-additions hopper, a mechanism for raising and lowering either the vacuum vessel or the ladle, and a five-stage steam-ejector system.

To degas a heat of steel by this method, the vacuum vessel is preheated internally to a refractory-surface temperature of about 1538° C (2800° F). Required alloy additions, such as ferrosilicon, aluminum, carbon and ferromanganese, are placed inside the alloy-additions hopper. The steel to be degassed is superheated from 17° to 83° C (30 to 150° F) above the tapping temperature of steel that is not to be degassed. The degree to which a heat of steel is superheated depends upon heat size, ability to preheat the vacuum vessel sufficiently, amount of alloys to be added, and expected duration of the degassing period.

The heat is tapped in an undeoxidized or partially deoxidized condition into a conventional ladle. The ladle is positioned beneath the vacuum vessel. The lower end of the snorkel tube is covered with a cone-shaped

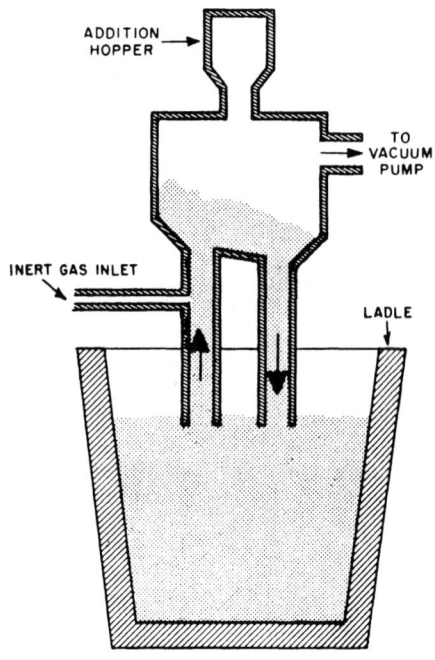

FIG. 19—10. Principle of operation of the R-H (Ruhrstahl-Heraeus) process. Argon gas injected into one extension or leg of the vessel causes molten steel to rise into the evacuated chamber with a boiling action that releases gases from the steel that then flows back into the ladle through the second leg. The recirculation of steel is continued until the desired degree of degassing is attained.

slag breaker. Two systems are used in this process: either the vacuum vessel is movable vertically or the ladle of steel is movable vertically. In either system, the lower end of the snorkel tube is immersed in the molten steel in the ladle and the chamber is evacuated to the desired low pressure. Molten steel is forced into the vacuum chamber by atmospheric pressure and is degassed. Then the snorkel tube is raised while keeping the lower end of the tube immersed in the steel, thus causing the degassed steel to flow back into ladle. This cycle is repeated 40 to 50 times as the pressure in the vacuum vessel gradually decreases to the range of 0.3 to 0.6 Torr and lower. During the later stages of degassing, the desired additions are made. After the steel is considered to be properly degassed, the vacuum vessel is restored to atmospheric pressure and the snorkel tube is withdrawn from the ladle. The ladle of degassed steel is then teemed or continuous-cast in the conventional manner.

Means are provided in some D-H installations to introduce argon and/or oxygen inside the snorkel in order to produce steels of lower hydrogen or carbon contents than those produced without the introduction of these gases.

b. R-H Process—Figure 19—10 illustrates the equipment used in the R-H process. It comprises a refractory-lined vacuum vessel with two legs or "snorkel" tubes at the bottom, an alloy-additions hopper, a mechanism for raising and lowering the vacuum vessel, a gas-inlet pipe on one of the legs, and a steam-ejector system. The vacuum vessel is preheated to high tem-

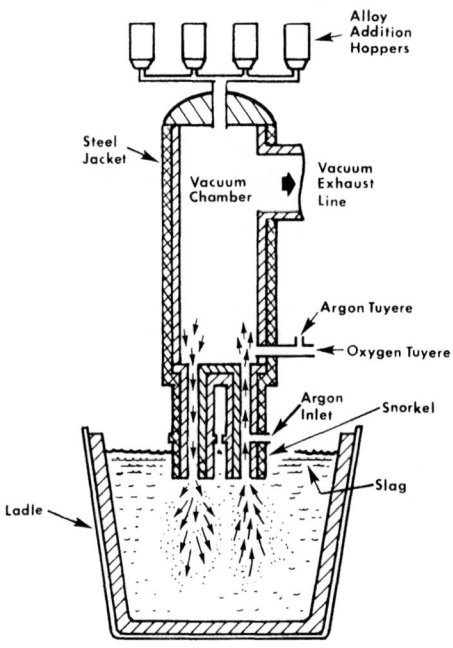

FIG. 19—11. Schematic diagram of the RH-OB vacuum degassing system.

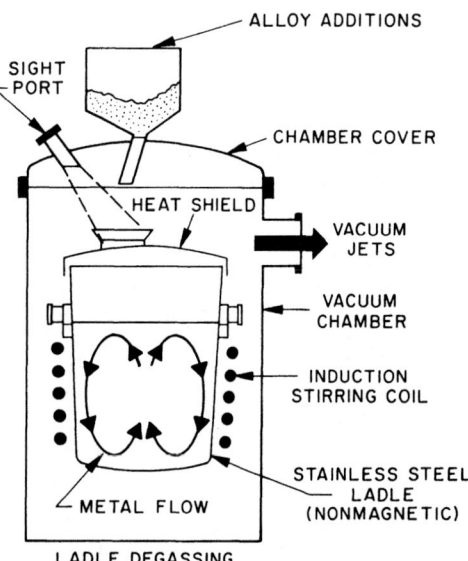

LADLE DEGASSING

FIG. 19—12. Schematic arrangement of equipment used in the induction-stirring ladle-degassing process.

peratures by either gas burners or electric heating rods to minimize steel temperature losses during degassing.

The steel to be degassed is superheated in the steelmaking furnace to a tapping temperature about 42° to 61° C (75° to 110° F) above that for steel that is not to be degassed. The steel is tapped into a regular ladle with little or no deoxidation, and the ladle is placed in the degassing station. The preheated vacuum vessel is lowered until at least about 150 mm (6 inches) of each leg is immersed in the molten steel. The steam ejectors are started and the vacuum chamber is evacuated to a low pressure. An inert gas is introduced into one of the legs and the lower density of the gas-steel mixture causes the steel to flow up that leg and into the vacuum chamber where it is degassed and then flows by gravity through the other leg back into the ladle. As degassing proceeds, the pressure inside the vacuum chamber gradually decreases to a level of about 0.3 to 0.6 Torr and lower. During the later stages of degassing, the deoxidizers and other alloys are added to the steel through the hopper. After the steel is considered to be properly degassed, the vacuum chamber is returned to atmospheric pressure, the vacuum chamber is raised and the ladle is moved to the pouring platform for conventional teeming of the steel.

c. RH-OB Process—Figure 19—11 shows a diagram of the equipment used in the RH-OB system. The RH-OB system is used primarily for the removal of hydrogen and a certain amount of carbon from the steel. The RH-OB process uses equipment nearly identical to that of the RH process: it comprises a refractory-lined vacuum vessel with two legs or "snorkel" tubes at the bottom, an alloy addition hopper, a mechanism for raising and lowering the vacuum vessel, and a steam-ejector system. The difference lies in the

gases injected into the system during the process of degassing. Whereas the RH system introduces an inert gas into one of the legs through a gas-inlet pipe to facilitate the flow of the metal, the RH-OB process also introduces oxygen, in addition to an inert gas, into the system. The inert gas aids in stirring the metal. The introduction of oxygen assists in the decarburization of the liquid steel.

VACUUM LADLE DEGASSING

This section comprises two processes: induction stirring and vacuum-oxygen decarburization.

Induction Stirring—In this process, illustrated schematically in Figure 19—12) induction coils are used to induce eddy currents in the molten steel to produce a stirring effect. To contain the steel during degassing, a special ladle with sections of the steel shell containing non-magnetic austenitic stainless steel, a high-alumina or basic refractory lining and about 1.5 metres (5 feet) of freeboard are used.

Steel to be degassed is superheated in the steelmaking furnace to between 39 degrees to 83 degrees C (70° to 150° F) above the temperature required for steel not to be degassed: higher superheat is required for heats that receive large alloy additions or require long degassing time at low pressures.

Prior to tapping, the furnace slag is removed from the heat in the electic-arc steelmaking furnace, as slag carry over into the ladle is undesirable. The steel is then tapped into the special ladle that has been preheated. If the heat has not yet been deslagged, it can be removed from the ladle by a ladle slag skimmer. As in the degassing processes using gas stirring (discussed below under Vacuum Degassing with Supplemental Reheating: Gas Stirring - Arc Reheating), furnace slags are undesirable in the ladle. Little or no deoxidizing of the steel is done before degassing. The ladle is then placed inside the induction coil positioned inside the vacuum tank, a heat radiation and splash shield is placed on top of the ladle, the cover is placed on the vacuum tank, and the steam ejectors are started to evacuate the tank to about 0.1 Torr in about 10 to 25 minutes.

During degassing, the induction coils are energized to stir the steel. At the end of the degassing period, the alloys are added (chiefly ferrosilicon, aluminum and carbon, which are placed in the hopper above the tank prior to degassing), and stirring is continued for about two minutes to mix the additions with the molten steel. The vacuum tank is then returned to atmospheric pressure, the tank cover and radiation shield are removed, and an inert synthetic slag is put on the surface of the steel to minimize heat losses and reoxidation. The ladle is then removed from the tank and teemed into molds either in air or under an inert atmosphere. (In the United States, this method is also known as the Republic-Stokes process.)

The less common variation of the ladle-degassing method mentioned above is the use of a flanged-top ladle equipped with porous plugs and a cover that fits on top of the ladle, forming a vacuum seal with the ladle flange. This enables the steel shell of the ladle and the ladle cover to serve as the vacuum enclosure. For this method, a heat of steel is tapped into the ladle. The cover is then placed on top of the ladle and the heat is stirred by admitting argon through the porous plugs while the space above the heat is evacuated to low pressures. During vacuum degassing, deoxidizers and ferroalloys may be added to the heat. After degassing, the space above the heat of steel is returned to atmospheric pressure, the cover is removed, an insulating synthetic slag may be added, and the steel is teemed in the conventional manner. The minimization of furnace slag and provision of adequate ladle freeboard also applies to this method of degassing.

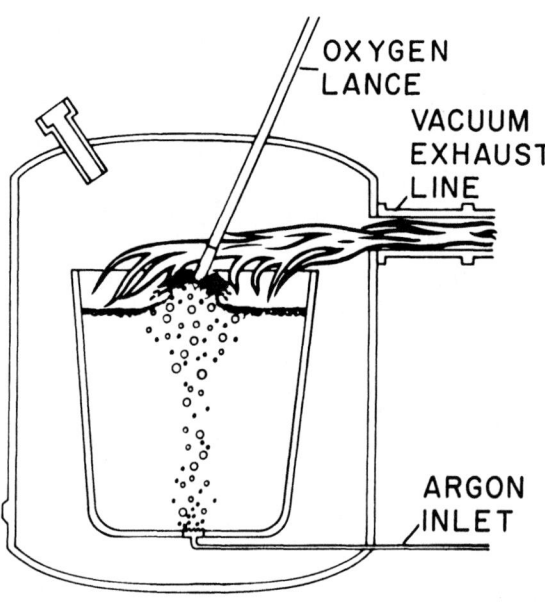

FIG 19—13. Schematic diagram of the VOD process.

Vacuum-Oxygen Decarburization (VOD)—This process was designed for the more economical production of stainless steel, as was the AOD process discussed below under Nonvacuum Ladle or Vessel Refining. A heat of ferrochromium, ferrosilicon, stainless-steel scrap, burnt lime and fluorspar is charged and melted, then heated to the desired tapping temperature. It is then tapped into a preheated basic-lined ladle equipped with porous plugs for blowing argon. Next the ladle is placed inside a vacuum chamber and the system is evacuated to low pressures while oxygen is blown through a lance above the bath of steel and argon is blown through the porous plugs in the bottom of the ladle, as shown in Figure 19—13. After the oxygen-blowing period, further decarburization and deoxidation can be accomplished by use of additional vacuum time: if deoxidation alone is desired, silicon and/or aluminum is added under vacuum. After the vacuum treatment, the chamber is restored to atmospheric pressure, the cover is removed and the steel is sampled and analyzed. If necessary, trim additions of ferroalloys are added. A desulphurizing slag can also be added to the heat after the vacuum treatment and argon stirring. About 50 per cent of the initial sulphur can be removed by this process. Induction stirring can be used to stir the steel instead of argon blowing.

VACUUM DEGASSING WITH SUPPLEMENTAL REHEATING

In this section the following processes will be discussed:

a. Ladle refining furnace (Finkl-Mohr and ASEA-SKF);
b. Vacuum-arc remelting (VAR);
c. Modified vacuum induction (Therm-I-Vac).

a. Ladle Refining Furnace—The processes discussed in this section are all performed in ladle refining furnaces where it is possible to add heat to the secondary steelmaking process. A ladle refining furnace has the ability to reheat the liquid steel. This allows the steelmakers to add greater amounts of alloys to the liquid steel. The furnaces' ability to provide heat during refining by arc-reheating gives the steelmakers the capability to desulphurize, deoxidize and perform metallurgical operations with synthetic slag. All of the ladle refining processes discussed below take advantage of these increased possibilities. Heat is supplied to the processes through electrodes, a process called arc reheating.

Ladle refining furnaces can either reheat or reheat and vacuum degas. The capabilities of the ladle refining furnaces which only reheat perform the following secondary steelmaking functions as well as or better than any other secondary steelmaking process: composition control, temperature control, deoxidation and microcleanliness. Ladle refining furnaces that also vacuum degas, in addition to sharing the above assets, perform desulphurization as well as or better than any other secondary steelmaking process.

There are several different companies which have developed individual variations to the ladle refining

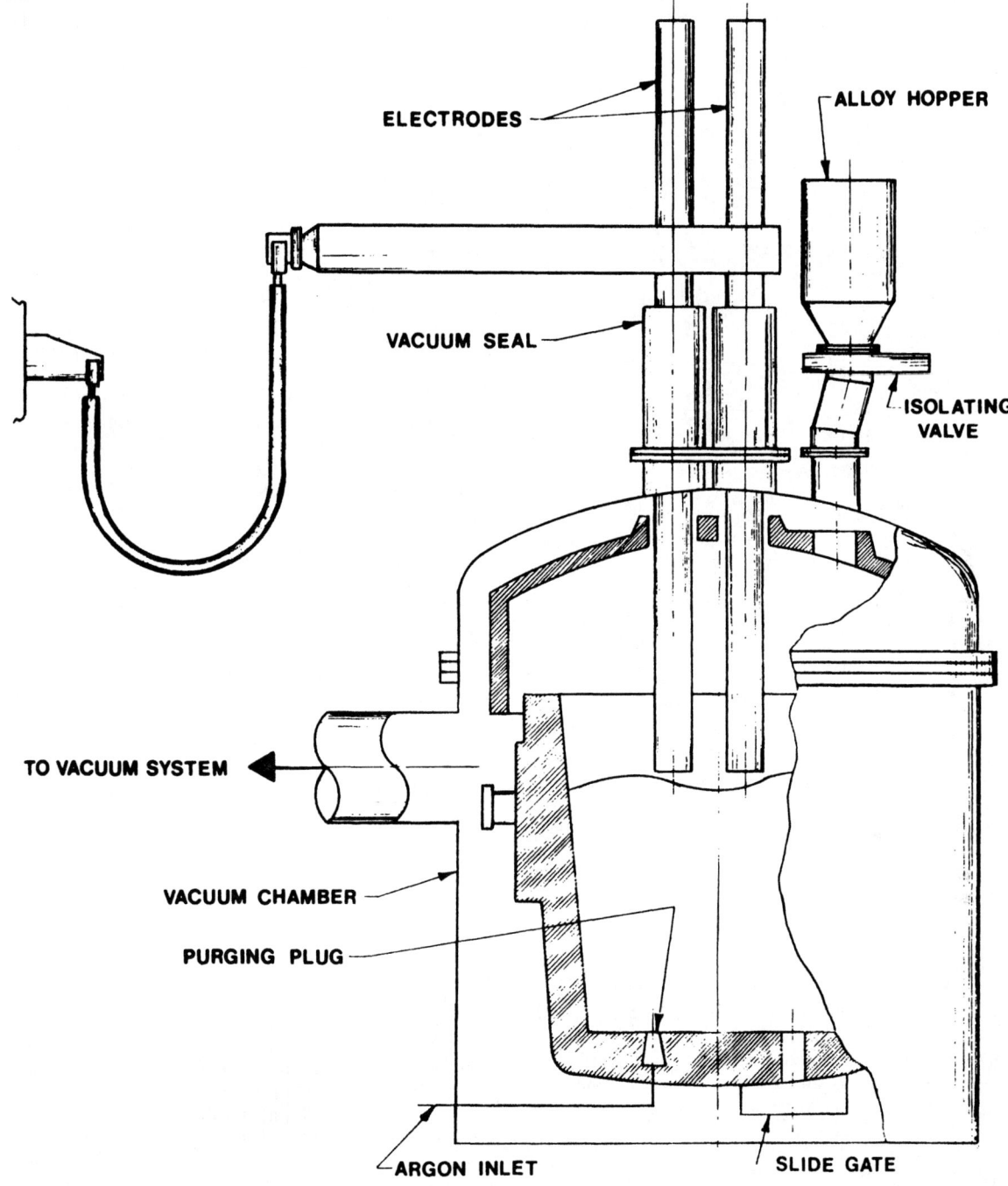

FIG. 19—14. Schematic arrangement of equipment used in the gas-stirring arc-reheating process. (Courtesy, Mohr Vacuum Process Division, John Mohr and Sons.)

furnace processes, such as Finkl-Mohr, ASEA-SKF, Surface/Heurtey, Daido and Lectromelt processes.

Gas Stirring-Arc Reheating and Induction Stirring-Arc Reheating

Gas Stirring-Arc Reheating—This system, also known as the Finkl-Mohr VAD degassing system, is designed for desulphurization and additional steelmaking refine-ment purposes, as well as degassing of the liquid steel.

This process (See Figure 19—14) is very similar to a ladle vacuum degassing process which is described below, except that heat is supplied to the steel before or after the degassing operation by electric arcs provided by electrodes inserted through the top of the vacuum-tank cover.

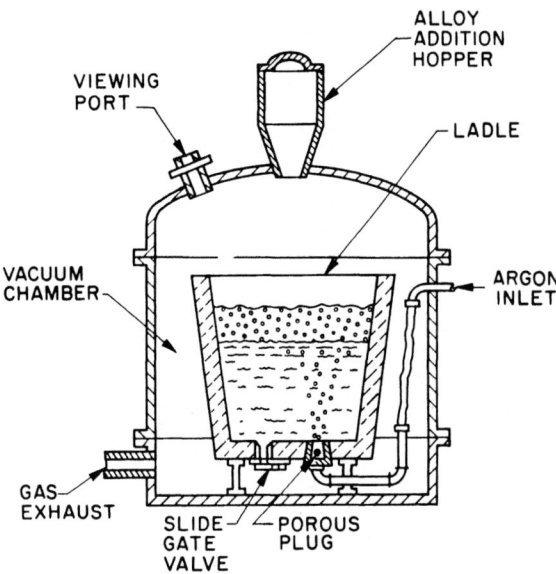

FIG. 19—15. Schematic arrangement of equipment for using an inert gas to stir molten steel during ladle degassing.

An inert gas, usually argon, is used to stir molten steel in this ladle degassing process. Figure 19—15 is a diagrammatic sketch of the equipment used. A special ladle with a slide-gate mechanism and porous refractory plugs in the bottom or sides is used, in conjunction with a vacuum tank that encloses the ladle during the degassing operation. A gas-inlet system introduces the inert gas through the tank wall and the porous plugs. The ladle is provided with enough freeboard above the molten steel to contain it during the violent action of the degassing operation. The vacuum tank has a removable cover that has a refractory heat shield. An alloy-additions hopper is on top of the ladle cover. An exhaust line connects the base of the vacuum tank with a steam-ejector system.

Steel to be degassed by this method is typically superheated in the steelmaking furnace about 55 degrees to 83 degrees C (100° to 150° F) above the temperature for tapping steel that is not to be degassed. The steel is tapped open, undeoxidized, into the special ladle containing porous plugs and gas-inlet pipes. As much slag as possible is held back in the furnace during tapping; further, as much slag as possible is skimmed off the steel in the special ladle. Furnace slag is undesirable because of phosphorus reversion and the iron-oxide reaction with the steel during vacuum degassing. The ladle is positioned above the vacuum tank and a flexible hose is connected between a pipe that passes through the wall of the tank and the inlet pipe on the ladle. This completes the path for the inert gas from outside the tank to the porous plug in the ladle lining. The flexible connection permits the ladle to be lowered into the vacuum tank, after which the cover of the tank is lowered into position. Inert gas is introduced into the steel as the steam ejectors are started. As the pressure in the tank decreases, the agitation of the metal in the ladle becomes more vigorous. This vigorous action continues until the pressure decreases to less than 0.10 Torr with continued pumping, at which point the metal is considered to be degassed. Deoxidizers are then added to the steel, with stirring continued until thorough mixing is achieved. The vacuum tank is then returned to atmospheric pressure, the tank cover is removed, an insulating synthetic slag is sometimes added on top of the heat, a crane removes the ladle of degassed steel from the tank, and the steel is teemed in the conventional manner. Depending upon the grade of steel and size of the heat, the steel may be exposed to low pressure from 10 to 25 minutes.

Induction Stirring-Arc Reheating—This system is commonly known as the ASEA-SKF ladle refining furnace process. The equipment can be arranged in different ways: in one method suggested for large heat-size units, it consists of a ladle-furnace, a mobile induction coil stand, a vacuum cover with an exhaust line, a steam-ejector system, and a cover fitted with three carbon electrodes. Sections of the steel shell of the ladle-furnace are made of austenitic non-magnetic stainless steel. At the top of the ladle-furnace is a seal to provide a fit with the vacuum cover. The ladle-furnace is unique in that it serves as a tapping and teeming ladle, a heating furnace, and a vacuum vessel.

In preparation for treatment by this method, the slag is removed from the bath in the steelmaking furnace and the steel is tapped into the ladle-furnace without superheating and without the addition of strong deoxidizers. In this process, also, furnace slags in the ladle-furnace are undesirable; therefore, in some installations, provisions are made to skim the heat in the ladle-furnace. The ladle-furnace then is placed inside the induction coil on the mobile stand and moved to the reheating station where the cover equipped with electrodes is lowered into position. The arcs are struck and the reheating and refining period begins. Fluxes are added to make a basic slag and alloys are added to meet the chemical-composition specifications for the steel. The process is also used for desulphurizing the steel to very low levels, less than 0.005 per cent sulphur content. Heat lost during the processing is restored, and the steel is then ready for degassing. At the vacuum degassing station, the vacuum cover is put in position and the steam ejectors gradually evacuate the ladle-furnace to a pressure of less than 1.0 Torr, usually to an end pressure of 0.2 Torr. During the degassing process, the induction coil is energized to stir the steel. Because of the vigorous action during degassing, sufficient freeboard must be provided to contain the steel. Near the end of the degassing period, aluminum and silicon are added to the steel. For a 27-metric-ton (30-net-ton) heat, the degassing period takes about 10 to 15 minutes.

The system is then restored to atmospheric pressure, the vacuum cover is removed, and the ladle is moved to the pouring platform where the steel is either teemed into molds or continuously cast.

b. VAR Process—In the vacuum arc remelting (VAR) process (See Figure 19—16), a steel electrode having a chemical composition about the same as that of the desired product and usually in the as-cast state is drip-melted into a water-cooled copper mold at a pressure

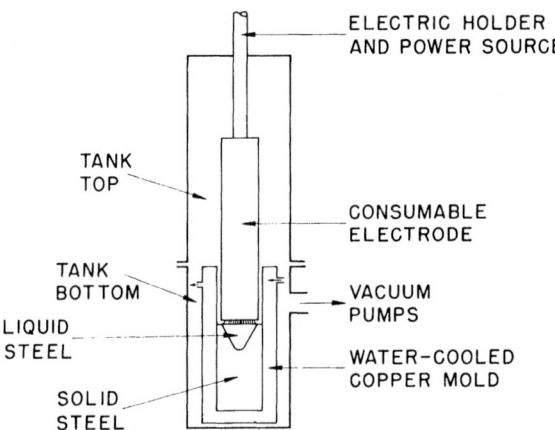

FIG. 19—16. Schematic diagram of a consumable-electrode arc furnace used in the vacuum-arc remelting (VAR) process.

not exceeding 0.1 Torr. The melting rate varies from somewhat over 100 kilograms (a few hundred pounds) per hour for small ingots (diameters of 500 mm or 20 inches) to as high as about 1135 kilograms per hour (2500 pounds per hour) for 1520-mm (60-inch) diameter ingots. Because of the high arc temperature and the small pool of liquid metal, the process can produce very sound ingots of dense crystal structure, low hydrogen and oxygen contents, and with minimal chemical and non-metallic segregation. Ingot sizes usually range from about 500 to 1520 mm (20 to 60 inches) in diameter. Because of the high cost of such a refining process, applications of these steels are limited to specialty products.

c. Modified Vacuum Induction—The Therm-I-Vac equipment can be used either as a stream-degassing

unit or for induction-furnace melting of steel under low pressure. The system can be used for cold charging steel in the induction furnace (See Figure 19—17), melting and refining the metal under low pressure, and teeming the steel into ingots while still under low pressure. However, the system can be used for hot charging molten steel into the induction furnace under low pressure (stream degassing takes place during this operation), reheating the steel in the induction furnace to make up for the heat lost during degassing, adding alloys for final adjustment of chemical composition, and teeming the heat under low pressure.

Figure 19—17 shows schematically how the unit is set up for using a hot charge (liquid steel from the steelmaking furnace) for degassing and refining. With ingot molds of the proper size and number positioned on cars in the mold tunnel, the tunnel door is closed, the refractory collar for directing the pouring stream into the induction furnace is placed inside the tank above the furnace, the tank is covered and the opening in the tank cover is sealed with an aluminum disc. The steam-ejector system is started and the system is evacuated to pressures between 0.2 and 1.5 Torr as required.

The heat of steel from the steelmaking furnace, in an oxidized or partly deoxidized condition, is tapped into a special tapping ladle that has a vacuum sealing flange at the bottom and around the nozzle region. Superheating of the steel is not required in the steelmaking operation, because the steel can be reheated in the induction furnace after stream degassing. The tapping ladle is placed on top of the the vacuum tank, the stopper rod is raised and molten steel flows from the ladle, melts the aluminum disc seal in the ladle cover, and passes through the refractory collar into the induction furnace. If the pressure in the vacuum tank is about 0.2 Torr at the start of the operation, it will rise to about

FIG. 19—17. Cut-away schematic diagram showing how a Therm-I-Vac unit can be used for stream degassing followed by vacuum casting of the degassed steel. Part of the vacuum tank and vacuum mold tunnel have been cut away to show arrangement of equipment enclosed in them.

0.5 Torr and higher, depending upon the pouring rate and the grade of steel being degassed. Degassing continues with the molten steel in the induction furnace. A valve in the tank cover is closed to maintain the vacuum. The ladle is then removed and replaced with an alloy-additions hopper, the valve is opened and the alloying elements are added. The temperature of the steel is raised to the desired level and then teemed by tilting the induction furnace to pour the steel through a pouring box into the ingot molds while still under vacuum. After teeming, the tank and mold tunnel are restored to atmospheric pressure.

NONVACUUM LADLE OR VESSEL REFINING

This section is devoted to secondary steelmaking processes that are conducted at atmospheric pressure rather than in a vacuum and which receive no supplemental heating during the refining process. The processes discussed include:

a. Argon bubbling, CAB (Capped Argon Bubbling) and CAS (Composition Adjustment by Sealed Argon Bubbling);
b. AOD (argon-oxygen decarburizing);
c. Electro-slag remelting;
d. Ladle injection.

Argon Bubbling—Argon stirring, trimming and rinsing: Argon bubbling practices are currently used for several purposes including rapid and uniform mixing of alloys, temperature homogenization, adjustment of chemical composition, and partial removal of nonmetallic inclusions.

Heats of steel are argon-stirred, -trimmed, or -rinsed in one of two ways (see Figure 19—2):

1. by lowering a refractory-protected lance into a ladle of steel to within about 300 mm (12 inches) above the bottom and blowing argon through the lance;

2. by blowing argon through porous refractory plugs inserted in the bottom or side wall of the refractory ladle lining.

Many of the secondary steelmaking processes make use of argon bubbling treatments to supply energy to the steel bath to promote bulk movement of the liquid steel for chemical and thermal homogeniety; to enhance the flotation of nonmetallics from the steel; or to promote intimate slag and metal mixing for refining operations (i.e. desulphurization). For this reason proper gas control and monitoring equipment must be available and the specific intent of the gas-bubbling procedures employed must be understood.

Concurrent with the development and refinement of the secondary steelmaking processes in the middle to late 1970's, the practice of injecting argon through a gas-permeable refractory plug mounted in the bottom of the steel ladle was being used in Canadian steel plants for improving the thermal and chemical homogeneity of steels.

The injection of an inert gas into or under steel offers a simple and inexpensive method of treatment. The benefits to be gained from inert gas injection are:

1. to promote the flotation of nonmetallic inclusions;

2. to provide temperature homogenization;
3. to provide chemical homogenization of the molten steel;
4. to promote slag and metal mixing for steel refining, such as deoxidation and desulphurization of the steel.

Argon stirring is the most vigorous of the argon bubbling treatments, injecting the highest flow of argon through the liquid steel, from 5 - 10 standard cubic feet per minute (scfm). It is normally performed at the time the heat of steel is tapped. The treatment is used primarily to mix for temperature and chemical homogenization and to promote intimate slag and metal mixing for refining operations as mentioned above. The rapid gas flow is vigorous enough to break through the layer of synthetic slag resting on the surface of the liquid steel, opening up an "eye" in the surface slag in order to introduce ferroalloys and deoxidizers into the steel for close control of chemical composition. The rapid flow rate of the argon is also intended to produce new heat-radiating surfaces in order to increase the rate of temperature loss. Since steel temperature is critical in continuous-casting operations, argon stirring is used by some steelmakers to minimize the steep temperature gradients in a ladle of steel and to chill the steel by radiant heat losses.

Precautions must be taken by tapping techniques or by installation of taphole equipment to minimize the carryover of the oxidizing furnace slag, because intimate mixing of that slag and the steel during argon stirring can deleteriously affect steel quality and chemical composition of the steel.

During some argon-stirring processes, a refractory-lined snorkel is lowered into the steel so that the steel inside the snorkel will be completely free of furnace slag. Deoxidizers and ferroalloys can then be added without any contact with furnace slag. The CAS process mentioned above uses the refractory-lined snorkel.

Argon rinsing usually follows argon stirring and is employed to promote the flotation of nonmetallics out of the bulk of the liquid metal up into the slag. The gas flow rate is much gentler, less than 5 scfm. Any flow rate more rapid would open up the slag and expose the steel to the air. The gas flow creates a plume of upward rising bubbles by which the nonmetallic substances are carried up. During argon rinsing, the gentle flow rate of the argon also prevents new heat-radiating surfaces from forming on the slag covering. The argon rinsing procedure usually takes between five and 20 minutes.

If argon trimming is used it will occur between argon stirring and argon rinsing. Trimming means adjusting the composition of the steel through ferroalloy additions, such as chromium. Since the alloys are lighter than steel they will float unless some agitation forces the material to mix with the steel. During argon trimming, just enough gas is introduced to prevent the alloys from being lost in the top slag layer. The amount of gas flow depends on the size of the ladle and the amount of material to be added. The flow rate can vary between less than 10 to as little as 5 scfm. Argon trimming takes between three and five minutes to complete.

There are two distinctive processes which achieve their objectives through special handling of the argon bubbling treatments. Argon bubbling from the bottom of the ladle was implemented into two secondary steelmaking processes by Nippon Steel Corporation: the Capped Argon Bubbling process (CAB) and the Composition Adjustment by Sealed Argon Bubbling process (CAS). Both of these processes provided an improved method for making controlled additions to the ladle as well as the capablility for improvements in steel refining and in the cleanliness quality of steels.

The CAB and CAS processes introduce argon into a ladle of steel as discussed above with these differences:

The CAB process is a refining treatment which uses a conventional ladle with a cover (See Figure 19—18). A synthetic refining slag is added to a heat of steel following the tapping process. This synthetic slag acts as a sponge designed to absorb the undesirable nonmetallics. The introduction of argon into the covered ladle stirs the liquid metal vigorously; in the argon treatments discussed above the argon flow was regulated according to the amount of agitation desired to keep the covering slag intact to prevent contact with the air. The purpose of the CAB process is to keep the steel protected from the air by a cap, and to stir the molten steel vigorously in order to mix the slag with the steel in order that the steel may react with the slag thus cleaning and purifying it. The CAB process enables steelmakers to achieve excellent composition control, deoxidation, desulphurization, microcleanliness.

The CAS process is shown schematically in Figure 19—19. This process uses a refractory-coated snorkel which is lowered into a ladle of steel during argon stirring so that the steel inside the snorkel will be completely free of furnace slag. This allows steelmakers to add deoxidizers and ferroalloys in an inert atmosphere without any contact with furnace slag.

CAS is an excellent secondary steelmaking process for achieving composition control of the steel.

Argon-Oxygen Decarburization (AOD)—This process was originally designed for the more economical production of stainless steels. As shown in Figure 19—20, an electric-arc furnace is charged with high-carbon ferrochromium, ferrosilicon, stainless-steel scrap, burnt lime and fluorspar. After the heat is melted and the desired temperature is obtained the heat is tapped, deslagged, weighed and poured into the AOD vessel (Figure 19—21). There the heat is decarburized from an initial level of 0.5 to 1.0 per cent carbon to about 0.03 per cent. At first, the oxygen-argon blowing ratio is typically 3:1 during which time, the steel is blown to a carbon level of about 0.12 per cent, then the ratio between argon and oxygen is modified to 1:1. During the last part of the process the oxygen-argon ratio is 1:3.

During the blowing, fluxes are added to the furnace. Immediately after the decarburization blow, ferrosilicon is added and the heat is argon-stirred for a short period of time. The furnace is then turned down and the heat is sampled for chemical composition and deslagged. Additional alloys are added if adjustments are necessary, and the heat is tapped into a ladle and poured into ingot molds or a continuous-casting machine. Almost all the chromium is recovered, and the steel can be desulphurized to a level of less than 0.005 per cent sulphur content. In addition, because of the purging effect of the gases, both hydrogen and nitrogen can be removed to low levels. Depending upon the grades of steel being refined, hydrogen content can be less than 2 ppm and nitrogen content less than 0.005 per cent.

c. Electroslag Remelting—In the electroslag-remelting (ESR) process (See Figure 19—22), one or

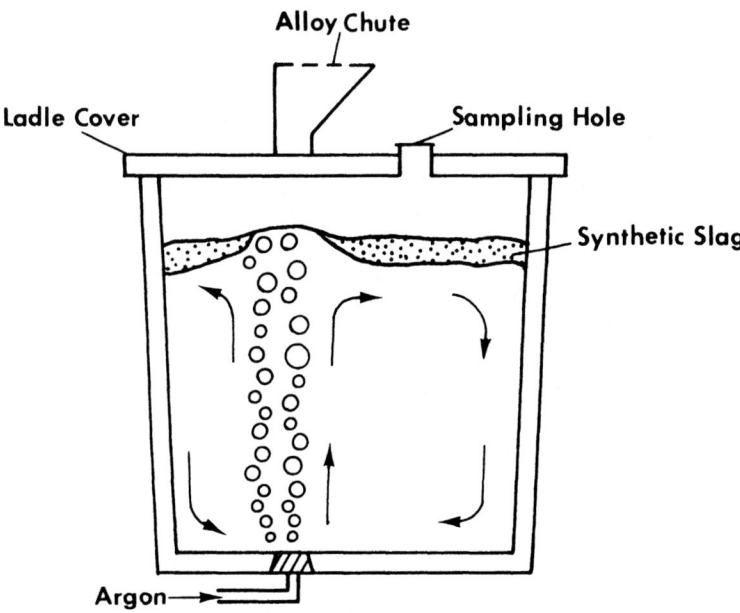

FIG. 19—18. Schematic of Capped Argon Bubbling System(CAB).

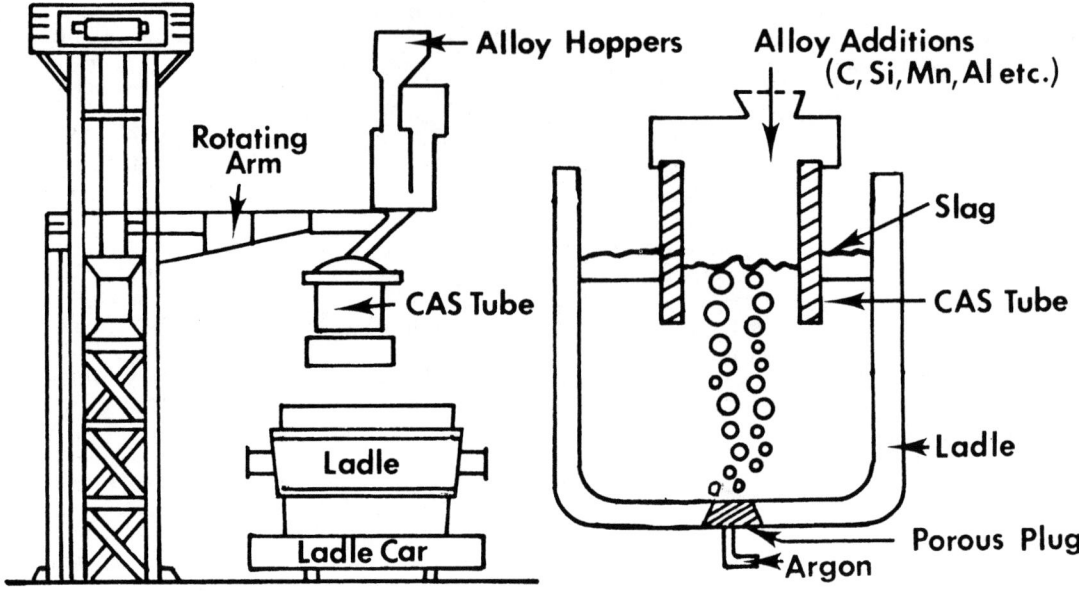

FIG. 19—19. Schematic diagram of the Composition Adjustment by Sealed Argon Bubbling System (CAS).

more steel electrodes of about the desired chemical composition are drip-melted through molten slag into a water-cooled copper mold at atmospheric pressure. The melting rate for this process is somewhat greater than that for the VAR process, otherwise the two processes are similar. Ingot sizes range from about 500 to 1520 mm (20 to 60 inches) in diameter for round ingots; rectangular and hollow-shaped ingots are also produced. The ESR process cannot eliminate hydrogen as the VAR process is able to do, but it has the following capabilities:

1. Multiple electrodes can be melted into a single mold;
2. Spacing between mold wall and electrodes is not critical;
3. Ingot surface quality is excellent, requiring little or no conditioning;
4. Steel can be desulphurized to 0.002 per cent sulphur content;
5. Round, square and rectangular shaped ingots can be produced;
6. Larger size and weight ingots can be produced.

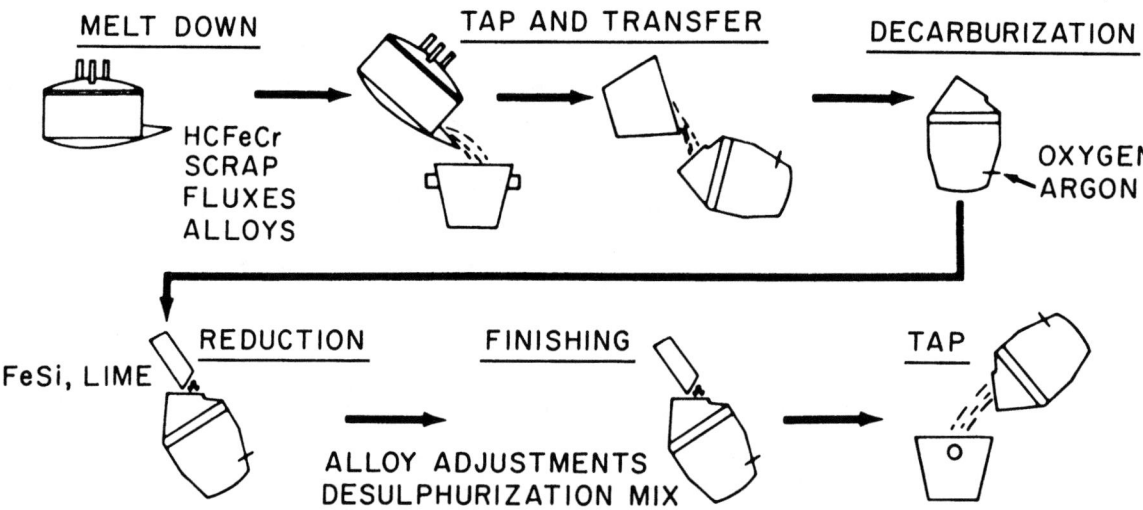

FIG. 19—20. Processing steps for refining stainless steel by the AOD process.

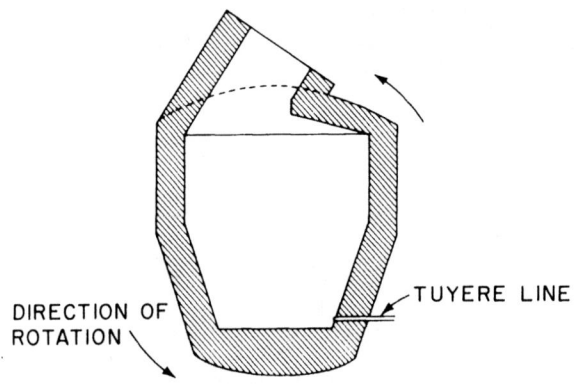

FIG. 19—21. Schematic cross-section of a 100-ton AOD furnace.

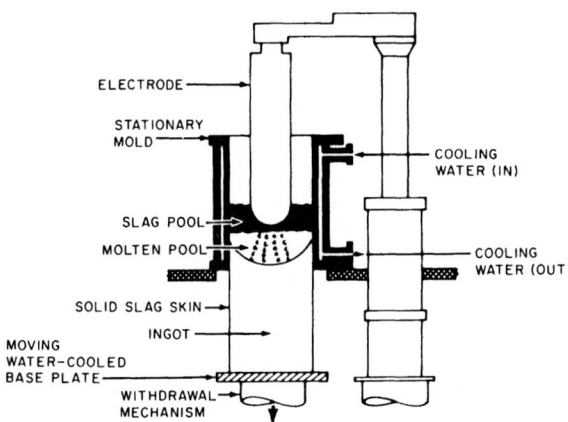

FIG. 19—22. Sketch showing principle of operation of the electroslag-remelting (ESR) process.

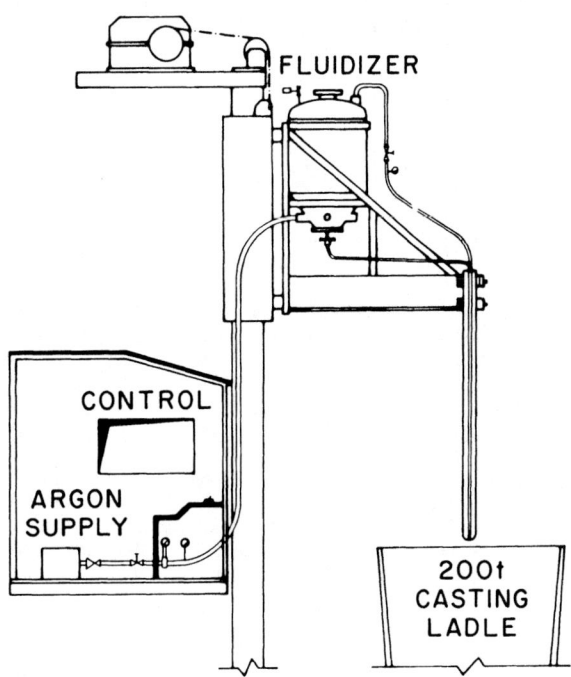

FIG. 19—23. The lance injection desulphurizing process.

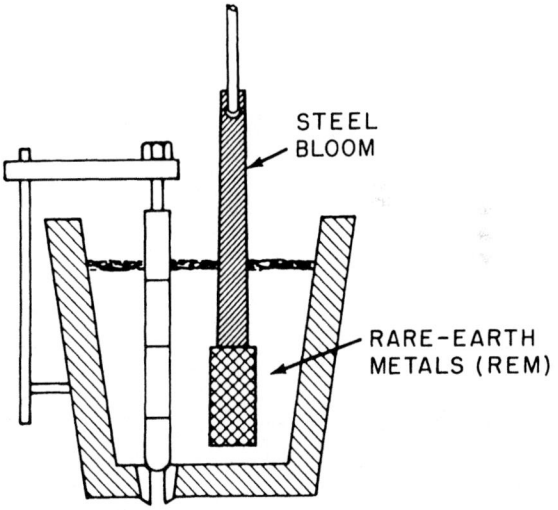

FIG. 19—24. Plunging method for adding rare-earth metals to a ladle of molten steel.

Similar to the VAR process, the ESR process is costly and limited to speciality products.

d. Ladle Injection—Ladle injection is a process by which deoxidizers, special materials and rare earth metals are added to ladles of steel. The process is the best of all secondary steelmaking processes for achieving excellent inclusion morphology in the steel. The injection can be done through powder or canister injection (See Figure 19—23). The lance powder injection is excellent for deoxidation and microcleanliness.

Ladle injections serve as an effective production method for adding rare-earth metals (REM) to molten steel. One of the methods involves preparing a large bloom as shown in Figure 19—24, fastened with predetermined amounts of REM at the lower end of the bloom assembly. A heat of steel is tapped into a ladle, alloyed, and deoxidized. Then the bloom with the REM container is lowered vertically into the ladle of steel. A small amount of magnesium may be added to the REM to provide the agitation needed for uniformly distributing the REM throughout the heat.

REM are very effective in changing the morphology of the sulphide inclusions from the stringer type to the globular sulphides and this change results in a more isotropic steel. The addition of REM has a minor role of desulphurizing the steel.

Bibliography

AIME Committee, "Electric furnace steelmaking," Vol. II, 1963, 206-210.

Barrick, J. R., Vacuum degassing of alloy and stainless steels. Trans. Vacuum Metallurgy Conf., 1966, 370-383.

Capuano, A., J. K. Preston and J. G. Mravec, Vacuum degassing of steel at Timken Steel and Tube Div. Dortmund-Horder unit and operation. Iron and Steel Engr., Nov. 1966, 85-92.

Chandler, H. E., A look at American steel technology. Metal Progress, Oct. 1978, 40-61.

Church, C. P., T. M. Krebs and J. P. Rowe, The application of vacuum degassing to bearing steel. Jour. of Metals, vol. 18, January 1966, 62-68.

Cordier, J. A. and J. Chipman, Activity of sulphur in liquid Fe-Ni alloys. Jour. of Metals, Aug. 1955.

Duckworth, W. E. and R. J. Wooding, Vacuum arc or electroslag melting? Trans. Vacuum Metallurgy Conf., 1968, 470-496.

Elliott, John F., Molly Gleiser and V. Ramakrishna, Thermochemistry for steelmaking, Vol. II, 1963, 525, 526, 548, 549.

Finkel, C. W., "Degassing — then and now", Iron and Steelmaker, AIME, December, 1981.

Fogarty, J. E., Carbon deoxidation and observations on induction-stirred ladle degassing. Trans. Vacuum Metallurgy Conf., 1964, 432-453.

Fogelman, E. L., H. W. Wilt and R. M. Smailer, Development of operating and metallurgical practices for Lukens' 150-ton degassing unit. Jour. of Metals, vol. 18, May 1966, 623-627.

Ford, J. M., and A. C. Mager, The vacuum tap degassing process. Trans. Vacuum Metallurgy Conf., 1964, 454-464.

Forster, E., et al, Apparatus for treating a melt. U. S. Pat. No. 3,891,196, June 24, 1975.

Giove, J. L. and R. S. Mulhauser, Effect of various deoxidation practices on vacuum-degassed open-hearth steel. Trans. Vacuum Metallurgy Conf., 1966, 397-414.

Goto, N. and K. Nagata, New Development of Steelmaking Equipment in Japan. Trans. ISIJ, vol. 21, 1981,446-453.

Gross, J. H. and R. D. Stout, Steels for hydrospace. Ocean Engineering, vol. 4, May 1969, 395-413.

Hallemier, W. and E. Steier, "Vacuum Degassing Processes and Their Industrial Importance", Steel Times International, Dec. 1980.

Hutnik, A. W. and J. B. Hemphill, Production processing and properties of low-sulphur steels. TMS AIME Open-Hearth Proc., vol. 57, 1974, 358-371.

Konkol, P. J. and M. F. Baldy, Effects of S, V, Nb and rolling practice on shelf energy of C-Mn plate steel. Metals Technology, vol. 57, July 1974, 358-371.

Kuwano, T., S. Maruhashi and Y. Aorjama, Decarburization of molten high chromium steel under reduced pressure. Trans. Iron and Steel Inst. of Japan, vol. 15, 1975, No. 7, 353-360.

Leach, J. C. C., Vacuum degassing and secondary refining of steel. Iron and Steel, April 1971, 105-114.

Mikuleckey, J. H., Production and testing of low sulphur steel at Texas Works. ISS AIME Steelmaking Proc., vol. 60, 1977, 338-344.

Nippon Steel Corp. Engineering Data Brochure No. QT 212489, 1974, 47-74.

Orehoski, M. A., Vacuum degassing methods used in the steel industry. Trans. Vacuum Metallurgy Conf., 1966, 321-337.

Orehoski, M. A., and J. N. Hornak, Vacuum casting of steel. AIME Proc., Super duty steels, 1958, 235, 266.

Orehoski, M. A., A review of vacuum degassing in the steel industry. Trans. Vacuum Metallurgy Conf., 1967, 131-158.

Orehoski, M. A., J. W. Bales and D. A. Venseret, Rim-stabilized steel made by molten-aluminum injection method. TMS AIME Open-Hearth Proc., vol. 57, 1974, 386-398.

Orehoski, Michael A., Improved crystal structure in cast steel ingots. Jour. of Metals, May 1969, 41-48.

Orehoski, M. A., Secondary steelmaking. Proc. of The Metal Soc. Conf., London, May 5-6, 1977. Metal Publication 190, 14-26.

Perry, T. E., Vacuum degassing; a review and a progress report. Blast Furnace and Steel Plant, Nov. 1965, 1017-1025.

"Proceedings of the Third International Conference on Refining of Iron and Steel by Powder Injection", Part I and Part II, Lulea, Sweden, June, 1983.

"Proceedings of the Seventh International Conference on Vacuum Metallurgy", The Iron and Steel Institute of Japan, Novemeber 1982.

Ridgeon, J. M., L. Pochon, R. T. Gross and V. Sharma, Comparison of hot-workability of electroslag and vacuum arc melted nickel- and cobalt-base alloys. Trans. Vacuum Metallurgy Conf., 1968, 525-552.

Schemp, E. G., and R. J. Taylor, Continuous monitoring of oxygen content during D-H vacuum deoxidation. Trans. Vacuum Metallurgy Conf., 1965, 342-354.

Schlatter, R., and A. Simkovich, Metallurgical and operational considerations of melting in the Therm-I-Vac furnace. Trans. Vacuum Metallurgy Conf., 1966, 338-369.

Schlatter, R., Progress in vacuum melting of aerospace alloys. Trans. Vacuum Metallurgy Conf., 1968, 333-354.

Sherman, C. W., H. I. Elvander and J. Chipman, Thermodynamic properties of sulphur in molten iron-sulphur alloys. Study of up to 4.8% S in molten Fe. Trans. AIME, vol. 188, 1950, 334-340.

Suzuki, Y. and T. Kuwabara, "Secondary Steelmaking: Review of Present Situation in Japan", Ironmaking and Steelmaking, No. 2, 1978.

Tiberg, M., U. Kalling and C. R. Eliason, ASEA-SKF steel refining process. Trans. Vacuum Metallurgy Conf., 1967, 159-168.

Tivelius, B., and T. Sohlgren, Secondary steelmaking by ASEA-SKF and the TN process. ISS-AIME, Steelmaking Proc., vol. 61, 1978, 154-171.

Troutman, J. W., Ingot structure control of VAR high temperature alloys. Trans. Vacuum Metallurgy Conf., 1967, 599-614.

Turkdogan, E. T., Interaction of solutes in liquid and solid solution in iron. Jour. Iron and Steel Inst., vol. 182, 1956, 66-73.

Ward, R. G., An introduction to the physical chemistry of iron and steel making. 1961, 101-105, Edward Arnold, London.

Watanabe, S., H. Watanabe, K. Asano and T. Saeki, Some chemical engineering aspects of R-H degassing process. Trans. Vacuum Metallurgy Conf., 1967, 183-190.

Yeo, K. B., and R. A. Borowski, Optimization of slag composition and operating temperatures for attaining improved AOD lining life. Iron and Steel Magazine, Nov. 1975, 36-41.

CHAPTER 20

Production of Semi-Finished Steel By Ingot Casting and Rolling

Introductory—This chapter starts with the liquid steel after steelmaking and ladle refining and follows it through a series of steps involving solidification to ingots (Section 1); rolling to semi-finished form of blooms, slabs, and billets (Section 2); and finally conditioning and cooling to prime semi-finished steel (Section 3).

STEEL SOLIDIFICATION

INGOT SOLIDIFICATION

Ingots—After a heat of steel is properly refined either in an oxygen-steelmaking furnace, an open-hearth furnace, or an electric furnace, the liquid steel is tapped into a refractory-lined open-topped vessel called a **steel ladle**. Alloying materials and deoxidizers may be added during the tapping of a heat. The steel ladle has an off-center opening in its bottom, equipped with a nozzle. Some ladles are equipped with a stopper-rod assembly and a mechanism called the ladle rigging for raising and lowering the stopper rod vertically to open and close the bottom hole. Other ladles employ a **sliding gate** mechanism by which the flow of steel from the ladle can be controlled externally by sliding a refractory plate with an opening to align the opening in the plate with the opening in the ladle bottom to permit steel to flow; flow can be stopped by sliding the plate so that the solid part of the plate covers the bottom opening. The ladle is moved by an overhead crane to a pouring platform where the steel is then **poured** or **teemed** into a series of molds of the desired dimensions (Figure **20**—1). The steel solidifies in each of the molds to form a casting called an **ingot.** During the course of solidification and cooling, the surface of an ingot is colder than its interior. In fact, for some types of steel, the centers of ingots are still molten during the subsequent stripping operation, in which the molds are removed from the ingots (Figure **20**—2). For other types they are permitted to stand for a period of time to ensure solidification before leaving the teeming area.

The stripped ingots are placed in a tightly covered soaking pit. The "track time" between stripping and inserting the ingots into the pits should be minimized to conserve energy. The soaking pit is equipped with fuel burners to supply heat to the pit when necessary. There, the ingots are heated to the desired temperature for rolling and held a suitable time at that temperature ("soaked") so as to equalize the temperature throughout the cross-section of the ingots. Modern soaking pits are, in reality, special heating furnaces, as described in Chapter 24. However, in early steel-processing practices, the soaking pits functioned differently. It was the custom at that time to strip the ingots from the molds as soon as possible after pouring and to place them in tightly covered holes or pits in the ground, where the heat from the interior (sometimes molten) of the ingot was conveyed to the relatively colder surface This procedure not only equalized temperature throughout the ingots, but also supplied heat to the pits so that, with careful manipulation, ingots could be heated and maintained at the proper rolling temperature. This early process was called **soaking;** hence the designation soaking pits.

Following these reheating operations, ingots are rolled in primary mills to semifinished steel shapes (slabs, blooms, billets, etc.). The more modern technology of continuous casting (see Chapter 21) replaces all these ingot casting, heating, and rolling operations by casting semifinished steel directly. Continuous casting is increasing rapidly; however, today in the U.S. (and in the world), most steel is still cast by the ingot route.

Ingot Characteristics—An ideal ingot would be one

FIG. 20—1. Overall view of the operation of teeming molten steel into ingot molds.

that was homogeneous both physically and chemically. It would have a fine, equiaxed crystal structure, and would be free of chemical segregation, nonmetallic inclusions, and cavities. Unfortunately, the natural laws that govern the solidification of liquid metal operate against attainment of the ideal condition and, instead, ingots develop within their interiors the well-known phenomena of pipe, blowholes, chemical segregation, nonmetallic segregation, columnar crystal structure, and internal fissures. Added to these manifestations of internal non-uniformity are detrimental surface occurrences such as ingot cracks and scabs. For a proper understanding of such phenomena, a brief description of the mechanism of ingot solidification is warranted.

Types of Ingot Molds—Ingot molds that are in common use are tall box-like containers made of cast iron and weigh from about 1 to 1.5 times as much as the ingots that are cast in them. The mold cavity for receiving molten steel is usually tapered from the top to the bottom of the mold, primarily to facilitate stripping of the ingot. As shown in Figure 20—3, the taper gives rise to the two principal types of molds: **big-end-down**, and **big-end-up**. The big-end-down molds are further classified as **open-top** and **bottle-top**; the big-end-up molds as **open-bottom, closed-bottom,** and **plug-bottom.** Some big-end-up ingot molds may have double tapers, the more severe taper being in the top section. The mold **stool** serves as the bottom closure for the mold cavity in all big-end-down molds and in the open-bottom big-end-up molds. The mold itself provides the bottom closure in the closed-bottom big-end-up mold. In the plug-bottom big-end-up mold, the interior is constricted at the bottom to a small, circular opening

FIG. 20—2. This illustration shows an ingot mold suspended in the jaws of the stripper above a slab ingot from which it has just been removed.

that is closed with a refractory or metal plug prior to casting (see Figure 20—3). Although erosion from impingement of the pouring stream is expected to be confined to the plug in the plug-bottom mold, some erosion of the mold occurs because the plug is too small in diameter to completely protect the bottom. The small opening in the bottom originally was intended to facilitate the use of a plunger to loosen ingots that might stick in the molds, but today most plants do not use a bottom plunger.

The inner walls of molds may be **plain sided, cambered, corrugated,** or **fluted** (Figure 20—4). The purpose of corrugating and fluting is to minimize surface cracking of some types of ingots during solidification and cooling by increasing mold-wall area, which has the effect of increasing initial ingot-skin thickness by promoting faster cooling. Corrugated molds are more common than fluted molds. Most corrugations vary from about 13 to 19 mm (0.5 to 0.75 inch) inches high, with distance between centers varying from about 75 to 150 mm (3.0 to 6.0 inches).

In addition to these molds, which are all filled by a stream of liquid steel poured into the mold cavity from above "top-pouring", "bottom-poured" ingot molds also exist, in which the steel is fed into the bottom of the mold through refractory lined feeders. In the USA, top-pouring is much more widely used, but the use of bottom-pouring is increasing, especially for high quality steels.

There is a close relationship between ingot shape and slabbing mill yield. Thus, for ingot weights below 3½ tons, the best yields are achieved with closed-bottom semi-circular or near semi-circular designs. For ingots over 3½ tons produced in closed-bottom molds

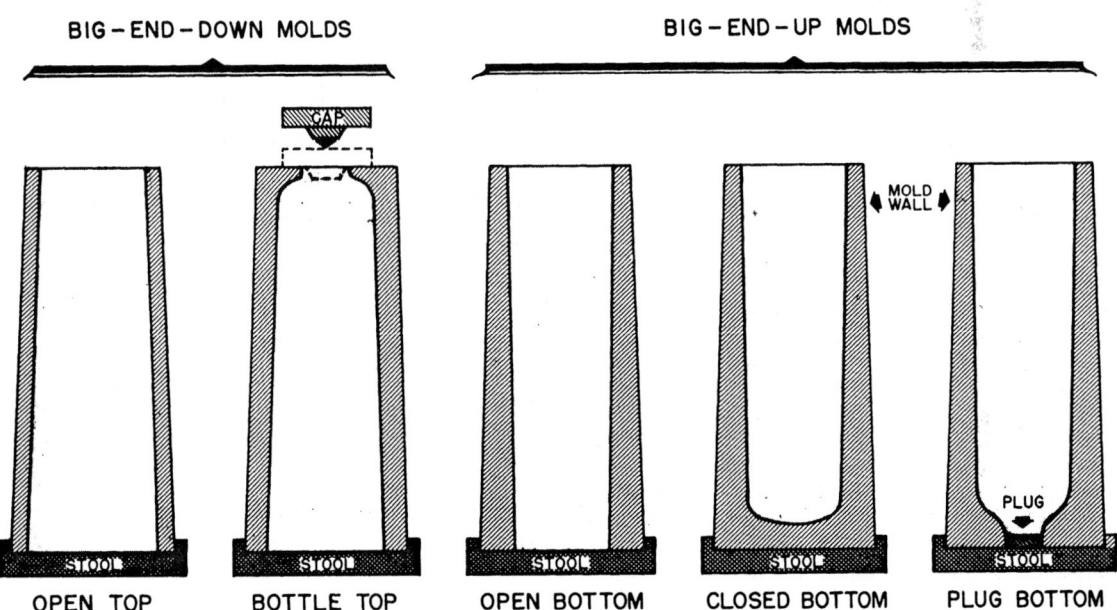

FIG. 20—3. Cross-sections (not to scale) of the five principal types of ingot molds. Molds usually are cast from molten pig iron directly from the blast furnace.

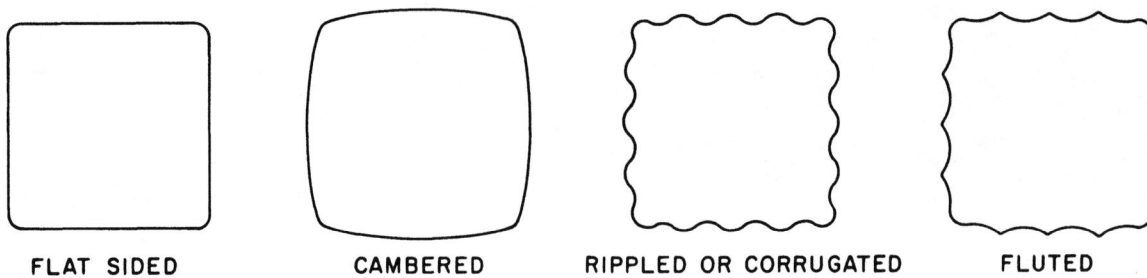

FLAT SIDED   CAMBERED   RIPPLED OR CORRUGATED   FLUTED

FIG. 20—4: Typical cross-sections of ingot molds.

and also for open-end molds, the optimum bottom design is a full volume pyramid possessing a volume less than 2 per cent, a depth of about 2½ inches and an angle of about 25 degrees. For top quality products, ingots with this cross sectional geometry will yield bottom-end losses of less than 2½ per cent.

Types of Steel—As the mold is being filled with molten steel, the metal next to the mold walls and mold stool is chilled by contact with the cold surfaces and solidifies in these regions to form an ingot **shell** or ingot **skin.** Early during solidification, the ingot skin contracts as it cools and forms an **air gap** between itself and the mold wall; this gap reduces the rate at which heat can be transferred from the steel to the mold and thence to the atmosphere. Also, as solidification proceeds, the thermal gradients become less steep. The thickness of the ingot skin (frozen zone) increases rapidly at first but slows down greatly as solidification proceeds.

The solubility of gases in molten steel decreases with decreasing temperatures, especially when the steel changes from the liquid phase to the solid phase. During the solidification of ingots, the gases are liberated in amounts dependent upon the amount of gases originally present in the molten steel. Oxygen is the chief gas that is involved. It reacts with carbon in the steel and produces carbon monoxide that is evolved from the steel. The addition of deoxidizing agents to the liquid steel decreases the amount of dissolved oxygen, and the degree of deoxidation establishes four types of steel—killed, semikilled, capped, and rimmed—to be discussed later.

Time for Solidification of Ingots—The rate at which heat is extracted from an ingot and, hence, the rate of solidification, is affected by many factors, some of which are the thickness, shape, and temperature of the mold; the amount of superheat of the the liquid steel; the cross-section of the ingot; the type of steel; and the chemical composition of the steel. Figure 20—5 shows the idealized solidification pattern of a 813-mm by 813-mm (32-inch by 32-inch) killed-steel ingot. The lines marked 20, 40, 60, etc., indicate the progress of ingot solidification for the corresponding number of minutes that had elapsed after pouring. The location of these lines was established on the basis of data obtained by casting a series of identical ingots and then dumping each ingot after progressive predetermined intervals to pour out the remaining liquid steel. The solidified shells were then removed from the molds and split vertically for examination and study. The ingots used in the experiment had the following chemical composition:

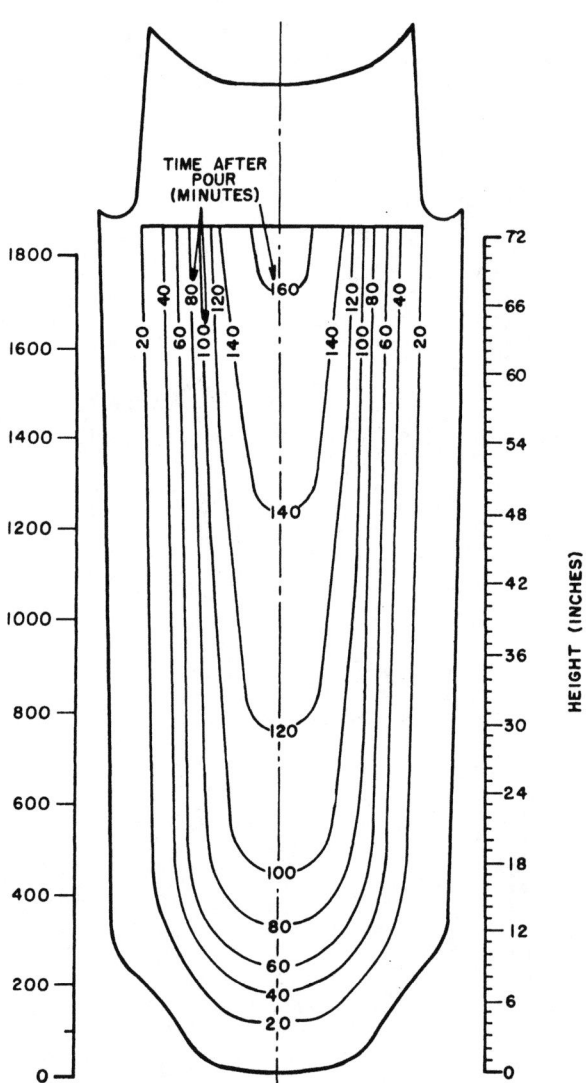

FIG. 20—5. Solidification pattern of 813-mm by 813-mm (32-inch by 32-inch) hot-topped big-end-up ingot of killed steel.

Element	Per Cent
C	0.83
Mn	0.77
P	0.014
S	0.024
Si	0.18
Ni	2.08
Cr	0.15

Recent work in Japan (1980) has shown the feasibility of rolling ingots that have not completely solidified during soaking. Interestingly, such a practice results in improved homogeneity of the workpieces as well as large reductions in the fuel used by the soaking pits.

The relationships for determining the comparative rates of ingot solidification for various sizes and shapes are complex and outside the scope of this book.

Range of Ingot Sizes—Ingots may range in weight from well over 270 metric tons (300 net tons) each for large forging ingots to only a few hundred kilograms (several hundred pounds) each for specialty-steel ingots (e.g., tool-steel ingots). Slab ingots range in weight from 9 to 36 metric tons (10 to 40 net tons), with many of them in the neighborhood of 18 metric tons (20 net tons) each. Ingot shape and weight are selected to meet the requirements of the product to be made and the rolling or other equipment that is to be used for hot working.

INTERNAL STRUCTURE OF INGOTS

Introduction to Types of Ingot Structure—When molten steel cools to the temperature range in which it begins to solidify, the solubility of gases dissolved in the steel decreases and the excess gases are expelled from the metal. Of greater importance, the chemical equilibrium between carbon and oxygen changes with decreasing temperatures, so that the two elements react to form carbon monoxide that is evolved as the system attempts to attain a new equilibrium. Molten steel does not solidify at one definite temperature but over a temperature range, so that the gases evolved from still-liquid portions may be trapped at solid-liquid interfaces of the remaining liquid with previously solidified metal to produce **blowholes.**

The amount of gases, chiefly oxygen, dissolved in liquid steel and the amount of gases released during solidification determine the types of ingots; **killed, semi-killed, capped,** and **rimmed.** The amount of oxygen dissolved in molten steel is dependent upon the carbon content of the steel, upon the type and amount of deoxidizers added to the steel, and the chemical composition of the steel.

Figure 20—6 illustrates diagrammatically eight typical conditions of commercial ingots, cast in identical bottle-top molds, in relation to the degree of suppression of gas evolution. The dotted line indicates the height to which the steel originally was poured in each ingot mold. The ingot structures range from that of a fully-killed or dead-killed ingot (No. 1) to that of a violently rimming ingot (No. 8). The differences between these structures are the result of the differences in the amount of gas evolved by these ingots as they solidified.

The fully-killed ingot (No. 1) evolved no gas, its top was slightly concave, and directly below the top was an

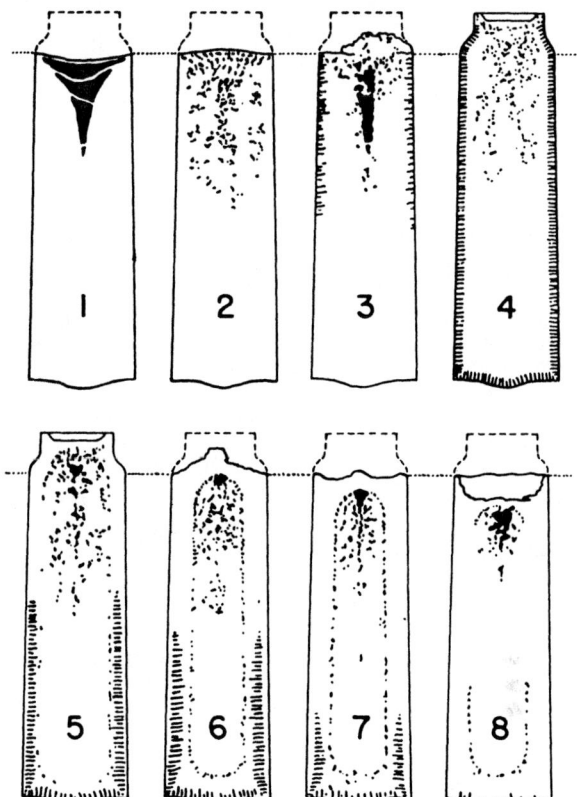

FIG. 20—6. Series of ingot structures.

intermittently-bridged shrinkage cavity that is commonly called **pipe.** While fully killed steels are commonly poured in big-end-up molds that have refractory-type hot tops so as to confine the pipe cavity entirely to the hot-top portion that is later discarded, ingot No. 1 has been included here for comparative purposes. When fully-killed steels are poured in big-end-down, open-top molds, refractory sideboards (insulating or exothermic) are commonly used as replacements for hot tops. Normally, mold additions of deoxidizers are not made to produce killed steel.

A typical semikilled ingot is shown as ingot No. 2. In this ingot, only a slight amount of carbon monoxide was evolved; however, the resulting blowholes developed slowly but were sufficient in volume to compensate fully for the shrinkage encountered during solidification. Ferrostatic pressure (hydraulic pressure exerted by liquid steel due to gravity) prevented the formation of blowholes in the lower half of the ingot. The pressure caused by the trapped gases in the blowholes was sufficient to bulge the surface of the ingot to produce a domed top.

Ingot No. 3 evolved more gas than ingot No. 2 during solidification. The resulting blowholes had a greater volume than that required to compensate for shrinkage resulting from solidification. Some of the blowholes formed very close to the side surface in the top half of the ingot. Blowholes are undesirable so close to the ingot surface since they may result in surface defects (seams) upon subsequent heating and rolling, as discussed later under "Blowholes." Also, the gas pressure ruptured the initially frozen top surface of the ingot and forced liquid steel up through the rupture where it froze; this phenomenon is called **bleeding.** Excessive bleeding causes pipe cavities or spongy surface on product rolled from such an ingot.

Ingot No. 4 evolved so much gas that the top ingot surface could not solidify immediately after pouring. Instead, numerous honeycomb blowholes formed very close to the side surface of the ingot, extending from top to bottom The evolution of gas caused the steel to rise after pouring and produced a boiling action that is commonly called **rimming action.** This action was stopped by a metal cap secured to the top of the mold.

Ingot No. 5 represents a typical capped ingot. It evolved so much gas that the resulting strong upward currents along the sides in the upper half of the ingot swept away the gas bubbles that otherwise would have formed blowholes. Even in the lower half of the ingot, the blowholes could not form until the gas evolution had moderated somewhat. The result was that a thick solid skin formed first that was then followed by the zone containing the honeycomb blowholes. An ingot of this type would not have the interiors of its blowholes exposed to oxidation by scaling of the ingot surface during heating and soaking. Because this ingot No. 5 had fewer blowholes than did ingot No. 4, the steel rose less rapidly to the cap at the top of the mold.

Ingot No. 6 is a rimmed ingot, as are also ingots No. 7 and No. 8. In ingot No. 6, the evolution of gas, while greater than in ingot No. 5, was insufficient to prevent the honeycomb blowholes from exceeding in volume the amount required to offset solidification shrinkage. Therefore, the top surface of the ingot rose slightly as it froze in from the sides of the mold.

Ingot No. 7 represents a typical rimmed ingot in which gas evolution was so strong that the formation of blowholes was confined to only the lower quarter of the ingot. The apparent increase in volume due to blowholes offset the shrinkage that occurred during solidification. As a result, the top of the ingot did not rise or fall appreciably during solidification.

Ingot No. 8 illustrates a violently rimming ingot, typical of low-metalloid steel. Honeycomb blowholes could not form and the top surface of the ingot fell markedly during solidification.

The foregoing eight ingot structures were selected merely to illustrate a series of cast structures ranging from a fully killed steel to a fully rimmed steel. Included in the series were the four main types of ingots that are produced commercially: killed steel (No. 1), semi-killed steel (No. 2), capped steel (No. 5), and rimmed steel (No. 7).

As was mentioned earlier, oxygen is the principal gas dissolved in steel that makes it possible to produce the various types of ingots. It reacts as follows with carbon during cooling and solidification:

$$O + C = CO \text{ (gas)}.$$

The reaction to the right of the equation may be explained as follows: at the tapping temperature, the oxygen and carbon contents of the liquid steel are essentially in equilibrium; however, as the metal cools, the equilibrium is changed and the reaction proceeds toward the right of the equation in an attempt to restore the chemical balance of the system. Because cooling in the mold is continuous, a new state of equilibrium is not attained, and gases continue to be evolved. The last gases to be evolved may not be able to rise in the ingot, and may collect as bubbles to form blowholes.

Blowhole Formation—The mechanism of formation of primary blowholes is illustrated in Figures 20—7, 20—8 and 20—9. In Figure 20—7, bubbles of gas first form at the solid-liquid interface (a) because of the decrease in steel temperature. If the gas evolution is fast and if the liquid is moving upward rapidly along the interface, the bubbles are swept away. If the gas evolution is slow and if the liquid is not moving rapidly, the bubbles will grow as solidification proceeds, as indicated in (b) and (c). Figure 20—8 illustrates that, if the bubbles grow slightly faster than the advancing solidification front, the protruding bubble will break off periodically as shown in (d_1), (e_1) and (g_1). Figure 20—9 shows that, if the bubble growth is accompanied by upward motion of liquid, the bubble will be swept away carrying some of the gas surrounded by solid steel as shown in (d_2), (e_2), (f_2) and (g_2). In a properly rimmed ingot, the rapidly moving metal sweeps the bubbles from the solid interface and no primary blowholes are formed in the top portion of the ingot and during the early stages of solidification. But in the bottom of the ingot, the action of the liquid is less pronounced, and the ferrostatic head suppresses the liberation of the gases, so that primary blowholes are formed. The primary blowholes will continue to grow but at a slower rate because there is less oxygen available in the steel until there is enough pressure resulting from the ingot top freezing or capping to suppress the release of gases temporarily. As solidification continues, there is enough oxygen remaining in the liquid to cause an evolution of gas despite the high internal pressure in the ingot; hence spherical secondary blowholes form. The internal pressure is decreased somewhat by the cooling of the steel.

Since the amount of oxygen dissolved in liquid steel decreases with increasing carbon content (excluding the effect of deoxidizers), it becomes apparent that rimmed or capped ingots, that require the evolution of large amounts of gas, cannot be produced if too much carbon is present. The practical upper limit of carbon content for such steels is 0.30 per cent. Killed and semikilled ingots can be produced in steels of low carbon and high oxygen contents by adding deoxidizers to the liquid steel to react with and remove the oxygen. However, in such low-carbon steels, the required large amounts of deoxidizers to be added not only would add to the cost of the steel, but also may produce an excessive number of nonmetallic inclusions representing the products of the deoxidation reactions. Therefore, there are often practical advantages in producing the lower carbon steels by rimmed or capped practice, and the higher carbon steels by semikilled or killed practice.

In all except killed-steel ingots, the evolution of gas produces cavities of roughly cylindrical shape (skin or honeycomb blowholes) or of spherical shape (located deeper in the ingot). Except for the ones located within several inches of the top of the ingot, such blowholes

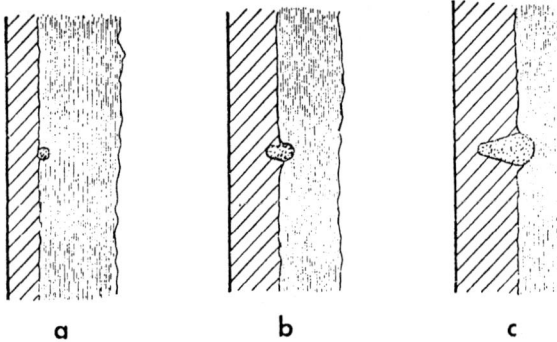

FIG. 20—7. Formation of primary blowholes in rimmed-steel ingot under conditions described in text.

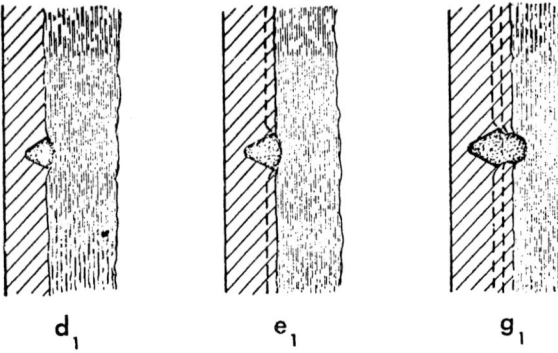

FIG. 20—8. Formation of primary blowholes in rimmed-steel ingots under conditions described in text.

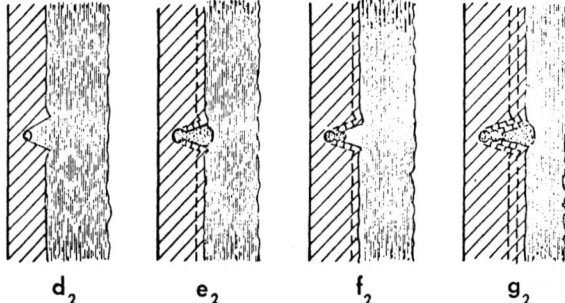

FIG. 20—9. Formation of primary blowholes in rimmed-steel ingots under conditions described in text.

tend to have interiors free of an oxide coating and clean enough to weld easily and completely during rolling. However, if the blowholes extend to the surface of an ingot, or lie at such shallow depth beneath the surface as to become exposed by scaling of the ingot surface during heating in the soaking pits, they can become oxidized and will not weld; instead, they may produce numerous seams (sometimes termed **brush seams**) in the rolled product. Properly made ingots, therefore, will have gas evolution during solidification so controlled that there will be a skin of adequate thickness over those blowholes closest to the surface. The fact that blowholes serve a useful purpose in diminishing or preventing the formation of pipe and improving ingot yield already has been mentioned.

Pipe—The shrinkage cavity, or pipe, located in the upper central portion of the ingot, is largest and most deeply located in the two extremes of ingot structure represented by ingots Nos. 1 and 8 in Figure 20—6. Less extreme structures such as No. 2 (semikilled) or No. 7 (strongly rimming) exhibit this tendency to a lesser extent, while the product of an ingot of intermediate structure such as No. 5 (capped) will be practically free from pipe after rolling. Big-end-down killed ingots (poured without a hot top) often have the lower, unoxidized portion of the pipe cavity below the bridges clean enough to be welded completely shut by pressure and deformation of the steel during rolling. This is particularly true for steels of higher carbon content or for steels that are extensively reduced during rolling, as in the lighter-gage flat-rolled products. A satisfactory yield of sound rolled product often can be obtained with such steels without taking special steps to prevent the formation of pipe.

If assurance of complete freedom from pipe is required, it is accomplished best in killed-steel ingots by making them big-end-up with a hot top, as shown in Figure 20—10 (No. 1). This figure also illustrates the extent of pipe in hot-topped and non-hot-topped ingots of the big-end-down and big-end-up types. The refractory material with which the hot top is constructed or lined absorbs heat less rapidly than the cast iron of the mold, so that the top of the ingot remains molten until after the remainder of the ingot has solidified, thus furnishing an overlying pool of liquid steel that feeds down into the portions of the ingot below the hot top to overcome the shrinkage due to solidification. By using big-end-up molds, this feeding is made still more effective.

To improve the feeding of metal by hot tops, especially during the late stages of solidification, efforts are directed toward keeping the metal pool in hot tops liquid as long as possible. Several methods are used to do this. One method is to use a highly insulating refractory material in the hot top. Another method is to use exothermic materials as part of the hot top and as a covering over the top of the steel. When killed steels

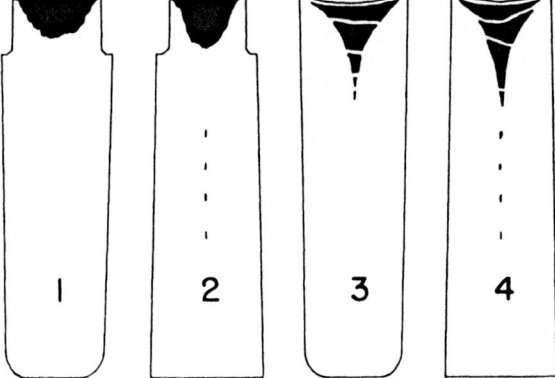

FIG. 20—10. Types of killed ingots.
1. Big-end-up, hot-topped.
2. Big-end-down, hot-topped.
3. Big-end-up, not hot-topped.
4 Big-end-down, not hot-topped.

are poured in big-end-down ingot molds, the more common practice is to use insulating or exothermic sideboards. Still another method is to employ an electric arc to provide heat to the top of a non-hot-topped ingot.

Segregation—The amount of segregation found in an ingot depends upon several factors, some of which are the chemical composition of the steel, the type of ingot (killed, semikilled, capped, or rimmed), and the ingot size. A detailed explanation of segregation, dendrite crystal formation, and solidification rates of ingots is outside the scope of this discussion. In general, the metal that solidifies very rapidly close to the mold wall (the **chill zone**) has about the same chemical composition as the liquid metal entering the mold. However, as the rate of solidification decreases, the mechanism of solidification is such that dendrite crystals of purer metal solidify first; that is to say, the first crystals to form contain less carbon, manganese, phosphorus, sulphur, and other elements than the liquid steel from which they formed, and the remaining liquid is enriched by these elements that are continually being rejected in the crystallization process. Thus, the last material to solidify contains the largest amount in total of the rejected elements. Segregation is frequently expressed as a departure from the average chemical composition. Thus, when the content of an element is greater than the average, the segregation is termed **positive segregation;** when the content is less than the average, it is termed **negative segregation.**

Some elements in steel tend to segregate more readily than others. Sulphur segregates to the greatest extent. The following elements also segregate, but to a lesser degree, and in descending order: phosphorus, carbon, silicon, and manganese. The tendency for elements to segregate while an ingot is solidifying increases with increased time for solidification, so that large ingots exhibit more severe segregation than do small ingots.

Also, when comparing ingots of the same cross-section, movement of liquid steel by convection currents or turbulence due to gas evolution in the steel in a mold during solidification increases the tendency of elements to segregate. Therefore, killed steels are less segregated then semikilled; and the semikilled less segregated than capped or rimmed steels. In a rimmed ingot, the rimmed zone exhibits negative segregation and the core zone exhibits positive segregation. The boundary between the rim and core zones of a rimmed ingot is very sharp and these zones are so different with respect to chemical composition that they resemble different steels.

There are certain other special aspects of segregation in killed steel that are of interest, but can only be mentioned here; these include **axial porosity** (associated with the " **'V' segregate**" along the central axis of an ingot) and **ingot pattern** that may be due to the ingot being disturbed while solidifying, or to the type of segregation referred to as "**inverted 'V' segregate**." Alternatively, these are known as " **'A' segregates.**"

Columnar Structure—Steel after solidification is a crystalline material. The first molten metal to contact the comparatively cold mold wall freezes (solidifies) rapidly, with a structure characterized by small and randomly oriented crystals that form a chill zone about 12 mm (½ inch) thick. After this initial zone of randomly oriented crystals has formed, large crystals (**dendrites**) that are characterized by a branching structure develop. Growth of the individual dendrites occurs principally along their longitudinal axes perpendicular to the surfaces of the ingot, and these large elongated crystals may extend all the way to the center of the ingot. An ingot possessing a preponderance of these large elongated crystals is referred to as having a **columnar structure** and, if the structure is exaggerated in extent, it is referred to as **ingotism.** Ingots exhibiting ingotism tend to crack excessively during rolling unless light drafts (small reductions in cross-sectional area per pass) are employed for the first few passes in the rolling mill. In most ingots, however, columnar structure gives way, toward the center of the ingot, to rather large, equiaxed, randomly-oriented crystals that also are dendritic in character. The relative proportion of columnar and equiaxed dendritic crystallization appears to be dependent upon many variables, among which are: composition of the steel, mold temperature, pouring temperature, and gas content of the steel.

Methods for moving the liquid steel in a mold during the initial time of solidification by various mechanisms such as pneumatic oscillation have been successful in decreasing and sometimes eliminating the zone of columnar crystals but such practices are the exception rather than the rule.

The success of these mechanisms in minimizing and sometimes eliminating the formation of columnar crystals can be explained as follows:

Essentially, all such induced movements of the liquid steel ingot core must achieve two objectives if columnar growth is to be terminated and equiaxed growth achieved. These two objectives are (1) the removal of all superheat in the liquid core (superheat is defined as the temperature of the liquid above its liquidus temperature.) Thus, the removal of all superheat means the liquid core will be at the liquidus temperature of the steel; and (2) the generation of nuclii fragments in the liquid core, either by mechanically breaking or remelting the columnar dendrite tips.

Long columnar crystals, especially in higher alloy grades that tend to resist plastic flow at hot-rolling temperatures, are undesirable because of their poor cohesive strength which causes the steel to rupture along the periphery of the columnar crystals during the hot-forming operation.

Internal Fissures—Tensile stresses in the interior of an ingot, arising during heating, cooling, or rolling, may produce internal fissures or internal **bursts,** sometimes of a very large size. If these do not extend to the surface, they may weld completely during the rolling operation, provided that the amount of hot work (percentage of reduction) is sufficient.

Ingot Cracks—These defects occur as both transverse and longitudinal ruptures in the ingot wall, and normally are observed first while the ingot is being rolled on the primary mill, although some are apparent on the surface of the ingot itself, especially if it becomes cold. Ingot cracking has been the subject of numerous investigations, and some of the causes have been brought to light, although many still remain obscure. Excessively high pouring temperature (a temperature considerably above the solidification temperature) has been established as one definite cause of ingot cracking. During solidification a dendritic crystalline structure is developed in the ingot and interdendritic zones of weakness are formed which extend from the ingot surface toward the center. The larger the dendrites, the more pronounced are these zones. Lower pouring temperatures will help eliminate cracking from this origin by limiting the size while increasing the total number of individual dendrites. **Folds** due to surging of the molten metal in the mold form discontinuities in the ingot wall that lead to transverse ingot cracking. This type of defect has been minimized by use of mold coatings and improved mold design. Steels of the 0.15 to 0.25 per cent carbon grades, especially the fine-grained killed types, have the greatest tendency toward transverse cracking. Generally, the higher carbon steels have the least tendency toward this transverse rupturing. **Hanging** of hot-topped ingots by fins forming over the edge of the mold wall usually produces a transverse crack approximately 150 mm (6 inches) below the hot-top junction, easily recognizable and termed a **hanger crack.** Hanger cracks from this source can be prevented by use of properly designed hot-tops. Plain-sided molds are more prone to produce or at least accentuate transverse cracking than the fluted type. Longitudinal cracks generally are related to the flute or corner design of the mold.

Other and more obscure causes and corrective measures for ingot cracking are a constant subject for study by practically all steel-mill personnel.

Nonmetallic Inclusions—All steel ingots contain nonmetallic inclusions that consist of oxidized material and lesser amounts of sulphides in various combinations and mixtures with each other. They are derived chiefly from the oxidizing reactions of the refining processes and the deoxidizing materials added to the steel in the furnace, ladle, or molds. Some may result from erosion of ladle and other refractories during pouring or chemical reaction of these refractories with the steels.

Scabs—In top-poured ingots, the pouring stream first strikes the stool or the mold bottom, and splashes violently against the lower part of the mold walls. Many of these splashes adhere and solidify, forming a continuous layer on the lower portion of the mold walls. This splashing diminishes as a pool of liquid metal forms in the bottom of the mold. The adhering splashes cool rapidly, and their surfaces oxidize. If the cooling and oxidation have progressed too far by the time the liquid steel in the mold rises past them, they will not be incorporated into the ingot, but will remain as adhering and imperfectly bonded **scabs** on the surfaces of the ingot. If thin, scabs may be oxidized away by scale formation in the soaking pit. If thick, they remain and produce a similar defect on the rolled product. As the continuous layer of splashes cools, its upper edges tend to bend inward and, as the rising liquid steel overflows them, to become enfolded. Horizontal ingot cracks called **butt cracks** often occur below and parallel to such **folds,** and the folds themselves can produce surface laminations or seams in rolled product.

The defects associated with pouring splashes can be reduced by filling the mold more rapidly, so that the rising level of liquid steel covers the splashes before they can cool and oxidize. This is done by using larger or multiple nozzles, which practice, however, leads to various mechanical difficulties if carried to extremes. Some steel plants use metal "boots" or stovepipe type sheet-metal cylinders set on end on the stool to contain the initial splash; after the pouring stream is opened full, these rapidly melt into the liquid metal. Bottom pouring also will minimize these defects, since the molten steel enters the mold from a runner through an opening in the mold bottom and there is little splashing as compared with top-pouring practice.

Mold Coatings—Another method of reducing the effects of splashing, and thereby improving the surface of ingots, is to coat the inside of the molds with a substance that will volatilize and tend to repel splashes. Some coatings such as tar and powdered pitch are effective as splash repellents but are no longer being used because of environmental and health hazards. Mold temperatures are important in applying mold coatings. If they are too hot, the coating will not adhere to the mold walls and, if the coating materials are carbonaceous, they will decompose and the resulting residual charred film has no beneficial effect. If they are too cold, the coatings are extremely heavy, and the excessive evolution of gases when in contact with the liquid steel gives rise to subsurface pinholes and blowholes in the ingot surface. Refractory-base mold coatings are also used to a limited extent.

Practices have been developed to eliminate mold coatings by using, instead, clean molds and optimum pouring rates that minimize surface imperfections.

PRODUCTION OF DIFFERENT INGOT TYPES

The foregoing discussion has shown that the final structure of an ingot is determined almost entirely by the degree to which the steel from which it was cast has been deoxidized. The several types of steel require different steelmaking and deoxidation practices, which are described briefly in the following summary of the principal steps involved.

Rimmed Steels—For rimmed steels, proper rimming action in the molds has been described as necessary to produce the surface conditions and ingot structure desired. Slag control is aimed at adjusting the lime-silica ratio and iron-oxide content of the slag to give the desired level of oxidation of the bath of metal when the heat is ready to tap. The exact procedures followed depend upon whether the steel has a carbon content in the higher ranges (0.12 to 0.15 per cent), in the lower ranges (0.06 to 0.10 per cent), or under 0.06 per cent.

Rimmed steel usually is tapped without additions of deoxidizers to the steel in the furnace, and with only small additions to the molten steel in the ladle, in order to have sufficient oxygen present to give the desired gas evolution by reacting in the mold with carbon. Ferromanganese may be added to the furnace before tapping in open-hearth or electric-furnace operations, or to the ladle, but it is usual to make the addition in the ladle. Aluminum, ferrotitanium, or other deoxidizers in small amounts may be added in the ladle, if needed. This type of steel, when properly made, can be cast into ingots having a minimum of pipe and a good surface, though they are subject to segregation. When the metal in the ingot mold begins to solidify, there is a brisk evolution of gas, resulting in an outer ingot skin of relatively clean metal. For many applications, particularly where the surface of the product is most important, this steel is used to a considerable extent.

The thickness of the outer skin and the absence of blowholes and oxides within it depend upon the skill of the steelmakers. When the temperature and the oxygen content of the steel as it is poured from the ladle are within the most desirable limits, the desired evolution of carbon monoxide from the steel in the molds is obtained. The rimming action of the first-cast ingot is observed, and if an increase or decrease in the rimming action is desired, this is obtained for subsequent ingots by making small adjustments in the amount of shot aluminum or gas-evolving materials depending upon whether the oxygen level is, respectively, too high or too low. If the steel is over-deoxidized (oxygen content too low), the rimming action will be incomplete because gas evolution is too small in volume and slow in starting. The use of hot molds will also suppress rimming action. If the steel is too high in oxygen content, there is an absence of blowholes, the level of liquid steel falls and the incidence of pipe increases.

Capped Steels—Capped-steel practice is a variation of rimmed-steel practice. The steel is poured into big-end-down bottle-top molds in which the constricted top or mouth of the mold facilitates the capping operation. The rimming action is allowed to begin normally, but is then terminated at the end of a minute or more by sealing the mold with a cast-iron cap. The addition of only a small amount of shot aluminum during pouring insures that the steel will rise to press against the cap. The oxygen level of the steel as poured into the mold is preferred to be not more, and possibly slightly less, than the level desired for rimmed steel. The capped ingot has a thin rim zone that is relatively free from blowholes, and a core zone that has less segregation than that for a rimmed-steel ingot of the same volume. In steels with a carbon content greater than 0.15 per cent, the capped-ingot practice is used with advantage. Steels of this type are applied to sheet, strip, skelp, tin plate, wire, and bars. Ingots are also "splash capped." For this practice, a rimming type of steel is poured into bottle-top molds just into the shoulder section of the molds, is allowed to rim for a predetermined period of time ranging from a few to several minutes, then the mold is filled into the neck section and immediately capped.

Semikilled Steel—Semikilled steel is deoxidized less than killed steel, and there is enough oxygen present in the molten steel to react with carbon and form gas after the steel is poured into molds. Semikilled steel finds wide application in structural shapes, plates, and bar products. This steel generally has a carbon content within the range of 0.15 to 0.30 per cent. The usual practice is to bring the carbon content of the steel in the furnace to the desired carbon content for tapping. Ferromanganese may be added to the furnace in the case of an open-hearth or electric furnace, to the ladle, or to both. Carbon, ferrosilicon, and aluminum may be added to the ladle. Usually, most deoxidation is done in

the ladle, so that only a few grams (ounces) of aluminum per ton of steel will be required as a mold addition to produce the desired level of deoxidation so that gas bubbles are trapped in the upper portion of the ingot during solidification.

Killed Steel—The term "killed" indicates that steel has been deoxidized sufficiently for it to lie perfectly quiet when poured into an ingot mold. There is no evolution of gas in the mold, and the top surface of the ingot solidifies with relative rapidity. Killed steel generally is used when a homogeneous structure is required in the finished steel. Alloy steels, forging steels, and steels for carburizing are of this type, when the essential quality is soundness. In general, all steels with more than 0.30 per cent carbon content are killed.

In making killed steel in the open-hearth, the usual steelmaking-furnace practice is to "catch the heat coming down," that is, to decrease the carbon content of the bath to the desired level and then either to **block the heat** (deoxidize it) by adding high-silicon pig iron (15 to 25 per cent silicon), 50 per cent ferrosilicon, or silico-manganese, or to tap the heat without blocking and depend upon ladle deoxidation. Blocking lowers the oxygen content of the liquid metal to prevent further oxidation of carbon; it also serves to protect alloying elements that are susceptible to oxidation and, consequently, are added after the heat has been blocked.

At the final part of the finishing period, the carbon will have been worked down until it is at a level within the range required for tapping and pouring. The phosphorus and sulphur contents should be below the specified maximum, the manganese usually will be below the minimum required, and the bath temperature should be proper for the composition and grade of the steel being produced. The steel is then ready for whatever ferroalloys need to be added.

The decision as to whether a ferroalloy addition is to be made to the furnace or to the ladle is determined largely by the susceptibility of the ferroalloy to oxidation. Manganese may be added to the furnace or to the ladle, or divided between them, but the additions to the ladle must not be so large as to chill the metal too much. The furnace additions are chosen and the timing of addition set so that maximum elimination of the solid oxides formed will take place by their floating up through the metal to the slag before the metal is tapped from the furnace. After tapping, other deoxidizing additions may be made to the steel as it runs into the ladle. These additions complete the deoxidation to the desired degree up to the pouring into molds. These ladle additions are usually ferrosilicon, aluminum, or some special alloys (calcium-silicon is an example) containing elements that have a strong affinity for oxygen. Additions containing such elements as manganese and silicon furnish part of these elements required to meet the chemical-composition specifications. Additions of deoxidizers may be made to the molds, depending upon the type of steel; however, this is rarely practiced except by some plants which may add aluminum to the last three or four ingots to make up for any possible loss to the oxidizing slag floating on the metal in the ladle.

For the pneumatic (BOP, Q-BOP) processes, the common practice is also to "catch the heat coming down" in carbon content, and if the heat is at the desired carbon, phosphorus, sulphur and temperature levels at turndown, the heat is tapped into a ladle where almost all additions are made during tapping and only small additions are sometimes made in the molds during teeming. Tapping temperatures are adjusted upward as the amount of ladle additions (except for silicon) are increased. When temperature losses are excessive, exothermic materials are used as ladle additions.

In some BOP shops, the heats are blown to low carbon contents (less than 0.10 per cent), tapped, and ladle-refined where the steel is recarbonized to high-carbon contents (more than 0.30 per cent).

Most killed steels are cast in hot-topped big-end-up molds. However, for economic reasons, some killed steels are cast in big-end-down molds with either insulating or exothermic sideboards.

SECTION 2

ROLLING OF BLOOMS, SLABS AND BILLETS

Introductory—There is no widely accepted precise definition for the terms bloom, slab, or billet, and local applications of the terms are used somewhat on a traditional basis. Distinctions are made according to general appearance, influenced by overall size and the proportions of the three linear dimensions and also by intended use. The distinction between **blooms** and **billets** is principally a distinction of size, billets being smaller than blooms in cross-sectional area, and both having a length several times greater than their maximum cross-sectional dimension. The distinction between **blooms** and **slabs** is principally one of cross-sectional dimensional proportion, blooms tending to be square, or nearly square, and slabs being always oblong and tending to be relatively wide, thin, and (until recently) of short length. There are many exceptions, and there are special names for pieces intended for special uses.

For example, any piece to be rolled into a plate is called a slab, regardless of size or of dimension proportion. Likewise, any piece produced on a billet mill is termed a billet, regardless of shape and size, with the exception of round billets, such as **tube rounds**. Blooms in short lengths are sometimes called **blanks** or **blocks**, and special-shape blooms for structural sections frequently are called blanks regardless of length.

A crude guide using only the cross-sectional characteristics as the distinguishing features which may serve in place of definitions is shown schematically in Figure 20—11.

Until recently, steel in the form of blooms, slabs and billets was produced chiefly by hot rolling ingots to produce blooms and slabs. In this section we first discuss the **primary mills** used for hot rolling ingots. Following this is a discussion of **blooming** and **slabbing mills**. General features of **roll pass design** are discussed, followed by a discussion of **billet mills**. Some blooms and slabs were (and still are) produced by other means of hot working, such as forging by hammering or pressing. These are discussed briefly in Chapter 22 and covered in more detail in Chapter 29.

Since the first blooming mill was built in Dowlais, Wales in 1866, there has been continual improvement. In the early 1940's, research and development resulted in the perfection of methods for the **continuous casting** of molten steel directly into the form of slabs and billets, by-passing the ingot stage and the necessity for hot-rolling operations formerly required to produce such products. Another process known as **bottom**

pressure casting is also employed to produce slabs and billets directly from molten steel. These processes are discussed in detail in Chapter 21.

Blooms, slabs and billets may develop a variety of defects arising from heating, rolling and casting that may have to be removed to prevent their affecting the surface quality of finished products made from them. These defects usually are detected and removed from the blooms, slabs and billets after they have cooled to

TYPICAL CROSS-SECTION
AND
DIMENSIONAL CHARACTERISTICS*

SLAB

ALWAYS OBLONG
MOSTLY 50 TO 230 mm (2 TO 9 IN.) THICK
MOSTLY 610 TO 1520 mm (24 TO 60 IN.) WIDE

BLOOM

SQUARE OR SLIGHTLY OBLONG
MOSTLY IN THE RANGE 150 mm BY 150 mm (6 IN. BY 6 IN.)
TO 300 mm BY 300 mm (12 IN. BY 12 IN.)

BILLET

MOSTLY SQUARE
MOSTLY IN THE RANGE 50 mm BY 50 mm (2 IN. BY 2 IN.)
TO 125 mm BY 125 mm (5 IN. BY 5 IN.)

* DIMENSIONS USUALLY GIVEN TO NEAREST ROUND NUMBER.
ALL CORNERS ARE ROUNDED, AS SHOWN.

FIG. 20—11. Comparison of the relative shapes and sizes of rolled steel governing nomenclature of products of primary and billet mills. (Cast sections produced by continuous or bottom pressure casting methods are similarly designated when of the same general proportions and dimensions as their rolled counterparts.)

ordinary temperatures. These cold steel products are referred to as **semi-finished steel**. After inspection and removal of defects by operations known by the collective name of **conditioning** (see Section 3), the semi-finished products are converted into finished products by reheating and further hot working by rolling or forging coupled, in some instances, with cold-working operations that follow the secondary hot working.

As will be observed in the subsequent discussion, there are some important variations from the sequence of events described above; however, these steps are the ones usually followed in the production of the bulk of the steel products discussed in this book.

PRIMARY MILL ACTIVITIES

It is possible, and often quite economical, to roll ingots directly through the bloom, slab, or billet stage into more refined and even finished steel products. The early stages in the production of useful objects from ingots consist of a series of operations, whereby the ingot cross-section is reduced to a square, oblong, round, or other simple shape, all having rounded corners, and of dimensions approximate to nominal specified size. The length of the ingot is increased, corresponding to the decrease in cross-section. The concluding operations cut a relatively short length called **crop** from each end of the rolled piece as scrap, and cut the remaining long piece, if necessary, into multiples to suit the required lengths or weights for subsequent operations.

These operations may be done in one mill of varying numbers of stands from one to about twenty, in a continuous operation, frequently without any reheating. Large tonnages of standard rails, wide-flange beams, and plates, and a lesser proportion of wide hot-rolled coiled strip are produced regularly by this practice from ingots of medium to large size. A few very small plants follow a similar practice in rolling small ingots, such as 102-mm to 152-mm (4-inch to 6-inch) squares, directly into wire rods, concrete-reinforcing bars and other bar products.

However, most of the ingot tonnage is rolled into blooms, slabs, or billets in one mill, following which they are cooled, stored, and eventually rolled in other mills or forged. The reasons for this more common practice are primarily economic; sometimes size or shape of product and certain quality requirements peculiar to the final article to be manufactured determine the steel-rolling method.

A variety of names for mills rolling ingots has come into common use to differentiate between them in reference to the particular kind of product intended or generally produced, the basic mechanical design features, or the general layout of the mill. For a long time, the term **blooming mill** was used rather commonly for all such mills, but it has come to be restricted by many to a mill producing blooms, and the term **primary mill** has gained acceptance as a generic term to cover both blooms and slabs. Each of these mills is known by a combination of size and descriptive name. This serves to provide a rough mental picture of the mill; for example, 1370-mm (54-inch) blooming mill, or 1090-mm (43-inch) three-high mill, or 1145-mm (45-inch) slabbing

mill. The size here is based on the center-to-center spacing of the pinions and hence roughly approximates the center-to-center spacing or the diameter of the work rolls.

The composite name refers specifically to the roll stand, or to the first roll stand if the mill is a multiple-stand one. However, the name is used as an abbreviation for the entire group of facilities needed to produce blooms, slabs, or billets and for the organization operating them, known more properly as a mill department.

The activities of all primary mills are the same, and it follows that the facilities for carrying out these activities are fundamentally the same in all mills; they differ in detail according to the particular requirements of each mill. The primary function is the conversion of a steel casting (the ingot) into rolled steel product; the secondary function of the rolling operation is to produce this rolled steel product in pieces of the desired cross-sectional dimensions and weights. Auxiliary operations include cropping, conditioning, cutting to length, and the collection and assorting of crops, roll scale and other by-products that are subsequently delivered to the steel-producing and blast-furnace departments.

The conversion of the steel casting into rolled steel product is carried out in a sequence of thermal and mechanical operations:

- (a) Heating the ingots (see Section 1 and Chapter 24).
- (b) Breaking up the coarse crystalline structure of the ingot into a refined structure by heavy rolling pressure and recrystallization during hot working.
- (c) Closing solidification voids by the same means as in (b).
- (d) Cutting off such portions of steel as are metallurgically (both physically and chemically) unsuited for the intended final purpose.
- (e) Cutting off test specimens.
- (f) Cooling to atmospheric temperature those products not destined for immediate further hot working.

The secondary function is performed in conjunction with operations (b), (c), and (d), above, since they can be carried out with the same facilities. The essential steps in performing the secondary function are:

- (a) Positioning the rolls, which have been shaped to produce the desired cross-section.
- (b) Cutting the rolled piece to measured length or to specified weight per piece.
- (c) Weighing individual pieces, or weighing and measuring test specimens.

The auxiliary operations pertain to the rolled product and the by-products:

- (a) Conditioning the steel products by removing injurious surface imperfections, removing portions containing injurious internal defects, and by correcting physical conditions by straightening bent pieces, cutting overlength pieces to proper length, and so on.
- (b) Collecting identified crops, other ferrous scrap, roll scale, and cinder and delivering these by-

products to steel-producing and blast-furnace departments.

In order to carry out these activities, all primary mills have the same general kind of facilities and all are operated in the same general manner. No two are exactly alike, although there are a few instances of close similarity. The differences among mills are due to the sizes of ingots and product, the particular kind of product, production rate considerations, the date when the mill was built or subsequently altered, and the most economical form of fuel and power available at the time of building. Mills differ in size, in details of design, in the auxiliary equipment provided, and in the arrangement with relation to other facilities, especially those producing the ingots and those rerolling the products of the primary mills.

In a large number of mill arrangements, the primary mill supplies hot steel products either directly or through reheating furnaces to other mills producing semi-finished products such as billets and coiled strip, or finished products such as rails or structural shapes. The number of such mills supplied by one primary mill is from one to three in various tandem and parallel combinations. When the still hot product coming from the primary mill is immediately rolled in a succeeding mill, without reheating, this is known as direct rolling. At this time, rolling capacities must be considered. In the ideal case, the primary mill will just keep the succeeding mill or mills busy, without loss of time in either the primary mill or the succeeding mills. The arrangement of the mills determines the material-handling facilities required and the size and location of the mill buildings.

The basic operation in a primary mill is the gradual compression of the steel ingot between the surfaces of two rotating rolls, and the progression of the ingot through the space between the rolls. The physical properties of the ingot prohibit making the total required deformation of the steel in one pass through the rolls, so that a number of passes in sequence are always necessary. There are several ways in which a sequence of passes can be made, and some particular way is selected by choosing one which is suited to the quality of material to be produced and also likely to be the least expensive. This choice determines the general type of mill. The elements of pressure, motion, weight, and time, together with the sizes of ingots and sizes of products, plus the quantity of product desired in a given time period, determine the size and design of the mill.

Each type of mill has been developed to meet a definite need, and, in performing this one task, it is superior to other types. However, mills are expensive to install and are capable of long life, while changing conditions in the steel industry sometimes result in a number of changes in the needs to be met by a primary mill during the period of its useful life. This results in adapting existing mills to new needs, frequently with alterations to some of the facilities. Many primary mills roll ingots and produce products that were not contemplated when the mills were built.

The general characteristics of each of the types of primary mills are outlined in the following section.

BLOOMING AND SLABBING MILLS

Two-High Reversing Mill—In the phrase "two-high reversing mill," the term "two-high" refers to the fact that the mill consists of two rolls, one over the other, as in Figure 20—12. "Reversing" means that after the piece has gone through the rolls in the direction of the first pass, the rolls are brought to a standstill and then caused to rotate in the reverse direction shown in Pass No. 2, so as to impose the next reduction on the piece (the next pass), and so on until the piece has been reduced by the desired amount. This type of mill has maximum flexibility in size range of ingot and product, as well as wide range of adjustment in the amount and rate of steel deformation in any pass.

For the same size ingot rolled to the same size product, the two-high reversing mill has a lower production rate, in terms of tons per unit of time, than any of the other mills. It was due to this relatively low output rate that the other types of mills were devised. However, the two-high reversing mill remains the principal type in use, exceeding in numbers all other types combined, by a wide margin. Its chief characteristic, flexibility, is so valuable that this type of mill is combined frequently

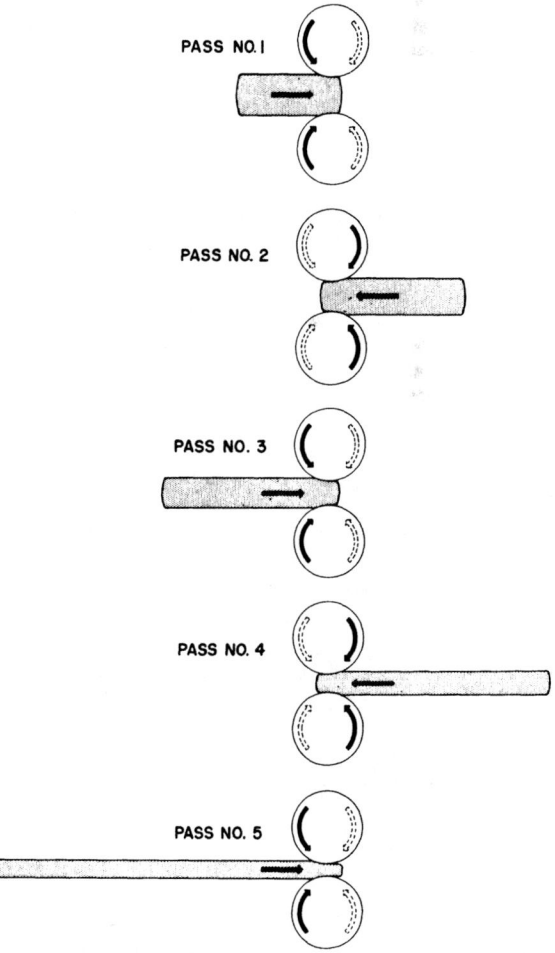

FIG. 20—12. Diagrammatic representation of the sequence of rolling operations involved in reducing an ingot to a slab on a reversing two-high mill.

with other types to compensate for low production rate, the extreme being two two-high mills of different sizes arranged in tandem. Many such combinations are in operation at this time.

While there is no limit to the size or weight of ingot for which this type of mill is used, there is an economic limit to the length of piece which should be rolled from one ingot on one mill. It is determined by two factors: inertia of moving parts of equipment, and size of building to house the mill. The ingot is a relatively short casting, mostly from five to nine feet long. Mill parlance refers to length of ingots as their height, because they are poured vertically and this dimension is height at first, but the mill operates on the ingot when the latter is in a horizontal position. In the early passes, the length of the piece is so short that one to three revolutions of the rolls is sufficient to complete a pass, and the rolling speed is preferably very slow. Reversals can be rapid, power loss in reversing is relatively slight, and impact among moving parts is light. As the piece becomes longer in successive passes, the rolling speeds must be increased if the steel is not to lose too much heat. At high speed, the motion in one direction is only a few seconds in duration, but reversals at high speed consume excessive power and subject all the moving parts in the mill tables as well as in the mill and the mill drive to severe shock, even with the best electrical controls.

Two-high reversing mills are of three forms: bloom-ing mill, high-lift blooming mill, and universal slabbing mill. The predominate form, known simply as a **bloom-ing mill**, is provided with one pair of horizontal rolls in which several grooved passes provide the means of controlling the shape of the piece during rolling, and particularly of working the corners of the piece. This form of mill exists in a wide range of sizes and is designed to roll ingots of square or nearly square cross-section with a maximum thickness of about 865-mm (34 inches). It can edge vertically, in a grooved pass, a piece of about 1020-mm (40-inches) maximum width. An example of an 1170-mm (46-inch) blooming mill is shown in Figure 20—13.

A variation of this form, designed for edging pieces up to about 1980-mm (78-inches) wide resting on their narrow edges, is known as a **high-lift blooming mill** and occasionally as a **blooming and slabbing mill**. It differs from the more conventional form in being provided with higher mill housings to permit greater elevation of the top roll for the edging passes on wide pieces. The roll body is usually longer and has fewer grooved passes, and the motors which operate the screws for controlling top-roll elevation are usually larger in order to provide faster raising and lowering of the top roll. This latter feature is provided since edging passes are few in number but require excessive time unless the top-roll travel is accomplished in no longer than that required for the manipulator to turn the ingot 90 degrees. This form of mill can be provided with interchangeable sets

FIG. 20—13. An 1170-mm (46-inch) two-high reversing blooming mill.

FIG. 20—14. A 1345-mm (53-inch) high-lift, two-high, reversing blooming and slabbing mill.

of rolls so that it can produce not only wide slabs but also blooms and, for this reason, beginning around 1940 it came to be rather widely installed. It has the greatest range in sizes of ingots and products of all three forms. An example of a 1345-mm (53-inch) high-lift blooming and slabbing mill is shown in Figure 20—14.

A third form, actually a special-purpose mill limited to the production of wide slab sections, is the **universal slabbing mill**, often referred to simply as a **slab mill.** It is designed to increase the production rate for wide slabs by avoiding expenditure of the time required for

vertical edging passes in a blooming mill. The mill is provided with a pair of vertical rolls in addition to the pair of horizontal rolls, the vertical rolls performing the edging. In this arrangement, no grooves are used in either pair of rolls. This lack of grooves prevents support of the ingot corners during rolling, and also limits the mill to production of slab sections, since it is incapable of avoiding consistently the inadvertent production of a diamond-shaped product in square or nearly square sections. The slab product rolled on universal mills has relatively sharp corners compared to that of

FIG. 20—15. An 1145-mm (45-inch) universal slabbing mill.

the other mills. Figure 20—15 illustrates an 1145-mm (45-inch) universal slabbing mill.

Rolling a certain size ingot to a certain size bloom is done at the same rate in the conventional and the high-lift blooming mills. The installation cost of the latter is slightly higher. Rolling the same size ingot to the same size slab is done at a somewhat faster rate in the universal slab mill than in the high-lift blooming mill, but the installation cost of the universal slab mill is substantially higher due to the cost of the vertical roll assembly and

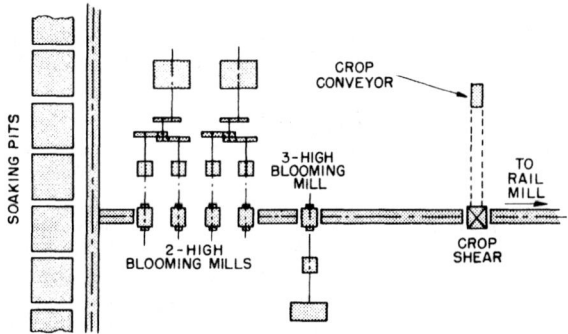

FIG. 20—16. Diagram of blooming mills in tandem.

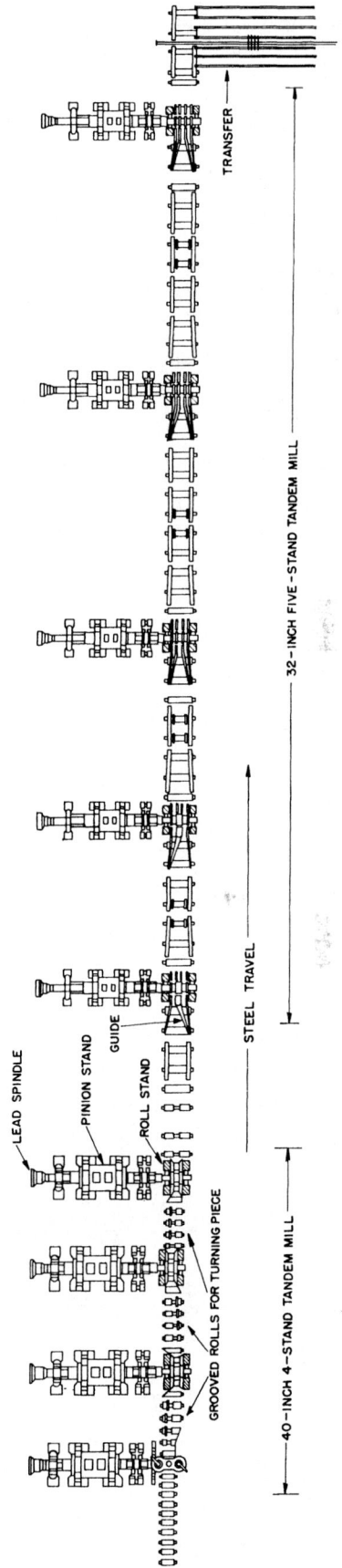

FIG. 20—17. Schematic arrangement of mill producing billets, narrow slabs, and small blooms.

the additional prime mover to drive it. Before the high-lift mill was developed, the universal mill was used to roll wide, heavy slabs for plate mills, and it has been built in sizes to produce widths to about 1980-mm (78-inches), maximum.

Two-High Tandem Mill—This mill consists of several single stands, each containing one pair of rolls, spaced one following another at such distances as to permit the rolled piece to be free between stands. Normally, the piece is rolled one pass in each stand, as shown in Figures 20—16 and 20—17.

Because each stand rolls only one pass, there is no time lost in reversing the travel of the piece, with the result that this type of mill has the maximum output rate of all types. In addition, each stand can be designed for the almost ideal draft and rotating speed for each of the successive rolling passes. This is a great advantage in the early passes where the weakening effect of the ingot crystal structure exists. Opposed to these three very desirable features is the narrow limit of ingot and product cross-sections which the mill can roll without sacrificing the advantages by excessive idle time to change rolls and by compromise on roll-pass designs. For these reasons, the tandem mill is used as a roughing mill to supply hot steel to succeeding billet or narrow slabbing mills.

Due to the large amount of equipment in a tandem mill, it is the most expensive mill to build for a given ingot-to-product reduction but, when scheduled to full capacity, its operating cost per ton of product is lower than that of any other type mill rolling the same-sized ingot to the same product. A tandem mill can roll any size of bloom or slab for which it is designed, but its application has been chiefly to the medium and small sizes, since these were the only sizes in which sufficient orders could be obtained to provide the quantity of production to match the capacity of a tandem mill designed to roll them.

Three-High Mill—The three-high rolling mill first came into use in 1853 under a British patent granted to R. Roden of the Abersychen Iron Works. The first three-roll mill in the U.S. was developed in 1857 by John Fritz for the rolling of bars and rods at the Cambria Works, Pittsburgh. In 1863, Bernard Lauth designed a significantly improved three-high mill. These mills all had advantages over earlier mills, for on the newer three-high mills no reversing of rolls was necessary: the middle roll was driven in one direction. The workpiece was then rolled in one direction by passage below the driven roll and in the opposite direction by passage above it.

This type of mill consists of one stand containing three horizontal rolls, one above the other, each of which has grooved passes, and an elevating table on each side of the mill so that the piece being rolled can be passed alternately between the bottom and middle rolls and between the top and middle rolls. In order to keep the roll dimensions and mill and table dimensions from being prohibitive in size, a compromise pass design is used wherein the grooves in the middle roll are used for rolling both with the top and with the bottom roll. This method limits the reduction possible in the second pass through each middle-roll groove to less than that possible in single passes. The rotating speed of the mill is also a compromise, since each pass should be progressively faster than the one preceding, but, with all passes in one set of rolls, that is impossible. The speed selected is usually too fast for the early passes and too slow for the last ones. There are normally five or seven passes, so that it is often necessary to employ heavy ragging in at least the first two passes to prevent slippage of the steel both in entering the pass and during rolling.

Because the rolls rotate in one direction only and at constant speed, a relatively simple and inexpensive constant-speed prime mover, aided by a flywheel, is used as the main drive. No power is lost overcoming inertia as there is no reversing. Raising and lowering the tables to guide the piece into successive passes can be accomplished in less time than the corresponding manipulations in a reversing mill. For these reasons, a three-high mill is less expensive to build and has a higher output rate than a two-high reversing mill rolling the same ingot to the same product.

Opposed to these favorable features are certain disadvantages. Like the tandem mill, the three-high mill with its fixed passes is rather inflexible with regard to size of ingot cross-section and product. Roll changing results in serious operating delays. Then, too, its usual rather fast rolling speed in the first few passes makes it less desirable generally for the rolling of ingots in these passes.

The three-high mill is best adapted to a place as an intermediate stand (Figure 20—18) where it can be supplied with bloomed-down ingots (ingots which have received a few heavy reduction passes), or large blooms, and where it can supply smaller hot blooms of one or two sizes directly to a following mill. Its rather simple design permits a rolling schedule among three mills in sequence which can be balanced within reasonable limits by alterations in the pass design to adjust the entering bloom and the product bloom sizes to control the production rate of the preceding and following mills.

It finds a use as an ingot-rolling mill in those plants whose standard ingot sizes are very small, on the order of 355-mm (14-inch) square or less in cross-section, and whose total ingot-producing capacity is also quite modest. In these applications, the three-high mill functions more as a billet mill or as a roughing stand for a directly connected finishing mill.

Operating Units Comprising a Blooming Mill—The physical layout will be described here in some detail for rolling ingots on a two-high reversing mill. Essentially the same auxiliary equipment is used in rolling on other types of primary mills. The main parts of the rolling mill itself are shown diagrammatically in Figure 20—19.

Two people are required to operate the controls of an electric-motor-driven two-high reversing mill, the **roller** and the **manipulator,** whose functions are defined by their titles; their **pulpit,** or control station, is placed on a bridge over one of the mill tables. (A steam-engine-driven mill requires the **engineer,** in the pulpit, to operate the engine controls.)

The rolling sequence begins when a **pot car** or **ingot buggy** (Figure 20—20) transfers a heated ingot from the soaking pits to the **ingot receiving table,** which delivers it to the **mill approach table.** The receiving table, in some mills, is equipped with an **ingot turner** so the ingot may be rolled butt-end first (Figure 20—21). This practice aids the shear operator in cutting the proper discard from the butt end of the rolled piece.

The mill approach table transports the ingot to the

FIG. 20—18. A 1020-mm (40-inch) three-high blooming mill, forming an intermediate stand that rolls roughed down ingots (from another mill stand) to blooms that are supplied to the 710-mm (28-inch) billet mill in the far left background. (Courtesy, Wean United, Inc.

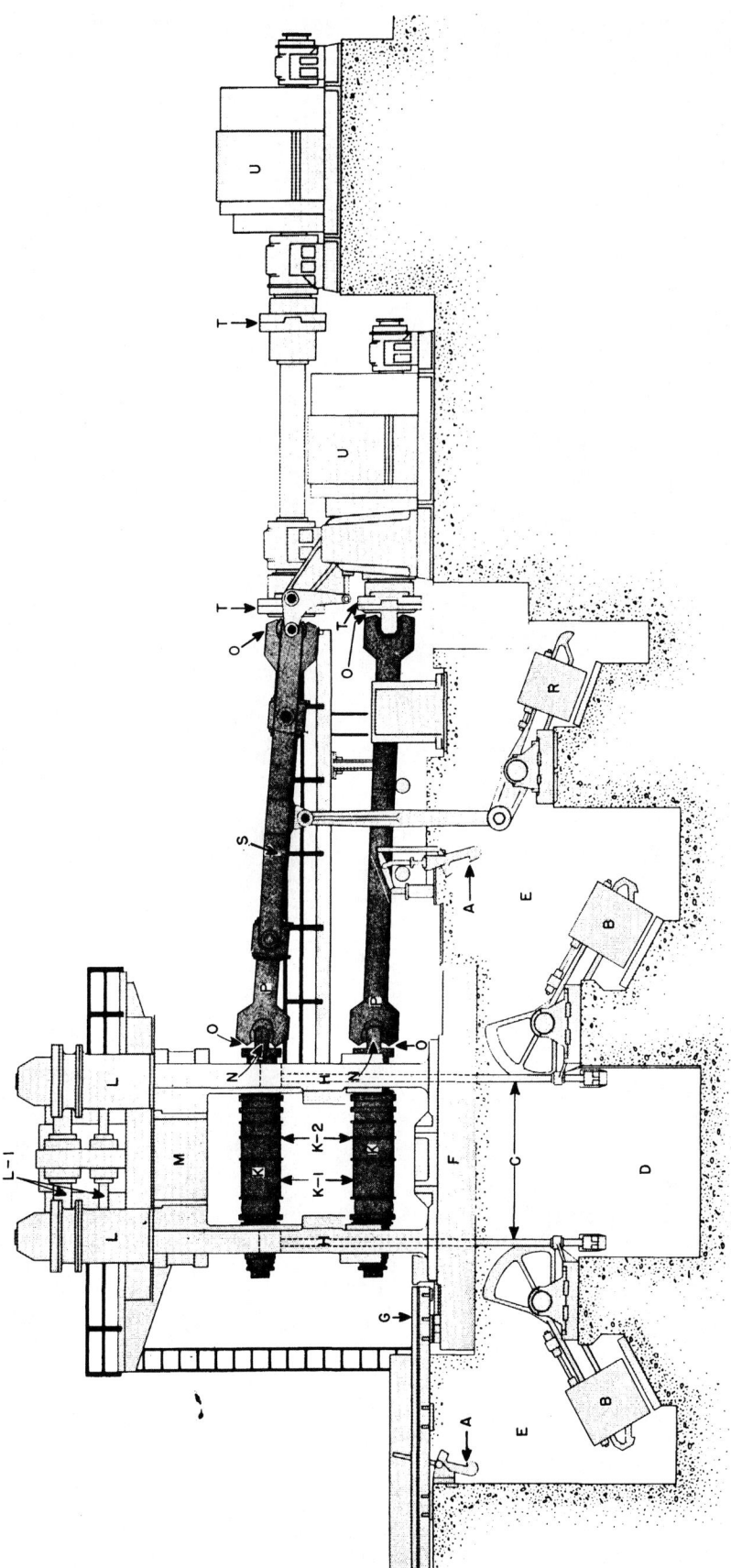

FIG. 20—19. Diagrammatic sketch of a high-lift, two-high, reversing blooming mill, with the main constructional details indicated. (Manipulator, roll tables and control elements not shown.)

LEGEND

A. Counterweight latches.
B. Roll-balancing counterweights.
C. Steelyard rods
D. Scale pit.
E. Counterweight pits.
F. Housing shoe.
G. Roll-change rig rail.

H. Housings.
K. Rolls.
    K-1. Bullhead pass.
    K-2. Edging passes and collars.
L. Screwdown drives.
    L-I. Crossover shafts.
M. Housing separator.

N. Wobblers.
O. Universal couplings.
P. Main-drive spindles.
R. Spindle counterweight.
S. Top-spindle carrier.
T. Motor couplings.
U. Main drive motors.

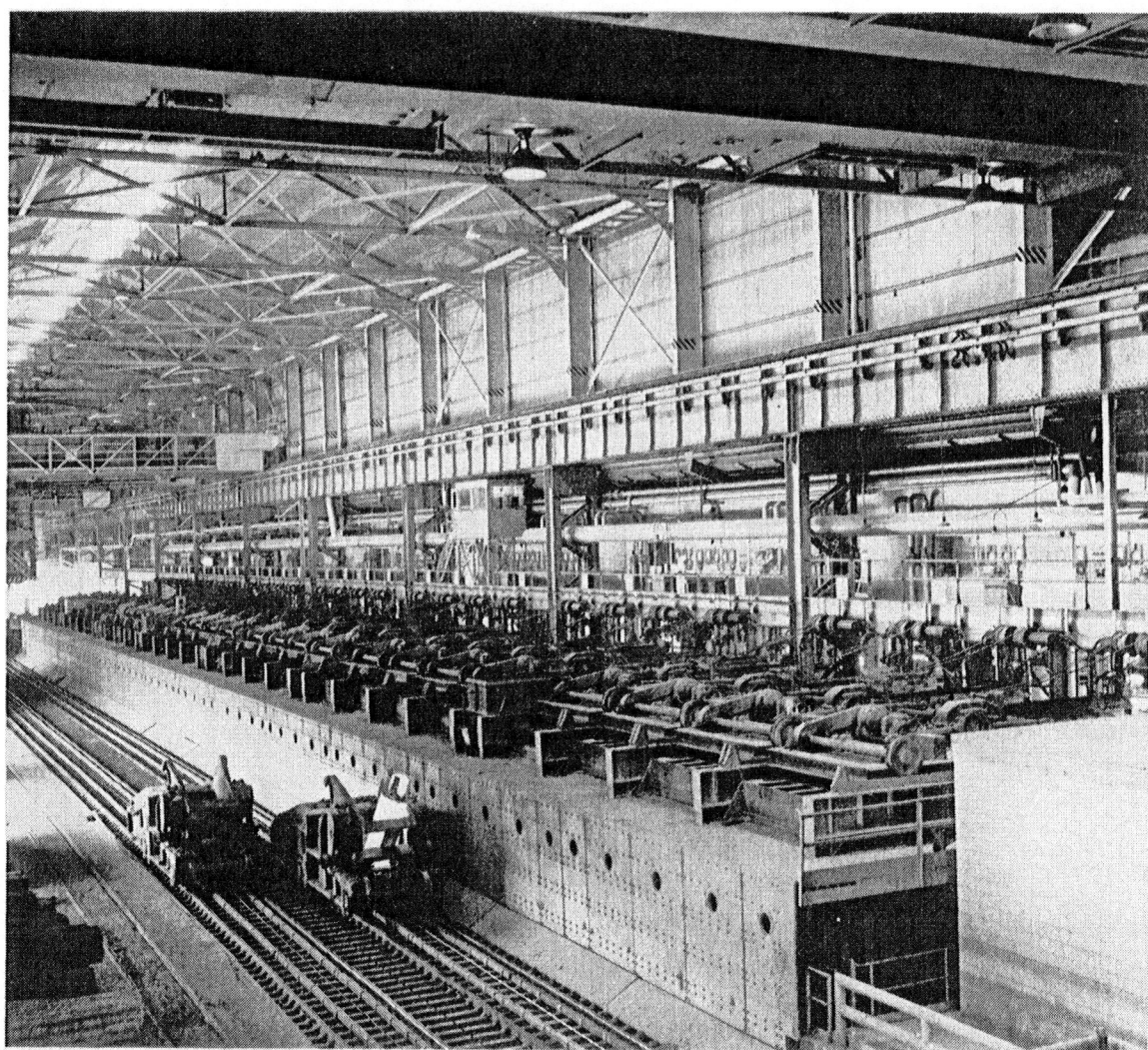

FIG. 20—20. Ingot buggy (pot car) transferring an ingot heated in the soaking pits in the background to the ingot-receiving table of an 1170-mm (46-inch) blooming and slabbing mill. Second pot car (empty) is on parallel track.

front mill table or roller table in preparation for rolling. It can move pieces at speeds of 1 to 2 metres per second (about 200 to 400 feet per minute) in modern mills. A few mills have a scale in the ingot receiving table or mill approach table for weighing the ingots before rolling (Figure 20—21).

During the complete rolling cycle, the hot steel is transported by reversible, live rollers in the mill tables. These rollers are subjected to high temperatures, heavy loads and impact. Forged steel is generally used in making the rollers, which are mounted in roller bearings having a circulating lubricating system. Rollers are usually driven electrically through a line shaft by miter gears; in some cases, they may be driven separately. As the pieces move along the tables, or are turned over by the manipulating equipment between passes, any loose scale falls between the rolls into a trough or depression beneath the tables and is removed to a settling pit with the aid of a stream of water.

Front and back roller tables alternately feed and re-

ceive the piece during each pass through the rolling mill, and mechanical units called **manipulators** rotate the piece through 90 degrees as required and move it from pass to pass. Some of the precautions to be observed in rolling are discussed later in this chapter.

The **shear approach tables** carry the rolled product from the back roller table of the mill to the **shear.** These tables have side guards that line the piece at right angles to the shear so that square cuts can be made. Some mills have a **hot-scarfing machine** in the mill delivery table or in the shear approach table for "skinning" the surface of the hot piece. This removes, in part, some of the surface defects.

Next, **shearing** operations begin. The shear, generally called the **crop shear,** usually is located some 30 to 60 metres (100 to 200 feet) from the rolls, and in line with the mill. Live rollers transport the rolled pieces between the mill and the shear. The primary function of the crop shear is to remove sufficient material from the front and back ends (corresponding to the top and bottom of the ingot, or vice versa), so that the sheared

FIG. 20—21. Ingot being turned 180 degrees at the ingot-receiving table of an 1170-mm (46-inch) blooming and slabbing mill. The ingot turner incorporates a scale for weighing ingots.

piece remaining will meet chemical and metallurgical specifications. Secondary functions are to cut the remaining piece into desired semi-finished lengths and to shear test pieces.

Shears may be operated with electric or hydraulic power. A typical slab shear has a 2030-mm (80-inch) wide knife, and a cutting-force capacity of 1090 metric tons (1200 net tons); it can make straight cuts up to 200-mm thick and 1520-mm wide (8-inches thick and 60-inches wide), at a rate of 10 cuts per minute. It is driven by two 260-kW (350 horsepower) air-cooled motors. The top knife comes down to act as a gag and hold the piece while the bottom knife moves upward to make the cut. This principle of operation has caused this type of shear to be designated as a **down-and-up-cut shear.**

The **after-shear tables** which receive the cut product are designed according to the type of shear. Modern down-cut shears have a table that can move vertically to compensate for the action of the material during shearing. Either kind of shear (down-and-up-cut or down-cut) may have a shuttle motion for the after-shear table, which permits it to move back so the scrap may drop onto a conveyor beneath the shear. A **crop pusher** on the front side of the shear facilitates pushing the last crop end from the shear. Most after-shear tables have adjustable guards for lining up the last piece to insure a straight cut.

The **shear gage** measures the length of pieces to be cut. It is operated by a motor-driven screw. The gage head stops the bloom or slab at the desired point between the shear knives.

Sheared-off discards are moved by the **crop conveyor** to transportation facilities for return to the steelmaking

departments where they are charged into the furnaces as scrap.

Located adjacent to and operating with the after shear tables are mechanical stamping devices or platforms for hand stampers to mark pieces for exact identification. A scale also usually is provided for weighing sheared pieces.

The **transfers** move sheared product from point to point, and may be of the continuous chain type with pusher dogs that engage and move the pieces over skid bars, or of the reciprocating-beam type which moves pieces progressively from one set of supports to another. Transfer buggies or cars also are used to move finished products from the runout table onto the pilers or transfer. Mechanical pilers separate the product into desired classifications as to weight, size of piece and type of cut, or other characteristics. Where the type of product permits, the sheared pieces may be moved by roller table from the shear to a piler, from which the piles may be removed by a "C" hook on an overhead crane, or moved by rollers onto a transfer car.

Combination of Conventional Type Mills—There are many possible configurations and combinations of conventional type mills. Examples include a **tandem** and **three-high mill** in tandem, and **four-stand** and **five-stand mills** in tandem. A primary mill consisting of two two-high reversing mills in tandem is briefly considered here. It was designed to feed a train of mills capable of rolling product ranging in size from 44-mm (1¾-inch) square billets up to and including 1270-mm (50-inch) wide slabs in whatever mix of sizes might be required at any given time. The installation consists of stripping facilities, soaking pits, a 1170-mm (46-inch) blooming and slabbing mill, a 915-mm (36-inch) blooming mill, and a four-stand continuous 535-mm (21-inch) billet mill, all in tandem and enclosed in one building (see Figure 20—22). A scrap- and scale-handling area and a shipping building were included, along with slow-cooling facilities and auxiliary equipment designed to handle the wide range of mill products. A wet-scale-handling flume system covers all areas under the entire mill and auxiliary equipment, including hot beds. An information-broadcasting and data-collecting system provide necessary rolling information and collect desired control data throughout the mill.

Stripping facilities consist of two stripping cranes, ground stripper, and a stool-coating station. The soaking pits comprise seven banks of one-way top-fired pits, fired with high-velocity multi-directional burners. Two cable-driven ingot buggies (Figure 20—20) transfer ingots from the pits to the mill receiving tables. Following delivery to the receiving table, the ingot passes through the overhead ingot turner that weighs and turns the ingot 180 degrees, when desired, and returns it to the mill approach table (Figure 20—21).

The 1170-mm by 2795-mm (46-inch by 110-inch) blooming and slabbing mill (Figure 20—23) is a high-lift reversing mill with individual roll drives. The mill is equipped with manipulating fingers on the front side only, with positioning side guards on each side of the mill stand. Table rolls adjacent to the mill are individually driven, and two feed rollers are incorporated into the housing on both the entry and delivery side of the mill.

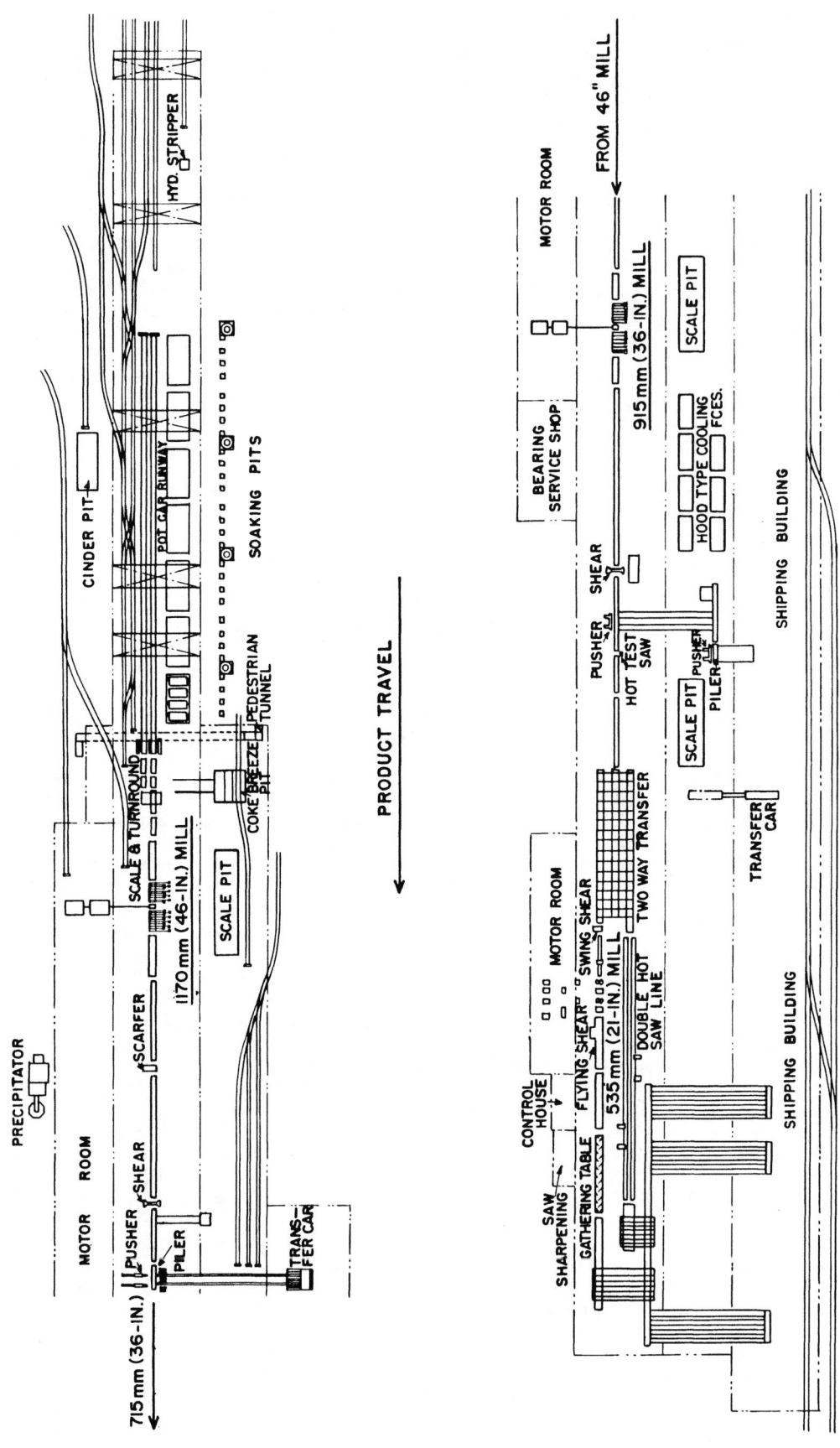

FIG. 20—22. Schematic arrangement (not to scale) of two two-high reversing mills in tandem, followed by a 4-stand 535-mm (21-inch) billet mill. (Mill tables for both mills are in same straight line: drawing has been divided for convenience in printing.)

FIG. 20—23. 1170-mm (46-inch) two-high reversing high-lift blooming and slabbing mill.

Product sizes from the 1170-mm (46-inch) mill include: slabs up to 1320-mm (52-inches) wide with a minimum thickness of 89-mm (3½-inches); blooms up to 510-mm (20-inches) square; and various breakdown sizes for subsequent rolling in the 915-mm (36-inch) mill. Product down to about 200-mm (8-inches) square is finished on this mill when the 915-mm (36-inch) mill is rolling to the smaller sections to balance output of the two mills.

A four-sided scarfer following the 1170-mm (46-inch) mill is capable of scarfing the complete range of products from this mill on the edges only, top and bottom only, or on all four sides.

The 1170-mm-mill (46-inch-mill) shear is an electrically driven up-and-down-cut shear capable of parting 3870 square centimetres (600 square inches) of mild steel and 2580 square centimetres (400 square inches) of stainless or high-carbon grades. An elaborate crop-handling system maintains proper segregation of steels of the wide variety of compositions rolled on these mills. Revert scrap is loaded into charging boxes at the mill.

The slab piler and transfer, next in line, is the outlet for slabs and larger bloom sizes finished on the 1170-mm (46-inch) mill. The transfer car travels to the shipping bay where it is unloaded by a crane equipped with a "C" hook. A pneumatic impact stamping machine, remotely controlled by the piler-transfer operator, is used to mark rolled pieces with heat number, ingot number, and cut letter.

The 915-mm by 1980-mm (36-inch by 78-inch) two-high reversing blooming mill (Figure 20—24) is next in line. Products rolled on this mill range from 98-mm (3⅞-inch) square billets to 380-mm (15-inch) square blooms, in sheared lengths from 1.5 to 6 metres (5 to 20 feet). Positioning sideguards are on both the entering and the delivery sides of the mill.

The shear following the 915-mm (36-inch) mill is electrically driven through an air-operated clutch and flywheel combination. The relatively small amount of revert scrap accumulated at this point is handled in self-dumping boxes, with a diverter in the scrap chute to guide the scrap to the desired box.

The sheared product is then transferred by crane and "C" hook either to slow cooling furnaces or to the shipping building.

The slow-cooling furnaces for controlled cooling of rolled products are gas-fired furnaces with removable hoods, capable of holding 54-metric ton (60-net ton) loads.

Just beyond the 915-mm (36-inch) mill transfer is a 1830-mm (72-inch) hot saw for securing metallurgical-test samples up to 380-mm by 380-mm (15 inches by 15 inches) in cross section. Blooms are routed into the saw by an air-operated diverter and returned to the transfer after test pieces have been cut.

FIG. 20—24. 915-mm (36-inch) two-high reversing blooming mill, showing control pulpit in right background.

Product placed on the 535-mm (21-inch)-mill transfer table may be moved in one direction to the hot-saw line equipped with one stationary and one movable 1525-mm (60-inch) hot saw with suitable stops and gages for cutting product from the 915-mm (36-inch) mill that requires hot-sawed ends; sections up to 180-mm by 180-mm (7-inches by 7-inches) square can be sawed into lengths of about 2.5 to 10.5 metres (8 to 34 feet), following which it is transferred to the hot-bed run-in table and discharged onto one of the three hot beds. In the other direction, the 535-mm (21-inch)-mill transfer table moves blooms to the 535-mm (21-inch)-mill run-in table. As the pieces approach the 535-mm (21-inch) mill, a swing shear with a section limitation of 230 square centimetres (36 square inches) is used to crop their front ends. Immediately preceding the first stand of the 535-mm (21-inch) mill is an air-operated billet turner that turns pieces 45 degrees to position them for proper entry into the first diamond pass of the mill.

The two two-high reversing mill has proven to be very versatile, and may easily be configured to supply adjacent continuous billet mills. In this way, a wide range of sizes of square and round billets and narrow slabs can be rolled at high production rates.

ROLL PASS DESIGN FOR PRIMARY MILLS

The design of primary rolling mills is based on the deformation to be given to the workpiece, the size of material to be handled, and the length, weight, and shape of the sections to be produced. It is necessary in all cases to have all parts of the roll stand designed to survive the severe service conditions imposed by the inherent nature of the operations performed.

The **foundations** for primary mill stands are proportioned to the size and weight of the mill, and are built to withstand any tendency to settle or distort. Foundations are anchored to bedrock by piling to increase their ability to withstand the shocks and blows common to primary rolling operations. In counterweight-balanced mills, such as that shown in the diagram represented by Figure 20—19, foundation design is complicated by the space needed for this mechanism. Design of foundations for two-high continuous mills is simplified by the type of housing and roll arrangement used. Three-high mill foundations must provide room under and adjacent to the mill for the mechanisms of lifting and manipulating devices that lift and turn pieces between passes.

Foundation design also must provide for the draining of water, oil, and greases to low spots, or sumps, from which they can be removed by pump or siphon if natural drainage cannot be provided. Passageways giving access to the equipment beneath the mill, sluice ways for handling scale that drops off pieces during rolling, and electric-cable tunnels also must be provided in the foundations of a modern mill.

Design of Mill Roll Stands—The essential parts of

rolling mills are discussed in Chapter 23, and need not be discussed again here for the particular case of the primary mills. The point should be made however, that the component parts and auxiliary equipment for primary mills are extremely rugged and of very large size as compared to most other mills. Mention might be made of the fact that, in primary mills, two-high tandem and three-high mills generally are constructed with open-top housings, while two-high reversing mills, high-lift reversing mills, and universal mills are built with closed-top housings.

Front and back feed or housing rollers are required to move the ingot or piece between the mill tables and the horizontal rolls of the mill, because housings of two high reversing, high-lift two-high reversing, and universal mills are so bulky in design and the tables for these mills are laid out in such a way that there is an open space between the last roller of the mill tables and the horizontal rolls of the mills. (Post and table construction of two-high continuous and three-high mills make it possible in these cases only for these rollers to be eliminated from their design.) These feed rollers have separate electric motor drives in modern mills. They must be designed with heavy bodies and necks to withstand pressure and impact, and should have carefully designed, well-lubricated bearings that are protected from scale and water.

In universal-type primary mills, one of the feed rollers must be omitted between the vertical rolls, since these rolls must move in and out with respect to the center of the table. This one roller is replaced by a short dummy roller. If a piece "stalls" on this dead roller, the vertical rolls must be run in to contact the piece to move it onto the live table rollers on either side of the vertical rolls.

Roll Design—Roll sizes for blooming-slabbing operations are determined by the type and size of mill and the product being rolled. In most two-high reversing mills, roll diameters are usually about 50- to 125-mm (2 to 5-inches) less than the size of the mill (based on the pinion drive). Rolls for a two-high tandem mill usually are short and heavy, since usually only one pass is made in each roll stand and diamond passes predominate.

In most three-high mills, the line diameters of the work rolls are different, the bottom roll being the largest, with the middle roll next in size and the top roll the smallest in diameter. These rolls are all of the same length body and neck.

Roll Positioning—There are several different methods for adjusting rolls on two-high mills. One is a simple method of using a rider bearing on each of the two bottom-roll necks. Two top-roll carrier bearings rest on liners on the bottom rider bearing. Wedges are placed between these liners to vary the distance between the rolls. Another method of adjustment consists of supporting the top roll by springs from the housing cap. Turning of the housing screw compresses the springs, and thus the space between the rolls is varied. Still another method of adjustment in which the top roll remains stationary and the bottom roll is moved up or down is used in a good many mills. This is accomplished in the following manner: a cast-steel screw box in which the threads have been babbitted to prevent excessive wear is placed on the sill of each housing. A screw bolt is placed in each of these, the outside one containing a left-hand thread while the inside one has a right-hand thread. The two screws are joined together by a cast coupling. Above the screw bolts, and resting upon them and the screw boxes, cast-steel wedges are mounted containing the same number of babbitted threads as the screw box. The bottom chocks are wedge-shaped and rest on the screw wedge. Since the two screws are coupled together, turning of the screw on the outside of the mill causes both screw wedges to move in or out and, due to the wedge shape of the bottom chock, the bottom roll is raised or lowered. This method of adjustment is used on many continuous mills.

Billet mills also make use of an electric screw-down to adjust rolls. With this type of adjustment, the top roll is supported by two hydraulic cylinders through steel-yard rods in each housing, reaching up to the top-roll carrier bearings. The top roll is held down by screws reaching through the cap to the breaker blocks on the rider bearings. An electric motor mounted on the housing cap actuates the two housing screws. More recently, in place of or in addition to screws, hydraulic roll positioning systems are being employed.

Effect of Pass Design on Rolling Procedures—Drawings of rolls for various types of blooming mills are shown in Figures 20—25 and 20—26. Figure 20—25A shows rolls designed to produce a great number of sizes of semi-finished products; this is typical of general practice. Since a large number of sizes must be produced on one set of rolls, it might be supposed that it would be well to roll on flat rolls without grooves. However, such procedure would produce blooms that were far from square. Grooves are provided, therefore, to produce the most popular sizes.

It is best practice to get the ingot into a box (grooved) pass as soon as possible in rolling, since any great amount of reduction of a piece without protection on the sides would cause the steel to crack. The spreading of the metal in these grooves causes wear of the rolls. Therefore, the sides of the grooves are tapered so that, with the dressing of the rolls, the pass may be made to approach its original width. The taper is made as great as possible for this purpose. However, if too much taper is allowed, the piece being rolled will have a tendency to turn down as it emerges from between the rolls. Tapers up to 15 degrees per side have been used effectively.

Figure 20—25B illustrates rolls designed for a more specific purpose, while allowing a certain degree of flexibility. The largest proportion of product made on this set of rolls consists of 100-mm by 100-mm (4-inch by 4-inch) square billets. To use these rolls, it is necessary to provide grooved table rollers for the diamond passes. This also restricts manipulation and, consequently, full use of the roll body.

Roll contours for a three-high, single-purpose mill are shown in Figure 20—26. In this case, the ingot size, bloom size, and number of passes is fixed. On this type of mill, much heavier reductions are given in the early passes than would be used in general practice on the two-high mill using the rolls of Figure 20—25A. With the rolls of Figure 20—26, edging occurs after each two passes. Since the ingot is protected on all corners and spreading is somewhat suppressed, and edging is done so frequently, it is possible to roll with the heavy reductions required to produce a bloom from an ingot in so few passes.

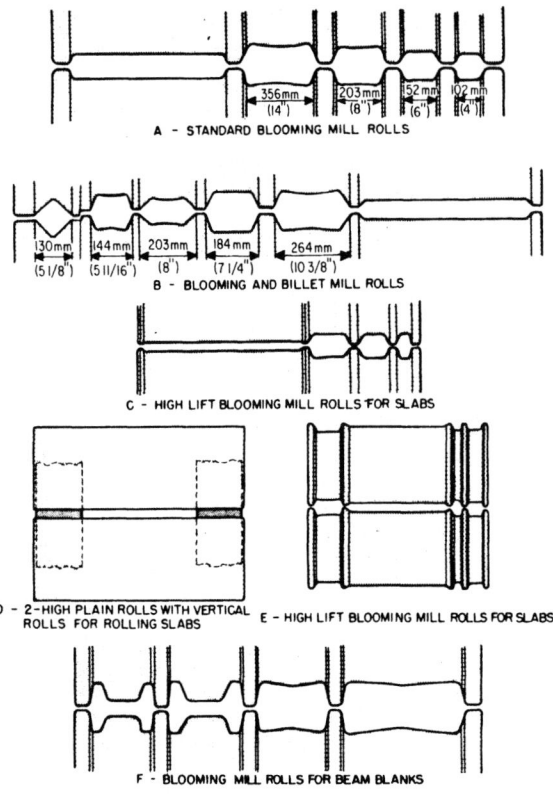

A - STANDARD BLOOMING MILL ROLLS

B - BLOOMING AND BILLET MILL ROLLS

C - HIGH LIFT BLOOMING MILL ROLLS FOR SLABS

D - 2-HIGH PLAIN ROLLS WITH VERTICAL ROLLS FOR ROLLING SLABS

E - HIGH LIFT BLOOMING MILL ROLLS FOR SLABS

F - BLOOMING MILL ROLLS FOR BEAM BLANKS

FIG. 20—25. Schematic representations of rolls for various types of blooming mills. In Sketches A, B, C and F, only part of each roll is shown to indicate the shapes of the pass openings. Sketches D and E show elevations of entire roll bodies.

Figure 20—25F illustrates the pass arrangement for the production of shaped blooms called beam blanks. This type of roll, like the blooming-mill rolls of Figure 20—25A, usually lacks the necessary space for the most desirable arrangement of passes. Generally, in the rolling of such blanks, the ingot is reduced to a smaller rectangle or square. Proportioning of web and flange must be made. For larger beam sizes, however, a beam-shaped ingot is used. This permits a more uniform reduction that results in a better flange built up in rolling.

Figures 20—25C and E show two sets of high-lift blooming-mill rolls. While the Method D of employing vertical and horizontal rolls is fastest for producing slabs, use of the horizontal rolls for edging, as in Figures 20—25C and E, is advantageous from the standpoint of surface of the semi-finished product of the mill. The edging pass, using horizontal rolls with the piece resting on its narrow edge, cracks the scale from the wide faces of the slab (which are in the vertical position) permitting the scale to fall off the piece and between the table rollers into the sluice-way beneath the mill. With the flat-rolling method using both vertical and horizontal rolls, the piece is not turned and some means must be employed to blow off the cracked scale; various ways are discussed later.

Convexity of Passes—The convex shape (belly) of some passes will be noted in the illustrations; this convexity serves a double purpose. The first purpose is to prevent the formation of fins. If a straight-sided piece is edged in a succeeding pass, spreading causes the metal to squeeze out between the collars of the roll, as shown by the dotted lines in Figure 20—27A. Convexity of the passes produces a shape with concave sides, allowing considerably more spread in succeeding passes before a fin can occur. Fins, when a piece is turned 90 degrees then rolled in a succeeding pass, are rolled over and produce what is known as a **lap,** and no amount of further rolling can eliminate this defect. It is impossible for fins to roll back into the parent metal because an

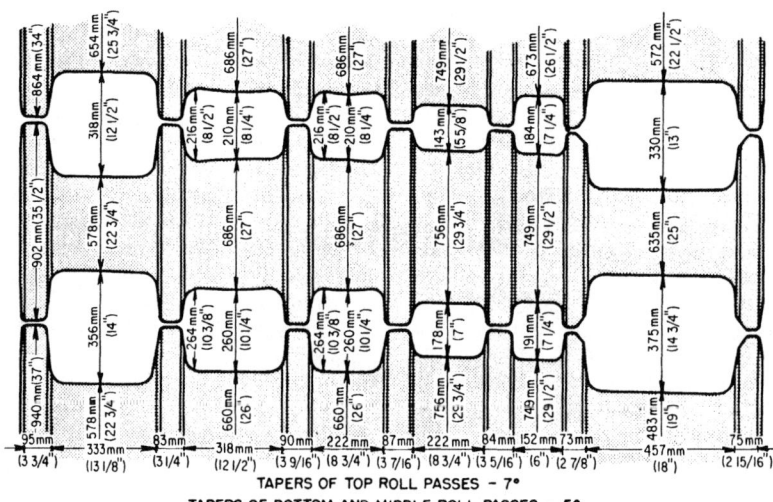

TAPERS OF TOP ROLL PASSES - 7°
TAPERS OF BOTTOM AND MIDDLE ROLL PASSES - 5°

FIG. 20—26. Schematic representation of rolls for a three-high, single-purpose mill. Only part of each of the top and bottom rolls is shown, sufficient to indicate the shape and size of the pass openings. The diameters of the rolls in the various passes, however, are given to fix the size of the top and bottom rolls. Dimensions are in millimetres with equivalent inches in parentheses.

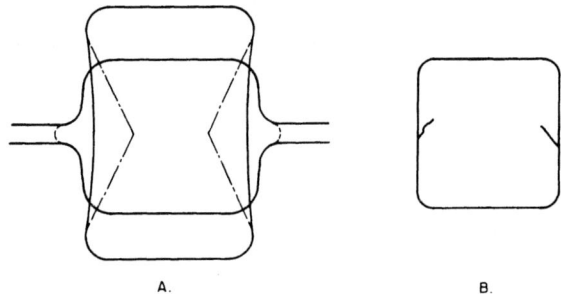

FIG. 20—27. (Left) Schematic representation of the effect of concavity of ingot sides in preventing the formation of fins. The large round-cornered shape is that of the piece entering the pass; the smaller shape represents the cross-section of the piece in the pass, with dotted lines indicating fins formed in the openings between the rolls if the ingot is straight-sided or has insufficient concavity. The dot-dash lines indicate extreme concavity which, as indicated at right, also will cause laps during rolling.

oxide forms on them which prevents clean metal-to-metal contact between the fin and the parent metal that would be essential to welding the two integrally together. On the other hand, laps (or seams) may result if the sides of a piece are excessively concave (Figure 20—27). Laps must be chipped, scarfed or ground out before the bloom or billet can be rolled on finishing mills for further conversion.

The permissible amount of convexity is dependent upon the draft. It might seem that the convexity should be made very great to eliminate all possibility of fin formation. However, if the convexity is made too great, as illustrated by the dot-dash lines of Figure 20—27A, the reduction, in penetrating to the center, will not be able to push out the metal in the center and folding will occur that ultimately will form laps, as indicated in Figure 20—27B.

It will be seen from Figure 20—25F, showing the beam-blank roll set, that the box passes have a deep convexity; this is desirable in this case because of the shaping effect which is helpful in producing the shaped bloom.

The second purpose of convexity is to provide a convenient place for the ragging placed on primary mill rolls. This will be discussed shortly.

Depth of Passes—The depth of any pass other than the bullhead pass in blooming-mill rolls is governed by two factors: (1) the maximum draft to be taken on any piece rolled in that pass, and (2) the minimum thickness to be produced in that pass. The relationship between the depth of any pass and the bullhead or barrel pass diameter must be such that the bottom of the bullhead pass is not elevated too high above the table rollers. If the bottom of the bullhead or barrel pass is too high above the table rollers, there would be excessive impact from the piece each time it passed through the mill and was discharged onto the rollers. Passes should be made as deep as possible, consistent with these considerations, however, because deep coverage of the bloom helps to guide the piece and also aids in preventing the formation of fins.

High-lift reversing mills require rolls of considerable length; most mills of this type have rolls 2540- to 3050-mm (100- to 120-inches) in body length. This is necessary to provide sufficient width for the grooved edging passes, along with a barrel or bullhead pass at least 150-mm (6-inches) in excess of the maximum width of slab to be rolled. Mills of this type may operate in conjunction with a separate vertical roll stand having relatively short, heavy rolls, located at the end of the mill runout table, to give a square edge to the finished product.

Universal-mill roll sizes are dependent directly upon the width of material to be rolled, along with the desired product size. In a typical mill, the bodies of the horizontal rolls are 1145-mm (45-inches) in diameter and 2030-mm (80-inches) in length.

Ragging, or grooves cut into blooming-mill rolls to prevent slippage when using heavy drafts, should be shallow and parallel to the axis of the rolls. Otherwise, it may prove injurious to the surface of the semi-finished steel by causing slivers and laps. Sometimes, ragging will be designed to protrude from the roll, instead of consisting of grooves cut into it. In general, the less ragging the better.

Roll-Opening Indicators—On mills where the distance between the rolls is varied at the will of the operator to suit the particular rolling operation, some means must be provided for indicating visually the position of (i.e., the distance or opening between) the rolls. The conventional type of indicator was a large, calibrated dial with clock-like hands. The shaft that drives the hands is connected mechanically by rods, gears and cables to some rotating part or shaft on the screw-down mechanism, so that the position of the hands on the dial changes in unison with the movement of the rolls. The dial usually is mounted at the top of the mill, where it can be seen from the pulpit. In more modern mills, electronic digital indicators would be more common.

Roll-Changing Devices—Rolls in open-top housings are changed by removing the housing cap and picking out the rolls separately or collectively with crane hooks.

With closed-top housings, roll changing is effected through a window of the housing by a counterweighted porter bar, by a sleeve using the ingoing roll as a counterweight, or by a "C" hook. Any of these methods requires the use of an overhead crane to remove the rolls one at a time. The rolls are pulled out of the housing by attaching a socket fitted over the protruding roll necks.

In modern mills, a **roll-changing rig** removes worn rolls and installs new ones. It is set in a permanent mounting, level with the housing sill, and consists of either a motor-operated crosshead mounted on a rail frame and driven by a rack and pinion or hydraulic cylinder and rod. It withdraws both top and bottom rolls, with their respective bearings, one at a time. The old set-up then is removed from the rig, on which the new set-up is placed and positioned in the mill housings.

Cooling Water—During operation of a blooming or slabbing mill, a liberal supply of cooling water should be distributed carefully and uniformly over the rolls. If possible, the rolls always should be warm, never chilled, since the sudden contact of the surface of a hot ingot with a cold roll will cause too rapid a differential expansion of the roll surface, and cracks called **fire cracks** may develop. Water should be turned off when the mill is not rolling. If water is kept flowing when the mill is not rolling, the rolls should be kept turning to avoid uneven cooling which is one of the most common sources of cracks in rolls. Rolling without water at first and then putting water on the heated rolls is also a cause of cracking.

In some mills, great care is taken to warm up rolls before they are used in primary rolling operations. Rolls are sometimes warmed by steam coils in a special pit before use. This practice minimizes fire cracking and roll breakage.

Manipulators—The rate of rolling steel in the blooming or slabbing mill depends upon the auxiliary equipment to a marked degree, as well as on the speed of the rolls. The importance of the auxiliaries is demonstrated by consideration of the fact that, in normal good rolling practice, a piece being rolled is in actual contact with the rolls only 25 to 40 per cent of the total rolling time. The remaining time is consumed in handling or manipulating the piece, emphasizing the importance of rugged, efficient equipment for performing this work. Transfer of the work from one point to another and away from the mill stand and between the different work stations is performed by the various live-roller arrangements discussed earlier.

Turning of the piece between passes, moving it sideways on the roller tables from one pass to another and, when necessary, straightening it, is the function of the manipulators. Usually, they are built to have both horizontal and vertical motion and, in most mills, are operated by electric or hydraulic power.

Location of blooming-mill and slabbing-mill manipulators depends upon the type of mill. In most three-high mills, the manipulators are located under the table, on the entering side of the rolls, the fingers coming up between the rollers to engage the piece as the table is lowered. The principle of operation is shown in Figure 20—28, as applied to a three-high billet mill. In other mills, they are located on side guards, stroking laterally over the tables on one or both sides of the mill. These side guards are equipped with retractable manipulator fingers or arms, which have a vertical or near vertical stroke and serve to lift the piece by a corner in the process of turning it 90 degrees.

BILLET MILLS

Development of the Billet Mill—It already has been described how, as the technique of pouring larger and larger ingots progressed, it became necessary to use blooming mills to roll blooms of sizes which could be handled by separate finishing mills, either directly, or after reheating. A further step in the trend became necessary as ingot sizes increased still further in that reversing blooming mills are not suitable for producing billets in sizes down to 98-mm by 98-mm (3⅞-inches by 3⅞-inches). Hence, an intermediate mill between the blooming and finishing mills is required for rolling billets.

Blooms from 150-mm by 150-mm to 250-mm by 250-mm (6-inches by 6-inches to 10-inches by 10-

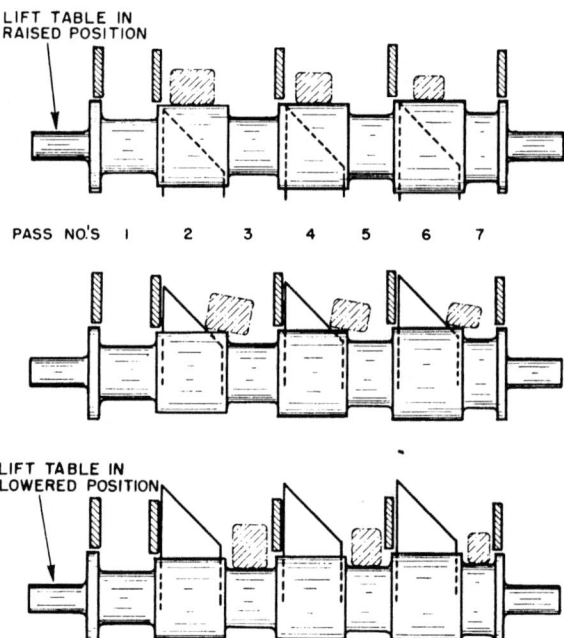

LIFT TABLE IN RAISED POSITION

PASS NO'S   1   2   3   4   5   6   7

LIFT TABLE IN LOWERED POSITION

Fig. 20—28. Diagram showing the action of the fingers of a stationary-type manipulator for advancing the piece from pass to pass, while simultaneously turning it 90 degrees. The fingers remain stationary, and their action is performed by the motion imparted to the piece as the table is lowered from the raised position. The middle diagram represents the lift table in intermediate position.

inches) usually are taken to the billet mill from the blooming mill directly without reheating. The ends of the bloom are cropped, and it is often necessary to cut the remainder of the bloom into two or more pieces, as will be explained later. Hot scarfing also may be performed at this time. These operations must be performed rapidly to retain as much heat as possible in the bloom to keep its temperature high enough for good rolling in the billet mill. In this connection, billet mills capable of handling large blooms not only save time on the blooming mill, but also receive hotter steel, because the smaller surface area per unit of weight exposed by pieces of heavy cross-section lessens heat losses.

Although most billets are rolled directly from blooms without reheating, there are some plants that provide reheating furnaces between the blooming and billet mills (Figure 20—29). The older blooming and billet mills generally did not provide for reheating since the grades of steel then produced could be rolled through a greater temperature range without harmful effects. Since the more critical grades of steel, with their restricted range of rolling temperatures, have become more widely used, some of the newer primary-mill installations provide reheating facilities between the blooming mill and the billet mill. These mills are arranged sometimes so that the ordinary grades of steel can be sent direct to the billet mill without reheating.

Since the billet section is a simple one and the requirements in the way of accuracy as to finish and dimensions of section are not exacting, the first requirement of the billet mill is that it be heavy enough to handle fairly large blooms, and speedy enough to reduce the piece to the desired size before it becomes too cold. Most billets, however, must be straight, square, and free from surface defects. Billets that are twisted, bent, or not square, will not charge into or push through pusher-type reheating furnaces properly, causing pile-ups in the furnace; this is due to the fact that billets lie on the hearth side by side, and depend, for their movement through the furnace, on pushing against the exposed side of the last billet charged to move all the billets forward through the heating chamber. Poor surface qualities due to the above problems cause excessive conditioning or even losses due to rejections.

In this section of the chapter, only the rolling of billets will be discussed. The rolling of sheet bar and skelp will not be considered, for, although these are both considered as semi-finished products, their production on billet mills has been abandoned almost completely in favor of using products of the continuous hot and cold strip mills as raw materials for the production of sheets, welded pipe, and tubes.

Three-High Billet Mills—Billet mills may be of several types. One is the three-high mill with lifting or tilting tables. This type of mill consists of three rolls mounted one above the other in a single roll housing. Billets are rolled in one direction between the bottom and middle rolls, and through the middle and top rolls in the opposite direction. The lifting or tilting tables move the billet to the two different pass levels. Mills of this type have fixed drafts, and only a few limited sizes of billets can be rolled with a given combination of rolls. To produce various sizes of billets on this type of mill would necessitate numerous roll changes, which are costly.

Figures 20—30 and 20—31 show the layout for a three-high billet mill and the rolls used in the mill. Alternate passes are used in the top and bottom rolls while every pass is used in the middle roll. On this mill, four rolls comprise a set, consisting of a top, bottom, and two identical middle rolls although only three rolls are used at one time. When the middle roll becomes worn to the extent that it must be replaced, the mill is changed; that is, the rolls are removed from the housing and the worn top roll is placed in the bottom position, the fourth roll or second middle is placed in the middle position and the worn bottom roll is placed in the top position. This presents a whole series of new passes and permits the same tonnage to be produced with the set of four rolls that otherwise would require two sets of three, or six rolls. Since the piece is rolled in one direction between the bottom and middle rolls and in the opposite direction between the middle and top rolls, the use of lifting or tilting tables is necessary to transfer the piece between the bottom and top pass lines. The table on the side of the mill from which the bloom enters (the side nearer the blooming mill) is called the front table, while that on the opposite side of the stand is called the rear or back table. Manipulators must be used in conjunction with these tables to turn the piece between passes and move it into line with the next pass.

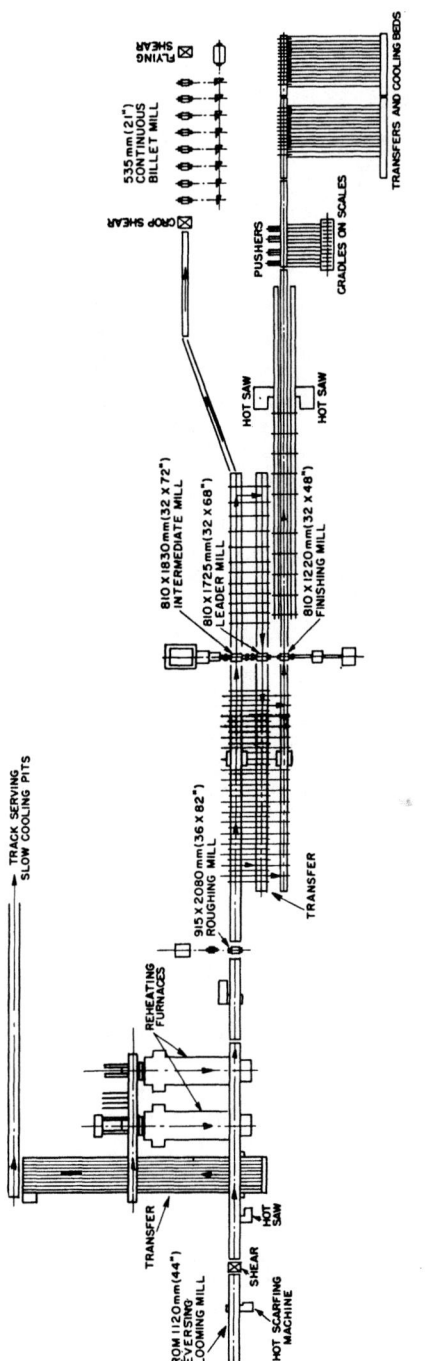

FIG. 20—29. Schematic arrangement of a cross-country billet mill.

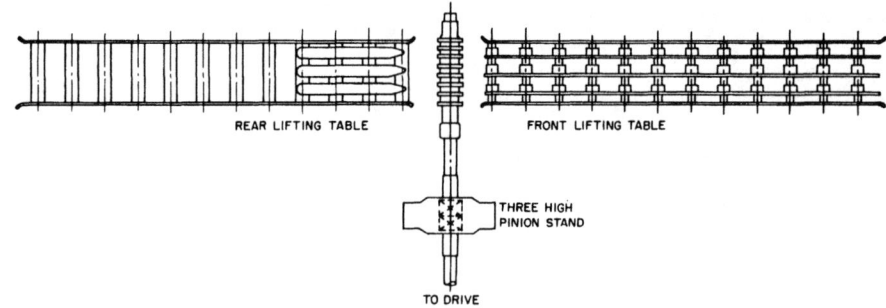

REAR LIFTING TABLE   FRONT LIFTING TABLE

THREE HIGH
PINION STAND

TO DRIVE

FIG. 20—30. Schematic layout for a three-high billet mill, using the rolls shown mounted in their housings in Figure 20—31.

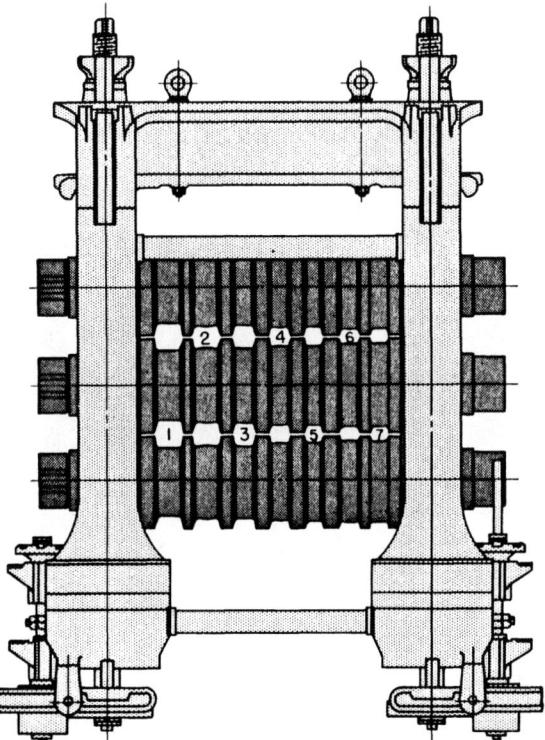

FIG. 20—31. The rolls (mounted in the housings) of a three-high billet mill. Numerals in the pass openings indicate the sequence of passes used in rolling (see also Figure 20—32).

Both tables on a three-high mill are raised and lowered as a unit. The lift type table is more common and will be described along with only one of the many possible arrangements of manipulators.

Figure 20—28 shows diagrammatically a three-high mill lift table in raised, intermediate and lowered position. The front table contains twelve cast-steel rollers, each of which has five collars for turning the billets. These collars create four grooves extending from end to end of the table. The rollers are driven by an electric motor. There are side guards on the edges of the table and at the front end are side guards for putting the bloom into the proper pass. The front table is equipped with a stationary manipulator for advancing

the billets from pass to pass which consists of fingers bolted to a pedestal on the foundation of the mill. The fingers do not reach above the level of the roll passes when the table is elevated and the billets run out on the collars of the rolls. When the table is dropped, the billets encounter the stationary fingers and slide down into the grooves in position for the next pass. The rear table is operated through the same shaft as the front table but owing to the fact that it not only must raise the billet from the bottom pass line to the top pass line, but also must advance them one pass toward the outside, it has to travel through an arc in rising to bring the billet in line with the next pass. This is done by causing the table to slide toward the next pass as it is raised, by the use of pull-over rods attached to pedestals on the proper side of the scale pit. When lowered, the table slides back into place again. This table consists of twelve cast-steel rollers driven by a motor similar to the one used on the front table. Three heavy cast-steel side guards between the four passes that are used on the bottom roll divide the table into four grooves. There is a manipulator in the first groove that consists of five forged-steel fingers. The upward motion of the table draws the fingers with it and causes them to turn the piece and, as stated before, the sideward motion of the table advances the billet to the next pass. This manipulator lies below the table when the billet is delivered from the bottom roll and only turns the billets 90 degrees from the first to the second pass.

Since the piece being rolled must be entirely out of the rolls for the lift table to transfer it between the top and bottom passes, its length must be limited to that which the tables will accommodate. This requires cutting the blooms from the blooming mill into the correct lengths so that when they are elongated, due to draft in the three-high mill, they can be handled properly on the tables. Cutting the blooms into pieces reduces the yield considerably as the ends of each piece usually are cropped after rolling.

The roll housings generally are cast steel, although some mills use cast-iron housings. The middle roll is not adjusted, but the top and bottom rolls are adjusted toward the middle. Cast-steel guides and side guards are held in guide cages or on rest bars which are bolted to the receiving and delivery sides of the housings. The pinions preferably are made with dou-

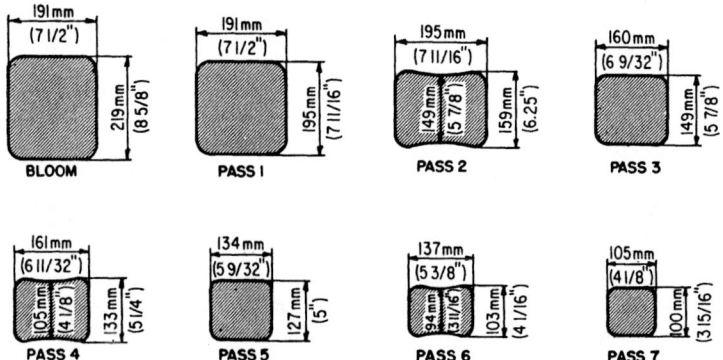

FIG. 20—32. Cross-sections of the pieces out of each pass when billets are rolled on the mill shown diagrammatically in Figures 20—30 and 20—31.

ble helical or herringbone teeth. The three-high mill was originally driven by a steam engine at constant speed. Later, electric motors were used. Methods of calculating power requirements for mills are discussed in Chapter 23.

The three-high billet mill shown in Figures 20—30 and 20—31 rolls four sizes of billets, which are subsequently reduced further in size on a smaller billet mill. As noted previously, the blooms are turned only once on the rear table and from one to three times on the front table, depending on the number of passes taken in the mill.

Figure 20—32 shows the shape of the piece out of each of these roll passes. The rolls used are cast alloy steel, typically 770-mm (30⅜-inches) in diameter with 1915-mm (76½-inch) roll body. Box passes, as shown in Figure 20—31, are used generally on three-high mills for rolling billets. Slightly greater reduction can be taken using a diamond and square series but, since in passes of this shape the piece is rolled on the diagonal, the overall height of the piece is greater, requiring a deeper groove in the roll than for box passes. Such deeper grooves weaken the roll and increase the rolling or contact angle. Generally speaking, billets produced in box passes do not have as uniform diagonals as those rolled in diamond passes. If the box pass is narrow enough to restrict spreading, the action of the steel attempting to spread wears the sides of the pass and it rapidly becomes too wide. When the pass wears wide, the stock entering the pass will "float" to opposite sides in the top and bottom of the pass and, hence, produce a **diamond** billet. The stock cannot be adjusted very easily to control this condition since the three-high mill is a fixed-draft mill, and any adjustment for one pass affects all passes. Another reason for not using too heavy reductions on the three-high billet mill is that the rolling or contact angle must be small enough to assure the entry of the piece in the roll pass. When several pieces are rolled at once, the failure of one to enter the pass causes the others to be held up until it enters. This delay results in loss of temperature by the steel, with its resultant harmful effects; it also reduces output, since it can affect the production of the blooming mill as well as any mills following the three-high mill.

Billets from a three-high mill usually are cut by sta-

tionary shears or, in the larger sizes and for special purposes, by hot saws. This type of mill usually is used where only a few sizes of billets are produced, or in conjunction with smaller billet mills that further reduce the billets from the three-high mill.

An innovation in the three-high type mill is the so-called **Jumping Mill** developed in 1964 by Morgardshammar A.B. of Sweden. This mill retains the advantages of three-high mill stands while avoiding the problems associated with tilting the mill tables. The mill stand design (Figure 20—33), features rolls up to 650 mm (about 26 inches) in diameter and up to 2150 mm (about 85 inches) in barrel length. It permits the installation of adequate guide equipment for all types of rolling while the fixed tables on both sides of the mill stand enable all types of manipulating equipment to be used. The housings raise and lower while the tables are stationary.

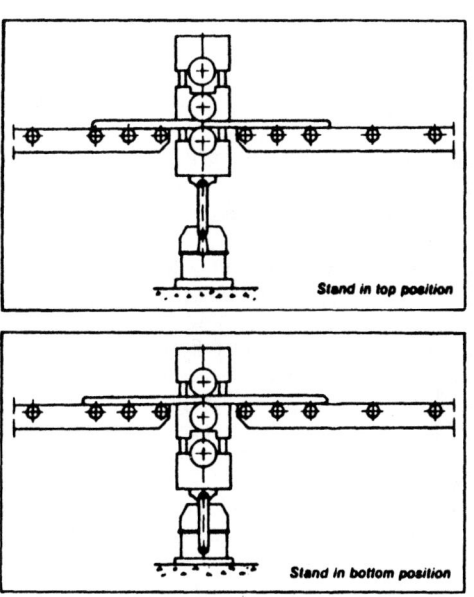

FIG. 20—33. Jumping three-high mill stand (Courtesy, Morgardshammar AB, Sweden).

The three-high jumping mill is used primarily for the rolling of billets from small ingots or continuously cast blooms, as a rougher for bar and rod mills and as a finishing stand for larger bars [down to 25 mm] (about one inch in diameter).

Cross-Country Billet Mills—Another design of mill is the cross-country mill. This type of mill is composed of several stands of rolls, so arranged that the piece to be rolled is never in more than one stand at the same time. The roll stands may be placed side by side and the piece transferred to the various roll tables, the direction of rolling being reversed in each stand; they may be arranged with two or more stands rolling in one direction, with the piece transferred to roll tables and then rolled through several other stands in the opposite direction, and so on. This type of mill is faster than the three-high, but as in the three-high, the piece from the blooming mill must be cut into several lengths before entering the mill. The cross-country mill is much more flexible than the three-high billet mill in that, in order to roll a complete range of billet sizes, it not only can take the product of the blooming mill during a greater percentage of the operating time, but also can roll various sizes of billets with only one complement of rolls throughout the mill, with only a few minutes delay necessitated by changing guides for the various passes. Quite frequently, production on the blooming mill is not affected by roll changes on the cross-country mill, except when they occur in the first two stands. Some cross-country mills are so arranged that a single pass is taken in each stand while others are composed of a combination of single-pass and multiple-pass stands.

In contrast to the three-high mill, the finishing pass in the cross-country mill can be faster than the first passes, provided separate drives or gear ratios are used. Since the piece being rolled is only in one stand of rolls at a time on a cross-country mill, tables must be provided to carry the piece from stand to stand. Some mills are provided with diagonal tables to direct the piece from one train to another, while other mills provide transfers at the end of each line of stands for moving the piece over to the next train line. The roll passes for a cross-country mill can be any shape required for good rolling practice and quality of product. The piece can be turned as desired between stands by manipulators, guides, transfers, turn-up rolls and other devices.

Advantages of Cross-Country Mills—The production of any mill is governed by the speed of the finishing pass and the percentage of the operating time during which it can be kept full. In the cross-country mill, the speed of the finishing pass can be greater than in the three-high mill, but pieces have to be spaced far enough apart in going through the mill so that each table is clear of one piece before receiving the following one. It should be possible to keep the finishing pass of this type of mill full for more than half the time (as opposed to the three-high mill) and this, together with the higher speed possible, enables such a mill to reach higher production rates than the three-high mill.

Cross-country mills, where the piece can be turned at will after every pass, and particularly those mills where the direction of rolling is reversed after every pass, are admirably suited for the rolling of quality steels.

Figure **20—29** shows the layout of a cross-country billet mill comprising part of an integrated installation designed for the production of electric-furnace alloy steels. An 1120-mm (44-inch) blooming mill supplies blooms to a 915-mm (36-inch) two-high reversing roughing mill which in turn feeds three 810-mm (32-inch) two-high stands. The original intent in the utilization of the mill was to roll hot-topped ingots, 635-mm (25-inches) square, weighing 5440 kilograms (12 000 pounds), and open-type steels from 585-mm by 635-mm (23-inch by 25-inch) ingots weighing 5215 kilograms (11 500 pounds). The 1120-mm (44-inch) reversing mill produces blooms, for subsequent rolling into billets, bars, etc., in the following sizes: 175 by 175-mm, 205 by 205-mm, 230 by 230-mm, 265 by 265-mm, 305 by 305-mm and 330 by 380-mm (7 by 7-inch, 8 by 8-inch, 9 by 9-inch, 10½ by 10½-inch, 12 by 12-inch, and 13 by 15-inch), depending on the finished size of the rerolled product.

Delivery tables, transfers and reheating furnaces of the 1120-mm (44-inch) blooming mill are so arranged that the blooms may be disposed of in any one of three ways. They may be sent without reheating directly to the 915-mm (36-inch) roughing mill and then on to the three 810-mm (32-inch) intermediate mills and the 535-mm (21-inch) continous billet mill. They may be kicked off on a transfer and then directed through one of two continuous reheating furnaces from which they are discharged onto the same table that leads directly from the 1120-mm (44-inch) mill to the 915-mm (36-inch) mill, or they may be removed from the transfer table and placed in slow-cooling pits after which they are conditioned and shipped or are charged cold into the reheating furnaces and re-rolled on the billet mills.

The 915-mm (36-inch) two-high reversing mill is equipped with 915 by 2080-mm (36-inch by 82-inch) rolls, laid out to use either 5 or 7 passes in roughing blooms down in preparation for rolling in the 810-mm (32-inch) mill. It is possible, however, to vary this practice by coordinating the work of this mill with that of the 1120-mm (44-inch) mill, which is the practice usually followed. The three 810-mm (32-inch) two-high stands are located 68.6 metres (225 feet) beyond the 915-mm (36-inch) two-high reversing roughing mill. The three stands are placed side by side. The first, or so-called intermediate stand, contains 810-mm by 1830-mm (32-inch by 72-inch) rolls; the second or leader stand has 810-mm by 1725-mm (32-inch by 68-inch) rolls; and the third or finishing stand has 810-mm by 1220-mm (32-inch by 48-inch) rolls.

It will be noted that a 535-mm (21-inch) continuous mill composed of eight horizontal stands is located beyond the first stand of the 810-mm (32-inch) cross-country mill. A 155-mm (6⅛-inch) bloom from the first stand of the 810-mm (32-inch) mill can be rolled into square billets ranging in size from 100-mm by 100-mm to 50-mm by 50-mm (4-inches by 4-inches to 2-inches by 2-inches) on the 535-mm (21-inch) mill. The bloom is cropped before entering this latter mill, and the finished billets are cut to length by a steam-operated flying shear.

It has been mentioned previously that some plants are equipped to reheat blooms between the blooming and billet mills. Figure 20—29 illustrates a mill arranged in such a way.

Continuous Billet Mill—The continuous mill consists of a series of roll stands, arranged one after the other so that the piece to be rolled enters the first stand and travels through the mill, taking but one pass in each stand of rolls and emerging from the last set as a finished product. These stands may be all horizontal, and may include one or more vertical edgers, or the stands may be alternately horizontal and vertical. In the continuous mill, where the piece is being rolled in several different stands simultaneously, the peripheral roll speeds must be such that the elongation which occurs as a result of reduction in cross-sectional areas of the piece is taken care of by increasing the speed of each successive roll pass. On mills where all of the various stands are driven through gears by one motor or engine, the elongation is compensated for in the original design of the mill by choosing gear ratios that drive each set of rolls at a higher speed than the preceding set. The diagram of such a mill appears as part of Figure 20—29, where it is designated a 535-mm (21-inch) continuous billet mill. Any deviation from the originally designed elongation must be compensated for by varying roll diameters and thus changing peripheral speed. The rolls usually are installed in sets and are all dressed together. However, excessive wear on one set of rolls requires excessive dressing on the remaining rolls in the train. Various means have been adopted to gain longer roll life. Some plants will carry extra rolls for those stands that receive the most wear, and thus save excessive dressing on the remainder of the train, while others change the rolls around from stand to stand. As an example of the latter methods, a 65-mm (2½-inch) square in No. 4 stand can be changed to a 75-mm (3-inch) square without reducing roll diameter, and used on No. 2 stand.

The use of individual drives for each stand in a continuous mill is a great improvement over the single-drive type. With this drive, roll diameters need not be matched and the speeds of the individual motors can be regulated to give the correct peripheral speed for each set of rolls. The speed ratio on continuous mills must be maintained closely to prevent the piece from pushing or pulling between stands.

Since the piece being rolled in a continuous mill is in several stands simultaneously, the piece cannot be turned between passes in successive horizontal stands but, rather, it must be twisted. (The necessity for twisting, however, is eliminated in mills composed of alternate horizontal and vertical stands.) Various methods have been devised for obtaining this twist, one of which incorporates the use of twist delivery guides. These guides are iron or steel castings, and are designed so that the piece is twisted gradually by the action of the guide which is contacted by two surfaces of the piece being rolled; the other surfaces have clearance in the guide. The twist in the guide is designed so that the piece being rolled is rotated to the correct degree for entry into the succeeding roll pass. Another method uses friction-driven twist rolls,

mounted in housings, between the various stands of work rolls. Some plants use twist delivery guides that are designed with a light over-twist, to start the twisting action, in conjunction with twist rolls. The twist rolls relieve the twisting slightly, thus reducing the pressure on the twist guide and minimizing guide scratches on the product. Two different types of twist rolls are shown in Figure 20—34. A third method accomplishes the twist as shown in Figure 20—35. With this latter method, the passes are cut in rolls obliquely, and the difference in diameters in the pass causes the twisting action. This method, however, requires the use of twist guides to insure the proper twist, since temperatures and the chemical compositions of steels being rolled are not always constant and can cause a variation in the degree of twist. This method of twisting causes excessive roll-pass wear.

High output is one of the chief advantages of the continuous mill. Scrap losses are low, due to the fact that blooms of any length can be rolled, making it unnecessary to cut the bloom after it leaves the blooming mill, except to discard pipe or any other flaws that might he present. Flying shears are placed after the finishing stand on a continuous billet mill, and are synchronized with the speed of the stand on which the billet is finished.

Sometimes, however, breaks occur in the steel due to the twisting of the billet; also, when the billet is

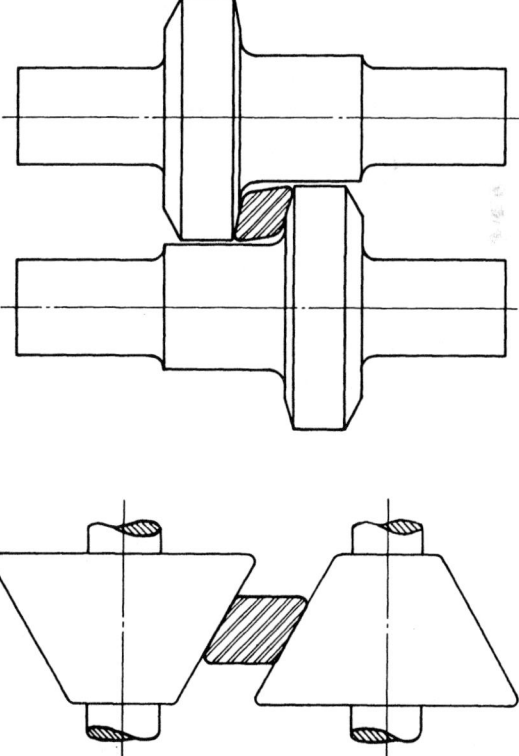

FIG. 20—34. Sketches illustrating two types of friction driven twist rolls, as used on a continuous billet mill to turn pieces 90 degrees between passes. Twist rolls such as those shown in the bottom diagram are called "cone-type" twist rolls.

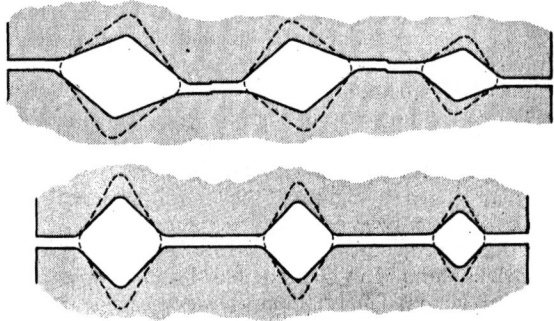

FIG. 20—35. Diagram of oblique roll passes designed to provide a twisting action on billets between stands when being rolled on a continuous mill. In each pass, the dotted line shows the rolled shape which is received from the preceding pass. The solid line shows the new shape produced, which consequently appears dotted in the next pass.

deflected from one pass line to another, it is more subject to scratching by the guides. These defects increase the amount of conditioning required by the steel. The continuous mill composed of alternate horizontal and vertical stands, which eliminate twisting and arranged so that it is never necessary to deflect the bar from one pass to another, produces a product with much better surface quality.

Figure 20—36 shows a schematic diagram of the roll stands and twist rolls for a continuous billet mill; Figure 20—37, the passes used on a 355-mm (14-inch) mill for rolling two different sizes of billets.

The **Six-Stand Continuous Mill** shown as part of Figure 20—38 embodies many interesting features uncommon to billet mills. A detailed description is provided here.

Designed to produce a wide variety of semi-finished products from 230-mm by 205-mm (9-inch by 8-inch) blooms, the mill is equipped to roll 125-mm by 125-mm (5-inch by 5-inch), 145-mm by 145-mm (5¾-inch by 5¾-inch) square billets, and 140-mm (5½-inch), 150-mm (6-inch) and 175-mm (6⅞-inch) tube rounds, on the same roll and guide set-up, with the

ability to alternate quickly on any of the above sizes by push-button-operated controls (Fig. 20—39).

The mill consists of three vertical roll stands numbered Nos. 1, 3, and 5, and three horizontal roll stands numbered Nos. 2, 4 and 6, set in tandem on 3-metre (10-foot) centers (Figure 20—40). Suitable gear reduction sets provide for a finishing speed of 1.75 to 3.50 metres per second (346 to 692 feet per minute) at No. 6 stand. All vertical mills are of sufficiently rugged construction to enable drafting equal to the horizontal stands. The main vertical housing containing two 760- by 1220-mm (30-inch by 48-inch) rolls is a self-contained, integral unit and resembles a horizontal stand laid over on its side. A fixed outer roll held securely by hydraulic pressure against the housing seat, and a movable inner roll hydraulically balanced against motor-driven screws, plus anti-friction roll-neck bearings and thrust units, comprise the roll-adjusting mechanism. Screw speed is 2.54-mm per second (6-inches per minute) and screw travel is sufficient to permit operating the rolls with the distance between their centers set at any point between 910- and 690-mm (35¾- and 27¼-inches).

Power for the vertical mill rolls is transmitted from the main gear drive (Figure 20—41) through a horizontal shaft to an intermediate bevel-gear drive, through a shaft inclined at 40 degrees to a combination bevel-gear drive and fully-enclosed 760-mm (30-inch) pinion stand. Universal couplings and spindles equipped with anti-friction thrust bearings, suspended and attached beneath the main mill housing, complete the drive. The main housing is enclosed within a secondary housing supported on the shoe plates. The secondary housing carries the screw-up mechanism and guiding slides for controlling the vertical movement of the main housing with respect to the horizontal-mill pass line. The motor-driven screw-up mechanism operates by push-button control at 4.5-mm per second (10.5-inches per minute), and provides for a maximum vertical movement of the main housing of 915-mm (36-inches). Thus, a set of vertical rolls may contain several passes within this range, each of which may be aligned properly with the horizontal mills. The male coupling end, transmitting mo-

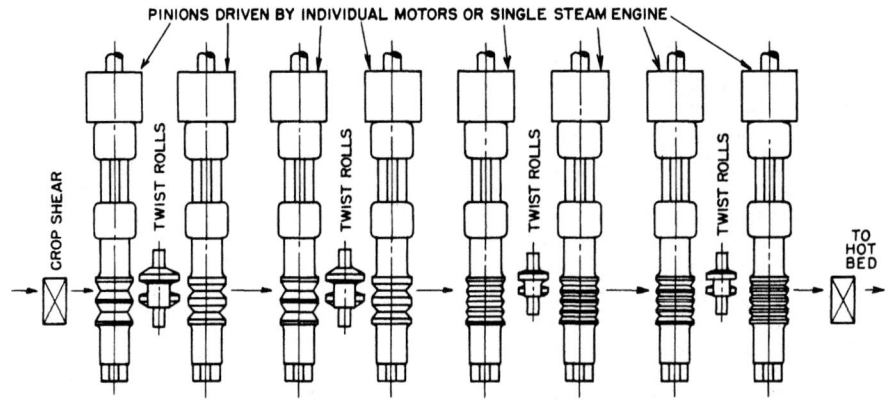

FIG. 20—36. Schematic arrangement of the roll stands and twist rolls for an eight-stand continuous billet mill.

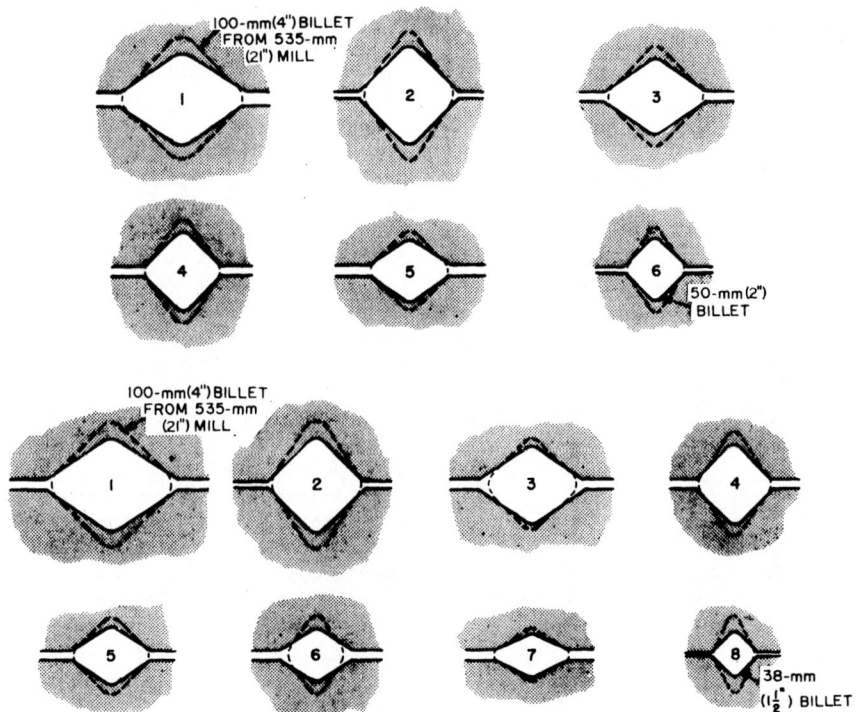

FIG. 20—37. (Above) Six successive roll passes employed in the production of 50 mm (2 inch) square billets from 100 mm (4 inch) square starting material on a 355 mm (14 inch) billet mill. (Below) Method of rolling 38-mm (1½-inch) square billets on the same mill, using the same starting material, in eight passes. In each numbered pass, the dotted line shows the rolled shape which is received from the preceding pass. The solid line shows the new shape produced, which appears dotted in the next numbered pass.

tive power from 760-mm (36-inch) pinions through universal couplings to the rolls, is forged in one piece with the spline shaft that telescopes with the pinions. This arrangement permits maintaining the drive connection during upward movement of the rolls. The three vertical mills are of similar design, having all parts interchangeable, including intermediate bevel-gear drive. Rolls with complete bearings are changed in pairs, using double roll-changing hooks.

Vertical-mill roll and guide equipment is of such design that changes are unnecessary when rolling any of the previously mentioned sizes. The useful proportion of rolling time on the mill is thus increased, since it is necessary to change rolls only when worn or in the event of possible breakage. The entry guide unit is attached to the secondary housing in a fixed, pre-leveled position, to accommodate the upper pass in the rolls. The remaining passes may be utilized by adjusting the screw-up mechanism upwards, thus positioning the desired pass in front of the guide.

Product is guided from the delivery side of the vertical roll set to the entry guides of the horizontal set by a covered cross-over trough. Upon changing rolls, the entry guide retracts clear of the rolls by a hydraulically operated mechanism. The cross-over troughs are removed from the pass line with an overhead crane. The rolling operations of the three vertical mills in conjunction with the three horizontal stands will be described subsequently.

The three two-high horizontal roll stands (Nos. 2, 4

and 6) resemble strip-mill stands in their construction features, having been designed for stability, smoothness of operation, accurate roll settings, and ease in roll changing. Housings are of cast-steel, closed-top construction. The top roll is hydraulically counter balanced against motor-driven screws. The screw-down permits the screws to function independently or in unison. The bottom roll is fixed, with provision on the outward chuck (alternatively called a chock) for endwise adjustment. The top-roll outward chuck is fixed, while both drive-side chucks are free to "float" in the housing. The 710-mm (28-inch) diameter by 1220-mm (48-inch) body length rolls are mounted on anti-friction bearings of the radial thrust type, all preassembled with fully-enclosed chucks. Rolls thus can be quickly changed by use of a "C" hook similar to those used in changing strip-mill rolls. Horizontal roll stands Nos. 2 and 4 each are driven independently through a double reduction-gear drive and a 685-mm (27-inch) pinion stand. No. 6 horizontal roll stand is driven through a single reduction gear and 685-mm (27-inch) pinion stand. Universal couplings and spindles mounted on carriers connect the pinion stands with the rolls.

A feature essential to the flexibility of operation of this mill is the endwise movement provided for the horizontal roll and pinion stands. To utilize all the grooves provided in the horizontal rolls for various product sizes, provision was made for rapid alignment of horizontal-roll grooves with respect to vertical-roll

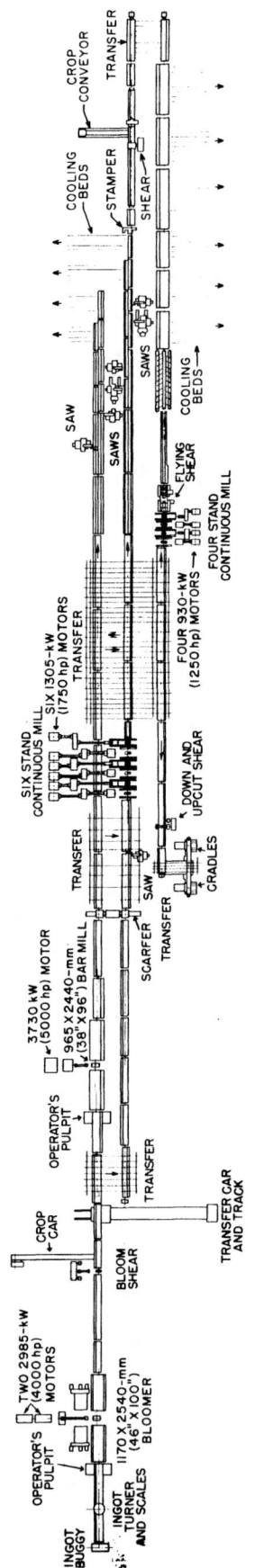

FIG. 20—38. Diagrammatic layout (not to scale) of a blooming, bar and billet mill. Flow of material through this mill is illustrated schematically in Figure 20—39, which also indicates typical sizes produced.

grooves to maintain a straight line of product travel throughout the desired grooves of all six stands. This was accomplished by mounting the roll stand, spindle-carrier base, and pinion stand as a unit on the bed plate. By using the bed as sliding ways, the entire unit can be moved endwise horizontally 1065-mm (42-inches) through power supplied by a 355-mm (14-inch) diameter hydraulic cylinder operating at 8550 kPa (1250 pounds per square inch). To maintain the drive connection during such movement, the main drive shaft telescopes a spline-type flexible coupling at the pinion stand. Hydraulically operated quick-release clamps anchor the assembly to the bed plate, obviating the use of foot bolts.

As in the case of the vertical mills, the roll and guide equipment of the horizontal stands was designed to roll all of the previously mentioned sizes without roll or guide changes. Entry and delivery guides are mounted as integral units on sliding rest bars which, in turn, are supported by the main rest bar attached securely to the roll housings. All main rest bars are adjustable vertically by jack screws mounted on the housings. Rolls must be changed after excessive wear occurs. To minimize roll-changing time and increase efficiency, all guide equipment is mounted on a retractable mechanism operated hydraulically, which makes it possible to move the guides clear of the housing window to permit rapid and simple removal or insertion of the roll-and-chuck assembly.

The run-out table from the six-stand mill delivers all product to transfer equipment designed to handle bar lengths up to nearly 49 metres (160 feet). Rounds may be delivered directly to any one of four saw lines, each equipped with a stationary saw for cropping and cutting to required lengths. Each of the saw lines converges to its own independent chain-conveyor cooling bed. Thus, tube-round product from the 965-mm (38-inch) mill and the six-stand mill for square billet conversion are transferred to a roller line that carries them to shearing equipment and inspection and storage. For further reduction to smaller billets, 125-mm by 125-mm (5-inch by 5-inch) square billets are delivered from the six-stand mill and transferred to a roller line directly preceding the four-stand continuous mill.

Processing Prior to Conditioning—The product from the blooming mill must be cropped to discard pipe before entering the billet mill and the finished product from the billet mill must be cut to length. The bloom-crop **shears,** however, usually are considered part of the blooming mill rather than the billet mill. When the primary mill set-up consists of several billet mills fed one from the other, and all or part by the same blooming mill, crop shears usually are placed before each unit. Crop shears assure a clean end entering the mill and also enable cutting of the stock to the right length to accommodate the various mill tables. These crop shears usually are stationary, and may be steam, hydraulically or electrically driven.

Billets of large cross-section generally are cut by **stationary shears.** Such billets are usually short enough to be handled on tables and transfers, and there are comparatively few cuts to the finished piece. These shears cut one or more pieces at the same time. The length of cut is determined by an adjustable stop on

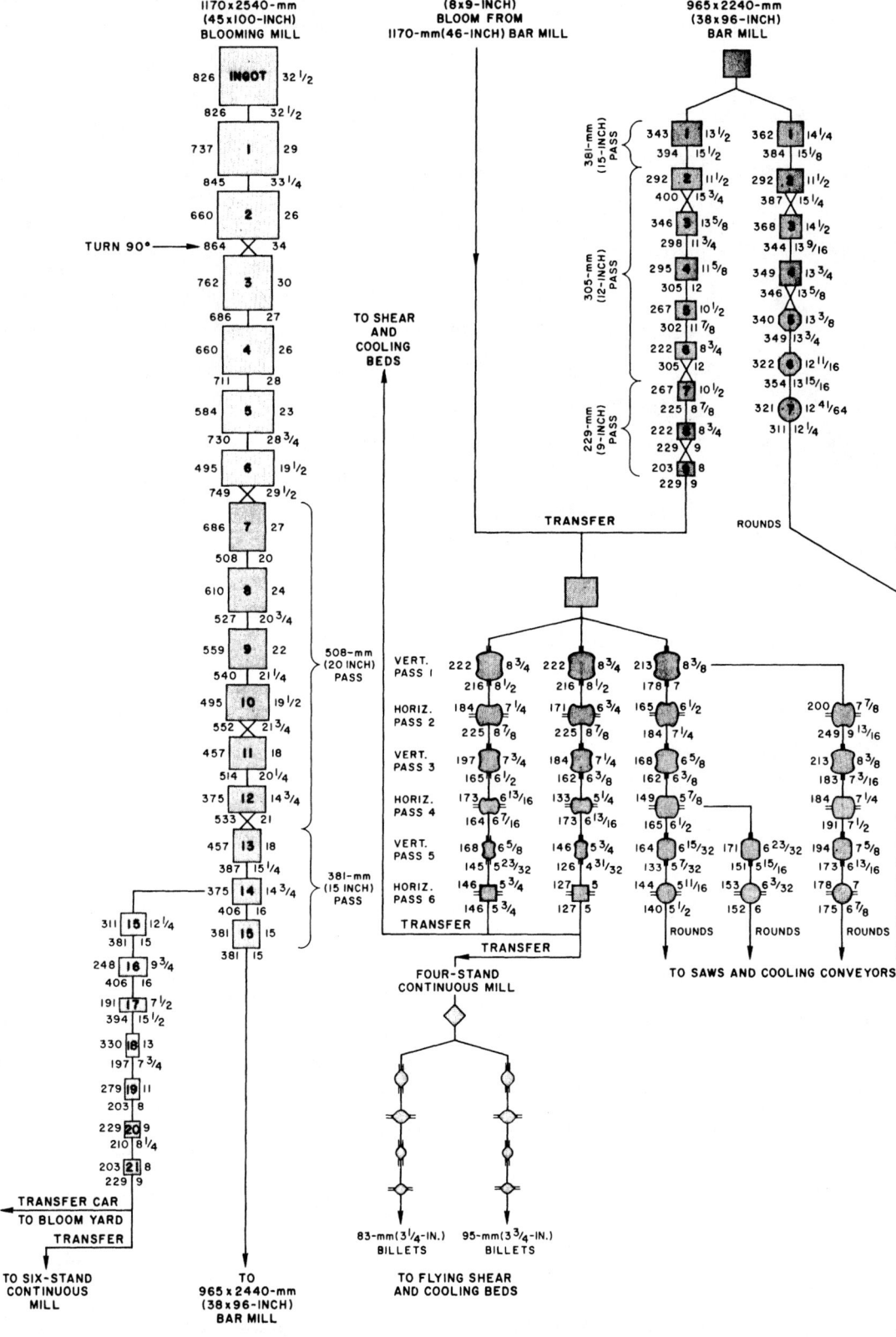

FIG. 20—39. Flow chart summarizing typical series of operations possible due to the extremely flexible arrangement of a blooming, bar and billet mill. Dimensions of each pass opening to the nearest millimetre are given at the left of the schematic representation (not to scale) of the pass, the upper number representing the height and the lower number of width of the pass. Corresponding dimensions in inches are given at the right of each pass.

FIG. 20—40. Six-stand continuous billet mill.

the shear table and the shears usually can cut accurately to within 6 mm (¼-inch) of the desired length.

Another form of shear is the gang shear in which several shears are placed in line in such a manner that any length of billet may be sheared by moving the position of individual shears within the limits of the first and last shear in line. They may be made to cut in unison, or separately, as desired.

The **flying shear,** as opposed to the stationary shear, cuts the piece as it travels in a continuous operation. These shears are used for the smaller sizes up to about 100 mm by 100 mm (4 in. by 4 in.) that would require excessively long tables and transfers and would, of course, consume too much time in handling, if cut on stationary shears. This is obvious when it is realized that a 7260-kilogram (8000-pound) ingot, rolled into a 38-mm by 38-mm (1½-inch by 1½-inch) billet, would be over 305 metres (1000 feet) long. By the use of flying shears, this piece can be cut effectively into shorter lengths as it leaves the finishing stand.

Flying shears are of two types: 1) coupling shear in which the drive motor runs continuously, the shear mechanism being engaged only when cutting is neccessary and 2) motor or start-stop shear in which the motor is operated only when cutting is to be undertaken. The development of flying shears has progressed rapidly in recent years and they can now be used at mill speeds up to 4000 fpm.

Billets also may be cut to length by saws or flame cutters to eliminate distorted ends produced by shears. Sawing is usually slower and more expensive than shearing, and saws are more costly to maintain.

Flame cutting entails a comparatively small initial investment, but is costly in that it is slower than shearing and requires the use of relatively expensive gases.

Prior to conditioning it is essential that every billet have proper **identification.** Every billet should be stamped with its heat number, and some orders require that top and bottom cuts from the ingot be identified. The steel must be followed closely through the mill, especially in those mills composed of several units, to make certain that the correct steel is coming for which the stampers already have received instructions. This tracing of the steel can become quite complicated at times, and must be adhered to rigidly in order to prevent costly errors in identification. Some mills identify all products by the use of special hand or pneumatic hammers having suitable heads for containing the stamps. Heat numbers can be machine stamped on billets after leaving the finishing stand by the use of a wheel, suitably mounted and containing the proper marking devices or stamps, contacting the steel. These wheels can be held in a stationary housing, so arranged that the height above the table can be varied for different sized billets, or they may be moved by a manually controlled hydraulic cylinder to contact each piece as it passes over the table. Usually, heat numbers only are applied in this manner. Some mills nail metal tags that are prepared ahead of time with the necessary information to the ends of billets, blooms or slabs.

After leaving the shears, billets are delivered usually to **cooling beds,** or **hot beds** as they are sometimes called, where they are cooled before shipping. In many mills, the above mentioned hand stamping is done on billets on the hot bed. The hot beds are at right angles to the direction of delivery from the mill and are wide enough to handle the longest billet length regularly produced on the mill. These beds are

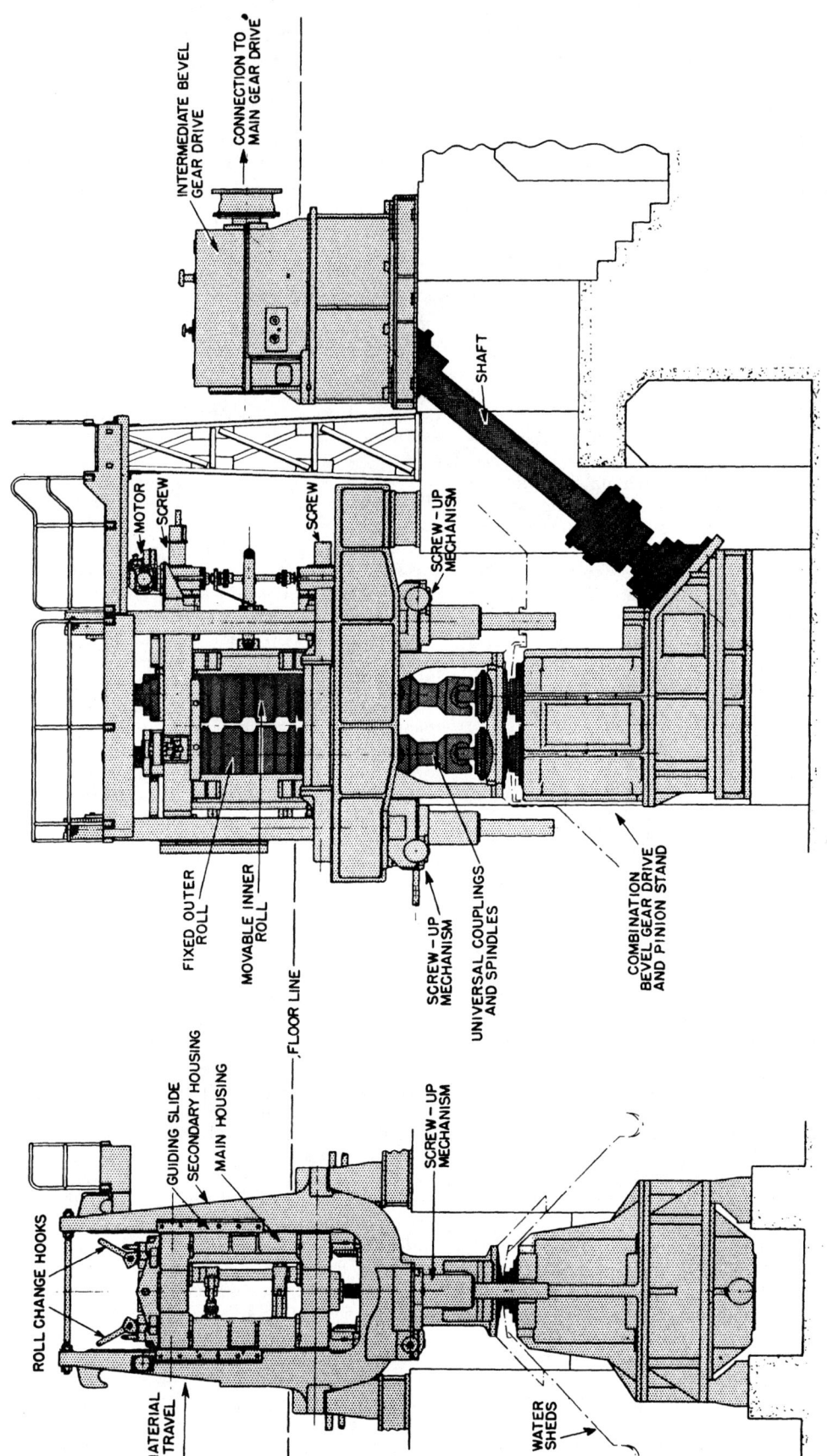

FIG. 20—41. Schematic arrangement of the principal parts of one of the vertical stands of a six-stand continuous billet mill.

composed of skids placed close enough together to support the billet, but not too close to permit air to circulate around them. The skids may be composed of rails or of a series of specially designed castings. The castings give much better results than rails since they do not bend so readily with prolonged use. Some mills use a turnover type cooling bed, equipped with V-shaped groves to hold the billet straight. The billet

is walked across the bed and is turned 90° each time it progresses, resulting in uniform cooling and very straight product. After cooling, the billets are pushed off the hot beds into **cradles,** from where they can be loaded easily into cars for shipment, or transferred to the conditioning area where they are handled as described in Section 3 of this chapter. Many mills now remove the billets directly from the bed with magnets instead of from cradles.

SECTION 3

CONDITIONING AND COOLING OF SEMI-FINISHED STEEL

Semi-finished steel is previously hot-worked cold steel subjected subsequently to further hot deformation. When steel is rolled from ingot to semi-finished forms, such as blooms, billets and slabs, a variety of ingot defects and some defects arising during heating and rolling may be carried through to appear on the surface of the semi-finished product. It often is necessary to identify and remove these defects before the steel is converted further into finished products by forging, rolling or other hot forming operations. The identification of these defects, known as **inspection,** and the removal of these defects, known as **conditioning,** will be the subjects of the earlier parts of the present section; a discussion of the principles and practice of controlled cooling of certain grades and sizes of semi-finished products from rolling temperatures will be presented later in this section.

TYPES OF DEFECTS AND INSPECTION

Surface defects on semi-finished products may be caused by defects on or in the ingot, by improper handling after pouring, or by variations in heating and rolling practice. Ingot defects, such as ingot cracks, scabs, and seams, caused during the earlier solidification stages (Figures 20—42, 20—43 and 20—44) were previously introduced in Section 1.

Those surface defects which originate at the soaking pits, and on the primary mill and other mills rolling semi-finished products are:

1. **Cinder Patch**—This defect is the result of pick-up of material from soaking-pit bottoms, and generally has the appearance of a very scabby bottom. Cinder patch can be overcome to a great extent by proper attention to bottom making. Some plants charge hot-top ingots of the big-end-up type upside down in the pits, thus causing the cinder patch to be confined to the hot-top or discard portion of the ingot.

2. **Burned Steel**—This defect is very often the result of flame impingement on the surface of ingots, usually on their corners, as they are heated in the soaking pits, causing penetrating oxidation at the grain boundaries with a resultant tearing or rupturing of the ingot on the primary mill. A burned bloom is shown in Figure 20—45. Burned steel seldom can be salvaged and normally must be scrapped. Some steels are more sensitive to burning than others.

3. **Laps**—Laps are the result of overfilling in the mill

passes that causes fins or projections which turn down as the material rolls through succeeding stands in the mill train.

A lapped billet is shown in Figure 20—46. Laps usually are deep and the product often cannot be salvaged economically.

The foregoing list of defects is not all-inclusive, but presents merely the more important causes of surface defects that require removal in conditioning. There are numerous other defects, such as: twist, guide marks, collar cut, diamond, bent, crop ends, and so on, the names of which are self-explanatory. Most of these will receive attention at various points in the text.

Inspection—Conditioning is that function of steel manufacture which, by removal of defects, renders steel more suitable for subsequent hot- and/or cold-forming operations. Of primary importance is the detection and proper removal of the defects where required. It is not always necessary (and, strictly speaking, probably not possible) to remove all of the imperfections.

What, if anything, is removed is determined by the prospective subsequent operations and final end use of the product. The following paragraph illustrates the method for determining the conditioning practice in a plant producing semi-finished steel for bar products.

Using an arbitrary classification, product end uses can be grouped to include under one class, semi-finished products which require the removal of all surface defects visible after pickling or skinning. As an example of this class of material, if the steel is intended for alloy-steel bars of high hardenability and the bars are further quenched in oil after reheating, it is obvious that any light seams may result in quench cracks, so that all defects should be removed from the semi-finished steel. In a second example, where the material is to be rolled or forged to some finished form, but not heat-treated, oxidation during the reheating for rolling or forging will eliminate, in many cases, slight surface imperfections, and semi-finished products for this use could be placed in another class for which removal of only those defects visually detectable without pickling would be required. If material is to receive considerable work in further hot rolling or cold machining, even medium-sized seams can be acceptable, offering another classification which might be defined as a class of material requiring the removal of only major defects

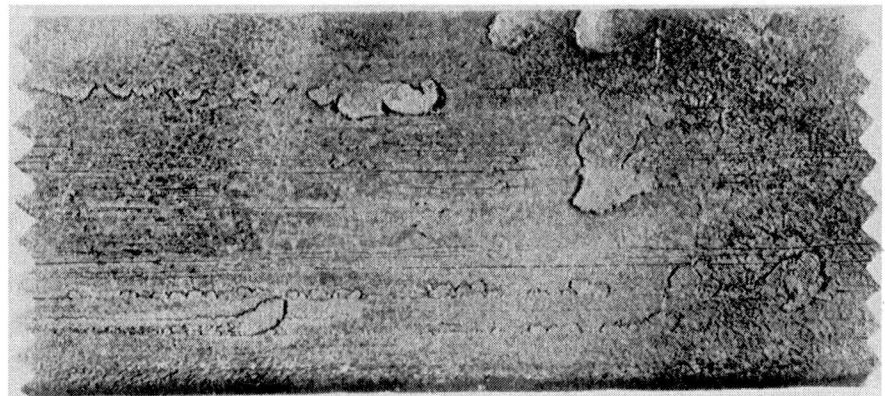

FIG. 20—42. Scabby surface of a bloom, showing the effects of roll collars on the pre-existing scabby surface on the ingot from which it was rolled.

FIG. 20—43. Deep seam on the surface of a semi-finished rolled product, originating with an ingot crack.

FIG. 20—44. Clustered seams on the surface of a semi-finished rolled product, pickled after rolling to facilitate detection of defects.

such as scabs, ingot cracks, laps, and deep seams. Finally, the remainder of the product can be placed in a class from which only those defects need be removed such as large scabs or ingot cracks, deep laps, and burned steel which might be damaging to either or both the final product or the operation (such as subsequent hot rolling where large scabs might cause cob-

bles on the mill, tear out the guides or cause other damage). Thus, it can be seen that the basis for inspection is predicated both on the end use of the product as well as upon the operations employed to shape the steel into its final form.

Pickling for Inspection—Pickling, as applied to surface inspection and conditioning, consists of immersing

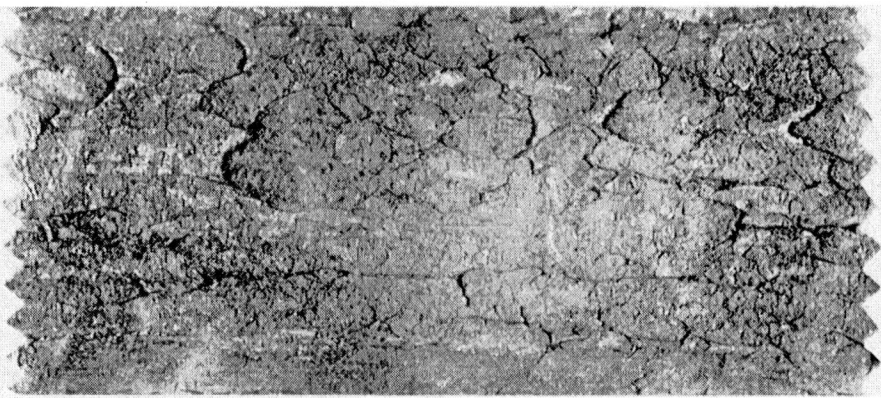

FIG. 20—45. Surface of a "burned" steel bloom.

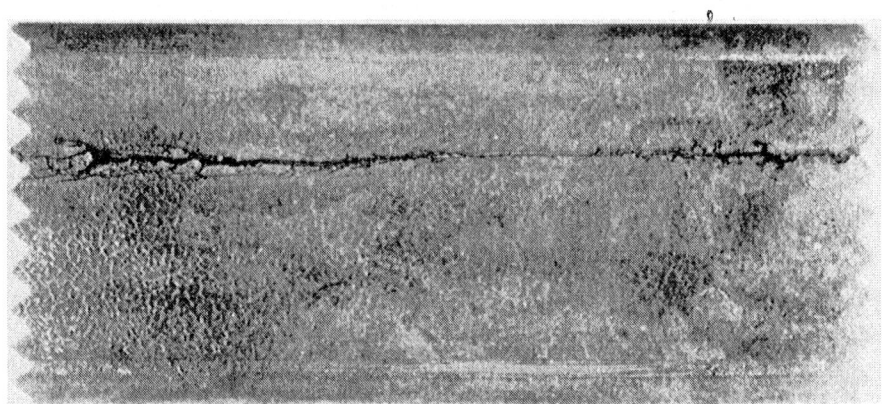

FIG. 20—46. A lap on the surface of a rolled steel billet.

the steel in a chemical solution for the purpose of re-moving the scale so that defects will be exposed. The usual solution for ordinary steels is composed of 5 to 8 per cent sulphuric acid in water, used at temperatures between 65° and 82°C (150° and 180°F). A good, clean surface is desired for optimum results in inspection, and to attain this, relatively close control of tempera-ture and acid concentration is necessary. A thorough rinse after pickling is required, either by immersion in clean, hot water or washing with high pressure nozzles. The time required for pickling varies with the grade of steel; for example, a 10-ton lift of blooms of plain-carbon steel may require pickling for around 45 min-utes, while the same quantity of constructional alloy steel may require treatment for from 1 to 2½ hours.

Pickling generally has been associated with hand chipping, as a dual operation. In those plants where hand scarfing has replaced hand chipping, the pickling practice has been abandoned, too. Modern practice for the same specifications is: for rectangular sections, to skin by scarfing and then spot scarf where needed; for round sections, to peel mechanically and then spot scarf. Magnetic-particle inspection (Magnaflux), fluo-rescent particle (Magnaglo) and eddy-current methods of defect inspection have become very popular. They are described in Chapter 51.

As discussed in Chapters 30 and 33, **grit blasting** is used for removing oxides from the surfaces of mer-chant bars and hot-rolled strip in some steel plants. However, it has not been adopted generally for clean-ing semi-finished steel prior to inspection.

CONDITIONING OF DEFECTS

Hand Chipping—Until a few years prior to World War I, the practice of extensive conditioning of semi-finished steel was almost unknown except for seamless tubes and special forgings. For most purposes, the use of a cold chisel and hand hammer was about all the work that was done, and this only to knock off such things as large slivers or scabs that might tear out the guides in rolling on a secondary mill. For the most part, steel was either accepted or scrapped in the semi-finished form. It is not to be assumed from this that a good deal of unusable material was produced, for in those days, a great deal of stock-removal in machining was accepted as commonplace and the tolerances for finished work were much wider. Hand chipping was the only method of steel preparation used to any great extent until the early 1930's.

Hand chipping consists of using a pneumatic ham-mer with a cold chisel placed in the barrel, so that the

FIG. 20—47. Chipper using a pneumatic hammer equipped with a cold chisel for removing selected portions of steel billet surface.

FIG. 20—48. Planer-type machine for removing surface imperfections from surfaces of semi-finished steel.

reciprocating action of the piston on the chisel causes the chisel to cut into the steel and remove the surface to the desired extent. Figure 20—47 shows a close-up view of a chipper at work.

Machine Chipping—There are available several large machines built specifically for the purpose of removing imperfections from the surface of steel blooms and billets. These machines fall into two general types: the planer and the milling machine.

The **planer type**, illustrated in Figure 20—48, is constructed with a large table which can be moved longitudinally and a tool holder which can be moved to any position in a plane vertical to the table motion. The steel is clamped onto the table by mechanical or hy-

draulic vise-type jaws. The table then is operated to move in a horizontal direction, back and forth in a reversing motion, so that the steel passes under the tool. The tool can be adjusted while the machine is in operation. The distance of travel and the depth of cut are governed by the operator. Speeds are varied in accordance with the hardness of the steel and the depth of cut desired. In this manner, any length of cut or repeated cuts in depth can be made, so that either complete skinning or partial surface removal can be accomplished.

The **milling machine**, typified in Figure 20—49, is constructed with a multiple set of tools in a revolving head. The steel is clamped in a stationary position, and

FIG. 20—49. Milling machine adapted to the removal of surface imperfections from semi-finished steel.

FIG. 20—50. A scarfing torch in use removing surface imperfections from a rolled steel slab.

the tool head can be moved in a horizontal as well as a vertical direction so that surface removal can be restricted to selected spots or the piece being conditioned can be skinned completely. The entire frame of the machine carrying the tool is capable of traveling the entire length of the pieces processed.

Scarfing—Scarfing in the steel mill consists of surface removal by the use of oxygen torches (Figures 20—50, 20—51 and 20—52). The oxygen rapidly oxidizes the steel surface, generating elevated temperatures that cause the oxidized product to become liquid. The process can be carried out on hot steel between stages of rolling, either by hand or by hot-scarfing machines.

When hot-scarfing machines are used, any further scarfing is done on cold steel in the conditioning area by the use of the hand torch. In principle, the tip of the scarfing torch is similar to the regular oxygen-gas torches used for the burning of steel. As early as 1919, surface-defect removal on semi-finished steel was attempted by hand scarfing, but because of lack of knowledge of proper torch design and the difficulties encountered with the fuel gas, the results obtained were unsatisfactory. During the 1920's, some attempts at specialized uses of torches met with success, but it was not until 1929 that a determined effort was made to procure fuels and equipment to perform satisfactorily

FIG. 20—51. View of a hot scarfer in operation at a slab mill.

the function of defect removal from the surface of semi-finished steels. Promotion of the chemical reaction that forms the melted iron oxide is not the only function of the oxygen; its kinetic energy must also force the liquid, oxidized metal from the path of the torch.

Hand Scarfing of Cold Steel—As might be expected, the first hand torches, though successful, were cumbersome and slow, but since 1929 many outstanding improvements have been effected so that the bulk of semi-finished steel now is conditioned by the scarfing method. So far, only oxygen has been mentioned, but fuel gas, too, is necessary to the functioning of the scarfing torch. The gas acts as a preheating agent to elevate a spot on the steel surface to such a temperature that the oxygen and the steel will begin to combine chemically. Once this preheating has been accomplished, the heat of reaction between oxygen and iron produces sufficient heat in front of the torch tip to maintain the reaction and fuel gas is no longer a necessary part of the operation. An interesting development with respect to this spot pre-heating is the so-called **starting rod,** which is a small rod of steel that extends into the flame and is heated until a small drop of melted rod falls onto the steel surface, thus instantly kindling the reaction. The melting of this steel rod is a matter of a

fraction of a second as opposed to the 6 or 8 seconds required to heat sufficiently a spot on the steel surface with the flame alone. Many different types of torches and many different kinds of fuel gases are used.

Perhaps the largest tonnage of all conditioned steel is processed by hand scarfing. Scarfing can be used for both small and large sizes down to about 50 mm by 50 mm (2 inches by 2 inches), below which a bowing effect is produced.

A factor present in scarfing cold steel which is not present in other types of conditioning previously described might be termed the metallurgy of scarfing. When a relatively small portion of a large piece of steel, such as the area adjacent to a cut made by the scarfing torch, is heated to a high temperature and the source of heat is suddenly removed, a **quenching effect** is produced by the rapid extraction of heat from the hot area by conduction of heat into the surrounding relatively cold areas. This rapid cooling of a heated portion of steel from a temperature above its critical temperature range causes it to harden, to a degree proportionate to its carbon content and to some extent also to the hardenability of the particular steel being scarfed. In plain low-carbon steel, this effect is not noticeable, but as the content of carbon or alloying elements increases, the effect becomes more and more severe, producing, in

FIG. 20—52. Overall view of a bloom-turning machine showing a bloom on the supporting arms being scarfed.

the extreme case, a hard, martensitic layer on the surface where scarfing has been performed.

This hardened surface will crack upon cooling if no preventive measures are taken. Scarfing cracks on a billet are shown in Figure 20—53. To prevent scarfing cracks, preheating before scarfing to temperatures between 150° and 260°C (300° and 500°F) normally will suffice, although in highly critical grades, such as UNS G15216 (formerly AISI or SAE 52100) containing 1.0 per cent carbon and 1.3-1.6 per cent chromium), postheating for stress relief after scarfing also should be employed. To eliminate completely the hardened surface, a normalizing or annealing heat treatment is nec-

essary. Ordinary brick pits, gas fired, can be used for routine preheating before scarfing, where a sufficient tonnage of sensitive steels is involved. Conventional car-bottom or continuous heat-treating furnaces are employed for complete softening of the steel by heat treatment. The necessity for proper preheating of some steels cannot be over-emphasized, since scarfing cracks cannot be tolerated in finished products.

Mechanical Scarfing of Hot Steel—The **mechanical hot scarfer** (Figure 20—51) originated about 1930. It is installed directly in the mill line, and is composed of a number of scarfing torches so designed that they form a pass on the mill. The hot-scarfing machines may be

FIG. 20—53. Surface cracks resulting from scarfing to remove surface imperfections from a rolled steel billet.

placed immediately after the bloom shears or, when multiple units are fed from the same blooming mill, they can be installed after the roughing or the first billet mill.

The torches are used in exactly the same manner as the hand torch. A starting rod is not used because the high temperature of the product being rolled is enough to obtain quick starting action. The working speed of the machine must be coordinated with the mill so that no delay in rolling production is encountered.

Only a moderate amount of metal can be removed but very satisfactory results can be obtained in conditioning some products. The machine desurfacing process can remove defects within 3 mm (0.09-inch) of the surface, which is usually sufficient to remove such defects as rolled seams, light scabs, checks, etc. It is not always economically advisable to cut at depths necessary to remove the deepest defects. General practice in those mills using the process is to desurface each grade of steel to the depth which strikes the right economic balance between loss in yield due to removal of metal and processing costs on one hand, and savings in lessened rejection loss on the other hand. This depth will vary with different materials, but the metal loss is generally 2 to 3 per cent of the product. This equipment is not confined to four-side removal, but has been employed for two-side removal (of slab edges and the like). Such machines can be designed to meet the needs of individual mills.

Four-side scarfing machines are now commonly used for the simultaneous conditioning of all sides of cold and hot workpieces. A recent innovation in scarfing was the development of automated spot scarfing machines in Japan, 1979, capable of spotting or skinning the full top surface and one edge of a slab in a single pass. The machine is operated from visually pre-inspected and marked defects. A further development was the use of lasers for automatic spot scarfing.

Grinding—Conditioning by grinding also is in general use. The technique has come into use because of environmental and metallurgical problems associated with conventional scarfing. For semi-finished steels, two types of grinding machines are employed. One type is the **stationary or swing-frame grinder**. This machine consists of a large circular grinding stone, up to 610 mm (24 inches) in diameter, electrically driven. The machine is suspended so that it can be raised or lowered readily to suit the product being ground. When the material is positioned properly and held firmly, the operator grasps the handle bars and, by body pressure, forces the revolving stone against that portion of the surface to be removed, moving the stone with a slight back and forth horizontal movement until the defect is gone (Figure 20—54).

The second type of grinding machine has a moving table or rollers on which the piece being ground is supported and on which it can be moved back and forth in a horizontal direction. The grinding wheel is mounted on a carriage that permits it to be moved back and forth over the piece in a direction at right angles to the motion of the piece on the table or rollers. Motions of the wheel and table or rollers are controlled by an operator in an adjacent pulpit.

The grinding wheel can be considered as a revolving cylinder carrying many sharp tools, which are the abrasive grains. When revolving at a proper speed, the grains cut very small chips from the material being ground. It is therefore, the cumulative effect of the action of a very large number of cutting points which produces the result. Grinding cracks can result from the heat generated by grinding unless the proper wheels are used; this occurrence is related to the quenching effect discussed under "Hand Scarfing of Cold Steel."

Grinding to remove defects from semi-finished steel formerly was considered to be a more or less specialized process, particularly adaptable to stainless steel and other products which cannot be scarfed or chipped. There has been an increasing trend toward the use of grinding for conditioning all grades of semi-finished steel, particularly billets, especially in conjunction with ultrasonic and other sophisticated inspection techniques (see, for example, Chapter 51).

Material Handling—Developments in the material handling field are being employed in increasing numbers to simplify the handling and stocking of material for conditioning. The use of **straddle carriers** (Figure 20—55), in some plants which have large conveniently located vacant areas, has revolutionized the transporting of semi-finished material, permitting flexibility and simplification of stocking areas, improved scheduling and better control of inventory. **Bloom-turning machines** (Figure 20—52) safely and efficiently position material automatically for hand scarfing, permitting increased tonnage to be conditioned. These machines are designed to be loaded and unloaded by fork lift trucks, eliminating the need for cranes.

CONTROLLED COOLING OF SEMI-FINISHED PRODUCTS

Blooms, billets and slabs are rolled at temperatures well above the critical temperature range of the steel and thus, in cooling to atmospheric temperature after rolling, must pass through its transformation range. Depending upon the chemical composition of the steel and the size of product rolled from it, it may be necessary to retard the rate of cooling by artificial means for two reasons:

(1) To prevent the formation of small internal ruptures, called flakes.

(2) To minimize development of internal stresses.

Some steels are very sensitive to rapid changes in temperature, and this **thermal sensitivity** requires that great care be exercised during both heating and cooling to avoid physical damage to the steel due to high internal stresses.

Nature and Prevention of Flakes—Flakes are small internal ruptures that usually occur some distance away from the end of a piece and often midway from the surface to the center of a section. Although there is not complete accord as to the relative importance of the various factors contributing to flaking, it generally is considered that hydrogen dissolved in the molten steel makes it more susceptible to flakes and that proper retardation of cooling from forging or rolling temperatures effectively will prevent their formation. The development of methods for degassing molten

FIG. 20—54. Two swing grinders in operation conditioning the surface of a steel slab.

FIG. 20—55. Straddle carrier in right foreground facilitates the transporting and stocking of semi-finished material.

steel provided an effective means for preventing flaking (see Chapter 19).

Development of Controlled-Cooling Practices—The earliest method employed for retarding the rate of cooling after forging or rolling was to bury the steel, as it came from the press or mill, in ashes or some other insulating material such as sand or slag. The purpose of this was not to prevent flakes, but to retard the cooling sufficiently to lessen the hardness developed by the steel on cooling and to minimize the development of internal stresses. Rates of cooling varied greatly, depending upon the type of insulating material used and the depth to which the steel was covered. No positive control of cooling rates was possible with such crude methods, and the time required before the steel safely could be removed usually was two to ten days. Another method of slow cooling involved the use of unfired pits, lined with brick to provide insulation. After the steel was placed in these pits, they were covered with insulated lids, and the charge cooled at a rate dependent upon the mass of the charge and the rate at which heat was lost by radiation. In another method, steel boxes were inverted over piles or bundles of hot steel as it was delivered from the mill. Because of the space and long time required, and the difficulties involved in handling, the burying processes proved entirely inadequate as increased tonnages of alloy steel came to be produced. Consequently, it became imperative to develop more suitable means for slow cooling.

In recent times, use has been made of automatically controlled furnaces, insulated and heated railway gondola cars with covers, and heavy pits equipped to control temperatures during slow cooling. With such equipment, it has been possible to take advantage of cooling cycles of considerably shorter duration with better control of results. When using equipment that makes possible close control of rate of cooling, the temperature of the steel can be lowered rapidly and arrested at the desired temperature which improved knowledge of heat-treating procedures has established as effective for each particular type of steel. After a suitable holding time, measured in hours instead of days (as was the case when the steel was simply buried in ashes), the steel can be removed and air cooled in the open, permitting prompt placement of the next charge in the furnace or pit.

Much experimental work has been done to determine the proper holding temperature for the different grades of steel. Figure 20—56 shows the cooling curves for UNS G 43400 and G 15216 (formerly AISI or SAE 4340 and 52100) steels superimposed on the isothermal transformation curves for these steels. The latter curves, formerly called "S" curves, show the time required to complete transformation at various temperature levels, and a more detailed description of their derivation and use will be found in a later chapter. The objective in selecting the holding time for each of these grades was to determine what temperature would require the minimum holding time to obtain transformation and still give a product free from flakes. In the case of the 52100 steel, 650°C (1200°F) was selected as the temperature from which the billets could be air cooled. For the 4340 steel, holding at 650°C (1200°F) was not feasible, as excessive time would be required for transformation. Since 315°C (600°F) corresponded to the minimum time of holding, this was the temperature selected. Transformation in this range would result in an appreciable amount of internal stress which might result in flakes on cooling in air to atmospheric temperature. Hence, instead of air cooling immediately after the end of the holding period, this particular steel should be heated instead to a temperature just below the critical range to relieve these stresses after which it may be cooled.

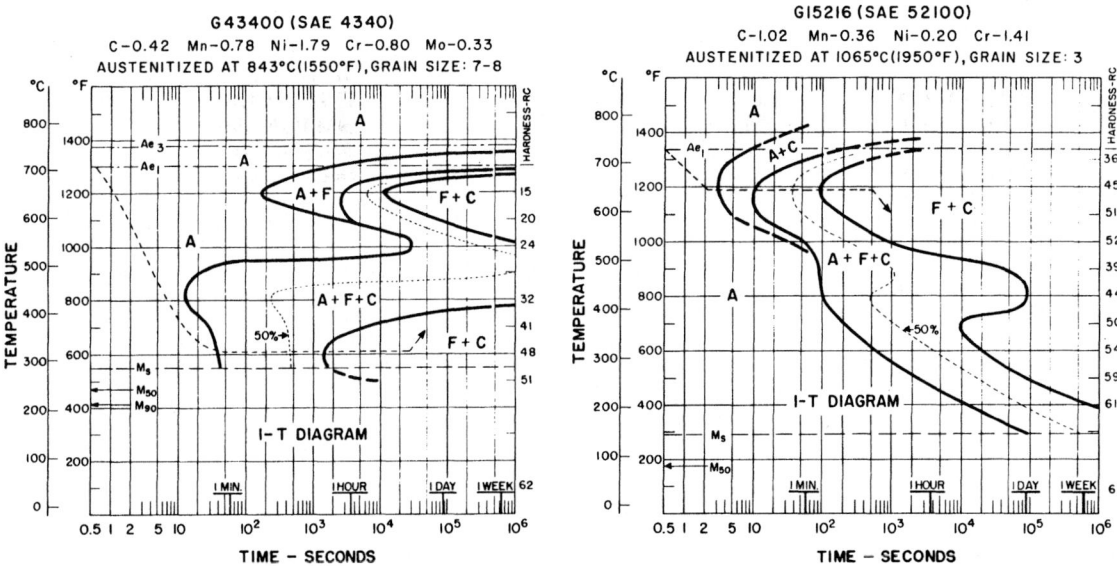

FIG. 20—56. Cooling curves superimposed on isothermal transformation diagrams of two steels, showing cooling cycles employed to effect transformation in a minimum of time while still avoiding flake formation.

These two types of cycles are applicable to a large number of different grades of steel, and the holding temperature for a particular grade can be determined from the proper isothermal transformation diagram. In general, the higher holding temperatures followed by air cooling can be used for the carbon and lower alloy grades of steel; the use of lower holding temperatures, followed by immediate reheating to just below the critical temperature range, is applicable to the more highly alloyed deep-hardening steels.

Bibliography

Am. Iron and Steel Institute, Steel Products Manual. Alloy Steel: Semifinished; Hot Rolled and Cold Finished Bars. The Institute, New York, 1964.

Am. Iron and Steel Institute, Steel Products Manual. Carbon Steel: Semifinished for Forging; Hot Rolled and Cold Finished Bars. The Institute, New York, 1964.

Bergstrand K.G., and P. Nilsson, Hot Surface Inspection by a New Eddy-Current Technique. Iron and Steel Engineer Year Book, 1980, Vol. 57, 75-77.

Beynon, R.E., Roll Design and Mill Layout. Assn. of Iron and Steel Engrs., Pittsburgh Pennsylvania, 1956, 35-49.

Brittain, Marshall and H. G., Frostick, The blooming mill programmed. Iron and Steel Engineer Year Book 38, No. 3 1961, 215-222.

Burden, M. & J. Hunt, A Mechanism for the Columnar to Equiaxed Transition In Castings and Ingots. Met. Trans., 1975, Vol. 6A.

Cardwell, G. P. and W. J. Reilly, Features of Lukens' 140-inch roughing and slabbing mill. Iron and Steel Engineer Year Book 40, No. 3, 1963, 253-263.

Clifford, T. D., Operational features of 45-inch slab mill and 80-inch hot strip mill. Iron and Steel Engineer Year Book 35, No. 9, 1958, 696-699.

Committee of Hot Rolled and Cold Finished Bar Producers. Cold Finished Steel Bar Handbook. American Iron and Steel Institute, 1000 16th St. N.W., Washington, D.C. 20036. 1968.

Cramer, R. E. and E. C. Bast, Production of flakes by treating molten steel with hydrogen and the time of cooling necessary to prevent their formation. Am. Society for Metals, Trans. 27, 433-445; discussion, 446-457 (1939).

Curtin, J. J., Design of a mill for rolling semifinished products. Am. Iron and Steel Institute Yearbook, 1947, 495-509.

Daniels, F.C.T. and D.L. Eynon, Bloom and billet mills and their rolls. Assn. of Iron and Steel Engineers, Proc., 1944, 301-313.

Fisher, K., Macrosegregation in Ingots and Castings, D. Phil. Thesis, Oxford University, 1979.

Fleming, D. H., Powder cutting and scarfing. Welding Engineer 32, 66-70 (1947).

Flemings, M.C., The Control of Macrosegregation in Ingots. J.Scand.Met., 1975, Vol.5.

Flemings, M.C., Solidification Processing, McGraw Hill, 1974.

Gray, A. C., and J. G. Mitchell, Operating and mechanical features of new 46-inch by 110-inch high-lift blooming mill at Stelco. Iron and Steel Engineer Year Book 35, No. 8, 1958, 598-606.

Henderson, G. A., Bethlehem Steel Co.'s 45-inch slabbing mill at Lackawanna Plant. Iron and Steel Engineer Year Book 40, No. 2, 1963, 73-88.

Hintz, O., and R.W. Blackbourn, Recent Developments in Machine Scarfing of Continuous Cast and Rolled Steel. Iron and Steel Engineer Year Book, 1978, 62-66.

Hultgren, A., Flakes or hair cracks in chromium steel, with a discussion on shattered zones and transverse fissures in rails. Iron and Steel Inst., J. 111, 113-148; discussion and correspondence, 149-167 (1925).

Ikushima, H., T. Hirasaw, I. Nakauschi, Y. Settai and Y. Yamagishi, Plasticine Model Tests to Determine Effects of Rolling and Ingot Geometry Variables on Bottom Crop Losses of Slab Ingots. Iron-making and Steelmaking No. 3, 1977, 176-180.

Jacobi, H., Casting and Solidification of Steel, CEC Steel Research Report EUR 5861, 1978, 94.

Johnson, E. R., S. W. Poole and J. A. Rosa, Flaking in alloy steels. Am. Inst. Mining and Metallurgical Engineers, Open Hearth Proc. 27, 1944, 358-377.

Kingsley, P. S., Prevention of flakes in steel forging billets. Metal Progress 47, 1945, 699-703.

Kol, J., Technical Advances in the Equipment for the New Slabbing Mill and Hot Wide Strip Mill at IJmuiden. Stahl Eisen 91, No. 10, May 13, 1971, 547-548. (BISITS Translation BISI 9549)

Laidlaw, J. L. and J. H. Walshaw, Algoma's 46-inch blooming and plate mill. Iron and Steel Engineer Year Book 38, No. 1, 1961, 19-29.

Lavette, R. F., Republic's 44-36-32 in. mills at Chicago. Assn. Iron and Steel Engineers, Proc., 1948, 717-721.

Lindberg, V. H., and J. E. Duffy, Automated structural rolling facilities at South Works of U.S. Steel. Iron and Steel Engineer Year Book 38, No. 11, 1961, 927-939.

Linde Air Products Co., Oxy-acetylene handbook. New York, The Company, 1943.

Marrs, R. E. and P. E. Perrone, Operating experience with the automated slabbing mill. Iron and Steel Engineer Year Book 40, No. 4, 1963, 280-291.

Miller, David H. and Dr. Terence E. Dancy, Continuous casting—past, present and future. Iron and Steel Engineer Year Book 40, No. 5, 1963, 381-393.

Montgomery, A., Jr. and J. K. Magee, New primary mills at Duquesne Works, United States Steel Corp. Iron and Steel Engineer Year Book 40, No. 6, 1963, 450-456.

Moran, P., P. Waterworth & I. Davies, Macrosegregation in Killed Steel Ingots, in Solidification Technology in the Foundry and Casthouse, Metals Society, London, 1981.

Nozaki, N., N. Jokei, Y. Matsumori, M. Kawasaki, and H. Shiraishi, On the Rolling of Rimmed Steel Ingots in an Unsolidified State. Proc. of Int. Conference on Steel Rolling, Iron and Steel Institute of Japan 1, 1980, 141-150.

Reid, Warren and R. H. Wright, Electrical features of universal slabbing and hot strip finishing mills at Fairless. Iron and Steel Engineer Year Book 32, No. 1, 1955, 85-93.

Roberts, William L., Cold Rolling of Steel. 1978. New York, Marcel Dekker, Inc.

Roberts, William L., Hot Rolling of Steel. 1983. New York, Marcel Dekker, Inc.

Samson, D.H. and K.R. Canfor, Dofasco's 88-Inch Slabbing Mill. Iron and Steel Engineer Year Book, 1974, 57-62.

Schlesinger, Kurt, Selection and economy of equipment for blooming and slabbing mills. Iron and Steel Engineer Year Book 34, No. 7, 1957, 597-608.

Shiraiwa, T. Hiroshima, and Y. Tamura, Development of Automatic Conditioning Systems. The Sumitomo Search No. 17, May 1977, 66-72.

Smith, S.J., Influence of Ingot Shape on Billet Yield. Journal of Iron and Steel Institute, March 1970, 247-254.

United States Steel Corporation, Isothermal transformation diagrams, 3rd ed. 1963. Pittsburgh, The Corporation.

Wray, P.J., Predicted Volume Change Behavior Accompanying the Solidification of Binary Alloys, Metall. Trans., 1976, Vol 7B, 639-646.

CHAPTER 21

Continuous Casting of Semi-Finished Steel Products

SECTION 1

INTRODUCTION

Continuous casting, a relatively recent development in the steel industry, has had a dramatic impact on steel production throughout the world. Of all technologies developed and applied in the steel industry, continuous casting has had the greatest effect on improving the efficiency of material utilization. In continuous casting, the process yield is more than 95 per cent compared with approximately 80 per cent for the production of semi-finished products made by rolling ingots in slabbing or blooming/billet mill facilities. In addition, significant quality benefits have also been achieved.

Large continuous slab and bloom casting machines have been widely installed in existing steel plants to replace ingot casting, as well as in plant expansions and large new integrated facilities. Smaller continuous billet casters, together with ultra-high powered electric arc furnaces, were the critical technology development which created the phenomenon of the mini-mill—a small capacity plant (originally a few hundred thousand tons per year) serving local markets.

PAST PRODUCTION AND PRESENT TRENDS

Continuous casting of semi-finished steel shapes (slabs, blooms, billets, rounds and other special sections) has increased markedly from the early 1960's when the operation of a significant number of production facilities first began. For example, it was predicted, in a 1981 study by the Institute for Iron and Steel Studies, that in 1983 continuous casting capacity in the United States would be 37.7 million metric tons (41.5 million net tons) per year including both presently installed and committed facilities. By comparison, in 1981 the production of continuously cast steel was 23.0 million metric tons (25.3 million net tons) which represented approximately 21.1 per cent of the total raw steel made. A classification of existing and projected continuous casting capacity by integrated, specialty, and scrap-based steelmakers is shown in Figure 21—1.

Production of continuously cast steel in the United States, Japan, Western Europe and the rest of the world is summarized in Table 21—I. This table shows both the significant increase in the quantity of continuously cast steel from 1965 to 1981 for the four regions as well as the increase in the percentage of raw steel continuously cast. These data, illustrated graphically in Figures 21—2 and 21—3, show that both the amount and percentage of continuously cast steel in Japan and Western Europe exceed, significantly, those of the United States and the rest of the world. The change in the total world production of raw steel and the proportion of continuously cast steel, which was approximately 30 per cent in 1981, is shown in Figure 21—4.

Growth in continuous casting is also depicted in Figure 21—5 which shows the total number of machines in operation worldwide since the late 1960's. This illustration, compiled by Concast Incorporated, shows the rapid adoption of both the large capacity slab and bloom machines as well as the smaller-capacity billet machines.

HISTORICAL DEVELOPMENTS

Although commercial production of continuously cast steel in significant quantities started in the early 1960's, the first design concepts for the continuous casting of metals occurred in the mid-1800's in the United States and in Europe. The following summary includes both historical events that occurred as well as the major developments leading to the operation of modern continuous casting facilities.

1840 First patent granted for continuous casting lead tubing to G.E. Sellers (U.S.A.).

1843 Patent granted for continuous casting of lead tubing which recognized the necessity of moving the mandrel to prevent adhesion of the cast material (J. Lang, U.S.A.).

1846 H. Bessemer's patent for the production of tin foil and sheet lead, and experimental production of sheet glass using water-cooled rotating rolls (England).

1856 Continuous casting of malleable iron using ro-

Table 21—I. World Wide Growth of Continuously Cast Steel by Region—1965 to 1981.
(Millions of Metric Tons)

	Region of World[1]											
	United States			Japan			Western Europe			Rest of World		
Year	Total Raw Steel	CC*	CC, % of Total	Total Raw Steel	CC*	CC, % of Total	Total Raw Steel	CC*	CC, % of Total	Total Raw Steel	CC*	CC, % of Total
1965	119	0.3	0.2	41	0.0	0.0	130	0.4	0.3	136	0.3	0.2
1970	119	4.5	3.8	93	5.2	5.6	161	8.1	5.0	220	7.7	3.5
1975	106	9.7	9.1	102	31.8	31.1	155	27.1	17.5	278	15.9	5.7
1976	116	12.2	10.5	107	37.6	35.0	164	34.0	20.8	286	20.6	7.2
1977	114	14.3	12.5	102	41.8	40.8	155	40.7	26.2	301	23.7	7.9
1978	124	18.9	15.2	102	47.2	46.2	164	48.5	29.6	322	29.8	9.2
1979	123	20.9	17.0	112	58.1	52.0	174	55.3	31.8	333	32.7	9.8
1980	101	20.6	20.3	111	66.3	59.5	162	62.9	38.9	337	35.5	10.6
1981	109	23.0	21.1	102	71.8	70.0	159	70.2	44.2	334	38.0	11.4

*CC = Continuous-Cast
[1]Based on IISI data.

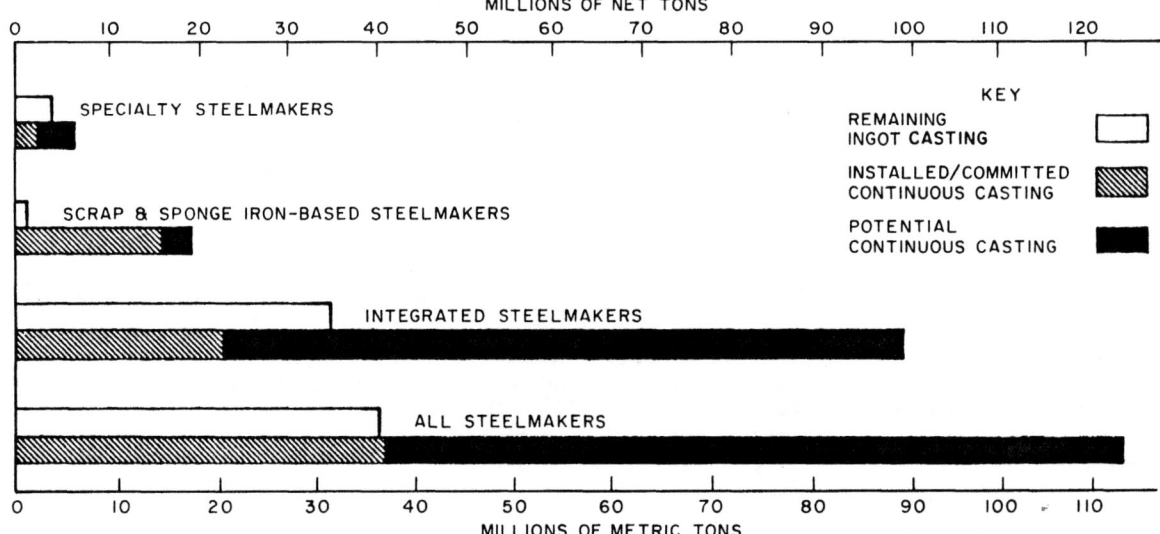

FIG. 21—1. Existing and projected continuous casting capacity
in the United States by type of steel producer—1981.

tating rolls announced and patent granted (H. Bessemer, England).

1872 Concept of continuous casting with moving molds (W. Wilkinson and E. Taylor, England).

1886 Basic principles of vertical casting of steel (B. Athea, U.S.A.). Production unit reported in operation until 1910.

1889 Vertical-type casting machine designed similar to those currently in operation (M. Daelen, Germany).

1912 Advantages of mold oscillation in the direction of the strand recognized with horizontal casting (A. H. Pehrson, Sweden).

1915 Automatic control of metal flow into mold depending on metal level in mold considered (G. Mellan).

1921 Continuous relative motion between strand and mold proposed (C. W. van Ranst).

1933 S. Junghans, a leader in continuous casting, proposed a nonharmonic mold oscillation which

would not influence heat transfer between the solidifying section and mold (Germany).

First industrial continuous casting plant built in Germany by S. Junghans for the production of brass sections. The facility embodied a vertical design with an open-ended mold and had a production rate of 1700 metric tons/month.

1935 Brass plates continuously cast by Scovill Manufacturing Co., using casting rolls until 1937 (U.S.A.).

1938 Semi-horizontal machine built for the casting of steel based on the Goldobin principle of moving molds (USSR).

1943 Early pilot plant for casting steel installed by Junghans (Germany).

1946/ Pilot plants for casting steel installed by Babcock
1947 & Wilcox (U.S.A.), Allegheny Ludlum (U.S.A.), Low Moor (England), BISRA (England), Bohler Bros. (Austria) and Steel Tube Works (Amagasaki, Japan). Melting capacity varied from 200 lb

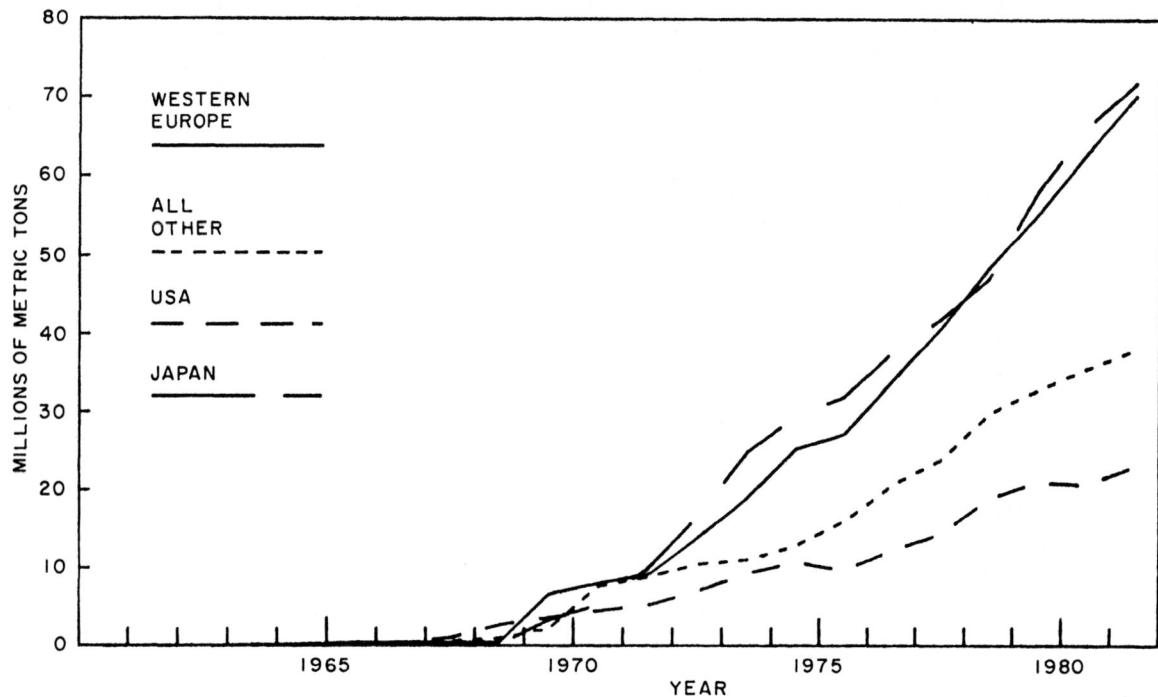

FIG. 21—2. World wide growth in production of continuously cast steel by region through 1981.

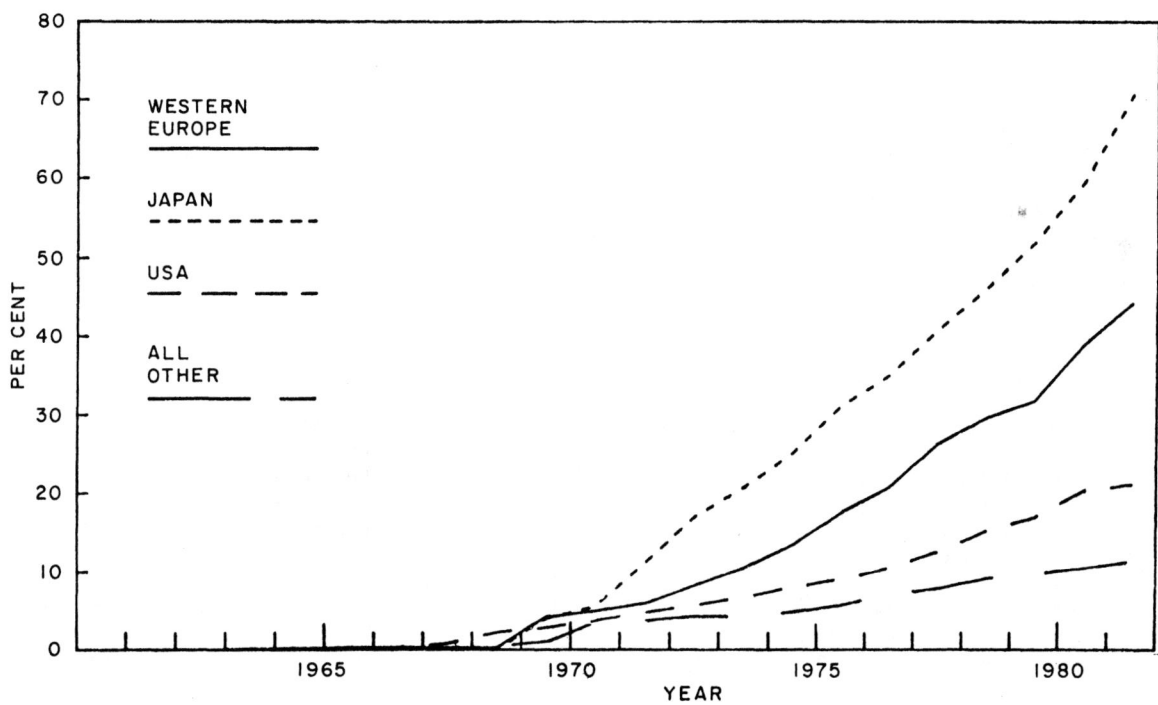

FIG. 21—3. World wide growth in percentage of steel continuously cast by region through 1981.

to approximately 7.5 metric tons.

1951 First slab caster for the production of stainless steel, installed at Krashy Oktubri, USSR. The caster produced section sizes up to 800 mm x 180 mm (32 in. x 7 in.) and had a rated capacity of 36 000 metric tons/year.

1952 First billet caster for the production of carbon and low alloy steel brought into operation at Barrow, England. Initially a single-strand machine for casting 50 to 100-mm sq. billets. Twin-

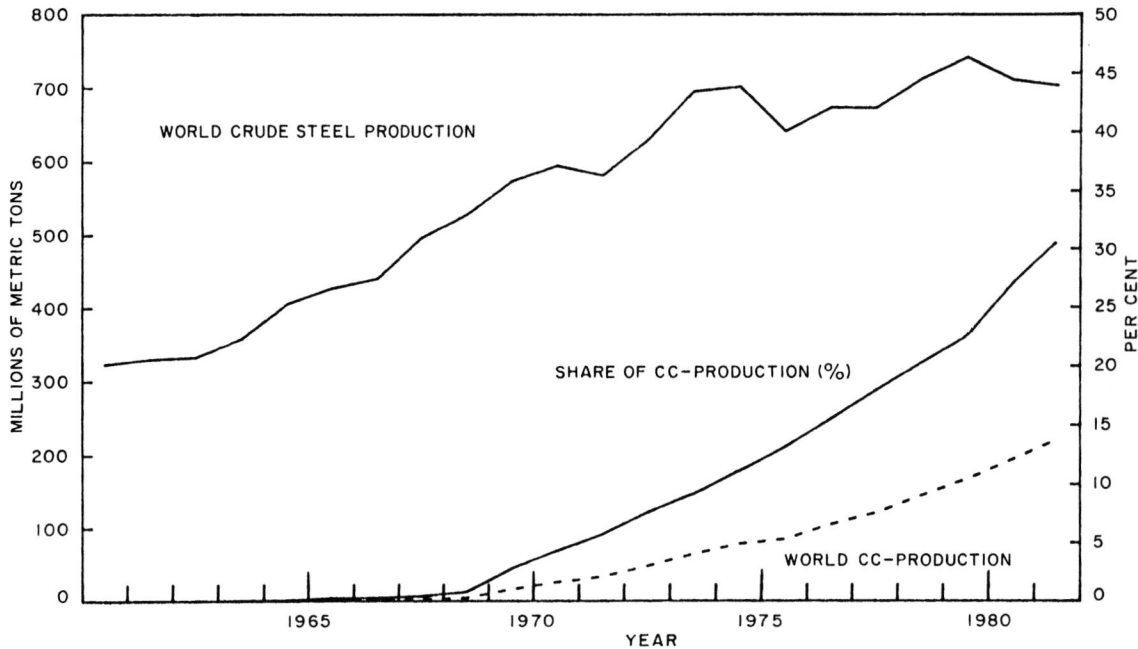

FIG. 21—4. World production of crude steel versus continuously cast production from major steel producing countries.

strand operation started in 1958.

Patent issued describing a curved casting machine (O. Schaeber, Germany).

First electromagnetic stirrer designed for continuous casting at Mannesmann (Junghans and Schaeber, Germany).

1954 First slab caster for the production of stainless steel slabs in North America started up by Atlas Steels, Canada. Slabs up to 622 mm x 165 mm (24½ in. x 6½ in.) were cast from 27.3 and 45.5 metric-ton heats (30 and 50 net ton).

First 4-strand unit installed for casting rounds by Mannesmann, Germany. (Production status not achieved until 1973).

1956 Cutting strands to length in the horizontal position following bending instead of in the vertical permitted a significant reduction in the height of the casting machine (Barrow, England).

Patent filed for a curved mold (E. Schneckenburger and C. Kung, Switzerland).

1958 First 8-strand billet caster installed by Societa per l'Industria e l'Elettricita, Italy.

1961 Start-up of first large vertical-type slab casting machine with bending of the strand for a horizontal discharge by Dillinger Steelworks, West Germany. The single-strand machine casts large low carbon steel slabs up to 1520 mm x 200 mm (60 in. x 8 in.) from 30 metric-ton (33 net-ton) heats.

1963 Start-up of first curved mold billet caster at Von Moos' Eisenwerke, Switzerland.

A world survey indicated a total of 61 machines in operation with 44 under construction.

1964 First new steelworks which depended 100 percent on continuous casting brought on stream by

Shelton Iron and Steel, England. Casting facilities consisted of four machines (11 strands total) which produced medium to large carbon and low alloy blooms from 140 mm (5½ in.) sq. to 622 mm x 432 mm (24½ in. x 17 in.) as well as slabs from 56 metric-ton (62 net-ton) heats.

Start-up of the first curved mold machine for large slabs, up to 1600 x 250 mm (63 x 9.8 in.) at Dillinger Steelworks, West Germany. This design permitted a 50 percent reduction in machine height in comparison with a vertical-type machine.

1965 Production casting of rounds on a 4-strand machine at Eschweiler Bergwerks (Germany).

1967 High productivity slab caster with an annual capacity of 1.5 million metric tons (1.65 million net tons) commissioned at U.S. Steel's Gary Works with in-line rolling to change slab widths.

1968 First 4-strand slab caster with curved mold installed in North America at National Steel's Weirton Division for tinplate applications.

1968/ Four slab casters placed in operation by
1969 McLouth Steel with combined annual capacity of 2.2 million metric tons (2.4 million net tons).

First continuous centrifugal caster by Vallourec and Creusot-Loire (France).

1970/ Recent developments adopted:
1983 • Rapid ladle and tundish changing equipment to improve productivity and yield through sequence casting.

• Longitudinal slitting of slabs to minimize mold changing and improve productivity.

• Variable width adjustable molds to minimize mold changing and improve productivity.

• Multi-point bending of strand to improve

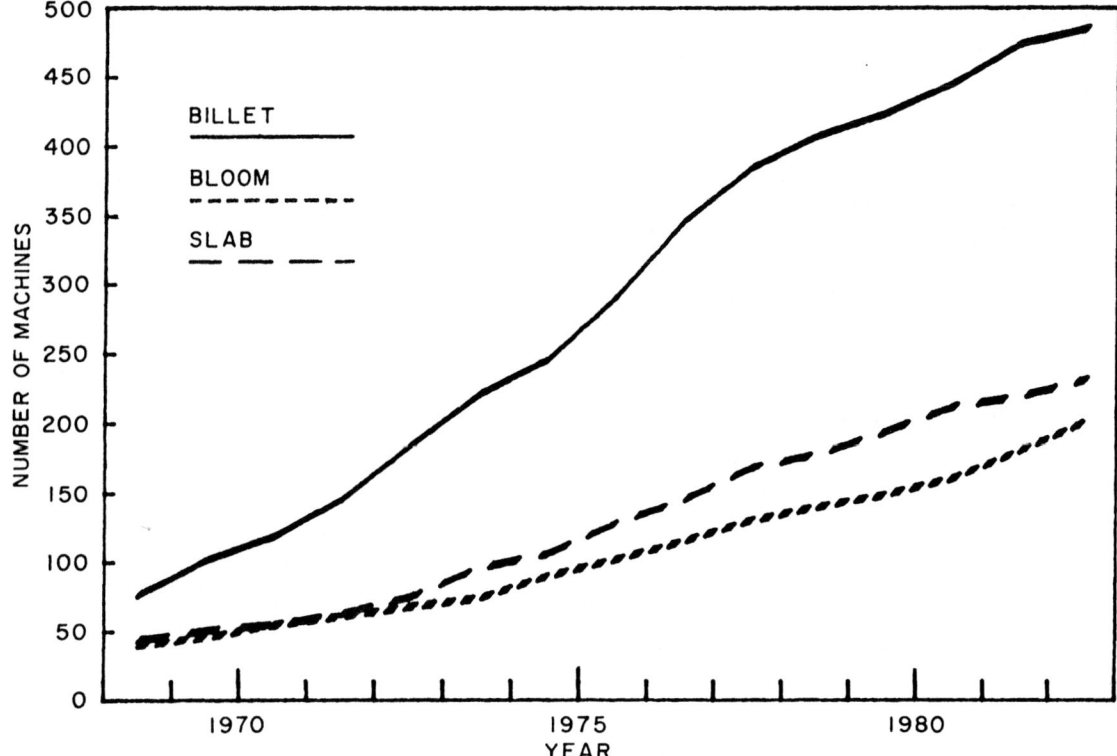

FIG. 21—5. Continuous casting machines in operation world-wide through 1981.

quality.

- Mist cooling using air-atomized water to improve both cooling efficiency and homogeneity and, consequently, product quality.
- Divided or split molds to improve productivity.
- Total shrouding of metal streams from ladle to tundish and from tundish to mold to improve quality.
- Electromagnetic stirring in and below the mold to improve quality.
- Integrated computer control of complete casting process.

BENEFITS OF CONTINUOUS CASTING

Sequence of Operations—Prior to the development of continuous casting, ingots provided the only starting material in wrought-steel products. The typical sequence of operations from the steelmaking furnace to the rolling mills was:

(1) Tapping liquid steel from a steelmaking furnace into a ladle.
(2) Transferring ladle to pouring platform and teeming liquid steel into ingot molds.
(3) Transferring filled molds to stripping area for ingot removal.
(4) Transferring and charging ingots into soaking pits and heating to rolling temperature.
(5) Removal of heated ingots from soaking pits and transfer to primary mill for rolling into semi-finished shapes.

(6) Transferring semi-finished shapes to subsequent rolling mills.

Using continuous casting, the following much shorter sequence of operations is required:

(1) Tapping liquid steel from a steelmaking furnace into a ladle.
(2) Transferring the ladle to a casting platform and continuously casting liquid steel into semi-finished shapes.
(3) Transferring the semi-finished shapes to rolling mills.

The benefits derived from the shorter sequence of operations provided the main impetus for the adoption of continuously casting: increased yield; improved product quality; energy savings; less pollution; and reduced costs.

Yield—Increased yield from liquid steel in the ladle to the semi-finished rolled shape results from a reduction in scrap generation in three areas: the primary rolling mill; the pouring operation; and ingot heating. The major contribution to the improved yield is the absence of crop losses corresponding to the ingot top and bottom location when an ingot is rolled in the primary mill. Reduction in yield losses associated with the pouring operation include "short" ingots, ingot butts and general pit scrap. Scaling losses associated with ingot heating in the soaking pit are also avoided.

Quality—Metallurgical quality improvements include less variability in chemical composition and solidification characteristics. In addition to improved segregation characteristics of carbon, sulfur and alloying elements across the section of a continuously cast

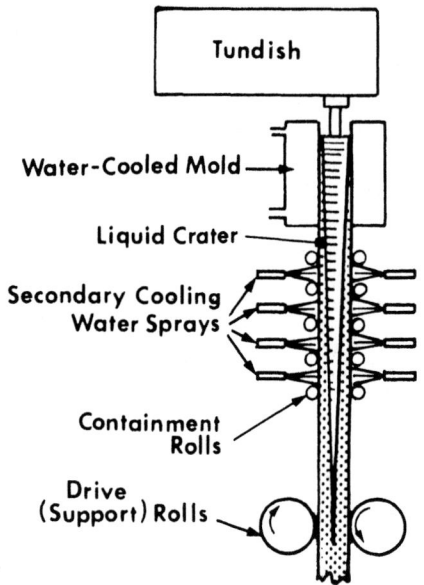

Tundish

Water-Cooled Mold

Liquid Crater

Secondary Cooling
Water Sprays

Containment
Rolls

Drive
(Support) Rolls

FIG. 21—6. Major components of a continuous casting machine.

shape, there is also less variability along the length of the cast shape. (In casting a heat into ingots there are a multitude of individual ingots each with their associated vertical segregation and structural variability, whereas a continuous cast strand is not only as one ingot but also an ingot which has less variability in a vertical direction.) In modern continuous casting, the surface quality of the cast shape is superior to that of a semi-finished rolled shape with respect to surface defects such as seams and scabs, and, consequently, conditioning requirements and yield losses are minimized. A majority of continuously cast steels can be further processed without any conditioning. Thus, an improved, more uniform finished product can be obtained with fewer internal and surface defects.

Energy—Energy savings are achieved with continuous casting because of the elimination of the energy-consuming steps in the ingot process. These include fuel consumption in soaking pits and the electric power requirements for operating the primary rolling mills. Energy is also indirectly saved through the increased yield which requires the production of less raw steel for a comparable quantity of semi-finished product. In addition to these savings, a practice in which hot continuously cast shapes are charged directly into a heating furnace in the finishing mills is receiving attention. Thus the sensible heat of the cast product is conserved.

Pollution—Continuous casting reduces pollution through the elimination of ingot-processing facilities such as soaking pits.

Costs—Both capital and operating costs are reduced with the installation of continuous casting in comparison with ingot processing. Capital assets savings are attributable to the elimination of the additional equipment required for ingot processing. Operating cost savings are primarily the result of lower manpower requirements and higher yields.

OVERVIEW OF CONTINUOUS CASTING PROCESS

The main components of a casting machine are depicted schematically in Figure 21—6 with a cross-section of a slab caster shown in Figure 21—7. Essentially, a casting machine consists of: a liquid metal reservoir and distribution system (a tundish); a water-cooled mold; secondary cooling zones in assocation with a containment section; bending rolls; a straightener; cutting equipment; and a runout table to cooling beds or directly to a product transfer area.

A casting machine can have a number of casting strands each of which is associated with an independent mold, secondary spray water cooling zone, containment section, etc. The number of strands depends principally on the shape being cast (slab, bloom, billet etc.) and the heat size.

To start the casting process, a dummy bar (which is connected to an external mechanical withdrawal system) is inserted in the mold and positioned so that the top of the dummy bar closes the bottom of the mold. Liquid steel is delivered in a ladle to the casting floor where it is poured at a controlled rate into the tundish. Liquid metal flows through nozzles in the bottom of the tundish and fills the mold. When the liquid metal level in the mold reaches a predetermined position, withdrawal of the dummy bar is initiated. The withdrawal speed of the dummy bar is preset based on the casting speed required or the metal flow rate from the tundish. When the dummy bar head, which is now attached to the solidified shape being cast, reaches a certain position in the withdrawal system, it is mechanically disconnected and the dummy bar removed. The solidified cast shape continues through the withdrawal system to the cutting equipment.

Solidification of the liquid steel starts in the water-cooled mold and continues progressively as the strand moves through the casting machine. Freezing begins at the liquid steel meniscus level in the mold forming a shell in contact with the walls of the mold. The distance from the meniscus level to the point of complete solidification within the machine is called the metallurgical length. Obviously, the point of complete solidification must occur ahead of the cutting point and in many casters is ahead of the straightener. Casting conditions are established such that the strength of the solidified steel shell leaving the mold is sufficient to withstand the ferrostatic pressure of the liquid steel in the mold. To prevent sticking of the solidified shell to the mold wall, the mold is oscillated in a vertical direction. Friction between the shell and mold is minimized by the introduction of mold lubricants such as oils or fluxes which form a fluid slag.

Below the mold, additional heat is removed from the strand in the secondary cooling zones and solidification is completed. The secondary cooling zones consist of a series of water sprays. Flow rates are closely controlled to obtain optimum cooling rates and strand surface temperatures. Support roll units are provided to contain the strand to avoid transverse movement and to prevent bulging of the hot solidifying shell from internal ferrostatic pressure. The strand cooling and containment systems are designed, as is the mold, to prevent external and internal defects in the cast section

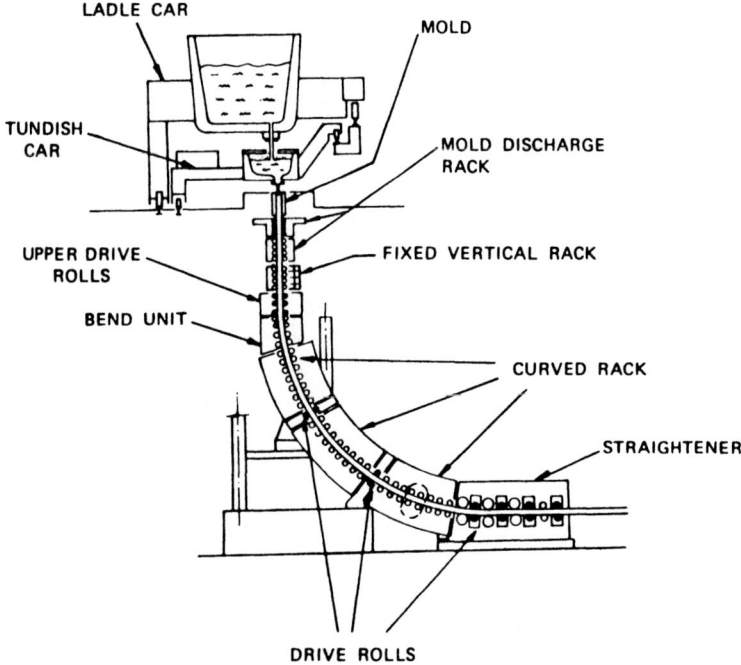

Figure 1. SECTION SCHEMATIC OF A U. S. STEEL CASTING MACHINE

FIG. 21—7. Cross section of a slab caster.

and insure final product quality.

The secondary cooling and containment area is followed, on certain types of machine, by a bending unit and a straightener which is present in all machines.

After straightening, the cast section is cut to the desired length either by torches or shears. The hot cut lengths are then either conveyed by a runout roller table to cooling beds or grouped and transferred directly to subsequent hot and cold-rolling operations.

TYPES OF CASTING MACHINES

Casting machines can be classified into four main groups depending on the section shape produced: billet; bloom; round; and slab. In some cases, overlaps occur where the molds on a particular machine can be changed to cast other shapes; for example, billets or blooms, blooms or small slabs, and blooms or rounds. In addition, machines exist where special shapes, such as rectangles and 'dogbone' structural sections can be cast as well as billets or blooms.

Billet machines, which cast section sizes up to approximately 5 inches square, are multi-strand machines that are widely used in the mini-mill sector of the industry but only to a relatively limited extent in fully integrated plants. This has occurred because of practical considerations which are related to the heat size, casting rate per strand (tons/minute) and casting time. In general, casting times are limited to approximately one hour for each heat because of heat losses in the

ladle. It is practical, for example, to cast a 50 net-ton heat on a 2-strand machine or a 100 net-ton heat on a 4-strand machine. However, the number of strands required for casting heat sizes in excess of 200 net tons, which are common in integrated steel plants, becomes impractical.

Bloom casters have been more widely installed by integrated plants because the casting rate for the larger section size is higher than for billet sizes and, consequently, larger heat sizes can be cast with relatively fewer strands. Bloom section sizes cast can vary, for example, from 7 in. sq., cast on a 6-strand machine from 150 net-ton heats, up to as large as 14.6 in. x 23.6 in., cast on a 3-strand machine from 180 net-ton heats.

The installation of machines for casting rounds, principally for seamless tube production, has been relatively slow. Although a 4-strand caster at Escheweiler Bergwerks was producing 125 and 210-mm (4.9 and 8.3 in.) diameter rounds from a 30 metric-ton (33 net ton) heat in 1965, potential surface cracking problems delayed the introduction of round casting in the U.S.A. Jones & Laughlin Steel Corp. modified an existing 6-strand billet/bloom caster in 1981 to produce 152-mm (6-in.) diameter rounds and in 1983 both U.S. Steel and Babcock & Wilcox installed bloom/round machines. The 640,000 metric tons (700,000 net tons) per year U.S. Steel machine at Lorain (described in Section 4) is a 6-strand caster producing up to 232-mm (9¼ in.) diameter rounds with the Babcock & Wilcox 4-strand machine producing 140-mm (5.5-in.) diameter rounds.

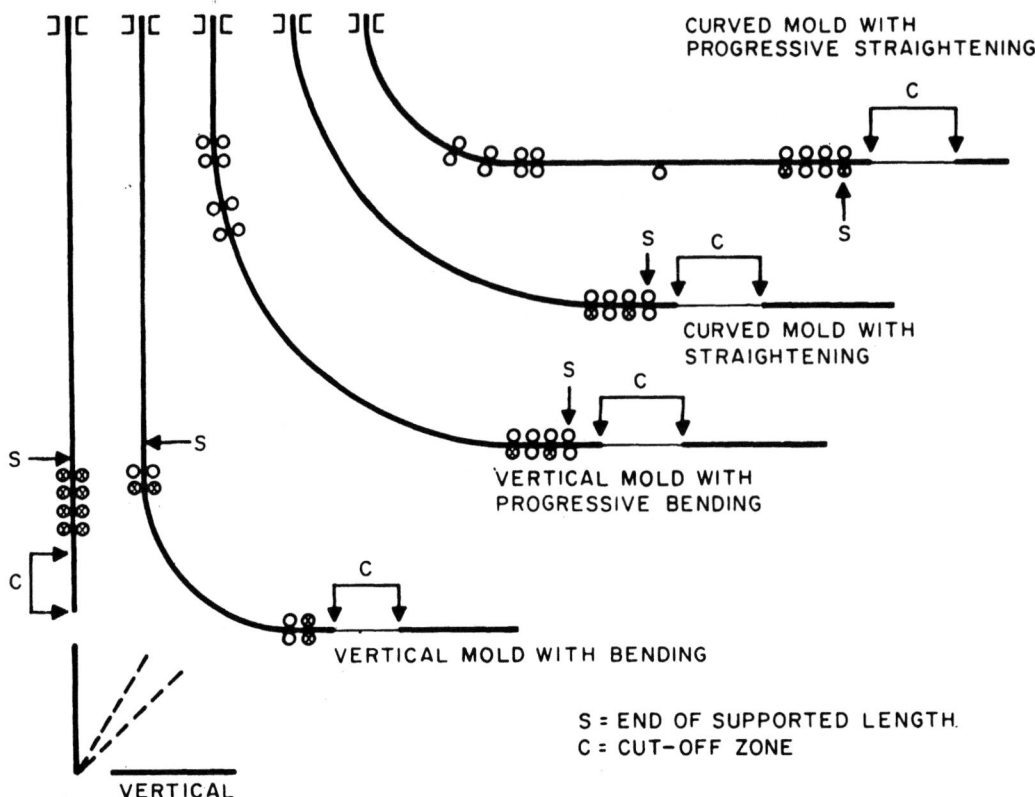

FIG. 21—8. Development of casting machines.

140-mm (5.5-in.) diameter rounds.

There are a large number of slab casters throughout the world which, although operated principally in integrated steel plants, are also used for producing stainless and specialty steel. These machines are generally high-production units with rated annual capacities of up to 1.4 million metric tons (1.5 million net tons) and above. They are usually either single or twin-strand machines casting large heat sizes up to 270 metric tons (300 net tons). The section sizes cast are also large; e.g., 2160 mm x 305 mm (85 in. x 12 in.) for plate applications. A wide variety of low carbon, low alloy, alloy and stainless steel grades are cast for sheet, strip, plate and specialty applications. Installations in the U.S.A. in 1983 included: a twin-strand, 1.6 million metric-ton (1.8 million net-ton) per year caster by Republic Steel which produces slab sizes up to 1829 mm x 254 mm (72 in. x 10 in.); a 2-strand (2.2 million net tons)/year machine by Wheeling-Pittsburgh, casting slab sizes up to 2032 mm x 254 mm (80 in. x 10 in.); and a two-machine, 2.8 million metric ton (3.12 million net ton) per year casting complex by Jones & Laughlin Steel Corp. (described in Section 4).

In addition to these four main groups of casting machines, a few continuous casting machines have been installed that cast special shapes such as beam blanks which are subsequently rolled into I beams.

One of the major objectives in the design of continuous casting machines has been to reduce the capital cost of the installation while at the same time maintaining or improving the quality of the cast product. This objective has been achieved by a progressive reduction in the height of the machine which has resulted in a reduction in the size of the supporting structure, building height and foundation. It has led to the development of five principal types of casting machines which are essentially applicable to all section shapes cast whether billets, blooms, slabs, etc. Chronologically, these types, illustrated schematically in Figure 21—8, are:

- Vertical machine with a straight mold and cutoff in the vertical position.
- Vertical machine with a straight mold, single-point bending and straightening.
- Vertical machine with a straight mold, progressive bending and straightening.
- Bow type machine with curved mold and straightening.
- Bow type machine with curved mold and progressive straightening.

With the introduction of the newer designs there has been an increasing adoption of the bow-type machines with curved molds for slab casters and to a lesser extent for billet and bloom machines.

The choice between these types of casting machines depends on a complex optimization of the specific facility requirements for caster productivity, product quality and machine complexity, and cost. Curved machines are usually simpler to build (i.e., lower cost) and maintain than vertical with bending machines, as the bender is eliminated. However, for some grades of steel, (for example, plate grades), quality and casting speed limitations were previously more restrictive on these curved machines. Recently, technical developments such as "clean" steel practices and electromagnetic stirring have been applied to curved machines to overcome these restrictions. In general, the complexity of the casting process and machine varies greatly between the type of product being cast (e.g., billet, bloom, or slab). This is due both to the thermomechanical characteristics of these cast sections,

(billet sections are self-supporting in the secondary cooling zone, slabs are not) and to the different applications of the cast product.

Generally, billet casters have tended to be simple in design, with open-pouring streams, limited automatic controls, and no roll support in the secondary cooling zone. Conversely, slab casters are complex and use the total range of subsystems described subsequently, including total stream shrouding, computer controls, and total roll containment throughout the machine. Bloom casters are intermediate between these two extremes.

SECTION 2

CASTING PRACTICES, EQUIPMENT AND OPERATION

STEELMAKING

Steelmaking practices for continuously cast steels are, with certain exceptions, similar to those employed for ingot steels whether produced in the electric furnace or basic oxygen convertors. There are two major exceptions: (1) temperature control; and (2) deoxidation practice.

Temperature Control—Temperature control is more critical than in ingot production. The tapping temperature is generally higher to compensate for heat losses associated with the increased transfer time to a caster and must be maintained within closer limits to avoid mold "breakouts", if the temperature is too high, or premature freezing in the tundish nozzles, if the temperature is too low. Casting temperature can also affect the crystallization structure of the cast product. Optimum structures are developed with low superheats which should be uniform throughout the entire cast. To meet this objective, temperature homogenization practices are employed. One practice widely employed is to stir the metal in the ladle by the injection of a small quantity of argon through porous plugs located in the bottom the ladle, or a lance which is lowered into the ladle.

Deoxidation—Continuously cast steel must be fully deoxidized (killed) to prevent the formation of blowholes or pinholes at or close to the surface of the cast product which cause seams in subsequent rolling operations. Depending on the grade of steel and product applications, either of two practices are employed: (1) silicon deoxidation with a small addition of aluminum for coarse grain steels; and (2) aluminum deoxidation for fine grained steel. Silicon-killed steels are easier to cast than aluminum killed steels because deposits of alumina in the tundish nozzle, which cause nozzle blockage, are avoided. (Special techniques to prevent the nozzle deposition of alumina are discussed later). For high-quality products, it is becoming a common practice to employ a ladle refining practice prior to casting (see Chapter 19).

LIQUID METAL FLOW CONTROL

At the casting machine, two steps are involved in

transferring the liquid steel from the ladle to the molds. Liquid steel is first fed continuously or semi-continuously from the ladle to the tundish which, in turn, distributes the liquid steel through nozzles in a continuous flow to the individual molds, (Figure 21-9). Metal flow through the ladle nozzle into the tundish is controlled by a stopper rod mechanism or by hydraulically or electrically controlled slide-gate systems. (The latter systems are rapidly receiving acceptance because of greater control capabilities and reliability.)

Tundish Design—The tundish is essentially a rectangular box with nozzles located along the bottom. In addition to serving as a metal distributor to the molds, it is designed to provide a number of important functions:

- Metal flow patterns which enhance the stability of the metal streams entering the casting mold.
- Relatively constant metal height above the tundish nozzle to produce a constant discharge rate which, in turn, leads to uniform casting speeds.
- A metal reservoir to facilitate in casting a sequence of heats, the changeover of an empty ladle for a full ladle without interrupting the flow of metal to the molds.
- Facilitate the separation of inclusions and slag entering the tundish from the ladle. Metal residence time is a critical parameter in meeting this condi-

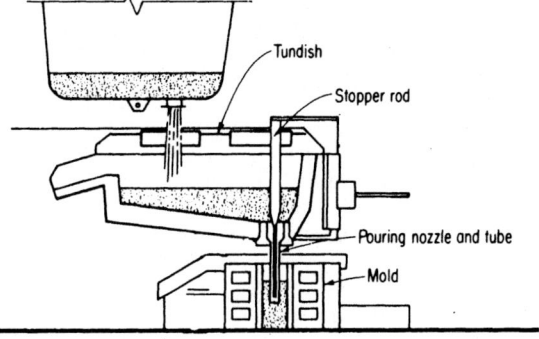

FIG. 21—9. Liquid steel flows from the ladle into the tundish and from the tundish into the mold.

tion, i.e., the time required for a unit volume to steel to flow through the tundish.

There are many types and shapes of tundish (see Chapter 2). One common tundish design for multi-strand billet and bloom casters is a trough shape with a pouring box offset at the midpoint; for slab casters the tundish is a short box or tub shape. The pouring stream from the ladle is directed downward to a position in the tundish bottom which is protected with a wear-resistant pouring pad. This position is usually as far as possible from the tundish nozzle to minimize turbulence. In other locations, the tundish is lined with refractory bricks or boards. Weirs and dams are used as flow-control devices which both increase the residence time as well as reduce the detrimental effects of turbulence on the metal surface, the metal streams entering the mold and dead zones.

Tundishes are usually preheated prior to casting to minimize heat losses from the liquid steel during the initial stages of casting and thus avoid metal freezing, particularly in the critical nozzle areas. Tundish covers are also used to reduce radiant heat losses throughout the casting operation.

. **Tundish Nozzles**—Two basic types of tundish nozzles are used: (1) a metering or open nozzle and; (2) a stopper rod-controlled nozzle. Metering nozzles, a simpler system, have been generally employed in billet and small bloom casters, producing silicon-killed steels. Metal discharge rate is controlled by the bore of the nozzle and the ferrostatic pressure (metal height in the tundish) above the nozzle. Different bores are selected depending on the section size cast and casting speed required. Stopper rod-controlled nozzles are used for casting slabs and large sections when aluminum-killed steels are produced. In this application, metal discharge rate through the nozzle is controlled manually or automatically by the setting of the stopper head in relation to the nozzle opening. Originally, over-sized nozzles were used for casting aluminum-killed steels: as alumina buildup occurred, the stopper head was raised to compensate for a reduction in flow rate.

Modern developments in deoxidation practice together with the use of argon bubbling through the stopper head and nozzle units have minimized the alumina buildup problem. Another development in controlling metal flow from the tundish is the application of slide gate systems which are similar to those employed on ladles. These gate systems can also provide the capability for changing nozzles during casting as well as changing nozzle size.

Liquid Metal Shrouding—Major changes have occurred in the practices adopted for transferring liquid steel from the ladle to the tundish and from the tundish to the mold since the introduction of continuous casting. These changes were brought about by problems experienced in casting certain types of steel (i.e. aluminum fine grain steels) as well as the need to meet both the surface requirements and cleanliness of high-quality products for critical applications.

There are basically two approaches to handling liquid steel: open stream casting and closed stream or shrouded casting. In open stream casting the liquid metal flows directly, through the air, from the ladle to the tundish or from the tundish to the mold. Under these conditions the unprotected metal stream picks up oxygen (and some nitrogen) from the air and deleterious inclusions are formed in the liquid steel. These inclusions are transferred into the casting mold where they are either retained within the cast section or float to the surface of the liquid steel. Those present on the liquid steel surface are subsequently trapped in the solidifying shell and either result in surface defects on the product in rolling or a catastrophic break in the shell below the mold. In addition to the direct formation of inclusions in the exposed steel stream, air entrained in the stream can also react with liquid steel both in the mold and tundish.

To avoid these problems shrouded-stream casting is employed. Emphasis was first placed on shrouding the metal stream between the tundish and mold because of severity of the problem. However, ladle to tundish stream shrouding is now widely employed, especially in slab casting of aluminum-killed steels where the prevention of alumina inclusions is of paramount importance. There are two basic types of shrouding with numerous variations and combinations: (1) gas shrouding; and (2) refractory tube shrouding.

Gas shrouding is frequently used in casting small sections on billet machines (i.e. 4-in. sq.) because of operating difficulties experienced with refractory tubes: there is insufficient space to introduce a tube without encountering metal freezing between the mold wall and tube. There is a variety of designs including: the Pollard steel tube shroud in which gas is introduced at the mid point of the tube at low velocity and exits between the tube and nozzle, and between the tube and mold; complete enclosure between the tundish and mold using a flexible coupling; trucated pyramidal enclosures; and a liquid nitrogen curtain (Figure 21-10.) Nitrogen or argon is used as the protection gas. Gas shrouding alone is not commonly used for preventing oxidation of the ladle to tundish stream. However, one design in use employs a circular ring which is attached to the ladle at one end and is sealed by a sand seal at the other end when the ladle is lowered toward the tundish thus forming an enclosed box: the box is then pressurized with argon.

Refractory tube shrouds are commonly used for casting aluminum-killed steel. They are used both between the ladle and tundish, and tundish and mold. One end of the tube is attached to the ladle (or tundish) with the other end immersed in the steel when the tundish (or mold) is filled with metal. Refractory tubes are usually made of fused silica or alumina graphite.

The mechanical design of the refractory tube is important, especially at the exit end which is immersed in the steel. One type is a straight-through design. Another type, generally used in the mold, has a multi-port (opening) design, such as a bifurcated tube with the bottom of the tube closed and two side openings located near the bottom of the tube. This type of shroud avoids deep penetration of the pouring stream into the crater of the solidifying strand and modifies the flow pattern in the mold. Thus, the inclusions in the pouring stream are not entrapped in the solidifying section but rise to the surface of the liquid metal and are removed with the slag formed by the mold powder.

In many plants, the design of the shroud attachment

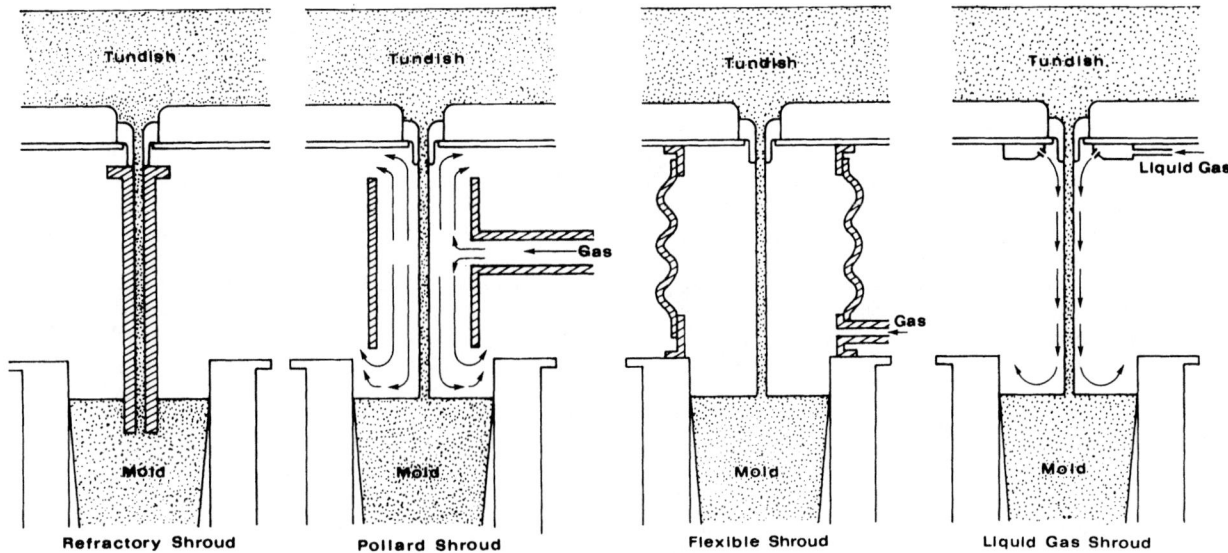

FIG. 21—10. Configurations for shrouding from ladle to mold.

includes the capability for replacing a worn shroud so that a long sequence of heats can be cast without interruption.

At some plants, argon is introduced into the refractory tube to avoid aspiration of air through pores and joints which is caused by the venturi effect of a moving metal stream.

MOLD DESIGN AND OPERATIONS

The primary function of the mold system is to contain and start solidification of the liquid steel in such a manner that both the exiting partly solidified steel section and final cast product are in the proper physical and chemical states. These include: shape (overall configuration and shell thickness); temperature distribution; and internal and surface quality (i.e. structure, chemical uniformity together with an absence of cracks, porosity and non-metallic inclusions).

Mold Design—The mold is constructed as an open-ended box structure which contains an inner lining fabricated from a copper alloy which serves as the interface with the steel being cast and provides the desired shape to the cast section. The liner is rigidly connected to an outer steel supporting structure. There are small water passages between the inner liner and supporting structure for the mold cooling water which absorbs heat from the solidifying steel in contact with the liner.

There are two types of mold designs; tubular molds and plate molds. Tubular molds conventionally consist of a one-piece copper lining which usually has relatively thin walls and is restricted to smaller billet and bloom casters. Plate molds consist of a 4-piece copper lining attached to steel plates. In some plate mold designs opposite pair of plates can be adjusted in position to provide different section sizes. For example, slab width can be changed by positioning the narrow-face plates, and the slab thickness changed by altering the size of the narrow-face plates. The plate mold is inherently more adaptable than the fixed-configuration tubular mold. In addition to permitting size changes, changes can also be made to the mold taper (to com-

pensate for different shrinkage characteristics of different steel grades) as well as ease of fabrication and reconditioning.

Although the material of construction of the inner lining is usually a high purity cold-rolled copper, copper with small amounts of silver is commonly used to obtain increased elevated-temperature strength. The working surface of the liner is often plated with chromium or nickel to provide a harder working surface and also to avoid copper pickup on the surface of the cast strand.

During the casting operation, the copper liner is subjected to distortion (a change in the internal dimensions of the mold). It is caused mainly by mold wear and mold deformation due to thermal and mechanical strains. For example, one type of distortion produces a reverse taper, caused by mold wear at the exit end of the mold, which can adversely affect product quality. Deformation due to thermal strains is particularly important. Two common causes are thermal expansion due to non-uniform heating of the mold wall, and restraint of the free expansion of the copper liner by the mold-support system. The resulting thermal strains and stresses may be sufficient to cause yielding and permanent deformation, especially at the meniscus level where the yield strength of the copper is reduced because the highest temperatures in the mold are encountered at this position.

Heat Transfer Conditions in the Mold—Heat transfer in the mold is critical and complex. The predominant transverse heat transfer can be considered as a flow of heat energy through a series of thermal resistances, from the high-temperature source of liquid steel core in the mold to the sink of cooling water of the mold-cooling system. It includes:

(1) Heat transfer in the solidifying casting.
(2) Heat transfer from steel shell surface (skin) to inner copper-lining surface.
(3) Heat transfer through copper lining.
(4) Heat transfer from outer copper-lining surface to mold-cooling water.

Heat transfer in the solidifying casting occurs in a complex way since the heat to be extracted originates from enthalpy changes in the steel strand both from temperature decreases and phase changes. The former is referred to as sensible heat change and the latter as latent heat. Moreover, phase changes involve not only the changes between solid phases, such as the δ-to-γ phase transformation, but also the conditions produced by the solidification of an alloy. For example, a "mushy zone" exists between the liquidus and solidus temperatures which depends on the carbon content of the steel. In addition, the thermal resistance increases as the shell thickness increases from the meniscus to the bottom of the mold. Heat transfer in this region is by conduction.

Heat transfer from the solidified steel surface to the mold is probably the least understood and most complex of the heat-transfer steps and involves mainly two mechanisms of heat transfer; conduction and radiation. The salient feature of this heat-transfer step is the shrinkage of the solidifying steel (which is a function of steel grade and caster operating conditions) and the resulting tendency for an air gap to form between the steel shell and the mold surface.

The formation of the air gap is complex and may vary both in the transverse and longitudinal direction. Thus, it has a variable effect on the heat-transfer mechanism and the magnitude of heat flux. For example, as the air gap is formed, the heat transfer proceeds mainly from conduction to radiation with a resulting decrease in heat flux. In general, this heat-transfer step represents the largest thermal resistance of all of the four steps, especially with respect to heat transfer through the copper lining and from the latter to the mold-cooling water.

This heat-transfer step is the controlling step in the mold; the entire pattern of heat removal in the mold is dependent on the dynamics of gap formation. In general, gap width tends to increase with increasing distance from the meniscus as the steel shell solidifies and shrinks away from the mold surface. In addition, as the shell thickness increases with distance from the meniscus, it tends to withstand the opposing bulging effect of the ferrostatic pressure to reduce the gap.

Heat transfer at the copper inner surface is further complicated by the effects of mold lubrication. Mold lubricants can be divided into two categories; oil lubricants and mold fluxes or powders. Oil lubricants (used with open pouring) tend to wet the copper mold and permit greater heat transfer at the upper part of the mold. Fluxes (used with refractory shroud pouring) also result in greater heat transfer.

Another factor influencing heat transfer at this mold surface is the mold taper, which tends to increase heat transfer because it opposes the effect of gap formation.

In general, the local heat flux down the mold length reaches a maximum value at or just below the liquid steel meniscus, and decreases down the mold length. The average heat flux for the whole mold increases with increasing casting speed.

Heat transfer through the copper lining is by conduction. It is dependent on the thermal conductivity of the copper and its thickness; the greater the thickness, the higher the hot-face temperature of the copper lining.

Heat transfer to the cooling water from the liner outer surface is accomplished by forced convection. Although the bulk temperature of the cooling water, typically about 40°C (90°F), is usually below its saturation temperature at a given water pressure, boiling is still possible at local regions at the mold outer surface if the local temperature of this surface is sufficiently high for water vapor bubbles to nucleate at the surface, pass to the colder bulk cooling water, and condense. This effect increases heat transfer. Nucleate boiling can result in cycling of the temperature field through the copper mold (both at the cold face and the hot face) and can result in deleterious product quality. Boiling can be suppressed by increasing the water velocity in the cooling system or by raising the water pressure. Incipient boiling is more likely in billet molds, which have higher cold-face temperatures than slab molds because of their thinner wall thicknesses. Typical values for cold-face temperature are in the range of 150°C (302°F) for billet molds and 100°C (212°F) for slab molds.

Cooling Water System—Control of heat transfer in the mold is accomplished by a forced-convection cooling-water system, which must be designed to accommodate the high heat-transfer rates that result from the solidification process. The primary functions of the system are to provide:

(1) The proper volume of water for heat extraction at the required water temperature, pressure and quality.

(2) The proper flow velocity of water uniformly through the passages around the perimeter of the mold liner.

In general, the cooling water enters at the mold bottom, passes vertically through a series of parallel water channels located between the outer mold wall and a steel containment jacket, and exits at the top of the mold. Typically, a pressurized recirculating closed-loop system is employed.

The cooling-water system is designed so that the rate of water flow is sufficient to absorb the heat from the strand without an excessive increase in bulk water temperature. A large increase in temperature could result in a decrease in heat-transfer effectiveness and higher mold temperatures. For this same reason, the inlet-water temperature to the mold should also not be excessive; a proper mold water pressure is also required. For example, as discussed previously, higher water pressures tend to suppress boiling but excessively high pressures may cause mechanical mold deformation.

Water quality is an important factor with regard to scale deposition on the mold liner. Scale deposition can be a serious problem because it causes an additional thermal resistance at the mold-cooling water interface that increases the mold-wall temperature leading to adverse effects such as vapor generation and a reduction in strength of the copper liner. The type and amount of scale formed is mainly dependent on the temperature and velocity of the cooling water, the cold-face temperature of the mold, and the type of water treatment.

To achieve the proper flow velocity, the cooling system is designed such that the velocity is high enough to produce an effective heat-transfer coefficient at the mold-cooling water interface. Too low a flow velocity will produce a higher thermal resistance at this interface, which may lead to boiling and its adverse effects. In general, the higher the cooling-water velocity, the lower the mold temperature. The cooling system should also be designed to maintain the required flow velocity distribution uniformly around the mold and to maximize the area of the faces that are directly water-cooled. Uniform flow distribution can be achieved by the proper geometrical design of the water passages with the use of headers and baffle plates.

Monitoring the operating parameters of the mold cooling system provides an assessment of the casting process. For example, with a constant cooling-water flow rate, the heat removed from a mold face will be directly related to the difference between the inlet and outlet water temperature, ΔT. Thus an excessively large ΔT may indicate an abnormally low flow rate for one or more mold faces, whereas an excessively small ΔT may indicate an abnormally large scale buildup for one or more mold faces. An unequal ΔT for opposite faces may result from an unsymmetrical pouring stream mold distortion, or from strand misalignment.

Mold Oscillation—To reduce mold-strand adhesion and the risk of breakouts, (in which liquid steel breaks through the thin solidified shell either in or below the mold) the mold is oscillated and lubricated. Oscillation may be accomplished by motor-driven cams which support and reciprocate the mold; other mechanical means, such as levers and cranks and the use of hydraulic actuation, are also employed.

The purpose of mold oscillation is to prevent high friction and sticking of the casting skin to the working face of the mold which may cause tearing of the skin. Sticking can occur as the strand moves down the mold and tensile forces are developed in the solidifying skin due to friction at the mold-strand interface (which can be further enhanced by increasing ferrostatic pressure). If these tensile forces exceed the cohesive forces of the solidifying steel, the skin will tear and a breakout may occur. Sticking can be exacerbated by local rough areas in the mold such as gouges.

Mold oscillating cycles are many and varied with respect to frequency, amplitude and form. Many oscillation systems are designed so that the cycle can be changed when different section sizes on steel grades are cast on the same machine. However, there is one feature that has been adopted, almost without exception, which applies a negative strip to the solidifying shell. Negative strip is obtained by designing the "down stroke" of the cycle such that the mold moves faster than the withdrawal speed of the section being cast. Under these conditions, compressive stresses are developed in the solidifying shell which tend to seal surface fissures and porosity and thus enhance the strength of the shell. During the "up stroke" portion of the cycle, the mold is very rapidly returned to the starting position and the cycle then repeated. Thus the shape of the oscillating cycle is non-symmetrical with respect to time.

Mold Lubricants—Mold oscillation alone is insufficient to prevent skin ruptures and the the use of mold lubricants is essential. Mold lubricants can be divided into two groups: (1) liquids; and (2) solids.

Liquid oil lubricants include those of mineral, vegetable, animal and synthetic origin. Rapeseed oil was commonly used but is being replaced by semi-refined vegetable oils. Because of the casting environment, the oil lubricants require high-temperature properties, such as a high flash point, so that they can effectively lubricate the mold surface in contact with the steel. The oil is continually injected through a series of small holes or slots in the upper portion of the mold above the steel meniscus to form a thin continuous film over the surface of the mold walls. Oils are principally used in billet or bloom machines casting silicon-killed steels.

Solid lubricants (mold fluxes or mold powders) are widely used with submerged refractory tube shrouds in casting aluminum-killed steels on slab and bloom casters. These powders serve not only as lubricants but also provide other functions:

(1) Enhanced heat transfer at the strand-mold interface.
(2) Protection of the liquid metal surface in the mold from reoxidation by surrounding air.
(3) Thermal insulation of the liquid metal surface to prevent unwanted solidification, particularly at the wall-meniscus interface and at the submerged shroud.
(4) Absorb non-metallic inclusions which float to the liquid surface.

Mold powder is added to the surface of the liquid steel shortly after the start of casting either manually by rakes or by mechanical feeders. Powder in contact with the liquid steel melts forming a liquid slag which then infiltrates between the mold wall and surface of the solidifying steel. Additional powder is added continually to replace that removed on the surface of the cast section. Lubrication by mold powders is a complex phenomenon and depends not only on flux properties such as viscosity, but also on the operating conditions, such as steel grade, casting speed, and oscillation condition.

In addition to viscosity, which is dependent on the silica and alumina content of the powder, the melting point or crystallization temperature characteristics of the powder are also important. Very "fluid" slags with low viscosities and low crystallization temperatures tend to provide the most effective heat transfer in the mold.

Additional characteristics affecting the other functional requirements of powders include: a minimal iron oxide content, for example, to protect the liquid steel surface from reoxidation; and a low density which, together with graphitic carbon to retard sintering, fusibility and melting, enhances the thermal insulation capabilities.

Mold powders consist of a mixture of materials of which SiO_2-CaO-Al_2O_3-Na_2O-CaF_2 is the basic component with varying amounts of carbon and other compounds. They can be broadly divided into fly-ash based powders; synthetic powders; and prefused, fritted or granulated powders.

SECONDARY COOLING, STRAND CONTAINMENT AND WITHDRAWAL

In modern slab casting machines, secondary cooling,

strand containment and withdrawal form a closely integrated and interlocked system which also includes strand bending and straightening. In the older designs of billet and bloom casting machines, there was a greater functional as well as physical separation of the components of this part of the casting operation. For the purposes of this discussion the concepts employed in the design and operation of modern slab casting machines will be considered.

Secondary cooling and the containment/withdrawal system extends from the bottom of the mold through complete solidification of the strand to the cut-off operations. The system is designed to produce a final cast section which has the proper shape, and internal and surface quality. To accomplish these results the solidifying section leaving the mold is cooled in a series of spray zones and contained and withdrawn by a series of roll assemblies until the solidified cast section reaches the cut-off machine and horizontal runout table.

Secondary Cooling—The secondary cooling system is normally divided into a series of zones to control the cooling rate as the strand progresses through the machine. This system, conventionally, consists of water sprays which are directed at the strand surface through openings between the containment rolls. Recently, air-water "mist" sprays (discussed later) have been employed which provide more uniform cooling.

Heat Transfer in Secondary Cooling—The main heat-transfer functions of the spray-water system are to provide:

- The proper amount of water to obtain complete solidification under the contraints of the casting operation, i.e., steel grade, casting speed, etc.
- The capability to regulate the thermal conditions of the strand from below the mold to the cut-off operation, i.e., strand surface temperature and thermal gradients in the strand.
- Auxiliary functions such as cooling of the containment rolls.

It is necessary to control both the temperature levels and thermal gradients in the strand to avoid the occurrence of surface and internal defects such as improper shape and cracks. At high temperature, the strength properties of the steel shell play a critical role in the ability of the shell to withstand the external and internal forces that are imposed by the casting operation. The primary forces are those exerted by the ferrostatic pressure of the liquid core and the traction of the withdrawal operation. In particular, the ductility of steel close to the solidus temperature is low and the shell is susceptible to crack formation. It is important to control temperature gradients because thermal strains can be caused which exceed the strength of the steel resulting in cracks. Excessive thermal strains result from changes in the heat-extraction rate by either over-or-under-cooling. The latter conditions can occur by reheating, which is produced when spray cooling is terminated improperly and the strand reheats by heat transfer from the interior with an increase in temperature before decaying by radiation heat transfer to the environment. Under these conditions, excessive strains and cracks can result. This effect can be reduced by extending and varying the water-spray cooling opera-

tion to provide a smooth transition with the radiation-cooling area.

Thus, in the design of a secondary cooling system, the thermal conditions along the strand must be established which satisfy the product integrity and quality. For example, the surface temperatures along the strand are specified. They are generally in the range of 1200° to 700°C (2190° to 1290°F). Based on this information the cooling rates along the strand are determined from heat-transfer equations. Important parameters in these calculations include the convection heat transfer coefficient of the water sprays and the water flux (the amount of water per unit area of surface contact). The type of spray nozzle, nozzle position with respect to the strand surface, number of nozzles and water pressure are selected to provide the required water flux and distribution throughout the secondary cooling sector. Multiple nozzles are typically used at each level along the strand which have an overlapping pattern.

Generally a series of cooling zones is established along the strand, each of which has the same nozzle configurations and heat-transfer characteristics. Since the required cooling rates decrease along the length of the strand, its water flux in successive zones decreases.

During operation, changes in the water flux are made to compensate for changes in casting conditions such as casting speed, strand surface temperature, cooling-water temperature and steel grade.

The spray-water system is typically a recirculating system.

Strand Containment—The strand is contained by a series of retaining rolls which extend across its two opposite faces of the cast sections in a horizontal direction: edge rolls may also be positioned across the other pair of faces in a direction perpendicular to the casting direction to further enhance containment. The basic functions of the mechanical strand containment and withdrawal equipment, which forms an integral part of its secondary cooling system, are: (1) to support and guide the strand from the mold exit to the cut-off operations; and (2) to drive the strand at a controlled speed through the caster. In both of these functions, the final objective is to minimize the mechanical stress and strains incurred during the process.

For illustrative purposes, a casting machine design which consists of a vertical discharge from the mold to horizontal delivery prior to the cut-off operations is discussed. In this typical case there is a series of rolls or guides arranged vertically below the mold, followed by a series of rolls arranged in a curve (which provides a transition to the horizontal) and a series of rolls in a horizontal plane before the cut-off equipment. Each series of rolls may be segmented and contain different diameter rolls and roll spacings to meet the conditions existing at that location.

Strand support involves the restraint of the solidifying steel shape which consists of a solid steel shell with a liquid core. The ferrostatic pressure, created by the height of liquid steel present, tends to bulge the steel especially in the upper levels just below the mold where the solidified shell thickness is small (Figure 21-11). Bulging at this location would not only cause product defects such as internal cracks but also cause a skin rupture and a breakout. Bulging is controlled by an

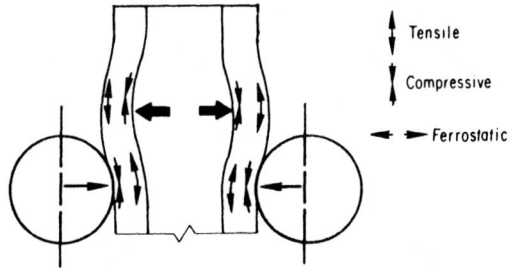

FIG. 21—11. Stresses in the solidifying skin due to ferrostatic pressure.

appropriate roll spacing which, in general, is closest just below the mold and progressively increases in the lower levels of the machine as the skin thickness increases. All four faces of the strand are usually supported below the mold with only two faces supported at the lower levels. In addition to ferrostatic pressure and skin thickness, roll spacing is also based on strand surface temperature and the grade of steel cast.

Strand Bending and Straightening—In addition to contaning the strand, the series of rolls which guide the strand through a prescribed arc from the vertical to the horizontal plane must be strong enough to withstand the bending reaction forces. During bending, the outer radius of the solid shell is placed in tension and the inner radius in compression. The resulting strain, which is a function of the radius of the arc and the strength of the particular grade of steel being cast, can be critical; excessive strain in the outer radius will result in metal failure and surface defects (cracks). To minimize the occurance of surface defects but, at the same time, maintain a minimum effective arc radius, triple-point bending has recently been introduced (i.e., three arcs, with progressively smaller radii).

A multi-roll straightner is installed following the completion of bending which, as the name implies, straightens the strand and completes the transition from the vertical to horizontal phase. During straightening the strand is "unbent" which reverses the tension and compression forces in the horizontal faces of the strand.

Strand Withdrawal—The strand is drawn through the different parts of the casting machine by drive rolls which can be located in the vertical, curved and horizontal roll sections. This multiple drive-roll system is designed, wherever possible, to produce compression forces in the surface of the strand to enhance the surface quality. Thus, the objective is to "push" the strand through the casting machine, as opposed to "pulling" the strand with the attendent tensile stresses which tend to produce surface defects. In addition, the use of multiple sets of drive rolls distributes the required traction force along the length of the strand and consequently reduces the deleterious effects of tensile forces. The proper placing of drive rolls can also reduce adverse bending and straightening strains by exerting an offsetting compression force, i.e., by placing drive rolls before a set of bending rolls. In all cases, the pressure exerted by the drive rolls to grip the strand must not be excessive; excessive pressure will deform the shape of the section being cast.

Following straightening, the strand is conveyed on roller tables to a cut off machine where the section is cut to the desired length. There are two types of cut-off machines: oxygen torches and mechanical shears. Oxygen torches are employed for large sections such as slabs and blooms. Billets are either cut by torches or shears. The cast product is then either grouped and transported directly to the finishing mills or, in the case of billets, to cooling beds which are predominantly of the walking beam type to maintain product straightness.

SECTION 3

PRODUCTIVITY, PRODUCT QUALITY AND AUTOMATION

In the relatively short time span since the commercial application of continuous casting in the early 1960's, there have been a wide variety of new process developments directed at improving productivity and product quality. These developments include new machine design concepts, metallurgical practices, and the application of process control and automation by computer systems. The main driving force behind these developments has been the recognition that substantial yield and energy savings are possible which have a dramatic effect on operating costs. Through these developments major quality improvements have been obtained such that the product is fully equivalent to and exceeds that of ingot steel. Today, essentially all grades of steel, including the highest qualities for critical applications, can be efficiently produced by continuous casting.

PRODUCTIVITY

Productivity improvements have been directed at decreasing the caster downtime and thus increasing the time that the machine is actually casting (utilization time) while maintaining the ability to produce the variety of product sizes and steel grades required by the finishing mills and ultimately by the customer. There are five major factors which contribute to downtime that have been addressed: (1) machine set-up time following the completion of a cast; (2) mold changing for casting different section sizes; (3) casting machine or strand stoppage because of failures such as strand breakouts, tundish nozzles blocked by frozen metal or inclusion build up, and uncontrolled flow of metal from the ladle (e.g., a running stopper); (4) out-of-specification heat composition and temperature and (5) machine maintenance. In addition to improved steelmaking control practices and techniques, the influence of these factors has been reduced by the development of new operating concepts and equipment designs. The major changes in operating concepts

FIG. 21—12. Slab slitting with two independent torch carriages.

include:

- Sequence casting to reduce machine set-up time.
- Slab slitting to reduce the frequency of mold changes and to reduce mold inventory.
- Variable-width adjustable molds to reduce mold changing time.
- Divided or split molds to reduce mold changing time and mold inventory and to increase casting rate (tons per hour per strand).
- Top-fed dummy bar to reduce set-up time.
- Hot charging and direct rolling.

Sequence Casting—Casting machine set-up, after the completion of a cast, is time consuming since it involves feeding the dummy bar through the entire length of the casting machine into the mold cavity and packing the dummy bar head to prevent leakage between the mold wall and head. Sequence casting was developed to reduce the frequency of setting the dummy bar by casting a series of heats in succession without interrupting the casting process.

This practice has been widely adopted and it is not uncommon in slab casting for a series of 40 heats to be cast successively representing several thousand tons; strings of several hundreds of heats, tens of thousands of tons have been cast. However, this practice also demands precise heat scheduling, high-machine reliability and the ability to rapidly change ladles (within one or two minutes), tundishes and refractory tube shrouds between the ladle and tundish and between the tundish and mold. Special purpose equipment has been designed to provide this capability. One example is the use of rotating ladle turrets which can have a single rigid slewing arm carrying two ladles, or two individual slewing arms, carry one ladle each. In addition, some designs include a ladle lifting mechanism and weighing equipment. Another example is rapid-change tundish cars which also have lifting devices to facilitate shroud changing.

Slab Slitting—One problem experienced in casting a long series of heats successively is that of scheduling the different slab sizes required by the finishing mills for different customer applications. Rather than interrupting a string of heats to change the mold size, which in itself represents a loss of casting time, a practice has evolved in which a small number of "master" slab sizes are cast with the slab product being slit longitudinally in a separate operation using mechanized oxy-natural gas torches (Figure 21-12). At one facility, for example, only two master slab sizes are produced: 2490 or 2640 mm wide by 240 mm thick (98 or 104 in. wide by 9½ in. thick) which are slit into multiple slab widths from 620 to 1880 mm (24½ to 74 in.). This facility has a capacity of 155,000 metric tons (170,000 net tons) per month. Sophisticated computer-assisted programs have been developed to minimize both the potential yield loss and inventory, as well as meet the scheduling requirements.

Adjustable Mold Width—To minimize both the time required to change a mold as well as the mold inventory, stepless-width adjustable molds were first developed for slab casting which could be adjusted without the mold being removed from the casting machine. The adjustment could be made either manually, electro-mechanically or hydraulically while the previously cast slab was being removed from the machine (Figure 21-13). More recently, as an alternative to slab slitting, the slab width can be changed during the actual casting operation. In one design, the mold taper can be adjusted by using different gear ratios for moving the top and bottom of the narrow mold faces.

Divided Molds—Another development which increases the productivity of a casting machine is the use of mold inserts. For example, in a new slab caster (described in Section 6) a 305 mm (12-in.) permanent divider in the center of the mold permits the casting of two narrow slabs simultaneously on a single strand us-

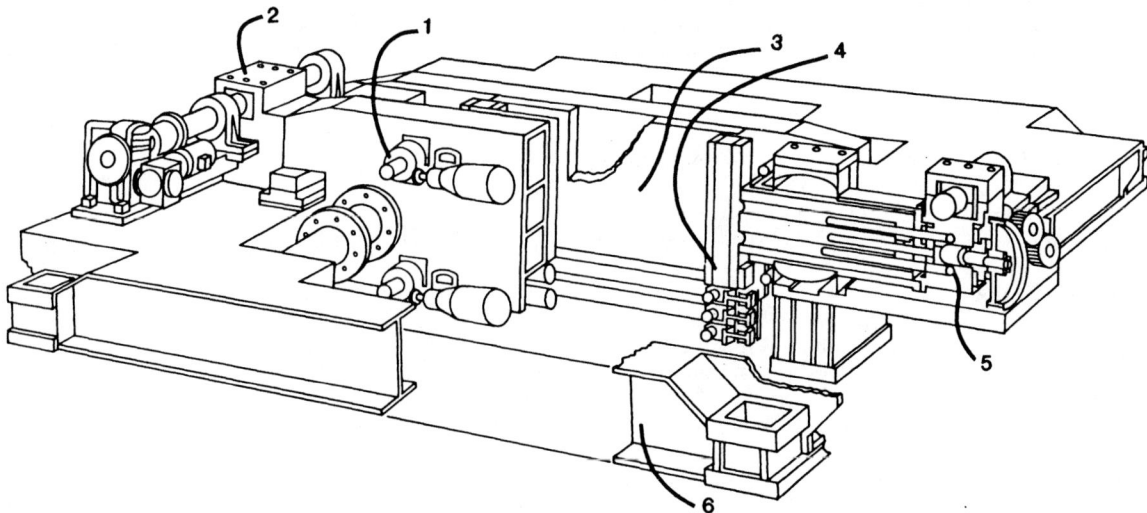

FIG. 21—13. Principal elements of variable width mold at Gary Works, U.S. Steel Corp.: 1—load-cell drive and jack; 2—taper mechanism control; 3—removable water-cooled copper broadface; 4—removable water-cooled copper end wall; 5—width adjusting mechanism; and 6—superstructure.

ing common containment and withdrawal units.

Top-fed Dummy Bar—Several systems have been developed for inserting the dummy bar into the mold. The shortest set-up time is obtained by inserting the dummy bar vertically through the top of the mold (as opposed to the older methods where it is inserted through the entire machine into the bottom of the mold). With the top-fed design, the dummy bar can be inserted before the previously cast strand has passed through the machine. In addition to this advantage, the top-fed dummy bar is also shorter than the more conventional bottom-fed type.

Hot Charging and Direct Rolling—Although the practice of hot charging a semi-finished shape into the reheating furnace of the finishing mills is not necessarily a productivity improvement attributable to continuous casting, it is, nevertheless, receiving wide attention because of the potential fuel savings.

In the early development of continuous casting the product was cooled to ambient temperature, inspected for defects and, if necessary, conditioned to remove the surface defects (a practice that is comparable to that used for many ingot-rolled, semi-finished products.) The product was then reheated and further processed in the finishing mills which involves an appreciable consumption of energy. By charging hot continuously cast product into the finishing mills, the sensible heat of the product is utilized with significant energy savings. This practice may avoid reheating altogether or require some intermediate reheating. However, it demands close coordination between the caster and finishing area. It also demands excellent surface quality because on-line hot inspection and conditioning of the cast material is not yet fully developed.

PRODUCT QUALITY

The quality of continuously cast steel is dependent on the steelmaking and casting practices employed. It is affected by the interaction of chemical and physical factors which must be closely controlled to obtain the full potential of the process.

Typical defects experienced in continuous casting have included:

- Surface: Deformed cross-section (including concavity and convexity)
  Cracks (longitudinal and transverse)
  Laps, scale and entrapped inclusions and slag
  Oscillation marks
- Sub-surface: Pinholes and blowholes
  Inclusions
  Cracks
- Internal: Cracks (central, diagonal and half-way)
  Porosity
  Inclusions
  Segregation

Crack formation is related to a wide variety of physical causes (discussed previously in Section 2). Techniques employed to eliminate or reduce the occurrence of external and internal cracks include:

(1) Surface cracks. Mold and secondary cooling, mold lubrication, mold coatings, mold wear control, machine alignment and casting speed.

(2) Internal cracks (and porosity). Machine type, machine alignment, electromagnetic stirring, in-line reductions, multi-point straightening, compression casting, liquid steel temperature and casting speed.

Laps and scabs are related to casting speed control and the integrity of the pouring stream between the tundish and mold. Oscillation marks are a function of the steel grade cast and the type of mold oscillation.

Pinholes and blowholes are controlled by deoxidation and tundish stream shrouding. Center line segregation has been minimized by low casting temperature, electromagnetic stirring and casting speed.

The frequency of inclusions, whether at the surface, sub-surface or in the interior of the cast sections, has been progressively reduced through improvements, for example, in steelmaking, deoxidation and shrouding practices, and equipment design. These improvements form an integtral part of a continuing effort to further upgrade the quality of cast products.

The most significant recent developments in improving product quality include: (1) the concept of "clean" steels; (2) the application of electromagnetic stirring; and (3) air-mist cooling to further reduce the incidence of surface cracks. One of the primary objectives, in the case of flat-rolled steels, is to produce a cast surface which does not require conditioning prior to further processing.

"Clean" Steels—The concept of "clean" steels is multifaceted and involves a number of techniques and practices designed to: minimize the number of inclusions formed during deoxidation; avoid inclusion formation by reoxidation; minimize inclusion pick-up from refractories; and facilitate the separation and removal of inclusions. Additional important objectives are the ability to desulfurize steel to low sulfur levels (including sulfide shape control) and oxide inclusion modification (primarily in aluminum-killed steels) to minimize inclusion deposition in the tundish nozzle. The full spectrum of practices (several of which have been discussed in the preceeding section) extend from furnace tapping through fluid flow control in the mold. They are:

- Slag-free tapping
- Secondary refining in the ladle (ladle metallurgy)
- Complete shrouding between the ladle, tundish and mold
- Tundish design
- Refractory selection for ladle and tundish
- Mold powders
- Electromagnetic stirring

Slag-free tapping techniques have been designed to retain steelmaking slag in the steelmaking vessel until the steel has been tapped into the ladle. The absence of slag in the ladle avoids a loss of deoxidation elements, and permits a closer control of deoxidation additions as well as avoiding a source of inclusions. It also provides a closer control in secondary refining operations such as desulfurization.

Secondary refining (see Chapter 19) serves a number of functions depending on the type of processes installed. They can include: temperature homogenization (by mixing) as well as the addition of heat; vacuum treatment (oxygen and hydrogen removal) to minimize the number of inclusions formed when solid deoxidizers are subsequently added; the addition of deoxidizers (either as bulk additions or via wire feeders); addition of oxide-inclusion modifiers, such as calcium wire; and the addition of desulfurizing agents and sulphide-shape modifiers.

Shrouding between the ladle, tundish and mold has been discussed previously. Refractory shrouds, tundish, nozzles and stopper heads have been designed to permit the introduction of inert gases to both prevent air infiltration as well as to minimize the deposition of aluminum-type inclusions on the walls of these components and subsequent blockage (Figure 21-14).

Large capacity, deep tundishes are used to increase the steel residence time to facilitate inclusion removal as well as to provide a "buffer" in sequence casting. Dams and weirs have also been incorporated to assist

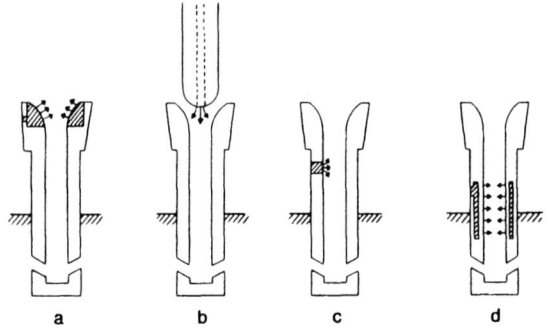

Fig. 21—14. Shrouding techniques using inert gas to prevent nozzle clogging by alumina clusters. A-porous ring, b-stopper, c-micropose, d-slit.

inclusion removal and to improve fluid-flow conditions to the nozzles.

High-quality refractories are used in the ladle and tundish to avoid inclusion pick-up from erosion, to extend the service life and, in the use of vacuum treatment, avoid chemical breakdown of the refractory constituents. High-alumina and basic refractories are being used to eliminate the pick-up of oxygen from the silica in fireclay refractories.

Mold powders, discussed previously, are used in casting aluminum-killed steel to absorb nonmetallic inclusions from steel in the mold, to enhance heat transfer to the mold wall, protect the liquid steel from reoxidation, and as a thermal insulation on the liquid steel surface.

Electromagnetic Stirring—The potential benefits to be derived from electromagnetic stirring of liquid steel during solidification are receiving wide attention. Improvements reported include:

- Internal quality (reduced segregation, cracking and porosity) through a preferred solidification structure.
- Sub-surface and internal cleanliness through a modified metal flow pattern.
- Reduced criticality of casting parameters (temperature and casting speed).
- Increased productivity through increased casting speeds.

Electromagnetic stirrers were initially installed on billet casters to reduce centerline segregation. This was achieved by a change in solidification structure: the area of the central equiaxed crystal zone was increased with a corresponding decrease in the area of the outer columnar crystal zone. Subsequently, the other improvements listed were recognized. In a recent survey, it was reported that over 100 stirrers are in operation, of which over 60 are on billet and bloom machines with approximately 40 on slab casters.

There are two basic types of stirrers, (rotary and linear) which can be installed either in or below the mold. In a rotary system installed in the mold of a billet caster, a rotating magnetic field produced by the coils imparts a circular motion to the liquid steel. The centrifugal force developed results in a sound skin, with the lighter phases (i.e., inclusions) moving towards the center. The central equiaxed zone is enlarged because

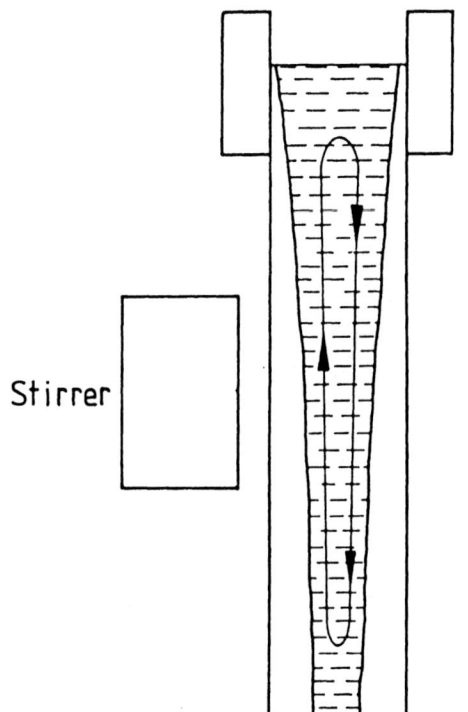

FIG. 21—15. Electromagnetic stirrer on a billet or bloom caster.

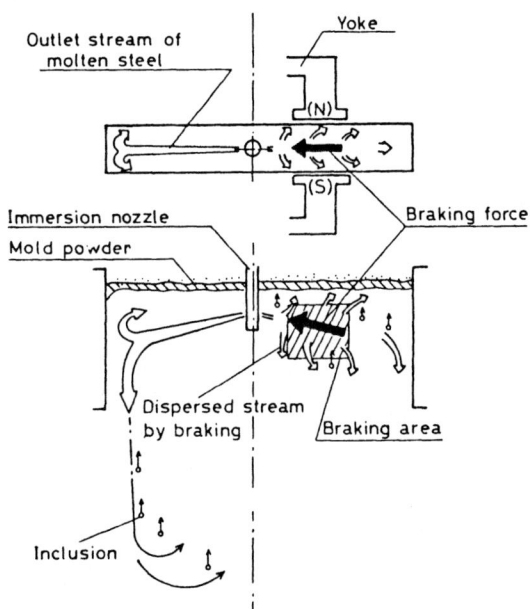

FIG. 21—16. Flow patterns of pouring streams in continuous casting molds. (Left) Conventional system (Right) Electromagnetic brake system.

the rotational flow promotes the fracturing of the tips of the columnar dendrites which serve as nucleii for equiaxed crystal formation in the central zone.

With the linear system, electromagnetic coils are installed along the side of a strand (below the mold) which produce a vertical circulation pattern (Figure

21-15). The increase in the central equiaxed crystal zone is obtained by a similar mechanism as that obtained by the rotary stirrer. Inclusions, which are normally concentrated in a band close to the upper surface in curved mold machines, are more uniformly distributed.

In another application of electromagnetic principles on a slab caster, a magnetic field is used as a brake to modify flow patterns within the mold in certain areas and to create flow patterns in others, with a subsequent improvement both in internal cleanliness and surface quality. This effect is achieved through the interaction of a moving steel stream in a stationary magnetic field. The metal stream moving through the magnetic field produces induced currents which, together with the stationary field, creates forces which brake the steel streams (Figure 21-16). In addition, steel between the streams and the poles of the electromagnet is accelerated which provides a strong stirring action. Thus, the velocity of a metal stream exiting the ports of a refractory tube shroud is reduced as well as the depth of penetration into the liquid crater. Under these conditions, inclusion concentrations are reduced and a more uniform shell growth occurs around the periphery of the mold which lessens the possibility of surface defects.

Mist Cooling—Conventional water sprays in the secondary cooling section can aggravate the occurrence of surface cracks initiated in the mold because the cooling is relatively uneven both in the longitudinal and transverse direction. Local overcooling can also occur, for example, in a slab casting machine from water trapped by a containment roll and the slab surface. A reduction in spray intensity by the use of smaller nozzles ("mild" spray cooling) is difficult to achieve because smaller nozzles are easily blocked.

Cooling by an air-water mist is being adopted with improved cooling characteristics being reported: heat transfer is improved, resulting in significantly lower water volumes (steam is constantly removed by the compressed air); and greater cooling uniformity which reduces both longitudinal temperature variation (minimal water retention at the containment rolls) as well as slab-edge-to-slab edge temperature variations. Air-water mist, by cooling the strand more uniformly, reduces the thermal stresses which can enlarge surface defects into cracks during casting.

The mist is produced by an atomized nozzle in which cooling water and compressed air are premixed; the mist is discharged through a slit outlet from a pressure chamber. Nozzle blockage is minimized because the discharge area is approximately 100 times larger than nozzles used for "mild" spray cooling.

AUTOMATION AND COMPUTER CONTROL

Automation and computer control of a continuous-caster operation is a rapidly evolving development which has a beneficial effect on productivity and product quality.

One type of computer control system consists of individual subsystems called first level controls. These subsystems may then be integrated by process data communication lines between these subsystems, between the subsystems and a supervisory computer (called a second level control), and to other plant operations. The supervisory computer provides mainly the function of coordinating all the data communication.

In this caster-operation control system, measurements are made of selected operating parameters to be controlled and regulated. Appropriate control changes are then effected, if required.

One main control function is automatic casting. This function may be effected by control and regulation of operation parameters, such as the control of liquid steel level in the tundish by regulating the steel flow from the ladle. Maintaining control of this level assists in the control of the liquid level in the mold, which may be performed by regulating the flow from the tundish (Figure 21-17). Maintaining control of the liquid level in the mold by means of thermocouples, radiation detectors, eddy current meters, etc, permits control of the caster speed by regulating the speed of the drive motors in the roll containment system. In addition, controlling the loads on these motors provides for the proper load distribution to minimize the occurrence of deleterious tension stresses in the strand skin while casting.

Another major function is the control of the cooling rate distribution in the secondary cooling system. This function is usually effected by controlling the strand

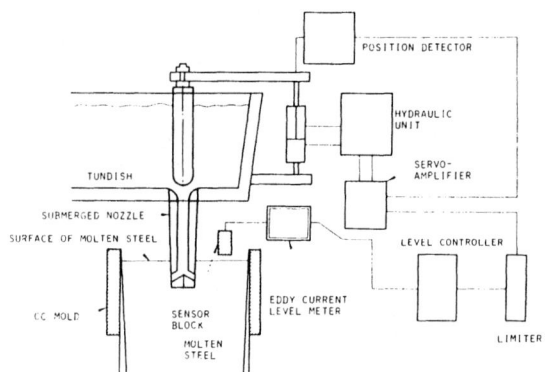

FIG. 21—17. Mold level control system.

surface temperature (measured by thermal radiation detectors or thermocouples) by regulating the water-spray flow distribution. Normally, both the casting speed and steel grade bring about changes in the cooling-control settings. Other functions that may be controlled are: mold powder additions to the liquid steel surface in the mold; cutting the cast steel shapes on the run-out tables to the proper length; and strand marking with the proper identification.

SECTION 4

EXAMPLES OF CONTINUOUS CASTING INSTALLATIONS

A MODERN SLAB CASTER

In 1983, Jones & Laughlin Steel Corporation started operating two single-strand slab casters at the Indiana Harbor Works. This casting facility has an annual capacity of 2,800,000 metric tons (3,100,000 net tons), making it the highest production caster in North America. Steel for the caster is made in a two-furnace, 260 metric ton (285 net-ton) heat size, BOF shop using desulfurized hot metal.

The liquid steel from the BOF is adjusted in composition and temperature at two duplicate treatment stations before it is sent to the caster. Equipment is provided for desulphurization and inclusion shape control using lime, lime/calcium silicon powder or a lime/magnesium premix. These powders are added to the steel through a replaceable refractory snorkel.

The caster facility consists of two independent single-strand machines No. 1 and No. 2, each supplied by its own ladle turret. The ladle turrets are located on 30-metre (98 ft) centerlines with a centerline distance between the two strands of 21 metres (68 ft 10 in.). Casting machine No. 1 (Figure 21-18) is designed to either cast twin-mold slabs from 710 to 1067 mm (28 to 42 in.) wide or single-mold slabs from 1015 to 1980 mm (40 to 78 in.) wide. The other machine, No. 2, is designed only for single-mold casting in the 1015 to 1980 mm (40 to 78 in.) width range. Slab thickness is 250 mm (10 in.).

Both the single and twin molds are equipped with a

remotely adjustable width changing feature that permits the slab width to be changed in 50-mm (2 in.) increments without stopping the cast (Figure 21-19).

The basic design data are summarized in Table 21—II.

The molds are equipped with a Cobalt 60 radioactive source-type mold level control system which, in conjunction with three plate throttling slide gates on the tundish, controls steel level in the mold to ±5 mm (0.2 in.) while keeping casting speed constant.

Casting machine No. 2 is equipped with a top-fed dummy-bar system consisting of a wind-up winch dummy-bar disconnect system and a dummy-bar handling car mounted on the casting floor. This design reduces machine turnaround time to 40 minutes.

The steel grades cast include low-carbon aluminum-killed steels, medium-carbon aluminum-killed steels, medium carbon silicon-killed steels, D&I tinplate, motor lamination steels and columbium-bearing high-strength low-alloy steels.

One of the objectives of the facility is to produce slabs of superior quality with a targeted goal of having over 95 percent of the slabs produced requiring no conditioning and the resultant ability to "hot charge" 45 percent of caster production to the hot strip mill reheat furnaces. The liquid steel stream from the ladle to the tundish is shrouded with argon and/or nitrogen to minimize oxidation of the stream. Immersion nozzles are used between the tundish and mold to fur-

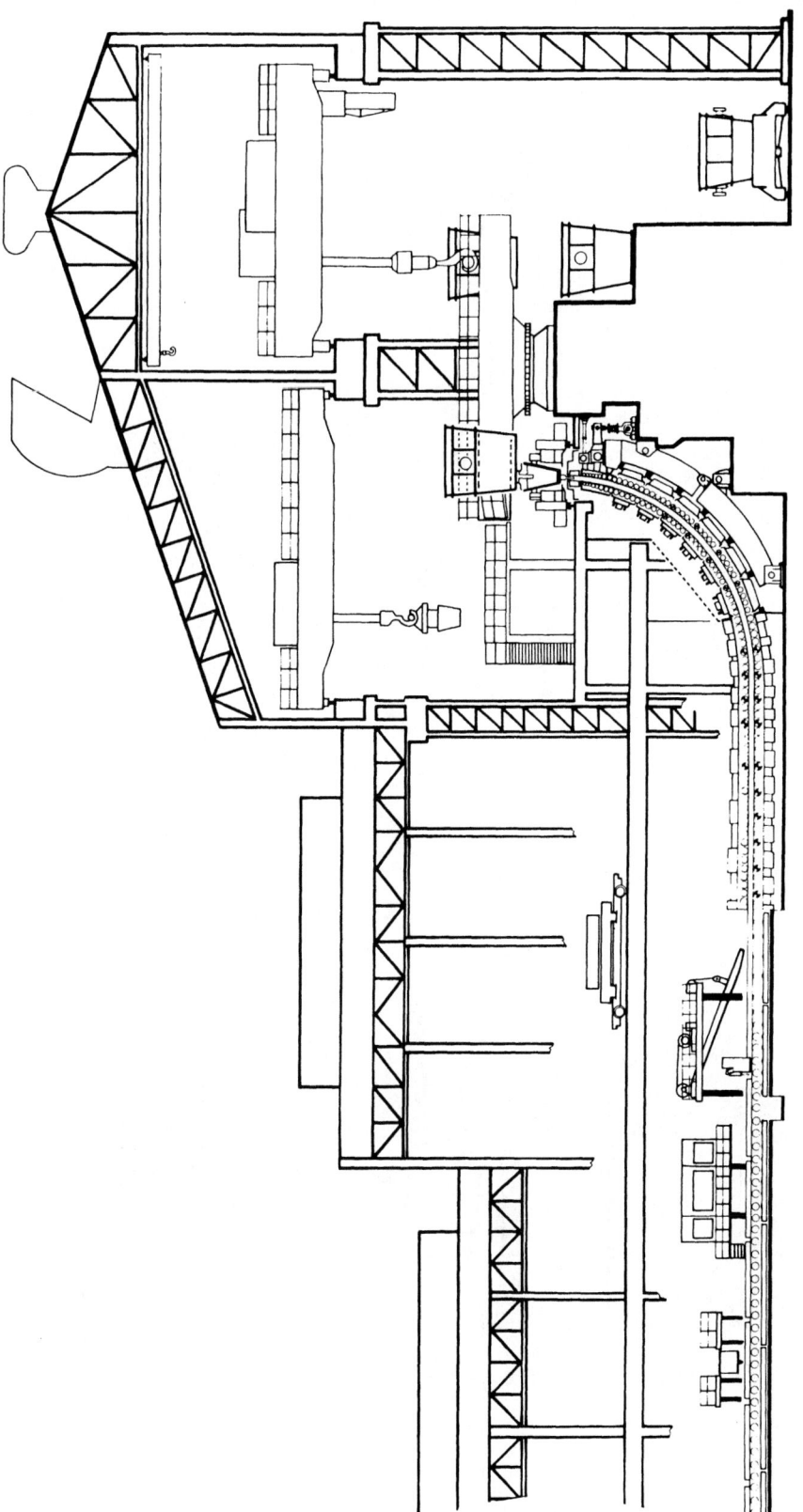

FIG. 21—18. Cross section of a Jones & Laughlin slab caster at Indiana Harbor Works.

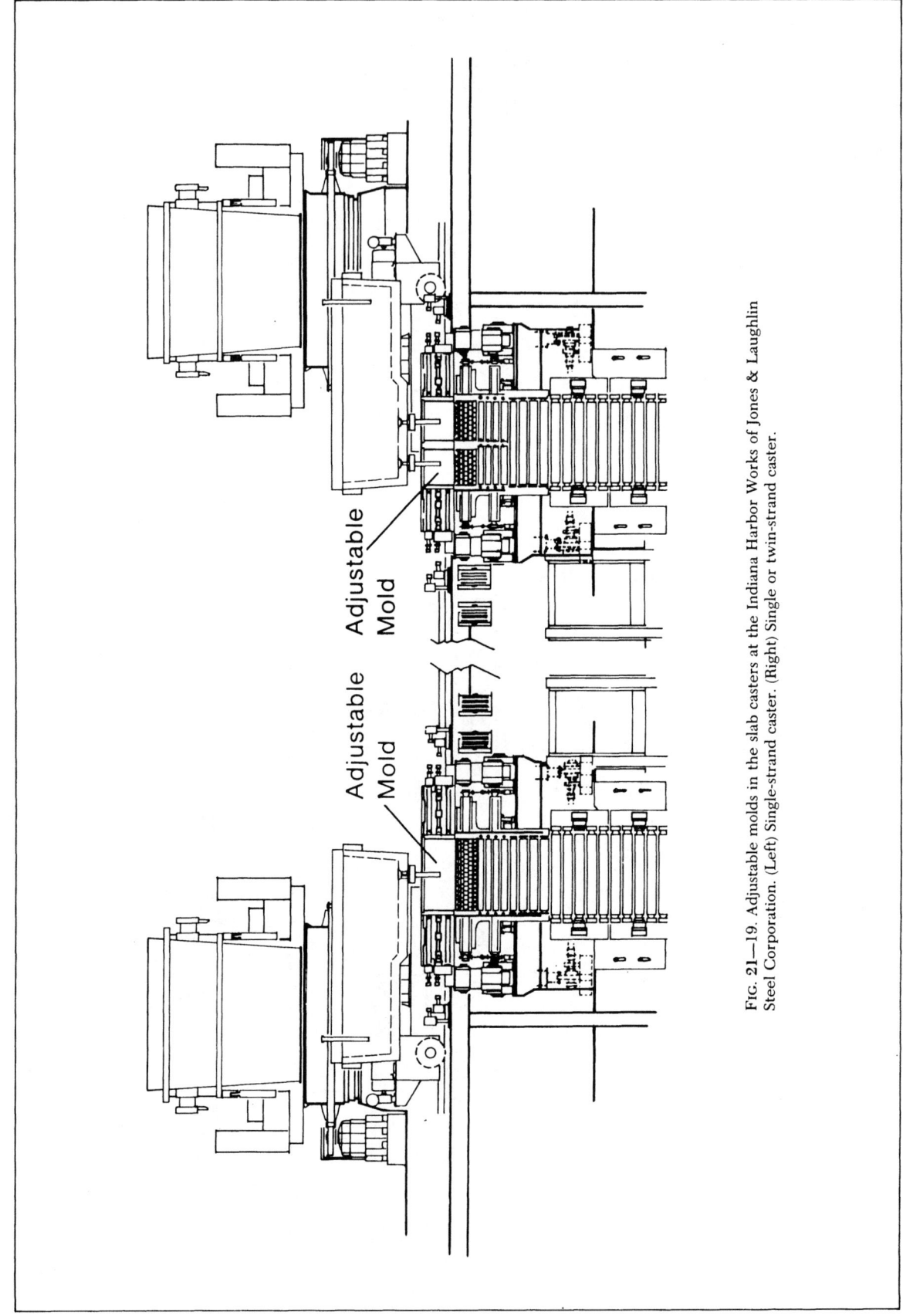

FIG. 21—19. Adjustable molds in the slab casters at the Indiana Harbor Works of Jones & Laughlin Steel Corporation. (Left) Single-strand caster. (Right) Single or twin-strand caster.

TABLE 21-II
Slab Casting Facility Characteristics

	Caster No. 1	Caster No. 2
Caster type	Curved Mold	Curved Mold
Number of strands	1 or 2	1
Casting radius, metres (ft)	12.2 (40)	12.2 (40)
Section sizes, mm (in.)	710 to 1067 wide x 250 (28 to 47 wide x 10)	1015 to 1980 wide x 250 (40 to 78 wide x 10)
Slab length, metres (in.)	5.5 to 11.3 (219 to 444)	5.5 to 11.3 (219 to 444)
Heat size, metric tons (net tons)	260 (285)	260 (285)
Tundish capacity, metric tons (net tons)	43 (47)	43 (47)
Ladle stream shrouding	Argon and nitrogen	Argon and nitrogen
Tundish stream shrouding:	Immersion nozzle	Immersion nozzle
Mold length, mm (in.)	900 (35.4)	900 (35.4)
Mold liner	Silver bearing copper with chrome plating	Silver bearing copper with chrome plating
Metallurgical length, metres (ft)	31.025 (101.8)	39.859 (130.8)
Casting speed (max.), metres/min (in./min)	1.4 (55)	1.8 (71)
Straightener rolls	32 pairs (14 driven rolls)	49 pairs (20 driven rolls)

Total mold cooling water, m³/s (gpm)	0.31 (4940)
Total internal machine cooling water, m³/s (gpm)	0.77 (12 200)
Total spray cooling water, m³/s (gpm)	0.64 (10 100)

ther protect the liquid steel from oxidation.

Grids and rolls in the first cooling zone support the slab upon its exiting from the mold. Narrow face width adjustment is accomplished by hydraulic motor-driven screws.

The roller containment section of each caster consists of six segments, each containing five roll pairs. Roll diameters range from 317 mm (12.5 in.) to 390 mm (15.3 in.). Roll diameters and spacing were designed to maintain a maximum slab bulging strain of 0.5 percent at maximum casting speed to avoid internal cracking. Roll gap sleds are installed in the dummy bars of both machines and provide a radio signal output to permit on-line monitoring of machine alignment during every dummy bar insertion. Segment change is carried out by a changer car running on a track paralleling the roller apron frame. It is equipped with a stop system capable of positioning and holding the car for segment removal.

The straightener/withdrawal unit provides the force required to insert the dummy bar, withdraw the slab at controlled speeds, and unbend the slab from the curved cast radius to the horizontal plane. Three-point unbending is used with radii of 12, 15 and 25 metres (40, 48.6 and 82 ft). The number of unbending points and radii were determined to minimize surface strain and avoid crack formation.

After leaving the straightener, the strands are cut by oxygen/gas torch cut-off machines into preset lengths of from 5.5 to 11.3 metres (219 to 444 in.). The slabs are then automatically weighed and marked with an identification number, destination code and conditioning code by a paint spray gun in accordance with data signals from a computer.

The casting process is controlled by two minicomputers, one a hot standby. These computers control the secondary spray cooling zones, slab tracking, on-line width change initiation, slab-length cutting optimization, monitoring of casting abnormalities and data logging.

A MODERN ROUNDS CASTER

Operation of a six-strand rounds caster was initiated at the Lorain Works of United States Steel Corporation in 1983, Table 21—III. Molten steel is provided by a 200 metric-ton (220 net-ton) BOF shop. The composition and temperature of the molten steel are adjusted in a capped argon bubbling (CAB) treatment facility.

The molten steel is delivered on a transfer car into the ladle handling bay (Figure 21-20). An overhead crane raises the ladle from the transfer car and places it on the ladle turret. The same handling equipment is used to return ladles to the BOF shop.

The ladle turret is used to position the steel ladle above the tundish and casting machine. By rotational movement, the ladle turret moves the ladle into position from its home point in the ladle handling bay into the casting bay. Exact positioning above the tundish is accomplished by vertical movement capabilities of the turret (Figure 21-21).

The casting bay contains the casting platform and casting machine. This bay is serviced by a floor-operated overhead crane used for preparing and maintaining the casting equipment. The tundish is mounted on a car which travels perpendicularly to the casting direction. The tundish is equipped with six stopper rods (one stopper rod for each strand), which are controlled by radiation level detectors in each mold for controlling the flow of steel into the mold.

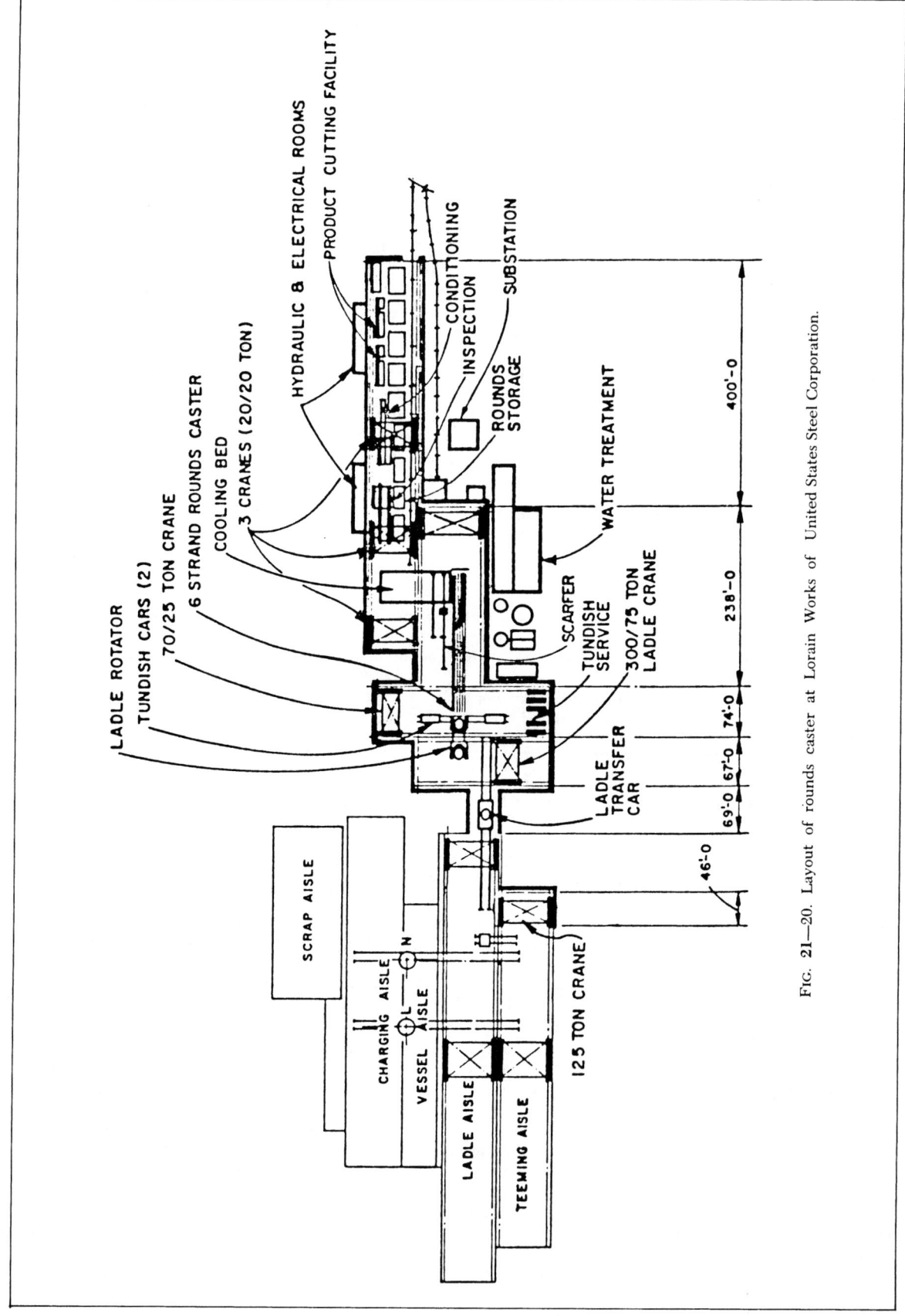

FIG. 21—20. Layout of rounds caster at Lorain Works of United States Steel Corporation.

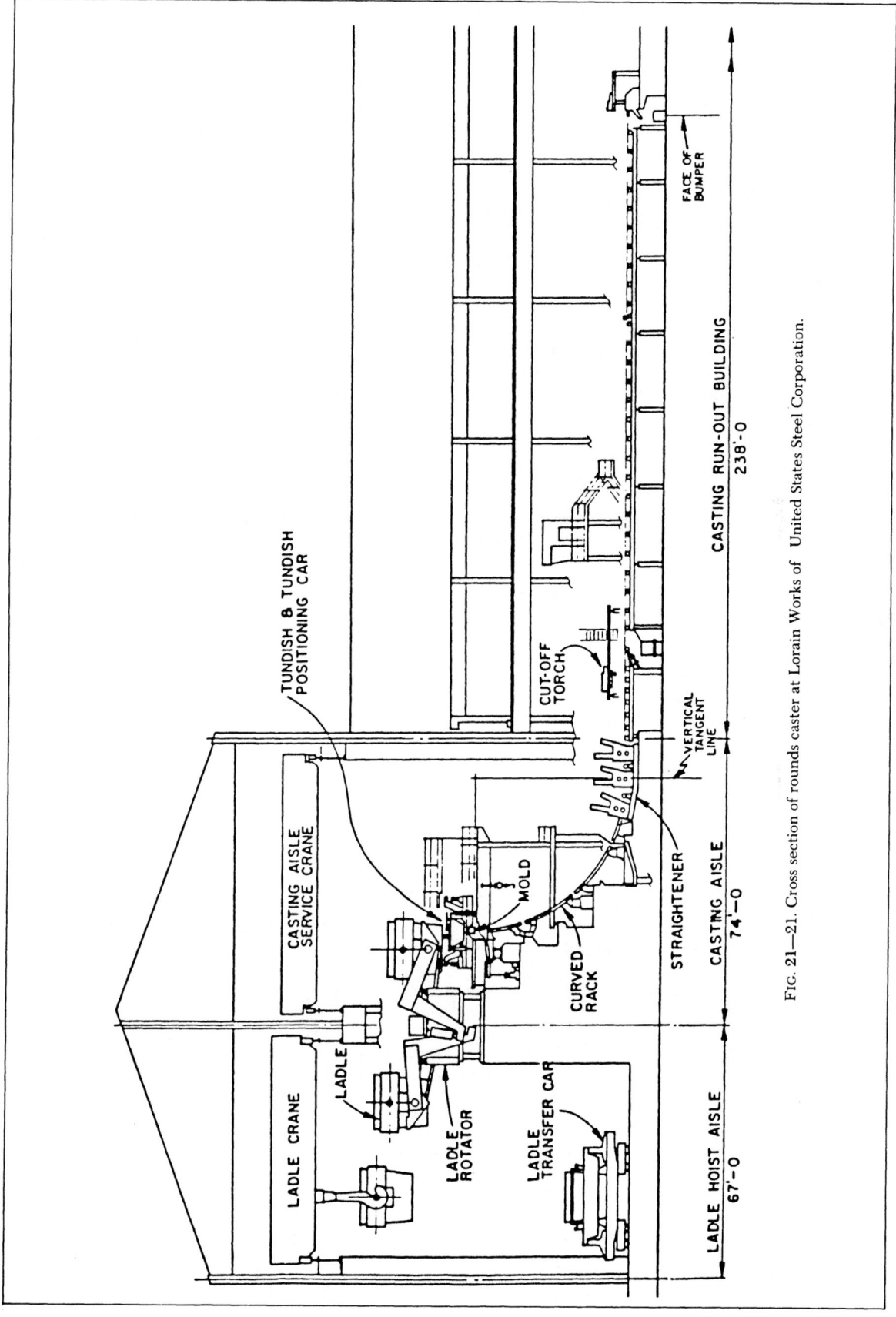

Fig. 21—21. Cross section of rounds caster at Lorain Works of United States Steel Corporation.

TABLE 21-III
Rounds Caster Design Specifications

Annual design capacity	500 000 metric tons (550 000 net tons)
Heat size	200 metric tons (220 net tons)
Number of strands	6
Products cast	round billets, 152, 178 and 235 mm. dia. (6, 7 and 9.25 in. dia.)
Product length	8.8 to 12.2 metres
	(29 to 40 feet)
Typical grades cast	N-80 Q and T J-55 — K-55 A-106 GRB
Casting radius	12.2 metres (40 ft)
Metallurgical length	19.8 metres (65 ft)
Casting time	80 minutes/ladle
Casting speed	1.2 to 2.9 metres/min (49 to 115 in./min)
Heats/day	9 to 12

A curved, water-cooled mold is used. Exiting the mold, the solidifying strand is conveyed through a spray chamber by containment rolls. Exhaust fans remove the steam created during strand cooling. When the strand reaches the withdrawal and straightening machine, the strand is completely solidified.

The runout bay is adjacent to the casting bay and contains runout tables, a dummy bar receiving device, torch cutting machines, stamping machines, and inspection and conditioning equipment. After straightening, the strand is cut into preset lengths by the torch cutting machines (Figure 21-22). The cut-to-length rounds are conveyed on roller tables up to a fixed stop. At that point, an identification number is stamped into the face of the rounds. The rounds are then cross transferred from the end stop area to the cooling beds.

An inspection area adjacent to the cooling beds is provided for surface quality and surface slag/inclusion inspection. Based on quality control requirements, rounds are transferred into the inspection area using a table roller system, inspected and transferred back into sequence on the cooling bed.

The final point before shipment is cutting the rounds into smaller pieces on a torch cutting machine and loading into rail cars for shipment. Overhead cranes are used to move product through the shipping area.

The main casting facility is controlled by a programmable logic controller (PLC) with the marking machines and the mold liquid metal level control monitored by PLC's communicating directly with the main PLC. The process control of the cooling system is a computer-directed microprocessor. Interactive graphic screen displays are used for operator control. All logic and control algorithms are capable of being performed at each control level for dynamic process updating. All of the process control processors communicate with a supervisory Level II minicomputer over a data highway which ties the casting plant into the steelmaking and planning computer. Process data storage and retrieval are included.

SECTION 5

ALTERNATIVE CONTINUOUS CASTING PROCESSES

The rapid adoption of vertical continuous casting by the steel industry, with the large improvements in processing yields and energy savings, generated further interest in the development of alternative continuous casting processes which would provide additional benefits. The objectives of these alternative processes are, in general, directed at improvements in specific areas, such as the reduction of capital costs, the casting of small section sizes and product quality. In most cases, they incorporate many of the design concepts used in vertical casting.

There are four major processes at various stages of development and application:
- Horizontal casting
- Rotary continuous casting
- Wheel-belt casting
- Strip casting

HORIZONTAL CASTING

Horizontal continuous casting is a new process in the steel industry with the first commercial facilities being installed in the late-1970's. However, its concept is not new: one of the earliest patents was granted to Sellers, in 1840, for the horizontal casting of lead tubing. It was first employed for the production of nonferrous metals and cast iron in the 1930's using graphite molds. Problems which had to be solved before the process could be used for casting steel involved the following areas: mold design and materials of construction; oscillation and section withdrawal; and the design and composition of the seal between the tundish and mold.

In 1966, Davy-Loewy Ltd., U.K., initiated a development program for casting steel and, in 1968, installed a pilot plant for producing 100-mm (3.9-in.) sections. One of the earliest production units, which started up in 1970, was designed and operated by General Motors Corp. in the U.S.A. A number of other facilities were installed in Europe, Japan and the USSR. The first commercial machine for casting stainless product in the U.S.A. was installed in 1983 by Armco in Baltimore. The Armco caster is a 2-strand machine for producing 100 to 150 mm (4 to 6-in.) rounds and squares from

FIG. 21—22. Cutting machines and runout tables of rounds caster.

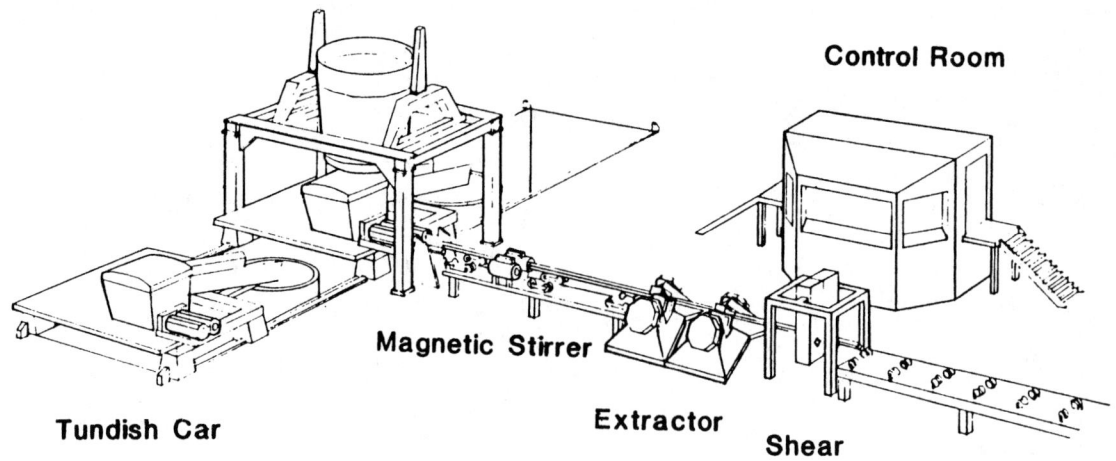

FIG. 21—23. Horizontal continuous casting machine.

heat sizes up to 41 metric tons (45 net tons).

Benefit of Process—The advantages of the process are reported to include:

- Lower capital cost compared with vertical casting.
- Low overall height, smaller space requirements and installation in existing buildings.
- Low ferrostatic pressure which minimizes strand bulging.
- Absence of stresses induced by strand bending and straightening which permits the casting of crack-sensitive grades.
- Absence of metal reoxidation between tundish and mold (especially important in casting small sections).
- Cost effectiveness in small-capacity plants producing ordinary and special steel from small heats.
- Small number of operating personnel.

The production capability is illustrated by the following example reported by Mannesmann Demag. The annual capacity of a 2-strand facility casting 120-mm (4.7-in.) billets from 25 metric-ton (27.5 net-ton) heats at a casting speed of 2.5 metres per min (98 in./min) is approximately 60 000 metric tons (66 000 net tons). The same capacity can be obtained with one strand, if the casting time is extended from 45 to 95 minutes by tundish heating.

Description of Process—The basic components of a typical casting machine are a tundish, mold, spray zone, strand withdrawal unit, cut-off unit and a runout table which are installed in a common, horizontal passline (Figure 21-23). The tundish and mold are stationary.

Steel is poured from a ladle into a tundish and flows horizontally into a stationary mold via a refractory connection on the side of the tundish. Partially solidified in the mold, the shell is withdrawn, in a pull-and-pause or a pull-and-push cyclic motion, through a secondary cooling zone where solidification is completed. The solidified section is then cut to length, cooled and transferred to the finishing mills. Two of the key elements in the system are the refractory connection between the tundish and mold, and the design and control of the

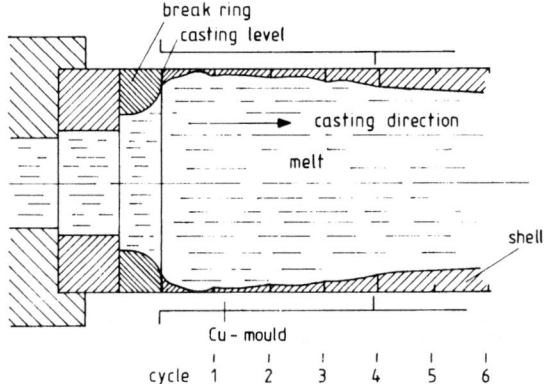

FIG. 21—24. Detail of junction between the tundish and mold in a horizontal caster showing the break ring and solidification sequence.

withdrawal unit.

The tundish is basically similar to that employed on a conventional vertical caster except for its opening at the bottom of one of the sidewalls. The mold is attached and sealed to the tundish. The junction between the tundish and mold consists of a refractory connection block (which is set in the tundish) and a high-quality refractory nozzle such as zirconia. In some installations, slide gates have been installed between the tundish and mold.

Mold—Mold components in one design include a break ring, which mates with the refractory nozzle, and water-cooled copper sections (Figure 21-24). The break ring, an essential part of horizontal casting, establishes the start of metal solidification. Metal freezes initially at the end of the break ring and along the mold wall. After a preset time (the pause portion of the withdrawal cycle) the solidified steel is withdrawn allowing fresh liquid metal to enter the cavity created between the end of the solidified shell and break ring. The cycle is then repeated. Thus, the break ring must be constructed of materials that facilitate the separation of the solidified metal without fracturing the shell, as well as possessing high thermal shock and abrasion resistance. Boron nitrides are generally used for this application.

There are two basic types of molds: a long mold which is made either entirely of copper or a combination of copper and graphite; and a short copper mold followed by multiple, movable, spring or hydraulically loaded after-coolers made of various combinations of materials including copper and graphite.

Withdrawal Unit—The function of the withdrawal unit is to provide an oscillation of the solidifying strand relative to the mold together with the withdrawal of the strand. It must be capable of accelerating and decelerating large masses of steel at speeds of several metres per minute with precision and reproducibly. Some units employ drive rolls, with the system controlled by a microprocessor; others use a hydraulically-operated system of clamps.

A variety of oscillating cycles are used which depend on the design of the machine, the particular steel grade and product being cast, and the operating conditions. One type of cycle can include strand withdrawal followed by a pause to allow new metal to solidify in the break ring-mold location (pull-and-pause). Another type involves strand withdrawal followed by a small reverse movement to place the shell under compression and consequently improve surface integrity (pull-and-push). Cycle selection, which includes frequency, anplitude and form, is important in achieving the optimum surface quality and avoiding breakouts.

Product—Typical product sizes cast commercially range from 40 to 250 mm (1.5 to 10 in.) both as rounds and squares; wires and rods from 3 to 25 mm have been cast. The potential exists for casting other shapes and sizes including slabs. Steel grades produced include structural, high carbon, ball bearing, high alloy, tool and stainless.

Steel cleanliness is reported to be superior to that of ingot steels because of the sealed connection between the tundish and mold, together with the ability to shroud the tundish. The macrostructure is similar to product cast on vertical casting machines. It is also reported that center segregation can be easily controlled by placing electomagnetic stirrers along the strand.

ROTARY CONTINUOUS CASTING

Rotary continuous casting of rounds was developed in the late 1960's to overcome some of the difficulties experienced at that time in producing high quality rounds for seamless tube in conventional vertical casters. The process is similar to conventional vertical casting except that the mold, guide rolls, and withdrawal system rotate together with the solidifying section. The rotating solidified cast section is then cut-to-length in the vertical position by a circular saw and subsequently tilted in the horizontal position in a basket and transferred to cooling beds. The secondary cooling sprays are stationary.

The processes was developed by Vallourec and Creusot-Loine in France and was first installed in two French plants: (Acieries et Usines Metallurgiques de Decazville and Societe des Acieries d'Anzin). In 1974, a 2-strand unit was placed in operation by the MacSteel

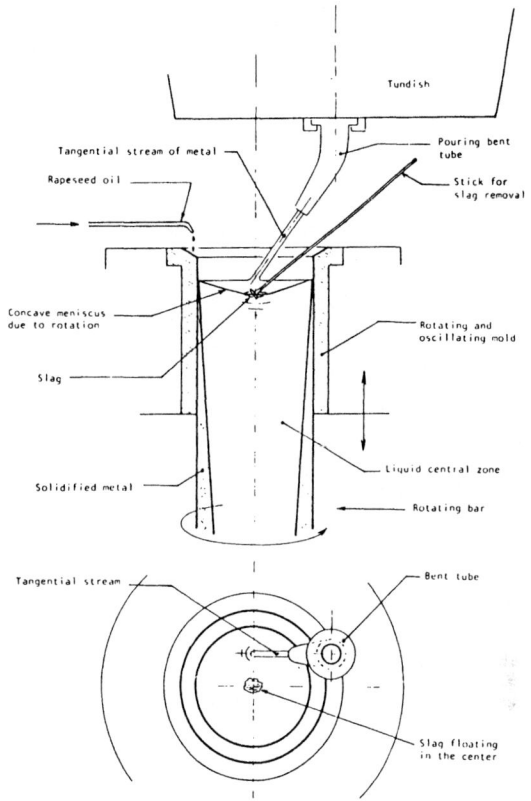

FIG. 21—25. Vallourec-Creusot Loire rotary continuous casting machine showing tangential entry of metal stream into mold and rod for inclusion removal.

Division of Michigan Seamless Tube Company for producing 105 to 180-mm (4 to 7¼ in.) rounds from 36 metric-ton (40 net-ton) heats. A 4-strand unit was also commissioned by NKK at its Keihin works in 1974 for the production of 120 to 240 mm (4.7 to 9.5 in.) rounds from 73 metric-ton (90 net-ton) heats.

Process Benefits—The simultaneous rotation of the strand and mold (30 to 120 rpm) is reported to have a beneficial effect on the quality of the product.

The centrifugal forces developed by the rotary action produce an intimate contact between the steel and the mold wall which results in uniform, high heat transfer rates during the initial stages of solidification. Thus, a relatively short mold can be used and shape uniformity is enhanced. In addition, the lighter inclusions and slag particles move towards the center of the section where they accumulate at a central vortex and can be removed manually with a metal rod (Figure 21-25). Occluded gases are also eliminated in a similar manner which minimizes the formation of pinholes and blowholes.

Secondary cooling is also more uniform because, although the water spray distribution is not uniform, the surface of the strand moves uniformly through the sprays. Thus, the induced thermal stresses are uniform and both the tendency to cracking and ovality are reduced.

Other improvements reported include a more extensive central equiaxed crystal structure under a wide range of operating conditions with less tendency to central segregation.

Products—Plain carbon and alloy seamless tube rounds produced by rotary casting have been rolled directly into tube with rejection levels similar to and generally lower than those obtained with ingot-rolled conditioned rounds. Other applications include rounds for rod and wire as well as forged parts.

WHEEL-BELT CASTING

The objectives in developing wheel-belt casters are similar in many respects to those of horizontal casting: the economic production of billets on a relatively small scale by compact, low cost, highly efficient casting machines.

The characteristic feature of wheel-belt casters is a traveling mold which has the potential for high casting speeds and, therefore, high productivity. One design is illustrated in Figure 21-26. The mold cavity is formed by enclosing a groove in the periphery of a vertical rotary wheel with an endless moving belt. Both the wheel and belt are water cooled. Contact between the belt and wheel starts on an upper part of the wheel, continues around the circumference of the wheel, and is withdrawn at a location which depends on the particular machine design. Molten metal enters at the upper position from a tundish nozzle, travels with the mold, with the solidified section or shell removed from the wheel at a location on the circumference which also depends on machine design.

Unlike conventional vertical or horizontal casting, the billet surface is free of oscillation marks because the cast steel travels with the mold. In addition, inclusion distribution is reported to be more uniform than in some conventional continuous casters because of the different geometry of the mold.

Commercial production has only recently started and the process must be considered, therefore, to be in the development stage. Examples are two machines

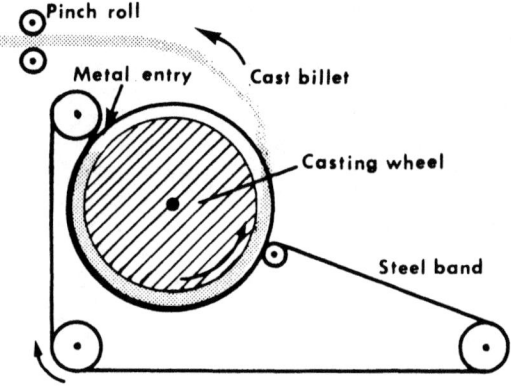

FIG. 21—26. Southwire Company wheel-belt caster showing "over-the-top" removal of cast billet.

designed by Southwire Company in the U.S.A. and Hitachi Ltd., in Japan.

The Southwire design concept (Figure 21—26) includes direct rolling. It is based on the successful casting and rolling of aluminum and copper rod. Casting speeds of 9 to 14 metres/min (30 to 45 ft/min) were obtained on a pilot machine producing 3110 and 5220-sq mm (4.8 and 8.1-sq in.) steel billets. A production unit for casting steel billets has been constructed. It has a wheel diameter of 3.7 metres (12 ft). An "over-the-top" billet extractor system is employed. Energy savings of 32 percent are estimated based on the ability to cast and roll product directly to rod (without reheating) in comparison with conventional continuous casting and rolling.

The Hitachi wheel-belt caster is the first commercial machine in operation (Figure 21—27). It has a casting speed of 3 to 6 metres/min. (10 to 20 ft/min) and a production rate of 30 to 60 metric tons (33 to 66 net tons) per hour per strand. The heat size cast is 27 metric tons (30 net tons). A trapezoidal-shaped section is

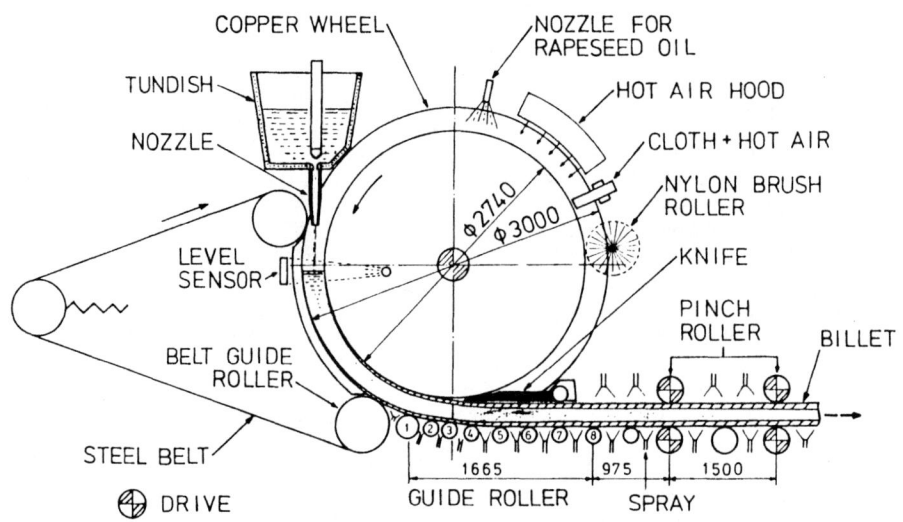

FIG. 21—27. This Hitachi wheel-belt caster has been in successful operation since June 1979.

produced, 190 mm (7.5 in.) at the bottom, 160 mm (6.3 in.) at the top and 130 mm (5.1 in.) high, which is rolled to a 130 mm (5.1 in.) square billet. The wheel is 3 metres (9.8 ft) in diameter and is enclosed with a low carbon steel belt, 2.6 mm (0.10 in.) thick. Both the wheel and belt are water cooled. The cast billet, partially solidified in an arc shape, is progressively straightened near the bottom of the wheel to ensure that a critical strain rate, which produces internal cracks, is not exceeded. After straightening, the billet passes through pinch rolls, a secondary cooling zone and a soaking zone to homogenize the internal temperature. The billet is then rolled in a vertical mill followed by a horizontal mill. The entire system is computer controlled. Finished products are flats for construction and other applications.

STRIP CASTING

Strip casting is an ambituous concept with the object of casting liquid steel directly into strip of similar thickness to hot band produced from hot strip mills. By eliminating the hot strip mill operation, together with the slab reheating phase, significant operating cost savings would be achieved together with higher processing yields in comparison with the production of hot band from either slabs rolled from ingots or cast on conventional vertical continuous casters.

In the 1960's, the Jones & Laughlin Steel Corporation was one company that investigated the concept of strip casting with the construction of a pilot plant machine. Strip was produced by casting steel on the internal periphery of a rotating drum. However, the development was not continued beyond the pilot plant stage.

In 1983, added impetus to develop strip casting was provided by the U.S. Department of Energy who also recognized both its potential energy and production cost savings. Proposals for a cooperative agreement to perform research in strip/sheet casting technology and pilot plant-scale demonstration were solicited by the Department of Energy. The object of the new technology is to cast steel and strip materials at "near to final product thickness 0.25 to 4.75 mm (0.010 to 0.187 in.) and widths up to 1830 mm (72 in.) which meet or exceed current product quality standards." A sum of $7 million was to be made available for this program in 1984 and with anticipation that additional funds of up to $30 million would be appropriated for the completion of the project in approximately five years. Provisions in the agreement include the repayment of federal funds from the commercial sale, lease, manufacture and use of the process.

In reviewing recent research and development work on steel stripmaking processes the Department of Energy reported that increasingly thinner sections had been cast with a present capability (in 1983) limited to 25 mm (1 in.) thickness. The Department of Energy also reported that, although there were several U.S.

and foreign patents for the continuous production of amorphous steel strip, which were limited to the production of thickness less than 0.25 mm (0.010 in.), the resulting structures differed from conventional steel strip. Thus, the gap between 0.25 and 25 mm (0.010 and 1 in.) represented an opportunity to further develop the technology for this strip production with substantial market potential.

Bibliography

A Study of the Continuous Casting of Steel. Part 1—Past Trends, Current Technology and Statistical Survey. Part 2—Future Impact of the Process. International Iron and Steel Institute, Committee on Technology, Brussels, 1977.

Continuous Casting. Vol. 1—Chemical and Physical Interactions During Transfer Operations. Vol. 2—Heat Flow, Solidification and Crack Formation. Vol. 3—The Application of Electromagnetic Stirring (EMS) in the Continuous Casting of Steel. Iron and Steel Society of AIME, Warrendale, 1984.

Definitions and Causes of Continuous Casting Defects. Publication No. 106. The Iron and Steel Institute, London, 1967.

Gueussier, A. L., et al., "Specific Aspects of Rotary Continuous Casting," AISE Year Book, 1982, pp. 67-73.

Haissig, M., "Horizontal Continuous Casting: A Technology for the Future," Iron and Steel Engineer, Vol. 61, No. 6, 1984, pp. 65-71.

Ishihara, S., "Latest Advances in Bulk Steelmaking Technology," Proceedings of 15th Annual Conference, Toronto, 1981 International Iron and Steel Institute, Brussels, pp. 224-250.

Kemeny, F., et al., "Fluid Flow Studies in the Tundish of a Slab Caster," Proceedings of 2nd Process Technology Conference, ISS-AIME, Vol. 2, February 1981, pp. 232-245.

Kruger, B., et al., "Continuous Casting of Steel: The Process for Quality Improvement," Iron and Steel Engineer, Vol. 61, No. 3, 1984, pp. 45-52.

Little, J., Van Oosten, M., and McLean, A. "Factors Affecting the Reoxidation of Molten Steel During Continous Casting," Canadian Metallurgical Quarterly, Vol. 7, No. 4, 1968, pp. 235-246.

McBride, D. L. and Dancy, T. E. (editors), Continuous Casting. Proc. of Tech. Sessions of Iron and Steel Div., AIME Detroit, Oct. 1961. Interscience Publishers, New York, 1962.

Miller, D. H. and Dancy, T. E., "Continuous Casting—Past, Present and Future", AISE Year Book, 1963, pp. 381-393.

Mizikar, E. A., "Spray Cooling Investigation for Continuous Casting of Billets and Blooms," AISE Year Book, 1970, pp. 299-306.

Niyama, E., et al., "Belt-Wheel Type Continuous Caster for Steel Billets," AISE Year Book, 1981, pp. 304-308.

Sakala, J. A., and Williams, Jr., R. W., "Installation and Start-Up of Jones & Laughlin's Continuous Slab Caster at Indiana Harbor Works," Iron and Steel Engineer, Vol. 61, No. 9, 1984.

Samways, N. L., Pollard, B. R. and Fedenko, D. J., "Gas Shrouding of Strand Cast Steel at Jones & Laughlin Steel Corporation," NOH-BOS Conf. Proc. AIME, Vol. 57, 1974, pp. 71-82.

Schumacher, T. F., Zugates, T. B., and DeLisio, A. J., "Design Considerations for Continuous Caster Water Systems," Iron and Steel Engineer, Vol. 61, No. 2, 1984, pp. 34-41.

CHAPTER 22

Plastic Working of Steel

SECTION 1

INTRODUCTION

Plastic working of metal is the **permanent deformation** accomplished by applying mechanical forces to a metal surface. The primary objective of such working is usually the production of a specific shape or size (**mechanical shaping**), although in some cases it may be the improvement of certain physical properties of the metal (**thermomechanical treating**). Often these two objectives can be attained simultaneously.

The study of plastic deformation has been approached from two major viewpoints. The one, here called **microscopic,** is concerned with a physical explanation of plasticity. It considers such questions as the relation of plastic behavior to the crystal structure and the interatomic forces, factors which are important in the design of materials with improved plastic properties. The other, here called **macroscopic,** is concerned more with a phenomenological explanation of plasticity. It considers such questions as the relation of plastic behavior to applied stresses, temperature and rate of deformation, factors which are important in the design of metal-forming processes, and in the design of structures and machines.

The study of plastic deformation within each of these two major viewpoints may also be conveniently, if somewhat artificially, subdivided into areas called **cold working** and **hot working.** In hot working, the forces required to deform the metal are very sensitive to the rate of application of loads and to temperature variations, but the basic strength of the metal after the deformation is essentially unchanged. In cold working, on the other hand, the forces are relatively insensitive to the rate of application of loads and to temperature variations, but the basic strength of the worked metal is permanently increased.

In a book of this kind, it is possible to consider only the major aspects of this complex and extensive subject. An effort is made, however, to show the basic principles and present capabilities and limitations of both the microscopic and macroscopic studies of plastic deformation. An attempt is then made to show how these studies can be applied to the design of metal-working processes, and the more important hot- and cold-working methods are summarized. Many of these methods will be described in detail in later chapters. Finally, there is a discussion of thermomechanical treatment of steels that emphasizes the most widely-used of such treatments, control rolling, to illustrate the effects on microstructure and properties of control of reheating, deformation and cooling.

SECTION 2

MICROSCOPIC NATURE OF PLASTICITY

Metals generally consist of regions called **crystals** or **grains** where the atoms are arranged in more or less regular, geometrical patterns, bounded by transition regions of irregular pattern called **grain boundaries.** The average diameter of the grains varies greatly with mechanical and thermal treatment but is usually in the range 0.25 mm to 0.025 mm (0.01 to 0.001 in.). The geometrical pattern or **crystal lattice** consists primarily of repeated fundamental groups of atoms called **unit cells.** The unit cell of pure a-iron, for example, is shown in Figure 22—1. It is called body-centered cubic.

Plastic deformation of metals occurs by several processes including grain rotation. By far the most predominant process, however, is **slip** of adjacent planes of atoms within the crystals. This slip takes place only on definite crystallographic planes, usually those of most dense atomic packing, and only in definite crystallographic directions, also usually those of most dense packing, within these planes. A **slip plane** and **slip direction** of most dense packing for the body-centered cubic structure are illustrated in Figure 22—1. Such a plane and direction constitute a **slip system.** Each particular crystal has its own characteristic slip systems. At least 48 are known for a-iron.

A force applied to the surface of a crystal is transmitted by interatomic reactions across each internal

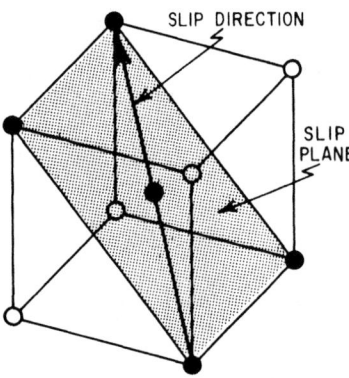

FIG. 22—1. Illustration of body-centered cubic cell and a common slip system.

crystallographic plane. The force transmitted across a particular plane can, in general, be resolved into a component normal to the plane and a component tangent to the plane. The tangent component can in turn be resolved further into components corresponding to definite crystallographic directions in the plane. In this manner, a definite force component can be associated with each particular slip system. Many experiments have shown that slip in a particular system will occur if this force exceeds some critical value on a unit area of the slip plane. This is known as the **critical shear stress law.** The critical value, or **yield stress,** is essentially independent of the normal force on the slip plane, but it does depend on the particular type of slip system. When a system of forces is applied to the surface of a polycrystalline metal, slip first occurs in that crystal and that system where the critical shear stress is first attained. Complex plastic deformation occurs, however, by slip in several systems within each of many crystals.

Early efforts to predict the magnitude of the yield stress in perfect metal crystals, using laws describing interatomic forces, showed that **simultaneous slip** of an entire layer of atoms could occur when the shear stress in the plane was about 6.89 x 10⁷ MPa (10⁷ psi). Experiments showed, however, that the actual yield stress was several orders of magnitude lower. This apparent discrepancy was resolved when it was realized that metal crystals contain many imperfections in structure, called **dislocations,** which permit **consecutive slip** of one line or loop of atoms at a time in such a manner that the required shear stress is greatly reduced. Figure **22—2** shows a single dislocation line, BC, in an otherwise perfect simple cubic lattice. In a typical metal crystal there may be 10⁷ to 10⁹ such lines crossing a square inch of area everywhere in the crystal. One of the characteristics of these dislocations is that they create internal forces between the atoms. These have been likened to the internal stresses that would appear in a perfectly elastic body by making a cut, such as ABC in Figure 22—2, displacing the upper portion one atom distance, b, and then cementing the cut together. In fact, an entire **mathematical theory of dislocations** has been developed which equates these lattice imperfections to displaced surfaces within elastic bodies.

As mentioned above, the observed yield stress in a typical metal crystal is several orders of magnitude be-

low the theoretical value for a perfect crystal. The yield stress of some iron whiskers, which are nearly dislocation free, however, is very high (about 6.89 x 10⁶ MPa or 10⁶ psi) approaching the theoretical value for a perfect crystal. The theoretical stress required to move a single dislocation, on the other hand, is very small, in fact much smaller that the yield stress of a typical metal crystal. This is explained by the fact that other imperfections in the typical crystal, i.e., vacancies, interstitial and substitutional foreign atoms, grain boundaries, or other dislocations, cause additional internal stress fields which oppose the motion of any dislocation. Finally then, a very simple picture of the effect of imperfec-

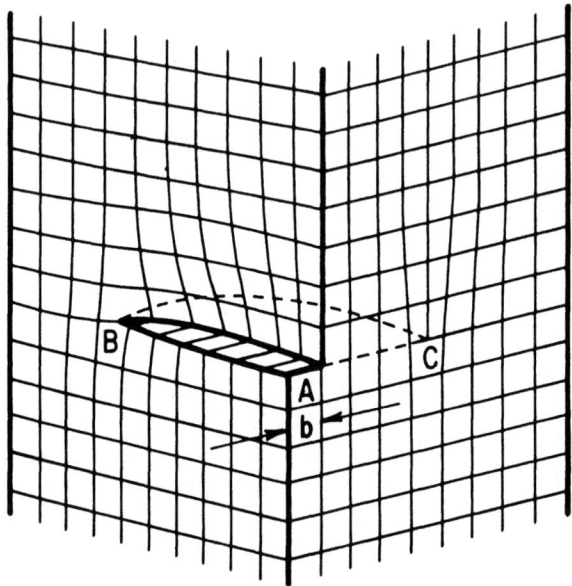

FIG. 22—2. Dislocation in a simple cubic lattice.

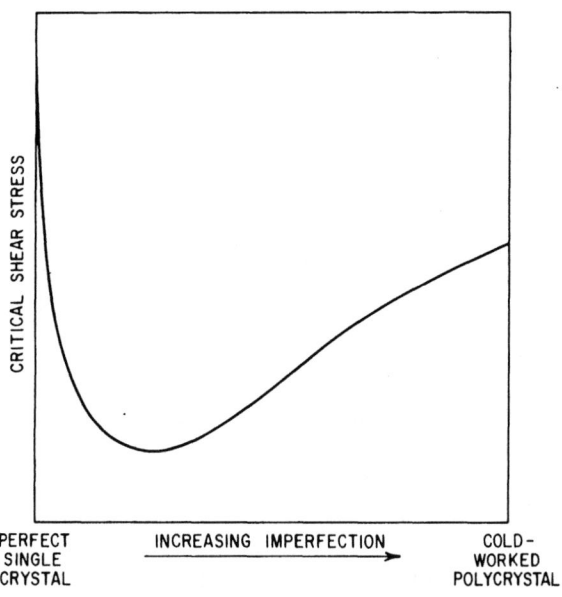

FIG. 22—3. Effect of imperfection on critical shear stress.

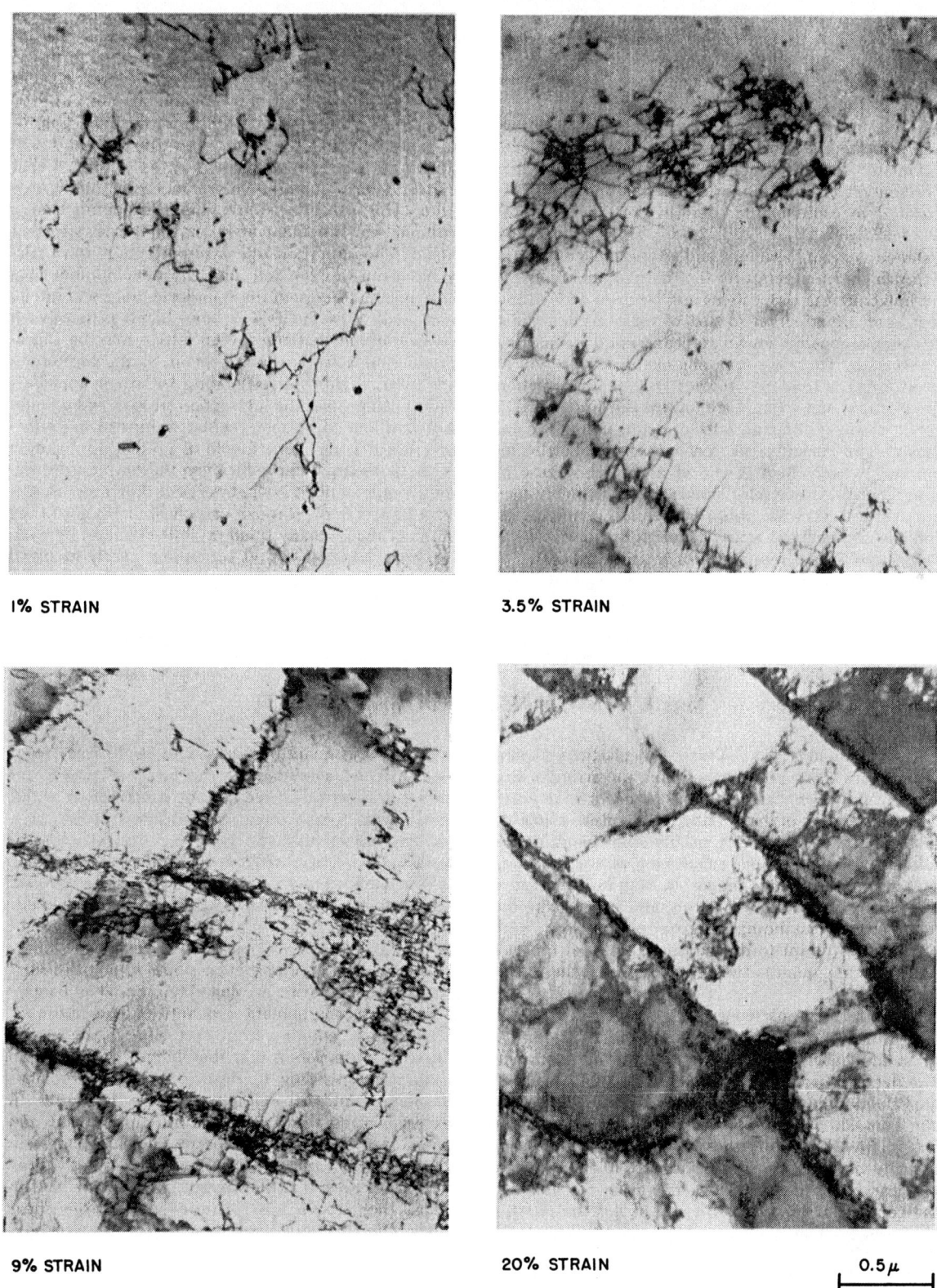

1% STRAIN

3.5% STRAIN

9% STRAIN

20% STRAIN

0.5 μ

FIG. 22—4. Dislocation tangles in iron deformed at 25°C. (a) 1% strain, (b) 3.5% strain, (c) 9% strain, (d) 20% strain. (Transmission electron micrographs, approximately 25 000×.)

tions on yield strength can be constructed as shown in Figure 22—3. A perfect crystal has a very high yield strength, one with only a few imperfections has a very low yield strength, but as the density of imperfections increases, the yield strength again becomes large.

As plastic deformation proceeds beyond yielding, new imperfections are formed and old ones may be annihilated. At low temperatures, the overall effect is the creation of more imperfections, raising the stress required to cause further deformation. (See Figures 22—4a, 22—4b, 22—4c and 22—4d). This phenomenon is known as **strain hardening** and is typical of **cold working**. At higher temperatures the thermal energy assists the movement of dislocations and tends to reduce the degree of imperfection to that of the annealed state. This phenomenon is known as **recovery** and is typical of **hot working**. Thus, two opposing factors are present in hot working, a tendency toward greater imperfection due to strain and a tendency toward less imperfection due to recovery. At high rates of deformation, the annealing is insufficient to overcome the strain hardening and the overall effect is one of a greater degree of imperfection. At low rates, however, the recovery may completely cancel the strain hardening, giving an essentially constant flow stress. Whereas the flow stress at low (cold-working) temperatures is quite insensitive to the rate of deformation, the flow stress at high (hot-working) temperatures is very dependent on this same factor. Thus, although it requires a relatively high stress to hot roll a steel bar, a much smaller stress may be required to cause creep at hot-rolling temperatures.

The above discussion may leave the impression that microscopic plasticity is now a closed chapter in physical science. This would be highly misleading. One of the gages by which to measure the true state of development of a branch of science is the ability of that science to make quantitative predictions within its own realm. This usually requires the development of a **mathematical statement of theory** of the laws of that science. The important role of imperfections in plasticity has been emphasized, and the development of a mathematical theory of these imperfections was briefly mentioned. This theory has been based primarily on the assumption that the crystal lattice may be represented as an isotropic, homogeneous elastic continuum with internal slips corresponding to lattice imperfections. It can predict the interaction stresses and equilibrium positions of certain regular groups of imperfections, but it is not yet capable of predicting a simple tensile stress-strain curve for even the simplest dislocation arrangement in a single crystal. For most metals the situation is even more complicated because they consist of many, many small crystals bounded by complex grain boundaries and containing highly involved dislocation groupings as illustrated in the electron micrographs of iron shown in Figure 22—4.

SECTION 3

MACROSCOPIC NATURE OF PLASTICITY

As mentioned above, microscopic plasticity begins with certain observations concerning the arrangement of atoms in the crystal lattice and proceeds to develop an understanding of the detailed mechanisms of plastic flow. Macroscopic plasticity, on the other hand, begins with certain observations concerning plastic deformation of polycrystalline metals in simple mechanical tests, such as simple tension tests, and proceeds to develop an understanding of gross plastic flow. Both viewpoints attempt to develop mathematical theories which enable quantitative predictions in their own realms.

In the macroscopic viewpoint, the metal is thought of as a **continuum** possessing properties like density, stress and velocity at all points within its outer surface. The detailed properties of the lattice structure and its imperfections including grain boundaries are smoothed out. Thus, although this viewpoint is particularly suited for situations involving the simultaneous deformation of many grains, it is completely unable to deal with the physical details of plastic flow. The great advantage of the continuum concept is that it enables the physical quantities, such as density, stress and velocity, to be treated as continuous functions, thus opening up the possibility of employing the large body of mathematics based on such functions. Even the microscopic viewpoint involving dislocation theory finds it necessary to adopt the continuum concept outside of the dislocation lines, i.e., in the more regular portions of the lattice.

Two quantities that play a central role in continuum theory are the quantities **stress** and **strain rate.** They describe in an average way the forces between the atoms in the crystal lattice and the deformation of the lattice, respectively. Stress, as described before, is a quantity representing the force per unit area on an internal plane. Obviously, however, the stress at a point in a continuum cannot be described by a single number because it varies for different planes through the point. Strain rate represents the relative velocities of neighboring internal planes. Thus, it too varies for the different planes through a particular point. Quantities such as stress and strain rate are called **tensors.** They may be thought of as mathematical quantities like numbers, but with more complex properties. One of the special properties of tensors is that they have three **principal values,** corresponding to three perpendicular planes through a point, that uniquely describe the tensor. In the case of stress these are called principal stresses and they correspond to three perpendicular planes on which there are no tangential or shear forces. Three such planes always exist through every point, and in general they are the only planes which have no shear stresses. Thus, the principal stresses are normal stresses, here denoted as σ_1, σ_2, and σ_3 such that $\sigma_1 \geq \sigma_2 \geq \sigma_3$, where tensile stresses are considered positive. The strain rate tensor also contains three principal values. These correspond to three perpendicular planes through each point across which there is simple stretching or compressing but no distortion or shearing. They are denoted ξ_1, ξ_2 and ξ_3, where stretching is considered positive.

The theory of macroscopic plasticity depends very largely on three macroscopic observations concerning stress and strain rate. The first of these is that *the volume remains essentially constant during gross plastic deformation*. This means that the sum of the principal strain rates, stretches and compressions, is zero, i.e.,

$$\xi_1 + \xi_2 + \xi_3 = 0.$$

This result is not really surprising since, as remarked above, the main mechanism of plastic flow is consecutive slip of adjacent rows of atoms, a mechanism that requires no volume change.

A second basic observation is that *plastic flow or yielding occurs at a point only after the maximum shear stress in some direction on some plane attains the critical value, k*. It is known from a study of the stress tensor that the greatest shear stress occurs on the two planes which are half way between the planes of greatest and least principal stress. Its magnitude is one-half the difference between the greatest and least principal values. Thus, the second observation is that, for gross plastic flow to occur,

$$\frac{\sigma_1 - \sigma_3}{2} = k$$

A third basic observation of plastic flow is that *the principal shear stresses are coincident with and proportional to the principal shear strain rates*. This may be written as

$$\frac{\sigma_1 - \sigma_3}{\xi_1 - \xi_3} = \frac{\sigma_2 - \sigma_1}{\xi_2 - \xi_1}$$

Since shear straining is similar to slip, this observation is similar to saying that the slip in a given direction is proportional to the shear stress in that direction, a statement that seems reasonable on the basis of microscopic plasticity.

As mentioned in the previous section, plastic flow in a single crystal requires the attainment of a critical shear stress in particular crystallographic directions. If one thinks of gross plastic flow as being the summation of many slips involving many crystals with all possible orientations, it is not surprising that the macroscopic critical shear stress is independent of crystal orientation.

This third basic observation of gross plastic flow is that *the directions of greatest shear strain rate coincide with the directions of greatest shear stress*. This observation requires a more complex mathematical statement than those above and such treatment will not be given here. It is noted, however, that since shear straining is similar to slip, the observation is similar to saying that the most slip occurs in the directions of greatest shearing stress, a statement which certainly seems reasonable on the basis of microscopic plasticity.

The mathematical statement of the above three macroscopic observations forms the foundation of macroscopic or *continuum theory of plasticity*. One additional statement is necessary concerning k, the critical shear stress. For cold working, it is generally found to be an increasing function of prior work (**work hardening**), but not of strain rate. For hot working it is a function of the prior thermomechanical history as well as the current temperature and strain rate. This is all in accord with the microscopic observations concerning the formation and annihilation of imperfections.

Before leaving this discussion of general plasticity it should be mentioned that the two viewpoints described here are extremes, the one specifically taking into account every crystal imperfection and the other averaging out even the individual grains. Several intermediate viewpoints are also possible. For example, one that has been considered is to use the properties of imperfections but to consider them continuously distributed over the lattice. Another is to consider the grains as individual continua treated with the same methods now applied to the polycrystalline body. Since all viewpoints are based upon experimental observation, they are mutually consistent, and are expected eventually to be merged into a single unified theory of plasticity.

SECTION 4

GENERAL PRINCIPLES OF METAL WORKING

In considering the application of the principles of plastic deformation to the design of metal-working processes, it is necessary to define the factors that can limit such deformation. These may be broadly classified into two types: **instability**, which is the creation of undesirable types of deformation due to small, but usually unavoidable, irregularities in the metal or in the load application; and, **fracture**, which is the creation of new surfaces such as holes, cracks, or actual separation into two or more parts.

Instabilities can be divided into two general types; those associated with compressive stresses and called **buckling**, and those associated with tensile stresses and called **necking**. The two types can be illustrated by considering a cup-drawing process such as shown in

Figure 22—5. If the flange is too thin or insufficiently supported, the compressive circumferential stresses may cause buckling or wrinkling. If the walls of the cup are too thin, the tensile stresses in the axial direction may cause local thinning or necking. Instability is a very complex problem and not completely understood. It can be said, however, that *factors which tend to spread plastic deformation reduce the likelihood of instability*. These may be material properties such as work hardening, or they may be mechanical design factors such as shape and method of load application.

Fracture may be considered as the formation of holes or cracks due to tensile stresses which overcome interatomic forces. Since the local tensile stresses are highest at internal discontinuities, and the interatomic

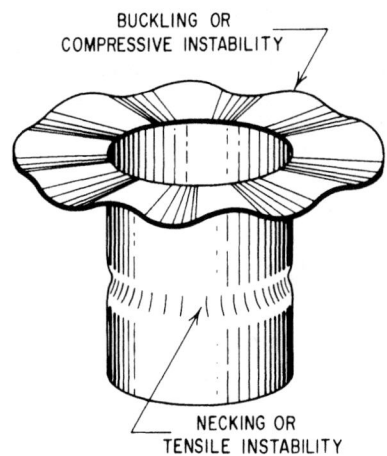

FIG. 22—5. Examples of forming instabilities in a deep-drawn cup.

resistance weakest at these same places, holes and cracks generally initiate at grain or phase boundaries. Macroscopic observations show that both the initiation and the growth of these cracks and holes are aided by tensile stresses. This leads to an important principle of metalworking; namely, that *the strains which can be achieved without internal damage are increased by compressive forces.* This is why larger reductions are possible in forging operations than in stretching operations, in single-pass rolling than in single-pass drawing.

In addition to avoiding instability and fracture, a metal working operation may be designed to minimize forces or to increase efficiency. Since plastic flow depends upon achieving a critical shear stress, *forming forces are minimized in tensile-compressive operations;* that is, those which involve both tensile and compressive stresses. This is because the difference between the principal stresses, and hence the maximum shear stress, is large in these cases while the absolute magnitudes of the stresses remain small. It explains, for example, why roll-separating forces are reduced by the application of strip tensions. *Efficiency is enhanced by minimizing tool surface friction and by avoiding redundant work; that is, work that does not contribute to the final shape.* A drawing process is inherently more efficient than a rolling process because, unlike rolling, it does not de-

pend on tool surface friction to transmit the work. Redundant work is quite difficult to avoid completely, but can be kept small by lubrication and die shaping.

Another factor which might be important in the design of metal-working operations is the achievement of some particular properties in the product. Thus, for example, it might be desirable to produce a product that has directional properties, or **anisotropy.** The rules for accomplishing this are not yet completely clear. It also might be desirable to produce a product that has no residual stresses, or one that has certain prescribed residual stresses. *Residual stresses can be minimized by hot working or by die designing such that every portion of the metal goes through the same deformation history.* Figure 22—6 shows an example of an ideal wire-drawing die that forces each element of the wire to go through the same history as every other element. This die also avoids redundant work since all of the deformation done on each portion of the metal goes toward the shape of the final product.

Many other factors may enter into the design of a specific metal-working operation. In particular, some knowledge of the frictional conditions between the worked material and the forming tools is often essential for a complete design. The above examples, however, are believed to be representative of the major design factors involving the principles of plastic deformation, and to illustrate how a knowledge of these principles can be useful in designing such operations.

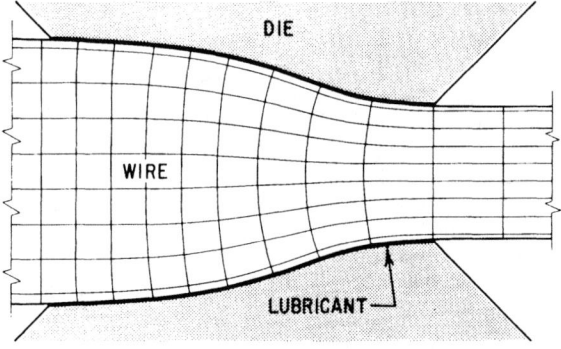

FIG. 22—6. Wire drawing with maximum efficiency and product uniformity.

SECTION 5

THEORETICAL ANALYSIS OF METAL-WORKING PROCESSES

The theoretical analysis of metal-working processes requires the mathematical solution of the equations of plasticity theory (Section 3) for boundary conditions representing work-tool geometry and motion as well as tool-interface friction. In general, these equations form a highly nonlinear set of partial differential equations. Thus, until the advent of high-speed digital computers with their tremendous capacity for processing numerical data, it has not been possible to analyze metal-

working processes without drastic simplifying assumptions such as the neglect of elastic, work-hardening, and recovery effects. Now, however, computer-based methods for performing detailed theoretical analyses are being developed. Present methods are limited primarily to two-dimensional cold-working processes such as wire drawing and cold rolling, but these are being extended to include three-dimensional and hot-working processes.

Perhaps the most versatile of the computer-based computational methods is the finite-element method. In this method the material to be processed is mathematically divided into many regions or elements each of which is assumed to deform very simply, for example by uniform strain, but collectively can deform in a very complex manner. Figure 22—7 shows the deformation of a finite-element mesh in a strip-drawing process. The finite-element method can be used to predict forming loads, local stresses and deformations, and residual stresses. Thus, it has the potential not only for analyzing existing processes, but also for designing processes which, for example, might minimize forming loads, avoid excessive tensile stresses and associated

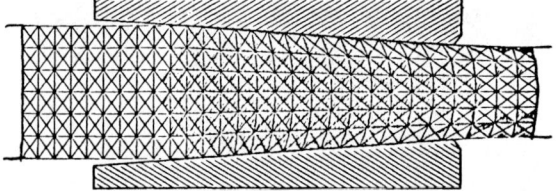

FIG. 22—7. Theoretical distortion of a finite-element mesh in a strip-drawing process.

microcracks, or deliberately induce favorable residual stresses.

SECTION 6

PRINCIPAL METHODS FOR HOT WORKING

The principal methods of forming steel by hot working are hammering, pressing, rolling, piercing and extrusion. Hot working by the first two methods is called forging.

Forging may be performed under hammers, in mechanical presses and upsetters or by a method known as roll forging. Pressing generally includes the manufacture of forged articles in hydraulic presses. Extrusion usually is performed in hydraulic presses which force the hot steel through a die. Rolling is performed in rolling mills of a variety of types described in detail in various other chapters. The present discussion will be limited to general descriptions of the various types of hot-working equipment and some of the principles of their operation.

The two principal reasons for deforming steel at elevated temperatures (hot working) are to reduce the forming loads through the reduction of the resistance of the steel to deformation, and to develop preferred metallurgical structures for strength and ductility of the formed steel.

Hammering—Hammering was the first method employed by man in shaping metals. The first forging was done by hand hammers wielded by workmen.

The first known power hammer, called a **tilt hammer**, was built in England. It was driven by water power and consisted of a beam of wood, hinged at one end and provided with an iron hammer head or die at the opposite end. At an intermediate point between the hinged end and the free end carrying the hammer head, cams on a revolving shaft driven by a water wheel alternately raised the free end and allowed it to fall upon an anvil or die fixed upon a suitable foundation. This was a crude tool compared to the steam hammers now used.

The first steam hammer was built in France during 1842. It consisted of a two-piece frame, constructed so as to support a vertical steam cylinder, fitted with a piston and piston rod, directly over a die or anvil. To the piston rod of the steam cylinder was attached a **tup** or hammer head. By admitting steam to the cylinder below the piston, the hammer was raised for any desirable length of stroke and then allowed to drop upon

the work piece supported on the anvil or bottom die.

In order to increase the striking force of the steam hammer above that derived from gravity alone, there was developed the **double-acting steam hammer,** in which steam can be admitted above the piston also and employed both on the downward stroke as well as for lifting the tup. The first double-acting steam hammer was built at Midvale, Pennsylvania, in 1888.

A variety of other types of forging equipment employing the impact principle for forming hot steel have been developed, one of which is the impact forging machine used in the manufacture of steel axles described in Chapter 29. Descriptions of some other types are available in reference works listed at the end of this chapter and in recent literature.

Pressing—The hydraulic forging press is an English invention dating from the year 1861. It was introduced into the United States about 1887. It consists (see Figure 22—8) essentially of a hydraulic cylinder supported by two pairs of steel columns which are anchored to a single base casting of great weight and strength. The piston or ram of the cylinder points vertically downward and carries the upper forging die, which is directly above a stationary die resting on the base casting to which the columns are attached. By admitting water under high pressure to the cylinder at its top, the ram carrying the upper die is forced down upon the material to be forged, which rests upon the lower forging die. Small auxiliary cylinders lift the ram after each application of pressure. The manufacture of large steel forgings by this method is described in Chapter 39.

The pressure, which must be very high if the forging press is to do effective work, is increased gradually and maintained until the metal yields. In practice, it has been found that the lowest pressure that can be effective in shaping steel at a full forging heat is about 1.65 MPa or 1.2 net tons per square inch, but the pressures employed in actual work often will exceed about 18 MPa or 13 net tons per square inch.

Piercing—The first stage of the forming of seamless tubing is the piercing of hot solid billets. In piercing, a billet is forced over a piercer point between skewed rolls. The resulting shell is reworked by rolling or

FIG. 22—8. Hydraulic press in operation, forging a massive ingot supported on a porter bar which, in conjunction with the link-chain support, permits manipulation of the ingot as desired.

drawing over a plug. The hot forming of seamless tubing is covered extensively in Chapter 32.

Extrusion—The hot-extrusion process consists of enclosing a piece of metal, heated to forging temperature, in a chamber called a "container" having a die at one end with an opening of the shape of the desired finished section, and applying pressure to the metal through the opposite end of the container. The metal is forced through the opening, the shape of which it assumes in cross-section, as the metal flows plastically under the great pressures used.

The equipment and methods for carrying out one type of hot extrusion are described in detail in Chapter 32.

Mechanical Forging—Many hot forgings are produced on mechanical presses. In machines of this type, pressure is applied to a vertical ram (carrying the upper forging die) through a connecting rod from a crankshaft. The heated work piece rests on the bottom die. The stroke of such a press is limited to the "throw" of the crankshaft.

Upsetting—A special type of mechanical press is the upsetting machine, in which the piece to be shaped is clamped between two dies with vertical faces and shaped by the action of a tool on a ram operated by a crankshaft. The ram of the upsetting machine operates with a horizontal, instead of vertical, stroke.

Hot Rolling—Of all the known methods of shaping steel, that of rolling, as introduced by Henry Cort in 1783, has come to be employed the most extensively. Though Cort is credited rightly with being the "father of modern rolling," because of his successful development of mills employing grooved rolls, the use of this principle in shaping metal antedates his mill by many years. There are records, for example, to show that in

the year 1553, rolls were employed in France to produce sheet of uniform thickness for the stamping of gold and silver coin. In Sweden, rolls were employed to produce certain sections prior to the year 1751, and even at that time assertion was made that as many as twenty times more bars could be reduced in a given time than could be shaped under the tilt hammer of those days. From the days of Cort to the present time, the rolling mill has passed through a rapid process of development, not only in the size, power and productive capacity of mills, but also in their design and in the increasing variety of shapes of sections rolled.

Mechanical Principle and Effects of Rolling—The process of shaping steel by rolling consists essentially of passing the material between two rolls revolving at the same peripheral speed and in opposite directions, i.e., clockwise and counterclockwise, and spaced so that the distance between them is somewhat less than the height of the section entering them (Figure 22—9). These rolls can either be flat (as illustrated) or contoured for the hot rolling of rods or shapes (see Chapter 27). Under these conditions, the rolls grip the piece of metal and deliver it, reduced in cross-sectional area and increased in length. The extent of sideways or lateral spreading (called **spread**) is found to depend mainly upon the amount of reduction and the shape of the cross section entering the rolls; thus in rolling plates of considerable width, the actual total spread is independent of the width, and actually may be less than that resulting from the first pass in the reduction of small, square billets, especially if the percentage reduction in cross-sectional area of the latter is great.

The turning of the rolls in contact with the work introduces a frictional force which acts along the arcs AB and A'B' of Figure 22—9, and is proportional to,

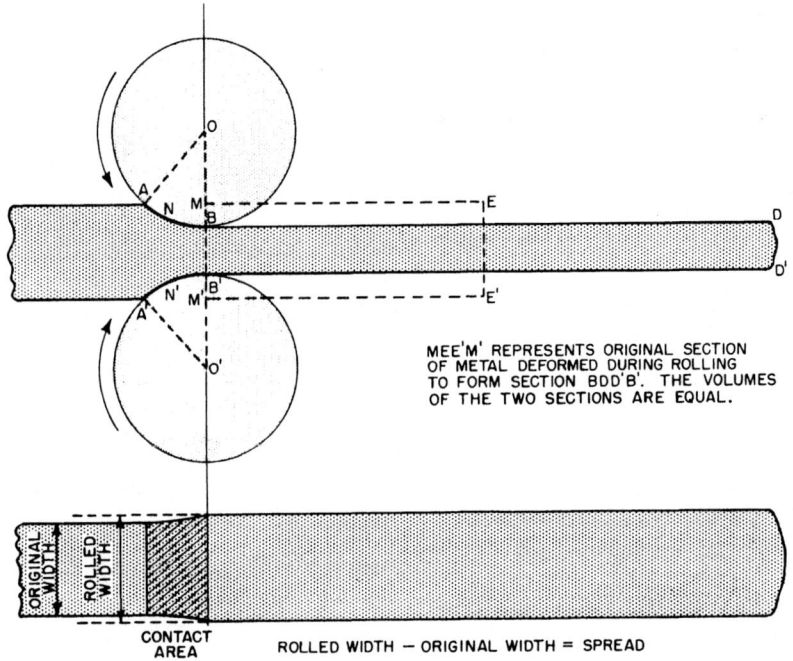

MEE'M' REPRESENTS ORIGINAL SECTION
OF METAL DEFORMED DURING ROLLING
TO FORM SECTION BDD'B'. THE VOLUMES
OF THE TWO SECTIONS ARE EQUAL.

ROLLED WIDTH — ORIGINAL WIDTH = SPREAD

FIG. 22—9. Diagram illustrating the action of plain rolls upon a piece of hot steel of originally square cross-section.

among other factors, the pressure between the rolls and the piece. This pulls the work into the opening between the rolls, against the wedging action of the tapered section entering the rolls. The piece is delivered at a higher speed than the roll-surface speed; it enters at a velocity lower than the roll-surface speed. The ratio of the speed with which the work leaves the rolls to the surface speed of the rolls themselves is called the **forward slip.** Evidently, there must be some point between A and B and A' and B' where the speed of the bar is equal to the roll-surface speed. This point, indicated on the drawing by N and N', is called the **neutral point,** which coincides with the point of maximum pressure in many instances. The arc AB is called the **contact arc,** and its included angle AOB the contact angle or **rolling angle.** When this angle is the maximum at which the piece will enter (without pushing), it is called the **angle of bite.** The area of steel under the contact arc is called the **contact area,** which is projected to show the spreading that may occur. The force exerted against the rolls by the piece is called the **separating force.**

Effect of Work Temperature—The effects of rolling with respect to changes in physical dimensions of the piece are influenced very markedly by the temperature of the piece being rolled with respect to both degree and uniformity of heating. Reduction of the resistance to deformation imparted to steel by relatively slight increases in temperature results in a reduction in the amount of power required for rolling and increases the ease with which hot steel can be made to flow plastically in the desired directions. Chemical composition of a particular steel and the nature of the rolling operation to be performed may limit, respectively, the maximum and minimum rolling temperatures that are

applicable. Control-rolling practices, combining certain per cent reductions at specific temperatures, are being used in modern plate mills to achieve improved toughness in hot-rolled steels.

Effects of Roll Diameter—Small-diameter rolls require less force than rolls of larger diameter to effect a given reduction. Advantage is taken of this fact in four-high and other mills employing small-diameter work rolls backed up by heavier rolls that prevent the smaller rolls from bending. Smaller separating force is encountered with small-diameter rolls for two reasons: first, the area of contact is less so that, with a given pressure, the total required is less; and second, the required average pressure is less because the smaller area of contact reduces the total frictional forces.

Miscellaneous Hot-Working Methods—Rotary swaging is performed to taper the end of bars, wires and tubes. The machine in which the work is done has two or four shaped dies, in opposing pairs suitably mounted in a rotating ring. As the ring rotates, alternate pairs of dies are forced against the metal being swaged, the resultant pressure shaping the piece. Such a machine is illustrated in Chapter 32.

In **hot spinning,** which is limited to shapes symmetrical about the spinning axis, the heated piece to be shaped is mounted in a lathe or similar machine that can rotate it rapidly. A tool is then brought to bear against the spinning piece and, by manipulating the position and pressure of the tool, the work piece can be shaped. For example, bowl-shaped sections, such as flanged and dished heads, can be formed from flat, circular plates by spinning.

Many steel parts are formed hot from plates and sheets by **hot deep-drawing** operations that would be impracticable if the material were at room temperature. Roof ribs for railroad box bars, one-piece gates for hopper cars, deep bowls, etc., are typical. The heated steel is formed in hydraulic or mechanical presses equipped with forming dies that produce the desired shapes. Another example is the manufacture of closed-end cylinders by a combination of hot cupping and drawing operations, described in Chapter 32.

COMPARISON OF METHODS FOR HOT WORKING

It is a very difficult matter to make a fair comparison between rolling and forging, or even between hammer and press forging. Each method has a field of its own with rather well-defined boundaries. Many irregular shapes are so intricate in design that rolling or extruding them is out of the question, and such shapes must be formed under the hammer or in the press. Certain crankshafts or a claw-hammer head serve as examples of these classes of shapes, which can be produced by no other form of mechanical working than by the forging method.

The hammer and the press are both slower and more expensive to operate than rolls. Special care must be given to all phases of the forging operation, including

heating for forging. It is, perhaps, this meticulous care required to produce high-quality forgings that has given rise in some quarters to the belief that forged articles always are superior to rolled articles. Through-thickness ductility is generally better in thick forged sections than in rolled product of similar size. Assuming that a given section can be produced by either rolling or forging, and that an equivalent amount of attention is given to all details of both processes with regard to heating, proper speed and amount of reduction, etc., the quality of rolled material can be equal to that of forged product.

With the complex stress-strain relationships involved in plastic deformation in hot working, and the basic differences in the nature of the major hot-working processes, direct comparison is difficult. However, studies have shown that the effects of rolling and pressing are comparable, provided equipment for carrying out both processes is of comparable capacity and the same sized pieces are being worked.

Rolls have the distinct advantage of speed of production where the shapes involved are of a nature suited to rolling. There is one field of operation in which rolling, hammering and pressing all can be applied; this is the shaping of blooms and billets from ingots.

SECTION 7

PRINCIPAL METHODS FOR COLD WORKING

Cold working, generally applied to bars, wires, tubes, sheet and strip, is a process of reducing the cross-sectional area by cold rolling, cold drawing, cold extrusion, or cold forging. Cold working is employed to obtain the following effects: improved mechanical properties, better machinability, special size accuracy, bright surface, and the production of thinner gages than hot work can accomplish economically. The present brief discussion will be supplemented in later chapters by discussion of methods and effects of cold working various steel-mill products.

Cold Rolling—Cold working by cold rolling consists of passing unheated, previously hot-rolled bars, sheets or strip (cleaned of scale) through a set of rolls, often many times, until the final size is obtained. Methods and effects of cold rolling wide strip are discussed in Chapter 34.

Cold Drawing—In this process a bar, rod, wire or

tube, after being cleaned, is pulled through a die having an opening smaller than the entering piece to reduce the latter to the required size (see Chapter 31).

Cold Extrusion—The cold extrusion of steel is carried out in a manner similar to the hot-extrusion process, with two main exceptions: (1) The steel is at room temperature, and (2) the surface of the piece is treated by some chemical process such as bonderizing to assist in reducing friction between the steel and the container wall and die, in conjunction with special lubricants.

Cold Forging—To obtain superior mechanical properties and tight dimensional tolerances with minimal machining and consumption of energy, the process of cold forging is now being used for the forming of small steel parts. As with extrusion, cold forging is similar to hot forging but requires the use of special lubrication techniques and tooling design.

SECTION 8

THERMOMECHANICAL TREATMENT

Thermomechanical treatment consists of combining controlled amounts of plastic deformation with the heat treatment cycle to achieve improvements in mechanical properties beyond those attainable by the usual rolling practices alone or rolling followed by a separate heat treatment. The tensile strength, of

course, is increased at the same time as the yield strength (not necessarily to the same degree) and other properties such as ductility, toughness, creep resistance and fatigue life can be improved.

In the past, thermomechanical treatments (TMT), as applied to alloy steels, have been classified as: deformation before transformation (ausforming, marworking, austentempering, control rolling); deformation during transformation (isoforming, intensified control rolling); or deformation after transformation (flow tempering, warm working). From the point of view of current and future commercial application, as applied to C-Mn steels, which comprise over 80 per cent of current tonnage, the most important process is that of control rolling. In TMT, the deformation/transformation sequence is only one of several stages of the processing that must be controlled in order to optimize the as-rolled properties of sheet and plate. For example, grain coarsening must be limited during reheating, recrystallization must be controlled during the rolling, cooling rate through the transformation must be limited to control the size and type of transformation products, and post-transformation cooling must be regulated to control precipitation processes. These objectives are realized through control of both processing and compositional variables (e.g., through microalloy additions).

Grain Coarsening During Reheating—Grain coarsening in C-Mn steels between 900 and 1200° C follows a well behaved continuous grain-growth law of the type $D = kt^a \exp(-\beta/T)$ where D is the average grain diameter, t is time, T is temperature, and k, α, and β are constants. Grain coarsening in microalloyed steels in the range 900 to 1300° C is, however, characterized by temperature regions of continuous growth at high and low temperatures between which is a rather narrow temperature range in which abnormal growth occurs that produces a duplex distribution of austenite grain sizes. In extreme cases the duplex structure may be difficult to remove during subsequent hot rolling. Accordingly, current practice is to restrict reheat temperatures to ranges that result in uniform structures while avoiding temperatures so low as to create excessive mill forces during subsequent rolling.

The reheat temperature necessary to achieve a uniformly fine initial grain structure in the slab varies with type and amount of microalloy carbide and/or nitride forming elements, since the abnormal grain growth is controlled by grain-boundary/particle interactions. Boundary pinning occurs when a critical fraction of the grain-boundary area is occupied by particles. Thus, pinning is favored by high volume fractions of fine particles that resist dissolution and growth.

These principles offer reasonable explanations for the grain-coarsening behaviors of microalloyed steels. Vanadium-bearing steels generally resist coarsening only up to 1000 to 1050° C, above which VCN is sufficiently soluble in austenite to begin ripening. Nb-bearing steels resist grain coarsening to temperatures up to 1100 to 1150° C because NbCN is less soluble in austenite than is VCN. Steels containing a fine dispersion of the very stable TiN can resist grain coarsening to temperatures above 1200° C because TiN has such an extremely low solubility in austenite that particle ripening is inhibited at lower temperatures.

Grain Refinement During Hot Rolling—If the reheated grain structure has been coarsened, refinement can be effected by repeated recrystallization during hot rolling if the process is adequately controlled. Three parameters are particularly important—the rolling temperature, the reduction per pass, and the num-

ber of recrystallization passes. At temperatures above 1000° C, rapid recrystallization occurs in most low-carbon, low-alloy steels. In tandem hot-strip rolling, where reductions per pass of 25 to 30 per cent are common, recrystallization may be dynamic (coincident with deformation); under these conditions the recrystallized grain size after several passes is quite fine ($\sim 20\ \mu$m). At the opposite extreme, heavy slab rolling on a reversing mill frequently begins with reductions of only 5 per cent per pass. Such low reductions can actually produce coarsening of the grains by strain-assisted growth during the time interval between the early passes. This trend can be reversed if there are sufficient subsequent passes with higher reductions, as is common in constant-draft schedules. Thus roughing schedules of 8 to 10 passes with reductions in the last passes approaching 20 per cent will produce, by multiple recrystallization, austenite grains of 30 to 40 μm diameter. However, final, light-reduction shaping passes must be avoided since they can again cause blown grains, thereby destroying the grain refinement produced by the preceding passes.

The general rule seems to be that any adjustment of the rolling parameters that increases the stress to which the work piece is subjected will result in finer recrystallized grains, presumably by increasing the density of recrystallization nuclei. Thus large reductions per pass are desirable, as are low rolling temperatures in the recrystallization range.

Plain C-Mn steels continue to recrystallize down to temperatures near 800° C and thus produce quite fine recrystallized austenite of 20 to 30 μm diameter on most reversing mills. However, austenite grains of such a size may not transform, under typical mill cooling rates, to ferrite that is sufficiently fine to meet modern strength/toughness requirements. Consequently, it is desirable to deform and flatten the refined austenite grains by rolling below the recrystallization range in order to increase the density and proximity of the austenite grain boundaries that act as ferrite nucleation sites. Altering the recrystallization temperature range of a steel is one of the roles of microalloy additions.

The recrystallization-stop temperature of C-Mn steels can be increased by the formation of microalloy carbides and/or nitrides that pin subboundaries and grain boundaries. Thus, by selection of the proper type and concentration of microalloy, recrystallization and refinement of austenite grains can be achieved in the first stage of hot rolling (roughing) between about 1100 and 1000° C, and flattening of these refined grains can then be effected in a second stage of rolling (finishing) between about 900 and 700° C. Generally a 3 to 1 or 4 to 1 reduction is performed in the finish passes to assure flattening of the recrystallized 30 μm austenite grains to heights (the austenite grain-boundary separation in the through-thickness direction) of 8 to 10 μm. To avoid the mixed grain structures that result from partial recrystallization, a delay is generally introduced after the roughing passes to allow the plate to cool before finishing to a temperature below which only grain flattening occurs. Figure 22—10-A shows the relation between the finishing reduction and the refinement of ferrite grain size in the product as a result of the flattening of the austenite grains. The importance

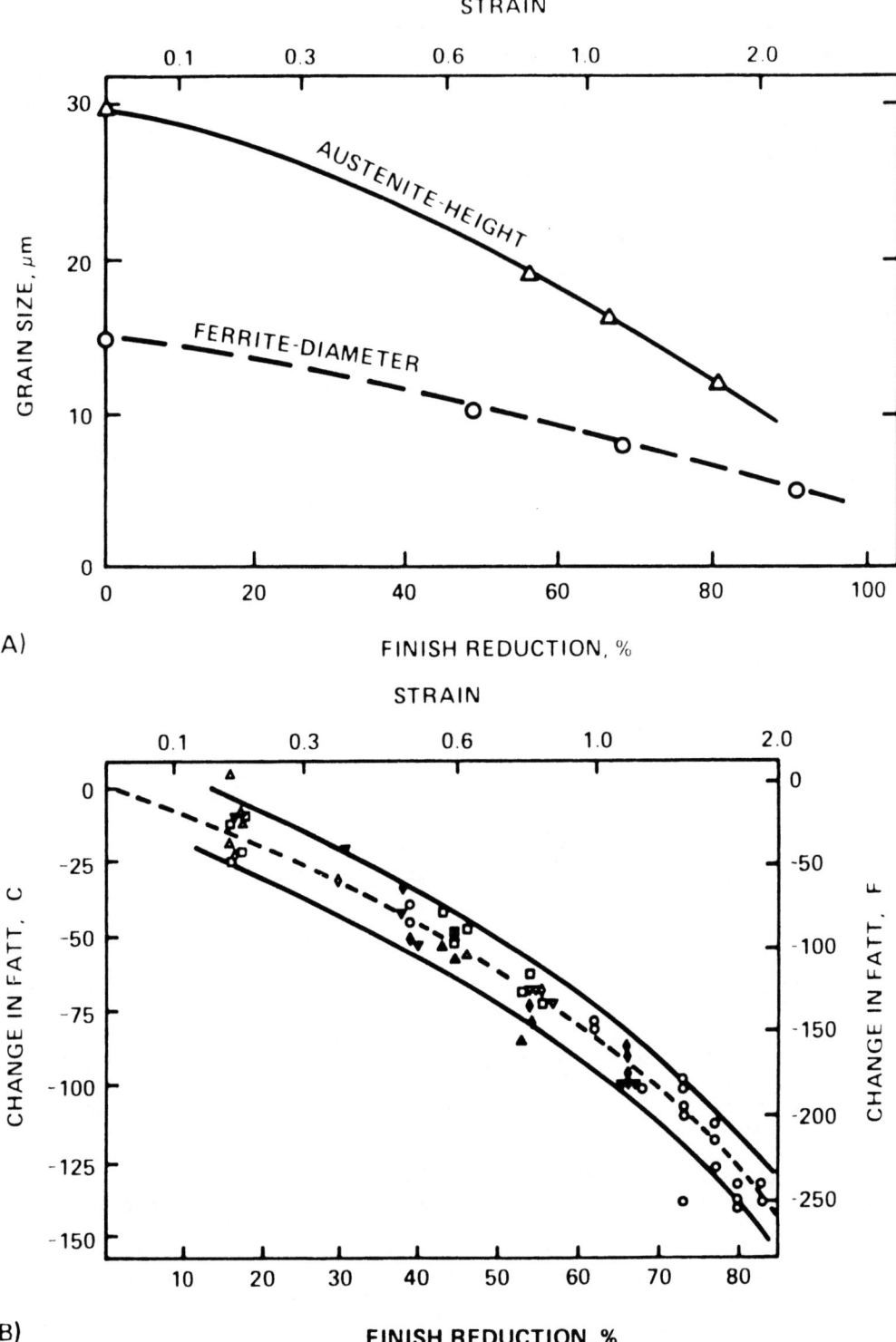

FIG. 22—10. Effect of amount of finishing reduction on: A) The
height of flattened austenite grains and the diameter of the
ferrite grains to which they transform; and B) The change in
the ductile-to-brittle transition temperature for a family of
microalloyed steels.

of this treatment to final toughness is shown in Figure 22—10-B.

To meet the requirements of rolling with lower reheat temperatures, large roughing drafts, and large total reduction at low finishing temperatures, the new generation of rolling mills must be larger and stronger. Four-high mills are an absolute minimum requirement, and considerable investigation of Sendzimir, planetary and other cluster mills is underway. In addition to the rolling itself there is considerable current effort in developing new technology for control of cooling through and after the transformation.

Controlled Transformation—Further strengthening of hot-rolled products may be achieved by intercritical rolling to produce work-hardened ferrite. There is, however, a limit to how far this process can be carried before the increase in work hardening (or dislocation density) begins to deteriorate ductility and toughness. With increasing intercritical rolling there is also an increase in the tendency for splits, or separations, to occur in the plane of the plate when it is loaded in tension in the through-thickness direction. The loss in through-thickness ductility (which has been variously attributed to stringered inclusions, impurity segregation to elongated austenite boundaries, textured ferrite, and weak interfaces between coarse and fine ferrite), may, depending on product specifications, also place a limit on the amount of intercritical rolling that can be performed.

Severe low-temperature rolling of unrecrystallized austenite (control rolling) markedly accelerates transformation thus raising the ferrite-start temperature (Ar$_3$) during the continuous cooling that attends hot rolling. This offsets to some extent the grain refinement achieved by control rolling. To counter this effect small additions of hardenability agents (molybdenum, boron) are used.

Alternatively, high-strength control-rolled product with less severe low-temperature rolling can be obtained by accelerated cooling through the transformation range. The faster cooling rate suppresses the Ar$_3$, thereby refining the transformation products and changing their morphology and distribution. Faster cooling also retains solute for subsequent precipitation strengthening of ferrite. If the austenite entering the transformation is uniformly fine, the transformation products will be fine polygonal ferrite-pearlite or acicular ferrite-fine bainite with good strength and toughness. Any regions of coarse austenite can, however, transform to large patches of upper bainite that will produce a deterioration in toughness. Consequently, accelerated cooling can be used only in combination with control rolling that is effective in refining the austenite.

Interruption of accelerated cooling below the Ar$_1$ followed by slow cooling in the ferrite region, as in coil cooling of strip, will allow full precipitation strengthening of the ferrite by those microalloy constituents that have been retained in solution by the rapid cooling through the transformation. Because accelerated cooling makes more efficient use of both hardenability agents and microalloys, leaner, more economical steels with lower carbon equivalent can be used to meet the requirements for strength, weldability, and ductility.

At the other extreme, deliberate slowing of the cooling rate through the transformation, or an intercritical hold, is being investigated for the production of as-hot-rolled dual-phase steels. The longer times in the intercritical region are necessary to promote sufficient partitioning of hardenability elements from the newly formed ferrite to the remaining austenite pools so that the austenite transforms to martensite or bainite on subsequent cooling.

Reactions in Ferrite—Control of subcritical cooling affects the mechanical properties of the ferrite. For example, reduction of coiling temperature, down to about 600° C, increases the strength of sheet products by limiting ferrite grain growth, by increasing the volume fraction of acicular ferrite, and, in microalloyed steels, by retaining solute to lower temperatures, thereby refining precipitate size and increasing its number density. Rapid cooling to too low a temperature, however, will suppress part or all of the precipitation reaction, thereby causing some decrease from maximum yield strength.

Bibliography

Adams, A. T., Wire drawing and cold working of steel. Sherwood Press, Cleveland, Ohio, 1936.

American Society for Metals, Metals handbook. The Society, Cleveland, Ohio, 1961.

Boulger, F. W., et al., Metal deformation processing (Vols. I, II and III). DMIC Report 208 (August 14, 1964), Report 226 (July 7, 1966), and Report 243 (July 10, 1967). Defense Documentation Center, Alexandria, Va.

Dieter, G. E., Jr., Mechanical metallurgy. McGraw-Hill Book Co., New York, N.Y., 1961.

Dove, A. B., ed., Steel wire handbook. The Wire Association, Inc., Stamford, Conn., 1965.

DeArdo, A. J., Ratz, G. A., Wray, P. J., Thermomechanical Processing of Microalloyed Austenite. TMS-AIME, Warrendale, Pa., 1982.

Grange, Raymond A., Microstructural alterations in iron and steel during hot working. Fundamentals of deformation processing (Backofen, Burke, Coffin, Reed and Weiss, eds.), Syracuse University Press, 1964, pages 299-320.

Hoffman, O., and G. Sachs, Introduction to the theory of plasticity for engineers. McGraw-Hill Book Co., New York, N.Y., 1953.

Leslie, W. C., The Physical Metallurgy of Steels. Hemisphere Publishing, Washington, D.C., 1981.

Lueg, W., Hot rolling of medium hard carbon steels. Stahl and Eisen 57, 1937. Translated by Henry Brutcher, Lansdowne, Pa.

Moses, Louis, The flow of metal between rolls. Iron and Steel Engineer 15, Feb. 1938.

Naujoks, W., Forging handbook. American Society for Metals, Cleveland, Ohio, 1939.

Pearson, C. E., and R. N. Parkins, The extrusion of metals. John Wiley and Sons, Inc., New York, N.Y., (1960).

Tanaka, T., Controlled Rolling of Steel Plate and Strip. Internat. Metals Rev., volume 26, 1981, pp. 185-212.

Trinks, W., Roll pass design, Vol. I, second ed. 1953, Vol. II, second edition 1934, and Supplement to roll pass design, Vols. I and II, first edition 1937. The Penton Publishing Company, Cleveland, Ohio.

Underwood, L. R., The rolling of metals: theory and practice. Chapman and Hall, London (1950).

Wistreich, J. G., Fundamentals of open-die forging. Proc. International Conf. on Research in Production Engineering, Pittsburgh, Pa. American Society of Mechanical Engineers (1963).

CHAPTER 23

Construction and Operation of Rolling Mills

SECTION 1

TYPES OF MILLS

General Classification—The three principal types of rolling mills used for the rolling of steel are referred to as **two-high, three-high,** and **four-high** mills, shown schematically in Figure 23—1. As the names indicate, the classification is based on the manner of arranging the rolls in the housings, a two-high stand consisting of two rolls, one above the other; a three-high mill has three rolls, and a four-high mill has four rolls, arranged similarly. When rolling is in one direction only on **two-high mills,** and the piece is returned over the top of the rolls to be rerolled in the next pass, the mill is known as **a pull-over** or **drag-over** mill. This type of mill formerly was used mainly for production of light sheets and tin plate; it still is used by merchant mills for rolling of tool and high-alloy steels. On **two-high reversing mills,** the direction of rotation of the rolls can be reversed, and rolling is alternately in opposite directions, with work done on the piece while traveling in each direction. The long mill tables of reversing mills make it possible to handle heavy pieces in long lengths that would be impractical to roll on ordinary two-high mills, or to handle on the lift tables of a three-high mill (see below). The reversing two-high type of mill occupies an important position in the industry and, with the use of manipulators, it is possible to produce on it slabs, blooms, plates, billets, rounds, and partially-formed sections suitable for later rolling into finished shapes on other mills. In all **three-high mills** each roll revolves continuously in one direction; the top and bottom rolls in the same direction and the middle roll in the opposite direction. The piece is lifted from the bottom pass to the return top pass by mechanically-operated lift tables, or by inclined approach tables. Usually the large top and bottom rolls are driven, while the smaller middle roll is friction driven. This latter roll is about two-thirds the size of the other two rolls, in order to permit removal through the housing windows. **Four-high mills** are used for rolling flat material, like sheets and plates, and represent a special type of two-high mill for both hot and cold rolling, in which large **backing-up rolls** are

employed to reinforce the smaller **working rolls:** either the working or back-up rolls may be driven. Four-high mills resist the tendency of long working rolls to deflect, and permit the use of small-diameter working rolls for producing wide plates, and hot- or cold-rolled strip and sheets of uniform gage. These mills often consist of a number of stands spaced closely together in one continuous line and are known then as **tandem mills;** the product passes in a straight line from one stand to the next. In **cluster mills** each of the two small working rolls is supported by two (or more) backing-up rolls. This latter type of mill is used for the rolling of thin sheets.

ARRANGEMENT OF MILLS

A single stand mill, which may be either two-, three- or four-high, and either reversing or non-reversing, represents the most common arrangement for rolling a wide range of products, including blooms, slabs, plates, sheets, and various sections. **Guide, loop,** and **cross-country** mills are made up of several two-high or three-high stands, or a combination of both, and are used for the rolling of bar sections. **Guide mills** are small hand mills consisting of several stands of rolls in a **train.** Mills in train have the rolls of separate stands in the same line, the rolls of one mill being driven from the end of the rolls of an adjacent stand. Guide mills take their name from the metal guides which support the piece in the correct position during its passage through the grooves of the various passes. For example, it is possible to roll from an oval section to a round in one pass, provided the oval is supported by metal guides. In many guide mills it is the practice of the catchers, in order to save time, to start the piece through each of the passes before it is through the preceding one, thus forming a loop, resulting in this arrangement being called a **looping mill.** The layout of a mill of this type is shown in Chapter 30. There originated in Belgium the plan for setting up an independent roughing stand pre-

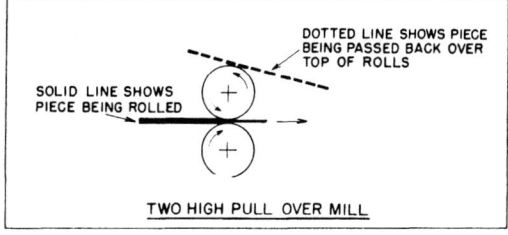

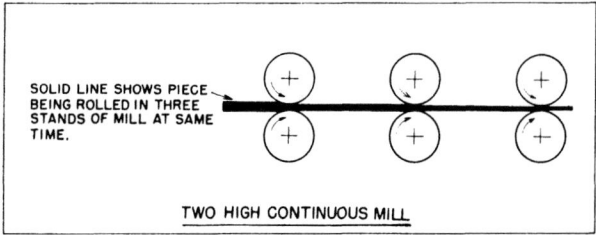

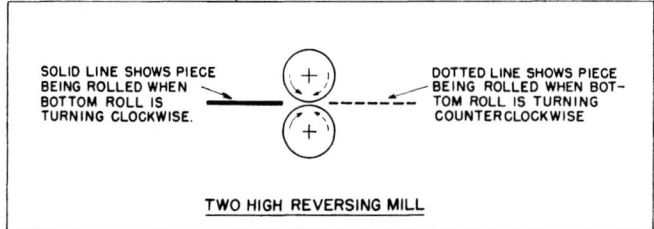

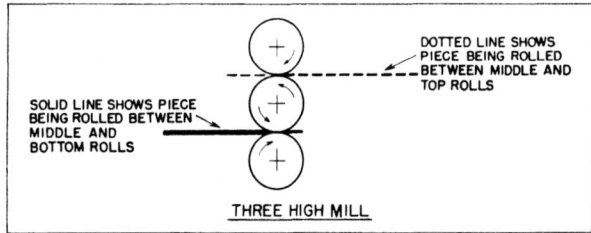

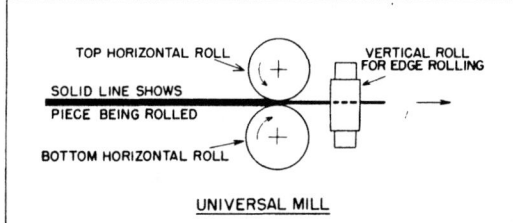

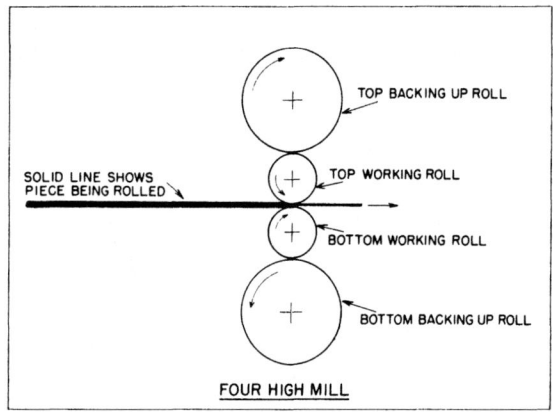

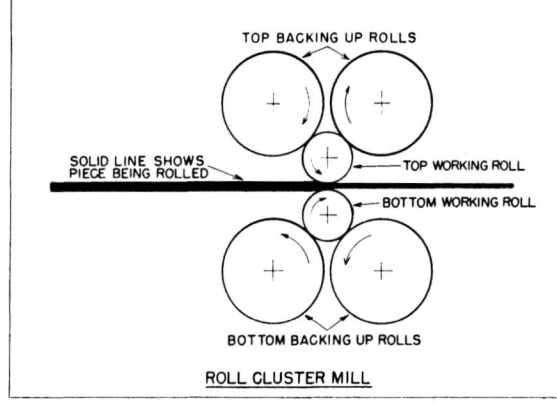

FIG. 23—1. Schematic representation of roll arrangements in the principal types of rolling mills.

ceding the finishing train of the looping mill. This arrangement became known as a **Belgian mill.** On looping mills, it was found that the loop could be made mechanically by a tube or horse-shoe type trough, called a **repeater,** and thus dispense with the hand catchers. Prior to the looping mill, the piece was rolled throughout the entire length in one pass before it could be entered in the next pass. The looping arrangement eliminates the temperature difficulties encountered with long lengths. The shapes produced range from simple rounds and squares to intricate special shapes, but must be relatively small in cross-section. The **cross-country mill** is so named because of the scattered loca-

tion of its roll stands, and was developed for rolling sections that, due to size or shape, are not adaptable to loop rolling. These mills involve the continuous idea, but the stands are placed so far apart that the piece must leave one set of rolls before entering the next (see Chapter 30). To save space and to avoid complicating the drives, the stands usually are arranged in two or more parallel lines, and the direction of travel of the piece is reversed during the rolling by employing transfer and skid tables. This arrangement results in a high-production mill of great flexibility, which may be used for a wide range of products, including structural shapes, rails, and splice bars. A **continuous mill** consists

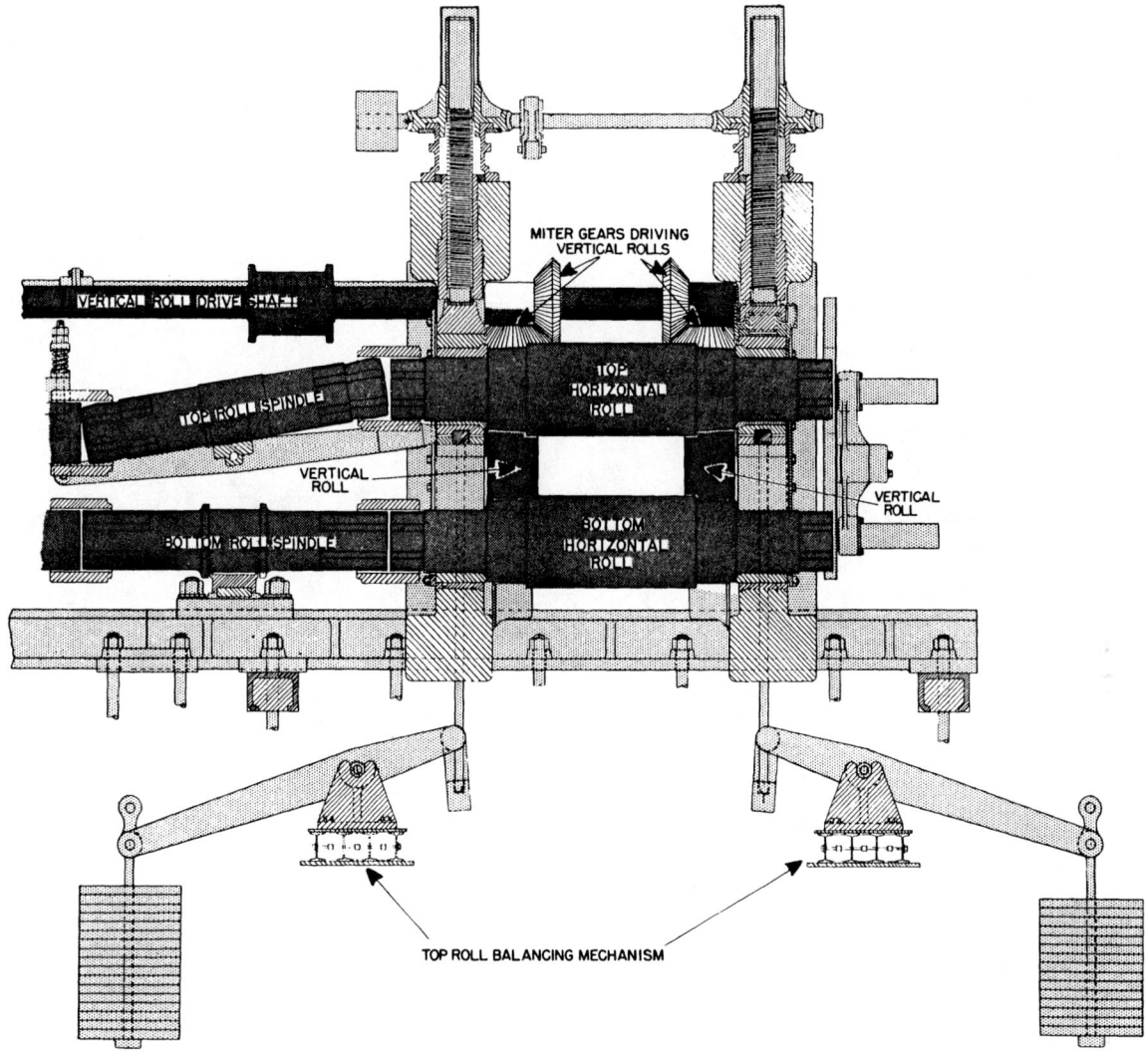

FIG. 23—2. Diagrammatic layout of the principal parts of a universal mill for rolling plates.

of several stands of rolls arranged in a straight line (in tandem), with each succeeding stand operating with roll surface speed greater than its predecessor. Such a mill is illustrated diagrammatically in Chapter 20. Reduction takes place in several passes at the same time until the piece emerges as a finished shape from the last roll stand. This type of mill is in very common usage for rolling strip, sheet, billets, bars, rods, etc. A **semi-continuous mill** comprises also a reversing roughing stand for reducing the piece prior to entering the continuous mill for reduction to the finished shape. This arrangement gives moderately high production with lower first cost than a continuous mill. **Combination mills** are those in which the roughing or major part of the reduction is performed in a continuous mill, and the shaping in a guide or looping mill.

SPECIALTY MILLS

The **universal mill** is a combination of horizontal and vertical rolls, usually mounted in the same roll stand (Figure 23—1). The mill is made up of two-high (and

occasionally three-high) horizontal rolls, with vertical roll sets on either or both sides of the horizontal stand. The vertical rolls also usually are driven. The direction of the piece is reversed after each pass in the mill. The universal mill is used to a limited extent also for plate product that requires rolled edges (see Figure 23—2). A special type of universal mill, known as the **Gray mill,** is well adapted for rolling beams and H-sections of great width and depth without taper on flanges (see Figure 23—3.) The horizontal rolls work on the web and flange thickness, while the idler vertical rolls in the same stand work simultaneously on flange thickness only. The roughing stands and intermediate stands are of the reversing type, and each has a separate stand of driven horizontal edging rolls which work on the flange height only. The finishing stand consists of the horizontal and vertical rolls in which the beams are given one pass only.

The **Wenstrom mill** is a similar modification of a universal plate mill, designed principally for rolling flats. Instead of acting upon the top and bottom and the two

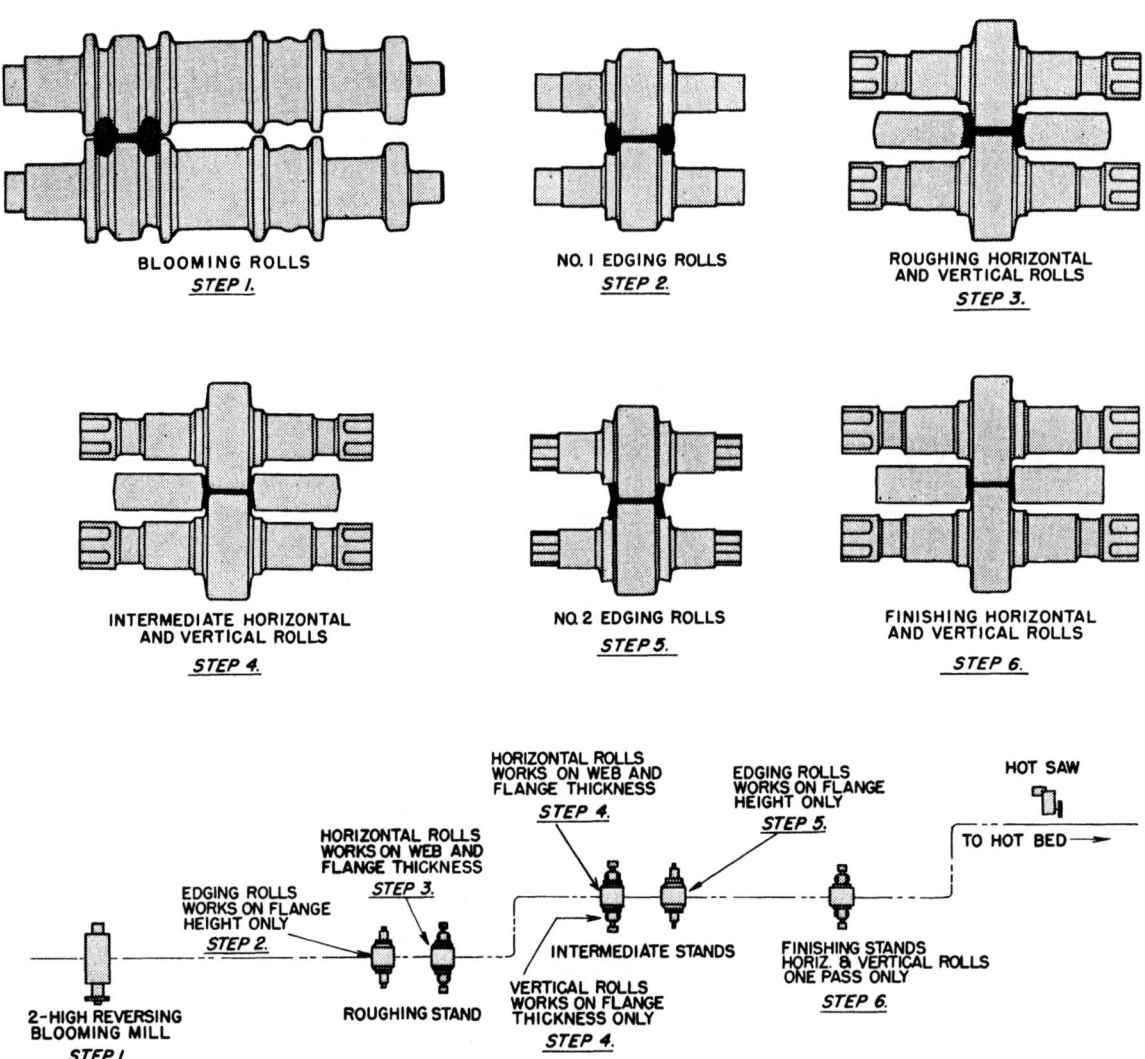

BLOOMING ROLLS
STEP 1.

NO. 1 EDGING ROLLS
STEP 2.

ROUGHING HORIZONTAL
AND VERTICAL ROLLS
STEP 3.

INTERMEDIATE HORIZONTAL
AND VERTICAL ROLLS
STEP 4.

NO. 2 EDGING ROLLS
STEP 5.

FINISHING HORIZONTAL
AND VERTICAL ROLLS
STEP 6.

HORIZONTAL ROLLS
WORKS ON WEB AND
FLANGE THICKNESS
STEP 4.

EDGING ROLLS
WORKS ON FLANGE
HEIGHT ONLY
STEP 5.

HOT SAW

TO HOT BED →

HORIZONTAL ROLLS
WORKS ON WEB AND
FLANGE THICKNESS
STEP 3.

EDGING ROLLS
WORKS ON FLANGE
HEIGHT ONLY
STEP 2.

INTERMEDIATE STANDS

FINISHING STANDS
HORIZ. & VERTICAL ROLLS
ONE PASS ONLY
STEP 6.

2-HIGH REVERSING
BLOOMING MILL
STEP 1.

ROUGHING STAND

VERTICAL ROLLS
WORKS ON FLANGE
THICKNESS ONLY
STEP 4.

FIG. 23—3. (Above) The rolls and their functions in forming an H section by rolling in two-high and wide-flange mills. (Below) Flow diagram showing layout of the mill utilizing the rolls shown above.

sides at different times, it does this simultaneously. The top roll can be adjusted vertically, and the bottom roll transversely, whereby pieces of different thickness and width can be produced with the same set of rolls. The **Sack** universal mill, designed principally for rolling **cruciform sections,** has horizontal and vertical rolls which act upon the piece simultaneously, the general arrangement being much like that of the Wenstrom mill. A somewhat similar principle is employed in the **Schoen mill** for rolling of railroad car wheels. whereby the tread and flange are rolled simultaneously with the web, while rotating the forged wheel blank in a vertical position. This is accomplished by a pair of driven web rolls, and an idler tread roll in simultaneous contact with the wheel blank. A pair of idler rim rolls controls the width of rim (see Chapter 28).

For rolling of billets, rods, and narrow slabs, a continuous mill called a **Morgan mill** may be used (see also Chapter 31). It consists of a series of horizontal roll stands arranged one after the other, so that the piece is

being rolled in a number of stands at the same time. The drive for each stand is through bevel gearing, with roll speed of each stand so proportioned that each set of rolls travels at a greater speed than necessary merely to compensate for the increase in speed due to elongation of the piece in the preceding pass, in order to keep the piece under tension at all times. Twist guides may be used to turn the piece 90° between passes. The **Garrett mill** (see also Chapter 31) is used for rolling small rods; it reduces the billet section in a roughing mill, which may be a looping train of three stands, or two or more stands arranged in tandem, followed by two trains of four or five stands each, lying end to end along two parallel lines in close proximity to each other, and so placed that the first intermediate pass is in line with the last roughing pass, and far enough away to give proper clearance for the pieces. Generally, catchers are employed on the oval side, while repeaters are used only on the opposite (or square) side of the intermediate and finishing trains. The Sendzimir **planetary hot-rolling**

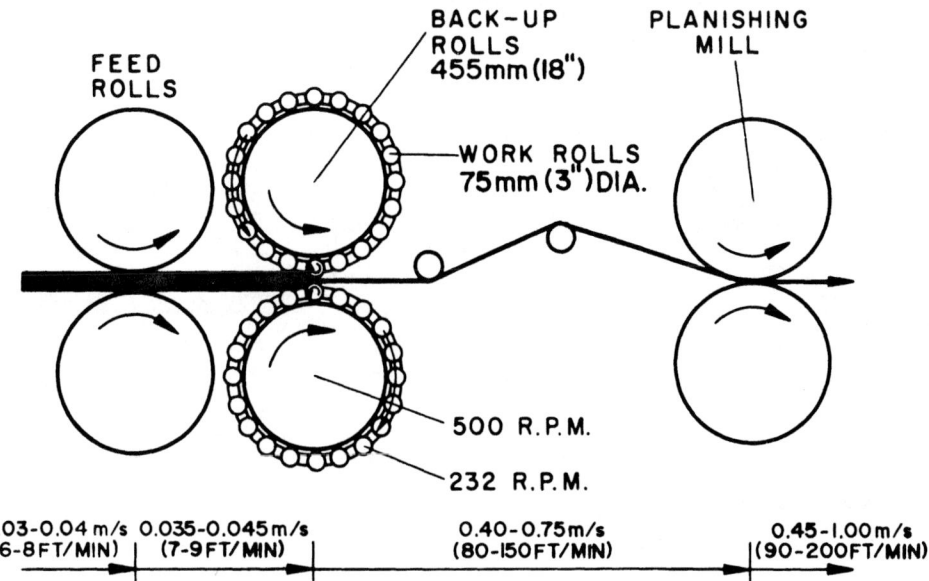

FIG. 23—4. Schematic representation of a planetary hot-rolling mill. Slab speed, roll speeds and product speeds were taken from a paper by H. W. Ward in the January 1958 issue of "Iron and Steel" and represent a typical set of operating conditions on an existing British mill.

mill (Figure 23—4) was developed to reduce slabs to coiled hot-rolled strip in a single pass. Such a mill consists of the planetary assemblies proper, and feed rolls (one pair preceded by a pair of pinch rolls or two pairs of feed rolls) which push the slab (taking a small reduction) through a guide into the planetary rolls where the main reduction takes place. Other means than rolls have been used for pushing slabs into the mill. The mill is followed by a two-high or four-high planishing mill and coiler. The planetary assemblies consist of two back-up rolls surrounded by a number of small work rolls that are are mounted in "cages" at their extremities. The cages are synchronized by external means so that each pair of opposed work rolls passes through the vertical center line of the mill at precisely the same time and so that the axes of the work rolls are always parallel to the axes of the back-up rolls. The angular velocity of the cage is somewhat less than half the angular velocity of the back-up roll. As shown in Figure 23—4 the cages rotate in the same direction as the corresponding back-up rolls, while the work rolls rotate in the opposite direction (i.e., the work rolls turn clockwise when their back-up roll and cage turn counterclockwise). Mills are built with different ratios between back-up roll and work-roll diameters, and with different numbers of work rolls per back-up roll (18, 20, 24, 26, and 30 work rolls per back-up roll have been employed on different existing mills). The rolling process is cyclic. Work rolls make contact with the unworked portion of the slab, then work downward gradually, making a rolling pass in the deformation zone and finally breaking contact with the material where it has reached the final thickness but not before the next pair of work rolls has contacted the slab. Generally speaking, the temperature of the strip is higher at the exit side of the planetary mill than the temperature of the slab at its entrance, due to the energy of plastic defor-

mation involved in the heavy reductions (as much as 98 per cent) that the mill is capable of effecting; this permits the rolling of slabs at lower entering temperatures than in conventional hot-strip mill practice. Mills of this type can be designed up to 2030 mm (80 inches) wide.

Sendzimir cold-rolling mills (frequently referred to as Z-mills) feature several roll arrangements, the predominant one of which is the 1-2-3-4 arrangement illustrated in Figure 23—5, which is the only one that will be discussed here. In this design, each work roll is supported throughout its entire length by two first-intermediate rolls that are, in turn, supported by three second-intermediate rolls (of which the outer ones are driven) which transfer the roll-separating forces to a rigid, one-piece cast-steel housing through four backing assemblies. The work rolls are driven by the four driven rolls through friction contact with the first intermediate rolls. Screwdown is controlled by rotation of the primary eccentric rings (see Figure 23—5) on the bearing shafts of the two top-center backing assemblies; hydraulic-cylinder-driven racks rotate the shafts by engaging pinion teeth mounted at both ends of each shaft. The two bottom-center bearing assemblies are similarly equipped to permit positioning of the lower work roll to balance the pass line. The two remaining pairs of eccentric shafts at the left and right of the diagram are also adjustable as pairs to maintain roll-opening capacity while compensating for roll wear. Depending upon the size of the mill and the roll arrangement, either mechanical or hydraulic means can be provided for crown adjustment by positioning individual secondary eccentric rings (see Figure 23—5) at the saddles. On most Sendzimir cold mills, such adjustment is possible during the operation of the mill. Lateral axial adjustment of the first-intermediate rolls (that are provided with minute crown or taper reliefs at opposite ends, as shown in Figure 23—5) enables the set-up of

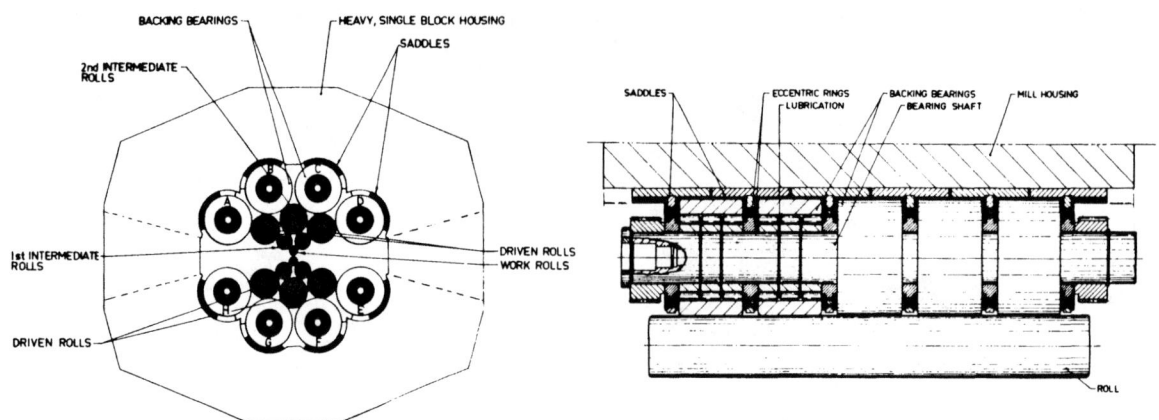

PRINCIPLE OF ROLL ARRANGEMENT AND SUPPORT
IN A SENDZIMIR MILL

TYPICAL BACKING ASSEMBLY OF A SENDZIMIR MILL

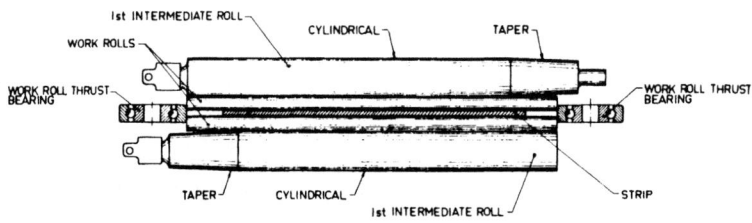

LATERAL ADJUSTMENT OF 1st INTERMEDIATE ROLLS ON A SENDZIMIR
MILL FOR STRIP SHAPE CORRECTION AND PROVIDING PRESSURE CONT-
ROL ALONG STRIP EDGES

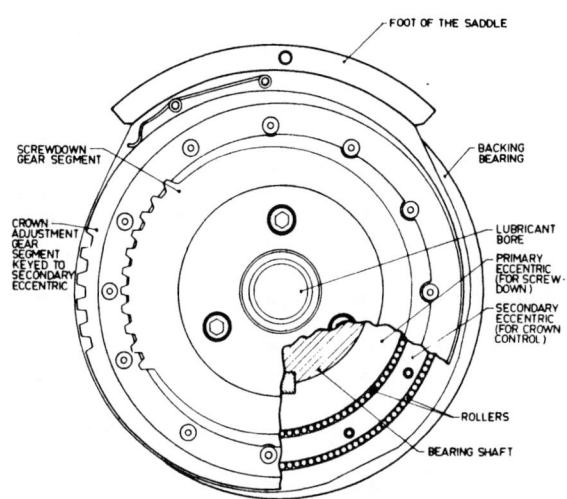

THE SENDZIMIR PRINCIPLE OF MILL SCREWDOWN AND
CROWN CONTROL
(END ELEVATION OF A BACKING ASSEMBLY WITH DOUBLE ECCENTRICS)

FIG. 23—5. Roll arrangement, details of backing assembly and lateral adjustment of first intermediate rolls on a Sendzimir cluster mill of the 1-2-3-4 type.

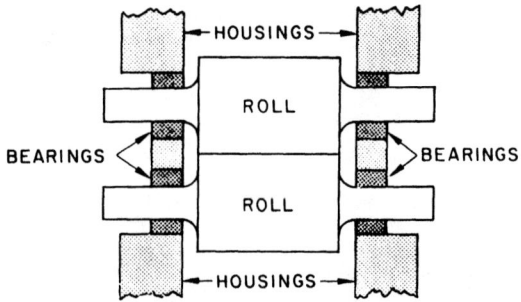

A. CONVENTIONAL ROLL MOUNTING

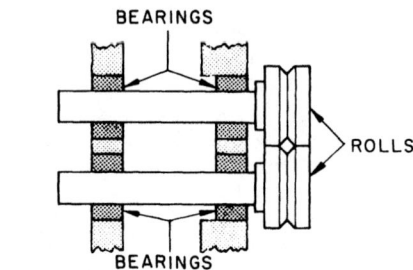

B. CANTILEVER ROLL MOUNTING

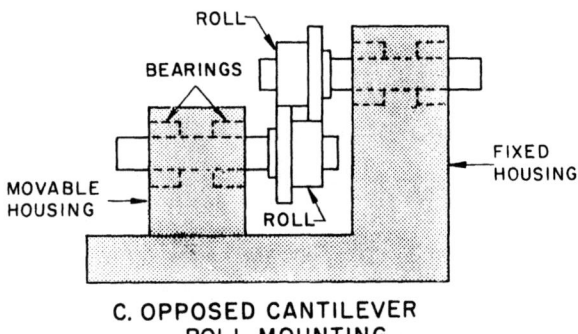

C. OPPOSED CANTILEVER
ROLL MOUNTING

FIG. 23—6. Schematic representations of three ways for mounting rolling-mill rolls in their housings.

the mill to be changed quickly for rolling strip of various widths, thicknesses, and hardnesses.

The **Unitemper mill** is used as one means for temper rolling or skin rolling of finished cold-reduced strip. This mill consists of two stands of two-high mills, but with the second stand above the first in the same mill housing, to provide closer spacing between the two stands. With the motor of the first stand acting as a generator, and the motor of the second stand acting as a motor, the strip can be elongated a controlled amount between passes with only a very slight reduction in each stand due to rolling.

Die rolling is the process of rolling a string of blanks, each of which has varying cross-sectional area produced by heavy reductions, and specified center to center length. When sheared to length the blanks are of identical shape. Products such as automobile axles and crank-shafts are produced satisfactorily by die rolling. The blanks are rolled with or without flashing, depending upon the particular product section.

For producing seamless tubing, the types of piercing mills include the **Mannesmann** or parallel-axis barrel-type roll piercing mill, the 60° **cone-roll** piercing mill, and the **Stiefel** or 180° disc mill. For the operating principles of these mills, refer to Chapter 32. A **continuous tube rolling mill,** built in 1948 by United States Steel Corporation, consists of nine tandem individually powered stands of two-high grooved rolls. The rolls in the consecutive stands have their axes at 90° with each other. A cylindrical mandrel extends entirely through the pierced billet and passes through the mill with the work piece. The reheated tubes are processed in a tension reducing mill consisting of sixteen two-high roll stands, without the use of a mandrel. For producing buttweld pipe the **Fretz-Moon tube mill** may be used. The skelp on leaving the furnace enters the first pair of rolls which form the piece downward, with the edges still apart. The second pair form the complete circle and bring the edges of skelp together, and the welding of the edges by their own heat takes place. The mill actually consists of three pairs of horizontal rolls and three pairs of vertical rolls. The last four pairs reduce and size the pipe, and furnish traction for pulling the skelp through the furnace and forming rolls.

Opportunity will be given in later sections to become better acquainted with the operating details of most of these mills. The types of rolls used in the more common mills, and their manufacture, will be discussed in Section 3 of this chapter.

Types of Roll Mountings—In general, there are three ways in which rolls may be supported in their housings. These methods are illustrated schematically in Figure 23—6.

The conventional method (Sketch A in Figure 23—6) is the manner in which rolls for two-high, three-high and four-high mills usually are mounted. The bearings, carried in fixtures supported by the mill housings, are at the ends of the rolls. The roll working surface, therefore, is between the bearings.

Sketch B of the illustration shows the cantilever roll-mounting method. The roll working surface is outside the bearings. Bearing size is limited by the diameter of the rolls. This type of mounting is employed on some mills using rolls of relatively small diameter (for example, continuous rod mills described in Chapter 31). The rolls are readily accessible for changing.

The opposed-cantilever roll-mounting principle is illustrated in Sketch C. The bottom-roll housing can be moved in and out in relation to the fixed housing to make the rolls easily accessible for changing and to permit work to be entered between and disengaged from the rolls without difficulty. This principle originally was employed in mills for rolling rings and tires for railway wheels. Bearing size is not limited by the diameter of the rolls with this particular type of mounting.

Mill-Size Description—Mills are classified by descriptive dimensions which tell the size of the mill, the arrangement of stands, and the product rolled. The source of the dimensions is different, depending upon the type of mill and its product.

In mills rolling plates, wide strip and other flat-rolled products, the descriptive dimension is the length of the roll body, except in universal plate or slab mills for which it is the maximum distance between vertical rolls. In mills rolling other products such as rails, blooms, structural shapes and billets, the descriptive distance is the distance between centers (pitch diameter) of the pinions, or, if no pinions are used, the approximate maximum diameter of the rolls. Thus, typical descriptions might be: a 4065-mm (160-inch) plate mill, a 2135-mm (84-inch) hot-strip mill, an 1145-mm (45-inch) universal slabbing mill, a 1090-mm (43-inch) blooming mill, a 355-mm (14-inch) structural mill, and so on.

SECTION 2

ROLLING-MILL ACCESSORIES

Many of the accessories of rolling mills are common to all types of mills, differing in design and operations to conform with the conditions in a particular mill. In addition to the rolls (described in Section 3 of this chapter), essential parts include the mill **drive, lead spindle, pinions** and their **housings, spindles** and **coupling boxes, chock bearings, screws** and **screw-down mechanism, edgers, front** and **back mill tables, manipulators** and **side guards, entering** and **delivering guides** and **roll-changing devices.** The newer mills may be equipped also with various control devices, such as pressure meters and automatic roll-setting devices. Many of these parts are discussed in the description of particular types of mills in later chapters. In the following, a discussion is presented of those parts essential to the operation of the rolls. To aid in locating some of the various parts described below, Figure 23—7 shows a cross-sectional diagram of a high-lift reversing blooming mill.

Lead Spindle—The lead spindle is used to connect the prime mover with the pinions, and may be of the universal type, either short-coupled, or long with carrier bearings, depending on the position of the motor in the layout. If short-coupled, standard flexible couplings can be used. The lead spindle is attached to the bottom pinion of two-high mills, and to the center pinion of three-high mills. If the spindles are of the universal type, they can be of generous proportions, as their diameter is not limited by the space available. Lubrication of all working surfaces is important. Grease usually is applied through a system of holes drilled in the jaws, and connected to various designs of spring-loaded cavities in the spindle body. Oil-mist lubrication may be used in some instances. The connection at either end of the spindle may be made by a **coupling box,** which is a hollow cylindrical casting corresponding in section to the wobbler, with one end fitting the spindle and the other the pinion (shown in Figure 23—13 in Section 3 of this chapter). The coupling box usually is made deliberately as the weak spot of the mill, either on the lead spindle or the spindles on the roll end. In the event of extreme overloading of the mill, the box breaks and disconnects the motor from the mill. The minimum length of spindle is slightly over twice the length of the coupling box that is used.

Mill Pinions—The pinions are gears serving to divide the power transmitted by the drive between the two or three rolls, driving the adjacent rolls in opposite directions. If twin-motor drives are used, no pinions are required, since the power is transmitted directly to

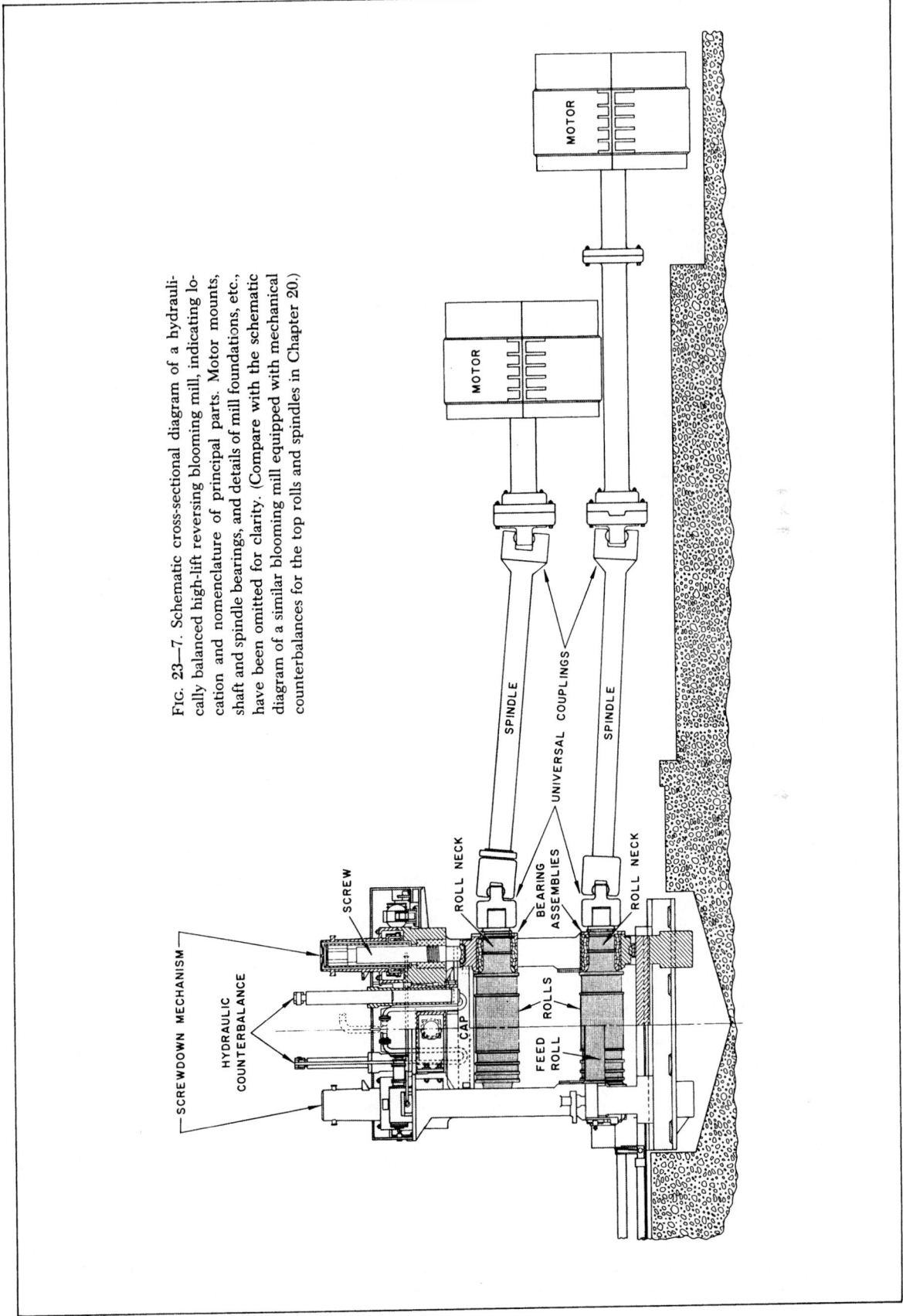

Fig. 23–7. Schematic cross-sectional diagram of a hydraulically balanced high-lift reversing blooming mill, indicating location and nomenclature of principal parts. Motor mounts, shaft and spindle bearings, and details of mill foundations, etc., have been omitted for clarity. (Compare with the schematic diagram of a similar blooming mill equipped with mechanical counterbalances for the top rolls and spindles in Chapter 20.)

each roll. The earlier pinions had either spur teeth or a divided face and staggered spur-type teeth, but the present practice is to use double helical teeth. The helical gears give a smoother drive, as some parts of the teeth are in contact at all times, making the transmission of power continuous. When rolling certain grades of materials, operation of the old type of pinions with spur or cast teeth was characterized by jarring of the meshing teeth. This mechanical jarring was transmitted to the rolls and produced marks on the product being rolled. Pinions are made of cast or forged steel. They are mounted in babbitted bearings and set in housings similar to those used for roll stands. The pinion stand must be of adequate strength to withstand the over-turning effect of the full and maximum torque of the drive motor. Housings should be sealed to eliminate dirt and scale, and forced lubrication should be provided for the pinions. Except in plate, direct-drive and universal mills, the distance from center to center of pinions determines the size of the mill. The pinions absorb about 6 per cent of the power transmitted

Spindles are used to connect the pinions with the rolls if the mill is not a direct-driven type; in the latter case the spindle is connected directly to the motors. Spindles are made of cast or forged steel and are fitted at each end with wobblers similar to those on the rolls (Figure 23—13), or with the universal couplings, depending upon the type of mill. The bottom spindle runs in spring-balanced carriers, and the top spindle is supported at its center of gravity by carrier bars balanced hydraulically or by counterweights. The spindles are supported by babbitt or composition bearings, and should operate as nearly level as possible to prevent excessive power loss. It is very difficult to operate with a spindle more than 15 degrees out of level. This angle may be kept within the desired limits by increasing the length of the spindles. If the angularity is greater than 6 degrees or 7 degrees, a universal coupling should be used, in which case the ends of the wobbler are cut from a section of a sphere to give them the rounded form necessary to permit them to work at different angles, and the spindle is supported by saddle bearings in a carrier which rises and falls with the top roll and holds the spindle in place.

Bearings—The bearings, which support the roll necks and align the rolls, may be of two general types: roller bearings or the older chock-type bearings. One type of roller bearing is illustrated in Figure 23—8.

Roller bearings first were used for rolling mills in Czechoslovakia in 1921 and in Sweden in 1922, where they were applied to small hot-rolling mills. In America, use of the roller bearing accompanied the development of the cold-reduction mill for strip and sheet, and its adoption has been extensive since 1932. The development of this type of bearing was slow on account of the difficulties that had to be overcome: (1) the roll necks had to be of adequate diameter to withstand the rolling loads; (2) the roll necks had to be kept free from wear; (3) the bearings had to permit quick roll changes; (4) the bearings had to permit adjustments of the roll laterally as well as vertically; (5) the bearings had to be self-adjusting to take care of slight changes in the shape of the rolls caused by heating and bending; (6) the bearings had to have ample carrying capacity; and (7)

in case of roll breakage, provisions had to be made to keep the bearings themselves from being damaged. Roller bearings are now common in new mills, although they have not replaced entirely the chock type of bearing, which possesses the advantage of simpler construction.

Modern roller bearings are showing extremely long life, with resulting lower cost per ton of product. Mills equipped with roller bearings may be expected to use from 15 to 20 per cent less power for the main drives than mills using bronze or conventional bearings. As roller bearings require no adjustment for bearing wear, greater tonnages can be rolled with less variation in section from piece to piece, permitting requirements to be met with greater ease and improved mill yield. Roller bearings also result in a general reduction in mill maintenance, and, as an automatic greasing system is used for lubrication, a clean mill operation is effected. Figure 23—8 (above) shows a horizontal section through the center-line of a bottom work-roll chock (sometimes called chuck) on a 4065-mm (160-inch) four-high reversing plate mill utilizing roller bearings.

Oil-Film Bearings—Figure 23—8 (below) shows a sectional view of the type of bearing known as an oil-film bearing, applicable to both primary and finishing rolling mills, either hot or cold. An alloy-steel forged sleeve, hardened and ground to a mirror finish, is keyed to the tapered roll neck. Surrounding the sleeve is a bushing made with a special alloy bearing metal accurately finished in the bore which mates with the sleeve to form the bearing surface that carries radial loads. The bushing is secured in the chock with a lock pin.

At the outboard end of the assembly, on one end of the roll, is a thrust bearing which in the larger-size bearings is of the roller type. The thrust bearing is independent of the radial bearing and can adjust itself to roll-neck deflection or other misalignments during mill operation. The bearing at the other end of the roll is free-floating, to compensate for dimensional and thermal expansion variables. A threaded ring, lock nut and end cover complete the bearing assembly. The lock nut has two functions: (1) when a bearing is being installed, it pushes the bearing firmly onto the roll neck and keeps it there; and (2) when a bearing is being removed, it acts as a mechanical puller to unseat the taper bore from the roll neck. A labyrinth seal at the inboard end of such bearings minimizes the possibility of contaminants entering the bearings.

Oil-film bearings have long life, a high load-bearing capacity and a low coefficient of friction. Their design makes bearing changes simple and rapid. They are adaptable to mills designed for contour control that employ work-roll or backup-roll bending as a means of strip-shape and/or gage control. They are also used on the new fast-response hydraulic-loaded type of mills—some with roll-gap sensing and possibly employing pre-stressed elements for control purposes. In recent years, the hydrostatic principle of lubrication has been applied to oil-film bearings for some rolling mills to provide very low starting torque under load.

Chock Bearings—The chocks usually are made in two parts, the bearing with the surface of its lining in contact with the roll neck, and the chock, which holds

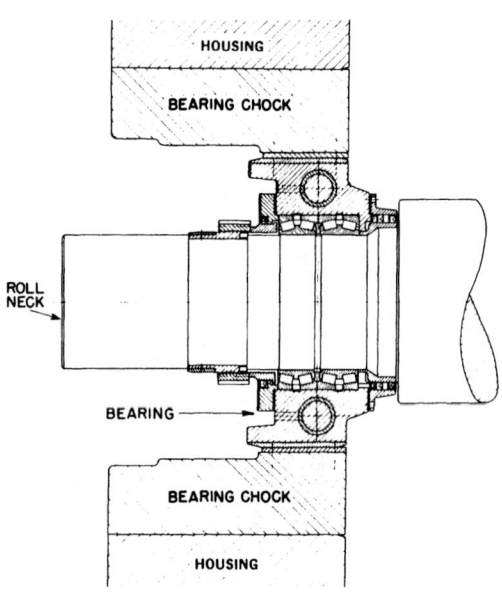

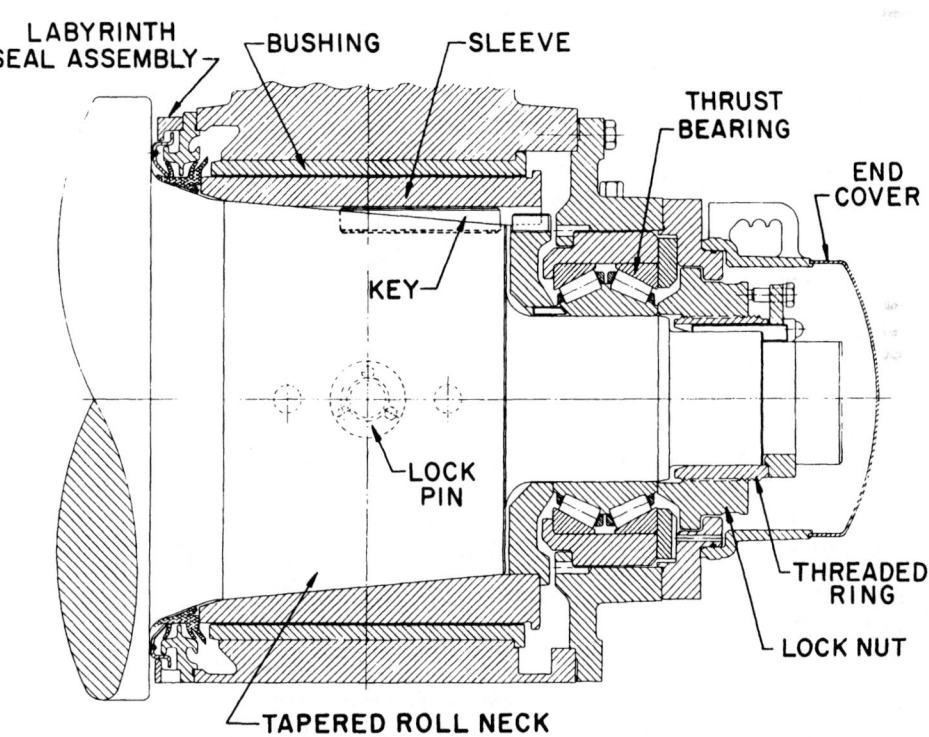

FIG. 23—8. Horizontal sections (not to same scale) through the centerlines of (above) a bottom workroll chock on a plate mill equipped with roller bearings, and (below) a roll-neck bearing of the oil-film type. (Illustration of oil-film bearing by courtesy of Morgan Construction Company.)

the bearing in place against the roll neck. The bearings may be lined with babbitt or may be made with babbitt, brass or bronze inserts or they may be nonmetallic bearings of the phenolic-resin type.

The alloys used in the bearings vary considerably. Table 23—I gives approximate compositions of some of the more common alloys used for bearings.

Table 23—I. Compositions of Typical Bearing Metals

Metal	Cu (%)	Zn (%)	Sn (%)	Sb (%)	Pb (%)
Red Brass	85	15	—	—	—
Yellow Brass	65	35	—	—	—
Bronze, No. 1	85	—	15	—	—
Bronze, No. 2	82	15	3	—	—
White Metals	—	—	10-15	12-20	65-80
Babbitts*	—	—	59-91	4-12	0.35-26

*As, 0.1 max.; Bi, 0.08 max.; Fe, 0.08 max.

The tin in babbitt bearing metals can be replaced satisfactorily in most instances by substituting lead-base babbitts for tin-base babbitts and making, in some cases, additions of silver or arsenic to increase the hardness of the alloy at elevated temperatures.

As bearings made of all of these metals are soft and not very strong, it is necessary to insert them in supporting castings which are set into the housings. These castings are box-like in shape, each one containing on one side a semicircular groove corresponding to, but larger than, the necks of the rolls. In order to reduce weight, the castings are cored out, and may be of either iron or steel.

Reference has been made above to the use of bearings of the phenolic resin type. These composition bearings of a laminated type operate with water as a lubricant and have replaced the bronze and babbitt types of bearings on a number of mills, including some blooming mills, plate mills, bar and billet mills, rod mills, strip mills, skelp mills, tube mills, three-high sheet mills, and structural mills. Bearings of hard wood, with water as a lubricant, also have been used successfully for certain mills. The advantages of the composition and wooden bearings include longer life, freedom from greasing, and reduced power requirements. Failure of this type of bearing in some mills may be due to one of several causes. The heat conductivity of such bearings is very low, and heat generated by friction must be removed continuously with adequate cooling water. Roll necks for these bearings must be smooth to prevent excessive bearing wear.

Arrangement of Chock Bearings—In a two-high mill, a chock is placed under each of the necks of the bottom roll and above each of the necks of the top roll. In case the top roll is adjustable, light bearings must be placed also under the necks of this roll for support; the upward thrust of the top-roll balancing device is exerted against these supporting bearings. The top-roll balance is for the purpose of keeping: (1) the top-roll necks in contact with the upper chock bearings; (2) the chocks tight up against the screw points; and (3) the screw-thread surfaces in contact. On heavy mills, the counterbalance may be of the overhead hydraulic-cylinder type connected to an accumulator system, or of the underneath counterweight-and-lever type; both types of counterbalances can be compared in Figures 23—7 and 20—19. On small mills, screw bolts extending through the housing serve the same purpose. The exposed half of the necks of the lower roll usually will be covered to protect them from scale, etc.

The arrangement of chocks in three-high mills is more difficult. The simplest way is to place double-groove chocks between the top and middle and the middle and bottom rolls, and then set them in the housings one above the other, so that all the adjusting made necessary by the wearing away of the bearings and the material of the rolls themselves may be made with large screws in the top of the housing that can be turned to move upward or downward. However, this arrangement causes the bottom bearing to wear down rapidly and increases the power required to drive the mill, owing to the additional friction induced on the bottom bearing by the weight of the upper two rolls and their chocks. This fault may be overcome in two ways: (1) by making the bottom roll fixed and supporting the weight of the upper two rolls and their chocks on the shoulders of the chocks themselves, the distance between rolls may then be regulated with shims, or "liners," by adding or removing the shims as the bearings wear down; (2) a better way, and the one most often employed in modern mills, is to make the middle roll fixed, in which case the bottom roll is raised or lowered by an adjusting wedge attached to a screw in the housing which permits it to be moved back and forth with a wrench from the outside of the housing.

In all mills, two-high as well as three-high, the top chocks are held down by two strong screws which work in threaded holes or nuts in the tops of the housings.

The functions of the chocks are not only to furnish vertical bearings for the rolls but also to prevent their movement laterally as well. This lateral displacement of the roll is prevented by the inner edge of the bearing which is formed to fit against the shoulder of the roll. Adjustment for wear in this direction is provided by adjusting screws which extend through the side of the housing and bear on the ends of the chocks. This lateral adjustment is a matter of great importance in rolling sections that require grooved rolls, the reason for which is self-evident.

Housings—There are two housings for each stand of rolls (see Figure 23—9.) They may be made either of iron or steel, the choice of materials depending upon the size of the mill and the strength required. Generally they are of annealed cast steel, but housings may be of welded steel construction, heavy plates or slabs. The housings are of an "O" or "U" form, each having an opening called the **window,** which serves as a receptacle for the bearings. Housings may be either **closed top** (O) or **open top** (U). In the former, the base, the two legs and the top are all in one piece, while in the latter, the top will form a separate part which can be removed. The base of the housing is cast with a projection on each side to form the **feet.** In the bottom of each foot is cut a groove which fits over a **shoe,** running parallel to the rolls. Suitably shaped bolts then serve to clamp the foot of the housing to the shoe, which is fastened firmly to the foundation by long bolts. This method permits the housing to be moved laterally, and facilitates the lining up of the mill. The tops of the two

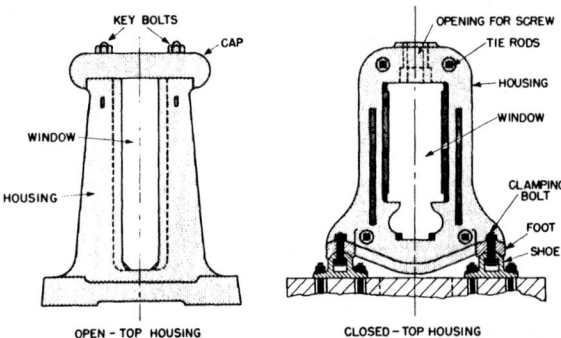

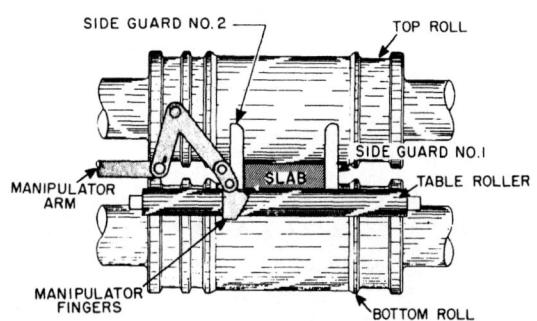

FIG. 23—9. Diagrams indicating principal parts and their names for (left) an open-top housing and (right) a closed-top housing.

housings in a set are prevented from spreading apart by suitable **tie rods** or, in the case of open-top housings, the top for both housings may be cast in one piece. Similarly, tie rods usually will be placed at the bottom. Recesses or other openings are cast in the inside face of each housing to receive the supports for the guards and guides, these supports being usually in the form of square **rest bars** which extend from housing to housing in front of the rolls. The immense pressure applied to the rolls between the top and bottom of the housing acts as a stretching force on the uprights of the housings; the degree to which the housings resist this force is an important factor in determining the reduction that can be effected in one pass and also the exactness with which the thickness of the piece is controlled. The uprights of the housings sometimes are referred to as **posts**.

A **screw down mechanism** is used on mills to position the top roll for each pass through the mill, except on continuous and three-high mills where fixed passes are used. The top roll is adjusted by **screws** which extend through the top of each housing. The screwdown bearings are of roller or antifriction type. The transmission of power was first accomplished by hydraulic pressure, but in modern mills the speed of the operation has been greatly increased by electric drives. On one recent installation, the maximum rate of raising or lowering the top roll is 0.25 m/s (50 ft/min.), compared with 0.010 or 0.015 m/s (2 or 3 ft/min.) on the older mills. On small mills where the adjustment is only occasional, the screws will be operated by hand with spanner bars. In all cases, compression of these screws is unavoidable and, combined with the stretch of the housings plus deflection and deformation of rolls and bearings, produces the **spring** of the mill, which in some cases is a surprisingly large amount.

Edgers or **edging rolls** are used to give a universal or rolled edge to the product. The edging unit may be a roll stand separate from the horizontal stands, but in the universal-type mill the edging unit usually is attached to the main-roll housings. The edging unit consists of two vertically-mounted rolls, each manipulated by screws of identical size, so that they move in unison in opposite directions with respect to the centerline of the mill. Each roll is operated through a screw-down mechanism like that used on the top horizontal roll of

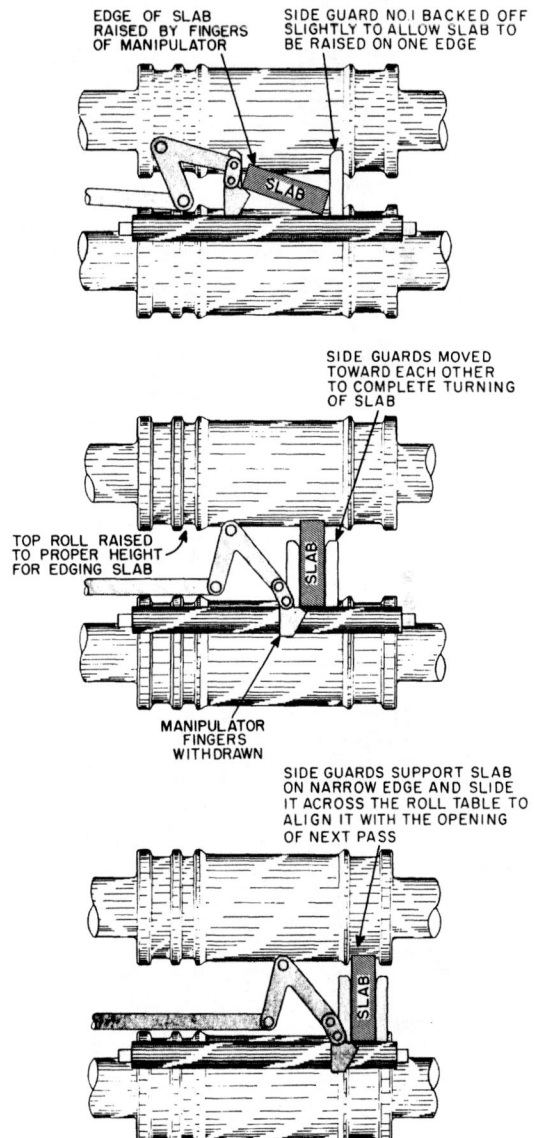

FIG. 23—10. Schematic representation of the actions of the principal parts of a manipulator in turning a slab 90 degrees. The mechanisms have been simplified and exaggerated in size for the sake of clarity.

FIG. 23—11. Blooming mill in operation, showing location of manipulator with respect to stand.

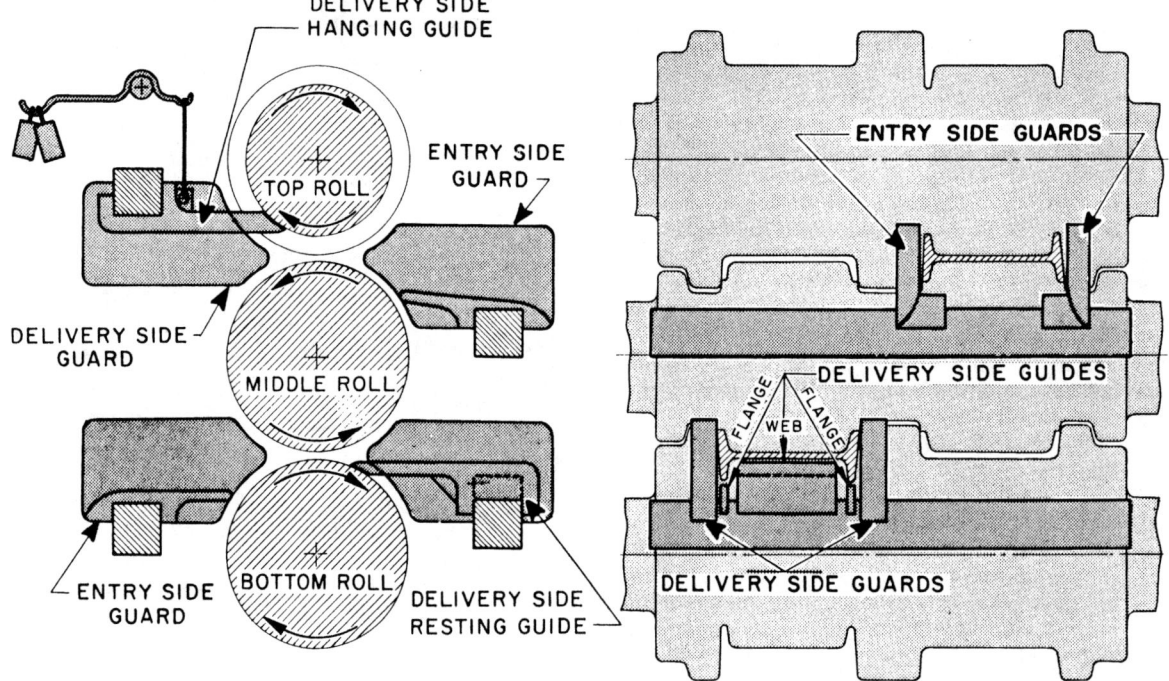

FIG. 23—12. Schematic representation of guides and guards applied to a three-high mill rolling a structural section.

a reversing or universal-type mill, with the exception that the motion is horizontal instead of vertical.

Front and back roller tables feed and receive the piece during each pass. Their speed should be matched closely to that of the rolls, and the width should be the same. On reversing mills these tables are subject to heavy usage due to impact from turning the piece, and should be of rugged construction. Forged steel generally is used today in making the rollers, which in new mills are mounted in roller bearings and equipped with automatic lubrication. In the older mills, bronze or babbitt bearings are lubricated by hand. The rollers usually are electrically driven by a line shaft through miter gears. On newer primary mills, front and back mill-table rolls and feed rolls are individually driven.

The **manipulator** consists of a pair of side guards, stroking laterally over the rollers of the front or back mill tables, usually on both sides of the mill (Figures 23—10 and 23—11). The manipulators operate to turn the piece between passes, to move it to another pass, and to straighten it when necessary. Most manipulators are constructed so that both horizontal and vertical motion is possible. In most mills, these units are operated by electric or hydraulic power. Depending upon the size of ingot or section to be manipulated, the height of side guards may run up to 1220 mm (48 inches). Lifting fingers, always on the front manipulator and sometimes on the back one also, are incorporated in the guard or guards toward the driver side. These have a vertical or nearly vertical stroke and serve to lift the piece by a corner in the process of turning it 90 degrees.

Guides and Guards—In order to prevent collaring and to insure that the piece enters and leaves the pass in the correct position, guides are employed. These guides vary in form and size to fit the conditions. In some cases, they are merely grooved fore-plates; in others, they are blunt-edged plates set up in front of the collars, dividing the space in front of the rolls into a series of pigeon holes; in large mills rolling heavy sections, they may take the form of grooved rollers; in the smaller mills like the guide mills, they are trumpet-shaped castings that fit close up to the roll and have exit openings to conform to the shape and size of the sec-

tion of the entering piece; in continuous mills they may be constructed so as to twist and thus turn the piece between two successive passes. Guides may be employed on both sides of a pass, in which case they are designated as **entering guides** and **delivery guides.** They are held in place by means of the rest bars previously mentioned in connection with the housings. **Guards** are devices employed mainly on the delivery side of the mill to control the direction of the piece after leaving the pass. Reversing and three-high mills are provided with guards on both sides of the mill (see Figure 23—12).

Roll-changing devices are dependent upon the construction of the mill housing. The rolls in an open-top housing are changed by removing the housing cap and picking out the rolls separately or collectively with crane slings. Roll changing in closed-top housings is accomplished through the window of the housing either: (1) by using a counterweighted porter bar, (2) by a sleeve utilizing the ingoing roll as a counter-weight, or (3) by a "C" hook. All of these methods require overhead crane service to remove the rolls one at a time. The rolls are pulled out of the housing by attaching a socket fitting over the protruding roll necks. In some large modern mills, a roll-changing rig is used to remove worn rolls and install new ones. The rig is placed in a permanent mounting, level with the housing sill, and consists of a rack-and-pinion motor-operated crosshead mounted on a rail frame. It withdraws both top and bottom rolls with their respective bearings, all at one time. The old setup is removed from the rig by the crane and the new rolls and bearings on the rig to be pushed back into the mill.

To minimize roll-changing time on modern hot and cold continuous mills, automatic roll changing devices are employed. Typically, a complete set of redressed rolls are set up on the operating side of the mill stands. During the roll-changing cycle, the rolls to be changed are mechanically withdrawn from the mill stands in line with the new roll sets. All roll sets are shifted to align the new roll sets with the mill stands. The new roll sets are mechanically inserted into the mill stands. The used rolls can be moved to the roll shop for dressing while the mill is productive.

SECTION 3

ROLLING-MILL ROLL DESIGN AND MANUFACTURE

Principal Parts of Rolls—Of the essential parts of the rolling mill, the rolls are of the greatest interest, as they control the reduction and shaping of the metal. There are three parts to a roll; namely, the **body,** or the part on which the rolling is done, the **necks** which support the body and take the rolling pressure, and the **wobblers,** where the driving force is applied through loose-fitting **spindles** and **boxes** which together form a sort of ingenious universal coupling. These parts are shown in Figure 23—13.

A plain-surface or cylindrically-bodied roll is used (in pairs) for rolling sheets and plates, while for bars and

shapes, grooves of suitable design are turned in the roll bodies. Such grooves are called **passes.**

Figure 23—14 shows two examples of grooved rolls:—a three-high set of rolls with 98 degree diamond passes, with the roughing-down operation from a billet indicated by the hatched lines, and a two-high set of rolls with grooves for rounds. Rolls are turned as required for a multitude of shapes. Figure 23—15 shows some of these grooves or passes and their nomenclature. The dotted lines over each pass show the cross-section of the piece leaving the preceding pass and entering the pass in question.

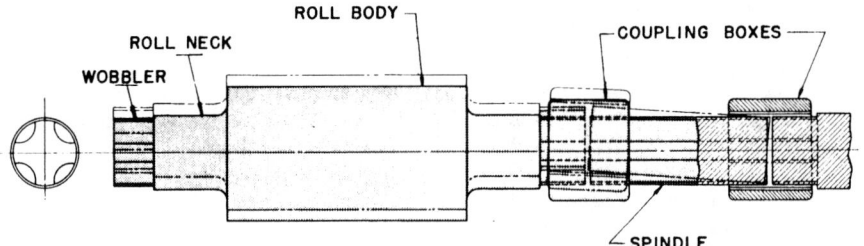

FIG. 23—13. Nomenclature of the parts of a rolling-mill roll and the units that connect it to the motive power driving the mill. The broken lines indicate position of the parts when the roll is raised.

Procedure in Designing—When a new section has been approved for rolling, a detailed drawing of the section is sent to the roll designer, together with instructions to proceed with the design.

In most cases, though not all, a cold-finishing **templet** of the section is made which is an exact cross section of the bar to be produced by rolling. This is the starting point in the design. The templet work is done by skilled artisans called **templet filers** who must work to very close tolerances. As the bars are finished hot, a certain amount of shrinkage takes place in cooling. To allow for this shrinkage it is necessary to prepare a hot-finishing templet to which the rolls are turned. The shrinkage varies with different finishing temperatures, but, on the average, is 15 mm per metre ($\frac{1}{64}$ inch to the inch).

The passes in the design are drawn very accurately from the finishing pass to the starting billet, bloom or ingot. Where it is necessary to use a definite size of billet or bloom, the design must be shaped so that the available billet or bloom will be satisfactory.

Plants having primary blooming mills have an advantage in that special sized billets, blooms, or shaped blanks may be made exactly to the size required.

In designing the passes, it is customary to draw each preceding pass, from the finishing pass to the initial pass, over the succeeding one, and the roll designer judges the correctness of the design by this method, drafting the passes in the form which his knowledge and experience has shown to be proper. This part of roll design has been referred to as intuitive ability in design and is a faculty not readily transferred to others.

When the templets have been prepared, drawings are made and roll castings ordered from the drawings. When the castings have been received, the rolls are turned on special lathes to match the drawings and to an exact fit of the templets. The roll-pass grooves may be made in some types of rolls (i.e., the rolls for rod mills) by crush grinding instead of cutting with a lathe tool. As will be shown later, rolls are cast from either iron or steel, with or without alloying elements, depending upon the use to which they are to be put.

The **pass guides** usually are designed by the roll department and are a very necessary part of any successful roll design. Many shaped sections are so cut in the rolls that they would have a tendency to wrap around the rolls unless stripped from the pass by guides. Entering and delivery pass guides also serve their purpose in guiding and delivering bars into the proper pass in a

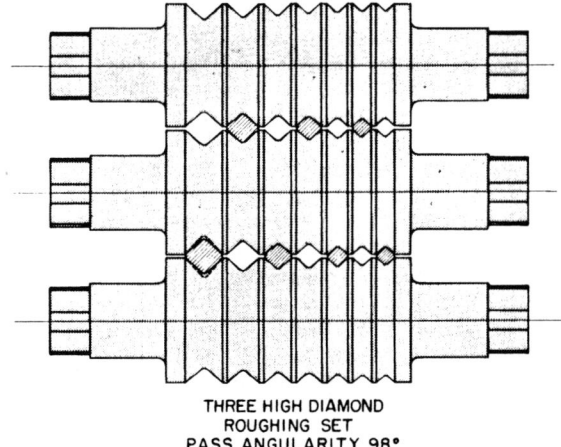

THREE HIGH DIAMOND
ROUGHING SET
PASS ANGULARITY 98°

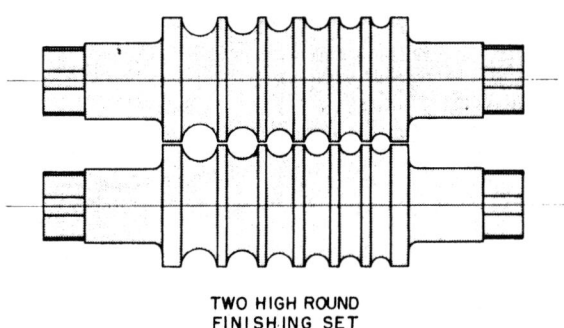

TWO HIGH ROUND
FINISHING SET

FIG. 23—14. (Above) Set of three-high rolls with diamond passes, showing the seven passes used in roughing down a billet to a round-cornered square section. (Below) Two-high set of rolls with grooves for finish rolling of rounds.

straight line. Figure 23—12 illustrates the guide principles.

Elements of Good Roll Design—In the designing of rolls there are many points to be observed if the design is to be satisfactory.

In determining the number of passes, one point to be remembered is that the fewer the passes, the smaller the roll expense. Drafts (amounts of reduction per pass), however, must be suited to roll diameters and

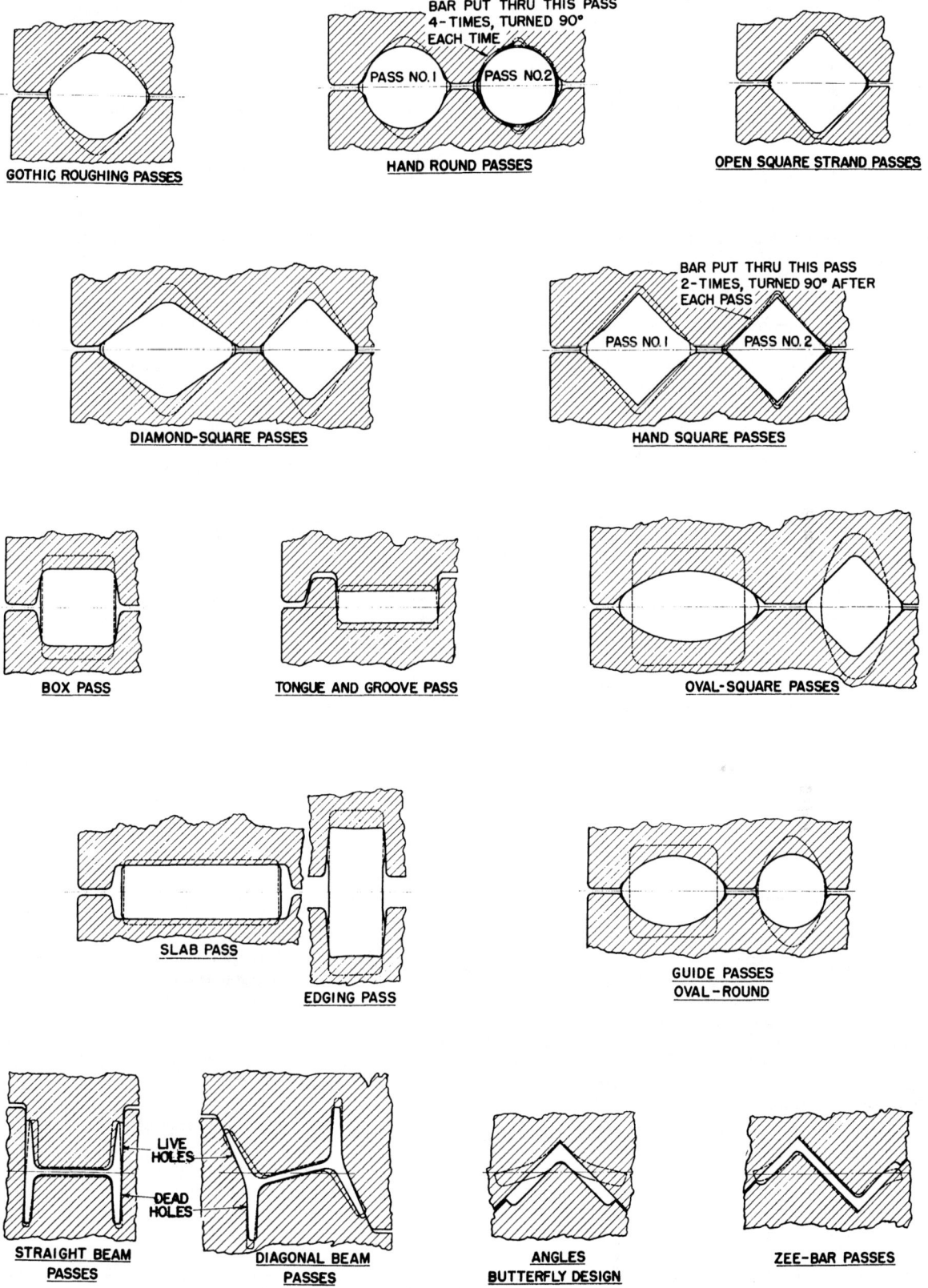

FIG. 23—15. Diagrammatic representation of some common types of roll grooves or passes. The cross-hatched portions are axial sections of the roll bodies. Dotted lines indicate cross section of the piece entering each pass.

plant heating capacity so that roll breakage will be small, and excessive use of power will be avoided.

The passes must be designed with liberal tapers on the various parts so that they may be dressed to templet with the minimum possible reduction in diameter.

A successful design also takes yield into consideration. A minimum of section variation, giving a bar uniform in size from the front to middle and back, insures good yield. Yield is defined as a measure of good product made from a given quantity of starting material.

The tonnage which may be obtained from the rolls also is important. The design which permits the greatest tonnage to be rolled before it is necessary to scrap the rolls because of their reduction in size, due to dressing to eliminate the effects of wear, gives the lowest roll cost.

In roll design, the fact that material being rolled elongates and also spreads in proportion to the draft must be given the most careful consideration, as this is the fundamental principle in rolling, and the proper directing of this elongation and spread is vital to the success of a roll design.

Spread of the various kinds of material must be taken into consideration, as there is a great deal of variation in spread of steels. Roll speeds must be considered in the design. High speeds in many cases restrict spread and low speeds increase it. Low speed permits heavy drafts, while high speed requires lighter drafts.

Temperature of the steel is also a factor. High temperatures permit heavy drafting, while low temperatures call for lighter draft.

Roll diameter also is important. A roll of large diameter is strong and permits heavy drafts without roll breakage, and also, because of the large diameter, the area of contact of roll on bar is large, permitting easy entrance. Rolls of small diameter require less power to drive, reduce spread and increase elongation, but are more easily broken, and heavy reductions cannot be taken readily on them as the area of contact between the work-piece and roll is small.

All these points and many more must be given full consideration in any successful roll design.

CASTING OF ROLLING-MILL ROLLS

The selection of the proper grade of roll to be used in any of the various stands of the rolling mills is most important. No longer is it adequate for the roll user to order rolls by hardness and carbon content. An understanding of roll microstructure is necessary if the user is to assure consistent roll properties from various roll suppliers for each specific rolling application. In general, primary mills such as blooming or slabbing mills require a roll in which strength is paramount. Such rolls are subject to tremendous shock and extreme pressures in the rolling of large ingot masses with heavy reductions. The heat from the ingots transmitted to the roll also has a tendency to cause surface or fire cracks through differential expansion of the surface, and the strength and toughness of the rolls must be such as to resist the further development of these cracks. Firecracking of cast-steel rolls is related directly to the carbon content; the lower the carbon content, the more resistant a cast-steel roll is to firecracking. Resistance to firecracking is better for rolls that have been normalized and tempered than for rolls that have been annealed. However, rolls that have been heat treated to obtain a ferritic matrix with a fine dispersion of spheroidized carbides and absence of grain-boundary carbide networks offer the best resistance to firecrack propagation. A composition of 0.85 per cent carbon, 1.00 per cent manganese, 1.00 per cent chromium and 0.25 per cent molybdenum is typical of blooming-mill rolls. The original plain-carbon steel roll containing 0.30 per cent to 1.50 per cent carbon, used in the past for such applications, now has been supplanted largely by the alloy-steel roll of greater strength and durability. Two additional type rolls gaining wide acceptance for primary-mill applications are the differentially hardened (D.H.) roll and the weld-overlayed roll. In both cases, the roll bodies are processed to improve wear and firecrack resistance. At the same time the roll-neck properties are such as to improve strength and ductility levels necessary to withstand both castastrophic and fatigue failure.

From the primary mills through the secondary and so on to the finishing mills, the required rolls generally become smaller and harder and any of a variety of roll grades may be used; carbon-steel or alloy-steel rolls for roughing, iron or alloy-steel rolls for intermediate work, and the various grades of iron-base rolls for the finishing stands. Specifically, however, each roll must be custom made to suit the requirements of the individual mill by a procedure within the limits of the manufacturer's equipment. For this reason, no two manufacturers use exactly the same procedure for the same roll and each must apply his own specifications for a given requirement.

One of the problems in determining the proper roll for a specific application is that mill conditions vary greatly. A mill that rolls finished product from ingots without intermediate conditioning and reheating requires different rolls than one where the product is allowed to go cold prior to reheating and finishing. The amount of cooling water used on the mill is also a factor in the selection of rolls. For these reasons, experience in roll applications on one mill will not always be successful if applied to another mill. Roll requirements must be developed for each application. This development can best be accomplished through an ongoing exchange of information between the roll user and the roll manufacturer. The success of this endeavor depends on the roll user's ability to; (1) maintain roll-performance records and (2) to keep abreast of changes in rolling practices.

Steel for casting rolls usually is produced in electric-arc furnaces lined with either silica (acid) or dolomite (basic) refractories. The use of a basic lining permits a double slag process to remove both sulphur and phosphorus. This in turn means that analytical control is not so dependent upon selection of good quality scrap charge material. However, an acid lining does offer savings in refractory costs and permits a higher residual silicon content in the molten steel. In both cases, good deoxidation practice in the ladle is necessary to produce a high-quality roll.

STEEL ROLLS

A problem in manufacturing cast-steel rolls is to achieve soundness or lack of pipe. This can be accomplished by the use of a sink-head of large volume (about 35 to 45 per cent of the volume of the finished roll) to act as a reservoir of molten metal. Metal from the sink-head feeds into the roll body as the latter freezes. The necessary volume of molten metal can be maintained in the sink-head by the use of an electric hot top which is similar to a small electric-arc furnace that is placed on top of the roll flask as soon as pouring of the roll is completed: heat generated by the arc keeps the metal in the sink-head molten.

The total sink-head volume can be reduced 50 per cent when an electric hot top is used. The electric hot top is used about eight hours for a 13.6-metric ton (15-net ton) roll to assure freedom from injurious pipe. Figure 23—16 shows an electric hot top.

The optimum casting shape for rolls from the standpoint of progressive solidification would be a cone with the point downward. Due, however, to the required design of the roll and the general rule that the drive end is best placed down, this optimum arrangement is often distorted or even inverted. In these cases provision must be made to control the freezing pattern by artificial means. This may be done in several ways, one of which is "padding" or adding extra thickness of metal to sections which are too small to solidify in the proper manner in relation to the balance of the casting. Increasing the rate of freezing by inserting heavy metal blocks, rings, or segments, called chills, in the mold, close to the surface contacted by the molten metal, is still another means for controlling the freezing pattern. Consideration must be given also to the fact that, in addition to increasing the freezing rate, chilling in the manner described also promotes formation of a dense, refined outer skin of the roll at the contact areas, which in most cases is very desirable. A balance must

FIG. 23—16. Electric hot top in position above a mold for casting a steel roll. The carbon electrodes at the left extend vertically into the sinkhead of the mold. After casting, electric arcs between the molten metal in the sinkhead and the electrodes keep the metal from solidifying.

FIG. 23—17. Illustrating use of a sweep in preparing a mold for casting a steel roll.

FIG. 23—18. Assembling the two halves of a mold for a steel roll.

be maintained between chilling, cast-diameter stock and usable roll-diameter life. Proper casting design will eliminate the presence of solidification/shrinkage defects on the rolling surfaces. Although rolls are made with meticulous care in both molding and pouring, particles of the mold sand frequently are entrapped accidentally on the surface and, in order that these or other surface defects will be removed properly during subsequent machining, rolls are cast slightly oversize. The extra stock thus cast on a roll must be of such thickness that, in those instances where heat treatment precedes the other processing, the tough, heat-treated skin will not be removed during subsequent machining.

The initial step in the manufacture of steel rolls is that of examining the blue print of the roll; the composition and heat treatment are determined from the requirements indicated. The pattern is designed after applying the metallurgical and mechanical considerations discussed above. This pattern is generally in the form of a **sweep**; that is, a flat board carved on one edge to conform to the longitudinal contour of the roll as modified for casting, with provision on the other edge for fastening the sweep securely to a spindle or axle (Figure 23—17). Then the plan of manufacture is given to the operating section.

The steel flasks or containers for roll molds are cylindrical in shape, and designed to separate into two longitudinal halves. Proper molding sand is first rammed into each half, after which the sweep mounted on the spindle is attached to the half flask and rotated, cutting the sand before it to the shape of the sweep contour. The same procedure is used on the second half-mold. Figure 23—17 shows a mold used for casting a large-diameter structural-mill roll. After molding, the half-molds are baked in drying ovens at temperatures somewhat above 200°C (400°F) to drive out moisture and bond the mold thoroughly. Figure 23—18 shows the

two halves of a mold being put together. Figure 23—19 shows the vertical section of a mold for an alloy-steel blooming-mill roll in position for pouring. Rolls are bottom poured and the metal enters the bottom neck

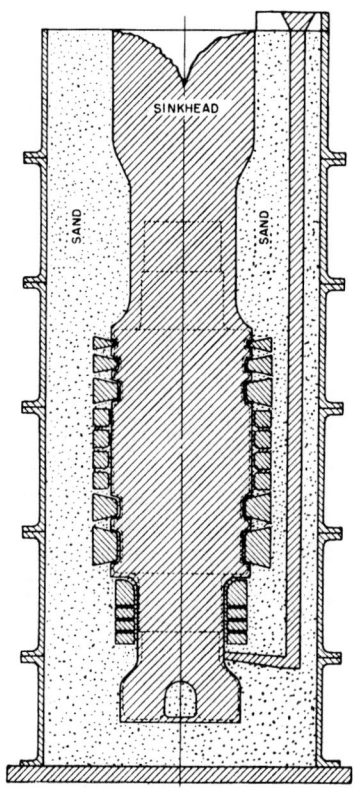

FIG. 23—19. Schematic section of a mold for an alloy-steel blooming-mill roll. Note location of chilling rings.

FIG. 23—20. Inserting chilling rings in a mold for a steel roll.

through a gate tangent to the periphery of the neck. This is to set up a swirling motion of the metal as it enters the mold. This centrifugal action concentrates dirt and foreign particles in the center of the rising molten metal in the roll mold, and as the metal continues to rise, they are carried to the top into the sink-head. The dotted lines seen in Figure 23—19 show the outline of the finished roll. The excess of metal at the top is either burned off before, or cut off during, the machining operation. The cavity in the sink-head is formed by feeding molten metal in the head to the roll body to compensate for shrinkage during solidification.

Figure 23—20 shows cooling or chilling rings being placed in a roll mold and Figure 23—21 shows the mold being smoothed over and near completion.

After a steel roll is shaken out of the mold, the adher-

FIG. 23—21. Smoothing over the face of one-half of a roll mold, after insertion of chilling rings.

ing sand is removed from its surface, the sink-head is taken off, and the roll is heat treated. The temperature of the roll at shake out is usually about 260° to 370°C (500° to 700°F), but some large rolls or rolls with large body diameters and small necks may be shaken out about 595°C (1100°F) and immediately charged into a heat-treating furnace that is at the same temperature as the roll. This latter practice, called hot shake out practice, reduces the possibility of internal tears in rolls by preventing the development of excessive stresses that would result from unusually large temperature differences that would otherwise occur due to the differences in cooling rate of different parts of the roll. Mathematical modeling techniques whereby roll solidification is simulated by a computer program have provided a great deal of valuable information concerning the magnitude and direction of thermal stresses in rolls after casting.

The initial heat treatment is annealing. Temperature increase to the holding temperature is controlled at from 30°C per hour to 10°C per hour (50°F per hour to 20°F per hour). Larger rolls and rolls with higher carbon contents require the slower heating rates. Holding time at holding temperature is one hour per 25 mm (per inch) of major thickness. Holding temperature varies with carbon content. The selection of the austenitizing temperature has a significant impact on the physical properties of the material. The effectiveness of the alloying elements is also established by the selection of the austenitizing temperature. In rolls containing above 1.50 per cent carbon, some grain-boundary carbide is present even with a holding temperature of 980°C (1800°F). While this structure enhances wearing characteristics, it is also brittle and likely to firecrack. For some applications, a second heat treatment is required consisting usually of normalizing followed by tempering.

After heat treatment, the roll goes to the roll lathe for removal of extra stock on the top neck and thorough testing of the roll body. Any defects which cannot be removed by the rough turning result in rejection and scrapping of the roll.

A special type of alloy-steel roll is the **built-up roll.** These are used for backing-up rolls on four-high plate and strip mills. The solid back-up rolls as used on large four-high mills are difficult to cast. The large volume of

the casting induces strain during cooling due to the unequal cooling between the roll surface and its center. Due to this unequal cooling, cracks are very apt to occur in the casting. In the built-up roll, the center arbor or mandrel is relatively small in diameter and easier to cast. It also cools more uniformly than a large-diameter solid roll, and is less liable to strain and cracking. The outer shell also is easy to manufacture and, due to its construction, can be heat treated to be much harder than the arbor. Figure 23—22 shows a built-up roll of special construction. In some cases, solid rolls which have been dressed down repeatedly in service are reduced further and fitted with sleeves and returned to service.

Another type of roll is the **forged-steel roll.** This roll is forged from an ingot of suitable size and, after being normalized in a heat-treating furnace, is turned close to the finished size. A hole is then bored longitudinally through the center of the roll to facilitate hardening by quenching after heating to the proper temperature. This type of roll is used mostly for the cold rolling of flat material. One of the favored compositions for a roll of this kind is: carbon, 0.85 per cent; manganese, 0.25 to 0.30 per cent; phosphorus and sulphur, below 0.05 per cent; silicon, 0.25 to 0.30 per cent; chromium, 1.60 to 2.50 per cent; molybdenum, 0.25 per cent; and a trace of vanadium.

IRON-BASE ROLLS

The iron-base roll differs from the steel-base roll principally in per cent of carbon; the steel roll containing from 0.30 to 2.50 per cent carbon, whereas iron rolls contain from 2.50 to 3.50 per cent. The material for iron rolls is usually melted in an electric coreless induction furnace. This results in little or no oxidation and low melt loss.

In the manufacture of iron rolls, a wide range of composition is employed to obtain the desired characteristics such as hardness, strength, resistance to spalling, and depth of dense, refined grain structure. Table 23—II summarizes the general range of elements for various types of rolls, and rolling-mill applications.

The normal charge for the melting furnace in the manufacture of iron rolls contains 25 per cent pig iron and 75 per cent worn-out rolls, heads and gate scrap. If lower carbon content is desired, a certain percentage

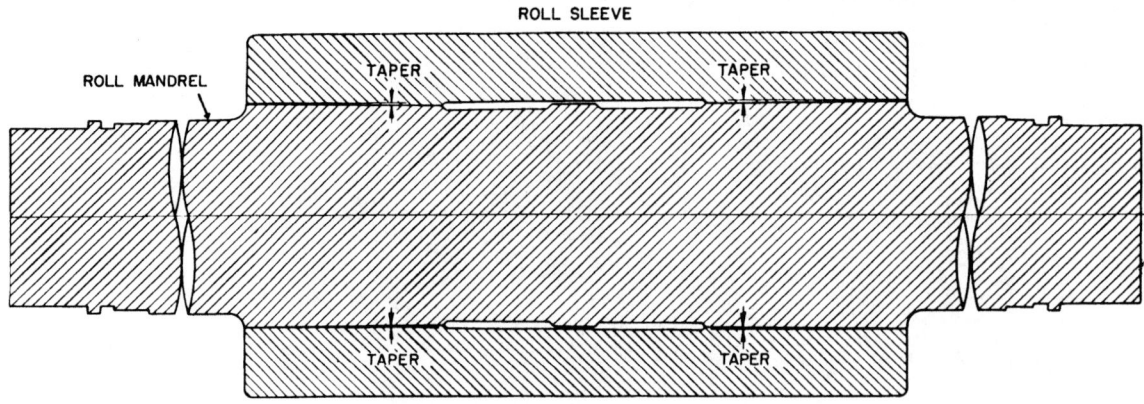

FIG. 23—22. Cross-section of a built-up roll.

Table 23—II. Types, Compositions and Uses of Iron Rolls

Type	Carbon (%)	Manganese (%)	Silicon (%)	Phosphorus (%)	Sulphur (%)	Chromium (%)	Nickel (%)	Molybdenum (%)	Scleroscope Hardness	Uses
Plain Chill	2.90-3.30	0.20-0.25	0.60-0.70	0.45	0.08	......	......	0.25	58-70	Sheet, Bar, Plate and Rod Mills.
Nickel Chill—Regular	2.90-3.30	0.18-0.25	0.40-0.60	0.35	0.08-0.12	0.25-0.50	2.50-3.00	0.25	65-70	Finishing; Bar, Rod, Skelp and Intermediate Stands of Some Strip Mills.
Nickel Chill—Hard	3.30	0.18-0.25	0.40-0.50	0.35	0.08-0.12	0.75-1.00	3.50-4.50	0.25	80	Finishing Stand; Strip and Band Mills.
Sand Iron (Low-Alloy Grain)	2.50-2.80	0.40-0.50	0.80-1.25	0.15-0.25	0.08-0.12	0.50	0.50	0.25	45-55	Roughing and Intermediate; Bar, Structural and Rail Mills.
Medium Alloy Grain	3.00-3.25	0.40-0.50	0.90-1.25	0.10-0.20	0.08	0.70-1.20	0.50-1.00	0.25	55-65	Intermediate, Finishing; Bar, Rod, Structural and Pipe and Skelp Mills.
High-Alloy Grain (Regular)	3.15-3.40	0.40-0.50	0.80-1.10	0.10-0.15	0.06	1.50-1.75	4.00-4.50	0.25	65-75	Intermediate Stand; Hot Strip Mills.
High-Alloy Grain (Hard)	3.40	0.90-1.50	0.80-1.00	0.10-0.15	0.06	1.50-2.00	4.50-5.00	0.25	80-90	Finishing Stand; Strip Mills and Cold Reduction Mills.
Ductile Iron	3.20-3.40	0.30-0.50	1.50-2.25	0.04-0.15	0.01-0.02	0.75 max.	1.50-2.50	0.25-0.50	40-65	Roughing or Intermediate Stands; Rod, Structural and Bar Mills.
High Chromium	2.60-3.30	0.40-1.20	0.40-0.10	0.01-0.05	0.01-0.05	13.00-18.00	0.40-1.70	0.50-3.00	65-90	Early Stands of Hot Strip Mills and Cold Reduction Mills.

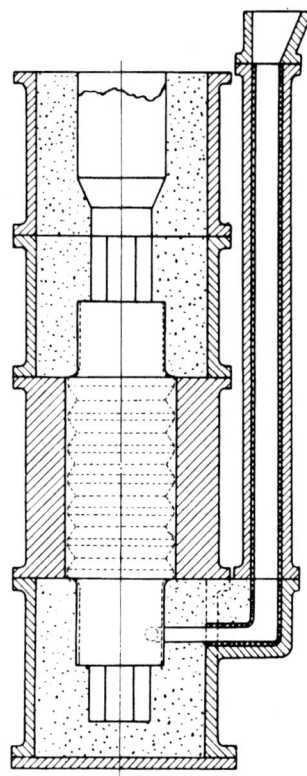

FIG. 23—23. Vertical section of mold for casting a plain-bodied roll, or a shape roll in which grooves are machined from the solid (dotted line).

of low-carbon steel scrap makes up a part of the charge. Alloys, such as nickel, chromium and molybdenum, if required, are also added with the charge.

When the iron is melted and sufficiently hot, two samples are taken, one for chemical analysis and the other for a test block which is cooled and broken. From the chemical analysis and the test coupon, the roll metallurgist determines the necessary additions of alloys for the type of roll being cast. The metal is tapped in a ladle and, when the correct temperature is reached (1330° to 1455°C corresponding to 2425° to 2650°F), poured into previously prepared roll molds. After cooling from one to four days, the rolls are shaken out, heat treated if necessary, and machined to specifications

Chill Rolls—Figure 23—23 shows a roll-mold arrangement for casting a plain-bodied roll, or a shape roll in which the grooves are machined from the solid. The wobblers and necks are cast in sand but the body of the roll is formed by heavy-walled cast-iron cylinders, known as chills. The purpose of these chills is to cool the molten iron quickly after it has been poured into the mold, while the necks and wobblers, being molded in sand, are not subjected to fast cooling. The carbon in molten iron is in solution. If the molten iron is allowed to cool slowly, the carbon in excess of 0.82 per cent separates as graphitic carbon and is distributed throughout the mass. But if iron of the proper composition is cooled rapidly from the liquid state, the

carbon combines with the iron to form cementite (Fe_3C), which is very hard and white in color. The depth of case on the body of the roll, induced by the fast cooling of the molten iron in contact with the chill mold, depends upon the chemical composition of the iron (see Chapter 40). As the mold loses its chilling effect, carbon starts to come out of solution to form a mottled area immediately below the hard case. This mottled area consists of a mixture of graphite, cementite and iron. Since the interior of the roll cools slowly, gray iron forms therein. Another factor controlling the depth of chill is the diameter of the roll being cast. As the diameter increases, the chilling rate is retarded due to greater cross-sectional area. On a 760-mm (30-inch) diameter roll, the following composition would give a clear chill about 19 mm (¾ inch) in depth, and a mottled area about 19 mm (¾ inch) deep before reaching the portion composed uniformly of gray iron:

Element	Per Cent
Carbon	3.00
Silicon	0.65
Phosphorus	0.40
Manganese	0.25
Sulphur	0.08

To improve the finish of the mill product and reduce spalling and wear, various alloys are added such as: nickel, chromium, molybdenum and vanadium in balanced proportions to control chill depth and roll structure.

Since the case or chill on the rolls is hard and brittle, there is a practical limit to the depth a roll can be chilled without causing failure through breakage. In general, chill specifications will vary between about 12.5 and 40 mm (about ½ and 1½ inches) in depth, depending upon the product to be rolled and the amount of diameter reduction through dressing expected to be permissible before the roll is scrapped.

Grain-Iron Rolls—For certain applications, such as the intermediate stands of some hot-strip, plate, rod and bar mills, where shock, temperature and extremely heavy loads frequently result in spalling, firecracking and breakage, chill rolls are not practical. This applies also to rolls for billet, rail and structural mills, with deep grooves or passes. To meet these requirements, an iron composition commonly referred to as **grain iron** is used. Sufficient alloys, such as chromium and nickel, are added to control hardness and to increase strength while silicon or other graphitizers are used to resist formation of a definite chill. While these rolls do not give the quality of finish obtained from a true chilled-type roll, they have deeper penetration of the refined structure with increased strength. The grain-type rolls become softer and the gray iron coarser progressing inward from surface to center, and this less desirable structure is encountered as the deeper passes are cut in a roll. To overcome this characteristic, rolls calling for deep working passes are molded similar to steel rolls, as shown in Figure 23—19, 23—20 and 23—23. The inserted iron rings promote a finer structure and better wearing metal in the passes.

Double Poured Rolls—High-alloy rolls (1.50 to 2.50 per cent chromium and 4.00 to 5.00 per cent nickel with 75 to 90 Scleroscope hardness) used in cold-

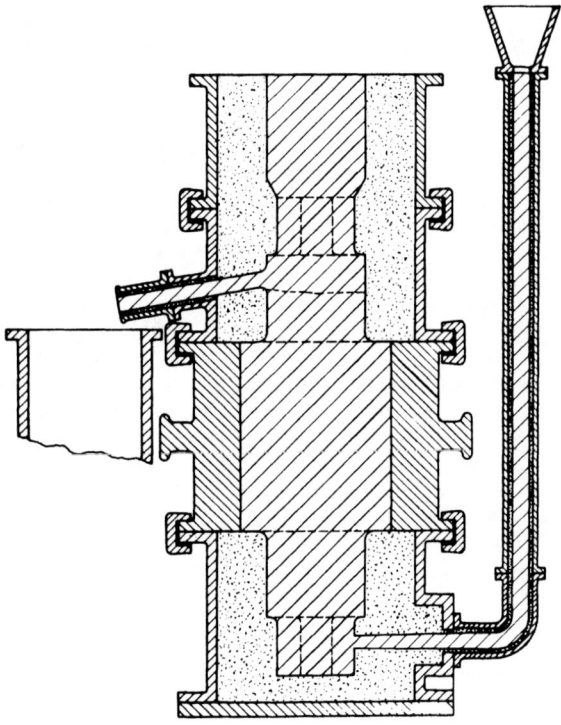

FIG. 23—24. Schematic representation of the vertical section of a mold for casting a composite iron roll.

reduction mills and finishing stands of some hot-strip mills, would, if cast solid, have hard, unmachinable necks. Casting strains are quite severe in this type of roll. To overcome these objectionable features, a different method of pouring has been developed as shown in Figure 23—24. The mold is made in the regular manner, except that a spout is connected to the top neck cavity a few inches above the body of the roll. Metal of high-alloy content is poured until the metal reaches the run-off spout. Pouring is stopped and a small amount of metal of a composition that produces much "softer" iron that the first high-alloy metal is poured down the runner to keep the ingate from freezing. After a predetermined time, pouring of the "soft" iron is continued, washing out through the spout connected to the top neck and into a "nugget pot" the still-molten hard roll metal that has not been chilled and solidified by the iron mold which forms the body of the roll. When sufficient "soft" iron has been introduced into the mold to wash out the still-fluid iron in the center of the roll, the pouring is stopped, and a plug inserted in the overflow spout. The same iron used to flush out the molten hard iron from the roll is introduced into the top neck either by pouring directly down through the neck or through a gate attached to the top neck just above the run-off spout.

A roll poured by this method will have an extremely hard fine-grained structure for a depth of 25 to 50 mm (1 to 2 inches). The necks will be machinable and the central part of the roll will be strong, the hard brittle iron having been replaced or diluted by the soft "flush

iron."

This type of roll can be produced by several alternate methods, two of the most recent being the sliding-gate technique and centrifugal casting, both of which are special processes for the manufacture of this roll type but are aimed at producing such a roll in a more economic manner. The development of these processes took place during the 1960's and early 1970's, and was particularly associated with the production of the high-chromium roll. The traditional displacement/dilution method referred to above becomes progressively less economic as the percentage of alloying elements in the shell material increases. The high-chromium roll, offering considerable advantages in the early stands of the finishing train in hot-strip mills and being used for certain applications in cold mills, is not well suited to production by the displacement/dilution method.

The **slide-gate process** utilizes a valve in the bottom of the roll mold to "drain" the high-alloy shell metal after sufficient material has solidified to provide the wear-resistant outer surface of the roll. Before oxidation of the inside surface of the solidified shell can occur, flush iron is added either from the top or bottom of the mold to form the "core" of the roll. The return of the high-alloy shell metal, undiluted and in the molten state, for subsequent re-use gives this method economic advantages with respect to both material utilization and energy consumption.

Centrifugal or **spin casting** uses the force generated from rotation of the roll mold to hold the high-alloy shell material in the appropriate position during solidification. The longitudinal axis of the roll mold can be in either the vertical or the horizontal plane. The latter is less popular for producing rolls because a degree of microstructural segregation can occur when using certain alloy systems. After pouring the precise quantity of shell material necessary the "flush" iron is added to the mold which has to be decelerated in order to avoid flinging metal out of the top of the mold. A flux may be added to prevent oxidation of the interior surface of the shell metal or the cone metal may be poured prior to complete solidification of the shell in order to facilitate the formation of a strong bond between the high-alloy shell and the central core of the roll. The much smaller volume of high-alloy material needed in this process gives considerable economic advantages arising from both energy and yield. In addition, the abrasion-resistant shell of the roll has improved metallurgical properties which are attributed to the greater density and high solidification rate achieved by this method.

Ductile-Iron Rolls—With the development of "ductile" or "nodular" cast iron, some manufacturers have produced rolls of this material which is made by the addition of magnesium or rare earth compounds to iron of restricted composition.

The remarkable strength and toughness of this iron results from the nodular shape of the free graphite in the structure as contrasted with the flake graphite common to gray iron. This iron, if properly made and heat treated, develops properties which approach the strength and ductility of steel. Spheroidal (nodular) graphite rolls are currently being produced with matrices of pearlite, bainite, and martensite in hardness ranges from 50 to 85 Shore Scleroscope.

MILL DRIVES AND POWER REQUIREMENTS

DEVELOPMENT OF MAIN MILL DRIVES

During the first part of the nineteenth century, rolling mills were driven by water wheels and low-pressure steam engines. The power was transmitted to the mills by direct mechanical connection, through suitable shafts, gears and couplings. Because of the low steam pressures used, short steam lines were necessary. As a result, many small boiler houses were built adjacent to the mills. Most of the engines were of the simple non-condensing type and this factor, together with the small, hand-fired boilers used, resulted in high-cost operation. Coal had to be delivered to the boiler houses, ashes disposed of, and steam lines maintained. In some plants, compound condensing engines were used, requiring a supply of cooling water and a sewer system. When the mills were shut down over the weekend, the steam lines had to be kept hot, as the contraction set up by cooling the line resulted in leaks at the joints. Nearly all of these early mills have been replaced, during the last fifty years, with modern electrically-driven mills that, while fewer in number than the mills they replaced, have a much greater combined capacity.

With the development of the electric generator and motor in the later part of the nineteenth century, a new and more efficient method of driving rolling mills became available. Instead of direct mechanical connection of the mills with the power source, generators could be placed in a convenient central location, and electric power could be transmitted over wires to motors attached to the mills. The generators could be driven either by gas engines, steam engines, or steam turbines.

Shortly after 1890, internal combustion engines were developed to operate on blast-furnace gas. This by-product fuel was available at all blast-furnace installations and at this period either was used for making steam or was wasted. The gas engine could use this fuel direct and thus eliminate the investment for a boiler house required for a steam engine. These gas engines were used as the prime movers for blast-furnace-blowing equipment and also for driving electric-power generators. This type of drive was selected for both the blowing units and the generators at the Gary Steel Works when this plant was designed in 1908. At that time, gas-engine-driven generators were available in sizes having outputs up to 2000 kW, and fifteen such units were installed. A few years later, the size was increased to 3000 kW, which equalled the largest generator driven by a reciprocating steam engine at that time. The gas-engine-driven generator installation at Gary was the first sizeable steel-mill power house in the United States, and made possible the use of 4500-kW (6000 horsepower) motors to drive the mills. At a later date, the maximum size of the gas-engine-driven generators was increased to 6000 kW and three units of this size were installed in the South Works, Chicago, Illinois.

While the development of the gas engine was progressing, another prime mover had entered the field—the steam turbine. Both the steam engine and the gas engine were slow-speed machines and, when used to drive electric generators, the speed limitation resulted in a physical limitation of the generators. The 5000-kW generators, which were about the largest built, had a diameter of over 9 metres (30 feet). On the other hand, the steam turbine is inherently a high-speed machine and the early generators of 5000-kW capacity that were driven by steam turbines required only a fraction of the space needed by other types of units.

Developments in boiler practice whereby steam pressures could be increased to 1725 kPa (250 psi) at first, then 3450 kPa (500 psi) and then, by 1948, over 13 800 kPa (2000 psi), all favored development of turbine designs and the size of turbine units increased until 275 000-kW generators are now common and 1 000 000-kW units are available. While these developments in electric-power-producing facilities were taking place, iron-silicon alloy steel in sheet form for transformer cores was being developed and improved and better insulating materials developed, with the result that larger quantities of electric power could be produced and, further, it could be transmitted economically over greater distances to the motors that used it.

Such improvement in both the generation and distribution of power was required before electrification of mills became possible.

Many steel plants, especially the smaller ones, are not self-contained in that they have no coke plant or blast furnaces. Without blast-furnace gas and coke-oven gas for fuel to produce steam for electric-power generation, such plants buy their electric-power supply from the power companies and, since they may impose large loads such as strip mills create, the power producer has the problem of providing sufficient generating and distribution capacity to serve such loads.

If a map showing the electric transmission lines in the United States is studied, it will be found that the various utility generating sources are interconnected by high-voltage lines to provide a pooled power supply. These interconnecting systems have many advantages, and are favorable to steel mills in that heavy peak loads caused by electric furnaces or strip mills do not necessarily have to be absorbed by a single generating station but, usually, are distributed among many units that are connected to the system. The growth of these interconnecting systems has followed the growth of electric-power consumption and today, with forecasted arrangements, it is possible to purchase power for all types of steel-mill electric loads.

The growth in generating capacity plus the interconnecting systems have made possible the large motor-driven installations in modern mills. Prior to about 1930, it often was a problem to obtain 6000 kW to drive a proposed blooming-mill reversing motor. Today, purchased-power contracts often cover the supply of power for loads of 300 000 kW, and additions to the supply are limited mostly by the time required to plan and install additional generating capacity. Now the utility industry response at a greenfield site may need 6 to 10 years.

In 1905, two 1120-kW (1500-horsepower), 230-volt, 100/125 rpm, direct-current motors were installed on the light-rail mill at the Edgar Thomson Works in Braddock, Pa. They were supplied with power from two 1000-kW, steam-engine-driven, direct-current generators in a nearby power house. In the same year, the first reversing direct-current main-drive motor was installed on the 762-mm (30-inch) universal-plate mill at South Works in Chicago. Power for this latter motor was furnished at 575 volts, direct current, from a 2200-volt, 25-cycle, motor-generator set.

At about the same time, another type of main drive, suggested by European developments, was installed on the light-rail mill at the South Works. This consisted of two 1120-kW (1500-horsepower), 2200-volt 25-cycle, 80/120 rpm, variable-speed, wound-rotor motors. The rotor windings of these motors were connected to the stator of a second wound-rotor motor on the same shaft; this is known as a cascade connection. By varying the resistance in the rotor circuit of this second motor, the speed of the main drive could be varied between 80 and 120 rpm.

About the same time (1908—1910), the Gary Works, designed to be the largest steel plant in the world, planned to use slow-speed, wound-rotor induction motors on the heavy rolling-mill drives. Some of these had windings that, by external contactors, could be connected to change the number of poles and thus provide what is known as a two-speed motor. Other than this, the main drive installations were designed to operate at constant speed, except for the variations that could be obtained by changing the resistance in the secondary circuits of the wound-rotor motors.

This type of drive was satisfactory for some types of mills but, for bar-rolling mills in particular, greater ranges of speed variation were needed and this led to the development of variable-speed controls. With these a-c systems, speed variations up to 50 per cent were made possible at the expense of complexity or poor efficiency.

The demand for wider speed ranges, better regulation and a more simplified control led to the development of large, variable-speed, direct-current motors, which are now preferred drives on practically all new installations.

While these developments were progressing, another problem confronted the steel industry—the alternating frequency of the power systems. In the period 1905—1910, practically all steel companies generated their own power and were not interconnected with utility systems. Slow-speed, direct-connected drives were needed, since high-efficiency gears such as now used were not available. For this purpose, 25-hertz motors offered an advantage and were adopted in the mills and, as late as 1910, all main drive motors in the steel industry had a 25-hertz source of power. Today, while there are some 25-hertz drives representing thousands of horsepower still operating, new installations using 25-hertz power have stopped and 60 hertz has become the standard frequency of the industry

within the United States.

POWER REQUIREMENTS IN THE STEEL INDUSTRY

According to a report of the American Iron and Steel Institute, the iron and steel industry used 57.5 billion kWh of electric power in 1981. Of this amount, about 14.8 per cent was generated in power houses in the industry and the remainder was purchased from utilities. With a raw-steel production of 109 590 000 metric tons (120 828 000 net tons), the average power consumption per ton of raw steel in 1981 was 524.6 kWh per metric ton (475.8 kWh per net ton).

Power Requirements for Various Operations in the Production of Steel—When new mills are contemplated in existing works or new plants planned, one of the questions to be answered is the quantity of electric power required to operate the new facilities.

Data on existing mills are the best source for this information. Actual power consumption by certain operations such as coke plants, blast furnaces and basic oxygen shops is found to be in accord with estimates based on established data from pre-existing similar units. Requirements of heavy mills such as blooming, slab and rail mills also can be estimated closely if the predicted output of product in tons per hour is matched by actual production. Finishing mills, however, require additional data as to the sections to be rolled and the pass design of the rolls.

In the total demand load of a plant, there are three loads that have the greatest effect on power peaks: those from the electric melting furnaces, the hot-strip mills and the cold-reduction mills. There are available data for the power requirements of all of these operations, and the total power required for new installations has to be estimated by using these data in conjunction with proposed operating schedules.

The following tabulation gives the average kWh consumption per ton of steel for some of the major operating units in a modern plant.

Operating Unit	Product Basis	kWh Consumption Per Metric Ton of Product	Per Net Ton of Product
Sinter Plant	Sinter	44	40
Coke Plant	Coke	38.5	35
Blast Furnaces	Hot Metal	27.5	25
Basic Oxygen Furnace	Raw Steel	33	30
Continuous Casting ..	Slabs/Billets	27.5-44	25-40
Blooming Mill	Blooms	33-44	30-40
Slabbing Mill	Slabs	35.2-49.5	32-45
Reversing Plate Mill .	Plates	110-132	100-120
Bar Mill	Bars	132-165	120-150
Hot-Strip Mill	Sheet/Strip	121-143	110-130
Cold-Reduction Mill .	Sheet/Strip	126.5-148.5	115-135

The preceding tabulation for mills covers the power used by the main drive motors, plus that used for auxiliaries such as tables, fans, lighting, and so on. On some mills, the auxiliary power amounts to as much per ton as the power used by the main drive.

The following tabulation indicates how power re-

Table 23—III. Estimate of Annual kWh Requirements for An Integrated Plant

Operating Unit*	Annual Production (Thousands of Tons)		kWh Consumption		Total kWh Consumption Per Year
	Metric Tons	Net Tons	Per Metric Ton	Per Net Ton	
Sinter Plant	1225	1350	44.0	40	54 000 000
Coke Oven and By-Products	1400	1540	38.5	43	66 300 000
Blast Furnaces	2600	2860	27.5	25	71 500 000
Oxygen Plant	182	200	649.0	590	118 000 000
Basic Oxygen Furnace	2725	3000	33.0	30	90 000 000
Continuous Caster	2545	2800	36.0	33	92 400 000
Reversing Plate Mill	545	600	121.0	110	66 000 000
Hot-Strip Mill	1725	1900	132.0	120	209 000 000
Hot-Strip-Mill Finishing	365	400	30.0	27	10 800 000
Pickling Line	1275	1400	20.0	18	25 200 000
Cold-Reduction Mill	1275	1400	137.5	125	175 000 000
Cold-Reduction Finishing (Cleaning, Tinning, Galvanizing, etc.)	1165	1280	137.5	125	160 000 000
Services/Utilities, including Material Handling, Pumps, Air Compressors, etc.					214 400 000
Total					1 352 600 000

*Includes auxiliaries.

quirements are increased as the finishing operations are continued. These figures are averages taken from the records of a modern strip mill whose end products are tin plate and sheets. Auxiliaries are not included except where noted.

Facility	kWh Consumption	
	Per Metric Ton	Per Net Ton
2032-mm (80-inch) Hot-Strip Mill*	121-143	110-130
Continuous Pickling	19.5	18
5-Stand Cold-Reduction Mill*	127-149	115-135
Continuous Annealing	22	20
Temper Mill	22-33	20-30
3-Stand Tandem Mill	38.5	35
Electrolytic Tinning	99-121	90-110
Electrolytic Cleaning	9.9	9
Galvanizing Line	33	30
Shearing/Slitting Lines	9.9-15.4	9-14

*Including auxiliaries.

In addition to the above operations, such a plant has pumps, shops, air compressors, yard lighting, environment loads and other miscellaneous loads that increase the total power consumption.

Estimation of the total electric-power requirements for a steel plant with a predicted demand is part of a study required to get a fuel balance for new installations. The demand of an installation or system is the load at the receiving terminals averaged over a specified time interval: for example, 15 minutes or 30 minutes. The study is also necessary to arrive at an estimated amount to be used in negotiating purchased-power contracts. Table 23—III is an example of the estimation of the annual kWh requirements for an integrated plant planned to produce 2 725 000 metric tons (3 000 000 net tons) of basic oxygen steel per year.

The example in Table 23—III represents an average

consumption of 112 700 kWh per month or approximately 496 kWh per metric ton (451 kWh per net ton). The average demand of this example is 154 000 kW. However, it is not the average load demand that has to be determined but, rather, the maximum 15-minute (or 30-minute or other interval) demand that determines capacity requirements from in-plant generation and/or purchased power. When load factors of the various facilities and overall diversity are considered, the maximum demand for this example would approximate 184 000 kW.

FACTORS AFFECTING SIZE AND TYPE OF MAIN-DRIVE MOTORS

The selection of the proper motor to be used on mill drives is a very important item in the design of a mill. Its size and characteristics must be based not only on tonnage requirements and rolling schedules, but also on displacement, reductions, temperatures, composition of product, speed of rolling, and finish. The accepted methods for calculating loads when rolling a product at a specific speed and temperature, with specified reductions per pass, are based on data taken from tests on existing mills and adjusted with adaptive programs in a computer.

The most common main-drive motor once used was on reversing primary mills. For these drives, the size and shape of the ingot was known, also the size of the finished bloom or slab, the composition and temperature of the steel, the roll size, and the efficiency of the machinery. It has been determined from existing reversing primary-mill units that the total power required is divided approximately into the following components:

Power Required for Rolling 59.00%
Friction of Pinions and Mill 5.90
Loss in Reversing Motor 10.84
Loss in Electrical Connections 0.75

Loss in Generators 11.44
Loss in Flywheel 1.32
Loss in Slip Regulator 2.69
Loss in Induction Motor 6.40
Loss in Exciters, Blowers, Etc. 1.66

Total 100.00%

The part of the total torque required of the motor shaft of a single-motor drive for a reversing blooming mill that is used in (1) deforming the metal being rolled; (2) overcoming the increase in roll-neck bearing friction during rolling; and (3) other mechanical losses, may be approximated from the following formula (see "The Modern Soaking Pit and Blooming Mill," by T. J. Ess and "Main Roll Drives for Blooming and Slabbing Mills," by R. H. Wright, both of which appeared in the 1943 Proc. of the Assn. of Iron and Steel Engineers):

(1a)

$$T = 275 \times C (A_1 - A_2) \times D\frac{A_2}{A_1},$$

where   T = Motor torque in pounds-feet
C = Factor for cubic inches of metal displaced (ordinate of Figure 23—25).
A_1 = Area of section in square inches before pass
A_2 = Area of section in square inches after pass
D = Diameter of roll at base of pass in inches.

This same formula, expressed in SI units, is as follows:

(1b)

$$T = 500 \times C (A_1 - A_2) \times D\frac{A_2}{A_1}$$

where: T = Motor torque in newton-metres
C = Factor for cubic metres of metal displaced (ordinate of Figure 23—25)
A_1 = Area of section in square metres before pass
A_2 = Area of section in square metres after pass
D = Diameter of roll at base of pass in metres

From these formulas, which apply only to the type of mill mentioned, the rolling schedule can be calculated pass by pass, and to the figures so obtained must be added the idling friction torque for the mill which may amount to 27 000 to 34 000 newton·metres (20 000 to 25 000 pounds-feet).

To obtain the maximum torque requirements for the motor, it is necessary to add the torque required for acceleration. The inertia of the mill parts and the armature for a 5220-kW (7000-hp) motor is about 179 000 kilogram-metres² (4 250 000 lb-ft²).

If the required increase in speed is 20 rpm and the time required to accomplish this change is 2 seconds, the torque required for acceleration is calculated from the formula:

(2a)

$$T_a = \frac{0.003255 \times WR^2 \times rpm}{t}$$

where: T_a = Torque required for acceleration (pounds-feet)

rpm = Speed change required
t = Time required for acceleration (seconds)
WR^2 = Inertia of moving parts (pounds-feet²)

Thus,

$$T_a = \frac{0.003255 \times 4\ 250\ 000 \times 20}{2} = 138\ 000 \text{ pounds-feet}$$

The same formula, expressed in SI units, is as follows:

$$T_a = \frac{0.1047 \times WR^2 \times rpm}{t}$$

T_a = Torque required for acceleration (newton-metres)
rpm = Speed change required
t = Time required for acceleration (seconds)
WR^2 = Inertia of moving parts (kilogram-metres²)

Thus,

$$T_a = \frac{0.1047 \times 179\ 000 \times 20}{2} = 187\ 000 \text{ newton-metres}$$

Another factor to be considered is selection of a motor of ample capacity to prevent excessive heating of the motor during operation (see Ess paper referred to in caption of Figure 23—25.)

Figure 23—26 shows a group óf curves for five different steels in which the resistance to deformation is plotted against temperature of the steel. Figure 23—27 shows similar curves for two grades of carbon steel, where rolling temperatures are plotted in relation to power consumption. From these curves it may be seen that a drop in temperature from 1250°C to 950°C (2282°F to 1742°F) will double the power requirements for rolling. This indicates the necessity for considering steel temperature when calculations of power for rolling are being made.

Another factor to be considered is the number of kilowatt-seconds per cubic metre, or horsepower-seconds per cubic inch, of metal displaced required during the various passes. Temperature again must be considered in this case as the bloom or slab cools off during the rolling operations. Figure 23—25 shows curves giving the specific power for motor torque for displacement of metal plus all increases in friction during the pass. From these curves, it will be noted that as the bloom becomes smaller the power requirements increase rapidly, especially after 25 per cent of the original area is reached. The constants "C" used in formulas (1) and (2) were taken from these charts.

Another form of these power curves is shown in Figure 23—28. These curves, for flat and structural sections, show that power requirements do not increase in direct proportion to per cent reductions, and that the smaller the section the larger the increase in power in terms of reduction.

The speed at which the piece is rolled affects the power requirements in two ways. Theoretically, the power requirements should be directly in proportion to speed; however, with increase in speed the temperature of the piece being rolled has less time in which to drop; therefore, the higher the rolling speed, the higher the finishing temperature and, consequently, the lower the power required by the mill. For this rea-

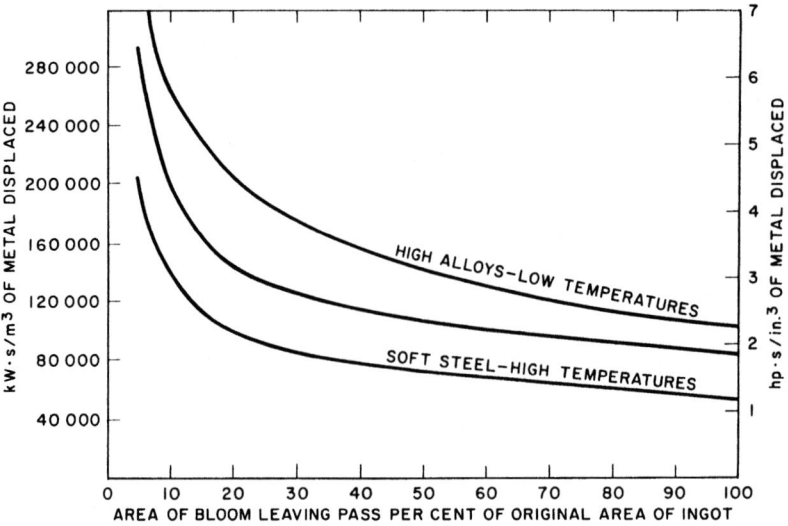

FIG. 23.—25. Curves showing amount of work for displacement of one cubic metre or of one cubic inch of metal including all increases in friction during the pass in a single-motor-driven reversing blooming mill. (See "The Modern Soaking Pit and Blooming Mill," by T. J. Ess, and "Main Roll Drives for Blooming and Slabbing Mills," by R. H. Wright: both in the 1943 Proceedings of the Association of Iron and Steel Engineers.)

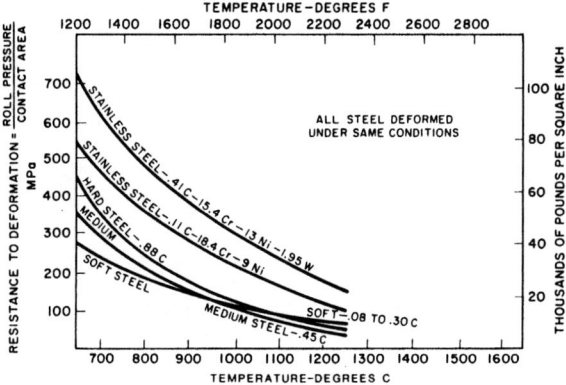

FIG. 23—26. Effect of temperature of steel on resistance to deformation. (After Trinks.)

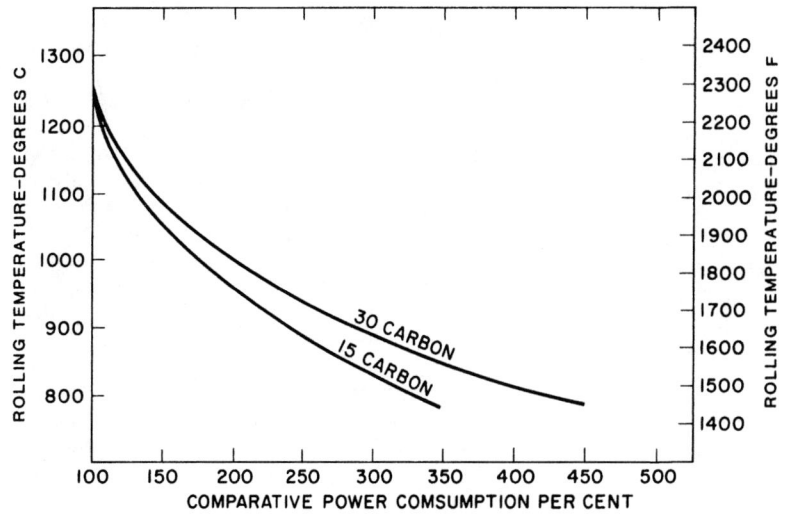

FIG. 23—27. Effect of temperature on power consumption during rolling steels containing 15 and 30 points (0.15 and 0.30%, respectively) of carbon. (Courtesy, Iron and Steel Engineer.)

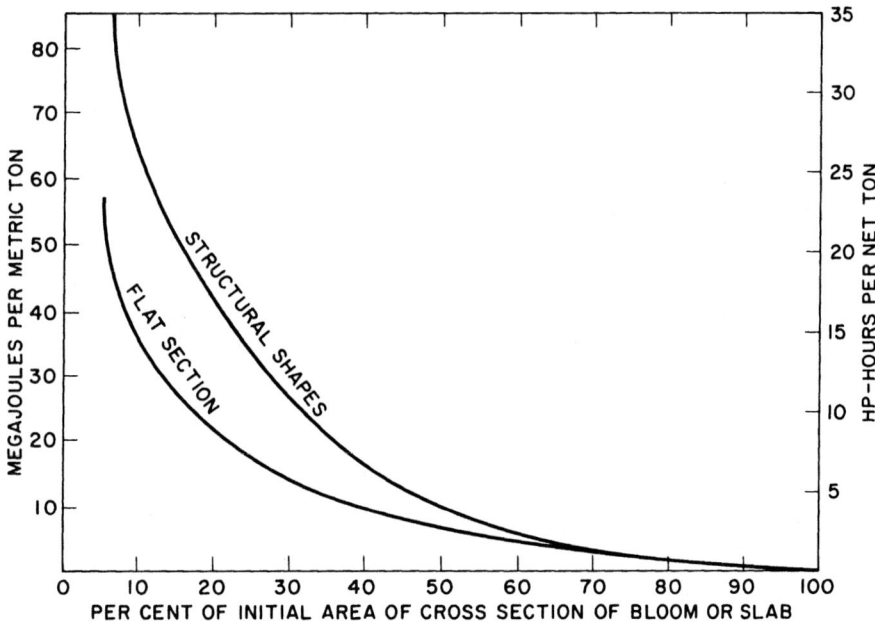

FIG. 23—28. Curves showing relation of megajoules per metric ton of steel rolled to per cent reduction of cross-section for shapes and flat sections.

son, it is sometimes possible to increase the speed and not increase the motor loading to any great extent.

In order to meet the surface-quality requirements on certain products, it is necessary to roll them in the finishing pass at relatively low temperatures—sometimes as low as 760°C (1400°F). On mills where this practice is to be used, the horsepower requirements for the motor should be determined on the basis of the reduction and temperature required for such sections.

TYPES OF MOTORS FOR MAIN DRIVES

There are four general classifications of motors that can be used on mill drives: synchronous, squirrel -cage, wound-rotor, and direct-current. All but the last named operate on alternating current. Each of these has characteristics which make it suitable for definite applications. The cost of each type also differs and is often the factor determining the type to be used.

The synchronous motor, if an exciter is not required, usually costs the least and, in certain speed ranges, may be used interchangeably with the squirrel-cage design. Wound-rotor motors cost more than either the synchronous or squirrel-cage, and direct-current motors are the most expensive of all. With the latter, a supply of direct current is needed, which necessitates the use of a motor-generator set or suitable rectifier to convert the alternating-current power supply of the plant to direct current, and these have to be considered as a part of the installation.

In the following paragraphs, short descriptions of the four general classifications of motors are given, along with suggestions for possible applications. It is assumed that the reader is familiar with the general design of such units; if not, details of their design and construction are available in the literature.

Synchronous Motors—The synchronous motor has the stator or stationary part wound with core and coils in the same manner as the squirrel-cage and wound-rotor motors. However, the rotating part or rotor has salient poles, positive and negative, wound for direct current. The power supply for these poles can be furnished by a small direct-current generator called the exciter. This may be mounted on the main motor shaft, be belt driven by same, or be driven by a separate motor in a motor-generator set. Many recent machines are built with brushless exciters which eliminate slip rings and brushes or they receive their separate excitation from a static power supply. In addition to the direct-current coils on the pole pieces, another winding which is buried in the pole faces is required. This is called the squirrel-cage winding or amortisser and controls the necessary starting and pull-in torques of the motor and also the uniformity of the accelerating torques between starting and pull-in. When driving reciprocating machinery, the squirrel-cage winding also develops a damping torque.

This amortisser winding consists of a number of round or rectangular rods or bars that pass through the pole shoes in the axial direction. The material used may be copper, brass, or bronze, of such electrical conductivity as to suit best the particular requirements of starting and pull-in torques. The ends of these rods or bars project sufficiently beyond the poles to allow making connections to the short-circuiting end-ring segments to which they are joined securely by brazing. In some cases, where both high starting torque and high pull-in torque are required, it is necessary to use double cage windings. These consist usually of two sets of damper bars but only one set of end connections.

To conduct the direct current to the field coils, collector rings mounted on the main shaft are required, the terminals for the field circuit being connected to these rings and the direct-current power supply from the exciter is connected to the brushes that contact the rings. The brushless exciter has a stationary field and ac output on the rotor. This ties directly to the motor field coils through a rotor mounted diode bridge and static starting control assembly.

In order to apply a synchronous motor correctly, it is necessary to give consideration to the starting torque, pull-in torque, pull-out torque, operating temperature, power factor and method of starting.

The time required for a synchronous motor to attain full speed depends upon the load, and the power required to pull into step from this speed depends upon the inertia of the revolving parts, so that the pull-in torque cannot be determined without knowing the inertia of the external load as well as the torque required to overcome it.

The permissible starting-power input will vary with different localities, being affected by local power company regulations, capacity of feeders, or the capacities of individual, isolated power stations.

These motors can be designed with considerable range of starting, pull-in and pull-out torques. Some applications require a starting torque of 150 to 200 per cent, a pull-in torque of 110 per cent, and a pull-out torque of 175 per cent.

A synchronous motor usually has a higher efficiency than that of a comparable induction motor, it may be less expensive in first cost, and it has the ability to correct power factor. They have been built in sizes up to 14 920 kW (20 000 horsepower) for mill drives. It must be remembered that they are a constant-speed motor and are locked in step with the frequency of the power system. For this reason, there is no cushion afforded by a drop in speed for a sudden peak load, and the pull-out torque must be high enough to handle the highest peak load encountered. Synchronous motors often are used to drive motor-generator sets, air compressors, pumps and other types of equipment where constant load conditions exist.

Wound-Rotor Induction Motors—This type of induction motor has a wide range of use in the steel industry. The difference between it and the squirrel-cage type is that the rotor winding consists of coils with the phase connections brought out to collecting rings mounted on the motor shaft. Connections can be made from these rings through external resistance boxes or a slip regulator; hence, the secondary resistance can be varied. By this means, the starting-power input and accelerating time can be controlled and, if the motor is used in connection with a flywheel, various divisions of the load between the motor and the wheel can be obtained.

These motors are available over the complete output and speed ranges required for mill drives, for operation on voltages up to 13 200. As in the case of squirrel-cage motors, the efficiency and power factor decrease with the base speeds.

Wound-rotor induction motors are standard for flywheel-equipped motor-generator sets, and usually have been designed for a speed of 360 rpm in medium sizes and 300 rpm for the largest sets. However, several large installations have been designed for 514 rpm.

One such unit consists of a 5970-kW (8000-hp) motor with a 59-metric ton, 160 400-kJ (65-net ton, 215 000 hp·s) flywheel; driving four 2500-kw and two 1750-kw direct-current generators.

Squirrel-Cage Motors—Squirrel-cage induction motors are the most simple with respect to design, and can be built in a wide range of sizes and torques. By varying the design resistance in the rotor winding, the starting torque can be low, normal, or high, as desired. It has become customary to start these motors "across the line" in the usual sizes, which necessitates strong bracing for the stator coils to withstand the powerful magnetic forces involved.

In the larger sizes, these motors have been designed for use in certain speed ranges in capacities up to 14 920 kW (20 000 horsepower), and for speeds as low as 100 rpm, operating on 60-hertz power.

This motor may have the lowest initial cost per installation, but its efficiency is less than that of the synchronous motor, and at the lower speeds the power factor is lowered. For instance, a 373-kW (500-horsepower), 60-hertz squirrel-cage motor has the following efficiencies and power factors at 900 compared with 514 rpm:

rpm	Efficiency (%)		
	½ Load	¾ Load	Full Load
900	91.1	92.9	93.1
514	91.3	92.3	92.3
	Power Factor		
	½ Load	¾ Load	Full Load
900	82.0	88.0	90.0
514	75.0	83.0	86.5

This type of motor is used on fans, pumps, compressors and shears, to name a few common applications, but is not used normally on main drives.

Direct-Current Motors—Due to the demand for wide speed ranges for mill drives, many large direct-current motors are being used for this purpose. They are designed with flat speed characteristics, to operate usually on 600 or 700 volts. The largest size, about 8960 kW (12 000 horsepower) has been used on reversing mills. These motors require forced ventilation due to the heat generated by operating conditions involving frequent reversals. The normal speed range is 2:1, but higher ranges can be obtained. On mill drives, it is not uncommon to have a maximum torque of 240 per cent of the full-load torque.

The trend toward the use of variable-speed direct-current motors on main drives began in the 1930's. The operating advantages of this type of drive offset the added cost. Table 23—IV, taken from annual summaries in the Association of Iron and Steel Engineers Yearbooks, shows this trend in motors (over 225 kW or 300 horsepower) sold by American motor builders to both domestic and foreign rolling mills in the steel and allied industries from 1949 to 1978, inclusive.

The speeds of synchronous and induction motors are a function of the number of poles in the motor windings. These speeds are determined by the formula:

Table 23—IV. Main-Drive Motors Installed 1949-1978*

| Year Installed | Alternating-Current Motors | | | | Direct-Current Motors | | |
| | Number Installed | Total Power Rating | | Number Installed | Total Power Rating | | |
		kW	hp		kW	hp
1949	11	24 990	33 500	78	119 345	160 100
1950	32	58 000	77 750	236	336 780	451 450
1951	30	26 070	34 950	174	226 970	304 250
1952	8	5 370	7 200	41	48 190	64 600
1953	8	21 410	28 700	67	69 040	92 550
1954	13	12 010	16 100	106	119 065	159 605
1955	30	27 900	37 400	290	350 550	469 909
1956	28	69 640	93 350	226	284 605	381 510
1957	19	22 680	30 400	179	202 615	271 605
1958	13	23 945	32 100	84	104 850	140 550
1959	14	41 665	55 850	226	358 990	461 220
1960	7	8 800	11 800	133	158 050	211 800
1961	11	10 705	14 350	191	223 740	299 920
1962	16	50 610	67 840	162	206 940	277 400
1963	18	50 690	67 950	246	347 650	466 022
1964	22	108 765	145 800	299	456 295	611 655
1965	15	64 195	86 050	204	295 985	396 760
1966	8	3 505	4 700	279	298 060	399 541
1967	8	43 340	58 100	154	226 135	303 130
1968	7	3 990	5 350	216	235 625	315 850
1969	2	1 240	1 730	112	150 375	201 575
1970	35	89 020	119 330	348	578 525	775 505
1971	10	27 185	36 440	187	288 560	386 810
1972	6	4 895	6 565	284	273 980	376 262
1973	12	7 855	10 530	99	84 435	113 185
1974	5	5 220	7 000	209	306 255	410 530
1975	—	—	—	40	52 830	70 820
1976	2	1 940	2 600	62	39 540	53 000
1977	12	6 340	8 500	56	46 440	62 250
1978	—	—	—	71	62 255	83 450
Totals	402 (7.4%)	821 975 (11.2%)	1 101 935 (11.2%)	5059 (92.6%)	6 552 675 (88.8%)	6 539 840 (88.8%)

*Total rated power of main-drive motors over 225kW (300-hp). Some of these are replacement motors and would not represent additions of total installed power ratings.

$$N = \frac{60 \times F}{\frac{1}{2}P}$$

where,

N = Speed (rpm)
F = Frequency
P = Number of poles

Table 23—V shows the possible speeds that can be obtained in both 25-hertz and 60-hertz motors. These are synchronous speeds and apply to synchronous motors. For induction motors, the full-load speeds are a few per cent less due to slip.

Although it is possible to build motors with more than 40 poles, they become very expensive, and the above table shows how it was advantageous to build the slow-speed, 25-hertz motors used in some of the early mills and why it would have been necessary to use gear sets if 60-hertz frequency had been specified.

PRINCIPLE AND APPLICATION OF FLYWHEELS

Many main-drive applications are on mills that have short-time peak rolling loads. Blooming mills, slabbing

Table 23—V. Synchronous Speeds of 25-hertz and 60-hertz Motors

| Number of Poles | Speed | |
	25-hertz (rpm)	60-hertz (rpm)
2	1500	3600
4	750	1800
6	500	1200
8	375	900
10	300	720
12	250	600
14	214	514
16	187½	450
18	166⅔	400
20	150	360
22	136	327
24	125	300
26	115	277
28	107	256
30	100	240
32	93.75	225
34	88.23	212
36	83⅓	200
40	75	180

mills, and reversing plate mills are in this classification. For such mills, it is often economical to use a wound-rotor induction motor having an adjustable secondary resistance in conjunction with a flywheel, to drive the generators that supply direct current to the main mill motors that are subject to peak loads. By this means, it is possible to select a motor having a capacity $\frac{1}{3}$ to $\frac{1}{2}$ of the value of the maximum load requirements of the mill, and so control its "slip" that the flywheel energy supplies the greater proportion of the energy to meet the peak loads. Some typical questions related to motors and flywheels of mill drives, and the methods of finding their answers, are given in the following examples.

1. Energy Stored in a Flywheel—Assume a large flywheel having a moment of inertia $(WR^2) = 15\,000\,000$ lb-ft² (that is, an effect of 15 000 000 lb at a radius of gyration of 1 ft, or 3 750 000 lb at a radius of 2 ft), and a wheel speed of 83 rpm. What is the total amount of energy stored in the wheel?

The energy (E) possessed by any body of weight W lb moving with a velocity of V ft per second is:

$$(1) \quad E = \frac{W \times V^2}{32.16 \times 2} \text{ ft-lb}$$

In this expression, 32.16 ft per second per second is the rate of acceleration due to gravity, usually expressed by the symbol, g. The rotating flywheel with a given WR^2 and running at N rpm may be considered as a body with weight W moving with the same velocity as the end of its radius of gyration R; this velocity in feet per second for the case under consideration is:

$$(2) \quad V = \frac{2 \times 3.1416 \times R \times N}{60}$$

Substituting this value of V in (1), it becomes:

$$(3) \quad E = \frac{WR^2 \times N^2}{5865} \text{ ft-lb}$$

Then, for the present example,

$$(4) \quad E = \frac{15\,000\,000 \times (83)^2}{5865} = 17\,500\,000 \text{ ft-lb}$$

Applying SI units of comparable magnitude to this same example, the moment of inertia is MR^2, where M is the mass of the wheel in kilograms and R is the radius of gyration in metres, and is equal to 632 000 kg·s.

The energy (E) possessed by any body of mass M kilograms moving with a velocity of V metres per second is:

$$(1) \quad E = \frac{1}{2}MV^2 \text{ newton·metres}$$

$$(2) \quad V = \frac{2 \times 3.1416 \times R \times N}{60} \text{ metres per second}$$

$$(3) \quad \text{becomes } E = \frac{MR^2 \times N^2}{182.4} \text{ newton·metres}$$

$$(4) \quad \text{becomes } E = \frac{632\,000 \times (83)^2}{182.4} = 23\,880\,000 \text{ newton·metres}$$

This energy is equal to the amount of work required to bring the flywheel from rest to a speed of 83 rpm.

Since 1 horsepower-second = 550 ft-lb

E = 31 900 horsepower-seconds

or, in SI units:

$$1 \text{ kWh} = 3\,600\,000 \text{ N·m}$$

$$E = 6.63 \text{ kWh}$$

From the foregoing formulas, it may be seen that the stored energy in a flywheel is proportional to the WR^2 of the wheel and to the second power of its speed.

2. Amount of Energy Available for Regulation—The energy calculated above is the total energy stored in the wheel at a certain speed. How much of this energy may be used for load equalization in case this flywheel is installed on a mill drive?

If it is assumed that a speed variation of more than 15 per cent is not permissible, the wheel speed could be varied from 83 rpm to 70.5 rpm. At 70.5 rpm, the stored energy in the wheel would be:

$$E_2 = \frac{15\,000\,000 \times (70.5)^2}{3\,230\,000} = 22\,950 \text{ horsepower-seconds}$$

or, using SI units:

$$E_2 = \frac{632\,000 \times (70.5)^2}{182.4 \times 3\,600\,000} = 4.78 \text{ kWh}$$

Thus the energy given up by the wheel in slowing down from 83 to 70.5 rpm would be:

31 900 − 22 950 = 8 950 horsepower-seconds, or, in SI units 6.63 − 4.78 = 1.85 kWh. This equals approximately 28 per cent of the total stored energy.

From this it is evident that the per cent of energy given up by the wheel is not directly proportional to the per cent of speed reduction. This relationship is shown in Figure 23—29.

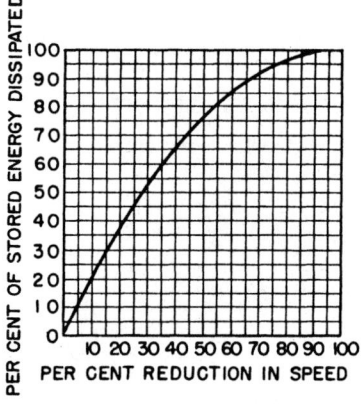

FIG. 23—29. Relationship of stored energy given up by a flywheel to per cent reduction of speed of a motor operating in conjunction with it. (Courtesy, Iron and Steel Engineer.)

VARIOUS MEANS FOR OBTAINING ADJUSTABLE SPEEDS

A. CONTROL OF TWO-SPEED A-C MOTORS

It is possible to design induction motors that will operate on two or more speeds with a single winding, the slower speed being one-half of the higher. This is done by bringing out the groups of phase leads to switches, by means of which the number of poles in the stator winding can be regrouped to give half the number of poles. Thus, a motor can be designed to operate at either 600 rpm or 300 rpm on 60-hertz power. Such a motor will cost 30 per cent to 40 per cent more than a standard motor, and would be a constant-torque machine, or, at the lower speed, would have only half the horsepower or kW rating. In addition to constant torque, they may be designed for constant power or variable torque.

A variation of this design can be built using two stator windings with which two, three, or four speeds can be obtained. In the past decade, a pole amplitude modulation form has been offered which provides many selections of two-speed control on a single winding. Ratios of 6:8, 8:10, 10:12 or wide ranges of 4:1 or 5:1 are offered.

In some of the older mills, two-speed, wound-rotor motors were installed so that the first passes on a reversing mill could be rolled at half speed; and, as the piece became longer, the contactors changed the winding connections to operate the motor at full speed and hence deliver its full power.

B. A-C MOTOR SPEED CONTROL BY SECONDARY RESISTANCE

When resistance is added to the secondary circuit of a wound-rotor induction motor, and a load imposed, the speed of the motor is reduced, and this drop in speed is called "slip." The slip increases with increasing load, hence, it is a function of both load and secondary resistance. For example, if with a given load and secondary resistance the motor has a slip of 10 per cent, which means it is operating at 90 per cent synchronous speed, and the load is doubled, the motor will then have 20 per cent slip and operate at 80 per cent synchronous speed. These statements illustrate why speed control by varying secondary resistance is not satisfactory for mill operations, as each load has a definite corresponding motor speed and as soon as the load changes, the speed changes.

This type of control, however, has wide application on flywheel sets where the secondary resistance is controlled by a slip regulator which can be set to control automatically the amount of slip. This, in turn, controls the relation between the momentum given up by the flywheel and the torque of the motor.

C. VARIABLE-SPEED CONTROLS FOR A-C MOTORS

Two common methods have been employed for controlling the speed of wound-rotor induction motors in mill drives. These are known as the Kraemer and the Scherbius systems. In the **Kraemer system,** slip rings of such a motor are connected to a rotary converter (plus auxiliaries) so that the slip energy of the main motor is transformed to d-c power, and the latter is either returned to the main drive shaft (to maintain constant power output), or converted to a-c power and returned to the line (to maintain constant torque). The **Scherbius system** employs an auxiliary regulating machine which converts the slip energy of the main drive motor into mechanical power on the shaft of that motor to maintain a constant power output. The main difference between the two systems is that, while the Scherbius system uses an a-c commutation machine to convert

slip energy of the motor to mechanical power on its shaft, the Kraemer system accomplishes this by using a rotary converter and a d-c motor. Both systems are largely of historical interest, so far as main drives for rolling mills are concerned, as none have been installed in this application since about 1930. As the energy crisis of the 1970's occurred, renewed interest in "slip power recovery" systems brought developments primarily in the static conversion areas.

Variable frequency speed control of any type a-c motor is feasible over a wide frequency range. (6 to 120 hertz) but in general has not been competitive in performance and first cost with the highly favored direct-current drives.

D. VARIABLE-SPEED CONTROLS FOR D-C MOTORS

Ward-Leonard Control—In the Ward-Leonard system of control the voltage of a d-c generator is varied from zero to full voltage and this power supply connected to a d-c motor or motors. The speed of a d-c motor with a given load and the shunt field kept constant is proportional to the impressed voltage. Hence, if the armature voltage is varied from zero to full voltage, the motor speed will vary from zero to base speed (full field). During this period, the motor exerts constant torque and operates as a constant-torque machine. This is the essence of the Ward-Leonard system of control, and is the method used to start large d-c motors as well as to control speeds up to the base speed. Such motors can be accelerated and decelerated very smoothly and rapidly without drawing an increased amount of power from the system, contrary to the case with a-c motors.

For speeds above the base speed, d-c motors in standard types have a 2:1 speed range. Higher speed ranges, up to 4:1, can be obtained in certain sizes and designs. An example of the higher extreme is the type of motor employed for driving the reel on cold-reduction mills. For speed control of such d-c motors above the base speed (usually 200 per cent of the base speed or higher), the Ward-Leonard control is not applicable because at base speed, the system is delivering full voltage that already is producing maximum armature current. Speed control of such d-c motors above base speed is accomplished by adding resistance in the field circuit that results in weakening of the shunt fields; during operation under these conditions the motor becomes essentially a constant-power machine.

The direction of rotation of d-c motors is controlled by reversing the direction of the flow of current irrespective of the method of speed control employed.

Relay and Continuous-Feedback Systems—During the past forty years, various electrical control schemes have been developed for maintaining constant speeds, speed relations between different motors, and divisions of load in a tandem train. These methods have the basic principle of sensing very small variations in some function such as speed, load, voltage, current or tension and amplifying them so that the regulating equipment has large quantities to work with in maintaining close control. These systems also eliminate many contactors by employing the principles of static power supplies and solid-state regulators, magnetic amplifiers or electronic equipment.

The choice of regulating system for mill drives requires a thorough knowledge of the requirements of the process and an appreciation of the proper function or combination of functions to be controlled to satisfy these requirements.

In general, there are many regulating systems that were used in a historical evolution. Rotating, magnetic-amplifier and electronic and solid-state regulators are examples of control elements utilizing the continuous-feedback principle.

Rotating Regulating Systems provided a high-power-amplification ratio and responded very quickly to a change in input power. These characteristics were obtained by a system of field and armature connections on a machine very similar to a conventional d-c generator.

The Magnetic-Amplifier Regulating System differs from the rotating-regulator system in that the control element is a static rather than a rotating device. This static power-control unit combines a saturable-core reactor in a self-saturated circuit.

The magnetic-amplifier regulating system has been applied in considerable numbers in the steel industry. It has been used for controlling tandem cold-reduction mills, reel drives for cold-reduction mills, speed control for high-speed side-trimming line tension reels, speed regulation for the bridle rolls on cleaning lines, tension-reel drives for processing lines, tensiometer-type tension regulators for skin-pass or temper mills and as a voltage regulator for ignitron rectifiers (see Harris reference at end of chapter).

Electronic regulators have been used for applications where the regulator has to "see," such as edge control for slitting and cleaning lines, and where very precise and fast-acting regulator components are required such as rod-mill droop regulators and electric flying-shear regulators. To obtain the required speed regulation, such as required on the tube and rod mills, combination electronic-rotating regulating equipment has been used.

Thyristors with Phase Control—Development of transistors and silicon rectifiers has resulted in continual improvements to automatic-control systems. Operational amplifiers and silicon-controlled rectifiers are now used almost exclusively, instead of magnetic or vacuum-tube amplifiers and rotating regulators, for new installations of automatic-control systems. This transition began in the early 1960's and continued to expand and be refined through the 1970's. In a modernization program involving automatic gage control, operational amplifiers and silicon-controlled rectifiers were used in the design of analog control systems for the main-drive equipment and auxiliary drives on the finishing stands of three wide continuous hot-strip mills of United States Steel Corporation. High-speed static switching devices were used in digital systems for screw-down-position control. Similar analog and digital control systems are used throughout the computer-controlled 2134-mm (84-inch) hot-strip mill at Gary Works; in fact, silicon-controlled rectifiers are used exclusively now in the industry instead of mercury-arc rectifiers or rotating generators, for the power supply to the main drives and screwdowns on the finishing stands and for various other auxiliary drives.

REVERSING-MILL DRIVES

On motor-driven primary mills, such as slabbing, blooming, or single-stand plate mills, it is customary to use a reversing mill drive. A few of the older mills were driven by steam engines which were retained for economic reasons. Since the installation of the first motor-driven reversing mill in 1915, very few engine drives have been installed and, today, only motor drives are given consideration.

For modern reversing-mill service, the main-drive d-c motor is of rugged design, with a welded frame of rolled steel, horizontally split. The pedestal for the bearing next to the mill is steel and is provided with a thrust bearing. These motors, being connected directly to the mill or pinion stand, necessarily must be of slow speed. One of the normal speed ratings is 0/40/80, which represents speed control from zero to 40 rpm, which is the constant-torque range, on voltage control, and from 40 to 80 rpm, which is the constant-horsepower range, on field control.

One of the original design problems on these motors was commutation and, in sizes above 2984-kw (4000 horsepower), they were built with two armatures on a single shaft. In recent years, however, the designs have been developed so that a 5970 kW (8000 horsepower), 0/50/100 motor has been built with a single armature. Sizes and speeds of some typical installations of these motors are shown in Table 23—VI.

This table indicates that the most popular size is about 5220 kW (7000 horsepower). Some of the characteristics of such a motor are:

Frame Diameter . . . . 5000 mm (197 inches)
Armature Diameter . . . . 3760 mm (148 inches)
Length of Shaft . . . . 6500 mm (295 inches)
Diameter of Shaft . . . . 815 mm (32 inches)
Weight of Complete Motor . 188 700 kg (416 000 lb)
Approximate Re-   50-0-50 rpm—2 seconds
versing Cycle . . . . 120-0-120 rpm—5 seconds

A motor of this type was installed on the reversing roughing stand of a 3350-mm (132-inch) semicontinuous plate mill of United States Steel Corporation. This 5220-kW (7000-horsepower) 0/25/60 rpm motor is really two 2610-kW (3500-horsepower) motors mounted as a single unit. The armatures are of unusual construction in that they have no conventional shaft. The spiders are bolted to stud flanged shafts at each end and the complete unit assembled at the mill. This design solved some of the shipping problems for the heavy pieces.

When the power requirements for a reversing mill exceed 5220 kilowatts (7000 horsepower), problems in motor and pinion design arise that suggest the use of two separate motors, one to drive the top roll and one to drive the bottom roll, connected together electrically. This is known as the twin drive (Figure 23—30). The first twin drives were installed at the 1370-mm (54-inch) blooming mill and the 1120-mm (44-inch) slabbing mill in United States Steel's South Works in 1928.

This type of drive has a lower inertia than the single-motor drive of comparable power, eliminates the pin-

Table 23—VI. Typical Installations of Reversing Mill Motors*

Company	Location	Mill Size		Motor Rating			Motor Voltage Rating
		mm	in.	kW	Horse-power	rpm	
Armco Steel Corp.	Butler, Pa.	1220	48	5220	7000	50-120	700
	Middletown, Ohio	1140	45	2-3730	2-5000	40-80	700
Bethlehem Steel Co	Steelton, Pa.	1120	44	4470	6000	40-120	600
	Sparrows Point, Maryland	1370	54	5220	7000	50-100	650
	Sparrows Point, Maryland	1015	40	5220	7000	50-100	650
Great Lakes Steel Corporation	Ecorse, Mich.	1170	46	5220	7000	40-100	700
Inland Steel Co.	Indiana Harbor, Indiana	1170	46	5220	7000	50-120	700
Jones & Laughlin Steel Corp.	Aliquippa, Pa.	1120	44	2-4470	2-6000	70-140	750
	Indiana Harbor, Indiana	1140	45	2-3730	2-5000	40-80	750
Republic Steel Corporation	Warren, Ohio	915	36	4850	6500	50-120	600
	Canton, Ohio	890	35	3730	5000	40-120	600
	Gadsen, Ala.	1015	40	5220	7000	50-120	700
United States Steel Corp.	Duquesne, Pa.	1170	46	2-3730	2-5000	40-80	750
	Homestead, Pa.	1120	44	5220	7000	50-120	700
	Homestead, Pa.	1140	45	2-3730	2-5000	40-80	700
	Gary, Indiana	1120	44	5970	8000	40-100	800
	South Chicago, Ill.	1370	54	2-3730	2-5000	40-80	700
	Fairfield, Ala.	1170	46	5220	7000	50-100	700
	Morrisville, Pa.	1140	45	2-4470	2-6000	40-80	700
Wheeling-Pittsburgh Steel Corporation	Steubenville, Ohio	1140	45	4470	6000	50-120	700

*Courtesy of "Iron and Steel Engineer."

ion stand, and represents about the same proportion of the cost of the complete mill as a single-motor drive. It has the advantage also of being able to operate with rolls of different diameters in the mill, and also to equalize loading of the two rolls.

One of the design problems involved in laying out a twin-motor drive is the determination of the length of the connecting spindles. The diameter of the main-drive armatures is a fixed distance for a given size and speed of motor, and this dimension, plus that of the field pole and the yoke, give the vertical distance between centerlines of the two drive shafts at the motor bearings. The diameter of the mill rolls, plus the maximum roll opening, set the condition at the mill end of the drive. The problem then is to use a length of spindle that results in a drive angle that will transmit the full torque without deflection or whipping. On some modern drives, each motor is rated at 3730 kW (5000 horsepower) 0/40/80 rpm, and has a maximum torque of 2 671 000 newton-metres (1 970 000 pounds-feet).

On the 54-inch reversing blooming mill referred to above, the maximum angle with the horizontal for the spindle was limited to four degrees.

On reversing slabbing mills, an edger drive usually is required, the motor of which must perform in relation to the main-drive motor or motors. This drive should not push a slab into the main rolls or exert a pulling action on it in opposition to the main motors. One means of preventing this is to space the edger far

enough away from the mill so that the slab is never in the edger and the mill at the same time. Several modern installations, however, have the edger as part of the main mill so that the slab can be edged and rolled simultaneously. Such an installation requires a coordinated motor and control design to distribute motor loads properly.

The Flywheel Motor-Generator Set—In most types of d-c reversing drives, the motor receives power from a flywheel motor-generator set (mercury-arc rectifiers and silicon-controlled rectifiers are used on some recent installations). Since the practical size for generators of this type is limited to about 4000 kW, there may be two or three generators on a continuous bed plate, driven by a wound-rotor induction motor, whose combined output will meet the mill requirements. The generator capacity is in the ratio of 1.15 to 1.35 of the motor capacity in order to meet the requirements of the maximum torque of the motor, plus losses. The induction motor driving the generator or generators, plus the flywheel, has a capacity equal to about the average net output of the drive motor (which may be only 50 per cent of the capacity as limited by heating effects), plus the electrical losses of the generators, slip regulator and the friction and windage losses of the flywheel.

In practice, these motor-generator sets are proportioned as follows:

FIG. 23—30. Twin-motor drive for an 1120-mm (44-inch) slabbing mill consisting of two 3730-kW (5000-horsepower) 40/80 rpm, direct-current motors.

Generator Capacity (kW)	Induction-Motor Rating	
	kilowatts	Horsepower
6 000	3 730	5 000
7 500	4 480	6 000
9 000	5 220	7 000
10 500	5 600	7 500
13 500	5 970	8 000

The flywheel is of hot-riveted, steel-plate construction and, as indicated in Table 23—VII, may be 4770 mm (15 feet) in diameter, weigh from 63 500 to 90 700 kilograms (140 000 to 200 000 pounds), and be driven at a peripheral speed of about 112 metres per second (22 000 feet per minute). A slip of about 15 per cent in speed, which will make available about 30 per cent of the flywheel energy, is the maximum desirable. To obtain the benefit of this stored flywheel energy, a slip regulator is connected in the secondary circuit of the wound-rotor induction motor. This device consists of three vertical electrodes with their lower ends immersed in a salt solution, and raising or lowering the electrodes inserts more or less resistance in the secondary circuit, which causes the motor to slow down or slip as the load is thrown on, permitting the flywheel energy to be used.

A torque motor that responds to the amount of current in the primary circuit of the induction motor acts to raise or lower the electrodes. These slip regulators generally are set to restrict the input to the induction motor at some value between 125 and 150 per cent of full-load rating. The regulator also is used as the starting resistor when the set is put in the line.

The drive motor of the motor-generator set often has another feature incorporated in its control scheme; that is, means for dynamic braking. This consists of a circuit that, by means of circuit breakers, can be connected to a d-c source, in order that after the a-c source is disconnected, these circuit breakers close and dc flows in two of the motor phase windings. The purpose of this is to shorten the time required for the set to drift to a standstill. Without the dynamic braking feature, the drift time for an average flywheel set may be as long as 30 to 40 minutes.

THREE-HIGH MILL DRIVES

Motors to drive three-high mills usually are of the low-speed, direct-connected, wound-rotor induction type. Often, the rotor is designed to take advantage of the flywheel effect. In other cases, a flywheel is mounted on the same shaft with the rotor. The motor speed is usually a compromise—slow enough to permit the slab or bloom to enter the mill, and fast enough to enable the mill to produce the specified tonnage.

Since the motor of a three-high mill operates continuously in one direction only, the proportion of total

Table 23—VII. Reversing Drive Motors*

Motor Rating						
kW	2798	3730	4737	5222	5968	7460
hp	3750	5000	6350	7000	8000	10 000
Motor Speed (rpm)	0-50-120	0-50-120	0-60-120	0-50-120	0-40-80	0-40-80
Continuous Rating Torque						
N • m	448 648	596 304	632 651	834 826	1 192 609	1 487 922
lb-ft	395 000	525 000	557 000	735 000	1 050 000	1 310 000
Maximum Torque, Field Breaker						
N • m	1 010 878	1 345 944	1 419 773	2 184 258	2 612 381	3 350 663
lb-ft	890 000	1 185 000	1 250 000	1 660 000	2 300 000	2 650 000
Maximum Torque, Main Breaker						
N • m	1 238 042	1 646 936	1 743 481	2 294 352	3 282 514	4 088 945
lb-ft	1 090 000	1 450 000	1 535 000	2 020 000	2 890 000	3 600 000
Generator Capacity, kW .	2 × 1800	2 × 2200	2 × 2700	2 × 3000 3 × 2500	2 × 3500 3 × 2500	3 × 3000
MG Set Drive						
kW	2238	2723	3245	3730-4476	3730-4476	4849-5595
hp	3000	3650	4350	5000-6000	5000-6000	6500-7500
MG Set Drive Speed, rpm	514	514	400	360	360	360
Flywheel Diameter						
mm	3962	3962	4420	4572	4572	4572
ft	13	13	14.5	15	15	15
Flywheel Weight						
kg	34 019	34 019	45 359	49 895	68 039	81 647
lb	75 000	75 000	100 000	110 000	150 000	180 000
Flywheel WR²						
kg • m²	69 531	69 531	116 307	134 848	181 202	219 129
lb-ft²	1 650 000	1 650 000	2 760 000	3 200 000	4 300 000	5 200 000
Flywheel Stored Energy						
MJ	100.7	100.7	99.9	96.9	130.5	156.6
hp-sec	135 000	135 000	134 000	130 000	175 000	210 000

*Courtesy of "Iron and Steel Engineer." SI values obtained by conversion of published fps units.

time the piece is in the rolls depends on the speed of (1) the screw-downs and (2) the tilting tables. There are some installations where d-c motors are used, which have the advantage of variable speed, and these mills can produce higher tonnages than comparable mills using constant-speed motors because of more favorable operating characteristics.

The chief applications for wound-rotor induction-motor drives have been on three-high plate and structural mills. The motors are controlled by slip regulators or notch-back control to limit the power peaks. The notch-back control consists of secondary grid resistors which are cut in or out in fixed steps by load relays and function in the same manner as the slip regulators. The chief difference is that the notch-back control resistance is in fixed steps while the slip regulator, with electrodes moving in and out of the salt solution, makes it possible to increase or decrease resistance in a continuous manner without steps, giving smoother operation.

CONTINUOUS-MILL DRIVES

Wide-Hot-Strip Mills—The major development of continuous-mill drives was associated with the wide strip mills.

The conventional layout for these mills has a long, narrow motor room paralleling the hot-strip mill, between the mill and the slab yard. The motor room houses all of the main-drive motors, the motor-generator sets or rectifiers, the gear-reduction units, control and switching equipment. Figure 23—31 shows a section of the motor room in one of the modern mills.

The drive motors for continuous mills follow a fairly uniform pattern, with the scale breaker and roughing stands being driven by synchronous motors or a wound-rotor motor with flywheel and slip regulator, and the finishing stands driven by variable-speed d-c motors.

The roughing stands usually are spaced so that the slab is in only one stand at a time. On the 2135-mm

Fig. 23—31. Interior partial view of a motor room showing some of the drive motors and auxiliaries for an 80-inch hot-strip mill.

(84-inch) hot-strip mill at United States Steel's Gary Works, the fourth and fifth roughing stands are operated in tandem. The fifth stand is driven by a synchronous motor and the fourth stand is driven by a d-c motor to provide variable speed control so that the two stands can be synchronized. For hot strip mills using reversing rougher stands, the drives would be variable speed d-c motors powered by thyristor rectifiers. An advantage of using reversing roughers is the reduction in overall length of both mill and building.

The basic data for calculating the horsepower requirements per stand for a continuous mill are based upon actual test. Composition of the steel, its temperature, speed of rolling, and width are variables in these data. In addition, allowance has to be made for the judgment that has to be exercised by the roller, who is given a certain size slab to be reduced in a number of stands to a product meeting close dimensional tolerances and high surface-quality standards. To do this, the theoretical reductions per pass may have to be varied, and it may be necessary to increase or decrease the draft on certain passes as compared with the calculated reductions.

The d-c finishing-mill motors are designed to operate on 600 to 700 volts. At this voltage, they will run at the base speed. These motors are started by the Ward-Leonard system, the armature voltage being increased gradually until the motors are at the base speed. If lower speeds are required, the d-c voltage is reduced, which also decreases the power output but does not change the torque. Higher speeds are obtained by weakening the d-c fields in the motors by the field rheostats or, in the more modern installations, by changing the field voltage by means of rotating or static control. The standard speed range for these motors on field control is 2:1 or more and they are designed to have approximately flat speed characteristics.

When operating, the motors must function as a unit and maintain their speed relationship regardless of load. Usually, a master control is installed so that, once this relationship is established, the speed of the stands can be raised or lowered as a unit. On recent installations, the interstand loopers are operated in the raised position when strip is in the mill, and the stand speeds are synchronized automatically by looper position regulators operating on the main-drive speed regulators. Table 23—VIII gives installed power, speeds, etc., for five modern hot-strip mills.

Tandem Cold-Reduction Mills—Another type of continuous mill is the cold-reduction tandem mill. Mills of this type have three, four, five or six stands, with a reel to wind up or coil the cold-reduced product. The

Table 23—VIIIa. Installed Motor Kilowatt Ratings, Speeds, Etc.—Hot-Strip Mills (SI Units)

	U. S. Steel Corp. Gary Works 2134-mm Mill	Great Lakes Steel Corp. Ecorse Works 2032-mm Mill	Inland Steel Co. Harbor Works 2032-mm Mill	Wheeling-Pittsburgh Steel Corp. 2032-mm Mill	Republic Steel Corp. Warren Works 1473-mm Mill
Vertical Scale Breaker—Roll Size (mm)	1168 × 508	1143 × 457	1143 × 356	1143 × 356	—
Motor kW, rpm, Type	1120, 257, Syn.	1120, 514, Syn.	943, 450, —	933, 257, Syn.	—
Gear Ratio, m/s	13.37/1, 1.118	30.0/1, 1.026	30.3/1, 0.887	—, 1.27	—
Horizontal Scale Breaker—Roll Size (mm)	1219 × 2134	—	1219 × 2032	—	1118 × 1473
Motor kW, rpm, Type	3360, 450, Syn.	—	3730, 300, Syn.	—	2610, 350, —
Gear Ratio, m/s	25.67/1, 1.12	—	17.4/1, 1.1	—	15.69/1, 1.30
No. 1 Rougher—Roll Size (mm)	1219 × 2134	1118 × 2057	1219 × 2032	1168 × 2032	—
Motor kW, rpm, Type	5250, 450, Syn.	3730, 450, Syn.	3730, 300, Syn.	3730, 950, Syn.	—
Gear Ratio, m/s	25.7/1, 1.12	25.8/1, 1.02	17.4/1, 1.1	23.88/1, 1.15	—
No. 2 Rougher—Roll Size (mm)	1219 × 2134	1118 & 1524 × 2032	1219 × 2032	1118 & 1524 × 2032	
Motor kW, rpm, Type	5250, 450, Syn.	7460, 450, Syn.	6715, 360. Syn.	3730, 450, Syn.	4-High Reversing Rougher
Gear Ratio, m/s	17.27/1, 1.66	25.8/1, 1.02	17.9/1, 1.32	23.7/1, 1.11	
No. 3 Rougher—Roll Size (mm)	1067 × 1524 × 2134	1118 & 1524 × 2032	965 & 1524 × 2032	1118 & 1524 × 2032	914 & 889 × 1422
Motor kW, rpm, Type	7460, 450, Syn.	7460, 450, Syn.	6715, 360, Syn.	7460, 450, Syn.	2-2985-kW, 40/100 rpm
Gear Ratio, m/s	12.37/1, 2.03	15.7/1, 1.67	10.4/1, 1.75	9.92/1, 2.8	1.9/4.8 m/s
No. 4 Rougher—Roll Size (mm)	1067 × 2134	1118 & 1524 × 2032	965 & 1524 × 2032	1118 & 1524 × 2032	—
Motor kW, rpm, Type	7833, 430, d-c	7460, 450, Syn.	6715, 360, Syn.	7460, 450, Syn.	—
Gear Ratio, m/s	6.79/1, 3.54	11.7/1, 2.25	5.75/1, 3.15	7.93/1, 3.33	—
No. 5 Rougher—Roll Size (mm)	1067 × 1524 × 2134	1118 & 1524 × 2032	965 & 1524 × 2032		—
Motor kW, rpm, Type	7460, 450, Syn.	7460, 450, Syn.	7460, 450, Syn.		—
Gear Ratio, m/s	6.79/1, 367	9.25/1, 2.25	5.75/1, 5.84		—
Scalebreaker Motor kW, rpm, Type	2-150/300, 420/840, d-c	375, 350, d-c	—	225, 300/600, d-c	—
Finishing Mill Roll Sizes (mm)	724 & 1524 × 2134	724 & 1524 × 2032	724 & 1524 × 2032	724 & 1524 × 2032	635 & 1372 × 1473
No. 1 Finishing Stand Motor kW and rpm	7833, 125/295	5225, 200/400	6714, 125/312	4476, 175/410	3730, 150/370
Gear Ratio and m/s	2.932/1, 1.62/3.81	7.91/1, 0.95/1.9	3.27/1, 1.96/3.62	5.42/1, 1.22/2.87	4.27/1, 1.16/238
No. 2 Finishing Stand Motor kW and rpm	7833, 125/295	5970, 125/275	6714, 125/312	4476, 175/410	4475, 100/230
Gear Ratio and m/s	1.84/1, 2.58/6.1	2.98/1, 1.6/3.45	1.96/1, 2.42/6	3.34/1, 1.99/4.65	1.951/1.7/3.9
No. 3 Finishing Stand Motor kW and rpm	7830, 125/295	5970, 125/275	6714, 125/312	4476, 175/410	4475, 100/230
Gear Ratio and m/s	1.29/1, 3.66/8.64	1.93/1, 2.47/5.39	1.402/1, 3.39/8.4	2.307/1, 2.88/6.73	1.35/1, 2.46/5.7
No. 4 Finishing Stand Motor kW and rpm	7833, 125/295	5970, 125/275	6714, 125/312	4476, 175/410	4475, 100/230
Gear Ratio and m/s	1/1, 4.73/11.18	1.33/1, 3.56/7.8	Direct, 4.7/11.8	1.586/1, 4.19/9.8	Direct, 3.3/7.7
No. 5 Finishing Stand Motor kW and rpm	7833, 200/430	5970, 125/275	6714, 125/312	4476, 175/410	4475, 125/300
Gear Ratio and m/s	1/1, 7.67/16.26	1/0.833, 4.7/10.4	1/1.402, 6.6/16.6	1.212/1, 5.98/12.8	Direct, 4.2/10
No. 6 Finishing Stand Motor kW and rpm	7833, 200/480	5970, 175/350	6714, 125/312	4476, 175/410	3730, 150, 370
Gear Ratio and m/s	1/1, 8.51/18.29	Direct, 6.6/13.2	1/1.672, 7.9/19.7	Direct, 6.64/15.5	Direct, 5.0/12.3
No. 7 Finishing Stand Motor kW and rpm	5222, 200/430	5225, 200/400			—
Gear Ratio and m/s	0.76/1, 9.9/21.34	Direct, 7.62/15.2			—
Finishing Motor d-c Supply by	Silicon Rectifiers	Mercury-Arc Rectifiers	Mercury-Arc Rectifiers	Motor-Generator Sets	Mercury-Arc Rectifiers

Table 23—VIIIb. Installed Horsepower, Speeds, Etc.—Hot-Strip Mills (fps Units)

	U.S. Steel Corp. Gary Works 84" Mill	Great Lakes Steel Corp. Ecorse Works 80" Mill	Inland Steel Co. Harbor Works 80" Mill	Wheeling-Pittsburgh Steel Corp. 80" Mill	Republic Steel Corp. Warren Works 58" Mill
Vertical Scale Breaker—Roll Size	46" × 20"	45" × 18"	45" × 20"	45" × 14"	—
Motor hp, rpm, Type	1500, 257, Syn.	1500, 514, Syn.	1250, 450, —	1250, 257, —	—
Gear Ratio, fpm	13.37/1, 220	30.0/1, 202	30.0/1, 175	—250	—
Horizontal Scale Breaker—Roll Size	48" × 84"	—	48" × 80"	—	44" × 58"
Motor hp, rpm, Type	4500, 450, Syn.	—	5000, 300, Syn.	—	3500, 350, —
Gear Ratio, fpm	25.67/1, 220	—	17.4/1, 216	—	15.69/1, 257
No. 1 Rougher—Roll Size	48" × 84"	44" × 81"	48" × 80"	46" × 80"	—
Motor hp, rpm, type	7000, 450, Syn.	5000, 450, Syn.	5000, 300, Syn.	5000, 450, Syn.	—
Gear Ratio, fpm	25.7/1, 220	25.8/1, 201	17.4/1, 216	23.88/1, 226	—
No. 2 Rougher—Roll Size	48" × 84"	44" & 60" × 80"	48" × 80"	44" & 60" × 80"	4-High Reversing Rougher
Motor hp, rpm, Type	7000, 450, Syn.	5000, 450, Syn.	9000, 360, Syn.	5000, 450, Syn.	
Gear Ratio, fpm	17.29/1, 327	25.8/1, 201	17.4/1, 260	23./1, 219	
No. 3 Rougher—Roll Size	42" × 60" × 84"	44" & 60" × 80"	38" × 60" × 80"		36" & 35" × 56"
Motor hp, rpm, Type	10 000, 450, Syn.	10 000, 450, Syn.	9000, 360, Syn.		2-4000 hp
Gear Ratio, fpm	12.37/1, 400	15.7/1, 330	10.4/1, 344		40/100 rpm
No. 4 Rougher—Roll Size	42" × 60" × 84"	44" & 60" × 80"	38" & 60" × 80	44" & 60" × 80"	377/942 fpm
Motor hp, rpm, Type	10 500, 220/430, d-c	10 000, 450, Syn.	9000, 360, Syn.	10 000, 450, Syn.	
Gear Ratio, fpm	6.79/1, 695	11.7/1, 442	5.75/1, 620	9.92/1, 522	
No. 5 Rougher—Roll Size	42" × 60" × 84"	44" & 60" × 80"	30" & 60" × 80"	44" & 60" × 80"	
Motor hp, rpm, Type	10 000, 450, Syn.	10 000, 450, Syn.	10 000, 450, Syn.	10 000, 450, Syn.	
Gear Ratio, fpm	6.79/1, 724	9.25/1, 561	5.75/1, 755	7.93/1, 655	
Scale Breaker Motor hp, rpm, Type	2-200/400, 420/840; d-c	550, 350/875 d-c		300, 300/600 d-c	
Finishing Mill Roll Sizes	28½" × 60" × 84"	28½" & 60" × 80"	28½" & 60" × 80"	28½" & 60" × 80"	25" & 54" × 58"
No. 1 Finishing Stand Motor hp & rpm	10 500; 125/295; d-c	7000, 200/400	9000, 125/312	6000, 175/410	5000, 150/370
Gear Ratio and fpm	2,932/1; 318/750	7.91/1, 188/376	3.27/1, 287/712	5.42/1, 241/565	4.27/1, 230/567
No. 2 Finishing Stand Motor hp & rpm	10 500, 125/295	8000, 125/275	9000, 125/312	6000, 175/410	6000, 100/230
Gear Ratio and fpm	1.84/1; 507/1200	2.98/1, 313/680	1.96/1, 477/1190	3.34/1, 391/915	1.95/1, 335/770
No. 3 Finishing Stand Motor hp & rpm	10 500, 125/295	8000, 125/275	9000, 125/312	6000, 175/410	6000, 100/230
Gear Ratio and fpm	1.29/1; 720/1200	1.93/1, 486/1060	1,402/1, 665/1660	2.307/1, 568/1325	1.35/1,485/1120
No. 4 Finishing Stand Motor hp & rpm	10 500, 125/295,	8000, 125/275	9000, 125/312	6000, 175/410	6000, 100/230
Gear Ratio and fpm	1/1; 930/2200	1.33/1, 700/1545	Direct, 934/2330	1.586/1, 824/1930	Direct, 655/1510
No. 5 Finishing Stand Motor hp & rpm	10 500, 200/430	8000, 125/275	9000, 125/312	6000, 175/410	6000, 125/300
Gear Ratio and fpm	1/1, 1490/3200	1/.833, 930/2050	1/1.402, 1308/3265	1.212/1, 1078/2525	Direct, 820/1970
No. 6 Finishing Stand Motor hp & rpm	10 500, 200/480	8000, 175/350	9000, 125/312	6000, 175/410	5000, 150/370
Gear Ratio and fpm	1/1, 1675/3600	Direct, 1300/2600	1/1.672, 1560/3884	Direct, 1307/3061	Direct, 982/2420
No. 7 Finishing Stand Motor hp & rpm	7000; 200/430	7000, 200/400			
Gear Ratio and fpm	0.76/1; 1950/4200	Direct, 1500/3000			
Finishing Motor d-c Supply by—	Silicon Rectifiers	Mer. Arc. Rectifiers	Mer. Arc. Rectifiers	Motor-Generator Sets	Mer. Arc. Rectifiers

Table 23—IX. Installed Motor Kilowatt Ratings and Generating Capacity—Tandem Cold-Reducing Mills (SI Units)*.

	U. S. Steel Corp. 6-Stand	U. S. Steel Corp. 6-Stand	Midwest Steel Co. 5-Stand	Wheeling-Pittsburgh Steel Corp. 5-Stand	U. S. Steel Corp. 5-Stand
Mill Size (mm)	584 & 1422 × 1320	584 & 1422 × 1320	584 & 1422 × 1320	533 & 1346 × 1219	584 & 1524 × 2032
#1 Stand Motor kW	1-1490 S.A.	2-1120 S.A.	2-1120 S.A.	1-1490 S.A.	2-1305 S.A.
Motor rpm	80/280	160/480	115/400	90/270	80/300
Gear Ratio & m/s	1/1, 2.4/8.5	1.686/1, 2.9/8.7	1.438/1, 2.45/8.5	Direct, 2.5/7.5	1.1, 2.45/9.1
#2 Stand Motor kW	2-1490 S.A.	2-1865 S.A.	2-1865 S.A.	1-4000 D.A.	2-2610 D.A.
Motor rpm	225/563	175/437	125/375	150/375	225/620
Gear Ratio & m/s	1.5967/1, 4.3/10.8	1.15/1, 4.67/11.68	1/1.11, 4.3/12.8	Direct, 4.2/10.5	1.512/1, 4.5/12.5
#3 Stand Motor kW	2-1865 S.A.	2-1865 S.A.	2-1865 S.A.	2-1490 S.A.	2-2610 D.A.
Motor rpm	225/530	175/437	125/375	150/375	225/620
Gear Ratio & m/s	1.0833/1, 6.35/15	1/1.22, 6.5/16.4	1/1.51, 5.8/17.4	1/1.51, 6.3/15.8	Direct, 6.86/18.9
#4 Stand Motor kW	2-2237 D.A.	2-2237 D.A.	2-2237 D.A.	2-1865 S.A.	2-2610 D.A.
Motor rpm	350/700	122/305	200/485	250/518	225/620
Gear Ratio & m/s	1.0882/1, 9.8/19.7	D.Bk.R.Dri., 9.1/22.7	1/1.51, 9.2/22.4	1/1.51, 10.6/21.8	1/1.293, 8.9/24.5
#5 Stand Motor kW	2-2237 D.A.	2-2237 D.A.	2-2610 T.A.	2-2237 D.A	2-2237 T.A.
Motor rpm	350/700	161/402	200/500	275/550	350/840
Gear Ratio & m/s	1/1.2828, 13.7/22.5	D.Bk.R.Dri., 12/30	½, 12.2/30.6	½, 15.4/30.7	Direct, 10.7/25.7
#6 Stand Motor kW	2-2610 D.A.	2-2610 D.A	—	—	—
Motor rpm	350/700	173/520	—	—	—
Gear Ratio & m/s	1/1.6904, 18.1/36.2	D.Bk.R.Dri., 12.9/38.7	—	—	—
Reel Motor kW	3-1566 T.A.	2-1120 T.A.	2-1342 T.A.	1-1120 T.A.	2-1044 D.A.
Motor rpm	300/1500	250/1250	200/1000	300/1120	260/1160
Gear Ratio	1/1	1/1.42	1/1.31	Direct	Direct
Total Kilowatts	23 937	26 100	22 073	16 778	24 832
	MG Set #1 / MG Set #2	MG Set #1 / MG Set #2	MG Set #1 / MG Set #2	MG Set #1 / MG Set #2	MG Set #1 / MG Set #2
MG Sets Motor kW	13 423 / 13 423	14 168 / 14 168	12 677 / 12 677	8 948 / 8 948	13 423 / 13 423
Generator kW #1	1600 / 2000	2000 / 2000	1440 / 1440	1600 / 2000	2800 / 2800
Generator kW #2	1600 / 2000	2400 / 2000	1440 / 1440	1600 / 2000	2800 / 2800
Generator kW #3	2400 / 2000	2400 / 2000	2000 / 1440	1600 / 1600	2800 / 2800
Generator kW #4	2400 / 2000	2400 / 2800	2000 / 1440	1600 / 1600	2800 / 2800
Generator kW #5	2400 / 2000	2400 / 2800	2400 / 2000	1600 / 1600	2800 / 2800
Generator kW #6	2400 / 1600	2400 / 2400	2400 / 2000	1250 /	2400 / 2400
Generator kW #7	/ 1600		2400		2400 / 2400

*S.A. indicates Single Armature, D.A. indicates Double Armature, T.A. indicates Triple Armature, D.Bk.R.Dri. indicates Direct Back-Up Roll Drive.

Table 23—X. Motor Drives for Modern Continuous Rod Mill

Stand No.		Motor Rating		Speed rpm
		kW	hp	
1	Roughing	375	500	250/500
2	Roughing	445	600	300/600
3	Roughing	520	700	400/800
4 & 5	Roughing	1120	1500	300/600
6 & 7	Intermediate	1120	1500	300/600
8 & 9	Intermediate	1490	2000	250/500
10	Intermediate	670	900	400/800
11	Intermediate	670	900	400/800
12A	Intermediate	375	500	250/500
12B	Intermediate	375	500	250/500
13A	Intermediate	375	500	250/500
13B	Intermediate	375	500	250/500
14-23	#1 Strand Finishing	2×930	2×1250	650/1115
14-23	#2 Strand Finishing	2×930	2×1250	650/1115
14-23	#3 Strand Finishing	2×930	2×1250	650/1115
14-23	#4 Strand Finishing	2×930	2×1250	650/1115

electrical problem here is more difficult than on the hot-strip mill, because tension maintained in the product between the stands must be controlled within close limits. The speeds on this type of mill have been gradually increased, several mills having a top finishing speed of over 35.6 metres per second (7000 feet per minute). Table 23—IX gives motor and generator data on some modern tandem cold-reduction mills.

Continuous Billet Mills—Modern continuous billet mills are provided with individual motor drives at each stand. On the 760-mm (30-inch) mill at one plant, each stand (three vertical and three horizontal) is driven through gearing by a 1305-kW (1750-horsepower), 300/600 rpm, 600-volt direct-current shunt motor. They are supplied with power from a motor generator set consisting of four 2000-kW generators driven by an 8355-kilowatt (11 200 horsepower) 514 rpm, 13 800-volt synchronous motor. The 535-mm (21-inch) billet mill at the same plant consists of two vertical and two horizontal stands, each individually driven by a 933-kilowatt (1250 horsepower), 400/600 rpm, 600-volt, direct-current shunt motor through gears. The power for these drives is supplied by a motor-generator set consisting of two 2000-kW generators driven by a 4180-kilowatt (5600-horsepower), 514 rpm, synchronous motor.

Continuous Bar Mills—Individual motor drives are also provided for the separate stands of modern continuous bar mills. At one 250-mm (10-inch) mill, there are 18 two-high stands, all in line. Ten have horizontal rolls and eight have vertical rolls. Each stand is driven by a 600-volt, direct-current motor, in sizes ranging from 300 to 520 kilowatts (400 to 700 horsepower). Power is supplied by a motor-generator set composed of three 2500-kW generators, one 150-kW exciter, and a 7835-kilowatt (10 500-horsepower), 514 rpm, 13 200-volt synchronous motor.

Continuous Rod Mills—Continuous rod mills consist of numerous stands arranged, for drive purposes, in groups with an individual motor driving the two or more stands in each group. The selection and distribution of motors in such a mill is dependent upon roll

speed, roll diameter, amount of reduction in each stand, the power required, and the degree of flexibility of control of the mill as an operating unit.

One straight-line continuous rod mill rolls 100-mm by 100-mm by 18.3-metre (4-in. by 4 in. by 60-ft) billets into coiled rods about 6 mm to 12 mm (¼-inch to ½-inch) in diameter. The delivery speed of this mill when producing 5.56-mm (⁷⁄₃₂-inch rod) is approximately 51 metres per second (10 000 ft per min.), and the coil weight is about 1450 kilograms (3200 lb). The motors with their speed ranges were selected as shown in Table 23—X.

Continuous Seamless Tube Mill—In the continuous seamless tube mill at Lorain Works of United States Steel Corporation (separate mills of which are described in Chapter 31), the piercing mill is driven by a 3360-kilowatt (4500-horsepower), 225 rpm, 13 800-volt, synchronous motor through a reduction gear having a ratio of 2.25:1. The nine continuous stands of the mandrel mill are individually driven by direct-current motors with a combined rated capacity of 6340 kilowatts (8500 horsepower). The product from the mandrel mill, after reheating, is further processed in either the stretch-reducing mill or the sizing mill. The stretch-reducing mill is made up of twelve two-high, overhung roll stands; each roll stand is individually driven by a 150-kilowatt (200-horsepower), 850/1700 rpm, 600-volt, shunt motor. The sizing mill consists of twelve overhung roll stands, each with an individual drive consisting of a 57-kilowatt (76-horsepower), 850-1700 rpm, shunt motor.

MOTOR-ROOM VENTILATION

The selection of an efficient and adequate ventilation system is a very important consideration in the design of motor and control rooms. Motors, generators, static power equipment, control panels and switchgear develop and emit considerable heat during operation. Therefore, it is necessary to provide sufficient cooling air to assure that the operating temperature of the equipment does not exceed the safe design level for the ambient conditions. Usually, this design level is for 40°C (104°F).

The following table indicates typical air volumes required for cooling per kilowatt of heat generated by each machine:

Air Volume Supplied Per kW Loss		Corresponding Rise in Air Temperature	
m³/min.	ft³/min.	°C	°F
2.83	100	17.5	31.5
3.12	110	15.9	28.6
3.4	120	19.6	26.2
3.68	130	13.4	24.2
5.66	200	8.7	15.7
8.5	300	5.8	10.5

Air cleanliness is of prime importance in an electrical-equipment ventilation system since open circuits, circuit grounds and short circuits can develop if a dirt build-up occurs. This is particularly true with some of the newer sophisticated static control systems. Usually, these printed circuit boards have a conformal coating to minimize the deleterious effects of humidity, corrosion and dirt. To provide the required ambient environment and maintain stable operation, it is often necessary to install critical pieces of sensitive control equipment in a clean, air-conditioned atmosphere at some constant temperature below 40°C (104°F), depending upon the control manufacturer's criteria. Controlled humidity is also sometimes necessary to maintain proper operating characteristics of the control components.

When categorized by flow direction, there are basically two main types of electrical-equipment-ventilating systems: (1) downdraft and (2) updraft. Both systems usually are designed to maintain a positive pressure in the motor room to reduce the infiltration of dirt. The desirable level of pressurization ranges from 25 to 62 pascals (0.10 to 0.25 in.) w.g. depending on motor-room size, construction characteristics, external wind load, and so on. Both downdraft and updraft systems can be arranged to recirculate some or all of the air which has been heated in passing over the electrical equipment. Depending on the outdoor temperature, such recirculation can conserve energy as well as reduce filtration requirements. Most motor-room ventilation systems also include metering fans in the basement of the motor room. These are used to assure a supply of clean ventilating air to the main-drive motors in the motor room as well as auxiliary drives such as screwdowns, manipulators and table motors which are located in the mill buildings.

Both the downdraft and updraft systems depend on high-integrity filtration equipment, usually of two-stage design, to trap air-borne dirt particles. One of the most practical and efficient systems currently being used by the steel industry utilizes an inertial separator in the first stage to remove the larger dirt particles (above 10 micrometres); high-efficiency cartridge filters of the strainer type are used in the secondary filter bank.

Electrostatic precipitators have been used for filtration in motor-room ventilation systems because of their high efficiency. They have the disadvantage of requiring very low velocity non-turbulent air flow. Careful maintenance is required to prevent short circuits which permit air to pass through the system unfiltered.

Roll-type fiberglass viscous-impingement filters are sometimes used for motor-room ventilation. However, these have a relatively lower filtration efficiency and depend on a viscous adhesive to trap air-borne particulates. These filters can be used in combination with a dry-type filter in a two-stage arrangement: the dry-type unit is designed to remove the smaller particles of dirt which are not caught by the viscous-impingement filters.

In the past ten years, a method of filtration which has shown much promise is the use of bag-type dust collectors for ambient-air filtration. Although slightly higher in first cost, this system has the advantage of extremely high efficiencies and low maintenance, and a life of 8 to 10 years for the filter media is not unusual. The development of pulse-jet bag collectors now permits use of air-to-cloth ratios of 20:1 for ambient-air filtration. This greatly reduces space requirements and permits the incorporation of bag collectors in factory-assembled package ventilation systems.

The finalized design of a motor and control room ventilation system is a reflection of many considerations, including: volume of air needed to remove heat generated by electrical equipment; minimum dust-particle size that can be tolerated for the equipment being ventilated; air pressure loss through the complete ventilating system; the positive pressure level required in the room to prevent infiltration of contaminants; maximum noise level that can be endured by personnel working in the room; and overall energy required to operate the ventilation system.

A careful engineering evaluation of each of the above parameters should result in the design of an efficient and an economical system.

AUXILIARY DRIVES

The auxiliary drives on mills are as important as the main drives, and require proper application and coordination of types and sizes of motors. Both a-c and d-c motors are used, the former being the accepted ones for driving fans, pumps, and, occasionally, shears and run-out tables. The majority of the motors, however, are mill-type d-c units, because of their ruggedness and the many operating characteristics that can be built into them. These mill-type motors were developed by the steel industry in cooperation with the electrical-equipment manufacturers for use in the rough service of driving mill tables, heavy-duty cranes, manipulators, etc. The ratings and principal dimensions have been standardized by the Association of Iron and Steel Engineers into fourteen frame sizes, ranging from 3.7 to 373 kW (5 to 500 horsepower), based on the 1-hour rating. They can be obtained with several designs of windings from straight series to shunt, and with various types of ventilation. The designs are very rugged, and the armatures have low inertia in comparison with general-purpose motors. Table 23—XI describes motors for various auxiliary-drive applications.

Table Rollers—The early strip-mill run-out table rollers were equipped with variable-speed a-c motors, one per roller, and speed variations were obtained by varying the frequency of the power supply by a variable-speed motor-generator set. Since 1937, the d-c mill-type motor has also been used for this application in shunt-wound and permanent-magnet field forms.

Schloemann rollers, a German design, in which the movable part of the table roller is the rotor of the motor and the stationary part wound to form the stator of an induction motor, have been used on some mills. When such rollers handle hot products, they must be water cooled. Speed variations are obtained by frequency control from a motor-generator set.

In modern mills, the rollers may be driven by individual, small direct-current motors of 2.2- to 3.0-kilowatt rating (3 to 4 horsepower), or they may be arranged in consecutive groups with each group driven separately by a direct-current motor of ample size through a line shaft and gearing. Variable-voltage control is used to regulate the speed of the table rollers. The current generally is supplied by group thyristor static power supplies.

Screw-Downs—Proper selection of motors for screw-down drives on the various mills is very important. In some mills, such as blooming and slabbing mills, a maxi-

Table 23—XI. Motor Sizes on Screws, Manipulators, Shears, Etc., on Some Blooming and Slabbing Mills°

Company and Size of Mill	Great Lakes 1170-mm (46-inch)	U.S. Steel Morrisville, Pa. 1065-mm (40-inch)	U.S. Steel Braddock, Pa. 1140-mm (45-inch)	Inland Steel 1170-mm (46-inch)	Jones & Laughlin 1770-mm (46-inch)
Screw Motors (No. and Size)	2-112 kW (2-150 hp)	2-112 kW (2-150 hp)	2-150 kW (2-200 hp)	2-112 kW (2-150 hp)	2-150 kW (2-200 hp)
Screw Motor Control	Variable Voltage	Variable Voltage	Variable Voltage	Variable Voltage	Variable Voltage
Manipulator Motors .... (No. and Size)	4-112 kW (4-150 hp)	4-75 kW (4-100 hp)	4-112 kW (4-150 hp)	4-112 kW (4-150 hp)	4-156 kW (4-200 hp)
Lift Fingers (No. and Size)	4-37 kW (4-50 hp)	2-75 kW (2-100 hp)	4-112 kW (4-150 hp)	4-112 kW (4-150 hp)	2-112 kW (2-150 hp)
Manipulator Control	Variable Voltage	Variable Voltage	Variable Voltage	Variable Voltage	Variable Voltage
Front Table Drive .... (No. and Size)	2-112 kW (2-150 hp)	2-75 kW (2-100 hp)	2-112 kW (2-150 hp)	2-112 kW (2-150 hp)	2-150 kW (2-200 hp)
Front Table Control	Variable Voltage	Variable Voltage	Variable Voltage	Variable Voltage	Variable Voltage
Back Table Drive .... (No. and Size)	2-112 kW (2-150 hp)	2-75 kW (2-100 hp)	2-112 kW (2-150 hp)	2-112 kW (2-150 hp)	2-150 kW (2-200 hp)
Back Table Control	Variable Voltage	Variable Voltage	Variable Voltage	Variable Voltage	Variable Voltage
Shear (maximum size cut)	1625 × 150 mm (64 in. × 6 in.)	405 × 560 mm (16 in. × 22 in.)	1525 × 230 mm (60 in. × 9 in.)	1015 × 205 mm (40 in. × 8 in.)	5160 cm² (800 sq. in.)
Shear Motors (No. and Size)	2-205 kW (2-275 hp)	2-185 kW (2-250 hp)	2-185 kW (2-250 hp)	2-205 kW (2-275 hp)	Hydraulic
Shear Motor Control	Variable Voltage	Variable Voltage	Variable Voltage	Variable Voltage	Pressure Pumps

° Courtesy, "Iron and Steel Engineer."
∞ All auxiliaries under variable-voltage control.

mum roll lift of 1675 mm (66 in.) may be required, with lifting speeds of up to 16.4 metres (645 in.) per minute. On strip mills, a maximum operating lift of only a few inches may be required, but the opening must be controlled within a few thousandths of an inch.

Regardless of the speed or distance involved, the screw-down motor and control must be capable of positioning the rolls within close limits. The inertia of the armature and brake wheel makes this a difficult problem, even with a compound-wound motor, and has resulted in the adoption of the Ward-Leonard system of control. This system has an added advantage that it limits the amount of torque that can be applied to the motor, thus preventing damage to the screws by jamming.

On primary mills, modern screw-down drives usually consist of two 230-volt, d-c motors of 75 to 150 kW (100 to 200 horsepower) rating. If operating under variable-voltage control, they are shunt motors with shunt brakes.

Manipulators and Side-Guards—Mill tests have shown that steel being rolled on primary mills is in the rolls only from 25 to 70 per cent of the total rolling time, the remainder of the time being required for screw-down, manipulator, and table operation. In a typical slabbing mill, the maximum opening of the side-guards is 2920 mm (115 inches), with an overtravel of about 1020 mm (40 inches) to permit changing of the table rollers. Each of the side-guards is driven by a motor rated at about 110 or 185 kilowatts (150 or 250 horsepower). Due to the heavy plugging service required for these devices, and the heavy moving mass that has to be controlled, modern installations are using shunt-wound, variable-speed motors with Ward-Leonard control for this service.

Lift fingers on manipulators are required to make 1220-mm (4-foot) lifts and complete the cycle in two seconds. They always must stop below the top of the rollers, and represent another application for variable-voltage control. Where magnetic control is used, compound-wound mill-type motors with reversing, plugging, dynamic-braking features, plus a jamming resistor and arranged to operate two motors in series with series brakes, should be specified.

Blooming-Mill Shears—Blooming-mill shears can be driven either electrically or hydraulically. Many of the older installations were driven by an induction motor with high secondary resistance to allow the fly-wheel to supply the peak power for the cut. The motor ran continually and the shear was actuated through a clutch. Some of the more recent shears are of a start-stop type, driven by two shunt-wound motors of about 150 to 225 kW (200 to 300 horsepower), each equipped with a shunt brake and variable-voltage control.

FUTURE DRIVES

This resume on drives for steel mills is indicative of the ever-changing design in electrical apparatus. It is no longer a problem to add loads of many thousands of kilowatts to existing systems. Motors have been improved in design and increased in rating so that they can be obtained for all types of service. Improved controls, too, have kept pace with motor improvements, so that it is possible to maintain close relationship between individual large motors on successive continuous-mill stands rolling product at rates above 30.5 metres per second (6000 feet per minute).

The present trend is toward still greater flexibility in drives. Recent continuous mills have been designed with separate thyristor power supplies for each drive motor; this allows the greatest range in speed.

Silicon-controlled rectifiers have continued to succeed mercury-arc rectifiers or motor-generator sets as a source of d-c power. With associated solid-state electronic control, this device can regulate voltage or speed within very close limits and increase overall efficiency.

The story of motor drives for mills is one of continual improvement in design to obtain higher speeds and closer regulation, to provide for more and better products from the mills. Drive efficiency has become a more important consideration but it will not make many changes in drive types.

SECTION 5

AUTOMATIC CONTROL OF ROLLING OPERATIONS

The continuing trend toward more powerful mill drives and higher rolling speeds developed as a result of efforts to obtain larger output of better quality product from each stage of the rolling processes. A necessary corollary of this trend has been the development of improved electrical equipment and controls, coupled with the application of automatic process-control systems. Automatic control of some or all of the functions of a rolling mill is resulting in more consistent operating practices that improve product quality and uniformity, improve productivity and yield while increasing the degree of utilization of the equipment.

With the adaptation of computers to automatic control of rolling mills, very sophisticated control systems have been and continually are being devised. This section will discuss first the basic strategies that are common to all process control applications and next the specific process control functions that are typically included in modern rolling mill computer control systems.

PRINCIPLES OF PROCESS-CONTROL SYSTEMS

The general form of a process-control system is one in which a multiplicity of inputs, such as the physical and chemical characteristics of the raw materials, the energy levels, the machine settings, and so forth are used to provide desired outputs such as product quality, productivity, and minimum cost, or some selected combination of these. In addition, there may be other outputs required; for example, processing information for supporting functions such as accounting and evaluation.

In most cases, the basic mode of process control is feed-forward control where the process machinery is preset, either manually or automatically, to produce a desired product based on measurements or estimates of the physical and chemical characteristics of the incoming materials. Feedforward control can be used alone to produce satisfactory results in many cases. However, when very accurate and consistent results are required, it is generally necessary to supplement the feedforward control system with feedback control. By this method, the actual output is compared to the desired output and the process settings are readjusted as necessary to eliminate any discrepancies. In a relatively simple system an operator may observe the value of an output, determine the extent and direction of any discrepancy from a desired value, and manually adjust some process variable or combination of variables to arrive at the desired output value. Many such systems have been made automatic by continually comparing the output to the desired value and making corrections automatically as required. These systems have been applied extensively to speed control, position control, and temperature control.

In a rolling operation, the roll opening and roll speeds of the various rolling stands are preset (set up) to produce a desired product thickness based on the thickness, width, and mechanical properties of the incoming material. This setup function is a good example of feedforward control.

Similarly, in most modern rolling operations, automatic gage control (AGC) is used to make compensating adjustments in the preset roll openings and roll speeds when and if the actual product thickness, as measured by an in-line thickness gage or as implied from roll separating force measurements, deviates from the desired thickness. AGC is a good example of feedback control.

COMPUTER CONTROL

As knowledge of the various manufacturing processes has increased through the years, so has the potential for precise control of those processes with attendant improvements in productivity and yield. To take advantage of that potential, process control systems have had to grow rapidly in sophistication and speed of response. The need has developed for control systems capable of making complex calculations and complicated logical decisions as materials move rapidly into and through various processing operations. The need has also developed to track materials through the manufacturing process and provide operations managers, business managers, and engineers with accurate and concise production and quality data to aid them in evaluating and improving machinery and procedures and in responding to customers needs and complaints. Modern computer control systems have evolved to satisfy these needs.

The principal components of a computer control system are:
1. Mathematical models that adequately describe the process.
2. Instrumentation to measure the required variables of the system.
3. Control equipment, including a digital computer, to perform the required functions for control of the system.

Process Models—A computer-controlled system can only follow orders, and it is necessary to instruct the computer as to what to do. This instruction is provided by programming the computer in accordance with mathematical formulations that describe the relationships between the process variables. The mathematical form of the relationships used will be different depending upon the specific application. They might include the differential equations derived from theoretical considerations, empirical equations developed from experimental data, statistical analysis, logical decisions, or some combination of these. Regardless of the approach used, there must be assurance that such treatment of the processing data will provide the desired degree of control. In addition to the computational instructions, the computer must also be instructed in the logic to be used; this is accomplished by programming the computer to establish the time sequence of events required, priorities of several possible control actions under certain circumstances and other decisions that must necessarily be made for proper process control.

Instrumentation—A computer-controlled system accepts the quantitative values of the many processing variables and executes its control function based on these values. A prime requisite of such a system is adequate instrumentation for translating a process variable from its physical or chemical units to a form suitable for use by the computer. Many instruments are presently available to provide rapid on-line measurements of such variables as width, thickness, position, force, temperature, and flow. Instruments that will measure other physical and chemical properties of both raw materials and finished products are available. However, these kinds of measurements generally involve the taking of a sample and subsequent analysis in an off-line laboratory. Because the decisions made by a computer control system are no better than the information with which it works, the importance of reliable instrumentation is obvious.

Control Equipment—The final component of a computer control system is the digital computer system, including hardware and software. The computer hardware includes a central processing unit (C.P.U.) which has the arithmetic and logical capability needed to run the mathematical models, a storage (memory) unit for accumulation of process measurement data and other needed information, and a computer interface to allow the C.P.U. and memory to communicate with the instrumentation, with process regulators like screw position regulators and speed regulators, with operators, and with other computers, including a business computer system.

The computer software consists of all the computer programs needed to accomplish the desired functions of the computer control system.

CONTROL OF PRIMARY ROLLING

Several reversing primary mills have been equipped with control systems in which all mill operations (main-drive acceleration and reversal, screw setting, manipulator and roll-table operations, and other functions) are performed automatically in accordance with instructions previously prepared on punched cards. The feedback in such systems is through the operator, who makes modifications to the programming on a relatively long-time basis as average performance dictates.

In some other cases, microcomputer-based systems are being used to calculate an appropriate drafting schedule for each ingot and then to automatically set the screws for each pass. In these systems, the operator controls the main drives, the manipulators, and the rolling table operations.

In recent years, relatively little effort has gone toward automating the primary mill itself. The major emphasis on computer controls for the primary rolling area has instead been directed toward improved combustion control in the soaking pits and toward coordinating the scheduling of ingots through the steelmaking/soaking pits/primary rolling complex.

PLATE MILL CONTROL

The most modern plate mill control systems include direct, automatic input of needed information (primary data) from a business computer system to the plate mill control system computer. The primary data consists of all information needed for rolling each slab, including slab dimensions and chemistry, ordered plate dimensions, release temperatures for controlled temperature rolling, and other rolling practice instructions. Plate mill control systems typically include a tracking function which follows the movement of slabs from pre-furnace weighing through the reheating, rolling, and finishing operations. The tracking function triggers the operation of process machinery and/or the accumulation of pertinent process measurements to coincide with the arrival of a slab at the various work stations (furnace dropout, roughing mill, finishing mill, leveler, etc.) throughout the plate mill. Typical computer control functions in the rolling area include:
1. Slab identification check—Each slab is weighed

prior to charging into a reheat furnace and scale weight is compared with primary data to verify that the correct slab is being charged.

2. Mill set-up and sequencing—The computer uses a mathematical model to calculate a drafting schedule (roughing stand and finishing stand) for each slab. The computer then sets the screws and sideguides automatically, controls sequencing and speeds of the mill rolls and tables, and turns descaling sprays on and off. The operator usually retains the responsibility for turning a slab for broadside rolling in the roughing stand; this is a relatively complicated operation and very difficult to automate with reliability. Rolling forces are measured during each rolling and pass and, if necessary, the computer model and/or the remainder of the drafting schedule are adjusted to compensate for any discrepancies between measured and expected rolling forces. Also, the final plate thickness is measured and, if necessary, the computer model is adjusted to compensate for any discrepancies between measured and desired thickness. This process of adjusting the computer model on the basis of measured results is called adaptive feedback; the idea is to automatically and continuously improve the accuracy of the results calculated by the model.

Noncontact sensors are now available to measure the width of slabs after the roughing mill. These measurements are compared with the target slab width and additional roughing mill passes are scheduled, if necessary, to make the target width. Control of plate shape (flatness) during the final finishing mill passes is important, but no sensor is available to date to measure plate flatness. Therefore, the finishing mill operator is required to assess plate flatness through visual observation and to initiate corrective action when he sees a problem. Two different approaches are in use for making corrections when a flatness problem is observed. In one case, the operator intervenes and modifies the computer-calculated drafting pattern to correct the problem. In the second case, the operator enters the nature of the problem, i.e., wavy edge, center buckle, etc., into the computer and the computer calculates a revised drafting pattern to compensate for and correct the flatness problem.

3. Controlled temperature rolling—Mill set-up and sequencing are automatic as in 2. However, in this case, the computer inserts hold time at specified points in the drafting sequence to let individual plates cool to a specified temperature prior to resuming rolling. Calculations and/or pyrometer measurements are used to track plate temperature. A typical requirement might be to hold a plate on the finishing mill tables and let it cool to 1 550° F before making the last three or four finishing passes.

4. Finishing mill automatic gage control (AGC)—Finishing mill roll separating forces are measured to detect plate thickness changes and screws are adjusted dynamically during one or more of the last finishing mill passes to minimize thickness

variability along the length of finished plates.

5. Printed logs—Various reports are printed to summarize pertinent information relative to operation and maintenance of the mill. These include production reports, engineering logs, delay logs, and alarm logs.

6. Business system reports—Pertinent production and quality data are accumulated and transmitted directly to a business computer system for inclusion in a corporate database. The database provides needed information for accounting and customer service activities.

In some cases, particularly in Japan, plate mill computer control systems have been expanded to include extensive automation of the slab yard and plate finishing operations as well as the rolling operation. These systems are adequately described in the technical literature and will not be discussed at length here.

HOT STRIP MILL CONTROL

In a modern hot strip mill, the temperature in each reheat furnace zone is individually regulated by a zone temperature controller, the rolling stands are all equipped with screw position regulators, and speed regulators are standard on the finishing stands. In addition, loopers, edgers, scalebreakers, and sideguides are all equipped with position regulators. In many cases, these regulating functions are implemented in conventional electronic equipment. However, in newer installations these functions have been programmed in microcomputers and minicomputers.

Overall control of a modern continuous hot strip mill is implemented in a supervisory computer system which uses mathematical models of the heating and rolling operations to calculate and send out appropriate setpoints for the mill regulators. These hot strip mill supervisory control systems represent the most extensive use of computer controls to date on any steel industry process. Specific functions that are typically included in the supervisory computer system are:

1. Automatic primary data input—All information needed for heating, rolling, and quenching each slab is entered into the supervisory computer directly from a business computer system.

2. Slab identification check—Each slab is weighed prior to charging into a reheat furnace and scale weight is compared with primary data to verify that the correct slab is being charged.

3. Tracking—The computer follows the movement of slabs through the reheat furnaces and mill. Send-out of appropriate regulator setpoints is triggered to coincide with the arrival of each slab at the various work stations along the mill (furnace zones, roughing stands, crop shear, finishing stands, runout table and coilers). Also, critical processing variables are measured and the measurements are accumulated in a coil record as each slab proceeds through the heating, rolling, and quenching operations.

4. Mill pacing—The rate at which slabs are extracted from the reheat furnaces is adjusted as required to match the furnace throughput rate to the rolling capacity of the mill. The objective is to main-

tain a desired, uniform tail-end to head-end time interval between bars at the first finishing stand. This insures that the available mill capacity is being fully utilized.

5. Slab heating control—Slabs are tracked through the reheat furnaces and furnace zone temperature setpoints are calculated to achieve target slab temperatures at the furnace exit with minimum fuel use.

6. Roughing mill set-up—Roll openings of the roughing stands, scale breakers, and edgers are preset to produce the proper slab thickness and width to facilitate finish rolling.

7. Crop shear control—The computer monitors signals from multiple hot metal detectors on the holding table and initiates operation of the crop shear at the proper time to crop a fixed length from the head-end and/or tail-end of each bar entering the finishing train.

8. Finishing mill set-up—Roll openings, speeds, and looper position setpoints are preset at all finishing stands for each incoming bar based on optimally distributing the total required reduction among the stands. The set-up calculations aim for a target crown and good shape (flatness) in the finished strip as well as target centerline thickness.

9. Automatic gage control—After threading of the finishing train, roll separating forces in the finishing stands are measured and compared with forces predicted by the finishing mill set-up model. Any difference between measured and predicted forces is interpreted as an error in strip thickness and screw positions and stand speeds are automatically adjusted to correct the thickness error. The finished strip centerline thickness is measured by an on-line gage after the last finishing stand and compared with the target thickness; the screws on one or more finishing stands are then automatically adjusted to eliminate any deviation from target thickness.

10. Coiler set-up—Coiler mandrel lead speed, tension, sideguides position, and other pertinent parameters are preset based on finished strip dimensions and properties.

11. Finishing temperature control—The finishing mill threading speed is preset to bring the head-end of the strip out of the finishing mill at the desired finishing temperature. After threading, the finishing mill acceleration is preset at the proper rate (as calculated) to maintain the desired finishing temperature throughout the full length of the coil. Water sprays between the finishing stands are often utilized for additional control of finishing temperature. In some cases, the finishing temperature is measured by a radiation pyrometer and the accelerating rate and/or interstand water sprays are modified as required to compensate for deviations between the measured and desired temperature.

12. Coiling temperature control—The number of runout table spray headers required to achieve the desired coiling temperature at the head-end of the coil is calculated and headers are turned on accordingly. After threading, additional headers are turned on (and off if necessary) in response to measured changes in strip speed and finishing temperature in order to maintain the coiling temperature within desired limits over the length of the coil. In some cases, the locations as well as the numbers of active headers are adjusted to control the strip cooling rate as well as the coiling temperature. Additionally, some systems utilize pyrometers located along the runout table for feedback control.

13. Printed logs—Various reports are printed to summarize pertinent information relative to operation and maintenance of the mill. These include production reports, engineering logs, delay logs, and alarm logs.

14. Business system reports—Pertinent production and quality data are accumulated and transmitted directly to a business computer system for inclusion in a corporate database. The database provides needed timely information for accounting and customer service activities.

Adaptive feedback is used extensively in hot strip mill computer control systems. Slab temperature measurements at the roughing mill are used to adjust the slab heating models. Measured roll separating forces in the finishing train and measured strip thickness at the head end of a coil are used to adjust the finishing mill set-up model. In some cases, a strip thickness gage continually traverses across the width of the finished strip and this measurement is combined with a centerline thickness measurement from a stationary gage to give a measure of strip crown. The strip crown measurement provides additional feedback for adapting the finishing mill set-up model. To date, visual observation is the only practical means for in-line assessment of strip shape (flatness) on a hot strip mill. Manually entered data describing the nature and magnitude of shape problems provides additional information for adapting the finishing mill set-up model. Coiling temperature measurements at the head end of the coil are used for adaptation of the coiling temperature control model.

CONTROL OF COLD-REDUCTION MILLS

Reversing Mills—Reversing mills for the cold reduction of steel strip can be provided with accurate controls for both front and back tensions to enable the operator to roll a wide variety of materials with comparative ease. Since tension control is most effective in controlling thickness of thinner and harder forms of strip, screwdown positioning is employed for gage control when rolling the thicker gages. For example, in the early passes of rolling a coil when the strip is relatively thick, adjustment of screwdown position is employed because the reel drive cannot exert enough torque to affect strip thickness to a significant degree. When the strip has become thinner and harder from progressive rolling, undesirably large screw movements would be necessary to effect the desired reduction. Thus, a system for automatic gage control on a reversing cold mill should combine screwdown positioning and tension control, preferably in such a manner that back tension would provide the major element of control, with screwdown positioning used to keep the back tension within a desired range. Built-in limiting devices control tension to within safe limits to prevent breakage of the strip. On heavy strip, the upper tension limit is quickly reached, and the screwdown control takes over; with light-gage strip, gage corrections are made almost entirely by controlling tension, and the screwdowns operate seldom if at all. Computer control has seen little application to date on reversing cold mills.

Tandem Mills—Modern tandem cold-reduction mills typically include a screw position regulator and speed regulator on each stand, interstand tension limit regulators, and reel tension regulators. In addition, some mills include means for automatic coil handling, automatic threading, and automatic slowdown for welds and tail ends. In most cases, these functions are implemented in conventional analog and digital electronic devices. In some newer installations, however, these functions have been programmed into microcomputers and minicomputers.

A supervisory computer control system is generally provided which uses mathematical models of the cold rolling operation to calculate and send out optimum setpoints to the regulating devices. The major functions included in the supervisory control systems are:

1. Mill set-up—Roll openings and speeds of all stands, sideguide positions, tension limit regulator setpoints, and reel tension setpoints are preset to produce a target finished strip thickness based on incoming coil thickness, width, and resistance to deformation. In some cases, the setpoints are selected from a limited number of precalculated drafting schedules. However, newer installations use a mathematical model to calculate a unique schedule for each coil. Roll separating force measurements and strip thickness measurements provide adaptive feedback for adjusting the setup model.

2. Oil film compensation—As the mill accelerates after threading, the screws are automatically backed off to compensate for oil-film buildup in the work roll bearings.

3. Tail-out compensation—The screws are run down automatically at each stand to compensate for the loss of tension when the tail-end of a coil clears each previous stand.

4. Automatic gage control—Several different forms of automatic gage control (AGC) are being used in tandem cold-rolling mills. In some cases, the AGC is implemented in the supervisory computer. In other cases, AGC is implemented either in a separate computer or in other types of electronic equipment.

One strategy that has been used for automatic gage control for a tandem cold-reduction mill employs screwdown positioning on the first stand for a coarse control element to minimize the effects of gage variation in incoming strip, using an x-ray gage after the first stand to compare actual thickness with the desired gage setting. An off-gage indication to the automatic gage regulator on the first stand actuates the screwdowns to correct the variation to within narrow limits. With a proper mill setup, the variations in thickness out of the first stand are reduced in the second, third, and fourth stands to improve the accuracy of

thickness of strip entering the last stand. A second x-ray gage following the last stand detects gage variations out of that stand, and any error in gage causes the speed of either the last stand or the last two stands to change automatically in the proper direction so as to alter interstand tension by an amount that will produce a final gage of the desired accuracy.

Some other installations of automatic gage control have employed variations in tension between the first two stands to maintain essentially constant gage out of the second stand. The speed of the first stand is varied in response to deviations in strip thickness measured by an x-ray gage following the second stand. The x-ray gage setpoint is automatically adjusted if finished thickness is off-gage or if there is an excessive amount of correction required from the gage-control system at the last stand. All interstand tensions are maintained within preset limits by automatic screwdown positioning.

Another design of an automatic gage-control system employs speed or tension control to maintain essentially constant gage on a volumetric basis. Entry speed and thickness are measured at one or more stands and, as the entry speed or thickness varies, stand speeds are automatically adjusted to maintain the correct relationship between entry volume and delivered volume at constant thickness. X-ray gages monitor the system and establish volumetric setpoints. All interstand tensions are maintained within preset limits by automatic screwdown positioning.

Like the plate mill and hot strip mill systems described earlier, modern cold-rolling computer control systems typically produce printed reports to facilitate operating and maintenance activities and also transmit pertinent production and quality data to a business computer system to facilitate accounting and customer service activities.

Bibliography

Allison, F. H., Jr. and C. E. Peterson, Modern manufacture and use of cast rolling mill rolls. Iron and Steel Engineer **31**, No. 12, pages 68-77, December 1954.

American Foundrymen's Association, Cast metals handbook (latest revision).

Anon., Hot Strip Mills in Japan: Tomorrow's Technology Today. 33 Met. Prod. **18**, No. 2, pages 50-53, February 1980.

Anon., Rod mill rolls formed by grinding. Iron and Steel Engineer **36**, No. 4, page 156, April 1959.

Antrim, M. B., The application of the mercury arc rectifier to large reversing mill drives. Iron and Steel Engineer **37**, No. 8, pages 71-80, August 1960.

Bailey, W., Manipulating equipment, guides, guards and strippers for rolling mills. Journal of the Iron and Steel Institute **173**, No. 10, pages 198-213, October 1953, Discussion **174**, No. 6, pages 249-255, June 1954.

Beadle, R. G., Characteristics of tandem mill drives. Iron and Steel Engineer **32**, No 5, pages 97-103, May 1955.

Bradd, A. A., Material and design defects in forged steel rolls, Iron and Steel Engineer **38**, No. 1, pages 85-98, January 1961.

Briggs, C. W., The metallurgy of steel castings. McGraw-Hill Book Company, Inc., New York (1946).

Brown, H. S. and A. P. Baines, Developments in electrical equipment for reversing plate mills. Journal of the Iron and Steel Institute **194**, No. 1, pages 225-240, February 1960.

Browning, E. H., Electrical drive systems for modern rolling mills. Blast Furnace and Steel Plant **44**, No. 3, pages 299-

308, March 1956.

Browning, E. H., Modern electrical systems in the steel industry—new concepts in design and maintenance. American Iron and Steel Institute Regional Technical Meetings 1959, pages 19-51.

Cox, H. N., and A. S. Norton, Computer setup and control of cold strip mills. Iron and Steel Engineer **45**, July 1968, pages 99-106.

Cramer, F. W., Twin motor drives for hot reversing mills. Iron Age **156**, No. 15, pages 58-61, October 11, 1945,.

Cullen, C. W., and George P. Petrus, Generation III Hot Strip Mill Automation System. Iron and Steel Engineer **56**, No. 11, pages 25-30, November 1979.

Darnell, R., J., Computer control on Inland's 80-inch hot strip mill, Yearly Proceedings, Assn. of Iron and Steel Engineers, 1968; pages 254-260.

Ess, T. J., Fairless Works. Iron and Steel Engineer **31**, No. 6, pages F62-F92, June 1954.

Fazan, B., D. Boubel, P. Ratte, J. Bouvard, and F. Weber, Optimum Computer Control of a Plate Mill. Iron and Steel Engineer, **57**, No. 11, pages 58-64, November 1980.

Fujii, S., Unmanned Slab and Coil Yard System for Hot Strip Mill. Automation in Mining, Mineral and Metal Processing, Montreal, Canada, pages 565-576, August 18-20, 1980.

Goss, R. D., and N. Toschi, Jr., Productivity Improvement Through Automation of a Bar Mill Facility. Iron and Steel Engineer **60**, No. 7, pages 39-40, July 1983.

Harris, W. R., Developments in electrical drives for steel mill applications. Iron and Steel Engineer **29**, No. 12, pages 69-74, December 1952.

Hollander, F., Reheating Processes and Modifications to Rolling Mill Operations for Energy Savings. Iron and Steel Engineer **62**, No. 6, pages 55-62, June 1983.

Holman, R. W., R. G. Beadle and W. E. Miller, Card programming control of rolling mills. Iron and Steel Engineer **35**, No. 6, pages 113-119, June 1958.

Hurme, E. A., Factors involved in the selection of direct connected and geared main roll drives. Iron and Steel Engineer **2**, No. 5, pages 189-222, May 1925.

Hyams, W., Automatic contour turning of large mill rolls. Iron and Steel Engineer **35**, No. 3, pages 82-89, March 1958.

Iida, N., Advanced Automation on the New Plate Mill at Mizushima Works. Iron and Steel Engineer, **55**, No. 10, pages 34-40, October 1978.

Ishikawa, S., K. Mori, and N. Aoki, Production Control System of Steel Making and Blooming Process. Sumitomo Search **24**, pages 103-121, November 1980.

Jones, R. and R. Hawley, Trends in rolling mill drive design. Iron and Steel Engineer **38**, No. 9, pages 121-128, September 1961.

Jordan, J. A., State of the Art Computer Control of the Carlam Hot Strip Mill. Iron and Steel Engineer **58**, No. 4, pages 33-40, April 1981.

Kasecky, J. J., and K. W. Roessing, Retrofitting Existing 4-Hi Reversing Cold Mills With Automatic Gage Control (Retroactive Coverage). American Iron and Steel Institute, Regional Technical Meeting, Pittsburgh, PA, November 7, 1973, page 24.

Kaufman, G. A. and A. S. Smith, Electrical design and operation of a modern 46-inch high lift slabbing-blooming mill. Iron and Steel Engineer **30**, No. 10, pages 61-68, October 1953.

Kenyon, A. F., Factors involved in the selection of electrical drive ratings for metal rolling mills. Iron and Steel Engineer **38**, No. 11, pages 121-129, November 1961.

Kimura, J., H. Nagase, and S. Matsuka, Computer Control System Applied to a Cold Reversing Mill. Iron and Steel Engineer **53**, No. 9, pages 37-42, September 1976.

Krummel, W. M., Hot strip mill electrical system design trends. Iron and Steel Engineer **38**, No. 5, pages 97-105, May 1961.

Larson, H. E., Mercury arc rectifiers for main roll drives. Iron and Steel Engineer **29**, No. 11, pages 61-73, November 1952.

Link, W. J., Plate mill computer control start-up and operation at Kaiser Steel. Yearly Proceedings, Assn. of Iron and Steel Engineers, 1969; pages 40-48.

Mayer, J. H., Mechanical design and operation of a modern 46-inch high lift slabbing-blooming mill. Iron and Steel Engineer **30**, No. 10, pages 55-63, October 1953.

Mignon, J., J. Vinciotti, and S. Wilmotte, New Development of the Sigma-Rho System for Computer Control in Wide and Narrow Hot Strip Mills. International Conference on Steel Rolling, Vol. I, Science and Technology of Flat Rolled Products, Tokyo, Japan, **29**, pages 399-409, Sept. 29-Oct. 4, 1980.

Miller, H. E., R. A. Smith, J. F. McCarthy, Homer Smith, Jr. and N. J. Hittinger, Control and automation of hot strip mills from the operator's viewpoint. Yearly Proceedings, Assn. of Iron and Steel Engineers, 1969; pages 481-489.

Miyazaki, Y., et al., Full Computer Control of Plate Mill—Outline of the Plate Mill Control System at Oita Works., Nippon Steel Corp. Tech. Rep. Overseas **14**, pages 120-137, 1979.

Morgan, M. H., Adjustable speed main roll drives. Iron and Steel Engineer **5**, No. 10, pages 429-448, No. 11, pages 467-489, October and November 1928.

Mulflur, W. H., Algoma's 106-inch wide hot strip mill. Yearly Proceedings, Assn. of Iron and Steel Engineers, 1965; pages 13-22.

Nilsson, A., Automatic Flatness Control System for Cold Rolling Mills. Iron and Steel Engineer, **56**, No.6, pages 55-60, June 1979.

Peck, P. E., Electrical drives for reversing hot mills. Journal of the Iron and Steel Institute **169**, No. 3 pages 309-323, March 1951: discussion **169**, No. 10, pages 169-174, 168, October 1951.

Perry, W. A. and W. B. Snyder, Performance of blooming mill auxiliary drives with Ward-Leonard control. Iron and Steel **16**, No. 3, pages 42-51, March 1939.

Peterson, E. C., Automation of bar and rod mills. Iron and Steel Engineer **44**, February 1967, pages 53-64.

Readal, G. J., T. A. Shuman, and K. H. Care, Computer Control for U.S. Steel's Gary Works' 84 In. Hot Strip Mill. Iron and Steel Engineer **59**, No.4, pages 52-56, April 1982.

Scheer, G. B., Rectifier applications on steel mill drives. Iron and Steel Engineer **29**, No. 10, pages 107-112, October 1952.

Sendzimir, M. G. and L. Zdanowicz, A mill for cold rolling metals to close tolerances. Iron and Steel Engineer **33**, No. 11, pages 65-70, November 1956.

Sendzimir, T., The planetary mill and its uses. Iron and Steel Engineer **35**, No. 1, pages 95-101, January 1958.

Snitkin, S. R., Process Control Computers at Altos Hornos del Mediterraneo Cold Rolling Facility. Iron and Steel Engineer, **58**, No. 8, pages 25-32, August 1981.

Snively, H. N., Electrical equipment for metal rolling mills. Blast Furnace and Steel Plant **50**, No. 10, pages 961-969; No. 11, pages 1086-1092; No. 12, pages 1173-1179, October, November and December 1962.

Snyder, W. B., Electrical equipment for merchant, bar and rod mills. Blast Furnace and Steel Plant **38**, No. 4, pages 427-437, April 1950.

Starling, C. W., An introduction to the theory and practice of flat rolling—8. Sheet Metal Industries **38**, No. 4, pages 247-252, April 1961.

Steel Founders' Society of America, Steel castings handbook (latest revision).

Steel Publication, Inc., Watkin's cyclopedia of the steel industry (ninth edition), Pittsburgh, Pa., 1963.

Takano, H., H. Matsubara, and K. Matsuda, Computer control of a plate mill. Yearly Proceedings, Assn. of Iron and Steel Engineers, 1968; pages 345-351.

Takano, H., and T. Miyabe, Steel Plate Production in Japan:

Tomorrow's Technology Today. 33 Met. Prod. 17, No. 2, pages 55-59, February 1979.

Takano, H., and T. Miyabe, Steel Plate Production in Japan: Tomorrow's Technology Today. 33 Met. Prod. 17 No. 3, pages 57-61, March 1979.

Thomas, C. C., Electric equipment for reversing hot strip mills. Iron and Steel Engineer 32, No. 12, pages 92-109, December 1955.

Thurman, A. L., and D. Hancke, A new approach to continuous mill drives. Iron and Steel Engineer 33, No. 2, pages 76-85, February 1956.

Umansky, L. A., Flywheels for steel mill drives. General Electric Review 26, No. 10, pages 688-707, October 1923.

Umansky, L. A., Adjustable speed drives for rolling mills. Iron and Steel Engineer 1, No. 9, pages 515-532, September 1924.

Vonada, E. E., Multiple generators for processing lines. Iron and Steel Engineer 29, No. 11, pages 77-86, November 1952.

Wallace, J. W., Integrated process control rolls steel more efficiently. Westinghouse Engineer, January 1969.

Wallace, J. W., Fundamentals of strip mill automatic gage control systems. Yearly Proceedings, Assn. of Iron and Steel Engineers, 1964; pages 753-762.

Wright, R. H., Modern reversing mill drives. Iron and Steel Engineer 6, No. 5, pages 210-217, May 1930.

Wright, R. H., Twin motor drives. Iron and Steel Engineer 8, No. 6, pages 246-250, June 1931.

Wright, R. H., Main roll drives for blooming and slabbing mills. Iron and Steel Engineer 20, No. 11, pages 55-61, November 1943.

Wright, R. H., Steel mill drives; past, present and future. Iron and Steel Engineer 34, No. 1, pages 76-81, January 1957.

Wusatowski, Z., Fundamentals of roll design for steel sections. Iron and Steel 36, No. 10, pages 499-503; No. 12, pages 596-599, October and December 1963.

Yabuuchi, K., Y. Noma, R. Shimizu, and S. Tanimoto, Unmanned Operation Technology for a Hot Strip Mill. Iron and Steel Engineer 59, No. 11, pages 31-37, November 1982.

CHAPTER 24

Heating Steel For Hot Working

SECTION 1

PRINCIPLES OF FURNACE DESIGN

Objectives and General Metallurgical Requirements—A heating furnace is utilized to raise the temperature of steel to prepare it for hot working (shaping). Heating furnaces may be divided into two general classes:

1. Soaking-pit furnaces
2. Reheating furnaces

The function of soaking pits and reheating furnaces is to raise the temperature of steel in the course of processing until it is sufficiently hot to be plastic enough for economic reduction by rolling or forging to the desired section.

Heating furnaces must be constructed of suitable materials to withstand the effects of the temperature levels at which they must operate. They must be provided with charging and discharging facilities which are adequate for the material size and handling rate, and with the proper means for heating the steel at the specified production rate. From the metallurgical standpoint, soaking-pit and reheating furnaces must be constructed to heat the steel uniformly and, by suitable temperature and combustion control instrumentation, hold it at the desired temperature for a specified length of time.

Basic Elements of Furnaces—There are many different designs for each of the general furnace classes noted above, but each design consists of certain common parts, as:

1. The heating chamber; an enclosure to contain the material and retain heat.
2. A hearth or support in the furnace for carrying the charge.
3. Facilities for the development of heat to raise and maintain furnace temperature.
4. Means for the distribution of heat and the removal of spent gases from the furnace.
5. Means for the introduction of work to be heated and removal of heated stock.

The enclosure to contain the material and heat generally is called the furnace proper. It is constructed of refractory material. The furnace hearth or support for the charged material may be constructed either of refractory or metallic material. Metallic supports generally are water cooled and insulated. Furnace hearths

are constructed to permit the charge either to remain in a fixed position in the furnace or to be moved during the heating process. An example of the first type of hearth is the conventional **batch-type** or "in-and-out" furnace. Examples of the second type, referred to generically as **continuous furnaces,** are found in **roller-hearth furnaces** in which the material moves as the series of rollers that constitute the hearth rotate, in **walking-beam furnaces** and **walking-hearth furnaces** in which material is moved in a controlled step-wise manner, in **pusher-type furnaces** in which a continuous line of material is pushed over skids, and in **rotary-hearth furnaces** with circular hearths that rotate in a horizontal plane. The combustion of fuel usually is employed to develop the required furnace temperature, but the conversion of electrical energy into heat is used also in some important heating-furnace operations, such as large-scale induction-heating facilities for heating slabs for hot working. The circulation of heat in fuel-fired soaking-pit and reheating furnaces is accomplished by natural convection and stack draft.

Facilities for the introduction and removal of heated stock vary with the type and size of furnace, the size and shape of the stock to be handled, and the general layout of the furnace and auxiliary facilities. Roller tables, conveyors, charging machines, overhead cranes, furnace pushers and pinch rolls are the principal kinds of equipment used for this service.

Furnace Size and Capacity—The size of a heating furnace usually is described by its hearth area. The hearth areas of the various types of the two general furnace classes differ greatly. Soaking pits may have a hearth area of from 9.3 square metres (100 square feet) to something over 27.9 square metres (300 square feet). Reheating furnaces may have hearths ranging in area from less than half a square metre (a few square feet), as in a small forge furnace, to over 370 square metres (4000 square feet) in large, continuous slab-heating furnaces. The productive capacity of a heating furnace often is related to its hearth area, and figures up to 1220 kilograms per square metre (250 pounds per square foot) of hearth area per hour are quoted for the various types. The high rate may apply to pit furnaces when heating already hot ingots to a higher temperature. The capacity of a furnace is determined primarily by the area of the surface of the material to be heated

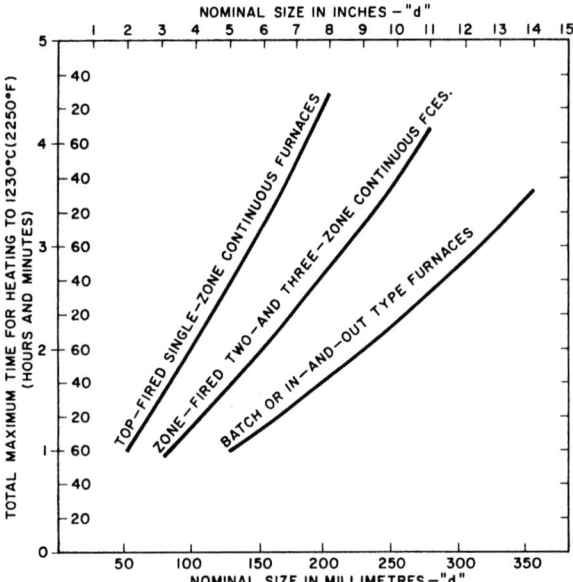

FIG. 24—1. Normal average time for heating varous sizes of square mild-steel (low-carbon) and medium-carbon steel billets from room temperature to rolling temperature of 1230°C (2250°F).

which is exposed to the furnace temperature, and the shape, thickness and composition of the material, its temperature and that of the furnace, and the emissivity of the heat source and of the material to be heated. The desired rate of heating is regulated by manually or automatically controlling the rate of heat input to the furnace.

High-carbon steels and heat-resisting alloys, particularly when charged cold, require longer heating cycles to attain uniformity of heating at the same temperature levels, as compared to low-carbon steels. Selection of the temperature and time required for properly heating various grades and sizes of steel is based on experience with given furnaces heating specific types of prod-

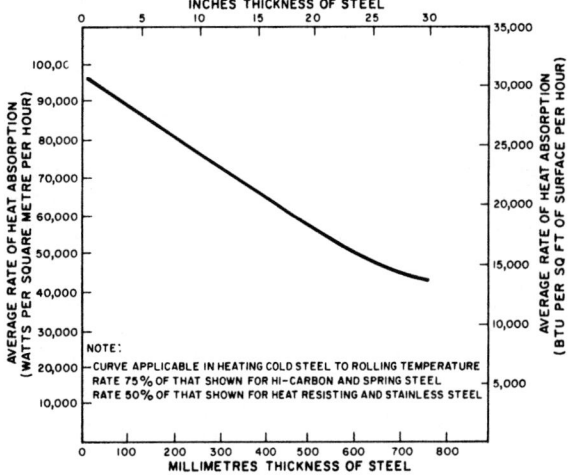

FIG. 24—2. Curve showing average rate at which heat is absorbed during an entire heating cycle by mild steel and medium-carbon steel in heating from room temperature to rolling temperature in reheating furnaces.

ucts and grades. Figure 24—1 shows the normal time for heating various sizes of cold mild steel and medium-carbon steel blooms and square billets to rolling temperature. The difference in the heating rates among the two furnace types is due to the difference in the number of exposed surfaces on the bloom or billet while it is in the furnace. The heating time is longest for a bloom or billet in a top-fired continuous furnace since heat is transferred to the product for the most part through the top surface only. In a two and three zone continuous furnace, the bloom or billet is heated from both the top and bottom surfaces during part of the time it is in the furnace. In the batch or 'in-and-out' furnace, the product rests on the hearth or a small pedestal and is heated on at least three sides. The heating time for high-carbon steel is about one-third longer and for heat-resisting grades about twice that shown.

The flow of the heat through a thick body of steel is relatively slow compared to surface absorption in high temperature furnaces and, therefore, caution must be exercised in regulating the supply of heat to prevent the surface from "sweating" (partial melting) while bringing the temperature inside the material up to the required level. Figure 24—2 shows the average rate of heat absorption for an entire cycle as generally practiced in heating mild or medium-carbon grades of steel from atmospheric to rolling temperature. During the early stages of the heating cycle, the heat-transfer rates are considerably above the average shown, while during the latter part of the cycle the rate is very low.

Furnace Type and Shape—There are many types of each of the two general classes of heating furnaces. The selection of type is determined by its suitability for heating economically particular grades, shapes and sizes of the material at the rate and to the temperature level desired. For instance, batch-type furnaces are especially suitable for heating blooms of mixed sizes and lengths in thicknesses over 200 mm (8 inches); continuous furnaces are used for heating slabs or billets for large orders of uniform length and thickness, and carbottom furnaces are used for heating large ingots for forging. The general shape of a furnace depends upon a number of factors, such as capacity desired, space available, auxiliary equipment and metallurgical requirements in heating. Refinements in furnace lines depend within rather wide limits on the kind of fuel used and on the grade and size of steel to be heated. The desired combustion space, temperature level requirements, uniformity of heating and fuel flow are major considerations in furnace design.

Thermal Efficiency—Thermal efficiency is defined in this case on a percentage basis as the amount of heat required to raise the temperature of the stock from its initial temperature to the temperature desired for hot working relative to the gross heat input to the furnace. The thermal efficiency of heating furnaces varies considerably because of differences in the temperature level of the heated stock and of the charged material, in the provision of heat recovery equipment such as regenerators and recuperators, in furnace insulation, in operating schedules, and in heating requirements. To cover the full range of all common types of heating furnaces, the thermal efficiency may fall anywhere between 5 per cent and 60 per cent. Large production-line furnaces, such as continuous furnaces equipped with recuperators and insulation, generally give 30 per cent to 40 per cent thermal efficiency over an average month's period of operation. Small shop furnaces, poorly loaded, with no insulation or heat-recovery facilities, have low thermal efficiency. The heat requirement of heated product from heating furnaces for the production line varies from 349 000 kilojoules to 5 230 000 kilojoules per metric ton (300 000 to 4 500 000 Btu per net ton). The lower figure is obtained when heating hot steel, the higher one with poorly-loaded furnaces heating cold steel.

The sensible heat lost in stack gases is the principal source of heat loss in a fuel-fired furnace. Other losses include the heat loss by conduction through furnace walls, hearth and roof; radiation through furnace openings; the heat absorbed by water-cooled furnace parts; and the latent heat and unburned combustibles in stack gases.

In Chapter 3, dealing with fuels, the means employed for reducing furnace losses and the salvage of heat in waste gases already have been reviewed. Regenerators were provided in many of the older batch furnaces which were operated at high-temperature levels. They provide, in addition to heat salvage, a reservoir of potential heat for equalizing the temperature of steel during the soaking period. Ordinarily, in regenerative batch-type furnaces, the actual time for firing fuel amounts to only 50 per cent of the total time from charge to charge; the balance of time is taken up by charging, drawing and soaking (soaking consists of holding the product in a furnace maintained at the desired temperature for a sufficient time to assure uniform distribution of heat in the product). Firing in regenerative furnaces is usually at a constant rate with intervals of soaking, which occur more frequently as the charge approaches rolling temperature.

The heat stored in thick furnace walls of ordinary firebrick at high hot-face temperatures is considerable. When the so-called **flywheel effect** of hot walls and regenerators is not desirable, such as in furnaces which occasionally must be cooled as quickly as possible to lower than usual operating temperatures to receive special alloy or high-carbon heats, a material saving in production, fuel economy and maintenance can be effected by constructing the walls and roof of insulating rather than of regular firebrick.

Recuperators generally are supplied for high-temperature heating with continuous reheating furnaces and soaking pits, and in some instances for batch-type furnaces. They are more desirable than regenerators when the control of atmosphere and a constant flow of fuel into the furnace is important. Often, they provide a lower preheat temperature of the combustion air than is obtainable with regenerators.

Materials of Construction—The temperature level carried in various parts of the furnace determines the kind and grade of construction materials that must be used. The hot end of continuous and regenerative batch furnaces, which employ burners or a fuel developing intensive combustion, may reach temperatures up to about 1535°C (2800°F). For this level, it is customary to construct the walls and roof of first-quality or superduty firebrick, plastic firebrick or ramming mixes and to water-cool any metallic parts which are exposed to high temperature. In high-temperature furnaces, there are applications for the use of special refractories. Hearths are constructed usually of refractories resistant to abrasion, slag attack or adherence to the steel being heated. Door jambs are made of refractories with non-spalling characteristics. Pier walls, such as those used in top- and bottom-fired continuous furnaces, use refractories with good hot-load-bearing properties. In pit furnaces, slag-resistant refractories are used in the lower wall areas to combat chemical attack by cinder.

Flue temperatures, especially following recuperators or regenerators, seldom exceed 870°C (1600°F), and for this application second-quality firebrick generally is used. Cast iron or heat-resisting steel is used for constructing dampers up to this temperature level; in flues subject to higher temperatures, the dampers are water-cooled.

Besides the refractories and metallic parts directly exposed to internal furnace temperature, many furnaces contain a number of other essential parts which must be given careful consideration due to the temperatures to which they will be subjected. Cast-iron doors and door frames usually are lined with refractory material for protection; roof hangers are cast of heat-resisting metal; furnace casings and steelwork usually are constructed of ordinary grades of carbon steel; regenerators generally utilize refractories; recuperators are constructed either of refractory or metal, depending upon the temperature level and combustion-air pressure at which they are to operate. Various grades of insulating material are used, dependent upon load, location and temperature level. Insulation for reheating-furnace walls, roof and hearth, not only provides a saving in fuel but also aids in maintaining a uniform temperature within the furnace, reduces stresses in furnace brick and steelwork, and improves working conditions around the furnaces.

SECTION 2

SOAKING-PIT FURNACES

Introductory—The function of the soaking pit is to provide uniform heating of ingots to the desired temperature with a minimum of over-heating of the surface. In most modern designs, this is accomplished with automatic controls. The normal range for heating ingots is between 1175° and 1345°C (2150° and 2450°F).

The proper temperature level varies with grades of steel and sizes of ingots and characteristics of the rolling mill. Low-speed mills with many passes require the higher level of heating for certain grades of steel. Soaking pits serve the dual function of heating and acting as a reservoir to correct irregularities in the flow of ingots

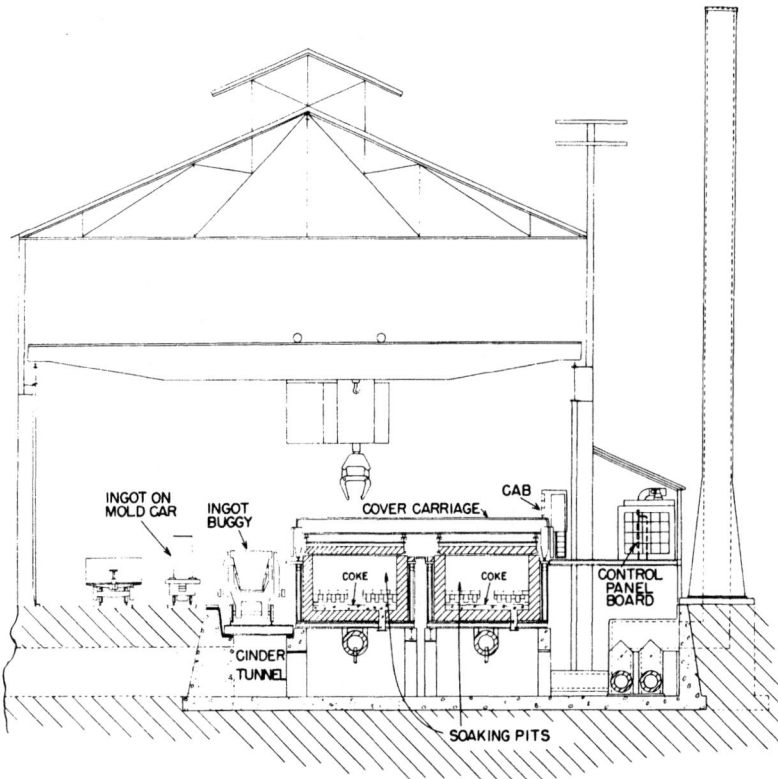

FIG. 24—3. Schematic cross-section through a soaking-pit furnace building. (Courtesy, Amsler-Morton Company.)

between the steelmelting shop and the primary rolling mills. Briefly, soaking pits are deep chambers, or furnaces, of square, rectangular or circular shape, into which ingots are placed in an upright position through an opening at the top (Figure 24—3). A removable cover closes the pit opening. A series of pits, installed usually in rows, are placed under cover of a building adjacent to the entering side of the blooming or slabbing mill to be served. The pits are spanned by one or more electrically operated traveling cranes equipped with a traveling hoist for charging the ingots into the pits and for lifting them out as they are needed by the mill. The lower end of this hoist is provided with adjustable tongs, by which an ingot may be grasped at the top and moved vertically a distance greater than the depth of the pit. This crane is used to transfer the ingots to the pits from the cars on which they were brought from the stripper, moving on tracks usually located along the side of the pits. For heated ingots, a pot car or ingot buggy is provided, which usually is propelled electrically along a track leading to the primary-mill tables, upon which it automatically deposits the ingots.

Types of Soaking-Pit Furnaces—There are several designs of soaking pits in operation today. The original soaking pits were nothing more than unfired holes or pits in the ground in which hot ingots were placed to 'soak-out' to a uniform temperature prior to rolling. Demands for high productivity and improved product quality required that burners be placed in these older soaking pits. The oldest of the so-called modern types of fired soaking pits is the regenerative pit. In this type, the ingots are heated by alternately burning the gas

through a port in the pit wall on one side, permitting the products of combustion to pass across the pit and out through the regenerator flues and stack to the atmosphere. The air, after each reversal, is passed through the hot regenerators to provide preheat for combustion of the fuel. If fuels of high calorific value,

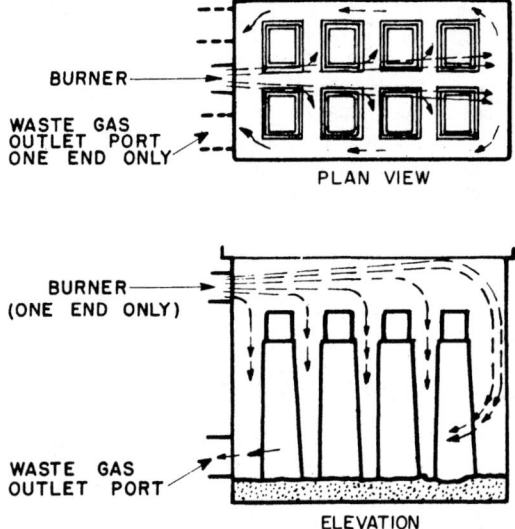

FIG. 24—4. Diagrammatic elevation and plan view illustrating principle of continuous firing and flow of hot gases in a "one-way top-fired" soaking pit. Hot gases from outlet ports pass to recuperators (not indicated).

e.g., fuel oil, natural gas, or coke-oven gas are burned, they are introduced either through pipes in the top of the checker chamber and directed toward the bridgewall, or through a well in the checker brickwork adjacent to the port bridgewall. In some installations, particularly when cold blast-furnace gas is used, burners have been installed in the rear walls of the checker chambers. Since the ports are located in the endwalls of the pit, the ingots are exposed to conditions of unequal heating on their opposite sides. To equalize ingot temperature, the practice of firing and dampering is generally followed. This practice provides uniformly-heated steel at a relatively-fast heating rate, but experience has shown that timely maintenance of seals around checker-chamber brickwork and flues is necessary to minimize unnecessary fuel consumption and added scale losses due to the infiltration of air through these openings or through faulty cover seals each time the pit is dampered. Difficulties from scaling and added fuel consumption are confined not only to regenerative-fired pits, but also to recuperative-fired pits if there are leaks in the recuperators.

One of the first major steps towards improvement in pit-furnace design was to provide sufficient space for the combustion of fuel. Pits of a continuous-fired design, known as **one-way fired pits** (Figure 24—4), equipped with recuperators, were designed to provide combustion space above the ingots, where the space available for combustion was not affected by **ingot coverage.** "Ingot coverage" is a term used to denote the tonnage or number of ingots charged into a pit. Instead of the horizontal flow of gases through the pit as in the regenerative type, the flow in this type is vertical in accordance with hydrostatic principles. Utilization of one-way top-fired pits has become very popular, and, when installed in banks of three or four with common-wall construction, such pits make possible a high ratio of pit-hearth area to building area.

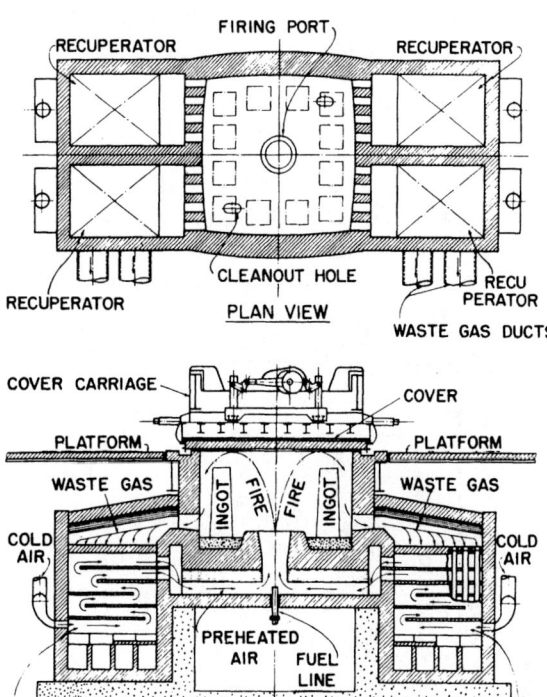

FIG. 24—5. Schematic elevation and plan view of a "vertically-fired" soaking pit, with the burner opening in the center of the pit bottom. (Courtesy, Amsler-Morton Company.)

In another type, designated as the **bottom center-fired** or **vertically-fired** pit a section of which is shown in Figure 24—5, the fuel is fired vertically through a port, centrally located in the bottom of the pit, around which the ingots are placed. As the products of com-

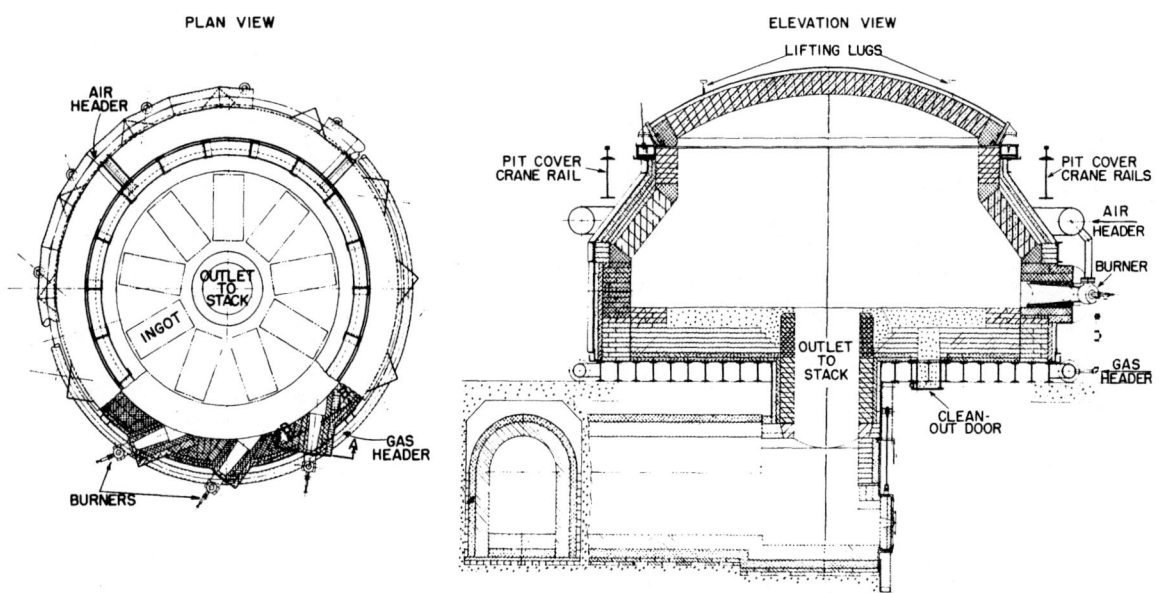

FIG. 24—6. Simplified section and plan view (not to same scale) of a tangentially-fired soaking pit. (Courtesy, Salem-Brosius Inc.)

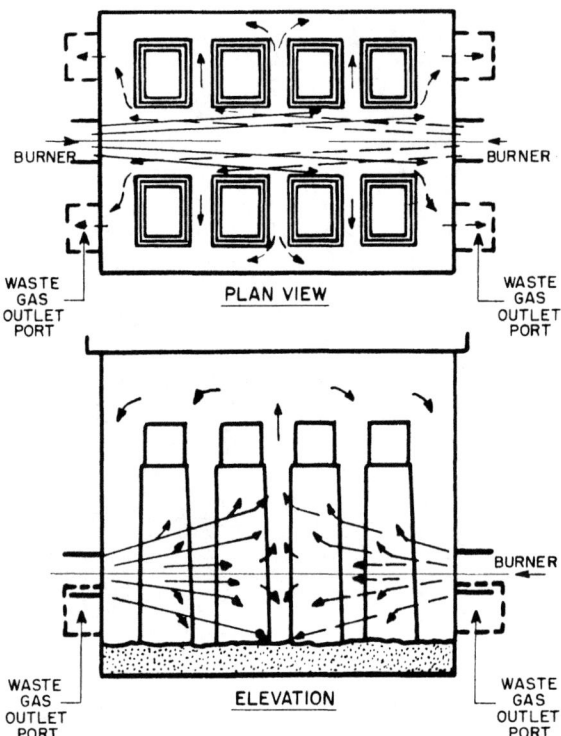

FIG. 24—7. Schematic diagram showing principle of firing and flow of gases in a side-fired or bottom two-way fired soaking pit. Gases from waste-gas ports go to a recuperator (not shown).

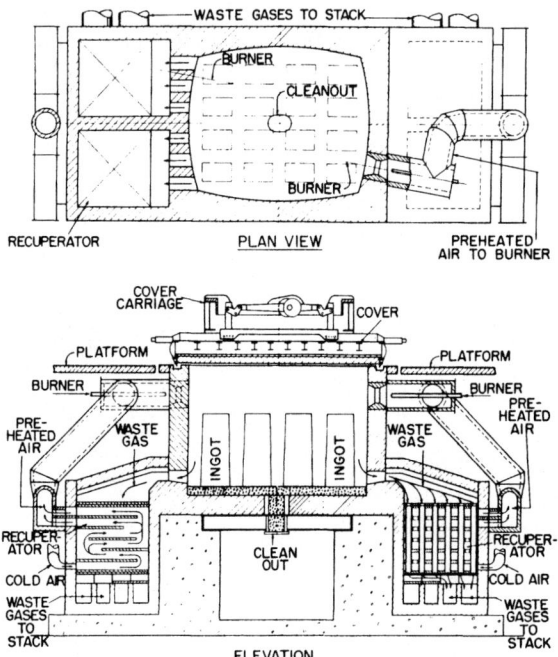

FIG. 24—8. Diagrammatic elevation and plan view of a top two-way-fired soaking pit. Left-hand section of plan view is through waste gas ports; right-hand section through one of the burners. Left-hand half of elevation is section through air passages of recuperator; right-hand, through waste-gas passages. (Courtesy, Amsler-Morton Company.)

bustion rise in the combustion zone, some of the spent gases which are moving downward around the ingots and next to the furnace walls are drawn into this inner column of hot rising gases, and the resulting good circulation equalizes furnace temperature. Flues are located along the bottom of opposing side walls of the pit to remove products of combustion and to aid in heat distribution. This design was introduced with recuperators and controls for carrying out program heating. Adequate provision for combustion of fuel and careful sizing of pits, of nearly square shape, to suit loading conditions are incorporated in the design.

Another type, which is called the **circular pit** (Figure 24—6) employs tangential firing from a series of recessed burners located in the lower periphery of inclined side walls, to permit unusually long travel of the gases and to induce recirculation of the spent gases before they leave the pits through a centrally-located exit port at the bottom of the pit. This design utilizes a method for tempering the flame and securing uniform pit temperatures through gas circulation. These pits normally are fired with a high-calorific fuel and cold air in premix or nozzle-mix type burners to obtain complete combustion of the fuel before the gases come in contact with the ingots.

Still another pit is a refinement of the continuous-fired pit furnace, known as the **bottom two-way fired pit** (Figure 24—7). In this design the pits are fired by burners located in opposite endwalls about 600 mm (2 feet) above the bottom of the pit. The waste-gas ports are located in the same endwalls at each of the four pit corners and the gases on each end go directly to a recuperator. Combustion of the fuel takes place in a centrally located aisle, at the sides of which the ingots are placed. The method of firing and the position of the burners and waste-gas ports provide turbulence to the flow of gases in the pit, and results in improved heating of the ingot bottoms.

The **top two-way fired pit** is a deep rectangular pit in which the fuel is fired from opposite ends into a combustion space above the ingots (Figure 24—8). The burners are set to fire horizontally at an angle to the centerline of the pit to obtain a swirling motion of the gases. Outlet ports are located in the sidewalls just above the cinder line. Long-flame burners are used generally to distribute properly heat from combustion above and between the ingots. The flow of gases is vertical, similar to that in the one-way fired pits. In this design, the method of firing allows the pit shape to be selected that will give the desired ingot-setting patterns; it may be built long and narrow or square. Curved (elliptical) sides and endwalls, now common to the design of most square and rectangular types of pits, are utilized also in these pits.

Electric soaking pits were developed to meet special requirements, such as the control of scaling and the maintenance of controlled atmospheres during the heating of stainless-steel and alloy-steel ingots. An electric soaking pit installation is very simple, consisting of the pit proper and electrical equipment. The pit itself consists of a rectangular steel casing heavily bound in rolled-steel sections and containing a refractory lining and insulation to form a closed unit. Internal brickwork divides the pit into a number of cells, each of which may hold one or more ingots, depending upon ingot size. The cells are provided with individual covers that are handled with a special cover-lifting machine. The heating elements are coke-filled refractory troughs that run the entire length of the pit through arched openings in the walls that separate the cells. Power is applied to the heating elements by electrodes that are led through the pit walls at each end. Temperature is controlled by adjusting the secondary voltage from transformers.

Auxiliary Facilities—The principal pit-furnace auxiliaries are pit covers, ingot pit cranes, facilities for ingot delivery to and from the pits and for cinder removal, along with the necessary instrumentation and controls for fuel and air supply, draft and temperature.

Pit covers are constructed in various ways, the essential parts being a metal framework to support the refractories and means for quickly removing the cover then replacing it upon the pit. In the older pits, the metal framework usually was made of cast iron, and the bricks were slightly arched for support. The covers were supported upon rollers or wheels which moved over tracks, originally by the use of hydraulic cylinders and later by electric motors. A depression in the track permitted the cover to drop, when in position directly over the pit, and seal the pit opening. This design, however, seldom provided a tight seal at all points and leakage of gases caused high cover maintenance. An improved design consists essentially of a heavy steel frame which supports a suspended arch of high-grade firebrick, or a rammed or cast plastic or fibrous refractory material held in position by refractory hangers. They are equipped with either individual cover carriage mechanisms for lifting the cover vertically to free it from the sand seal and then move the cover to the desired pit opening, or with a cover crane to effect similar movements. The cover crane spans a row of pits and serves several. The steel frame of the pit cover is protected on its under side by heat-resisting-alloy castings. These castings are provided with a lip which penetrates into the sand seal. The sand seal, located around the periphery of the top of the pit, is a gutter-shaped space filled with sand. The suspended arch of a modern cover usually is backed up with insulation to reduce radiation losses.

Ingot Pit Cranes—In modern plants, the ingots are handled from the ingot receiving cars to the pits and from the pits to the mill delivery ingot buggy by pit cranes. These cranes are of the conventional electric overhead design except for the trolley and position of the crane cab. Instead of the conventional hoist drum or drums, the trolley consists of the usual track wheels and a frame supporting a vertical housing in which travels a ram equipped at the bottom with mechanically operated tongs. The crane cab is attached to the lower end of the vertical housing and moves with the trolley. This arrangement of the crane cab is to provide maximum vision for the operator in order that the ingots may be placed in proper position in both pits and mill ingot buggy without damage to either. The pit crane is also used to handle coke breeze for bottom making.

Ingot receiving facilities consist usually of a track and ingot buggies. The ingot buggies which deliver the

ingots to the soaking-pit building are the same ones which held the molds for casting at the steel melting shop. The ingots usually are stripped of molds at the stripper, but in some cases where big-end-up ingots are used, the molds are only loosened at the stripper and are later removed by the ingot pit cranes, thus reducing heat losses enroute to pits. The layout of ingot-delivery tracks to avoid crossings and switches which interfere with other plant movements is very important. Minimizing delay in the delivery of ingots from the steelmaking furnace pouring platform through the stripper and the pit furnace building into the pit furnace not only conserves heat but provides better metallurgical control of quality and less conditioning cost of the mill product.

Mill delivery facilities consist usually of an electrically-operated ingot buggy or pot car which runs directly from the pits to the mill approach table. In one type of buggy, the contact of the buggy with a cam arrangement at the mill approach table tilts the cradle of the buggy and throws the ingot from its vertical position in the buggy to a horizontal position on the mill approach table. The other type of buggy, in order to avoid battering the mill tables with heavy ingots and the necessity of approaching the tables at a high rate of speed to throw the ingot onto the table, is a self-tipping type. This design has a motor for driving the track wheels, for tilting the cradle, and for turning the table rollers which deliver the ingot in a horizontal position to the approach table.

Cinder-Removal Facilities—Where coke breeze is used as a bottom-making material, the so-called **wet-bottom** practice, pit bottoms are made up with coke breeze to a depth of approximately 300 to 400 mm (12 inches to 16 inches). The wet-bottom practice derives its name from the fact that scale which falls off the ingots during heating combines with carbon in the coke breeze to form a mixture which melts at normal furnace operating temperatures. This liquid mixture, along with ashes from burned breeze, refractory, and other material, is removed through cinder holes, of which there are usually two, located in the bottom of the pit. In modern practice these holes, provided with gates at the outer end, discharge the cinder into a box located in the cinder alley under the pit bottom. The cinder box often is supported by a lift tractor or car which carries the box through the cinder tunnel to a hoist where it can be lifted to yard level and dumped into broad-gage cars for disposal. Some plants are equipped with an underground narrow-gage track system for moving the car to the cinder hoist. Bottoms for the older pit designs generally are made up each day; in modern pits they are made up only every 5 to 7 days under normal conditions.

The **dry-bottom** practice utilizes dolomite or magnesite as a 75-mm to 100-mm (3-inch to 4-inch) covering on the pit hearth. With a dry-bottom practice, bottom life is extended to several months as opposed to the 5- to 7-day life for the wet-bottom practice. To successfully exploit this advantage, the pit must be designed so that it has a relatively large cinder volume—that is, the space between the hearth floor and the waste-gas outlet port (or ports) must be considerably larger than for pits in which the wet-bottom practice is used. The reason for this is that the longer life of the dry bottom results in a larger accumulation of scale (cinder) from the ingots. At the high temperatures employed in the soaking pits, this scale, particularly when mixed with the hot-topping compounds which fall off ingots, is molten and, if the capacity of the cinder volume is insufficient, this molten mixture will run out of the waste-gas outlet port, tend to restrict gas flow and often cause damage to the waste-gas ducts and recuperators.

Removal of the cinder from a so-called dry-bottom pit is considerably more difficult than for a wet-bottom pit. Before the accumulation of the cinder (or scale) reaches the waste-gas outlet port, the pit is shut down, allowed to air cool for a short time, and then cooled more rapidly by the injection of a fairly large volume of water applied by hoses to the pit bottom. The water also serves to "break up" the bottom cinder and scale. As the temperature of the pit decreases, the volume of water is substantially reduced and, as soon as the pit is cool enough, cinder and scale are further broken up by mechanical equipment specially designed for the purpose.

The bottom-making operation, including the cooling of the "old" pit and the start-up of the "new" pit, requires 32 to 48 hours. Obviously, this lost time plus the cost of removal of the old bottom can only be made up by the longer pit life gained by the use of the dry-bottom practice. Generally, a minimum bottom life of about 90 days is considered to be "break-even" point for the dry-bottom practice.

Fuel, air and draft facilities of a soaking pit are extremely important as these control furnace temperature and atmosphere. In modern pits the quantity of fuel, the desired fuel-air ratio and the draft or pressure in the pits are controlled automatically. The rate of fuel input is controlled by temperature measurement to maintain some predetermined level. The air is proportioned to the amount of fuel fired at a ratio that will give a slight excess of air in the exhaust gas. The furnace draft or pressure is controlled by automatically raising or lowering the stack damper to maintain the desired furnace draft. Pit-furnace burners are of various designs to provide the type of flame most suitable for heating in each particular pit design. In the regenerative pit the port acts as a burner. Some pit designs utilize a long-flame burner in which the air and gas are not mixed intimately in order to develop slow combustion and a longer path of heat release. In others, the mixing is very rapid and a short nonluminous flame is developed. Some burners are designed to inspirate either the gas or air. Low-pressure inspirating burners usually employ gas at a low pressure to inspirate air for combustion. With high-pressure inspirating burners, either the gas or air is carried at sufficient pressure to inspirate the required volume of the other that is at low or atmospheric pressure. Soaking pits are provided with accurate means for measuring and controlling the air volume required for the combustion of fuel. Generally, motor-operated fans are used to supply the air. Fans are used for delivering either hot or cold air to the burners. Fans may be installed in some designs on either the cold or hot side of the recuperator. If on the cold side, the fan construction is simple and does not require special features such as alloy blades and an in-

sulated casing, as is the case with hot-air fans. Cold-air fans, however, subject the recuperator to a high pressure differential across its tubes and joints which sometimes causes excessive air losses and lowered recuperator efficiency.

Objectives in Soaking Pit Design—The principal requirements of pit furnaces are:

1. Uniform heating of all ingots in the pit with no localized overheating of any ingot.
2. Heating rate equal to the ability of the steel to safely absorb heat.
3. Sufficient holding capacity to accommodate the number of ingots representing either one heat or one-half of a heat without overcrowding.
4. Low operating cost.
5. Control of furnace atmosphere.
6. Ability to duplicate heating practices.
7. Minimum expenditure of ground space.
8. Minimum capital expenditure.

The first objective noted above is the most difficult to attain. Many different types of pits have been introduced to solve this problem. Uniform heating of all ingots in a pit with no localized overheating of any ingot permits attaining the highest possible temperature in the pit compatible with the ability of the steel to absorb heat without sweating the surface or injuring the steel from severe temperature strain. Optimum heating rates are obtained by holding the surface just below the sweating point. This permits the most rapid flow of heat by conduction into the body of the steel. Since the flow of heat from the surface decreases as the temperature of the center of the ingot rises, the rate of firing must be reduced as heating proceeds to prevent overheating the surface. Variations in firing rate introduce problems in pit furnace heat distribution, as the hottest part of the flame, or the length of travel of heat release of the fuel, changes with the fuel volume and air-fuel ratio. Experience with various pit types generally dictates the proper adjustments necessary for variation in firing rate. In some cases alterations in furnace draft is the remedy chosen. In many instances, when heating with regenerative pits, and occasionally with other types, adjustment for lack of uniform temperature distribution is secured by firing at a uniform rate for a period of time followed by a period during which the pit is dampered and no fuel is admitted. During the latter period, the steel is "soaked" to equalize the temperature between ingots in the pit and between the surface and the interior of individual ingots. This practice, known as **"firing and dampering,"** provides a product with a uniform temperature at high production rates even though uneven heating developed during the firing period.

In order to aid uniform heating, circulation of the gases in the pit furnace has been given considerable thought in modern design. Entrainment of cooler, spent gases with the hotter products of combustion, especially near the burner, dampens the tendency of the flame to develop a hot spot, lengthens the time for full release of heat from the fuel and aids in uniform heating. Other effective methods utilized for obtaining uniform temperature distributions are: (1) designing the burner to release the heat of combustion of the fuel over as great an area in the pit as possible to avoid concentration of heat, (2) developing the flame sufficiently distant from the ingot surface to control radiation within safe limits, (3) reducing flame emissivity by using non-luminous-flame type burners or the equivalent, and (4) use of higher pit pressure. A cause of uneven heating in pit furnaces is the fact that the top of an ingot is exposed to heat from all sides while the bottom rests on a relatively cold unexposed surface. It is necessary to compensate for this either by proper distribution in the flow of gases in the pit, or, as is done in some designs, by maintaining higher effective flame radiation at the pit bottom. In some cases, as noted above, uniform heating is secured only by "soaking." Since the corners of ingots are heated from two adjacent sides, these parts reach the desired temperature first and, therefore, prevent the carrying of an otherwise permissible higher furnace temperature during a considerable part of the heating time.

Fuel consumption and the maintenance of soaking pits have been reduced considerably by improved design, the installation of automatic controls and better materials of construction. Fuel-air ratio control for combustion and furnace atmosphere has reduced scale losses and the frequency of bottom making. Temperature measurements associated with automatic fuel-input control have made it possible to duplicate heating practices.

Soaking Pit Heating Practices—The time required to heat a charge of ingots in soaking pits generally is associated with transit time. Transit time is the elapsed time beginning with the 'start to pour' of ingots at the steel-producing source and ending with the charging of the first ingot of the heat into the soaking pit. The heat content of an ingot and the degree of solidification which has taken place when it is charged into the pit is related to transit time. As the transit time increases, the heat content of an ingot decreases and the amount which has solidified in the ingot increases, both of which affect the time required to heat an ingot to proper rolling temperatures in a soaking pit. Figure 24—9 illustrates the relationship between heating time and transit time. As indicated, there is an optimum transit time which corresponds to the minimum heating time. When the transit time is shorter than the optimum, the heat content of the ingot is high, but the ingot residence time in the soaking pit must be longer than the minimum to allow the ingot to completely solidify prior to rolling. When the transit time is longer than the optimum, the ingot is completely solidified by the time it is charged to the soaking pit. It follows that the heating time increases as colder steel is charged to the soaking pit. The relationship shown in Figure 24—9 can be approximated for a specific ingot size and operating conditions using a complex heat transfer analysis method. However, in practice, it has been common for soaking pit operations to use various rules of thumb for the relationship between heating time and transit time. For instance, the heating time required for ordinary carbon steels is approximately 1½ times the transit time when the track time is not excessive. The heating time of cold ingots is normally determined according to their size and usually requires from 8 to 12 hours, but, if the ingot is unusually large or of some particular type of steel which requires special treatment, it may re-

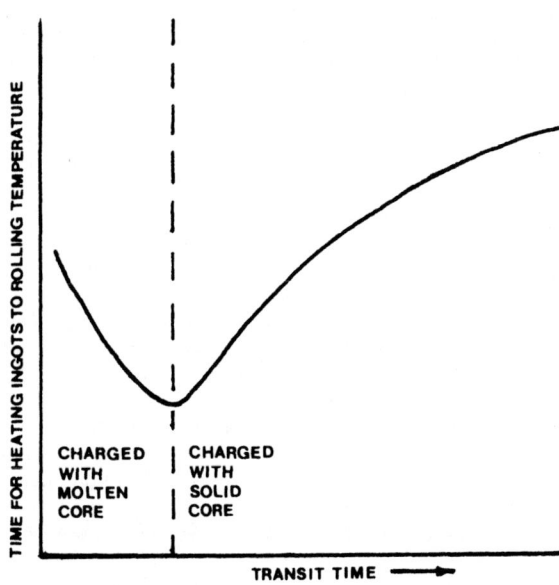

FIG. 24—9. Sketch illustrating the relationship between heating time and transit time in a soaking pit.

quire up to 18 hours.

In many soaking pit operations, the heater must use his experience and process information such as temperature-time and fuel input data to make a decision of when a pitload of ingots is ready-to-roll. Formerly, by careful observation of the ingots through sight holes in the cover and his knowledge of temperature requirements for each particular type of steel, he determined when the charge was ready for rolling. While this practice is still common with old regenerative-type pits, the heating in continuous-fired pits is governed largely by automatic controls which carry out the heating program desired.

In heating ordinary grades of hot carbon steel in modern pits with automatic control, the operator sets the control dial at the temperature level desired for drawing. A full, or maximum, head of fuel automatically is fired into the pits in the early stages of heating. The fuel input is cut back gradually as the heat soaks into the ingot until it is finally reduced to a minimum, just sufficient to cover pit radiation losses. When the fuel input has remained at this low level for some specific time, such as a half hour or more, as experience dictates, the ingots are drawn. With high-carbon steel and many alloy grades which are cold or have had a long transit time, it is customary to cool the pit down to the temperature of the charged steel before placing these grades in the pits. They are then heated rather slowly to some fixed temperature below the rolling temperature. After the ingots have been soaked at this lower temperature, most types of steel may be heated rapidly to rolling temperature. A common practice in heating certain alloy steels is known as **step heating.** Step heating consists of heating ingots to rolling temperature through several levels, each level being held for some specific time to permit equalization of temperature through the ingot. This procedure eliminates stresses due to severe temperature gradients in the ingots and provides slow, uniform heating. With step heating, it is customary to reduce the fuel-rate input considerably below the maximum setting permitted for heating ordinary types of steels.

Since one of the most significant factors affecting fuel consumption in a soaking pit is ingot transit time, the trend has been toward reducing transit time as much as possible within the guidelines of safety and product quality. An ingot with a short transit time contains a large amount of unsolidified steel in its center, which contains both sensible and latent heat. Heating practices which consist of a long dampering period to allow the ingot to soak out and solidify, followed by a brief firing period to heat its surface, have been developed to utilize the large reservoir of energy contained in these ingots and substantially reduce fuel consumption. This practice has gained wider use along with the application of computer process control in soaking pit operations. Process computers programmed with heat transfer algorithms can instantaneously compute the temperature and degree of solidification in an ingot as it is being transported from the melt shop and while it is being heated to make decisions regarding the best heating practice to be followed and to determine when a pitload of ingots is ready to roll.

The best practice for heating each type and size of steel may be duplicated with modern control facilities. The fuel-air ratio control affects the flame characteristics and the furnace atmosphere. Flame characteristics have an important bearing on temperature distribution in the pit, and furnace atmosphere affects scale formation and character of the scale. By controlling the furnace atmosphere to advantage, the frequency of bottom making may be reduced, and minor surface defects on the ingot (from cooling, pouring or deoxidation practice) may be corrected or prevented from becoming more serious.

Operating Statistics—The principal operating data, aside from quality of heating and maintenance, relate to production output and fuel consumption. For comparative purposes, production is based generally on "live" pit area since there are so many different sizes and types of pits. The live pit-hole area is the area available in the pits on which ingots can be placed for heating. With proper loading this coverage of live pit area amounts to approximately 35 to 40 per cent. In the case of ingots with short transit time, the hearth coverage can be slightly higher than 40 per cent. An average month's practice is heating 29 to 98 metric tons of ingots per hour per 100 square metres of live pit-hole area (30 to 100 net tons per hour per 1000 square feet). This wide variation is due principally to the temperature at which the ingot is charged and to the type of steel. Other important factors causing the wide spread in pit furnace productivity are differences in the size and length of ingots, irregularity in mill operations, melting-shop pouring schedules, and the portion of live pit area occupied by the ingots.

The amount of fuel consumed per metric ton of steel heated varies from approximately 233 000 kJ to 2 330 000 kJ per metric ton (200 000 Btu to 2 000 000 Btu per net ton). This variation also is due largely to variations in the temperature of the charged steel. Other factors contributing to fuel economy are careful design, proper use of insulation, reduction of stack-gas losses with recuperators and regenerators, proper installation of controls, utilization of the proper air-fuel ratio, utilization of the maximum percentage of hearth, controlled metallurgical practices, and regularity of melting shop and rolling-mill schedules.

SECTION 3

REHEATING FURNACES

Furnace Types—Reheating furnaces are divided into two general classes:
1. Batch type.
2. Continuous type, including pusher-type, rotary-hearth-type, walking-beam-type, walking-hearth-type and roller-hearth-type furnaces.

Batch furnaces are those in which the charged material remains in a fixed position on the hearth until heated to rolling temperature. Continuous furnaces are those in which the charged material moves through the furnace and is heated to rolling temperature as it progresses through the furnace. Batch furnaces are the older type and are used for heating all grades and sizes of steel. However, small billets seldom have been heated in this type since the introduction of continuous furnaces. Batch furnaces are fired with either gaseous or liquid fuel, with preheated or cold air for combustion. The air may be preheated by regenerators and the furnace firing reversed from one end to the other, as in open-hearth furnaces. Batch furnaces, in which the air is preheated by recuperators, are not reversed and fir-

ing is continuous from one or both ends, depending upon the location of the gas outlet port (Figure 24—10). The steel to be heated in a batch furnace commonly is charged and drawn through front doors by a charging machine. Batch furnaces vary in size from those with hearths of less than a square metre (only a few square feet), with a single access door, to those about 6 metres (20 feet) in depth by 15 metres (50 feet) long, with five or six doors.

Pusher-Type Furnaces—Continuous pusher-type furnaces were designed initially for heating billets and small bloom sections. The hearths were relatively short in length and were sloped downward longitudinally towards the discharge end to permit an easy movement of billets through the furnace. In early designs, the furnaces were fired by burners located at the discharge end and the billets were heated by the hot gases flowing through the furnace above the top surface of the steel toward the charging end. Pushers were used to push forward the charge of cold billets. The flow of gas and steel in the furnace were counter-current. The

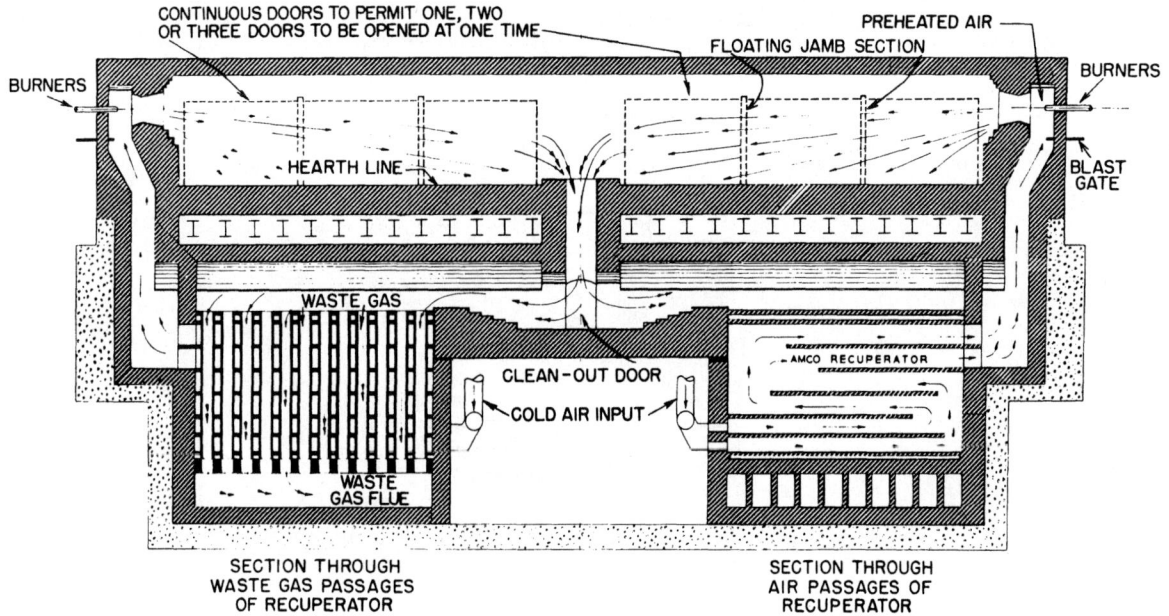

FIG. 24—10. Schematic longitudinal section through a recuperative batch-type reheating furnace. (Courtesy, Amsler-Morton Company.)

modern continuous pusher-type furnace has been altered in many respects from those of early design, although a large number of the older ones, particularly billet-heating furnaces, are still used. Longer furnaces generally are constructed now. Some have hearths about 24.5 to 32 metres (80 to 105 feet) long, with top and bottom firing, and contain preheating, heating and soaking zones. The hearths usually are constructed level. Recuperators are utilized to provide waste-heat recovery. A transition from the early designs to the modern five-zone slab heating furnace is illustrated in Figure 24—11.

The steel to be heated in a continuous furnace can be charged either from the end or through a side door. In either case, the steel is moved through the furnace by pushing the last piece charged with a pusher at the charging end. As each cold piece is pushed into the furnace against the continuous line of material, a heated piece is discharged by several methods, such as through an end door by gravity upon a roller table which feeds the mill, or pushed through a side door to the mill table by suitable manual or mechanical means or withdrawn through the end door by a mechanical extractor. Figure 24—12 shows a section through a modern triple-fired pusher-type furnace using counter current flow of gases and steel.

A distinctly different type of continuous reheating furnace is the rotary-hearth type, a cross-section of which is shown schematically in Figure 24—13. It is used frequently for heating rounds in tube mills and for heating short lengths of blooms or billets for forging. The rotary-hearth type permits the external walls and roof to remain stationary while the hearth section of the furnace revolves.

Walking-Beam-Type Furnaces—The general principal of operation of a walking-beam furnace is illustrated in Figure 24—14. The walking-beam concept for conveying materials is not new. However, in the case of furnaces, the idea at first was applied successfully only to furnaces that operated at maximum temperatures of about 1065°C (1950°F), compared with reheating furnaces that must heat steel to temperatures up to 1315°C (2400°F). The early furnaces employed alloy steel walking beams that were exposed directly to the heat of the furnace and were subject to heat corrosion and warping and did not retain enough strength at the higher temperatures. Difficulties were met in trying to seal the slots between the walking beams and the fixed portion of the hearth, so that control of furnace pressure and atmosphere presented problems.

Walking-beam furnaces that operated successfully in the higher ranges of temperature began to be used in Europe in the early 1950's, being first applied to the reheating of billets and blooms. They are now used here and abroad to reheat slabs as well as billets and blooms. The walking beam may consist of water-cooled steel members topped with refractories in such a manner that only the refractories are exposed directly to the heat of the furnace, as shown in Figure 24—15. Alternatively, the beams and supports may be constructed of water-cooled tubular sections (with "buttons" on the top surfaces to keep the hot steel from direct contact with the water-cooled tubes) as shown in Figure 24—16. Walking-beam furnaces can be designed for side or end charging and discharging. Either hydraulic or mechanical methods can be used to actuate the beams. Cross firing with side-wall burners above and below the stock being heated (Figure 24—16) and heating with radiant-type burners in the furnace roof (Figures 24—15 and 24—17) or in both the roof and below the stock have been employed.

ONE ZONE – END CHARGE, SIDE DISCHARGE

TWO ZONE – SIDE CHARGE, SIDE DISCHARGE

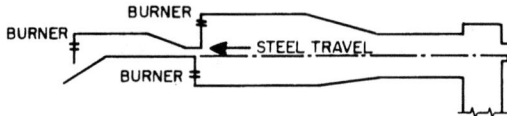

THREE ZONE FURNACE – END CHARGE, END DISCHARGE

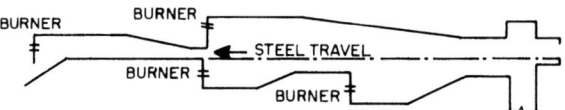

FOUR ZONE FURNACE – END CHARGE, END DISCHARGE

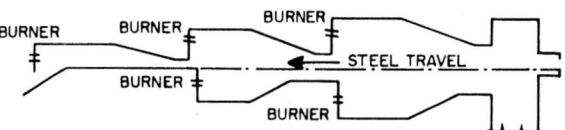

FIVE ZONE FURNACE – END CHARGE, END DISCHARGE

FIG. 24—11. Diagrammatic steps in the evolution of the modern five-zone pusher-type slab-heating furnace from earlier designs.

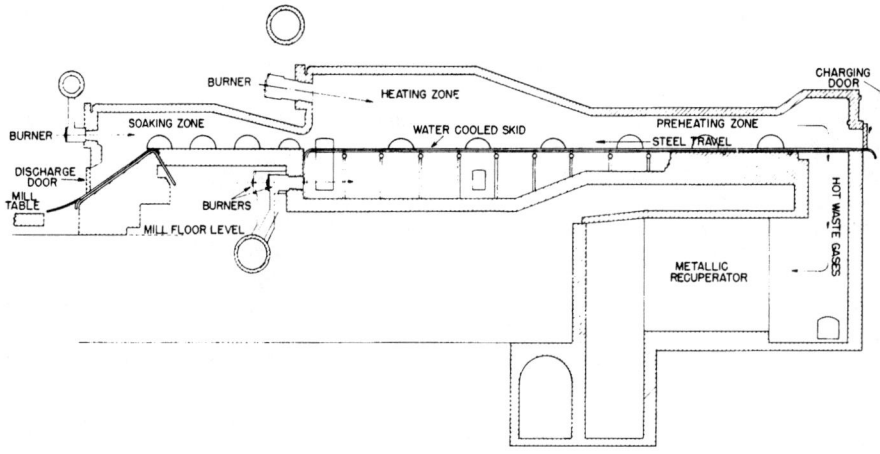

FIG. 24—12. Schematic longitudinal section through a three-zone counter-current fired pusher-type continuous reheating furnace. (Courtesy, Rust Furnace Company.)

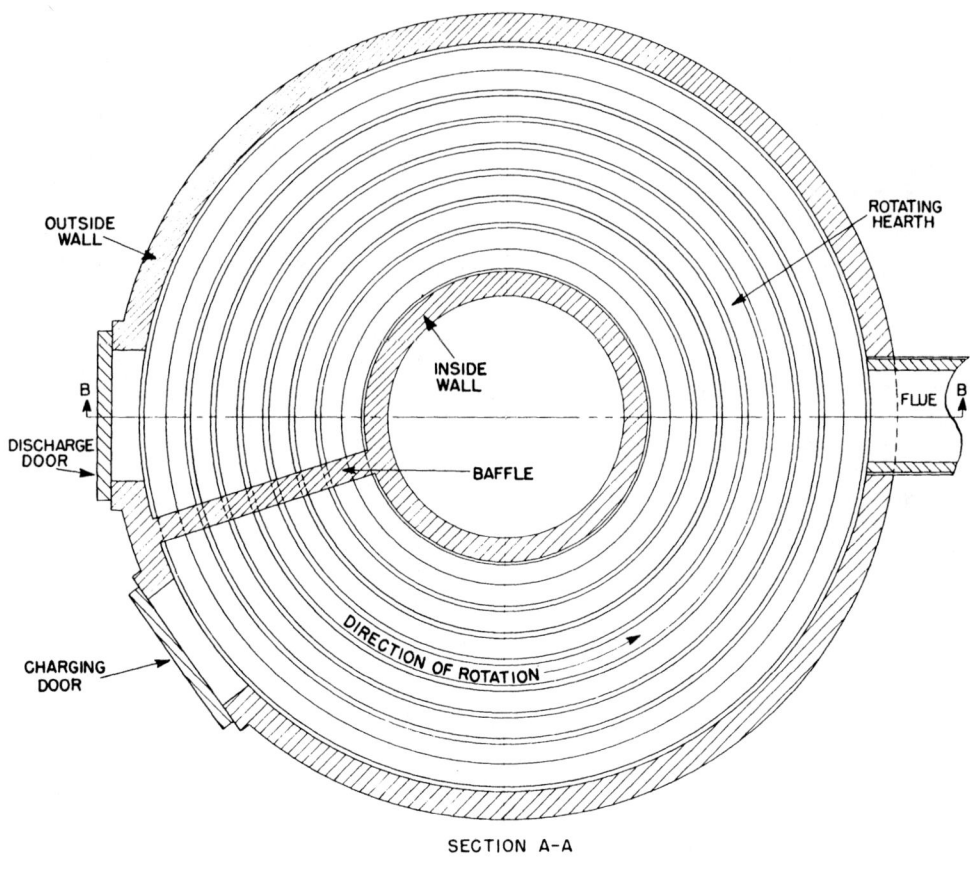

OUTSIDE
WALL

ROTATING
HEARTH

INSIDE
WALL

FLUE

B

B

DISCHARGE
DOOR

BAFFLE

DIRECTION OF ROTATION

CHARGING
DOOR

SECTION A-A

SECTION B-B

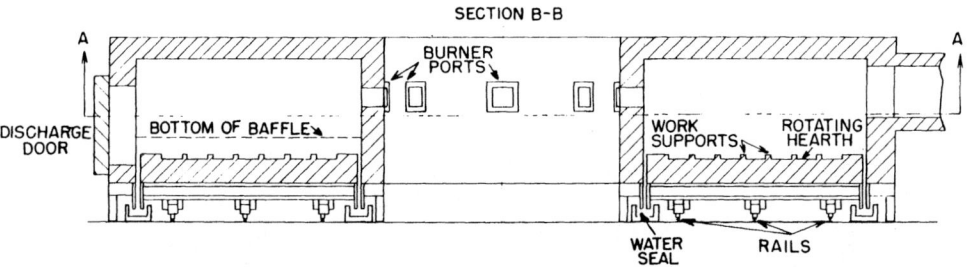

A

A

DISCHARGE
DOOR

BURNER
PORTS

BOTTOM OF BAFFLE

WORK
SUPPORTS

ROTATING
HEARTH

WATER
SEAL

RAILS

FIG. 24—13. Schematic arrangement of a relatively small continuous rotary-hearth heating furnace. The plan view (above) shows a section above the hearth near the bottom of the baffle. Larger furnaces of this type have burners firing through both inside and outside walls above the hearth, or very large furnaces (up to 100 feet in diameter) utilize multiple heating zones that can be fired either with roof-type burners or burners located in the vertical portion of saw-tooth roof construction.

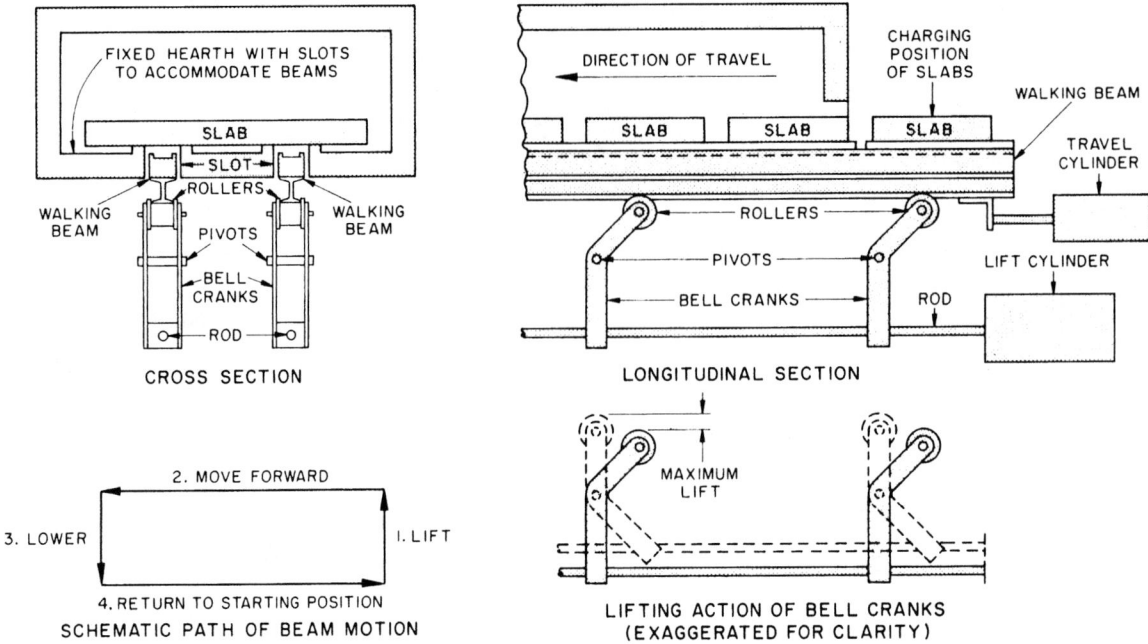

FIG. 24—14. Schematic diagram (not to scale) showing a method by which walking beams can operate to advance slabs along the hearth of a heating furnace. In actual furnaces, water seals beneath the hearth prevent escape of furnace gases (or air infiltration) through the openings between the walls of the slots in the hearth and the sides of the walking beams. In these diagrams, slabs at rest are supported on raised ridges on the hearth (upper left). The hydraulic lift cylinder (upper right) moves the rod to the right to actuate the bell cranks. Rollers on the upper arms of the bell cranks move in an arc to a higher position (lower right) raising the walking beam they support along with the slabs that are lifted above the hearth ridges. The hydraulic travel cylinder holds the beams in the proper horizontal position until the beams are raised: it then exerts a push to the left to move the beams with their load of slabs forward by a preset amount. The lift-cylinder rod then moves to the left to lower the slabs in their new advanced position onto the hearth ridges. The travel cylinder then pulls the walking beams back to their starting point. As shown at the lower left, the path of motion of the beams in this system is rectilinear: other modes of actuation can result in a circular or elliptical motion to effect the same result.

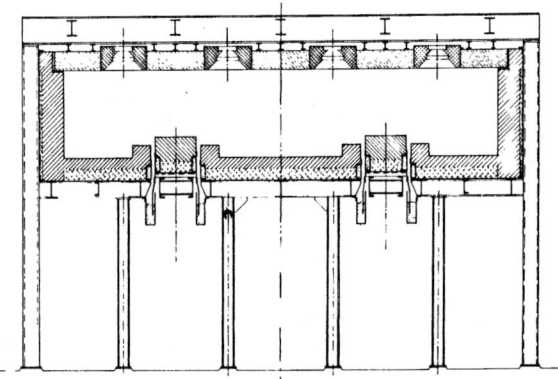

FIG. 24—15. Cross-section of a walking-beam furnace, heated by radiant burners in the roof. Note plates extending from sides of walking beams into water troughs beneath the furnace to seal the slots. (Courtesy, Salem-Brosius (Canada) Ltd.)

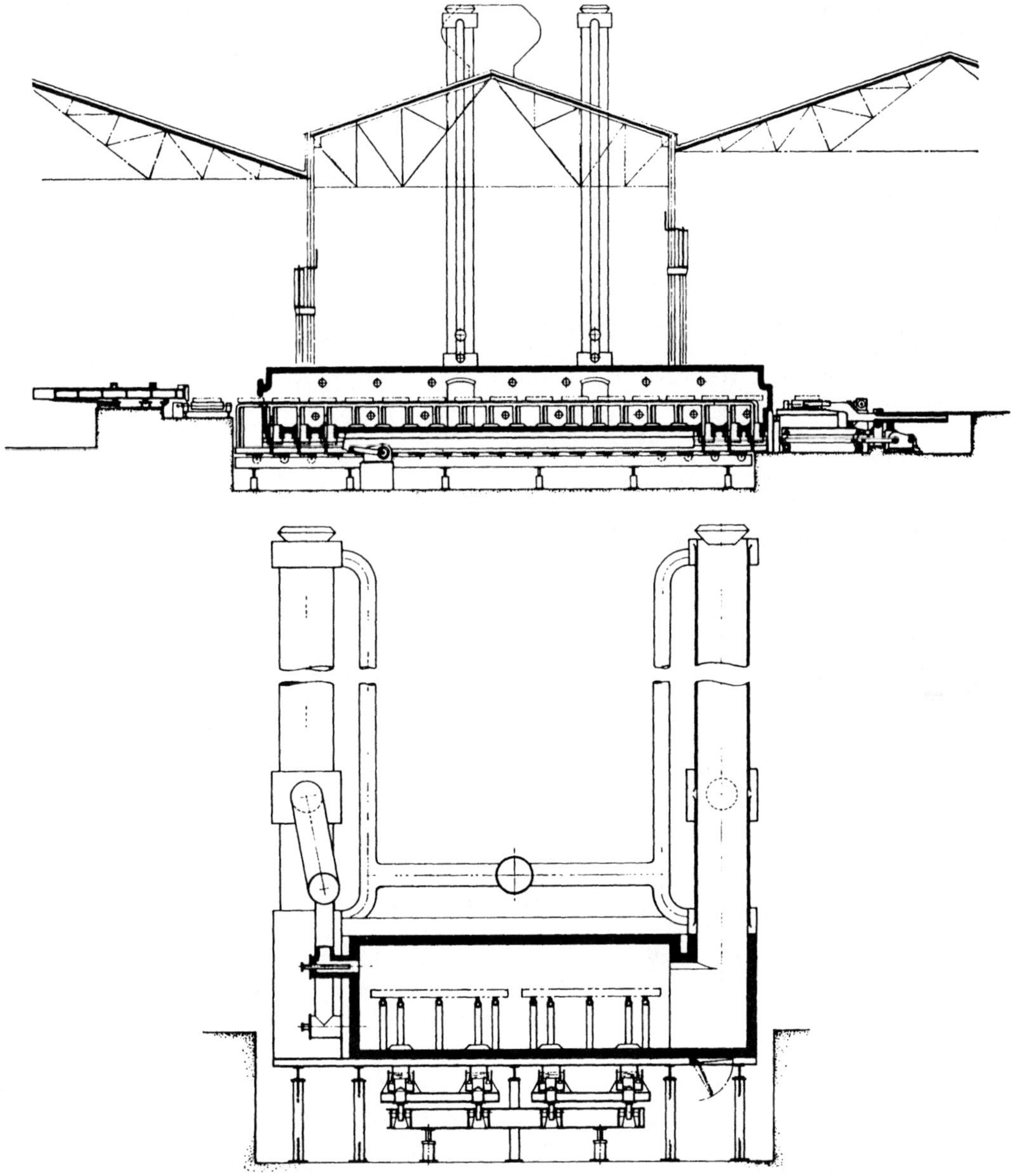

FIG. 24—16. (Above) Schematic longitudinal section of a reheating furnace equipped with walking-beam conveyor, showing pusher-type charging machine outside the furnace at left and slab-extracting mechanism at right. (Below) Cross-section (not to same scale) of the same furnace, showing how slabs are supported and moved by water-cooled tubular members. Note side-wall burners firing above and below slabs and stack-type recuperators for waste-heat recovery. (Courtesy, Surface Combustion Div., Midland-Ross Corp. and "Iron and Steel Engineer."

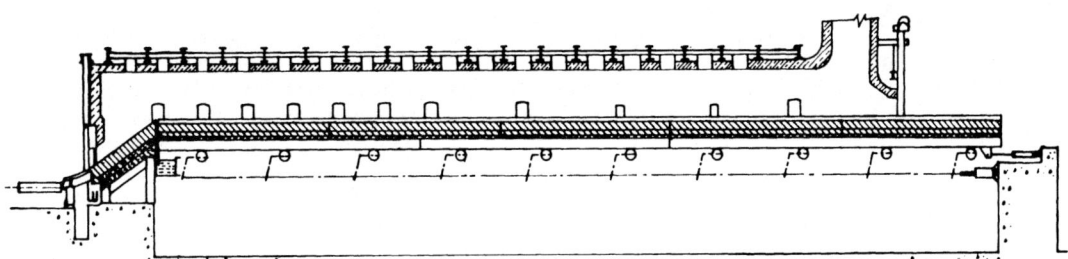

FIG. 24—17. Longitudinal section through a radiant-roof walking-beam furnace. (Courtesy, Salem-Brosius (Canada) Ltd.)

Walking-Hearth-Type Furnaces—In a walking-hearth furnace, travel of the work through the heating chamber follows the same general path as in the walking-beam furnace. The main difference in method of conveyance in these two furnace types is that, in the walking-hearth furnace, the work rests on fixed refractory piers. These piers extend through openings in the hearth and their tops are above the hearth surface during the time when the work is stationary in the furnace. The furnace gases can thus circulate between most of the bottom surface of the work and the hearth.

To advance the work toward the discharge end of the furnace, the hearth is raised vertically to first contact the work and then raise it a short distance above the piers. The hearth then moves forward a preset distance, stops, lowers the work onto its new position on the piers, continues to descend to its lowest position and then moves backward to its starting position toward the charging end of the furnace to await the next stroke.

Roller-Hearth Reheating Furnaces—Roller-hearth furnaces, although used in many types of heat-treating operations, have not been used extensively for reheating steel to hot-working temperatures in conventional steelmaking practices. However, the advent of continuous casting has prompted studies of how this type of furnace might be used to advantage in heating much longer slabs than would be practical in a pusher-type or walking-beam-type furnace.

Figure 24—18 shows a cross-section of a specially designed roller-hearth furnace for reheating slabs immediately after continuous casting. Slabs move through the furnace in the direction of their length. Main burners fire across the furnace above and below the slab to heat the top and bottom surfaces, and auxiliary burners fire through the sidewalls in line with the slab edges to supply additional heat for the edges and corners.

GENERAL CONSIDERATIONS IN FURNACE-TYPE SELECTION

Batch-Type Furnaces—Some of the particular advantages of batch-type furnaces are:

1. They provide means for heating steels of various types and sizes, which can be heated more properly in separate batches than when mixed with other types in the same furnace, especially when specific heating practices are required.

2. They are suitable as a reservoir for holding hot steel directly from the primary mill for later rolling in the finishing mills.

3. They can be operated to heat steel to temperatures above 1315°C (2400°F) more satisfactorily than a continuous furnace. Steel can be given a "wash heat," when desirable, without trouble from the pieces sticking together. (Steel is sometimes "washed" or heated until an earlier oxide scale is melted and a new scale formed, to reduce surface defects.)

Some of the important disadvantages of batch furnaces are:

1. High capital investment per unit of production.

2. Low hearth area efficiency. That is, the hearth in the conventional type is not utilized fully because of interference of door jambs and clearance required in-

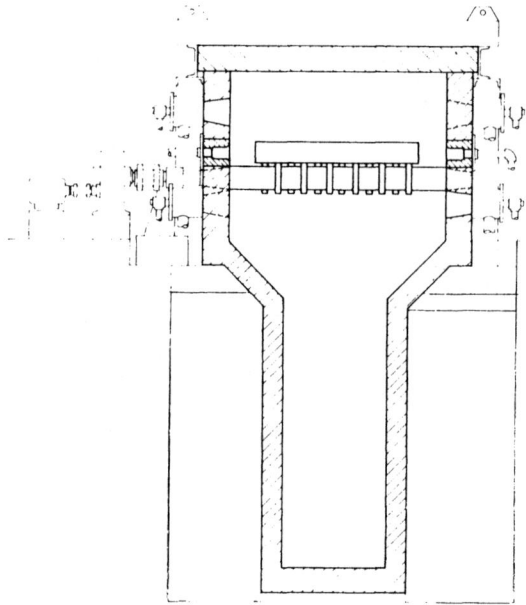

FIG. 24—18. Cross-section of a roller-hearth-type furnace for reheating continuously cast slabs immediately after casting to equalize temperature for rolling. Note over- and under-firing main burners and auxiliary edge-heating burners. (Courtesy, Surface Combustion Div., Midland-Ross Corp. and "Iron and Steel Engineer.")

side the furnace for handling the stock with charging machines.

3. High man-hours per ton of heated steel.

4. Lack of flexibility for heating steel slowly in the lower temperature range. Furnaces must be cooled to charge high-carbon steel and some alloy steels.

5. Length of pieces to be heated is limited by tendency to bend when they are removed.

Pusher-Type Furnaces—The advantages of the pusher-type continuous furnaces are:

1. High production per dollar investment.

2. Low man-hours per ton of steel heated.

3. Greater ease in charging and drawing steel.

4. High hearth area efficiency.

5. Better means for controlling the rate of heating at all temperature levels. Gradual rise in temperature permits charging all grades of cold steel without cooling furnace.

6. Less trouble from temperature inequalities between each succeeding piece drawn.

7. High production per square foot of ground space occupied.

8. Can be built for any reasonable length of piece to be heated, resulting in higher mill yield, because of fewer crop ends from lengths beyond batch-furnace possibilities.

Some of the important disadvantages of pusher-type continuous furnaces are:

1. Lack of flexibility for heating efficiently small orders of different lots of steel or thicknesses. Heating time must be increased to suit the heating cycle of the

piece, requiring lower heat transfer rate which may be detrimental to the heating of the other grades and sizes in the furnace.

2. Trouble from water-cooled skid maintenance, and from building up of scale on hearth, which results in piling of steel in furnace, due to climbing of pieces over each other, particularly if square contact is lost between adjacent pieces, when pressure is applied by the pusher.

3. Face of contacting surface of stock must be square to prevent piling.

4. Expensive to empty furnace at end of schedule.

5. Difficulty in pushing mixed sizes through furnace.

6. Where water-cooled skids are used, thickness of product is limited to 300 to 350 mm (12 to 14 inches).

7. Presence of visibly colder "stripes" on the hot steel caused by contact with water-cooled skids, unless adequate soaking time is provided.

Rotary-Hearth Furnaces—Some of the important advantages of rotary-hearth furnaces are:

1. They eliminate either the manual labor required for rolling rounds forward on horizontal or moderately sloped hearths, or the disadvantages of excessively sloped hearth in continuous furnaces.

2. They have better means for controlling the rate of heating at all temperature levels than batch-type furnaces.

Some of the important disadvantages of the rotary-hearth furnace are:

1. High capital cost per unit of production.

2. Low hearth area efficiency.

3. Large ground space required per unit of production.

4. Maintenance of hearth seals and wall refractories at the hearth level.

Walking-Beam Furnaces—Some of the important advantages of walking-beam furnaces are:

1. Pieces can be separated from one another on the hearth, eliminating stickers.

2. Pile-ups are reduced.

3. Furnace retention time is reduced.

4. Furnace can be emptied easily from either end by activating the beam mechanisms.

5. No line contact with water-cooled skids occurs, thus eliminating troublesome skid marks (also called "stripes" or "black marks") on the heated stock.

6. There is no rubbing or friction between the stock and the hearth, minimizing hearth wear and stock damage.

7. By selecting the proper number of walking beams, better hearth utilization can be obtained when charging mixed sizes.

8. The potential for the extension of overall furnace length to improve the utilization of furnace waste gases and reduce fuel consumption. A similar advantage is not available with other furnace types because of limitations on overall furnace length.

Some of the disadvantages of walking-beam furnaces are:

1. Complexity of the walking-beam system.

2. Inherently higher cost per ton of capacity.

3. Precautions must be taken to avoid problems from scale that drops off the stock being heated.

4. Maintenance of hearth seals and hearth refractory.

In general, the same advantages and disadvantages of the walking-beam furnace apply to walking-hearth furnaces.

Roller-Hearth Furnaces—Some advantages of the roller-hearth design for reheating furnaces are:

1. Ability to handle very long pieces.

2. Zone control is simpler when cross-firing is employed.

3. Stock suffers little mechanical damage.

4. Skid marks are avoided.

5. Furnace is self-emptying.

Some disadvantages of the roller-hearth furnace are:

1. High initial cost per unit of capacity.

2. Water cooling of the rollers results in high heat losses, unless adequately insulated.

3. Although relatively narrow, roller-hearth furnaces must be considerably longer than pusher-type or walking-beam furnaces of the same capacity.

Some features in both batch and continuous furnaces are worthy of note:

1. Regenerators or recuperators act as a reservoir of heat supply, which is especially valuable for efficient soaking of steel.

2. Continuous furnaces provided only with top firing require longer hearths for equal production than those with top and bottom firing, but, in the case of pusher-type furnaces, do not require a special soaking zone to eliminate cold spots on the work caused by contact with water-cooled skids.

3. Continuous furnaces with single-zone firing have higher scale losses and greater tendency to cause decarburization of high-carbon steel than the top-and-bottom fired furnaces, since the steel is in the furnace longer. Decarburization is caused primarily by hydrogen and water vapor combinations in furnace gases, and increases almost directly with the time the steel is at elevated temperatures. Free air and carbon dioxide in the furnace atmosphere cause decarburization to a lesser degree. The scaling of steel is practiced sometimes deliberately to remove the decarburized surface layer.

4. A level hearth in a continuous furnace eliminates the stack effect of hearths sloping upwards towards the charging end. This stack effect draws cold air into the furnaces at the hot end causing higher fuel consumption and scale losses.

5. Batch-type furnaces used for heating in a production line require supplementary furnaces for preheating certain grades of alloy and high-carbon steels for transfer into the hotter furnaces. The preheating zone of a continuous furnace makes this unnecessary.

6. Side-discharge continuous furnaces have less air infiltration at the hot end than end-door discharge furnaces. End-door discharge of the usual gravity type induces cold air into the furnace by the stack effect at the discharge section of the furnace. End-door discharge, however, is mechanically simpler for removing the heated stock; particularly slabs and heavy blooms.

Furnaces have been developed to raise the temperature of steel for hot shaping by exposing the piece to be heated to intense radiant energy from gas burners of special design, set very close to the path of the steel. This method of heating has been very effective in obtaining fast heating for small round sections. The

FIG. 24—19. Barrel-type furnaces in tandem for continuous reheating of pierced seamless tubes for sizing. Conveyor rolls between adjacent furnaces support the work in its travel. Temperatures are automatically controlled by radiation-type pyrometers.

barrel-type furnaces in tandem shown in Figure 22—19 are used for reheating pierced tubes for sizing in seamless-tube mills. This method has been applied to soaking pits and reheating furnaces.

Operating Statistics—The production rate of furnaces varies widely. Batch furnaces, as a general class, produce much less tonnage per furnace operating hour than the continuous type. While batch furnaces often are used for heating hot charges, as in structural and rail mills, continuous furnaces usually are used for heating cold steel. Batch furnaces heating cold charges for rolling mills are designed to heat from 4.5 or 5.5 metric tons up to 22.5 metric tons (5 or 6 net tons up to 25 net tons) per furnace hour. When heating a hot charge, the production is much higher, and the rate depends largely upon the temperature of the charged material and the size of the furnace.

The production of a continuous furnace depends principally upon the hearth size of the furnace and whether the material is heated from more than one side. Many continuous furnaces heat over 45 metric tons (50 net tons) of steel per furnace hour. A modern three-zone slab-heating continuous pusher-type furnace 6 metres (20 feet) wide, with an effective hearth length of 24.4 metres (80 feet), top and bottom fired, is capable of heating 100 metric tons (110 net tons) of slabs per hour to a rolling temperature of 1230°C (2250°F). Five-zone furnaces are capable of heating rates in excess of 270 metric tons (300 net tons) per hour. The productive rate of a furnace for a fixed thickness of steel is directly proportional to the surface exposed to the flame, other conditions remaining the same. For comparison of furnace performance, the relative number of pounds heated per effective square metre or square foot of hearth area per hour is used as a common factor. By this comparison, furnaces heat from 146 to 730 kilograms per square metre (30 to 150 pounds per square foot) of hearth area per furnace hour. The variation in heating capacity is due principally to the difference in the total surface of the steel exposed to heat in the furnace, which the hearth area does not measure. Other factors are the temperature of the heated product and the grade of steel. In many batch furnaces, the material receives heat from three sides and in continuous furnaces from two sides. Heating from more than one side decreases heating time and, therefore, increases production rates. The production and fuel rates of furnaces are affected by continuity of operation. A furnace will not provide high production rates and fuel economy if it is operated for only short intermittent periods. It also must be fully loaded to provide optimum production and fuel economy.

The fuel consumption of reheating furnaces varies from 349 000 to 5 230 000 kilojoules per metric ton (300 000 to 4 500 000 Btu per net ton). When hot steel is heated, very little additional heat may be required to attain rolling temperature. If the steel charged is at a temperature of 980°C (1800°F), only about 163 000 kilojoules per metric ton (140 000 Btu per net ton) must be added in the furnace to heat to 1230°C (2250°F) while with cold steel about 814 000 kilojoules per metric ton (700 000 Btu per net ton) must be added in the furnace. If the hot steel requires 581 000 kilojoules per metric ton (500 000 Btu per net ton) input to the furnace and only 163 000 kilojoules (140 000 Btu) are absorbed by the steel, the furnace fuel efficiency is

$$\frac{163\,300}{581\,000} \times 100, \text{ or 28 per cent.}$$

If cold steel is heated with a fuel consumption of 2 325 000 kilojoules per metric ton (2 000 000 Btu per net ton), as obtained in many modern furnaces, the furnace fuel efficiency is

$$\frac{814\,000}{2\,325\,000} \times 100 = 35 \text{ per cent.}$$

The best fuel efficiency is obtained in continuous furnaces that are relatively long, have hearths fully covered with steel, and are operated at a high rate of continuity. Reheat furnaces are being integrated into the continuous casting facilities for slabs, blooms, and billets. When these large pieces are charged hot into the reheat furnace, the result is a moderate fuel savings. Process control instrumentation has been developed to adjust separate zone temperatures to match the heating capacity of the furnace to the rolling capacity of the mill, thereby reducing fuel consumption.

In both batch and continuous furnaces, the length of gas travel and the velocity of the gases through the furnace have an important effect on fuel economy. An excessive velocity of furnace gases tends to increase the exit temperature of the gases leaving the furnace and, therefore, low fuel efficiency results. Fuel efficiency is improved by providing features in design that reduce the inherent furnace losses. The greatest losses usually are represented by the sensible heat carried away in the stack gases. Reduction of this loss can be obtained by the installation of recuperators, regenerators or waste-heat boilers, and by designing the furnace cross-sectional area for proper gas velocities. Automatic fuel-air ratio and furnace-pressure control reduces stack losses by reducing the volume of gases from excess air for combustion, or from air infiltration. Radiation losses are reduced by insulation of the furnace walls, roof and hearth. Losses from water-cooled elements of continuous furnaces can be reduced by eliminating any unnecessary exposure of them to hot furnace gases, and by use of refractory covering.

CHAPTER 25

Production of Steel Plates

SECTION 1

PLATE-MILL PRODUCTS

Plates are classified, by definition, according to certain size limitations to distinguish them from sheet, strip, and flat bars. According to this classification, plates generally are considered to be those flat, hot-rolled, finished products that come within the following dimensional and weight limitations:

Width		Thickness		Weight	
mm	in.	mm	in.	kg/m²	lb/ft²
Over 203 to 1219, incl.	Over 8 to 48, incl.	5.842 and thicker	0.2300 and thicker	46.97 and heavier	9.62 and heavier
Over 1219	Over 48	4.572 and thicker	0.1800 and thicker	26.72 and heavier	7.53 and heavier

There are a few exceptions to the above classification. Flat, hot-rolled, semifinished products such as slabs, sheet bar and skelp are not classed as plates, although their dimensions and weight may be within the foregoing limits. Also, the dimensional limitations for stainless-steel plates differ slightly from those listed above (see AISI Steel Products Manual entitled "Stainless and Heat-Resistant Steels").

Principal end uses for plates include such fabricated structural and plate products as bridges, trestles, storage tanks, pressure vessels and penstocks; railroad freight and passenger cars; shipbuilding; line pipe, industrial machinery and equipment; weldments and many special applications.

Plates are produced by hot rolling, either from reheated slabs or directly from ingots. The terms applied to plates to differentiate between the several kinds (without regard to the chemical composition of the steel) are based upon the types of mills used in their production, e.g., **sheared plate** or **sheared mill plates** when rolled between straight horizontal rolls and later trimmed on all edges. Mill edge is the normal edge produced by hot rolling between horizontal finishing rolls: **mill edge plates** have two mill edges and two trimmed ends. Plates are called **universal plates** or **universal mill plates** (abbreviated **U.M. plates**) when rolled simultaneously between both horizontal and vertical rolls and trimmed on the ends only; grooved rolls are sometimes substituted for the plain horizontal rolls

used for universal plates.

Carbon-steel **rolled floor plates** are also flat, hot-rolled, finished steel products that come under the plate classification. Floor plates are hot-finished in the final pass or passes between one or more pairs of rolls. One roll of each pair has a pattern cut into it so that one surface of the plate passing between the rolls is forced into the depressions on the pattern roll to form a raised figure at regular intervals on the surface of the plate. Individual floor-plate patterns are produced exclusively by each manufacturer, the patterns differing in both dimensions and appearance.

The majority of steel plates produced are rolled from carbon steels. In addition to carbon steels, the complete range of steel-plate production includes high-strength, alloy, and stainless steels.

The as-rolled product of plate mills is rectangular in form, and a considerable proportion is shipped as **rectangular plates. Circular** and **semi-circular plates** and **sketch plates,** including **rings,** are produced by shearing or gas cutting hot-rolled rectangular plates to specified shapes.

A considerable proportion of plate-steel production is subjected to some form of heat treatment prior to shipment from the mill, especially in the case of alloy-steel plates, to develop the desired mechanical properties. Heat treatments employed include: annealing, spheroidize annealing, stress relieving, normalizing, accelerated cooling, quenching, and tempering. The principles and purposes of carrying out these treatments are discussed in Chapter 41. Modern heat-treating installations designed specifically for plates are described in Sections 3 and 4 of this chapter.

The same rolling mills and auxiliaries are used to produce carbon, high-strength low-alloy and alloy-steel plates. The basic quality requirements for these steels are described in ASTM A 6, "General Requirements for Rolled Steel Plates, Shapes, Sheet Piling and Bars for Structural Use," and ASTM A 20, "General Requirements for Steel Plates for Pressure Vessels." The quality requirements for stainless-steel plates are described in ASTM A 488, "General Requirements for Flat-Rolled Stainless and Heat-Resisting Steel Plate, Sheet and Strip."

SECTION 2

PLATE-MILL OPERATIONS

The sequence of operations for plate mills is covered by the following general subdivisions:

1. Scarfing or slab preparation
2. Heating slabs for rolling
3. Descaling
4. Rolling
5. Leveling or flattening
6. Cooling
7. Shearing or cutting

Steps 2 through 7 in the production of steel plates will be discussed in the order in which they are listed above; Step 1 was discussed in Chapter 21.

HEATING SLABS FOR ROLLING

Batch-Type Heating Furnaces—Early plate-producing plants were laid out so that the slabs from the primary mills could be transported to the heating furnaces of the plate mill as hot charges. The continuous-type heating furnace had not yet been developed, and batch-type furnaces were used. Such layouts were motivated by the considerations: (1) that the small ingots then generally produced were not so susceptible as present-day large ingots to being the source of major surface imperfections that carried through to the finished product; (2) that economical, intermediate-conditioning processes had not yet been developed; and (3) that the economies of increased production rates from hot charges would more than offset the surface rejections which might be incurred.

Heating in a fuel-fired batch furnace is essentially a process of transmitting heat by radiation and convection from the combustibles to the top slab surface and by conduction through the thickness of the slab to the bottom surface and the hearth. Even where the temperatures of both the hearth and the charge are high in relation to the drawing temperature at the beginning of a cycle of charges and draws, the heating rates per unit area of furnace hearth are low in contrast with continuous slab-heating furnaces. When the cycle involves cold charges, the initial abstraction of heat from the hearth, which must be restored later in the cycle, slows the process to the point where continuous operation is impracticable, if the heating capacity was installed initially for hot charges. The process of heating from top to bottom through the thickness, in itself, suggests the existence of a temperature differential which can be minimized only by soaking or by turning slabs over partway through the heating cycle. Changes which have occurred in the industry over the years have created a trend in the direction of cold charges and have confronted many plants with the necessity of determining whether to operate mills on an intermittent basis, add heating capacity, or build new mills.

Continuous-Type Heating Furnaces—Most of the existing continuous slab-heating furnaces were designed for and are operated on cold charges. The slabs are charged into the low-temperature end of the furnace and progress toward the high-temperature end by being pushed progressively forward on water-cooled skids. Walking-beam furnaces are replacing push furnaces on most modern plate mills. Burners located above and below the skid level furnish heat by radiation and convection to both the top and the bottom surfaces of the slabs until they reach the high-temperature zone, which is, in effect, a soaking zone. Only top burners are provided for the high-temperature zone. The slabs are at approximate rolling temperatures when they arrive in this zone, and elimination of **skid marks** (visibly colder "stripes" on slabs caused by contact with the water-cooled skids) and other temperature non-uniformities, plus making up radiation losses to maintain the desired furnace temperature, are the principal heat requirements to be met.

Furnace Control—Furnace-control equipment is intended not only to effect fuel economies, but also to improve the quality of heating. Slab heating is improved if slabs are brought to the plastic state most satisfactory for rolling, with a minimum of temperature variation in each slab without any part ever attaining an excessive temperature, and with the production of a scale coating conducive to the best surface finish. The thickness and type of scale formation is dependent on temperature, time at temperature, composition of steel and availability of oxygen. The temperature that must be used is dependent upon the required plastic state of the heated steel; time of heating is controlled by the heating capacity of the furnace; therefore, control of the fuel-air ratio is the only flexible practical determinant of the types of scale that may be formed, which may vary from extremely adherent to loose. Automatic furnace-control equipment was not available when the bulk of existing batch furnaces was designed. Some were modified later to permit installation of controls. Several furnaces, such as at the Homestead Works, constructed during World War II, were designed for and equipped with automatic furnace controls. The preponderant number of batch furnaces still in operation are dependent on manual control. Slab-heating furnaces of the continuous type either had furnace controls included in the original design or they were added later, inasmuch as such furnaces are very adaptable to automatically controlled operation. Scale formation, consequently, can be controlled closely with respect to both thickness and type.

DESCALING

Prior to the introduction of the modern hydraulic pressure sprays, various expedients were resorted to for primary scale removal. Moistened burlap thrown on the slab at the entry side of the rolling mill embodied the same scale-removing principle of the more common use of salt. In both instances the materials were used to get moisture between the work rolls and the stock, where it vaporized instantaneously with explosive effects and removed the scale. Of all flat, hot-rolled products, hot-rolled sheet and strip must meet the most stringent surface finish requirements, and it was natural that scale-removal methods and equipment would first reach their maximum development in connection with those product classifications which were rolled on continuous mills. The trend of development moved in the direction of the use of a leading roll stand in which light drafts were taken for the initial breaking of the scale. Hydraulic sprays, impinging on both top and bottom surfaces, were placed on the delivery side of the scale breaker and on the entry side of the roughing stands and the first finishing stand. Hydraulic pressures were increased with various installations from the 4137 to 5516 kPa (600 to 800 psi) which prevailed for general plant hydraulic equipment, up to 10 342 kPa (1500 psi). It was found, however, that with proper heating, spray pressures of 6895 kPa (1000 psi) per square inch were adequate for the production of surface finishes to meet consistently the most stringent requirements. The continuous hot-strip mill method of scale removal was, therefore, adopted for new plate-mill installations.

PLATE ROLLING

Plate-Rolling Variables—The rolling of plates is subject to a number of variables. Control of temperature, if not properly effected, may cause variations in mechanical properties in steel from the same melt or lot of material. Inherent characteristics of equipment permit one mill to roll plates to closer limits than another with respect to thickness, weight, width, length, camber (the greatest deviation from a straight line along a longitudinal edge), and flatness of plate after mill finishing-pass delivery and ordinary roller-leveler treatment.

Other variables affecting rolling are the **mill spring** (the total looseness under load of all mechanical parts) between the roll necks and the housings, which makes it necessary to increase spring allowances for light mill construction as compared with heavier or more rigid mill construction for the same loading.

Bending of rolls and resultant **crown** in the plates are also variables which affect the accuracy of rolling. When the rolls are subjected to the separating force of a plate being rolled between them, they are equivalent to beams supported at the bearing centers and subjected to uniformly distributed load over the length. Uniformly loaded beams have maximum deflection in the center and a similar condition exists in the rolls. For the same drafts, small-diameter rolls are subject to less separating force than large-diameter rolls but the latter have greater resistance to deflection. The minimum roll deflection and plate crown are found in a mill with small-diameter work rolls backed up by large-diameter rolls. Crown (increase in thickness of the rolled center of the plate over its edges) is related to the amount of bending or deflection in the rolls. A minimum crown and uniform thickness of plates are desirable for many applications, especially forming operations.

Roll wear is an important factor in rolling plates. There is only one point in the arc of contact of the rolls with the stock at which the linear speed of the rolls and the stock is identical. As mentioned in Chapter 22, this is called the neutral point. There is backward slippage of the stock on the entry side and forward slippage on the delivery side of the neutral point. Slippage contributes to roll wear, and as the central portion of the roll

face comes in contact with the stock on all widths, it is subject to the most wear and gradually develops a hollow contour across the roll face, accompanied by a surface roughness.

The effect of roll wear on plate crown and surface finish can be visualized readily. An additional effect relates to plate flatness and becomes more pronounced with the thinner plate gages. Roller levelers operate on the principle of imposing a slight extension, through the bending-roll action, on each increment of constricted plate surface as it goes through each bending action in the leveler.

Because edges of rolled plate cool faster than the center, the desired shape for entry into the leveler, particularly for light gages, is one with a slight edge wave. A mill which is too full will deliver light-gage plates with center buckles that require excessive edge extensions in the leveler to secure flat delivery. Conversely, a hollow mill will deliver light-gage plates with more edge wave than can be compensated for by the limited extension of the central portions of the plate which can be accomplished in the leveler.

Temperature variation in the plate from the front end to the back is also a problem in rolling plates. Every roll pass involves a time interval for its accomplishment. In initial passes, when the stock is relatively thick and short, the time interval has no practical effect in creating a temperature differential from front to back. If the stock is being reduced to wide, light gage, and is elongated to long, thin lengths in the last several passes, the time interval will cause temperature differentials of such magnitude that the consequent gage differential must be anticipated in mill settings in order to retain the gage within tolerances. Differentials are minimized by shortening the time required for rolling.

LEVELING (FLATTENING)

The amount of flattening required by the plate after leaving the rolls generally increases with decreasing thickness of the plate. The effectiveness with which flattening can be accomplished by a leveler increases with decreasing roll diameters and spacing and with increasing temperatures. For light-gage plates, therefore, a roller leveler with small-diameter rolls and with close roll spacings should be located near the mill finishing stand.

Heavier-gage plates, at corresponding temperatures, require greater strength and rigidity of rolls than lighter-gage plates. The heavier gages usually are delivered from the rolls at higher finishing temperatures and require accelerated cooling prior to leveler entry or the leveler must be located at a greater distance from the finishing stand to permit proper cooling before leveler entry. A compromise in roll diameters and leveler location may be satisfactory if the range of gages produced does not include extremes.

Levelers with small-diameter bending rolls backed up by short, rigidly supported backing-up rolls are used for cold releveling. They provide the requisite combination of severe bending and roll rigidity.

COOLING

Plates delivered from the roller leveler must be

cooled uniformly to avoid localized stresses that again would set up permanent localized distortions. As more heat is given up by the plate on contact with another metallic surface than by exposure to the atmosphere, it is necessary that cooling conveyors be so constructed that only momentary, staggered contacts are made with the bottom surface of the plate and cooling of the bottom side is effected primarily by radiation, similarly to the top surface. This cooling condition should be maintained until temperatures are reached at which the plates are no longer susceptible to distortion as a result of non-uniform cooling.

SHEARING AND CUTTING

Shearing—Normally, plates are sheared above atmospheric temperature. The most important factor in shearing plates to the desired size is the proper allowance for shrinkage. The cooling rate of plates decreases as they approach atmospheric temperature, consequently, extremely long cooling lines would be required to permit plates to cool sufficiently for accurate shearing to be done without the necessity for making allowance for shrinkage.

Manual layout of plates for shearing introduces some degree of deviation from theoretical accuracy. Mechanical layout machines and calibrated dials integral with the shearing equipment reduce the frequency and extent of deviations. The fundamental characteristics of the shearing equipment itself influence the degree of accuracy that can be attained consistently.

Cutting—The chemical composition of the steel limits in a general way the maximum thickness of steel plate that can be produced by shearing.

Rectangular plates, circular and semi-circular plates, sketch plates, and rings all may require gas cutting or special cutting on machine-tool type equipment in certain combinations of chemical composition and thickness.

IDENTIFICATION, INSPECTION AND LOADING

Except for material in the lighter gages or narrower widths, each plate is marked by hand-stamping, painting or writing with chalk to show the heat number and any other necessary identification marks. It also is inspected for possible imperfections. If the order calls for special treatment such as annealing or other forms of heat-treatment, the plates are sent to the heat-treating department. Samples cut from the plates are usually subjected to the specified mechanical tests. If the plates meet the requirements, they then are loaded in railroad cars in accordance with standard loading practice.

SECTION 3

GENERAL TYPES OF PLATE MILLS

Plate-rolling mills are generally considered in two very broad design classifications. One type includes the **universal mills,** which are characterized by vertical rolls preceding and following the horizontal rolls. The horizontal and the vertical rolls are integrated into a single mill unit and work the stock simultaneously. The purpose of the vertical rolls is not only to work the edges of the stock in the process of reduction, but also to produce a rolled width in conformance with specified standard tolerances.

The second type includes the **sheared-plate mills,** some of which may include edge-working equipment. While some installations use the edging equipment for both edge working and approximate width sizing, final widths are attained by edge shearing. Sheared-plate mills, in turn, may be subdivided into the following mill types: (1) the single-stand two-high pull-over mill, (2) the single-stand two-high reversing mill, (3) the conventional single-stand three-high mill, (4) the single-stand four-high reversing mill, (5) the tandem mill, (6) the semi-continuous mill, and (7) the continuous mill.

A. TWO-HIGH PULL-OVER, TWO-HIGH SINGLE-STAND REVERSING AND THREE-HIGH PLATE MILLS

Two-High Pull-Over Mills—The two-high pull-over plate mills were essentially an adaptation of the then existing sheet mills to plate rolling. Prior to the development of mechanical passers, single-stand hot sheet mills were drafted in one direction only, and the sheet packs were returned for successive passes by being lifted manually on to the top roll and pulled over by hand to the starting or entry side with the aid of the tractive friction of the top roll. Roll settings were altered after each pass by hand operation of levers, which rotated the head screws. Plates rolled on these mills were limited in size and weight by the ability to handle manually. This restricted their utility and soon rendered these mills obsolete for producing finished plate. Two-high non-reversing mills still are in operation, however, in the form of single-pass units, such as scale breakers and roughers. In tandem arrangements in which the finishing-mill unit is either a three- or a four-high mill.

Two-High Single-Stand Reversing Mills—Hot-rolled sheet production by the pack method anteceded volume tonnage production, and recognition of the impracticability of handling heavy plates on two-high pull-over mills prompted the introduction of reversing engines for two-high mills for rolling plates. These mills, likewise, soon reached their practical limit of application, particularly for the production of wide, lighter-gage plate. The fact that all passes from slab to finished plate were made on the same set of rolls accelerated roll wear, which is always greatest in the central portion of the roll body and in itself imposes a restriction on the gage that can be finished satisfactorily. Roll deflection also is an important factor and, for the same roll diameter, increases with the body length for like reductions. An increase in the roll diameter to provide more strength and stiffness increases the separating force between the rolls for the same draft, because of the greater area of contact between the stock and the rolls. The lessening of roll deflection, therefore, progresses at a diminishing ratio with increasing roll diameters.

The art of casting massive rolls developed gradually. Increased roll separating forces for like reductions meant greater power requirements for rolling, as well as to overcome greater resistance to acceleration and deceleration. The design of reversing engines of increasing power also was a gradual development. Neither of these developments progressed sufficiently to meet expanding requirement for plates prior to the invention of the three-high mill.

Two-high, single-stand, reversing plate mills are obsolete in this country, and a detailed description of any installation would have historical value only.

Three-High Plate Mill—The invention of the three-high mill provided design features which, to a degree, overcame some of the principal limitations of the two-high reversing type. In the three-high mill, the top and the bottom rolls are of large diameter, whereas the middle roll is friction-driven and usually about two-thirds of the diameter of the top and bottom rolls. The top roll can be raised and lowered in the housing by power-operated screws, and the middle roll can be brought into contact alternately with the top and the bottom rolls. In making the bottom pass, the stock passes between the middle and the bottom roll while the top roll serves as a backup roll. The stock is raised on the delivery side by a tilting table for a return pass between the middle and the top roll, while the bottom roll serves as a backup roll. The sequence of alternate passes is continued until the stock is reduced to the desired finished plate thickness.

The middle roll is changed when combined roll wear produces a crowned plate which approaches the permissible tolerance limits. The replacement roll is itself crowned to compensate for the wear which already has taken place on the top and the bottom rolls. Successive replacements in the course of a week's rolling schedule are turned with progressively increased crowns to compensate for the continued wear of the top and the bottom rolls. During the weekly mill-repair shutdown, the top and the bottom rolls are either turned in place to their original contours or are replaced with newly dressed rolls. The cycle of replacement of middle rolls with progressively increased crowns is repeated in the following week.

The fact that one of the rolls in contact with the stock on each pass is of smaller diameter than is required for strength (the requisite strength being provided by the roll serving as a backup) reduces the total roll separating force for the drafts in contrast with a two-high mill. The principle, when applied to wide plate production, served to solve the problem of providing rolls of the required strength without being too massive. The fact that only one smaller-diameter roll needed to be replaced in the course of a rolling schedule for wear compensation eased the roll-changing problem. The three-high roll arrangement, with tilting tables on entry and delivery sides, eliminated the necessity for using reversing engines. The tilting tables

and the unidirectional main drives permitted a shortening of the rolling-time cycle and provided a temperature advantage in the finishing of light-gage plates. Continued trade demands for improved finishes, lighter gages, and closer and more uniform rolling to desired dimensions (width and thickness) for many applications, which the three-high mill could not meet consistently, hastened development of other plate-mill types.

The era of the predominance of three-high plate mills was coincident with the development of open-hearth furnaces in which furnace capacities were relatively small, ranging from 36 to 91 metric tons (40 to 100 tons). Ingots, with the exception of those poured for armor plates, were also relatively small and 1016-mm (40 inch) maximum width dimensions were about the largest produced. The three-high mills, however, were not as restricted in the production of plate-size ranges as might be inferred from the restricted ingot sizes. Slabs were conveyed to the mill in a broadside position and after the requisite number of passes were taken to attain plate widths, turning hooks attached to rigid masts were employed to turn the stock 90 degrees by utilizing table traction. The stock was then reduced to final thickness in successive passes. The types of mills described above are now considered obsolete and are not a major factor in modern quality plate production.

B. FOUR-HIGH REVERSING PLATE MILLS

The development of the four-high reversing plate mill further increased the advantages which the three-high mill possessed when compared with the two-high reversing mill. The backup-roll to work-roll diameter ratios were increased to over two-to-one as compared with the three-to-two ratios prevailing in three high mills. For like reductions, therefore, not only is the total roll separating force less, but also strength is provided for each pass by the backup rolls on both top and bottom sides. A four-high reversing plate mill was in operation in 1917 at the Coatesville, Pennsylvania, plant of the Lukens Steel Company; but the general adoption of the type for plate rolling was retarded until antifriction bearings for the rolls became available. Such bearings were developed for the hot-strip mills and the cold-reduction mills and their use was extended to four-high reversing plate mills. A concurrent development was the designing of multi-armature reversing motors for primary mills. The development of the latter motor types reduced the inertia effects of massive mill parts and their relation to acceleration and deceleration on reversals. The multi-armature motors replaced reversing steam engines as prime movers.

4064/5334-mm (160/210-inch) PLATE MILL "A"

This 4064/5334-mm high plate mill "A" is a single stand reversing mill and one of the largest in the world, Figure 25—1.

Slab Yard—All slabs for this plate mill are produced on a 1168-mm (46-inch) universal slabbing mill.

The slabs, after being conditioned either by hand scarfing or by machine scarfing at the conditioning area, are delivered to the 4064/5334-mm (160/210-inch) mill slab yard by one of three transfer cars. Two

Fig. 25—1. Schematic layout of 4064/5334-mm (160/210-inch) plate mill "A".

of the cars are arranged to transfer slabs directly from the conditioning area to the slab yard after machine scarfing. Three electric overhead traveling cranes, equipped with slab tongs, are used to place the slabs into stock for later transfer to the continuous-furnace assembly area or the in-and-out furnace charging area. At this point the slabs are identified with a serial number and placed on the slab-charging table.

Slab-Reheating Furnaces—Heating facilities consist of two double-row continuous-type furnaces and four batch-type in-and-out furnaces.

The continuous furnaces each have six water-cooled skids, with a pusher for each group of three skids. Slab sizes range from 102 to 305 mm (4 to 12 inches) thick and from 1829 to 3658 mm (72 to 144 inches) in length. A television receiver indicates to the pusher operator the slab position at the drop-end of the furnace. The furnaces are quadruple fired, with four heating zones. The steel enters the primary or preheating zone cold. As it is pushed through this zone, it is heated to between 425° and 650°C (800° and 1200°F). It then passes through the two heating zones (the first bottom-fired and the second top-and bottom-fired) where it is heated to its rolling temperature of approximately 1315°C (2400°F). In the final or soaking zone, slab temperatures are equalized with the slabs supported on a dry hearth.

The four in-and-out batch-type furnaces normally are used to heat slabs under 102 mm (4 inches) thick, over 305 mm (12 inches) thick, or less than 1829 mm (72 inches) in length, and slabs for small orders or other slabs requiring special heating. An electric overhead traveling "peel"-type crane charges and draws the slabs and places them directly on the mill-approach roll table.

Coke-oven gas with a heating value of 21 014 kJ/m³(535 Btu/ft³) is normally used for firing both the continuous and the in-and-out furnaces, with provision for firing fuel oil.

Three hood-type preheating furnaces are located in the slab yard, adjacent to the in-and-out furnaces.

Scalebreaker—The scalebreaker is a two-high mill with 1016-mm (40-inch) diameter grooved rolls. A slab measuring gage is located ahead of the scalebreaker to prevent overdrafting. Water sprays operate at 10 342 kPa (1500 psi) for descaling on both the entry and exit sides. In addition to scale breaking, this stand establishes the slab thickness for slabs entering the four-high mill.

Slab Turnaround—A slab turnaround with a mushroom-type lift is used for turning slabs 90 degrees before broadsiding at the four-high mill. The lift is operated by pressure supplied at 16 548 kPa (2400 psi) by the hydraulic roll-balance system, and has a double-acting cylinder for positive return. The rotating drive is powered by a non-reversing electric motor.

Four-High Reversing Stand—The four-high reversing stand where the actual rolling of plates is performed (Figure 25—2) has adjustable housings that permit operations as either a 4064-mm (160-inch) or a 5334-mm (210-inch) mill. This is accomplished by shifting the housings to the proper positions and installing rolls of the proper length. In order to move the housings, all mill rolls, bearings, feed rolls, spray headers and stripper plates are removed, and hydraulic

clamps holding the housings at the shoe-plate elevation are loosened. A center tapered pin that keys the upper portion of the housings together is lifted from position, and hydraulic pressure from the roll-balance system is applied to slide the housings into the new position. A four-high set of rolls of the proper length is then installed and the mill accessories are replaced.

Two 4476-kilowatt (6000-horsepower), 40/80 rpm, single-armature, direct-current motors drive the two work rolls with a separate motor driving each roll. To permit the same drive spindles to be used with rolls of either 4064- or 5334-mm (160- or 210-inch) length, the 4064-mm (160-inch) rolls have a 635-mm (25-inch) integral extension on the drive ends.

A roll-changing sled is used for changing both the work rolls and back-up rolls. The roll assemblies are retracted on the sled and there are disassembled and handled with an electric overhead traveling crane.

The back-up rolls are ground truly cylindrical and the work rolls are crowned 0.20 mm (0.008 inch) on the diameter when a complete set of four new rolls are installed. Work rolls are replaced from time to time without changing the back-up rolls until wear of the latter has become excessive. As wear on the back-up rolls increases, work-roll crowns are gradually increased to 0.41 mm (0.016 inch), after which new backup rolls are installed.

Oil-film bearings are employed on the back-up rolls, and anti-friction roller bearings on the work rolls.

Plate sizes produced by the mill are 4.76 mm to 381 mm (3/16 inch to 15 inches) in thickness, 762 mm to 5080 mm (30 inches to 200 inches) in width and 3.048 m to 21.336 m (10 feet to 70 feet) in length. Longer lengths may be rolled, depending upon slab weights, physical rolling characteristics and transfer-equipment problems for the particular plate size. Most slabs are reduced in thickness at the scalebreaker, turned 90 degrees on the mushroom-type lift, rolled to proper width on the four-high mill, turned 90 degrees on the spin table at the delivery side of the four high mill, and then rolled straightaway to gage. Some plate product requires additional turning during rolling to obtain optimum yield.

Scale removed from slabs at the scalebreaker is collected by dry-type shuttle conveyors up to the entry end of the scalebreaker stand; after this point, all scale is sluiced by water to conventional retention pits that are equipped with oil-skimming devices.

A rolling pulpit is located just to the entry side of the mill, approximately 9.144 m (30 feet) above floor level.

Tables—All mill tables are individually driven, with adjacent rollers driven from opposite sides of the table. The plate-spin table, comprised of 11 alternately tapered rolls, turns the plate by having adjacent rollers rotate in opposite directions.

Transfer Tables and Cooling Sprays—A long runout table is located between the mill and the levelers where an electric overhead traveling crane, equipped with a cradle, has access between the rolls to remove heavy-gage flame-cut product, over 254 mm (10 inches) thick, to a transfer buggy. This buggy travels to the south flame-cut building, where the product is removed by a "take-off table." Also located over these tables is a system of cooling sprays comprised of in-

Fig. 25—2. Delivery side of 4064/5334-mm (160/210-inch) plate mill "A". In this illustration, the mill is operating as a 4064-mm (160-inch) mill.

verted L-shaped pipes that are motorized to swing out of line in case of a cobble on the roller line. These sprays cool plates in two separate spray patterns, 3302 mm and 5334 mm (130 inches and 210 inches) in width, respectively.

The No. 1 flame-cut transfer cooling bed is located after the levelers for transfer of 76.2-mm to 254-mm (3-inch to 10-inch) thick flame-cut product into the south flame-cut building. The cooling bed is of the link-and-dog type and can handle plates up to 18.288 m (720 inches) in length. The plates are removed by electric overhead traveling cranes equipped with a spreader beam having either multiple "C" hooks or magnets, and are placed on cooling skids or directly on the brick floor when cooling in piles is required.

All other plates, under 76.2 mm (3 inches) thick, pass over the main cooling-bed transfer that will handle plate with a maximum length of 38.100 m (1500 inches). The transfer is of the chain-link, disappearing dog type, with a sprocket drive having five sections, with a cast-iron grid design that affords maximum support for 4.76-mm (3/16-inch) thick plate and excellent cooling for heavier gages.

Levelers—There are two levelers: the 4064-mm (160-inch) leveler which normally levels plates less than 3810 mm (150 inches) wide and the 5334-mm (210-inch) leveler for plates from 3810 mm (150 inches) wide. Since the 5334-mm (210-inch) leveler has three 229-mm (9-inch) back-up rolls for each leveler roll, it is also a good supplementary leveler for lighter-gage plates that are narrower than 3810 mm (150 inches) wide. Both levelers are of the retractable type with attached roll table. The levelers are designed to level hot plates from 4.8-mm to 50.8-mm (3/16-inch to 2-inch) thickness, and both have sufficient opening to pass plates up to 152.4 mm (6 inches) thick.

Plate-Inspection Turnovers—There are three plate-inspection turnovers of the multiple-arm scissors type that permit turning plates 180 degrees to permit complete bottom-side inspection.

The first turnover is part of the cooling-bed transfer where plates up to 38.100 m (1500 inches) long can be turned and inspected. Any surface imperfection can be marked for possible cutting out at the shears or, if the plate requires minor spot grinding, that particular surface may be left in the upward position. After this inspection, plates over 38.1 mm (1½ inches) thick are removed to the north flame-cut building by magnet-equipped electric overhead traveling cranes. All sheared plate is transferred on the roller line to the plate-marking tables.,

Two additional plate turnovers, designed to turn plates up to 18 288 mm (720 inches) in length are located in the finishing end to facilitate final inspection of plates.

Plate Marking—At the plate-marking tables, the layout crew checks the rough plate for pattern size, stencils necessary slab identification, customer requirements and shipping destination on the plate. Test specimens are also sketched out and stamped at this point.

The plate-marking machine traverses the full length of the table on rails, with an arm extending out over the entire roller line. Attached to the arm are three marking heads: two to mark the ordered width of the plates for side-shear cutting, and one as an auxiliary head to mark a guide line 610 mm (24 inches) from the left line) for the shearman's use. The machine is adjustable to mark plates from 4.8 mm to 38.1 mm (3/16 inch to 1½ inches) thick, up to 5334 mm (210 inches) wide and 38.100 m (125 feet) long.

Crop Shear—The crop shear can cut a 38.1 mm (1½ inch) carbon steel plate, 5334 mm (210 inches) wide. This shear (as are the dividing and side shears) is of the motor driven start-stop crank type with open throat and has a raked guillotine-type moving upper knife. The hold-down is independently mounted on the top knife. Usually, both front and back end crops are cut with this shear. It also performs any multiple cutting required by the rough plate length being greater than that of the transfer bed. After cropping, the plate is transferred 90 degrees across a transfer of the chain-link disappearing-dog type to the side-shear line.

Side Shears—Plate-positioning magnets mounted below the top of the table rollers position the front end of the plate under the No. 1 side-shear knife. A traverse pusher and a hold-down maintain plate alignment during shearing to assure a straight sheared edge.

The No. 2 side shear is equipped with almost identical features as the No. 1 side shear but, of course, is located on the opposite side of the shear line. The edge cut by the No. 1 side shear is pushed by a straight edge to provide a straight cut at the No. 2 side shear. No. 2 side shear is equipped with a motor-driven plate-width gage, used for remote indication of plate width. A digital readout is located in the No. 2 side-shear pulpit for the operator to set the desired sheared width. At this point, it can be determined how close to ordered plate size the shears have trimmed. With the extended travel of the plates during the cooling cycle, accurate width and length tolerances can be met.

Dividing Shear—The dividing shear has a dual function of dividing the plate into ordered lengths and of accurately cutting the ends of the plate. Plate-positioning magnets and movable edging rollers move and hold the plate edge snugly against the fixed side guard. There are two shear-gage stops, one used for short cutting and one used for long cutting, thus minimizing the time required for setting plate-cutting lengths.

Scrap Shears—Each product shear is equipped with an individual pallet-type scrap conveyor and scrap shear. All crop scrap is cut to open-hearth charging-box size, and transferred by chutes to a system of shuttle conveyors at basement level. The shuttle conveyors discharge to self-dumping boxes remote from the immediate work area for transfer to rail cars.

Test pieces are cut by the scrap shearman and transferred to the operating floor by apron-type conveyors.

Inspection and Piling—After shearing, each plate is weighed on a roller-type scale table of the load-cell type, gaged, and inspected for dimensional accuracy. The plates are then transferred to the piling tables in the shipping bay by two magnet-equipped transfer cranes. At this point, the plates either are loaded directly into cars or assembled and stored in piles for later loading.

Flame Cutting—The flame-cutting operation is performed in two adjacent building bays. The light flame-cutting area usually handles plates up to 76.2 mm (3 inches) thick, 5334 mm (210 inches) wide and of lengths up to 25.4 m (1000 inches). This burning bed consists of five motorized gantry-type burning machines, each equipped with five torches. The usual operation consists of one gantry machine making four parallel length cuts while alternate machines are making and cutting the end crops. Crop scrap is prepared simultaneously by hand torch. A car-bottom furnace is provided in this area for various heating, stress-relieving and normalizing operations.

The heavy flame-cutting bed is similarly equipped with three motorized gantry-type burning machines, and usually handles plates 101.6 mm to 381.0 mm (4 inches to 15 inches) thick, up to 4064 mm (160 inches) wide.

Flame-cut plates are inspected, weighed, and shipped by rail and truck in the immediate area.

Heat-Treating Facilities—A complete continuous heat-treating department has been provided to heat-treat certain grades of plates rolled on the 4064/5334-mm (160/210-inch) mill; the layout and equipment are described in Section 4 of this chapter.

Roll Shop—The roll shop has facilities for handling the rolls and bearings of the 4064/5334-mm (160/210-inch) mill and the 1168-mm (46-inch) slab mill that supplies the 4064/5334-mm (160/210-inch) mill. The equipment includes roll lathe, roll grinder, bearing extractors, degreaser, and shear-knife grinder, with adequate storage and working area served by an electric overhead traveling crane.

Lubrication—The greasing system for all table bearings, spindles, spindle carriers, jacks, slide joints, etc., is of the automatic "single-line" reversing type and contains eleven major pumping units.

Oil lubrication for the majority of the equipment, such as table line shafts, gears and bearings, leveler drives, screw-down drives, reduction gears, pinions, backup-roll bearings, and motor-room equipment, is accomplished by circulating systems of the single- and double-tank types.

4064-mm (160-inch) FOUR-HIGH PLATE MILL "B"

This 4064-mm (160-inch) four-high mill "B", Figure 25—3, is a two-stand reversing mill. It has a cold-slab storage yard that provides for stocking approximately 13 600 metric tons (15 000 net tons) of slabs in accordance with a predetermined rolling sequence. Six gas-fired preheating pits are located in this yard; each has a capacity of approximately 10 slabs weighing a total of 45 metric tons (50 net tons) and is capable of heating cold steel to a temperature of 650°C (1200°F) in 24 hours. Three gas-fired preheating hoods, each with a capacity of approximately 109 metric tons (120 net tons), can heat cold steel to a temperature of 870°C (1600°F) in 24 hours. Slabs are transferred to the reheating-furnace areas by two standard-gage shuttle-car transfers, powered by diesel-electric locomotives.

The slab-heating equipment consists of two continuous-type furnaces and two batch or in-and-out type furnaces. The batch-type furnaces are designed to heat slabs of a maximum size of 3658 mm (144 inches) long

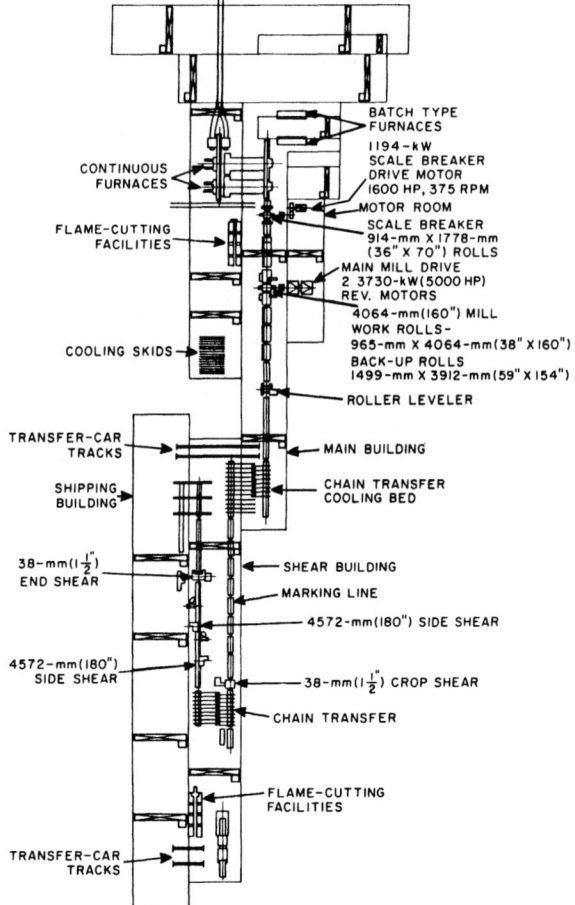

FIG. 25—3. Diagram (not to scale) of the physical layout of equipment comprising the 4064-mm (160-inch) mill "B".

by 521 mm (20½ inches) thick. They are two-zone-controlled, end-fired, recuperative-type furnaces, designed to operate on a mixture of coke-oven and natural gas and have a rated heating capacity of 13.6 metric tons (15 net tons) per hour from cold to 1230°C (2250°F). These furnaces are served by a charging crane which charges the cold slabs from the transfer car into the furnace and draws the heated slabs and delivers them onto the furnace delivery table.

The continuous-type furnaces are designed to heat slabs from approximately 75 to 305 mm (3 to 12 inches) thick, 1120 to 1525 mm (44 to 66 inches) wide and 1780 to 3050 mm (70 to 120 inches) long. They are conventional two-zone-controlled, triple-fired, end-charged-and-discharged, recuperative-type furnaces. The top and the bottom burners of the primary heating zone are designed to burn a mixture of coke-oven and natural gas as a primary fuel and also can burn fuel oil as an alternate fuel. Only the gas mixture is burned in the holding zone. The heating capacity of each furnace is approximately 63.5 metric tons (70 net tons) per hour from cold to 1230°C (2250°F) with full hearth coverage.

The continuous furnaces are served by an electric overhead traveling crane which takes the slabs from the transfer cars by magnet or slab-handling tongs, and

FIG. 25—4. General view of the charging end of continuous-type reheating furnaces supplying hot slabs to the 4064-mm (160-inch) plate mill "B".

FIG. 25—5. Close-up view of the scalebreaker stand of the 4064-mm (160-inch) plate mill "B".

places them on the furnace charging table. The slabs are fed through the furnaces by double rack-type pushers arranged to feed two rows of slabs through each furnace (Figure 25—4). Slabs are discharged from the delivery end, and slide down the furnace dropout skids to the furnace delivery table against spring-backed bumpers.

Heated slabs are conveyed on a roller table from the furnaces to a scalebreaker stand (Figure 25—5) which is a two-high mill with fluted alloy-steel rolls, operating in water-lubricated composition bearings.

The top roll is balanced by carrier bars supported from a yoke which is actuated by hydraulic jacks and accumulator to hold the top roll against the housing screw. The screws are operated by 37-kW (50 horse-power) motors through worm reduction gearing at a lineal vertical speed of nearly 280 mm (11 inches) per minute. The rolls are changed at approximately one-month intervals under normal operating conditions. Approximately four hours are required to make the roll change with the use of a "C" hook and an overhead crane.

The mill is driven through a double-reduction fly-wheel gear drive, a universal leading spindle, mill pinions and universal mill spindles, resulting in a lineal roll speed of 90 m (295.6 feet) per minute. A meter in the pulpit gives the operator a reading of the opening between rolls at all times. Drafts of 12.7 mm (½ inch) may be taken on this stand, and the passage of the slab is guided to and from the mill by twin adjustable side-guides. Primary descaling is completed on the delivery side of the mill as the slab is passed through top and bottom hydraulic sprays operating at a pressure of 10 342 kPa (1500 psi).

The distance between the 4064-mm (160-inch) reversing-mill stand (Figure 25—6) and the scale breaker is spanned by 12.2 m (40 feet) of 1829-mm (72-inch) wide table and 36.6 m (120 feet) of 4267-mm (168-inch) wide table. All table rollers in this mill are journalled in antifriction bearings. A lift-and-turn table device is located on the reversing-mill side of the junction between the narrow and the wide table sections. The turntable is elevated and lowered by oil-hydraulic pressure and turned by a motor-operated rack and pinion. This permits either straightaway or broadside slab entry into the reversing mill. When spreading is required in the first pass, the slab may be entered following the preceding plate without a turning delay.

The crowned (0.61-mm or 0.024-inch maximum) alloy-iron work rolls of the four-high reversing stand are equipped with four-row tapered roller bearings. The alloy-steel backup rolls employ oil-film bearings

FIG. 25—6. The 4064-mm (160-inch) reversing mill stand of the 4064-mm (160-inch) plate mill "B", showing powered side guards for squaring and centering slabs and holding plates during the rolling operation.

through which the roll-separating force is transmitted. Thrust is taken by two-row steep-angle tapered thrust bearings. The top rolls are balanced by hydraulic jacks mounted in the bottom backup chucks. An automatic force-feed lubricating system is connected with all wearing surfaces on the mill other than the oil-film bearings.

Three sets of work rolls are used in the course of a normal weekly rolling schedule. Approximately one hour is required for a work-roll change, which is accomplished by an overhead crane-suspended "C" hook, equipped with a motor-driven shifting device that positions the hook suspension over the center of gravity of the unloaded or loaded hook. The average time between complete roll changes, including the backup rolls, is three weeks. The complete roll change requires approximately ten hours with the backup rolls being moved in and out of the mill on a motor-driven slide.

The reversing motors are each directly connected to a work roll through universal spindles. The motor speed range is from 40 to 80 rpm which provides a roll surface speed of approximately 122 to 244 m (400 to 800 feet) per minute. The mill screws are operated by two 112-kilowatt (150-horsepower) mill-type motors, supplied by a variable-voltage motor-generator set. A pre-set automatic screw-down control, with optional manual control, is provided for this mill. Provision is made to set up one schedule while another is being run. The control is operated by the screw-down operator, who also reverses the mill drive and who can make manual adjustments to the screw settings by interrupting any automatic setting without affecting the accuracy of the following automatic stop. The mill is capable of producing the minimum gage of 4.8 mm (3/16 inch) to sheared widths of 3048 mm (120 inches). Auxiliary facility limitations restrict the sheared-plate gages to a 4.8-mm to 38.1-mm (3/16-inch to 1½-inch) range, widths to a 978-mm to 3658-mm (38½-inch to 144-inch) range and lengths to a 18.288-m (720-inch) maximum. Flame-cut plates range from 38.1-mm to 381-mm (1½-inch to 15-inch) thickness and a maximum of 3658 mm and 12 192 mm (144 inches and 480 inches) in width and length, respectively.

The front and the rear mill tables have tapered table rollers. Even-numbered rollers all are tapered in the same direction and driven as a unit. Odd-numbered rollers are tapered in the opposite direction and also are driven as a unit. By rotating the alternate roller units in opposite directions, it is possible to turn a slab 90 degrees as desired. Heavy, powered sideguards are on each side of the mill for squaring and centering the slabs and holding the plates during the rolling operation. Hydraulic sprays on both sides of the mill, operating at a pressure of 9653 kPa (1400 psi), are utilized for top and bottom secondary scale removal.

The remainder of the distance between the four-high mill and the leveler is occupied by a conveying table. Top and bottom water sprays facilitate the cooling of heavier-gage plates prior to leveler entry and retard secondary scale formation. A section of the table on each side of the leveler is designed to permit plate removal by a "C" hook suspended from an overhead crane. The plates may be transferred to either one of two motor-driven transfer cars and subsequently

transferred to the cooling area of the adjacent flame-cutting building or to the controlled-cooling facilities located in the shipping yard.

The leveler has two entry pinch rolls and 6 top and 5 bottom bending rolls. The machine is driven by a 187-kilowatt (250 horsepower), 400 rpm motor, and positioning is accomplished through a selective screwdown assembly which is equipped with drum indicators located at all four corners of the leveler housing. The leveler can level structural-grade plates of 38.1-mm (1½-inch) thickness and can be raised to a maximum passage opening of 152 mm (6 inches).

From the leveler, the plates travel to a chain-transfer cooling bed. This grid-type bed delivers the plates from the mill building onto a disc-roller marking table located in the shear building. A motor-propelled automatic measuring and marking machine moves parallel with this table. Indexed longitudinal and transverse movements enable the operator to measure and mark off mechanically any desired size and shape of a rectangular plate, and scribe a reference line which is used by the side-shear operator in the trimming of the side scrap. Beyond the marking machine, all required identification stamping and painting of the plates and test coupons are performed manually prior to entry in the crop shear.

The crop shear or end shear is capable, in common with the main line shear units, of cutting 38.1-mm (1½-inch) gage to the maximum width produced. Powered mechanical manipulators are available for the square positioning of the plate. Oil hydraulic hold-down gags furnish positive clamping for the shearing stroke. The shear operates on the oil-hydraulic principle, and has a self-contained pump with motor drive. The crop shear is used for dividing plates that are too long to pass over a second transfer, and for front-end shearing of heavy, wide plates, which must be free of shear bow.

Plates from the crop shear are delivered either to another chain transfer or to a table section beyond the transfer, which is equipped with roll-off lifting arms and a side piling bed. This run-off and piling bed is utilized to divert flame-cut plates of such size and composition that will permit their conveyance to this point without the necessity of slow-cooling cycles. Plates to be finish-sheared are moved across the transfer to the transfer run-out tables, which double back through the shear building and are parallel to the marking line.

The main shear line consists of two side shears and one end shear. These shears are similar in design and construction to the crop shear, with the exception that the side shears have longer knives. The side shears are staggered on either side of the table. The side crop is sheared at the first side shear by positioning the plates to the reference line. A series of electromagnets with hydraulic vertical motion and screw drive transverse motion are utilized to position the plate so that the reference line is matched with a positioning line on the shear block. A mechanical width-gage guide is provided at the second side shear, which facilitates the positioning of the plate by working against the finished sheared edge; and its position in relation to the shear knife is recorded on a dial located on the shear housing. The end shear is a duplicate of the crop shear and is equipped with squaring pushers located on the entry side, and two motorized gage stops on the delivery side, which gage plates mechanically to predetermined lengths. One stop is used for short and the other for long lengths in the 1524-mm to 18 288-mm (60-inch to 720-inch) sheared length range.

Crop and end shear auxiliary equipment consists of motor-driven scrap shears. A butterfly chute arrangement diverts test coupons, "block sheared," to a test box in which they are periodically delivered to a test-cutting shear for final handling. The test-cutting shear is also motor-driven. The side shears are provided with motor-driven scrap shears to which side scrap is fed by conveyors located below the table level, and the cut scrap is guided to removal buckets through gravity chutes.

Plates are delivered from the end shear to a scale table, where they are weighed individually. They then proceed to a table section spanned by two transverse, selective, electromagnetic-type piling cranes. One crane can handle a 9144-mm (360-inch) plate length, and the two cranes working together can handle a 18 288 mm (720 inch) plate length. Lifting force of the magnets can be regulated at will to provide a wide range in the lift combinations. The cranes transfer the plates into the shipping building and deposit and pile them on gravity conveyor tables. The piles are moved by a conveyor-chain dog into positions from which they can be picked up by the overhead shipping cranes.

The shipping building is equipped with a shipping track extending its entire length. A connecting spur track of an 8-car capacity is located in an adjoining lean-to-section of the building. This track is not serviced by overhead cranes and is used for preparing cars which require special blocking, either prior to or after loading. The shipping building is serviced by four electric overhead traveling cranes of the double-hoist type. This building also is equipped with controlled-cooling hoods. Two additional inlets into this yard, other than the piling cranes just mentioned, are provided in the form of motor-driven transfer cars located at the extreme ends of the building. One car delivers directly from the mill building, and the other car delivers from the flame-cutting unit located in the end of the shear building.

The flame-cutting unit, located at the end of the shear building, is equipped with a flame planer, three pantographs, and sketch-cutting machines. The flame planer is designed primarily for side and end trimming as well as splitting rectangular-shaped plates. The pantograph machines are designed primarily for intricate shape cutting, although they can be used efficiently for cutting rectangular shapes. Magnetic tracing mechanisms, traveling over templates or operated manually over sketch drawings, guide the torch through the course desired.

Another flame-cutting unit is located in a building parallel and adjacent to the mill building. A continuation of this building merges with the end of the continuous-furnace building, and the two cranes servicing this unit operate on a common runway with the crane servicing the continuous furnaces. This unit is equipped with three pantograph sketch-cutting machines, two large heat-treating ovens, and a heat-treating pit. The heat-treating facilities are used extensively in preheating, stress-relieving, normalizing and the controlled cooling of flame-cut products requiring these treatments. The heat-treating ovens also can be used for preheating slabs prior to charging in the slab-heating furnaces when such practice is necessary. Cooling skids are provided in this area for heavy-gage plates which are transferred hot from the mill by a transfer car previously mentioned.

The greasing system for all table bearings, spindles, spindle carriers, jacks, slide joints, etc., is of the automatic "single-line" reversing type and contains eleven major pumping units.

Oil lubrication for the majority of the equipment, such as table line shafts, gears and bearings, leveler drives, screw-down drives, reduction gears, pinions, backup-roll bearings, and motor-room equipment, is accomplished by circulating systems of the single- and double-tank types.

4064-mm (160-inch) FOUR-HIGH PLATE MILL "C"

The 4064-mm (160-inch) four-high plate mill "C", which became operational in May 1970, is a 2-stand reversing mill. Figure 25—7 shows the mill layout, illustrating the product flow. Slabs are received in the slab yards and then flow to the reheat furnaces, through the plate mill, heat treating and then on to shipping at the far end of the building.

The range of slabs heated is 88.9 to 610 mm (3½ to 24 in.) thick by 1829 to 7620 mm (72 to 300 in.) long by 1016 to 1981 mm (40 to 78 in.) wide in the continuous furnaces. The in-and-out furnaces handle slabs up to 610 mm (24 in.) thick. All four heating furnaces are fired with natural gas.

Preprogrammed logic equipment permits semi-automatic as well as manual operation for rapid and uniform rolling of plate.

The two-high reversing rougher has 1346-mm (53-in.) diameter rolls with 4064-mm (160-in.) body length. The four-high reversing finisher has 901-mm (39-in.) diameter, 4064-mm (160-in.) body length work rolls with 1829-mm (72-in.) diameter, 3912-mm (154-in.) body back-up rolls.

The two-high work rolls and four-high back-up rolls run in tapered-neck, sleeve-type, oil-film bearings. The work rolls of the four-high mill run in antifriction, tapered roller bearings. This being a sheared-plate mill, vertical rolls are not used.

A descaling box is located between the continuous furnaces and the roughing mill. Also, a descaling spray station is located on each side of the roughing mill and at the entry to the four-high finishing mill. A descaling pressure of 11 030 kPa (1600 psi) is used at all points. The descaling-box header can be positioned automatically or manually by the two-high-mill operator.

The next major piece of equipment is the plate leveler, which is designed to handle 4.8 to 50.8 mm (³⁄₁₆ to 2 in.) thick by 3988-mm (157-in.) wide plate. The work rolls of the leveler are 229-mm (9-in.) diameter with 260-mm (10¼-in.) back-up rolls. All are driven through duplex double-reduction speed reducers and a pinion stand by spindle and gear-type couplings.

The guides on the entry side of the leveler consist of a fixed bottom apron and a movable top entry guide.

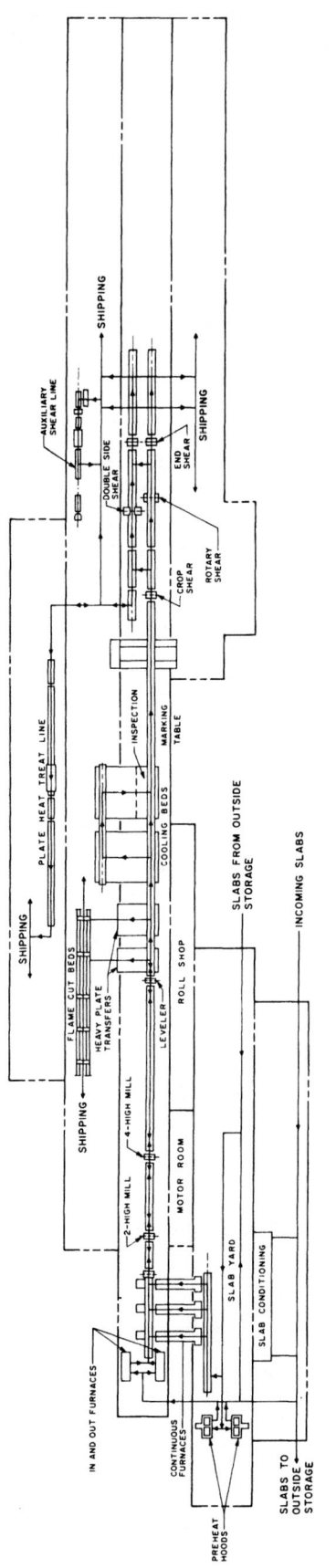

FIG. 25—7. Layout of 4064-mm (160-in) plate mill "C", showing product flow.

The delivery side incorporates only a fixed bottom apron.

Maximum opening of the leveler rolls is 457 mm (8 in.)

Each of the two chain-type cooling beds has a grid-plate turnover for inspecting both sides of plates.

Heat-Treatment Facilities—For heat treating, both a continuous line handling 4.8 through 102-mm (³⁄₁₆ through 4-in.) thick plates (carbon and alloy grades) and a car-bottom furnace are available. The continuous line has a roller-hearth hardening furnace, roller quench unit and a roller-hearth tempering furnace.

The length of the continuous hardening furnace is 60.66 metres (199 feet), with 137 burners using natural gas; this unit has 10 zones of control. The continuous tempering furnace is 73.15 metres (240 feet) long with 184 natural-gas burners, also with 10 zones of control. The temperature maintained in the continuous hardening furnace is controlled between 760° and 1010° C (1400° and 1850° F). Temperature of the continuous tempering furnace is controlled between 400° to 730° C (750° to 1350° F).

The roller-pressure quenching unit has three pumps, each with a capacity of 37 850 litres per minute (10 000 gpm). Two are in operation at all times, thus delivering 75 700 litres per minute (20 000 gpm) of water to the elevated storage tank. The low-pressure sections are fed by gravity from the elevated tank, delivering 291 500 litres per minute (77 000 gpm) of water at approximately 586 kPa (85 psi).

The plate is moved directly from the hardening furnace into and through the quenching unit. Quenching begins the instant the plate enters the unit, minimizing heat loss. The plate is held firmly in the unit by top and bottom rollers, the plate being kept in motion throughout the quenching cycle. The rolls minimize distortion. The entire plate surface is exposed to spray from headers located both top and bottom, between the rolls, thus assuring uniform quenching.

The car-bottom furnace, for stress relieving, normalizing, annealing and tempering plate, maintains temperatures between 150° and 1205° C (300° and 2200° F), with five zones of control. Plates up to 3810 mm wide by 18.29 metres long (150 in. wide by 60 feet long) can be heat treated in this facility.

Shearing and Cutting Units—Shearing equipment includes a rotary shear with an operating speed of 0.66 to 1.32 metres per second (130 to 260 fpm). Other shearing equipment includes: a double side shear that makes synchronized parallel cuts, a 19.1-mm (¾-in.) end shear, a 38.1-mm (1½-in.) end shear, a crop shear and a rocker-type dividing shear. In addition to the shearing equipment, special shapes can be produced with a circle shear and shape flame cutting equipment located up the line near the cooling beds. Also, conventional flame cutting equipment for producing rectangular plates is available in this same area. In the shearing aisle, plate is handled by two piler cranes equipped with magnets which transfer plate to piling tables located in the adjacent bays.

Lubrication—For lubrication, each stand uses recirculating oil of 2000 SUS viscosity and centralized grease systems. The bearings served by the oil system are those on the leveler, back-up rolls on the four-high mill and work rolls on the two-high mill. The grease system lubricates bearings on the four-high mill work rolls, roller tables and shears.

C. TANDEM MILLS

Many of the original single-stand plate mills have been supplemented by an additional stand to form a tandem plate mill. The various tandem-mill arrangements represent a wide variety of mill unit combinations, which achieve two principal objectives. When the total work of reducing slab to plate is divided between two mill stands, satisfactory surface finish and shape of rolls can be maintained for longer periods between roll changes. Secondly, since the work is divided between two units operating simultaneously, the required time interval for the reduction of a slab to a plate is reduced and the overall capacity of the unit is increased correspondingly.

A wide variety of tandem arrangements exists because they represent modifications of original rather than new installations, and because they were accomplished by the maximum utilization to existing equipment, the minimum expenditure for new equipment and a minimum of alteration to auxiliary facilities. The tandem arrangements, consequently, may include a two-high reversing rougher with a three-high finisher; a two-high reversing rougher with a four-high finisher; a three-high rougher with a three-high finisher; or a three-high rougher with a four-high finisher.

The two principal advantages of the tandem mills over the corresponding single-stand mills have been stated. No specific installation has mill or auxiliary equipment of engineering interest which is not covered in decriptions of more standardized arrangements.

D. SEMI-CONTINUOUS AND CONTINUOUS MILLS

These plate mills constitute the plate-mill groupings which include multipass, reversing, roughing units for the semi-continuous mills, and non-reversing roughing units for the continuous mills, coupled with two or more single-pass continuous units in which the plate is reduced simultaneously. Two-, three- and four-high stands with or without scale breakers, broadside stands, squeezers, and edgers are used as roughing units, while four-high stands are used as finishing units. The semi-continuous mill arrangement, although requiring a larger capital investment, has a number of operating advantages over both the single-stand and the tandem-mill types. The total reduction work is divided between individual stands to an even greater extent than in the case of the tandem mills. The roll wear of individual stands is, therefore, less than that of the prior mill types. The total time increment for reduction from slab to plate also is less, and the tonnage capacity per unit of time correspondingly is greater. The decreased time element permits sheets as well as plates to be rolled on these units.

THE 2540-mm (100-inch) SEMI-CONTINUOUS PLATE MILL "D"

The 2540-mm (100-inch) Semi-Continuous Plate Mill "D"—This semi-continuous plate mill (Figure 25—8 and 25—9) has a slab-storage yard with a capacity of 16 300 metric tons (18 000 net tons) stocked in rolling sequence. Two gas-fired, insulated, fabricated-steel preheaters, each having a capacity of 63.5 metric tons (70 net tons), are located in this yard. Slabs are placed in lifts on a yard table which delivers them to a magazine feeder located in the slab-storage yard which serves as an intermediate unit between the yard and the furnace tables. These tables, as well as all others in this mill, are equipped with anti-friction bearings.

The feeder raises each pile of slabs to the level of the furnace charging table and discharges them one at a time onto the furnace table by an endless chain pusher. The slabs then are positioned in front of the furnace doors for charging by two pushers which can be operated independently or in unison for each furnace. The

FIG. 25—8. General view of the 2540-mm (100-inch) semi-continuous plate mill "D".

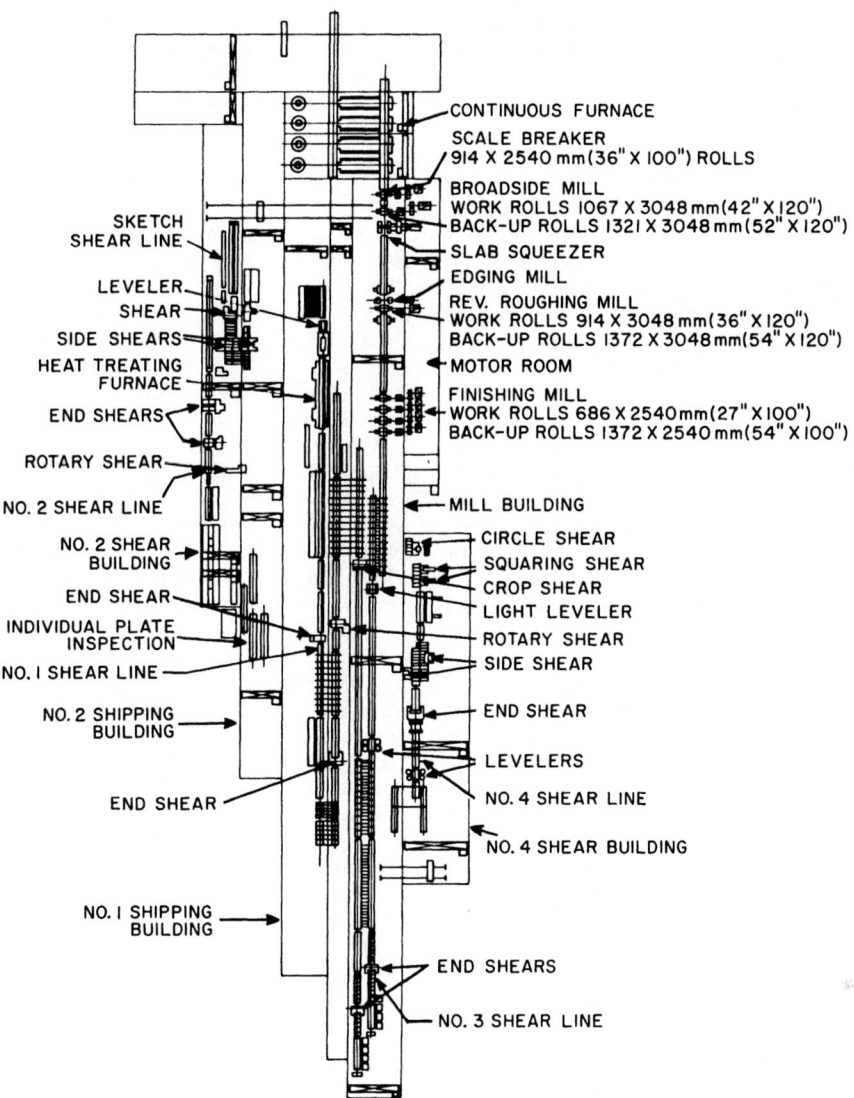

FIG. 25—9. Schematic arrangement (not to scale) of the production units comprising the 2540-mm (100-inch) semi-continuous plate mill "D".

mill is served by four conventional, continuous, triple-zoned furnaces with non-metallic recuperators. A mixture of natural and coke-oven gas is the primary fuel, although fuel oil can be burned as an alternate. The furnaces are equipped with combustion and temperature controls. They have heated, on a continuous basis, about 62 metric tons (68 net tons) an hour per furnace from cold to 1250° C (2250° F). Slabs charged vary from 610 mm to 1524 mm (24 inches to 60 inches) in width, 82.6 mm to 305 mm (3¼ inches to 12 inches) in thickness, and 1651 mm to 2337 mm (65 inches to 92 inches) in length.

The slabs slide down the furnace dropout against inclined spring-loaded bumpers or an improved shockless bumper onto the furnace tables. A skew table with movable side guards permits straightaway entry or may be used as a turnaround for broadside entry to

the scale breaker. This is a two-high stand with smooth rolls, operating in bronze-insert babbitt bearings. The mill is driven by a 746-kilowatt (1000-horsepower), 500-rpm motor, transmitting power through a reduction gear and pinion stand. Top and bottom descaling sprays operating at 6895 kPa (1000 psi) are located at the delivery side of the scale breaker and at the entry sides of the reversing rougher and the first finisher.

The distance between scale-breaker and broadside stand is spanned by a roller table into which is incorporated an electrically-operated lift-and-turn platen at the broadside entry. The broadside mill is a four-high nonreversing stand. Both work and backup rolls operate in bronze-insert babbitt bearings. The mill is driven by a 3357-kilowatt (4500-horsepower), 370-rpm, ac motor, transmitting power through a 22-to-1 gear ratio and conventional pinions to give a lineal speed of 0.94

metres per second (185 feet per minute) at the work-roll face. Movable side guards on the entry side permit centering of the slab. Pushers on entry and delivery sides facilitate entry and return for another spreading pass if required.

A duplicate of the turn-around on the entry side is located on the delivery side in the table leading to the slab squeezer. The squeezer is a two-ram width-sizing machine. The rams are supported above the mill table rollers and are positioned for the squeezing stroke in conformance with the plate widths being rolled. A 373-kilowatt (500-horsepower) motor, through gearing and mechanical linkage, actuates the forging rams. A hydraulic hold-down prevents bowing of the slab as the edges are subjected to the squeezing action.

On the entry side of the four-high reversing roughing mill are movable side guards which guide the slab into a vertical edging mill. The vertical rolls have a 1016-mm (40-inch) diameter and a 279-mm (11-inch) face with a 5-degree downward taper. They are driven by a 448-kilowatt (600-horsepower), 125 to 406-rpm, variable-speed motor.

The reversing roughing mill is connected directly through mill pinions to a 5222-kilowatt (7000-horse-power), 40 to 80-rpm, variable-speed motor which provides a lineal speed range at the work-roll face of 1.91 to 3.82 metres per second (376 to 752 feet per minute). Both work and backup rolls operate in antifriction roller bearings. Heavy ram-type guides on entry and delivery sides and a turn-around on the delivery side permit the finish rolling of plates wider than could be finished through the continuous stands.

The four, four-high, continuous finishing stands are exact duplicates of each other. Both work and backup rolls operate in anti-friction roller bearings. The finishing stands designated as No. 4 to No. 7 are each driven by a 3730-kilowatt (5000-horsepower) motor. No. 4 to No. 6 stands have a speed range of 110-250 rpm and No. 7 has a speed range of 125-265 rpm. Reducing-gear ratios for No. 4 to No. 7 are, respectively, 2.32, 1.77, 1.46 and 1.35. Corresponding lineal speed ranges at the work-roll faces are 1.69 to 3.86, 2.24 to 5.08, 2.69 to 6.10 and 3.33 to 7.04 metres per second (333 to 760, 440 to 1000, 530 to 1200 and 655 to 1386 feet per minute). Pull back chutes with a variable opening up to a 2540-mm (100-inch) maximum guide the rolled plate through the individual stands. Steam top-side blow-offs are provided between stands in addition to the hydraulic sprays at No. 4-stand entry.

The scale-breaker top roll and the top work and backup roll of all the other stands are balanced by hydraulic jacks and accumulators. Overhead cranes and counter-balances are used to change scale-breaker rolls and the work rolls of the broadside and the finishing stands. A roll-changing rig is used to change the reversing-rougher work rolls, and the broadside, the reversing-rougher, and the finishing-stand backup rolls.

The first section of the run-out table is equipped with a series of cooling sprays. The far portion of the table is a part of a transfer, over the cooling grids of which the plates are moved laterally by a rope-and-carriage transfer. Located immediately beyond the transfer is the light leveler with 17 bending rolls. A cooling table with individually-driven disk rollers spans the distance

between the light and the heavy levelers. The heavy leveler is an 11-bending-roll machine. After leveling, the plates may progress through either one of two alternate routes, to the No. 3 shear unit or toward the rotary-shear line.

No. 3 Shear Unit—The No. 3 shear unit consists of two similar shear lines, each one of which is a continuation of the dual transfer table. Each line consists of a shear approach table, a motor-driven guillotine shear, a powered gage and gage table, and a "kick-off" table with a stacking bed. A scrap shear serves both end shears.

As all the kick-off tables and the stacking beds in this mill are of similar design, a brief description of one will suffice for all. The kick-off mechanism consists of a series of arms which, during a plate delivery from the shear, are positioned between and below the table rollers. The arms are of a double-bar design with small idler rollers free to rotate in the space between the bars. The stacking-bed ends of the arms are keyed to a pivot shaft which, on being rotated through a partial arc, will lift the arms correspondingly through a partial arc and permit the plates to slide down along the idler rollers.

The stacking beds consist essentially of a series of double beams between which movable stops may be shifted and secured in a position to correspond with the widths of plates to be received. A series of beams is fastened to a bed frame and constitutes a stacking-bed section. The sections can be raised individually or in unison to receive the first plate of a stack and are then lowered in consecutive increments as the stack is built up. The stacks are removed from the beds by overhead cranes equipped with sheet carriers or spreaders.

The No. 3 shear unit is utilized for cutting sheet and light plate gages up to 9.5 mm (⅜ inch). The mill is capable of rolling material to thicknesses of about 2.5-mm (0.10-inch) minimum, 1830 mm (7.2 inches) wide. Sheets and plates sheared on the No. 3 unit generally are cut into multiples of the ordered lengths in order to keep ahead of the mill rolling rates. The mill has rolled a maximum of 1938 metric tons (2137 net tons) in an 8-hour period. The multiple-length, side-untrimmed sheets and plates are transferred to the No. 4 shear unit by a transfer car or by placing them on a gravity-feed table, which moves them under a magnetic depiler.

No. 4 Shear Unit—The No. 4 shear unit is housed in a building adjacent to and parallel with the mill building on the motor-room side. Two gravity conveyor tables, one located in the shear building and one in the mill building, move the stacks under a magnetic depiler. The stacks are depiled, and the plates are placed singly on the approach table of a backed-up roller leveler. The leveler has two pinch rolls, nine bending rolls and eleven backup rolls. A marking mechanism attached to the delivery side of the leveler scribes the ordered width on the plates.

The plates progress from the leveler to an end shear equipped with a powered gage which can be set in the range of 1219 mm to 12 954 mm (48 inches to 510 inches). Delivery from the shear gage table is made to the caster bed of a left-hand side-trimming shear and from there to the caster bed of a right-hand shear of otherwise identical design. A scale table and a kick-off table with an 18.3-metre (60-foot) stacking bed complete the main shear line.

Beyond the main shear line are located a squaring shear and a circle shear serviced by gravity conveyors and a jib hoist. The circle shear can cut 203-mm to 2438-mm (8-inch to 96-inch) diameter circles from 2.4-mm to 9.5-mm ($\frac{3}{32}$-inch to $\frac{3}{8}$-inch), 0.30-carbon steel at a cutting speed of 0.25 to 0.51 metres per second (50 to 100 feet per minute). A scrap shear serves both the resquaring and the circle shears, and a similar shear serves each of the side shears.

Rotary Shear Line—No. 1 Shear Unit—Heavier-gage plates, after moving across the dual transfer, continue their travel in a reverse direction to the mill rolling direction, over spool-type marking and inspection tables toward a crop shear. The distance traveled by a plate from the finishing stand to this crop shear is approximately 366 metres (1200 feet). The plates are cropped and are divided into multiples of the ordered lengths at this point and progress to a roller-chain lift transfer located immediately beyond the shear. Required painting and stamping identifications are applied as the plates travel over this transfer.

Sketch plates and other plates which are beyond the capacity of the rotary shear are diverted from the delivery side of the transfer by a table extension and stacking beds. They are moved by overhead crane and transfer car to the No. 2 shear unit for cutting. Plates within the capacity of the rotary shear resume travel in the rolling direction toward the unit.

The rotary-shear approach table is equipped with magnetic manipulators to position the plates for the shear entry. The shear is driven by a 224-kilowatt (300-horsepower) motor. Scrap from each side is guided from the main cutters through chutes to the scrap cutters located below the large knives. The scrap is cut by a rotary, eccentric motion and is dropped through chutes into disposal buckets. The shear has a capacity for cutting 508-mm to 2286-mm (20-inch to 90-inch) widths, 2.38-mm to 19.05-mm ($\frac{3}{32}$-inch to $\frac{3}{4}$-inch) gage, of 0.30-carbon material at a cutting speed of 0.40 to 1.20 metres per second (79 to 237 feet per minute). The cutters may be set to 1.59-mm ($\frac{1}{16}$-inch) increments within the width range.

The trimmed plate may be sent to either one of two end shears for final cutting to ordered length. The near or No. 2 end shear, located in the shipping building, is reached by traversing a rope-and-carriage transfer located immediately beyond the rotary-shear delivery table. The far or No. 3 end shear approach table is immediately beyond the transfer referred to, and is reached by direct, continuous travel from the rotary shear. Both shears can handle material 19.05-mm ($\frac{3}{4}$-inch) thick by 2540 mm (100 inches) wide, and each is driven by a 112-kilowatt (150-horsepower) motor. They are followed by gage tables equipped with motor-operated, tilting, and traveling plate stop-gages with a 1220-mm to 18 288-mm (48-inch to 720-inch) travel range from the knife edge. The gage carriage travel speed is 0.25 metre per second (50 feet per minute). Each end shear is serviced by a scrap shear.

The sheared plates from the No. 3 end shear are transferred into the No. 1 shipping building by a roller-chain lift transfer. Travel direction is reversed to move over a scale table with an automatic weight-recording device and continue on to a kick-off table and stacking bed. Plates from No. 2 end shear travel over a similar scale table and on to a kick-off table and stacking bed. The kick-off table connects with the conveyor feed table of the continuous normalizing furnace.

Heat-Treating Furnace—Plates to be heat treated are placed in stacks on a gravity feed table which moves them under an unpiler. The unpiler is a motor-driven traverse bridge and hoist with five selective magnetic lifters having a total capacity of ten tons. It operates on a structural framework runway. The runway extends over a gravity conveying table which also connects the No. 1 shear line kick-off table with the furnace.

The furnace is of the straight conveyor type, divided into five zones. All zones, with the exception of the entrance or heating zone, are equipped for recirculation to improve temperature uniformity. There is an alloy baffle between the heating and the recirculating zones. The furnace is 39.6 metres (130 feet) long, 2.7 metres (9 feet) wide, and the roof is 864 mm (2 feet, 10 inches) above the conveyor. The conveyor consists of sprocket-driven rolls. A variable speed range with a maximum of 0.38 metre per second (75 feet per minute) is provided. Pressure burners for all zones, burning a mixture of coke oven and natural gas, with temperature and fuel-air ratio controls, provide the heat input. A stack at each end of the furnace provides natural draft.

Plates emerging from the furnace may be air cooled for normalizing or quenched in a quench platen. The quenched plates are subsequently tempered in the same furnace at a suitable lower tempering temperature. The backed-up leveler is a duplicate of the one in No. 4 shear unit. Plates from the leveler traverse a cooling bed and are stacked in conventional stacking beds at the far side of the traverse.

No. 2 Shear Unit—The No. 2 shear unit has two main shear lines, consisting of a rotary shear line housed entirely in the No. 2 shear building, and the sketch shear line, which has the feeding table and two shears in the shear building and one shear, scale table, and stacking beds in the No. 2 shipping building.

The sketch shear line, as its designation implies, is used to shear irregularly shaped plates. It also is utilized to shear structural-grade, rectangular plates heavier than 19.05-mm ($\frac{3}{4}$-inch) gage, and alloy plates of thinner gage which, because of composition, are beyond the cutting capacity of the No. 1 shear line. Such plates diverted from the No. 1 shear line beyond the No. 2 transfer and moved to the No. 2 shear unit by transfer car, as previously described.

Plate stacks for the sketch shear line are placed on a gravity feed table by an overhead crane. An overhead, selective, magnetic unpiler puts the plates singly on a marking and layout, five-chain conveyor table. Both sketch and rectangular plates are laid out manually on this table. The first shearing unit in the line, an end shear, cuts the plates to length. The plates may be turned 180 degrees on the caster beds immediately beyond and returned, if necessary, to have both ends front-end cut for the elimination of shear bow. A continued, manually propelled movement over caster beds brings the plate to a side shear for edge trimming. A lateral movement into the shipping building brings the plate to an opposite hand but otherwise identical shear for the trimming of the opposite edge. Further manual propulsion opposite to the original travel direction moves the plate back to powered traction on the scale table and a kick-off table and stacking beds. Two scrap shears serve the three major shearing units in this line.

When the mill rolling rate on 9.53-mm to 19.05-mm (⅜-inch to ¾-inch) gage plates exceeds the capacity of the No. 1 shear line, the excess multiple-length plates are diverted at the same point and are transferred to the No. 2 shear unit in the same manner as sketch plates. The plates are spread on a chain-conveyor marking table and move over an approach table to a rotary side-trimming shear. In general design features, the shear is similar to the one in the No. 1 shear line. The capacity ranges from 762 mm to 2438 mm (30 inches to 96 inches) in width and 2.38 mm to 19.05 mm (³⁄₃₂ inch to ¾ inch) in gage at a cutting speed of 0.25 metre per second (50 feet per minute). The scrap cutters are of the revolving-drum type. A duplicate of the end shear in the sketch shear line serves the rotary shear. It is followed by a back shear table, a scale table and a kick-off table with stacking beds. A guillotine scrap shear serves the end shear.

Auxiliary shearing equipment in No. 2 shear unit consists of a resquaring shear, and a test-cutting shear. The squaring shear is a motor-driven unit. The motor-driven circle shear has a capacity ranging from 508-mm to 3810-mm (20-inch to 150-inch) diameter, 2.38-mm to 31.75-mm (³⁄₃₂-inch to 1¼-inch) gage, in structural grades at a cutting speed of 0.28 metre per second (56 feet per minute). The test-cutting shear has a capacity equivalent to cutting 508-mm by 20.64-mm (20-inch by ¹³⁄₁₆-inch) thick plates.

An individual plate inspection unit is made available by feeding from the No. 2 shear building and repiling in the No. 2 shipping building. It includes two parallel sections of gravity table for plate entry and two parallel sections for plate delivery. Each section employs pipe rollers operating in roller bearings. Located between the gravity table lines is a motor-driven roller table over which individual plates are moved and are tilted for inspection by eight lifting arms to an angle of 75 degrees. The unit is served by one unpiling unit and one piling unit. Each traverse and hoist, with five selective magnetic lifters, operates on a structural framework.

The greasing system for all table bearings, spindles, spindle carriers, jacks, slide joints, etc., is of the automatic "single-line" reversing type and contains eleven major pumping units.

Oil lubrication for the majority of the equipment, such as table line shafts, gears and bearings, leveler drives, screw-down drives, reduction gears, pinions, backup-roll bearings, and motor-room equipment, is accomplished by circulating systems of the single- and double-tank types.

The 2438-mm (96-inch) FOUR-HIGH CONTINUOUS PLATE MILL "E"

The four-high continuous plate mill which will be

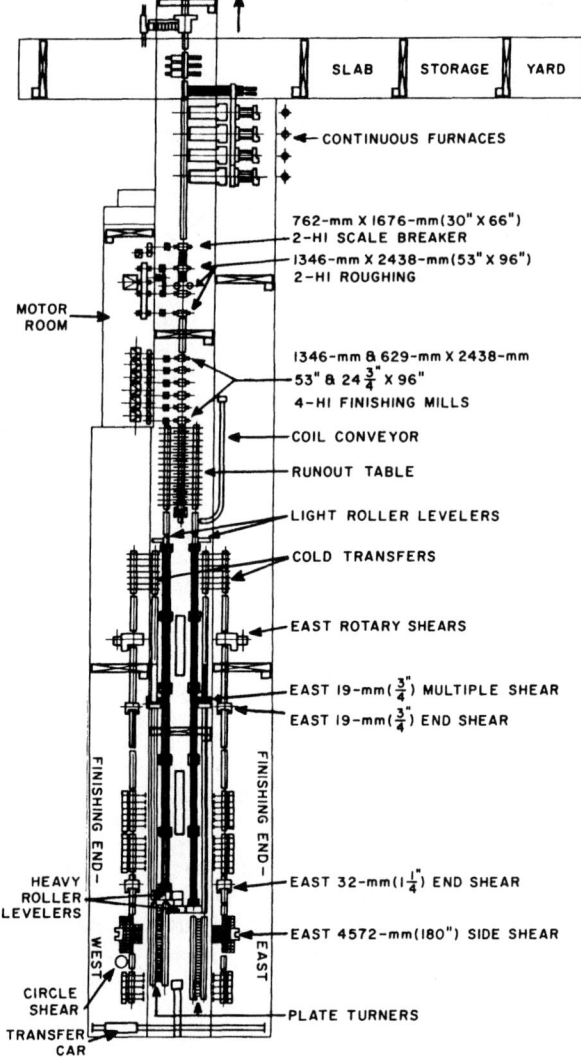

FIG. 25—10. Diagram (not to scale) of the layout of the mills and auxiliary equipment comprising the 2438-mm (96-inch) four-high continuous plate mill "E".

described in detail was placed in operation in the early part of 1931 and, therefore, represents an early stage in the development of the wide-four-high continuous mills. The wide, four-high, continuous hot-strip mills, which have been placed in operation in the intervening period and which are described in another section of this book, are more representative of the potential productive capacities, the surface finish, and the width and gage uniformity which can be achieved with a modern mill of this type designed primarily for plate production.

The 2438-mm (96-inch) continuous plate mill "E", Figure 25—10, receives the bulk of its charge in the form of cold slabs from a multi-purpose slab-yard building running at 90 degrees to the mill center line and serviced by four overhead cranes. The 2438-mm (96-inch) plate mill and its supplying unit, the 1118-mm (44-inch) slab mill, have a common and continuous center line. The slab mill shear delivery and piler tables form a continuous table line with the 2438-mm (96-inch) mill furnace and mill approach tables so that hot slabs may be table-conveyed directly from the slab-mill shears into the 2438-mm (96-inch) mill. A pusher and chain transfer adjacent to No. 1 reheating furnace permits the diversion of hot slabs from the mill table and their transfer to the charging ends of the reheating furnaces. It also is utilized for the recharging of furnace kick-outs.

The slab mill has a rated capacity of 127 000 metric tons (140 000 net tons) of product per month. Virtually all the slab-mill production is removed from the run-out tables by three sectional pushers and pilers located in the slab yard. Slabs in ingot stacks are removed from the pilers by overhead cranes.

All 2438-mm (96-inch) plate mill "E" slabs are placed in the central portion of the slab yard, adjacent to the pilers, for cooling. After cooling, they are conditioned, and piled in rolling sequence in an area near the pilers. This area has a storage capacity of 4535 metric tons (5000 net tons). A portion of the slab tonnage produced for other applications and cooled in the central area also is conditioned here and is loaded out on standard-gage tracks. Cool slabs in excess of the yard conditioning capacity are sent to other yards for conditioning.

The plate-mill furnace-charging tables extend into the slab yard. The slab charger is a side unpiler consisting of a skidded deck with a vertical screw motion and a pusher operating transversely to the table travel direction and the table-top elevation. Slab stacks are placed on the unpiler deck in its down position, and the slabs are pushed singly on the charging table as the unpiler is moved upward in slab-thickness increments. The slabs are positioned at the furnace charging doors and are charged into the furnaces by double-row pushers which can be operated singly or in unison.

The heating equipment for this mill consists of four continuous-type, two-zone, triple-fired slab furnaces equipped to burn natural gas or fuel oil through all burners. One of the originally installed furnaces is manually controlled, is not equipped with either recuperators or regenerators and is only used for standby service. When this furnace is used, it is single-row charged with slabs of 2104-mm (83-inch) maximum length. Three furnaces are equipped with automatic heating controls and metallic recuperators. Slabs charged into these furnaces vary from 89 mm to 165 mm (3½ to 6½ inches) in thickness, from 711 mm to 1524 mm (28 to 60 inches) in width, from 1270 mm to 2104 mm (50 to 83 inches) in length for a double-row charge and up to a maximum of 4521 mm (178 inches) for a single-row charge. Each of these furnaces has rated heating capacity of 45 metric tons (50 net tons) of slabs per hour.

The two-high scale-breaker stand is driven by a 458-kilowatt (600-horsepower), 488-rpm motor, driving through a gear reducer and pinions. The steel rolls operate in bronze-insert, babbitt bearings. The top roll is counter-weight balanced. Drafts on this stand are limited to 9.5-mm (⅜-inch) maximum. Top and bottom hydraulic sprays operating at 6895 kPa (1000 psi) are located on the delivery side of the mill.

The roughing train consists of three duplicate two-high stands. All three stands are driven by a 4576-kilowatt (6000-horsepower), 370-rpm motor, driving through a special gear-reduction set with two flywheels on the high-speed shaft and through conventional pinion stands. No. 1, No. 2, and No. 3 stands operate, respectively, at 8.5, 10.1 and 15.15 rpm, corresponding to lineal surface speeds of 0.59, 0.70 and 1.05 metres per second (116, 138 and 206 feet per minute). Only one slab can be undergoing reduction in the roughing train at any time. The cast-steel rolls for these stands operate in bronze-insert, babbitt bearings. The top roll is balanced hydraulically. The screwdowns are operated by two motors tied in with a magnetic clutch so that they may be operated in unison for draft settings or individually for roll alignment. High-pressure hydraulic sprays are located at No. 1 and No. 3 delivery and at No. 4 entry.

Motor-driven sectional side guards on the skew roller table between the scale breaker and No. 1 stand may be positioned to permit either a straightaway entry to No. 1 stand or to form a turning pivot for a 90-degree turn and a broadside entry to No. 1 stand. Therefore, No. 1 stand may be utilized as either a straightaway or a broadside mill. A rack-type carriage pusher operating on a structural framework above the table provides for a square entry for the broadside pass. Powered side guides at the entry of each roughing stand, as well as the first finishing stand, permit centered entry.

A similar arrangement of side guards and skew table on the delivery side of No. 1 stand makes possible either continued straightaway progress of the slab or a 90-degree turn after a broadside pass.

The finishing train consists of six duplicate four-high stands. Each mill is driven by a 2611-kilowatt (3500-horsepower), 165-330-rpm motor through a gear-reduction unit, conventional pinion stand and wobbler-type spindles. The speeds of the work rolls of the various stands in rpm, m/s and ft/min are listed below:

Stand No.	Speed Range		
	rpm	m/s	ft/min
4 Finishing	20 to 49	0.65 to 1.60	128 to 315
5 Finishing	30 to 72	0.98 to 2.35	193 to 463
6 Finishing	39 to 94	1.27 to 3.05	250 to 600
7 Finishing	49 to 118	1.60 to 3.86	315 to 760
8 Finishing	54.5 to 132	1.78 to 4.32	350 to 850
9 Finishing	60 to 146	1.96 to 4.78	385 to 940

Grain-iron work rolls are used in No. 4 and No. 5 stands, and chilled iron work rolls in the remainder of the stands, with steel backup rolls. Both work and backup rolls operate in antifriction roller bearings. The top backup and work roll are balanced hydraulically as an assembly, with the work roll being held against the backup roll by two spring suspension take-up rods positioned through yoke extensions. Draft settings are made by motors driving the screwdowns through worm reduction gearing. The two motors for each stand are connected by magnetic clutches and are operated in unison for draft settings and singly, when the stock is in the mill, for camber correction.

The spacings between the mill stands are each occupied by a two-sectional retractable table which in a normal operating position presents a continuous, smooth iron liner surface to plate travel. Adjustable, powered side guides center the plates for admission at the point of entry, and top and bottom stripper-guide assemblies are attached to the delivery section. A looping roll normally is positioned below the table surface and is raised in an arc by pneumatic cylinders when it is necessary to take up stock slack while speed adjustments are being made. Prior to a roll change, the two table sections are retracted so that the liners and the supports of one section overlap the other in the center of the spacing, and both the entry- and the delivery-guide assemblies are free of the rolls.

An overhead-crane-suspended sleeve, into which a work roll is fitted as a counterweight, is used for work-roll changes. The frequency of work-roll changes is extremely variable and is dependent on the particular roll stand as well as on the preponderance of product gages rolled. In the rolling of sheet gages, the leader and the finisher may require changing at four-hour intervals. When the schedule includes only plate rolling and the preponderance of the tonnage is in the heavier plate gages, the work rolls in No. 4 and No. 5 stands may last out the weekly schedule or require changing only once to meet wide-plate rolling requirements in the latter part of the week. Work-roll changes are made in 25 minutes.

A cast counterweight secured on an integral arm and sleeve which fits over the roll-neck extension is used for the changing of the backup rolls, whereas a sleeve and another roll as a balancer are used for changing the roughing rolls. The frequency of backup-roll changes is also variable and dependent on the roll-stand position and gages rolled. Since one roll is changed per week, the average time interval amounts to 12 weeks. No. 3 roughing rolls are changed each week, No. 2 at weekly or by-weekly intervals dependent on schedules, and No. 1 at 3 to 4-week intervals. Backup-roll changes require an average of 2½ hours per roll, and roughing rolls require 3½ hours a set.

Product is delivered from the finishing stand onto a central runout table, consisting of individually-driven disk rollers which protrude through openings in cast alloy-iron aprons. Plates are stopped on the runout table by raising the lifting aprons above the table rollers. The plates are moved laterally over iron gridwork by a cable-carriage dog transfer to either one of two duplicate, parallel finishing lines.

The leveler-approach tables convey the plates to the light levelers, the first processing units in each line. They are located immediately beyond the hot transfer and have seventeen bending rolls. Delivery from the leveler is made to the first of five 38.1-m (125-foot) sections of sprocket-driven chain spool conveyors. These conveyors span the distance to the heavy levelers and also serve as cooling and top surface inspection tables. The heavy leveler is a 13-roll machine with seven top and six bottom, 356-mm (14-inch) diameter by 2540-mm (100-inch) long, rolls. From the leveler the plates are discharged on a 40.2-m (132-foot) combination table and turnover device. The turnover consists of two series of arms keyed at pivot shafts located between two parallel roller tables. In an idle position both sets of arms are below the roller-top levels of the two tables. When the plate is delivered from the

leveler, the arms are moved toward each other in an arc, with the sending arms passing through a greater arc. This results in a transfer of the plate to the receiving arms as they approach a verticle position. The lowering of the arms transfers the plate to the adjacent parallel table which reverses travel direction toward the multiple shear.

Approach tables for the multiple shears serve as a bottom surface inspection and marking table. The first identification is made by chalking on the plate surface. Multiple lengths to be sheared are also marked on the plate with allowances for necessary tests. The multiple shear is a 19-mm by 2540-mm (¾-inch by 100-inch), motor-driven, open-throat, downcut shear. A chain scrap conveyor is common to the multiple and the adjacent end shear and carries the scrap from both to an alligator scrap shear.

Continued table travel brings the multiple plates to a cold transfer. This is a chain lift transfer. During the crosswise travel, physical identification of product and test pieces is completed by manual painting and stamping. Routing information for placement in the shipping area or direct car loading also is indicated on the plates.

Travel in the rolling direction is resumed on the approach table to the rotary shear. The finish-shear lines are located in the shipping buildings which are adjacent and parallel to the mill building. Three magnetic manipulators traveling in a direction transverse to that of the table are available for positioning the plates for the rotary-shear entry. An electromagnet, traveling with the table direction, holds the plate in a fixed lateral position. Cutting capacity of the rotary shears ranges from 3.2-mm (⅛-inch) to 19-mm (¾-inch) gauge and 508-mm to 2438-mm (20 inches to 96 inches) in width at a cutting speed of 0.40 to 0.80 metres per second (79 to 158 feet per minute). Width settings may be made in 1.6-mm (1/16-inch) increments. Side trimmings are guided to rotary scrap cutters, cut into short lengths and dropped into a hold for removal by magnet.

Two table sections that serve as delivery and entry tables for the rotary and the end shears respectively have powered side guards to guide the delivery from the rotary shears and facilitate square entry into the end shears. The end shears are duplicates of the multiple shear. A motor-operated lift plate gage with a travel speed of 0.25-metre per second (50 feet per minute) can be set from 1.8 metre (6-foot) to an 18.3-metre (60-foot) distance from the shear knife. The delivery table is followed by a scale table and two sections of 19.5-metre (64-foot) push-off tables and side pilers. The side pushers travel in transverse ways across the tables to stack the plates in the pivoted side pilers, the table ends of which are lowered in conformance with the height of the plate stack.

Plates heavier than 19-mm (¾-inch) and up to 32-mm (1¼-inch) gage are laid out manually on the push-off tables and continue over short approach tables to end shears. The shears are conventional motor-driven downcut type followed by a short depressing table and a delivery table. The plates then are moved manually over caster beds to a side shear. The plates are turned 180 degrees on the caster beds and moved back to the side shear to trim the opposite edge.

Sidetrim scrap is moved manually to an alligator scrap shear for cutting into short lengths.

A circle shear is located beyond the side shear of the west shear line. The shear has a cutting capacity of 3-mm to 19-mm (⅛-inch to ¾-inch) gage and 406-mm to 2133-mm (16-inch to 84-inch) diameter at a cutting speed of 0.10 to 0.30 metre per second (20 to 60 feet per minute). A transfer car at the extreme end of the building permits the transfer of circles and plates between the various buildings.

The greasing system for all table bearings, spindles, spindle carriers, jacks, slide joints, etc., is of the automatic "single-line" reversing type and contains eleven major pumping units.

Oil lubrication for the majority of the equipment, such as table line shafts, gears and bearings, leveler drives, screw-down drives, reduction gears, pinions, backup-roll bearings, and motor-room equipment, is accomplished by circulating systems of the single- and double-tank types.

E. UNIVERSAL PLATE MILLS

Universal plate mills are single-stand units which roll plates to a width within standard tolerances. Although universal plate mills producing widths up to 1524-mm (60-inches) have been in operation, the bulk of the installations has been of 1219-mm (48-inch) width and under. The number of installations and the capacity has decreased rather than increased in the past several decades. The production of deep wide-flange beams displaced a large tonnage of fabricated beams and columns that formerly had utilized universal plates for the web and flange portions. Consequently, universal plate mills in operation today date back several decades to their installations dates; and the unit to be described is typical in spite of its early construction date.

THE 762-mm (30-inch) UNIVERSAL PLATE MILL "F"

The 762-mm (30-inch) universal plate mill "F", Figure 25—11 was erected in 1907 to produce rolled-edge plates ranging from 4.8 mm (³⁄₁₆ inch) to 152.4 mm (6 inches) in thickness and 152 mm (6 inches) to 762 mm (30 inches) in width. Slab sizes utilized to produce this range vary from 63.5 mm (2½ inches) to 292 mm (11½ inches) in thickness, 127 mm (5 inches) to 787 mm (31 inches) in width and 1397 mm (55 inches) to 3048 mm (120 inches) in length.

The conditioning and slab-storage yard serving this mill is under an open crane runway serviced by two cranes. The yard consists of four sections, for slab receipt, conditioning by hand scarfing, stock storage, and current charge storage. Charge slabs are piled in upright racks in the proper sequence. About 4080 metric tons (4500 net tons) can be stored in the charge and stock racks.

Reheating Furnaces—Two single-zone, end-fired, hydraulic-pusher-charged, continuous reheating furnaces service the mill. The furnaces are rated 18 metric tons (20 net tons) an hour per furnace for full hearth coverage and a cold charge. Heated slabs and billets slide down the dropout skids onto the mill approach table.

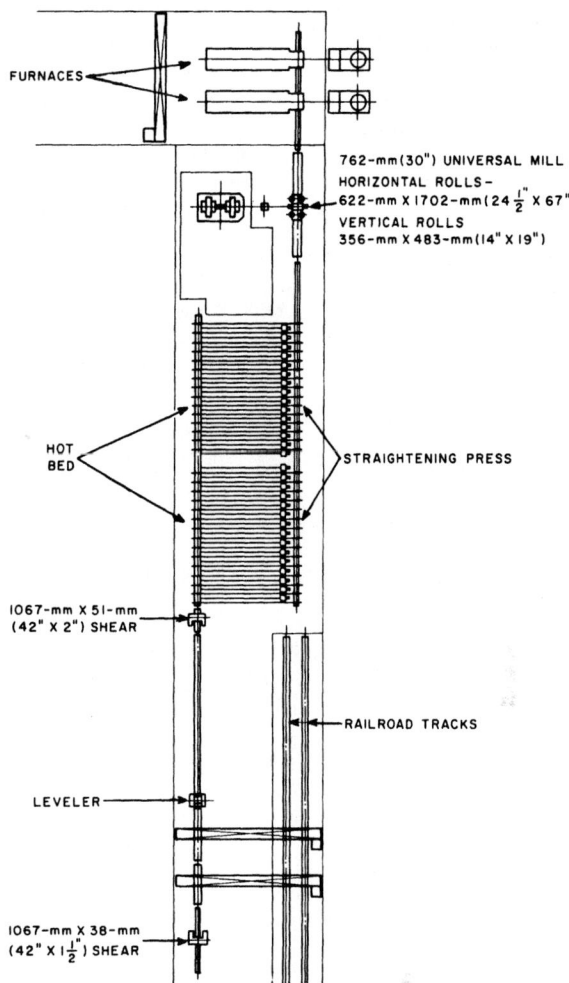

FIG. 25—11. Layout (not to scale) of the 762-mm (30-inch) universal plate mill "F".

762-mm (30-Inch) Universal Plate Mill Stand—The single-stand two-high reversing mill, with vertical edgers front and rear, has chilled alloy-iron horizontal rolls and chilled-iron vertical rolls.

The mill is driven by a 2984-kilowatt (4000-horsepower), 150-rpm reversing motor, the first reversing-mill motor to be built and placed into successful operation. In 1973, the mill was rebuilt to provide new mill housings designed for individually driven motor-driven vertical rolls and a new pinion stand, both complete with anti-friction bearings.

Rolling—Slabs intended for bars in the 152-mm to 203-mm (6-inch to 8-inch) width range are provided 25 mm (1 inch) narrower than the finished plate width. A slab 152 mm (6 inches) thick and 203 mm (8 inches) wide rolled down to 25-mm (1-inch) gage will finish 248 mm (9¾ inches) wide on this mill if it is allowed to spread freely. Therefore, on the basis of slab provision for a 25-mm (1-inch) gage, 203-mm (8-inch) wide bar, a 25-mm (1-inch) spread is allowed; and the edges actually are worked down 19 mm (¾ inch). Slabs provided for over 203-mm (8-inch) to 305-mm (12-inch) widths are 12.7 mm (½ inch) narrower than the finished size. Slab and finished widths are equivalent in the over 305-mm (12-inch) to 432-mm (17-inch) range. Slabs 12.7 mm (½-inch) over finished size are provided for the over 432-mm (17-inch) to 762-mm (30-inch) range. Edging drafts are taken alternately with the vertical rolls on the delivery side of the horizontal-roll drafting passes. As the horizontal-roll drafts are variable, vertical- and horizontal-roll diameters must be kept matched so that the peripheral speed of the vertical rolls exceeds that of the horizontal rolls. Main-roll changes average three per week for a normal schedule and require 55 minutes to complete. Vertical rolls are changed at three-week intervals and require 3½ hours to change both sets. Spray nozzles operate at about 13 800 kPa (2000 psi) pressure for descaling on both sides of the mill.

Leveler and Hot Bed—In 1973, a new hot-cold leveler was installed at the end of the mill runout table; designed to flatten hot plates up to 102 mm (4 inches) thick and cold plates up to 63.5 mm (2½ inches) thick. Most material is hot leveled at least four passes, timed with the rolling mill to maximize leveling. Water sprays are used to aid in heat dissipation at the leveler, thus lowering plate temperature into a range where less distortion results.

A new hot bed was installed in 1973, with integral screw clamp straighteners to handle lengths through 33.58 meters (1440 inches). The bed has closely spaced skids with chains to move plates across it without lifting.

An automated shear gage was also installed to handle lengths 1.83 through 33.58 metres (72 inches through 1440 inches).

SECTION 4

HEAT-TREATING FACILITIES FOR STEEL PLATES

Types of Heat Treatment—Depending upon the grade of steel and the intended end use of plates, they may be subjected to annealing, normalizing, stress relieving, spheroidize annealing, accelerated cooling, quenching, tempering, or certain combinations of some of these treatments.

While the purposes and principles of these heat treatments (and the types of equipment in which they can be performed) are discussed in detail in Chapter 40, each will be defined briefly here.

Annealing consists of a single thermal treatment, intended to place the steel in a suitable condition for fabrication. The steel is heated to a temperature in or near the critical temperature range and is cooled at a predetermined rate or cycle. Plates are generally annealed in "open" furnaces without atmosphere control.

Normalizing consists of heating the steel above its critical temperature range, and cooling in air. This treatment refines the grain size and improves uniformity of microstructure and properties of the hot-rolled plate.

Stress relieving consists of heating the steel to a temperature below the critical range to relieve stresses induced by flattening or other operations such as cold working, shearing, or gas cutting. It is not intended to alter the microstructure or mechanical properties significantly.

Spheroidize annealing is performed by prolonged heating of the steel in a controlled-atmosphere furnace at or near the lower critical point, followed by retarded cooling in the furnaces, to produce a lower hardness than can be obtained by regular annealing. The purpose is usually improvement of performance of the steel in cold forming.

Accelerated cooling is employed to improve resistance to impact (toughness) and refine the grain size of certain grades and thicknesses of plate. Such cooling is accomplished by fans to provide circulation of air during cooling, or by a water spray or dip.

Quenching consists of heating the steel to a suitable austenitizing temperature, holding at that temperature for a sufficient time to effect the desired change in crystalline structure, and immersing and cooling the steel in a suitable liquid medium that will depend upon the composition of the steel and its cross-section.

Tempering is carried out by preheating previously quenched or normalized steel to a predetermined temperature below the critical range, holding for a specified time at that temperature, and then cooling under suitable conditions to obtain the desired mechanical properties.

Furnaces for Heat Treating Plates—The size, thickness and grade of steel determine to some extent the type of furnace employed for the heat treatment of plates. In-and-out batch-type, car-bottom and semi-continuous roller-hearth furnaces are among the types employed.

PLATE HEAT-TREATING EQUIPMENT
FOR 4064 mm (160 inches) AND 2540 mm (100 inch) "D" PLATE MILLS.

With installations in the 4064-mm (160-inch) "B" and 2540-mm (100-inch) "D" plate mills, the Harvey Shop and the Forgings Division, carbon-, alloy-, and stainless-steel plates ranging from 2.5-mm (0.10 inch) to 152-mm (6 inches) thick and heavier can be heat treated. Facilities consist of the following:

Location	Maximum Thickness of Plate Treated	
	mm	inches
4064-mm (160-Inch) Mill "B"		
No. 1 Plate Treating Line	50 and under	2 and under
No. 2 Plate Treating Line	50 and under	2 and under
Stainless-Steel Furnace	50 and under	2 and under
2540-mm (100-Inch) Mill "D"		
Normalizing Furnace	19 and under	¾ and under
Harvey Shop		
Car-Bottom Furnaces	Over 50	Over 2
Forgings Division		
Car-Bottom Furnaces	Over 50	Over 2

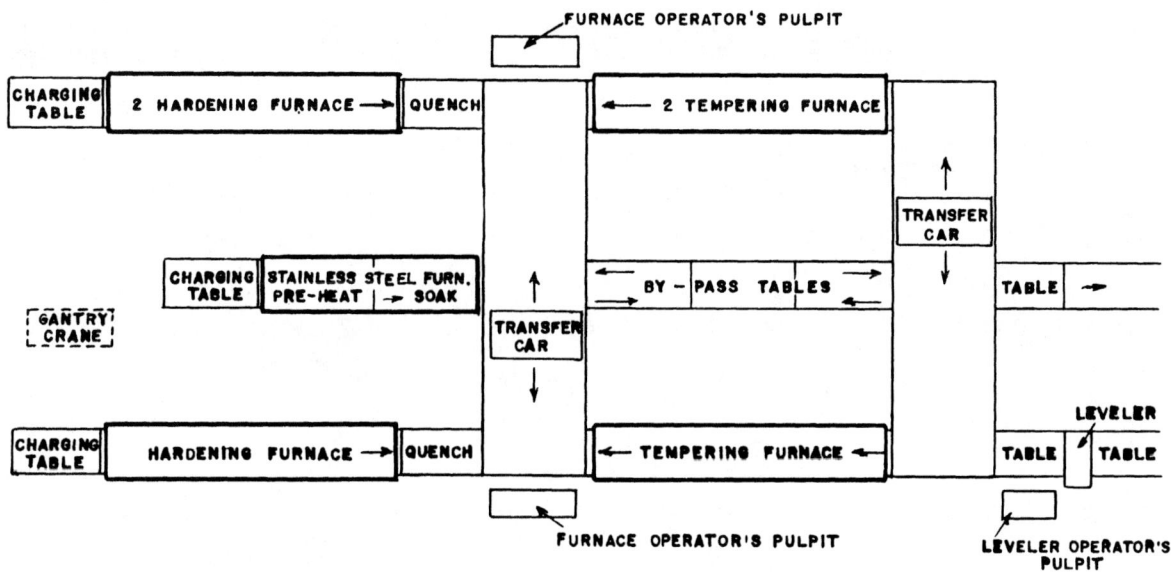

FIG. 25—12. Schematic layout (not to scale) of the plate heat-treating lines in the 4064-mm (160-inch) mill "B". (See also Figure 25—13.)

4064-mm (160-Inch) Mill "B" Heat-Treating Facilities—The heat-treating facilities of the 4064-mm (160-inch) plate mill comprise: (1) a preliminary inspection and conditioning area; (2) a classification and shipping building; and (3) the heat-treating and shearing building in which are located the No. 1 and No. 2 plate treating lines and the furnace for heat treating stainless-steel plates, arranged as shown in Figures 25—12 and 25—13. These facilities are designed to slow cool, inspect and heat treat plates up to 3962 mm (156 inches) wide and 13 056 mm (514 inches) long, and up to 50 mm (2 inches) thick, that have been produced by hot rolling on the 2540-mm (100-inch) and 4064-mm (160-inch) plate mills, and to shear all such plates that are no more than 25 mm (1 inch) in thickness.

Plates are delivered to the inspection building directly from the rolling mills. The inspection area has hoods for slow cooling of heavy plates. All plates are inspected top and bottom, with the assistance of a turnover rig. Those having defects revealed by the inspection are taken to grinding beds for conditioning. A gantry crane handles the product throughout the various operations. When inspection and any necessary conditioning are completed, the plates are loaded onto mobile trailers and delivered to the classification building by a diesel-powered tractor.

In the classification building, the plates are unloaded from the trailers, grouped for most efficient heat treatment, and then moved by gantry crane onto the charging table of the appropriate furnace in the heat treating and shearing building.

As shown in Figure 25—12, the No. 1 and No. 2 plate heat treating lines are situated on opposite sides of the heat treating and shearing building. The furnace for heat treating stainless steel is located between the two plate treating lines. Operating data for the various furnaces are given in Table 25—I.

Plates to be processed in the No. 1 and No. 2 plate treating lines are placed on the charging tables of the continuous hardening furnaces (Figure 25—14), whose function is limited to heating the plates to a hardening temperature. The individual plates are carried through these furnaces, by the motor-driven high-alloy steel rollers that form the hearths. Speed of travel of plates through a furnace is regulated according to the thickness of the plate, with thinner plates moving through more rapidly than thicker ones. Each plate heated for hardening is delivered to the water quench which is sized to accommodate the largest plates. A fast runout section of table at the delivery end of each hardening furnace rapidly moves heated plates into the quench, with only an anticipated 20 seconds of elapsed time until the water is applied.

When a plate is in position in the quench, the transfer rollers are lowered so that the plate rests on a fixed platen. The hydraulically operated top platen is then lowered until it comes in contact with the top surface of the plate. Both platens have ribbed surfaces extending across the width of the quench, and water pipes are located in the recesses between the ribs. With the plate in place and the platens exerting pressure over the entire surfaces of the plate, water sprays from the pipes play on the steel from above and below. Quenching time is approximately three minutes per 25 mm (1 inch) of plate thickness.

From the quench, the plate is rolled onto an electrically operated transfer car that delivers it to the by-pass roller tables, parallel to the tempering furnaces. The plate is then transferred by another car to the entry end of the tempering furnace in which it is to be processed. The tempering furnaces, like the hardening furnaces, are equipped with heat-resistant steel rollers that move the plates through the furnace and into the

FIG. 25—13. General view of the heat-treating lines in the 4064-mm (160-inch) mill "B", seen from the charging end. No. 1 plate-treating line is at the right, No. 2 line at the left, with the stainless-steel heat-treating furnace between. (See also Figure 25—12.)

Table 25—I. Operating Data for Furnaces in the 4064-mm (160-inch) Plate Mill "B".

Characteristic	No. 1 Hardening Furnace	No. 2 Hardening Furnace	No. 1 Tempering Furnace	No. 2 Tempering Furnace	Stainless-Steel Heat-Treating Furnace
Temperature Limits					
°C	675-955	675-955	400-675	400-675	675-1150
°F	1250-1750	1250-1750	750-1250	750-1250	1250-2100
Number of Temperature-Control Zones	6	11	10	12	3
Connected Fuel Input					
m³/h (at 0°C. 760 mm Hg)	1800	1800	1676	1837	1418
ft³/h (at 60°F. 30 in. Hg)	67 000	67 000	62 400	68 400	56 800
Average Fuel Consumption at Maximum Capacity					
m³/h (at 0°C, 760 mm Hg)	913	1074	833	1289	860
ft³/h (at 60°F, 30 in. Hg)	34 000	40 000	31 000	48 000	32 000
Roller speeds					
Charging					
m/s	0.0015 to 0.038	0.0025 to 0.064 or 0.102 to 0.406	0.381	0.0025 to 0.064 or 0.445 to 0.889	0.254
ft/min	0.3 to 7.5	0.5 to 12.5 or 20 to 80	75	0.5 to 12.5 or 87.5 to 175	50
Through Furnace					
m/s	0.0015 to 0.0038	0.025	0.0018 to 0.445	0.025	0.005 to 0.010*
ft/min	0.3 to 7.5	5.0	0.035 to 87.5	5.0	1 to 2*
Discharging					
m/s	0.889	0.445 or 0.889	0.889	0.445 or 0.889	0.889
ft/min	175	87.5 or 175	175	87.5 or 175	175

*Oscillating.

FIG. 25—14. Alloy-steel plate entering the hardening furnace of the No. 2 plate heat-treating line of the 4064-mm (160-inch) mill "B".

quench when treating plates of certain grades that must be rapidly cooled after tempering. Direction of travel of plates through the tempering furnaces can be reversed when treating plates that do not require water cooling after tempering.

Some carbon-steel plates are normalized. In this operation, the steel is heated to the proper temperature in a hardening furnace but, instead of being quenched, is cooled on roller tables in the open air.

Stainless-steel plates, as contrasted with others, are not tempered. They are heated in the stainless-steel furnace and then quenched, following which they are delivered to the roller leveler for flattening.

Plates that have been tempered or quenched and tempered are delivered by the transfer cars and bypass tables to the delivery end of the roller leveler. With the exception of stainless-steel plates, roller leveling is rarely necessary and the plates simply pass through the rollers of this machine to the cooling tables. A transfer car at the exit end of the cooling tables returns the plates to the classification building, where test coupons are removed and sent to the metallurgical laboratory for tests. Plates that conform to test requirements are returned by transfer car from the classification building to the approach table of the shear line. Individual plates pass through a turntable to the side shear where one side of each plate is sheared. The roller table is reversed to return each plate to the turntable where it is then turned end for end and returned to the shear

for shearing the other side. Plates continue down the roller table from the side shear to the end shear, where both ends are sheared. After shearing, plates are removed from the table and loaded for shipment. Plates over 25 mm (1 inch) thick cannot be sheared on this shear line and are sent to heavy shearing or gas-cutting facilities elsewhere in the plant.

2540-mm (100-Inch) Mill "D" Hardening-Tempering Furnace—This furnace is an integral part of the 2540-mm (100-inch) mill "D" facilities. It is located in the shipping building and plates to be heat treated can be transferred from the normal travel of product to its approach table. This combination semi-continuous hardening and tempering furnace is similar in design to the furnaces of the No. 1 and No. 2 plate treating lines of the 4064-mm (160-inch) mill "B". It is designed to heat treat plates and sheets up to a maximum of 2438 mm (96 inches) in width and 9.1 mm (30 feet) in length, with thicknesses ranging from 2.5 mm to 19 mm (0.10 inch to 0.75 inch). Operating temperature limits of the furnace are from 285° to 955° C (550° to 1750° F), and temperature-control zones make very uniform heating possible. Plates move through the furnace at speeds ranging from 0.0066 to 0.0660 metres per second (1.3 to 13 feet per minute), depending upon the thickness of the product being treated. Heating cycles are established by metallurgical requirements dictated by grade of steel, thickness of the plate, and the treatment desired. After heating to either hardening or tempering temperature, the plate can be conveyed into a platen-type water quench.

Car-Bottom Heat-Treating Furnaces—For heat-treating plates too thick for the continuous lines (product over about 50 mm or 2 inches and up to approximately 203 mm or 8 inches thick) large, direct top-and-bottom-fired, car-bottom furnaces with operating temperatures ranging between room temperature and 1060° C (1940° F) are utilized. Seven of these furnaces are located in the Harvey Shop and six in the Forgings Division. The Harvey Shop also is equipped with a 5.5-m by 18.3-m (18-foot by 60-foot) water-quench tank, 2.4 m (8 feet) deep with a working depth of 1.4 m (4½ feet). Stools, about 1 metre (3½ feet) high, are placed at the bottom of the quenching tank on which to lay the plates.

PLATE HEAT-TREATING EQUIPMENT FOR
4064/5334-mm (160/210-inch)
MILL "A"

A continuous plate heat-treating line operates in conjunction with the 4064/5334-mm (160-210-inch) plate mill "A". It is capable of heat treating plates up to 5080 mm (200 inches) wide and up to 76.2 mm (3 inches) thick, the actual combination of dimensions being related to the sizes of plate that can be rolled on the mill.

The plate heat-treating department was installed as a continuation of the South Flame Cut Building of the 4064/5334-mm (160/210-inch) mill "A" parallel to and extending beyond the Shearing Building (see Figure 25—1). It is equipped to normalize or quench and temper carbon-steel alloy-steel and high-strength low-alloy steel plates as required by the grade of steel and the mechanical properties desired in the final product. The arrangement of the equipment was shown in figure 25—1.

The gas-fired roller-hearth hardening furnaces and tempering furnaces are about 47.5 metres and 63.4 metres (156 feet and 208 feet) in length, respectively. Following the delivery end of each hardening furnace is the quench, equipped with platens that exert pressure against both surfaces of a plate while it is spray quenched with water. The roller leveler is a four-high unit capable of flattening alloy-steel plates up to about 19 mm (¾ inch) thick.

Electric overhead traveling cranes, transfer cranes, transfer cars, roller tables and other handling equipment facilitate movement of plates from one work station to another.

Plates thicker than about 25 mm (1 inch) are removed from the transfer table located beyond the roller levelers in the 4064/5334-mm (160/210-inch) mill "A" line (item 5 in figure 25—1) and transported by overhead crane to the slow-cool area of the heat-treating department. Alloy-steel plates are placed in unfired hoods for slow cooling and then inspected, checked for layout, and conditioned by grinding if necessary, prior to heat treating. Plates approximately 25 mm (1 inch) thick and lighter that do not require slow cooling progress across the cooling bed to the marking and crop shear line shown in Figure 25—1 to be inspected, checked for layout, cropped, and parted if necessary. The transfer crane of the heat-treating department, operated remotely by the crop-shear operator, transfers plates into the heat-treating area where they receive further inspection and test grinding.

The plates then are scheduled for treatment, stacked in schedule order in the single-leg gantry area by overhead crane, and placed individually by the gantry on the charging table of a roller-hearth hardening furnace. Plates to be quenched and tempered are heated progressively while proceeding through the furnace by continuous or indexing movement and, after attaining the proper temperature, are rapidly moved into the platen quench press where they are quenched by water sprays.

Normalized plates are heated in the same manner, but are "dummied" through the quench press, and then transported to the by-pass table by the transfer car adjacent to the press.

Quenched or normalized plates requiring tempering are moved from the by-pass table and charged into a tempering furnace by the transfer car that operates at the other end of the tempering furnace (adjacent to the leveler).

After tempering (or after heating, in the case of normalized plates), the plates are air-cooled on the by-pass table (some grades are quenched in water after tempering), transferred to the leveler (if necessary) and then to the layout and test-cutting tables where the plates are marked for shearing or flame-cutting and test samples are removed. All plates then are moved by overhead crane from the tables and stacked in the storage area pending results of metallurgical tests.

After test release, the heat-treated plates are moved by over-head crane to either the transfer-crane area or south flame-cut areas of the mill. Sheared plates are placed on the mill shear line by transfer crane, trimmed to finished size, and transferred to the shipping building for final inspection, conditioning and shipment. Flame-cut plates are cut to final size, inspected, conditioned and shipped.

CHAPTER 26

Railroad-Rail and Joint-Bar Production

SECTION 1

ROLLING OF RAILROAD RAILS

Dating from the invention of the steam locomotive, the railroad rail represents one of the first sections to be rolled. The railroad rail is a most vital part in railroad operations and represents a difficult section for the roll designer and roller. With the advancement in speed of travel and weight of loads carried, more and more has been required of the rail, until today no product is subject to more severe service conditions. Exposed to the weather at all times, it is subjected, under constantly varying conditions, to high compression and bending stresses, impact, vibration, friction and wear.

The railroad rail section should be designed to have the greatest possible bending strength, to provide an abundance of metal for wear, to present a wide base for fastening to the cross tie, and, for the sake of economy, it should be of the lightest section possible. The American Tee Rail best meets all of these requirements and represents the accepted design.

Historical Development of Rail Sections—The history of rail development is indicated in the sketches of Figure 26—1. The first running surfaces for the early railroad rolling stock comprised of strap rails comprised of cast-iron plates approximately 100 mm by 32 mm by 1.5 metres long (4 inches by 1¼ inches by 5 feet long), which were attached to a wooden base. The first strap rails were used around 1767. Various types of cast and malleable iron rails were used until about 1820 when the first iron rails were rolled. These were supported by cast iron holders, called chairs, attached to stone supports. In an effort to eliminate use of the expensive chair required for this type of iron rail, a rail with a wide and relatively heavy flange on the bottom was rolled in 1831. The difficulty of rolling the flange led to the better balanced Lock rail of 1837, the bull head and U-shape rails of 1844, and the pear head rail of 1845. Then came the compound rail of 1856 (not shown in Figure 26—1) and another form in 1858, which was the U-shape of 1844 with the lower parts closed in and welded to form a web. As neither of these forms proved serviceable, a demand for more metal in the head for wear forced a final return in 1858-1868 to the tee shape with wide thin flanges. Subsequent to 1858-1868 the quality of the steel, the design of the rails and rolls, and the equipment of the mills have improved continu-

ously. Present standard railroad rails are rolled in various sizes ranging in weight from 32.2 kilograms per metre (65 pounds per yard) up to and including 69.4 kilograms per metre (140 pounds per yard). Rails 29.8 kilograms per metre (60 pounds per yard) and lighter are classed as light rails. Representative chemical compositions for standard rails are shown in the American Railway Engineering Association (A.R.E.A.) tabulation given in Table 26—I. In addition, typical chemical compositions of various alloy and heat-treated rails produced for higher strength applications are shown in Table 26—II along with their mechanical properties.

Mills for Rolling Rails—There does not seem to be much accurate information available about the first mills which rolled rails. It is probable that existing mills designed to roll bars were utilized with such alterations and additions as were necessary. Credit for rolling the first steel rail in 1857 is given to the Dowlais Plant, Wales, while the credit for rolling the first steel rail in this country is given to Captain Ward's North Chicago Rolling Mills, where the first Bessemer steel rails were rolled experimentally in 1865 from blooms made of hammered ingots produced at Wyandotte, Michigan.

Rails were originally rolled on the pullover mill, and later on the reversing mill. In this country rails have been rolled for many years on the three-high mill, which was usually made up of a single train of three stands driven by one engine. Sixty-nine mills were reported to be rolling rails of various weights in 1874. During this great railroad-expansion period, rails were

Table 26—I. Representative Chemical Compositions for Rails

Constituents	Nominal Weight in Lb. per Yard	
	90/120	121 and Over
Carbon	0.67-0.80%	0.72-0.82%
Manganese .	0.70-1.00	0.75-1.05
Phosphorus, Max.	0.035	0.035
Sulphur, Max.	0.040	0.035
Silicon	0.10-0.35	0.10-0.35

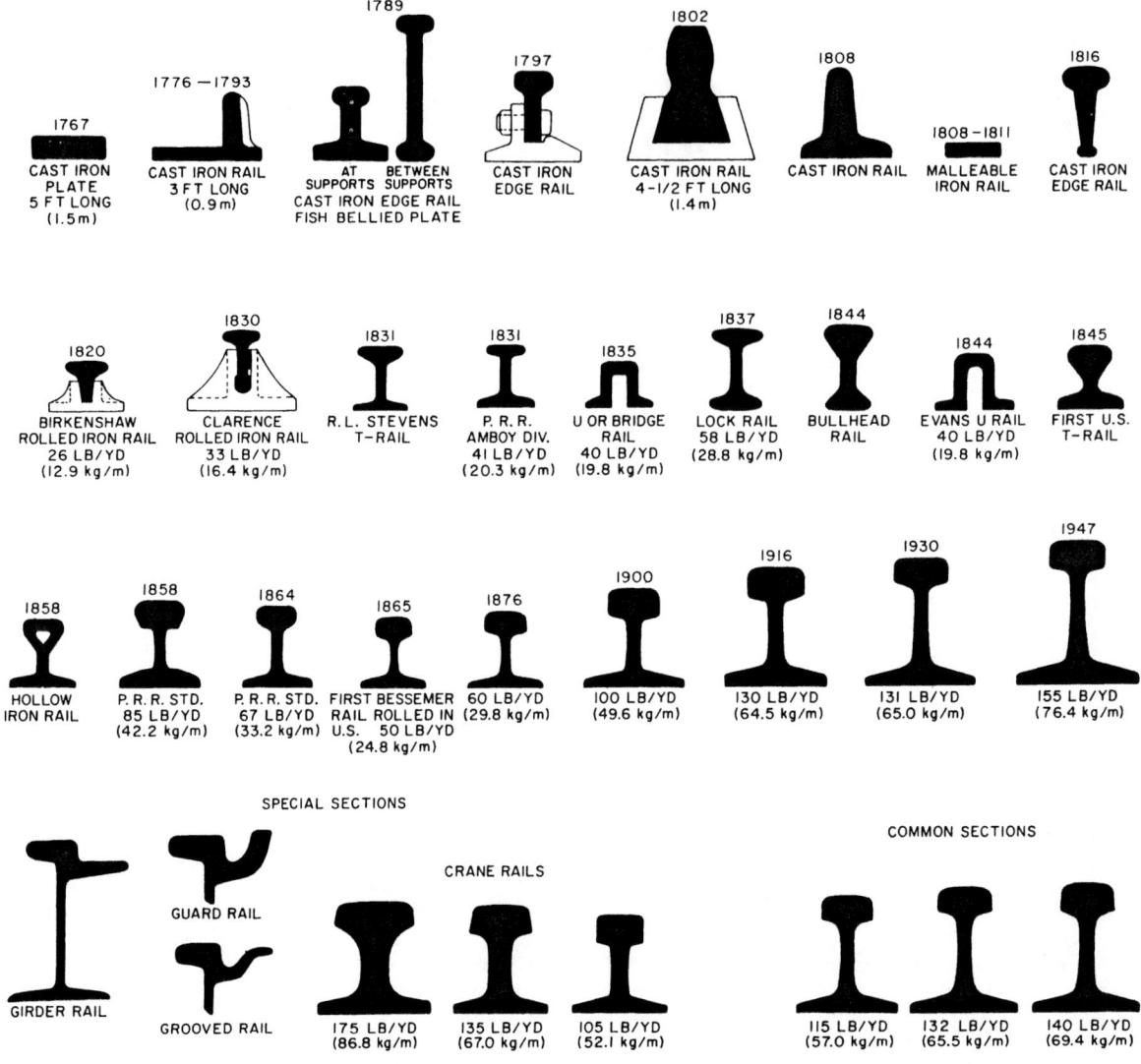

FIG. 26—1. Sketches of cross-sections of rails from the earliest periods of railroading until the present, showing the evolution of the modern railroad-rail design. Certain special rail sections are shown at the lower left.

in such demand that even this large number of mills could not supply the demand. These mills were scattered around the country with one as far west as Laramie, Wyoming. The tonnages produced on the old mills were small in comparison to present-day tonnages. Since about 1900, rail mills have undergone many changes because of the ever-increasing requirements of the railroads with respect to quality of the steel, size tolerances of the rail section, length of rail, freedom from internal and surface imperfections, and a variable demand as to quantity.

Rails are formed by three general methods of rolling: (1) the **tongue-and-groove, flat** or **slab-and-edging** method; (2) the **diagonal** or **angular** method; and (3) the **universal** method. Several of the rail mills in this country combine the first two methods, some of the passes being designed to form by the first method and the remaining passes to roll the section by the second method. In the tongue-and-groove, flat or slab-and-edging method, illustrated by the second roughing stand in Figure 26—2, the axis of symmetry of the rail coincides with the pitch line and is parallel to the train line of the rolls.

The diagonal or angular method of rolling is represented by the roughing stand shown in Figure 26—3. It differs from the slabbing method in that the shaping of the rail is begun with the first pass in the roughers and instead of first compressing the bloom to a smaller size and then forming the section partly through compression and partly by spreading, the process is one of compression from beginning to end.

The universal method for rolling rails is used in Japan, France, and recently in the United States. As described in Chapter 23, a universal rolling mill employs both horizontal and vertical rolls in one stand to effect shaping of the work piece. The use of universal rolling

Table 26—II. Typical chemical compositions and mechanical properties of standard and high strength rail steels

Manufacturer	Rail Type	Chemical Composition						
		C, %	Mn, %	Si, %	Cr, %	V, %	Cb, %	Mo, %
Standard AREA		0.80	0.90	0.20	–	–	–	–
C.F.&I.	Hi-Si	0.75	0.80	0.65	–	–	–	–
	Cr–Mo	0.78	0.84	0.21	0.74	–	–	0.18
Algoma	Cr	0.75	0.65	0.25	1.15	–	–	–
British Steel	Cr	0.75	1.25	0.35	1.15	–	–	–
Krupp	Cr–Si	0.70	1.05	0.75	1.00	–	–	–
Thyssen	Cr–Si–V	0.65	1.05	0.60	1.15	0.20	–	–
Klockner	Cr–Mo–V	0.65	0.80	0.30	1.00	0.10	–	0.10
Sydney	Mn–Cr–V	0.70	1.65	0.20	0.30	0.10	–	–
	Cr–Si–Cb	0.70	1.10	0.55	0.80	–	0.06	–
Brazil Steel	Si–Cb	0.74	1.30	0.80	–	–	0.03	–
Bethlehem	Through Hardened	0.80	0.90	0.20	–	–	–	–
U.S. Steel*	Head Hardened	0.80	0.90	0.20	–	–	–	–
Russia	Through Hardened	0.75	0.90	0.30	–	–	–	–
Nippon Steel*	Head Hardened	0.75	0.80	0.22	–	–	–	–
Nippon Kokan*	Head Hardened	0.75	0.83	0.22	–	–	–	–

Manufacturer	Rail Type	Mechanical Properties						
		0.2% Proof Stress		Ultimate Tensile Strength		Elong., %	Redn. of Area, %	Brinell Hardness Number
		N/mm²	ksi	N/mm²	ksi			
Standard AREA		510	73.9	920	133.4	11	18	255
C.F.&I.	Hi-Si	520	75.4	980	142.1	11	14	285
	Cr–Mo	807	117.0	1228	178.1	9	17	352
Algoma	Cr	650	94.3	1100	159.5	9	17	320
British Steel	Cr	690	100.1	1130	163.9	11	17	325
Krupp	Cr–Si	675	97.9	1140	165.3	12	20	315
Thyssen	Cr–Si–V	680	98.6	1130	163.9	12	20	320
Klockner	Cr–Mo–V	705	102.3	1145	166.1	12	20	325
Sydney	Mn–Cr–V	705	102.3	1035	150.1	12	18	325
	Cr–Si–Cb	705	102.3	1040	150.8	10	16	340
Brazil Steel	Si–Cb	645	93.5	1070	155.2	10	15	320
Bethlehem	Through Hardened	870	126.2	1220	176.9	13	30	365
U.S. Steel*	Head Hardened	910	132.0	1260	182.8	12	33	380
Russia	Through Hardened	820	118.9	1250	181.3	14	40	380
Nippon Steel*	Head Hardened	857	124.3	1231	178.5	12	34	370
Nippon Kokan*	Head Hardened	817	118.5	1192	172.9	14	37	365

*Tensile and hardness properties measured at 10 mm (0.39 in) and 5 mm (0.2 in) respectively from the top corners of the rail head.

(Courtesy of Vanitec [Vanadium International Technical Committee])

mills for rolling rails in combination with two-high reversing mills that have only horizontal rolls is shown schematically in Figure 26—4. The vertical work rolls of the universal stands in this rail mill are of relatively small diameter and are backed up by vertical rolls of larger diameter. It may be noted that the finishing stand of this mill (Pass No. 13 in Figure 26—4) employs only one vertical roll working on the base of the rail flange while the web, head and top of the flange are worked by the two horizontal rolls. The universal rolling method allows a greater degree of hot working of the head and flange of a rail by compression than can be obtained when horizontal rolls alone are used.

At the end of 1983, there were four mills in the United States that were producing railroad rails. These mills are far from standardized in layout. Some have a large number of stands, others only a few; some are two-high throughout, while others are three-high. One is universal. No one rail mill may be cited as an example typical of all, but a brief general description of a rail mill which rolls rails directly from ingots will illustrate operating conditions of a rail mill. Its layout is somewhat similar to the cross-country type of rolling mill.

The mill starts with four tandem stands of 40-inch, two-high rolls. In these stands, the ingot is given one pass per stand, and is turned after each pass. The passes

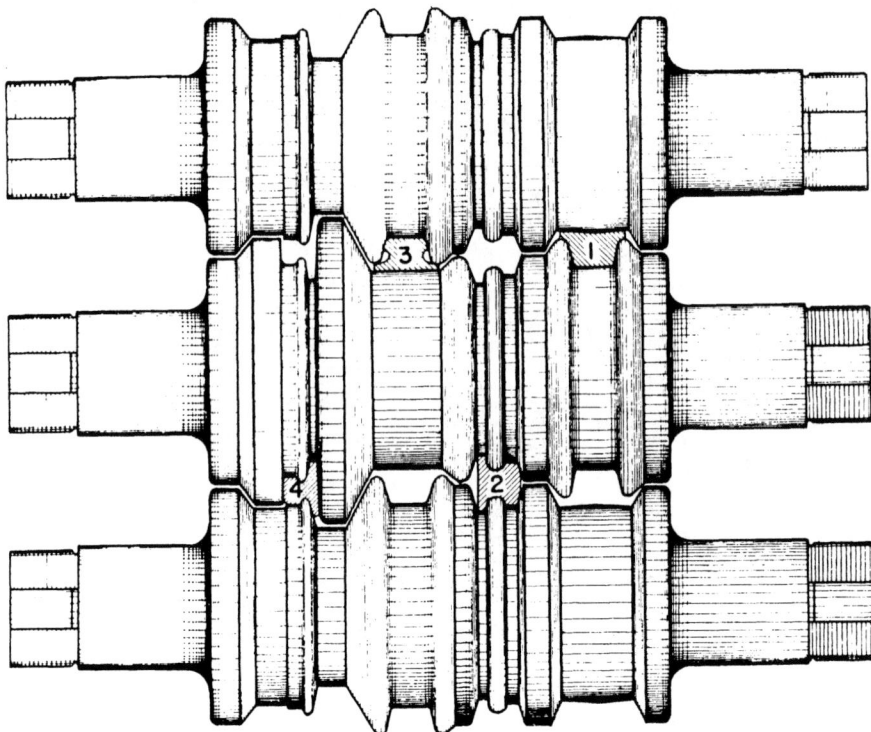

FIG. 26—2. Sketch of rolls used in second roughing stand of a rail mill rolling by the tongue-and-groove, flat or slab-and-edging method, showing shape produced in each of the four passes.

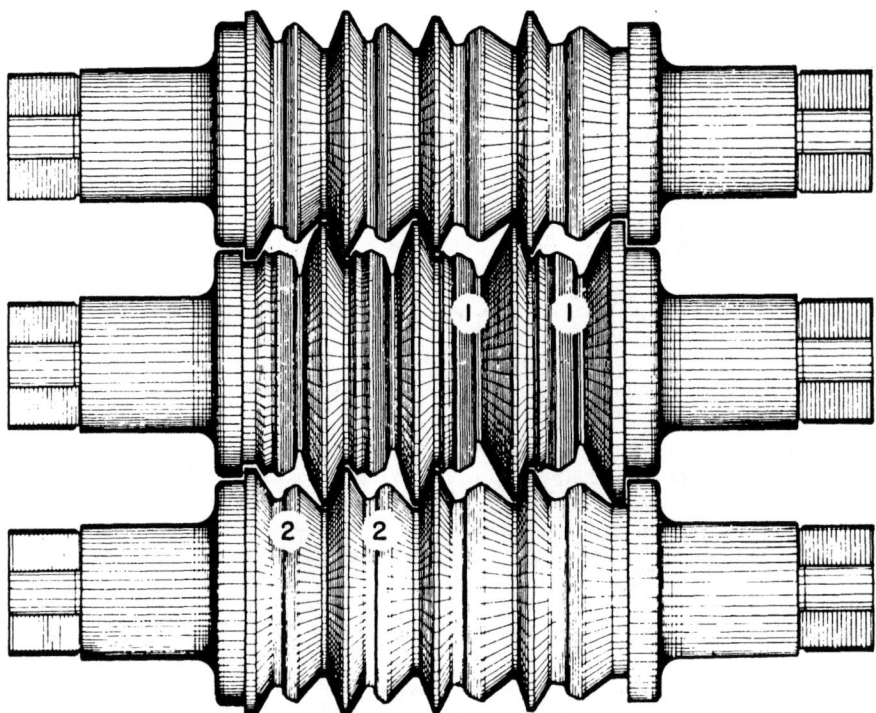

FIG. 26—3. Sketch of rolls used in the roughing stand of a rail mill rolling by the diagonal or angular method.

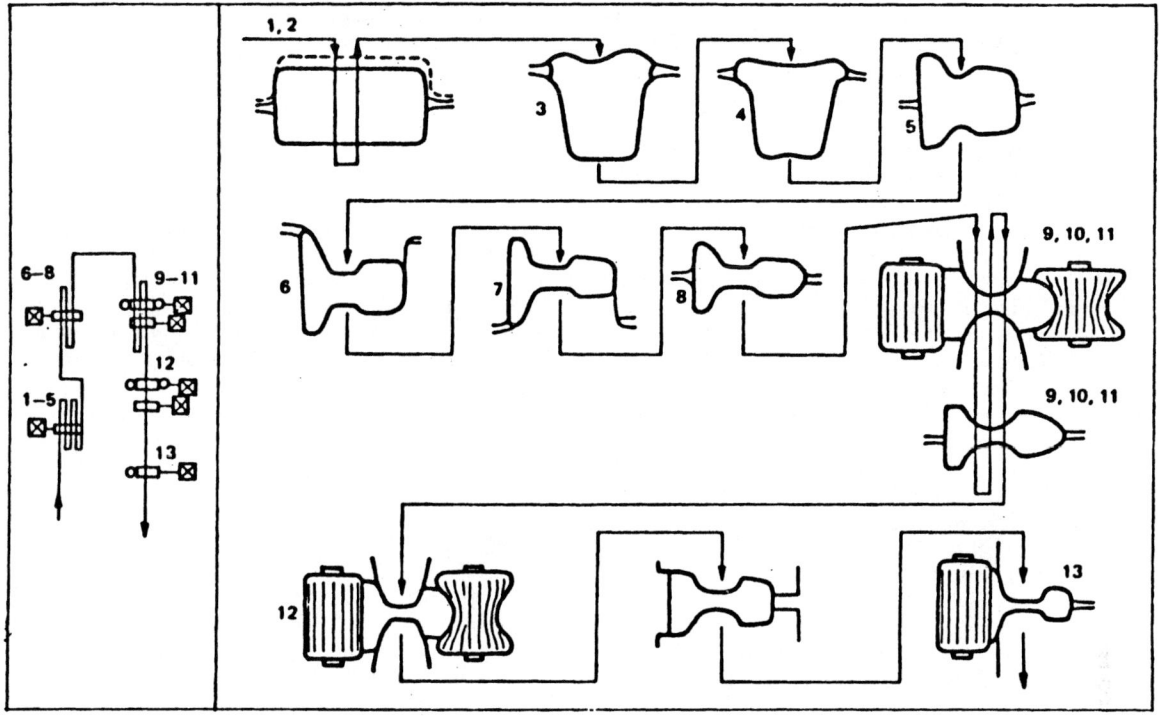

FIG. 26—4. Successive stages in the reduction of a bloom to a rail in thirteen passes in a rail mill composed of four two-high reversing stands and three universal stands arranged as shown in the plan at the left. (Courtesy, Nippon Steel Corporation.)

are of diamond, diamond-square, and box-pass design. After the initial passes, there is a 40-inch three-high mill with five box passes, the final pass being slightly shaped to give the hitherto rectangular bloom a form more suitable for subsequent rolling. Following this three-high blooming stand is the bloom shear and then the cross-country arrangement of seven stands of rolls arranged in two groups of three stands each and one separate stand. The leader stand has a vertical roll working on the head of the rail; the finishing stand one for the rail base.

DESIGNING THE ROLLS FOR RAILS

The first consideration in designing the rolls for rails is to produce a finished piece of the correct size and form. This objective can be accomplished only by directing and controlling the flowing, spreading and bending of the steel. The ease with which this forming is done depends on the plasticity of the metal, which in turn is affected by the temperature. With the speed of the rolls fixed, the temperature confined to a narrow range, and the grade of steel given, the only means of control remaining to the roll designer is the size and shape of the passes, and in part of these, at least, the size will be governed by the size of the bloom. In designing the passes, a good designer will endeavor to work the steel in such a manner that the quality of the product will be benefited, and no surface imperfections will be developed. The imperfections that should be carefully avoided are fins, laps, overfills, and underfills. Laps may result from fins or a collaring of the piece in the rolls; overfills, from worn rolls, poor or improper

design; and underfills either from poor design or incorrect adjustments of the rolls.

Stages of Reduction—The formation of the rail from the bloom may be considered as taking place in three steps or stages. The first stage, called the **roughing,** is merely one of preparation; in it a large amount of work is done, but this work is expended mainly in reducing the size of the section and elongating the piece. The intermediate stage proceeds with the forming of the rail and involves a combination of **slabber, dummy, former, edger** and **leader passes,** dependent on the mill layout. The **finishing pass** completes the formation of the rail.

The Section—No original designing of section is done by the roll designer. The first requirement in the rolling of a new section is that the roll designer be supplied with a drawing or print of the section, which must be accompanied with all the dimensions, preferably indicated on the print. The weight of rail desired or expected should also be given. Here the matter of dimensions is of extreme importance, for the designing of the templets cannot be started until each and every dimension required is given. These dimensions not only include linear measurements, such as height of rail, width and thickness of parts, but also the radius of all curves, amount of slope on inclined surfaces expressed in degrees or percentages, and the tolerances of dimensions.

Roll Preparation—With all the necessary information available, the first step taken by the roll designer is to prepare a drawing for the cold templet. This drawing is constructed on the axis of symmetry of the rail, which is the vertical line drawn through the center of the head, of the web, and of the flange. On this line the section of rail is symmetrically constructed to the dimensions given on the drawing, all the dimensions being made with extreme care and accuracy. Upon completion of this very accurate drawing, the area of the section is measured with a planimeter in order to check the weight of the section, after which the cold templet is prepared from either brass or steel.

The next step, which is really the first step in designing the roll passes, is the making of the hot templet. This templet is similar to the cold templet, but larger in size, as it represents the section of the rail at the finishing temperature of rolling.

From the hot templet the various passes are designed successively as the experience and judgment of the designer dictates. Roll designers may prepare the various pass templets by several methods. Usually the pass templets from a similar section are used as a guide, since many of the sections currently being rolled have been perfected through many modifications from initial designs for the passes. Frequently, the designer constructs each pass outline in a drawing showing the different passes superimposed upon each other. As a preliminary step toward designing passes back to the bloom from the finished templet, a table is usually prepared. This table will consist of various passes, cross-sectional area of the head, web, and base, per cent reduction, and spread in inches. The designer must constantly keep in mind the danger of forming fins. In order to avoid fins, the designer may arrange the passes so that each side of the piece alternately enters an open and a closed side of the groove when rolling by the tongue-and-groove, flat or slab-and-edging method. Even with this arrangement, fins would still be formed if the passes were not properly designed. To avoid the possibility of forming fins, two modifications of design may be used. In the leader pass, the corner of the head,

which is to come opposite the openings between the rolls in the finishing pass, is well rounded off, so that the spread or flow of the metal will be taken up in filling out this rounded corner and none will remain to be forced into the clearance. For the same reason, that half of the flange on the same side of the rail is left much shorter. This provision is made in many of the passes. Great care is necessary in distributing the reduction of each part to prevent the metal flowing away from parts where it is needed. For example, if a too great reduction in the web takes place in one pass, it will produce a flow of metal away from the head, causing the latter to be underfilled. The cause for much of the trouble of this kind lies in the different diameters of the pass, which cause a different roll-surface speed for head, web, and flange, and hence different rates of elongation. If the elongation produced through compression is at the point of less speed, the section will be imperfectly formed, or cracks will result. The accompanying illustration (Figure 26—5) will help in understanding this point.

The roll designer strives to keep the roughing passes of such shapes and sizes that the same set may be used for a large number of different rail sections. The pass contours for producing the 132-pound RE rail are illustrated in Figure 26—6.

Upon completion of the pass templets, including both male and female for the cold templet, they are sent to the tool shop where they are used as patterns in making a set of tools for turning the rolls for the section. As many as 24 different tools may be required for the last six passes of each size rail. After shaping these tools a little over-size, they are heat treated and then redressed to exact size before they are used to turn the rolls. When ready, templets and tools go to the roll shop, where the work of turning the rolls is done. Rail-mill rolls may be sand iron, alloy iron, or alloy steel rolls. Frequently more than one grade of rolls will be

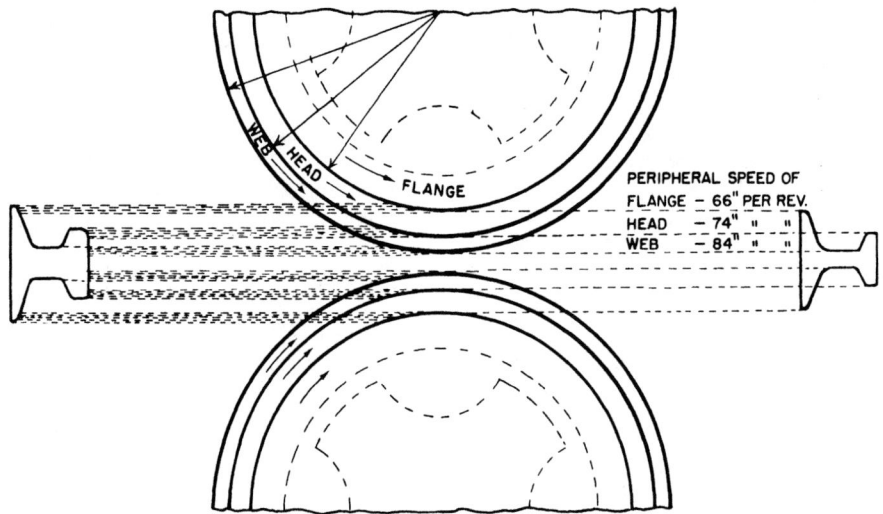

FIG. 26—5. Differences in peripheral speeds at various points in the roll pass contacting, respectively, the flange, head and web of a rail during rolling.

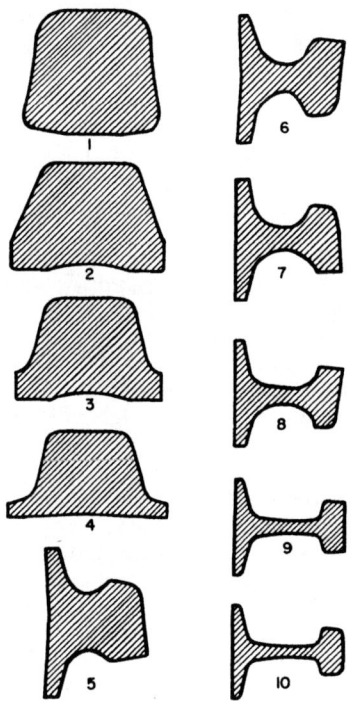

FIG. 26—6. Roll-pass contours for producing a 132-pound RE rail.

used on the same mill.

ROLLING PRACTICE

After the rolls have been properly turned, they are placed in the housings in their proper position and carefully lined up. A trial rolling with one bloom will then be made with the roller checking the piece closely to see that it goes through the mill in a satisfactory manner. The trial rail thus produced is gaged by templets which permit checking of base over-all, head width and contour, flange length and thickness, head radius and fish. (Fishing is a mill term for the dimensions of the theoretical contour of the under side of the head, the side of the web, and the upper side of the base.) By comparison with a master or cold templet the height and web thickness can be checked. If this section is found to be correct, the mill is then ready to begin the rolling. At frequent intervals, the rails are gaged and examined for surface imperfections such as collar marks, underfills, roll marks, overfills, guide marks, cracks, seams or pickups. One of the things that cannot be avoided in the rolling is the wear of the rolls which is aggravated by the slippage associated with the irregular section. The grooves tend to wear down rapidly, which will produce a "loose fish." "Rocking base" rails may result from worn rolls.

A.R.E.A. specifications for rails require that the cold templet shall conform to the specified section as shown in detail on the drawing. The rail section shall conform as accurately as possible to the templet or drawings. A variation of 0.4 mm (0.015 inch) less or 0.8 mm (0.030 inch) greater than the specified height is permitted. A variation of 0.9 mm (0.040 inch) in the width of either flange is permitted, but the variation in total width of base shall not exceed 1.2 mm (0.050 inch). No variation is permitted in dimensions affecting the fit of the joint bars, except that the fishing templet may stand out not to exceed 1.6 mm (1/16 inch) laterally.

Crane Rails—Crane rails, in three sections of 52.1, 67.0 and 86.8 kilograms per metre (105, 135 and 175 pounds per yard) are produced. These rails in general have a heavier head and web than the railroad rail in order to withstand the heavy loads imposed in service. These rails are rolled on rail mills and are made by the same manufacturing practices as railroad rails.

Light Rails—Light rails are rolled on several types of mills—bar mills, light structural and, in some cases, mills built and operated exclusively for light rails. Light rails are produced from standard rails or from billets. In the rerolling of light rails from standard rails, either passes of the edge and flat design or of the diagonal design may be used. The rolling of light rails from billets is accomplished in much the same manner as the rolling of standard rails. An average of nine passes is generally used, although in a few extreme cases as few as five passes have been used.

FINISHING OPERATIONS FOR STANDARD RAILS

Cutting and Cambering—Railroad rails in the United States have traditionally been cut to produce the standard 11.8-metre (39-foot) rail. However, recent trends are toward 23.6-metre (78-foot), and two producers in the United States now have this capability. The rails are cut to desired lengths by hot saws operated singly or in gang, mounted over the runout table beyond the finishing stand. In cutting the hot rails, proper allowance must be made for linear thermal contraction, which, for an 11.8-metre (39-foot) rail, is approximately 152 mm (6 inches). The exact amount of the contraction depends upon the temperature at which the rail is sawed. It is necessary to set the saws very accurately as the first rail from an ingot as well as the last rail from the same ingot must be within the length tolerance of plus or minus 11.1 mm (7/16 inch). Vertical alignment of the saws must also be accurate and the saws must be maintained in sharp condition in order to meet a 0.8-mm (1/32-inch) maximum off-square tolerance. Drop-test and the nick-and-break samples are obtained from the crop ends cut at the hot saws. After sawing, the rails pass under a stamping machine which marks the heat number, ingot number and position of the rail in the ingot. The latter is designated by letters beginning with A at the top of the ingot. For continuous cast rail, a numerical and/or alphabetical code establishing position in the cast heat and strand is established between the purchaser and manufacturer. Between the stamping machine and cooling beds is a cambering machine which consists of a set of horizontal rolls with a vertical roll on each side, all in one housing and set to bend the rail slightly so as to make the top surface of the rail convex from end to end. This bending is done to compensate for the camber produced in a straight rail as it cools on the cooling bed, this camber being caused by the different rates at which the head and the base cool. A scale located near the end of the delivery table is used for checking the weight of the rails as often as desired, before they are advanced to the cooling beds. Rail specifications state that a variation of 0.5 per cent from the calculated weight of section as applied to the entire order is permitted.

Marking and Branding—One of the mandatory requirements for standard railroad rails is that each rail shall be legibly marked for complete identification. So far as practicable, this branding is done by numbers, symbols and letters cut in the bottom finishing roll to give raised characters on one side of the web of the rail. Markings that cannot be thus rolled into the web are marked intaglio on the opposite side of the web by a hot stamping machine following the hot saws. The nature of the information required and the methods of marking are as follows:

Method of Marking	Kind of Mark	Nature of Information
Engraving on bottom finishing roll	Raised character	Weight and section number Method of hydrogen elimination Manufacturer's brand Year rolled Month rolled
Hot stamping	Intaglio	Heat number Rail letter Ingot number

Controlled Cooling—Rails must be free of shatter cracks caused by hydrogen. To accomplish this, hydrogen must be eliminated by vacuum treatment of the steel or by controlled cooling of rails or blooms. Controlled cooling of rails on a production basis was begun in 1937.

Present specifications for rails intended for railroad service require controlled cooling of the rails within the temperature range between 385° and 150°C (725° and 300°F) except when produced from vacuum degassed steel, in which case the rails may be air cooled. In mill operations, the rails are allowed to cool to under 535°C (1000°F) and then are placed in either stationary or movable insulated containers. The rails must be placed in the container within the temperature range between 535° and 385°C (1000° and 725°F), which is checked by a pyrometer. One or more thermocouples of the chromel-alumel type are placed between tiers of rails in the slow-cool container in order to obtain temperature readings during cooling. The containers are insulated to meet the specified cooling cycle of not dropping below 150°C (300°F) in 7 hours for rails 49.6 kilograms per metre (100 pounds per yard) in weight or heavier, from the time that the bottom tier is placed in the container, and 5 hours for rails of less than 49.6 kilograms per metre (100 pounds per yard) in weight. Rails must remain in the cooling container for a minimum of 10 hours. Complete records of the cooling cycle to 150°C (300°F) are maintained on each container.

Testing of Standard Rails—Testing of standard and higher strength railroad rails includes chemical, me-

chanical and internal tests. Two ladle test samples, representing the composition of one of the first three and one of the last three ingots applied, are obtained for chemical analysis. The drop test is relied upon to supply the impact test information required. This test is not required for continuous cast rail. Crop ends 1.2 to 1.8 metres (4 to 6 feet) long are cut from the top end of the "A" rail from one of the first three, one of the middle three, and one of the last three ingots of each heat. ("A" rails are those rolled from the upper portion of the ingot, "B," "C," "D," etc., rails being rolled from steel in successively lower parts of the ingot.) These specimens are placed upon supports and subjected to the impact of a tup weighing 907 kilograms (2000 pounds), falling from a height of 5.7 to 6.7 metres (19 to 22 feet), depending upon the weight of the section. For rails 52.6 kilograms per metre (106 pounds per yard) or less in weight, the supports are placed 0.91 metre (3 feet) apart and for rails over 52.6 kilograms per metre (106 pounds per yard) to 69.4 kilograms per metre (140 pounds per yard) this distance is increased to 1.2 metres (4 feet), with the tup striking the rail midway between the supports. If a specimen breaks, all the "A" rails of the heat are rejected and the "B" rails must be similarly tested. Failure of a "B" rail specimen causes all "B" rails to be rejected in addition to the "A" rails and similar tests must be made in "C" rails. Failure of a "C" rail specimen is cause for rejection of the whole heat.

The nick-and-break test is a fracture of the rail adjacent to the top end of the top rail of each ingot rolled. If the fracture of any test specimen exhibits seams, laminations, cavities or foreign matter, the top rail represented is closely examined on the top end and bolt holes of the finished rail for these imperfections. If the finished rail shows these imperfections, the rail must be cut back to sound metal and used as a short rail. Ultrasonic testing of rails for internal imperfections or pipe may be specified as a supplement by the purchaser or manufacturer. The purchaser shall choose the calibration standard. Rails shall be produced as specified by the purchaser to one of two levels of hardness:

	Standard Rail	High-Strength Rail
Brinell Hardness	248 minimum	321-388

The test is made on the top or side of the rail head of a piece of rail at least 15-cm (6-inches) long from a rail of each heat of steel. Heat-treated rails may be retreated if they fail to meet the hardness requirements.

FINISHING AND INSPECTION

Standard Rails—Finishing operations for standard rails include preliminary inspection upon unloading from slow-cool cars or containers, removal of saw burrs, straightening, drilling, grinding or milling of ends, beveling of heads when specified, inspection, classification, and painting. The purpose of the preliminary inspection is to check identification and lengths and to locate harmful surface imperfections and thus save the expense of finishing a rail that would be rejected. The removal of burrs made by the saws on the ends of the rails usually occurs prior to the straightening, although in some instances burr removal occurs after the straightening operation. The burrs are cut off with chisels and subsequently smoothed with a file or a grinder.

Roller Straightening—Rails, including the rail ends, must be straightened to stringent A.R.E.A. specifications to provide better welding, handling, and service. The most common method for straightening is the use of a roller-straightening machine, the principle of which is illustrated in Figure 26—7. Three manufacturers in the United States and more foreign manufacturers use roller straightening. If three rolls are arranged in a triangle, a piece passing through them will be continuously bent. This is a common method used to form cylinders from flat plates. The radius of the bend can be controlled by moving the bottom roll upward or downward; upward motion decreases the radius.

If a second triangle is formed by adding a fourth roll to the last two rolls of the first triangle, the piece will be bent in the opposite direction and will leave the second triangle with a known curvature, regardless of the original bend. By adding a fifth roll, this known curvature can be removed by another reverse bend and the piece will leave the roll assembly straight. In actual practice, a minimum of seven rolls (for example, four upper and three lower) is usually used to improve control. When such a machine is used for straightening rails, the bending rolls are shaped with grooves that restrain the rail to assure proper guidance to prevent uncontrolled sideways movement. Rails usually are passed through the machine with their heads up.

The foregoing applies to the straightening of a rail bent only in the vertical plane. If the rail is bent laterally also in a horizontal plane, it is possible to straighten it in both planes simultaneously in some machines by making the horizontal bending rolls adjustable along their axes so that a rail can be bent a controlled amount to the right or left in a horizontal plane as it passes through the successive rolls that are also bending the rail in a vertical plane. Other machines employ a second set of adjustable rolls with their axes vertical, following the first set of horizontal rolls, to remove distortions by successive reverse bending in a horizontal plane in the second set of rolls. Instead of the grooves used in horizontal rolls to retain the rail during bending, the vertical rolls have collars that contact the web of the rail.

Straightening on Gag Presses—Some rails are straightened on gag presses, each of which is provided with two bottom supports located in the table proper on which the rail rests, and a top block with die attachment which moves up and down with a fixed stroke between the supports. The stroke of the block is of such length as not to touch the rail by about two inches at its lowest point. On each side of the press at spaced intervals are located two or three stands, each with a roll, to allow bringing the rail to the press and manipulating the rail back and forth while it is being straightened. The die has a double face, each side of which slopes to the center line. The gag, a rectangular block of steel slightly beveled to fit the die, is inserted between the rail and the die. The die form, in combination with the different dimensions of the gag, makes it possible to control the amount of bend the rail receives

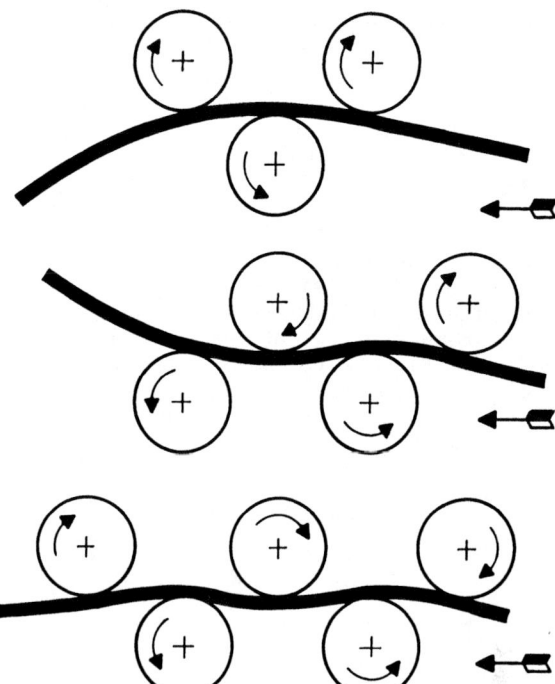

FIG. 26—7. Principle of straightening in a roller-type machine. (Courtesy, The Metals Society, London, England: see Dinwoodie listing in bibliography.)

and to adjust the press to the several rail dimensions. To straighten a rail, one man called a gagger is stationed in front of the machine and a second referred to as a straightener at the end. By sighting along the rail, the straightener at the end locates the areas of rail which require straightening and brings them under the blocks while the gagger before the machine, acting under directions from the straightener at the end, inserts the gag in such a way that the stroke of the machine will bend the rail enough to straighten it, which requires that the rail be bent beyond its elastic limit in order to give it a permanent set.

Drilling— Although most rails for mainline service are welded, some rails are still bolted. The rails for bolted track are next moved to the drilling machines, which are arranged in pairs and so spaced that when one machine has completed the drilling on one end of the rail, it is moved under the other machine which drills the holes in the opposite end. These machines are provided with three drilling spindles, the middle one of which is fixed, and the outside two adjustable, and may be made to drill from one to three holes at one time. After drilling, the rails are moved to inspection beds where they are walked, or inspected, ends checked, ends beveled when specified, classified, painted, and separated for loading. When rails are for welded track, the ends to be welded are not drilled.

Inspection—The finished rail inspection is made by mill and purchasers' inspectors to satisfy themselves that the rails are according to specifications. The rails are inspected twice, one time with the base up and the second time with the head up. The location of surface imperfections or rails not satisfactorily straightened are marked with chalk. If the surface imperfections are located near the end, that portion of the rail may be sawed off and the rail still applied on the order as a short of first grade, number one; shorts are accepted in limited quantities. If the surface imperfections are many or are near the center, the rail is either classed as a number two, or scrapped. Number two rails may contain slight imperfections which do not make them unfit for service. Number one rails must be free of injurious imperfections and flaws of all kinds. Rails which are not straight are transferred to a restraightening press for restraightening whereas rails required to be sawed are transferred to a recut unit which is equipped with slow-speed saws. Off-square ends are corrected either by use of end-milling machines or hand grinders.

Loading—Rails are classified into the following groups (each group is loaded in separate cars):

Classification	Painting on Rail Ends
No. 2 rails	White
"A" rails	Yellow
No. 1 rails of less than standard lengths	Green
Heat-treated rails	Orange
Alloy rails	Aluminum

The high-strength heat-treated or alloy rail must also be identified by a plate, stamp, or brand identifying the type.

Crane Rails—Crane rails are processed about the same as standard railroad rails. In addition, they may be heat treated as described later.

Light Rails—The finishing and inspection of light rails are different from the same operations for heavy rails. Because of the relatively small tonnage of most orders, light rails cannot be handled as separate heats and many light rails are rolled from billets of lower carbon content than heavy rails. Light rails are supplied in 4.6 and 9.1-metre (15- and 30-foot) lengths and usually straightened on a machine straightener. Bolt holes may be punched or drilled.

HEAT TREATMENT

Most rails, including alloy, are used in the as-rolled condition; however, some heat treatment may be used to produce rails with special characteristics, as in the following examples.

High-Strength Rails—Rails installed in curves where traffic density is high are subjected to severe wear. The abrasive action of wheels on the gage corner of the rail, and mashing out of the low rail, result in shortened rail life. The heads only of standard tee rails may be heat treated by a continuous process to give them higher hardness and improved service life. These heat-treated rails are called USS "Curvemaster" rails. The head only of the rail is heat treated from end to end by passing the rail lengthwise through an induction coil shaped like an inverted U to heat the head to austenitizing temperature. Heating is followed by an air quench. The air-quenched portion is tempered by residual heat in the rail. A special prebend and subsequent final water cooling result in a straight rail after treatment. Rail heads are hardened to between 321 and 388 Brinell hardness along the entire 11.9 metres (39 feet) of the rail (only No. 1 rails are so treated). Some rails are fully heat-treated for severe service. Others are alloyed to increase hardness and wear resistance as rolled.

End hardening is a heat treatment that may be given to the top end surface of a rail to make it more resistant to batter and wear. The heating of the rail end is accomplished by induction heaters. Compressed air is used for quenching. The hardened pattern covers the full width of the rail head and extends longitudinally 38 mm (1½-inch) minimum from the end of the rail. The effective hardness zone about 13 mm (½-inch) from the end of the rail is at least 6.4 mm (¼-inch) deep. Production-control testing on end-hardened rails includes the Brinell test on representative rails.

Crane Rails—Crane rail sections may be fully heat treated to between 321 and 388 Brinell hardness with very satisfactory results where heavy loads are involved. The rails are slowly heated during transverse movement through a furnace and, when heated above the critical temperature, are quenched in oil. After quenching, the rails are tempered at a temperature below the critical for a predetermined length of time. Heat-treated crane rails are used in heavy industry where severe mashing of the rail head results in short rail life. Both heat-treated and control-cooled crane rails are in common use.

SECTION 2

THE ROLLING OF RAIL-JOINT BARS

Types of Rail Joints—Paralleling the development of rails to present-day standards has been the development in rail-joint bars and new techniques, such as welding for joining rails. Through the years, rail-joint bars have been known by various names, such as splice bars, angle bars, and fish plates. A great number of rail-joint bar designs were in use around 1925, such as: the Duquesne rail joint; the 100 per cent rail joint; the Weber rail joint; the Hatfield rail joint; the Wolhaupter rail joint; the Barschall rail joint; the Bonzano rail joint; the Abbott rail joint; the Williams reinforced rail joint;

and the reinforced angle bar. Only a few of these rail joint bars are in use at the present time. In recent years there has been a trend toward standard designs of joint bars, and less use of joint bars because of welded rail.

Present Rail-Joint Bars—Even with one type of joint bar, there are many different designs, for the joint bar must fit accurately between the head and the base of the rail, and each change in these dimensions of the rail requires a change in the joint bar. The most popular rail-joint bar type at the present time is the short-toe joint bar of either the head-free or head-contact design.

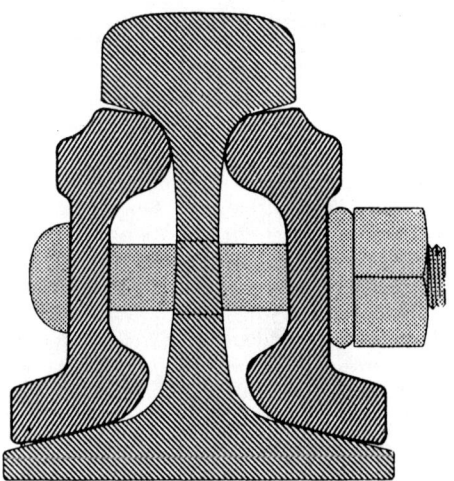

FIG. 26—8. Cross-sectional diagram of the head-free type of conventional rail-joint bar.

A sketch of the conventional rail-joint bar of the head-free type (with short toe) is shown in Figure 26—8. The head contact type of conventional rail-joint bar is identical with the sketch shown in Figure 26—8, except there is contact of the under surface of the head of the rail with the head of the joint bar. The head-free design is considered to have some advantage with respect to a lower rate of bolt-tension loss and lower resistance to expansion movement of rail ends. The principal advantage of the head-free design, however, is the fact that its use permits the desired thickening of the upper web and lengthening of the fillet radius. Nomenclature for the present conventional rail joint (with long toe) and the older type of continuous rail joint is shown in Figure 26—9. The continuous joint bar shown is in very limited use. This section is rolled with the lower base flared, the flare being pressed into a size to fit the rail base when the bar is punched. Other types of joint

bars, with or without reinforcements in the web, are in very limited use at the present time.

Problems in Rolling Rail-Joint Bars—Passes for rolling three typical joint bars are shown in Figure 26—10. Joint bars are rolled from billets or blooms with minimum carbon contents of generally 0.45 percent; however, two other grades are produced primarily for industrial and mine use with lower carbon content. The conventional joint bar and the joint bar with long toe present rolling problems because of their irregular section and lack of symmetry. In the conventional joint bar, the angles at which the section is rolled are limited by possibility of undercuts, and the shape of the passes in which the piece is necessarily reduced is favorable to the formation of laps and seams. If the joint has a depending flange, or long toe, these difficulties are multiplied, while the excessively protruding parts of a joint, such as in the continuous section (Figures 26—9 and 26—10) often prevent the piece from entering the pass properly, by striking the rolls first or being a trifle colder than the rest of the piece. For similar reasons, it is difficult to make guides that will properly handle such sections, and they are prone to become cobbled or caught in the rolls of the tables. The rolled section of joint bar is cut into convenient lengths for handling and sent to the joint bar shop for processing into finished joint bars.

FINISHING JOINT BARS

In general, rail-joint bars may be finished in one of three ways: First, all the operations of shearing to length, straightening, punching and slotting may be performed upon the cold pieces without heating in any way; the finished pieces then are referred to as **cold-worked joint bars.** Second, the bars may be heated, after shearing to length, and the work of punching and slotting be done while they are hot, after which they are allowed to cool in air. In this case they are called **hot-worked joint bars.** Third, instead of cooling the bars in the air after hot working they may be quenched by

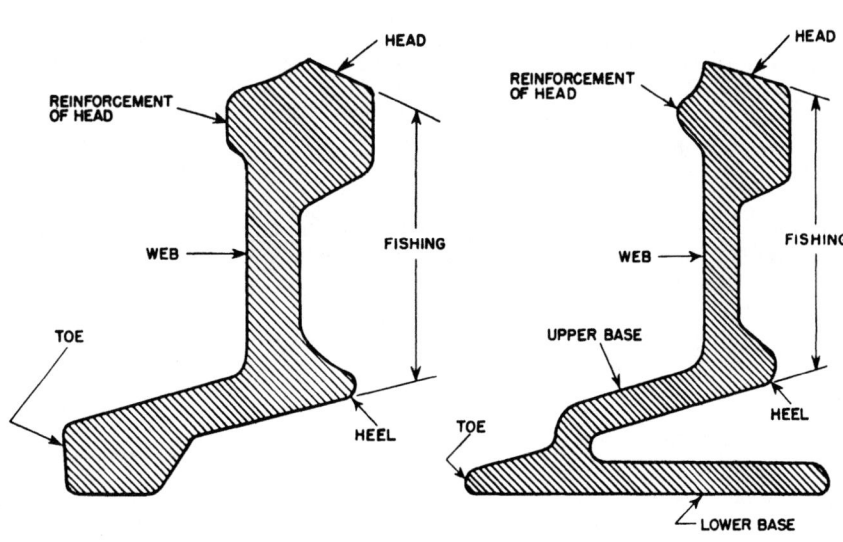

CONVENTIONAL JOINT BAR   CONTINUOUS RAIL JOINT

FIG. 26—9. Diagrammatic cross-sections of two types of rail-joint bars, with nomenclature of principal components.

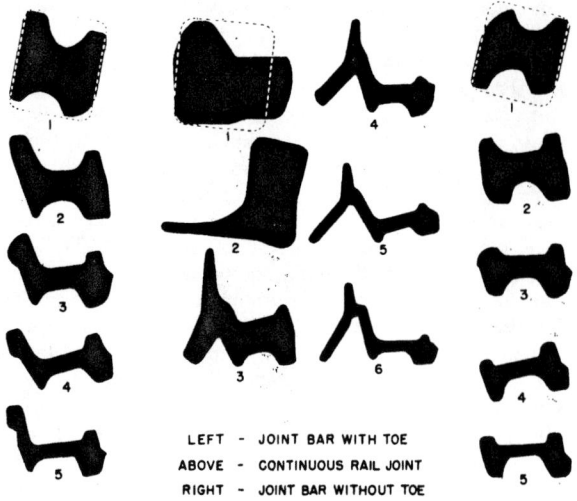

LEFT - JOINT BAR WITH TOE
ABOVE - CONTINUOUS RAIL JOINT
RIGHT - JOINT BAR WITHOUT TOE

FIG. 26—10. Successive roll passes for the production of three typical rail-joint bars.

set fusion weld. After welding is completed, the upset metal at the weld is scarfed or sheared off.

An electric flash-welding process that was developed in Europe has been adopted by most railroads in the United States for the fabrication of continuous rail. Rail ends to be joined are properly cleaned and then clamped in the heads or platens of a welding machine, and electrical current is applied. One of the platens is movable, and is repeatedly moved back and forth to make and break the electrical contact between the two rail ends. This "flashing" operation heats the rail ends to the plastic state, at which they are brought together under high pressure and an upset weld is produced the same as under the butt pressure method that employs gas for heating the rail ends. Upset metal is removed by shearing and grinding and the weld is tested to assure complete soundness.

immersing them in oil, when they are designated as **hot-worked** and **oil-quenched bars.** It will be observed that in this latter method, as in the other two methods, the bars are sheared to length cold, hot shearing being impractical, and hot or cold sawing too expensive. The hot-worked and oil-quenched joint bars possess such superior mechanical properties that practically all joint bars used on trunkline railroads are finished by this process.

WELDED RAIL JOINTS

The joining of rails by welding the joint is used by practically all railroads. The Thermit process, the electric-flash method, and the gas-heated butt-pressure method have all been used, with the electric-flash method being most generally accepted for railroad installations.

A typical example of the gas butt-pressure method for welding rail joints as practiced by one of the major railroads in this country is as follows: Rails to be welded are held together end to end and a saw cut is made at the junction of the two rails to produce parallel, clean and smooth end faces. After end preparation, the rails are placed end to end in the welding unit under high pressure. A gas-fired heating head with multiple orifices brings the two ends to a temperature resulting in plastic deformation from the pressure, giving an up-

Bibliography

American Iron and Steel Institute, Railway track materials, Steel Products Manual, January 1981.

American Railway Engineering Association, Specifications for steel rails—1979.

American Society for Testing Materials, Standard specifications for carbon steel tee rails, ASTM A 1.

American Society for Testing Materials, Standard specification for steel joint bars, low, medium, and high carbon (non-heat-treated), ASTM A 3.

American Society for Testing Materials, Rail steels—developments, processing, and use, ASTM STP 644, 1978.

Association of American Railroads Symposium, Memphis, Tennessee, November 1983, to be published.

Beynon, R. E., Rail mills and rail mill roll design, Iron and Steel Engineer, June 1946, 53-80.

Charbonnier, H., Straightening of rails in a modern installation at the Micheville plant of Sidelor, C.D.S. Circ. 1967(9), 2055-2062; BISI translation 6123.

Dinwoodie, W. A. J., Straightening of sections and rails, Journal Iron and Steel Institute, November 1955, 263-272; discussion August 1956, 423-432.

Frederick, C. O., and D. J. Pound, ed., Rail Technology, Technical Print Services, Ltd., Nottingham, U.K., 1983.

Kinoshita, K., H. Hayashida, M. Hattori and K. Isozumi, On reconstruction of rail mill and newly developed rails of Nippon Steel Corporation, Technical Report Overseas No. 3, June 1973, Nippon Steel Corporation, Tokyo, Japan.

Marich, S., Development of improved rail and wheel materials, Vanadium in Rail Steels, Proceedings of Seminar in Chicago, Vanitec, November 1979, 23.

Vanitec, Vanadium in Rail Steels, Proceedings of Seminar in Chicago, November 1979.

CHAPTER 27

Structural and Other Shapes

SECTION 1

EQUIPMENT FOR PRODUCING SHAPES

"Structural shapes" is the general term applied to rolled flanged shapes, having at least one dimension of their cross-section 76 mm (3 inches) or greater, which are used in the construction of bridges, buildings, ships, railroad rolling stock and for numerous other constructional purposes. In the past, the terms "shapes" and "sections" were used synonymously. The American Institute of Steel Construction and the producers have now adopted "shapes" as the standard designation.

Structural shapes are designated as wide-flange shapes, standard I-beams, channels, angles, tees and zees.

Other shapes include H-piles, sheet piling, tie plates, cross ties and those for special purposes.

The production of shapes, as enumerated above, involves a number of processes which are generally common to all of them. These processes include heating of the bloom, rolling to proper contour and dimensions, cutting while hot to lengths that can be handled, cooling to atmospheric temperature, straightening, cutting to ordered lengths, inspecting, and shipping.

The heating of the bloom for large sections is done in either of two types of furnaces, the in-and-out, or the continuous, which are described in Chapter 24. The in-and-out furnace is the more common of the two and serves nearly all of the older structural mills. A typical mill uses three furnaces of this type having hearth areas about 5½ metres by 11 metres (18 feet by 36 feet). The newer mills tend to use continuous furnaces because of the greater economy, one or two continuous furnaces being sufficient. Practical widths of this type furnace can accommodate blooms up to 9.1 metres (30 feet), and one furnace of proper length, designed according to the cross-section of blooms to be heated, can have sufficient capacity to supply a mill. To hold heat loss to a minimum, furnaces are usually located adjacent to a bloom-storage yard, or the delivery table from a blooming mill, or both, and at a minimum distance from the first stand of the mill on which shapes are to be rolled.

A typical mill for the production of structural sections is shown schematically in Figure 27—1. It has a two-high reversing breakdown stand in which the initial shaping is accomplished, followed by a group of three stands, arranged in train, where the rolling process is completed. The first of these three stands, known

as the rougher, is three-high. The middle stand, which is known as the intermediate, is also three-high, and continues the formation of the shapes to almost finished dimensions. The finishing stand, which is usually two-high, establishes the final shape of the rolled section. Under some conditions it is desirable to have a three-high stand for the finisher. The two-high breakdown stand is fitted with rolls of cast steel, either carbon or alloy, with pitch diameter typically 904 mm (36 inches), and body length 2032 mm (80 inches). The roughing, intermediate, and finishing stands, which are sometimes referred to collectively as the finishing mill,

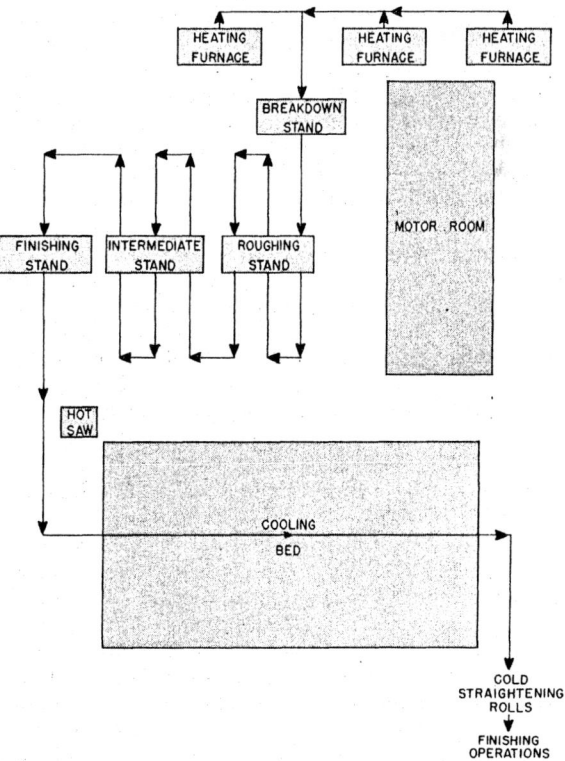

FIG. 27—1. Diagrammatic layout (not to scale) of a typical mill for rolling structural sections.

usually employ rolls with a pitch diameter between 711 mm (28 inches) and 838 mm (33 inches), and a body length of 1727 mm (68 inches). Cast rolls are used on these three stands, usually of carbon or alloy steel, or grain or sand iron (see Chapter 23), the selection being based on service requirements.

In modern mills all of these rolls are driven by electric motors, with direct-current variable-speed reversing motors being essential for the breakdown stands, and preferred for all stands. On breakdown mills, the motor drives through a flexible connection to a two-high set of pinions, and through spindles to the rolls. Most structural finishing mills drive the three stands from a single motor, through a single three-high set of pinions, the drive being carried by spindles from the pinions to the roughing rolls, and by other spindles from the roughing rolls to the intermediate rolls. Two more spindles, connected to the middle and bottom rolls of the intermediate stand, drive the finishing rolls. Power requirements vary, with 3730 kilowatts (5000 horsepower) being typical for breakdown stands, and 4475 kilowatts (6000 horsepower) typical for the finishing mill.

The breakdown stand is normally served by stationary roller tables equipped with mechanical manipulators. The shape is conveyed into, and received out of, the finishing mill passes by traveling, tilting roller tables.

The mill equipment for the production of wide-flange beams and H-piles is substantially different from that used in the production of other shapes. The distribution of metal in the web and flanges of wide-flange beams is such that beams of this type have maximum resistance to bending moment with minimum weight per unit length of beam, and the thickness of metal in the web is usually less than that in the flanges. On the other hand, the bearing piles, known as steel H-piles, while essentially wide-flange beams, are produced with an equal thickness of metal in the web and flanges to provide (1) ruggedness to resist driving forces; and (2) the same thickness of metal in all parts to provide uniform life under corrosive conditions, for the entire cross-section. Two or three groups of roll stands may be used, with each group containing more than one set of rolls. The rolls for these stands were shown in Figure 23—3 of Chapter 23. Figure 27—2 shows the schematic arrangement for a typical mill. This mill has a roughing group of stands which consist of a two-high roughing edging stand, closely associated with a roughing main stand, which has two horizontal main rolls and two friction-driven vertical rolls in a single vertical plane. The rolls in the roughing edging stand have a nominal diameter of 762 mm (30 inches) at the working zone and a body length of from 508 mm to 1219 mm (20 inches to 48 inches) varying according to the depth of section rolled. The roughing main rolls have a nominal body diameter of 1321 mm (52 inches) with a body length matching the edging rolls. Vertical rolls in the roughing stand, having a nominal diameter of 1067 mm (42 inches) and a face width of 457 mm (18 inches), are friction-driven by the shape about a roller bearing in their bore. The intermediate group of stands is identical in all of the above particulars to the roughing group, with the exception of the edging stand being on

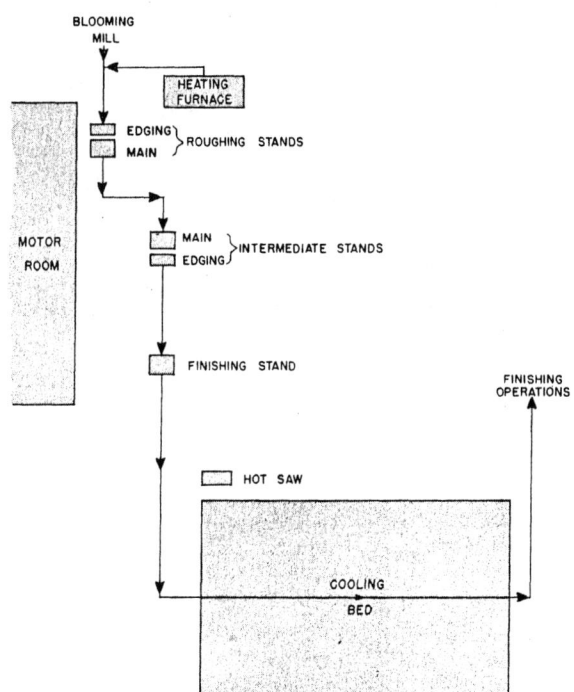

FIG. 27—2. Schematic arrangement of a typical wide-flange mill.

the opposite end of the main stand. The finishing stand resembles the roughing and intermediate main stands in that it has main and vertical rolls of the same size in like arrangement, but does not have edging rolls. The general arrangement of the mill is such that the hot shape from the blooming mill or reheating furnace enters first into the roughing edging stand and before clearing that stand enters the roughing main stand. The intermediate group of stands is placed a minimum distance to the side and one maximum shape length beyond the roughing group. The main intermediate stand is on the side toward the rougher and is closely followed by the intermediate edging stand, again arranged so that a single shape is in both stands of the group simultaneously (3.8-metre or 12½-foot centers). The finishing stand is in line with the intermediate group and 56.7 metres (186 feet) beyond the intermediate edging stand. Stationary roller tables are provided throughout with a short transfer mechanism to move the shape from the line of the roughing group of stands to the line of the intermediate group. All of these stands are driven by direct-current reversing motors, each individual stand having a separate motor, two-high pinions, and the necessary spindles. The roughing and intermediate main stands are driven by 5222-kilowatt (7000-horsepower) motors, and their respective edgers by 1492-kilowatt (2000-horsepower) motors. The finishing-stand motor is rated at 2984-kilowatt (4000-horsepower) power.

Rolled shapes are further processed with equipment which is substantially the same for shapes produced on the standard type of structural mill as on the wide-flange-beam mill.

An 864/1168-mm (34/46-inch) Structural Mill—One 864/1168-mm (34/46-inch) structural mill (Figure

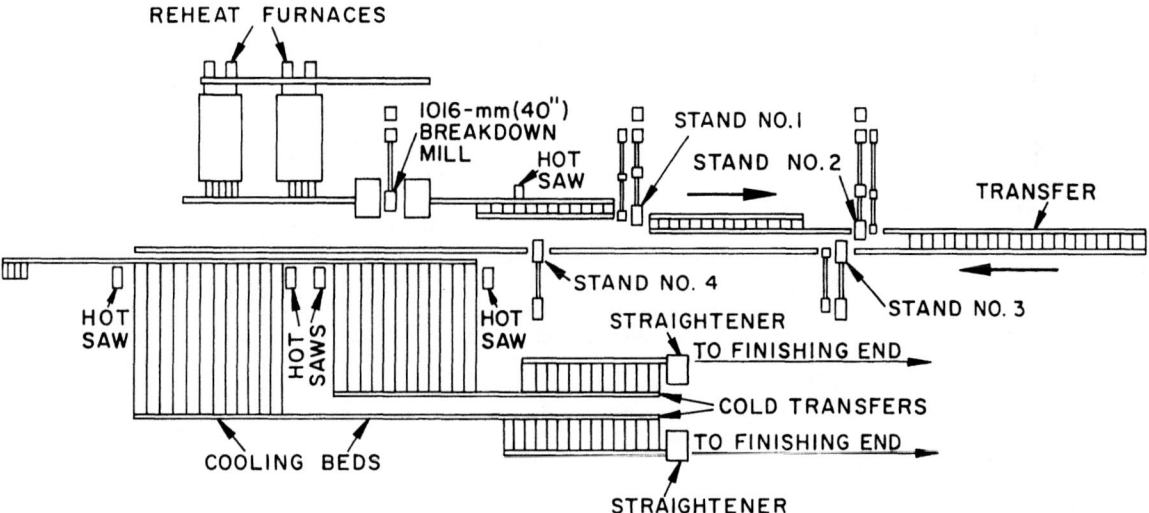

FIG. 27—3. Schematic arrangement of 864/1168-mm (34/46-inch) structural mill for rolling standard structural and wide-flange beam products. This sketch shows the mill set up with universal stands (with edgers for Stands Nos. 1, 2 and 3) for rolling wide-flange-beam products; the universal stands and edgers can be replaced with two-high horizontal-roll stands when rolling standard structural products.

27—3) is arranged in a double-line hairpin with two stands in each line. The first and second stands are the reversing type and the last two are single-pass unidirectional stands. When rolling wide-flange sections, Stands Nos. 1, 2 and 3 consist of a 1168-mm (46-inch) universal and edging combination, and Stand No. 4 (the finisher) is a single 1168-mm (46-inch) universal stand. When rolling other structural sections, universal stands Nos. 1 and 2 are replaced with 864 by 2032 mm (34 by 80-inch) two-high horizontal-roll stands, Stands Nos. 3 and 4 are replaced with 864 by 1727-mm (34 by 68-inch) two-high stands, and the edgers are replaced with movable table sections. Each stand is a complete unit arranged for clamping to the shoes and fitted for rapid coupling and uncoupling of oil lines and drives.

The first roughing stand is equipped with twin-motor individual-roll drives, but is also provided with a pinion stand which can be used for tying both rolls together, or in which shafts can be used to drive the rolls separately. This unique set-up is provided with a 2984-kilowatt (4000-horsepower) drive on each roll. The remainder of the mill stands are powered with single motors and pinions. The second stand has a 4476-kilowatt (6000-horsepower) motor; the third, a 2984-kilowatt (4000-horsepower) motor and the fourth, a 2238-kilowatt (3000-horsepower) motor. Each edger is driven by a 1119-kilowatt (1500-horsepower) motor through a pinion stand.

The mill was designed to operate with card-programmed control.

Hot Sawing—The removal of ends not filled to proper section, commonly called "crop ends," and cutting the hot shape into lengths which can be handled in further processing, is done with a hot saw. This equipment consists of a toothed circular saw blade

mounted on a sliding frame and driven at high speed by an electric motor. Blades range up to 1676 mm (66 inches) in diameter and copious water cooling helps maintain the cutting edges. The saw and its drive are moved on a sliding base at right angles to the hot shape so that the revolving blade cuts through the stationary shape.

Cooling—Cooling of the rolled shapes is accomplished by placing them on a cooling bed which is a steel structure, arranged to support a continuous layer of shapes, while providing for a maximum circulation of air upward around them. A mechanism is provided to pull the shapes sideways into place and to slide them across the bed onto the discharge table.

Straightening—Large shapes are straightened by roller-type straightening equipment, or a gag press. The former normally consists of seven or eight cast-iron or cast-steel rolls assembled in a single housing with a single drive, driving either part or all of the rolls. The rolls in the top row are spaced midway between the bottom rolls and may be moved vertically by screws. All rolls are arranged for axial adjustment. The gag press is a horizontally-reciprocating ram midway between two support points on a platen so arranged that it can be moved closer to or farther away from the ram to vary the amount of bend made in the shape.

Cold Cutting—Cold cutting to final length is accomplished by shearing or cold sawing. The shears consist of a stationary bottom blade over which the shape is positioned, and a top blade which is forced down on the top of the shape to cut it through, either with a single shearing action, or by punching a slug, or short piece, out of the shape. The cold saw is similar in design and action to the hot saw, but is fed at a much slower rate.

SECTION 2

ROLLING METHODS AND PROCEDURES

The practice of heating blooms for rolling into shapes varies with the grade of the steel, the size of the bloom, and the temperature at which the blooms are charged. A typical mill charges ordinary carbon-steel blooms received from the blooming mill at about 982°C (1800°F) and heats them to 1232°C (2250°F) in approximately 45 minutes. On a mill using three in-and-out type furnaces, normal operations find one furnace being charged, one heating, and the other being drawn, at a given time. Charging of the single layer of blooms in a furnace is begun as soon as the drawing has been completed. When steel at atmospheric temperature is charged, the heating to 1232°C (2250°F) requires about 2½ hours for the average size bloom. Handling the blooms into and out of the furnaces is accomplished by charging machines which grip the blooms on their sides. This necessitates a space between adjacent blooms of some 150 or 200 mm (6 or 8 inches) as clearance for the charging-machine tongs.

Rolling—Heated blooms are deposited by the charging machine on the breakdown-stand approach table, which conveys them to the rolls. The manipulators, on the entry side of the mill, are brought into play to align the bloom with the first pass and to turn it about its longitudinal axis, if necessary. When properly aligned, the table rollers are revolved to feed the bloom into the first pass of the rolls. The rolls for the breakdown-stand generally have three or more pass grooves, some of which may be rectangular blooming passes. The number of different shaped grooves in the rolls, and the number of passes made in each groove, vary with the section being rolled. After making the required number of passes through the first groove, the bloom is re-aligned for subsequent passes by the manipulators.

Blooms processed through the breakdown-stand progress over the stationary rollers of the delivery table to the traveling tilting table which carries the bloom into line with the first pass of the roughing stand. Since the finishing-mill rolls are not reversed during operation, only a single pass is made through each groove. Similar tables on the opposite side of the finishing mill receive the shape from the first groove and enter it into succeeding grooves. Roughing- and intermediate-stand rolls usually contain from two to five pass grooves, while finishing-stand rolls are generally limited to a single pass on a shape. Finishers usually have duplicate grooves for the same section, or provide grooves for a variety of sections.

The type of section resulting from the rolling process is determined by the shape of the various pass grooves through which the material progresses, with optional groups of pass shapes frequently being available for the production of a given section. In the case of standard beams, at least three different systems of passes may be used.

The most common pass shapes used in rolling standard beams are those of the straight-flanged method which are shown in Figure 27—4. This method owes its popularity to the large range of web thicknesses which can be produced from a single set of rolls, the ability to use the early passes for the production of channels as well as beams, and the small thrust loads transmitted by the rolls to the bearings. The versatility of the weights produced results from the relatively small total taper in the live or open flanges which permit rolls to be separated to produce heavier webs with a minimum thickening of the live flange. The most undesirable feature of this method is the relatively slight taper in the pass sides which reduces the gain on dressing and results in passes getting progressively wider and flanges relatively heavier when dressed, thereby limiting the life of the rolls.

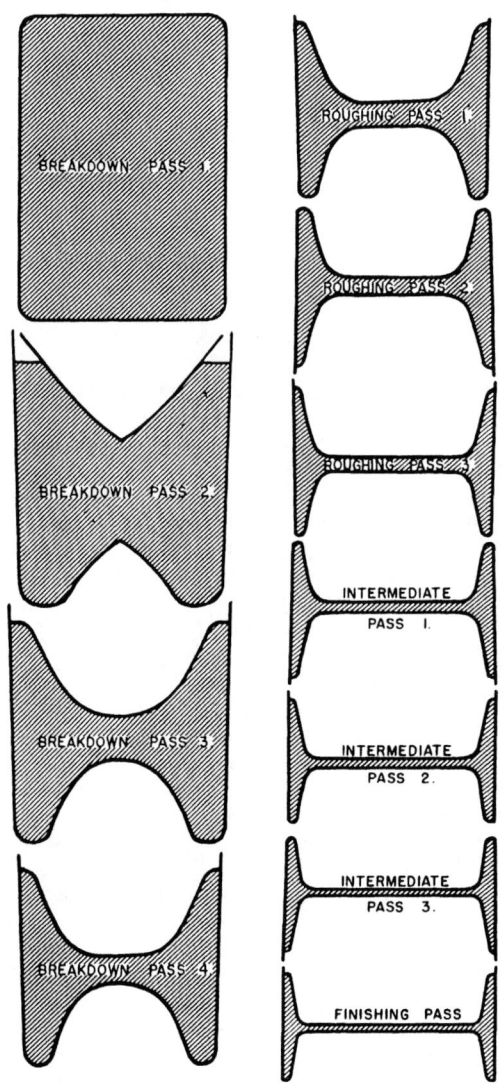

FIG. 27—4. Roll passes for rolling standard beams by straight-flanged method.

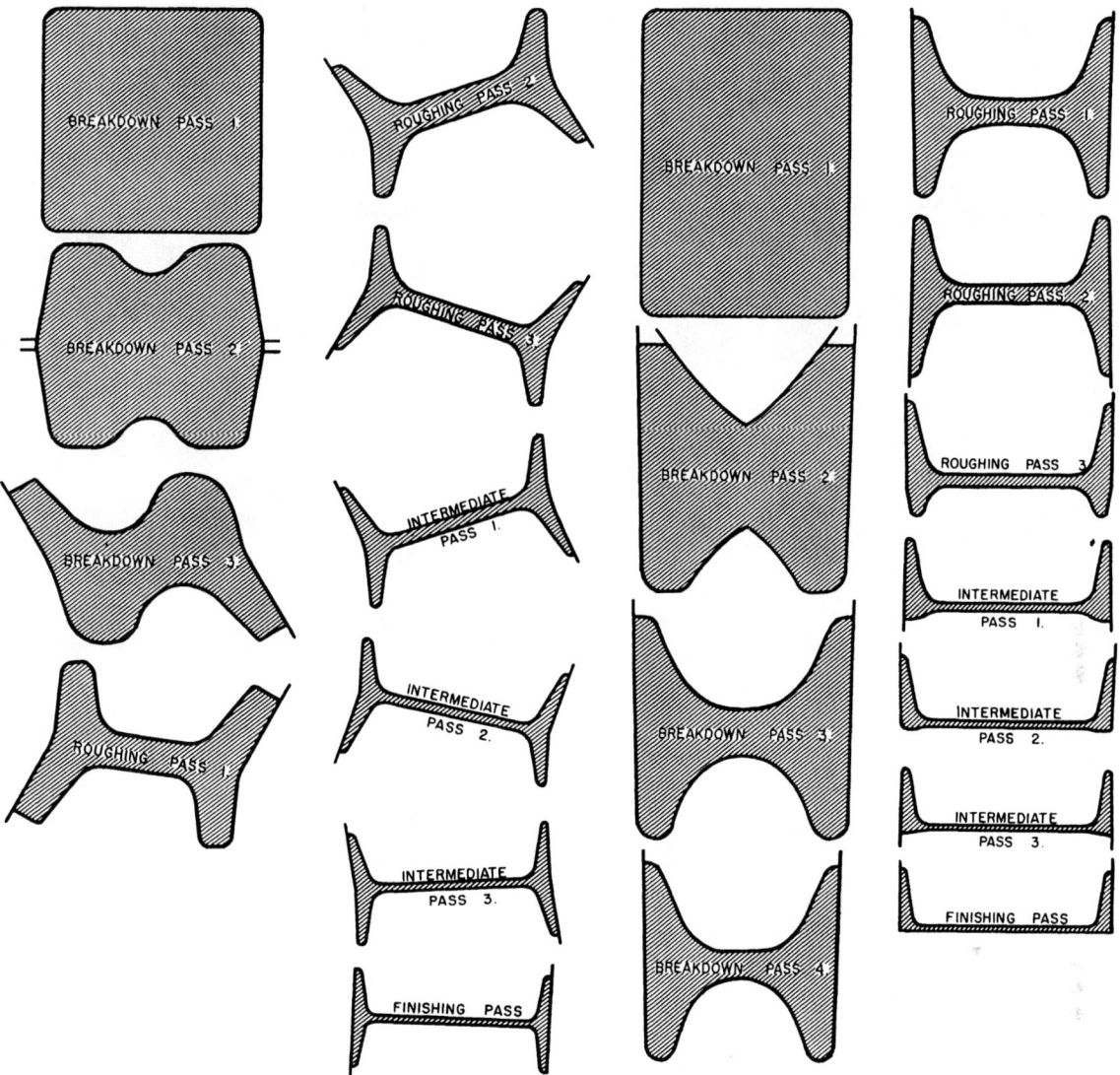

FIG. 27—5. Roll passes for rolling beams by the diagonal method.

FIG. 27—6. Roll passes for producing a channel, using early passes shared for rolling beams.

The butterfly method of rolling beams closely resembles the straight-flanged method. The outstanding difference is the fact that the live flanges are bent out to a much greater degree in all passes preceding the finishing pass. This results in greater ease of restoring the open flanges by dressing, but imposes a serious limitation on the range of web thicknesses that can be produced in a given set of passes.

Figure 27—5 illustrates the passes used in the diagonal method of rolling beams. This method makes good provision for the restoring of both flange thickness and pass width by dressing, but involves serious restrictions in the web thicknesses that can be satisfactorily produced from a single set of rolls, and vastly increases the thrust loads that are imposed on the bearings which support the rolls. Since thrust-bearing wear permits longitudinal motion of the rolls with corresponding changes to the thickness of the open flanges, rolling difficulties result.

The method of rolling a channel using early passes shared with a standard beam is illustrated in Figure 27—6. This plan results in a smaller roll inventory, good flange control in the production of a large range of weights, narrower passes and stronger collars than can be had if the butterfly method is used. Figure 27—7 shows the butterfly design. It has the advantage of producing channels which are filled to proper section very close to the ends of the rolled bars. However, this method of production results in relatively thin weak collars on the rolls and almost precludes the rolling of multiple weights. This is caused by the extreme thickening of the flanges in the early passes when the rolls are separated to produce a thicker web.

Angles also offer the roll designer a choice between a number of proven rolling methods. Those most generally approved today may be referred to as the butterfly and the flat-and-edge methods. Typical passes for the butterfly angle are shown in Figure 27—8. The

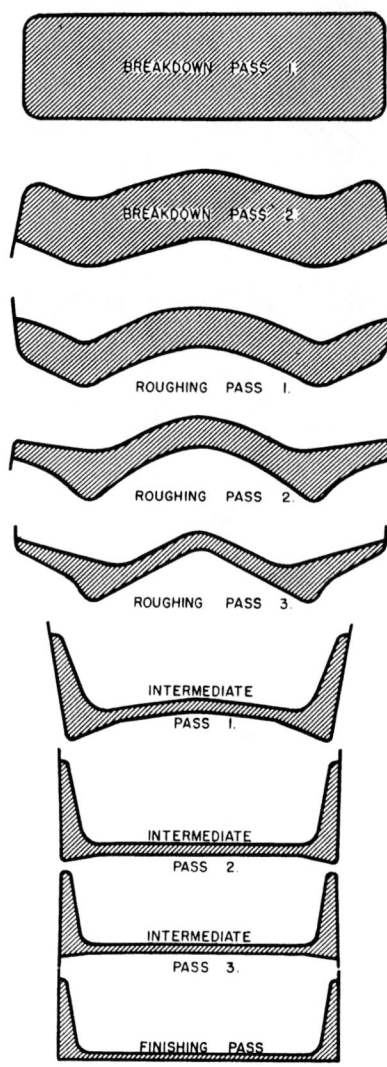

FIG. 27—7. Roll passes for rolling a channel by the butterfly method.

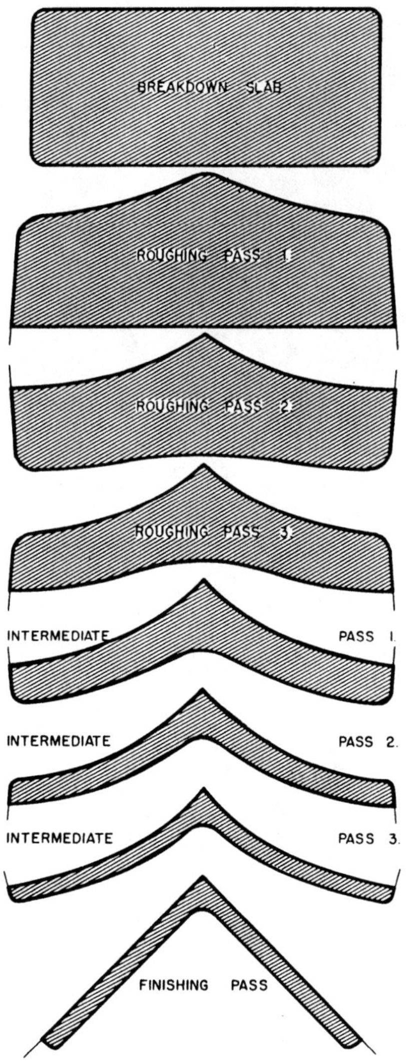

FIG. 27—8. Roll passes for rolling an angle by the butterfly method.

current popularity of this method is due to the relatively small crop loss on the shapes, the absence of vertical rolls or turning for edging, and the ease of controlling the bending, which is slight in any one pass. Undesirable features of the design are the lack of proportional thickening of the various parts of each leg when rolls are separated to make heavy weights, and somewhat poorer control of length of leg.

The flat-and-edge type of design can be seen in Figure 27—9. This method is popular on those mills having vertical edging rolls and is sometimes used where the shape is turned 90° for edging in a pass in horizontal rolls. Leg lengths are readily controlled in the edging passes and thicknesses remain uniform when rolls are set to produce heavy weights. These advantages are frequently offset by the difficulty in maintaining proper entry alignment of the section in the later passes where bending is drastic. Turning long lengths of hot, flexible

steel for edging, where necessary, is also difficult.

The special structural sections include shapes rolled rarely or for a single purpose. The passes used for the production of zee bars involve the same principles as the rolling of angles by the butterfly method, with each half of the pass resembling an angle pass, and with the two halves being fitted together reversed. The half center-sill section, used in making modern railroad cars, is a special type of zee bar, being irregular in both thickness of the members and flange lengths. Passes for its production are included here (Figure 27—10) because of the relatively great demand for the section.

Modern practice tends towards the production of tees by splitting a beam of proper size through its web to make two of the required sections. Some tees, however, are still rolled as such. The rolls are unduly complicated and the actual rolling is difficult. Where a choice is offered, the rolling of tees is usually avoided

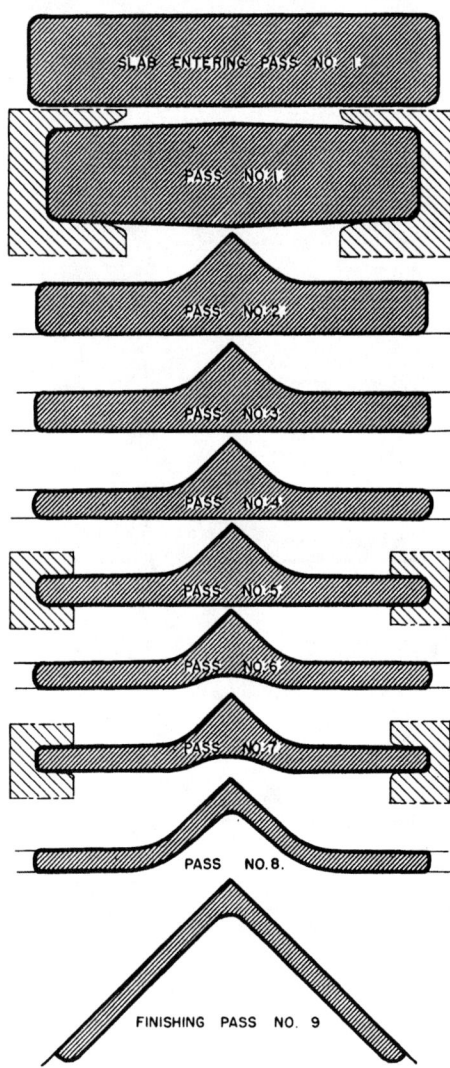

FIG. 27—9. Roll passes for producing an angle by the flat-and-edging method.

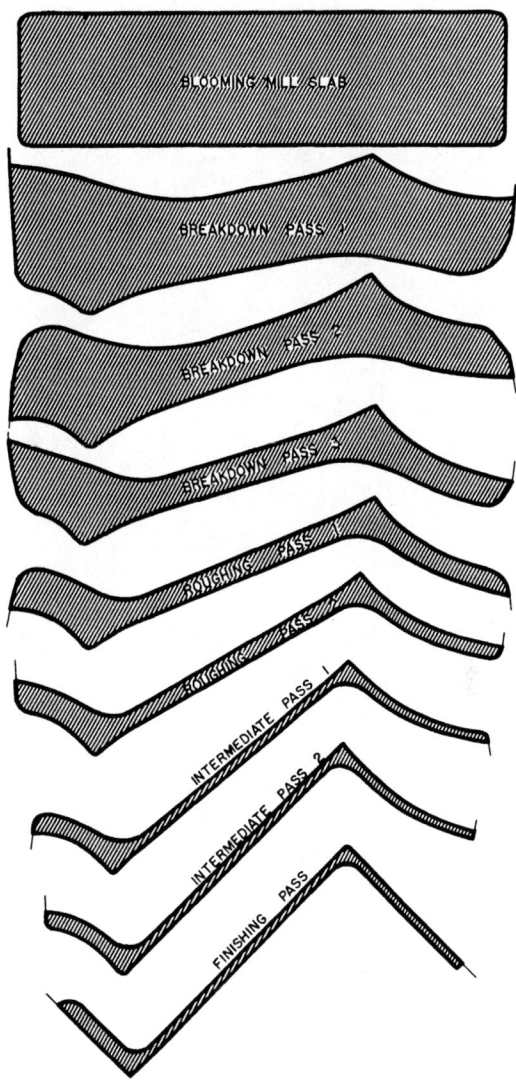

FIG. 27—10. Roll passes for the production of a half centersill section, a special type of zee bar.

because of the necessity of turning the section between passes to permit each of the two tapered flanges and the parallel stem to be worked alternately, so far as possible, in open and closed grooves.

One of the most interesting groups of sections produced upon standard structural mills is the sheet-piling group. (Steel sheet piling of domestic manufacture is produced to ASTM specifications A 690 and A 328.) Usually produced in three general types, known as straight web, arch web, and zee, these sections have been adapted as wall members in both permanent and temporary structures requiring strong vertical walls for lateral support, such as cofferdams, piers, dykes, breakwaters, etc. In the rolling of piling sections, the usual problems are complicated by the necessity of bending the flange (which has been rolled straight) to form the interlock. Precise control of the bend is necessary since proper clearance within the interlock must

be maintained and the resulting opening between flange tip and thumb must be within close limits. The particular arch web section for which passes are shown in Figure 27—11 accomplishes this bending of the flange in two steps in the leader and finishing passes. Attention should be given to the fact that the shaping of the section starts in the blooming mill. This is typical of most piling sections and large beams, and is the result of the large size of the sections coupled with their intricate shape which precludes getting enough passes into the limited space of the breakdown and finishing mills to properly form the section.

While the rolling procedure for a section on the standard type of structural mill does not normally employ the same pass groove for more than one reduction in the shape, multiple passes being limited to the breakdown stand and seldom exceeding three passes in a groove, the opposite situation exists on the wide flange

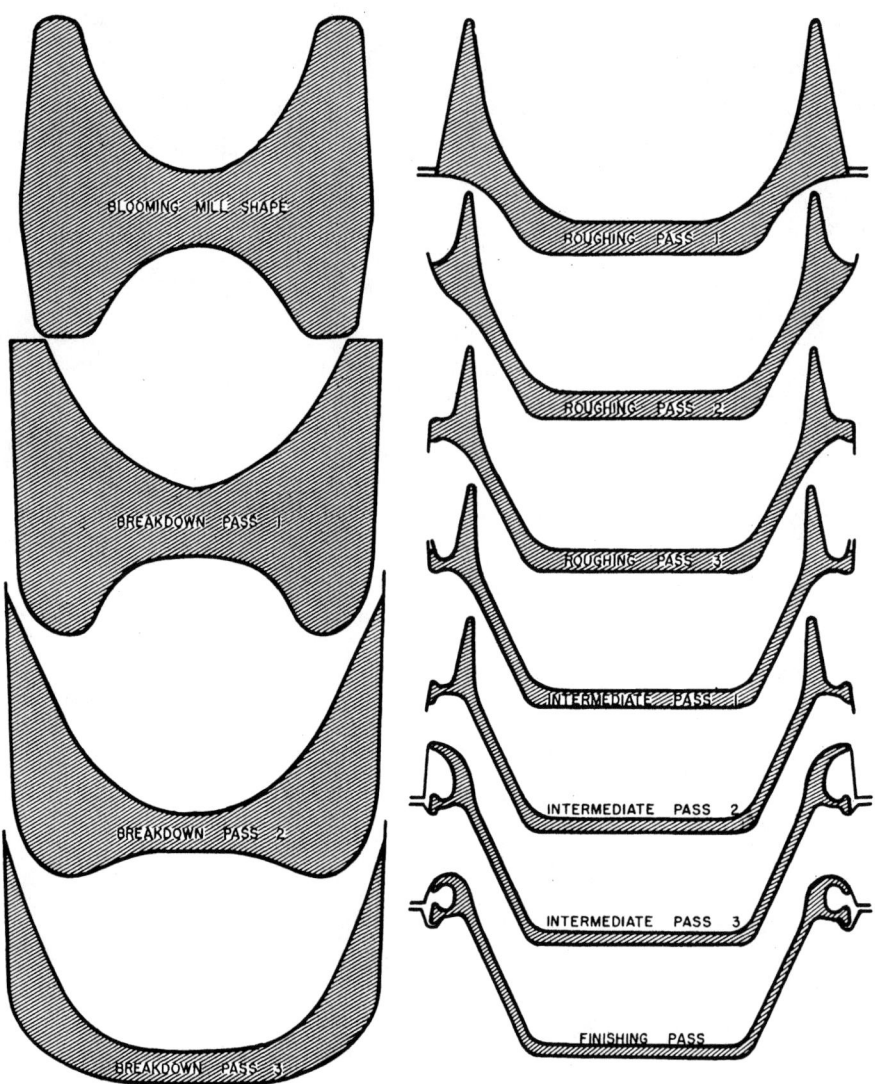

FIG. 27—11. Roll passes for rolling an arch web section of sheet piling, showing how flange is bent to accurate configuration in the leader and finishing passes.

beam mill. Here as many as fifteen successive passes are taken through the main and edging stands of the roughing group. In a typical layout, the main and vertical rolls of the roughing main stand and the rolls of the roughing edging stand are moved to new settings with respect to each other by an automatic screwdown control. With this mechanism, the operator sets in advance of rolling the reduction to be taken on each pass. Then, pass by pass, he advances the lever of the master control switch a notch, with each of the three sets of rolls of the roughing group moving closer together simultaneously to the new settings. The control equipment is so designed that each of the three sets of rolls can be made to move different amounts, thus permitting proportional rather than like reductions on flange and web.

The equipment of the intermediate group of stands is practically the same as the roughing equipment. As

in the rougher, fifteen passes are provided for in the control equipment. In practice, the passes are divided between the roughing and the intermediate stands in a proportion that gives equal duration of rolling cycle in each of the two stands. Since the shape at the roughing stand is relatively heavy and short, more passes are required there to balance the relatively few passes made on the elongated shape at the intermediate stands. Passes in actual use range from fifteen roughing and nine intermediate on sections requiring heavy overall work, to five roughing and three intermediate on the sections requiring a minimum of work; nine roughing and five intermediate passes represent the average condition. In all cases the intermediate passes are followed by a single finishing pass. In rolling wide-flange beams it is customary to roll a bloom which has, as nearly as possible, the same proportions as the finished beam. Each succeeding pass in the rolling cycle strives to maintain these proportions, resulting in reduction, pass by pass, being proportional. Since at no time after leaving the blooming mill are the flanges rolled in a closed groove, the loss in flange length is very small, and the rolling of beams with rather wide flanges is entirely possible.

Rolled material from either of the described structural mills is delivered by roller table to the hot saw. The blade of the saw is normally revolved in such a direction that the cutting edge is moving downward toward the shape. The action of the rapidly revolving toothed blade combines a mechanical sawing and melting of the steel, the chips when thrown clear frequently being molten. Cuts are usually made at the hot saw to remove the crop ends, to part the usable material into lengths that can be handled for further processing, and to provide short test pieces. Tests are cut for the guidance of the roller in correcting shape deviations and for various mechanical tests which are part of the inspection procedure.

Cooling is accomplished by the natural circulation of air about the shapes on the cooling bed. The rate of cooling can, to some extent, be controlled and is of relative importance. Cooling can be accelerated by maintaining space between shapes on the bed, and by providing ample space below the bed for the passage of cool air. Retarding of the cooling cycle results from putting the hot shapes on the bed in a solid touching layer. If cooling is too drastically retarded, the size of the hot bed must be increased or rolling delays will result. On the other hand, too rapid cooling may result in mechanical properties outside the ordered range. Non-uniform cooling results in warpage of the shapes as they cool. This must be avoided in symmetrical shapes, if possible. Most non-uniform shapes warp during cooling, sometimes to the point that they are difficult to convey to the straightening machine.

SECTION 3

FINISHING AND INSPECTION

When the shapes have been properly cooled they are conveyed again by roller table to the straightening machine. For the products rolled on the standard type of structural mill and the smaller wide-flange beams, a roller straightener is normally used. The first shape to be straightened is stopped in the machine while the rolls are adjusted laterally to it. The top rolls are then moved downward to deform the shape with each successive top roll making less deformation than the preceding one. This shape is backed out of the machine and then run through for most of its length. If not straight to the eye of the experienced operator it is again reversed in the machine. Further adjustment is made with this process being repeated until a straight shape is obtained. Subsequent shapes will require only slight alteration to the setting of the machine to compensate for wear and other variables. The larger wide-flange beams and some few standard mill products are straightened in the gag press. Here the process consists of advancing the stationary platen and the shape towards the reciprocating ram until a bend is accomplished in the shape. The operator has full control of the point on the shape at which the bend will be made, the amount of the bend, and by turning on the table, the direction of the bend. Bending strives only to offset initial crookedness.

Cutting to final length is accomplished by one of several means. Hot shearing or hot sawing is most commonly used, wherever the class of product permits. Cold sawing is the next most frequently used method. Flame cutting is available for extremely heavy sections and unusual applications. Where overall length must be accurate, as in the case of columns, the ends may be milled; this results not only in close length tolerance, but in a reduction in the ends' out-of-square allowance as well.

Before the products rolled on the various mills may be shipped, they are inspected for imperfections and tested for certain mechanical properties as may be specified. This is accomplished by two general methods. First, the test cuts taken at the hot saw are sent to the laboratory where tests are made to determine the mechanical properties of the material, such as tensile strength and yield point. The other method of inspection is visual inspection of the material on an inspection bed where the shapes are inspected for metallurgical features such as pipe, blister and scabby surface, and for dimensional variations such as length, straightness, and out-of-square condition. Dimensions such as web or flange thickness, flange height, weight per unit length, etc., are checked on tests taken at the hot saw by the mill inspector, and the roller may adjust the mill to correct improper conditions, if necessary. Special sections may require additional inspection. An example is the slot and thumb on sheet piling. Chemical composition of all material is listed as part of a record kept of all rolled products showing the number of the heat from which they were rolled and any other pertinent information.

The final steps are to separate and assemble for shipment. The material then is loaded in freight cars, river barge, or truck, and shipped to its destination.

CHAPTER 28

Production of Steel Wheels

Introduction—The different types of steel wheels are classified according to methods of manufacture as well as their intended use. Very light wheels such as automobile wheels and extremely large wheels generally are fabricated from several pieces that are joined by riveting or welding. Most wheels for railroad or industrial service are manufactured either by forming them from a solid block of metal by a sequence of hot forging and/or rolling operations, or by casting them directly from liquid metal into their final shape. Steel rims, shrink-fitted to cast-iron or cast-steel centers, are still in common use on railroads in many parts of the world. On the North American continent, however, solid one-piece wheels are used with rare exceptions. The use of cast-iron wheels on freight cars has been discontinued by the American railroads. This chapter describes the production of steel wheels by (1) forging and rolling and (2) casting in molds.

SECTION 1
STEEL WHEELS AND THEIR CLASSIFICATION

Parts of Steel Wheels—Steel wheels are used in heavy load and traction services. The size and contour of a specific wheel design is based on the load it must carry and the space limitations of the equipment on which it is to be used. The contour of a wheel is considered as being composed of five parts: namely, the **hub**, the **web** or **plate**, the **rim**, the **tread** and the **flange**. These parts, for the case of a railroad wheel, are illustrated in Figure 28—1. The hole in the hub into which the axle is fitted is called the **bore**. The tread is the outer surface of the rim on which the wheel rolls or makes contact with the rail. The flange keeps it from leaving the track. The **dishing**, or **coning**, refers to the offset of the hub with regard to the rim and the resulting slanting of the web section. The flange, the rim, the service to which they are applied, the methods of forming, the finish, and the finishing treatments are all bases for classifying wheels. Thus, they may have a **single**

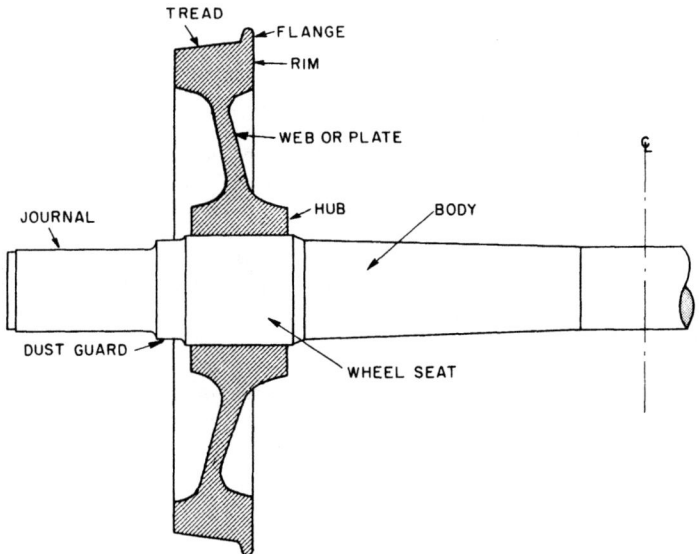

FIG. **28**—1. Sketch showing cross-section of wrought-steel railroad wheel on a roller-bearing raised-wheel-seat axle, giving nomenclature of the principal parts of each.

911

flange (a flange on the inside only) or a double flange (a flange on both sides) and a rim thickness that will permit worn treads and flanges to be machined one or more times to restore their original contours. The term one-wear wheels implies that only one major period of wear is generally obtained. However, with one-wear wheels of design rim thickness of 31.75 and 38.10 mm (1¼ and 1½ inches) it is often possible to machine the worn tread and flange to obtain additional life. Multiple-wear wheels have a rim thickness of 63.5 mm (2½ inches) or more, and may be machined two or more times to the original tread and flange contour. A two-wear wheel with a 50.8 mm (2-inch) thick rim also is produced; as the name implies, two wear periods are intended during its service life.

Classifications based upon the service to which the steel wheels are to be applied are defined as follows:

(1) **Industrial car wheels** are single-flange wheels designed for use under cars for industrial purposes, such as mine cars, railroad cars or buggies used in steel mills, furnace bottoms, transfer cars, and so on.

(2) **Industrial locomotive wheels** are single-flange wheels designed for use under electric and other types of locomotives used in mines, and in industrial plants.

(3) **Crane track wheels** may have either a single or a double flange, and are used under such equipment as traveling, gantry and bridge cranes; transfer and turntables; and floor-type charging and drawing machines.

(4) **Railroad freight-car wheels** are one-wear, two-wear, and multiple-wear wheels used under freight cars depending on their capacity and usage.

(5) **High-duty wheels** are usually multiple-wear wheels and are used under railroad, electric railway, and rapid-transit passenger cars, as well as under locomotives.

Steel wheels are made to meet the requirements of standard specifications including the type and kind of steel used, the design of the wheels, the dimensional tolerances, and the inspection and tests. While alloy steels are used in some circular shapes, steel wheels have generally been made of carbon steel.

Almost all wrought-steel wheels of 711-mm (28-inch) diameter and larger are produced by forging and rolling, whereas smaller wheels are forged only. Wheels may be machined on the tread, flange, rim faces and hub faces, while relatively few are machined all over. To promote soundness, wheels are not air cooled from forging or rolling temperatures, or after removal from the mold in the case of cast wheels made by the controlled pressure pouring process (described later), but are always controlled cooled to ambient temperatures.

Classes of Steel Wheels—Steel wheels are either manufactured to Association of American Railroads or American Society for Testing and Materials specifications. These specifications designate five classes of steel wheels. The five classes differ in chemical composition and hardness and were developed after extensive research followed by many studies of wheels in different types of railroad service. Each class is processed and treated to meet a specific service requirement and it is important that wheels of the proper class be used in the service for which they are intended.

Association of American Railroads specification M-107 and American Society for Testing and Materials

specification A 504 cover one class of untreated wheels (Class U) and four classes of heat-treated wheels (Classes L, A, B and C) grouped according to carbon content as follows:

Element	Per cent
Carbon Class U	0.65-0.77
Class L, not over	0.47
Class A	0.47-0.57
Class B	0.57-0.67
Class C	0.67-0.77
Manganese	0.60-0.85
Phosphorus, not over	0.05
Sulphur, not over	0.05
Silicon, not less than	0.15

The service for which the various classes are intended generally is as follows:

Class U—General service where an untreated wheel is satisfactory.

Class L—High-speed service with more severe braking conditions than other classes and light wheel loads.

Class A—High-speed service with severe braking conditions, but with moderate wheel loads.

Class B—High-speed service with severe braking conditions and heavier wheel loads.

Class C—(1) Service with light braking conditions and high wheel loads.

Class C—(2) Service with heavier braking conditions where off-tread brakes are employed.

Classes L, A, B and C are heat treated to the following hardness requirements:

	Brinell Hardness Number	
Class	Minimum	Maximum
L	197	277
A	255	321
B	277	341
C	321	363

The specifications state:

"The Brinell measurement shall be made on the front face of the rim with the edge of the impression not less than 4.8 mm (³⁄₁₆ inch) from the radius joining face and tread. Before making the impression, any decarburized metal shall be removed from the front face of the rim at the point chosen for measurement. The surface of the wheel rim shall be properly prepared to permit accurate determination of hardness."

These specifications require that heat-treated wheels be stamped with the class letter L, A, B or C, for wheels that have the rim quenched, as follows:

	Significance	
Mark	Class	Treatment
L	L	Rim Treated
A	A	Rim Treated
B	B	Rim Treated
C	C	Rim Treated

MANUFACTURE OF WROUGHT-STEEL WHEELS

Outline of Methods for Forming Wrought-Steel Wheels—All methods for forging and rolling wrought-steel wheels require a cylindrical, or nearly cylindrical, block of steel for the initial forming operations. Many years ago, the blanks were sheared from slabs, but this method located the central portion of the ingot diametrically across the wheel, so that the line of segregation terminated at opposite points in the tread. The objections to this method are obvious. Therefore, the procedure was changed, as shown in Figure 28—2, so that metal in the line of segregation, or central portion of the ingot, would be punched out in forming the bore.

To accomplish this, the steel blocks may be prepared either by hot shearing or cold cutting the blocks from a round bloom, or by cutting the blocks from ingots. After the blocks have been press forged to the general form of a wheel, called the **wheel blank,** the shaping is completed by rolling the blank in the wheel mill which may be one of two types designated, from the position in which the blank is held in the mill during rolling, as a vertical (Figure 28—3) or a horizontal mill (Figure 28—4). A vertical mill has the advantage that any scale that might form on the wheel plate will fall free during rolling and thus promote the production of wheels with smooth plate surfaces. The steps of a typical production process are described below.

Preparation of Blocks—Basic oxygen or basic electric furnace steel to be rolled into round blooms for wheels is cast in 645- by 645-mm (25⅜- by 25⅜-inch) hot-top molds with rippled or corrugated walls, and rolled on a 1370-mm (54 inch) mill into round blooms either 381 or 445 mm (15 or 17½ inches) in diameter, according to the weight of the wheels to be made. With proper allowance for discard, these rounds are hot sheared into blocks of the desired weight and dimensions. Blocks are

put into special, insulated containers and allowed to cool slowly to below 150°C (300°F). The cooling is thus controlled as a precaution against the possible formation of minute internal cracks in the blocks. Some

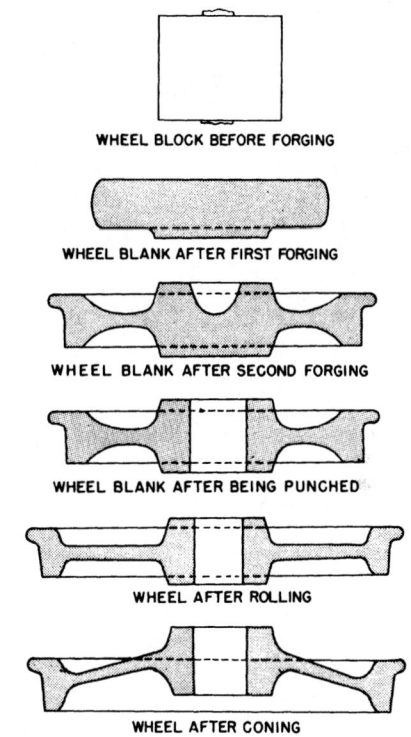

FIG. 28—2. Cross-section showing basic steps in the procedure of making a wrought-steel wheel by a combination of forging, rolling and pressing operations.

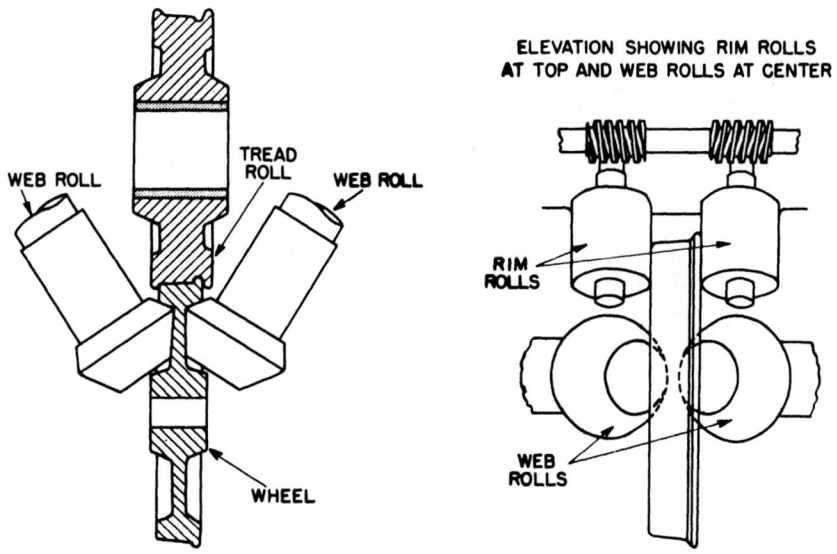

FIG. 28—3. Schematic arrangement of the rolls in a vertical mill for rolling wheels.

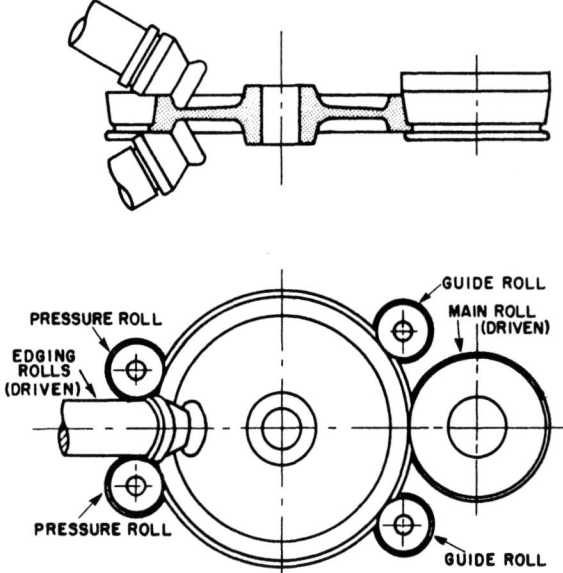

FIG. 28—4. Elevation (above) and plan (below), showing disposition of the rolls in a horizontal mill for rolling wheels.

round blooms are cold cut on special multiple-tool lathes after controlled cooling. The blooms are then inspected, surface conditioned, and fractured.

Heating and Forging—Wheel blocks are brought from the storage area in open gondola-type railroad cars and are unloaded with a magnet on an overhead crane onto a platform on which their individual identity is checked. Wheel blocks entering the production line are rolled onto a chain conveyor at the edge of the platform and passed through an automatic weighing station. Only blocks within the correct weight tolerances of the wheel design to be rolled proceed to the

heating furnace. Blocks that are too heavy or too light are automatically rejected. The blocks are charged in a rotary-hearth furnace for heating to forging temperature by automatic hydraulically actuated tongs that are program controlled, as shown in Figure 28—5. The furnace is about 19 metres (62½ feet) in diameter, has 5 zones, and is heated by natural-gas-fired roof burners. The first zone of the furnace is unfired and acts as a preheat zone utilizing the residual heat in the hearth. The furnace has five concentric file locations and each block is positioned at a specific spot so that its identity can be maintained. After approximately five hours in the furnace the blocks have been heated uniformly to forging temperature of 1175°C (2150°F). During heating, conditions inside the furnace are continuously monitored with the aid of two closed-circuit television cameras which sight through windows near the charging and discharging doors. The heated block is removed from the furnace by a set of hydraulically actuated tongs similar to those that charged it. Before forging, the block is automatically conveyed to a descaler where high-pressure water sprays remove the scale.

Both the first and second forging operations are performed on the same press which has a capacity of 9000 metric tons (10 000 net tons). This press has two sets of top and two sets of bottom dies which are moved horizontally by hydraulic rams that position the proper set of dies for each forging operation (Figure 28—6). In the first forging operation, the cylindrical block is upset forged into a disk with a centrally located projection on the bottom that will eventually be part of the hub of the wheel. The first forged blank is then ejected from the bottom die and is grasped and suspended by a pair of tongs while the change to the second-forging dies is made (Figure 28—6). The second-forging dies are moved into position, and the first forged blank is lowered into the bottom die for second forging. In second forging, the rough contour of the wheel is formed—

FIG. 28—5. View from heater's control pulpit, showing wheel block being charged into rotary-hearth furnace.

FIG. 28—6. Hot wheel block in position on lower die of a forging press at the start of the first operation in forging a wheel. The lower die and block are shifted beneath the upper die (right) for first forging.

shaping of the web, rim, flange, and hub is started. A view of the second forged blank, emerging from the press, is shown in Figure 28—7.

At some wheel producing plants, the first and second forging operations are performed on two separate presses, and the first forged blank is reheated before second forging and/or for rolling. In another method, all hot-forming operations are performed after a single heating.

Wheel-Rolling—The wheel-rolling process is a unique metal-forming operation because of the simultaneous use of the various rolls to form the different parts of this complicated geometric shape. Unlike at most other mills, the hub is not punched after second forging to accommodate a mandrel to hold the wheel during rolling. Eight rolls simultaneously hold and shape the wheel to the desired diameter and contour. These include a driven pressure roll that forms the tread and flange contour, two edging (idler) rolls that form the front and back faces of the rim, and two driven web rolls that shape the contour of the web or plate. Three additional rolls support the wheel at its periphery during rolling. All the rolls are movable and gradually change their position during rolling as the

wheel changes to the shape desired. The limits of movement for all rolls on this mill are preset with a template before rolling wheels of a given design so that the desired contour and dimensions are consistently maintained during rolling (Figure 28—8).

Stamping, Coning, and Hub Punching—After rolling, the wheel is transferred to the stamping machine in which its individual identification numbers and letters are hot stamped into the back face of the rim. As required by the Association of American Railroads, this identification includes the month and year of manufacture, a wheel serial number, specification class, manufacturer's brand and design designation.

Next the wheel is transferred to the coning press where the shape of the plate of the wheel is transformed from a flat disk to that of a truncated cone by close-fitting dies that contact the entire wheel with the exception of the tread (Figure 28—9). Punching of the hole in the hub is also carried out on this same press by a ram that is actuated immediately after the coning operation is completed.

Controlled Cooling and Heat Treatment—After coning and hub punching, which complete the hot-forming operations in making a wrought-steel wheel,

FIG. 28—7. Wheel blank raised from die after the second forging operation.

FIG. 28—8. Wheel being rolled in the rolling mill.

FIG. 28—9. Wrought-steel wheel after coning and hub punching.

the wheel is automatically transferred to a chain-link conveyor system on which it is transported in a vertical position on a hook which suspends it by the bore.

All wheels are transported by this conveyor system through a continuous controlled-cooling furnace which permits them to cool below the critical temperature and then maintains them at a temperature of about 535°C (1000°F) (Figure 28—10). Upon emerging from this furnace, wheels that are not to be heat treated (Class U wheels) are conveyed to a slow-cooling tunnel where they are automatically removed from the conveyor and stacked in piles on buggies which transport them through the tunnel (Figure 28—11). The wheels are slow cooled in this tunnel at a controlled rate to below 150°C (300°F).

Wheels that are to be heat treated are not controlled cooled but are conveyed directly into a tunnel-type continuous gas-fired furnace where they are reheated to a uniform temperature of about 870°C (1600°F). The wheels emerge from this furnace in groups of six and are conveyed to the quenching unit shown in Figure 28—12. Here the wheels are automatically lowered onto rollers in the quenching tank that support and rotate them in vertical position during quenching. The water level in the quenching tank is adjusted so that only the rim at its lowermost portion is submerged. Thus, only the rim of the wheel is quenched to harden it and thereby increase its resistance to wear. The time of quenching is varied depending on the rim thickness and the diameter of the wheel.

FIG. 28—10. Hot wheel entering the temperature-equalizing furnace on a continuous conveyor after leaving the coning press.

FIG. 28—11. Wheels entering the controlled-cooling tunnel.

After quenching, the wheels are conveyed through the tempering furnace where they are reheated and maintained at the proper tempering temperature. Upon emerging from the tempering furnace, the wheels are transferred through the slow-cooling tunnel described previously for slow-cooling untreated wheels and are slow cooled in the same manner.

Finishing—To obtain final dimensions that meet specified requirements, the tread, flange, rim faces, hub faces, and bore of the wheels are machined on modern lathes using carbide-tipped cutting tools (Figure 28—13).

After machining, the wheel passes down an inspection line, shown in Figure 28—14, where its surfaces are visually inspected and its different dimensions are measured. The wheel is checked for rim thickness, thickness and height of the flange, the radius of the flange throat, concentricity and diameter of the bore, and location of the hub with reference to rim. The hub wall thickness, rotundity, and plate thickness are also measured. The rims of the wheels then are automatically ultrasonically inspected to evaluate their internal quality. This is accomplished by transmitting high-frequency sound energy through the rim of the wheel from the front rim face with an ultrasonic transducer. A second high-frequency sound wave is transmitted through the rim from the tread and is oriented 90 degrees from the rim-face sound wave. In this test, the high-frequency sound waves travel through the rim of the wheel, strike the opposite surface, and rebound. The time and intensity of the ultrasonic waves that are reflected can be measured electronically, and changes

FIG. 28—12. Group of six wheels being quenched simultaneously during the "rim-toughening" heat-treatment process.

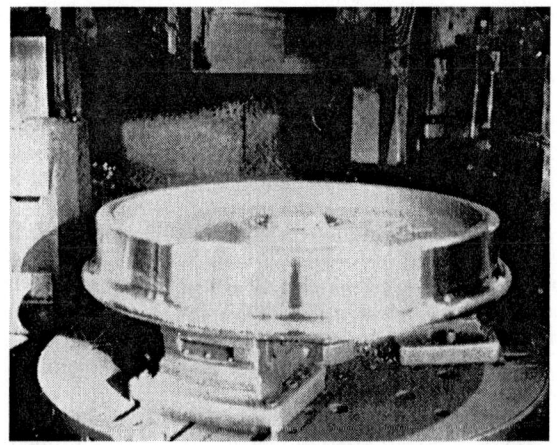

FIG. 28—13. Wheel after machining to finish dimensions on a multiple-station lathe equipped with tungsten-carbide-tipped tools.

in the sound characteristics are used to indicate the size and location of any internal discontinuities that may be present. Because small differences in internal structure can be detected by this technique, ultrasonic testing is considered an important inspection tool.

Each wheel is shot peened in the plate area to improve its fatigue strength. The circumference of the wheel is then measured and it is stencilled with the tape size (number indicating its circumference in eighths of an inch) and final inspected. The plate areas are then wet flourescent magnetic-particle inspected, supplementing visual inspection, to detect discontinuities which may be harmful to service life.

Forged Wheels and Circular Sections—This special category of mine-car wheels; crane-track wheels; industrial wheels; circular sections such as gear blanks and sheave wheels; and special wheels 711 mm (28 inches) in diameter and smaller are forged to their final rough dimensions by a sequence of closed-die forging operations. A separate 9000-metric-ton (10 000-net-ton) press is used for this product line. Inasmuch as most wheels have complex geometric shapes, few can be forged in a single press operation. Design of the forging dies must be such that the surface of the block will expand to fill out the dies completely without folding at any constricted part of the contour and forming a lap. Consequently, several progressive forging steps are usually necessary to form most wheel designs. The heat treatment, machining and inspection are similar to those for rolled wheels, but are performed in a separate special shop.

FIG. 28—14. Inspection line where wheels are visually inspected, gaged, and ultrasonically inspected.

SECTION 3
PRODUCTION OF CAST-STEEL WHEELS

General—According to the "Wheel and Axle Manual" of the Association of American Railroads (AAR), processes used by individual cast-steel wheel manufacturers differ in many details, but all have been developed to produce wheels that meet the requirements of the AAR specifications discussed in Section 1. Cast-steel wheels of various diameters are made in one-wear, two-wear and multiple-wear designs and are furnished in various heat-treated classes and carbon levels as covered in the specifications.

The metal for cast-steel wheels is melted in basic electric-arc furnaces. Melting stock for the furnace charge consists of scrap steel wheels, other steel scrap and foundry return metal. Practices discussed in Chapter 18 on electric-furnace steelmaking are employed to produce finished steel of the required quality and composition. The finished steel is brought to the proper temperature for subsequent casting operations, following which it is tapped from the furnace into a ladle. The steel is cast into molds to form wheels according to the practice of the individual foundry; two current practices are described below.

Controlled Pressure Pouring—One manufacturer produces wheels by a patented controlled pressure pouring process which enables wheels to be cast in graphite molds to final dimensions, accurate within thousandths of an inch and requiring no machining other than boring of the hub. (Another variation of the controlled pressure pouring process used to produce steel slabs was discussed in Chapter 21.) The principle of the process is shown in Figure 28—15.

The two-part molds are machined from graphite blocks. The upper part or cope contains the risers, which are lined with sand sleeves, and a central vertical opening through which a stopper assembly passes. The lower part or drag of the mold has a central opening through which molten metal enters the mold from the bottom.

The ladle of molten steel is placed in a pit that is closed by an air-tight cover. A ceramic tube called a snorkel projects from the center of the cover into the molten steel. The mold is positioned with the hole in the drag coinciding with the opening in the end of the snorkel. Pressure in the casting pit is raised by admitting compressed air, forcing molten steel upward through the snorkel into the mold. The flow of metal is

continued until the mold and risers are filled. The risers provide a reservoir of molten metal that can flow back into the mold to compensate for the contraction of the metal in the mold as it solidifies. The stopper rod is then lowered to prevent metal from leaving the mold, and pressure in the pit is restored to atmospheric pressure and the molten metal in the snorkel runs back into the ladle. The filled mold is removed and replaced by an empty mold and the operation is repeated.

Within a few minutes, the wheel is removed from the mold and proceeds through various processing steps according to the class of wheel being made. The mold is then cleaned, inspected and prepared for reuse by spraying the faces of the cope and drag with a protective wash and remaking riser sleeves and the stopper rod in the cope.

Wheels are removed from the molds at about 1095°C (2000°F). They are carried by conveyor to a continuous cooling furnace in which they cool to about 540°C (1000°F) in about one hour as they travel through the furnace. After slow cooling, the wheels have the riser stubs removed and each wheel is stamped with the appropriate identification data. Following this, the hub of each wheel is rough bored by an oxygen cutting torch.

Immediately after rough boring, wheels are charged into a rotary-hearth furnace for normalizing at 925°C (1700°F); this operation, including the heating and soaking periods, takes about 1½ hours. When the wheels are discharged from the normalizing furnace, they proceed down a roller conveyor.

Class U wheels pass from the roller conveyor to a slat-type conveyor, where insulating covers are placed in suspension over the wheels so that the hubs cool

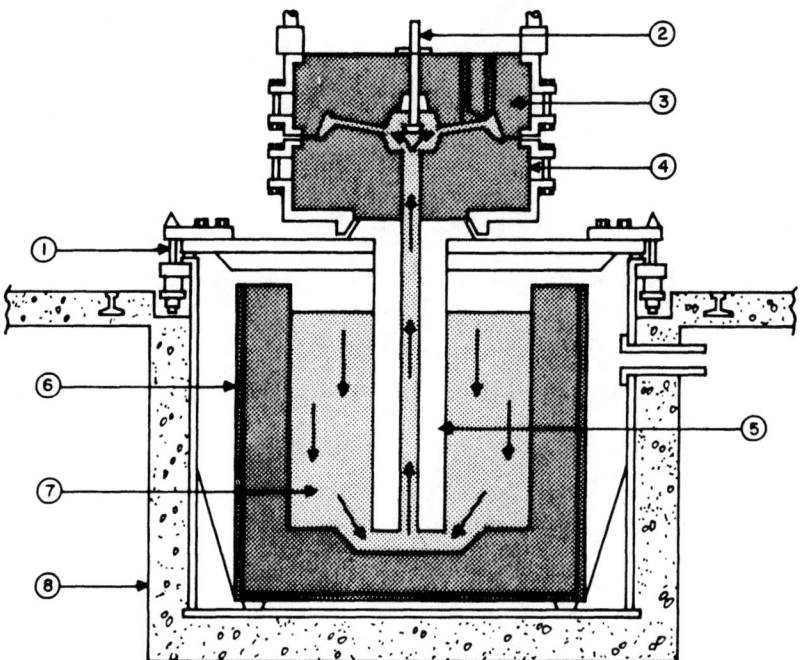

FIG. 28—15. Schematic arrangement of the equipment used in the controlled pressure pouring process for casting steel wheels. (1) Casting pit cover; (2) Stopper assembly; (3) Cope; (4) Drag; (5) Snorkel; (6) Ladle; (7) Molten steel; (8) Casting pit .

slowly. When Class U wheels have cooled below the transformation temperature (675°C or 1250°F), they are removed from the slat-type conveyor by an automatically operated crane and stacked on a storage conveyor where they are allowed to cool to room temperature before finishing and inspection.

Wheels of Classes L, A, B and C that require quenching of the rim, when discharged from the normalizing furnace, are transported by roller conveyor to the rim-quenching station and positioned in the quenching machine as rapidly as possible to minimize heat loss. A rotating spray ring with nozzles spaced equally around the inside of the ring is lowered over the wheel which is held in a horizontal position. Water sprayed from the ring cools the wheel tread uniformly. Quenching time varies with thickness of the wheel rim.

After quenching, wheels are transferred back to the roller conveyor which carries it to a continuous tempering furnace operating at about 480°C (900°F). Wheels travel through the furnace hung individually on hooks suspended from trolleys that travel on a monorail above the furnace. Tempering requires two hours. Following tempering, water is sprayed inside the bore of the wheels to produce a slight compressive stress.

Following normalizing of Class U wheels and rim-quenching and tempering of wheels of Classes L, A, B and C, all wheels are cleaned by shot blasting.

Each wheel is inspected for surface discontinuities by a wet magnetic-particle test, and the soundness of each wheel is checked by ultrasonic testing. Blemishes and surface discontinuities are removed by hand grinding, following which the wheels are re-inspected, shot peened to raise fatigue strength in high stress areas, and finish bored. Weighing and final inspection complete the process.

ACT Process—Another patented casting process for steel wheels utilizes top pouring into graphite molds with sand liners; this method was named the ACT (acronym for "advanced casting technique") process by its developers. The mold arrangement is illustrated in Figure 28—16.

An advantage of this process is that the same graphite molds can be used to produce single-wear, two-wear and multiple-wear wheels by changing only the pattern to accommodate more or less sand behind the rim and flange to obtain the required variation in rim thickness. Identification data for each wheel is formed in the sand portion of each mold and becomes an integral part of the casting at the beginning of the manufacturing process.

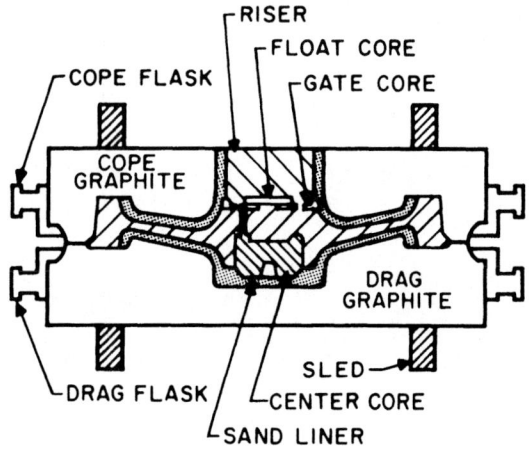

FIG. **28**—16. Schematic cross-section of a mold for casting steel wheels by the ACT process.

Top pouring is made possible by the unique gating arrangement and the fact that metal does not splash on or flow over the surface of the graphite until it reaches the tread. The gate core has a series of relatively small holes near its periphery and a larger hole in the center. The center hole is covered by the float core when the mold is assembled. Molten metal enters the mold through the smaller holes when pouring begins and, when the mold is full, the float core rises with the surface of the metal as the riser is filled, allowing molten metal to flow back through the central hole to compensate for shrinkage as the metal in the mold solidifies.

After the casting is removed from the mold, the riser is removed and the wheel is ready for heat treatment and finishing. All wheels are heat treated, either by normalizing alone or by normalizing followed by rim quenching and tempering as described above in the description of cast-wheel manufacture by the controlled pressure pouring process.

Following heat treatment, the tread surface, rim face, and the front hub face to which the riser was attached are machined, and the wheel hub is bored out to accommodate an axle of the required size.

Each wheel receives a visual and dimensional inspection and is also inspected over its entire surface by magnetic-particle testing. Each wheel is also ultrasonically inspected in a vertical position from both the rim face and tread surface.

CHAPTER 29

Production of Railroad Axles

Axles, which are a vital part of every wheeled vehicle, vary in size from those weighing several ounces to those that weigh several thousand pounds. The United States Steel Corporation is one of the principal manufacturers of axles used by the railroad industry. This includes axles for freight and passenger cars, diesel and electric locomotives, subway and rapid transit cars, and industrial applications. United States Steel's axle-producing facilities include the most modern innovations in axle forging, heat treatment, and finishing equipment.

The shapes of the three basic types of railroad axles are shown in Figure 29—1.

Composition and Heat Treatments—Axles are generally made of carbon steels of various classes, ranging from 0.40 to 0.59 per cent carbon and 0.60 to 0.90 per cent manganese, with phosphorus under 0.045 per cent and sulphur under 0.05 per cent.

The various classes of axle as designated by the Association of American Railroads indicate the size of the journal, i.e., the diameter and the length in inches.

Axles generally are ordered untreated or double normalized and tempered. Special axles have been heat treated as follows: quenched and tempered; or normalized, quenched and tempered, according to the size of the axle, the kind of steel, and the mechanical properties desired. The carbon content may vary depending upon the heat treatment involved.

Detailed requirements relating to chemical composition, microstructure, mechanical properties, ultrasonic testing, and dimensions of axles are included in Association of American Railroads Specification M-101.

Methods of Forming Axles—In general, axles are formed by a combination of rolling and forging operations, and the practices of different plants vary somewhat. The axle bloom, which is either round or square, may be rolled or continuous cast. The forming is completed by forging. In other plants, where it is necessary to start with cold steel after it has been rolled into blooms (which must correspond in dimensions and weight to the size of the axles for which they are intended), the various steps in the process of manufacture are as follows:

Inspection of the Blooms—The blooms are subjected to a rigid inspection before the steel is shipped to the axle plant. Those blooms that show any signs of pipe or insufficient discard at the blooming-mill shears are rejected. Surface imperfections, such as seams, slivers,

AAR Class	Size of Journal, inches (diameter x length)	
	mm	in.
A	95.3 x 177.8	3¾ x 7
B	108.0 x 203.2	4¼ x 8
C	127.0 x 228.6	5 x 9
D	139.7 x 254.0	5½ x 10
E	152.4 x 279.4	6 x 11
F	165.1 x 304.8	6½ x 12
G	177.8 x 364.8	7 x 12

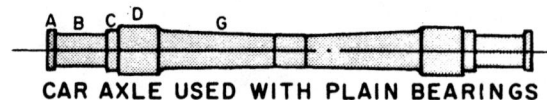

CAR AXLE USED WITH PLAIN BEARINGS

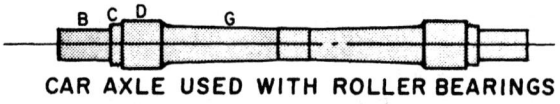

CAR AXLE USED WITH ROLLER BEARINGS

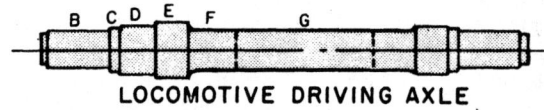

LOCOMOTIVE DRIVING AXLE

FIG. 29—1. Typical forms of steel axles. Legend: (A) End collar, (B) Journal, (C) Dustguard seat, (D) Wheel seat, (E) Gear seat, (F) Motor-support bearing, and (G) Body.

and surface cracks, are eliminated by surface-conditioning treatments. Blooms passing this inspection are shipped to the axle plant, where they are stored until needed.

The Axle Plant—The essential equipment of the plant usually includes furnaces for heating the blooms; forging machines; straighteners; double cutting-off facing and centering machines, as well as rough-turning and finishing lathes; and a complete heat-treating plant.

FIG. 29—2. General view of automatic axle impact-forging machine.

The layout of the plant provides for the most economical handling of the materials. The steel for axles, in the form of blooms, enters at one end of the plant and continues in one direction, progressing step by step through the various operations, until arrival at the other end of the plant as finished axles, ready for shipment.

Heating the Blooms—Proper heating of the blooms for forging requires that they be uniformly heated throughout and brought gradually to the forging temperature. The importance of heating slowly is obvious; rapid heating may cause the outside of the bloom to become somewhat hotter than the core, which condition would promote non-uniformity in grain structure and in the flow of the metal during the forging operations. In addition, high internal stresses would be developed. Slow heating gives the heat a chance to "soak" into the bloom and thus produce that uniformity in temperature from center to surface so necessary to secure a finished forging of the best quality. Reheating furnaces of the continuous type are used, and these are equipped with recording pyrometers. In addition, the temperature of the steel is periodically checked with an optical pyrometer as the bloom is drawn from the furnace. With this type of furnace, the bloom is placed in the furnace at the cold end and is slowly pushed or rolled toward the hot end. The bloom reaches forging temperature of 1175°C (2150°F) only a short time before it is drawn from the furnace.

The Forging Operation—After heating to the proper temperature for forging, the blooms are pushed out of the hot end of the heating furnaces onto a conveyor that carries them to the forging machine.

The Automatic Axle Forging Machine—At United States Steel's axle plant, a precision impact forging machine is used. The machine, shown in Figure 29—2, was the world's first impact forging machine designed to forge railroad axles.

With this innovative forging machine, an axle can be forged more rapidly and accurately than with the former steam hammer-manipulator combination. The steam hammer could deliver up to 120 blows per minute, while the impact forging machine delivers 310 blows per minute. The machine has four dies that are positioned 90° from each other, all in the same cylindrical forging head (Figure 29—3). They are mechanically actuated, and the operation of this precision forging machine is fully automated. Twenty-seven variable functions are programmed and controlled. A program change from one axle design to another can usually be made in less than 15 minutes.

Heated blooms are delivered from the furnace to the

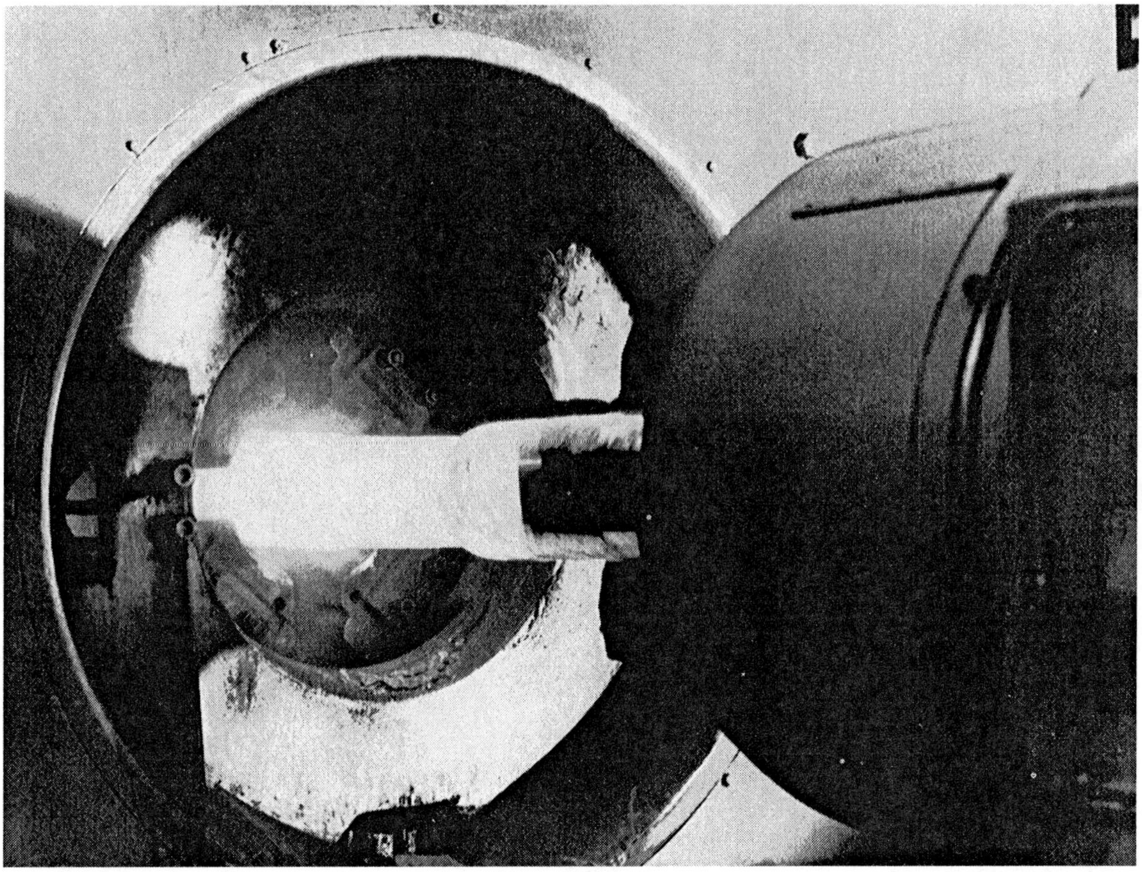

FIG. 29—3. Close-up view of axle impact-forging dies.

forging machine by a roller conveyor, and are automatically lifted by a pair of tongs that transfers them to one of the two forging manipulators on the machine.

FIG. 29—4. Flame cutting the ends of the as-forged axle.

During forging, the work piece is transferred back and forth through the forging dies by the manipulators. The beginning passes break down the square bloom into a round. The subsequent passes forge the journal ends, wheel seats, and center portion of the axle. The stroke of the forging dies automatically changes to control the diameter as forging progresses from the axle body to the wheel seat and to the journal of the axle. A template guides the taper in the axle body. Because the four forging dies operate simultaneously, the axle comes out straight and does not require subsequent straightening. Such a concentric forging is almost impossible to attain when forged on a regular steam hammer.

After forging, the axle moves by conveyor to the double-torch flame cutter where the axle is cut to approximate length (Figure 29—4). The serial number and heat lot number are then stamped on the end to maintain individual axle identification. Axles that are not to be heat treated are placed in insulated slow-cooling pits and control cooled. This step may be omitted if properly vacuum-degassed steel is used.

Heat Treatment—Axles to be heat treated are conveyed directly to the heat-treating line for double-normalizing and tempering. This unique line consists of three gas-fired furnaces that are automatically controlled. The axles are rolled through the sequence of

FIG. 29—5. One of the furnaces in the automated axle heat-treatment line.

FIG. 29—6. Two-stage lathe for end facing and rough machining the outside diameter of the journal ends.

Fig. 29—7. Three-stage lathe for finish machining the axle journals, dust guards, wheel seat, and axle body to specified dimensions.

furnaces and the intermediate cooling tables between them on skids. Continuous movement and rotation of the axles is accomplished by dogs that project between the axles from continuous chains that run the length of the line (Figure 29—5).

The axle enters the first normalizing furnace at about 535°C (1000°F), is heated to 900°C (1650°F), then air-cooled to 315°C (600°F). This cycle is repeated, in the second normalizing furnace by reheating to 815°C (1500°F) and again cooling in the air to 315°C (600°F). Finally, the axle moves into the tempering furnace where it is reheated to 775°C (1250°F), then discharged and placed in a slow-cooling pit for 32 to 48 hours to cool to ambient temperature, or it may be air cooled if properly vacuum degassed. The constant rotation provides uniform heating and cooling, which helps to minimize internal stresses, eliminate hard and soft spots, and improve grain structure and mechanical properites.

In order to conform to AAR requirements, 5 per cent of the axles are forged from an extra-long bloom with a "prolongation" on one end. After heat treatment, a test sample is removed from a "prolongation". It is machined to tension-test size, then tested for determina-tion of the yield point, tensile strength, elongation, and reduction of area. Each test represents no more than 70 heat-treated axles of the same size classification. The test axle is cold sawed to remove the "prolongation", and routed back into the normal production flow.

Machining—Axle machining begins at a 2-stage lathe shown in Figure 29—6. The first stage mills the ends of the axle to final length. The second stage turns the outside diameter of the journal ends to within 3.2 to 6.4 mm (⅛ to ¼ inch) of final diameter, and drills center holes in the ends. Serial and heat numbers must now be replaced, including mill designation and month of man-ufacture.

Typical final machining is done on a 3-stage lathe equipped with an automatic transfer handler (Figure 29—7). The first station of the lathe finish machines the journals, wheel seats and dust guards to specified di-mensions. The second station roughs out the center portion of the axle, while the third station finishes the center portion to final dimensions and to a surface fin-ish of 2.54-micrometre (100-microinch) average. The automatic transfer handler moves 4 axles at a time, putting one in place at each station on the machine, and discharging one onto the finish rack.

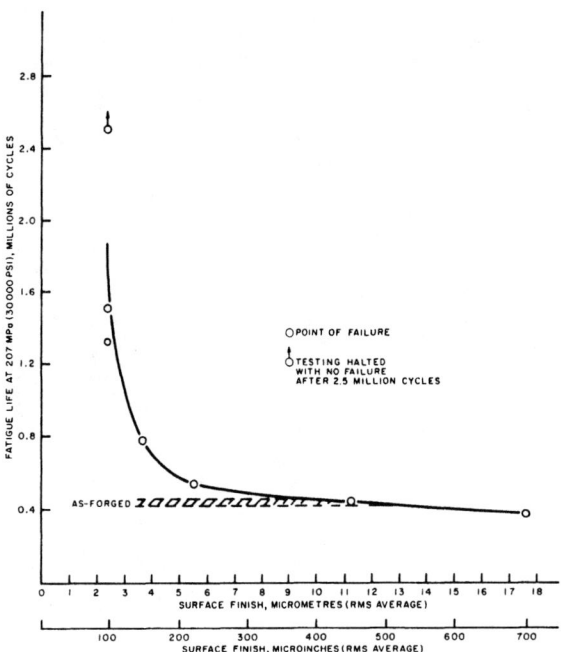

FIG. 29—8. Relationship of surface finish to fatigue life of axles.

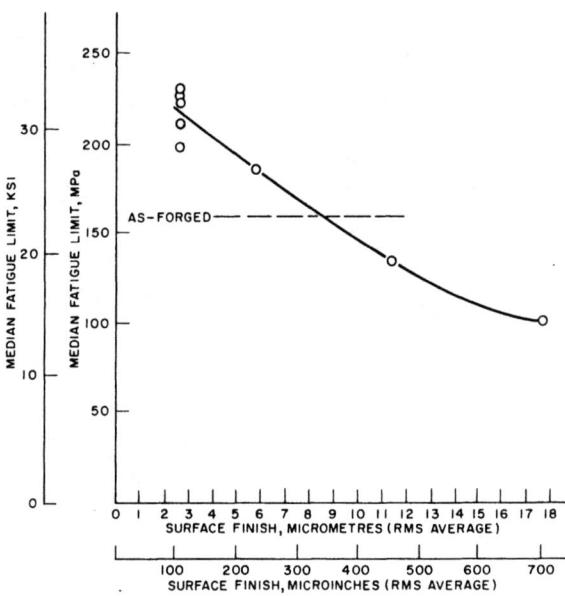

FIG. 29—9. Median fatigue life of Rayflex fatigue-test specimens with various surface finishes.

Final Inspection—Complete dimensional and visual surface inspection is made on each axle. Ultrasonic inspection is made on axles produced to Association of American Railroads Specification M-101 and ASTM Specification A21.

Axles with Improved Fatigue Strength—Although the fatigue strength of steel is generally proportionate to its tensile or ultimate strength, other factors which cause stress concentrations in a given part may be more important to resistance to fatigue failure. Surface finish or roughness is one of these factors.

Because of the trend by American railroads to build larger freight cars and to operate them at higher speeds, the possibility of failure of freight-car axles has increased. Until August 1965, freight-car axles were supplied with the surface of the body or center portion in the as-forged condition. The remaining portions were machined to accommodate the wheels, the dust guards, and the journal bearings. For freight-car axles to withstand the increased stresses imposed by changing operating conditions, the Association of American Railroads decided that the body portions should be machined to remove the as-forged surfaces.

In line with United States Steel's continuing efforts to improve the quality of its products, a research study was conducted to determine the type of surface finish on axle bodies that would provide optimum resistance to fatigue failure.

The results of the study indicated that axles having surface finishes with roughness values of 8.89 micrometres (350 microinches) or more would exhibit fatigue strengths no better than those of axles with as-forged surfaces. However, studies on the effect of surface finish in the range 8.89 to 2.54 micrometres (350 to 100 microinches) (root-mean-square or rms, average) showed that the finer the surface finish, the greater the fatigue strength (Figure 29—8). The surface-roughness measurements were made with commercial instruments. All measurements were made using automatic scanning with a rate of 7.62 mm (0.3 inch) per second. Roughness width cut off was set at 0.76 mm (0.03 inch). The surface-roughness measurements reported in the data are the average root-mean-square values. Although variations from the average for an individual axle generally do not exceed plus or minus 10 per cent of the average, some greater variations in roughness occasionally occur along the length of an axle body. This roughness is about 3.2 micrometres (125 microinches) root-mean-square, for an improved freight car axle.

For Rayflex specimens cut from the machined surfaces of axles with finishes of 2.54 micrometres or 100 microinches (rms, average) the fatigue strengths averaged about 40 per cent greater than those of specimens with as-forged surfaces, as judged by tests, some of which ran 10 million cycles without failure (Figure 29—9). (The Rayflex machine is a magnetic resonant frequency machine that subjects the specimen to bending stresses.)

Based on the results of these studies, U.S. Steel's axle-manufacturing plant is equipped with finishing lathes that produce axles with the optimum body surface finish of 2.54 micrometres (100 microinches) average, root-mean-square, or finer.

CHAPTER 30

Production of Bar Product

SECTION 1

BAR MILLS AND THEIR PRODUCTS

The name "**merchant mill**" arose in the early days of rolling, when bar mills carried a stock of their products from which merchants selected bars at their convenience. The name was carried over from that time and, until relatively recently, was still applied in the designation of modern bar mills, particularly those on which some of the small shaped sections were rolled. Other designations sometimes used for some early mills were "**hand bar mills**" and "**guide mills.**" The present designation for mills rolling bar product is simply **bar mills.**

In the early mills, rounds and sharp-cornered squares were rolled by hand; that is, the bars were prevented from turning over in the passes by a tongsman, who forcibly held the bars in the proper position in the grooves as they passed through the rolls. As a general rule, the finishing stand of rolls was two-high in these mills, though in rare instances a three-high stand was used. In the two-high mill, the bar delivering from a given pass was returned over the top roll for entering into the next pass, and so on until the section was finished. This type of rolling was arduous and the length of bar that could be rolled under these circumstances was limited to about 5 to 6 metres (16 to 20 feet). This limited length that could be rolled, in combination with the heavy labor involved, led to the adoption of the so-called guide mills, in which rounds and sharp-cornered squares were held in the passes by close-fitting entering guides. To form rounds, tongsmen entered squares or gothic squares into an oval pass and the resulting oval went through on edge into a round finishing groove. This new pass design rolled longer lengths than hand-rolling; it also kept the oval from turning over in the pass as it was rolled. Figure 30—1 illustrates both the hand and guide method of rolling rounds.

With these early mills as a start, the bar mill gradually became a more and more finished machine until, today, a modern bar mill can turn out in one day more tonnage than one of the pioneer mills could roll in a year.

Evolution of the Bar Mill—As indicated, the first bar mills consisted of one two-high stand of rolls. The number of stands was gradually increased and in time four to five stands in line (in train[*]), driven by the same engine, became the popular design. Various names were given to the stands of a mill of this kind, such as **roughing** for the first stand, **strand** for the second, **pony** or **leader** for the third, and **finishing** or **planishing** for the fourth. In a typical layout the roughing (first) stand was a three-high; the strand (second) stand, three-high; the leader was two-high in a three-high housing, and the finishing stand was two-high. The leader was in a three-high housing to permit driving the leader rolls through the top and middle strand rolls, with which they were in line, while the top finishing or planishing roll was driven through the bottom leader roll, and the bottom finishing roll was driven by a spindle from the bottom strand roll. This spindle passed through the three-high housing of the two-high leader in the space where the bottom roll would have been if the leader had three rolls. Such an arrangement permitted the leader and finishing stands to operate with opposite rolling directions. After the development of three-high stands, the number of stands for a bar mill was increased to as many as seven in line (in train).

The drawback to this type of mill was the limited roll speed at the roughing stand. Beyond a certain speed, the momentum of an emerging bar pushed the tongsman off his feet. This held up the rest of the train, which could roll at much faster speeds. In 1938, to solve this problem, new mills were built with separate roughing stands. A typical layout of this kind of mill is shown in Figure 30—2. In this design, the three-high roughing mill could be independently operated at a speed suitable for the best roughing practice, while the finishing train could be run at a much higher speed. This combination mill was referred to as a **Belgian mill,** as it was said to have originated in Belgium. The mill could roll longer lengths and finish them at a higher rate of speed and, consequently, could produce greater tonnage than earlier mills.

As the next innovation, the Belgians and Germans introduced the looping of bars. In this practice, the bars

---

[*] Rolling mills are "in train" when they are set relatively close together, side by side, and the rolls of one stand are driven by connecting them to those of an adjacent stand, as in Figure 30—1.

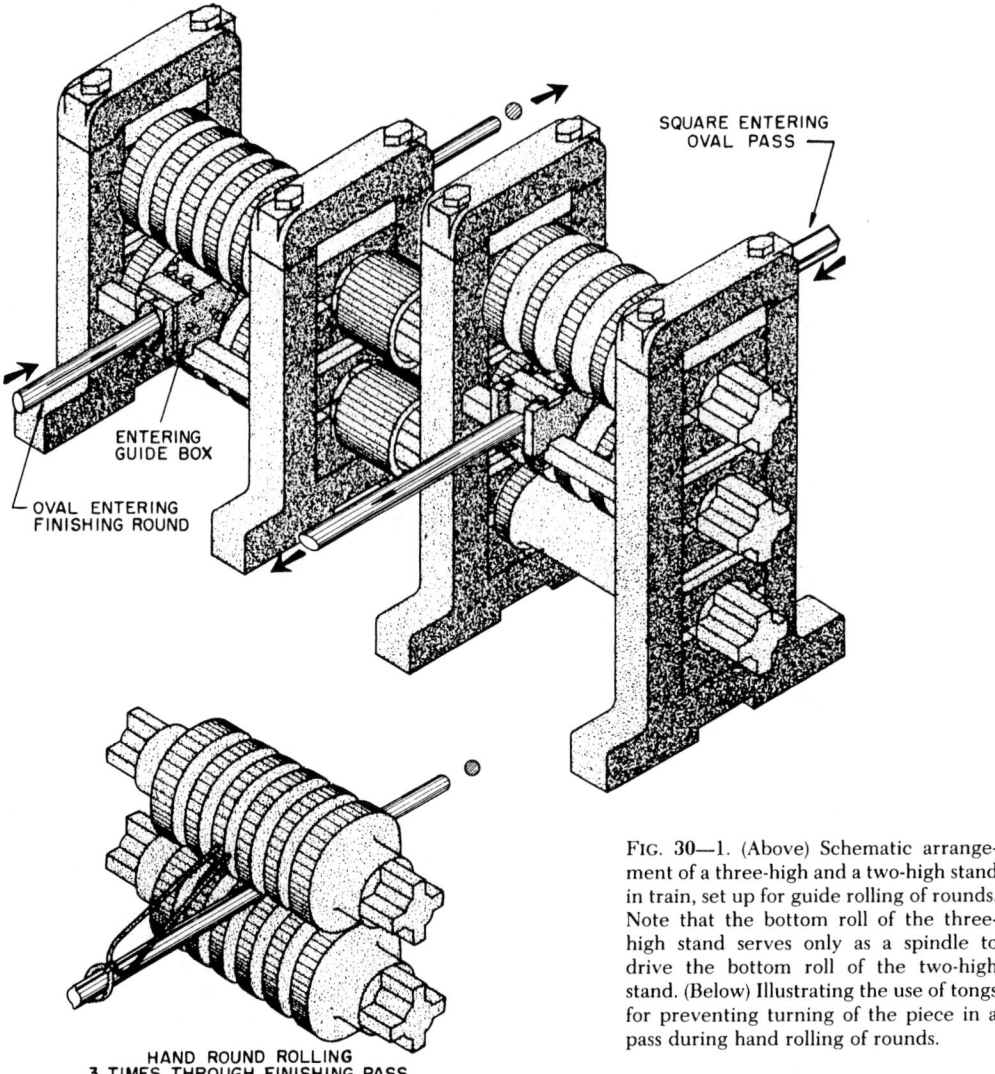

SQUARE ENTERING
OVAL PASS

ENTERING
GUIDE BOX

OVAL ENTERING
FINISHING ROUND

HAND ROUND ROLLING
3 TIMES THROUGH FINISHING PASS

FIG. 30—1. (Above) Schematic arrangement of a three-high and a two-high stand in train, set up for guide rolling of rounds. Note that the bottom roll of the three-high stand serves only as a spindle to drive the bottom roll of the two-high stand. (Below) Illustrating the use of tongs for preventing turning of the piece in a pass during hand rolling of rounds.

in the finishing train, when they reached a stage in their reduction at which they were relatively flexible, were caught by the tongsmen on the front end as they emerged from the rolls, pulled around in a half circle and entered into the next pass. As a result of this practice, rolling time for a bar was appreciably reduced and more bars could be rolled per hour, producing greater tonnages. Some of the more improved mills of this kind had two or three roughing stands in train, on the same housing shoes, which permitted still greater production because the roughing passes could be divided among the two or three stands.

A trough called a **repeater** eliminated the need for manual looping. Generally, the repeater directed the bar out of a square pass into a succeeding oval pass, while tongsmen directed the ovals into square passes. (An oval-to-square repeater must twist the bar so that the longer axis is vertical in the repeating trough; otherwise, the bar will jump out of the trough.) Repeaters for three-high stands differed from the others in that they conducted emerging bars back into the next

groove on the same roll stand, a task that could not be manually performed in such close quarters. These repeaters were used on the last two or three passes of the roughing operation, when the bars were small and long enough to provide the necessary flexibility. Figure 30—3 shows a three-high mill repeater in use on the roughing stand of a mill in which the finishing stands have been divided into two groups, the second group being run faster than the first for increased production. The finishing stands employ two-high repeaters for looping square into oval. Some mills in the United States use three-high repeaters of this type.

In 1882, mills began to employ the 'continuous roughing principle' by placing an eight-stand tandem continuous roughing mill and four finishing stands in line. Figure 30—4 shows this type of mill with two

---

° Rolling mills are said to be "in tandem" when they are arranged so that the rolled piece can progress from one stand to the next in a continuous straight path, as in the continuous roughing stands in Figures 30—4 and 30—5.

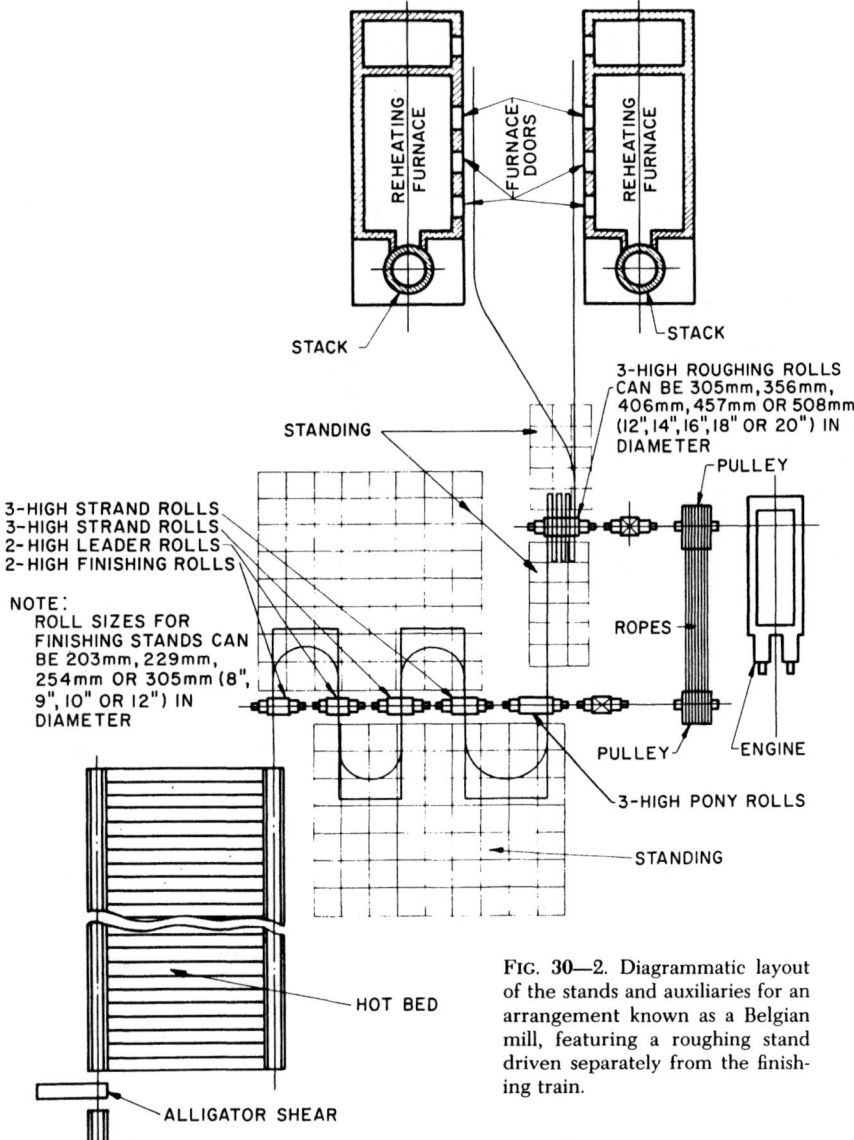

FIG. 30—2. Diagrammatic layout of the stands and auxiliaries for an arrangement known as a Belgian mill, featuring a roughing stand driven separately from the finishing train.

additional stands added for the rolling of smaller bars.

The early mills of the type shown in Figure 30—4 had some serious disadvantages. As discussed in Chapter 24, since the cross-sectional area of a piece is reduced during rolling, the piece leaves the rolls at a higher velocity than that at which it entered, due to its being elongated. The continuous roughing stands of the early mills had fixed speeds, with each successive stand running faster than the preceding one by an amount calculated to compensate for bar elongation. The fixed speeds made it imperative to have perfect balance between speed and elongation among the several stands, otherwise either the bars would be stretched or a loop would form between stands, either of which was undesirable. At the same time, the distances between stands were relatively short, causing heavy pressure on the delivery guides used to twist the bars 90 degrees for entrance into each succeeding stand. This heavy pressure often resulted in injurious scratching of the bars by the guides and these scratches, when rolled in, gave the bars a seamy appearance.

In the finishing mill the four roll stands, being in one line (in train), are driven at the same speed. As there is a reduction in cross-sectional area of the bar in each succeeding pass through the mill, and consequently a corresponding elongation of the bar, a successively longer loop forms between each stand. As a result, the back end of the bar becomes colder than the front end because of more contact with the floor and longer exposure to the air, a condition which introduces a variation in size of the ends of the bar as compared with the middle.

A number of new mill designs attempted to overcome these disadvantages. Some had the roughing line split into two or three groups so that the bar would run out to its full length between the groups. The finishing

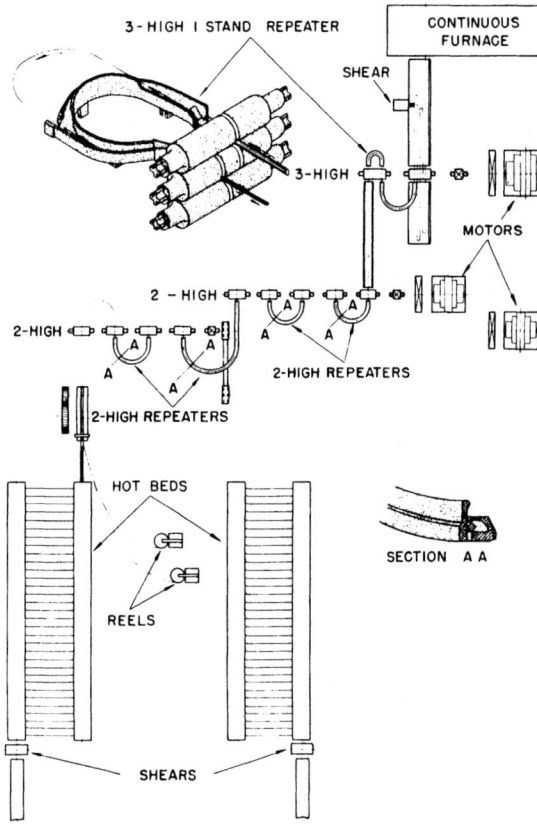

FIG. 30—3. Schematic arrangement of the stands and auxiliaries for a looping mill, employing repeaters of the type shown in the inset at the upper left on the three-high roughing stand as well as the two-high finishing stands. The three-high roughing stand is driven separately from the two trains of two-high finishing mill stands.

stands were also separated into groups of two or more roll stands, each group being separately driven and at a higher speed than the preceding one. Thus, between each group of rolls, the loop could be kept at a minimum, though they still formed between the stands which were in train. Space does not permit showing all of the arrangements devised to eliminate long loops.

A mill designed to overcome some of the disadvantages of these older mills is shown in Figure 30—5. In this mill, which has a double finishing train, the finishing stands have been divided into groups, with no more than two roll stands driven at the same speed. In each of these groups of two stands driven at the same speed, the second stand has rolls of larger diameter than the first in order to keep the loop between stands at a minimum. In a mill of this type the continuous roughing mill has sufficient capacity to supply both finishing trains, thus increasing production. Two billets can be sent through the roughing train, one for each finishing train.

Figure 30—6 shows a mill with a semi-continuous roughing arrangement in which the first two stands are driven by one motor and the bar runs free after each stand. These two stands are followed by four continuous stands driven by individual variable-speed motors.

The finishing unit is comprised of four stands, each one driven by an individual variable-speed motor at a successively increased speed. This mill, through the use of separate motors, permits excellent control over push and pull in the roughing stands, except in the first two, and, as the stands are a good distance apart, the heavy pressure on the guides experienced in the older close-coupled mill is relieved. In the finishing stands, the speed control provided by the separate motors permits close control of the length of the loop between any two passes. This mill was a considerable improvement over the older types.

Figure 30—7 shows a bar mill of the cross-country type which rolls a wide range of products. The mill has two hot-beds and finishes the large sizes out of stand No. 10. The smaller sizes are finished in stand No. 14 or No. 16. The mill is further distinguished by having a vertical roll stand immediately following stand No. 14 and continuous with it. The vertical stand provides the

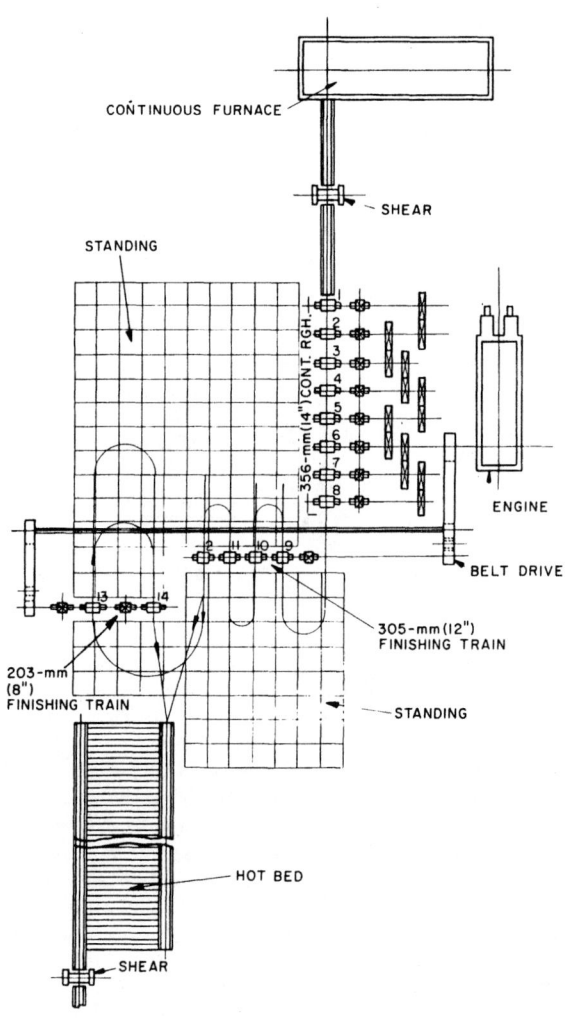

FIG. 30—4. Layout of the stands for a bar mill employing the continuous roughing principle, comprising an eight-stand tandem continuous roughing mill supplemented by a 305-mm (12-inch) four-stand and a 203-mm (8-inch) two-stand finishing train. All stands are two-high.

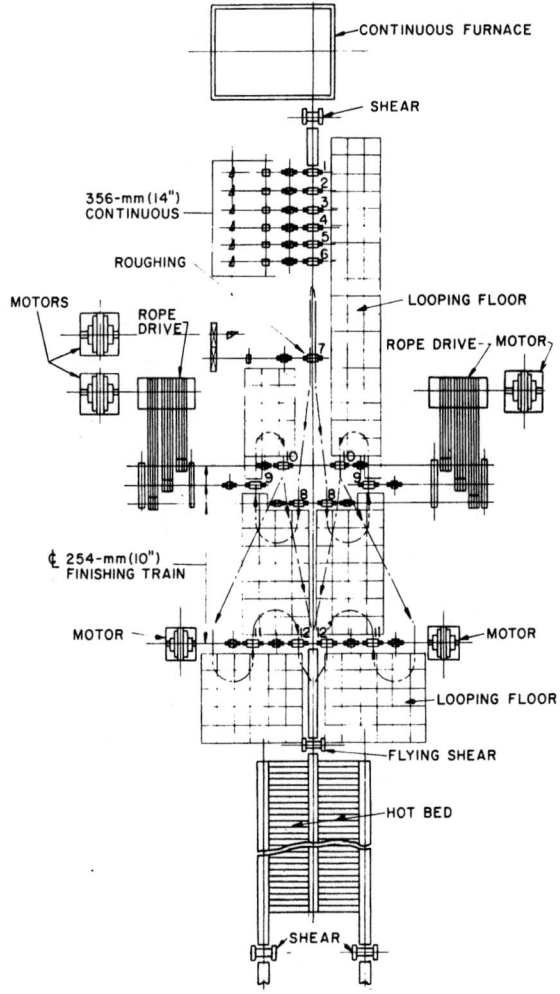

FIG. 30—5. Arrangement of a bar mill employing a six-stand tandem continuous roughing mill, followed by a double finishing train. All stands are two-high.

rolls, an arrangement which, curiously enough, was embodied in the first continuous mill built by Bedson in 1862. The use of alternate horizontal and vertical stands obviates the necessity for twisting the bars, and the twist guides, with their tendency to scratch bars, are eliminated. The stands of the mill are spaced far enough apart so that a slight loop can be formed between stands. Individual variable-speed drives for practically every stand permit regulation of speed relationships to take care of bar elongation between stands, and the slight loop eliminates all push and pull between stands.

The 10-inch and 12-inch continuous bar mills at Lorain Works of United States Steel (Figure 30—9) are modern examples of the use of horizontal and vertical stands in combination. A continuous billet mill at the same plant, described in Chapter 21 also uses alternating horizontal and vertical stands.

Lorain Works' Continuous 10-inch Bar Mill—The 10-inch continuous bar mill at Lorain provides for a total of 18 passes. Depending upon the product being

means for rolling off all over-sizes or shoulders formed in the horizontal stand, giving a more perfect round. So-called precision rounds, which are rounds rolled to approximately one-half standard tolerances, are finished through these two stands. The vertical stand is removed when finishing in stand No. 16 or when rolling to standard tolerances.

Figure 30—8 shows another cross-country design. This layout does not have any continuous stands and the bars run out free after each pass. The sections rolled on this mill range from 12.7-mm (½-inch) to 130-mm (5⅛-inch) rounds, squares in corresponding sizes, flats from about 25 to 230-mm (1 to 9 inches) in width and other products of comparable dimensions. The larger size products are finished from stand No. 9-A, which has 406-mm (16-inch) rolls, and the smaller sizes are finished from stand No. 12, which has 305-mm (12-inch) rolls. The distinguishing feature of the mill is that each stand has a variable-speed motor.

The most modern of all the bar mill designs is the continuous mill with alternate horizontal and vertical

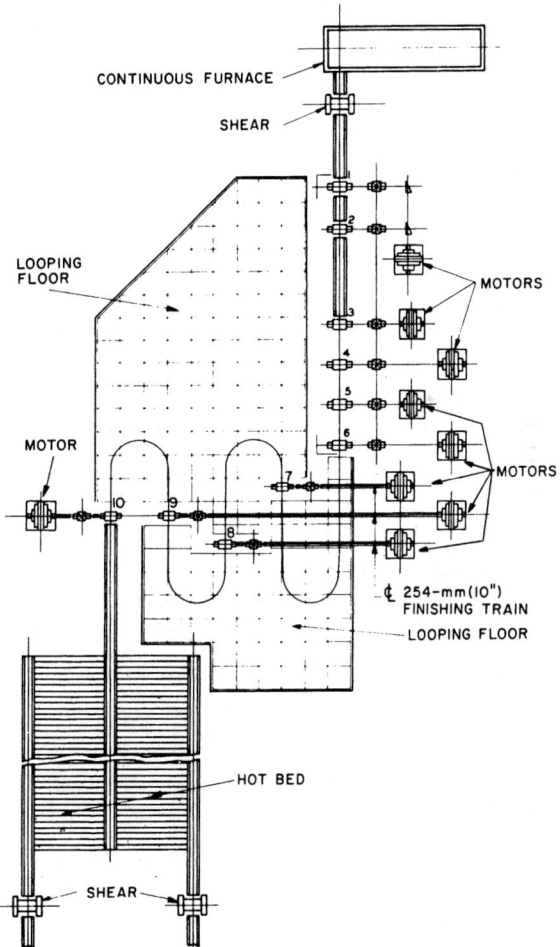

FIG. 30—6. Bar mill designed with a semi-continuous roughing arrangement followed by a four-stand continuous roughing mill, supplying roughed-down billets to the four stands of the finishing mill. All stands are two-high, and each is separately driven except for the two stands of the semi-continuous roughing set-up.

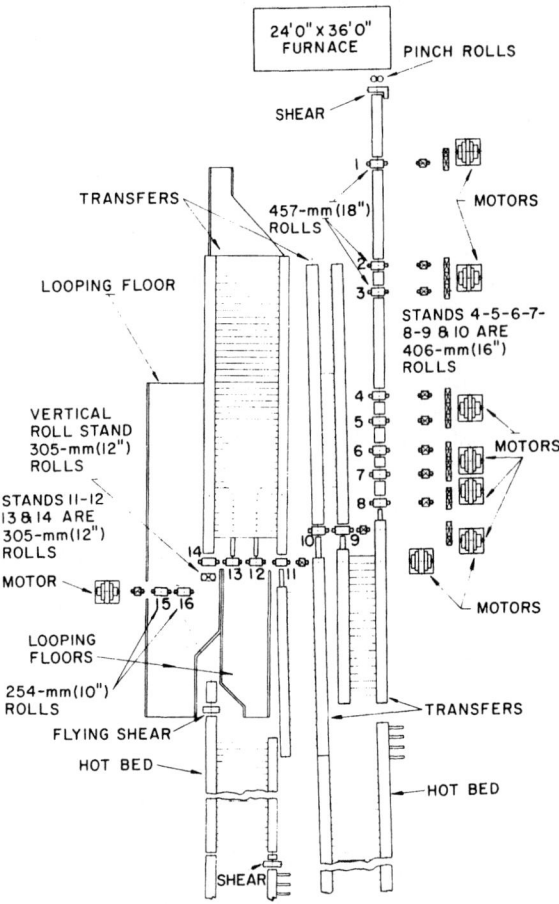

FIG. 30—7. Schematic layout of the stands and auxiliaries for a bar mill of the cross-country type, employing the continuous principle in the latter roughing stands.

Total main-drive power for this 10-inch mill is 11 695 kilowatts (15 675 horsepower), distributed as follows.

Stand or Shear Unit	Motor Rating		Motor Speed (rpm)
	kw	hp	
Billet Shear	56	75	720
No. 1 Vertical	187	250	350/1250
No. 2 Horizontal	187	250	350/1250
No. 3 Vertical	261	350	450/1400
No. 4 Horizontal	261	350	450/1400
No. 5 Horizontal	298	400	450/1350
No. 6 Vertical	298	400	450/1350
No. 7 Horizontal	373	500	450/1350
No. 8 Vertical	373	500	450/1350
Crop and Cobble Shear	2-224	2-300	500/1000
No. 9 Horizontal	373	500	450/1350
No. 10 Vertical	448	600	450/1125
No. 11 Horizontal	448	600	450/1125
No. 12 Vertical	448	600	450/1125
No. 13 Horizontal Right	597	800	450/1020
No. 13 Horizontal Left	597	800	450/1020
No. 14 Vertical Right	597	800	450/1020
No. 14 Vertical Left	597	800	450/1020
No. 15 Horizontal Right	597	800	450/1020
No. 15 Horizontal Left	597	800	450/1020
No. 16 Vertical Right	597	800	450/1020

rolled, billet sizes for the mill are 83-, 102- and 127-mm (3¼-, 4- and 5-inch) squares, 12.2 metres (40 feet) in length. The first section of the mill has twelve tandem stands. Stands Nos. 1, 3, 6, 8, 10 and 12 are vertical, and Stands Nos. 2, 4, 5, 7, 9 and 11 are horizontal. A crop and cobble shear is in line between Stands Nos. 8 and 9. Bars leaving the No. 12 Stand can be directed to either a right or left rolling line of six tandem stands each for continuous rolling to finished size. In each line, Stands Nos. 13, 15 and 17 are horizontal stands and Nos. 14, 16 and 18 are vertical stands. Product leaving Stand No. 18 in either line may be guided to pouring reels for coiling or over the hot runout table to the cooling beds for cut-length product.

The rolls for Stands Nos. 1 through 4 are 457 by 914 mm (18 by 36 inches); for Stands Nos. 5 through 8, 356 by 711 mm (14 by 18 inches) for Stands Nos. 9 through 12, 305 by 610 mm (12 by 24 inches) and for Stands Nos. 13 through 18, in the right and left rolling lines, 254 by 508 mm (10 by 20 inches). Roll-changing time at the mill line is minimized by roll cartridge units and by having the inner-housing assemblies of the stands hydraulically moved into and out of the housings on steel tracks.

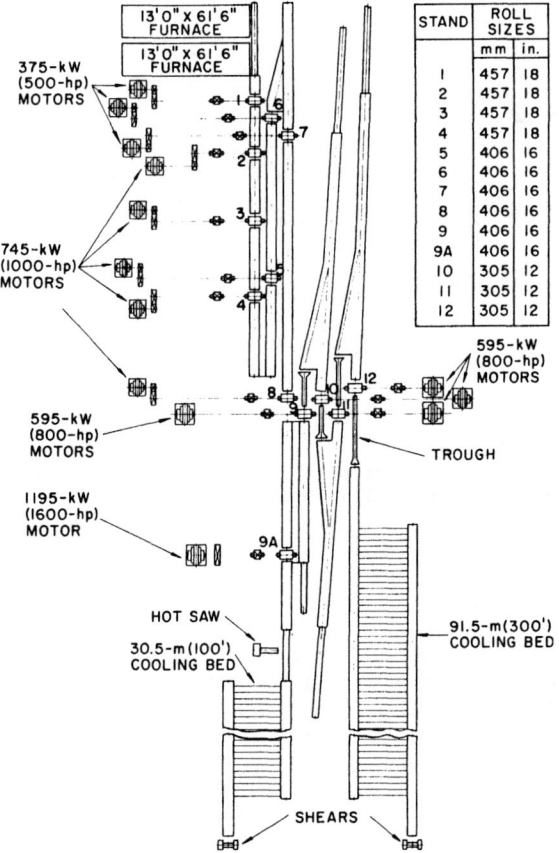

FIG. 30—8. Bar mill of the cross-country type differing from that of the preceding illustration in that no continuous stands are used and the bars run out free after each pass.

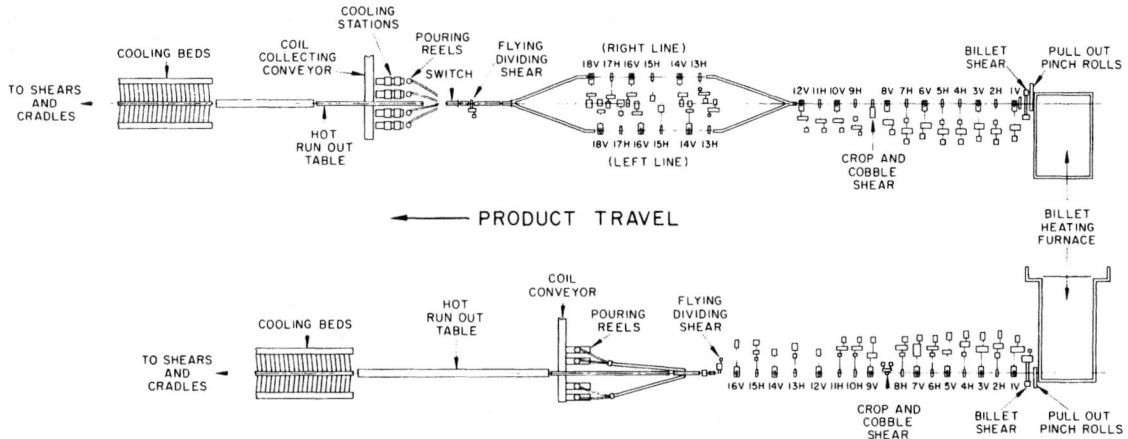

FIG. 30—9. (Above) Continuous 10-inch bar mill and (below) continuous 12-inch bar mill at the Lorain Works of United States Steel Corporation.

No. 16 Vertical Left	597	800	450/1020
No. 17 Horizontal Right	597	800	450/1020
No. 17 Horizontal Left	597	800	450/1020
No. 18 Vertical Right	597	800	450/1020
No. 18 Vertical Left	597	800	450/1020
Dividing Shear	75	100	515/1030

In addition to the installed power rating of the main drives on the 10-inch mill, the auxiliary drives represent a combined rating of 6530 kilowatts (8755 horsepower), for a total of 18 225 kilowatts (24 430 horsepower).

Product range capabilities of the 10-inch Lorain bar mill are as follows:

Rounds	9.5 to 25.4 mm	⅜″ to 1″ diam.
Squares	9.5 to 25.4 mm	⅜″ to 1″
Hexagons	9.5 to 25.4 mm	⅜″ to 1″
Flats	12.7 to 50.8 by	½″ to 2″ by
	5.2 to 31.8 mm	¹³⁄₆₄″ to 1¼″
Reinforcing bars	9.53 through	No. 3 through
	22.23 mm	No. 7
Cut lengths	1.5 to 27.4 metres	5′ to 90′
Coils	227 to 1361 kilograms (as rolled)	500 to 3300 lb (as rolled)
	4536 kilograms (welded and compacted)	10 000 lb (welded and compacted)

Lorain Works' Continuous 12-inch Bar Mill—The 12-inch continuous bar mill at United States Steel's Lorain Works is a 16-stand mill. Stands Nos. 1, 3, 5, 7, 9, 12, 14 and 16 are vertical stands; Stands Nos. 2, 4, 6, 8, 10, 11, 13 and 15 are horizontal stands. There is a crop and cobble shear in line between Stands Nos. 8 and 9. This mill has the ability to roll 127- and 146-mm (5- and 5¾-inch) square billets, 12.2 metres (40 feet) in length into products in the following size ranges:

Rounds

—Cut lengths	19.1 to 66.7 mm diam.	¾″ to 2⅝″ diam.
—Coils	19.1 to 38.1 mm diam.	¾″ to 1½″ diam.

Squares

—Cut lengths	19.1 to 50.8 mm square	¾″ to 2″ square
—Coils	19.1 to 31.8 mm square	¾″ to 1¼″ square

Hexagons

—Cut lengths	19.1 to 50.8 mm across flats	¾″ to 2″ across flats
—Coils	19.1 to 31.8 mm across flats	¾″ to 1¼″ across flats
Flats	50.8 to 101.6 by 5.2 to 38.1 mm thick	2″ to 4″ wide by ¹³⁄₆₄″ to 1½″ thick
Reinforcing bars	19.05 through 43.00 mm (nominal)	No. 6 through No. 14
Cut lengths	1.5 to 27.4 metres	5′ to 90′
Coils	454 to 1905 kilograms (as rolled)	1000 to 4200 lb (as rolled)

Roll size for Stands Nos. 1 through 4 is 457 by 914 mm (18 by 36 inches); for Stands Nos. 5 through 8, 406 by 762 mm (16 by 30 inches); for Stand No. 9, 356 by 711 mm (14 by 28 inches); for Stands Nos. 10 and 11, 356 by 660 mm (14 by 26 inches); for Stand No. 12, 356 by 711 mm (14 by 28 inches); and for Stands Nos. 13 through 16, 305 mm by 508 mm (12 by 20 inches).

Total installed power rating for this mill is 15 770 kilowatts (21 140 horsepower), with the main drives representing 10 650 kilowatts (14 275 horsepower).

	Motor Rating		Motor Speed
	kw	hp	(rpm)
Billet Shear	56	75	720
No. 1 Vertical	298	400	450/1350
No. 2 Horizontal	298	400	450/1350
No. 3 Vertical	448	600	450/1125
No. 4 Horizontal	448	600	450/1125
No. 5 Vertical	448	600	450/1125
No. 6 Horizontal	597	800	450/1020
No. 7 Vertical	597	800	450/1020
No. 8 Horizontal	597	800	450/1020

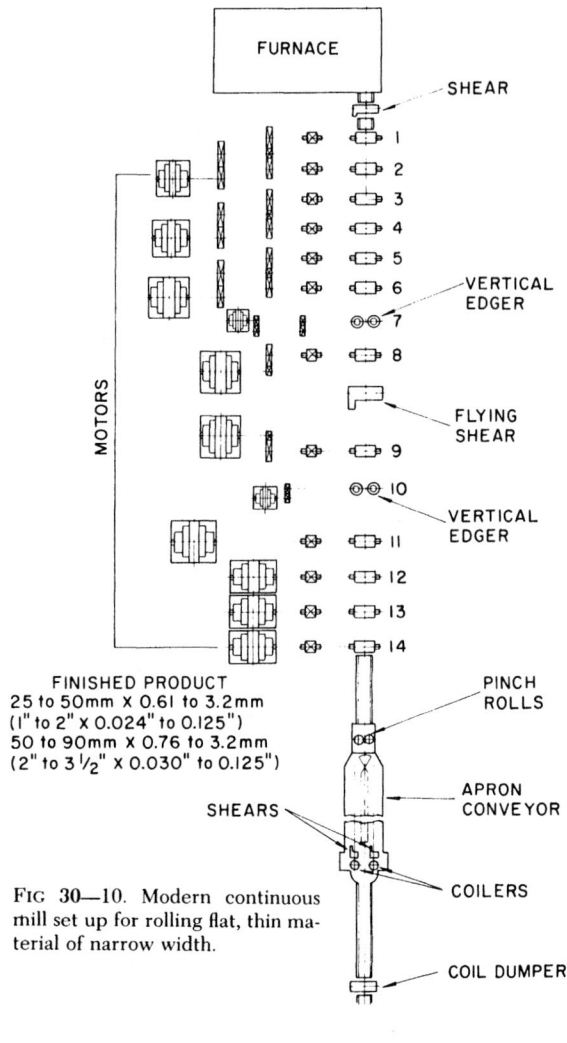

FIG 30—10. Modern continuous mill set up for rolling flat, thin material of narrow width.

Crop and Cobble Shear	2-224	2-300	500/1000
No. 9 Vertical	746	1000	300/900
No. 10 Horizontal	746	1000	300/900
No. 11 Horizontal	933	1250	250/750
No. 12 Vertical	933	1250	250/750
No. 13 Horizontal	746	1000	300/900
No. 14 Vertical	746	1000	300/900
No. 15 Horizontal	746	1000	300/900
No. 16 Vertical	746	1000	300/900
Dividing Shear	75	100	515/1030

Mills for Rolling Light, Narrow Flat Material—Flat thin material of narrow width is produced on various types of bar mills. This material is known as **narrow strip, band iron, hoop,** and **cotton tie.** When first produced, this type of product was rolled on mills similar to the one shown in Figure 30—2, but as mills of this type could not roll any large tonnage of such light weight material, better mills were soon developed for the purpose.

Figure 30—10 shows a modern mill designed to roll this thin material. This continuous mill produces comparatively large tonnages of hoop, cotton tie, and narrow strip. In mills of this kind, the very narrow widths

of the thin gauge materials are obtained by rolling a multiple width of the material and then slitting the rolled piece (after it is cold) into the desired widths. This permits the mill to maintain a rate of production that could not be realized if very narrow widths were to be rolled.

Rail-Slitting Mills—Rail-slitting mills were first developed when Bessemer rails were being rolled in large quantities and before open-hearth furnaces were common. Bessemer rail scrap was a drug on the market, as it could not be used in any quantity in the Bessemer vessel. The rail-slitting mills, a number of which are still in use, solved the problem of utilizing this scrap by rerolling it into other and useful sections.

Figure 30—11 illustrates a rail-slitting mill. These mills take a heated rail, usually 3 to 5 metres (10 to 16½ feet) long, and run it through a two-high stand of rolls with sharp collars which cut through and separate head, web, and base. Each one of these parts is then rolled into a bar of some kind directly after it is slit. The products include rounds, squares, fence posts, angles,

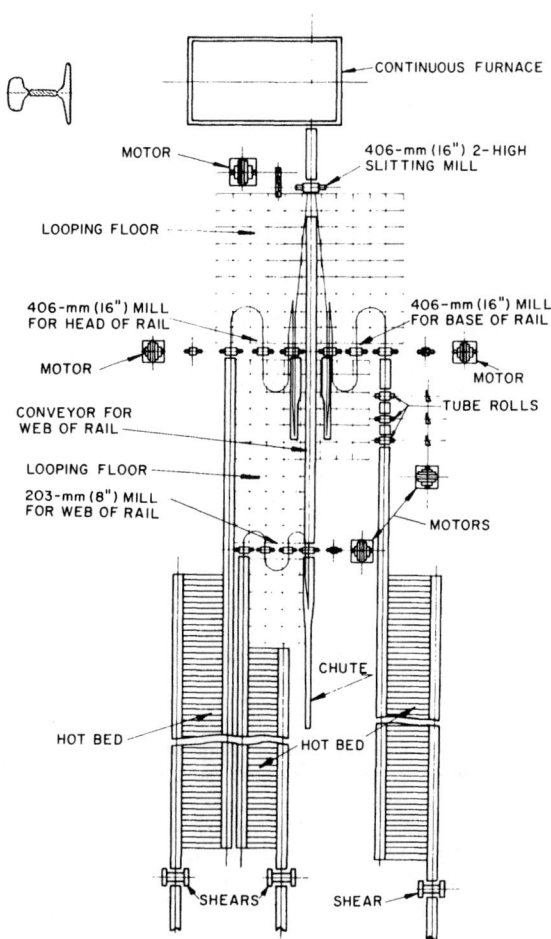

FIG. 30—11. Schematic arrangement of the slitting mill and roll stands of a rail-slitting mill. Inset at upper left shows how the head, web and base of the heated rail are separated in the slitting mill prior to rolling each, without reheating, into a bar. This particular mill has three continuous stands (marked "tube rolls") for forming a tube as described in the next.

concrete-reinforcing bars, flats, and a variety of other small sections. The continuous stands shown on the drawing are used to bend and form a flat, rolled from the base of the rail, into a cylinder which is then welded at the joint to make a tube or pipe.

Steel producers are using automation to increase the productivity and efficiency of their bar mills. Fully automated systems can produce round bar; hexagons, squares, and special shapes need some manual supervision. One completely automated system has a total of 22 passes. Of these, the first four are roughing stands; the next six comprise a first intermediate train; the next six make a second intermediate train; and the final six are finishing stands. The intermediate and finishing trains alternate between horizontal and vertical, ending with a vertical pass. The roughing train has four stands arranged in a vertical-horizontal-horizontal-vertical pattern. This arrangement facilitates scale removal.

Nine interstand loopers (in line from stands 13 to 22) relieve the tension on billets, and five interstand cooling boxes (built into the second intermediate and finishing trains) keep the temperature from becoming too high. Automatic gages on stands 21 and 22 assure that the bar is within specified tolerances or half-tolerances. A dividing shear on the last pass snips off the end of an emerging bar; this sample is used for spark testing, surface examination, diameter measurement, and Magnaglo testing.

The interstand cooling boxes deserve special note. The process of interstand cooling emerged in the early 1970's as a result of research into methods of lowering finishing temperatures. Research indicated that a lower finishing temperature produced bar with better mechanical and metallurgical properties. Two other techniques, slowing mill speed and greatly reducing dropout temperatures, proved to be unfeasible because they lowered mill productivity and caused motors to overload.

Interstand cooling produces a bar of uniform grain structure, with a thin, tight, easily pickled scale. Another advantage is that the process can easily be retrofit into an existing line. If space between stands is tight, cooling chambers can be installed throughout the mill layout so that the hot bar cools in stages. Typically, bar passes through cooling chambers at the points where it has a round cross-section. The chamber surrounds the hot bar with nozzles that spray water at points all the way around the bar's circumference. Moveable guides prevent the steel from damaging the nozzles, and filters prevent the nozzles from becoming clogged with pipe scale.

This mill has also undergone extensive computerization. From the time the plant's mainframe computer receives a customer's order until the bar is ready for shipping, the mill's ten peripheral computers control the mill's operations and run quality checks. These ten process control computers perform the following specific functions: coil weighing, transfer, storage, and handling; dividing shear control; furnace temperature control; cooling bed control; and straight bar production.

The mainframe computer takes care of inventory and production scheduling. When it receives a customer's order, the computer fits the order into the

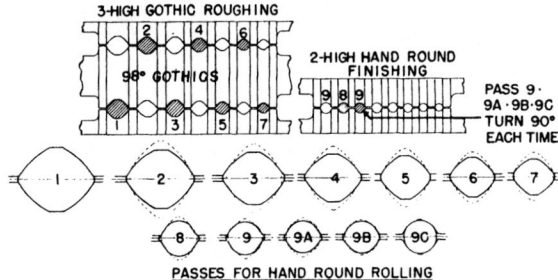

FIG. 30—12. Rolls and passes for rolling a hand round, using gothic passes in the roughing and hand-round grooves in the finishing rolls. The numbers of the sketched passes correspond to the cross-hatched and numbered pass openings between the rolls above.

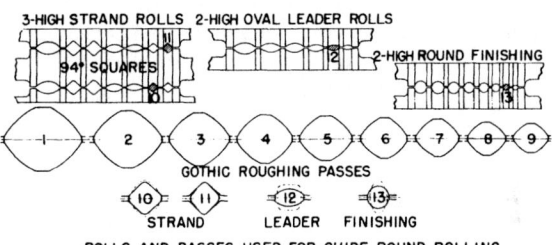

FIG. 30—13. Rolls and passes for rolling guide rounds. The roughing rolls containing the nine gothic roughing passes are not shown. Numbers on the sketched strand, leader and finishing passes correspond to the cross-hatched and numbered pass openings in the respective sets of rolls shown immediately above.

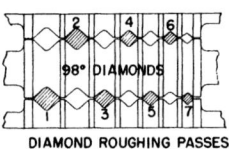

FIG. 30—14. Rolls with diamond roughing passes of the type which has largely superseded gothic roughing passes for rolling rounds by either the hand or guide method.

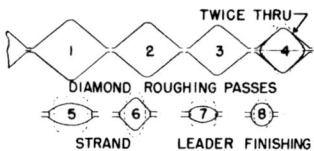

FIG. 30—15. Diamond roughing passes, oval and square strand passes, oval leader pass, and round finishing pass for guide-round rolling. Dotted lines indicate shape of piece entering pass, corresponding to shape of piece out of preceding pass turned 90 degrees.

mill's schedule, proposes a rolling sequence, and transmits the rolling sequence and any other important data to the process control computers. The main computer also keeps and updates an inventory of each billet storage area.

ROLL DESIGN FOR BAR MILLS

The first roll designs for bar mills were very simple. They consisted of box passes, gothic passes, tongue-and-groove passes for flats, and so-called hand-round grooves for rolling rounds.

Figure 30—12 shows a design of passes and a sketch of the rolls for rolling a hand-round, using gothic passes in the roughing and hand-round grooves in the finishing rolls. Figure 30—13 shows a typical series of passes for rolling guide-rounds, using gothic roughing passes, strand open-square passes, and oval and round finishing passes. These two methods were extensively used in the older bar and guide mills.

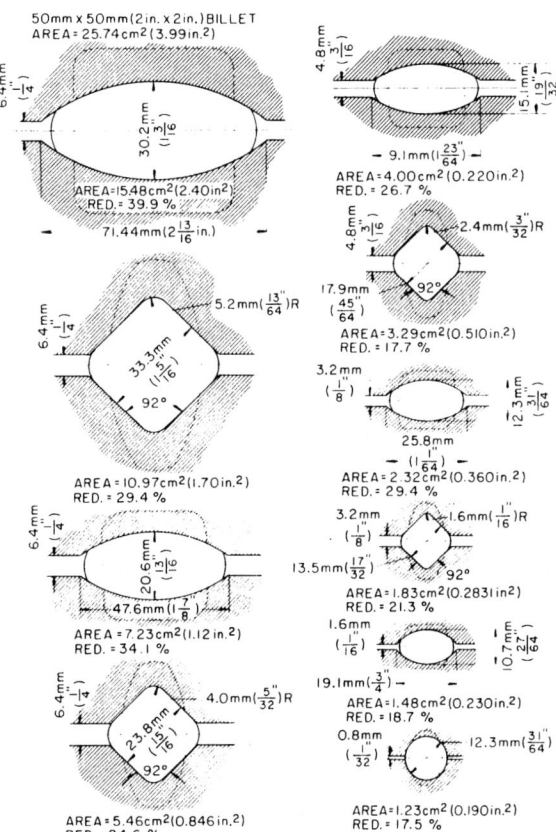

FIG. 30—16. Series of passes for rolling 50-mm (2-inch) square billet into a 12.3-mm ($^{31}/_{64}$ inch) round in ten passes, using an alternate oval and square reduction down to the finishing oval and round. Dotted lines indicate shape of piece entering pass, corresponding to shape of piece out of preceding pass turned 90 degrees. Per cent reduction of the piece in each pass has been calculated from the cross-sectional areas, as shown.

The gothic roughing rolls are not used to any great extent today. The design has been superseded in most hand mills by the diamond roughing set. Figure 30—14 shows one of these sets, which are also used for roughing passes for hand or guide rounds. Figure 30—15 shows a line of passes for a guide round using diamond roughing grooves, oval and square strand grooves, and oval and round in the leader and finishing passes. The three-high strand rolls shown in Figure 30—13 have a combination of oval and open-square passes, also diamond passes. The open-square passes referred to are of 94-degree angularity. To make a 90-degree square using these grooves it was customary to go twice through the same size groove.

The mill with the eight-stand continuous roughing arrangement and the looping finishing mills shown in Figure 30—4 used a 100- by 100-mm (4- by 4-inch) billet of not over 102 kilograms (225 pounds). The continuous stands were close together and a roll design was provided which would require the least possible twisting of the bars. The roll design on this mill provided for a 11.9-mm ($^{15}/_{32}$-inch) round in 14 passes. The first and second passes were both flat ones in box grooves. The bar then was twisted 90 degrees into a hexagon-shaped groove (pass No. 3) and then on into pass No. 4, which was a square in the diagonal position. By using this design, only one twist of the bar was required in the first four grooves. After pass No. 4, the bar was twisted 45 degrees to enter pass No. 5, which was an oval, and from there on the passes were alternately square and oval, the bars being twisted 90 degrees in the oval form and 45 degrees in the square form, before entering the next pass. The last pass, naturally, was round. Passes Nos. 9, 10, 11, 12, 13 and 14 were looped by hand on this mill.

As mentioned earlier, mills of this type rolling short 100-mm (4-inch) square billets, with their close coupled roughing stands and long loops, were not able to roll bars of close tolerance on the front and back ends of the bars. These disadvantages led to the adoption for modern continuous mills of 9.1-metre (30-foot) billets of 38- to 76-mm (1½- to 3-inch) squares. Rolling 9.1-metre (30-foot) billets produces bars more uniform in section throughout their lengths because the back end of such a billet is still in the furnace and retaining heat while the front end is emerging from the continuous roughing mill and, as the mills generally have good speed adjustment in the finishing stands, mill loops are kept at a minimum to insure uniformity of size throughout the rolled piece.

Figure 30—16 shows the pass design for rolling a 50- by 50-mm (2- by 2-inch) billet on a modern-type mill into a 12.3-mm ($^{31}/_{64}$-inch) round in 10 passes, using alternate oval and square reduction down to the finishing oval and round.

In the cross-country mill types, somewhat different roll designs are used. Figure 30—17 shows a roll design for the mill shown in Figure 30—8, which rolls many different grades of alloy steel, including stainless steel. The design shows the steps in reducing a 150-mm (6-inch) square billet to a 39.7-mm (1%6-inch) round in 12 passes. The first four passes are flat and edge, the flat passes being in plain rolls and the edging in box passes. The principal object in using these flat and edge passes

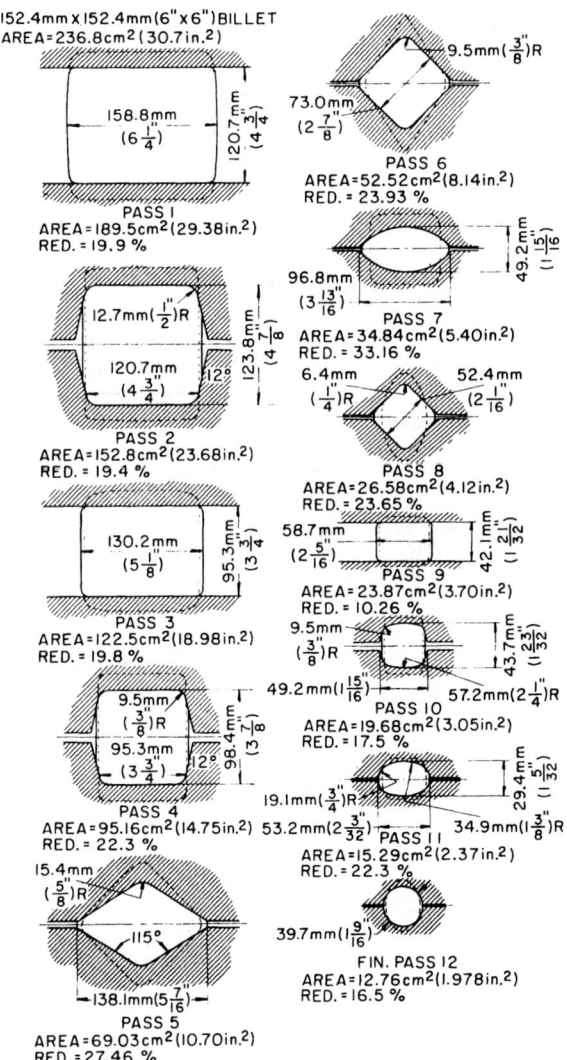

FIG. 30—17. Steps in the reduction of a 152-mm (6-inch) square billet to a 39.7-mm (1%16-inch) round in twelve passes on the cross-country mill shown in Figure 30—8. Dotted lines indicate shape of piece entering pass, corresponding to piece out of preceding pass after turning the required number of degrees for proper entry.

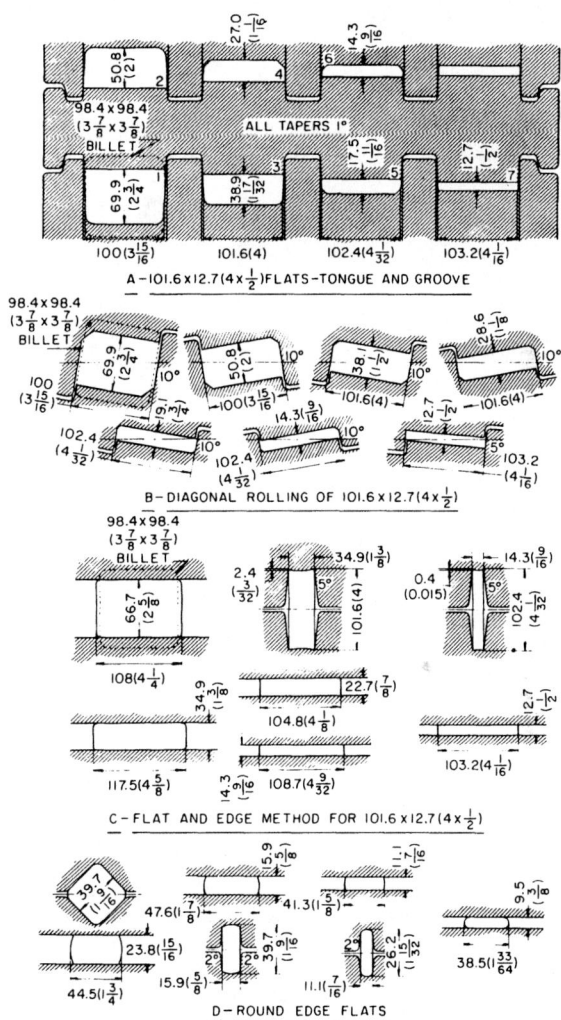

have a heavy reduction in order that passes Nos. 9 and 10 may be more lightly drafted. Pass No. 9 is a slabbing or flatting pass which reduces the 52.4-mm (2$\frac{1}{16}$-inch) square to a rectangle measuring 58.7 by 42.1-mm (2$\frac{5}{16}$ by 1$\frac{21}{32}$ inches). Pass No. 10 is called a "former" pass. The function of this pass is to round the upper and lower edges so that the following oval will be well-rounded on the sides, for when the oval is rounded at

is to break and free the bar of the scale that formed in the furnace during heating. In the first pass the roll pressure is on the top and bottom of the piece; the sides are free and, being compressed, crack the scale and allow it to drop off. In the next pass, the other two sides are free and the scale drops off here also. The next two passes repeat the same procedures. It is not practical to use this type of pass throughout the mill because box passes are not efficient in drafting in the smaller sizes, and for this reason the next two passes are diamond and square, a form of pass which effects substantial reduction in cross-sectional area and also has the merit of contacting all sides of the bar at once. In the next two passes, oval and square are used as it is desired to

Figure 30—18 shows three roll-pass designs for rolling square-edge flats, and one design for round-edge flats, as used on bar mills. Figure 30—18A illustrates a tongue-and-groove design used on a simple four-stand old-style bar mill. Figure 30—18B shows a diagonal design used on some mills, while Figure 30—18C shows the most popular design, one used on many modern bar mills, consisting of a flat and edge layout wherein the widths of the flats are regulated by edging grooves which at the same time give the section a square edge. This design permits the rolling of several sizes in the same grooves. Round-edge flats are commonly rolled as shown in Figure 30—18D.

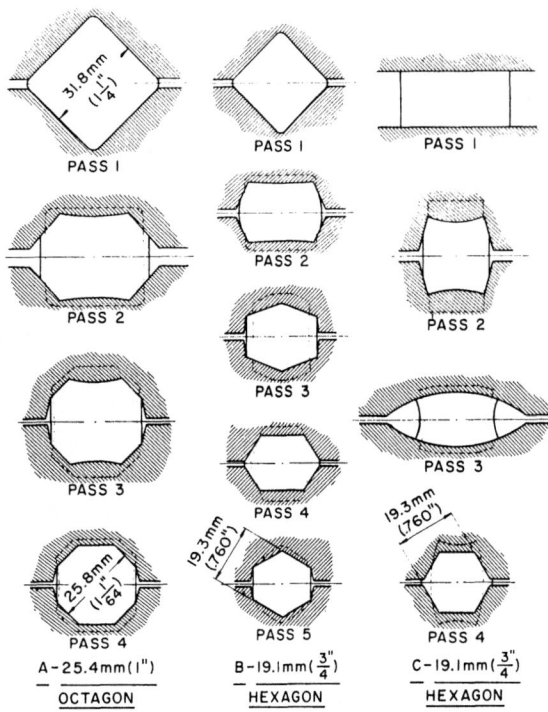

Fig. 30—19. Each vertical line of roll passes illustrates steps in rolling a different size of hexagon. Method C is somewhat uncommon. Dotted lines show shape of piece out of preceding pass and position on entering each succeeding pass.

the roll parting, scale will not readily adhere to it at this point. When scale is present, it is rolled into the finished round. In some mills, a square is used to enter the round oval and, when so used on some grades of alloy steel, scale sticks to the bar at the roll parting and is rolled into the bar.

One multi-roll process, commonly called RER (for round-edged rectangle), uses grooveless (plain-surfaced) cylindrical rolls. Either the rolls can alternate between horizontal or vertical positions, or the billet or bloom itself can be twisted 90 degrees between stands. Generally, twisting the billet is less efficient and may result in defects. RER rolls accept larger billets than conventional grooved rolls and still produce within the same product range. They also will reduce an existing billet to a smaller finished size than grooved rolls.

Round-Edge Rectangle rolling produces the following quality improvements: fewer laps and overfills; fewer defects caused by billet cracks or seams; the same product tolerance as that achieved by conventional rolling; and less downtime (since the same rolls are used for most billet sizes.)

Octagons are often produced as illustrated in Figure 30—19A, but the most popular design for hexagons is shown in Figure 30—19B. Figure 30—19C shows an uncommon hexagon design used in one mill.

Figure 30—20A shows a 25-mm (1-inch) sharp-cornered square design and Figure 30—20B a design used for rolling triangular file steel. Figure 30—20C

illustrates a half-oval design also used for file steel and Figure 30—20D an arrangement for rolling half-rounds.

Shapes such as angles, channels, small beams, tees, small agricultural shapes, window sash, etc. can be rolled on bar mills. Figure 30—21 shows a number of sections and steps in their formation from the billet or slab.

Rail-slitting mills, such as shown in Figure 30—11, roll a variety of bars and shapes. These mills, of course, must confine their product to material which can be rolled from steel having the chemical composition of rail steel. Figure 30—22 shows a number of typical bars and shapes produced from the various parts of a rail.

Hoop and cotton-tie passes on the older mills were made by the tongue-and-groove method. The tongue-and-groove passes restricted the spread of the bar somewhat and also regulated the width. A series of passes using tongue-and-groove in the next to the last three passes is shown in Figure 30—23A. A later design is shown in Figure 30—23B. In this design the flattening is done in plain rolls and the width is regulated by a vertical edging stand. These passes represent the design used in the first continuous hoop and cotton-tie mill of 1895. The passes used in the most modern mills are of this same general design.

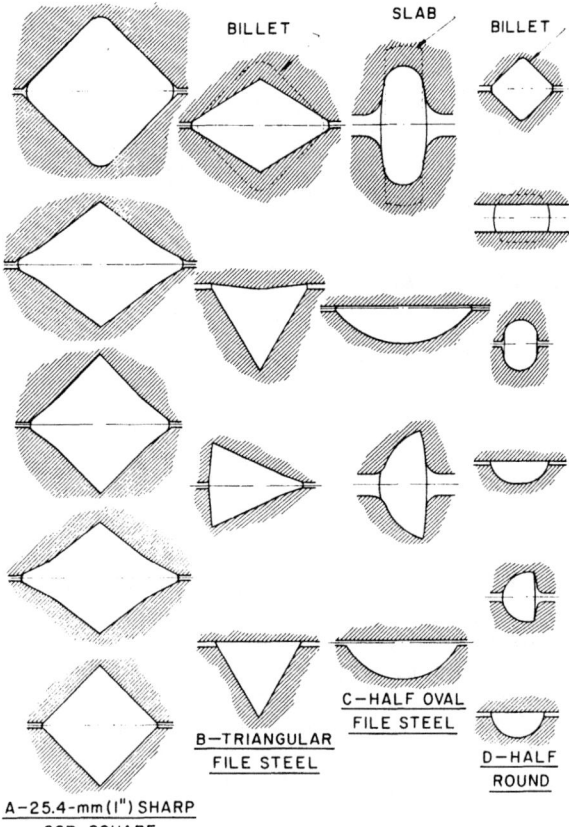

Fig. 30—20. Steps in rolling (A) 25-mm (1-inch) sharp-cornered squares, (B) triangular file steel, (C) half-oval file steel, and (D) half-rounds on a bar mill.

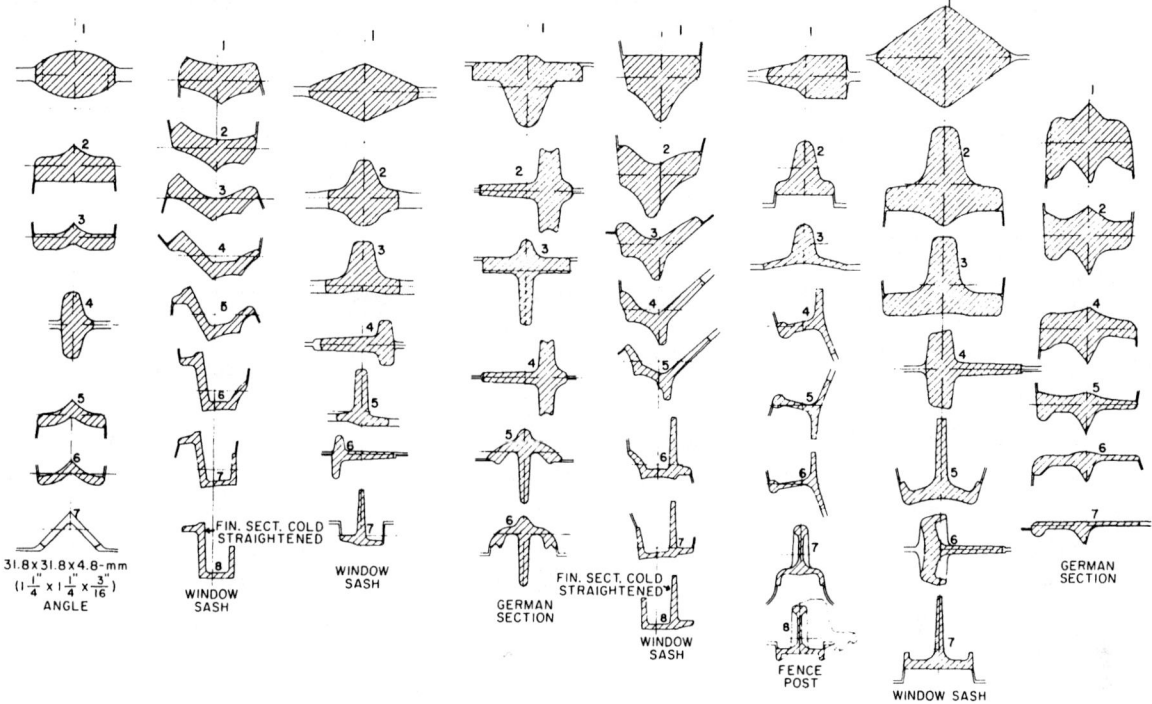

Fig. 30—21. Steps in rolling various small sections from billets or slabs on bar mills.

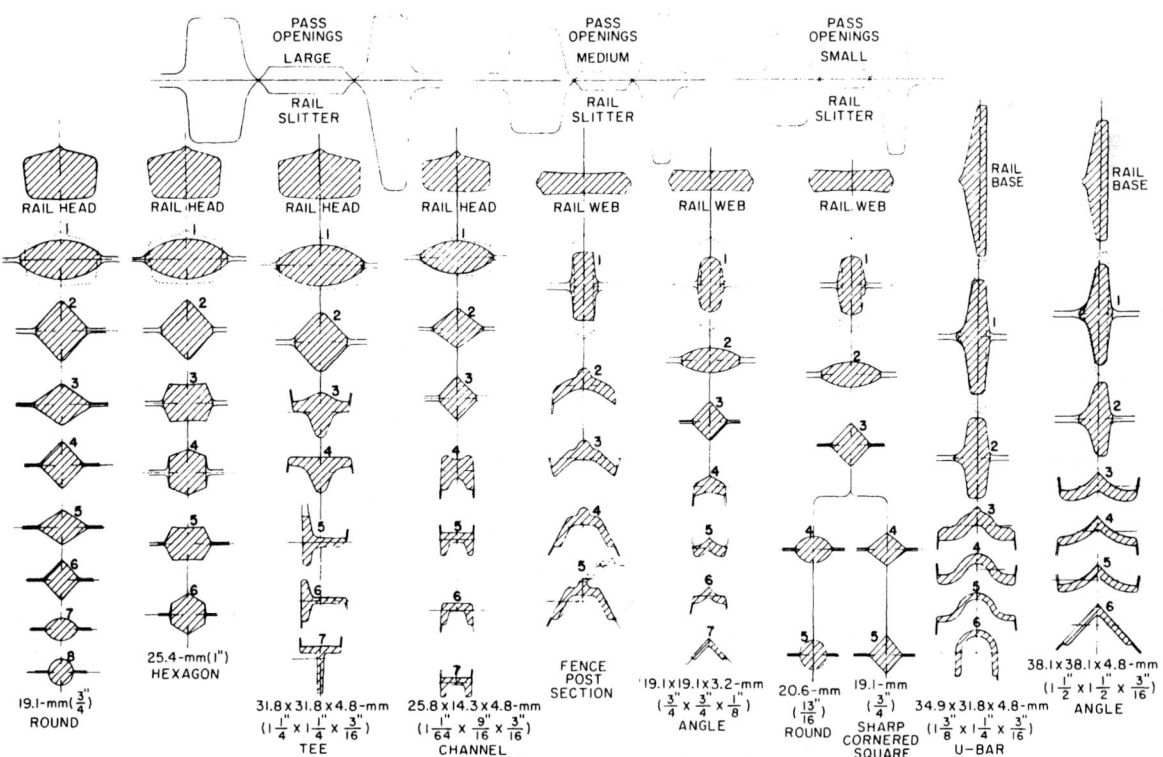

Fig. 30—22. Roll passes for producing various bars and shapes on a rail-slitting mill such as shown in Figure 30—11.

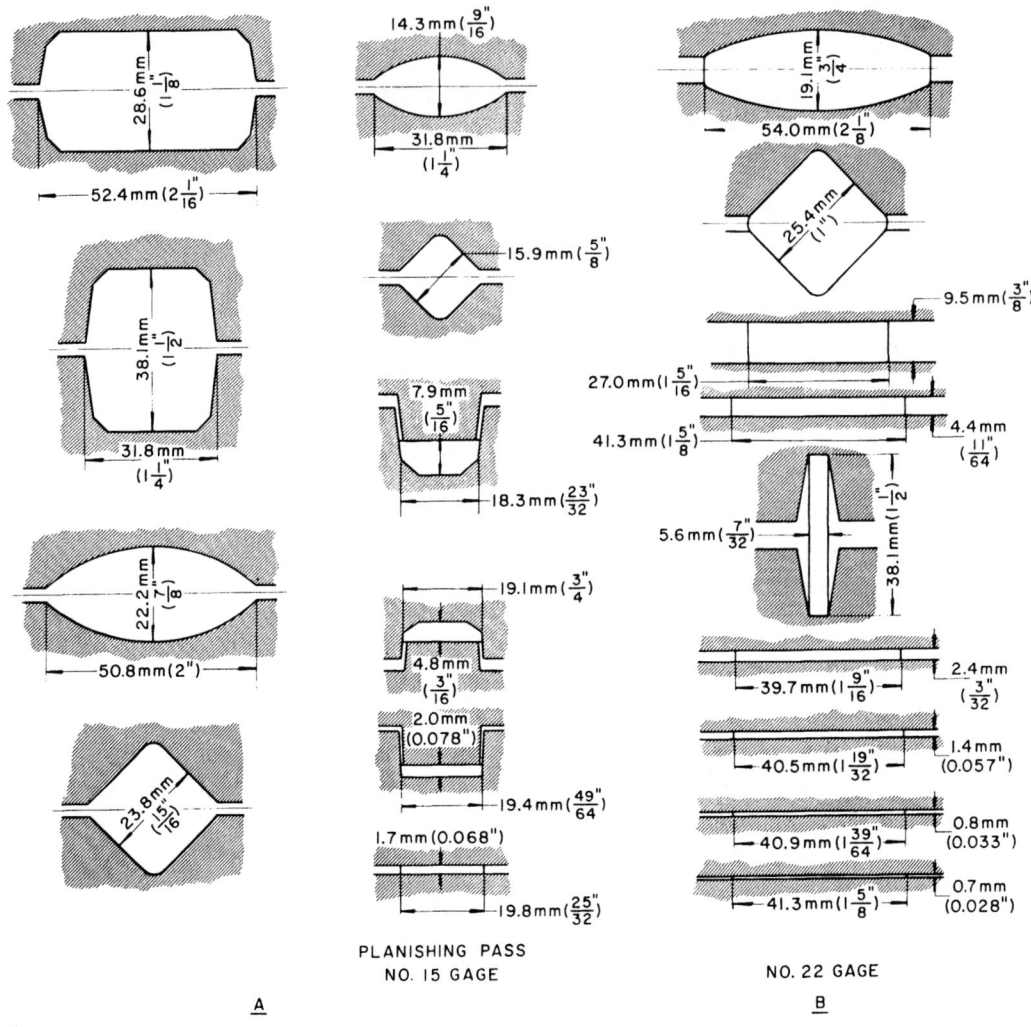

Fig. 30—23. Roll passes for producing hoop and cotton-tie on a bar mill.

SECTION 2

FINISHING AND SHIPPING OF BAR PRODUCT

Bar-Mill Finishing and Shipping Processing Functions—The processing functions required in the finishing and shipping of bar product are in accordance with customer specifications or mill practices. These processing functions include:

Cutting product to length.
Bundling or coiling the product on the rolling mill.
Straightening.
Cleaning (pickling/grit blasting).
Surface coating for corrosion protection or lubrication.
Heat treating.
Inspection.
Surface conditioning.

METHODS OF CUTTING PRODUCT TO LENGTH

The cutting of bar product in most plants is performed by one of the following methods or by combinations of these methods: (1) Cutting by mill shears, directly to ordered lengths if practicable, or in multiple lengths if necessary to facilitate mill delivery, the latter requiring subsequent recutting to a variety of shorter lengths in the finishing departments. (2) Recutting by a finishing-department shear to meet ordered requirements, or for salvaging of portions of the product which may have surface or end defects. (3) Hot sawing of large rounds or sections which are not adaptable to hot shearing. (4) High-speed friction sawing where

Fig. 30—24. Slow-speed saw for recutting steel bars.

recutting is necessary after the straightening operation on a product previously cut by a mill shear or hot saw. (5) Machine cutting, which consists of the use of an abrasive cut-off machine, a hack saw, a cracker shear, or a slow-speed saw (Figure 30—24), where the closest tolerance on end squareness is required, or where, by the nature of the product, recutting cannot be done satisfactorily by other methods. (6) Flame cutting, where the nature of the product and specifications permit the application of heat for rough cutting.

PROCESSING BAR COILS

There are some manufacturers, such as those that produce fasteners or cold-drawn bars, who find it more convenient and economical to obtain bars (either rounds, squares, hexagons, or flats) in coil form rather than in cut lengths. As most of the smaller bar and rod mills are not equipped with flying shears to crop the front and back ends of a coil, it is necessary to complete this operation in the finishing department of the mill. Typically, coils are conveyed from the mill coiling tubs to hooks, spaced 3 metres (10 feet) apart, which carry the coils to be cooled, inspected, sheared, and wired or banded. These conveyance lines are sometimes over 300 metres (1000 feet) in length. A shipping tag is attached to each coil to maintain identity.

By the time the coils have reached the end of the conveyor, they are cool enough to be loaded into the railroad cars for shipment. In some plants having high-speed mills, where cutting to length would seriously retard output of the mills, bars which finally are to be shipped in straight lengths first are rolled in coils; subsequently, in the finishing department, they are then uncoiled, straightened, and sheared to length on a processing line designed for the purpose.

STRAIGHTENING

The type of rolling-mill delivery equipment used has a very significant bearing on the straightness of bar product. Most modern mills have certain features in the design of the hot beds—straight edges, kick-off equipment, roller tables and guides—to minimize bending and kinking that result in the necessity for subsequent straightening. When rolling-mill design permits, straightening units may be installed in the hot-bed delivery roll train ahead of the shears. Such an installation generally eliminates straightening as a finishing operation and necessitates fewer material-handling steps prior to loading.

However, many bar products require straightening after hot shearing or sawing at the mill. Mechanical straightening may be necessary to assure that the product meets the required straightness tolerance. In the finishing departments this is usually accomplished by roll-type machines or by gag presses. The selection of the type of straightening equipment depends on the characteristics of the product and the end-use requirements. There are numerous commercial types of straightening machines, but they all are similar in principle and are fundamentally designed to process a specific shape within certain dimensional limitations. On this basis, straightening equipment can be divided into three main classifications by product shape:

(1) **Flat product.** This equipment includes multi-roll units for flats, squares, hexagons, etc,; and two-way roll units for straightening in planes at right angles in one pass.

(2) **Round products.** This equipment includes roll units which rotate the product through either a two-roll (cross-roll) or a multi-roll unit, and gag presses. Figure 30—25 illustrates a two-roll straightening machine.

(3) **Shapes and special sections.** This equipment includes both light and heavy multiple-roll units having grooves to fit each separate section, gag presses (both vertical and horizontal, with interchangeable guides and adjustable stroke), and roll units for shaping such products as small channels or U-bars which are hot rolled in an open form and then closed by cold forming before the straightening operation.

SIZING, TURNING AND CENTERLESS GRINDING

In the manufacture of bar products, the design and type of rolling mill or the characteristics of the steel frequently preclude the possibility of hot-rolling a round to within precise sectional or out-of-round limitations, or to roll a round with a surface suitable for subsequent fabrication requirements. When such requirements must be met, rounds are further processed in the finishing departments by such operations as sizing, turning, centerless grinding, or by combinations of the latter two. Bars to be processed by any one of these methods are hot-rolled over-size by an amount predetermined by experience.

Sizing is applied to bars when a moderate reduction in section or an improvement in out-of-roundness is desired. It is performed by passing the bar through a two-roll or cross-roll straightening machine in one or more passes as required. Sizing can impart a cold-drawn appearance to the surface of the bar if the rolls are new and properly adjusted; however, it is not generally used for the purpose of improving the surface appearance. As sizing elongates the bar, due to reduction of cross-sectional area, recutting to length must follow.

Turning improves the surface of a bar by removing undesirable defects that may be present in the surface. Turning is accomplished by passing the bar through a turning machine or lathe, using one or more passes depending upon the amount of material to be removed. There are several designs of turning machines used in producing bar products, which differ in the arrangement of the revolving heads or the manner in which the bars are supported and fed to the revolving heads. In all types, a suitable cutting oil is fed to the tool or tools in the revolving head by a pressure system. In one type of machine, two cutting heads are mounted very close together, in tandem, between two guide bearings. In another type, the revolving heads are a considerable distance apart, separated by guide bearings. Due to the arrangement of the revolving cutting heads, the former type is more adaptable for bars of short length, while the latter is more suitable for bars of long length. Turning machines are made in various sizes and can handle rounds in sizes up to and including 150 millimetres (6 inches) and in lengths up to 12.2 metres (40 feet). The amount of metal removed per pass is dependent on the type of steel, and the diameter and the length of the bar. Removal of about 3 millimetres (approximately ⅛ inch) on the diameter

per pass is considered normal for bars of medium or large diameter. Speeds of 305 to 380 millimetres per minute (12 to 15 inches per minute) are normal when removing this amount of metal.

In finishing some grades of alloy and stainless steels, surfaces and dimensions are required that can be met only by **centerless grinding.** Centerless grinding differs from turning in that a grinding wheel is used for removal of metal instead of a cutting tool, and more accurate dimensions and better surface finish are obtained. The centerless grinding machine is constructed so that the bar is supported under the greater portion of its length as well as under the grinding wheel, rather than at the ends. This design enables the machine to operate to close tolerances by elimination of axial thrust, always present when a bar is supported at the ends. A suitable coolant is used on the grinding wheel or wheels. Most commercial centerless grinding machines are constructed to handle round bars up to 100 millimetres (4 inches) in diameter, in ranges of approximately one inch in diameter. When the diameter of the bar must be reduced materially, it is more economical to remove a large portion of unwanted metal by the turning machine before centerless grinding. Standard centerless grinding machine tolerances are ±0.05 millimetres (±0.002 inches) on bar sizes up to 50 millimetres (2 inches) in diameter and ±0.08 millimetres (±0.003 inches) on larger sizes up to 100 millimetres (four inches) in diameter.

In some cases a highly polished bar is desired rather than the standard centerless ground finish. The pol-

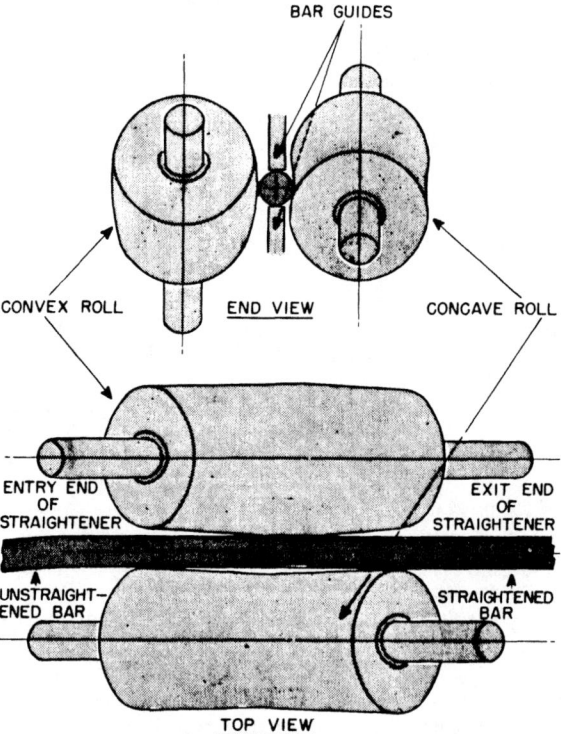

Fig. 30—25. Sketch showing end and top views of a typical two-roll unit for straightening rounds.

ished finish is produced by passing centerless ground bars through a polishing or lapping machine. This machine is similar to the standard centerless grinding machine except that the grinding wheels are composed of a fine grit and the bar-supporting apparatus is constructed to operate within very narrow limits of accuracy. Polishing or lapping tolerances are one-half of those for standard centerless grinding. In precise polishing or lapping of long bars, it is very difficult to keep within these tolerance limits because of the wear of the grinding wheel and the spring of the metal. For these reasons, more precise tolerance limits than those stated above are not considered practical for commercial products.

PICKLING

Stationary or Vat Pickling—Pickling is the term given the descaling process by which the hard black oxide formed on the surface of a bar during hot rolling is removed by chemical action. The removal of hot-rolled scale by pickling may be performed in order to: (1) prepare the surface of the bar for inspection, (2) prepare the product for ultimate end use. Pickling is also used, though to a much lesser degree, to remove the slow-forming red rust that develops on bar products after long exposure to air. Stationary or vat pickling in its simplest form consists of immersing the steel bars in a dilute acid bath, which is held at a predetermined temperature, and permitting them to remain stationary until the pickling action is completed. Most modern installations have improved the method by including means of keeping the bath in motion or agitation. Stationary or vat pickling is one of a number of pickling methods which are classified according to the manner in which the operations are conducted. Other methods of pickling and equipment are described elsewhere in the text.

There is considerable difference of opinion as to what actually takes place as the iron oxide is removed. One theory proposes that the scale or oxide layer is relatively less soluble than the iron underneath and that the acid solution, passing through tiny cracks or fissures in the scale (formed by differences between the cooling rates of the metal and scale), reacts chemically with the bonding structure between the underside of the scale and the true metallic surface of the bar. In this reaction hydrogen gas is liberated which, it is claimed, dislodges the scale particles. A second theory proposes that the acid solution attacks the iron oxide or scale, resulting in its actual dissolution in the pickling bath. A third theory combines the first and second.

Stationary pickling works on most types of steel bar, provided that the process takes all physical and chemical variables into consideration. Hot-rolled low-carbon steels are most easily pickled with a sulphuric-acid solution. Hot-rolled high-carbon steels or alloy steels, however, present difficulties. High-carbon steels discolor in the form of a black smudge of carbonaceous, insoluble material. On alloy steels, pickling does not uniformly remove the hot-rolled scale because of differences in chemical composition of the scale. Proper selection of acid in an effective acid concentration, adequate agitation, proper bath temperature and the inclusion of an effective inhibitor will successfully pickle high-carbon and alloy-steel bars. Very high-alloy steels, such as stainless steels, are pickled in baths containing sulphuric plus another acid, another acid alone, or by a combination of other acids. The choice of the acid combination and concentration depends upon the characteristics of the steel to be pickled and the finish desired. The most commonly used acid for pickling is sulphuric. Other acids used in this method are hydrochloric (muriatic), nitric, and phosphoric. The exception to the above is hot-rolled high-carbon or alloy steels which have been normalized or annealed prior to pickling. These steels must often be cleaned by other means because pickling may cause severe pitting.

The equipment for stationary or vat pickling consists of three tanks. The first tank is used for the dilute acid solution or pickling bath; the second contains only water, for washing the steel bars after immersion in the pickling bath; the third tank contains an alkaline liquid (lime water), for neutralizing any acid which remains on the bars after washing.

Tanks or vats containing the dilute acid solution must be water-tight and capable of resisting attack by the acids used in the pickling solution. They are constructed either of wood, concrete, brick or lined metal. Cypress wood, rubber, and acid-resistant brick or terra-cotta tile are the most efficient linings for tanks.

In addition to the three tanks required for pickling, bar-mill finishing-department equipment usually includes a steel tank, containing a rust-preventative oil into which the bars are dipped in order to provide adequate protection against subsequent rusting. Mills that produce bars that will be cold-drawn may need to coat the bar with lime, in which case another tank is required. Liming tanks are usually made of steel and provide for continuous agitation.

A unit's efficiency depends to a large extent on its ability to handle an assortment of similar pieces at once. The solution must reach the entire surface of each individual bar for the process to be efficient. For this reason, various designs of crates and racks have been developed, one of which is shown in Figure 30—26. The steel bar product to be pickled is placed in tiers on the crates or racks. In the tiers, the bars are kept apart by separators which are called **combs**. The materials used in the construction of the crates, racks, separators, and combs must be of a strong, acid-resisting material.

Coils are pickled by passing a C-hook or a chain through the open center of a group of coils, and attaching the C-hook or the ends of the chain to the hook of an overhead crane. Such a group of coils is referred to as a "lift," and the practice is refered to as chain-pickling. Monel metal or aluminum-bronze alloys are used in many parts and fixtures, including the chains.

Steam has been the most widely used for heating pickling baths, but, the submerged gas burner method, a relatively recent development, has become popular with many operators. Where steam is used, hand-operated or automatically controlled valves maintain the temperature. Also, the steam should not beat against the sides or the bottom of the tank, as this will quickly erode the surface and cause damage to the tank.

The submerged gas burner provides for burning gas

and air under automatic control in submerged burners. The hot waste gases from the burners are conducted into lead tubes which run the length of the tank on the bottom. This method heats the pickling bath quite efficiently, and the waste gases, which are forced out under pressure through holes located throughout the length of the tubes, provide the desired agitation of the bath. When this type of heating equipment is used, however, care must be taken to avoid accumulation of sludge and scale in the bottom of the tank. If the lead tubes become completely covered with sludge and scale, they are shielded from the pickling solution in the tank, become overheated, and are destroyed.

When a new pickling bath is to be prepared, the tank first is filled to approximately three-fourths capacity with water. Acid in an amount to provide the proper concentration is run into the tank, and then enough water is added to bring the solution to an operating level. The heat is then turned on and the bath is brought to the operating temperature. When placing products in the pickling bath, care must be taken to avoid any rapid or irregular movement that may splash acid solution over the sides of the tank, where it rapidly attacks the outer tank shell, concrete floors and tank foundations. It is very important that wash tanks be emptied frequently, as acid in damaging amounts builds up in the wash water and interferes with the proper function of the washing operation.

The surface of steel bars to be pickled must be free from oil or grease, as these materials protect the surface from the action of pickling solution. Any substance on the bars which may serve to contaminate or neutralize the acid solution must be removed before they are immersed in the pickling bath. After the steel bar product is washed and dipped into the neutralizing bath, it is further safeguarded against rusting by being thoroughly dried. Live steam and sometimes air, blown against the bars as they hang suspended in the rack, are used for this purpose.

Summarizing, the rate of pickling and the iron loss depend on: (1) acid concentration, (2) temperature, (3) time in bath, (4) percent of iron (ferrous) sulphate in the bath, (5) presence of inhibitors, and (6) agitation of the bath. Testing the pickling solution by titration will reveal the concentration of the acid bath, and will also show the rate of iron dissolution into the bath, thus giving a good indication of the efficiency of the pickling operation. A rapid increase of iron in the solution indicates that the true metal is being attacked severely, and that acid and good steel are being wasted. The bath should be dumped when the iron content of the pickling solution reaches approximately 60 grams per litre (0.5 pound per gallon).

Temperature of the Pickling Bath—Temperature greatly affects the action of the acid on scale. Raising the temperature of the pickling bath greatly increases the action. However, increasing the temperature of acid solution also increases the tendency of the acid solution to attack the steel itself. The resulting salts saturate the solution, and the efficiency of the process decreases very rapidly. On the other hand, lower acid concentrations and lower bath temperatures require more time per ton of steel pickled, but the quality of the work improves. Less metal is lost from the surface of the bar and less acid is consumed in the process.

An **inhibitor** is an agent added to the pickling solution for the purpose of protecting the exposed surface of the metal of the bars, by inhibiting or retarding acid attack upon the metal without affecting, to any appreciable degree, the pickling action which removes the scale. The inhibiting action is not understood clearly but is generally explained by the electrolytic theory of corrosion. Inhibitors show little change and lose little of their efficiency during the pickling operation. However, they may be broken down and their function destroyed by overheating.

Many different substances are used as inhibitors in acid pickling. They range from vegetable or animal

Fig. 30—26. Loaded pickling rack, showing how combs keep bars separated to allow free circulation of acid solution.

matter to complex synthetic organic chemicals. The prime requirements of an inhibitor are that it must **disperse** colloidally in the bath, prevent hydrogen evolution and not leave a smudge or film on the surface of the steel. Many inhibitors contain substances that cause foaming, and the floating layer or blanket of foam on the bath prevents the escape of acid with escaping gases and vapors. As inhibitors are expensive and only a small amount is needed for desired results, less costly foaming substances are added separately by many operators when increased foaming activity is desired.

Common difficulties encountered in pickling are over-pickling, under-pickling, smudge and pitting. Over-pickling may be defined as the etched appearance of the surface of a product caused by over-activity of the acid solution. Conversely, under-pickling may be defined as incomplete removal of the scale due to limited activity of the acid solution and/or the use of too low a pickling temperature. Smudge is a carbonaceous precipitate or stain which forms on bars of high carbon content. Pitting may be defined as the appearance of crater-like indentations on the surface of the bar, which may result from over-activity of the acid solution on the metal in areas where scale has been loosened mechanically or removed prior to the pickling operation, or as the result of an electrolytic action taking place in areas where there are concentrations of dense scale on the surface.

COATING

In addition to the three tanks required for the pickling operation, bar-mill finishing-department equipment usually includes a steel tank, containing a rust-preventive oil into which the bars are dipped in order to provide adequate protection against subsequent rusting.

As a coating of lime may be specified on bars, particularly those to be cold drawn, an additional tank for this purpose is generally part of the equipment. The lime coating serves as a lubricant in the cold-drawing process and reduces wear on the dies. It is applied after the washing and neutralizing stages of the pickling process. The tank for lime coating generally is constructed of steel and, as the lime should be kept in suspension in the water, a means of providing continuous agitation is usually a part of the installation. Pickling, washing, neutralizing, and liming tanks should be equipped with proper means of heating the bath and provision must be made for proper means of disposing of the waste solutions.

GRIT-BLASTING

Grit-blasting or blast-cleaning is a mechanical process used for removing scale and rust from bar products. It consists of eroding or abrading away the scale from the surface of the bar by impinging an abrasive substance like sand, aluminum oxide, or a metallic substance like cast-iron or steel shot. The abrasive material may be directed against the work by air under pressure or by a mechanical apparatus utilizing centrifugal force. All types of steel bar products and shapes can be cleaned successfully by this method. As a result of the difference in economy of operation between the grit-blasting and the pickling method, the former generally

is confined to the cleaning of one of the following types of products: (1) those which must have certain physical characteristics essential to subsequent processing; (2) those which cannot be cleaned satisfactorily by the pickling method after such thermal treatments as normalizing or annealing; (3) those with high alloy content whose scale cannot be removed satisfactorily by ordinary pickling methods.

The principal difference between the various types of grit-blasting equipment is in the means employed for throwing the grit or shot against the work. One type of grit-blasting machine makes use of centrifugal force generated by a rotor or impeller which rotates at a very high speed. Grit in the form of metal shot is fed from a hopper into the revolving rotor through an opening in the center of the rotor housing. The shot, on entering the revolving rotor, is picked up by the rotor vanes and is thrown by centrifugal force away from the center of the rotor toward its periphery. From the periphery, the shot is directed outward to impinge on the work at the angle desired for the best abrasive effect. Since this type of machine develops its abrasive characteristics on the basis of the velocity and the weight of the particles, lighter grit, such as sand, cannot be used satisfactorily.

Another type of machine employs compressed air or forced air from a blower for throwing the grit particles. There are several modifications of this principle, but in each type the grit is permitted to entrain in a fast-moving air stream and is directed upon the surface of the work through a hose having a nozzle designed for the purpose. The lighter grit is used most effectively with this type of equipment. The material used for, and the size of, the grit or shot in blast cleaning are dictated by the requirements for a given job and, as previously noted, by the type of equipment in which it is to be used.

From the standpoint of density, grit can be classified as either light or heavy. Light grit is a nonmetallic inorganic material with excellent abrasive characteristics. It is purchased according to particle size. Examples of light grit are the widely used natural abrasive known as Ottawa sand and the synthetic abrasive, aluminum oxide. Heavy grit is principally of the metallic type such as cast-iron shot. Like the light grit, it is purchased according to particle size. Metallic grit is the type most generally used on bar product.

The chief advantages of grit blasting are: (1) it leaves the surface of the bar with a bright metallic finish without any adhering scale; (2) it is capable of cleaning a number of types of products that cannot be cleaned successfully by pickling; (3) it does not produce such physical or chemical conditions as over-pickling, under-pickling, smudge or pitting which may attend the pickling process. The disadvantages of grit blasting are: (1) higher cost; (2) only a few bars can be processed at one time in the blast machine; (3) as the grit or shot builds up with the scale the efficiency of the cleaning effect is decreased; and (4) the bars may be more difficult to inspect for surface defects because of the peening action of the grit or shot.

Grit blasting does produce minute indentations on the surface of the product which are more beneficial in some subsequent fabrication operations than they are detrimental. The size of these indentations and their shape serve to produce a surface sheen or finish determined somewhat by the particle shape and size of the grit. The character of this sheen or finish is very important where cold-drawing operations are to be performed. As a grit-blasted steel surface is very susceptible to rusting, bars cleaned by this method must be protected immediately from moisture and alkaline or acid vapors. Also, when a bar is grit-blasted with cast-iron shot, particles of the shot have a tendency to adhere to the product and will cause rapid rusting or corrosion on some types of steel if they are not removed immediately. In such cases the grit-blasting operation must be followed by a light pickling operation.

BAR INSPECTION AND TESTING

Inspection—In order to produce satisfactory bar products it is important that an adequate quality control system be established and maintained. This system should include mechanical and metallurgical testing as required, and the inspection for surface or other defects. The procedures for making mechanical and metallurgical tests and the inspection of steel prior to rolling in the bar mills are covered in other chapters of this book. The inspection of bar product in the majority of plants is carried out along the same general lines and can be divided into two divisions: (1) mill inspection, and (2) finished product or final inspection. The duties and responsibilities of the inspection forces in these divisions must be thoroughly co-ordinated to accomplish the best results.

Mill inspection is performed during the rolling process and is the means of minimizing or preventing discrepancies at the source. The duties of the inspector at the mill are: (1) to check the identity of the steel being rolled, (2) to inspect the surfaces of the product to determine its suitability for further processing in the finishing departments or in its final intended use, (3) to check section and length of the product to determine its suitability for application on a particular order, and (4) to procure necessary samples for mechanical or metallurgical tests. The inspection of the finished bar products, on the basis of commercial standards or for dimensions, straightness, surface defects, and other items of form as may be required, is carried out upon completion of the finishing end operations at designated units.

The duties of the final inspector are in a measure a repetition of the duties of the mill inspector, with the additional responsibility of checking and posting heats. The checking and posting of heats is a method of reporting the amount of product accepted or rejected (per heat) on an order, with reasons for further processing or for rejections. The final inspector must lay particular emphasis on surface quality if the product has been conditioned and straightened. Various tools are required for inspecting bar products. A list of these tools and a brief description of their use follows: (1) micrometers and calipers, for measuring the section of rounds, flats, and squares; (2) snap gages, for precision rounds; (3) tape, for measuring length; (4) slide scales, for measuring width; (5) protractors, depth gages, radius gages and templets for shapes; and (6) the square edge for measuring off-squareness.

Surface defects, as well as other defects which cause the rejection of bar products, may be the result of steelmaking practices which carry through from the ingot or may be caused by the rolling mill equipment used to produce the product. A list follows giving the general mill terminology of the most regularly occurring types of mill defects and a brief description of each:

1. **Fins and overfills** are protrusions formed when the section is too large for the pass it is entering, or when proper allowance has not been made for lateral spreading in the rolls. Overfills are broad and less sharp than fins. As a rule, overfills occur more frequently than fins and in many cases are associated on the same bar with underfills.
2. **Underfills** are the reverse of overfills, that is, they are areas in which the section is incompletely filled. They are formed by permitting the bar to be rolled scant in certain dimensions. Underfills appear most frequently on rounds and channels.
3. **Slivers** are loose or torn segments of steel rolled into the surface of the bar. They may be caused by a bar shearing against a guide or collar, incorrect entry into a closed pass, or a tear from other mechanical causes. Sometimes slivers are present in the billet and are carried through to the hot-rolled bar or shape.
4. **Laps** can be said to be a rolled-over condition caused by a bar having been given a pass through the rolls after a sharp overfill or fin has been formed, causing the protrusion to be rolled into the surface of the product.
5. **Seams** are crevices in the steel that have been closed but are not welded. They are a type of a defect very difficult to detect on certain types of steel products. Seams are caused by blow holes and cracks in the original ingot, or by faulty methods of rolling in both semi-finishing and finishing mills.
6. **Fire cracks and roll marks** are impressions in the product, of varying degree and pattern, caused by mill rolls becoming overheated, and cracking or spalling.
7. **Scratches** are long nicks or indentations in the product caused by the surface or surfaces of the bar rubbing against sharp or pointed objects such as guides on the mill, chutes, "dead" conveyor rolls, chain hoists or other mechanical equipment.
8. **Rolled-in scale** is a defect in the surface caused by scale, formed during a previous heating, which has failed to be eliminated during the rolling operations.
9. **Buckle and kink** is a corrugated or wrinkled surface condition caused either by worn out pinions on a roll stand or uneven cooling beds. Buckle is an up-and-down wrinkle, whereas kink is a side wrinkle.
10. **Burned steel** appears as a rough area with checked or serrated edges. It is caused by steel being exposed to excessive temperature and is always scrapped.
11. **Camber** is the deviation of the side edge of a bar

from a straight line. It is caused by improper heating of the billet, uneven dimensions causing differential expansion or contraction, or improper alignment on the hot beds.

12. **Hook** is a short bend or curvature caused either by improperly adjusted delivery guides or by any obstruction which may halt momentarily the forward motion of the bar from one roll stand to another.

13. **Pipe** is a steel-making defect carried through from the ingot. The presence of pipe is detected as a small cavity located in the center of an end surface.

14. **Shear distortion** is a mashed or deformed end on a bar caused by defective or improperly adjusted shearing equipment.

15. **Twist** is a condition wherein the ends of a bar have been forced to rotate in relatively opposite directions about its longitudinal axis. It may be caused by excessive draft, faulty setting of delivery guides, or lack of uniform temperature in the bar.

Testing—Numerous tests are made during the finishing operations, the purposes of which are to reveal defects otherwise impossible to detect during surface inspection. The tests most commonly employed are described as follows:

1. **The pickling test** consists of immersing short pieces of product for several minutes in dilute sulphuric acid. The acid removes the scale from the bar and exposes to view such surface defects as may be covered or hidden by the scale.

2. **The upset test** consists of subjecting test pieces to severe compression under a hammer. The compression or upsetting action will force open any defects which could not be detected while the steel was in the as-rolled condition. A sound steel will be indicated by the absence of areas which open up.

3. To inspect finished bars for seams, **eddy-current testing** is being used widely: **ultrasonic test** methods also are employed. Although it has been used for finished-bar inspection, the **magnaflux test** currently is used more for inspecting semifinished products (billets and blooms), insofar as rolling operations are concerned. All three of these methods are discussed in detail in Chapter 51 on "Nondestructive Inspection of Steel."

4. **The file test** consists in removing the scale, by filing, on any surface area which may be suspected of containing a hidden defect. The file test is a quick method employed by inspectors for determining the extent and depth of seams in bars.

5. **The bend test** is made on certain classes of material to determine the soundness of internal structure and to denote the degree of ductility. The test consists merely in bending a standard test specimen through a certain specified arc. Examination of the bend will disclose surface defects.

6. **The etch test** is one which is used repeatedly in standard manufacturing and fabricating processes to determine the soundness of internal

structures. A test piece is cut from the desired location in a bar and from this test piece a specimen of the size required for etching is removed. The surface to be examined is ground and then dipped in a solution of hydrochloric, sulphuric, nitric or picric acid. For some products, solutions of iodine, copper ammonium sulphate, ferric chloride, cupric chloride or cupric sulphate are used. Directions for etching various products may be found in "Metals Handbook."

7. **The grit-blasting test** is used for much the same purpose as the pickling test. The grit removes the hot-rolled scale so that the surface of the test piece can be visibly inspected. It is not recommended for detecting fine seams on certain types of steel because of the peening action of the grit.

CONDITIONING METHODS AND EQUIPMENT

It has been noted that bars which are free from surface defects cannot always be produced. Some of these defects are common to all steel products, whereas others are more or less peculiar to bar product. In order that the product meet required quality standards, these defects must be removed by chipping or grinding. Their removal by either of these methods is known by the term "conditioning." Thorough inspection must precede conditioning. In a large number of cases pickling before inspection is necessary in order to reveal all the defects that may be present. The product requiring conditioning must be marked properly with a suitable chalk or crayon in the areas showing defects.

Conditioning is carried out by either chipping or grinding, depending upon the characteristics of the product. Conditioning by chipping is confined to soft and medium-hard steels and to those products on which subsequent fabrication procedure will permit the presence of grooves from chipping. Well-maintained pneumatic hammers operated by an adequate air supply, and chisels which have been properly heat-treated and dressed, are the principal equipment necessary for chipping. Conditioning by grinding is confined to those steels with a high hardness, or to bars or shapes whose contours would be changed by chipping grooves. The grinding operation is accomplished by: (1) pneumatic grinders, or (2) high-cycle electric grinders. The choice of grinder depends upon local conditions such as air and power supply, or upon the suitability of the machine to perform the required task. Each type of grinder possesses certain advantages and disadvantages. The pneumatic grinder has the advantage of being lighter in weight and has a minimum of moving parts, which facilitates repairs. Its major disadvantage is its inability to maintain full speed on all loads. It also must have an adequate air supply at 620 to 690 kPa (90 to 100 psi) without a large drop in pressure from source to point of use. The high-cycle electric grinder, because of its special construction, has the advantage of being able to maintain full speed at all loads. It has the disadvantages of being somewhat heavier to handle, of having many complicated working parts, and of requiring a frequency changer adjacent to the work. The grit, bond, shape and speed of the grinding wheels are dictated by the characteristics of the steel and by the shape of the product being worked.

Some mills have automated their billet grinding process. In these mills, the operator is enclosed in a booth and remotely controls the grinder. The grinding wheel is stationary; billets are pulled back and forth across the wheel. Each pass removes a section about 2 inches wide and about 0.08 inch deep. After grinding, the operator can spot-grind any remaining flaws.

In an effort to improve grinding wheels, manufacturers developed zirconia abrasives, which consist of a zirconia-alumina alloy mixed with binders. When the mixture is molded under high temperature and pressure, it produces a very dense wheel with almost no porosity. The zirconia wheel is especially effective because as it wears, it continually forms a fresh cutting surface that does not require dressing or replacement.

NARROW FLAT-ROLLED PRODUCTS

Band, Hoop and Cotton Tie—Band for use in commercial packaging is usually rolled in coil form in multiple width on the rolling mill, and then slit to narrow widths on multiple slitting units installed in the bar finishing departments. Edge conditioning is necessary to eliminate the sharpness of the slit edge, before final coiling and shipment. The usual sizes are 19.1 by 0.71, 19.1 by 0.89, 31.8 by 0.89, 31.8 by 1.27 and 50.8 by 1.27 millimetres (¾ by 0.028, ¾ by 0.035, 1¼ by 0.035, 1¼ by 0.050 and 2 by 0.050 inches). It is made from a steel in the 0.50 to 0.60 per cent carbon content range, with 0.80 per cent manganese and 0.15 to 0.30 per cent silicon.

Hoop—There are four general classifications of this type of product: (1) tight cooperage hoop for barrels to hold liquids, (2) slack barrel hoop for barrels to hold dry products, (3) tobacco barrel hogshead hoop, and (4) special hoop for special packages.

Hoop (except tight cooperage hoop) is made from steel in the 0.08 to 0.10 per cent carbon content range. Tight cooperage hoop is generally made from steel having a carbon content of 0.30 to 0.35 per cent.

Hoop is made either by slitting coiled strip, rolled in multiple width, into narrow coiled strip of the desired width; or, from narrow coiled strip with a hot-rolled or mill edge. The type and width of hoop being produced influences the choice of method used.

Hoop is produced in widths increasing in increments of 3.2 millimetres (⅛ of an inch), beginning with a minimum of 28.6 millimetres (1⅛ inches) and extending to and including 50.8 millimetres (2 inches). It is made in thicknesses between 0.64 to 1.24 millimetres (0.025 to 0.049 inches). It is prepared in cut lengths for hoops from 825.5 mm (2 feet 8½ inches) to 2591.8 mm (8 feet 6 inches) in circumference. Automatic machines are used in the fabrication of hoop (Figure 30—27). These machines are so designed that the strip from the coil, passing through a machine in a horizontal position, is first run through rolls in which a slight bend is made on the edge to be beaded. This is followed by a beading operation, done in forming dies that operate horizontally at approximately 400 strokes per minute. The beaded strip next moves into a combination shear and rivet-hole punching die, where it is sheared to a specified length, and where the rivet holes are punched. Hoop is produced as "curled hoop" or a "straight length." Curled hoop is made by a pinch-roll and curved guide-shoe arrangement that permits the hoop to take a circular form. A straight length hoop is produced merely by removing the curved guide shoe.

Cotton tie is a light, narrow, hot-rolled strip used, as the name implies, to bind bales of cotton, hemp, jute, etc. It is fabricated in a manner quite similar to hoop. After being finished in pinch rolls and a vibrator it is delivered onto apron conveyors. From the apron conveyors, the strip is coiled and delivered on a coil conveyor to cold shears where it is sheared to a length of approximately 3.7 metres (12 feet). It is then bundled by hand at which time buckles are inserted. The bundles are then dipped into a tank containing an asphalt base paint if desired. Cotton tie is shipped in bundles consisting of 30 ties and 30 buckles, either painted or unpainted. Two sizes of bundles are pro-

Fig. 30—27. Automatic hoop-forming machine in operation. Finished hoops are on stand in right foreground.

duced; the standard, weighing about 20.4 kilograms (45 pounds) of 23.8 mm wide and 1 mm thick ($^{15}\!/_{16}$ inch wide and 0.042 inch thick) cotton tie, and a special bundle, weighing approximately 27.2 kilograms (60 pounds), of 23.8 mm wide and 1.2 mm thick ($^{15}\!/_{16}$ inch wide and 0.049 inch thick) cotton tie.

CONCRETE REINFORCING BAR

Concrete reinforcing bar is a bar product consisting of plain rounds and deformed rounds. It is used to furnish tensile strength to concrete sections subject to bending loads and to furnish additional compressive strength in sections where unreinforced concrete would prove too bulky.

All types of concrete structures are commonly reinforced with either deformed or plain bars. Concrete reinforcing bars are usually deformed and this discussion will be confined to that type of bar. Deformed concrete reinforcing bars are bars in which the surface is provided with lugs or protrusions (called "deformations") which inhibit longitudinal movement of the bars relative to the surrounding concrete. The surface deformations (Figure 30—28) are hot formed in the final roll pass by passing the bars between rolls having patterns cut into them so that the surfaces of the bars are forced into the depressions in the rolls to form characteristic deformations. Deformed bars are produced in accordance with the specifications for minimum requirements for the deformations of deformed steel bars for concrete reinforcement. Table 30—I furnishes dimensional data on these bars.

Table 30—I. Deformed Concrete-Reinforcing Bar Designation Numbers, Unit Weights and Nominal Dimensions

Bar Designation Number*	Unit Weight		Diameter		Cross-Section Area		Perimeter	
	kg/m	lb/ft	mm	in.	cm²	in.²	mm	in.
3	0.560	0.376	9.53	0.375	0.71	0.11	29.92	1.178
4	0.994	0.668	12.70	0.500	1.29	0.20	39.90	1.571
5	1.552	1.043	15.88	0.625	2.00	0.31	49.86	1.963
6	2.235	1.502	19.05	0.750	2.84	0.44	59.84	2.356
7	3.042	2.044	22.23	0.875	3.87	0.60	69.82	2.749
8	3.973	2.670	25.40	1.000	5.10	0.79	79.81	3.142
9**	5.060	3.400	28.65	1.128	6.45	1.00	90.02	3.544
10**	6.404	4.303	32.26	1.270	8.19	1.27	101.35	3.990
11**	7.907	5.313	35.81	1.410	10.06	1.56	112.52	4.430
14	11.384	7.650	43.00	1.693	14.52	2.25	135.13	5.32
18	20.239	13.600	57.33	2.257	25.81	4.00	180.09	7.09

*Bar numbers are based on the number of eighths of an inch included in the nominal diameter of the bars. The nominal diameter of a deformed bar is equivalent to the diameter of a plain bar having the same weight per foot as the deformed bar.
**Bars of designation Nos. 9, 10, and 11 correspond to the former 25.4-mm (1-in.) square, 28.6-mm (1⅛-in.) square and 31.8-mm (1¼-in.) square sizes and are equivalent to those former standard bar sizes in weight and nominal cross-sectional areas.
Note: the above table including the footnotes is in agreement with U. S. Department of Commerce Simplified Practice Recommendation 26-50 covering Steel Reinforcing Bars.

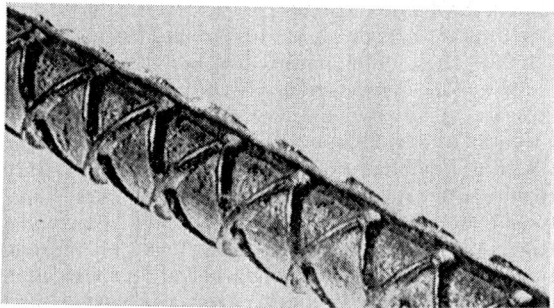

Fig. 30—28. Photograph of a deformed concrete-reinforcing bar, showing protrusions produced by rolling the bar between rolls having a pattern cut into them.

Deformed concrete reinforcing bars are produced to standard specifications for concrete reinforcement (ASTM designation A 615), rail-steel bars for concrete reinforcement (ASTM designation A 616) and axle-steel bars for concrete reinforcement (ASTM designation A 617). These classifications are largely self-explanatory. Two grades of steel are produced under ASTM specification A 615; namely, grades 40 and 60. Under ASTM A 616, two grades, namely, 50 and 60 are produced. Under ASTM 617, two grades are produced; namely, grades 40 and 60.

Referring to Table 30—I; bar sizes 3 to 18, inclusive, are produced to ASTM designation A 615.

Concrete reinforcing bars are shipped from the mills in straight lengths, either cut to design length in the mill shears or in long lengths to be recut for fabrication, as required in specific applications.

Engineers' and architects' designs and specifications are prepared in accordance with the Manual of Standard Practices of the Concrete Reinforcing Steel Institute. Bar fabricators furnish concrete reinforcing bars either straight and cut to the proper length or bent or curved in accordance with plans and specifications.

PACKAGING AND LOADING

The packaging of bar products in most plants is performed according to: (1) standard practices, or (2) special practices as required by the customer. Standard practices are controlled by such factors as: (1) the weight of the bundle or lift, (2) the means of binding or fastening the bundles or lifts, (3) the means of identification, (4) the means of protecting the product in transit, and (5) the geographical location of the customer. Special practices are predicated upon the type or capacity of handling and processing equipment used by the customer.

Either wiring or banding may be applied for binding or fastening bundles or lifts, depending upon the shape, size and length of the product. Special practices for binding or fastening bundles or lifts are applied only when specified. Wiring is used generally on large rounds and heavy sections, whereas bands are used most generally for small sections or for flat products. Standards have been established which specify the spacing and the number of wires or bands to be used for various sizes of bundles. Additional wiring or banding, either for domestic or export shipment, is handled as special practice. Special practice also applies when stack piling is required, as the number of pieces and the dimensions of the bundle are limited as well as the weight. Wiring and banding is accomplished by such special tools as stretchers and sealers.

Identification of bar products requires that each bundle or lift for domestic trade be identified by a standard manila or metal tag, approximately three inches by five inches, attached to one end. Where coils are involved the same type of tag is used on each coil. The information on the identification tag will vary somewhat in different plants but it generally contains a part or all of the following information: customer's name, destination, customer's order number, mill order number, part number or special markings, number of pieces, section, length, weight, heat number, bundle or lift number, and the grade of steel. On products for export, additional identification is required on each bundle or lift in the form of duplicate manila tags on opposite ends and a metal tag attached to the center of the bundle. This is necessary because of the possible loss of a single manila identification tag when the product is rehandled at railroad terminals, shipping docks, or aboard ship. In addition to the standard manila tag, as mentioned above, the identification of bar products may involve the die stamping of heat numbers on the ends of bars of certain dimensions. These specifications are as follows: heat numbers are stamped on one end of all bars (round, squares or hexagons) of 63.5 millimetres (2½ inches) or above, and on all flats 75 millimetres by 25 millimetres (3 inches by 1 inch) and above. When special practices are involved, bars are stamped with codes or symbols as specified. When product in sizes about 75 millimetres (3 inches) or over (or of equivalent cross-section) consigned for export is to be shipped loose, each piece is stenciled with the necessary information for proper identification.

Color marking may be specified on one end of each lift or bundle or may be specified on one end of each bar, depending on the product size. In special practices, either for export or domestic trade, special marking is applied only when required by the customer.

Bundles or lifts are loaded in open cars without protection from the weather. When shipping carload lots, bars and bar shapes are shipped loose or in large lifts in open cars. When special practices are required, lifts may be wrapped or shrouded individually, or the entire carload may be shrouded as a unit. Loading, like packaging, is divided into (1) standard practice, or (2) special practices as required by the customer. The gondola car is the standard means of transit, either for magnet, chain, or crane unloading. Standard practice for loading and bracing is conducted in accordance with the rules of the American Association of Railroads. Special practices as requested by the customer generally fall into one of the following categories: (1) modified methods of gondola-car loading for chain or crane unloading, or magnet unloading alone; (2) flat-car loading for chain or crane unloading, magnet unloading, or tractor unloading; (3) box-car loading for either tractor unloading or hand unloading; (4) double loads or sets for chain or crane unloading; (5) other methods of special blocking. Double loads are used principally for long lengths of structural material which exceeds the length of a single standard gondola car. Box-car loading generally is used on products which cannot be exposed to the weather and which require the best handling methods.

Material-Handling Equipment—Numerous and varied types of handling equipment are used in the finishing and shipping of bar products, depending upon the type of product and the arrangement and design of the finishing department equipment. The types of handling equipment most generally used are: (1) overhead electric cranes, (2) tractors (principally the lift or peel type), (3) transfer buggies, (4) jib cranes, (5) box-car loading machines, (6) conveyors, and (7) miscellaneous auxiliary apparatus such as coil hooks, spreaders, package lifters, chains, and slings.

SECTION 3

HEAT TREATING CARBON AND ALLOY BAR STOCK

Heat treating may be defined as an operation or combination of operations in which a metal or alloy in the solid state is heated and cooled, under controlled conditions, according to a predetermined schedule, to obtain desired properties.

The purpose of heat treatment is to develop the full effect of the various elements in the steel as related to desired properties, through structural or phase changes. "As-rolled" bars vary in hardness and microstructure in relation to the chemical composition of the steel and, therefore, usually require some form of heat treatment to obtain a physical condition best suited for the final product. Low- and medium-carbon bars often are used in the as-rolled condition, but higher-carbon steels and most alloy steels require heat treatment. This treatment consists of some form of annealing, normalizing, or quenching and tempering, or

a combination of any two or even three of these.

It has been found that the results of heat treating conform to certain definite principles, a detailed discussion of which will be found in Chapter 41. Application of these principles makes it possible to obtain transformation of a particular steel at a stipulated temperature by controlling the rate of cooling from temperatures above its critical range of temperature, and thus obtain desired mechanical properties. More detailed descriptions of how this is accomplished will be given in the following discussion of the various types of heat treating.

PROCESSES AND THEIR EFFECTS

Annealing—The term annealing is used rather loosely to describe several types of heat treatment which differ greatly in procedure yet all accomplish

TYPES OF ANNEALING

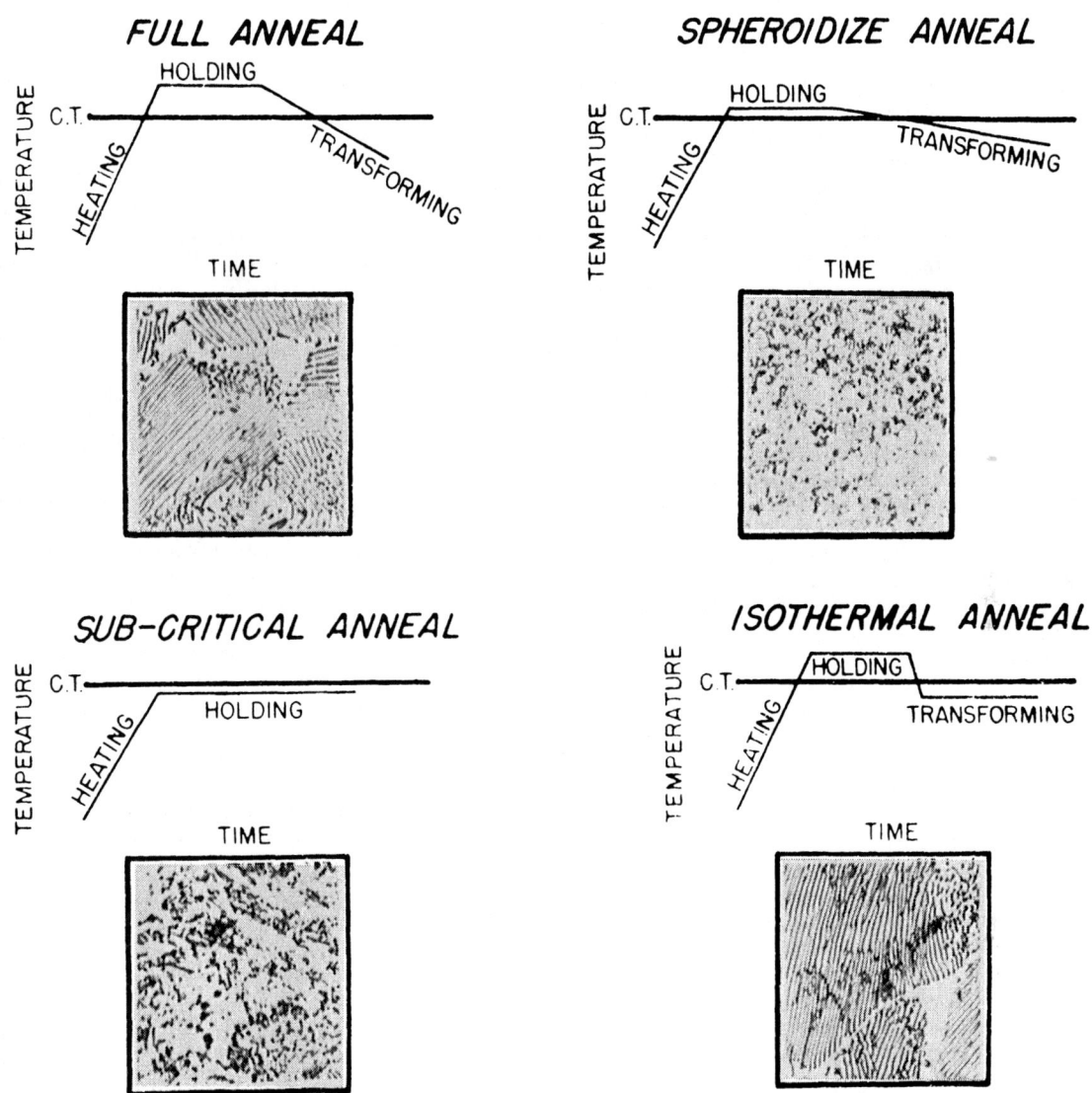

Fig. 30—29. Thermal cycles and resultant microstructures obtained from the four basically different types of production annealing.

one or more of the following effects:

1. Remove stresses
2. "Soften," by altering mechanical properties
3. Refine the grain structure
4. Produce a definite microstructure

In most commercial operations, more than one of these effects usually are obtained simultaneously, although only one may be desired specifically. Therefore, the selection of a specific annealing process is dependent on the particular predominant or overall effect desired and the grade of steel being processed.

Full Annealing—If, for example, it is desired to refine the grain structure and produce a lamellar pearlite, a full annealing cycle should be used. This consists

of heating the steel to a temperature above the transformation range, holding for one to two hours, and then cooling at a predetermined rate to obtain the desired mircrostructure (see Figure 30—29). Grain refinement is accomplished in this instance by the recrystallization of the steel in passing through the critical range both in heating and in cooling, as explained in greater detail in Chapter 41. The microstructure obtained in cooling any steel from above the critical temperature range is dependent both upon the temperature range in which transformation occurs and the time required for completion of transformation in that range. Thus, it is obvious that the rate at which any steel is cooled determines the final microstructure, since the degree of transformation will depend on the amount of time allowed for it to occur. Therefore, the slower the rate of cooling and the higher the temperature at which complete transformation occurs during full annealing, the coarser the pearlite will be with correspondingly lower hardness. Such treatment is performed usually on steel of 0.30 to 0.60 per cent carbon content which is to be machined.

Isothermal annealing is a type of full annealing in which the steel first is cooled to the temperature at which it is desired to have transformation occur, at a rate sufficiently rapid to prevent any structural change above that temperature. The steel then is held at the selected temperature for the time necessary to complete such transformation (see Figure 30—29). Thus it is possible, with this process, to obtain a more uniform microstructure than could be expected by continuous cooling. However, since it is necessary to drop the temperature rapidly to prevent any transformation above the desired temperature, there are definite limitations as to the mass that can be so treated. It is applicable, therefore, only to smaller sections and would not be suitable for large bars or large loads in batch-type furnaces, since it would be impossible to cool them at a rate sufficiently rapid to prevent some transformation.

A modified application of isothermal annealing is possible, however, in which the charge is heated in one furnace and transferred to another, which has been set at a temperature somewhat lower than the desired temperature of transformation, in order that the temperature of the charge will drop rapidly to that required. The selection of the temperature of the second furnace will be governed by the temperature to which the charge first is heated, the mass of the charge and the desired transformation temperature. Suitable handling equipment must be available to transfer the entire charge rapidly, since any undue delays might result in portions of the charge being cooled to too low a temperature. Continuous furnaces also are applicable to this type of cycle.

Process or Subcritical Annealing—Another type of annealing called process or subcritical annealing consists of heating the steel to a temperature just under the lower critical (Ac₁) and holding at this temperature for the proper time (usually 2 to 4 hours) followed by air cooling (see Figure 30—29). This type of annealing results in softening the steel due to a partial coagulation of the carbide to form spheroids or small globules of carbide. It is not suitable when a close control of hardness or structure is desired, because the prior structure of the steel determines to a marked degree the extent of spheroidization which will occur. For example, an originally coarse lamellar structure may show very little evidence of spheroidization after this treatment, whereas an originally fine lamellar or martensitic structure would show a marked degree of spheroidization. The treatment is, however, quite satisfactory for rendering bars more suitable for cold sawing or shearing, and is used to a great extent for these purposes. Since the temperature to which the bars are heated is somewhat lower than in a full anneal, there is less scaling, and warping can be controlled.

Spheroidization—Spheroidization is a type of annealing which causes practically all carbides in the steel to agglomerate in the form of small globules or spheroids. There may be a wide range of hardness with such a structure for any grade of steel since the size of the globules has a direct relation to hardness, i.e., the larger the globules the lower the hardness. Spheroidizing may be accomplished by heating to a temperature just below the lower critical and holding for a sufficient period of time. However, as pointed out in the discussion under subcritical annealing, the prior structure of the steel affects to a marked degree the final result of such a treatment. Therefore, if a well spheroidized uniform structure is desired, such a treatment is not suitable because too many uncontrollable variables are encountered. A more desirable and commonly used method for spheroidizing is to heat to a temperature just above the critical and cool very slowly (about 5.6° C or 10° F per hour) through the critical range (see Figure 30—29) or to heat to a temperature within the critical range but not above the upper critical and cool slowly.

This treatment is used for practically all steels containing over 0.60 per cent carbon that are to be machined or cold formed and for bearing steels such as G15216 (SAE 52100 or AISI E52100) prior to machining. For the latter, the size and distribution of the spheroidal carbide particles are extremely important since they all are not dissolved in subsequent heating for hardening but remain in the condition obtained in the annealing process. Small well-defined spheroids are the most desirable and coarse elongated carbides should be avoided at all times.

Normalizing—Normalizing is a process wherein the steel is heated to a temperature about 55° to 83° C (100° to 150° F) above the transformation range and cooled in air. The resulting structure will differ greatly for steels of different chemical composition, since no attempt is made to regulate the temperature at which transformation on cooling occurs. It is used for the purpose of producing a more uniform structure and removing the irregularities caused by high or low rolling or forging temperatures for large sections. It also is applicable to a wide extent in treating low-carbon steels of all sections to produce a more uniform structure, and is preferred to full annealing for these steels as the more rapid cooling minimizes banding (marked segregation of carbide and ferrite) which is very undesirable for machining.

Quenching and Tempering—The heat treating processes described so far have all been for the purpose of "softening" the steel by regulating the rate of cooling so that transformation occurs at relatively high temperatures. Such processes are necessary to render the steel suitable for further operations such as machining and cold cutting. However, this condition does not represent the optimum mechanical properties which can be developed in the steel, and a further treatment termed quenching and tempering is employed to develop these properties. When the amount of machining or cutting to be done is not great, steel sometimes is given this treatment in the bar form in preference to an annealing treatment.

The combination of quenching and tempering consists of first heating the steel above the critical range, and then cooling it rapidly by immersing it in a liquid cooling medium such as oil or water. If the rate of

HEAT, QUENCH AND TEMPER

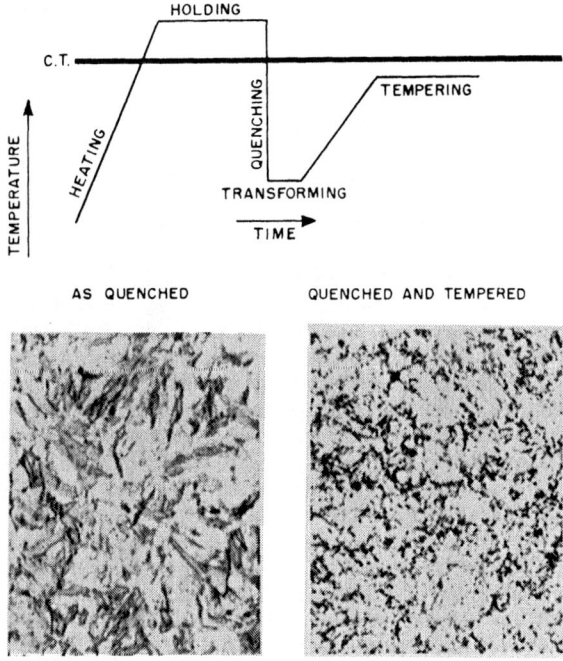

Fig. 30—30. Thermal cycle and microstructures obtained in (a) quenching and (b) quenching and tempering.

cooling is sufficiently rapid, transformation does not occur until the lower temperature ranges are reached with the resultant formation of martensite or martensite and bainite. These structures are much harder than the structures obtained by transformation at higher temperatures. Martensite, however, is quite brittle and would be unsuitable for most applications where the steel may be subjected to shock. Therefore, the steel is given a tempering treatment which consists of heating it to an intermediate temperature, very seldom higher than 650° C (1200° F) and usually somewhat lower. This treatment reduces the hardness by coagulation of the carbide and increases the toughness or shock resistance of the steel. A schematic illustration of quenching and tempering with the resultant microstructures after each operation is shown in Figure 30—30. It is possible to obtain any desired hardness within a wide range, with corresponding variations in strength and ductility, by selecting the proper tempering temperature. This can be seen readily by examining the chart in Figure 30—31.

Tempering is sometimes called **drawing.** Tempering is the preferred nomenclature. In ancient times and in the early days of relatively modern steelmaking practices, steel for swords, tools, etc. was said to be "tempered" after it only had been hardened. Since the hardened steel was too brittle, some of its "temper" was "drawn" by suitable heat treatment, giving rise to the term "drawing." At present, temper is considered as describing the final condition of hardened and tem-

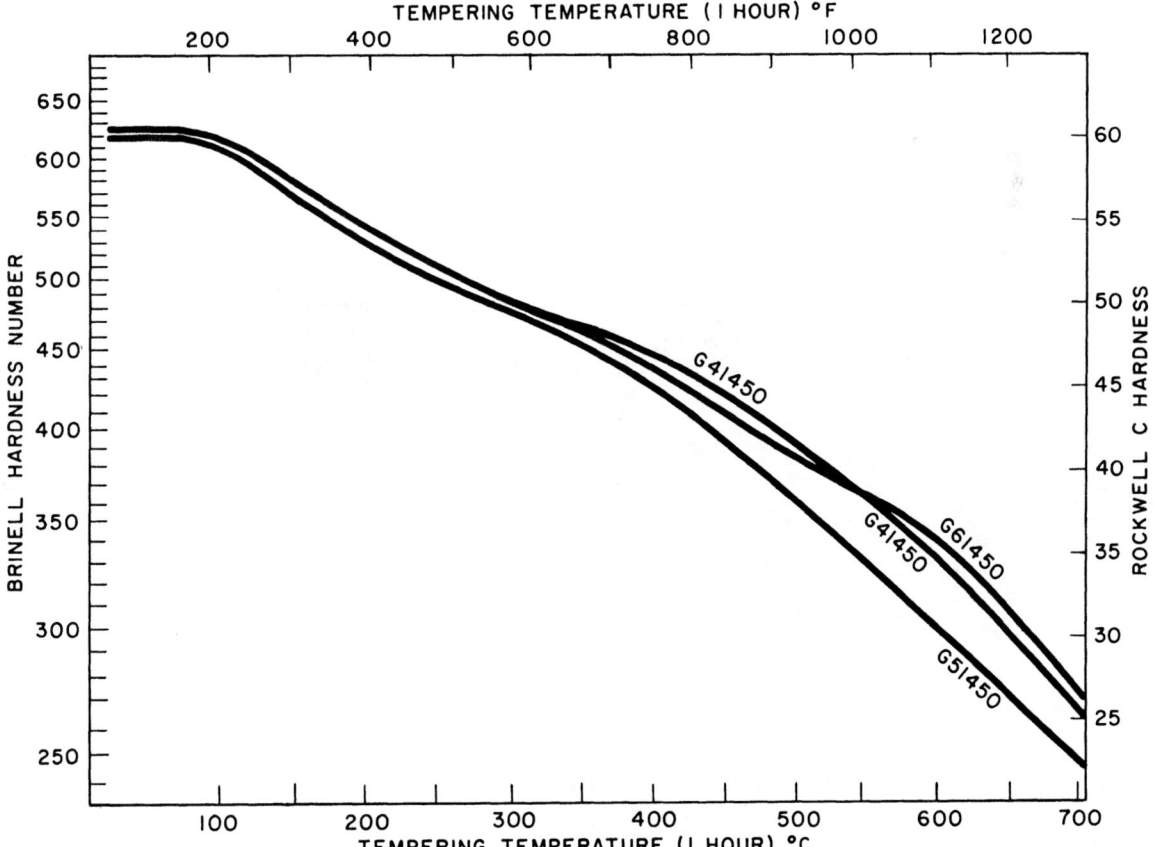

pered steel, and it is more logical to give the operation that regulates the final condition the name "tempering."

HEAT-TREATING PLANTS

Modern heat-treating plants vary widely with respect to type of furnaces, auxiliary equipment and general arrangement. No matter how widely they may differ in these general aspects, all must have adequate and reliable means for controlling and checking temperatures and have handling facilities suitable for transferring the heated steel from the furnace to the quenching tanks with a minimum of delay. These are factors which must be given careful consideration when planning new units or additions to existing units, and will be governed to a great extent by the type of product to be processed and the heat treatment to be given.

Another important factor to be considered is the provision of adequate space for accumulating and building up charges so that they may be ready to load into the furnace as soon as it is available. This lessens time lost in charging the furnaces and increases the efficiency of these units.

Furnaces—In general, furnaces which are suitable for annealing are not the most desirable for quenching and tempering, nor are the most desirable furnaces for quenching and tempering entirely suitable for annealing. This can be shown by analyzing the basic requirements for furnaces considered most suitable for these operations.

A furnace to be suitable for quenching and tempering must be capable of maintaining uniform temperatures over a rather wide range throughout the entire charge. It can be heated by either gas or electricity, and possibly oil, although the latter fuel is not too desirable at temperatures below 650° C (1200° F).

Since uniform temperatures must be maintained in furnaces to be used for quenching or tempering, and rapid cooling of the charge in the furnace is not required, such furnaces usually are insulated heavily to minimize heat losses due to radiation. Since only single-layer loads are charged for quenching, the height of the heating chambers need not be very great. This enables a more uniform temperature to be maintained throughout the charge since the differential in temperature between the top and bottom of the furnace is much less than would be the case in a larger heating chamber.

Uniformity of temperature is also very desirable in furnaces used for annealing, but the requirements are not necessarily as exacting as for furnaces used for quenching and tempering. However, it is essential that a rather wide range of cooling rates be attained. Therefore, furnaces that cool too slowly, such as the heavily insulated type referred to in the discussion on furnaces for quenching and tempering, are not suitable for very many annealing cycles. A furnace which will lose heat at a fairly rapid rate can be adapted to a wider range of cycles because various rates of cooling can be attained by control of the heat input. Thus, if the heat input is cut off entirely, the furnace and charge will cool rapidly; by firing at reduced rates, the heat losses from the furnace and charge can be balanced to any desired degree to control the net heat loss and, consequently, the over-all rate of cooling. Possible methods for increasing the rate of cooling in a heavily-insulated furnace would be to blow air into the furnace or to open it. This is not very desirable, however, because it is extremely difficult to regulate the temperature drop uniformly throughout the charge and erratic, non-

Fig. 30—32. Partially-loaded car-bottom of a heat-treating furnace, showing use of spacer bars between layers of the load, which is supported above the refractory hearth on heat-resisting alloy castings.

Fig. 30—33. Car-bottom batch-type heat-treating furnace with bell-type cover in operating position.

uniform results may be obtained. Furnaces have been developed which are equipped with large fans that circulate the furnace gases over water-cooled pipes, thus reducing the temperature to around 425° to 480°C (800° to 900°F). The partially-cooled gases are then forced into the furnace chamber at the bottom and up through the charge, causing a rapid drop in temperature. Such equipment has many potential advantages but care must be exercised in its use to avoid nonuniform cooling.

Since annealing cycles are usually rather long and tie up a furnace for some time, it is desirable that they be large enough to accommodate the maximum size load that can be cooled uniformly. For this reason, furnaces designed for annealing have a great deal more height in the heating chamber than those designed exclusively for quenching. Consequently, there is more danger of greater differential in temperature between the top and bottom of the chamber. This affects selection of the type of furnace to be used.

Control of Temperature—As stated previously, adequate and reliable control and checking of temperature is absolutely essential for all types of heat treating. Furnaces equipped with automatic temperature-control devices are generally more reliable and can be operated more economically than manually controlled units. Thermocouples to indicate and control temperature generally are located near the top of the heating

Fig. 30—34. The bell-type cover (upper left corner of illustration) of the heat-treating furnace of Figure 30—33 has been raised and hot load is being moved on furnace car bottom toward crane carrier arms.

Fig. 30—35. Hot steel bars, supported on crane carrier arms above the quenching tank in the foreground (see also Figure 30—36.)

chamber, as this location will be the first to reach the desired temperature on heating; placing the controlling thermocouples in this position prevents overheating. In the case of furnaces designed primarily for quenching and tempering, these couples will be close to the charge (which consists of only one layer of bars or other sections) so that additional couples should not be needed to assure uniform heating. However, in the case of furnaces designed primarily for annealing, in which the chamber height is much greater and in

which there is a correspondingly greater possibility of temperature differentials, provision should be made to insert couples in the charge also, being sure that some are located so that they indicate the temperature of the bottom layer. With couples so located, the temperature throughout the charge will be known and sufficient time can be allowed to attain uniformity.

Methods of Loading—As stated previously, loads to be quenched should consist of only one layer; if there is more than one layer the load would not be uniformly

Fig. 30—36. Looking down on a load of hot steel bars supported on carrier arms above quenching tank.

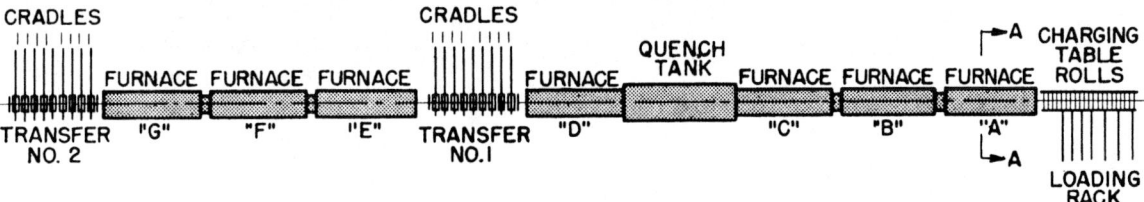

Fig. 30—37. Layout of the various units comprising a typical continuous heat-treating line for bar products. Work travels from right to left. See also Figures 30—38, 30—39 and 30—40.

quenched because the efficiency of this operation is dependent on the entire surface of the bar being in contact with the quenching medium. The same principles apply to normalizing, since uniform properties are dependent on attaining a uniform rate of cooling along the entire length of the bar, which can only be obtained by having the entire surface exposed to the air and not in contact with other hot bars.

Loads for annealing are generally more than one layer high but care should be taken to avoid having a load that is too compact, as this would prevent uniform cooling throughout the entire charge. A common practice is to separate the load into layers by the use of spacer bars, as shown in Figure 30—32. The load also may be separated lengthwise down the middle, leaving an open space approximately 150 millimetres (6 inches) wide. These methods of loading provide for better circulation of gases throughout the charge, thus enabling the operator to cool the entire charge at a reasonably uniform rate.

Auxiliary Equipment—Effective quenching is dependent on immersion of the steel in the quenching medium while the temperature is still above the critical range. If there is too great a delay in moving the load from the furnace to the quench tank, it may air cool below the critical range, especially at the ends of the bars. Thus, there is the danger of obtaining some products of transformation in the higher temperature ranges. To offset this hazard the quench tanks should be located close to the furnace and cranes or lifting devices should be provided which are capable of moving the entire load to the tanks at a rapid rate. By moving the entire load at one time, all bars will receive the same treatment. If only one bar or any portion of the load is quenched at a time, there is always danger that the last bars quenched will have cooled below the critical temperature before reaching the quenching tank.

Figure 30—33 shows a car-bottom batch-type furnace equipped to handle loads rapidly from the quench

Fig. 30—38. Hot steel bars in position on elevator roll-table section above quenching tank, ready to be lowered into the quenchant. This is furnace "C" in Figure 30—37.

Fig. 30—39. Perspective view of most of the actual continuous heat-treating line shown diagrammatically in Figure 30—37.

tank to the furnace. This is a rectangular bell-type furnace with a car bottom. To discharge the load, the furnace bell is raised (there are no doors on this furnace) and the car is moved out from under it, alongside the quenching tank which is in front and to one side (see Figure 30—34). The entire load then is picked up by a crane, equipped with carrier arms or bales as shown in Figures 30—35 and 30—36, which moves it quickly into the quenching medium. Agitation is obtained by alternate raising and lowering of the load in the tank through a travel distance of approximately 1.2 metres (4 feet). This particular unit is equipped with automatic controls so that the entire sequence of movements of the equipment is initiated by closing a single control switch. This eliminates delays which might be caused by improper timing in closing successive switches by the operator. The time required from the instant the bell of the furnace is lifted until the load is immersed in the quenching medium is one and one-half minutes.

A roller-hearth furnace with similar handling equipment is shown in Figures 30—37, 30—38, 30—39 and 30—40. In this furnace, the bars in single layers are placed on the charging table indicated in Figures 30—37 and 30—39. From here they progress through the three separate chambers of the furnace, where they are preheated to about 650°C (1200°F) in the first chamber, then progressively heated above the critical temperature range and soaked in the other two. Progress is not continuous through these chambers, the bars being held for a sufficient period of time in each chamber to be heated throughout to the prescribed temperature. The time required in each chamber, of course, is determined by the size and number of bars being treated. After the bars have been soaked properly in the last chamber, they are run out on the delivery table, which is suspended over the quenching tank (Figure 30—38), and immediately are immersed in the quenching medium. When bar temperature has fallen sufficiently, they are raised from the tank and run into

the succeeding furnace for stress relieving, after which they may be removed from the transfer table to continue through the remaining three furnaces for tempering.

Facilities should be available for quenching in either oil or water. The choice of quenching medium is to be determined by the grade of steel and the section, as explained in greater detail in Chapter 41. Facilities also should be provided for cooling the oil between the quench loads to maintain a uniform temperature of the quenching medium, because variations of an appreciable magnitude in this temperature will have a marked effect on results. Storage facilities for the oil also should be provided since a reserve supply is necessary to replenish that lost through dragout. One quenching tank can be used for either oil or water, the oil being pumped back to the storage tank when it is necessary to use water in the quenching tank. However, this is not the most desirable arrangement since there is always danger of contaminating the oil with water. It is more desirable to have separate tanks for oil and water when the size of the unit warrants such an installation.

SECTION A—A

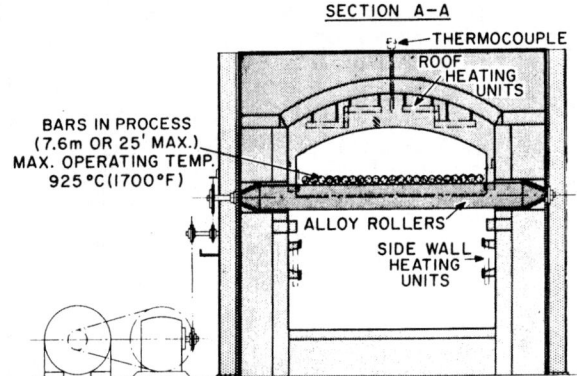

Fig. 30—40. Schematic cross-section of one of the high-temperature furnaces in the line shown in Figure 30—37.

CHAPTER 31

The Manufacture of Steel Wire and Steel Wire Products

HISTORY AND IMPORTANCE OF STEEL WIRE

Historical—In general, wire is a term that may be applied to any metallic shred, thread, or filament, or to any exceedingly slender rod or bar of metal having a uniform cross section. Considering the term in this sense, wire is of very ancient origin. Gold wire is mentioned in the Bible in connection with the sacerdotal robes of Aaron, and was used to form part of a necklace, found at Denderah, which bears the name of the Pharaoh who reigned in Egypt about 2750 B.C. Evidences of the use of wire by the Assyrians and Babylonians as far back as 1700 B.C. have also been found.

As to the methods of manufacture used by these ancient peoples, practically nothing is known. The specimens of ancient wire that have been found so far are flat, and probably were produced solely by hammering.

How, when or of what material round wire was first made is not known. No doubt it was first produced by hammering, but the difficulty of forming a fine round wire by this method probably caused the metalworkers, early in the history of the art, to seek some better method. An old Latin manuscript written by Theophilus sometime during the 8th and 9th centuries describes a method of making wire of an alloy of lead and tin, which was cast into an ingot, hammered into a long slender bar and drawn through holes in a wire-drawing plate.

The draw plate, itself, is described as "a piece of iron three or four fingers wide, smaller at the top and bottom, rather thin and pierced with three or four rows of holes through which wire may be drawn." However, there is reason to believe that the art of wire drawing was known and practiced before this time.

The chain armor used by the knights in the great crusades against the Saracens and Turks is thought by many to have been made of drawn wire. Other authentic records are available to show that wire drawing on a commercial scale was practiced in France in 1270, in Germany in 1350 and in England in 1465. The first

wire-drawing mill in this country was built in 1775 by Nathaniel Miles at Norwich, Conn. But, though three other mills, two of which were in Pennsylvania, were started during the next twenty years, none of these appear to have been very prosperous, for in 1820 practically no wire was being made in this country. In 1831, however, the industry was reestablished in this country by Ichabod Washburn of Worcester, Mass., who, with Benjamin Goddard, founded the firm of Washburn, Moen and Company. From that date, 1831, the American wire industry grew very rapidly.

Principal Uses of Steel Wire—The adaptability of wire to the manufacture of household articles of common use has expanded until it has become known as the steel product of 150 000 uses. These articles, including such items as pins, needles, brooms, strainers, hooks, egg beaters, toasters, etc., are almost indispensable to our comfort. Of the many inventions developed since 1820 that make use of large quantities of wire, only a few of the most important can be mentioned here. The first of these was wire rope. Rope made of bronze or brass wire is known to be of ancient origin, while iron wire rope was used in Europe as early as 1820. The first wire rope manufactured in this country was made in 1840 at Saxonburg, Pa., by John A. Roebling, founder of the firm of John A. Roebling's Sons Company. In 1844 the first telegraph line, extending from Washington to Baltimore, was completed by Professor Morse, but for more than ten years the growth of the telegraph was hampered because the wire essential in the construction of the lines could be produced only in short lengths, which required much time and labor for splicing.

The use of crinoline wire, a high-carbon heat-treated wire, for hoop skirts has been cited by some as a novel use of the product and as a great boon to the industry, but is better as an example of the unstable character of a business that depends on fashions. Starting about 1860, this business, for the next ten years, consumed about 1350 metric tons (1500 net tons) of wire annually, but suddenly collapsed with the passing of the hoop skirt in 1870. This was not to be the last time that feminine fashions would play a part in the wire industry, for in 1916 some 13 600 metric tons (15 000 net tons) of wire were being consumed annually in the manufacture of corsets and several hundred tons for hair pins, but these have largely disappeared as did the crinoline wire.

The collapse of the crinoline-wire business was replaced by more staple articles, for in 1868 the first patents were obtained on the use of wire for bale ties and barbed wire for fence. Both these articles came quickly into wide use, the latter article being especially useful in fencing the great unwooded lands west of the Mississippi River, which were rapidly being brought under cultivation about this time. By 1876 the development of the telephone had reached the practical stage, and began to create a demand for more wire of the same type used by the telegraph. It was in 1879 that the beneficial effect of sulphur on machinability was discovered, and the manufacture of high-sulphur screw stock was begun. Chas. H. Morgan also began the manufacture of coiled-wire springs in this year. In 1884 woven-wire fence was first made by machinery. This fence, of which there are now several types, fulfilled all needs that could not be met with barbed-wire or straight-wire fence and possessed all the advantages of any fence that could be constructed of wood, hence it was soon being used in all parts of the country. In 1851 the first American-made wire-nail machine was built by Thomas Morton; in 1875 the first steel-wire nails were made; and in 1888 the production of the wire nails exceeded that of the cut nail for the first time. Today, a great many more wire nails than cut nails are made, and about 10 per cent of all the steel wire produced is made into nails.

Early Method of Manufacture—The methods employed at the time of the introduction of wire drawing into this country were wholly inadequate to meet the demands that were soon to arise. As the Bessemer and open-hearth processes for making steel were then unknown, practically all wire was made from wrought iron, or the softer metals like copper, brass, bronze, etc., though crucible steel in softer tempers was also used. Practices varied somewhat in different localities, and detailed descriptions of the operations are lacking but, from information gleaned from different sources, the process and operations for drawing wrought-iron and steel wire appear to have been about as follows: Starting with a small square bar or billet, the metal was heated and hammered into as small a round as practicable, say 6.4 to 9.5 mm (¼ to ⅜ in.) in diameter and some 1.8 to 3.0 metres (6 to 10 ft.) in length. One end of this rod was tapered to a size a little less than the hole in the die through which the rod was to be drawn. The rod was then scoured bright with a piece of sandstone or other abrasive. In the case of iron and steel, it was early discovered that this cleaning of the rod was not only essential to good work but was well worth the trouble, for the hard scale, when not removed, soon scored the die. Then, the tapered end of the rod was inserted in the proper hole of a draw plate made of cast iron or hard steel, where it was grasped with a pair of

tongs attached to a lever, and the rod was drawn through the hole a few inches at a time, about 125 mm (5 in.) at each stroke being the common practice. The lever could be operated either by hand or by water power, and the rod, as it passed through the hole, was generally lubricated with butter or other grease, to facilitate the drawing and save the die as much as possible. After the first draft, the operation was repeated through progressively smaller holes in the draw plate until the wire was drawn to the size desired or until it became too hard and brittle for further drawing. If the wire was to be drawn to still finer sizes, it was annealed at a red heat, generally in an oven sealed off from the air, after which the thin coat of oxide thus formed was removed by tumbling with a mixture of sand and water, or by immersing the wire for some time in sour beer or fermented barley water. The machine for drawing this fine wire differed somewhat from that used for drawing the rod. The draw plate was mounted on a table or bench in front of a drum or block that could be revolved by a hand-operated crank and gear drive.

Improvements in the manufacture of iron and steel wire have kept pace with the demands. In connection with this statement, it should be noted that the basic principles of wire making have remained unchanged. As already indicated, the process consists essentially in drawing a rod or a slender piece of the metal through a tapered hole in some harder materials, or successively through a series of such holes. It involves the application of force in the simplest manner, a straight pull, and the use of one of the simplest tools, the wire-drawing die. The efficiency of the latter and the simplicity of both these elements have left little room for improvement, and many doubt the possibility of finding a substitute for the process that will do the work so well. The improvements that have been made pertain to the accessory equipment and to the complementary operations, rather than to the process of wire drawing itself. Thus, the need for long lengths was met not by any radical change in the wire-drawing process, but by changing the method of producing the wire rod. As long as the rod was made by hammering, it was too short to coil, or draw with a block; but when the need for long lengths began to be felt, efforts were made to roll the rod. About 1840, Washburn succeeded in adapting the wire-drawing block to coarse-wire drawing. This block was power driven, and provided a means not only of applying a greater pulling force continuously to the rod, but also of storing the wire as it was drawn through the die. This development made it desirable to use very long rods, and this demand was met first by the invention of the three-high mill about 1857, which was succeeded by the Morgan continuous mill about 1875 and the Garrett rod mill in 1882.

Another important development was the substitution of soft steel for wrought iron, made possible by the invention of the Bessemer and the open-hearth processes for steelmaking. This change was inaugurated in this country by the Washburn and Moen Manufacturing Company at Worcester, Mass., in 1871. Important improvements in methods for handling, cleaning, coating, or lubricating the rod, and in heat treating and finishing the wire, were also made.

ROLLING THE WIRE ROD

The Wire Rod—The smallest round sections of steel that can be produced by hot rolling are known as **wire rod.** Smaller diameters are achieved by cold working the hot-rolled rod by drawing it through a die to produce wire. Almost all wire rod is cold drawn into wire to improve the dimensions, surface smoothness or mechanical properties required for the finished products. A 5.5-mm (7/32-in.) diameter rod is the smallest size that is practicable to produce on a rolling mill and it is the most commonly produced size. Modern rod mills employing controlled cooling generally roll diameters up to about 12.7 mm (½-in.). Larger diameter wires are drawn either from rods rolled on special coarse rod mills or from hot-rolled bars.

Rod is wound into coils about 760 mm (30 in.) inside diameter and weigh from 450 to 2000 kg (1000 to 4400 lb). Each coil represents the rod made from a single billet. While the wire rod represents a finished product of the rolling mills, it constitutes the raw material used in the wire mill and should be considered as the first step in the making of wire. Rod is produced in a variety of chemical-composition grades and quality levels suitable for specific requirements of the finished wire. For each rod order, billets must be selected for proper chemical composition, surface conditioning and center soundness compatible with the intended wire application. The rod rolling, cooling and inspection practices are tailored to these requirements.

Development of Rod Mills—Two types of rod mills, employing fundamentally different principles of roll arrangements, began to develop separately in the late 1800's. The original **Belgian** or **looping mill** evolved into the **Garrett mill.** During the evolution and perfection of looping mills, the continuous mill was introduced. Both looping and continuous mills continued to make improvements and exploit their separate advantages. Hybrid mills appeared which combined features of both types of mills. Present rod mills are not simply direct descendants of the early continuous mills, but retain elements of both mill designs including the use of loops to eliminate tension between rolling trains.

The single-stand three-high mill was introduced for rod rolling of wrought iron about 1857. The first multiple-stand rolling was accomplished in single-train Belgian, or looping mills. The rolling stands were arranged side-by-side and all rolls were driven at the same speed by a common shaft parallel to the rolls. The rod made a loop of 180 degrees to enter stands in succession and formed an "S" pattern as the loops reversed through the stands. As a rod was being rolled, each loop continued to grow longer. This was the physical result of rod passing through successive stands with decreasing cross-sectional areas and traveling at the same speed. Rod accumulated ahead of each stand in ever increasing lengths of loops. The tail end of the rod would become too cold to finish if the length exceeded certain limits. Rod produced in single train Belgian mills was limited to lengths of about 90 m (300 ft) or 23 (50 lb) maximum.

The Continuous Rod Mill—An entirely new type of mill was patented in 1862 by George Bedson of Manchester, England. It was introduced in this country in 1869 when the Washburn and Moen Manufacturing Company erected a **Bedson mill** at their Grove Street Works, Worcester, Mass. The mill consisted of rolling stands arranged in tandem so that the bar passed through the mill in a continuous, straight line. Stands were alternately arranged with horizontal and vertical pairs of rolls which made a draft at right angles to that in each successive stand. This arrangement of rolls overcame the necessity of giving the piece a quarter turn between passes, as in the looping mill. Then by placing the stands close together, Bedson was able to drive the mill through two long shafts and a system of gears, whereby the rolls of each stand revolved faster than the preceding stand. This system of driving was to regulate the speed of each stand so that the peripheral speed of the rolls would be nearly the same as the linear speed of the bar, which increases after each pass due to the elongation. Looping of the rod between passes was therefore avoided. This mill met all expectations as to speed and length of rods rolled, but the rods could not be coiled and taken out of the way as rapidly as the mill could produce them. C. H. Morgan, then the general manager for Washburn and Moen, overcame this restriction of output by devising a power traction reel for coiling the rods.

While the Bedson mill represented a great advance in rod rolling, it possessed certain mechanical features which were objectionable. Its chief faults were found to be due to the vertical rolls. They were not easily kept in adjustment, and the shaft and gears for driving them were all beneath the mill floor. The scale and water from the rolls fell upon these gears and bearings and caused excessive wear, which, combined with the difficulty of getting to them under the floor, made them a source of much trouble. After a few years' experience with the Bedson mill, therefore, Morgan and his associates developed the twisting guide, by means of which the vertical rolls could be eliminated and horizontal rolls substituted for them. This twisting guide was a closed delivering guide in which the grooves were cut in a spiral, so that with this guide properly mounted after any pass the piece was forced through it and twisted a quarter turn before it entered the next pass. Besides overcoming most of the disadvantages of the Bedson mill, this Morgan plan presented the additional advantage of more than one rolling line, that is, of rolling two or more rods side by side at the same time and on the one set of rolls. This plan of increasing the capacity of the mill, however, increases the difficulty of keeping such mills in adjustment, and was not taken advantage of until some years after the erection of Morgan's first mill. Even with a single rolling line, the development of the mill to a practical basis of operation presented some difficult problems. It is evident that such a mill requires very fine adjustment of the draft, roll diameter, and speed of rotation, for these factors

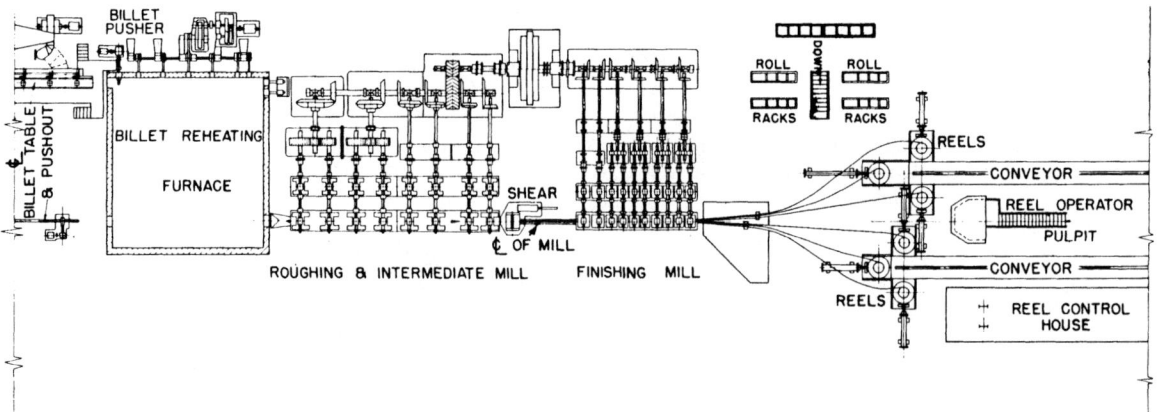

FIG. 31—1. Plan and layout of continuous rod mill for 2-inch billets, including reels and part of the rod bundle conveyor.

are brought into close relation. The relations of all these factors can be determined by proper mathematical calculations, but as the draft fixed by the grooves in the rolls is affected by the adjustment of the rolls, the temperature of the bar, composition of the steel, and changes with the wearing of the grooves and bearings, much difficulty was at first experienced in getting the mill adjusted so that the piece would not loop or jam between the different stands of rolls. This was particularly true in the case of the last few roll stands of the high-speed finishing mill. This difficulty was finally solved by increasing the speed of the rolls enough to keep the rod under slight tension at all times, and by making the bottom as well as the top roll adjustable.

Layouts for Continuous Mills—Until recently, continuous rod mills were designed to roll billets from 45 to 100 mm (1¾ to 4 in.) square and about 9 m (30 ft) in length. To work these sections into a 5.5 mm (⁷⁄₃₂-in.) rod without unduly increasing the danger of developing rolling imperfections requires from 16 to 23 passes. For straight continuous rolling, the stands in these mills were divided into roughing, intermediate and finishing trains (Figure 31—1). The roughing set was placed close to the discharge door of the heating furnace, which was usually of the side-discharge type. This arrangement makes it possible to push one end of the billet into the roughing set while most of the billet remains inside the furnace. Immediately following the last intermediate stand is a flying shear which crops the distorted cold front end before the piece enters the finishing mill. To provide space for shearing and allow a little slack between the two groups of rolls, the finishing train is placed 6 to 9 m (20 to 30 ft) beyond the intermediate. Rod leaving the finishing stand travels through pipe guides to the reeling machines. There are usually two reels for each rolling line or strand. The figure illustrates a two-line, 18-stand continuous mill capable of rolling small billets to 5.5 mm (⁷⁄₃₂-in.) rod at perhaps 20 metres per second (4000 feet per minute). The compact arrangement promotes uniformity of rolling temperature and lends itself to driving the mill with a single motor or steam engine or with two motors, one for each group of rolls.

The Garrett Mill—After the first Morgan mill was built about 1875, the Belgian looping mills were disad-

vantaged as to speed, length of rod, and tonnage produced. The Belgian mills of that era charged as 23 kg (50 lb) maximum billet weight, about 50 mm (2 in.) square. These billets had to be rolled from 100 mm (4 in.) blooms on a three-high billet mill. To counter the competitive advantages of the continuous rod mills, William Garrett, who was plant superintendent for the Cleveland Rolling Mill Company, conceived a plan to modify the looping mill to roll longer lengths of rod direct from 100 mm (4 in.) billets without reheating. Since a large section loses heat less rapidly than a small one, he reasoned that the 100 mm (4 in.) billet could be rolled in the roughing mill to a size suitable for the rod mill and fast enough to retain heat for finishing. It could not be exposed too long in the looping mill or the tail end would become cold. So he combined the three-high billet mill with the rod mill, and divided the rod mill into three groups or trains of rolls. These were arranged in echelons and driven at progressively increasing rates so that large sections were roughed slowly and the finishing train operated at the highest speed. Not only could a given length of rod be rolled in less time, but the size of the loops could be controlled and their length reduced. Garrett erected his first mill in 1882 and was able to more than double the output of the older looping mills.

Garrett was able to add some improvements to his mill including the use of **repeaters.** These are semicircular troughs which guide the front end of the piece from one pass into the next. Repeaters are open at the top and once the loop has formed, it is free to rise out of the trough and enlarge to correspond to the elongation, or otherwise adjust to any difference in speeds of the two passes. Repeaters eliminated the need for catchers to feed the piece into the next pass. In the case of the earlier looping rod mills, however, repeaters were used only on the side of the mill producing square sections. Twist guides would stand an oval on edge in the continuous mills but it took many years to perfect oval repeaters for looping mills which did not result in excessive cobbles or scrap. Another improvement introduced by Garrett, who was the first to employ the scheme, was the practice of finishing more than one rod at a time. The roughing train was able to break down the short billets much faster than the finishing

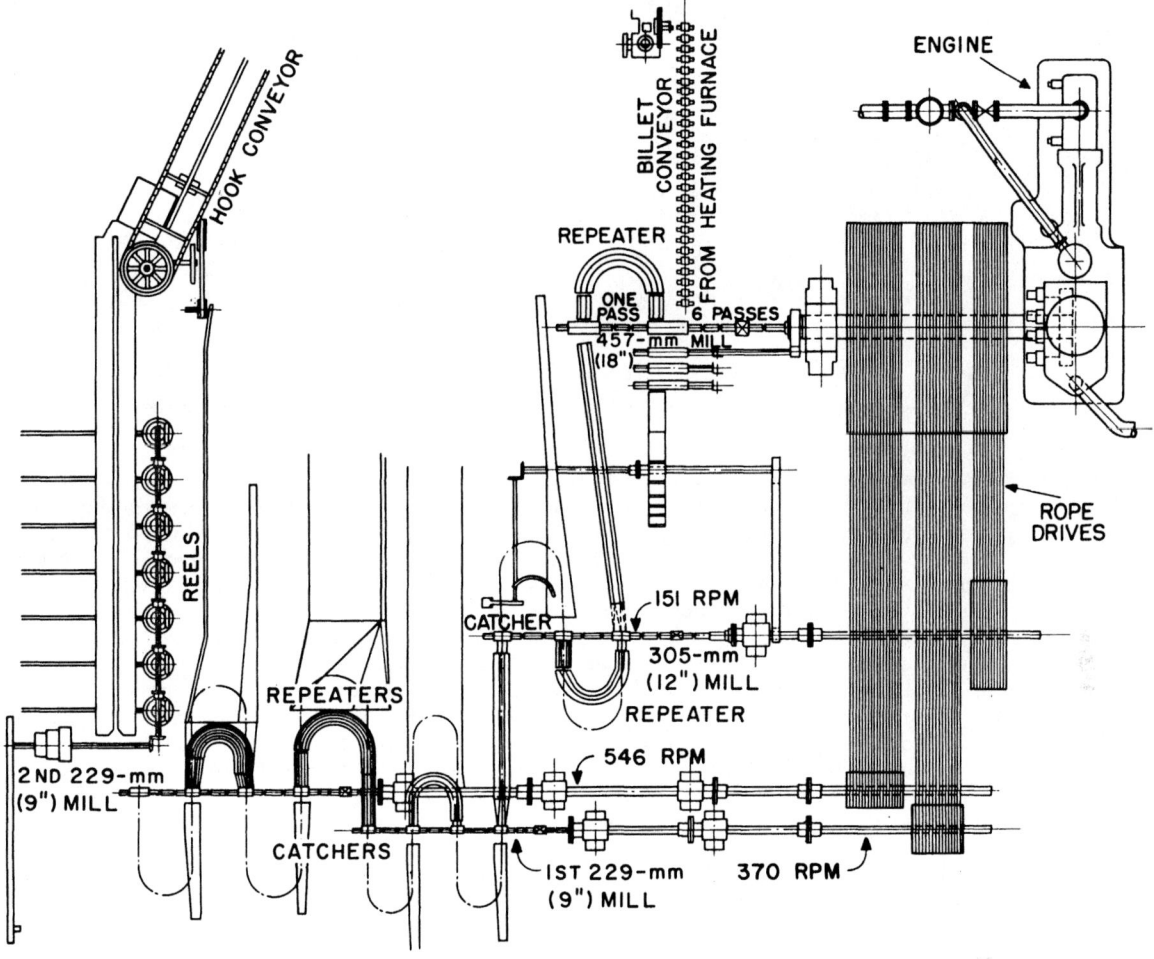

FIG. 31—2. General layout of Garrett rod mill.

train could roll the rod in a single strand, even when the latter was run at the maximum speed of 550 rpm. By adopting power reels to collect and coil the rods, and by cutting an additional pass in each finishing mill roll, Garrett was able to finish two strands simultaneously and double the output of the mill. Later Garrett mills rolled four, and even six strands through the finishing train.

Layouts of Garrett Mills—The arrangement of the roll trains in later looping mills was subjected to considerable variation. A typical layout is shown in Figure 31—2. As originally planned by Garrett, the three-high billet mill was placed near the heating furnaces. The roughing mill was placed at a great enough distance from the billet mill to permit clearance from the last pass on the three-high mill. Then came the intermediate and finishing trains with their looping floors sloping in opposite directions from the rolls to facilitate the enlargement of the loops. These trains usually consisted of four or five stands each, lying end to end along two parallel lines closely together. This layout illustrates 18 passes through 13 stands and requires five catchers.

Continuous and Looping Mills Compared—Virtually all rod mills erected recently are basically continuous

mills but they often combine features of the looping mills, such as repeaters, and include some of the innovations devised for the Garrett mills. The major advantage of the looping mills was the simple power drive and gearing system. The loops eliminated tension in the piece between stands and permitted flexibility in adjusting the relative speeds of the rolling trains. Although looping mills can accommodate large 100 mm (4 in.) billets, they are limited to a billet weight of about 80 kg (175 lb) for small rods because the tail end of the rod can cool to 800°C (1480°F) or lower. Power requirements become excessive, roll breakage is possible, and the variation in size can be objectionable when looping mill rod is rolled in long lengths. Because the continuous mill cools the rod less during rolling and uses less hand labor for catching, it permits the use of a greater number of roll stands and passes. The additional passes allow smaller reductions with better shaped passes and lessen the possibility of overfill. Garrett mills are generally restricted in the present market to specialty mills which emphasize the rolling of tool steels and exotic alloys. Looping mills have the flexibility to roll a broad range of grades with different reheating-temperature requirements and rolling characteristics. Small produc-

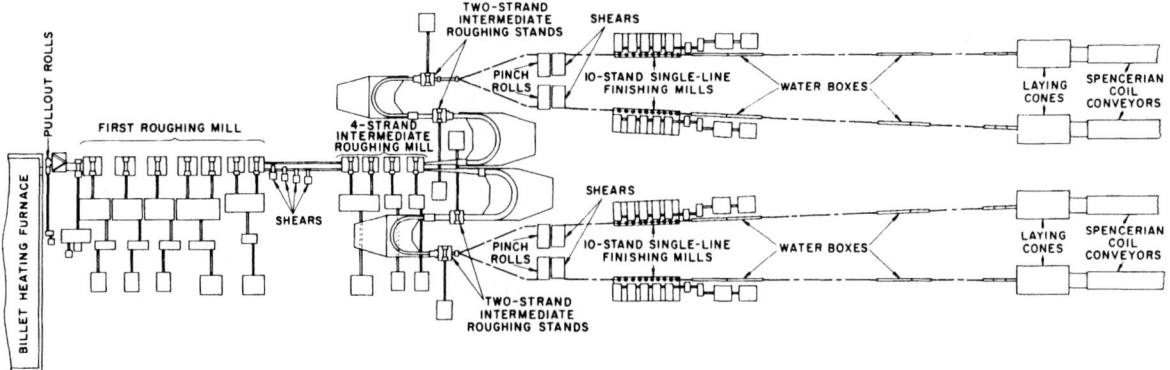

FIG. 31—3. General arrangement of a high-speed rod mill at United States Steel Corporation.

tion runs present few problems and size control is adequate. Small coil weights and relatively high rod conversion costs are not necessarily prohibitive on these items.

Modern Rod Mills—After many years of gradual refinement of rod-mill designs, two revolutionary improvements were introduced simultaneously and have become standard equipment in ferrous rod mills in the free world since 1964: the twist-free finishing mill and controlled cooling. Numerous benefits justify capital expenditure for this equipment but the demand for increased production rates and heavier rod-coil weights accelerated their acceptance by the industry.

The twist-free mill is a single-line, continuous ten-stand finishing train which uses small tungsten-carbide rolls set in pairs with an angle of 45 degrees to horizontal. Each pair is set 90 degrees to the previous stand. High rolling speeds are achieved by rotating the axis of the rolls instead of physically twisting the rod after the oval passes. The conventional mill of the former era that used static twist guides induced several types of imperfections, especially at high speeds. Wide-faced rolls cut with multiple grooves were used to finish multiple strands on the same finishing mill. Accelerating these massive steel rolls to high speed contributed to bursting from centrifugal force. Size adjustments could not be made to each strand individually because they all finished on the same rolls. Changes in one line to compensate for roll wear or temperature variations invariably affected other lines and close control of tolerance was difficult.

Another dissatisfaction with the conventional rod mills of the former era was that coil weights of small diameter rods ranged from about 100 to 450 kg (200 to 1000 lb). Such small coils posed problems in handling and shipping and required numerous welds. Billet-reheating furnaces could be lengthened to accommodate larger billets but massive coils collected in reeling chambers would retain heat for a long time. The heat and weight could bend a "C" hook of a hook-type conveyor and tall coils would tend to spill if transported on a drag conveyor. Large coils of high-carbon grades would cool slowly and transform into a metallurgically undesirable structure of predominantly coarse pearlite. Improvements incorporated into the design of new-generation rod mills enhance rolling speed, coil weight

and microstructure. This design will be illustrated by tracing the flow of material through one of these mills.

The rod mill illustrated in Figure 31—3 started production in February, 1969. It is typical of those mills designed by the Morgan Construction Company that incorporated the Morgan no-twist finishing mill and Stelmor cooling process in their design. The 4-line mill consists of 23 stands which convert 1400 kg (3100 lb) billets into 5.5 mm (7/32-in.) diameter rods at speeds averaging 56 m/s (11 000 fpm). Sizes up to 12.7 mm are delivered at 14 m/s (2800 fpm). As of 1980, there were 13 such rod mills in the country with a total of 31 lines. As a rule-of-thumb, each line produces roughly 16.3 metric tons (180 net tons) per turn.

The mill uses only one billet size of 100 mm square x 27 m (4 in. square x 60 ft) and about a one-month supply of billets can be stored in the yard. Billets arrive in special flat-bed cars and are stacked in bins by cranes with magnets. To fill the rolling schedule, billets of a specified heat and ingot section are selected from the bins and transferred by crane to separating skids. A series of walking beams and conveyors moves the billets into position for charging into the billet-reheating furnace. The billet-reheating furnace is a 3-zone, top-fired, cross-push furnace rated at 135 metric tons (150 net tons) per hour. A ram charger pushes two billets at a time into the side of the furnace. The cross pusher advances the billets down the sloping hearth from the preheat zone to the heating and soaking zones. When the billets reach a temperature of 1260°C (2250°F), a "peel" bar pushes the billets out of the side door. A switch plate guides the billets into one of the four passes of stand No. 1. A shear is located on each line just ahead of the first stand. In the event of a cobble, it can be used to cut the billet and return the remainder to the furnace. The shear can also be used to crop bad ends or divide the billet for rolling trial bars to facilitate size changes.

The first roughing mill consists of five roughing stands with short, rugged rolls designed to prevent roll deflection and two intermediate roughing stands. The first three roughing stands are individually driven by 373, 448 and 522 kW (500, 600 and 700 hp) motors, and the last two are driven in tandem from one 1110 kW (1500 hp) motor. The steel sections are rolled into a box, oval, square, oval and square pattern, in that order, and stationary guides twist the steel 90 degrees after each oval pass. After the bar has left the roughing stands, it enters the four-strand intermediate mill which consists of stands No. 6 through No. 11. The first two intermediate mill stands immediately following the roughing stands are driven by a common 1119 kW (1500 hp) motor. The housings of all intermediate stands are slightly smaller than the roughing stands and, should a roll break in any of the intermediates, the entire stand can be removed and immediately replaced by a spare stand. Following No. 7 stand, the steel passes a set of crop and cobble shears and is then rolled through four more 4-strand intermediate mill stands. Stands No. 8 and No. 9 are driven by one 1492 kW (2000 hp) motor and stands No. 10 and No. 11 are individually driven with 671 kW (900 hp) motors on each stand.

As the steel leaves stand No. 11, it enters the first of two 180-degree repeaters. It is at this point that the four strands, which have traveled in a straight path from stands No. 1 through No. 11, change direction and also divide into two strands which loop left to stand No. 12A and two strands which loop to the right to stand No. 12B. Upon exiting from the No. 12 stands, the pairs of strands reverse in another 180-degree repeater to restore the original direction of travel. Two strand rolling in stands Nos. 13A and B then presents a round section to the finishing mill. Continuous mills without repeaters may induce tension and distortion in the rod. The use of repeaters eliminates tension at the end of the intermediate mill. Unlike bar mills, modern continuous rod mills make use of tension between finishing mill stands to help reduce the steel section to small diameters. The tension must be controlled to avoid dimensional variations and the use of repeaters ahead of the finishing mill contributes to this control. The use of repeaters also reduces the length of the mill and lowers the cost of the building.

As the strands leave No. 13 stand, they separate into four individual lines and pass through the second set of crop and cobble shears and continue on to the uploopers for speed control. The uplooper consists of a roller mounted on a pivoted arm which is initially lowered below the mill pass line. The uplooper is raised after the head end of the piece has entered the finishing mill. This causes a deflection of the piece and a small loop is formed. A photoelectric scanner senses the height of the loop and provides a signal to the speed regulator. The speed of the finishing mill is adjusted so that the loop height remains approximately the same. The high speeds of new finishing mills make it necessary to have this automatic speed regulation to avoid tension or excess slack in the line ahead of the finishing mill. Uploopers, or sideloopers as they were arranged in subsequent mills, are not as efficient as repeaters in isolating the finishing mill from deviations of delivery speed from the intermediate mill but are used in place of repeaters in some continuous mills.

The four single-strand finishing mills have ten pairs of rolls which are close-coupled into one unit driven by two 932 kW (1250 hp) motors in tandem. The rolls are angled alternately at 45 degrees above and 45 degrees below the horizontal so that the rod does not need to twist. The first two stands are 200 mm (8 in.) in diame-

ter and the balance are nominally 150 mm (6 in.) in diameter. All finishing-mill rolls are solid tungsten carbide with one or two passes cut into each roll. The small rolls are simple to change manually, accelerate rapidly and operate at delivery speeds averaging 56 m/s (11 000 fpm) with small-diameter rod. Cobbles and dimensional tolerance are easily controlled. The hardness of the carbide composition extends roll life and minimizes roll flexing, both of which contribute to dimensional accuracy.

Upon leaving the finishing mill, the cooling rate of the rod is controlled from hot-rolling temperatures to obtain preselected metallurgical properties. The rod passes through water boxes on its way to the laying cone. Inside each water box, the rod is conducted through pipes with annular nozzles which apply water under high pressure to the rod to reduce the temperature. To prevent the formation of martensite on the rod surface, resulting from residual cooling water, high pressure air wipes are often positioned at the exit end of each water box. The temperature is reduced 110 to 220°C (200 to 400°F) in about 0.6 seconds, in the case of small rod, to establish a proper laying-cone temperature. The laying temperature is normally controlled by varying the water pressure. The mill has two water boxes on each line. The distance between the two was engineered to permit the equalization of temperature through the rod cross section before the rod enters the second box.

The rod travels horizontally to the laying cone where it is passed between a driven wheel guide and a roller chain which directs the rod vertically downward through a short, curved "laying" pipe. The bottom end of the pipe revolves rapidly and forms the rod into rings which overlap in spencerian form as they fall onto the conveyor. The conveyor carries the rings from the laying cone to the reforming tubs 47 m (153 ft) away. The rings are supported by two roller link chains, spaced 560 mm (22 in.) apart, which transport the rod over four 9 m (30 ft) plenum chambers. Each plenum chamber has a separate fan, independently controlled, which can blow air upward through the rings at selected velocity. The air is channeled through vane openings in the cast plates immediately below the pass line. A greater amount of air is directed toward the edges where the rings touch each other, to provide a uniform cooling rate from the center-line of the conveyor to the edges. Initial heat loss is principally by radiation but as the rod turns black and the pearlite transformation occurs, cooling is accomplished primarily through convection heat loss to the air blast. Body cooling mechanisms are controlled by adjusting the speed of the conveyor, and hence the spacing of the rings, and settings for the four fans on each line.

At the end of the conveyor, the rings fall into a reforming chamber and a coil forms in the eye-vertical position. The coil is pushed into a cradle mounted on a car known as a downender where it is rotated to an eye horizontal position and moved onto a "C" hook. Trimming, sampling and inspection of each coil is facilitated by a "power-and-free" hook conveyor system which automatically releases hooks from the traveling conveyor chain and allows a stationary coil inspection without stopping the line. The coils then proceed to the four compactor-banders and the unloading station. A total of 104 hooks can accumulate between the downenders and the banding station so that mechanical malfunctions at that point do not necessitate curtailing rolling operations.

Operation of the mill from the water boxes through the downenders is controlled by the Stelmor pulpit operator. Pulpit instrumentation includes gauges and recording charts for laying-cone temperature which is measured by an infra-red pyrometer with peak picking circuitry located near the laying cone. The pulpit operator adjusts water pressure in the two boxes to establish the laying temperature. Gauges also indicate pressure in the plenum chambers and the per cent openings of vanes at the fan air intakes. The operator can turn fan motors on or off and vary the air flow by adjusting the vane openings. Mill-delivery and conveyor-chain speeds are indicated in pulpit gauges and the operator will increase conveyor speed for faster cooling or for small rod diameters which are delivered from the finishing roll stands at higher speeds.

Stelmor Practices—Laying-cone temperatures are usually set between 790 and 900°C (1450 and 1650°F) for plain carbon steels in a typical Stelmor operation. This temperature is the primary determinant of grain size and scale characteristics and also influences the microstructure. Wire-mill techniques for removing scale from rod may be broadly categorized as either chemical or mechanical. A thin scale is best for acid pickling because less time and acid consumption are required. A thick scale will extend shelf life of rod for outside storage but the scale itself ultimately represents a yield loss. A low-temperature practice not only provides a weight of scale under 0.5 per cent, but it promotes the formation of wustite (FeO) which is more soluble in acid than Fe_3O_4 or Fe_2O_3. There are several varieties of mechanical descaling but those systems which involve bending the rod, with or without subsequent water jets, air blast or mechanical scouring, appear to function most effectively with a relatively thick scale. A laying temperature of 870°C (1600°F) or higher ensures a scale weight of about 0.6 per cent on small-diameter rod which is adequate for a variety of descaling devices.

A high-temperature laying cone practice is usually applied to low-carbon grades to provide a large ferritic grain size and low tensile strength. In contrast, a high temperature laying cone practice increases the tensile strength of high-carbon grades by promoting more pearlite. The cooling rate, which is governed by the amount of air blast and speed of the conveyor, is designed to be fast enough to provide a uniform structure of fine pearlite. The objective of the rolling practice for high carbon steel rod is to balance the laying temperature and cooling rate to achieve the maximum amount of fine pearlite without producing traces of undesirable low-temperature transformation products. Excessive coarse pearlite and free ferrite from slow cooling can be as serious a microstructural deficiency as the formation of martensite. Selecting the optimum cooling cycle is complicated by variations in steelmaking practices such as deoxidation methods, residual element control and steelmaking and casting method. Each rod mill must specify slightly different cooling practices to achieve a suitable structure. Rather rapid cooling of high-carbon grades will produce a structure similar to patenting and permit the direct drawing of many grades into wire without additional rod heat treatment. A moderate cooling rate is prescribed for steels containing high carbon and manganese combinations or high residuals and some alloy grades. Medium-carbon grades are low enough in hardenability that Stelmor mills usually cool them as rapidly as possible from high temperature. This tends to be true, to a lesser extent, for large diameter rods which cool slower because of the greater mass per unit of length. High-carbon rods 12.7 mm (½-in.) in diameter average 93 MPa (13.5 ksi) lower tensile strength than 5.5 mm (7/32 in.) rod. A typical tensile prediction equation for rapidly cooled, small-diameter medium-carbon and high-carbon rod is:

$$\text{Tensile Strength (MPa)} = 1027 \text{ C} + 186 \text{ Mn} + 221$$
$$\text{or}$$
$$\text{Tensile Strength (ksi)} = 149 \text{ C} + 27 \text{ Mn} + 32$$

Medium-carbon grades for cold-heading applications require a much different treatment. Cold heading requires maximum proeutectoid ferrite for upsetting rather than fine pearlite for drawability. This is achieved by slow cooling from low laying-cone temperatures. The laying-cone temperature, air blast and conveyor speed are manipulated in various combinations to achieve a wide range of microstructures and physical properties and retain reasonable uniformity even in heavy-weight coils. Between 1969 and 1974, U.S. Steel started ten Stelmor lines at its rod mills in Fairless, Pa.; South Works, Ill.; and Pittsburg, Calif. The major microstructural benefits from these first-generation Stelmor mills accrue to the users of high-carbon rod. Controlled cooling in this type of mill is confined to a range of relatively rapid cooling. Conveyor chains and structural members generally limit cooling rates to a minimum of about 5.6°C (10°F) per second, or about the cooling rate of air patented rod. Facility changes have been engineered to now permit slower cooling rates in order to improve the ductility of low- and medium-carbon steels.

Rod Slow-Cooling Facilities—Modernization of the rod mill at the Lorain-Cuyahoga Works was undertaken in 1975 to provide a versatile mill capable of slowing cooling rates down to 0.08°C (0.15°F) per second for a meaningful period of time, but retaining the uniformity and fast cooling of the regular Stelmor process. Requirements for the new facility evolved from investigations conducted at the mills and the company's Technical Center. The unique cooling conveyor on the Cuyahoga Plant's No. 1 mill is 66 m (217 ft) long. The loops are supported by 114 mm (4.5 in.) diameter stainless-steel rolls which transport the rod in the speed range of 3 to 58 m/min. (10 to 190 fpm). There are six plenum chambers along the conveyor and each is served by an independently controlled fan. The first three 9.1 m (30 ft) zones are equipped with furnace covers having gas-fired radiant tubes. The stainless-steel covers are insulated and retard cooling by minimizing radiation and convection heat loss. Heat can be supplied by the muffle tubes, or, if necessary, room-temperature air can be forced through the tubes to lower the furnace temperature. Through this com-

FIG. 31—4. Cooling conveyor of the Cuyahoga rod mill with the covers raised.

bined action, a controlled, preset temperature is maintained under the covers. The covers are hinged on one side so that they can be raised to a vertical position when not in use. Figure 31—4 shows the conveyor with the covers raised. The fourth zone has insulating covers without the radiant tubes. A transverse section through the conveyor of the rod mill is shown in Figure 31—5.

The options presented by this arrangement are numerous, but most cooling cycles can be reduced to three general categories:

1. Spiracool is a relatively rapid cooling cycle utilizing the Stelmor process using forced air and without covers or externally supplied heat. It is used at U. S. Steel rod mills at Cuyahoga and elsewhere.

2. Spiracool "Soft" is the designation of a slow cooling rate achieved by slowing the conveyor to about 4.6 m/min. (15 fpm) and packing the loops on top of one another to a height 50 to 120 mm (2 to 5 in.). It utilizes natural air cooling without forced air, covers or externally supplied heat.

3. Spiracool "Extra Soft" includes cooling practices utilizing the furnace covers and using a conveyor speed of 4.6 m/min. (15 fpm) or less. The initial 27.4-m (90-ft) portion of the conveyor is divided into three zones of equal length enclosed by heated covers. It is controlled, zone-by-zone, at prescribed, preset temperatures.

The regular Spiracool, or rapid controlled-cooling

process, is generally applied to high-carbon steel to produce a microstructure of predominantly fine pearlite suitable for direct drawing. This practice provides a rod product having scale characteristics and physical properties similar to other controlled-cooled mills. The uniformity of mechanical properties is improved by the use of austenitic conveyor rolls and the high speed of the conveyor. The rolls provide constantly changing support points for the loops; whereas the use of conveyor chains in other mills tends to disrupt the air flow and cause non-uniform cooling at the points of contact between the chain and the rod surface. Regular Spiracool practices for low-carbon rod at the Cuyahoga mill include cooling rates as low as 0.8°C (1.5°F) per second. This cooling is achieved by using conveyor speeds of 7.5 m/min. (25 fpm) with the cooling fans turned off. Rimmed G10080 (AISI or SAE 1008) steel rods produced in this manner will average 345 MPa (50 ksi) tensile strength and 77 per cent reduction of area compared to values for regular Stelmor mills of 372 MPa (54 ksi) and 71 per cent reduction of area. Tensile values for rods cooled in this manner are close to those of conventionally reeled rod, but the uniformity of mechanical properties and scale is superior.

Improvements in mechanical properties are achieved in some low carbon grades with the Spiracool "Soft" practice. Cooling rates on the order of 0.3 to

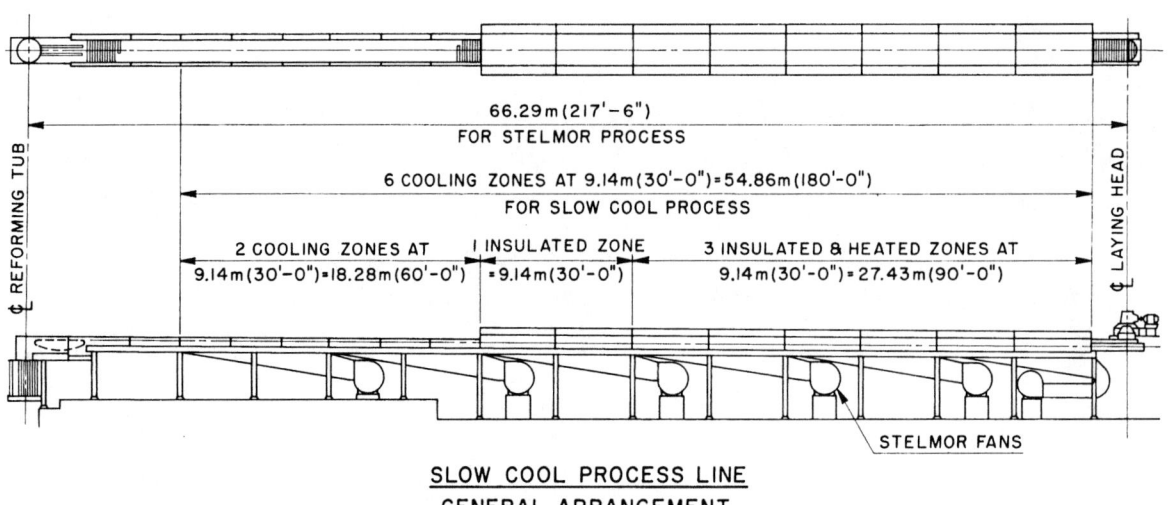

SLOW COOL PROCESS LINE
GENERAL ARRANGEMENT

Fig. 31—5. Transverse section through the cooling conveyor of the Cuyahoga Works rod mill. (Courtesy, Morgan Construction Company.)

0.5°C (0.5 to 0.9°F) per second are produced by slowing the conveyor to about 4.6 m/min. (15 fpm). Certain grades and applications benefit from this operation including Cold Heading Quality, low-carbon welding rod and medium-carbon machining grades.

The slowest cooling rates, down to 0.08 to 0.3°C (0.15 to 0.5°F) per second, are obtained from the Spiracool "Extra Soft" operation. The furnace covers are lowered along the first four zones and temperatures are automatically controlled by thermocouples in the first three zones. Generally, ascending temperatures from about 480°C (900°F) in zone 1 to 590°C (1100°F) in zone 3

Table 31—I. A Comparison of Typical Mechanical Properties Obtained in Rods Cooled by USS Spiracool "Extra Soft" and Regular USS Spiracool Procedures.

Steel Grade*	USS Spiracool "Extra Soft" Rods			USS Spiracool Regular Rods		
	Tensile Strength		Reduction of Area Per Cent	Tensile Strength		Reduction of Area Per Cent
	MPa	ksi		MPa	ksi	
G10170 Capped	421	61.0	65	434	63.0	66
G10180 Capped	441	64.0	69	455	66.0	66
G10180 Killed	462	67.0	68	517	75.0	67
G10220 Capped	496	72.0	69	552	80.0	64
G10220 Killed	514	74.5	65	563	82.0	63
G10300 Coarse Grain	555	80.5	58	641	93.0	55
G10380 Coarse Grain	607	88.0	52	758	110.0	47
G10B210 Q-Temp	503	73.0	61	572	83.0	55
G86200	621	90.0	56	800	116.0	39

*UNS Grade numbers correspond, respectively, to AISI or SAE 1017, 1018, 1022, 1030, 1038, 10B21 and 8620.

provide the desired cooling rate in the steel and minimize the temperature gradient from coil edges to centers. Positive thermal control of zone temperatures through the use of radiant tubes is, nevertheless, afforded by the system. Several medium-carbon and alloy grades produce the lowest strength and maximum ductility when slow cooling is conducted through a temperature range of 650 to 700°C (1200 to 1300°F). Optimum properties are obtained in lower carbon rods by transforming at higher temperatures. At 3 m/min. (10 fpm), the rod remains in that portion of the furnace where temperatures are controllable for a duration of nine minutes. This is sufficient time to assure the completion of transformation in low- and medium-carbon grades and alloys including G40370 and G86200 (AISI or SAE 4037 and 8620) before the rod leaves zone 3. Widely differing rod-cooling rates, tailored to the grade and application, can be obtained on these novel conveyors but maximum ductility and lowest strength are achieved by the Spiracool "Extra Soft" procedure.

A comparison of typical mechanical properties of rods of several steel grades cooled by Spiracool "Extra Soft" and regular Spiracool procedures is shown in Table 31—I.

SECTION 3

CLASSIFICATION OF WIRE RODS

CARBON STEEL RODS

Carbon steel rods are produced in various qualities; each quality requires a specific combination of steel melting; billet rolling; billet conditioning; rod cooling; and rod inspection to meet the needs of its end application.

Standard Quality rods, which range to 0.23 per cent maximum carbon, are used to produce wire for a wide variety of products including welded fabric, fence, nails, chain, wire forms, paper clips, fine wires such as fuse wire, and common wire such as tie wire.

Cold Finished Bar Quality rods are rolled from steel grades intended for producing straightened and cut bars with good machining properties.

Medium High Carbon Quality and **High Carbon Quality** rods use steel grades with greater than 0.23 per cent carbon content. Wire drawn from these rods is used to produce such products as "U" bolts, concrete snap ties, mechanical springs, tire bead, strand, upholstery springs, and automotive zig zag springs.

Welding Quality rods call for a close control of steel composition; in fact, this quality represents the special case where the low carbon rimmed or capped steel used to produce it must meet product chemical-composition limits.

Cold Heading Quality rods are intended for the production of wire which will be rolled, extruded, or headed. Billet conditioning for seam removal prior to rolling the rod is critical in this product.

ALLOY STEEL RODS

Alloy steel rods are used to produce alloy wire and cold finished bars which will be heat treated after fabrication or alloy wire which will become filler metal in weldments. Specific alloy rod qualities are as follows:

Aircraft Quality rods are intended for use in highly stressed parts of aircraft. Electric furnace steel and magnetic particle inspection are mandatory requirements when producing this quality.

Bearing Quality rods are used in the production of antifriction bearings. This quality is usually rolled in alloy carburizing grades and the high carbon chromium alloys. Steelmaking practices usually include vacuum treatment.

Cold Heading Quality alloy rods are used to produce alloy wire which will be cold or hot formed into high strength fasteners and components.

Welding Quality alloy rods are used to produce alloy wire for electric arc welding or for building up surfaces subjected to considerable wear.

SECTION 4

OUTLINE OF WIRE-DRAWING PROCESSES

The rolling of the rods is a fairly well standardized process. The primary object of rod rollings is, of course, to put the steel into such shape that it can be most efficiently cold drawn into wire. Using steel of the required composition, the rod is rolled to the desired size and shape by following a standard practice on a standard rod rolling mill. From this point, however, the method for manufacture into wire depends entirely upon the end use to which the wire is to be put. These uses are many and varied, and the wire drawing practices necessary to provide the wide ranges of required wire characteristics vary accordingly. Obviously, it would be impossible to cover all the practices in detail here. However, a brief summary of the various operations necessary to process the rod to finished wire will be outlined.

Preparing the Rod for Drawing—In order to prepare the rod for drawing, it is first given some necessary preliminary treatments. In general, these consist of cleaning and coating following a heat treatment when required.

The heat-treating process is usually one of the following: patenting, annealing, or normalizing. These processes will be described in more detail later. Low-carbon rods which are to be drawn into wire and high-carbon or alloy rods rolled on control-cooled rod mills and intended for production of wire for many applications usually do not require any heat treatment and are cleaned and coated after rolling.

In the process of rolling, the rods acquire a mill scale or oxide on the surface. Oftentimes, it is necessary to store the rods for some time before drawing. In storage they may pick up either some additional oxide in the form of rust or just plain dirt. All mill scale, oxide, or dirt must be removed before the rods are drawn into wire. This is accomplished by mechanical means or by placing the rods in a solution of hot diluted sulphuric or hydrochloric acid for from ten to thirty minutes. The action of the acid loosens the scale and frees the rod of all rust and dirt, leaving the surface of the rods clean. The rods are then removed from the acid and given a thorough rinsing in a spray of high-pressure water. This removes all the acid from the surface and from between the various strands.

The next step is to give the rods a suitable coating, usually lime, borax, or phosphate. The purpose of the coating is threefold: first, to protect the surface of the cleaned rods from rusting in the atmosphere; second, to neutralize any traces of acid left from the cleaning; and, third, to serve as a carrier for the lubricant used in drawing the rod to wire.

For many years, it was customary to coat rods which were to be drawn into ordinary finished wire with a sull coating. This is accomplished by placing the cleaned rods on a rack or a traveling conveyor and spraying them with a fine mist-like spray of water. The rods are then dipped in a tank containing hot milk of lime. In recent years the practice of sull coating has been discarded, and the rods are coated with the milk of lime directly after the water rinse. Sull coating is used now only on a few kinds of wire; in some instances for manufacture of cold-headed bolts.

Borax is another coating which has been used quite extensively in recent years with good results. After the coating is applied, the rods are baked in an oven or other type of baker to dry the coating. This function will be described later in more detail.

Drawing the Rod—After the rods are properly cleaned, coated, and baked, they are delivered to the wire-drawing equipment. Today, most wire which is drawn three or more drafts is produced on continuous machines. Wire which is to be drawn one or two drafts is produced on single or double-deck motor blocks. Wire 15.9 mm (⅝-inch) and coarser is drawn on horizontal bull blocks. However, there is still a considerable amount of wire produced on the so-called wire drawing

frame. This frame, in the type most generally used for drawing rods, supports a single **die** and the power driven **block** for drawing the rod through the die, also a **drawbar** for drawing the first few feet of the rod. One end of the rod is now pointed, or tapered, so that it may be threaded through the die hole, which is somewhat smaller than the rod in section. Next the die holder, on the entering side of the die, is filled with a specially prepared grease, or some other suitable lubricant, so that in passing through the die the rod must first pass through the lubricant. The pointed end of the rod is now inserted through the proper die hole, where it is grasped by tongs attached to the drawbar, and drawn through far enough to be attached to the draw block. When this block is started revolving, it coils the wire about itself and thus continuously draws the rod through the die, thereby bringing about a fixed decrease in its sectional area and a proportional increase in its length.

Draft, Drawing and Process Wire—The amount of the reduction in the sectional area, as in the case of rolling, is expressed in per cent of the original area and is known as the **draft,** while the operation itself is called **drawing.** In general the draft on the rod varies from 10 to 45 per cent according to the kind of wire to be made. As soon as a rod has been given a draft, it is thereafter designated as a wire, though many more drafts and various other treatments may be necessary to work it into wire of the size, finish and strength desired. In wire-drawing plants, any wire which, following the initial drawing from the rod, is to receive further work or treatment before it is finished is designated as **process wire.** After the first draft from the rod, process wire may be finished in various ways.

Dry Drawing and Wet Drawing—There are two processes for drawing wire. These are designated as dry drawing and wet drawing. Mechanically, the processes are the same; that is, the wire is drawn through a die and wound up on a block. The difference in the process is in the coating applied to the wire and the lubricant used. All wet-drawn wire is first given one or more dry drafts from the rod. This process wire is then thoroughly cleaned, rinsed, and immersed in a dilute solution of copper or tin sulphate or a mixture of both of these salts. A chemical reaction takes place which results in the deposition upon the wire of a thin metallic coating from the solution used. After coating, the wire is usually kept under water to protect it from the influence of the atmosphere.

In subsequent wet drawing, the pay-off reel containing the wire is placed in a tub of water. To this water a special type of soap is usually added to act as a lubricant. Special coating solutions have been developed which are applied to the wire after the copper or copper-tin solution treatment. These coatings protect the surface from the atmosphere and are also quick drying, so it is possible to store the wire without keeping it under water and also to permit drawing from a dry reel. In drawing, the copper or copper-tin solution imparts a characteristic metallic color to the wire which is known commercially as **coppered wire** or **liquor-finished wire.** The latter term designates the brass-colored, straw-colored or bright "white liquor finish" (depending upon the percentage of tin in solution).

PROCESSES AND EQUIPMENT FOR PREPARING RODS AND WIRE FOR DRAWING

Importance of Cleaning—All of the hard, brittle oxide commonly called mill scale, which forms on hot-rolled rods or on heat-treated rods, and also the slow-forming red rust which forms on long exposure to the air, must be entirely removed before drawing. If these oxides remain on the surface of the rods, they result in very rapid wear on the wire-drawing dies and also cause scratched and off-gage wire. Proper cleaning and coating is largely responsible for success in the wire-drawing operation.

Method of Cleaning—As already stated, the method generally employed for cleaning the material preparatory to drawing consists in dipping the coils into a vat of hot, dilute sulphuric or hydrochloric acid. The action by which sulphuric acid removes the scale is for the most part mechanical rather than chemical, for ferrous-ferric oxide, Fe_3O_4, which is the chief constituent of the scale, is but slightly soluble in sulphuric acid. The acid, however, is able to penetrate to the metal beneath the scale, where it reacts with the iron forming iron sulphate, a soluble neutral salt, and liberating a mixture of gases, mainly composed of hydrogen. This action results in loosening and detaching the scale from the surface of the metal, when it sinks to the bottom of the vat where it accumulates and must be removed at frequent intervals. Unlike the black scale, the red rust (which is a hydrated sesquioxide of iron of the general formula $XFe_2O_3 \cdot YH_2O$ or a double compound of ferric oxide and ferric hydroxide, $Fe_2O_3 \cdot 2Fe(OH)_3$, is readily soluble in sulphuric acid, and under the conditions of pickling this action results in the formation of iron sulphate, also. This salt remains in solution, and as concentrated acid is added from time to time to replace that neutralized, the solution eventually becomes saturated to such an extent that it is no longer fit for pickling. The exhausted solution must then be replaced.

When hydrochloric acid is used for removing scale from steel rods or heat-treated wire prior to drawing, the acid removes the scale by chemical attack and converts the various oxides of iron in the scale to ferrous chloride which is water soluble. Generally, hydrochloric acid will remove scale more effectively than sulphuric acid; however, its use requires more rigid control of the maximum temperature, concentration of acid and concentration of the ferrous chloride to minimize objectionable fuming. Economic factors restrict wider use of this acid for rod and wire cleaning or pickling.

Scale removal also can be effected by mechanical means such as shot blast cleaning with abrasive particles, or by reverse bending over sheaves (Figure 31—6 and 31—7) to fracture and remove the hard, brittle scale from the rod surface. The diameter of the sheaves must be carefully selected to provide sufficient outer fiber strain to fracture the scale without introducing excessive work hardening of the rod. The use of either of these two methods is restricted to in-process mate-

rial or to wire whose surface finish is not of primary importance. To produce the highest quality rod surface for continuous, in-line scale removal, mechanical scale cracking may be combined with high-current density electropickling in a sulfuric acid electrolyte.

Manner of Handling the Material—When the rods are removed from the hook conveyor in the rod mill, they are loaded into standard-gauge gondola cars. The coiled rods are loaded in an upright position to facilitate unloading with a so-called "C" hook (Figures 31—8 and 31—9).

From the rod mill, the rods are transported to the rod storage where they are unloaded by an overhead crane. The crane is equipped with the "C" hook referred to above, which has a capacity of two to three tons. Facilities for storing rods consist of large bins or racks.

In modern integrated mills, the rods may be moved directly from the rod storage to the cleaning crane by the rod-storage crane. In other mills, the rods are set down in racks by the rod-storage crane and then moved to the cleaning crane by ram tractor or on rod trucks. In either case, the rod-storage crane sets up the exact amount of rods which the cleaning crane can handle in one lift, so that there is a minimum of handling.

The "C" hook or hairpin hook, as it is sometimes called, is a specially designed device for handling coiled rods and wire. It is designed to lift as a unit the number of coiled rod bundles which the cleaning crane was designed to handle in one lift. There are two types, the open-end type and the latch type. Both are made by using a section of extra-heavy pipe, or I-beam, which is long enough to span the cleaning tank. This pipe has a metal eye attached at the approximate center, or balance point, into which the crane hook is inserted for lifting. From the pipe, a very heavy hanger is suspended. In the open type (Figure 31—8), this hanger is in the form of a large letter "C" with the bottom section straightened horizontally and made long enough to lift the proper number of coils.

In the latch type (Figure 31—9), the hanger is made more in the shape of the letter "L", and the horizontal section is pivoted to the vertical section which allows for a slight movement of the cross piece. At the opposite end of the pipe, a second hanger is suspended. This hanger is pivoted to the pipe allowing it to swing freely. When the hook has been threaded through a lift of rods, this second hanger is swung down and latched to the free end of the cross piece. This gives support to both ends of the cross piece. All of the parts which come in direct contact with the acid are made of phosphor bronze, aluminum bronze, monel alloy or other acid-resisting material.

A hook with its load is referred to as a "pin" or "stem" of rods or wire.

Types of Cranes—There are two types of cranes generally used in cleaning, namely, the circle crane and

FIG. 31—6. Four-wheel mechanical descaler for removing scale from steel rods prior to drawing. In passing over the series of sheaves, the rod is bent 180 degrees in four different directions. The raised cover is lowered when the machine is in operation. (Courtesy, Morgan Construction Company.)

FIG. 31—7. Mechanical descaler for removal of scale from wire rods prior to drawing by bending the steel in various directions as it passes continuously over the sheaves. The hinged cover is closed when the machine is in operation. (Courtesy, Vaughn Machinery Division, Wean Industries, Inc.)

the overhead electric traveling crane. One type of circle crane consists of an upright standard set in bearings to allow the upright standard to turn 360°. A horizontal boom is placed on the upper part of the standard. This boom or arm is built long enough to reach the center of the cleaning and coating tanks which are placed in an arc around the base of the standard. Power for lifting the material to be cleaned is supplied by a steam piston which is placed on the boom or standard. Power for turning the standard and boom is supplied by placing a ring gear around the upright standard. This ring gear is then driven by an electric motor which is connected by the proper gearing and shafting.

Use of the circle type of crane requires the cleaning and coating tanks to be placed in an arc around the upright standard. The number of tanks is then limited by the radius of the boom.

When the electric overhead traveling crane is used, the cleaning and coating tanks may be placed in a straight line or sometimes two lines are used.

The gantry crane is another type of electric crane in use. This crane requires all the tanks to be in one line as the crane has no cross travel, and the crane hook moves in relatively the same horizontal line at all times. This type of crane has value where there is a large volume of work and few varieties of product.

Construction of Tanks—The rectangular acid tanks are constructed of welded steel plate, and are usually

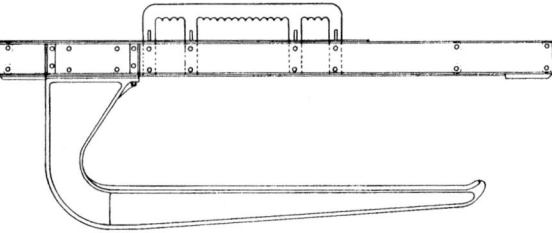

FIG. 31—8. Open-type "C" hook for handling coiled rods and wire.

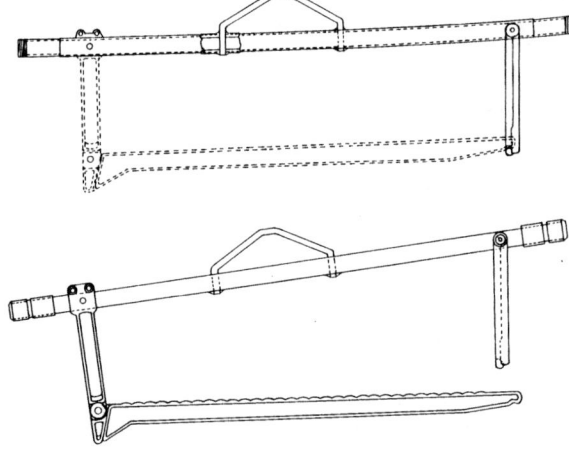

FIG. 31—9. Latch-type "C" hook for handling coiled rods and wire.

FIG. 31—10. General view inside a straight-line cleaning house. Rods can be cleaned in acid, coated with borax, lime, or sulphur, dried and baked in one continuous operation.

large enough to hold two stems of rods. The bottom is reinforced with channels and angles, and the interior is lined with a 4.76-mm (³⁄₁₆-in.) thick membrane of rubber or neoprene to protect the steel from the acid. Inside the rubber lining a layer of acid-proof brick is placed completely covering the sidewalls and bottom of the tank. The brick lining is bonded to the rubber or neoprene with acid-resisting sulphur-base cement. Acid-resisting synthetic-resin cement is used in the brick joints. All unprotected exterior surfaces of the tank are covered with a black vinyl-resin-base acid-resisting paint.

Water-rinse tanks are also usually rectangular and constructed of welded steel plate. Such tanks usually are of one-stem capacity. All surfaces are coated with a black vinyl-resin-base acid-resisting paint. The upper section of a water-rinse tank is provided with a series of

high-pressure water nozzles, while the lower part of the tank contains fresh running water for rinsing.

Lime tanks are also rectangular and constructed of welded steel plate, and usually have a capacity of one stem.

Arrangement of Tanks—As already indicated, when a circle crane is employed, the number of tanks is limited by the radius of the boom of the crane. When an overhead traveling crane is used, the number of tanks is unlimited. Figures 31—10 and 31—11 show typical arrangements for so-called straight-line cleaning. In general, the acid tanks are used first in the cycle, then the water-rinse tank, and finally the coating tanks. Sull coating today is in very slight demand, but usually a small section is provided where the rods can be treated for formation of this coating.

Concentration of Acid—There are three factors

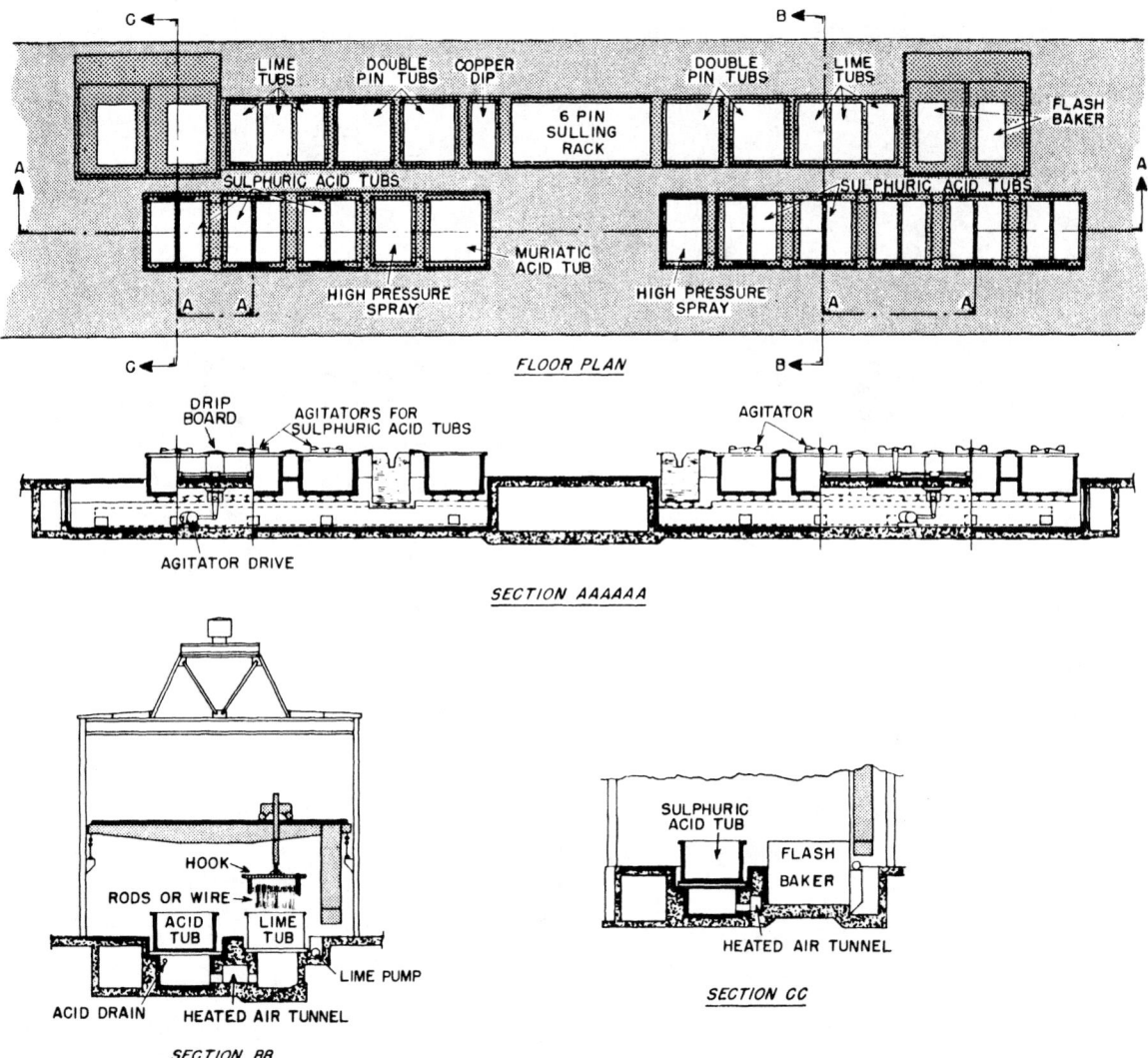

FIG. 31—11. Typical arrangement for so-called straight-line cleaning of rods in preparation for drawing.

which affect the cleaning operation, namely, time, temperature, and concentration of acid. In operation, the tanks are first partially filled with water. Then the acid is added in the required amount and the steam for heating the bath is turned on. By the time the tanks are filled by the addition of more water, the steam has raised the temperature to the required point, usually about 82°C (180°F) for sulphuric acid and 60°C (140°F) for hydrochloric acid, and the acid concentration has been adjusted to the required amount.

Certain kinds of steel require longer time for cleaning than others. Low-carbon rods which may be finished at a high temperature on the rod mill will have a very heavy scale. Also, rod coils which are very compact, as is usually true with the coarser gages, are more difficult to clean. Rods made of high-carbon steels react much faster to the acid, and the time of cleaning is shorter. This is particularly true of high- carbon patented rods which have the scale broken up by the

sheaves on the take-up frame of the patenting furnace. The acid concentration, then, is varied according to the kind of steel being treated, the low-carbon heavily-scaled material requiring a solution having a higher concentration of acid, with the higher-carbon steels and patented material requiring a lower concentration.

Temperature for Cleaning—The control of temperature in the cleaning operation is very important since the rapidity of the reaction between acid and steel is greatly affected by temperature. It is known with sulphuric acid that the reaction is 100 times as fast at 88°C (190°F) as at room temperature. Too high a temperature, therefore, can be very wasteful. It causes rapid and high usage of the acid, develops unnecessary fumes which, in turn, can cause excessive corrosive action on the steelwork of the building. What is more important, it causes pitting action on the surface of the steel being treated. The tanks in most modern cleaning houses are equipped with thermostatic controls so that a uniform pickling-solution temperature is maintained.

Inhibitors are used to aid in preventing pitting or over-cleaning. These usually have a nitrogenous hydrocarbon base. The theory of the use of inhibitors is that when the scale has been removed from the steel, the inhibitor forms a protective film on the cleaned surface and minimizes any additional attack by the acid.

Time of Cleaning—The time required for cleaning will vary according to the amount of scale to be removed and the type of steel being treated. This may vary from as little as 10 minutes for high-carbon patented rods to 30 or 35 minutes for low-carbon heavily-scaled rods. Another factor which affects the time of cleaning is the change in specific gravity of the acid solution. As the work proceeds and more and more scale is removed, the iron content of the bath builds up. This is reflected in a change in specific gravity of the solution, which is usually measured by a hydrometer that provides a Baumé reading. As the Baumé figure increases, the action of the acid is retarded and the cleaning process gradually slows down. When the Baumé reading reaches about 20, the acid in the tanks is discarded, after which the tanks are washed out and refilled. In modern cleaning houses, where the tanks hold from 18.9 to 26.5 cubic metres (5000 to 7000 gallons), it is customary to have one extra acid tank. This tank is filled and heated to temperature, ready to operate when one of the other tanks is being emptied of its spent acid. By rotating the use of the tanks in this way, continuous operation can be maintained.

Rinsing—It is very important to have the cleaned rods or wire thoroughly rinsed after the acid cleaning. The cleaned material is dipped into a tank of fresh running water. There is a series of high-pressure sprays placed around and at the top of the water tank. When the cleaned rods are dipped in the water, a switch automatically turns on a high-pressure pump. As the crane moves the rods up and down, the high-pressure sprays wash them thoroughly in an attempt to remove all traces of acid or acid residue.

Coatings—After the rods are thoroughly cleaned and rinsed, it is necessary to apply a coating. The purpose of the coating is three-fold: First, it prevents oxidation or rusting of the surface; second, it neutralizes any traces of acid which may be left on the steel; third, it acts as a carrier for the lubricant used in drawing. For dry drawing, dry slaked lime or hydrated lime have been found to be the best materials for coating. Lime is a low-cost material and is easily applied by dipping the cleaned rods into a tank of hot milk of lime kept at a temperature between 88° and 93°C (190° and 200°F). Borax is another coating which has been used for dry drawing in recent years. It is a little more costly than lime, but it offers some advantages. It dries very quickly, and it does not flake off or form a dust in the wire-drawing room, thereby making conditions much cleaner in the mill. Where borax is used, a separate tank is provided and the consistency of the solution is adjusted according to the number of drafts the wire is to be given.

Recently, phosphate coating has become quite popular, primarily for wire to be subjected to cold heading or cold extrusion, where a chemical rather than a mechanical bond between the coating and the metal is

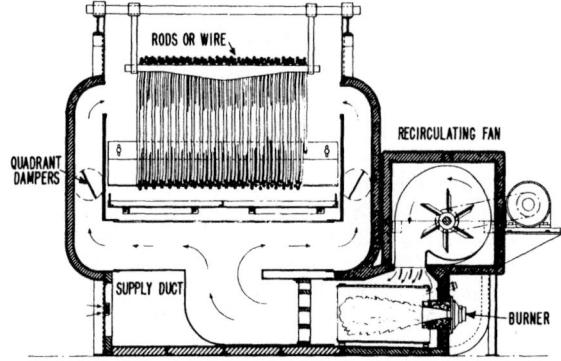

FIG. 31—12. Cross-section of a flash baker for drying rods and wire after cleaning.

desirable. Either a zinc-phosphate or an iron-phosphate solution may be used, and both are applied by immersing the wire in the heated solution (at about 82°C or 180°F) for from 2 to 10 minutes depending upon the weight of coating desired. After suitable hot- and cold-water rinsing, the phosphate coating is neutralized with lime or some other suitable agent, baked, and drawn through the normal dry-drawing lubricants (calcium- or aluminum-stearate and lime mixtures) or coated with a neutral sodium tallow soap. This type of coating, when heavy enough, gives excellent results for heavy cold extrusions and long shelf life.

Baking—After the rods or wires are properly cleaned and coated, they must be thoroughly dried before drawing. The drying or baking is done in ovens or bakers. In the cleaning process, hydrogen is liberated by action of the acid upon the steel, and sometimes this hydrogen is absorbed by the steel. This can cause the drawn wire to be brittle, a condition known as "acid brittleness." The baking process has two purposes: one, to dry the coating so that it will function properly in dry drawing and, two, to remove any hydrogen that may have been absorbed by the steel.

There are several types of bakers. The **flash type** is the most popular. This is a baker of about the size and shape of a lime tub. One pin of rods is placed in the baker for from five to fifteen minutes, depending upon the weight of coating to be dried. The temperature is usually kept between 230° and 315°C (450° and 600°F). The **tunnel type** is another baker used. This is a long, rectangular oven with a roll or chain conveyor running through its length. In operation, the rods are placed on the conveyor and pass slowly through the baker. The time may be one or two hours at temperatures from 205° to 260°C (400° to 500°F). **Compartment-type** bakers are still used to a limited extent. In these bakers, the rods are loaded onto steel buggies which are pushed into the oven on tracks running throughout its length. The temperature is usually between 120° and 160°C (250° and 325°F) which requires a longer baking time. This may be from three to eight hours. The flash and tunnel bakers are fired with oil or gas and are equipped with high-velocity fans which rapidly recirculate the heated air. Figure 31—12 shows a sketch of a flash baker.

WIRE-DRAWING EQUIPMENT

DIES

The wire-drawing die is one of the most efficient tools used in industry. It has no moving parts; it does not remove any of the metal; yet it uniformly reduces the cross-sectional area of the steel and at the same time improves the finish and physical properties.

Over the years many different materials have been used for wire-drawing dies. Chilled iron, steel plates, alloy steel, all have been used, but in the late 1920's tungsten carbide was developed and tried out. This material was an immediate success, and in a comparatively few years it replaced all other materials except the diamond. Diamond dies are still used for very fine sizes of high-carbon and alloy-steel wire but even for these the tungsten-carbide die can be used.

Tungsten carbide and diamonds are very hard materials and have great wearing characteristics but do not have very great resistance to impact. They must have some outside reinforcing when put to use. This is accomplished by pressing a small section of tungsten carbide, called a "nib," into a cylindrical steel casing, as shown diagrammatically in Figure 31—13.

Diamond dies are frequently used when accuracy and uniformity of section are required in the finer sizes of wire, especially in the process of continuous wet drawing. In the construction of diamond dies, a diminutive flat crystal of diamond is securely fastened in a center opening of a small circular metal disc, then a hole of required diameter is worked through this diamond by special drills and diamond dust. Diamond is one of the hardest substances known and has great wearing qualities. Long lengths of fine wire, can, in consequence, be drawn through these dies with little or no change of sectional area due to wear of the die.

Die Contours—The shape of the die has been found to be very important. There are four distinct zones in the die as shown in Figure 31—14. The first zone, on the entering side of the die, is somewhat larger in diameter than the rod or wire to be drawn; its purpose is to afford room for the die lubricant that adheres to the rod or wire to be drawn into the hole. This is called the **bell and entering angle** and gradually tapers into the second zone. The second zone is called the **approach angle** and is the section where most of the actual reduction takes place. The next zone is called the **bearing zone** and it may have a very slight angle of taper. The exit zone or **back relief** is in the form of a countersinking of the back part of the hole. This is done as a strengthening to prevent the circular edge of the hole from breaking away.

THE BLOCK

Wire was first drawn in very short lengths and was merely pulled through the die in straight pieces. As the lengths grew longer, some means of storage became necessary. The wire-drawing block serves this purpose. The block is a steel casting in the form of a cylinder the sides of which have a slight taper. The base of the cast

cylinder is solid and there are enough cross members inside the casting to give adequate support to the sides. At the base, there is also a flanged section extending horizontally outward around the cylinder. The juncture of this horizontal flange and the vertical sides of the block is machined to a definite radius. This is called the **fillet** of the block. In the center of the block, a vertical cylindrical hole is left to allow for the drive shaft to be keyed to the block.

In operation, the wire is threaded through the die and attached to the block by a vise or clamp. The block revolves and pulls the wire through the die. As the wire is wound around the fillet of the block it is forced upward. At each revolution of the block the wire moves upward approximately one diameter of the wire being drawn. The sides of the block may be 150 or 200 mm (6 or 8 inches) high. At the top of the block four vertical pins are placed. As the wire feeds upwards and reaches the top of the block, it is stored around these four pins. The result is a uniformly wound, compact coil of wire.

Double-Deck Blocks—This is an arrangement in

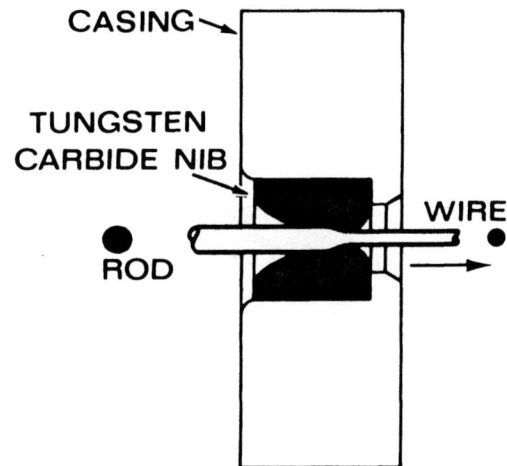

FIG. 31—13. Schematic cross-section (not to scale) of a wire-drawing die employing a nib of sintered carbide mounted in a circular steel holder.

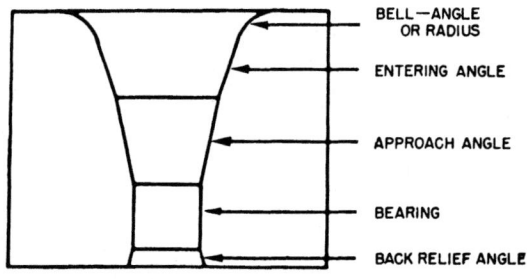

FIG. 31—14. Enlarged cross-section of die nib showing names of parts.

FIG. 31—15. Closeup view of a double-deck block being used as the first and second drafts on a non-cumulative type of continuous wire-drawing machine. (Courtesy of Vaughn Machinery Div., Wean Industries, Inc.)

which one block is placed above the other, both being mounted on a common spindle. In some cases the wire is drawn first on the upper block and then on the lower block, while in others the reverse is true. The first block is made smaller in diameter than the second block with the ratio of the circumferences of the two blocks determined by the amount of second draft, with the attendant increase in length per unit weight. In drawing, the wire is pulled through one die and wound around the first block several times, then it passes around a sheave and into the second die and onto the second block. Thus, two drafts are drawn on the one spindle.

Sizes from 4.76 to 3.18 mm (³⁄₁₆ to ⅛ in.) may be drawn by the double-block method. Due to the heat developed, the use of double-deck blocks is usually limited to the drawing of low-carbon wire.

Double-deck blocks may be used in drawing frames, motor blocks, or as the initial drafts of continuous machines, as described later in this section. Figure 31—15 shows a close-up view of a double-deck block used as the first and second drafts of a non-cumulative continuous wire-drawing machine. The first draft is on the upper block and the second draft on the lower, larger block.

CAST

To facilitate handling of wire, and to prevent tangling as the wire is paid off for use, it is necessary to control the **cast.** The term cast includes two characteristics. One is the **free-circle diameter,** which is the diameter of a ring of wire cut from the coil an placed on a flat surface. The free-circle diameter should be about ten per cent larger than the diameter of the block on which the wire was drawn. The other characteristic is **side cast,** which is the pitch of the helix which a ring of wire assumes when suspended at its midpoint. The term "dead cast" means that the ends of the ring of wire meet, or that the pitch is zero. In actual practice, a slight side cast, no greater than about ten per cent of the diameter of the block on which the wire was drawn, is usually present. The side cast should be in a

positive direction, that is, the free end of the wire should tend to spring up from the top of the coil, rather than down into the coil.

Cast can be controlled by changing the position of the wire-drawing die. Most modern wire-drawing machines have adjustable die boxes to facilitate control of cast. In some wire-drawing machines, straighteners are provided for this purpose. These straighteners consist of a sheave for leading the wire from the block into several rolls which can be adjusted to control cast, and a sheave for leading the wire back onto the block. In use, the wire passes through the die, onto the block for several convolutions, then through the straightener, and back onto the block.

While it is most important to control cast in the final draft, it is also desirable to maintain good control in all drafts of continuous machines, especially in high-carbon wire, since the residual stresses imposed by radical change of cast in the final draft may affect the forming characteristics of the wire.

DRAWING MACHINES

There are several types of drawing machines. These may be grouped as follows: drawing frames, bull blocks, motor blocks, multiple-draft machines, fine-wire machines, turk's-head shaped-wire drawing machines and drawbenches. These will be briefly described.

Drawing Frames—Before the development of continuous machines, all wire was drawn on what are known as frames, and a small percentage of wire still is drawn by this method. A frame is equipped with several wire-drawing blocks, all capable of being driven from a common shaft, either singly or in unison. The horizontal shaft extends the length of the machine, usually below floor level. From this shaft, vertical spindles are geared, and at the top of each spindle a drawing block is mounted. A separate clutch for each block permits the individual blocks to be started, operated or stopped without affecting the others on the frame. The blocks are usually 559 or 660 mm (22 or 26 in.) in diameter and the spindles set at about 1.5 m (5 ft) centers. The blocks are about 635 to 762 mm (25 to 30 in.) above the floor level and all gearing and other moving parts are enclosed. Each block has a die holder or die box which also holds the drawing lubricant. Each block is also equipped with a pull-out mechanism, which is geared to the main shaft. In operation, the rod is placed on a pay-off reel and one end is pointed. This pointed end is then threaded through the die. The pointed end is then grasped by a pair of pincer jaws which are attached to the pull-out mechanism. About a metre (3 ft) of wire is then pulled through the die. The pincer jaws are released and the wire is attached to the block by a clamp on the side of the block. The block is then started by attaching it to the spindle by actuating the jaw clutch which is operated by a foot lever. When the coil has been drawn, the block is declutched and the coil is removed by an overhead hoist. The coil is then placed on another pay-off reel and the process is repeated on another block until the required size is produced.

Obviously, there are disadvantages to this method. Each coil must be handled separately for each draft. All

FIG. 31—16. Heavy machines employing horizontal blocks called "bull blocks" for drawing coarse (12.7 to 25.4 mm or ½ to 1 in.) sizes of wire.

FIG. 31—17. Motor block, with vertical spindles, for drawing wires.

blocks on a given frame turn at the same speed and as the diameter of the wire is reduced, the length of the wire and consequently, the running time increase, and more blocks must be provided for drawing wire to the smaller sizes in order to keep pace with the blocks drawing coarser wire through the early drafts.

Bull Blocks—Sizes 12.7 to 25.4 mm (½ to 1 in.) are usually drawn on horizontal blocks called bull blocks. These are very heavy machines built to pull these coarse sizes (Figure 31—16.) They are driven by an individual variable-speed motor. The blocks are usually 914 mm (36 in.) in diameter. The horizontal block makes it easier and safer to handle coils of drawn wire in these coarse sizes.

Motor Blocks—These are also driven by an individual motor, but the blocks have a vertical spindle (Figure 31—17). The size range drawn on these machines is usually 4.76 to 12.70 mm (³⁄₁₆ to ½ in.). The blocks may be 660, 760 or 915 mm (26, 30 or 36 in.) in diameter. Motor blocks may also be equipped with double-deck blocks.

Multiple-Draft Machines—Wire which requires three drafts or more from the rod is usually drawn on continuous machines. The tungsten carbide die, due to its long-wearing characteristics, makes it practical and economical to use continuous machines for the so-called multiple-draft work. There are two general types used in multiple-draft dry drawing. These may be described as cumulative or non-cumulative machines.

Cumulative Machines—In one type of cumulative

FIG. 31—18. Overall view of a cumulative-type wire-drawing machine. (Courtesy of Morgan Construction Company.)

machine, the wire is drawn on a conventional-type block and is allowed to build up around the block pins. At the top of the pins a ring, free to rotate around the block, is placed. The first end of the wire is threaded through a loop on this ring and then is fed over a dome and down through the center shaft of the block, which is hollow. From there it is led around sheaves and into the next die and onto the next block. Figure 31—18 shows this type of cumulative machine and Figure 31—19 shows a close-up of the blocks. In another type of cumulative machine, the wire passes from the ring over a sheave which is mounted above the center of the block and then down around another sheave and into the next die.

In yet another type of cumulative machine, each block has two segments, separated by a ring, free to rotate around the block. The wire is drawn through a die onto the lower segment. The first end of the wire is passed over a sheave on the ring. This changes the direction of coiling of the wire on the upper segment of the block. From the top of the upper segment the wire passes over sheaves to the die box of the next draft. An over-all view of this type of machine, known as a "double-block" machine, is shown in Figure 31—20.

Non-Cumulative Machines—In this type of machine, the wire is drawn through the die and wrapped around

a block in the conventional manner. This block will be about 150 to 200 mm (6 to 8 in.) in height. After a number of turns on the block, the wire passes around the sheave. This sheave is mounted at the end of an arm which is attached to a rheostat, which in turn controls the speed of the block. From this sheave, the wire passes to another sheave and into the next die. Each block of this type of machine has its own individual motor. The drafting is usually laid out to meet the nominal speeds for which the machine is designed. If, however, this drafting should get out of line and one block draws more wire than the preceding block supplies, the arm moves forward and actuates the rheostat so that the speed of the preceding block is increased. By this means the production of each block is kept in balance with all the other blocks and there is no build-up on the blocks.

Both types of machines can be equipped with means for cooling the wire between each draft. In drawing medium high-carbon and high-carbon wire (0.30 carbon and over) it has been found that the wire will become brittle if the heat developed in drawing is not removed between drafts. Both types of machines use a blast of air directed at the wire to cool it. The non-cumulative machine and some cumulative machines also employ a water spray inside the block. Figure

FIG. 31—19. Close-up view of the blocks of a cumulative-type continuous wire-drawing machine. (Courtesy of Morgan Construction Company.)

FIG. 31—20. Double-block cumulative 7-draft continuous wire-drawing machine. (Courtesy of Morgan Construction Company.)

FIG. 31—21. General view of a non-cumulative type of continuous wire-drawing machine. (Courtesy of Vaughn Machinery Div., Wean Industries, Inc.)

FIG. 31—23. General view of a circular-tandem type of fine-wire drawing machine.

31—21 shows the non-cumulative machine.

Fine-Wire Machines—Sizes finer than 0.76 mm (0.030 in.) are usually drawn by the wet process on specially designed machines. There are several types of these machines. The **tandem type** uses horizontal spindles and the drawing blocks are vertical. The wire is drawn through the first die and given a few wraps around the block and passes directly into the next die.

FIG. 31—22. General view of a cone-type fine-wire drawing machine.

The **step-cone type** uses horizontal spindles and has several blocks of different diameters mounted on each spindle, giving the appearance of a cone. The coarser wire is drawn on the smaller block and passes around an idler sheave and into the next die and onto the next larger block. The diameters of the blocks are designed to take care of the normal elongation of the wire as determined by the amount of reduction of diameter in each draft. Another type is the **circular tandem type** in which the blocks are arranged in a circle, with a die between each block. Each block is geared to draw enough faster than the preceding block to take up the increase in length. The finishing block is usually eight inches in diameter, but many of these machines are equipped to take up the wire on spools.

In wet drawing, the drawing lubricant is a soap solution diluted with water. This solution is pumped from a tank on the machine or from a central tank supplying several machines. The solution is piped directly to a nozzle which sprays the lubricant into the bell of the drawing die. There is enough surplus to spray the drawing block also. In this manner the die, the wire, and the drawing block are kept well lubricated. The surplus lubricant falls to the bottom of the machine where it drains out to the supply tank or into a filter system for recirculation. Figure 31—22 shows a cone-type machine and Figure 31—23 shows the circular tandem type.

"Turks-Head" Shaped-Wire Drawing Machine— Common four-sided shaped wire, such as squares, oblongs or keystone shape, are frequently produced on

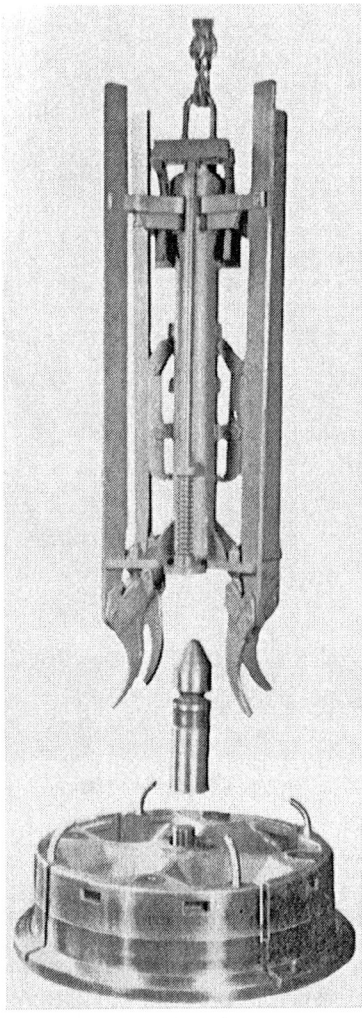 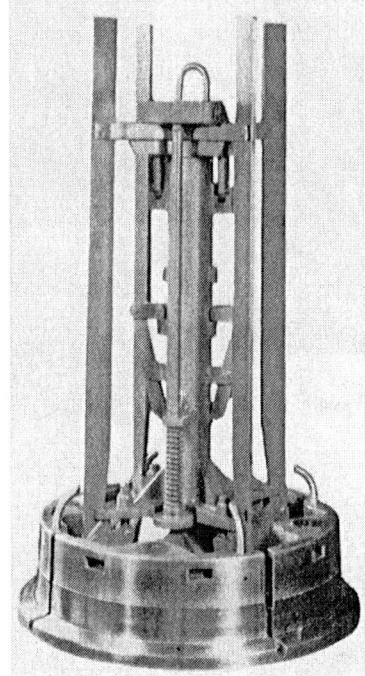 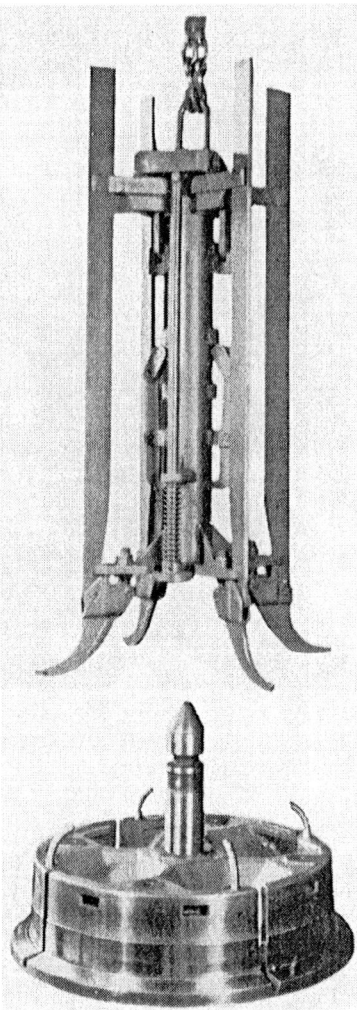

FIG. 31—24. (Left) Riding stripper for large bundles suspended over a wire-drawing block. (Center) Riding stripper in operating position on block. (Right) Stripper with coil-supporting fingers extended as it would be raised from the block after a coil of the desired weight had been formed. By collapsing the stripper to the position shown in the illustration at the left, the coil is released. (Courtesy, Vaughn Machinery Division, Wean Industries, Inc.)

cold-rolling equipment known as a "Turks-head" machine. This consists of four hardened-steel rolls set in planes at right angles to each other. The narrow face of the rolls, as set in the framework, is adjustable on the same plane so that the assembly of the overlapping roll edges facing each other will project the contour of the opening so formed, into the desired shape of the cross-section of the wire to be made. The process wire, of a size somewhat larger than the finished size desired, is pointed and pulled through the Turks-head, being thus rolled to shape and size, after which it is coiled on a regular wire-drawing take-up block.

Drawbench—A drawbench is a mechanism used to give a single draft to heavy material, which is afterwards usually straightened and cut to a definite length. It handles the largest sizes drawn and is especially adapted for drawing shapes, screw stock and small shafting. The machine itself consists of a horizontal framework 15.2 to 30.5 m (50 to 100 ft) long, along the

center line of which runs a heavy roller chain driven by heavy sprocket wheels. The die through which the material is drawn is located at the opposite end of the frame from the drive. A carriage mounted on wheels, and arranged to travel along the upper surface of the frame or bench, has suitable jaws for gripping the material and pulling it through the die, and also a heavy hook for connection with the roller chain. This equipment is similar to that used for drawing seamless tubes (see Chapter 32).

Machines have been developed recently which have three sets of grippers, operating in sequence. The first set grips the material and pulls it through the die for a length of about 250 mm (19 inches). As the first set of grippers nears the end of its stroke, the second set grips the material and continues to pull it through the die, and then the third set of grippers similarly advances the material through the die. By this time, the first set of grippers has returned to its starting position, and the cycle is repeated. Thus it is possible to draw any length desired, without providing the chain and other machinery employed by the conventional drawbench.

TERMINAL EQUIPMENT

Developments which affect the terminal equipment

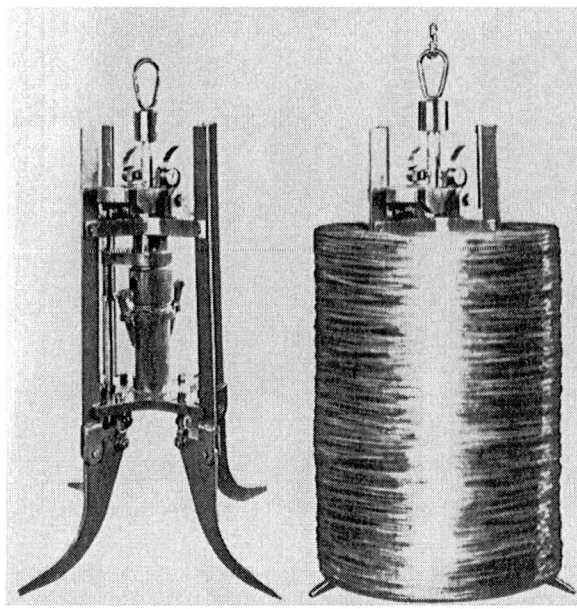

FIG. 31—25. (Left) Riding stripper for wire-drawing block. (Right) Bundle of drawn wire accumulated on riding stripper. Stripper can be collapsed to release bundle. (Courtesy, Morgan Construction Company.)

on wiredrawing machines include riding strippers, stationary or so-called "dead" blocks, and spooling takeups. The riding stripper (Figures 31—24 and 31—25) is a device mounted on the finishing block of the wire drawing machine which permits the accumulation of 454 to 1361 kg (1000 to 3000 lb) of wire depending on size of wire and block diameter. When the desired coil weight is produced the riding stripper is lifted over a tubular coil carrier and collapsed in such a manner as to allow the wire to fall onto the coil carrier.

The stationary block or "dead" block, as the name implies, is a device for accumulating a heavy weight coil of wire onto a block which does not revolve (Figures 31—26, 31—27 and 31—28). This is accomplished by wrapping the wire onto the block by a die or casting device carried on a rotating arm or disc.

For certain types of wire, spooling equipment is provided, which takes the wire directly from the last drawing block and winds it onto spools or reels. This type of takeup is widely used in plants which produce wire rope, where the wire is placed on the spools to be used in stranding equipment. Figure 31—29 shows steel wire being directly taken up on a spooler in tandem with a wire-drawing machine, and Figure 31—30 shows a closeup of the spooler, capable of producing 454- to 907-kilogram (1000- to 2000-pound) spools of wire.

AUXILIARY EQUIPMENT

Pay-Off Reels—For rods 8.7 mm ($^{11}/_{32}$ in.) and smaller, a flipper-type reel is used. This consists of two horizontal arms placed one above the other, both extending from a vertical standard. The arms are hinged so that they may be moved up or down. In use the arms

FIG. 31—26. Stationary block conversion unit integrally mounted on a continuous wire-drawing machine, showing transfer sheave, die holder and casting rolls. (See also Figure 31—27). (Courtesy, Vaughn Machinery Division, Wean Industries, Inc.)

FIG. 31—27. Wire accumulating on stationary block feeds off the block onto tubular guides that form a bundle of the drawn wire on a coil carrier. (Courtesy, Vaughn Machinery Division, Wean Industries, Inc.)

FIG. 31—28. Stationary drawing block for 15.9-mm (⅝-in.) wire, showing how rod passes over sheave and through a die (both mounted on the large rotating vertical disc) and is wrapped on the stationary block. Wire accumulating on one end of the block pushes coils off the opposite end and over a guiding "horn" to form a bundle. (Courtesy, Morgan Construction Company).

FIG. 31—29. Steel wire being taken up directly on a spooler (lower right hand corner) in tandem with a wire-drawing machine. (Courtesy of Morgan Construction Company.)

are collapsed and the rod coil is suspended on the upper arm. Both arms are then moved to the horizontal position and the rod is held vertically. In this manner each convolution of the rod coil can be paid off or

FIG. 31—30. Closeup view of the spooler in Figure 31—29, capable of producing 454- to 907-kilogram (1000- to 2000-pound) spools of wire. (Courtesy of Morgan Construction Company.)

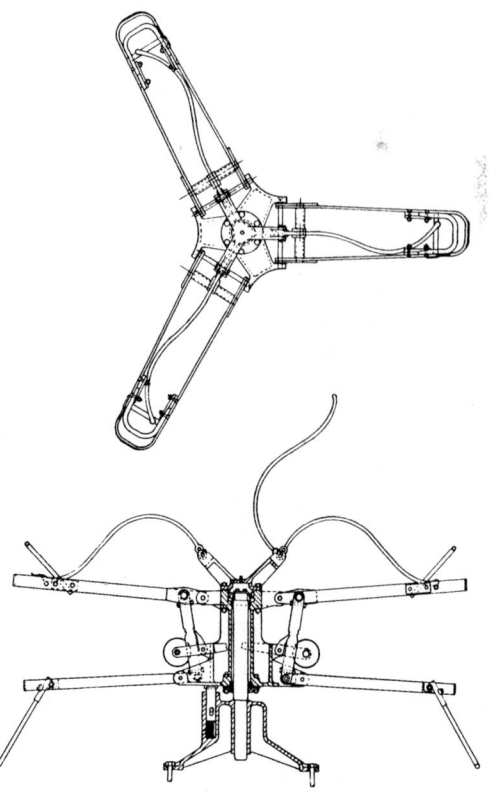

FIG. 31—31. Plan and elevation sketch of a three-arm pay-off reel. (Courtesy of Morgan Construction Co.)

FIG. 31—32. Stem of rods from the cleaning house being loaded by tractor into the stem-type pay-off.

flipped and pulled up to the die. Some reels are made with three arms set at 120°. While one coil is paying off, the other arms are loaded and the ends are welded so there is no stoppage necessary between rod coils. Figure 31—31 shows a 3-arm reel. Another method is the stem payoff, Figures 31—32 and 31—33. By this method the full stem of rods from the cleaning operation is placed on a special holder. A long bar or U-shaped rod is placed through the center of the coils and acts as a holdback to keep the convolutions paying off uniformly. The coils are welded as in the flipper pay-off.

Sometimes revolving reels or dead payoff equipment is used to pay rod or process wire into wire-drawing equipment.

Safety Stop—All machines and frames are equipped with stops to cut off the power quickly in case of trouble. Between the pay off device and the first die a snarl stop is placed which will stop the machine or block in case of a tangle or snarl before the rod is broken. All machines are equipped with bars around all working parts. By merely pushing against these bars the machine can be stopped. The motors are also equipped with dynamic brakes so that there is immediate response to the stop signal and there is no tendency for the machine to "coast."

Welders—For continuous drawing, electric butt welders or flash welders are used. These have been developed to perform very well on high- and low-carbon steel. For welding high-carbon steel, the welders are equipped with annealing jaws to give the steel the proper structure for drawing after the weld is made. Grinders are also provided to quickly remove the weld "burr" or upset.

Pointers—Several methods are used for pointing the rods or wire for the initial threading of the dies. The fine sizes are literally pulled apart and the "necking down" forms a point. The intermediate sizes are pointed by means of a "roll pointer" as shown in Figure 31—34. This consists of a pair of oscillating rolls with grooves of varying diameters. The rod end is worked down to smaller sizes by successive rolling in the

FIG. 31—33. Battery of wire-drawing machines (left) fed by stem pay-offs in the foreground.

FIG. 31—34. "Roll pointer" for pointing rods or wire for initial threading through the dies. (Courtesy of Morgan Construction Co.)

grooves of decreasing size. The larger sizes are swaged or hammered on rotary swaging machines or are machined by specially designed cutting machines.

SPECIAL FINISHING OPERATIONS

Straightening and Cutting Wire—With the exception of the largest sizes, all wire, both round and shape, is drawn on a block and, therefore, at the finished size, is still in the form of a coil. But for certain purposes it is desired to have the wire furnished in straight lengths, rather than in coils. This straightening and cutting work is usually done on some type of machine.

Rotary Straighteners—For common round wire, automatic machines are employed for straightening and cutting the wire. These machines are required to perform three operations simultaneously or in very rapid succession, namely, pull the wire forward, straighten it, and cut it into the lengths desired. The mechanism for straightening the wire is known as a **head** or **flyer**. It contains a number of staggered dies which bend the wire slightly, as it passes through, in reverse directions so as to remove the bends and kinks and leave it straight. The wire to be straightened is automatically fed through the flyer and onto an apron, where it actuates a mechanically or electrically driven cutting tool which cuts the wire into uniform pieces of the length desired.

Roll Straighteners—For shapes and flats the **roll straightener** is used. This machine consists of sets of vertical and horizontal rolls which can be so adjusted that pressure can be brought to bear on the sides, top and bottom of the wire. On these machines the first sets of rolls put a considerable bend in the material to remove the kinks, and succeeding sets of rolls reverse this bend just enough to leave the wire practically straight. Squares, rectangles and hexagons, narrow flats that are not too thin, most shape wires, and also very fine wires, can be straightened on this type of machine provided the machine is properly designed for them.

SECTION 7

WIRE-DRAWING PROCESSES AND OPERATIONS

Results of Cold Drawing—The results attainable by the wire-drawing process may be summarized as follows:

1. Metal may be elongated and reduced in section to an extent not attainable by other methods.

2. A greater degree of accuracy as to size and section can be attained than is possible by other methods excepting cold rolling, which is not applicable to common sizes of wire.

3. A uniformly smooth and highly polished surface can be produced.

4. The process serves as a test of the quality of the metal. The fact that a wire has satisfactorily withstood the drawing operation may be taken as an indication that the metal was originally sound and free from imperfections which might impair its serviceability.

5. Finally, the process affects the mechanical properties of the metal. As noted elsewhere, this makes it possible, by employing this process in conjunction with heat treatment, to produce many wires having different mechanical properties from the same steel rod.

Effect of Drawing Upon Mechanical Properties—Wire drawing, like any cold working of metals such as iron or copper, will increase the hardness, stiffness, tensile strength and elastic limit. The ductility, as indicated by the elongation and reduction of area, will be correspondingly decreased. The extent of these changes in mechanical properties is not always directly proportional to the amount of drafting or cold work done upon the metal, as it is affected by various factors such as the total amount of drafting, number of drafts, per cent reduction per draft and the type of material itself. However, for each set of conditions, the change in mechanical properties has been determined, and the processing necessary to produce the required grade of wire may be regulated accordingly. The gain in tensile strength in drawing of various carbon steels under typical conditions is shown in Figure 31—35.

The Cause of These Changes—These changes in the characteristics of the metal brought about by cold working are to be attributed to the changes in grain structure such working produces. In a hot-rolled rod or

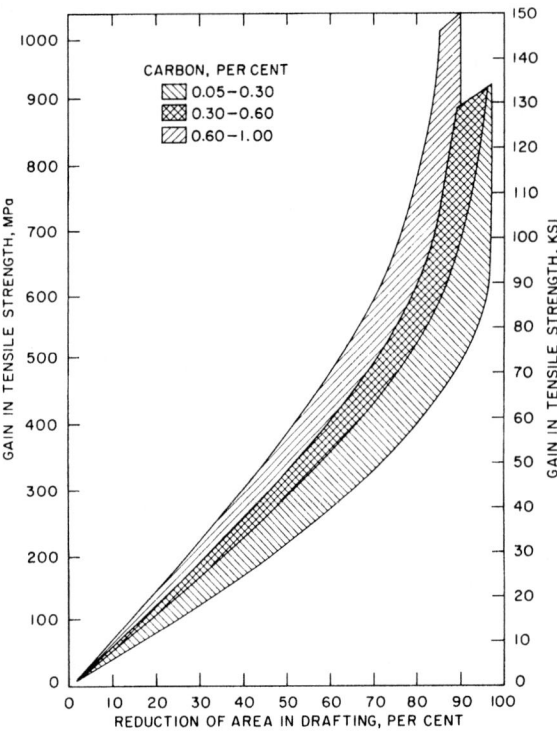

FIG. 31—35. Gain in tensile strength on drafting steels of carbon contents from 0.05 to 1.00 per cent.

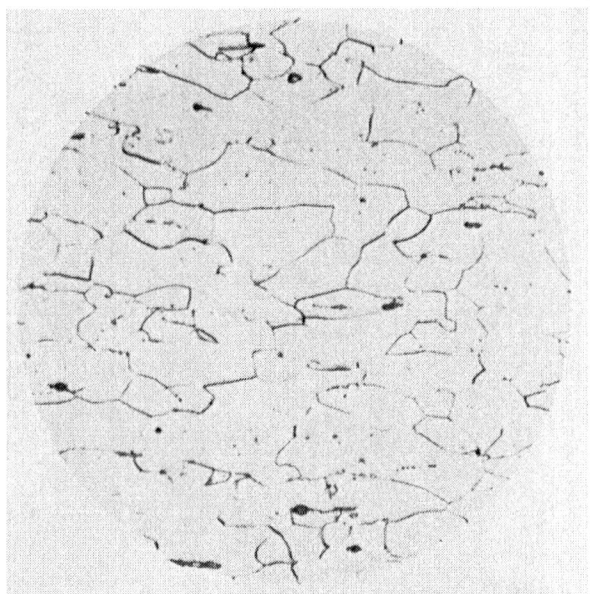

FIG. 31—37. Structure of 0.06 carbon-steel wire after one draft. Longitudinal section. Magnification: 500X.

in an annealed wire the grains have a polygonal form and are arranged about as shown in Figure 31—36. The microstructure of an annealed low-carbon steel wire is shown in the photomicrographs of Figures 31—39 and 31—40. All photomicrographs shown represent longitudinal sections, parallel to the direction of drafting.

Figure 31—36 exhibits grains in the steel before drafting. Figure 31—37 shows the condition of the grains after the material has had one draft. Here it can be seen that, as the wire is being elongated while passing through the die, the grains in the steel actually become elongated, also. The grains elongate in the direction of the drafting and become correspondingly narrower at right angles to that direction. Figure 31—38 shows wire after having had a number of drafts. Here the grains have been elongated to a considerable extent. Due to the stretching of the wire in drawing and the

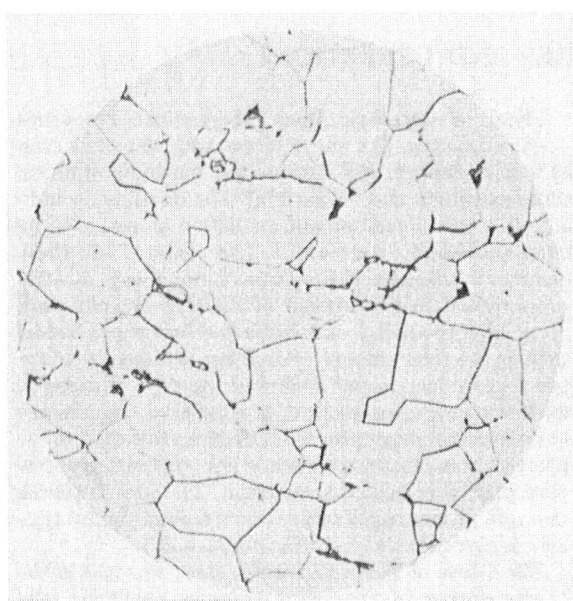

FIG. 31—36. Full-annealed 0.06 carbon-steel wire. Longitudinal section. Magnification: 500X.

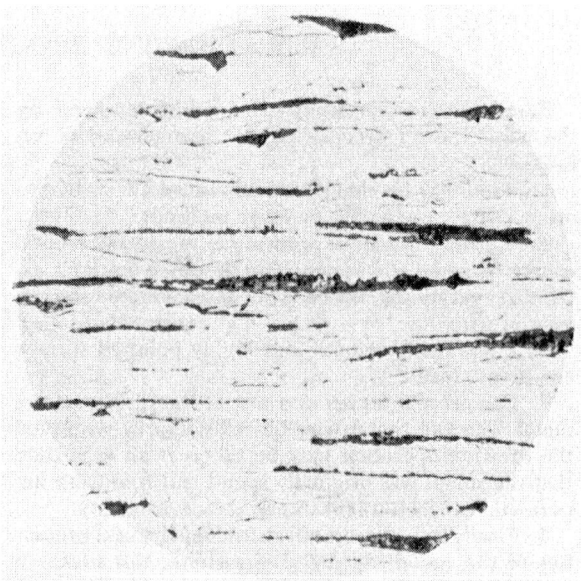

FIG. 31—38. Longitudinal section of hard-drawn 0.15 carbon-steel wire. Magnification: 500X.

crushing effect produced by the pressure exerted by the die in all directions towards the center of the wire, the structure has become so altered that it is difficult, if not impossible, to locate the exact boundaries of the individual grains in their strained and distorted condition.

Heating Effect in Wire Drawing—The plastic deformation involved in drawing wire will induce internal heat, and this rise in temperature of the wire will be dependent on the composition or hardness of the steel and the amount of cold work or reduction of cross-sectional area of the wire. In the ordinary dry drawing of frame or individual drafting, the wire will cool somewhat between successive drafts, but on continuous drafting, the heating of the wire may be cumulative resulting in temperatures frequently detrimental to the quality of the wire. This rise in temperature may be restricted by several methods, such as air or water cooling of the wire-drawing capstans or blocks and the use of water-cooled dies, the latter being more of a secondary rather than a direct effect. The effect of the wire-drawing temperatures is not as detrimental to low-carbon steel wires as to high-carbon wires. The toughness and uniformity are adversely affected, and in drawing high-carbon wire of high tensile-strength requirement, by either single or continuous drafting, the temperature must be controlled and kept as low as possible. The efficiency of lubricants used in the wire-drawing die, and carried on the wire in continuous drafting, is adversely affected by excessive wire-drawing temperatures. Without the cooling devices, the present high wire-drawing speeds could not be attained and die life would be considerably reduced.

Limitations of Drawing—Steel may be drawn several drafts further than the state illustrated by Figure 31—38, but when wire has been put into this highly strained condition by drawing, it loses its ductility to such a degree that it is not practicable to submit it to further drawing, and if required to be reduced in section still more, it becomes necessary first to restore the grains to the form and arrangement characteristic of the unworked (not strained) condition as illustrated by Figures 31—39 and 31—40. This changing back of the grains to a non-distorted form simultaneously restores the properties that existed in the wire before it was subjected to drawing. The processes employed for restoring the grain to strain-free formations are annealing and patenting. The effects of these processes will be discussed in the section on "Heat-Treatment of Wire." Enough having been said to explain the reasons for them, some of the chief features of wire-drawing practice will now be considered.

DRY DRAWING

Low-Carbon Wire—Most low-carbon wire is drawn directly from the hot-rolled rod. In producing sizes coarser than 4.76 mm (3/16 in.) in diameter, it is normal practice to roll the rod 1.59 mm (1/16 in.) larger than the required finished-wire diameter. in one draft.

Sizes 12.70 to 25.40 mm (½ to 1 in.) are drawn on horizontal bull blocks.

Sizes 4.76 to 12.70 mm (3/16 in. to ½ in.) generally are drawn on single-spindle motor blocks.

Wires having diameters smaller than 4.76 mm (3/16 in.) are drawn from a 5.56 mm (7/32 in.) diameter hot-rolled rod, which is the smallest hot-rolled rod size that it is commercially feasible to produce.

Low-carbon wire finer than 4.76 mm down to 3.56 mm (3/16 down to 0.140 in.) is drawn usually on double-

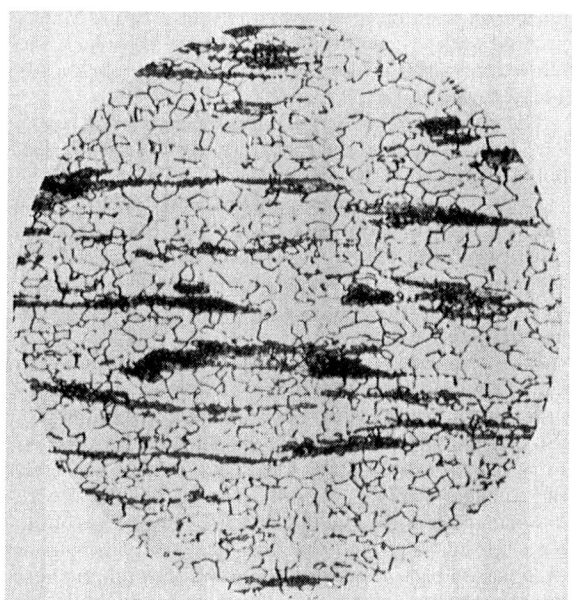

FIG. 31—39. Structure of 0.15 carbon-steel wire after short-time-cycle sub-critical anneal. Longitudinal section. Magnification: 500X.

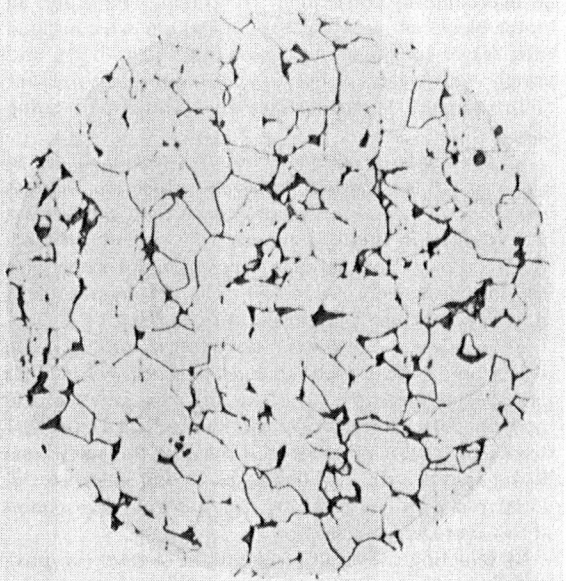

FIG. 31—40. Structure of 0.15 carbon-steel wire, full annealed after hard drawing. Longitudinal section. Magnification: 500X.

deck motor blocks. Sizes finer than 3.56 mm (0.140 in.) are drawn on multiple-spindle machines that, as previously described, may have 3 to 10 spindles.

Under favorable conditions, low-carbon wire can be drawn to reduce the cross-sectional area 95 per cent in a series of drafts without intervening heat treatment, starting with a hot-rolled rod of 5.56 mm (7/32 in.) diameter.

High-Carbon and Specialty Wire—High-carbon and other specialty wire are also produced by the dry-drawing method. As the designation implies, these wires are intended for use in applications of a particular special nature. It is important for the wire manufacturer to know the intended use of the wire and how it is to be fabricated or processed in attaining its end-use form, so that the wire-making practice can be adjusted to produce wire that will perform satisfactorily during the customer's manufacturing operations and in service.

In most cases, the wire is drawn from a patented or controlled-cooled rod. The patenting process, described in more detail later, is a special heat treatment used only on high-carbon steel and peculiar to the wire industry alone. It imparts uniformity of microstructure, hardness and strength to steel so treated, thereby overcoming any non-uniformity that may have resulted from the hot-rolling operation by which the rod was produced.

After patenting (if required), cleaning, coating and baking of the rod, the drawing operation on high-carbon wire is similar to that of drawing low-carbon wire, except that the reduction in cross-sectional area per draft is smaller for high-carbon wire and the cooling of the wire between drafts becomes an important factor in controlling the properties of the finished wire. If the wire is not cooled properly between drafts, the cumulative build-up of heat in the wire has an embrittling effect in the finished product that cannot be overcome by subsequent treatment. Therefore, all motor blocks and wire-drawing machines are equipped with water-cooled or air-cooled blocks, or both, and usually with water-cooled dies, as required to prevent embrittlement of the wire by the cumulative heating effect.

As previously mentioned, the cross-sectional area of a low-carbon hot-rolled rod may be reduced as much as 95 per cent in a series of drafts without intervening heat treatment but, as a general rule, the total reduction in cross-sectional area in drawing high-carbon wire from a rod does not exceed 75 to 85 per cent before a process patenting operation is performed.

Sometimes, a hot-rolled, unpatented rod may be drawn one or two drafts to a process size before it is patented and drawn to finished size. This practice promotes greater uniformity in the finished wire over that produced from a patented rod because the roundness of the cold-drawn wire before patenting is improved, compared with the relatively non-uniform cross-section of the average hot-rolled rod.

By selecting the right combination of steel composition, heat treatment, and amount of cold work, the wire producer can manufacture wire that can meet almost any of the many and varied requirements of the wire consuming industries.

WET DRAWING

The wet-drawing process is used to produce wire for decorative purposes or for special applications where extra-clean finish is required. In order to produce satisfactory wet-drawn wire, good cleaning and coating facilities and practices are required. Wet drawing is more costly than dry drawing; therefore, it is customary to dry-draw wire to a process size and then clean and coat the process wire prior to wet drawing to finished size.

Cleaning of the dry-drawn process wire prior to coating and wet drawing is usually effected by immersing the wire in a cold hydrochloric-acid or hot sulphuric-acid solution, or both. The objective is to remove all of the surface film resulting from dry drawing, so that a clean metallic surface is ready for plating.

The plating solution consists of copper and tin sulphates in a liquid solution made by mixing with cold water and a small amount of sulphuric acid. By adjusting the proportions of copper sulphate and tin sulphate in the batch, various shades of color ranging from a typical reddish copper color to a nearly metallic-white tin finish can be obtained as required.

After cleaning, the coils are spread out on a stem or hair-pin hook and dipped into the plating solution. The plating is accomplished. by deposition, and the thickness of coating and its uniformity are dependent upon duration of immersion and degree of agitation of the bath. After plating, the wire is washed thoroughly in fresh, cold water. At this point, the wire is either dipped in a special hot coating solution and baked, or placed in barrels and kept under water until it is delivered to the wire-drawing machines.

In principle, the wet-drawing process is identical with dry drawing. The equipment and actual practices, however, are entirely different. As a rule, lighter drafts (around 15 per cent) are taken.

With the older "underwater" method, the wire is placed on a submerged "wet reel" in a tub of water, pointed and pulled through a die onto the block. No lubricant is used in the die box; however, lubricant is added to the water.

The more modern method of using coated and baked wire paid off from a "dry reel" requires the use of additional lubricant in the die box.

For single-draft work, a single-spindle motor block may be used as well as frames. Block diameters are usually 660, 560, 406 or, in some cases, 305 mm (26, 22, 16 or, in some cases, 12 in.). Finished wire sizes range from 0.88 mm (0.035 in.) diameter to 19.05 mm (0.75-in.) diameter.

Wet Drawing—Multiple Drafts—This is done on continuous machines that differ from those used for dry drawing. In general, they are cone-type machines with the drawing blocks mounted adjacent to each other on the same horizontal shaft. Naturally, the first block has the smallest diameter. After a sufficient length of wire has been pulled through the first die, a few wraps of the wire are taken around the first block and the wire is then passed back around a sheave and through the next die onto the next block, and so on. The second and each succeeding block is made sufficiently larger in diameter than its immediate predecessor to take care of the increased length due to the decrease in diameter of the wire after each draft. After the required size is reached, the wire is gathered on a 203 mm (8 in.) diameter block.

Wet-drawing machines are usually capable of mak-

ing 10 to 14 drafts, and are completely enclosed, with a mixture of lubricant in water being sprayed continuously on the blocks and dies. In some cases, the blocks and dies are completely submerged in a water-and-

Table 31—II. Standard Size Tolerances for Wire.

Size Tolerances for Uncoated Coarse Round Wire in Coils

Size		Tolerance, Plus and Minus	
mm	in.	mm	in.
12.70 and larger	0.500 and larger	0.08	0.003
Under 12.70 to 1.93, incl.	Under 0.500 to 0.076, incl.	0.05	0.002
Under 1.93 to 0.89, incl.	Under 0.076 to 0.035, incl.	0.03	0.001

Maximum out of round is customarily one half the total gage tolerance.

Size Tolerances for Coarse Round Galvanized Wire in Coils

Tolerances, Plus and Minus

Size		Regular Coating		Type 1 Coating		Type 2 Coating		Type 3 or Class A Coating		Class B and C Coating	
mm	in.	mm	in.	mm	in.	mm	in.	mm	in.	mm	in.
12.70 to 6.38	0.500 to 0.251	0.08	0.003	0.08	0.003	0.10	0.004	0.13	0.005	0.15	0.006
6.35 to 3.76	0.250 to 0.148	0.08	0.003	0.08	0.003	0.08	0.003	0.10	0.004	0.13	0.005
3.73 to 1.93	0.147 to 0.076	0.08	0.003	0.08	0.003	0.08	0.003	0.10	0.004	0.10	0.004
1.91 to 0.89	0.075 to 0.035	0.05	0.002	0.05	0.002	0.05	0.002	0.05	0.002	0.08	0.003

Size Tolerances for Uncoated Fine Round Wire in Coils*

Size		Tolerances, Plus and Minus	
mm	in.	mm	in.
1.588/0.884	0.0625/0.0348	0.025	0.001
0.881/0.688	0.0347/0.0271	0.020	0.0008
0.686/0.508	0.0270/0.0200	0.015	0.0006
0.505/0.384	0.0199/0.0151	0.013	0.0005
0.381/0.257	0.0150/0.0101	0.010	0.0004
0.254/0.152	0.0100/0.0060	0.008	0.0003
0.150/0.112	0.0059/0.0044	0.005	0.0002

Size Tolerances for Galvanized Fine Round Wire in Coils*

Size		Standard Tolerances							
		Type 1 Coating				Type 3 Coating			
		Plus		Minus		Plus		Minus	
mm	in.	mm	in.	mm	in.	mm	in.	mm	in.
1.588/0.884	0.0625/0.0348	0.038	0.0015	0.038	0.0015	0.051	0.002	0.051	0.002
0.881/0.688	0.0347/0.0271	0.038	0.0015	0.025	0.0010	0.051	0.002	0.025	0.001
0.686/0.508	0.0270/0.0200	0.033	0.0013	0.013	0.0005	0.038	0.0015	0.025	0.001
0.505/0.384	0.0199/0.0151	0.025	0.0010	0.013	0.0005	—	—	—	—
0.381/0.257	0.0150/0.0101	0.020	0.0008	0.010	0.0004	—	—	—	—
0.254/0.152	0.0100/0.0060	0/013	0.0005	0.008	0.0003	—	—	—	—

*These tolerances do not apply to special wires which have been annealed as a separate operation following cold drawing and immediately prior to galvanizing.

lubricant solution that is constantly recirculated by a pump from a reservoir adjacent to the machine.

Drawing Limits and Tolerances—Although every effort is made to draw the wire as true to the required shape and size as possible, exactness in these respects is impossible under commercial conditions, and all wire produced under such conditions will vary somewhat in diameter and section. These variations may be due to a varying degree of hardness in the metal being drawn, to wearing of the die, or, in the case of wet-drawn wires, to unsuitable or imperfect coatings. Some tolerance is necessary therefore, but it is important that these tolerances be kept within certain limits defined by the use to which the wire is to be put. The tolerances used by the wire industry are divided into four groups of allowable variations in the diameter of the common grades of round wire, other than special grades, and designated as **standard, semi-special, special,** and **extra special** drawing. Each of these sets of tolerances contains the variations allowable, which are adjusted to the size of the wire as indicated in Table 31—II. The **standard** tolerances are those adopted by the American Iron and Steel Institute (AISI) and are given in its Steel Products Manual, Wire and Rods, Carbon Steel and Wire and Rods, Alloy Steel.

All wire is gaged carefully in two or three places around the circumference, just after the first end of the coil is pulled through the die and at the finished end to insure that it is of the correct size and shape. Wire ordered by millimetres, by gage number, by decimal, or by fraction of an inch, is gaged with a micrometer gage.

INSPECTION AND TESTING

Importance of Inspection—Of equal importance with the various processes in making wire is the thorough system of inspection and testing to which the material is subjected during its progress through the mill and after finishing. Beginning with the hot-rolled rod, the material undergoes frequent and careful inspection during manufacture and is subjected to whatever tests are necessary to determine its fitness for the service required.

Final Tests on Wires—Wire for shipment to customers is regularly inspected and tested for size, finish, and ductility. For ductility, small sizes are tested by a kink test and medium sizes by a mandrel bend test. Certain grades of wire are subjected to a machine test for determining tensile strength and elongation, and some of it may be given severe torsion tests. All telephone and telegraph stock is tested also for electrical conductivity. Cold-heading and cold-forging wire is subjected to upset testing and/or etch and file tests for surface-quality evaluation.

IMPERFECTIONS IN WIRE

After the wire has been drawn, it may be found unsuitable for the purpose for which it was intended. Process control tests and final inspection are established to detect and discard such wire. The chief sources of trouble can usually be placed under one of the following headings: off size or shape, internal imperfections, surface imperfections, and improper mechanical properties.

Size and Shape—The reasons for wires being off size or shape have been already given. While effort is made to draw the wire to the diameter ordered, the tolerances may be too great to permit use of the wire in an exacting application. Where extreme dimensional accuracy is needed, proper specifications should be established before the wire is drawn so that product can be made to suit the special requirements.

Internal Imperfections—Pipe and segregation are the most common types of internal imperfections, and are detected by deep etching of a cross-section of the wire, or by making a nick-and-break test.

Pipe and/or segregation cause uneven tensile strength, brittleness and what is known as **cuppy wire.** Cuppiness (cup and cone fracture) may also be caused by over-drafting.

Surface Imperfections—Sometimes the finish of the wire does not meet requirements. If the finish is not satisfactory it may be due to poor cleaning, to the use of improper coatings or other causes. In the case of liquor-finished wires, especially, the color may not be of the particular shade desired, or it may vary too much in different bundles. When these features are of special importance, they should be stated on the order, so that special attention may be given to meeting requirements in this respect. Other surface imperfections in wire are scratches, slivers, and seams. Scratches are caused by using a die in poor condition, or they may result from improper lubrication or from pieces of metal or other gritty substances being drawn into the die with the wire. Slivers, as the name indicates, are sharp pointed projections of metal that rise from the surface of the wire. They may be caused from any one of a number of sources, including the ingot and rolling practice. Seams are longitudinal cracks in the surface of the wire and are generally traceable to billet preparation, conditioning, and rod rolling. In general, seams are relatively shallow and are detected by upset testing, etching, magnetic-particle inspection or eddy-current testing as appropriate to the application of the wire.

MECHANICAL PROPERTIES

The mechanical properties of any wire will depend upon the chemical composition and quality of the steel as well as the exact nature of the drafting practice and of the heat treatment it has undergone. The particular application or method of fabrication determines the grade of wire necessary. A finished wire taken at random will not be suitable for all purposes, as the strength and ductility may vary from a 345 Mpa (50 ksi) tensile strength for an annealed low-carbon tie wire to 3310 MPa (480 ksi) tensile strength for a high-carbon music wire. If the wire manufacturer is fully informed both as to the desired properties of the wire and its application, the requirements can be met by proper selection of composition, heat treatment and predetermined amount of hardening by cold working in wire drawing. Many products such as upholstery springs, rope, telegraph wire, etc., are recognized in the industry as regular grades of wire for standard applications, being covered by standard specifications, and may be ordered as such.

HEAT TREATMENT OF WIRE

Heat-treating processes used in the wire industry include annealing, patenting, and oil tempering. These play such an important role in the industry and the methods of applying these processes have become so highly specialized that, although their general principles and their application to other branches of the steel industry have been thoroughly discussed elsewhere in this book, a brief account of their specific application to the wire industry is included here.

ANNEALING

Annealing may be separated into two types: process or subcritical annealing, and full annealing. These will be discussed separately below.

Process Annealing to Soften Hard-Drawn Wires—In the previous section, it was pointed out that each successive draft in drawing a wire has a certain hardening effect upon the metal; that the grains become elongated in the direction of drafting; that, as the drafts are repeated and the wire becomes progressively harder, a point is reached at which the wire will break if subjected to further drafting; and that well before such a condition of brittleness is reached, it is necessary to heat-treat the wire in order to put it in a condition that will permit further processing without injury. For reasons which have already been stated or will be made apparent in the discussions to follow, high-carbon wires are generally patented, while low-carbon wires are annealed.

For softening wires in process of manufacture, process annealing at sub-critical temperatures is usually employed. This type of annealing is used mainly to restore the ductility of hard-drawn low-carbon wires at process size so that they may be drawn to finer sizes, although it is also employed to adjust the mechanical properties of some wires at finished size as required by their application. To accomplish these results, it is only necessary to heat the wire to below the lower critical range, as a temperature between 535° and 675°C (1000° and 1250°F) is sufficient to soften the steel to such an extent that the material is almost as ductile as it was before any drafting was done. The reason for this is illustrated by comparison of the photomicrographs of wire in Figures 31—36 through 31—40. Figure 31—36 shows the form, arrangement, and size of the grains in full-annealed low-carbon steel. In this condition, this steel exhibits a tensile strength of approximately 330 MPa (48 000 psi) and an elongation of approximately 33 per cent in 250 mm (10 in.). Figure 31—37 shows the structure after a single draft; the grains are somewhat elongated in the direction of drafting. Figure 31—38, shows the structure of the wire after it has had about five drafts, and it will be observed that the grains have been generally elongated into a stringer-like or fibrous structure, greatly changed from the original equiaxed grains of Figure 31—36. This wire would have a tensile strength of about 860 MPa (125 000 psi) and an elongation of about one per cent. At an elevated temperature, such as 535°C (1000° F) the grains recrystallize, thereby

forming new grains. The greater the amount of plastic deformation, the lower the recrystallization temperature. Such incipient recrystallization or forming of new grains is indicated in Figure 31—41. Grain size increases as the temperature of annealing and the time the wire is held at temperature increase. The grain size obtained after long time annealing and recrystallization at 650°C (1200°F) is similar to that of the original structure, although certain remnants of the structure produced by drawing still remain, with some residual directional properties. The mechanical properties obtained by annealing hard-drawn low-carbon wires at various temperatures up to 650°C (1200°F) are shown in the graphs of Figure 31—42.

Sizes of Grains—While the preceding paragraph shows the effect of grain distortion upon the mechanical properties of steel, nothing is said about the effect of grain size on these properties, and in all annealing, as well as other forms of heat treatment, the matter of grain size is of considerable importance. Large grains lower the tensile properties of the metal somewhat, while small grains make it not only stronger but tougher. The reasons for this, as well as the effects of cold working, are to be found in the structure of the grains themselves. Structurally, each grain is in reality a small crystal in which the atoms are arranged in regular rows and "layers," as described elsewhere in this book. Crystals deform through movement of adjacent parts along definite planes, called **slip planes,** whose location and direction within the crystal are determined by the positions of the atoms within the space lattice. Anything that interferes with this movement will make the metal more resistant to deformation, while anything that promotes slip will lower its resistance to deformation. Small grains offer greater interference to slip than large grains, because the slip planes in different grains lie at different angles, and in any one of such small grains the slip cannot progress far before it is stopped by an adjacent grain. Each grain offers considerable resistance to the force causing the slip, because the direction of possible slip movement may differ greatly from the direction of the applied force. In a small section, such as a wire, where the ratio of the diameter of the wire itself to that of the crystals is relatively small, the effect of large grains on lowering the tensile properties is likely to be very noticeable. The increase of strength due to drawing may also be explained from the standpoint of slip interference, for, as the grains are distorted, the slip interference becomes greater and greater until most of the grains have been drawn into parallel stringers and there are no more planes of easy slip on which further motion can take place. In this condition, the force required to pull the wire apart should approach the force of attraction of the molecules or atoms themselves, since the wire no longer can adjust itself to the load by deforming, but it never does because some of the grains begin to break before the others. The resulting ferrite grain size is referred to as "structural grain size," and is not to be confused with austenitic grain size.

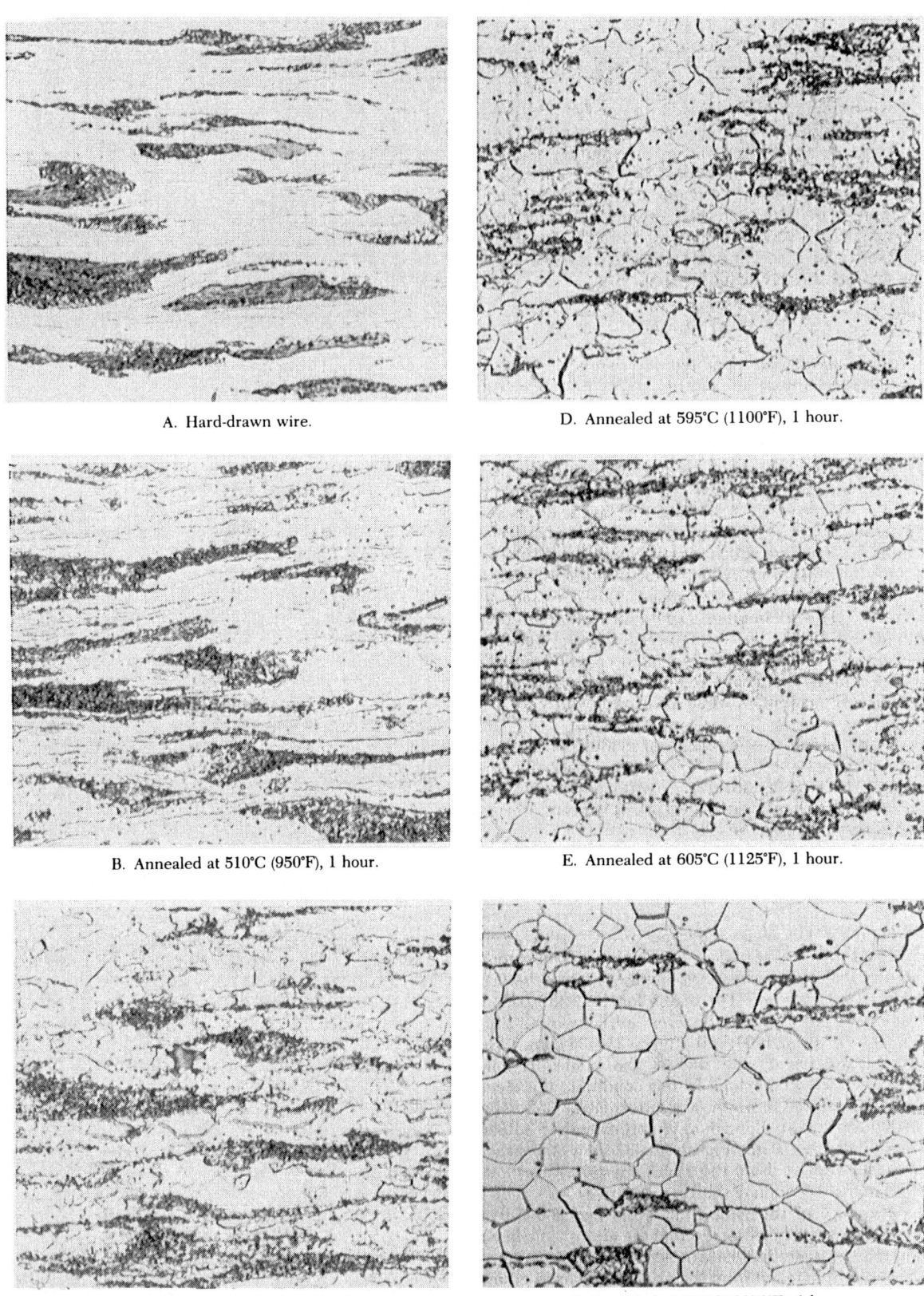

A. Hard-drawn wire.

D. Annealed at 595°C (1100°F), 1 hour.

B. Annealed at 510°C (950°F), 1 hour.

E. Annealed at 605°C (1125°F), 1 hour.

C. Annealed at 540°C (1000°F), 1 hour.

F. Annealed at 620°C (1150°F), 1 hour.

FIG. 31—41. Series of six photomicrographs illustrating the changes in microstructure of a hard-drawn low-carbon steel wire after annealing at various temperatures. All longitudinal sections magnification; 1000X.

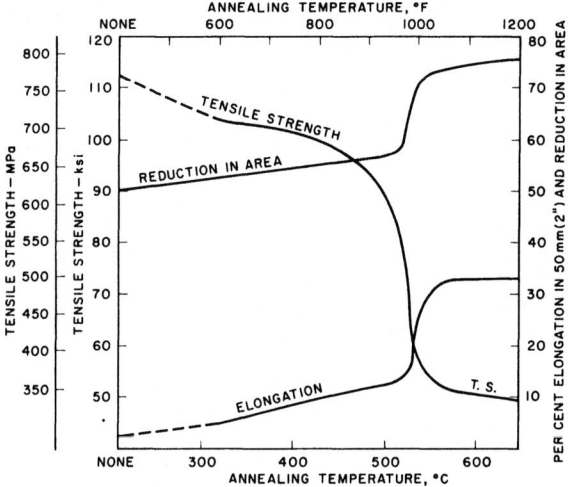

FIG. 31—42. Curves showing changes in mechanical properties of hard-drawn low-carbon steel wire after annealing at various temperatures up to 650°C (1200°F).

The conditions for process annealing involve three factors, namely, strain, time, and temperature. Without strain no change in the crystalline structure takes place on heating below the critical range. As to the effect of the amount of the strain in causing grain growth, or rearrangement, below the critical range, in low-carbon steel, very slight straining followed by subcritical annealing produces no grain growth; moderate straining (approximately 15 per cent), such as might be effected by a single light draft, and annealing cause the grains to grow to very large size resulting in "orange peel" surface if the wire is deformed; over-straining, or severe distortion, and annealing at the same temperature and for the same length of time produce grains of small size, comparable with the structures shown in Figures 31—36 and 31—37. The condition of strain producing large grains is seldom met with in process annealing of wire, because the grains of such wires are usually distorted far beyond that point. However, when a consumer expects to anneal wire, this fact should be made known to the manufacturer; it will then be possible, except in a few cases, to supply wire that will not form large grains under any annealing conditions.

Time and Temperature for Annealing—From what has already been said, and the evidences supplied by Figure 31—41, it is evident that, for the best results in annealing, careful regulation and adjustment of the time and temperature are necessary. In all types of annealing, sufficient time must be allowed for the whole charge to reach the required temperature, and in low-temperature annealing additional time is required for the development of the structure desired. Process annealing requires careful adjustment of time and temperature in order to avoid the formation of large grains, and, at the same time, eliminate the effects of the cold working, because at a low temperature the atoms or molecules move at a slower rate, and more time is required for their adjustment than at higher temperatures. As to temperature, the fact that full control and careful regulation of this factor in all

annealing is necessary to satisfactory results cannot be too highly emphasized. If the temperature is too low, it may fail entirely to accomplish the results desired in process annealing. In some cases adjustment of the temperature to the size of the wire is also necessary. Fine wire in coils, for example, is generally annealed at as low a temperature as practicable, in order to avoid the sticking of the wire in the coils. After these remarks it is scarcely necessary to say that accurate and carefully standardized pyrometers form an essential part of any equipment for the proper annealing of wire.

Annealing for Definite Structures—This type of annealing is applied to the higher carbon steel wires, i.e., those with a carbon content of 0.35 per cent or more, in order to obtain certain definite structures, and is similar to the treatment given carbon tool steels and other high-carbon steels in order to make them easy to machine. If steel having a pearlitic (lamellar) structure is heated to 705°C (1300°F), and is held at that temperature for a sufficiently long period of time, the iron carbides slowly coalesce, changing from a platelike form to a granular, globular, or spheroidal form. By properly regulating the temperature of the treatment, structures intermediate between these two extremes and best suited to the purpose for which the wire is to be used may be obtained with temperatures in excess of 720°C (1330°F) yielding a correspondingly greater percentage of pearlite as the temperature increases. These structures are sought in wires mainly to enhance their workability in processing or their formability in manufacturing the end products. Close regulation of the temperature is imperative in this type of annealing, for in heating the steel to a point so near the critical range, a difference 28°C (50°F) in the temperature often means the difference between obtaining the spheriodized carbide structure sought and the lamellar (pearlitic) structure which it is desired to efface.

The tensile strength (at various carbon levels) of wire full-annealed or recrystallized by annealing above the upper critical point, followed by slow cooling, is shown in Figure 31—43, as well as that of wire subjected to annealing in a molten lead bath (one form of a "subcritical anneal" involving a short time cycle of heating and cooling). It should be noted that it is possible for sub-critical annealing to produce approximately the same tensile strength values obtained by full annealing, providing the long time cycle of heating and cooling used in the former treatment is observed. A comparison of the effects of full annealing on recrystallization and grain size with short time subcritical annealing of hard-drawn wire is indicated by the photomicrographs of Figures 31—39 and 31—40. These differences in structure are reflected in the lower tensile properties of full-annealed wire shown in Figure 31—43.

Methods of Annealing Wire—While it is possible, theoretically at least, to anneal wire in any apparatus in which the proper conditions of time and temperature can be maintained, the efficient handling of the material calls for, and some of the finishes and conditions to be met demand, equipment specially designed for the purpose in mind. These conditions, in general, call for differences in the time of annealing, and differences in the methods of applying the heat and usually require

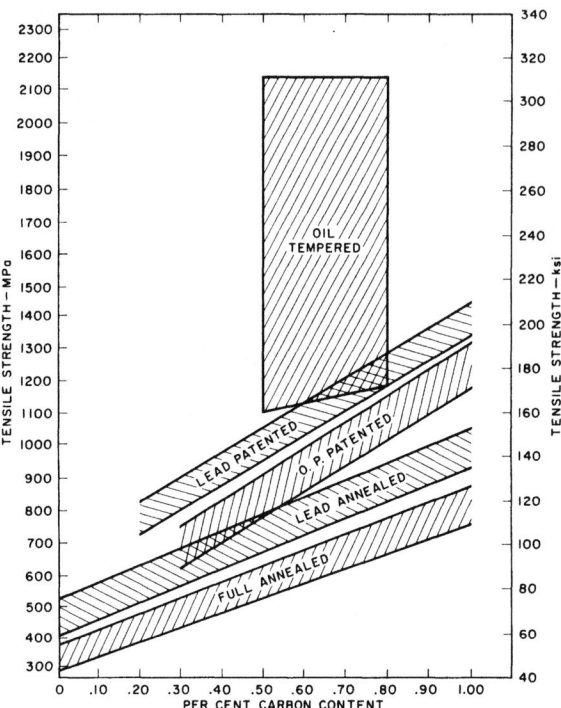

FIG. 31—43. Curves showing approximate effect of carbon content and heat treatment on tensile strength of steel wire.

that the material be protected against oxidation. To meet these conditions, several different methods requiring entirely different equipment for each are employed: these are known as **controlled-atmosphere annealing, salt-bath annealing, continuous lead annealing,** and **tube annealing.**

Controlled-Atmosphere Annealing—Controlled atmosphere annealing is the most recently adopted method of annealing used by the wire industry. Both batch-type and continuous-type furnaces are employed. The protective gas or controlled atmosphere maintained in the heating chambers or muffles of these furnaces is prepared by various methods such as the partial combustion of natural gas to produce a "neutral" atmosphere gas (consisting principally of nitrogen with controlled amounts of methane, carbon monoxide and carbon dioxide). The atmospheres are formulated to prevent oxidation and decarburization of the steel rods or wire undergoing heat treatment. Both types of furnaces are heated by gas-fired radiant tubes. Further details about controlled-atmosphere furnaces, the means for heating them, and the generation of controlled atmospheres will be found in the section of Chapter 41 dealing with heat-treating furnaces.

Furnaces referred to as bell-type or cover-type are employed for **batch-type annealing** of rods or wire under a controlled atmosphere. Coils of rods or wire are "threaded" on vertical stems placed on a lattice-work cast-steel base. A stainless-steel inner cover or hood is lowered over the charge. A high-velocity fan in the base circulates controlled atmosphere within the inner cover during the annealing cycle. The bell, with

radiant-heating tubes installed on its inner surfaces, is then lowered over the entire base and inner cover that encloses the charge.

Radiant-tube-heated continuous furnaces for annealing rods and wire in coils under a controlled atmosphere are of the pusher type. The furnaces are zone-controlled to provide connecting zones for heating, soaking and cooling of the charge. The wire or rods are loaded onto mesh-type trays that are pushed, one following another, into the entry end of the furnace. As a tray is pushed into the entry end, a tray of annealed wire is pushed out of the furnace at the exit end, the thrust being transmitted from tray to tray through the continuous line of trays in the furnace. Escape of controlled-atmosphere gas is minimized by a vestibule with inner and outer doors at each end of the furnace.

For annealing rods and wire, the continuous-type furnace is believed to possess some advantages over batch-type furnaces, notably: shorter annealing cycles and better uniformity of heating of the coils. However, the efficient operation of a continuous furnace requires that large tonnages of similar material be available for heat treatment to minimize changes in the operating cycle of the furnace, and operation of the furnace should be uninterrupted for the longest possible periods, since down time resulting from cycle changes or lack of or delays of stock to be annealed can increase operating costs considerably.

When annealing hot-rolled rods by either the batch or continuous method, the rods should be cleaned and lightly coated with lime, since scale or rust will cause decarburization of the rods by reacting with carbon in the steel in spite of the presence of a controlled atmosphere.

Salt-Bath Annealing—Molten salt-bath annealing is sometimes practiced for process annealing of common sizes of wire. The wire, in coils, is immersed for one-half hour to one hour in gas-fired pots containing molten salt which is held at some predetermined temperature between 535° and 705°C (1000° and 1300°F), depending on steel composition and structure or physical properties desired. Advantages over other methods are that small amounts of wire may be quickly annealed at closely controlled temperatures without scaling the surface of the wire. Such annealing is somewhat limited as to its application in wire processing.

Continuous lead annealing consists merely in drawing the wire through a bath of molten lead heated to the proper temperature. The molten lead is contained in a shallow rectangular steel pan, some 250 to 380 mm (10 to 15 inches) deep, 0.9 to 1.2 metres (3 to 4 feet) wide, and 4.6 to 7.6 metres (15 to 25 feet) long, the exact dimensions depending upon conditions. (Sometimes two pans are used, the first known as the cold pan and the second as the hot pan, and the wire is drawn through each in succession.) The pan is heated from below usually by gases from an oil or gas fire and is supported by a brick setting adapted to the method of heating. In practice several strands of wire are drawn through the bath in parallel. For this purpose the coils of wire to be annealed are placed on free-running reels before the annealing furnace, and the wire from each coil is drawn through the bath by a take-up block placed at a convenient distance from the other end.

For keeping the wires immersed in the molten lead, devices of various forms, known as **sinkers,** are used. To obtain bright annealing by this process, it is necessary to cool the wires without their being in contact with the air. Consequently, when such a finish is desired, the wires from the hot lead bath are conducted directly into long tubes which are kept as nearly sealed as possible.

The advantages of lead annealing are obvious. In addition to protecting the material from the air, the bath of molten metal is readily maintained at a uniform temperature throughout its mass, and its temperature can be accurately ascertained at any time by a pyrometer. Also, since the process is continuous, it eliminates much handling of the material.

The principal use of lead annealing is in connection with galvanizing plants, where it is employed to anneal process wire. In these plants, layouts are provided that permit the wire to be annealed, cooled, cleaned, washed and dried and galvanized or tinned, all in one continuous operation. In continuous annealing and cleaning, when the wire is not to be galvanized or tinned, the wire, after passing through the lead bath, is conducted through crushed coal or sand banked at the end of the lead pan; this bank of coal or sand is known as a **header.** The wire then passes successively through a water bath, a weak acid bath, a cold water bath, and a hot water bath, thence to the take-up blocks, which are located several yards beyond the hot water bath to permit the wire to dry. This elimination of the customary process of dip-cleaning with cranes is very desirable in the case of process wire.

PATENTING

Patenting is a heat treatment applied to rods and wire generally having a carbon content of 0.40 per cent and higher, the term being peculiar to the wire industry. The object of patenting is to obtain a structure which combines high tensile strength with high ductility and thus impart to the wire the ability to withstand heavy drafting to produce the desired finished sizes possessing a combination of high tensile strength and good toughness.

Methods of Patenting—Patenting is always conducted as a continuous process and consists in first heating the material to a point well above the upper critical temperature, then cooling through the critical temperature at a comparatively rapid rate to a predetermined temperature level at which the transformation will yield the desired microstructure (usually fine pearlite) and mechanical properties. In practice, there are various ways of carrying out the treatment. Thus, the wire may be heated by passing it through alloy-steel tubes arranged in an open muffle or in an open flame without tubes and be cooled by pulling it from the furnace into the open air, which method is now referred to as "O.P." (old process or air) patenting. In a second method, the wire may be heated as in the first method, then cooled by passing it into a lead bath held at a relatively low temperature; this process is known as the "M.H." or **metallic hardening process.** The wire, in the third process called "D.L." or double-lead patenting, can be heated in a bath of very hot lead and cooled

in another bath of lead at a lower temperature. In this process the temperatures of both baths can be readily controlled and accurately measured. These features make it possible to obtain desired structure even in wires of highest carbon content, a condition not easily attainable in "O.P." patenting. The third method is advantageous in that less scale is formed than in the other two methods. In the wire industry, both the metallic-hardening process and the double-lead process are generally referred to as "lead patenting."

Because of environmental considerations associated with the use of lead as a heat-treating medium, lead is being replaced with either molten salts or with fluidized beds. Although neither of these media has heat transfer coefficients as high as lead, the resulting mechanical properties of the heat treated rod or wire are comparable and satisfactory for most applications. Because the fluidized bed uses inert particles for transferring heat, the particles do not react with the wire or rod surface and are therefore easily removed by air wiping.

Another method of patenting involves the use of electric direct-resistance heating and quenching in a molten alloy metal bath. A recent development particularly applicable to patenting very high carbon and hypereutectoid steels involves a double cascade quenching of the rod or wire from the austenitizing temperature. In the first cascade quenching chamber the temperature of the material is reduced to below the Ar_1 temperature zone. The quenchant used is preferably a eutectic $NaNO_3$-$NaNO_2$ composition maintained at a temperature within the range of 205° to 370°C (400° to 700°F). After traveling through the first cascade quenching chamber, the material passes through a second cascade quench in which the temperature is maintained between 400° and 565°C (750° and 1050°F). From the second cascade quenching chamber, the material passes into a holding furnace in which the temperature of the steel is held at the level required to develop the desired properties and for a sufficient time so that complete transformation occurs.

Many modern rod mills are equipped with controlled cooling facilities which impart to the rod the advantages of a patented structure. In some cases, the separate patenting treatment (especially O.P.) of the rod can be eliminated.

Properties of Patented Wires—The structure obtained by patenting is extremely tough and possesses the best characteristics for drawing to high tensile strengths. As indicated in Figure 31—43, the tensile strengths of air- or lead-patented rods or wire are considerably higher than the same steels in the annealed condition. The lead patenting process is definitely required in the production of any exceedingly high strength and tough wire such as music wire. As an illustration, by properly patenting and drawing 0.75-carbon steel, a wire is produced having a tensile strength of 2590 MPa (375 000 psi) or over. In spite of the great amount of drafting required to raise the tensile strength to this point, such wire will be tough enough to wrap around itself (i.e., can be wound around a pin of the same diameter as the wire) or can be hammered flat to one-half its original thickness without cracking.

OIL TEMPERING

In other branches of the steel industry, hardening and tempering are usually considered as separate operations, but in wire-making they are more often conducted in a single continuous operation known as **oil tempering**. As in the case of tools and other products of steel, the object in oil tempering wire is to control the hardness, or "temper," in order to adapt the material to the use for which it is intended. Oil tempering is applied to a great variety of wires, such as overhead-door spring wire, sewer-auger wire, and wires for measuring tapes, clock and watch springs, automobile-engine valve springs, door-check springs, etc. Wires for such purposes are generally treated in wire form by a continuous process, but many kinds of springs, tools, and miscellaneous special products, are heat-treated after the wire has been formed into the article desired, when oil-tempered wire with the desired mechanical properties cannot be formed to the required shape.

Methods of Hardening and Tempering Wire—For greatest efficiency in hardening and tempering, it is necessary to adapt the method of treatment to the nature of the material and the size and form of the article. Therefore, because of the great variety of steels and of articles produced by the wire industry, various methods of treatment involving the use of various kinds of apparatus and of various heating and cooling media, are employed. Thus, for hardening, the material may be heated in muffle, tube or electric furnaces, or in molten lead, and be quenched in oil, or some other medium. For tempering, the material may be reheated in open furnaces, in baths of molten lead or hot oil, or by bringing it in contact with hot sand or with hot plates or discs of iron or steel. In continuous oil tempering, the heating for hardening may be accomplished in a manner similar to that for patenting, and the quenching may be accomplished by passing it quickly into a tank of oil; in the tempering operation, the material may then be drawn through a suitable bath, such as molten lead or a bed of hot fluidized alumina sand, the wire being in continuous motion throughout the process.

Properties of Oil-Tempered Wire—The mechanical properties of oil-tempered wire can be varied over a considerable range, as indicated in Figure 31—43. This is accomplished by varying the chemical composition, principally the carbon content, and by varying the temperature of the tempering medium. For applications such as crimping for screen cloth or for cold rolling, temperatures up to about 677°C (1250°F) are used, resulting in low tensile strength and high ductility. For such applications as highly stressed mechanical springs, overhead-door springs and sewer augers, temperatures as low as about 438°C (820°F) are used, resulting in high tensile strength. The decreased ductility which is naturally associated with the higher strength is not detrimental in applications which do not involve severe deformation of the oil-tempered wire. Within a given class of applications, small-diameter wires are generally produced to higher tensile strength than large-diameter wire.

SECTION 9

PROTECTIVE METALLIC COATINGS

Kinds of Coatings—Probably more than one-third of all the steel wire drawn is given some kind of metallic coating, either for decorative or protective purposes. This phase of the industry has reached huge proportions and is of great importance. As already indicated, appreciably large quantities of the medium and larger sizes of wire are given a light coating or "liquor finish," varying in color from that of copper through the various shades of brass to the color of tin, for decorative purposes only, and practically all soft fine wires have to be given this coating for lubricating purposes. This coating, which may consist of copper or tin or a combination of the two, is very thin and has but little value as a protection against corrosion. As is well known, all of the common grades of steel will rust sooner or later when exposed to atmospheric conditions unless their surface is covered with some substance that will protect it from moisture and air. To afford such protection, it is the practice to galvanize steel wire, i.e., coat it with zinc, or to tin it, or to coat it with aluminum, or, for certain purposes, to give it a coating of paint, lacquer or japan. While it is also possible to coat steel wire with copper, lead, nickel, and the precious metals, these metals are seldom used for coating steel wire, except in a few particular cases, so they need not receive any further consideration here. The remainder of this section will, therefore, be devoted to galvanizing, tinning, and aluminum coating.

WIRE GALVANIZING

Advantages of Galvanizing—As a protective coating, zinc offers several advantages over other metals: (a) Zinc is electropositive to iron, which means that if the steel of a galvanized wire should be exposed to corrosive influences, owing to defective or damaged coating, the zinc will corrode first, and its presence will protect the iron from corrosion until after a considerable area has been exposed. Just the opposite reactions might occur with other metals, such as tin, which are electronegative to iron. The presence of tin alone under the same conditions would hasten the corrosion of the iron, but, for specific purposes, tin is added in small amounts to zinc for galvanizing to improve adherence of the metallic coating, and the size and appearance of the spangle, without affecting corrosion resistance. (b) Zinc may be obtained in a comparatively pure state at a relatively low cost. (c) With zinc, it is easy to obtain a hard, smooth coating with relatively good resistance to abrasion. (d) The color of the zinc coating is satisfactory for general purposes.

Methods of Galvanizing—There are two common

methods in use for galvanizing wire, known as (1) the hot galvanizing process and (2) the electrogalvanizing process, but the second is the less widely used. In both processes, some 30 (more or less) parallel strands of wire pass first through certain preparatory processes, thence through the galvanizing bath to take-up frames on which each strand of wire is separately coiled, all these operations being made by attaching the last end of each coil to the first end of the succeeding one. As the wires pass through the various baths required to clean and coat them, they are submerged by suitable forms of sinkers (either of the stationary or rotary type) or, in the case of electrogalvanizing, under contact fingers in the electrolytic or plating solution.

Processes Preliminary to Hot Galvanizing—In many cases, wire to be galvanized must be annealed to remove the effects of cold working, and, in order to minimize handling of the wire, it is common practice to do this work in conjunction with the galvanizing operations by one of the continuous annealing processes. For this purpose, as well as for burning off the wire drawing lubricant, a molten-lead pan in front of the cleaning and galvanizing apparatus has generally been used. Environmental considerations now dictate the use of alternate means such as molten salts or a fluidized bed. For those galvanized products that require only stress relieving, heating in the zinc pan is sufficient and a pre-heat treatment is not required. Lubricants, however, must then be removed from the wire surface by an alternate operation such as electroalkaline cleaning. Since the rate of cooling in process annealing has little effect upon the physical properties of the wire, the wires are cooled in air, or, if the space is limited, low-carbon wires will be cooled by conducting them from the annealing furnace into a vat of water. Following annealing, the next step is cleaning, for which purpose the wire is conducted into a bath of hot muriatic (hydrochloric) acid at predetermined concentration. This acid is used instead of sulphuric, which is the acid commonly employed for cleaning in the galvanizing of sheets and tubes, because it acts much more quickly than sulphuric, and is much more effective in removing traces of lime remaining from the drawing operations. The iron chloride formed by the action of the acid, as well as any particles of loosely adhering scale, must next be removed and these objectives are accomplished by passing the wire through a bath of hot water.

While the wire is now perfectly clean, it must also be perfectly dry before it is brought into contact with the molten zinc, but in drying, any exposure to the air, which in practice it is impossible to avoid, results in the formation of a light coat of oxide, or rust. This difficulty is overcome by the use of a flux. In galvanizing sheets, tubes, and various wire products, this flux consists of fuxed ammonium chloride, commonly known as sal ammoniac, which lies upon the surface of the zinc bath, but for the continuous galvanizing of wire, better results are obtained by passing the cleaned and thoroughly washed wires directly through a hot solution of zinc chloride or zinc ammonium chloride flux at predetermined concentration or Baumé; and then through or over a dryer into the molten zinc. The wire thus becomes coated with a thin film of zinc chloride, which tends to protect it from oxidation during drying and also removes any traces of rust that may be formed.

Apparatus for Hot Galvanizing—From the dryer, the wires are drawn at once into the molten zinc, or **spelter,** as it is commonly called. This molten metal is contained in a **spelter pan,** which is usually made of boiler plate and is supported by a brick setting of suitable construction for firing with the most satisfactory fuel available. Figure 31—44 shows the general form of the pan and the furnace commonly used. The dimensions of these pans are subject to much variation, and depend upon several conditions. Thus, the width will vary according to the number of strands it is desired to galvanize at once, which is usually about 30. The length is dependent upon the size of the wire, the speed of travel, and the thickness of the coating desired. Pans designed for galvanizing coarse sizes of wire may reach a length of 9.1 metres (30 feet). The depth of the pan must be sufficient to prevent the wires from coming into contact with the **dross** which settles and collects upon the bottom. This dross is an alloy of iron and zinc containing from 3 to 7 per cent iron, which is solid at the temperature of molten spelter, forms a pastelike mixture, and is very harmful to the coating. As molten zinc oxidizes rapidly, the pan is provided with some form of covering medium, which rests upon the molten spelter and protects it from the air. Periodically, the surface of the zinc metal must be skimmed to remove the buildup of oxides. The mixture of zinc and zinc oxide thus obtained is known as **zinc skimmings;** these skimmings, together with the dross, represent a considerable loss, for the proportion of these waste products to the total zinc used is relatively large, and, although both are refined, the cost of recovering the zinc they contain is relatively high.

Wiping the Wire—As the wires emerge from the galvanizing pan, some of the zinc they carry remains in the molten state for a brief period, and, unless prevented from doing so, tends to flow downward on the surface of each wire, making the coating rough and uneven. Such a coating makes the wire hard to handle and renders it unsuitable for many purposes, especially when it is to be fabricated by machinery. It is evident that in order to form a smooth, evenly distributed coating, the surplus zinc must be removed while it is still in a fluid state. This is accomplished by passing the wire, just after it emerges from the zinc bath, through devices known as **wipes, headers,** or **gas knives.** In the case of wipes, the parts in contact with the wire are made of asbestos, and the type most commonly used is known as the **split wipe.** It consists essentially of two balls, or molds, of asbestos fibers held together by some suitable binding material, or stranded asbestos rope of suitable diameter formed into wiping pads. In service, these two balls are pressed lightly against and about the wire, and are held securely in place by cup-shaped holders. The aim in the split wipe is to make it possible for the operators to separate the two halves in order to permit the splices joining two coils of wire to pass, as otherwise the wipe is torn out entirely or badly damaged. Headers or gas knives may be employed for wiping only when heavier coatings are desired. A header may consist of crushed charcoal of definite mesh or pebbles in combination with an atmosphere such as CH_4-H_2S. For extra heavy coatings, a circular gas knife

FIG. 31—44. Overall view of the apparatus employed in the continuous hot-dip galvanizing of steel wire.

using a controlled pressure of N_2 may also be used to control the coating thickness.

Cooling the Coated Wire—After passing the wipes, the coating of the wire is completed, but the manner in which it is cooled should also be given some consideration. If the wire is allowed to cool naturally in the air, the coating, because of the formation of a thin film of zinc oxide on the surface, will be dull and lusterless in appearance, known as air-cooled galvanized wire, but if the coating is cooled suddenly, as by immersing the wire in cold water as soon as possible after leaving the wipes, it will have a bright lustrous surface. However, since the real value of the coating as a protection against corrosion depends upon the thickness of the coat and the completeness with which it excludes moisture from the iron, it is clear that the luster, which is a characteristic of suddenly cooled coatings, is no indication of the quality of the coating, and since this luster can be preserved only for a short time in any natural atmosphere, its attractive appearance is of little value. For heavier coatings, the zinc should be rapidly solidified to prevent coating sag and to minimize the formation of a brittle FeZn alloy.

Coiling the Wire—The wire, which is delivered to the galvanizing department in coils, is drawn through the annealing, cleaning and galvanizing apparatus by blocks, which form it into coiled bundles again (Figure 31—44). The blocks commonly used for this purpose are of the so-called continuous type, and are similar to a wire-drawing block except that their spindles are mounted in a horizontal position. With these blocks, bundles of wire may be removed at any time without stopping the blocks, so that the speed of travel of the wire through the annealing and galvanizing baths may be uniformly maintained. Many modern galvanizing

lines are equipped with dead blocks as described previously under "Drawing Machines—Terminal Equipment," or with inverted live blocks to produce heavyweight continuous coils which are accumulated on tubular coil carriers.

Some Features of the Operations for Hot Galvanizing—To the novice the galvanizing of wire may appear to be a simple process. The wires being drawn through the various pieces of equipment and recoiled on the blocks apparently require little attention from the operator. As a matter of fact, however, the process demands considerable experience and skill in order to obtain uniformly good results. For example, the firing of such long shallow pans requires not only a properly designed furnace, but constant care and observation on the part of the operator to keep the temperature of the spelter uniform and constant and to avoid injury to the pan. Similar statements apply also to the annealing and cleaning equipment. Some of the factors that determine the quality of the zinc coat are as follows:

The quality of the zinc coat, by which is meant its effectiveness in protecting the wire from corrosion and its adaptability to the intended use of the wire, depends upon the completeness with which the zinc covers the wire, the adhesion, the thickness, uniformity, physical properties, structure, and chemical composition of the coat. The completeness with which the wire is covered by the zinc depends upon the thoroughness of cleaning and fluxing. Attention has already been called to the effects of the manner of cooling the wire as it is leaving the spelter bath. It should also be noted that the coating may be injured in fabricating the wire. If the cooled wire is bent at sharp angles or otherwise severely deformed, the coating may be cracked or peeled off, especially if the brittle alloy layer

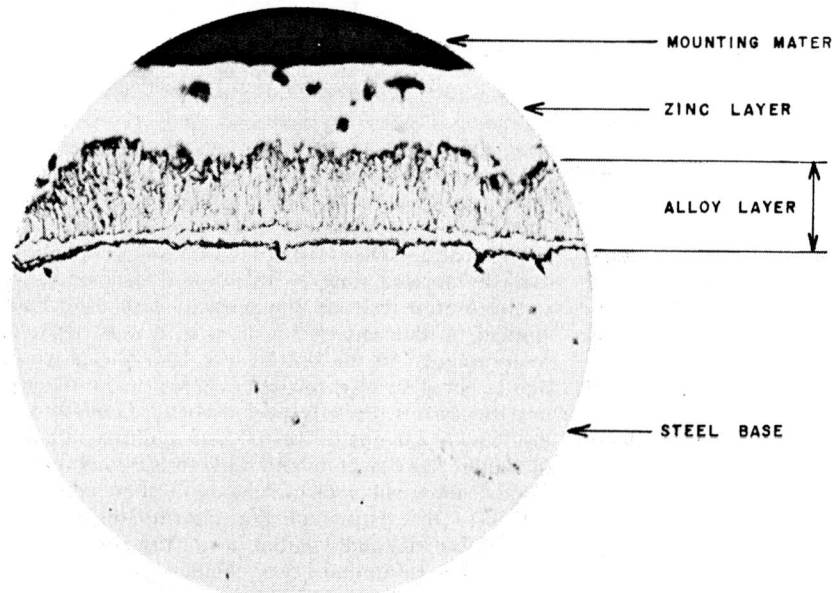

MOUNTING MATERIAL

ZINC LAYER

ALLOY LAYER

STEEL BASE

FIG. 31—45. Structure of the zinc coating on a steel wire galvanized by immersion in molten zinc. Magnification: 250X.

is thick and the coating is rough. Small additions of aluminum to the spelter and rapid cooling of the coated wire improve the ductility and adhesion of the coating. As to the thickness of the coat and the uniformity with which the spelter is distributed over any given wire and adherence of the zinc coating, these features depend upon the type of steel used, thoroughness of cleaning, the methods of wiping, the temperature and composition of the molten spelter, and the time of immersion of the wire in the spelter bath. Up to certain limits, the thickness of the alloy layer increases as the temperature of the bath increases and as the time of immersion is prolonged, thereby increasing the coating weight. The time of immersion is controlled by the length of the spelter pan and the speed of travel of the wire through the bath. In this connection it should be pointed out that the thickest coat is not always the best, for, as the thickness of the coat increases, the coat becomes rougher and its tendency to peel or crack also increases. Although alloying the spelter will decrease the thickness of the brittle alloy layer, this may decrease the resistance of the zinc to corrosion, and for certain products the spelter is kept as pure as possible. The composition of the spelter is checked by chemical analysis, and close limits are fixed for all the impurities it may contain.

The Structure of the Zinc Coat—The structure of the zinc coating as obtained on hot-dipped galvanized wire is shown in the photomicrograph of Figure 31—45. It will be seen that the spelter coating is made up of three different layers. The lower portion of the photomicrograph represents a section of the steel base, and the zinc layer represents the outer layer of spelter, which may be considered pure zinc. The alloy layer is made up of two different alloys of iron and zinc, the one next the steel base containing slightly more iron in solution than the other. Formerly it was thought that when iron is brought in contact with molten spelter, two compounds, having the formulas $FeZn_3$ and $FeZn_7$ are always formed. More recent studies indicate that the constitution of these coating layers is not precisely represented by these formulas. Assuming for the present that the formulas are substantially correct, however, the alloy next to the steel base consists of the compound $FeZn_3$, while the remainder of the alloy layer is composed of the compound $FeZn_7$. The dark lines separating different areas in the photomicrographs are not cracks, but the result of grain boundary effect in etching, similar to that observed in developing the structure of pure metals.

Electrogalvanizing—Electrogalvanizing is the name applied to the process of covering any metal with a coating of zinc by means of an electric current. It is sometimes termed "cold" or "wet" galvanizing to distinguish it from the more common method of hot galvanizing. For galvanizing flat wire and also strip steel, this cold process has certain advantages over hot galvanizing. For example, in hot galvanizing such materials, the edges, especially of strip steel, tend to destroy wipes, and special care is required to secure a smooth uniform coating, whereas the thickness of the zinc coating may be readily controlled in electrogalvanizing. The most important advantage in electrogalvanizing is in the application of very heavy coatings; these heavy coatings can be applied to closer weight specifications and considerably smoother than hot-dip coated wire. The thickness of an electrodeposited coating depends primarily on two factors, namely, current density and time. By current density is meant the number of amperes flowing per unit of surface of the metal exposed in the electrolyte. The coating, when well applied, adheres firmly to the metal and will stand severe bending without flaking off. But many features of the process require special attention, because the quality of the coat depends upon the condition of the electrolyte, the cleanliness of the surface, and on general working conditions.

The equipment for electrogalvanizing consists of a long shallow horizontal vat, known as the plating vat, which is usually made of steel lined with rubber or a synthetic mixture. Attached to one outer side of this vat are the positive bus bars, or feeders, connected to the plating generator, while on the other side are similar negative feeders. As considerable time is required for suitable electroplating, the vats are usually from 30.5 to 61.0 metres (100 to 200 feet) long in order that the speed of the wires or strips through the vats can be maintained at a reasonable rate for production and still remain submerged in the electrolyte long enough to receive the desired coating. Since the working voltage is low and the amperage high, the conductors consist of heavy copper bars. In addition, there are needed a low-voltage direct-current supply source, and means for making and storing fresh electrolyte.

Operation of the Process—The vat is nearly filled with the plating liquid, called the electrolyte, consisting of a solution of some zinc salt such as zinc sulphate. This solution must be continually agitated to maintain a uniform density, and it must also be kept clean and at a fairly constant and suitable temperature. There are two methods of supplying the zinc content to the electrolyte and these distinguish the process as being a **soluble anode process** or an **insoluble anode process**. In the case of the former slabs of zinc are submerged in the electrolyte and are electrically attached to the positive bus bar and thus constitute soluble anodes which supply the zinc to the solution. In the insoluble anode process, special lead-alloy bars properly connected to the positive bus bars constitute the anodes by which the plating current is introduced into the bath. The zinc is supplied to the plating bath from an outside source where metallic zinc is dissolved in sulphuric acid or the zinc is obtained by direct leaching from zinc ore and the solutions thus obtained are subjected to required controls for purity, concentrations, etc. Alternating with the anodes are strong tungsten contact bars which are placed across the vat, immersed up to 100 mm (4 in.) below the surface of the electrolyte, and connected with the negative bus bars outside the vat. These contact bars support the wires being plated, which collectively form the **cathode.** The wires to be electrogalvanized are generally annealed, acid pickled, electropickled, then rinsed before they pass through the plating vat. From the vat the wires pass through or over some simple form of wipe and onto takeup blocks placed far enough away from the wipe to permit the wires to dry in air.

Factors in Controlling the Thickness of the Coat—Theoretically, about 1.1 kilograms (2.5 pounds) of zinc can be deposited per hour per 1000 amperes of current. In practice, the amount of zinc deposited will be somewhat less, as no electrochemical operation can be carried on at 100 per cent efficiency. The practical maximum current density varies and must be determined experimentally for each set of working conditions. Having determined this and also the efficiency to be expected, it will be possible to calculate the required total current needed to deposit a required amount of zinc in a given time. The current flowing from the anodes through the electrolyte into the cathode wires causes the zinc to be deposited on the wires. Since the zinc ions traveling through the electrolyte actually carry the current, the amount deposited will be directly proportional to the current strength and to the time of flow. Since the former factor is fixed by the capacity designed into the unit, the latter must be increased if a greater weight of coating is desired.

Tests for Galvanized Coatings—The whole galvanizing process is placed in the hands of experienced operators, who keep constant watch over each phase of the process and at frequent intervals make certain tests to determine the fitness of the material for the purpose for which it is to be used. With respect to the galvanized coatings, these tests are both physical and chemical. The more common of the physical tests are known as the **button test** and the **mandrel test,** each being applied to different classes of wire to meet different requirements. In the button test, the wire is wound tightly about its own diameter, while in the mandrel test the wire is similarly wound about a smooth mandrel having a diameter that is some multiple of its own diameter. The object in both these tests is to determine if the coating will crack or flake under these conditions and, if so, to what extent. For determining the thickness of the coat and whether or not the zinc has been uniformly distributed, two chemical tests are employed. One of these, called the **antimony hydrochloride test,** is a laboratory test used for determing the exact quantity of zinc per unit of surface galvanized. This is the best known and most generally used chemical test for galvanized coatings. The other, known as the **Preece** or **copper-sulphate test,** when made under properly standardized conditions, will determine whether or not the coating is uniform in thickness and whether the thinnest portion will dissolve in a certain solution of copper sulphate in one, two, three, four or more minutes. The terms "one minute wire," "two minute wire," etc., are based on this test, and mean that the coating will successfully withstand immersion in a standard neutral copper sulphate solution for one, two, or more minutes.

WIRE TINNING

Because of the high cost of tin, and of the difficulty of entirely avoiding pin holes in tinning steel wire, only a small amount of steel wire is tinned, and that largely for decorative purposes or to prepare the wire for assembly by soldering. As already explained, tin is cathodic to iron. Consequently, in the case of steel wire coated with tin and exposed to the atmosphere, the destruction of the steel base will be accelerated by the presence of tin when corrosion begins. The tinning process is very similar in principle to the hot galvanizing process and need not be described at length. The wire has to be thoroughly cleaned and fluxed before entering the bath of tin, and as it emerges from this bath it passes through a wipe of wicking wound about the wire, after which it goes to the take-up blocks, either with or without an intermediate water cooling. The tin must be maintained at a fairly constant temperature of 260° to 285°C (500° to 550°F), for if the bath temperature is too low, the coating will be rough and uneven, and if too high, it will be discolored by the yellowish tin oxide formed on the surface.

ALUMINUM COATINGS

In recent years aluminum coatings have been applied to steel products for resistance to corrosion in marine, industrial and the usual environments. The aluminum is applied to steel wires by the hot-dip process, where the cold-drawn wires are cleaned prior to immersion into a molten aluminum bath. Two processes most commonly used today are: 1) the use of a fluoride flux prior to immersion in the aluminum bath; or, 2) after cleaning, preheating the wires to the bath temperature by induction heating, in a reducing atmosphere such as cracked ammonia. Aluminum forms alloys with steel in a manner similar to zinc, and the process must be carefully controlled; otherwise, a brittle alloy layer will form which will cause cracking and spalling during subsequent forming operations. The aluminum bath is often alloyed with silicon, which

lowers the melting point of the bath and decreases the thickness of the brittle Fe-Al alloy layer.

The aluminum bath is operated at a temperature of approximately 660° to 675°C (1225° to 1250°F). This temperature is sufficient to soften carbon steels with a corresponding loss in tensile strength. Aluminum-coated wire can be cold drawn after coating to improve or raise the tensile strength. Also, by drawing after coating, very bright finishes can be produced which make the wire suitable for many decorative applications.

Wire which can be coated with aluminum for corrosion protection includes: field-fence wire, barbed wire, chain-link-fence wire, netting wire, tie wire, strand wire, telephone wire and A.C.S.R. core wire.

SECTION 10

CLASSIFICATION OF STEEL WIRE

Bases for Classification—Steel wire products are classified in various ways, as indicated in Table 31—III. From the standpoint of a manufacturer, the factors of greatest interest are the kind and grades of steel; the size, shape and mechanical properties of the wire; and the methods of drawing and treatment. In commercial transactions, classifications based upon mechanical properties of the wire, its treatment and finish, and its commercial applications are of most interest. The following brief descriptions are intended to assist the reader to a better understanding of the classes.

Table 31—III. Kinds and Classes of Steel Wire

Classification, according to:

Method of making the steel—Basic oxygen, open hearth, and electric-furnace processes.

Compositions of steel—Carbon, alloy, or stainless steel. Carbon steel may be divided into four classes, as follows:

Low carbon includes grades of steel to 0.15 per cent maximum carbon content e.g., UNS G10060, G10080, G10100, G10120, G11080, G11100, corresponding to former AISI or SAE grades 1006, 1008, 1010, 1012, 1108, 1110); medium low carbon includes grades of steel with a maximum carbon content exceeding 0.15 per cent to and including 0.23 percent (e.g., UNS G10130, G10150, G10180, G10220, corresponding to former AISI or SAE grades 1013, 1015, 1018, 1022); medium high carbon includes grades of steel with a maximum carbon content exceeding 0.23 per cent to and including 0.44 per cent (e.g., UNS G10250, G10300, G10380, G15410, corresponding to former AISI or SAE grades 1025, 1030, 1038, 1541); high carbon includes grades of steel with a maximum carbon content over 0.44 per cent (e.g., UNS G10450, G10500, G10650, G15660, G15720, G1086, corresponding to former AISI or SAE grades 1045, 1050, 1065, 1566, 1572,

1086.)

Shape—Round, half-round, oval, half-oval, square, rectangular, hexagonal, crescent, triangular, etc.

Method of drawing—Dry-drawn, wet-drawn, single-draft drawn and continuous drawn.

Size—Coarse wire, fine wire.

Treatment and Finish—Bright (hard dry-drawn), bright soft (annealed in process), normalized (pre-normalized), spheroidized, extra smooth clean bright, sull coated, annealed, black annealed (pot annealed), strand annealed (lead annealed), bright annealed, lime bright annealed, patented, oil tempered, liquor finished (coppered or bright coppered wire), galvanized, tinned, aluminum coated.

Uses or commercial applications—Armor wire, bale-tie wire, bolt wire, scrapless-nut wire, recessed-head screw wire, broom wire, cold-heading or cold-extrusion wire, concrete reinforcement fence wire, music wire, piano wire, poultry netting, rope wire, valve-spring wire, upholstery-spring wire, mechanical-spring wire, telephone wire, telegraph wire, tire wire, welding wire, wool wire (for steel wool), special wires, miscellaneous wires.

Kinds and Composition of Steel Used for Wire—The steel for wire is produced by all the modern processes, including the basic oxygen, basic open-hearth, and basic electric-arc-furnace processes. Plain carbon steels chiefly are produced by the basic oxygen processes; and open-hearth processess; stainless steels and alloy steels for certain special-purpose wires are made in basic electric-arc furnaces. The alloy steels known as constructional alloy steels (see Chapter 43) generally are made in basic oxygen and basic open-hearth furnaces.

By properly selecting and handling the plain carbon steels in drawing and heat treating, it is possible to develop a wide range of mechanical properties. Ordinarily, sulphur and phosphorus are kept within the usual limits for each grade of steel, while the silicon, manganese, and carbon contents are varied according to the mechanical properties desired. Occasionally, the sulphur or sulphur and phosphorus contents will be increased, or lead, bismuth, or tellurium may be added to improve the cutting qualities (machinability) of the steel.

In recent years, there has been a constantly increasing demand for wire possessing properties that cannot be obtained by cold working and heat treating plain carbon steel. This demand has been met by the addition of alloying elements, such as nickel, chromium, manganese, silicon, vanadium, molybdenum or combinations of them. Many of these alloy steels can be produced satisfactorily by the basic oxygen and basic open-hearth processes. For those that cannot and for the stainless steels containing large percentages of chromium and nickel, electric furnaces of the arc type are commonly used and, to a decreasing extent, the induction type of electric furnace is sometimes used. Each of these processes produces steel having the best properties for certain purposes or kinds of wire.

Wire Shapes—While wire is ordinarily thought of as being round, it may have any one of an infinite number of sectional shapes, as required by the particular use for which it is desired. After the ordinary round wire of circular section, the most common shapes are square, hexagon, octagon, oval, half-oval, half-round, triangular and flat. Besides these regular, or symmetrical shapes, wire is also made in various odd and irregular shapes for specific purposes.

The American Iron and Steel Institute (AISI) Steel Products Manual entitled "Carbon Steel Flat Wire" defines a flat wire as a cold-rolled product, with a prepared edge, rectangular in shape, 12.7 mm (½ inch) or less in width, under 6.4 mm (¼ inch) in thickness. "Low-carbon steel flat wire is generally produced from hot-rolled rods or specially prepared round wire, by one or more cold-rolling operations, primarily for the purpose of obtaining the size and section desired and for improving surface finish, dimensional accuracy and varying mechanical properties. Low-carbon steel flat wire can also be produced by slitting cold-rolled flat steel to the desired width. The width-to-thickness ratio together with the specified type of edge generally determine the process which will be necessary to produce a specific flat wire item." Those edges, finishes, and tempers obtainable in flat wire are similar to those furnished in cold-rolled strip.

Sizes of Wire—The practice of expressing wire size by gage numbers, formerly widely used, has recently been giving way to expression in absolute units, or decimal parts thereof. The absolute unit for expressing diameter in SI is the millimetre. In this country the absolute unit has been the inch, and the diameter is obtained by a micrometer capable of making measurements accurate to at least one hundredth of a millimetre or one thousandth of an inch. As stated in Chapter 53, "Gage Numbers", there are several different gages which are sometimes used to designate the size of wire.

The size limits for the products commonly known as wire range range from approximately 0.127 mm (0.0050 in.) to under 25.4 mm (1 in.) for round sections, and from a few hundredths of a millimetre (a few thousandths of an inch) to approximately 12.5 mm (½ in.) for square sections. Larger rounds and squares, and all sizes of hexagonal and octagonal shapes are commonly known as bars.

As wire may include a wide range of sizes, its classification has been governed somewhat by manufacturing equipment and methods of handling. It normally is separated into two broad groups known as **coarse** and **fine** wire. The fine-wire classification is usually recognized as including 1.59 mm (0.062 in.) and smaller wire, normally produced in approximately 200-mm (8 in.) diameter coils. Coarse wire includes sizes 0.884 mm (0.0348 in.) and coarser, normally drawn on 406- 559-, 660- or 762-mm (16-, 22-, 26- or 30-inch) blocks. The distinction is not clearly drawn for 1.59 to 0.884 mm (0.062 to 0.0348 in), inclusive, as this range is commonly regarded as coarse wire for some end uses, and as fine wire when made by a different manufacturing practice for other end uses. The most commonly used wire sizes are those drawn from a 5.5 mm (7/32 in.) rod to 4.11 mm to 0.89 mm (0.162 to 0.035 in.), inclusive.

Surface Finishes of Wire—A variety of finishes on the wire may be obtained by controlled processing in manufacture. The more common finishes are usually designated as bright, black annealed, liquor finished, coppered, tinned, galvanized, aluminum coated, and painted or enameled. For certain specific uses, a drawn aluminum coated or a drawn galvanized finish is sometimes produced, by cold drawing the wire one or more drafts through a die after coating. The bright finish is obtained by dry drawing, liquor finish by wet drawing, black annealing finishes by oxidation of the surface in heat treatment, while the tinned, galvanized and painted finishes are produced by subjecting the wire to a coating process as a final operation. The methods used in producing these different finishes will be developed more fully in the description to follow.

Mechanical Properties—In the wire industry, tensile strength is usually used to stipulate a required level of mechanical properties. The tensile-strength test can be performed rapidly, and is less subject to testing error or variation due to heterogeneity of the steel, residual stress, etc., than other tests such as hardness tests. It is also possible to derive other information from the tension test, such as yield strength and per cent elongation or reduction of area, which give an indication of the ductility of the wire.

Tensile strength is affected by the amount of carbon, manganese, and phosphorus, or other alloying elements present in the steel, the amount of cold drawing following the last heat treatment, and the type of heat treatment given the wire before or after drawing.

By properly correlating all these factors—composition, drawing and heat treatment—the mechanical properties of steel wire may be varied over a wide range of hardness, strength, toughness, ductility, or softness. As an illustration of the possibilities in varying these properties, there may be cited the strength of steel wires, which are regularly made with tensile strengths varying from 345 MPa to as high as 3450 MPa (50 000 to 500 000 psi) for the smaller sizes of high-carbon wire. It must be remembered, however, that generally whatever is done to strengthen the wire will decrease its ductility and vice versa.

TYPICAL FINISHED WIRES FOR MANUFACTURING PURPOSES

The object of this section is to list according to their common trade names some of the different kinds and grades of wire used for manufacturing purposes and indicate some of the uses of each.

The applications of the various types of wire require different characteristics in the finished wire as to physical properties, finish, gage tolerance, electrical conductivity and many others. These requirements are met by designing manufacturing practices which employ selection of the proper grade and quality of steel, various amounts of cold work (drawing), various types of heat treatments, or combinations of both. All of these practices are carefully worked out and are designed to provide wire which will adequately meet consumer requirements and at the same time make the most efficient use of producing equipment by the wire manufacturer.

Standard Quality Wire is produced from Standard Quality rods which use carbon steel grades having up to 0.23 per cent carbon maximum. This quality represents the largest tonnage of wire in today's market.

The finer sizes of Standard Quality wire are used to manufacture paper clips, bag ties, fly-screen cloth, staples, fuse wire, bookbinder wire, fine brads and nails, etc.

In coarser sizes, Standard Quality wire is used for such high-tonnage items as nails, fence, welded mesh, coat hangers, wire forms, barbed wire, refrigerator-condenser wire, and baling wire.

The versatility of Standard Quality wire is demonstrated by its availability in a clean finish for painting and resistance welding; a smooth finish for electroplating; annealed at finish size for tying wire or redrawing purposes; zinc coated for appearance or corrosion resistance; tin coated for wire which contacts food products or for minimal corrosion resistance; aluminum coated at finish size for maximum corrosion resistance.

Medium High Carbon Quality wire is produced from carbon steels with carbon ranging from 0.24 per cent to 0.44 per cent maximum. The primary use of medium high carbon wire is to supply a product of higher tensile strength than that available in the Standard Quality grades either as drawn from the parent rod or after a heat treatment. Automotive "U" bolts, concrete snap ties, and concrete nails are typical applications of this wire.

Cold Forming Wire and **Recessed Head Screw Wire** are drawn from Cold Heading Quality rods, with a surface free of harmful surface imperfections.

Since this wire will be upset, extruded, or rolled, it is imperative that the wire be supplied in a condition of ductility suitable for processing into its end application. Low-carbon wire for cap screws, wood screws, solid rivets, steel balls, and machine screws can be headed as drawn from the parent rod; medium high carbon wire in grades such as UNS G10350, G10380, and G15410, corresponding to former AISI or SAE grades 1035, 1038, and 1541 usually require that the parent rods be annealed prior to wire drawing to head the same fasteners.

For wire to produce fasteners with washer faces, recessed heads, extra large heads, or components where backward extrusions are involved, a thermal treatment in the process of wire drawing or a thermal treatment at finished wire size is required due to the high ductility needed to cold form such parts.

Along with a suitable ductility, cold forming wire is drawn with specific surface finishes suitable for plating, extrusion, open die work or solid die work.

A possible third requisite for cold forming wire is that it be supplied in a steel grade capable of meeting a customer's heat-treatment requirements. Steel grades and austenitic grain size are usually designated by specifications governing the finished part.

Free Machining Wire and **Cold Finished Bars**—These commodities are discussed in Chapter 52, "Machinability of Carbon and Alloy Steels."

Welding Wire—Various steels, both carbon and alloy, are used to produce welding wire for the manufacture of electrodes and rods for electric and gas welding, and these range from 0.05 carbon to about 1.0 per cent carbon. While considerable gas welding is still employed, by far the greater proportion of welding operations are performed by the electric-arc process. Wire for gas welding is copper coated and used in straight lengths, composition being governed by properties required in the weld metal.

Electric-arc welding is divided into two classifications: manual welding and automatic welding.

Wire for manual welding is furnished in straightened and cut lengths with surface as free from wire-drawing lubricant as possible to enable electrode manufacturers to apply a flux-coating by an extrusion method. Welding steels are made especially for this application, and selection of heats or portions of heats is mandatory to conform to required composition limits and ranges. Low sulphur, not exceeding 0.035 per cent, is generally preferred. Most of the grades of steel made for welding electrodes are of the rimming type, with carbon in the neighborhood of 0.10 to 0.15 per cent. However, there is also some demand for rimming-type steel with 0.31 per cent carbon as the maximum of the range. Killed steels of 0.55 to 0.65 per cent carbon and 0.90 to 1.10 per cent carbon are also used to some extent, along with some alloy grades for special purposes such as resistance to abrasion. Composition of the steel is governed by composition of the flux coating, the combination being designed to provide certain welding characteristics and specific properties in the deposit metal.

The wire thus serves as the core of the coated electrode and is expected to provide good welding characteristics and a dense, sound deposit. Wire sizes in popular demand are 3.18, 3.97, 4.76 and 6.35 mm ($\frac{1}{8}$, $\frac{5}{32}$, $\frac{3}{16}$ and $\frac{1}{4}$ in.). Tolerances for diameter, length, camber, and burrs in the straightened and cut wire are listed in the AISI Steel Products Manual "Wire and Rods, Carbon Steel and Wire and Rods, Alloy Steel."

The flux coating on the electrode provides a blanket which surrounds the arc during welding and serves as a protection against oxidation from the atmosphere. In automatic electric-arc welding, there are two main processes: submerged-arc and inert-gas. In the former, the welding is shielded by a blanket of fusible material, usually referred to as flux which is fed on to the work. In the latter, shielding is obtained from inert gas such as argon or helium. Both automatic and semi-automatic equipment is used for each process.

Wire compositions vary for submerged-arc welding according to requirements in the deposit metal. The wire is produced with a bright copper finish to improve the contact surfaces and so facilitate the introduction of the relatively high currents used. It is important that the surface of the wire be as free as possible from wire-drawing lubricant which would tend to clog the nozzle through which the wire is fed. Wire for submerged-arc welding is furnished in layer-wound coils with card-board cores. Dimensions of these coils are designed for insertion into a reel from which the wire is fed through the welding machine. Wire sizes in demand range from about 2.38 mm (³⁄₃₂ in.) up to about 7.94 mm (⁵⁄₁₆ in.).

The inert-gas welding process calls for smaller wires ranging from 1.14 mm (0.045 in.) to around 3.18 mm (0.125 in.). Wire for this application is furnished with a bright, clean copper coating for the same reason as mentioned above and is wound on spools of the dimensions necessary for insertion in the welding equipment. The required wire compositions vary according to requirements of consumers.

Steel Wool Wire—The mechanics of the steel wool cutting process and the special quality characteristics that have to be imparted to the final product require a steel wire with an unusual combination of metallurgical and mechanical properties. The wire must have unique machinability that makes it amenable to cutting different grades of steel wool, from the very fine to the very coarse. The steel wire must cut into steel wool without fragmenting or "dusting" and the fibers produced must be dimensionally uniform for a particular grade; at the same time, the fibers need to have sufficient continuity, cohesiveness, and strength to permit uninterrupted traverse of the steel wool fibers from the individual cutting tools to a take-up reel. The steel wool pads must be resilient with good resistance to crumbling and have a required degree of hardness and abrasiveness to serve their intended purpose.

The surface quality of the wire is extremely important especially in the cutting of the fine wool grades. The cutting blades, having as many as 360 cutting points per 25 mm (1 in.) are very delicate and sensitive to damage even by the impact of relatively small surface imperfections. Also, even light surface seams may result in undesirable needle-like slivers being cut into the steel wool.

The properties required to satisfy all of these conditions are built into the wire by an appropriate combination of chemical composition, billet conditioning, thermal treatment, and cold work. A specially developed steel grade for this application is essentially a relatively low carbon steel that is nitrogenized and rephosphorized. The silicon and sulphur are controlled at appropriate low levels, and the aluminum content is maintained at an absolute minimum. The thermal treatment of the rods in the hot rolling or patenting

must be carefully monitored to produce a suitable microstructure with a high degree of uniformity. The wire is drawn from a predetermined rod size calculated to give the wire an approximate average tensile of about 965 to 1030 MPa (140 000 to 150 000 psi) depending on the customers' individual requirements.

Steel wool wire is normally supplied in a wire diameter range of about 2.62 to 2.85 mm (0.103 to 0.112 in.).

Liquor-Finished Fine Weaving Wire—Some of this grade of wire, which may be used for weaving household window-screen cloth, is made in very fine sizes, 0.30 mm and 0.26 mm (0.0118 and 0.0104 in.) being the common sizes. Wire for this purpose is usually made from low-carbon steel and to reduce the wire from a 5.5 mm (⁷⁄₃₂ in.) rod to such fine sizes requires annealing in process and 22 to 24 drafts, depending on composition and requirements.

Brush Wire—Untempered brush wire is used for bristles in wire brushes. There are three grades: Low-Carbon Brush Wire; Scratch Brush Wire and High-Strength Brush Wire. Brush wire is supplied in the following finishes: straw liquor finish, steel bright, coppered, tinned, galvanized, or cadmium coated. The Low-Carbon Brush Wire is usually produced in sizes under 0.20 mm (0.008 in.) from low-carbon steel to an approximate tensile strength of 965 MPa (140 000 psi). Scratch Brush Wire is generally produced in sizes 0.89 mm (0.035 in.) and under from carbon steel of approximately 0.45 to 0.60 per cent carbon content, in tensile strengths from 1585 to 2000 MPa (230 000 to 290 000 psi). High Strength Brush Wire is usually produced in sizes 0.89 mm (0.035 in.) and under from steel of approximately 0.55 to 0.75 per cent carbon content in tensile strengths ranging from 2068 to 2620 MPa (300 000 to 380 000 psi).

High Carbon Screen Wire—The special properties of this grade of wire were developed to accommodate the particular requirements needed in the fabrication of industrial screens used in classifying and screening various minerals and aggregates. Because of the variety of materials involved and the diversity in the design of the equipments and screens used in this operation, several grades of screen wire have evolved to satisfy the fabricating and service requirements of the ultimate product.

In the fabrication of the screens, the most critical operation is the crimping which may be performed by roll crimping or by the use of a press with suitable jig fixtures. The severity of this formation depends on the depth of the crimp, the bending radius, and the degree of restraint to the metal flow during the formation of the crimp. The ductility of the wire is, therefore, the controlling property and all other properties of the wire must be designed to be compatible with the minimum ductility required for the particular crimp formation. The most important service requirements are strength and abrasion resistance. In general, therefore, the highest tensile strength wire, of the highest carbon content which can be crimped and woven, represents the most ideally suited material for this application. To meet the many combinations of tensile strength and ductility, high carbon screen wire is produced as hard drawn and oil tempered to several tensile-strength levels and for progressively more severe crimping, this grade is also available thermally treated in process, lead tempered, and spheroidize-annealed at finish size. The sizes produced range from 0.89 to 15.88 mm (0.035 to 0.625 in.).

Upholstery Spring Wire—Upholstery spring wire is used in the manufacture of spring elements for mattresses, box springs, foldaway beds, cots, furniture, cushions, and automobile seats. This category also includes high-carbon and medium-high-carbon wire used to produce the structural and assembly elements used in the construction of the various upholstery units. The diversity of grades developed for each particular application includes automatic coiling and knotting quality for springs used in bedding and furniture; sinuous furniture type wire for the corrugated springs in furniture; zig-zag spring wire for corrugated springs in automobile seats; border and brace wire for structural components; and automatic lacing quality wire or severe crimping and clinching wire for assembling the various upholstery units.

In the manufacture of this wire, the mechanical properties, which are unique to each application, are imparted to the wire by first carefully selecting a steel of appropriate chemical composition. The wire is drawn from a thermally treated rod or wire of a predetermined diameter calculated to impart the necessary amount of cold work to attain the required tensile strength and ductility in the finished wire. The post-pickling coatings applied to the thermally treated rod or wire and the lubricants used in wire drawing are carefully selected to produce a wire finish that is most conducive to the particular type of fabrication used in the manufacture of the various types of springs and construction components. Upholstery spring wire can also be produced with a galvanized coating applied at the finish size or drawn one draft after galvanizing. The tensile strengths of the galvanized upholstery spring wire are usually about 5 per cent to 10 per cent lower than the bright hard drawn wire. The approximate size ranges available are 0.89 to 5.72 mm (0.035 to 0.225 in.) in the bright wire and 0.89 to 4.11 mm (0.035 to 0.162 in.) in the galvanized grades.

Hard Drawn Mechanical Spring Wire—In terms of quantities used, this grade of spring wire ranks first. It is generally used in springs subject to static loads or relatively infrequent stress cycles. The mechanical properties are controlled by careful selection of plain carbon steels of appropriate composition which are then subjected to a precise thermal treatment at one or more of the wire drawing levels and by a predetermined amount of cold working which is controlled by the calculated selection of the exact rod or process wire size. With the proper balance of these practice elements, this grade of wire is made available in several tensile-strength classes with compatible ductility and toughness levels.

Springs from this grade of wire are produced by cold forming followed by low-temperature tempering to relieve cold forming stresses and to increase the elastic limit of the wire. Hard drawn mechanical spring wire may also be produced with a galvanized coating for use where exposure to corrosion is a consideration.

Oil Tempered Mechanical Spring Wire—Like hard drawn mechanical spring wire, this grade is also intended for springs with static loads or relatively infrequent stress cycles. The manufacture of oil tempered wire is similar to the manufacture of hard drawn mechanical spring wire with regard to the selection of the appropriate steel composition and the incorporation of a suitable thermal treatment at the proper wire drawing level. However, the total amount of cold working as a control of the final mechanical properties is less critical because the principal control, which distinguishes these two grades, is in the final oil tempering heat treatment. This heat treatment comprises the last step in the processing of this product and consists of passing the individual strands of wire through a continuous sequence of austenitizing, oil quenching, and lead tempering.

Because of the superior uniformity attained by this strand type heat treatment and because of the straightness that is characteristic of oil tempered wire, a more precise control of the spring dimensions can be maintained with a minimum of adjustment in the coiling equipment. This process has the added advantage of making appreciably higher tensile strength available in the coarser sizes than are offered in hard drawn mechanical spring wire. Oil tempering, moreover, imparts a high elastic limit to the wire and, for some applications, the springs produced therefrom may not require a tempering heat treatment after coiling.

High Carbon Spring Wire—This spring wire category is intended for the fabrication of special springs, spring elements, or wire components that cannot be formed from hard drawn or oil tempered spring wire because of the severity of the formations involved, or where the mechanical properties specified for the final product are attainable only by heat treating after forming. Since the properties of the fabricated component are controlled by the final heat treatment, this wire is generally ordered to a specified chemical composition. The mechanical properties required in the as received wire will depend on the particular fabricating process and the severity of the formations to be imparted to the wire. Accordingly, this wire is produced as hard drawn, spheroidize-annealed in process, or spheroidize-annealed at finish size.

Valve Spring Wire—In early automobile engines, operating at speeds around 2200 rpm, any good commercial-quality hard-drawn or oil-tempered spring wire would make a satisfactory valve spring. However, modern engines, designed for higher speeds require valve springs made from special wire. Higher-horsepower motors require greater energy storage in the springs. At higher speeds, the effect of inertia in the valve-spring system becomes much more pronounced, and the spring designer must employ the highest design stresses possible to obtain the desired load without excessive weight in the valve springs. High speeds also tend to introduce considerable amounts of dynamic stress, in addition to the high static stress mentioned above. The designer usually attempts to keep these dynamic stresses to a minimum by adding dampener coils to the spring or by using a separate dampener spring inside the main valve spring. If the stresses encountered in service exceed the fatigue limit of the material, or if significant surface or internal imperfections that can cause localized concentrations of stresses are present, fatigue failures may take place, often in the first few hundred miles of operation.

By careful selection of materials at all stages of wire manufacture, the use of the best processing equipment and practices available, and very rigorous inspection, all the requirements of valve spring wire for modern high-speed high-power engines can be met. Some of the precautions taken with this material are: (1) Use of steel of exceptional cleanliness; (2) Generous cropping from both top and bottom of the ingot; (3) Inspection and conditioning of billets; (4) Special rod-rolling equipment and practices; (5) Extra heat treatments from rod to finished wire; (6) Careful control of processing to produce high, uniform tensile strength and coating satisfactory for uniform spring coiling; (7) Special tests, including tensile-strength, twist tests, and surface inspection by acid etch and examination under binocular microscope on both ends of every bundle; and (8) Special tests and investigations to determine fatigue limit of wire samples or test springs coiled from the wire.

Most of the valve springs produced in this country are made from wire that has been oil tempered at finished size. However, some percentage of valve spring wire has been used in the hard-drawn condition, with tensile strength built up by a combination of process heat treatment and cold drawing. Wire diameters usually range from 4.11 to 4.88 mm (0.162 to 0.192 in.) and tensile strength from 1450 to 1725 MPa (210 000 to 250 000 psi).

Music Wire—Music wire is the term applied to wire having the physical requisites necessary for use in musical string instruments. Although commonly known as music wires, these wires are generally referred to as piano wire; harp wire and mandolin wire, depending upon the purpose for which it is used. In conjunction with the manufacture of music wire, there is a grade of wire manufactured known as music spring wire. This wire is not quite as high in quality as piano wire and is used for high-grade, cold-wound, high-tensile-strength springs.

Piano Wire—Piano wire has rightly been called the "specialty of specialties." It represents the highest attainment in the art of wire manufacture. The wire possesses the highest tensile strength of any form in which carbon steel is used for purely stress-resisting purposes. It is made from the finest quality of steel, having a carbon content of from 0.80 to 0.95 per cent. Different grades and sizes require varying drafting and heat-treating practices to develop the proper tensile strength and toughness.

In addition to the mechanical properties required, piano wire must possess acoustic properties. Attainment of this requirement is dependent upon the accuracy of the size, the soundness of the steel and the finish of the wire. Furthermore, the tension required for different pitches and lengths means that high elastic limit with uniform toughness must be developed to a maximum for the tensile strength to permit forming the wire into loops or eyes by the bass-string manufacturers. The bass strings must also be flattened locally on the ends to form an anchor for the covering wire. Tensile strengths for the highest grades of piano wire are given in Table 31—IV, from which data it should be noted that the higher strengths are obtained in the finer sizes.

Table 31—IV. Properties of Piano Wire.

Kind of Wire	Diameter		Tensile Strength	
	mm	in.	MPa	ksi
Bass Strings	0.80 to 1.70	0.035 to 0.067	2000 to 2400	290 to 351
Treble Strings	0.74 to 1.24	0.029 to 0.049	2200 to 2690	319 to 390

Bronze Finish Tire Bead Wire—Tire wire is used in the beads of automobile tires. This product receives special care at every stage of its manufacture to be sure it will meet requirements, which are very exacting. It must be round and true to gage, and possess the proper tensile strength, torsion and elongation qualities. Some consumers demand only a minimum breaking weight while others specify both a maximum and a minimum. A wire having the high tensile strength demanded (approximately 1900 MPa or 275 000 psi, minimum), that will withstand a minimum of 58 twists in 200 mm (8 in.), must be free from injurious steel imperfections such as pipe, seams, and segregation.

Another very important requirement for wire used in tire construction is that the wire surface must be extremely clean and properly plated to assure that it will develop a strong adherent bond with the rubber. Minimum rubber adhesion values and procedures for measuring them are specified by all consumers of this product. To assure compliance, special attention is directed to the preparation of the wire surface before bronze plating and to the precise control of the bronze plating and post coating process conditions.

Brass-Plated Tire Cord—Tire cord fabric is used in the belt construction of automobile, truck, off-the-road and airplane tires. The cord is processed from high carbon rod by a series of heat-treating and drafting operations. The wire is finally patent heat treated and brass plated at process size, then wet drawn to finish size, and finally bunched or stamped into finished cord. The brass plate is applied both as an aid in wire drawing and also to provide the necessary adhesion of the cord to rubber during the vulcanization process. A typical cord construction would be a 1 × 4 construction in which 4 individual 0.0098-inch (0.25 mm) diameter wires are bunched into a single cord.

Rope Wire—Rope wire is made from high-carbon steel produced by the basic oxygen, open-hearth, or electric-furnace processes. Carbon content ranges from 0.45 to 0.80 per cent. The various grades, namely; Mild Plow, Plow, Improved Plow Steel, and Extra Improved Plow Steel require very exacting processing to produce wire to meet the tensile-strength and torsion-test requirements. At least one and perhaps three patenting heat treatments at various stages of reduction are necessary to produce the various sizes of wire down to 0.13 mm (0.005-in.) diameter.

Rope wire is generally made in any of the following finishes:

   A. Bright—Dry drawn in coarser sizes, wet drawn in finer sizes.
   B. Drawn galvanized.

C. Galvanized at finished size.

Bridge wire is the term commonly applied to special rope wire used in the construction of cable suspension bridges, in which the main cables consist of parallel wires compacted to act as a unit. These bridges are remarkable for their graceful beauty, the absence of heavy superstructures, and their extraordinary long spans. Some of the largest bridges in the world are of this type, including the George Washington Bridge in New York, the Golden Gate and San Francisco-Oakland Bay Bridges in California, the Mackinac Bridge in Michigan, and the Narrows Bridge in New York.

Interesting data on the Narrows Bridge are: main span, approximately 1.3 km (0.8 mile), the longest and heaviest span in existence; total length, including two side spans of 370 m (1215 ft) each, the Staten Island toll plaza and the Brooklyn approach ramps, approximately 4.2 km (2.6 miles). The bridge has two decks, each accommodating six lanes for vehicular traffic, with navigational clearance of 69.5 m (228 ft) at the center of the main span. Width of the bridge (center to center of outer cables) is about 34 m (112 ft). Total width of the suspended structure is 35 m (115 ft); the towers are somewhat wider. The roadway structure is suspended from the four main cables, two on each side of the deck. Each main cable is about 915 mm (36 in.) in diameter and is composed of 26 108 cold-drawn extra galvanized wires, each having a diameter of 4.98 mm (0.196 in.).

The total length of the bridge cable wire is 229 300 km (142 500 miles) miles, totaling nearly 35 000 metric tons (38 500 net tons); the 1048 suspender ropes total 119 km (74 miles) in length.

Bridge wire requires steel of high quality and extreme care in every step of its manufacture from melt to finished wire. Specifications for the finished wire have required: (1) The wire to be heavily coated with zinc to protect it from corrosion—galvanized to withstand 5 immersions in the standard Preece test. (2) A minimum average tensile strength of 1550 MPa (225 000 psi). (3) A minimum elongation of 4 per cent in 250 mm (10 in.). (4) A coating that will permit wrapping without peeling about a round mandrel of a diameter equal to 1½ times the diameter of the wire.

At one time, specifications required the steel to be made by the acid open-hearth or the crucible processes, but it is now generally accepted that steel equally as satisfactory or better can be made by the basic oxygen process or the basic open-hearth process. The wire is made from selected high-carbon steel, the composition of which is held within close limits. The composition of the steel used heretofore has usually been held within the following limits: carbon, 0.75 to 0.85 per cent; manganese, 0.55 to 0.75 per cent; phosphorus, under 0.03 per cent; sulphur, under 0.03 per cent; silicon, 0.15 to 0.30 per cent, with maximum limits for other elements that may be present in small amounts. Some of the more important requirements of the manufacturing practice are listed as follows: The manufacturing processes in the production of bridge wire are very closely controlled in order to produce the required mechanical properties of high tensile strength, toughness, and resistance to fatigue. The

bright wire is produced most commonly in two wire sizes, 4.88 mm (0.192 in.) and 4.11 mm (0.162 in.) the former being more frequently used. The bright wire is usually hot galvanized. In order to prevent undue bending, and to supply wire which is straight, as required for the subsequent spinning of the cables, the wire is finished in 1524 mm (60 in.) diameter coils. Coils are coupled together to form long, continuous lengths of wire, and are shipped to the bridge site on large reels, from which the wire is unreeled during the process of spinning the cables.

A recent development is the preassembly of bridge wire into parallel wire strands, and shipping these strands to the bridge site on large reels from which they are unreeled in the process of fabricating the main cables of the bridge.

Tinned armature-binding wire is another specialty wire. It is made in several grades, which differ in tensile strength. The specifications require a product having high elastic limit and high elongation combined with toughness. To obtain this combination, important essentials are: first, steel of the proper composition; second, proper heat treatment; third, a uniform drafting practice; fourth, a tinning temperature which will give a good finish and will aid in obtaining the desired physical properties in the finished wire.

Metal-stitching wire was developed for fastening non-stressed parts in automotive assembly operations and is now also used in the assembly of non-stressed aircraft structures as well as in a multitude of other types of assemblies. In this application, the wire is fed from a spool through a stitching machine which cuts the wire to length, forms it into a staple and drives the legs of the staple through the material to be fastened. In most cases, the legs are folded back or clinched after penetrating the two or more layers being assembled. This wire is frequently used to unite metal to metal or metal and non-metal parts, such as plastics, rubber, fiber, felt or plywood. The tensile strength of wire for these applications may range from 1380 to 2275 MPa (200 000 to 330 000 psi), depending on the thickness and type of metal required to be penetrated by the fasteners. For this purpose, it is evident that the wire must be very uniform in temper and possess great toughness. Metal-stitching wire is made with several different finishes, according to whether a protective or decorative finish is desired.

Tying Wire—In strapping systems that use round steel strapping rather than flat steel strapping, the tying material is generally a regular galvanized wire. The wire for this application must meet a specified tensile strength, have sufficient ductility to withstand the severe twisting in the knotting operation, and be elastic enough to withstand some sustained stress in tension. Although this product can be manufactured to a broad range of tensile strength levels, general usage and experience have narrowed the number of common grades to four or five tensile strength levels with specified minima ranging between 795 and 1100 MPa (115 000 and 160 000 psi). The actual tensile-strength ranges for the various sizes and grades are usually stated in terms of the minimum and maximum breaking weights.

To attain these tensile strengths and the concomitant

elongations specified, generally between 5 per cent and 8 per cent in a 250 mm (10-in. length), this wire is drawn from steels with carbon content ranging between 0.40 and 0.75 per cent. The wire is drawn from a thermally treated rod or wire and is annealed in the process of galvanizing to effect the final adjustment in the tensile strength and elongation.

In addition to the high tensile grades noted above, galvanized tying wire is also produced from low carbon steels to a tensile-strength range of approximately 414 to 620 MPa (60 000 to 90 000 psi). The size range, encompassing all grades of galvanized tying wire, extends from 1.04 to 4.11 mm (0.041 to 0.162 in.).

ACSR Steel Core Wire—Aluminum cables for the transmission of electricity, commonly known as ACSR (Aluminum Conductors, Steel Reinforced), are strengthened with a core of high-tensile-strength wire. The steel reinforcement core, which may be either a single wire or a group of concentric-lay strands, provides 55 per cent to 60 per cent of the strength of the total ACSR construction. As a deterrent to corrosion, the wire is galvanized or aluminum coated in accordance with rigidly prescribed minimum requirements. Specifications for this specialty wire also include stipulations for the minimum tensile strength; minimum yield strength at one per cent extension under load; minimum elongation in a 250 mm (10 in.) gage length; and compliance to a specified toughness test. The approximate size range of this wire grade is 1.27 to 4.83 mm (0.050 to 0.190 in.). The unit coil quantity is usually specified in nominal lengths with special length tolerances to assure minimum lengths to conform with distances to be spanned between supports.

The wire is produced from a thermally treated rod or wire of a carefully selected chemical composition to assure compliance to the tensile and yield strength requirements. Welds are not permitted at finish size but may be made at or prior to the thermally treated size prior to final drawing.

Alloy Spring Wire—Alloy steels have been designed to expand the scope of spring performance beyond that attainable with plain carbon steels. Alloy steel spring wire is used to manufacture springs for special applications that require performance at elevated temperatures, or on springs that demand a higher degree of surface or depth hardening to withstand higher stresses. Alloy steels are also used where it is important to maximize the resistance of the spring to permanent set under conditions that impose continuous spring stressing.

Alloy steel springs are produced either from an oil tempered wire or from an untempered wire that is heat treated after hot or cold forming. The most common spring alloy steels are UNS G61500, G92540, G92600 and G51600 steels, formerly identified as AISI or SAE 6150, 9254, 9260 and 5160 steels. Alloy steel spring wire is produced to surface quality requirements that are more restrictive than those that are commercially standard for plain carbon spring wire grades. The wire is drawn from an annealed or thermally treated rod or process wire.

STAINLESS-STEEL WIRE

Stainless-steel wire has become an established com-

modity for many purposes because of its utilitarian nature, its corrosion- and heat-resisting characteristics, and its ability to withstand a variety of forming operations. Among the established products are cold-heading wires for the manufacture of bolts, rivet pins, and screws; tinned non-magnetic armature-binding wire for use in highly-powered motors for Diesel engines, and other motors which require a high-tensile non-magnetic wire; welding wires furnished for all methods of welding; spring wires to be used where resistance to chemical solutions and atmospheres is a necessity; weaving wires for a wide range of screens from very fine mesh flour sifters to coal and coke screens and continuous high-temperature conveyor belts; rope wires for specialized purposes such as mine sweeping, aircraft-control cables and strands for yachts and ships. The unusual properties of stainless steel which permit a lasting bright finish to be developed, combined with its utilitarian nature, give it great intrinsic value when used as display racks, dishwasher and refrigerator racks.

While UNS S43000, S41000, S30200, S30400 and S30500 steels (formerly identified as AISI Types 430, 410, 302, 304 and 305) may be considered the more popular grades, almost the entire list is drawn to wire for a great variety of uses.

The processing and drawing of these steels are similar, in general, to that of the carbon-steel wires. They will differ somewhat in individual practices because of the inherent nature of the alloys, such as resistance to some acids and activation by others, all of which combine to make the coating and lubricating problems in drawing stainless-steel wires one of meticulous control. Also, because of their chemical compositions, the stainless steels require different heat treatments than carbon steels.

FLAT WIRE

Flat wire is ordinarily cold rolled from a drawn round wire, properly annealed and treated to permit additional reduction in rolling and to produce a reasonably bright surface. Natural round, smooth edges are produced by this method of manufacture. Flat wire can be made up to 12.7 mm (½ in.) wide, with ratio of width to thickness being in accord with good manufacturing practices. Normally, flat wire is best produced in sizes up to 9.53 mm (⅜ in.) width with roughly a 5-to-1 ratio in thickness. However, this ratio can in some instances be as high as 10-to-1 or 15-to-1. Flat wire is available or can be made to include soft or medium or hard rolled tempers. This material has a variety of uses; some typical applications are in window-shade roller springs (Curtain Spring Wire), leaf-type feeler (thickness) gauges (Feeler Gauge Steel), electricians' tools for "fishing" wires between walls and through conduits (Fish Tape Wire) and staples for many fastening operations involving stapling machines (Flat-Preformed Staple Wire). Flat wire can be formed into flat wound springs. It also is used in miscellaneous products where spot welding is required in assemblies. UNS S30200 and S43000 steels (former AISI Types 302 and 430) are grades of stainless steel which have been most commonly furnished in the form of flat wire, although flat wire can be produced from many of the other grades.

SECTION 12

SOME FABRICATED STEEL-WIRE PRODUCTS

Importance of Fabricated Wire Products—Steel wire is fabricated into thousands of different kinds of articles, which are used for a great variety of purposes. In a list of such articles will be found many items of common use, such as automobile and bicycle spokes, hoops, rivets, bolts, chains, buckles, cotterpins, sifting screens, wire netting, wire cloth, and a host of others, each of which consumes large amounts of wire. Since description of the fabrication of all these articles would require a great deal of space, only a few of the more important commodities are briefly discussed here.

WIRE NAILS

Wire nails are produced in machines which use the same processing steps as their prototype which was invented just before the turn of this century. The steps involved are: hammering the head; feeding additional wire below the head; cutting and forming the point; ejecting the finished nail.

The finished nails are tumbled on themselves in rotating drums to remove particles of head flash and the whiskers which often remain on the cut and pointed ends. This same drum may contain a medium (such as sawdust) which effects cleaning and polishing of the nails during tumbling, otherwise the tumbled nails can be transferred to units which clean the nails with solvents or vapor degreasers.

After tumbling and cleaning, the nails may be given a subsequent processing such as deforming (ring shanking or screw shanking), painting, resin coating, or galvanizing.

Common nails such as those purchased for hand hammering are inspected during processing for visible defects only, such as bent shanks, mashed heads or points, and long or short lengths.

During the 1950's, the appearance of automatic nailing machines brought about radical changes in the nail industry. Nails now required closer dimensional control for hopper feeding and collating. Nail makers responded to this challenge by installing equipment capable of achieving closer tolerances and in conjunction with revised inspection methods they are now supplying the needs of automatic nailers with nails bearing the description of "Special Inspection Specifications Nails".

WIRE FENCE

An increasing amount of steel wire is used for fence purposes. Today an enormous tonnage of galvanized steel wire is fabricated into fence.

Woven-Wire Fence—There are a great many types of woven-wire fence, varying in style or design, and each may be made up in many different sizes. In a general way, the various styles resemble one another in that they have several **horizontal** or **lateral wires** which are secured in position with vertical or diagonal **stay wires,** the former being stronger and stiffer than the latter and provided at frequent intervals with **tension curves,** to take care of expansion and contraction due to temperature changes. The crossing of stay wires, with the horizontals, form the meshes, which may be quite large, as in cattle fences, or very small, as in poultry netting, and may be of any one of four forms, namely, triangular, rectangular, hexagonal, or diamond-shaped. They are also fastened together in various ways. In the **cut-stay fence** there is a short piece of stay wire for each space, having its ends twisted about the laterals. The stays may be electrically welded to them, or the two wires may be woven together. In any case, the work is done by specially constructed and rather complex machines from which the coils of fence will emerge all ready for the market (Figure 31—46). In the making of these fences, a most rigid system of inspection and tests is maintained, not only in the drawing and galvanizing of the wire, but also in the weaving room, in order to turn out as perfect a product as possible. These same fences made of extra-strong wire constitute exceptionally good reinforcement for concrete work. The leading fence manufacturers also produce a line of gates, special fittings, steel posts, and other articles used in fencing.

Barbed-Wire Fence—In one type of this fence, two wires, usually galvanized, and known as **line wires,** are twisted together, and, at regular and frequent intervals of 75 to 100 mm (3 to 6 in.) either two or four **barbs,** which may be round, flat or oval in section, are wound about one or both of these line wires. The barbs are diagonally cut so as to produce a long sharp point extending at right angles to the line wires. Here again, a great variety of styles and sizes of fence are made by fast-running and rather complex machines. The bulk of barbed wire, however, has the two line wires of 2.51 mm (0.099 in.) wire, while the barbs are usually made of 2.03 mm (0.080 in.) wire. The fence is furnished to the market in 400-m (80-rod) or in "catch-weight," spools.

Another type of barbed wire, developed in recent years, employs a single 2.32 mm (0.092 in.) galvanized steel wire for the line wire. This line wire is crimped throughout its length to form a continuous series of "tension curves" that guard against overstretching and allow the wire to expand and contract lengthwise with changes in temperature without loss of tension. Four-point barbs are wound about the line wire at regular spaced intervals. The wire is marketed in 400-m (80-rod) packs.

Chain-Link Fence—Chain-link fence fabric is a fencing material made from wire, helically wound and interwoven in such a manner as to provide a continuous mesh without knots or ties, except for knuckling of, or twisting of, the ends of the wire to form the selvage (finished edge) of the fabric. This results in a continuous fabric having approximately uniform square openings, having parallel sides and horizontal and vertical diagonals of approximately uniform dimensions.

FIG. 31—46. Battery of machines producing various types of woven wire fence.

The fabric may be produced from aluminum-coated wire, from extra galvanized steel wire, from wiped galvanized steel wire covered with an extruded or powder coated plastic, or from uncoated steel wire with the fabric galvanized after weaving.

Wire sizes from 3.05 to 4.88 mm (0.120 to 0.192 in.), incl., mesh openings of 45 to 54 mm (1¾ to 2⅛ in.), incl., and heights of 915 to 3660 mm (36 to 144 in.), incl., are normally furnished.

The leading fence manufacturers also produce posts, rails, gates, and other accessories required for fence installation, and often install the fence.

CONSTRUCTION PRODUCTS

Although concrete offers great resistance to compressive stress, it is lacking in tensile and bending strength unless reinforced by some material which possesses these characteristics to a marked degree. In this respect steel excels, particularly cold drawn steel, because of its high tensile strength and high yield strength.

REINFORCED CONCRETE

Plain round steel wire, produced to ASTM A82 or deformed steel wire produced to ASTM A496 are employed for reinforcement to control crack growth. The deformed surface of the latter enhances the anchoring or bonding of the wire in the concrete.

Accurate spacing of the longitudinal and transverse members is important to attain the full reinforcing value of the wire. Therefore, the wire is usually fabricated into a wire fabric. Longitudinal and transverse wires are fixed in position by a machine designed to space them accurately and weld them together at the contact points by electric-resistance welding. ASTM

A185 covers welded steel wire fabric produced from plain round wire (A82) and A497 covers welded steel wire fabric produced from deformed steel wire (A496). In addition to insuring accurate placement of the wires, fabricating into welded wire fabric provides a positive anchor in the concrete at each welded intersection, and aids in handling the reinforcement material. It is supplied in flat sheets or rolls of desired length and width.

Electric-welded wire fabric reinforcement has a variety of applications. It is used to reinforce concrete roads, buildings, dams, etc. It is used in such precast concrete products as pipe, posts, and slabs. As a matter of fact, every type of concrete construction should be reinforced.

PRESTRESSED CONCRETE

Prestressing of concrete is the introduction of desirable compressive forces into a concrete member. These compressive forces are designed to offset or neutralize any subsequent tensile forces which occur when the concrete member is loaded. Prestressing permits a concrete member to withstand tensile forces without cracking. Because prestressing steel must sustain very high tensile stresses for long periods of time without excessive stress relaxation, very high tensile strength products are used. Prestressed concrete structures usually also contain reinforcing wire or wire fabric.

Prestressing is accomplished by one of two methods:

(a) **Pretensioning.** Steel wire or strand is tensioned and then concrete mix is poured around the steel. When the concrete has attained full strength, the wire or strand is released from its tensioning apparatus, and the tensile forces of the steel induce equal and opposite compressive forces in the concrete. The bond between the concrete and steel continues to hold the steel in tension and the concrete in compression.

(b) **Postensioning.** Postensioning is a method of prestressing in which tension is not applied to the steel members until after the concrete has attained sufficient strength to carry the compressive forces imposed upon it. This can be accomplished in several ways. Tensioning members may be introduced through holes formed in the concrete, or they may be located entirely outside of the concrete section. The most common method is to enclose steel in flexible metal conduits and place it in forms before depositing concrete. In any of these methods, there is no bond between steel and concrete prior to applying tension to the steel; instead, mechanical anchorage is utilized to transfer prestressing force to the concrete section. After tensioning, bond is usually provided by pressure-grouting spaces around the wire. Grouting protects the steel from corrosion and also improves the performance of the member at high overloads.

Seven-wire stress-relieved strand, produced to ASTM A416 is generally used in pretensioning, in such members as bridge beams, piling, deck and roof trusses, etc., in which a sufficient number of similar members is required to justify the building of a tensioning bed, and the members are of such size that they can be transported from the tensioning plant to the construction site.

Individual stress-relieved wire, produced to ASTM A421, is generally used in postensioning, when a limited number of similar members, unusually large members, special design, or other considerations make construction on the job site more economical.

In certain cases, hard drawn high carbon wire, not stress relieved, is employed. One of these is prestressed concrete pressure pipe, used for conveying liquids under pressure. It is made by forming a steel plate into a liquid-tight cylinder with a welded seam. A layer of centrifugally cast concrete is deposited inside the steel shell. When this has attained the desired strength, the assembly consisting of concrete core and steel cylinder is wrapped under tension with a helix of wire produced to ASTM A648. After wrapping, the entire pipe is covered with a mortar coating to protect the wire from corrosion. With suitable design, compressive prestressing forces act to overcome the bursting pressure caused by liquid carried in the pipe, and no tension will occur in the concrete pipe. Another similar application of non-stress relieved wire is in prestressed concrete tanks.

For certain prestressed concrete applications, special products such as two- or three-wire strands or deformed wires of various diameters are employed in either pretensioning or postensioning.

CHAPTER 32

Manufacture of Steel Tubular Products

SECTION 1

HISTORY AND CLASSIFICATION OF STEEL TUBULAR PRODUCTS

The growth of the steel tubular industry was due to many factors, chief of which is the diversity of uses to which steel tubular products may be and are applied. The mere mentioning of the names, oil, gas, air, water, plumbing, heating, ammonia, and boiler tubes is sufficient to call to mind industries almost wholly dependent upon steel pipe or tubing. In addition to the many pressure-retaining applications, considerable quantities of pipe and tubing are used in structural applications because of favorable strength-to-weight ratio.

DEVELOPMENT OF THE BUTT-WELD PROCESS

About the year 1815, William Murdock, a Scottish inventor, introduced at London the use of coal gas for lighting purposes. For conveying this gas, Murdock collected old musket barrels and screwed them together to form continuous tubes. The popularity of this lighting system created a demand for tubes, and stimulated inventors to seek some means of producing the tubes more rapidly and at a lower cost. The first to succeed in this undertaking was James Russell, who filed patent papers describing his process as "an improvement in the manufacture of tubes for gas and other purposes" in 1824. In his method, the tube was formed by butting the white-hot edges of a bent plate together. The initial welding was done with a tilt-hammer provided with round grooves in the head, and the rough tube thus formed was finished by reheating it and passing it through a round groove in a rolling mill and over a mandrel which was supported in the pass, or opening, between these rolls. The next year, however, Russell's work was overshadowed by the invention of Cornelius Whitehouse, who succeeded in forming a commercially perfect tube by merely drawing the flat plate, heated to a proper temperature, through a "bell" or die. This invention, which became the basis of one of the butt-weld processes used until recently, made it possible to produce tubes of superior quality much more cheaply than before. Shortly after this invention, about 1832,

the first shop for making butt-weld pipe in the United States was established in Philadelphia by Morris, Tasker and Morris. Four years later, this firm built the works afterwards known as the Pascal Iron Works. Following their success, other plants appeared in Eastern Pennsylvania, Eastern New York, New Jersey and Massachusetts, but no plants were built west of the Allegheny Mountains until after 1860. The idea that pipe could be butt-welded continuously was conceived by John Moon in 1911. Later, with S. F. Fretz, Jr., he built the first experimental mill. This equipment proved successful and in 1921 and 1922 the Fretz-Moon Tube Company was formed and continuously butt-welded pipe was made on a production basis. Modern continuous butt-welding methods are discussed in Section 2 of this chapter.

DEVELOPMENT OF SEAMLESS TUBES

Concurrent with these developments in making welded tubes, inventors turned their attention to the production of seamless tubes. At first they attempted to duplicate with iron or steel the method (extrusion) used to produce tubes of lead and some other metals, namely, forcing the hot metal through an orifice formed by a mandrel or punch located and supported in the center of a circular die. In 1836 such a process was patented by Hanson in England, but this method proved impracticable. Two other methods were brought out in 1840 and 1845. They involved the cupping of a plate or the piercing of a round billet in a press, and subsequently elongating the rough tube thus formed by drawing or rolling. While practicable, these methods were costly, and for fifty years the use of seamless tubes was restricted on that account. The modern developments beginning about 1890 are described in Section 5.

In the piercing process, developed around 1840, a round hole was first made along the central axis of a round billet which was then rolled and drawn over mandrels to form a tube. The mill for lengthening such hollow billets was first patented in England by Church and Harlon about 1841. The rolling mill and the draw bench afforded simple and comparatively cheap methods of elongating the hollow billet, but the development of a method for piercing it provided a difficult matter. The oldest and simplest method was to heat the billet to a high forging temperature and hydraulically force a punch through its center while in this hot state. As it was essential that the hole be exactly concentric with the billet throughout its entire length, a feat that is hard to accomplish with a billet more than a foot or so in length, recourse was had to drilling a small hole in the cold billet, then heating the billet and enlarging the hole by piercing in a press. About 1888 a patent was granted for a process whereby a small ingot was cast about a core of refractory material, which hollow ingot was to be treated as described for hollow billets.

In the cupping process, which was first used about 1845, a circular sheet or plate was forced by successive operations through several pairs of conical dies, each pair being 'eeper and more nearly cylindrical than the previous one, until the plate took the form of a tube, or cylinder with one end closed. This method has been supplanted by the roll-forge process described in Section 8.

The Mannesmann machine, employing the principle of helical rolling for piercing round billets for making seamless tubes, was patented in 1886; its design and operation will be discussed later (in Section 5). The Pilger process for making seamless tubes was developed later but, for economic reasons, was employed only to a limited extent in this country and will not be discussed in detail here; it utilized a rolling mill with a pair of grooved rolls with shaped passes that exerted a forging force on the piece as it was forced into the roll pass rather than drawn into the pass by the action of the rolls. Other processes employing special machines including the Stiefel disc piercer, the Assel mill, and the Diescher mill, and special techniques such as plug rolling, double piercing, rotary rolling and reeling and the press-piercing mill (PPM) have been introduced over the years to yield additional improvements to the processing of seamless tubes.

Innovations by Briggs and Riverside Iron—Besides these developments in methods of manufacture, two other events should be mentioned because of their far-reaching effects on the industry as a whole. About 1862 Robert Briggs, then superintendent of the Pascal Iron Works, formulated the dimension of pipe (tube) threads, and compiled a table giving the nominal sizes, the exact diameters and the number of threads per inch for all sizes of pipe and tubes up to NPS 10*. These formulae and tables were subsequently adopted as standard for the manufacture of all tubes and pipes up to 15 inches (381 millimetres) in size and are widely known as "Briggs Standards," but the name now officially adopted is ANSI (American National Standards Institute) B2.1 Pipe Threads (except Dry Seal). In 1887 the Riverside Iron Works, Wheeling, W. Va., began making butt- and lap-welded pipe of soft Bessemer steel. Up to that time, wrought iron had been the only

---

*See subsection on "Sizes of Pipe" later in this section.

Table 32—I. Steel Tubular Products Classified According to Use.

1. Standard Pipe
   a. Standard Weight Pipe (Black and Galvanized)
   b. Extra Strong Pipe (Black and Galvanized)
   c. Double Extra Strong Pipe (Black and Galvanized)
2. Special Light Weight Pipe
3. Conduit Pipe
4. Line Pipe
   a. Gas Lines
   b. Oil Lines
   c. Water Mains
   d. Slurry Pipe
5. Pressure Piping
6. Structural Pipe
   a. Railings
   b. Fence Pipe
   c. Columns
   d. Bridge and Roof Trusses
7. Pipe Piles
8. Sprinkler Pipe
9. Welding Fittings Pipe
10. Water-Well Pipe
11. Oil-Country Tubular Products
    a. Casing
    b. Tubing
    c. Drill Pipe
12. Pressure Tubes
    a. Boiler Tubes
    b. Condenser Tubes
    c. Superheater Tubes
    d. Economizer Tubes
13. Still Tubes
14. Mechanical Tubes

---

material used for welded pipe and tubes. Riverside proved that steel was not only equal to wrought iron for this purpose, but actually superior to it in many respects.

Classification of Steel Tubular Products—The many uses to which steel tubular products have been applied have led to a great variety of products and to the use of a large number of more or less descriptive terms in designating the products used for different purposes. Use, therefore, may form a basis for classifying steel tubular products, as shown in Table 32—I.

Another classification is based on methods of manufacture. On this basis all steel tubular products may be classified under the two main headings of welded and seamless, with subclasses, as shown in Table 32—II.

Methods of Manufacturing Welded Tubular Products—A butt-welded pipe or tube is made from hot-rolled skelp with square or slightly beveled edges, the width and thickness of the skelp being selected to suit the various sizes to be made. The coiled skelp is uncoiled, heated, and fed through forming and welding rolls where the edges of the skelp are pressed together at a high temperature to form a weld.

In the electric-weld process, hot-rolled strip or plate, of a gage corresponding to the thickness of the wall of the pipe desired, but of an overall width slightly greater than its circumference, is first edge-trimmed to obtain parallelism and accurate width. To produce fusion-welded pipe, the plate is then bent into cylindrical shape with the beveled edges abutting to form a

Table 32—II. Steel Tubular Products Classified According to Methods of Manufacture.

A. Welding Process

I. Butt-Weld Process. This process is used in the manufacture of NPS ⅛ to NPS 4 pipe.

II. Electric-Weld Processes.
   a. Electric-Resistance Weld.
      This process uses the resistance of the material itself to generate the heat required for welding.
   b. Arc Welding Process.
      The heat required for welding is generated by electric arcs, which may be shielded, and may use added material in the weld.

B. Seamless Process

I. Piercing Processes—Roll Piercing. This process is used to make pipe and tubes up to 660 mm (26 inches) in diameter.
   a. Hot Finished.
      (to 660 mm or 26 inches in diameter).
   b. Cold Drawn.
      (any size up to 305 mm (12 inches) outside diameter).

II. Press Piercing Mill (PPM). This process is applied in the manufacture of pipe 89 to 245 mm (3½ to 9⅝ inches) outside diameter.

III. Extrusion Process. Used primarily for small sizes of stainless or high-alloy pipe.

"V" into which the electrode is melted. Fusion welding, which is particularly adapted to large-diameter pipe, is a term used to distinguish this method from electric-resistance welding used in the manufacture of smaller sizes of tubing. By the latter process, union of the seam is effected by the application of pressure and heat, the heat being generated by the resistance to current flow (either transformed or induced) across the seam during the welding.

Seamless Tubular Products—Seamless tubular products are made by four basic methods:

(1) Rotary piercing of a solid round bar or billet, followed by various methods of refining to produce the wall thickness and size required.

(2) Piercing a bloom or section of steel in a vertical press, leaving one end closed, and then further processing to the required size by the roll-forge process.

(3) Extruding a short, large-diameter round in a hydraulic extrusion press.

(4) Press piercing of a solid square billet followed by processing methods such as rotary elongating, mandrel mill, and stretch reduction.

Pipe—Applied in a general sense, pipe is a term used to designate any long hollow body used for conducting gases or liquids, and may be of clay, cement, wood, lead, brass, cast iron, or steel. Restricted to the steel industry, the term is one that is applied to all tubular products intended for the purposes for which such products are ordinarily used, as for conducting water, fuel, gas, steam, air, oil, etc. The term "wrought" distinguishes forged iron or steel pipe from cast-iron pipe.

Sizes of Pipe—Pipe sizes are frequently referred to as **nominal pipe size**. Up to and including 12 inches (304.8 millimetres) nominal pipe size there are wide differences between the nominal pipe sizes and the actual outside diameters. In nominal pipe sizes 14 inches (355.6 millimetres) and larger the actual outside diameter corresponds to the nominal pipe size.

The nominal pipe sizes 10 inches (254 millimetres) and under were for the most part established in the British tube trade between 1820 and 1840. The sizes were designated, roughly, according to their internal diameters.

Robert Briggs, about 1862, while superintendent of the Pascal Iron Works, formulated the nominal dimensions of pipe and pipe threads up to and including 10 inches (254 millimetres). These dimensions spread broadly and became widely known as "Briggs Standards". It is interesting to note that the perfect thread length in the Briggs Standards was a function of the outside diameter while the wall thickness (today's standard weight) was calculated to yield a thickness of the pipe under the root of the thread at the end of the pipe equal to 0.0175 D + 0.025 inch where D = outside diameter in inches.

Details on Briggs' Standards may be found in Paper No. 1842, "American Practices in Warming Buildings by Steam", presented before the British Institute of Civil Engineers by Robert Briggs. The paper is contained in the Institute Proceedings, Vol. LXX1, Session 1882-1883, Part 1. The substance of the paper is quoted in the report of the Committee on Standard Pipe and Pipe Threads to the American Society of Mechanical Engineers at the seventh annual meeting and is published in Vol. VIII, Paper No. 226, of their proceedings. The paper was accepted by the American Society, December 29, 1886.

Recent actions have been taken by American National Standards Institute in collaboration with the American Society for Testing and Materials (ANSI/ASTM) to substitute the dimensionless designator NPS for such traditional terms as "nominal diameter", "size", and "nominal size". The designation NPS 2, for example, will replace 2 inches (50.8 millimetres) nominal diameter. The NPS designator will be used in this text.

SECTION 2

BUTT-WELD PROCESS (CONTINUOUS)

The continuous butt-weld process is a true continuous process starting with coiled skelp and ending with finished pipe. Figure 32—1 shows an arrangement that is typical of many of the continuous butt-weld mills in use today.

Production of Skelp—The skelp mill described below is a typical mill designed primarily to roll skelp to meet the requirements of a continuous butt-weld pipe mill. This mill normally rolls skelp in coils up to 1420 millimetres (56 inches) in outside diameter, weighing

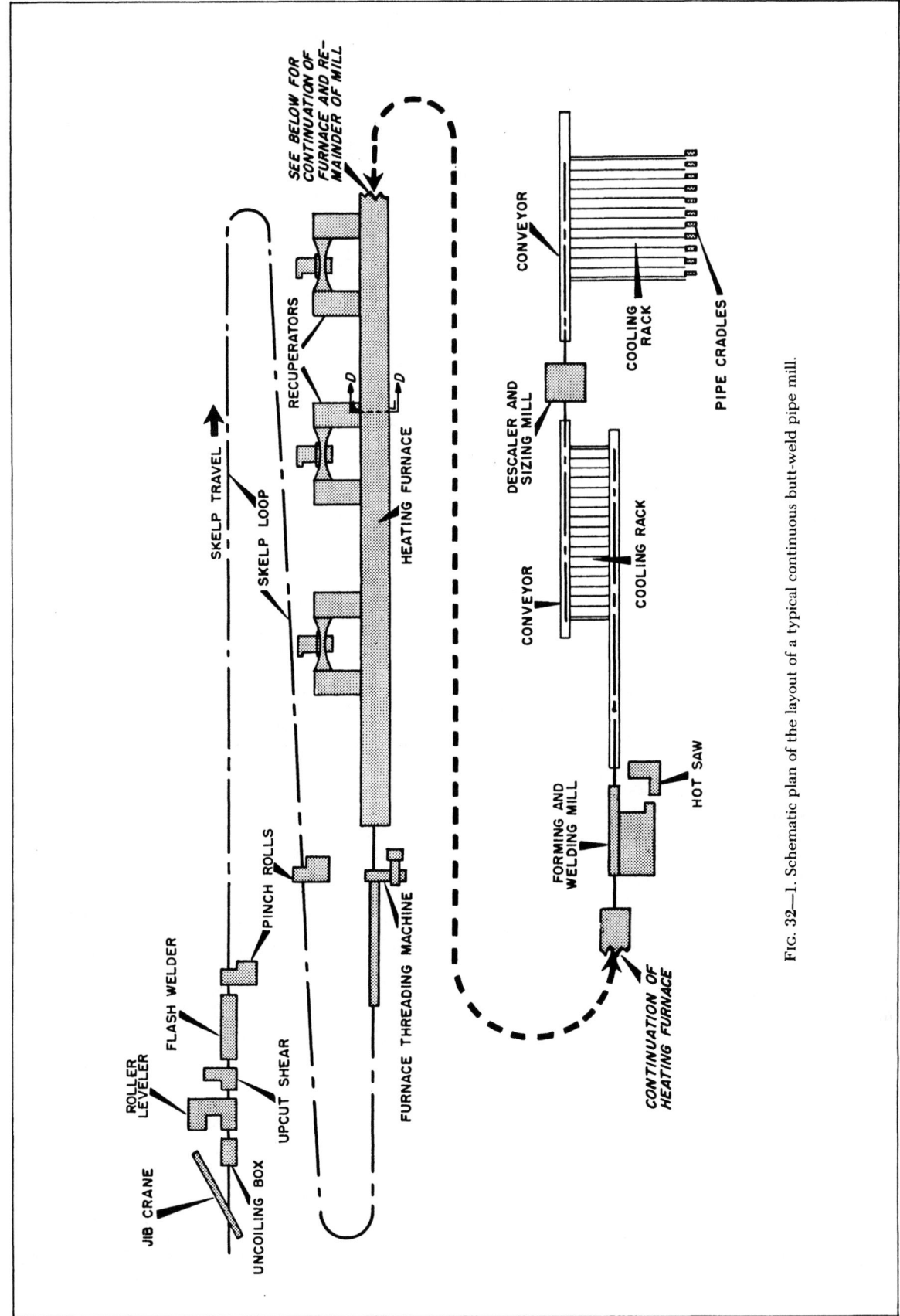

FIG. 32—1. Schematic plan of the layout of a typical continuous butt-weld pipe mill.

up to 9.8 kilograms per millimetre of width (550 pounds per inch of width) at the average rate of approximately 635 metric tons (700 net tons) per 8-hour turn on a monthly basis, from slabs 11 to 12 metres (36 to 40 feet) in length. Typical widths and thickness of skelp for the production of pipe of various sizes and wall thicknesses are as follows.

Skelp Width		Skelp Thickness		
(mm)	(in.)	(mm)	(in.)	Pipe Size
222	8¼	3.0	0.120	NPS ½
222	8¼	4.2	0.165	NPS ½
241	9½	3.2	0.125	NPS ¾
241	9½	4.2	0.165	NPS ¾
280	11	3.8	0.150	NPS 1
280	11	4.8	0.190	NPS 1
320	12½	3.8	0.150	NPS 1¼, NPS 1½
320	12½	5.5	0.215	NPS 1¼, NPS 1½
350	13¼	3.9	0.155	NPS 2
350	13¼	5.7	0.225	NPS 2
375	14¾	5.0	0.195	NPS 2½
375	14¾	5.2	0.205	NPS 2½
375	14¾	7.0	0.275	NPS 2½
445	17½	4.6	0.180	NPS 3, NPS 3½, NPS 4
445	17½	5.2	0.205	NPS 3, NPS 3½, NPS 4
445	17½	5.5	0.215	NPS 3, NPS 3½, NPS 4
445	17½	7.6	0.300	NPS 3, NPS 3½, NPS 4
445	17½	8.3	0.327	NPS 3, NPS 3½, NPS 4

The incoming slabs are delivered to the open slab-storage yard on specially built standard-gauge slab cars.

The slabs are removed from the cars in 3-metre (10-foot) wide tiers by a 27 metric-ton (30-net-ton) electric overhead traveling crane equipped with two rectangular magnets carried on a rotating trolley. This makes it possible to pile the slabs in alternate tiers, rotated 90 degrees to secure piling stability, and provides for the stocking of unusually large piles of slabs separated into their proper sizes and grades of steel. Slab-separating skids 11.5 metres (38 feet) wide by 7.5 metres (25 feet) long and the furnace-charging conveyor are located in the slab yard. The separating skids consist of both stationary and movable skids arranged to provide a cascade effect that separates and aligns individual slabs and deposits them singly onto the 33.5 metre (110 foot) long furnace-charging conveyor. In conjunction with the charging conveyor is a chain-driven ram-type charging car for pushing slabs into the furnace.

The slab-heating furnace (Figure 32—2) is of the side-charge and side-discharge, zone-controlled, double-fired type. The effective heating length of the furnace is 18 metres (60 feet), its inside width is 13 metres (43 feet) and its maximum heating capacity is 109 metric tons (120 net tons) per hour. The furnace has a suspended roof throughout its length, and the hearth is sloped upward from the point of discharge. Zone-controlled heating, utilizing primary and secondary heating zones, is employed. After a slab has passed through the primary heating zone, it enters the secondary zone where heating to rolling temperature is completed and temperature throughout the slab is equalized. The furnace is top-fired in both zones, and is equipped to burn either coke-oven gas or oil. Combustion air is preheated in a refractory-tile recuperator

FIG. 32—2. General view of furnace for heating slabs for rolling in a skelp mill.

located beneath the furnace. Located on the charging
end of the furnace is the slab cross pusher that operates
at right angles to the furnace-charging conveyor and
pushes slabs off the conveyor and advances them
through the furnace. The cross pusher has a 1070 mm
(42 inch) stroke and a maximum speed of 3.04 strokes
per minute and is mounted on its own independent
foundation that is not attached in any way to the furnace structure.

A pushout bar and a pair of pull-out rolls remove
heated slabs from the furnace. The pushout bar assembly, located entirely outside the furnace, consists of the
pusher bar and a pair of driven rolls; the bar is clamped
between the rolls and is driven by friction. The complete unit can be shifted by a hydraulic mechanism to
align the bar with slabs in the furnace. The bar has a
maximum speed of 1.05 metres per second (206 feet
per minute) and a maximum travel of 14.5 metres (48
feet). A pair of pullout rolls located outside the furnace-
discharge door aids the push-out bar in delivering slabs
to the mill-approach table; the bottom roll of the pair is
driven by an adjustable-speed motor so that the speed
of delivery of slabs from the furnace can be matched
approximately to the speed of the mill-approach table.

The mill-approach table, which conveys heated slabs
to the mill, is approximately 20 metres (65 feet) long
and has a speed range of 0.16 to 0.48 metres per second
(32 to 96 feet per minute). A hydraulically operated
slab turnover and discard permits slabs to be presented
to the mill with the heavy scale on the bottom or to be
removed from the approach table when desired.

The general arrangement of the skelp mill proper is
shown in Figure 32—3; a general view of the mill in
operation is shown in Figure 32—4. Table 32—III provides data with regard to the capacity of the main
drives and other features of the mill.

The mill proper consists of ten 485 mm (19 inch)
closed-top roll stands, one 405 mm (16 inch) scale-
breaking edging stand, and four 305 mm (12 inch) edging stands. The ten horizontal stands are of heavy cast-
steel construction with the windows equipped with
liners. All horizontal rolls are carried by oil-film-type
sleeve bearings. The top rolls are carried on hydraulically supported rails that extend through the front and
back housing windows, enabling the supports to serve
also as guide rails when changing rolls. The bottom
roll-neck bearing on the operating side of each mill is
located axially by a parallelogram adjusting device that
not only clamps the roll to prevent axial movement,
but also provides for limited axial adjustment if required. The bottom rolls may be adjusted vertically by
a screw and wedges accessible from the operating side
of each mill. The first five horizontal roll stands are
equipped with hand-operated screwdowns since, in
normal operation, very little screw adjustment is required on these stands. The last five horizontal roll
stands are equipped with single 11 kilowatt (15 horsepower) motor-driven screwdowns, so arranged that the
top screws may be adjusted in parallel or individually.

The heavy-duty 16-inch scale-breaking edging stand
is located ahead of No. 1 horizontal-mill stand and, to
maintain close control of product width, the 305 mm
(12 inch) edging stands are located ahead of horizontal
stands Nos. 4, 6, 8, and 10 as shown in Figure 32—3.

A rotary cobble shear, situated between horizontal
stands Nos. 5 and 6, disposes of strip leaving the roughing stands in the event of a cobble occurring in the
finishing portion of the mill.

Air-operated loopers are provided between stands
Nos. 5 and 6, 7 and 8, 8 and 9, and 9 and 10 to assist in
width control.

High-pressure-water descaling boxes are provided
both ahead of and following the edging scalebreaker to
chill the scale on the slabs entering the edger and to
remove it following the edger.

The runout table of the mill is approximately 24.5
metres (80 feet) in length and contains 39 individually
driven rollers on 610 mm (24 inch) centers. This table
has a speed range to match the mill delivery speed of
2.66 to 7.98 metres per second (523 to 1570 feet per
minute).

To turn the finished-rolled skelp on edge, polished
cast-iron twist troughs are provided, preceded by a horizontal pinch-roll unit and followed by a vertical pinch-
roll unit and a strip switch. The strip switch, automatically operated by a photoelectric system, is a
mechanical guide trough that is swiveled about its entry end while its delivery end is free to move to direct
the skelp alternately to either of two vibrators that feed
the skelp in serpentine pattern onto an apron conveyor. From the strip switch through to the unloading
of coils in to storage, dual facilities have been provided
to handle the mill output.

The serpentine-formed skelp is fed directly onto the
apron-type conveyor, which is 34 metres (110 feet) long
and 2.1 metres (7 feet) wide between side guards, at a
speed of from 0.27 to 0.82 metres per second (54 to 162
feet per minute) (Figure 32—5). The conveyor will hold
a maximum of about 490 metres (1600 feet) of skelp.
The skelp is held on the conveyor until the leading end
is trimmed square and then fed by an adjustable-speed
vertical pinch roll into the reels for coiling. The trimming shears, of the vertical alligator type, can cut skelp
up to 460 mm (18 inches).

Two vertical-type skelp reels receive the skelp delivered from the apron conveyors. The reel centers are
of the collapsible type and, upon completion of a coil,
are collapsed automatically and withdrawn downward
from the coil. Centers are designed to form coils with
an inside diameter of 510 mm (20 inches). The reels
have a speed range of 122 to 244 rpm. Each reel has a
coil pushoff for pushing the coils onto the coil
conveyors. This method of coiling avoids telescoping
and produces coils much less subject to edge damage in
handling to storage or from storage to subsequent processes.

The coil conveyors are of the double roller-chain
type, with the coils riding on the chains on 1.8 metre (6
foot) centers. The conveyors are approximately 27
metres (90 feet) long and have a maximum speed of
0.08 metres per second (15.5 feet per minute). Scales in
each conveyor weigh the coils.

To facilitate the handling of coils to storage in 13.6
metric ton (15-net-ton) lifts by either the 22.7 metric
ton (25-net-ton) electric overhead traveling coil-storage
crane with rotating trolley or by a 13.6 metric ton (15-
net-ton) capacity self-propelled ram truck, there are a
coil upender and coil assembly conveyor after each of

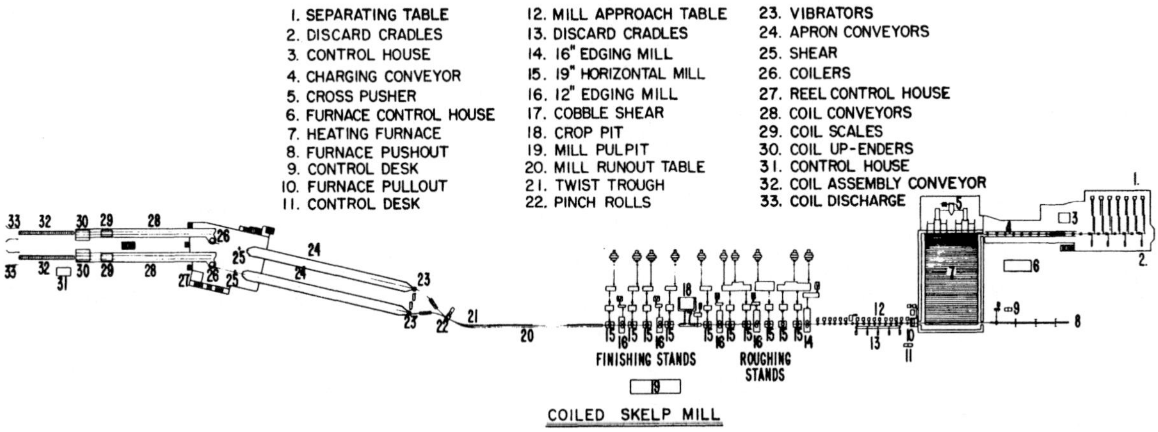
1. SEPARATING TABLE
2. DISCARD CRADLES
3. CONTROL HOUSE
4. CHARGING CONVEYOR
5. CROSS PUSHER
6. FURNACE CONTROL HOUSE
7. HEATING FURNACE
8. FURNACE PUSHOUT
9. CONTROL DESK
10. FURNACE PULLOUT
11. CONTROL DESK
12. MILL APPROACH TABLE
13. DISCARD CRADLES
14. 16" EDGING MILL
15. 19" HORIZONTAL MILL
16. 12" EDGING MILL
17. COBBLE SHEAR
18. CROP PIT
19. MILL PULPIT
20. MILL RUNOUT TABLE
21. TWIST TROUGH
22. PINCH ROLLS
23. VIBRATORS
24. APRON CONVEYORS
25. SHEAR
26. COILERS
27. REEL CONTROL HOUSE
28. COIL CONVEYORS
29. COIL SCALES
30. COIL UP-ENDERS
31. CONTROL HOUSE
32. COIL ASSEMBLY CONVEYOR
33. COIL DISCHARGE
FINISHING STANDS
ROUGHING STANDS
COILED SKELP MILL

FIG. 32—3. General layout of a skelp mill for producing coiled skelp. (See also Figure 32—4.)

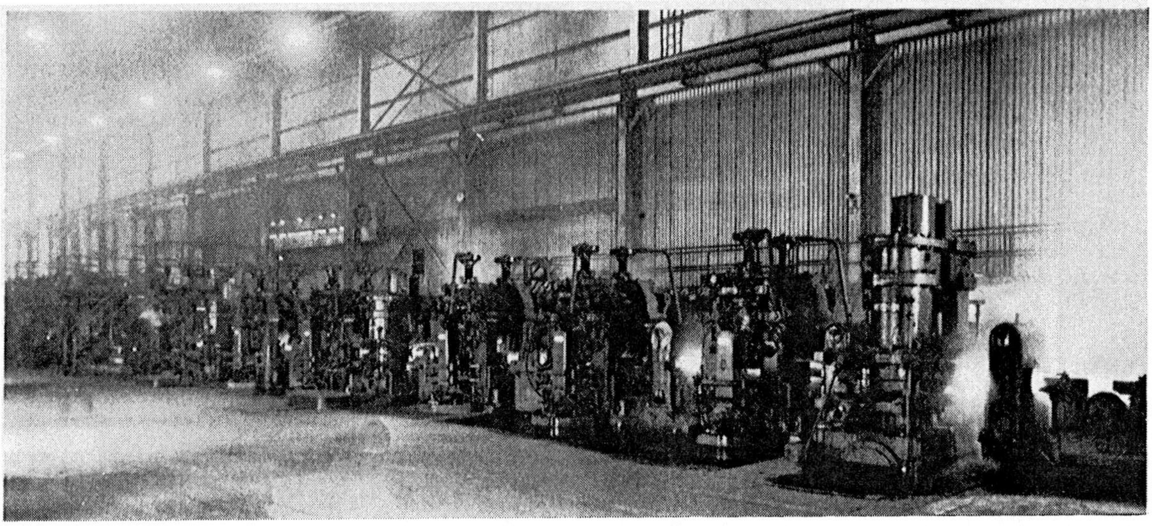

FIG. 32—4. General view of skelp mill rolling coiled skelp for use in the continuous butt-weld process for pipe. Product travel is from the vertical 405-mm (16-inch) edging mill at the right toward the left of the illustration. (See also Figure 32—3.)

Table 32—III. Drive and Mill-Speed Data for Typical Skelp Mill

Stand Number	Motor kW	Motor hp	Motor Speed Range (rpm)	Total Gear Reduction	Roll Speed (rpm)	Roll-Pass Diameter (mm)	Roll-Pass Diameter (in.)	Delivery Speed m/s	Delivery Speed (ft per min)
Edger 1	149	200	225/1200	51.77	4.3 to 23.2	318-381	12½-15	0.08 to 0.46	14 to 91
No. 1	448	600	225/900	33.95	6.6 to 26.5	440-484	17 5/16-19 1/16	0.15 to 0.67	30 to 132
No. 2	448	600	225/900	25.54	8.8 to 35.2	440-484	17 5/16-19 1/16	0.20 to 0.89	40 to 176
No. 3	597	800	262.5/875	17.68	14.8 to 49.5	200-498	7⅞-19⅝	0.35 to 1.29	69 to 254
Edger 2	112	150	225/1200	9.50	23.8 to 126.8	229-305	9-12	0.28 to 2.02	56 to 398
No. 4	597	800	262.5/875	11.56	22.7 to 75.7	470-505	18½-19⅞	0.55 to 2.00	108 to 394
No. 5	597	800	262.5/875	7.85	33.4 to 111.5	465-510	18 5/16-20 1/16	0.81 to 2.98	160 to 586
Edger 3	112	150	225/1200	4.83	46.8 to 249.6	229-279	9-11	0.56 to 3.65	110 to 719
No. 6	597	800	262.5/875	5.69	46.1 to 154	437-481	17 3/16-18 15/16	1.05 to 3.88	207 to 764
No. 7	597	800	262.5/875	4.74	55.4 to 184.8	440-484	17 5/16-19 1/16	1.28 to 4.68	251 to 922
Edger 4	112	150	225/1200	3.67	60.8 to 324.4	305-330	12-13	2.97 to 5.61	191 to 1104
No. 8	597	800	262.5/875	4.07	64.6 to 215.2	441-486	17⅜-19⅛	1.49 to 5.47	294 to 1077
No. 9	597	800	262.5/875	3.54	74.1 to 247	441-486	17⅜-19⅛	1.71 to 6.28	337 to 1237
Edger 5	112	150	225/1200	2.85	78.6 to 419.2	305-330	12-13	1.25 to 7.25	247 to 1427
No. 10	597	800	262.5/875	3.19	82.2 to 274.4	445-489	17½-19¼	1.92 to 7.02	377 to 1381

FIG. 32—5. Finish-rolled skelp fed in serpentine pattern onto apron conveyor that carries it to reel.

the conveyors that carry coils from the reels. The upender is hydraulically operated and raises the coils to a vertical position and pushes them onto a 40-foot long assembly conveyor where the coils are assembled into approximately 13.6 metric ton (15-net-ton) lifts.

The coiled skelp is stored in a large building in which are located the discharge end of the apron conveyors, the two coiling reels, the two coil conveyors, two coil upenders and assembly conveyors. All floor areas within the building are paved and suitable for storing the coils three high. The building is served by a standard-gauge railroad siding. A transfer car is provided for delivering stored coils as required to the continuous-weld pipe mills.

Production of Continuous Butt-Welded Pipe—In a modern continuous-weld mill, coiled skelp is formed into high-quality butt-welded pipe at speeds of 0.76 to 5.59 metres per second (150 to 1100 feet per minute). The automatically controlled equipment assures a uniformity of wall thickness and diameter of finished product never before attained. The continuous mills produce pipe in the full range of sizes from NPS ½ to NPS 4 nominal diameter from only a few different widths of skelp.

Coils of skelp from storage are loaded on the arms of

a turntable that provides a continuous supply to the mill. The coils are removed from the turntable, one at a time, and placed on a cradle car. Each coil is rotated on the cradle car until it is in the right position for the coil peeler, which opens the coil and bends the end of the skelp so that it can be fed into the pinch rolls of the uncoiler-leveler. When the skelp is started through the leveler, the coil is transferred to an expanding mandrel for payoff into the mill. Figure 32—6 illustrates one type of coil-handling equipment at the entry end of a continuous butt-weld mill.

An upcut shear squares the leading end of each coil and the trailing end of the coil preceding it so that the ends can be joined by flash welding. A flash welder is designed to position and clamp the two coil ends, heat them, and force them together to form an upset weld. After welding, a drawcut planer removes the flash formed in the welding operation. The principle of operation of a flash welder is shown in Figure 32—7.

Pinch rolls and the uncoiler-leveler push the strand of skelp away from the welder. The steel is guided to the floor and formed into a large storage loop that provides skelp for continuous mill operation while end welds are being made. Four magnet-roll units help to feed skelp into the storage loop. At the end of the stor-

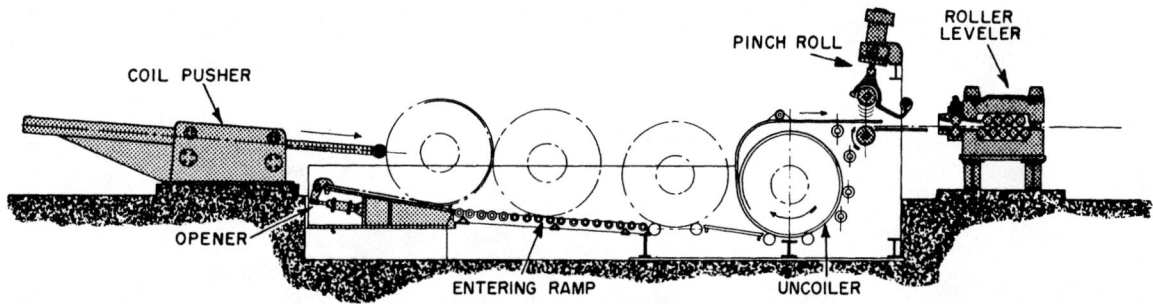

FIG. 32—6. Schematic elevation of coiled-skelp handling equipment at the entry end of a continuous butt-weld pipe mill.

age loop, a horizontal pinch-roll unit moves skelp into a secondary loop, then feeds it into the preheater or upper main furnace chamber (Figure 32—8.) Pinch-roll speed is varied by automatic controls that maintain the extent of the secondary loop within set limits.

The preheat chamber is heated by waste gases from the main furnace. The skelp is fed into the preheater where, after traveling some distance, it reverses direc-

tion and makes a second pass through the preheater. The preheated skelp then passes over a large sheave and into the main furnace, which is fired by 434 gas burners and is divided into four separately controlled heating zones. The furnace and preheater are capable of heating 63.5 metric tons of 5.7 mm by 445 mm skelp per hour (70 net tons of 0.225 inch by 17½ inch skelp per hour). A schematic section of one type of skelp-

A. TRAILING END OF SKELP STRIP IS HELD STATIONARY BY HOLDING JAWS, WHILE LEADING END OF INCOMING STRIP IS BROUGHT A- GAINST GAGE BAR.

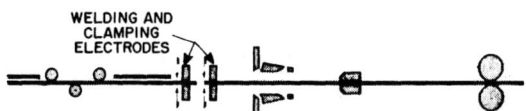

B. WELDING AND CLAMPING ELECTRODES GRIP BOTH ENDS OF SKELP. GAGE BAR IS REMOVED.

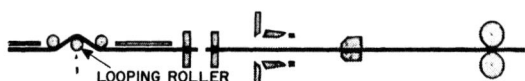

C. LOOPING ROLLER MOVES UP TO ALLOW SLACK IN SKELP FOR NEXT OPERATION.

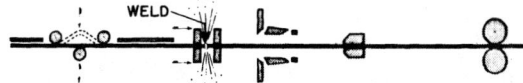

D. POWER IS APPLIED TO ELECTRODES. LEFT HAND ELECTRODES MOVE TO RIGHT BRINGING ELECTRIFIED ENDS TOGETHER FORM- ING ARC, UPSETTING AND WELDING.

E. WELDING AND CLAMPING ELECTRODES RELEASE. SKELP IS PULLED THROUGH FLASH STRIPPER. STRIPPER JAWS RELEASE AND PINCH ROLLS PULL WELDED AND CLEANED SKELP TO CONTINUOUS LINE.

FIG. 32—7. Welding cycle for a typical flash-type skelp welder.

FIG. 32—8. Skelp from storage loop in right foreground being fed into furnace extending into center background that heats the skelp for continuous butt welding. The roller leveler and flash-welding machine are visible at the left of this illustration.

heating furnace is shown in Figure 32—9.

When starting operations or under other circumstances when it is necessary to "thread" skelp through the furnace, a furnace-threading machine installed in the line (Figure 32—10) is employed. This machine has driven pinch rolls, one convex and one concave, to give the skelp a dished shape that imparts sufficient stiffness to permit it to be pushed through the furnace. As soon as the skelp is entirely through the furnace and is being pulled by the forming and welding unit, the threading rolls are opened up and become inoperative until the mill must again be threaded.

An air blast cleans the surfaces of the hot skelp before it enters the forming, welding, and reducing stands.

The functions performed in the forming and welding mill are shown schematically in Figure 32—11. Skelp passes through the forming stand that forces it into an arc of about 270 degrees. Then it goes through a welding horn and into the welding stand, where the edges are pressed firmly together. As shown in the diagram, a nozzle following the welding horn supplies oxygen to the edges of the skelp to further heat them before being pressed together. The last five stands of the welding mill provide for reduction of diameter and resultant change in wall thickness. An overall view of a forming and welding mill is shown in Figure 32—12.

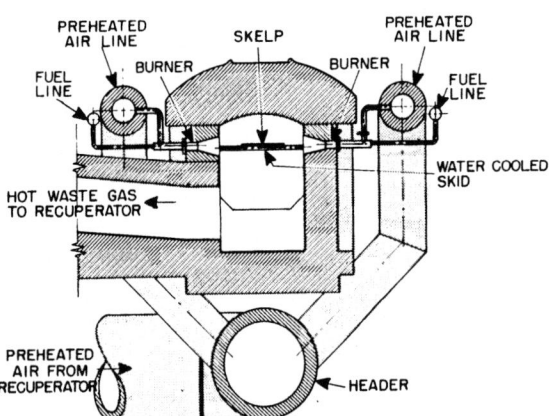

FIG. 32—9. Cross-sectional diagram showing design of a furnace used on a continuous butt-weld pipe mill for heating the edges of the skelp prior to welding. (This sketch corresponds to Section D-D of Figure 32—1.)

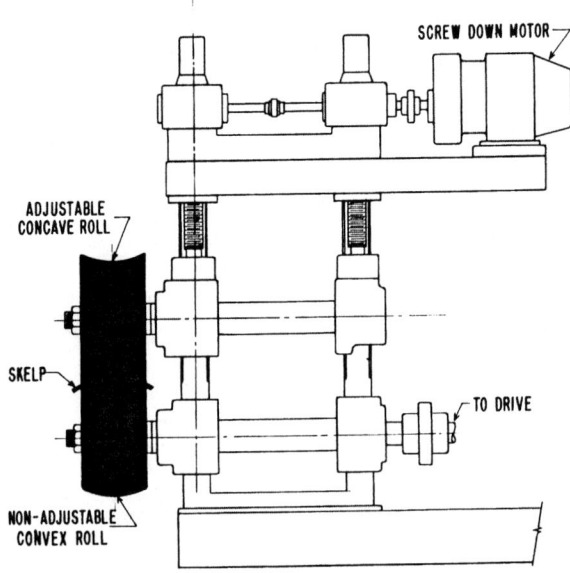

FIG. 32—10. Schematic elevation of a furnace threading machine.

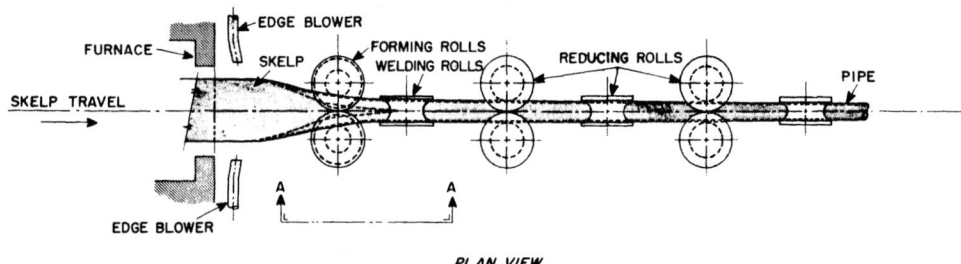

PLAN VIEW

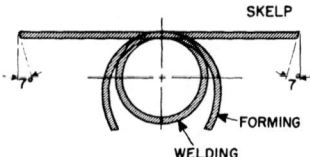

SKELP AT VARIOUS STAGES IN THE FORMATION OF PIPE
(NOTE BEVELED EDGES OF SKELP)

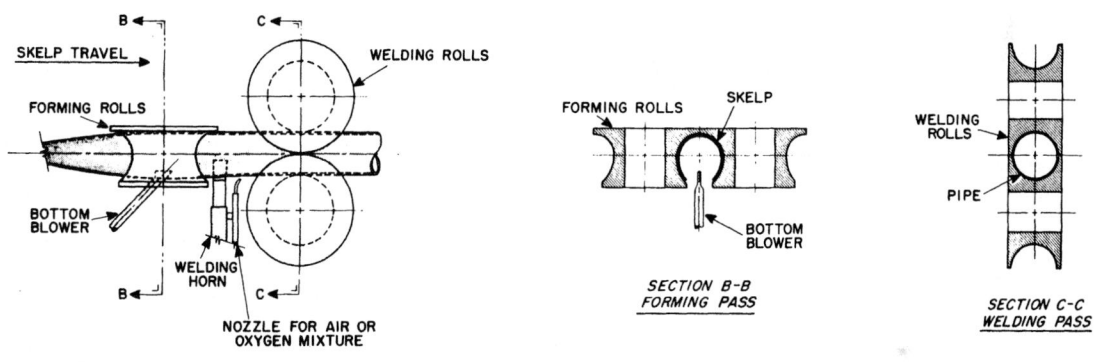

FIG. 32—11. Diagram depicting schematically the operations performed in a continuous forming and welding mill.

FIG. 32—12. Overall view of forming and welding stands of continuous butt-weld mill.

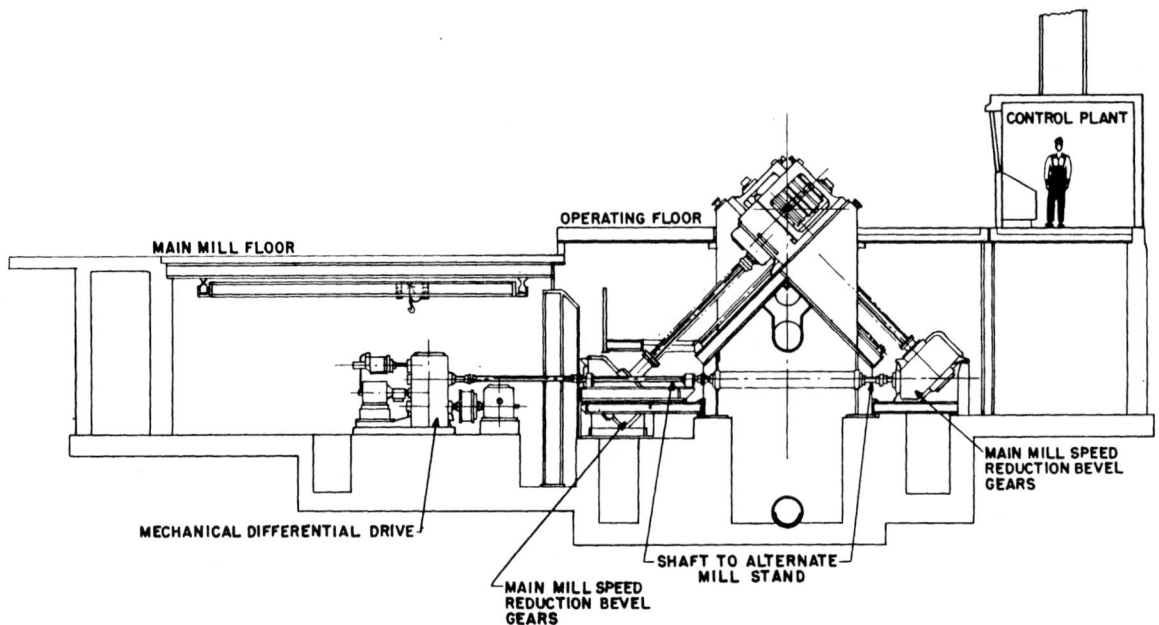

FIG. 32—13. Schematic cross-section of two typical stands of a stretch-reducing mill, illustrating components of the differential drive system that ensures a constant interstand speed ratio.

The stretch mill, a development of United States Steel, has been used since the early 1930's for simultaneous reduction of diameter and wall thickness by maintaining tension between stands. Although it was first applied to stretch lengths of furnace-welded pipe to raise productivity in the smaller sizes, it is more applicable to continuous mills. By this method, and without the use of supporting mandrels, the wall thickness of a tube can be maintained or even decreased while the diameter is being reduced. Figures 32—13 and 32—14 illustrate some of the principles of design of a stretch-reducing mill; further details are given in Sec-

FIG. 32—14. Close-up view of two active stands (right) of a stretch-reducing mill. Two stands at left have been retracted. Welded pipe leaving active stands passes through tubular guide in the left foreground.

tion 5 under "Continuous Seamless Process." Multiple-pass rolls permit rolling of six pipe sizes without changing rolls. This minimizes downtime for size changes and roll wear.

Roll shafts of the multiple-pass stretch-reducing mill are at 45 degrees from horizontal, and roll axes on successive stands are perpendicular. Each stand is mounted on a steel sled that houses the pinion stand; sleds are moved by hydraulic cylinders and positioned by adjustable stops. Hydraulic clamping mechanisms hold the sleds in position.

The drive system ensures compatible speeds for the last six stands of the forming and welding mill and all seven stands of the stretch-reducing mill. Each stand has a controllable variation of 30 per cent above or below its normal speed with an accuracy of 0.1 per cent.

All forming, welding, and reducing stands are driven by a 746-kilowatt (1000-horsepower) direct-current motor through a common lineshaft. Except for No. 1, stands receive power through differential gear units that are connected to the main lineshaft by right-angle gear boxes. The differential gearing permits variation or adjustment of stand speeds without changing the speed of the lineshaft. Such variation or adjustment is accomplished by rotating the differential gearing independently. Rotation of the gearing in one direction increases output-shaft speed, and rotation in the opposite direction decreases output-shaft speed. A hydraulic motor drives the differential gearing in the desired direction at the desired speed.

The hydraulic motors are operated by fluid from variable-displacement pumps that receive power directly from the differential-input gears. Control of the hydraulic-pump discharge and the direction of fluid flow between the hydraulic pump and motor permits mill rolls to run at an infinite number of speeds—30 per cent above or below the predetermined base speed.

Hydraulic tachometers automatically control and regulate a metering device that governs oil flow from the hydraulic pump to the hydraulic differential drive motor.

If skelp thickness varies and off-gage pipe is delivered to the stretch-reducing mill, over-all elongation can be distributed among the stands by available controls to bring wall thickness to its standard value.

A rotary flying saw cuts the continuous pipe into predetermined lengths at maximum mill-delivery speeds. Hydraulic saw-drive equipment is separately powered but controlled by an extension of the main mill shaft. Controls, using the lineshaft speed as a reference, cause the saw cycle to coincide with the product speed.

Cut lengths are reduced to the required hot size on a three-stand sizing mill, delivered by a kickout mechanism to a middle-cut saw, and dropped on a cooling bed. Crop saws at opposite ends of the cooling beds trim the pipe ends.

Pipe passes from the cooling bed to a water bosh tank for fast cooling; then it is taken to conveyors that feed rotary straighteners in the finishing bay.

The pipe is then ready for finishing operations.

ELECTRIC-RESISTANCE-WELDED TUBULAR PRODUCTS

The art and processing equipment for manufacturing electric-resistance-welded tubular products, sometimes referred to as ERW tubing, has improved considerably during the recent past. These improvements have widened the range of available ERW tubular products from small diameters and thin walls to permit the manufacture of tubing up to 510 mm (20 inches) in diameter with wall thicknesses as thin as 2.4 mm (0.094 inch) and as thick as 9.5 mm (0.375 inch). The development of nondestructive testing methods such as ultrasonic, eddy current, and fringe flux has been a major factor towards promoting ERW tubular product acceptance. These nondestructive testing devices are normally placed directly into the welding-mill line to check steel quality and also weld integrity. The steel product used as starting material for making ERW tubing is generally strip, sheet or plate in coil form (some mills are designed to use uncoiled plate) produced from either rimmed or killed steel; either hot rolled or cold rolled; with pickled, grit blasted, or mill surface.

Figure 32—15 shows an electric-resistance-weld pipe mill capable of producing pipe from NPS 8 to NPS 20 and 24.5 metres (80 feet) in length with wall thicknesses as desired between 2.4 and 10.2 mm (0.094 and 0.400 inch).

Steps in Manufacturing Electric-Resistance-Welded Tubing—The sequence of operations required in the fabrication of electric-resistance-welded tubing are: slitting (when multiple-width strip is used), forming, welding, sizing, cutting, and finishing.

Slitting—Frequently, tubing is manufactured direct from single-width strip: i.e., strip the width of which will equal the perimeter of the tubing to be welded. However, many manufacturers of tubular goods purchase wide-width coils which can be slit into the desired widths.

The wide coils are loaded onto a ramp from which they are permitted to roll, as required, onto a charging buggy which moves the coils to the pay-off reel of the slitting machine. The buggy is provided with an elevator by which it is possible to center the coil on the reel before the retaining bands on the coil are removed. The reel, which is motor driven, is rotated to slowly unwind the strip which is threaded into a set of breaker rolls that flatten it sufficiently to facilitate threading through pinch rolls into the slitter knives. These knives, which may take the form of tool-steel discs, are mounted on arbors above and below the strip and are spaced with rings to slit to the desired width. From the slitter, the strips pass to the recoiler where they are

FIG. 32—15. Overall view of an electric-resistance weld (ERW) mill for producing line pipe in a wide range of diameters and wall thicknesses, up to 24.5 metres (80 feet) in length. Flat-rolled starting material enters the line at the lower right. Welded pipe is discharged at the end of the line at the upper left. Sequence of operations is described in the text.

wound between thin, steel disc spacers. The narrow strips of scrap from the edges of the wide strip pass over the recoiler onto the scrap winder.

After the strip has been threaded completely through the line, the recoiler and slitter motors are adjusted to pull it through the slitter knives with any desired tension. The slitter, recoiler, scrap winder, breaker rolls and pay-off reel are not synchronized with the others as they are used only when threading the strip.

When the entire coil has been slit and wound on the recoiler mandrel, steel bands are placed around each of the narrow coils. An unloading buggy is elevated under the coils which are then pushed off hydraulically onto the buggy. The banded coils are weighed, and are then ready for the welding mill.

Forming—Although the range of sizes handled by any mill is limited, the process of forming and welding the tubing is generally similar in mills of this type.

Coils are fed either directly into forming rolls or into a "looper" to permit continuous forming of strip welded end-to-end in the smaller sizes. The strip first passes through an edge trimmer where the desired width is established and the edge is made smooth and clean for good welding.

ERW mills normally make use of three types of rolls to progressively form the flat steel section into the round form prior to welding. These three types of rolls are: (1) Breakdown or forming rolls, (2) Idler vertical closing rolls, and (3) Fin pass rolls, as shown schematically in Figure 32—16.

In the cases of the breakdown and fin pass sections, the rolls are horizontal and are driven either by universal line shaft or by individual drives to allow perfect speed control of each stand. Breakdown rolls provide the initial shaping of the strip towards the round form. These are followed or interspersed by idler rolls which further close and guide the strip into the fin pass rolls. The fin pass rolls provide perfect guidance into the welding section and in addition coin the strip edges to provide the precise circumference.

Welding the Tube—After forming has occurred, the open tube passes directly into the welding section of the mill. Here the tube is held in squeeze rolls at the

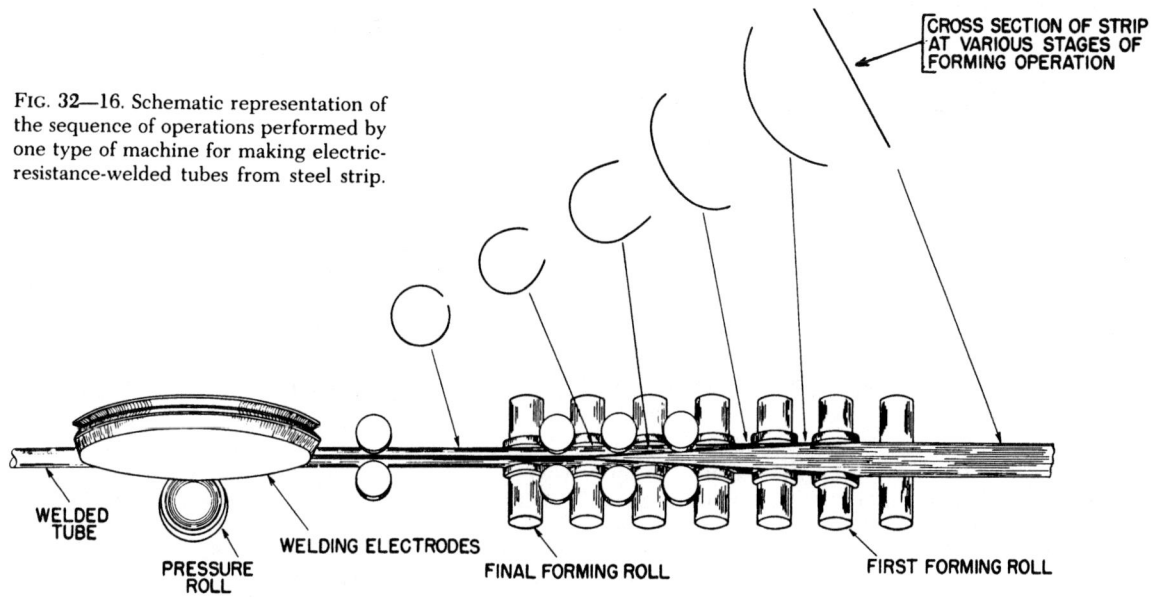

FIG. 32—16. Schematic representation of the sequence of operations performed by one type of machine for making electric-resistance-welded tubes from steel strip.

CROSS SECTION OF STRIP AT VARIOUS STAGES OF FORMING OPERATION

WELDED TUBE

PRESSURE ROLL

WELDING ELECTRODES

FINAL FORMING ROLL

FIRST FORMING ROLL

FIG. 32—17. View of the welding operation in the pipe mill shown in Figure 32—15. Two concave rolls (only one is visible here) force the edges of the joint together while sliding contacts supply 480 000-hertz (480 000 cycles per second) power to generate heat for welding.

correct pressure to provide the desired weld as the edges are heated at this point to welding temperature.

The heat for welding is provided either by low-frequency power through electrode wheels (Figure 32—16), or radio-frequency power through sliding contacts or coil induction. Typical radio-frequency power for welding is supplied at 450 000 hertz (cycles per second); higher or lower frequencies may be used (Figure 32—17).

The welded tube then passes under a cutting tool which removes the outside flash resulting from the pressure during welding. Inside flash is likewise removed by cutting tools in this area. After removing the flash, the tube is subjected to proper post-weld treatment as metallurgically required; e.g., such treatment may involve sub-critical annealing or normalizing of the welded seam or normalizing of the full cross-section of the tube.

Sizing the Welded Tube—After cooling, the tube is sized to obtain a round finished product of the desired

FIG. 32—18. Flying cut-off for the pipe mill illustrated in Figure 32—15. Parting tools (or discs) are used in this machine for cutting pipe. Pipe can be seen emerging from concave rolls of the last stand of the sizing mill that precedes the cut-off. Product travels from right to left in this view.

diameter. The sizing mill consists of several driven horizontal rolls and several idle vertical rolls. One sizing-mill design permits straightening to occur here.

Finishing—After the tube leaves the sizing mill, it is cut to determined lengths by suitable cut-off machinery. Cutting may be performed by discs, parting tools, shears, or cold saws which travel with the tube while cutting. Figure 32—18 illustrates the flying cut-off for the pipe mill shown in Figure 32—15.

The cut tubes are then transferred to the finishing floor where they are straightened, if required, using standard facilities either continuous or of the gag-press design.

Also on the finishing floor, special cutting and end finishing is performed when needed. After finishing, tubes are inspected and packed for shipment.

SECTION 4

ELECTRIC-WELDED LARGE-DIAMETER PIPE

Applications of the Process—Large-diameter pipe in sizes beyond the practical limits of the seamless process is fabricated by electric welding. This type of pipe, which is employed for water lines, gas mains, oil lines, tanks, pressure vessels, etc., may be made from rolled-steel plate of any weldable quality. The size of the pipe which can be made by this process is practically unlimited, since, when the circumference of the desired pipe exceeds the plate width capacity of the rolling mill, two or more plates may be welded together longitudinally to provide the necessary width or, if only a short length is desired, the plate may be bent lengthwise, permitting it to be formed with only one longitudinal weld. Where long-length, large-diameter pipe is required, the desired length may be made by welding together two or more pieces circumferentially.

Steps in the Manufacture of Electric-Welded Pipe—The sequence of operations required to make plates into pipe by the electric-weld process are shearing, planing, crimping, bending, welding, expanding and finishing.

Shearing and Planing—Plates employed in the manufacture of electric-welded pipe are rolled on either a plate or strip mill as described elsewhere in this book. The plate is transferred to the edge-planing machine where it is aligned so that the two edges will be parallel and square with the ends after planing. A clamping bar, hydraulically operated, holds the plate during the edging operation. Along the full length of both sides of the table, lead screws drive carriages carrying a series of cutting tools which trim the edges of the plate.

The series of cutting tools are arranged so that no single tool will be required to remove more than a reasonable amount of stock from the rough edges of the rolled or sheared plate.

Crimping—Forming plate into the circular shape required for pipe is usually performed in three operations. The first operation, called crimping, consists of bending the edges of the plate so as to avoid a flat surface near the longitudinal seam of the pipe.

Crimping may be performed in a large crimping press which deforms the edges of the plate for a distance of approximately 150 mm (6 inches) in a hydraulically operated press. Crimping may also be performed by crimping rolls which roll the edges to the desired radius as the plate is drawn through the roll pass.

Bending—The crimped plate is then conveyed to what is called the "U"-ing machine (Figure 32—19). In this operation, the plate is centered over a series of parallel rocker-type dies which lie along the axis of the plate. A large "U"-shaped die, which is as long as the longest length of plate fabricated and which is operated by a press of at least 1800 metric tons (2000 net tons) capacity is moved down on the plate, forcing it between the dies which automatically conform themselves to the operation and assist in forming the plate into the "U" shape. The plate is then transferred to what is called the "O"-ing machine. This machine consists of two semi-circular dies which are as long as the plate to be formed. Rollers mounted on vertical spindles prevent the "U"-shaped plate from falling and keep it in correct alignment as it enters the "O"-ing machine. The "U"-shaped plate rests in the bottom die, and the top die, operated by hydraulic press, having a 16 300-metric-ton (18 000-net-ton) capacity or more as required is forced down, deforming the plate until it is the shape of an almost closed circle which is then ready for welding.

FIG. 32—19. The "U"-ing machine forming the crimped plate in a "U" shape.

FIG. 32—20. Pipe beginning to emerge from the outside welding machine.

Welding—Welding may be performed by any one of the numerous methods. However, the most common is the submerged-arc method. In this method, coalescence is produced by heating with an electric arc or arcs between a bare-metal electrode or electrodes and the work. The following description refers to the use of a single electrode but two and even three electrodes are used in some operations. The welding is shielded by a blanket of granular, fusible material or flux on the work. Pressure is not used and filler metal is obtained from the electrode and sometimes from a supplementary welding rod.

It is extremely important that the gap in the formed pipe be properly positioned for outside welding. Two steel archways supporting a longitudinal guide are located over the conveyor approaching the welding machine. As the pipe moves over the conveyor, the guide enters and continues along the gap, guiding it into the welding machine. Figure 32—20 shows a pipe emerging from an outside-welding machine.

The welding operation is started by striking an arc beneath the flux on the work. The heat produced melts the surrounding flux so that it forms a sub-surface conductive pool which is kept fluid by the continuous flow of current. The end of the electrode and the work piece directly under it become molten and molten filler metal is deposited from the metal electrode onto the work. The molten filler metal displaces the flux and forms the weld.

The specially designed welding heads used for this process perform the triple function of progressively depositing the flux along the joint, feeding the electrode, and transmitting welding current to the electrode. The flux is supplied from a hopper connected to the head by tubing. The bare electrode is fed into the welding head from a coil mounted on a reel.

After the pipe is welded on the outside, the weld is inspected and conditioned on the ends when required. At this time, square pieces of metal are welded to the pipe at the ends of the weld to enable the inside-welding machine to start at the end of the pipe.

The automatic machine which does the internal welding is similar in design to the outside welder, and is mounted on the end of a long cantilever arm and the pipe is drawn over this arm by a carriage.

After welding, the scaly deposit left from the flux must be cleaned out. This may be accomplished by running the pipe over a series of rollers and, at the same time, allowing the entry of a cantilevered tube which is attached to a vacuum system; another tube-cleaning method causes the tube to be rotated in a fixture at such an angle as to cause the flux to fall from the tube by gravity. The weld is then carefully inspected and defects corrected with semiautomatic welding equipment.

In some mills, primarily to facilitate guidance in making the weld, the inside weld is made first. In this case, the unwelded "can" is closed by a series of hy-

FIG. 32—21. Hydraulic expanding machine for final sizing of diameter of electric-welded large-diameter pipe. The ribbed sections are retaining jackets.

draulically operated clamps in a tacking press and tabs are affixed across the closed seam by semiautomatic welding. If necessary, additional tacks can be made along the lengths of the can between clamping bands.

The pipe is then transferred to the inside welder, which again hydraulically clamps the pipe with the weld down, and the welder boom carrying the welder heads strikes arcs on the tab and proceeds through the pipe, maintaining guidance by following a key placed on the plate edge before forming.

The pipe is then transferred to an outside-diameter welder where the OD weld is made by a traveling welding carriage.

Sizing or Expanding—The final pipe diameter is obtained either by hydraulically expanding the welded shell against a retaining jacket or by mechanically expanding it over an inside mandrel.

In hydraulic expansion, the pipe is placed in an expanding machine (Figure 32—21) and mandrels are forced into each end to expand the pipe to the required diameter at the ends only Retainer jackets then encircle the body of the pipe which is filled with water and expanded by hydraulic pressure to the limits of the jackets. The pressure is then reduced and serves as a hydrostatic test.

In mechanical expansion, a segmented and expandable cylindrical element is inserted in the pipe and expanded to the desired final inside diameter of the pipe. The section being expanded is usually only a fraction of the length of the pipe. At least two operations normally are required to completely expand the pipe, the work being accomplished in two stations. Separate hydraulic or other non-destructive tests are required with mechanical expansion.

The expansion by either method sizes, rounds and straightens the pipe and provides a good test of the weld, while cold working the material to provide higher mechanical properties.

Attention is continuously given to non-destructive inspection of weld quality (see Chapter 51). This is accomplished by X-ray examination of the weld at both ends of the pipe at suitable intervals. In addition, the entire longitudinal weld can be subjected to X-ray examination by viewing a fluoroscopic screen followed by photographic X-ray inspection of any points wherein fluoroscopic examination suggests further inspection is in order. Ultrasonic inspection is also used along with X-ray examination of the longitudinal welds, as discussed in Chapter 51.

Finishing—Following the rounding-up or expansion operation, the pipe is placed in special machines which face the ends. This operation ensures that the ends will be smooth and accurately within a plane at right angles to the longitudinal axis of the pipe. If the pipe is being prepared for welded joints, the ends are beveled in this operation.

SECTION 5

SEAMLESS STEEL TUBULAR PRODUCTS

In the decade preceding the twentieth century, the bicycle became a very popular vehicle, and the growth of the bicycle industry created a demand for high-class seamless tubes in very large quantities. The early methods of seamless tube manufacture were slow and tedious, and orders aggregating millions of feet of tubing for bicycle construction could not be filled. In 1895 the first American seamless tube plant using the principle of rotary piercing was constructed in Ellwood City, Pennsylvania, and operated successfully for many years in the manufacture of bicycle tubing. The equipment at this plant consisted of a Stiefel disc piercer, a Pilger rolling mill, and cold-drawing benches on which the tubes were finished.

After 1900 the automotive industry began to grow and soon caused a great decrease in the manufacture of bicycles, with a consequent loss of seamless-tube business. On the other hand, the great increase in the use of automobiles created an immense demand for motor fuels and lubricants, which greatly stimulated expansion in the oil industry and thus developed a new and enormous market for seamless tubes. In the transition period, the seamless-tube manufacturer turned to the boiler industry, and the next step in the development of the seamless process related to the production of all forms of stationary and locomotive-boiler tubes. The seamless tube soon displaced the wrought-iron and lap-welded steel tubes which had previously been standard in boilers.

After World War I, the discovery of enormous oil pools with flush production created a demand for pipe for wells and transmission lines. The later discovery, after 1920, of huge natural-gas reservoirs in remote districts from which the gas could be transported to consuming markets only in steel pipelines resulted in another large market for seamless pipe. The seamless tube mills prior to 1920 were limited to producing tubes with a maximum diameter of 230 mm (9 inches) and a maximum length of 9 metres (30 feet).

In the year 1925, the United States Steel Corporation developed the so-called double-piercing process, which consisted of piercing a heavy-walled shell and then expanding this in a second piercing operation to a larger diameter tube with a lighter wall. This development made possible the production of tubes up to 16 inches in diameter. The Pilger process had also reached a production stage at this time, but a thorough study of this method caused its rejection for American manufacturing conditions. Further study of the double-piercing and expanding process later led to the rotary-rolling process, by which it was possible to produce seamless tubes up to 660 mm (26 inches) in diameter and in lengths exceeding 12.2 metres (40 feet).

Scope and Requirements of Seamless Tube Products—The expansion of the seamless-tube industry described above was also greatly influenced by the adaptation of the piercing process to the many different types of steels which were developed during this period. Steels melted by many processes can now be successfully converted into seamless tubes. In general, killed and semi-killed steels made by open-hearth, electric-furnace and basic oxygen processes are used.

Because of the severity of the forging operation involved in piercing, the steels used for seamless tubes must have good characteristics with respect to both surface and internal soundness. A sound, dense cross-section, free from center porosity or ingot pattern, is the most satisfactory for seamless tubes. For this reason steels of the thoroughly killed types and in some cases semi-killed steels are to be preferred to strongly rim-

ming steels. Although rimming steels have been successfully converted into seamless tubes, the results are not always uniform, either because of a poor inside surface, due to internal porosity, or because of high losses from external seams, caused by surface defects. Metallurgical developments in recent years have contributed greatly to the improvement of steels for seamless tubes.

As a result, the seamless process has been extended to include practically all of the regular and alloy grades of steel. At the present time all of the ordinary carbon steels, even those containing as much as one per cent of carbon, are processed in commercial quantities. Some of the high-sulphur and leaded steels developed for machining are also manufactured into seamless tubing.

All of the constructional alloy steels, such as those listed in the specifications of the American Iron and Steel Institute and the Society of Automotive Engineers, are available in tubing.

Over the years, many steels of special compositions that impart resistance to heat and corrosion have been developed. Seamless tubes are now satisfactorily made from most of these steels. These include the chromium steels containing from 1 to 30 per cent chromium and numerous other alloyed steels containing chromium with additions of such elements as molybdenum, nickel, managenese, columbium (niobium), silicon, and titanium. In general, it can be stated that all the alloys which are ferritic or pearlitic in structure are satisfactory for piercing. The alloy steels which are generally of the austenitic type can be successfully pierced if the austenite remains stable at forging temperatures. A few of the austenitic steels which have a considerable proportion of ferrite at forging temperatures are difficult to pierce. For example, an alloy such as the 25 per cent chromium-10 per cent nickel, which is partially ferritic, may be difficult to pierce, whereas an alloy with 25 per cent chromium and 20 per cent nickel can be pierced satisfactorily. However, seamless tubes of alloys containing free ferrite may be produced by the extrusion process.

Seamless pipe with a variety of special properties has been developed to meet the needs of the oil industry. In this field the pipe employed for rotary drilling must possess great torsional strength and high resistance to fatigue stresses. An idea of the stresses to which the pipe is exposed in drilling service may be gained from the realization that the power for the rotation of the drill bit on the lower end of the pipe is applied at the upper extremity, which, in some cases, is 14.5 kilometres (9 miles) distant. The string of drill pipe itself may weigh over 400 000 kilograms (900 000 pounds), and rotate at 200 rpm or more, so that only material of the highest strength and toughness will perform satisfactorily. Drill pipe is usually made with heavy upset ends for attaching tool joints and the full lengths of pipe are normalized or quenched and tempered on the completion of the forging operations.

The casing for such deep oil wells requires seamless pipe having a high resistance to collapse to withstand the high external pressures. Casing has been set to depths of over 10 058 metres (33 000 feet) under very high hydrostatic heads and depths of 12 190 metres (40 000 feet) are contemplated. Other classes of seamless pipe and specialty tubing made for the oil industry include oil-well tubing, pump tubing, and line pipe.

The oil-refining, chemical and high-pressure-steam industries also demand special seamless pipe. Satisfactory alloy-steel tubing and pipe are made to withstand temperatures up to 650°C (1200°F) coincident with pressures as high as 34 500 kilopascals (5000 psi). These industries also use pressure pipe lines at temperatures as low as −268°C (−450°F), and steel pipe which is tough and strong at these low temperatures is now produced in considerable quantities.

Sequence of Seamless-Pipe-Mill Operations—The sequence of operations in seamless-pipe mills varies slightly from mill to mill. At the Lorain Works of United States Steel it follows three general patterns determined by the size of the pipe to be produced. The round billets are first uniformly heated to piercing temperature in special gas- or liquid-fuel-fired furnaces. They are then processed through the various operations in the following order, depending upon the diameter of tube to be produced:

Pipe Size—NPS 2 to NPS 4
    Piercing Mill
    Multi-Stand Mandrel Mill
    Reheating Furnace

    Sizing Mill, or
    Stretch-Reducing Mill
Pipe Size—NPS 3½ to NPS 9⅝
    Press Piercing Mill
    Rotary Elongator
    Multiple-Stand Pipe Mill
    Reheating Furnace
    Stretch-Reducing Mill
Pipe Size—NPS 5½ to NPS 16
    First Piercing Mill
    Second Piercing Mill
    Reheating Furnace
    Plug Rolling Mill
    Reeling Machine
    Sizing Mill
Pipe Size—NPS 14 to NPS 26
    First Piercing Mill
    Second Piercing Mill
    Reheating Furnace
    Plug Rolling Mill
    Reheating Furnace
    Rotary Rolling Mill
    Reeling Machine
    Reheating Furnace
    Sizing Mill

FIG. 32—22. Mannesmann piercer in operation.

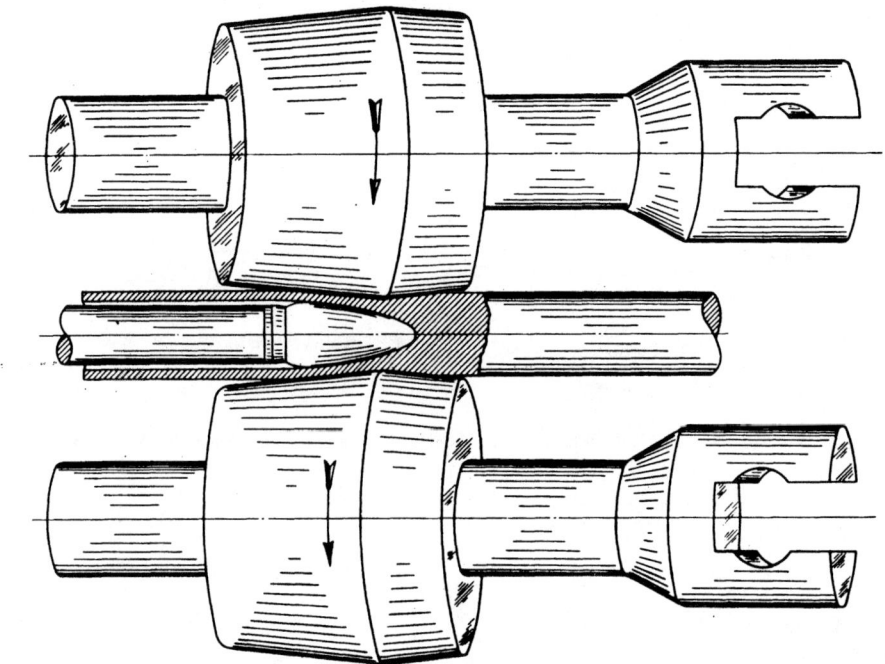

FIG. 32—23. Sketch illustrating action of different parts of Mannesmann piercer in the piercing of a solid billet.

The operation of each of the above units will be described in detail in the remainder of this section. In the overlapping sizes of the range NPS 14 to NPS 16, the lighter-wall pipe is rotary rolled, the heavier-wall pipe by-passing the second reheating operation and the rotary-rolling mill.

The Mannesmann Machine for piercing round billets for making seamless tubes (Figure 32—22) was patented in 1885. In this machine, the principle of helical rolling is employed. The two steel rolls, which bring into play the forces used to produce the cavity in the work piece, are positioned side by side and have their axes inclined at opposite angles 6 degrees to 12 degrees with the horizontal centerline of the mill (Figure 32—23). These rolls measure from 510 to 760 mm (20 to 30 inches) in length and from 810 to 1220 mm (32 to 48 inches) in diameter. The roll surfaces are contoured so that, in the horizontal plane through the centerline of the pass, the space between the rolls converges toward the delivery side for a length of from 130 mm to 380 mm (5 inches to 15 inches) to a minimum, called the gorge, and then diverges to form the pass outlet. The converging and diverging angles formed by the roll surfaces vary from 4 degrees to 12 degrees with 7 degrees as the usual standard. The shafts of these rolls are mounted in bearings which can be adjusted laterally in the housing to permit the space between the rolls to be properly set for the size of work piece being rolled. The inlet end of each roll shaft is fitted with a universal coupling which, through long spindles, connects with a common reduction gear powered by an electric motor. The size of this motor and reduction unit depends on the size range to be produced on the mill, varying between 559 and 3730 kilowatts (750 and

5000 horsepower). These motors are designed for a pullout torque of 300 per cent. The rolls are cooled by water sprays. The elevation of the centerline of the pass is determined by two guides, one of which is mounted above and the other below the center of the mill in the space between the rolls.

Between these guides in the pass outlet a projectile-shaped piercing mandrel is held in position on the end of a water-cooled mandrel-support bar, located on the delivery side of the mill. The opposite end of this bar is mounted in a thrust bearing which is carried in a reciprocating carriage that is latched stationary during the piercing operation. The pointed end of the piercing mandrel extends just beyond the gorge toward the entering side of the rolls.

The Operation of Piercing—A solid round bar or billet of the proper length and diameter to make the size and weight of tube desired, is heated uniformly to the usual temperature for rolling light sections, which is 1200° to 1260°C (2200° to 2300°F). With the rolls revolving at constant speed of 4.06 to 5.59 surface metres per second (800 to 1100 surface feet per minute), the heated billet is transferred to a horizontal trough, which positions the axis of the billet on the inlet side of the mill coincident with the centerline of the pass.

The heated billet is pushed forward into the space between the rolls, which has been adjusted so that the gorge is approximately 90 percent of the diameter of the billet. As soon as the leading end of the billet has contacted the rolls, the force of the pusher is removed. Because of the obliquity of the roll axes, the motion imparted to the billet between the rolls is one of rotation and axial advance. When the leading end of the billet has advanced to the gorge, it encounters the nose or pointed end of the piercing mandrel. The grip of the

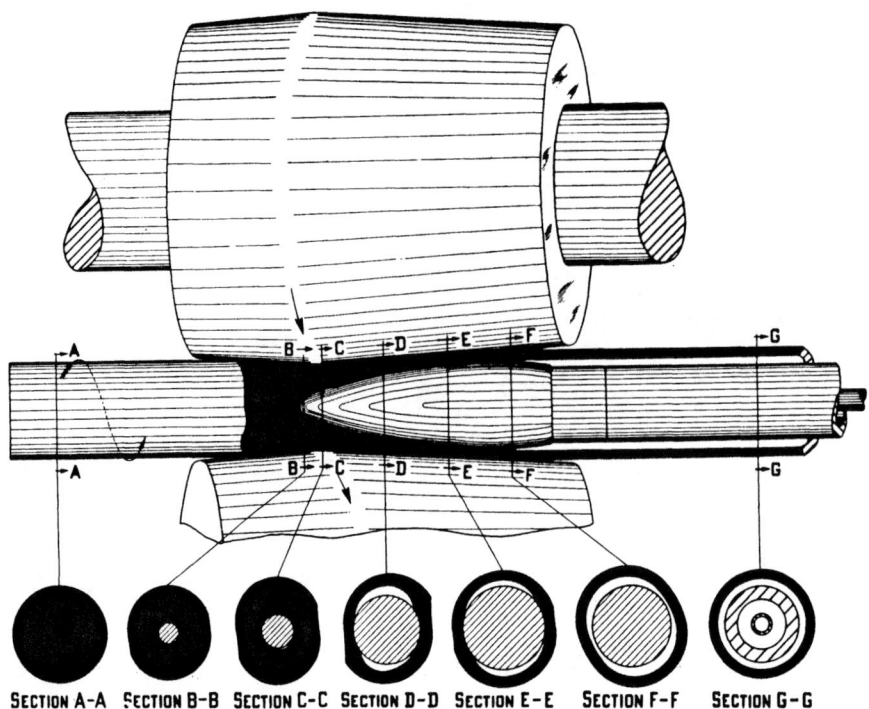

SECTION A-A   SECTION B-B   SECTION C-C   SECTION D-D   SECTION E-E   SECTION F-F   SECTION G-G

FIG. 32—24. Sketches illustrating action of rotary-piercing mill on the round billet.

rolls is sufficient to continue the advance of the work piece against the retarding effect imposed by the piercing mandrel. When the rearward end of the billet is rolled clear of the piercer mandrel, the thrust-bearing carriage is unlatched and the mandrel withdrawn from the billet, now a hollow shell, which is then conveyed to a reheating furnace in preparation for further fabrication into a finished tube. These shells are produced in lengths up to 8 metres (26 feet) in less than a minute.

The Action of the Rolls—It is evident that the forward motion of the billet is caused by the inclination of the axes of the rolls. How these two rolls, by exerting pressure only on a surface of the billet, are able to force the metal over the mandrel to form a tube from a solid billet is not so readily grasped. It is to be especially noted that the mandrel is not forced through the metal, but that the rolls cause the metal to flow over and about the mandrel. To bring about this result, the rolls must first draw metal away from the center of the billet, which action tends to form a central hole, or cavity, for the entrance of the piercer point. The truth of this statement is evident from the fact that a small, but somewhat irregular, hole may be formed in a billet without the use of the piercer point. In practice, the end of the piercer mandrel is placed sufficiently forward to prevent the formation of a cavity ahead of the point. Advantage is taken only of the tendency of the rolls to form this cavity.

Principle Involved in Forming the Cavity—Indeed, such a hole can be opened up in the center of any solid cylindrically shaped plastic body by rolling it between, even, two flat surfaces. Steel workers, particularly hammermen, are familiar with the fact that, if a piece of steel in the form of a round be pressed or hammered

into an oval form several times in succession, a rupture will occur in the center that will extend longitudinally through the middle of the bar. The reason for the formation of this rupture is plainly due to the fact that when pressure is applied to the round bar at diametrically opposite points sufficient to deform it, making one diameter shorter and that at right angles to it longer, the spreading of the metal, which takes place along the long diameter and in opposite directions, sets up a lateral tension that may cause its particles to be drawn away from the center (Figure 32—24).

Flow of the Metal in Piercing—As the billet, which is in a plastic state, enters the mill, the rolls grasp it at diametrically opposite points on its circumference. As they draw the billet forward in the converging portion of the pass, they continue to compress it at these opposite points and, since the billet is being revolved rapidly, these points are continually changing. As the compressive rolling or cross-rolling proceeds from the point of initial contact of the piercer rolls to the gorge, the diameter of the billet is reduced and the section is changed from a circle to an oval with the long diameter in a vertical position. Since the billet is rotating, the central portion is acted on by all of the forces which are applied around its circumference during successive contacts between the billet and rolls. If a sufficient reduction is effected in this manner, a cavity will be formed in the center of the billet even without the presence of a piercer mandrel. In practice, this cavity is not permitted to form in advance of the nose of the mandrel, since the rough surface of the self-formed cavity might not permit the inner surface of the shell to be subsequently rolled smooth by the action of the mandrel. It is for this reason that the end of the piercer

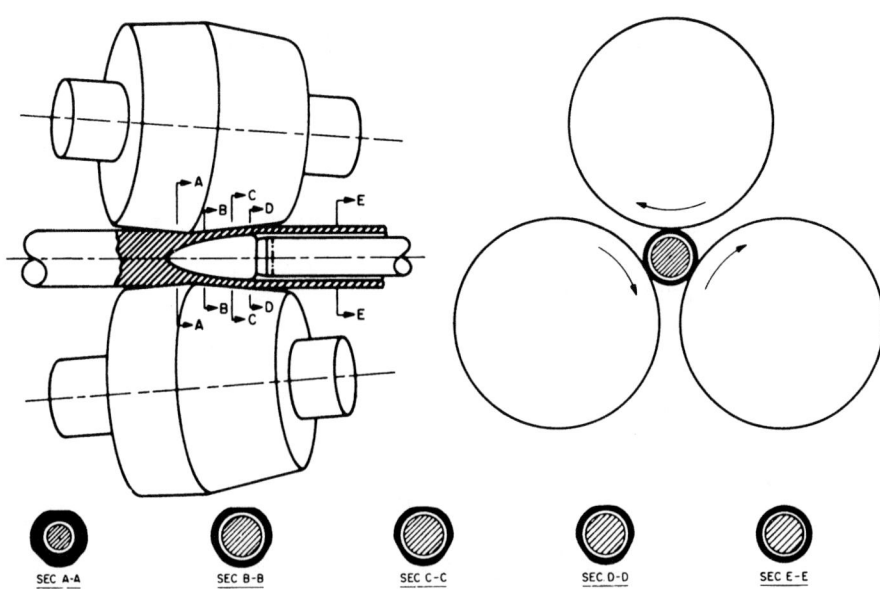

FIG. 32—25. Schematic diagram showing principle of operation of a three-roll piercing machine.

mandrel is positioned in advance of the gorge so that it will actually effect the opening in the billet, being assisted in its function by the cross-rolling action. To avoid the difficulty of accurately centering the mandrel, the forward end of the billet is centered to ensure that the point of the mandrel will penetrate the billet at or very near its axis. Centering is not absolutely necessary; however, it assists in starting the end of the mandrel in the center of the billet and reduces wear on the mandrel. If the end of the piercer mandrel is positioned too far in advance of the gorge, the cross-roll action on the center of the billet will not be great enough to reduce the resistance sufficiently to permit the grip of the rolls to advance the billet over the mandrel.

Once the end of the mandrel has penetrated the axial center of the billet, the piercer mandrel serves as a third roll so that, with the properly designed pass, the metal of the work piece is helically rolled over the piercing mandrel (rather than extruded) to produce the hollow shell. The grip of the rolls on the hollow in the diverging portion of the pass, due to the obliquity of the roll axes, tends to draw the billet forward as the reduction in wall thickness is being made, and, as a consequence, tends to increase the length of the pierced hollow shell at the expense of diameter.

Double Piercing—It is to be especially noted that the amount of metal displaced in piercing increases considerably as the OD size (outside diameter) is increased. In 1925, United States Steel Corporation developed the process known as double piercing. In this process the solid billet is first pierced to a comparatively heavy-walled shell, after which, without reheating, it is put through a second piercing mill. The second mill further reduces the wall thickness and increases the diameter and length of the piece. This practice has extended the permissible diameter range of the automatic-mill method of producing seamless tubes by dividing the

requisite work of piercing in the two stages.

Three-Roll Piercing—The three-roll piercing process (Figure 32—25) differs from the Mannesmann process in that three (instead of two) contoured rolls are triangularly disposed around the centerline of the pass and angularly disposed longitudinally. The use of three rolls, rather than two rolls and two shoes which constitute the closed pass on the Mannesmann piercer, provides a more concentric tube with improved inside surface and freedom from stationary guides or shoes.

The Plug Rolling Mill is a motor-driven, non-reversing, single two-high stand, which resembles a reversing bar mill (Figures 32—26 and 32—27). There are, however, several differences between the two. Instead of roll tables, the mill is equipped on the entering side with a movable trough and pusher and on the delivery side with a stationary guide table, and mandrel-bar support. The guides on the latter table, two or more in number, are of the double-bell type and are mounted on cross beams and lined up one behind the other directly in back of the grooves in the rolls. The water-cooled mandrel bar, which is anchored in the support at the rear of the table, projects through a series of these guides with its opposite end terminating about 6.4 mm (¼ inch) short of the vertical centerline of the rolls. The free end of the mandrel bar provides support for the mandrel or plug during the working cycle. The work rolls, which are from 560 to 965 mm (22 to 38 inches) in diameter, depending upon the size of the mill, have several semi-circular grooves machined in their surface. With the rolls in position one above the other, the opening formed by the groove is not a true circle but is slightly oval with the long axis in a horizontal plane. This flare of the groove at the roll surface is provided to prevent the edge of the groove from shearing the work piece. In general, only one mandrel bar and one groove are used in the rolling of a given size tube. The tube is passed through this roll

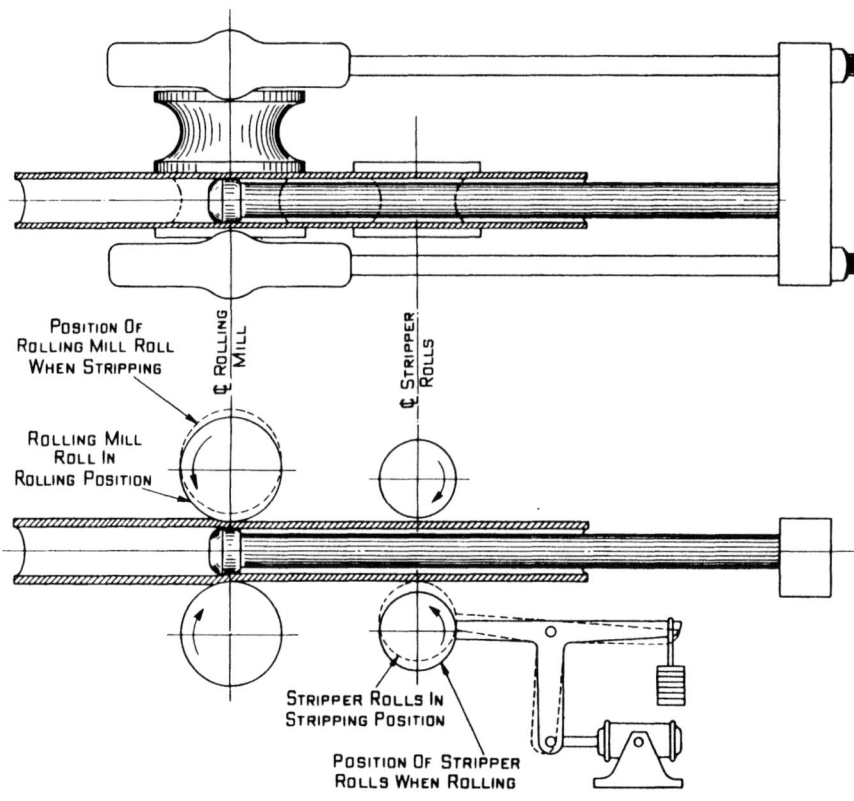

POSITION OF
ROLLING MILL ROLL
WHEN STRIPPING

ROLLING MILL
ROLL IN
ROLLING POSITION

STRIPPER ROLLS IN
STRIPPING POSITION

POSITION OF STRIPPER
ROLLS WHEN ROLLING

FIG. 32—26. Schematic plan view (above) and elevation (below) of a plug rolling mill.

FIG. 32—27. Plug rolling mill.

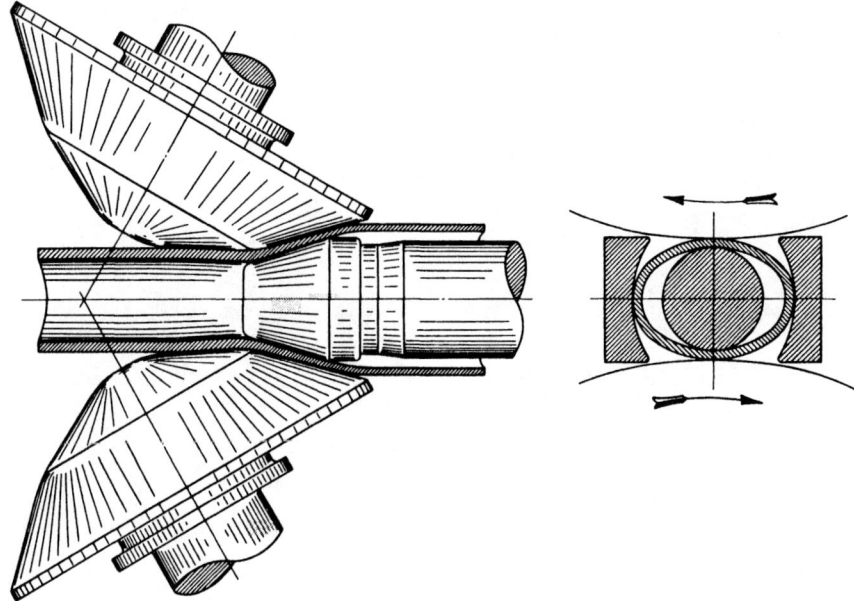

FIG. 32—28. Rotary-rolling mill.

stand twice, being rotated through 90 degrees between the passes so that the entire surface receives an equal and similar treatment in the slightly ovalled groove. To permit the tube to be stripped rapidly from the mandrel bar, the top roll is supported by counterweights and arranged to be elevated mechanically for a rapid opening of the pass. Stripper rolls, located just to the rear of the main rolls, are grooved to correspond with the main rolls. The lower stripper roll is also arranged for mechanical movement for rapid closing of the pass. These stripper rolls rotate in a direction opposite to that of the main rolls and function only when the top main roll is in its elevated or open pass position.

The Operation of Plug Rolling—The pierced shell, except for the smaller pipe sizes, is reheated after the piercing operations. With the pierced shell lying in the feed or delivery trough, an alloy-steel mandrel or plug is attached to the end of the bar, the bar holding the plug at the correct position in the roll groove. The plug is somewhat larger in diameter than the support bar in order to provide clearance between the inside of the tube and the support bar. In order to start the shell over the plug and permit the rolls to secure a good bite upon it, the shell is shoved into the pass with considerable force by a compressed-air-operated ram or pusher. Once started, the force of friction due to the pressure exerted by the revolving rolls is sufficient to draw the shell rapidly over the plug, slightly reducing its diameter and wall thickness and increasing its length. As soon as the shell has passed through the groove, the mandrel is removed from the bar. The top work roll is elevated approximately 38 mm (1½ inches). The lower stripper roll is then elevated to raise the tube clear of the bottom work roll and to grip it in the grooves of the stripper rolls, which return it to the entering side of the mill. Another mandrel is then placed on the bar, and the tube is rotated through an angle of 90 degrees. The top work roll and the lower stripper roll are returned to their original position, and the pusher again enters

the tube in the pass. As soon as the tube has passed through the groove for the second working pass, it is returned again to the entering side of the mill, from which it is discharged for further fabrication. In this way the wall of the tube, supported by the plug on the inside and subjected to the action of the rolls on the outside, is reduced in thickness to the gage desired. The pierced billet is proportionately lengthened and slightly reduced in outside diameter. The wall reduction normally made in the plug mill is approximately 3.2 to 6.4 mm (⅛ to ¼ inch). After plug rolling, the tube has a uniform wall of the desired thickness throughout but is slightly out of round or oval shaped, not perfectly straight, and still at a bright-red heat.

Rotary Rolling—The large demand for pipe between 405 and 915 mm (16 and 36 inches) in outside diameter for the transportation of natural gas for long distances raised a serious question as to the best manner of manufacturing such pipe. The lap-weld processes then used for making such pipe sizes were both costly and slow and also unsuited for the manufacture of lengths over 12.2 metres (40 feet). It is also not feasible to roll pipe over 405 mm (16 inches) in diameter on the automatic rolling mill, and this feature made it questionable whether seamless pipe in these sizes could be economically made. The rotary-rolling mill was developed by United States Steel Corporation, which has made possible the production of pipe as large as 660 mm (26 inches) in outside diameter in lengths up to 13.7 metres (45 feet) and with wall thicknesses as light as 7.1 mm (0.281 inch). Other large sizes are produced with wall thicknesses as light as 6.4 mm (0.250 inches).

The rotary-rolling mill is a modification and enlargement of the cone-type piercing mill. The shafts of this mill, which drive the 1880-mm or 74-inch (diameter) conical rolls, are in separated horizontal planes and are at an angle of 60 degrees with the axis of the pipe being rolled. A diagram of this mill is shown in Figure 32—28; an actual mill in operation is shown in Figure 32—29.

FIG. 32—29. Rotary-rolling mill in operation.

Each shaft is powered with a 1120-kilowatt (1500-horsepower), 200-500-rpm, d-c motor, which provides peripheral roll speeds of 4.06 to 12.19 metres per second (800 to 2400 feet per minute). In operation, the conical rolls grip and spin the pipe, feeding it forward over a large tapered mandrel, thereby effecting a decrease in the wall thickness of the pipe and an increase in the diameter. The length of the tube is substantially unchanged by the operation. The rolling action is simi-

lar to that which takes place in a tire- or ring-rolling machine, except that in the rotary-rolling mill a forward helical advance is imparted to the tube, which is supported on the inside by the tapered mandrel.

The Reeling Machine (Figure 32—30) is similar in construction and operation to the Mannesmann piercer except that the rolls, which are about 760 mm (30 inches) long and 865 mm (34 inches) in diameter, are almost cylindrical in form. The rolls are adjusted later-

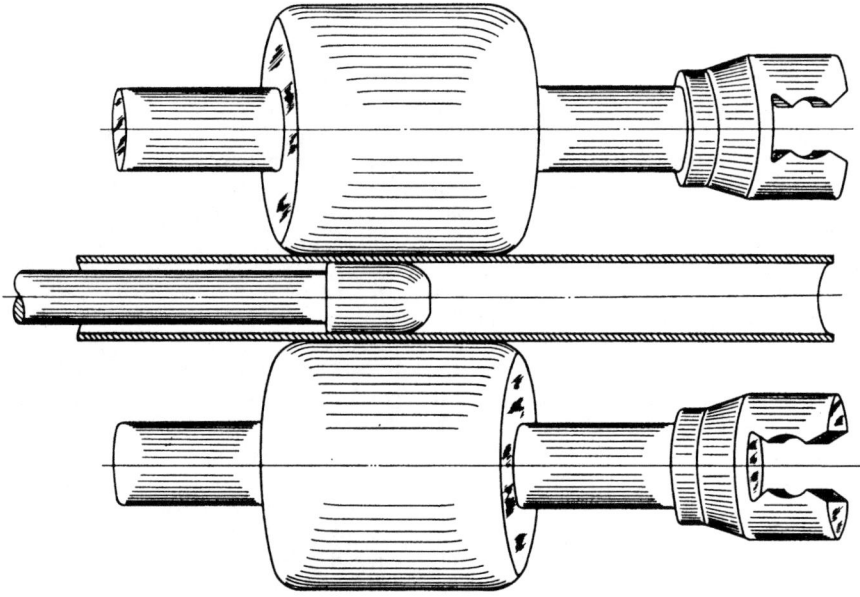

FIG. 32—30. Reeling machine.

ally in the same manner as that described for the Mannesmann piercer and are separated by a space a little less than the diameter of the tube to be reeled. The rolls are motor-driven and are geared together to revolve in the same direction at a surface speed of approximately 4.57 metres per second (900 feet per minute). In operation, a cylindrical mandrel, which is placed between the rolls, is supported on the delivery end by a water-cooled mandrel-support bar. Like the piercing mill, the opposite end of this bar is connected to a thrust bearing carried in a reciprocating carriage, locked stationary during the reeling operation. On the inlet side of the mill a conveyor carries the tube through stationary guides to contact with the rolls. Since the rolls are revolving in the same direction and with axes oppositely inclined, they cause the tube to revolve and helically advance over the mandrel. The elevation of the mandrel and tube during reeling is maintained in the proper horizontal position by stationary guides mounted between the reeler rolls above and below the pass. Owing to the fact that the total space between the reeler rolls and the mandrel is a few hundredths of a millimeter (thousandths of an inch) less than twice the wall thickness of the entering tube, a slight reduction in the thickness of the wall is effected during the reeling operation. This slight reduction made in the reeling operation has the effect of burnishing the inside and outside surfaces of the tube and slightly increasing its diameter. The function of the reeler is, therefore, to round up and to burnish the inside and outside surfaces of the tube delivered from previous fabricating operations.

Sizing the Tube—The manner of sizing the reeled tube depends upon the diameter of the pipe that is being produced. For sizes from approximately 140 mm (5½ inches) and over, the sizing process consists merely of passing the tube, reheated in some cases, through two or three stands of sizing rolls the grooves of which are slightly smaller than the reeled tube. The diameter reduction effected is to ensure uniform size and roundness throughout the length of the tube.

Seamless mills producing pipe from 90 mm to 180 mm (3½ to approximately 7 inches) outside diameter usually are equipped with a seven-stand gear-driven reducing mill where reductions up to 12.7 mm (½ inch) in diameter can be made. Such a reduction in diameter provides greater flexibility in operation and minimizes the change time of the mill by lessening the number of steel sizes and high-mill roll changes required for the size range of the mill.

Since it is not economical to pierce, roll and reel tubes of small diameter, the production of hot-finished tubes less than about 75 mm (3 inches) in diameter requires a reducing and sizing process for which a special machine is employed. This machine is similar to a continuous rolling mill. It consists of 8 to 16 stands of two-high grooved rolls about 305 mm (12 inches) in diameter, arranged on the continuous plan and set about 610 mm (24 inches) apart, center to center. Instead of standing vertically, the housings for these rolls are inclined 45 degrees, so the adjacent stands lie at right angles to each other and the loci of the centers of the pass openings formed by the grooves, which gradually decrease in size from the first to the last, are in

the same straight line. The grooves in the initial stands are slightly oval in shape. However, the grooves in the last two stands on the delivery end of the mill are preferably round. As the tube from the reeling machine is too cold to be reduced in diameter, it is passed endwise into a long reheating furnace located at the entering end of the reducing mill where it is heated to a uniform temperature just below the scale-forming point. It is then pushed by a mechanical pusher directly into the first stand of the reducing mill, being drawn continuously through the successive stands in which it is elongated and reduced to the outside diameter desired. The smallest size to which tubes generally are reduced by this process is 38 mm (1½ inches). The preferred maximum diameter reduction per stand in this type of reducing mill is 3 per cent. Since the relationship of wall change and elongation varies with the ratio of diameter to wall thickness (D/T), a medium condition is assured and the diameter of the rolls in successive stands is chosen to provide a linear speed as close as possible to the natural flow of the metal.

An improvement to the reducing mill is the stretch-reducing mill described later in connection with the continuous seamless process.

Spray-Quenched Deep Well Casing—When higher strength casing than that produced by low alloy additions or by normalizing is desired for deep oil wells, carbon-manganese steel pipe is heat treated by a special process. In this process, the hot-rolled pipe is heat treated in a continuous unit by water quenching from temperatures of approximately 845°C (1550°F) and tempering at 480° to 705°C (900° to 1300°F). A suitable furnace is used for heating the pipe. During heating and quenching, the pipe is rotated. The furnace is normally gas-fired, but could take the form of a continuous induction heating unit. After heating to 845°C (1550°F), the pipe is quenched as it passes through a specially designed water-spray ring containing nozzles which spray water on the complete periphery of the pipe.

After quenching, the steel pipe is placed in a tempering furnace and heated to 480° to 705°C (900° to 1300°F) depending upon the final properties desired. The pipe is rotated during heating in the tempering furnace to maintain straightness and uniformity of heating. The continuous tempering furnace is fired with gas and the pipe temperature is checked with a radiation-type pyrometer as it leaves the furnace.

When the pipe leaves the tempering furnace, it immediately passes through a set of five stands of sizing rolls. Due to the formation of martensite during quenching, an increase in volume of the steel occurs which increases the outside diameter of the pipe, and introduces a limited amount of ovality and out-of-straightness. The sizing operation ensures uniform size in the quenched and tempered product; however, the severity of the sizing pass must be limited.

Test results show that a carbon-manganese steel with average hot-rolled mechanical properties of 441 MPa (64 000 psi) yield strength, 689 MPa (100 000 psi) ultimate strength and 30 per cent elongation will, after spray quenching and tempering, exhibit average mechanical properties of 848 MPa (123 000 psi) yield strength, 938 MPa (136 000 psi) inch ultimate strength and 23 per cent elongation. A marked increase in collapse resistance also results from this heat-treating process. For example, 178 mm (7 inch) OD casing with a wall thickness of 10.4 mm (0.408 inch) will have its collapse resistance increased from 44 MPa (6400 psi) in hot-rolled pipe to 85 MPa (12 290 psi) after heat treating—an increase of 92 per cent.

Experimentation has shown that cold rotary straightening can have a deleterious effect on mechanical properties, which is caused by the Dauschinger effect. If the cold work associated with cold straightening is great enough, the yield strength of a pipe or tube can be reduced. Furthermore, casing pipe will also suffer a reduction in collapsing resistance because of this effect. As a consequence, many seamless mills are now using hot rotary straightening, which is done at temperatures of about 700°F (330°C) and higher, but in the case of quench and temper product does not exceed the tempering temperature. Tubular products processed in this manner do not exhibit any reduction in mechanical properties.

The Continuous Seamless Process—Some seamless mills utilize equipment entirely different than that used in the conventional seamless process. The rolling mill and reelers of the conventional mill are replaced by a continuous rolling mill (Figure 32—31) with nine tandem individually powered stands of two-high grooved rolls. Figure 32—32 illustrates the method of reduction employed by this continuous rolling mill. The rolls in the consecutive stands have their axes at 90 degrees to each other and are driven by motors which provide a total of 6340 kilowatts (8500 horsepower). The pipe mill requires an internal mandrel against which the work piece is rolled to reduce wall thickness. This cylindrical mandrel extends entirely through the pierced billet and passes through the mill with the work piece. In the first two roll stands, the diameter of the pierced billet is reduced so that the inner surface is in substantial contact with the mandrel bar. Each of the next two stands makes a reduction in wall over a portion of the circumference, the two jointly completing the first increment of reduction. The next two stands, the fifth and sixth in this mill, make a similar complete reduction but of somewhat less magnitude. The next two succeeding stands (7 and 8) are designed to effect a very slight reduction, the purpose of which is to planish the tube surface. The shape of the tube which has been oval in the preceding stands is changed to approximately circular section in the ninth stand. The rounding up operation effected by this stand frees the inner surface of the tube from the mandrel bar to facilitate withdrawal of the mandrel.

In the operation of the mill, after a billet has been pierced by a conventional Mannesmann piercing mill, a lubricated mandrel, considerably longer than the pierced shell, is inserted and both pass through the rolling mill. The tube and mandrel are then kicked out of the pass line to a stripper which mechanically removes the mandrel. The rolled tube is then further processed by one of two methods depending on the desired product.

After withdrawal of the mandrel, the rolled tubes are reheated before being processed in either a sizing mill or a tension reducing or "stretch" mill. The stretch mill (Figure 32—33) which is similar in construction to the continuous rolling mill, consists of twelve two-high roll stands with the individual stands powered by 150 kilowatt (200 horsepower) variable-speed motors. Tension reducing is unique in that without the use of a supporting mandrel the wall thickness is diminished while the

FIG. 32—31. Over-all view of a continuous seamless-pipe mill, known as a mandrel mill.

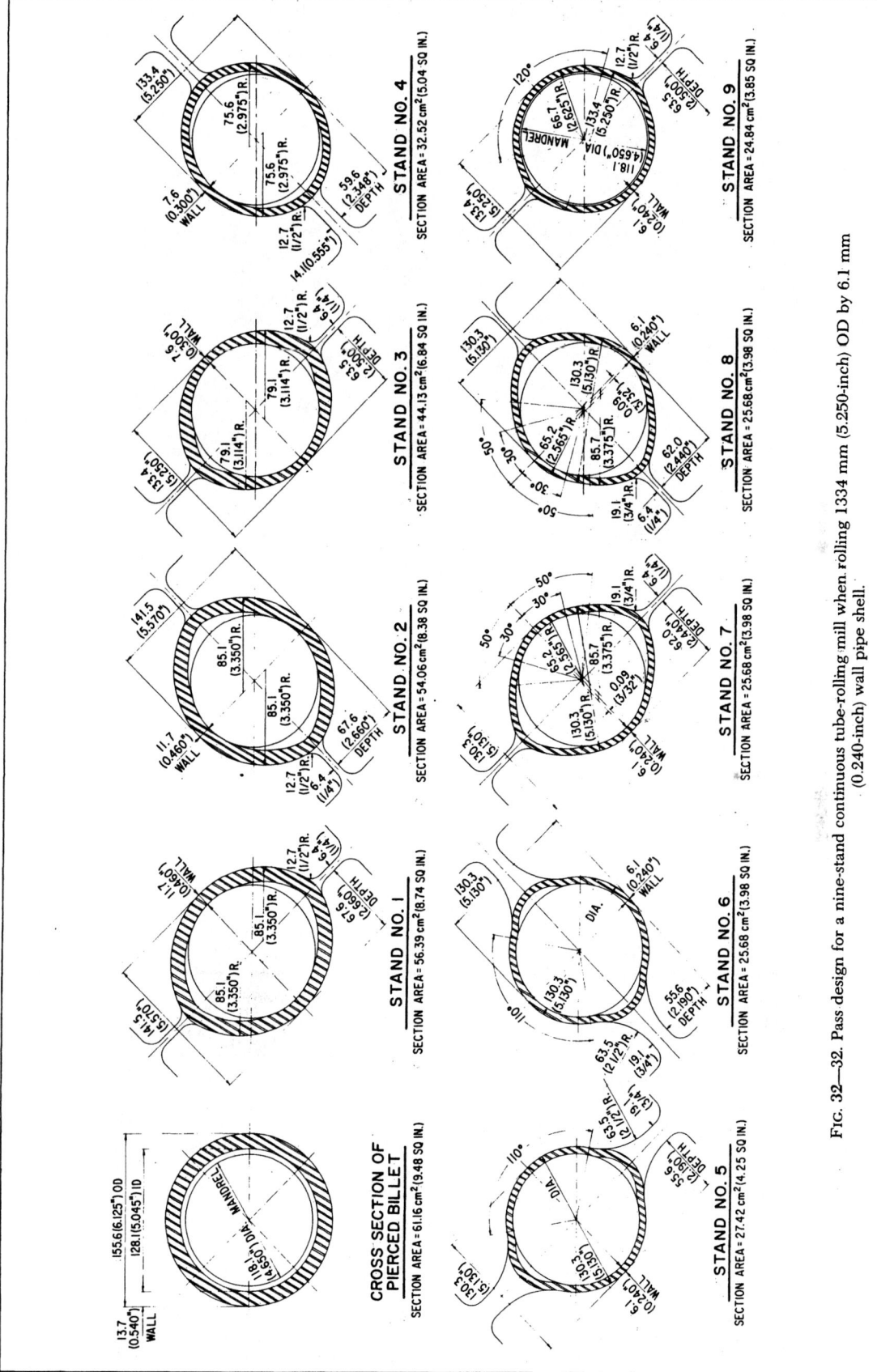

FIG. 32—32. Pass design for a nine-stand continuous tube-rolling mill when rolling 1334 mm (5.250-inch) OD by 6.1 mm (0.240-inch) wall pipe shell.

FIG. 32—33. Over-all view of 12-stand stretch-reducing mill, with tube emerging from tube-reheating furnace in the background.

diameter is reduced. This operation differs from the conventional reducing mill in which the wall thickness is increased as the diameter is reduced. In the tension reducing mill, the tension forces to which the tube is subjected between roll stands are not only effective in reducing wall thickness of the tube but the reduction in diameter performed in each stand can be tripled over conventional practice.

A typical schedule of roll-pass design for making 60.3 mm (2⅜ inch) outside diameter by 4.8 mm (0.190 inch) wall thickness by 42.7 metre (140 foot) length from a shell size of 131.0 mm (5.156 inch) outside diameter by

5.7 mm (0.225 inch) wall thickness by 16.5 metre (54 foot) length is shown in Table 32—IV for 356 mm (14 inch) diameter rolls.

Having both a sizing and stretch mill permits continuous production since it is possible to divert tubes to either unit, depending upon the desired size.

The pierced billet which is 149.2 mm (5⅞ inches) in outside diameter by 11.4 mm (0.450 inch) wall thickness and 6.7 metres (22 feet) long, pierced from a solid round that is 139.7 mm (5½ inches) in diameter and 2.44 metres (8 feet) long, will emerge from the continuous reducing mill as a tube with 127 mm (5 inches)

Table 32—IV. Schedule of Roll-Pass Design for A Tension Reducing Mill with 356-mm (14-Inch) Diameter Rolls.

Stand	Height of Groove		Cutter Diameter		Mean Diameter		Groove Depth		Reduction per Pass (%)	Roll Speed (rpm)
	mm	in.	mm	in.	mm	in.	mm	in.		
1	117.6	4.630	141.2	5.560	126.2	4.970	58.4	2.300	3.5	115
2	109.5	4.310	131.3	5.170	117.4	4.620	54.4	2.140	7.0	123
3	101.9	4.010	120.1	4.730	108.5	4.270	50.6	1.990	7.5	132
4	93.8	3.692	109.7	4.320	99.6	3.920	46.5	1.831	8.1	154
5	85.4	3.362	90.1	3.900	90.4	3.560	42.3	1.666	9.1	167
6	77.8	3.062	89.4	3.520	82.0	3.230	38.5	1.516	9.1	181
7	71.1	2.798	81.0	3.190	74.7	2.940	35.2	1.384	9.0	204
8	64.6	2.542	73.7	2.900	67.9	2.672	31.9	1.256	9.0	240
9	62.1	2.445	67.1	2.640	63.9	2.514	30.7	1.208	6.0	265
10	61.3	2.414	63.5	2.500	62.2	2.447	30.3	1.192	2.7	275
11	61.3	2.414	61.3	2.414	61.3	2.414	30.3	1.192	1.3	275
12	61.3	2.414	61.3	2.414	61.3	2.414	30.3	1.192	0	275

outside diameter with 5.7 mm (0.225 inch) wall and 16.5 metres (54 feet) long. This tube, after passing through the stretch-reducing mill, if rolled into 60.3 mm (2⅜ inch) outside diameter tubing with 4.8 mm (0.190 inch) wall, will be 42.7 metres (140 feet) long.

The long tubes are cut into two sections by a rotary saw on the cooling table. The half sections then are cut into predetermined lengths by high-speed rotary-blade cutters before going to the finishing floors.

Seamless Fabricating Practices—In the discussion of the various phases of seamless-tube manufacture in this section, the role of each unit was described. It was also indicated that the number of operations is dependent upon the size of the tube to be produced. To further clarify these discussions, the following examples, which typify the variations employed in the manufacture of seamless tubes, are given to develop the operations step by step.

In producing a 60.3 mm (2⅜ inch) OD by 3.9 mm (0.154 inch) wall single-length hot-rolled tube, a solid billet 83 mm (3¼ inches) in diameter and weighing approximately 37 kilograms (82 pounds) is pierced to a shell 87 mm (3⁷⁄₁₆ inches) OD with 5.5 mm (0.215 inch) wall, about 3.4 metres (11 feet) long. This pierced shell is passed, without reheating, to the plug rolling mill where it is plug-rolled in an 82.6 mm (3¼ inch) groove to produce a tube 82.6 mm (3¼ inches) OD with a 3.6 mm (0.140 inch) wall and 5.3 metres (17 feet, 6 inches) long. The plug-rolled shell is then reeled to approximately 87 mm (3⁷⁄₁₆ inches) OD, 3.6 mm (0.140 inch) wall and about 5 metres (16 feet, 6 inches) in length. After reeling, the tube is passed through a reheating furnace and into the reducing-sizing mill from which it emerges 60.3 mm (2⅜ inch) OD with 3.9 mm (0.154 inch) wall and approximately 6.8 metres (22 feet, 3 inches) long. From this tube, the crop-ends, and any test pieces that may be required, are cut in the finishing operations described in a later section.

In producing a double-length hot-rolled tube 219 mm (8⅝ inch) OD with 7 mm (0.277 inch) wall, a solid billet 210 mm (8¼ inches) in diameter, 2 metres (6 feet, 5 inches) long, weighing 529 kilograms (1166 pounds), is pierced in the first piercer to a shell 203 mm (8 inches) OD, wall thickness of 32 mm (1¼ inches) and 3.9 metres (12 feet, 10 inches) long. Without reheating, this shell is further processed in the second piercer to a shell 219 mm (8⅝ inches) OD, with an 11.9 mm (0.470 inch) wall, and approximately 8.5 metres (28 feet) long. The shell from the second piercer, after reheating, is plug-rolled in a 214 mm (8⁷⁄₁₆ inch) groove to 214 mm (8⁷⁄₁₆ inches) OD, 5.8 mm (0.227 inch) wall and 14.3 metres (46 feet, 9 inches) long. The plug-rolled tube is transferred directly to the reeling machine which produces a tube 222.3 mm (8¾ inches) OD, 7 mm (0.277 inch) wall, and 13.3 metres (45 feet, 5 inches) long. The tube then receives two or more passes in the two-high sizing mill from which it emerges 219 mm (8⅝ inch) OD, with a 7 mm (0.277 inch) wall and about 14 metres (46 feet) long, ready for end cropping and the other finishing operations.

To produce 660 mm (26 inch) OD, 7.7 mm (0.303 inch) wall double-length hot-rolled tubes, a solid billet 311 mm (12¼ inches) in diameter, 3 metres (10 feet) long and weighing 1818 kilograms (4007 pounds) is pierced to a shell 355 mm (14 inches) in outside diameter, 41.7 mm (1.640 inch) wall, and 5.6 metres (18 feet, 3 inches) long in the first piercer. This shell, on the same heat, is entered in the second piercer where it is rolled to 430 mm (17 inches) OD, 19 mm (0.750 inch) wall, and 9.2 metres (30 feet, 2 inches) long. After reheating, the shell is plug-rolled to form a tube 426 mm (16¾ inches) OD, 12.7 mm (0.500 inch) wall, and 13.6 metres (14 feet, 9 inches) long. The plug-rolled shell is reheated a second time, after which it is rotary rolled to 667 mm (26¼ inches) OD, 7.7 mm (0.303 inch) wall, and 13.9 metres (45 feet, 8 inches) long. Without reheating, it is then reeled to 673 mm (26½ inches) OD, 7.7 mm (0.303 inch) wall, 13.8 metres (45 feet, 3 inches) long. The tube, after reeling, is again reheated in the tunnel-type reheating furnace, after which it passes through two stands of two-high sizing-mill rolls, forming a hot-rolled tube 660 mm (26 inches) OD, 7.7 mm (0.303 inch) wall, and 14 metres (46 feet) long, ready for the finishing operations.

SECTION 6

PRODUCTION OF SEAMLESS PIPE BY PRESS-PIERCING MILL

The press-piercing mill (PPM) is composed of three basic elements: a roll stand with a round pass between a pair of driven rolls; a billet pusher; and a fixed plug located between the two rolls. The billet, enveloped in a 4-sided guide, is forced against the plug by the combined action of the pusher and the driven rolls. The material deformation by this process is mainly compressive with low elongation (maximum 1.2) which avoids subjecting the cast material to high tensile stresses.

The advantages of this process include:
- The use of continuous-cast square billets.

- Elimination of internal porosity and segregation common to continuous-cast blooms. This is the result of the combined action of the three elements of the PPM which, together, cause radial displacement and heavy internal surface compaction by the plug. It is made possible by the opposing external surface compaction of the wall by the mill rolls while under the heavy longitudinal compressive force exerted by the pusher.

- Transformation from solid to hollow under compression. This method improves the physical quality of the cast material, thereby rendering harm-

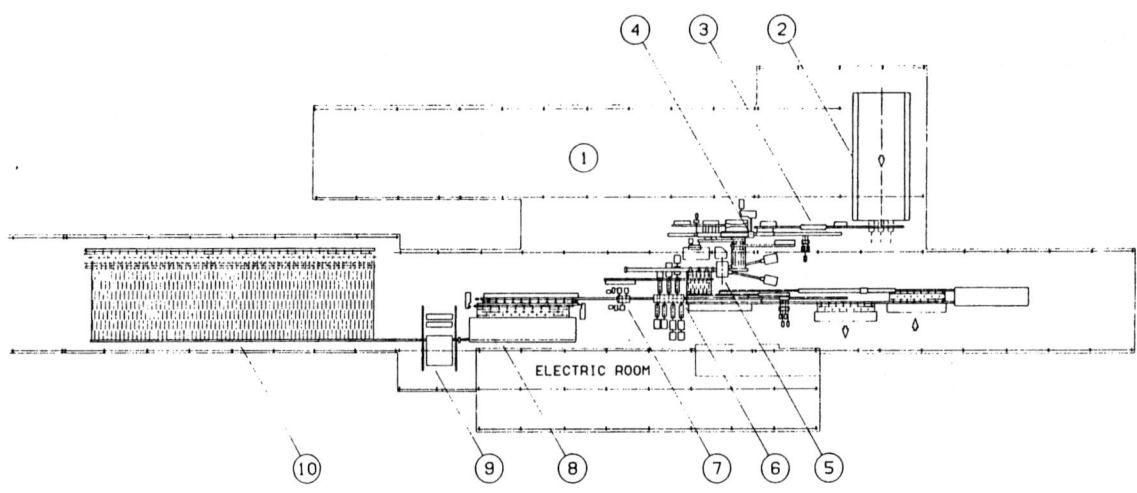

FIG. 32—34. U.S. Steel 245-mm (9⅝-in.) seamless mill: 1-billet conditioning and cutting; 2-walking beam furnace; 3-descaler; 4-press piercing mill; 5-cross roll equalizer; 6-retained mandrel mill; 7-extracting mill; 8-reheat furnace; 9-stretch reducing mill; 10-cooling bed.

less the effects of cross rolling in the elongation process which follows.

- Ability to pierce high-alloy steel grades. Both 5 per cent-Cr and 13 per cent-Cr grades have already been processed.
- High piercing ratios. Ratios of length to ID in excess of 35 have been achieved with good wall concentricity.

Three press piercing mills were in operation and two more were under construction in 1983.

The retained mandrel mill (MPM) is an extrapolation of the conventional floating mandrel mill. Pipe is rolled in a series of stands over a mandrel which moves at a constant controlled speed. The extraction of the tube from the mandrel is made in-line by means of a separate extracting mill which imparts a continuous pull on the tube.

The merits of this rolling method are:

- Excellent quality on the internal surface of the pipe.
- Good wall tolerance over the total pipe length which is achieved by the adoption of a closed-pass design. This design can be used because of the uniform flow of material which is a result of a controlled mandrel speed.
- Heavy wall thickness capability because the tube is continuously pulled from the mandrel in-line.
- Long pipe capability. Lengths of 50 metres (164 feet) are feasible.
- High alloy steel grade capability resulting from the compressive piercing action by the PPM and the closed-pass roll design of the MPM.
- Low power consumption. The closed-pass roll design eliminates the waste of power resulting from redundant rolling and alternating displacement of over-filled stock at the roll parting line which is common to free mandrel mills.
- High yield as a result of the uniform wall thickness and long length.

- Adaptability. The process is applicable to mills of relatively low production requirements as well as those of high output, eg., a 245-mm (9⅝-in.) mill can produce 800 000 to 900 000 tons per year.
- Process flexibility. The process is flexible with respect to the use of various feed stocks and different types of final sizing equipment. Continuous-cast square billet feed stock may be fed to the MPM via the press piercing mill and rotary equalizer or round continuous-cast or rolled billets may be fed via a conventional cross roll piercing mill. The process section can proceed from the MPM and extractor mill either to a sizing mill or to a stretch reducing mill or, as at some installations, both options are available.

A PPM-MPM mill was installed by U. S. Steel at its Fairfield, Alabama plant. The mill commenced operations in 1983.

The facility at Fairfield will produce 545 000 net tons per year of OCTG pipe in the range from 89 to 245 mm (3½ to 9⅝ inches) in all grades specified in the API standards as well as alloy steels (Figure 32—34). The major equipment includes: a walking beam furnace, descaler, press piercing mill, elongator/equalizer, retained mandrel mill, mandrel retaining and circulating system, extracting mill, stretch reducing mill and cooling bed.

All of the equipment is arranged in two parallel longitudinal bays 19.8 and 35 metres (65 and 115 ft) wide. This mill is built on two levels with the hot mill equipment installed at the upper floor level.

The walking beam furnace has an initial capacity of 200 net tons/hr and a final capacity of 240 net tons/hr. Billets are heated to approximately 1280°C (2340°F).

The descaler, located after the furnace, operates with 182.8 kg/sq cm (2600 psi) water to clean the scale from the billet prior to rolling. Descaling rings are also provided immediately before the MPM and the stretch reducing mill.

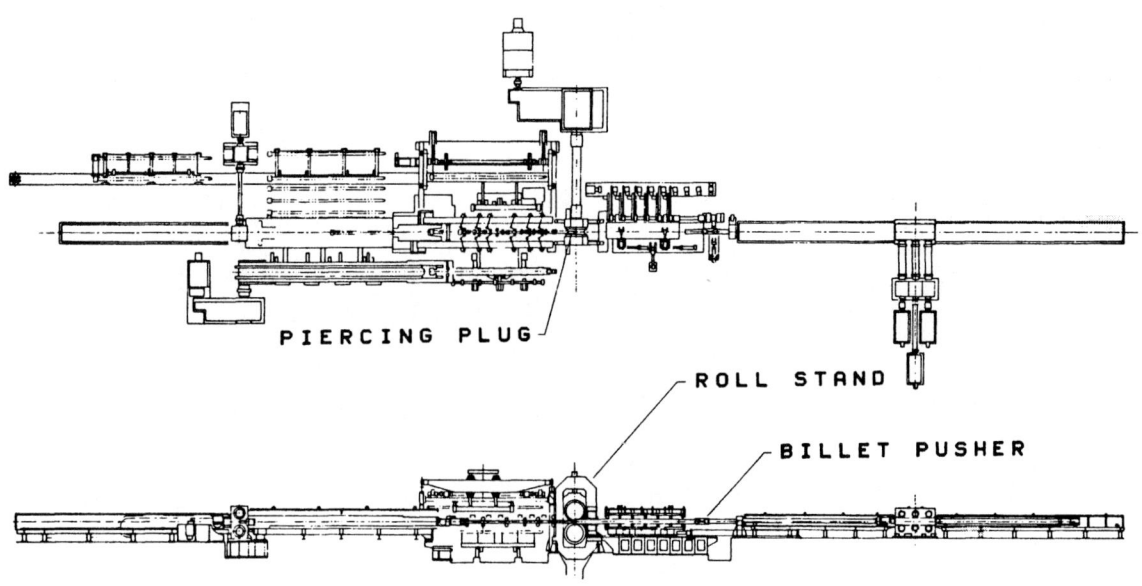

FIG. 32—35. Press piercing mill.

The press-piercing mill (Figure 32—35) consists of a 2 high-roll stand with 1300-mm (51.18-in.) diameter rolls, with installed power of 950 kW (1275 hp). Maximum rolling speed is 0.5 metres/s (98.4 fpm). A rack and pinion mechanism drives the billet pusher and a similar system controls the forward and backward strokes of the piercer plug bar. The machine is equipped with a plug bar circulating system which provides sufficient time for cooling, lubricating and inspecting the plug and bar. The circulating system allows the plug bar to be extracted from the hollow, off-line, by means of a mechanical stripper.

Roll changing on the press-piercing mill is accomplished by extracting from the main roll stand an inner cartridge by means of a hydraulic cylinder. The inner cartridge completely contains the operating rolls and chocks. The new roll and chock assembly has been built up off-line in a second cartridge which is placed adjacent to the roll stand for rapid exchange with the operating unit at the conclusion of rolling. The mill entry guides are also rapidly exchanged in a similar manner.

The elongator/equalizer which receives the hollow bloom from the PPM is a cross roll machine with the rolls arranged in the vertical plane, one above the other. The roll diameter is 1100 mm (43.30 in.) with continuous angle adjustment up to 10°. Maximum axial speed is 0.8 metres/s (157.5 fpm). The machine is equipped with adjustable, fixed lateral shoes. Rolls are directly driven by individual twin motors with total installed power of 9000 kW (12 065 hp). On the outlet side, four bar steadiers guide the plug bar and shell during rolling. Plug bar extraction is made in-line by the mechanically driven thrust block carriage. As with the other installations, deoxidizing powder is sprayed into the shell during the transfer to the mandrel preinserting station.

The retained mandrel mill is made up of seven stands arranged at 45°. The roll diameter for the first three stands is 780 mm (30.71 in.) and 660 mm (25.98 in.) for the last four stands. Maximum tube outlet speed will be 3.7 metres/s (728.3 fpm) with a maximum tube length of 35 metres (114.8 ft.) The mandrel working section length is 16 metres (52.5 ft). The motors are inclined at 45° to accommodate the high torque requirements through parallel gearing and have a total installed power of 15 400 kW (20 643 hp). Mandrel support stands are provided.

A new method has been adopted for roll changing in the 45° roll stands. In place of the conventional exchange of the entire roll stand, only the chocks and rolls are changed. Each pair of used chocks and rolls is hydraulically pushed upward and out of the fixed roll housings on rails which align with those of a pivoting carriage. A simple translation of the carriage presents an alternate carriage on which has previously been placed the new pair of rolls and chocks. The new set is hydraulically lowered into the roll stand. In this manner, sets of rolls and chocks may be simultaneously exchanged. The time for the complete tool change is approximately 20 minutes.

The motion of the retained mandrel during rolling, as well as for insertion prior to rolling and extraction after rolling, is controlled by a double rack and multiple pinion driving system. After rolling, the mandrels return to air and water cooling stations from which they first pass through a lubrication chamber and then to the hollow shell preinsertion station. The preinsertion feature has been adopted for all of the new installations to improve cycle time. Two benches are provided to charge and discharge new and old sets of mandrels. A furnace is used to preheat the new mandrels before being lubricated and placed in service.

The extracting mill has three stands arranged at 45°.

TABLE 32-V Press piercing mill material flow

PROCESS	SECTION DIMENSIONS	
FEEDSTOCK FEEDSTOCK		
PPM	Square	267 mm (10.5 inches)
	Weight	1360 to 2540 kg (3000 to 5600 pounds)
PIERCER EQUALIZER	OD	316 mm (12.44 inches)
	Wall thickness	79 mm (3.11 inches)
	Piercing ratio, max	35
MPM	OD	310 mm (12.20 inches)
	Wall thickness	27.25 to 51.25 mm (1.083 to 2.018 inches)
	Elongation, max	2.43
EXTRACTOR	OD	261 mm (10.275 inches)
	Wall thickness	6.0 to 30.5 mm (0.236 to 1.200 inches)
	Length, max	35 metres (114.8 feet)
	Elongation, max	5.05
SIZING SRM	OD	252 mm (9.92 inches)
	OD	88.9 to 244.5 mm (3.500 to 9.625 inches)

The roll diameter is 720 mm (28.35 in.) and the total installed power is 900 kW (1206 hp).

Following the extractor, two saws are provided to cut the crop ends of pipe. The pipe is delivered to a holding furnace from which it is discharged at approximately 900°C (1650°F) for the subsequent stretch reducing operation.

The stretch reducing mill consists of 24 individually driven stands containing three internally driven rolls and roll diameters of 500 mm (19.70 in.) for the large stands and 430 mm (16.93 in.) for the small stands. The total installed power is 5480 kW (8686 hp). Tool changing is by a special overhead crane equipped with two side by side independently controlled load beams which carry the sets of roll stands.

The basic rolling diagram is shown in Table 32—V.

SECTION 7

COLD-DRAWN, OR COLD-FINISHED, TUBES

While the ordinary requirements for pipe and tubing can be met by the hot-rolling process just described, there are many requirements that demand greater accuracy, higher physical properties, better surfaces, thinner walls and smaller diameters than can be produced by hot-working methods. This demand is met by cold drawing the hot-rolled tubes as a finishing operation. This phase of tube production is analogous to the drawing of wire and the cold rolling of sheets described in other chapters.

In the early history of seamless tube production, cold drawing was employed in almost every instance as a finishing operation, but, as the art of hot rolling developed, the necessary finish and dimensional requirements have been met to a greater and greater degree in the hot-rolling operation. However, a substantial proportion of tubing is still cold drawn for the following reasons:

1. To produce tubes with thinner walls than can be hot rolled.
2. To produce tubes with smaller diameters.
3. To produce tubes longer than can be hot rolled in certain sizes.
4. To secure better surface finishes.
5. To obtain closer dimensional tolerances.
6. To increase certain mechanical properties, such as tensile strength.
7. To produce shapes other than round.
8. To produce tubes with varying diameters and wall thicknesses from end to end.
9. To make small lots of tubing of odd sizes and gages that do not justify a hot mill run.

The practical minimum tube diameter produced by hot rolling is 33.3 mm (1⅜ inches), and 2.1 mm (0.083 inch) is the thinnest wall of commercial hot-rolled tube, which is available in the smaller diameters only. The diameter and wall thickness range of cold-drawn tubing produced by the United States Steel Corporation is from 12.7 mm (½ inch) OD by 0.9 mm (0.035 inch) wall to 219 mm (8⅝ inch) by 50 mm (2 inch) wall.

The fact that hypodermic needles are seamless steel tubes that are cold drawn from hot-rolled tubing in small specialty plants conveys the idea of how far seamless tubing can be reduced by cold drawing.

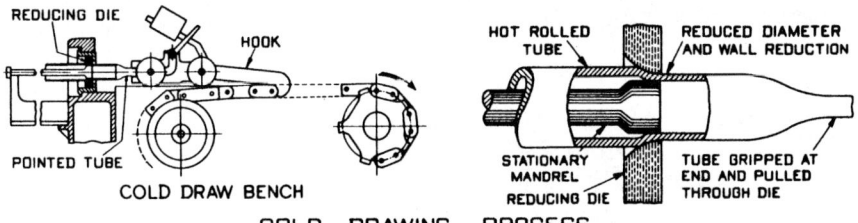

COLD DRAWING PROCESS

FIG. 32—36. Cold drawing a seamless steel tube.

Principle of Cold Drawing—Tube drawing is essentially the same as wire drawing, as explained in preceding chapters with two important exceptions: viz., the inside diameter of the tube must be supported while it is passing through the die to effect wall reduction and control the size of the hole, and comparatively short lengths are involved as in bar drawing. The process consists of pulling the tube through a die, the hole of which is smaller than the outside diameter of the tube being drawn, and at the same time supporting its inside surface by a mandrel anchored on the end of a rod so that it remains in the plane of the die during the drawing operation. Figure 32—36 shows the operation. The mandrel may be omitted if it is not necessary to make a reduction in the wall thickness, or if the dimensions and surface of the inside are not important. A modification of this method consists of drawing on a bar rather than over the mandrel, in which method the bar travels through the die with the tube and must be removed later. The resistance of the metal to passage through the restricted space between the die and the mandrel exceeds the yield strength of the metal at this section, thereby resulting in plastic flow. As a result of the reduced metal section leaving the die, the velocity is increased, the amount of increase being dependent upon the cross-sectional reduction. It is evident that the reduction or draft may be increased only up to a certain limit, depending on the ultimate strength of the section leaving the die, because, if the resistance and balancing pull exceeded the stress, the section leaving the die would break. In drawing tubes over a stationary mandrel, the maximum practical sectional area reduction does not exceed 40 per cent per pass, while, in drawing tubes on a bar which is free to move with the tube, the area reduction can be 50 per cent. As in wire drawing, all of the metal passing through the die is subjected to stress almost up to its breaking strength and is thus given a test of its physical fitness to withstand high stresses in its ultimate service. Any flaw or defect of consequence is brought to light under this severe treatment.

The Draw Bench—A cold-draw bench for tubes consists of a heavy steel frame or bench, in the middle of which is located a die head for holding the die. At one end of the bench is located an adjustable holder to anchor the mandrel rod. At its other end a shaft is mounted carrying a sprocket wheel over which passes a heavy, endless, square-linked chain. This chain lies in a trough on top of the bench, which extends from the sprocket wheel to the die head, where the chain passes around an idler and returns underneath the bench to the sprocket wheel. The sprocket wheel is driven by a variable speed motor through suitable reduction gear-

ing. A carriage called a plyer runs on tracks on the top of the bench and over the chain that lies in the trough between the tracks. This plyer is equipped on one end with the jaws to grip the tube and on the other end with a hook to engage a link of the draw chain. The plyer is connected by cable to a motor-actuated drum by which means it is returned to the die head after drawing a tube. The jaws grip the reduced or pointed end of the tube which projects through the die about 150 mm (6 inches). The closing of the jaws is effected by the motion of the hook in dropping into engagement with the chain. The whole action of gripping the tube and engaging the chain is automatic, once the operator pushes a button to return the plyer to gripping position. The mandrel-rod anchor is equipped with an air cylinder to push the mandrel into operating position inside the die after the pointed end of the tube is inserted in the die. Draw benches for small tubes are equipped with two mandrels with their supporting rods, so that a tube is being loaded on one mandrel while another tube is being drawn off the other. A motor-driven indexing mechanism places the anchor in the pass line so that the mandrel rod is in perfect alignment with the die and plyer jaws prior to the beginning of the draw. A friction-roller mechanism automatically loads the tubes on the mandrel rods. All controls for the draw benches are grouped at the operator's position near the die head so he does not have to move away from that position when operating the bench.

The total length of a bench is about 24.5 to 30.5 metres (80 to 100 feet). The capacity of draw benches may be from 222 400 to 1 780 000 newtons (50 000 to 400 000 pounds) pulling power. Chain speed may vary from 0.10 to 0.76 metre per second (20 to 150 feet per minute), and is automatically controlled so that the tube is started through the die at a slow speed, and, as soon as it is fairly started, the speed increases to the predetermined drawing rate. Some smaller benches have been equipped to vibrate the mandrel at an ultrasonic frequency. Reduced power for drawing, increased drawing speeds and improved surface conditions are claimed to have been achieved by this method. Dies are made with a conical outer surface which fits in a holder mounted in the die head. Dies up to about a 75-mm (3-inch) hole size are made with tungsten-carbide inserts or nibs. Larger dies are made from hardened tool steel, chrome-plated on the bearing surface. Mandrels are chrome-plated, hardened tool steel and are either made in the form of a bar from 150 to 300 mm (6 to 12 inches) long with one end upset to form the working surface and the other end tapped for connection to the mandrel rod (see Figure 32—36) or, in the case of larger mandrels (2 inches and over),

FIG. 32—37. Rotary tube pointer and feeder mechanism. Tube-end heating furnace is at left.

they are made in disc form with a central hole for engagement with the mandrel rod. In the drawing-on-the-bar method, hardened and ground bars of a diameter to correspond with the inside diameter of the

FIG. 32—38. One step in the pickling operation for tubes.

drawn tube and somewhat longer than the drawn tube are used. As the bar-removal operation requires some time, three or more bars constitute a set, which permits the drawing of a tube during the interval required for inserting a bar in the next tube and extracting the bar from the tube previously drawn.

Preliminaries to Cold Drawing—The hot-rolled tubes after cooling are **pointed** on one end. This pointing consists of reducing the outside diameter, for a distance of about 150 mm (6 inches), sufficiently to permit the reduced portion to enter the hole in the draw die freely, so that the jaws of the plyer can grip this end of the tube. If more than one cold-draw pass is to be given the tube, the point is made slightly under the final die size, if possible. Where large diameter reductions are made on small tubes, a point that would enter the final die may be too small and weak to stand the earlier reduction, in which case the original point may be further reduced after a few passes or, in the case of very small tubes, the pointed end may have to be cut off and a new point made on the reduced diameter and wall section. This procedure may have to be repeated several times in drawing very small tubing. Tubes with diameters 63.5 to 75 mm (2½ to 3 inches) and over are usually **open pointed** with a reduced section 50 or 75 mm (2 or 3 inches) long, which has a rather sharp shoulder or offset joining the two diameters. A pulling pin with a cylindrical head is inserted in the tube so that the head engages the shoulder of the point and the stem of the pin projects through the die far enough to enter the plyer jaws. Pointing is done on rotary swagers or steam or air hammers after the end of the tube has been heated to a forging heat (see Figure 32—37). Tubes of certain grades of steel receive an annealing heat treatment, prior to cold drawing. This treatment, which is usually confined to high-carbon and alloy steels of the air-hardening type, is necessary to obtain additional ductility. All tubes are pickled in dilute acid solutions to remove scale and oxides from the outside and inside surfaces. The pickling practice is quite similar to that employed in preparing wire for drawing or sheets for cold rolling. The pickle tubs are 12 to 21 metres (40 to 70 feet) long, about 1.2 metres (4 feet) wide, and 1.7 metres (5½ feet) deep (Figure 32—38). Steam for heating and agitating the bath is led into the bottom of the tubs and discharged through jets or perforated acid-resisting pipes. A wash tub of similar construction is located near the pickle tubs, and after the tubes are pickled free of scale they are dipped in the wash tub to remove the acid and any sludge or loose scale. Facilities are provided for applying lubricant for cold drawing. One lubricating method involves application of a phosphate coating followed by coating with a lubricant normally consisting of a proprietary soap. Throughout these operations the tubes are handled in bundles up to 9 metric tons (10 net tons) in weight. From the drying oven the tubes are then sent to the cold-draw benches or to stock piles near the benches.

The Cold-Drawing Operations—One cold-draw pass produces a cold-drawn tube of close dimensions, good surface, and of any mechanical property within the usual limits of cold-worked steel. Additional passes may be necessary to secure: (1) thinner walls, (2) better surface finishes, (3) smaller diameters, or (4) longer lengths. The bundle of pointed, pickled, inspected and lubricated tubes is laid on a table at the chain end of the bench, with open ends toward the mandrel-anchor end of the bench. The mandrel on its supporting rod is held so that the powered pinch rolls, into which a tube has been entered, drive the tube over the mandrel and rod until the mandrel is just back of the pointed portion. The operator then swings the tube into the pass line and operates the control for the push-up which advances the tube and mandrel to drawing position. The cable-driven plyer-return mechanism moves the plyer carriage toward the die head, in a direction opposite to that in which the draw chain is moving, until the carriage contacts a limit which releases the cable mechanism and permits the hook, which has been elevated, to drop into engagement with the draw chain. The engagement of the hook with the chain permits the grip jaws to close on the projecting point, and the motion of the carriage draws the tube through the die. When the open end of the tube clears the die, the plyer hook automatically disengages from the chain. Another tube is loaded on the mandrel and the cycle is repeated as before. When, after a series of draw passes, the inside diameter of the tube becomes too small to accommodate a substantial mandrel, the tube may be further reduced in both its outside and inside diameters by sinking it through a die without a mandrel or bar on the inside. Such a reduction results in a thickening of the tube wall which is uniform and predictable; however, the inside diameter cannot be held to the close tolerances produced with the use of a mandrel or bar support.

Annealing and Redrawing—While a large proportion of tubes receive only one cold-draw pass, many require a number of passes for reasons previously noted. Because cold drawing hardens and reduces the ductility of tubes, it is necessary to anneal them after each cold-drawing operation. Before further cold-drawing, the annealed tubes must be pickled and lubricated as previously described. All tubes, except unannealed mechanical tubes, receive a final anneal or heat treatment after the last cold-draw pass. Many tubes receive a special normalizing treatment before the last pass in order to obtain the proper grain structure in the finished tube; this annealing is performed in either continuous tunnel or car-bottom batch furnaces fired with gas. The **continuous furnaces** are provided with heat-resisting driven rolls spaced nearly a metre (about 3 feet) apart, on which the tubes are carried through the furnace at a predetermined rate depending on the tube section, annealing temperature, time at temperature, etc. (see Figure 32—39. The **car-bottom batch furnaces** are arranged in a battery and are served by a special charging crane. Less than two minutes is required to discharge a 4.5 metric-ton (5-net-ton) batch of annealed tubes and recharge the furnace with a new batch. The furnaces will accommodate tubes up to 15 metres (50 feet) long (see Figure 32—40). Both types of furnaces are fully equipped with recording pyrometers. Each furnace is divided into four zones, and the temperature is automatically controlled in each zone. When extreme softness or freedom from scale or both are requested, the tubes are enclosed in a heat-resisting sheet-metal box and annealed in a car-bottom batch furnace. The final anneal or heat treatment is varied to produce tubing with the desired mechanical properties. Table 32—VI illustrates the effect of this heat treatment on AISI 1015 steel. The mechanical properties shown are representative expected values for all but the soft-annealed condition, where they reflect the

FIG. 32—39. Continuous annealing furnace.

FIG. 32—40. Car-bottom annealing furnace.

approximate softest condition expected for this steel.

Finishing Operations on Cold-Drawn Annealed Tubes—After the final heat treatment, the cold-drawn tubes are finished in preparation for shipment. The principal finishing operations consist of straightening, cutting, inspecting and testing. Straightening is performed on various types of straighteners, viz., press, rotary, continuous and post. The rotary type consists of rolls set with axes oblique with the pass line (see Figure 32—41). These rolls are somewhat smaller in the center than at the ends to afford a line contact with the tube, which passes between the two driven rolls on one side and three idle rolls on the other. The tube is helically advanced by the rolls which are adjusted to bend the tube progressively as it moves through the machine. Initial bends in the tube are removed and a straight tube is produced. Care must be exercised to have the roll setting adjusted accurately, especially on light-walled tubing, to prevent crushing the tube. Some very

thin-walled tubes can be straightened successfully only on a post set in the floor carrying grooved blocks which support the tube while it is sprung into straightness by hand. Where exact straightness is required, as in some mechanical tubes, gag press straighteners are used. Operators of these presses develop great skill in giving the tubes the proper deflections between the supporting dies so that the desired amount of permanent set remains. A proving table, on which the tube rotates, is equipped with dial gages, which provide a quick and accurate means of determining straightness. Each tube is given a visual inspection for surface defects, both inside and outside. Pressure tubing is hydrostatically tested to internal pressures in excess of service pressures that may be encountered. The usual range of test pressure is from 6895 to 34 470 kPa (1000 to 5000 psi). On pressure tubing, manipulation tests on coupons consist of flattening, expanding, flanging and crushing. All tubes are carefully measured for outside diameter, size

Table 32—VI. Approximate Mechanical Properties of Hot-Rolled, and Cold-Drawn Low-Carbon
Steel Tubing. Steel Specification AISI 1015

	Yield Strength		Ultimate Strength		Elongation (Per Cent)	Hardness	
	MPa	psi	MPa	psi		Rockwell	Brinell
Hot Rolled	228	33 000	379	55 000	40	B-64	107
Normalized	241	35 000	344	50 000	40	B-57	97
Soft Annealed	207	30 000	331	48 000	40	B-50	...
Medium Annealed	276	40 000	448	65 000	30	B-73	128
Finish Annealed	379	55 000	517	75 000	20	B-81	149
As-Drawn	448	65 000	552	80 000	15	B-84	159

Fig. 32—41. Rotary-type straightener. (Courtesy, Mackintosh-Hemphill Division, E. W. Bliss Company.)

and wall thickness. A protective coating of oil, or rust preventive, is usually applied before shipment. Light-gage tubes are boxed and small-diameter tubes are bundled to prevent injury during shipment.

Mechanical Tubing—Tubing for mechanical purposes is made in a wide range of sizes extending from 12.7 mm (½ inch) OD to 219 mm (8⅝ inches) OD and in many wall thicknesses. Mechanical tubing has wide application and can be found in airplanes, automobiles, agricultural machinery, electrical equipment, household equipment, etc. Specific uses include ball- and roller-bearing races, gravity conveyor rolls, bushings, separators, hydraulic cylinders and hoists, oil-well pumps, bicycle frames, metal furniture, etc.

Pressure Tubing—Tubing used to withstand internal or external gas, steam or fluid pressure in refining, chemical, or evaporator apparatus is designated as pressure tubing and is usually furnished in the full-annealed state to assure ductility under service conditions. The most common applications of pressure tubing are in boilers, condensers, heat exchangers, evaporators, cracking stills, refrigerators, and air-conditioning apparatus. Many of these applications can be met by hot-rolled tubing, but the smaller-size, light-walled tubes can be produced only by the cold-drawing process or by the tube-reducing process.

Dimensional Tolerances of Cold-Drawn Mechanical Tubing—While dies for cold drawing mechanical tubing are held to the same exact sizes as wire- and bar-drawing dies, the fact that thin-walled sections will spring out of round while a solid section cannot, accounts for the somewhat greater tolerances required in tubing than with solid sections.

Surface Finishes—The various surface finishes in both hot-rolled and cold-drawn tubing are as follows:

1. Hot Finished
   Hollow-forged billets and hot-rolled tubing.
2. Normalized
   Hot-rolled or cold-drawn tubing.
3. Soft Annealed
   Hot-rolled or cold-drawn tubing.
4. Medium Annealed
   Cold-drawn tubing only.
5. Finish Annealed
   Cold-drawn tubing only.
   (Continued on next page)
6. Hard Drawn (Unannealed)
   Cold-drawn tubing only.
7. Bright Annealed
   Cold-drawn tubing only.
8. Pickled
   Hot-rolled or cold-drawn tubing.

1. **Hot Finished**—Hollow-forged billets have a surface appearance much smoother than hammered or pressed forgings, but may show a slight marking that will clean up with very little stock removal. Hot-rolled tubing has a surface finish comparable to plates or sheets of equal thickness. The thin-walled tubing, because of the high reduction in the rolling operation and the low finishing temperatures, will have a better surface than tubing with heavy walls. A light, tightly-adhering mill scale, blue-black in color, is found on both outside and inside surfaces of hot-rolled tubing.

2. **Normalized**—Whether hot-rolled or cold-drawn, all normalized tubing will be coated with scale, the

thickness of which depends upon the thickness of section and grade of steel. Thin-walled tubing can be brought to temperature and cooled rapidly, thus avoiding long exposure to oxidizing atmospheres at high temperatures, while heavy sections require a longer exposure. Scale formation is directly proportional to time and temperature. Some alloy steels, notably those containing chromium or nickel in small percentages (1 to 5), usually are more heavily scaled when normalized, due to the time-temperature effect.

3. **Soft Annealed**—Soft annealing, as commonly applied to pressure tubing, or mechanical tubing that is to be manipulated cold, leaves a light scale from reddish brown to blue-black in color that is comparable in thickness to the scale on hot-finished tubes, but is usually less tightly adhering and of a more porous nature. This surface may at times resemble that caused by rusting in storage or in transit. Oil used for protective coating may be absorbed by the oxide film, and the tubing may have the appearance of not being properly oiled, when fully protected against normal atmospheric corrosion.

4. **Medium Annealed**—Due to the lower temperature employed, medium annealed tubing is very slightly scaled and the loose scale can usually be rubbed off easily, leaving a black oxidized surface.

5. **Finish Annealed**—As the temperature at which tubing is finish annealed produces an oxide film of blue color, the tubing has a blue-black appearance and the smooth surface produced in the cold-drawing operation is not disturbed. The oxide film offers a slight protection against local rusting or discoloration.

6. **Hard Drawn (Unannealed)**—As the tubing has no heat treatment after the cold-drawing operation, the surface is more or less bright, depending on the number of passes through the die and on the nature of the lubricant used in cold drawing. Normally, thin-gage tubing will have a smoother and more uniform surface than heavy-walled tubing.

7. **Bright Annealed**—Annealed cold-drawn tubing may be furnished with a scale-free surface, when so specified, by annealing the material in a controlled atmosphere or bright annealing furnace.

8. **Pickled Finish**—Where mill scale or scale from heat treatment is objectionable, tubing can be furnished with pickled surfaces, both inside and outside. This is often desirable when the tubing is to be machined, especially in automatic machines using formed cutters, as the tool life is materially increased. Pickling also permits close surface inspection.

SECTION 8

THE PIERCE AND ROLL-FORGE PROCESS

Application of the Process—A considerable quantity of seamless tubular product is made by the process of piercing and drawing, starting with a bloom of steel of the required composition. The process is used primarily where special alloy steels, combined with nonstandard pipe diameters and wall thicknesses, are the requirements for the particular product. The large use of this product at United States Steel is in the manufacture of cylinders for containing high-pressure gases. The continued increase in the use of steel cylinders for the storage and/or transportation of hydrogen, helium, and oxygen—as well as many unusual gases—has necessitated the use of steels of much higher alloy grade and of much greater wall thicknesses than are employed in the manufacture of the more-standardized types of seamless pipe and tubing, and has revived the original piercing process to meet the needs for special products in the high-pressure field. Walls up to 152 mm (6 inches) thick and in diameters up to 660 mm (26 inches) are not unusual for pressure vessels for the containment of gases at pressures up to as high as 103 400 kPa (15 000 ksi).

Chief Details of the Pierce and Roll-Forge Process—The piercing and roll-forge process requires skill and experience, although its principles are simple and easily understood (Figure 32—42). Depending upon the desired end result, the starting material is a solid steel billet ranging in size from 355 to 635 mm (14 to 25 inches) square. The billet is heated to a forging temperature of about 1260°C (2300°F) and is performed into a solid cylindrical form by pushing through the two roll stands of the push bench. The billets are then pierced and formed into a closed-end hollow on a vertical press. After reheating, the pierced billet is positioned on a mandrel of the proper size at the roller push bench. The mandrel pushes the forging through opposing concave rolls to reduce the diameter while increasing the length. The hot forging is returned to its original position, rotated, and the operation repeated until the specified size is achieved. When the roll forge operation is complete, the mandrel is withdrawn and the forging is cooled. The cupped, closed end is removed and the open end is squared by flame cutting. The resulting seamless roll-forged pipe is then ready for further processing.

Forming Cylinders—Where a high-pressure vessel is to be produced, one end of a tube produced in the manner just described is heated and hammer forged to convert it into a hemispherical closed end. The other end is then heated and forged to impart a conical shape without closing the end. The forging operation on either end is also known as **swaging**. Swaging is performed on an air- or steam-operated forging hammer. Hemispherical or conical-shaped cavities in two mating dies, one mounted on the solid bed of the hammer and one on the ram, impart the desired shape to the end being swaged. A series of properly tooled forging hammers with suitable reheating furnaces spaced between them allows for a succession of swaging operations to be performed to produce the desired end result.

Following the completion of all forging operations and heat treating, the open formed end section is ma-

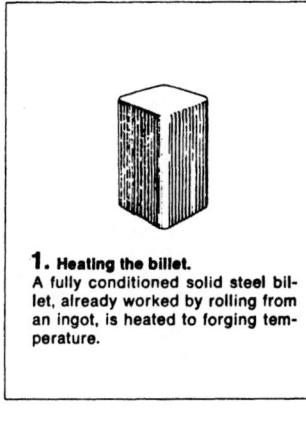

1. Heating the billet.
A fully conditioned solid steel billet, already worked by rolling from an ingot, is heated to forging temperature.

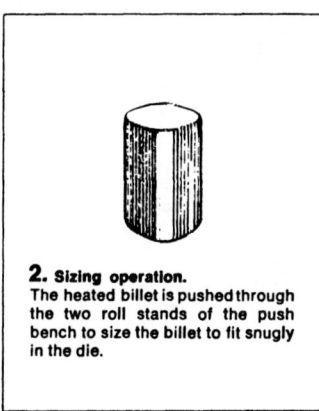

2. Sizing operation.
The heated billet is pushed through the two roll stands of the push bench to size the billet to fit snugly in the die.

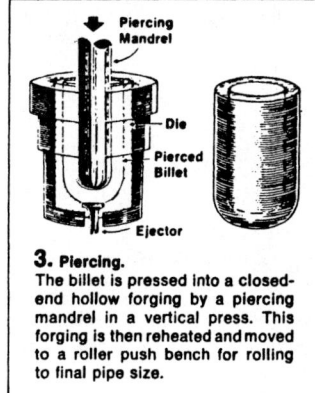

3. Piercing.
The billet is pressed into a closed-end hollow forging by a piercing mandrel in a vertical press. This forging is then reheated and moved to a roller push bench for rolling to final pipe size.

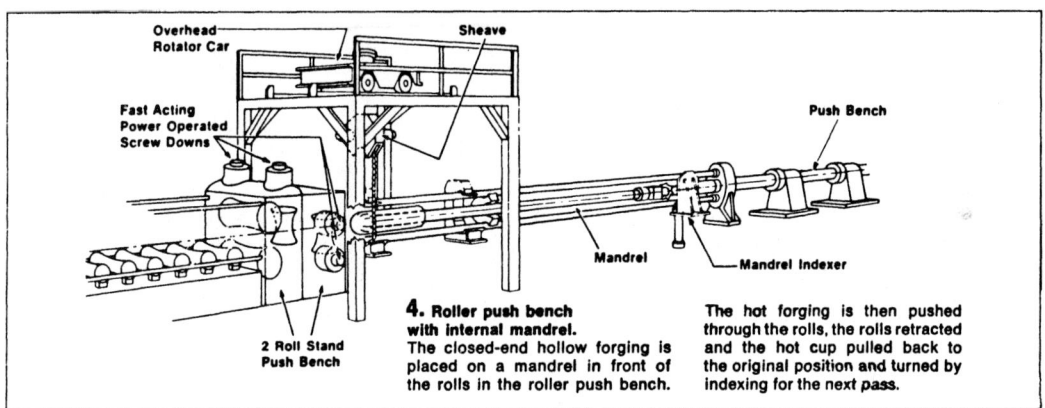

4. Roller push bench with internal mandrel.
The closed-end hollow forging is placed on a mandrel in front of the rolls in the roller push bench.

The hot forging is then pushed through the rolls, the rolls retracted and the hot cup pulled back to the original position and turned by indexing for the next pass.

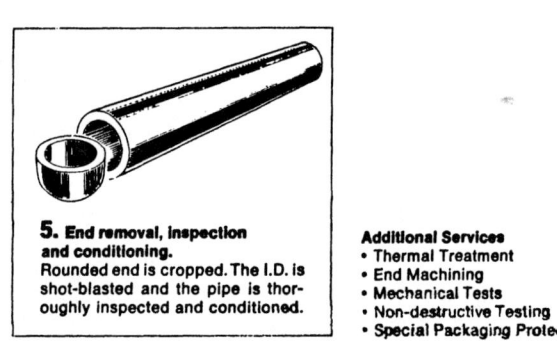

5. End removal, inspection and conditioning.
Rounded end is cropped. The I.D. is shot-blasted and the pipe is thoroughly inspected and conditioned.

Additional Services
• Thermal Treatment
• End Machining
• Mechanical Tests
• Non-destructive Testing
• Special Packaging Protection

FIG. 32—42. Steps in the manufacture of seamless pipe by the pierce and roll-forge process.

chined and threaded for the required detail of piping or fixtures that will be assembled in the cylinder end.

To assure the degree of internal cleanliness required by the individual gases to be stored, the interior surface of the cylinder is first shot blasted to remove all scale that developed in the forging or heat-treating processes. In many cases, complicated chemical washing is also done to assure what is sometimes called "hospital cleanliness" for the interior of the cylinder. Special interior coatings of either the air-drying or baked types are often applied. The exterior of the vessel is shot blasted to ensure a clean metal surface for the application of exterior coatings. Pipe fittings and flanges with seal welded connections are often required, following

the installation of which the entire assembly is tested with air at operating pressure to test the leak-proof qualities of the unit. Another function of inspection to fill the requirements of ASME or ICC specifications calls for hydrostatic testing of all cylinders made. Non-destructive testing beyond hydrostatic testing consists of ultrasonic testing of the entire vessel area; this becomes more necessary with increased service-pressure requirements.

The described method of producing pressure cylinders is used for vessels of 254 mm (10 inch) or greater diameter and of 12.7 mm (½ inch) or greater wall thickness. The operation called **spinning** is used to produce vessels of smaller diameters and lighter walls. The

FIG. 32—43. Ends of a tube at different stages of the spinning process, showing the progress of the operation from the tube on the left to the closed cylinder on the right.

spinning operation (Figure 32—43) is performed by placing the pipe section in the hollow spindle of a lathe-type machine and rotating the piece at relatively high speeds (600 to 1200 rpm), the piece having been heated to a dull red color prior to insertion in the hollow spindle. A tool holder is mounted on a carriage that permits longitudinal and cross feeding of a blunt spin-

ning tool. The tool is fed against the work piece, moving from the outside diameter, and the sweep of the spinning tool is controlled to form the necks and ends of the cylinders. Sufficient heat is developed from friction of the tool against the work-piece to form a weld in the case of closing the end of a cylinder.

SECTION 9

FINISHING OPERATIONS

Pipe made by the hot-rolling process described in previous sections must be subjected to many mechanical operations, such as straightening; end cropping; plain-end machining for various mechanical uses; chamfering, reaming and facing for welded line pipe; and chamfering and reaming for threaded line pipe or casing. These are some of the characteristic operations classified as "finishing" and each requires its own special technique.

Straightening—Pipe from the hot mills must be straightened either by continuous "in-line" processing through a skewed-roll machine, or by a press. Both units use the principle of supporting a pipe at two points while applying a deflecting force to the opposite side of the pipe at a point midway between the supports in a direction opposite to the bend.

In one type of continuous roll straightener, two pairs of skewed power rolls rotate and advance the pipe and serve as the fixed fulcrum points mentioned above. A single skewed roll located midway between the sets is positioned ahead of the axial centerline to exert the deflecting force. Details of this general type of straightener vary. In some machines, power is applied to both rolls of each of the driver pairs; in others, to only one. Some use a pair of rolls in the deflection set; others, only one, with or without power. In one case, the deflection rolls are formed into a cluster that holds the pipe over approximately one-half of the pipe circumference. Some straighteners employ as many as seven rolls, with three rolls arranged in a cluster at both the entry and exit ends and the deflection roll between the clusters.

In all straighteners, including the press, the deflection is selected that will straighten bent or bowed pipe, but will not exceed the elastic limit of the pipe that is straight before entering the machine.

Inspection and Cutting—When seamless pipe leaves the straightening rolls, it is delivered to a table where it is thoroughly inspected for straightness, size and

external and internal surface defects such as seams, pits, etc. The pipe is then delivered to the cutting-off machine where the crop ends are removed and the ends are cut smooth and perpendicular to the pipe axis, beveled for welding, or if the pipe is to be threaded, the ends are chamfered to aid in starting the threading dies. The ends are then reamed to remove any burrs, and the wall thickness of the cut end is measured to establish that the pipe has the proper uniform wall thickness. Depending upon the class of pipe, a number of cut-off portions may be subjected to a flattening test to determine the ductility of the steel in the individual pipe.

Pipe Joints—A pipe joint is a means of connecting two or more lengths of pipe so as to permit transportation of liquids or gases under leak-proof conditions, or to permit the use of long lengths of pipe for mechanical or structural purposes. Generally speaking, joints may be divided into three classes:

(1) Joints with threaded ends for couplings or flanges.
(2) Special connectors or couplings for use with plain-end or flanged-end pipe; as, for example, Dresser, Victaulic, Vanstone, and similar joints.
(3) Welded joints.

Joints with Threads and Couplings—Pipe to be used for oil and gas-well casing is mostly finished to American Petroleum Institute (API) specifications which specify that all sizes from 114.3 to 508 mm (4½ to 20 inch) outside diameter, inclusive, shall be threaded on both ends with round top and bottom sixty-degree threads. API tubing is also provided with this type of thread whereas API drill pipe is furnished with upset ends for attachment of rotary connections by welding. Much tubing and casing today is also available in special proprietary type end finishes utilizing both upset and non-upset tubular ends. Threaded API line pipe requires a modified Briggs thread. An increasing proportion of casing that is to be subjected to severe ten-

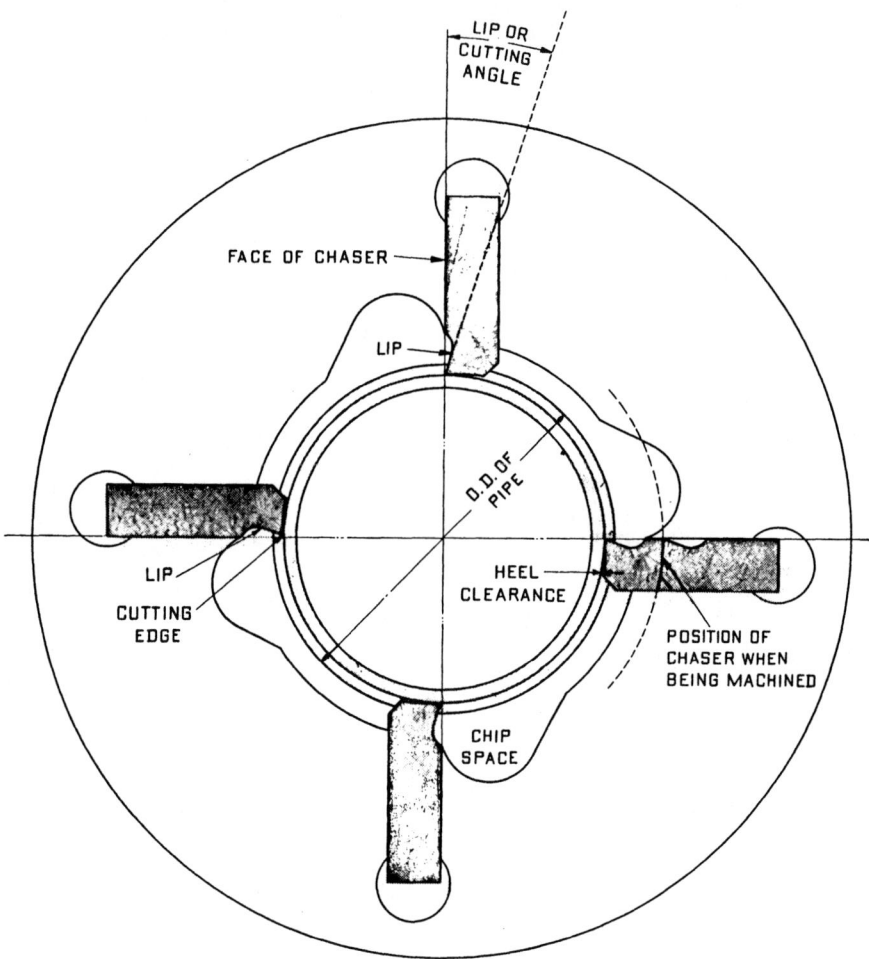

FIG. 32—44. Threading die showing lip, lip angle, chip space, clearance and cutting edge.

sile stresses (as in deep wells) is threaded with the buttress casing thread that was developed by the United States Steel Corporation. This makes a joint that is practically equal in tension to the pipe that it joins. Standard pipe that is to be fitted with couplings or flanges is manufactured to the applicable parts of the American Standards Association Threading Specifications. The great increase in the loads and pressures to which many types of threaded joints have become subject in recent years has resulted in additional efforts to improve physical properties and threading practice. Thread form, thread depth, lead, and taper are maintained within the limits of the rigid tolerances given in the specifications of the American Petroleum Institute and other organizations, so that coupling threads will mate properly with the pipe threads when made up to the power-tight position.

Threading Pipe—To secure good threaded joints it is necessary to have clean, smoothly cut threads of the proper taper and pitch, and to secure such threads it is necessary to have threading dies made with full consideration for the following: lip, chip space, clearance, lead, lubricants or cutting oils, and, for power machines, number of chasers.

Lip—Figure 32—44 illustrates clearly what is meant by lip on a chaser. The lip forms a slanted cutting edge which promotes curling of the chips and gives an easy cutting action, similar to that of a properly ground lathe tool, instead of the pushing-off effect caused by chasers which have no lip; it also permits a higher cutting speed. The angle to which the lip should be ground depends upon the kind of material to be threaded and the style and condition of the chasers and chaser holder.

Chip Space—Chip space is the space required in the die holder in front of the chasers to prevent the accumulation or packing up of chips. If sufficient chip space is not allowed, the chips will rapidly pack in front of the chaser, causing rough, torn threads, and creating a tendency on the part of the chaser to pick up stickers. The best design for this chip space provides an even curve for the chips to follow, with the back of each chaser well supported.

Clearance—Clearance is the space between the threads of the chaser and the threads on the pipe at a given distance from the cutting edge ("heel clearance" in Figure 32—44.) This clearance is secured by die manufacturers in various ways. A simple method for getting clearance in the type of die known as "cutting-edge-on-center" or "center cut," consists of setting the

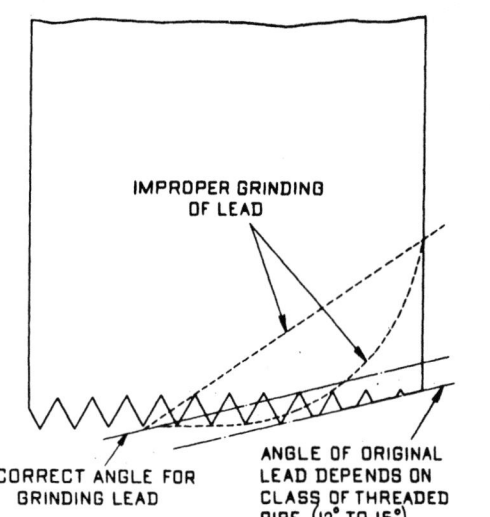

IMPROPER GRINDING
OF LEAD

CORRECT ANGLE FOR
GRINDING LEAD

ANGLE OF ORIGINAL
LEAD DEPENDS ON
CLASS OF THREADED
PIPE (12° TO 16°)

FIG. 32—45. Sketch showing correct angle for regrinding lead of chasers.

chasers for machining with their cutting edge tangent with a larger circle than they are set for cutting threads. Clearance may be obtained on the "stock-on-center" type of chasers by machining in the same manner as "cutting-edge-on-center" chasers, with the exception that their cutting edge is tangent to a smaller circle than when they are set for cutting threads. Stock-on-center chasers can also obtain their clearance by setting the chasers ahead of the centerline when they are being machined.

Lead or Throat—Lead is the angle which is machined or ground on the first three threads, more or less, of each chaser to enable the die to start on the pipe, and also to distribute the work of making the first cut over a number of threads. The lead may be machined or, as is more common, it may be ground after the chasers are tempered. The proper amount of lead is about three threads. As the heaviest cutting is done by the lead, this section of the chaser should have a slightly greater clearance angle than the rest of the threads, but care must be used to see that this angle is not excessive. Excess lead clearance will cause the die to feed too fast, and the half threads cut by the lead are consequently damaged by the full teeth of the chasers (see Figure 32—45.)

Number of Chasers—To get good results in threading at one cut, the die head should have a suitable number of chasers. The number is determined by the size of the pipe. In some cases as many as eighteen chasers are required. The number necessarily depends upon the design, size, and operative principle of the die; hence, no exact rule can be laid down for universal acceptance. When an insufficient number of chasers is used, the die will chatter and cut a rough thread.

Dies—Dies usually are designed with the chasers evenly spaced and arranged either with the stock on center or with the cutting edge on center, as shown in Figure 32—44, in which case the face of the chaser is in the same plane as the central axis. A more recent development places the chasers at an angle of 24 degrees with a radial or center line and spaces them at varying angles around the pipe. An odd number of chasers is used. This has several advantages among which is the steadying effect the unequally spaced chasers have on the die, thereby reducing chatter. Sharpening is simplified since the rake or lip angle is set at 24 degrees by the angle of the chaser with the center line and sharpening consists of merely grinding parallel to the face of the chaser.

Chaser teeth are usually designed to have the even-numbered chasers cut one flank of the thread and the odd-numbered chasers cut the other flank. This is done so that the threads of any individual chaser will not cut both flanks and results in a smoother thread.

Lubricant—Care should be taken to provide the proper quality of threading lubricant as the best die made will not produce good results with poor or insufficient lubricant. Poor lubricants are destructive to dies, and more power is required to cut a thread when they are used.

Gaging Pipe Threads—To keep pipe-threading practice at a high degree of accuracy, the mills maintain complete sets of standard gages with which the pipe-thread dimensions may be measured. For each size of pipe threaded, a master plug gage is kept at the mills, and these gages are returned periodically to the National Bureau of Standards for examination of accuracy. Except for the pipe-thread vanish angle, the threaded master plug gage represents a theoretically correct pipe thread as to pitch diameter, lead, taper, and thread form. The ring gage is the transfer medium, also certified for accuracy by the Bureau of Standards, and represents a theoretically correct coupling thread with reference to lead, taper, and thread form. The ring gage is threaded and sized so that it will screw onto the master plug gage a predetermined distance, known as the hand-tight position. On American Petroleum Institute gages, the hand-tight position is expressed as **standoff,** which is the distance from a scribed mark or notch in the master plug gage to the face of the ring gage. The standoff represents the advance from the hand-tight to power-tight position which has been determined as necessary to obtain a leak-proof and otherwise efficient joint. The working plug and ring gages used to verify the accuracy of the couplings and threaded pipe, respectively, are in turn compared with the master plug and ring gages, so that the makeup from hand-tight to power-tight position in the joint will be uniform and within the specification tolerance limits. In addition to plug and ring gages, the mills are provided with various other types of high-precision gaging instruments, each one serving its particular function in indicating the accuracy of thread depth, thread form or angle, taper of threads in a specified length, pitch diameter, and lead or pitch. A more complete description of thread inspection and gaging instruments may be had by reference to American Petroleum Institute Standards 5-B.

Coupling Forgings—All threaded couplings are made of seamless steel forgings of a grade of steel at least equal to that of the threaded pipe with which they are used. Seamless coupling forgings, also called blanks, are cut from lengths of seamless pipe, pierced and rolled to the required coupling diameter and wall thickness.

Finishing Steel Couplings—The blanks are stamped with the necessary identifying marks. They are then placed in machines which true up the ends, taper the bore to conform accurately with the taper of the thread and, if required, recess or chamfer the ends internally. This is all performed in the same operation to assure alignment of the two ends of the finished coupling. The preparation of the coupling blanks is of great importance and care is taken to see that all operations associated with their preparation are properly carried out. The prepared blanks are tapped on automatic tapping machines. After tapping, all couplings are inspected for pitch, taper and thread depth with precision instruments specially designed for this type of work. Size or pitch diameter is checked on hardened and ground, threaded gages.

To prevent galling of the threads, the couplings are given phosphate coatings or are electroplated with either zinc or tin, depending upon the type of thread and/or the yield strength of the coupling. As the first step in the zinc plating process, the couplings are first washed with a soda solution to remove the oil adhering from the threading operation. Thoroughly cleaned on the inside, they are placed, several at a time, in a specially constructed plating tank filled with a zinc solution. In this tank the couplings are supported so that they collectively form the cathode, while zinc poles project into the couplings to form the anode. Upon the passage of direct current through the apparatus, the zinc is deposited upon the inside of the coupling as a firmly adhering coat, the thickness of which depends upon current density and time. Since the coat will not adhere firmly if the current density is too great, this factor is limited and maintained to give a tight coat, and thickness is controlled by time.

As an additional safeguard against galling, to facilitate the tightening up of the couplings on pipe and also to prevent rusting of the threads in service, special thread compounds such as A.P.I. Modified, consisting of various metallic powders suspended in non-drying greases, have been developed for application to the pipe threads.

Testing the Pipe—From the threading machines, or from the cutting-off machines if the pipe is plain end or beveled for welding, the pipe is moved over a final inspection table, where each length is carefully inspected for surface imperfections, end finish, size, etc., to the hydrostatic-testing machine. If the pipe is threaded, the threads are lubricated with thread compound after which the couplings are screwed up to the established hand-tight position, examined for stand-off, and then brought up to power-tight position by a power screwing-on machine. The pipe is then filled with water and an internal hydrostatic pressure is applied. This pressure may be from 2758 to as high as 103 422 kPa (400 to as high as 15 000 psi), depending upon the kind of pipe, the size, and the service for which it is to be used. For the hydrostatic test a specially designed machine is provided, which consists of a bench mounted on one end with a water-tight head connected with a hydraulic line and with a similar head, made adjustable to suit different lengths of pipe, mounted on the other. A number of clamps for supporting the pipe and an air hammer for tapping the

pipe during the test are mounted between these heads. When being tested, welded pipe is placed between the heads, the supporting clamps are applied, and the adjustable head is moved forward to seal the ends of the pipe tightly with packing rings or gaskets. Water is admitted until the pipe is full and no air pockets remain. The pressure line valve is then opened until the gage indicates the specified test pressure, which is maintained for five seconds, during which time the ERW pipe is jarred automatically with the air hammer. (Seamless pipe does not require tapping.) The face of the air hammer is brought to bear upon and vibrate against the pipe, which is thereby subject to impact and vibration while under maximum internal stress. The pipe is then unclamped from the testing machine, and one end is elevated while the water flows out, carrying with it flakes of detached scale.

Because of the enormous head pressures encountered in testing pipe, the higher test pressures are usually applied by what is known as the "field testing" method, so called because it was originally used with portable equipment. The extremely high pressures are employed to test casing and tubing to be used in deep wells where very high working pressures are encountered. In one type of tester, a cap is screwed onto the field end pipe threads and a plug, through which the water enters the pipe, is screwed into the coupling on the other end of the pipe. Since there are no restraining forces on the ends of the pipe, the internal pressures acting on the cap and plug hydrostatically test the pipe joint in tension. Because of the very high pressures involved in this test, safety precautions are an important factor when installing and operating this equipment.

In another type of field-testing unit, the end closures seal on the outside diameter; it is used for testing electric-welded tubular products.

A very important factor in setting oil-well casing is the amount of external pressure (collapse) exerted on the casing. This pressure is usually directly proportional to the depth of the well and together with the bursting pressure and joint strength determines the grade of steel and wall thickness of casing required in deep wells. The resistance to collapse is proportional to the ratio between the outside diameter and the wall thickness of the casing which is usually expressed as D/t. It also varies with mechanical properties, especially the yield strength, below the range of elastic failure.

In order to assure a satisfactory level of collapse resistance, representative specimens of casing are subject to external hydraulic pressure under laboratory controlled conditions. After outside diameter and wall measurements are recorded, the ends are sealed either with portable leakproof heads or by welded plugs and the specimen is inserted in a heavy wall forged jacket which is sealed and closed with heavy bayonet-type heads. After filling the jacket and bleeding out all entrapped air, hydraulic pressure is applied to the outside of the pipe by a high-pressure electrically driven reciprocating pump until failure of the specimen occurs through collapse of the section. The hydraulic pressure applied during the test is indicated on a calibrated mercury pressure gauge. Open-end tests are made by sealing the specimen using suitable packing rings through

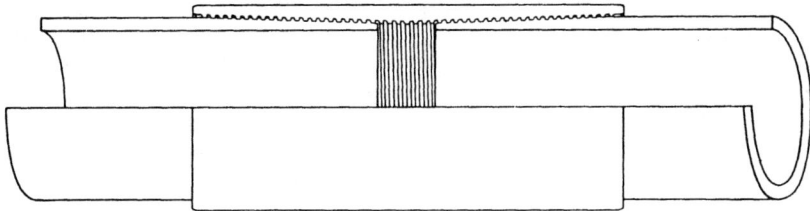

FIG. 32—46. API buttress thread on seamless casing.

openings in the heads of the collapse jacket. In this method, the length of the test specimen is fixed by the length of the collapse jacket and consequently the ratio of the length of specimen to its diameter varies with each size. The method of applying and registering the pressure during the test is the same as with the closed end test.

The test results are recorded and used as a process control, and as part of the experimental work which is constantly being done to improve casing quality and manufacturing method.

Oiling—Each length of pipe as it leaves the testing pump is measured, and in some cases weighed, and this information together with the necessary identification marks is stenciled on the pipe, which is then given a coating of protective oil as it passes through a spray machine. This oil is a hardening transparent oil that leaves a lacquer finish. Sizes NPS 1½ and smaller are identified by stamping the necessary information on metal tags rather than stenciling, and these sizes, after oiling, are bundled to facilitate handling and shipment.

Types and Uses of Joints—The joints shown in Figures 32—46 through 32—55 represent those best known and most commonly used.

Upsetting—For severe service, it is often necessary to provide additional strength in the joint, and for this reason the ends of the pipe are upset before cutting the threads. To accomplish this upsetting, the end of the pipe is heated to forging temperature, then inserted endwise between two semi-circular dies of the upsetting machine. These dies clamp the pipe from the outside, while a mandrel, carrying a collar of the exact size of the outside diameter of the upset end, is inserted into the pipe. As the mandrel advances, the collar comes in contact with the end of the pipe and pushes the hot metal back to fill the ring-like space between the mandrel and the die. By changing the design of the dies, the upsetting may be controlled to displace the extra thickness either to the inside or to the outside of the pipe. Figure 32—49 and 32—54 show

external upset tubing.

API Buttress Thread Casing—This joint (Figure 32—46) was developed to satisfy the petroleum industry's need for a casing joint which will safely and economically support the weight of casing designed for deep wells. The high tensile strength of the buttress-thread joint is due largely to the combined effect of the coupling threads completely engaging the casing thread throughout their entire length, including the vanishing threads, and the three-degree flank angle of the threads which support the weight of the casing in the well. Used on the API casing grades and with full inside clearance, 114.3 mm (4½ inch) through 508 mm (20 inch) OD casing combination strings of multiple weight can be designed for 6100 metre (20 000 foot) depths and corresponding approximately 68 950 kPa (10 000 psi) bottom hole rock pressure.

API Long Round-Thread Casing—API casing with long couplings is manufactured in sizes 114.3 to 508 mm (4½ to 20 inches) outside diameter. The general outline of the joint is as shown in Figure 32—47.

API Short Round-Thread Casing—This casing (Figure 32—48) was designed for oil-well depths of 1525 to 2135 metres (5000 to 7000 feet) with a safety factor of 2.0 against failure in tension based on minimum physical properties. However, this joint may be used in the bottom sections of longer strings of casing, thus resulting in some reduction to the cost of the string. The API Short Round-Thread Casing is furnished in sizes 114.3 to 508 mm (4½ to 20 inches) outside diameter.

API Line Pipe—API line pipe is threaded to American Petroleum Institute Standards 5B, and the same rigid control of threading tolerances maintained as for API casing and all other classes of threaded joints. Until recently, the essential differences between line pipe and standard pipe joints were in the coupling diameters and length and also in the thread length, each of these dimensions being greater in API line pipe than in standard pipe. The American Petroleum Institute has since adopted the American National Standards Insti-

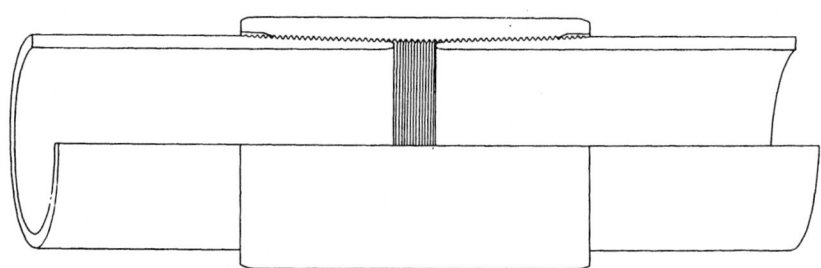

FIG. 32—47. API long round-thread casing.

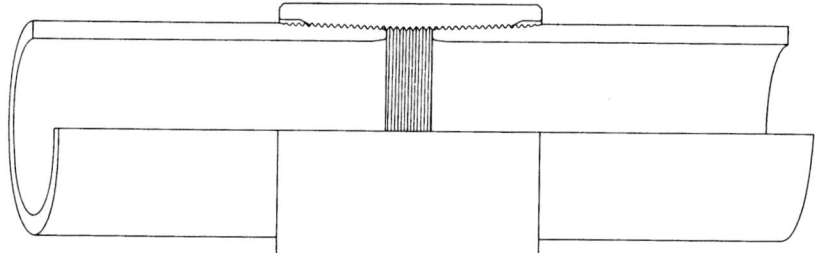

FIG. 32—48. API short round-thread casing.

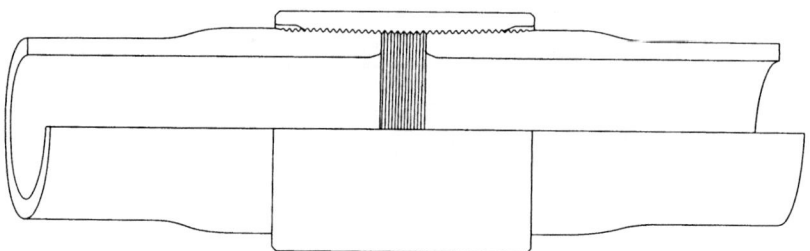

FIG. 32—49. API external upset tubing.

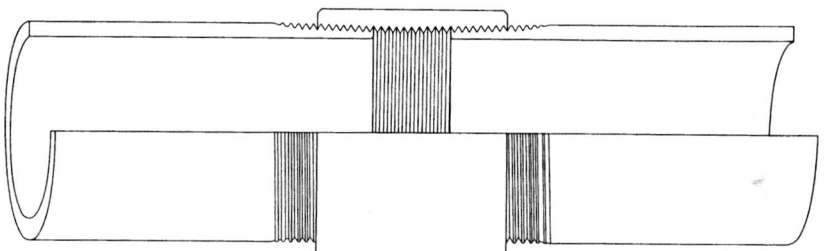

FIG. 32—50. Standard pipe.

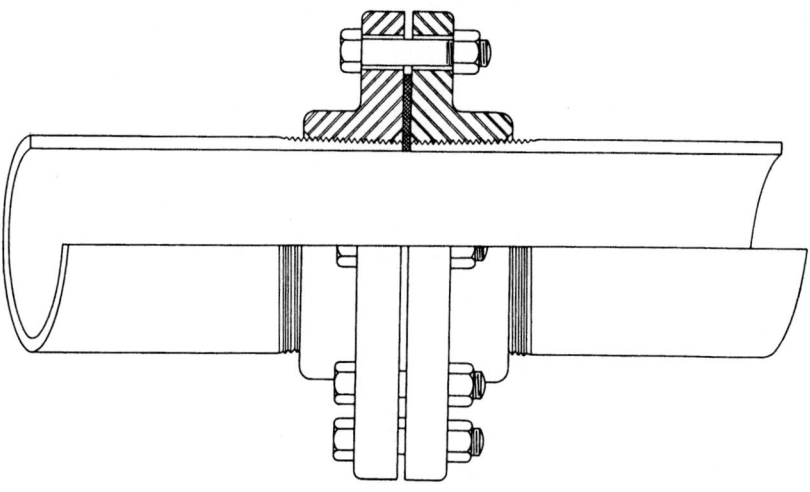

FIG. 32—51. Flanged joint.

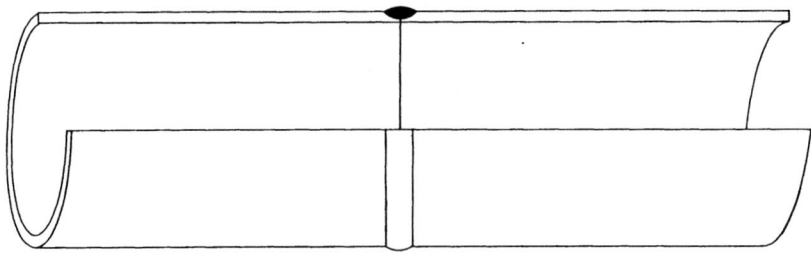

FIG. 32—52. Beveled end for welding.

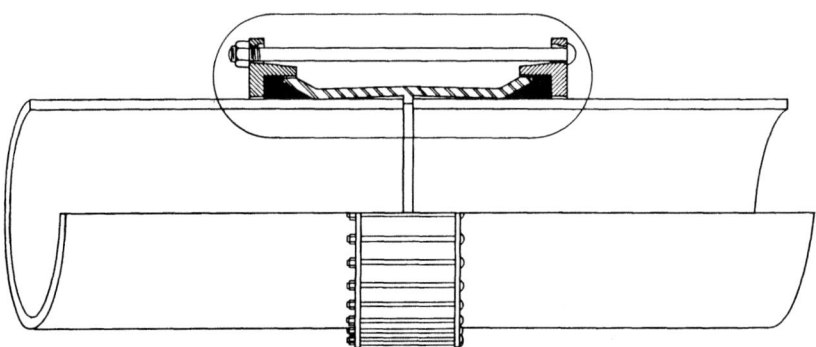

FIG. 32—53. Plain-end coupling.

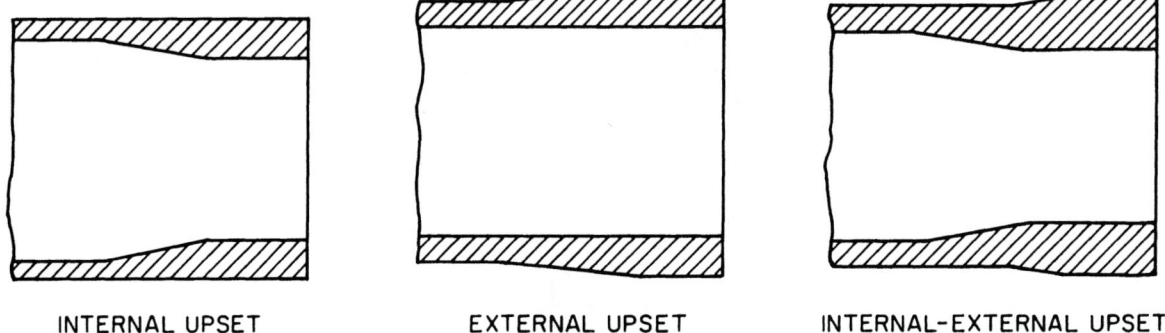

INTERNAL UPSET   EXTERNAL UPSET   INTERNAL-EXTERNAL UPSET

FIG. 32—54. Types of ends available on API upset drill pipe for weld-on tool joints.

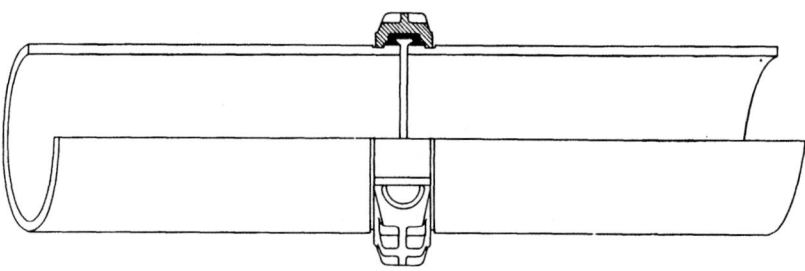

FIG. 32—55. Victaulic-type joint.

tute thread lengths for line pipe, thus eliminating the longer line-pipe threads. Line-pipe and standard-pipe couplings may now be used on pipe threaded to either the API or ANSI specifications. The threaded line-pipe joint, however, is subject to higher test pressures since, in actual service, line pipe is generally subject to far higher working pressures than is standard pipe, the latter being more suitable for low-pressure work, such as piping for plumbing and sprinkler systems.

API External Upset Tubing—API external upset tubing (Figure 32—49) is desirable for deeper wells and for wells of any depth that are pumped. Both ends of the tubing are externally upset in order that the metal area at the root of the first exposed thread, when the joint is properly made up, may be at least equal to the metal area in the body of the pipe and thus of equal strength in tension. This tubing can therefore better absorb the dynamic stresses induced by pumping as the possibility of a fatigue failure in the threaded section is reduced to a minimum.

API Non-Upset Tubing—API non-upset tubing is generally used in open-flow wells and in shallow wells where tubing is pulled from the well infrequently and is a medium for pumping or otherwise raising oil to the well surface and thence to storage tanks. The API non-upset tubing joint is similar in general outline to the API casing joint (Figure 32—47.)

API Upset Drill Pipe—API upset drill pipe is manufactured with either an internal upset and external upset, or a combination internal-external upset for weld-on tool joints (see Figure 32—54). In the past, flash welding was a common practice, but at present most tool joints are friction welded to the drill pipe. The upset dimensions and configurations, as shown in API Specification 5A and API Specification 5AX, were chosen to accommodate the various bores of tool joints and to maintain a satisfactory cross-section in the weld zone after final machining of the assembly. The assembled drill pipe acts as a rotating stem for drill bits in rotary drilling and as a conveyance for air or drilling mud which cools the bit and conveys cuttings to the surface.

Standard Pipe—The standard pipe joint (Figure 32—50), while primarily intended for low-pressure steam, gas, and water lines, as found in buildings and industrial plants, is also used for structural purposes, as in hand railing, scaffolding, etc. This joint has been superseded for long transportation lines by the API line-pipe joint (Figure 32—48) and welded line-pipe joint (Figure 32—52).

Flanged Joints—While only one type of flanged joint (a threaded-flange joint) is illustrated in Figure 32—51, there are a number of other types of flanged joints that may be more suitably adapted to specific problems, and the fact that they are not described or illustrated does not in any way detract from their recognized merits.

Pipe to be fitted with standard screwed flanges is threaded to ANSI B2.1 (Briggs threads). The flanges for use with steel pipe are generally manufactured from forged or alloy steel, depending on the temperature and pressure to which they are to be subjected. Joints of this type (Figure 32—51) are used for power plants, refineries, etc., where working pressures of 10 340 kPa (1500 psi) and temperatures of 540°C (1000°F) are quite

common in regular practice.

Dresser-Type Joint (Plain-End Coupling)—This joint (Figure 32—53) has been frequently used for transportation of high-pressure natural gas, also for low-pressure oil and water lines. It is easily and rapidly assembled and is made leakproof by tightening of bolts, thus exerting pressure on gaskets and in turn on the pipe perimeter. The two gaskets are of rubber, or other suitable material, depending on the temperature to which the joint is to be subjected, and also upon the corrosive action of the fluid being carried in the pipe. The joint is designed to accommodate contraction or expansion due to temperature changes encountered in the line, and will also take care of a certain amount of misalignment. The ends of the pipe for Dresser-type couplings are gaged to specified diameter and roundness tolerances to insure proper assembly in the field.

Victaulic-Type Joint—Like the Dresser-type joint, the Victaulic-type joint (Figure 32—55) is quickly assembled and designed to take up expansion or contraction due to normal temperature changes. The joint will also absorb a certain amount of angular misalignment. As shown in Figure 32—55, the pipe ends are grooved by machining or roller-grooving to provide a seat for the coupling. The joint is assembled by first lubricating the gasket exterior including the lips and/or the pipe ends and housing interiors, and then slipping the gasket over one pipe end, after which the second pipe end is brought into position and the gasket positioned to cover both ends of pipe. The metal housing, normally made in two sections, is mounted over the ring gasket so that it fits into the grooves and is made leakproof by tightening the bolts, thus forcing the gasket against the pipe face. As shown in the illustration, the gasket is so formed that pressure inside the pipe tends to increase the resistance to leakage.

Galvanizing—The hot-dip process of galvanizing is widely used for applying zinc coating to steel pipe for protection from corrosion. The pipe is first thoroughly cleaned by washing in a water solution of caustic soda to remove all traces of oil, paint and grease. It is then pickled in a hot, dilute sulphuric-acid solution (containing an inhibiting agent) to remove all rust and mill scale, after which it is washed free of adhering acid by immersing in a bath of fresh, clean water. The fluxing operation then follows and is accomplished by immersing the pipe in a hot solution of zinc ammonium chloride until the temperature of the pipe approximates that of the solution. Upon removal from the fluxing tank, the pipe is thoroughly drained, then placed on the charging table of the galvanizing kettle (Figure 32—56) where it is introduced into a bath of molten zinc maintained at a temperature range of 440° to 460°C (820° to 860°F). The length of time the pipe is immersed in the molten zinc bath varies with the pipe size. Excessive immersion times are to be avoided to minimize the thickness of the brittle alloy layer (see Chapter 38).

The pipe is withdrawn on an inclined plane from the galvanizing kettle by a magnetic roller. As the pipe is withdrawn, it passes through an air-wipe ring which removes the excess zinc from the outside surface of the pipe. The pipe is then conveyed to a station where superheated steam is blown through the pipe to re-

FIG. 32—56. Continuous pipe-galvanizing line. (1) Pipe on charging chains with sinker arms. (2) Galvanizing pot. (3) Magnet-roll conveyor. (4) Roll-out station. (5) Quench tank. (6) Inspection table.

move excess zinc from the inside surface. The wiping and blowing result in a smooth surface and uniform weight of coating. The coated pipe is then cooled rapidly by quenching to arrest the growth of the alloy layer and to maintain a bright surface luster.

Value of Zinc Coating for Pipe—Though zinc coating is readily soluble in dilute acids, strong alkalies and some mineral-salt solutions, galvanizing affords effective protection against ordinary atmospheric corrosion, because the zinc, when exposed to the air, immediately reacts with oxygen to form zinc oxide. This reaction progresses only to a limited extent, however, for the oxide remains as a thin film on the surface of the zinc coating, and protects the zinc from further oxidation. Since this oxide is insoluble in water and since the film adheres rather tenaciously to the zinc, it protects the underlying coating of zinc against all ordinary types of weathering, and as long as the zinc remains intact, the steel beneath is secure from corrosive action. If the zinc coating is broken, by bending or abrasion, and the underlying steel is exposed to corrosive influences, the zinc will still afford considerable protection. Being elec-

tropositive to the steel, the zinc is dissolved instead of the steel as long as any zinc remains closely adjacent to the exposed steel. In underground piping, where the pipe is in direct contact with the earth, galvanizing is not generally suitable as protection against corrosion, because it may be quickly destroyed either by certain acids or by certain alkali substances in the soil.

Bibliography

Am. Institute of Mining and Metallurgical Engineers, Tube producing practice (Institute of Metals Div. Symposium Series, vol. 4) N.Y., The Institute, 1951.
    (Schroeder, J W., The metallurgical factors affecting the production of seamless pipe, p. 57-68)

American Iron and Steel Institute, Carbon Steel Pipe, Structural Tubing, Line Pipe and Oil Country Tubular Goods, April 1982.

American Iron and Steel Institute, Committee of Steel Pipe Producers, The Handbook of Steel Pipe, latest edition.

Bray, T. J., Manufacture of welded steel tubing. Engineers Society of Western Pennsylvania Proceedings, 1888.

Camp, J. M. and C. B. Francis, The making, shaping and treating of steel; 4th ed. Pgh., Carnegie Steel Co., 1925 (The lap-weld process, pp. 1043-1059).

Hill, William J. and Ricci, Mario, Recent applications of press piercing mills and retained mandrel mills in North America. Iron and Steel Engineer (August 1983).

Herb, C. O., Steel tubing made by resistance welding. Machinery 44, 1-5 (Sept. 1937).

Sanders, E. N., Progress in steel pipe manufacture with particular reference to seamless pipe. Am. Iron and Steel Institute Yearbook, 1947, p. 446-454; Discussion, 454-458.

Seamless Steel Pressure Vessels, United States Steel, ADUSS 46-2560-02, December 1972.

Thirty-inch pipeline. Steel 120, 74-75 (March 24, 1947).

Wilder, A. B., When stronger line pipe is needed . . . pipeliners will get it. Oil and Gas Journal 54, 130-133 (May 9, 1955).

Wright, E. C. and S. Findlater, Manufacture of seamless steel pipe in the plants of the National Tube Co., Iron and Steel Institute Journal 138, 109 P-124P (1938).

CHAPTER 33

The Manufacture of Hot-Strip Mill Products

SECTION 1

CLASSIFICATION OF FLAT-ROLLED STEEL PRODUCTS

The products of the hot-strip mill are classed among flat-rolled steel products. About half of the rolled-steel products now made in the United States are flat-rolled material. Flat-rolled steel products (including sheet, strip, tin plate, black plate, flat bars, slabs, plates, skelp and hoop) may be distinguished from other forms of rolled steel in two general ways. First, flat-rolled steel is produced on rolls with smooth faces in contrast to the cut or grooved roll faces employed in the manufacture of shapes and, second, in flat-rolled products the ratio of width to thickness is generally high as distinguished from other rolled products. The ranges of dimensions are wide, varying in thickness from 0.13 millimetre (0.005 inch) in light strip to 381 millimetres (15 inches) in heavy plates, and in width from 4.76 millimetres ($\frac{3}{16}$ inch) in narrow strip to 5182 millimetres (204 inches) in wide plates.

Sheet, strip, black plate and tin plate comprise about three-fourths of the total tonnage of all flat-rolled steel products. Shipments of various grades of these commodities from mills in the United States in 1982, according to the Annual Statistical Report of the American Iron and Steel Institute for that year, were as follows:

	Thousands of Metric Tons	Thousands of Net Tons
Hot-Rolled Sheet	8 210	9 052
Hot-Rolled Strip	452	498
Cold-Rolled Sheet	10 097	11 132
Cold-Rolled Strip	657	724
Metallic-Coated Sheet and Strip		
Galvanized		
Hot-Dipped and Electrolytic .	4 870	5 369
Tin Plate		
Electrolytic and Hot-Dipped .	2 737	3 018
Tin-Free Steel	833	918
All Other	689	760
Black Plate	289	319
Electrical Sheet and Strip	404	445

Flat-rolled steel products fall into two major categories: hot rolled and cold rolled. Hot-rolled products are reduced to final thickness by heating and rolling at elevated temperature. Hot rolling on a continuous hot-strip mill usually begins at about 1205°C (2200°F). In ordinary practice, hot rolling is usually completed within the range 815 to 955°C (1500 to 1750°F). By the use of water sprays or laminar jets, the hot-rolled strip is cooled to a specific temperature, usually within the range 510 to 730°C (950 to 1350°F), before it enters the coilers. Cold-rolled products are really "cold finished" products, since much of the reduction from ingot to final thickness is, of course, done while the product is hot, in a manner similar to that employed for hot-rolled products. Cold rolling is carried out on products which have not been heated immediately prior to the cold-rolling operation in which they are reduced to final thickness. However, the temperature of the steel is raised due to frictional effects of rolling. The temperature of the steel in coils immediately after cold reduction has been measured, and coil temperatures ranging from 120 to 230°C (250 to 450°F) have been recorded; the temperature of the steel in the actual nip of the rolls is probably higher than this, but is quickly lowered by the coolants used in rolling. The important distinction is that cold-rolled products receive enough cold reduction in the final rolling operation to affect the surface and mechanical properties of the finished product.

Flat-rolled steel products include both semi-finished and finished materials. Among the **semi-finished** products are **slabs** and **flat bars**. An important semi-finished flat-rolled product of the hot-strip mill, which will be discussed later in this chapter, is **hot-rolled breakdowns for cold reduction, in coils**.

The chief hot-rolled **finished** flat products are divided into four major groups, namely, **bars, plate, hot-rolled strip,** and **hot-rolled sheet**. Dimensions, particu-

Table 33-I. Dimensional Limitations of Sheet.

Thickness Range		Width Range	
(mm)	(In.)	(mm)	(In.)
5.839 to 4.571	0.2299 to 0.1800	Over 304.8 to 1219.2, incl.	Over 12 to 48, incl.
4.570 to 1.140	0.1799 to 0.0449	Over 304.8 to over 1219.2	Over 12 to over 48

Table 33-II. Dimensional Ranges of Strip.

Width		Thickness	
(mm)	(In.)	(mm)	(In.)
Up to 88.9, incl.	Up to 3½, incl.	0.648 to 5.156, incl.	0.0255 to 0.2030, incl.
Over 88.9 to 152.4, incl.	Over 3½ to 6, incl.	0.874 to 5.156, incl.	0.0344 to 0.2030, incl.
Over 152.4 to 304.8, incl.	Over 6 to 12, incl.	1.140 to 5.839, incl.	0.0449 to 0.2299, incl.

larly thickness and width, are the principal bases of classification.

In this chapter, finished, flat-rolled carbon steel produced within the dimensional limitations in Table 33—I is considered to be **sheet,** whether in cut lengths or in coils. Modern hot-strip mills are capable of rolling and coiling material that is considerably thicker than the 5.839-mm (0.2299-inch) limit given in Table 33—I for sheet. However, to be classified as sheet, the product dimensions must be within the above ranges.

By common custom, finished flat hot-rolled carbon-steel **strip** is produced in the dimensional ranges shown in Table 33—II.

Size limitations of bars and plates are discussed in Chapters 30 and 25, respectively.

The chief cold-rolled products, all classified as **finished,** are divided into four major groups, namely, **bars, strip, sheet,** and **black plate.** The dimensional bases for differentiating between the three latter commodities are discussed in Chapter 34; dimensional limitations for bars are given in Chapter 30. Continuous hot-strip mills roll product in very long lengths, up to over a thousand metres (several thousand feet), according to the width and thickness of the strip and the size and equipment of the mill. The product of a continuous hot-strip mill is generally produced in coil form, although it can be cut to specified lengths after rolling, either directly on the mill or as a subsequent operation.

The distinction between the hot-rolled and the cold-rolled classes of these commodities lies, as mentioned above, in the methods used to attain finished thickness. The cold-reduction process, however, applied to the hot-rolled and pickled steel imparts, after proper heat treating and finishing operations, greatly superior surface and mechanical properties to the hot-rolled counterpart of each commodity. The starting material for the cold-reduction process consists of the semi-finished product of the hot-strip mill designated as hot-rolled breakdowns in coil form.

Further sub-classification of sheet, strip and black plate is necessary to approach an understanding of the diversity of characteristics which enables steel in these forms to be applied to so many important uses. As one example, most black plate actually is not used as such, but is coated with tin to produce tin plate of many varieties for many uses, including the common "tin" can. Such group subdivisions are based on steel type, product treatment, characteristic properties, and final use, and will be discussed after description of the manufacturing methods used.

SECTION 2

HISTORICAL DEVELOPMENT

The mills used for rolling flat-rolled steel products include the following, named in the order of their development:
(1) The two-high mill for hot rolling sheets in packs.
(2) The two-high mill for rolling sheared plates.
(3) The universal mill for rolling plates.
(4) The three-high or Lauth mill for rolling sheared plates.
(5) The continuous or tandem hot-strip mill for rolling sheets, strip and hot-rolled breakdowns for cold reduction in coils.
(6) The cold-reduction (cold-rolling) mill for sheet and strip.
(7) The continuous sheared plate mill.

The plate mills mentioned are described in Chapter 25.

The method for rolling sheet on two-high, single-stand mills originated between 1720 and 1728 and antedates all other methods for rolling iron and steel. Up to about 1890, the finished flat-rolled steel products could be classed as sheet, plates and bars, although many thin products, such as pipe skelp, were rolled on the merchant-bar mills. The bar mills also rolled thinner sections, and about this time there developed such a demand for very thin flats that special mills were built to supply the material. The differentiation first took place in the narrower widths, ranging from 15.9 to 76.2 mm (⅝ to 3 inches), and from 1.65 to 0.71 mm (0.065 to 0.028 inch) in thickness. The higher grades of this material were designated as **hoop;** a common grade was known as **cotton tie.** In 1890, **bands** 371 by 3.6 mm (14⅝ by 0.14 inch) were rolled at Warren, Ohio and, in 1892, a mill designed to roll sheets in tandem rolls was built in Austria. In 1893, a mill at Bridgeport, Connecticut produced thin sections up to 178 mm (7 inches) in width and, in 1895, a semi-continuous mill was built there to produce thin hot-rolled products up to 254 mm (10 inches) wide. In rolling on these first mills it was found that the limits as to width and thickness bore a certain relation to each other. In the Bridgeport mill, the ratios of width to thickness of product varied between 100 and 160. From these beginnings, the widths within the same range of thicknesses were increased at intervals by various producing mills until, in 1920, steel 559 mm (22 inches) wide and 2.67 mm (0.105 inch) thick was rolled successfully at Weirton, West Virginia. Some of the product of the successful strip mills subsequently was pickled and cold rolled, when it was designated as **cold-rolled strip steel,** but much of it was used as rolled and was known as **band steel,** or **hot-rolled strip.** In 1920, therefore, flat-rolled products were classed commercially as sheet, plates, bars, bands and

hoop, although the term "strip" commonly was applied to light products which were somewhat narrower than heavier sheet rolled on pack mills.

Up to 1920, it was customary to observe a certain width-to-thickness ratio in the hot rolling of strip, and the maximum ratio for successful hot rolling was considered at that time to be 250. In 1923, however, a continuous hot-rolling mill at Ashland, Kentucky, began rolling much wider strip, the widest and thinnest being 914 mm (36 inches) by 1.65 mm (0.065 inch). This mill can be considered the forerunner of the modern continuous wide hot-strip mills described later in this chapter.

In 1926, at Butler, Pennsylvania, the first mill was built to combine successfully use of the following principles: (a) four-high finishing stands, (b) control of the direction of travel of steel through the pass line of the tandem finishing mills by progressively decreasing the product crown in successive mill passes, and (c) hot-coiling equipment at the discharge end of the mill. This installation was the first of the modern wide continuous hot-strip mills as known today.

SECTION 3

SOURCES AND TYPES OF STEEL FOR SHEET, STRIP AND TIN PLATE

Chemical Compositions—Steel compositions used for the manufacture of thin, flat steel products range from so-called "pure iron," in which the sum of all elements other than iron in the product is less than one-third of one per cent of the total weight, to the high-alloy stainless and heat-resisting steels composed of as much as 50 per cent alloying additions. Over 75 per cent of the sheet, strip and tin plate tonnage rolled, however, is made from steel compositions within the following ranges (based on ladle analyses):

Element	Per Cent
Carbon	0.03 to 0.12
Manganese	0.20 to 0.60
Phosphorus	0.04 maximum
Other elements	Low as possible

This general range of compositions provides the best combination of rollability during manufacture and formability in most of the applications for which these products are used. Such compositions, too, are well suited for the production of rimmed steel, which is. preferred for flat products because of the superiority of its ingot surface. Deviations from this basic composition range are deliberately employed to obtain specific desired properties in the steel, according to principles discussed later. Within the basic composition range, however, most steel plants further subdivide the indicated ranges for individual elements to fit particular production conditions or consumers' special needs. The end result of such adjustments of composition will differ from plant to plant; accordingly, most consumers' requirements are expressed best in terms of suitability for particular applications or of desired properties, with

composition restricted only where a direct relationship between composition and performance is known.

Sulphur, silicon, copper, nickel and chromium generally are considered as the "other elements" of the basic composition given above. Except in the steels where they are added deliberately to produce alloy steels with definite properties, these elements offer no advantages and, when present in greater than certain amounts may even be detrimental to the rolling or fabricating properties of steels for sheet, strip and tin plate. An effort is made to keep sulphur content below 0.30 per cent, chromium content below 0.05 per cent, and copper and nickel contents below 0.15 per cent. Silicon content naturally falls under 0.02 per cent in the rimmed and capped steels popularly used for sheets, strip and tin plate, but is present in amounts up to 0.15 per cent when this element alone is used as the deoxidizing agent in the manufacture of steel in the range of the basic low-carbon composition. Other elements seldom are found in undesirable amounts, although unusual local conditions affecting the scrap or ore supply may result in the presence of enough molybdenum or tin, or both, in the steel to cause it to be somewhat harder in the finished condition than would be the case if these elements were absent or present in only very small amounts.

Steelmaking Processes—The steel for sheet, strip and tin plate is made in this country by the basic oxygen, open-hearth or electric-furnace process, each being used where it is best suited to produce steel having the desired composition and properties. Acid Bessemer steel formerly was used also, for some limited applications. Stainless and some other alloy steels are melted in the electric-arc or induction furnace for conversion to sheet and strip, the processes in these cases being chosen for their ability to produce alloyed grades of steel with minimum loss of alloying elements and steels with fewer non-metallic inclusions. However, about 95 per cent of the steel for sheet, strip and tin plate is being produced in basic oxygen and basic open-hearth furnaces.

Rimmed, capped, semi-killed and killed steels all are used for conversion to thin, flat steel products. Rimmed steels comprise more than half of the sheet, strip and tin plate tonnage made, since steel of the basic low-carbon composition given above, when properly refined, tapped and teemed, provides a natural rimming steel that can be cast into ingots with sound surfaces and possesses a high degree of cleanliness and ductility. Mechanically-capped steel retains most of the surface qualities of rimmed steel and provides more uniformity of hardness throughout the cross-sections of rolled products, while increasing the yield of sound steel obtained from each ingot. This modification of rimmed steel by mechanical capping is, therefore, of importance in producing steel for such an application as tin plate where a controlled degree of uniform stiffness is desirable in the end product. Aluminum capping or top killing also is employed; in this practice, the rimming action is stopped after having progressed to the desired point by adding aluminum to the molten steel in the top of the mold. The aluminum killed ("special killed" or "fine-grained") extra-deep-drawing steel is a highly specialized modification of the low-carbon type, having virtual freedom from age-hardening and, hence, unique suitability for some types of drawing operations.

Rimmed, capped and even the "special killed" low-carbon steels are cast into ingot molds without hot tops. "Semi-killed" grades, usually made to possess somewhat higher carbon and manganese contents, still allow sufficient deoxidation control (with aluminum or silicon) of gas evolution and shrinkage to provide a sound ingot from an open-topped mold. Steels having high contents of carbon or alloying elements, those fully killed to attain certain desired end properties, and occasionally the "special killed" grade are hot-topped to eliminate the formation of "pipe" resulting from the shrinkage characteristic of the solidification of such grades. The stainless steels are the best known hot-topped steels converted to thin flat-rolled products.

Slabs—Slabs are the raw material for the modern continuous hot-strip mill. A slab is defined as a rectangular steel section having a minimum thickness of 38.1 mm (1½ inches) and minimum width not less than twice the thickness. Slabs (including continuously cast slabs) are generally provided in thicknesses of 50.8 to 304.8 mm (2 to 12 inches), widths of 304.8 to 2032 mm (12 to 80 inches), and lengths of 1524 to 12 192 mm (60 to 480 inches), depending on strip-mill requirements. They must be accurate enough in dimensions and sound enough in structure to permit conversion in subsequent rolling operations with a minimum of difficulty, and their edges and surfaces must be free of injurious defects which would carry through to the finished product.

Slabs may be produced by continuous casting, by bottom-pressure pouring or by hot rolling of ingots.

Rolling Slabs from Ingots—Two methods are prac-

ticed for converting steel from ingot to slab form and then to hot-rolled sheet, strip or breakdowns for cold reduction. By one procedure, the ingots may be heated in soaking pits and rolled on a blooming or slabbing mill to slabs of the required width and thickness, then sheared to length and immediately passed along to the hot-strip mill for final reduction to desired thickness, without reheating. The second and more generally used method is similar to the foregoing, except that the slabs are allowed to cool after being sheared to length at the slab-producing mill. The cooled slabs then are laid out for inspection to locate visible surface and edge defects which are marked for conditioning by the scarfing process. The conditioned slabs then are charged into reheating furnaces at the continuous hot-strip mill.

The first or single-heating method described above results in substantially lower fuel costs and minimizes handling and conditioning expenses, but does not provide maximum flexibility in scheduling hot-strip mills rolling a widely varying product mix or a substantial portion of small orders. The reheating method (the second described above) has proven more advantageous as it provides full flexibility of hot-strip mill scheduling, permits closer metallurgical control of steel-rolling temperatures and minimizes injurious steel-surface defects resulting from defective slab areas.

In accordance with the latter practice, conversion of ingots to slabs is effected by the following typical steps: After stripping, the ingots are charged in the soaking pits of a blooming or slabbing mill and brought to a uniform temperature, approximately 1315°C (2400°F) for the low-carbon steel grades which comprise the bulk of this tonnage. They then are removed by crane-borne tongs to ingot buggies that convey them to the entry roll table of the slabbing or blooming mill. Some mills are equipped with turntables at this point which automatically record ingot weight and place the ingots butt first with respect to the rolling stands and shears; this practice is advantageous as it provides a close check on ingot weight and efficient control of end scrap. Ingots then are passed along a roller table to the reducing stand, consisting of a slabbing mill or a blooming mill of the single stand, reversing type. The slabbing mill is equipped with both horizontal and vertical rolls that work on all four sides of the ingot simultaneously, while the blooming mill operates with horizontal rolls only. Greater tonnages and wide slab sections can be produced on the slabbing mills as a result of this difference in design. After reduction to the prescribed slab thickness and width, the elongated slab is advanced toward a shear at the end of the mill roller table. Some mills, at a point between the reducing stand and the shears, are equipped with automatic flame scarfing equipment for the purpose of removing all but the worst edge or surface defects from the slab. The shears cut the slab product to the designated lengths, cropping sufficient scrap from the two ends of the slab, corresponding to what originally was the top and bottom of the ingot, respectively, to ensure elimination of pipe, porosity, mechanical end laps, slag deposits, and so on.

Immediately after hot shearing, the slabs are hot stamped with identification markings, such as heat number, ingot number and cut number. After shearing and stamping, the slabs are piled on cooling beds and permitted to cool to a workable temperature, then laid out individually for inspection and marking of surface and edge defects such as scabs, ingot cracks, spongy surface, breaks, tears, and so on which, if not removed, would result in surface slivers, scabs, skin laminations or cracked edges on the finished strip. The defects are removed by scarfing with an oxy-acetylene torch or, on stainless steels, by grinding with abrasive wheels or powder scarfing. The slabs then are repiled, and each is painted with identifying information normally including heat, ingot and cut numbers, thickness, width, weight and code letters or numerals representing chemical composition and steelmaking process.

The finished slabs finally are transported to the storage yard of the hot-strip mill, where they are stacked in orderly fashion according to size and steel grade to facilitate their selection and charging into magazines feeding the reheating furnaces of the hot-strip mill.

Continuous Casting of Slabs—Slabs may be cast by the continuous methods described in Chapter 21. Different-sized molds may be used to cast slabs of the desired sizes or, to minimize the number of required mold sizes, the steel may be cast into molds of only a few sizes, and the castings hot rolled continuously to size in special rolling equipment immediately following the casting machine.

Most continuously cast slabs, after cutting to length, inspection and marking for identification, are handled in the same manner as slabs rolled from ingots as described above. Some continuous castings, however, after in-line heating to restore and equalize temperature, pass directly into continuous hot-rolling mills for reduction to strip form.

Bottom-Pressure Pouring of Slabs—In this process, a special casting technique is employed, with molten steel being forced at a controlled rate into molds of the desired size where it solidifies in the form of a slab.

SECTION 4
CONTINUOUS HOT-STRIP MILLS

Development and Output—The terms "strip" and "sheet," as applied to the finished products, have definite reference to width and gage limitations, as shown in Section 1 of this chapter, and the term "hot-rolled breakdowns for cold reduction" defines a semifinished product for subsequent rolling by another process. These distinctions do not exist when the terms are used in connection with a continuous mill rolling such products; thus, a mill rolling continuous lengths of strip, sheet, or breakdowns commonly is known as a "strip mill." The history of development of continuous hot-strip mills was summarized in Section 2 of this chapter.

As of December 1981, forty-one wide continuous hot-strip mills were in operation in the United States, their combined annual capacity being approximately 64.1 million metric tons (70.7 million net tons). As a group, the wide strip mills roll flat steel in thicknesses of 1.19 to 12.7 mm (0.047 to 0.500 inch), widths of 610 to 2438 mm (24 to 96 inches), with coil weights up to 43 000 kilograms (95 000 pounds). Each mill has its own limitations as to sizes of finished product though, as a general rule, no mill exceeds a product width-to-thickness ratio of 1000:1.

General Arrangement of Modern Mills—The modern wide hot-strip mill has become quite standardized in its general layout. Slabs are heated in two or more continuous reheating furnaces. A typical rolling train will consist of a roughing scalebreaker, four four-high roughing stands, a finishing scalebreaker, and six four-high finishing stands. Some recently built wide hot-strip mills have five roughing stands and seven finishing stands. Driven table rolls convey the steel from furnace to mill and also from stand to stand. If the mill is to produce strip, sheets or breakdowns of greater width than the maximum width of slab available, the first rougher or roughing stand is a broadside mill in which the width of the slab is increased in a single pass by cross rolling. In this case, turntables for manipulating the slab must precede and follow this stand. A slab

squeezer also follows the broadside mill. The remaining roughing stands usually are provided with integral vertical edgers in front of each stand. Separating the roughing train from the finishing train is a holding table, while the finishing end is a closely grouped tandem train composed of the finishing scale-breaker and six (or seven) finishing stands.

High-pressure hydraulic sprays to remove scale from the hot slab are located after the two scale-breakers and perhaps at several roughing stands. Water is supplied to the spray nozzles by suitable high-pressure pumps.

As the steel proceeds from the mill, it is carried over a long table called the runout table, consisting of individually driven rollers. On the runout table, water is applied to the top and bottom surfaces of the strip by water sprays or laminar jets to reduce the strip temperature to a prescribed value. Two or more coilers are located at the end of the runout table. Additional tables may be installed parallel to the central runout table, with suitable transfers for moving material to them; this equipment is used principally when the heavier gages are being rolled on some strip mills.

The most commonly used hot-mill arrangement just described, employing continuous roughing and continuous finishing trains, provides high rolling capacity and rapid steel travel with little loss of heat, but entails a high installation cost and a fixed number of passes, with some loss of flexibility in making rapid changes in the mill setup when the size of product to be rolled is changed. An alternate arrangement, used in several instances, employs a reversing roughing mill and a continuous finishing train; this arrangement has a lower original installation cost, requires less floor space, and is flexible with regard to the number of passes available, but at a sacrifice in capacity and operating cost. Another modification of the conventional mill (the single-heating practice already described) provides for rolling the slab directly from the blooming operation, utilizing retained heat and by-passing the slab-reheating furnace; this practice saves fuel but sacrifices flexibility of scheduling rolling operations on the continuous hot mill.

Still another arrangement for continuous hot rolling of steel calls for the use of a reversing hot mill for finishing. This mill may have a conventional roughing train or single reversing stand for roughing, followed by the single reversing finishing mill. The reversing finishing mill has a pair of pinch rolls and a paddle-type coiler located on both sides (entry and exit) of the mill. The coilers operate inside of small heating furnaces which keep the steel hot and permit the finishing rolling operation to be carried out by repeated reversing of the finishing stand. This type of mill entails low initial cost and is highly satisfactory for the production of small orders or the rolling of alloy steels.

Control of Finished Product Quality—When an order is accepted for production by a particular mill, the first step taken is to determine the proper grade of steel and the size and surface quality of the slab necessary to make the order. The order is then grouped with others and scheduled in its proper rolling sequence. The principal factors taken into consideration at this important stage are rolling width, gage, and steel composition.

The next step in production requiring control is the operation of heating the slabs to rolling temperature. The slabs must be heated uniformly throughout, and also must have a uniform "scale jacket" that will "clean up" readily in rolling. Many rolling delays and mechanical difficulties are a direct result of poor heating practice; the steel may not be "soaked" sufficiently, may be too hot, too cold, or unduly hard to clean.

The third step is to rough down the slab to a predetermined intermediate thickness. As the slab leaves the last roughing stand, it should be flat, straight, free of furnace scale, true to width and of a cross-section suitable for further reduction on the finishing stands. The first rolling pass on the slab is done on a scale-breaker which is followed immediately with a high-pressure hydraulic spray to facilitate removal of the furnace scale. In addition, there are usually one or two more descaling sprays following the second or third roughing stands and numerous steam and air sprays available to remove any further scale that may be loosened during rolling or edging. Proper use of the broadside mill, slab squeezer and the vertical edgers normally will guarantee the uniform width essential for all subsequent operations.

Next, the finishing train must be operated with careful regulation to obtain a finished hot-rolled product of prime quality. As mentioned later in this chapter and as discussed in Chapter 23 on "Rolling Mills," various automatic-control elements have been incorporated in mill design to assist the operators in producing strip of uniformly high quality. Surface, gage, width, finishing temperature and cross-sectional contour of the product are all required to meet given standards depending upon the subsequent treatment or ultimate use of the material in question. As an example, metallurgical requirements may dictate a definite finishing temperature for a particular gage and width to be rolled. Time on the holding table prior to coiling, number of

descaling sprays used during rolling in the finishing stands, speed of the finishing train, and method of drafting, all affect the finishing temperature and must be controlled in order to meet requirements. Defects in the surface of the rolled steel, if not evident in the rough slab, usually can be traced to defects in the surfaces of the work rolls on the finishing stands and are corrected readily by substituting newly-surfaced work rolls. The principal factors affecting the overall dimensional accuracy of the finished hot-rolled product include contour of the work rolls and back-up rolls as installed, changes in the contour of the work rolls and back-up rolls due to intermittent heating and cooling, method of drafting (i.e., amount of reduction in successive passes), and rolling sequence of various gages and widths. Also involved in the occurrence of gage (thickness) variations in hot-rolled strip are: the difference in the speed of the strip leaving the last roughing stand and the entry speed of the strip entering the first finishing stand, which results in temperature variations along the length of the strip; and variations in the tension applied to the strip between stands. Most of these factors that affect product quality are amenable to control by automatic means to assist the mill operators in achieving the best results from the rolling operation. Some of the control principles are discussed in Chapter 23 on "Rolling Mills."

Figure 33—1 shows typical reductions per pass in the finishing train of a continuous wide hot-strip mill.

The final step in rolling an order on a strip mill is disposition of the hot-rolled product. The greater portion of hot-rolled flat material is handled by the hot-coiling method; this includes the semi-finished product designated as hot-rolled break-downs in coils for subsequent cold reduction, as well as hot-rolled sheet in coils which may be shipped as such or transferred to the finishing department where they are uncoiled and processed into the form of flat cut sheet. The essential requirements of the coiler are to receive the material at mill speeds and coil it tightly without excessive tension, telescoping, scratching or marking and, finally, discharge the finished coil quickly without damage.

A 2135-mm (84-INCH) CONTINUOUS HOT-STRIP MILL

The 2135-mm (84-inch) continuous hot-strip mill at the Gary Works of United States Steel Corporation serves as an example of a modern mill of this type. Operation of the mill, from slab-heating-furnace entry to coil delivery, is controlled by computer, including such functions as automatic setting of roll openings, automatic gage control, and data recording. Maximum operating speed is 20.32 metres per second (4000 feet per minute).

The total installed horsepower of the electric motors for the mill is 147 700 kilowatts (198 000 horsepower). The main-drive motors represent an aggregate of 92-100 kilowatts (123 500 horsepower), divided between synchronous alternating-current motors totaling 29 800 kilowatts (40 000 horsepower) and direct-current motors totaling 62 300 kilowatts (83 500 horsepower). Auxiliary alternating-current and direct-current drives total 21 300 and 14 500 kilowatts (28 555 and 19 375 horsepower), respectively; miscellaneous drives for

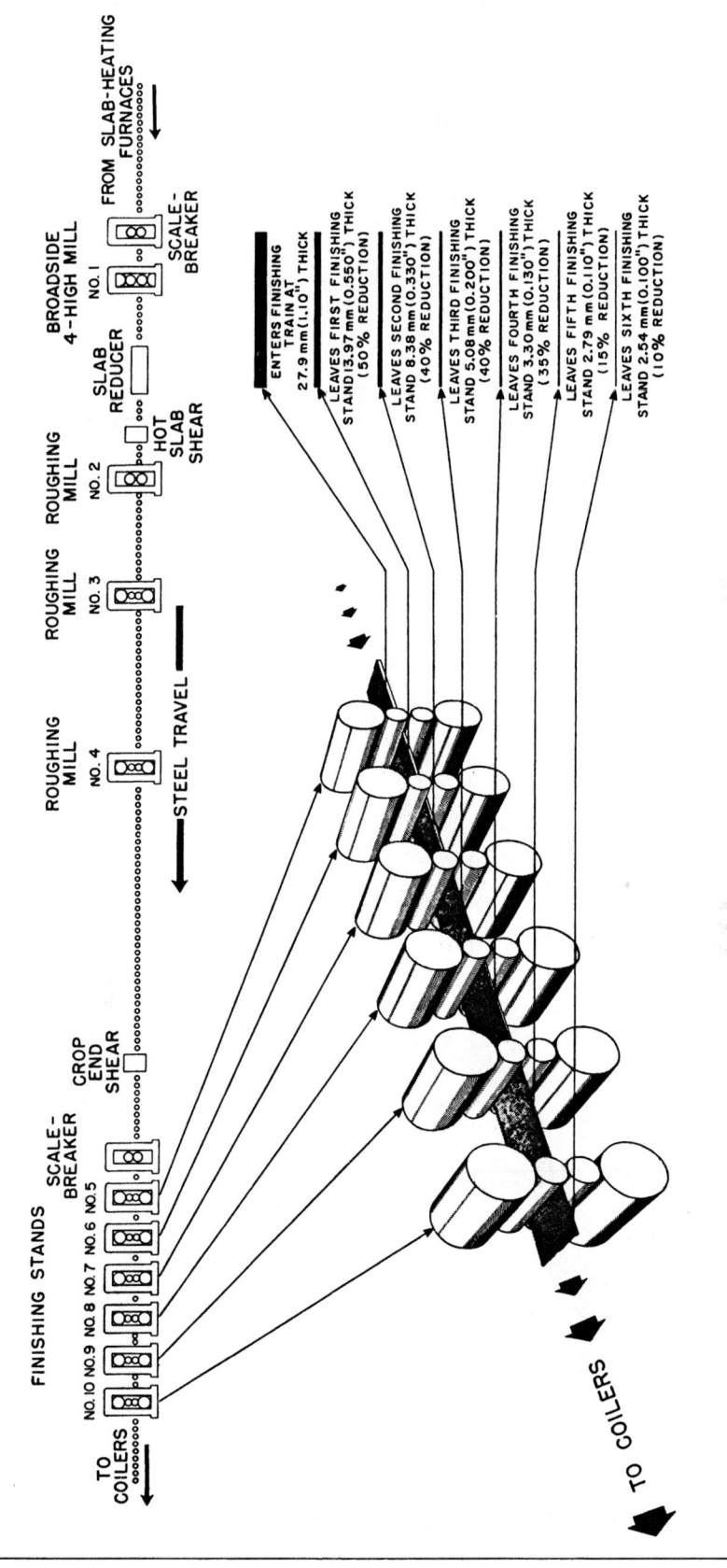

HOT STRIP ROLLING MILL

TYPICAL REDUCTIONS PER PASS IN FINISHING STANDS

(THIS DRAWING IS ENTIRELY SCHEMATIC AND NOT TO SCALE)

FROM SLAB-HEATING FURNACES

BROADSIDE 4-HIGH MILL NO. 1

SCALE-BREAKER

SLAB REDUCER

ROUGHING MILL NO. 2

HOT SLAB SHEAR

ROUGHING MILL NO. 3

ROUGHING MILL NO. 4

STEEL TRAVEL

FINISHING STANDS

SCALE-BREAKER

CROP END SHEAR

NO. 10 NO. 9 NO. 8 NO. 7 NO. 6 NO. 5

TO COILERS

ENTERS FINISHING TRAIN AT 27.9 mm (1.10") THICK

LEAVES FIRST FINISHING STAND 13.97 mm (0.550") THICK (50% REDUCTION)

LEAVES SECOND FINISHING STAND 8.38 mm (0.330") THICK (40% REDUCTION)

LEAVES THIRD FINISHING STAND 5.08 mm (0.200") THICK (40% REDUCTION)

LEAVES FOURTH FINISHING STAND 3.30 mm (0.130") THICK (35% REDUCTION)

LEAVES FIFTH FINISHING STAND 2.79 mm (0.110") THICK (15% REDUCTION)

LEAVES SIXTH FINISHING STAND 2.54 mm (0.100") THICK (10% REDUCTION)

TO COILERS

FIG. 33—1. Typical reductions per pass in the finishing stands of a hot-strip rolling mill equipped with four roughing stands and six finishing stands.

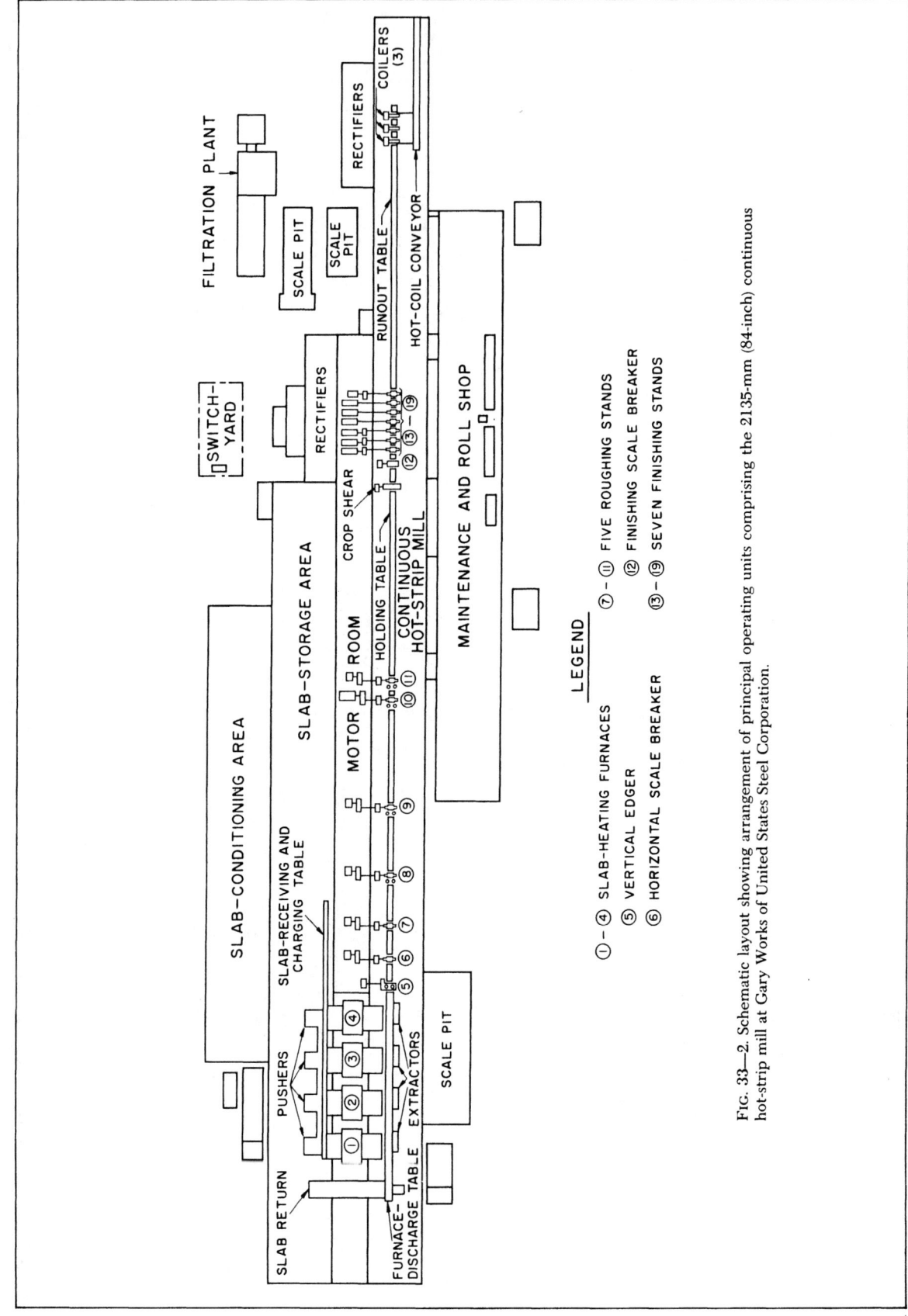

FIG. 33—2. Schematic layout showing arrangement of principal operating units comprising the 2135-mm (84-inch) continuous hot-strip mill at Gary Works of United States Steel Corporation.

cranes, pumps and other units total 19 800 kilowatts (26 570 horsepower). All direct-current power is supplied by silicon-controlled rectifier (SCR) units.

The mill produces hot-rolled strip (in coil form only) in the thickness range of 1.19 to 12.7 mm (0.047 to 0.500 inch), in widths from 457 mm (18 inches) minimum to 1943 mm (76½ inches) maximum. Coils have an internal diameter of 724 mm (28½ inches), and external diameters ranging from 914 mm (36 inches) minimum to 2134 mm (84 inches) maximum. Maximum coil weight is 33 100 kilograms (73 000 pounds). Provision has been made for future production of coils with 813-mm (32-inch) internal diameter and a maximum weight of 42 000 kilograms (94 000 pounds).

General layout of the mill is shown schematically in Figure 33—2.

Slab Conditioning and Storage Areas—Slabs for the mill range from 101.6 to 254 mm (4 to 10 inches) in thickness and 457 to 1943 mm (18 to 76½ inches) in width. Maximum length is 12.19 metres (40 feet) for 229-mm (9-inch) slabs and 10.7 metres (35 feet) for 254-mm (10-inch) thick slabs.

The slab-conditioning area is 38.1 metres (125 feet) wide and nearly 274.3 metres (900 feet) long, and is served by two electric overhead traveling cranes of 113-metric-ton (125-net ton) main-hoist capacity with 54-metric-ton (60-net-ton) auxiliary hoists, along with two gantry-type cranes that are each of 54-metric-ton (60-net-ton) capacity. The slab-storage area, 38.1 metres (125 feet) wide and 480.7 metres (1577 feet) long, has a capacity for storing 63 500 metric tons (70 000 net tons) of slabs and is served by two 113-metric-ton (125-net-ton) electric overhead traveling

cranes with 54-metric-ton (60-net-ton) auxiliary hoists and two 54-metric-ton (60-net-ton) capacity gantry-type cranes.

Slab-Heating Furnaces—A slab scheduled for rolling is placed by crane on one of three hydraulically operated slab-lowering devices of the mechanical-counterweight type that lowers the slab gently onto the slab-receiving table that is in a continuous line with the furnace-charging table. The combined length of these two tables is 158.5 metres (520 feet). The receiving and charging tables transport the slab to a position before the pushers of the proper furnace, with the long dimension of the slab at right angles to the direction of travel through the furnace.

Each heating furnace has a double pusher of the rack-and-pinion type, with a 4.4-metre (14-foot, 7-inch) stroke (Figure 33—3). Each individual pusher is operated by a 205-kilowatt (275-horsepower) direct-current motor through a worm reducer driving a single-reduction reducer, employing an air-operated clutch.

At the delivery or discharge end of each furnace, individual slabs are raised and carried out of the furnace by lifting arms called extractors that lower each slab gently onto the furnace-discharge table (Figure 33—4).

The four slab-heating furnaces, two of which are shown in Figure 33—4, all are of the five-zone recuperative continuous type. Each furnace is 32 metres (105 feet) long and wide enough to accommodate a single row of 12.2-metre (40-foot) long slabs or a double row of 4.9-metre (16-foot) to 6.1-metre (20-foot) long slabs. Each furnace is able to heat approximately 305 metric tons (336 net tons) of steel slabs per hour to a maximum

FIG. 33—3. One of two rack-and-pinion-type pushers at the charging end of one of the four slab-heating furnaces of the 2135-mm (84-inch) continuous hot-strip mill at Gary Works of United States Steel Corporation.

FIG. 33—4. Heated slab being withdrawn from one of the heating furnaces of the 2135-mm (84-inch) continuous hot-strip mill at the Gary Works of United States Steel Corporation.

FIG. 33—5. General view of No. 4 and No. 5 roughing stands of the 2135-mm (84-inch) continuous hot-strip mill at Gary Works of United States Steel Corporation.

of 1315°C (2400°F). The furnaces may be fired with natural gas, coke-oven gas or oil, and are equipped with automatic combustion and temperature controls. The maximum fuel rate per furnace is 844-million kilojoules (800-million Btu) per hour. Each furnace has two recuperators for preheating the combustion air that is supplied by motor-driven fans. Flue gas is conducted through underground flues to individual venturi ejector-type stacks, each equipped with a fan driven by a 670-kilowatt (900-horsepower) electric motor, and can be diverted through two waste-heat boilers.

Slabs move through the first 24.4 metres (80 feet) of the furnace on six water-cooled skids which permit heating both tops and bottoms of the slabs. The last 7.6 metres (25 feet) of travel is through the soaking zone that has a solid flat bottom where the slabs are brought to uniform temperature throughout.

Slabs are withdrawn from the furnaces at temperatures ranging from 1095 to 1315°C (2000 to 2400°F), depending upon the grade of steel and intended thickness of the finished product. They are carried to the entry end of the continuous hot-strip mill by the 109.7-metre (360-foot) long furnace-discharge roller table; this table is reversible so that, when necessary, slabs can be returned to storage by a transfer car operating on tracks parallel to the centerlines of the furnaces (Figure 33—2).

Roll Stands—Major operating units of the 2135-mm (84-inch) continuous hot-strip mill are, in succession, a two-roll vertical edger, a two-high roughing scalebreaker, two two-high roughing stands (the second having a vertical edger), three four-high roughing stands with vertical edgers (Figure 33—5), seven four-high finishing stands (Figure 33—6) and three downcoilers. Information relating to the main drives,

roll sizes, speeds, and so on, of these units is summarized in Table 33—III. The building housing the hot-strip mill is 32 metres (105 feet) wide and nearly 762 metres (2500 feet) in length. It is served by three 90-metric-ton (100-net-ton) electric overhead traveling cranes with 54-metric-ton (60-net-ton) auxiliary hoists.

The vertical edgers that operate in conjunction with the last four of the five roughing stands are each driven by a 112/224-kilowatt (150/300-horsepower) direct-current motor. The two roll-traverse motors that position the vertical rolls sideways are each a 32-kilowatt (35-horsepower) direct-current motor.

All of the rolling stands of the continuous hot-strip mill are equipped with hydraulic roll-balancing equipment.

Motor-driven side guides are provided on the approach tables at the entry sides of the vertical edger and roughing scalebreaker, and between the second and third roughing stands, third and fourth roughing stands, and following the fifth roughing stand.

A width gage and an X-ray thickness gage follow the fifth roughing stand.

The holding table following the roughing stands is equipped with three reject-cobble pushers of the rack-and-pinion type for removing rejected rolled material from the table. A cobble shear of the downcut type is provided for cutting up rejected material.

A rotary-type crop shear is located at the end of the holding table, adjacent to the finishing descaler that immediately precedes the first of the seven finishing stands. This shear is equipped with retractable guides.

Motor-driven entry guides are provided for each of the seven finishing stands. Hydraulically operated delivery guides are located between successive finishing stands and following the seventh finishing stand.

FIG. 33—6. The seven four-high finishing stands of the 2135-mm (84-inch) continuous hot-strip mill at Gary Works of United States Steel Corporation. The runout table with laminar-flow cooling sprays is in the right foreground.

Table 33 — III. Main-Drive Power Ratings, Roll Sizes, Etc. of the 2135-mm (84-Inch) Hot Strip Mill at Gary Works of United States Steel Corporation.

Stand	Work Rolls mm	Work Rolls in.	Back-Up Rolls mm	Back-Up Rolls in.	Distance m	Distance ft & in.	kW	hp	Voltage	rpm	Type	Gear Ratio	m/s	ft per min	Roll rpm	Screwdown No.	Screwdown kW	Screwdown hp
Vertical Edger	1163	46					1119	1500	4160	257	a-c sync.	13.367:1	1.17	231	19.22	2	37/75	50/100
Roughing Scale-Breaker	1219	48			16.815	55' 2"	3357	4500	13800	450	a-c sync.	25.676:1	1.12	220	17.52	2	56/112	75/150
Roughing Stands																		
No. 1	1219	48			18.745	61' 6"	5222	7000	13800	450	a-c sync.	25.7:1	1.12	220	17.50	2	56/112	75/150
No. 2[a]	1219	48			29.312	96' 2"	5222	7000	13800	450	a-c sync.	17.293:1	1.66	327	26.08	2	56/112	75/150
No. 3[a]	1067	42	1524	60	41.504	136' 2"	7460	10000	13800	450	a-c sync.	12.375:1	2.03	399	36.3	2	75	100
No. 4[a]	1067	42	1524	60	64.364	211' 2"	7833	10500	700	200/430	d-c	6.793:1	1.64/3.54	323/696	29.4/63.3	2	75	100
No. 5[a]	1067	42	1524	60	10.262	33' 8"	7460	10000	13800	450	a-c sync.	6.793:1	3.69	727	66.2	2	75	100
Finishing Stands																		
No. 1	724	28.5	1524	60	138.608	454' 9"	7833	10500	700	125/295	d-c	2.932:1	1.62/3.81	318/750	42.6/100.6	2	75	100
No. 2	724	28.5	1524	60	5.486	18' 0"	7833	10500	700	125/295	d-c	1.841:1	2.54/6.00	500/1181	67.8/160	2	75	100
No. 3	724	28.5	1524	60	5.486	18' 0"	7833	10500	700	125/295	d-c	1.2996:1	3.64/8.60	716/1693	96/227	2	75	100
No. 4	724	28.5	1524	60	5.486	18' 0"	7833	10500	700	125/295	d-c	Direct drive	4.73/11.18	932/2201	125/295	2	75	100
No. 5	724	28.5	1524	60	5.486	18' 0"	7833	10500	700	200/430	d-c	Direct drive	7.58/16.30	1492/3208	200/430	2	75	100
No. 6	724	28.5	1524	60	5.486	18' 0"	7833	10500	700	200/480	d-c	Direct drive	7.58/18.19	1492/3581	200/480	2	75	100
No. 7	724	28.5	1524	60	5.486	18' 0"	5222	7000	700	200/430	d-c	Direct drive				2	75	100
Coilers (3)							746 (ea.)	1000 (ea.)	500	400/1200	d-c	1:1.314	9.93/21.41	1954/4215	262/565	2	75	100

(a) A vertical edger is provided on the entry side of this stand; rolls are 838-mm (33 inches) in diameter with a 381-mm (15-inch) face, driven by a 112/224-kilowatt (150/300-horsepower) direct-current motor.

(b) In addition to the main-drive motors listed in the body of the table (totaling 92 130 kilowatts or 123 500 horsepower), there are auxiliary direct-current drive motors totaling 14 455 kilowatts (19 375 horsepower); auxiliary alternating-current drive motors totaling 21 150 kilowatts (28 355 horsepower); and miscellaneous drives for cranes, pumps, etc., totaling 19 820 kilowatts (26 570 horsepower) for a grand total of 147 700 kilowatts (198 000 horsepower).

(c) These are all direct-current motors.

A looping roll is provided between successive finishing stands to control tension on the strip between stands.

A hydraulic roll-contour control device is installed on each of the last three finishing stands to assist in controlling edge-to-edge uniformity of strip thickness.

Two X-ray gages (used alternately) and a width-measuring gage follow the last finishing stand.

All four-high stands of this continuous hot-strip mill (both roughing and finishing) have equipment for automatic changing of the work rolls. These rolls are handled on hydraulically operated side-shift cars. The back-up rolls for the four-high stands are handled on sled-type transfer units during roll changing.

Oil-film-type roll-neck bearings are used on the back-up rolls of all ten four-high stands in the mill. The work rolls of these stands are equipped with anti-friction-type roller bearings.

Runout Table—The 139.0-m (456-foot) long runout table following the finishing stands is provided with laminar-flow cooling jets above and below the rollers for rapid cooling of the rolled strip to the desired temperature for coiling. This table also carries the strip to the coilers.

Coilers—Following the runout table are three coilers, each driven by a 746-kilowatt (1000-horse-power), 500-volt direct-current motor. Expanded diam-

eter of the mandrels of these identical coilers is 724 mm (28½ inches). The three blocker rolls of each coiler are 559 mm (22 inches) in diameter. These coilers can produce coils having a 2134-mm (84-inch) maximum outside diameter, weighing a maximum of 33 100 kilograms (73 000 pounds). By substituting a 762-mm (30-inch) mandrel, coils of 42 600 kilograms (94 000 pounds) can be handled. Air-operated pushers remove the coils from the mandrels onto coil tilters (Figure 33—7) which place coils on end onto the hot-coil conveyor.

The hot-coil conveyor that carries finished coils away from the mill is 1189 metres (3900 feet) long and operates at a lineal speed of 0.15 metre per second (30 feet per minute).

Computer Controls—The computer controlled areas of the mill are the reheat furnaces, roughing mill, finishing mill, and runout-table cooling. Each area has individual controls which are coordinated with all other controls to produce a prime coil with the desired target thickness, width, and temperature. The following specific control functions are employed.

Furnace tracking function monitors the slabs from furnace entry to furnace exit. Location and identity information are available for use by the reheat furnace control and downstream functions. Furnace heating data are also tracked as generated.

FIG. 33—7. Coil of hot-rolled strip being removed from a coiler on the 2135-mm (84-inch) continuous hot-strip mill at Gary Works of United States Steel Corporation. (This is not a production coil; its condition is the result of having been used repeatedly in testing the equipment during installation and adjustment.)

Reheat furnace control calculates temperature setpoint references for the various furnace zones to achieve the target slab temperature at the furnace exit.

Mill tracking maintains the identity and location of all slabs as they progress from the reheat furnaces to the coilers. Mill tracking provides the signals to initiate sendouts to the mill, scans variables, and changes displays as the slab progresses through the mill.

Mill pacing maintains the gap between slabs at the finishing mill entry. Control signals are generated for extraction of slabs from reheat furnaces on the basis of maximum production rates available in all mill areas. Predicted rolling times for future slabs are transmitted to the reheat furnace control for estimating the heating time available for each slab.

Production logging records data necessary for analysis and evaluation of mill operation and production quality through coil quality logs, mill delay logs, turn summary logs, and statistical quality information logs.

The roughing mill setup distributes the total required reduction from entry slab thickness to rougher delivery thickness in accordance with desired drafting practices and mill mechanical and electrical constraints. The computer calculates the screw settings for all roughing stands and the speed reference for the close-coupled d-c rougher.

The edger setup establishes finished strip width by calculating edger drafts to achieve the desired slab width prior to transfer to the finishing mill. The computer also calculates the references for the vertical scale breaker, sideguides, and the width gage at the exit of the roughing mill.

The finishing mill setup calculates the screw settings, stand speeds, and interstand looper tension references required to deliver on-gage strip at the exit of the mill.

The finishing temperature setup determines the mill threading speed required to bring the head end of the strip out of the finishing mill at the required finishing temperature.

The finishing temperature control calculates the acceleration rate required to maintain the desired finishing temperature throughout the full length of the coil and modifies this rate in response to finishing temperature measurements. This control also initiates mill slowdown to achieve the desired tail and exit speed.

Shape setup determines the final load distribution array of stand forces in the finishing mill to produce a target strip crown. When the target crown cannot be met because of a constraint on flatness, the target crown is modified so that a flat product is always rolled.

Coiling temperature setup selects the number of laminar flow spray headers to be turned on to achieve the desired head end coiling temperature. Coiling temperature control turns sprays on/off in response to changes in finishing temperature and strip speed, to maintain coiling temperature within desired limits over the length of the strip.

Lubricating Systems—Six separate circulating-oil systems automatically serve the oil-lubrication needs of the 2135-mm (84-inch) continuous hot-strip mill. Three of these systems supply lubricant to the oil-film bearings on the back-up rolls of the four-high mills; the other three systems meet the general need for automatic oil lubrication of other operating units.

Grease requirements are met by several centralized greasing systems strategically located throughout the mill.

Motor Room—The motor room is 21.3 metres (70 feet) wide and nearly 396 metres (1300 feet) in length. It is served by an electric overhead traveling crane of 72 metric tons (80 net tons) capacity. The main mill drives with their reduction gears, control equipment, switchgear, and so on, are located in this room. The room has a recirculating-type ventilating system that controls the pressure and temperature of the air in the room and is equipped to filter all make-up air.

METALLURGY OF HOT-ROLLED COILS

Wide hot-rolled coils from a modern continuous mill may be used in the as-rolled condition, in which case it is referred to as "hot-rolled sheet," with or without the application of such auxiliary treatments as pickling, shearing and flattening. When produced as hot-rolled breakdowns for cold reduction, it is pickled in coil form, cold reduced as much as 90 per cent of its original thickness, heat treated and further processed to cold-rolled sheet, strip, black plate, or the various coated sheet-mill and tin-mill products. In any case the metallurgical requirements of the great bulk of the product are relatively simple and lend themselves to best operating conditions on the hot-strip mill and subsequent processing units.

The last hot-rolling operation (in the last finishing stand) should be conducted above the upper critical temperature on virtually all continuous hot-mill flat-rolled products. Such a practice permits the rolled steel to pass through a phase transformation after all hot work is finished and produces a uniformly fine, equiaxed ferritic grain throughout all portions of the steel. For the low-carbon steels generally used, proper finishing temperatures will have been attained when the apparent product temperature emerging from the last rolling stand is over 845°C (1550°F). This finishing temperature is practical over most thicknesses rolled on most modern mills at normal maximum rolling speeds.

If part of the hot rolling is conducted on steel which already has transformed partially to ferrite, the deformed ferrite grains usually will recrystallize and form patches or layers of abnormally coarse grains during the self-anneal induced by coiling at the usual temperatures of 650° to 730°C (1200° to 1350°F). Such a structure is more likely to occur at the surface of the product, which is colder than the interior during rolling. Very thin hot-rolled material, inadvertently finished far below the upper critical and coiled too cold to self-anneal, may retain microstructural evidence of hot-working. Neither condition is suitable for some types of severe drawing applications; both may be corrected by normalizing the sheet.

A special case occurs in the steels of the so-called "pure-iron" or "enameling sheet" compositions, in which the sum of the carbon and manganese contents may be well under 0.10 per cent. Such compositions often exhibit a hot-short temperature range between 900° to 1035°C (1650° to 1900°F), and normal hot rolling in that range may produce deep cracks on the edge of the product. Accordingly, it is the practice on many mills to complete the roughing operations above the hot-short range, to allow the steel to cool through the range by holding it on the conveyor table between the last roughing stand and the finishing train, and to resume rolling by passing the product into the finishing train below the hot-short range. By this practice it is impossible to finish above the upper critical temperature of these steels.

The runout table following the last rolling stand of most hot-strip mills is long enough and equipped with enough quenching sprays to cool the maximum thickness of rolled product up to 335°C (600°F) below the finishing temperature before coiling. In addition, some mills have auxiliary tables or holding beds which allow single-thickness cooling to a take-off temperature of 260°C (500°F) or lower. The cooling practice employed largely determines the metallurgical properties of the steel, its suitability for further processing and its final applicability to the intended use.

On hot-rolled products properly finished above the upper critical temperature, a uniform microstructure has been established and the runout cooling practice determines the carbide characteristics and, to some extent, the ferrite grain size. The self-annealing effect of a large mass of steel coiled at around 730°C (1350°F)

produces considerable carbide agglomeration, a coarse ferrite grain and a soft, ductile sheet. Coiling around 650°C (1200°F) yields a fine, dispersed spheroidal carbide in a finer ferrite matrix, resulting in a somewhat harder sheet, which still retains excellent ductility. Even more drastic quenching produces various transformation states of carbide, down to and including martensite. For most low-carbon steel made either for use as hot-rolled sheet or as breakdowns for subsequent cold reduction, coiling temperatures of 650° to 705°C (1200° to 1300°F) are employed; this range provides optimum uniformity of mechanical properties without excessive scale formation or over-annealing. Steels of 0.15 to 0.30 per cent carbon content and alloy steels often are coiled at lower temperatures to attain higher strength levels.

As most heat treatment after cold reduction is carried out below the lower critical point, the cold-reduced, box-annealed microstructure usually bears a relationship to the microstructure of the hot-rolled material. In the case of the aluminum-killed, cold-rolled deep-drawing sheet, coiling temperatures under 650°C (1200°F) are employed when rolling breakdowns on the continuous hot mill to provide the best drawing properties in the finished cold-reduced, box-annealed sheet.

SECTION 5

OXIDE REMOVAL

(Pickling and Shot Blasting)

Necessity for Removal—The presence of oxide (scale) on the surface of strip, sheet, or breakdowns, is objectionable when they are to be processed further. For example, the oxide must be removed and a clean surface provided if satisfactory results are to be obtained from the hot-rolled sheet or strip in any operation involving deformation of the material. If the sheet is for drawing applications, removal of the oxide is essential, as its presence on the steel surface tends to shorten die life, cause irregular drawing conditions and destroy surface smoothness of the finished product. Oxide removal is also necessary if the sheet or strip is to be used for further processing involving coating in order to permit proper alloying or adherence of metallic coatings and satisfactory adherence when a non-metallic coating or paint is used.

In the production of cold-reduced steel sheet and strip, it is necessary that the oxide resulting during hot rolling the steel slab to breakdown form be removed completely before cold reduction to prevent lack of uniformity and eliminate surface irregularities.

Types of Oxide—The term "oxide" as used here refers generally to the chemical compounds of iron and oxygen formed on the surface of the steel by exposure to air while the metal is at an elevated temperature. "Scale" is specifically the oxidized surface of steel produced during heating for working and during hot working of steel. Hence, the oxide produced on steel surfaces in hot-rolling processes is known as **mill scale**. Chemical compounds thus formed are iron oxides FeO,

Fe_2O_3 and Fe_3O_4.

The mechanism whereby mill scale is formed generally is considered to be of a dynamic nature, whereby alternate formation and reduction of the higher oxides of iron occur. Fe_2O_3 is formed first and then reduced successively to Fe_3O_4 and FeO by the availability of iron. Additional Fe_2O_3 is formed at the atmosphere-surface interface and the process becomes continuous; the final result is a scale composed of layers richest in oxygen at the scale surface and richest in iron at the metal surface. At elevated temperatures, FeO, the layer next to the steel, constitutes about 85 per cent of the scale thickness, Fe_3O_4 about 10 to 15 per cent and Fe_2O_3 about 0.5 to 2 per cent. There is evidence, too, of a molecular or ionic diffusion process involving oxygen moving inward and iron moving outward through the scale. During slow cooling of coils of hot-rolled strip, most of the FeO is transformed to Fe and Fe_3O_4 and the latter oxide is predominant after cooling.

The rate of oxide formation is dependent on the temperature, composition and physical characteristics of the steel, and temperature, character and rate of flow of the atmosphere, as well as the length of time the steel is exposed to oxidizing conditions. Figure 33—8 shows graphically the oxide development at temperatures of 535° to 1205°C (1000° to 2200°F) for a low-carbon steel exposed for two hours to atmospheres of oxygen, dry air, carbon dioxide and water vapor. It is noted that rate of scaling increases uniformly with temperature when water and carbon dioxide atmospheres contact the steel surfaces and that interruptions occur in the rate of scaling when oxygen and dry air are employed. These interruptions are attributed to blistering of the scale which leads to pitting of the undersurface.

Figure 33—9, in which scale loss is plotted against time for samples of low-carbon steel exposed to carbon dioxide, air and oxygen atmospheres, shows increasing loss for an increasing time at a given temperature. Figures 33—8 and 33—9 also show relative oxidizing activity of the four atmospheres, with carbon dioxide the least oxidizing and dry air, oxygen and water vapor more strongly oxidizing in the order named.

PRINCIPLES OF PICKLING

Pickling is the process of chemically removing oxides and scale from the surface of a metal by the action of water solutions of inorganic acids. While pickling is only one of several methods of removing undesirable surface oxides, this process is used most widely in the manufacture of sheet and tin mill products, due to comparatively low operating costs and ease of operation. Considerable variation in type of pickling solution, operation and equipment is found in the industry. Among the types of pickling equipment may be mentioned the batch picklers, modified batch, semi-continuous and continuous picklers.

The reaction occurring when steel or iron materials are immersed in dilute inorganic acid solutions includes the solution of metal as a salt of the acid and the evolution of hydrogen. Steel pickled in dilute sulphuric-acid solutions is an example of this reaction, with the end products of reaction being ferrous sulphate and hydrogen. Adherent films of oxides are undermined by the acid attack through the previous scale on the base metal. FeO is not dissolved as readily as the steel, but does have higher reaction rates than Fe_2O_3 and Fe_3O_4, both of the latter being soluble very slowly in the acid. When oxides are dissolved in acids, the ferrous salt and water are products of reaction. Ferric sulphate is formed first and then is reduced to ferrous sulphate by the free hydrogen. Sulphuric-acid baths rarely contain significant amounts of ferric sulphate since this compound is unstable in the presence of reducing agents. Certain metals, such as copper, chromium and nickel, retard the rate of pickling when they occur in the steel base, since the scale bearing these alloying metals inhibits acid attack. Silicon and aluminum form refractory-type oxides, which in turn lower the solubility rate of the oxide in the acid.

The rate of pickling is affected by numerous variables, including the aforementioned steel-base constituents and type of adherence of oxide to be removed. Solution temperature and concentration, ferrous sulphate concentration, agitation, time of immersion and presence of inhibitors influence the rate of acid attack. While the rate of pickling increases in direct proportion to the concentration of the acid from zero to 25 per cent by weight, the influence of temperature is much more pronounced. For example, in 15 per cent sulphuric acid an increase in temperature over the range 21°C to 99°C (70° to 210°F) doubles the pickling rate for each rise of 8° or 11°C (15° or 20°F) in temperature. Rate of solution of iron at 82°C (180°F) is

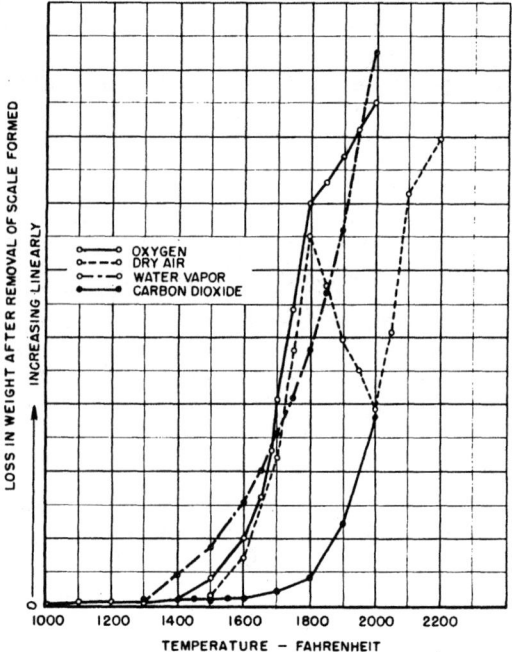

FIG. 33—8. Comparative effects of temperature and atmosphere on the scaling of plain carbon steels, all exposed for a constant time.

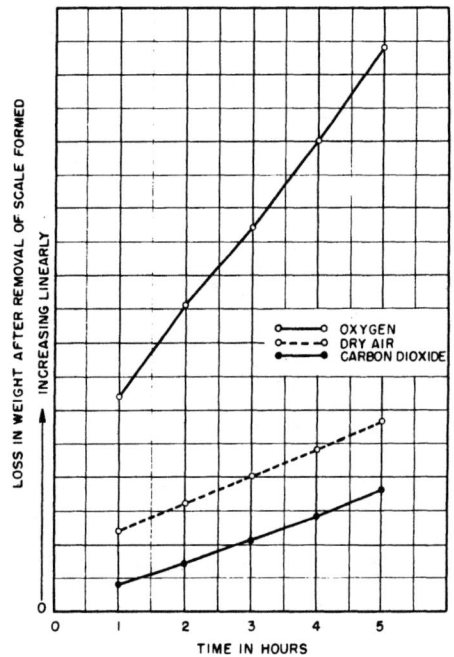

FIG. 33—9. Comparative effects of time of exposure and nature of the atmosphere on the scaling of plain carbon steels at 825°C (1520°F).

about five times the rate at room temperature. The trend in recent years in batch pickling of hot-rolled steels is to maintain temperatures at 65° to 79°C (150° to 175°F) if possible and to increase or decrease pickling activity with adjustment of the concentration, thus affording some savings in fuel for heating and avoiding decomposition of the inhibitor. Acid concentration is varied over a relatively wide range in the industry, dependent on the amount of pickling required to prepare the surface for the succeeding process. Much higher concentrations are used in continuous pickling methods, as the desired surface must be secured in the shortest possible time; hence, temperatures are maintained at 93° to 104°C (200° to 220°F) and concentrations at 12 per cent to 25 per cent. These specific examples pertain to the commonly-used sulphuric-acid baths. The required concentration is also a function of the kind of acid used; for example, hydrochloric acid can be used in concentrations of 5 per cent to 50 per cent. The maximum concentration of hydrochloric acid in continuous picklers is about 15 per cent in the fourth, or exit, tank section.

The retarding or inhibiting effect of ferrous sulphate is recognized widely and provisions are made in every pickling operation for adequate control of the salt build-up in the bath. Continued use of the bath without replenishment results ultimately in complete ineffectiveness of the solution. It is usually considered good practice to permit build-up of ferrous sulphate to 25 per cent and work the bath until the free acid content is reduced to less than 5 per cent.

Agitation of either the work or the bath saves time, metal and acid and is a practice used widely throughout the industry. Several methods have been adopted, such as raising and lowering the work in the bath, agitating the bath with plungers or circulatory systems or passing the work through the bath horizontally.

With hydrochloric acid, much higher concentrations of the ferrous salt can be tolerated without retarding the pickling process. Pickling rate is increased by increases in acid concentration and/or temperature.

Inhibitors—The effect of inhibitors has been known and used commercially for years. An inhibitor is any substance added to a solution that inhibits or lessens acid attack on the steel itself, while permitting preferential attack on the iron oxides. Originally such substances as wheat bran were used but in recent years complex synthetic organic chemicals have been manufactured for this purpose. The principal requirements of an inhibitor are that it must be effective in very low concentrations, its effectiveness must be constant at all bath temperatures and concentrations, it must not leave an oily or otherwise harmful film on the steel, and it must be economical for use on the basis of cost per ton pickled. Many inhibitors promote foaming to restrict acid and heat losses from the bath so blanketed.

Wetting agents, while often constituting a supplementary ingredient to inhibitors, in most cases are somewhat of an inhibiting agent in themselves and are used principally to improve rinsing and wetting of the surfaces. Wetting agents are organic compounds that lower the interfacial tension between the steel and the liquid. Some improvement in pickling rate is believed to result from their use and improvement in rinsing is definite.

Hydrochloric-Acid Pickling—Starting in 1964, numerous steel plants have changed from sulphuric to hydrochloric acid for pickling. The first major steel

plant to change did so to gain needed additional capacity from their old continuous picklers. However, subsequently, the relative cost of the two acids changed drastically, with a scarcity of sulphur increasing the price of sulphuric acid, while generation by the chemical industry of increasing quantities of hydrochloric acid as a by-product in the production of vinyl and polyvinyl chlorides by the ethylene chlorination process reduced the price of hydrochloric acid. As a result, approximately two-thirds of the steel pickled for sheet and tin mill products now is pickled in hydrochloric acid.

In addition to the economic advantage which accelerated the change from sulphuric to hydrochloric acid, and which is the major reason for it, users have identified the following benefits:

(1) Faster pickling. Hydrochloric-acid pickling is 2.5 to 3.5 times as fast as sulphuric-acid pickling at comparable concentrations and temperatures, except in the case of some steels such as silicon steel or some high-strength low-alloy steels where the difference is less significant.

(2) Cleaner pickling. Hydrochloric acid will dissolve the higher oxides on the surface of hot-rolled strip and FeO under these oxides with or without breaking of the scale. It will also dissolve and remove smut, or the rolled-in finely divided iron and/or magnetic iron oxide. The brighter hot-rolled and pickled sheet is preferred by some customers. Also, contamination of the rolling solution in subsequent cold-reduction operations is reduced.

(3) Lower acid consumption and greater utilization of the acid. As continuous-pickling-line speeds have been increased, increasing acid-consumption rates per ton of steel pickled have occurred at those mills using sulphuric acid, with consumptions varying between 15 and 22.5 kilograms (30 and 45 pounds per net ton) of acid per ton of steel pickled with sulphuric acid. By comparison, hydrochloric-acid consumption is about 5 to 7 kilograms of acid per metric ton of steel pickled (10 to 14 pounds per net ton). Also, hydrochloric acid, because of its fast reaction time, can be almost completely utilized. Approximately 88 per cent of hydrochloric acid is actually consumed in continuous pickling whereas only 53 per cent of equivalent sulphuric acid is consumed. Spent hydrochloric acid normally contains about 0.5 per cent of acid, whereas spent sulphuric acid will contain about 8 per cent acid. This reduces by one-third or more the volume of spent acid requiring disposal.

(4) Less steam consumption and reduced quantities of waste pickle liquor. Because hydrochloric acid pickles steel effectively at 71° to 77°C (160° to 170°F) as compared to 99° to 104°C (210° to 220°F) used for sulphuric-acid pickling baths, steam consumption is reduced about 40 per cent. This, in turn, reduces the steam condensate in the bath and contributes to the reduction of the amount of waste pickle liquor produced and the extent of the operations required for its disposal. However, operation with hydrochloric acid at about the same temperatures as used with sulphuric acid does permit use of much lower concentrations to achieve similar pickling rates.

(5) Scale breaking and temper rolling are necessary

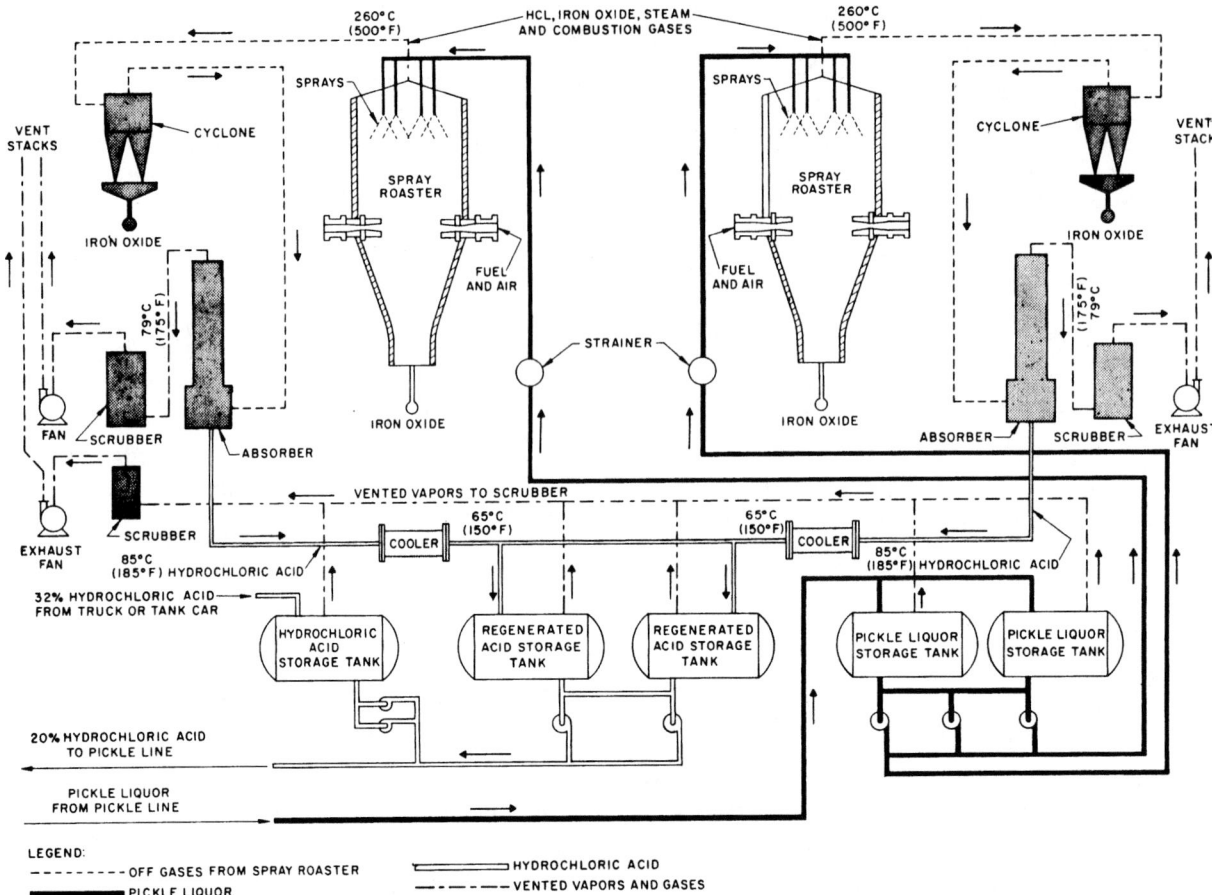

FIG. 33—10. Simplified flow sheet of a process for regenerating spent hydrochloric acid by treatment of pickle liquor from a continuous pickling line. Pickle liquor is sprayed into the combustion chamber of the spray roaster. Most of the iron oxide formed in the reaction is recovered from the bottom of the roaster; iron oxide in the off gases is removed in a cyclone. Hydrochloric acid is recovered from the gases in an absorber and returned to the pickling line. (Courtesy, Dravo Corporation.)

for high-speed pickling in sulphuric acid, but they also increase the attack of the acid on the base metal, particularly in the event of a line stoppage, resulting in overpickling. Overpickled strip results in lower yield and can cause difficulties in the cold-reduction operation.

(6) Greater versatility. For hot-rolled and pickled steel sheet products, it is frequently desirable to coil hotter on the hot-strip mill to achieve a softer steel with greater formability. However, at coiling temperatures over 650°C (1200°F), the amount of scale formed increases rapidly, as shown in Figure 33—8. Hydrochloric acid will remove this extra scale at pickling speeds much greater than are possible with sulphuric acid. In addition, temper rolling to break the scale imparts enough cold work to the steel to suppress the yield-point elongation (approximately 3 per cent), and also promotes strain aging which increases the hardness of the steel by 3 to 5 points on the Rockwell B scale. For hot-rolled sheet for severe forming operations, temper rolling in the pickling operation can be omitted and good pickling still achieved with hydrochloric acid.

The only significant disadvantage of hydrochloric acid is its volatility which is greater than that of

sulphuric acid. The volatility of hydrochloric acid varies as a logarithmic function of temperature, concentration, and the amount of ferrous chloride in the bath. Close control of these variables must be maintained in order to avoid hydrochloric-acid losses to the exhaust system. In addition, tank covers and seals must be maintained in good condition to avoid acid fume and mist leakage that will cause discoloration of product stored in the area and corrosion of equipment and buildings.

Spent Hydrochloric-Acid Disposal—Because the reaction products of the neutralization of hydrochloric acid waste pickle liquor are soluble in the treated liquor instead of being precipitated as in the case of the neutralization of sulphuric acid waste pickle liquor with lime, disposal of spent hydrochloric acid pickling solutions was recognized early as a problem.

One early solution to this problem was the regeneration of the spent acid by removing the iron and recycling the hydrochloric acid (Figures 33—10 and 33—11). This process has been successfully applied in some plants in the United States, Canada, Japan and Europe.

FIG. 33—11. Exterior view of an installation for regeneration of spent hydrochloric acid by treatment of pickle liquor, utilizing the process illustrated in Figure 33—10. (Courtesy, The Steel Company of Canada, Ltd.)

Another means of spent-acid disposal is the use of a deep well where geological conditions will permit, such as in the midwestern steel-producing district of the United States. The spent acid is filtered to remove solids larger than specific micrometre (micron) sizes and is then forced by gravity or pumped into underground strata about 1200 to 2100 metres (4000 to 7000 feet) below the surface.

Alternative methods of regeneration and neutralization and solids separation of spent hydrochloric acid are being tried here and abroad. To date, all effective methods have involved extensive facilities requiring large investments.

CONTINUOUS PICKLING LINES

With the advent of continuous cold-reduction mills, it was necessary to design and develop suitable equipment to remove the oxides resulting from the continuous hot-rolling operation and prepare the hot-rolled breakdowns for cold reduction in coil form. This operation is performed in a continuous pickling line (Figure 33—12 through 33—18). The primary function of a

continuous pickling line, as of other pickling processes, is the removal of oxide from the steel surface. This serves to promote maximum reduction with a minimum of power, to assure good roll life in the cold-reduction mills and to secure the increased surface density possible with cold work. Modern continuous pickling lines operate at speeds as high as 6.1 metres per second (1200 feet per minute).

The thickness of the oxide varies considerably on steel rolled on the hot-strip mill. Loose coiling permits greater atmospheric penetration into the wraps, with corresponding heavier oxide formation on the edge areas. Flexing of the steel in passing through the pickling-line uncoiler and temper mill breaks this scale or oxide film and permits more rapid attack by the acid bath.

The continuous pickler has other advantages or supplementary functions. The product of the hot-strip mill is subject to fluting (formation of creases when the steel is bent or otherwise deformed) due to lack of springiness. Continuous pickling lines usually are equipped with suitable apparatus for cold working the material so that severe local strains are eliminated and fluting

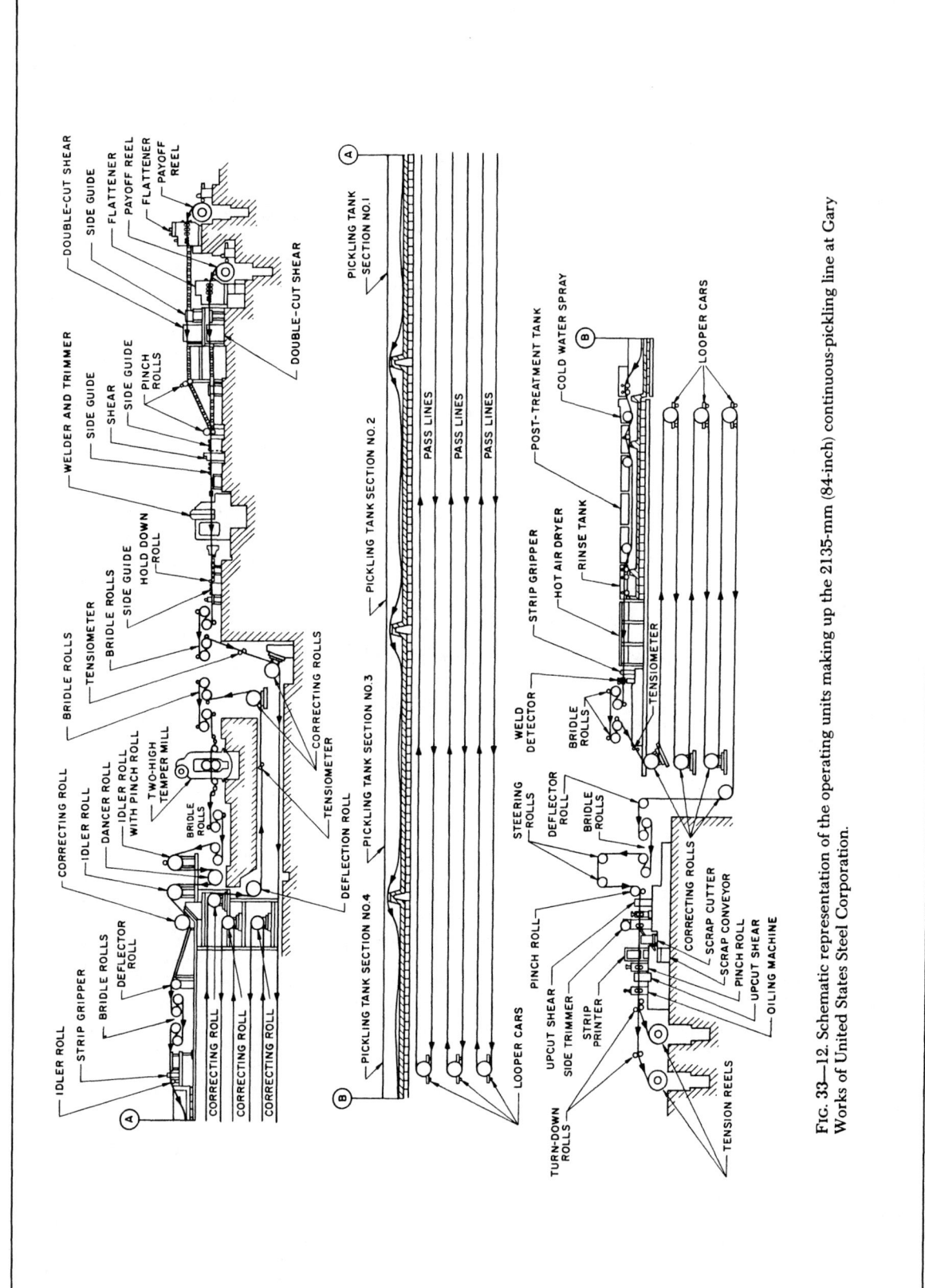

Fig. 33—12. Schematic representation of the operating units making up the 2135-mm (84-inch) continuous-pickling line at Gary Works of United States Steel Corporation.

FIG. 33—13. General view of the entrance and of a 2135-mm (84-inch) continuous-pickling line, showing strip entering the line from the payoff reel. The two coils in the left foreground are in position to be moved onto the reel. The coil visible at a lower level behind these two coils is in position to be moved onto the second of the two payoff reels with which the line is equipped (see also Figure 33—12). (Courtesy, Aetna-Standard Engineering Company)

largely is prevented. Another advantage is that they permit individual lengths to be joined into a single coil containing multiple lengths, often necessary because of coil length limitations of the continuous hot-strip mill. The result is a much longer continuous coil for the cold-reduction mills. The pickling line permits inspection of the steel for defects and suitability for the next operation, and finally oiling of the steel as a protection against rusting and as an aid to cold reduction.

At the coil entry end of a typical continuous pickling line are facilities for handling and charging coiled product into the line. These usually consist of conveyors on which the coils are placed in proper sequence by overhead cranes, upenders in cases where the coil is delivered with the axis vertical, and a motor-driven integrated buggy and hoist for placing the coil in the uncoiling or pay-off equipment. The primary cold-working equipment, called a "processor," or "flattener," integral with the uncoiling equipment, consists of a mandrel on which the coil is placed, a hold-down roll, and a series of smaller diameter rolls. After the coil is charged on the mandrel and the lead-end entered into the small diameter rolls, the hold-down roll is brought down and pressure applied to the material. This action alternately flexes the steel around the rolls, thus effectively "breaking" the surface scale into nu-

merous fine cracks, and increasing the available sub-oxide area for pickle attack. This flexing also cold works the steel enough to eliminate, in large part, the fluting tendencies of the hot-rolled steel. The group of small driven rolls immediately following the hold-down or breaker roll applies tension to the steel and also serves to straighten and flatten it. A stationary shear is located after the processor for the cropping and squaring of the coil ends for butt-welding or stitching.

While most of the pickling lines in the industry include stitchers for fastening coil ends for continuously processing hot-rolled product, many also have installed flash butt-welding as supplementary equipment. The main advantage of this method of joining coil ends is that it provides a joint which can be cold reduced, whereas the lapped and stitched joint cannot.

Following welding, the flash, or excess metal resulting from the upsetting action of the welder, is trimmed off by a cutter designed for the purpose. The looping pit is next in line and provides a continuous storage space for material to compensate for short delays at the charging end and to permit a uniform rate of travel through the acid tanks. The looping pit is usually 3 to 3.7 metres (10 to 12 feet) deep and 6.1 to 12.2 metres (20 to 40 feet) long, depending on equipment speed. Construction is generally of concrete with adjustable

FIG. 33—14. Another view, in the opposite direction, of the entry end of the 2135-mm (84-inch) continuous-pickling line shown in Figure 33—13 (see also Figure 33—12). (Courtesy, Aetna-Standard Engineering Company)

wood side guides for retaining the lapped-over material and preventing twisting. Water is kept in the pit to minimize scratching and increase wetting action in the first pickling tank.

In some continuous pickling lines, an auxiliary or secondary scalebreaker follows the looping pit, to break the scale even further than was achieved in the processor at the entry end, and thus increase the speed at which the line can be operated and still produce satisfactorily pickled strip. The secondary scalebreaker may be a machine similar to the entry-end processor, or it may be a two-high temper mill preceded and followed by a tension bridle at the entry and exit sides of the mill. Use of a two-high temper mill results in extension of the strip on the order of 3 per cent and increases its hardness 3 to 5 points on the Rockwell B scale.

The pickling zone consists of several individual acid-proof tanks located in a series, comprising an effective immersion length of about 76 to 91 metres (250 to 300 feet). While most lines have from three to five tanks, each about 21 to 24 metres (70 to 80 feet) long, some

modern lines have only one long tank, divided by weirs into four or five sections, thereby increasing effective immersion depth about 10 per cent to 15 per cent. The inside dimensions of these tanks have been more or less standardized at four feet in depth and about one foot wider than the maximum product width. A steel shell is used for support with layers of rubber bonded to the steel and the rubber is protected from abrasion by a lining of about 229 mm (9 inches) of silica-base acid-proof brick. For operating temperatures in excess of 93°C (200°F), a bakelite-base cement generally is used for bonding. In modern high-speed lines operating at 93° to 104°C (200° to 220°F), the brick facing gradually is eroded away, so that replacement is required after several years of operation. Occasionally, small leaks in the rubber lining and the steel tank require patching. However, if care is taken to prevent acid attack on the outside of the tanks, the tank assemblies may be regarded as permanent for the life of the line.

A few continuous-pickling lines have been built in which the pickling section consists of a multiple-pass

FIG. 33—15. Flash welder (right center) in operation on a 2135-mm (84-inch) continuous-pickling line. The pickling tanks in this view extend into the far upper left. Operating units between the welder and the first pickling tank can be identified in Figure 33—12. (Courtesy, Aetna-Standard Engineering Company)

vertical tower in which the moving steel strip is sprayed with a heated solution of hydrochloric acid. (Figure 33—18).

Following the acid tanks are rinsing tanks consisting of a cold-water spray rinse and a hot-water tank. The cold water rinses the acid carry-over from the steel. The hot-water rinse is a tank with an effective product immersion length of 4.5 to 6 metres (15 to 20 feet). This tank completes the rinsing and by warming the steel, promotes flash drying prior to entering the succeeding set of pinch rolls. Situated between the final rinse tank and the pinch rolls are one, two or three banks of hot-air dryers operating at low pressures. Pinch rolls at the exit end of the pickling tanks control the speed of product travel and, in conjunction with the pinch rolls which provide back tension at the entry end of the line, help to maintain the proper loops in the tanks.

The delivery end of the continuous pickling line has, in the order listed, a looping pit, pinch rolls, shear, oiler, recoiler and suitable supplementary equipment for conveying the finished product from the line. The pinch rolls preceding the shear are located so that product delivery to the shear is facilitated. Stitches are removed at this point, as well as short sections which inspection has shown to be inferior quality. Some lines are provided also with rotary side trimmers at the entry end or, more commonly, at the delivery end.

The description just given fits the majority of continuous picklers operating today, except for the fact that two-thirds of them now use hydrochloric instead of sulphuric acid. However, the most recent generation of continuous picklers are super lines which have changes and additional features as follows (see also Figures 33—12 through 33—17):

(1) High speed. Speeds of 5 metres per second (1000 feet per minute) or higher are maintained through the pickling tanks, with 10 metres per second (2000 feet per minute) speed on the entry and 6 metres per sec-

FIG. 33—16. Equipment at the exit end of No. 4 pickling tank of a 2135-mm (84-inch) continuous-pickling line. The line of pickling tanks extends into the far upper right of this view. The several operating units can be identified by reference to Figure 33—12. (Courtesy, Aetna-Standard Engineering Company)

ond (1200 feet per minute) on the exit end.

(2) Dual uncoilers and recoilers able to handle coils up to 55.3 kilograms per metre of width (1220 pounds per inch of width) and 2032 mm (80 inches) in outside diameter are provided.

(3) Entry and exit horizontal loopers in multiple tiers under the pickling tanks (Figure 33—12) permit continuous operation through the pickling tanks while the entry end is stopped for welding coils together, or while the exit end is stopped for making a transfer from one mandrel recoiler to the second one.

(4) One 110-metre (360-foot) pickling tank with three weirs forms four pickling sections and minimizes the number of tank ends and openings. Strip lifters are provided and some lines have strip threading devices in the tank covers. Water seals are provided for the tank covers and a fume exhaust and fume scrubber remove emitted gases and vapors. Automatic titration, control, and feed of acid are provided with the acid feed into either No. 3 or No. 4 (exit) tank sections. Double wringer rolls and air or water sprays are pro-

vided at the exit end of the pickling tank to minimize carryover into the subsequent cold- and hot-water spray and dip rinse tanks.

(5) Six large-diameter four-roll pulling and tension bridles are used instead of the smaller pulling pinch rolls. Twelve steering rolls with tracking sensors guide the strip through the line.

(6) Leaving the exit looper tiers, the strip passes through a strip-inspection device which permits inspection of both surfaces at high speed. Strip thickness is measured at both the entry and exit ends of the line and can be displayed on a recorder.

(7) Dual side trimmers and scrap choppers are provided to permit knife changes with a minimum of line delay.

(8) Television is used to monitor critical sections of the line and an inter-communications system connects operating stations.

Acid and water control on all present continuous picklers is important for quality and economic reasons. On the new super high-speed lines, pickling at rates up

FIG. 33—17. General view of equipment at the exit end of a 2135-mm (84-inch) continuous-pickling line, showing coil discharge conveyor at upper left, with coil-storage area in the far background (see also Figure 33—13). (Courtesy, Aetna-Standard Engineering Company)

to 270 metric tons (300 net tons) per hour and using hydrochloric acid, the acid and water control is very important. To assist the operator, automatic acid controls are available which monitor the HCl content in one or more tank sections of the line and automatically add acid to maintain a preset concentration. In a typical line, the acid is added in No. 3 or No. 4 tank section. The pickling solution then cascades counter strip travel through shallow channels cut on each side of the weirs between tank sections. Weir heights are decreased 25.4 mm (1 inch) per weir from the exit toward the entry end of the tank, and in a typical line would be 1016, 991 and 965 mm (40, 39 and 38 inches) in height.

Ferrous chloride is monitored by an automatic titration instrument and water is added as necessary to maintain the concentration between 15 and 18 per cent. A typical line controls this in No. 2 tank section where the build-up of ferrous chloride was found to be the highest.

Typical acid consumption is 5.5 kilograms of 100 per cent HCl acid per metric ton (11 pounds per net ton) of steel pickled. Waste pickle liquor averages 50 litres per metric ton (12 gallons per net ton) of steel pickled.

Maintenance of coil identity through the pickling operation is essential. Schedules are made out showing the coil number and other unit identification opposite a pickling sequence number. The coils are then charged into the line in the sequence listed on the schedule and unit identity is reassigned to the coil in the same sequence following pickling. Product delivered to the cold-reduction mills carries complete identity as to coil number, heat number, width and gage. Means of transporting the pickled coil to the storage area ahead of the cold-reduction mill varies from plant to plant, depending on the plant layout, nature of transportation available and the most economical method. Occasionally, long gravity-type conveyors are installed to deliver coils to a central point, from which a tractor or overhead crane stores the coils in areaways adjoining the cold-reduction mills. Slat-type conveyor-tractors also are employed and where the distances involved become greater, rail transportation is resorted to.

Inspection of the raw pickled product is carried on continuously at the exit end of the pickling lines. Each coil is inspected for surface and edge quality, width and gage. Some of the defects commonly causing rejection or diversion are as follows: slivers, cracked edges, laminations, off gage, off width, roll marks, underpickling, overpickling, handling damage and pitting.

Underpickling results when the steel has not had sufficient time in the pickling tanks to become free of adherent scale and occurs when acid concentration, solution temperatures and line speed are not balanced properly. Variations in the oxide and composition of the steel are also factors in underpickled product, as well as such factors as coiling temperature off the hot-strip mill and inadequate amount of cold working

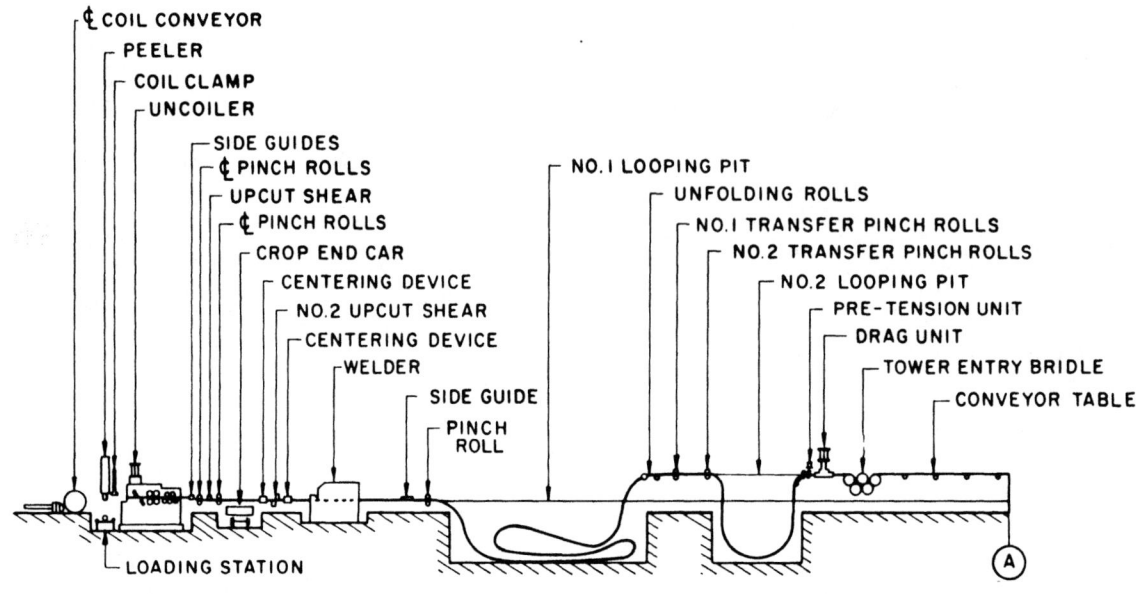

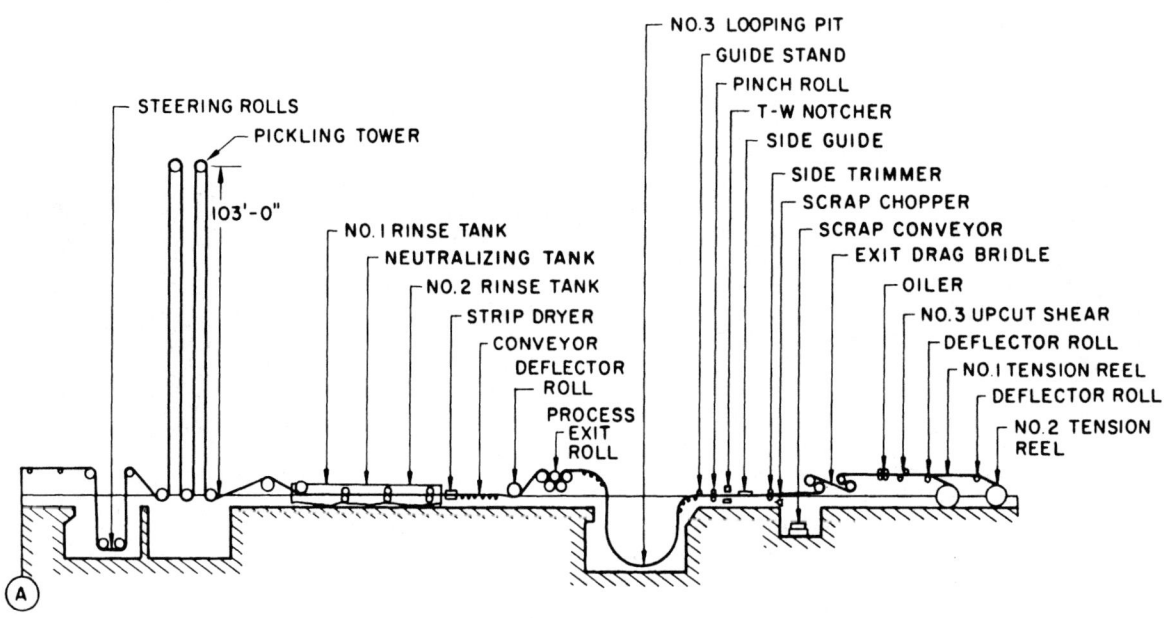

FIG. 33—18. Schematic representation of a continuous-pickling line, utilizing a two-pass vertical tower in which moving steel strip is sprayed with a heated solution of hydrochloric acid. (Courtesy, The Steel Company of Canada, Ltd.)

through the processor. Overpickling results from line delays which permit sections of the steel to remain in the acid too long. The presence of an inhibitor reduces iron loss, but when an inhibitor is not used, iron loss during a short delay period appreciably reduces thickness of the steel and raises the hazard of hydrogen embrittlement. Pitting is related to overpickling, the presence of non-metallic inclusions near the steel surface and to rolled-in scale, slag or a refractory substance. While overpickling is not common in continuous pickling operations, its occurrence does have a very serious effect on cold-reduction performance and surface appearance of the finished product. Product damage from handling or improper equipment adjustment can render the steel unsuitable for further processing.

Prior to recoiling, the pickled steel passes between a set of oiling rolls which cover both surfaces with a small amount of oil. The type of oil used to lubricate the steel, and protect it from rusting during storage and from scratching during handling, is determined by the type of lubricating system on the cold-reduction mill unit. Hence, palm oil diluted with light mineral oil, is applied to the steel at the pickling lines when a straight palm oil or a solution containing palm oil is used on the cold-reduction mill. Finally, the pickled and oiled product is recoiled on a conventional up-type or down-type coiler.

BATCH PICKLING

Pickling of steel sheets and other light-gage sheared lengths is performed with specialized equipment in which a batch of sheets are processed together. In general, there are two basic methods of batch pickling and several modifications of them, but the employment of agitation as a means of increasing pickling rate is common to both. One method is to employ a large wooden or acid-resistant concrete tank installed in the floor so that the top of the tank rises about 610 mm (2 feet) above floor level. The tank is divided into two sections by a wooden partition extending between about 150 and 250 mm (6 to 10-inches) from the tank bottom; one section is smaller than the other. A large float is inserted in the smaller compartment and the material to be pickled is placed in the larger side. Forcing the float down into the bath displaces acid solution in the smaller section and raises the level of the acid solution in the other compartment. The wood float is attached to a gear-driven crank, making the motion quite rapid, and effectively agitating the bath. Sheets are placed in a near vertical position on an acid-resistant metal rack, and pins inserted between every few sheets to facilitate exposure of all surfaces. The rack is then placed in the bath by an overhead crane and allowed to remain for several minutes. After sufficient pickling, the sheets are rinsed in a cold-water tank to wash off any iron-salt residue. Acid concentration is varied from 4 to as high as 12 per cent depending on the amount of surface etch required. Temperatures usually are maintained between 65° and 88°C (150° and 190°F).

In the second general type of batch picklers, the work is agitated instead of the bath. Equipment consists of a minimum of two, and more often three, tanks recessed into the floor and located at 90 degree steps to

each other on about a 4.5-metre (15-foot) radius. At the center is a vertical mast topped by a four-armed spider from which the holding racks are hung. The pickling sequence begins at the loading station, followed by a high-concentration acid bath and, in a two-tank system, succeeded by rinsing and unloading stations. In the three-tank system, the first pickle tank is followed by either a more dilute acid solution or a cold-water rinse, followed in turn by a hot-water rinse. Racks must be unloaded at the fourth station prior to loading. A variation of this method has the tanks installed in tandem, with the first tank containing the highly concentrated acid solution, followed by a very dilute acid or cold-water rinse and finally with a hot-water rinse. The racks are suspended from a sectionalized conveyor system which plunges the work in the bath, the section immediately above the tanks being separated from the return rack conveyor off to the side of the pickle tanks. In this manner, racks are never removed from the overhead conveyor system, except for maintenance, and the product flow assumes a more continuous aspect. When pickled sheets are furnished to the customer for use in that condition, it is necessary to force-dry them to minimize rusting. Equipment for drying consists of a short hydrochloric-acid bath, water spray rinse, high-volume V-type dryers, and pilers. Roller leveling frequently is included to obtain standard mill flatness and a rotary brush scrubber may be incorporated in the drying sequence.

Adjustments of the pickling bath for batch-type picklers is accomplished by adding raw acid after the water has been brought to the correct level. Sulphuric acid is used for low- and medium-carbon steels in concentrations ranging from 1½ per cent to 4 per cent for tin plate and from 4 per cent to 8 per cent for sheet products, with the temperature maintained between 65° and 88°C (150° and 190°F). During working of the solution the acid is depleted and fresh acid must be added periodically as indicated by suitable tests for free acid and ferrous sulphate. Inhibitors generally are used so that iron loss is low and the hazard of hydrogen embrittlement reduced to a minimum.

After pickling certain steel containing small amounts of copper or nickel, a "smut" or dark film appears on the sheet surface. As a rule, the film contains relatively high percentages of the alloying elements and a dilute solution of hydrochloric acid must be used to attain the necessary cleanliness.

SHOT BLASTING

There are two cleaning lines, actually called "pickle lines," in the United States that use grit abrasive blasting as the basic means for removing scale from strip. The entry and exit ends of these lines are identical with standard continuous pickling lines such as those already described under "Continuous Pickling." In the processing section, abrasive cleaning machines designed to remove all of the scale are substituted for the first two or three acid tanks of a standard line. One or two full-length pickling tanks follow the abrasive cleaning equipment to remove any fine particles of abrasive or scale that might have been left on the strip surface, and to remove some of the sharp ridges of steel caused by the thrown abrasive. These lines operate at relatively slow speed, but effectively clean strip passed through them.

In 1966, an abrasive-blast process called "No Acid Descaling" for cleaning hot-rolled strip and sheet for cold reduction was introduced. This process cleans in two stages with round steel shot and steel grit. The first blasting is designed to remove substantially all of the mill scale by the mass velocity impact shattering and abrasive action of the round steel shot, or more than 95 per cent of the mill scale. The second blasting with the steel grit is designed to remove the more tenacious imbedded scale, particles of work-hardened metal and amorphous iron from the surface of the steel by a cutting and scouring action of the angular sharp-edged steel grit. The steel strip passes horizontally through the apparatus.

A scale-free strip is presumably produced suitable for cold reduction with a surface finish of 3 to 3.3 micrometres (120 to 130 microinches) and surface hardness increased 4 or 5 points on the Rockwell B scale over the original hardness.

While this modification of shot blasting avoids the problem of waste pickle-liquor disposal and presumably has a total cleaning action, it is essentially a slow-speed, low-production type of operation, with the recognized problems of shot blasters. These include the high cost of shot and grit, the adjustments of blasting wheels and shot-flow rates required for changes in width, and the masking effect that occurs at high shot-flow rates as speed is increased. One unit operating in Canada has two stations, each of which consists of three four-wheel units to give a reported speed of about 1.5 to 2.5 metres per second (300 to 500 feet per minute).

SECTION 6

FINISHING OF HOT-STRIP MILL PRODUCTS

Hot-rolled product is sold in several forms, including as-rolled hot bands, or as various hot-rolled sheet products after appropriate finishing operations. Hot-rolled bands are normally supplied with no further processing except to correct irregularities in coiling, if necessary, or to secure tests from representative coils if the specification so dictates. Ends of coils are not cut back to conform to established width or thickness tolerances (except where testing is required). Hot-rolled sheet product in coils or cut lengths is provided with edges trimmed or untrimmed (mill edge) as specified, and may have additional processing including pickling and oiling, flattening, temper rolling, etc. Coil ends are cut back to assure conformance to established thickness and width tolerances.

Hot-rolled pickled and oiled product is processed through continuous pickling lines as described in the preceding section. Modern high speed pickling lines incorporate flexing rolls and/or temper rolling to facilitate the pickling process which also enhances the flatness somewhat over that of the as-rolled product. These lines are equipped with in-line side trimming and oiling, and therefore pickled coils are usually supplied as pickled and oiled, cut edge product, although dry (unoiled) or mill edge can be provided if required. In addition to processing coils to be sold as hot-rolled pickled and oiled product, these lines also supply all of the material to be further processed into cold-reduced flat-rolled product.

A hot-rolled strip finishing department is usually equipped with units for further processing of hot-rolled, or hot-rolled pickled and oiled coils for the designated finished product. These units include facilities to recoil, level, temper roll, side trim or slit, and/or shear to length as required for the product ordered.

Temper Rolling (Skin Passing) —Temper rolling or skin passing is a relatively light cold rolling operation (1 to 4 per cent reduction in thickness) used primarily to improve flatness, and to minimize surface disturbances such as coil breaks and fluting during forming, and to

alter the mechanical properties of the material.

Roller Leveling —Roller leveling is a flattening operation where the setup or sheet is passed between an upper and lower set of rolls offset and meshed as illustrated in Figure 33—19, causing the strip to be alternately flexed up and down. The rolls are adjusted so that the amount of flexing decreases as the strip travels towards the exit end of the leveler, with the final rolls performing chiefly a straightening operation. The material is slightly cold worked in the roller leveling operation, which retards fluting somewhat, but without much other effect on mechanical properties or thicknesses of the material.

Recoilers—Recoilers or coil preparation lines are used to remove off gage material from the head and tail of hot-rolled coils, and may also be equipped with side trimming and/or slitting capabilities. For slitting, circular slitting knives are packed on the slitter arbors with accurate spacers in between to establish the width of the slit material and to maintain the proper cutting clearances between the mating knife edges. Slitting lines may be either driven lines or pull-through lines, where the exit coiler provides all the power for slitting

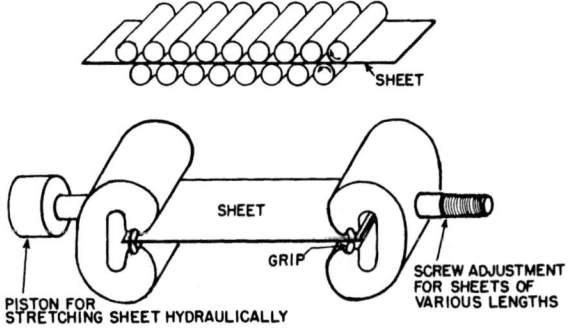

FIG. 33—19. (Above) Schematic representation of the disposition of rolls in a roller leveler. (Below) Diagrammatic sketch of the principle of operation of a stretcher leveler.

by pulling the strip through the slitting knives. Driven lines are powered by the slitting arbors and generally provide a wider range of product thickness capabilities. Slitters (and sometimes recoilers) generally have roller leveling equipment for improved flatness.

Shear Lines—Shear lines, or cut to length lines, consist of an uncoiler plus flattening equipment (leveler and/or temper mill) followed by a flying shear (light gages) or upcut shear and gaging table (heavier gages). Some heavy gage lines are capable of shearing plate thickness product hot-rolled in coil form.

Combination Lines—A number of processing lines have been designed as combination lines capable of performing several functions in one tandem line. These frequently include an uncoiler, leveling or flexing rolls, and a temper mill with a recoil mounted as in the temper mill unit. The strip can also be bridged over the coiler direct to a shearing facility in the line. This is usually followed by an additional roller leveling unit for further flattening capabilities after shearing to cut sheets. Temper mills installed in a shear line or combination line must be preceded and followed by a tension bridle.

Bibliography

Am. Iron and Steel Institute, Steel Products Manuals:
"Carbon Steel: Plates; Structural Sections; Rolled Floor Plates; Steel Sheet Piling."
"Carbon Steel: Semifinished for Forging; Hot-Rolled and Cold-Finished Bars; Hot-Rolled Deformed Concrete Reinforcing Bars."
"Carbon Steel Sheets."
"Cold-Rolled Carbon Steel Strip."
"Tin Mill Products."

American Society for Metals, Metals Handbook, Ninth edition, Vol. 5, Surface Cleaning, Finishing, and Coating, 1982, pp. 59-96.

Am. Society for Testing and Materials, ASTM Standards, Part I, Ferrous metals specifications, The Society, Philadelphia, Pa., (latest edition).

Angle, John E., Maximum production on hot strip mills. Am. Iron and Steel Institute Regional Technical Meetings, 1953; pp. 29-44.

Anon., Continuous strip pickle lines (mild steel) installed in North America since 1951. 33 Magazine 4, No. 5, May 1966, pp. 84-85.

Assn. Iron and Steel Engineers, The modern strip mill. Pittsburgh, Pa., 1941.

Auman, P. M., Griffiths, D. K., and Hill, D. R., Hot-Strip-Mill Runout Table Processing, Iron and Steel Engineer Year Book, (1967), pp. 678-685.

Cook, J. W., Hot-Strip-Mill Roughing Train Arrangements and Power Selection, Iron and Steel Engineer, (Dec. 1979), pp. 25-31.

Cullen, C. W. and Petrus, G. P., Generation III Hot-Strip-Mill Automation System, Iron and Steel Engineer, (Nov. 1979), pp 25-30.

Eibe, W. W., A New Economical Concept of Producing Hot-Strip and Plate, Iron and Steel Engineer, (Sept. 1981), pp. 25-36.

Ess, T. J., The Hot Strip Mill—Generation II, American Iron and Steel Engineer, 1970, Pittsburgh, Pa.

Gehman, Hugh, Pickling acids and inhibitors. Wire and Wire Products 41, No. 10, Oct. 1966, pp. 1609-10, 1612, 1703.

Greenberger, J. S., Developments in scalebreaking and continuous pickling lines. Iron and Steel Engineer 33, May 1956, pp. 69-79.

Held, J. F., Kruger, H. J., and Bucher, J. H., Effect of Hot-Rolling Practice on the Structure and Properties of Low-Carbon Steels, Mechanical Working and Steel Processing VIII, AIME, (1970), pp. 340-372.

Hindson, R. D., Metallurgical factors in the hot working and cleaning of hot rolled strip and their influence on the cold reduced annealed product. Proc. International symposium on annealing of low-carbon steel, pp. 9-17. Lee Wilson Engineering Co., Cleveland, Ohio, 1958.

Hudson, R. M., and Warning, C. J., Effectiveness of Organic Compounds as Pickling Inhibitors in Hydrochloric and Sulfuric Acids, Metal Finishing, Vol. 64, No. 10, (1966), pp. 58-61, 63.

Hudson, R. M., and Warning, C. J., Factors Influencing the Pickling Rate of Hot-Rolled Low-Carbon Steel in Sulfuric and Hydrochloric Acids, Metal Finishing, Vol. 78, No. 6, (1980), pp 21-28.

Hudson, R. M., and Warning, C. J., Effect of Strip Velocity on Pickling Rate of Hot-Rolled Steel in Hydrochloric Acid, Journal of Metals, Vol. 34, No. 2, (1982), pp. 65-70.

Hudson, R. M., and Warning, C. J., Effect of Strip Velocity on Pickling Rate of Hot-Rolled Strip in Sulfuric Acid Solutions, Metal Finishing, Vol. 82, No. 3, (1984).

Iron and Steel Institute (British), Special Report No. 67, Production of wide steel strip (1960).

Kenyon, A. F., The Reversing Hot-Strip Mill, Iron and Steel Engineer Year Book, (1953), pp. 410-418.

LeBas, J., Factors affecting the pickling rate of hot rolled steel strip. BHP Technical Bulletin, 2, No. 2, Sept. 1958, pp. 10-16. Broken Hill Proprietary Co., Ltd., Melbourne, Australia.

Lowey, John R., Hydrochloric acid pickling (tower pickler). Blast Furnace and Steel Plant 53, No. 10, Oct. 1965, pp. 948-952.

Lyle, David, The reversing hot strip mill—its place in the steel industry. Iron and Steel Engineer 33, April 1956, pp. 83-90.

Martin, E. D., Continuous strip pickling. Am. Iron and Steel Institute Yearbook, 1948.

Miltenberger, R. S., The use of hydrochloric acid in conventional pickling facilities. Blast Furnace and Steel Plant 53, No. 9, Sept. 1965, pp. 833-836.

Morgan, E. R., Dancy, T. E., and Korchynsky, M., Improved High-Strength, Low-Alloy Steels Through Hot-Strip-Mill Controlled Cooling, AISI Year Book, (1965), pp. 281-304.

Ride, John S., Analysis of operational factors derived from hot strip mill tests. Iron and Steel Engineer 37, Nov. 1960, pp. 77-90.

Roberts, W. L., Hot Rolling of Steel, Marcel Dekker, Inc., (1983), New York, N. Y.

Treasure, R. W., Continuous strip steel pickling. J. Iron and Steel Institute (British), 162, No. 2. (June 1949), pp. 201-212.

CHAPTER 34

Manufacture Of Cold-Reduced Flat-Rolled Products

SECTION 1

COLD-FINISHED FLAT-ROLLED PRODUCTS

The principal finished flat-rolled products are **flat bars, cold-rolled strip, cold-rolled sheets, and black plate**. With the exception of black plate (a product made of low-carbon steel), these products may be made from carbon, alloy, or stainless steels as required by the intended end use. The cold-finished flat-rolled products differ from each other principally in dimensions. In certain size and thickness ranges, surface finish and edge finish may determine their classification, as shown in Table 34—I.

While the several principal types of cold-finished flat-rolled products will be discussed briefly in this section, the remainder of the present chapter will be devoted principally to the manufacture of the carbon-steel flat-rolled products designated as cold-rolled sheets and black plate, produced by the cold reduction of hot-rolled coils sometimes referred to as "breakdowns," or

"hot bands," after the coils have been descaled by continuous pickling.

As discussed in other chapters, cold-rolled carbon-steel sheets are the base material for such coated products as terne coated sheets (Chapter 37), galvanized sheets (Chapter 38) and aluminum-coated sheets. Most black plate is subsequently coated with tin to produce tin plate (Chapter 36). A special variety of cold-rolled sheet is produced for porcelain enameling; some characteristics of the steel for enameling sheets were discussed in Chapter 33. Electrical (silicon-steel) sheets are discussed in Chapter 46.

COLD-FINISHED FLAT BARS

Cold-finished flat bars are commonly produced in the following dimensional ranges: 6.35 mm (¼-inch) and

Table 34—I. Product Classification By Size of Flat Cold-Rolled Carbon Steel.

Width		Thickness		
(mm)	(in.)	6.35 mm (0.2500 in.) and thicker	6.34 mm (0.2499 in.) to 0.361 mm (0.0142 in.)	0.358 mm (0.0141 in.) and thinner
To 304.8, incl.	To 12, incl.	Bar	Strip[1,2]	Strip[1]
50.8 to 304.8, incl.	2 to 12, incl.	Bar	Sheet[3]	Strip
Over 304.8 to 608, incl.	Over 12 to 23¹⁵⁄₁₆, incl.	Strip[2]	Strip[2]	Strip
Over 304.8 to 608, incl.	Over 12 to 23¹⁵⁄₁₆, incl.	Sheet[4]	Sheet[4]	Black Plate[4]
Over 608	Over 23¹⁵⁄₁₆	Sheet	Sheet	Black Plate

[1]When the width is greater than the thickness with a maximum width of 12.7 mm (½ inch) and a cross-sectional area not exceeding 32.28 mm² (0.05 in.²), and the material has rolled or prepared edges, it is classified as **flat wire**.
[2]When a particular temper is defined in ASTM specification A 109, or a special edge, or special finish is specified, or when single-strand rolling is specified in widths under 609.6 mm (24-in.).
[3]Cold-rolled sheet coils and cut lengths, slit from wider coils with No. 3 edge (only) and in thicknesses 0.36 to 2.09 mm (0.0142 to 0.0821 inch) incl., carbon 0.20 per cent maximum.
[4]When no special temper, edge or finish (other than Dull or Luster) is specified, or when single-strand rolling widths under 609.6 mm (24 in.) is not specified or required.

over in thickness, and up to 304.8 mm (12-inches) in width. These dimensions prevent the convenient use of cold drawing for the production of many such bars, and, therefore, cold rolling is employed.

Hot-rolled bars of any suitable cross-section are cleaned of scale by pickling or other means, and then passed repeatedly through a set of driven rolls that reduce the section with each pass until the piece is worked down to the required thickness and surface finish. The edges of thin, flat sections are sometimes milled square after cold rolling. The cold working of the metal by this rolling process imparts to the bars a smooth surface finish and accurate thickness, and also increases the tensile strength, yield strength, and hardness, with a corresponding decrease in ductility. It is possible, by selecting the proper combinations of steel composition, degree of cold working, and heat treatment, to produce bars that will have mechanical properties best suited for intended uses.

Flat bars normally are furnished in straight lengths but, in sizes up to 14.3-mm by 15.9-mm (9/16- by 5/8-inch) or others having cross-sectional areas not more than 193.5 mm² (0.30 in.²) they may be supplied in coil form.

COLD-ROLLED CARBON-STEEL STRIP

Cold-rolled carbon-steel strip is manufactured in a variety of finishes, tempers, and edges, all depending upon the end use. By common custom, cold-rolled carbon-steel strip is made in a width range of over 12.7 to 608-mm (1/2-inch to 23 15/16-inches) and up to 6.34-mm (0.2499 inch) thick. It has a carbon content of 0.25 per cent maximum; material of this form containing more than this amount of carbon is considered as cold-rolled carbon spring steel which is discussed later in this section.

Cold-rolled strip is produced in coils on any of the conventional reversing mills, tandem mills, or by single-stand rolling. While the strip may be supplied in coils, it can be furnished in cut lengths by the straighten-and-cut process.

Before cold rolling, the mill scale is removed from the hot-rolled strip, usually by pickling. From this point, the strip may be cold reduced to final thickness or to some intermediate gage where it is annealed and further cold reduced to obtain the desired temper and gage.

Temper—Many degrees of temper are possible in the manufacture of cold-rolled strip by controlling the combinations of cold rolling and annealing. However, many years of use have brought certain ranges of temper into common usage, and these have come to be recognized by number, as follows:

Temper	Maximum Carbon Content (Per Cent)*	Rockwell Hardness	
		Minimum	Maximum**
No. 1 (Hard Temper)	0.25	{B-84ᵃ {B-90ᵇ	
No. 2 (Half Hard)	0.25	B-70	B-85
No. 3 (Quarter Hard)	0.25	B-60	B-75
No. 4 (Skin Rolled)	0.15	—	B-65
No. 5 (Dead Soft)	0.15	—	B-55

*Ladle analysis
**Approximate
ᵃFor thickness 1.78 mm (0.070 inch, and greater)
ᵇFor thickness less than 1.78 mm (0.070 inch)

No. 1 Hard Temper is a very stiff springy strip intended only for flat work where no bending is required.

No. 2 Half Hard Temper is less stiff than No. 1, but is intended for limited cold forming and will only withstand 90 degree bends made across the direction of rolling around a radius equal to the thickness.

No. 3 Quarter Hard Temper is intended for limited bending and cold forming and can be bent 90 degrees in the direction of rolling around a radius equal to the thickness and 180 degrees across the direction of rolling over its own thickness.

No. 4 Skin-Rolled Temper is intended for cold forming such as bending flat upon itself in any direction and for deep drawing. The purpose of the skin pass is to prevent the formation of stretcher strains.

No. 5 Dead Soft Temper or annealed temper is intended for severe cold forming and deep drawing where the formation of stretcher strains is not objectionable.

Three finishes have come to be recognized as standard within the industry, again by common usage.

No. 1 Dull Finish does not have any luster and is actually made rough intentionally by rolling on rolls roughened either mechanically or chemically. This finish is suitable where paint adherence is desired, or in deep drawing since the lubricant will stick to it.

No. 2 Regular Bright Finish is cold rolled on rolls having moderately smooth finish. It is suitable for many requirements, although it is not generally applicable to plating.

No. 3 Best Bright Finish is the highest luster finish produced by cold rolling and is particularly suited for electroplating.

Cold-rolled carbon-steel strip in No. 1 Dull Finish, No. 2 Regular Bright Finish, and No. 3 Best Bright Finish is also available with "rolled-in" designs produced through the use of work rolls with the desired pattern engraved into the roll by conventional photographic-etching techniques.

Six types of edges have become recognized as standard.

No. 1 Edge is a prepared edge of a specified contour (round, square, or beveled) which is produced when a very accurate width or edge finish is required.

No. 2 Edge is a natural mill edge carried throughout the cold rolling from the hot mill without additional processing of the edge.

No. 3 Edge is an approximately square edge produced by slitting.

No. 4 Edge is a rounded edge produced by edge rolling either the natural edge of the hot-rolled strip or the slit-edge strip. This edge is produced when an approximate round edge is desired and when the finish of the edge is not important.

No. 5 Edge is an approximately square edge produced by rolling or filing of a slit edge to remove the burr.

No. 6 Edge is a square edge produced by edge rolling the natural edge of the hot-rolled strip or slit-edge strip when the width tolerance and finish required are not so exacting as for No. 1 Edge.

STAINLESS COLD-ROLLED STRIP STEEL

Stainless cold-rolled strip steel is manufactured from hot-rolled annealed and pickled strip by cold rolling in mills equipped with ground or ground and polished rolls, depending upon the surface finish requirements. Strip steel is normally supplied up to and including 608-mm (23^{15}/$_{16}$-inches) wide in a great many of the more common types of stainless steels.

A range of mechanical properties can be provided by taking advantage of the possibilities offered by combining the influences of chemical composition, cold work (cold rolling), and heat treatment. These are varied according to the mechanical properties desired. Mechanical properties developed in this manner are referred to as tempers and are associated with the capacity each temper level possesses in regard to resistance to cold deformation. In the straight chromium grade (UNS S40000 or former AISI 400 series), only three tempers are generally recognized by the industry, viz., full hard (Rockwell C 20 minimum), No. 4 Temper (approximately Rockwell B 80-90), and No. 5 Temper (approximately Rockwell B 83 maximum). In the chromium-nickel steels (UNS S30000 or former AISI 300 series), No. 4 Temper will show approximately 80-90 Rockwell B and No. 5 approximately Rockwell B 83 maximum. In addition to these soft tempers, a variety of hard tempers are available classed as ¼, ½, and ¾ hard and full hard. These are based on minimum values for tensile strength or yield strength or both.

Finishes—Various surface finishes are possible in cold-rolled stainless strip steel although there are gradations. For example, a dull finish without luster can be produced by rolling on rolls roughened by chemical or mechanical means. A luster finish can be produced by rolling on rolls having a moderately high finish. For the very best or high luster, all of the treatments must be carefully done and the rolling performed on highly-polished rolls.

Basically only two finishes are recognized in the stainless steel strip industry:

No. 1 Finish, which is cold-rolled, annealed, and pickled, and

No. 2 Finish, which is cold-rolled, annealed, pickled, and rerolled.

Because of the difference in the alloy contents between the straight chromium and the chromium-nickel steels, it is possible to anneal the former so that a bright finish results, while the annealed and pickled chromium-nickel steel will be dull. In the chromium-nickel steels, No. 1 Finish is classed as dull, while the hard tempers are relatively bright as will be the No. 2 Finish, No. 4 Temper.

Since the annealed and pickled straight chromium steels result in a bright No. 1 Finish, it is possible to change the processes to enhance this finish in making a No. 2 so that it is even brighter. An especially bright finish is supplied by a process known as bright annealing, in which stainless-steel strip is annealed in a controlled atmosphere. The special atmospheric environment does not allow scale formation; therefore, the

annealed strip is extremely bright and reflective and does not require pickling. Dull finishes are also available in these grades where the requirements are necessary.

There are three edges available in stainless strip depending upon the width and thickness:

No. 1 Edge is a rolled edge, either round or square as specified, and is recommended when a very uniform width is required. It is limited to strip of approximately 127-mm (5-inches) in width and under.

No. 3 Edge is an approximately square edge produced by slitting. This edge is not burr free. Width is not a limiting factor in furnishing this edge.

No. 5 Edge is an approximately square edge produced by rolling or filing for the primary purpose of removing burr originating in the slitting operation. Width is not a limiting factor in furnishing this edge.

COLD-ROLLED CARBON SPRING STEEL

Cold-rolled carbon spring steel is furnished either untempered or hardened and tempered in a variety of finishes, tempers, and edges. It is produced in coils in a manner similar to cold-rolled strip and it can also be supplied in cut lengths by the straighten-and-cut process.

Tempered cold-rolled carbon spring steel is customarily produced with a No. 1 edge. Untempered cold-rolled carbon spring steel is customarily produced with edge Nos. 1 to 6, inclusive, as described above for cold-rolled strip.

Untempered cold-rolled carbon spring steel is commonly furnished with a No. 2 finish as described above for cold-rolled strip. The following types of finishes are commonly produced on hardened and tempered cold-rolled spring steel:

Black Tempered
Scaleless Tempered
Bright Tempered
Tempered and Polished
Tempered, Polished and
Colored (Blue or Straw)

Temper—As with cold-rolled carbon steel strip, many degrees of temper are possible in the manufacture of cold-rolled carbon spring steel. Untempered spring steel is commonly furnished to the temper designations of Hard, Intermediate, and Annealed.

Hard temper cold-rolled carbon spring steel is a very stiff, springy product intended for flat work and not for cold forming. It is furnished to a minimum Rockwell hardness value of B-98.

Intermediate temper cold-rolled carbon spring steel is intended for applications where hardness ranges are required, the maximum being higher than that customarily obtained by annealing at finish thickness. The Rockwell hardness ranges commonly specified vary according to the maximum of the carbon range involved, the maximum of the required hardness range, and the thickness of the material.

Soft temper cold-rolled carbon spring steel is intended for moderately severe cold forming requiring low hardness and is produced to a specified maximum

Table 34—II. Expected Rockwell Hardness for Different Ranges of Carbon Content and Thickness of Soft Type Cold Rolled Spring Steel.

Carbon Content (Maximum of Range- Per Cent)	1.02 mm (0.040 in.) and Over ("B" Scale)	Under 1.02 mm to 0.64 mm (Under 0.040 in. to 0.025 in.,) Incl. ("30T" Scale)	Under 0.64 mm (0.025 in.) ("15T" Scale)
0.30	74	67	84
0.35	76	68	84
0.40	78	70	85
0.45	80	71	85
0.50	82	72	86
0.55	84	73	87
0.60	85	74	87
0.65	87	75	88
0.70	88	76	88
0.75	89	76	88
0.80	90	77	89
0.85	91	77	89
0.90	92	78	89
0.95 and over	92	78	90

Rockwell hardness value only. This type is produced by performing the final annealing at finish thickness. The Rockwell hardness values indicated in Table 34—II are commonly expected for the indicated ranges of carbon content and thickness.

Hardened and tempered cold-rolled carbon spring steel customarily has a carbon content over 0.60 per cent. It is commonly produced to the Rockwell hardness ranges given in Table 34—III, dependent upon maximum per cent carbon content of the range and strip thickness.

COLD-ROLLED CARBON-STEEL SHEETS

Cold-rolled carbon-steel sheets are produced as either coils or cut lengths, within the dimensional limitations shown in Table 34—I. Their manufacture involves two distinct stages: (1) reduction of the product to the desired thickness (gage) and (2) the necessary finishing operations. Cold-rolled sheets are produced by the cold reduction of hot-rolled coils from the hot-strip mill.

BLACK PLATE

Black plate was the term used originally to designate small thin sheets produced by hand hammering. The term continued to be used when hot pack rolling was developed, and today the name is still retained for a product manufactured by the cold reduction of hot-rolled coils from the hot-strip mill. As indicated in Table 34—I, black plate is produced in thicknesses of 0.358 mm (0.0141 inch or 29 gage) and thinner. Because of its classification as a tin-mill product, the uncoated black plate is specified in the same manner as tin plate; namely, by weight per base box (see Chapters 36 and 53).

Black plate originated as the starting material for the manufacture of tin plate, and by far the greatest part of all that is produced is still used for that purpose, as described in Chapter 36. The modern product manufactured by the cold-reduction method and sold uncoated with any metal is cleaned, continuously annealed or box annealed, and temper rolled after cold reduction to make it suitable for intended purposes. Heat treatment, and chemical or other surface treatments, may be applied. The product is available in coil form or as cut sheets.

Two general types of finish are commonly produced on black plate: dull finish and bright finish. Dull finish is produced by rolling the plate on rolls roughened by chemical or mechanical means; bright finish is a semiluster finish produced by rolling on rolls having a moderately smooth surface. By varying the possible combinations of steel composition, degree of mechanical (cold) working and heat treatment, it is possible to produce black plate in a range of tempers to suit particular applications; these tempers are described in Chapter 36 on "Manufacture of Tin-Mill Products."

Table 34—III. Rockwell Hardness Range Limits for Hardened and Tempered Cold Rolled Carbon Spring Steel for Various Ranges of Carbon Content and Strip Thickness.

Thickness		Rockwell Scale	Per Cent Carbon, Maximum						
Millimetres	Inch		0.75	0.80	0.85	0.90	0.95	1.00	1.05
			Limits of Rockwell Hardness Range*						
Over 2.92 to 3.18, incl.	Over 0.115 to 0.125, incl.	C	32-44	37-45	38-46	39-47	40-48	41-49	42-50
Over 2.54 to 2.92, incl.	Over 0.100 to 0.115, incl.	C	33-45	38-46	39-47	40-48	41-49	42-50	43-51
Over 2.16 to 2.54, incl.	Over 0.085 to 0.100, incl.	C	34-46	39-47	40-48	41-49	42-50	43-51	33-52
Over 1.78 to 2.16, incl.	Over 0.070 to 0.085, incl.	C	35-47	40-48	41-49	42-50	43-51	44-52	45-53
Over 1.40 to 1.78, incl.	Over 0.055 to 0.070, incl.	C	36-48	41-49	42-50	43-51	44-52	45-53	46-54
Over 0.89 to 1.40, incl.	Over 0.035 to 0.055, incl.	C	37-49	42-50	42-51	44-52	45-53	46-54	47-55
Over 0.38 to 0.89, incl.	Over 0.015 to 0.035, incl.	30N	57-68	62-69	63-70	64-71	64.5-71.5	65-72	66-73
Up to 0.38	Up to 0.015	15N	78-84	80.5-84.5	81-85	81.5-85.5	82-86	82.5-86.5	83-87

*By common usage, a Rockwell hardness range is the arithmetical difference between two limits. It is customary to specify Rockwell range requirements within the above limits for each grade of hardened and tempered cold rolled carbon spring steel in accordance with the following:

Rockwell Hardness Scale	Specified Range
C	Any 4 points
30N	Any 4 points
15N	Any 3 points

PRINCIPLES OF COLD REDUCTION

Cold rolling is a generic term applied to the operation of passing unheated metal through rolls for the purpose of reducing its thickness; producing a smooth, dense surface; and, with or without subsequent heat treatment, developing controlled mechanical properties. Any single one or combination of these three effects may be the reason for cold rolling of a particular product.

Cold reduction is a process in which the thickness of the starting material can be reduced by relatively large amounts in each pass through a single-stand reversing cold mill or in a series of passes through a tandem cold mill. Thus, in the production of most cold-rolled sheets, cold-rolled strip, and black plate, the cold-reduction process is employed to reduce the thickness of the starting material (hot-rolled coils) between about 25 and 90 per cent. After cleaning and annealing, a large proportion of such products is subjected to a cold-rolling operation referred to as "temper rolling" that reduces the thickness of the material from less than one to only a few per cent in the process of imparting the desired mechanical properties and surface characteristics to the final product.

The introduction of continuous methods for cold reduction of steel was the most significant step in the development of present-day rolling facilities for sheets and tin plate. As in the case of the hot-strip mill, cold reduction of wide hot-rolled coils was developed from the rolling of narrow products. The original purpose of cold rolling was, however, to achieve the desired surface and mechanical properties, and reduction in thickness was of incidental importance. Cold rolling of strip probably originated in Germany early in the nineteenth century as a process for cold rolling high-carbon wire to produce a flattened cross-section. Similar practices were adopted thereafter in this country. Low-carbon cold-rolled strip, the forerunner of today's cold-reduced strip, sheets and tin plate, was first made in this country at the Stanley Works at New Britain, Connecticut. For years, the raw material for such strip was produced on merchant bar mills. Later, cold-reduction equipment developed rapidly, concurrent with the introduction of narrow continuous hot-strip mills and their evolution to the modern wide mills. Today, the narrow and wide cold-reduction mills provide the major outlet for hot-rolled coils.

In the cold reduction of hot-rolled coils, the prime objective is reduction in thickness of the material, since the modern hot-strip mill cannot reduce the steel to gages thinner than about 1.24 mm (0.049 inch or 18 gage). The development of the cold-reduction process for wide products has been even more rapid than that of the continuous hot-strip mill. The first efforts toward cold reduction of wide hot-rolled coils were made in the 1920's and the process changed rapidly from single-stand non-reversing mills through single-stand reversing mills and then to the present-day high-speed tandem mills. Using tin plate as an example of the rate of change in the methods for reducing very thin flat-rolled

steel, it is found that in 1936, 24 per cent of the black plate for this country's tin plate production was cold reduced, with the balance made by the old hot pack-rolling method. By 1939, 75 per cent of all tin plate was made by the cold-reduction process and this figure reached 100 per cent by 1943.

Sequence of Operations in Cold Reduction—After hot-rolling, the hot-rolled coils are uncoiled, passed through a continuous pickler, dried, oiled, and re-coiled. The oil serves as a protection against rusting and as a lubricant during cold reduction. There are several types of cold-reduction mills which vary in design from single reversing two-high, four-high or multiple-roll units to continuous four-high stands with up to six units in tandem. In tandem rolling, the product is given one pass through four, five or six stands (Figures 34—1, 34—2, 34—3, and 34—4), each contributing to thickness reduction and each driving the material being rolled, usually at speeds so synchronized that the steel is under tension at each of the stages between the payoff reel, the various sets of work rolls and the re-coil reel. Unlike hot-rolling, no scale is formed, but much greater pressures and driving forces are required to effect a given reduction in thickness.

For any given pass in the cold-reduction process, the resultant of the compressive forces of the rolls on the steel and the tensional forces along its length between the reels and rolls must exceed the elastic limit of the steel to produce permanent deformation.

The proper reduction of thickness which should be made at each pass of a reversing mill or in rolling on each stand of a tandem mill can be expressed only in very general terms. The work load should be distributed as uniformly as possible at the various stages without falling very much below the maximum capacity of any stage. The maximum is determined in each case by several factors, of which the most important are mill design, power available, steel width and total reduction to be taken, steel lubrication, steel cross-sectional contour, steel hardness, steel tension, steel surface, roll diameter and roll surface. Generally, the lowest percentage of reduction is in the last reduction pass to permit better control of the flatness, gage, and finish of the product. On the conventional wide four-high tandem mills used to cold reduce most sheet and tin plate stock, the individual pass reduction will range generally between 25 and 45 per cent on all stands but the last, where it will fall between 10 and 30 per cent. The resultant total reduction on these mills from hot-rolled coils to finished product will usually run 45 to 75 per cent for most sheet gages and 80 to 90 per cent for most tin plate.

Heavy reductions at high speed on any of the various types of mills generate considerable heat and not only raise the temperature of the product but also that of the rolls. The heat generated must be dissipated by a system of flood lubrication in which a tallow based oil or a mixture of oils is directed in small streams or jets against the roll bodies and the surface of the steel. Some mills

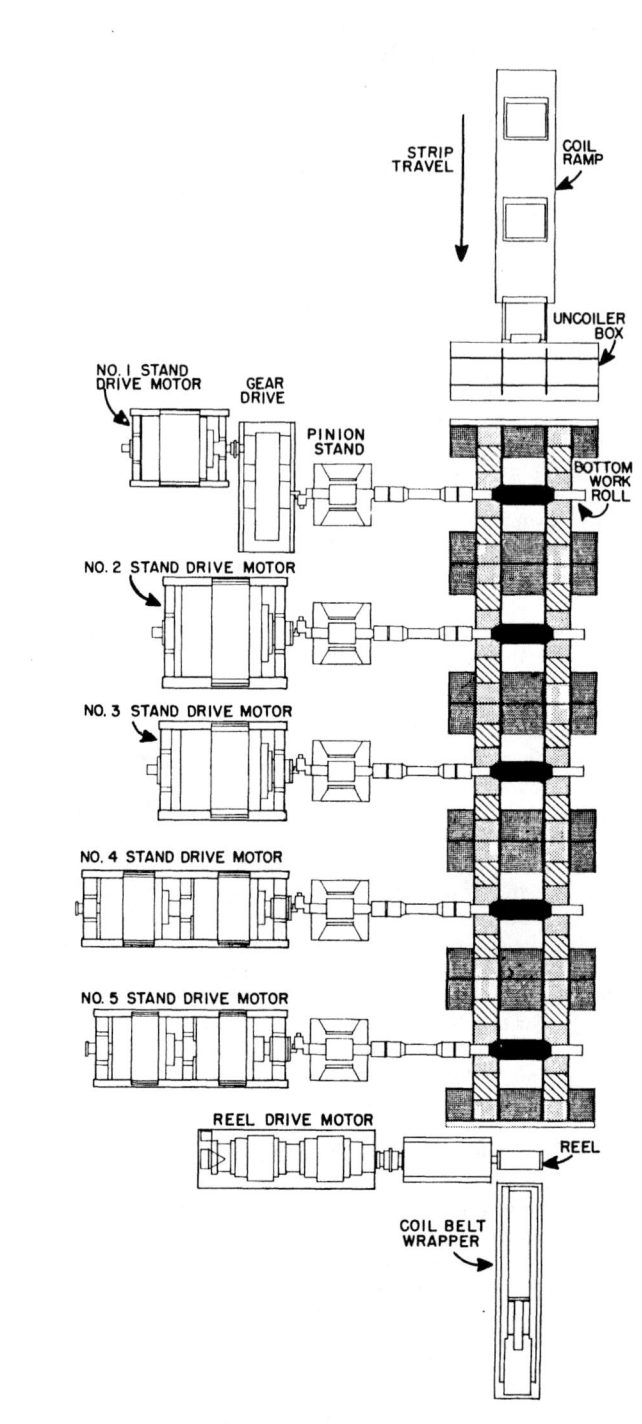

Fig. 34—1. Schematic arrangement of the mill stands and auxiliary equipment comprising a five-stand tandem cold-
reduction mill. The shaded sections of the drawing represent
the mill housings of the five four-high mill stands. Only one
work roll of each stand is indicated in black; no other rolls are
shown.

Fig. 34—2. A 1372-mm (54-inch) high-speed, four-stand tandem cold-reduction mill. The delivery end of this mill is at the left of the illustration.

Fig. 34—3. A 1219-mm (48-inch), high-speed, five-stand tandem cold-reduction mill. Product being rolled travels through the mill from right to left in this view.

Fig. 34—4. A 1321-mm (52-inch), six-stand tandem cold-reduction mill, used in the production of coils of steel for subsequent coating with tin.

use palm oil on the steel and high-pressure water on the rolls. In any event, the resultant steel temperature generally runs between 65° and 120°C (150° and 250°F). On high-speed tandem mills rolling tin plate stock, the temperature of the steel may become as high as 205°C (400°F).

Numerous automatic-control devices have been installed to assist the mill operators in maintaining the optimum rolling conditions to achieve maximum quality of the rolled product. Computer control for setup and optimum operation of the mills is now common place.

SECTION 3

ROLL ARRANGEMENT FOR COLD REDUCTION

In reducing the thickness of steel, the action of a pair of rolls may be visualized as being somewhat similar to that of two very blunt knife edges in cutting partially through a stationary piece across which the knife edges are opposed under pressure. The smaller the roll diameter or the sharper the knife edges, the less pressure is required to do the necessary work. Other factors being equal, the force imposed across a set of work rolls to obtain a given reduction in thickness of steel varies with the square root of the roll diameter. To minimize the power requirements and mill size, it is therefore, advantageous to use work rolls of as small diameter as possible, consistent with being able to remove the heat generated.

The steel being rolled is necessarily narrower than the length of the work rolls. In a previously used two-high mill, these rolls were supported only at their ends and rolling pressure was applied there. In view of this, it is obvious that rolls of some minimum diameter no longer will be sufficiently rigid to withstand the necessary pressures without bending. In extreme cases, this bending along the length of the roll might be sufficiently great to cause the rolls to touch each other at the ends. The tendency, of course, is for the bending to increase as the mill width (the roll length) becomes greater.

In order to overcome the difficulties arising from bending of work rolls of small diameter without sacrificing the advantage of the small diameter, the modern mill backs up the work rolls with rolls of larger diameter. These more massive rolls resist the bending force of the screw-down pressures applied to their ends, and the pressures are transmitted to the work rolls along their entire length. For example, four-high cold-reduction mills in the 1067- to 2489-mm (42- to 98-inch) width range, which produce nearly all cold-rolled sheets and tin plate, will have two work rolls of 406- to 610-mm (16- to 24-inch) diameter, each backed up by a roll of 1067- to 1422-mm (42- to 56-inch) diameter. The driving power is applied at one end of each of the work rolls or back-up rolls of such a mill and screwdown pressure at both ends of the back-up rolls.

While the widespread use of such conventional four-high mills, in single stands or, more commonly, in tandem arrangements, has long ago proved their economic suitability for producing most of the cold-rolled flat steel products, many departures from this type have been tried to obtain even smaller work-roll diameters, particularly in the production of specialty sheets such as stainless and silicon sheets. In the four-high Steckel mill, none of the rolls are driven and the steel is pulled back and forth between the 50- to 125-mm (2- to 5-inch) diameter work rolls by reversing power-driven reels. The back-up rolls in this case are 610 to 914 mm (24 to 36 inches) in diameter. Most Steckel installations are confined to the rolling of narrow strip. Other mills, known as the "six-high" or the "cluster" types, use two or more backup rolls for each work roll. The currently most successful of the multiple-roll class is the Sendzimir mill, with two work rolls of about 25- to 64-mm (1- to 2½-inch) diameter. Each work roll is backed up by two rolls of twice that diameter, each of which in turn is backed up by two segmented rolls of even larger diameter. The intermediate backup rolls are driven and the outer segmented rolls, which are actually rows of bearings on common shafts, provide a caster-like support action and permit the application of

screw pressure by rotating these shafts eccentrically (Chapter 23). A light, wholly-enclosed housing is possible and the entire roll system is immersed in lubricant. This mill appears to be well-adapted to taking heavy reductions on steels of the harder grades such as stainless and high-silicon steels. Six-high mills with much larger roll diameters are now in use in high-speed tandem mills.

Even with the rigidity provided by massive backup rolls, work rolls of the common four-high mills will flex or bend somewhat under the rolling pressures used; this tendency is greater in the wider mills. Accordingly, on mills rolling sheets and tin plate, work rolls are "crowned," that is, ground with roll diameters which increase very gradually from end to center. This convexity, which may consist of a diameter differential of 0.25 mm (0.010 inch) on the widest mills, compensates for normal roll flexure. On mills about 1016 mm (40 inches) wide, the crown may be about 0.03 mm (0.001 inch) or less, and on narrow cold-reduction mills it is usually unnecessary. The work rolls themselves are made of forged, hardened steel, surface ground to a high finish on specially-designed precision lathes. Work rolls for the last stand of a tandem mill for sheets generally are roughened slightly by shot-blasting to impress a matte pattern on the sheet as an aid in preventing sticking of stacked sheets in annealing. Work rolls for narrow cold-reduced strip, which requires a much brighter surface than sheets, are highly polished. Backup rolls on wider mills are generally made of cast steel and are always surface ground. They can be shaped "flat" (no crown) or with up to 0.25-mm (0.010-inch) convexity.

Instrumentation for measuring the shape and profile of strip being rolled is now available. Automatic control of these variables is accomplished by work roll benching, inflatable crown backup rolls, and shifting of intermediate rolls on a 6-high mill.

SECTION 4

TYPICAL MILL LAYOUTS

Four-High Tandem Mills—The bulk of the cold-reduced flat steel made in this country is rolled on four-, five- or six-stand four-high tandem mills.

Most tin plate falls in the thickness range of 0.20 to 0.36 mm (0.008 to 0.014 inch) and requires 80 to 90 per cent cold reduction from the thinnest hot-rolled coils available in the width range of 610 to 914 mm (24 to 36 inches). Tin plate, therefore, has been rolled on five-stand mills around 1067 to 1219 mm (42 to 48 inches) wide, but the use of six-stand mills for tin-mill products is common.

Most cold-rolled sheets for automotive bodies, agricultural implements, architectural use, furniture and household equipment are required in thicknesses of 0.64 to 1.65 mm (0.025 to 0.065 inch) and widths of 762 to 1829 mm (30 to 72 inches). Material for these applications, most of which is given 45 to 75 per cent cold reduction, is rolled on four- or five-stand mills producing material ranging from 1372 to 2489 mm (54 to 98 inches) in width. Four-stand mills, usually 1219 to 1422

mm (48 to 56 inches) wide, may be used to roll tin plate but are best adapted to produce 0.38 to 0.76 mm (0.015 to 0.030 inch) thick, 610 to 1194 mm (24 to 47 inches) wide sheets for such applications as roofing, signs and containers. Maximum delivered cold-reduced strip speeds on such tandem mills may run as high as 5.08 metres per second (1000 feet per minute) on a three-stand, 15.24 metres per second (3000 feet per minute) on a four-stand, 30.48 metres per second (6000 feet per minute) on a five-stand, and higher than 35.56 metres per second (7000 feet per minute) on a six-stand mill. The average operating speeds vary greatly throughout the industry.

All such mills have uncoiling reels, cradles, or boxes from which the coil is fed into the actual roll train, coil ends being started usually by pinch rolls. At the discharge end the finished coils are recoiled on a mandrel, being started by a belt wrapper (a continuous driven fabric belt bearing against the mandrel), which guides the head end of the cold-reduced strip around the mandrel and is withdrawn when the re-coiling has been so started.

Mill housings are massive castings, similar to those on finishing stands of a hot-strip mill. On most tandem mills, both work rolls of each stand are driven through a pinion arrangement by a single motor, up to 3730-kilowatt (5000-horsepower) motors being used, although some of the latest mills have individual motors for each work roll or backup roll on some stands. For the first stand or stands, in which the rolling speed is relatively low, motors are geared down; direct drive may be used for one or more of the remaining stands.

A typical example of a four-high tandem mill is the 1320-mm (52 inch) six-stand tandem mill shown in Figure 34—4. This mill rolls a pickled, 21 800-kilogram (48 000-pound) maximum hot-rolled breakdown in coil form. Two rolling set-ups are used: one for sheet-mill product and the other for tin-plate product. The hot-rolled product entering the mill is 2.79-mm (0.110-inch) maximum thickness with a delivery thickness of 0.91 mm (0.0359 inch or 20 gage) and 46 inches wide up to 15.24 metres per second (3000 feet per minute) maximum mill speed for sheet product. For tin-plate product, the mill set-up delivers cold-reduced product 0.10 mm (0.004 inch) thick and 1067 mm (42 inches) wide at 35.56 metres per second (7000 feet per minute) maximum. The six stands, plus the reel, require a total of 9960 kilowatts (13 350 horsepower) for driving. From the moment when the head of the coil has been threaded through the stands until the mill reaches top speed, approximately 150 per cent of motor rated torque is applied to work rolls of each stand. This represents approximately 1 metre per second (200 feet per minute) acceleration and is adjustable in the control room. After steady rolling state is attained, the torque requirement reduces to approximately 80 per cent of motor rating depending upon product being rolled. The main mill motors on stands Nos. 4, 5 and 6 are of the double-armature type, with the reel being of triple-armature design. Such motors provide a marked reduction in accelerating time and improved control at high speeds due to their lower inertia. The work rolls of stands Nos. 1, 2 and 3 range from 495 mm (19½ inches) to a maximum of 546 mm (21½ inches) with stands Nos. 1 and 2 using cast-iron rolls. Work rolls for stands Nos. 4, 5 and 6 have roll diameters ranging from 546 to 597 mm (21½ to 23½ inches) of forged steel. Stand No. 3 also uses forged steel work rolls. All stands have back-up roll diameters ranging from 1270 mm (50 inches)

through 1422 mm (56 inches). Work rolls are equipped with roller bearings while the back-up rolls are equipped with oil-film bearings. The cold-reduced strip tension reel at the delivery end is designed to pick up the end of the strip coming from the mill, grip it, and put it under tension without sudden or undue stress. The reel also winds the cold-reduced strip under constant tension, slowing down automatically without altering the tension as the coil increases in diameter. Such control is necessary to prevent tearing of the very thin, hard, cold-reduced strip. The reel is of the collapsible type and is easily stripped by a hydraulic plunger. A control computer is included in the motor-control circuitry which automatically sets the finish gage, screw positions for all stands along with mill speed and interstand tension limits. In addition, the computer automatically records a log of product for operating and accounting management use.

Four-High Reversing Mills—A typical four-high reversing mill consists of a single stand with reels located on each side of the mill. The mill itself is essentially the same in design and arrangement as the individual stands of a tandem mill. In the reversing mill, the steel must be passed back and forth until the required reduction is obtained. On the entry side of the mill, means are provided for uncoiling and feeding the coil through the mill to the tension reel on the delivery side. After the first pass, the tail end of the coil coming from the uncoiler is gripped by the second tension reel on the entry side of the mill. In each pass, the reel serving as the payoff unit is operated as a generator, providing back-tension to minimize rolling friction and feeding of current into the drive-reel motor. On the last pass, the tail end of the coil is released from the unwinding tension reel, completely wound on the winding reel and stripped in a manner similar to the action on the delivery reel of the tandem mill. For the rolling of tin plate product, a reversing mill usually requires five passes; for relatively heavy sheet product, usually three passes. All passes are considerably slower than the delivery pass of a tandem mill. From this it can be seen that a reversing mill is inherently flexible but cannot compete with a modern tandem mill from a production or cost standpoint where large tonnages are involved. Low installation costs in comparison to tandem mills make reversing mills popular for the production of specialty items that vary widely in dimensions and are ordered in small tonnages for each specification.

SECTION 5

DISPOSITION OF PRODUCT

Cold-Reduced Product for Strip—Disposition of the cold-reduced hard product depends upon prior processing and ultimate use. Much narrow high-finish strip is sold as cold reduced, for "full-hard" applications, on which heavy cold reductions have been used; or as intermediate tempers, in which the cold-reduced strip has been reduced a predetermined percentage from hot-rolled coils or intermediate annealed gage to develop specified mechanical properties.

Cold-Reduced Product for Sheets—In the manufacture of cold-rolled sheets, practically all coils have been side trimmed to width on the continuous pickling line before cold reduction, and the coil is conveyed directly to the annealing department from the cold-reduction mill. After annealing, most coils for sheets are temper rolled. The greater part of cold-rolled sheet product is

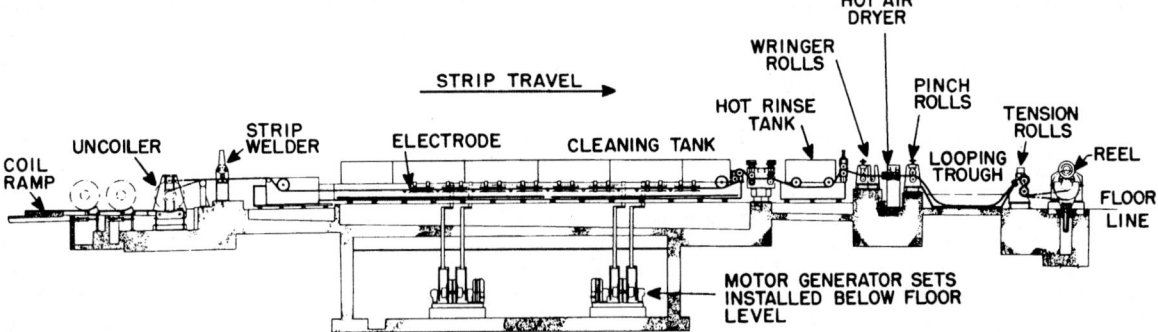

Fig. 34—5. Schematic arrangement of the equipment comprising a typical horizontal electrolytic cleaning line for processing cold-reduced steel.

shipped in coil form. However, some coils are taken to continuous sheet-shearing lines, where they are sheared into sheets of ordered lengths. The latter may be subjected to one or more of the other finishing operations discussed in Section 10 of this chapter. Some coils may be slit into narrower widths as also outlined in Section 10.

Cold-Reduced Product for Tin Plate—Cold-reduced coils for tin plate are uncoiled, cleaned and recoiled (Figure 34—5), to remove surface dirt and oil from the rolling operation, prior to annealing. For material that is to be box annealed, the cleaning is a separate operation. Material that is to be continuously annealed is uncoiled and passed through a continuous cleaning operation that is an integral part of the continuous annealing line before it enters the heat-treating section of the line; after continuous cleaning and annealing, the product is recoiled.

Annealed coils for tin plate generally are temper rolled, side-trimmed and recoiled and then delivered to the continuous electrolytic-tinning lines. In the manufacture of double-reduced tin plate (see Chapter 36) annealed coils may be rolled in a two-stand or three-stand cold-reduction mill instead of being temper rolled prior to electrolytic tinning.

SECTION 6

CLEANING OF COLD-REDUCED STEEL

In order to produce satisfactory tin plate, tin foil steel, terne plate, or hot-dip or electrogalvanized sheet, it is necessary to clean the cold-reduced steel to remove the lubricant used in cold reduction since the lubricant left on the steel will decompose during annealing, and leave undesirable residues of carbonaceous material on the annealed product. Until recently, it had not been common practice in North America to clean cold-rolled sheets prior to box annealing; however, the practice is becoming more widely practiced. For continuous annealed cold-rolled sheets, cleaning stations are an integral part of the continuous annealing and processing lines.

Without exception, cold-reduced strip cleaners currently employ hot alkaline detergent solutions to remove the rolling-mill oils or solutions.

In most modern alkaline cleaners electrolysis is also used to facilitate the cleaning operation. For electrolysis the sheet is connected to a power source which provides direct current. There are two principal types of electrolytic alkaline cleaning operations which are differentiated by the method of connecting the sheet/strip to the power source as follows:

Conductor-Roll Cleaning—On vertical cleaning lines the electrical connection to the strip is made via conductor rolls. If the current enters the solution from the strip, the strip is said to be **anodic** (positive) and grids must be provided in the solution adjacent to the strip as an exit path for the current. Those grids (negative) are also connected to the same DC power supply by busses or cables that convey the electric current back to the power supply. If the direction of the current is reversed and the current leaves the solution via the strip and conductor roll, the strip is said to be **cathodic.** Thus the polarity refers to the direction the electric current flows in the strip and controls the type of chemical reaction that takes place.

If the strip is anodic, oxygen gas produced on the strip, and the volume of gas produced is controlled by the quantity of electric current that flows. If the strip is cathodic, hydrogen gas will be produced on the strip instead of oxygen but for the same current flow the volume of hydrogen generated will be twice the volume of oxygen that would be generated at the anode. Because the gas is generated at the steel surface, it helps scrub the grease and oil off the surface. Obviously the larger volume of gas and the smaller size of a hydrogen molecule as compared with an oxygen molecule will produce a more effective scrubbing action. For this reason cathodic polarity is usually more desirable in alkaline cleaning. In some instances, however, the cleaner polarity is anodic. For example, when a tin-free steel coil (coated with chromium oxide) must be recoated, the cleaner must be anodic to strip off the old coating so that a new coating can be applied in the plating section of the line. The metal stripped from these recoated coils dissolves in the cleaner solution and will plate out on subsequent coils if the polarity is made cathodic.

Grid-to-Grid Cleaning—Some coating lines and almost all cleaning lines used before box-annealing operations have horizontal cleaners. In this case the current enters the strip as it passes between sets of horizontal grids which are connected to the power supplies. This arrangement precludes the selection of only one cleaning polarity because the polarity of the strip reverses as the strip passes between successive sets of grids. During operation some of the grids become coated with an insulating film of the soils that have been removed, and an increasingly higher voltage is required to maintain the desired current density. When this happens, the polarity should be reversed to clean the dirty grids. The voltage should then drop back to the previous operating value.

Alkaline cleaners usually consist of sodium hydroxide, sodium carbonate, and/or various silicates and phosphates. The cleaners are generally classified as highly silicated (18 to 28% SiO_2), medium silicated (6 to 11% SiO_2), or slightly silicated (2 to 3% SiO_2). The nonsilicated cleaners can be classified by their phosphate content: high phosphate (8 to 12% P_2O_5), medium phosphate (4 to 7% P_2O_5), or nonphosphated. Other agents may also be added to the cleaning solution; for example, wetting agents optimize contact between the soiled surface and the cleaner and emulsifiers combine with oil removed from the surface and thereby facilitate removal of the oil from the cleaning bank.

For good cleaning efficiency the concentration of the cleaning solution is usually maintained between 15 and 45 grams/litre (2 to 6 oz./gal.) when added as a powdered cleaner. Consequently, the cleaners are analyzed regularly and additions made to maintain the required concentration. Current density and cleaning time are important factors that influence cleaning efficiency. High current densities are more effective in cleaning because they release more gas (oxygen or hydrogen depending on polarity) and therefore produce more scrubbing action. A current density of at least 100 amperes per square foot (or 11 amperes/square decimeter) is usually used for normal action. Thorough rinsing of the steel after cleaning is essential to completely remove all residual cleaning solution.

SECTION 7

HEAT TREATMENT OF COLD-REDUCED STEEL

Purposes and Types of Heat Treatment—As is true of other steel products, sheets, strip and tin plate are heat treated primarily to effect changes in mechanical properties which render the material suitable for the intended purpose. Other heat-treating objectives, involving relatively small tonnages of flat-rolled steel but highly important to the specialized commodities to which they apply, include: solution of chromium carbides to attain maximum corrosion resistance of austenitic stainless steels; development of optimum magnetic properties and establishment of the thin insulating oxide on silicon-bearing electrical sheets; and dispersion or spheroidization of carbides to influence later heat-treating characteristics of alloy and high-carbon sheets.

Except for the very small proportion of "full-hard" cold-reduced strip and sheets used in the as-cold-reduced state to take advantage of the high strengths developed by cold reduction, some form of heat treatment is applied as a separate operation to all cold-reduced flat products to restore the ductility lost in cold reduction. In the case of hot-rolled sheets and strip made on a modern continuous hot-strip mill, supplementary heat treatment is usually unnecessary because rolling practices are used which include definite finishing and coiling temperatures as the steel leaves the last pass of the mill; the resultant "mill heat treatment" can be varied to provide mechanical properties in the as-rolled state which are satisfactory for most uses. Supplementary heat treatment of hot-rolled coils is used for a few specialized products. For example, for a specific class of high strength sheets, called "dual phase" steels because they consist of two phases (ferrite and martensite), the as-hot rolled coils are heat treated to obtain a unique combination of strength and formability. For these steels, the control of the steel composition, heat treatment temperature and cooling is critical to obtaining the required mechanical properties. This section will consider only those types of heat treatment conducted as separate operations.

Heat treatment of cold-reduced sheets, strip, and tin plate may be divided into: (a) "batch" operations such as conventional or "tight coil" box annealing, in which a large, stationary mass of steel is subjected to a long heat-treating cycle by varying the temperature within the furnace that surrounds it; and **open-coil annealing,** in which a tight coil is first rewound with a suitable spacer between each wrap of the coil to permit circulation of the furnace atmosphere between individual wraps to hasten and improve uniformity of heating and to permit the use of special atmospheres as described later; and (b) "continuous" operations that include **continuous** or **strand annealing,** and **normalizing,** in which a single thickness of cold-reduced sheet or strip is passed through a furnace in a relatively short time and is subjected to a heat-treating cycle determined by the temperature distribution within the furnace and the dimensions and rate of travel of the steel.

Virtually all box-annealing practices on tight coils slowly raise the steel to a temperature level at or below the lower critical temperature, as discussed in Section 9. Such a cycle exposes the steel to high temperatures for long time periods, which provides full recrystallization of the cold-reduced steel and results in the softest possible finished product; accordingly, box annealing is the principal batch annealing heat treatment applied to cold-reduced steel. In recent years computer-control practices for box annealing have been developed and are now widely used; the firing (heating) rates, temperature-time cycles and cooling rates are closely controlled to minimize energy usage and yet provide the required mechanical properties throughout the coils. Most of these computer controls are applied to stationary coils that are stacked in a furnace. However, special procedures have also been developed for moving the tightly wrapped coils through furnaces having closely controlled temperature zones.

As usually practiced in the heat treatment of cold-rolled carbon-steel sheets, open-coil annealing is carried out in the same general type of equipment as conventional box annealing, with some modifications. However, the space between wraps of the open coils permits more rapid and uniform heating and cooling as well as some special treatments that are impractical with conventional tight coils. For example, it is possible to almost completely decarburize coils of steel for special applications. It is also possible to alter the composition of the steel in other ways, such as by increasing the carbon content (carburizing), or by adding chromium to its surface (chromizing), or by adding nitrogen (nitriding) to change either the mechanical or physical properties of the steel or both. Spacing between wraps for open-coil annealing has been effected by rewinding tight coils on a second mandrel with formed-wire spacer between wraps.

Depending on the nature of the steel and the results to be obtained, in continuous heat treatment the steel may be heated quickly to a maximum temperature at or somewhat above the lower critical temperature (continuous or strand annealing), or it may be heated slightly above the upper critical temperature (normalizing). In any case, the time at temperature is only a few minutes and the cooling rate is fast as compared to the hours-long cooling cycle of a box-annealed charge. For many years continuous annealing was used, and continues to be used, for the production of coated sheets and tin mill products where formability is not a critical requirement. But until recently continuous annealing was not used for cold-rolled low-carbon steel sheets because the continuous annealed sheets were usually less ductile and more susceptible to strain aging than box annealed sheets. However, in recent years metallurgical developments (new steel compositions and modified hot-rolling practices) have allowed the production of cold-rolled low-carbon steel sheets on newly designed continuous annealing lines so that the resulting mechanical properties are comparable to those of box annealed sheet. These new continuous annealing lines usually incorporate a separate overaging zone to minimize the susceptibility to strain aging and often incorporate in-line temper rolling. The inclusion of several operations (such as electrolytic cleaning and temper rolling) into the new continuous annealing lines has resulted in improved productivity. In addition, because of the better control of temperature, which results from annealing a single strand (in contrast to annealing a large coil in batch operations), the uniformity of properties throughout the coil is usually better than for batch annealed product. These advantages together with the capability of new continuous annealing technology to improve the flatness and surface cleanliness of the sheet have accounted for a wider use of continuous annealing in recent years for the production of uncoated cold-rolled sheets. Often these products are subsequently processed through coating lines (such as electrolytic galvanizing lines) to produce one or two-side coated products.

While the great bulk of flat-rolled steel products can be heat treated adequately at temperatures around 675°C (1250°F), a few because of their sluggish recrystallization tendencies after cold reduction, or their need for relatively quick cooling from the austenitic state to attain optimum properties, must be heated to temperatures ranging from 760° to 1205°C (1400° to 2200°F). In such cases, continuous heat treatment is the most convenient method, as it avoids the hazard of pressure welding ("sticking") inherent in large masses of cold-reduced products held for hours at such temperatures. Continuous heat treatment is the only way to attain the quick cooling necessary to hold chromium carbides in solution in austenitic steels such as UNS S30400 (formerly AISI Type 304) or to develop high strength levels in alloy sheets of the UNS G41300 (formerly AISI 4130) type. It permits maximum reaction between a decarburizing atmosphere and the steel being treated (in single thicknesses) for such products as silicon-bearing sheets, whose magnetic properties require very low carbon contents. The controlled recrystallization possible in continuous heat treatment makes a fine-grained microstructure easy to obtain when such is desired, for example, in high-strength cold-rolled sheets.

SECTION 8

EFFECTS OF HEAT TREATMENT ON MICROSTRUCTURE

Box Annealing—Prior to cold reduction, low-carbon rimmed steel in the form of hot-rolled coils has more or less equiaxed ferritic microstructure, with the carbides visible as pearlite or cementite (depending on whether the product was coiled cold or hot); it is relatively free of internal stresses, particularly if coiled hot and so "self-annealed." Cold reduction, however, elongates the grains up to ten-fold, greatly distorts the crystal lattice, and induces heavy internal stresses; the resultant product is very hard, with little ductility. During heat-treatment, this severely cold-worked microstructure will recrystallize at temperatures below those required for transformation. If such recrystallization is allowed to continue to completion by holding the steel at the proper temperature for sufficient time, the resultant structure again will consist of clearly-defined equiaxed ferrite grains and dispersed carbides with undistorted lattices and the steel again will be soft and ductile. In whatever state they were in the hot-rolled product, the carbides will have formed cementite, either small scattered spheroids (from hot-rolled breakdowns coiled cold) or coarse agglomerates (from hot-rolled coils coiled hot). But it should be noted that the form of the carbides in cold-rolled and annealed sheet not only depends on the hot-mill coiling temperature, but also on the amount of cold reduction and on the annealing temperature. For example, when coarse carbides result from high coiling temperatures, these usually break up to smaller, angular carbides when heavy cold reductions are used (such as the 80 to 90% levels used for tin mill products). Also, coarse carbides can result from annealing above the lower critical temperature, especially where long annealing times are involved (such as in batch annealing operations).

Assuming sufficient time at the annealing temperature, steel given a heavy cold reduction will begin to recrystallize at a lower temperature, will complete recrystallization more quickly, and finish with a finer ferrite grain, than steel given a light cold reduction. This is because the former material is more distorted before annealing and so has more centers of nucleation and higher localized stresses to induce the crystalline realignment. Similarly, those variations in practice during the hot-rolling operation which affect grain size of the product will affect similarly the microstructure of the cold-reduced, box-annealed steel, within the limits determined by the steel grade, the degree of cold reduction and the annealing practice.

Recrystallization begins at each nucleation center with a return from distorted to "normal" atom alignment and is propagated by absorption of the surrounding distorted material into that alignment until the growth stops, establishing a single ferrite grain. Grain formation generally is stopped by the advancing fronts of differently-oriented adjacent grains. Within the practical range of recrystallization temperatures, however, the tendency of adjacent grains to assume the same lattice alignment, and to merge to form a larger grain, increases with increasing temperature; the maximum annealing temperature at which the steel is held for a significant time, therefore, determines the finished grain size for a given steel grade, hot-rolling practice and degree of cold reduction.

Beyond the period permitting full recrystallization—one to four hours at maximum temperature is sufficient for common steel grades and mill practices—time at subcritical temperature has relatively little effect on the grain size of the common steels. Extended "soaking" used to be employed in the belief that lower hardnesses result. However, in modern box annealing practices soaking is either minimized or eliminated to conserve energy and optimize productivity.

A box-annealing practice developed to attain a given end result in grain size and mechanical properties, therefore, must balance several factors, some of them outside the annealing operation itself and invariable to any effective degree. Fortunately, minor variations in annealing practice have little effect on results and effective variations aimed at attaining lowest steel hardness values are not difficult to make or control.

Continuous Annealing—Continuous annealing of light-gage cold-reduced steel in a reducing atmosphere is widely used for tin mill products and hot-dip galvanized sheets. In addition, continuous annealing is finding increased use for the production of cold-rolled sheet which may be used as such or subsequently coated in electrolytic plating lines. The nature of the continuous annealing process differs slightly from one product to another, which has metallurgical and equipment consequences.

For tin mill products a single strand of steel, cold reduced to the thin section (about 0.01 inch) used for tin plate, travels at high speed through a heating zone having a controlled atmosphere, where it is brought to a temperature just above the lower critical in a very short time, recrystallizes almost instantaneously, passes through a cooling zone and emerges into the air cold enough to avoid oxidation. Such an operation provides fully-recrystallized steel, ductile in spite of its relative hardness and nearly free of directionality in mechanical properties. The extremely short time at temperature is effective because recrystallization has been suppressed by the rapid temperature rise; the resultant increase in energy level at all potential nucleation centers causes the microstructure to recrystallize rapidly once the process begins.

The resulting product is usually harder than corresponding box annealed product because of the finer grain size and because of a larger amount of carbon and nitrogen retained in ferrite solid solution owing to the rapid cooling rate. For hot-dip, galvanized sheets the heating rate is not as high as for continuous annealing of tin-mill products primarily because of the thicker sections 0.4 to 1.8 mm (0.015 to 0.070 inch). However, the recrystallization is still rapid and fine grain sizes usually result. The cooling rate is usually controlled so that the sheet enters a pot of molten zinc at about the same temperature as the zinc (about 455°C or 850°F). However, the cooling is not delayed sufficiently to allow complete precipitation of carbon from solid solution and as a result the in-line annealed product is susceptible to carbon strain aging as well as nitrogen strain aging in non-aluminum-killed steels. Like tin mill products in-line, continuous annealed hot-dip galvanized sheet is harder than box-annealed product. Yet for all but the most difficult-to-form parts (for which pre-box annealed sheet is used), the properties are entirely satisfactory. For cold-rolled sheets more of the applications have historically required higher formability than for hot-dip galvanized sheets. For this reason the continuous annealing process has been modified to provide an overaging zone after the recrystallization annealing section. Overaging at temperatures of about 288° to 425°C (550° to 800°F) allows for a more complete precipitation of carbon from ferrite solid solution and hence increases the ductility (an important factor for good forming) and reduces the tendency for strain aging. The time required for carbon precipitation during overaging depends on the rate of cooling from the primary annealing temperature (usually 730°C to 830°C) (1346°F to 1526°F). Rapid cooling rates (such as obtained by water quenching) require shorter overaging times (1 to 1¼ minutes) compared with slow cooling rates (gas jet cooling) which require about 3 to 5 minutes. The extent of carbon precipitation during overaging also depends on the composition of the steel especially the carbon content; for relatively high carbon contents (above 0.025 per cent) the precipitation of carbon is usually more complete than for very low carbon steels (about 0.01 per cent) because the presence of carbides in the higher carbon steels acts as sights for precipitation of carbon. In box annealed aluminum-killed steels, excellent deep drawability is obtained by forming preferred crystallographic textures by ensuring precipitation of aluminum nitride coincident with recrystallization. However, in continuous annealing there is insufficient time for precipitation to coincide with the rapid recrystallization. Therefore, for continuous annealed cold-rolled sheets, the steels are usually coiled hot during hot rolling (in contrast to the low coiling temperatures used for box annealed product) to precipitate the aluminum nitride and coarsen the carbides. Such practices, in combination with controlling the steel composition and the annealing cycle, allow good formability (ductility and crystallographic texture) to be developed in continuous annealed cold-rolled sheets. However, the properties are still not quite as good as box annealed deep drawing quality (DDQ) sheet, also referred to as drawing quality special killed (DQSK sheet), unless radically different steels such as titanium or columbium treated interstitial-free steels are used. Therefore, for the most part continuous annealing of cold-rolled sheets is used primarily for commercial quality (CQ) and drawing quality (DQ) sheet grades. Nevertheless, much research is underway to develop continuous annealed DDQ sheet that is fully competitive with box annealed DDQ product, and it is likely that continuous annealing will be used to produce DDQ sheet in the future. Continuous annealing is particularly useful for the production of high-strength cold-rolled sheets. The rapid annealing and excellent control of temperature results in higher, more uniform strengths than by box annealing. In addition, the rapid cooling allows transformation to unique structures such as the ferrite/martensite mixtures of duel-phase steels.

Normalizing—Compared with box and continuous annealing, only small tonnages of sheet and tin mill products are continuous normalized. Normalizing involves heating the cold reduced sheet above its upper critical temperature (which depends on the steel composition but usually is around 910°C to 980°C (1700°F to 1800°F) and cooling at a rate which permits the formation of the required ferrite grain size. Although continuous normalizing as a means of heat treating sheet and tin mill products has declined in the last 10 to 15 years, it is still used for three principal product areas:

- low metalloid "irons" or steels such as those used for porcelain enameling applications. Because of low carbon and manganese contents these steels often have high oxygen contents and hence contain numerous fine oxide particles. These fine oxides often prevent complete recrystallization during conventional box annealing and consequently the steels are continuous normalized instead; the higher temperatures ensure the required recrystallization.

- high-strength alloy-steel sheets. These steels also tend to be sluggish in recrystallization and even if complete recrystallization is obtained by box annealing, mixed ferrite grain sizes and coarse pearlite colonies are often obtained which do not provide optimum strengths. Consequently, continuous normalizing is used for these steels to produce uniform fine ferrite grain sizes, fine pearlite and hence suitable strengths.

- high-strength martensitic sheets. These steels have very high strength levels (tensile strengths of 150 to 200 ksi) and are produced by continuous normalizing to form austenite at the high temperature followed by rapid cooling to form high-strength martensite instead of low-strength ferrite. Some modern continuous annealing lines have the temperature capability for continuous normalizing, but the composition of the steel and the cooling rate are also critical factors to obtaining structures containing large proportions of martensite (70 to 100 per cent).

SECTION 9

HEAT-TREATING EQUIPMENT AND PRACTICES

Because of the wide range of flat-rolled products that receive heat treatment in their course of manufacture, and because the design of equipment is influenced by operators' preferences and the recommendations of those who design and build furnace equipment, heat-treating equipment and practices will be discussed here in general terms.

BOX ANNEALING

Box-Annealing Equipment—Box-annealing equipment consists of annealing bases on which to place the steel charge, furnaces to apply the heat, and, generally,

inner covers which fit over the charge in the furnace and contain the protective atmosphere that prevents oxidation of the steel (Figure 34—6). Each of these basic units may vary considerably in design, with little or much auxiliary equipment; in any steel plant several sets of units are grouped together into an annealing department, which is serviced as a whole by tracks, cars, tractors, cranes, and atmosphere preparation equipment.

In most box-annealing equipment the bases are stationary and the portable furnaces are lowered by crane onto the loaded base and attached to fuel and control connections for the annealing operation. To attain

Fig. 34—6. A 450-metric-ton (500-net-ton) capacity annealing furnace for coils. The furnace is being lowered by crane over the base loaded with eight stacks of coils. The burner tubes may be seen extending across the interior of the furnace. Note guide posts at the corners of the base.

maximum furnace utilization, two, three, or four bases and an equivalent number of inner covers are provided for each furnace. Thus, no furnace time is wasted while the bases are being loaded or cooled to handling temperature.

An alternative form of box annealing uses stationary furnaces through which the coils are moved progressively so that the annealing cycle is controlled closely and the energy use is optimized.

Annealing bases are usually rectangular, although circular bases (and furnaces) are not uncommonly used for coil annealing of cold-reduced products. A base consists of a shallow tray of cast iron or, more commonly, refractory-lined steel. For sealing the open down-end of the inner cover, the bottom of the base may be covered with a layer of sand or the sand may be contained in a trough around the periphery of the base. Where expensive protective atmospheres are used, such as cracked ammonia (to protect high-carbon steel against decarburization), the trough into which the edge of the inner cover fits is filled with oil, low-melting alloy, or insulating material for perfect sealing.

For annealing coils, rectangular bases have two to eight raised "stools," each capped with an annular plate and generally containing a motor-driven fan for atmosphere circulation and better heat transfer. In addition, most bases are fitted with thermocouple inlets and atmosphere inlets and outlets. The dimensions of a base are, of course, determined by the length and width of the furnace used on it.

Inner covers are commonly open-bottomed thin-walled steel cylinders to fit cylindrical charges of stacked coils. These shells commonly are formed and welded either from low-carbon steel or stainless-steel sheets. They may be strengthened by beading or corrugating the walls and the carbon-steel covers may be coated with aluminum oxide, sodium silicate or other protective compounds to reduce their oxidation rate.

Box-annealing furnaces are stationary or, more commonly, portable. **Stationary furnaces,** rectangular in shape, may be built singly or in batteries of two or four. Each furnace consists of a rectangular steel frame lined with refractory materials. Charging ends are fitted with hinged or counter-balanced doors, the latter traveling in vertical tracks. Floor tracks extend from the loading area into the furnace hearth and one or two charged bases are pushed into the furnace on cast-iron balls or carried in on a charging machine. Such furnaces may be fired with coal, oil, or gas.

Portable annealing furnaces consist of a structural-steel frame covered with steel plate and lined with refractory material. The rectangular furnace generally used varies greatly in size and in annealing load weight, depending on mill requirements, handling facilities and effectiveness of design in providing temperature uniformity.

Such furnaces commonly are gas-fired, although any fluid fuel may be used and electrical resistance heating has been tried. In some cases, the burners fire into "radiant tubes" of stainless steel. These tubes are 75 to 150 mm (3 to 6 inches) in diameter and as long as the furnace size and desired temperature distribution dictate. They may run vertically or horizontally along the inner walls of the furnace or, in large furnaces for annealing coils, may span the distance between the walls for better temperature distribution. The tubes vent the products of combustion on the outside of the furnace.

Heating of the inner cover thus occurs by radiation. Proper tube location and the flexibility possible in having each tube served by an individual burner permit the necessary uniformity of heating of the large masses of steel contained in one annealing charge. For minor adjustments, burners can be controlled individually. They are manifolded together, however, and the fuel flow to the individual burners of one furnace is usually determined by the main fuel valve of the furnace, which is controlled automatically by predetermined tube-temperature control settings and by charge-temperature controls.

In common use today are tubeless furnaces, in which the burner discharges onto a steel heat shield which prevents flame impingement on the inner cover but permits circulation of the hot combustion gases between the furnace wall and the inner cover. Convection plays an important part in heating the inner cover in this design. The decision to use a tubeless furnace is influenced by the relative importance of fuel economy, freedom from radiant tube costs, higher inner-cover costs and different temperature-distribution characteristics.

Box-Annealing Practices—To begin an annealing operation in portable furnace equipment, coils are loaded on the base stools in stacks about 2.4 to 3.7 metres (8 to 12 feet) high; loading is done by traveling overhead cranes equipped with hooks, slings, retractable racks or magnets. Thermocouples are inserted in standard locations in the charge. Individual cylindrical "ash can" inner covers are lowered over each stack of coils and settled in the sand seal. The furnace is then lowered onto the base, fuel line and thermocouple connections are made, the flow of deoxidizing gas to purge the air from the space under the inner cover is begun, and the burners are ignited. Base fans, if available, are turned on to effect high-speed circulation of the atmosphere in the inner cover.

The subject of reducing-gas protective atmospheres is treated more fully elsewhere in this volume. An atmosphere that is used less now than formerly consists of the products of partial combustion of a fuel gas in a limited volume of air, these products then being treated to remove most of the resultant water vapor, dirt and carbon particles, and in some cases further treated to remove sulphur dioxide and carbon dioxide. The resultant mixture of 75 to 85 per cent nitrogen plus varying percentages of hydrogen, carbon monoxide, carbon dioxide and methane protects the bright steel surface from visible oxidation throughout the annealing cycle. In most common use for annealing of sheet and tin mill coils to better promote surface cleanliness required for these products are dry mixtures of hydrogen and nitrogen with hydrogen contents from 5 to 15 percent. Such mixtures are frequently prepared by blending nitrogen from an air liquefaction plant with hydrogen that may be produced on site by steam reforming of methane. Other effective atmospheres include dissociated ammonia and high-purity nitrogen.

During the heating-up period, fuel consumption is maintained at a constant rate until the gas stream temperature, determined by a thermocouple connected to a recording-controlling instrument, reaches a predetermined level.

A simplistic view of box annealing is that each portion of the charge is heated slowly but uniformly to a specified temperature at which the steel is "soaked" for a fixed time to ensure complete softening. However, modern box annealing practices recognize that a charge of several coils of steel is not heated uniformly throughout to the same temperature at the same time, that is, there is no specific "soak" time/temperature. The geometry of a coil, the stacking within a furnace, and the heat-transfer characteristics of coils and furnace prevent uniform temperature distribution in the coil. However, there is a "cold spot" that is about one third of the way in from the inside of the coil and at midwidth of the coil, and a "hot spot" that is usually at the top outer edge of the coil. The annealing cycle is defined usually by a minimum temperature to be achieved at the cold spot and a maximum temperature not to be exceeded at the top outer edge. In current practice, the heating portion of the annealing cycle for each charge is determined by computer calculation of the requirements to properly anneal the most difficult-to-anneal coil in the charge. Factors considered in these calculations for each coil in the charge are coil weight, coil width, sheet thickness, and the coil location in the charge. The computer defines the required gas-stream temperature and the total heating time. After cooling, the furnace is removed to begin a cycle on another base, and the charge is allowed to cool (still in a protective atmosphere under the inner cover) to about 150°C (300°F), when it can be exposed to air without oxidizing. The cooling period takes at least as long as the combined total time of heating and soaking.

The temperature and times specified for annealing practices vary greatly from plant to plant, even in the manufacture of the universally similar dead-soft low-carbon sheet. This is true because of inherent differences in the cold-reduced steel to be annealed and differences in furnace size, design, and load size. A stack of coils consists of a hollow cylinder with a 406 to 762 mm (16 to 30 inch) inside diameter and a 914 to 2134 mm (36 to 84 inch) outer diameter, receiving inner-cover radiant heat perpendicular to the curved planes of the layers of steel comprising the coils. This handicap can be overcome in part by blowing the atmosphere at high speed through the spaces inside and outside the coil stack, and so transferring some of the heat from the inner cover to the inner diameter of the coils. Another means is to separate each pair of coils in a stack with "convector plates," resembling flattened doughnuts with passages for the hot circulating atmosphere between top and bottom surfaces. Thus, some heat is transmitted to the ends of each coil, from which it has an easier path to follow.

Regardless of the wide differences necessary in specified control cycles, the coldest spot of an annealing charge, whether or not actually accessible to a thermocouple under production conditions, will be found to have been annealed for 10 to 20 hours at a temperature in the range of 660° to 690°C (1225° to 1275°F) in virtually all plants making deep-drawing sheets, with the proper formula within these limits being determined by the prior history of the cold-reduced steel. Depending on furnace and load characteristics, the hottest part of the charge will have been at 690° to 730°C (1275° to 1350°F) for 20 to 50 hours, and the furnace will have been under fire for 30 to 90 hours.

Such temperature differences are more critical in annealing tin plate, where higher hardness levels are wanted. Partial protection against over-annealing exists in this case because the heavy cold reductions used limit the grain size to a relatively fine, and therefore relatively hard, structure, regardless of any annealing temperature employed below the transformation-temperature range. Such heavy cold reduction also induces full recrystallization at a lower temperature than is necessary on steel thicknesses given less cold reduction, and tin plate can be annealed fully in the slightly lower temperature range necessary to attain proper hardness.

For products other than low-carbon cold-reduced sheets, strip or tin plate, annealing practices can vary considerably from the outline given. Box-annealing furnaces are used occasionally at temperatures as low as 540°C (1000°F) to stress relieve, rather than recrystallize, certain specialty steels, and temperatures up to 1095°C (2000°F) have been employed on other specialties.

Open-Coil Annealing—As has been stated earlier, open-coil annealing minimizes the difficulties inherent in box annealing in regard to uniformity and rapidity of heating. Open-coil annealing can employ protective atmospheres to prevent scaling, decarburization, or carburization of the charge equally as well as box annealing; however, by proper selection of the atmosphere, the open-coil technique can also be used to deliberately change the chemical composition of the steel, or to obtain desirable surface effects in a controllable manner. Obviously, time-temperature relationships as well as atmosphere controls are factors in obtaining the desired results from such special uses of the open-coil technique.

CONTINUOUS ANNEALING

From an engineering standpoint, the continuous-annealing operation is made practical by building the heating and cooling zones as towers and increasing their effective length by threading the steel back and forth around rolls at the top and bottom of the towers. This principle is illustrated in Figure 34—7, which shows schematically the path of full-width tin-plate stock passing through the line that consists, in the direction of strip travel, of a double payoff reel; shears for squaring the end of coils for welding; a welding unit; an alkaline electrolytic cleaner with a brush scrubber, water rinse and drying unit; and an entry looping tower that can "store" 305 metres (1000 feet) of wide strip and from which strip is taken to maintain constant line speed when a weld is being made. This line was designed to operate at a speed of 7.6 metres per second (1500 feet per minute). Following the entry-end equipment just enumerated is the furnace proper that consists of a gas-fired heating zone wherein the strip makes ten passes; an electrically heated holding zone of ten passes; an electrically heated slow-cooling zone of six

passes; a water-jacketed fast-cooling zone of thirty passes; a final cooling zone of four passes; another looping tower; and two recoilers. Each pass consists of 15.2 metres (50 feet) of strip steel, and all passes are vertical. The strip is heated to approximately 730° C (1350° F) in the heating zone in 20 seconds and maintained at this temperature for 20 seconds in the holding zone. Upon leaving the holding zone the strip is cooled to about 540° C (1000° F) in 12 seconds in the slow-cooling zone and thence to about 115° C (240° F) in 60 seconds in the fast-cooling zone. From its entry into the heating zone until it leaves the fast-cooling zone, the strip is protected from oxidation by a gas atmosphere containing 95 per cent nitrogen and 5 per cent hydrogen. Subsequent cooling is done in 8 seconds in air. At an operating speed of 7.62 metres per second (1500 feet per minute) less than two minutes (112 seconds) elapse from the time a given section of strip enters the heating zone to the time it leaves the fast-cooling zone.

The equipment used in the continuous annealing of hot dip galvanized sheet is described more fully in the chapter on galvanized products. The continuous hot dip galvanizing line serves three principal functions: 1) the reduction of surface oxides by heating in a controlled reducing atmosphere which prepares the steel surface for subsequent coating with molten zinc, 2) heating steel either to just above the molten zinc temperature or to higher temperatures (for in-line recrystallization annealed product), and 3) controlling cooling of the sheet to the zinc-pot temperature and below so that the required coating thickness and surface condition can be obtained.

The continuous annealing of cold-rolled sheets is accomplished on lines that usually incorporate other processing steps in addition to annealing. For this reason the lines are often referred to as continuous annealing and processing lines. These lines usually include electrolytic cleaning of the sheet, preheating and heating sections, controlled cooling (by either gas jet cooling, water quenching, or gas-water mist cooling) overaging, pickling and rinsing (to remove oxide films), temper rolling and oiling. The lines which are usually 600 to 750 feet long are designed to handle sheet coils of 0.4 to 2.0 mm (0.015 to 0.079 inch) thickness and 24 to 65 inches wide at entry/delivery speeds of 600 to 900 feet per minute. The lines are usually vertical and the number of loops within the furnace and overaging sections depends on the thickness of the sheet and the annealing temperature and time required. Heating is usually accomplished by radiant tube burners operating on coke oven gas, by electric heating or both. A reducing furnace atmosphere is provided by a gas mixture of nitrogen and up to about 5 percent hydrogen. After a preheat zone which uses recirculated spent gases from the combusion of coke oven gas, the sheet passes into the main heating (annealing) furnace and is raised to temperatures in the range 730°C to 830°C (1345°F to 1526°F) and held for 1 to 2 minutes. (The choice of annealing temperature and time depends on the grade and quality of the steel being produced. Usually higher temperatures and longer times are used for higher quality forming grades). After leaving the heating furnace the sheet is cooled in one of several ways: gas jet cooling (cooling rates of about 5° to

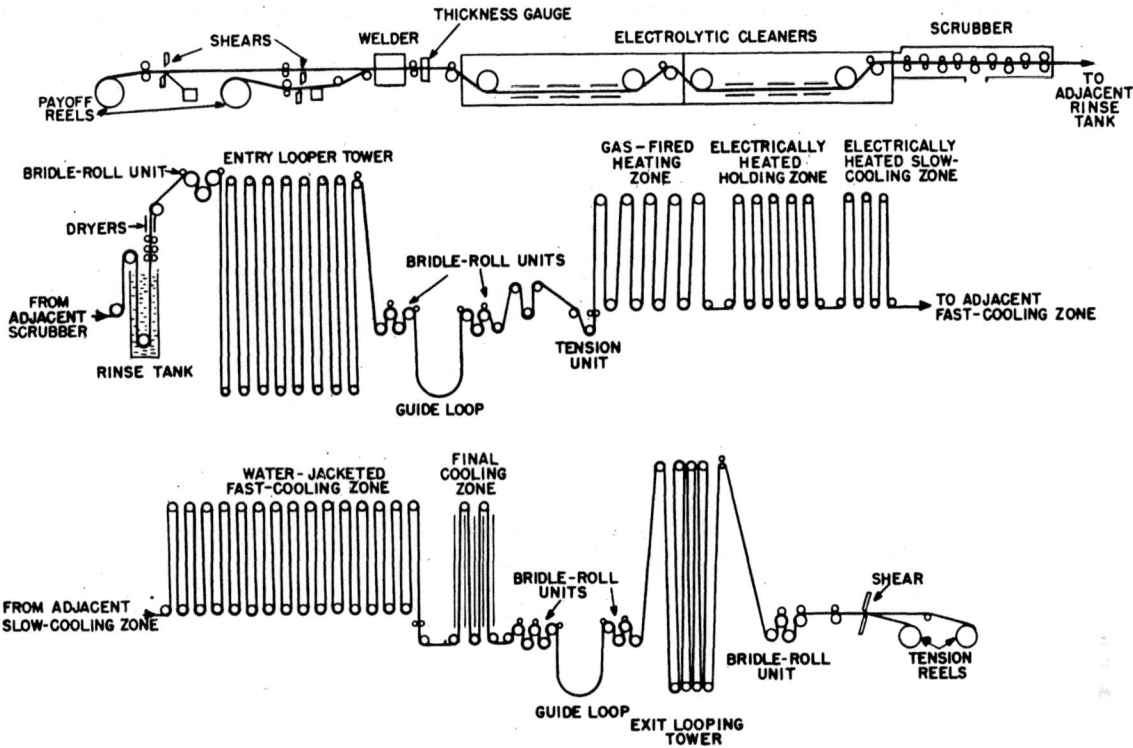

Fig. 34—7. Schematic diagram (not to scale) of a continous-annealing line used for tin mill products and designed to operate at a strip speed of 7.62 metres per second (1500 feet per minute).

50°C/sec.), quenching in boiling water (cooling rates of 25° to 150°C/sec.), gas-water spray cooling (cooling rates of 80° to 300°C/sec.), cooling by contact with water cooled rolls (cooling rates 80° to 150°C/sec.) or quenching in cold (room temperature) water (cooling rates of above 1000°C/sec.). When cold-water quenching is used the cooling process cannot be interrupted at the overaging temperature (about 288° to 425°C or 550° to 800°F), hence the sheet must be reheated to the overaging temperature. Reheating consumes more energy than those processes which use other cooling techniques that allow interruption at the overaging temperature. However, the overaging time required usually increases as the cooling rate (from the heating/annealing zone) is decreased, hence the length of the overaging furnace increases as cooling rate decreases. The diameter of the rolls in the overaging furnace is critical for avoiding high surface stresses on the sheet as it passes over the rolls during each looping pass. High surface stresses can result in deterioration of the ductility of the sheet. Therefore, roll diameters between 1200 and 1600 mm (47 to 63 inches) are used in the overaging section. Modern continuous annealing lines have sophisticated computer control procedures for monitoring/controlling sheet tracking, the sheet tension, line speed, and sheet temperature. The high degree of computer control optimizes the uniformity of properties and flatness in continuous annealed product.

Normalizing —Whereas, in the past, normalizing was carried out by passing one or several sheets through multi-zone (usually 3 zones) conveyor type furnaces, today virtually all normalizing is carried out by continuously heating and cooling the strands from coils. Several different facilities are used which depend on the product being normalized.

Many modern hot-dip galvanizing lines have the capability for continuous normalizing which is often required for alloy-steel sheets. The furnaces can be of the horizontal type (single or multi layer) or vertical type. For alloy steels and enameling "irons" (steels), continuous normalizing is accomplished on horizontal type, roller-hearth lines which have many rolls throughout the furnace to support the sheet. In contrast, free loop or catenary type furnaces are designed to continuously normalize a sheet that passes through the furnace without contacting rolls inside the furnace (the support and guide rolls are immediately before and after the furnace). Therefore the heating zones tend to be short (about 6 to 15 meters or 20 to 50 feet).

Such furnaces may have pickling or other descaling equipment at the exit end to remove, in the same operation, the surface oxides formed on the steel during normalizing. Catenary furnaces so equipped, but without a cooling zone, are widely used for the heat treatment of stainless steel, as the temperatures of 1040° to 1205°C (1900° to 2200°F) used on the austenitic grades would result in short life for rollers used in the furnace. Steam or water quenching facilities are usually provided at the exit end of furnaces for rapidly cooling the stainless steel.

TEMPER ROLLING

The purpose of temper rolling as applied to cold-reduced flat-rolled products depends upon the type of product. In most sheet products, the main purpose is to suppress the yield-point elongation that is present in the as-annealed state for most steels. If not suppressed the yield-point elongation results in surface markings called "stretcher-strain markings" or "luder lines" on lightly deformed panels as are often experienced in complex stampings. For parts that are visible to the end user, such as painted automobile body panels, the stretcher-strain markings result in an undesirable surface appearance. For sheets, temper rolling is equally important for imparting the necessary surface finish (as determined by surface roughness and peaks per inch). Temper rolling is also used to improve the shape and flatness of sheets and tin mill products. For tin mill products especially, temper rolling is used to develop the proper stiffness or temper by cold working the steel in controlled amounts. For high-strength cold-rolled sheets, heavy temper rolling (1.5 to 3.0 percent extension) is sometimes used to raise the strength of the as-annealed sheet to a specified minimum yield-strength level. However, the heavier the temper rolling the greater the deterioration in sheet ductility. For electrical sheets, especially cold-rolled motor laminations, heavy temper rolling (2 to 8 percent extension) is used to improve magnetic properties after the stamping and decarburization-annealing steps are carried out by the electrical equipment manufacturer. In addition, temper rolling tends to improve the flatness of annealed strip, to develop desired mechanical properties, and to impart the desired surface finish to the finished product.

The mechanical and stiffness properties imparted to steel by temper rolling are related to the degree of reduction in thickness effected by the cold working. Since an increment of length is added to each unit length that passes through the mill due to the reduction in thickness with the width remaining essentially the same, increase in length, referred to as **extension**, is used as the criterion for determining the relative reduction in thickness. The temper rolling of sheet coils on a temper mill involves back tension on the product between the uncoiling reel and the mill, and forward tension between the mill and the recoiling reel. Since coil temper mills operate at relatively high speeds, extension of the strip due to the combination of tension and roll pressure is controlled by automatic measuring and control devices.

Sheet temper mills are of both two-high and four-high types and consist of either single-stand or two-stand four-high or six-high types. Single stand four-high or six-high temper mills are used primarily for sheet products where relatively small extensions are used to suppress yield-point elongation, impart the required surface finish, and improve the flatness. Most modern temper mills are equipped with automatic flatness-monitoring devices. On many mills the flatness information is fed back to the temper mill to provide automatic control of the shape/flatness. Two-stand four-high coil temper mills for sheets are used to impart improved flatness to the product. Two-stand tandem four-high temper mills are also used to produce higher strength sheet and tinplate products and for electrical sheets.

The finish of the rolled product is controlled by using rolls having a variety of surface finishes developed to impart the desired finish to the product. Roll finishes range from ground and polished rolls to impart a bright finish, to shot-blasted or electric-discharged textured rolls that produce a dull, velvety finish on the steel surface.

Mechanical properties imparted by temper rolling vary with the amount of extension (which, as stated above, is proportional to the reduction in thickness). Sheets intended for deep-drawing applications receive about 0.5 to 1 per cent extension, which is sufficient to suppress the formation of stretcher strains during forming without significantly impairing ductility. Sheets having lesser ductility requirements are given about 1 to 1.5 per cent extension.

SHEARING, SIDE TRIMMING, SLITTING AND LEVELING

Shearing to Length—Much less shearing is being done in sheet and tin mills than formerly, because most of the tonnage is being shipped in coil form. There are two general types of shears: (a) those for making cuts across the width of the strip, and (b) those known as side-trimmers or slitters that make continuous cuts along the length of a moving strip.

Practically all sheets produced by shearing from coils of cold-reduced material are cut to length on continuous-shearing lines employing **flying shears** that may be of either the guillotine type or the rotary type. As the name "flying shear" implies, both types perform the cutting operation on strip passing through the shearing line at some pre-set speed.

The guillotine-type flying shear operates on the same principle as a stationary-type guillotine shear except that it is mounted in a movable housing that can move in the same direction and at the same speed as the strip while the knives perform the cutting operation. After each cut, the housing moves back to its original position in preparation for the next cut. Flying guillotine shears are used in lines capable of cutting strip moving at speeds up to 1.8 metres per second (350 feet per minute).

Rotary-type flying shears consist of two horizontal cylinders, mounted one above the other in a housing, each carrying a knife that is parallel to the axes of the cylinders and that extends a suitable distance from the cylinder face. The cylinders can be operated so that the knife edges are brought together at intervals to achieve the proper length of cut. Lines employing rotary-type flying shears can operate at strip speeds up to 5.1 metres per second (1000 feet per minute).

The speed advantage of the rotary shear is best realized on lines employed to cut large numbers of sheets of the same size. When small lots of sheets of different lengths are to be cut, it is possible to cut the same number of sheets per day on a line employing a guillotine-type flying shear.

Side Trimming and Slitting —As was stated in Chapter 33, most continuous-pickling lines have a side trimmer installed at the exit end. These trimmers employ mating circular knives, mounted on arbors, to remove continuously the desired amount from both edges of the strip, thereby establishing accurate and uniform width and producing parallel and reasonably smooth edges. Side trimmers have also been installed on continuous-tinning lines and continuous-annealing lines. Many sheet-shearing lines, however, do not have side trimmers because practically all sheet-mill strip is side trimmed at the continuous-pickling line and may not need a second trimming. Side trimming of strip for electrolytic tin plate may be done in separate lines called coil-preparation lines or, in some cases, by trimmers incorporated in the tinning line itself or in the cutting-up line that follows the continuous-tinning line.

The principle of operation of the machines used for slitting was described in Section 6 of Chapter 33.

Leveling—Roller-leveling and stretcher-leveling of cold-reduced product for sheets are performed on equipment similar to that described in Section 6 of Chapter 33, and for the same purpose—to improve flatness of individual sheets to the degree required for the intended application.

Semi-continuous stretcher leveling (tension leveling) of cold-rolled strip product is made practical by building bridles on each side of the leveling unit and providing a pay-off reel, welder, and recoiler to make the operation semi-continuous. The principle is illustrated in Figure 34—8, which shows schematically the units in a leveling line for full-width sheet stock.

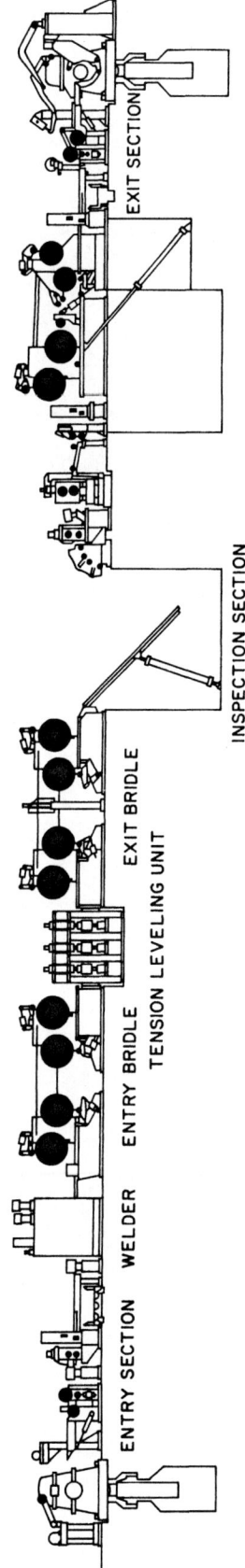

Figure 34—8. Schematic diagram (not to scale) of a semi-continuous tension leveling line.

Bibliography

Am. Iron and Steel Inst., Steel products manuals (Carbon Steel; Semifinished for Forging; Hot-Rolled and Cold-Finished Bars; Hot-rolled Deformed Concrete Reinforcing Bars. Also, Carbon Steel Sheets. Also, Cold-Rolled Carbon-Steel Strip. Also, Tin Mill Products). The Institute, Washington, D.C.

American Soc. for Testing and Materials, ASTM Standards, Part 3. Steel Plate, Sheet Strip and Wire; Metallic Coated Products ASTM 1981, Philadelphia, Pa.

Blickwede, D. J., "Sheet Steel—Micrometallurgy by the Millions," 43rd Campbell Memorial Lecture, Trans. ASM Vol. 61, 1968; also in "Source Book on Forming of Steel Sheet," ASM, Metals Park, Ohio, 1976.

Bramfitt, B. L. and P. L. Mangonon, eds. "Metallurgy of Continuous Annealed Sheet Steel." The Metallurgical Society of AIME, Warrendale, Pa, 1982.

Leslie, William C., "The Physical Metallurgy of Steels." McGraw-Hill Series in Materials Science and Engineering. McGraw-Hill, 1981.

Nakamura, T., K. Kobayashi, K. Terai, T. Nishimura, M. Fujioka, T. Nanba, "Recent Trends in Steel Strip and Processing Lines," Nippon Steel Technical Report No. 20, December, 1982, pp. 131-151.

"Recent Developments in Annealing," The Iron and Steel Institute Special Report 79, 1963. The Iron and Steel Institute, England.

Roberts, William L., "The Cold Rolling of Steel". Marcel-Dekker Inc. New York, 1978.

CHAPTER 35

Corrosion and Protective Coatings

SECTION 1

CAUSES OF CORROSION

The Mechanism of Corrosion—The corrosion of the common metals in usual environments is an electrochemical phenomenon. That is, it is associated with the flow of electric currents over finite distances. Electric currents associated with corrosion have been detected in numerous cases, and in a limited number of instances the amount of corrosion occurring has been accounted for quantitatively by the amount of electric current which passed.

Knowledge that corrosion is electrochemical is important, since it assists in the development of methods for combating corrosion. For instance, it is obvious that, in order for electrochemical corrosion to occur, there must be differences in potential between different areas of the corroding surface. Such differences can be caused by the use of dissimilar metals or alloys in contact with each other. However, differences in potential can be caused by heterogeneities of any kind in the metal surface or in the environment contacting the metal. Some of the most important of these heterogeneities will be discussed below.

FACTORS WHICH AFFECT CORROSION RATE

The fundamental reason why metals corrode is that the corrosion products are more stable than the metals themselves. Thermodynamically speaking, in order for metals to corrode there must be a decrease in free energy associated with the formation of the appropriate corrosion product from the metal. This free energy decrease is the **driving force** of the corrosion reaction. However, the magnitude of this driving force gives little information regarding the rate at which corrosion will occur. The rate of corrosion is determined by other factors which will be described later in this chapter.

Metal going into solution involves a loss of electrons and is by definition an anodic reaction. The area where this takes place is called an anodic area. Examples of anodic reactions are:

$$Fe \rightarrow Fe^{++} + 2 \text{ electrons}$$
$$Zn \rightarrow Zn^{++} + 2 \text{ electrons}$$

The reactions of either water or oxygen with electrons from the metal surface are by definition cathodic reactions. The area where these reactions occur is called the cathodic area.

The two most important cathodic processes in aqueous corrosion are the hydrogen-evolution reaction

$$H_3O^+ + e \rightarrow H_{ads} + H_2O$$
$$H_{ads} + H_{ads} \rightarrow H_2(gas)$$

and the oxygen-reduction reaction

$$O_2 + 2H_2O + 4e \rightarrow 4OH$$

Because metals are good conductors, the electrochemical reactions are short-circuited on a corroding metal surface. The anodic and cathodic reactions polarize toward each other until the potential difference between the two areas is just sufficient to pass the resultant corrosion current through the net resistance of the corrosion cell. This can be illustrated on a potential-current diagram as shown in Figure 35—1.

The anodic current on the anode area is equal to the cathodic current on the cathode areas which is equal to the corrosion current. The corrosion current is related directly to rate of metal loss by Faraday's law. The total

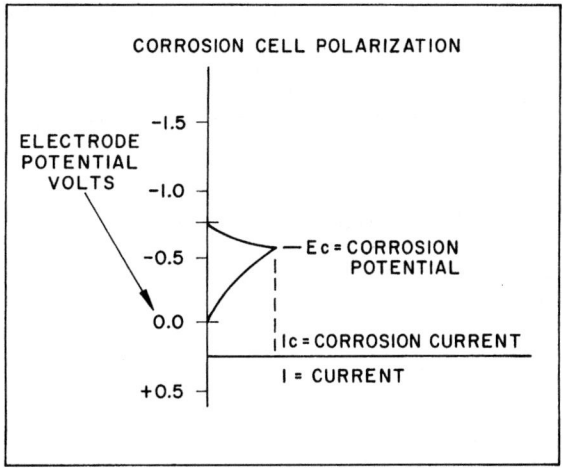

FIG. 35—1. Potential-current diagram.

current on the anodic and cathodic areas always is equal, but the current density on the two areas usually is not the same. This is because the two areas generally are not equal in size. Pitting-type attack occurs when the anodic areas are small (compared with the cathodic areas) and stationary. General attack occurs when the anodic areas are large, or if movement of the local cells on the surface occurs.

Moisture—The presence of liquid or gaseous water is the factor of most importance in stimulating normal types of corrosion (although in special cases the presence of some moisture may retard corrosion). The reason for the customary stimulation in corrosion rate caused by the presence of moisture is that moisture generally increases the electrical conductivity of the environment contacting the metal surface. Since corrosion is commonly electrochemical in nature, an increase in electrical conductivity of the environment will permit flow of larger electrical currents and, therefore, result in higher corrosion rates for given potential differences from point to point on the metal surface. Examples of the effect of moisture in stimulating corrosion are so well known that it is hardly necessary to elaborate greatly on this point. It has been established that even unalloyed steels will remain uncorroded if they are exposed to air with a relative humidity less than about 30 per cent. At higher humidities appreciable rusting will occur. Similarly, contact between steel and dry cloth or paper causes no serious attack of the steel, whereas, contact with damp cloth or paper will.

Salts—Neutral salts can stimulate corrosion in the presence of moisture by either or both of two mechanisms. They increase the electrical conductivity of the solution and thus increase corrosion currents. In addition, certain salts can form complexes with the metal corrosion products, thus increasing the solubility of the metal ion. This also stimulates corrosion. Local differences in salt concentration in a liquid which contacts a metal surface can cause severe localized attack as a result of the formation of concentration cells. These will be discussed in greater detail.

Acids—There are two broad classes of acids:—(1) oxidizing acids such as concentrated nitric acid and (2) the non-oxidizing acids such as hydrochloric acid. The non-oxidizing acids stimulate corrosion by permitting the more rapid evolution of hydrogen as a result of reduction of the hydrogen overvoltage. Oxidizing acids may or may not be corrosive, depending on whether they form thin protective films on the metal surface. The so-called **passivity** of steel in concentrated nitric acid is the result of the formation of a thin, insoluble film on the steel surface upon contact with nitric acid above a certain concentration. Very concentrated sulphuric acid also forms a protective film upon contact with steel. Under usual service conditions, an increase in the acidity of a solution generally tends to increase its corrosivity.

Alkalis—For most ferrous materials, an increase in the alkalinity of a solution generally will tend to reduce the total amount of corrosion, although it may increase the intensity of attack at local areas. Very strong caustic solutions, particularly at elevated temperatures, or molten caustic materials also can be corrosive to a serious extent on account of the amphoteric nature of iron.

Oxygen and Oxidizing Compounds—Free oxygen and many oxidizing compounds have a complex effect on the corrosion of steel. In aqueous solutions they stimulate the total amount of corrosion but tend to restrict the area which is attacked. On the other hand, if sufficient oxygen or oxidizing compound is present in the solution, attack will be prevented entirely. This inhibiting action of oxygen and oxidizing compounds is of most importance in the case of stainless steels, while the accelerating action of dissolved oxygen is of great importance in the corrosion of unalloyed steels by natural waters and many chemical solutions.

The behavior of free oxygen and oxidizing compounds in general can be described in electrochemical terms. These materials stimulate the cathodic reaction by depolarization, but they tend to retard the anodic reaction by forming films on the surface of the anode. Thus, their behavior is complex.

In gaseous exposures at elevated temperatures, the presence of oxygen in the atmosphere generally results in increased rates of scaling or oxidation.

High Temperatures—In general, the higher the temperature of the atmosphere, the faster corrosion will proceed. There are many exceptions, since change in temperature can affect simultaneously several factors which will influence corrosion rates. For example, raising the temperature of an aqueous solution exposed to air may either increase or decrease the rate of attack on metallic surfaces contacting the solution. Some of the effects of raising the temperature of a solution can be described in electrochemical terms. Increase in temperature increases the conductivity of the solution and also tends to decrease cathodic polarization. Both of these factors tend to stimulate corrosion. However, raising the temperature will decrease the concentration of oxygen dissolved in the solution which can cause more continuous films to be formed on anodic areas. Both of these factors tend to reduce corrosion rates. In practice, increase in the temperature of solutions freely exposed to the air usually results first in an increase in the corrosion rate, but as temperatures above about 80°C (180°F) are exceeded, the corrosion rate will decrease up to the boiling point.

Increase in temperature generally increases corrosion by gases, although here too special effects can come into play so that for some regions of temperature variation, relatively small increases in temperature will actually reduce corrosion rates. One of the most obvious examples of this is where the gas contains water vapor. At temperatures below the dew point, liquid water condenses from the gas and corrosion may be rapid. At somewhat more elevated temperatures, the humidity is less than 100 per cent, so liquid water does not form and the corrosion rate is much lower.

Galvanic Action—When dissimilar metals are placed in electrical contact and exposed to a conducting solution, generally corrosion of one member of the combination is accelerated while that of the other member is retarded. The metal of the combination which has the most active solution potential under the particular conditions of exposure is anodic and sends electric current through the solution to the cathodic metal. The direction of the current flow determines the member of the couple that will suffer accelerated or **galvanic corro-**

Table 35—I. Solution Potentials of Metals or Alloys in
A Solution Containing 53 g/1 NaCl plus 3 g/1 H_2O_2.

Metal or Alloy	Potential (Volts)[a]
Magnesium	-1.73
Zinc	-1.10
Aluminum	-0.84
Iron	-0.63
Copper	-0.20
Stainless Steel (18-18)	-0.15

[a]Measured against a 0.1 N calomel half-cell.

sion. The magnitude of the current determines the corrosion rate. The electromotive force series is a rough guide in predicting whether contact between two dissimilar metals in an aqueous environment will result in accelerated corrosion of one of them. However, a more dependable guide is obtained by measuring the potentials of the metals or alloys of interest in the aqueous environment to which they will be exposed in service. For many natural environments, measurement of potentials in solutions of sodium chloride or sodium chloride plus hydrogen peroxide gives more valuable information on probable galvanic behavior than does the emf series. Data of this type are given in Table 35—I.

From this table, it could be predicted that in most natural aqueous environments contact between zinc and aluminum would result in accelerated corrosion of the zinc and cathodic protection of the aluminum. This behavior actually occurs in sea water and in many natural fresh waters.

However, the emf series that appears in chemical handbooks lists aluminum as being anodic to zinc and, therefore, suggests that contact between the two metals would result in accelerated corrosion of the aluminum and protection of the zinc. As was mentioned above, this generally does not occur when the two metals are exposed in contact in most natural aqueous environments.

Thus, the emf series is not a dependable guide to galvanic effects. Tables such as Table 35—I are better but not infallible for several reasons. For example, that table indicates that magnesium is strongly anodic to

Table 35—II. Effect of Relative Areas on Galvanic Corrosion in Flowing Sea Water.

	Corrosion Rate[a] (mg/dm² per day)
Uncoupled Carbon Steel	0.0106
Carbon Steel coupled to 12 per cent Cr Steel of ⅛ the area	0.0150
Carbon Steel coupled to 12 per cent Cr Steel of equal area	0.0240
Carbon Steel coupled to 12 per cent Cr Steel 8 times the area	0.1200

[a]The specimens were exposed for 108 days. Corrosion of the 12 per cent Cr steel was negligible.

aluminum. It reasonably might be predicted that contact between magnesium and aluminum in sea water would result in accelerated corrosion of the magnesium, which occurs in practice, and in cathodic protection of the aluminum, which does not occur in practice. Instead, the aluminum member of the couple also will suffer accelerated corrosion. The reason for the severe attack on the aluminum, in spite of its being the cathode of the couple, is that aluminum is an amphoteric metal and is corroded readily by the alkali formed at its surface as a result of electrolysis of sea water by current generated by contact with the magnesium.

In other special cases, the cathodic member of a couple may suffer accelerated attack although, in general, it is protected.

The relative areas of the two dissimilar metals that are in contact also affect the amount of corrosion that occurs on the anodic metal. In general, the larger the surface area of the cathodic metal, or the smaller the surface of the anodic metal, the more severe the corrosion of the anodic metal (Table 35—II). In other words, it is the anodic current density, assuming the same duration of the exposure, that determines the extent of the accelerated corrosion of the anodic metal.

Since changes in practically all environmental factors: temperature, pressure, solution conductivity, solution velocity and the like, affect anodic current density, the knowledge that there is some specific potential difference between two dissimilar metals is not sufficient information to predict the extent of galvanic action, although it will predict accurately its direction.

Of special interest is the fact that some metals and alloys polarize rapidly at low cathodic current densities. The austenitic stainless steels are representative of this group of materials. For this reason, contact between anodic metals, such as aluminum, and stainless steel in aqueous environments does not result in as severe galvanic attack of the aluminum as does contact between aluminum and copper.

Whenever it is feasible to avoid contact between dissimilar metals, where they will be exposed for prolonged periods of time to conducting solutions, it is good practice to do so. However, whenever it is necessary to use dissimilar metals under such conditions, then consideration should be given in advance of actual construction to use designs which will alleviate undesirable galvanic effects.

There are at least six ways of alleviating galvanic corrosion when dissimilar metals are involved:

1. Select metals which do not generate appreciable corrosion currents under the conditions of use.
2. Electrically insulate the dissimilar metals from each other.
3. Add a "waster" section of a third metal which is anodic to the other two and which will cathodically protect both of the original dissimilar metals.
4. Add an appropriate corrosion inhibitor to the contacting solution.
5. Cover the surface of the cathodic member of the couple with an insulating coating.
6. Make the anodic member of the couple extra thick and design it so that easy replacement is possible.

It should be pointed out that dissimilar metals can be

used in contact with each other with impunity if they rarely or never are wetted by conducting solutions and even in structures exposed outdoors to the atmosphere, provided:

1. The atmosphere is not contaminated with conducting materials such as salt spray; and
2. The structures are designed to drain freely so that rain water or dew does not contact the dissimilar metals for prolonged periods of time.

It also should be pointed out that apparent single potentials of different metals change markedly upon exposure to different chemical solutions. For example, metal A can be anodic to metal B in one solution, cathodic to metal B in a second solution, and have about the same potential as metal B in a third solution. Thus, although measurement of apparent solution potential in sodium chloride solutions is a reasonably good criterion of galvanic action in natural environments, it is no better than the emf series in predicting galvanic behavior in solutions that differ widely from sodium chloride solutions in characteristics.

Stray Currents—Stray direct current electricity is a common cause for corrosion of underground steel pipe lines and other buried steel structures. It also is encountered in harbor structures of steel, steel-hulled vessels while tied up at docks, and in lake or river structures. In some cases, it has caused severe corrosion in chemical plants and other plants where large amounts of electrically conductive liquids are handled. Stray current causes more rapid attack than almost any other commonly encountered source of corrosion. In fact, wherever very rapid attack of a metallic structure buried in the ground, immersed in water or exposed to conducting liquids is encountered, stray currents should be investigated as one of the most likely causes.

Alternating current electricity is much less likely to cause severe corrosion of steel structures unless it is of very low frequency (say one cycle per second).

Concentration Cells—When one portion of a metal surface is exposed to an electrolytically conducting medium which differs in any way from the electrolytically conducting medium that contacts another portion of the same metal surface, selective corrosion of the portion of the surface contacting one of the two types of media is likely to occur.[6] For example, suppose a steel tank is partially filled with a salt solution. If, for any reason, the concentration of salt is different in the portion of solution near the bottom of the tank than it is toward the top of the tank, comparatively more severe corrosion of the steel contacting the more dilute solution is likely to occur. This is true even though no corrosion would occur if the tank were full of solution having a uniform concentration from top to bottom. Corrosion in this case is caused by **concentration cell action**.

A special type of concentration cell action is caused by local differences in oxygen concentration of the liquid. For example, let us suppose that liquid saturated with oxygen is circulating freely in contact with most of the inside surfaces of a steel vessel. However, there is a narrow channel or fissure in some part of the surface where stagnant liquid is trapped. The oxygen content of this stagnant liquid will generally be less than that of the freely circulating liquid in contact with it. Thus, an oxygen concentration cell is set up which causes more severe corrosion of the metal surface in contact with the portion of the liquid having the *lower* oxygen content.

Oxygen concentration cells are a very common cause of corrosion under service conditions. The severe localized corrosion at joints and crevices, and on surfaces in contact with wet insulating materials, is largely the result of oxygen concentration cells.

Stress—Applied or residual stresses, either static or dynamic, can greatly accelerate corrosion.[1] Generally speaking, the acceleration of corrosion by static stress is greatest in environments which do not cause appreciable general corrosion. In fact, stress corrosion often occurs under conditions which would cause almost negligible attack in the absence of stress. A particularly dangerous form of stress corrosion is **stress corrosion cracking**. Under some specific sets of conditions, a stressed part of almost any metal or alloy may suddenly crack, although there may have been no appreciable corrosive attack evident prior to the sudden development of the stress corrosion crack. Stress corrosion cracks rarely occur when metallic parts are stressed appreciably below the yield strength. Since design stresses for most stressed assemblies are normally kept well below the yield strength, it is normally not the stresses for which the structure was designed that cause stress corrosion cracking. Instead, it is generally the residual or so called "internal" stresses that cause this type of failure. These residual stresses are induced during the fabrication of the structure, by bending, welding and other fabricating procedures. The magnitudes of such stresses are frequently not known to the design engineer and are consequently not considered by him in designing the structure.

Cyclic stresses also frequently accelerate corrosion. In cases where the frequency of the alternation in cyclic stresses is high, failures accompanied by corrosion are said to be caused by **corrosion fatigue**. Cyclic stresses, especially if of high frequency, normally cause more rapid failures when accompanied by corrosion than do static stresses of the same magnitude.

Abrasion, Erosion and Cavitation—Surface effects such as abrasion or wear, erosion, impingement of liquid at high velocity, or cavitation effects caused by the collapse of gas or vapor bubbles at the metal surface can all contribute to intensified corrosion damage in specific cases.

Other Surface Effects—Practically any local difference at the metal surface can cause corrosion under suitable conditions. Thus, a scratch in the metal surface can cause corrosion under suitable conditions. Thus, a scratch in the metal surface can form a weak point at which localized attack may occur. Local differences in temperature, velocity of liquid flow, degree of surface roughness and even amount of illumination can all cause localized corrosion in specific cases. In fact, any heterogeneity in environment, metal surface or, as will be indicated subsequently, in the metal itself, can cause localized attack.

Metallurgical Factors—As mentioned just above, any heterogeneities in the metal or alloy itself can give rise

---

[6]References are at end of chapter.

to localized corrosion. For most commercial metals or alloys these "internal" effects are normally a minor factor in influencing corrosion under service conditions. Generally, environmental factors are of much greater importance. This is very fortunate since it is impossible to avoid internal heterogeneities in commercial metals and alloys. All such commercial materials contain inclusions or minor segregations. Whenever localized attack develops in service, it is common practice to put the blame on these small particles of different phases. Actually, very careful control of the uniformity of the corroding environment and of the metal surface is normally required to reveal the effect of inclusions on cor-rosion. In special cases, with specific metals or alloys, metallurgical factors can be important. For example, in the case of 18 per cent chromium—8 per cent nickel stainless steels, it is well known that slow cooling from the austenitizing temperature, or reheating in a critical temperature zone for an appropriate time can cause the precipitation of carbides at grain boundaries, thus sensitizing the alloy to intergranular attack by certain media. In this case, the metallurgical condition is of great importance in controlling corrosion behavior. However, the fact remains that, in general, the metal-lurgical condition of most commercial materials is not of great importance in affecting corrosion rates.

SECTION 2

METHODS OF PREVENTING CORROSION

A knowledge of the factors which accelerate corro-sion is of value to most people only as it aids them to understand and to guard against corrosion failures in service. A knowledge of appropriate methods of pre-venting corrosion is important for the same reason.

Material Selection—The most obvious method of preventing corrosion is to build the structure of a mate-rial which is unaffected by the service. Unfortunately, it is not always feasible to do this. The most inert mate-rials may be too expensive or otherwise unsuited for the article to be built. Generally, the engineer must make a compromise. He cannot afford to use the most corrosion-resistant material but instead must compro-mise on a material that has the lowest combined initial cost plus maintenance costs for some selected period of time. The more accurately the engineer knows the cor-rosion behavior of the various competitive materials under the desired service conditions, the more accu-rately can he select the most economical material to use. As a guide to proper selection, nothing has yet replaced previous service use. Since small variations in service conditions can sometimes affect corrosion rates greatly, even previous service use is not infallible. Nevertheless, it is the most trustworthy criterion avail-able.

If it is desired to select material for equipment re-quired for some new process or chemical, there will be no previous background of experience. In such a case, the engineer must be guided by knowledge of the be-havior of various materials when used as equipment for similar processes. Better yet, a pilot plant or small-scale service test can be made using a material or materials of construction deemed likely to be satisfactory. The selection of materials for these small-scale tests can be based on laboratory tests or published information. In conducting the laboratory tests, it should be kept in mind that the closer these tests can be designed to simulate actual practice, the more confidently can the results be used to predict satisfactory performance.

Selection of a suitable material of construction may eliminate the need for using any other form of corro-sion prevention. However, it is frequently more eco-nomical to use some less resistant but cheaper material and to employ one or even several protective meas-ures. In most cases it is not economical to use the most resistant material of construction, but instead to use the cheapest material that will do a satisfactory job.

Appropriate Design—It is frequently overlooked that small changes in design can make it possible to use cheaper materials of construction. For instance, it might be feasible to use carbon-steel pipe as a vapor line for handling gaseous chemicals if the line were insulated to prevent condensation. If the insulation is omitted, such severe corrosion of the steel might occur that it would be necessary to employ stainless steel or non-ferrous pipe in order to reduce corrosion to a toler-able value.[4]

Similarly, the design of a processing tank might be such that liquids could lodge in crevices, pockets, joints or other dead spaces. Severe corrosion is likely to occur in such areas as a result of concentration cell formation. By altering the design to eliminate these areas, the same material of construction can be used with greatly reduced corrosion damage. In the case of tanks or proc-essing vessels for chemicals, the most severe corrosion often occurs on the *outside* of the vessel, not on the inside as might be anticipated. Such equipment should be designed so that moisture will not be trapped in external joints between the vessel wall and the sup-porting members, between the bottom of the vessel and the supporting wood or concrete base, or in absorptive thermal insulation which contacts the exter-nal surfaces of the vessel. Moisture in these locations, either condensed moisture or tap water, can cause se-vere corrosion. To prevent corrosion, the design may be as important as the selection of the materials of construction.

Protective Coatings—There are a large number of different types of protective coatings. They can be classified in different ways. Based on their characteris-tics, it is convenient to classify them as:

1. Anodic coatings
2. Cathodic coatings
3. Inert coatings
4. Inhibitive coatings

When a coated metal article is exposed to an electro-lyte, if there are discontinuities in the coating, several possibilities present themselves. An electric current may flow from the coating through the solution (elec-trolyte) to the base metal. If this happens, the coating is anodic to the base metal. Furthermore, if the current density at the exposed area (or areas) of the base metal is of the correct magnitude, corrosion of the base metal will be prevented. Thus, anodic coatings tend to prevent corrosion of exposed areas of the base metal by sending electric current to them through any contacting film or layer of an electrolytically conducting medium.

In contrast to this, cathodic coatings tend to stimulate corrosion at exposed areas of the base metal under similar conditions of exposure.

Coatings showing the most pronounced anodic or cathodic behavior are the metallic coatings. Non-metallic coatings, especially oxide or sulphide coatings under some conditions of exposure, act as cathodic coatings. However, there is no definite evidence of non-metallic coatings acting as anodic coatings. It should also be pointed out that the same metallic coating on the same base metal can behave as an anodic coating under one set of exposure conditions, as a cathodic coating under other conditions, and as an inert coating or even as an inhibitive coating under still different conditions.

Tin coatings on steel form a good example of this variation in behavior under different environmental conditions. When exposed outdoors, in sea water, in most natural waters, or even to many food products in the presence of air, tin is cathodic to exposed areas of the steel base. However, when exposed to nearly air-free food products, tin is generally definitely anodic to steel. Presumably, if the exposure were to food products containing some critical amount of oxygen, tin would be neither anodic nor cathodic to steel. That is, it might then be classed as an inert coating. Again, under different conditions, the accumulation of dissolved tin compounds from the coating in the food product might reduce greatly the corrosiveness of the product to steel. In this last case, the tin compounds would serve as corrosion inhibitors, and the tin coating could be classed as inhibitive.

It should also be pointed out that, although cathodic coatings *tend* to stimulate corrosion of exposed areas of the base metal, this does not always mean that increased attack of these areas will, in fact, occur. For example, if the cathodic coating is thick, and if the exposed areas of base metal are small, attack can be stifled by plugging of the small pores in the coating with corrosion products from the base metal.

Inorganic coatings sometimes are inert, sometimes cathodic and sometimes inhibitive. Organic coatings are generally either inert or inhibitive. Either inorganic or organic coatings which contain water-soluble chromates generally function as inhibitive coatings in most natural environments. It is obvious that the inhibitive value of such coatings is greatest when there is only limited opportunity for leaching of the soluble inhibitor to occur. For example, it would be expected that the inhibitive action of a coating containing a soluble chromate would be more in evidence if the coating were on the interior of a tank which contained only a small amount of stagnant water than if it were on the interior of a pipe through which unrecirculated water was passing continually.

Organic coatings also can function in a manner which does not permit their classification in one of the four simple groups mentioned above. For example, in special cases, organic coatings when exposed to liquid media can serve as semipermeable membranes. Then, by osmosis, dialysis or electro-dialysis, liquid can be transferred through the organic coating to the metal-coating interface. The composition of the liquid which collects at the interface may differ markedly from that of the liquid on the outside of the coating. It may be either more or less corrosive to the base metal than the parent liquid. This behavior is still obscure but it is known to exist.

Treatment of Environment—Corrosion can sometimes be prevented either by adding something to the corrosive medium or by removing some corrosive agent from the medium.[3] For example, a certain tap water may be highly corrosive to the steel tank in which it is stored. By adding a corrosion inhibitor, such as sodium chromate, to the water the attack may be prevented. Alternatively, by removing dissolved oxygen from the water, the corrosion rate may be greatly reduced.

Corrosion prevention by treating the environment is normally employed when there is only a limited amount of the corrosive material. Thus, it is more widely used for waters which are recirculated than for waters which flow continuously from the source without recirculation.

Chromates are by far the most versatile corrosion inhibitors, although phosphates, silicates and various complex organic compounds also are used. The mechanism of inhibitor action differs for the various inhibitors and types of uses. In some cases inhibitors function simply by forming a protective film or layer on the metal surface. In others, they retard one or more of the electrolytic processes necessary for corrosion to occur.

Cathodic Protection[2]—It has already been mentioned that anodic or active coatings have the property of sending an electric current through an electrolyte to exposed areas of cathodic metals. This current flow tends to prevent corrosion of the cathodic metal. Protection in this manner by current flow from any source is termed **cathodic protection**, since the metal being protected is made the cathode of an electrolytic cell.

Sometimes a coating metal is used to provide cathodic protection. Cathodic current from pieces of an anodic metal in electrical contact with the article to be protected and also exposed to the electrolyte can be used. These are termed "galvanic anodes." Alternatively, direct current from a storage battery or from a generator, or rectified alternating current from a power line, can be used. The only essential is that a sufficient cathodic current density must be maintained at all areas of the protected article which contact the corrosive solution.

Determining the magnitude of the current density just sufficient to prevent corrosion under various types of service conditions is a difficult problem. Fortunately, for most steel structures, it is not necessary to know the limiting current density with great accuracy. *Any* cathodic current density which is applied to a given area will reduce corrosion. Furthermore, if the limiting current density is exceeded, no harm is done to the structure. This means, simply, that protection is costing more than necessary because some of the current is being wasted.

Cathodic protection is a very effective way for preventing corrosion by most types of electrically conducting media. Also it is relatively safe, since as long as all the current flows to the structure, corrosion will be reduced even if protection is not complete. It is relatively easy to apply cathodic protection to small, geometrically simple structures. However, skilled cathodic protection engineers are required to develop efficient cathodic protection for large or geometrically complicated structures.

At the present time, cathodic protection is applied widely to the steel hulls of marine vessels, to the interiors of small and large steel water tanks, to the exteriors of buried steel pipe lines and to a variety of types of chemical equipment. Another important application of this method is the use of magnesium anodes to prevent corrosion of glass-lined hot-water tanks. It should also be mentioned again that the usefulness of tin plate for food containers and of galvanized sheet for roofing, siding and the like, depends upon cathodic protection of the steel base by the anodic coating metal.

Another method of corrosion control is anodic protection, which consists of applying an anodic current to passivate the metal surface. The method is being used on a limited basis primarily in the chemical and paper industries for prevention of corrosion in chemical-process environments. Tests usually are required to show that any specific metal/environment combination is amenable to the use of this method.

Periodic Cleaning—It is frequently overlooked that periodic cleaning may greatly reduce corrosion damage. The fundamental reason why cleaning is beneficial is that it removes moist layers of solid matter from the metal surface. Corrosion products may be hygroscopic and generally are water sorptive. Thus, a metal surface coated with a heavy layer of corrosion products and exposed to the atmosphere will be wet for a considerably greater proportion of the time than will a similar clean surface. In like manner, layers of dust or soot of many types stimulate corrosion. Some materials hold moisture, others contain soluble products which in themselves are highly corrosive. It is only rarely that clinging debris is protective.

Coatings Combining Decorative and Protective Properties—Coatings may be not only protective but also decorative and, in fact, more often than not the feature of appearance is of prime importance. For example, galvanized sheets often have been required to exhibit a pleasing spangled surface. The presence of this spangle or bright appearance has been associated with the highest quality product although tests have shown no longer actual service life than that of galvanized sheets of duller luster with equivalent weight of coating. Thus, dependent on the required appearance of the ultimate product, a variety of bright or dull coatings can be manufactured. In the field of organic coatings, a variety of hues may be obtained by a variation of the pigments or stains used in the coating mixture.

SECTION 3

PREPARATION OF STEEL SURFACES FOR PROTECTIVE COATINGS

Importance—Correct surface preparation is the primary and most important requirement for satisfactory performance of protective coatings on steel. Without a properly cleaned surface, even the most expensive coatings will fail to adhere or to prevent rusting of the steel base. It is axiomatic that coating performance is proportional to the degree of surface preparation. However, the economics of the particular coating requirement must be considered. In painting, for example, complete descaling by blast cleaning or pickling, followed by a suitable pretreatment, is best. Because maximum paint life is obtained when paint is applied to rust-free and scale-free surfaces, such thorough cleaning, while expensive, is most economical when painted surfaces are exposed to severely corrosive environments. For withstanding corrosion in mild atmospheres, wire brushing of steel surfaces before painting is adequate and economical, since intact mill scale in the absence of electrolytes is a good base for paint. When lesser degress of surface cleaning are warranted, the protective coating must be capable of developing good adhesion to the prepared surface. Good adhesion is the primary requisite of a protective coating, and good adhesion is obtained only by appropriate surface preparation. A comprehensive discussion of and specifications for mechanical and chemical surface preparation are given in the "Steel Structures Painting Manual," Volumes I and II.[12]

Mill Scale—Mill scale is a relatively thick layer of iron oxide that is easily cracked and does not adhere well to steel. It is cathodic to the underlying metal and must be removed to prevent failure of coatings that are exposed to severe environments.

Rust—Rust is perhaps the worst surface contaminant: it prevents satisfactory adhesion of the protective coating to the metal during application and, if covered over, leads to premature failure of the coating in service. Depending upon the history of its formation, rust may contain chemical constituents that can accelerate corrosion.

Oil, Grease, Soil—These must be removed because they prevent satisfactory wetting of the surface by almost all protective coatings and lead to poor adhesion.

Solvent Cleaning—Solvents clean metal surfaces by dissolving and diluting foreign matter such as oil, grease, soil, and drawing and cutting compounds. Oil or grease can be removed by wiping or scrubbing the surface with rags or brushes wetted with a solvent followed by a final wiping with clean solvent and clean rags or brushes. The steel also may be completely immersed in the solvent, or solvent sprays may be used, or the steel can be subjected to vapor degreasing in equipment in which vaporized solvent condenses on the surfaces to be cleaned. Solvents used include mineral spirits, naphthas, and certain chlorinated hydrocarbons.

Alkaline Cleaning—This method is used where mineral and animal fats and oils must be removed. Pressurized spray application of alkaline solutions of cleaners at their optimum concentrations and operating temperatures are preferred, but dipping with solution agitation will occasionally be satisfactory. The use of electrolytic cleaning may be advisable for large-scale production or where this method yields a cleaner product. Caustic soda, soda ash, alkaline silicates, and phosphates are common alkaline cleaning agents. Sometimes the addition of wetting agents to the cleaning bath will facilitate cleaning.

Hand- or Power-Tool Cleaning—These methods include brushing (wire or bristle), scraping, chipping, sanding, impact-tool cleaning, and grinding. They are used widely on larger items in the shop or in the field, particularly such items as tanks and bridges. They do not completely remove rust or mill scale, and may be used in combination with solvent cleaning to remove oil and grease. Therefore, these methods are limited in their effectiveness and must be used with protective coatings that are compatible with the resultant surfaces.

Flame Cleaning—Special oxy-acetylene torches are used to partially remove scale and rust from the surfaces of heavy steel sections. The surface immediately after cleaning is warm and, therefore, in good condition for painting, but only partial removal of rust and scale is accomplished.

Blast Cleaning—Abrasives such as sand, steel or iron grit (crushed shot), or shot are impinged at high velocity against the surfaces to be cleaned, either by compressed air in a nozzle-type blast-cleaning apparatus or by centrifugal force in rotary-type blast-cleaning machines. The surface is roughened by these cleaning methods, and the degree of roughness must be regulated by selection of the proper type, size, shape or speed of abrasive so that the anchor-pattern profile of the surface is satisfactory for the intended protective coating. The degree of cleanliness achieved ranges from (1) white-metal blast cleaning, in which no visible residues remain on the surface and the surface is uniform in color; (2) near-white blast cleaning, in which traces of scale and rust remain on the cleaned surface; (3) commercial blast cleaning, which is done at a considerably faster rate than (1) and (2) and, consequently, leaves considerably more scale and rust on the surface than (2); to (4) brush-off cleaning, which is a very fast cleaning that removes any loose material from the surface and abrades it only slightly.

Pickling—In large-scale operations, pickling in a hot solution of sulphuric acid (H_2SO_4) is a common method for removing scale and rust. The time of pickling, and concentration and temperature of the solution, will be varied depending upon the type of scale to be removed or the type of steel being pickled.

In continuous pickling of strip and in other operations where the size and shape of the material being pickled does not vary greatly, electrolytic pickling may be employed at room temperature. The article to be pickled is made the anode if some actual solution of the steel is required, or the cathode where only the blast action of hydrogen evolved at the cathode is necessary. Cold muriatic (hydrochloric acid, HCl) will pickle steel very efficiently, and is being used increasingly for large-scale pickling operations. Other acids are not used as commonly as sulphuric and hydrochloric acids, although the use of phosphoric acid (H_3PO_4) is increasing in those cleaning operations where a slight phosphate coating is desired on the cleaned steel.

Special Methods—Many protective coatings require scrupulously clean surfaces or special surface-preparation procedures for their successful application. A series of cleaning procedures may be necessary; for example, preparation of steel for electrolytic tinning, hot-dip galvanizing, aluminum coating, terne coating, and porcelain enameling require detailed procedures, many of which are covered elsewhere in this book. Metallizing requires blast cleaning to the white-metal condition and a rough surface. Thick coatings such as hot bituminous melts are applied advantageously to rough surfaces. Decorative finishes such as the coatings used in the automotive industry and electroplated finishes for appliances require smooth surfaces.

Protective coatings often are applied as temporary coatings after cleaning to protect the cleaned surfaces until the product receives its final coating; some temporary coatings must be removed by degreasing or other means before the final coating operation.

SECTION 4

METALLIC PROTECTIVE COATINGS

The most commonly used metallic coatings for steel include tin, zinc, terne metal (lead plus tin), aluminum-zinc alloy, nickel, chromium, cadmium, copper, aluminum, bronze, brass, silver, gold, and lead. As mentioned previously, a metallic coating may be anodic or cathodic to the metal to which it is applied. If anodic, it is "sacrificial" or less noble than the base. If cathodic, it is more noble than the base and its protective value is due to its own relative chemical inactivity in the environment to which it is exposed.

A rough indication of the activity of the metals may be obtained from the aforementioned electromotive series, a classification of metals in an order of electrode potential referred to the standard hydrogen electrode at a temperature of 25°C (77°F) (see Table 35—III). This table often is used mistakenly as if the order of metals were invariable, each metal displacing from solution or protecting from corrosion those below it. That this is often not true is shown in the second column of the table, where the values of potential in a normal salt

Table 35—III. Comparison of the Electromotive Series and Solution Potentials in Sodium-Chloride Solution of The Common Metals and a Few Alloys.

	Potential in Volts	
Metals or Alloys	From emf Series (Normal hydrogen scale)	In 1 N (5.85%) NaCl containing 0.3% H_2O_2(0.1N) Calomel Scale
Magnesium, Mg	−2.37	−1.73
Aluminum, Al	−1.66	−0.85
Zinc, Zn	−0.76	−1.00
Chromium, Cr⁺⁺	−0.74	....
Iron, Fe⁺	−0.44	−0.63
Cadmium, Cd	−0.40	−0.82
Cobalt, Co	−0.28	....
Nickel, Ni	−0.25	−0.07
Tin, Sn⁺⁺	−0.14	−0.49
Lead, Pb⁺⁺	−0.13	−0.55
Hydrogen, H_2	0.00	....
Copper, Cu⁺⁺	+0.34	−0.20
Silver, Ag	+0.80	−0.08
Gold, Au⁺⁺	+1.50	....
Brass (60-40)	....	−0.28
Stainless Steel (18-8)	....	−0.15
Monel Metal	....	−0.10
Inconel	....	−0.40

solution are given. From the electrochemical series it would be anticipated that the corrosion of zinc would be retarded by contact with aluminum, while the potential measurements in salt solution indicate that the zinc should protect aluminum. In sea water and many natural waters, this protection of aluminum by zinc actually occurs. The electrochemical series is useful since metals near the top are generally protective of those near the bottom of the table. When, however, two metals differ little in potential, i.e., are close together in the first column of Table 35—III, one cannot predict which will protect the other, without testing. In many instances the results of potential measurements in a salt solution (second column of Table 35—III) are a better guide as to the ability of one metal to protect the other under natural conditions of exposure than is the electromotive series. Many factors, such as the environment to which it is exposed, the magnitude of current generated, the relative area of metals exposed, the texture of the metal surface and the inherent tendency of the metal to form an insoluble protective film notably affect the corrosion of metals themselves or their corrosion rate when used as anodic coatings for steel.

Table 35—IV illustrates the variable effect of environment or exposure conditions on the solution potentials of several common metals. It will be noted that the relative potentials of these metals vary when they are exposed to different solutions. Thus, no one table of solution potential values can indicate the electrochemical behavior of the different metals under all conditions of use. Actual tests under conditions similar to those of service are required before it is possible to make accurate predictions. The electrochemical behavior of metallic coatings on steel under conditions of atmospheric corrosion is now fairly well established. Zinc is anodic

to steel under most exposure conditions and will prevent corrosion even at small discontinuities in the coating. The behavior of aluminum coatings is more complex. In some environments they tend to protect exposed steel areas but in others there is no evidence of galvanic protection.

Tin, terne metal, nickel, copper, silver, gold and lead are all cathodic to steel under most conditions of atmospheric exposure and, if used as coatings for steel, will tend to accelerate corrosion at pores, scratches, and pinholes. In some cases the corrosion products formed in the areas where steel is exposed will stifle corrosion. This is particularly true for heavy coatings of lead, tin, terne metal and aluminum. From the above it can be concluded that protection against corrosion is not a simple consideration, but an extremely complex one.

METHODS OF APPLYING METALLIC COATINGS

Metallic coatings are applied to steel surfaces by the following methods:

Hot Dip Processes—The steel article to be coated is immersed, after thorough cleaning, in a molten bath of the metal forming the coating. Zinc, tin, terne metal, aluminum, aluminum-zinc alloy, and lead are applied commercially in this manner and are discussed fully elsewhere (see especially Chapters 36, 37 and 38).

Metal Spraying—This method, introduced about 1910, can be used with most of the common metals such as aluminum, copper, lead, nickel, tin and zinc, and alloys such as brass, bronze, babbitt metal, monel metal and stainless steel. The coating metal usually is drawn into wire and fed into a specially constructed spray gun. This gun is operated with compressed air and a fuel gas. It is small and compact although it contains an air-gas mixing chamber, a special nozzle for burning the mixture and melting the wire, an outer compressed-air nozzle concentric with the inner nozzle, and an air turbine driving knurled rolls which draw the wire from its spool and feed it through the inner nozzle. The gases at the nozzle are ignited, the wire is melted as it is fed to the nozzle and is projected against the surface to be coated at a speed of over 152 metres (500 feet) per second. Although the particles of molten metal are cooled instantly to a temperature of about 27°C (80°F), the impact causes them to adhere firmly to the steel surface, provided it has been cleaned thor-

Table 35—IV. Solution Potentials of Several Metals in Various Solutions (all 1 molar in concentration). Referred to 0.1 N calomel half-cell*

Metal	Sodium Chloride	Sodium Chromate	Nitric Acid	Sodium Hydroxide
Magnesium	−1.72	−0.96	−1.49	−1.47
Aluminum	−0.86	−0.71	−0.49	−1.50
Zinc	−1.15	−0.67	−1.06	−1.51
Iron	−0.72	−0.16	−0.58	−0.22

* From: Light Metals for the Cathodic Protection of Steel Structures, by R. B. Mears and C. D. Brown; Corrosion, Vol. 1, No. 2, September, 1945, National Association of Corrosion Engineers.

oughly, as by machining or by sand or shot blasting. Metal spraying is used for building up surfaces and sometimes for the application of thin coatings as a protection against corrosion.

Metal Cementation—The metals zinc, chromium, aluminum and silicon are successfully applied in this manner, in which the protecting metal is alloyed into the surface layers of the steel.

In *sherardizing*, practiced since about 1900, the parts to be coated, usually small articles such as nails, are thoroughly cleaned by pickling or sand blasting; packed in metal drums with fine zinc dust, usually containing 5 to 8 per cent of zinc oxide; and heated for several hours at between 345° and 400°C (650° and 750°F), the drums being slowly rotated in the furnace during the heating. The coating is thin and consists of intermetallic compounds of iron and zinc, ranging from an iron-rich alloy next to the steel base to almost pure zinc at the surface, but it affords good protection against atmospheric corrosion. The process was invented by Sherard Cowper-Coles.

Chromizing is a cementation process analogous to sherardizing. The parts to be treated are packed in a container with a mixture of 55 parts of chromium or ferrochromium powder and 45 parts of alumina by weight. They are then heated "in vacuo" or in a protective atmosphere (preferably hydrogen) at 1300° to 1405°C (2370° to 2560°F) for three or four hours, although a shorter time and lower temperature may be used when less penetration is desired. Chromizing also can be accomplished by a gaseous method in which the part to be treated is enclosed in a hydrogen atmosphere with chromium or ferrochromium powder and an ammonium halide. A gaseous chromium halide is formed at elevated temperatures. This reacts with the steel and releases chromium that diffuses into the steel surface. The halide carrier then recombines with chromium from the chromium or ferrochromium powder and the process is repeated. Coatings formed by this process are generally about 0.1 mm (0.004 inch) thick, and may contain about 40 per cent of chromium at the surface.

In *calorizing*, developed by General Electric Company about 1925 to 1930, the thoroughly cleaned steel articles are packed in steel drums containing a mixture of aluminum, aluminum oxide, and a small amount of ammonium chloride. A reducing gas is passed into the drum, which is rotated in the furnace and heated for about 5 hours at between 945° and 955°C (1730° and 1750°F). The resulting coating is said to be a solid solution of aluminum in iron, richest in aluminum at the outer surface, and is used principally to protect the steel from oxidation at elevated temperatures, as in pyrometer tubes, superheater tubes and oil-refinery equipment.

Ihrigizing is a special type of siliconizing, or impregnation of the surface of low-carbon steels with silicon. In this process, the surface of low-carbon steel, freed of sand and heavy scale, is packed with silicon carbide or ferrosilicon mixed with mill scale (iron oxide), heated to a temperature of 705°C (1300°F) or higher and exposed to the action of chlorine for two hours or more depending on the temperature used and depth of case desired. At this temperature the chlorine reacts with the carbon or the iron of the silicon-bearing substance, leaving the silicon in nascent form to combine with the iron in the steel. The siliconized layer, usually 0.13 to 2.54 mm (0.005 to 0.1 inch) thick as desired, is very hard and resistant to corrosion by nonoxidizing acids, such as hydrochloric and sulphuric acids, to wear and oxidation at temperatures up to 870°C (1600°F), and is capable of absorbing and retaining substantial amounts of oil. In usual practice, the silicon content of the case remains practically constant for the first millimetre (0.04 inch), varying from about 14 per cent at the surface to 12 per cent at 1.3 mm (0.050 inch) below, then decreases gradually in the next 0.5 to 0.6 mm (0.020 to 0.025 inch) to the silicon content of the core.

In *corronizing*, developed about 1938 by Standard Steel Spring Company of Coraopolis, Pennsylvania, the steel is electroplated with nickel and subsequently with zinc or nickel-zinc alloy (U.S. Patent No. 2,419,231). The plated steel may be heated to about 400°C (750°F) to form a nickel-zinc alloy, if zinc is the final coating.

Metal Cladding—*Copper cladding* processes give bimetal products. Usually those containing steel consist of an inner steel core covered with a heavy layer of copper. In the usual process the steel core, with a clean surface, is mounted in a covered mold and heated out of contact with air to a temperature slightly above the melting point of the copper, which then is cast about it. Other methods consist of dipping the solid steel core into a bath of molten copper, or of depositing the copper electrolytically. Starting material for copper-clad steel wire is made by forcing a steel rod into a closely fitting copper tube. Semi-finished products prepared by any of these methods may be heated to around 925°C (1700°F) and hot rolled, then finished by cold rolling or drawing, as in forming copperclad wire. The wire is used widely for electrical conductors, combining the strength of steel with the high conductivity of copper. Bundy-weld steel tubing is hydrogen-welded, copper-coated, rolled steel tubing. It is used for gasoline and oil lines in automobiles and for refrigerator coils. Copper-clad sheet steel was produced during World War II for the fabrication of copper-jacketed bullets.

Aluminum cladding is accomplished best by rolling flat steel almost to gage, cleaning it thoroughly and either placing it between two sheets of aluminum and cold rolling, or heating to between 315° and 400°C (600° and 750°F) and rolling. The latter method results in a better bonding of the aluminum with the steel. Subsequent annealing above 535°C (1000°F) causes the aluminum to unite with the iron forming the very brittle $FeAl_3$. With carbon steel containing above 0.25 per cent silicon, the temperature of this reaction is raised above that for regular carbon steel so that the coated strip or sheet may be annealed at a somewhat higher temperature after cold rolling without becoming brittle.

Stainless cladding may be accomplished by (1) electrowelding stainless steel onto the carbon steel (2) casting the stainless steel around a solid carbon-steel slab or (3) placing a slab of carbon steel between two plates of stainless steel and hot rolling them. In the last mentioned method, fluxes or metals have been used to facilitate bonding but are not necessary if both steels

Table 35—V. Electroplating Baths.

Kind of Coating	Type of Solution	Typical Composition of Baths			pH	Temperature		amperes		volts
		Constituents	grams, plus water to make 1 litre	ounces, plus water to make 1 gal		°C	°F	per dm²	per ft²	
Nickel	Watts	nickel sulphate[1] nickel chloride[2] boric acid	330 45 38	44 6 5	1.5-4.5	45-65	113-149	2.5-10	23-93	6-12
	Bright	nickel sulphate[1] nickel chloride[2] boric acid and 0.1 to 1.0 per cent addition agents	330 45 38	44 6 5	3-4.5	50-65	122-149	2.5-10	23-93	6-12
	Hard	nickel sulphate[1] ammonium chloride boric acid	180 25 30	24 3.3 4	5.6-5.9	43-60	109-140	2.5-5	23-46	6-12
Chromium	Bright	chromic acid sulphuric acid[3]	250 2.5	33 0.33	—	39-49	100-120	7.5-17.5	75-175	6-12
	Hard	chromic acid sulphuric acid[3]	250 2.5	35 0.33	—	52-63	125-145	31-62	288-576	6-12
Zinc	Acid	zinc sulphate[4] zinc chloride[5] boric acid	180 14 12	32 2 4	2.5-4.5	25-40	77-104	0.5-7.5	5-70	4-10
	Neutral chloride	zinc chloride[5] potassium chloride	71 207	10 28	4.8-5.8	21-35	69-94	2-40	19-37	1-5
	Cyanide	zinc cyanide sodium cyanide sodium hydroxide	61 42 79	8 6 10.5	>13.0	20-40	68-104	2.5-10	23-93	6-15

(Continued on next page)

Table 35—V (Continued)

Kind of Coating	Type of Solution	Typical Composition of Baths		pH	Temperature		amperes		volts
		Constituents	grams, plus water to make 1 litre / ounces, plus water to make 1 gal			°C	°F	per dm²	per ft²	
Cadmium	—	cadmium oxide	23-34 / 3-4.5	13.0	20-30	68-86	0.5-5	5-46	6-12
		sodium cyanide	75-90 / 10-12							
		plus brighteners								
Tin	Alkaline	sodium stannate	100 / 13	—	80-85	140-185	0.5-3	5-28	4-6
		sodium hydroxide	10 / 1.3							
	Acid	stannous sulphate	60 / 8	—	20-30	68-86	3-10	28-93	6-12
		phenol sulphonic acid	60 / 8							
		ethoxylated naphthol sulphonic acid	2 / 0.25							
Copper	Cyanide	copper cyanide	15 / 2	12.0-12.6	41-60	106-140	1.0-3.2	9-30	6
		sodium cyanide	23 / 3							
		sodium carbonate[7]	15 / 2							
	Acid	copper sulphate[6]	188 / 25	—	90-109	32-43	34-50	4-6	
		sulphuric acid[3]	74 / 10							
Brass	—	copper cyanide	26.2 / 3.5	10.3-11.0	27-35	81-95	1.0	9	2-3
		zinc cyanide	11.3 / 1.5							
		sodium cyanide	45 / 6							

[1]NiSO$_4$•6H$_2$O
[2]NiCl$_2$•6H$_2$O
[3]H$_2$SO$_4$—100%
[4]ZnSO$_4$•H$_2$O
[5]ZnCl$_2$
[6]CuSO$_4$•5H$_2$O
[7]Na$_2$CO$_3$ (anhydrous)

are cleaned thoroughly before making the "sandwich." Welding is usually done around the perimeter of the slab. Considerable care is necessary in the preparation of such duplex material to avoid formation of blisters.

Fusion Welding of Coatings may be accomplished in different ways, as by depositing weld metal under a slag covering by the electric-arc method, or by fusing the surfaces of two bodies of metal in contact by passing a current of sufficiently high density. These initial steps are followed by heating and forming operations carried out in the usual manner, as by rolling. Similarly, weld metal may be deposited upon metal of another kind to afford greater resistance to abrasion, such as manganese-nickel steel welding rod used to face excavating and similar tools.

Electroplating—This process is an old art, practiced not only to protect the base metal from corrosion but also for decorative purposes and, more recently, to protect the base metal from wear by friction or abrasion. Metals used for coatings include cadmium, chromium, copper, gold, tin, lead, nickel, silver and zinc, and alloys such as brass, bronze and lead-tins as well as cobalt-tungsten, tungsten-nickel, nickel-zinc and cadmium-tin alloys. It will be noted that with the exception of zinc, which is anodic to steel, i.e., protects by sacrificing itself, nearly all electroplated coatings are cathodic to steel and provide protection through surface coverage. The decorative coatings commonly used vary in thickness, according to the life required. Durability usually depends upon the properties of the coating, especially adhesion and porosity. The severity of conditions of exposure, particularly with reference to acidic gases in the atmosphere, also affects the service life. In plating, the preparation of the base metal is necessary to obtain

good adherence but also because surface preparation has much to do with the final appearance, since the intermediate or final coating is frequently buffed to a high lustre. Therefore, the base metal should be smooth if polishing and buffing costs are to be kept to a minimum. The decorative coatings most commonly applied are nickel, cadmium, nickel followed by chromium, or copper followed by nickel or by nickel and chromium. To protect hard steel and iron surfaces from wear or abrasion, coatings of chromium or alloy coatings of tungsten and cobalt or nickel and tungsten are sometimes used although some of these may not always be applied by electroplating. A few typical plating baths with operating data as commonly used by job-shop platers are shown in Table 35—V.

Miscellaneous Metallic Coating—*(1) Cathode Sputtering.* When relatively high voltage is applied between two electrodes in a partial vacuum, inducing a glow discharge, the cathode disintegrates and the metal thus removed can be deposited in a thin film on near-by objects within the chamber. In suitably designed chambers, objects may be arranged with respect to the cathode so that they will receive a uniform, thin coating of metal. The process is particularly suitable for the metallization of electrically non-conducting materials such as fabrics and phonographic recording waxes.

(2) Evaporation or Condensation. This process, closely related to cathode sputtering, is of more recent origin in practical application. The metal vapor is produced by thermal instead of electrical means. A coiled filament of platinum or tungsten in a higher vacuum than for cathode sputtering is a convenient heat source. The process is usually confined to the deposition of pure metals.

SECTION 5

SURFACE CONVERSION COATING

Steel often is treated in various ways by heating to form a uniform blue or black coating of oxide which, although not thoroughly protective, is pleasing in appearance and, especially if coated with oil, wax or other clear protective coating, is much more resistant to corrosion than steel not so treated. "Blue annealed" plate now produced on continuous plate mills is one example of this finish. **Steam-blued** and **air-blued** finishes on thin sheets used for common stove-pipe stock are others and the **gun metal** finish applied to gun barrels is another. Strip steel is satisfactorily colored by passing through heated sand. From three to ten minutes treatment at 345°C (650°F) will produce a rich blue color. Highly polished steel may be blued by placing it in a bed of hot

charcoal about two feet deep. The lower part of the charcoal is in a state of incandescence whereas the upper layers are lower in temperature and suitable for the development of oxide colors. After development of the desired shade of blue, the article is rubbed vigorously with waste or cloth dipped in raw sperm oil.

The **gun-metal** or **carbonia** finish used on rifles, shotguns and revolvers, as well as on many other metal parts, is obtained by placing the steel loosely in a retort with a small amount of charred bone and heating to 370° to 425°C (700° to 800°F).

After the articles are thoroughly oxidized, the temperature is dropped to about 345°C (650°F) and a mixture of bone and carbonia oil are added, after which heating is continued for several hours. On removal, the articles are dipped in sperm oil or tumbled in oily cork to develop a uniform, black finish. If a lower temperature is necessary to prevent excessive softening of the steel, a longer time is usually required and the color obtained is less permanent than at the higher temperatures. Articles which have been first nitrided, when treated by the gun-metal or carbonia process, take a pleasing, rust-resistant finish and retain their surface hardness since coloring temperatures do not temper nitrided articles.

Barffing is a process somewhat analogous to steam bluing of sheets and black plate. The steel articles are cleaned, placed in air-tight ovens and then are heated to a dull red heat. Super-heated steam at 414 to 689 kPa (60 to 100 psi) pressure is introduced and a slate-blue coating is obtained consisting largely of magnetic oxide of iron. The coating is of considerable depth and is quite durable especially when oil or wax coatings are used as the final application. The **Bower-Barff process** is similar except that, after the steam treatment, the steel articles are cooled to 150°C (300°F), dipped in hot linseed oil and kept at 150°C (300°F), until the oil becomes oxidized. The process has been modified by introducing benzene with the steam to shorten the treatment and produce a heavier coating.

SECTION 6

CHEMICAL TREATMENT OF STEEL SURFACES

Black, blue or brown finishes on steel also may be produced in various shades by a wide variety of chemical treatments. Molten-salt baths produce effective colors on clean, polished steel, but often the temperature is so high that a change in hardness and other mechanical properties can result. The method may be high in cost because of drag-out salt adhering to the metal on removal from the molten-salt bath.

Niter Baths—Molten mixtures of sodium and potassium nitrates are effective bluing agents in the absence of rust. Manganese dioxide is generally added to aid in the production of good colors. Potassium nitrate is used when a bath operating at low temperature is employed, although the sodium nitrate alone can be used without affecting the quality of the work. The temperatures used vary from 315° to 535°C (600° to 1000°F). If a lower temperature is necessary, a black color may be obtained by using a 40 per cent aqueous solution of sodium hydroxide to which about 5 per cent each of sodium and potassium nitrates are added. This solution operates at 120° to 140°C (250° to 285°F). The colored articles are usually finished by immersing in hot oil followed by wiping and polishing.

Polished steel is **oil blackened** by packing in a carburizing box with spent carburizing compound, excluding air, and heating to 650°C (1200°F) for about one and one-half hours, and then quenching in oil. Variations of this process at lower temperatures may be effected by heating the articles rapidly to 535° to 650°C (1000° to 1200°F) in air, and quenching in oil. Small parts are colored by introducing them into a rotary furnace retort operating at about 400°C (750°F). A small amount of linseed or fish oil is added to the charge and the parts rotated for three to ten minutes after this addition, after which parts are cooled in air and dipped in a rust-retarding oil.

Articles which have been quenched in oil may be placed, without removal of oil, in a rotary, unperforated-drum retort, heated to 260° to 345°C (500° to 650°F), maintained for proper tempering time according to color desired. The longer the time the deeper and more desirable will the black color be. The retort is allowed to cool to 260°C (500°F) then articles are removed and tumbled in a slightly-oiled granular cork to brighten. This method gives a combination tempering and oil-blackening treatment.

Browning of steel is accomplished by a wide variety of processes. The thoroughly cleaned steel is coated by spraying, brushing or dipping with two coats of a browning solution which generally consists of a mixture of metallic salts, acids, alcohol or water. The coating is allowed to dry, heated to 60° to 80°C (140° to 175°F), and then placed in a humidity room or chamber at the same temperature used for preheating, where it is allowed to rust. It is then washed in boiling water for 15 minutes, dried and cleaned with a wire brush or fiber wheel to remove loose particles of rust. Three more rustings with intermediate cleanings are applied after which the browned surfaces are coated with a rust-preventive oil.

Solutions for the coloring of steel by chemicals are very numerous and often contain lead, iron, mercury, antimony or copper salts in combination with sulphur or selenium compounds. Usually the colors obtained by the relatively cold chemical methods are not as brilliant as those from heat tinting in air or steam, or the product from salt baths. Nevertheless, certain new chemical treatments have been widely adopted, not because of their superior appearance, but because they are beneficial in bonding paints and lacquers, especially when these are baked. Such coatings have a very durable, final finish and one in which corrosion due to porosity of or imperfections in the paint film is minimized. The most important of these are the phosphate treatments whereby a thin adherent coating consisting largely of iron phosphate and zinc phosphate is applied to the steel. Also, thicker coatings of zinc phosphate are frequently used to aid in deep drawing operations. There are in addition a number of proprietary or commercial coatings that may be used for the above purpose.

SECTION 7

CHEMICAL TREATMENT OF METALLIC COATINGS

Chemical treatments may be applied to the surface of metallic coatings such as zinc, tin and aluminum to increase their durability or facilitate the application and adherence of enamels, paints and the like.

Zinc-coated (galvanized) steel is usually treated to retard the formation of white corrosion products that are very detrimental to the appearance of the galvanized sheet. Treatments applied for this purpose by the producer of galvanized products are usually washes with dilute water solutions of water glass (sodium silicate) and sodium dichromate, chromic acid, and mixtures of phosphoric and chromic acids. Most of these treatments do not significantly alter the bright appearance of the galvanized sheet. Some colored chromate treatments are applied when severe storage conditions are anticipated. If paint adherence is of primary importance, then proprietary phosphate treatments such as "Bonderite" can be used. Many other proprietary treatments may be used for special purposes.

Surface treatments commonly used for tin plate are described elsewhere in this book (see Chapter 36). In general, these are usually rapid electrochemical treatments in dilute chromate or phosphate-chromate solutions which have the dual function of stabilizing the surface against oxidation and/or discoloration and of imparting good lacquer and enamel adherence.

Aluminum-coated steel may, if desired, be treated by methods much similar to those discussed above. There are many patents describing treatments for aluminum that may also be applicable to aluminum-coated steel.

SECTION 8

VITREOUS-ENAMEL COATINGS

These coatings consist of a layer of glass fused to the properly prepared steel base and, thus, are quite different from enamels of organic origin which will be discussed later. To adjust the properties of the finished articles to the ultimate uses, wide variations in the composition of this glass are required but, in general, it must adhere well to the steel base and possess a coefficient of expansion adjusted to that of the base metal. Good adherence is achieved by incorporating in the ground coat (primer) enamel certain oxides, usually cobalt oxide. Adjustment of coefficient of expansion is accomplished by a variety of compositions which are compounded by fusing together quartz and feldspar, with fluxes such as borax, fluorspar, cryolite, soda ash, sodium nitrate and litharge. Opacifiers such as oxides of titanium are usually added when the glass is ground to a fine powder. This is generally accomplished in a pebble or ball mill.

Ground coat application is made to the thoroughly cleaned sheet steel article by immersing in a water suspension or "slip" or pulverized enamel ingredients. The prior cleaning may consist of degreasing, pickling in acid, rinsing in a neutralizing bath, sometimes followed by a nickel-solution dip to improve enamel adherence and behavior during firing, after which the articles are washed and dried. After application, the slip is allowed to drain, dried, heated (**fired or burned**) at as high a temperature as 815° to 870°C (1500° to 1600°F) for 1 to 4 minutes and cooled to room temperature. This fired ground coat offers sufficient protection to the steel base and may be used alone. However, since the "slip" usually contains cobalt oxide, the resulting coating is dark blue and may not be suitable for all purposes, hence a finish coat may be applied.

Finish Coats are applied when a light color or additional protection such as acid resistance is desired. The fired ground-coated article is sprayed with a slip of the required finish-coat composition, dried, fired for 1 to 3 minutes and cooled. The operation may be repeated several times using the same or different slips and many attractive color combinations can be obtained if desired.

Single-Finish Coats—Much experimental work has been directed toward the development of a suitable sheet steel on which the finish enamel coat can be applied directly. This work has led to the production of sheet steels of very low carbon content that are suitable for this purpose. These steels are being used in commercial production with satisfactory results and, because of the economics associated with single-coat enameling, are finding increasing use.

Low-temperature vitreous enamels—To provide a coating where the high gloss or decorative enamels are not required, ceramic coatings have been developed which can be applied by spraying and firing at a temperature of as low as 535°C (1000°F). These have found a limited field of use. The composition is said to consist of an alkaline aluminum silicate.

SECTION 9

MISCELLANEOUS INORGANIC COATINGS

Cements—These coatings differ from vitreous enamel in that they are not always fused to the steel, although the constituents consist largely of finely ground, vitrified products. Cracking vessels in the oil industry are sometimes protected by a mixture of furnace cement and sand to which short fiber asbestos and water glass are added. In this instance curing is facilitated by heating to about 480°C (900°F). The interior of cast-iron or steel pipe and steel tanks may be coated with cement to resist corrosive waters, salt solutions, oil having a high sulphur content and the like. Concrete coatings are used on the exterior of pipe when it is buried in extremely moist or corrosive soils. This concrete is a rich mixture approximating two parts sand and one part portland cement and may be two to four inches in thickness. The alkalinity of cements usually inhibits corrosive attack of the steel to which they are applied.

Core Plate—These coatings, which are discussed in greater detail under silicon or electrical-steel sheets (Chapter 46), may sometimes be inorganic in nature and are usually applied to silicon-steel sheets used for transformer laminations to improve the insulating properties. Other core plates used for a similar purpose or for insulating motor laminations are organic-varnish coatings.

Metal Powders in Inorganic Vehicles—Metal powders may be incorporated with inorganic silicates or phosphates to produce a protective coating. For example, zinc dust is used in inorganic zinc-rich silicate paints that can provide sacrificial protection to steel surfaces.

SECTION 10

ORGANIC COATINGS

Steel requires an abundance of both oxygen and water to rust; organic coatings can prevent corrosion by: (1) interposing a barrier between the steel and corrosive media; (2) inhibitive action; or (3) restricting flow of galvanic currents. They also may serve as decorative or functional coatings to obtain color, reflectivity, antiskid properties or fire retardancy. Painting, for the sake of corrosion protection only, cannot be considered economically warranted under mild conditions represented by low humidity and absence of corrosive media, but painting is often justified under such conditions on the basis of good appearance. Where protective organic coatings must be applied to combat corrosive influences, a coating may be selected that will provide good appearance as well as protection without adding materially to the cost.

Important factors that can be controlled to improve the performance of organic coatings are:
   (a) Designing to minimize corrosion and paint failure and to facilitate painting.
   (b) Using a degree of surface preparation compatible with the intended service and paint scheme.
   (c) Applying paint in a manner ensuring maximum life commensurate with practical difficulties.
   (d) Using paints properly formulated and capable of performing the required service.
Painting should be considered as a complete system that includes surface preparation, pretreatment, primer, intermediate coat or coats, finish coat, and method of application.

The type of organic coating or painting system used obviously depends upon the steel product and its intended use. The painting of steel structures with suitable specifications for surface preparations, pretreatments, paint application and paints is fully covered in the Steel Structures Painting Manual.[8] Production-line procedures generally used in product finishing on sheet or strip products differ radically from those for structural steel. Production finishing procedures are adequately described in the literature.[9,10] Different metallic substrates require special coatings or pretreatments. Galvanized steel requires zinc-dust paints, cement-in-oil paints, wash primers or phosphatizing pretreatments for satisfactory adhesion to be achieved.[11] Terne-coated steel may have a film of residual oil which must be removed; otherwise, the selected coating must be capable of wetting the terne surface through the residual oil. Aluminum-coated steel requires pretreatment to achieve good paint adhesion. A wide variety of inks, varnishes, lacquers and pigmented organic coatings are used on tin plate for caps, closures, cans, and lithography. These coatings must be compatible with the surface conditions imposed by the chemical treatments used on tin plate.

Organic coatings are commonly known as paints, varnishes, enamels, and lacquers. The pigments for organic coatings may be of the inorganic or the organic type and are used to impart color, increase film density, and protect the organic binder from the effects of sunlight. Inhibitive pigments are often used in primers to enhance the corrosion resistance afforded the steel by the coating. Paints are mixtures of pigments with drying oils, usually linseed or tung oils (which dry by oxidation), with varnish vehicles, or with synthetic resins or polymers. Varnishes are solutions of resinous materials in oils or volatile liquids which dry by evaporation or oxidation. Enamels are varnish or synthetic resin solutions to which pigments have been added. Lacquers comprise solutions of shellac, resins, cellulose derivatives or various polymerization products in suitable solvents, all of which dry by evaporation. Some of these paints are also available as dispersions or solutions in water and are used where solvent vapor evolution from painted surfaces is objectionable.

The subject of organic coatings is too extensive and complex for a thorough discussion here. The importance of such coatings in the protection of steel is, however, too great to dismiss without a brief discussion of some of the types of coatings that have been developed. This field is in a constant state of flux and changes due to the rapid adoption of newly developed materials.

Synthetic Resins—The introduction of the synthetic alkyd resins a comparatively few decades ago led to drastic changes in organic-coating technology. While paints mixed with drying oils continue to be used in large quantities with satisfactory performance, most organic coatings are based on synthetic-resin vehicles. Alkyd resins are used in many of the maintenance paints and in product finishes because of their hardness, gloss, and color retention. Phenolic resins are used in combination with oils in varnishes and paints, especially to withstand immersion in water, high humidity, and condensation: these resins are used extensively in linings and tin-plate coatings. Production finishes utilize resins that require baking such as the ureas and melamines usually blended with other synthetics to achieve high levels of hardness, abrasion resistance, and durability in service. Vinyl resins, usually copolymers designed to have specific properties, serve well in wet or corrosive environments: they possess resistance to many chemicals and may be applied as solution coatings, plastisols, or organic dispersions, and are also being used as latices in water-base coatings. Epoxy resins, cured with amine or polyamide reactants at ambient temperatures or by baking, resist many chemicals and are widely used in maintenance paints and product finishes; these resins, like the vinyls, alone or combined with other resins, are used as chemical-resistant drum and tank linings and for tin-plate containers and closures. Combinations of epoxy resins with coal tar form epoxy coatings of good chemical resistance and high film build (productive of thick films) at moderate cost. Acrylic resin coatings are regularly used as lacquers and thermosetting finishes on automobiles, appliances, and siding; the acrylics have good weather durability and make good clear coatings for outdoor use.

Pigments—Pigments impart color, hiding power, reflectance, rust inhibition, abrasion resistance, anti-skid characteristics, and fire resistance, and reduce water-vapor transmission and prolong coating life by screening out ultraviolet rays from sunlight. Red-lead and zinc-chromate have been used in rust-preventive paints; basic lead silico-chromate is a newer pigment with good rust-inhibiting characteristics. Use of lead compounds are being eliminated for health reasons.

Zinc dust performs well in rust-preventing paints when incorporated with zinc oxide; these paints are particularly useful for painting galvanized steel. Zinc-dust pigment provides sacrificial protection to steel when the pigment is used in zinc-rich paints (paints heavily loaded with zinc dust). Inert pigments, such as those containing titanium dioxide or iron oxides, are used for intermediate and finish coats, or for primers when rust inhibition is not required or is provided by other pigments. Extenders are low cost pigments that are used to increase the pigment volume content of a paint at moderate cost.

Bituminous Coatings—These are based upon bituminous resins—coal tars, asphalts, or asphaltums. The resins are applied as solution coatings of low build (productive of relatively thin films), as mastics that have increased viscosity and high build (productive of thick films) due to incorporation of fillers, or as hot melts called enamels. Water-emulsion dispersions of these resins are also used. The bituminous resins perform well underground and in contact with water; they do not have good weather durability when exposed to sunlight.

Application and Drying of Organic Finishes—Since the bond between metal and coating is the weak point in most paint systems for metal, the proper application is extremely important. Prior to application of a coating, thorough cleaning of the surface is necessary. Since no paint system is entirely impermeable to moisture, coating durability is increased by pretreatment of the steel base with inhibitive washes, many of which are proprietary in nature. The application of such treatments prior to painting is practiced extensively in fabricating steel articles.

The older and more common methods of application of finishes by brushing or dipping are still widely used, but, in most factories, conveyor systems for handling parts permit the use of roller-coating methods for flat products and spraying for more complicated or formed shapes. These operations are usually performed mechanically and are often followed by a closely controlled baking operation. Banks of electric infra-red lamps, or gas-fired or electrically heated ovens, may be used for the baking operation. In some instances the heat from baking merely serves to drive off excess solvents. In other cases, reactions or oxidation of the coating required to produce a durable finish are also accomplished: here, the composition of the oven atmosphere is important. Baked coatings are usually harder, tougher, and often more durable than those that are air-dried, and quite often the coating mixture for an air-dry coating is different from one that requires baking. With the more general use of continuous coating methods, baking is a common practice in quantity production.

An entire industry to coat steel products in coils has developed in recent years because of the demand for prepainted strip steel in coils to make such articles as roofing and siding. A typical "continuous strip coating" line will receive a coil of steel at one end of the line, uncoil it, pass the strip through a five-stage cleaner and pretreater, dry the strip, apply one or two coats of paint to top and bottom of the strip by roller coater, bake the coated strip in long, catenary-loop or floating-strip ovens, cool, and recoil the coated product, all in one continuous operation.

Temporary Organic Coatings—For temporary protection in shipment or storage, steel is usually coated with oils known as slushing oils. Most often, mineral oils are mixed with inhibiting or polar compounds which deter rust formation. Mineral oils alone are sometimes used where corrosive conditions are not severe. Slushing oils are generally removed subsequent to fabrication and prior to application of a permanent organic coating. If the temporary protective film is not removed, troubles are generally encountered due to its incompatibility with the permanent organic coating. For severe conditions such as outdoor storage or overseas shipment, rustproofing compositions such as heavy greases or waxes compounded with inhibitors are applied. When conditions are unusually severe and the value of the steel product justifies the cost, plastic films are applied that can be removed by stripping. This latter expedient may be used for overseas transportation and storage of expensive equipment.

References

1. Fundamental aspects of stress-corrosion cracking. National Association of Corrosion Engineers (1969).
2. Morgan, J. H., Cathodic protection. Leonard Hill, Ltd., 1959.
3. Char, T. L. Rama and D. K. Padma, Corrosion inhibitors in industry. Trans. Inst. Chem. Eng. 47, 1969, PT 177.
4. Mears, R. B., The prevention of corrosion by use of appropriate design. Australasian Corrosion Engineering 6, No. 12, 1962.
5. Burns, R. M., and W. W. Bradley, Protective coatings for metals, 2nd ed. New York, Reinhold Publishing Co., 1955.
6. Mears, R. B., and R. H. Brown, Causes of corrosion currents. Industrial and Engineering Chemistry 33, 1001-1010 (1941).
7. Lowenheim, F. A., Modern electroplating, New York, John Wiley and Sons, Inc., Third Edition, 1974.
8. Keane, J. D., Editor, Steel Structures Painting Manual, Vol. I, Good painting practice; Vol. II, Systems and specifications, Second Edition. Steel Structures Painting Council, 4400 Fifth Ave., Pittsburgh, Pa. 15213.
9. Mattielo, J. J., Protective and decorative coatings. New York, John Wiley and Sons, Inc., 1941.
10. Payne, H. F., Organic coating technology. New York, John Wiley and Sons, Inc., 1954.
11. Bigos, J., H. H. Greene and G. R. Hoover, Results of the AISI research project on the paintability of galvanized steel using trade-sales paints. New York, American Iron and Steel Inst., Contribution to the metallurgy of steel.

General Bibliography

Champion, F. A., Corrosion testing procedures, 2nd ed., New York, John Wiley and Sons, Inc., 1963.
Evans, U. R., The corrosion and oxidation of metals. London, Edward Arnold, Ltd., 1960.
Fontana, M. G., and N. D. Greene, Corrosion engineering. New York, McGraw-Hill Book Co., 1967.
La Que, F. L., and H. R. Copson, Corrosion resistance of metals and alloys (American Chemical Society monograph series). New York, Reinhold Publishing Co., 1963.
Romanoff, M., Underground corrosion. U. S. Bureau of Standards Circular 579, 1957.
Shreir, L. L., Corrosion, Vols. 1 and 2, 2nd ed. London, England, Newnes-Butterworths, 1976.
Speller, F. N., Corrosion: causes and prevention; 3rd ed. New York, McGraw-Hill Book Co., 1951.
Tomashov, N. D., Theory of corrosion and protection of metals. New York, MacMillan Co., 1966.
Uhlig, H. H. (ed.), Corrosion handbook. New York, John Wiley and Sons, Inc., 1948.
Uhlig, H. H., Corrosion and corrosion control, 2nd ed. New York, John Wiley and Sons, Inc., 1971.

CHAPTER 36

Manufacture of Tin-Mill Products

SECTION 1

TIN-MILL PRODUCT TERMINOLOGY

SECTION 1

TIN-MILL PRODUCT TERMINOLOGY

Tin-mill products originate with flat-rolled, low-carbon steel in relatively thin gages. They comprise **black plate, tin-free steel (electrolytic chromium-coated steel) and tin plate.** Of these, tin plate is the most important commercially, accounting in 1981 for nearly 70 per cent of the combined tonnage of all tin-mill products. Tin-free steel was introduced in 1966 as a substitute for tin plate and, by 1981, represented about 19 per cent of the tonnage of tin-mill products. Black plate accounts for most of the balance.

Tin-mill products are specified according to the following Standard Specifications of the American Society for Testing and Materials: A 623 for general requirements for tin-mill products (up to 302.7 kilograms per SITA*, equivalent to 135-lb. base weight); A 624 for single-reduced electrolytic tin plate; A 625 for single-reduced black plate; A 626 for double-reduced electrolytic tin plate; A 657 for single- and double-reduced tin-mill product electrolytic chromium coated; and A 599 for electrolytically tin-coated cold-rolled steel sheet (heavier than 302.7 kilograms per SITA, equivalent to 135-lb. base weight).

Black Plate—Although originally designating thin steel plates produced by hand hammering, the term "black plate" has persisted and now defines the product of the cold-reduction method in thicknesses of 0.358 mm (0.0141 inch) and lighter (No. 29 gage and under). Some of the details of rolling and heat treating such light-gage product were discussed in Chapter 34.

"Black" plate does not have a black appearance. Present methods of manufacture generally result in a flat-rolled product having the typical appearance of clean steel. The appearance of black plate may be affected, however, by modifications of annealing practice, or of processing methods that affect the roughness of the surface texture.

*SITA is an acronym derived from the name of the unit called Systeme International Tin Plate Area, and represents an area of 100 square metres.

Black plate as such is produced in the form of either cut sheets or coils, and is used for fabricating a variety of items including containers, trays and toys. When coated with suitable organic coatings, it exhibits considerable resistance to corrosion, whereas uncoated single cold-reduced black plate is extremely susceptible to rusting and precautions to prevent condensation of moisture must be taken during shipment and warehousing. All manufacturers are equipped to apply thin films of protective oils to minimize rusting. Some mills are equipped to produce chemically treated steel (CTS) or "full-finish" black plate, which is black plate given a protective chemical treatment to enhance rust resistance and adhesion of organic coatings. In addition, there has been limited use of black plate as produced by cold reduction after annealing. This product, termed "QAR" for "quality as rolled," was used in manufacture of beverage cans. The rolling oil is not removed in this process and, as a result, the product has considerable resistance to oxidation during storage. Special lacquers tolerant of heavy oil films are required in its use. All black plates usually require a protective organic coating applied by the user. Precautions to prevent condensation of moisture are necessary during shipment and warehousing of all three types—oiled, CTS and QAR black plate.

The most important use of black plate is in the manufacture of tin-free steel and tinplate, as described hereafter.

Tin-Free Steel—Tin-free steel, also referred to as electrolytic chromium-coated steel, may be described as black plate additionally processed and electrolytically plated, for example, with 0.00762 micrometre (0.3 microinch) of metallic chromium plus a chromium-oxide film generally ranging in weight from 3.2 to 21.5 milligrams per square metre (0.3 to 2.0 milligrams per square foot), determined as trivalent chromium. Although having rust resistance superior to that of black plate, tin-free steel must be lacquered on both surfaces to be used for containers. Tin-free steel has excellent adhesion with organic coatings and shows high resistance to undercutting of lacquers, which makes this product suitable for packing certain food products and beverages. The main applications at present are for beer and soft-drink three-piece cans and ends as well as draw-redraw cans, ends for food cans, and caps and crowns for glass containers.

Tin Plate—Tin plate may be described as black plate additionally processed and coated on both sides with commercially pure tin. The wide-spread use of this major steel-mill product arises from its combination of the strength of steel with the protective properties and solderability of tin. When coated by the hot-dip process, the tin plate was termed **coke tin plate** or **charcoal tin plate. When coated by the electrolytic process, it is termed electrolytic tin plate.** In the United States, there is now no commercial production by the hot-dip process.

Hot-Dipped Tin Plate—Tin plate formerly was produced by the hot-dip tinning of thin plates that had been hammered or rolled from bars of puddled iron. The pig iron used in puddling might have been made in blast furnaces using charcoal or coke as fuel. Plates made from "charcoal iron" were considered a more ductile and higher grade product; hence, tin plate made by coating charcoal-iron plates with tin (charcoal tin plate) was regarded as a product of higher quality than tin plate with a coke-iron base (coke tin plate). At present, the designation "charcoal tin plate" merely indicates plate with a relatively heavy tin coating as compared with "coke tin plate" as described below, and has no significance so far as quality of the steel base is concerned.

Eventually, the thin sheets (black plate) for hot-dipped tin plate were rolled from sheet bars on single-stand sheet and tin plate hot mills. Both acid-Bessemer and basic open-hearth steels were used.

Since the 1930's, black plate for hot-dip tinning was produced by the hot-strip rolling and cold-reduction methods described in Chapters 33 and 34 and later in Section 4 of this chapter, the coiled product being cut into single sheets prior to the tinning operation.
Section 4 of this chapter, the coiled product being cut into single sheets prior to the tinning operation.

Various grades of hot-dipped tin plate were produced, the terminology indicating in a general way the weight of tin coating or, more exactly, the amount of tin used to produce the given unit or base box of plate. Coke tin plate has always designated the plate produced with the lowest amount of tin and formerly was

Table 36—I. Tin-Coating Weights of Standard Grades of Hot-Dipped Tin Plate.

Class Designation	Minimum Average Tin-Coating Weight Test Value	
	g/m²	lb/bb
Common Cokes*	19.07	0.85
Standard Cokes	23.56	1.05
Best Cokes	26.70	1.19
Kanners Special Cokes	31.42	1.40
1A Charcoal*	40.39	1.80
2A Charcoal	51.61	2.30

*The terms "coke" and "charcoal" stem from the old practice of making hot-dipped tin plate from coke iron and charcoal iron. Today, they merely refer to different tin-coating-weight levels.

called **Cokes** or **Common Cokes.** In ascending weight of coating, the various grades were **Common Cokes, Standard Cokes, Best Cokes, and Kanners Special. Charcoal tin plate** carries still heavier coatings including 1A and 2A. (See Table 36-I).

Electrolytic tin plate is available in both melted and matte (unmelted) finish. Electrolytic tin plate with a melted finish has bright luster and is similar in appearance to hot-dipped tin plate; whereas matte finish tin plate is lacking in luster. Originally, electrolytic tin plate was produced only from black plate that had been annealed and temper-rolled, but the ever-increasing trend toward stronger, lighter-weight tin plate has resulted in the development of double reduced tin plate, which is produced from black plate that receives a second cold reduction after annealing (see Section 4 of this chapter). Electrolytic tin plate can be produced as coils or cut sheets.

Theoretically, electrolytic tin plate can be produced in any coating weight. The commercial grades of electrolytic tin plate available at present include coating weights of 2.2, 5.6, 7.7, 11.2, 16.8 and 22.4 grams of tin per square metre (0.10, 0.25, 0.35, 0.50, 0.75 and 1.00 pound of tin per base box), identified by the numerals 10, 25, 35, 50, 75 and 100 respectively. In addition, differentially coated tin plate that has different tin coating weights on opposite surfaces is available in coating weights of 6.7-4.5, 11.2-5.6, 16.8-5.6, 22.4-5.6, 30.3-5.6 and 22.4-11.2 grams of tin per square metre (0.30-0.20, 0.50-0.25, 0.75-0.25, 1.00-0.25, 1.35-0.25 and 1.00-0.50 pound of tin per base box). These are also identified by numerals as, for example 50-25. The 7.7 and 6.7-4.5 gram per square metre (0.35 and 0.30-0.20

Table 36—II. Nominal Thicknesses and Weights of Tinplate Commonly Produced.

Thickness, Millimetres	Equivalent Weight, kg/SITA	Weight, lb/bb	Equivalent Thickness, Inches
0.14	109.9	50	0.00550
0.15	117.8		
0.16	125.6	55	0.00605
0.17	133.4	60	0.00660
0.18	141.3	65	0.00715
0.19	149.2		
0.20	157.0	70	0.00770
0.21	164.8	75	0.00825
0.22	172.7		
0.23	180.6	80	0.00880
0.24	188.4	85	0.00935
0.25	196.2		
0.26	204.1	90	0.00990
0.27	212.0	95	0.01045
0.28	219.8		
0.29	227.6	100	0.01100
0.30	235.5		
0.31	243.4	107	0.01177
0.32	251.2	112	0.01232
0.33	259.0		
0.34	266.9	118	0.01298
0.35	274.8		
0.36	282.6	128	0.01408
0.37	290.5		
0.38	298.3		
		135	0.01485

pound per base box) weights were introduced primarily for the sake of economy in the production of two-piece containers by the drawing and ironing (D&I) operation. The D&I can is a relatively recent innovation in container manufacture and replaces the conventional three-piece soldered, welded, and adhesive-bonded cans for certain applications, principally for packaging beer and carbonated beverages.

The differentially coated tin plate was developed to conserve tin and to lower container costs, with the heavier-coated surface being employed as the inside surface of the container where greater protection is required. The lighter-coated surface was formerly identified by making it less lustrous than the heavier-coated surface either by roughening the steel surface during temper rolling or by anodizing the plated surface in a suitable electrolyte just before melting. Identification of the coating grade is accomplished by the use of chemicals to print prescribed patterns on one or the other surface prior to melting the tin coating.

Symbols and Definitions of Units of Sale—Tin plate is now sold both on a weight per unit area basis and on a thickness basis. The old unit of area is the **base box**, equal to the area of 112 sheets, 14 by 20 inches, or 31 360 square inches (217.78 square feet). The new unit of thickness is the millimetre and the corresponding area designation for pricing and weight or area calculations is the SITA (Systeme International Tin Plate Area) which equals 100 square metres or 4.9426 base boxes. The nominal weights produced for the two systems differ slightly as shown in Table 36—II.

In the early eighteenth century in England, 355.6-by 508-mm (14-inch by 20-inch) plates were packaged in lots of 112 sheets to make 50.8 kilograms (one hundredweight or 112 pounds). When the plate was lighter or heavier, it was identified by suitable symbols as discussed in Chapter 53 under "The Tin Plate Gage." A **package** of tin plate still consists of 112 sheets.

The use of symbols to designate gage was displaced by the use of **base weights** (or **basis weights**) expressed in pounds per base box that also indicated the approximate thickness of the tin-mill product (tin plate, black plate, and tin-free steel). With the recent introduction of the metric system (SI) in production, gage is the basic unit for specification and area becomes only a derived value for commercial usage.

SECTION 2

OCCURRENCE, MINING, AND REFINING OF TIN

Tin, though one of the common metals, is the most sparsely distributed metal in common use. The deposits that produce probably 85 per cent of the world's supply are in Malaysia, Indonesia, Bolivia, Thailand, China, Zaire, and Nigeria; the remaining 15 per cent is accounted for by the deposits in Australia, in the centuries-old mines of Cornwall in the United Kingdom, and in the Union of South Africa and India.

The most abundant source of tin is the oxide, **cassiterite, or tin stone,** and most of the world's supply is derived from alluvial deposits in river beds. The cassiterite originally occurred in veins and lodes in highly acid igneous rocks. The ore may be found also as primary veins in metamorphosed sedimentary rocks. Cassiterite has a relatively high specific gravity (6.4 to 7.1).

Alluvial deposits of tin stone occur associated with gravel, which may be bonded with clay and covered with more or less overburden consisting of soil, clay, etc. This type of deposit, found in Malaysia, Indonesia, Thailand, China and Nigeria, is worked by various types of dredging or hydraulic mining. Screening and washing of the deposit with running water causes a breaking up of the gravel and a carrying away of the lighter pebbles while the heavier tin stone is retained and recovered.

Tin ore in Bolivia and Cornwall is practically all vein tin. Here the tin ore is found in lodes and beds, in older rocks such as granite, gneiss and mica schist, associated with a large proportion, perhaps 95-99 per cent, of gangue consisting principally of silica and siliceous minerals, metallic minerals like the sulphides of iron, copper, lead and zinc, iron oxides and wolframite (a tungstate of iron and manganese). The ore first is crushed to pass approximately a twenty-mesh screen; the tin-bearing minerals of the ore are separated by washing and recovered in settling tanks, or on slime tables.

Regardless of which method is used to concentrate the ore, the concentrates then are roasted to remove sulphur and arsenic. This operation in turn is followed by a second washing, although in some cases the roasted ore is treated in magnetic separators to remove iron oxide and wolframite prior to washing. The tin stone is then ready for the smelter. Reduction of the tin ore is accomplished either by the use of a reverberatory furnace or a blast furnace. The crude tin thus obtained is refined by a liquating operation in which advantage is taken of the low melting point of tin. The impure metal is heated on the inclined bed of a furnace to a temperature just above its melting point. Comparatively pure tin trickles down to a basin below, leaving the higher melting-point impurities on the bed of the furnace. The low melting-point impurities, lead and bismuth, are removed by either or both of two oxidizing methods. The first of these is an operation called **boiling,** in which sticks of green wood are immersed in the molten metal and undergo destructive distillation to produce bubbling by the steam and gases generated to agitate the molten metal bath. Different portions of the metal thus are exposed to air and are oxidized. The oxidized impurities float on top of the molten metal to form **dross** that is skimmed from the surface of the bath. Similarly, pouring ladlefuls of molten metal from a height into the bath permits the oxidation of impurities. This latter operation is called **tossing.** Drosses may be resmelted to recover the tin they contain.

Table 36—III. Chemical Composition of Commercial Grades of Tin.

Element	Chemical Composition, Per Cent						
	AAA	AA	A	B	C	D	E
Tin, min	99.98	99.95	99.80	99.80	99.65	99.50	99.00
Antimony, max	0.008	0.02	0.04	—	—	—	—
Arsenic, max	0.0005	0.01	0.05	0.05	—	—	—
Bismuth, max	0.001	0.01	0.015	—	—	—	—
Cadmium, max	0.001	0.001	0.001	—	—	—	—
Copper, max	0.002	0.02	0.04	—	—	—	—
Iron, max	0.005	0.01	0.015	—	—	—	—
Lead, max	0.010	0.02	0.05	—	—	—	—
Nickel—cobalt, max	0.005	0.01	0.01	—	—	—	—
Sulphur, max	0.002	0.01	0.01	—	—	—	—
Zinc, max	0.001	0.005	0.005	—	—	—	—

In more modern extraction methods, ore concentrates are calcined with or without additions intended to facilitate further processing. The iron and other impurities then are removed by leaching the ores in solutions of hot hydrochloric acid. Leaching is followed by filtering of the chloride solution. After washing with dilute acid, the residues are of such purity that subsequent treatment in reverberatory furnaces will yield a metal containing at least 99.80 per cent tin. It is possible to obtain metal of 99.98 per cent tin content by electrolytic refining. The grades of tin listed in Table 36—III are published by the American Society for Testing and Materials as a guide to the maximum impurities that normally are found in brands of tin that fall into these grades. Grade A tin is usually used in the production of electrolytic tin plate.

Properties and Uses of Tin—Tin has a silver color with a slight bluish tinge, a brilliant luster, a structure which is distinctly crystalline, and is soft and malleable at ordinary temperatures. Other physical properties of tin are:

Atomic Weight: 118.7 (isotopes with masses 112 to 124)

Atomic Number: 50

Density: 7300 kg/m³ at 15°C (59°F)

Specific Volume: 0.1395 at 20°C (68°F)

Hardness on Mohs' Scale: 1.8

Tensile Strength: about 15.2 MPa (2200 psi) with elongation of 86 per cent

Melting Point: 231.9°C (449.4°F)

Boiling Point: 2270°C (4120°F)

Tin is alloyed with other metals to make bronze, Britannia metal, pewter, solder or white bearing metal. Pure tin or lead-tin alloy, rolled very thin, is known as tin foil. Tin amalgam is used in making mirrors, and tin condenser tubes are used in laboratory stills. Tin, in conjunction with an acid, is used as a reducing agent.

Stannic oxide is used as a polishing powder, and as an opacifying agent in glasses, glazes and enamels. The chlorides are used as mordants in weighting silk and in dyeing. The largest use of tin at the present time is in the manufacture of tin plate.

SECTION 3

USES AND IMPORTANCE OF TIN-MILL PRODUCTS

Tin-mill products represent some of the major items produced by the steel industry in the United States. In normal years, the tonnage has represented approximately 7 per cent of the total steel products shipped. Statistics compiled by the American Metal Market show that production increased steadily from about 308 000 metric tons (340 000 net tons) in 1900 to about 2 490 000 metric tons (2 750 000 net tons) in 1940. By 1948, new facilities had increased production to about 3 600 000 metric tons (4 000 000 net tons). In 1981, according to the Annual Statistical Report of the American Iron and Steel Institute for that year, shipments were about 4 400 000 metric tons (4 933 000 net tons). The lower rate of growth of the tinplate industry reflected the increased competitive pressure from competitive materials, such as aluminum, glass, and plastic.

The importance of tin plate to the food industry is well recognized and its widespread utilization attests to the unique properties of this product in which are com-

bined the strength of steel and the corrosion resistance of tin. Tin plate is fabricated readily on three-piece can manufacturing equipment and in the production of two-piece cans by draw and ironing, or draw-redraw processes. It has a pleasing appearance. It is relatively inexpensive and is non-toxic. Because of these and other properties, tin plate has been found to be the ideal material for fabricating food containers (tin cans), aerosol and beverage cans, crown caps and other bottle caps or closures, for kitchen utensils such as baking pans, for various drawn or fabricated parts in radios, and for such articles as electrical equipment and toys.

The largest use of tin plate is for containers, and many of the improvements in its manufacture have been the result of research directed toward meeting the requirements of the container-manufacturing industry. Tin cans are used not only for food and beverages, but also for paints, oils, tobacco, insecticides and proprietary drugs.

PROCESSING OF STEEL FOR TIN PLATE

Types of Steel Used—Most of the steel used for the production of tin plate is made by the basic oxygen and open-hearth steelmaking processes.

The steels utilized for tin-plate production are carbon steels having a maximum carbon content of 0.15 per cent. Until the advent of continuous casting, the preponderance of tin-plate steels were produced as either rimmed or mechanically capped ingot steels. The rimmed steels are utilized in general for the softer and deep-drawing tin-plate requirements, whereas the capped steels are used for those applications which require a harder tin plate or are less critical with regard to stamping or forming. The capped steels for tin plate have the definite advantage of being relatively more uniform throughout in grain size, cleanliness and chemical composition than rimmed steels.

Continuous-cast steel is now replacing ingot-cast steel for tinplate manufacture. Most continuous-cast steel is produced using aluminum deoxidation practices, although a small amount of continuous-cast steel is produced using deoxidation with manganese, silicon, and aluminum. Continuous-cast steels have excellent surface quality, improved uniformity of composition and properties, and superior steel cleanliness compared with ingot-cast steel.

Of special importance in the manufacture of tin plate is the selection of a steel base of the proper composition for the job. For particular applications, as determined by temper requirements and pack corrosivity, the desired composition ranges will be selected as discussed in Section 8. The acid fruit products represent in general the most corrosive media. For these applications, the phosphorus, silicon, copper and "tramp" elements in the steel are held to low limits, and nitrogen is added to impart additional strength if desired.

Strand-cast steel has been examined in corrosion pack tests with a variety of foods in both plain (unlacquered) and lacquered containers and has performed as well as ingot-cast steels. Its performance in can fabrication is excellent.

EQUIPMENT AND PRACTICE

The sequence of operations in the manufacture of tin plate is as follows: slabs are heated and hot-rolled to coil form on the hot-strip mill. The coils are continuously pickled and taken to the cold-reduction mills where they are reduced to the desired final tin-plate gages or to gages which may be as much as twice that required for the finished products. The cold-reduced material is cleaned, annealed, and then either temper-rolled (up to about 2 per cent extension) or cold-reduced (up to 40 per cent) in coil form in preparation for the tinning operations described in Section 5 of this chapter.

As preceding sections of this book discuss and describe in detail the equipment and operation of the continuous hot-strip mill, the continuous pickler, the cold-reduction mills, and all other equipment necessary for the processing of black plate up to the actual coating operation, the following explanations will be limited to general control measures designed to produce the most suitable product.

Continuous Hot Rolling—In continuous hot rolling, the slabs of steel, ranging from 100 to 250 mm (4 to 10 inches) in original thickness, are hot-rolled at an elevated temperature to a single continuous length which is coiled. This coiled product may be as thin as 1.65 mm (0.065 inch) or as thick as 3.18 mm (0.125 inch), depending upon the end use of the product and the desired thickness of the product after cold reduction; it will generally be between 1.78 and 2.54 mm (0.070 and 0.100 inch) for the bulk of the tin plate produced in this country. Factors such as type of hot-rolled surface desired, ease with which oxide can be removed prior to the cold-reduction operation, desired hardness of the hot-rolled product and resultant mechanical properties, grain size and corrosion resistance of the final product, all must be considered in establishing an optimum hot-rolling practice. In general, the 1.78- to 2.54-mm (0.070- to 0.100-inch) thick hot-rolled product is at a temperature between 815° and 870°C (1500° and 1600°F) at the exit side of the last finishing stand. Similarly, this product going into the coiler generally will be in the range of temperature between 595° and 675°C (1100° and 1250°F).

Continuous Pickling—For the production of tin plate, it is sufficient to say that the primary functions of the continuous pickler are to remove uniformly all of the scale and oxide from the surface of the hot-rolled steel and subsequently to oil the pickled product. These operations are necessary prior to taking the very heavy cold reductions necessary to obtain the light gages required for tin plate and to ensure the clean surface finally necessary for the tinning operation.

Cold Reduction—The importance of the cold-reduction operation to the over-all quality of tin plate cannot be over-emphasized. To a great extent, this operation determines the gage uniformity and the surface quality and flatness of the final tin plate product. These three properties require particularly close control because of the demands of the very sensitive container-lacquering operations and the high-speed automatic can-making operations. To attain the 80 to 90 per cent cold reduction used on the steels for tin plate manufacture, nearly all plants in this country use five-stand or six-stand four-high tandem cold-reduction mills with delivery speeds of 10.2 to 38.1 metres per second (2000 to 7500 feet per minute). The design and operation of these mills are described in Chapter 34. It should be mentioned here that the mills used are generally from 1065 to 1420 mm (42 to 56 inches) wide to produce the usual 610- to 915-mm (24 to 36-inch) widths. The cold-reduction operation is accomplished with the aid of suitable lubricant which must be removed from the strip before annealing so that a bright clean strip will be available for tinning. Depending upon the type of annealing utilized, cleaning may be a separate operation or in line with the annealing furnace. Electrolytic

Table 36—IV. Commonly Produced Surface Roughness Grades for Tin Mill Products

	Desig-nation	Approximate Arithmetic Average (A-A) Roughness, Aim Range, Microinches	Characteristic
Shot Blasted Roll	5B	30-50	Standard blasted surface for uncoated black plate and a few tin plate requirements.
	5C	40-60	Principally a base for unmelted electrolytic tin plate for crowns and closures.
	5D	70-100	Primarily for unmelted electrolytic tin plate for D&I cans.
Ground Roll	7A	5 or less	A lustrous smooth finish for uncoated black plate intended for electro-plating.
	7B	7-15	A smooth finish which may contain fine grid lines for special applications.
	7C	12-22	A smooth finish with grit lines. This is the standard finish for tin mill products.

Table 36—V. Typical Mechanical Characteristics of Tin Mill Products

	Temper Desig-nation	Aim Rockwell Hardness Range, 30-T	Approximate Longitudinal Yield Strength, ksi	Application Characteristics
Single-Reduced Products	T-1	46-52	25-42	Soft for drawing.
	T-2	50-56	34-46	Moderate drawing where some stiffness is required.
	T-3	54-60	40-52	Shallow drawing, general purpose with fair degree of stiffness to minimize fluting.
	T-4 CA	58-64	48-60	General purpose where increased stiffness is required.
	T-5 CA	62-68	56-68	Increased stiffness to resist buckling.
Double Cold-Reduced Products	DR-8	73-Mean	75-85	Relatively hard with limited ductility and highly directional mechnical properties.
	DR-9 CA	76-Mean	85-105	Higher strength than DR-8.

cleaning is done in alkaline solutions, with strip polarity and current density varied as required.

Annealing—The cold-reduced product is quite hard, having a Rockwell (30-T scale) hardness of approximately 80 to 85, and must be softened by annealing. Batch-type (box annealing) and continuous-annealing furnaces are used.

In the **box-annealing process,** coils of steel for tin plate are annealed in a protective atmosphere within the steel-temperature range of 620° to 675°C (1150° to 1250°F), for periods of 4 to 12 hours. While the heating cycle in relation to the metalloid content of the steel largely determines grain size of the finished product and exerts a strong influence of the mechanical properties of the finished tin plate, the cooling cycle plays an important part as regards surface; a clear, bright, steel surface, substantially free of oxides, can be obtained if a reducing atmosphere (usually simple mixtures of hydrogen and nitrogen) is maintained around the coils until their temperature is below about 120°C (250°F).

In the **continuous-annealing process,** the steel is heated in protective atmospheres to about 650°C (1200°F) in a fraction of a minute at strip speeds up to 7.6 metres per second (1500 feet per minute) in older furnaces and at speeds of 10.2 metres per second (2000 feet per minute) in newer furnaces. Continuously annealed strip is inherently harder and stiffer than box-annealed strip, but possesses good formability; thus, continuous annealing offers a method for obtaining tin plate of higher hardness without change in steel composition.

Temper Rolling—The function of the temper mills in the manufacture of tin plate is three fold: to impart the desired surface finish to the product, to improve its flatness and to obtain the desired metallurgical properties such as temper and freedom from fluting tendencies.

The surface finish is controlled largely by the smoothness of the exit work rolls. Finishes may be produced by grinding, shot-blasting or electro-discharge texturing. Smooth-ground rolls impart a bright, dense surface finish whereas rough-ground or shot-blasted rolls impart a rougher finish. All grades except blasted (No. 5) finish are finished with smooth-ground rolls. The commonly supplied finishes are described in Table 36—IV.

Flatness is obtained by proper adjustment of such factors as roll contour, finish, pressure and strip tension.

For the softest tempers, temper rolling (also called **skin passing**) is done on a conventional four-high single-stand coil temper mill although, with care, these tempers can be produced on two-stand tandem temper mills. The objective here is to flatten the strip and impart the proper surface finish in one pass with a minimum of hardening effect due to cold working of the product, the thickness reduction being held to less than one per cent. For the harder tempers, a two-stand four-high tandem temper mill is employed, or the strip is given two passes through single-stand mills. On such material, temper rolling has the added function of increasing the steel hardness, and thickness reductions of about two per cent are common. As an aid to temper rolling, the work-roll surface of the first mill stand is relatively rough, being rough-ground or shot blasted, and the work rolls of the second stand or finishing mill are ground smooth. Thus, superficial hardening, arising from second-stand smoothing of the rough strip surface imparted in the first stand, is added to the full-section hardening of the strip under rolling pressures and reel tension. Modifications of temper mills for tin plate include means for applying high strip tensions and for developing heavier drafts in the first stand, as mentioned in the section on finishing. In addition, there are installations where flattening is achieved by stretcher leveling and no gage reduction is produced by rolling.

For double-reduced tin plate, the temper-rolling step is replaced by a cold-rolling operation in which the annealed black-plate coils are reduced 30 to 40 per cent on a four-high tandem cold-reduction mill of either two or three stands. The resultant product has high hardness, yield strength and tensile strength but possesses sufficient formability for a wide range of container applications. Tin-mill product temper designations and mechanical properties are listed in Table 36—V.

After temper rolling or cold rolling, the coils of black plate may be handled in one of several ways. They may be side trimmed and packaged or side trimmed, sheared and packaged for shipment. However, black-plate coils for electrolytic tinning or chromium plating (tin-free steel) are taken to a side-trimming and recoiling line for preparation, or sent direct to plating units with in-line side trimmers found in most new installations.

SECTION 5

ELECTROLYTIC TIN PLATE AND TIN-FREE STEEL

Introduction—Prior to 1937, all tin plate produced commercially was manufactured by the hot-dipping process. While the electrodeposition of tin on steel had been a known process for many years, its application to the production of tin-plated sheets could not be made to compete economically with the hot-dipping process due to much higher equipment and labor costs. However, with the introduction to the sheet and tin plate industry of continuous cold-reduction mills in the early 1930's, the possibility of continuous high-speed electrotinning became obvious. As early as 1935, small experimental units were designed and constructed capable of continuously electroplating tin on steel strip moving at relatively high speeds. From the results of these studies, electrolytic tin plate appeared on the market as a commercial item in 1937.

The early development of electrolytic tin plate in the United States was given a major impetus by the requirements of the dry package market for a light-coated product which could not be produced by the hot-dip process. The precipitous necessity to conserve vital tin during World War II resulted in a phenomenal expansion of this development so that by 1948, over half of the tin plate produced was electrolytic tinplate. Now, in this country, all tin plate is being produced by electrolytic processes.

The superiority in tin-coating uniformity which electrolytic tin plate exhibits, as compared to hot-dipped tin plate, together with the close control of tin-coating weight obtainable, has resulted in large savings of tin.

Basic Principles of Electrotinning—To the English scientist, Michael Faraday, belongs the credit for placing electrochemistry on a quantitative and orderly basis. In 1833, he postulated certain laws which today bear his name and which can be summarized as follows:

1. In any electrolysis, the quantities of materials liberated at the electrodes bear a direct relationship to the quantity of electricity passed through the system.

2. An equal number of equivalents of substances are set free by the same quantity of electricity.

It can be seen from a study of these laws that tin plating can be controlled when the electrode reactions are known. Electrotinning can be accomplished with acid or alkaline electrolytes.

In the **acid processes** the anode reaction consists of direct oxidation from metallic tin to the bivalent stannous ion with the liberation of two electrons as follows:

$$Sn^\circ = Sn^{++} + 2e$$

The reaction is usually 100 per cent efficient, thus making the quantity of tin driven into solution directly proportional to the electrical energy used. The reverse reaction takes place at the cathode,

$$Sn^{++} + 2e = Sn^\circ$$

thus causing metallic tin to be deposited on the cathode which, in electrolytic tinning lines is the moving steel strip. The cathode reaction is usually less than 100 per cent efficient. The use of organic addition agents in acid baths appears to be essential to the production of dense adherent tin deposits. While such agents may lower the operating cathode efficiency somewhat, the effect is slight.

There are several successful acid baths used for electrotinning, among them being the phenolsulphonic acid bath, the fluoride-chloride bath, the fluoborate bath and the sulphate bath. In each of these the electrode reactions are the same. Additions of acid are commonly made to raise the conductivity of the electrolyte and thus lower the plating-power consumption.

The electrode reactions which take place in the **alkaline sodium-stannate baths** are somewhat complex. The over-all effects are well known, however, thus permitting satisfactory control of the deposition. The net reaction of the anode is to oxidize tin from its metallic state to the quadrivalent stannic condition as follows:

$$Sn^\circ = Sn^{++++} + 4e$$

It will be noted that four electrons must be liberated to drive an atom of tin into solution in this process as compared to two electrons in the acid process. This means that twice as much electric current is required to dissolve one pound of tin. It must be emphasized that the anode reaction shown above is an over-all effect since the actual mechanism at the anode indicates the formation of intermediate tin compounds during the transfer process. This complexity of the tin-anode reaction in the alkaline stannate process limits its operating current density and thus necessitates the use of large plating areas to achieve high operating speeds.

At the cathode (steel strip) the tin is reduced again to the metallic condition with the absorption of four electrons per atom:

$$Sn^{++++} + 4e = Sn^\circ$$

When operating temperatures are kept above 95°C (200°F), the electrode efficiencies of the alkaline processes are quite close to 100 per cent.

COMMERCIAL ELECTROLYTIC TINNING EQUIPMENT

The equipment used for the production of electrolytic tin plate in no way resembles the more common electroplating equipment seen in most plating shops. The complex and bulky machinery required to handle the heavy steel coils and the fast moving steel strip itself is much more expensive and usually takes up more space than the plating unit proper.

Most commercial tinplating lines in the world use the FERROSTAN process. The process used in these lines is based on the use of a sulphonic-acid electrolyte in which tin is reduced from the stannous state (Sn^{+2}). A schematic drawing of a typical FERROSTAN line is shown in Figure 36—1, together with lines using competitive processes. An actual installation of a line similar to that shown at the top of Figure 36—1 is depicted in Figures 36—2 and 36—4 through 36—7. In the FERROSTAN process, a coil of properly prepared steel (black plate) is placed on an uncoiler, fed into the welder where it is attached to the tail end of the preceding coil, and passed through the looping tower and into the drag bridle. The looping tower or strip accumulator contains enough strip to permit continuous operation of the line while a new coil is being welded. After passing through the drag bridle, which provides the desired strip tension to permit tracking the strip through the line, the strip usually is electrolytically cleaned in an alkaline solution to remove grease, oil and dirt. It is then rinsed and electrolytically pickled to remove oxides, rinsed with water, and passed into the plating tanks. By adjusting the number of plating passes, speed and/or the current density, the desired tin-coating weight is obtained. The strip then goes through a drag-out-recovery tank where most of the excess electrolyte is removed and returned to the plating tanks through the drag-out-recovery system. The strip is then dried. In the reflow tower, the coating is melted, using either resistance or induction electrical heating as preferred, and the strip is water quenched. Following quenching, the tin plate is chemically or electrochemi-cally treated to improve storage stability and lacquering properties. To reduce abrasion damage, the strip is oiled. It is then recoiled or sheared into sheets as desired.

These FERROSTAN lines are compact in construction because of the use of vertical plating and treating tanks; hence, building costs are minimal for high-capacity lines. A line may be designed to plate steel strip of almost any gage desired, although the thinnest strip that is commonly plated is 0.155 mm (0.0061 inch) and the heaviest is 0.63 mm (0.025 inch). All steel tempers and both conventional (single-reduced) and 2 CR (double-reduced) strip may be plated. Any tin-coating weight required may be deposited; the normal tin-coating-weight range is 2.25 to 30 grams per square metre (0.10 to 1.35 pounds per base box). The amount of tin on each surface can be controlled so that each surface carries a different weight of coating. Suitable means are available for identifying these differentially coated surfaces.

As normally constructed, lines of this type will permit the use of many different chemical treatments, the choice of which depends upon the desired end use of the tin plate being processed. The tin coating may be melted or not as desired. Oiling is electrostatic with any of the conventionally used tin-plate oils. The strip may be recoiled or sheared in line. Lines can be built to handle any strip width and shear cut to any length required.

Regardless of the type of plating unit used, the steel strip for manufacturing electrolytic tin plate is similarly prepared. The strip itself is manufactured as described elsewhere; that is, it is hot rolled from slabs, continuously pickled, cold reduced, electrolytically cleaned, box or continuously annealed, and either temper rolled or cold reduced. In annealing, a protective reducing atmosphere is used to ensure a uniform, bright surface and the temper rolling operation is carefully controlled to maintain this condition. The coiled temper-rolled or cold-reduced strip is usually delivered to the side trimming units (Figure 36—3) where it is uncoiled, run through pairs of rotary knives and recoiled. The rotary knives are adjustable in such a manner that the strip can be very accurately side trimmed to the ordered width. It is general practice to adjust the slab-selection and hot-rolling practice to obtain about one inch protective over-width on the process strip which is removed at these slitters. In some instances, this side trimming of strip is done after tinning in a separate shearing operation. In many plants where the strip is trimmed before coating, mash seam-welding equipment is included in the side trimmers to permit welding several coil units together and thus provide a continuous section of steel strip for plating, up to 10 or 11 kilometres (6 or 7 miles) long. In other plants, the side-trimming operation is performed in the electrolytic tinning unit itself.

ENTRY-END EQUIPMENT

The entry end of an electrolytic line is usually so designed as to provide two **uncoilers** in line (Figure 36—4). This permits the operator to "pay off" from one uncoiler while charging a coil into the other. Electro-

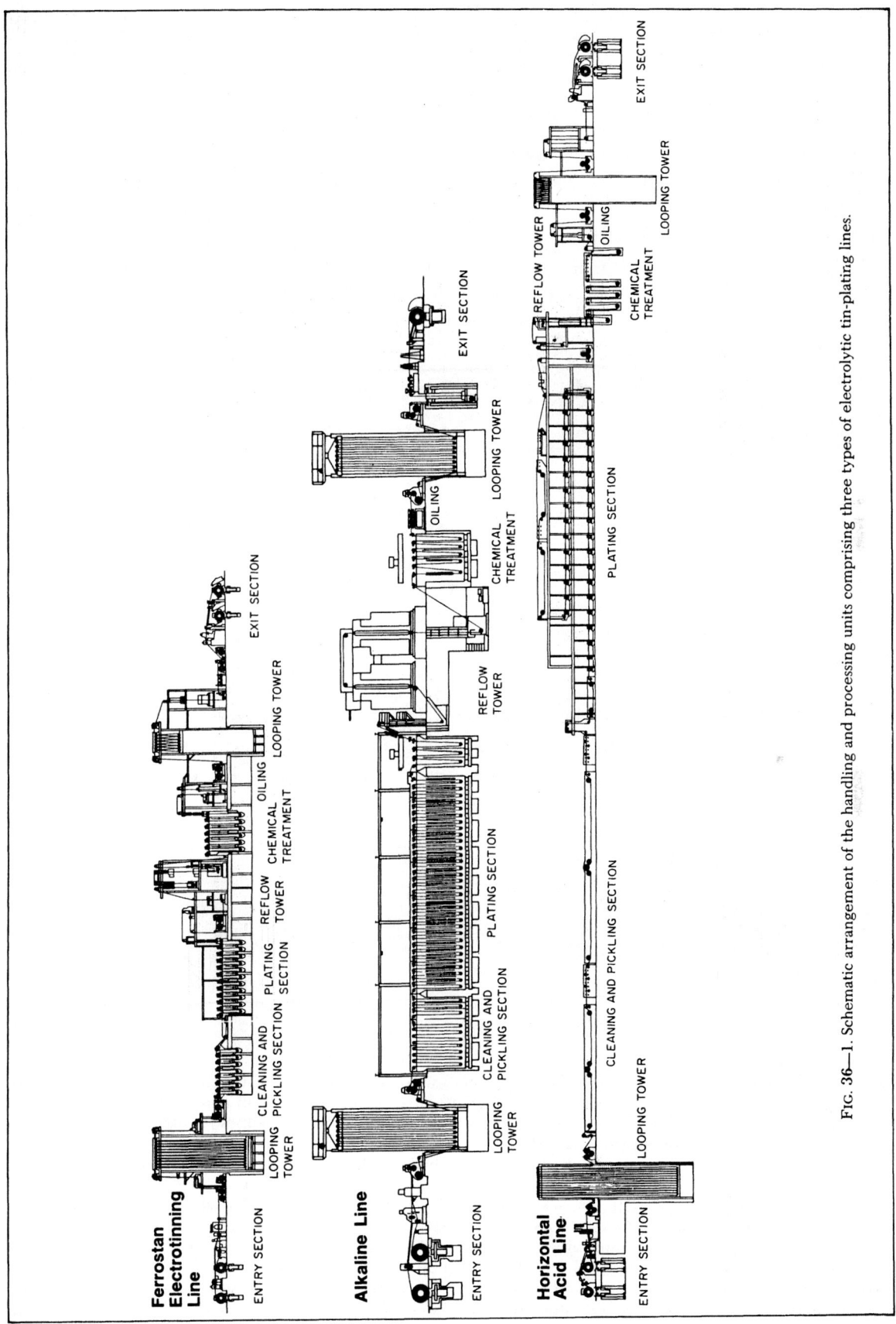

FIG. 36—1. Schematic arrangement of the handling and processing units comprising three types of electrolytic tin-plating lines.

FIG. 36—2. Close-up view of the plating unit of a sulphonic acid electrolytic tinning line, with melting tower immediately behind in right center background.

FIG. 36—3. Side-trimming unit that prepares cold-reduced, annealed and temper-rolled coils of flat steel for electrolytic tinning by removing excess material from side edges to produce strip of exact desired width. An air-operated shear following the side trimmer cuts off the crop ends of coils preparatory to seam welding of successive coils together to form larger single coils weighing up to 13 600 kilograms (30-000 pounds).

lytic tinning units do not require any special type of unreeling equipment and either the conventional cone or expanding types may be used. All of the auxiliary uncoiler equipment such as brakes, forward and reversing drives, hydraulic lifts, and strippers is usually to be found on all units.

In preparing a coil for processing, the lead edge of the strip is manually engaged in a set of small **pinch rolls** which can be opened and closed by air pressure and which are usually motor driven. The function of these rolls is to permit the operator to advance the lead edge of a new coil into the welding assembly. This

welding assembly consists primarily of an up-cut shear, a mash seam welder, and a set of large pinch rolls.

It is desirable to maintain a high strip speed in the plating baths, so facilities are provided to join fresh coils to the strip without reducing line speed. As the coil in process is being unrolled, the operator takes care that the maximum amount of strip is contained in the **looper** located just after the **entry bridle** (Figure 36—1). When the last several wraps of the coil unwind from the uncoiler in use, the operator stops the entry bridle, "trues-up" the tail end of the coil with the **shear,** moves this tail end into welding position with the head end of

FIG. 36—4. Entry end of an electrolytic tinning line, with one uncoated coil being paid off into the line and another in reserve position. The trailing end of the coil being fed will be welded to the leading end of the reserve coil (after squaring in the shear) to provide a continuous feed to the line.

FIG. 36—5. Tin anode being placed in the plating tank of an electrolytic tinning line, with other anodes lined up awaiting placement.

FIG. 36—6. General view of an electrolytic tinning line, looking toward the entry end and showing the melting tower in the center foreground.

FIG. 36—7. Discharge end of an electrolytic tinning line, showing sheared-plate pilers (left).

the new coil, welds the two together and starts the entry bridle. All this must be done before the strip previously stored in the looping unit has been completely used. The entry bridle is then run at some speed greater than line operating speed to refill the loop, at which point automatic electrical devices slow these rolls back to synchronism with the rest of the unit. Since these looping units are usually designed to accumulate 90 to 150 metres (300 to 500 feet) of strip, it is obvious that the welder operator must act rapidly in order to keep a line operating at speeds no less than 2.0 to 3.0 metres per second (400 to 600 feet per minute). These loopers may be of the "tower" type or of the "pit" type. A **tower-type looper,** as its name indicates, is constructed predominantly above ground and usually consists of fixed and movable sets of rolls over which the strip is passed. One set of rolls moves down or up, depending on whether strip is being expended from or accumulated in the equipment. For high-speed lines (up to 9.4 metres per second or 1850 ft per min.), these tower-type units may accumulate in excess of 300 metres (1000 feet) of strip. The **pit-type looper** consists of a deep pit in which hangs a long catenary of strip.

MAIN PROCESS SECTION

From the looper the strip enters the main process section of the line. In the acid lines, the process section usually consists of a dynamic tension device (often called a **drag bridle** or **tension bridle**), an alkaline **electrolytic cleaner**, a rinsing unit, a pickler, another **rinsing unit**, a plating unit, a third **rinsing unit**, a fusion unit, a quench tank, a **chemical-treating unit**, a fourth **rinsing unit**, a drying unit, an **oiling unit**, a **drive** or **pull-through bridle**, and finally a set of **recoilers** or a **shear** or both. The alkaline lines have the same sequence of units except that the alkaline cleaner is usually not part of such lines inasmuch as the alkaline plating bath itself does sufficient cleaning.

Tension Bridle—The function of the **tension bridle** is to produce sufficient drag on the strip to maintain a positive strip tension throughout the line. It consists of a series of rolls, some of which may be pinch rolls, through which the strip passes out of the looper. These rolls are usually geared together and to a generator. By controlling the field voltage on this generator and "shorting out" the armature through a resistance, a controllable drag can be applied to the strip through the geared rolls.

Cleaning and Pickling Units—In the acid-electrolyte units, the strip passes from the drag bridle to the alkaline **electrolytic cleaners**. In some lines these cleaners have horizontal electrodes as described in the section on the electrocleaning of cold-reduced strip. In other lines, vertical units are used. Current densities in these units normally vary from 538 to 3230 amperes per square metre (50 to 300 amperes per square foot). However, in a recently developed process, current densities several times greater than these are employed, resulting in both more effective cleaning or oxide removal in pickling and also permitting much faster line operation. The units developed for the latter process also employ horizontal passes and thus are not subject to the speed limitations imposed by the deflector rolls of vertical passes. Although the few such units installed in conventional processing units and currently in use operate at conventional speeds, the potential for operation at speeds two to three times normal exists with suitable ancillary equipment modification. The hot cleaning solutions are alkaline detergents. Strip polarity may be either anodic or cathodic or combinations of the two.

The strip passes from the alkaline cleaner into a **rinsing unit**. Its function is to remove all alkali from the strip in preparation for the pickling operation. This rinsing unit is usually comprised of water sprays playing on both sides of the strip and, in some cases, of rotary bristle brushes which rotate vigorously against the strip. There is a trend toward the use of high-pressure water sprays and the elimination of brushes in such units.

The conventional **strip-pickling units**, which are used on electrolytic tinning lines, may be of the hot immersion type or of the cold electrolytic type. The immersion type usually consists of a large rubber- or brick-lined tank through which the strip passes vertically or horizontally. These tanks are filled with hot sulphuric acid of a strength varying up to 12 per cent and the pickling time is regulated by the operating strip speed. The electrolytic picklers are usually small units, as the control of pickling is maintained by regulation of the electric current. These units generally are built similar to the alkaline cleaning tanks and the electrical circuits are also similar. After pickling, the strip is again rinsed in a unit similar to the one used after the alkaline cleaner and enters the plating tank.

Plating Tanks—As explained earlier, the main difference in the various electrolytic tinning units lies in the type of electrolyte used. The unit used in **phenolsulphonic-acid lines** is designed for operation at high current densities and consists of several vertical compartments in each of which the strip passes over metal contact rolls and down into the electrolyte between banks of tin anodes. (Another compartment in line with the plating tanks and identical in appearance plays no part in the plating operation but merely collects the solution dragged out of the plating system.) Thus, the current can be considered to pass from the tin anodes through the solution to the strip and up the strip to the metal deflector rolls which act as the negative contact of the system. By such a circuit, the tin is deposited from the solution onto the strip and is also equally driven into solution from the anodes. The anodes consist of tin bars which, for example, may be 75 mm by 50 mm (3 inches by 2 inches) in cross section and nearly 2 metres (approximately 6 feet) long, weighing about 45 kilograms (100 pounds) each (Figure 36—5). The life of the individual anode depends on the quantity of electric current passing through it. The electrolyte is constantly recirculated through the plating tanks after passing through a settling tank and several heat exchangers. The temperature of the bath is generally maintained at about 38°C to 49°C (100°F to 120°F). All of this equipment must of necessity be constructed of corrosion-resistant materials and care must be taken to provide sufficient insulation in the system to prevent electric-current leakage. Generating equipment capable of developing 100,000 amperes at 24 volts has been a typical installation. The total amperage required is of course dependent upon strip speed and tin-coating weight desired. A line designed for 11.7 metres per second (2300 feet per minute) has been built.

Alkaline-type plating tanks, while essentially of the same basic design, require much greater floor space than the acid type. Inasmuch as alkaline stannate baths are operated at temperatures in excess of 95°C (200°F), no recirculation of the electrolyte for cooling is necessary. These plating tanks are fabricated as a single large unit with contact rolls placed at the top of the tanks and rubber deflector rolls placed in the bottom. The tin anodes used in the alkaline units are usually very large slabs of tin hanging under and between the contact rolls. These slabs are large enough to allow several weeks of operation before replacement is necessary. Alkaline lines have been provided with up to 90,000 amperes at 10 volts, with current densities of up to 645 amperes per square metre (60 amperes per square foot) permitting strip speed up to 3 metres per second (600 feet per minute) on half-pound per base box coatings. As indicated earlier, strip speed and tin-coating weight requirements will dictate the amperage requirements.

The third type of unit, employing a **halogen-type** electrolyte, consists of a series of small cells, each with its own circulation system, contact roll and anode bank. These tanks are so designed that the strip is barely immersed in the electrolyte and is plated on the bottom side only. After passing through a number of these units, the strip is deflected upward and backward so that the original top of the strip now becomes the bottom. It then passes through another series of similar plating cells until the desired amount of tin is deposited on this side of the strip. The halogen-type electrolyte used exclusively in these units is constantly recirculated through the cells, cooled and filtered during its circulation. The tin anodes used in the individual cells resemble regular pigs of tin and rest on side supports just under the strip pass line. Halogen lines have been designed for strip speeds greater than 1 metre per second (about 2000 feet per minute). Generator capacity exceeds 100,000 amperes at a voltage of from 8 to 10 volts. The current density used is approximately 3230 amperes per square metre (approximately 300 amperes per square foot).

Drag-Out Control—The plated strip, regardless of plating process, is now freed from the dragout and rinsed in pure water or condensate. The electrolyte which is dragged into this wash water is all or in part returned to the plating tank. To accomplish this, dragout recovery systems varying from complex recirculation and evaporation systems to simple counter-rinsing with partial recovery are used. Too much emphasis cannot be placed on the necessity for efficiently recovering the dragged-out electrolyte since solution losses from this source can be enormous at high speeds, reaching as much as 113.5 litres (30 gallons) per hour.

Fusion Units—The tin coating, as it emerges from the plating bath, is gray-white and semi-lustrous. This as-deposited surface is satisfactory for some applications, but for food can or other applications where bare tin plate is used a bright lustrous surface is desired. This is obtained by melting and quenching the electrodeposited tin which gives it the brilliant luster typical of hot-dipped plate (Figure 36—6).

Although the melting point of tin is 231.9°C (449.4°F), it is necessary to heat the strip to temperatures somewhat higher than this to assure complete melting and uniform appearance. The upper temperature limit, however, is restricted by the possibility of strip distortion (poor shape), surface oxidation, excessive iron-tin alloy formation, and coalescence.

There are three types of units in which this tin fusion is accomplished.

One of these melting units makes use of **electrical**

resistance heating. The strip is run in a verticle loop between contact rolls, the second of which may be immersed in water. These two contact rolls form the terminals of an alternating-current circuit in which the strip is the closing resistance. Thus, by regulation of the current flow through the strip (or applied voltage) it is possible to bring the plated strip up to 232°C to 235°C (450°F to 455°F) just prior to passage into the water.

A second type of melting unit utilizes **electrical high-frequency induction heating** for melting. In this unit, the strip passes down and through water-cooled copper coils on which is impressed a high-frequency voltage. The induced eddy currents in the tinplate strip cause it to heat up with resulting fusion of the tin coating. Control is again exercised by voltage variations on the induction-coil terminals.

The third type of melting unit in commercial use is gas-heated. It employs the principle of **radiant heating** and is equipped with special ceramic burners that radiate controlled amounts of heat to the strip. Control on such units is exercised by lateral movement of the burner banks, closer to or farther away from the strip, depending on whether more or less heat is required. Further limited control is available by regulation of the gas supply to the unit. As in the other units, the strip is quenched directly after fusion of the tin is accomplished.

The fused and quenched coating is now given a **filming treatment** which may be either chemical or electrochemical, as described later in this section. After such treatment, the tinned strip is **rinsed** with clean water and is dried either by blasts of hot air or high-pressure steam.

Oil-Film Application—Unlike hot-dipped tin plate, the electrolytic tin plate is not oily as it emerges from the coating operation; hence, it is necessary to deposit a controlled film of lubricant on the product in order to improve its handling properties in succeeding operations. The lubricant used is usually either di- (2-ethylhexyl) sebacate, (commonly referred to as dioctyl sebacate or DOS) or acetyltributyl citrate (ATBC). For tin-free steel production, butyl stearate (BSO) is used. The oiling method now most widely used is the **electrostatic process** which makes use of a high potential between the strip and a fixed electrode which creates a powerful electrostatic field around the moving strip. Into this electrostatic field is allowed to rise a mist of ATBC or DOS which is deflected onto the strip by the proper adjustment of strip polarity. Despite the relative complexity of the equipment, these oiling units operate very satisfactorily and economically.

Pull-Through or Drive Bridle—The strip next enters the unit which supplies tractive power to the strip to pull it entirely through the electrolytic line. This piece of equipment is called the "pull-through" or drive bridle. It actually does the pulling of the strip through the process section and is the basic unit with which the plating current and both entry and delivery ends are synchronized. The equipment itself is usually quite similar to the tension bridle described elsewhere in this section. However, instead of a drag generator being geared to the rolls, they are coupled with a powerful motor which is sufficiently large to pull the strip through the whole process section at high speeds. A tachometer generator attached to this motor provides the impulse to regulate the plating and melting currents in required relation to strip speed.

Helper Drives—In addition to the main drive bridle, other helper drive bridles may be located at various points in the line to minimize strip tension. Also, most of the rolls are driven with helper motors to aid strip propulsion.

DELIVERY-END EQUIPMENT

Alkaline tinning lines are provided with large **loopers** and a single recoiler at the delivery end into which the strip passes from the drive bridle. Such an assembly performs in a manner similar to that of the entry end previously described in that the looper acts as a strip accumulator. Acid lines are provided with two recoilers, a special shear and guides and belt wrappers which make it possible to cut the strip and wind it on the second recoiler at high speeds. A number of electrolytic plating lines are provided with both quick-change coilers and flying shears. However, the coil of coated product may be sent to the conventional **shearing units** where it is sheared to size, oiled (if not so treated in the electrolytic-plating line), assorted, counted, and piled.

In those units where **flying shears** only are included in the equipment assembly, no loopers are necessary. Instead, a very small catenary is maintained immediately in front of the shear in order that perfect guiding into the shear knives can be accomplished. Failure to guide properly into a shear results in miscutting and "out-of-square" sheets. The flying shears themselves are conventional units for light-gage-strip shearing, as described elsewhere in this chapter. There are some modifications, however, which should be mentioned. The most important ones concern the adaptation of these units for careful inspection, classification and counting of the product (Figure 36—7).

The flying shears are equipped with four piling stations and a system of conveyor belts which allows the deposition of sheets in any of these pilers at the will of the operator.

In lines containing four **pilers,** all sheets containing perforations are accumulated in the first piler. Sheets containing bad surface defects and product which is outside the accepted gage tolerances are accumulated in the second piler. The third piler is used to receive sheets while the fourth piler is changed to pile a new bundle. The accumulated sheets in the third piler are assorted on an auxiliary unit and the prime sheets are applied on the order. The fourth or end piler is the so-called **prime piler.** As the **sheet counter** is located between the third and this prime piler, the product

delivered into it is usually counted into ten-, twelve-, or fifteen-package bundles (1120, 1344 or 1680 sheets) ready for packaging and shipping.

The methods of inspection and classification of electrolytic tin plate on these flying shears are rather ingenious. Located somewhere after the melting unit is a noncontacting **thickness gage.** When the strip is too thick or too thin to meet specifications, the gage actuates an electronic memory device which rejects those particularly heavy or light sheets at the second piler. Likewise, a device sometimes called a **pin-hole detector** utilizes a photoelectric cell to continuously scan the coated strip and cause sheets with perforations to be deflected into the first piler.

Trained inspectors are located at the pull-through bridle and the shear, respectively; they operate contact buttons which allow them to deflect at will any defective product they detect into the second or third piler (depending on product classification) and pass prime product into the fourth piler where the attendant sees that it is properly counted and that the shear operator is kept advised of the product flatness.

Although most lines are equipped with x-ray or beta-ray gages for continuous determination of tin-coating weight, laboratory determinations are made regularly to determine the tin-coating and oil-film weights on the finished product. The lacquerability of the tin plate is judged by specialists. Various laboratory controls of cleaning, pickling and plating solution characteristics are necessary.

Commercial Electrolytic Chromium-Plating Equipment—The equipment used to produce tin-free steel (TFS or electrolytic chromium-coated steel) is very similar to that existing for electrolytic tin plate. In fact, in the early development of TFS, commerical production was achieved by utilizing the chemical-treatment tanks of the electrolytic-tin-plate line for the plating electrolyte and by-passing the tin-plating tanks. In some instances, "combination" lines exist where production of either electrolytic plate or TFS can be accomplished.

Two processes were developed to produce TFS. One process is a "one-step" process wherein metallic chromium and chromium oxide are deposited from the same solution. A competitive two-step process requires two sets of tanks, one for a solution from which metallic chromium is deposited and a second set from which the surface oxide layer is deposited. In both cases, a lubricating oil, butyl stearate, is applied by electrostatic spray as is the case in tin-plate production. The conventional tin-plate oils DOS or ATBC do not have sufficient lubricity for the more abrasive chromium-oxide surface, and the butyl stearate oil (BSO) is required to facilitate handling and shearing without excessive surface scratching and abrasion.

SECTION 6

CANMAKING

The technology of canmaking is discussed in detail in several of the publications listed in the bibliography at the end of this chapter. Generally, there is no one can

correct for every application. Can designers seek to achieve an optimum combination of material properties, including strength and thickness; coating type and

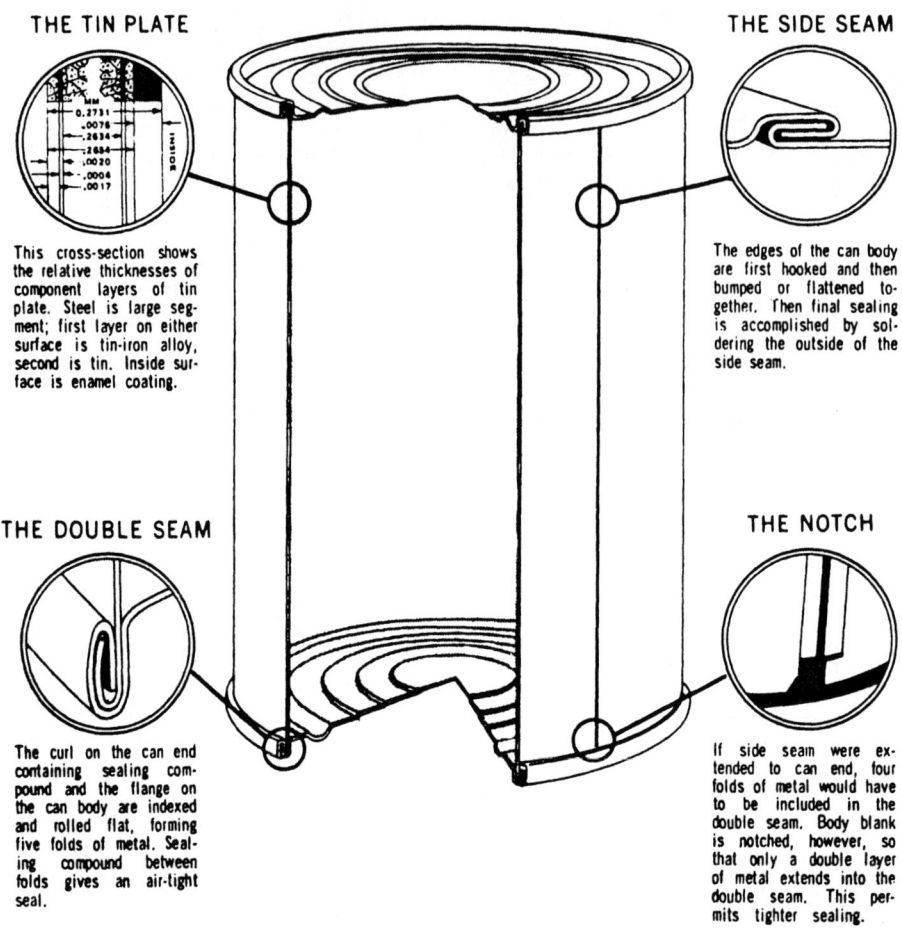

THE TIN PLATE

This cross-section shows the relative thicknesses of component layers of tin plate. Steel is large segment; first layer on either surface is tin-iron alloy, second is tin. Inside surface is enamel coating.

THE SIDE SEAM

The edges of the can body are first hooked and then bumped or flattened together. Then final sealing is accomplished by soldering the outside of the side seam.

THE DOUBLE SEAM

The curl on the can end containing sealing compound and the flange on the can body are indexed and rolled flat, forming five folds of metal. Sealing compound between folds gives an air-tight seal.

THE NOTCH

If side seam were extended to can end, four folds of metal would have to be included in the double seam. Body blank is notched, however, so that only a double layer of metal extends into the double seam. This permits tighter sealing.

3-PIECE WELDED OR CEMENTED CANS

Many 3-piece cans are produced by welding or use of an adhesive instead of a soldered side seam. In these instances the edges of body cylinder are simply lapped and then welded or bonded with an adhesive. The notch is unnecessary in welded and cemented cans.

FIG. 36—8. Architecture of the enameled sanitary tin can.

coating weight; the appropriate lacquer required for the material being packed; and the manufacturing technology to produce the lowest cost can for the application.

Cans are described as being three-piece, that is, having two ends and a body, or two-piece, having a seamless body with an integral end and a conventional end for closing.

THREE-PIECE CANS

Three-piece cans consist of a body and two ends shown in Figures 36—8 and 36—9. The body side seam can be accomplished by soldering, cementing, or welding. The body blank is lacquered and decorated prior to canmaking with only a protective stripe of lacquer applied over the side seam after joining.

Soldering Method—Soldering requires that the body blank must be notched, hooked, formed, lock seamed, soldered and flanged. A minimum tin coating weight of 5.6 grams of tin per square metre (0.25 pounds of tin per base box) is required to assure solderability.

Traditionally, soldering has been the primary method of side seaming tin plate containers. Concern over possible lead contamination from solder (2% tin/98% lead) has caused can manufacturers to voluntarily seek alternative lead-free methods of can manufacturing. Cementing, welding and seamless construction are viable alternatives.

Cementing Method—Cementing requires a rectangular body blank that is not notched or hooked. It is

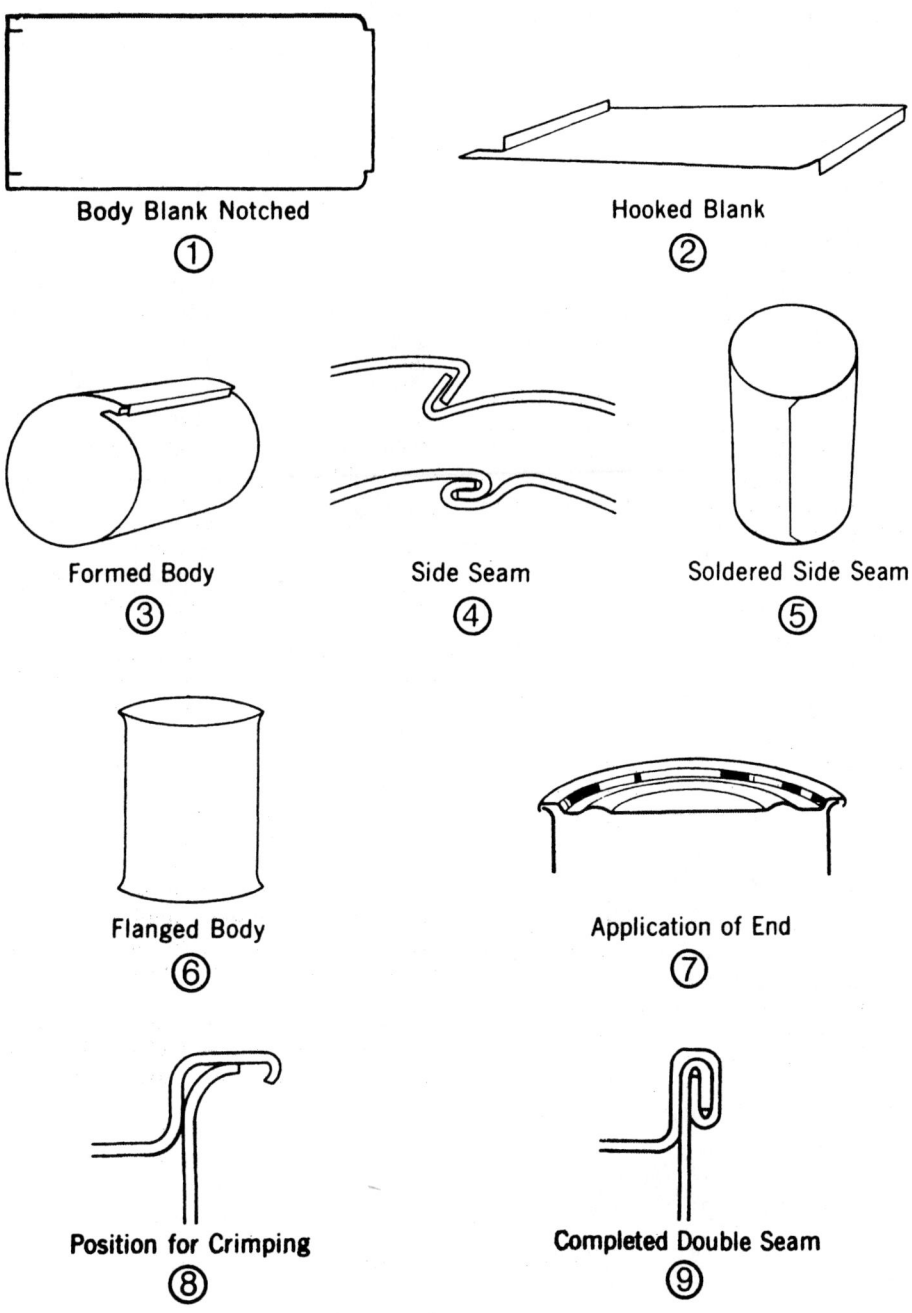

Body Blank Notched
①

Hooked Blank
②

Formed Body
③

Side Seam
④

Soldered Side Seam
⑤

Flanged Body
⑥

Application of End
⑦

Position for Crimping
⑧

Completed Double Seam
⑨

FIG. 36—9. Steps in·can fabrication (3-piece soldered can).

simply formed with an overlap (approximately 0.2 inch) to permit adhesive bonding, and flanged. Both electrolytic chromium-coated black plate and tin coating weights less than 5.6 grams of tin per square metre (0.25 pounds of tin per base box) can be cemented to reduce metal cost below that required for the soldered can. The cementing method has had limited use, which has been primarily for beverage containers.

Welding Method—Welding requires a rectangular body blank that is accurately sized and squared. It is then formed with a narrow overlap (approximately 1/64 inch), welded and flanged. Base metal parameters

affecting weldability and welding performance are carbon level, temper, consistency of physical properties, shape and gage. Welder variables include consistency of weld overlap and alignment, welding speed and current. Tin coating weights less than 5.6 grams of tin per square metre (0.25 pounds of tin per base box) can provide proper welds. Electrolytic chromium-coated black plate can be reliably welded only if the chrome surface is mechanically removed by grinding or scraping from the weld zone. Welding is quickly becoming the dominant method for producing three-piece sanitary cans.

SEAMLESS CANS

Drawn and Ironed Method—Drawn and ironed (D&I) cans are now used almost exclusively in the beer and carbonated beverage fields. Much of their popularity can be attributed to the 20 percent metal savings over the three-piece can.

In addition to this cost savings, D&I steel cans have advantages in:
1. Strength and abuse resistance.
2. Mobility and magnetic-handling capabilities during manufacture and filling.
3. Accurate volumetric capacity.
4. Suitability for magnetic collection and separation in solid waste.

Although material specifications for D&I containers vary, the most common consist of tin plate in thicknesses of 0.24 to 0.30 millimetres (85 pounds to 103 pounds base weight) with tin coating weights of 2.2 to 5.6 grams of tin per square metre (0.10 to 0.25 pounds of tin per base box). The newest technology permits the use of blackplate with special lubricants employed during can forming.

Manufacturing starts when a coil of tin plate is unwound, lubricated and fed into a multiple cupping press. Cuppers usually form seven cups per stroke at speeds in excess of 100 strokes per minute. The cups are gravity fed into the ironers where they are relubricated, redrawn and ironed into can bodies. In the ironing operation the cup is elongated and the sidewall reduced in thickness by a horizontal punch forcing it through two or three coaxial dies where the clearance between the punch and dies is less than the thickness of the incoming metal. The final thickness of the sidewall is 0.09 to 0.10 millimetres (0.0035 to 0.0038 inch) and the shell is at least 127 millimetres (five inches) high. The bright reflective shell is then trimmed to a uniform height and thoroughly washed. The bright container provides an excellent background for lithography since the smooth flow of decoration is not interrupted by a side seam.

Outside decoration usually is applied immediately after washing. Usually a white base coat is applied to the can and cured in a pin-type oven. The label is then printed over the base coat, and this is also cured in a pin oven.

Next is the important operation of inside coating. As the can is rotated at high speed, a lacquer is sprayed over the inside surface. This lacquer film will protect the can from the product it is to hold. At the same time the outside bottom is sprayed for corrosion protection. These coatings are then cured in the internal bake oven.

Final metal-forming operations are necking, which reduces the diameter of the top of the shell, followed by spin flanging so that an end may be attached after filling. Completed cans are tested for leaks.

Thin walled D&I beverage cans depend on the internal pressure associated with product carbonation to attain the desired can rigidity for stacking and warehousing. However, D&I steel cans are also used to pack food products. These are heavier walled cans with sidewall beads to withstand the pressures and vacuums of food processing.

Draw and Redraw Method—The draw and redraw canmaking method includes cutting a blank from previously lacquered material, drawing the blank through a die to form a cup, and redrawing the cup one or two times without thinning the sidewall to form a can of desired height and diameter. Subsequently the bottom profile is formed and excess material is trimmed from the flange. Blanking and cupping are generally done in one press, whereas redrawing, profiling and trimming are done in another press.

The draw and redraw method achieves its metal cost advantage from substitution of chromium-coated black plate for the more expensive tin plate (although tin plate is used for some applications) and from the greatly reduced rate of manufacturing spoilage. The use of double-reduced material can result in further savings.

A major use for the draw and redraw steel can is the packaging of fruits, vegetables, baby foods, and tuna; and soups and prepared foods dispensed from vending machines.

SECTION 7

METALLURGICAL ASPECTS

General—A fuller appreciation of the metallurgical aspects of tin plate may be had by considering it as a nine-layer sandwich (Figure 36—10), the layers of which consist of:

1. Oil layer
2. Oxide film
3. Free tin
4. Tin-iron alloy
5. Steel base
6. Tin-iron alloy
7. Free tin
8. Oxide film
9. Oil layer

Adaptability for a specific purpose may depend on the properties of a given layer. To illustrate for a pressed part, such as a toy, the drawing characteristics of the steel base may be the controlling factor. In the case of electrolytic tin plate, the chemical or electrochemical filming treatment gives an oxide layer found beneficial for adhesion of lacquers and enamels as well as helping storage stability. Similarly, tin plate must have an oil film adequate both to promote good "feeding" in automatic equipment and to prevent scratching during fabrication and yet not so heavy as to cause difficulty during roller coating with lacquers or printing inks. Thus, each layer requires adequate control for some of the applications. Generally, since the manufacture of tin plate is on a mass production basis, it is most economical to standardize on the optimum quality of each layer for total tin-mill production.

The Steel Base—The metallurgical controls of the

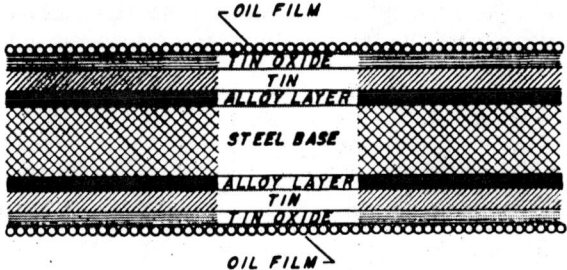

FIG. 36—10. Schematic enlarged cross-section of a sheet of tin plate, showing approximate relative thicknesses of the various "layers." The approximate thickness of the individual layers is as follows:

Layer	Thickness	
	micrometres	inches
Oil film	0.00254	10^{-7} (0.0000001)
Oxide film	0.00254	10^{-7} (0.0000001)
Tin	0.254	10^{-5} (0.00001)
Alloy layer	0.0254	10^{-6} (0.000001)
Steel base	254	10^{-2} (0.01)

steel base are similar to those employed for the production of sheet and strip and, hence, will not be elaborated on further. It is sufficient to point out that mechanical properties of the finished tin plate depend on the composition of the steel, the heat treatments and the rolling operations. The steel used is a low-carbon mild steel and the usual hardening agents are looked upon with disfavor because of their detrimental effect on corrosion resistance, as well as cost.

Strength, or **temper**, is controlled by regulating the amount of temper rolling or cold rolling prior to plating. The tin-plate industry has adopted the Superficial Rockwell Hardness test as a control measure of the temper of the strip. This test is described elsewhere.

As the major portion of tin plate is used for containers, the requirements of this application dictate most of the metallurgical considerations in processing of the steel base. As will be pointed out more fully in a later paragraph, the corrosion resistance of tin plate to food products is an important consideration—and in this respect the production of steel sheet for tin plate differs from the production of sheets for other applications. Many factors affect the corrosion resistance of tin plate, but quality of the steel base is probably the most important. Besides the composition, the various processes, such as hot-strip rolling, cold reduction and annealing, are subjected to even closer control than would be necessary from the standpoint of mechanical properties alone.

The various cleaning and pickling operations, previously described, are also closely controlled so as to prepare a suitable surface for subsequent coating with tin.

The Tin-Iron Alloy Layer—When a clean steel surface comes in contact with molten tin, a reaction takes place with the formation of a tin-iron alloy layer intimately bound to the steel surface. This alloy layer is quite thin (about 0.0635 to 0.1270 micrometres or 2.5 to 5 millionths of an inch), in electrolytic tin plate.

The metallurgy of tin-iron alloy has received considerable attention by various investigators. The phase diagram for the iron-tin system in Figure 36—11 reveals that there are probably at least three compounds of tin

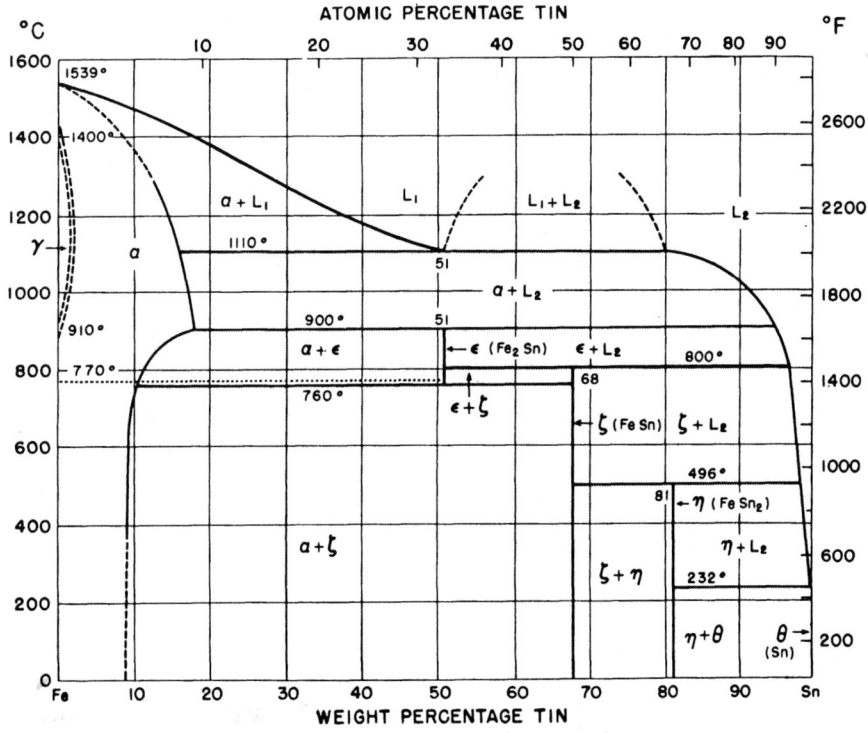

FIG. 36—11. Equilibrium diagram for the iron-tin system.

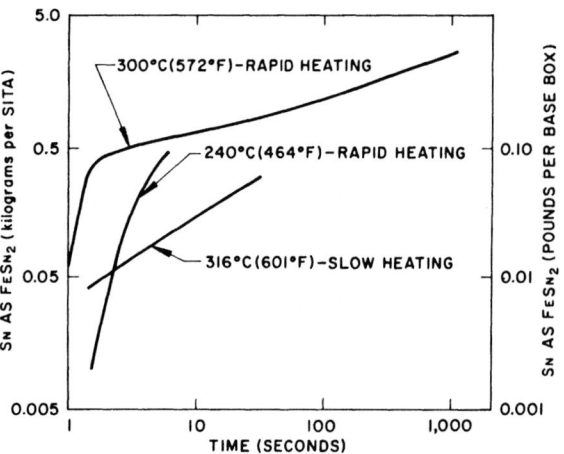

FIG. 36—12. Tin-iron alloy growth with slow and rapid heating rates in constant-temperature baths.

and iron, as follows:

1. Fe_2Sn, referred to as the epsilon (ϵ) phase, which is stable between 760°C (1400°F) and 900°C (1652°F) but can, in the presence of sufficient tin, react at 800°C (1472°F) to form FeSn or zeta (ζ) phase.

2. FeSn, or zeta (ζ) phase, which is stable at all temperatures below 800°C (1472°F) but which reacts with excess tin below 496°C (925°F) to form the compound $FeSn_2$, known as the eta (η) phase.

3. $FeSn_2$, or the eta (η) phase, which is stable below 496°C (925°F) and does not further react with tin.

Inasmuch as tinning operations are always carried out at temperatures considerably below 496°C (925°F), it follows that the alloy layer should contain both eta ($FeSn_2$) and zeta (FeSn) phases. The results of X-ray diffraction studies and chemical analyses indicate that the compound $FeSn_2$ predominates.

The growth of the alloy layer on tin plate has been studied extensively for the range of conditions encountered in commercial tinning operations. It has been demonstrated that the amount of tin-iron alloy formed is a function of time, temperature and heating rate. Typical growth curves are shown in Figure 36—12. During commercial flow brightening of electrolytic tin plate, the alloy forms at a rate similar to the initial rate at 300°C (572°F) with rapid heating.

The thickness of alloy can be determined by electrostripping, by chemical stripping, by magnetic testing, or by metallographic cross-sectioning. The most commonly used is the electrochemical stripping method. A 25.8 square-centimetre (4-square-inch) sample is attached to a suitable holder and immersed in a 1 N solution of hydrochloric acid containing a carbon electrode and a silver—silver-chloride reference electrode. The tin coating is stripped anodically at a constant pre-determined stripping current. The potential difference between the specimen and the reference electrode is continuously measured by a potentiometer recorder during the stripping operation. The slope of this potential-time curve will change as the stripping

proceeds from the surface, to the alloy layer, to the base metal. With proper cathode-anode geometry, the changes in slope occurring when the potential changes at the several interfaces are clearly discernible. The distances between the so-called end points, or points where the slope changes (measured parallel to the time axis), are measures of the thickness of the free tin and alloy layers.

The alloy layer on tin plate has a very definite pattern according to the manufacturing practice. Hot-dipped tin plate produced with the use of zinc-chloride flux has a distinctive mottle shown in Figure 36—13. This pattern is the result of variation in thickness and crystal size of the alloy, the dark areas being thinner and finer grained than the light areas (Figures 36—14 and 36—15). Electrolytic tin plate (melted) has a smooth fine-grained alloy layer similar to the dark areas in coke tin plate alloy layers (Figures 36—16 and 36—17). The arrangement of the individual crystallites of the alloy layer on electrolytic tin plate depends on the orientation of the underlying grains of steel. For this reason, the alloy layer differs in its growth pattern from one local area to another (Figure 36—17). Matte finish (unmelted) electrolytic tin plate does not have an alloy layer detectable by ordinary metallographic or chemical procedures. Heating matte finish plate at temperatures below the melting point of tin will result in formation of tin-iron alloy, the amount formed being dependent on the time and temperature.

The thickness and continuity of the tin-iron alloy layer are important with respect to tin-plate quality. An increase of thickness of the alloy layer in 5.6 grams per square metre (0.25 pound per base box) electrolytic tin plate results in a decrease of thickness of the tin layer and may result in soldering difficulties. The continuity of the alloy layer is important from the corrosion standpoint in that the alloy layer acts as a barrier between the steel and the tin coating as discussed in Section 8.

Physically, the compound $FeSn_2$ is very hard and brittle and of itself cannot stand much bending. In commercial tin plate the alloy layer is very thin and is covered with a much thicker layer of ductile tin so that even sharp bending does not expose much of the steel base because the tin will bridge gaps in the alloy layer. With thick layers of $FeSn_2$, bending will cause breaks through the tin.

The Tin Layer—As seen in the photomicrographs, the tin layer is considerably thicker than the alloy layer. It is customary to estimate the thickness of the tin layer in terms of weight of tin per unit of area, rather than in terms of thickness measurements. Thus, electrolytic tin plate will have on the average 22.4 grams of tin per square metre (1.00 pound per base box) spread over both surfaces of 20.23 square metres (31 360 square inches) of steel. Assuming that 1.8 grams of tin per square metre (0.08 pound of tin per base box) is present as iron-tin alloy layer, 20.6 grams of tin per square metre (0.92 pound of tin per base box) is present as free tin. The weight of coating is usually determined by testing methods incorporating electrolytic methods for dissolving the tin followed by standard volumetric titrations. The use of X-ray or beta-ray gages for continuous determination of tin-coating weight has also

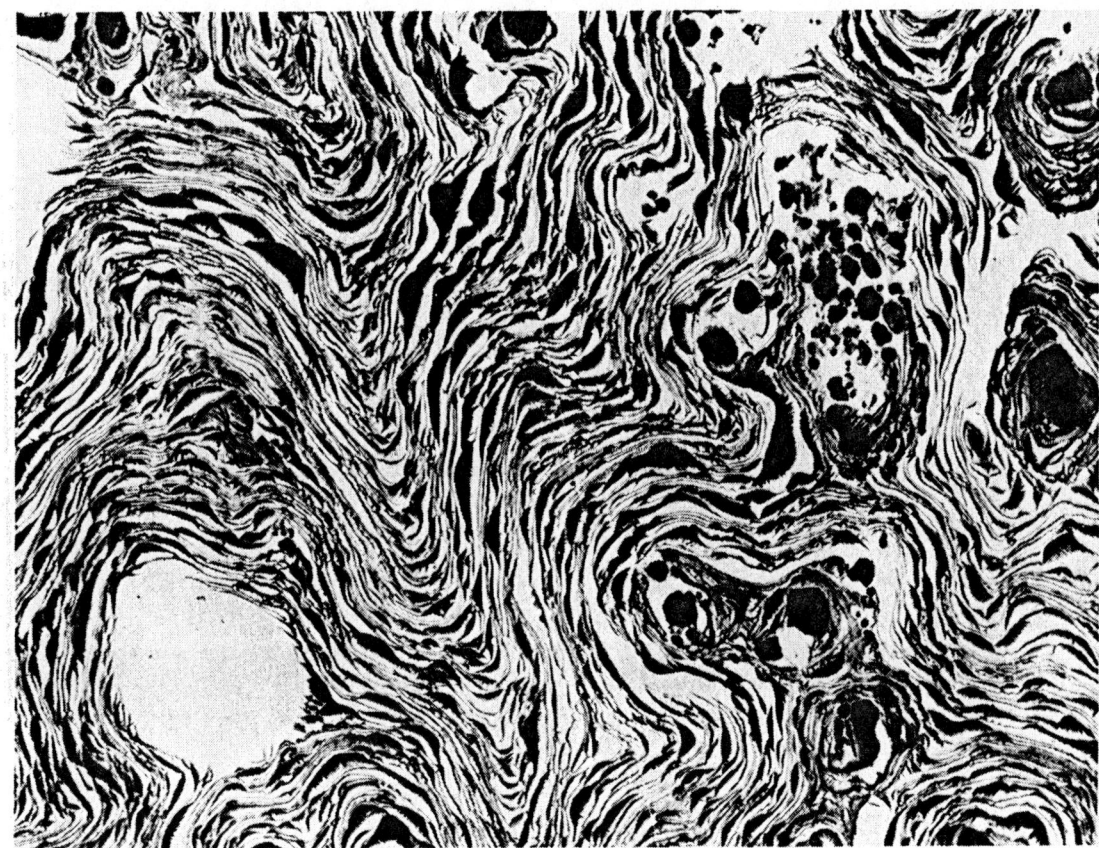

Fig. 36—13. Flux pattern on the surface of the steel base (interface of iron-tin alloy and steel) of commercial coke tin plate having a coating weight of 33 grams per square metre (1.49 pounds per base box), after removal of both the tin and the iron-tin alloy. Magnification: 3X.

gained favor and most electrolytic-tinning lines are equipped with such gages for control purposes. Generally speaking, the tin plate from each producing unit is sampled periodically for tin-coating weight. Methods for determining tin-coating weights are presented in Specification A 630 of the American Society for Testing and Materials (ASTM).

Electrolytic tin plates are produced to much closer tolerances with respect to tin-coating weight than are possible for the hot-dipped tin plates, and the spread in tin-coating weight is usually within 10 per cent of the average. Electrolytic coatings are capable of fairly precise control, variations being due to fluctuations in current flow or variations in distance of the strip (cathode) from the individual tin anodes.

Tin plate usually has a bright, lustrous, mirror-like surface. The luster varies somewhat with the surface finish of the steel base. Rough base finish will result in poorer luster than smooth base finish. The luster is also affected by the thickness of the tin coating. Heavy tin coatings, as on charcoal tin plates, have bright luster irrespective of base-metal roughness. The smoothness of the tin deposit on electrolytic plate also affects the luster.

Mild etching of the tin coating reveals the tin-crystal pattern which may take a variety of shapes. This pattern is evident on the interior of all non-lacquered cans which have contained fruits or vegetables and can also be seen on unetched surfaces when viewed under polarized light. The size of the tin crystal can only be partially controlled by the rate of quenching. Slow cooling such as is present in hot-dip tinning produces a massive crystal whereas rapid cooling, as by water quenching in the electrolytic plating lines, produces smaller crystals. This is illustrated in Figure 36—18. By control of the temperature of the quench water, larger crystals can be obtained on electrolytic tin plate. Cleanliness of the steel strip surface prior to electrotinning affects the tin crystal size with larger tin crystals resulting from cleaner strip. Electrolytic tin plate with large tin crystals tends to show better internal corrosion resistance when packed with the citrus fruits.

The tin coating, even when very heavy, is not continuous and microscopic areas of steel and alloy are exposed through the tin. The areas where steel is exposed are known as pores. The porosity of tin plate has been the subject of considerable research and a number of methods for indicating the degree of porosity have been developed; details of the procedures are summarized in International Tin Research Institute Bulletin, No. 662. The total area of iron exposed in pores is reported to be only about 0.7 square millimetre per square metre of surface. The satisfactory commerical service of tin plate for fruit containers shows that de-

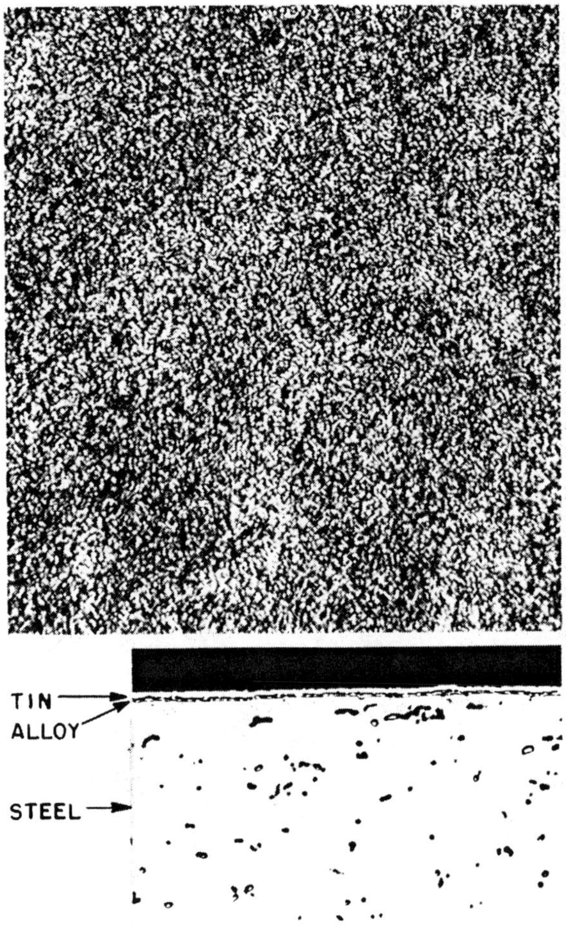

FIG. 36—14. (Above) Representative surface of iron-tin alloy layer on hot-dipped tin plate in dark areas, representing the interface between the iron-tin alloy and the tin of the coating, after removal of the tin portion of the coating. (Below) Representative cross-section through a dark area similar to that shown above, etched to show the tin coating and the iron-tin alloy layer. Magnification of both photomicrographs: 1000X.

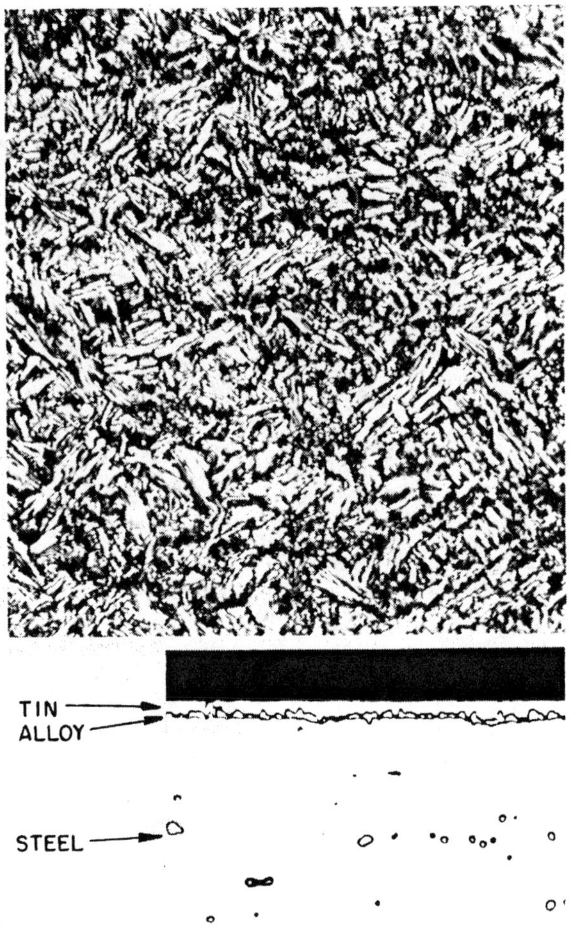

FIG. 36—15. (Above) Representative surface of iron-tin alloy layer on hot-dipped tin plate in light areas, representing the interface between the iron-tin alloy and the tin of the coating, after removal of the tin portion of the coating. (Below) Representative cross-section through a light area similar to that shown above, etched to show the tin coating and the iron-tin alloy layer. Magnification of both photomicrographs: 1000X.

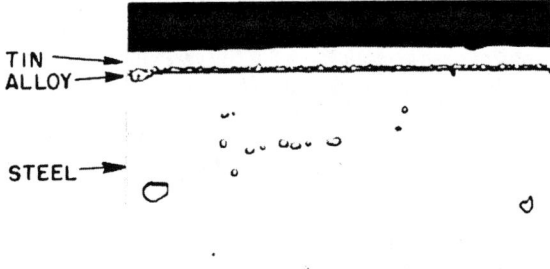

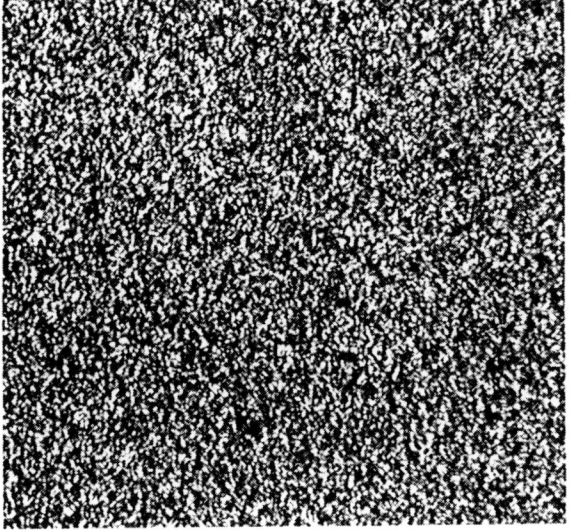

FIG. 36—16. (Above) Cross-section representative of commercial melted electrolytic tin plate (coating weight 12.6 grams per square metre or 0.56 lb. per base box), etched to show iron-tin alloy layer. (Right) Unetched surface of iron-tin alloy layer (interface between iron-tin alloy layer and tin coating) after removal of the tin portion of the coating shown above. The surface of the iron-tin alloy has a pebbly appearance, indicating that it is not of uniform thickness. Magnification: 1000X for both photomicrographs.

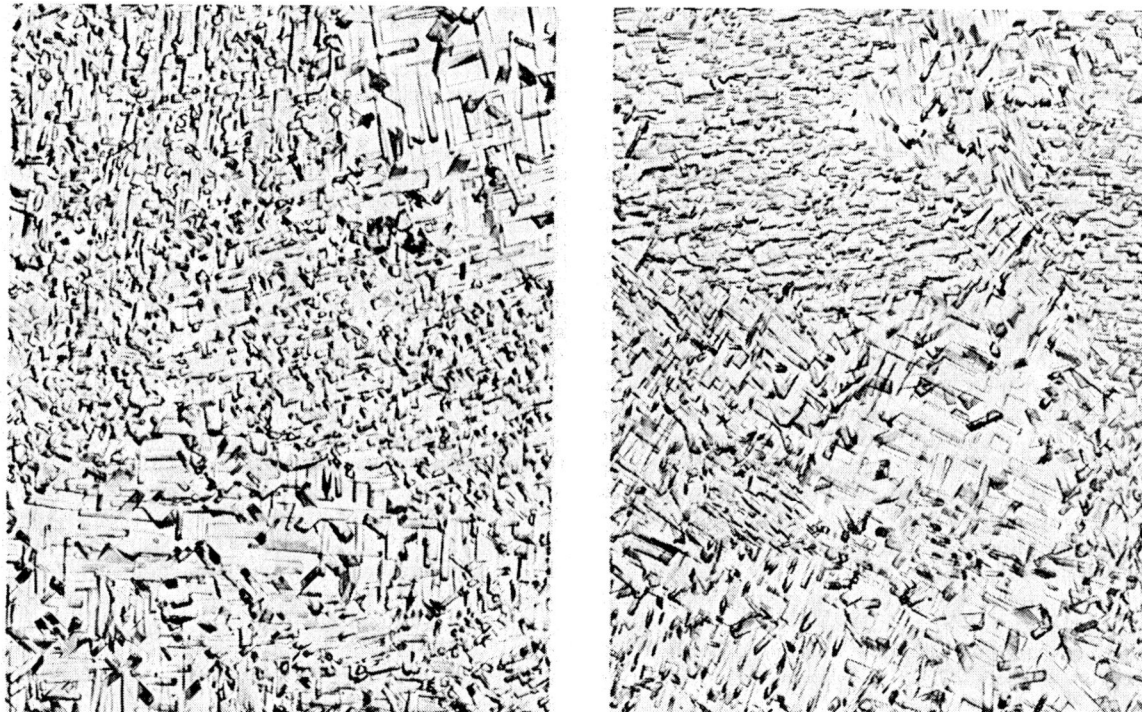

FIG. 36—17. Electron micrographs of carbon replicas of alloy layer on samples of electrolytic tin plate. Tin removed electrolytically with NaOH. Magnification: 14 000X, reduced ½ in reproduction.

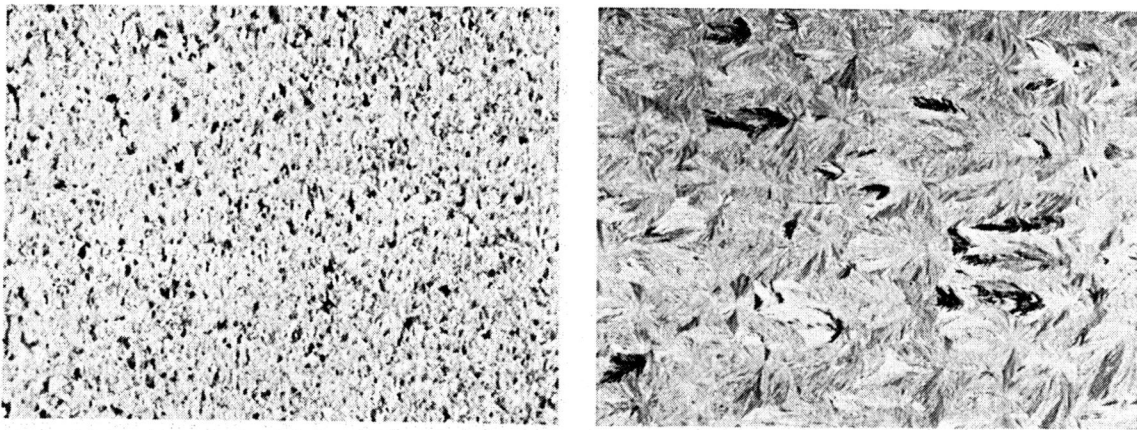

FIG. 36—18. Effect of cooling rate on the comparative size of tin crystals produced on tin plate. Small grain size at left resulted from rapid cooling (water quenching); larger grain size at right resulted from cooling at a slower rate (quenching in oil).

spite these minute pores, the tin coating has excellent protective properties. This will be discussed more fully under resistance to corrosion.

The Oxide Layer—An extremely thin oxide film is formed on tin plate upon contact with air, and although the film exerts some protective action, it can grow to the point where it becomes visible as a result of the development of interference colors beginning with yellow, then blue, and purple under very severe conditions of exposure. (The oxide formed on pure tin at temperatures as low as 75°C (167°F) has been identified by electron diffraction patterns as alpha stannous oxide.) The growth of the oxide film may be expedited by lacquer-baking operations or by warehouse storage in

areas of high temperature and humidity. The Gulf Coast and West Coast are among the most critical storage areas in this country.

The visible oxide film (yellow discoloration) is objectionable not only from the appearance standpoint but also because it can cause poor adhesion of lacquer and lithographing ink and adversely affect solderability. Thus, it is necessary that the tin-plate surface be stabilized to inhibit oxide-film growth. Stabilization of tin-plate surfaces by chemical and electrochemical treatments in solutions or various chemicals has been studied extensively, and as a result a number of treatments have been developed. The protective treatment utilized must be readily applied and controlled, and compatible with the lacquers involved, without being visible or affecting container performance or solderability. Because of these broad restrictions, no individual treatment is satisfactory for all applications.

The commercial solutions most commonly used for the treatment of tin-plate surfaces contain chromates. One of the most convenient is a chromic-acid treatment which was developed during the early days of the substitution of electrolytic tin plate for hot-dipped tin plate and is still in limited use today. The treatment involves the immersion of the tin plate in a hot (82°C or 180°F) solution of chromic acid. Chromic-acid-treated tin plate possesses satisfactory lacquerability and exhibits improved resistance to discoloration over untreated tin plate but may not be suitable for prolonged storage. Electrochemical treatments in dichromate-phosphate solutions in which the tin plate is successively the cathode and the anode have also been utilized. Such treatments yield lacquerability and storage-stability properties similar to those obtained with the chromic-acid treatment.

Electrochemical treatments in solutions containing only dichromate effectively inhibit oxide growth on tin plate for long storage periods and generally exhibit satisfactory lacquerability. Furthermore, they possess improved resistance to sulphide staining.

A cathodic treatment in sodium dichromate is now used extensively for a majority of applications. Referred to as CDC (cathodic dichromate), product given

this treatment has excellent storage stability and lacquer performace. However, the protective oxide film may decrease the ease of soldering and also inhibit the dissolution of the tin required for satisfactory appearance of certain foods. As a result, a treatment with the same solution but without electrolytic current has been found to yield a less protective film with better performance in the applications noted. This treatment, described as SDCD (sodium dichromate dip) does not have long-term storage stability, however.

The Oil Film and Its Application—It has been found necessary for tin plate to have a thin film of oil to permit feeding of sheets to fabricating equipment and to prevent scratching and abrasion during fabrication on automatic equipment. Too much oil, however, causes trouble if the tin plate is to be lacquered, because globules of oil immiscible in the lacquer film produce thin spots and uncoated areas. Satisfactory oil films of di-(2-ethylhexyl) sebacate (DOS) usually are about 0.005 to 0.15 gram per square metre (about 0.10 to 0.30 gram per base box) in weight. This means an oil film less than ten molecular layers in thickness. To illustrate the minute thickness of this oil film, imagine spreading one teaspoon of salad oil uniformly over a wall about 3 metres (10 feet) high and 610 metres (1000 feet) long; this is equivalent to 0.005 gram per square metre (0.10 gram per base box).

As mentioned earlier, electrostatic methods are used almost entirely to apply oil to electrolytic tin plate.

SECTION 8

CORROSION RESISTANCE

To a large degree, the successful application of tin plate for containers depends on its corrosion resistance. Food products react with the tin coating and the base metal but the rate of reaction is generally sufficiently slow to permit fairly long storage or shelf life. The type of corrosion that takes place depends on the nature of the corrosive media. Thus, the corrosion problem can be divided into:

A. Atmospheric corrosion—rusting.
B. Discoloration of the interior of the container, as by etching due to food acids, or blackening by food sulphides.
C. Hydrogen-producing corrosion by contents of cans giving rise to **swells** or **hydrogen springers** in which the pressure of the evolved gas bulges the ends of the cans, thereby making them unmerchantable. In some cases, the cans may perforate.

Atmospheric Corrosion—Tin plate is very durable in dry air but all tin plate will rust eventually in the atmosphere, especially when moisture is present. The amount of external rusting depends, in large measure, on the porosity of the tin coating and the resulting area of steel base exposed. Under conditions involving moisture and oxygen, the exposed iron behaves as an anode and the tin as a cathode (see that part of Chapter 35 dealing with "Theory of Corrosion"). Increasing the thickness of the tin coating reduces the rusting, by re-

ducing porosity. Rust resistance of tin plate can be somewhat controlled also by the nature of the tin-oxide film and the presence of oil. The composition of the steel base also has some bearing on rust resistance under certain conditions.

Discoloration of the Interior of Cans—When foods rich in sulphur-containing proteins are packed in plain tin cans (that is, without lacquer), it is generally noted that the inside of the can is stained purple, brown or black. This stain is tin sulphide and is in no way harmful. The usual method of preventing this type of corrosion is by lacquering, especially with zinc-oxide-impregnated lacquers. The sulphur reacts with the zinc oxide to form zinc sulphide, which is white and not noticed in the can. Sulphide blackening may also be prevented by a surface-filming treatment as mentioned under the section discussing the tin-oxide layer in an earlier part of this chapter.

With some condensed soups such as green pea and bean, sulphide staining of the plain can bodies can be eliminated by a high-current-density surface filming treatment. It also has been found that with condensed soups such as vegetable and tomato which are severe detinners, the high-current-density surface filming treatment greatly minimizes the detinning action.

In some mildly corrosive foods such as evaporated milk, localized detinning or dark staining may occur if the tin-oxide film is not soluble in the food product.

This localized detinning may be prevented by electro-chemically treating the tin plate in a solution of sodium carbonate.

The etching of the tin plate by food products is no indication of spoilage. All food products react with tin, some very slowly and others more rapidly. Spinach and other greens, rhubarb and squash are rapid detinners, and are frequently packed in lacquered containers.

Hydrogen-Producing Corrosion—The usefulness of tin plate for food-container manufacture depends largely on its resistance toward the formation of hydrogen "swells" or perforations.

The corrosion of tin-plate containers by food products is a complex problem and the state of our knowledge is still largely empirical. There are many factors affecting the rate at which tin plate corrodes. Major factors include:

1. Type of food product.
2. Packing procedure used in canning.
   a. The initial vacuum.
   b. The headspace volume.
   c. Use of inhibitors.
3. The storage conditions.

1. Food Product—Food can be roughly divided into three classes with respect to corrosiveness:

Most Corrosive—This group includes the highly colored fruits and berries. They are generally packed in lacquered containers because tin has a reducing action on the anthocyanin pigments which results in bleaching of the color. The acidity and pH (intensity of acidity) of the food product are no criteria of its corrosivity. In fact, corrosivity can often be reduced by adding an organic acid, as by adding lemon juice to dried prunes in syrup.

Mildly Corrosive—This group includes the bland fruits such as peaches, pears, apricots and the citrus juices from grapefruit and oranges. The mildly corrosive fruits are generally packed in plain (not lacquered) tin cans.

Slightly Corrosive—This group includes the vegetables and meat products which normally do not produce hydrogen springers in the time required to merchandise the products.

2. Food-Packaging Procedure—The canning procedure is carefully controlled to provide maximum shelf life. Cans are sealed under a sufficient vacuum to assure maximum exclusion of oxygen because the presence of oxygen within the can markedly accelerates corrosion. The vacuum is obtained either thermally or mechanically. In the former procedure the cans are sealed at as high a temperature as feasible, normally above 74°C (165°F). At this temperature, water has an appreciable vapor pressure so that the atmosphere immediately above the liquid level of the can is largely water vapor which displaces the air that would normally be present. Upon cooling the can, the water vapor condenses, the contents shrink, increasing the headspace volume somewhat, and a vacuum is produced within the can. Mechanical methods of vacuumizing involve evacuating the chamber of the sealing machine in which the cans are sealed by a vacuum pump, or sucking the air out of the headspace of the filled can with a jet of steam passing over the end of the container at the moment of attaching the lid. Mechanical vacuumizing eliminates preheating of the contents.

The headspace volume also affects the time required for springer formation by acting as a hydrogen reservoir. Overfilled cans require much less hydrogen to dissipate the vacuum and build up a pressure within the can than do cans with normal headspace. Underfilled cans, however, are prohibited by Federal statutes.

3. Storage Conditions—The temperature of storage of packed cans profoundly affects the corrosion rate as temperature affects the rate of most chemical reactions. The higher the storage temperature the shorter is the pack life. Cans of fruit which would normally last four years at 21°C (70°F) might last only a year or less at 38°C (100°F) sustained storage temperature.

Characteristics of Tin Plate Affecting Its Corrosion—The resistance of tin plate to corrosion by food varies considerably according to the methods of manufacture.

Fundamentally, the corrosion of a tin can is the corrosion of the steel-tin couple as modified by the tin-iron alloy layer. In the region of the pores in the tin coating, local electrolytic cells are formed when the tin plate is in contact with an electrolyte. Depending upon the environment, either the steel exposed through the pores in the tin coating or the surrounding tin may behave as the anode. Generally speaking, tin is anodic to steel under the normal conditions existing in a can closed so as to have a minimum amount of oxygen present, and the protection afforded the steel by the tin as affected by the tin-iron alloy layer is responsible for the excellent performance of tin-plate containers. The amount of protection that the tin provides to steel is illustrated by Figure 36—19, which compares the prune-pack performance of cans with plain black-plate can ends with black-plate ends protected with tin wire.

Extensive study by numerous investigators has helped to clarify the mechanism by which various foods corrode tin plate. Much of the work involved accelerated-corrosion tests with prunes in water as the test medium but other work has used accelerated tests with grapefruit juice. This work has been concerned with the role of the steel base, the tin-iron alloy layer, and the tin coating, as well as with evolution and diffusion of hydrogen in tin-plate containers. Obviously, the nature of the food product packed in tin-plate containers is quite important.

In acid food products such as prunes or prune juice which contain depolarizers, all tin plates detin at about the same rate, and increased corrosion resistance can be obtained in unlacquered cans by increased tin-coating weights (see Figure 36—20). Hydrogen is not evolved in these cans until insufficient tin remains to cathodically protect the steel base. In other acid food products such as grapefruit juice, it has been observed that the rate of tin solution is a function of the particular lot of tin plate involved and that as tin goes into solution hydrogen is evolved (see Figure 36—21). Thus, in these instances, increasing the tin-coating weight beyond a certain limit will not result in improved corrosion resistance. The relationships between tin, evolved hydrogen, and iron in prunes and in grapefruit are shown in Figures 36—22, and 36—23.

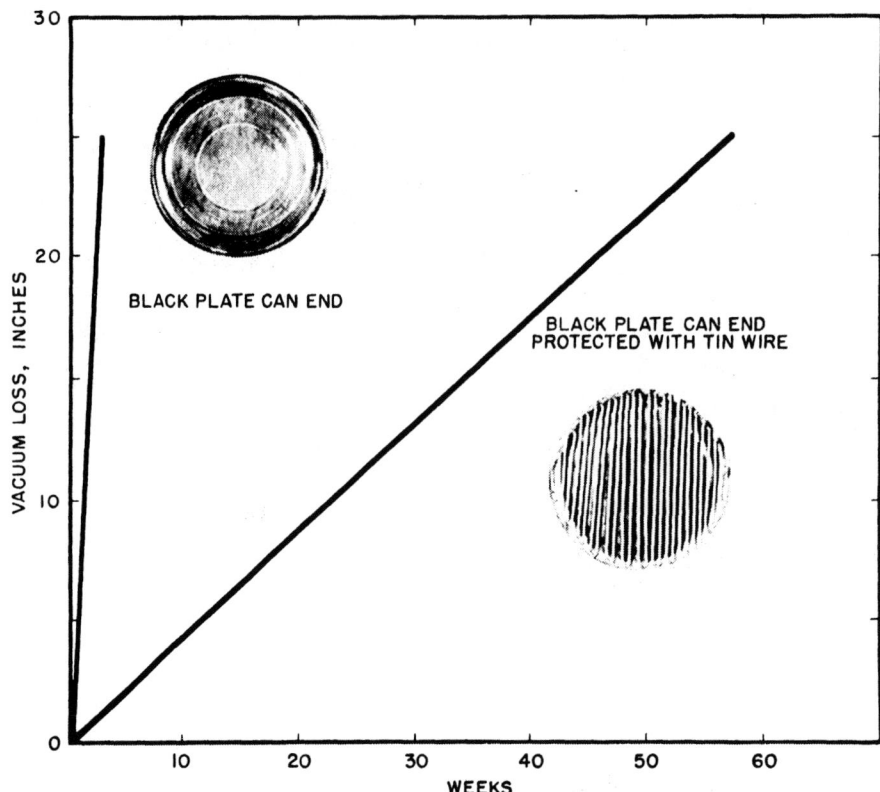

FIG. 36—19. Chart comparing results of prune-pack test on (upper left) can with black-plate end, and (lower right) similar can with black-plate end protected with tin wire, illustrating cathodic protection of steel by tin under the conditions of the test.

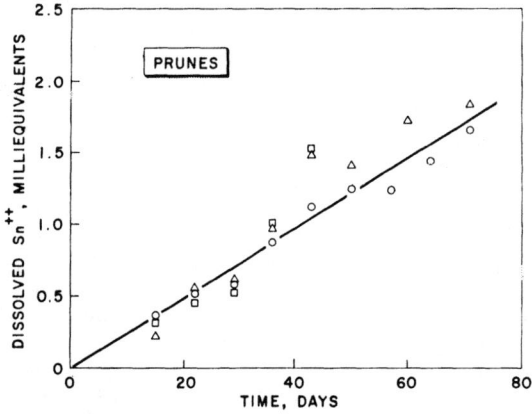

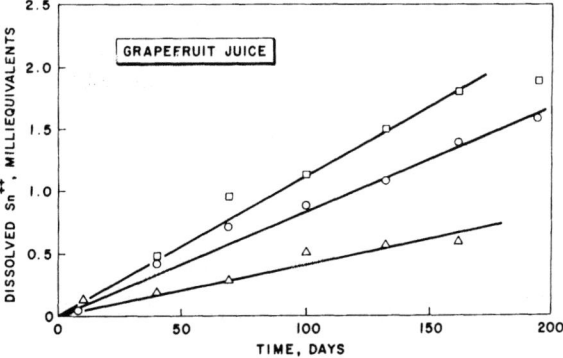

FIG. 36—21. Tin-dissolution curves for various tin plates represented by containers packed with grapefruit juice.

FIG. 36—20. Tin-dissolution curves for various tin plates represented by containers packed with prunes.

Analyses of various food products stored in unlacquered cans show as much as 120 parts per million of tin after about a year's storage at room temperature. Tin pickup from lacquered cans is very much lower, averaging 10 to 30 parts per million. The iron content of food in cans which have a good vacuum even after prolonged storage is invariably low, less than 20 parts per million.

Considerable attention has been given to the devel-

opment of accelerated tests to aid in predicting the corrosion performance of tin plate in food packs. A number of these have proved sufficiently useful in this regard that they are now utilized by the steel and can-manufacturing industries. However, none of the tests is completely satisfactory.

Referred to as "special property tests," these are chemical or electrochemical methods known as alloy-couple test (ACT), iron-solution-value test (ISV) and pickle-lag test. These, together with a fourth test indicating tin-crystal size, comprise a group, conformation

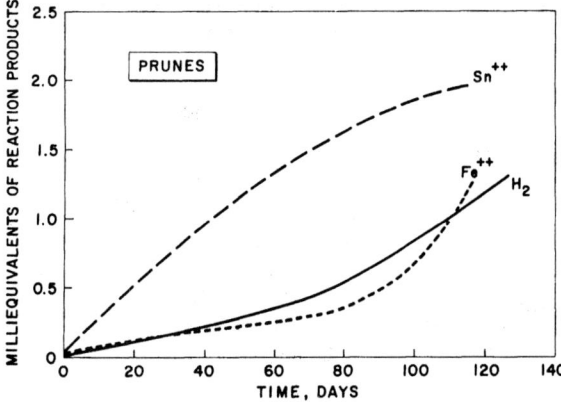

FIG. 36—22. Reaction products for corrosion of tin-plate containers packed with prunes.

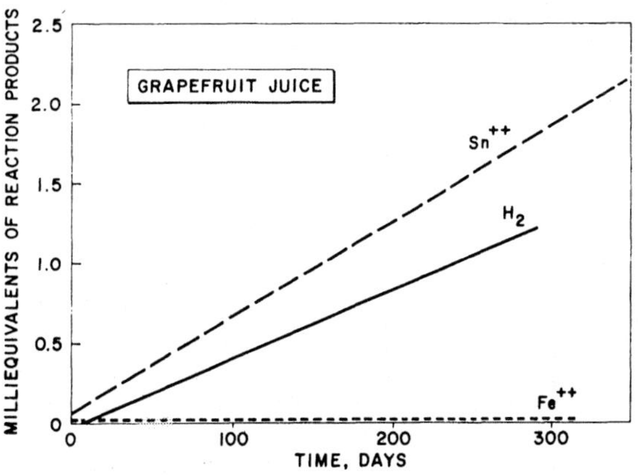

FIG. 36—23. Reaction products for corrosion of tin-plate containers packed with grapefruit juice.

to which represents Type K quality. Tin plate for corrosive applications may be purchased to this quality as a specification. It must be noted, however, that conformance to these specifications does not ensure a specified pack life.

Characteristics of the Steel Base Affecting the Corrosion of Tin Plate—The composition and methods of manufacture of the steel base are probably as important in controlling the corrosion resistance of the resulting tin plate as are the canning factors.

In the 1930's, the results of experimental packs of corrosive fruits in cans from many lots of tin plate demonstrated that phosphorus, copper and silicon may have pronounced effects on corrosion resistance, whereas, within the limits usually found in tin-plate steel, the effects of carbon and sulphur were not of commercial significance. Phosphorus and copper had a profound effect on the service life of enameled (lacquered) cans whereas silicon had a significant effect with plain cans. Increasing amounts of phosphorus and silicon were found to be detrimental to the service life of containers, while copper increased or decreased the corrosion resistance, depending on the product packed. Therefore, tin plate made from a steel base low in phosphorus (0.015 per cent maximum), copper (0.06 per cent maximum) and silicon (residual) was considered desirable for containers for the strongly corrosive products.

Container manufacturers order tin plate according to the type of food product packed. The following steel grades are in general use:

Type L—Cold-reduced basic oxygen process or open-hearth steel containing:

Element	Per Cent (Max.)
Carbon	0.013
Manganese	0.60
Phosphorus	0.015
Sulphur	0.05
Silicon	0.02
Copper	0.06
Chromium	0.06
Nickel	0.04
Molybdenum	0.05

This is used for the most corrosive foods.

Type MR—Cold-reduced basic oxygen process or open-hearth steel similar to Type L except that the phosphorus limit is increased to 0.02 per cent, the copper limit is increased to 0.20 per cent and no tolerances are specified for the other residual elements. This grade is used for mildly corrosive products.

Type D—Cold-reduced basic oxygen process or open-hearth steel aluminum-killed for applications involving difficult drawing, containing:

Element	Per Cent
Carbon	0.03-0.07
Manganese	0.25-0.40
Phosphorus	0.015 max.
Sulphur	0.035 max.
Silicon	0.02 max.
Copper	0.06 max.
Nickel	0.05 max.
Chromium	0.05 max.

Steel made to these grade specifications may be strand-cast or ingot-cast for tin plate applications. When strand-cast steel is furnished, the silicon maximum may be increased to 0.08 per cent unless expressly prohibited by the purchaser.

Extensive research conducted over many years has shown that the manner of preparation of the steel strip for tinning is as significant in controlling ultimate corrosion resistance as is composition of the steel. Control of steelmaking, ingot-conversion, and rolling practices so as to minimize surface defects and discontinuities as well as maintenance of pickling, cleaning and annealing conditions leading to strip-surface cleanliness are all-important in achieving high corrosion quality. The required practices can be maintained consistently in commercial operations and the special-property tests previously mentioned are now available for monitoring the resultant product.

In essence, the problem of achieving good corrosion resistance of tin plate resolves itself into methods of slowing down the attack on the steel base. This can be accomplished by:

1. Producing good-quality steel base.
2. Applying heavier tin coatings.
3. Coating with lacquer in certain cases.

Lacquered Tin Plate—Lacquered cans are normally fabricated from tin plate and tin-free steel that have been lacquered in sheet form. In the United States, the trade refers to the lacquers as enamels. The lacquers vary in composition but are either organic solutions of resins or mixtures of resins and vegetable oils, the resins being either naturally occurring or synthetic. They may be clear or pigmented. Waterborne and high-solids lacquer systems are finding increasing acceptance as a means of avoiding the pollution problems associated with solvent-based lacquer.

Lacquered cans are normally used to prevent the bleaching of highly colored fruits or to prevent the sulphide staining caused by sulphur-bearing foods. Considerable success in the reduction of hydrogen swells has, however, attended the use of lacquered cans for certain food products. During World War II, considerable quantities of 11.2 grams per square metre (0.50 pound per base box) electrolytic tin plate, and even black plate, protected by lacquers, gave satisfactory service life for meat and vegetable products. The use of lacquered 5.6 grams per square metre (0.25 pound per base box) electrolytic tin plate has increased substantially, so that currently it accounts for a major portion of cans. Tin-free steel is lacquered on both surfaces for normal food-packaging applications.

The corrosion process in lacquered cans is considerably more complex than in plain cans. The type of lacquer and methods of application may be as important as the quality of the tin plate used. Generally, improved service life may be obtained by more continuous coverage, as by spray or flush lacquering the fabricated cans. The conventional metal-decorating process, however, consists of roller coating with lacquers and if a decoration is used, it is applied by offset lithography on the flat sheet.

Cracks in the lacquer coating expose minute areas of tin and iron. At these areas the tin tends to protect the iron, but because of the limited amount of tin exposed, it is soon used up and corrosion of the tin plate may be found in areas which have no apparent discontinuities in the lacquer film. Indeed, lacquered cans of fruit which have abnormally long service life frequently show pronounced etching of the tin beneath the lacquer, indicating that lacquer films may be acting as semiporous membranes and the corroding tin beneath the lacquer is protecting the iron cathodically.

Trial packs are used to determine whether pack life can be improved by the use of a specific lacquer on the tin plate. As yet there are no laboratory tests which will determine the suitability of a specific lacquer system for a specific food product for a specific surface condition. Because certain formulation constituents may make a lacquer incompatible with certain surface-treatment films, specific lacquers must be evaluated on the surface for which they are being considered.

Bibliography

American Iron and Steel Institute, Steel products manual: Tin mill products. Washington, D.C., 1982.

Britton, S. C., Tin versus corrosion, International Tin Research Institute, Publication No. 510.

Food Processors Institute, Proceedings, Shelf Life: A key to sharpening your competitive edge, Washington, D.C., 1981.

Ibid, Proceedings, The A-to-Z of Container Corrosion: Another path to productivity, 1982.

Gonzalez, O. D., P. H. Josephic, and R. A. Oriani, The undercutting of organic lacquers on steel, Journal of the Electrochemical Society, Vol. 121, No. 1: 29-34 (January 1974).

Hoare, W. E., E. S. Hedges, B. T. K. Barry, The Technology of Tinplate, Saint Martin's Press, N.Y, 1965.

International Tin Council, Study on consumption, United States of America, tin in tinplate, London, 1983.

International Tin Research Institute, Guide to tinplate, Publication No. 622.

International Tin Research Institute, First international tinplate conference, Publication No. 530, London, 1976.

International Tin Research Institute, Second international tinplate conference, Publication No. 600, London, 1980.

Mannheim, C. and N. Passy, Internal corrosion and shelf-life of food cans and methods of evaluation, CRC Critical Reviews In Food Science and Nutrition, Vol. 17, No. 4: 371-407 (1983).

Uchida, H., O. Yanabu, A. Horiguchi, and H. Sato, Chromium plated steel strip for canmaking, Sixth International Conference on Electrodeposition and Metal Finishing: 138-145 (1964).

Yonezaki, S., Research and development on tin-free steel sheet, Sheet Metal Industries, Vol. 48, No.1: 25-35 (1971).

CHAPTER 37

Terne Coated Sheet

Terne coated sheet is any sheet steel that has been coated by immersion in a bath of molten terne metal. Terne metal is an alloy of lead and tin. Sheet steel coated with terne metal is duller in appearance than sheet steel coated with tin alone, and it is this feature of the product that gave rise to the name "terne"—meaning "dull".

Formerly, there were two classes of terne-coated products: **long terne sheet** (sometimes called simply **long ternes**) which was a hot-dipped sheet-mill product that fell within the dimensional ranges covered by the uncoated-sheet commodities; and **short ternes** (also called **terne plate**) which was a hot-dipped tin-mill product in black-plate gages (thickness less than 0.36 mm (0.014 inch). Short-ternes (terne plate) are no longer produced.

In commercial practice, terne coated sheet is seldom manufactured thinner than 0.36 mm (0.014 inch), thicker than 3.2 mm (0.125 inch), or wider than 1372 mm ·(54 inches). Terne coated sheet occupies a relatively minor position tonnagewise in the sheet business, but offers a combination of properties that make it very suitable for such applications as air cleaners and fuel tanks. Terne coated sheet is produced both in cut lengths and in continuous coils.

Terne Coatings—The composition and weight of terne coatings were drastically restricted in 1941, when the normal supplies of tin were cut off by World War II. Prior to that time, it was customary to coat with terne metal containing approximately 80 per cent lead and 20 per cent tin, although this composition was not mandatory and could be varied for specific applications. Since the object of terne coating is to apply to the steel base an inexpensive, corrosion-resistant coating of lead, the percentage of tin used need only be sufficient to obtain a smooth, continuous coating. Lead alone does not alloy with iron, so that it is necessary to incorporate

a certain amount of another element, in this case tin, which alloys readily with the steel base and forms a solid solution with the lead. It was formerly thought impractical to coat with terne metal compositions containing much less than 15 per cent tin. Experience gained in the late 1960's has proved otherwise. Terne coatings of entirely satisfactory quality are now regularly produced with a tin content of 8 per cent. It is to be noted, however, that decreasing the amount of tin in the coating alloy necessitates higher pot temperatures and lessens the alloying or "wetting" properties of the terne metal. To improve wetting, small additions of other alloying elements such as antimony may be used to replace some of the tin that was removed and improved fluxes may be used. ASTM A 308 permits tin contents as low as 3 per cent.

Terne coated sheet is ordered to the coating designations and corresponding minimum coating-weight test limits given in Table 37—I, specified in grams per square metre or ounces per square foot total coating on both sides. The designation of terne coating weights in terms of ounces per square foot followed the practice used in designating coating weights of galvanized sheet, as terne coated sheet formerly was made commonly on equipment located in galvanizing shops. The LT01 coating is furnished unless a special coating weight is ordered. Terne coated sheet with LT01 coating is a well-coated product but is not subject to any minimum coating test.

The thickness and nature of the coating are the most important factors governing the corrosion resistance of terne coated sheet. Both lead and tin are highly corrosion resistant, and their combinations as used on terne coatings also resist corrosion well. However, both lead and tin are cathodic to iron under most conditions of exposure, so that terne metal actually will accelerate corrosive action on the steel base when any portion of

Table 37—I. Coating Designation and Minimum Coating Test Limits for Terne Coated Sheets.

Coating Designation	Corresponds to Old Coating Class Now Obsolete (oz/ft²)	Minimum Coating Check Limits			
		Triple Spot Test Total Both Sides		Single Spot Test Total Both Sides	
		(g/m²)	(oz/ft²)	(g/m²)	(oz/ft²)
LT01	Commercial	no check		no check	
LT25	0.35	76	0.25	61	0.20
LT35	0.45	107	0.35	76	0.25
LT40	0.55	122	0.40	92	0.30
LT55	0.75	168	0.55	122	0.40

the steel surface becomes exposed to attack. Since a light coating has a greater tendency to form pinholes or other discontinuities than does a heavier coating, exposure of the base metal is minimized and corrosion resistance is increased as the thickness of the coating increases. However, even the most heavily coated terne sheet should be well-painted when maximum protection against corrosion is desired.

Composition and Preparation of Steel Base—The steel base of terne coated sheet may be provided on either a "commercial quality" or a "drawing quality" basis. In the former case, rimmed, capped, or semi-killed steel is generally used. In the latter, specially selected rimmed or aluminum-killed steels are provided, because of their superior ductility. All of these steels usually fall within the following composition limits:

Carbon (%)	Manganese (%)	Phosphorus (%)	Sulphur (%)	Silicon (%)
0.03-0.10	0.25-0.50	0.025 max.	0.050 max.	0.020 max.

Selection of the composition and treatment of the steel to be used for the base of terne coated sheet is governed by the end use of the products to be fabricated from it and hence must take into consideration a number of factors, important among which are the customers' fabricating practices. Factors normally considered in addition to the usual requirements for flatness, finish, and weight of coating, are such forming hazards as breakage in deep-drawing operations, fluting, stretcher strains, and the conditions under which the material is to be used. In this latter connection, for example, a copper content of 0.20 per cent minimum is usually recommended for applications involving corrosion hazards, such as in roofing.

The treatment accorded the steel base used in manufacture of terne coated sheet depends upon the desired combination of gage, surface finish, and physical properties. It is basically identical with the treatment which would be given a cold-rolled product manufactured for the same type of customers' fabricating operations. This material is rolled on continuous cold-reduction mills.

The continuous cold-reduction process involves the use of lubricating solutions containing oil, the removal of which from the surface of the steel is of vital importance in producing a base metal which is to be terne coated. Any oil which remains on the sheet or strip after cooling will decompose when heated during box-annealing prior to coating. The relatively inert, carbonaceous residue which results can interfere seriously with the application of a smooth, continuous coating of terne metal, since these residues are not attacked to any appreciable degree by the pickling solutions commonly employed for descaling and cleaning the steel prior to terne coating. As a result, they prevent the terne metal from "wetting" or alloying with the steel base, causing rough coating or entirely bare areas. Some manufacturers make use of an electrolytic cleaning operation prior to annealing in order to prevent the formation of the "carbon smudge;" others cold reduce the base metal in thin mixtures of volatile water-soluble oil.

Cold-reduced product is usually box annealed since it is more economical than normalizing and the steel may be kept scale-free by using a deoxidizing atmosphere, thereby reducing the iron loss due to pickling. To further enhance the surface quality, special box-annealing practices have been developed to provide clean strip surface. A temper rolling operation may be performed after annealing, but may be omitted on certain gages and applications wherein it is known that the customer's fabrication involves a deep-drawing operation in which fluting or straining of the base metal is not objectionable.

The cold-reduced and box-annealed (and perhaps temper rolled) material is cleaned and pickled, as necessary, either prior to or in the terne-coating line. Hydrogen embrittlement of the base metal is a hazard and close control of the pickling operation is required to avoid deleterious effects.

Originally, terne coating was developed as a process using single sheets. However, this has been replaced in the United States by the semi-continuous and fully continuous coil-coating process which provides higher productivity, superior coating quality and production of both sheets and coils. In adapting the lines to operate with coils, the basic steps of the single-sheet process were retained until recently when further modifications were made in coating-weight control systems.

SINGLE-SHEET COATING PROCESSES

Although single-sheet coating processes have been supplanted in the United States by the coil-coating processes described later, they will be discussed briefly here to establish the principles common to both types of processes.

For applying terne metal to single sheets in the manufacture of terne coated sheet, two methods—a flux process and a combination flux-oil process—are used. The chief objective of both of these methods is to bring the properly prepared surface of the steel base into contact with the molten terne metal, and to remove the coated sheets from the pot, under conditions that assure a uniform distribution of the coatings.

The **flux process** employs a flux of molten (or a water solution of) zinc chloride, or a solution of zinc chloride in hydrochloric acid, to remove any oxides of iron that are present and also effect a rapid drying of the sheets.

The process is carried out in a "rigging" or machine which carries the sheet through the several steps of the process as follows: The pickled sheets are fed through a hydrochloric acid wash tank to remove the residues which might have formed on the sheets subsequent to pickling. After removing excess acid from the sheet, it is passed into the flux box, where it is cleaned and dried. The sheet then passes downward through the molten terne-metal bath operated at 325° to 360°C (620° to 680°F), then upward through a bath of oil (palm oil, fish oils, mineral oils, or combinations thereof) floating on top of the molten metal, being conveyed through the oil by the oil rolls. The oil bath on the exit side of the metal bath, in addition to protecting the surface of the molten metal bath from oxidation, tends to keep the coating on the sheet in a molten condition and thus allows the oil rolls to control the distribution and thickness of the coating. As the coated sheets leave the oil bath they are conducted to the conveying system of the cooling and cleaning train.

In the **combination process,** a coating pot using the flux process is followed by a second coating pot of the same type but with oil substituted for the zinc chloride. This type of arrangement permits complete control over the coatings, particularly those in the heavy weights. Since most terne coated sheet is ordered in the lighter coating weights, the bulk of the material produced is coated on a single unit by the flux process.

After the coating operation, the freshly coated sheet is cooled with forced-air blasts sufficiently to permit cleaning without injury to the soft coating.

Regardless of the coating process used, the terne sheet emerges from the coating operation covered with oil which has to be removed by using both liquid cleaners and dry branning equipment. The liquids are solvents such as mineral spirits. The branning operation consists of wiping off the oil with cloth-covered rolls and bran middlings or hardwood sawdust. Flannel-covered rolls at the exit from the branner exert a smoothing action on the surface when a bright finish is desired. Terne coated sheet is produced with a "dry" or "bright" finish, and an "oil" finish. Sheets leaving the cleaning equipment practically free of oil are designated as dry finish, but if they carry an appreciable amount of oil, or are subsequently coated with oil, they are designated as oil finish. When oil finish is specified, oiling rolls can be inserted in the line to apply the desired amount of oil on the surface. Upon completion of cleaning, the sheets are ready for inspection by the inspectors stationed at the end of the line.

CONTINUOUS-STRIP PRODUCTION

The principles of the coil-coating (continuous strip) process are identical to those of the single-sheet process and, indeed, the operations were quite similar up until about 1972 when the successful use of gas-wiping knives eliminated the oil-bath system for controlling coating weights; the principle of the gas-wiping knife is described in the discussion of continuous galvanizing in Chapter 38. The majority of commercial lines continue to use the flux process for preparing the steel prior to coating. However, very recently a line using the anneal-and-coat process widely used in continuous-strip galvanizing started operation in one country, and a coil line in which the terne coating is deposited electrolytically on the sheet started operation in another country.

The process of continuous terne coating has several unique advantages. The first of these is the elimination of the intermittent operation characteristic of sheet pots, whereby a space of from 0.3 to 0.9 metre (1 to 3 feet) was left between successive sheets as they passed through the line. The next distinct advantage is the uniformity of the coating that can be realized by having a continuous flow of the base metal through the coating rigging with no break in the operation. The single-sheet method resulted in a heavier coating on the front and back ends, "entry" and "list" ends respectively, so that continuous coating makes more efficient use of a given quantity of coating metal. A disadvantage of the continuous process lies in the fact that it is less adaptable to producing small, varied types of orders since it is most efficient when used to produce coils of one coating weight. However, this disadvantage has

been overcome by using the aforementioned gas-wiping metering system.

Continuous lines for the coating of strip steel with terne metal may be of the semi-continuous or the fully-continuous type. The semi-continuous line has only one payoff reel and one tension reel, which necessitates threading each individual coil through the line. The fully continuous line is equipped with double payoff reels, welder, and looping towers or pits, which permit a continuous flow of strip through the coating unit. One arrangement of a fully continuous line is shown in Figure 37—1.

The equipment contained in a continuous coating line may vary from shop to shop depending on numerous operating conditions and the preference of the manufacturers concerned. Essentially, the equipment and process involved are as follows: A coil holder, followed by a payoff reel, feeds the strip into a pinch-roll unit, which in turn is followed by a squaring shear and a welder if the process is fully continuous. When necessary, oil and grease are removed from the steel with a conventional electrolytic alkaline cleaning unit with a suitable rinsing and scrubbing facility. The strip then enters an acid pickling tank, which may consist of a single-dip tank, wherein normal chemical pickling is effected, or which may be an electrolytic pickling unit. Subsequently, the steel is passed through (1) an aqueous zinc-ammonium-chloride solution, (2) a molten flux bath floating on the surface at the entry of the terne-metal bath, and (3) the molten terne metal. The weight and distribution of the coating metal is metered either by (1) an oil-roll system (which covers the exit surface of the terne-metal bath) or (2) the recently developed gas-knife system. Above the metering system in a vertical tower, the molten coating on the steel is solidified using forced-air cooling. If the oil-roll system is used, the oil is removed after cooling using the conventional liquid cleaner and dry branning. With the gas-knife system, the oil is eliminated so that cleaning and branning are not necessary, and the steel is conducted to the delivery looping tower, the roller leveling and the recoiling or shearing section of the line.

INSPECTION AND TESTING

The inspection of terne coated sheet normally takes place during coiling or piling into lifts. All but a small percentage of the sheets coated are **primes,** i.e., in compliance with the requirements of the customers' orders. However, even with the best practice and control of processing variables, a certain percentage of the sheets from the terne pots may be classified as **rejects. Scrap sheets** are those totally unsuitable for reconditioning or salvaging. If desired, the coating from scrap sheets may be reclaimed by heating them in palm oil to the proper temperature, and allowing time for most of the coating to run off and collect in the bottom of the container.

Rejected terne coated sheet may be classified into three general types, depending on the basic cause for the non-compliance with specification. These types are as follows:

(1) Those due to an imperfection in the steel base.
(2) Those due to faulty processing of the material · prior to the coating operation.
(3) Those due to faulty coating practice.

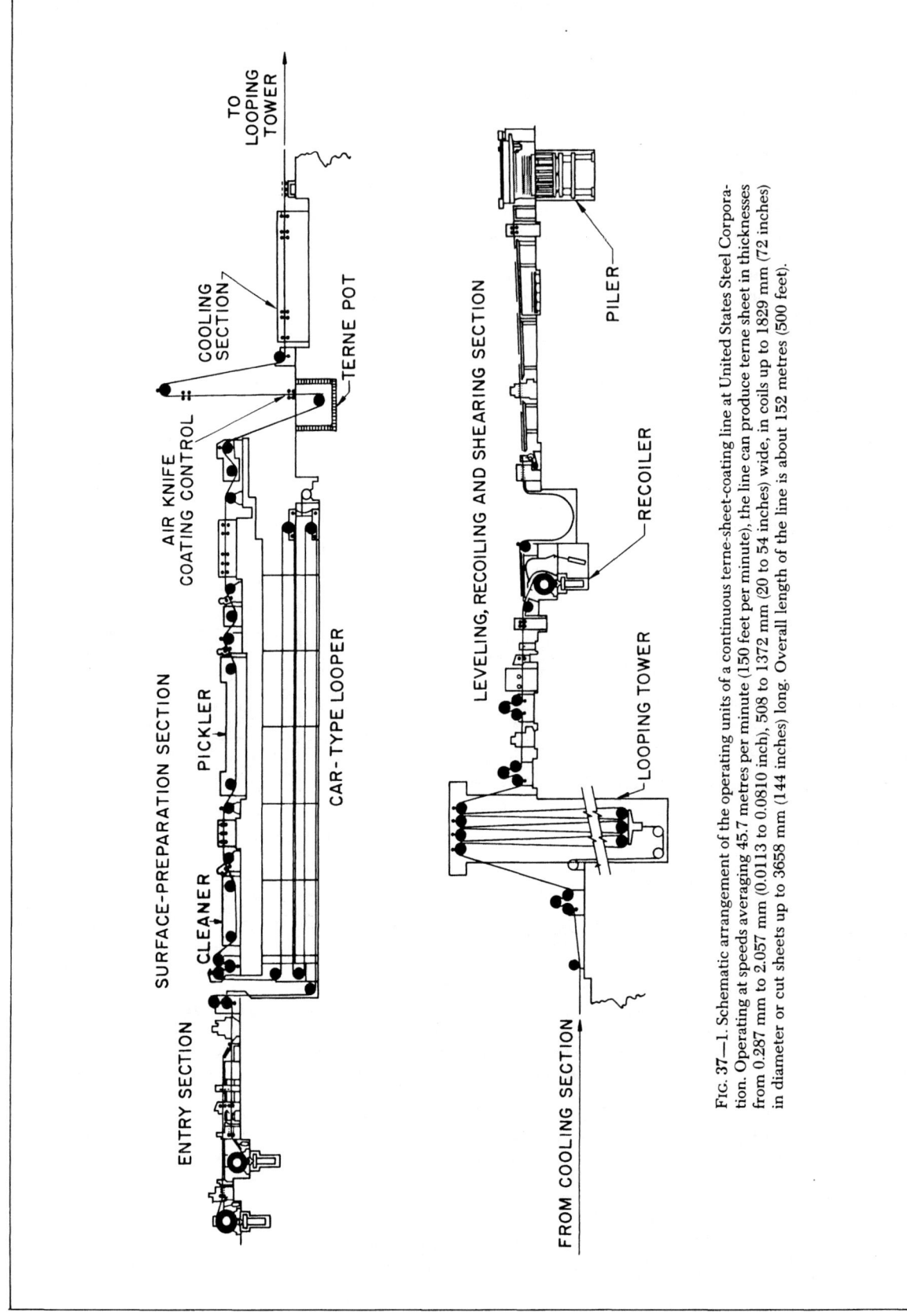

FIG. 37—1. Schematic arrangement of the operating units of a continuous terne-sheet-coating line at United States Steel Corporation. Operating at speeds averaging 45.7 metres per minute (150 feet per minute), the line can produce terne sheet in thicknesses from 0.287 mm to 2.057 mm (0.0113 to 0.0810 inch), 508 to 1372 mm (20 to 54 inches) wide, in coils up to 1829 mm (72 inches) in diameter or cut sheets up to 3658 mm (144 inches) long. Overall length of the line is about 152 metres (500 feet).

The major imperfections encountered in categories (1) and (2) are blisters, seams (the latter being designated sometimes as "skin laminations" or "slivers"), rolled-in scale, pits, gouges, scratches and underpickled areas as described for cold-rolled sheets in the American Iron and Steel Institute Manual for "Carbon Sheet Steel." These may be eliminated by making corrections in preceding processing operations. Another imperfection falling into the second category is the previously mentioned "rough coating," which occurs when the terne metal fails to "wet" or alloy with the steel base because of a film of reduced iron or carbonaceous residue which has not been satisfactorily removed in pickling. In this connection, it is to be noted that the cleanliness of the base metal is much more critical in the coating of terne sheet than in the case of hot-dip zinc coatings.

Imperfections comprehended in category (3) include such items as uncoated areas, oil stains, branner marks, and other minor items which may be controlled by the terne-line operator. In addition to the inspection for the above enumerated imperfections, the sheets are also checked for gage, size and flatness by the inspectors and are stamped with the appropriate identification marks.

The tests to which terne coated sheet is subjected after completion of the coating operation may be classed as chemical, microscopic and mechanical. To determine the weight of terne coating on the sheet and to determine conformance to specifications it is customary to use the **triple spot test method.** For this test, three specimens exactly 57.15 mm (2.25 inches) square, or discs of equivalent area, are cut from each test sheet, one being cut from the center and the other two from near but not closer than 50.8 mm (2 inches) from the sides, in order to avoid the heavier coating on these areas. After each specimen has been weighed on an analytical balance, the coating is dissolved by immersion in an appropriate solution. The specimen is then washed, dried and reweighed, the difference in weight in grams being numerically equal to the weight of coating in ounces per square foot (one ounce per square foot equals approximately 305 grams per square metre). From time to time, and particularly when certain specifications governing the composition of the terne coating must be met, the coating on a test sample is chemically analyzed to determine the percentages of tin and lead present.

A **microscopic examination** of a properly polished and etched cross-section may be made to determine the ferritic grain size of the sheet base. Since the adherence of the coating to the base metal is never a problem in the production of terne sheet, adherence testing is not required as it is for galvanized sheet. **Mechanical property tests** are used primarly as a guide in determining the ductility of the coated sheet and its suitability for the intended application. These mechanical tests include Rockwell hardness, Olsen cup, bend, fluting and tension test. Hardness tests are normally made on the base metal after the coating has been removed since the soft coating would give erroneous values if the hardness impression were made through it. Besides providing a measure of the ductility, the Olsen test is a valuable tool for detecting hydrogen embrittlement of the base metal. In the Olsen cup test, the material normally "necks" at the periphery of the base of the cup, but when absorbed hydrogen has embrittled the base metal, a ragged fracture will occur on the side of the cup or, in extreme cases, at the base of the cup. Fluting tests are made by forming a suitable test piece into a cylinder and observing the presence or absence of flutes or creases.

APPLICATIONS

The major portion of the terne coated sheet production is used in the manufacture of gasoline tanks for the tractor, truck and automotive industries. Automotive gas tanks are usually drawn in two halves, top and bottom, and seam welded around the perimeter. Truck and tractor tanks are sometimes drawn in the same manner, but are also fabricated by forming the sheets into cylinders by using bending rolls and then lockseaming or soldering the ends. The use of terne coated sheet for these applications is dictated by the resistance of the coating to corrosion by gasoline, water and air. Other automotive parts manufactured from terne coated sheet include water distributor tubes, radiator parts, oil pans and air cleaners. The latter are considered a minor corrosion hazard but manufacturers of air cleaners prefer to draw the shapes involved from long terne sheet because of the lubricating value of the coating in deep-drawing operations. Other uses are for roofing, hand fire extinguishers, outboard-motor gasoline tanks, portable gasoline cans, firedoors, laboratory furniture, radio and TV chassis, condenser and capacitor cans and burial caskets.

Bibliography
American Iron and Steel Institute, Steel products manual, carbon sheet steel. The Institute, New York, 1979.
American Society for Metals, Metals Handbook, Ninth Edition, Volume 1. The Society, Metals Park, Ohio, 1978.
American Society for Testing and Materials, Annual book of standards (Designation A 306). The Society, Philadelphia, 1978.

CHAPTER 38

Production of
Galvanized Sheet and Strip

SECTION 1

GENERAL

Production and Uses of Galvanized Sheet and Strip—It is important to provide protection against corrosion for steel articles having a light section to extend the service life of the articles. Coating the steel with zinc is a very effective and economical means of accomplishing this end. Zinc coatings are commonly applied by dipping or passing the article to be coated through a molten bath of the metal. This operation is termed "galvanizing," "hot galvanizing," or "hot-dip galvanizing" to distinguish it from zinc electroplating processes which are termed "cold" or "electrogalvanizing."

The present discussion relates to (1) **single-sheet galvanizing** which is the hot-dip galvanizing of cut-length sheets by passing them one by one and in close succession through the molten zinc, (2) **continuous (strip) hot-dip galvanizing,** in which material in coiled form from the rolling mills is uncoiled and passed continuously through the galvanizing line, continuity of operation being achieved by joining the trailing end of one coil to the leading end of the next, and (3) an **electrogalvanizing process** for the production of one-side and two-side galvanized sheet.

The word "strip," as used in this chapter, denotes the *physical form* of the material in process, i.e., a continuous strand of steel of sheet gage and width. Commercially, the meaning of the term "strip" has the limitations outlined in Chapter 33.

Of all the common metals used for protective coatings, zinc is the most cost effective. Zinc and tin are by far the most widely used coating metals for flat rolled steel. In 1980, 379 227 metric tons (418 111 net tons) of slab zinc were consumed by the galvanizing industry in the United States. Of this amount, 220 744 metric tons (243 378 net tons) were used for the galvanizing of sheet and strip. The balance of the zinc was used largely in wire galvanizing, hot-dip galvanizing of construction materials such as guard rails, pipe, and heavy sections (used, for example, in transmission towers), and hot-dip galvanizing of prefabricated articles such as wash tubs, garbage cans, and pails. The reported shipments of galvanized sheet and strip in the United States during 1980 amounted to over 4.7 mil-

lion metric tons (5.25 million net tons), about 90 per cent of which was produced by hot-dipping (about 411 027 metric tons or 456 697 net tons were produced by electrogalvanizing processes).

Of domestic galvanized-sheet production in 1980, about 34.2 per cent was supplied to the construction and contractors'-products industries, and about 30.1 per cent was supplied to steel service centers and distributors. Galvanized sheets were also supplied to the automotive (19.6 per cent); appliance (4.3 per cent); agricultural (4.6 per cent); electrical-equipment (1.8 per cent); machinery and industrial-equipment (0.9 per cent); and container, packaging and shipping-materials (0.6 per cent) industries.

Specific uses for galvanized sheets include: automotive parts such as quarter panels, door inners and outers, rocker panels, trunk lids, underbody parts, and mufflers; agricultural roofing and siding; roofing and siding for industrial buildings; residential siding; culverts; eave troughs; conductor pipe; air ducts; heating furnaces; air-conditioning equipment; metal lath; corner beading; switch boxes; chimney flues; metal awnings; garbage cans; ash cans; pails; tubs; refrigerators; deep-freeze units; kitchen equipment; laundry tubs; automatic washers and dryers; storage tanks; beverage coolers; well casings; poultry equipment; manure spreaders; hay loaders; tractors; and wagons.

FACTORS INFLUENCING EFFECTIVENESS OF GALVANIZED COATINGS

The effectiveness of a protective coating depends to a considerable extent upon the nature of the environment in which it is used. In general, galvanized coatings are subjected to atmospheric corrosion (such as rain and airborne contaminants), liquid corrosion (such as road splash), and, less frequently, to soil corrosion (such as acids in the soil). Their effectiveness in the atmosphere is strongly influenced by the amount of acidic contaminants present. Thus, in the relatively pure air of rural districts, the life of zinc coatings can be considerably longer than that of the same coatings in

Table 38—I. Galvanized Sheet Gage Numbers with Equivalent Unit Weights

Galvanized Sheet Gage Number	Grams Per Square Metre	Ounces Per Square Foot	Pounds Per Square Foot	Pounds Per Square Inch	Thickness Equivalents	
					(Inches)	(mm)
8	34 329	112.5	7.03125	0.048828	0.1681	4.270
9	31 278	102.5	6.40625	0.044488	0.1532	3.891
10	28 226	92.5	5.78125	0.040148	0.1382	3.510
11	25 175	82.5	5.15625	0.035807	0.1233	3.132
12	22 123	72.5	4.53125	0.031467	0.1084	2.753
13	19 072	62.5	3.90625	0.027127	0.0934	2.372
14	16 020	52.5	3.28125	0.022786	0.0785	1.994
15	14 495	47.5	2.96875	0.020616	0.0710	1.803
16	12 969	42.5	2.65625	0.018446	0.0635	1.613
17	11 748	38.5	2.40625	0.016710	0.0575	1.461
18	10 528	34.5	2.15625	0.014974	0.0516	1.311
19	9 307	30.5	1.90625	0.013238	0.0456	1.158
20	8 086	26.5	1.65625	0.011502	0.0396	1.006
21	7 476	24.5	1.53125	0.010634	0.0366	0.930
22	6 866	22.5	1.40625	0.0097656	0.0336	0.853
23	6 256	20.5	1.28125	0.0088976	0.0306	0.777
24	5 645	18.5	1.15625	0.0080295	0.0276	0.701
25	5 035	16.5	1.03125	0.0071615	0.0247	0.627
26	4 425	14.5	0.90625	0.0062934	0.0217	0.551
27	4 120	13.5	0.84375	0.0058594	0.0202	0.513
28	3 814	12.5	0.78125	0.0054253	0.0187	0.475
29	3 509	11.5	0.71875	0.0049913	0.0172	0.437
30	3 204	10.5	0.65625	0.0045573	0.0157	0.399
31	2 899	9.5	0.59375	0.0041233	0.0142	0.361
32	2 746	9.0	0.56250	0.0039062	0.0134	0.340

industrial areas.

For any specific set of exposure conditions, the thickness of a zinc coating is the most important factor in determining its service life. The amount of zinc on a galvanized sheet is stated in terms of grams per square metre (or ounces per square foot), including the coating on both sides. Coatings applied to sheets by the hot-dip process range, in general, from 122 to 839 grams per square metre (0.4 to 2.75 ounces per square foot). A coating weight of 839 grams per square metre can be expected to have approximately seven times the life of the 122 grams per square metre (0.4 ounce per square

Table 38—II. Standard Designations and Weights of Coating (Total Both Sides) for Galvanized Sheet*

Type	Coating Designation	Previous Coating Class	Minimum Check Limits			
			Triple-Spot Test		Single-Spot Test	
			Oz./Sq Ft (of sheet)	g/m²	Oz./Sq Ft (of sheet)	g/m²
Regular	G 235	2.75	2.35	717	2.00	610
Regular	G 210	2.50	2.10	640	1.80	549
Regular	G 185	2.25	1.85	564	1.60	488
Regular	G 165	2.00	1.65	503	1.40	427
Regular	G 140	1.75	1.40	427	1.20	366
Regular	G 115	1.50	1.15	351	1.00	305
Regular	G 90	1.25 Commercial	0.90	275	0.80	244
Regular	G 60	Light Commercial	0.60	183	0.50	152
Regular	G 30	—	0.30	92	0.25	76
Regular	G 01	—	No Min.	—	No Min.	—
Alloyed	A 60	—	0.60	183	0.50	152
Alloyed	A 40	—	0.40	122	0.30	91
Alloyed	A 25	—	0.25	76	0.20	61
Alloyed	A 01	—	No Min.	—	No Min.	—

NOTE: The coating designation number is the term by which this product is specified. The weight of coating in grams per square metre or oz. per sq. ft. of sheet refers to the total coating on both surfaces. Because of the many variables and changing conditions that are characteristic of continuous galvanizing, the weight of zinc coating is not always evenly divided between the two surfaces of a galvanized sheet; neither is the zinc coating evenly distributed from edge to edge. However, it can normally be expected that not less than 40 per cent of the single-spot check limit will be found on either surface.

*From: ASTM Specification A 525 published by the American Society for Testing and Materials, Philadelphia, Pa. The latest version of the complete specification should be consulted for further details.

foot) coating under similar exposure conditions.

Localized areas of heavy coating are not useful in service because a part formed from a coated sheet will fail at the spot with the thinnest coating and thus be of no value. For this reason, uniformity of coating is also an important consideration. In addition, a uniform coating makes for the most economical use of a given amount of zinc. As a practical matter, a perfectly uniform coating is not produced in a commercial hot-dipping operation; however, no area or spot of a hot-dipped sheet may carry a lesser weight of coating than the trade-specified minimum for the grade. These minimums vary with the different coating-weight levels available and are specified by coating designation; see "Coating Weight and Gage Requirements" below.

One of the advantages of zinc as a protective coating is its anodic relationship to iron in the electromotive series. Due to this relationship, the protection afforded is extended to small areas of exposed steel adjacent to coated areas. Thus, lack of coating at a sheared edge or minor damage (scratches) to the coating do not seriously impair its service life if the uncoated areas are not too large.

COATING WEIGHT AND GAGE REQUIREMENTS

Galvanized sheets are manufactured to several general specifications, the requirements of which have been found by usage and test to be the most relevant for the applications involved.

The so-called G 90 Coating Designation (formerly called Commercial Coating Class) generally includes those coatings that are produced on sheets for direct weather-exposure applications. Galvanized Formed Roofing and Siding and Galvanized Flat Sheets for Roofing and Siding are produced to ASTM Specification "Steel Roofing Sheets" A 361 in G 90 coating designation. This designation specifies 275 grams per square metre (0.90 ounce per square foot) minimum by triple spot test, and 244 grams per square metre (0.80 ounce per square foot), minimum by single-spot test. (Tests for galvanized sheets are described in Section 6 of this chapter.) Galvanized Flat Sheets for other than roofing and siding products, but for an intended equivalent corrosion hazard, may be produced to ASTM Specification "Steel Sheets, Zinc Coated (Galvanized) by the Hot-Dip Process" A 525 and designated to have a similar G 90 coating.

In some cases, galvanized sheet is produced with a thinner coating for applications in which corrosion resistance is not the only important consideration, for example, to reduce spot-welding problems. This product is made to ASTM Specification A 525 and designated to have a lighter G 60 coating, with minimums

specified as 183 grams per square metre (0.60 ounce per square foot) by triple-spot test and 153 grams per square metre (0.50 ounce per square foot) by single-spot test.

Heavier than G 90 coatings are supplied when required for specific applications. The Zinc Institute "Seal of Quality" roofing sheet is specified as 610 to 687 grams per square metre (2.00 to 2.25 ounces per square foot) coating class with a minimum 549 grams per square metre (1.8 ounces per square foot) as checked by triple-spot test and a minimum 458 grams per square metre (1.50 ounces per square foot) as checked by single-spot test (see Section 6). Heavier coatings are often used for culvert sheets which, according to ASTM Specification A 444, are required to carry a minimum 610 grams per square metre (2.00 ounces per square foot) coating when tested by the triple-spot-test method.

Galvanized sheet gage numbers and their weight equivalents are given in Table 38—I; thickness equivalents are also shown in this table. Table 38—II shows the coating designation in grams per square metre or ounces per square foot and the specified minimums for coatings heavier than G 90 galvanized coatings.

GENERAL QUALITY DESIGNATIONS

Sheets of G 90 coating designation are suitable for bending and moderate forming requirements. The coating and steel must be capable of withstanding standard bend tests (see Section 6). The adherence of the coating must be sufficient to withstand bending without flaking when tested in accordance with ASTM Specifications A 525 and A 526.

For superior resistance of the base metal to atmospheric corrosion, copper bearing steel or low alloy steel containing small amounts of chromium and/or other elements can be used as base metals for hot-dip galvanizing.

FORMED SHEETS FOR ROOFING AND SIDING

Roofing and siding are produced either flat or in the form of "Corrugated Sheets," "V-Crimped Roofing," "Corrugated Roll Roofing," or regular "Roll Roofing." In addition, corrugations produced by roll forming and based on rectangular and trapezoidal profiles were developed in the 1960's; the design of these profiles provides a greater covering span of lighter gage sheets having much better rigidity and structural integrity for modern metal building designs.

V-Crimped sheets, which were formerly widely used, have largely been replaced by rectangular and trapezoidal sheets produced with modern roll-forming equipment. Limited amounts of V-Crimped roofing sheets are still produced on older equipment.

SECTION 2

METALLURGICAL FEATURES OF THE HOT-DIP GALVANIZING PROCESSES

Hot-dip coatings are produced on steel objects by passing the properly prepared base metal through a bath of molten coating metal. Zinc, aluminum, terne

metal (a lead-tin alloy), and zinc-aluminum alloy are commonly used as coating metals. As in all coating processes, the surface of the base metal must be clean to insure a continuous coating. This is even more critical in aluminum or zinc-aluminum alloy coating operations, for example, than it is for galvanizing. In addition, successful hot-dipping requires the surface of the molten bath to be kept clean, particularly at the entrance to the bath. This is because oxide films of the coating metal at this point may be picked up by the surface of the entering base metal and interfere with the wetting of the steel by the molten coating metal.

The surface of the steel should not be contaminated with any material; the contaminant must be removed by acid pickling, alkaline cleaning or other preparatory treatment prior to introduction in the zinc bath. To further facilitate the production of a continuous adherent zinc coating, (1) suitable fluxes carried on the surface of the bath and/or applied to the surface of the steel just prior to entry into the bath are used; or, in the case of operations which include an annealing or other heat-treating step immediately ahead of the galvanizing pot, (2) the heated strip is kept under the protection of a reducing furnace atmosphere, until it passes below the surface of the molten zinc.

For many years, galvanized sheets were produced by passing single sheets of steel in close succession through a coating pot; the operations are described in detail in Section 3 of this chapter. However, with the development, after World War II, of wide strip mills capable of rolling to sheet gages, there has been a general transition to processing sheet products in strip form. In such operations, the steel is unwound from coils and passes as a continuous strand through the processing equipment. The product may then be recoiled or cut into individual sheets as desired at the end of the line. Processing in strip form is referred to as "continuous processing" to distinguish it from similar operations involving the treatment of individual sheets. Continuous processing permits many operations formerly conducted on separate pieces of equipment to be combined into one continuous processing line. In addition,

certain problems inherent in operations on individual sheets are precluded when the steel is processed in strip form. For example, in the hot-dip coating of sheets, the necessary gap between successive sheets causes a considerable variation in the weight of coating applied to the leading and trailing ends of the sheet. The absence of this gap in operations on material in strip form therefore results in a more uniform coating and consequently affords better use of a given quantity of the coating metal. Continuous-type operations, however, require a very high investment in equipment. A detailed description of various continuous hot-dip galvanizing processes is given in Section 4 of this chapter.

At and above its melting point, zinc readily forms intermetallic alloy compounds with iron. In hot-dip galvanizing, these alloys form at the interface between the base metal and the coating metal and ultimately constitute an important portion of the finished coating, about 10—25 per cent of the total coating thickness. After solidification, the coating consists of an outer layer composed primarily of the metal from the coating bath, and inner layers, generally termed "alloy layer," consisting of intermetallic phases of iron and zinc, with the iron content of the alloy layer higher in proximity with the steel base.

The outer layer of a hot-dip galvanized coating solidifies as a cast crystalline structure, starting at scattered nuclei and developing into a more or less regular "frost-flower" or "spangled" pattern (Figures 38—1 and 38—2). The size of the spangles is influenced by the composition, surface texture, and prior treatment of the base metal, by the rate of cooling until the coating solidifies, by the composition of the coating bath, and by the coating weight. The zinc bath composition is the most important determinant of spangle shape and spangle boundary configuration.

The alloy layer serves as the metallurgical bond between the outer layer of relatively pure zinc and the

FIG. 38—1. Typical spangles on a sheet galvanized with aluminum-free spelter. Average coating weight: 336 grams per square metre (1.10 ounces per square foot). Actual size, unetched.

FIG. 38—2. Typical spangles on a sheet continuously galvanized with aluminum-bearing (nominal 0.15 per cent Al) spelter. Average coating weight: 403 grams per square metre (1.32 ounces per square foot). Actual size, unetched.

steel base. Complete coverage of an alloy layer on the steel surface is necessary, but an excessively thick layer of the brittle intermetallic-alloy compounds could cause poor coating adherence.

Adding aluminum to the zinc bath in small amounts (0.10 to 0.25 per cent) suppresses the alloy formation and produces coatings of improved adherence. Even with a proper aluminum content, however, a clean

steel surface leading to a metallurgical bond with good integrity is essential for good coating adherence.

Various investigators have studied the equilibrium relationship between iron and zinc at different temperatures. The constitution diagram as developed by J. Schramm (1938) is shown in Figure 38—3. The composition and lattice structure of each phase indicated on the diagram are shown in Table 38—III.

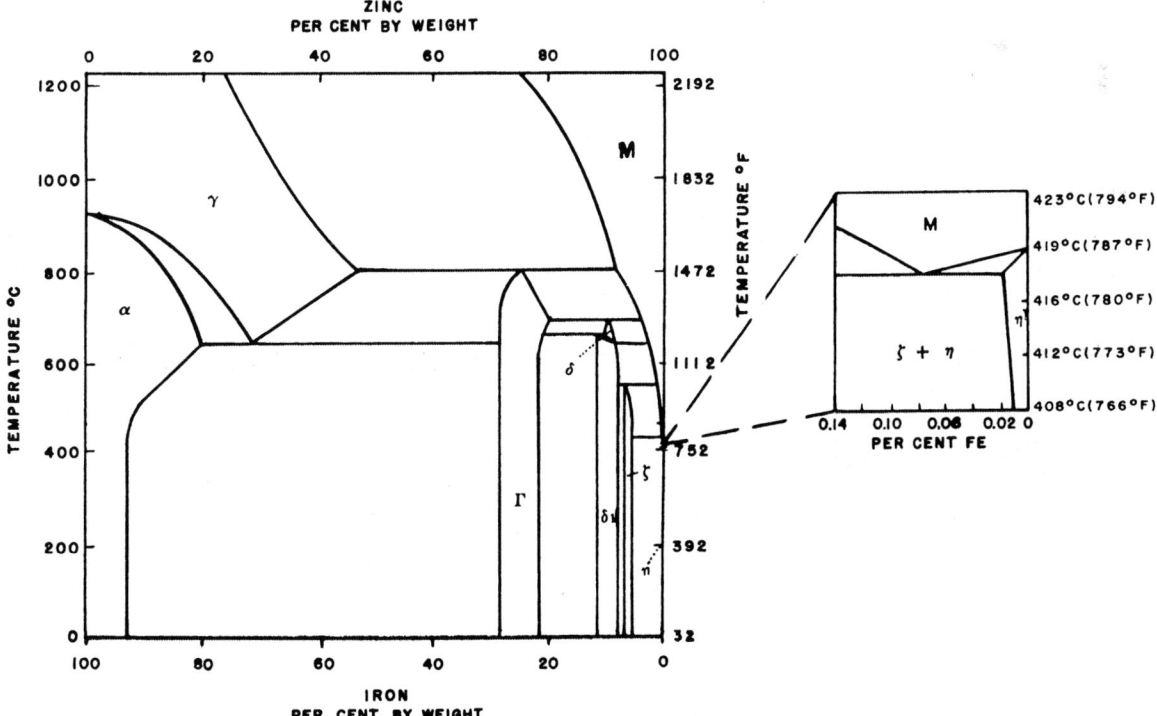

FIG. 38—3. The iron-zinc constitution diagram, according to J. Schramm.

Table 38—III. Phases in the Iron-Zinc Constitution Diagram

| Phases | X-Ray Formula | Limits of Composition, (Per Cent Fe) | | Space Lattice |
		Atomic	Wt.	
Alpha (α) Fe	FeZn		80-100	Body-Centered Cubic
Gamma (γ) Fe	FeZn		55-100	Face-Centered Cubic
Capital Gamma (Γ)	Fe_5Zn_{21} or Fe_3Zn_{10}	23.2-31.3	20.5-28	Body- Centered Cubic
Delta₁ (δ_1)	$FeZn_7$	8.1-13.2	7-11.5	Hexagonal
Delta (δ)	$FeZn_7$	8.1-11.5	7-10	Monoclinic
Zeta (ζ)	$FeZn_{13}$	7.2-7.4	6-6.2	Hexagonal
Eta (η)	Zn		Max. 0.003	Close Packed

Intermetallic compounds present in hot-dipped galvanized coatings produced under a variety of conditions are indicated in Figure 38—4 through 38—7. This series of illustrations shows that the relative amounts of the various constituents of the alloy layer can be varied by varying time and temperature of immersion. The composition of the steel also, particularly the silicon content, can have a noticeable influence on the alloy layer. Aluminum has the predominant effect on the character of the alloy layer. Figure 38—8 is a photomicrograph at 1000 diameters showing the cross-section of a galvanized coating produced with aluminum-free spelter.

Figure 38—9 is a similar photomicrograph of a continuously galvanized coating produced in aluminum-bearing (nominal 0.15 per cent Al) spelter. In the latter instance, almost all the coating is the eta phase of zinc.

COATING METAL USED IN HOT-DIP GALVANIZING

There are various common grades of zinc used for hot-dip galvanizing. However, the predominant grade is Prime Western zinc, which contains relatively low levels of lead (0.05 - 0.30 per cent). Generally, a low-

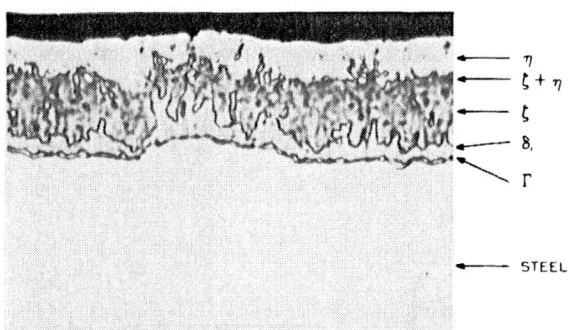

FIG. 38—4. Photomicrograph of commercial sheet galvanized coating. Average coating weight, 192 grams per square metre (0.63 oz. per sq ft). Magnification: 1000X.

FIG. 38—5. Photomicrograph of experimental galvanized coating formed by immersing sheet steel for one minute in molten zinc at 450°C (840°F). Magnification: 500X.

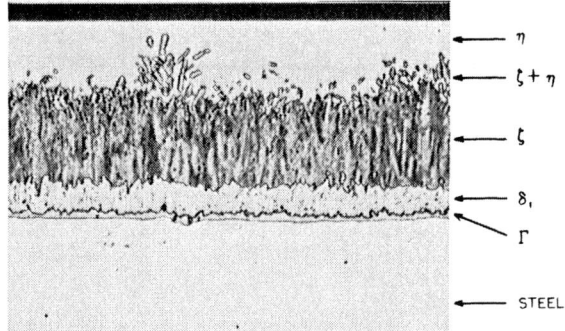

FIG. 38—6. Photomicrograph of experimental galvanized coating formed by immersing sheet steel for five minutes in molten zinc at 450°C (840°F). Magnification: 500X.

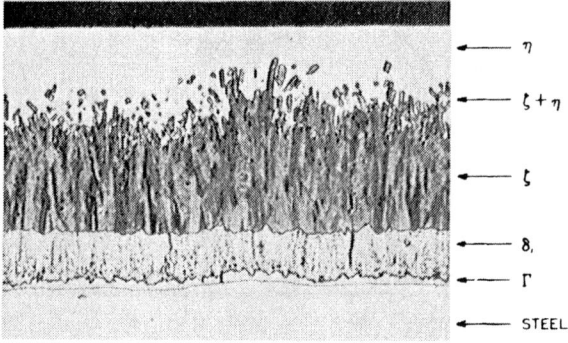

FIG. 38—7. Photomicrograph of experimental galvanized coating formed by immersing sheet steel for ten minutes in molten zinc at 450°C (840°F). Magnification: 500X.

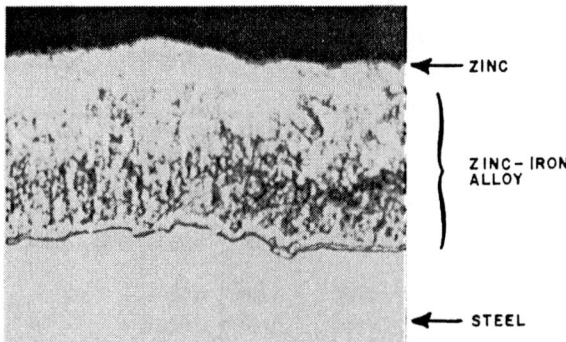

FIG. 38—8. Photomicrograph showing thickness of the alloy layer in the coating on a commercial sheet galvanized with aluminum-free spelter. Average coating weight 345 grams per square metre (1.13 oz. per sq ft). Magnification: 1000X.

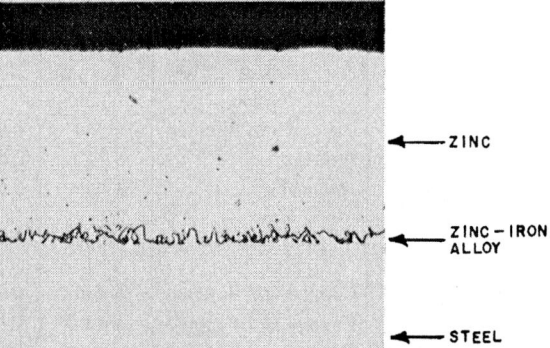

FIG. 38—9. Photomicrograph showing thickness of the alloy layer in a continuously galvanized coating produced with aluminum-bearing spelter (nominal 0.15 per cent Al). Average coating weight 357 grams per square metre (1.17 oz. per sq ft). Magnification: 1000X.

lead coating is desirable for hot-dip galvanizing applications that require low solidification shrinkage (low spangle relief) at spangle boundaries. In addition to the pronounced effect of lead on spangle characteristics, other elements, such as antimony, cadmium, and tin, have a pronounced effect on the zinc spangle. However, these elements are generally not used in hot-dip galvanizing.

The primary addition made to all hot-dip continuous galvanizing baths is aluminum ranging from 0.10 - 0.25 per cent. Generally, zinc baths are maintained at about 0.15 per cent for most hot-dip galvanizing processes; slightly lower levels (0.10 - 0.12 per cent) are used in flux processes (such as Process C described in Section 4 of this chapter) because aluminum in the bath can react with the flux to form $AlCl_3$. The main purpose of aluminum in the bath is to minimize the thickness of the intermetallic layer and thus provide coating adherence sufficient to prevent the coating from flaking in severe forming operations in which the coating is highly strained. The aluminum restricts alloy later growth by preferentially reacting with the steel (as it enters the bath) to form a thin surface layer of $FeAl_3$ and/or Fe_2Al_5. The alloy layer growth is inhibited because the interdiffusion (mutual diffusion) of iron and zinc is retarded by the presence of the iron-aluminum layers. With long residence times in the bath, the iron aluminum layers will be converted to iron-aluminum-zinc compounds through which diffusion will become very rapid.

A small amount of **iron** is always present in a hot-dip galvanizing bath. The excess iron (above the solubility limit) gradually settles to the bottom of the pot as **dross,** $(FeZn_{13})$. However, when excess amounts of aluminum are maintained in the galvanizing bath, the iron may react and take the form of an iron-zinc-aluminum alloy. This alloy, because of a lower density, will gradually rise to the surface of the bath (**top skimmings**), and can then be easily removed.

STEELS USED FOR HOT-DIP GALVANIZING

Hot-dip galvanized coatings may be successfully applied to a variety of ferrous metals, ranging from cast iron to low-carbon and alloy steels. In the coating of sheet and strip, the bulk of the production is from steel made by the low-carbon basic oxygen process. Various grades of low-carbon and low-alloy steels are processed on continuous galvanizing lines. The low-carbon grades include rimmed, capped, aluminum special-killed, silicon semi-killed, rephosphorized, renitrogenized, rimkilled process, and continuous cast steels. For a given quality of sheet, more than one composition of low-carbon steel may be used to achieve equivalent mechanical properties. For example, Structural Quality A 446 Grade D sheet may be produced at local plant option using any of the following three steels: rephosphorized, renitrogenized, or silicon-semikilled. Different qualities of galvanized sheet and strip are produced using selected annealing cycles and steels to obtain a wide range of mechanical properties. The four predominant types of galvanized products are described below.

(1) Commercial Quality (CQ) sheet and strip are produced chiefly with rimmed, capped, and continuous cast steels. CQ sheets are excellent for bending, forming, and moderate drawing, and can be ordered to ASTM Specification A 526.

(2) Lockforming Quality (LFQ) sheet and strip are generally produced with rimmed steel. LFQ sheet and strip are excellent for continuous forming in lock seams and can be ordered to ASTM Specification A 527.

(3) Drawing Quality (DQ) sheet and strip are generally produced with select rimmed or killed steels. The most severe drawing applications may require the use of the post-box-annealing treatment after galvanizing. The production of Drawing Quality steel involves the exercise of close control in the selection of steel and processing to assure performance of the material in press drawing operations. Drawing Quality sheets are made from strand-cast (continuous-cast) and rimmed or killed steel and are described in ASTM Specification A 528.

(4) Drawing Quality Special Killed (DQSK) sheet and strip are produced with aluminum special-killed steel. The most severe DQSK applications may

Table 38—IV. Composition of Steels Used for Galvanizing

Quality	Steel Grade	ASTM Spec and Grade	Composition, per cent (maximum or range)					
			C	Mn	P	S	Si	Other
Commercial (CQ)	Capped and Rimmed	A 526	0.05/0.15	0.60	0.035	0.035	0.010	—
Lockforming (LFQ)	Rimmed	A 527	0.05/0.10	0.50	0.035	0.035	0.010	—
Drawing (DQ)	Rimmed	A 528	0.03/0.10	0.50	0.025	0.020	0.010	—
Drawing Special Killed (DQSN)	Aluminum Special Killed	A 642	0.03/0.10	0.50	0.025	0.020	0.010	0.03/0.08 Al
Structural (Physical) PQ	Capped and Rimmed	A 446A	0.05/0.15	0.50	0.025	0.035	0.020	—
	Capped and Rimmed	A 446B	0.10/0.15	0.60	0.025	0.035	0.020	—
	Capped and Rimmed	A 446C	0.10/0.15	0.60	0.025	0.035	0.020	—
	Rephosphorized	A 446D	0.10/0.15	0.60	0.09/0.15	0.035	0.020	—
	Renitrogenized	A 446D	0.20/0.25	0.40/0.90	0.025	0.035	0.020	0.009/0.015N
	Silicon Semi-Killed	A 446D	0.15/0.25	0.60/0.80	0.025	0.035	0.10	—
	Capped	A 446E	0.05/0.10	0.50	0.025	0.035	0.020	—

require the use of a post-box-annealing treatment after galvanizing. For stringent drawing and forming applications, DQSK sheets can be ordered to ASTM Specification A 642. The limitations described above for "Drawing Quality" also apply to this grade of material.

In addition, recent steelmaking technology permits production of very low carbon steels (0.005 - 0.020 per cent). These steels are increasingly being used to produce galvanized sheets for critical formability applications.

Compositions of steels used for galvanizing are shown in Table 38—IV.

MILL TREATMENT OF STEEL PRIOR TO HOT-DIP GALVANIZING

Various processing practices have been used to produce steel for **single sheet hot-dip-galvanizing**. Continuous-hot-rolled sheets were either hot rolled in strip form or hot rolled and pickled in strip form before they were sheared to cut lengths for the galvanizing operation. Cold-reduced sheets were either hot-rolled, pickled, cold-reduced, and recoiled; or they were hot-rolled, pickled, cold-reduced, recoiled, and sheared to cut-length sheets before box annealing. Alkaline cleaning was often used after cold reduction to remove rolling lubricants. Material that was box annealed in coils was temper rolled and sheared before pickling for galvanizing, and material that was box annealed in sheet form was temper rolled before pickling for galvanizing.

For **continuous galvanizing,** coils are also prepared in a number of ways. Thus, they may be hot rolled and pickled; hot rolled, pickled and cold reduced; or hot rolled, pickled, cold reduced, box annealed and temper rolled before galvanizing. Again, alkaline cleaning may be used after cold reduction to remove rolling lubricants. For more detail on the pickling process, see Chapter 33, "Manufacture of Hot-Strip Mill Products."

SPECIAL FINISHES

In addition to the **regular spangled** finish produced in sheet or strip galvanizing, provision may be made in

the processing facilities to accommodate the production of special finishes. When the relatively large size of regular spangles, resulting from normal solidification of the zinc naturally cooled with ambient air, is not desired, a **minimized-spangled** sheet can be produced by treating the molten zinc on the strip with atomized water, air water, or air-steam spray which restricts the size of the spangles. Also, a recent technological advance allows the impingement of fine zinc powder onto the molten zinc coating in the cooling run-out tower to self-nucleate a very fine-spangled or non-spangled coating.

Galvannealed coatings, which are composed entirely of zinc-iron intermetallic alloys, possess a gray matte finish of relatively low reflectivity. These sheets may be produced by passing the zinc-coated sheet, after it leaves the coating pot, through a heated chamber to alloy the zinc completely with the base metal. They have generally good paint-adherence properties without additional surface preparation and can withstand moderate forming. In addition, galvannealed coatings are more weldable than are galvanized coatings.

Differentially coated sheets are produced with a specific weight on one surface and a significantly lighter weight on the other surface. The lighter coating may be composed entirely of iron-zinc alloy, if specified. If this is the case, the side with the heavier coating is considered "galvanized" while that with the lighter coating is called "alloyed". To meet the trade demand for **one-side coated sheets** for automotive body parts, sheets with the equivalent of a G 90 zinc coating on one surface and no coating on the other surface are being produced commercially by both the hot-dip and the electrolytic processes.

Sheets given a phosphate pre-treatment may be produced by processing in line or as a separate supplementary processing operation to apply a thin film composed largely of zinc phosphate on the sheet surfaces. The purpose of this pre-treatment is to prepare the steel for painting. Phosphatized sheets have outstanding paint-adherence properties, although some finishes may require special primers. In addition to phosphatized sheets which are meant for applications involving painting, large tonnages of galvanized sheet in coil form are continuously painted on coil coating lines for applications involving high finish such as metal buildings or appliances.

Temper rolling may be employed to produce an "Extra Smooth" sheet for applications involving critical end uses when galvanized sheets are required to have a higher degree of smoothness than is obtained with regular as-coated commercial quality, or drawing quality product. Examples are for exposed auto body applications and for critical painted sheet applications. Also, in-line surface conditioning may be used to produce a smooth, non-spangled, high-luster surface.

Galvanized sheets shipped and stored in stacks of cut-length sheets or coils are subject to white, grey, or black stains when water is allowed to remain in contact with surfaces in the interiors of stacks and coils for any appreciable period of time. This condition is usually designated as "humid-" or "wet-storage stains." Various chemical films and oil films have been found to be effective in retarding the formation of wet storage stains, but, even with filming treatments, wet storage stain is a hazard when sheets or coils are allowed to remain wet. Common sources of moisture may be rain water, melted snow, or condensation, the latter resulting from temperature changes.

Good packaging, handling and adequate warehousing facilities have been found to be essential requirements for preserving the original finish on galvanized steel until the material is used.

SECTION 3

HOT-DIP SINGLE SHEET GALVANIZING

The hot-dip sheet-galvanizing process is used very little in the United States. However, the sheet-galvanizing process is used in other countries. Following is a brief description of how the process was carried out in this country.

Pickling for Sheet Galvanizing—Pickling for sheet galvanizing was usually conducted as a batch operation in stationary tubs provided with an agitator. This operation sometimes was conducted as a continuous process in equipment provided with a sheet conveyor and means for electrolytic acceleration.

Very light pickling requiring only a short-time exposure to the pickling solution was suitable for products, such as roofing and siding, that required little mechanical deformation. Deeper penetration of the base metal was generally necessary when forming requirements were severe.

Following the pickling operation, a thorough water rinse, with or without an alkaline neutralizing dip, was employed to remove iron salts and residual acid. A prolonged immersion in boiling water was sometimes used to minimize hydrogen embrittlement.

Equipment for Sheet Galvanizing—The equipment used for sheet galvanizing (Figure 38—10) consisted of mechanical facilities for transporting cut-length sheets successively through acid washing, fluxing, hot-dipping

FIG. 38—10. A sheet galvanizing line in operation, showing the hooded setting which contained the galvanizing machine immersed in the pot containing molten zinc, the spangle conveyor, and control station.

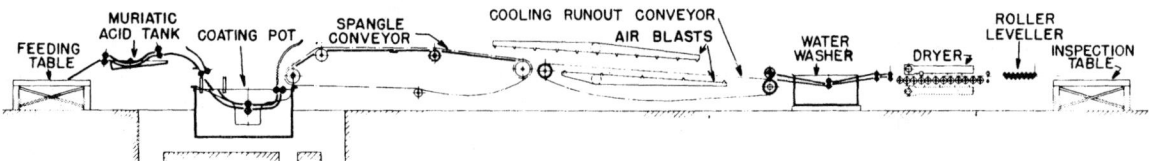

FIG. 38—11. Schematic side elevation of a conventional sheet galvanizing line.

and cooling operations. The coating bath itself was contained in a heated low-carbon steel vessel or **pot.** A framework or **rigging** with suitable **entry feed rolls, sheet guides,** driven **bottom pinch rolls,** and driven **exit rolls,** was suspended in the bath in such a manner as to completely submerge all but the entry rolls, part of the exit rolls, and part of the supporting framework. A baffled section or **flux box** at the entry end contained a floating fused-chloride **flux** prepared from sal ammoniac.

General Arrangement and Operation of a Sheet-Galvanizing Line—A typical sheet-galvanizing line consisted of a **feeding table,** a muriatic **acid tank,** a coating **pot,** a **spangle and cooling conveyor,** a **water rinse tank,** a **dryer,** a **roller leveler** and an **inspection table** (see Figure 38—11). In operating the line, pickled sheets were either fed dry from a feeding table or wet from water boshes directly into a set of driven rubber pinch rolls. The sheets passed forward through a shallow muriatic-acid bath and a second set of rubber pinch rolls into steel intake rolls that forced them downward into the flux box of the coating pot. The sheets then passed into the coating bath downward to the driven bottom rolls, and upward through exit rolls, the latter being located at the bath surface. As they emerged from the exit rolls the sheets may have been blasted

with air, sulphur-dioxide fumes, or powdered sal ammoniac to produce the surface finish desired. To control the amount of coating applied, the level of the bath may have been adjusted or the pressure against the exit rolls may have been regulated.

Endless belts for conveying coated sheets away from a galvanizing pot were generally made of coarse woven-wire netting. The meshes of the netting contacted the sheet in such a manner as to cause rapid solidification at regularly distributed points. These provided nuclei for crystallization and served, to some extent, as a means for spangle control. A cooling runout conveyor, usually provided with air blasts, followed the spangle conveyor in a typical installation. Cooled sheets from the conveyor were roller leveled in line before inspecting and piling. In some coating lines, a water-rinsing and drying operation preceded the leveler. At the delivery end of the coating line, the sheets were inspected, counted, stenciled, and stacked into lifts, after which they may have been branded and bundled or otherwise prepared for shipment.

If a special finish, such as a phosphatized finish, were applied to the galvanized sheets, the sheets were then delivered from the galvanizing line to further processing units before shipment.

SECTION 4

CONTINUOUS (STRIP) HOT-DIP GALVANIZING

General Arrangement and Operation of Continuous Galvanizing Lines—Processing steel in the form of a continuous strand unwound from coils requires more elaborate equipment than does processing cut-length sheets, but grouping several manufacturing steps in one operation makes possible more economical production and better control of product quality. Thus, in the early 1980's, cut-length steel sheet is rarely galvanized using hot-dip processes.

Continuous galvanizing lines of several designs have been in commercial use in the early 1980's. The four most widely used designs are:

(1) Anneal with flame cleaning (Process A)
(2) Anneal with liquid cleaning (Process B) (See Figures 38—12 and 38—13)
(3) No anneal with liquid cleaning and flux (Process C)
(4) Anneal with no liquid cleaning and no flux (Process D)

Although different steel producers may use variations on these processes, each process involves five basic

steps:

(1) Cleaning, to prepare the surface for coating
(2) Annealing, to soften the steel for good formability
(3) Coating with zinc, to make the steel resist corrosion
(4) Chemical treating, to protect the coating from storage stains
(5) Working, to insure uniform forming

Most or all of these steps are carried out "in line" in the galvanizing process.

This section will describe the general arrangement of each of the four types of continuous galvanizing lines.

Cleaning— The cleaning step prepares the steel surface to be galvanized for coating. In the entry section, the steel first passes to the cleaning equipment. The steel strip is fed from the uncoilers (pay-off reels) through pinch rolls and a shear to an electric-resistance welder where the tail end of the coil being processed is attached to the head end of the new coil. Following the welder, the strip passes through a bridle to the accumulator which takes up strip to permit the entry end of

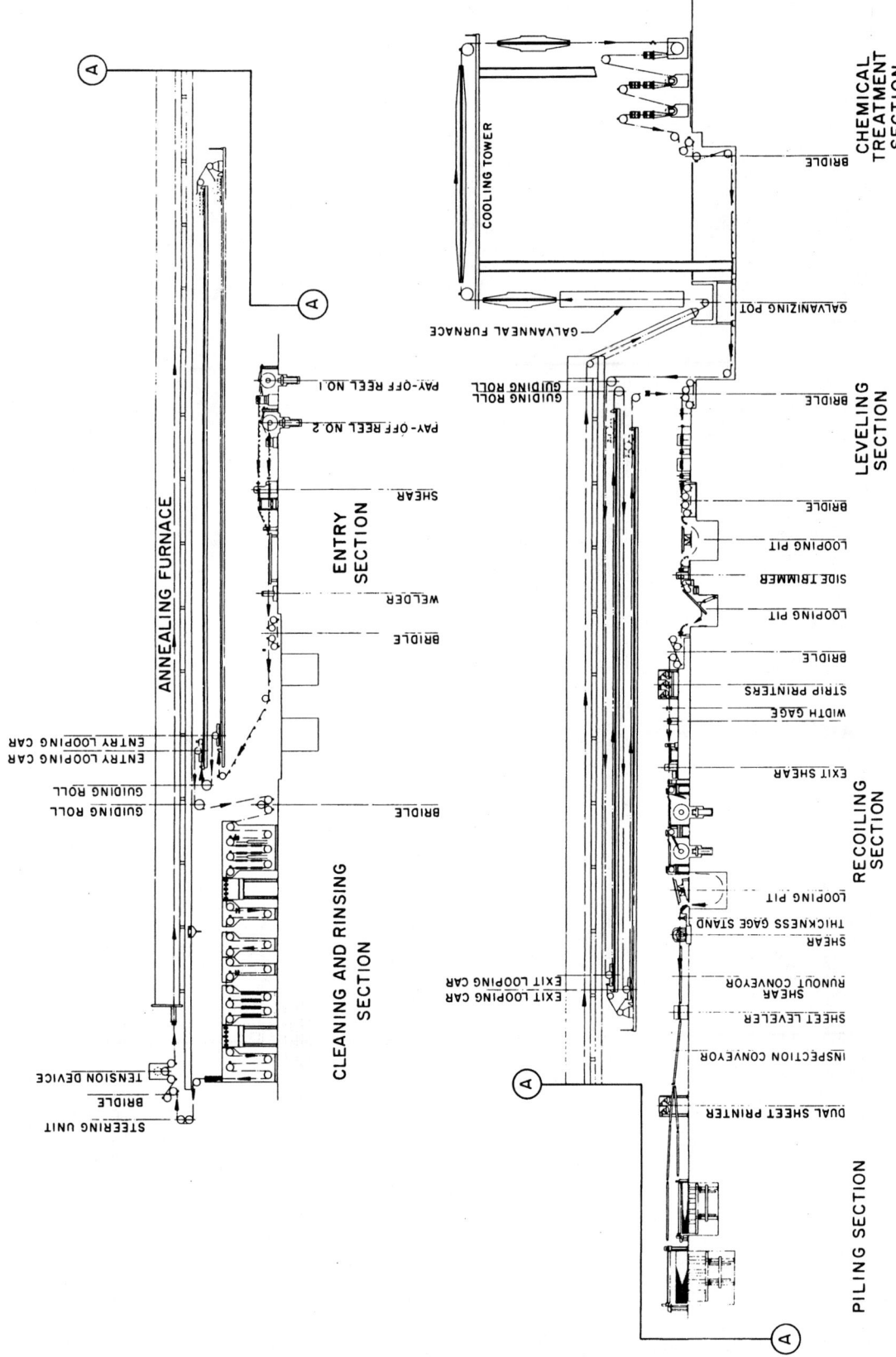

FIG. 38—12. Schematic elevation of a modern continuous-galvanizing line is an example of process A or B. Strip in coiled form enters the line at the pay-off reels in the upper part of the drawing, and may be either coiled or cut into sheets after coating on the finishing-end equipment in the lower part of the drawing.

Fig. 38—13. General view of the continuous galvanizing line shown schematically in Figure 38—12.

the line to be stopped for the joining operations without affecting operation of the process section of the line. Instead of car-type accumulators (loopers), which are most commonly used, some lines employ looping pits or towers.

The strip leaves the looper and enters the cleaning section. In Process A, the strip is flame oxidized to remove surface oils. In Process B, cleaning consists of electrolytic cleaning tanks, scrubbers, hot-water rinse tanks, and hot-air drying. Cleaning in Process C consists of acid pickling to remove any surface residues, an alkaline dip to remove surface oils, and flux to make the steel reactive with the molten zinc. In Process D, a reducing-flame furnace cleans the steel of oil and simultaneously heats the steel to a preheat (full-hard products only) or subcritical annealing temperature. Thus, cleaning is not a discrete step in the process, but rather, is accomplished at the same time as annealing.

Annealing— Steel is annealed before hot-dipping to soften it for good formability. In Processes A, B, and D, annealing is carried out in the galvanizing line in a non-oxidizing atmosphere to make the surface reactive with molten zinc. The cleaned strip passes through steering rolls, a bridle, and tensioning rolls into the annealing furnace. The furnace consists of heating and holding sections that are heated by gas-fired radiant tubes, electric resistance elements, or direct-fired burners. The steel then moves into the cooling sections where it is cooled to about 482°C or 900°F. The heat-treated strip passes from the furnace through a discharge chute into the coating bath without being exposed to air.

In Process C, the steel must be box annealed before it enters the line. The annealing furnace is obviated, and cleaning, pickling and fluxing do not require a protective atmosphere; the fluxed strip passes directly into the coating bath.

The remaining three steps in hot-dip galvanizing are common to all four processes.

Coating with Zinc— In this step, strip steel is coated with zinc to protect it from corrosion in service. The coating step is the most important part of the process, and is essentially the same for all types of continuous hot-dip galvanizing lines. Here, the strip is directed around sink rolls and out of the molten zinc bath. Bath residence times range from one to ten seconds.

The coating bath is usually contained in a refractory-lined, induction-heated furnace. An induction-heated pre-melt furnace supplies molten zinc to the coating bath. Some lines use electric or gas-fired melting and coating units.

After leaving the coating bath, coating weight (thickness) on the strip is controlled by the use of a "gas knife," which consists of a stream of gas, usually air or steam, directed at both sides of the strip as it emerges from the coating bath. The wiping gas is generated in blowers or compressors and is then directed to a main header. From the header, the wiping gas is further directed through a rectangular slit in the form of a compact uniform stream. The pressure of the gas and the positioning of the knife relative to the strip surface are controlled to give the desired weight of coating for the speed used. In special cases, nitrogen gas is used to produce a smoother finish.

After passing through the gas knives, strip can be galvannealed in a vertical gas-fired furnace. In this optional step, the coating is converted from zinc to a zinc-iron alloy. The coated strip, after passing through this part of the line, then travels through the cooling tower and a water-quench tank.

Chemical Treating— Strip that has been galvanized can be treated chemically to protect the surface from humid- or wet-storage stains. For most orders, the galvanized sheets are chemically treated to resist staining. Various passivation treatments are used to make the galvanized surface less reactive with moisture. All of these treatments are based on protection with chromium compounds. The treatments are produced by dipping the galvanized sheet in or spraying it with water-base solutions containing hexavalent chromium compounds in the form of chromium salts, chromic acid, or combinations of the two. The solution also contains film formers, activators or catalysts, and wetting agents. The treatment provides both a physical barrier film and a conversion film that inhibits corrosion of the zinc surface.

Chemical treating is not always the best way to protect steel from humid-storage stains. In some cases, oiling is used for customers who plan (after removing the oil) to paint the sheet; for these customers, a chemical treatment on the surface might detract from the paint's adhesion. However, if the passivation and prepaint treatments have been designed to be compatible, normal chemical treating can be used in the galvanizing line.

Coating thickness is continuously monitored on the line using radiation-type gages. Use of the coating-thickness gage in conjunction with the gas knives, a computer, and data logger provides a completely automatic, closed-loop control of the gas-knife operation with significant savings in zinc-utilization costs. The strip moves through an exit accumulator that permits the process section to continue operating at times when the delivery section must be stopped.

Working— Sheet steel in the as-annealed condition exhibits "discontinuous yielding." That is, as the sheet is bent, formed, or passed around small rolls resulting in stresses greater than the yield stress, the steel does not plastically deform uniformly. Instead, it deforms in localized areas resulting in surface imperfections known as stretcher strains, coil breaks, "Lüders lines," flutes, or chevrons. Cold working the steel sheet by leveling or temper rolling eliminates (for a finite period of time) the discontinuous behavior at the yield stress, and thus eliminates the breaking tendency. It does this by introducing into the steel "mobile dislocations," features of the crystalline lattice, which facilitate uniform forming behavior. With increasing storage or "aging" time after cold working, these dislocations become pinned by reacting with carbon and nitrogen in the steel, and discontinuous yielding behavior returns. Thus, the amount of cold working must be sufficient to maintain uniform forming behavior for the aging time before the customer forms the product. However, the amount of cold working should not be greater than necessary, because this cold working also reduces the ductility of the steel which could result in the steel fracturing during forming.

Levelers in continuous-galvanizing lines cold work the galvanized-sheet product to provide uniform forming. These levelers also serve to flatten any shape deficiencies (wavy edges, crossbow, center buckles, etc.) in the galvanized sheet.

A number of different machines have been devised to level the strip. All of these develop the necessary deformation stresses with bending, tension or a combination of tension and bending. While simple roller leveling uses simple bending stresses, stretcher leveling is achieved by using two sets of bridle rolls to apply tension to the strip. In stretch-bend levelers, a number of strip deflector rolls are located between the bridles to introduce a combination of bending and tensile stresses into the strip.

In addition to the levelers, the delivery section includes a printer, a width gage, an oiling unit, and tension reels for coiled product. Coils may be a maximum of 1829 mm (72 inches) in diameter, weighing up to about 28 000 kilograms (62 000 pounds).

If the product is to be sheared into sheets, it passes over the coiling stations through a thickness gage into a flying rotary shear. The sheared sheets pass through a sheet leveler, inspection conveyor, and a printer to the sheet pilers. Sheets may be sheared to any desired length between 1016 mm (40 inches) minimum and 6096 mm (240 inches) maximum.

Processes A, B, and D, as described above for galvanizing, can also be used to coat steel with Galvalume or aluminum. Terne coated sheets are produced by a flux process, similar to Process C above.

SECTION 5

ELECTROGALVANIZING PROCESSES

The process of electrogalvanizing contributes a small but growing portion of each year's total galvanized steel output. While overall reported shipments for galvanized steel dropped almost 8 per cent during 1977-1980, shipments of electrogalvanized steel increased almost 75 per cent. This increase reflects demand in the automobile industry for a product with greater corrosion resistance.

In the process of electrogalvanizing, the furnaces, galvanizing pot, and cooling tower of the hot-dip process are replaced by a series of electrolytic cells through which steel strip passes. Electrical current, in each of the cells, flows through a zinc solution from anode to conductor roll, bonding zinc to the steel strip. Anodes are of two kinds: soluble anodes made of zinc slab supplying zinc for the solution; insoluble anodes made of lead, lead alloys, or platinized titanium. Processes using soluble anodes require replacement of the anodes as they are depleted; processes using insoluble anodes require continuous replenishment of the solution with zinc-bearing chemicals, such as zinc sulfate. The off-line production of zinc sulfate by mixing zinc with sulfuric acid releases large quantities of hydrogen gas. A zinc chloride solution is used with the soluble anode process.

Because electrogalvanizing bonds zinc to base metal electrochemically, the coating which results is unlike the zinc coatings that are produced by hot-dip methods. Electrogalvanized coatings lack the spangled surface characteristic of hot-dip sheet; in addition, their smoothness and uniformity permit lighter-weight zinc and zinc-alloy coatings that are reliable, even in very thin applications. Electrogalvanizing also permits zinc-coating of any grade of steel, including grades adversely affected by the high temperatures of the hot-dip method. (Table 38-V.)

Through the 1960's, the construction industry provided the chief market for electrogalvanized sheet. However, in the early 1970's, automobile manufacturers in the United States expressed interest in obtaining one-side galvanized sheet for use in body parts exposed to road salt and other corrosive agents. In 1980, several steel companies supply one-side galvanized sheet made by various proprietary continuous-strip processes, including both hot-dip coating and electrogalvanizing. The coated side of one-side sheet is used for underbody surfaces requiring greatest corrosion resistance, and the uncoated side provides the appearance of cold-rolled steel and good paintability for exterior surfaces. Automobile industry forecasts indicate a future demand for two-side galvanized sheet, with thick zinc coatings for the underbody and thin coatings of zinc or iron-zinc alloy for exterior surfaces.

World-wide, there are three major electrogalvanizing systems in use in 1984. Each features a distinctly different type of electrolytic cell: horizontal, vertical, and radial. The oldest operating electrogalvanizing lines, dating from the early 1940's, use horizontal cells. Steel strip passes through the cells, with anodes placed above and below it. The Nippon Steel Company Anode-Center Injection Cell (ACIC) is a more recently developed horizontal-cell system. Using insoluble anodes and a zinc sulfate bath, the ACIC system injects electrolytic solution into the cell at high velocity,

Table 38—V. Electrolytic Zinc Coating and Minimum Coating Test Limits*

Coating Class	Minimum Check Limit Triple Spot				Minimum Check Limit Single Spot			
	Decimal Equivalents		Coating Weight		Decimal Equivalents		Coating Weight	
	One Side		Total Both Sides		One Side		Total Both Sides	
	in.	mm	oz/ft²	g/m²	in.	mm	oz/ft²	g/m²
A	none	none	none	none	none	none	none	none
B	0.000065	0.00165	0.075	22.9	0.000060	0.00152	0.070	21.3
C	0.000140	0.00356	0.165	50.4	0.000125	0.00318	0.150	45.8

*From ASTM Specification A 591, published by the American Society for the Testing and Materials, Philadelphia, Pa.

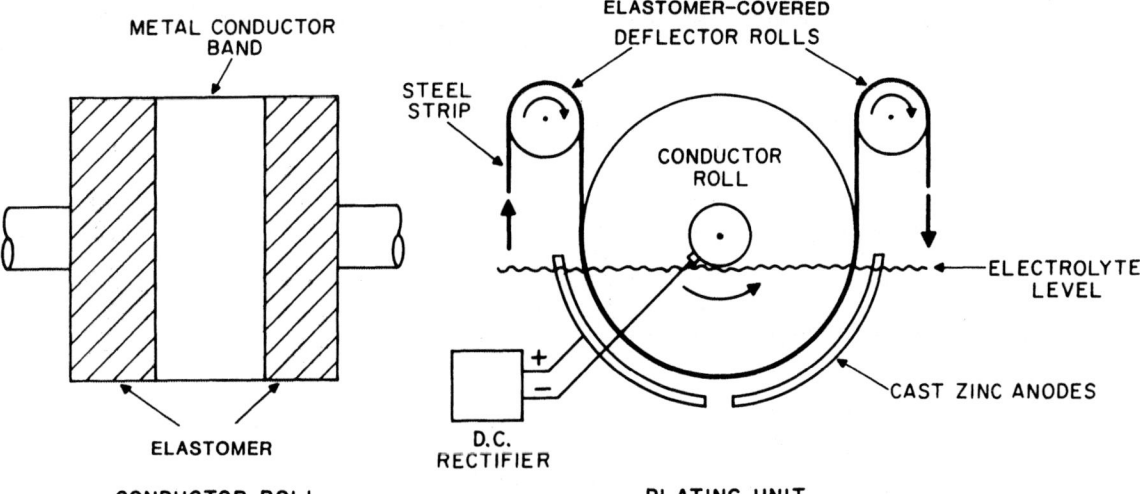

FIG. 38—14. Schematic arrangement of conductor roll and plating unit for USS CAROSEL one-side electrogalvanizing process.

counter to the direction of the strip. The system has been used to produce primarily light coatings, for one-side and two-side coated strip.

Electrogalvanizing lines using vertical cells are fewer in number. In this system, steel strip passed up and down between pairs of vertical anodes in a series of adjacent cells. The upward movement from each cell reduces "drag-through," the pulling of solution from the cell; however, the looping path taken by the strip as it dips into each of the cells requires a complex arrangement of conductor rolls, sink rolls, and hold-down rolls. Producing chiefly two-side coatings, vertical-cell systems in operation use either soluble or insoluble anodes with a zinc sulfate electrolyte.

The radial cell system, developed by United States Steel, achieves efficient power usage by effectively reducing the distance between anodes and strip. Developed for the production of one-side sheet, the Consumable Anode Radial One-Side Electrogalvanizing (CAROSEL) process uses a zinc chloride electrolyte. The CAROSEL process is also capable of two-side coating and permits different coatings to be applied on each side of the strip.

In the CAROSEL process, the steel strip is passed around large-diameter conductor rolls rotating partially submerged in the electrolyte, Figure 38—14. The electrical conducting surface of the rolls covers only the center portion, with the remainder of the roll surface being an elastomer. With the strip passing tightly around the rolls, the edges are sealed against the elastomer portion thereby preventing zinc deposition on the surface in contact with the roll.

A schematic of a CAROSEL line is shown in Figure 38—15. The steel is fed off one of the two uncoilers and welded to the trailing end of the previous coil. During the welding operation, the stored strip in the horizontal looping unit is utilized to minimize the need to slow down the line. Any slowdown required is computer controlled and is gradual. After welding, the loop car is over-speeded to refill it. The strip passes through the looping unit where guiding systems keep it centered, and then passes into the main tension bridle. It is then cleaned in a high-current-density unit utilizing a hot commercial alkaline cleaning solution. Current densities of about 130 amperes/dm² (1200 amperes/ft²) are used. Following rinsing, the strip is dip-pickled in hydrochloric acid and rinsed again.

The strip then enters the first of 18 plating cells. The conductor rolls, described previously, are 2.4 m (8 ft) in diameter. The total plating pass length is about 67 m (220 ft) for all 18 rolls. The strip moves from one plating roll to the next over elastomer-covered deflector rolls and is plated on both the down and up passes. The plating current is 28 000 amperes per pass, or a total of 1 008 000 amperes for the 18 plating cells (36 passes). The maximum plating voltage is 12 volts. Plating current density varies with strip width, because the maximum available current is used; it ranges from about 98 amperes/dm² (915 amperes/ft²) for the widest strip to about 179 amperes/dm² (1664 amperes/ft²) for the narrowest strip. Strip widths on this line range from 914 to 1664 mm (36 to 65.5 in.). Line speed is computer con-

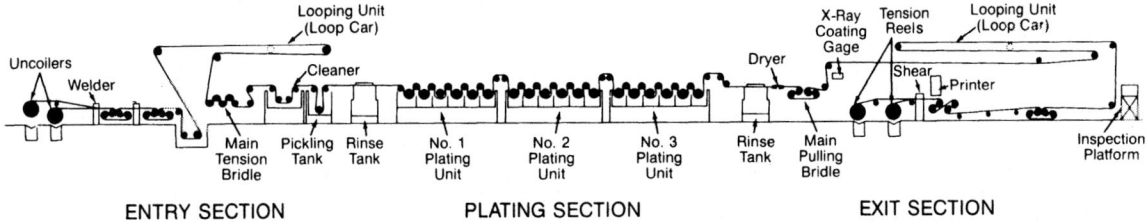

FIG. 38—15. Schematic arrangement of a one-side electrogalvanizing line at a United States Steel plant.

trolled to maximize production with available plating current. Maximum line speed is 183 m/min (600 ft/min).

The zinc for electrolyte replenishment is provided from cast zinc anodes supported on conducting bridges. The anodes are moved across the bridges at a rate determined by the line control computer. Approximately one half of the anode is consumed as it moves across the bridge. It is then removed and remelted for casting into new anodes.

After leaving the plating section, the strip is rinsed and dried and then enters the main pulling bridle. An x-ray coating gage continuously monitors the coating thickness and provides a feedback signal to the computer for line speed adjustments to maintain the specified thickness. The average coating thickness normally applied is 14.9 μm (0.00059 in.) with a minimum spot of 13.7 μm (0.00054 in.). This corresponds to the minimum coating thickness on G 90 hot-dip-galvanized sheet.

Following the x-ray gage, the strip enters the exit looping unit and allows coil transfer on the tension reels. The strip then passes through a visual inspection station, the exit bridles, a printer, and a snip shear, and is recoiled for shipment.

SECTION 6

TESTING GALVANIZED SHEETS

The **weight of coating** on a galvanized sheet is determined by weighing a 57.15-mm (2¼-inch) square test piece or a 64.49-mm (2.539-inch) diameter circular-disc test piece, dissolving the coating in an aqueous hydrochloric acid solution containing antimony chloride as an inhibitor, and reweighing the washed and dried test piece. The weight loss in grams is numerically equal to the coating weight in ounces per square foot of sheet; the weight loss in grams multiplied by 305.15 is numerically equal to the coating weight in grams per square metre of sheet. Triple-spot tests selected from the center and both edge locations are used for continuous hot-dip galvanized sheet product. It is customary to obtain edge spot tests 50.8 mm (2 inches) inward from the actual edge of the sheet or strip. A test to check the weight of coating is run on all hot-dip galvanized steel.

Five standard forming tests are used to evaluate the coating's adherence. They involve different formations of the steel that are characteristic of different applications. These tests are the bend test, bead test, Olsen cup test, lockseam test, and Gardner impact test. All involve deforming steel in a particular way to determine how well the coating adheres to the base metal. Flaking shows that the coating is not adhering well to the base metal; crazing (cracking) indicates that fractures are occurring down through the coating to the steel along characteristic weak lines in the zinc.

Bend tests are sometimes used for determining coating adherence and base-metal formability, especially in heavy-gage galvanized sheets. The bend, which may be made in a vise or punch press, is generally a 180° angle around a specified number of sheet thicknesses of the same gage as the test piece. The number of sheet thicknesses is identified by 0T for none, 1T for one, and so on. This test is run on all hot-dip galvanized sheet product. In some cases, a strip of cellophane tape is placed along the fold. The tape is pulled away to see if the coating adheres or comes away on the tape.

The **bead test** is commonly used for testing coating adherence, especially on light-gage material. A bead is used to deform the steel such that the sample contains a continuous ridge. The depth of the bead varies depending on the sheet thickness of the sample; lighter gage steel must be deformed with a deeper ridge for equivalent strain. The width, depth, and radii of the beads also vary somewhat depending upon the beading machine employed in making the test, although it is normal for the individual operator to have established definite standards for his testing operation. In conducting the bead test, the impressions are made to cross each other and run straight across the width and edge of the test piece.

The **Olsen cup test** can be used to obtain an indication of the drawing properties of the base metal as well as the adherence of the coating. In this test, male and female dies are used to deform the steel into a cup shape. The force draws the metal into the die slowly so the coated sheet stretches to conform to the shape of the die. The depth of the cup is measured when the steel fractures, and the coating is then examined.

The **lockseam test** is used extensively in the sheet-galvanizing industry to test coating adherence. This test involves bending the steel to form an 'S' shape and assessing the adherence of the coating along the apexes. This test is important because lockseams are commonly found in galvanized sheets formed for certain applications (such as buckets).

The **Gardner impact test** is similar to the Olsen cup test in that the steel is deformed to a cup-shape. In the Gardner impact test, however, a projectile is dropped from a particular distance to dent the steel to various depths. Thus, the strain rate in this test is considerably greater than in the Olsen cup test. The impact is measured in inch-pounds, and the coating adherence is determined by assessing flaking or crazing on the convex side of the cup.

Other tests are used to test various properties of the base metal and coating. Tensile properties of the base metal are determined in a test that measures (1) the yield point; (2) the tensile strength before localized necking occurs; and (3) the percentage of elongation in a 50.8 mm (2 inch) gage length.

The **Rockwell 'B' test** determines the hardness of the base metal. The procedure involves making indentations on pre-cut samples of metal. A standard machine pushes a ball or diamond tip down to hold the sample. Then a certain load is applied and the machine measures how deeply into the steel the tip goes. This test is run on all hot-dip galvanized sheet product, usually after stripping the coating.

Ferrite grain size is measured to determine the steel structure which has a significant effect on the level of strength and formability of the steel. Various other microstructural features such as carbide morphology and inclusions also can be determined by standard metallographic procedures.

Corrosion of galvanized sheets is best assessed by testing the galvanized sheet product in an environment that simulates the service conditions to which the sheet will be exposed. For example, many companies have atmospheric exposure test racks at locations that represent rural, industrial, marine, etc. environments. Also, galvanized samples may be mounted under automobiles to determine corrosion rate in chloride-containing road salt splash. Various accelerated tests may also be used, but these tests have greater or lesser value depending on their ability to simulate service conditions in a particular application. Corrosion of galvanized sheet is described in detail in Chapter 35, "Corrosion and Protective Coatings."

Selected Bibliography

Historical Interest

Bablik, H., Bending conditions of galvanizing. Iron Age **125**, 1452-54 (1930).

Cowper-Coles, S., Method of hot-dip coating metallic objects. U. S. Pat. No. 979,931 (1910).

Daniels, E. J., The attack on mild steel in hot galvanizing. Institute of Metals Journal **56**, 81-96 (1931).

Richards, J. W., Use of aluminum in galvanizing. U.S. Pat. No. 456,204 (1891).

General Interest

Am. Iron and Steel Institute, Contribution to the Metallurgy of Steel: Paintability of galvanized steel. N. Y., The Institute (Oct. 1966).

Am. Iron and Steel Institute, 1983 Annual Statistical Report. N. Y., The Institute, 1983.

Beachum, E. P., Useful properties of galvanized sheets. AISI Regional (Buffalo, N. Y.) Meeting, Sept. 1962.

Edwards, H., Technique of sheet galvanizing by hot dip process. Sheet Metal Industries 22 1546-52, 1725-30, 1914-22, 2096-2103 (1945).

Marshall, W. E., Process of coating metal articles with molten metal and of preparing metal articles for hot coating. U.S. Pat. No. 2,310,451 (1943).

Special Technical Interest

Am. Iron and Steel Institute, Steel products manual: carbon sheet steel, N. Y., The Institute, 1974.

Am. Society for Metals, Metals handbook, 9th ed. Metals Park, Ohio, The Society, 1978. (See H. Geduld. 'Zinc Plating,' pp. 244-255.)

Am. Society for Testing and Materials, ASTM Standards (volumes on ferrous metals, latest editions) covering galvanized products. Philadelphia, The Society.

Bablik, H., Galvanizing (hot-dip). 3rd ed., London, E. and F. N. Spon, 1950.

Burns, R. M., and W. W. Bradley, Protective coatings for metals, 3rd ed., N. Y. Reinhold, 1967.

Cone, Carroll, Continuous Strip Galvanizing. Iron and Steel Engineer Year Book, 1962.

Higgs, R., M. Komp, and E. Oles, Jr., One-Side-Electrogalvanized Steel Sheet for Automotive Applications. SAE Technical Papers Series, Society of Automotive Engineers, 1980.

Lowenheim, Frederick, ed., Modern Electroplating, 3rd ed., New York, John Wiley and Sons, 1974.

Lynn, H. W., Continuous galvanizing of strip steel. Iron Age **164**, 96-100 (July 21, 1949).

Mauger, A. J. and A. H. Ward, Metallic coating alloy. U.S. Pat No. 2,360,784 (1944).

Mathewson, C. H., Zinc—science and technology of the metal, its alloys, and compounds. Am. Chemical Society Monograph Series, 1964.

Oganowski, K., Continuous galvanizing by the Sendzimir process. Metal Finishing **48**, 63-68 (Oct. 1950).

Radtke, S. F., MacKinney, J. J , and Freytag, N. A., How to resistance weld galvanized steel. Welding Design and Fabrication **38**, No. 2, 39-45 and No. 3, 37-40 (1965).

Rowland, D. H., Metallography of hot-dipped galvanized coatings. Am. Society for Metals Trans. **40**, 983-1011 (1948).

Tama, M., Induction furnaces for galvanizing. Proceedings of Galvanizers Committee, Zinc Institute **42**, 15-26 (1960).

Turner, G. C., Modern continuous galvanizing lines. Yearly Proceedings of Association of Iron and Steel Engineers, 245-252 (1963). Pittsburgh, Pa., The Association.

Ward, A. H., Continuous galvanizing—a development program. Iron Age **164**, 74-79, 154 (Oct. 13, 1949).

Ward, A. H., Continuous processing ferrous strip or sheet material. U.S. Pat. No. 2,588,439 (1952).

White, F. G., Developments in galvanizing. Products Finishing **14**, 48, 50, 52, 56 (May 1950).

CHAPTER 39

Manufacture of Heavy Press Forgings

This chapter will be confined to discussion of the production of steel forgings of large size in hydraulic presses utilizing open dies, and will be based generally on the practices employed at one plant of United States Steel Corporation. Capacities of the hydraulic forging presses at this plant are 88 964, 35 586 and 17 793 kilonewtons (10 000, 4000 and 2000 net tons force). Auxiliary equipment includes furnaces of ample capacity for heating and reheating ingots and forgings, complete heat-treating facilities, and machine tools designed to handle the massive forgings that are produced.

The principle of operation of the hydraulic press was outlined in Chapter 22, along with a comparison of forging with other methods for the hot-working of steel. Although steel can be hot-worked in various ways, perhaps no other method of hot-working can surpass the forging process for producing the best combination of properties.

Open-die forging may be defined as the hot-working of steel between flat or contoured dies. This hot work produces the following advantageous effects: refinement of the relatively coarse crystal structure inherent in the as-cast ingot; performance of sufficient work on the ingot to obtain the desired mechanical and metallurgical properties; and production of a sound, homogeneous mass of steel of the desired size and shape. Typical large forgings produced by the methods to be described are used for generator shafts, steam-turbine rotors, hydroelectric shafts, marine propulsion shafting, ring- and disc-type nuclear components, work rolls, sleeves, die blocks, and miscellaneous mill-equipment parts. Forged billets, blooms and slabs for further working in customers' plants are also produced.

HEATING FOR FORGING

Rate of Heating—At this plant, the majority of large forgings are produced from ingots from 1320 to 3405 mm (52 to 134 inches) in diameter and weighing up to 366 500 kilograms (808 000 pounds). Such large masses of steel require careful control of heating practices that must be varied according to the chemical composition, size, and prior thermal history of the ingots. These same factors govern the several reheating operations usually involved in the production of forgings of large section. The general aims in the control of heating operations are the attainment of a uniform temperature

throughout the ingot or reheated forging, and establishment of heating rates that will achieve proper degree and uniformity of temperature in the shortest practical time. Practical considerations related to time and rate of heating are the minimizing of the amount of scaling and decarburization of the steel surfaces.

Furnaces of the direct-fired car-bottom type (Figure 39—1) are popular for heating large ingots and forgings. Any furnace employed for this work should be equipped with suitable instrumentation for the accurate measurement, control, and recording of temperature.

Only general rules can be prescribed for heating large sections. Large ingots or forgings should be heated slowly and uniformly. The rate of heating fairly well establishes the length of time necessary for the steel to attain forging temperature throughout its mass, and a rate should be selected that avoids excessive temperature differentials between the inside and outside of the mass. The temperature of the interior lags behind that of the exterior during a large part of the heating cycle, and a period of time near the end of the cycle must elapse after the exterior has attained forging temperature for sufficient heat to be conducted into the interior to raise it to the proper temperature. The slower the heating rate, the shorter will be the time for such temperature equalization to take place. **Step heating** may be employed; that is, the steel may be held at one or more temperature levels below forging temperature and allowed to equalize before proceeding to a higher temperature level. It has been found that after its temperature has been equalized at a point slightly above the upper critical temperature (about 800° C or 1475° F), steel can be heated at a rate of 22° to 33° C (40° to 60° F) per hour until forging temperature is attained. This cycle results in heating times corresponding to approximately ¾ to 1 hour per 25 mm (1 inch) of diameter or thickness of the ingot or forging. In general, carbon steels containing over 0.50 per cent carbon and alloy steels require slower rates of heating than carbon steels of lower than 0.50 per cent carbon content.

Forging Temperature—Forging temperature is selected to provide the best condition for hot-working a given steel. Although the final properties of a finished forging are established largely by heat treatments applied subsequent to hot-working, the temperature at which hot-working is completed influences, to varying

FIG. 39—1. Overhead crane removing a 1955-mm (77-inch) diameter fluted ingot from the hearth of a car-bottom heating furnace after a 70-hour heating cycle to prepare it for forging.

degrees depending on grade, what heat treatments are necessary as well as the final mechanical properties of the steel. In general, lower finishing temperatures result in a finer-grained microstructure when forging is completed. The finer-grained structures respond better to heat treatment than do coarser-grained structures. However, the finishing temperature must be kept high enough to prevent the occurrence of forging bursts (internal ruptures) that may result from severe stresses induced by working large masses of steel at too low a temperature.

Soaking times at the forging temperature are also important, particularly in the hot working of the austenitic grades of steel. The times should be sufficiently long to accomplish the achievement of uniform temperature throughout the steel, but not so long as to promote excessive grain growth of the austenite. Since the austenitic stainless steels do not undergo a phase transformation during heat treatment, the final grain size of a forging is determined by the hot-working process. Poor control of forging temperatures or soaking times can result in an extremely coarse grain size in a stainless-steel forging, which then cannot be evaluated by ultrasonic inspection due to the tendency for the coarse grain structure to attenuate or scatter the sound wave.

Caution must be exercised to avoid **overheating** and **burning** of the steel. The safe upper limit of the hot-working range is a suitable temperature interval below the melting point of the lowest-melting constituent of a steel. **Burning** consists of heating a steel to a high temperature in an oxidizing atmosphere so that actual fusion and oxidation occur at the austenite grain boundaries, causing hot shortness that results in badly torn surfaces and internal ruptures during hot working. Burned steel cannot be salvaged. **Overheating** has less obvious effects, caused by heating to a high temperature but not sufficiently high as to cause burning. The effects of slight overheating can be removed by subsequent hot working, but more severe overheating can cause low ductility and low toughness in forgings tested after final heat treatment.

HANDLING EQUIPMENT

Special equipment is required for handling the heavy masses of steel represented by forging ingots and the forgings themselves.

Electric overhead traveling cranes with special lifting devices are employed in charging ingots and forgings onto and from the hearths of car-bottom heating furnaces, and for transporting them to and from the forging presses (Figure 39—1).

An electric crane at the forging press carries a **turning gear** suspended from the main hoist. The turning gear consists of a frame carrying a drum that can be rotated by an electric motor through gearing. An endless chain called a **sling**, constructed of flat links and

FIG. 39—2. Hydraulic press in operation, forging a massive rectangular ingot supported on a porter bar which, in conjunction with the link-chain sling and turning gear suspended from an overhead crane, permits manipulation of the ingot as desired.

pins, passes over the drum and moves with it (Figure 39—2).

A device called a **porter bar** has a hollow end shaped to fit the sinkhead of the ingot being forged. The load represented by the ingot and porter bar is balanced by placing the sling at the center of gravity of the combined load (Figure 39—2); the sling has to be moved from time to time to preserve balance as the dimensions of the forging are changed.

Faster and more accurate handling of the hot steel during forging is accomplished by the use of machines called **manipulators.** These machines are equipped with powerful tongs at the end of a horizontal arm that can be moved from side to side, raised or lowered, and rotated about its longitudinal axis. The manipulator shown in Figure 39—3 operates on tracks in the floor in front of a 88 964 kilonewton (10 000-net-ton force) hydraulic press, and has a capacity for handling pieces weighing up to 68 metric tons (75 net tons). Smaller manipulators are designed to operate on wheels that have resilient or solid tires.

OPEN DIES FOR FORGING

The dies used in open-die forging are of three basic types, shown schematically in Figure 39—4. They are known as **flat dies, V dies,** and **swage dies.** For hollow forgings or ring forgings, a **mandrel** or **expanding bar** is inserted in a hole in the piece to be forged, and forging is carried out by utilizing the mandrel or bar as the bottom die.

PRINCIPAL FORGING OPERATIONS

Superior quality, toughness and strength characteristics have gained for forgings their reputation for dependability under the most severe service conditions. The consecutive operations employed in producing heavy press forgings by open-die forging are carefully planned and executed in proper order so as to arrive at the final contour (in which the sections may vary greatly) while at the same time achieving the proper degree of grain refinement and internal soundness.

Initial working of an ingot is usually referred to as **cogging** (Figure 39—5), and removes the flutes, ripples or corrugations that were formed on the ingot as it solidified in a contoured mold and which are intended to prevent cracking of the ingot surface during solidification and cooling. Light drafts (small reductions) are taken all over the ingot until the surface irregularities

ALLIANCE

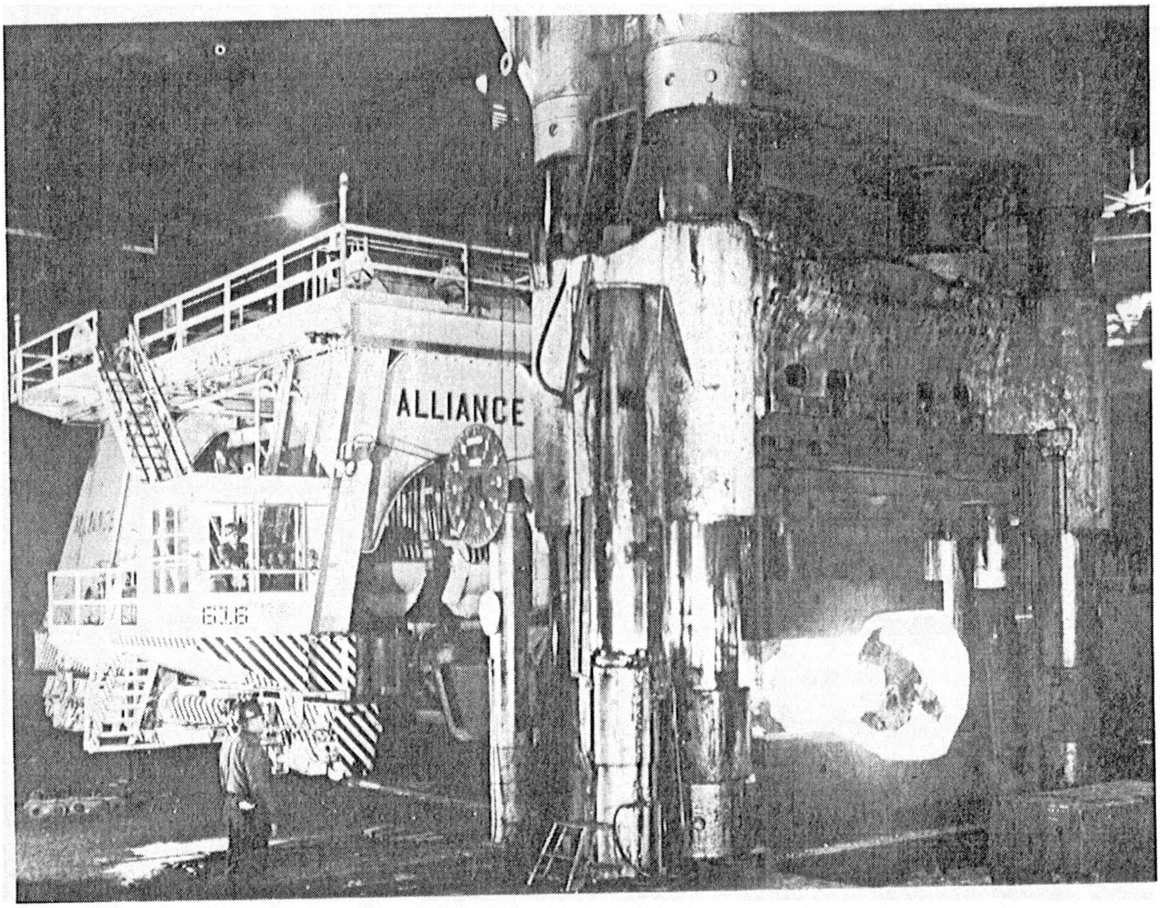

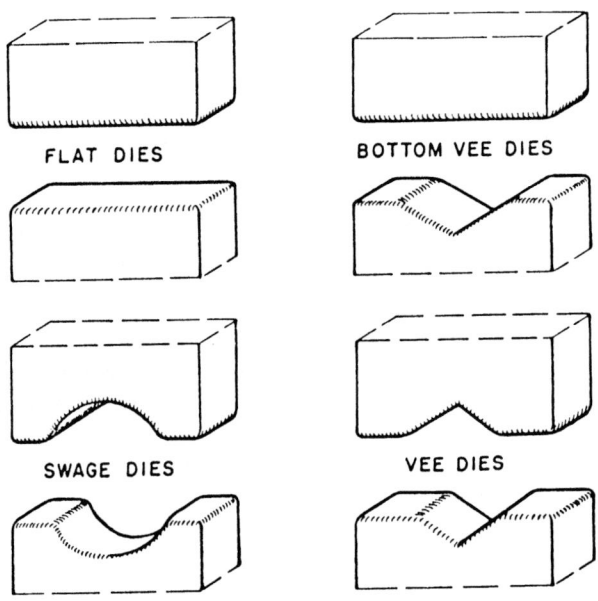

FIG. 39—3. Heavy duty manipulator holding an ingot in position while a 88 964 kilonewton (10 000-net-ton force) press squeezes the hot steel into the rough shape of the finished product.

FLAT DIES
BOTTOM VEE DIES
SWAGE DIES
VEE DIES

FIG. 39—4. Basic shapes of dies for open-die forging.

are smoothed. Heavier drafts are then taken and working continues, usually to convert the cross-section of an originally round ingot to an octagon shape, then to a square, back to an octagon, and so on. This **straight-down** or **setting-down** type of forging (Figure 39—6), also called **drawing out** and **forging solid**, is used on products such as blooms, rounds, and shafting having constant or variable section size, in which the work flows in a longitudinal direction. For variable section sizes or step-downs, **marking knives** or **veeing tools** are used to mark off the necessary volume of metal for any particular section (as shown later in Figure 39—11).

Upsetting (sometimes called **pancaking**) is employed to alter the geometry of a piece. For example, forging pressure applied to the ends of a cylindrical piece will shorten its axis while increasing its diameter. Upsetting also produces a benefit that results from circumferential flow that induces the best condition for parts subjected to either tangential or radial stresses, or both. Upsetting is used to produce disc-type forgings and is sometimes employed as an intermediate operation in the production of steam-turbine rotors. In this instance, the ingot is forged to a bloom of predetermined size,

usually octagonal in shape, and a slug of sufficient length is sheared. The slug is upended and forged to reduce its length, with the percentage of upsetting being based on the grade and size. Further forging consists of reworking along the length, marking, and contouring. Benefits derived from this sequence of operations include working in all directions to enhance soundness and reduce the directionality of properties. In upsetting, it is necessary to keep the length of the slug within certain limits (usually not more than 2½ times the octagon size) to prevent kinking or bending during upsetting. In some types of upsetting operations, hollow, cylindrical die-like tools called **bolsters** may be employed to retain part of the slug and prevent change in its dimensions while the upsetting force is applied to deform the rest of the slug. A simple application of a bolster is shown schematically in Figure 39—7.

Piercing and **punching** are performed by forcing a solid punch into hot steel to form a cavity. Piercing is employed to make a blind cavity by displacement without removal of metal. Punching produces a hole that extends through the entire section and both displaces and removes metal in the form of a slug. Sometimes a

FIG. 39—5. Large corrugated ingot in the process of being "cogged" or reduced to an octagonal shape during the first stage of forging.

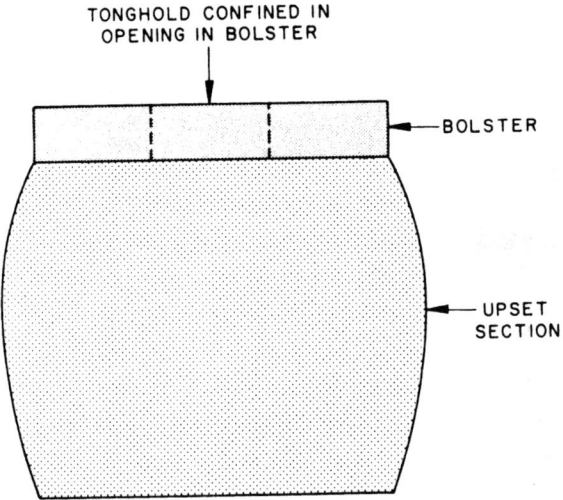

FIG. 39—6. Straight-down forging of a 225-metric-ton (250-net-ton) bloom between flat dies on a hydraulic forging press.

TONGHOLD CONFINED IN
OPENING IN BOLSTER

BOLSTER

UPSET
SECTION

FIG. 39—7. Schematic representation of the use of a bolster to maintain a previously forged tonghold on a piece during upsetting.

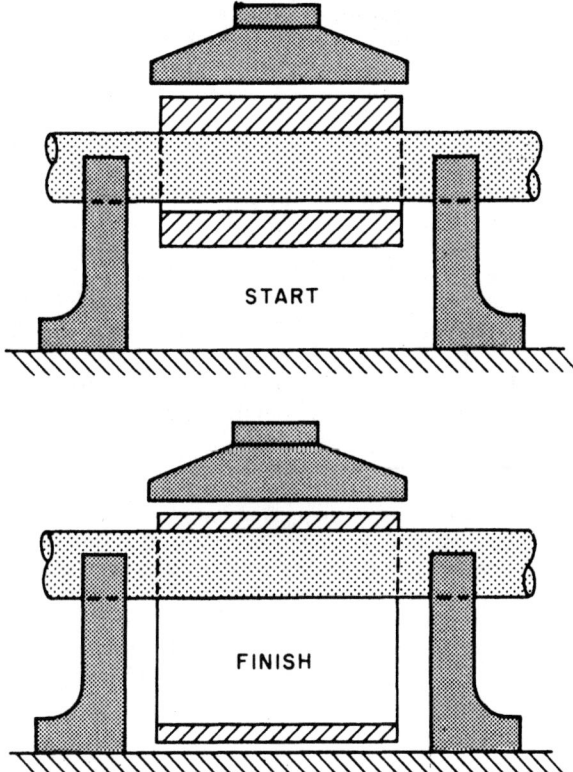

FIG. 39—8. Sectional diagrammatic representation of the expanding operation for simultaneously increasing the inside and outside diameters of a hollow cylindrical forging while reducing wall thickness. The expanding bar acts as a bottom die.

hollow punch is used to remove some of the central metal as a core; this operation is generally termed **hot trepanning** or **hot trephining.**

Expanding is a special process for increasing the diameter of hollow forgings. It can be used for finishing hollow forgings or to increase the size of the hole prior to finish forging on a mandrel. It is also employed in the forging of rings. For expanding a forging of considerable length, a top tool with a narrow face parallel to the expanding bar is used (Figure 39—8), to keep the length-wise elongation of the piece to a minimum while reducing the thickness. As shown in Figure 39—8, the bottom die is replaced by an **expanding bar** or mandrel that passes through the opening in the forging and rests on supports beyond the forging. When pressure is applied, the thickness of the material between the top tool and the bar is reduced, displacement of the metal resulting in an increase in the circumference of the forging. By successive incremental movement of the work piece followed by pressing, the wall thickness can be reduced uniformly while attaining the desired inside and outside diameters. Uniformity of temperature throughout the piece is important for successful control of expanding, and a bar of sufficient strength to withstand the bending moment is essential. Figure 39—9 illustrates the forging of a large ring on an expanding bar.

The process of **hollow forging on a mandrel** differs somewhat from the expanding process, in that the mandrel establishes the inside diameter as the piece is forged with pressure from opposed top and bottom tools. As shown in Figure 39—10, the hollow work piece is fitted to a mandrel of the desired size that is supported on both ends to position the work between

FIG. 39—9. Forging a large steel ring in a hydraulic press, using an expanding bar.

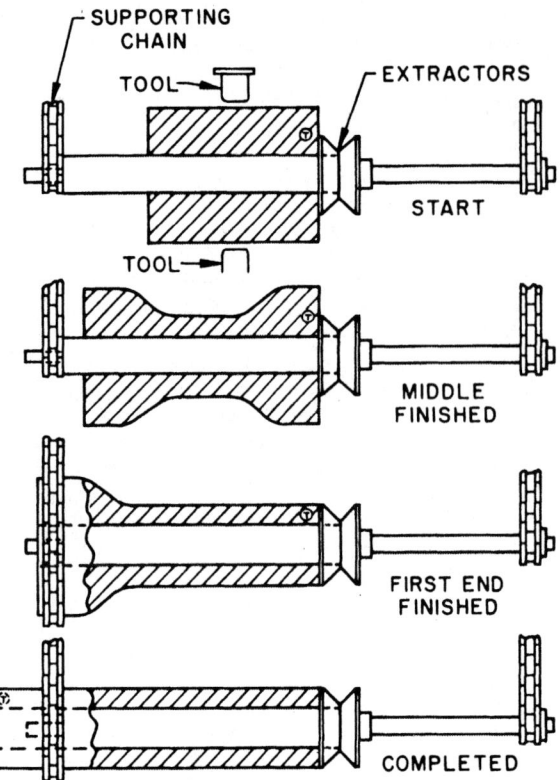

FIG 39—10. Schematic diagram showing steps involved in forging a hollow cylinder on a mandrel. The mandrel establishes the inside diameter of the forging as pressure is applied by the top and bottom tools.

the top and bottom dies. Narrow-faced tools are used in this type of work to cause the metal to flow lengthwise of the piece. The mandrel is usually tapered slightly to facilitate its removal from the finished forging.

Closing in is an operation on a hollow forging, using flat, tapered, curved, or formed dies, to partially close the end, or reduce some other portion of a forging; as, for example, in forming the hemispherical ends of a boiler drum. One or more local reheatings of the part of the forging being worked are usually involved.

Slabbing consists of forging an ingot to a large, heavy slab or plate section that is beyond the capabilities of rolling facilities.

EXAMPLES OF FORGING PROCEDURE

Figures 39—8, 39—9 and 39—10 have shown the principles employed in making hollow forgings such as cylinders and rings. These expanding and elongating principles are used singly or in combination to forge boiler drums, chemical reactors, pressure vessels, roll sleeves and many other products.

Figure 39—11 shows the steps in producing a sleeve for a cold-reduction-mill back-up roll. Such sleeves generally are made from a high-carbon (0.65 to 0.75 per cent) nickel-chromium-molybdenum-vanadium steel. This particular sleeve has a 1084 mm (42⅝-inch) outside diameter, a 711 mm (28-inch) inside diameter, and a 1089 mm (42⅞-inch) body length. Three sleeves of

this size are produced from a 1219 mm (48-inch) diameter ingot weighing 33 000 kilograms (72 800 pounds). The round, corrugated ingot is first forged to a 1118 mm (44-inch) octagon, then sheared into three pieces, each 1067 mm (42 inches) in length. The pieces are reheated and each is upset forged to 914 mm (36 inches) in length, after which a 419 mm (16½-inch) diameter hole is punched through the longitudinal axis of each. After reheating, each piece is forged on a 406 mm (16-inch) diameter expanding bar to increase the hole to 508 mm (20-inch) diameter. Each expanded piece is again reheated and forged on a 495 mm (19½-inch) diameter bar to a 997 mm (39¼-inch) octagon and the hole is enlarged to 533 mm (21 inches). Another reheating is followed by forging the ends of each piece to a level contour. After a final reheating, the sleeves are each forged on a 483 mm (19-inch) diameter bar to 660 mm (26 inches) in inside diameter, 1133 mm (44⅝ inches) in outside diameter, and approximately 1346 mm (53 inches) in length. After this final forging operation, the sleeves are slowly cooled, then heat treated to secure the proper microstructure for subsequent hardening by water quenching and tempering, following rough machining.

A Ni-Mo-V generator-rotor shaft with a forge weight

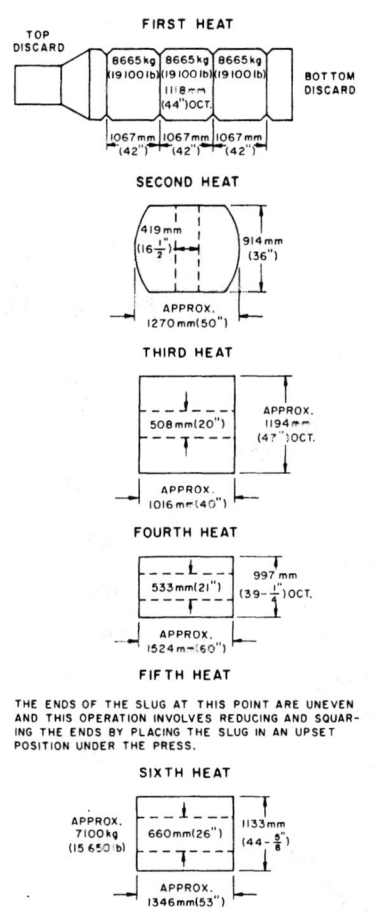

FIG. 39—11. Steps involved in forging a sleeve for a backup roll for a cold-reduction mill.

of 68 to 73 metric tons (75 to 80 net tons) and a rough-machined weight of 50 to 54 metric tons (55 to 60 net tons) would be forged in three heats from a 154 metric ton (170 net ton), 2413 mm (95-inch) diameter ingot. On the first forge heat, the corrugated ingot body is forged to approximately a 1397 mm (55-inch) octagon, the sinkhead is sheared off, and a chuckhold forged on the top end. On the second heat, the ingot is further reduced to approximately an 1194 mm (47-inch) octagon, marked and "veed" to define the body and journal steps, and at this stage the bottom discard is sheared off. On the third and final heat, the various steps from body to journals are forged to size, a final forge pass made on the body, and the top chuckhold sheared off. The appropriate slow cooling, heat treating, and machining operations are carried out before the shaft is shipped.

COOLING AFTER FORGING

It is important that large forgings be cooled after hot-working is finished in a manner that will prevent the formation of thermal **bursts** or **ruptures** caused by internal stresses related to differences in rate of cooling of different parts of the forging, or **flakes** that are attributed to gases (particularly hydrogen) absorbed by the liquid metal during steelmaking. As described in Chapter 19, vacuum stream degassing of molten steel to lower the hydrogen content to safe limits will effectively prevent formation of flakes, even in very large ingots, and consumable-electrode melting and vacuum melting (see Chapter 18) can achieve the same end in smaller ingots. However, prevention of damaging internal stresses can only be achieved through closely controlled cooling. For applications where minimum residual stresses are necessary, very slow cooling rates and long tempering times are advisable.

HEAT TREATMENT OF FORGINGS

Few heavy forgings are shipped without some form of heat treatment. Here, horizontal car-bottom furnaces and vertical pit-type furnaces are used, depending upon the shape and weight of the forgings undergoing heat treatment.

Typical heat-treating operations include:
1. Annealing (various cycles).
2. Normalizing (with optional accelerated air-cooling) and tempering.
3. Water or oil quenching followed by tempering, utilizing tanks for quenching.
4. Water-spray quenching, followed by tempering.
5. Stress relieving.
6. Various combinations of the above.

Car-Bottom Furnaces—The plant has thirteen car-bottom furnaces for the heat treatment of forgings. These are direct top-and-bottom fired, using natural gas as fuel, and have a maximum operating temperature of 1060°C (1940°F).

Vertical Furnaces—Seven vertical pit-type furnaces are available for heat treating large and lengthy shafting, steam-turbine rotors, generator shafts, and similar types of forgings.

The principal use of these furnaces is for grain-refinement treatments and stress relieving of rotors and shafts. A part of these furnaces is below ground level, as shown in Figures 39—12 and 39—13. All seven of the furnaces are charged from the top by an overhead crane. Work rests on a special hearth casting at the bottom of each furnace and is stabilized by pins inserted through sides of the furnace to contact the forging, except in one furnace that has a rotating hearth that revolves the charge about its longitudinal axis throughout the heating cycle. Natural-gas-fired burners fire tangentially into the heating chambers to prevent flames from impinging directly on the charges and to circulate the products of combustion in a manner that promotes temperature uniformity in the furnaces.

A typical heating cycle consists of a heat-up period at a controlled rate of temperature rise, a soak period, another heating at a controlled rate to a higher temperature, another soak period, and then a controlled cooling period. Because of the slow cooling rates involved, it is usually necessary to provide a small amount of heat to maintain the desired rate of cooling. When the natural cooling rate is slower than that desired, the gas is shut off and the proportioning controller operates the air valve only, thus increasing the cooling rate. A complete heat-treating cycle can last from one to eight days.

Quenching Facilities—Some forgings to be quenched in liquid media (water or oil) require a more rapid cooling rate than air-cooling in the furnaces can provide. For this purpose, a vertical quench chamber (Figure 39—14) is located near a battery of the pit-type furnaces. Forgings are positioned vertically and held securely in the base that can be rotated continuously during the quenching operation. A variety of cooling media can be employed, such as water spray, fog spray, a combination fog-and-air spray, or an air quench.

For liquid-media quenching, the plant has tanks of ample dimensions, equipped for forced circulation.

TESTING AND INSPECTION OF FORGINGS

The forgings produced are manufactured in accordance with various customer or industry standards which specify desired material characteristics and properties. In addition to the usual dimensional considerations, the specified requirements may be many or few; and may, for example, relate to chemical composition, mechanical properties, heat treatment and machined-surface finishes. Non-destructive tests, such as ultrasonic and/or magnetic-particle testing, may be specified to determine the presence of internal or surface discontinuities.

Mechanical-Property Tests—Various tests are employed to measure the mechanical properties of heat-treated forgings and include:
1. Tension tests
2. Impact tests (Charpy V-notch, drop weight)
3. Hardness tests (Brinell, Rockwell, Shore Scleroscope)
4. Bend tests
5. Corrosion bend tests

The above tests are discussed in more detail elsewhere in this book.

When tension, impact or bend tests are specified, provision must be made for testing at locations within

FIG. 39—12. Forging positioned over one of a battery of four pit-type heat-treating furnaces (see also Figure 39—13).

FIG. 39—13. General view of battery of pit-type furnaces (right) and vertical quench chamber (left) for the heat treatment of large forgings. Generator-shaft forging is shown rotating about its vertical axis within the quench chamber, the doors of which normally are closed during operation but which were left open for this photograph.

the forging which will not result in destruction of the forging, or from excess test-metal prolongations added to the forging configuration. Location and size of the added test-metal prolongations are determined by the orientation of the test-specimen axis. On most forgings, the tests are oriented parallel to the direction in which the forging is most drawn out or extended by forging, i.e., on shaft forgings, test specimens are parallel to the

FIG. 39—14. Generator-shaft forging being lowered onto holding fixture on rotating base of quench chamber. Structure at floor level in left foreground is an upending cradle that raises long forgings from a horizontal position to a vertical position to facilitate handling by crane. Sliding vertical doors close front of chamber during quenching.

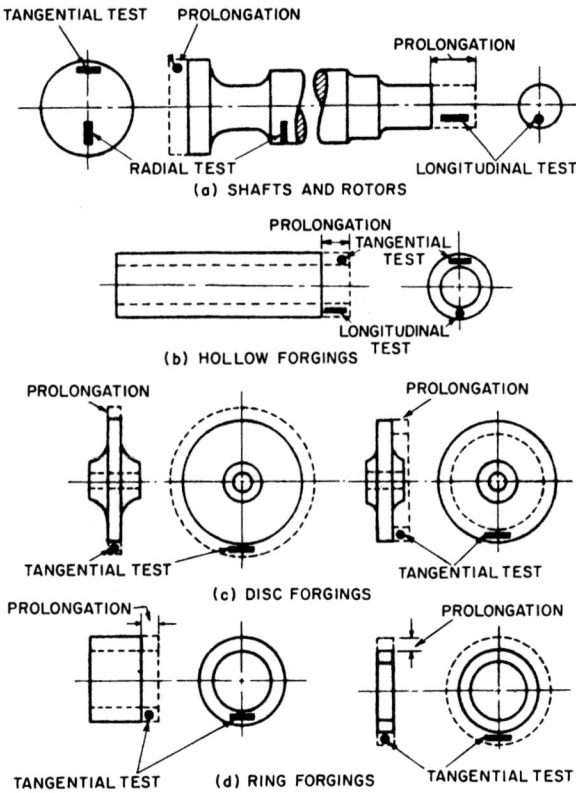

TANGENTIAL TEST   PROLONGATION

PROLONGATION

RADIAL TEST

LONGITUDINAL TEST

(a) SHAFTS AND ROTORS

PROLONGATION

TANGENTIAL TEST

LONGITUDINAL TEST

(b) HOLLOW FORGINGS

PROLONGATION   PROLONGATION

TANGENTIAL TEST   TANGENTIAL TEST

(c) DISC FORGINGS

PROLONGATION   PROLONGATION

TANGENTIAL TEST   (d) RING FORGINGS   TANGENTIAL TEST

FIG. 39—15. Location of test specimens in forgings. (Adapted from the "Open Die Forging Manual.")

axis of the forging; on disc-type or ring forgings, test specimens are oriented tangential to a circle drawn using the axis of the forging as a center. Steam-turbine and generator forgings which are subjected to high rotational stresses in service require adequate transverse properties and are tested in a direction perpendicular and radial to the forging axis. Typical test plans for various types of forgings are shown in Figure 39—15. Actual removal of test specimens from the forging can be accomplished in a number of ways, such as sawing and trepanning, or from blocks removed by an electrochemical milling machine.

Hardness tests are made in prepared areas of the forgings at locations required by the material specification. Sufficient metal is removed from the surface of the specific test site to eliminate decarburization and other surface irregularities.

Non-Destructive Testing—The internal and surface quality of forgings is verified by one or more of the following methods which are covered in more detail in other chapters.

1. Visual inspection

2. Ultrasonic inspection
3. Liquid-penetrant inspection
4. Magnetic-particle inspection.

Surfaces of forgings are inspected for imperfections by visual means which may be supplemented with either liquid-penetrant or magnetic-particle test methods. Optical periscopes may also be employed to inspect inaccessible areas such as the bores of shaft-type forgings.

Internal quality of forgings is evaluated by ultrasonic methods using a longitudinal wave, and, when required, shear-wave techniques. The inspections are performed using 0.5, 1.0, 2.25 or 5.0 MHz transducers with the instrument calibrated to display a specific back-reflection response from the forging itself or from an artificial flaw (flat-bottom hole) placed in the forging or in a compatible external test block. Distance-amplitude corrections may also be employed to evaluate flaws at various positions within a long test distance. A typical distance-amplitude curve is shown in Figure 39—16.

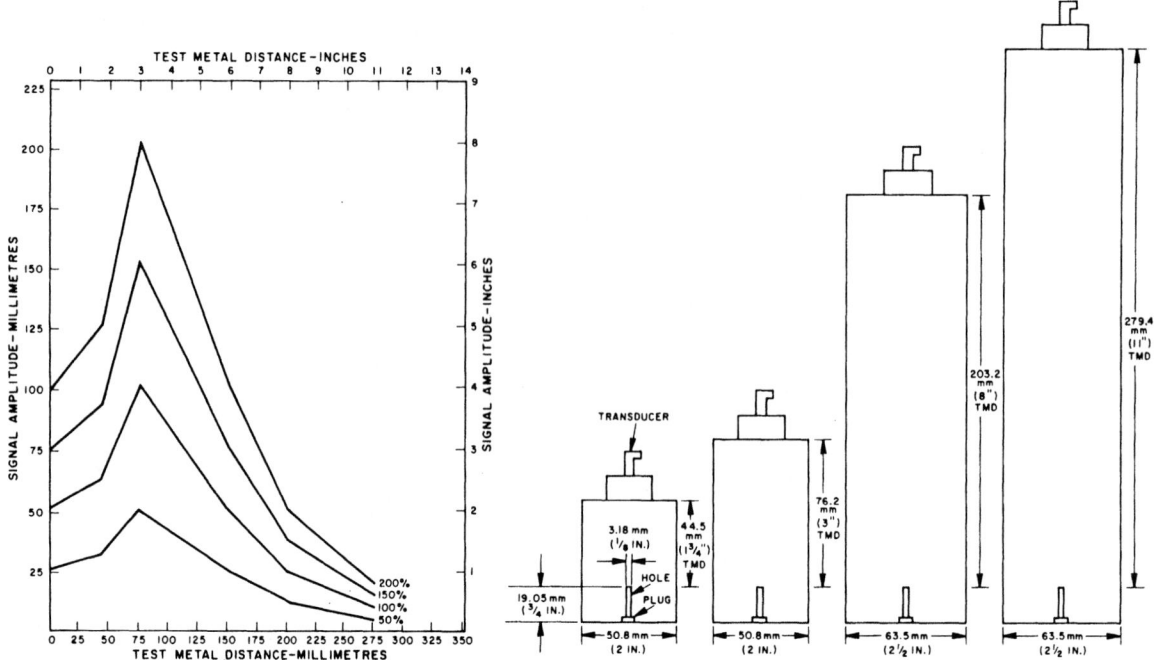

FIG. 39—16. Flaw-size estimation using distance-amplitude correction. (Left) Distance-amplitude curve for 2.25-MHz 1-inch-diameter crystal in normalized low-carbon nickel-chromium-molybdenum steel; 100 per cent amplitude is equivalent to 17.81 mm² or 0.0276 in.² (4.76-mm or 3/16-inch-diameter) reflecting area. (Right) Schematic sections of cylindrical sonic test blocks with different test-metal distance (TMD). Flat-bottom holes are 3.18 mm (⅛ in.) in diameter, 19.05 mm (¾-in.) deep, plugged at open end.

CHAPTER 40

Castings – Steel And Iron

SECTION 1

STEEL CASTINGS

Casting Compared with Other Forms of Shaping Steel—The process of making steel castings consists simply of pouring metal into a mold which is of the desired shape, dimensionally accurate and of sufficient stability to permit the metal to solidify in the exact shape of the mold cavity. Intricate and complicated castings of practically any desired shape or size, and for almost any particular application, can be made in this manner. The versatility of the foundry trade commands recognition throughout the industrial world, and design engineers in every field of endeavor have come to rely upon the utility of steel castings.

Mechanically, steel castings are considered inferior to wrought-steel products, and it is true that wrought steels do exhibit higher mechanical properties than cast steels, especially when the former are tested in the direction of rolling or forging. Moreover, cast structures, unless designed in strict conformity with the natural characteristics of metal solidification, sometimes contain internal defects or surface imperfections which may seriously affect serviceability and otherwise render them less dependable than a wrought-steel product. However, the casting of steel is the most direct method of producing a given shape and, for this reason, the method provides the basis for a key industry.

Tonnagewise, the steel-casting or steel-foundry industry would appear to be of minor importance, approximately two per cent of the total steel production per year being in the form of castings. However, few industries enjoy so prominent a role in the industrial and domestic development of the world as the foundry industry. Many articles, especially turbine shells, valve bodies, and machine parts are made by casting, not by choice, but because they cannot be made as readily by other processes. For example, valve bodies, exhaust manifolds, pump casings, and turbine diaphragms could be fabricated by other processes or from wrought products only with extreme difficulty and at a serious economic disadvantage. Also, cast structures are generally considered to be more rigid than their wrought counterparts.

Steel castings vary in weight from a few ounces to hundreds of tons and cover a multitude of designs and services. Latch castings for airplane cowls weigh less than 113 grams (4 ounces), while one of the largest castings ever produced was a 218-metric-ton (240-net-ton) casting for an armor press of 311 375 kilonewtons (35 000 net tons force) capacity. Castings for steel-mill service alone cover a very large field and include housings, gears, charging boxes, guides, blast-furnace bells, hoppers, and cinder pots. Rolls for certain types of rolling mills are made of cast steel; because of the special foundry techniques employed in roll casting, the manufacture of rolling-mill rolls by casting is discussed separately and at length in Chapter 23. The transportation industry relies upon castings for railroad and marine use, such as couplings, journal boxes, bolsters, frames, brake shoes, cylinders, housings, valves, crankshafts, engine beds and steam chests. These are but a few of the examples in everyday use and hundreds of other applications could be mentioned serving the chemical, petroleum, mining, excavating, agricultural, ceramic and construction industries.

Composition and Mechanical Properties of Cast Steels—Cast steels are available in both carbon and alloy grades. Carbon content usually is limited to 0.20 per cent minimum, and the silicon content is between 0.35 per cent minimum and 0.80 per cent maximum. The carbon content is restricted to 0.20 per cent minimum since this is the minimum carbon content that can be melted in an acid-lined open-hearth or electric-arc furnace without excessive slag-line or bottom erosion. The silicon level must be higher than for rolled or forged products since the solidification rate of castings poured in sand molds is much slower than for ingots poured in cast-iron molds. The slower solidification rate increases the possibility of gas evolution and blowholes, so the molten steel must be higher in silicon content to reduce the possibility of gas evolution. Aluminum also is added in most cases to further reduce the possibility of blowholes.

Various alloy-steel castings are available with tensile strengths and yield points as high as 1207 and 1000 MPa (175 000 and 145 000 psi), respectively. The grade of alloy castings most used, however, has a minimum tensile strength of 621 MPa (90 000 psi) and a yield point of 414 MPa (60 000 psi) minimum.

When carbon-steel castings are to be readily weldable, the carbon content is restricted to 0.35 per cent maximum. Alloy-steel castings of any carbon content, however, require extensive preheat and postheat practices to be successfully welded.

Tables **40—I** and **40—II** give the chemical compositions and corresponding mechanical properties of some carbon- and alloy-steel castings for a variety of applications.

Table 40—I. Chemical Compositions of Some Low-Carbon and Alloy Steels for Castings (Per Cent).[a],[b],[c]

Grade	C	Mn	Si	P	S	Mo	Ni	Cr	Other
	ASTM Designation A 27—81a Mild- to Medium-Strength Carbon-Steel Castings for General Applications
N-1	0.25[d]	0.75[d]	0.80	0.05	0.06				
N-2	0.35[d]	0.60[d]	0.80	0.05	0.06				
U–60–30	0.25[d]	0.75[d]	0.80	0.05	0.06				
60–30	0.30[d]	0.60[d]	0.80	0.05	0.06				
65–35	0.30[d]	0.70[d]	0.80	0.05	0.06				
70–36	0.35[d]	0.70[d]	0.80	0.05	0.06				
70–40	0.25[d]	1.20[d]	0.80	0.05	0.06				
	ASTM Designation A 216—77 Carbon-Steel Castings Suitable for Fusion Welding for High-Temperature Service
WCA	0.25[e]	0.70[e]	0.60	0.04	0.045	0.25*	0.50*	0.40*	Cu:0.50*; V:0.03*
WCB	0.30	1.00	0.60	0.04	0.045	0.25*	0.50*	0.40*	Cu:0.50*; V:0.03*
WCC	0.25[f]	1.20[f]	0.60	0.04	0.045	0.25*	0.50*	0.40*	Cu:0.50*; V:0.03*
	ASTM Designation A217—81 Alloy-Steel Castings for Pressure-Containing Parts Suitable for High-Temperature Service
WC1	0.25	0.50–0.80	0.60	0.04	0.045	0.45–0.65	0.50*	0.35*	Cu:0.50*; W:0.10*
WC4	0.20	0.50–0.80	0.60	0.04	0.045	0.45–0.65			Cu:0.50*; W:0.10*
WC5	0.20	0.40–0.70	0.60	0.04	0.045	0.90–1.20	0.60–1.00	0.50–0.90	Cu:0.50*; W:0.10*
WC6	0.20	0.50–0.80	0.60	0.04	0.045	0.45–0.65		1.00–1.50	Cu:0.50; Ni:0.50*; W:0.10*
WC9	0.18	0.40–0.70	0.60	0.04	0.045	0.90–1.20		2.00–2.75	Cu:0.50*; Ni:0.50*; W:0.10*
WC11	0.15–0.21	0.50–0.80	0.75	0.020	0.015	0.45–0.65		1.00–1.60	A1:0.01; Cu:0.35*; Ni:0.50*; V:0.03*
C5	0.20	0.40–0.70	0.75	0.04	0.045	0.45–0.65		4.00–6.50	Cu:0.50*; Ni:0.50*; W:0.30*
C12	0.20	0.35–0.65	1.00	0.04	0.045	0.90–1.20		8.00–10.00	Cu:0.50*; Ni:0.50*; W:0.10*
CA–15	0.15	1.00	1.50	0.040	0.040	0.50	1.00	11.5–14.0	
	ASTM Designation A 356—77 Heavy-Walled Carbon- and Low-Alloy Steel Castings for Steam Turbines
1	0.35[g]	0.70[g]	0.60	0.035	0.030				
2	0.25[g]	0.70[g]	0.60	0.035	0.030	0.45–0.65			
5	0.25[g]	0.70[g]	0.60	0.035	0.030	0.45–0.65		0.40–0.70	
6	0.20	0.50–0.80	0.60	0.035	0.030	0.45–0.65		1.00–1.50	
8	0.13–0.20	0.50–0.90	0.20–0.60	0.035	0.030	0.90–1.20		1.00–1.50	V:0.05–0.015
9	0.13–0.20	0.50–0.90	0.20–0.60	0.035	0.030	0.90–1.20		1.00–1.50	V:0.20–0.35
10	0.20	0.50–0.80	0.60	0.035	0.030	0.90–1.20		2.00–2.75	
	ASTM Designation A 389—77a Alloy-Steel Castings Specially Heat Treated for Pressure-Containing Parts Suitable for High-Temperature Service
C23	0.20	0.30–0.80	0.60	0.04	0.045	0.45–0.65		1.00–1.50	V:0.15–0.25
C24	0.20	0.30–0.80	0.60	0.04	0.045	0.90–1.20		0.80–1.25	V:0.15–0.25

[a] These compositions for the indicated ASTM Designations are excerpted from ASTM Standard Specifications in ASTM Standards published by the American Society for Testing and Materials, 1916 Race St., Philadelphia, Pa. 19103, and represent only part of each specification. The appropriate ASTM Standard Specification for each of the designations should be consulted for complete specifications.

[b] All values maximum, except where range is specified.

(Continued on next page)

(c) See Table **40**—II for mechanical properties of heat-treated steels of these compositions.

(d) For each reduction of 0.01 per cent carbon below the maximum specified, an increase of 0.04 per cent manganese above the maximum specified is permitted to a maximum of 1.40 per cent for grade 70-40 and 1.00 per cent for the other grades covered by ASTM A 27-81a.

(e) For each reduction of 0.01 per cent below the specified maximum carbon content, an increase of 0.04 per cent manganese above the specified maximum is permitted up to a maximum of 1.10 per cent.

(f) For each reduction of 0.01 per cent below the specified maximum carbon content, an increase of 0.04 per cent manganese above the specified maximum is permitted to a maximum of 1.40 per cent.

(g) For each 0.01 per cent reduction in carbon content below the maximum specified, an increase of 0.04 percentage points of manganese over the maximum specified for that element is permitted up to 1.00 per cent.

*Only as residual element: not added intentionally.

Table 40—II. Mechanical Properties of Some Cast Steels[a],[b],[c]

Grade	Tensile Strength[e] MPa	ksi	Yield Point MPa	ksi	Elongation in 50 mm or 2 in. (%)	Reduction of Area (%)	Heat Treatment[d]
ASTM Designation A 27—81a Mild- to Medium-Strength Carbon-Steel Castings for General Application							
U–60–30	415	60	205	30	22	30	Excepting castings of the N–1 and
60–30	415	60	205	30	24	35	U–60–30 grades which may be furnished
65–35	450	65	240	35	24	35	as-cast, castings of all grades in this
70–36	485	70	250	36	22	30	designation are heat treated by full
70–40	485	70	275	40	22	30	annealing, normalizing, normalizing and tempering, or quenching and tempering, as required by their design and chemical composition.
ASTM Designation A 216—77 Carbon-Steel Castings Suitable for Fusion Welding for High-Temperature Service							
WCA	415 to 585	60 to 85	205	30	24	35	Castings of all grades in this designa-
WCB	485 to 655	70 to 95	250	36	22	35	tion are heat treated by annealing, nor-
WCC	485 to 655	70 to 95	275	40	22	35	malizing, or normalizing and tempering, as required by their design and chemical composition.
ASTM Designation A 217—81 Alloy-Steel Castings for Pressure-Containing Parts Suitable for High-Temperature Service							
WC1	450 to 620	65 to 90	240	35	24	35	Castings of all grades in this designa-
WC4, WC5,							tion are heat treated by normalizing
WC6, WC9	485 to 655	70 to 95	275	40	20	35	and tempering.
WC11	550 to 725	80 to 105	345	50	18	45	
C5, C12	620 to 795	90 to 115	415	60	18	35	
CA15	620 to 795	90 to 115	450	65	18	30	
ASTM Designation A 356—77 Heavy-Walled Carbon- and Low-Alloy Steel Castings for Steam Turbines							
1	485	70	250	36	20	35	Castings of all grades in this designa-
2	450	65	240	35	22	35	tion are heat treated by normalizing,
5	485	70	275	40	22	35	tempering and stress relieving.
6	485	70	310	45	22	35	
8	550	80	345	50	18	45	
9	585	85	415	60	15	45	
10	585	85	380	55	20	35	
ASTM Designation A 389—77a Alloy-Steel Castings Specially Heat-treated for Pressure-Containing Parts Suitable for High-Temperature Service							
C23	483	70	276	40	18	35	Castings of both grades in this designa-
C24	552	80	345	50	15	35	tion are heat treated by normalizing and tempering.

(Continued on next page)

MAKING STEEL FOR CASTINGS

Three types of furnaces are employed by foundries for melting steel. These are: (1) the electric-arc furnace; (2) the electric-induction furnace; and (3) the open-hearth furnace. Induction furnaces are used only for high-alloy, corrosion- or heat-resistant castings, so their use is limited. The present trend is to replace open-hearth melting furnaces with electric-arc furnaces. Electric-arc furnaces are a more economical source of molten steel, and the quality of electric-furnace steel for castings is generally better.

Electric-arc furnaces in foundries vary in size and the normal weight of charge runs from about 1 to 27 metric tons (1 to 30 net tons). Open-hearth furnaces in foundry practice usually have capacities between about 9 and 91 metric tons (10 and 100 net tons).

In general, furnace charges consist of purchased scrap in the form of billets, shearings, flashings, punchings, plates and turnings; also, scrap from the foundry itself in the form of gates, heads, and scrapped castings. High-grade pig iron, low in phosphorus and sulphur, is used in the open-hearth furnace charges, and frequently in the electric-furnace charges when a high-carbon melt is required. Ferroalloys and slag-making materials are used in about the same proportions in the melt as for steel-ingot production. However, the casting-from-melt yield is considerably lower than that generally experienced with ingots. For every ton of steel castings produced, at least three tons of raw materials are consumed, including scrap steel, pig iron, fuel oil, limestone, sand, organic and clay binders, and miscellaneous materials. This includes the mold-making materials.

The greater portion of the steel for castings is now produced in electric furnaces, but an appreciable amount is still melted in open-hearth furnaces. The basic open-hearth furnace offers one distinct advantage in permitting partial removal of sulphur and phosphorus, thereby making possible the use of less costly scrap. However, in localities where scrap low in sulphur and phosphorus is abundant, the acid open-hearth process commonly is used for various reasons.

The most important advantage of the acid open hearth is that the rate of carbon drop can be closely controlled; if no ore is added, the heat will remain at a given carbon level. This characteristic permits close control of composition of the finished steel, and permits a heat to be held in the furnace for as long as 30 minutes if molds are not available.

Another advantage of an acid furnace is that the bottom and banks are not harmed by the sand that often adheres to the heads and gates that are part of each furnace charge. This sand would be extremely erosive to a basic furnace bottom.

Hydrogen content of acid open-hearth steel is lower than that of basic open-hearth steel by about two or three parts per million, and the recoveries of ferroalloys are higher and more consistent for acid open-hearth steel.

Much has been written concerning the operation of open-hearth furnaces, and elsewhere in this book the subject is treated in some detail. Suffice it to say that the melting process in foundry practice is very similar to that practiced in the steel mills, the chief difference being that greater amounts of deoxidizers are required in foundry practice to produce sound castings, free from porosity. It is necessary to add silicon and manganese in amounts in excess of 0.35 and 0.60 per cent, respectively. The 0.35 per cent minimum silicon content is necessary to assure castings free from blowholes (dead-killed steels) and the 0.60 per cent of manganese is necessary to keep "hot tearing" to a minimum. Also, aluminum and alloys of calcium, manganese, silicon and zirconium commonly are added to insure complete deoxidation.

Electric-arc furnaces offer several distinct advantages over the open-hearth furnaces in the foundry. The modern arc furnace, with its top-charge mechanism, improved switchgear and automatic control, is an extremely flexible unit and rapidly is replacing the open-hearth furnace in the foundry. Electric furnaces permit a wider range of scrap selection than the open hearth, can be charged quickly and without the aid of a charging machine, operation requires fewer men, and the power costs are less than fuel costs in many localities. The furnaces can be shut down and allowed to cool to room temperature with less damage. The types of steels that can be produced in electric furnaces are unlimited. Highly-alloyed steels can be made as well as the plain carbon steels and electric furnaces are more flexible in the size and type of heats. This facilitates foundry planning and affects customer relations by the ability to make quicker deliveries.

Electric furnaces are lined with acid or basic refractories as available scrap or product characteristics may warrant. High-manganese steels are always melted in basic-lined furnaces because the slag from such steels is very destructive to acid refractories. Where low-sulphur and low-phosphorus scrap is not available, basic slags must be used to remove these elements, hence basic-lined furnaces are used. However, the major part of the electric-steel-casting tonnage is made by the acid-electric-furnace process. Many new electric-arc furnaces are equipped with water-cooled sidewall and roof panels. With the sidewall panels, 80 per cent of the sidewall refractory lining can be eliminated: it is not uncommon for sidewall panels to have a life of 1500 or more heats. Water-cooled roofs eliminate up to 95 per cent of the refractory roof lining with only the replaceable delta section being lined.

The generally accepted method of making steel in an acid electric-arc furnace is to melt in at least 20 points (0.20 per cent) above the ordered carbon content, and to lower the carbon content by the use of an oxygen lance. No oxygen is used for 20 minutes prior to adding alloys or deoxidizers to the bath. Precise control of tapping temperatures is achieved by the use of platinum—platinum-rhodium thermocouples.

Temperature control is most important in producing castings, since a heat that is too hot will cause excessive sand "burn-in" on the casting surface and may cause excessive grain growth that cannot be refined by heat treatment.

There is greater variation in the melting operation of basic furnaces than acid furnaces. However, the three principal methods employed in the basic furnace are: (1) the **dead-melt process,** in which a reducing slag is made up as quickly as possible and maintained throughout the heat; (2) the **double-slag process** in which the first oxidizing slag is removed and replaced with a white, lime finishing slag; and (3) the **single-slag method** with an oxidizing slag which may or may not be made reducing before tapping. Each method has its own specific advantages, depending upon the type of steel being made.

MOLDING FOR CASTING STEEL

Patterns and Molds for Steel Castings—The construction of the **pattern** is perhaps the most important single factor in the production of a casting. Not only must the pattern be dimensionally accurate, but full consideration must be given to making it meet the requirements of the foundry equipment and technique. The patterns for steel castings are usually made about 15.5 to 21.0 millimetres per metre (³⁄₁₆ to ¼ inch per foot) larger than the dimensions shown on the drawing to compensate for metal shrinkage, because steel castings cooling from the liquid state to room temperature contract approximately 21 millimetres per metre (¼ inch per foot) depending upon the chemical composition of the steel and the size and design of the casting. It follows, therefore, that to make dimensionally accurate castings provision must be made for metal shrinkage. Patternmakers make such adjustments by **shrink rules** or **patternmakers' rules** which are graduated to compensate for the necessary shrink allowance. Other details of pattern construction, such as allowance for **draft** to facilitate removal of the pattern from the molding sand, **padding** for feed purposes, avoidance of sharp changes in metal sections, and elimination of sharp edges, corners, and reentrant angles, all require full consideration.

There are several types or classes of patterns, each fulfilling a specific need. Patterns may be made of wood, metal or plastic, as required, and used in conjunction with hand-molding or machine-molding methods, depending upon the number of castings to be made and the degree of precision required. Descriptions of patterns and molding machines and procedures will be found in the books listed at the end of this chapter.

The materials used for making molds vary to a great extent, not only from foundry to foundry, but within the same foundry. The size and type of castings, the composition and temperature of the metal, pouring methods and foundry technique, sand mixing and reclaiming facilities, and location of the foundry exercise a profound influence upon the type of molding material. Some foundries prefer to make molds of crude or **bank sand.** For special work **calcined ganister** or **chamotte** is sometimes favored.

Patents have been issued for many molding media, such as those used in the **Fischer process** (calcined aluminous-clay grog), or in the **Randupson process** (sand and cement). Highly refractory **silica sands** of known particle size, mixed with weighed or measured quantities of various types of **clays, resins, dextrins, vegetable oils,** and water to develop desired molding characteristics are used. **Plaster, plastic** and **wax patterns** have been used for special purposes, such as making an experimental casting. Wax patterns are used extensively for castings weighing a few ounces and a special operation known as the "lost wax" or "precision casting" process is based upon the use of such patterns; this process is described in more detail later in this chapter.

Other molding materials include **mold and core washes** which are sprayed, or swabbed, on the mold surfaces to make a smooth mold and resist metal penetration, **gaggers** or reinforcing rods which are placed around the pattern in the sand to add strength to the mold, and **chills** of various types are used to promote directional solidification and reduce the effect of severe temperature gradients caused by sharp changes in metal sections of the casting. **External chills** are placed directly on the pattern and the chill is flush with the mold wall when the pattern is removed from the sand. **Internal chills** are placed in the mold after the pattern has been removed. Internal chills are often more effective than external chills, but they must be positioned with the utmost discretion. **Chaplets** and **stem anchors** are used within the mold cavity to support an internal core. **Nails** frequently are inserted in the mold surface to prevent metal penetration and sand erosion. Rolled-steel, cast-steel or wooden frames, commonly called **flasks,** are used to hold the molding sand around the pattern.

Flasks are made in two sections—the lower half is called the **drag** section; the upper half is called the **cope** section. Occasionally, there is need for a third section which is placed between the cope and the drag. This is known as the **cheek** section. The bottom of the flask is termed the **bottom board** or **bottom plate.** It is important that the cope and drag sections match properly. With small or medium flasks, this is accomplished by hardened pins placed on the outside of the drag section: with larger flasks, this matching is performed with guides, peep-sights, match blocks and many other ingenious devices. The flask sections and the bottom board are held together by "C" clamps or wooden or

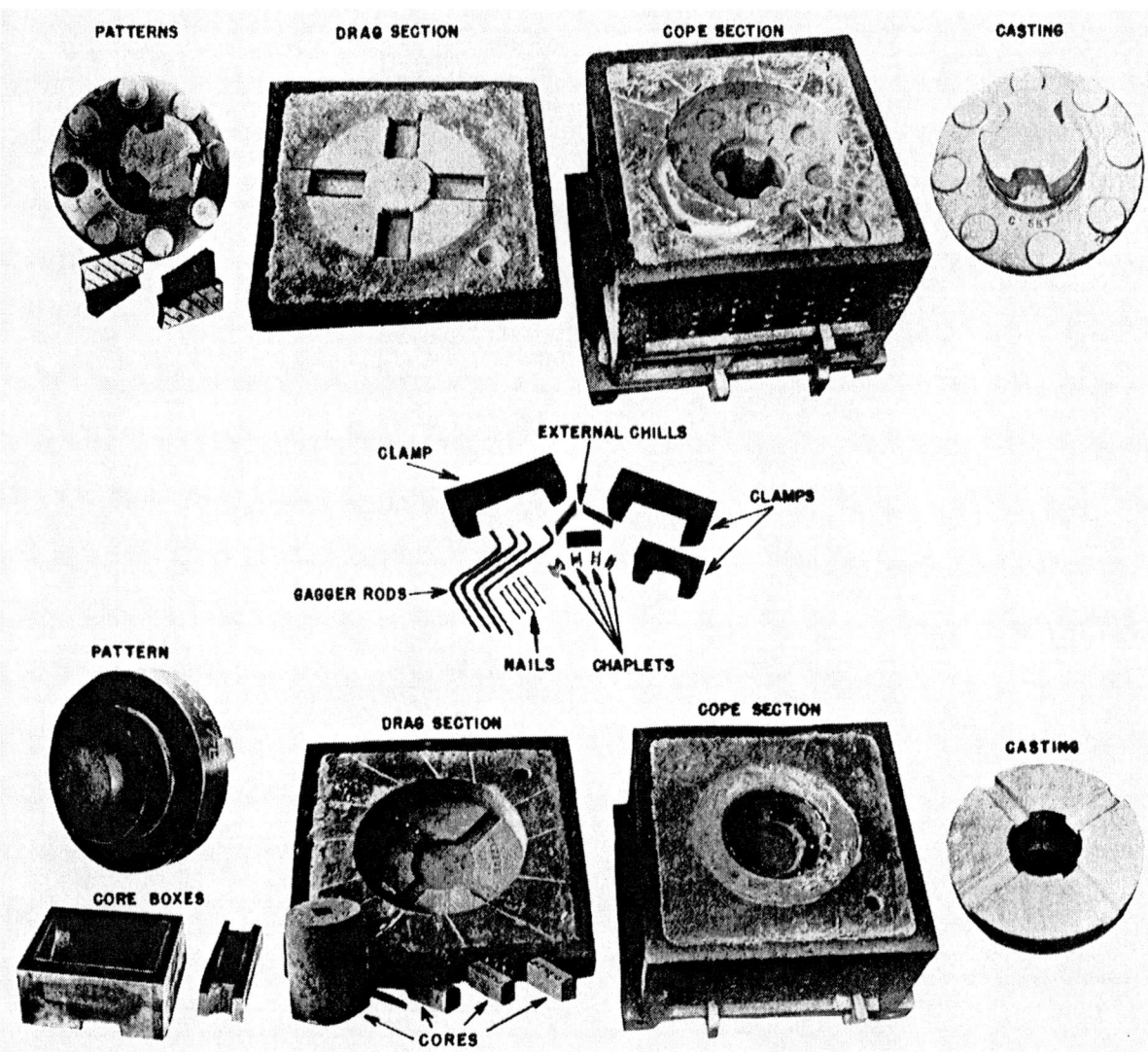

FIG. 40—1. Parts and accessories for preparing typical molds for two relatively small ferrous castings.

steel wedges. Figure 40—1 shows patterns, cope and drag sections of two different molds, and accessories required for one of the molds, and the finished castings produced with this equipment.

Making the Mold—The size and shape of the casting are the controlling factors in deciding how the molten steel should be introduced into the mold and where to locate **gates, risers** and **vents.** A flask is selected which is sufficiently larger than the pattern to provide room between the pattern and the flask wall to accommodate at least about 100 mm (several inches) of sand and the gate system. The flask also must be sufficiently large to permit placement of the **feedheads** or **risers** which are attached to the pattern and leave spaces in the finished mold to serve as reservoirs for molten metal that supply extra metal to feed the voids formed by shrinkage as the metal cools and passes from the liquid to the solid state.

In preparing the mold for a typical steel casting of relatively small size, the bottom board or bottom plate is clamped securely to the flange of the drag section of the mold as shown in Figure 40—1. The pattern is set on the bottom plate in such a position that sand may be rammed over the top of the pattern. The amount of sand must be sufficient to prevent metal runout, to develop the required strength to resist ferrostatic pressure, and to permit handling of the mold. **Facing sand** is riddled or sifted over the surface of the pattern to a depth of 25 mm (1 in.) and packed in any pockets and around the corners of the pattern. **Heap sand,** which is nothing more than used facing and core sands, sometimes re-bonded, is rammed into the flask to a depth of about 100 mm (4 inches). Pneumatic air hammers may be used for this purpose, and many foundries use sandslingers. More heap sand is added to a depth of 150 to 200 mm (6 to 8 in.), or until the flask is full and, after ramming, the excess sand is removed by a straight-edge. This operation is known as **striking off.** A bottom plate is clamped over the top of the flask and the entire flask section is rotated or turned 180°. The first bottom board is now removed and the cope section of the pattern is placed on the drag section. The two parts of the pattern are matched properly by dowel pins fitted into holes located in the face of the **parting line,** joint line, or split sections of the pattern. The cope section of the flask, which is reinforced with cross bars, is set on the drag section and fitted properly, as described. The cross bars in the cope section also serve to support gagger rods which reinforce the sand, to prevent possible distortion of the mold cavity by pressure of the metal entering the mold.

A finely ground sand, known as **parting sand,** is dusted on the face of the drag section to prevent cohesion of the sand in the cope section with the sand in the drag section. **Riser patterns** are placed at the desired locations on the casting pattern and a **sprue stick or gate tile** is placed upright on the sand surface of the drag near the point where the metal is to enter the mold cavity. Facing sand is then riddled over the pattern and packed firmly by hand. The gagger rods are placed and heap sand rammed into the flask in the same manner as described for the drag section. After ramming is complete, the riser patterns and sprue stick, if used, are removed. The cope section is removed from the drag and the cope and drag patterns drawn from the sand by **lifting screws.** Pneumatic **vibrators** are sometimes attached to the patterns for the purpose of freeing the pattern from the sand. The mold is smoothed off and patched, and rough corners and edges are rounded. The gate is cut in the drag section from the base of the sprue stick to the mold cavity. **Cores,** and chaplets, if required, are properly placed as shown by core prints and markings on the pattern. Provision for elimination of mold and core gases is made by jabbing a rod through the mold wall or by scratching **vents** across the drag section of the mold at the parting line. Internal chills, if required, are placed at this time. External chills are placed on the pattern surface prior to adding the facing sand.

If the casting is to be what is known as a **green sand casting,** the cope section of the mold is placed on the drag section and clamped securely to the bottom board. A runner cup, a sand mold having an internal shape similar to a funnel, is placed directly over the gate, and the mold is ready to pour.

If the casting is to be what is known as a **dry sand casting,** the mold cavity is sprayed or swabbed with a mold wash, and placed in a mold oven before the cores are set. Operating temperatures of **mold drying ovens** vary from 150°C to 425°C (300°F to 800°F) and the time of drying may vary from 4 hours to 72 hours, depending upon the type of molding sand, the size of the mold, and the drying characteristics of the oven. After drying, cores and chaplets, if required, are placed and ceramic fiber rope or putty is placed on the joint surface for a seal. The cope section then is fitted properly on the drag section and, after placing the runner cup, the mold is ready to receive the molten metal.

Several other sand molds in popular usage, especially in jobbing foundries, are the furan no-bake and the carbon dioxide/sodium silicate type molds. These molds are made of sand mixes which dry and harden without oven treatment. Minimum ramming is required in producing these molds, molds can be made without flasks, and production rates are relatively high. These molds have higher strength and hardness and can produce castings to closer tolerances than green sand molds. Cores are also being produced of these mixes.

Machine Molding—Fundamentally, machine molding methods differ little from the process previously described for the manual or floor molding operation, the chief difference being that the ramming of the sand and the removal of the pattern from the sand are performed by machine. Also, there are details of pattern construction, such as integral gates and, frequently, attached risers, which eliminate some of the work formerly done by hand. However, the placement of cores, patching and finishing of the mold still remain as the chief function of the molder.

A molding method known as "shell molding" lends itself to the production of molds by machine methods. This method of molding is described briefly in Section 2 of this chapter. It is especially adapted to production of large numbers of small repetitive castings.

Cored Molds for Hollow Castings—Many castings are designed with overhanging flanges, ribs, bolt holes, bosses and hollowed-out sections. Such castings, because of their irregular shape, cannot be produced simply by making a mold, and the foundryman must resort to the use of cores to meet the demands of the design. A core is nothing more than a solid shape made of sand. Sand is rammed either by hand or machine, or blown into a **core box.** When the core box is filled with sand and the excess sand "struck off" with a straight edge, it is turned over and the box lifted from the core thus formed. The ramming operation includes placement of reinforcing rods, or wire, for strengthening purposes. Vent rods are jabbed through the core to permit escape of gases. Sometimes wax wire is rammed in with the sand and, upon baking, the wax wire melts, thereby forming a passageway through which mold gases escape. As soon as the core is vented and sprayed with core wash, it is placed in an oven to bake. The baked core should be hard, strong, and smooth, but it should be sufficiently collapsible at high temperature in order not to cause the casting to "hot tear," i.e., pull apart due to contraction as the metal cools. Appendages attached to the molder's pattern permit the cores to be placed at the proper locations in the mold. These appendages are known as **core prints** and have the same dimensions as the inside dimensions of the core box. The core prints are of sufficient length to provide a good bearing surface to support the core. It is sometimes necessary to tie the cores in the cope section of the mold or to use chaplets and stem anchors to provide additional support.

Cores are made by hand and by various types of molding machines, including sandslinger and rollover machines. For high production work, core blowers are used extensively. Core baking and drying operations require considerable attention and much use is being made of dielectric heating methods, as well as automatically controlled oil- and gas-fired ovens. Perhaps one of the greatest problems in the foundry is the selection and development of core sands to fill the specific needs of the various types of castings. Such problems are so varied and of such importance that specialists are employed in many of the foundries to study sands and sand compositions.

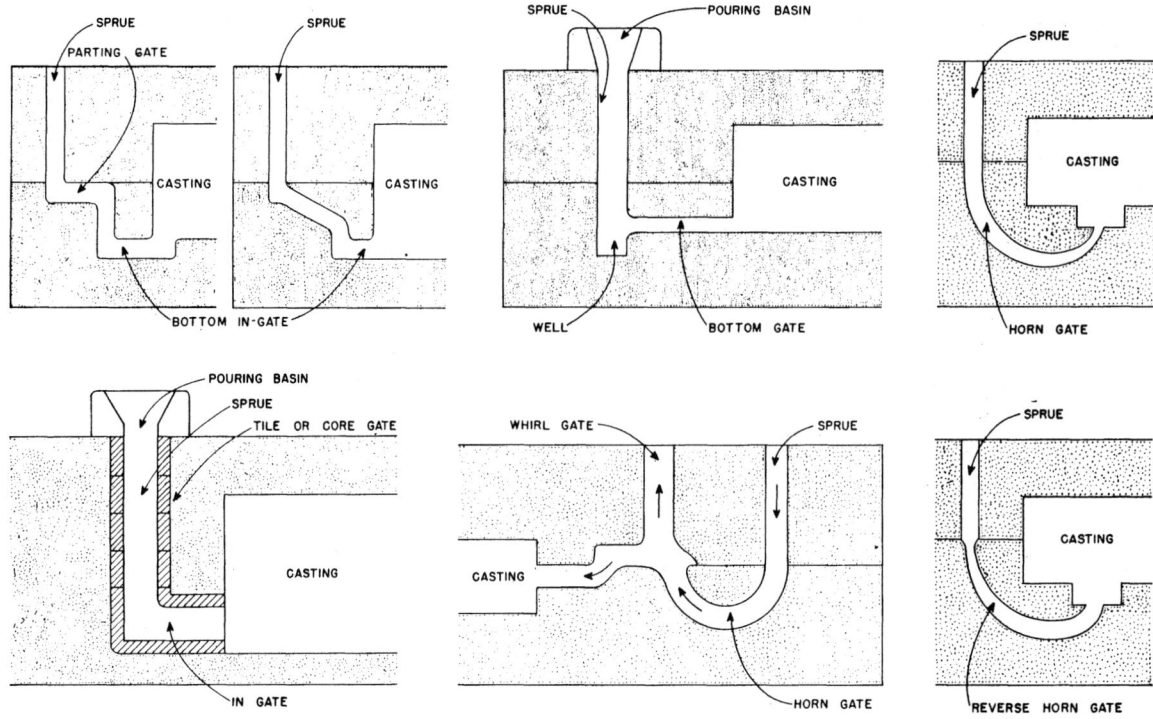

FIG. 40—2. Schematic vertical sections through various molds, illustrating types of gates frequently employed (After Briggs).

Gates, Risers and Vents—The gate, as mentioned previously, is the channel or passageway through which the metal flows in filling the mold. The height, the cross-sectional area, and the shape of the gate, as well as the point at which it enters the mold cavity, are all important factors in any gate system. Many defects in steel castings, such as imbedded sand, cracks, shrinkage cavities, core failures, and misrun castings are attributed directly to poor gating practice. A gate, in order to function properly, must permit the flow of the metal into the mold with the least amount of turbulence. It must carry sufficient volume of metal with enough pressure to fill the mold quickly, and must be arranged to permit proper distribution of temperature gradients.

There are four general types of gates, (1) the **bottom gate**, (2) the **parting gate**, (3) the **top gate**, and (4) the **step gate** which is a combination of the three aforementioned types. The bottom gate is used most commonly for large floor-molded castings. The parting gate is favored for smaller work and is used on practically all machine-molded castings. There are many modifications of each of the four types, such as **horn gates, pencil** or **finger gates, swirl** or **whirl gates, skimmer gates, shower gates, ring gates** and **strainer gates.** Figure 40—2 shows types frequently used.

There are two types of risers; (1) the **open riser**, and (2) the **blind riser**. Open risers are attached to the surface of the casting at some location on the cope side and extend through the sand to the surface of the mold. The size of the riser depends upon the size of the section to be fed and upon the design of the casting. In general, the cross section of the neck of the riser should be equal to the section thickness of the casting at the point of attachment. The body of the riser should be

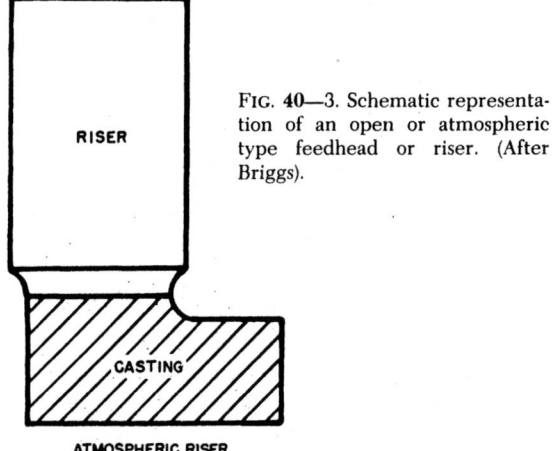

FIG. 40—3. Schematic representation of an open or atmospheric type feedhead or riser. (After Briggs).

slightly larger than the neck and the height at least 1.5 times its diameter or width (see Figure 40—3.)

Blind risers usually are attached to the drag side of the casting and are covered completely with sand. The neck thickness should be 1½ to 2 times the sectional thickness of the casting at the point of attachment. The diameter of the riser body should be 2 to 2½ times the sectional thickness at the point of attachment. The height of the riser should be about 1.5 times the diameter and the topmost surface rounded off or dome shaped. In order to insure proper feeding action of blind risers, a sand core or carbon rod should be inserted in the top section. Also a vent or **pop riser** should extend from the top of the riser to the surface of the mold. The use of the core in blind risers is patented and

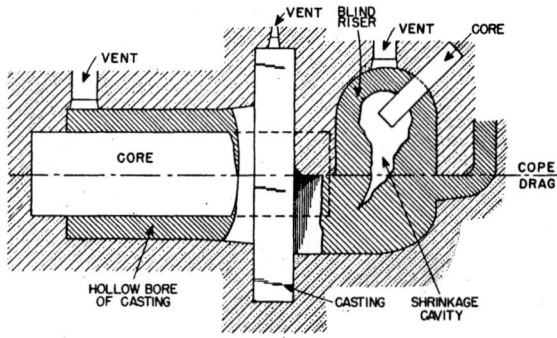

FIG. 40—4. Vertical section of a mold employing a Williams core in conjunction with a blind riser. (After Briggs).

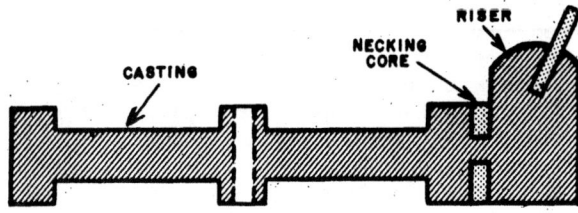

FIG. 40—5. Washburn necking core used to neck down the riser to facilitate removal, applied to a blind riser using a Williams core. (After Briggs).

is known as the **Williams method.** The core serves to admit atmospheric pressure, thereby increasing feed efficiently. Figure **40—4** is a sketch of a Williams core on a blind riser.

Risers placed on steel castings present a serious problem to foundrymen from the standpoint of cleaning, because all risers must be removed prior to shipment. For this reason the **Washburn core** or **necked-down riser** is finding considerable use in conjunction with both open risers and blind risers. The necking core is nothing more than a core through which a small opening connects the riser to the casting. The thickness of the core and the size of the opening depend upon the size of the riser required. The necking core has the advantage that risers can be removed easily by sledge hammers, whereas oxyacetylene torches are necessary to remove conventional risers. Figure **40—5** illustrates use of a necking core with a blind riser.

During the teeming process many mold gases are formed by volatilization and burning of sand binders and by the expansion of the cool air within the mold cavity. Provision for the exhaust of such gases is made by jabbing wires through the mold to the surface or by the use of very small risers. The small holes thus formed are called vents and usually are located over points of the mold cavity where trapped gas might be expected, such as pockets and high points, or at flanges where a blind riser is used in place of an open riser. The vent rod preferably should be flat in shape rather than round.

STEEL CASTING AND FINISHING OPERATIONS

Casting—There are two classes of castings, static

castings, and **centrifugal castings.** The discussion, so far, has dealt only with static castings, i.e., castings which depend upon proper molding practice, atmospheric pressure and gravity to form castings free from internal shrinkage cavities and other defects. The second class, centrifugal castings, as the name implies, makes use of centrifugal action to perform the function of gravity in static casting. There are two types of centrifugal castings—the horizontal type and the vertical type. In the horizontal type, the mold rotates on a horizontal axis. This method is used principally for making long, concentric, hollow castings with uniform wall thickness, where risers are not required. Tubing, pipe, gun barrels, bushings, sleeves, etc., are typical castings made by this process. The vertical method, employing rotation about a vertical axis, is essentially a pressure-casting method and depends, in part at least, on gravity, atmospheric pressure, and proper location of riser or risers to produce sound castings. Gears, piston rings, impellers, propellers, turbine diaphragms, etc., with sections too thin to be cast by static methods or where exceptional feeding problems arise which cannot be surmounted in any other way, are made by the vertical centrifugal process. There are several advantages to be gained from centrifugal casting methods: (1) castings are said to be sounder and have fewer inclusions, (2) fewer cleaning problems, and (3) a 10 to 30 per cent increase in yield.

Shaking Out, Cleaning, Finishing and Testing—After the castings have been poured and sufficient time allowed for solidification and cooling, they are **shaken out,** i.e., the castings are removed from the flask and sand. This operation is performed by placing the entire flask on a vibrating grid called a **shakeout machine.** The clamps are removed from the flask sections and the vibration of the machine causes the sand to fall loose and free of the castings. The castings are then removed to the cleaning room where they are subjected to blast cleaning or put through a tumbling barrel, where the abrasive action of sand particles or metallic shot impinging upon the walls of the castings loosens and removes adhering sand. Risers and gates are removed by oxyacetylene torch, electric arc, abrasive or friction saws, or by sledge hammers. Fins and surface defects found during preliminary inspection are removed as far as possible by chipping and burning. The powder-injection acetylene torch has been applied successfully to the removal of burnt-on sand. Another recent improvement in removing gates, fins, risers, and surface imperfections is by the carbon-arc—air process. This method employs a copper-clad electrode with which an arc is struck against the casting, while a stream of compressed air directed behind the arc flushes away the slag created by the "burning" or oxidation of the metal of the casting liquefied by the heat of the arc. This process offers several advantages over conventional chipping and grinding practice, particularly on carbon-steel castings, and permits removal of risers and gates to the pattern line. The riser necks, gate stubs, and slight surface imperfections are dressed by grinding wheels of various types and sizes. Repairs by welding also are performed sometimes, depending upon the carbon and alloy content of the steel. Castings containing more than 0.30 per cent carbon or of equivalent hardness that require repairing by welding are welded prior to heat treatment. After heat treatment, the castings are subjected to final surface-finishing operations—chipping, grinding, or sand blasting—as required. Finally, they are inspected for surface defects and dimensional accuracy. Inspection may utilize X-ray, gamma-ray, magnetic-particle, fluorescent-penetrant, or ultrasonic methods.

HEAT TREATMENT OF STEEL CASTINGS

In the "as cast" or "green" state, steel castings are relatively brittle and possess poor mechanical properties. It follows, therefore, that in order to render them serviceable they must be subjected to a heat treatment which will refine the grain and break up the dendritic structure, relieve internal stresses, and develop the desired physical and mechanical properties. The heat treatment applied to steel castings depends upon the chemical composition, section size, design, grain size, and the desired mechanical properties. Because of the variations in size, section and design of castings, their heat treatment requires considerable care. However, steel castings respond to heat treatment similarly to wrought steel products and, for all practical purposes, the same rules apply, with care being exercised in placing castings in the furnace in such a way that there is minimum of danger of warpage, distortion or cracks which may cause them to be unfit for service. The common heat treatments applied to steel castings are as follows:

Annealing—The castings are placed in a furnace and heated slowly to a temperature slightly above the Ac_3 point, usually 900°C (1650°F) for carbon steels. The castings are held at that temperature 0.4 hour per centimetre of thickness (1 hour per inch of thickness) of the heaviest section and cooled slowly in the furnace. Such treatment relieves casting stresses, refines the grain, and serves to eliminate the dendritic structure. Annealing raises the tensile and yield strength and increases ductility. It also improves machinability, especially of high-carbon steels.

Normalizing—The normalizing treatment is similar to the annealing process, except that the castings are removed from the furnace at the end of the soaking or holding time and allowed to cool in still air. The normalizing treatment produces a harder steel with higher yield and tensile strength than the annealed product, with ductility value approximately the same. However, internal stresses are not removed to the same extent as in the annealing process. Double normalizing is often employed to produce a more uniform grain structure and such treatment improves the ductility. To remove internal stresses or to reduce the hardness of normalized steels, a **draw** or **tempering** treatment is used frequently, i.e., heating the casting to a temperature below the critical range and allowing it to cool in the furnace. Tempering temperatures for steel castings range from 260° to 675°C (500° to 1250°F).

Quenching and Tempering—The quenching and tempering treatment is confined principally to high-carbon and alloy-steel castings where high strength and resistance to impact and/or abrasion is required. The general practice is to anneal or normalize the casting,

reheat and quench. The tempering treatment should follow immediately, because internal stresses set up by the quench may cause the castings to crack. Sometimes, it is necessary to use a time quench to eliminate cracking, in which case the casting is immersed in the quenching bath for a predetermined time interval, removed, and subjected to the tempering treatment immediately. Such procedure must be controlled closely; otherwise, wide variations in hardness values may result. The end-quench hardenability test is of great value in determining quenching and tempering procedure for steel castings.

Some alloy steels, particularly straight manganese steel in the range of 1.00 per cent up to 2.00 per cent, are susceptible to temper brittleness. Such steels have low impact and ductility values when cooled slowly from the tempering temperature. To avoid this, it is sometimes necessary to cool quickly from the tempering temperature by quenching or air cooling.

Flame Hardening—It is sometimes desirable that castings be differentially hardened, i.e., parts of the casting subject to extreme wear or abrasion should be harder than another part of the same casting which requires a machine operation. A pinion gear, for example, should have hard, wear-resistant teeth with a machinable bore. In such cases, the castings first are annealed or normalized and tempered and then only the surfaces to be hardened are heated to the hardening temperature by a torch or induction-heating apparatus. The heated parts are quenched in water. A time quench often is employed to prevent critical internal stresses. Sections can be hardened up to 6.4 mm (¼-inch) in depth by the flame-hardening process.

HEAT- AND CORROSION-RESISTANT STEEL CASTINGS

Highly Alloyed Steels—Normally, high-alloy steel castings contain at least 12.0 per cent of alloying elements, such as chromium, nickel, cobalt, copper, molybdenum and tungsten, although 4 to 6 per cent chromium steel is generally considered in this class. Various alloy combinations have been developed by consumers and manufacturers for specific applications where conditions of heat and corrosion are critical. There is no clear line of demarcation between heat-resistant castings and corrosion-resistant castings since corrosive media, in the form of fumes and exhaust vapors, usually accompany high temperatures. Moreover, the so-called stainless or corrosion-resistant castings possess excellent mechanical properties at high temperatures. It is not uncommon, therefore, to see castings of a composition designed to resist corrosion being used for heat-resistant applications, and vice versa.

Typical Applications—Heat- and corrosion-resistant castings serve a very broad field. Mine pumps, impellers, fans, valves, valve trim, tubing, sleeve nuts, return bends, agitators, retorts, stills, vessels, digesters, thimbles, nozzles, and centrifuges, are but a few of the castings being used in chemical plants for corrosive applications. The explosives manufacturers, oil and petroleum refineries, paper mill plants, nylon and fabric manufacturers, food and medicinal producers, and the mining and ore-concentrating industries use many corrosion-resistant alloy castings. Heat-resisting alloy castings are used in practically every installation where temperatures exceed 650°C (1200°F) or where alternate heating and cooling cycles are employed; typical examples are chain conveyors, blower tubes, carburizing trays, recuperators, heat exchangers, hearth plates, grate bars, dampers, retorts, rolls, furnace doors, skid bars, carrier blades, muffles, safety sleeves, fractionating towers, normalizing shafts, guides, piercer points, and tube supports.

In the past, the optimum chemical compositions were determined largely by trial and error methods, but in recent years scientific studies and tests conducted by the Alloy Casting Institute and other interested parties have done much toward the standardization and proper use of the various grades of high-alloy steels. Table 40—III shows the compositions of some of the important grades of heat- and corrosion-resistant steels used for castings: Table 40—IV gives their mechanical properties.

Melting—Steels for heat- and corrosion-resistant castings are melted in arc-type electric furnaces and high-frequency electric induction furnaces. The direct-arc electric furnace is used more extensively, but modern developments in induction melting equipment make it an extremely desirable unit for melting alloy steels for metallurgical, as well as operating, reasons. Furnace linings may be acid or basic and, in the case of the induction furnace, the lining may be neutral. The

basic-lined furnace has the advantage of greater alloy efficiency and makes possible the use of chrome ore and nickel oxide for chromium and nickel additions. Also, it is possible to oxidize and reduce heats containing alloy scrap without serious loss of expensive alloys. On the other hand, acid furnaces require less skill to operate, they are faster, and the refractory costs are lower. Melting costs of induction furnaces compare favorably with acid- or basic-lined arc furnaces, and chemical composition and metal temperatures can be controlled within closer limits. Moreover, alloy recoveries are said to be higher. The induction furnace is limited to the dead-melt process but with present developments in flushing inert gases, such as argon and nitrogen, through the bath to remove dissolved gases, this does not pose a serious disadvantage.

Regardless of the type of melting unit, strict metallurgical control is necessary to produce high-alloy steels. Care must be exercised in the purchase of ferroalloys and other raw materials, and scrap must be segregated according to composition and used with utmost discretion.

Casting—High-alloy castings are made in static molds and centrifugal molds of the horizontal or vertical type. By far the greater tonnage is produced in static molds by conventional molding methods. However, a large number of horizontal centrifugal castings are made in the form of tubing, retorts, rolls and rollers. For work of this type, the centrifugal method is

Table 40—III. Chemical Compositions of Corrosion- and Heat-Resistant Iron-Chromium and Iron-Chromium-Nickel Alloys for Castings (Per Cent).[a]

Grade	C max	Mn max	Si max	P max[d]	S max[d]	Cr	Ni	Mo	Nb (Cb)	Se	Cu
\multicolumn{12}{c}{ASTM Designation A 743—81a}
\multicolumn{12}{c}{Corrosion-Resistant Iron-Chromium and Iron-Chromium-Nickel Alloy Castings for General Applications}
CF–8	0.08	1.50	2.00	0.04	0.04	18.0–21.0	8.0–11.0	–	–	–	–
CG–12	0.12	1.50	2.00	0.04	0.04	20.0–23.0	10.0–13.0	–	–	–	–
CF–20	0.20	1.50	2.00	0.04	0.04	18.0–21.0	8.0–11.0	–	–	–	–
CF–8M	0.08	1.50	2.00	0.04	0.04	18.0–21.0	9.0–12.0	2.0–3.0	–	–	–
CF–8C	0.08	1.50	2.00	0.04	0.04	18.0–21.0	9.0–12.0	–	(e)	–	–
CF–16F[b]	0.16	1.50	2.00	0.04[b]	0.04[b]	18.0–21.0	9.0–12.0	(b)	–	(b)	–
CH–20[c]	0.20[c]	1.50	2.00	0.04	0.04	22.0–26.0	12.0–15.0	–	–	–	–
CK–20	0.20	2.00	2.00	0.04	0.04	23.0–27.0	19.0–22.0	–	–	–	–
CE–30	0.30	1.50	2.00	0.04	0.04	26.0–30.0	8.0–11.0	–	–	–	–
CA–15	0.15	1.00	1.50	0.04	0.04	11.5–14.0	1.00 max	0.50 max	–	–	–
CA–15M	0.15	1.00	0.65	0.04	0.04	11.5–14.0	1.00 max	0.15–1.0	–	–	–
CB–30	0.30	1.00	1.50	0.04	0.04	18.0–21.0	2.00 max	–	–	–	(f)
CC–50	0.50	1.00	1.50	0.04	0.04	26.0–30.0	4.00 max	–	–	–	–
CA–40	0.20–0.40	1.00	1.50	0.04	0.04	11.5–14.0	1.0 max	0.5 max	–	–	–
CF–3	0.03[g]	1.50	2.00	0.04	0.04	17.0–21.0	8.0–12.0	–	–	–	–
CF–3M	0.03[g]	1.50	1.50	0.04	0.04	17.0–21.0	9.0–13.0	2.0–3.0	–	–	–
CG–8M	0.08	1.50	1.50	0.04	0.04	18.0–21.0	9.0–13.0	2.0–3.0	–	–	–
CN–7M	0.07	1.50	1.50	0.04	0.04	19.0–22.0	27.3–30.5	2.0–3.0	–	–	3.0–4.0
CN–7MS	0.07	1.00	2.50–3.50	0.04	0.03	18.0–20.0	22.0–25.0	2.5–3.0	–	–	1.5–2.0
CW–12M[h]	0.12	1.00	1.50	0.04	0.03	15.5–20.0	remainder	16.0–20.0	–	–	–
CY–40[i]	0.40	1.50	3.00	0.03	0.03	14.0–17.0	remainder	–	–	–	–
CA–6NM	0.06	1.00	1.00	0.04	0.03	11.5–14.0	3.5–4.5	0.4–1.0	–	–	–
CD–4MCu	0.04	1.00	1.00	0.04	0.04	24.5–26.5	4.75–6.0	1.75–2.25	–	–	2.75–3.25
CA–6N	0.06	0.50	1.00	0.02	0.02	10.5–12.5	6.0–8.0	–	–	–	–

(Continued on next page)

Table 40—III (Continued)

ASTM Designation A 297—81
Heat-Resistant Iron-Chromium and Iron-Chromium-Nickel Castings for General Applications

Grade	C max	Mn max	Si max	P max	S max	Cr	Ni	Mo[j]	Nb (Cb)	Se	Cu
HF	0.20–0.40	2.00	2.00	0.04	0.04	18.0–23.0	8.0–12.0	0.50	–	–	–
HH	0.20–0.50	2.00	2.00	0.04	0.04	24.0–28.0	11.0–14.0	0.50	–	–	–
HI	0.20–0.50	2.00	2.00	0.04	0.04	26.0–30.0	14.0–18.0	0.50	–	–	–
HK	0.20–0.60	2.00	2.00	0.04	0.04	24.0–28.0	18.0–22.0	0.50	–	–	–
HE	0.20–0.50	2.00	2.00	0.04	0.04	26.0–30.0	8.0–11.0	0.50	–	–	–
HT	0.35–0.75	2.00	2.50	0.04	0.04	15.0–19.0	33.0–37.0	0.50	–	–	–
HU	0.35–0.75	2.00	2.50	0.04	0.04	17.0–21.0	37.0–41.0	0.50	–	–	–
HW	0.35–0.75	2.00	2.50	0.04	0.04	10.0–14.0	58.0–62.0	0.50	–	–	–
HX	0.35–0.75	2.00	2.50	0.04	0.04	15.0–19.0	64.0–68.0	0.50	–	–	–
HC	0.50 max	1.00	2.00	0.04	0.04	26.0–30.0	4.00 max	0.50	–	–	–
HD	0.50 max	1.50	2.00	0.04	0.04	26.0–30.0	4.0–7.0	0.50	–	–	–
HL	0.20–0.60	2.00	2.00	0.04	0.04	28.0–32.0	18.0–22.0	0.50	–	–	–
HN	0.20–0.50	2.00	2.00	0.04	0.04	19.0–23.0	23.0–27.0	0.50	–	–	–
HP	0.35–0.75	2.00	2.50	0.04	0.04	24–28	33–37	0.50	–	–	–

[a] These compositions for the indicated ASTM Designations are excerpted from ASTM Standard Specifications in ASTM Standards published by the American Society for Testing and Materials, 1916 Race St., Philadelphia, Pa. 19103, and represent part of each specification. The appropriate ASTM Standard Specification for each of the designations should be consulted for complete specifications. (See Table 40—IV for mechanical properties of steels of these compositions and Table 40—V for heat-treatment requirements.)

[b] For free-machining properties the composition of Grade CF-16F may contain suitable combinations of selenium, phosphorus, and molybdenum (Grade CF-16F) or of sulphur and molybdenum (Grade CF-16Fa) as follows:

Selenium, phosphorus and molybdenum:	
Selenium, %	0.20–0.35
Phosphorus, max, %	0.17
Molybdenum, max, %	1.50
Sulphur and molybdenum:	
Sulphur, %	0.20-0.40
Molybdenum, %	0.40-0.80

Other combinations of elements for free-machining properties may be agreed upon by the manufacturer and the purchaser.

[c] For the more severe general corrosive conditions, and when so specified, the carbon content shall not exceed 0.10%. This low-carbon grade shall be designated as Grade CH-10.

[d] Chemical analysis is not normally required for the elements, phosphorus, sulphur, and molybdenum, but if they are present in amounts over those stated they may be cause for rejection.

[e] Grade CF-8C shall have a columbium content of not less than 8 times the carbon content and not more than 1.0%. If a columbium-plus-tantalum alloy in the approximate Cb:Ta ratio of 3:1 is used for stabilizing this grade, the total columbium-plus-tantalum content shall not be less than 9 times the carbon content and shall not exceed 1.1%.

[f] For Grade CB-30 a copper content of 0.90 to 1.20% is optional.

[g] For purposes of determining conformance with this specification, the observed or calculated value for carbon content shall be rounded to the nearest 0.01% in accordance with the rounding method of Recommended Practice E 29.

[h] Also contains 5.25 max W, 0.40 max V, 2.50 max Co and 7.50 max Fe.

[i] Maximum Fe content, 11%.

[j] Castings having a specific molybdenum range agreed upon by the manufacturer and the purchaser may also be furnished under these specifications.

of great advantage since casting yield approaches 90 per cent, no cores are required, and the molds may be made of sand or, in some instances, cast iron sprayed with a highly refractory material. The vertical centrifugal method of casting is used only when static methods fail and the castings cannot be made any other way. Impellers containing thin vanes, for example, or castings designed in such a manner that they cannot be properly fed, are made by the vertical centrifugal method.

Molding—Fundamentally, the molding procedure for high-alloy castings is the same as that employed for carbon-steel castings. However, greater skill and more precision is required on the part of the molder because excess stock on castings necessitates expensive machining operations. Every effort must be expended to produce clean, smooth, sound castings. The molding sand must be rigidly controlled and the particle size and bonding must be such that details of the pattern are reproduced accurately. The risers and gates should be located so as to avoid internal defects and center-line shrinkage, for such defects in castings exposed to high temperatures or corrosive media cause premature failures. The use of internal chills should be avoided, but if the casting problem cannot be surmounted in any other way, the chill material should be of the same chemical composition as the metal in the casting. Likewise, stem anchors and chaplets should be used sparingly and with care. Chill nails should never be used on high-alloy castings because contamination of the sur-

Table 40—IV. Mechanical Properties of Some Corrosion- and Heat-Resistant Iron-Chromium and Iron-Chromium-Nickel Alloys for Castings.[a]

Grade	Tensile Strength (min.) MPa	ksi	Yield Strength (min.) MPa	ksi	Elongation in 50 mm or 2 in. min. (%)	Reduction of Area min. (%)
ASTM Designation A 743—81a						
Corrosion-Resistant Iron-Chromium and Iron-Chromium-Nickel Alloy Castings for General Applications						
CF-8	485[b]	70[b]	205[b]	30[b]	35	–
CG-12	485	70	195	28	35	–
CF-20	485	70	205	30	30	–
CF-8M	485	70	205	30	30	–
CF-8C	485	70	205	30	30	–
CF-16 and CF-16Fa	485	70	205	30	25	–
CH-20 and CH-10	485	70	205	30	30	–
CK-20	450	65	195	28	30	–
CE-30	550	80	275	40	10	–
CA-15 and CA-15M	620	90	450	65	18	30
CB-30	450	65	205	30	–	–
CC-50	380	55	–	–	–	–
CA-40	690	100	485	70	15	25
CF-3	485	70	205	30	35	–
CF-3M	485	70	205	30	30	–
CG-8M	520	75	240	35	25	–
CN-7M	425	62	170	25	35	–
CN-7MS	485	70	205	30	35	–
CW-12M	495	72	315	46	4.0	–
CY-40	485	70	195	28	30	–
CA-6NM	755	110	550	80	15	35
CD-4MCu	690	100	485	70	16	–
CA-6N	965	140	930	135	15	50
ASTM Designation A 297—81						
Heat-Resistant Iron-Chromium and Iron-Chromium-Nickel Alloy Castings for General Applications						
HF	485	70	240	35	25	
HH	515	75	240	35	10	
HI	485	70	240	35	10	
HK	450	65	240	35	10	
HE	585	85	275	40	9	
HT	450	65	–	–	4	
HU	450	65	–	–	4	
HW	415	60	–	–	–	
HX	415	60	–	–	–	
HC	380	55	–	–	–	
HD	515	75	240	35	8	
HL	450	65	240	35	10	
HN	435	63	–	–	8	
HP	430	62.5	235	34	4.5	

[a] These mechanical properties for the indicated ASTM Designations are excerpted from ASTM Standard Specifications published by the American Society for Testing and Materials, 1916 Race St., Philadelphia, Pa. 19103, and represent part of each specification. The appropriate ASTM Standard Specifications for each of the designations should be consulted for complete specifications. (See Table 40—III for chemical compositions and Table 40—V for heat-treatment requirements for developing the properties of the corrosion-resistant steels listed here. Properties of the heat-resistant steels are for the as-cast condition.)

[b] For low-ferrite or nonmagnetic castings of this grade, the following values shall apply: tensile strength, min., 450 MPa (65 ksi); yield point, min., 195 MPa (28 ksi).

face of the casting by the steel nail head may initiate corrosion and ultimately cause failure.

Finishing Operations—Cleaning and finishing operations for high-alloy steel castings differ but slightly from ordinary steel castings. Shot blasts or other abrasive cleaning devices must contain high-alloy shot or what is called high-alloy sand, since steel shot causes the castings to corrode. Risers and gates are removed by burning with an electric arc, although abrasive or friction saws are sometimes used for this purpose. The riser and gate pads are removed completely by grinding and all surface imperfections must be smoothed off by grinding. Frequently, alloy castings are ground to template. Casting repair by welding is limited strictly to minor defects and when welding is permitted, a rod of the same chemical composition as the casting generally is used. Heat treatments vary according to compositions, and range from carbide-solution treatments for corrosion-resistant castings such as 18 Cr-8 Ni, to precipitation-hardening treatments for the high-chromium, low-nickel-molybdenum steels as shown in Table 40—V. Steels containing 11.5 to 14.0 per cent chromium respond to the conventional hardening and tempering treatments. Heat-resistant alloys, such as 24 Cr-12 Ni, are sometimes treated to improve machin-ability, in which case they are heated to approximately 870°C (1600°F) and slowly cooled. The purpose of the treatment is to precipitate carbides.

Before shipment, castings are checked thoroughly for dimensional inaccuracies and surface imperfections and finally are cleaned in a sand blast or similar equipment.

Methods of Sampling and Testing—Sampling and preparation of samples for chemical tests have become standardized fairly well in all foundries. In general, test specimens for chemical analyses are poured from the ladle at about the mid-point of the teeming operation so that the sample is representative of the entire heat. The sample is cleaned and drilled; the first 6.4 mm (¼-inch) of drillings is thrown away as insurance against contamination. The chemical composition is then determined from subsequent drillings according to proven analytical procedures. Heat numbers are stamped on the castings and test specimens, thereby making possible the identification of the chemical composition of the castings. Most foundry laboratories are equipped to determine completely and accurately the compositions of steels, ferroalloys, sands, and various raw materials.

Mechanical properties of cast steels are determined

Table 40—V. Heat-Treatment Requirements for Developing Acceptable Corrosion Resistance in Iron-Chromium and Iron-Chromium-Nickel Alloy Castings.[a]

Grade	Heat Treatment
CF-8, CG-8M, CG-12, CF-20, CF-8M, CF-8C, CF-16F, CF-16Fa	Heat to 1040°C (1900°F) minimum, hold for sufficient time to heat casting to temperature, quench in water or rapid cool by other means so as to develop acceptable corrosion resistance.
CH-20, CE-30, CK-20	Heat to 1093°C (2000°F) minimum, hold for sufficient time to heat casting to temperature, quench in water or rapid cool by other means so as to develop acceptable corrosion resistance.
CA-15, CA-15M, CA-40	(1) Heat to 955°C (1750°F) minimum, air cool and temper at 595°C (1100°F) minimum, or (2) Anneal at 790°C (1450°F) minimum.
CB-30, CC-50	(1) Heat to 790°C (1450°F) minimum, and air cool, or (2) Heat to 790°C (1450°F) minimum, and furnace cool.
CF-3, CF-3M	(1) Heat to 1040°C (1900°F) minimum, hold for sufficient time to heat casting to temperature, and cool rapidly so as to develop acceptable corrosion resistance, or (2) As cast if corrosion resistance is acceptable.
CN-7M, CN-7MS	Heat to 1120°C (2050°F) minimum, hold for sufficient time to heat casting to temperature, quench in water or rapid cool by other means so as to develop acceptable corrosion resistance.
CY-40	As cast.
CW-12M	As agreed upon by the manufacturer and the purchaser so as to develop acceptable corrosion resistance.
CA-6NM	Heat to 955°C (1750°F) minimum, air cool to 95°C (200°F) maximum, and final temper between 565°C (1050°F) and 620°C (1150°F).
CD-4MCu	Heat to 1120°C (2050°F), hold for sufficient time to heat casting uniformly to temperature, furnace cool to 1040°C (1900°F), hold for a minimum of 15 min., and quench in water or rapid cool by other means so as to develop acceptable corrosion resistance.
CA-6N	Heat to 1040°C (1900°F), air cool, reheat to 815°C (1500°F), air cool, and age at 425°C (800°F), holding at each temperature sufficient time to heat casting uniformly to temperature.

[a] Excerpted from ASTM Standard Designation A 743—81a, published by the American Society for Testing and Materials, 1916 Race St., Philadelphia, Pa. 19103, which should be consulted for complete specification.

from **test coupons** or **test lugs** cast integrally with the castings. At times it is necessary, by reason of the size or design of the casting, to make separate test coupons which are identified properly with the castings they represent by heat numbers stamped on the castings and coupons. The test bars always are heat treated with the castings, machined and tested in accordance with standard mechanical-testing procedure. Test specimens cut from castings sometimes show inferior mechanical properties when compared to standard coupon specimens. This is due principally to the mass effect, although in some instances it may be attributed to improper foundry technique, such as inadequate feeding action. Generally, results of mechanical tests as performed in the foundries may be considered indicative of the quality of the castings, but it should be recognized that excellent mechanical properties shown by tests do not preclude the possibility of poor mechanical properties in a casting.

Many foundries make use of modern non-destructive testing equipment, such as X-ray and gamma-ray apparatus. Also, an ultrasonic method for detecting internal defects and magnetic particle testing for detecting surface imperfections, such as cracks and small sand inclusions, is used extensively. In the case of high-alloy steels of the austenitic type which are non-magnetic, a method which makes use of a fluorescent penetrant and black light, is used to indicate cracks. Destructive tests in which castings are machined to complete destruction are employed frequently to determine the internal soundness of castings. Such tests are used as control measures and serve to establish foundry technique for subsequent castings. The microscope has been used in foundries for many years as a control tool to establish heat-treating practices and for other metallurgical investigations.

PRECISION STEEL CASTINGS

The **lost wax** or **investment molding** process of making castings is often referred to as "precision casting." Castings weighing one pound or less can be made to tolerances of 0.13 to 0.25 mm (0.005 to 0.010 inch); considerably larger castings have been made successfully by this method. The process is restricted largely to high-alloy steel parts which cannot be machined readily and which cannot be formed by hot or cold working methods. Such parts include valve parts; turbine blades, buckets and nozzles; molds and dies for the plastic and ceramic industries; small gears for timers; hobs, milling cutters, magnets, jewelry, surgical and dental tools, etc. Figure 40—6 shows some typical precision castings.

The process consists of making a pattern of free-machining steel, around which a mold is formed of an alloy of low melting point. The mold thus formed is used for making wax patterns. The wax patterns are then used to make up molds similar to those used in the conventional molding process, except that risers are not required and the molding sand is bonded with an ethyl silicate, making a very hard and highly refractory mold when dried. The wax patterns within the mold are not lifted out of the sand but are removed by melting with steam or hot air; hence the name "lost wax." The molds then are inverted over small induction furnaces containing the metal to be cast and clamped to the furnace. The furnace is rotated 180 degrees and the molten metal forced into the molds by air pressure, by centrifugal force, or by vacuum. The castings finally are cleaned and finished just as in conventional casting practice.

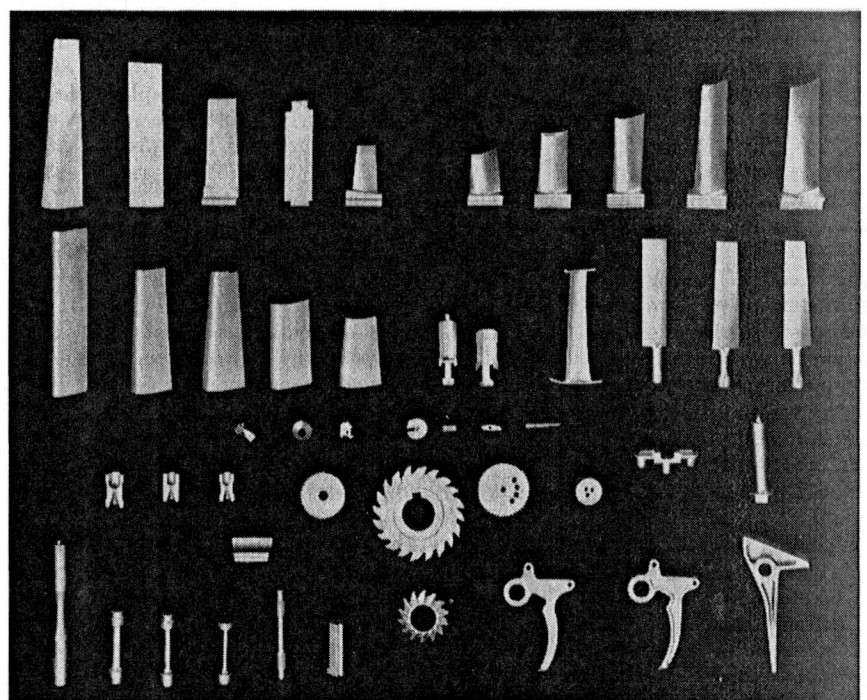

FIG. 40—6. Typical castings produced by the lost wax or investment molding process for precision casting of materials not readily machinable that cannot be formed by hot or cold working. (Courtesy, Westinghouse Electric Corporation.)

IRON CASTINGS

Pig Iron for Castings—Iron castings are of innumerable kinds and uses, roughly grouped as **chilled-iron castings, gray-iron castings, alloyed-iron castings, malleable castings** and **nodular-iron castings.** In general, castings are made by mixing and melting together different grades of pig iron; different grades of pig iron and foundry scrap; different grades of pig iron, foundry scrap and steel scrap; or different grades of pig iron, foundry scrap, steel scrap and ferroalloys or other metals. In rare instances, molten iron direct from the blast furnace is run to a mixer, then to an electric furnace in which its composition is adjusted. In all remelting operations, the pig iron undergoes some change. Hence, physical properties of the pig iron itself are held subordinate to chemical composition in iron for remelting. However, by selecting iron of different grades and controlling the rate of cooling, the widest variations in mechanical properties may be obtained, from extreme hardness and brittleness with low impact resistance to extreme softness with, however, considerable strength and enhanced toughness. Thus, without the use of alloys and by selecting different malleable and foundry pig irons and controlling the rate of cooling, the following ranges in properties may be obtained from gray irons:

Brinell Hardness Number . . . . 100 to 500
Tensile Strength . 69 to 414 MPa (10 000 to 60 000 psi)
Deflection (Tranverse) . 1.0 to 9.1 mm (0.04 to 0.36 in.)
Modulus of Elasticity . . . . 82 738 MPa to 199 949 MPa
(12 000 000 to 29 000 000 psi)

Other important properties in iron for castings are:
a. **Fluidity,** at time of casting, which depends upon composition and temperature. The melting and freezing points of pig iron and cast irons of eutectic and hypoeutectic composition, for a given phosphorus content, vary inversely with the carbon and silicon, from 1088°C (1990°F) with 4.40 per cent carbon and 0.6 per cent silicon to 1250°C (2280°F) with 3.56 per cent carbon and 2.40 per cent silicon, the melting point varying inversely with the combined carbon for a given low (under one per cent) silicon content.
b. **Shrinkage,** which is the net result of contractions and expansions in cooling to atmospheric temperature, ie., through the point where solidification begins, through the solidification range, and in cooling from this range to atmospheric temperature. Shrinkage depends upon the temperature of the iron as it enters the mold, the composition of the iron, the rate of cooling, and subsequent heat treatments, and it varies from 2.60 to 10.41 mm per metre (1/32 to 1/8 inch per foot).
The net volume change on cooling from the liquid state is incurred stepwise and is complicated by structural and other changes that take place, somewhat in the following sequence and at the indicated approximate temperatures.
   1. Contraction in the liquid state—Tapping temperature to solidification (say 1205°C, corresponding to 2200°F).
   2. Contraction, liquid-to-solid, solidification—Temperature nearly constant at freezing, largest change.
   3. Contraction of austenite and ledeburite—1205°C (2200°F) to 1120°C (2050°F).
   4. Expansion due to graphitization—1120°C (2050°F) to 1065°C (1950°F)—large.
   5. Contraction due to cooling—1065°C (1950°F) to 720°C (1325°F).
   6. Expansion—austenite to pearlite, gamma to alpha iron—720°C (1325°F).
   7. Contraction due to cooling—720°C (1325°F) to room temperature.

c. **Growth** is the tendency of castings to increase in volume after repeated heatings to temperatures between 455° and 900°C (850° and 1650°F) in the absence of stress. It is usually measured in linear units and expressed in per cent. It varies from essentially zero to several per cent (there is an example of a reported extreme of 50 per cent) according to the time of heating or number of heating cycles and type of iron. For ordinary castings, 5 per cent is extreme, while a growth of 0.002 centimetre per linear centimetre (0.002 inch per linear inch) is common, and depends upon composition and prior heat treatment. The causes of growth are believed to be (1) graphitization and (2) penetration of active gases into the discontinuities of the coarser-grained irons.
d. **Creep** is the tendency to increase in length under stress (tension) at elevated temperatures above 370°C (700°F), the stress being but a fraction of that required to break a specimen of the material in a short-time tension test at room temperature. As it is expressed in different ways, one example is cited for an ordinary gray iron. In short-time tension tests, the breaking loads were 236 MPa (34 200 psi) at 21°C (70°F), and 242 MPa (35 100 psi) at 370°C (700°F); at a testing temperature of 370°C (700°F) the test broke in 9 days under a load of 124 MPa (18 000 psi) and in 90 days under a load of 62 MPa (9 000 psi). Creep is lessened by certain alloy additions.
e. **Porosity, density, and closeness of grain** are designations referring primarily to macroscopic structure, particularly of heavier sections, as revealed by fractured surfaces. A porous structure is conducive to weakness, and is indicative of a condition bordering on **unsoundness,** which term is applied when the structure shows more clearly visible blowholes or gas pockets, or small cavities due to bleeding or shrinkage. The term "density" is sometimes used loosely to designate grain size rather than mass per unit weight; thus, iron showing a fine-grained fracture is said to be a "dense" iron. Porosity may be due to segregation of low-melting constituents, to minute slaglike inclusions, or oxides that react with carbon to form gas while the casting is solidifying.

f. **Machinability** (the capability of being machined) may be viewed from various standpoints. Most frequently it is considered to be the characteristic (of the metal being machined) which causes more or less wear on the cutting tool; less frequently the definition involves the power required for the machining operation; often it refers to the degree of smoothness obtainable on the machined surface. With due precautions to keep the casting free of sand or dirt, machinability is controlled through the composition and rate of cooling, but often must be sacrificed for some more essential property, such as strength or toughness.

g. **Graphitization,** a phenomenon common to pig iron and cast iron, refers to the decomposition of iron carbide or, in any event, the rejection of elemental carbon in a casting after solidification has taken place, the carbon being liberated in the form of graphite which is usually found existing as minute, flaky particles disseminated throughout the casting. This property is controlled basically through composition of the iron, and is very important in castings, particularly malleable castings in which, upon

reheating, carbon is rejected as nodules. As will be shown later, the formation of graphite is promoted by slow cooling, hindered or prevented by rapid cooling, and, if suppressed, can be brought about by subsequent heat treatment.

The effects of the different elements upon the properties of pig iron and cast iron will now be discussed.

IRON COMPOSITION vs. PROPERTIES

Forms of Carbon in Pig Iron—The factors influencing the carbon content of pig iron were discussed in Chapter 15. In considering the effect of temperature on molten iron, it is necessary to keep in mind that the effective temperatures are inevitably above a certain minimum of about 1140°C (2085°F), marking the freezing point of the iron-carbon eutectic, Referring to the iron-carbon equilibrium diagram (see Index) and Figure 40—7, it will be observed that iron at 1300°C (2370°F) may absorb about 5.0 per cent carbon. Being a saturated solution, **cementite** (Fe_3C, 93.33 per cent Fe and 6.67 per cent C) crystallizes and separates as the liquid cools to 1130° to 1140°C (2065° to 2085°F), at which temperature we have the eutectic (lowest freez-

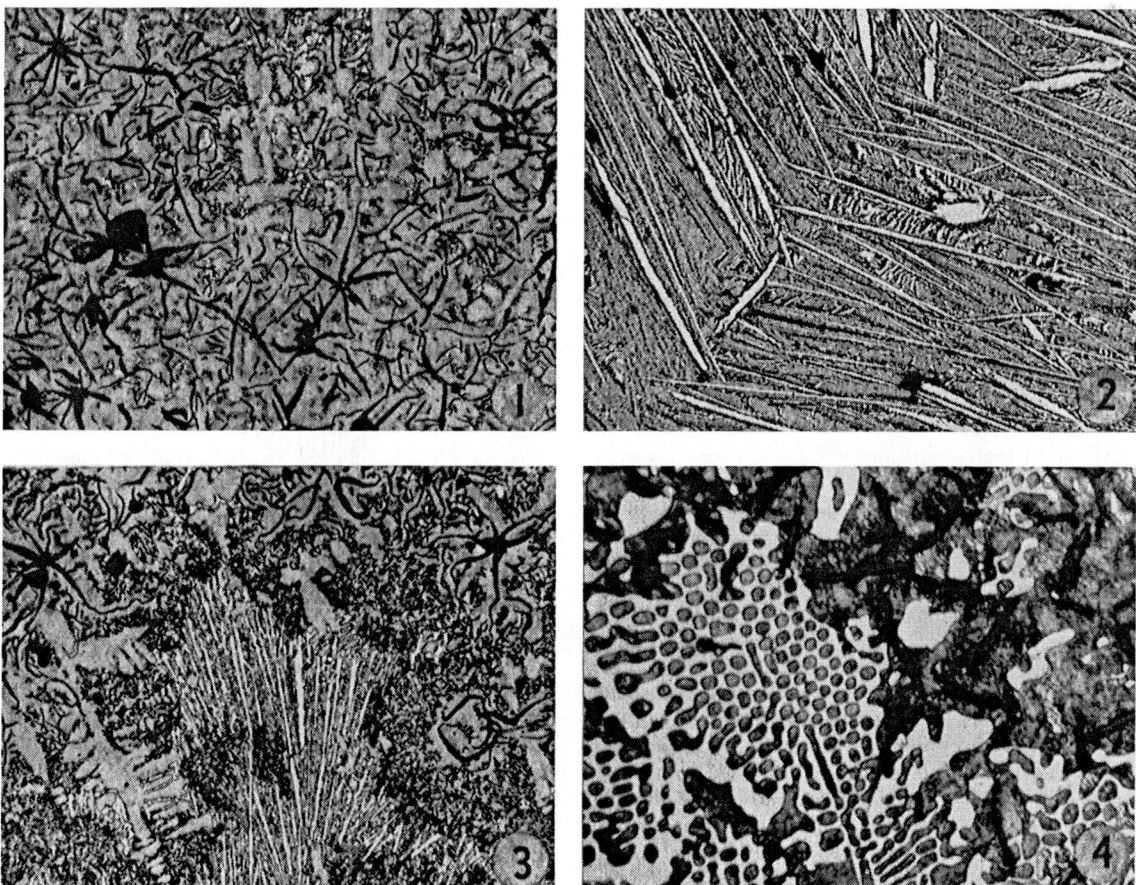

FIG. 40—7. Photomicrographs of typical microstructural constituents of cast iron as influenced by heat treatment and cooling rate. Sample 1: Annealed specimen, showing graphite flakes in a fine pearlite matrix. Sample 2: "Chilled" sample showing ledeburite containing long needles and irregularly shaped areas of iron carbide (cementite; Fe_3C). Sample 3: Annealed specimen; region near chilled surface, showing areas of "chilled" ledeburite and graphite in a fine pearlite matrix. Sample 4: Annealed specimen; showing ledeburite and graphite flakes in a fine pearlite matrix. White constituent is cementite. All samples etched with picral. Magnifications: Samples 1, 2 and 3, 100×; Sample 4, 1000×.

ing) liquid, called **ledeburite,** containing 4.3 per cent carbon. Upon freezing, this eutectic liquid becomes a two-phase solid, one phase of which is composed of primary austenite (47.7 per cent) containing 1.7 per cent carbon in solution (equivalent to 12.1 per cent Fe_3C) and the other (52.3 per cent) of cementite. In the freezing of the eutectic, no drop in temperature occurs until heat equivalent to the heat of fusion of this eutectic is abstracted. Then, if the cooling is extremely rapid, practically all the carbon remains in these combined forms and the metal is extremely hard and brittle. But the saturated liquid solution may, in rapid cooling, produce a supersaturated solid solution. As another complication, the iron changes its allotropic form at about 710°C (1310°F) losing its power to hold carbon in solution, and if then the cooling is slow, some of the already formed cementite breaks down, or decomposes, to precipitate free carbon, which assumes the graphitic form, and usually is distributed as tiny flakes throughout the metal. The remainder of the carbon remains as combined carbon, part of which may be present as free or "proeutectoid" cementite, and some as **pearlite,** a fine lamellar aggregate of iron and cementite (so-called from its resemblance to mother-of-pearl). This last component is a microscopically laminated structure composed of alternate layers of nearly pure iron, called **ferrite,** and cementite, in the proportion of about 7 parts ferrite to 1 part of cementite (about 0.80 per cent carbon). The proportion is by no means constant. The presence of pearlite in cast iron usually strengthens it without producing too great an embrittling effect.

Influence of Silicon—The silicon content of the iron is second only to its carbon content in regard to its effectiveness as a means of controlling the properties of the castings. As it is increased above 3.5 per cent, it makes the iron matrix more and more brittle, forming silvery iron, so that it is generally held to amounts between 0.5 and 3.0 per cent. Since it tends to throw carbon out of solution, as explained in the discussion of carbon in pig iron in Chapter 15, it is used as a "softener" in gray-iron castings, as a graphitizing agent in malleable castings or castings to be heat treated, and to regulate depth of chill in chill castings. With sulphur under 0.05 per cent, even one per cent silicon makes it difficult to obtain a **chill** (an external zone of hard cementitic iron without appreciable graphite). Below one per cent, the chilling properties are roughly inversely proportional to the silicon present. Its effect in increasing the rate of graphitizing the combined carbon in white iron for malleable castings is remarkable. In a certain casting, for example, raising the silicon from 0.7 per cent to 1.0 per cent shortens the annealing cycle from 180 hours to 72 hours. Silicon above 2 per cent hardens the matrix of iron solid solution to such an extent that machinability is adversely affected. Silicon added to white iron until it just turns gray increases the toughness by changing massive Fe_3C to pearlite and graphite, but more weakens it by forcing the pearlite ratio below 0.80 per cent carbon and increasing the graphite. Silicon tends to decrease shrinkage, to prevent blowholes, but increases the tendency to growth; silicon is oxidized in the cupola or air furnace and excessive amounts thus may favor a dirty casting, due to the entrapment of silicate-type

inclusions. Silicon, from 1.5 up to 4.5 per cent, increases the resistance of the iron to atmospheric and acid corrosion, and more than 10 per cent greatly protects the metal from all forms of oxidation and from chemical attack.

Effects of Manganese—As to whether high manganese content has an overall good or a bad effect on cast iron, there is much difference of opinion, some considering it almost as a cure for all troubles and others condemning it as a source of much trouble, especially in chilled castings. While it tends to hold carbon in solution, iron in which chill is produced by increasing the manganese content alone is "soft" and tends to spall. In *moderate* amounts, it is said to prevent cracking of the surface and also spalling to some extent, especially in chilled rolls, and it may harden the chill, if other conditions are right. It does tend to decrease blowholes, increase fluidity, and to neutralize the effect of the sulphur present. The amount used in castings varies from 0.1 to 2.0 per cent, 0.5 to 0.7 per cent being most common. In malleable castings, it is added in proportion to the sulphur, and according to the formula, $Mn = 2S + 0.15$. Held to these proportions, it will be found ultimately as MnS and in the cementite, and will not prevent graphitization of the cementite while larger amounts of manganese tend to stabilize cementite and increase growth. In high-grade gray-iron castings, the per cent manganese is held between 5 and 7 times the sulphur.

Influence of Sulphur—Sulphur except for special purposes (eg., enhanced machinability), is undesirable in steel. In castings, it is varied from about 0.04 per cent to 0.20 per cent, though in irons for foundries it will seldom exceed 0.1 per cent. Iron melted in cupolas always takes on sulphur from the coke, the percentage sometimes being doubled. Sulphur with iron forms sulphide, which is soluble in the liquid metal and has a melting point that is lower than the other constituents of the iron. This sulphide in iron used for castings has a three-fold influence. First, it tends to hold the carbon in combined condition, hence, can be used to increase the depth of chill in chilled castings, but in malleable castings and other castings that are to be heat treated, it thus retards graphitization, if it is not fully neutralized with manganese, and may be very undesirable. Chill produced with sulphur may be very brittle; low-silicon iron containing between 0.2 and 0.4 per cent sulphur often cracks when cooled rapidly. Second, its low melting point causes it to segregate as the iron solidifies, thereby causing the condition in castings known as bleeding. Third, it increases the shrinkage of the iron to a marked degree, thus increasing the difficulty of making accurate castings and increasing the tendency to form cracks, which are a result of the high shrinkage.

Influence of Phosphorus—Since compounds of phosphorus present in the materials charged into the blast furnace are completely reduced, all the phosphorus in the raw materials is ultimately found in the metal. Therefore, its content must be regulated by proper selection of raw materials. High phosphorus content causes a slight brittleness in pig iron and markedly reduces the total carbon content of the iron. **Ferrophosphorus** containing about 24 per cent phosphorus is almost carbonless and melts between 1230° and 1265°C (2245° and 2310°F). The melting point range for the 17 to 19 per cent grade is 1330° to 1370°C (2426° to 2498°F). Lesser amounts permit a proportionate increase of carbon, so that the total carbon in an iron containing 0.20 per cent phosphorus may be as high as 4.25 per cent. In this respect its action is not selective, since the ratio of combined to graphitic carbon is not affected. Phosphorus is known to form a compound, Fe_3P, with iron, containing 15.6 per cent phosphorus, but it apparently is able to combine or alloy with it in any proportion up to 25 per cent.

In pig iron and cast iron, Stead found that phosphorus forms a ternary eutectic solution containing 91.19 per cent iron, 6.89 per cent phosphorus, and 1.92 per cent carbon, to which Sauveur gave the name **steadite**. The amount formed in pig iron depends upon the phosphorus present, and since steadite has a low fusion point, the influence of phosphorus is to lower the freezing *range* of the iron. Another effect is to decrease the pearlite, so that more than one per cent phosphorus decreases the strength of the casting rapidly, the maximum strength being obtained with 0.25 to 0.40 per cent. For uniformly thin castings, such as stove plate and sanitary ware that is to be enameled, from 0.55 to 0.75 per cent is used with marked benefit to obtained the necessary fluidity and good mechanical properties. But in castings with thin and thick parts, such as engine blocks, these percentages of phosphorus make the thick portions porous, so that not more than 0.30 per cent can be permitted. High phosphorus also increases the shrinkage, particularly of the heavier parts, and increases the harmful stress-producing shrinkage (called "draw" shrinkage) and those making highest type castings generally agree that phosphorus above 0.30 per cent tends to produce unsound, porous, and brittle castings. Since the required fluidity can be had by raising the temperature, better results are obtained by lowering the phosphorus to 0.12 to 0.20 per cent and increasing the temperature. Castings thus made have a closer grain and machine better than those made with higher phosphorus content. Therefore, this range is recommended as the best for general work, including machinery castings of various kinds. In general, the resistance of unalloyed cast iron to corrosion decreases as the phosphorus content is increased above 0.05 per cent.

Effects of Chromium—Chromium occurs only occasionally, and in traces, in pig iron made from ordinary iron ore, but it now is added commonly to iron for high-grade castings in amounts from 0.1 to 3.5 per cent. It is a carbide-forming element, hence it holds the carbon in the combined state, opposes graphitization, decreases the tendency for growth, and increases the hardness of the matrix. In moderate amounts of 0.1 to 0.5 per cent, it increases the strength, but greater additions are made with some increase of brittleness. In high-grade castings, it is used mainly to increase the resistance to wear. It is used also with the other elements, particularly nickel and copper, to obtain resistance to corrosion and growth. Additions of chromium are made to the iron in the ladle, as solid or preferably molten ferrochromium. Its effects were earlier studied and reported by Hadfield in 1892 (see: Iron and Steel Institute, Vol. II, 1892).

Influence of Nickel—Nickel, like chromium, is rare in pig iron made from ordinary iron ores; but its use in high-grade castings is common, the amount added to gray-iron castings varying from 0.10 to 2.50 per cent, and to special alloy castings from 5 to 15 per cent. Nickel dissolves in the iron, and, like silicon, promotes graphitization; but, unlike silicon, it does not graphitize eutectoid cementite (that is, the portion of cementite that goes to make up pearlite) and it causes a reduction in the size of the graphite plates, giving a "closer grain." Therefore, in the smaller percentages, it is added to toughen the iron, prevent formation of massive carbides, and increase machinability. For example, one per cent added to gray iron, with carbon reduced from 3.50 to 3.00 per cent and silicon from 1.50 to 1.00 per cent, doubles the tensile strength, increases the Brinell hardness from 175 to 225, and gives an iron that is readily machinable. From 5 to 6 per cent nickel hardens the iron and may result in a martensitic matrix, the Brinell hardness being 250 to 280. A maximum of about 360 Brinell is obtained with 10 to 12 per cent nickel; while larger amounts, about 12 to 16 per cent, produce a distinct type wholly austenitic (except for cementite or graphite) and much softer, the Brinell hardness being as low as 130 with 18 per cent nickel. In these larger amounts, it is used in conjunction with chromium, with copper and chromium, or with silicon and chromium, to produce various special corrosion- and heat-resistant irons. In cupola practice, nickel is added in the form of shot to the metal in the ladle; in air- or electric-furnace practice, it may be added as pig nickel to the charge in the furnace, or in the form of shot in the ladle. Commercial forms of nickel used for this purpose contain some carbon, which should be considered if it is added in large amounts.

Influence of Copper—Copper may occur in pig iron, if it is present in the charge. Usually, it is absent. It is added to castings in various proportions from 0.10 to 2.00 per cent in gray-iron castings and up to 7.00 per cent in special alloy-iron castings. In the smaller amounts, its effects are similar to nickel, decreasing slightly the combined carbon, increasing the strength, preventing formation of massive carbides, improving machinability, and in addition, increasing the fluidity and decreasing the shrinkage slightly. Its solubility in iron is limited, but is increased by nickel, in conjunction with which it is used up to 6.5 to 7.0 per cent to produce corrosion- and heat-resistant castings. As to its influence on graphitization, one per cent is equivalent to 0.1 to 0.2 per cent silicon. In malleabilizing, it may be, therefore, both a hindrance and a help, tending to make the iron mottled in casting but shortening the annealing time. The strength of malleable iron containing about one per cent copper can be increased 13.8 to 55.2 MPa (2 000 to 8 000 psi) by precipitation hardening, a treatment usually carried out by heating for one hour at 750°C (1380°F) and reheating for 3 hours at 500°C (930°F).

Effects of Molybdenum—Molybdenum is rarely found in pig iron, but is added in the ladle as **ferromolybdenum** in cupola practice, or in the furnace as ferromolybdenum or **calcium molybdate** or **molybdenum oxide** in air- and electric-furnace practice. It strengthens and toughens the metal, and is added in amounts of 0.30 to 1.25 per cent to increase strength, hardness, and resistance to shock. It is credited also with preventing cracking. It does not decrease the shrinkage on solidification, but does decrease slightly the contraction following solidification. Its effect upon graphitization is slight, and it appears not to affect the amount of combined carbon, but does increase the depth of chill when added in amounts of 0.25 to 0.60 per cent. It increases resistance to wear, hinders graphitization somewhat, has little effect on growth, improves the properties at high service temperatures, and promotes uniformity in mechanical properties as between large and small sections.

Effects of Titanium and Aluminum—Both of these elements have been added to cast irons, but only titanium is used regularly. Both are reported to promote graphitization, titanium decreasing the size of the graphite flakes. A little titanium, 0.05 to 0.10 per cent, is common in pig iron, and its effect appears to be generally beneficial. The addition of titanium to cast iron is reported to impart greater toughness and resistance to wear.

Influence of Vanadium—Vanadium has been added to cast iron in amounts of 0.10 to 0.15 per cent. It is an expensive addition, and its chief effect seems to be that of opposing graphitization.

Effects of Special Additives—Various chemical elements other than those discussed above are added to iron for casting to effect changes in microstructure, improve mechanical properties, and so on. Such elements may be added singly or in various combinations. To control chilling tendencies and to achieve optimum size, shape and distribution of the graphite in cast-iron structures, a nucleating addition is generally made to the molten iron as near as possible to pouring. Known as an inoculant, this addition usually consists of 75 per cent ferrosilicon and/or other commercial inoculants consisting of various amounts of silicon, calcium, aluminum, barium or other nucleating elements. When the molten iron is treated with an appropriate "nodulizing" mix of magnesium and/or cerium alloys in addition to the inoculant, there is produced what is designated as "nodular iron"; this practice will be described later in this chapter under the heading of "Kinds and Uses of Iron Castings."

IRON-FOUNDRY MELTING METHODS

The chapter on pig iron describes the manufacture of iron used in the production of iron castings. The pig iron for castings may be melted in one of several types of furnaces, the cupola, the air furnace or the electric furnace, though the cupola is the principal source of molten metal for iron castings. In the case of certain alloy and high-test iron castings, the metal is frequently duplexed; ie., melted in a cupola and further processed in an electric furnace. The open-hearth furnace also has been used for melting iron for castings, but its use for this purpose is limited.

The Cupola—The cupola resembles a miniature blast furnace. It differs primarily in that pig iron and steel scrap replace iron ore in the charge. The cupola is lined with fireclay or firestone refractories. Since, in most installations, no water cooling is utilized in the melting zone, the lining has to be repaired with plastic-fireclay patching between periods of operations. In the new installations where water-cooling of the melting zone is employed, such frequent repair of the lining is not required. The charging door is located on the side of the cupola near the top. The cupola is supported by legs, permitting the use of a drop bottom which facilitates the removal of the remaining burden after the last charge has been tapped. Intermittent operation is the general rule, but continuous operation is possible.

The charge is composed of coke, steel scrap, iron scrap and pig iron in alternate layers of metal and coke. Sufficient limestone is added to flux the ash from the coke and form the slag. The ratio of coke to metallics varies, depending on the melting point of the metallic charge. Ordinarily, the coke will be about 8 to 10 per cent of the weight of the metallic charge. It is kept as low as possible for the sake of economy and to exclude sulphur and some phosphorus absorption by the metal.

During melting, the coke burns as air is introduced at a 4.31 to 8.62 kPa (10 to 20 ounces per square inch) pressure through the tuyeres. This melts the metallic charge and some of the manganese combines with the sulphur, forming manganese sulphide which goes into the slag. Some manganese and silicon are oxidized by the blast and the loss is proportional to the amount initially present. Carbon may be increased or reduced, depending on the initial amount present in the metallic charge. It may be increased by absorption from the coke or oxidized by the blast. Phosphorus is little affected, but sulphur is absorbed from the coke. Prior to casting, the slag is removed from the slag-off hole which is located just below the tuyeres. The molten metal is then tapped through a hole located at the bottom level of the furnace. The depth between these two tapping holes and the inside diameter of the furnace governs the capacity of the cupola.

The Air Furnace is a type of reverberatory furnace somewhat similar to the puddling furnace described in Chapter 1. It has a fireplace at one end, the stack at the other end, and between them a hearth covered by a roof sloping toward the stack. A cross or "bridge" wall near the stack (the flue bridge), and another next to the fireplace (the fire bridge), together with the lining (usually of silica sand) form a rectangular basin which holds the charge. A removable-bung type of roof is used to permit charging large pieces through the top with a crane. Coal, fuel oil or gas are used as fuel, the liquid or gaseous fuels being preferred. When coal is used, the fireplace is constructed and manipulated to serve as a crude gas producer. About 15 per cent of the carbon, 30 per cent of the silicon, 45 per cent of the manganese, and 1 to 2 per cent of the iron of a pig-iron charge are oxidized in the air furnace, the exact amounts varying with the composition of the charge and the oxidizing conditions of the flame.

The Electric Furnace—A description of the arc-type electric furnace, which is used to a limited extent for melting iron for iron castings, is given in Chapter 18. Induction-type electric furnaces have become increasingly popular for melting gray and alloy irons: these furnaces are also described in Chapter 18.

KINDS AND USES OF IRON CASTINGS

One of the principal reasons for using iron castings in applications at elevated temperatures is the fact that cast iron is resistant to warping and cracking in the presence of heat. At ordinary temperatures, cast iron—through proper control of composition and molding and casting methods—can be made to exhibit such valuable properties as; high hardness and wear resistance, good damping capacity and corrosion resistance, while lending itself to the design of castings characterized by intricacy of shape and/or massiveness. The various kinds of iron castings are: **Chilled-iron castings, malleable castings, alloyed-iron castings, gray-iron castings, and nodular-iron castings.** Among the innumerable uses may be mentioned pipe, rolls, permanent molds, ingot molds and stools, sanitary ware, engine blocks and all kinds of machinery parts. A few of the various kinds of iron castings, with their compositions, properties and uses, are given in Tables 40—VI and 40—VII. Further details may be found in "Cast Metals Handbook" published by the American Foundrymen's Association.

Chilled-Iron Castings are extremely hard on the surface. They are made by melting iron of certain compositions, and casting the molten metal in such a way that the parts to be hardened will be solidified on contact with a metal or graphite block capable of abstracting heat rapidly and thus causing quick cooling. Chilled-iron castings are made of ordinary low-silicon iron and of irons alloyed, usually, with nickel and chromium. The chilled surface is very hard, and when fractured, shows white for a distance beneath the chilled surface varying with the rate of cooling and the composition of the iron; hence, such irons are commonly spoken of as chilled iron. The hardness and the white fracture are due to the fact that all the carbon, in the clear chill at least, is in combined form, the rapid cooling preventing the separation of graphite. In heavy sections, the clear chill will extend only a short distance, 3.2 to about 50 mm (⅛ to 2 inches) from the surface, where it merges into mottled, then into a gray appearance, all these structures being the result of varying the rate of cooling. Such castings are used for rolls and various other articles which require a hard, wear-resisting surface. All three types of furnaces are used for melting, depending upon the kind of iron and other factors.

Malleable Castings, while not strictly malleable, are soft and can be bent without breaking. They are of two kinds, known as **white heart,** or European, and **black heart,** or American, these terms indicating differences in the process and the products and countries of origin. As an initial step in making malleable castings, a charge consisting of malleable grades of pig iron (10 to 15 per cent), steel scrap (35 to 40 per cent), and cast-iron scrap (45 to 55 per cent), consisting of feeders, runners, sprues, and defective castings, may be melted to give metal containing 2.25 to 3.00 per cent carbon, 0.30 to 0.50 per cent manganese, 0.05 to 0.08 per cent phosphorus, 0.06 to 0.11 per cent sulphur, and 0.60 to 1.15 per cent silicon, the exact composition, particularly in regard to silicon, being varied according to the thickness of the section being made. If the carbon is under

2 per cent, graphitization in annealing is slow and if the carbon is near 4 per cent, graphite may be formed in casting. To prevent the latter occurrence, the silicon is lowered as the carbon is increased in a given casting, if the iron must be white all the way through after casting, with all the carbon in the combined form. When the iron has melted, it is tapped and cast in well-prepared green-sand molds, made about 20.8 mm per metre (¼ inch per foot) oversize to allow for total shrinkage, or liquid-to-solid contraction. When cool, the castings are removed from the molds and, unless finish is of no consideration, cleaned by tumbling, sand blasting, pickling, or hand scouring. The castings next are packed carefully in annealing boxes or pots with an oxidizing agent, such as roll scale or crushed furnace slag for white-heart castings, or with nonoxidizing materials such as blast-furnace slag, fine sand, etc., for black-heart castings. The packed castings then are placed in an annealing furnace and heated gradually to about 900°C (1650°F) for white-heart castings or to near 871°C (1600°F) for black-heart castings. The former are held at this temperature 3 to 5 days, during which time the iron carbide is eliminated almost completely by a process of migration and surface oxidation of the carbon to CO or CO_2, leaving a metal similar to soft steel but of much coarser grain and less ductile. Since the migration of the carbon is very slow, only thin castings can be decarburized successfully or "malleableized" by this process. Black-heart castings are treated by an "annealing cycle," requiring about 30 hours for heating to temperature, about 45 hours holding at temperature, 30 to 35 hours for cooling to and holding at 650°C (1205°F) and 5 hours for cooling to handling temperature. In this cycle, the combined carbon is completely graphitized, causing the casting to grow slightly and leaving the iron as ferrite. But instead of forming plates, the graphite is dispersed in a very finely divided form known as temper carbon, which, under the microscope, is seen as black spots distributed in haphazard fashion throughout the metal except near the surface of the casting where the metal may be partly or wholly decarburized. Thoroughly malleableized castings have a yield point of about 186 MPa (27 000 psi), and a tensile strength of about 310 MPa (45 000 psi). Some standard specifications require a yield point of more than 221 MPa (32 000 psi) (224 MPa or 32 500 psi and 241 MPa or 35 000 psi minimum for two grades), and a tensile strength of more than 345 MPa (50 000 psi) (345 to 365 MPa or 50 000 to 53 000 psi minimum for two grades). The modulus of elasticity is about 172 370 MPa (25 000 000 psi). As to ductility, standard specifications require a minimum elongation of 10 to 18 per cent in 2 inches. Castings that have been malleableized are immune to growth.

Alloyed Castings—Alloyed irons are used most extensively for applications where resistance to wear, to heat (including growth), and to corrosion, along with the high strength of the alloyed iron, rigidity, "damping" of vibrations and amenability to heat treatment are of prime importance. They are produced and used extensively by the steel industry but are used most widely by the automotive industry for purposes where the above properties are a requirement. The alloying elements used are silicon, nickel, chromium,

Table 40—VI. Composition Ranges (Percent) and Uses of Some Varieties of Iron Castings.[a]

Iron Types and Uses	Si	S	P	Mn	T.C. (*)	G.C. (*)	C.C. (*)	Cu	Ni	Cr	Mo	Other Elements
Gray Iron—Ordinary	1.25-2.40	0.12 max	0.20 max	0.60-0.70	3.00-3.35	–	–	–	–	–	–	–
Cast Iron Pipe	As required	0.10 max	0.90 max	0.40-0.60	–	–	about 3.00	–	–	–	–	–
Automotive Cylinder Blocks Type A	2.00-2.20	0.09-0.12	0.14-0.18	0.70-0.85	3.25-3.40	2.00-2.75	0.55-0.65	–	0.75-0.85	0.25-0.35	–	–
Type B	2.35-2.40	0.15 max	0.18 max	0.55-0.75	3.25-3.40	–	0.50-0.65	0.75-1.00	–	0.20-0.30	–	–
Type C	2.10-2.50	0.10 max	0.20 max	0.50-.090	3.00-3.40	–	0.60-0.80	–	–	0.20-0.40	–	–
Type D	1.80-2.00	0.12-0.15	0.16-0.20	0.70-0.90	3.20-3.40	2.60-2.80	0.60-0.80	–	–	0.15-0.20	0.15-0.20	–
Crank Cases	1.90-2.10	0.09-0.12	0.14-0.18	0.70-0.85	3.20-3.40	2.70-2.80	0.50-0.60	–	1.00-1.10	–	–	–
Crank Shafts Type A	2.20-2.40	0.15 max	0.18 max	0.65-0.90	3.20-3.40	–	0.70-0.90	–	0.40 max	0.80-0.90	0.40-0.50	–
Type B	2.20-2.50	0.08 max	0.08 max	0.90-1.00	2.60-2.80	–	0.60-0.75	–	0.75-1.00	0.10-0.20	0.70-1.25	–
Piston Rings	1.90-2.25	0.10 max	0.35-0.75	0.50-0.70	3.40-3.60	2.70-3.10	0.45-0.75	–	0.90-1.10	–	0.20-0.30	–
Dies—Forming	1.40-1.50	0.10 max	0.18 max	0.60-0.80	3.20-3.50	–	–	–	0.90-1.00	0.60-0.70	0.80-0.90	–
Dies—Forming and Bending	2.00-2.40	0.10 max	0.20 max	0.55-0.70	2.60-2.90	1.90-2.10	0.65-0.80	–	1.40-1.70	0.35-0.50	0.40-0.60	–
High-Alloy Irons Ni-Resist—Typical	1.10-2.00	0.06 –	0.04-0.30	0.60-1.50	2.40-2.90	1.60-1.80	0.60-0.75	5.00-7.30	14.50-16.50	1.50-4.00	–	–
Silal—Typical	5.70	0.06	0.30	0.70	2.40	2.30	0.08	–	–	–	–	–
Nicor—Silal	4.50-5.90	0.04	0.04	0.60-0.70	1.80	1.50	0.30	–	17.70-18.70	2.10-2.60	–	–
Abrasion-Resistant White Irons Ni Cr HC	0.8 max	0.15 max	0.30 max	1.3 max	3.0-2.6	–	–	–	2.3-3.0	1.4-4.0	1.0 max	–
25% Cr	1.0 max	0.06 max	0.10 max	0.5-1.5	2.3-3.0	–	–	1.2 max	1.5 max	23.0-28.0	1.5 max	–
Nodular Iron Ferritic	2.50-2.80			0.30 max	3.35-3.65							Mg:0.04-0.08

[a] See Table 40—VII for mechanical properties of the irons in this table.
* T.C. = Total carbon content; G. C. = Graphitic carbon; C. C. = Combined carbon content.

molybdenum, copper and titanium, in amounts varying from a few tenths to 20 per cent or more. They may be classed as: (1) low-alloyed gray-iron castings, (2) high-alloyed gray-iron castings, and (3) austenitic alloy castings, the latter containing sufficient alloying elements to hold the iron in the austenitic condition. These are used for resistance to corrosion, both atmospheric and chemical; for resistance to heat, including oxidation and growth; and for their high thermal coefficient of expansion. In the higher alloy group of irons, a popular type for certain extremely high-wear, low-impact applications are the medium and high-alloy white irons. These irons, usually of relatively lower silicon and medium-to-high levels of chromium, consist of carbidic-martensitic structures which are extremely hard and abrasion-resistant. A standard specification for these irons is ASTM A 532. Many of these irons are patented compositions and are sold under various trade names, such as Ni-resist, Nihard, Causal metal, Silal, Nicrosilal, etc.

Gray-Iron Castings—Gray-iron castings are made of pig iron, of mixtures of pig iron and steel, or of mixtures

Table 40—VII. Mechanical Properties and Uses of Some Varieties of Iron Castings.[a]

Iron Types and Uses	Brinell Hard. No.	Tensile Strength		Yield Strength		Elonga- tion (%)
		MPa	ksi	MPa	ksi	
Gray Iron—Ordinary	187–235	207–310	30–45	–	–	–
Cast-Iron Pipe		138–172	20–25	–	–	–
Automotive						
Cylinder Blocks—Type A	187–196	228–248	33–36	–	–	–
Cylinder Blocks—Type B	192–220	262–290	38–42	–	–	–
Cylinder Blocks—Type C	187–240	241–310	35–45	–	–	–
Cylinder Blocks—Type D	212–231	262–283	38–41	–	–	–
Crank Cases	240–270	–	–	–	–	–
Crank Shafts—Type A	263–300	331–359	48–52	–	–	–
Crank Shafts—Type B	220–240	414–552	60–80	–	–	–
Dies—Forming	200–220	–	–	–	–	–
Dies—Forming & Bending	–	379–448	55–65	–	–	–
High-Alloy Irons						
NiResist—Typical	120–170	138–241	20–35	–	–	–
Silal—Typical	–	228	33	–	–	–
Nicor-Silal	–	159	23	–	–	–
Abrasion-Resistant/White Irons						
NiCrHC	550 min.	–	–	–	–	–
25% Cr	450 min.	–	–	–	–	–
Nodular Iron						
Ferritic	–	448	65	310	45	12.0
Pearlitic	–	552	80	379	55	6.0

[a]Chemical compositions of the irons listed in this table are given in Table 40—VI; ranges in properties relate to ranges in composition.

of pig iron, steel and other metals in smaller amounts, and have been referred to as semi-steel, high-test iron, and alloy iron. They are frequently sold under trade names, such as Meehanite, Gunite, Ermalite, Ferrosteel, Guniron, etc. Chemically, gray-iron castings include a large number of metals covering a wide range in composition, with carbon varying from 2 to 4 per cent, and silicon from 0.5 to 3 per cent, with small amounts of Ni, Cr, Mo, and Cu frequently added. Grouped according to uses, they include (1) pipe-foundry castings, (2) sanitary ware, (3) automotive castings, (4) locomotive castings, (5) light machinery, (6) heavy machinery, (7) miscellaneous shapes.

Mention has been made of the use of cast iron for ingot molds and stools. If available, molten iron direct from the blast furnace may be used instead of cupola iron because of its lower cost and high percentage of graphitic carbon. In some cases where iron direct from the blast furnace is not available, cupola iron is used for this purpose; however, studies invariably indicate that blast-furnace iron produces a mold with longer life than those produced with more refined cupola iron. The composition of the iron, pouring temperature and their relation to ingot-mold and stool life require considerable study; however, the consensus of mold makers in this country indicates that the composition should be about as follows:

Silicon	1.25 to 1.75%
Phosphorus	0.120 to 0.140%
Sulphur	0.035 to 0.050%
Manganese	0.75 to 1.25%

A casting temperature of 1260° to 1285°C (2300° to 2350°F) is desirable.

Ingot-Mold Casting—Ingot molds are made either by foundries belonging to steel producers or by outside manufacturers who generally obtain blast-furnace iron from blast furnaces belonging to nearby steel producers. Cupola iron is used by some mold manufacturers. Molds are normally made with green-sand forms, but there is a trend toward the use of air-setting sand ("no-bake") which does not require the use of energy-consuming ovens or similar equipment. Some molds are now being manufactured using insulating boards to replace sand.

Several sand-lined forms are used in mold manufacture. A large sectional casting or fabrication, called a flask, is placed over a pattern having the outside contour of a mold. Sand is introduced into the space between the flask and pattern to form a sand replica of the outer mold wall. This form is often called a "cheek".

A hollow pattern called a core-box, the inside surface of which has the inside contour of a mold, is placed over a pedestal-type casting or fabrication called an arbor or core-barrel. Sand is introduced between the core-box and the arbor to form the replica of the inner mold wall (actually the ingot contour). This is called a core. A third piece of equipment, which forms the bottom of a mold, is called a drag. It is usually a casting with a depressed top surface into which a layer of sand (either green or "no-bake") is hand-applied. At some operations, the drag may be permanently attached to the core-barrel. The sand on all three of the above forms is then coated with a "blacking" which is dried before use.

The dry, coated forms are then normally assembled by placing the core into the drag, and then placing the cheek concentrically over and around the core. When all mold sets are assembled, blast-furnace iron is bottom poured into each through a down-gate which has been provided in one corner or a face during the preparation of the cheek. The metal flows down the gate and enters just above the bottom of the mold through an in-gate. When the metal level reaches the top of the mold, precipitated iron graphite (kish) is skimmed off to provide a smooth top surface.

After a prescribed solidification period, the drag, arbor and flask are removed, leaving the sand around the mold. After a cooling cycle the remaining sand is removed and the mold is conditioned by chipping, grinding and/or machining (top, bottom). The mold is then ready for use. However, for optimum mold life, a mold should be allowed to remain unused for a period long enough to relieve casting stresses.

Pipe-foundry castings include cast-iron pipe and fittings. A great part of cast-iron pipe now is cast centrifugally from ordinary iron containing less than 0.1 per cent sulphur, under 0.9 per cent phosphorus, with carbon and silicon controlled to give the required mechanical properties. When not centrifugally cast, the molten metal is poured into cored dry-sand molds supported in a verticle position, those 457 mm (18 inches) or more in diameter being cast with the hub end down.

Nodular-Iron Castings—Nodular iron, also called ductile iron and spheroidal graphite iron, was introduced around 1948; it has been used for castings having sections from 3.18 up to 1016 mm (⅛ inch up to 40 inches) thick. It is produced by treating with cerium or magnesium alloys molten iron that normally would produce a soft, weak, gray iron casting. The addition of these special alloys results in castings which have the carbon present in spheroidal form. Castings so made have relatively high strength and better ductility than ordinary gray iron. Pearlitic nodular irons have a tensile strength in excess of 552 MPa (80 000 psi), with an elongation of at least 3 per cent, while ferritic grades, having tensile strengths of over 414 MPa (60 000 psi), will show an elongation of from 10 per cent to 25 per cent. Iron having a sulphur content below about 0.10 per cent is required for the process. A sulphur content of 0.05 per cent would be a reasonable aim, with 0.03 per cent preferred for economy to obtain the maximum efficiency from the additions. Several types of matrix structures can be developed by alloying or heat treatment; the pearlitic and ferritic matrices were mentioned above. A standard specification for nodular iron (ductile iron) is ASTM A 536. As this type of iron has been discussed extensively in the literature, no further details will be given here.

IRON-FOUNDRY MOLDING AND CASTING PRACTICE

Molding practice for iron castings is somewhat similar to that already described for steel castings in Section 1 of this chapter, the chief differences being in the preparation and types of sand and the placement and size of gates and risers, the latter being modified by reason of the lower shrinkage in cast iron.

The scope of this book does not permit more than a brief description of the casting of iron castings. The metal is cast in one of six types of molds: namely, (1) **green-sand molds,** made of moist sand which is rammed about a pattern (usually of wood) in a "flask" of wood or iron, and the metal is poured with the mold in the condition as rammed; (2) **dry-sand molds,** made up like green-sand molds, but dried before they receive the metal; (3) **loam molds,** made up of loam (a kind of low-grade sand) which, for heavy castings, is backed with brick and faced with other more refractory material; (4) **permanent molds** or **semi-permanent molds,** which have become more and more popular for certain applications (5) **shell molds** and (6) **chemically bonded** or "no-bake" molds.

The permanent mold is a cast-iron or graphite mold into which the molten iron is poured. The semi-permanent mold is built up of cast iron and sand, the latter having to be replaced after use. In both the latter cases, the molds are prepared with a graphite coating and warmed to 65° to 93°C (150° to 200°F) before the hot metal is poured into the mold. Shell-molding techniques are described in numerous recently published articles and papers and will not be discussed in detail here; briefly, shell molds are made by applying a coating of sand mixed with a synthetic resin or other suitable binder to a prepared pattern and then "curing" the coating by heating to form a solid shell that can then be stripped from the pattern. Shell molds are used in the production of steel and non-ferrous castings, as well as iron castings.

TESTING OF CAST IRON

The tests most commonly employed for gray cast iron are the tension test, transverse load and deflection (measured bend) test, and hardness (Brinell and file) tests. In making tension tests, standard test specimens are machined from a standard cast test bar, and "pulled" in a tensile-testing machine until the piece breaks, the load calculated to MPa (or pounds per square inch) being taken as the tensile strength. In short-time tension tests, most grades of cast iron show no point corresponding to the yield point of steel and very little, if any, elongation or reduction of area. (Nodular iron is the exception, requiring both yield strength and elongation determination from a steel-type test specimen.) The form of specimen used for making tension tests on gray iron is somewhat different from that used for testing steel. There are three standard sizes; viz., 12.83 mm (0.505 inch), 19.05 mm (0.750 inch), and 31.75 mm (1.25 inches) in diameter (at the center), the diameter being varied with the thickness (and design) of the castings they represent. As standard specifications do not require the measurement of elongation but only the tensile strength, specifications covering the lengths of specimens merely state that the affected test length shall not be less than the diameter. The tensile strength varies: (a) with the diameter of the test bar, being higher for bars of smaller diameter and lower for bars of greater diameter; (b) with the temperature above 120°C (250°F), being almost constant up to 315°C

(600°F) and decreasing rapidly above 425°C (800°F); and (c) with the time after casting, most castings increasing in strength with age. This change is attributed to relief of casting strains, which may be relieved also by tumbling and by heat treatment. The relief of strain by aging is called **seasoning.**

Transverse testing consists of placing a standard bar upon supports 12, 18, or 24 inches apart, then either applying a specified load and noting the deflection, which is measured in inches, or of loading until deflection occurs and then gradually increasing the load till the specimen breaks. With the latter procedure, the **modulus of rupture** is found from the equation $M R = \frac{2.546LS}{D^3}$ where M R = modulus of rupture, L = distance between supports in inches, S = the breaking load, and D = the diameter of the test bar.

In the Brinell hardness test, a special machine is used to apply a load of 3000 kg to a steel ball 10 mm in diameter, resting on a filed-smooth surface of iron to be tested. The Brinell number is taken as the quotient of the load divided by the area of the impression made by the ball. Thus, small numbers up to 100 indicate softness, while high numbers (400 to 600) indicate great hardness.

Bibliography

American Foundrymen's Society, Cast Metals Handbook (1957 Edition).

American Society for Metals, Source Book on Ductile Iron (1977).

American Society for Testing and Materials, ASTM Standards, Vol. 01.02, Ferrous castings; ferro-alloys (1984).

C. W. Briggs, The Metallurgy of Steel Castings, McGraw-Hill Book Co., New York (1946).

Iron Castings Handbook, edited by C. F. Walton and T. J. Opar, Iron Castings Society, Inc. (1981).

Steel Founders' Society of America, Steel Castings Handbook (1980 Edition: also, Steel Casting Design Engineering Data Files).

CHAPTER 41

Principles of Heat Treatment of Steel

SECTION 1

METALLOGRAPHY

The Importance of Heat Treatment—The outstanding advantage of steel as an engineering material is its versatility. This is illustrated by the wide variety of products already described and remaining to be described in this book. The reader has seen and will see that through heat treatment one or another of the characteristics of steel may be enhanced, and it is generally true that much of the versatility of steel arises from the fact that its properties can be controlled and changed at will by heat treatment. Thus, if steel is to be formed into some intricate shape, it can be made very soft and ductile by heat treatment; if, on the other hand, it is to resist wear, it can be heat treated to a very hard, wear-resisting condition.

This important faculty which steel possesses of being amenable to control of its properties by heat treatment arises largely from the fact that these properties, in turn, reflect primarily the **constitution** of the steel, that is, the nature, distribution, and amounts of its **metallographic constituents**, as distinct from its chemical composition. The metallographic constituents are controlled by changes in alloy composition and heat treatment. Thus, steel's versatility largely reflects the very wide range of constitutional variations that can be brought about by heat treatment.

The Science of Heat Treatment—The science of heat treatment, then, deals with the factors and mechanisms involved in the control of the constitution of steel by chemical composition in conjunction with heating and cooling and the relationships between the constitution and properties of steel. Many, although by no means all, of these constitutional changes can be followed by microstructural studies, or a broader type of study known as **metallography,** and many of the known relationships between constitution and properties are in terms of microscopic structures, which, with the advent of the electron microscope, are fine indeed. However, although the microscope has certainly been the most important single tool used in the development of the science of heat treatment, many other methods and techniques have been utilized, including dilatometric studies, magnetic measurements, thermal analyses, X-ray diffraction and electron diffraction.

THE CONSTITUENTS OF STEEL

The two constituents of steel, the amount and distribution of which primarily control the properties, are iron and iron carbide. Thus, iron and carbon are the most important elements in steel. Most plain carbon steels will also contain manganese, silicon, phosphorus, sulphur, oxygen and traces of nitrogen, hydrogen and other chemical elements such as aluminum and copper. These elements, however, as will be shown later, may modify to a certain extent the major effects of the constitution in respect to iron and iron carbide but the distribution of these two constituents is always the predominating influence. This is largely true even of medium-alloy steels, which may contain considerable percentages of such elements as nickel, chromium, molybdenum, vanadium or titanium. The properties of such steels are still dependent primarily upon the distribution and amount of iron carbide in the steel. The major effect of the alloying elements is to help in the control of this distribution, although the properties may be modified somewhat by solution of the alloying element in the iron or by its combination with the carbide phase.

High purity iron has very low strength. Iron containing less than two parts per million (0.0002 per cent) of oxygen, nitrogen, hydrogen and carbon has been made in the laboratory by a technique known as zone melting. Yield strengths as low as 48 MPa (7ksi) and tensile strengths of 165 MPa (24 ksi) have been measured on samples of this iron.

Ferrite—The metallographic name for iron in steel is ferrite and the microstructural appearance of ferrite in a low-carbon steel in which it is the predominant constituent is shown in Figure 41—1. It appears as polyhedral grains, distinguishable from one another by the boundaries which have been etched out by the etching reagent and by differences in orientation which cause different etching behaviors. The reagent most commonly used to bring out the ferrite grain boundaries is a dilute solution of nitric acid in ethyl alcohol. Figure 41—2 shows the appearance of ferrite in a slowly-cooled steel of somewhat higher carbon content in which the ferrite appears as a white or light-etching

Table 41—I Allotropic Forms of Iron

Allotropic Forms	Crystallographic Form	Temperature Range
Alpha	Body Centered Cubic	Up to 910°C (1670°F)
Gamma	Face Centered Cubic	910-1403°C (1670°-2557°F)
Delta	Body Centered Cubic	1403-1535°C (2557°-2795°F)

network surrounding bodies of "pearlite," another very common constituent to be described shortly

In pure iron-carbon alloys, the ferrite consists of iron with a trace of carbon in solution, but in steels it may also have considerable amounts of alloying elements such as manganese, silicon or nickel dissolved in it.

As is shown in Table 41—I, iron exists in two allotropic forms, body-centered-cubic and face-centered-cubic. The atomic arrangement in crystals of the two allotropic forms of iron is shown in Figure 41—3. Ferrite is the allotropic form of iron with a body-centered cubic crystal structure. When ferrite is present at low temperature, this allotropic form is known as **alpha iron.** When ferrite is present at high temperature, this allotropic form is known as **delta iron.** The other allotropic form of iron is known as **gamma iron:** it has a face-centered cubic structure and is usually present in the temperature range of 910° to 1403°C (1670° to 2557°F). The metallographic name for gamma iron is austenite: this constituent will be discussed shortly.

Cementite—Cementite is the metallographic term for iron carbide in steel. This is the form in which carbon occurs in steels and the proportions of iron and carbon correspond to the chemical formula Fe_3C. Cementite thus consists of 6.67 per cent carbon and 93.33 per cent iron. Little is known about its properties except that it is very hard and brittle. It is the hardest

constituent of plain carbon steel and will scratch glass and feldspar but not quartz. It has about two-thirds the induction of pure iron in a strong magnetic field. The atomic structure of cementite is illustrated in Figure 41—4.

Its metallographic appearance in the grain boundaries of a slowly-cooled, relatively high carbon steel is shown in Figure 41—5. In this case, the cementite appears as a brilliant white network outlining pearlite colonies or as needles interspersed with the pearlite. Figure 41—6 shows the metallographic appearance of cementite in a steel which has been heated to a temperature just below that at which austenite first forms. This heating has caused the cementite to coalesce into spheroidal particles and the illustration shows these spheroidal particles in a matrix of ferrite. This form of cementite is known as **spheroidized cementite,** or the whole structure as "spheroidite."

Austenite—Although austenite is not ordinarily a constituent of steel after it has been cooled, it seems appropriate to describe it at this point since it is the important high-temperature phase, the decomposition of which on cooling forms the room-temperature constituents which are being discussed. It is a homogeneous phase, consisting of a solid solution of carbon in the gamma form of iron. It is formed when steel is heated to a temperature above the upper critical point. The

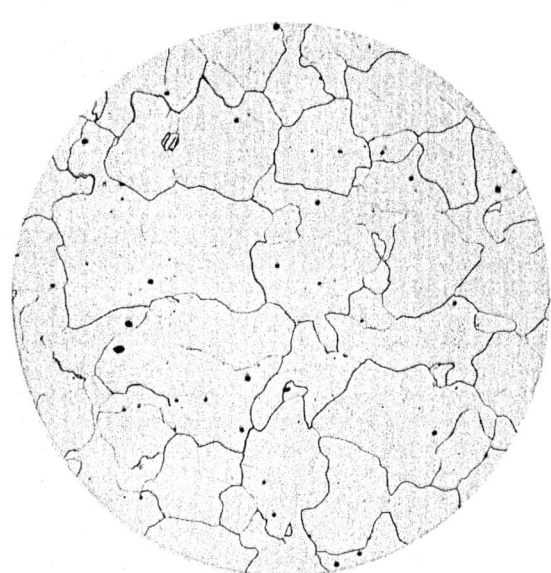

FIG. 41—1. The microstructure of ferrite. Magnification: 100X (This sample has a coarse grain size.)

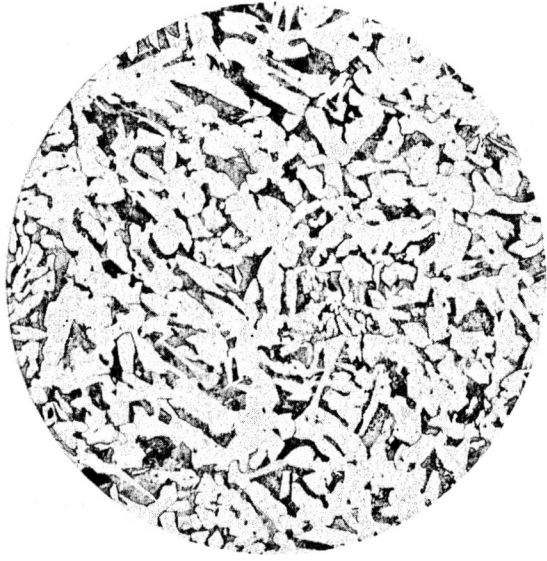

FIG. 41—2. The microstructure of slowly cooled hypoeutectoid steel showing ferrite and pearlite. Magnification: 100X. (See Figure 41—8 showing lamellar structure of pearlite at 1000X.)

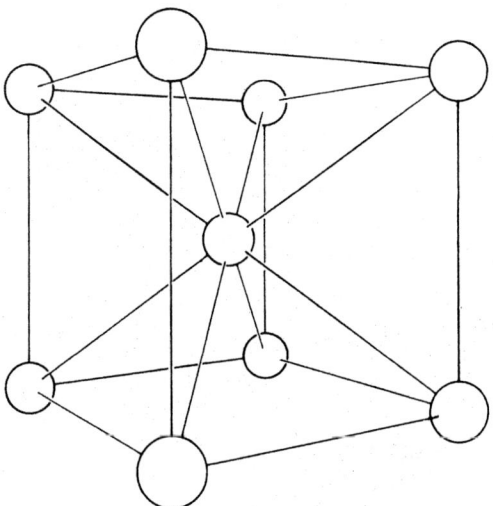 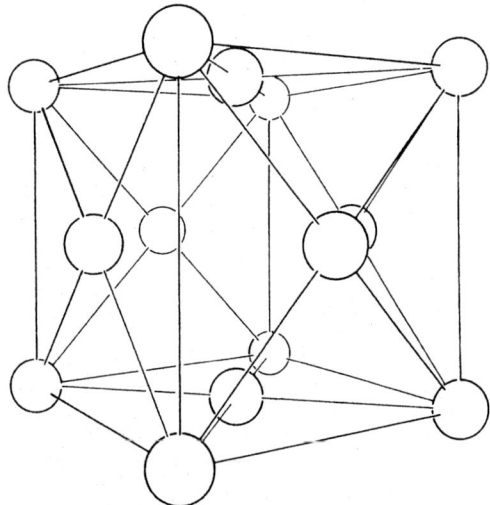

FIG. 41—3. Crystalline structure of the allotropic forms of iron. Each white sphere represents the relative position of an atom in a "unit cube" of: (left) alpha iron ferrite and delta iron, which have the body-centered cubic form and (right) gamma iron caustenite), which possesses the face-centered cubic form.

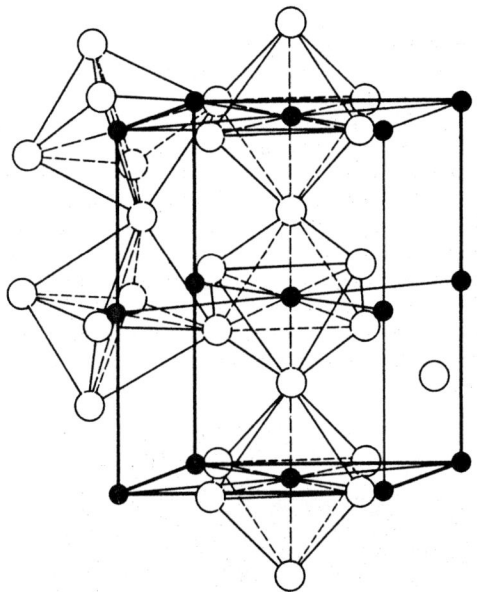 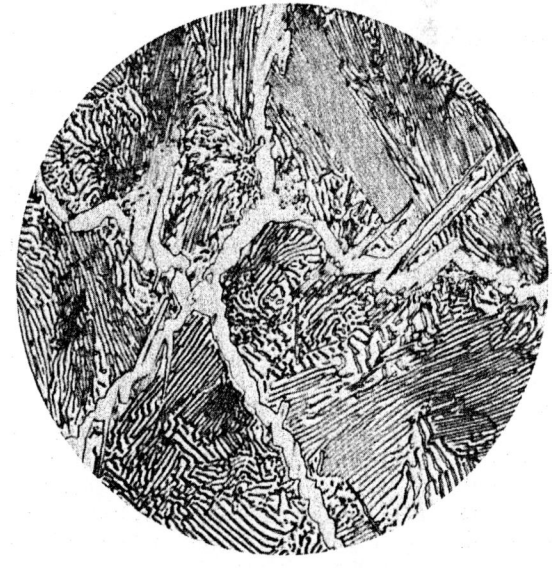

FIG. 41—4. The atomic structure of cementite. Positions of carbon atoms are indicated by solid circles, positions of iron atoms by open circles. (Hendricks, S. B., Zeitschrift fur Kristallographic. Vol. 74 (1930), 534-545.)

FIG. 41—5. The microstructure of slowly cooled, high-carbon steel showing pearlite with cementite in the grain boundaries Magnification: 1000X.

limiting temperatures for the formation of austenite vary with composition and will be discussed in connection with the iron-carbon diagram. The metallographic appearance of austenite is shown in Figure 41—7, which represents a sample of an alloy steel which has been very rapidly cooled from the temperature range at which the austenite is stable, the steel having sufficient alloy content to make possible the retention of austenite structure at room temperature.

As indicated earlier, the atomic structure of austenite is that of gamma iron—face-centered cubic. The atomic

spacing varies with the carbon content.

Pearlite—When a plain carbon steel of approximately 0.80 per cent carbon is cooled slowly from the temperature range at which austenite is stable, all of the ferrite and cementite precipitate together in a characteristic lamellar structure known as pearlite. This structure is illustrated in Figure 41—8. It is generally similar in its characteristics to an eutectic structure but since it is formed from a solid solution rather than from a liquid phase, it is known as an **eutectoid structure. Ledeburite** (discussed later) is a true eutectic: its

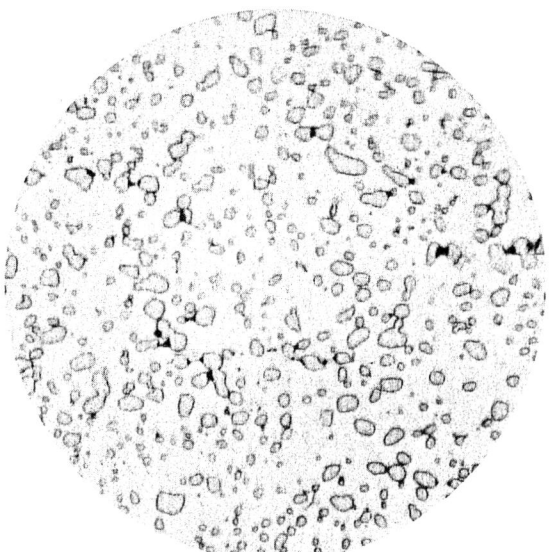

FIG. 41—6. Micrograph showing spheroidized cementite in matrix of ferrite. Magnification: 1000X.

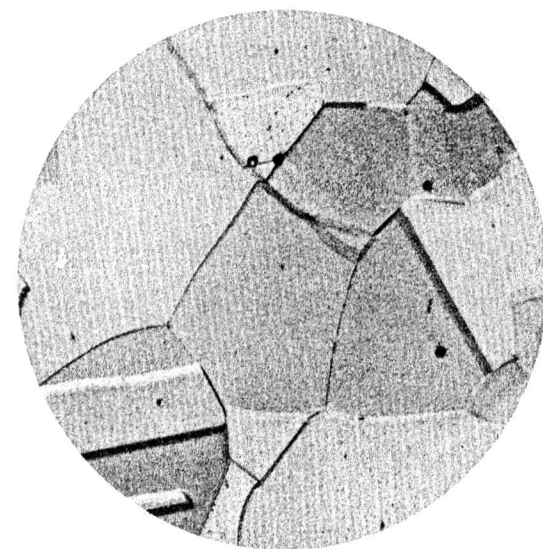

FIG. 41—7. The microstructure of austenite. Magnification: 500X.

structure is shown in Figure **41**—10.

At carbon contents above and below 0.80 per cent, pearlite of about 0.80 per cent carbon is likewise formed on slow cooling, but the excess ferrite or cementite first precipitates, usually as a grain boundary network, but occasionally also along cleavage planes of the austenite. This excess ferrite or cementite rejected by the cooling austenite is known as a **proeutectoid** constituent. The carbon content of a slowly-cooled steel can be estimated from the relative amounts of the pearlite and proeutectoid constituents in the microstructure.

The Iron-Carbon Equilibrium Diagram—The iron-carbon equilibrium diagram furnishes a "map" showing the ranges of compositions and temperatures within which the various phases are stable and the boundaries at which phase changes occur. Although heat treatment is largely concerned with a controlled departure from equilibrium, this diagram represents the limiting conditions and is basic to an understanding of heat-treating principles (Figure **41**—9).

The diagram, covering the temperature range from 600° C (1112° F) to the melting point of iron and carbon contents of from 0 to 5 per cent, represents the equilibrium conditions for the entire range of steels and cast irons in both the liquid and solid states. Use of this diagram in ensuing discussions involves two constituents, **ledeburite** and **graphite,** which have not been mentioned up to this point. Although these are not ordinarily constituents of steels, their characteristics will be briefly discussed at this point.

Ledeburite—Ledeburite is the metallographic term for the iron-iron carbide eutectic, containing 4.27 per cent carbon. This eutectic is a constituent of iron-carbon alloys containing more than 2.01 per cent carbon and for this reason the dividing line between steels and cast iron is customarily set at 2.0 per cent carbon. The metallographic appearance of ledeburite in cast

iron is shown in Figure **41**—10.

Graphite—Cementite is metastable and decomposes into iron and graphite. Thus, in most slowly-cooled cast irons, graphite is an equilibrium constituent at room temperature. The gray appearance of the fracture of such slowly-cooled cast irons reflects the presence of the graphite and the metallographic appearance of graphite in gray cast iron is shown in Figure **41**—11. Graphite may, under certain conditions, be a constituent of steels and the metallographic appearance of graphite in a low-carbon steel which has been subjected to a prolonged heating at a temperature below that at which austenite is formed is illustrated in Figure **41**—12.

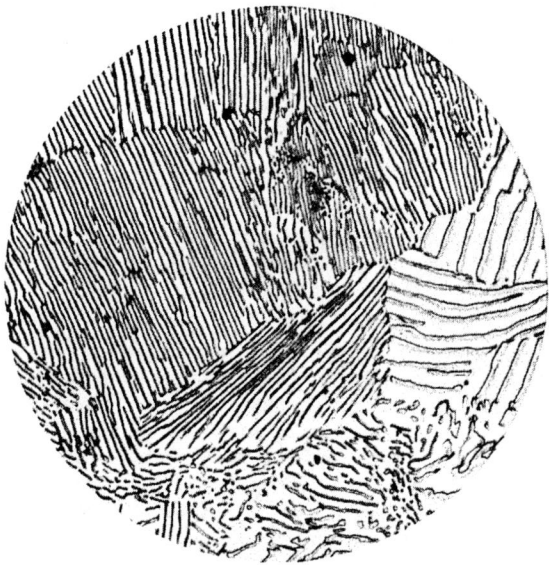

FIG. 41—8. The microstructure of pearlite. Magnification: 1000X.

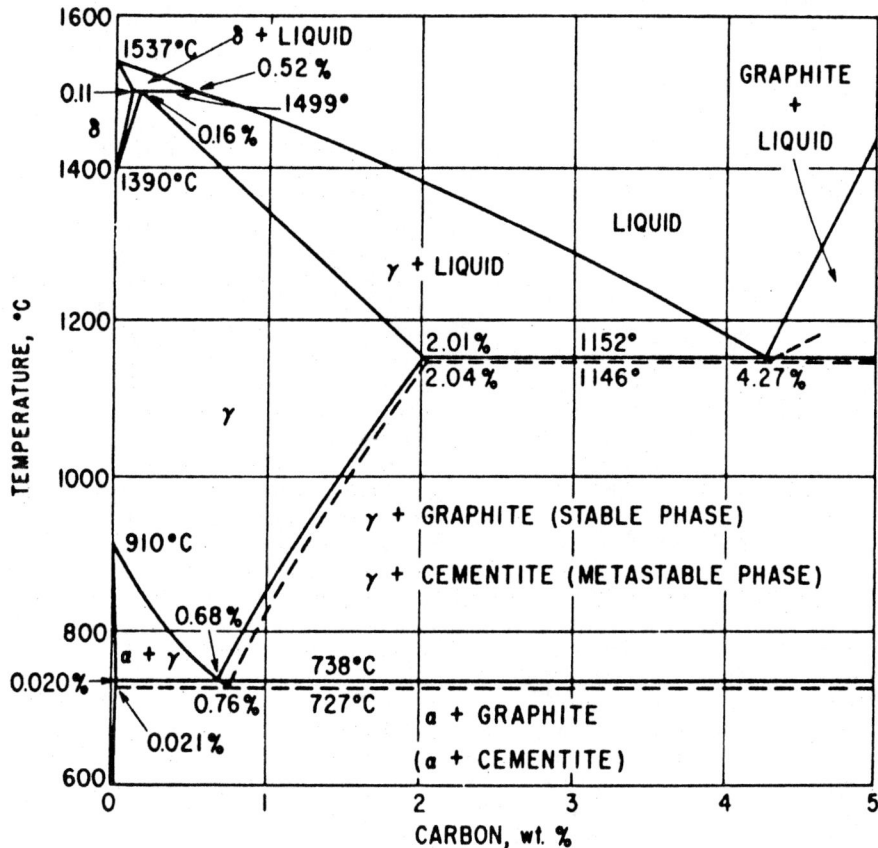

FIG. 41—9. The iron-carbon equilibrium diagram, for carbon contents up to 5 per cent.

The Iron-Iron Carbide Equilibrium Diagram for Steels—The portion of the iron-iron carbide diagram of concern in connection with the heat treatment of steel is that part extending from 0 to 2.01 per cent carbon. The general features of this diagram (Figure 41—9) will

be discussed and its application to heat treatment will be illustrated by considering the changes occurring on heating and cooling steels of selected carbon contents as depicted by the diagram.

Critical Temperatures—In Table **41**—I, iron is listed

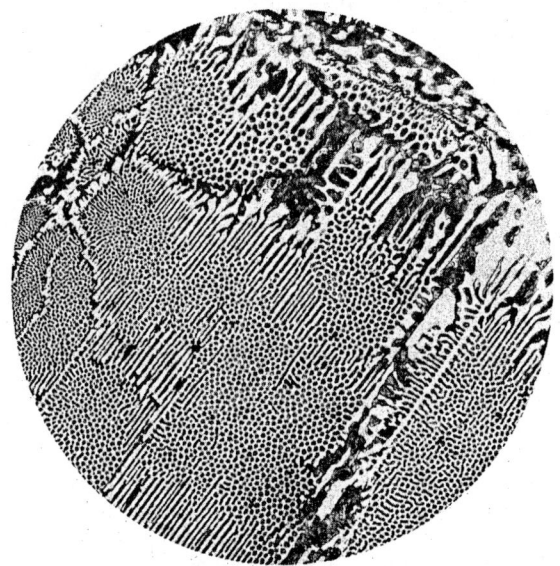

FIG. 41—10. The microstructure of ledeburite. Magnification: 150X.

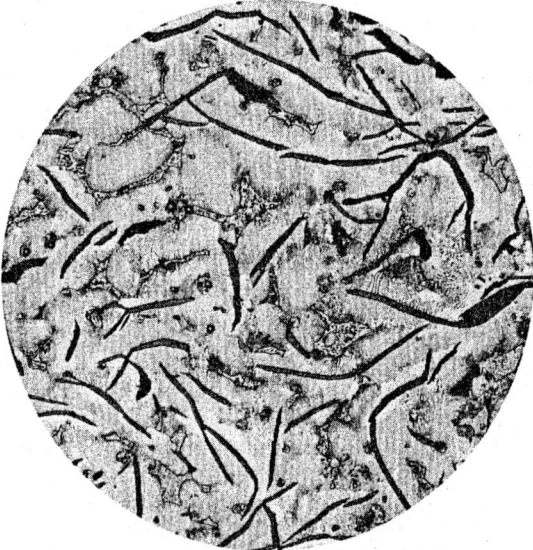

FIG. 41—11. The microstructure of gray cast iron showing graphite flakes. Magnification: 100X.

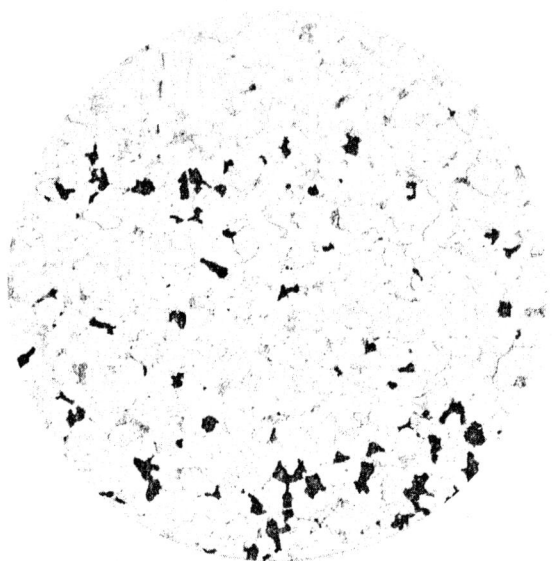

FIG. 41—12. Micrograph showing graphite particles in low-carbon steel. Magnification: 100X.

as occurring in two allotropic forms, alpha or delta (the latter at very high temperature) and gamma. The temperatures at which these phase changes occur are known as critical temperatures and the boundaries in Figure 41—9 show how these temperatures are affected by composition. For pure iron, these temperatures are 910° C (1670° F) for the alpha-gamma phase change and 1390° C (2534° F) for the gamma-delta phase change. The critical temperatures on heating are designated as A_c (from the French "chauffage," meaning "heating") temperatures, while those on cooling are designated as A_r (from the French "refroidissement," meaning "cooling") temperatures. Although, in principle, the transformations A_c and A_r result from the same temperature of equilibrium, in practice, the A_r temperatures are lower than the corresponding A_c temperatures, falling as the cooling is more rapid; even with very slow heating or cooling these temperatures do not coincide and, therefore, the subscript "e" (for equilibrium) is used in the designation of the critical temperatures in the equilibrium diagram. The meaning and significance of the critical temperatures will be clarified in the following discussion of the changes occurring on heating and cooling iron-carbon alloys as depicted by the equilibrium diagram.

Changes Occurring on Heating and Cooling Pure Iron—The only changes occurring on heating or cooling pure iron are the reversible changes, (1) at about 910° C (1670° F) from body-centered alpha iron to face-centered gamma iron and (2) from the face-centered gamma iron to body-centered delta iron at about 1390° C (2534° F).

Changes Occurring on Heating and Cooling Hypoeutectoid Steels—Hypoeutectoid steels are those which contain less than the eutectoid percentage of carbon (about 0.80 per cent). The diagram shows the equilibrium constituents are ferrite and pearlite, the relative amounts of each depending upon the carbon content. The diagram also shows that at 600° C

(1112° F) the ferrite may hold in stable solution about 0.007 per cent carbon. Up to 727° C (1340° F), the solubility of carbon in the ferrite increases until at this temperature, the ferrite contains about 0.025 per cent carbon. The first phase change on heating (if the steel contains above 0.025 per cent carbon) occurs at 727° C (1340° F) and this temperature is therefore designated as the A_1 critical temperature. On heating just above this temperature, the pearlite (ferrite and cementite) all changes to austenite. Some proeutectoid ferrite, however, remains unchanged. As temperature rises farther above A_1, the austenite dissolves more and more of the surrounding proeutectoid ferrite, becoming lower and lower in carbon, until at the A_3 temperature, the last of the proeutectoid ferrite has been absorbed into the austenite having the same average carbon content as the steel.

On slow cooling the reverse changes occur. The austenite first rejects ferrite (generally at grain boundaries) on cooling below A_3 and becomes progressively richer in carbon, until, just above the A_1 (eutectoid) temperature, it is substantially of eutectoid composition. On cooling below A_1, this eutectoid austenite changes to pearlite so that the final product after cooling below A_1 is a mixture of ferrite and pearlite, the relative proportions of each constituent depending upon the carbon content.

Changes Occurring on Heating and Cooling Eutectoid Steels—Since no excess ferrite or cementite is present in eutectoid steel, the only change occurring on slow cooling or heating is the reversible change from pearlite to austenite at the eutectoid temperature. Thus, in the case of eutectoid steels, the A_3 and A_1 temperatures coincide and this eutectoid composition and temperature is designated as the $A_{3\cdot 1}$ point.

Changes Occurring on Heating and Cooling Hypereutectoid Steels—The behavior on heating and cooling hypereutectoid steels (steels containing more than 0.80 per cent carbon) is similar to that of hypoeutectoid steels except that the excess constituent is cementite rather than ferrite, so that on heating above A_1, the austenite gradually dissolves the excess cementite until at the A_{cm} temperature all of the proeutectoid cementite has been dissolved and austenite of the same carbon content as the steel is formed. Similarly, on cooling below A_3, cementite precipitates and the carbon content of the austenite approaches the eutectoid composition. On cooling below A_1, this eutectoid austenite changes to pearlite and the room temperature constitution is, therefore, pearlite and proeutectoid cementite.

The A_2 Formerly Designated Critical Temperature—Early iron-carbon equilibrium diagrams indicated a critical temperature at about 768° C (1414° F). It has since been found that the behavior at this temperature differs from those at A_1 and A_3 in that it does not involve a phase change. In the neighborhood of 768° C (1414° F) and up to about 790° C (1454° F) there is a gradual magnetic change, ferrite being ferromagnetic below this temperature range, and paramagnetic above. The change is also accompanied by a heat effect. This A_2 change is of little or no significance in regard to the heat treatment of steel.

The Effect of Alloys on the Equilibrium Dia-

gram—The iron-carbon diagram may, of course, be profoundly altered by alloying elements, and, therefore, its application should be limited to plain carbon and low-alloy steels. The most important general effects of the alloying elements may be listed as follows:

1. The number of phases which may be in equilibrium is no longer limited to two as in the iron-carbon diagram.

2. The temperature and composition range, with respect to carbon, over which austenite is stable may be increased or reduced.

3. The eutectoid temperature and composition may be changed.

The alloying elements may be divided generally into two classes in respect to the second effect: those which enlarge the austenite field and those which reduce it. The elements which enlarge this field include manganese, nickel, cobalt, copper, carbon and nitrogen. Because of this characteristic, elements of this type are descriptively known as **austenite formers.** Figure 41—13, which shows the 13 per cent manganese section of the iron-carbon-manganese equilibrium diagram, is illustrative of the effect of elements of this type. The large field in which austenite is stable, the lowering of the eutectoid temperature and carbon content, and the three-phase field in which alpha iron, austenite and carbides exist in equilibrium should be noted.

The commoner elements which decrease the "size" of the austenite field include chromium, silicon, molybdenum, tungsten, vanadium, tin, niobium (columbium), phosphorus, aluminum and titanium. Such elements are known as **ferrite formers.** Figure 41—14 showing the effect of molybdenum on the composition and temperature range over which austenite is stable will serve to illustrate the effect of alloys of this type. These steels will likewise have the three-phase zone in which austenite, ferrite and carbides will be in equilibrium at temperatures below the austenite field.

The effect of the elements on the eutectoid temperature and composition is summarized in Figure 41—15. It will be noted that manganese and nickel lower eutectoid temperature while the other elements generally raise it. All of the elements seem to lower the eutectoid carbon content.

Shown below are empirical formulae proposed by Andrews for calculating the Ac_1 and Ac_3 temperatures for steels containing less than 0.6 per cent carbon and less than 5 per cent of each of the other alloying elements. These formulae permit one to estimate the influence of carbon and alloy content on the transformation temperatures. Direct experimental determination of the transformation temperatures of a given alloy steel by dilatometric techniques is preferable.

$$Ac_1°C = 723 - 10.7\ Mn - 16.9\ Ni + 29.1\ Si + 16.9\ Cr + 6.38\ W + 290\ As$$

$$Ac_3°C = 910 - 203\sqrt{C} - 15.2\ Ni + 44.7\ Si + 104\ V + 31.5\ Mo + 13.1\ W - 30\ Mn - 11\ Cr - 20\ Cu + 700\ P + 120\ As$$

The symbols represent the weight per cent of the element indicated. The coefficients for phosphorus and arsenic are very approximate.

Grain Size—As described above, when a piece of steel is heated above the critical temperature, the ferrite and carbide react with one another to form austenite. The austenite is a crystalline phase differing distinctly from either the ferrite or carbide from which it is formed. Like any metal composed of a solid solution it exists in the form of polyhedral grains. The reaction which forms austenite begins at a number of points in the interface of the carbide and ferrite. Each of the little islands of austenite grows until finally it reaches its similarly growing neighbors. As the temperature above the critical increases, further grain growth occurs, presumably by encroachment of a grain into adjacent grains. The final austenite grain size is, therefore, a function of the temperature above the critical to which it is heated. This grain growth may, however, be inhibited by carbides which dissolve slowly or remain undissolved in the austenite or by a suitable dispersion of non-metallic inclusions. Hot working refines the coarse grain size formed by heating to the relatively high temperature used in forging or rolling and the grain size of hot-worked products is determined largely by the finishing temperature, that is, the temperature at which the final stage of the hot-working process is carried out.

Grain Size and Properties—The coarseness of the ferritic and pearlitic "grains" in the cooled steel is related to the grain size of the austenite prior to its transformation and the properties of the product are profoundly influenced by its grain size. The general effects of the austenite grain size are summarized in Table 41—II.

Grain size is determined by making measurements with the aid of a metallographic microscope. Three basic procedures for estimating grain size are described in ASTM Designation E112 and are known as the com-

Table 41—II. Trends in Heat-Treated Products

Property	Course-Grain Austenite	Fine-Grain Austenite
Hardenability	Deeper Hardening	Shallower Hardening
Toughness	Less Tough	Tougher
Distortion	More Distortion	Less Distortion
Quench Cracking	More Prevalent	Less Prevalent
Internal Stress	Higher	Lower
For Annealed or Normalized Products		
Machinability (Rough)	Better	Inferior
Machinability (Fine Finish)	Inferior	Better

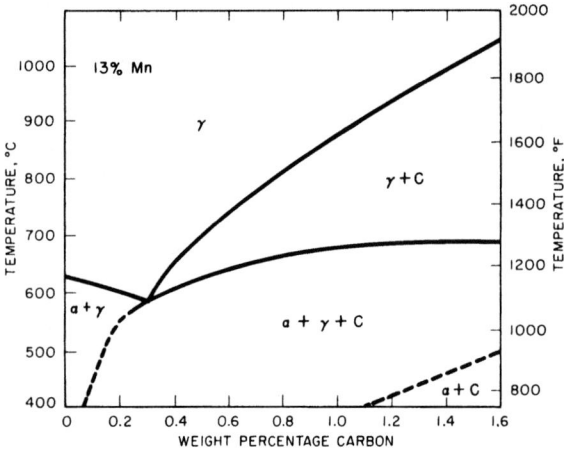

FIG. 41—13. Section of the Fe-C-Mn equilibrium diagram at 13 per cent manganese content.

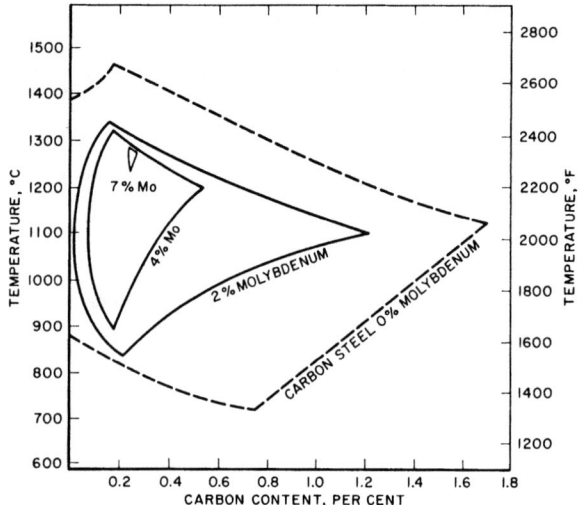

FIG. 41—14. The effect of molybdenum on the composition and temperature range over which austenite is stable. (After Bain.)

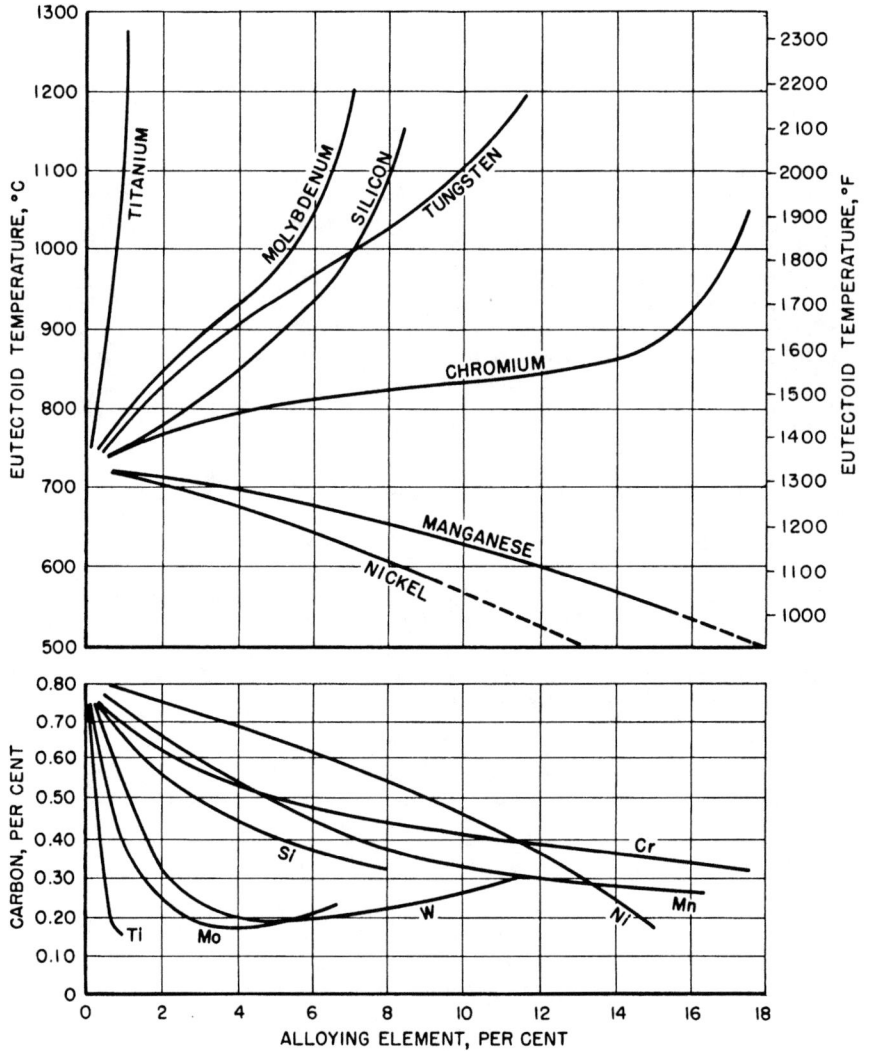

FIG. 41—15. Eutectoid composition and temperature as influenced by several alloying elements. (After Bain.)

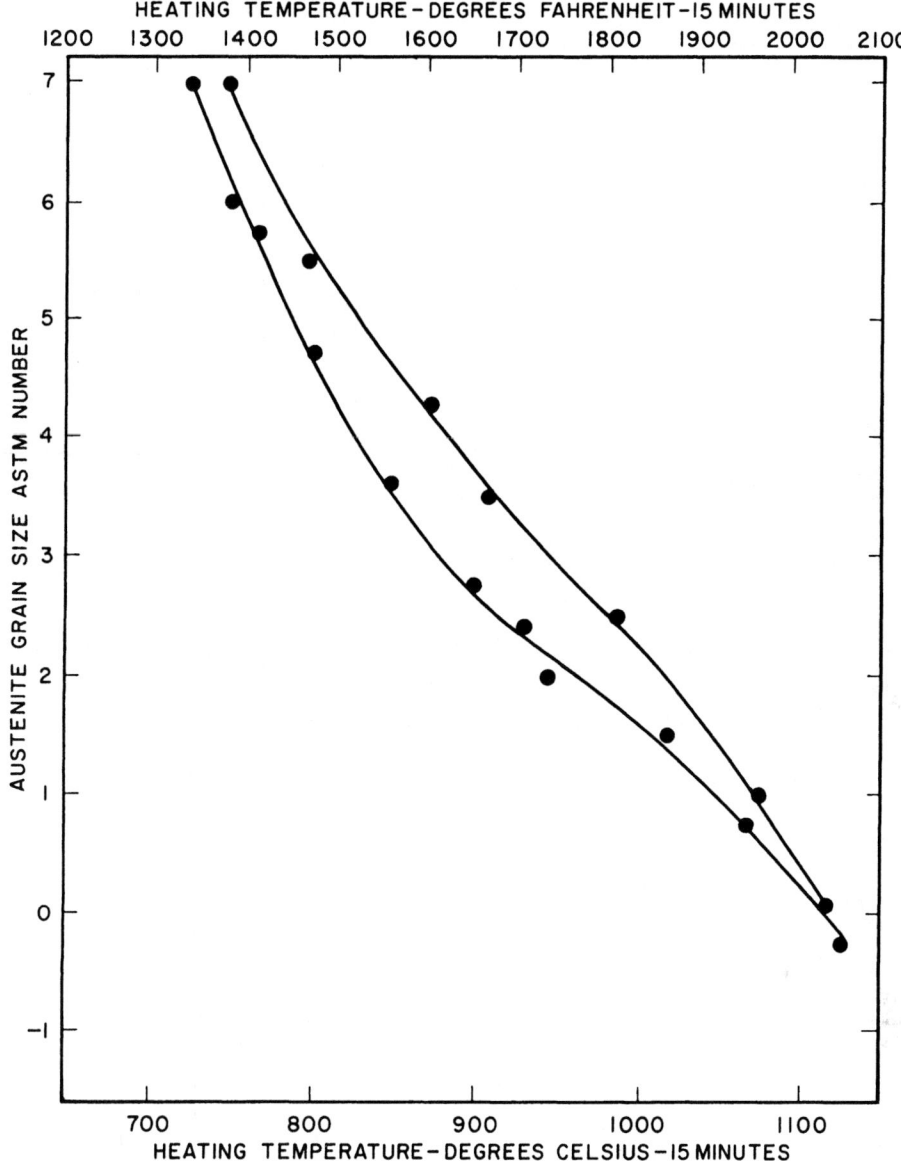

FIG. 41—16. Grain size as a function of austenitizing temperature for inherently coarse-grained steels. (After Bain.)

parison procedure, the intercept (or Heyn) procedure and the planimetric (or Jeffries) procedure. The grain size is commonly given as the ASTM grain size number. The ASTM grain size number is an arbitrary exponential number that refers to the mean number of grains per square inch at a magnification of 100X according to the following relation:

Mean number of grains per square inch at $100X = 2^{n-1}$ where "n" is the ASTM grain-size number.

General relations between the ASTM grain size number and other parameters characterizing the grains are shown in Table 41—III.

Fine- and Coarse-Grain Steels—As was mentioned previously, austenitic-grain growth may be inhibited by undissolved carbides or by a suitable distribution of non-metallic inclusions. Steels of this type are commonly referred to as inherently fine-grained or simply as fine-grained steels, while steels which are free from these grain growth inhibitors are known as coarse-grained steels.

The general pattern of grain coarsening in steels of the coarse- and fine-grained types on heating above the critical temperature is illustrated in Figures 41—16 and 41—17. It will be noted that the coarse-grained steel coarsens gradually and consistently as the temperature is increased, while the fine-grained steel coarsens only slightly, if at all, until a certain temperature is reached, above which an abrupt coarsening occurs. This temperature is known as the **coarsening temperature.** It should further be noted that either type of steel may be heat treated so as to be either fine or coarse grained, and as

Table 41—III Micro-Grain Size Relationships.

ASTM Micro-Grain Size Number	Calculated "Diameter" of Average Grain		Average Intercept Distance[b]		Calculated Area of Average Grain Section		Average Number of Grains per cu mm	Nominal Grains per sq. mm at 1 ×	Nominal Grains per sq. in. at 100 ×
	mm	in.	mm	in.	sq. mm	sq. in.			
		$\times 10^{-3}$		$\times 10^{-3}$	$\times 10^{-3}$	$\times 10^{-3}$			
00[a]	0.508	20.0	0.451	17.8	258	400	7.63	3.88	0.250
0	0.359	14.1	0.319	12.6	129	200	21.6	7.75	0.50
0.5	0 302	11.9	0.268	10.6	91.2	141	36.3	11.0	0.707
1.0	0.254	10.0	0.226	8.88	64.5	100	61.0	15.5	1.0
....	0.250	9.84	0.222	8.74	62.5	96.9	64.0	16.0	1.03
1.5	0.214	8.41	0.190	7.47	45.6	70.7	103	21.9	1.41
....	0.200	7.87	0.178	6.99	40.0	62.0	125	25.0	1.61
....	0.180	7.09	0.160	6.29	32.4	50.2	171	30.9	1.99
2.0	0.180	7.07	0.160	6.28	32.3	50.0	172.3	31.0	2.0
2.5	0.151	5.95	0.134	5.30	22.8	35.4	290	43.8	2.83
....	0.150	5.91	0.133	5.24	22.5	34.9	296	44.4	2.87
3.0	0.127	5.00	0.113	4.44	16.1	25.0	488	62.0	4.0
....	0.120	4.72	0.107	4.20	14.4	22.3	578.9	69.4	4.48
3.5	0.107	4.20	0.0948	3.73	11.4	17.7	821	87.7	5.66
....	0.900	3.54	0.0799	3.15	8.10	12.6	1 370	123	7.97
4.0	0.0898	3.54	0.0797	3.14	8.06	12.5	1 380	124	8.0
4.5	0.076	2.97	0.0671	2.64	5.70	8.84	2 320	175	11.3
....	0.070	2.76	0.0622	2.45	4.90	7.59	2 920	204	13.2
5.0	0.064	2.50	0.0564	2.22	4.03	6.25	3 910	248	16.0
....	0.060	2.36	0.0533	2.10	3.60	5.58	4 630	278	17.9
5.5	0.0534	2.10	0.0474	1.87	2.85	4.42	6 570	351	22.6
....	0.050	1.97	0.0444	1.75	2.50	3.88	8 000	400	25.8
6.0	0.045	1.77	0.0399	1.57	2.02	3.13	11 000	496	32.0
....	0.040	1.58	0.0355	1.40	1.60	2.48	15 600	625	40.3
6.5	0.038	1.49	0.0335	1.32	1.43	2.21	18 600	701	45.3
....	0.035	1.38	0.0311	1.22	1.23	1.90	23 000	816	52.7
7.0	0.032	1.25	0.0282	1.11	1.01	1.56	31 000	992	64.0
....	0.030	1.18	0.0267	1.05	0.90	1.40	37 000	1 110	71.7
7.5	0.027	1.05	0.0237	0.933	0.713	1.10	52 500	1 400	90.5
....	0.025	0.984	0.0222	0.874	0.625	0.969	64 000	1 600	103
8.0	0.0224	0.884	0.0199	0.785	0.504	0.781	88 400	1 980	128
....	0.0200	0.787	0.0178	0.699	0.40	0.620	125 000	2 500	161
8.5	0.0189	0.743	0.0168	0.660	0.356	0.552	149 000	2 810	181
9.0	0.0159	0.625	0.0141	0.555	0.252	0.391	250 000	3 970	256
....	0.0150	0.591	0.0133	0.524	0.225	0.349	296 000	4 440	287
9.5	0.0134	0.526	0.0119	0.467	0.178	0.276	420 000	5 610	362
10.0	0.0112	0.442	0.00997	0.392	0.126	0.195	707 000	7 940	512
....	0.0100	0.394	0.00888	0.350	0.10	0.155	1.00×10^6	10 000	645
10.5	0.00944	0.372	0.00838	0.330	0.089	0.138	1.19×10^6	11 200	724
....	0.00900	0.354	0.00799	0.315	0.081	0.126	1.37×10^6	12 300	797
....	0.00800	0.315	0.00710	0.280	0.064	0.0992	1.95×10^6	15 600	1 010
11.0	0.00794	0.313	0.00705	0 278	0.063	0.0977	2.00×10^6	15 900	1 020
....	0.00700	0.276	0.00622	0.245	0.049	0.0760	2.92×10^6	20 400	1 320
11.5	0.00667	0.263	0.00593	0.233	0.045	0.0691	3.36×10^6	22 400	1 450
....	0.00600	0.236	0.00533	0.210	0.036	0.0558	4.63×10^6	27 800	1 790
12.0	0.00561	0.221	0.00498	0.196	0.031	0.0488	5.66×10^6	31 700	2 050
....	0.00500	0.197	0.00444	0.175	0.025	0.0388	8.00×10^6	40 000	2 580
12.5	0.00472	0.186	0.00419	0.165	0.022	0.0345	9.51×10^6	44 900	2 900
....	0.00400	0.158	0.00355	0.140	0.0160	0.0248	15.62×10^6	62 500	4 030
13.0	0.00397	0.156	0.00352	0.139	0.0158	0.0244	16.0×10^6	63 500	4 100
13.5	0.00334	0.131	0.00296	0.117	0.011	0.0173	26.9×10^6	89 800	5 800
....	0.00300	0.118	0.00266	0.105	0.009	0.0140	37.0×10^6	111 000	7 170
14.0	0.00281	0.111	0.00249	0.0981	0.0079	0.0122	45.2×10^6	127 000	8 200
....	0.00250	0.098	0.00222	0.0874	0.00625	0.00969	64.0×10^6	160 000	10 300

[a]The use of 00 is recommended instead of "—1" or "minus 1" to avoid confusion.
[b]Value of Heyn intercept for equiaxed grains.
[c]From: ASTM Standards Designation E-112-81, Vol. 03.03, 1984, Metallography; Nondestructive Testing published by American Society for Testing and Materials, Philadelphia, Pa.

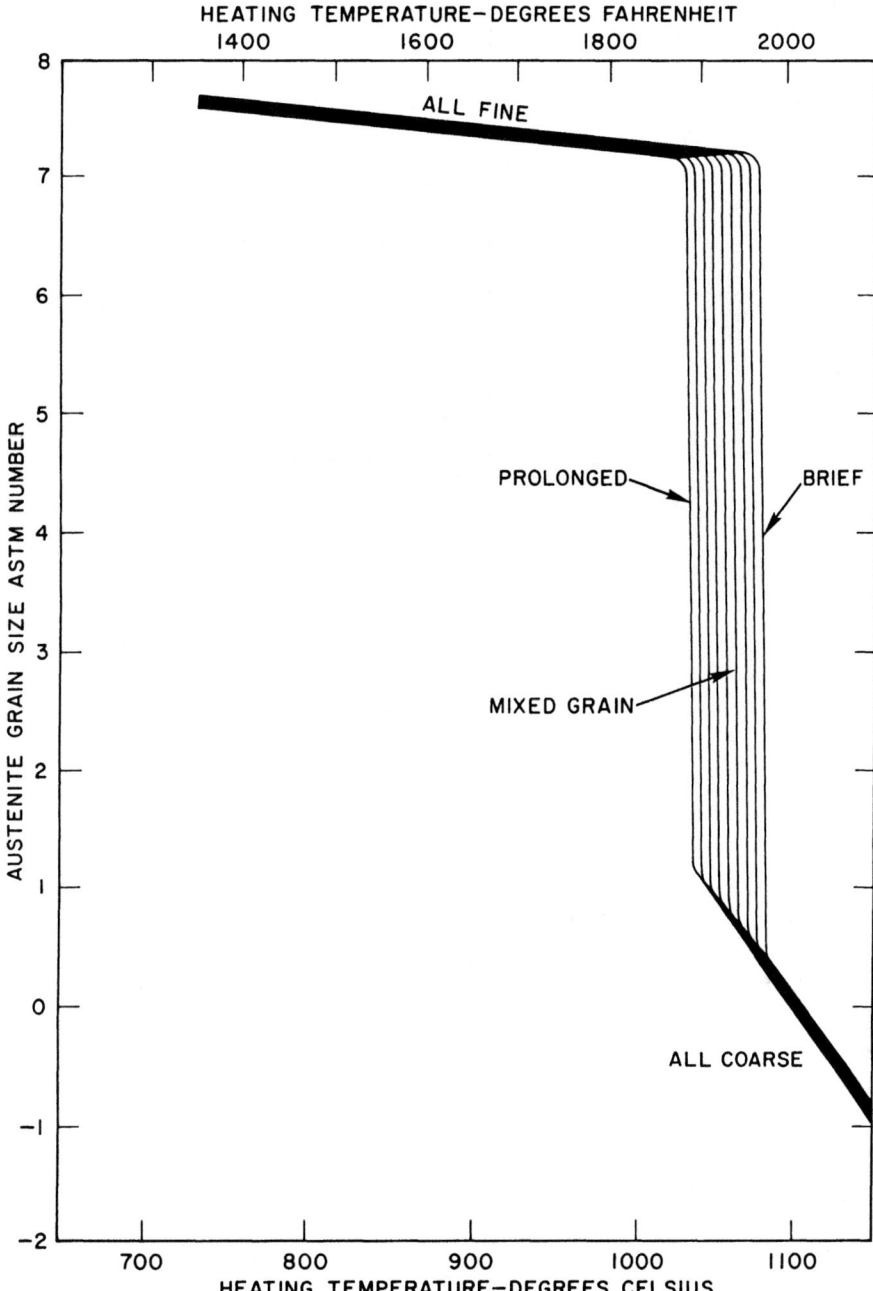

FIG. **41—17**. Grain size as a function of austenitizing temperature for an inherently fine-grained steel (schematic). (After Bain.)

a matter of fact, at temperatures above its coarsening temperature, the fine-grained steel will usually exhibit a coarser grain size than the coarse-grained steel at the same temperature.

The usual method of making steels which remain fine grained at 925° C (1700° F) (ASTM-E19-46) involves the judicious use of aluminum deoxidation. The inhibiting agent in such steels is generally conjectured to be a submicroscopic dispersion of aluminum nitride, or, perhaps at times, aluminum oxide.

THE TRANSFORMATION OF AUSTENITE

Thus far, this chapter has been concerned with equilibrium conditions or conditions resulting from slow cooling or heating. Under such conditions, it has been shown that austenite transforms to pearlite when it is cooled below the A_1 critical temperature, and at a temperature only a little below the Ae_1 temperature. When more rapidly cooled, however, this transformation is depressed and does not occur until a lower tem-

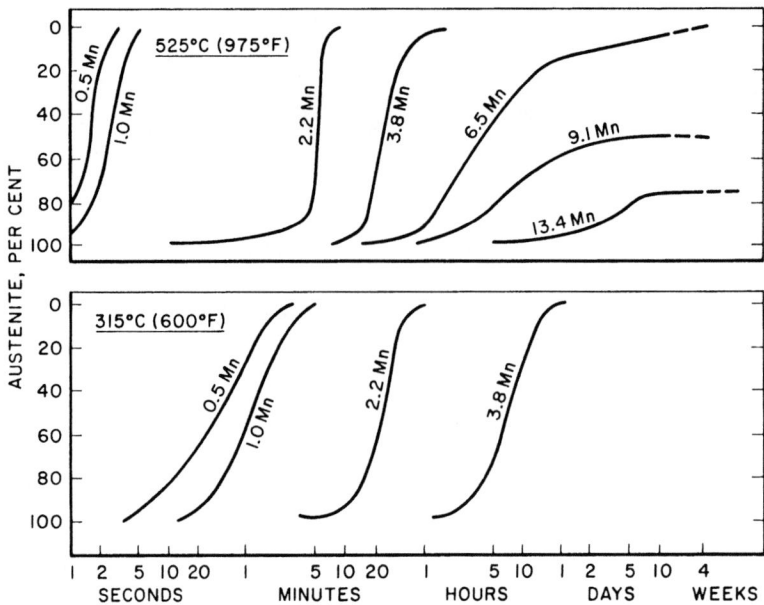

FIG. 41—18. Transformation behavior at a single temperature in a series of manganese steels of about 0.55% carbon content. (After Bain.)

perature is reached. The faster the cooling rate, the lower is the temperature at which transformation occurs. Furthermore, the nature of the ferrite-carbide aggregate formed when the austenite transforms varies markedly with the temperature of transformation and the properties are found to vary correspondingly. Thus, heat treatment is seen to involve a controlled supercooling of austenite, and in order to take full advantage of the wide range of structures and properties which this permits, a knowledge of the transformation behavior of austenite and the properties of the resulting aggregates is essential.

Isothermal Transformation Diagrams—The transformation behavior of austenite can best be studied by a technique developed at the United States Steel Corporation Fundamental Research Laboratory in 1930. This involves studying the transformation behavior at a series of temperatures below A_1, by quenching small samples to the desired temperature in a liquid bath, allowing them to transform isothermally and following the progress of the transformation metallographically or by dilatometric measurements. This procedure not only gives a picture of the rates of transformation at the various temperatures, but also furnishes information as to the metallographic structures characteristic of the various temperatures of transformation and permits a determination of the properties of these individual microstructures.

The general pattern of this transformation behavior at a single temperature is shown in Figure 41—18. It will be seen that there is first a period of time before any transformation starts. This period is sometimes spoken of as an **incubation period.** The incubation period presumably represents the period at the start of transformation when the volume of transformed phase around each nucleus is increasing very slowly. The transformation accelerates as it progresses so that the

fastest transformation rate corresponds to about 50 per cent transformation. The rate then slows down again and the transformation goes to completion rather slowly. This general pattern is characteristic, but the rates themselves will vary with the temperature of transformation and, as will be shown later, with the composition and grain size of the austenite.

It has become customary to present data obtained in this manner as a plot, with the times required for the beginning and completion (and a few other stages) of the transformation as the abscissa and temperature as the ordinate. Such a diagram is known as an **isothermal transformation diagram.**

An isothermal transformation diagram for a eutectoid plain carbon steel, together with the hardness values and microstructures characteristic of transformation at the various temperature levels, is shown in Figure 41—19. It will be noted that the diagram has a **nose** or temperature of most rapid transformation at above 535° C (1000° F). The transformation rate at temperatures near the A_1 is very slow and it is likewise relatively slow at a lower temperature range.

Transformation to Pearlite—Transformation over the temperature range of about 705° to 535° C (1300° to 1000° F), in carbon and low-alloy steels forms pearlitic microstructures and the characteristic lamellar appearance of these structures is apparent in the photomicrographs. It will also be noted that, as the transformation temperature decreases, the lamellae become more closely spaced, so that as transformed at 535° C (1000° F) they can hardly be resolved by the light microscope. The much greater resolving power of the electron microscope does, however, permit resolution of the closely spaced lamellae of fine pearlite, as illustrated in the electron micrograph of Figure 41—20, showing the microstructure of fine pearlite formed at 595° C (1100° F). The hardness is also seen to increase as the lamellar

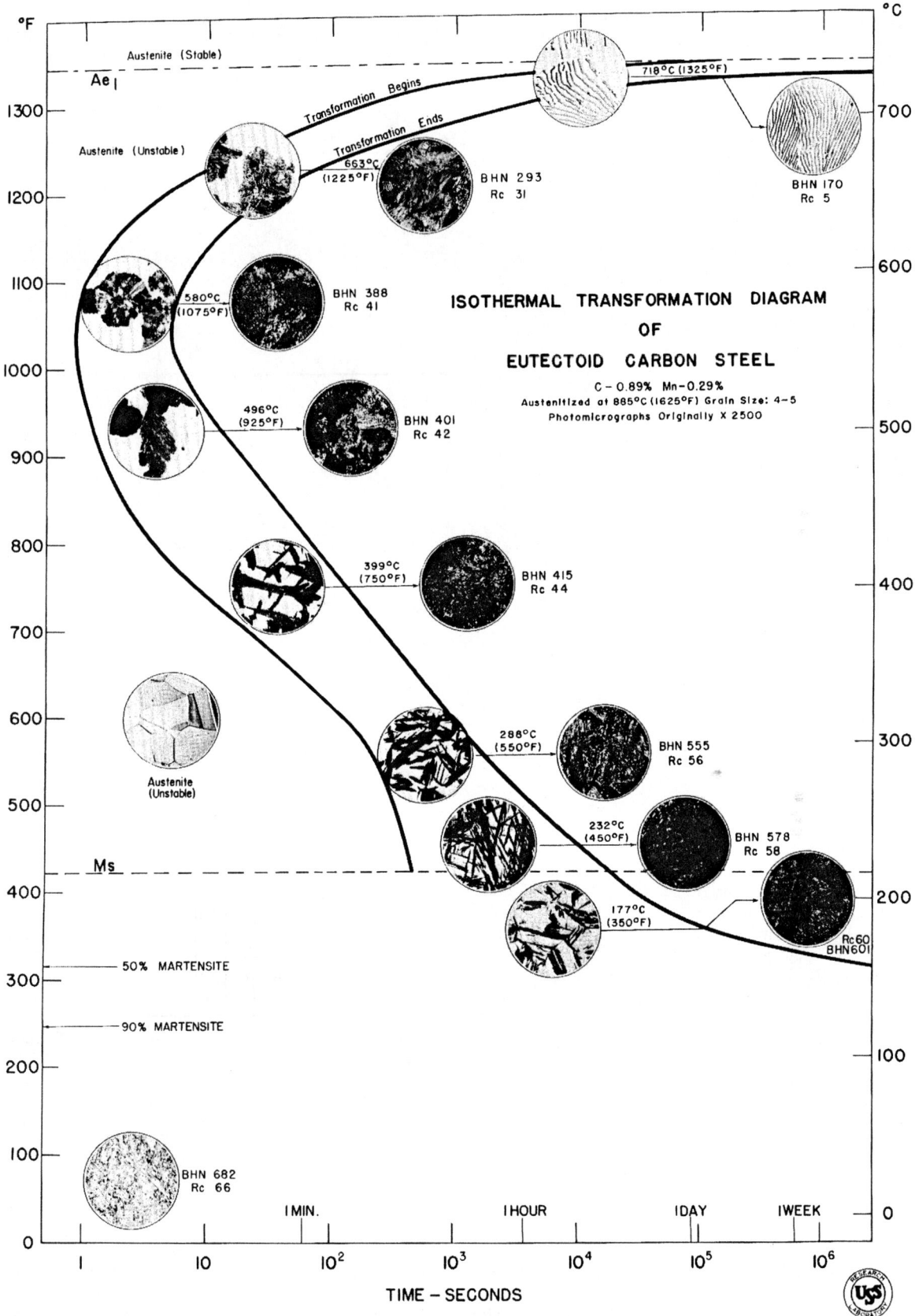

FIG. 41—19. Isothermal transformation diagram for a plain carbon eutectoid steel.

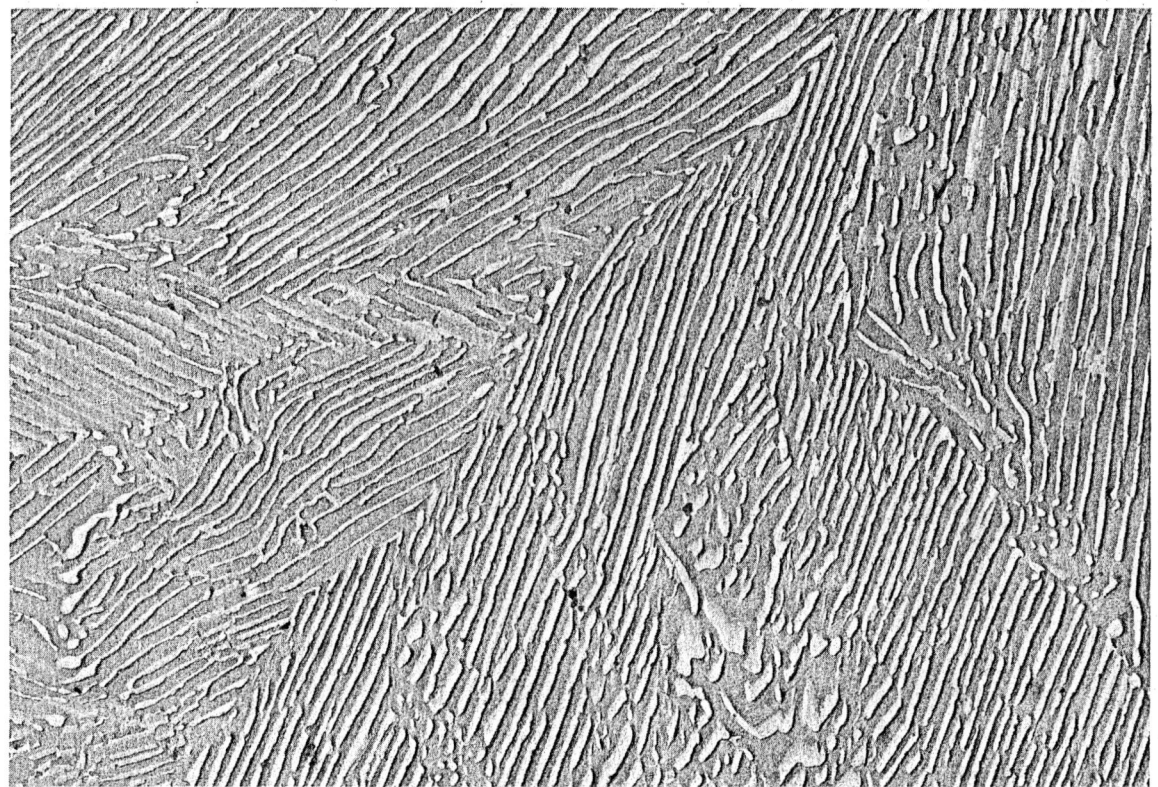

FIG. 41—20. Electron micrograph showing microstructure of fine pearlite formed at 575° C (100° F). Magnification: 15 000X.

spacing becomes smaller.

Transformation to Bainite—Transformation to bainite occurs over the temperature range of about 535° to 230° C (1000° to 450° F). The bainitic microstructures differ markedly from pearlitic in that they are acicular in nature. Here again, the hardness increases as the transformation temperature decreases, through the bainite formed at the highest possible temperature is often softer than pearlite formed at a still higher temperature. The distribution of carbide in bainite formed at low temperatures is very fine and as a result lower bainite is harder and stronger than upper bainite. The details of the bainitic microstructures forming in this temperature range are, as with fine pearlite, generally irresolvable with the light microscope, but are readily resolved with the electron microscope. Electron micrographs of the bainitic microstructures formed on complete transformation of a eutectoid steel at 455°, 370° and 260° C (850°, 700° and 500° F) are shown in Figures 41—21, 41—22 and 41—23.

Transformation to Martensite—Transformation to martensite, which in this steel occurs at temperatures below 230° C (450° F), differs from transformation to pearlite or bainite in that it is not time dependent, but occurs almost instantly during cooling and the percentage of transformation is dependent only on the temperature to which it is cooled. Thus, in this steel, transformation to martensite will start on cooling to 230° C (450° F), designated as the M_s temperature, will be 50 per cent complete on cooling to about 150° C

(300° F), and will be essentially completed at about 95° C (200° F) (designated as the M_f temperature). The microstructure of martensite is likewise acicular but it is generally lighter etching than bainite. It is the hardest of the transformation products of austenite. It is possible to form a little martensite at, say, 425° F and then to cause bainite to form thereafter isothermally.

MICROSTRUCTURAL AND MECHANICAL PROPERTIES

The dependence of the properties of steel upon its constitution has been emphasized in this chapter and, as would be expected, the properties of steel vary with the temperature at which the austenite transforms in accordance with the corresponding microstructural changes.

The microstructures discussed above fall into three general classes: pearlite, bainite, and martensite. In discussing the relationships between microstructure and properties, a fourth class of microstructure, tempered martensite, must be considered. This is the structure formed when martensite is reheated to a subcritical temperature after quenching. Its microscopic appearance is illustrated in the electron micrograph of Figure 41—24. The general effect of tempering martensite is to precipitate and agglomerate the carbide particles, so that tempered martensite microstructures consist of carbide particles dispersed in a ferrite matrix. The steel in this illustration has been tempered at a

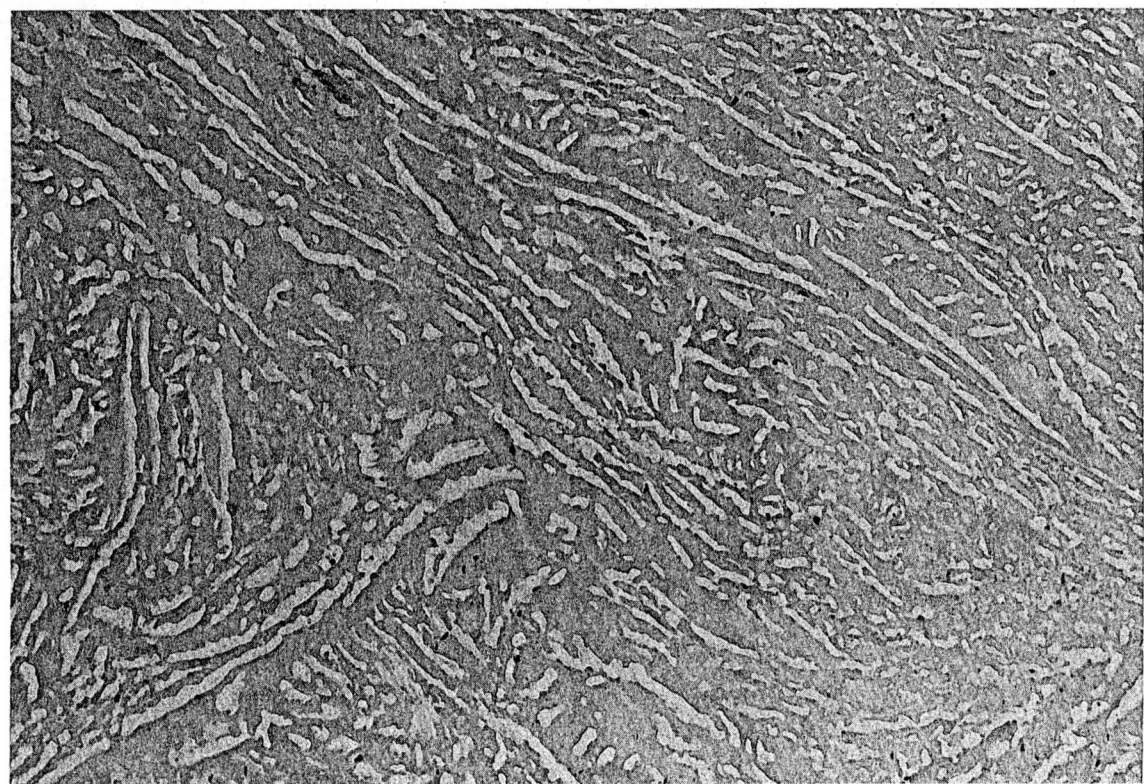

FIG. 41—21. Electron micrograph of bainitic microstructure formed on complete transformation of a eutectoid steel at 455° C (850° F). Magnification: 15 000X.

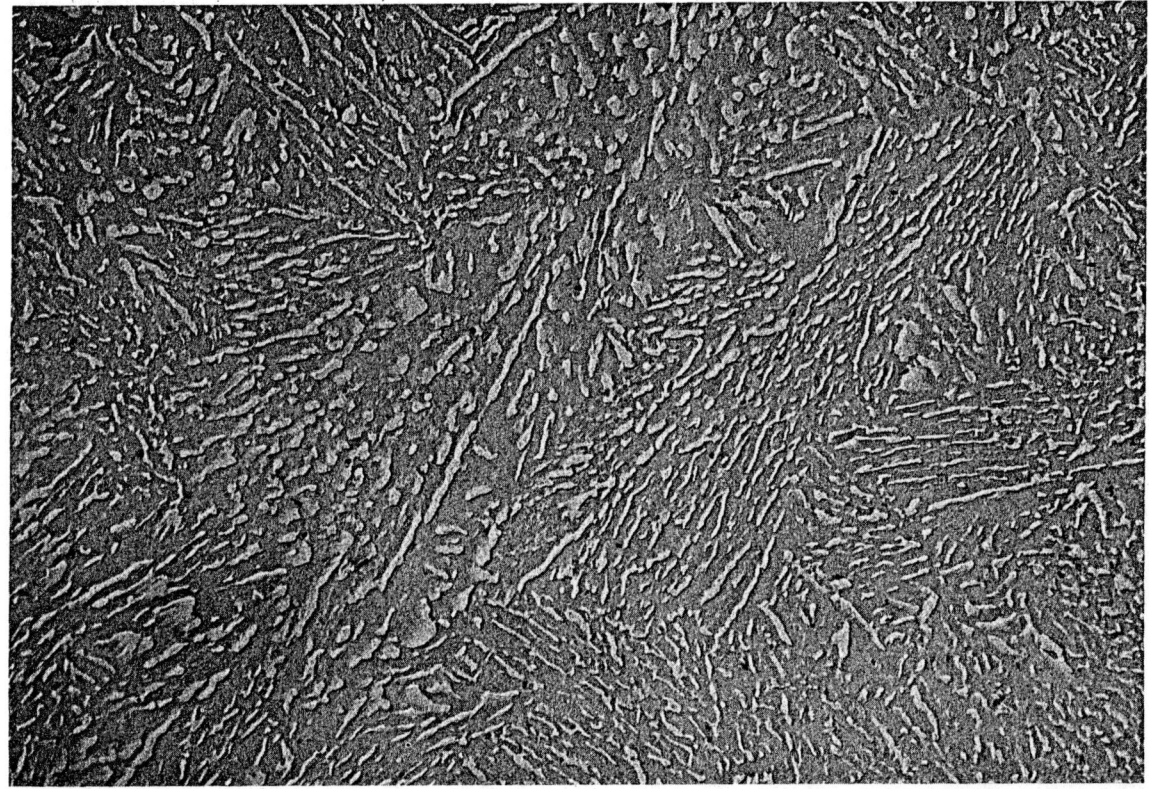

FIG. 41—22. Electron micrograph of bainitic microstructure formed on complete transformation of a eutectoid steel at 370° C (700° F). Magnification: 15 000X.

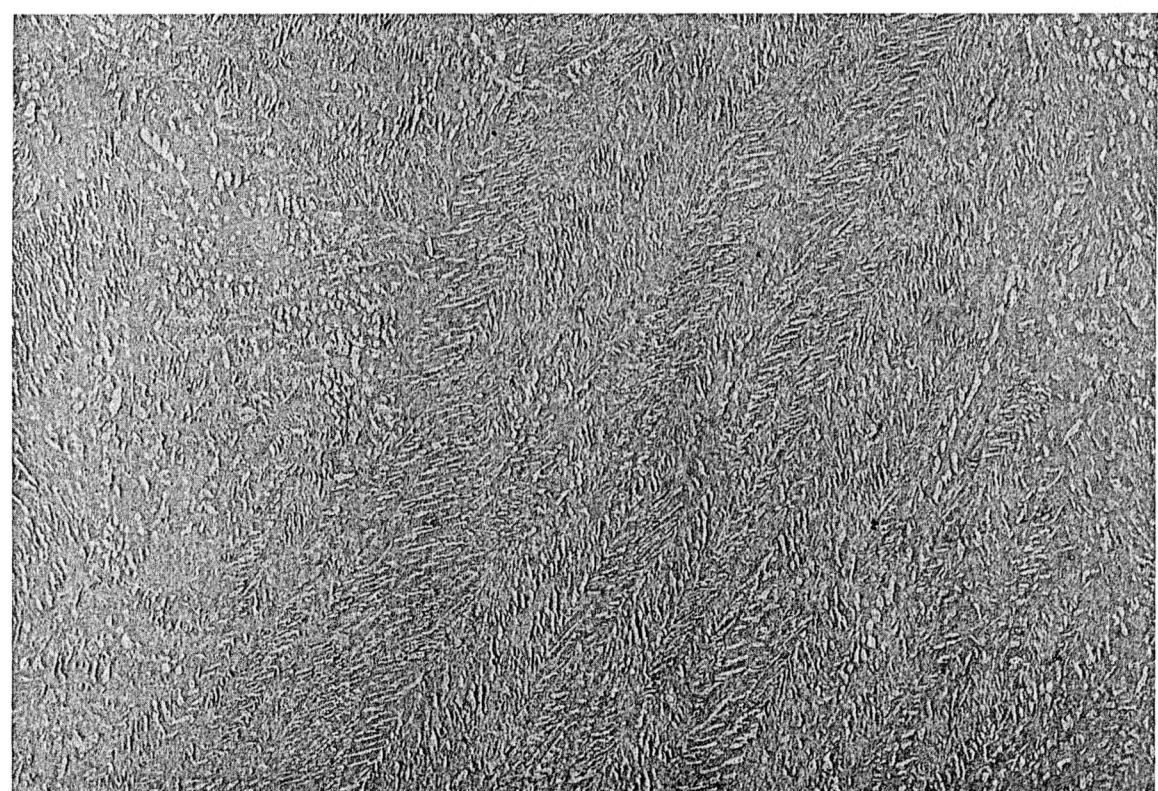

FIG. **41—23.** Electron micrograph of bainitic microstructure formed on complete transformation of a eutectoid steel at 260° C (500° F). Magnification: 15 000X.

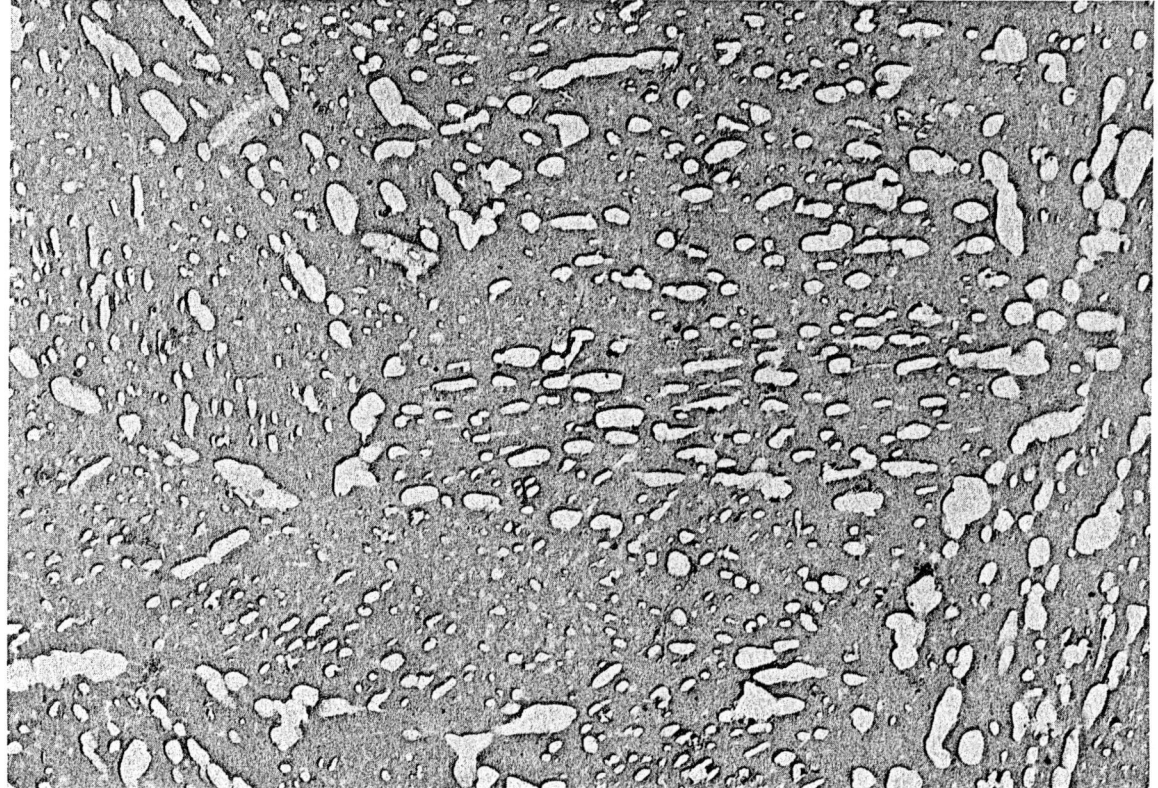

FIG. **41—24.** Electron micrograph of the structure of tempered martensite. Tempering temperature. 595° C (1100° F). Magnification: 15 000X.

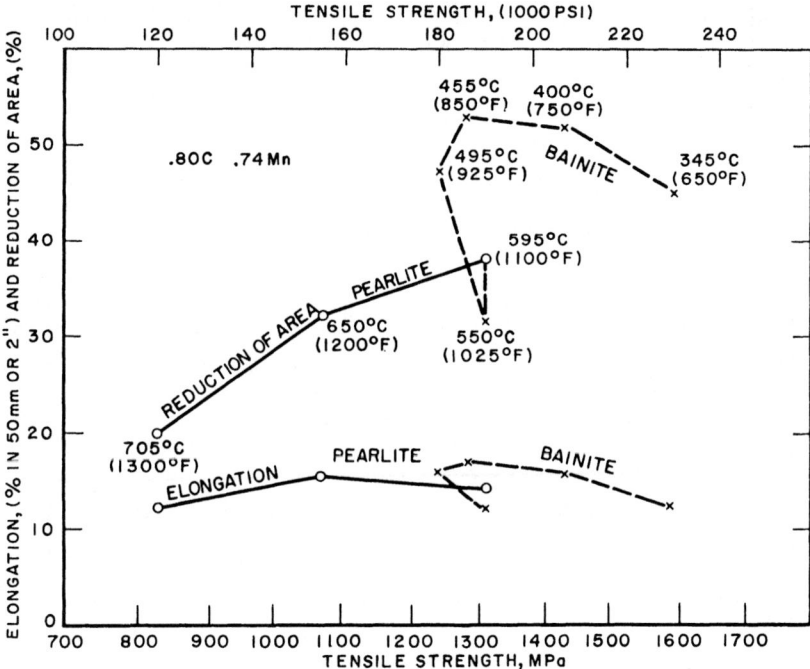

FIG. 41—25. The properties of pearlite and bainite in a eutectoid steel.

relatively high temperature and some tendency for the particles to assume a spheroidal form can be noted. This microstructure, consisting of spheroidized or partially spheroidized carbides in a ferrite matrix is, as might be expected, a very favorable one in respect to ductility.

Each of these types of microstructures has characteristic properties typical of the class but the properties of pearlite and bainite also each vary quite widely with transformation temperature.

Properties of Pearlite—In any steel, the pearlites are, as a class, softer than the bainites or martensites. In general, even though softer, they are less ductile than the lower temperature bainites and for a given hardness they will be far less ductile than tempered martensite. As the transformation temperature decreases within the pearlite range, the interlamellar spacing decreases, as was described above, and these "fine" pearlites, formed near the nose of the isothermal diagram, are both harder and more ductile than the

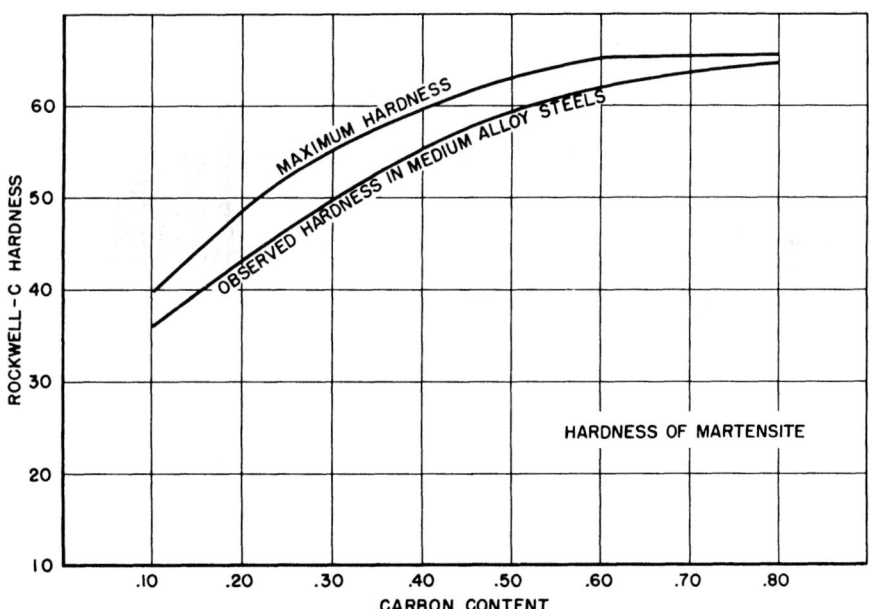

FIG. 41—26. The hardness of martensite as a function of carbon content.

"coarse" pearlites, formed at higher temperatures. Thus, although, as a class, pearlite tends to be soft and not exceedingly ductile, its hardness and toughness both increase markedly with decreasing transformation temperatures.

Properties of Bainite—In a given steel, bainitic microstructures will generally be found to be both harder and tougher than pearlite, although the hardness will be lower than that of martensite. Within the class,

as with pearlite, the properties generally improve as the transformation temperature decreases and "lower" bainite will compare favorably with tempered martensite at the same hardness. "Upper" bainite, on the other hand, may be somewhat deficient in toughness as compared with fine pearlite at the same hardness.

The properties of pearlite and bainite in a eutectoid steel are summarized in Figure 41—25.

Properties of Martensite—Martensite is a supersatu-

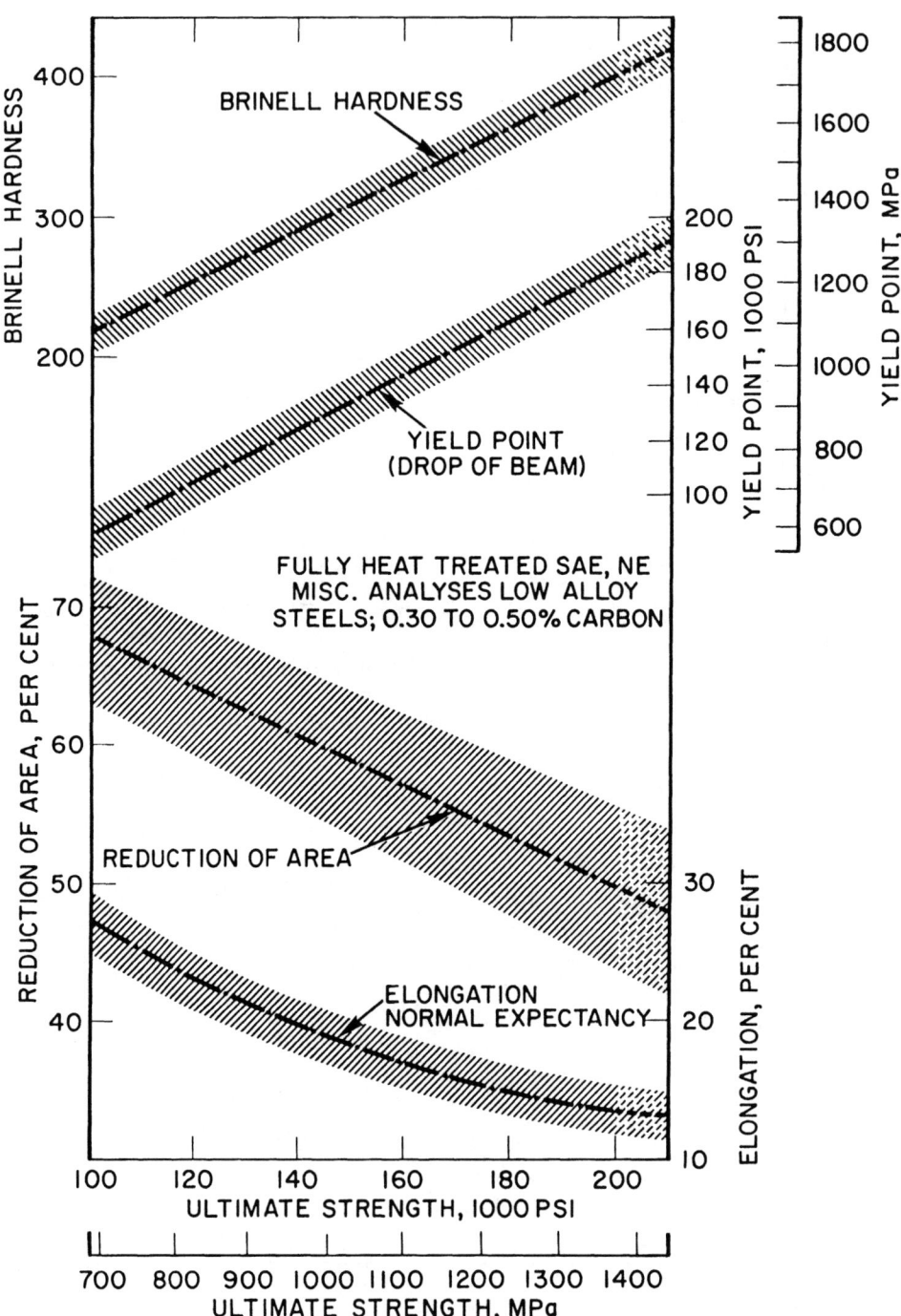

FIG. 41—27. The properties of tempered martensite.

rated solid solution of carbon in alpha iron. Martensite exhibits an acicular structure. It is the hardest and, except in the low-carbon martensites discussed in Chapter 42 on "Alloy Steels," the most brittle of the microstructures obtainable in a given steel. The hardness of martensite as a function of carbon content is shown in Figure 41—26. The hardness of martensite, at a given carbon content, varies somewhat with the cooling rate and this figure shows both the maximum hardness obtainable with very rapid cooling and the average hardness values which might be expected in practice.

Although, for some applications, particularly those involving wear resistance, the high hardness of martensite is desirable in spite of the accompanying brittleness, the principal importance of this microstructure is as the starting material for tempered martensite structures, which latter have definitely superior properties.

Properties of Tempered Martensite—Tempered martensitic structures are, as a class, characterized by relatively high toughness at any strength level. Their properties are illustrated in Figure 41—27. This chart designates, within plus or minus 10 per cent, the usual mechanical properties of any steel with this microstructure, regardless of composition. Because of its high ductility at a given hardness, this is the structure that is aimed for in heat treating for toughness by quenching and tempering.

FACTORS AFFECTING TRANSFORMATION RATES

The major factors affecting the rates of transformation of austenite are its composition, grain size, and homogeneity. In general, increasing carbon and alloy content tend to decrease transformation rates. Increasing the grain size of the austenite likewise tends to decrease transformation rates.

Effect of Carbon Content—The effect on the transformation rates of decreasing the carbon content of a plain carbon steel is illustrated by comparison of Figure 41—28, which shows the isothermal transformation diagram for a 0.35 per cent carbon steel, with that for the eutectoid steel shown in Figure 41—19. It will be noted that the effect of lowering the carbon content has been to shift the lines of the diagram to the left, that is, toward more rapid transformation rates. This diagram differs from that for the eutectoid steel also in that the transformation to pearlite is preceded by a precipitation of ferrite and the diagram, therefore, shows a line designating the time for the initiation of this ferrite precipitation at the temperature levels wherein this separation precedes the formation of pearlite.

Effects of Alloys—Figure 41—29 shows an isothermal transformation diagram for a 0.35 per cent carbon, 1.85 per cent manganese steel. Comparing this with the lower manganese steel (Figure 41—28), it will be noted that the entire curve has been displaced to the right; that is, transformation at all temperature levels starts later and is slower to go to completion. This is characteristic of the effect of alloys in solution in the austenite; in general, increased alloy content delays the start of the transformation and increases the time for its completion.

Although alloy additions tend in general to delay the start of transformation and to increase the time for its completion, they differ greatly, nevertheless, in both the magnitude and the nature of their effects. Figure 41—30 represents the isothermal transformation diagram for a 0.33 per cent carbon, 0.45 per cent manganese, 1.97 per cent chromium steel. By comparison with the plain carbon steel (Figure 41—28), it will be noted that the effect of the chromium has been, not only to move the curve to the right, but also to change the shape of the curve. The time for beginning of transformation in the pearlite region has been greatly increased, while that for the beginning in the bainite region has been only moderately increased. Thus the diagram now has two "noses" (or time minima), one in the temperature region of transformation to pearlite and the other in the bainite region.

Figure 41—31 represents the isothermal transformation diagram for a more-complex alloy steel. This steel (SAE 4340) contains 0.42 per cent carbon, 0.78 per cent manganese, 1.79 per cent nickel, 0.80 per cent chromium and 0.33 per cent molybdenum. It will be noted that the effect of the addition of moderate amounts of these several alloying elements has been to displace the curve even farther to the right than that of the 2 per cent chromium steel (Figure 41—30). This is characteristic of the effect of alloys; relatively small amounts of several alloying elements are more effective in decreasing transformation rates than are larger amounts of a single alloy, i.e., more retarding than if they were merely additive.

Summarizing the effects of alloying elements on transformation behavior, it can be seen that the general effect of increasing the alloy content is to delay both the start and the completion of transformation and that the effect of alloy additions is cumulative. The effects of alloying elements, however, differ greatly both in magnitude and in specific effects on transformation in different temperature regions, so that a precise prediction of the effect of a given alloy combination is not yet quite possible.

If cooling rates are rapid enough to avoid transformation to pearlite or bainite, transformation to martensite will initiate at a given temperature and will progress as the temperature falls. Empirical formulae have been derived for estimating from chemical composition the temperature at which martensite starts to form (M_s temperature). Very little work has been done concerning the effect of composition on the temperature where the transformation of austenite is virtually complete (M_f temperature). Elements that lower M_s probably lower M_f even more, thus widening the temperature range of austenite formation. Three typical formulae for calculating the approximate M_s temperature of alloy steels are:

$M_s(°C) = 500 - 300C - 33Mn - 17Ni - 22Cr - 11Si - 11Mo$ (Nehrenberg);

$M_s(°C) = 561 - 474C - 33Mn - 17Ni - 17Cr - 22Mo$ (Stevens and Haynes);

$M_s(°C) = 539 - 423C - 30.4Mn - 17.7Ni - 12.1Cr - 11Si - 7.5Mo$ (Andrews).

These formulae hold for steels austenitized 50° to 100°C above the A_3 temperature for times sufficient to assure that all the alloying elements are in solid solution

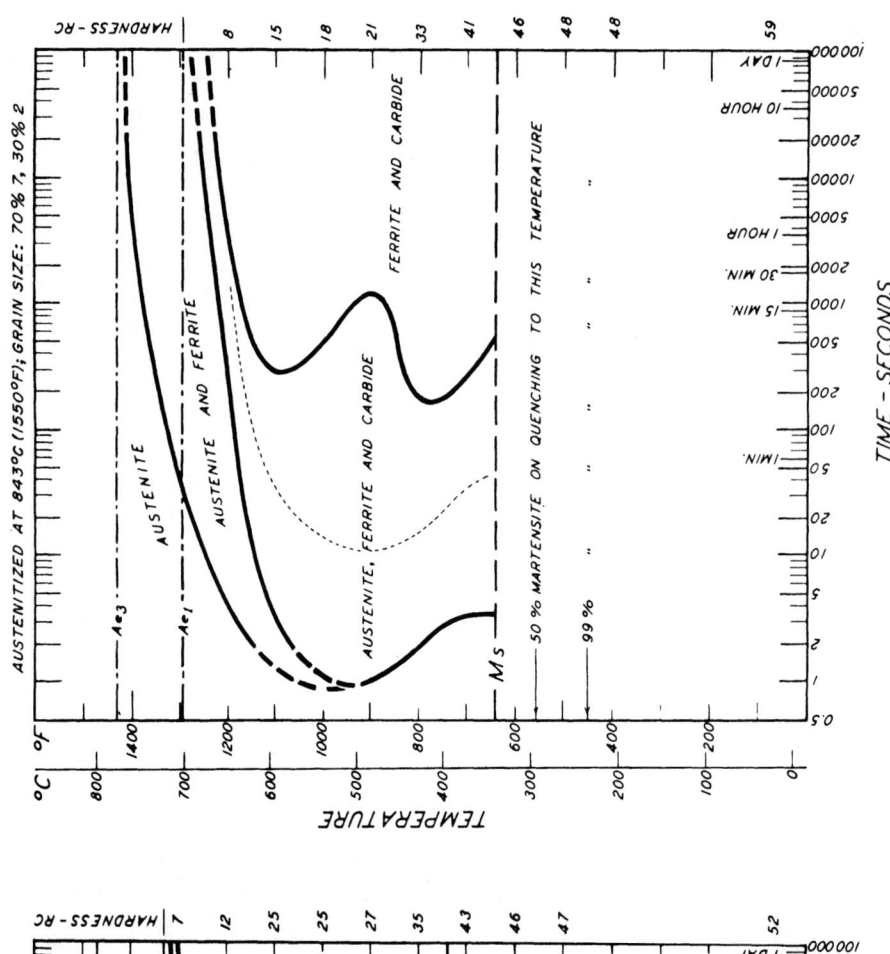

FIG. **41**—29. Isothermal transformation diagram for a 0.35% carbon, 1.85% manganese steel.

FIG. **41**—28. Isothermal transformation diagram for a 0.35% carbon, 0.37% manganese, plain carbon steel.

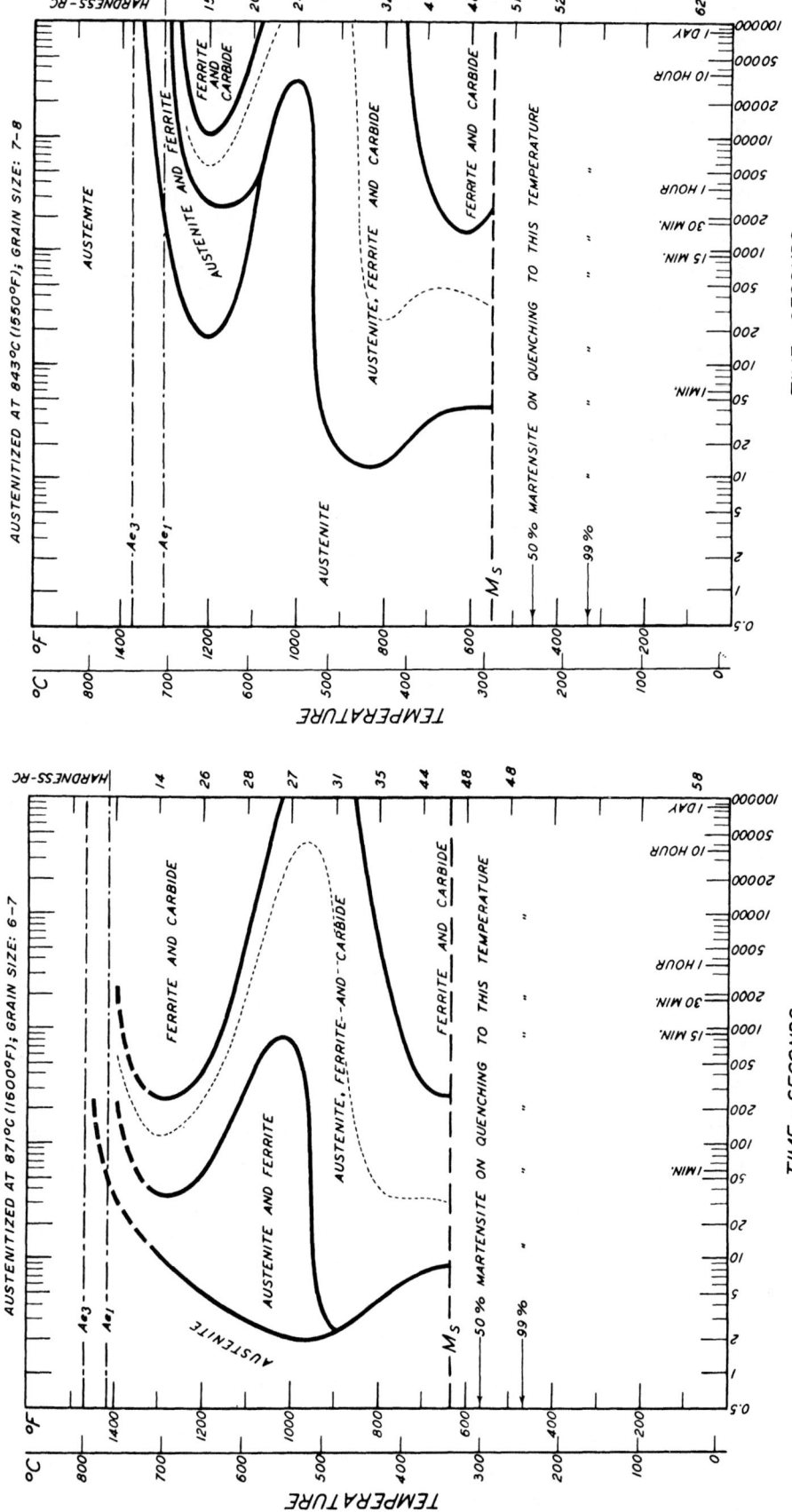

FIG. 41—31. Isothermal transformation diagram for a G43400 (AISI or SAE 4340) steel, containing 0.42% carbon, 0.78% manganese, 1.79% nickel, 0.80% chromium and 0.33% molybdenum.

FIG. 41—30. Isothermal transformation diagram for a 0.33% carbon, 0.45% manganese, 1.97% chromium steel.

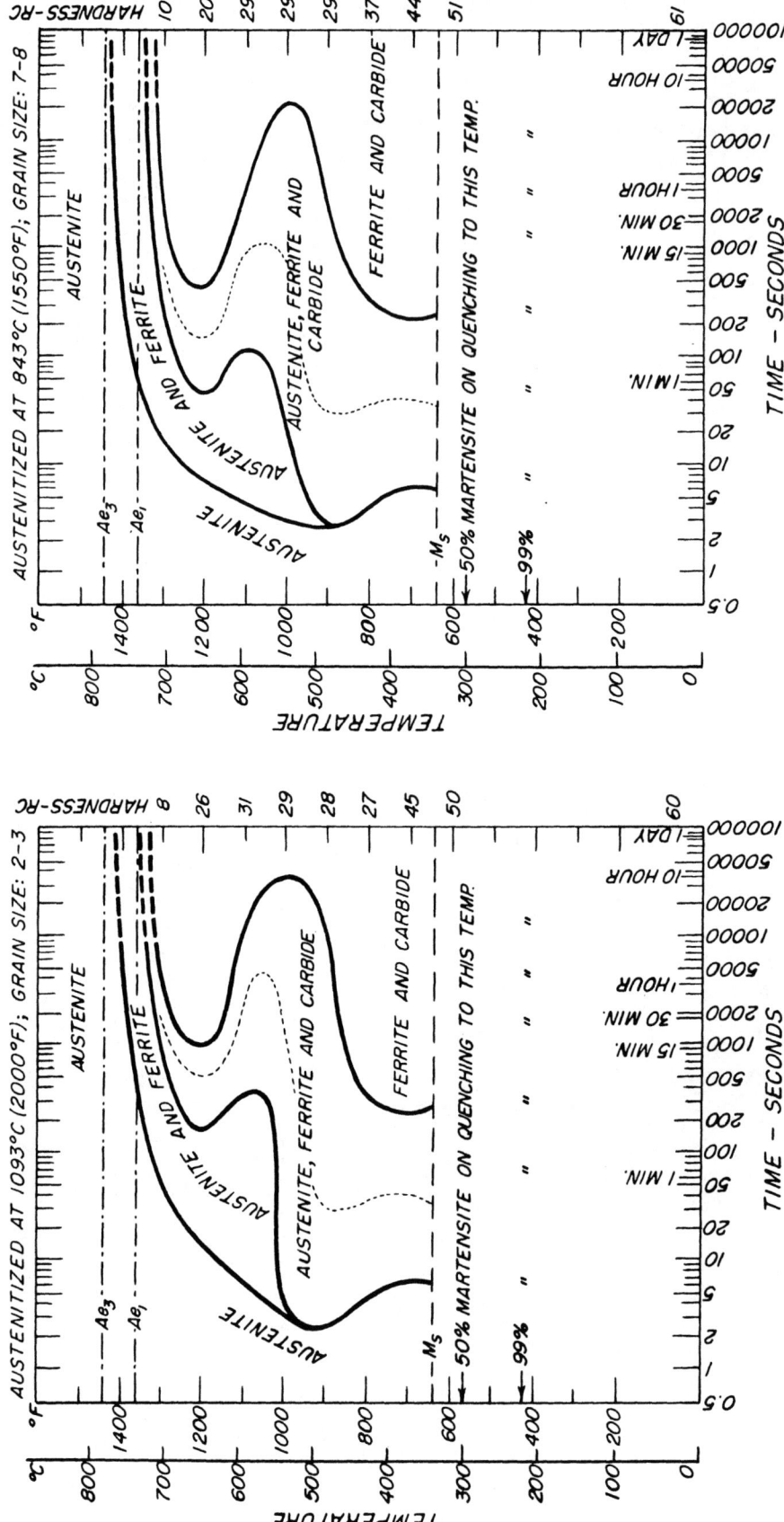

Fig. **41**—32. Comparison of the effect of grain size on the isothermal transformation of a C41400 (AISI or SAE 4140) alloy steel, containing 0.37% carbon, 0.77% manganese, 0.98% chromium and 0.21% molybdenum.

in the austenite. The steels should contain less than 0.6 per cent C and less than 5 per cent of each of the alloying elements. Direct experimental determination of the M_s temperature by dilatometric techniques is preferred to calculation.

Effect of Grain Size—The effect of increasing the grain size of the austenite is similar to that of alloys; it delays both the start and completion of the transformation. This is illustrated by Figure 41—32, which shows the isothermal transformation diagrams for the same alloy steel with both fine- and coarse-grained austenite.

Effect of Homogeneity of Austenite—The general effect of inhomogeneous austenite will be to speed up the start of transformation. This occurs because the initial transformation will occur in the portions of the austenite which are "leaner" in alloy. In addition, undissolved carbides may act as nuclei for transformation, thereby hastening the start of transformation.

TRANSFORMATION ON CONTINUOUS COOLING

The preceding section has described the manner in which the microstructure and, therefore, the properties of steel vary with the temperature of transformation, and has shown how the isothermal transformation behavior governing these microstructural changes can be studied and depicted as isothermal transformation diagrams. The factors affecting transformation characteristics have been enumerated and the nature of their effects described. The composition of the steel, particularly in respect to the alloying elements, has been shown to be the major factor, and the effects of austenite grain size and homogeneity have also been described. Thus, the basic information about the transformation behavior of a steel is fully described by the isothermal transformation diagram.

This basic information tells what structure is formed at each reaction temperature, if the cooling is interrupted so that the reaction goes to completion at that temperature. The information is equally useful for interpreting behaviors when the cooling proceeds directly without interruption, as is the case in the industrial processes of annealing, normalizing and quenching. In these industrial processes, the time at a

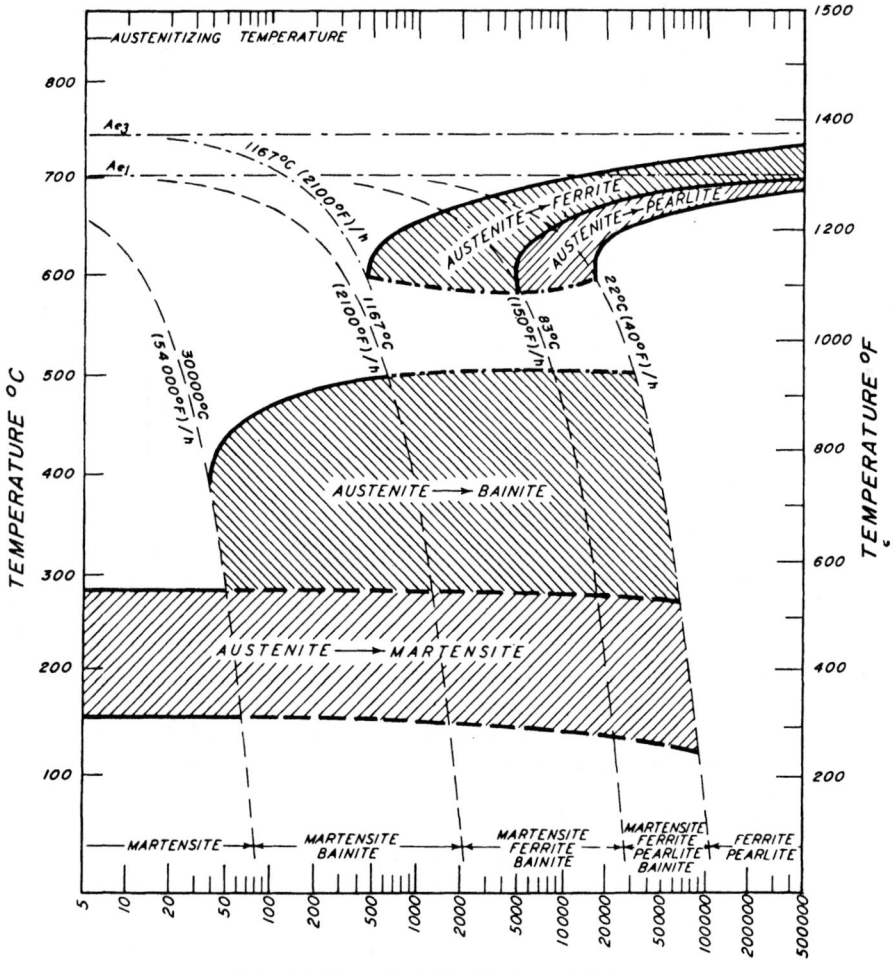

FIG. 41—33. Continuous-cooling transformation diagram for a G43400 (AISI or SAE 4340)-type alloy steel, with superimposed cooling curves illustrating the manner in which transformation behavior during continuous cooling governs final microstructures.

single temperature is generally insufficient for the reactions to go to completion at such a single temperature; instead, the end structure consists of an association of microstructures which individually were formed at successively lower temperatures as the piece cooled. But the tendency to form the several structures is still explained by the isothermal diagram.

The final microstructure after continuous cooling will obviously depend upon the times spent at the various transformation temperature ranges through which the piece is cooled. The transformation behavior on continuous cooling thus represents an integration of these times and this integration can be carried out by the method developed by Grange and Kiefer at the U. S. Steel Fundamental Research Laboratory. By this method, a continuous-cooling diagram generally similar to the isothermal transformation diagram, but depicting the transformation behavior on cooling at a series of constant cooling rates, can be constructed.

Such a diagram for an alloy steel is shown in Figure 41—33. This continuous-cooling diagram lies below and to the right of the corresponding isothermal diagram. That is, transformation on continuous cooling will start at a lower temperature and after a longer time than the intersection of the cooling curve and the isothermal diagram would predict, and this displacement is a function of the cooling rate, being larger as the cooling rate increases.

In order to illustrate the manner in which the transformation behavior on continuous cooling will govern the final microstructures, several cooling-rate curves have been superimposed on this diagram. A consideration of the changes occurring during these various cooling cycles will serve to illustrate the manner in which diagrams of this nature can be correlated with heat-

treating practice and used to predict the resulting microstructure.

Considering first the relatively slow cooling rate (less than about 20°C or 40°F per hour), the steel will be cooled through the regions in which transformation to ferrite and pearlite will occur and these constituents, ferrite and pearlite, will, therefore, make up the final microstructure. This cooling rate corresponds to a slow furnace cool, such as might be used in annealing.

At a somewhat faster cooling rate (about 20° to 85°C or 40° to 150°F per hour), such as might be obtained on normalizing a large forging, the ferrite, pearlite, bainite, and martensite fields will all be traversed and final microstructure will contain all of these constituents.

At cooling rates of about 85° to 1165°C (150° to 2100°F) per hour, the pearlite field will be missed entirely and the resulting microstructures will consist of ferrite, bainite, and martensite. This, therefore, is the microstructure to be expected on normalizing small or moderate sections of this steel.

Finally, on cooling at rates of from 1165° to 30 000°C (2100° to 54 000°F) per hour, the microstructure will be free of proeutectoid ferrite and will consist largely of bainite with a small amount of martensite present. A cooling rate of at least 30 000°C (54 000°F) per hour is necessary to obtain the fully martensitic structure desired as a starting point for tempered martensite on quenching and tempering this steel.

Thus, the final microstructure, and therefore, the properties of a steel, are dependent upon the transformation behavior of the austenite and on the cooling conditions, and can be predicted if these factors are known, or can be governed by controlling either or both of these factors.

SECTION 2

HARDENABILITY

The one attribute of a steel which is certainly of the greatest significance to the heat treater is its capacity for hardening, commonly referred to as its **hardenability**. This attribute has a two-fold significance; it is important, not only in relation to the attainment of a higher hardness or strength level by heat treatment, but also in relation to the attainment of a high degree of toughness through heat treatment to a desirable microstructure, usually tempered martensite or lower bainite. As a matter of fact, the attainment of toughness is the most important, since the attainment of a certain high strength level may often have little significance unless accompanied by a sufficient toughness to meet service requirements.

It should be clearly understood that hardenability refers to the **depth of hardening** or to the size of piece which can be hardened under given cooling conditions and not to the maximum hardness that can be obtained in a given steel. This maximum hardness, as previously described, is dependent almost entirely upon the carbon content (Figure 41—26), while the hardenability (depth of hardening) is dependent upon the carbon and

alloy content and the grain size of the austenite.

Relationship of Hardenability to Transformation Rates—In the preceding discussion, it was shown that, in general, the hardness of steel increases as the transformation temperature decreases. It was also shown that the lower-temperature transformation products, lower bainite and martensite, when tempered, exhibit superior properties in respect to ductility and toughness at a given strength level. It is apparent that, in order to realize the superior properties of these low-temperature transformation products, prior transformation at a higher temperature to softer products must, insofar as possible, be prevented. This means that the steel must be cooled through these high-temperature transformation ranges at a rapid enough rate that transformation does not occur, even at the nose of the transformation diagram. This rate, which will just permit transformation to martensite without any prior transformation at a higher temperature is known as the **critical cooling rate** for martensite, and furnishes one method for expressing hardenability. It can be readily ascertained from the continuous cooling diagram. For example, in the steel of Figure 41—33, the critical cooling rate for martensite is 30 000°C (54 000°F) per hour or 8.33°C (15°F) per second.

How Hardenability Is Expressed and Measured—Although the critical cooling rate can be used to express hardenability, it has the disadvantage that, in practice, cooling rates are ordinarily not constant, but vary during the cooling cycle. This is particularly true of liquid quenching, in which case, the cooling rate is always slower as the temperature of the cooling medium is approached, and is also greatly affected by the presence of a vapor phase in the earlier part of the quenching cycle. Furthermore, as already mentioned, hardenability refers to depth of hardening.

In order to facilitate the application of hardenability measurements to practice, it is, therefore, customary to express hardenability in terms of depth of hardening in a standardized quench. The quenching condition used in this expression is a hypothetical one, in which the surface of the piece is assumed to come instantly to the temperature of the quenching medium. This is known as an ideal quench, and the diameter of a round which will just quench to the desired microstructure, or corresponding hardness value at the center, in an ideal quench is known as the **ideal diameter** (symbol D_I).

Since the cooling rate relationships between the ideal quench and other quenching conditions are known, hardenability values in terms of ideal diameter can be used to predict the size of round which will harden in any quench, the characteristics of which are known, or similarly, if the diameter which will just harden to the center in a standardized quench is known, this can be converted into the ideal diameter value used to express hardenability.

The most direct method of measuring hardenability in terms of ideal diameter is by quenching a cylinder series. In this method, a series of bar sizes are quenched under identical conditions. These bars should have a length at least four times the diameter. They are then sectioned, etched, and cross-section hardness measurements made. The depth of hardening of each of the bars is determined by the point at which the etching characteristics change, which corresponds to a microstructure of 50 per cent martensite, or by the corresponding hardness value. This microstructure of 50 per cent martensite is a very commonly used criterion of hardenability because of the ease with which it may be located. The diameter of the bar in this series which just hardens to the center is noted and this is known as the **critical diameter** (D) for the series.

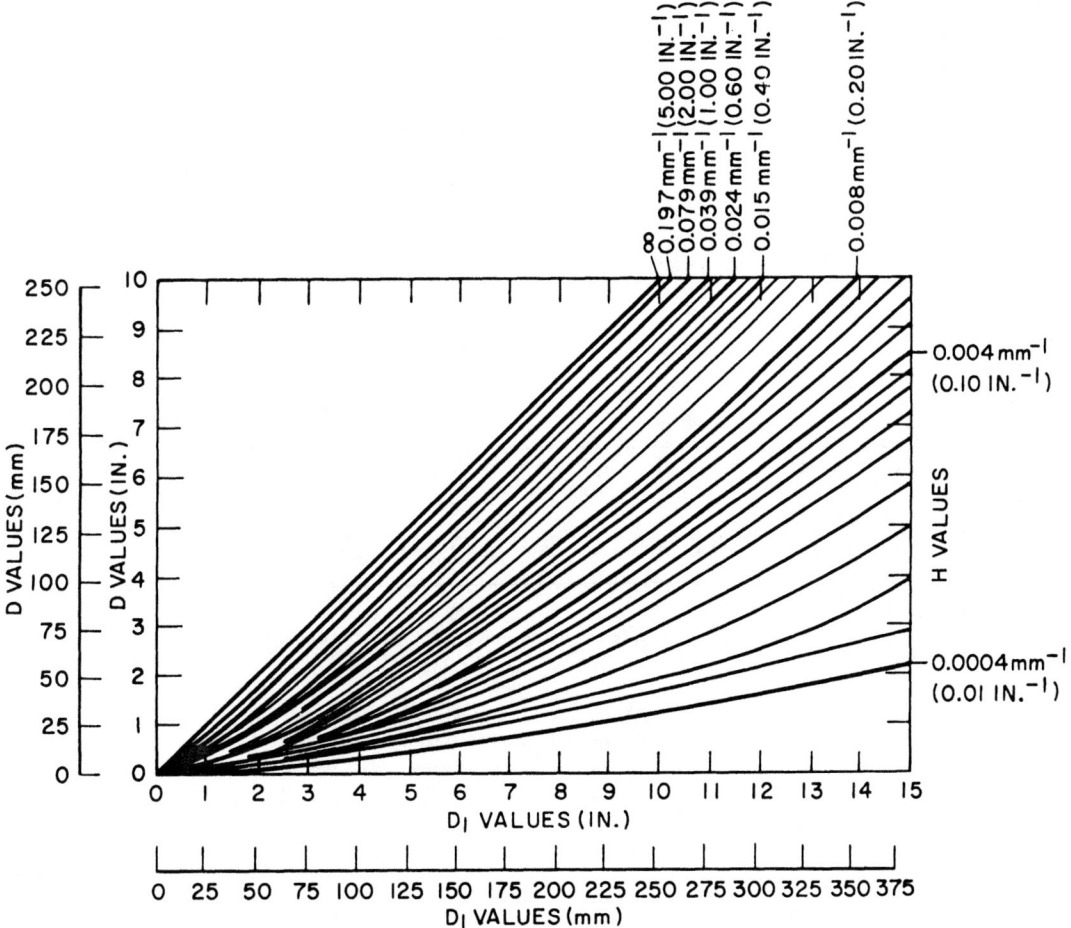

FIG. 41—34. Relationships among ideal diameter, critical diameter and severity of quench. See also Figure 40—35.

As mentioned above, this critical diameter value can be translated into the fundamental terms of ideal diameter (D_I) by the charts of Figures 41—34 and 41—35. In order to make this conversion, however, it is necessary to evaluate the factor expressing the **severity of the quench** (H factor). Typical values of this H coefficient are tabulated in Table 41—IV.

The cylinder series method, just described, is the most direct method of measuring hardenability, but because of numerous advantages, the **end-quench test**, developed by Jominy and Boegehold, is the hardenability test which is now by far the most generally accepted and used. In this test, a cylindrical specimen 25.4 mm (1 inch) in diameter and 101.6 mm (4 inches) long is heated to the desired hardening temperature and quenched in a fixture by a stream of water impinging upon only one end. The bar is then ground on two opposite sides to a minimum depth of 0.38 mm (0.015 inch) below the surface and hardness measurements made at 1.59 mm ($\frac{1}{16}$ inch) intervals along the length of the specimen. The hardenability is expressed as a curve of hardness versus distance from the quenched end of the specimen. Figure 41—36 illustrates the type of quenching fixture used for this test and a typical end-quench hardenability curve is shown in Figure 41—37. Standard procedures for this test have been established by the American Society for Testing and Materials and the Society of Automotive Engineers and the reader is referred to the publications of these societies for the details of the testing procedures.

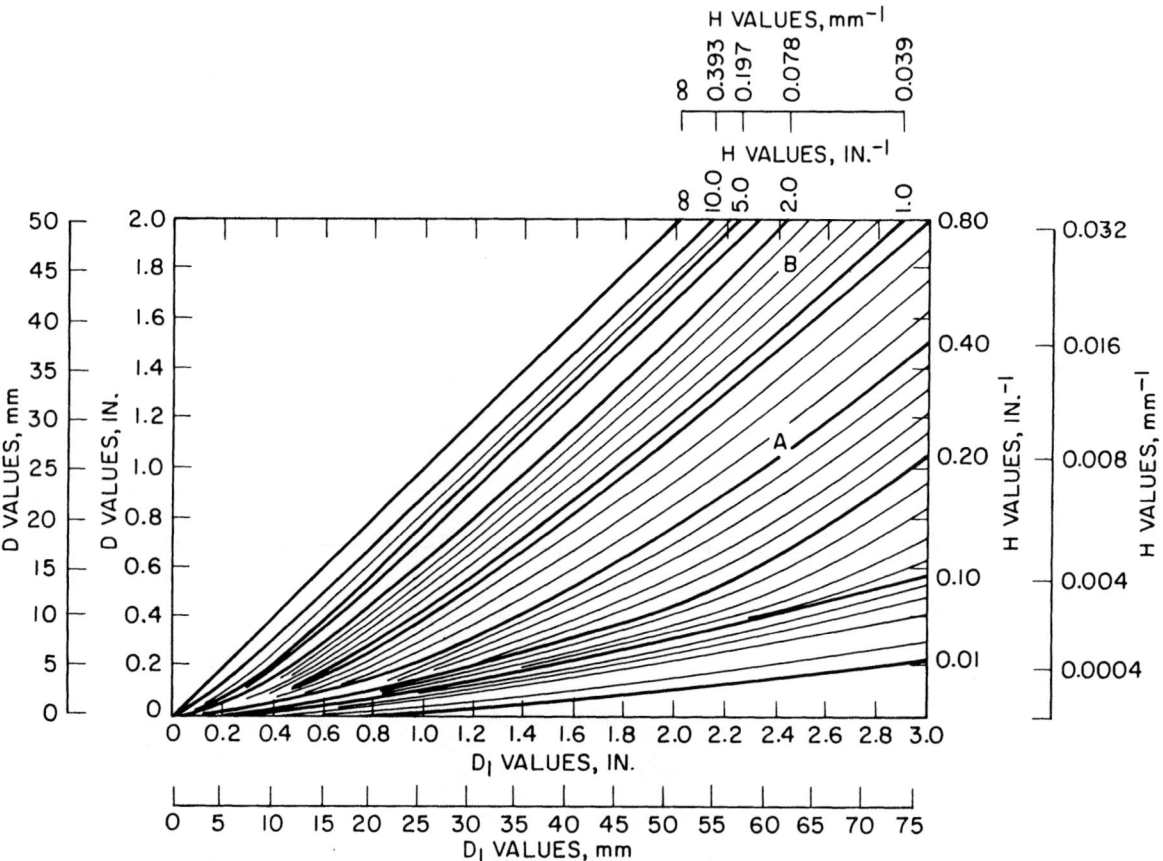

FIG. 41—35. Enlargement of the portion of Figure 41—34 for D values between 0 and 2.0, and D_I values from 0 to 3.0.

Table 41—IV. Typical Values of the H Coefficient Designating Severity of Quench (H Value)

Agitation	Oil		Water		Brine	
	mm⁻¹	in.⁻¹	mm⁻¹	in.⁻¹	mm⁻¹	in.⁻¹
None	0.0098-0.0118	0.25-0.30	0.0354-0.0394	0.9-1.0	0.0787	2
Mild	0.0118-0.0138	0.30-0.35	0.0394-0.0433	1.0-1.1	0.0787-0.0866	2.0-2.2
Moderate	0.0138-0.0157	0.35-0.40	0.0472-0.0512	1.2-1.3		
Good	0.0157-0.0197	0.40-0.50	0.0551-0.0591	1.4-1.5		
Strong	0.0197-0.0315	0.50-0.80	0.0630-0.0787	1.6-2.0		
Violent	0.0315-0.0433	0.80-1.1	0.1575	4.0	0.1969	5.0

FIG. 41—36. Quenching fixture for end-quench test.

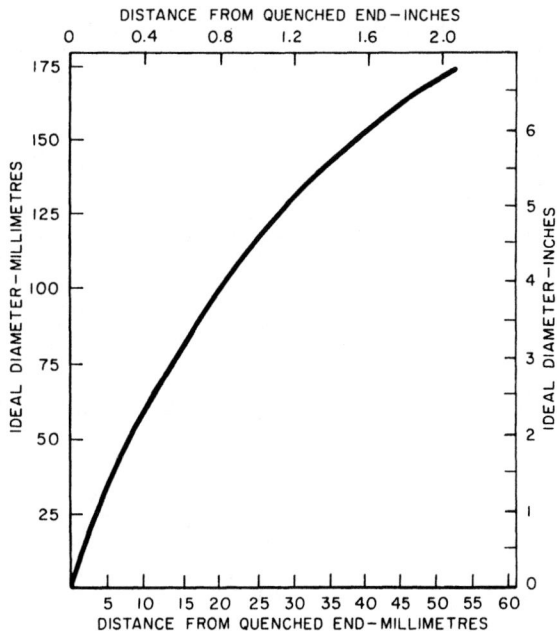

FIG. 41—38. Curve for converting distance from quenched end corresponding to the desired microstructure (or hardness) in the end-quench test to hardenability values in terms of ideal diameter. (After Carney)

This test furnishes a method of applying a continuous series of varying cooling rates to a single specimen, and, since these rates are known the results can be converted to hardenability values in terms of ideal diameter. The curve used for this conversion is shown in Figure 41—38. To use this curve, the distance along the end-quench bar to the desired microstructure, or corresponding hardness value, is noted and the ideal diameter corresponding to this distance is read from the curve. This ideal diameter value may then be con-

verted into terms of bar size which can be hardened under any given quenching conditions, by the methods described above.

Hardenability and Heat Treatment—It has been emphasized in the preceding sections of this chapter

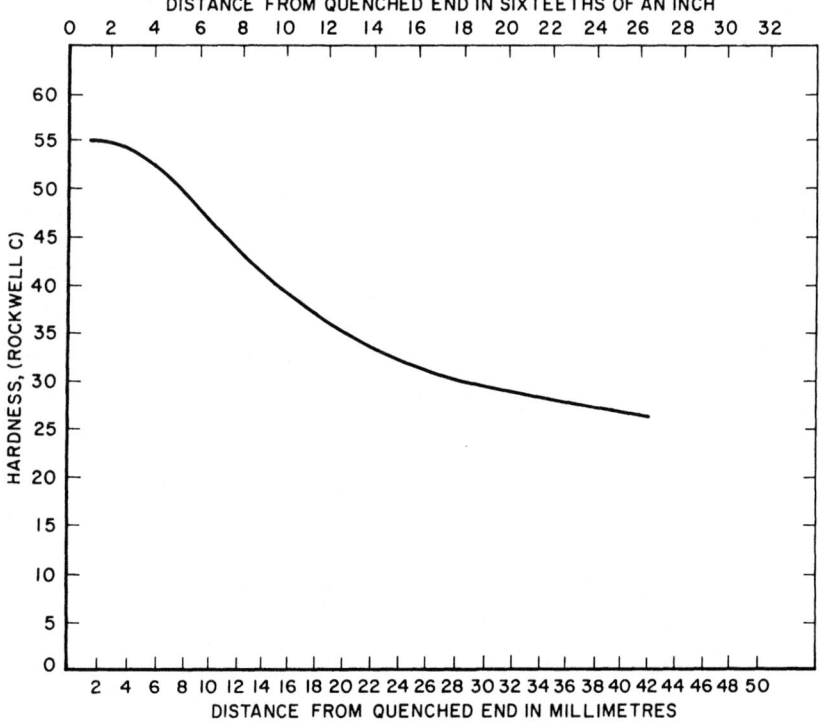

FIG. 41—37. Typical end-quenched hardenability curve.

that the most desirable microstructural constituents from the standpoint of strength and toughness, are those involving transformation at the lower temperature levels,—lower bainite and tempered martensite. In order to obtain these desirable structures, the transformation rates must be slow enough, or in other words, the hardenability must be high enough, to prevent prior transformation at a high temperature during the cooling cycle. The results of hardenability measurements serve to establish the limiting conditions in terms of cooling rates or quenching practices necessary to meet this requirement. Similarly, if the heat-treating practice and cooling conditions have been established and evaluated, the hardenability necessary to obtain the desired microstructure may be determined by the methods described above.

Thus, it is seen that, in general, the suitability of a steel for a given heat treatment practice or the suitability of a heat treatment practice for a given steel is determined largely by its hardenability.

SECTION 3

HEAT-TREATMENT PROCEDURES

QUENCHING AND TEMPERING

The desirable properties of tempered martensitic microstructures have been emphasized in this chapter. Quenching and tempering is the heat treatment commonly used to obtain such microstructures and, therefore, represents the final heat treatment ordinarily used to obtain optimum properties in heat-treated materials.

This method is depicted diagrammatically in Figure 41—39. It involves a continuous cooling from the austenitizing temperature through the martensite transformation temperature range at a rate rapid enough to prevent any transformation at temperatures above the M_s temperature, followed by tempering to the desired hardness or strength level.

Heating—The first step in this heat treatment, as in most of the heat treatments to be described, is the heating of the material to a temperature at which austenite is formed. The actual austenitizing temperature should, in general, be such that all carbides are in solution in order that full advantage may be taken of the hardenability effects of the alloying elements, although in some cases, (for example, in G15216 (AISI E52100 or SAE 52100) steel used in bearings) it may be desirable to leave some undissolved carbides. The temperature should not, however, be so high that pronounced grain growth occurs. The piece should be held at the austenitizing temperature long enough to dissolve carbides but, again, not long enough for excessive growth to occur.

Too rapid a heating rate may set up high stresses, particularly if irregular sections are involved, and is, therefore, generally undesirable. A heating time of one hour per 25 millimetres (one hour per inch) of section is commonly employed, and this is a safe rule. In numerous cases, however, much more rapid heating rates may be employed. In such cases, the safety of the practice must generally be determined by experiment. The available heating rate will, of course, be determined by the mass of the material being heated and the rate at which it can absorb heat, the temperature to which it is desired to heat, and the temperature and heat-transfer characteristics of the heating medium. In general, heating rates will be faster the higher the temperature, and the times will vary with the square of the thickness or diameter. Salt or liquid baths will have generally higher heat-transfer coefficients and, therefore, will heat more rapidly than furnaces in which the heating is in air. Since the heating rate is a function of the difference in temperature between the piece and the heating medium, rapid heating may be obtained by using a heating medium at a temperature well above the desired austenitizing temperature and removing the piece when this temperature is reached. Advantage is taken of this principle in continuous-furnace practice in which the temperature of the furnace is kept well above the desired temperature and the passage through the furnace regulated so that the piece being treated will reach the desired temperature at the outgoing end of the furnace. Temperature control is, however, uncertain in such treatment. Flame hardening, in which rapid heating is obtained by the actual impingement of a high-temperature flame on the surface of the piece being treated is also based on this principle. These rapid heating practices are the exception, however, and the usual and safe practice is a relatively slow and uniform heating to the austenitizing temperature, followed by a holding period at that temperature long enough to insure that the piece is at a uniform temperature throughout.

Unless special precautions are taken, heating will usually result in a certain amount of oxidation or **scaling** and may also result in **decarburization**. Both scaling and decarburization are usually undesirable. Scaling represents a loss of metal, mars the surface finish and may prevent rapid extraction of heat in quenching. Decarburization results in a soft surface and may seriously affect the fatigue life. The processes do not however, necessarily proceed together. For this reason, a slightly oxidizing atmosphere is often desirable when freedom from decarburization is important. Since the amount of scaling is largely determined by the time and temperature of the heating operation, austenitizing temperatures and times should be as low as is consistent with the principles described above in order to minimize scaling. Scaling is materially reduced by the presence of 4 per cent or more of carbon monoxide in the furnace atmosphere.

Special measures are necessary if complete freedom from scaling or decarburization is necessary. These measures include heating in a muffle containing reduc-

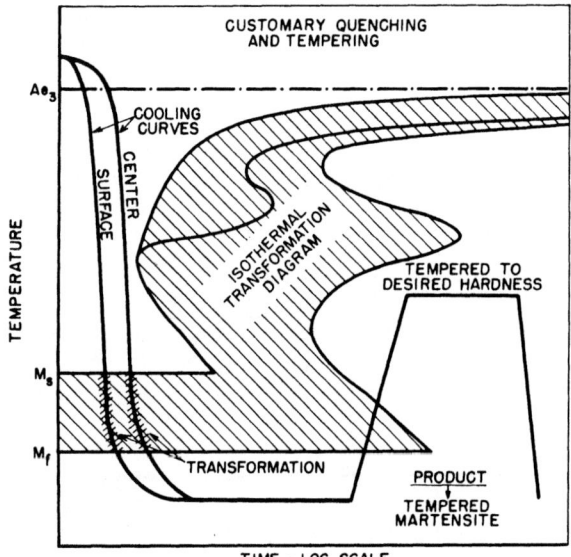

FIG. 41—39. Schematic transformation diagram for quenching and tempering.

ing gases such as carbon monoxide or methane and hydrogen mixtures, packing in cast-iron chips or in a mixture of charcoal and sodium carbonate or heating in neutral salt or lead baths. All of these methods, however, have limitations and disadvantages and require special precautions to insure their success. The composition of the gases used in controlled atmosphere heating varies with the temperature and the composition of the steel, and must be carefully balanced so that neither carburization nor decarburization occurs. At the higher temperatures packing mixtures such as charcoal and sodium carbonate may also lead to carburization and their use is frequently very inconvenient. Salt or lead baths may become contaminated with oxides through contact with the atmosphere and these may accelerate decarburization.

Quenching—The primary purpose of quenching is, as described above, to cool the piece rapidly enough that no transformation occurs at temperatures above the martensite range. The first requisite of a quenching medium is, therefore, a sufficient cooling rate to accomplish this result. The necessary cooling rate is, in turn, determined by the size and hardenability of the piece being quenched, so that the choice of a quenching medium is primarily determined by these factors. The temperature gradient set up by the quenching operation results in relatively high thermal and transformation stresses which are usually, although not always, undesirable since they may lead to cracking or distortion. In order to minimize these stresses, the quenching rate should not be much in excess of that dictated by the size and hardenability of the piece.

The quenching media most commonly used are water, oils, or brine. The relative severity of quench of these media is indicated in Table 41—IV of this chapter. As indicated by this table, brine quenching is the most severe, although when thoroughly agitated, as by a submerged pressure-spray, water approaches it in

severity. Oil is considerably less drastic, although its cooling rate may likewise be markedly increased by a proper and sufficient agitation.

Agitation of the quenching medium is important both because of acceleration of the cooling rate and because of the more uniform cooling obtained. Such agitation may be obtained from judiciously placed propellers, from pumps or from pressure sprays.

The severity of water quenching varies with the temperature of the quenching bath, hot water being quite markedly slower than cold water. This is presumably because of the large amounts of steam which are formed in quenching into hot water and which cling to the work and surround it with "gas pockets." The cooling rate in hot water is, however, not only slower, but less uniform and this lack of uniformity may lead to distortion or even cracking. The increased cooling rate in brine is also presumably due to its increased boiling point which diminishes the chance of gas envelopes forming around the work. The cooling rate of oil quenches tends to increase somewhat with a moderate increase in temperature, presumably because of the decreased viscosity at the higher temperature.

Tempering—The martensite formed by quenching is very hard and very brittle and, as described above, its formation leaves high residual stresses. The purpose of tempering is to relieve these stresses and to improve the ductility, which it does at the expense of strength or hardness. The operation consists of heating at temperatures below the lower critical temperature (A_1). The stress relief, the lowering of strength (hardness) and the recovery of ductility are brought about through precipitation of carbide from the supersaturated unstable alpha-iron solid solution (martensite) and through diffusion and coalescence of the carbide as the tempering operation proceeds.

An empirical method has been developed by Grange, Hribal and Porter (see bibliography) for calculating the hardness of tempered martensite in low-to-medium carbon alloy steels from the chemical composition of the steel, the tempering temperature and the tempering time. The method consists of adding to the hardness of a plain carbon steel base the increment of hardness ($\triangle$HV) attributable to each alloying element present. The hardness of the base and the increments in hardness for each alloying element are taken from a set of master curves developed for one hour tempering at different tempering temperatures. The effect of time at temperature is calculated separately. For example, Figure 41—40 gives an example of the effect of carbon content on the hardness of martensite tempered for 1 hour at different temperatures. Figure 41—41 shows the $\triangle$HV values that must be added to hardness of the base steel when different amounts of Mn, Si, P, Ni, Cr, Mo, and V are present, and the steel is tempered for 1 hour at 538°C (1000°F).

The effect of tempering on the residual stresses is illustrated by Figure 41—42. It will be noted that a considerable stress relief has occurred in tempering at 150°C (300°F) and that tempering at 480°C (900°F) has lowered the stresses to a quite low value.

A typical illustration of the effect of tempering on notch toughness measured by the notch impact test is shown in Figure 41—43. It will be noted that the

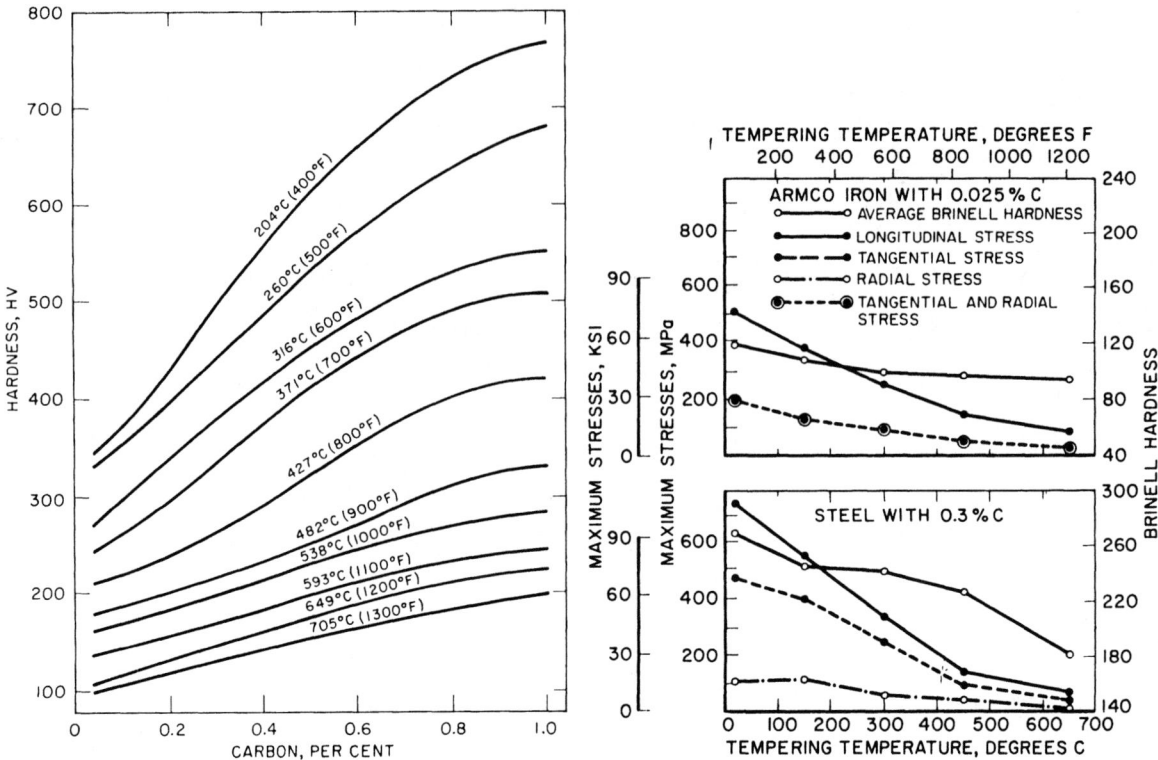

FIG. 41—40. Hardness of tempered martensite in iron-carbon alloys (base steel).

FIG. 41—42. The effect of tempering on residual stresses, in quenched cylinders. Buhler, Buchholtz and Schulz: Archiv fur das Eisenhuttenwesen. Vol. 5. 1932: pages 413-418.

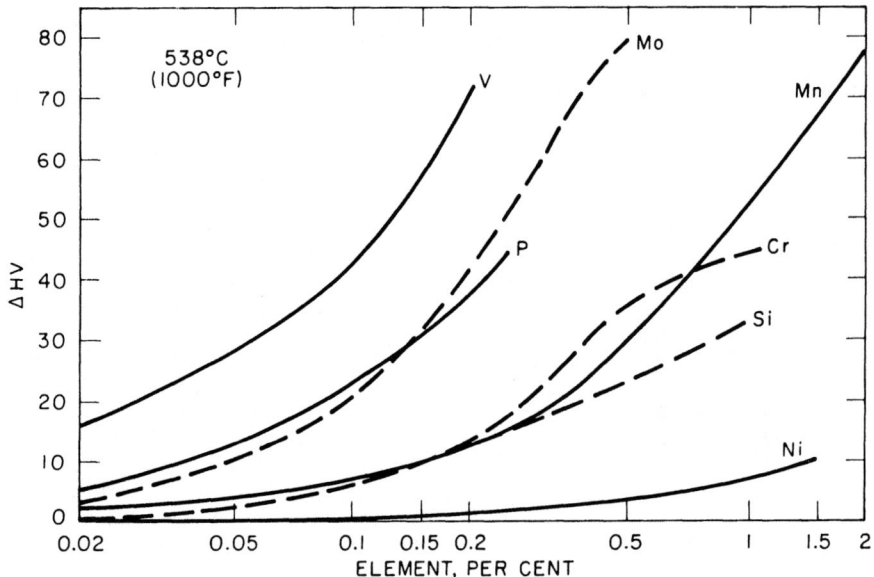

FIG. 41—41. Effect of elements on the hardness of martensite tempered at 538° C (1000° F) for 1 hour.

toughness first increases on tempering at temperatures up to 205°C (400°F), then decreases on tempering at temperatures between 205° and 315°C (400° and 600°F), and finally increases rapidly in tempering at temperatures of 425°C (800°F) and above. This is a characteristic behavior and, in general, the temperature range 230° to 315°C (450° to 600°F) should be avoided in tempering.

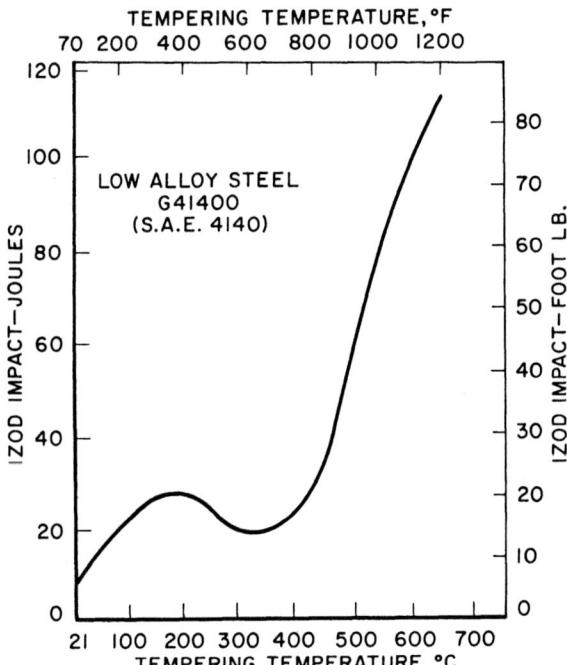

TEMPERING TEMPERATURE, °F

FIG. 41—43. The effect of tempering temperature on notch toughness of low-alloy steel.

In order to minimize cracking, the tempering operation should immediately follow the quench. Allowing fully-quenched pieces to stand overnight before tempering is liable to result in a large proportion of cracked work.

The tempering of martensite results in a contraction and if the heating is not uniform, stresses will be set up by this unequal contraction which will cause distortion or even cracking. Similarly, too rapid a heating for tempering may be dangerous because of the sharp temperature gradient set up between the surface and interior of the piece. Recirculating-air furnaces are ideal for obtaining the uniform heating desired for tempering and are very commonly employed for this purpose. Oil or salt baths are very commonly used for low-temperature tempering and are generally safe, in spite of their rapid heating rate, since the temperature differential is low. Lead or salt baths may be used for higher tempering temperatures if the pieces to be tempered are not too large or irregular so that the heating stresses may be kept at a safe level.

Some steels exhibit a loss of toughness on slow cooling from temperatures of about 535°C (1000°F) and above (the phenomenon known as "temper brittleness" which will be discussed further in another chapter) and therefore, a rapid cooling after tempering is generally desirable in these cases.

MARTEMPERING

As discussed above, the transformation to martensite, occurring during the rapid cooling through the martensite temperature range with the accompanying sharp temperature gradient, results in high stresses. A modified quenching procedure, known as martem-

pering, which was developed by B. F. Shepherd, is helpful in lowering these stresses after quenching. This method is illustrated diagrammatically in Figure 41—44. In practice, it is ordinarily carried out by quenching the piece into a molten-salt bath at a temperature just above the M_s temperature, holding in this bath long enough to permit the piece to acquire the temperature throughout, and then air cooling to room temperature. Transformation to martensite then occurs during the relatively slow air cooling and, since the temperature gradient characteristic of the conventional quench is absent, the stresses set up by the transformation are much lower than in conventional quenching and tempering. Along with these lower stresses goes, of course, a much greater freedom from distortion and cracking. After martempering, the piece may be tempered to the desired strength level. Martempering has been applied to the heat treatment of tools, bearings, dies, etc. in which difficulty was encountered with quench cracking or distortion when heat treated by conventional quenching and tempering.

AUSTEMPERING

As discussed above, the properties of lower bainite are generally similar in respect to strength and somewhat superior in ductility to those of tempered martensite. Austempering, which is an isothermal heat treatment to lower bainite, therefore, offers an alternative method of heat treatment for obtaining optimum strength and ductility.

The austempering treatment is illustrated diagrammatically in Figure 41—45. It involves quenching to the desired temperature in the lower bainite region, usually in molten salt, and holding at this temperature until transformation is complete. It is the usual practice to hold for a time twice as long as that indicated by the isothermal transformation diagram to insure complete transformation of segregated areas. The piece may be quenched or air cooled to room temperature after transformation is complete and may be tempered to a lower hardness level if desired.

Austempering has the tremendous advantage over conventional quenching and tempering that the bainite transformation takes place isothermally at a relatively high temperature so that the transformation stresses are very low, with a resultant absolute minimum of distortion and a practically complete assurance that quench cracking will not occur.

Austempering, on the other hand, has the disadvantage, which it shares with martempering, that, because of the slower cooling rates of the molten salt baths as compared with the usual water or oil quenches a higher hardenability steel is required to prevent high temperature transformation during the cooling to the bainite temperature. Along with these higher hardenabilities also go longer times for complete transformation to bainite so that austempering may be considerably more time consuming than martempering or conventional quenching and tempering.

This hardenability limitation may be overcome to a certain extent by the introduction of a prequench in water or oil to a temperature just below the M_s temperature, so that some martensite transformation occurs

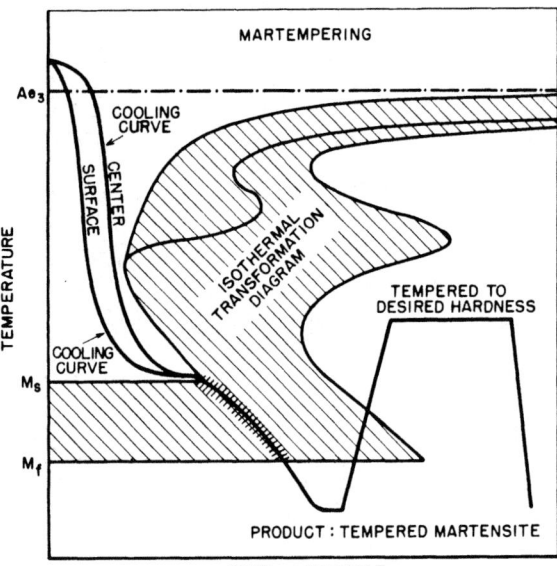

FIG. 41—44. Schematic transformation diagram for martempering.

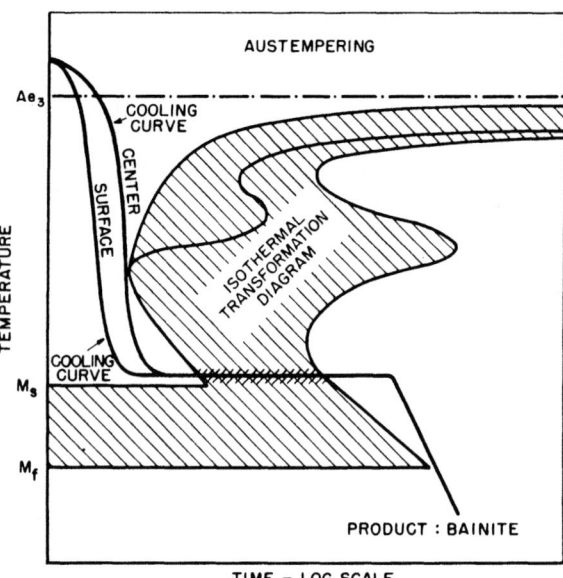

FIG. 41—45. Schematic transformation diagram for austempering.

prior to the final holding at the bainite transformation. The final product is then a mixture of tempered martensite and bainite and steel with this microstructure has good properties.

Largely because of this hardenability limitation, austempering has found its widest application in the heat treatment of plain high-carbon steels in small section sizes, such as sheet, strip and wire products. It is, however, also, being used for the heat treatment of alloy steels and cast irons for applications in which it is essential that distortion be held to a minimum.

NORMALIZING

Normalizing involves reheating the steel above its critical temperature (Ac_3) and air cooling. It has two primary purposes: to refine the grain, and to obtain a carbide size and distribution which will be more favorable for carbide solution on subsequent heat treatment than the as-rolled structure.

The as-rolled grain size depends principally upon the finishing temperature in the rolling operation. This is subject to wide variations and there is, therefore, a corresponding wide variation in the grain size of the as-rolled products. The normalizing operation, as the name implies, serves to refine a coarse grain size resulting from a high finishing temperature and to establish a uniform, relatively fine-grained microstructure.

In alloy steels, particularly if they have been slow cooled after rolling, the carbides in the as-rolled condition tend to be rather large and massive. These large carbides are difficult to dissolve on subsequent austenitizing treatments. This carbide size, likewise, will be subject to wide variations, depending on the rolling and slow-cooling practice. Here again, normalizing tends to establish a more uniform and finer carbide particle size which will facilitate subsequent heat treatment to a more uniform final product.

The usual practice is to normalize from about 35° to 65°C (100° to 150°F) above the critical temperature, but for some alloy steels with carbides that are soluble only with difficulty, considerably higher temperatures may be used to obtain carbide solution. Heating, in general, should be slow enough to insure uniform temperatures and low thermal stresses. It is now a very common practice to carry out this operation in continuous furnaces. Continuous normalizing is particularly well adapted to sheet and strip because it may be heated quickly, but it is also used for plates and bars. The heating operation may, however, be carried out in any type of furnace which will permit uniform heating and accurate temperature control.

ANNEALING

The principal purposes of annealing are to relieve cooling stresses induced by cold or hot working, and to soften the steel so as to improve its machinability or formability. It may involve only a subcritical heating to relieve stresses, to recrystallize cold-worked material, or to spheroidize the carbides or it may involve heating above the critical temperature with subsequent transformation to pearlite or directly to a spheroidized structure on cooling.

Full Anneal—As discussed above, the most favorable microstructure for machinability in the low- or medium-carbon steels is coarse pearlite. The customary heat treatment to develop this microstructure is a full anneal, illustrated diagrammatically in Figure 41—46. It consists of austenitizing at a relatively high temperature so that full carbide solution is obtained, followed by a slow cooling so that transformation occurs only and completely in the high-temperature end of the pearlite range. This is a simple heat treatment and is reliable for most steels. It is, however, rather time consuming since it involves a slow cooling over the entire temperature range from the austenitizing temperature

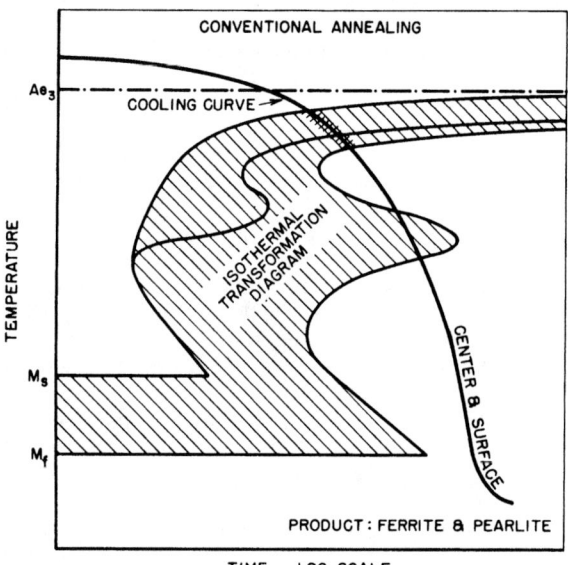

FIG. 41—46. Schematic transformation diagram for full annealing.

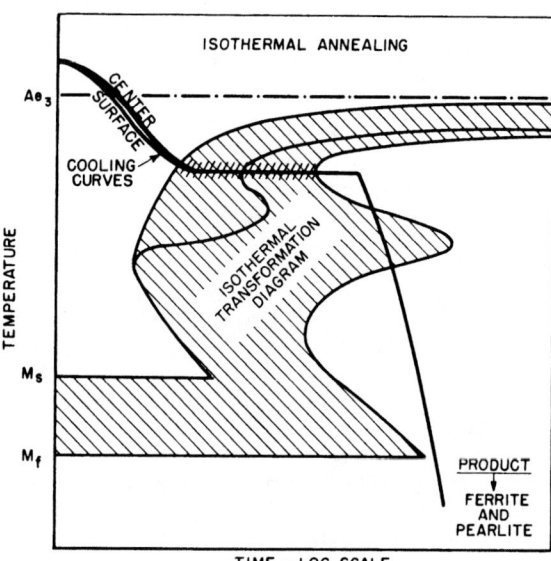

FIG. 41—47. Schematic transformation diagram for isothermal annealing.

to a temperature well below that at which transformation is complete.

Isothermal Annealing—This annealing to coarse pearlite can, of course, be carried out isothermally by cooling to the proper temperature for transformation to coarse pearlite, and holding at this temperature until transformation is complete in a manner similar to the austempering procedure. This method is illustrated diagrammatically in Figure 41—47. Such an isothermal-annealing cycle may make possible a very considerable time saving over the conventional full-annealing treatment described above. Neither the time from the austenitizing temperature to the transformation temperature, or from the transformation temperature to room temperature is critical and these may be speeded up as much as is desired or is practical. Furthermore, if the extreme softness of the coarsest pearlite is not necessary, the transformation may be carried out at the "nose" of the curve where the transformation goes to completion most rapidly and the operation thereby further expedited; the pearlite is much finer and the hardness is higher.

Isothermal annealing is most practical for applications in which full advantage may be taken of the rapid cooling to the transformation temperature and from this temperature down to room temperature. Thus for small parts which can be conveniently handled in salt or lead baths, this isothermal annealing makes possible large time savings as compared with the conventional slow furnace cooling. It is also very conveniently adapted to continuous heat treatment, and continuous annealing by this method is commonly referred to as "cycle annealing." This is usually carried out in an especially designed furnace, incorporating an air-blast chamber in order to cool rapidly from the high-heat stages for austenitizing down to the lower-temperature stages in which the transformation to pearlite occurs. This permits an accelerated cooling down to the transformation temperature. On the other hand, the

method offers no particular advantage for applications such as the batch annealing of large furnace loads in which the rate of cooling to the center of the load may be so slow as to preclude any rapid cooling to the transformation temperature. For such applications, the conventional full-annealing method usually offers a better assurance of obtaining the desired microstructure and properties.

Spheroidize Annealing—Coarse pearlite microstructures are too hard for optimum machinability and formability in the medium and higher carbon steels, and such steels are, therefore, customarily annealed to develop spheroidized microstructures. This may be accomplished by tempering the as-rolled, slow-cooled or normalized materials at a temperature just below the lower critical temperature. Such an operation is known as "sub-critical annealing." Full spheroidization of the carbides by this method may require long holding times at the sub-critical temperature and the method may, therefore, be slow, but it is a simple heat treatment and may frequently be more convenient than annealing above the critical temperature.

It has been found, however, that the procedures described above for annealing to produce pearlite, can, with some modifications, be applied to annealing methods that will result in spheroidized microstructures. If free carbide remains after the austenitizing treatment, transformation (in the temperature range at which coarse pearlite would ordinarily form) will proceed to spheroidized rather than to pearlitic microstructures. Thus, heat treatment to form spheroidized microstructures can be carried out in a manner completely analogous to heat treatment to form pearlite, except for the use of lower austenitizing temperatures. Spheroidize annealing may thus involve a slow cooling similar to the full-annealing treatment to produce pearlite or it may be an isothermal heat treatment similar to the isothermal annealing to form pearlite. An austenitizing temperature not more than about 35°C (100°F) above the lower critical temperature is customarily used for this super-critical annealing to produce spheroidized microstructures.

Process Annealing—Process annealing is the term used to describe the sub-critical annealing of cold-worked materials. It customarily involves heating at a temperature high enough to cause recrystallization of the cold-worked structure and to soften the steel.

An important example of process annealing is the box annealing of cold-rolled low-carbon sheet steel. The coils or sheets are packed in a large box which is sealed to protect them from oxidation. This annealing is usually carried out at temperatures from about 595° to 705°C (1100° to 1300°F). The heating and holding at temperature usually takes about 24 hours after which the charge is cooled slowly in the box, the entire process taking about 40 hours.

Box annealing of coils can be conducted by a modification of the process known as "open-coil" annealing. In this modification, a nylon string or steel wire is introduced during coiling between each layer of the coiled, cold-worked strip. This separation of the layers of the coils permits a much more effective and uniform heating of the interior portions of the coils during box annealing than is possible with the usual tightly wound coils. In addition, it permits a free access of the protective atmosphere to the coil surfaces, and this procedure is readily adapted to treatments in which it is desired to change the steel composition by interactions with the furnace atmosphere.

To reduce processing time, provide better uniformity and save energy, process annealing is often done in a continuous operation in which the cold-worked material is annealed as it moves through heating and cooling zones. Coils of sheet steel are continuously annealed by uncoiling the sheet and passing it through a controlled-atmosphere furnace. The furnace consists of heating and cooling zones built as towers in which the effective length of the sheet being treated is increased by threading the sheet back and forth around rolls at the top and bottom of the towers. Wire is continuously annealed in coils in controlled-atmosphere furnaces or in salt baths. Wire is continuously annealed in strands by drawing the wire through a bath of molten lead.

Processes and equipment for batch and continuous annealing of sheet are described in detail in Chapter 34. Annealing of wire is described in Chapter 31.

SPECIAL HEAT-TREATING PROCESSES

Dual-Phase Heat Treatments—When steels are heated into the intercritical region (between the A_1 and A_3 temperature), a microstructure consisting of a mixture of ferrite and austenite is produced. As a result, the austenite areas are high in carbon and thus have higher hardenability than the steel would have if it were heated above the A_3 temperature and were fully austenitic. If the cooling rate from the intercritical region is fast enough, the austenitic areas will transform to martensite and the resulting microstructure will be a mixture of ferrite and martensite.

Low-carbon hot- or cold-rolled sheet steels have been developed that are heat-treated by intercritical continuous annealing and exhibit a dual-phase microstructure consisting of ferrite with 15 to 20 per cent martensite present as small isolated areas uniformly dispersed throughout. The alloy content of these relatively low-alloy sheet steels is adjusted to provide sufficient hardenability in the austenite areas so that they will transform to martensite on air or accelerated cooling.

Sheets with this dual-phase microstructure exhibit a continuous stress-strain curve (no yield point or Lüder's extension) with low 0.2 per cent offset yield strength (over 340 MPa or 50 ksi), high tensile strength (over 660 MPa or 95 ksi) and superior ductility. Dual-phase sheet steels provide a more uniform distribution of strain during forming than that provided by steels strengthened by conventional methods and allow relatively complex high-strength parts such as automotive bumpers, wheels, control arms and steering-column assemblies to be produced successfully.

Grain-Refining Heat Treatment—The beneficial effect of finer grain size in increasing strength and toughness has led to widespread use of steels killed with aluminum or containing other grain-refining elements, such as vanadium or niobium (columbium). Grain refinement can also be obtained, and usually carried further, by a special heat treatment developed and patented by United States Steel Corporation. The grain-refining effect of austenite transformation, wherein each austenite grain transforms to many grains of ferrite or martensite, is utilized by the treatment, which comprises two or more rapid austenitizing-quenching cycles. Progressive grain refinement occurs with each successive, properly designed cycle, at least up to the point where substantially finer grains are obtained than is possible in any particular steel by the conventional, relatively slow, single austenitize-and-quench cycle. The grain size is progressively refined by each cycle with the result that an ultrafine austenite grain size of No. 12 ASTM can be developed after the fourth cycle. The number of cycles may be varied depending on the response of a particular steel. Cooling rate in the last cycle may be varied to develop the desired austenite-transformation product or, more commonly, the steel is quenched to martensite and subsequently tempered in the conventional way to optimum strength level for the intended use.

The grain refinement resulting from this rapid austenitizing brings about an increase in strength with little or no loss in ductility and notch toughness and with a marked improvement in ductile-to-brittle transition temperature.

Sub-Zero Treatments—In many instances, particularly in the higher carbon and relatively high alloy steels used for applications involving wear, or for tool steels, the M_f temperature may be so low that considerable quantities of austenite remain untransformed at room temperature or at the temperature of the quenching medium. This retained austenite is generally undesirable. It may result in significantly lower hardness in the heat-treated article with a resultant decrease in wear resistance. In addition, austenite present after quenching may transform during tempering to undesirable transformation products, or it may remain untransformed during tempering and transform to martensite during cooling, with resultant embrittlement. Retained austenite in the heat-treated article may, furthermore, transform to martensite during service, with a resulting danger of cracking, distortion or embrittlement.

This problem may often be overcome by cooling the steel to sub-zero temperatures immediately after quenching. The austenite which remains untransformed at the temperature of the quenching medium will transform on cooling to a lower temperature and complete transformation of the austenite to martensite will occur when the M_f temperature is reached. Dry ice (solid carbon dioxide) which can cool to about –62°C (–80°F) and liquid nitrogen which cools to about –184°C (–300°F) can be used as coolants in this process, but the use of refrigerated containers, into which the pieces may be placed on a batch basis after quenching, is preferable. A temperature of –73°C (–100°F) in such containers will generally suffice to obtain the desired transformation to martensite in most of the steels in which this problem exists.

SURFACE-HARDENING HEAT TREATMENTS

For many applications, particularly those involving wear, a combination of a hard, wear-resisting surface and a strong, fracture-resistant core is desired. For such applications, dual heat treatments, the first of which establishes the desired properties in the core, and the second of which imparts a high hardness to the surface layers are commonly used. The first step of such processes will be essentially similar to those described above, since the usual objective for the core properties is an optimum combination of strength and toughness. The surface hardening step may involve simply a rapid reheating so that only the surface layers are heated above the A_3 temperature followed by a quench to transform the austenite to martensite, or may involve a change in chemical composition in the surface layers.

Decremental Hardening—The process of increasing surface hardness without a change in the chemical composition of the surface involves heating the piece under conditions such that a large difference in temperature exists between the piece and the heating medium so that a steep temperature gradient is developed in the piece. This heating is continued until the A_3 temperature is reached at the desired depth below the surface at which a high hardness is desired, and the piece is then quenched so that the austenite transforms to martensite. The rapid heating required to establish the requisite temperature gradient can be accomplished by heating in a furnace, maintained at a high temperature, or, in the process referred to as **flame hardening**, by the direct impingement of a high-temperature flame. Special furnaces, in which the piece is heated by radiation from ceramic discs which are heated to incandescence by gas flames, permit a more rapid heating rate than is possible with the usual muffle furnaces, because of their more effective heat transfer, and are frequently used in decremental hardening processes. Alternatively, the requisite temperature gradient and rapid heating may be obtained by induction heating, and this type of decremental hardening is referred to as **induction hardening**. In the latter process, the depth of hardening is a function of the frequency, with higher frequencies resulting in higher surface temperatures and steeper temperature gradients.

In these decremental hardening processes, a steep temperature gradient is desirable in order to minimize the size of the zone which may be heated to temperatures between the A_1 and A_3 temperatures and which will, therefore, have an undesirable microstructure when quenched. A steep temperature gradient is also desirable in order to minimize the time during which the surface portion will be at a high temperature. High surface temperatures will result in grain coarsening and increase the possibility of retained austenite in the quenched structure, and both of these undesirable effects will be more pronounced as the time at the high temperature increases. It is apparent, therefore, that these decremental hardening processes require a close control of the rapid heating practice.

The hardenability requirements for decremental hardening are obviously less stringent than those for through hardening, as only the surface portions need to be rapidly cooled, and heat flow in cooling occurs both into the cold core and the quenching medium. Thus, for applications in which surface wear is of primary concern, the process offers a method of producing a product of superior properties from steels of relatively low alloy content and cost.

The rapid heating and cooling and the resulting high temperature gradients characteristic of these processes, as well as the differential expansion which occurs when the surface-hardened portions transform to martensite, generally lead to a relatively high level of surface residual stresses in decrementally hardened products. The resulting surface residual stresses will, however, generally be compressive, if the depth of the hardened portion is not excessive, so that, in addition to the increased resistance to wear, the surface compressive stress developed in decrementally hardened products may markedly increase their resistance to fatigue. The stresses resulting from this process may, on the other hand, lead to a serious distortion in the heat-treated product and, in some instances, it has been found necessary to pre-form such products in order to obtain a dimensionally accurate product after heat treatment.

Case Hardening—The surface hardening processes which involve a change in chemical composition of the surface portion are referred to as **case-hardening processes**. These include carburizing, in which the carbon content of the surface portions is locally increased; nitriding, in which the nitrogen content of the surface portions is increased; and carbonitriding, in which both the carbon and nitrogen contents are increased.

Carburizing—In the carburizing process, a high-carbon surface layer is imparted to low-carbon steel by heating it in contact with carbonaceous material or in a carbon-rich atmosphere. On quenching after carburizing, the high-carbon "case" becomes very hard, while the low-carbon core remains comparatively soft. The result is a product similar to that obtained by decremental hardening, but with a higher hardness in the case portion because of its high carbon content.

Carburizing can be done by packing the steel in boxes with carbonaceous solids, sealing to exclude the atmosphere, and heating to about 927°C (1700°F) for a period of time depending upon the depth of case desired. A time of about 8 hours at 927°C (1700°F) results in a case depth of about 1.6 millimetres 1/16 inch). This process is known as **pack carburizing.**

Carburizing may also be carried out by heating the steel in direct contact with carburizing gases, which is known as **gas carburizing,** or shallow cases may be imparted by heating in molten salt baths containing cyanides, which is known as **liquid carburizing.** In gas carburizing, the common practice is to use a heated retort into which mixtures of methane and carbon monoxide are continuously introduced, and in which the work is tumbled by continuously rotating the retort. This process is faster and is capable of better control than pack carburizing. In addition, with gas carburizing, the carburizing cycle can be followed by a diffusion cycle in which no carburizing gas is admitted to the retort, and thereby a lower surface carbon content and a better gradation of carbon content in the case can be obtained. The carburizing gas may be produced in a separate gas generator, or natural gas (which consists largely of methane) may be used.

Carbonitriding—Carbonitriding is a modification of the gas carburizing process in which both carburizing gases and nitrogen from dissociated ammonia are introduced into the retort during the carburizing cycle. The resulting case contains both higher carbon content and iron nitride, both of which contribute to the hardness so that a higher hardness case can be obtained, or the desired hardness can be obtained at lower carbon contents and with shorter cycles than for gas carburizing.

These advantages of carbonitriding have led to an increasing use of this process in recent years.

Liquid Carburizing—In liquid carburizing, a light, hard case, which is also a mixture of carbides and nitrides, is obtained by immersing the steel in a molten salt bath containing about 30 per cent sodium cyanide at 870°C (1600°F) for periods of about 1/2 to 1 hour. This results in a case depth of about 0.25 millimetres (0.010 inch).

Heat Treatment of Carburized Parts—Since carburized articles have a high-carbon case and a low-carbon core, the proper austenitizing temperature must be a compromise between these two different steels. Years ago, heat-treaters used a double quench; the first from a high temperature to refine the grain of the core and assure complete austenitization, and the second from a lower temperature to dissolve enough carbide from the case without disturbing the grain refinement of the core. In the 1930's, however, the "fine-grained" steels were introduced; by addition of controlled amounts of aluminum, these steels could be held for long carburizing periods at 927°C (1700°F) without grain growth in either case or core, and quenching from this temperature could be done with minimum distortion and maximum toughness, especially in the core. It is, therefore, common practice today to direct-quench from gas-carburizing retorts or liquid-carburizing baths. Pack-carburized articles are usually air-cooled to permit their extraction from the pack, after which they are reheated, held for a short time to allow sufficient carbide solution, and oil quenched. These reheatings should be done in controlled-atmosphere or salt-bath furnaces to avoid decarburization. The final tempering, regardless of austenitizing practice, is at a low temperature of 93° to 204°C (200° to 400°F) to maintain a high hardness in the case. Certain medium-alloy carburized steels such as G33120 (AISI or SAE 3312) steel may require a sub-zero treatment, as discussed earlier in this chapter, or intermediate tempering to minimize the amount of retained austenite in the finished parts.

Nitriding—The nitrogen case-hardening process which is termed "nitriding" consists of subjecting machined and heat-treated parts to the action of a nitrogenous medium, commonly ammonia gas, at temperatures of about 510°C (950°F) to 538°C (1000°F). Almost any quenched and tempered steel, if the tempering temperature is above 538°C (1000°F), may be nitrided with little or no distortion or change in surface finish. Nitrided cases are ordinarily shallow, with a case depth of 0.25 to 0.38 millimetre (0.010 to 0.015 inch) being obtained in 48 hours at 524°C (975°F). They are, however, very hard, with surface hardnesses of 700 to 1200 Vickers, and this hardness is retained, even after reheating to temperatures up to 482°C (900°F). This high hardness, together with the fact that the treatment does not result in scaling, distortion or discoloration, and that no further heat treatment is required, makes this process a very useful method in increasing the wear resistance of steel products. It has been found that chromium and aluminum are desirable in steels for nitriding and alloy steels containing these elements, which are especially adapted to nitriding, have been developed. About 1 per cent of aluminum will produce a very high surface hardness of about 1200 Vickers compared with the 700 Vickers in steels not containing this element.

The outer layer of a conventionally nitrided case contains a shallow (0.05 millimetre or 0.002 inch) but very brittle "white layer" consisting of iron and other nitrides.

Where specifications or performance requirements will not permit the presence of white layer, the layer is removed by costly finish grinding or pickling operations. However, grain-boundary network damage, which frequently accompanies white-layer formation, cannot be removed and becomes a serious problem in applications requiring good fatigue resistance. Attempts to limit white-layer formation and network damage during conventional nitriding are unsuccessful because of a lack of control of cyanide generated during nitriding. The cyanide is formed at nitriding temperatures by reaction of the nitriding atmosphere with carbon on the surface of carbon-containing materials, such as the work load, support fixtures, retort, and so on. Though present only in the parts-per-million range, any accumulation of cyanide will seriously retard the nitriding rate.

United States Steel Corporation has developed a process of nitriding known as **bright nitriding.** The bright-nitriding process is a gas-phase nitrogenation treatment for producing case-hardened ferrous-alloy parts without the presence of white-layer formation or grain-boundary network damage. Regardless of work load, a specific case or hardness profile can be predicted and consistently and reliably reproduced. This process is applicable to most alloys which can be nitrided by conventional gas or salt-bath treatments.

The bright-nitriding process eliminates the cyanide problem. The nitriding atmosphere is purified by cycling it through a scrubber that removes the cyanide impurities and returns the nitriding atmosphere to the furnace retort for reuse. Only enough fresh gas is added to the system to maintain the nitrogen activity required to sustain the nitriding rate. As a result, process-gas consumption in the bright-nitriding process is less than one-half that required for conventional nitriding in an installation of equivalent capacity.

The United States Steel process uses a binary mixture of ammonia and hydrogen which eliminates the need for ammonia dissociators required in conventional gas nitriding which use a ternary mixture of ammonia, hydrogen and nitrogen. The use of an all-metal system for surfaces exposed to the nitriding gas permits precise control of the binary gas mixture as well as the moisture content for optimum nitriding conditions.

SECTION 4

HEAT-TREATING FURNACES

General Design Requirements—Heat-treating furnaces are grouped into either batch or continuous furnaces. There are many different types in each group, and only general design features can be discussed here. Furnaces for specific applications are described in other chapters that deal with the manufacture of such steel products as bars, plates, flat-rolled material, heavy forgings, wire, etc.

The simplest furnaces are the direct-fired batch type with manual controls. The more elaborate installations used for large production lines are continuous furnaces with automatic program control. In a number of installations, specific facilities for controlling the atmosphere in the working chamber are provided to obtain the desired surface condition. The most common heat treatments performed in furnaces are annealing, normalizing, spheroidizing, hardening, tempering, carburizing and stress relieving. Heat-treating furnaces are seldom designed for temperatures in excess of 1095°C (2000°F), and generally are operated in the 425° to 870°C (800° to 1600°F) range. They are usually well insulated and built tight to prevent air infiltration or a loss of special atmosphere gas. Attention in design is directed toward procuring uniform temperature distribution in the working chamber of the furnace. The position and method of heat application, and the circulation of gases in the furnace, are of major consequence in this. In annealing furnaces, means for controlling both the rate of heating and cooling of the stock usually are provided. Since the required heating and cooling rates of different types of steel vary, it is necessary to provide flexible means for controlling these functions. Insulating firebrick is used generally in heat-treating furnace construction, due to its low heat-storage capacity, to permit heating and cooling the furnace quickly. For intermittent furnace operation this is particularly vital. Attention is directed in design to spacing the charge in order to attain the most efficient flow of heat around the stock. In coil-annealing furnaces, and in furnaces for heat treating other material, special facilities often are provided to improve the circulation of special atmosphere gases during heating and cooling. In all heat-treating layouts, special consideration is given to providing sufficient furnace capacity to maintain the desired time-temperature relation of the treatment. Furnaces forced beyond their normal capacity usually yield an erratic and non-uniform product.

For handling batch loads of material to be heated, quenched, and tempered, quench tanks and cranes should be located in such a way that little time is lost in getting the material from the furnace into the quenching medium. Furnaces also must be arranged so that, after quenching, a second furnace is available for taking the charge promptly for further treatment. Usually three furnaces are provided for operations of this type: two for heating and one for tempering.

In large-scale heat-treating operations, the layout of facilities is very important. The furnaces must be arranged to suit an orderly flow of material through the shop. Adequate space for temporary storage and handling is necessary for both the raw and the finished material. A sufficient number of cranes or other stock-handling facilities of the proper capacity must be provided to eliminate bottlenecks or interference with prescribed furnace cycles. A central station for the preparation of atmosphere gas generally is provided in the larger furnace layouts where particular attention to steel surface is required.

In heat-treating furnaces, many furnace parts, such as conveyors or rollers in continuous furnaces, radiant tubes for indirect firing and covers in coil and pack-annealing furnaces, are of metallic construction since the temperature seldom exceeds the 425° to 870°C (800° to 1600°F) range. Special alloy materials are utilized to reduce the maintenance of these parts to a minimum. In selecting the furnaces for a heat-treating plant, careful consideration must be given to the type of product to be heated, to the kind of treatment to be performed, and to the production rate required. The ensuing sections explain pertinent factors in design and describe the application of various furnace types.

Method of Heat Application—The character of the material to be heated and the type of treatment to be performed have an important bearing on the choice of method of heat application. Heat-transfer laws govern the flow of heat to the steel in heat-treating as in other heating furnaces. The surface of the material absorbs heat transmitted to it by radiation or by convection or both, and this heat is transferred through the body of the material by conduction. In heat-treating furnaces, the transfer of heat by convection is relatively more significant than in furnaces operated at a higher temperature level. In heat treating steel, the rate of heat transfer to the surface is usually low in order that each individual piece, as well as all pieces in the furnace, may be brought up uniformly to the required temperature level. However, in some furnaces which utilize induction heating or radiant-type open-flame burners, uniform heating can be accomplished rapidly with high heat-transfer rates.

Gaseous fuel and electric power are the two main sources of heat used in heat-treating furnaces. In some cases, fuel oil has been substituted for gas due to a shortage of the latter or for economic reasons. In gas heating, a number of variations in method of firing are used. These variations may be separated into two general classes **direct** and **indirect firing.** Direct firing is used more generally. This method permits the products of combustion of the fuel to circulate about the material to be heated. In direct-fired furnaces, open burners may be used either in the furnace proper or they may be installed outside the work-heating chamber in the path of an external fan which circulates large volumes of hot gases through the furnace. Temperatures attainable in this type of furnace are limited by the materials of which the fan is constructed. The latter modification is used in **convection-type** furnaces. In indirect-fired furnaces the products of combustion do not enter the work-heating chamber. Indirect firing is used in **muffle** furnaces, an example of which is shown in Figure 41—48.

Another common application of indirect firing is obtained with radiant tubes, an example of which is shown in the furnace in Figure 41—49. Furnaces using direct firing are relatively lower in operating cost and original capital investment. Furnaces using indirect firing generally are selected where the control of furnace atmosphere is of particular importance.

Electric power is used as a source of heat in many heat-treating furnaces due to the ease it affords for controlling temperature, to its suitability for use with protective-gas atmosphere and to its cleanliness. Resistance-type heating elements, which either are imbedded in the furnace refractory lining or suspended from heat-resisting hangers, radiate heat to the furnace charge. The resistance units are positioned in the furnace to permit uniform heating. Furnaces with zone heating sometimes use electrical resistance units because of their easy adaptation to the control of temperature levels. Another method for heating electrically is by induction. Induction heating is done by passing a high-frequency alternating current through a coil surrounding the material to be heated. The coils are shaped to suit the material to be heated. The rapidly alternating electrical field, in which the material to be heated is held, causes the steel to heat very rapidly, due to eddy currents and hysteresis. Due to the rapidity of the process, the duration of heating is very critical, necessitating precise control.

Atmosphere Control—The effect of a heat-treating operation on the surface condition of the work pieces is influenced by the time of heating, the temperature level maintained, and the atmosphere surrounding the material. Figure 41—50 shows graphically the relative amount of scale formed in a batch reheating furnace for variations of these conditions. While the temperature level in heat-treating furnaces is somewhat lower than that shown in the figure, the time is generally longer. By using the proper atmosphere in the working

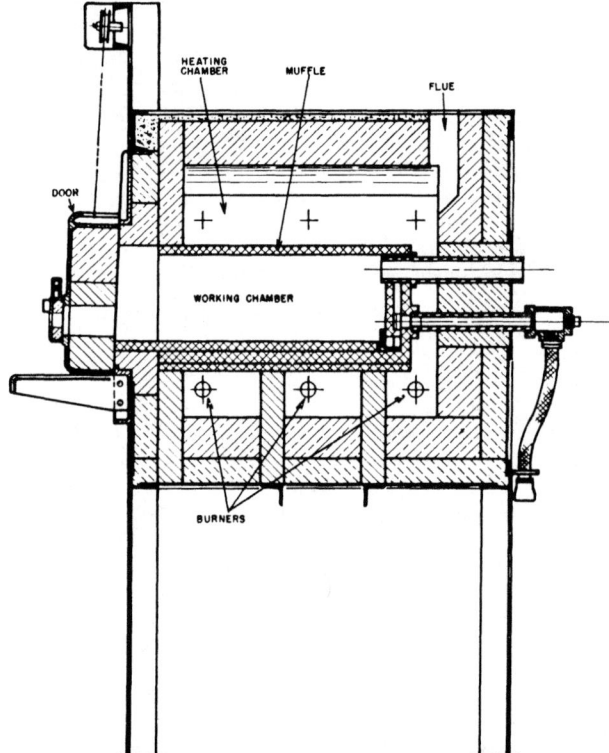

FIG. 41—48. Diagrammatic cross-section of an indirect-fired muffle type furnace equipped for use of a controlled atmosphere in the muffle. Openings in rear provide for insertion of control thermocouple and entrance of prepared atmosphere gas. (Courtesy, Surface Combustion Div., Midland-Ross Corp.)

chamber, a clean scale-free surface is obtained. Such a surface is required for most sheet and strip material and other important steel products, such as wire and tubes. The three gases most injurious to surface condition are oxygen, carbon dioxide, and water vapor. The effect of each of these, as well as effects from other gases, follows:

Oxygen reacts with the iron of the steel to produce iron oxide. For this reason it must be excluded entirely for bright annealing. It also reacts with the carbon in steel to lower the carbon content of its surface; that is, it decarburizes the steel. In some types of controlled annealing, oxygen of the air may be caused to scale the steel faster than it decarburizes it. This results in a product having a decarburized surface layer of minimum thickness and the scale, being flaky, is easily removed.

Nitrogen, in the molecular state, is entirely passive to iron and it is entirely satisfactory for bright annealing low-carbon steels. If pure and very dry, it will be passive to high-carbon steel, but the presence of even slight traces of moisture will cause decarburization.

Carbon dioxide and carbon monoxide are considered together since the ratio of their concentration in the atmosphere plays an important part in their action on the steel surface. As an example, if the ratio of carbon dioxide to carbon monoxide is 0.6 or higher at 815°C (1500°F), the atmosphere will scale steel readily. If the

FIG. 41—49. Radiant-tube fired cover-type furnace employed in heat treating coils of flat rolled products enclosed in inner covers under which controlled atmosphere gas is circulated. (Courtesy, Surface Combustion Div., Midland-Ross Corp.)

ratio is reduced to 0.4, or lower, the atmosphere will no longer scale steel but will remain decarburizing to a 1 per cent carbon steel. For low-carbon steels, a ratio of carbon monoxide to carbon dioxide on the order of about two to one will be in equilibrium with the steel, and a gas of this composition is used to advantage for producing bright-annealed sheets. A higher carbon-monoxide content actually will carburize the steel.

Hydrogen is highly reducing to iron oxide, and therefore, opposes the formation of a heavy, flaky scale and when present in the products of combustion, promotes formation of a tight scale that is hard to remove. At certain temperatures it is absorbed by the steel and is likely to result in embrittlement, more particularly in high-carbon steel. If the hydrogen is dry, it has no scaling effect on high-carbon steel at elevated temperatures, but it does cause considerable decarburization. A common prepared atmosphere which is used for bright-annealing work is composed of 75 per cent hydrogen and 25 per cent nitrogen. It is formed by cracking anhydrous ammonia.

Water vapor is oxidizing to iron and combines with carbon of steel to form carbon monoxide and hydrogen within the temperature range of the "water-gas reaction." It is reactive to a steel surface at temperatures even as low as 205° to 370°C (400° to 700°F), and is thus often the cause of formation of a blue oxide during the cooling cycle.

Hydrocarbons, more specifically methane, are carburizing gases. They are subject to thermal decomposition at annealing temperature, liberating hydrogen and depositing soot on the steel.

The most commonly prepared gases for control of atmosphere are formed by the partial combustion of hydrocarbon gases, contained in such fuels as coke-oven gas, natural gas, propane, or butane. Manufacturers of converters for the preparation of gases describe the various kinds under trade names, such as "DX" gas, "Drycolene," etc. The first step in making such a gas is to burn a mixture of the fuel gas with air. This provides a gas high in nitrogen, but containing other undesirable gases including water vapor, which must be

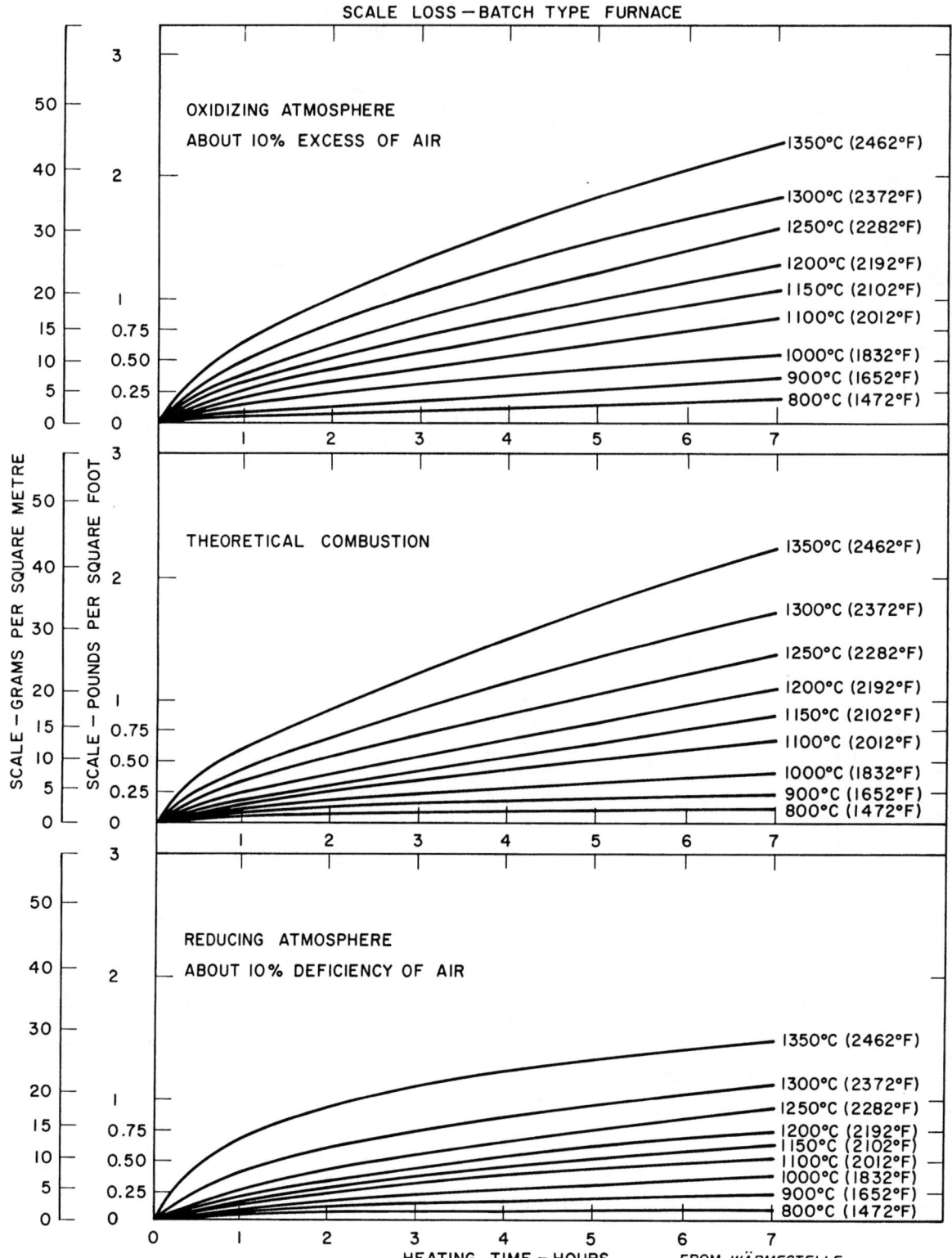

FIG. 41—50. Graphical representation of relative amounts of scale formed on steel heated in a batch-type reheating furnace with variations in time of heating, temperature level, and heating-chamber atmosphere. (From "Warmestelle.")

removed. Other atmosphere gases are prepared by (1) cracking a non-combustible mixture of air and gas with a catalyst at high temperature, (2) by cracking anhydrous ammonia at high temperature, and (3) by passing air through a heated retort filled with charcoal.

Furnaces using controlled atmosphere have a number of construction features not incorporated in ordinary furnaces. These features are essential to prevent loss of gas and to minimize the entrance of air into the furnace which would upset the control established. Casings for furnaces with controlled atmosphere are welded gas-tight. Batch-type furnaces are provided with sand seals. Continuous furnaces charged and discharged from the ends sometimes have flame curtains to burn out the oxygen from any possible air infiltration, or the furnace may be operated under sufficient pressure to prevent air infiltration. In the latter case, a small loss of the prepared atmosphere usually occurs through the small unavoidable openings in the furnace. Doors, where required, are fitted snugly by sloping fronts with ground surfaces or by wedging devices or clamps. Modern continuous pusher-type furnaces for special heat treatment of bar and wire coils utilize a vestibule and inner-door arrangement at the charge and discharge ends of the furnace. This permits complete purging of air as material is charged and discharged from the furnace

Batch-Type Furnaces—The five principal general types of batch furnaces are described below:

1. **Box furnaces** are constructed with a solid hearth. They are shaped, as their name implies, similar to a box and are charged through door openings by tongs or some mechanical charger. The furnace hearth may vary from less than half a square metre to nearly 3 square metres (a few square feet to over 30 square feet). Heating may be done by direct or indirect fuel firing or by electricity. Muffle and semi-muffle type construction often is employed when control of atmosphere is required. This type of furnace is used frequently for individual-piece or small-lot heat treating, for laboratory test and shop work, and for general production work on a small scale. Box furnaces have been constructed for convection heating either with a fan underneath the roof or with one external to the furnace for recirculation. Furnaces of this type are used for annealing, normalizing, tempering and carburizing.

2. **The car-bottom furnace** consists of a furnace shell equipped with burners or heating units with the hearth built upon a separate car which runs in and out of the furnace shell to charge and unload the furnace. The car usually is moved into and out of the furnace by a toothed rack attached to the bottom of the car and a stationary pinion actuated by an electric motor, the car itself resting on rollers or wheels that move over a two-rail track. The doors of the furnace are of the vertically lifting type, full width of the furnace, and are hydraulically or electrically operated. In order that the entire surface of the charge may be exposed to heat of the same intensity and to aid circulation, the charge is supported above the floor of the car bottom by heat-resisting alloy castings or on refractory piers. The car bottom is made to fit the furnace closely and the escape of hot gases around it is prevented by sand seals. Car-bottom furnaces have been constructed to process charges from a few tons to several hundred tons. They are used for heat treating of axles, bars, heavy plates, castings and miscellaneous shapes.

For operations involving heating, quenching and tempering, it is desirable that the quenching tank be located in close proximity to the furnace to enable the charge to be placed in the tank in the shortest possible time. In some installations, less than a minute is required to transfer the charge from a closed furnace to the quenching tank.

Car-bottom furnaces may be direct or indirect fired, and various designs have been developed to improve heat distribution in the working chamber. Electric heating also is employed in some car-bottom furnaces. Car-bottom furnaces sometimes are constructed of two chambers side by side, with a common division wall to facilitate annealing and tempering operations. In some installations, an auxiliary cooling system employing blowers is provided to accelerate cooling. Some car-bottom furnaces are known as **elevator furnaces** where the car is rolled under the furnace shell and then raised into the furnace by a motor-driven lifting mechanism. Those in which the shell is lowered over the car, as shown in Figure 41—51, are used to provide a more complete sand or water seal than is obtainable with conventional car-bottom furnaces.

3. **The bell-type furnace** has a removable shell or cover. The furnace usually is used for processing material which requires special surface protection from oxidation or decarburization. The furnace shell is removed by a crane and set aside while the hearth of the furnace is charged. The shell is then replaced, as was shown in Figure 41—49. Furnaces of this type, used for annealing sheet, strip, rod and wire, usually are called **box annealing, pack annealing, coil annealing,** or **cover annealing furnaces.** In these, the material is stacked on a permanent base or **stand,** a light **inner cover** is placed over the stack, sealed with sand at the bottom and provided with a constant supply of prepared gas atmosphere, and then the portable heating unit is lowered over the assembly. The heating covers are square, rectangular or cylindrically shaped. Loads vary from about 32 to 363 metric tons (35 to 400 net tons) per charge, distributed on one to eight stands per base. In most instances, a number of bases and inner covers are provided with one or more covers for heating. After heating of each charge is completed on a base, the heating cover is moved to another base, leaving the charge protected by the atmosphere under the inner cover which is left in place. The covers may be direct fired or equipped with radiant tubes for indirect firing, or they may be heated by electrical resistance units. The heating elements are attached to the inside of the heating cover, which is built of steel and lined with a refractory insulating material and braced substantially in order that it can be moved from base to base with an overhead crane. Many inner covers are made of heat-resisting alloys. All are sealed to the base at their bottom edges with a powdered refractory. The heating time in a cover annealing furnace for coils of sheet or tin plate range from 24 to 44 hours for the larger sized furnaces, depending upon the length of soaking period required. The soaking period usually is about 4 to 12 hours. In furnaces of about 135 to 270

metric tons (150 to 300 net tons) capacity, the average production is about 5 metric tons (5.5 net tons) per hour, the fuel consumption about 1 163 000 kilojoules per metric ton (1 000 000 Btu per net ton), the maximum fuel-burning capacity about 3 517 000 watts (12 000 000 Btu per hour), and the atmosphere-gas consumption about 34 cubic metres (1200 cubic feet) per hour. In pack or box annealing furnaces, natural gas or some inert gas is used to surround the charge; the circulation of gases inside the inner cover is by natural or forced convection. In cover furnaces for annealing coils, separators are placed between each coil to aid in distribution of the gas inside the inner cover. Circulation of this gas in a number of modern installations is forced. The fan is located in the base below each stand. The trend in annealing sheet and tin plate has been towards the greater use of coils and heating by forced convection rather than by the former pack method of annealing, in order to obtain higher production and improved uniformity of heating.

4. **Pit furnaces** are furnaces of cylindrical or rectangular shape in which the material is charged and withdrawn through an opening in the furnace top. The larger furnaces are installed usually with at least part of their work chambers below floor level, while many of the smaller and shallower furnaces rest on the working floor, for convenience in handling material. The material to be processed can be suspended by a fixture, loaded into a basket and set into the working chamber, or, as in the case of large forgings, be supported on a suitable base in the furnace. Pit furnaces employ either direct firing or electrical heating, in either case with natural or forced circulation. They may or may not be equipped with special facilities for atmosphere control. Pit furnaces are used for normalizing, hardening, annealing, tempering, and carburizing.

5. **A salt-bath or lead-bath furnace** is another type of heat-treating furnace. It is designed to hold a bath of molten salt or lead in which the material is immersed for treatment. These furnaces are usually small pot-like affairs used in batch operations, but some large furnaces have been constructed of rectangular shape with depths of over 4.5 metres (15 feet) to suit the shape and size of the material to be handled, with conveyors or other means for carrying out continuous operations. They are equipped usually with a hinged cover or a ventilating hood for minimizing fumes. Such furnaces are used to obtain uniform temperature distribution and close temperature control of the work piece. The bath is heated and maintained at proper temperature either by electrical resistance or by combustion of a fuel. Furnaces with a molten bath for heat treating are called **pot furnaces** when the bath is contained in a pot or crucible constructed of a heat- and corrosion-resistant metal, usually externally heated by suitable burners. Other bath-type furnaces may be heated by electric current passing through the (salt) bath between immersed electrodes, or by immersed resistance coils or fuel-fired tubes.

Continuous Furnaces—In continuous furnaces, the

FIG. 41—51. Charge of mixed sizes of steel bars, supported above hearth by cast-alloy fixtures, ready to be rolled under bell-type furnace body, which then will be lowered over charge. Toothed rack above floor at lower left is driven by a pinion to move car. (Courtesy, Surface Combustion Div., Midland-Ross Corp.)

material moves through the furnace, and two basic types of these are recognized. In one, the furnace is circular with a rotating hearth which carries the charge. The walls and roof are stationary, and the furnace enclosure is made by contacting the walls with the periphery of the moving hearth through a sand or liquid seal. In the other type, the furnace is composed of a single, long straight chamber, or series of chambers, through which the material is moved. Differentiation of continuous furnace types may be made according to the way the material is moved, such as the **rotary-hearth**, the **roller-hearth**, the **pusher**, the **conveyor**, the **walking-beam**, the **tunnel**, the **continuous-strand** and the **monorail** types. Modern production methods, dealing with ever increasing tonnages of material of identical size and treatment, favor the continuous-furnace type best suited to the nature of material to be heat treated.

Continuous furnaces are designed with and without auxiliary equipment for atmosphere control. Heat may be applied by direct or indirect firing or electrically. They are especially suited for zone heating and cooling. A brief description of continuous furnace types and their application is given below.

1. **Rotary-hearth furnaces** are used generally for heating pieces that are to be handled individually. Typical applications are the heating of gears, shells, cylinders, billets, etc., that are to be fixture-quenched or handled individually for scale-free hardening without decarburization, for normalizing or drawing. This furnace type is used also for heating smaller parts loaded in lightweight trays, and for pack carburizing. Charging and discharging are accomplished at the same location. Rotary-hearth furnaces are built in a wide range of hearth sizes, to heat from less than 200 kilograms to over 50 metric tons (a few hundred pounds up to 60 net tons per hour). A typical rotary-hearth furnace for heat treating steel is shown in Figure 41—52.

2. **Roller-hearth furnaces** are high production, continuous-type units, especially suited for uniform treatment of large orders of identical material. This type of furnace is used widely for bright annealing of tubes, stampings, drawn parts, etc.; for normalizing, annealing, hardening and tempering steel bars; for annealing malleable iron, small steel and iron castings, and forgings; and for normalizing flat-rolled products. Roller-hearth furnaces are constructed as a single furnace or as a line of furnaces for zone heating and cooling, and sometimes have an intermediate section with a tank for quenching.

In some modern furnaces used for continuously treating short lengths of sheet, disc rollers made of heat-resisting alloys with polished surfaces are utilized to reduce the cooling effect of full contact with the ordinary type of roller and to avoid scratching of the piece. The discs are staggered and mounted upon water-cooled shafts, which are driven by variable-speed motors either through a chain and sprocket system or shafts and gears. A gas-fired normalizing furnace with automatic pyrometric control is shown in Figure 41—53. Furnaces of this type are built up to 2540 millimetres (100 inches) in width and vary from about 35 to over 60 metres (120 to 200 feet) in length, the larger ones having capacities for normalizing up to 270 metric tons (300 net tons) per 24-hour day. Sheets undergoing treatment in roller-hearth furnaces may be protected further from contact with rollers of whatever type by the use of **rider sheets**, which are placed on the rollers and support the work. **Cover sheets** on top of the work also may be used to further increase protection. Rider and cover sheets may be used repeatedly before they must be scrapped, since they generally are made of alloy steel.

3. **Pusher-type furnaces** are of two general types. In one type the parts are pushed against each other, as in the continuous reheating furnace. In the other type, the parts are loaded in trays or other types of carriers which are pushed through the furnace.

4. **Conveyor-type furnaces** are constructed similarly to roller-hearth furnaces except that belt conveyors are

FIG. 41—52. Rotary-hearth furnace for heat-treating steel. Hearth is rotated by a chain-and-sprocket drive, seen in the foreground. (Courtesy, Surface Combustion Div., Midland-Ross Corp.)

used to carry the material through the furnace. They are suitable for accurately heat treating small miscellaneous pieces which would not ride properly on a roller hearth. Belt conveyors are made of alloy material of sufficient strength to carry the load and are resistant to heat, oxidation, corrosion and abrasion. A number of different designs of conveyors are utilized to satisfactorily meet the requirements. Some conveyors consist of several individual chains held on constant centers by spacer bars provided with suitable projecting lugs upon which the load, such as sheet or plate cut to length, rests. Many other belts are constructed of open mesh or woven chain to permit free circulation of hot furnace gases or protective atmosphere, while another construction utilizes pans or trays connected to a roller chain to carry the material. The production rate or heating cycle in this furnace type is controlled both by the temperature setting and by varying the speed of the conveyor.

5. **Walking-beam furnaces** employ a special mechanism within the furnace, known as a "walking beam" to move the material through the furnace. The walking beam consists of a number of alloy supports or beams which are arranged in rows of two or more beams to the row, throughout the length of the furnace. The beams are staggered with the one immediately ahead, and are placed in longitudinal slots in the furnace hearth. They are attached from below to toggles or cams that intermittently raise the beams, move them forward, and then lower them, thus depositing the material on the beams ahead. By this step action the material is moved through the furnace at the desired rate. This furnace type is used commonly for tubes, bars, structural shapes or similar material. The principle of the walking-beam furnace is also discussed and illustrated in Chapter 22.

6. In **tunnel-type furnaces,** the stock to be heated is placed upon cars which then are pushed or pulled slowly through the furnace. This furnace type was used at one time for continuous box annealing of sheets. Furnaces used for this purpose sometimes reach over 90 metres (300 feet) in length and were built in the form of a long tunnel, not necessarily straight, nor with a level bottom. The sheets were piled upon a base and sealed in a box for annealing. The box then was placed upon a small car which was pushed into the furnace. As the furnace was full of these cars, one car was removed as each car entered the furnace.

7. **Continuous strand-type furnaces** have been developed to reduce the extra handling and the long heating and cooling periods required in annealing sheet and tin plate in coil form. Heat treating uncoiled strip provides better control of the time-temperature requirements for the entire piece and therefore, a more uniform product. A coil can be processed in a matter of minutes or a few hours compared to the long cycle required in a batch furnace. Another special advantage of this furnace type is that other operations, such as cleaning, coating, temper-rolling and levelling may be combined with the heat-treating process to avoid extra handling and the expense of duplicated handling equipment for separate lines. For this reason the modern continuous strand annealing lines for sheet are often called continuous annealing and processing lines. For example, modern continuous annealing lines for sheet and tin mill products often include electrolytic cleaning prior to the heating furnace, pickling and rinsing (to remove the thin oxide layers formed during heating and cooling), temper rolling and oiling. A feature of these modern continuous annealing lines is the presence of a low temperature (288°C or 425°C or 550°F to 800°F) reheating furnace, after the main heating furnace, which is used to overage the sheet and thereby improve the ductility and reduce the strain aging susceptibility of the sheet. Continuous strand-type furnaces are constructed either as horizontal or vertical units. Furnaces of the latter type of construction are sometimes referred to as **tower-type furnaces** and are used primarily to conserve floor space. Furnaces of this type utilize either electric or radiant-tube heating or both.

Figure 41—54 shows a tower-type furnace for annealing cold-reduced material at one stage in the manufacture of tin plate.

The horizontal type of continuous-strand furnace sometimes utilizes catenary suspension of the uncoiled strip, where neither rolls nor any other type of support are used throughout the heating zone. The heating zone of these furnaces may be from about 6 to 15 metres (20 to 50 feet) in length. The preheating and

FIG. 41—53. Continuous disc-roller-hearth type of normalizing furnace, divided lengthwise into zones each having individual automatic temperature control.

FIG. 41—54. Tower-type furnace in a continuous annealing line in a tin mill. The furnace is surrounded by structural-steel operating platforms, and is heated by a combination of gas-fired and electric heating units.

cooling zones usually are constructed shorter than in the conveyor type and for some kinds of work are omitted entirely.

8. In **overhead monorail furnaces,** the material undergoing heating is suspended from rods that serve as hangers or even may be welded to the suspension rods. The suspension rods are attached at their upper ends to the carriers that operate on the monorail. If the rods are welded to the work pieces, they are removed after the assembly leaves the heating furnace.

Bibliography

American Society for Metals, Metals Handbook, 9th edition, Vol. 4.

American Society for Metals, Atlas of Isothermal transformation and Continuous Transformation Diagrams, 1977.

American Society for Testing and Materials, ASTM Designation F 44-80, Vol. 03.03.

American Society for Testing and Materials, ASTM Designation E 112-81, Vol. 03.03.

Bain, E. C., Functions of the alloying elements in steel, published by the American Society for Metals, 1939. (Second edition by E. C. Bain and H. W. Paxton, 1961.)

Bramfitt, B. L. and P. L. Magnanon, Metallurgy of continuous annealed sheet steel; Proceedings of the Metallurgical Society of AIME, published by AIME, Warrendale, Pa., 1982.

Doane, D. V. and J. S. Kirkaldy, (Editors) Hardenability concepts and applications to steel, TMS-AIME, 1978.

Grange, R. A., C. R. Hribal and L. F. Porter, Hardness of tempered martensite in carbon and low-alloy steels, Metallurgical Transactions, S. A., 1977, pages 1775-1785.

Krauss, G., Principles of heat treatment, published by American Society for Metals, 1980.

Siebert, C. A., D. V. Doane, and D. H. Breen, The hardenability of steels, ASM, 1977.

CHAPTER 42

Carbon Steels

SECTION 1

CLASSIFICATION AND APPLICATION

The plain carbon steels represent the most important group of engineering materials known. They represent by far the major percentage of steel production and the widest diversity of application of any of the engineering materials. These applications are so diversified that anything like a complete listing, or even a classification on the basis of application, is not feasible. Some of the more important classes of application have, however, been discussed in this book. These include castings, forgings, tubular products, plates, sheet and strip, wire and wire products, structural shapes, bars, and such railway items as rails, wheels and axles. Although a complete classification by application is impossible, plain carbon steels may be described by their method of manufacture as, for example, basic open-hearth, basic oxygen, acid open-hearth or basic electric-furnace steels. This method of classification may be extended to include the deoxidation practice employed; for example, steels may be rimmed, capped, semi-killed or fully killed. A further method of classification for plain carbon steels recognizes steel produced by teeming into ingots and then rolling into blooms, slabs or billets, as opposed to strand-cast steels (also termed continuous cast) wherein the molten steel is cast directly into blooms, slabs or billets. Steels made by any of the steelmaking processes can, in general, be used for similar end applications.

The plain carbon steels also may be classified on the basis of carbon content as hypoeutectoid or hypereutectoid steels; the hypoeutectoid steels are those in which the carbon content is below the eutectoid value of about 0.80 per cent, and the hypereutectoid steels those with carbon contents above this value.

The properties of plain carbon steels depend chiefly on their carbon content and microstructure. As discussed in Section 2 of this chapter, in addition to carbon these steels always contain some manganese, silicon, phosphorus and sulphur; minor amounts of other elements also may be present. Plain carbon steels, by American Iron and Steel Institute definition, may contain up to 1.65 per cent manganese, 0.60 per cent silicon and 0.60 per cent copper in addition to much smaller amounts of other elements. Composition ranges for plain carbon steels have been established jointly by the Society of Automotive Engineers (SAE) and the American Iron and Steel Institute (AISI), and these are listed in Tables 42—I, 42—II and 42—III where they are identified by their UNS (Unified Numbering System) designations as well as the SAE and AISI designations formerly used. The designation numbers assigned to the individual steels have a significance based on the system outlined in Section 2 of Chapter 43 on alloy steels.

SECTION 2

FACTORS AFFECTING CARBON-STEEL PROPERTIES

The principal factors affecting the properties of the plain carbon steels are the carbon content and the microstructure. Manganese is a solid-solution strengthening element in steel and is effective in increasing hardenability in heat-treated applications. The general relationships between microstructure and properties, and the factors governing microstructure, have been discussed in the preceding chapter on heat treatment, and need not be repeated here. Most of the plain carbon steels are, however, used without a final heat treatment and the factors affecting the microstructure and thereby the properties in such as-rolled or as-forged products will be emphasized in this chapter.

In addition to the predominant effects of carbon content and microstructure, the properties of plain carbon steels may be modified by the effects of residual ele-

ments other than the carbon, manganese, silicon, phosphorus and sulphur which are always present. These incidental elements are usually picked up from the scrap, from the deoxidizers, or from the furnace refractories. The properties of carbon steel may also be affected by the presence of gases, especially oxygen, nitrogen and hydrogen and their reaction products. The gas content is largely dependent upon the melting, deoxidizing and pouring practice. The final properties of the plain carbon steels are, therefore, influenced by the steelmaking practice used in their production.

Thus, the factors governing the properties of a plain carbon steel are primarily its carbon content and microstructure, with the microstructure being determined largely by the composition and the final rolling, forging or heat-treating operation, and secondarily by

Table 42—I. Carbon-Steel Compositions Applicable Only to Semifinished Products for Forging, to Hot-Rolled and Cold-Finished Bars, to Wire Rods, and to Seamless Tubing.[1],[2],[4]

Unified Numbering System Designation	Chemical Composition Limits (Ladle Analyses), Per Cent[3]				Corresponding SAE or AISI Number
	C	Mn	P, max.	S, max.	
G10050	0.06 max.	0.35 max.	0.040	0.040	1005
G10060	0.08 max.	0.25-0.40	0.040	0.050	1006
G10080	0.10 max.	0.30-0.50	0.040	0.050	1008
G10100	0.08-0.13	0.30-0.60	0.040	0.050	1010
G10120	0.10-0.15	0.30-0.60	0.040	0.050	1012
G10130	0.11-0.16	0.50-0.80	0.040	0.050	1013[5]
G10150	0.13-0.18	0.30-0.60	0.040	0.050	1015
G10160	0.13-0.18	0.60-0.90	0.040	0.050	1016
G10170	0.15-0.20	0.30-0.60	0.040	0.050	1017
G10180	0.15-0.20	0.60-0.90	0.040	0.050	1018
G10190	0.15-0.20	0.70-1.00	0.040	0.050	1019
G10200	0.18-0.23	0.30-0.60	0.040	0.050	1020
G10210	0.18-0.23	0.60-0.90	0.040	0.050	1021
G10220	0.18-0.23	0.70-1.00	0.040	0.050	1022
G10230	0.20-0.25	0.30-0.60	0.040	0.050	1023
G10250	0.22-0.28	0.30-0.60	0.040	0.050	1025
G10260	0.22-0.28	0.60-0.90	0.040	0.050	1026
G10290	0.25-0.31	0.60-0.90	0.040	0.050	1029
G10300	0.28-0.34	0.60-0.90	0.040	0.050	1030
G10350	0.32-0.38	0.60-0.90	0.040	0.050	1035
G10370	0.32-0.38	0.70-1.00	0.040	0.050	1037
G10380	0.35-0.42	0.60-0.90	0.040	0.050	1038
G10390	0.37-0.44	0.70-1.00	0.040	0.050	1039
G10400	0.37-0.44	0.60-0.90	0.040	0.050	1040
G10420	0.40-0.47	0.60-0.90	0.040	0.050	1042
G10430	0.40-0.47	0.70-1.00	0.040	0.050	1043
G10440	0.43-0.50	0.30-0.60	0.040	0.050	1044
G10450	0.43-0.50	0.60-0.90	0.040	0.050	1045
G10460	0.43-0.50	0.70-1.00	0.040	0.050	1046
G10490	0.46-0.53	0.60-0.90	0.040	0.050	1049
G10500	0.48-0.55	0.60-0.90	0.040	0.050	1050
G10530	0.48-0.55	0.70-1.00	0.040	0.050	1053
G10550	0.50-0.60	0.60-0.90	0.040	0.050	1055
G10590	0.55-0.60	0.50-0.80	0.040	0.050	1059
G10600	0.55-0.65	0.60-0.90	0.040	0.050	1060
G10640	0.60-0.70	0.50-0.80	0.040	0.050	1064
G10650	0.60-0.70	0.60-0.90	0.040	0.050	1065
G10690	0.65-0.75	0.40-0.70	0.040	0.050	1069
G10700	0.65-0.75	0.60-0.90	0.040	0.050	1070
G10740	0.70-0.80	0.50-0.80	0.040	0.050	1074
G10750	0.70-0.80	0.40-0.70	0.040	0.050	1075
G10780	0.72-0.85	0.30-0.60	0.040	0.050	1078
G10800	0.75-0.88	0.60-0.90	0.040	0.050	1080
G10840	0.80-0.93	0.60-0.90	0.040	0.050	1084
G10850	0.80-0.93	0.70-1.00	0.040	0.050	1085
G10860	0.80-0.93	0.30-0.50	0.040	0.050	1086
G10900	0.85-0.98	0.60-0.90	0.040	0.050	1090
G10950	0.90-1.03	0.30-0.50	0.040	0.050	1095

NOTE: LEAD. When lead is required as an added element to a standard steel, a range of 0.15 to 0.35 per cent, inclusive, generally is used. Such a steel is identified in UNS by changing the final zero in the designation to "4" and in the SAE or AISI systems by inserting the letter "L" between the second and third numerals of the grade number: for example, UNS G10450 (SAE 1045) becomes G10454 (SAE 10L45).

[1] Based on SAE Standard J403 in 1982 SAE Handbook, published by the Society of Automotive Engineers, the latest edition of which should be consulted for complete details regarding these steels.

[2] Certain SAE grades had slightly wider ranges for carbon and manganese when producing steels for structural shapes, plates, strip, sheet and welded tubing. There were no corresponding AISI numbers for steels having these broader ranges of composition.

[3] These steels may be produced by the basic oxygen, the basic open-hearth, or the basic electric steelmaking process. Where silicon is required, the following limits and ranges are commonly used: for steel designations up to but excluding UNS G10150 (SAE or AISI 1015), 0.15 per cent max.; for UNS G10150 to G10250 (SAE or AISI 1015 to 1025), 0.10 per cent max. or ranges of 0.10 to 0.20 per cent or 0.15 to 0.30 per cent; for over UNS G10250 (SAE or AISI 1025), ranges of 0.10 to 0.20 per cent or 0.15 to 0.30 per cent.

[4] The SAE and AISI grades originally numbered 1024, 1027, 1036, 1041, 1047, 1048, 1051, 1052, 1061 and 1066, with manganese maximum in excess of 1.00 per cent, were renumbered 1524, 1527, 1536, etc., respectively, and formed part of the SAE and AISI 1500 series shown in Table 42—III where the steels are identified also by their UNS designations.

[5] SAE only.

Table 42—II. Free-Cutting (Rephosphorized and Resulphurized, and Resulphurized) Carbon-Steel Compositions Applicable Only to Semifinished Products for Forging, Hot-Rolled and Cold-Finished Bars, Wire Rods and Seamless Tubing.[1],[2],[3]

Unified Numbering System Number	Chemical Composition Limits (Ladle Analyses), Per Cent					SAE or AISI Number
	C	Mn	P	S	Pb	
REPHOSPHORIZED AND RESULPHURIZED STEELS						
G12110	0.13 max.	0.60-0.90	0.07-0.12	0.10-0.15		1211
G12120	0.13 max.	0.70-1.00	0.07-0.12	0.16-0.23		1212
G12130	0.13 max.	0.70-1.00	0.07-0.12	0.24-0.33		1213
G12150	0.09 max.	0.75-1.05	0.04-0.09	0.26-0.35		1215
G12144[3]	0.15 max.	0.85-1.15	0.04-0.09	0.26-0.35	0.15-0.35	12L14
RESULPHURIZED STEELS						
G11080[4]	0.08-0.13	0.50-0.80	0.040 max.	0.08-0.13		1108[4]
G11100	0.08-0.13	0.30-0.60	0.040 max.	0.08-0.13		1110
G11170	0.14-0.20	1.00-1.30	0.040 max.	0.08-0.13		1117
G11180	0.14-0.20	1.30-1.60	0.040 max.	0.08-0.13		1118
G11370	0.32-0.39	1.35-1.65	0.040 max.	0.08-0.13		1137
G11390	0.35-0.43	1.35-1.65	0.040 max.	0.13-0.20		1139
G11400	0.37-0.44	0.70-1.00	0.040 max.	0.08-0.13		1140
G11410	0.37-0.45	1.35-1.65	0.040 max.	0.08-0.13		1141
G11440	0.40-0.48	1.35-1.65	0.040 max.	0.24-0.33		1144
G11460	0.42-0.49	0.70-1.00	0.040 max.	0.08-0.13		1146
G11510	0.48-0.55	0.70-1.00	0.040 max.	0.08-0.13		1151

[1] Based on SAE Standard J403 in 1982 SAE Handbook, published by the Society of Automotive Engineers, the latest edition of which should be consulted for complete details relating to these steels.

[2] These steels may be produced by the basic oxygen, the basic open-hearth, or the basic electric steelmaking process.

[3] When lead is required as an added element to other grades of steel, a range of 0.15 to 0.35 per cent, inclusive, generally is used. Such a steel is identified in UNS by changing the last numeral of the designation to 4 (for example, when lead is added to G11170, the leaded steel is designated G11174): while in the SAE and AISI systems, a steel to which lead was added was identified by inserting the letter "L" between the second and third numerals of the number of the unleaded grade (for example, 1117 became 11L17).

[4] SAE only.

NOTE: UNS grades G11080, G11090, G11100, G12110, G12120, G12130 and G12144 (SAE or AISI 1108, 1109, 1110, 1211, 1212, 1213 and 12L14) customarily are furnished without specified silicon content. When silicon is required, the following limits and ranges commonly are used: for UNS G11080, G11090 and G11100 (AISI or SAE 1108, 1109 or 1110), 0.10 per cent max.; for UNS G11160 (SAE or AISI 1116) and over, 0.10 per cent max., or ranges 0.10-0.20 per cent, or 0.15-0.30 per cent.

Table 42—III. High-Manganese Carbon-Steel Compositions Applicable Only to Semifinished Products for Forging, to Hot-Rolled and Cold-Finished Bars, to Wire Rods, and to Seamless Tubing.[1],[2],[3]

Unified Numbering System Number					SAE or AISI Number	
	C	Mn	P, max.	S, max.		
G15130	0.10-0.16	1.10-1.40	0.040	0.050	1513	1513
G15220	0.18-0.24	1.10-1.40	0.040	0.050	1522	1522
G15240	0.19-0.25	1.35-1.65	0.040	0.050	1524	1524
G15260	0.22-0.29	1.10-1.40	0.040	0.050	1526	1526
G15270	0.22-0.29	1.20-1.50	0.040	0.050	1527	1527
G15360	0.30-0.37	1.20-1.50	0.040	0.050	1536	[4]
G15410	0.36-0.44	1.35-1.65	0.040	0.050	1541	1541
G15480	0.44-0.52	1.10-1.40	0.040	0.050	1548	1548
G15510	0.45-0.56	0.85-1.15	0.040	0.050	1551	1551
G15520	0.47-0.55	1.20-1.50	0.040	0.050	1552	1552
G15610	0.55-0.65	0.75-1.05	0.040	0.050	1561	1561
G15660	0.60-0.71	0.85-1.15	0.040	0.050	1566	1566

[1] Based on SAE Standard J403 in 1982 SAE Handbook, published by the Society of Automotive Engineers, the latest edition of which should be consulted for complete details relating to these steels.

[2] These steels may be produced by the basic oxygen, the basic open-hearth, or the basic electric steelmaking process.

[3] Where silicon is required, the following limits and ranges are commonly used: for steel designations up to and including G15240 (SAE 1524 or AISI 1524), 0.10 per cent max. or ranges of 0.10 to 0.20 per cent or 0.15 to 0.30 per cent.

[4] SAE only.

the residual alloy, non-metallic and gas content of the steel which, in turn, depend upon the steelmaking practice.

Carbon Content and Properties—The average mechanical properties of as-rolled 25-mm (1-inch) bars of carbon steels, as a function of carbon content, are shown in Figure 42—1. These values are based on statistical analyses made by several investigators. This figure is illustrative of the general effect of carbon content when the microstructure and grain size are held reasonably constant. These data are also representative for hot-rolled sheet and strip products. It will be seen that the hardness, tensile strength and yield strength increase with increasing carbon content, while the elongation, reduction of area, and Charpy impact values decrease sharply.

Effect of Microstructure and Grain Size—The general relationships between microstructure and properties have been discussed in Chapter 41 on heat treatment. The carbon steels, being of relatively low hardenability, generally have ferrite-pearlite or pearlite-cementite microstructures in the cast, rolled or forged conditions. The constituents of the hypoeutectoid steels are, therefore, ferrite and pearlite, and of the hypereutectoid steels, cementite and pearlite. As described in the previous chapter, the properties of such steels are dependent primarily upon the relative proportion of these constituents, interlamellar spacing of the pearlite and the grain size. Both the hardness and the ductility increase as the interlamellar spacing or the pearlite-transformation temperature decreases, and the ductility increases with decreasing grain size. The effect of grain size was discussed in some detail in the preceding chapter. The effect of the interlamellar spacing of the pearlite was also discussed and illustrated for a eutectoid steel. The effect of this variable on tensile strength is further illustrated in Figure 42—2, which shows the approximate relationship between tensile strength and carbon content for a series of plain carbon steels isothermally transformed to fine and coarse pearlite microstructures. The line for the 650°C (1200°F) transformation product in this illustration is generally similar to, although slightly above, the tensile strength line for as-rolled bars in Figure 42—1, indicating that these as-rolled bars have transformed during cooling at temperatures in the vicinity of 650°C (1200°F)

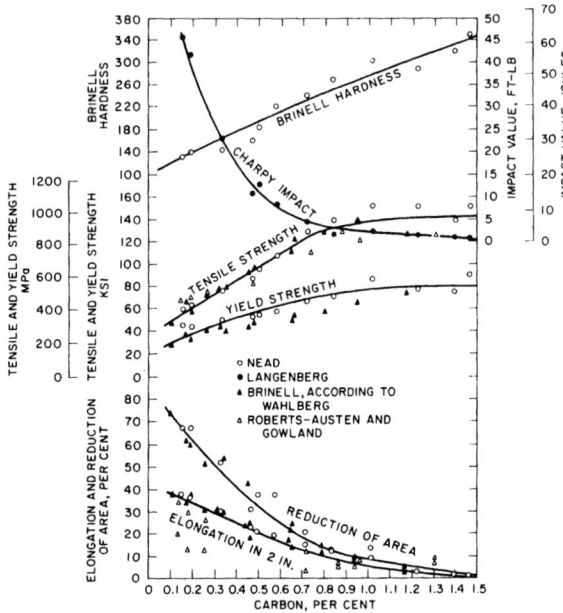

Fig. **42—1**. Variations in average mechanical properties of as-rolled, 25-mm (1-inch) diameter bars of plain carbon steels, as a function of carbon content.

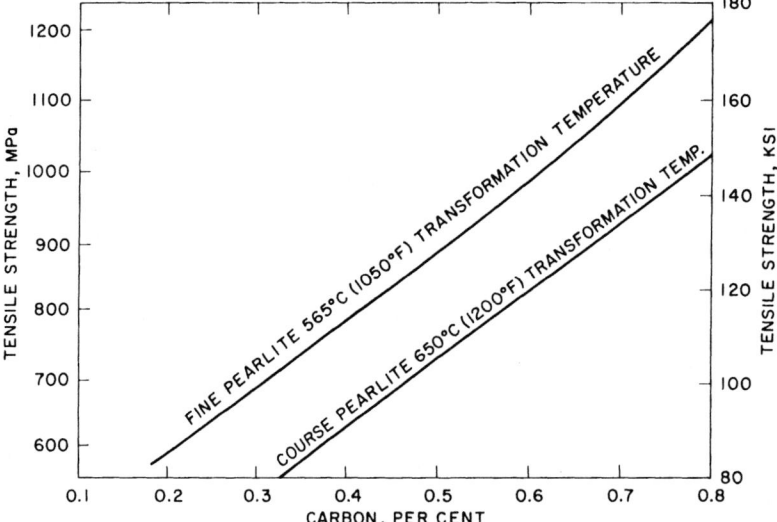

Fig. **42—2**. Relationship between tensile strength and carbon content of a series of plain carbon steels isothermally transformed to fine and coarse pearlitic microstructures.

SECTION 3

FACTORS AFFECTING MICROSTRUCTURE AND GRAIN SIZE

Composition—As explained in Chapter 41 on heat treatment, the microstructure of steel is determined by the temperature range in which transformation of the austenite takes place on cooling. This, in turn, is determined by the cooling rate employed and the transformation rate of the steel. This latter factor is dependent largely upon the composition; thus, for a given cooling rate after rolling, for instance, the resulting microstructure is largely dependent upon the composition. The composition will, of course, similarly control the microstructure for given cooling conditions in cast, as-rolled, or heat-treated carbon steels.

The austenite-transformation behavior in carbon steel is determined almost entirely by the carbon and manganese content; the effects of phosphorus and sulphur are almost negligible, and the silicon contents are normally so low that they are likewise ineffective. The carbon content is ordinarily chosen in accordance with the strength level desired and the manganese content then selected in order to produce suitable microstructure and properties at this carbon level under the given cooling conditions.

Microstructure of Cast Steels—The microstructure of as-cast steels is, of course, determined by the composition and cooling conditions in the same manner as in wrought steels. Cast steels are usually very coarse grained since the austenite forms at a high temperature, and the pearlite is usually coarse since the cooling through the critical range, particularly if the casting is cooled in the mold, is usually quite slow. In hypoeutectoid steels, ferrite is precipitated ordinarily at the original austenite boundaries during the cooling. In hypereutectoid steels, cementite is similarly precipitated. Such mixtures of ferrite or cementite and coarse-grained coarse pearlite have, as would be expected, poor properties both in respect to strength and ductility, and heat treatment is usually necessary to obtain suitable microstructures and properties in cast steels.

The dendritic segregation occurring during the solidification of steel castings also results in an irregular microstructure and correspondingly poor properties, and the homogenization of this segregated structure is another function of the heat treatment of cast steels.

A typical microstructure of an as-cast carbon steel is shown in Figure 42—3.

Effects of Hot Working—Many carbon steels are used in the form of as-rolled finished sections and the microstructure and properties of these sections are determined largely by the composition, rolling practice and cooling conditions after rolling. The rolling or hot working of these sections is ordinarily carried out in the temperature range at which the steel is austenitic and has four major effects, as follows:

1. Considerable homogenization that tends to eliminate dendritic segregation occurs during the heating for rolling.

2. The dendritic structure is broken up during rolling.

3. Recrystallization occurs during rolling so that the final austenitic grain size is determined by the temperature at which the last passes are made (the finishing temperature).

4. Grain structure and inclusions are reoriented in

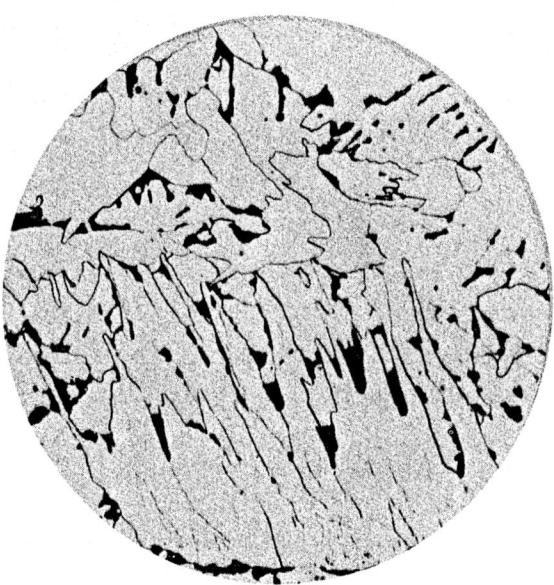

Fig. 42—3. As-cast microstructure of 0.20 per cent carbon steel. Nital etch; magnification: 200X.

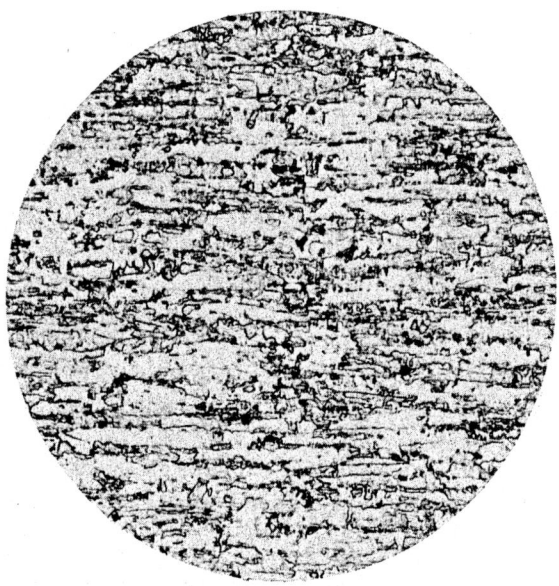

Fig. 42—4. Microstructure of full-hard cold-reduced black plate (85 per cent reduction). Nital etch; magnification: 200X.

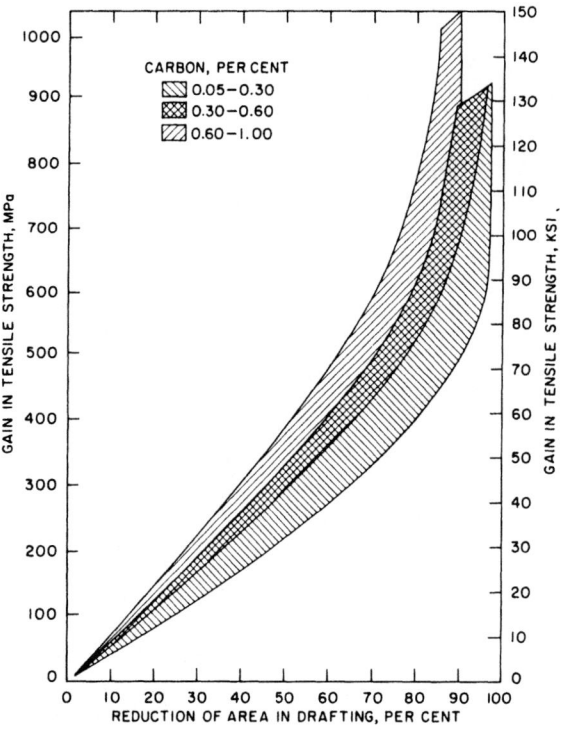

Fig. 42—5. Increase of tensile strength of plain carbon steel with increasing amounts of cold working.

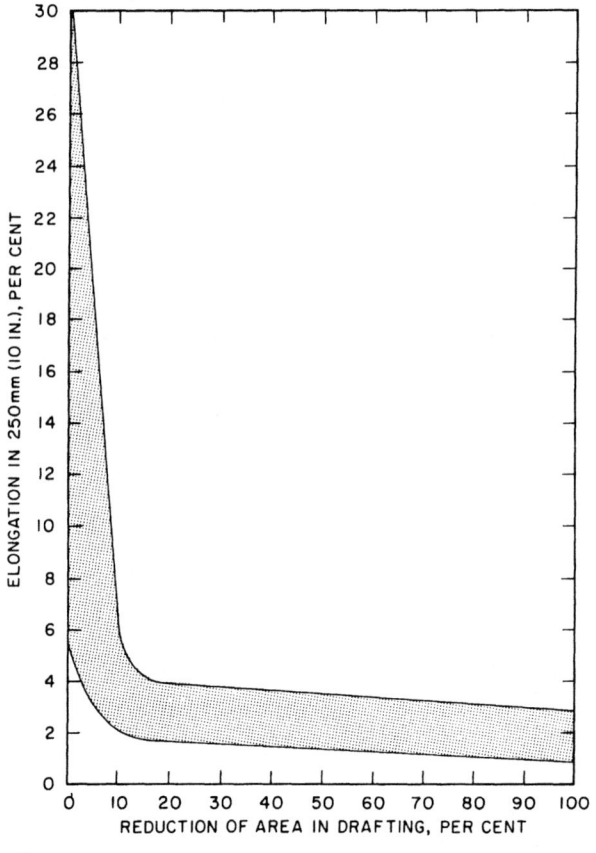

Fig. 42—6. Effect of cold working on the ductility of plain carbon steel.

the rolling direction so that the final ductility in the rolling direction is markedly improved.

Thus, homogeneity and grain size of the austenite are largely determined by the rolling practice. The distribution of the ferrite or cementite and the nature of the pearlite are, however, as has been explained earlier, determined by the cooling rate after rolling. Since the usual practice is air cooling, the final microstructure and, therefore, the properties of these as-rolled sections will be principally dependent on the composition and section size.

Effects of Cold Working—The manufacture of wire, sheet and strip, and tubular products often involves a cold-working operation and the general effects of cold working will, therefore, be discussed in this chapter. Some products, particularly wire, are used in the cold-worked condition. The effects of this cold work are essentially destroyed by annealing.

A typical microstructure of a heavily cold-worked steel is shown in Figure 42—4. The elongated and generally distorted microstructure is characteristic. The most pronounced effect of this cold work is an increase in strength and hardness and a decrease in ductility as represented by elongation and reduction of area. Effects of cold working on tensile strength and elongation are shown in Figures 42—5 and 42—6. Upon reheating cold-worked steel to the recrystallization temperature (400°C or 750°F) or above, depending upon composition, amount of cold work and other variables, the original microstructure and properties may be restored. The annealing of cold-worked steels (process annealing) has been discussed in the chapter on heat treatment.

SECTION 4

HEAT TREATMENT OF CARBON STEELS

Although the majority of carbon steels are used without a final heat treatment, heat treatment may be employed to improve the microstructure and properties for specific applications. The principles of these heat treatments have been discussed in Chapter 41 and many of the heat-treating practices have been described in detail in the chapters on the various carbon-

steel products. As mentioned earlier, the heat treatment of *cast* carbon steels improves the properties of the material by breaking up the dendritic structure and refining the grain size and microstructure. These treatments usually involve normalizing the castings at a high temperature to homogenize the dendritic structure, followed by annealing at a lower temperature for grain refinement. The heat treatments employed in processing *wrought* steel products are described in the following paragraphs.

Annealing—Annealing is practiced for applications requiring better machinability or formability than would be obtained with the as-rolled microstructure. This is usually a full anneal to form coarse pearlite, although a sub-critical anneal or spheroidizing treatment is also utilized for steels that are to be subjected to severe cold-forming processes. Process annealing to obtain optimum formability in cold-rolled strip, sheet and tubing is, of course, a universal practice.

Normalizing—The grain size of hot-rolled products is, as described above, largely dependent upon the finishing temperature in rolling and this is difficult to control. Therefore, a final normalizing treatment from a relatively low austenitizing temperature may be used to establish a fine uniform grain size for critical applications in respect to ductility or toughness.

Quenching and Tempering—Quenching and tempering is a common practice used to raise the strength levels of plain carbon steels in combination with optimum ductility. Because of the relatively low hardenability of these steels, this type of treatment falls generally into two classifications, as follows:

1. Heat treatment to produce essentially tempered martensite for optimum properties. Hardenability restrictions generally limit the application of this type of treatment to section sizes of not more than 9.5 to 12.5 mm (⅜ to ½ inch). The treatment is commonly used for small tools, fasteners, sheet and strip, seamless tubes, rods, etc., within this size range.

2. Heat treatment to form fine pearlite. Quite large sections of plain carbon steels may be quenched and tempered to produce fine pearlite microstructures, thereby making available the improved strength and ductility associated with this microstructure compared with the properties of the coarse pearlite of the usual as-rolled or normalized products.

Austempering—Austempering is used to improve the toughness of certain steels of high hardness by producing bainite instead of the more crack-sensitive martensite. Thin sections, 5-mm (0.2-inch) and below, of carbon steels are particularly suitable for austempering, since the times for transformation to bainite are relatively short.

Boron-Containing Steels—Boron-containing carbon steels provide hardenability characteristics that are required for a wide variety of heat-treated parts. Medium-carbon steels containing boron are used in many applications, often in place of more costly alloy steels. As indicated in Chapter 43 on alloy steels, very small quantities of boron, as little 0.0005 per cent, increase hardenability markedly. Although such small quantities have practically no effect on the strength of hot-rolled or annealed carbon steels, boron has been used effectively in low-carbon steels supplied for cold-headed or cold-extruded products which are subsequently heat-treated. For such applications, United States Steel developed a series of boron-treated carbon (and alloy) steels, identified by the trademark "Q-Temp", which are low-carbon grades designed to replace higher-carbon steels for parts heat treated to hardness as high as 40 Rockwell "C". Because of their low carbon content, Q-Temp steels, in many instances, do not require annealing prior to cold heading or forward extruding.

Generally similar grades of boron-containing low-carbon and medium-carbon steels are also supplied as hot-rolled sheets to provide a low-cost replacement for high-carbon and low-alloy sheets and strip. The lower carbon boron steels have better cold-forming characteristics and can be heat treated to equivalent hardness and improved toughness for a wide variety of applications such as tools, machinery components, and fasteners.

SECTION 5

AGING IN CARBON STEELS

Quench Aging—Aging in carbon steels falls into two categories: quench aging and strain aging. Quench aging is a consequence of the precipitation of carbon, nitrogen, or both, from supersaturated solid solution in ferrite. As shown in Figure 42—7, the solubilities of these two elements, which occupy interstitial positions in the alpha-iron lattice, decrease sharply with decreasing temperature, to vanishingly small values at room temperature. Because carbon steels contain much more carbon than nitrogen, carbon is the principal agent in quench aging. If carbon is retained in solid solution by rapid cooling, it can precipitate subsequently at ambient or slightly elevated temperatures, as finely dispersed particles of iron carbide. The consequences of such precipitation are manifested in in-creased hardness, increased yield and tensile strength, decreased elongation and reduction of area in a tension test, increased temperature of transition from ductile to brittle fracture in a notch-impact test, an increase in coercive force and a decrease in magnetic permeability.

The general nature of the hardness increase from quench aging is shown in Figure 42—8. It will be noted that aging above room temperature results in a more rapid hardness increase, but that the maximum hardness attained is lower than in steel aged at room temperature. Aging for periods beyond that corresponding to the maximum hardness results in a decrease in hardness, a phenomenon called **overaging**. During the process of precipitation, carbide particles nucleate,

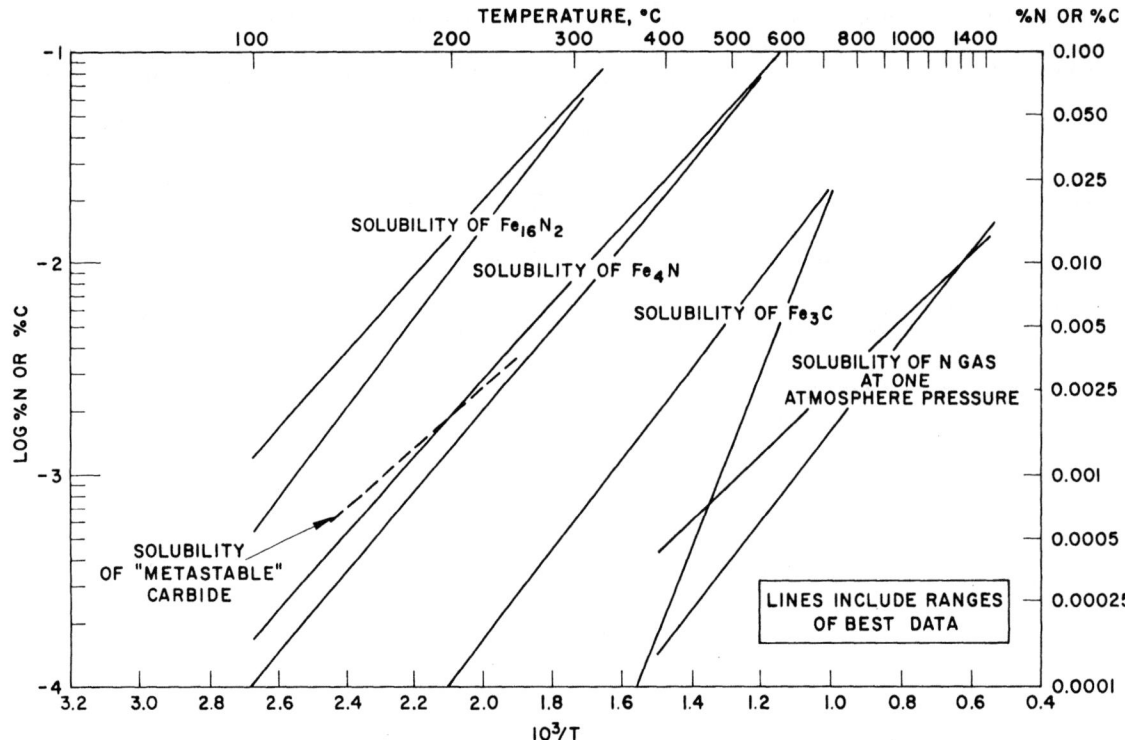

Fig. 42—7. Solubility of nitrogen, iron nitrides and iron carbides in ferrite. (After Leslie and Keh: see bibliography.)

grow, and increase in number. At a certain combination of particle size and interparticle spacing, the effect on various mechanical properties is at a maximum. Further aging causes the particles to coalesce and to become more widely separated, and the effects on mechanical properties decrease. The effects of aging on tensile properties follow the same pattern as the changes in hardness, as indicated in Figure 42—9.

The coercivity and permeability of a low-carbon steel can also be strongly affected by the precipitation of carbides. However, the increase in coercivity and the decrease in permeability reach their maxima after much longer periods than are required for the maximum increase in strength. Examples of this can be seen by comparing Figures 42—8 and 42—9 with Figure 42—10. The precipitate particle size having the greatest effect on magnetic properties is much larger than the particle size having the greatest effect on mechanical properties.

The elimination of quench aging in steels is simple in

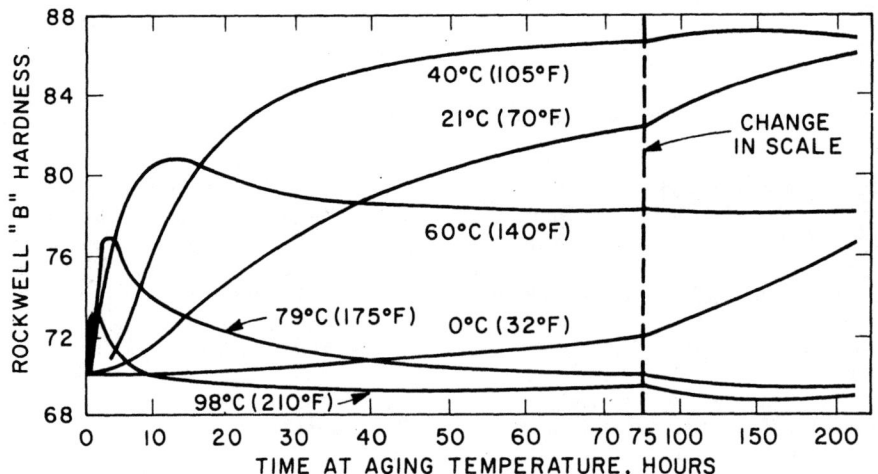

Fig. 42—8. Changes in hardness of 0.06 per cent carbon steel quenched from 720°C (1325°F) after aging at indicated temperatures. (From "Metals Handbook," 1948 Edition; American Society for Metals.)

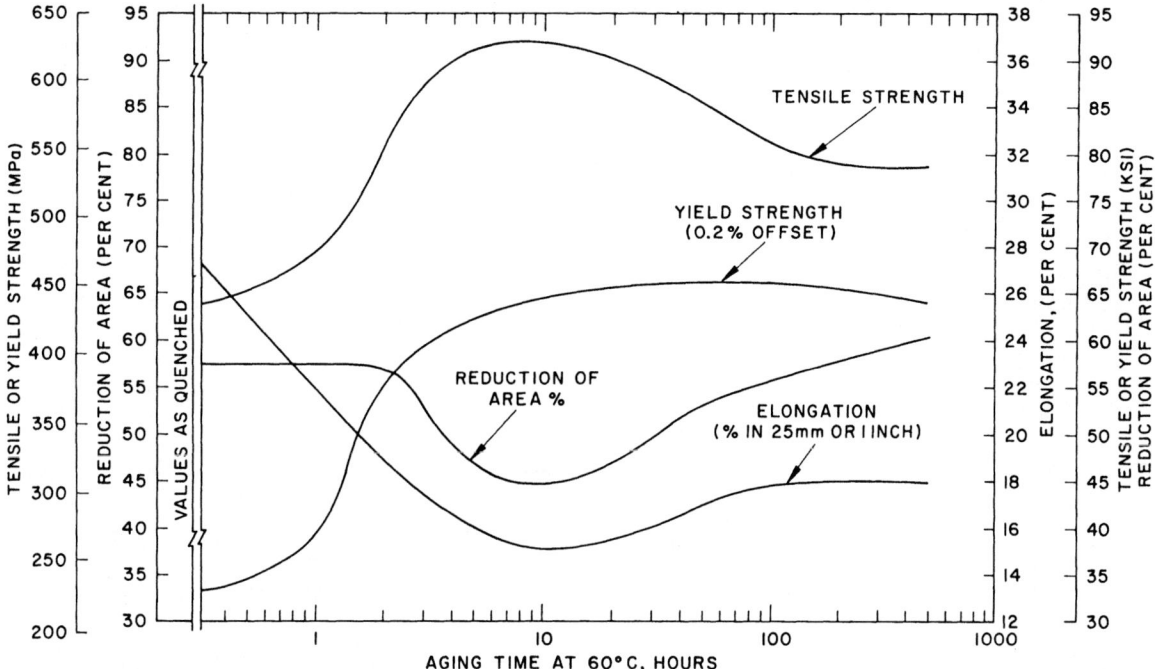

Fig. 42—9. Changes in tensile properties during the aging of 0.03 per cent carbon rimmed steel, after quenching from 725°C (1335°F). (After Leslie and Keh: see bibliography.)

principle. The carbon and nitrogen in supersaturated solid solution in the final product must be minimized, since the extent of aging is proportional to the concentration of these elements in solid solution. The customary and most economical procedure for achieving this goal is to cool the steel relatively slowly, especially through the temperature range from 540° to 315°C (1000° to 600°F). A matter of a few minutes in this range is usually sufficient to eliminate subsequent quench aging. Quench aging also can be eliminated by removing nearly all of the carbon and nitrogen by a decarburiza-

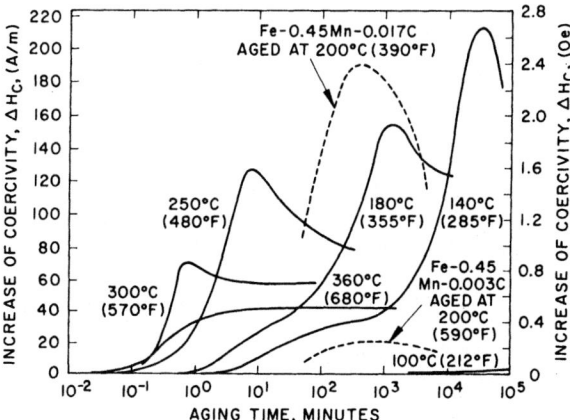

Fig. 42—10. Dashed lines compare change of coercivity of an Fe-Mn-C alloy (0.45 per cent Mn, 0.017 per cent C) aged at 200°C (390°F) with that of the same alloy decarburized to 0.003 per cent C. Solid lines show changes in coercivity of an Fe-C alloy (0.019 per cent C) with time at various temperatures. (After Leslie and Keh: see bibliography.)

tion treatment in the solid state, or by forming stable carbides and nitrides by alloying the steel with elements such as titanium and zirconium. The former procedure is generally considered impractical. The latter is currently in use.

Strain Aging—Strain aging differs from quench aging in that plastic deformation is necessary before the aging process can begin. Also, unlike quench aging, a supersaturated solution of carbon or nitrogen in ferrite is not essential for strain aging. However, if such supersaturation is present, the rate and extent of strain aging will be enhanced. Strain aging and quench aging can occur simultaneously. Because of its greater solubility in iron at low temperatures as compared to carbon, nitrogen is the element usually responsible for strain aging in steels, but carbon is effective whenever the aging temperature is high (above about 100°C or 212°F) or whenever the concentration of carbon in solution is high.

The cause of strain aging is the segregation of interstitial solute atoms (carbon and nitrogen) to the strain fields of dislocations in the alpha-iron lattice. Within these fields, the regular periodicity of the lattice is disrupted, and the interstitial sites can be larger than normal. Within the dislocation strain fields, nitrogen and carbon atoms can migrate from nitride or carbide particles to disclocation sites. When this occurs, the dislocations are fixed in place. Subsequent plastic deformation requires a higher stress.

Most of the manifestations of strain aging are similar to those of quench aging—increased hardness, yield and tensile strength, reduced ductility and increased notch-impact transition temperature. In addition, strain aging has another troublesome characteristic. Low-carbon sheet steel often is temper rolled to elimi-

ΔY = CHANGE IN YIELD STRESS DUE TO STRAIN AGING
ΔU = CHANGE IN ULTIMATE TENSILE STRENGTH DUE TO STRAIN AGING
ΔE = CHANGE IN ELONGATION DUE TO STRAIN AGING

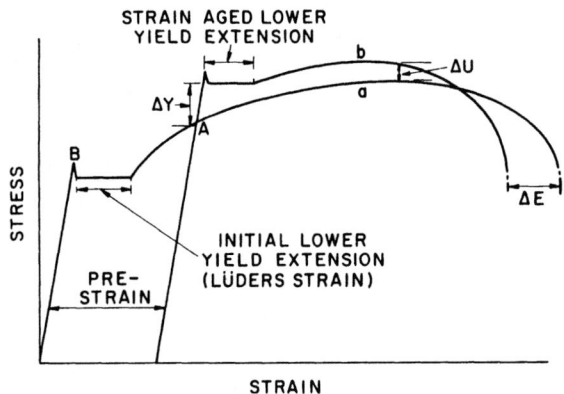

Fig. 42—11. Stress-strain curve for annealed or normalized low-carbon steel strained to point A, unloaded, and then immediately restrained (Curve a) and after aging (Curve b). (After Baird: see bibliography.)

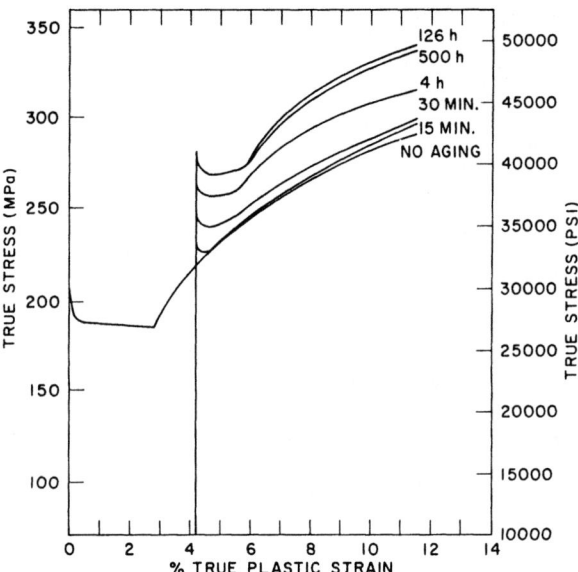

Fig. 42—12. Strain aging of a 0.03 per cent carbon rimmed steel at 60°C (140°F).

nate the abrupt yield point elongation characteristic of this material. After temper rolling, the sheet can be formed with uniform yielding and related smooth contours. If aging takes place after temper rolling, the abrupt yield point returns, and the sheet is then susceptible to discontinuous yielding, "fluting," and "stretcher strains" on subsequent deformation. Flexing the sheet by effective roller leveling just prior to forming will minimize this susceptibility.

The most common measure of strain aging is the increase in yield strength, ΔY in Figure 42—11. Strain

Table 42—IV. Aging Times at Several Temperatures Required to Produce Approximately Equal Aging Effects.

0°C (32°F)	21°C (70°F)	100°C (212°F)	120°C (250°F)	150°C (300°F)
1 year	6 months	4 hours	1 hour	10 min.
6 months	3 months	2 hours	30 min.	5 min.
3 months	6 weeks	1 hour	15 min.	2½ min.
1 month	2 weeks	20 min.	5 min.	
1 week	4 days	5 min.		
3 days	36 hours	2 min.		

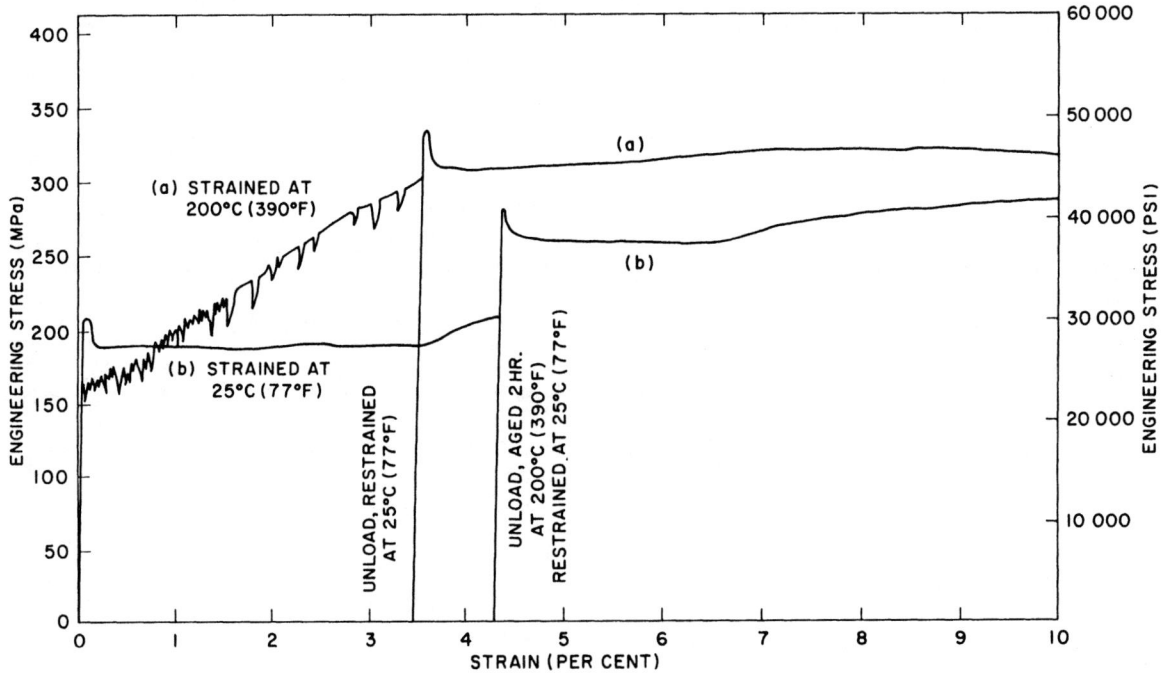

Fig. 42—13. (a) Dynamic strain aging versus (b) static strain aging of 0.03 per cent carbon rimmed steel at 200°C (390°F). (After Leslie and Keh: see bibliography.)

aging differs from quench aging in that there is little or no overaging, as can be seen from Figure 42—12. The rate at which strain aging can occur depends upon the concentration and the rate of diffusion of carbon and nitrogen at the aging temperature. As the aging temperature increases, the rate of aging increases sharply. Table 42—IV gives the aging times at several temperatures to produce approximately equal aging effects.

In using this table, it should be remembered that a low-carbon rimmed steel will be fully aged after 1 to 2 hours at 100°C (212°F).

If a specimen is strained at a temperature (150° to 350°C, or 300° to 660°F) at which carbon and nitrogen can diffuse to dislocations during the duration of the test, strain aging will occur during the test. This phenomenon is called **dynamic strain aging,** and less frequently, **blue brittleness.** It is characterized by sharply serrated stress-strain curves, as illustrated in Figure 42—13. This process can be used to strengthen steels economically. Straining at about 200°C (390°F) is more effective in strengthening than straining at room temperature and then aging.

It is extremely difficult to eliminate strain aging by removing nitrogen and carbon from steel, although the extent of aging can be reduced greatly by this procedure. The most common technique for minimizing strain aging is to add aluminum to the steel in amounts such that about 0.05 per cent aluminum is retained in the product. With proper treatment, the aluminum combines with nitrogen to form AlN. The removal of nitrogen from solution effectively prevents strain aging at room temperature. If the aging temperature is raised to about 100°C (212°F) or higher, the steel will strain age because of the action of carbon. Other alloying elements which form stable nitrides can be used in place of aluminum.

The quench aging and strain aging of steels can be effective techniques for strengthening in applications where ductility and toughness are of secondary importance, and the knowledgeable application of aging to achieve strengthening is expected to increase in the future.

SECTION 6

EFFECT OF RESIDUAL ELEMENTS

In addition to the carbon, manganese, phosphorus, sulphur and silicon which are always present, carbon steels may contain small amounts of other elements. These include gases, such as hydrogen, oxygen or nitrogen which are introduced during the steelmaking process, nickel, copper, molybdenum, chromium and tin which may be present in the scrap, and aluminum, titanium, vanadium or zirconium which may be introduced during the deoxidation process.

The effects of nitrogen have been discussed in the section on aging. In addition to its effect on aging, nitrogen has a strong strengthening effect in carbon steels and is added in low-carbon nitrogenized hot-rolled and cold-rolled sheet steels to provide high-strength grades with moderately good formability at low cost.

In steel, hydrogen has a definite embrittling effect, the mechanism of which is not well understood. Although hydrogen may diffuse out of steel at room temperature if the section is small enough and sufficient time is allowed, tension tests on carbon steels containing hydrogen will show low ductility if made soon after rolling. After sufficient time at room temperature, the ductility will again increase, with shorter times being required at slightly elevated temperatures. This effect is, of course, more pronounced in larger section sizes because of the longer time required for the diffusion of hydrogen to the surface. Hydrogen contents of more than about 0.0005 per cent will give rise to this effect and such contents are common in as-rolled steels that have been cast into ingots in the conventional way. It is possible to reduce the content of hydrogen (and other gases) in steel ingots by employing the vacuum degas-

sing processes, such as those described in Chapter 20.

Hydrogen, in excess of about five parts per million (0.0005 per cent), also plays an important role in the phenomenon known as **flaking** which is manifested as internal cracks or bursts, usually occurring during the cooling from rolling or forging. The phenomenon is more pronounced in heavy sections and in the higher carbon steels. In carbon steels, flaking may be prevented by slow cooling after rolling or forging. This slow-cooling operation presumably permits the hydrogen to diffuse out of the steel and thereby minimizes the susceptibility to flaking. Such a controlled slow-cooling operation after rolling is now standard practice in the manufacture of rails, and this practice has practically eliminated the occurrence of flaking and resultant occasional "transverse fissure" failures.

The alloying elements, such as nickel, chromium, molybdenum and copper, which may be introduced in the scrap will, of course, increase the hardenability of carbon steels, although, since the percentages are ordinarily low, this effect will usually not be large. It may, however, change the heat-treating characteristics and for applications in which ductility is important, such as steels for deep drawing, the increased hardness from these residual elements may be serious.

Tin in relatively low amounts is harmful in steels for deep drawing, but for most applications the effect of tin in the amounts ordinarily present is negligible.

Aluminum, as described above, is generally desirable since it acts as a grain refiner and tends to decrease the susceptibility to strain aging. It has the disadvantage, however, that it tends to promote graphitization and is, therefore, undesirable in steels to be used for high-temperature applications. The other elements which may be introduced as deoxidizers, titanium or zirconium, are, unless intentionally added, ordinarily present in such small amounts as to be generally ineffective.

Bibliography

American Society for Metals, Metals handbook (1948 and 1961 eds.). The Society, Metals Park, Ohio.

Baird, J. D., Strain aging of steel—a critical review; Part I: Practical aspects. Iron and Steel 36, May 1963, 186-192.

Leslie, W. C., and A. S. Keh, Aging of flat-rolled steel products as investigated by electron microscopy. Mechanical Working of Steel II, T. G. Bradbury, Ed. Published for AIME in New York by Gordon and Breach Science Publishers, Inc., 1965.

Sisco, Frank T., Alloys of iron and carbon: Vol. II—properties. Published for the Engineering Foundation by McGraw-Hill Book Co., Inc., New York, 1937.

CHAPTER 43
Constructional Alloy and Ultraservice Steels

SECTION 1
FUNCTIONS OF THE ALLOYING ELEMENTS

Alloy steels may be defined as those steels which owe their enhanced properties to the presence of one or more special elements or to the presence of larger proportions of elements such as manganese and silicon than are ordinarily present in carbon steel.

This chapter covers the **constructional alloy steels** identified in the Unified Numbering System (UNS) by designations between G13xxx and G94xxx, inclusive, which formerly were (and often still are) referred to as SAE-AISI alloy steels. In addition, **quenched-and-tempered low-carbon constructional alloy steels, quenched-and-tempered low-carbon ultraservice steels** and **maraging steels** are discussed.

Other major classifications of steels with special properties developed by the addition of alloying elements include: high-strength low-alloy steels, alloy tool steels, stainless steels, heat-resistant steels and electrical steels (silicon steels). Each of these five types of alloy steels is discussed in other chapters.

As stated above, alloying elements are added to steel to enhance its properties. In the broadest sense, alloy steels may contain up to approximately 50 per cent of alloying elements, and the enhancement of properties may be a specific and direct function of the alloying elements, as in the instances of the increased corrosion resistance of the high-chromium steels and the enhanced electrical properties of the silicon steels. In the narrower and more technical sense, however, the term "alloy steels" refers to the heat-treatable alloy constructional and automotive steels which contain from about one to three or four per cent alloying elements. The American Iron and Steel Institute definition of alloy steel is as follows: "By common custom steel is considered to be alloy steel when the maximum of the range given for the content of alloying elements exceeds one or more of the following limits: manganese, 1.65 per cent; silicon, 0.60 per cent; copper, 0.60 per cent; or in which a definite range or a definite minimum quantity of any of the following elements is specified or required within the limits of the recognized field of constructional alloy steels: aluminum, chromium up to 3.99 per cent, cobalt, columbium, molybdenum, nickel, titanium, tungsten, vanadium, zirconium, or any other alloying element added to obtain a desired alloying

effect." It may be noted that steels that contain 4.00 or more per cent of chromium are included by convention among the special types of alloy steels known as stainless steels (Chapter 47).

The constructional alloy steels originally standardized and classified jointly by the American Iron and Steel Institute and the Society of Automotive Engineers represent by far the largest tonnage of alloy steels. The composition of these steels as published in Part I of the 1982 "SAE Handbook" is shown in Table 43—I, where they are identified by their UNS designations as well as their former SAE or AISI numbers. Small quantities of certain elements are present in alloy steels which are not specified or required. These elements are considered as incidental and may be present to the following maximum amounts: copper, 0.35 per cent; nickel, 0.25 per cent; chromium, 0.20 per cent; and molybdenum, 0.06 per cent.

As was emphasized in the chapter on heat treatment, the mechanical properties of steel are dependent upon its microstructure. In the constructional (SAE-AISI) alloy steels the effect of the alloying is indirect; i.e., through their influence on the microstructure of the material. The alloy contents of these alloy steels make it possible to attain desirable microstructures and corresponding desirable properties over a very much wider range of sizes and sections than is possible with the carbon steels.

Hardenability—The mechanism by which the alloying elements affect the microstructure obtained with a given heat treatment is discussed in the chapter on heat treatment. It is shown that the alloying elements in general decrease the rates of transformation of austenite at sub-critical temperatures, thereby facilitating the attainment of low-temperature transformation to martensite or lower bainite when these are the end products desired, without prior transformation to unwanted higher temperature products. It was pointed out that this function of the alloying elements could be evaluated and expressed in terms of the property known as hardenability. Alloying elements thus control microstructure through their effect on hardenability, and this hardenability effect is by far their most important function.

It was also shown in the chapter on heat treatment that the properties of steel were characteristic of the microstructure rather than the composition, with tempered martensite being the most desirable microstructure in respect to strength and toughness. Thus, alloy steels of equal hardenabilities, but utilizing different combinations of alloying elements, are generally interchangeable for heat treatment to produce this microstructure. This principle permits an intelligent choice of alloy combinations which, for reasons of econ-

Table 43—I. Alloy Steel Compositions[a,d,e,f]

UNS No.	SAE No.	C	Mn	P	S	Si	Ni	Cr	Mo	V	Corresponding AISI No.
		Ladle Chemical Composition Limits, %									
G13300	1330	0.28-0.33	1.60-1.90	0.035	0.040	0.15-0.35	—	—	—	—	1330
G13350	1335	0.33-0.38	1.60-1.90	0.035	0.040	0.15-0.35	—	—	—	—	1335
G13400	1340	0.38-0.43	1.60-1.90	0.035	0.040	0.15-0.35	—	—	—	—	1340
G13450	1345	0.43-0.48	1.60-1.90	0.035	0.040	0.15-0.35	—	—	—	—	1345
G40230	4023	0.20-0.25	0.70-0.90	0.035	0.040	0.15-0.35	—	—	0.20-0.30	—	4023
G40240	4024	0.20-0.25	0.70-0.90	0.035	0.035-0.050	0.15-0.35	—	—	0.20-0.30	—	4024
G40270	4027	0.25-0.30	0.70-0.90	0.035	0.040	0.15-0.35	—	—	0.20-0.30	—	4027
G40280	4028	0.25-0.30	0.70-0.90	0.035	0.035-0.050	0.15-0.35	—	—	0.20-0.30	—	4028
G40320	4032	0.30-0.35	0.70-0.90	0.035	0.040	0.15-0.35	—	—	0.20-0.30	—	—
G40370	4037	0.35-0.40	0.70-0.90	0.035	0.040	0.15-0.35	—	—	0.20-0.30	—	4037
G40420	4042	0.40-0.45	0.70-0.90	0.035	0.040	0.15-0.35	—	—	0.20-0.30	—	—
G40470	4047	0.45-0.50	0.70-0.90	0.035	0.040	0.15-0.35	—	—	0.20-0.30	—	4047
G41180	4118	0.18-0.23	0.70-0.90	0.035	0.040	0.15-0.35	—	0.40-0.60	0.08-0.15	—	4118
G41300	4130	0.28-0.33	0.40-0.60	0.035	0.040	0.15-0.35	—	0.80-1.10	0.15-0.25	—	4130
G41350	4135	0.33-0.38	0.70-0.90	0.035	0.040	0.15-0.35	—	0.80-1.10	0.15-0.25	—	—
G41370	4137	0.35-0.40	0.70-0.90	0.035	0.040	0.15-0.35	—	0.80-1.10	0.15-0.25	—	4137
G41400	4140	0.38-0.43	0.75-1.00	0.035	0.040	0.15-0.35	—	0.80-1.10	0.15-0.25	—	4140
G41420	4142	0.40-0.45	0.75-1.00	0.035	0.040	0.15-0.35	—	0.80-1.10	0.15-0.25	—	4142
G41450	4145	0.43-0.48	0.75-1.00	0.035	0.040	0.15-0.35	—	0.80-1.10	0.15-0.25	—	4145
G41470	4147	0.45-0.50	0.75-1.00	0.035	0.040	0.15-0.35	—	0.80-1.10	0.15-0.25	—	4147
G41500	4150	0.48-0.53	0.75-1.00	0.035	0.040	0.15-0.35	—	0.80-1.10	0.15-0.25	—	4150
G41610	4161	0.56-0.64	0.75-1.00	0.035	0.040	0.15-0.35	—	0.70-0.90	0.25-0.35	—	4161
G43200	4320	0.17-0.22	0.45-0.65	0.035	0.040	0.15-0.35	1.65-2.00	0.40-0.60	0.20-0.30	—	4320
G43400	4340	0.38-0.43	0.60-0.80	0.035	0.040	0.15-0.35	1.65-2.00	0.70-0.90	0.20-0.30	—	4340
G43406	E4340[b]	0.38-0.43	0.65-0.85	0.025	0.025	0.15-0.35	1.65-2.00	0.70-0.90	0.20-0.30	—	E4340
G44220	4422	0.20-0.25	0.70-0.90	0.035	0.040	0.15-0.35	—	—	0.35-0.45	—	—
G44270	4427	0.24-0.29	0.70-0.90	0.035	0.040	0.15-0.35	—	—	0.35-0.45	—	—
G46150	4615	0.13-0.18	0.45-0.65	0.035	0.040	0.15-0.35	1.65-2.00	—	0.20-0.30	—	4615
G46170	4617	0.15-0.20	0.45-0.65	0.035	0.040	0.15-0.35	1.65-2.00	—	0.20-0.30	—	—
G46200	4620	0.17-0.22	0.45-0.65	0.035	0.040	0.15-0.35	1.65-2.00	—	0.20-0.30	—	4620
G46260	4626	0.24-0.29	0.45-0.65	0.035	0.04 max.	0.15-0.35	0.70-1.00	—	0.15-0.25	—	4626
G47180	4718	0.16-0.21	0.70-0.90	—	—	—	0.90-1.20	0.35-0.55	0.30-0.40	—	4718
G47200	4720	0.17-0.22	0.50-0.70	0.035	0.040	0.15-0.35	0.90-1.20	0.35-0.55	0.15-0.25	—	4720
G48150	4815	0.13-0.18	0.40-0.60	0.035	0.040	0.15-0.35	3.25-3.75	—	0.20-0.30	—	4815
G48170	4817	0.15-0.20	0.40-0.60	0.035	0.040	0.15-0.35	3.25-3.75	—	0.20-0.30	—	4817
G48200	4820	0.18-0.23	0.50-0.70	0.035	0.040	0.15-0.35	3.25-3.75	—	0.20-0.30	—	4820
G50401	50B40[c]	0.38-0.43	0.75-1.00	0.035	0.040	0.15-0.35	—	0.40-0.60	—	—	—
G50441	50B44[c]	0.43-0.48	0.75-1.00	0.035	0.040	0.15-0.35	—	0.40-0.60	—	—	50B44
G50460	5046	0.43-0.48	0.75-1.00	0.035	0.040	0.15-0.35	—	0.20-0.35	—	—	—
G50461	50B46[c]	0.44-0.49	0.75-1.00	0.035	0.040	0.15-0.35	—	0.20-0.35	—	—	50B46
G50501	50B50[c]	0.48-0.53	0.75-1.00	0.035	0.040	0.15-0.35	—	0.40-0.60	—	—	50B50
G50600	5060	0.56-0.64	0.75-1.00	0.035	0.040	0.15-0.35	—	0.40-0.60	—	—	—
G50601	50B60[c]	0.56-0.64	0.75-1.00	0.035	0.040	0.15-0.35	—	0.40-0.60	—	—	50B60
G51150	5115	0.13-0.18	0.70-0.90	0.035	0.040	0.15-0.35	—	0.70-0.90	—	—	—
G51170	5117	0.15-0.20	0.70-0.90	0.040	0.040	0.15-0.35	—	0.70-0.90	—	—	5117
G51200	5120	0.17-0.22	0.70-0.90	0.035	0.040	0.15-0.35	—	0.70-0.90	—	—	5120
G51300	5130	0.28-0.33	0.70-0.90	0.035	0.040	0.15-0.35	—	0.80-1.10	—	—	5130
G51320	5132	0.30-0.35	0.60-0.80	0.035	0.040	0.15-0.35	—	0.75-1.00	—	—	5132
G51350	5135	0.33-0.38	0.60-0.80	0.035	0.040	0.15-0.35	—	0.80-1.05	—	—	5135
G51400	5140	0.38-0.43	0.70-0.90	0.035	0.040	0.15-0.35	—	0.70-0.90	—	—	5140
G51470	5147	0.46-0.51	0.70-0.95	0.035	0.040	0.15-0.35	—	0.85-1.15	—	—	5147
G51500	5150	0.48-0.53	0.70-0.90	0.035	0.040	0.15-0.35	—	0.70-0.90	—	—	5150
G51550	5155	0.51-0.59	0.70-0.90	0.035	0.040	0.15-0.35	—	0.70-0.90	—	—	5155
G51600	5160	0.56-0.64	0.75-1.00	0.035	0.040	0.15-0.35	—	0.70-0.90	—	—	5160
G51601	51B60[c]	0.56-0.64	0.75-1.00	0.035	0.040	0.15-0.35	—	0.70-0.90	—	—	51B60

(Continued on next page)

Table 43—I (Continued)

UNS No.	SAE No.	Ladle Chemical Composition Limits, %									Corresponding AISI No.
		C	Mn	P	S	Si	Ni	Cr	Mo	V	
G50986	50100[b]	0.98-1.10	0.25-0.45	0.025	0.025	0.15-0.35	—	0.40-0.60	—	—	—
G51986	51100[b]	0.98-1.10	0.25-0.45	0.025	0.025	0.15-0.35	—	0.90-1.15	—	—	E51100
G52986	52100[b]	0.98-1.10	0.25-0.45	0.025	0.025	0.15-0.35	—	1.30-1.60	—	—	E52100
G61180	6118	0.16-0.21	0.50-0.70	0.035	0.040	0.15-0.35	—	0.50-0.70	—	0.10-0.15	6118
G61500	6150	0.48-0.53	0.70-0.90	0.035	0.040	0.15-0.35	—	0.80-1.10	—	0.15 min	6150
G81150	8115	0.13-0.18	0.70-0.90	0.035	0.040	0.15-0.35	0.20-0.40	0.30-0.50	0.08-0.15	—	8115
G81451	81B45[c]	0.43-0.48	0.75-1.00	0.035	0.040	0.15-0.35	0.20-0.40	0.35-0.55	0.08-0.15	—	81B45
G86150	8615	0.13-0.18	0.70-0.90	0.035	0.040	0.15-0.35	0.40-0.70	0.40-0.60	0.15-0.25	—	8615
G86170	8617	0.15-0.20	0.70-0.90	0.035	0.040	0.15-0.35	0.40-0.70	0.40-0.60	0.15-0.25	—	8617
G86200	8620	0.18-0.23	0.70-0.90	0.035	0.040	0.15-0.35	0.40-0.70	0.40-0.60	0.15-0.25	—	8620
G86220	8622	0.20-0.25	0.70-0.90	0.035	0.040	0.15-0.35	0.40-0.70	0.40-0.60	0.15-0.25	—	8622
G86250	8625	0.23-0.28	0.70-0.90	0.035	0.040	0.15-0.35	0.40-0.70	0.40-0.60	0.15-0.25	—	8625
G86270	8627	0.25-0.30	0.70-0.90	0.035	0.040	0.15-0.35	0.40-0.70	0.40-0.60	0.15-0.25	—	8627
G86300	8630	0.28-0.33	0.70-0.90	0.035	0.040	0.15-0.35	0.40-0.70	0.40-0.60	0.15-0.25	—	8630
G86370	8637	0.35-0.40	0.75-1.00	0.035	0.040	0.15-0.35	0.40-0.70	0.40-0.60	0.15-0.25	—	8637
G86400	8640	0.38-0.43	0.75-1.00	0.035	0.040	0.15-0.35	0.40-0.70	0.40-0.60	0.15-0.25	—	8640
G86420	8642	0.40-0.45	0.75-1.00	0.035	0.040	0.15-0.35	0.40-0.70	0.40-0.60	0.15-0.25	—	8642
G86450	8645	0.43-0.48	0.75-1.00	0.035	0.040	0.15-0.35	0.40-0.70	0.40-0.60	0.15-0.25	—	8645
G86451	86B45[c]	0.43-0.48	0.75-1.00	0.035	0.040	0.15-0.35	0.40-0.70	0.40-0.60	0.15-0.25	—	—
G86500	8650	0.48-0.53	0.75-1.00	0.035	0.040	0.15-0.35	0.40-0.70	0.40-0.60	0.15-0.25	—	—
G86550	8655	0.51-0.59	0.75-1.00	0.035	0.040	0.15-0.35	0.40-0.70	0.40-0.60	0.15-0.25	—	8655
G86600	8660	0.56-0.64	0.75-1.00	0.035	0.040	0.15-0.35	0.40-0.70	0.40-0.60	0.15-0.25	—	—
G87200	8720	0.18-0.23	0.70-0.90	0.035	0.040	0.15-0.35	0.40-0.70	0.40-0.60	0.20-0.30	—	8720
G87400	8740	0.38-0.43	0.75-1.00	0.035	0.040	0.15-0.35	0.40-0.70	0.40-0.60	0.20-0.30	—	8740
G88220	8822	0.20-0.25	0.75-1.00	0.035	0.040	0.15-0.35	0.40-0.70	0.40-0.60	0.30-0.40	—	8822
G92540	9254	0.51-0.59	0.60-0.80	0.035	0.040	1.20-1.60	—	0.60-0.80	—	—	—
G92600	9260	0.56-0.64	0.75-1.00	0.035	0.040	1.80-2.20	—	—	—	—	9260
G93106	9310[b]	0.08-0.13	0.45-0.65	0.025	0.025	0.15-0.35	3.00-3.50	1.00-1.40	0.08-0.15	—	—
G94151	94B15[c]	0.13-0.18	0.75-1.00	0.035	0.040	0.15-0.35	0.30-0.60	0.30-0.50	0.08-0.15	—	—
G94171	94B17[c]	0.15-0.20	0.75-1.00	0.035	0.040	0.15-0.35	0.30-0.60	0.30-0.50	0.08-0.15	—	94B17
G94301	94B30[c]	0.28-0.33	0.75-1.00	0.035	0.040	0.15-0.35	0.30-0.60	0.30-0.50	0.08-0.15	—	94B30

[a] For standard variations in composition limits, see Table 4 of SAE J409. Small quantities of certain elements which are not specified or required may be found in alloy steels. These elements are to be considered as incidental and are acceptable to the following maximum amount: copper to 0.35%, nickel to 0.25%, chromium to 0.20%, and molybdenum to 0.06%.

[b] Electric-furnace steel.

[c] Boron content is 0.0005 to 0.003%.

[d] Based on SAE Standard J404 in the 1982 SAE Handbook, published by the Society of Automotive Engineers, the latest edition of which should be consulted for complete details relating to these steels.

[e] Applicable to blooms, slabs, billets and hot-rolled and cold-rolled bars.

[f] These steels may be produced by the basic oxygen, the basic open-hearth, or the basic electric steelmaking process.

NOTE: These tables are subject to change from time to time, with new steels sometimes added, other steels eliminated, and compositions of retained steels occasionally altered. The reader is referred to current publications of the Society of Automotive Engineers and the American Iron and Steel Institute for latest published information.

omy or availability, are best suited for particular applications.

Effects of the Alloying Elements on Hardenability—The Multiplying Factor Principle—The effects of the alloys on hardenability may be quantatively evaluated by hardenability measurements on a series of steels in which a single alloying element is the only variable. This method is illustrated by Figure 43—1 which shows the hardenabilities of two series of steels in terms of ideal diameter for the microstructure of 50 per cent martensite. These series were made by additions of phosphorus to successive ingots so that in each series the composition was constant except for the phosphorus. It will be noted that the hardenability increases regularly with increasing phosphorus content, and that

the rate of increase is more rapid for the steel of the higher base hardenability. In order to obtain a numerical evaluation of the effect of phosphorus on hardenability, the hardenabilities of the steels containing phosphorus may be divided by the base hardenability of the steel containing no phosphorus. This value expressing the effect of the element on hardenability is known as a **multiplying factor,** and the multiplying factors for phosphorus, derived from the series in Figure 43—1, are plotted in Figure 43—2. It will be seen that a steel with 0.020 per cent phosphorus will have 1.05 times the hardenability of a steel with no phosphorus, while a steel with 0.100 per cent phosphorus would have roughly 1.25 times the base hardenability. Also, with this information, the effect of increasing the

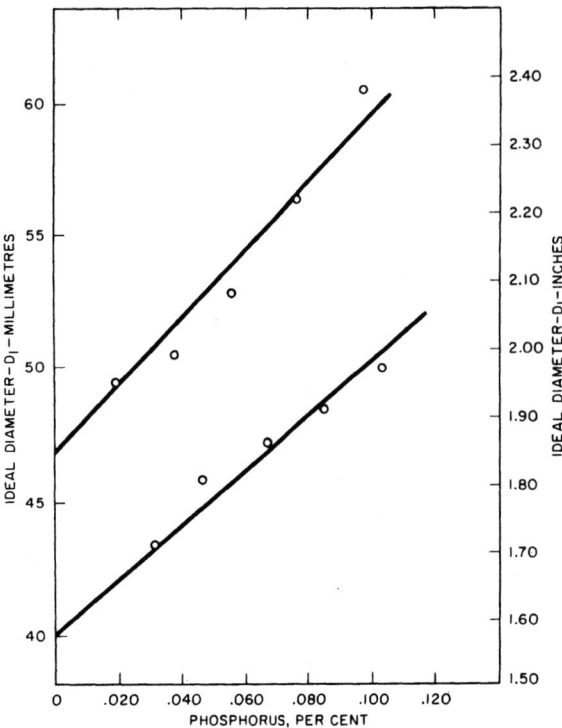

FIG. 43—1. Hardenability as a function of phosphorus content in two series of steels.

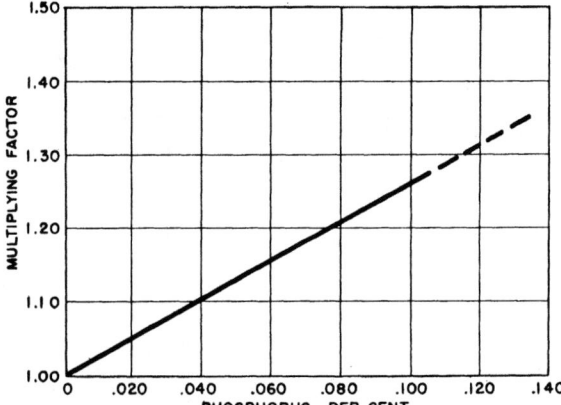

FIG. 43—2. Multiplying factors for phosphorus.

phosphorus content from 0.020 to 0.100 per cent can be evaluated by multiplying the hardenability of the 0.020 per cent phosphorus steel by

$$\frac{1.27}{1.05} \text{ or } 1.21$$

The pioneer work of this nature was done by Grossmann in 1941. Hardenability factors for many of the common alloying elements, as well as hardenability values for pure iron-carbon alloys and the effect of grain size, were evaluated at that time. These factors

have, in some cases, been modified by the work of other investigators and the values as published by the American Iron and Steel Institute are shown for alloying elements in Figure 43—3. Grain size also affects hardenability as shown in Figure 43—4.

Grossmann and his associates further found that the cumulative effects of alloying elements on hardenability could be evaluated by multiplying the base hardenability of the iron-carbon alloy progressively by the multiplying factors for the elements. Thus the chart of Figure 43—3 enables one to calculate the hardenability in terms of 50 per cent martensite microstructures of a given alloy combination.

The element boron markedly increases the hardenability of steel, but its effect differs from that of the other alloying elements since additions over a certain optimum amount produce no further increase in hardenability and may even cause a decrease. Experience has shown that the optimum boron content can range from 0.0005 to 0.005 per cent, depending on the nitrogen and oxygen content and on the presence of elements such as aluminum, titanium, zirconium and vanadium which can be added before the boron to tie up

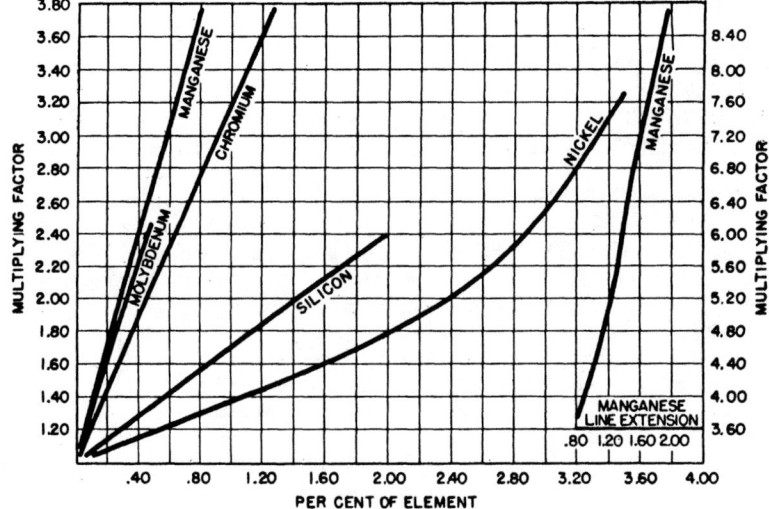

FIG. 43—3. Multiplying factors for a variety of alloying elements (American Iron and Steel Institute).

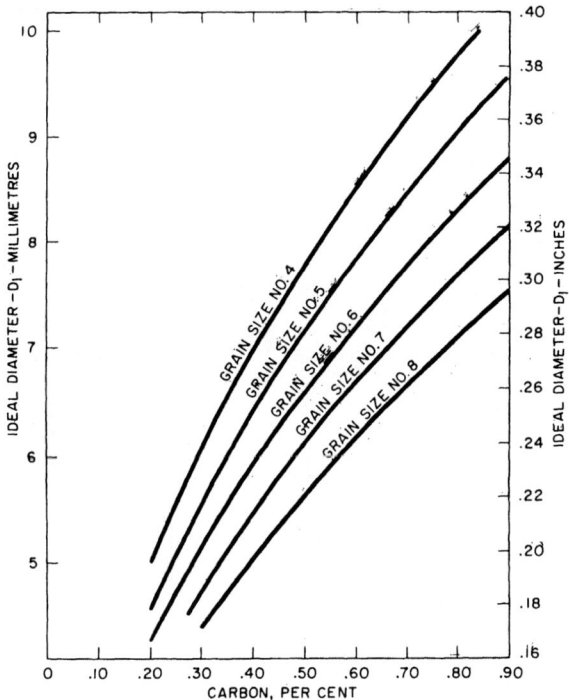

FIG. 43—4. Effect of grain size on the hardenability of pure iron-carbon alloys, expressed as ideal critical diameter, D_I.

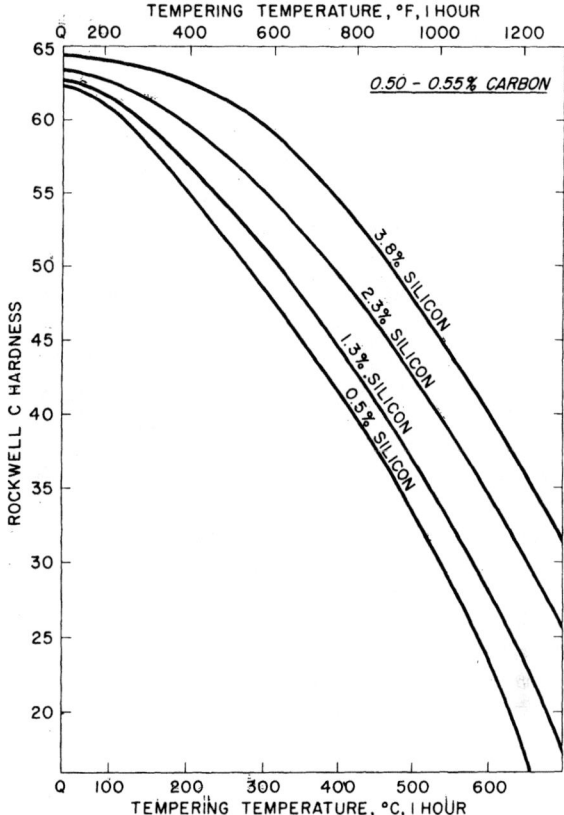

FIG. 43—5. The effect of silicon on tempering rate of 0.50-0.55% carbon steel, tempered for 1 hour at indicated temperatures.

the nitrogen and oxygen in the steel. The multiplying factor for boron depends primarily on base composition of the steel and varies from approximately 3.0 for plain low-carbon steels to nearly 1.0 (no effect of boron) for high-carbon heavily alloyed steels. In hypereutectoid steels, boron accelerates carbide precipitation, and this has a negative effect on hardenability.

The multiplying factor principle is of importance not only as a means of predicting the approximate hardenability of a steel from its composition, but also because it shows that, in general, the addition of relatively small amounts of several alloying elements is more effective in increasing hardenability than a relatively large amount of a single element.

Effects of the Alloys on Tempering—While the martensitic microstructure is the product of proper quenching, it must be realized that martensite itself is brittle, and that in order to realize the properties generally sought in machine parts, this martensite must be reheated or tempered. It is essential to an understanding of alloy steels, therefore, to consider the roles played by the alloying elements in this tempering process, and the manner in which they will affect the behavior on tempering.

The primary purpose of tempering is to impart a degree of plasticity or toughness to the steel to alleviate the brittleness of the martensite. Although this process may, and usually does, soften the steel, this softening is only incidental to the very important increase in toughness. This increase in toughness after tempering reflects two effects of tempering: (1) the relief of internal stresses set up by the quenching operation; and (2) precipitation, coalescence and spheroidization of iron and

alloy carbides, resulting in a microstructure of greater plasticity.

The addition of alloying elements which increase hardenability may also be very helpful in decreasing the magnitude of the internal stresses resulting from the quench, because they will permit the attainment of a martensitic microstructure with a less drastic quench. For this reason, the use of any alloy steel and a mild quench for an application requiring high hardness, and therefore, a low tempering temperature with an accompanying relatively low degree of stress relief, may be very advantageous.

The alloying elements will, however, have a direct and significant effect upon the second behavior, that of precipitation and coalescence of the carbides. In general, the effect of alloying elements will be to slow up the processes of precipitation and coalescence. This means that an alloy steel will customarily require higher tempering temperatures, or longer times at temperature, to obtain a given hardness, or strength level.

The effects of some of the individual alloying elements on the tempering rate are illustrated in Figure 43—5 for silicon, Figure 43—6 for chromium, Figure 43—7 for molybdenum and Figure 43—8 for vanadium. These charts show the hardness of tempered martensite in these steels after tempering one hour at the indicated tempering temperature.

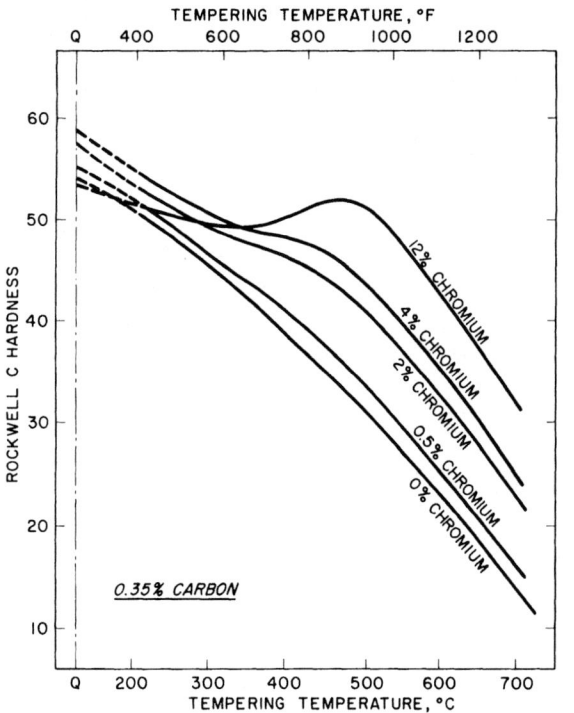

FIG. 43—6. The effect of chromium on tempering rate of 0.35% carbon steel, tempered for 1 hour at indicated temperatures.

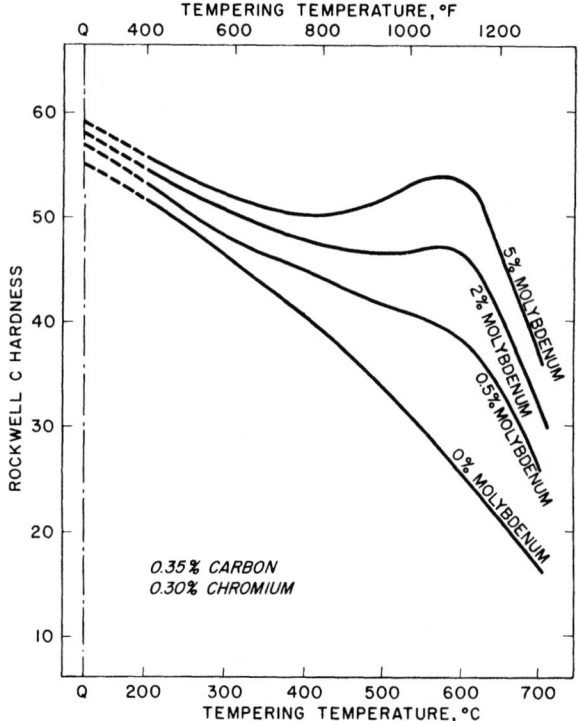

FIG. 43—7. The effect of molybdenum on tempering rate of 0.35% carbon, 0.30% chromium steel, tempered for 1 hour at the indicated temperatures.

The effects of nickel and manganese, as in the case of silicon, while they are significant, are quite moderate, and the hardness changes are nearly a direct function of the tempering temperature. This type of behavior is characteristic of alloys which dissolve largely in the ferrite phase, and do not tend to form carbides.

Boron in the amount used to increase hardenability has no perceptible effect on hardness changes during tempering.

The **carbide-forming elements,** such as chromium, molybdenum or vanadium, however, have very marked effects on the tempering behavior. Elements of this type not only raise the tempering temperature to obtain a given hardness, but with the higher percentages of these elements, the rate of softening is no longer a continuous function of the tempering temperature. In steels of this type, such as the 0.5 per cent molybdenum steel of Figure 43—7, there is a tempering temperature range in which the softening is retarded or, with still higher alloy content as in the 2.0 per cent molybdenum steel, in which the hardness first decreases, then increases slightly before continuing to decline with increasing tempering temperature.

In order that a carbide forming element may manifest its full effect upon the tempering behavior, it must dissolve in the austenite at the austenitizing temperature. This is illustrated by Figure 43—9, which shows the tempering behavior of a chrome-molybdenum-vanadium steel after quenching from 815°, 980° and 1150°C (1500°, 1800° and 2100°F). It will be noted that the secondary hardening behavior is very marked in the steels quenched from the higher temperature, but

is almost absent in the steel quenched from 815°C (1500°F) in which a considerable proportion of the carbides are undissolved.

This secondary hardening effect is also evident in studies of the effect of time at a given tempering temperature on the hardness of alloy steels with carbide forming elements. As an illustration of this, Figure 43—10 shows the manner in which the hardness of a 2.0 per cent molybdenum steel varies with time at several different tempering temperatures. It will be noted that at the highest tempering temperature 650°C (1200°F), the secondary hardening effect occurs at times of from 10 seconds to 10 minutes, while at the lowest tempering temperature 350°C (660°F), there is no indication of this effect in 1000 hours.

This phenomenon can best be explained on the basis of a delayed precipitation of alloy carbide. Because of the relatively small number of alloy atoms in comparison to the iron atoms, and because of the slow diffusion rate of the alloying elements, the first precipitate to form on tempering will certainly be iron carbide, and the initially rapid drop in hardness represents the coalescence of these iron-carbide particles. However, with longer times and particularly with higher temperatures at which the diffusion rate of the alloys becomes more rapid, some alloy carbide will precipitate, and since this occurs after the spheroidization of the iron carbide has progressed to a considerable extent, these fine particles will result in a reversal of the softening action. With relatively low alloy content, this may be manifested as only a decrease in the rate of softening while with high

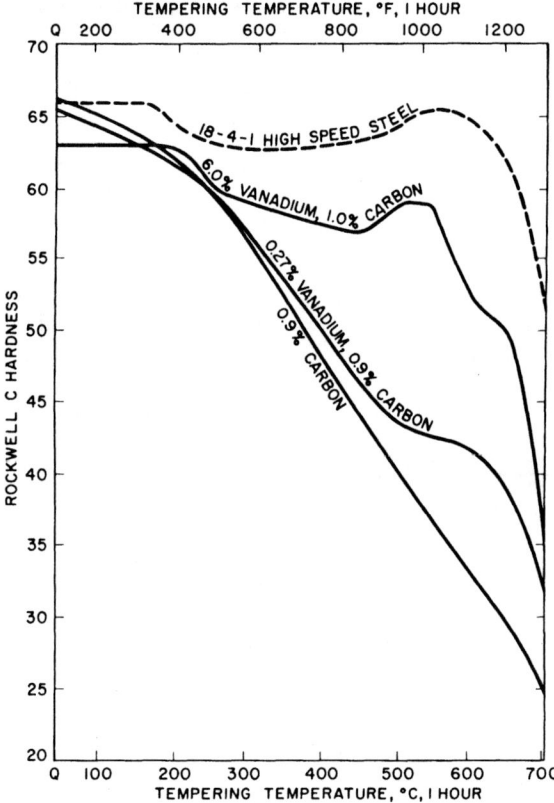

FIG. 43—8. The effect of vanadium on tempering rate of steels tempered for 1 hour at the indicated temperatures.

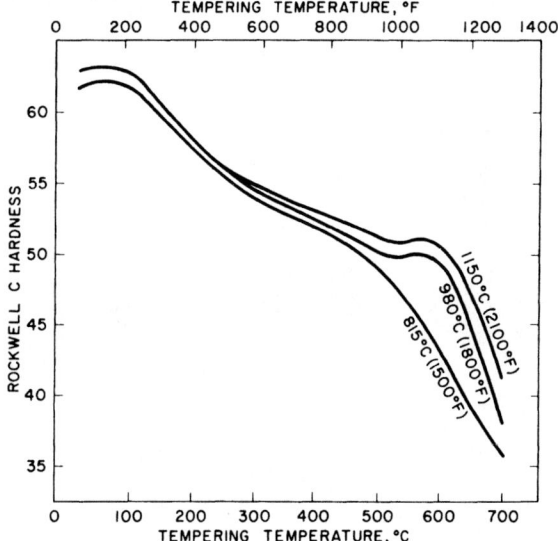

FIG. 43—9. The effect of carbide solution on tempering rate of a Cr-Mo-V steel quenched from 1150°, 980° and 815°C (2100°, 1800° and 1500°F).

alloy content, an actual increase in hardness may occur as this secondary precipitation proceeds.

In the case of modern alloy tool steels which are of such a composition as to consist of a matrix of considerable plasticity with a dispersion of rather sizeable undissolved carbides after heat treatment, full advantage is taken of this secondary hardening effect. The composition and heat treatment of these steels is such that, although a number of the alloy carbides remain undissolved, enough are taken into solution to bring about a marked resistance to softening at temperatures up to 595°C (1100°F). Such tools can then be used for high-speed machining at relatively high operating temperature without softening.

It might further be mentioned that this effect of the alloying elements, and particularly the carbide-forming elements, on tempering may be reflected in an increased toughness of high-alloy steels. We have seen that the toughness of tempered martensite results from both the relief of internal stresses, and the formation of a desirable carbide dispersion. The higher tempering temperatures for a given strength level, characteristic of these high-alloy steels, will permit a greater degree of stress relief with an accompanying increase in toughness. Furthermore, as we have seen, a given hardness level in the tempered alloy steels reflects not only the state of dispersion and spheroidization of the iron

carbides, but also that of the alloy carbides, and thus the spheriodization of the iron carbides must progress further for a given hardness to offset the secondary hardening effect of the alloy carbides. This more completely spheroidized microstructure would also favor plasticity, particularly at hardness levels at which a moderate coalescence of the alloy-carbide dispersion has taken place.

The increase in plasticity on tempering has so far been considered as though it were a continuous effect, with the steel softening and becoming more ductile continuously as the tempering temperature is increased. However, this is not altogether true, as many steels exhibit a minimum toughness on tempering at temperatures between about 230° and 370°C (450° and 700°F). This behavior is illustrated in Figure 43—11, which shows the impact values of several alloy steels as a function of tempering temperature. This phenomenon is not fully understood, nor can the effects of the alloying elements on this behavior be evaluated. However, it should be realized that, in general, tempering temperatures in this range (230°-370°C or 450°-700°F) should be avoided wherever possible.

Temper brittleness in alloy steels is another common example of a discontinuous increase in plasticity on tempering. This phenomenon is manifested as a loss of toughness on slow cooling after tempering at temperatures of 595°C (1100°F) or above, or on tempering in the temperature range of approximately 455° to 595°C (850° to 1100°F). Thus, a steel which is susceptible to this type of embrittlement may lose much of its ductility as indicated by a notched-bar impact test on slow cooling from a tempering temperature of 630°C (1150°F), although the same steel, in relatively light section sizes, will be very tough if it is quenched from the same tempering temperature, and this expedient of quenching from the tempering temperature is a com-

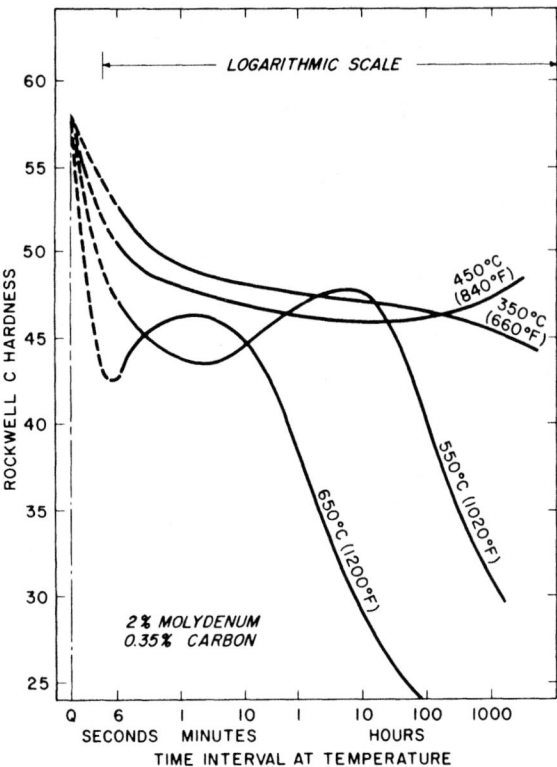

FIG. 43—10. The effect of time of tempering on the secondary hardening behavior in tempering.

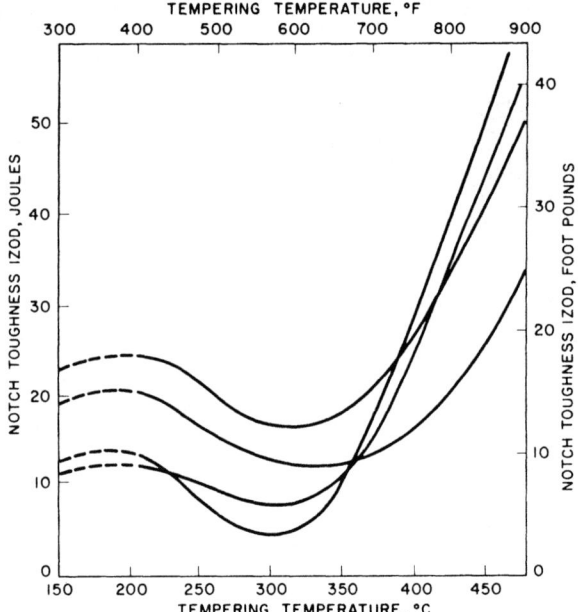

FIG. 43—11. The loss of notch toughness in the Izod test in several alloy steels on tempering at about 315°C (600°F).

mon practice to insure freedom from this embrittlement. However, in such steels, embrittlement will also occur on tempering at 455° to 565°C (850° to 1050°F), particularly if the tempering times are protracted and quenching from the tempering temperature will, in such cases, never completely restore the toughness. This phenomenon also is not completely understood, although the behavior suggests that something which dissolves at temperatures of 595°C (1100°F) and above precipitates in a damaging form at the lower temperatures, either during slow cooling or on reheating to these temperatures. High manganese, phosphorus, silicon, and chromium contents appear to accentuate this behavior, and molybdenum seems to have a retarding effect. The presence of minor amounts of trace elements such as antimony, arsenic and tin have also been found to increase the susceptibility to this type of temper brittleness.

OTHER FUNCTIONS OF THE ALLOYING ELEMENTS

The primary function of the alloying elements in alloy steels has been seen to be that of enhancing the properties through control of the microstructure, particularly in conjunction with suitable heat treatments, and this function has been discussed in considerable detail. However, the alloying elements may exert other useful influences, particularly in steels which may be classed as special purpose steels. These functions will now be briefly discussed.

Ferrite Strengthening—Alloying elements dissolved in pure iron will increase its hardness, and this furnishes a method of increasing the strength of steels in the unhardened state. This ferrite-strengthening function of the alloys is, thus, independent of the effect of the alloys on microstructure, and may be utilized to increase the strength of steels which essentially receive no heat treatment, except for the cooling after the hot-working operation. This hardening effect is, of course, small as compared with that obtainable by changes in the dispersion of the carbide. This is illustrated by Figure 43—12, which shows the hardness of chromium steels as a function of chromium content in a series of steels which have been slowly cooled, in which the ferrite-strengthening effect will be predominant, as compared with a more rapidly cooled series in which the effects of microstructural changes will be predominant. Each of the alloying elements will exert its individual effect in ferrite strengthening. The relative effectiveness of some of the alloying elements in this respect is indicated by Figure 43—13. Significant increases in strength may be obtained by the use of several elements in a single composition. In general, the higher strength levels obtained by this method of ferrite strengthening will be accompanied by a relatively small loss in plasticity, as compared with the considerable loss in ductility accompanying the hardness increases resulting from microstructural changes.

As discussed in the chapter on the low-alloy high-strength steels, these ferrite-strengthening effects have been utilized to their fullest extent in such steels.

Corrosion Resistance—Another function of some of the alloying elements in steel is to increase its resistance to corrosive attack. This function is also utilized in the low-alloy high-strength steels, in which chromium, copper, and phosphorus have been found very effec-

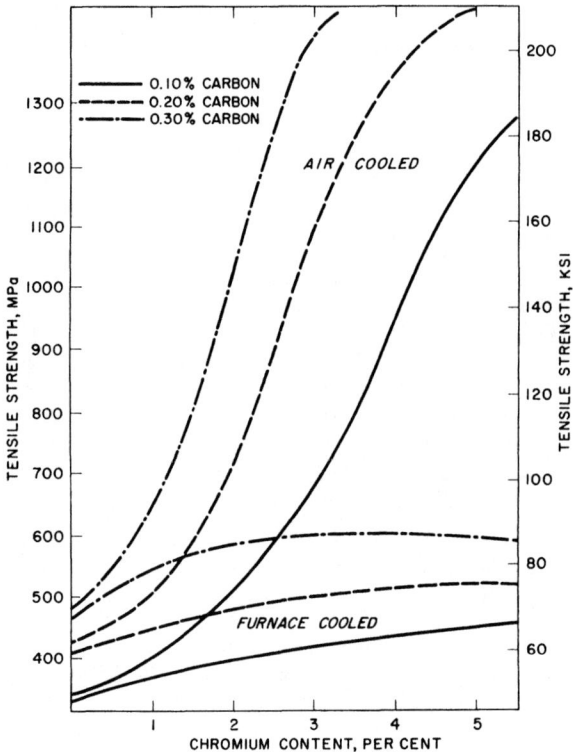

FIG. 43—12. The minor effect of chromium in furnace-cooled steels compared with its strong effect as a strengthener through its influence upon structure in air-cooled steels.

life from corrosion.

Chromium in high percentages imparts an extraordinary corrosion resistance to steel, and the stainless steels are largely based on this effect of chromium. This effect is discussed in detail in Chapter 47 on the stainless steels and need not be elaborated on here.

Abrasion Resistance—The compositions of alloy tool steels are such that at the heat-treating temperature many of the alloy carbides remain undissolved in the austenite. These hard, refractory carbide particles serve to increase the abrasion resistance of the steel, and this represents another function of the alloying elements. The elements commonly used for this purpose are tungsten, molybdenum, vanadium, titanium and chromium.

The austenitic manganese steels represent another class of abrasion-resistant steels in which the function of the element is to stabilize austenite in order to produce austenitic steels which harden on cold working.

Magnetic (Electrical) Characteristics—The addition of alloying elements may greatly modify the characteristics of steel used for electrical equipment. Improved electrical characteristics for a desired application may be obtained by utilizing these effects as discussed in Chapter 46.

One example of this function of the alloying elements is represented by the silicon steels used for transformer cores. These steels, containing silicon up to 5 per cent, have a greatly increased electrical resistivity and, as annealed, high permeability. When used as transformer cores, these properties result in greatly reduced core losses.

The magnet steels, for permanent magnets, are another example of alloy electrical steels. The outstanding property of these steels is their retentivity or ability to retain magnetism. Cobalt, chromium, and tungsten are the alloying elements commonly used to enhance this characteristic.

tive in increasing resistance to atmospheric corrosion, thereby permitting the use of the lighter sections made possible by their higher strengths, without decreased

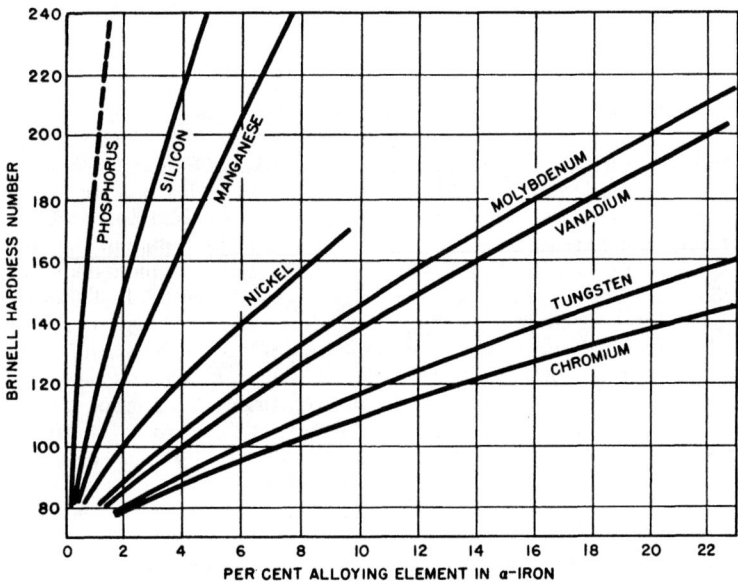

FIG. 43—13. The relative effectiveness of the alloying elements as ferrite strengtheners.

THE CONSTRUCTIONAL (SAE AND AISI) ALLOY STEELS

Classification and Standardization of the Constructional Alloy (SAE and AISI) Steels—The alloy steels most commonly used for heat-treated parts originally were classified by the American Iron and Steel Institute and the Society of Automotive Engineers. The composition ranges of these steels have been listed in Table 43—I.

Theses steels were identified in the AISI classification by a numerical index system that was partially descriptive of the composition. The first digit indicated the type to which the steel belonged; thus "1" indicated a carbon steel; "2" indicated a nickel steel; "3" indicated a nickel-chromium steel. In the case of the simple alloy steels, the second number usually indicated the percentage of the predominant alloying element. Usually the last two or three digits indicated the average carbon content in "points," or hundredths of a per cent. Thus, "2340" indicated a nickel steel of approximately 3 per cent nickel (3.25 to 3.75) and 0.40 per cent carbon (0.35 to 0.45).

The basic numerals for the various types of AISI steels (including plain-carbon steels) were:

Series Designation	Types
10xx	Nonresulphurized carbon steel grades
11xx	Resulphurized carbon steel grades
12xx	Rephosphorized and resulphurized carbon steel grades
13xx	Manganese 1.75 per cent
15xx	Manganese over 1.00 to 1.65 per cent
23xx	Nickel 3.50 per cent
25xx	Nickel 5.00 per cent
31xx	Nickel 1.25 per cent—Chromium 0.65 per cent
33xx	Nickel 3.50 per cent—Chromium 1.55 per cent
40xx	Molybdenum 0.25 per cent
41xx	Chromium 0.50 or 0.95 per cent—Molybdenum 0.12 or 0.20 per cent
43xx	Nickel 1.80 per cent—Chromium 0.50 to 0.80 per cent—Molybdenum 0.25 per cent
44xx	Molybdenum 0.40 or 0.53 per cent
46xx	Nickel 1.55 or 1.80 per cent—Molybdenum 0.20 or 0.25 per cent
47xx	Nickel 1.05 per cent—Chromium 0.45 per cent—Molybdenum 0.20 per cent
48xx	Nickel 3.50 per cent—Molybdenum 0.25 per cent
50xx	Chromium 0.28 or 0.40 per cent
51xx	Chromium 0.80, 0.90, 0.95, 1.00 or 1.05 per cent
5xxxx	Carbon 1.00 per cent—Chromium 0.50, 1.00 or 1.45 per cent
61xx	Chromium 0.80 or 0.95 per cent—Vanadium 0.10 per cent or 0.15 per cent min.
81xx	Nickel 0.30 per cent—Chromium 0.40 per

	cent—Molybdenum 0.12 per cent
86xx	Nickel 0.55 per cent—Chromium 0.50 or 0.65 per cent—Molybdenum 0.20 per cent
87xx	Nickel 0.55 per cent—Chromium 0.50 per cent—Molybdenum 0.25 per cent
88xx	Nickel 0.55 per cent—Chromium 0.50 per cent—Molybdenum 0.35 per cent
92xx	Manganese 0.85 per cent—Silicon 2.00 per cent
93xx	Nickel 3.25 per cent—Chromium 1.20 per cent—Molybdenum 0.12 per cent
B	denoted Boron Steel, as in 51B60 and others.
BV	denoted Boron Vanadium Steel, as in TS 43BV12 and TS 43BV14.
L	denoted leaded steel, as in 10L18.

The basic numbering system adopted by the Society of Automotive Engineers was quite similar, differing only in minor details.

Needless to say, this list, representing as it did a standardization and simplification of thousands of alloy-steel compositions, was a very valuable aid to the specification and choice of alloy steels for various applications. Many of these steels were developed for specific applications, and their continued satisfactory performance has resulted in a considerable degree of standardization of application among these compositions.

The Unified Numbering System for Metals and Alloys—In 1967 the Society of Automotive Engineers (SAE) and the American Society for Testing and Materials (ASTM) began to explore the possibility of developing a unified numbering system for both ferrous and non-ferrous metals and alloys to replace the numerous and uncoordinated designation systems that had been developed independently over a 60-year period by producers and users of metals and alloys and societies and trade associations concerned with numerical systems for classifying and identifying these materials.

A project sponsored jointly by SAE and ASTM, under a contract issued to SAE by the U. S. Army Materials and Mechanics Research Center in May 1969, established the feasibility of setting up a unified numbering system and proposed how such a system could be developed. In the course of the feasibility study, major trade associations, producers and users of metals and alloys were consulted. In 1972, SAE and ASTM established an advisory board that developed and refined the proposed system—again in consultation with interested groups and individuals—and completed a recommended practice for numbering metals and alloys in March 1974. The advisory board subsequently coordinated the establishment of specific designations for over 1000 steel, stainless steel, tool steel, superalloy, aluminum, copper, cobalt, nickel and rare-earth alloys.

During the period 1974-1977, further refinements were made in the basic numbering system, secondary systems were developed for 26 additional categories of metals and UNS numbers were assigned to alloys of each category. The results of all the foregoing activities are summarized in "Unified Numbering System for Metals and Alloys (SAE HS1086a and ASTM DS-56A), Second Edition," published in September 1977 by SAE.

The Unified Numbering System (UNS) established 15 series of numbers for metals and alloys (the series applying to ferrous metals and alloys are listed in Table 43-II). Each number consists of a single-letter prefix followed by five digits. In most cases, the letter is suggestive of the family of metals identified: for example, G is the prefix for the UNS numbers replacing the numbers formerly used for the AISI and SAE carbon and constructional alloy steels, S relates to the heat- and corrosion-resistant (stainless) steels and H relates to the H steels of the formerly used AISI classification. Where feasible, the numbers from formerly existing systems were incorporated into the UNS numbers: for example, the carbon steel formerly identified as SAE or AISI 1020 is covered by UNS G10200.

Two important aspects of the Unified Numbering System should be noted: (1) While identification numbers from former systems are incorporated, where feasible, into UNS numbers, the arbitrary assignment of UNS numbers derived unofficially from former numbers should be avoided: the proper trade association (SAE, AISI or ASTM) or its publications should be consulted to determine the proper UNS number; and (2) a UNS number is not a specification in itself: it is only a part of a unified system for identifying metals and alloys for which controlling limits have been established elsewhere.

Table 43—II. Unified Numbering System (UNS) Series Numbers for Ferrous Metals and Alloys.[a]

Series Numbers	Metals
D00001-D99999	Specified mechanical properties steels
F00001-F99999	Cast irons
G00001-G99999	AISI and SAE carbon and alloy steels (except tool steels)
H00001-H99999	AISI H-steels
J00001-J99999	Cast steels (except tool steels)
K00001-K99999	Miscellaneous steels and ferrous alloys
S00001-S99999	Heat and corrosion resistant (stainless) steels
T00001-T99999	Tool steels

[a] Based on "Numbering Metals and Alloys (SAE J1086 and ASTM E 527)—SAE and ASTM Recommended Practice" in Part I of the 1982 SAE Handbook.

Applications of the Constructional (SAE and AISI) Alloy Steels—The low-carbon steels (0.10-0.25 per cent carbon) in this classification are designated as carburizing steels and they are applied almost exclusively to carburized parts. However, the choice of a steel within this group is determined largely by the core properties desired for the specific application. The lower alloy combinations, such as UNS G40230, G41180 or G50150 (SAE 4023, 4118 or 5015) are used where somewhat better core properties than those obtainable with the plain carbon compositions such as UNS G10180 or G11170 (SAE 1018 or 1117) are desired. They have the further advantage of being hardenable in oil in moderate sections, and therefore can be heat treated with less distortion than the types requiring water quenching. The higher manganese and sulphur steels are used where superior machinability is required. Typical ap-

plications of these low-alloy carburizing grades would be for the production of cam shafts, wrist pins, clutch fingers, and other automotive parts.

The higher alloy carburizing steels, such as UNS G43200, G46200, G48150, G51200, G61180, G86200, G93100 or G94B170 (SAE 4320, 4620, 4815, 5120, 6118, 8620, 9310 or 94B17) are used where superior case hardness or core properties are desired. The choice of steels within this group is determined primarily by the hardenability necessary to obtain the desired core properties under the given conditions of section size and heat treatment. The nickel-molybdenum (UNS G46200 or SAE 4620) plain chromium (UNS G51200 or SAE 5120), and low nickel-chromium-molybdenum (UNS G86200 or SAE 8620) steels can be used for automotive gears, universal joints, small hand tools, piston pins, bearings, and similar parts of moderate section for relatively severe service. The higher alloy steels UNS G48150, G93100 or G94B170 (SAE 4815, 9310 or 94B17) are used for severe service applications or heavy sections. Typical applications of these steels are aircraft-engine parts, truck transmissions and differentials, rotary rock-bit cutters and large antifriction bearings.

Similarly, the choice of the higher carbon alloy steels is based largely on the hardenability requirements of the specific applications. This will, of course, be a function of the heat treatment and section size. Intricate sections or higher carbon materials (over 0.40 per cent carbon) which must be oil quenched to prevent danger of quench cracking may frequently require higher alloy compositions than the simpler sections or low-carbon materials which can be safely heat treated under more drastic quenching conditions.

As with the carburizing steels, the lower alloy higher carbon steels, such as the manganese (UNS G13300-13450 or SAE 1330-45), plain molybdenum (UNS G40370-40470 or SAE 4037-4047), plain chromium (UNS G51300-51350 or SAE 5130-50), the low nickel-chromium-molybdenum (UNS G86300-86500 or SAE 8630-50), or chromium-molybdenum UNS G41300 or 41500 (SAE 4130 or 4150) steels are used for applications involving relatively small sections, but which are subject to severe service conditions, or in larger sections which may not necessitate optimum properties, but in which advantage is taken of the weight saving derived from the higher strength of the alloy steels. Typical applications can be the use of the manganese steels for high-strength bolts, molybdenum steels and chromium steels for automotive steering parts and low nickel-chromium-molybdenum steels for small machinery axles and shafts. These lower alloy steels are also widely used for high-quality small tools.

The higher alloy constructional (SAE-AISI) steels, such as the UNS G43370-G43400 (SAE 4337-40) or UNS G86B450 (SAE 86B45) compositions are used for heavy sections or for parts subject to particularly severe service conditions or for which very mild quenches must be used to prevent distortion. Typical uses would be for relatively heavy aircraft or truck parts or for ordnance materials.

In addition to the more or less general uses described above, some of the constructional (SAE-AISI) alloy steels have quite specialized applications. Thus the UNS G15216 (SAE 52100) steels are used almost exclusively for ball-bearing applications, and the chromium steels UNS G51550 and G51600 (SAE 5155 and 5160) were developed for and are used almost entirely for spring-steel applications.

Hardenabilities of Constructional Alloy Steels—Hardenability has been stressed as the most important function of the alloying elements in these steels, and the above discussion of their applications has shown that the choice of the alloy steel to be used for a given application is largely based on its hardenability. Realizing the importance of hardenability, the American Iron and Steel Institute, together with the Society of Automotive Engineers, have established minimum and maximum end-quench hardenability curves, known as **hardenability bands,** for most of these alloy steels.

These bands or hardenability limits, typified by Figure 43—14, are based on the analysis of data collected from hundreds of heats of each grade of steel. Such information permits these steels to be sold on the basis of hardenability: steels sold to a hardenability specification are known as "H" steels.

Since, for application purposes, the minimum hardenability values for a given steel are usually the most pertinent, the minimum hardenability limits for numerous "H" steels are shown in Figures 43—15 and 43—16. Figure 43—15 also compares the minimum hardenability limits for a group of steels of different alloy content but having essentially the same carbon content. Figure 43—16 shows the effect on the minimum hardenability limit of increasing carbon content in G86000 (SAE 8600) series (nickel-chromium-molybdenum) steels.

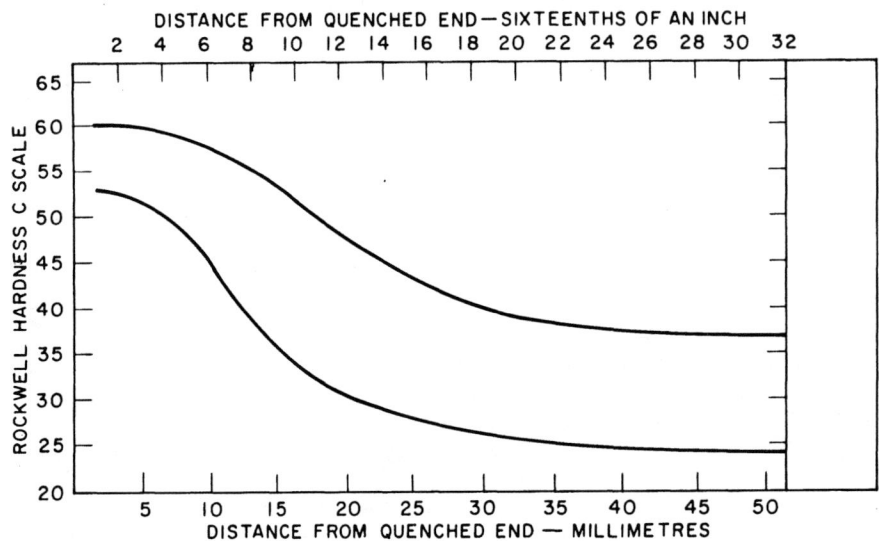

FIG. 43—14. Typical hardenability band for UNS H41400 (SAE 4140H) steel.

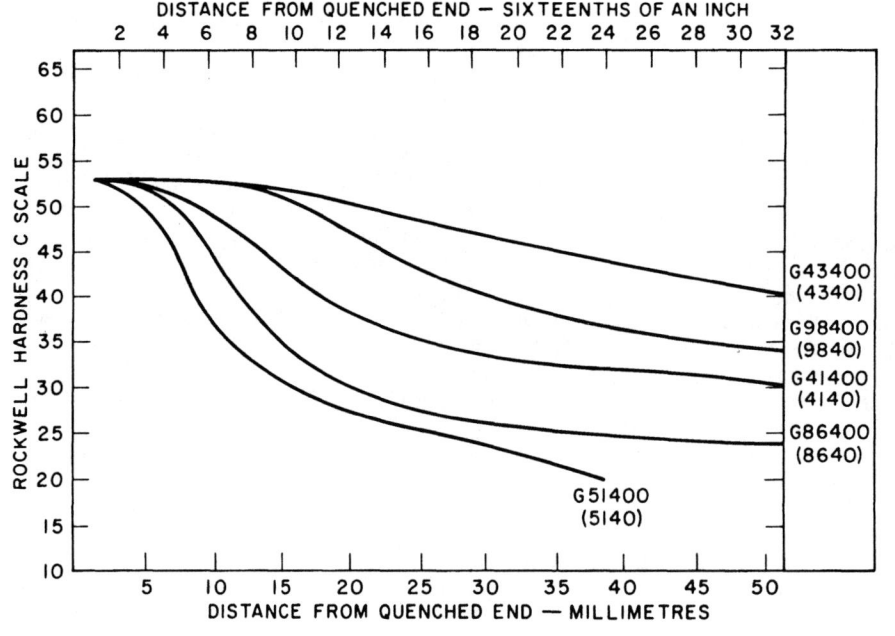

FIG. 43—15. Minimum hardenability limits for UNS H41400, H43400, H51400, H86400 and H98400 (SAE 4140H, 4340H, 5140H, 8640H, and 9840H), comparing minimum hardenability for steels of different alloy content but of the same carbon content.

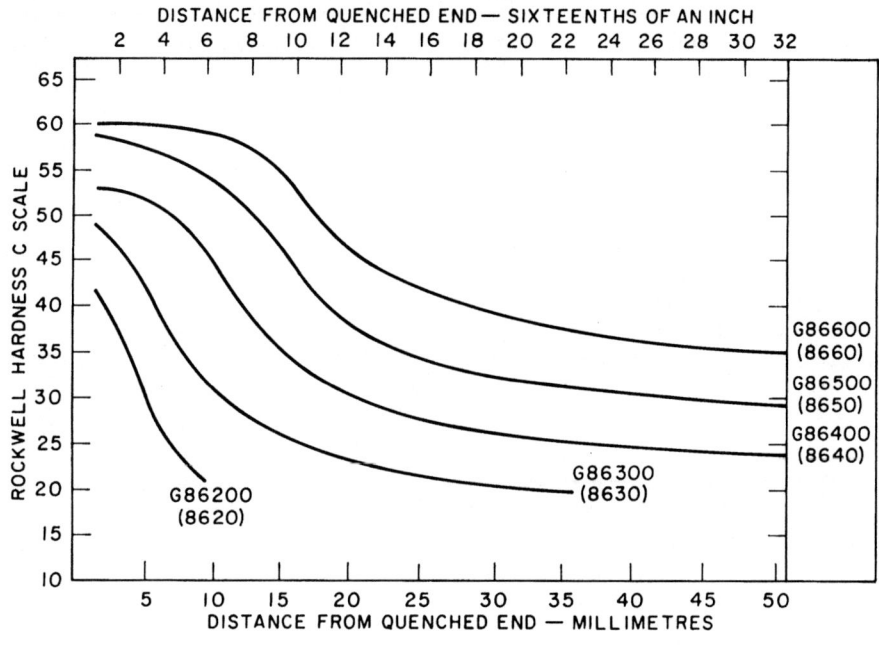

DISTANCE FROM QUENCHED END— SIXTEENTHS OF AN INCH

FIG. 43—16. Minimum hardenability limits of a group of steels of the UNS H86000 (SAE 8600) (nickel-chromium - molybdenum) series, showing the increase in minimum hardenability with increasing carbon content.

SECTION 3

QUENCHED AND TEMPERED LOW-CARBON CONSTRUCTIONAL ALLOY STEELS

A class of quenched and tempered low-carbon constructional alloy steels has been very extensively used in a wide variety of applications such as pressure vessels, earth-moving and mining equipment and as major members of large steel structures. Their outstanding attributes and their usefulness for a wide variety of applications warrant their consideration as a separate and important class of constructional alloy steels. These steels were not included in the SAE-AISI classification of alloy steels.

As a general class, these steels are referred to as **low-carbon martensitic steels,** to differentiate them from those constructional alloy steels of higher carbon content that develop high-carbon martensite upon quenching. They are characterized by a relatively high strength, with minimum yield strengths of 690 MPa (100 ksi), toughness at temperatures down to –45°C (–50°F), and weldability, with welded joints showing full joint efficiency when welded with suitable practices and low-hydrogen electrodes. They are used in the forms of plates, sheets, bars, structural shapes, or forgings.

The compositions of some steels of this type are shown in Table 43—III, and the minimum specified mechanical properties of these steels are shown in Table 43—IV.

The steels are all low-carbon steels with carbon contents in the 0.10 to 0.23 per cent range, and this low carbon content is a very important factor in the toughness and weldability of these steels. The hardenability of each steel is such that transformation on quenching in sizes within the specified thickness range occurs predominantly at low temperatures to lower bainite or martensite. The alloying elements used to impart this

requisite hardenability have been chosen with economy in mind so that the outstanding attributes of these steels can be realized at a relatively low cost. The very significant hardenability effect of a small percentage of boron is utilized in some of these steels and also, in some instances, different grades are available, with the alloy content and hardenability being varied in accordance with the section size in which it is to be applied. Thus, advantage can often be taken of the higher strength and superior toughness of some of these alloy steels at a lower cost than that for plain carbon steels which would be used in heavier sections at their lower strengths. The atmospheric-corrosion resistance of these alloy steels is, furthermore, from two to six times that of plain carbon steels. Although the high strength and toughness of these steels reflects primarily their low carbon content and their efficient use of the hardenability function of the alloying elements, the additional function of the alloying elements of controlling the tempering behavior is also utilized in some of these steels. In these steels, small additions of strong carbide-forming elements such as vanadium or titanium are used to permit retention of the high strength of the steels on tempering at relatively high temperatures.

The low-carbon martensitic steels are also used for cold-heading and cold-forging applications. The steels are furnished as semifinished product, which is cold-headed or cold-forged into parts such as fasteners or pins and heat treated to the desired properties. The low carbon content of these steels results in improved formability and facilitates cold forging, as well as permitting the development of high strength and toughness in the finished heat-treated product for many applications.

Table 43—III. Composition Ranges of Quenched and Tempered Constructional Alloy Steels (Per Cent).*

UNS	Designation (USS)	Producer	C	Mn	Si	Ni	Cr	Mo	V	Ti	B	Cu
K11576	"T-1" Steel	U.S. Steel	0.10 / 0.20	0.60 / 1.00	0.15 / 0.35	0.70 / 1.00	0.40 / 0.65	0.40 / 0.60	0.03 / 0.08	—	0.0005 / 0.006	0.15 / 0.50
K11630	"T-1" type A	U.S. Steel	0.12 / 0.21	0.70 / 1.00	0.20 / 0.35	—	0.40 / 0.65	0.15 / 0.25	0.03 / 0.08	0.01 / 0.03	0.0005 / 0.005	—
K11646	"T-1" type B	U.S. Steel	0.12 / 0.21	0.95 / 1.30	0.20 / 0.35	0.30 / 0.70	0.40 / 0.65	0.20 / 0.30	0.03 / 0.08	—	0.0005 / 0.005	—
	"T-1" type C	U.S. Steel	0.14 / 0.21	0.95 / 1.30	0.15 / 0.35	1.20 / 1.50	1.00 / 1.50	0.40 / 0.60	0.03 / 0.08	—	—	—

*Similar steels are listed in ASTM Specification A 517, published by the American Society for Testing and Materials.

Table 43—IV. Longitudinal Mechanical Properties of Quenched and Tempered Constructional Alloy Steels.*

Designation***		Minimum Yield Strength		Tensile Strength		Minimum Elongation (in 51mm or 2 in.)	Minimum Reduction of Area (%)	Impact Resistance** at–45°C (–50°F)	
UNS	U. S. Steel	MPa	ksi	MPa	ksi			joules	ft-lb
K11576	"T-1" Steel	690	100	760/900	110/130	18.0	50.0	27	20
K11630	"T-1" type A	690	100	760/900	110/130	18.0	50.0	20	15
K11646	"T-1" type B	690	100	760/900	110/130	18.0	50.0	20	15
	"T-1" type C	690	100	760/900	110/130	18.0	50.0	47	35

*For 25-mm (1-inch) thick plates. Properties vary with thickness and compositions are selected from ranges in Table 43—III to give desired properties for a given thickness.
**Charpy V-notch specimens.
***Similar steels are listed in ASTM Specification A 517, published by the American Society for Testing and Materials.

SECTION 4

QUENCHED AND TEMPERED LOW-CARBON ULTRASERVICE ALLOY, ARMOR, AND MARAGING STEELS

The need for high-performance materials with higher strength-to-weight ratios for critical military needs, for hydrospace explorations, and for aerospace applications has led to the development of quenched and tempered ultraservice steels. Although these steels are similar in many respects to the low-carbon constructional alloy steels described in the previous section, the significantly higher notch toughness at yield strengths up to 965 MPa (140 ksi) results from recent process improvements and material developments and distinguishes the ultraservice steels from the constructional alloy steels.

The compositions of several of the ultraservice steels are shown in Table 43—V and the minimum specified mechanical properties of these steels in several product forms are shown in Table 43—VI.

Because these steels may be used in large welded structures and these structures may be subjected to unusually high loads, the steels must exhibit excellent weldability and toughness. The composition of each of the ultraservice steels is a balance between the carbon content and the alloying elements necessary to maintain adequate hardenability and toughness without sacrificing weldability. The importance of nickel to strength and toughness of the ultraservice steels may be seen by examinations of the compositions and mechanical properties in the tables. For the 552 MPa (80 ksi) minimum yield strength of HY-80 steel, a 2.00 per cent minimum nickel content is adequate but for the 896 MPa (130 ksi) yield strength HY-130 steel, a 4.75 per cent minimum nickel content is required. Another important aspect in attaining good toughness, particularly at the higher strength level, is the necessity of a low sulphur content. It has also been observed that the nitrogen and oxygen content must be low to insure attainment of the high level of toughness. Although the strength and toughness requirements shown in Table 43—VI are important criteria for the ultraservice steels, such steels must also have a high resistance to fatigue and corrosion to meet the rigorous require-

Table 43—V. Composition Ranges of Quenched and Tempered Ultraservice Steels (Per Cent)*.

Designation	C	Mn	P	S	Si	Ni	Cr	Mo	V	Specifications Issued for
HY-80	0.12 0.18	0.10 0.40	0.020	0.020	0.15 0.35	2.00 3.25	1.00 1.80	0.20 0.60	—	Plates, extrusions, rolled shapes, bars, forgings, castings.
HY-100	0.12 0.20	0.10 0.40	0.020	0.020	0.15 0.35	2.25 3.50	1.00 1.80	0.20 0.60	—	Plates, extrusions, rolled shapes, bars, castings.
ASTM A 543/ A 543M (UNS K42338)	0.23	0.40	0.035	0.040	0.20 0.35	2.60 3.25	1.50 2.00	0.45 0.60	0.03	Plates.
HY-130	0.12	0.60 0.90	0.010	0.010	0.15 0.35	4.75 5.25	0.40 0.70	0.30 0.65	0.05 0.10	Plates.
ASTM A 579-77 Grade 12 (UNS K51255)	0.12	0.60 0.90	0.010	0.010	0.35	5.25	0.70	0.65	0.10	Forgings.

*Maximum unless range is shown.

Table 43—VI. Mechanical Properties of Quenched and Tempered Ultraservice Steels.*

Designation	Yield Strength		Tensile Strength		Elongation (%)	Reduction of Area (%)	Charpy V-Notch Energy Absorption	
	MPa	ksi	MPa	ksi			joules	ft-lb
For 25-mm (1-inch) Thick Plates								
HY-80	550/685	80/99.5	Information only		20 in 50 mm (2 in.)	50	47 at –85°C 81 at –18°C	35 at –120°F 60 at 0°F
HY-100	690/830	100/120	Information only		18 in 50 mm (2 in.)	45	41 at–85°C 75 at –18°C	30 at –120°F 55 at 0°F
ASTM A 543/ A 543M Class 1 (UNS K42338)	585	85	725/860	105/125	14 in 50 mm (2 in.)	Not specified	Negotiable	Negotiable
Class 2 (UNS K42339)	690	100	795/930	115/135	14 in 50 mm (2 in.)	Not specified		
Class 3 (UNS K42340)	485	70	620/795	90/115	16 in 50 mm (2 in.)	Not specified		
HY-130	895/1000	130/145	Information only		15 in 50 mm (2 in.)	50	81 at –18°C and at room temperature	60 at 0°F and at room temperature
For Forgings Up to 100 mm (4 inches) Thick								
ASTM A 579-77 Grade 12 (UNS K51255)	965	140	1035	150	13	40	68 at room temperature	50 at room temperature

*Minimum unless range is specified.
**Tentative designation used by the United States Navy.

ments for hydrospace operations for which they originally were developed.

Traditionally, some of these steels such as the HY steels have been viewed as separate and distinct and are commonly referred to as "armor," having originally been developed for military applications. In addition to the applications for military, hydrospace and aerospace needs, the ultraservice steels may find uses in the advancing technology for pressure vessels and thick-walled cylinders. Furthermore, ultraservice steels with even higher yield strengths are in use, such as the AF1410 steel, a 10 Ni-Cr-Mo-Co steel with a minimum yield strength of 1240 MPa (180 ksi), and the maraging steels, usually 10 to 18 per cent nickel, age-hardenable, martensitic steels containing substantial amounts of elements such as chromium, molybdenum, and cobalt, with yield strengths in the range 1035 to 2070 MPa (150 to 300 ksi).

Bibliography

American Society for Metals, Metals handbook, 1982 edition.

Bain, E. C., and Paxton, Harold W., Functions of the alloying elements in steel. American Society for Metals, Cleveland, Ohio, 2nd. ed., 1961.

Leslie, W. C., and Keh, A. S., Aging of flat-rolled steel products as investigated by electron microscopy. Mechanical Working of Steel II, T. G. Bradbury, ed. Published for AIME in New York by Gordon and Breach Science Publishers, Inc., 1965.

CHAPTER 44

Alloy Tool Steels

Compositions and Applications—The principal functions of the alloying elements in tool steels are to increase hardenability; to form hard, wear-resisting alloy carbides; and to increase resistance to softening on tempering. The alloy tool steels may be roughly classified according to the extent of their utilization of these three functions, as follows:

1. **Relatively low-alloy tool steels.** These are of higher hardenability than the plain carbon tool steels in order that they may be hardened in heavier sections or with less drastic quenches and thereby less distortion.

2. **Intermediate alloy tool steels.** These steels usually contain elements such as tungsten, molybdenum or vanadium, which form hard, wear-resisting carbides.

3. **High-speed tool steels.** These contain large amounts of the carbide-forming elements which serve not only to furnish wear-resisting carbides but also to promote secondary hardening and thereby to increase resistance to softening at elevated temperature.

High-speed steels are used for applications requiring long life at relatively high operating temperatures such as for heavy cuts or high-speed machining. The intermediate alloy types are used for finishing operations in which extreme wear resistance and the ability to retain a smooth cutting edge on light cuts are necessary. Steels of the first class are general purpose tool steels, the choice of which is based primarily on section size, permissible distortion, intricacy of design, and the hardness and toughness requirements of the application, all of which are, to a considerable extent, functions of the hardenability. The higher hardenability steels are used in cases where a low "movement" (change of dimension) in hardening is required, since relatively slow oil or even air quenches may be used. These steels are also capable of hardening from relatively low quenching temperatures, which also tends to decrease distortion and danger of quench cracking. Within this class, the higher carbon steels are used for applications requiring high resistance to wear or abrasion, and the lower carbon steels where resistance to shock or impact is important.

Typical constituents of commonly used tool steels (including some plain carbon steels) are given in Table 44—I. More detailed information on the composition, treatments and applications of tool steels may be found in the reference material in the bibliography at the end of the chapter.

The AISI developed the method of identification shown in Table 44—I which grouped the commonly used tool steels under seven major headings. Each group was identified by a letter symbol and contained individual types identified by suffix numbers following the letter symbol; these numbers and letter symbols have been replaced by numbers from the Unified Numbering System (UNS) that also are given in the table. The percentages of elements shown are for identification purposes only. Steels of the same type, made by different producers, may differ in mean composition from the values listed and may contain elements that are not listed.

HEAT TREATMENT OF ALLOY TOOL STEELS

General—The general principles of heat treatment, as described in Chapter 41, naturally apply to the heat treatment of the alloy tool steels. The alloy tool steels are generally high carbon, and many are relatively high-alloy steels so that their heat treatment involves special precautions to avoid distortion, cracking, and decarburization. Heating must be conducted at a slow rate to minimize thermal stresses, and relatively low austenitizing temperatures are used to minimize distortion and cracking. It is a common practice to carry out the heating in two stages; a preliminary preheat to an intermediate temperature preceding the heating to the final temperatures. Decarburization is usually particularly harmful in tool steels, and preventive practices such as use of controlled atmospheres, packing in cast-iron chips, or heating in neutral liquid baths are commonly used.

Because of their sensitivity to cracking and the dangers of distortion, relatively mild quenches are commonly used in the heat treatment of alloy tool steels, and many of them are of high enough hardenability to permit air quenching.

Because the residual stresses are high in these high-carbon steels after quenching, the stress-relieving function of the tempering operation is of particular importance. Tempering to relieve these stresses and toughen the steels is therefore an essential part of the heat-treating operation and should immediately follow the quench. Because high hardness is usually desired, the tempering temperatures are generally low, 120° to

Table 44—I. Classification of Commonly Used Tool Steels[a]

| Type Designation | | Identifying Elements, Per Cent | | | | | | | | | |
UNS	AISI	C	Mn	Si	W	Mo	Cr	V	Co	Ni	Al
HIGH-SPEED TOOL STEELS											
Molybdenum Types											
T11301	M1	0.85[d]	—	—	1.50	8.50	4.00	1.00	—	—	—
T11302	M2	0.85; 1.00[d]	—	—	6.00	5.00	4.00	2.00	—	—	—
T11313	M3, Class 1	1.05	—	—	6.00	5.00	4.00	2.40	—	—	—
T11323	M3, Class 2	1.20	—	—	6.00	5.00	4.00	3.00	—	—	—
T11304	M4	1.30	—	—	5.50	4.50	4.00	4.00	—	—	—
T11306	M6	0.80	—	—	4.00	5.00	4.00	1.50	12.00	—	—
T11307	M7	1.00	—	—	1.75	8.75	4.00	2.00	—	—	—
T11310	M10	0.85; 1.00[d]	—	—	—	8.00	4.00	2.00	—	—	—
T11333	M33	0.90	—	—	1.50	9.50	4.00	1.15	8.00	—	—
T11334	M34	0.90	—	—	2.00	8.00	4.00	2.00	8.00	—	—
T11336	M36	0.85	—	—	6.00	5.00	4.00	2.00	8.00	—	—
T11341	M41	1.10	—	—	6.75	3.75	4.25	2.00	5.00	—	—
T11342	M42	1.10	—	—	1.50	9.50	3.75	1.15	8.00	—	—
T11346	M46	1.25	—	—	2.00	8.25	4.00	3.20	8.25	—	—
Tungsten Types											
T12001	T1	0.75[d]	—	—	18.00	—	4.00	1.00	—	—	—
T12004	T4	0.75	—	—	18.00	—	4.00	1.00	5.00	—	—
T12005	T5	0.80	—	—	18.00	—	4.00	2.00	8.00	—	—
T12006	T6	0.80	—	—	20.00	—	4.50	1.50	12.00	—	—
T12008	T8	0.75	—	—	14.00	—	4.00	2.00	5.00	—	—
T12015	T15	1.50	—	—	12.00	—	4.00	5.00	5.00	—	—
HOT-WORK TOOL STEELS											
Chromium Types											
T20810	H10	0.40	—	—	—	2.50	3.25	0.40	—	—	—
T20811	H11	0.35	—	—	—	1.50	5.00	0.40	—	—	—
T20812	H12	0.35	—	—	1.50	1.50	5.00	0.40	—	—	—
T20813	H13	0.35	—	—	—	1.50	5.00	1.00	—	—	—
T20814	H14	0.40	—	—	5.00	—	5.00	—	—	—	—
T20819	H19	0.40	—	—	4.25	—	4.25	2.00	4.25	—	—

(Continued on next page)

Table 44—I. (Continued)

Table 44—I. Classification of Commonly Used Tool Steels(a)

Type Designation			Identifying Elements, Per Cent								
UNS	AISI	C	Mn	Si	W	Mo	Cr	V	Co	Ni	Al
Tungsten Types											
T20821	H21	0.35	—	—	9.00	—	3.50	—	—	—	—
T20822	H22	0.35	—	—	11.00	—	2.00	—	—	—	—
T20823	H23	0.30	—	—	12.00	—	12.00	—	—	—	—
T20824	H24	0.45	—	—	15.00	—	3.00	—	—	—	—
T20826	H26	0.50	—	—	18.00	—	4.00	1.00	—	—	—
Molybdenum Types											
T20841	H41	0.65	—	—	1.50	8.00	4.00	1.00	—	—	—
T20842	H42	0.60	—	—	6.00	5.00	4.00	2.00	—	—	—
T20843	H43	0.55	—	—	—	8.00	4.00	2.00	—	—	—

COLD-WORK TOOL STEELS
High-Carbon, High-Chromium Types

UNS	AISI	C	Mn	Si	W	Mo	Cr	V	Co	Ni	Al
T30402	D2	1.50	—	—	—	1.00	12.00	1.00	—	—	—
T30403	D3	2.25	—	—	—	—	12.00	—	—	—	—
T30404	D4	2.25	—	—	—	1.00	12.00	—	—	—	—
T30405	D5	1.50	—	—	—	1.00	12.00	—	3.00	—	—
T30407	D7	2.35	—	—	—	1.00	12.00	4.00	—	—	—

Medium-Alloy, Air-Hardening Types

UNS	AISI	C	Mn	Si	W	Mo	Cr	V	Co	Ni	Al
T30102	A2	1.00	—	—	—	1.00	5.00	—	—	—	—
T30104	A4	1.00	2.00	—	—	1.00	1.00	—	—	—	—
T30106	A6	0.70	2.00	—	—	1.25	1.00	—	—	—	—
T30107	A7	2.25	—	—	1.00(b)	1.00	5.25	4.75	—	—	—
T30108	A8	0.55	—	—	1.25	1.25	5.00	—	—	—	—
T30109	A9	0.50	—	—	—	1.40	5.00	1.00	—	1.50	—
T30110	A10(e)	1.35	1.80	1.25	—	1.50	—	—	—	1.80	—

Oil-Hardening Types

UNS	AISI	C	Mn	Si	W	Mo	Cr	V	Co	Ni	Al
T31501	O1	0.90	1.00	—	0.50	—	0.50	—	—	—	—
T31502	O2	0.90	1.60	—	—	—	—	—	—	—	—
T31506(e)	O6(e)	1.45	0.80	1.00	—	0.25	—	—	—	—	—
T31507	O7	1.20	—	—	1.75	—	0.75	—	—	—	—

(Continued on next page)

Table 44—I. (Concluded)

Table 44—I. Classification of Commonly Used Tool Steels(a)

| Type Designation | | Identifying Elements, Per Cent | | | | | | | | | |
UNS	AISI	C	Mn	Si	W	Mo	Cr	V	Co	Ni	Al
SHOCK-RESISTING TOOL STEELS											
T41901	S1	0.50	—	—	2.50	—	1.50	—	—	—	—
T41902	S2	0.50	—	1.00	—	0.50	—	—	—	—	—
T41904	S4	0.55	0.80	2.00	—	—	—	—	—	—	—
T41905	S5	0.55	0.80	2.00	—	0.40	—	—	—	—	—
T41906	S6	0.45	1.40	2.25	—	0.40	1.50	—	—	—	—
T41907	S7	0.50	—	—	—	1.40	3.25	—	—	—	—
MOLD STEELS											
T51606	P6	0.10	—	—	—	—	1.50	—	—	3.50	—
T51620	P20	0.35	—	—	—	0.40	1.70	—	—	—	—
T51621	P21	0.20	—	—	—	—	—	—	—	4.00	1.20
SPECIAL-PURPOSE TOOL STEELS Low-Alloy Types											
T61202	L2	0.50-1.10(c)	—	—	—	—	1.00	0.20	—	—	—
T61206	L6	0.70	—	—	—	0.25(b)	0.75	—	—	1.50	—
WATER-HARDENING TOOL STEELS											
T72301	W1	0.60-1.40(c)	—	—	—	—	—	—	—	—	—
T72302	W2	0.60-1.40(c)	—	—	—	—	—	0.25	—	—	—
T72305	W5	1.10	—	—	—	—	0.50	—	—	—	—

(a) Based on American Iron and Steel Institute Steel Products Manual, "Tool Steels," published September, 1981.
(b) Optional.
(c) Various carbon contents are available.
(d) Other carbon contents may be available.
(e) Contains free graphite in the microstructure to improve machinability.
NOTE 1: Some of the types can be produced with a sulphur addition to improve machinability.
NOTE 2: Some less commonly used tool steels, formerly included in the above classification, are listed in the manual referred to above.

230° C (250° to 450° F), although, where resistance to shock and impact is important, higher temperatures may have to be used. Tempering in the temperature range of 260° to 315° C (500° to 600° F) should generally be avoided.

High-Speed Steel—As mentioned above, the high-speed steels differ from the lower alloy tool steels, not only by the presence of higher percentages of carbide-forming elements, but also by the fact that the secondary hardening effects of these elements impart a high resistance to softening at elevated temperature. These steels require a special heat treatment in order that their unique properties may be fully realized. In outline, this procedure consists of heating to a high temperature of 1175° to 1315° C (2150° to 2400° F) to obtain solution of a substantial percentage of the alloy carbides, quenching to room temperature, at which stage a considerable amount of austenite is retained, tempering at 535° to 620° C (1000° to 1150° F), and again cooling to room temperature. During tempering, alloy carbides are precipitated, resulting in marked secondary hardening and a reduction of alloy content in the retained austenite, which then transforms to martensite on cooling to room temperature and results in a still greater hardness increase. It is often desirable to temper a second time to temper the martensite formed on cooling from the original tempering.

To prevent excessive grain growth and decarburization, the steels are held at high quenching temperature for only a few minutes before quenching. Steels are customarily preheated to between 760° and 870° C (1400° and 1600° F) before transferring to the high-heat furnace. This serves the dual purpose of eliminating the severe thermal shock of placing the cold tool into the high-temperature furnace, and of decreasing the decarburization because of the shorter time of exposure to the high temperature. The use of controlled-atmosphere furnaces, or of neutral liquid baths for the high-heat treatment is also very desirable, and in many cases essential, to minimize decarburization. The time in the high-heat furnace will, of course, vary with the heating rate and size and shape of the piece. Typical hardening temperatures are 1230° to 1285° C (2250° to 2350° F) for the tungsten types, 1175° to 1230° C (2150° to 2250° F) for the molybdenum types, and 1275° to 1315° C (2325° to 2400° F) for the cobalt types.

Quenching may be in air, oil, or in liquid baths. Air cooling has the disadvantage of the formation of a tightly adherent scale during cooling. Oil quenching, while it facilitates the removal of this scale, results in higher stresses. These may be minimized, however, by removing the piece from the oil at the flash point and then air cooling. By this method, the tools will be air cooled through the temperature range in which transformation to martensite occurs. The third method, which is very commonly practiced to ensure low quenching stresses, is to quench into a liquid bath at 535° C (1000° F), hold until equalized, and then slowly air cool to between 95° and 150° C (200° and 300° F) and immediately temper. This procedure is particularly applicable to hardening intricate tools without undue distortion or cracking.

Tempering may be carried out in salt baths, lead baths, or circulating-air furnaces. The latter are particularly desirable because of their adaptability to close control of temperature and uniformity of heating the work. The rapid heating of high-speed tool steels immersed in lead baths is undesirable since it may set up stresses which would lead to cracking. The maximum hardness will usually be developed at temperatures of 535° to 595° C (1000 to 1100° F), and the holding times in this temperature range usually are from 1 to 4 hours. The specific time and temperature will vary with the hardness and toughness desired.

As mentioned above, a second tempering at a relatively low temperature (315° to 345° C or 600° to 650° F) will temper the martensite formed on cooling from the first tempering operation and will increase the toughness of these steels without causing appreciable softening.

Bibliography

American Iron and Steel Institute, Steel products manual, Tool steels. Washington, D.C., September 1981.

American Society for Metals, Metals handbook. Cleveland, Ohio (Current Edition).

Bain, E. C., Functions of the alloying elements in steel. American Society for Metals, Cleveland, Ohio, 1939 (revised by E. C. Bain and Harold W. Paxton, 1961).

Grossmann, M. A., and Bain, E. C., High-speed steels. John Wiley and Sons, Inc., New York, 1931.

Roberts, G. A., and Cary, R. A., Tool steels. American Society for Metals, Cleveland, Ohio, 1980.

CHAPTER 45

High-Strength Low-Alloy Steels

Introduction—The generally accepted designation for the steels that are the subject of this chapter is "high-strength low-alloy steels" or, for convenience, HSLA steels.

HSLA steels are a group of steels intended for general structural or miscellaneous applications and that have specified minimum yield points above about 275 MPa (40 000 psi) and as high as 1035 MPa (150 000 psi). These steels typically contain small amounts of alloying elements to achieve their strength in the hot-rolled or heat-treated conditions. The steels are often sold as proprietary grades as well as to society specifications. The large number of proprietary steels accounts for the numerous grades in some society specifications such as American Society for Testing and Materials (ASTM) A 588. Producers sometimes have different methods of achieving the desired mechanical properties. There are hundreds of brand names of HSLA steels, some of which have properties not yet covered by specifications. Some of the typical specifications that do cover the HSLA steels are ASTM A 572, A 588, A 607, A 633, A 656, A 715, A 808, and Society of Automotive Engineers (SAE) J 1392. Some specifications, like the latter four listed, cover a number of different strength levels, and some of the specific steels are available in a broad range of products, such as sheet, plate, bars, and shapes. Sometimes, however, because of special processing practices required to achieve the required properties, the type of product is limited to the type of mill capable of achieving these properties.

Historical Background—Although a steel containing chromium was specified for certain members of the Eads Bridge, erected at St. Louis between 1867 and 1874, the steel most widely used for construction before 1900 was mild carbon steel having a tensile strength of about 415 MPa (60 000 psi). In 1902, the design engineers of the Queensboro Bridge, that was to span the East River in New York City, requested a stronger steel so that the number and size of supporting members could be reduced. Carnegie Steel Company, now part of the United States Steel Corporation, supplied 3.25 per cent nickel steel for this application. This steel also was used in the stiffening trusses of the Manhattan Bridge in 1906. Although it was satisfactory for riveted structures from the standpoint of strength, this material was relatively expensive and economical only for structures in which reduction in weight or size of members was a necessity. Another steel, offering higher strength than mild carbon steel and containing about 1.00 per cent silicon and 0.25 per cent carbon, was used for the hull plates in the S. S. Mauretania in 1907. Although this latter steel was less expensive than the nickel steel mentioned above, its use was ultimately discontinued because of many difficulties that were encountered in its application. Another steel, stronger than mild carbon steel, that contained only a nominal amount of silicon and depended for its strength on its high carbon content (usually over 0.30 per cent), was first used in 1915 in a bridge spanning the Ohio River at Metropolis, Illinois. This grade, under the designation "structural silicon steel," was one of the most widely used materials for riveted structures.

In 1927, the American Bridge Company, now a division of United States Steel Corporation, used a 1.60 per cent manganese steel for the lower chord members of the Kill van Kull Bridge connecting Staten Island with the mainland at Bayonne, New Jersey.

Similar developments had taken place in other countries. Engineers in Great Britain, attempting through weight reduction to effect economies in ocean freights and handling charges, had used carbon steels containing generous amounts of silicon and manganese. On the Continent similar steels were used, and early in 1933 "The Flying Dutchman," the first lightweight streamlined train, was built in Germany of a high-strength steel (ST. 52) that contained additions of silicon, manganese and copper.

The earliest of the present-day HSLA steels was "COR-TEN" brand, a steel with excellent atmospheric corrosion resistance, which was introduced by United States Steel in 1933. This so-called 'weathering' steel has been improved continuously to meet the increasingly severe demands of fabricators—an experience common with most new steels.

Although the strengthening effects of niobium (columbium) and vanadium in steels were recognized in the 1930's and 1940's, it was not until the late fifties and early sixties that HSLA steels were fully introduced as a structural material. The interest in those steels was mainly due to their high strength along with suitable weldability and formability; the atmospheric corrosion resistance of such steels is usually the same as that of carbon steel.

An important and rapidly growing new requirement for HSLA light-gage hot-rolled and cold-rolled sheets started to develop about 1972 for automobile bumpers, bumper supports, wheels, chassis and frame members, and exposed and unexposed body panels as a result of mandated fuel savings and damageability requirements. This has led to the continuing development of a new class of low-carbon (0.15 per cent maximum) microalloyed HSLA steels to meet the requirements of automotive production, which include a very high level of formability and excellent resistance spot weldability.

Table 45—I. Chemical Composition Ranges of Representative HSLA Sheet Steels

Element	Composition, max, % Cast or Heat (formerly Ladle) Analysis
Carbon	0.15
Manganese	1.65
Phosphorus	0.025
Sulfur	0.035

Types, by Added Elements, %

Type 1:	
Titanium, min	0.05
Silicon, max	0.10
Type 2:	
Vanadium, min	0.02
Silicon,[a] max	0.60
Nitrogen,[a] min	0.005
Type 3:	
Columbium, min	0.005
Vanadium,[a] max	0.08
Silicon,[a] max	0.60
Nitrogen,[a] max	0.020
Type 4:	
Zirconium, min	0.05
Silicon, max	0.90
Chromium,[a] max	0.80
Titanium,[a] max	0.10
Boron,[a] max	0.0025
Columbium[b]	0.005-0.06
Type 5:	
Columbium,[c] min	0.03
Molybdenum,[c] min	0.20
Silicon, max	0.30
Type 6:	
Columbium	0.005-0.10
Silicon, max	0.90
Type 7:	
Columbium or vanadium or both, min	0.005
Silicon, max	0.60
Nitrogen, max	0.020
Type 8:	
Columbium	0.005-0.15
Zirconium, min	0.05
Type 9:	
Columbium, min	0.01
Vanadium,[a] min	0.05
Silicon,[a] max	0.60

[a]Not added to Grades 50 and 60.
[b]Might not be added to Grade 50.
[c]Available as Grade 80 only.

The composition ranges of some of the steels used to produce 345, 415, 485, and 550 MPa (50, 60, 70, and 80 ksi) minimum yield strength sheet are shown in Table 45—I (ASTM A 715). Both uncoated and coated (for example, galvanized) HSLA hot- and cold-rolled sheets are used in various automotive applications.

FUNDAMENTAL CHARACTERISTICS

To be of interest as construction materials, HSLA steels must have characteristics and properties that result in economies to the user when the steels are applied properly. They should be considerably stronger, and in many instances tougher, than structural carbon steel. Also, they must have sufficient ductility, formability and weldability to be fabricated successfully by customary shop methods. In addition, improved resistance to corrosion often is required so that equal or longer service life in a thinner section is obtained in comparison to that of a structural-carbon-steel member. The combination of important characteristics that an all-purpose HSLA steel should possess, and the methods by which these characteristics are commonly determined, are shown in Table 45—II. All of the HSLA steels have higher strength than most as-rolled or normalized structural carbon steels, and most are readily weldable using good shop or field practices and have good formability commensurate with their strength level. Their corrosion resistance depends upon composition, and their notch toughness depends upon composition as well as the kind of processing they have received.

Table 45—II. Important Characteristics of An All-Purpose HSLA Steel.

Property	Method of Determination
High yield strength	Tension test.
Good formability	Bend test, ductility in tension test, and fabrication performance.
Good weldability	Standardized weldment performance tests.
Good atmospheric corrosion resistance	Weight loss in exposure-rack tests and useful life judged from service performance.
Suitable toughness under adverse conditions	Energy absorption in notched-bar impact test.

Strength—A comparison of the tensile properties of a typical HSLA steel plate with those of structural carbon steel is shown in Table 45—III. The yield point of a constructional member determines the stress to which a structure may be subjected without permanent deformation. Therefore, allowable stresses of each structure are based upon this important property. The minimum yield point of a typical structural carbon steel (ASTM A 36 steel) is 250 MPa (36 000 psi); that of a typical HSLA steel is 345 MPa (50 000 psi). Hence, on the basis of the proportionality of their yield points, the allowable stress employed with the HSLA steel may be increased to 1.4 times that used with the structural carbon steel. A greater increase can be realized with the use of HSLA steels of higher strength. The use of higher allowable stresses generally permits reduction in the size of a structural member, and this results in a decrease in weight. It should be noted that, in members where buckling can occur, the allowable stresses must be modified to the extent necessary to assure stability. Frequently, HSLA steels are substituted for structural carbon steel without change in section, the sole purpose being to produce a stronger and more durable structure with no increase in weight. Savings in weight are of utmost importance in mobile structures when it permits these structures to carry greater payload or to be transported using less energy.

Table 45—III. Comparative Tensile Properties for Typical HSLA and Carbon Plate and Structural Steels.

Property	ASTM A 441, A 572 Grade 50 and A 588 HSLA Steels	ASTM A 36 Structural Carbon Steel
Yield Point Minimum, MPa (psi)	345 (50 000)	250 (36 000)
Tensile Strength, MPa (psi)	480 (70 000) minimum	400 to 550 (58 000 to 80 000)
Elongation Minimum, Per cent in 200 mm (8 in.)	18	20

Higher strength is frequently obtained by the addition of small amounts of niobium (columbium) or vanadium, or titanium. These elements provide economical strengthening by precipitation hardening. Other alloying elements that are introduced for various purposes also provide strengthening. In hot-rolled sheet product, closely controlled hot rolling and coiling practices are employed to develop the desired uniform strength. For cold-rolled sheets, special annealing and temper-rolling practices are used to provide additional strengthening while maintaining adequate formability.

A recent development is the introduction of certain HSLA sheet steels which, by intercritical annealing and rapid cooling, produce a mixed microstructure (or a dual-phase microstructure) of martensite and ferrite. Such sheet product has excellent formability and typically exhibits 310 to 345 MPa (45 000 to 50 000 psi) yield point which is increased to 550 MPa (80 000 psi) or greater by straining in the press forming of automotive parts.

Formability—HSLA steels must have suitable properties so that they may be hot or cold worked readily and economically into various commodities for engineering structures. These operations, and others such as shearing, punching, and machining, can generally be performed on HSLA steels about the same as on structural carbon steels. Despite their high yield points, many HSLA steels can be formed satisfactorily in the same press brakes, draw benches, presses, and other equipment used for cold forming structural carbon steels, even when these forming operations are quite severe, though some die modifications may be necessary.

There are some inherent differences between the cold-forming characteristics of HSLA steels and those of structural carbon steels. First, more force is required to produce a given amount of permanent set in a HSLA section than in a structural carbon steel section of the same dimensions. Second, a somewhat greater allowance for springback should be provided when forming the HSLA steels.

Experience has shown that more liberal radii of bend must be used with HSLA steels than with structural carbon steel for successful cold forming unless the HSLA steels are treated for inclusion-shape control.

The HSLA automotive sheet steels, such as shown in Table 45—I and meeting ASTM A 715, have good formability for their strength levels because of their low-carbon contents. In addition, many are produced with very-low-sulphur contents and/or inclusion-shape control to provide improved transverse ductility.

Weldability—Since welding often is employed in fabricating structural steel, it is important that HSLA steels for these applications be readily weldable by all the widely used arc-welding processes in sheet and strip thicknesses. It is equally important that the welds in fabricated structures have the required strength and ductility to withstand the most adverse conditions anticipated in the contemplated service. The development of the present-day HSLA steels has paralleled the growth of the various welding processes, and particular care was exercised to make certain that these steels possessed suitable welding characteristics. Most of these steels are considered to be readily weldable by conventional processes when good shop or field practices are employed.

For shielded-metal-arc welding of HSLA steels having minimum yield point up to about 345 MPa (50 000 psi), E 60 or E 70 group mild-steel covered electrodes are generally satisfactory. E 70 grade electrodes are suggested for steel grades having somewhat higher minimum yield points. For heavier sections and for grades that have higher carbon and manganese contents, preheating and/or low-hydrogen type electrodes such as E 7016 or E 7018 are required. For some HSLA steels, only low-hydrogen practices should be used for all thicknesses. Low-alloy-steel electrodes are generally required for steels with minimum yield points higher than about 415 MPa (60 000 psi) or when specific characteristics, such as enhanced corrosion resistance, are required in the weld metal. Electrodes or electrode-flux combinations that provide filler metal similar to that of the suggested electrodes for shielded-metal-arc welding are recommended for submerged-arc, gas-metal-arc and flux-cored-arc welding.

In the automotive HSLA sheet steels, good spot weldability is achieved through the use of low-carbon contents which are generally restricted to about 0.13 per cent or less.

Corrosion Resistance—When HSLA steels are employed, it is desirable to take advantage of their strength by employing thinner sections, not only to save weight when this is desirable but also to make the steel selection as economical as possible. However, adequate consideration must always be given to corrosion, and the thinner the steel section, the more important is the prevention of corrosion. Corrosion prevention on any structure is ordinarily achieved by the application of paint coatings on suitably prepared surfaces and by maintenance of the protective coatings.

One class of HSLA steels, such as the COR-TEN steels, has much-improved resistance to atmospheric corrosion which not only results in improved paint performance but in some cases provides steels that, with suitable design precautions, can be used exposed to the atmosphere in the bare or unpainted condition.

No material is equally resistant to all the corrosive conditions to which it might conceivably be exposed. A large number of corrosion tests, particularly atmospheric rack tests, have been conducted under many conditions of exposure, with both painted and bare samples, for the purpose of determining performance. Performance is evaluated by measurement of the trends in weight loss due to corrosion (after descaling) after various time periods.

The atmospheric-corrosion resistance of HSLA steels varies with the combination and content of those alloying elements most effective in building up this resistance. The elements providing increased atmospheric corrosion resistance are copper, phosphorus, silicon, chromium, nickel, and molybdenum. Some of the HSLA steels (see Table 45—I) contain alloy additions that appreciably increase their resistance to atmospheric corrosion. Several steels of this type, notably those meeting ASTM Specifications A 242 and A 588, possess approximately four times the atmospheric-corrosion resistance of structural carbon steel having a low copper content. The superior atmospheric-corrosion resistance which many of the HSLA steels have shown in rack tests has been confirmed by their performances in many different kinds of service.

This superior resistance to atmospheric corrosion has led to a new concept in the design of structures such as buildings, bridges and towers. Such structures are now being built using exposed members of appropriate HSLA steels, for example the COR-TEN steels, in the bare condition. When properly exposed to the atmosphere, such bare steels develop a tight protective oxide coating during the first several months of weathering. Thereafter, there is little additional corrosion of the steel. Bare steel is sometimes specified by architects because they desire the unique appearance of the weathered oxide on the surface. At other times, bare HSLA steel is specified because the resultant elimination of the need for painting provides marked economies during the life of the structure. The first major use of bare HSLA steel in an architectural application was in the Deere and Company buildings that were constructed in Moline, Illinois, during the early 1960's (see Figure 45—1). In these buildings, all exterior walls and associated members were made from COR-TEN A steel and were not painted or otherwise coated.

When using this class of corrosion-resistant HSLA steels in the bare condition, design considerations must be taken into account so that no surfaces remain continuously wet. Also, careful attention must be given the specific atmospheric environment to be sure the corrosion rate will be acceptable under the existing conditions. For example, high concentrations of strong chemical or industrial fumes are not desirable. A thor-

Fig. 45—1. Deere and Company building in Moline, Illinois. The first major architectural use of unpainted HSLA steel for structural members exposed to the weather. Architect: Eero Saarinen and Associates.

ough evaluation of conditions, and even exposure tests, may be necessary to assess whether use of the bare steel in certain environments is acceptable.

Figure 45—2 shows time corrosion curves for structural carbon steel and COR-TEN A steel exposed to an industrial atmosphere in the United States. As previously mentioned, in other exposure tests it has been shown that paint coatings applied to the corrosion-resistant HSLA steels exhibit a longer service life than when applied to structural carbon steels. Figure 45—3 illustrates an instance of superior paint life on COR-TEN steel as compared to that on carbon steel containing copper. Protective coatings are available that will provide long life on the corrosion-resistant steels in aggressive or continuously wet environments that would be too severe for the use of bare steel. Appropriate inspection and maintenance to assure satisfactory performance are recommended as on any structure.

Much has been written concerning the details and results of corrosion tests and a number of references are included in the bibliography at the end of this chapter for the use of those who wish to pursue the subject further.

Notch Toughness—HSLA steel grades are designed to have adequate notch toughness for their intended structural application, but generally are not supplied to a minimum notch-toughness criterion. The suitability of the notch toughness of specific HSLA grades may be based on established service performance alone or in combination with the results of impact tests on notched specimens. Some HSLA steels are produced with exceptionally good notch toughness to meet the stringent requirements of certain applications. For example, controlled hot-rolling practices are now used commonly in the production of HSLA plate to be fabricated into welded line pipe; such pipe are required to meet notch-toughness specifications established by the American Petroleum Institute (API). As another example, FIFTY-N and SIXTY-N steels are produced in the normalized condition which, coupled with selected

compositions, results in ductile-to-brittle transition temperatures below –60°C (–75°F) in plate thicknesses up to 75 mm (3 in.). The ASTM A 572 and A 588 grades must meet impact requirements at –12 to 21° C (10 to 70°F) when used for main tension members in highway bridges.

EFFECT OF CHEMICAL COMPOSITION ON PROPERTIES AND CHARACTERISTICS

In the development of chemical compositions to obtain the desired properties in HSLA steels, it was, of course, imperative that strength be given first consideration. Since increased strength can be obtained with various combinations of alloying elements, a number of different compositions have been produced which offer interesting combinations of other properties and characteristics in addition to the required minimum strength. The compositions of a representative group of current HSLA steels are shown in Table 45—I, earlier in this chapter.

The mechanical properties of all steels, including the HSLA grades, are determined primarily by their microstructures. HSLA steels generally have ferrite-pearlite microstructures and their properties are affected by changes in the microstructure in the same way as is described for carbon steel in Chapter 42. That is, strength can be increased by increasing the amount of pearlite, increasing the fineness of the structure, and increasing the amount of dispersed phases (precipitation hardening). As strength is increased by these microstructural changes, notch toughness is usually impaired in proportion to the strength increase with the exception that a finer microstructure (finer grain size) is accompanied by an increase in notch toughness. A finer grain size in as-rolled steel is usually obtained by controlled rolling at low hot-rolling temperatures. When grain size is reduced, as in controlled rolling, its favorable effect on notch toughness can override a deleterious effect from, say, precipitation hardening. Thus,

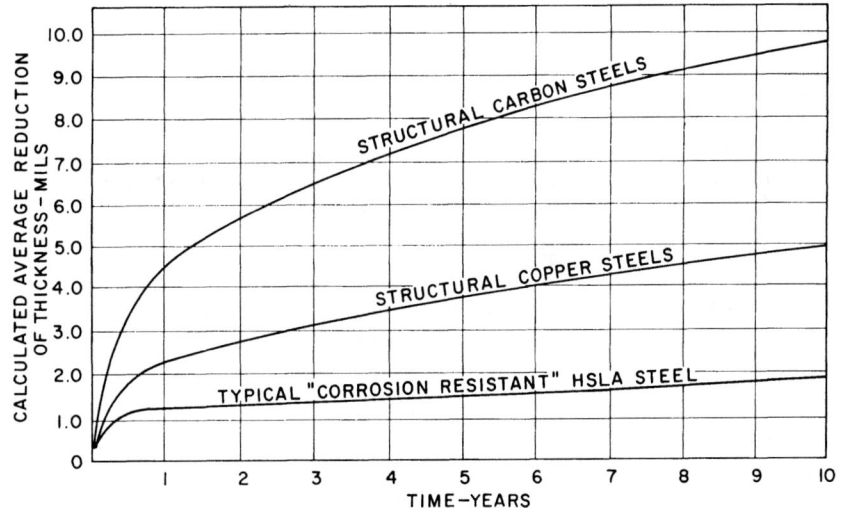

FIG. 45—2. Comparative corrosion of steel specimens exposed to an industrial atmosphere. (1 mil = 25.4 micrometres.)

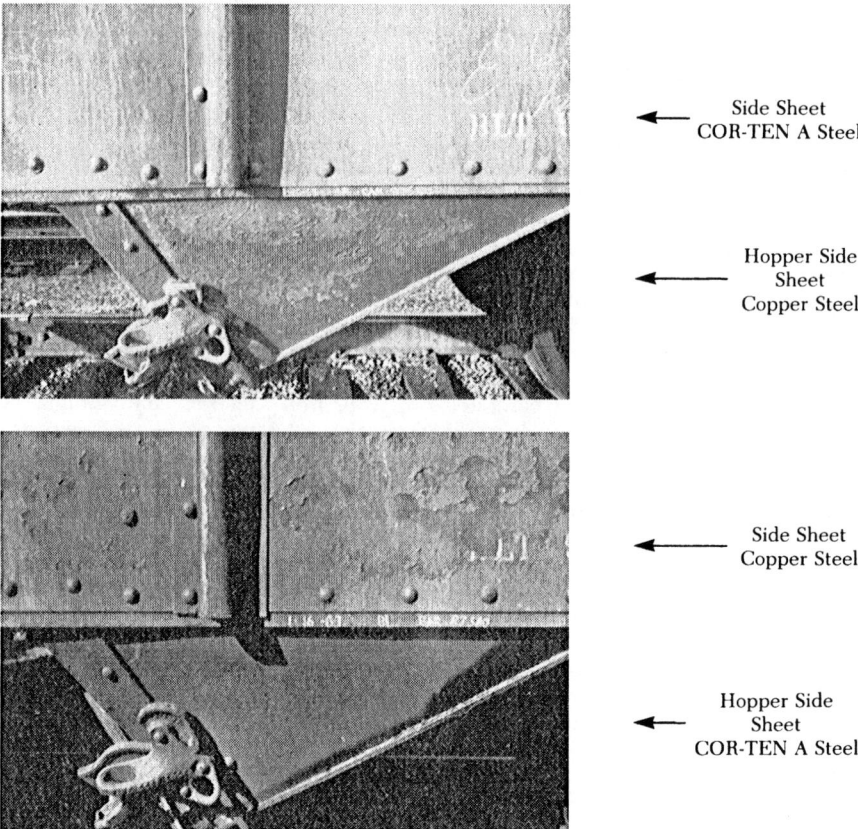

Side Sheet
COR-TEN A Steel

← Hopper Side
Sheet
Copper Steel

← Side Sheet
Copper Steel

← Hopper Side
Sheet
COR-TEN A Steel

FIG. 45—3. Appearance, after 12 years service, of painted hopper side sheets on opposite sides of the same railroad car. Note the deterioration of paint on the copper-steel side sheet in the bottom illustration.

the controlled rolling of niobium-treated (columbium-treated) steel can result in an excellent combination of strength and toughness. In addition to microstructural effects, certain alloying elements provide solid-solution strengthening of ferrite as is described for alloy steels in Chapter 43. The atmospheric-corrosion characteristics of HSLA steels are determined by chemical composition per se.

Carbon—Carbon is one of the more economical strengthening elements. The effect of this element on steel properties has been discussed in previous chapters.

It is generally desirable to keep the carbon content below certain maximum values when metal-arc welding is involved to avoid embrittlement of the heat-affected zone adjacent to the weld. HSLA steels may have increased hardenability due to the presence of certain alloying elements so that the tendency for martensite to form in the heat-affected zone is increased. The higher the carbon content, the harder the martensite-type structure will be in the event that it does form. Thus, for a given carbon content, low-hydrogen electrodes and/or a preheat may be required for a HSLA steel but not for a plain carbon steel. For the same reasons, a preheat and/or a postheat may be required when flame cutting HSLA steel.

Manganese—HSLA steels generally have higher manganese contents than structural carbon steels. Man-ganese increases the strength and, at the same time, improves the notch toughness. Because of these favorable effects, it is frequently present in HSLA steels at levels of 1 per cent or more. In steels for welding applications, this element should be kept below some maximum value that depends on the over-all composition but mainly on the carbon content.

Phosphorus—An important group of present-day atmospheric corrosion-resistant HSLA steels has phosphorus contents in the range of 0.07 to 0.15 per cent. Also, when small amounts of copper are present in the steel, the effect of phosphorus is greatly enhanced so that a given amount of phosphorus and copper together provides a greater beneficial effect on corrosion resistance than that produced by the corresponding amount of either of the individual elements.

Additions of phosphorus increase the strength properties of steel by entering into solid solution in the ferrite, but this increase may be accompanied by a decrease in ductility. Phosphorus formerly was considered to cause embrittlement in steels when present in amounts exceeding about 0.10 per cent. It was found, however, that the embrittling effect of phosphorus is influenced markedly by the carbon content and that this effect is not so pronounced in steels with carbon contents below about 0.15 per cent. Several grades of hot-and cold-rolled sheet with phosphorus additions have been proposed for automotive or structural applications.

Copper—Many present-day HSLA structural steels may contain copper in amounts ranging from 0.20 to 1.50 per cent. Copper, in the concentrations used, is by far the most potent of all the common alloying elements in improving atmospheric-corrosion resistance. Copper is especially effective in amounts up to about 0.35 per cent in regular carbon steel. Continued improvement may be obtained up to about 1.00 per cent copper, but the effect is not nearly so marked as with additions up to about the first 0.35 per cent copper.

Copper increases the strength properties of low- and medium-carbon steels to a moderate extent by ferrite strengthening with only a slight accompanying decrease in ductility. Copper up to at least 0.75 per cent is considered to have little effect on notch toughness or welding performance. Steels containing over about 0.60 per cent copper are capable of producing precipitation hardening of the ferrite. Steels containing about 0.50 per cent or more of copper frequently exhibit "hot shortness" during hot working, so that cracks or extremely roughened surfaces, sometimes referred to as "checking," may develop during hot deformation. The occurrence of these undesirable surface conditions can be minimized by careful control of oxidation during heating and taking care not to overheat for hot working. Also, the addition of nickel in an amount equal to at least one-half the copper content is very beneficial to the surface quality of steels containing copper.

Vanadium—Vanadium is a widely used strengthening agent in HSLA steels. This element, in amounts up to about 0.12 per cent, provides increased strength without impairing the weldability or markedly reducing the notch toughness of as-rolled steels. For this reason, vanadium is an essential ingredient in most specifications for as-rolled HSLA steel shapes and plates such as A 441, A 572, and A 588.

Vanadium-bearing steels are strengthened by precipitation hardening of the ferrite and by refining the ferrite grain size. Any refinement of grain size due to vanadium in hot-rolled steel is dependent on thermomechanical processing variables. The refinement will not generally take place unless fairly low finishing temperatures are employed. A manganese content of about 1.0 per cent or higher provides more effective precipitation hardening and grain refinement, and nitrogen contents of about 0.010 per cent and higher are particularly helpful in promoting vanadium-nitride precipitation.

Niobium (Columbium)—The use of niobium as a strengthening agent in HSLA steels became commercially important arount 1957 when economic availability of this element was markedly increased. Small additions of this element provide significant increases in the yield point and in the tensile strength of carbon steel. For example, the addition of 0.02 per cent of niobium has been found to provide an increase of about 70 to 105 MPa (10 000 to 15 000 psi) in yield point in hot-rolled 0.20 per cent carbon steel. However, the increased strength is accompanied by a marked impairment of notch toughness unless special rolling practices are used. Such practices involve the use of lower-than-normal temperatures for the final reduction passes. Therefore, hot-rolled niobium-bearing HSLA steels

generally are limited to the lighter plate thicknesses or to sheet that can be processed economically by the special rolling practices.

As is the case with vanadium, niobium provides strengthening by both precipitation hardening and ferrite grain refinement, and the effectiveness of these mechanisms is influenced by thermal processing variables and the presence of other alloy elements in the steel.

Nitrogen—The strength level of HSLA steels is affected by nitrogen content. Typical amounts of nitrogen (about 0.003 to 0.012 per cent depending upon steelmaking practice) provide definite contributions to strength, and when the nitrogen content is altered, such as by a change in steelmaking practice, adjustments in composition may be required to meet a specified strength level. Additions of nitrogen in amounts up to about 0.02 per cent to plain carbon steels have been used as a very economical way to obtain strengths typical of HSLA steels. In carbon steel, this practice generally is limited to light-gage as-rolled products because the increased strength may be accompanied by an impairment in notch toughness. There is generally no impairment when the steel contains a strong nitride former and the steel is properly heat-treated such as with USS SIXTY-N steel (A 633 Grade E). However, nitrogen additions to HSLA steels containing vanadium have become commercially important because such additions enhance the precipitation-hardening mechanism. Precipitation hardening usually is accompanied by a loss in notch toughness which may be overcome by the use of a lower carbon content and/or processing practices that result in a finer ferrite grain size. Use of vanadium-nitrogen strengthening to obtain a given strength level at a lower carbon content can provide improvements in weldability.

Nickel—Nickel is added in amounts up to about 1.0 per cent in some HSLA steels. This element provides a moderate increase in strength by solution hardening of the ferrite. It also provides enhancement of the atmospheric corrosion resistance, and when present in combination with copper and/or phosphorus, also increases the corrosion resistance in steels subject to wetting by sea water. The strengthening obtained by nickel generally is accompanied by a slight improvement in notch toughness. As mentioned previously, nickel often is added to copper-bearing steels to overcome hot shortness.

Other Elements—In addition to the elements discussed separately above, other alloying elements are added to HSLA steels in varying amounts. Among these are silicon, chromium, molybdenum, titanium, zirconium, and rare earths. Silicon provides a moderate increase in strength due to ferrite hardening but enhances the atmospheric-corrosion resistance. Chromium usually is added to obtain improved atmospheric-corrosion resistance. The use of molybdenum in normally hot-rolled or normalized HSLA steel plate and structurals may be limited because its potent effect on hardenability may cause some low-temperature transformation products (bainites and/or martensite) to form which generally impart poor toughness to the steel. The detrimental effects of these low-temperature transformation products are offset or eliminated in certain molybdenum-containing steels that are quenched and tempered or that are rolled by special controlled-rolling practices on a plate or hot-strip mill. The effects of titanium are similar to those of vanadium and niobium, but because of its strong deoxidizing characteristics, it is only useful in fully killed steels. Zirconium, titanium, and rare earths are added to killed HSLA steels to obtain improvements in inclusion characteristics, mainly, rounding of the sulphides in order to improve transverse bend properties or formability.

Many articles and papers have been published that deal with the properties, characteristics, and fabrication of the steels of various compositions that are termed HSLA steels, and a number of references that cover the subject in greater detail will be found at the end of this chapter.

APPLICATIONS

HSLA steels can be used advantageously in any structural application where the greater strength can be utilized either to decrease the weight or increase the durability of the structure. These steels find application in all recognized market classifications. The earliest large field of application was in the construction of transportation equipment. Since 1934, one of the leading grades of corrosion-resistant HSLA steels has been used in the construction of about one-half million railroad freight and passenger cars. Previously, the main emphasis in railroad freight-car usage was on the savings in operating costs obtained by using HSLA steels in somewhat reduced thicknesses to decrease the dead weight of cars or to increase their capacity without increasing the dead weight. The emphasis has shifted in some applications, however, from weight reduction to obtaining stronger, more-durable equipment. A big benefit obtained from using the corrosion-resistant HSLA steels in railroad cars has been reduced maintenance associated with reduced corrosion.

Weight savings has been a major consideration for using HSLA steels in automobiles and trucks in order to obtain better fuel economy. The use of HSLA steels in mobile equipment is expected to greatly expand as standards for fuel economy become more stringent while at the same time safety standards and environmental controls serve to increase car weight.

In bridges, designers are giving increased recognition to the importance of reducing dead weight. One solution has been the use of HSLA steels, particularly for bridges involving long spans in which a reduction of weight at the center permits additional savings in the weight of supporting members. Federal and State Highway Code agencies have approved grades such as ASTM A 588 and A 572 (Grades 42, 45, and 50) for welded-steel bridges. HSLA steels also lend themselves to economical tower construction, where the properties permit the use of sections smaller than would be required in structural carbon steel. This advantage is important in tall television towers, where forces resulting from wind resistance are lessened by use of smaller sections, and in transmission towers where lighter weight is a substantial advantage in reducing freight and handling costs.

HSLA steels are being used for columns in high-rise buildings. Judicious use of such steels in place of, and in combination with, structural carbon steel can result in substantial cost savings and an increase in usable floor area. HSLA steels also are being used to advantage in the framing members of industrial and farm buildings.

COR-TEN steel and other corrosion-resistant HSLA steels have been used for exposed members of buildings because the architects desired the appearance of the tightly adherent oxide coating that forms on this steel. Also, because of the superior resistance to atmospheric corrosion of this steel, it is being used in the "bare" condition for towers and bridges to eliminate the cost of maintenance painting.

The weight of portable containers for liquefied petroleum gas, such as used to supply gas for domestic and other low-capacity heating requirements, has been reduced appreciably by the use of HSLA steels, making them easier and less costly to handle and ship. Almost all such portable containers are now made of HSLA steel.

A few of the many other applications for HSLA steels include: the inner bottoms, floors, tanks, and hatch covers of ore boats; coal bunkers; street-lighting poles; portable oil-drilling rigs; jet-blast fences; cable reels; automobile bumpers; pole-line hardware; air-conditioning equipment; oil-storage tanks; stokers; agricultural-machinery parts; earth-moving equipment; military and domestic shipping containers; and air-preheater tubes.

Bibliography

American Iron and Steel Institute, High Strength Sheet Steel Source Guide, SG-603D, AISI, 1981.

American Iron and Steel Institute, Welding in the Automotive Industry, SG-81-5, AISI,1981.

American Society for Metals, Corrosion of metals, Cleveland, Ohio, The Society (1946).

American Society for Metals, Metals Handbook, 9th ed. Vol. 1, Cleveland, Ohio, The Society, 1978. (High-Strength Structural and High-Strength Low-Alloy Steels, 403-420).

American Society for Testing and Materials, Annual Book of ASTM Standards; Parts 3 and 4, Philadelphia, Pa., The Society, latest edition.

American Welding Society, Welding Handbook; Section 4, 7th Edition, New York, The Society (1976).

Architectural Record (Editors), The steel that will weather naturally, 132, 148-150 (1962).

Austin, J. B., Trends in the metallurgy of low-alloy high-yield strength steels, American Society for Testing and Materials, 1963 Gillett Memorial lecture.

Bain, E. C. and Paxton, H. W., Alloying Elements in Steel, Second Edition, ASM, Cleveland, Ohio, 1961.

Brockenbrough, R. L. and Johnston, B. G., Steel Design Manual. Pittsburgh, U. S. Steel Corporation, (1981) (ADUSS 27-3400-04).

Coburn, S. K., Gilliland, G. W., and Pohlman, J. C., Bare steel structures—a new concept. Electrical Engineering 82, 666-672 (1963).

Copson, H. R. and Larrabee, C. P., Extra durability of paint on low-alloy steels. ASTM Bull. no. 242, 68-74 (1959).

Corrosion, Edited by L. L. Shreir, Newnes and Butterworths, London, (1976).

Formable HSLA and Dual Phase Steels, Conference Proceedings, The Metallurgical Society of AIME, New York, N. Y., 1981.

Frost, R. W., The static and dynamic behavior of hybrid steel beams. Society of Automotive Engineers, Paper 769A, October 1963.

Haaijer, G., Economy of high-strength steel structural members. American Society of Civil Engineers, Transactions 128; part 2, 820-847 (1963).

Horton, J. B., The Rusting of Low-Alloy Steels in the Atmosphere, AISI Regional Technical Meeting, Pittsburgh, Novermber 1965.

Kelly, B. J , Corrosion of railroad hopper car body sheets. Corrosion 7, 196-201 (1951).

LaQue, F. L. and Boylan, J. A., Effect of composition of steel on performance of organic coatings in atmospheric exposure. Corrosion 9, 237-241 (1953).

Larrabee, C. P., Corrosion resistant steels for marine application. Corrosion 14, 501-504 (1958).

Larrabee, C. P., Mechanism of atmospheric corrosion of ferrous metals. Corrosion 15, 526-529 (1959).

Larrabee, C. P., Corrosion resistance of high-strength low-alloy steels as influenced by compositon and environment. Corrosion 9, 259-271 (1953).

Microalloying 75, Proceedings of an International Symposium on HSLA Steels, New York, N.Y. (1977).

Processing and Properties of Low Carbon Steels, Conference Proceedings, The Metallurgical Society of AIME, New York, N. Y., (1973).

Schmitt, R. J. and Gallagher, W. P., Unpainted High-Strength Low-Alloy Steel for Architectural Applications, Materials Protection, 8 (12), 70-77 (1969).

Schmitt, R. J. and Mathay, W. L., Performance of Low-Alloy Steels in Chemical Plant Environments, Materials Protection, 6 (9) 37-42 (1967).

Society of Automotive Engineers, SAE Handbook, New York, The Society (Recommended Practice J 1392).

Structure and Properties of Dual-Phase Steels, Conference Proceedings, The Metallurgical Society of AIME, New York N. Y., (1979).

Technology and Applications of HSLA Steels, Conference Proceedings, American Society for Metals, Metals Park, Ohio (1984)

Thermomechanical Processing of Microalloyed Austenite, Conference Proceedings, The Metallurgical Society of AIME, New York, N. Y., (1982).

Uhlig, H. H., Corrosion and Corrosion Control, John Wiley & Sons, Inc., New York, N. Y. (1971).

CHAPTER 46

Electrical Sheet Steel

SECTION 1

INTRODUCTION

The use of silicon as an alloy to improve the magnetic quality of steel was patented by Sir Robert Hadfield in England around 1900 and introduced in the United States several years later. The silicon-bearing sheet steels that have evolved since the Hadfield discovery have made possible the development of more efficient and more powerful electrical devices and have played an extremely important role in the rapid growth of the electrical-power industry. In recent years, the quality of carbon steel made specifically for magnetic applications has been improved also and now the nonsilicon grades are included in the family of electrical sheet steels and command a large share of the market.

Annual consumption of electrical sheet in the United States, both carbon-type and silicon-bearing grades, averaged an estimated 1.3 million metric tons (1.4 million net tons) during the five-year period 1979 through 1983.

The usual basis of sale for electrical sheet is a magnetic property called **core loss**. This quantity, expressed in watts per kilogram in SI and in watts per pound in the customary system, is determined by an Epstein test and is a measure of the efficiency of the steel's performance in conducting an alternating magnetic field, as is the purpose of the core of electrical devices. Core loss can be defined very simply as the electrical energy that is expended in the core steel without contributing to the work of the device. Most of this energy is dissipated as heat but some portion is consumed in magnetizing the steel. Most importantly, in practice, close control of the amount of this heat loss is necessary to avoid damage to the materials which insulate the magnetic core from the electrical circuitry. This loss is also an important operating cost factor.

Silicon in electrical sheet steel has an effect upon both **eddy current loss** and **hysteresis loss**, the two components of total core loss. These and other technical terms are defined in Section 8 at the end of this chapter. Increasing silicon content increases the electrical resistivity of the steel and thereby reduces the eddy-current loss which is that portion of core loss attributable to induced electrical currents circulating in the steel when it is subjected to an alternating magnetic field. The hysteresis component is also reduced as silicon content is raised. In this case, silicon reduces the magnetic reluctance and lessens the amount of energy expended in magnetizing the core steel. However, there are practical limitations to the amount of silicon that can be added to commercially available steels because the higher silicon alloys are inherently brittle at ambient temperatures. Therefore, the composition of cold-reduced grades is restricted to a maximum of approximately 3 per cent whereas, hot-rolled grades may contain up to about 5 per cent silicon. The hot-rolled grades are no longer commercially produced in the United States.

In addition to silicon content, other factors are influential in magnetic performance. Steel thickness affects eddy-current losses, while internal stress, crystallographic texture and impurities are important with regard to hysteresis characteristics. Best performance is generally associated with thinner gage, higher silicon, freedom from stress, appropriate crystallographic texture and minimum impurities.

In this chapter, reference is made only to steels produced in sheet form, coils and cut lengths, for the core sections of electrical equipment. These qualities are commercially available as cold-reduced product in relatively thin gages, 0.23 mm to 0.99 mm (0.009-in. to 0.039-in.), and containing up to about 3 per cent silicon.

SECTION 2

CLASSIFICATION AND APPLICATION OF ELECTRICAL SHEETS

Electrical sheet grades are divided into two general classifications, (1) oriented steels and (2) nonoriented steels. The oriented steels are given mill treatments designed to yield exceptionally good magnetic properties in the rolling, or lengthwise, direction of the steel. Nonoriented grades are made with a mill treatment that yields a grain structure, or texture, of a random nature and, therefore, the magnetic properties in the rolling direction of the steel are not significantly better than those in the transverse direction.

Subdivisions of the oriented category of electrical sheet include semi-processed, fully processed and a low-stress type. The semi-processed type has limited commercial availability and requires high-temperature annealing (approximately 1175°C or 2150°F) in pure dry hydrogen by the purchaser to develop appropriate magnetic characteristics. Fully processed also requires careful annealing in core or lamination form but in a stress-relief anneal at about 790°C (1450°F). The stress-free type of oriented steel has been developed specifically for those applications where annealing prior to core assembly is not intended.

Published core-loss maximums, shown in Table 46—I for oriented grades, recognize two products. One is a regular fully-processed quality and the other is a high-permeability steel. The high-permeability product is sold on the basis of core-loss test results at a flux density of 1.7 teslas (17 000 gausses) only and is expected to have permeability at 795.8 A/m (10 oersteds) field strength higher than 1880. The regular quality can be ordered to either 1.7 or 1.5 teslas (17 000 or 15 000 gausses) core-loss limits. In all cases, the magnetic properties are measured on all lengthwise Epstein samples that have been annealed prior to test.

The nonoriented grades of electrical sheet are generally subdivided into silicon-bearing and carbon grades. The silicon-bearing grades are divided further into fully processed and semi-processed types depending upon whether or not the maximum core-loss guarantee is based on the Epstein test sample being in an "as sheared" condition for test or after a "quality development anneal" (QDA). The quality development anneal simulates customer lamination annealing and applies to the semi-processed type which is intended for those applications where laminations will be annealed. Carbon grades are similar to the semi-processed silicon-bearing grades in that magnetic quality is frequently judged on the basis of Epstein test samples given a QDA.

Published core-loss maximums are shown in Table 46—II for the non oriented, fully processed types and in Table 46—III for semi-processed silicon-bearing qualities. Core losses at 1.5 teslas (15 000 gausses) are listed, but there is an option, in most cases, of purchasing on the basis of 1.0 teslas (10 000 gausses) limits. In all cases, the core-loss determinations are based on half lengthwise, half transverse Epstein samples, "as sheared" for fully processed and after "quality development anneal" for semi-processed.

The carbon grades are used in the mill-processed condition as well as after lamination annealing. Core losses are usually most important when annealing in

Table 46—I. Maximum Core Loss Limits for Oriented Electrical Sheet of the Fully Processed and High Permeability Types*.

| ASTM Grade Designation | Normal Trade Designation | Thickness | | Maximum Core Loss Limits for Indicated Flux Density (B)* | | | |
| | | | | 1.5 teslas | 1.7 teslas (15 000 gausses) | | (17 000 gausses) |
		mm	in.	W/kg	W/lb	W/kg	W/lb
A. FULLY PROCESSED							
—	M-3	0.25	0.010	1.08	0.49	—	—
27G058	M-4	0.28	0.011	1.17	0.53	—	—
30G058	M-5	0.30	0.012	1.28	0.58	—	—
35G066	M-6	0.36	0.014	1.46	0.66	—	—
—	M-3	0.25	0.010	—	—	1.63	0.74
27H076	M-4	0.28	0.011	—	—	1.68	0.76
30H083	M-5	0.30	0.012	—	—	1.83	0.83
35H094	M-6	0.36	0.014	—	—	2.07	0.94
B. HIGH PERMEABILITY							
—	M-0H	0.28	0.011	—	—	1.32	0.60
—	M-1H	0.28	0.011	—	—	1.39	0.63
—	M-2H	0.28	0.011	—	—	1.46	0.66
—	M-0H	0.30	0.012	—	—	1.39	0.63
27P066	M-1H	0.30	0.012	—	—	1.46	0.66
30P070	M-2H	0.30	0.012	—	—	1.54	0.70
—	M-3H	0.30	0.012	—	—	1.63	0.74
—	M-2H	0.36	0.014	—	—	1.59	0.72
30P076	M-3H	0.36	0.014	—	—	1.68	0.76
—	M-4H	0.36	0.014	—	—	1.79	0.81

*All lengthwise Epstein samples annealed at 788°C (1450°F) prior to test at 60 Hz.

Table 46—II. Maximum Core Loss Limits for Nonoriented Electrical Sheet of the Fully Processed Type.[a]

Normal Trade Designation	0.36-mm (0.014-in.) Gage			0.47-mm (0.0185-in.) Gage			0.64-mm (0.025-in.) Gage		
	ASTM Designation	Core Loss		ASTM Designation	Core Loss		ASTM Designation	Core Loss	
		W/kg	W/lb		W/kg	W/lb		W/kg	W/lb
M-15	36F145	3.20	1.45	47F168	3.70	1.68	—	—	—
M-19	36F158	3.48	1.58	47F174	3.84	1.74	64F208	4.59	2.08
M-22	36F168	3.70	1.68	47F185	4.08	1.85	64F218	4.81	2.18
M-27	36F180	3.97	1.80	47F190	4.19	1.90	64F225	4.96	2.25
M-36	36F190	4.19	1.90	47F205	4.52	2.05	64F240	5.29	2.40
M-43	—	—	—	47F230	5.07	2.30	64F270	5.95	2.70
M-45	—	—	—	47F290	6.39	2.90	64F340	7.50	3.40
M-47	—	—	—	47F380	8.38	3.80	64F470	10.36	4.70
Stator B[b]	—	—	—	47F450	9.92	4.50	64F550	12.13	5.50
Con-Core[b]	—	—	—	—	10.91	4.95	—	13.78	6.25

[a] Half lengthwise and half transverse 30 by 305 mm (1.18- by 12.01-in.) Epstein samples tested at 60 Hz in the as sheared condition. Flux density (B) = 1.5 teslas (15 000 gausses).
[b] United States Steel Corporation grade designations.

Table 46—III. Maximum Core Loss Limits for Nonoriented Electrical Sheet of the Semi-Processed Types.[a]

Normal Trade Designation	0.47-mm (0.0185-in.) Gage			0.64-mm (0.025-in.) Gage		
	ASTM Designation	Maximum Core Loss Limit		ASTM Designation	Maximum Core Loss Limit	
		W/kg	W/lb		W/kg	W/lb
M-27	47S178	3.92	1.78	64S194	4.28	1.94
M-36	47S188	4.15	1.88	64S213	4.70	2.13
M-43	47S200	4.41	2.00	64S230	5.07	2.30
M-45	47S230	5.07	2.30	64S260	5.73	2.60

[a] Half lengthwise and half transverse 30 by 305 mm (1.18 by 12.01 in.) Epstein samples tested at 60 Hz after a quality development decarburizing anneal at 843°C (1550°F). Flux density (B) = 1.5 teslas (15 000 gausses).

lamination form is required; therefore, core-loss maximums have been established only for those qualities where customer annealing is intended. The Epstein samples are annealed at 788°C (1450°F) in a decarburizing atmosphere. Maximum losses are shown in Table 46—IV.

Designers of electrical equipment work with many parameters, including the grade of core steel, steel thickness and surface insulation. Compromises with many cost considerations, lamination dimensions, space requirements, winding arrangements, and so on, result in one manufacturer selecting one specific grade for

Table 46—IV. Maximum Core Loss Limits for Carbon Electrical Sheet.[a]

Actual Thickness		Maximum Core Loss			
		ASTM Type 2-S		USS Q-Core	
mm	in.	W/kg	W/lb	W/kg	W/lb
0.46	0.018	7.94	3.60	6.48	2.95
0.48	0.019	8.27	3.75	6.84	3.10
0.51	0.020	8.60	3.90	7.17	3.25
0.53	0.021	9.04	4.10	7.50	3.40
0.56	0.022	9.48	4.30	7.94	3.60
0.58	0.023	9.92	4.50	8.38	3.80
0.61	0.024	10.36	4.70	8.60	3.90
0.64	0.025	10.80	4.90	9.04	4.10
0.66	0.026	11.25	5.10	9.26	4.20
0.69	0.027	11.80	5.35	9.70	4.40
0.71	0.028	12.35	5.60	10.14	4.60

[a] Half lengthwise and half transverse 30 by 305 mm (1.18 by 12.01 in.) Epstein samples tested at 60 Hz after decarburizing quality development anneal at 788°C (1450°F). Flux density (B) = 1.5 teslas (15 000 gausses).

the core steel and another choosing an entirely different grade to build an almost identical unit. It is, therefore, very difficult to always relate a specific electrical sheet grade to a specific electrical device. However, there are some generalities.

Applications for oriented electrical sheet are of the high-efficiency equipment types where designs can effectively utilize the pronounced directional properties. Included are power transformers, distribution transformers, large generators and a wide variety of small transformers. Core configurations are predominantly sheared flat laminations and wound cores but also include segmental laminations for large generators, and some "E" and "I" types.

Nonoriented electrical sheet have broad usage in all types of rotating equipment, dry-type network transformers, small transformers, magnetos, relays and saturable reactors. Higher silicon, lower core-loss grades are specified when efficient operating characteristics are most important. The lower silicon, high core-loss grades are specified when efficiency concerns are less important than cost factors.

The carbon electrical sheet grades are widely used in smaller rotating equipment; subfractional, fractional and small- to medium-horsepower motors. Many of the applications are high volume, intermittent-duty motors for washing machines, refrigerators, furnace blowers, fans and other appliances. Carbon steels are also used in ballast transformers, doorbell transformers and hand power tools.

SECTION 3

PROCESSING OF ELECTRICAL SHEET

Oriented—These grades are made from steel containing about 3 per cent silicon which is refined to yield very low carbon and sulphur contents in the finished product. The selection of the charge for the heats, and the melting and refining are under very close controls to insure uniform chemical composition within narrow limits. The alloying addition of silicon in the form of ferrosilicon is made to the molten steel in the ladle along with other additions which may be sulphur, antimony, boron or aluminum. These latter four additions or equivalents are extremely important to development of the desired crystallographic texture. The molten steel is carefully refined, for example, by an argon-oxygen decarburization (AOD) practice and then stirred with an inert gas (argon) in the ladle to mix the ladle additions and flush out contaminants prior to continuous or ingot casting. Teem stream protection is desirable. Ingots are converted to slabs by normal hot-rolling practices but both continuous-cast and rolled slabs require controlled cooling to avoid fracture.

Slabs are reheated to an extremely high temperature for hot rolling on the continuous hot-strip mill in order to attain the desired rolling and coiling temperatures, and control precipitation of grain-boundary inhibitors. After hot rolling, the product is normalized, wheelabrated, pickled and oiled for cold reduction. Subsequent steps in the processing of regularly oriented electrical sheet include two cold reductions, three continuous anneals, two coating operations and a high-temperature batch-anneal treatment. The more essential components of a typical mill processing after normalizing the hot-rolled band include:

Production Step	Purpose
Cold Reduction	reduce from band gage to twice the finish thickness
Continuous Annealing	recrystallize the cold-reduced structure
Cold Reduction	to finish thickness
Continuous Annealing	recrystallization and decarburization
Coating	separating medium to prevent stickers and enhance sulphur removal during high-temperature annealing
Annealing at 1150°C (2100°F) in H_2 (Batch Furnace)	desulphurization and to develop crystallographic texture, grain growth and "glass film" insulation
Continuous Annealing	flattening and stress relief
Core Plating	C-5 for surface insulation

Steel processed according to this full treatment is intended for core construction that requires good flatness. For applications such as wound cores, where curvature, or coil set, is not detrimental, the oriented steel is used directly after the high-temperature batch annealing without being given the final flattening and coating treatment.

High-Permeability Oriented—The exceptionally high permeability-oriented, or super-oriented, grade requires even more precise control of mill processing than the regular grades. These super grades are made with a cold reduction of 80 to 90 per cent after the initial normalizing step. Special emphasis is given to (1) cooling after normalizing, (2) minimizing temperature buildup during cold reduction, and (3) avoiding excessive oxidation during the decarburizing continuous annealing. The formation of the magnesium silicate "glass" film during high-temperature batch annealing is also closely controlled. The final step is application of special surface coating for the purpose of developing stresses at the steel-surface/coating interface.

Nonoriented—Mill processing treatments for the nonoriented electrical sheets are not so extensive as for oriented, but careful control of each step is also essential to the proper development of a random-type crystallographic texture. Magnetic quality is dependent upon carbon and sulphur content, cleanliness, degree of stress and, of course, silicon and aluminum content. Restrictions upon each of these factors become increasingly important, as the core loss objective is to attain lower and lower losses.

Current domestic production of all electrical sheet grades is by the cold-reduction method. The inherent brittleness of silicon steel at room temperature limits production to steels containing less than about 3.00 per cent silicon in contrast to the now obsolete practice of hot rolling on hand mills which could readily handle sheets containing up to about 5 per cent silicon—for many years the amount of silicon used in the best transformer grade. Therefore, the production of the better grades of cold-reduced nonoriented today must involve refinements of several processing steps to compensate for the inability to roll steels of the higher levels of silicon.

Nonoriented electrical steel can be melted in electric, open hearth, basic oxygen or Q-BOP furnaces and requires a low-carbon steel with oxidation and impurities as low as possible. Silicon and aluminum are additions to the molten steel in the ladle. Degasification, decarburization and in some instances desulphurization of the molten steel are important. Following casting, ingot or continuous, the slabs are hot rolled to band gage and further processed.

For the higher silicon grades, the hot-rolled coils are frequently given a continuous normalizing treatment followed by pickling and oiling prior to cold reduction. The cold-reduction operation may be accomplished on tandem mills, reversing or cluster mills. The cold-reduced coils are then heat treated by continuous annealing under conditions that are formulated to yield the desired magnetic and mechanical properties. In some instances, semiprocessed is made utilizing a batch anneal-temper roll practice. Fully processed steel must be continuously annealed as the final operation in order to maintain the residual stress level in the finished steel as low as possible. Exact temperatures and atmospheres used in continuous annealing vary considerably depending upon steel composition and the desired final properties.

Carbon—The nonsilicon-bearing grades of electrical sheet are produced from low-carbon steel, cold-reduced to the thinner gages that are typical of electrical sheet. A rephosphorized steel is frequently used to enhance the stiffness and provide better mechanical characteristics for lamination stamping. Following cold reduction, this steel is annealed, batch-type or continuous, and then temper rolled. Temper rolling, in the case of steels having magnetic property limitations, is extremely important to the magnetic quality that can be developed in customer's decarburizing lamination annealing. Optimum practice is to wet temper roll to about 5 to 7 per cent extension as compared to the normal practice for flatness alone of about ¾ to 1 per cent. Two-stand temper mills provide the capability for attaining the high extensions and at the same time produce good flatness.

The carbon grades are provided with a matte finish to minimize the tendency for laminations to stick together during annealing by the customer. A light matte finish measures about 0.89 micrometres (35 microinches) AA on the profilometer while a rough matte is about 1.78 micrometres (70 microinches). The roughened surface also promotes decarburization, provides for more uniform surface-oxide development under the most frequently-used lamination annealing practices, and facilitates handling by reducing the cohesion of laminations wet with punching lubricants.

SECTION 4

CORE PLATE COATINGS

The cores of electrical equipment are constructed of thin gage laminations to limit the flow of eddy currents within the core assembly and thereby reduce the watts loss attributed to core steel. Eddy currents not only flow within the cross section of individual laminations but will also flow from lamination to lamination throughout the core if there is inadequate insulation between the lamination surfaces. In many cases, the normal mill surface oxide offers sufficient interlamination resistance, but when dimensions are large and pressures in a core assembly are high, it may be necessary to increase the resistance.

Core plates are used to increase the interlamination resistance beyond that obtained with the normal surface oxide resulting from mill or customer processing. Coreplating refers to coating the sheet or lamination with a thin layer of varnish or inorganic material which has good electrical resistance. Core plates are usually applied by roller coating on both surfaces of the steel and then baking, assuring a stable, uniform coating of the proper physical properties and interlamination resistance. Coatings are more effective when applied to steel having a smooth surface texture.

Coating thickness is controlled depending upon insulation requirements but is limited by the curing facilities and the effect upon stacking factor. Heavier coatings will provide greater insulation but, if too thick, will decrease stacking factor, thereby reducing the weight of steel that will fit in a given stack length and having an adverse effect upon electrical performance. Heavier coatings are also subject to undercuring which results in a soft, tacky condition that may not exhibit satisfactory insulating properties at the pressures and operating temperatures occurring within the core assembly. Two layers of coating are sometimes more effective than a single heavy coat.

Core plates also have an effect upon the life of materials used in the construction of lamination dies and this characteristic of the steel is usually referred to as die life, punchability, or punching characteristic. For example, Core Plate C-3, an organic type coating, acts as a very good lubricant and effectively reduces the wear on the cutting edges of lamination dies, particularly when the higher silicon content electrical sheet grades are used. Core Plate C-5, an inorganic coating, is generally less effective than Core Plate C-3 as a die lubricant and may not yield much improvement over uncoated steel as far as die wear is concerned.

When the organic-type core plate is used in applications involving welding of stator cores or aluminum die casting of rotors, a thin coating is desirable. The thin, well-cured coating contains less volatile material, and therefore, the possibilities of blowholes and voids in die-cast aluminum and weldments are minimized. The inorganic coatings are better suited for welding and die-casting operations.

Of the two major types of core plate in general use Core Plate C-3 is typical of an organic type, while Core Plate C-5 describes an inorganic coating, as follows:

Core Plate C-3—A high-grade varnish intended for applications where good insulating properties and excellent punching characteristics are desired. This coating is suitable for most core operating temperature conditions but will not withstand lamination annealing.

Core Plate C-5—An inorganic coating, usually a complex iron-phosphate layer, that is heat and oil resistant and will withstand the usual lamination annealing conditions without significant impairment of insulating characteristics. Punchability of C-5 is, in most cases, not as good as C-3.

Curves illustrating the interlamination resistance of a typical Core Plate C-3 and the effect of pressure on the surface insulation are shown in Figure 46—1. The

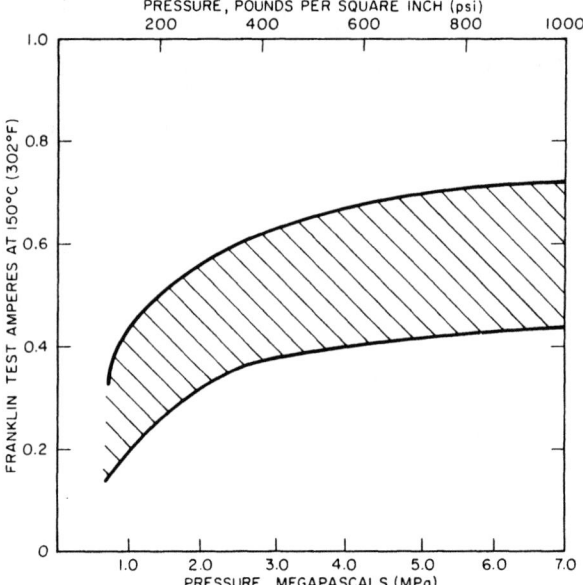

FIG. 46—1. Typical interlamination resistance (Core Plate C-3).

method of measurement is described in ASTM Standard Test A 717-81. In this method, frequently called the Franklin test, an ammeter reading is easily used as the indicator of insulation—a high current results from poor insulation, while a low current reading signifies a good level of resistance.

SECTION 5

FACTORS AFFECTING MAGNETIC PROPERTIES

Among the many factors that affect the magnetic properties of electrical sheet, the most important are: (1) composition, (2) internal stress, and (3) crystallographic texture.

Composition—Practically all elements other than silicon and aluminum, when added to iron, adversely affect the magnetic properties desired in these "soft" magnetic materials. The absolute influence of each element is masked by the effects of other elements, but it is generally agreed that carbon is the most detrimental, followed in order by sulphur, oxygen, and nitrogen. Manganese and phosphorus apparently have little effect on magnetic properties, at least in the quantities normally present in commercial steels. Consideration must be given to the state or form in which the impurity is present, however, as this may greatly alter its effect on magnetic properties. For example, widely-dispersed, fine particles of an impurity are more harmful than an agglomeration of the same impurity into a few relatively large particles.

Internal Stress—Three of the more important magnetic properties, permeability, coercive force, and hysteresis loss, are adversely affected by internal stress. For that reason, the objective for attaining best possible

magnetic quality is to produce a lamination that is essentially stress-free, either by proper mill processing or by annealing in the finished form. There are two main sources of internal stress: (1) impurities that cause distortions in the crystal lattice, and (2) mechanical stresses introduced during the rolling, coiling and leveling operations. Precautions are taken throughout the processing to reduce impurities such as carbon, sulphur, nitrogen, and oxygen to the lowest possible level to avoid internal stresses from this source. Mechanical stresses are minimized in the fully-processed grades by annealing at high temperatures to completely remove the stresses introduced during working. In semi-processed grades, internal stresses may be intentionally introduced, for example, by temper rolling with the objective of effecting a better response to customer annealing operations. In this case, the lamination producer develops the stress-free condition.

Crystallographic Texture—Most magnetic properties are markedly affected by crystal orientation. That is, such properties are better in one of the three principal directions of the crystal than they are in the other two directions. This directionality of magnetic properties can be undesirable in many applications such as in ro-

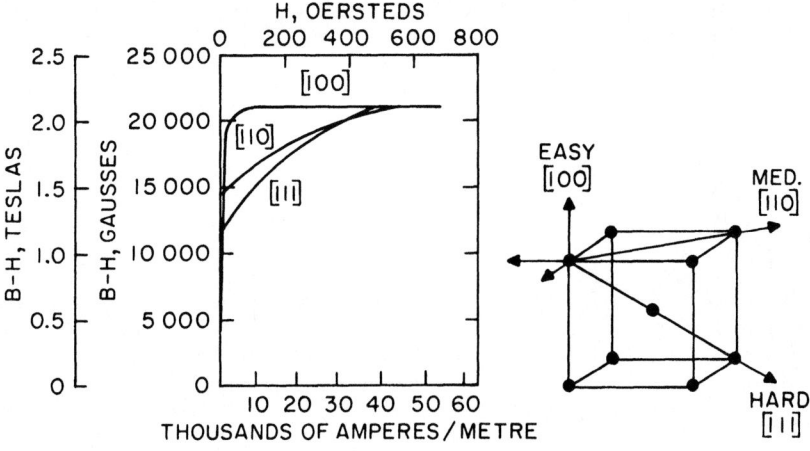

FIG. **46**—2. Effect of orientation on the magnetic properties of a crystal, showing relative ease with which the cubes comprising the iron-silicon space lattice can be magnetized in different directions.

tating machinery, but it has definite advantages in other applications. The cores of distribution and power transformers can be wound or constructed from laminations cut to take advantage of such directionality. Consequently, the manufacturers of oriented electrical sheet strive to develop this characteristic to a high degree.

The processing of oriented steel results in most of the grains being so arranged that edges of the unit cubes comprising each grain are aligned parallel to the rolling direction in a cube-on-edge position with face diagonals aligned in the transverse direction. Because each cube is most easily magnetized along its edge, the [100] direction, the magnetic properties of oriented sheets are best in the rolling direction. As shown in Figure **46**—2, the face diagonal, [110] direction, of each cube is more difficult to magnetize than the cube edge, and the cube diagonal, [111] direction, is the most difficult to magnetize. Thus, the magnetic properties are best in the

rolling direction, poorer at 90 degrees to the rolling direction, and poorest at 55 degrees.

Core loss at a flux density (B) of 1.5 teslas (15 000 gausses) in the rolling direction is approximately 2½ times better than the transverse loss, while permeability in the lengthwise direction can be 50 times better.

Other Effects of Silicon in Low-Carbon Steel—In addition to improving magnetic properties by decreasing eddy-current and hysteresis losses and by increasing permeability, silicon also has the following metallographic, physical, and mechanical effects.

(1) The silicon in low-carbon steel restricts the formation of the gamma (austenite) phase; so that in excess of 2¼ per cent silicon and with 0.01 to 0.02 per cent carbon or less, the alloy is ferritic from room temperature to the melting point. The addition of even as little as 0.05 per cent carbon, however, causes some gamma iron to be present up to at least 5 per cent silicon, as shown in Figure **46**—3.

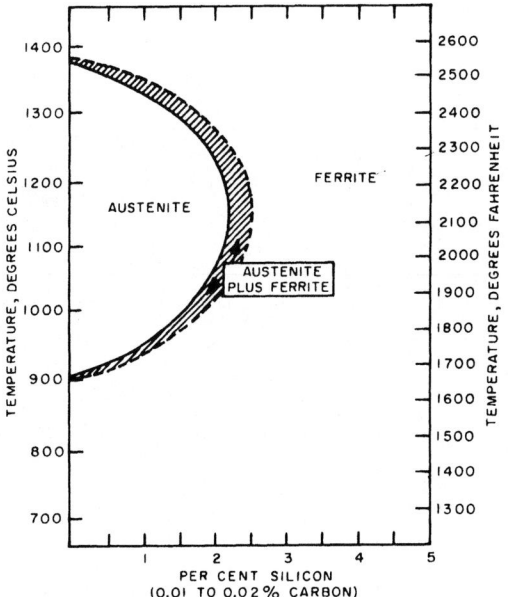

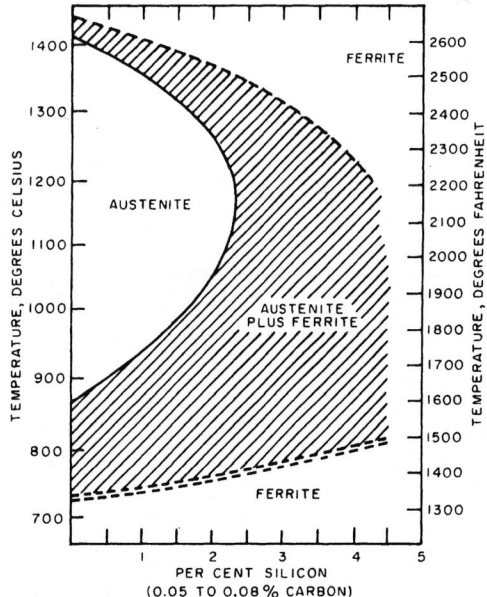

FIG. **46**—3. (Left) Section of the ternary iron-silicon-carbon diagram at the 0.01 to 0.02 per cent carbon level; (right) same at the 0.05 to 0.08 per cent carbon level.

Table 46—V. Typical Mechanical Properties of Some Electrical Sheet Grades.

ASTM Core Loss Type	Common Trade Designation	Normal Thickness		Approximate Alloy Content (Si + Al)%	Electrical Resistivity (μΩcm)	Yield Point*		Tensile Strength*		% Elong. in 50 mm (2 in.)	Apparent Shear Strength		Rockwell Hardness B
		mm	in.			MPa	ksi	MPa	ksi		MPa	ksi	
ORIENTED													
35G066	M-6	0.35	0.0138	3.0	50	345	50	379	35	18	503	73	77
NONORIENTED FULLY PROCESSED													
36F158	M-19	0.36	0.014	3.3	52	365	53	503	73	21	490	71	78
36F180	M-27	0.36	0.014	2.8	49	365	53	503	73	22	483	70	77
47F205	M-36	0.47	0.0185	2.65	49	365	53	496	72	24	469	68	75
64F270	M-43	0.64	0.025	2.35	39	345	50	476	69	28	427	62	75
64F360	M-45	0.64	0.025	1.85	35	345	50	476	69	30	421	61	74
64F490	M-47	0.64	0.025	1.05	23	352	51	434	63	27	393	57	65
NONORIENTED SEMI-PROCESSED													
47S188	M-36	0.47	0.0185	2.65	49	400	58	517	75	24	476	69	75
64S230	M-43	0.64	0.025	2.35	39	441	64	545	79	20	434	63	82
64S280	M-45	0.64	0.025	1.85	35	462	67	517	75	27	434	63	80
NONORIENTED CARBON													
Type 2S	Type 2S	0.61	0.024	0.15	16	379	55	434	63	24	345	50	62
Type 1	Type 1	0.61	0.024	0.15	14	269	39	372	54	33	324	47	49

*Average of lengthwise and transverse test specimens.

(2) The magnetic transformation (Curie temperature) is depressed about 9°C (16°F) for each per cent of silicon up to about 4 per cent silicon. At a silicon content of about 4 per cent, the magnetic transformation occurs at approximately 730°C (1346°F).

(3) Silicon increases the electrical resistivity of iron 11.4 microhm—cm for each added per cent of silicon.

(4) The addition of silicon reduces the density of the resulting alloy.

(5) The addition of silicon to a low-carbon steel decreases the tendency for the material to age (impairment of magnetic properties with increased time and temperature), provided other factors are normal and properly controlled.

(6) Brittleness, or lack of ductility, increases as the percentage of silicon increases.

Effect of Silicon on Mechanical Properties— Although no appreciable drawing operations are involved in fabricating articles from electrical sheet, the material must have good punching and shearing qualities and must be reasonably flat so that motor, generator, and transformer laminations may be punched or sheared without difficulty. Typical mechanical properties of the various silicon-steel grades are listed in Table 46—V.

In general, as the silicon content increases, elongation decreases, and hardness, yield point, and tensile strength increase. However, differences in processing treatments may alter these trends to some extent.

SECTION 6

LAMINATION AND CORE ANNEALING

Oriented—Oriented electrical sheet are used in electrical equipment of highest efficiency and, consequently, final annealing in lamination or core form is desirable to assure best possible magnetic quality. This heat treatment is essentially a stress-relief anneal because processing by the mill has thoroughly refined the product and developed the proper oriented-crystallographic texture. The primary precaution when annealing oriented grades is to avoid contamination from carbon and oxygen. Temperatures may range from 788°C to 816°C (1450°F to 1500°F) for a soaking period of sufficient time to attain thorough heat penetration of the steel.

Suitable stress relief can be accomplished in various types of furnaces—continuous batch, continuous strand, or batch type. Whatever the method, it is essential that the finished laminations or core be in a stress-free condition; flat if the product is used in the flat, or in a proper core configuration if the application is a wound core. Heating and cooling rates must be controlled to avoid distortions that can occur from excessive temperature gradients. Oxidation, if not kept to an absolute minimum, will produce distortions along wound core edges from the growth of excessive oxide layers.

Optimum furnace-atmosphere conditions depend upon the duration of the stress-relief process and include vacuum, air and dry nitrogen in addition to mixtures of various gases. Air atmospheres are used in continuous-strand annealing where the exposure at high temperature is only a matter of minutes. Batch-annealing cycles are of long duration, and as a consequence, the steel charge is highly subject to contamination from oxygen and carbon possibly present either in the protective atmosphere or in the furnace environment, including trays, plates, brickwork or inner covers.

Nonoriented—Fully processed grades are normally used without benefit of an anneal after stamping, but in some cases, a stress-relief anneal is desirable to restore magnetic characteristics which have been adversely affected by stresses imparted during punching, shearing or handling operations.

Semi-processed grades of nonoriented carbon and silicon-bearing steel require proper annealing in lamination or core form to develop fully the magnetic properties which have been only partially developed in mill processing. Semi-processed steels are made specifically for applications where annealing is intended and frequently the magnetic quality in the as-shipped condition can be quite poor.

The type of annealing equipment employed normally depends more on capacity requirements than on some special capability of a particular furnace design. Each furnace and furnace design will have its own peculiarities and, consequently, specific time-temperature and atmospheric conditions necessary to attain desired end results in each case will probably require some experimentation under actual operating conditions. However, there are basic objectives common to all, so the following general observations on annealing practices are offered.

Laminations are invariably coated with a lubricant that is applied during blanking and punching operations to increase die life. These lubricants should be removed prior to annealing because they will contaminate the furnace atmosphere and may cause welding of the lamination surfaces (stickers) or even carburization of the steel. Punching lubricant is sometimes removed by a degreasing operation; however, in most cases, lubricant is removed in the annealing cycle by a process called "burn-off". This step involves heating the laminations in air to 425°C to 480°C (800°F to 900°F) for sufficient time to volatilize oils entrapped throughout the stack of laminations.

Burn-off can be accomplished in a separate furnace section or as the first stage of a continuous-annealing furnace prior to entering the controlled-atmosphere zones. In the case of batch annealing, where the stacks of laminations are usually enclosed by an inner cover, the volatilized lubricants can be removed during the lower-temperature portion (to 480°C or 900°F) of the cycle by venting and using forced air or atmosphere gas for thorough evacuation.

Because heating rates for the steels used for electrical-equipment laminations generally are not critical, they are determined by the capacities of the annealing equipment. Cooling rates are sometimes critical and depend principally on the amount of distortion permitted by the application. Large laminations, for example, could become severely distorted under conditions which are entirely satisfactory for smaller laminations. Generally, a cooling rate not exceeding 150°C (300°F) per hour from soak temperature to about 540°C (1000°F) is adequate for all but the most special applications.

In many applications, a surface oxide on laminations is highly desirable to improve interlamination resistance in the assembled core. The most favored oxide is a "blue oxide" which can be developed in conjunction with lamination annealing by subjecting the steel to an air or steam blast as it cools through the 540°C to 425°C (1000°F to 800°F) temperature range. Usually these oxides are thin, discontinuous and widely variable in color, but despite their nonuniformity are quite acceptable for many smaller units. Where a uniformly heavy oxide is desired, it can be formed with dry steam at a temperature of 425°C to 540°C (800°F to 1000°F), in equipment that is separate from the regular annealing treatment.

The use of appropriate enclosures or inner covers can afford the necessary protection for the steel during batch-annealing cycles which are necessarily long, 12 to 24 hours, depending on the amount of steel undergoing heat treatment. The charge is held at the prescribed annealing soaking temperature for a sufficient time to ensure that all parts of the charge attain this temperature.

Fully Processed—When it is desired to anneal laminations of fully processed material to relieve the various stresses imparted during lamination production operations, soak temperatures as low as 705°C (1300°F) may be used. Adequate decarburization and grain size have already been achieved during mill heat treatment, and therefore, the prime objective of further heat treatment is to relieve stresses.

Deterioration of magnetic properties can occur with excessive oxidation at high temperature; therefore, it is important that a dry protective atmosphere be used when annealing the fully processed steels, particularly in batch-type facilities. Atmospheres that may be used would consist of a dry nitrogen-hydrogen combination with 5 per cent minimum hydrogen, or an exothermic atmosphere made by incomplete combustion of natural gas that has had the excess moisture removed.

Semi-Processed Silicon—Higher temperatures, 788°C to 843°C (1450°F to 1550°F), are employed when annealing semi-processed material so that the desired decarburization, grain growth, and optimum magnetic properties will be obtained.

The higher temperatures involved in the annealing of semi-processed steels may contribute to the sticking of laminations, and therefore, a matte finish is usually supplied to minimize this tendency of laminations to weld together. The use of a controlled atmosphere is required not only to minimize sticking but to prevent overoxidation, and to effect the desired decarburization.

Controlled atmospheres that may be used have as essential ingredients hydrogen, nitrogen and water vapor. One of the most common atmospheres used is an exothermic atmosphere generated by incomplete combustion of natural gas. What is optimum may vary among manufacturers, but experience has shown that an air-to-gas ratio in the exothermic generator of approximately 7.5 to 1 with a dew point of 18°C to 27°C (65°F to 80°F) will yield favorable results. Atmospheres obtained by mixing pure nitrogen and hydrogen to obtain 20 per cent minimum hydrogen, and the introduction of moisture to achieve a hydrogen-to-water ratio of 5 to 1, will also produce favorable results.

Semi-Processed Carbon—Grades of carbon electrical sheet, with some exceptions, are annealed in lamination form to develop their magnetic-property potential.

The selection of a controlled atmosphere is a very important consideration because the degree of decarburization will determine to a large extent the magnetic quality of the annealed laminations. Permeability and magnetic aging characteristics are highly influenced by carbon level. For example, when the steel has been decarburized below 0.004 per cent carbon, the permeability will be as high as commercially possible. A quality development anneal utilizing an atmosphere consisting of 15 per cent hydrogen, 85 percent nitrogen, with a dew point of 21°C to 29°C (70°F to 85°F) will allow the steel to develop the desired excellent level of magnetic quality. Most production furnaces, however, are equipped with exothermic generators which provide atmospheres composed of nitrogen, hydrogen, carbon monoxide, carbon dioxide and moisture. There is considerable leeway in the choice of combinations, but the most important factor is moisture—the dew point should be maintained in the 16°C to 27°C (60°F to 80°F) range. Air-gas ratios can be varied depending upon the degree of oxidation desired on the laminations, but a reasonable atmosphere-gas generator setting is 7 parts air to 1 part gas, which yields approximately 10 per cent hydrogen, 8 per cent carbon monoxide, and 7 per cent carbon dioxide.

The annealing or soaking temperature for the carbon grades is critical because the objective is to attain good grain growth and decarburization without causing laminations to stick together. Because of the absence of silicon, these steel types tend to stick at relatively low temperatures, and considerable care must be taken in the development of annealing conditions applicable to a specific furnace and the selection of a compatible surface texture or coating. The annealing temperature must be high enough to effect magnetic response but low enough to keep laminations from sticking together. In general, the annealing or soaking temperature must be in the range about 730°C to 790°C (1350°F to 1450°F) for a period of time sufficient for all portions of the charge to be at the annealing temperature for approximately one hour. During this period, carbon removal will progress to a very low level under appropriate atmospheric conditions and grain growth will occur.

FUTURE DEVELOPMENT IN ELECTRICAL SHEETS

Grain Oriented Sheets—Incremental improvements of grain oriented sheet have occurred in recent years and will continue to occur in the immediate future. These improvements include rolling thinner sheet and hence improving core loss. Such improvements occur when improved steelmaking and/or rolling facilities are brought to bear. Reduction of interaction between domain walls in stacked laminations by reducing domain size near the sheet surface has resulted in a recent improvement in core loss. Novel means of obtaining this improvement such as stress coatings and limiting surface grain size are presently under various stages of development.

Amorphous ferromagnetic alloys produced from a base alloy of 80 atomic per cent iron and 20 atomic per cent boron show great promise as a substitute for grain oriented sheets in power distribution transformers. These alloys are presently entering the quality audio equipment market in speakers, transformers and in tape reading heads. Present problems with redesign of equipment to use the sheet, handling of the very thin (0.001-inch thick) very hard sheet, and a cost penalty have limited the use of amorphous electrical sheets despite their almost order of magnitude improvement in core loss over conventional grain oriented sheets.

Nonoriented and Low Carbon Steel Electrical Sheets—Recent improvements in these products have been associated with advances in steelmaking such as the ability to limit C, S, and N contents to values less than 20 parts per million, to produce inclusion free steels, and to improve product uniformity through continuous casting. Presently, the differences between the magnetic quality of the poorer grades of nonoriented electrical sheets and the better grades of carbon steel electrical sheets are beginning to diminish. This results in a confusion concerning both quality and cost effectiveness of the various sheet products. The cost of the low carbon electrical sheets, that are primarily produced in the large sheet mills of integrated steelmaking facilities, has been less than the silicon bearing electrical sheets traditionally produced in specialty steel mills. As market forces lead to improvements in carbon steel sheets, they are being made with larger percentages of resistivity-enhancing alloys such as silicon and aluminum. Hence, not only are the properties of these two groups of steels converging, so is their composition. It appears that this trend will continue and the traditional distinction between nonoriented electrical sheets and low carbon electrical sheets will become more nebulous.

DEFINITIONS OF TERMS AND METHODS OF TESTING

The more important characteristics and terms used in the evaluation of electrical sheet steels are defined, as follows:

Aging Coefficient—The percentage change in a specific magnetic property resulting from a specified aging treatment. The aging treatments are usually:
 a. 100 hours at 150°C (300°F)
 b. 600 hours at 100°C (212°F)

Aging, Magnetic—The deterioration of magnetic properties of a magnetic material resulting from metallurgical change. This term applies whether the change occurs due to a continued normal or accelerated aging procedure.

Core Loss, Total (P_c)—The power expended as heat within a magnetic circuit (core) when there is a cyclically alternating induction. Total power includes eddy current and hysteresis losses and is usually expressed as watts per kilogram (watts per pound in the customary system) for a specific induction and frequency. The Epstein test for core loss is described in ASTM Standard A 343, Test for Alternating-Current Magnetic Properties of Materials at Power Frequencies Using Wattmeter-Ammeter-Voltmeter Method and 25 cm Epstein Frame.

Coercive Force (H_c)—The d-c magnetizing force that must be applied in a direction opposite to the residual induction to reduce the magnetic induction to zero.

Core Plate—Thin coatings applied to the surface of electrical sheets or laminations to increase interlamination resistance and thereby limit the flow of eddy currents from one lamination to another. Core plates may be organic or inorganic materials.

Eddy-Current Loss (P_e)—That part of the total core loss which is due to induced electrical currents circulating in a magnetic material. Eddy-current loss can be calculated from Maxwell's equation, or it can be determined by a graphical method of separating core loss into its components.

Hysteresis Loss (P_h)—That part of total core loss which is proportional to the energy loss per cycle resulting from magnetic hysteresis.

Interlamination Resistance—The term applied to the electrical resistance measured perpendicular to the lamination plane. It indicates the effectiveness of surface oxides or core-plate coatings on the laminations in reducing interlamination (eddy-current) losses. The standard methods of test include ASTM A 717, Test for Surface Insulation Resistivity of Single Strip Specimens, and ASTM A 718, Test for Surface Insulation Resistivity of Multi-Strip Specimens.

Lamination Factor or Space Factor—The ratio of the volume of a stack of laminations under a given pressure to that of the solid material of the same mass, assuming a definite density based on the chemical composition. Thus, the factor indicates the deficiency of effective steel volume due to the surface roughness and lack of flatness of the laminations, or to the presence of oxides and core-plate coatings on the surface of the laminations. Lamination factor is determined according to ASTM A 719, Test for Lamination Factor of Magnetic Materials.

Magnetic Flux or Flux Density—The number of lines of magnetic flux per unit area at right angles to the direction of the flux. When a magnetic core having a closed magnetic circuit is magnetized by current flowing in the windings of the coil which enclose the core, magnetic lines of force are generated which are designated as magnetic flux. The total magnetic flux in the core, designated as ϕ (phi), divided by the cross-sectional area of the core in square centimetres, gives the flux density (B), in lines per square centimetre, or gausses. The recommended SI unit for flux density is the tesla: one tesla equals 10 000 gausses.

Magnetizing Force—The magnetomotive force per unit of core length, and is designated by the letter H. When the length of the core has been expressed in centimetres, the unit of magnetizing force or field strength has been the oersted; the recommended SI unit is amperes per metre (A/m); one A/m equals 79.577 oersteds. Another unit of magnetizing force sometimes used has been ampere turns per inch, which is 2.02 times greater than an oersted.

Permeability (μ)—Ratio of magnetic induction (B) and magnetizing force (H) for a magnetic material. This ratio is a measure of the ease with which a material can be magnetized and how much better it is as a path for magnetic fields than air (permeability of 1).

Permeability, Impedance (μ_z)—Ratio of the measured peak value of magnetic induction to the value of the apparent magnetizing force, H_z, calculated from the measured rms value of the exciting current.

Permeability, Inductance (μ_L)—An a-c permeability evaluated in terms of core and circuit geometry and the measured parallel inductance of the path that is considered to carry only the magnetizing current (bridge method).

Permeability, Peak (μ_p)—The ratio of the measured peak value of magnetic induction to the peak value of magnetizing force (H_p) calculated from the measured peak value of the exciting current.

Residual Induction (B_r)—The induction remaining in a material when the magnetizing force has been reduced to zero.

Resistivity, Electrical (ρ)—The electrical resistance between opposite faces of a cube of unit dimensions. It is often referred to as specific resistance or volume resistivity. The standard test is described in ASTM A 712, Test for Electrical Resistivity of Soft Magnetic Alloys.

Saturation Induction—The maximum intrinsic induction possible in a magnetic material. Further increases in magnetic force over that necessary to achieve this saturation flux density will not cause the generation of any additional flux within the material. Saturation occurs at a flux density of about 2.15 teslas (21 500 gausses) in low-silicon steels and about 1.95 teslas (19 500 gausses) in high-silicon steels.

References

American Society for Testing and Materials.
  A 345-75 Standard specification for flat-rolled electrical steels for magnetic applications.
  A 340-77 Standard definitions of terms, symbols and conversion factors relating to magnetic testing.
American Iron and Steel Institute, Washington, DC, Steel products manual: "Flat rolled electrical steel."
Barrett, C. S., Structure of metals. New York, McGraw-Hill Book Company, Inc., (1943).
Bozorth, R. M., Ferromagnetism. New York, D. Van Nostrand Company, Inc., (1951).
Greiner, E. S.; Marsh, J. S.; Stoughton, B.; Alloys of Iron and Silicon. New York, McGraw-Hill Book Company, Inc., (1933).
Kneller, E., Ferromagnetismus. Berlin, Springer-Verlag, 1962.
United States Steel Corporation, Nonoriented sheet steel for magnetic applications. 1978.
United States Steel Corporation, Oriented electrical steel sheets. 1964.

CHAPTER 47

Stainless Steels

SECTION 1

GENERAL

As the name implies, stainless steels are more resistant to rusting and staining than are plain carbon and lower alloy steels. This superior corrosion resistance is brought about by addition of the element chromium to alloys of iron and carbon. Although other elements such as copper, aluminum, and silicon, nickel and molybdenum, also increase the corrosion resistance of steel, they are limited in their usefulness in the absence of chromium. Hence, the discussion in this chapter will be limited to the iron-chromium and iron-chromium-nickel steels in which chromium is the major element for conferring corrosion resistance.

The minimum amount of chromium necessary to confer this superior corrosion resistance depends upon the corroding agent. The American Iron and Steel Institute has chosen 4 per cent chromium as the dividing line between "alloy" and "stainless" steel and for this discussion, the AISI view will be adopted. Selected listings of the standard types of stainless steels are shown in Tables 47—I, 47—II, 47—III and 47—IV. These listings were obtained from the book "Unified Numbering System (UNS) for Metals and Alloys," published by the Society of Automotive Engineers in January 1975. Most of the stainless steels listed in the tables are available in the main product forms such as plates, bars, shapes, wire, sheet, strip and tubes. A list of products and sizes available for each type of stainless steel is given in the AISI Steel Products Manual entitled "Stainless and Heat-Resisting Steels" published by the American Iron and Steel Institute in 1974, and in a supplement with the same title published in 1979. One excellent reference on stainless steels is Handbook of Stainless Steels, edited by D. Peckner and I. M. Bernstein.

SECTION 2

CONSTITUTION

The iron-chromium phase diagram is represented best by Figure 47—1, most of which is taken from Elliott's "Constitution of Binary Alloys." In this figure, the upper temperature portions of the diagram are well accepted. However, the boundaries of the sigma regions and the alpha plus alpha prime regions are still not settled and probably never will be settled because of the extremely long reaction times involved in achieving equilibrium. In fact, no "critical experiment" has demonstrated that sigma phase will decompose to alpha and alpha prime. However, the diagram proposed is certainly adequate in explaining almost all of the observed phenomena that occur in the iron-chromium system. Certainly, this diagram removes the mystery which for many years surrounded the cause for the 475°C (885°F) embrittlement of these alloys.

Iron-Chromium-Nickel System—As might be expected, the uncertainty which exists in the boundaries of the iron-chromium system is carried over into the iron-chromium-nickel system and for the same reason, namely, the sluggishness of these alloys in attaining equilibrium conditions.

Figures 47—2, 47—3 and 47—4 illustrate the accepted diagram for the liquidus surface, the solidus surface and the 1300°C (2372°F) isotherm, respectively. For lower temperatures, the best diagrams available are those shown in Figures 47—5, 47—6 and 47—7.

Iron-Chromium-Carbon System—In Figures 47—8, 47—9 and 47—10 are shown constant sections at 0.05, 0.10 and 0.20 per cent carbon, respectively, through the iron-chromium-carbon ternary diagram. As might be expected, carbon has a marked effect with respect to increasing the stability range of the gamma phase. This effect is apparent from a comparison of Figure 47—8 at 0.05 per cent carbon with Figure 47—10 at 0.20 per cent carbon.

Iron-Chromium-Manganese-Nitrogen System—Shortages of nickel that occurred during World War II and

Table 47—I. Chemical Compositions of Wrought Chromium-Nickel Austenitic Stainless Steels.*
(Not Hardenable by Thermal Treatment)

UNS Number	C max.	Mn max.	Si max.	P max.	S max.	Cr Range	Ni Range	Other Elements	AISI Type Number
S20100	0.15	5.5–7.5	1.00	0.060	0.030	16.00–18.00	3.50–5.0	N, 0.25 max.	201
S20200	0.15	7.5–10.00	1.00	0.060	0.030	17.00–19.00	4.00–6.00	N, 0.25 max.	202
S30100	0.15	2.00	1.00	0.045	0.030	16.00–18.00	6.00–8.00	—	301
S30200	0.15	2.00	1.00	0.045	0.030	17.00–19.00	8.00–10.00	—	302
S30215	0.15	2.00	2.00–3.00	0.045	0.030	17.00–19.00	8.00–10.00	—	302B
S30300	0.15	2.00	1.00	0.20	0.15 min.	17.00–19.00	8.00–10.00	Mo, 0.60 max.[b]	303
S30323	0.15	2.00	1.00	0.20	0.06	17.00–19.00	8.00–10.00	Se, 0.15 min.	303Se
S30400	0.08	2.00	1.00	0.045	0.030	18.00–20.00	8.00–10.50	—	304
S30403	0.03	2.00	1.00	0.045	0.030	18.00–20.00	8.00–12.00	—	304L
S30409	0.10	2.00	1.00	0.045	0.030	18.00–20.00	8.00–10.50	—	304H
S30500	0.12	2.00	1 00	0.045	0.030	17.00–19.00	10.50–13.00	—	305
S30800	0.08	2.00	1.00	0.045	0.030	19.00–21.00	10.00–12.00	—	308
S30900	0.20	2.00	1.00	0.045	0.030	22.00–24.00	12.00–15.00	—	309
S30908	0.08	2.00	1.00	0.045	0.030	22.00–24.00	12.00–15.00	—	309S
S31000	0.25	2.00	1.50	0.045	0.030	24.00–26.00	19.00–22.00	—	310
S31008	0.08	2.00	1.50	0.045	0.030	2?.00–26.00	19.00–22.00	—	310S
S31400	0.25	2.00	1.50–3.00	0.045	0.030	23.00–26.00	19.00–22.00	—	314
S31600	0.08	2.00	1.00	0.045	0.030	16.00–18.00	10.00–14.00	Mo, 2.00–3.00	316
S31609	0.10	2.00	1.00	0.045	0.030	16.00–18.00	10.00–14.00	Mo, 2.00–3.00	316H
S31603	0.03	2.00	1.00	0.045	0.030	16.00–18.00	10.00–14.00	Mo, 2.00–3.00	316L
S31700	0.08	2.00	1.00	0.045	0.030	18.00–20.00	11.00–15.00	Mo, 3.00–4.00	317
S32100	0.08	2.00	1.00	0.045	0.030	17.00–19.00	9.00–12.00	Ti, 5 × C min.	321
S32109	0.10	2.00	1.00	0.045	0.030	17.00–19.00	9.00–12.00	Ti, 5 × C min.	321H
N08330	0.15	2.00	1.50[a]	0.045	0.030	17.00–20.00	34.00–37.00	—	—
S34700	0.08	2.00	1.00	0.045	0.030	17.00–19.00	9.00–13.00	Cb + Ta, 10 × C min.	347
S34709	0.10	2.00	1.00	0.045	0.030	17.00–19.00	9.00–13.00	Cb + Ta, 10 × C min.	347H
S34800	0.08	2.00	1.00	0.045	0.030	17.00–19.00	9.00–13.00	Cb + Ta, 10 × C min.; Ta, 0.10 max.; Co, 0.20 max.	348
S34809	0.10	2.00	1.00	0.045	0.030	17.00–19.00	9.00–13.00	Cb + Ta, 10 × C min.; Ta, 0.10 max.; Co, 0.20 max.	348H
S21400	0.12	14.50–16.00	1.00	0.045	0.030	17.00–18.50	0.75 max.	N, 0.35–0.50	USS TENELON[c]
S38100	0.07	2.00	1.80–2.20	0.010	0.020	17.00–19.00	17.50–18.50	—	XM15[d]

*Based on information in "Unified Numbering System for Metals and Alloys" published by the Society of Automotive Engineers in January 1975, and the AISI Steel Products Manual entitled "Stainless and Heat-Resisting Steels" published by the American Iron and Steel Institute in December 1974 and March 1979 supplement.

[a]To minimize carbon or nitrogen pick-up 0.75–1.50 Si is recommended for high-temperature application involving carbon or nitrogen atmosphere.

[b]At producer's option; reported only when intentionally added.

[c]Proprietary grade: no AISI type number assigned.

[d]USS 18-18-2.

Table 47—II. Chemical Compositions of Wrought Martensitic Chromium Stainless Steels.*
(Hardenable by Thermal Treatment)

UNS Number	C max.	Mn max.	Si max.	P max.	S max.	Cr Range	Ni Range	Other Elements	AISI Type Number
S40300	0.15	1.00	0.50	0.040	0.030	11.50–13.00	—	—	403
S41000	0.15	1.00	1.00	0.040	0.030	11.50–13.50	—	—	410
S41400	0.15	1.00	1.00	0.040	0.030	11.50–13.50	1.25–2.50	—	414

(Continued on next page)

Table 47—II. (Continued)

UNS Number	C max.	Mn max.	Si max.	P max.	S max.	Cr Range	Ni Range	Other Elements	AISI Type Number
S44002	0.60–0.75	1.00	1.00	0.040	0.030	16.00–18.00	—	Mo, 0.75 max.	440A
S44003	0.75–0.95	1.00	1.00	0.040	0.030	16.00–18.00	—	Mo, 0.75 max.	440B
S44004	0.95–1.20	1.00	1.00	0.040	0.030	16.00–18.00	—	Mo, 0.75 max.	440C
S50100	Over 0.10	1.00	1.00	0.040	0.030	4.00– 6.00	—	Mo, 0.40–0.65	501
S50300	0.15	1.00	1.00	0.040	0.040	6.00– 8.00	—	Mo, 0.45–0.65	503
S50400	0.15	1.00	1.00	0.040	0.040	8.00–10.00	—	Mo, 0.90–1.10	504

*Based on information in "Unified Numbering System for Metals and Alloys" published by the Society of Automotive Engineers in January 1975, and the AISI Steel Products Manual entitled "Stainless and Heat-Resisting Steels" published by the American Iron and Steel Institute in December 1974 and March 1979 supplement.

[a]At producer's option; reported only when intentionally added.

Table 47—III. Chemical Compositions of Wrought Ferritic Chromium Stainless Steels.*
(Not Hardenable by Thermal Treatment)

UNS Number	C max.	Mn max.	Si max.	P max.	S max.	Cr Range	Ni Range	Other Elements	AISI Type Number
S40500[a]	0.08	1.00	1.00	0.040	0.030	11.50–14.50	—	Al, 0.10–0.30	405
S40900	0.08	1.00	1.00	0.045	0.045	10.50–11.75	0.50 max.	Ti, 6 × C or	—
S43000	0.12	1.00	1.00	0.040	0.030	16.00–18.00	—	0.75 max.	430
S43020	0.12	1.25	1.00	0.06	0.15 min.	16.00–18.00	—	Mo, 0.60 max.[b]	430F
S43023	0.12	1.25	1.00	0.06	0.06	16.00–18.00	—	Se, 0.15 min.	430F Se
S43400	0.12	1.00	1.00	0.040	0.030	16.00–18.00	—	Mo, 0.75–1.25	—
S43600	0.12	1.00	1.00	0.040	0.030	16.00–18.00	—	Mo, 0.75–1.25; Cb+Ta, 5 × C min. or 0.70 max.	—
S44200	0.20	1.00	1.00	0.04	0.035	18.00–23.00	—	—	—
S44600	0.20	1.50	1.00	0.04	0.030	23.00–27.00	—	N, 0.25 max.	446

*Based on information in "Unified Numbering System for Metals and Alloys" published by the Society of Automotive Engineers in January 1975, and the AISI Steel Products Manual entitled "Stainless and Heat-Resisting Steels" published by the American Iron and Steel Institute in December 1974 and March 1979 supplement.

[a]Essentially non-hardenable by heat treatment.
[b]At producer's option; reported only when intentionally added.

Table 47—IV. Chemical Compositions of Wrought Precipitation Hardenable Stainless Steels.*

UNS Number	C max.	Mn max.	Si max.	P max.	S max.	Cr Range	Ni Range	Mo Range	Other Elements
S13800	0.05	0.10	0.10	0.01	0.008	12.25–13.25	7.50–8.50	2.00–2.50	Al, 0.90–1.35; N, 0.010
S15500	0.07	1.00	1.00	0.040	0 030	14.00–15.50	3.50–5.50	—	Cu, 2.50–4.50; Cb Ta, 0.15–0.45
S17400	0.07	1.00	1.00	0.040	0.030	15.50–17.50	3.00–5.00	—	Cu, 3.00–5.00; Cb Ta, 0.15–0.45
S17700	0.09	1.00	1.00	0.040	0.040	16.00–18.00	6.50–7.75	—	Al, 0.75–1.50

*Based on information in "Unified Numbering System for Metals and Alloys" published by the Society of Automotive Engineers in January 1975.

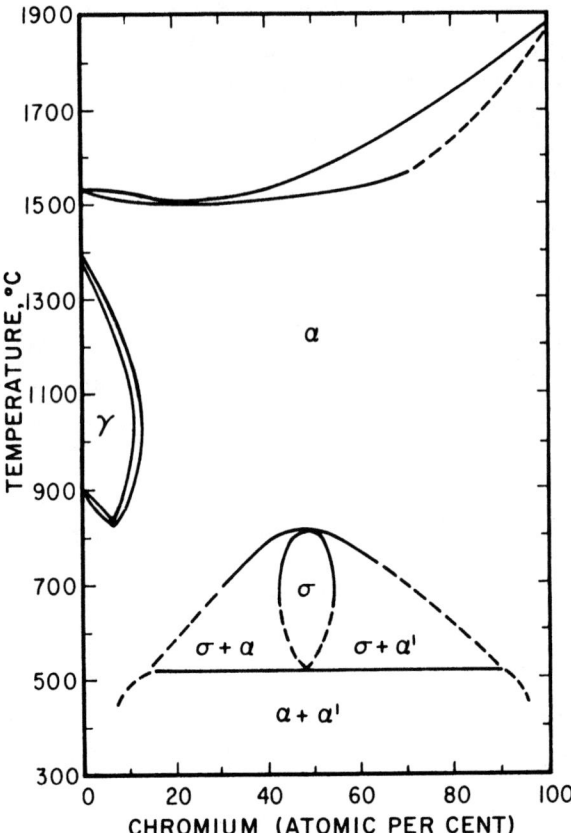

FIG. 47—1. The iron-chromium equilibrium diagram. (Based largely on Elliott's "Constitution of Binary Alloys' listed in the bibliography at the end of this chapter.)

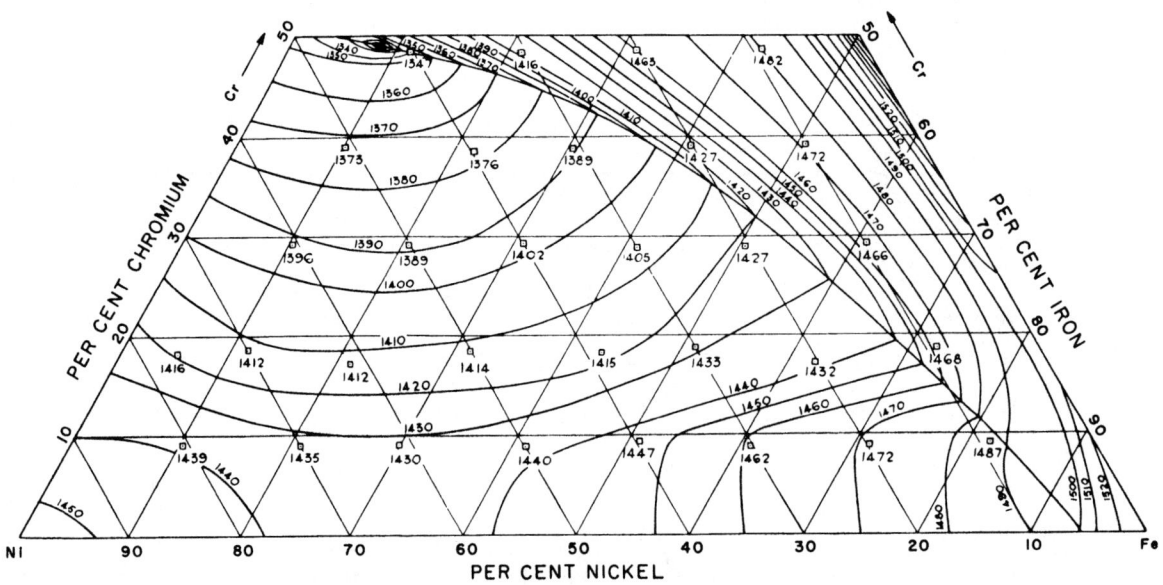

FIG. 47—2. The liquidus surface of the iron-chromium-nickel system. Temperatures are in °C. (After Jenkins, et al.: see bibliography.)

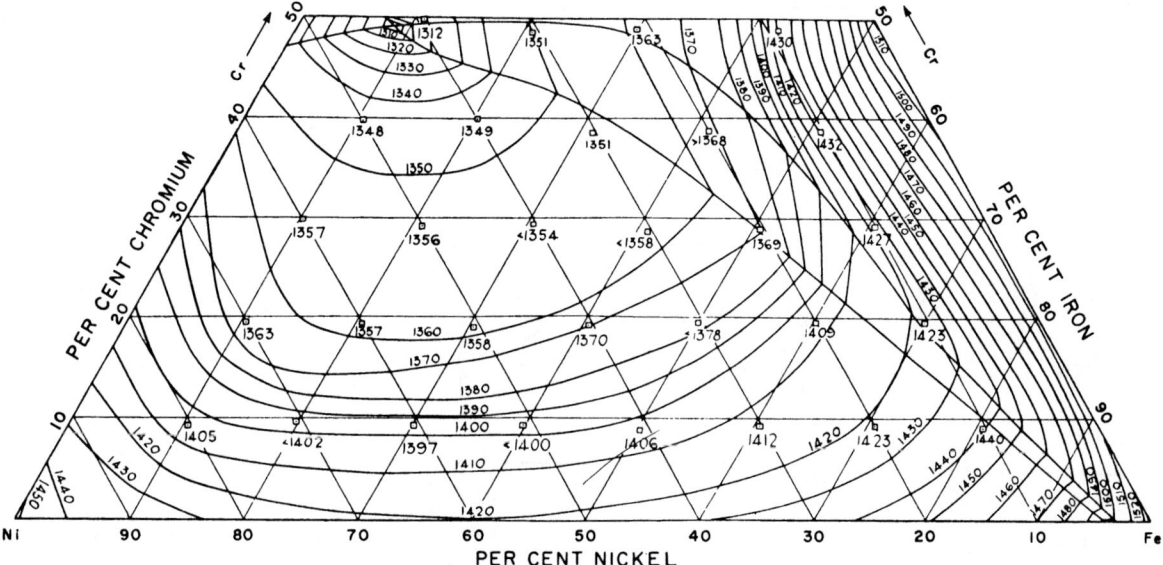

FIG. 47—3. The solidus surface of the iron-chromium-nickel system. Temperatures are in °C. (After Jenkins, et al.: see bibliography.)

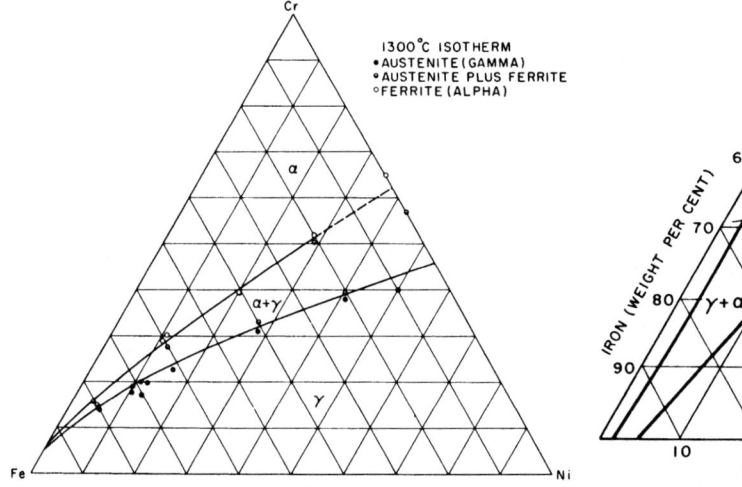

FIG. 47—4. 1300°C (2372°F) isotherm of the iron-chromium-nickel system, determined metallographically. (After Price and Grant: see bibliography.)

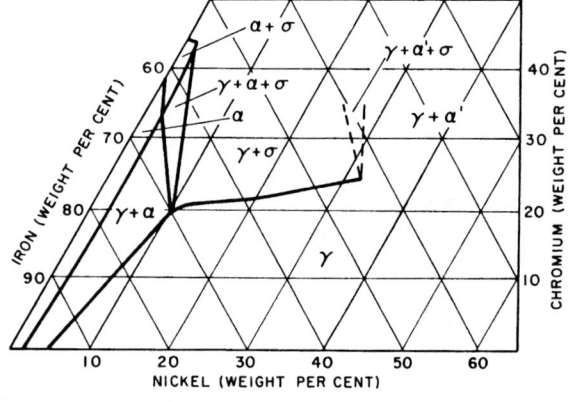

FIG. 47—6. Constitution diagram derived for 800°C (1472°F). (After Rees, Burns and Cook: see bibliography.)

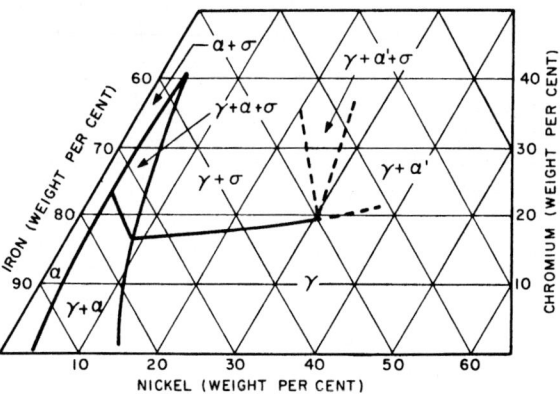

FIG. 47—5. Constitution diagram derived for 650°C (1202°F). (After Rees, Burns and Cook: see bibliography.)

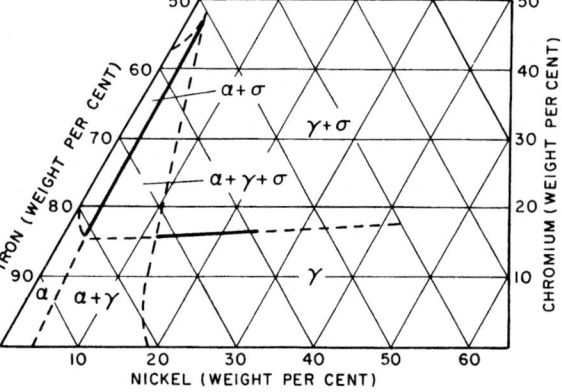

FIG. 47—7. Constitution diagram derived for 550°C (1022°F). (After Rees, Burns and Cook: see bibliography.)

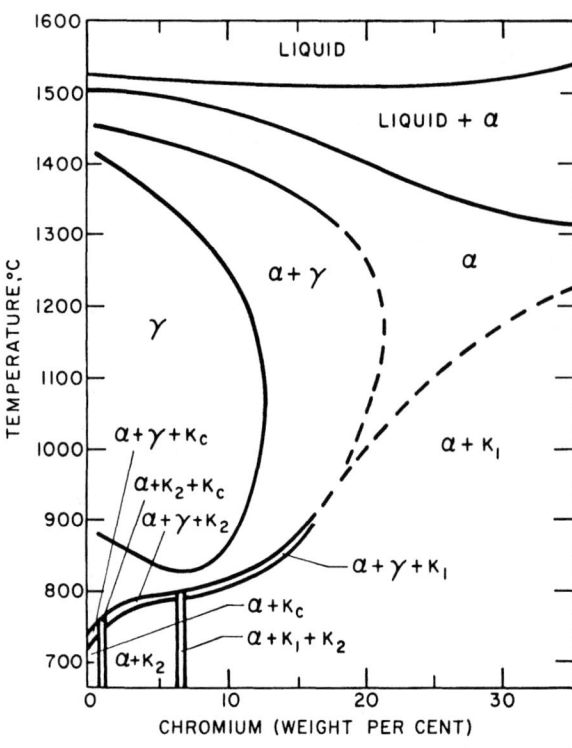

FIG. 47—8. Constant section at 0.05 per cent carbon through iron-chromium-carbon diagram. $K_c = Fe_3C$; $K_1 = (Fe,Cr)_{23}C_6$; $K_2 = (Fe, Cr)_7C_3$. (After Bungardt, Kunze and Horne: see bibliography.)

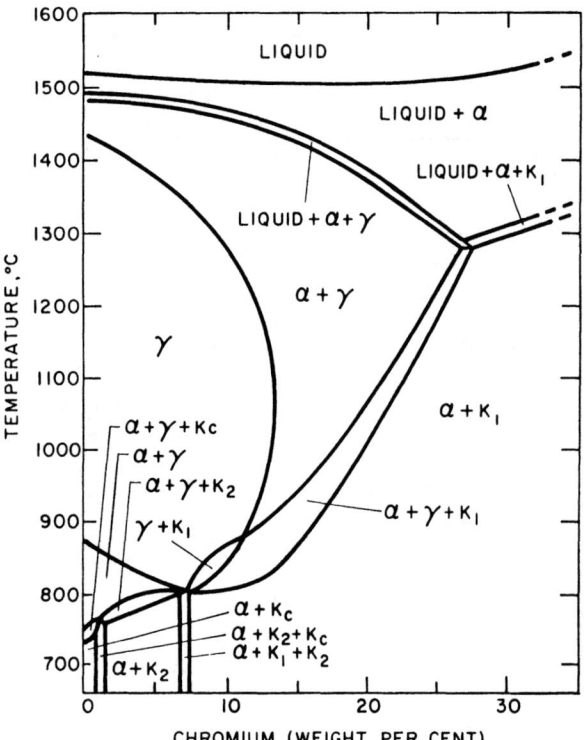

FIG. 47—9. Constant section at 0.10 per cent carbon through iron-chromium-carbon diagram. $K_c = Fe_3C$; $K_1 = (Fe,Cr)_{23}C_6$; $K_2 = (Fe,Cr)_7C_3$. (After Bungardt, Kunze and Horne: see bibliography.)

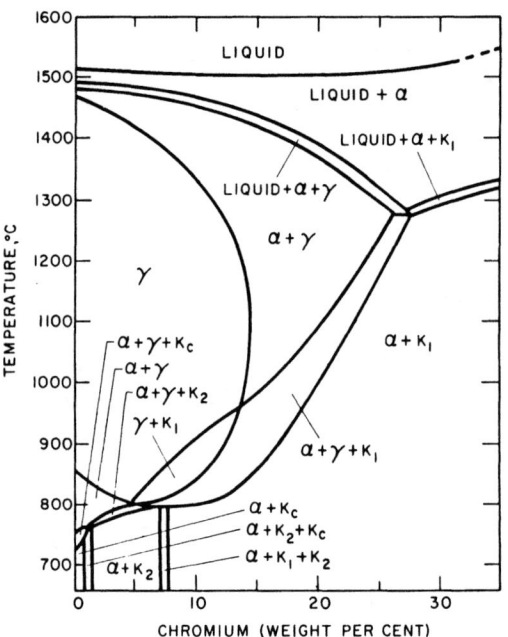

FIG. 47—10. Constant section at 0.20 per cent carbon through iron-chromium-carbon diagram. $K_c = Fe_3C$; $K_1 = (Fe,Cr)_{23}C_6$; $K_2 = (Fe,Cr)_7C_3$. (After Bungardt, Kunze and Horne: see bibliography.)

once again during the Korean conflict prompted a considerable amount of research in both the United States and Germany on substitutes for nickel in the austenitic stainless steels. Only a few elements (other than nickel) expand the austenite region in the diagram of the iron-chromium system. The principal elements are carbon, nitrogen, manganese, copper and cobalt. None of these elements as a single addition is completely satisfactory. For example, carbon in amounts necessary to form completely austenitic structures has a detrimental effect on ductility and corrosion resistance; nitrogen cannot be added in sufficiently large quantities to achieve the desired effect; manganese, even in amounts above 25 per cent, will not form completely austenitic structures in alloys containing over 15 per cent chromium; copper has a detrimental effect on hot ductility; and cobalt is only slightly effective and is quite expensive.

Research in United States Steel Corporation indicated that additions of manganese to the iron-chromium system increased the solid and liquid solubility of nitrogen to the extent that additions of nitrogen sufficient to make completely austenitic structures were possible.

These results are summarized in Figures 47—11, 47—12, 47—13 and 47—14, which are so-called structure diagrams. These diagrams show the influence of chromium, manganese and nitrogen on the austenite—austenite-plus-ferrite boundary at 1260°C (2300°F) and at nickel-content levels of zero, 1, 2 and 3 per cent. The temperature of 1260°C (2300°F) represents a maximum hot-working temperature and austenite formed at this temperature will not transform on cooling to room temperature.

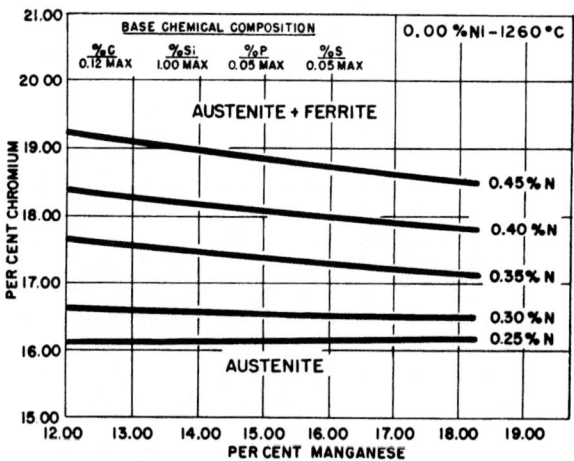

FIG. 47—11. Structure diagram for 0.00 per cent nickel, 0.25-0.45 per cent nitrogen steels after heating for one hour at 1260°C (2300°F) and water quenching.

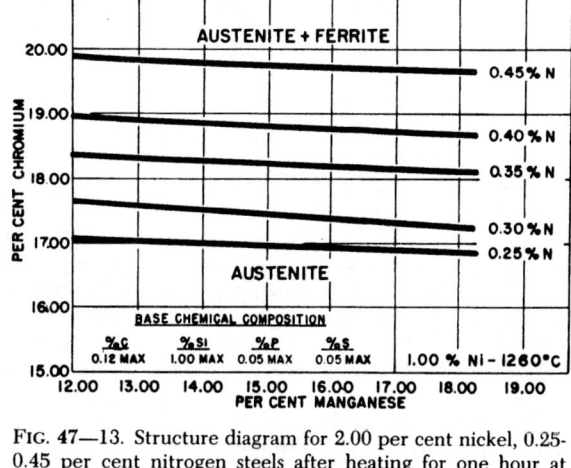

FIG. 47—13. Structure diagram for 2.00 per cent nickel, 0.25-0.45 per cent nitrogen steels after heating for one hour at 1260°C (2300°F) and water quenching.

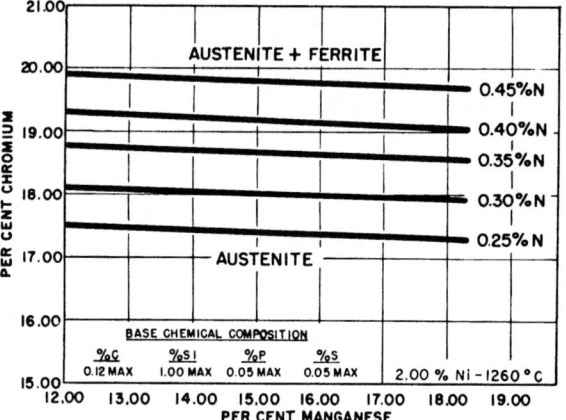

FIG. 47—12. Structure diagram for 1.00 per cent nickel, 0.25-0.45 per cent nitrogen steels after heating for one hour at 1260°C (2300°F) and water quenching.

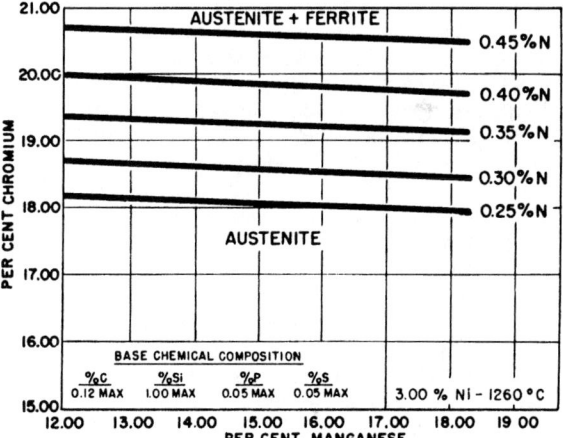

FIG. 47—14. Structure diagram for 3.00 per cent nickel, 0.25-0.45 per cent nitrogen steels after heating for one hour at 1260°C (2300°F) and water quenching.

SECTION 3

MANUFACTURE AND FABRICATION

For the purpose of general discussion, the stainless steels are generally grouped into three classes:

1. **Martensitic**—The martensitic stainless steels include: (A) those iron-chromium alloys that lie within the gamma loop and thus are hardenable by heat treatment (see Table 47—II); and (B) the **precipitation-hardening stainless steels** that are iron-chromium nickel alloys to which are added controlled amounts of titanium, niobium, aluminum and/or copper to achieve a precipitation-hardening reaction. The precipitation-hardening reaction can occur in stainless steels having any crystallographic structure. However, the most important precipitation-hardening stainless steels have a martensitic structure. These steels undergo an austenite-to-martensite transformation before a precipitation-hardening reaction (see Table 47—IV for selected grades).

2. **Ferritic**—Those iron-chromium alloys that are largely ferritic and not hardenable by heat treatment (ignoring the 475°C or 885°F embrittlement): commercial grades of such steels are listed Table 47—III.

3. **Austenitic**—The iron-chromium-nickel alloys not hardenable by heat treatment and predominantly austenitic as commercially heat treated. Other austenitic stainless steels include those in which all or part of the nickel of the iron-chromium-

nickel type of steel is replaced by manganese and nitrogen in proper amounts, such as USS TENELON and UNS S20100 and S20200 (AISI Types 201 and 202). Table 47—I lists commercial austenitic stainless steels.

Melting and Casting—Most of the stainless steels produced in the world are melted in an electric-furnace or BOP and refined by the argon-oxygen decarburization (AOD) process described in Chapter 19, or another suitable ladle-metallurgical process such as vacuum-argon-decarburization. Although a substantial tonnage of the stainless steel produced is cast into ingots and hot-rolled into semi-finished forms, the use of continuous casting for semi-finished products is growing rapidly and, except in the case of several low-volume producers, will replace ingot casting and primary rolling.

Hot Working—General—Before discussing the details of any particular process used for the production of stainless steel, a few general remarks will be made concerning the hot- and cold-working characteristics of these materials. These general remarks, describing the salient differences in behavior between the carbon steels and the stainless steels, will apply to any hot- or cold-working operation to which the stainless steels may be subjected. All of the stainless steels have lower thermal conductivity at temperatures below about 815°C (1500°F) than the carbon and low-alloy steels and, accordingly, precautions must be taken when heating below 815°C, or surface burning will result. Above 815°C, stainless steels can be heated the same as carbon steels. Also, for most of the stainless grades, the temperature ranges for optimum hot-working characteristics are narrower than those for the carbon steels and, hence, closer temperature control is necessary when hot working the stainless steels.

The martensitic stainless steels can be forged, pierced and rolled. However, because these steels are air hardening, they must be slow cooled or annealed after rolling before any subsequent operation such as conditioning or cold working. Their high "as-rolled" hardness also results in low toughness which must be taken into account when handling the hot-rolled product.

The ferritic stainless steels also can be forged, pierced and rolled. These steels are very soft when hot, thus they are easily marked by quides or rolls, and spread considerably during hot rolling. Over-heating these grades causes excessive grain growth, which makes the materials susceptible to tears and cracks. Additions of nitrogen have helped somewhat in preventing grain growth. To refine the grain size, finishing temperatures are kept as low as possible.

The austenitic stainless steels are generally stronger than ferritic steels at rolling temperature and, consequently, require more power for deformation. Like the ferritic steels, the austenitic steels are susceptible to grain growth and overheating should be avoided. Low finishing temperatures are not practicable because of the power required. During the heating of these nickel-bearing austenitic steels, special precautions are taken to keep the sulphur content of the furnace or soaking pit atmospheres at a minimum because these steels, after being heated in atmospheres containing sulphur, tend to tear and crack during rolling. Apparently, the sulphur in the atmosphere combines with the nickel in the steel to form nickel sulphide. This reaction usually occurs at the grain boundaries of the metal and, because the nickel sulphide is liquid at the rolling temperatures, the steels so attacked are weak and easily break apart during rolling. The presence of delta ferrite in the microstructure is considered to be detrimental. For example, those steels such as 18-8 Mo (UNS S31600 or AISI Type 316) 18-8 Cb (UNS S34700 or AISI Type 347) 18-8 Ti (UNS S32100 or AISI Type 321) and 25-12 (UNS S30900 or AISI Type 309) show poor hot-working characteristics which are blamed on the presence of delta ferrite. The explanation for the poor working characteristics is that the difference in plasticity between the soft ferrite and the tough austenite causes ruptures.

Table 47—V lists the suggested forging and annealing temperature ranges for some standard stainless steels.

The Rolling of Stainless-Steel Ingots to Blooms and Slabs—The equipment used for the heating and rolling of stainless-steel ingots is the same as that used for carbon-steel ingots. However, as previously mentioned, close temperature control and avoidance of sulphur contamination are precautions which should be followed when heating the stainless steels. Also, the stainless steels require more conditioning than the carbon steels. The bloom and slab products are completely conditioned.

Rolling of Billets—The blooms used for the production of billets are completely conditioned prior to heating for rolling. As was true for the rolling of ingots, the rolling of stainless-steel blooms to billets is performed on the same equipment used for carbon steels and the usual precautions of close temperature control and avoidance of sulphur contamination are taken.

After rolling, the air-hardenable martensitic grades must be cooled slowly in order to soften the material. This practice prevents thermal cracking during subsequent conditioning.

Rolling of Stainless-Steel Plates—The equipment used for the heating and rolling of stainless-steel plates is the same as that used for the heating and rolling of carbon-steel plates. However, because the austenitic stainless steels are very stiff at elevated temperatures, they require more power for rolling. Consequently, the amount of reduction per pass is smaller for the austenitic grades. Also, these steels spread less than do the ordinary steels and due allowances are made for this lack of spread in order that the resulting plate widths will satisfy dimensional requirements.

After rolling, the stainless-steel plates are annealed and pickled. As might be expected, the annealing temperatures employed depend upon the composition of the material, and the specific annealing temperatures used for the standard grades are listed in Table 47—V. The pickling procedure used for stainless steel varies from plant to plant. One installation consists of a 10 per cent sulphuric acid bath operated at 60° to 70°C (140° to 160°F) and a 10 per cent nitric acid, 4 per cent hydrofluoric acid bath operated at 55° to 65°C (130° to 150°F). The first bath softens and loosens the scale but will not remove it completely; the second solution will remove the scale loosened by the first solution.

Table 47—V. Forging and Annealing Temperatures for Some Stainless Steels.

Grade		Preheating Range(a)		Begin Forging		Finish Forging		Annealing Temperature		Rate of Cooling
UNS Number	AISI Number	°C	°F	°C	°F	°C	°F	°C	°F	
AUSTENITIC STEELS										
S20100	201	815–870	1500–1600	1150–1260	2100–2300	925 min.	1700 min.	1035–1095	1900–2000	Rapid
S20200	202	815–870	1500–1600	1150–1260	2100–2300	925 min.	1700 min.	1035–1095	1900–2000	Rapid
S30200	302	815–870	1500–1600	1150–1260	2100–2300	925 min.	1700 min.	1035–1095	1900–2000	Rapid
S30403	304L	815–870	1500–1600	1150–1260	2100–2300	925 min.	1700 min.	1010–1120	1850–2050	Air
S30300(b)	303(b)	815–870	1500–1600	1150–1260	2100–2300	925 min.	1700 min.	1035–1095	1900–2000	Rapid
S34700	347	815–870	1500–1600	1150–1260	2100–2300	925 min.	1700 min.	1010–1120(c)	1850–2050(c)	Air
S32100(b)	321(b)	815–870	1500–1600	1150–1260	2100–2300	925 min.	1700 min.	955–1120(c)	1750–2050(c)	Air
S31600	316	815–870	1500–1600	1150–1260	2100–2300	925 min.	1700 min.	1065–1120	1950–2050	Rapid
S31603	316L	815–870	1500–1600	1150–1260	2100–2300	925 min.	1700 min.	1035–1120	1900–2050	Air
S30900	309	815–870	1500–1600	1095–1230	2000–2250	980 min.	1800 min.	1095–1150	2000–2100	Rapid
S31000	310	815–870	1500–1600	1095–1230	2000–2250	980 min.	1800 min.	1095–1150	2000–2100	Rapid
S38100(f)	–	815–870	1500–1600	1150–1260	2100–2300	980 min.	1800 min.	1120 min.	2050 min.	Rapid
S21400(g)	–	815–870	1500–1600	1205–1260	2200–2300	980 min.	1800 min.	980–1035	1800–1900	Rapid
FERRITIC STEELS										
S40500	405	760–815	1400–1500	1065–1120	1950–2050	Under 815	Under 1500	730–785	1350–1450	Air
S40900	409	760–815	1400–1500	1065–1120	1950–2050	Under 815	Under 1500	730–785	1350–1450	Air
S43000	430	760–815	1400–1500	1035–1120	1900–2050(d)	Under 815	Under 1500	760–815	1400–1500	Air
S44600	446	760–815	1400–1500	1065–1120	1950–2050(d)	Under 785	Under 1450	845–900	1550–1650	Rapid
MARTENSITIC STEELS										
S41000	410	760–815	1400–1500	1095–1205	2000–2200	Under 815	Under 1500	845–870(e)	1550–1600(e)	Slow
S41600	416	760–815	1400–1500	1095–1205	2000–2200	Under 815	Under 1500	845–870(e)	1550–1600(e)	Slow

(a) Allow preheating time about twice that required for plain carbon steel of equivalent section.
(b) Not recommended for extremely severe forging operations.
(c) Annealing on low side of range provides maximum carbide stability with some sacrifice in optimum mechanical properties. Annealing on high side of range provides optimum mechanical properties with some loss in carbide stability.
(d) Long soaking at forging temperatures should be avoided to guard against excessive grain growth.
(e) Cool 28°C (50°F) per hour maximum to 595°C (1100°F).
(f) USS 18-18-2.
(g) USS TENELON.

Another procedure for the pickling of stainless steels is to use molten salts consisting of sodium hydroxide to which is added some agent such as sodium hydride. These molten-salt processes are rapid and efficient and have replaced many acid pickling installations.

After annealing and pickling, the stainless plates are sheared to size, and are then suitable for shipment.

Rolling of Stainless-Steel Bars—The billets used for the rolling of stainless-steel bars are conditioned as the surface requires it. Martensitic steels must not exceed 275 Brinell hardness prior to conditioning, and if this hardness is exceeded, the billet must be annealed before grinding to prevent thermal cracking which might occur during the grinding operation or during heating for rolling. Prior to heating for rolling, the ends of the billets are pointed by a scarfing torch to prevent the splitting of the ends of the bar and also to decrease slippage when entering the mill.

During the rolling of the ferritic grades, which spread easily, spread control is important and is accomplished by providing a billet size slightly less in cross-section than that which would be used for carbon or alloy steels. This smaller size permits lighter initial drafting.

On delivery from the rolling mill, the austenitic and ferritic stainless-steel bars are rapidly (air) cooled, while cooling of the martensitic stainless-steel bars is deliberately retarded. Sections of the martensitic stainless-steel bars 50 millimetres (2 inches) and over are cooled slowly in covered pits, while those under about 50 millimetres (2 inches) have cooling retarded by packing on the hot bed.

After annealing and pickling, stainless-steel bars are straightened on standard equipment and may be shipped in this condition or finished by centerless grinding or cold drawing.

Rolling of Stainless-Steel Sheet and Strip—In this section, the discussion of the rolling of stainless-steel sheet and strip will be confined to the continuous method by which the largest tonnage of stainless steels is produced.

The slab or billet used for the production of stainless-steel sheet or strip is completely conditioned. As is true for the other products, stainless-steel sheet and strip is rolled on the same equipment as that used for carbon-steel sheet and strip.

After rolling and annealing, all of the stainless steels are descaled, usually by pickling in acids. For the straight-chromium grades of stainless steel containing up to about 21 per cent chromium, the hot-rolled sheet or strip, in coil form, is batch annealed at subcritical temperatures. On the other hand, the straight-chromium grades containing over about 21 per cent chromium, and the austenitic grades, are annealed on a continuous unit and quenched from the annealing temperature. Often, for the austenitic steels, this annealing is performed in an oxidizing atmosphere which, by producing a heavy scale, "burns off" the defects and thus reduces the amount of conditioning necessary at some later stage. The quenching practice used depends upon the thickness of the material. For thick materials, high-pressure water sprays are used, but for thin materials, cooling in air is sufficient. All of the stainless steels are descaled on continuous units, usually arranged in tandem with a continuous annealing unit.

A typical continuous descaling or pickling installation (see Figure 47—15) consists of two 10.7-metre-long (35-foot-long) tanks containing approximately 15 per cent hydrochloric acid (HCl) at 71°C (160°F) followed by a tank of similar size containing about 4 per cent hydrofluoric acid (HF) and 10 per cent nitric acid (HNO_3) at 65° to 75°C (150° to 170°F). In some installations, low- or high-current-density electrolytic-pickling facilities using either cold sulphuric acid (H_2SO_4) or nitric acid (HNO_3) are substituted for the first two tanks.

Molten-salt descaling processes are also used com-

Fig. 47—15. Stainless-steel strip entering a continuous pickling line.

mercially on continuous lines. Normally a light acid pickle, usually hot nitric acid, follows the descaling treatment. These processes provide scale removal on all grades of stainless steel without metal loss and result in a smoother surface than is obtainable from acid pickling. The method requires more heat input than pickling, and care must be taken to prevent the introduction of any water.

If, after the first annealing and pickling operation, some of the surface defects produced during hot rolling are still present, the material may be reannealed (and pickled) to "burn off" these defects or the defects can be removed in a continuous coil-grinding operation.

Cold Working—General—With the exception of the high-carbon hardenable steels, all of the stainless steels can be cold worked. However, certain precautions must be taken.

The ferritic stainless steels, especially those containing over 20 per cent chromium, are extremely notch sensitive at room temperature and care must be taken to avoid notches, otherwise considerable breakage will result. However, between 205°C (400°F) and 315°C (600°F) the steels are tough, and difficult cold-working operations are successfully accomplished by working the material in this temperature range.

Cold work causes some austenitic stainless steels to transform partially to a low-carbon martensite. This transformation, plus the effect of the strain hardening caused by the cold work itself, causes such austenitic steels to have a high rate of work hardening. More power is required to work these steels.

To obtain lighter gages and improvements in surface, grain size and mechanical properties, the material is cold rolled in coil form. Usually, the cold rolling of stainless-steel sheet and strip is performed on a reversing mill, although a tandem mill may be used. Depending upon the final thickness desired, an intermediate anneal may or may not be used. This anneal, like the first anneal, may be performed on a continuous annealing and pickling line or it may be performed in a continuous bright-annealing furnace. For best results, stainless steel is bright annealed in pure hydrogen having a dew point of –73°C (–100°F) or lower. Generally, the surface of the bright-annealed product is smoother (and brighter) than that of the conventionally annealed and pickled product.

After annealing and pickling, the surface of the cold-rolled material has what is called a No. 2-D (dull, cold-rolled) finish for sheet or a No. 1 cold-rolled finish for strip. In this condition, the material either may be sheared to desired lengths and shipped, or may remain in coil form and be subjected to further processing. If a brighter finish is desired, the material is rolled on a temper mill, the finish resulting from this process is called a No. 2-B (bright, cold-rolled) finish for sheet or a No. 2 cold-rolled finish for strip. As previously mentioned, bright-annealed material has a higher finish (brighter surface) than material that has been conventionally annealed and pickled. Even higher finishes are obtained by mechanical polishing either as sheets or as coils. Flattening, or leveling, is accomplished on standard units used for this purpose.

Of special interest is the work-hardening characteristic of the austenitic stainless steels which permits these grades to be produced to tensile strengths as high as 1380 MPa (200 000 psi). The composition best suited for the production of these high strengths is UNS S30100 (AISI Type 301) containing 17 per cent chromium, 7 per cent nickel. Usually, this grade is supplied to four standard minimum tensile strength levels of 860, 1035, 1205 and 1275 MPa (125 000, 150 000, 175 000, and 185 000 psi). Materials having these respective tensile strengths are designated commercially as having ¼, ½, ¾ and full-hard temper. The amount of cold reduction necessary to produce these strengths depends upon the composition of the material.

HEAT TREATMENT

Heat Treatment of Iron-Chromium Stainless Steels—From 5 per cent chromium to 12 per cent chromium, using 0.15 per cent carbon as a base, the steels are hardenable by the austenite-to-martensite transformation. As would be expected from the high chromium contents, the hardenability of these steels is high and increases with increasing chromium and carbon.

However, most of the applications for the 5 to 9 per cent chromium steels require the metal to be in the most ductile condition, and the "softening" practice employed consists of normalizing and tempering or one of the various annealing cycles. Ductility is required because these steels, before they are placed into service, are subjected to various cold-fabrication practices.

Depending upon the intended application, the 12 per cent chromium steels are used in the fully-hardened and tempered as well as the annealed condition. These materials may be either air-cooled or oil quenched from the hardening temperature. Oil quenching produces slightly higher hardnesses but air cooling is employed in order to minimize the danger of cracking or warping. Annealing is usually accomplished by heating above the lower critical temperature and slow cooling, but also may be accomplished by subcritical annealing.

A group of steels containing about 0.10 per cent of carbon and between 15 and 21 per cent of chromium deserves special mention. Such steels, when air cooled from about 815°C (1500°F), can contain up to 50 per cent of martensite and, in this sense, they can be considered as martensitic stainless steels. However, because the steels are used almost exclusively for their corrosion-resistant properties in applications that require optimum formability, a ferrite-carbide microstructure is desired. This microstructure is achieved by an annealing treatment that consists of heating in the range 760° to 815°C (1400° to 1500°F) for a sufficiently long time to spheroidize the carbide phase and then slow cooling to room temperature.

The higher chromium steels (over about 21 per cent chromium) can be considered as being completely ferritic. Annealing would then seem to be a simple process of heating at a recommended temperature for a reasonable time and slow cooling. However, two difficulties arise. First, the alloys are single phase except for carbide; hence, no grain refinement by a phase transformation is possible. A large grain size once formed will be retained on cooling to room temperature. Only by cold work and recrystallization can the grain size be reduced. The fact that the high-chromium steels are inherently notch sensitive makes the effect of grain coarsening even worse. Additions of nitrogen have been used to obtain a finer grain size. The nitrogen causes small pools of austenite, which inhibit grain growth, to form at the heat-treating temperature.

The second difficulty is that embrittlement occurs when the steels are heated in or slowly cooled through the temperature range of 425° to 760°C (800° to 1400°F). The embrittlement, which is actually an age-hardening phenomenon, is caused by the precipitation of a body-centered cubic phase of iron and chromium containing 70 to 80 per cent chromium.

To avoid grain growth and embrittlement, the high-chromium steels are annealed by heating in the temperature range 760° to 925°C (1400° to 1700°F) and cooling rapidly.

Iron-Chromium-Nickel Stainless Steels—The austenitic stainless steels also are considered to be single phase although this belief is erroneous for two reasons.

First, in ordinary 18—8, the austenite is not thermodynamically stable at room temperature. By the means of plastic deformation at or below room temperature, metastable austenite can be transformed, at least partially, to martensite.

Second, the carbide phase, unfortunately, cannot be ignored. In Figure **47**—16 it will be observed that the carbide solubility changes abruptly with temperature. Therefore, during slow cooling a carbide precipitation occurs and these carbides, rich in chromium, precipitate at the grain boundaries. At the temperature where the precipitation occurs, chromium diffusion from the matrix is not rapid enough to replenish the chromium taken out of the immediate vicinity of the carbide and, consequently, this region is low in chromium. Because chromium is the element largely responsible for the excellent corrosion resistance, the region adjacent to the carbide becomes lower in corrosion resistance and the material is susceptible to intergranular corrosion in some environments.

The austenitic steels are, therefore, heat treated by an anneal at a temperature high enough to effect carbide solution but low enough to minimize grain growth, and then cooled to room temperature rapidly enough to keep the carbides in solution.

Such a treatment is not always possible, especially where these steels are welded in the field, and modifications of the austenitic grades have been developed. These modified steels contain titanium or columbium (niobium) which combine with the carbon and eliminate intergranular chromium carbide precipitation and susceptibility to intergranular corrosion. Titanium in amounts of five times the carbon and columbium in amounts of ten times the carbon are considered to be sufficient, although the actual amounts necessarily depend upon the grain size and composition (other than carbon content) of the material (see also "Intergranular Corrosion" in Section 5 of this chapter).

The relationship of these variables to the amount of titanium required has been quantitatively evaluated, and a suitable formula developed.

The titanium or columbium grades are sometimes given a stabilizing treatment at 871°C (1600°F) to insure complete chemical combination of carbon with titanium or columbium.

Another solution to the problems encountered in field welding and/or stress relieving is the use of austenitic stainless steels containing 0.03 per cent maximum of carbon, such as UNS S30403 and S31603 (AISI Types 304L and 316L). Although these steels are not "completely" immune to susceptibility to intergranular corrosion in the sense that they cannot be heated for prolonged periods of time in the sensitizing temperature range, they are satisfactory for almost all applications requiring welding and stress relieving.

Precipitation-Hardenable Stainless Steels—the precipitation-hardenable stainless steels are martensitic stainless steels whose strength is enhanced by a precipitation-hardening (or age-hardening) reaction. In the first three steels listed in Table **47**—IV, this transformation occurs on cooling (generally air-cooling is satisfactory) from the solution-annealing temperature which is in the range of 926° to 1036°C (1700° to

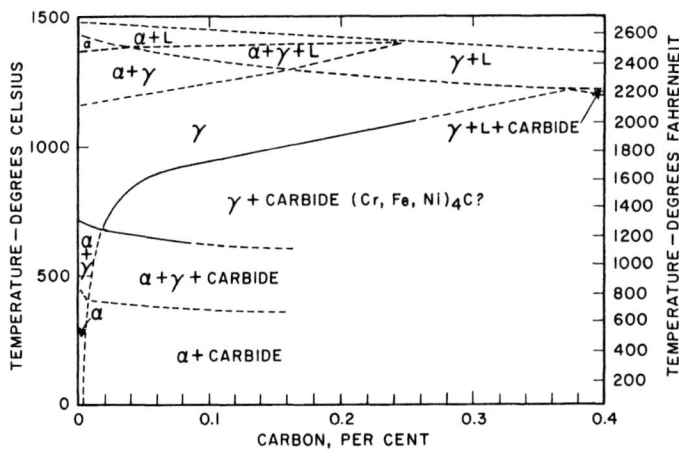

FIG. **47**—16. The effect of carbon on the constitution of stainless steels containing 18 per cent chromium and 8 per cent nickel.

1900°F). In the fourth steel, Table **47**—IV, cooling from the solution-annealing temperature of 1036°C (1900°F) results in a microstructure that is essentially austenite. The transformation of this austenite is accomplished by one of three treatments:

a. Cold rolling or cold drawing.

b. Reheating to 760°C (1400°F) and air-cooling to 16°C (60°F).

c. Reheating to 936°C (1700°F). Refrigerating to minus 73°C (minus 100°F).

The precipitation-hardening treatment consists of heating in the range 482° to 565°C (900° to 1050°F).

SECTION 4

MECHANICAL PROPERTIES

A. Martensitic Stainless Steels

Metallurgical Characteristics—Metallurgically, the martensitic stainless steels may be considered as high-chromium counterparts of the quenched and tempered carbon and alloy steels. Therefore, an understanding of these steels is based upon an interpretation of the effect of chromium on several metallurgical aspects such as equilibrium (phase) diagrams, hardenability, martensitic transformation and tempering characteristics.

The first characteristic to be attributed to chromium is that, as it is added to a 0.10 per cent carbon steel, it restricts the temperature range over which the austenite phase is stable under equilibrium conditions. In this sense, chromium has been called a "ferrite former." Figure **47**—9 illustrates this effect of chromium. The most important phase to be considered in this diagram is the austenite (gamma) phase because only those steels which form austenite on heating will transform to martensite on cooling.

The second effect of chromium is that of increasing hardenability. Figures **47**—17 and **47**—18 illustrate time-temperature-transformation (TTT) diagrams for a 0.35 per cent plain carbon steel and a 12 per cent chromium (0.1 per cent carbon) steel, respectively. The important effect of chromium to be noted is the shifting of the "nose" of the TTT curve to the right (toward longer times), thus causing the austenite-martensite transformation to occur at cooling rates slower than would be possible in the absence of chromium. Also shown in these figures are the results of end-quenched hardenability studies using the standard Jominy or modified Jominy specimens which further portray the effect of chromium on increasing hardenability.

Figures **47**—17 and **47**—18 also illustrate a third effect of chromium, namely that of decreasing the temperature for the start of martensite transformation (M_s temperature). The 0.35 per cent carbon steel has a M_s temperature of approximately 400°C (750°F); whereas the 12 per cent chromium steel has a M_s temperature of approximately 350°C (660°F).

The final important effect of chromium is that of increasing resistance to tempering. A convenient method of comparing tempering curves is to use a curve based on the combined time-temperature parameter developed originally by Hollomon and Jaffe. This method permits a wide range of different tempering conditions to be illustrated on a single curve. However, the user of such a method should be warned that with steels having strong secondary hardening characteristics, it is necessary to check the main features of

the tempering of such steels with isothermal tempering curves. Figure **47**—19 illustrates the tempering of a 0.94 per cent carbon steel, and Figure **47**—20 is a tempering curve for a 0.14 per cent carbon, 12 per cent chromium steel. As will be noted, the plain carbon steel begins to soften immediately (at low values of the parameter); whereas the 12 per cent chromium steel does not begin to soften appreciably until higher values of the parameter are obtained. Based on tempering times of approximately one hour, the softening of the 12 per cent chromium steel does not begin until a temperature in excess of about 480°C (900°F) is achieved. Moreover, these steels may undergo a secondary hardening following tempering at temperatures of 400° to 455°C (750 to 850°F). This is illustrated better by the isothermal tempering curves of Figure **47**—21, than from the parameter type curve of Figure **47**—20.

By way of summary, the physical metallurgy of the martensitic stainless steels is essentially the metallurgy of the effect of chromium as follows:

1. Chromium decreases the range of austenite stability under equilibrium conditions.

2. Chromium increases the hardenability.

3. Chromium decreases the temperature for beginning of martensite transformation.

4. Chromium increases resistance to tempering. Furthermore, the precipitation of chromium carbides during tempering causes secondary hardening that can be enhanced further by additions of suitable alloying elements.

Room-Temperature Properties—Although the martensitic stainless steels generally are used in the quenched and tempered condition so as to take advantage of the aforementioned effects of chromium, their annealed tensile properties are of significance particularly to a fabricator who must form these steels into suitable shapes before their full utility can be realized.

Table **47**—VI lists the representative properties of the martensitic stainless steels in the annealed condition. UNS S41000, S41600, and S41623 can be annealed to produce ferrite and spheroidal carbide microstructures; hence, the yield and tensile strength of these steels are similar to those of plain carbon and lower alloy steels. The other martensitic stainless steels have yield and tensile strengths significantly higher than the carbon and alloy steels of the same carbon content, because the softest microstructure that can be achieved in these steels is overtempered martensite. Particular attention should be given to the relatively high strengths and lower ductilities of the steels con-

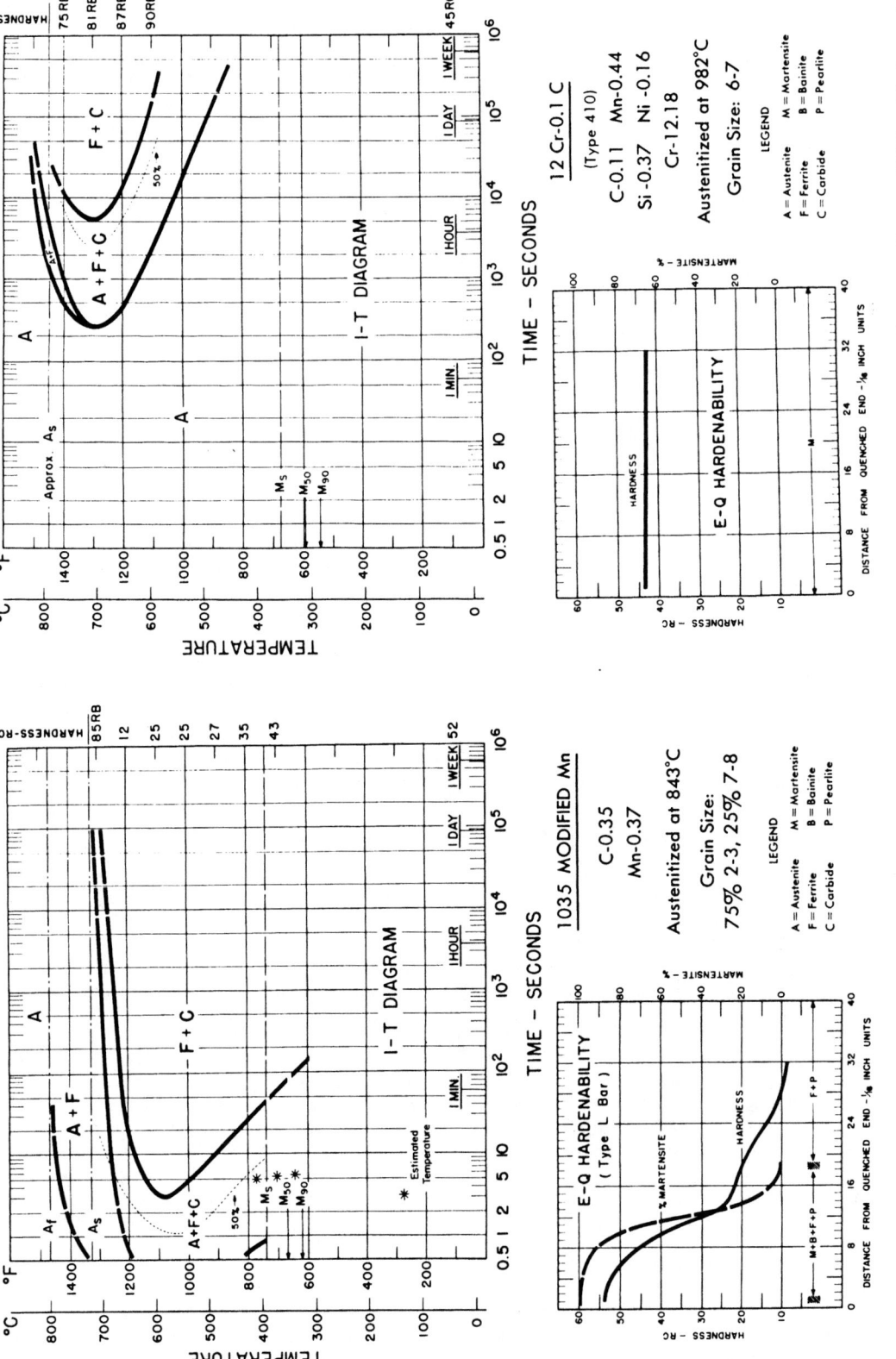

Fig. 47—18. Time-temperature-transformation diagram (TTT or IT diagram) and end-quench hardenability diagram for a 0.10 per cent carbon, 12 per cent chromium (S41000 or AISI Type 410) stainless steel.

Fig. 47—17. Time-temperature-transformation diagram (TTT or IT diagram) and end-quench hardenability diagram for 0.35 per cent carbon steel.

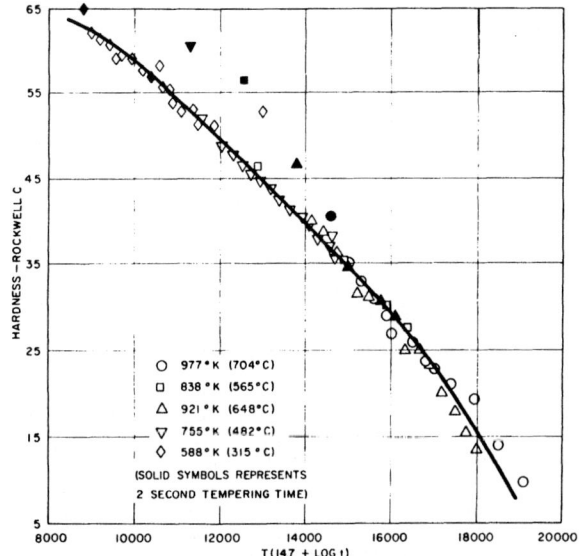

FIG. 47—19. Hardness vs. time-temperature parameter for tempering 0.94 per cent carbon steel. Structure before tempering: martensite plus retained austenite. Treatment before tempering: 815°C (1500°F), 10 min., brine quench. T = temperature in °K; t=time in seconds. (After Hollomon and Jaffe: see bibliography.)

taining nickel (UNS S41400 and S43100 or AISI Types 414 and 431). These higher strengths reflect the propensity of nickel to lower the temperature for austenite formation on heating, thus restricting the maximum

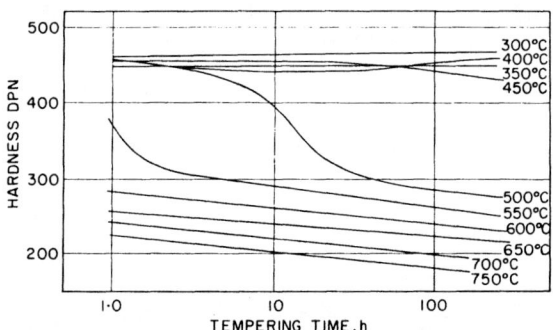

FIG. 47—21. Isothermal tempering curves for 0.14 per cent carbon, 12 per cent chromium steel. (After Irvine, Crowe and Pickering: see bibliography.)

tempering temperature for achieving softening. The solid solution hardening effect of carbon can be seen by comparing the strengths of UNS S41000 or AISI Type 410 (0.1 per cent carbon), UNS S42000 or AISI Type 420 (0.2 per cent carbon) and UNS S44000 or AISI Type 440 (0.60 to 1.0 per cent carbon).

Because of the previously mentioned effects of chromium on increasing hardenability and resistance to tempering, the martensitic stainless steels are capable of being heat treated to a variety of desirable mechanical properties. Figure 47—22 illustrates the mechanical properties obtainable from tempering a 12 per cent chromium, 0.1 per cent carbon steel. Enhancement of these already fine mechanical properties together with even further increased resistance to tempering can be

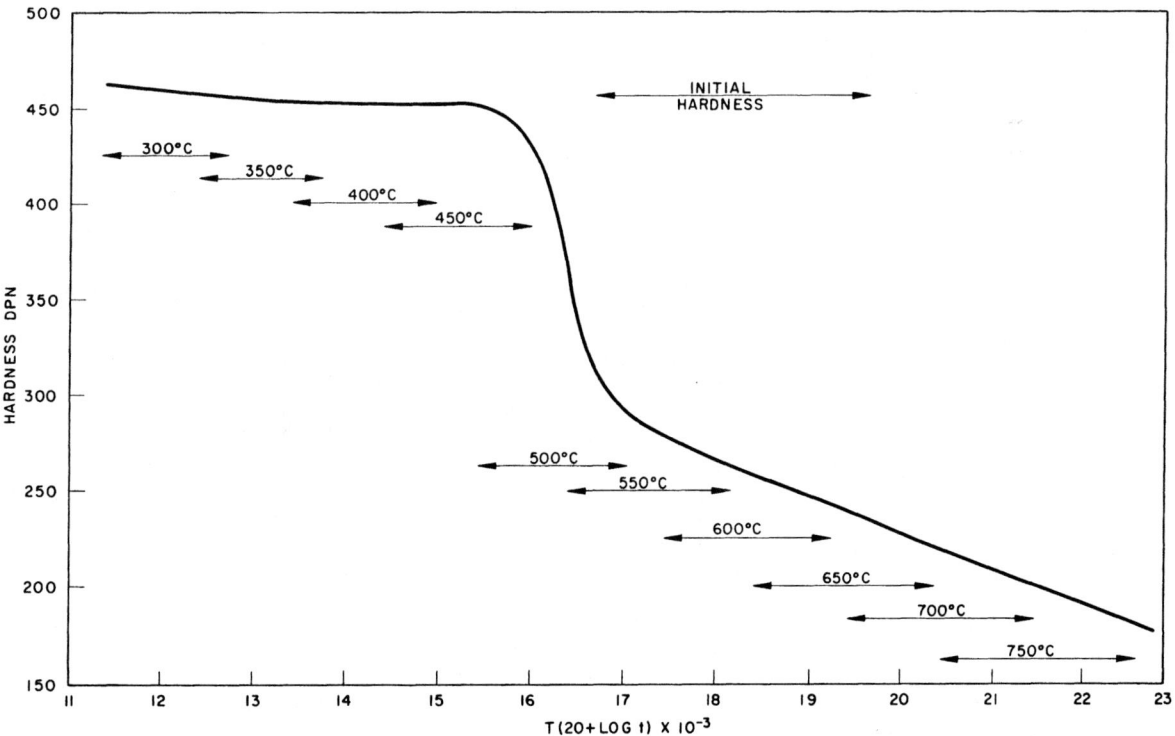

FIG. 47—20. Tempering curve for 0.14 per cent carbon, 12 per cent chromium steel. (After Irvine, Crowe and Pickering: see bibliography.)

Table 47—VI. Representative Mechanical Properties of Annealed Martensitic Stainless Steels (Wrought Products).

Designation		Yield Strength (0.2% offset)		Tensile Strength		Elongation (% in 50 mm or 2 in.)	Reduction of Area (%)
UNS	AISI Type Number	MPa	ksi	MPa	ksi		
S41000	410						
	Sheet	310	45	483	70	25	
	Bar	276	40	517	75	35	70
	Plate	241	35	483	70	30	
S41400	414	621	90	793	115	20	60
S41600,	416,						
S41623	416Se	276	40	517	75	30	60
S42000	420	345	50	655	95	25	55
S43100	431	655	95	862	125	20	55
S44002	440A	414	60	724	105	20	45
S44003	440B	427	62	738	107	18	35
S44004	440C	448	65	756	110	14	25

obtained by judicious selection of alloying elements, keeping in mind the strict limitations imposed by the already narrow temperature range of stability of austenite and problems of retained austenite.

Toughness characteristics, that is, the ability to deform plastically under the influence of a notch, also are important and a typical illustration of the effect of tempering temperature on the Charpy V-notch impact strength (one measure of toughness) of martensitic stainless steels is shown in Figure 47—23. This figure also gives some idea of the effect of chemical composition on impact (or toughness) characteristics. Considering the bottom curve of the top portion of Figure 47—23 (the 12 per cent chromium, 2 per cent nickel steel) the impact strength first increases with an increase in tempering temperature. This increase is probably due to a stress-relief effect of tempering temperature. Above a temperature of about 230°C (550°F), however, the impact strength decreases until it reaches a minimum which varies between 455° to 550°C (850° to 1025°F) depending upon the specific chemical composition of the material. This decrease in impact strength, or toughness, corresponds to the secondary hardness peak in the tempering curve. On heating to higher temperatures, of course, the impact strength increases markedly but, at these higher temperatures, the hardness and tensile strength also decrease rapidly.

Figure 47—23 also illustrates that additions of alloy-

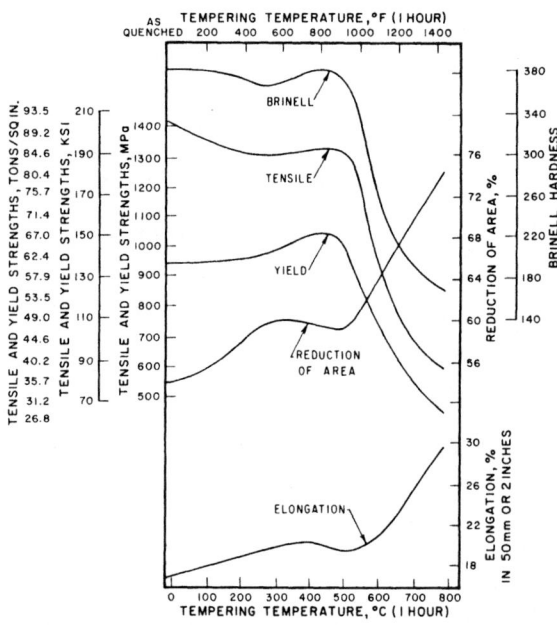

FIG. 47—22. Effect of tempering temperature on the mechanical properties of 12 per cent chromium S41000 (AISI Type 410) stainless steel.

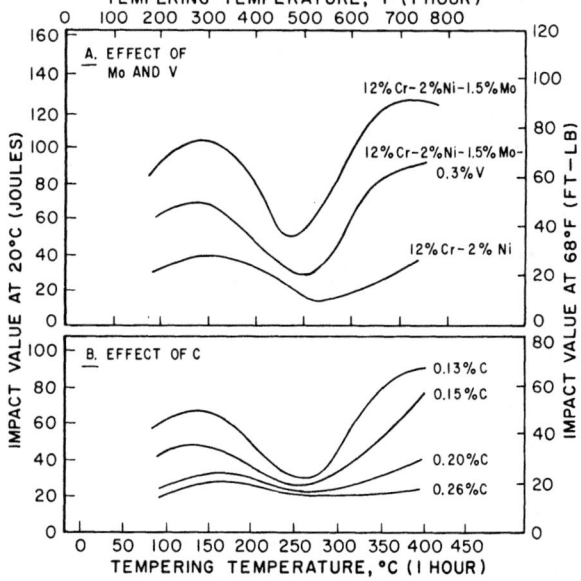

FIG. 47—23. Effect of tempering on impact properties (Charpy V-notch) of 12 per cent chromium steels (After Irvine, Crowe and Pickering: see bibliography.)

ing elements, particularly vanadium and molybdenum, tend to alleviate the low impact strength in the range of 455° to 550°C (850° to 1025°F). However, this effect is merely to raise the impact strength level but not to influence the depth of the "trough" in this range. The detrimental effect of carbon on lowering the impact strength at all temperatures is to be noted.

Small amounts of ferrite in the structure of the martensitic stainless steels significantly enhances their impact strength (or toughness). Fortunately, by suitable selection of chemical composition and heat treatment, these small amounts of ferrite can be achieved with but a minimum sacrifice of strength. On the other hand, the presence of ferrite in a martensitic stainless steel may be deleterious to the fatigue strength.

Elevated-Temperature Properties—The propensity of chromium to increase the resistance to tempering makes for an enhanced elevated-temperature strength of the martensitic stainless steels as compared to low-alloy and carbon steels. The elevated-temperature properties of the martensitic stainless steels are described in Chapter 48.

Cryogenic Properties—The martensitic stainless steels have impact transition temperatures that are, at best, only slightly below room temperature. Therefore, because the impact strength or toughness of these steels is low at temperatures below –18°C (0°F), they are not considered for structural applications at cryogenic temperatures.

B. Ferritic Stainless Steels

Metallurgical Characteristics—Referring to Figure 47—9, the ferritic stainless steels classically are defined as those steels whose carbon and chromium contents fall to the right of the gamma loop in this figure and their structure consists essentially of the alpha phase (ferrite) at all temperatures up to the melting temperature. Surprisingly, however, none of the so-called ferritic steels meet this definition. For example, the 17 per cent chromium steel, UNS S43000 or AISI Type 430 which is the most popular of the ferritic stainless steels, will develop from about 30 to 50 per cent austenite in its microstructure on heating to temperatures above about 800°C (1500°F). Therefore, during air cooling, this austenite will transform to martensite so that often in the hot-rolled condition the microstructure of the 17 per cent chromium steel will consist of a mixture of ferrite and martensite. Even the higher chromium ferritic steels such as 27 per cent chromium, UNS S44600 or AISI Type 446, with its normal amounts of carbon (0.08 per cent), manganese (0.75 per cent) and nitrogen (0.15 per cent), will develop a small amount of austenite in its microstructure when heated to high temperatures (above about 1090°C or 2000°F). Consequently, for the purposes of this chapter, the ferritic stainless steels are defined as those steels that contain in excess of about 11 per cent chromium and are used in a condition in which their final microstructure consists essentially of ferrite (plus carbides).

Largely because of difficulties associated with welding, coupled with their relatively low toughness, the ferritic steels (UNS S43000 to UNS S44600 in Table 47—III) have not found wide acceptance as structural materials. The welding difficulties are associated with the large grain size which cannot be refined by any thermal treatment. Their toughness in the annealed condition decreases as the chromium content increases. For example, the 17 per cent chromium steel has a Charpy V-notch impact transition temperature that is near room temperature and the 27 per cent chromium, UNS S44600 or AISI Type 446, stainless steel has a Charpy V-notch impact transition temperature of about 150°C (300°F). Of course, these impact values are for steels having the normal amounts of impurities, such as carbon, nitrogen, phosphorus, sulphur, silicon and manganese. An exception to the low toughness characteristics of the ferritic stainless steels is found in the 12 per cent chromium—aluminum UNS S40500 or AISI Type 405 stainless steel, which has exhibited toughness satisfactorily even in the heat-affected zone of welds; and, consequently, has found use as a structural material for elevated-temperature service in many chemical and petroleum-refining applications.

UNS S40900 or AISI Type 409 is a general purpose non-hardenable constructional stainless steel. It is primarily intended for automotive exhaust systems and other light structural applications.

Iron—chromium alloys, and particularly alloys having very low carbon and nitrogen contents, have excellent toughness. With the AOD refining process, the low-carbon and low-nitrogen contents necessary for good toughness are commercially attainable. Many of these nonstandard ferritic stainless steels are discussed in the "Source Book of Ferritic Stainless Steels" by R.A. Lula (see bibliography).

Room-Temperature Properties—Table 47—VII lists the room-temperature tensile properties of the annealed ferritic stainless steels. These steels exhibit yield and tensile strengths slightly in excess of those for low-carbon steel, but with slightly lower ductility as measured by elongation and reduction of area.

The ferritic stainless steels have two shortcomings, one of which, called roping (or ridging), can be overcome; the other, which cannot be overcome, is the so-called 475°C (885°F) embrittlement.

The phenomenon of ridging (or roping) is a defect that is manifest in the form of surface corrugations or ridges that result from a forming operation. The ridges are always in the direction of the final cold rolling of the product regardless of the direction in which the material is stretched. An illustration of this phenomenon as encountered in the deep drawing of an automobile hub cap is shown in Figure 47—24.

Investigations have shown that ridging is caused by development of preferred textures in the material, following the cold-reduction and annealing operations. The degree of ridging can be minimized by employing high temperature (870°C or 1600°F) box-annealing treatments or high-temperature normalizing treatments. Also, additions of the carbide- and nitride-forming elements, columbium (niobium) and titanium, have been beneficial.

The so-called 475°C (885°F) embrittlement is actually a precipitation hardening phenomenon which occurs when the ferritic steels are heated in the range of about 370° to 540°C (700° to about 1000°F). The precipitation hardening that occurs during the heating operation drastically reduces ductility and toughness. The precip-

Table 47—VII. Representative Room-Temperature Tensile Properties of Annealed Ferritic Stainless Steels.

UNS Number	AISI Type Number	Product Form	Yield Strength*		Tensile Strength		Elongation (Per Cent in 50 mm or 2 in.)	Reduction of Area (Per Cent)
			MPa	psi	MPa	psi		
S40500	405	Sheet	276	40 000	448	65 000	25.0	—
S40900	409	Sheet and strip	241	35 000	448	65 000	25.0	—
S43000	430	Sheet and strip	345	50 000	517	75 000	25.0	—
S44200	—	Bars	310	45 000	552	80 000	20.0	40.0
S44600	446	Sheet and strip	345	50 000	552	80 000	20.0	—
S44600	446	Bars	345	50 000	552	80 000	25.0	45.0

*0.2 per cent offset.

itate causing these changes in properties has been identified as a chromium-rich ferrite. Although additions of alloying elements (except those that will change the structure of the material from ferrite to austenite) have no effect insofar as minimizing the embrittlement, ma-

terial that has been "embrittled" can be restored to its original ductility and toughness by heating at temperatures above about 595°C (1100°F).

Elevated-Temperature Properties—The elevated-temperature properties of the ferritic stainless steels are described in Chapter 48. Although these steels are the weakest of the stainless steels at elevated temperatures, they have found applications particularly where resistance to oxidation as well as carburization and sulphidation is required.

The aforementioned 475°C (885°F) embrittlement, together with the tendency for ferritic stainless steels to form sigma phase when heated for long periods of time in the range of 540°C (1000°F) to about 760°C (1400°F), further limits the application of the ferritic stainless steels for elevated-temperature service.

Cryogenic Properties—Because the ferritic stainless steels have impact transition temperatures at or above room temperature, they are not considered to be suitable for low-temperature applications.

C. Austenitic Stainless Steels

Metallurgical Characteristics—The austenitic stainless steels may be placed into two classifications based on their microstructure; namely, stable austenitic and metastable austenitic. The stable austenitic steels are those whose microstructure remains austenitic even after considerable amounts of cold work. The metastable austenitic steels are those whose austenitic microstructure transforms to an acicular or martensitic-type structure during cold work (or plastic deformation). The difference between the two types of steels is illustrated best by the stress-strain behaviors of these steels which is shown in Figure 47—25. UNS S30400 (AISI Type 304), which is representative of a stable austenitic steel, exhibits a normal stress-strain curve for the austenitic structure. The "parabolic" shape of this curve indicates that strain hardening occurs throughout the duration of the application of the stress. On the other hand, the stress-strain relationship for UNS S30100

FIG. 47—24. Photograph of deep-drawn automobile hub cap of 17 per cent chromium UNS S43000 (AISI Type 430) stainless steel, illustrating "ridging" (or "roping").

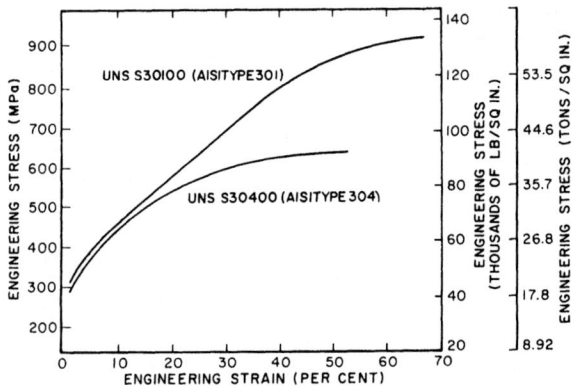

FIG. 47—25. Engineering stress and strain curves for UNS S30100 and S30400 (AISI Types 301 and 304) austenitic stainless steels.

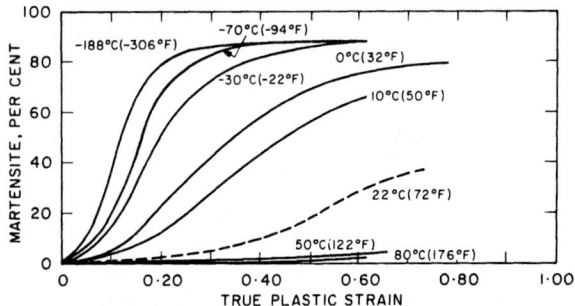

FIG. 47—26. Formation of martensite by plastic tensile strain at various deformation temperatures. (After Angel: see bibliography.)

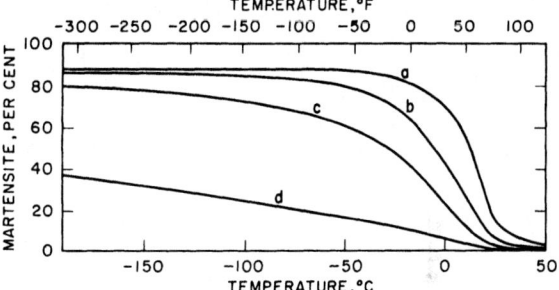

FIG. 47—27. Temperature dependence of martensite formation at true plastic strains of (a) 0.50, (b) 0.30, (c) 0.20, (d) 0.10. (After Angel: see bibliography.)

(AISI Type 301), which is representative of a metastable austenitic steel, indicates an accelerated rate of strain hardening after an initial increment of between 10 and 15 per cent plastic strain. This increase in strain hardening exhibited by the metastable steels is a direct result of the instability of the austenite; that is, formation of martensite due to plastic straining.

The importance of martensite, produced by deformation, to the mechanical properties of the austenitic stainless steels and the relationship of the amount of martensite to the chemical composition of the steel, and particularly to the temperature at which plastic deformation occurs, has been the subject of many investigations. Figure 47—26 illustrates the effect of deformation (as measured by true strain in a tension test) and temperature on the amount of martensite in a given steel. These curves show a striking resemblance to the curves for martensite formation in carbon steels where time rather than true plastic strain is the independent variable. The strong influence of temperature is shown in Figures 47—26 and 47—27, which show that above a certain temperature no martensite will form regardless of the amount of plastic deformation. The effect of chemical composition has been investigated and the results indicated that increasing amounts of all the elements studied lower the temperature of martensite formation; that is, increase the stability of the austenite.

Table 47—VIII lists the room-temperature tensile properties and impact strengths of a selected group of austenitic stainless steels in the annealed condition. The effect of chemical composition and phase change (martensitic transformation) during plastic strain on these properties is immediately obvious. For example, yield strength which is generally considered to be the stress near the termination of the elastic strain is not affected by any effects due to plastic strain. The compositional effects on yield strengths are thus simply solid-solution hardening effects, and inasmuch as the greatest contributions to solid-solution hardening are made by the interstitial elements, particularly carbon and nitrogen, those steels having the higher carbon and nitrogen contents should have higher yield strengths. Indeed, this proves to be the case, for as noted in Table

47—VIII, UNS S20100 and UNS S20200 (AISI Type Numbers 201 and 202, respectively) steels having the highest carbon and nitrogen contents have the highest yield strengths.

The demarcation between the stable and metastable steels is sharply manifested in tensile strengths as can be found by comparing the tensile strengths of the metastable UNS S20100, S20200 and S30100 (AISI Types 201, 202 and 301) with those of the stable UNS S30400 and S31000 (AISI Types 304 and 310). The metastable steels, of course, have the higher tensile strengths. The ductility of the austenitic stainless steels as represented by elongation and reduction of area is excellent, indicating that these steels can successfully undergo severe deformation. Also as noted, their toughness as represented by Charpy V-notch impact strength is excellent. The modulus of elasticity of the austenitic stainless steels is slightly less than that of the plain carbon and alloy steels, suggesting that for a given stress the austenitic steels will undergo slightly more elastic deflection.

The high rate of strain hardening discussed previously permits the austenitic stainless steels to be cold worked to exceptionally high yield and tensile strengths. Moreover, at these high yield and tensile strengths, a significant amount of ductility and toughness remains even after cold working. Thus, in the cold-worked condition, the austenitic stainless steels are truly high-strength, corrosion-resistant structural materials. Obviously, however, because this high strength is imparted by cold working, fabrication oper-

**Table 47—VIII. Representative Room-Temperature Tensile Properties and
Impact Strengths of Annealed Austenitic Stainless Steels.**

Steel*	Product Form	Yield Strength**		Tensile Strength		Elongation (Per Cent in 50 mm or 2 in.)	Reduction of Area, Per Cent	Charpy V-Notch Impact Strength (min.)		Modulus of Elasticity	
		MPa	psi	MPa	psi			J	ft-lb	MPa	psi
S20100 (Type 201)	Sheet and strip	310	45 000	655	95 000	40.0	—	—	—	197×10^3	28.6×10^6
S20200 (Type 202)	Sheet and strip	310	45 000	621	90 000	40.0	—	—	—	—	—
S30100 (Type 301)	Sheet and strip	276	40 000	758	110 000	60.0	—	—	—	193×10^3	28.0×10^6
S30400 (Type 304)	Sheet and strip	290	42 000	580	84 000	55.0	—	—	—	193×10^3	28.0×10^6
S30400 (Type 304)	Plate and bar	241	35 000	565–586	82–85 000	60.0	70.0	149	110	193×10^3	28.0×10^6
S31000 (Type 310)	Plate	310	45 000	655	95 000	50.0	65.0	122	90	200×10^3	29.0×10^6
S38100*** (XM15)	Plate	207	30 000	517	75 000	40.0	—	135	100	—	—

*AISI Type Number is in parentheses beneath UNS Number.
**0.2 per cent offset.
***USS 18-8-2

ations that require heating (such as welding) destroy the effects of the cold work. However, by ingenious welding techniques, together with ingenious design of welded joints, the cold-worked high-strength stainless steels can be joined and full utility made of the high strength imparted by cold working. Practical examples of structures made from high-strength, cold-worked austenitic stainless steels include railway passenger cars, truck trailers and the Atlas missile.

Figure 47—28, 47—29, 47—30 and 47—31 illustrate the effect of cold work on the tensile properties of UNS S20200, S30100, S30500 and S31000 (AISI Types 202, 301, 305 and 310, respectively) austenitic stainless steels, respectively. For a given amount of cold work, the metastable steels, UNS S20200 and S30100 (AISI Types 202 and 301, respectively) (Figures 47—28 and 47—29 respectively) attain a higher yield strength and tensile strength as well as a higher elongation than the stable austenitic stainless steels, Types UNS S30500 and S31000 (AISI Types 305 and 310, respectively) (Figures 47—30 and 47—31, respectively). Another example of the effect of composition on the properties of the cold-rolled austenitic stainless steels is illustrated in Figure 47—32. Again the distinction between the stable and metastable austenitic stainless steels is obvious.

Within a given grade of austenitic stainless steel, composition influences the properties in the cold-worked (as well as in the annealed) condition. The effect of composition as might be expected is a dual one:

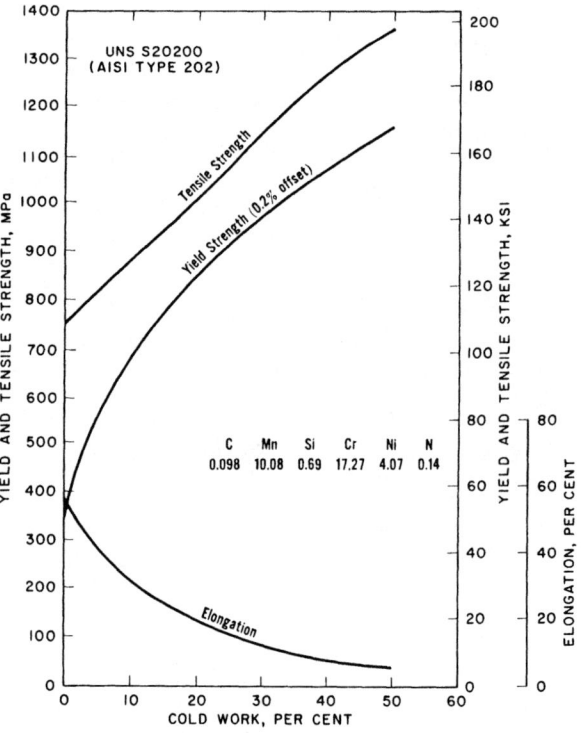

FIG. 47—28. Effect of cold working on the mechanical properties of 18 per cent chromium, 5 per cent nickel, 8 per cent manganese, UNS S20200 (AISI Type 202) stainless steel. (Courtesy, International Nickel Company, Inc.)

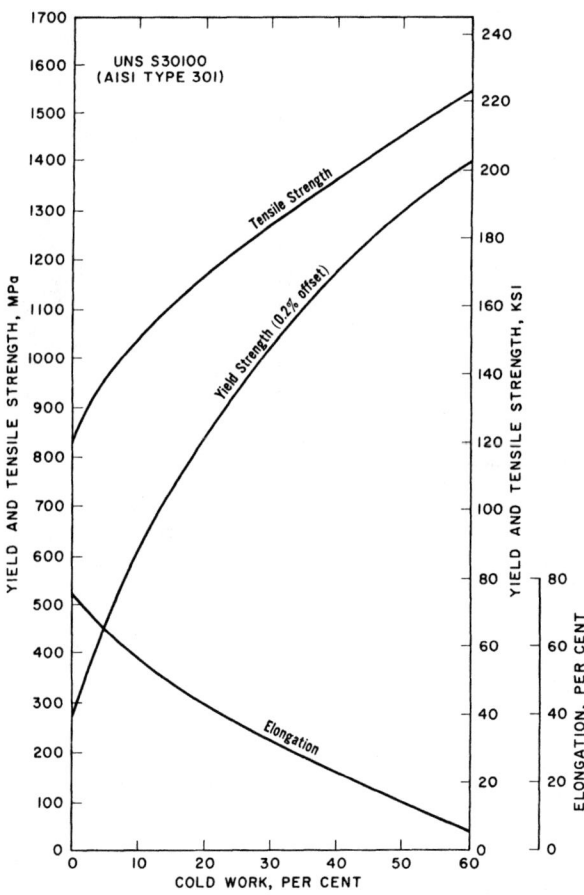

FIG. 47—29. Effect of cold working on the mechanical properties of 17 per cent chromium, 7 per cent nickel, UNS S30100 (AISI Type 301) stainless steel. (Courtesy, International Nickel Company, Inc.)

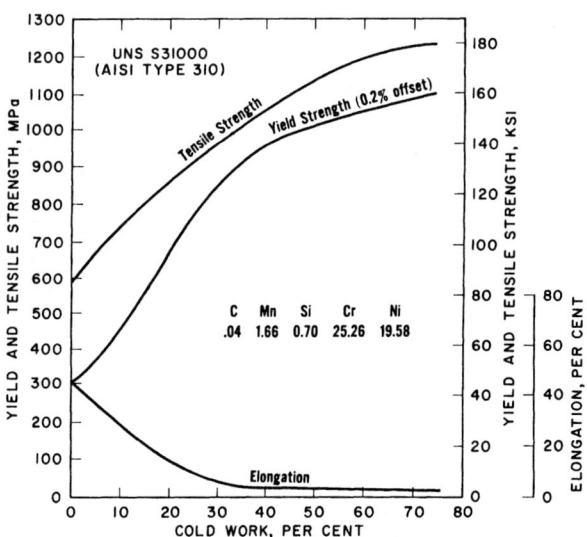

FIG. 47—31. Effect of cold working on the mechanical properties of 25 per cent chromium, 20 per cent nickel, UNS S31000 (AISI Type 310) stainless steel. (Courtesy, International Nickel Company, Inc.)

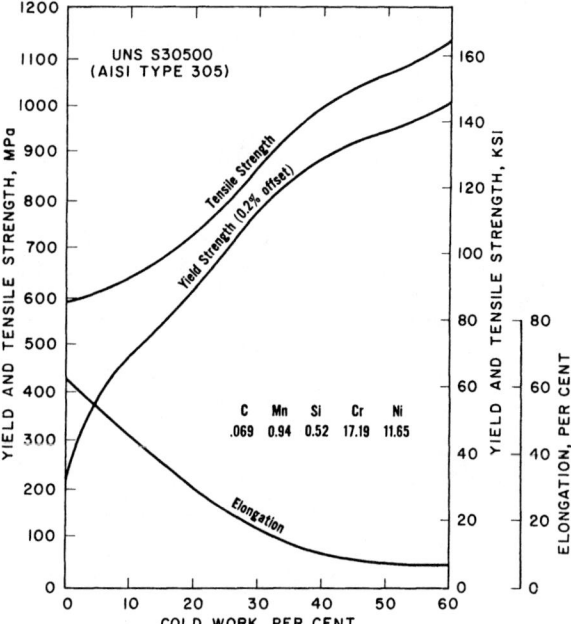

FIG. 47—30. Effect of cold working on the mechanical properties of 18 per cent chromium, 12 per cent nickel, UNS S30500 (AISI Type 305) stainless steel. (Courtesy, International Nickel Company, Inc.)

namely, its influence on transformation (or stability) of austenite, and its influence on solid-solution hardening. Because additions of almost all elements to a stainless steel tend to make its austenite more stable, this influence of composition is complex. For example, initial additions of a given element may have but little effect on stabilizing the austenitic structure, but because of its high solid-solution hardening effect, its addition will result in higher tensile and yield strengths and possibly lower elongation. On the other hand, above a certain amount the influence of this element may be principally that of stabilizing the austenite and thus it will tend to result in a lower tensile and yield strength (and possibly higher elongation). In other words, the austenite-stabilizing effect may offset the solid-solution hardening effect.

Perhaps the most important variable affecting the cold-worked strength of the austenitic stainless steels is that of the temperature at which the cold work is performed. Metallurgically, this effect can be described in terms of the M_d temperature (the temperature below which martensite will form from metastable austenite during deformation). If cold rolling is performed below the M_d temperature, plastic straining will cause the austenite to transform to martensite and significant increases in strength will occur. However, if the cold working is performed at about the M_d temperature, no martensite will form and the strain hardening will be proportionately less. For the so-called "metastable" austenitic stainless steels, the M_d temperature is above room temperature, thus cold-rolling operations conducted at room temperature result in large increases in strength. However, if the temperature of cold rolling is increased to about 205°C (400°F), the strain hardening

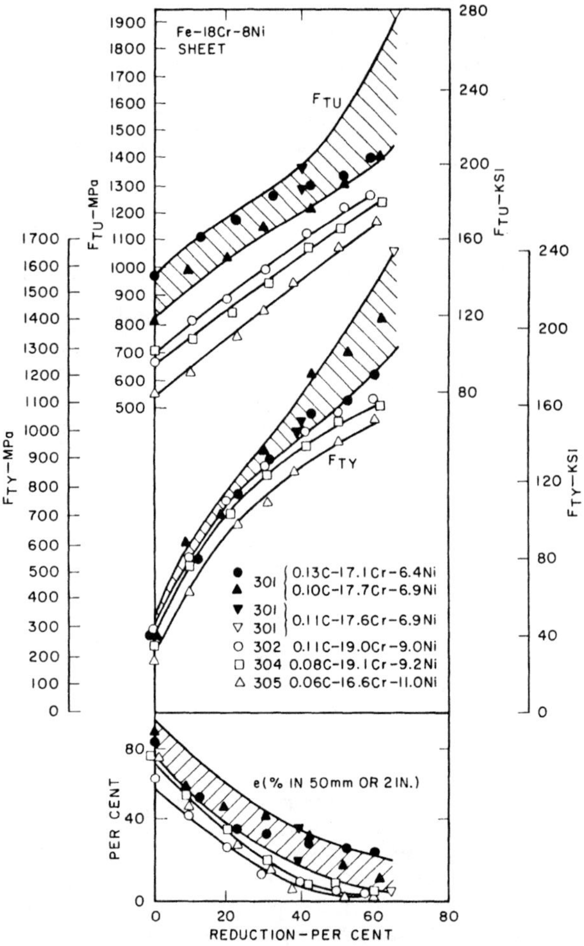

FIG. 47—32. Effect of cold rolling and chemical composition on the tensile properties of austenitic stainless steels. The AISI Type 301, 302, 304 and 305 steels in the list of constituents correspond respectively to UNS S30100, S30200, S30400 and S30500 steels. (From: "Aerospace Structural Metals Handbook," compiled by Air Force Materials Laboratory: see bibliography.)

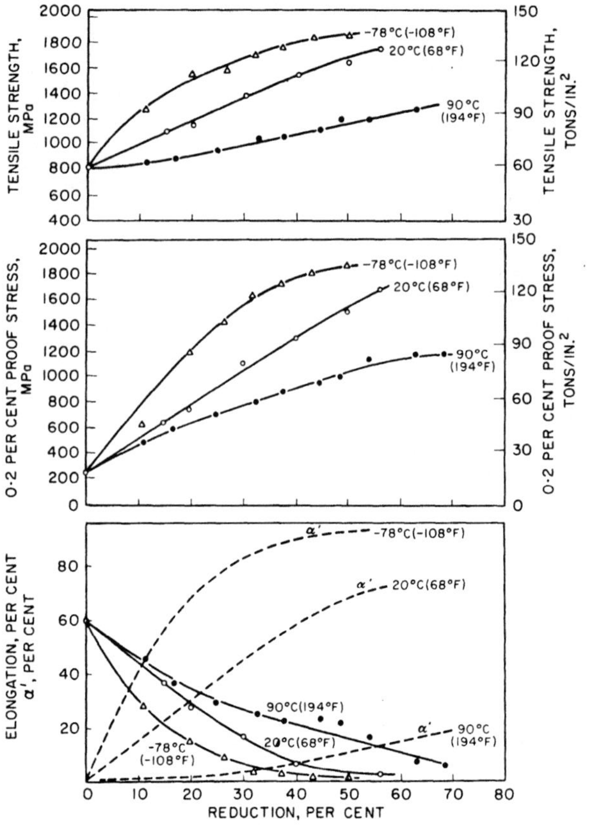

FIG. 47—33. Effect of temperature of rolling on the properties and martensite content of 17 per cent chromium, 7 per cent nickel, UNS S30100 (AISI Type 301) stainless steel. (After Llewellyn and Murray: see bibliography.)

decreases appreciably. Also, because the rate of strain hardening is greatly diminished at these so-called "warm" temperatures, the power required for a given amount of deformation is also significantly less. Thus, when severe deformation is required and power limitations prevent these severe deformations, advantage can be gained by performing these deformations at a slightly elevated temperature. The effect of temperature of rolling on the properties of UNS S30100 (AISI Type 301) is illustrated in Figure 47—33.

As mentioned previously, the toughness of the austenitic stainless steels, even after cold reduction at a high tensile strength, is still excellent. Figure 47—34 illustrates the effect of cold reduction on the toughness as measured by notch tensile strength of UNS S30100 (AISI Type 301) stainless steel. As will be noted, the notch tensile strength of this material in the longitudinal direction with a notch having a severity of K = 18.7 MN·m $^{-\frac{3}{2}}$ (17 ksi $\sqrt{\text{in.}}$) is the same or slightly higher than the unnotched tensile strength of the ma-

terial at cold reductions of between 30 and 60 per cent (tensile strengths of 1170 to 1515 MPa or 170 000 to 220 000 pounds per square inch). In fact, the notch toughness in the transverse direction, although less

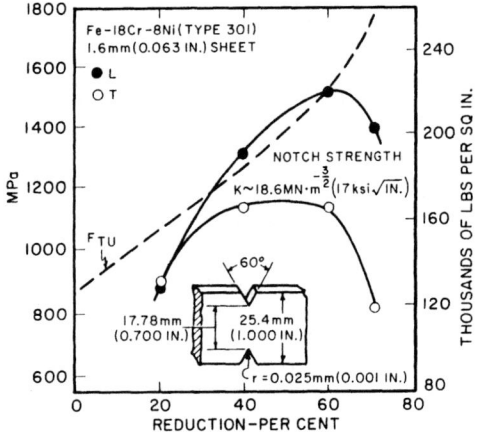

FIG. 47—34. Effect of cold rolling and test direction on the notch strength of UNS S30100 (AISI Type 301) stainless steel sheet. (From: "Aerospace Structural Materials Handbook," compiled by Air Force Materials Laboratory: see bibliography.)

Table 47—IX. Typical Room-Temperature Properties of the Precipitation-Hardenable Stainless Steels in the Solution-Annealed Condition.

UNS Number	Product Form	Yield Strength		Tensile Strength		Elongation (Per Cent in 50 mm or 2 in.)	Reduction of Area (Per Cent)
		MPa	psi	MPa	psi		
S13800	Bar	827	120 000	1103	160 000	17	65
S15500	Bar	1000	145 000	1103	160 000	15	55
S17400	Bar	1000	145 000	1103	160 000	15	55
S17700	Sheet	276	40 000	896	130 000	35	—

than the unnotched tensile strength, is still quite adequate.

Elevated-Temperature Properties—The austenitic stainless steels possess the highest elevated-temperature strength of any of the steels. These properties are described in Chapter 48.

Cryogenic Properties—The cryogenic properties of the austenitic stainless steels, which are excellent, are described in Chapter 49.

D. Precipitation-Hardenable Stainless Steels

Although, as stated previously, the precipitation-hardenable stainless steels essentially are martensitic stainless steels, they have several unique characteristics in addition to having to undergo an austenite-to-martensite transformation. One is that these steels must contain appreciable quantities of nickel because the precipitate responsible for the age-hardening is a nickel-aluminum and/or a nickel-titanium compound. A second is that the intensity of the age-hardening reaction is determined by the amounts of either titanium or aluminum present in these steels. The chemical compositions of some precipitation-hardenable stainless steels are given in Table 47—IV.

Room-Temperature Properties—Table 47—IX lists representative room-temperature tensile properties of these steels in the solution-annealed condition. With the exception of the fourth steel (S17700), these properties are typical of low-carbon martensite. The fourth steel, because of its predominantly austenitic structure in the solution-annealed condition, has a lower yield strength and correspondingly higher ductility. Its tensile strength approaches that of the other three steels because its austenitic structure is transformed to martensite due to the plastic flow that occurs in the tension testing.

In the precipitation-hardened condition, these steels exhibit yield strengths in the range 724 MPa (105 000 psi) to 1448 MPa (210 000 psi), and corresponding tensile strengths of 1000 MPa (145 000 psi) to 1551 MPa (225 000 psi) with ductilities in the range of 10 per cent to 20 per cent in 50 mm (2 inches). These are excellent properties for high-strength structural applications, and the fact that they are obtained by heat treatment requires the use of suitable pre- and post-heat techniques to assure retention of these properties in welded structures. For some applications, full advantage of the highest strengths may not be achieved because at these strengths the steels are susceptible to stress-corrosion cracking (or hydrogen embrittlement) and are somewhat notch-sensitive.

SECTION 5

CORROSION RESISTANCE

As was mentioned previously, the corrosion resistance of the stainless steels generally increases with increasing chromium content. There has been some speculation as to why chromium should impart stainlessness to steel. The popular concept is that when sufficient chromium is present, a thin, tight oxide is formed on the surface and this oxide prevents any further oxidation or corrosion. Environments that are oxidizing in nature strengthen this film while reducing environments tend to break down the film and cause the steel to corrode.

Voluminous data have been published on the corrosion resistance of specific grades of stainless steel in specific environments. These data represent both controlled laboratory tests and actual service records. A discussion of these data follows.

CORROSION RESISTANCE IN VARIOUS ENVIRONMENTS

Atmospheric Corrosion—The atmospheric-corrosion resistance of the stainless steels is primarily a function of the chloride content of the atmosphere. Therefore, proximity of the ocean or other sources of chloride contamination is of major importance. Other factors, such as sulphur contamination, which are important in determining the atmospheric-corrosion performance of other metals, are not particularly important with the stainless steels. The amount of rainfall is important only insofar as it affects the concentration of chlorides at the steel surface.

In rural atmospheres, UNS S41000 and S43000 (AISI Types 410 and 430) stainless steels and the austenitic stainless steels will give indefinite service without significant change in appearance. Thus, the selection of stainless steel for an application involving rural exposure can be based on cost, availability of materials, mechanical properties, ease of fabrication, and appearance.

In industrial atmospheres free of chloride contamination, UNS S43000 (AISI Type 430) stainless steel and the austenitic stainless steels will provide long-time service and will remain essentially free of rust staining. A film of dirt can be expected to form on the surface but when this film is removed the stainless steel will be found to be unattacked and to have retained its original bright appearance. The presence of chlorides in an industrial atmosphere can result in attack of the stainless steel. Austenitic stainless steels exposed in New York City for periods up to 26 years were essentially unaffected, but the same steels exposed in Niagara Falls near chemical plants producing chlorine and hydrochloric acid were attacked to varying degrees in much shorter periods.

In marine-type atmospheres, UNS S41000 and S43000 (AISI Types 410 and 430) stainless steels will develop a thin rust film in a short period of time but there will be no appreciable dimensional change. The austenitic stainless steels such as UNS S30100, S30200 or S30400 (AISI Types 301, 302 or 304) may develop some rust staining when exposed in marine environments. The staining is usually superficial and can be removed easily. UNS S31600 (AISI Type 316) stainless steel, which contains molybdenum, is essentially resistant to rust staining in marine environments.

Aside from atmospheric conditions, two additional factors that affect the atmospheric-corrosion behavior of stainless steels are surface condition and fabrication procedures. The degree of polish will influence the resistance of stainless steels to tarnishing and staining in chloride-containing environments. Dull finishes (No. 2 and 2D) are most susceptible to staining, whereas ordinary commercial finishes (2B and No. 4) are less susceptible to staining. The removal of dirt, rust and stain also are influenced by the surface finish. Dirt and corrosion products are removed easily from highly finished steels but are more difficult to remove from dull-finished surfaces. More frequent cleaning is necessary for dull-finished stainless steel if the original appearance is to be preserved.

Stainless steels are used widely in outdoor architectural applications. Information on the performance of stainless steels in some of these applications was gathered by a task group of ASTM Committee A-10 which inspected various buildings periodically and has reported their findings in the references listed under the American Society for Testing and Materials in the bibliography at the end of this chapter.

Fresh Water—Fresh water can be defined as water from any river, lake, pond or well that is not classified as acid, salty or brackish.

The corrosivity of fresh water is influenced by the pH, oxygen content and scale-forming tendencies of the water. In scale-forming (hard) waters, corrosivity is determined largely by the amount and type of scale formed on the metal surface; this scale formation is a function of the minerals present and the temperature.

In nonscaling (soft) waters, which are generally more corrosive than hard waters, corrosivity can be reduced by raising the pH or lowering the oxygen content. Where this is not possible, it is necessary to rely on suitable water treatment or to employ corrosion-resistant materials such as the stainless steels.

UNS S41000 (AISI Type 410) stainless steel is appreciably more resistant to corrosion by fresh waters than carbon steel and has given excellent performance in many fresh-water applications. It is used extensively to obtain high strength and corrosion resistance in the construction of docks and dams, for example. It should be recognized, however, that under some circumstances S41000 may be susceptible to moderate pitting attack in fresh waters. The pitting can be prevented completely by cathodic protection, which will result when the steel is in contact with an appreciable area of carbon steel. UNS S43000 (AISI Type 430) stainless steel and the austenitic stainless steels are almost completely resistant to corrosion by fresh waters at ambient temperatures.

The stainless steels also show excellent resistance to cavitation damage in fresh water. To illustrate this, tests on towboats operating on the Monongahela River were conducted by attaching plates of UNS S43000, S20100, S20200, S30200, S30400 (AISI Types 430, 201, 202, 302, 304) and USS TENELON stainless steels to areas of the hull where carbon steel was subject to severe cavitation damage. After seven years of exposure, no damage was noted on any of the stainless steels. Other tests have indicated that USS TENELON stainless steel is substantially more resistant to cavitation than other stainless steels.

Acid Water—The term "acid water" refers to natural waters that are contaminated by leachings from ore and coal and, as a result of their more acidic characteristics, are considerably more corrosive than neutral fresh water. Such acid waters usually contain appreciable quantities of free sulphuric acid because of the leaching action of the water on the sulphides contained in coal or ore. In addition, these waters can contain large quantities of ferric sulphate, which can have a pronounced effect on the corrosion of carbon steel.

Carbon-steel equipment exposed to acid waters usually corrodes rapidly. The results of tests conducted with various materials exposed to acid river water indicate that the austenitic stainless steels are highly resistant to corrosion by this environment.

The superior corrosion performance of the austenitic stainless steels in fresh and acid river waters and particularly the fact that their corrosion film presents only a minor barrier to heat transfer has resulted in a widespread use of stainless-steel tubes in heat exchanger applications.

Salt Water—A characteristic of corrosion by salt water is that it frequently takes the form of pitting. In the case of the stainless steels this is largely due to the fact that salt water induces local breakdown of the passive film to which these steels owe their ability to resist corrosion. Another cause of pitting of these steels is the fact that barnacles and other marine organisms which attach themselves to steel equipment are capable of causing oxygen concentration cells. Once established, these cells are very active and produce a considerable amount of corrosion and pitting. Under conditions of high flow rates in salt water, such as in the case of pump impellers, attack of the austenitic stainless steels is usually negligible.

In condensers employing stainless-steel tubes it is necessary to maintain a water velocity in excess of 1.5 metres (5 feet) per second to minimize a buildup of marine organisms and other solids in the tubes. Also, in the construction of stainless-steel equipment to handle salt water, it is good design practice to eliminate crevices and to employ heavy wall components.

Soil—Metals buried in soils are subject to a heterogeneous set of conditions that may vary from time to time, depending upon the weather and other circumstances. The most extensive and systematic testing of the corrosion of metals in soils has been done by the National Bureau of Standards (NBS).

The NBS Circular 579 entitled "Underground Corrosion" shows that, in general, the austenitic stainless steels have excellent resistance to most soils, whereas UNS S41000 and S43000 (AISI Types 410 and 430) stainless steels are subject to severe pitting in many soils. UNS S31600 (AISI Type 316) stainless steel is completely resistant to pitting in all the soils tested.

Nitric Acid—Ferritic stainless steels containing 14 per cent or more chromium and the austenitic stainless steels exhibit excellent resistance to nitric acid. UNS S43000 (AISI Type 430) stainless steel has been widely used for process equipment in nitric acid plants. However, UNS S30400 (AISI Type 304) has largely replaced UNS S43000 (AISI Type 430) for this service because of its generally better formability and weldability.

The corrosion resistance of the other austenitic stainless steels in nitric acid is similar to that of UNS S30400 (AISI Type 304). UNS S43000 (AISI Type 430) stainless steel generally shows slightly higher corrosion rates than UNS S30400 (AISI Type 304) and is more adversely affected by higher temperatures and concentrations. Corrosion-resistance ratings for Types UNS S43000, S30400 and S31600 (AISI types 430, 304 and 316) stainless steels in nitric acid are shown in Table 47—X.

Hot nitric acid will cause intergranular corrosion of both austenitic and ferritic stainless steel if the steels are improperly heat treated. The subject of intergranular corrosion will be discussed in detail later. At this point it is merely noted that this type of corrosion may be prevented by proper heat treatment or by the use of grades of steel that are resistant to this form of attack.

Sulphuric Acid—The standard grades of stainless steel are seldom used in sulphuric acid solutions because of their narrow range of usefulness. At room temperature, UNS S31600 (AISI Type 316) stainless steel (which is the most resistant of the standard grades in this acid) is resistant at acid concentrations less than 15 per cent or greater than 85 per cent. In the higher concentration ranges, however, carbon steel can normally be used. The martensitic and ferritic stainless steels generally are not resistant to sulphuric acid solutions.

As in the case of nitric acid, sulphuric acid can cause intergranular corrosion of stainless steels if they are not properly heat treated. For welded structures that can-

Table 47—X. Corrosion-Resistance Ratings of Stainless Steels in Nitric Acid.

Concentration of HNO₃ (Per Cent)	Tempera-ture (°C)	(°F)	Ratings* UNS S43000 (AISI Type 430)	UNS S30400 (AISI Type 304)	UNS S31600 (AISI Type 316)
10	80	175	B	A	A
20	50	125	B	A	A
30	50	125	B	A	A
30	boiling		B	A	A
50	50	125	B	A	A
50	boiling		D	A	A
70	50	125	B	A	A
70	boiling		D	B	B
90	50	125	D	B	B
90	boiling		D	D	D
100	100	212	D	D	D

*A = Corrosion rates less than 127 micrometres (5 mils) per year.
 B = Corrosion rates 127 to 508 micrometres (5 to 20 mils) per year.
 C = Corrosion rates 508 to 1270 micrometres (20 to 50 mils) per year.
 D = Corrosion rates greater than 1270 micrometres (50 mils) per year.

not be heat-treated after welding, the low-carbon grades UNS S30403 or S31603 (AISI Types 304L or 316L) or stabilized grades UNS S32100 or S34700 (AISI Types 321 or 347) should be used.

Phosphoric Acid—The austenitic stainless steels possess good resistance to phosphoric acid solutions, and they are widely used for equipment in the production and handling of this acid. These steels have useful resistance at temperatures up to 107°C (225°F) for most concentrations. The "superphosphoric acids" (greater than 100 per cent H_3PO_4) are satisfactorily handled in UNS S31600 (AISI Type 316) stainless-steel equipment at temperatures up to about 95°C (200°F).

A note of caution should be injected here, in that trace impurities of fluoride or chloride salts are sometimes present in phosphoric acid made by the "wet process." The presence of these halides in the acid can have a deleterious effect on the corrosion resistance of the stainless steels.

The martensitic and ferritic stainless steels are appreciably less resistant to phosphoric acid than are the austenitic grades, and are not normally used in this acid.

Hydrochloric Acid—Hydrochloric acid solutions of all concentrations rapidly attack stainless steels, even at room temperature. Use of stainless steels in this acid is therefore not feasible.

Other Inorganic Acids—The austenitic stainless steels generally show good resistance to boric, carbonic, chloric, and chromic acids at most concentrations and temperatures, except in the case of 100 per cent chloric acid. UNS S41000 and S43000 (AISI Types 410 and 430) stainless steels are considerably less corrosion resistant than the austenitic grades in chromic acid, but show relatively good resistance to boric and carbonic acids.

Acetic Acid—The austenitic stainless steels generally show excellent resistance to acetic acid, whereas the martensitic and ferritic grades would not be considered adequate for most applications. The austenitic stainless steels are completely resistant to all concentrations of acetic acid at room temperature. At higher temperatures, UNS S31600 and S31700 (AISI Types 316 and 317) show greater resistance than the other austenitic grades.

Formic Acid—Formic acid at room temperature can be safely handled with any of the austenitic stainless steels. When hot, however, this acid can rapidly attack the nonmolybdenum-bearing grades, and UNS S31600 or S31700 (AISI Types 316 and 317) stainless steel is required. Formic acid is quite corrosive to the martensitic and ferritic stainless steels at all temperatures.

Lactic Acid—UNS S30400 (AISI Type 304) stainless steel can be used for storing lactic acid at temperatures up to about 38°C (100°F). At higher temperatures, the nonmolybdenum-bearing grades of austenitic steel are subject to pitting, and UNS S31600 or S31700 (AISI Type 316 or 317) is preferred. The martensitic and ferritic stainless steels have generally poor corrosion resistance in this acid.

Oxalic Acid—In general, the stainless steels have good resistance to this acid at room temperature up to concentrations of at least 50 per cent. At higher temperatures, however, oxalic acid solutions are quite corrosive to all the stainless steels, as is the 100 per cent acid at room temperature.

Alkalis—The stainless steels generally show excellent resistance to weak bases such as ammonium hydroxide. In strong bases such as sodium and potassium hydroxides, the austenitic stainless steels have good resistance in concentrations up to about 50 per cent and at temperatures up to about 105°C (220°F). At higher concentrations and temperatures, corrosion rates can become appreciable. At temperatures above the atmospheric boiling point (and at slightly lower temperatures near 50 per cent concentration), stress-corrosion cracking of austenitic stainless steels has been reported.

Salt Solutions—The stainless steels generally show excellent resistance to salt solutions, except for halide solutions under certain conditions. In acid salts, their performance is naturally governed to some extent by the specific acid formed by hydrolysis of the salt. At high temperatures in acid salt solutions, the austenitic molybdenum-bearing grades of stainless steel (UNS S31600 and S31700, corresponding to AISI Types 316 and 317) are usually superior to the other grades.

When stainless steels are to be used in halide solutions, particularly chloride solutions, it should be recognized that even though corrosion rates are usually low, pitting and/or stress-corrosion cracking may occur under certain conditions. Although there are many instances where stainless steels are used with excellent results in the presence of chlorides (for example, in food processing equipment and in flowing sea water at relatively low temperatures), each application must be considered on an individual basis. Whether or not pitting or stress-corrosion cracking will occur depends on numerous factors in the environment and in the design and operation of the equipment.

CORROSION PHENOMENA

Pitting Corrosion—As mentioned earlier, the stainless steels owe their excellent corrosion resistance to the formation of an invisible oxide film that forms on the steel surface and renders it passive. The passive film may form either as a result of the steel reacting with oxygen when exposed to the atmosphere or as a result of contact with other oxygen-containing environments. If the passive film is destroyed, corrosion of the stainless steels may develop. In many cases the passive film is destroyed only in localized areas on the metal surface and the corrosion that occurs manifests itself in the form of tiny holes or pits. The affected area or pit becomes anodic to the surrounding passive surface and an electrolytic cell is established. The subsequent development of these active anodic areas into deep pits is a direct result of the flow of current between the small anodic area and the large cathodic area. In many cases, particularly in chloride environments, the pits show undercutting.

Factors That Promote Pitting—1. Pitting is most likely to occur in the presence of chloride ions combined with such depolarizers as oxidizing salts. An oxidizing environment is usually necessary for preservation of passivity with accompanying high corrosion resistance, but unfortunately, it is also a condition for occurrence of pitting. The oxidizer can often act as a depolarizer for passive-active cells established by breakdown of passivity at a specific point or area. The chloride ion in particular can accomplish this breakdown. (Stainless steels are subject to pitting in chloride solutions if the redox potential of the solution is more noble than the value of the critical potential (V_c) for pitting, which is close to 0.2 volts noble (hydrogen scale). Where the solution redox potential is more active than V_c, however, the same steel will be immune to pitting).

2. Aerated neutral or near-neutral chlorides can pit stainless steels. Pitting is less pronounced in rapidly moving aerated solutions than in partially aerated stagnant solutions because the flow of liquid carries away corrosion products which would otherwise accumulate at crevices or cracks. It also insures uniform passivity through free access of dissolved oxygen.

3. Pitting rate increases with temperature. For example, in 4 to 10 per cent sodium chloride solutions, a maximum weight loss produced by pitting is reached at 90°C (195°F); for more dilute solutions this maximum occurs at still higher temperatures.

Methods to Avoid Pitting—1. Avoid concentration of halogen ions.

2. Insure uniform oxygen or oxidizing solutions, agitate solutions, and avoid pockets of stagnant liquid.

3. Either increase oxygen concentration or eliminate it. Increasing oxidizing capacity of the solution augments passivity and resistance to attack. On the other hand, elimination of oxygen avoids passive-active cells as, for example, in salt solutions.

4. Increase pH. As compared with neutral or acid chlorides, appreciably alkaline chloride solutions cause fewer pits or none at all. (The hydroxyl ion acts as an inhibitor.)

5. Operate at lowest temperature possible.

6. Add passivators to the corrosive medium. A small concentration of nitrate or chromate is effective in many media. (The inhibiting ions preferentially adsorb on the metal surface, thereby preventing adsorption and subsequent attack by the chloride ion).

7. Apply cathodic protection. There is evidence that stainless steels protected cathodically by galvanic coupling to mild steel, aluminum, or zinc do not pit in sea water. (Note that it is not necessary to polarize the stainless steel to a potential more active than its open circuit value. It suffices to polarize (using sacrificial anodes or small impressed currents) only to a potential more active than the critical potential, V_c.)

Of the stainless steels, the austenitic grades containing 2 to 4 per cent molybdenum exhibit the best resistance to pitting attack. Examples of specific corrosive media in which the use of molybdenum-containing austenitic stainless steels markedly reduces pitting attack or general corrosion are sodium chloride solutions, sea water, and sulphurous, sulphuric, phosphoric, and formic acids.

Intergranular Corrosion—The unstabilized grades of austenitic stainless steels (those that do not contain titanium or columbium) containing more than 0.03 per cent carbon are subject to intergranular corrosion in certain environments if improperly heat treated. The damage occurs when these steels are heated in the range 425° to 815°C (800° to 1500°F), or are slowly cooled through this range. Such heat treatments cause precipitation of chromium carbides at grain boundaries (sensitization), and the resulting depletion of chromium in immediately adjacent areas renders those areas susceptible to corrosive attack. Sensitization can occur during welding, causing subsequent localized corrosion in the heat-affected zone of the weld.

The most commonly used method of checking for sensitization is the Huey test, in which specimens of the steel are exposed to boiling 65 per cent nitric acid for five consecutive 48-hour periods, and the weight loss determined for each period. It is usually specified that the average corrosion rate for the five test periods should not exceed 0.05 mm (0.002 inch) per month. Intergranular corrosion of welded structures of austenitic stainless steel can be prevented by the following methods:

1. Use of the low-carbon grades UNS S30403 or S31603 (AISI Types 304L or 316L) or the stabilized grades UNS S32100 or S34700 (AISI Types 321 or 347). Use of these grades prevents a damaging amount of chromium carbide from precipitating during welding.

2. If the finished part is small enough to furnace-anneal after fabrication, the part may be annealed at 1040° to 1150°C (1900° to 2100°F) to dissolve the chromium carbides and rapidly cooled through the 425° to 815°C (800° to 1500°F) range to prevent re-precipitation.

Although it is not as well known, intergranular corrosion can also occur in welded ferritic stainless steels in certain media. This is believed to be caused by a straining of the metal lattice by carbide or nitride precipitates when the steel is rapidly cooled from above 925°C (1700°F). Stress relieving at 650° to 815°C (1200° to 1500°F) after welding removes the stresses and restores the corrosion resistance. Addition of titanium to UNS

S43000 (AISI Type 430) stainless steel in amounts greater than eight times the carbon content greatly reduces the intergranular attack of the as-welded steel in some media. The titanium addition is not effective in concentrated nitric acid, however.

Stress-Corrosion Cracking—Stress-corrosion cracking is defined as the combined action of static stress and corrosion which leads to cracking or embrittlement of a metal. Only tensile stresses cause this type of failure. Practically all metals and alloys (with the exception of very pure metals) are subject to stress-corrosion cracking in certain environments. There is some difference of opinion as to whether certain metal failures are properly termed "stress-corrosion cracking" or "hydrogen embrittlement" (for example, cracking of high-strength steels in hydrogen sulphide). For the purpose of this discussion, all such environmentally induced failures are included under the general category of stress-corrosion cracking.

Hardened (quenched and tempered) martensitic stainless steels are susceptible to stress-corrosion cracking in environments containing chlorides, hot caustics or nitrates, or hydrogen sulphide. For austenitic stainless steels, concentrated chloride and caustic solutions are the major agents causing stress-corrosion cracking. Several other environments have been reported to cause stress-corrosion cracking in martensitic and austenitic stainless steels. However, it should be noted that in many of these environments the cracking may have been caused by the presence of impurities (chlorides, for example).

Sensitized austenitic stainless steel is susceptible to an intergranular form of stress-corrosion cracking. If the sensitization is severe and/or the stress level is high, this form of cracking may occur in what otherwise are considered to be mild environments. Under no circumstances should sensitized austenitic stainless steels be considered for stressed applications until sufficient testing has been performed to assure that the environment or environments to be encountered will not cause intergranular stress-corrosion cracking. Obviously, this same procedure of testing beforehand is a wise one when any form of stress-corrosion cracking is suspected.

The circumstances under which stress-corrosion cracking failures occur are usually quite complex. For example, the stress involved is normally not the operating stress alone, but a combination of this and residual stresses in the metal due to fabrication, welding, or heat treatment. Many times this situation can be alleviated by stress-relieving the equipment after fabrication. Also, as indicated above, the corrosive agent that causes the cracking is often only an impurity in the product being handled. The amount of the corrosive agent present may not be great enough to cause cracking in the bulk solution, but a localized concentration of the agent in crevices or in the splash zone above the liquid may cause failure.

Although several general measures for preventing stress-corrosion cracking have been reported, the best preventive measure is simply to use a material that is resistant to stress-corrosion cracking in the environment in question. Thus, in hot chloride environments AISI Type XMI5 (USS 18-18-2) or the ferritic stainless steels should be considered. In hydrogen sulphide environments, ferritic and austenitic stainless steels are usually suitable, but hardened martensitic stainless steels should not be used.

SECTION 6

SUMMARY

The stainless steels are alloys of (1) iron and chromium, (2) iron, chromium, and nickel, (3) iron, chromium, nickel and manganese, and (4) iron, chromium, manganese and nitrogen. Occasionally, small amounts of certain other elements are added in order to enhance corrosion resistance and mechanical properties or to immunize the steels to the action of certain harmful impurities. The inherently slow reaction rates of the stainless steels have hampered the establishment of precise equilibrium diagrams; however, the diagrams now in existence permit at least qualitative conclusions to be drawn regarding the structure of these steels. In regard to corrosion resistance, the chromium content seems to be the controlling variable and the effect of chromium may be enhanced by additions of molybdenum, nickel, and other elements. The mechanical properties of the stainless steels, like those of the plain carbon and lower alloy steels, are functions of the structure and composition of the material. Thus, the austenitic steels possess the best impact properties at low temperatures and the best strength at elevated temperatures while the martensitic steels possess the highest hardness at room temperature. Therefore, the stainless steels, by being available in a variety of structures, exhibit a range of mechanical properties which, combined with their excellent corrosion resistance, makes these steels highly versatile from the standpoint of design.

Bibliography

Adcock, F., The chromium-iron constitution diagram. Journal Iron and Steel Inst. **124** (1931), 99-149.

Air Force Materials Lab., Research and Technology Div., Air Force Systems Command, Wright-Patterson Air Force Base, Ohio, Aerospace structural materials handbook, Vol. I (Ferrous alloys), 1963. (Project 7381, Task No. 73810.)

American Iron and Steel Inst., Steel products manual—stainless and heat resisting steels, Dec. 1974.

American Iron and Steel Inst., Steel products manual supplement—stainless and heat resisting steels, March 1979.

American Society for Metals, Metals handbook (Vol. I, 8th ed.). Metals Park, Ohio, The Society (1961), 409.

American Society for Testing and Materials, Report of inspection of corrosion resistant steels in architectural and structural applications. Appendices to ASTM Committee A-10 Reports, Proc. ASTM **61**, 1961, 188; and **65**, 1965, 145.

Andersen, A. G. H. and E. R. Jette, X-ray investigation of the iron-chromium-silicon phase diagram. Trans. ASM **24** (1936), 375-419.

Angel, Tryggve, Formation of martensite in austenitic stainless steels. Journal Iron and Steel Inst. **177**, Part I (May 1954), 165-174.

Anon., Three-roll piercing developed by TI. Iron and Steel **40**, Feb. 1967, 55-57.

Armstrong, P. A. E., Method of welding alloy steels and product thereof. U. S. Pat. No. 1,997,538, April 9, 1935.

Armstrong, P. A. E., and Raymond R. Rogers, Composite ferrous bodies, U. S. Pat. No. 2,044,742, June 16, 1936.

Bain, E. C., The nature of alloys of iron and chromium. Trans. ASST, 9, January 1926, 9-32.

Bain, E. C., and W. E. Griffiths, An introduction to the iron-chromium-nickel alloys. Trans. AIME **75**, (1927), 166-213.

Barclay, W. F., The mechanisms of deformation and work hardening in AISI Type 301 stainless steel. ASTM Special Technical Publication No. 369, 26-29. Philadelphia, Pa., the Society, 1965.

Bates, J. F., Effect of stress on corrosion. Ind. and Eng. Chem. **58**, No. 2 (Feb. 1966), 19-29.

Bates, J. F., and A. W. Loginow, Principles of stress corrosion cracking as related to steels. Corrosion **20**, No. 5 (June, 1964), 189t-197t.

Bechtoldt, C. J., and H. C. Vacher, Phase diagram study of alloys in the iron-chromium-molybdenum-nickel system. Journal of Research of the National Bureau of Standards **58**, No. 1 (Jan. 1957), 7-19.

Bell, T. F., and D. G. Bowser, Process for the manufacture of stainless steel. U. S. Pat. No. 3,198,624, August 3, 1965.

Bergman, Gunnar, and David P. Shoemaker, The determination of the crystal structure of the sigma phase in the iron-chromium and iron-molybdenum systems. Acta Cryst. **7** (1954), 857-865.

Binder, W. O., and H. R. Spendelow, The influence of chromium on the mechanical properties of plain chromium steels. Trans ASM **43** (1951), 759-772.

Bloom, F. K., W. C. Clarke, Jr., and P. A. Jennings, Relation of structure of stainless steel to hot ductility. Metal Progress **59**, Feb. 1951, 250-256.

Bloom, F. K., and E. E. Denhard, Jr., High temperature ductility of boron-treated steel. AIME, Journal of Metals **13**, Dec. 1961, 908-911.

Bungardt, K., E. Kunze, and E. Horne, Untersuchungen uber den Aufbau des Systems Eisen-Chrom-Kohlenstoff (Structure of the system Fe-Cr-C). Archiv fur das Eisenhuttenwesen **29**, March 1958, 193-203.

Carney, D. J., Nickel free and low nickel austenitic stainless steels. Regional Technical Meetings AISI, 1955, 103-113.

Carney, D. J., and B. R. Queneau, Solidification of stainless steel ingots. AIME Proc. Electric Furnace Steelmaking Conf. **14**, 1956, 116-123.

Chao, Hung-Chi, The mechanism of ridging in ferritic stainless steels. ASM Transactions Quarterly **60**, No. 1 (March 1967), 37-50.

Chelius, Edward J., Production of stainless steels, U. S. Pat. No. 2,226,967, Dec. 31, 1940.

Christian, J. L., and L. D. Girton, The effects of cold rolling on the mechanical properties of Type 310 stainless steel at room and cryogenic temperatures. Trans. ASM **57**, 1964, 199-207.

Christian, R. R., The importance of corrosion testing in selecting process equipment. TAPPI **47**, No. 9 (Sept. 1964) 124A-129A.

Cook, A. J., and B. R. Brown, Constitution of iron-nickel-chromium alloys at 550—880 C. Journal Iron and Steel Inst. **171**, August 1952, 345-353.

Cook, A. J., and F. W. Jones, The brittle constituent of the iron-chromium system (sigma phase). Journal Iron and Steel Inst. **148**, No. II, 1943, 217-226.

Elliot, John F., and Molly Gleisen, Thermochemistry for steelmaking, Vol. 1, 1960. Reading, Mass., Addison-Wesley.

Elliot, Rodney P., Constitution of binary alloys, 1st supplement. New York, McGraw-Hill, 1965.

Fisher, R. M., E. J. Dulis and K. G. Carroll, Identification of the precipitate accompanying 885 F embrittlement in chromium steels. AIME Trans. **197** (1953), 690-695.

Fontana, M. G., Corrosion of Durimet 20 and Carpenter 20 by nitric acid. Ind. and Engrg. Chem. **45** (July 1953), 91A-92A, 94A.

Form, G. W., and W. M. Baldwin, Jr., The influence of strain rate and temperature on the ductility of austenitic stainless steel Trans. ASM **48**, 1955, 474-485.

Franks, R., W. O. Binder and J. Thompson, Austenitic chromium-manganese-nickel steels containing nitrogen. Trans. ASM **47**, 1955, 231-266.

Garofolo, F., P. R. Malenock and G. V. Smith, The influence of temperature on the elastic constants of some commercial steels. ASTM Special Technical Publication No. 129, 1952, 10-30.

Hamilton, Jack L., Matti H. Pakkala and Raymond Smith, Method of welding carbon steel to stainless steel. U. S. Pat. No. 2,704,833, March 29, 1955.

Handbook of Stainless Steels, (Peckner, D. and Bernstein, I.M., eds.) New York, McGraw Hill, 1977.

Hansen, M., Constitution of binary alloys. New York, McGraw-Hill Book Co., 1958.

Hattersby, B., and W. Hume-Rothery, Constitution of certain austenitic steels, Journal Iron and Steel Institute **204**, July 1966, 683-701.

Hawkins, G. A., J. T. Agnew and H. L. Solberg, The corrosion of alloy steels by high-temperature steam. Trans. ASME **66**, 1944, 291-295.

Healy, G. W. and D. C. Hilty, Significance of oxygen blowing rate in stainless steel melting. AIME Proc. Electric Furnace Steelmaking Conf. **13**, 1955, 187-191.

Hellawell, A., and W. Hume-Rothery, The constitution of alloys with transition elements of the first long period. Phil. Trans., Royal Soc., London **249**, 1957, 417-459.

Hilty, D. C., H. P. Rossbach and Walter Crafts, Observations of stainless steel melting practice. Journal Iron and Steel Institute **180**, June 1955, 116-128.

Hochmah, J., Properties of vacuum melted steel containing 25 per cent chromium. Revue de Metallurgie **48**, 1951, 734-758.

Hollomon, J. H., and L. D. Jaffe, Time-temperature relations in tempering steel. Trans AIME **162**, 1945, 223-249.

International Nickel Co., Inc., The, Bulletin B, Corrosion resistance of the austenitic chromium-nickel stainless steel in atmospheric environments. New York, The Company, 1963.

Irvine, K. J., D. J. Crowe and F. B. Pickering, The physical metallurgy of 12 per cent chromium steels. Journal Iron and Steel Institute **195**, August 1960, 386-405.

Irvine, K. J., D. T. Llewellyn and F. B. Pickering, Controlled transformation stainless steels. Journal Iron and Steel Institute **192**, Part 3, July 1959, 218-238.

Jenkins, C. H. M., E. H. Bucknall, C. R. Austin and G. A. Mellor, Some alloys for use at high temperatures, Part IV—The constitution of the alloys of nickel, chromium and iron. Journal Iron and Steel Institute **136**, 1937, 187-222.

Jette, Eric R., and Frank Foote, The Fe-Cr alloy system. Metals and Alloys **7**, 1936, 207-210.

Jones, J. D., and W. Hume-Rothery, Constitution of certain austenitic stainless steels with particular reference to the effect of aluminum. Journal Iron and Steel Institute **204**, January 1966, 1-7.

Kadinov, E. I., and S. I. Khitrik, Basic factors governing the melting loss of chromium in high-chromium steel baths during oxygen lancing, Izvest. VUZ-Chern. Met. **5**, No. 10 (October 1962), 50-57.

Kawakami, Kiminari, Production of alloy steels in the oxygen converter. 33/The Magazine of Metals Producing **4**, June 1966, 88-98, July 1966, 81-98.

Kinzel, A. B., and Walter Crafts, Alloys of iron and chromium—vol. I, chapter II, The constitution of iron-chromium alloys. New York, McGraw-Hill, 1937.

Kinzel, A. B., and Russell Franks, Alloys of iron and chromium—vol. II—High chromium. New York, McGraw-Hill, 1940.

Koster, W., The iron corner of the iron-manganese-chromium system. Archiv fur das Eisenhuttenwesen **7**, 1933-34, 687-688.

Krauss, G., Jr., and B. L. Averbach, Retained austenite in precipitation hardening stainless steels. Trans. ASM **52**, 1960, 434-450.

Krebs, T. M., and No. Soltys, A comparison of the creep-rupture strengths of austenitic stainless steels of the 18-8 series. Joint International Conf. on Creep, Institute of Mechanical Engineers, London, Paper 34, 1963.

Kubaschewski, O., Knowledge of the thermochemistry of refractory metals and its application in prediction of engineering structures. From "Refractory Metals" (ed. by N. E. Promisel), London, Pergamon Press, 1964, 191-204. (AGARD Conf., Oslo, 1963.)

Kubaschewski, O., and G. Heymer, The thermodynamics of the chromium-iron system. Acta Metallurgica **8**, July 1960, 416-423.

Langenberg, F. C., G. Pestel and C. R. Honeycutt, Grain refinement of steel ingots by solidification in a moving electromagnetic field. Trans. AIME **221**, No. 5, October 1961, 993-1001.

Larrabee, C. P., and W. L. Mathay, Controlling corrosion in coal-chemical plants. Corrosion **14**, 1958, 445t.

Llewellyn, D. T., and J. D. Murray, Cold worked stainless steels. Special Report 86, High alloy steels, The Iron and Steel Institute, 1964, 197-212.

Ludwigson, D. C., A study of the plastic behavior of metastable austenitic stainless steels. Doctorate thesis, University of Pittsburgh, 1966.

Ludwigson, D. C., and H. S. Link, Further studies on the formation of sigma in 12 to 16 per cent chromium steels. ASTM Special Technical Publication No. 369. Philadelphia, The Society, 1965.

Lula, R.A., Source book of ferritic stainless steels, American Society for Metals, Metals Park, 1982.

Marsh, John S., Sigma phase in the iron-chromium system. Metal Progress **35**, March 1939, 269-272.

McCabe, C. L., R. G. Hudson and H. W. Paxton, Activity measurements in the system iron-chromium. AIME Trans. **212**, 1958, 102-105.

Martin, W. A., Metallurgical development of weld overlays for cavitation resistance. Ontario Hydro Research Quarterly, Second Quarter 1963, 7-14.

Mears, R. B., The prevention of corrosion by use of appropriate design. Australasian Corrosion Engineer **6**, No. 12, Dec. 1962, 9-21.

Miller, R. F., and J. J. Heger, The strength of wrought steels at elevated temperatures. ASTM Special Technical Publication No. 100. Philadelphia, The Society, 1950.

Monypenny, J. G. H., Stainless iron and steel. New York, John Wiley and Sons, 1926.

Moshkevich, E. I., R. D. Mininzon, V. F. Smolyakov and M. F. Sorokina, Improving the ductility of OKh23N18 and Kh23N18 steel. Stal in English No. 8, August 1964, 645-647.

National Association of Corrosion Engineers, A report of Task Group T-5A-3, Corrosion by acetic acid. Corrosion **13**, No. 11, 1957, 79-88.

National Association of Corrosion Engineers, Second status report of Group T-4A-3 on methods and materials for ground rods. Materials Protection **4**, No. 12, Dec. 1965, 75-84.

Nelson, G. A. (Compiler), Corrosion data survey, 1960 edition. Shell Development Co.

Phelps, E. H., and D. C. Vreeland, Corrosion of austenitic stainless steels in sulfuric acid. Corrosion **13**, Oct. 1967, 619t-624t.

Pickering, F. B., Physical metallurgy of stainless steel developments. International Metals Review, Dec. 1976, 42.

Post, C. B., D. G. Schoffstall and H. O. Beaver, Hot workability of stainless steel improved by adding cerium and lanthanum. AIME Journal of Metals 3, Nov. 1951, 973-975, 976A-976B, 977A-977B.

Price, P. E., and N. J. Grant, 1300 C Isotherm in the system iron-chromium-nickel. Trans. AIME 215, 1959, 635-637.

Puzicha, W., and A. Kirsch, Influence of rate of extension upon tensile strength and elongation of austenitic steels. Manuscript 43, Kaiser-Wilhelm-Institut fur Eisenforschung, Clausthal-Zellerfield, June 1944.

Rees, W. P., B. D. Burns and A. J. Cook, Constitution of iron-nickel chromium alloys at 650 C to 800 C. Journal Iron and Steel Institute 162, 1949, 325-336.

Rohrig, J. A., R. M. Van Duzer and C. H. Fellows, High-temperature steam corrosion studies at Detroit. Trans. ASME 66, 1944, 277-290.

Romanoff, M., Underground corrosion. Circular 579, U. S. Dept. of Commerce, National Bureau of Standards, 1957, 49-52.

Schafmeister, P., and R. Ergagn, Constitutional diagram of the iron-chromium-nickel system with a particular attention to the brittle phase appearing after long annealing. Archiv fur das Eisenhuttenwesen 12, 1939, 459-464.

Schmitt, R. J., Behavior of carbon and stainless steels in acid waters. Corrosion 14, 1958, 445t.

Schwartzberg, F. R., S. H. Osgood, R. D. Keys and T. F. Kiefer, Cryogenics materials data handbook. U. S. Department of Commerce, National Bureau of Standards, 1965.

Sejournet, Jacques, Extrusion of steel using glass as a lubricant. Development of the Sejournet process. Revue de Metallurgie 53, No. 12, 1956, 897-914.

Shaffer, T. F., Field tests show corrosion resistance of stainless steels in liquid fertilizer service. Materials Protection 2, No. 8, Aug. 1963, 8-17.

Simmons, W. F., and H. C. Cross, The elevated temperature properties of stainless steels. ASTM Special Technical Publication No. 124, Philadelphia, The Society, 1952.

Simmons, W. F., and H. C. Cross, Elevated temperature properties of carbon steels. ASTM Special Technical Publication No. 180. Philadelphia, The Society, 1955.

Simmons, W. F., and Howard C. Cross, Report on elevated temperature properties of chromium steels. ASTM Special Technical Publication No. 228. Philadelphia, The Society, 1958.

Smith, R., E. H. Wyche and W. W. Gorr, A precipitation hardening stainless steel of the 18 per cent chromium, 8 per cent nickel type. Trans. AIME 167, 1946, 313-345.

Smithells, Colin J., The metals reference book, Vol. 2, fourth ed. Plenum Press, 1967.

Society of Automotive Engineers, Unified numbering system (UNS) for metals and alloys, January 1975.

Sticha, E. A., Structural stability of commercial wrought austenitic steels for power plant piping to 1450F. Proc., American Power Conf. 22, 1960, 288.

Swandby, R. K., Corrosion charts; guides to materials selection. Chem. Engr. 69, Nov. 12, 1962, 186-201.

Uhlig, H H., The corrosion handbook. New York, John Wiley and Sons, 1958, 165-168.

United States Steel Corporation, The atlas of isothermal transformation diagrams—supplement. Pittsburgh, Pa., The Corporation, 1953.

United States Steel Corporation, Fabrication of USS Stainless steel. Pittsburgh, Pa., The Corporation, 1969.

Wagstaff, R. S., G. E. Stock and G. N. Layne. Continuous casting of stainless steels at Atlas Steels Quebec plant. Iron and Steel Engineer 43, July 1966, 71-76.

Warren, K. A., And R. P. Reed, Tensile and impact properties of selected materials from 20 to 300K. Monograph 63, U. S. Dept. of Commerce, National Bureau of Standards, 1963.

Watson, J. F., and J. L. Christian, Mechanical properties of high strength Type 301 stainless steel at 70, –320 and –423 F in the base metal and welded joint configuration. ASTM Special Technical Publication No. 287, 136-149. Philadelphia, The Society, 1961.

Wever, F., and W. Jellinghaus, Das dreistoffsystem eisen-chromnickel (The iron-chromium-nickel system). Mitt. Kaiser-Wilhelm-Institut fur Eisenforschung 13, 1931, 93-108.

Williams, R. O., Further studies of the iron-chromium system. Trans. AIME 212, August 1958, 497-502.

Wilson, L. H., and T. S. Fitch, A discussion of the controlled pouring process for producing slabs and certain cast products. AISI Regional Technical Meeting, September 30, 1964.

Wolosin, Edward, The melting of 18-8 stainless steel. AIME Proc. of Elec. Furnace Steelmaking Conf. 19, 1961, 144-149.

Woodburn, J., G. R. Lohman and R. J. Nylen, Pressure pouring steel slabs. AIME Journal of Metals 16, Dec. 1964, 967-971.

CHAPTER 48

Steels for Elevated-Temperature Service

SECTION 1

CLASSES OF STEEL

The designation "elevated-temperature service" covers various applications such as steam boilers, vessels for hydrogenating coal or oil, heat-treating furnaces, and fittings for combustion engines. For these applications, there are a variety of steels available to meet the needs of industry. This chapter is intended as an introduction to selection and use of steels for elevated-temperature service and contains an outline of the considerations that are involved in steel selection. This outline is followed by a general discussion of the properties of steels that are relevant to elevated-temperature service. Oxidation and corrosion resistance are treated as separate items. Finally, an overview is presented of the general characteristics of steels used for elevated-temperature service.

Steel Selection—The selection of a steel for a particular application requires a consideration of the service conditions, reliability requirements, cost, and material availability. For most applications. weldability is also a major consideration.

As the first step in steel selection, the environment, stresses, temperatures, deformation, and frequency and rate of change of stress that are likely to be encountered must be identified. The stresses imposed by direct loading, such as internal pressure, are normally easy to identify and calculate. However, secondary stresses such as those due to thermal gradients or to nonuniform heating and cooling can be relatively large and difficult to assess. These secondary stresses may be of sufficient magnitude to affect the service performance of the steel. If the loading and service temperatures are cyclic, thermal and mechanical fatigue stresses must also be considered. As a general rule, the design of a component should minimize constraint that prevents the component from accommodating the thermal expansion associated with elevated-temperature service. The accommodation of thermal strain is especially important under cyclic thermal conditions, including start-up, operation, and shutdown.

The intended environment and service temperature determine both the candidate steels and the strength characteristics that are important. With regard to the strength characteristics, design stresses are usually limited by yield and tensile strength up to moderate temperatures (about 425°C to 535°C, corresponding to 800°F to 1000°F, depending on the steel); at higher temperatures, the design stresses are limited by the creep and creep-rupture strength. The specific design stress criteria determine the transition temperature at which the design stress is limited by tensile properties to that limited by the creep and creep-rupture properties. Using the design stress criteria of the ASME Boiler and Pressure Vessel Code, Unfired Pressure Vessels, Section VIII, Division I, the transition starts at about 425°C (800°F) for carbon steel, 480°C (900°F) for the Cr-Mo steels, and 535°C (1000°F) for the Cr-Ni austenitic stainless steels.

In addition to the stress and temperature conditions, the environment that the steel will encounter during its service history is a primary consideration. As was the case for the stresses, it is important to consider both the environment during service and the environment present during shutdown. For example, acid condensate formed during shutdown may be the limiting material consideration. Because of its complexity, a separate section in this chapter is devoted to the oxidation and corrosion-resistance considerations.

Once the stress, temperature, and environment are identified, the reliability requirements should be assessed. The reliability requirements influence the steel selection and the design stress criteria selected for the intended application. For example, the selection of a steel for a critical nuclear component requires that the selected steel have the highest degree of reliability. In such a case, the cost of the steel is an insignificant factor. On the other hand, the selection of a steel for a heat-treating furnace grate may be such that the selected steel can have much less reliability if the consequence of failure involves only a production delay. The selection of a material for the furnace grate would involve cost considerations.

For a plate-steel application where a high degree of reliability is required, such steels are purchased to "pressure-vessel" quality as defined by the ASTM A 20 specification. The purchaser may impose property requirements beyond those stipulated where additional reliability considerations so dictate. For plate-steel applications not requiring pressure-vessel quality, "structural" quality steel produced in accordance with ASTM A 6 may be considered. The structural-quality steels differ from pressure-vessel steels in that they are generally not heat treated, and they are generally produced to relatively wide composition ranges. In addition, the testing and inspection requirements are less stringent than for pressure-vessel-quality steels.

With regard to availability, carbon steel (0.30 per cent maximum), 2¼Cr-1Mo, and UNS S30400 (AISI Type 304) stainless steel are the most readily available steels for elevated-temperature service. It should also be noted that with the exception of carbon steel and some low-alloy steels, most high-temperature steels are not available in structural shapes.

The weldability aspects of material selection are not covered in this book. However, in elevated-temperature service, like most other types of service, good weldability is essential to satisfactory performance.

Ideally, the weld-metal and heat-affected zone properties should match those of the base metal. However, the very nature of most welding processes results in inhomogeneity in composition and microstructure in the weld metal and results in altered microstructures in the heat-affected zone. Thus, it is rare that matching properties are realized. Careful consideration of weld-metal elevated-temperature properties and weld-joint location and geometry are essential. References dealing with the properties of weldments are listed in the bibliography.

Properties to be Considered—Shown in Table 48—I are the major properties to be considered in the selection of a steel for elevated- temperature service. The property of primary importance varies with application. For example, for turbine blading, the steel must be resistant to creep; for a quench basket, the steel must be resistant to thermal shock; and for a furnace baffle, the steel must be resistant to oxidation.

The first property, oxidation and corrosion resistance, listed in Table 48—I, is treated in a subsequent section in this chapter. As is evident from the earlier discussion in this chapter, the strength property of interest depends on the conditions of stress and temperature imposed during service. For static loading, the yield and tensile strengths are of primary interest at moderate temperatures. At high temperatures, creep and creep-rupture strengths are used as a basis of de-

sign. Extensive data compilations summarizing tensile, creep, and creep-rupture properties of steels have been prepared by G. V. Smith; these compilations are listed in the bibliography. ASME code case N 47 contains detailed fatigue and creep data for 2¼Cr-1Mo, alloy 800H and AISI types 304 and 316 stainless steel. For service involving cyclic thermal and mechanical stresses, the elevated-temperature fatigue properties should be considered. In some applications, both fatigue and creep must be considered. For example, a power plant that is only used to supply power during periods of peak demand would involve both creep and fatigue. Data reported by Curran address the problem of fatigue creep interaction. Sources of fatigue data are listed in the bibliography.

The next property, ductility, is often measured by the elongation and reduction of area observed in tension and creep-rupture tests. Data on the ductility of many steels used for elevated-temperature service are presented in the Smith compilations already identified. In general, steels that exhibit high ductility are more tolerant of "upset" service conditions and nonuniform heating and cooling during service than steels exhibiting low ductility. However, steels that exhibit high strength may exhibit low ductility. Thus, a balance between strength and ductility may be required in steel selection. It is often observed that steels which exhibit low ductility (under 10 per cent) in the creep-rupture tests will exhibit reduced notched-bar creep-rupture strength. Data on the notched-bar creep-rupture strength are very limited; hence, ductility values are often the only basis for assessing the notched-bar creep-rupture properties of a steel. In general, as the strength of a steel increases, its possible susceptibility to reduced notched-bar creep-rupture strength increases. Data illustrating this dependence of notched-bar creep-rupture strength are documented in the paper by Sticha listed in the bibliography.

As was indicated earlier, the notch toughness is a consideration during start-up and shutdown conditions. The general question of notch toughness is treated in other portions of this book. During service, the notch toughness of the steel may change due to microstructural changes, including segregation of alloying elements to grain boundaries. This aspect of the properties is treated under thermal stability.

With regard to thermal stability, service may alter the steel properties. The extent to which the properties change is determined by the initial microstructure, alloy content, strain, and service conditions of temperature and time. Carbon steel may undergo a change in properties as a result of graphitization and strain aging. The detrimental effect of graphitization is usually associated with strain. Failures associated with graphitization often occur in the vicinity of a weld where the strain and thermal cycle associated with welding result in the formation, during elevated-temperature service, of a nearly continuous graphite network that results in low strength. Strain aging results in an increase in tensile strength and a reduction in the notch toughness of the steel.

For alloy steels containing a minimum of about 0.5 per cent chromium, graphitization is usually suppressed. In addition to strain aging, alloy steels, includ-

---

Table 48—I. Properties to be Considered in the Selection of a Steel for Elevated-Temperature Service

1. Oxidation and corrosion resistance
2. Strength at elevated temperature
   a) Yield and tensile strength of base metal and of weld metal
   b) Creep and creep-rupture strength
   c) Fatigue strength
3. Ductility
4. Notch toughness, especially at start-up and shutdown temperatures
5. Thermal stability
6. Thermal expansion and conductivity

ing the high-strength low-alloy steels, may undergo temper embrittlement. Temper embrittlement refers to the loss of toughness that occurs with exposure in the range of about 345 to 565°C (about 650 to 1050°F) without an appreciable change in strength or hardness. The degree of temper embrittlement is affected by microstructure (martensite being more susceptible than a ferrite/pearlite aggregate) and the level of such elements as As, Sb, Sn, and P.

The ferritic or martensitic stainless steels containing chromium above about 12 per cent can undergo a significant loss of notch toughness when exposed at about 480°C or 900°F (475°C or 885°F embrittlement). In addition to this embrittlement, the latter steels may also undergo a loss in toughness as the result of sigma formation (see the paper by Ludwigson and Link listed in the bibliography). The austenitic stainless steels may also undergo a loss of toughness as a result of either sigma formation or carbide precipitation or both.

The final properties, thermal expansion and thermal conductivity, are of interest because of their influence on thermal stresses. Data on the thermal expansion and thermal conductivity are presented in a subsequent section.

SECTION 2
OXIDATION AND CORROSION RESISTANCE

As mentioned above, the property of corrosion resistance is of primary importance, since the metal must not deteriorate excessively during service at elevated temperatures. One of the simplest forms of corrosion, and one frequently encountered, is oxidation of the metal which occurs by a process of diffusion of oxygen inward and of alloying elements outward. In plain carbon steel, the amount of oxidation in air is negligible below about 535°C (1000°F). Above this temperature, the rate of oxidation of carbon steel increases rapidly.

The most important alloying element for increasing the oxidation resistance of carbon steel above 535°C (1000°F) is chromium. This element appears to oxidize preferentially to iron. It forms a tightly adherent layer of chromium-rich oxide on the surface of the metal, retarding the inward diffusion of oxygen and inhibiting further oxidation. Other elements such as silicon and aluminum also increase the oxidation resistance, particularly when added to a steel containing chromium. These elements likewise have a greater affinity for oxygen than does iron, and are thus preferentially oxidized.

The rate of oxidation decreases as the oxide layer becomes thicker and additional protective layers are formed. The nature of the progress of oxidation with time at 593°C (1100°F) is shown in Figure 48—1 for carbon steel and 5 per cent chromium steel containing molybdenum. Initially, both steels oxidize rapidly, but as the protective chromium-rich oxide forms on the 5 per cent chromium steel its rate of oxidation decreases. Extensive data on the oxidation resistance of various steels in an air atmosphere have been cited in the literature. The results of a comprehensive series of such air oxidation tests are summarized in Figure 48—2.

In these tests, the amount of oxidation was measured by the gain in weight in 1000 hours. Several interesting points immediately become apparent. As the temperature is increased about 535°C (1000°F) the amount of oxidation in the plain carbon and molybdenum steels increases and the addition of 2 per cent to 3 per cent chromium moderately improves the oxidation resistance in air. The 5 per cent chromium steels are somewhat better but their oxidation resistance decreases above 650°C (1200°F). Additions of 9 per cent to 12 per cent chromium considerably improve the oxidation resistance, these materials showing little oxidation in

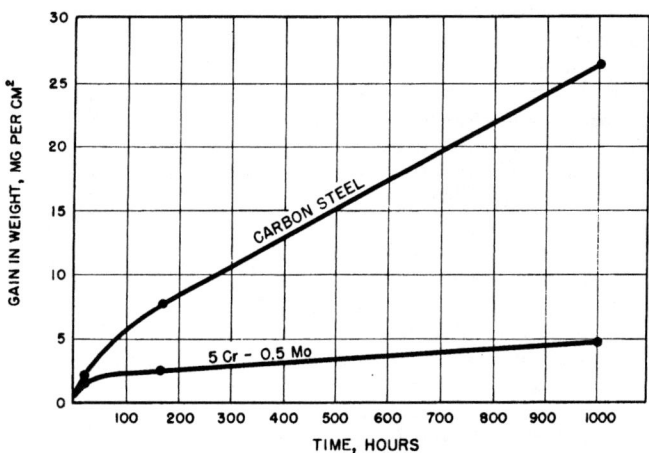

FIG. 48—1. Oxidation of plain carbon and 5 per cent chromium—0.5 per cent molybdenum steel at 593°C (1100°F).

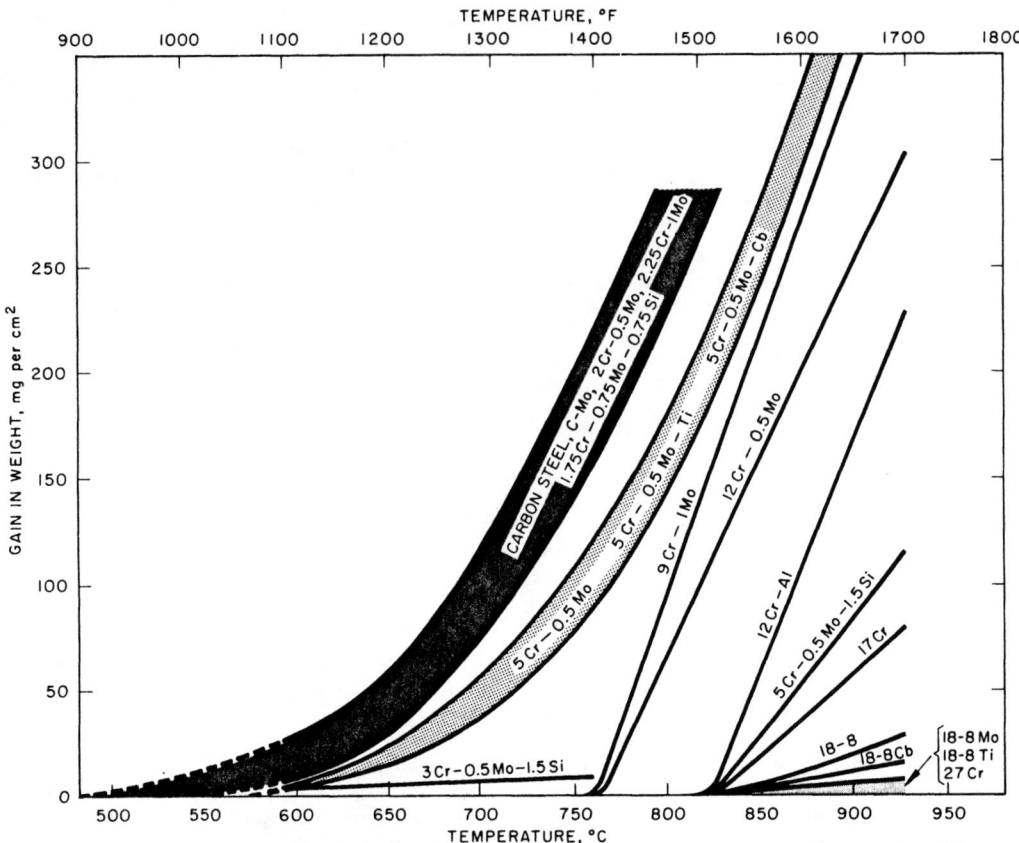

FIG. 48—2. Amount of oxidation of carbon, low-alloy, and stainless steels in 1000 hours in air at temperatures from 593° to 925°C (1100° to 1700°F).

1000 hours below 760°C (1400°F). The increase in oxidation resistance in going from 5 per cent to 9 per cent chromium is noticeable. It should also be noted that additions up to 1.5 per cent silicon greatly improve the oxidation resistance of the chromium steels. The silicon additions allow a decrease in the amount of chromium required for protection at a given temperature.

Effect of Various Atmospheres—The corrosion of steels at elevated temperatures in air is not necessarily indicative of their performance in other atmospheres. However, it is generally true that the corrosion resistance of steel increases with its chromium content. The precise behavior must, however, be established under the conditions in which the material will be used in service. Based on service experience in various atmospheres as well as on laboratory tests, a list has been prepared, Table 48—II, showing the suggested maximum service temperature for various steels in oxidizing gases. A plot of these data, Figure 48—3, shows that the chromium content required for freedom from oxidation is roughly proportional to the temperature.

Steam—In general, the relative corrosion resistance of various steels to steam is the same as it is in air. As shown in Figure 48—4, the chromium content of steel is a primary factor in determining its resistance to high-temperature steam. In this test carbon steel was resistant to the steam up to about 580°C (1075°F), whereas the UNS S34700 (AISI Type 347) stainless steel (18-8

Table 48—II. Maximum Temperature Without Excessive Oxidation.

Alloy Nominal Analysis	Designation		Maximum Temperature Without Excessive Scaling	
	UNS	AISI Type Number	°C	°F
1 Cr, ½ Mo		—	565	1050
2¼ Cr, 1 Mo		—	580	1075
3 Cr, 1 Mo		—	595	1100
5 Cr-Mo	S50200	502	620	1150
9 Cr		—	650	1200
12 Cr-Al	S40500	405	815	1500
12 Cr	S41000	410	705	1300
12 Cr	S42000	420	650	1200
17 Cr	S44002	440	760	1400
17 Cr	S43000	430	845	1550
27 Cr	S44600	446	1095	2000
18-8	S30200	302	900	1650
18-8	S30300	303	900	1650
18-8	S30400	304	900	1650
25-12	S30900	309	1035	1900
25-20	S31000	310	1095	2000
18-8 Mo	S31600	316	900	1650
18-8 Ti	S32100	321	900	1650
18-8 Cb	S34700–S34800	347-348	900	1650

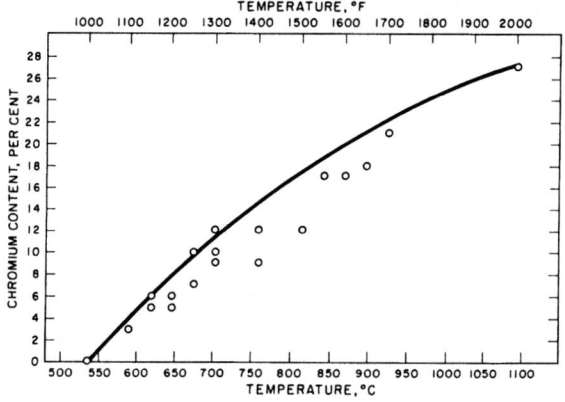

FIG. 48—3. Maximum amount of chromium necessary for freedom from oxidation at temperatures from 535° to 1095°C (1000° to 2000°F).

Cb) was resistant up to about 910°C (1670°F). Results of long-time exposure tests in steam at 595°C (1100°F) are shown in Figure 48—5, and it is apparent that the stainless steels generally show excellent resistance to high-temperature steam. The profound effect of chromium content of the steel is again evident in Figure 48—5.

The addition of silicon to steel, although beneficial in improving oxidation resistance in air, has essentially no beneficial effect in steam.

Flue Gases—Depending on the fuel source, flue gases may contain varying amounts of sulphur compounds along with the carbonaceous combustion products. The corrosion of steel in clean flue gas at temperatures above the dewpoint of the gas is usually not materially different from that in air if the gas sulphur content is low and if the gas is oxidizing. If the sulphur

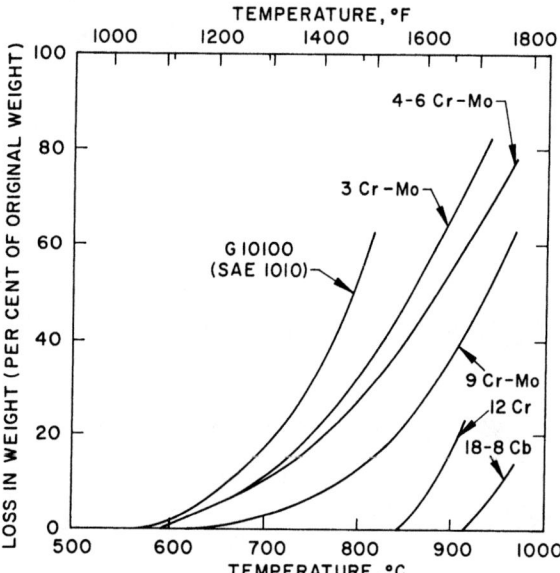

FIG. 48—4. Corrosion of steel bars in contact with steam for 500 hours at various temperatures. (After Hawkins, et al.; see bibliography.)

content of an oxidizing flue gas is high, steels containing more than about 12 per cent chromium are preferred, because of the good sulphidation resistance of chromium. Chromium-nickel steels also may be used under these same flue gas conditions with no loss in the sulphidation resistance characteristic of the chromium.

The presence of sulphur in flue gas under reducing conditions creates a much more aggressive environment than when the flue gas is oxidizing. In such flue-gas environments, chromium steels and chromium-

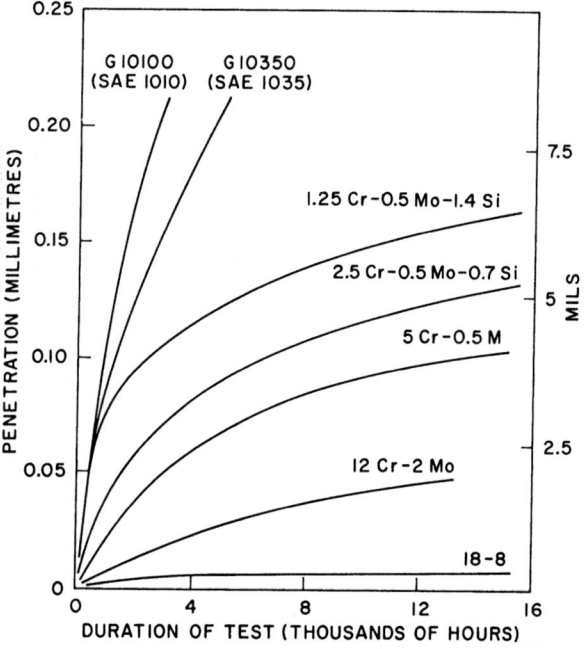

FIG. 48—5. Effect of long-time exposure to steam at 595°C (1100°F) on steel bars. (After Rohrig, et al.; see bibliography.)

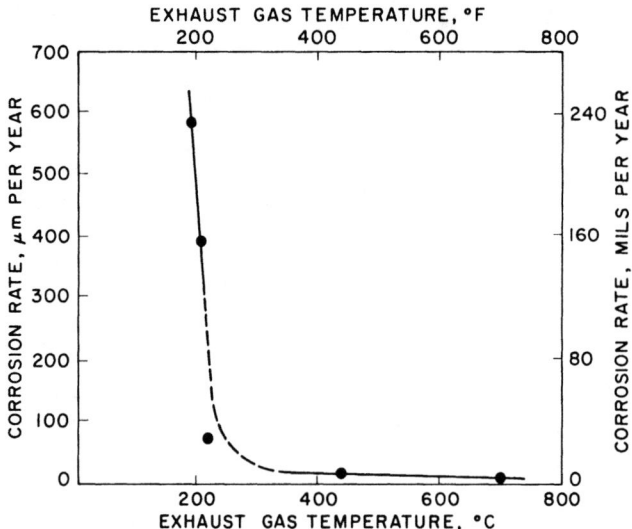

FIG. 48—6. Effect of temperature on corrosion rate of carbon steel in sulphur-bearing exhaust gas.

aluminum steels both containing more than 12 per cent chromium are usually the preferred materials, because the sulphidic scales on these steels are protective.

A complication that can arise in flue-gas systems where coal or residual fuel oils are used is the presence of corrosive ash deposits. These fuel ashes sometimes contain low-melting constituents such as vanadium oxides and alkali-metal sulphates which can flux with the protective oxides on the stainless steels, forming molten slags and causing catastrophic rates of corrosion. This type of attack normally occurs at temperatures above about 650°C (1200°F).

Another problem in sulphur-bearing flue gases is condensation corrosion. If the metal temperature in the system drops below the acid dewpoint of the gas (which in some cases can be greater than 150°C or 300°F), hot acidic condensates form which are quite corrosive to all steels. The effect of temperature on the corrosion rate of carbon steel in a sulphur-bearing exhaust gas is shown in Figure 48—6; the stainless steels also show high rates of attack under these conditions. The best solution to this problem is to insulate the system to prevent condensation.

High-Temperature Hydrogen—The stainless steels, because of their high chromium content, are resistant to high-temperature hydrogen attack. This type of at-

Table 48—III. Corrosion-Resistance Ratings for Stainless Steels in Various Gases at Elevated Temperatures.

Gas	Temperature (°C)	Temperature (°F)	UNS S41000 (AISI Type 410)	UNS S43000 (AISI Type 430)	UNS S30400 (AISI Type 304)	UNS S31600 (AISI Type 316)
Ammonia	315	600	A	A	A	A
Carbon Dioxide	370	700	A	A	A	A
	815	1500	—	A	A	A
Carbon Monoxide	370	700	A	A	A	A
Chlorine	100	212	—	A	B	B
	205	400	—	—	B	B
	260	500	—	—	D	—
Fluorine	205	400	D	A	A	A
	260	500	—	B	D	—
	315	600	—	D	—	—
Hydrogen Sulphide	315	600	B	B	B	B
Sulphur Dioxide	300	575	—	A	A	A
	370	700	B	B	B	B
	650	1200	B	B	B	B

*A—Corrosion rates less than 125 micrometres (5 mils) per year.
 B—Corrosion rates 125 to 500 micrometres (5 to 20 mils) per year.
 C—Corrosion rates 500 to 1250 micrometres (20 to 50 mils) per year.
 D—Corrosion rates greater than 1250 micrometres (50 mils) per year.

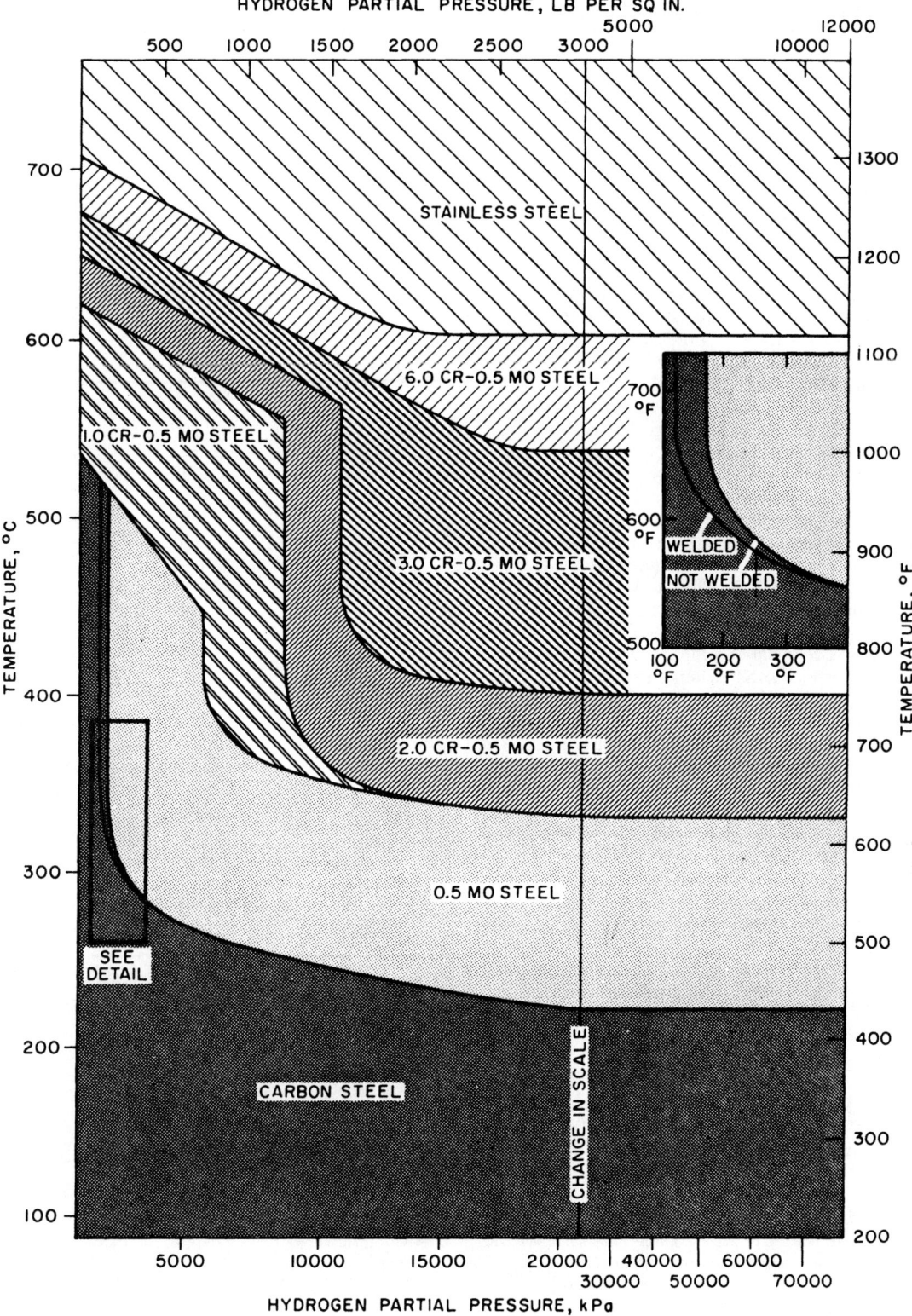

FIG. **48**—7. Safe operating zones for steel in hydrogen service. The stainless steels referred to in the upper part of this diagram include the steels for high-temperature service that are discussed in this chapter. Temperatures in insert are in °F. (After Nelson; see bibliography.)

tack is caused by diffusion of hydrogen into the steel where it reacts with carbides to form methane. Chromium, being a carbide-stabilizing element in steel, provides resistance to high-temperature hydrogen. The amount of chromium required is dependent on the temperature and the partial pressure of hydrogen in the gas. Safe operating limits for various steels in hydro-

gen have been published by Nelson (see bibliography); these are shown in Figure 48—7.

Other Gases—Corrosion-resistance ratings for stainless steels in several other gases at elevated temperatures are shown in Table 48—III. As seen in the table, the stainless steels show satisfactory resistance to most of these gases at the temperatures indicated.

SECTION 3

CHARACTERISTICS OF STEELS FOR ELEVATED-TEMPERATURE SERVICE

Now that some of the principal factors in selecting steels for elevated-temperature service and the properties of steel that must be considered for such service have been reviewed, the characteristics of the steels that are presently available for elevated-temperature service will be briefly discussed. These steels generally can be classified into three categories: carbon steels, alloy steels, and stainless and heat-resistant steels.

Carbon steel, the most widely used steel, is suitable where corrosion or oxidation is relatively mild. It is used for such applications as tubes in condensers, heat exchangers, boilers, superheaters, and stills. Depending on operating conditions, the maximum service temperature may be 400° to 535°C (750° to 1000°F).

The elevated-temperature properties of carbon steel are highly variable and are influenced by a number of factors such as deoxidation practice, chemical composition, and processing. Small amounts of residual elements can also affect the properties of carbon steel. For example, the effect of molybdenum, as determined by Glen and Barr, on the 100 000-hour rupture strength of carbon steel is shown in Figure 48—8.

The alloy steels that are used for elevated-temperature service generally contain one or more of the following elements: molybdenum, chromium, and silicon. The molybdenum is added principally to give higher strength; chromium is added to suppress graphitization and to give improved oxidation resistance, and silicon is added to give a further improvement in oxidation resistance.

The Cr-Mo steels, such as 2¼Cr-1Mo steel, are used extensively for a variety of high-temperature applications. The 2¼Cr-1Mo steel is available in a variety of product forms including heavy (up to about 405 millimetres or 16 inches thick) plate that can be produced to different strength levels by heat treatment. Table 48—IV shows the effect of heat treatment on the yield strength of 2¼Cr-1Mo steel at 27°, 204° and 427°C (80°, 400° and 800°F). As expected, the effect of heat treatment is greatest at 27°C (80°F) and least at 427°C (800°F). It is also observed that the creep-rupture properties of 2¼Cr-1Mo steel vary with the room-temperature tensile properties. Table 48—V shows the variation in the 100 000-hour rupture strength of 2¼Cr-1Mo steel produced to various tensile-strength values. The dependence of creep-rupture strength on tensile strength is large at 427° and 482°C (800° and 900°F), but small at 538°C (1000°F). It should be recognized that, as the strength level increases in 2¼Cr-1Mo steel (and other steels), the steel is more likely to show

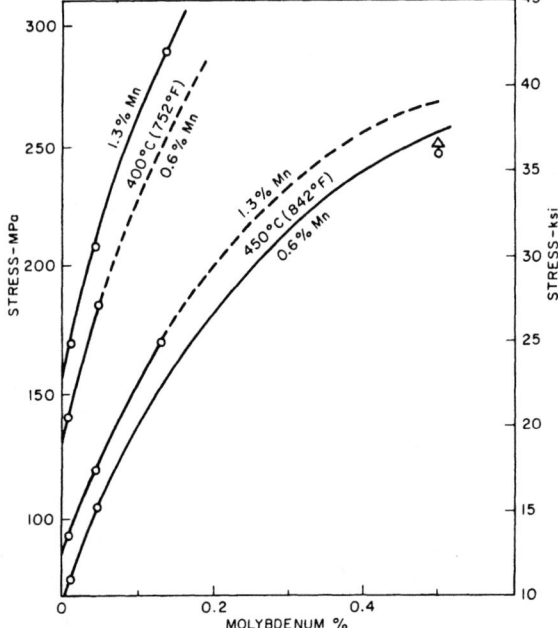

FIG. 48—8. Effect of molybdenum on rupture strength (100 000 hours) of carbon-manganese steels at 400°C (752°F) and 450°C (842°F). After Glen and Barr; see bibliography.

Table 48—IV. The Effect of Heat Treatment on the Yield Strength of 2¼Cr-1Mo Steel

	Test Temperature					
	27°C (80°F)		204°C (400°F)		427°C (800°F)	
Treatment	MPa	ksi	MPa	ksi	MPa	ksi
Annealed	207	30.0	185	26.9	184	26.7
Normalized and Tempered	310	45.0	276	40.0	241	34.9
Quenched and Tempered	586	85.0	538	78.0	445	64.6

Values from tables of yield strength, ASME Boiler and Pressure Vessel Code, 1980.

a susceptibility to temper embrittlement and/or reduced notched-bar strength.

The stainless and heat-resistant steels can be classified into three general groups: the ferritic steels, the

Table 48—V. Correlation Between Room-Temperature Tensile Strength and 100 000-Hour Rupture Strength for 2¼Cr-1Mo Steel

27°C (80°F) Tensile Strength		100 000 Rupture Strength, at					
		427°C (800°F)		482°C (900°F)		538°C (1000°F)	
MPa	ksi	MPa	ksi	MPa	ksi	MPa	ksi
655	95	365	53.0	262	38.0	186	27.0
724	105	445	64.5	302	43.8	193	28.0
793	115	524	76.0	334	48.5	200	29.0

Data from Smith, ASTM Data Series DS 6S2.

austenitic steels, and the precipitation-hardenable steels. The ferritic steels, which include the martensitic stainless steels, may be considered those that contain between about 5 and 27 per cent chromium as their principal alloying element. The 12 per cent chromium stainless steel, UNS S41000 (AISI Type 410), is a popular steel for elevated-temperature service up to 705°C (1300°F), where good resistance to corrosion and oxidation is required. However, it is seldom used over 650°C (1200°F), where high-temperature strength is a requirement. It is widely used in the quenched-and-tempered condition for steam-turbine blading and is also popular for oil-refinery vessels, tubes, and catalytic-processing units. Extensive use has been made of a 12 per cent Cr-0.30 per cent Ti steel for the catalytic converter of

automobiles. This application requires good oxidation resistance at temperatures up to about 760°C (1400°F). The 17 per cent chromium stainless steel, UNS S43000 (AISI Type 430), is used in applications that require oxidation and corrosion resistance up to 815°C (1500°F). Where elevated-temperature strength is a requirement, the use of this composition is limited because of its relatively low creep strength. Its limited ductility before and after welding also restricts its use in some applications. It is ductile between about 400° and 595°C (750° and 1100°F), but will be brittle when it is cooled to ambient temperature after prolonged heating in this range. The brittleness may be eliminated by reheating to about 760°C (1400°F). The 27 per cent chromium stainless steel, UNS S44600 (AISI Type 446), which has relatively low elevated-temperature strength, is used between 870°C to 1095°C (1600° to 2000°F) in applications where the most severe oxidation is encountered. It is also subject to the same embrittling phenomena as UNS S43000 (AISI Type 430) steel. The major application of UNS S44600 (AISI Type 446) steel is in such items as furnace parts, soot blowers, and thermocouple protection tubes, where stresses are negligible and support is good.

The austenitic stainless steels are essentially alloys of iron, chromium, and nickel. These steels as a class are the strongest steels for service above about 535°C (1000°F). This is shown in Figure 48—9, which is a comparison between the stress to produce a creep rate of

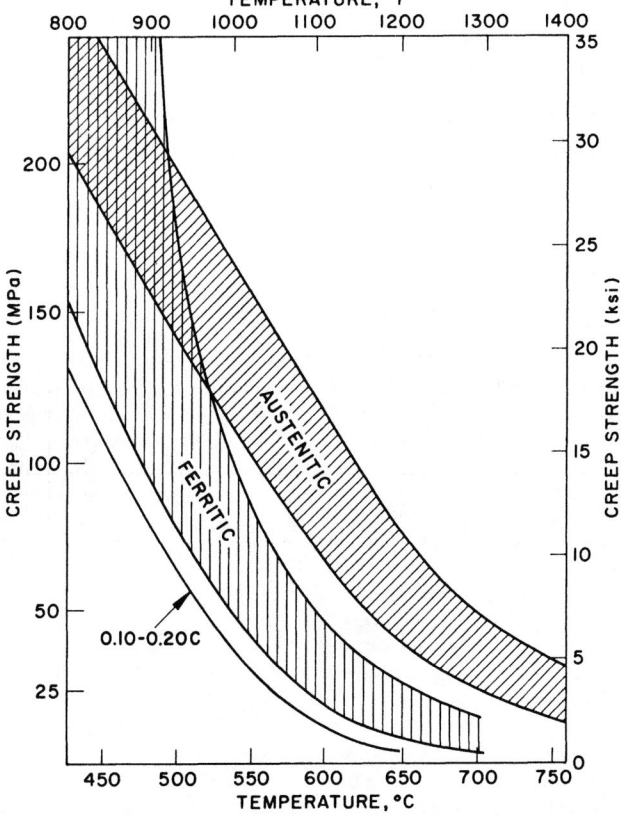

FIG. 48—9. Creep strength of various steels between 425° and 760°C (800° and 1400°F).

Table 48—VI. Principal Types and Nominal Compositions of Austenitic Stainless Steels

UNS Number	AISI Type Number	Nominal composition, Per Cent					
		Cr	Ni	Mo	Si	Cb	Ti
S30400	304	19	10				
S31600	316	17	12	2.5			
S32100	321	18	11				0.5
S34700	347	18	11			0.8	
S30900	309	23	13				
S31000	310	25	20		2		
S31400	314	25	20		2		

0.0001 per cent per hour between 425° and 760°C (800° and 1400°F) in carbon steel, ferritic stainless steels, and the austenitic stainless steels. The principal types and nominal compositions of the austenitic steels that are used for elevated-temperature service are shown in Table 48—VI. The first of these, UNS S30400 (AISI Type 304), is the most common grade of austenitic chromium-nickel steels, which as a group are used for handling many corrosive materials or resisting very severe oxidation. UNS S30400 (AISI Type 304) steel has excellent resistance to corrosion and oxidation, has high creep strength, and is frequently used at temperatures up to 815°C (1500°F). Where strength is not a major factor, it will resist progressive oxidation up to about 900°C (1650°F). S30400 (Type 304) steel is being used successfully and economically in high-temperature service in such applications as high-pressure steam-pipe and boiler tubes, radiant super-heaters, and oil-refinery tubes. It can be considered to be the "work horse" of the austenitic stainless steels for elevated-temperature service. UNS S32100 and S34700 (AISI Types 321 and 347) stainless steels are similar to UNS S30400 (AISI Type 304) except that titanium and columbium, respectively, have been added to these steels. The titanium and columbium additions combine with carbon and minimize intergranular corrosion that may occur in certain media after welding. However, the use of columbium (or titanium) does not ensure complete immunity to sensitization and subsequent intergranular attack when the steel is exposed for long times in the sensitization range of 425° to 815°C (800° to 1500°F). UNS S32100 and S34700 (AISI Types 321 and 347) stainless steels are widely used for such purposes as radiant super-heater tubes, internal-combustion exhaust pipes, and other similar high-temperature applications. UNS S31600 (AISI Type 316) stainless steel which contains molybdenum is used for high-strength service up to about 815°C (1500°F) and it will resist oxidation up to about 900°C (1650°F). However, above this temperature, in still air, the molybdenum will form an oxide that will volatilize and result in rapid oxidation of the steel. For service above 870°C (1600°F), UNS S30900, S31000 and S31400 (AISI Types 309, 310 and 314) stainless steels, which contain about 23 to 25 per cent chromium, are used. These steels have the best high-temperature strength of the austenitic stainless steels at these temperatures, and because of their chro-

mium contents, they can be used in applications where extreme corrosion or oxidation is encountered.

The precipitation-hardenable stainless steels have been developed during the past two decades. These steels offer moderate to good corrosion resistance, good fabricability, and relatively high strength at room and elevated temperatures. The precipitation-hardenable stainless steels can be classified as either martensitic, semiaustenitic or austenitic. At normal austenitizing temperatures, these steels are predominantly austenitic, but the austenite, depending on composition and heat treatment, may transform to martensite on cooling or during mechanical deformation. Classification, therefore, is on the basis of austenite stability.

The martensitic types undergo transformation of austenite to martensite on cooling to room temperature. This transformation results in the partial hardening of the matrix. The martensite that forms is usually thought to be highly supersaturated with certain solute elements that precipitate in the form of second phases during tempering or aging between about 425° to 535°C (800° to 1000°F) and produce additional hardening. The martensitic precipitation-hardenable stainless steels, which exhibit room-temperature yield strengths between about 1240 and 1515 MPa (180 000 and 220 000 pounds per square inch)—depending on heat treatment—can be regarded as having useful strength up to about 315°C (600°F).

The composition of the semiaustenitic types is such that after being annealed at temperatures near 1065°C (1950°F), they remain austenitic on cooling, being readily fabricable in this form. Subsequent heat treatment at 760°C (1400°F) or at 925°C (1700°F) depletes the austenite of chromium and carbon to the extent that martensite forms on cooling to room temperature or -73°C (-100°F), respectively. Mechanical deformation may also produce transformation. Final hardening is effected during the tempering or aging treatment between about 425° and 535°C (800° and 1000°F) as for the martensitic types. These steels in the hardened form have about the same strength as the martensitic precipitation-hardenable steels and, similarly, are useful up to about 315°C (600°F).

The austenitic precipitation-hardenable stainless steels differ from the types just considered in that they remain austenitic on cooling to room temperature. A moderate condition of supersaturation is developed in these steels during cooling as a result of the decreasing solubility of solute elements with decreasing temperature. Precipitates form near 705°C (1300°F) and these precipitates strengthen the austenitic matrix.

SPECIAL PROPERTIES

In addition to strength, stability and corrosion resistance, other properties occasionally have to be considered in designing for elevated-temperature service; some of these properties are thermal conductivity, thermal expansion, and modulus of elasticity.

Thermal Conductivity—Of importance at elevated temperature is the thermal conductivity of the material. Average data for various steels are shown in Figure 48—10. It will be noted that the addition of alloying elements decreases the thermal conductivity of carbon

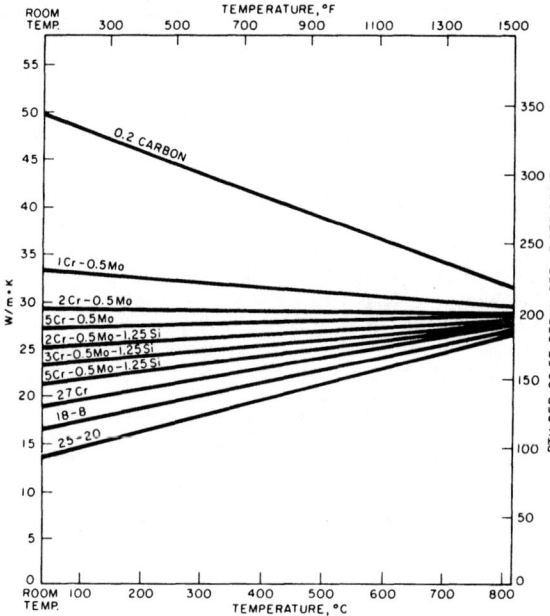

FIG. 48—10. Thermal conductivity of various steels at temperatures between room temperature and 815°C (1500°F).

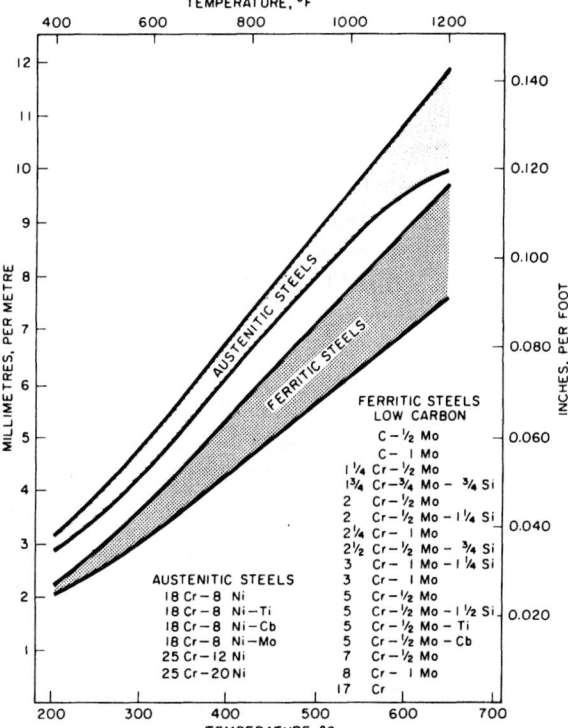

FIG. 48—11. Influence of temperature on linear thermal expansion. The diagram shows the change of length, in metres per metre and inches per foot, occurring in carbon, low-alloy, and stainless steels as they are heated from room temperature to any temperature between 205° and 650°C (400° and 1200°F). For the ferritic steels, the thermal expansion decreases with increasing chromium content, following the order of the steels as listed on the diagram. In the austenitic steels the thermal expansion is larger than in the carbon steel.

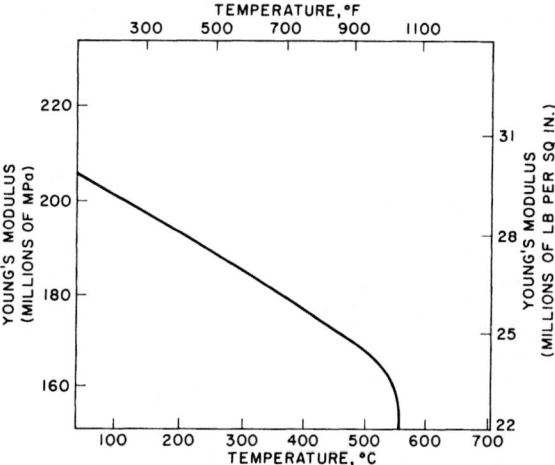

FIG. 48—12. Variation of tensile modulus of elasticity of ferritic alloy and stainless steels with temperature. (After Garofalo, et al.; see bibliography.)

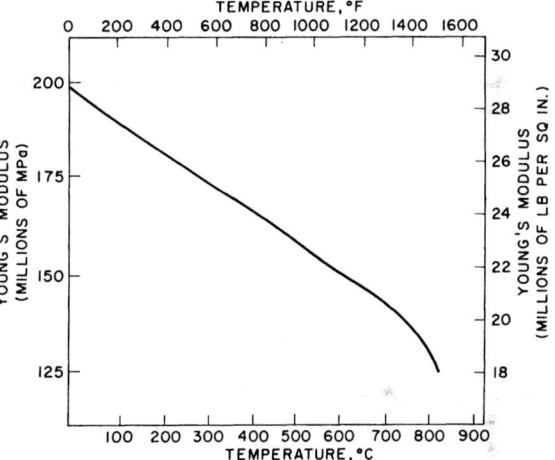

FIG. 48—13. Variation of tensile modulus of elasticity of austenitic stainless steels with temperature. (After Garofalo, et al.; see bibliography.)

steel, and that the difference among the thermal conductivities of the various steels decreases with increasing temperature. The thermal conductivities of the austenitic 18 per cent chromium, 8 per cent nickel steels and of the 25 per cent chromium, 20 per cent nickel steels, are the two lowest on the chart.

Thermal Expansion—In designing apparatus for use at elevated temperature, allowance must be made for the thermal coefficient of expansion of the component materials.

The linear thermal expansion (increase in length) in going from room temperature to any elevated temperature up to 650°C (1200°F) is shown in Figure 48—11.

The steels are listed in the order in which they occur on these bands. It is seen that the austenitic stainless steels have a higher coefficient of expansion than the ferritic steels.

The tensile modulus of elasticity of the alloy ferritic steels is close to 207 000 MPa (30 000 000 psi) at room temperature. It decreases linearly to about 172 000 MPa (25 000 000 psi) at 480°C (900°F) and then begins to drop at an increasing rate at higher temperatures, as shown in Figure 48—12.

The modulus of elasticity for austenitic stainless steels is about 193 000 MPa (28 000 000 psi) at room temperature. Again, there is a linear decrease, to roughly 138 000 MPa (20 000 000 psi) at 705°C (1300°F), before a rapid drop begins, as shown in Figure 48—13.

Elastic moduli are, for practical purposes, independent of composition for a given class of steel; e.g., the moduli of UNS S30400, S31600, S34700, S31000, etc. (AISI Type 304, 316, 347, 310, etc.), all classed as austenitic, are similar. The same is true for the ferritic or martensitic types of steel.

The value of the elastic modulus is affected by the method of testing. Static methods that measure deflection under load increments generally give results lower by about 13 800 MPa (2 000 000 psi) than do so-called dynamic methods. The dynamic methods are based on induced mechanical vibration or ultrasonic pulses. They probably approach the true elastic properties of the steel more closely than do static tests, but the static values are of greater interest in structural design.

Caution should be used in employing elastic moduli for design at elevated temperatures, where creep (continuous non-reversible plastic deformation with time under load) may occur.

Bibliography

American Society for Testing Materials, "Tables of data on chemical compositions, physical and mechanical properties of wrought corrosion resisting and heat resisting chromium-nickel steels." Philadelphia, The Society (Dec. 1942).

American Society for Testing and Materials, Report on the elevated-temperature properties of chromium steel (12-27 per cent). Special Technical Publication No. 228, July 1958.

American Society for Testing and Materials, Supplemental report on the elevated-temperature properties of chromium-molybdenum steels. Data Series Publication DS 6-S1, June 1966.

American Society for Testing and Materials, Supplemental report on the elevated-temperature properties of chromium-molybdenum steels (an evaluation of 2¼ Cr-1 Mo steel). Data Series Publication DS 6-S2, March 1971.

American Society for Testing and Materials, Evaluation of the elevated-temperature tensile and creep-rupture properties of C-Mo, Mn-Mo, and Mn-Mo-Ni steels. Data Series Publication DS 47, Nov. 1971.

American Society for Testing and Materials, Elevated-temperature static properties of wrought carbon steel. Special Technical Publication No. 503, March 1972.

American Society for Testing and Materials, Evaluation of the elevated-temperature tensile and creep-rupture properties of ½ Cr-½ Mo, 1 Cr-½ Mo and 1¼ Cr-½ Mo Si steels. Data Series Publication DS 50, Sept. 1973.

American Society of Mechanical Engineers, Tables ACS-2 and AHA-1, ASME Boiler and Pressure Vessel Code, Pressure Vessels, Section VIII, Division 2, 1980.

American Society of Mechanical Engineers, Properties of steel weldments for elevated-temperature pressure containing applications, 1978.

American Society of Mechanical Engineers, Properties of weldments at elevated temperatures, 1968.

American Society of Mechanical Engineers, Code Case N 47-21, Code Cases Nuclear Components, 1983.

Curran, R. M., and Wundt, B. M., Interpretive report on notched and unnotched creep fatigue interspersion tests in Cr-Mo-V, 2¼Cr-1Mo, and Type 304 stainless steel. Ductility and toughness considerations in elevated temperature service, American Society of Mechanical Engineers, 1978.

Day, M. J., and Smith, G. V., Iron alloy scaling, Ind. Eng. Chem. 35, 1943, 1098-1103.

Garofalo, F., Malenock, P. R., and Smith, G. V., Influence of temperature on the elastic constants of some commercial steels. ASTM Special Technical Publication No. 129, Symposium on Elastic Constants, 1952, 10.

Glen, J., High-temperature steels in steam-plant practice. Murex Review I, No. 6, 1950.

Glen, J., and Barr, R. R., Effect of molybdenum on the high temperature rupture strength of carbon steel. British Iron and Steel Institute Special Report No. 97, High temperature properties of steels, Paper No. 37, 1967.

Hawkins, G. A., Agnew, J. T., Solberg, H. L., The corrosion of alloy steels by high-temperature steam, Trans. ASME 66, 1944, 291-295.

Heindlhoffer, K. and Larsen, B. M., Rates of scale formation on iron and a few alloys, Trans. ASST 21, 1933, 865-895.

Heindlhoffer, K. and Larsen, B. M., Resistance to scaling at elevated temperature, in "The Book of Stainless Steels" (2nd Ed) by E. E. Thum, 548-556. Cleveland, Ohio, American Society for Metals, 1935.

Ludwigson, D. C., and Link, H. S., Further studies on the formation of sigma in 12 to 16 percent chromium steels. American Society for Testing and Materials Special Technical Publication 369, Advances in technology of stainless steel, 1965.

Mears, R. B., The prevention of corrosion by use of appropriate design, Australasian Corrosion Engineering, 6, 1962.

Miller, R. F., Benz, W. G., and Day, M. J., Creep-strength stability of microstructure, and oxidation resistance of Cr-Mo and 18 Cr-8 Ni steels, Trans. ASM, 32, 1944, 381-407.

Morris, L. A., Corrosion resistance of stainless steels at elevated temperatures. ASM Engineering Quarterly, 8, May 1968, 30-47.

Nelson, G. A., Corrosion data survey. (1960 Ed)., Shell Development Company, 15.

Rohrig, I. A., Van Duzer, R. M., and Fellows, C.H., High-temperature-steam corrosion studies at Detroit, Trans. ASME, 66, May, 1944, 277-290.

Simmons, W. F., and Cross, H. C., Report on the elevated-temperature properties of carbon steel. American Society for Testing and Materials Special Technical Publication No. 180, Sept. 1955.

Simmons, W. F., and Van Echo, John A., Report on the elevated-temperature properties of stainless steels. American Society for Testing and Materials Data Series Publication DS 5-S2, Dec. 1965.

Smith, G. V., An evaluation of the yield, tensile, creep and rupture strengths of wrought 304, 316, 321 and 347 stainless steels at elevated temperatures. American Society for Testing and Materials Data Series Publication DS 5-S2, Feb. 1969.

Smith, G. V., An evaluation of the elevated-temperature tensile and creep-rupture properties of wrought carbon steel. American Society for Testing and Materials Data Series DS 11-S1, Jan. 1970.

Smith, G. V., Evaluation of the elevated-temperature tensile and creep-rupture properties of 3 to 9 per cent chromium-molybdenum steels. Data Series Publication DS 58, Oct. 1975.

Smith, G. V., An evaluation of the elevated-temperature properties of wrought carbon steel. American Society for Testing and Materials, Data Series DS 11S1, January 1970.

Smith, G. V., Elevated-temperature static properties of wrought carbon steel. American Society for Testing and Materials Special Technical Publication 503, March 1972.

Smith, G. V., Elevated-temperature tensile and creep-rupture
properties of 12 to 27 percent chromium steels. American
Society for Testing and Materials Special Technical Publication DS 59, August 1980.

Smith, G. V., An evaluation of the yield, tensile, and rupture
strengths of wrought 304, 316, 321, and 347 stainless steels
at elevated temperatures. American Society for Testing and
Materials Data Series 5S2, February 1969.

Smith, G. V., Evaluation of the elevated-temperature tensile
and creep-rupture properties of ½Cr-½Mo, 1Cr-½Mo, and
1¼Cr-½Mo-Si steels. American Society for Testing and Materials Data Series DS 50, 1973.

Smith, G. V., Evaluations of the elevated temperature tensile
and creep-rupture properties of C-Mo, Mn-Mo, and Mn-
Mo-Ni steels. American Society for Testing and Materials
Data Series DS 47, November 1971.

Smith, G. V., Supplemental report on the elevated-temperature properties of chromium-molybdenum steels (an
evaluation of 2¼Cr-1Mo steel.) American Society for Testing and Materials Data Series DS 6-S2, March, 1971.

Sticha, E. A., Notched stress-rupture data for quenched and
tempered 2¼Cr-1Mo steel. 2¼ Chrome 1 Molybdenum
Steel in Pressure Vessels and Piping, published by American
Society of Mechanical Engineers, 1971.

Van Duzer, Jr., R. M., and McCutchan, A., High-temperature-
steam experience at Detroit, Trans. ASME, 61, July 1939,
383-398.

CHAPTER 49

Steels for Low-Temperature and Cryogenic Service

SECTION 1

INTRODUCTION

The fundamental concepts of low-temperature and cryogenic technology are used by almost every major industry. Cryogenic applications in space, in the oxygen process for the manufacture of steel, in the chemical-process industries, in the medical field, and the applications of cryogenic processes to the petroleum, natural gas, glass, cement, food, and electronic industries give the cryogenic field a firm foundation for future growth. One of the major benefits of cryogenic technology results from the fact that many gases transform to a liquid at temperatures below ambient. Because of the enormous reduction in volume resulting from liquefaction of gas, with one cubic foot of liquefied gas being equivalent to many hundreds of cubic feet of gas volume at normal pressure and temperature, handling of liquefied gases requires less container space with accompanying saving in material, transportation, fabrication and erection costs.

In terms of volume, the steel and chemical-processing industries represent the largest single commercial consumers for the products from cryogenic processes. The low-temperature and cryogenic products of commercial importance include refrigerated propane and anhydrous ammonia, carbon dioxide, nitrous oxide, ethane, ethylene, methane, oxygen, nitrogen, argon, chlorine, hydrogen and helium. The chemical formulas and boiling points of these and the other products normally associated with low-temperature and cryogenic applications are shown in Table 49—I, along with the designations of classes of steels used in equipment for service at the boiling-point temperatures of the products.

The terms "low-temperature" and "cryogenic" may be defined for present purposes as involving temperatures to $-100°C$ ($-150°F$) and $-273°C$ ($-459°F$), respectively, as discussed in the following section.

SECTION II

MATERIALS AND PROPERTIES

The selection of metals for low-temperature applications must be based on many mechanical properties, including the familiar yield and tensile strength, fatigue limit, ductility and toughness.

The "low-temperature" steels are those especially suited for extremely cold climates and for the handling of relatively "warm" (to $-100°C$ or $-150°F$) liquefied gases such as propane, anhydrous ammonia, carbon dioxide and ethane.

"Cryogenic" steels, such as 9 per cent nickel steel, and the austenitic stainless steels, are capable of retaining toughness in applications involving the storing and handling of liquefied methane, oxygen, nitrogen, argon, hydrogen and helium to $-273°C$ ($-459°F$).

Tables 49—II and 49—III give the UNS, ASTM or

AISI designation and the nominal compositions relating to carbon, alloy and stainless steels for these applications. The mechanical properties having major significance for applications in low-temperature and cryogenic service are also shown in Tables 49—II and 49—III.

Although these mechanical properties are important in considering a candidate steel for low-temperature or cryogenic applications, the final selection may be made on the basis of other properties. Low heat conductivity, low coefficient of thermal expansion, low emissivity and cleanliness are properties that can be used to advantage in storage vessels, vacuum-transfer lines and other components of low-temperature or cryogenic systems. The physical properties of major significance for these

1377

applications are shown in Tables **49**—IV and **49**—V. Of all the metals useful in construction for low-temperature applications, steels remain the most popular because they are the most efficient, most readily available, most versatile and most economical.

A. NOTCH TOUGHNESS

In general, most engineering structures are subject to stress concentrations under service conditions due to mechanical notches resulting from inadequate design and/or fabricating practices, or from the micro-

Table 49—I. Boiling Points of Gases and a List of Steels for Service at Boiling Points.

Commodity	Chemical Formula	Approximate Boiling Point at 101.325 kPa (1 atm)				Steels Normally Considered for Service at Boiling Point (Note 1 below)
		°C	K	°F	°R	
Butane	C_4H_{10}	−0.6	272.5	30.9	490.6	ASTM A 333, Grades 1, 6 and 7
Sulphur Dioxide	SO_2	−10.0	263.1	14.0	473.7	(UNS K03008, K03006 and K21903)
Isobutane	$(CH_3)_2C_2H_4$	−10.2	262.9	13.6	473.3	ASTM A 334, Grades 1, 6 and 7
Methyl Chloride	CH_3Cl	−23.7	249.4	−10.7	449.0	(UNS K03008, K03006 and K21903)
Fluorocarbon Refrigerant$_{12}$	CCl_2F_2	−30.0	243.1	−22.0	437.7	ASTM A 516[b] (UNS K01800, K02100, K02403, K02700)
Ammonia	NH_3	−33.3	239.8	−27.9	431.8	ASTM A 537[c] (USS CHAR-PAC, UNS K02400)
Fluorocarbon Refrigerant$_{22}$	$CHClF_2$	−10.6	232.5	−41.0	418.7	ASTM A 662, Grade A ASTM A 734, Type A
Ketene	C_2H_2O	−41.0	232.1	−41.8	417.9	ASTM A 736, Classes 2 & 3; ASTM A 782
Propane	C_3H_8	−42.3	230.8	−44.1	415.6	ASTM A 517 (USS "T-1", UNS K11576)
Propylene	C_3H_6	−47.0	226.1	−52.6	407.1	ASTM A 203 (2¼% Ni steel; K21703, K22103)
Hydrogen Sulphide	H_2S	−59.6	213.5	−75.3	384.4	ASTM A 203 (3½% Ni steel; UNS K31718, K32018)
Carbon Dioxide[a]	CO_2	−78.5	194.6	−109.3	350.4	ASTM A 645 (5% Ni steel; UNS K41583)
Acetylene	C_2H_2	−84.0	189.1	−119.2	340.5	ASTM A 333, Grade 3 (UNS K31918)
Ethane	C_2H_6	−83.3	184.8	−126.9	332.8	ASTM A 334, Grade 3 (UNS K31918)
Nitrous Oxide	N_2O	−89.5	183.6	−129.1	330.6	Stainless steels (UNS S30000 or AISI 300 series)
Ethylene	C_2H_4	−103.8	169.3	−154.8	304.9	ASTM A 645 (5% Ni steel, UNS K41583)
Xenon	Xe	−109.1	164.0	−164.4	295.3	ASTM A 553, Grades A and B[d]
Ozone	O_3	−111.9	161.3	−169.4	290.3	(9% Ni steel; UNS K81340 and 8% Ni steel, UNS K71340)
Krypton	Kr	−151.8	121.3	−241.2	218.5	ASTM A 353 (9% Ni steel, UNS K81340)
Methane	CH_4	−161.4	111.7	−258.5	201.2	ASTM A 333, Grade 8 (9% Ni steel,
Oxygen	O_2	−183.0	90.1	−297.4	162.3	UNS K81340)
Argon	Ar	−185.7	87.4	−302.3	157.4	ASTM A 334, Grade 8 (9% Ni steel,
Fluorine	F_2	−187.0	86.0	−304.6	155.1	UNS K81340)
Carbon Monoxide	CO	−192.0	81.1	−313.6	146.1	
Nitrogen	N_2	−195.8	77.3	−320.4	139.3	Stainless steels (UNS S30000 or AISI 300 series)
Neon	Ne	−245.9	27.2	−410.6	49.1	ASTM A 213 (austenitic stainless steel UNS S30400, AISI Type 304)
Tritium	T_2	−248.0	25.1	−414.4	45.3	ASTM A 240 (austenitic stainless steel UNS S30400, AISI Type 304)
Deuterium	D_2	−249.5	23.6	−417.1	42.6	ASTM A 269 (austenitic stainless steel UNS S30400, AISI Type 304)
Hydrogen	H_2	−252.7	20.4	−422.9	36.8	ASTM A 312 (austenitic stainless steels
Helium	He^4	−268.9	4.2	−452.1	7.6	listed in NOTE 2 below)
Helium	He^3	−269.9	3.2	−453.8	5.9	

[a]Sublimes.
[b]Normalized
[c]With modifications.
[d]Grade B to −170°C (−275°F).

NOTE 1: The steels listed in each grouping are intended as a general guide to steel selection. The steels listed in each grouping may not be suitable for all the gases listed under the corresponding commodity grouping. Steels intended for low temperature and cryogenic applications should be purchased to specific notch toughness requirements.

NOTE 2: ASTM A 312 relates to the following austenitic stainless steels: UNS S30400, S30403, S30409, S30451, S30900, S31000, S31600, S31603, S31609, S31651, S31700, S32100, S32109, S34700, S34709, S34800, S34809, S38100, S20910 and S24000, corresponding to AISI Types 304, 304L, 304H, 304N, 309, 310, 316, 316L, 316H, 316N, 317, 321, 321H, 347, 347H, 348, 348H, XM15, XM19 and XM29.

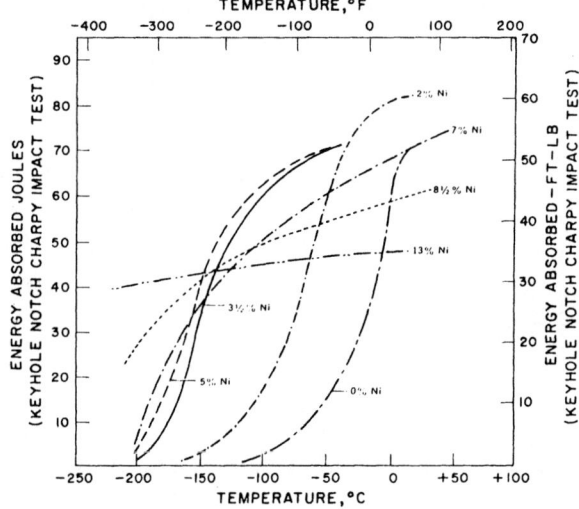

FIG. 49—1. Effect of nickel on notch toughness of steel as a function of temperature. (Courtesy, International Nickel Company, Inc.)

structural variations inherent in polycrystalline metals.

Notch toughness is a property of steel reflected in its resistance to brittle failure under conditions of high stress concentration, such as impact loading in the presence of a notch. This property is a prime requisite of metals intended for low-temperature application, and is vital to selection of materials for storage and

transportation of liquefied gases.

The notch toughness of most ferritic steels decreases with decreasing temperatures, and this factor is of critical importance in consideration of these materials for cryogenic applications. The notch toughness of ferritic steel can be improved by grain refinement, by additions of nickel, and/or by heat treatment.

In the alloy steels, nickel is the most common of the alloying elements used to lower the transition temperature. The general effect of nickel on notch toughness is shown as a function of temperature in Figure 49—1.

A steel extensively used for a low-temperature service, particularly in the handling of liquid propane at −42°C (− 44°F) and for other applications down to −60°C (−75°F) is low-carbon 2¼ per cent nickel steel (ASTM A 203, Grades A and B). Grade B has the widest range of applications because of its higher ASME maximum allowable design stress (121 MPa or 17 500 psi). Grade A has an allowable stress of 112 MPa (16 250 psi). These steels also are covered by SA-203 of the ASME Boiler and Pressure Vessel Code. Typical Charpy V-notch properties of 13-mm (½-inch) 2¼ per cent nickel steel plate are shown in Figure 49—2. Pipe and tubing are covered by ASTM A 333, Grade 7 and A 334, Grade 7.

Low-carbon 3½ per cent nickel steel has had wide use in land-based facilities for the storage of liquefied gases down to −100°C (−150°F). It frequently is specified for tankage and piping to handle liquid propane, carbon dioxide, acetylene, ethane and ethylene. It is governed by ASTM Specifications A 203, Grades D and

Table 49—II. Typical Specification Numbers, Nominal Composition and Minimum Tensile Properties for Cryogenic and Low-Temperature Carbon and Alloy Steels.

Designation		ASTM Spec. No. and Grade	Nominal Composition, Per Cent					Tensile Strength		Yield Strength		Elongation (Per Cent in 50 mm) or 2 in.)	Lowest Usual Service Temperature	
UNS	USS		C	Mn	Si	Ni	Cr	MPa	ksi	MPa	ksi		(°C)	(°F)
Carbon Steels														
K02400[a]	USS CHAR-PAC[a]	A 537 Cl1	0.18	1.15	0.35	–	–	483	70	345	50	–	−45	−50
K02400[b]	USS CHAR-PAC[b]	A 537 Cl2	0.18	1.15	0.35	–	–	552	80	414	60	23		
K01800		A 516	0.16	0.75	0.25	–	–	379	55	207	30	28	−45	−50
K02100		A 516	0.19	0.75	0.25	–	–	414	60	221	32	26		
K02403		A 516	0.22	1.10	0.25	–	–	448	65	241	35	24		
K02700		A 516	0.25	1.10	0.25	–	–	483	70	262	38	22		
Alloy Steels														
K11576	USS "T-1"	A 517-F	0.15	0.85	0.25	0.85	0.55[c]	862	125	793	115	18	−45	−50
K21703	USS 2¼% Ni	A 203-A	0.20	0.70	0.25	2.30	–	448	65	255	37	25	−59	−75
K22103		A 203-B	0.23	0.70	0.25	2.30	–	483	70	276	40	23		
K31718	USS 3½% Ni	A 203-D	0.17	0.70	0.25	3.50	–	448	65	255	37	24	−101	−150
K32018		A 203-E	0.20	0.70	0.25	3.50	–	483	70	276	40	22		
K71340	USS 8% Ni	A 553-B	0.10	0.50	0.25	8.00	–	690	100	586	85	22	−170	−275
K81340	USS 9% Ni	A 353	0.10	0.50	0.25	9.00	–	690	100	517	75	22	−195	−320
K81340		A 553-A	0.10	0.50	0.25	9.00	–	690	100	586	85	22	−195	−320

[a]Normalized.
[b]Quenched and tempered
[c]Also contains 0.50 Mo, 0.50 V, 0.25 Cu, and 0.003 B.

NOTE: These steels are included in other ASTM and ASME specifications for a variety of product forms such as seamless and welded tubing (ASTM A 333 and A 334).

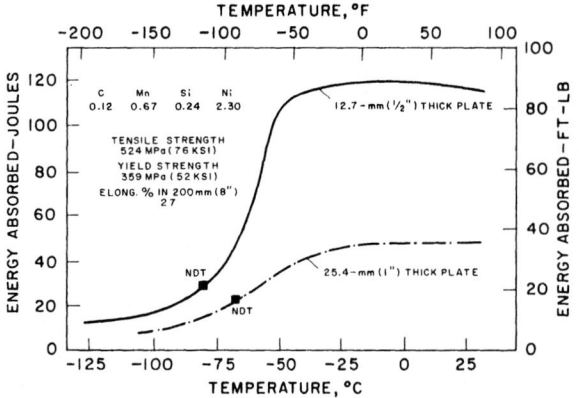

FIG. 49—2. Typical V-notch Charpy transition curves and NDT temperatures for normalized 2¼ per cent nickel steel plate. (Source: International Nickel Company, Inc.)

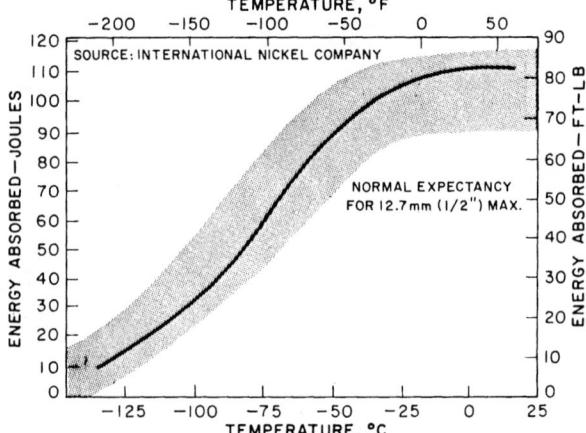

FIG 49—3. Charpy V-notch impact values for 3½ per cent nickel steel (normalized, tempered, 13-mm (½-inch) thick plate).

E; A 333, Grade 3; and A 334, Grade 3. Normal expectancy Charpy impact properties of normalized 13-mm (½-inch) thick 3½ per cent nickel steel plate are shown in Figure 49—3.

Low-carbon 5, 8 and 9 per cent nickel steels are ferritic steels, specifically developed to meet the requirements for cryogenic applications in the atmospheric-gas equipment field. ASTM Specification A 353 covers normalized and tempered 9 per cent nickel steel, ASTM Specification A 553 covers the quenched and tempered 9 per cent nickel steel (Class 1), and 8 per cent nickel steel (Class 2), and ASTM Specification A 645 covers the quenched, tempered, and reversion—annealed 5 per cent nickel steel. Nine per cent nickel steel pipe and tubing are covered, respectively, by ASTM Specifications A 333, Grade 8 and

Table 49—III. Nominal Compositions and Minimum Tensile Properties for Cryogenic Stainless Steels in Plate Form.

Designation			Nominal Composition, Per Cent					Tensile Strength		Yield Strength, 0.2 Per Cent Offset		Elongation (Per Cent in 50 mm or 2 in.)	Lowest Usual Service Temperature	
UNS	USS	AISI Type No.	C	Mn	Si	Ni	Cr	MPa	ksi	Mpa	ksi		(°C)	(°F)
S20100	USS 17-4-6	201	0.10	6.0	0.60	4.5	17.0 [1]	655	95	310	45	40	-195	-320
S20200	USS 18-5-8	202	0.10	9.0	0.60	5.0	18.0 [2]	620	90	310	45	40	-195	-320
S30200	USS 18-8	302	0.10	1.50	0.60	9.0	18.0	517	75	205	30	40	-270	-452
S30400	USS 18-8S	304	0.06	1.50	0.60	10.0	19.0	517	75	207	30	40	-270	-452
S30403	USS 18-8L	304L	0.02	1.50	0.60	10.0	19.0	483	70	172	25	40	-270	-452
S30500	USS 18-8FS	305	0.10	1.50	0.60	11.5	18.0	517	75	207	30	40	-270	-452
S30900	USS 25-12	309	0.06	1.50	0.60	13.5	23.0	517	75	207	30	35	-270	-452
S30908	USS 25-12S	309S	0.06	1.50	0.60	13.5	23.0	517	75	207	30	40	-270	-452
S31000	USS 25-20	310	0.10	1.50	0.60	20.0	25.0	517	75	207	30	35	-270	-452
S31600	USS 18-8Mo	316	0.06	1.50	0.60	12.0	17.0 [3]	517	75	207	30	40	-270	-452
S31603	USS 18-8MoL	316L	0.02	1.50	0.60	12.0	17.0 [4]	483	70	172	25	40	-270	-452
S32100	USS 18-8Ti	321	0.06	1.50	0.60	10.5	18.0 [5]	517	75	207	30	40	-270	-452
S34700	USS 18-8CbTa	347	0.06	1.50	0.60	11.0	18.0 [6]	517	75	207	30	40	-270	-452
S34800	USS 18-8Cb	348	0.06	1.50	0.60	11.0	18.0 [7]	517	75	207	30	40	-270	-452
S38100	USS 18-18-2	XM15	0.06	1.50	2.0	18.0	18.0	517	75	207	30	40	-270	-452

[1] Also contains 0.20 N.
[2] Also contains 0.20 N.
[3] Also contains 2.0 to 3.0 Mo.
[4] Also contains 2.0 to 3.0 Mo.
[5] Also contains Ti = 5 × C min.
[6] Also contains Cb + Ta = 10 × Cmin.
[7] Also contains Cb + Ta = 10 × C min., Ta = 0.10 max., Co = 0.20 max.

NOTE: The austenitic stainless steels described in this table are included in many ASTM and ASME specifications for a variety of product forms such as plate, sheet, bar, forgings, and seamless and welded tubing. Examples of the ASTM specifications include: A 412, A 240, A 213, A 269, and A 312. Examples of the ASME specifications include: SA 240, SA 213, SA 269, and SA 312.

Table 49—IV. Physical Properties of Carbon and Alloy Steels for Low-Temperature and Cryogenic Service[a]

Property	Carbon Steels			Alloy Steels			
	ASTM A 537[b]	ASTM A 333[c] A 334[c]	ASTM A 516[d]	ASTM A517[e]	ASTM A 203[f] A 333[g] A 334[g]	ASTM A 203[h] A 333[i] A 334[i]	ASTM A 353[j] A 333[k] A 334[k]
Density							
kg/m³		7833	7833				
lb/in.³		0.283	0.283				
Electric Resistivity							
$\mu\Omega\cdot$mm		97.1	97.1				
$\mu\Omega\cdot$in		3.82	3.82				
Specific Heat Capacity							
kj/kg·K							
–100°C to 27°C					0.335	0.334	0.368
At 20°C		0.461	0.461				
27°C to 370°C							0.498
27°C to 535°C					0.603	0.615	
Btu/lb.°F							
–150°F to 80°F					0.080	0.0798	0.0878
At 68°F		0.11	0.11				
80°F to 700°F							0.119
80°F to 1000°F					0.144	0.147	
Thermal Conductivity (k value)							
W/m·K							
–195°C							13.17
–100°C						30.86	24.37
–45°C					35.76		
20°C		50.33	50.33		38.50	36.48	27.25
95°C					40.38	38.93	30.14
Btu·in/ft²·h·°F							
–320°F							91.3
–150°F						214	169
–50°F					248		
68°F		349	349		267	253	189
200°F					280	270	209
Linear Expansion Coefficient							
K⁻¹							
At –185°C							7.2×10^{-6}
–185°C to –17°C							9.5×10^{-6}
–45°C to 65°C	11.7×10^{-6}			11.7×10^{-6}			
–17°C to 95°C					11.16×10^{-6}	11.07×10^{-6}	
at 20°C		11.75×10^{-6}	11.75×10^{-6}				
21°C to 705°C					13.93×10^{-6}		
in./in.·°F							
At –300°F							4.0×10^{-6}
–300°F to 0°F							5.3×10^{-6}
–300°F to 200°F							5.6×10^{-6}
–50 to 150°F	6.5×10^{-6}			6.5×10^{-6}			
0°F to 200°F					6.2×10^{-6}	6.15×10^{-6}	
At 68°F		6.53×10^{-6}	6.53×10^{-6}				
70°F to 1300°F					7.74×10^{-6}		5.8×10^{-6}

[a]Properties determined at room temperature unless otherwise indicated.
[b]With modifications (UNS K02400, USS CHAR-PAC).
[c]Grades 1 (UNS K03008) and 6 (UNS K03006).
[d]UNS K01800, K02100, K02403 and K02700.
[e]Grade F (UNS K11576, USS "T-1" steel).
[f]Grade A (UNS K21703, USS 2¼% Ni).
[g]Grade 7 (UNS K21903, USS 2¼% Ni).
[h]Grade D (UNS K31718, USS 3½% Ni).
[i]Grade 3 (UNS K31918, USS 3½% Ni).
[j]9% Ni Steel (UNS K81340, USS 9% Ni).
[k]Grade 8 (UNS K81340, USS 9% Ni).

A 334, Grade 8. Nine per cent nickel steel is suitable for service temperatures to −195°C (−320°F), whereas 5 and 8 per cent nickel steels are suitable for temperatures to −170°C (−275°F).

The ASME Boiler and Pressure Vessel Code approved the use of 5, 8, and 9 per cent nickel steels in construction of vessels without the necessity of postweld heat treatment in thickness to 50 mm (2 inches), inclusive. Nine per cent nickel steel has been used in the construction of various types of equipment for the manufacture and storage of oxygen, nitrogen, argon, liquid methane; and low-temperature separation of helium and, to a smaller extent, a number of other commodities. Five per cent nickel steel has been used in the construction of liquid-ethylene tanks for stationary storage. Charpy V-notch impact values for 9

Table 49—Va. Physical Properties of Stainless Steels for Low-Temperature and Cryogenic Service.[a],[b]

Property	UNS S20100 (AISI Type 201)	UNS S20200 (AISI Type 202)	UNS S30100 (AISI Type 301)	UNS S30200 (AISI Type 302)	UNS S30400, S30403 (AISI Types 304, 304L)	UNS S30500 (AISI Type 305)	UNS S30900, 30908 (AISI Types 309, 309S)
Density							
kg/m³	7750	7750	8027	8027	8027	8027	8027
lb/in.³	0.28	0.28	0.29	0.29	0.29	0.29	0.29
Electric Resistivity							
μΩ·mm	691	691	721	721	721	721	782
μΩ·in.	27.2	27,2	28.4	28.4	28.4	28.4	30.8
Magnetic Permeability							
Annealed					1.003		1.003
10% Cold Worked					1.10		
Heavily Cold Worked					7.0		
Specific Heat Capacity							
kJ/kg·K							
At −195°C			0.155				
At −100°C			0.368				
0°C to 100°C	0.50	0.50	0.50	0.50	0.50	0.50	0.50
Btu/lb·°F							
At −320°F			0.037				
At −150°F			0.088				
32°F to 212°F	0.12	0.12	0.12	0.12	0.12	0.12	0.12
Thermal Conductivity (k value)							
W/m·K							
At −145°C					8.13		
At −105°C					13.0		
At 21°C					17.0		
At 100°C	16.3	16.3	16.3	16.3		16.3	15.6
At 315°C					18.7		
Btu·in./ft²·h·°F							
At −230°F					56.4		
At −155°F					90.0		
At 70°F					117.6		
At 212°F	113	113	113	113		113	108
At 600°F					130.0		
Linear Expansion Coefficient							
K⁻¹							
−185°C to 21°C					13.3×10^{-6}		
−130°C to 21°C					13.9×10^{-6}		
−73°C to 21°C					14.3×10^{-6}		
0°C to 100°C	16.6×10^{-6}	16.6×10^{-6}	16.9×10^{-6}	17.3×10^{-6}	17.3×10^{-6}	17.3×10^{-6}	14.9×10^{-6}
0°C to 315°C			17.1×10^{-6}	17.8×10^{-6}	17.8×10^{-6}	17.8×10^{-6}	16.7×10^{-6}
in./in.·°F							
−300°F to 70°F					7.4×10^{-6}		
−200°F to 70°F					7.7×10^{-6}		
−100°F to 70°F					8.2×10^{-6}		
32°F to 212°F	9.2×10^{-6}	9.2×10^{-6}	9.4×10^{-6}	9.6×10^{-6}	9.6×10^{-6}	9.6×10^{-6}	8.3×10^{-6}
32°F to 600°F			9.5×10^{-6}	9.9×10^{-6}	9.9×10^{-6}	9.9×10^{-6}	9.3×10^{-6}

[a]See Table 49—Vb for properties of additional steels.
[b]Properties determined at room temperature unless otherwise indicated.

per cent nickel steel in both the double-normalized and tempered and the quenched and tempered condition are shown in Figure 49—4, and those for quenched and tempered 8 per cent nickel steel are shown in Figure 49—5.

Fracture toughness data from various types of specimens from plate and weldments of 2¼-, 3½-, 5-, 8-, and 9 per cent nickel steels are available in Welding Research Council Bulletins Nos. 205 (May 1975) and 278 (June 1982) and in the Welding Journal, July 1981

(pp 113-s to 120-s) and March 1982 (pp 94-s to 96-s).

Austenitic stainless steels (which are nonhardenable by heat treatment) are excellent materials over the entire range of cryogenic applications because of their excellent notch toughness and high ductility at low temperatures and because of their excellent corrosion resistance. Table 49—VI gives the low-temperature impact properties of various annealed austenitic stainless steels. As can be seen, these steels remain tough at liquid-hydrogen and liquid-helium temperatures.

Table 49—Vb. Physical Properties of Stainless Steels for Low-Temperature and Cryogenic Service.[a],[b]

Property	UNS S31000 (AISI Type 310)	UNS S31008 (AISI Type 310S)	UNS S31600, S31603 (AISI Types 316, 316L)	UNS S32100 (AISI Type 321)	UNS S34700, S34800 (AISI Types 347, 348)	UNS S38100 (AISI Type XM15)[c]
Density						
kg/m³	8027	8027	8027	8027	8027	7750
lb/in.³	0.29	0.29	0.29	0.29	0.29	0.28
Electric Resistivity						
$\mu\Omega \cdot$mm	782	782	742	721	726	865
$\mu\Omega \cdot$in.	30.8	30.8	29.2	28.4	28.6	34.1
Magnetic Permeability						
Annealed	1.003		1.003		1.02	1.003
10% Cold Worked						
Heavily Cold Worked						1.01
Specific Heat Capacity						
kJ/kg·K						
At –195°C						
At –100°C						
0°C to 100°C	0.50	0.50	0.50	0.50	0.50	
Btu/lb·°F						
At –320°F						
At –150°F						
32°F to 212°F	0.12	0.12	0.12	0.12	0.12	
Thermal Conductivity (k value)						
W/m·K						
At –145°C						
At –105°C						
At 21°C						
At 100°C	10.64	10.64	16.3	16.2	16.2	
At 316°C						
Btu·in./ft²·h·°F						
At –230°F						
At –155°F						
At 70°F						
At 212°F	73.8	73.8	113	112	112	
Linear Expansion Coefficient						
K⁻¹						
–185°C to 21°C						
–130°C to 21°C						
–72°C to 21°C						
0°C to 100°C	15.8 × 10⁻⁶	15.8 × 10⁻⁶	16.0 × 10⁻⁶	16.7 × 10⁻⁶	16.7 × 10⁻⁶	13.7 × 10⁻⁶
0°C to 315°C	16.2 × 10⁻⁶	15.8 × 10⁻⁶	16.2 × 10⁻⁶	17.1 × 10⁻⁶	17.1 × 10⁻⁶	15.8 × 10⁻⁶
in./in.·°F						
–300°F to 70°F						
–200°F to 70°F						
–100°F to 70°F						
32°F to 212°F	8.8 × 10⁻⁶	8.8 × 10⁻⁶	8.9 × 10⁻⁶	9.3 × 10⁻⁶	9.3 × 10⁻⁶	7.6 × 10⁻⁶
32°F to 600°F	9.0 × 10⁻⁶	8.8 × 10⁻⁶	9.0 × 10⁻⁶	9.5 × 10⁻⁶	9.5 × 10⁻⁶	8.8 × 10⁻⁶

[a]See Table 49—Va for properties of additional steels.
[b]Properties determined at room temperature unless otherwise indicated.
[c]USS 18-18-2.

QUENCHED AND TEMPERED
9.5-mm(3/8") PLATES (3/4-SIZE SPECIMENS)

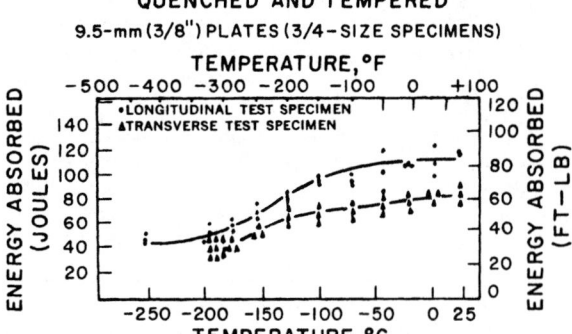

12.7-mm(1/2") PLATES (FULL-SIZE SPECIMENS)

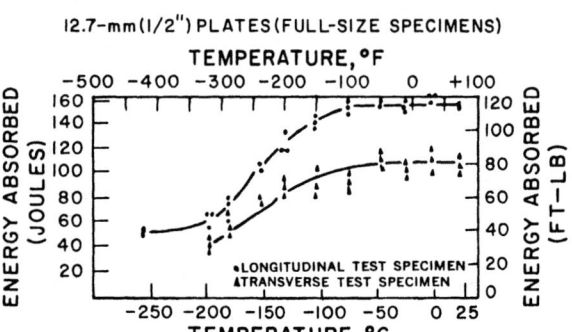

DOUBLE NORMALIZED AND TEMPERED
9.5-mm(3/8") PLATES (3/4-SIZE SPECIMENS)

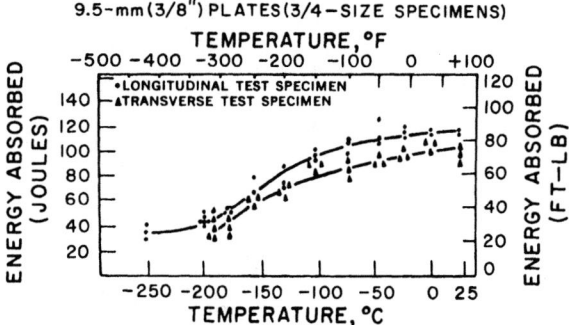

12.7-mm(1/2") PLATES (FULL-SIZE SPECIMENS)

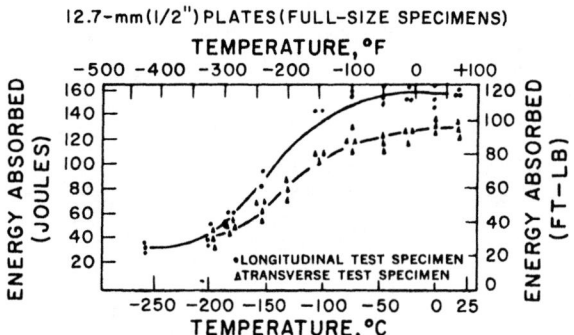

FIG. 49—4. Results of Charpy V-notch impact tests for 9 per cent nickel steel.

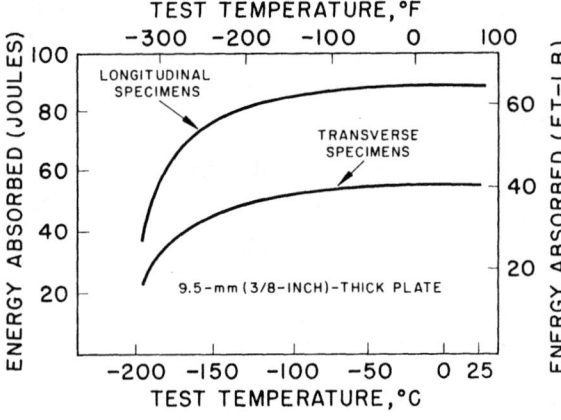

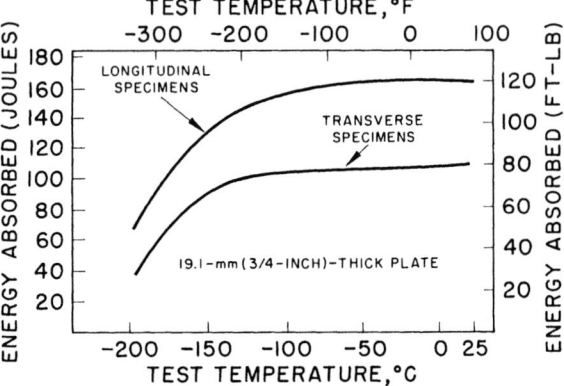

FIG. 49—5. Results of Charpy V-notch impact tests on 9.5- and 19-mm (⅜- and ¾-inch) thick plates of 8 per cent nickel steel. Specimens from the 19-mm plate were ¾-size (7.5 mm).

Table 49—VI. Room- and Low-Temperature Toughness of Several Annealed Austenitic Stainless Steels.

Designation			Temperature		Energy Absorbed*	
UNS	AISI Type	Product Size	(°C)	(°F)	(joules)	(ft-lb)
S20100	201	19.1-mm (¾-in.) Plate	27	80	298	220
			−195	−320	83	61
S20200	202	19.1-mm (¾-in.) Plate	27	80	298	220
			−195	−320	76	56
			−254	−425	66	49
S30400	304	12.7-mm (½-in.) Plate	27	80	209	154
			−195	−320	118	87
			−254	−425	122	90
S30403	304L	89.9-mm (3½-in.) Plate	27	80	160	118
			−195	−320	91	67
			−254	−425	91	67
S31000	310	89.9-mm (3½-in.) Plate	27	80	193	142
			−195	−320	121	89
			−254	−425	117	86
S34700	347	89.9-mm (3½-in.) Plate	27	80	163	120
			−195	−320	89	66
			−254	−425	77	57

*Charpy V-notch test.

Table 49—VII. Summary of Some Specifications for Steels for Low-Temperature and Cryogenic Service.

Designation ASTM	UNS	Tensile Strength MPa	Tensile Strength ksi	Minimum Yield Point MPa	Minimum Yield Point ksi	ASME Allowable Stress[(e)] MPa	ASME Allowable Stress[(e)] psi	Minimum Elongation (Per Cent) in 50 mm (2 in.)	Minimum Elongation (Per Cent) in 200 mm (8 in.)	Lowest Usual Service Temperature[(a)] (°C)	Lowest Usual Service Temperature[(a)] (°F)	Applications
A 537[(b)] (USS CHAR-PAC) Normalized	K02400	482–620	70–90	345	50	121	17 500		19	−45	−50	Welded pressure vessels and storage tanks, when weight and strength are not critical. Refrigeration; transport equipment.
Quenched and tempered	K02400	552–690	80–100	414[(c)]	60[(c)]	207	30 000	23		−45	−50	
A 516[(d)] Grade 55	K01800	379–448	55–65	207	30	90.7	13 150	28	24	−45	−50	Welded pressure vessels and storage tanks, when weight and strength are not critical. Refrigeration; transport equipment.
Grade 60	K02100	414–496	60–72	221	32	103	15 000	26	22	−45	−50	
Grade 65	K02403	448–531	65–77	241	35	112	16 250	24	20	−45	−50	
Grade 70	K02700	483–586	70–85	262	38	121	17 500	22	18	−45	−50	
A 517 Grade F (USS "T-1")	K11576	793–931	115–135	690[(c)]	100[(c)]	198	28 750[(e)]	18		−45	−50	Highly stressed pressure vessels. Tank trucks for handling LP gases.
A 203 (2¼% Ni) Grade A	K21703	448	65	255	37	112	16 250	25	21	−59	−75	Tanks, vessels and piping for liquid propane.
Grade B	K22103	483	70	276	40	121	17 500	23	19	−59	−75	
A 203 (3½% Ni) Grade D	K31718	448	65	255	37	112	16 250	25	21	−101	−150	Land-based storage of liquid propane, acetylene, carbon dioxide, ethane and ethylene.
Grade E	K32018	483	70	276	40	121	17 500	23	19	−101	−150	
A 553 Class 1 (9% Ni)	K81340	690–827	100–120	586[(c)]	85[(c)]	164	23 750	22		−195	−320	Large tonnage, oxygen producing equipment. Transportation and storage of methane, oxygen, nitrogen and argon.
Class 2 (8% Ni)	K71340	690–827	100–120	586[(c)]	85[(c)]	164	23 750	22		−171	−275	

(Continued on next page)

Table 49—VII. (Continued)

Designation		Tensile Strength		Minimum Yield Point		ASME Allowable Stress[e]		Minimum Elongation (Per Cent)		Lowest Usual Service Temperature[a]		Applications
ASTM	UNS	MPa	ksi	MPa	ksi	MPa	psi	in 50 mm (2 in.)	in 200 mm (8 in.)	(°C)	(°F)	
A 353 (9% Ni) A 240	K81340	690–827	100–120	517[c]	75[c]	164	23 750	22		-195	-320	In petrochemical, nuclear, missile and other areas where purity of product is essential. Hauling liquid hydrogen.
Type 304 Stainless	S30400	517	75	207[c]	30[c]	129	18 750			-269	-452	
Type 304L Stainless	S30403	483	70	172[c]	25[c]	121	17 500			-269	-452	

[a]Service temperatures may be lower if requirements of ASME Section VIII, Par. U.G. 84, are met.
[b]With modifications.
[c]Yield strength (see Tables 49-II and 49-III).
[d]Normalized
[e]For service applications -20 to +650°F, except A 240 (-20 to +100°F).

Ferritic and martensitic stainless steels generally are not recommended for cryogenic use.

B. YIELD AND TENSILE STRENGTH

Normally, such conventional properties as tensile, yield and fatigue strengths increase whereas ductility and toughness decrease at low temperatures. The modulus of elasticity of metals also generally increases at low temperatures. The design of code-regulated, low-temperature equipment generally is based on tensile properties of material at ambient (room) temperature.

However, the ASME Boiler and Pressure Vessel Code does provide alternative rules for pressure vessels constructed of 5, 8, and 9 per cent nickel steels (part ULT) and of Type 304 austenite stainless steel (Code Case 1909) which take advantage of the higher yield and tensile strengths of these steels at low temperatures. Similarly, in certain military and space applications (non-code), the higher yield and tensile strengths of austenitic stainless steels at low temperatures have been used to advantage to reduce weight and cost. Data for guidance purposes are given in Table 49—VII and Figures 49—6 and 49—7.

C. THERMAL PROPERTIES

Thermal Expansion and Contraction—The large dimensional changes between ambient and operating temperatures in the cryogenic field can lead to a number of problems in both stationary equipment and machinery unless proper design consideration is given to thermal expansion and contraction.

This factor is important in sandwich construction of cryogenic vessels and in double-wall vessels that use different materials. In some cases, thermal expansion might influence the decision to make the inner and outer tanks of a cryogenic vessel of the same material.

In most non-ferrous alloys, the coefficient of thermal expansion is considerably higher than that for steel, as shown in Figure 49—8. Large dimensional changes can result in dangerous stresses as a result of high coefficients of expansion, and allowance should be made for them in designs. For cryogenic service, the coefficients of expansion might be more aptly reported as coefficients of contraction, since it is the contraction during the cooling-down operation that creates the major problems in design.

Thermal properties of some steels are shown in Tables 49—IV and 49—V. Figures 49—9 and 49—10 give the thermal expansion properties of UNS K81340 and S30400 (9 per cent nickel and AISI Type 304) stainless steels.

Thermal Conductivity—Thermal conductivity is of considerable importance in the selection of materials for low-temperature applications. For support, piping, sensing lines, and so on, design engineers prefer a material with low thermal conductivity. A high conductivity reduces thermal gradients during the initial cool-down but results in much higher heat gains through the piping when the equipment is in service. The levels of thermal conductivity of steel are satisfactory for the vast majority of cryogenic equipment (Tables 49—IV and 49—V).

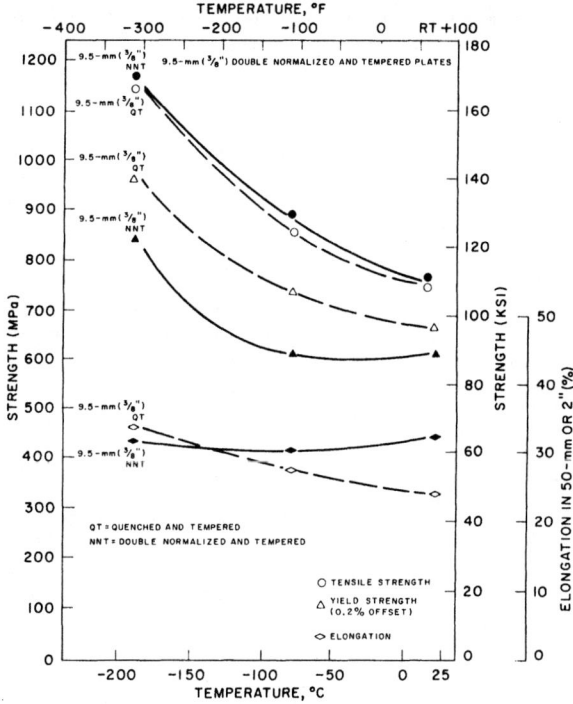

FIG. 49—6. Effect of low temperature on strength and elongation of 9 per cent nickel steel.

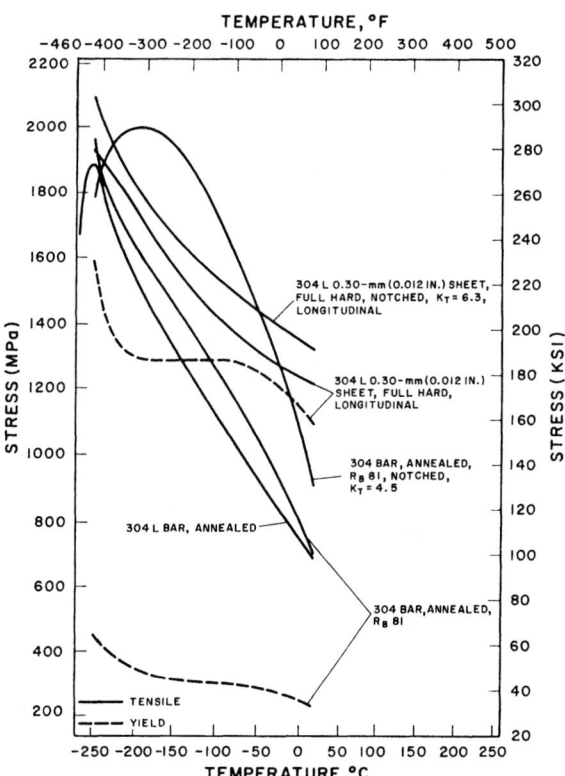

FIG. 49—7. Effect of temperature on the strength of UNS S30400 (AISI Type 304) stainless steel.

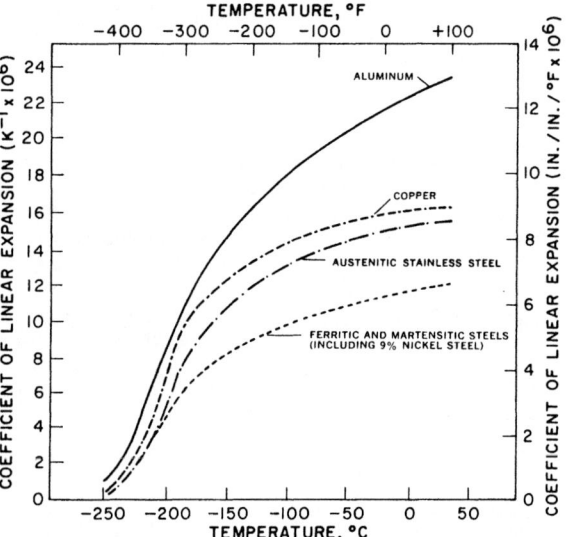

FIG. 49—8. Coefficient of linear thermal expansion for several materials.

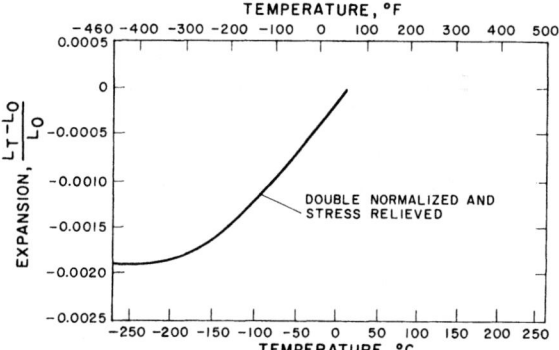

FIG. 49—9. Thermal expansion of 9 per cent nickel steel. L_0 = length at room temperature (20°C or 68°F); L_T = length at testing temperature.

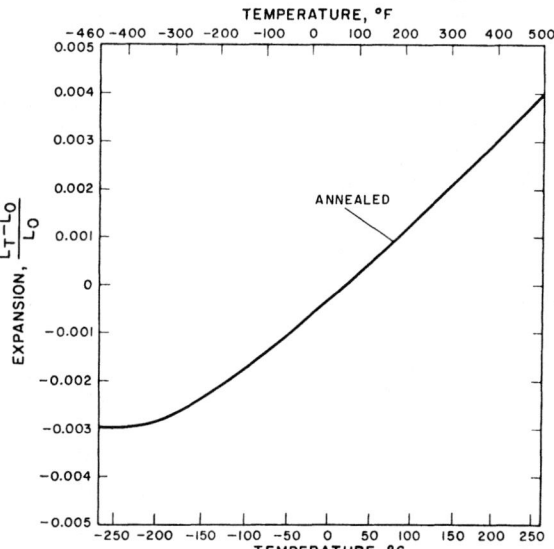

FIG. 49—10. Thermal expansion of UNS S30400 (AISI Type 304) stainless steel. L_0 = length at room temperature (20°C or 68°F); L_T = length at testing temperature.

D. CORROSION RESISTANCE

Although the selection of materials of construction for low-temperature service is based necessarily on mechanical properties, the corrosion resistance of the materials also must be considered. Corrosion resistance is important for assuring adequate service life and also for minimizing product contamination.

The austenitic stainless steels, of course, afford excellent corrosion resistance in many chemical media, including both strong oxidizing acids and alkalis, and are practically immune to atmospheric corrosion in most locations. However, moist chloride environments can cause breakdown of the passive film on stainless steels, resulting in some staining and possible pitting. The alloy steels used for low-temperature service, such as 2½, 3½ and 9 per cent nickel steels, although completely resistant to nearly all cryogenic liquids, are inherently less corrosion resistant than the stainless steels, and some corrosion is to be expected with these materials in the atmosphere or in chemical solutions.

Fortunately, because of the low reactivity of most chemicals at low temperatures, corrosion problems resulting from materials in contact with cryogenic liquids have been relatively few. The nickel steels give adequate service from a corrosion standpoint in nearly all low-temperature applications. Because of their excellent resistance to atmospheric corrosion when the equipment is not in use, austenitic stainless steels also are used for systems in which contamination of the cryogenic liquid by solid particles is critical.

CHAPTER 50

Mechanical Testing

SECTION 1

INTRODUCTION

The manner in which metals react to applied forces is important in almost all practical applications. Steels, for example, are used in a multitude of applications involving their ability to withstand service loadings without permanently deforming or rupturing. Of equal importance, on the other hand, is the ability of steel to undergo large permanent deformations, thus permitting the formation of useful shapes under the application of the proper forces. Even though the final service application of a metal part may not in any way involve load-carrying ability, it is almost certain that at some stage in fabrication the manner of reaction of the metal to applied forces has come into play. **Mechanical properties** are the characteristic response of a material to applied forces; the **methods of measurement** of these properties and of evaluating their significance form the subject matter of this chapter.

It is both logical and useful to think of mechanical properties as falling into two broad categories; strength and ductility. **Strength** properties are related to the ability of the material to resist applied forces, while **ductility** is a measure of the ability to undergo permanent changes of shape without rupturing. Many mechanical characteristics or behaviors depend on both strength and ductility, frequently in a complex manner, as illustrated so well by that characteristic of materials known as **toughness,** or the ability to absorb energy. Even such complex behavior can be better understood, however, by considering it in terms of the separate contributions of strength and ductility.

The mechanical reactions of metals to applied forces are extremely diversified, depending upon the exact nature of the forces and the conditions under which they are applied. It is essential that cognizance be taken of this fact in attempting to devise or select tests which will permit a prediction of service performance. The final answer to the question of suitability for service can, in general, only be found in an actual service test. In most cases, however, actual service tests are impractical and simpler tests must be found, especially where frequent inspection of material is essential. In devising simpler tests, it is important that the type of forces involved in the particular service application of interest be known, i.e., whether the loading is tension, compression, bending, twisting, shear, etc. It is also im-

portant to know whether the loading is static or dynamic, and if dynamic, it is important to know the nature of the rate of application and variation of the loading. The temperature at which the loading is applied also may be of critical importance. If these features of the service conditions are known, one can be guided in the selection of tests most likely to provide a reliable evaluation of the suitability of a material for a particular application. One criterion of test validity with respect to service would be, then, the extent to which the test simulates the actual application. For example suppose that a large structural member is required to carry a certain compressive load at a temperature of 260°C (500°F). A test of the actual member cannot be performed because of the large size. An alternate course of action is to select from the material in question a small specimen suited to available testing equipment and to make a compression test at 260°C (500°F). The type of loading and the service temperature are thereby duplicated, and it is likely that the test results will be a reliable guide in selecting a material suitable for the application.

Another criterion by which a test may be selected depends on whether or not the test measures the same ultimate fundamental property on which successful performance in service depends. For example, in bending, the minimum bend radius obtainable with a particular material depends on the ability of the material to elongate over very short gage lengths on the tension fibers of the bend. Since the reduction of area in the tension test may be interpreted also as a measure of ability to elongate over very short gage lengths, it has been possible in some instances to relate minimum bend radii to reduction of area. The test is thus measuring a property which is directly responsible for the behavior in the application.

There exist a great many applications of metals, however, which are characterized by such complex conditions of loading that a directly applicable test may not be readily found. Because of this circumstance, the selection of mechanical tests for a particular instance is based primarily on experience. It may be known that, over a period of years, many lots of a certain grade of steel having properties falling within a certain range have performed satisfactorily in service. It can be expected, then, that new lots of this grade of steel having the same mechanical properties will perform satisfactorily in the same application.

A good example of a test selected on the basis of experience is offered by the extensive use of hardness testing in the inspection of deep-drawing sheets. It is known from experience that, in order for a sheet to be formed satisfactorily into a particular part, its hardness must lie within a certain range. Although hardness in itself can hardly be considered the property which controls the performance, it reflects the critical properties to a degree which makes it extremely useful as a control test, particularly in view of the ease with which hardness tests can be made. Testing on this basis, however, should only be considered as a means of increasing the probability of selecting satisfactory material. Unless the actual characteristics which determine the performance are known and are actually being tested for, an arbitrary test, even though backed by considerable experience, may at some time fail to discriminate between good and poor material.

In the field of the mechanical behavior of metals, there is a strong trend away from arbitrary tests and toward a more exact evaluation of the fundamental properties which are actually called upon in particular service applications. Experimental stress analysis, particularly the use of electrical-resistance strain-gage techniques, have permitted great advances to be made in more thorough evaluation of service requirements. Once the mechanics of an application have been analyzed and the critical fundamental material characteristics have been ascertained, the task of selecting a suitable test is greatly simplified.

Many steel products are sold to mechanical property specifications and much of the testing carried out by the steel producer is for the purpose of ascertaining whether or not a given specification is being met. If the product is a widely used one, a standard specification may be set up by a body such as the American Society for Testing and Materials (ASTM) in order to assist the user in specifying material for a certain application.

Such specifications may provide fundamental design figures for the engineer, or in many cases may be based on previous experience which indicates that, if a material has certain properties, it will serve a particular purpose satisfactorily.

As already indicated, mechanical testing may be carried out as a means of quality control, even though the product may not be sold to a mechanical-property specification. In such instances, the experience of the producer dictates the test most suitable for the purpose. The particular test chosen may depend to a great extent upon the ease with which it can be made if frequent inspection or extensive sampling is needed.

Another important function of mechanical testing is to determine the causes for failure in service. If the material is shown to be at fault, a substitution of another material or the elimination of the deficiency of the first material can be guided by judicious mechanical tests. If no fault can be found with the material, a change in design may be indicated.

Still another function of mechanical testing results from its great usefulness in the development of new and improved products. Even in applications so complex in nature that a final test in service may be required, it is usually possible, by judicious mechanical tests, to establish trends and select the materials most likely to perform satisfactorily.

In the subsequent sections of this chapter, the mechanical tests most generally applied to steel will be discussed. Attention will be focused primarily upon the tension test, the hardness tests, the notch-toughness tests, the fatigue test, and the creep and rupture tests. In addition, certain miscellaneous tests sometimes made on steel will be briefly described. Emphasis will be placed on the significance of the various mechanical properties to be discussed, as well as on the precautions necessary to obtain reliable test results. Test procedures approved and recommended by the ASTM will be indicated for those tests for which standards or tentative standards have been established.

SECTION 2

THE TENSION TEST

Because considerable research effort currently is being expended in the field of mechanical behavior of materials, the field is changing rapidly. New tests are being developed; new ways of performing old tests are being formulated; new interpretations of old tests are being made. For these reasons, the reader is urged to peruse the literature for changes that may have occurred in the various subjects discussed in this chapter since it was written; this is especially true in the areas of notch-toughness testing and fatigue testing.

Because of the large amount of information which can be derived from it, the tension test is undoubtedly the most generally useful of all the mechanical tests applied to steel. The versatility of the tension test lies in the fact that it permits both strength and ductility

properties to be measured. The strength properties measured in the tension test are directly useful in design, whereas the ductility properties provide some indication of the extent to which changes in shape can be brought about by plastic forming, or an indication as to whether sufficient ductility is present for the intended service. Since the tension test is used quite extensively in the steel industry, a rather complete description of the test and of the properties measured will be presented.

Testing Machines—The machines employed in tension testing consist essentially of a load-producing mechanism and a load-measuring mechanism. The most elementary method of producing a tensile load simply involves the suspension of dead weights from the specimen to be loaded. This procedure is in general not practicable, however, because of the inconvenience of handling the large weights which would be required. Dead-weight loading, sometimes with the force multiplied by a lever, is used in certain instances, however, where it is desired to subject a specimen to a fixed load for a period of considerable duration. Creep and rupture testing, which is discussed in a later section, is perhaps the best example of the use of dead-weight loading for testing purposes.

The more commonly used testing machines employ either a mechanical system of loading actuated by screws or a hydraulic system in which the load is applied through a hydraulic ram. In both types of machines, there is a fixed crosshead and a moving crosshead through which the tensile force is applied. The moving crosshead in the screw machine is motivated either by threaded columns (screws) rotating in stationary nuts or by rotating nuts, depending upon the design of the machine. Various speeds of crosshead movement are available through changes of the gear combination in the drive or by variable-speed drives or both. In the hydraulic machine, the moving crosshead is powered by a hydraulic ram. A continuous variation in the rate of crosshead movement can be obtained; however, the slow rate of a hydraulic machine can usually be adjusted to a rate slower than most mechanical machines with variable-speed drive.

The mechanical or screw type of testing machines were formerly characterized by the load-measuring mechanism employed. The oldest machines used a mechanism in which the load is transmitted through a system of levers acting on fulcrums of hardened steel to a beam carrying a moveable weight or poise. Because of the characteristic load-measuring mechanism, this type of machine is generally referred to as a "beam-and-poise" or "lever" machine. One of the greatest objections to the lever lies in the manual operation of the weighing mechanism and the attendant human factor introduced into the test. To circumvent this objection, machines were developed in which a pendulum counteracts the machine force and the movement of which provides a continuous and automatic indication of load on a dial. The more recent machines use the electronic devices discussed later.

Hydraulic testing machines are often classified as lapped-ram type or packed-ram type. In the lapped-ram type, the piston and cylinder are lapped to an extremely close fit so that oil leakage is negligible and the movement is almost frictionless. Because of the extremely low friction, the oil pressure in the cylinder can be used directly as a measure of load on the machine. This pressure is usually measured with an electronic pressure transducer, a Bourdon-tube gage, or, in older machines, a secondary piston device; there are numerous variations in the ways that these devices are used.

The piston and cylinder of packed-ram type machines do not have tolerances as close as do the lapped-ram machines and so a packing is required to prevent leakage of oil from the system. Although current technology uses a series of seals and wipers, instead of hemp packing as was formerly used, the terms packing and packed-ram remain. Since the packing causes friction

forces that vary in magnitude with motion of the piston, the possibility of using a direct pressure measurement to provide an accurate indication of load is precluded. Therefore, some type of device that is independent of the hydraulic pressure must be used to measure the applied force. This device is often an electronic load cell or other device such as the Tate-Emery hydraulic capsule. Because of the large number of Tate-Emery capsules that are still in use, they are described below.

The load on the machine acts on the capsule which is part of a closed hydraulic system. The capsule is a very rigid assembly containing an oil layer of only about 0.76-mm or 0.030-inch thickness. As the load is applied, a pressure is developed in the capsule and transmitted to a Bourdon tube or to an electronic pressure transducer. Changes in the load cause motion of the tube due to volumetric changes in the capsule. In the Tate-Emery null-method design (that is, one in which the indicating system does no work), this motion is resisted by a pneumatic servometer so that the tube end does not move. The pneumatic servometer is the driving force, and has ample power not only to overcome any friction in the dial indicating system, but also to operate other accessories.

Electronic devices are employed for load measurement in most mechanical and hydraulic machines of recent design. One method uses electrical-resistance strain gages to measure the changes in dimensions of an elastic member through which the load is applied. The system is calibrated so that the changes in resistance of the strain gages provide an indication of the load. Another method uses linearly variable differential transformers to sense elastic motion of torque bars caused by the applied load. Different load ranges can be obtained by appropriate selection of the dimensions of the elastic member. A third method uses electronic pressure transducers installed directly in the hydraulic lines of hydraulic machines or in the load-capsule line of capsule weighing machines of the type shown in Figure 50—1.

Generally, a number of load-measuring ranges are provided so that the most sensitive and accurate range can be chosen for a particular test, that is, the range having its top limit just above the maximum load expected in the test. It is important to know the accuracy of the load-measuring mechanism, and frequent calibration should be carried out to insure that the accuracy remains within the desired limits. With modern load-weighing mechanisms, accuracies of one-half of one per cent of the indicated load over all but the lowest 10 per cent of the range are usually guaranteed. For most purposes, one per cent accuracy is considered satisfactory, but even at this level of accuracy, frequent calibration is desirable. (ASTM Standard E 4 recommends that machines in constant use be verified at intervals of twelve months.) If the errors are found to exceed the desired limits, the load-measuring mechanism should be adjusted until further calibration indicates the accuracy to be within the desired range.

In the ASTM Standard E 4, three methods for the verification of testing machines are described. These include: (a) verification of standard weights of known mass (only applicable in the case of machines weighing

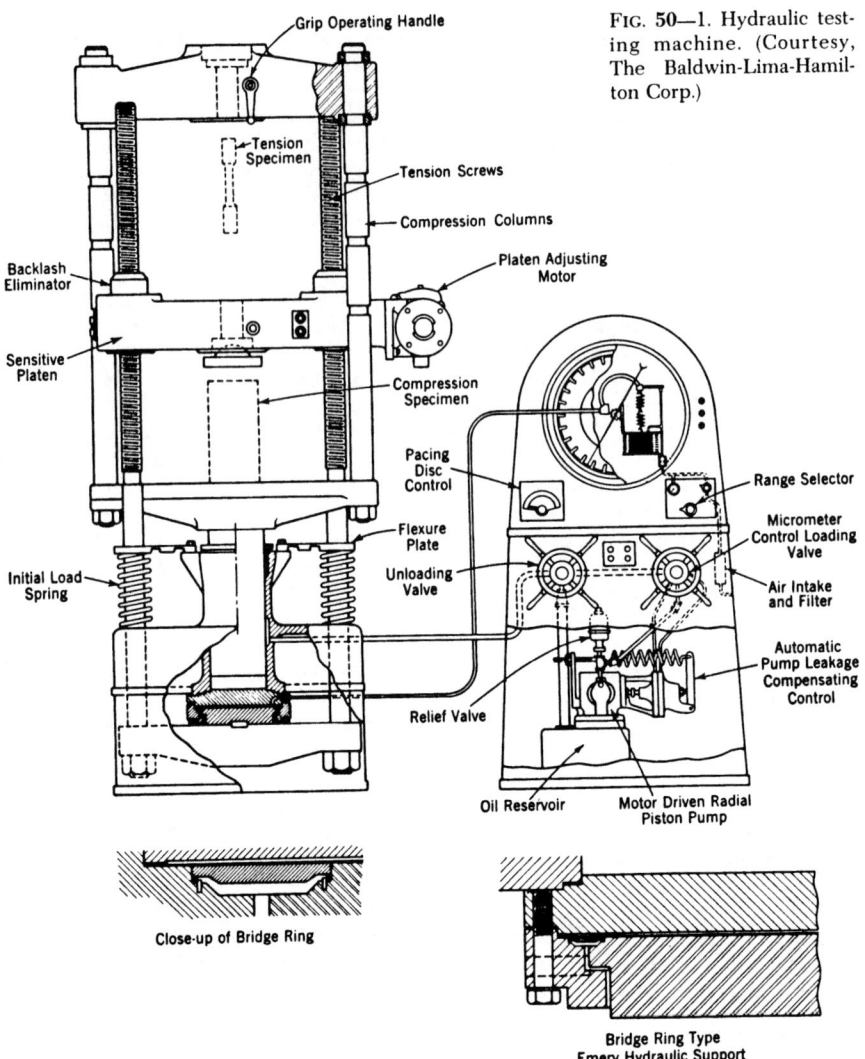

FIG. **50**—1. Hydraulic testing machine. (Courtesy, The Baldwin-Lima-Hamilton Corp.)

Close-up of Bridge Ring

Bridge Ring Type
Emery Hydraulic Support

a downward force), (b) verification by standardized proving levers, and (c) verification by elastic calibration devices. The first two of these methods make use of standard weights, in one case applying the weight directly to the machine, and in the other case making use of levers to multiply the forces that can be applied by the standard weights. Although both methods are quite accurate, they are obviously very inconvenient, and the feasible range of load capacities which can be covered does not encompass the larger machines.

The most widely used method of calibration makes use of elastic calibration devices. A frequently employed device of this type is the so-called proving ring, shown in Figure 50—2. Such a ring is subjected to a series of loads in the machine to be calibrated, and the deflection of the ring is measured by a micrometer which is an integral part of the proving ring. Proving rings are available in a variety of ranges up to 136 000 kilograms (1 000 000 lb) and are calibrated and certified by the National Bureau of Standards.

Calibration agencies are also using electronic load

cells for verifying testing machines. The accuracy, repeatability and stability of these devices have proven to be within required limits (see ASTM Method E 74). Since these devices can be easily designed for either tension loading or compression loading, or both, and are generally physically smaller and lighter in weight than proving rings, they are being used in increasing numbers.

Extensometers—Certain aspects of tension testing require the use of some device for the measurement of the extension of the specimen; such a device is called an extensometer. There are a very large number of types of extensometers available and no attempt will be made here to describe the features and specific applications of all the various types. The characteristics of a few of the more widely used types will be discussed briefly in order to illustrate the general nature of commercially available extensometers.

Extensometers can be considered to fall into two main groups, depending on the range of extensions which can be covered. On the one hand, there are the

FIG. 50—2. Proving ring, compression type. (Courtesy, Morehouse Machine Company.)

very accurate, high-sensitivity extensometers used for the measurement of minute extensions. These extensometers are characterized by very small ranges; i.e., the total extension which can be measured is quite small. On the other hand, there are the long-range extensometers designed to measure extensions up to the instant of rupture of the specimen. It is evident that, if long range is desired, sensitivity must be sacrificed; while it is equally true that sensitivity can be gained only at the expense of range. Extensometers can be either of the direct-reading (indicating) or recording type. The indicating extensometers make use of a dial gage to measure the movement of some element of the extensometer. A great many combinations of range and sensitivity can be obtained in dial-gage extensometers, the choice depending on the strain range over which observations are to be made.

The optical types of extensometers provide a high degree of sensitivity and accuracy for extremely small extensions and are of the indicating type. These extensometers are used in elastic-strain measurements and for very precise indications of the beginning of plastic yielding. Such an extensometer is the Tuckerman extensometer. This extensometer is attached to the specimen so that a fixed knife edge and a moveable knife edge are in contact with the specimen. The movable knife edge consists of a lozenge, one face of which is polished to a mirror finish. A light beam is focused on the lozenge, and as the lozenge rotates as a result of extension of the specimen, the reflected beam is displaced relative to the incident beam. A specially calibrated collimator provides the light source and measures the deflection of the reflected beam. Extensions as small as 5 one-hundred thousandths (0.00005) of a millimetre per millimetre (two millionths of an inch per inch) can be measured on a gage length of 50 millimetres (2 inches).

Generally, the great sensitivity of the optical extensometers is not required and other more convenient types of low-range extensometers are employed.

Bonded electrical-resistance strain gages are sometimes used in the determination of stress-strain curves, although their main application is in experimental stress analysis. Essentially, this type of gage consists of a grid of etched foil or of fine wire which is suitably bonded directly to the specimen, usually by special adhesives. As the specimen is strained, the grid undergoes similar strains which produce a change in its cross section. This change in cross-section results in a change in electrical resistance that is proportional to the amount of strain and can be readily measured. The range of strain of most electrical-resistance gages is about two per cent, but special gages having considerably higher ranges are available. When suitable techniques are employed, the sensitivity and accuracy of electrical-resistance gages is at least as good as the best optical strain gages.

Recording extensometers are now in wide use and offer the advantage of an automatically plotted stress-strain curve from which determinations of various material characteristics can be made, as well as the advantage of a permanent record of the test. High-sensitivity short-range recording extensometers can be obtained with an accuracy of better than 0.0025 millimetre per millimetre of indicated strain (0.0001 inch per inch). of indicated strain (see ASTM Standard E 83), and with a range of up to 0.5 to 1.0 millimetre (0.02 inch to 0.04 inch). Extensometers with sensitivities up to 0.0005 millimetre (0.00002 inch) can be obtained.

Two of the currently popular types of recording extensometers are the LVDT or linearly variable differential transformer type that operates on a magnetic principle and the resistance strain-gage type that is described below. In the LVDT type extensometer, a lever system is arranged in such a way that extenion of the specimen results in a movement of an iron core in a small magnetic coil, thus changing the inductance of the coil. A similar magnetic coil is contained in the recorder, and is automatically kept in balance with the extensometer coil by a servomotor that also actuates the recorder. By a different arrangement of the lever system in this type of recording extensometer, the range can be greatly extended and such extensometers, known as total-elongation extensometers, are commercially available. The magnification ranges of this extensometer vary from 5 to 20 as compared to 250 to 1000 for the more sensitive, short-range extensometer. Resistance strain gage extensometers are also available in which a thin member, with an electrical-resistance strain gage mounted on it, is elastically bent as the specimen deforms. The output of the strain gage activates a recorder. The movement of any stress-strain (or usually, more properly, load-elongation) recorder can be calibrated to indicate the extension directly. Extensometers of this type can be furnished with magnification ranges similar to the LVDT type.

Another type of long-range recording extensometer which has seen considerable application is the wedge type. In this extensometer, a wedge of known taper is pulled through a slot in the extensometer which opens as the specimen elongates. The movement of the wedge is used to rotate the recorder drum and since the taper of the wedge is known, the strain can be calculated. The magnification obtainable with this ex-

tensometer depends upon the degree of the taper in the wedge.

For strain measurements in and just beyond the elastic range, it is desirable to use extensometers which incorporate an averaging mechanism. Unless extreme precautions are taken, axiality of loading will not be obtained and a non-uniform strain distribution will result. By averaging the extension along two opposite fibers of the specimen, a closer approximation to the extension which would have occurred had the loading been axial can be obtained.

Specimens—Certain standard tension specimens have been adopted and are recommended by the American Society for Testing and Materials. The shapes and dimensions of the most frequently used specimens are shown in Figure 50—3. The rectangular cross-section specimen with 8-inch gage length is used for tests of plates and structural sections. The rectangular cross-section specimen with 2-inch gage length generally is used in sheet-metal testing. The standard circular cross-section specimen with 2-inch gage length is shown in Figure 50—4. The details of the ends of this type of specimen vary widely, and depend on the types of grips employed. If the section from which the specimen is to be taken is too small to permit the procurement of a full-size specimen, smaller specimens can be used if the dimensions are kept in geometric proportion. It is especially important that the gage length for measuring elongation be proportioned to the nominal diameter of the specimen. One-inch and 1.4-inch gage-length specimens are frequently used and have diameters of 0.252 inch and 0.357 inch respectively. The nominal diameters of 0.252, 0.357 and 0.505 inch for the specimens having gage lengths of 1 inch, 1.4 inches, and 2 inches, respectively, were selected to simplify calculation of loads in pounds per square inch from actual loads. These diameters provide specimens having respective cross-sectional areas of very nearly 0.05, 0.1, and 0.2 sq. in. Other specimens for special products can be found in the ASTM Standards. For certain types of products, such as small bars, tests may be made on the full section, in which case the only preparation necessary is the cutting of the specimen to length.

As mentioned above, if the specimen size is changed for whatever reason, the dimensions of the new specimen should be proportional to those of the former specimen. It is especially important that the gage length be proportional to the cross-sectional area if the elongation values obtained from different size specimens are to be directly compared. For the circular cross-section specimens shown in Figure 50—4, the ratio

$$\frac{\text{gage length}}{\sqrt{\text{area}}}$$

is equal to 4.47; in Europe a ratio of 5.65 or 11.3 is often used even for specimens with rectangular cross-sections. (For specimens with circular cross-sections, a ratio of 4.47 means that the gage length is 4.0 times the diameter, while a ratio of 5.56 means that the gage length is 5.0 times the diameter.) This is further discussed in the subsection on elongation.

A number of general precautions should be observed

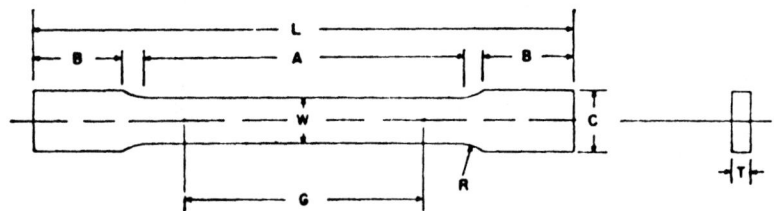

DIMENSIONS

	Standard Specimens				Subsize Specimen	
	Plate-Type, 1½-in. Wide		Sheet-Type, ½-in. Wide		¼-in. Wide	
	in.	mm	in.	mm	in.	mm
G—Gage length (Notes 1 and 2)	8.00 ± 0.01	200 ± 0.25	2.000 ± 0.005	50.0 ± 0.10	1.000 ± 0.003	25.0 ± 0.08
W—Width (Notes 3, 4, and 5)	1½ + ⅛ −¼	40 + 3 −6	0.500 ± 0.010	12.5 ± 0.25	0.250 ± 0.002	6.25 ± 0.05
T— Thickness (Note 6)			thickness of material			
R—Radius of fillet, min	½	13	½	13	¼	6
L—Over-all length, min (Notes 2 and 7)	18	450	8	200	4	100
A—Length of reduced section, min	9	225	2¼	60	1¼	32
B—Length of grip section, min (Note 8)	3	75	2	50	1¼	32
C—Width of grip section, approximate (Notes 4, 9, and 10)	2	50	¾	20	⅜	10

NOTE 1—For the 1½-in. (40-mm) wide specimen, punch marks for measuring elongation after fracture shall be made on the flat or on the edge of the specimen and within the reduced section. Either a set of nine or more punch marks 1 in. (25 mm) apart, or one or more pairs of punch marks 8 in. (200 mm) apart may be used.

NOTE 2—When elongation measurements of 1½-in. (40-mm) wide specimens are not required, a gage length (G) of 2.000 in. ± 0.005 in. (50.0 mm ± 0.10 mm) with all other dimensions similar to the plate-type specimen may be used.

NOTE 3—For the three sizes of specimens, the ends of the reduced section shall not differ in width by more than 0.004, 0.002 or 0.001 in. (0.10, 0.05 or 0.025 mm), respectively. Also, there may be a gradual decrease in width from the ends to the center, but the width at either end shall not be more than 0.015 in., 0.005 in., or 0.003 in. (0.40, 0.10 or 0.08 mm) respectively, larger than the width at the center.

NOTE 4—For each of the three sizes of specimens, narrower widths (W and C) may be used when necessary. In such cases the width of the reduced section should be as large as the width of the material being tested permits; however, unless stated specifically, the requirements for elongation in a product specification shall not apply when the narrower specimens are used. If the width of the material is less than W, the sides may be parallel throughout the length of the specimen.

NOTE 5—The specimen may be modified by making the sides parallel throughout the length of the specimen, the width and tolerances being the same as those specified above. When necessary a narrower specimen may be used, in which case the width should be as great as the width of the material being tested permits. If the width is 1½ in. (38 mm) or less, the sides may be parallel throughout the length of the specimen.

NOTE 6—The dimension T is the thickness of the test specimen as provided for in the applicable material specifications. Minimum nominal thickness of 1½-in. (40-mm) wide specimens shall be 3⁄16 in. (5 mm), except as permitted by the product specification. Maximum nominal thickness of ½-in. (12.5-mm) and ¼-in. (6-mm) wide specimens shall be ¾ in. (19 mm) and ¼ in. (6 mm), respectively.

NOTE 7—To aid in obtaining axial loading during testing of ¼-in. (6-mm) wide specimens, the over-all length should be as the material will permit.

NOTE 8—It is desirable, if possible, to make the length of the grip section large enough to allow the specimen to extend into the grips a distance equal to two thirds or more of the length of the grips. If the thickness of ½-in. (13-mm) wide specimens is over ⅜ in. (10 mm), longer grips and correspondingly longer grip sections of the specimen may be necessary to prevent failure in the grip section.

NOTE 9—For standard sheet-type specimens and subsize specimens the ends of the specimen shall be symmetrical with the center line of the reduced section within 0.01 and 0.005 in. (0.25 and 0.13 mm), respectively. However, for steel if the ends of the ½-in. (12.5-mm) wide specimen are symmetrical within 0.05 in. (1.0 mm) a specimen may be considered satisfactory for all but referee testing.

NOTE 10—For standard plate-type specimens the ends of the specimen shall be symmetrical with the center line of the reduced section within 0.25 in. (6.35 mm) except for referee testing in which case the ends of the specimen shall be symmetrical with the center line of the reduced section within 0.10 in. (2.5 mm).

FIG. 50—3. Rectangular tension-test specimens. (From ANSI/ASTM Standard A 370-77.)

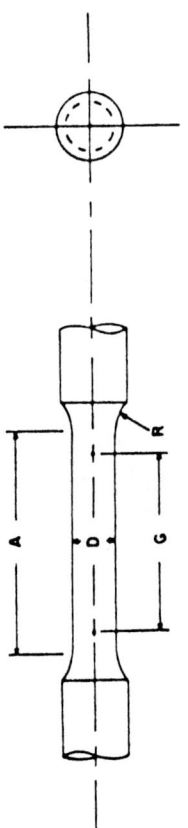

DIMENSIONS

Nominal Diameter	Standard Specimen		Small-Size Specimens Proportional to Standard							
	in.	mm	in.	mm	in.	mm	in.	mm	in.	mm
	0.500	12.5	0.350	8.75	0.250	6.25	0.160	4.00	0.113	2.50
G—Gage length	2.00 ± 0.005	50.0 ± 0.10	1.400 ± 0.005	35.0 ± 0.10	1.000 ± 0.005	25.0 ± 0.10	0.640 ± .005	16.0 ± 0.10	0.450 ± 0.005	10.0 ± 0.10
D—Diameter (Note 1)	0.500 ± 0.010	12.5 ± 0.25	0.350 ± 0.007	8.75 ± 0.18	0.250 ± 0.005	6.25 ± 0.12	0.160 ± 0.003	4.00 ± 0.08	0.113 ± 0.002	2.50 ± 0.05
R—Radius of fillet, min	3/8	10	1/4	6	3/16	5	5/32	4	3/32	2
A—Length of reduced section, min (Note 2)	2 1/4	60	1 3/4	45	1 1/4	32	3/4	20	5/8	16

NOTE 1—The reduced section may have a gradual taper from the ends toward the center, with the ends not more than 1 percent larger in diameter than the center (controlling dimension).

NOTE 2—If desired, the length of the reduced section may be increased to accommodate an extensometer of any convenient gage length. Reference marks for the measurement of elongation should, nevertheless, be spaced at the indicated gage length.

NOTE 3—The gage length and fillets shall be as shown, but the ends may be of any form to fit the holders of the testing machine in such a way that the load shall be axial (Fig. 9 of ASTM A 370—77). If the ends are to be held in wedge grips it is desirable, if possible, to make the length of the grip section great enough to allow the specimen to extend into the grips a distance equal to two thirds or more of length of the grips. (Fig. 9 is part of the original standard and is not reproduced here.)

NOTE 4—On the round specimens in Figs. 5 and 6 of ASTM A 370—77, the gage lengths are equal to four times the nominal diameter. In some product specifications other specimens may be provided for, but unless the 4-to-1 ratio is maintained within dimensional tolerances, the elongation values may not be comparable with those obtained from the standard test specimen. (Figs. 5 and 6 are part of the original standard and are not reproduced here.)

NOTE 5—The use of specimens smaller than 0.250-in. (6.25 mm) diameter shall be restricted to cases when the material to be tested is of insufficient size to obtain larger specimens or when all parties agree to their use for acceptance testing. Smaller specimens, require suitable equipment and greater skill in both machining and testing.

NOTE 6—Five sizes of specimens often used have diameters of approximately 0.505, 0.357, 0.252, 0.160, and 0.113 in., the reason being to permit easy calculations of stress from loads, since the corresponding cross sectional areas are equal or close to 0.200, 0.100, 0.0500, 0.0200, and 0.0100 in.², respectively. Thus, when the actual diameters agree with these values, the stress (or strengths) may be computed using the simple multiplying factors 5, 10, 20, 50 and 100, respectively. (The metric equivalents of these fixed diameters do not result in correspondingly convenient cross sectional areas and multiplying factors.)

FIG. 50—4. Standard 0.500-in. (12.5-mm) round tension-test specimen with 2-in. (50-mm) gage length and examples of small-size specimens proportional to the standard specimen. (From ANSI/ASTM Standard A 370-77.)

in the procurement and preparation of tension-test specimens. It is important that heating and cold working of the specimen be kept to a minimum during procurement and preparation if reliable results are to be obtained. When specimen blanks are flame cut from plates, for example, allowance should be made for machining off all metal affected by the heat introduced during flame cutting. Cold shearing or punching of specimen blanks should be performed with care, since specimens improperly prepared in this manner may become cold worked and thus not be representative of the material being sampled. Care should also be taken to insure that the test specimen is straight and flat. An initial bow or curvature in a tension specimen will result in distortion of the elastic loading line and may affect the stress level at which initial plastic yielding occurs. Removal of initial bow by a straightening operation such as bending will result in a distortion of the elastic loading line and will affect the stress level at which the initial plastic yielding occurs. In machining tension-test specimens, precautions should be taken to insure that the various portions of the specimen are symmetrical with respect to the loading axis of the specimen. If such precautions are not observed, the specimen will be loaded eccentrically, and bending stresses will be set up. Such eccentric loading will affect the elastic loading line and may influence the initial yielding behavior.

Grips used in tension testing vary considerably, depending upon the type of specimen being used. For many tests, the so-called wedge grips, which are serrated and simply grip the specimen by a transverse pressure which builds up as the specimen is loaded, are satisfactory. However, if the specimen is rather short, eccentric loading may become serious, necessitating the use of machined specimen ends and of grips with spherical bearings, an arrangement which to some extent provides self-alignment. Eccentric loading of plate or sheet-type specimens may be caused by the method used in preparing the test specimen; for example, shear or punch drag, or beads of metal deposited on the edges of the grip ends when burning out the specimen. A sketch of typical spherical seated grips is shown in Figure 50—5. For extremely accurate work, adjustable grips may be used in which the load is transmitted through a small hardened-steel ball and which permit a shift of the specimen with respect to the load application points in order to obtain a very high degree of axiality.

The Tension Test and the Properties which are Determined—In conducting a tension test, the specimen is introduced into the machine in such a manner that a load can be applied along the specimen axis, and the applied load is then gradually increased until the specimen breaks. An important factor which is generally controlled within certain predetermined limits in tension testing is the speed of testing. Some of the tensile properties, especially the yield point of low-carbon steels, are markedly affected by rate of straining, and it is important to specify and control the speed of testing within certain definite limits. These limits should be chosen in a way as to prevent more than a certain percentage variation in the property being measured from arising as a result of variations in the speed of

testing. Speed of testing may be specified in terms of rate of stressing the specimen, rate of crosshead movement, or rate of straining of the specimen. The most reliable of the three methods of specifying speed of testing is the rate of straining of the specimen, since any variations in properties as a result of variations of speed of testing arise directly from variations of strain rate in the specimen itself. In some instances the rate of crosshead movement may be closely enough related to rate of strain in the specimen that it can be used satisfactorily for control purposes.

Before considering the determination of specific properties in the tension test, it is necessary to define stress and strain, since these concepts are used in the definitions of the various tensile properties. The definitions of stress and strain by the American Society for Testing and Materials are as follows:

Stress—The intensity at a point in a body of the internal forces or components of force that act on a given plane through the point. Stress is measured in force per unit area using such units as pounds-force per square inch, kilograms-force per square millimetre, etc. However, the recommended unit in SI is the megapascal (MPa).

Strain—The unit change, due to force, in the size or shape of a body referred to its original size or shape. Strain is a non-dimensional quantity, but it is frequently expressed in inches per inch, centimetre per centime-

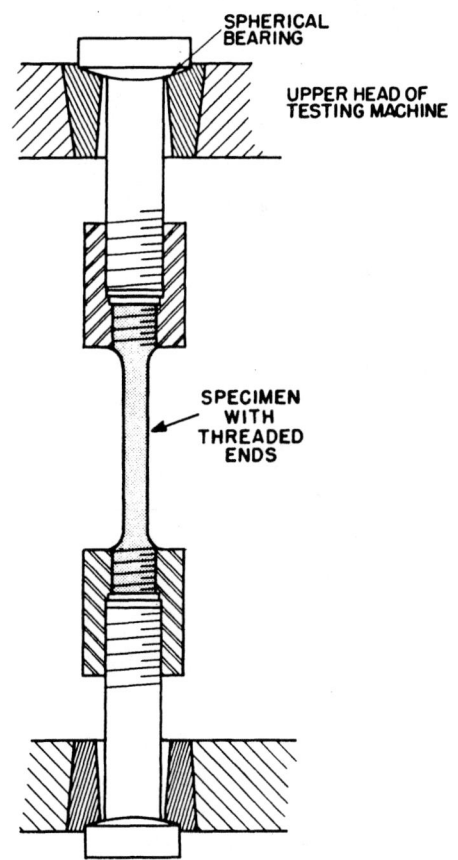

FIG. **50—5**. Sketch of self-aligning tension grips.

tre, etc.

In discussing the various properties which are determined in tension tests, these properties will be classified according to whether some aspect of strength or ductility is being measured. Further details of testing procedure will be mentioned as they apply to the measurement of specific properties.

A. STRENGTH PROPERTIES

Modulus of Elasticity (Young's Modulus)—When a metal is subjected to load, there is an initial range of loading in which no permanent deformation of the specimen occurs; i.e., if the load is removed at any value within this range, the specimen will return completely to its original dimensions. Furthermore, within this range of loading, which is designated as the **elastic range**, the strain produced is directly proportional to the applied stress; i.e., the strain produced by 138 MPa (20 000 psi) stress will be twice that produced by 69 MPa (10 000 psi). The law of proportionality between stress and strain in the elastic range is known as **Hooke's Law.** The **modulus of elasticity** is defined by ASTM as "the ratio of stress to corresponding strain for stresses below the proportional limit" and can be obtained from the slope of a plot of stress against strain within the elastic range. Because the stress-strain relations of many materials including some steels do not conform to Hooke's Law throughout the elastic range, but deviate therefrom even at stresses well below the elastic limit, four further methods of defining modulus of elasticity are illustrated in Figure 50—6: (a) **initial tangent modulus**; that is the slope of the stress-strain curve at the origin (line Os in sketch a): (b) **tangent modulus**; that is the slope of the stress-strain curve at

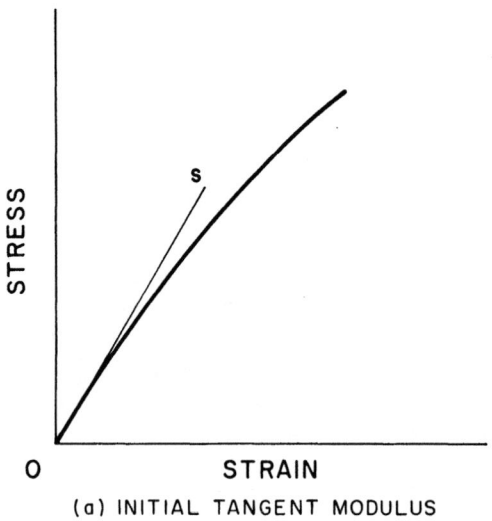

(a) INITIAL TANGENT MODULUS

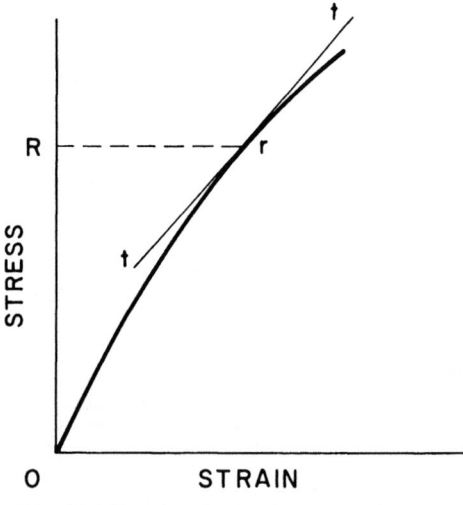

(b) TANGENT MODULUS AT ANY STRESS R

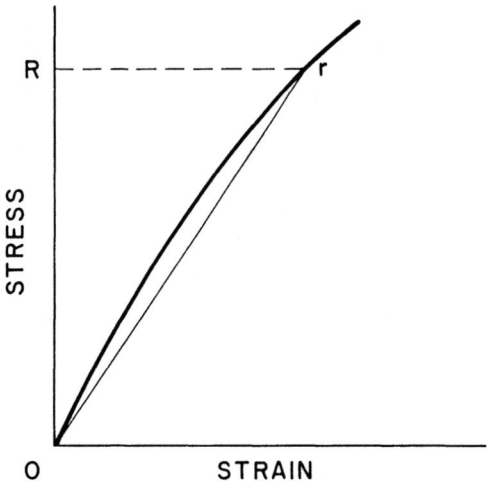

(c) SECANT MODULUS BETWEEN THE ORIGIN
AND ANY STRESS R

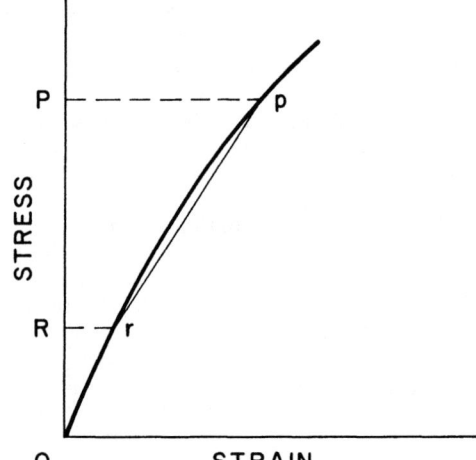

(d) CHORD MODULUS BETWEEN ANY TWO
STRESSES P AND R

Fig. **50**—6. Stress-strain diagrams showing straight lines corresponding to (a) initial tangent modulus, (b) tangent modulus, (c) secant modulus and (d) chord modulus. (From ANSI/ASTM Standard, E111-61, Reapproved 1978.)

any given stress R (line tt in sketch b): (c) **secant modulus;** that is, the slope of the secant drawn from the origin to any specified point r (usually a specified stress value) on the stress-strain curve (line Or in sketch c): and (d) **chord modulus;** that is, the slope of the chord drawn between any two specified points, r and p (usually specified stress values) on the stress-strain curve (line pr in sketch d). The modulus for ordinary steels is usually taken as 200 000 MPa (30 000 000 psi) in design work. Since the modulus does not vary much from steel to steel, it is not determined except in special instances where a more accurate value may be needed, or for example, where the effect of temperature on the modulus must be ascertained. Very great accuracy of load measurement and strain measurement is necessary for reliable modulus determinations. Furthermore, special precautions must be exercised to procure specimens free from residual stress and to insure very accurate specimen alignment.

Elastic Limit—In practice, the elastic limit is determined by subjecting a specimen carrying a strain-measuring device (extensometer) to a series of loading steps in which the maximum load applied is gradually increased, the load being released completely at each step. A load will finally be reached upon release of which the specimen will fail to return to its original length: this load is the elastic limit. The size of the load increments used and the sensitivity of the extensometer used will, of course, affect the value obtained, and, consequently, this property is seldom determined.

Proportional Limit—The proportional limit represents an aspect of elastic behavior similar to the elastic limit, the principal difference lying in the method of determination. The straight-line proportionality between stress and strain in the elastic range has already been discussed. It is the upper limit of the range of proportionality that defines the proportional limit. In other words, the proportional limit is the greatest stress which a material is capable of developing without a deviation from the law of proportionality (Hooke's Law).

In practice, the proportional limit is determined from a plot of stress against strain, being taken as the stress at the first visible departure from the straight line drawn through the points in the elastic range. Since the departure from linearity is in general quite gradual, the determined value will depend on the accuracy and sensitivity of the strain-measuring device employed in the test. The experimentally determined value for a given material will be found to decrease as the sensitivity of extensometer used is increased, that is, as the ability to detect smaller and smaller strain increments is increased. As a reference, see R. L. Templin paper listed in the bibliography at the end of this chapter. Because of these uncertainties, proportional limit is very seldom employed in specifications.

Yield Strength—As the tensile load on a specimen is increased through the elastic range, a stress will be reached at which the specimen will begin to deform in a plastic manner; i.e., it will undergo a permanent set which is not recoverable upon release of load. In the design of structural members to be subjected to static loads, it is generally necessary to design so that the service loads do not cause large deformations, since the

usefulness of the structure may be destroyed. It is for this reason that the portion of the tension test concerned with the onset of plastic yielding is of extreme importance. It has already been shown that in many materials the very first stages of plastic yielding are very difficult to detect, and that the stresses corresponding to the apparent beginning of yielding depend on the sensitivity of the strain measuring instrument used. It has become customary, therefore, to refer to the stress at which a material exhibits a specified limiting permanent set as the yield strength. The choice of the limiting amount of offset is to some extent arbitrary, but insofar as possible should be based on that amount of plastic yielding that would be considered damaging in a statically loaded member of a structure. Generally, yield strength is based on a 0.2 per cent permanent set.

Assuming that a 0.2 per cent permanent set has been chosen, the following example will illustrate the so-called **offset method** which is commonly used in determinations of yield strength. In Figure 50—7, the early portion of a stress-strain diagram extended up to about one per cent elongation is shown. Let the origin be designated as O and the elastic line and its extension as OA. Now, on the strain axis, lay off OB = 0.2 per cent, and construct BC parallel to OA. The stress corresponding to the point at which BC intersects the stress-strain curve represents the 0.2 per cent offset yield strength. The offset method is based on the observed fact that if the load is released at X, the specimen will recover along BX until at zero load, the permanent set OB remains. The procedure for any other amount of offset is identical with the exception that the offset OB will correspond to the new amount of permanent set chosen as the basis of the yield strength. The amount of offset used should always be reported with yield-strength values.

If a large number of tests, as for example in production control testing, are to be made on a material for which the stress-strain characteristics are known from experience, a shortened procedure for determination of yield strength may be used, which is known as the **extension-under-load method.** This method is based on the fact that the total extension, i.e., the sum of the elastic and plastic extensions, corresponding to the amount of permanent set chosen as a basis for the yield strength determination will be known within certain fairly narrow limits. Since for a given material the stress corresponding to the yield strength will not ordinarily vary more than a certain percentage, say 10 per cent, the elastic strain present at the offset on which the yield strength is based also will not vary more than 10 per cent. This 10 per cent variation in elastic strain will be a small fraction of the total strain at 0.2 per cent offset. For this reason, the yield strength can satisfactorily be determined at a total extension which is based on the permanent set chosen plus a mean expected value of elastic strain. An extensometer reading to 0.0025 mm per mm (0.0001 in. per in.) should be used for the total-strain method.

It should be pointed out that the elastic limit and the proportional limit can be considered as special cases of yield strength, the permanent set or offset corresponding to the least permanent strain detectable with the

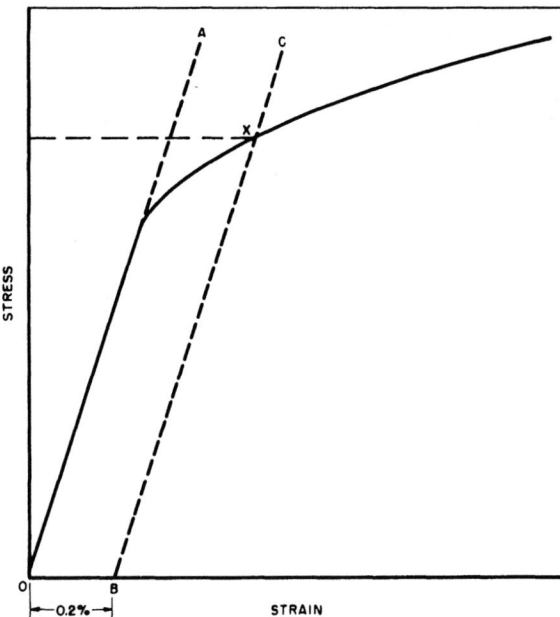

FIG. 50—7. Diagram illustrating offset method for determining yield strength.

instruments used.

Yield Point—It is only for those materials that with increasing stress show a gradual departure from elastic behavior and thus exhibit a stress-strain curve such as shown in Figure 50—7 that it is necessary to define the onset of plastic yielding in terms of yield strength. Many steels exhibit a rather abrupt yielding and may show an initial increase of strain without any appreciable increase of stress when yielding occurs. Such materials are said to exhibit a yield point, the yield point being defined as "the first stress in a material, less than the maximum attainable stress, at which an increase in strain occurs without an increase in stress." This definition of yield point is that presented in the ASTM Standards and forms the basis for yield-point specifications. As will be pointed out, the definition does not provide a complete description of the yield-point behavior.

Since the yield-point phenomenon is complex, an understanding of its general features is essential for the proper planning and execution of yield-point determinations. The degree to which the yield point is manifested varies widely, depending upon the grade of steel being tested and upon the thermal and mechanical history of the steel. In some steels, the yield point may appear as little more than a "jog" in the stress-strain curve, while in low-carbon deep-drawing steel whose final treatment has been box annealing, the yield-point behavior may extend over an elongation of several per cent.

The yield-point behavior can most easily be pictured by a description of the sequence of events during the yielding of a steel which shows a very pronounced yield point. If such a steel is stressed in tension, the load rises as the specimen is strained in the elastic region, and suddenly falls when the first yielding occurs. After this initial drop, with continuing elongation, the load fluctuates about some fairly constant value for a time and

then begins to rise again. The maximum stress before the sudden drop is known as the **upper yield point,** and the average value of the relatively constant stress level that follows is known as the **lower yield point.** The amount of extension which occurs between the upper yield point and the point at which the load begins to rise steadily again is called the **yield-point elongation.** The yielding process just described is very heterogeneous, different portions of the specimen successively undergoing yielding. After the first drop in load, locally depressed areas can be seen to form and to grow over the entire specimen as straining proceeds. These locally deformed areas are usually visible in the form of surface irregularities or strain markings which are referred to by a variety of terms including Lüders' lines, Hartmann lines, stretcher strains, "worms," and others. The formation of such strain markings has also been referred to as the Piobert effect, since the Frenchman Piobert was one of the first to describe the phenomenon. As the specimen is strained through the yield-point elongation range, which may amount to as much as 10 per cent, the strain markings continue to grow and gradually merge until the entire specimen has yielded. From this point on, the specimen deforms in a homogeneous manner, as opposed to the highly localized mode of deformation occurring during the yield-point elongation. Aggregate features of the yielding behavior just described are well illustrated by the stress-strain curve shown in Figure 50—8 for an annealed-last, rimmed, deep-drawing steel. Generally, the heterogeneous yielding process will be much less pronounced as, for example, is illustrated by the stress-strain curve shown in Figure 50—9 for a quenched and tempered bar of G43400 (AISI 4340) steel.

Heterogeneous yielding is associated with the presence of carbon and nitrogen in solid solution in ferrite. The carbon and nitrogen atoms are so situated that they exert an anchoring effect against the onset of plastic flow. When a sufficiently high stress is reached, the anchoring effect is suddenly overcome, and a yield point is observed.

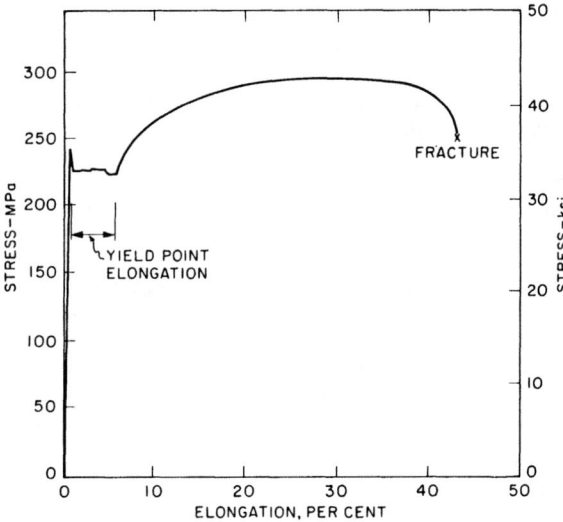

FIG. 50—8. Stress-strain curve for box-annealed rimmed deep-drawing steel.

One of the outstanding features of the yield-point behavior is the extreme sensitivity of the upper yield point to specimen preparation and testing conditions. Initial yielding can be greatly affected by non-axial loading, sharp fillets in the specimen, cold working of the specimen during procurement or preparation, the presence of residual stresses in the specimen, and the strain rate during the test. Any of these factors may be sufficiently effective to obscure the upper yield point entirely, and if the material being tested does not exhibit appreciable yield-point elongation, the influence of the aforementioned factors may eliminate all evidence of the yield point.

A number of methods of determining the yield point are recognized by the American Society for Testing and Materials and are discussed in ASTM A 370. These include the drop-of-the-beam or halt-of-the-pointer (halt-in-gage) and autographic-diagram methods for material having a sharp-kneed stress-strain diagram; and the total-extension-under-load (total-strain) method for materials that do not exhibit a well-defined proportionate deformation that characterizes a yield point as measured by the drop-of-the-beam, halt-of-the-pointer or autographic-diagram methods. The foregoing methods for the determination of yield point are described in the following quotations from ASTM A 370:

"**Drop of the Beam or Halt of the Pointer Method.**—In this method an increasing load is applied to the specimen at a uniform rate. When a lever and poise machine is used, the operator keeps the beam in balance by running out the poise at approximately a steady rate. When the yield point of the material is reached, the increase of the load stops, but the operator runs the poise a trifle beyond the balance position, and the beam of the machine drops for a brief but appreciable interval of time. When a machine equipped with a load-indicating dial is used there is a halt or hesitation of the load-indicating pointer corresponding to the drop of the beam. The load at the 'drop of the beam' or the 'halt of the pointer' is noted and the corresponding stress is recorded as the yield point." [Unless the material has a well defined yield point and some yield point elongation, this method is subjected to interpretation as to when the pointer halts and therefore, should be used with caution.]

"**Autographic Diagram Method.**—When a sharp-kneed stress-strain diagram is obtained by an autographic recording device, the stress corresponding to the top of the knee or the point at which the curve drops is to be taken as the yield point" (Figure **50**—10 left).

"**Total Extension Under Load Method**—When testing material for yield point and the test specimens may not exhibit a well-defined disproportionate deformation that characterizes a yield point as measured by the drop of the beam, halt of the pointer, or autographic diagram methods described above, a value equivalent to the yield point in its practical significance may be determined by the following method and may be recorded as yield point: Attach a Class C or better extensometer (an extensometer having maximum error of indicated strain of 0.001) to the specimen. When the load producing a specified extension is reached, record the stress corresponding to the load as the yield point,

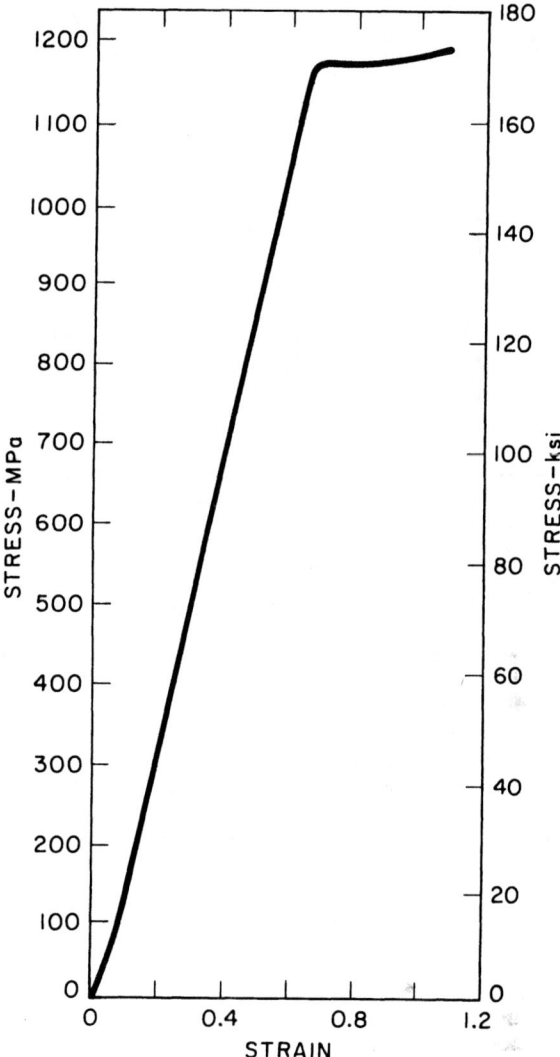

FIG. 50—9. Stress-strain curve for G43400 (AISI 4340) steel, quenched and tempered at 480°C (900°F).

and remove the extensometer." (Fig. **50**—10 right). ASTM A 370 further states that, for steels with a yield point specified not over 550 MPa (80 ksi), an appropriate value for the specified extension under load is 0.127 mm per mm of gage length (0.005 inch per inch of gage length) and that for values above 550 MPa (80 ksi) this method is not valid unless the limiting total extension is increased; for these latter materials yield strength should preferably be specified. Automatic devices are available that determine the load at the specified total extension without plotting a stress-strain curve and, according to ASTM A 370, may be used if their accuracy has been demonstrated.

Tensile Strength—As the specimen is strained past the yield point, the load rises. For ductile materials, the load passes through a maximum and fracture eventually occurs, as shown in the schematic load-extension diagram of Figure **50**—11. Some less-ductile materials may fracture while the load is still increasing, that is, without passing through a maximum. The tensile

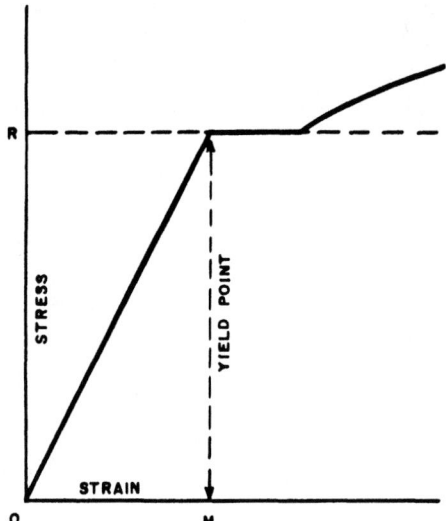
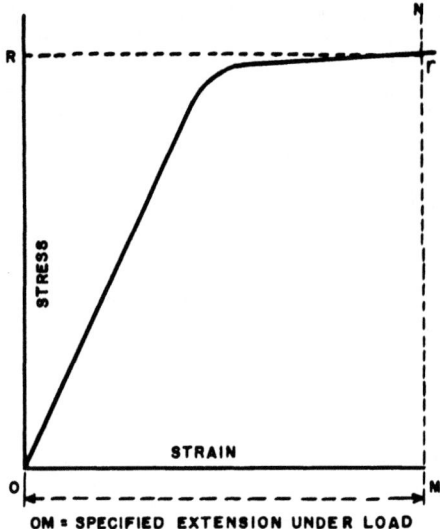

FIG. 50—10. (Left) Stress-strain curve showing yield point corresponding with top of knee. (Right) Stress-strain diagram showing yield point by extension under load method.

strength, according to the ASTM, "shall be calculated by dividing the maximum load carried by the specimen during a tension test by the original cross-sectional area of the specimen."

As already mentioned, speed of testing can affect the tension properties of a material. Considerable research has been conducted on the effects of speed of testing, and it has been generally concluded that the yield point is affected to a much greater extent than the tensile strength. It has been shown that a tenfold increase in rate of pulling increased the yield point of a 0.12 per cent carbon steel by about 49.6 MPa (7200 psi) in a 50-mm (two-inch) gage length specimen. The effect on tensile strength in the range of speeds investigated was considered to be negligible. These observations have the practical significance that, in general, the control of the speed of testing is more critical in the region of the yield point than in the later stages of the test.

B. DUCTILITY PROPERTIES

Elongation—One aspect of the ductility of a material which is generally determined in the tension test is the elongation which the material is capable of undergoing before the occurrence of fracture. The elongation is measured over some arbitrarily chosen gage length which is laid out on the test. The gage length chosen depends upon the specimen being tested, but in the United States is usually 50 mm or 200 mm (2 inches or 8 inches) for specimens with rectangular cross sections (Figure 50—3), and four times the diameter for specimens with circular cross sections (Figure 50—4). As mentioned previously, in Europe, the ratio of

$$\frac{\text{gage length}}{\sqrt{\text{Area}}} = 5.65 \text{ is used,}$$

for round specimens; the gage length is then equal to five times the diameter. After the specimen is broken, the two fractured portions are fitted together and the new distance measured, usually to the nearest 0.25 mm (0.01 inch). The percentage elongation is then calculated in the following manner:

Per Cent Elongation =
$$\frac{\text{New length} - \text{Original length}}{\text{Original Length}} \times 100$$

The original length refers to the initial gage length chosen and the new length refers to the length to which the initial gage length has been extended during the test.

The elongation at fracture of a ductile metal is not distributed uniformly along the length of the specimen. This nonuniform distribution is a consequence of **neck-**

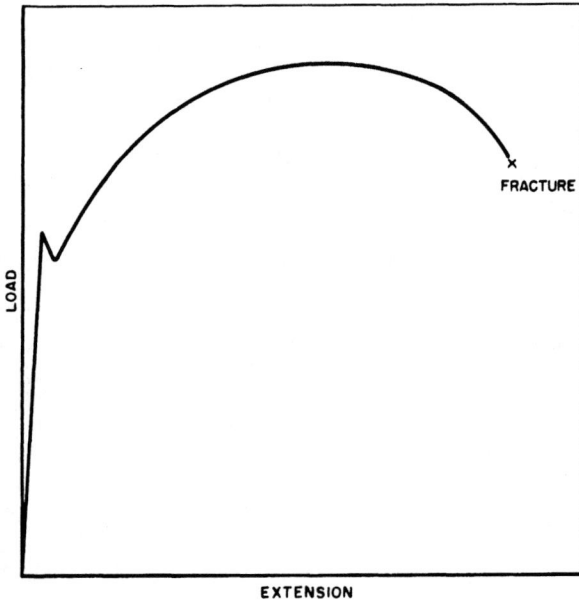

FIG. 50—11. Typical load-extension diagram.

ing down, the local elongation being greatest in the necked down region of the specimen. This behavior is illustrated in Figure 50—12, which shows the distribution of elongation over various gage lengths along a fractured specimen. Up to the maximum load, the specimen elongates in a uniform manner; at maximum load, however, the strain becomes localized and the specimen necks down. The portions of the specimen sufficiently removed from the necking zone then cease to elongate. It can be seen, therefore, that the total elongation measured at fracture over some arbitrary gage length is a sum of two components: the elongation up to maximum load (uniform elongation) and the elongation after maximum load (local elongation). The elongation will obviously vary, therefore, with the gage length over which it is measured, being greater the smaller the gage length. This variation of elongation with initial gage length is illustrated in Figure 50—13 for the same specimen represented in Figure 50—12.

When comparing elongations obtained from different sizes of specimens, it is essential that geometric similarity be observed in the comparison. In other words, the ratio of gage length to cross sectional dimensions must be held constant if comparable results are to be obtained. **Barba's Law of Similarity** states that for tension tests on different sizes of specimens from the same material, the same elongation values are to be expected only if the gage lengths are maintained in proportion to the square root of the cross-sectional area of the specimens. A relationship that is used to convert elongation obtained using a specimen having a ratio of gage length to square root of the area other than 4.47 to that which probably would have been obtained if the ratio had been 4.47 is presented in ASTM Method A 370; the conversion given is limited to carbon, carbon-manganese, molybdenum and chromium-molybdenum steels within the tensile-strength range of 275 to 585 MPa (40 to 85 ksi) and in the hot-rolled, in the hot-rolled and normalized, or in the annealed condition with or without tempering.

Reduction of Area—The reduction of area provides a measure of the ultimate local ductility of a material up to the instant of rupture. In determining the reduction of area, the fractured tensile specimen is fitted together, and the dimensions at the minimum cross-section are measured. From the original and final areas, the percentage reduction of area is calculated in the following manner:

Per Cent Reduction of Area =
$$\frac{\text{Original Area}-\text{Final Area}}{\text{Original Area}} \times 100$$

Per cent reduction of area is most accurately determined from round specimens. Since considerable restraint to plastic flow occurs at the corners of rectangular specimens, the outline of the fractured surface is not rectangular and, hence the final area is more difficult to determine.

C. SIGNIFICANCE OF THE TENSION TEST

In addition to providing engineering design data, the tension test is used to a very great extent as an empirical test for evaluating the suitability of materials for

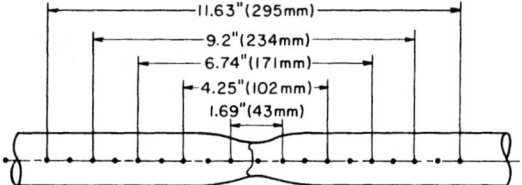

FIG. 50—12. Distribution of elongation along fractured tension specimen. (Original spacing between gage marks, ½ inch.) (From "The Mechanical Testing of Metals and Alloys," by P. Field Foster, page 76. Published by Sir Isaac Pitman & Sons, Ltd., London, 1948.)

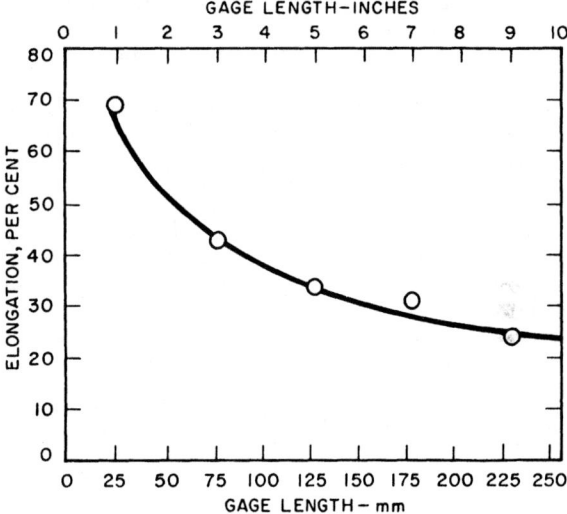

FIG. 50—13. Per cent elongation as function of gage length for fractured tension specimen. (From "The Mechanical Testing of Metals and Alloys," by P. Field Foster, page 77. Published by Sir Isaac Pitman & Sons, Ltd., London, 1948.)

particular mechanical applications. The test is interpreted in such instances on the basis of a wealth of practical experience, which permits an engineering opinion of the probable performance of a material to be drawn according to the tensile properties of materials known to have performed satisfactorily in the past. Many tensile-property specifications are drawn up in this manner and may not be used directly for actual design purposes.

Traditionally tensile strength has been used widely as a basis for design. At present, there is a marked trend to base the engineering design of structures to be subjected to static loads upon yield point or yield strength, with the application of safety factors based on engineering judgment. The use of the yield point or yield strength as a basis for design is, of course, based on the premise that any appreciable over-all yielding of the structure will destroy its usefulness. Tensile ductility specifications are almost universally based on practical experience as to what amount of ductility in the tension test has accompanied adequate ductility in service.

With regard to the use of the tensile yield strength or yield point in design, the question may well be raised as to how such a property, determined in a simple unidirectional tensile loading, can be used in the design of structures which are to be subjected to complex loads and complex states of stress. Any state of combined stresses can always be broken up into three so-called principal normal stresses which act in three mutually perpendicular directions. These three principal stresses can be designated as S_1, S_2, and S_3, the subscripts indicating the order of algebraic magnitude; i.e., S_1 is the algebraically greatest principal stress. Some examples of simple stress combinations are listed below and are illustrated in Figure 50—14. Tensile stresses are ordinarily designated as positive, while compressive stresses are designated as negative.

Simple tension:	S_1 positive; $S_2 = S_3 = 0$
Simple compression:	$S_1 = S_2 = 0$; S_3 negative
Balanced biaxial tension:	$S_1 = S_2$, both being positive; $S_3 = 0$
Balanced triaxial tension:	$S_1 = S_2 = S_3$, all positive
Torsion or twisting:	$S_1 = -S_3$, S_1 positive and S_3 negative; $S_2 = 0$

Two principal theories of initial yielding are used in predicting the strength under combined stresses; the **critical shear stress theory** and the **critical shear strain energy theory**. Shear stresses are those stresses which tend to cause one part of a body to slip over another part, as opposed to normal stresses which act in such a way as to tend to separate the body along a plane normal to the stress direction. It is recognized that shear stresses are the important stresses in controlling plastic flow. The maximum shear stress theory of yielding states that plastic action will begin when the maximum shear stress reaches some critical value characteristic of the material, the maximum shear stress being given by half the difference between the greatest and the least principal normal stresses. If the yield point in the tension test is designated as S_0, the maximum shear stress yielding criterion can be stated as:

$$\text{Critical shear stress} = \frac{S_1 - S_3}{2} = \frac{S_0}{2}$$

As an example of the use of this relationship, the value of the greatest principal stress at yielding will be calculated for torsion, where the state of stress is: $S_1 = -S_3$; $S_2 = 0$. For purposes of illustration, it is assumed that the yield point in simple tension, S_0, in SI units is 690 MPa. It is then found that:

$$\frac{S_1 - S_3}{2} = \frac{690}{2}$$
$$S_1 - (-S_1) = 690$$
$$S_1 = 345$$

In fps units, it could be assumed that the yield point in simple tension, S_0, is 100 000 lb per sq in. It is then found that:

$$\frac{S_1 - S_3}{2} = \frac{100\,000}{2}$$
$$S_1 - (-S_1) = 100\,000$$
$$S_1 = 50\,000$$

This solution indicates that a bar which is twisted will yield at half the value of the greatest principal stress,

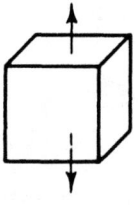
SIMPLE TENSION

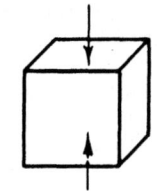
SIMPLE COMPRESSION

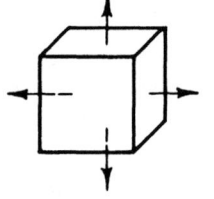
BALANCED BIAXIAL TENSION

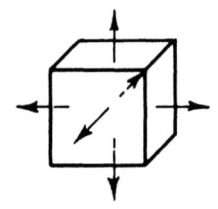
BALANCED TRIAXIAL TENSION

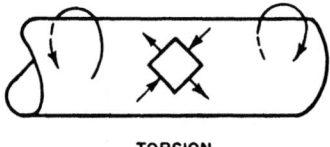
TORSION

FIG. 50—14. Illustration of simple states of stress.

S_1, at which a bar pulled in tension will yield.

The other important theory of yielding is the shear strain energy theory. As a body is loaded in the elastic range, elastic strain energy is stored up, just as when a spring is stretched. The total elastic energy stored in a body consists of two parts: that resulting from a change in volume of the body and that resulting from a change in shape of the body. The latter portion is known as the distortion energy or the shear strain energy. According to the shear strain energy criterion for initial yielding, yielding will occur when the shear strain energy reaches a critical value, which is dependent upon the material. This criterion can be expressed in terms of the three principal stresses as follows:

$$(S_1 - S_2)^2 + (S_2 - S_3)^2 + (S_3 - S_1)^2 = 2S_0^2,$$

where S_0 is again the yield point in simple tension. If the problem of the torsion of a bar of 689 MPa (100 ksi) yield point material is again considered, it will be found that the value of S_1 at yielding is 397 MPa (57.7 ksi) as compared to the value of 345 MPa (50 ksi) predicted by the maximum shear stress theory, a difference amounting to 15.4 per cent. In actual experiments conducted to test the two theories, the predictions of the shear strain energy theory are generally found to be in better agreement with the experimental results. The maximum shear stress theory is sufficiently accurate for many purposes, the difference between the predictions of the two theories never amounting to more than the 15.4 per cent cited for the example of torsion.

Two special cases exist, one in which the stress in one of the three principal directions is equal to zero, plane stress, and the other in which the strain in one of the principal directions is equal to zero, known as plane strain. Examples of the plane stress condition are that of a thin sheet loaded in tension in its own plane, and in thin rotating disks. Examples of the plane strain condition are a thick plate loaded in tension in its own plane (this is further discussed in the fracture-mechanics portion of the section on notch toughness), and thick walled tubes deformed plastically under internal pressure.

Another direction in which the significance of the tension test has been extended is the development of the concept of the so-called **true stress-strain curve.** Ludwik, a German investigator who was one of the outstanding figures in the development of better understanding of the mechanical behavior of materials, pointed out around 1910 that the definitions of stress and strain, as commonly used in materials testing and in defining engineering properties, are in some respects lacking in fundamental significance. Only in relatively recent years, however, have the concepts of Ludwik come to be widely used in this country. The principal criticism of the customary treatment of tensile data lies in the definitions of stress and strain, these quantities generally being referred to the initial dimensions of the specimen throughout a test. Such a procedure is obviously in error and has no real physical significance except at the very beginning of the test.

As a tensile specimen is stretched, its cross-sectional area diminishes progressively. Only at the beginning of the test, therefore, can stress be based on the original area without introducing appreciable, and, indeed, serious error. **True stress,** then, is defined as the instantaneous stress acting on the specimen and is computed by dividing the instantaneous load by the actual cross-sectional area at the instant that particular load is acting.

Since the area will have changed a negligible amount at initial yielding, the usual definitions of yield point and yield strength are satisfactory. Since the change in cross-sectional area at the maximum load is usually considerable, however, it must be concluded that the tensile strength is not a real stress and has no fundamental physical significance. Its principal use is for relative comparison purposes and in testing for uniformity of product.

The usual definition of strain, in which strain is referred to initial dimensions, is likewise in error. The inaccuracy of referring elongation increments to an original gage length can be visualized in the following manner. If a bar 250 millimetres (10 inches) long is stretched 25 millimetres (one inch), the elongation expressed as a percentage is 10 per cent. Now suppose that the bar, initially 250 millimetres (10 inches) long, has been stretched to a length of 2500 millimetres (100 inches) and is then stretched 25 millimetres (one inch) further. The last 25 millimetres (one inch) of stretch referred to the original length would represent an

elongation of 10 per cent. If, however, we base the elongation produced by the last 25 millimetres (one inch) increment of stretch on the length of the specimen just prior to that increment, the elongation would be considered to be 1 per cent. Obviously, the latter procedure provides a truer physical description of the stretching process than referring the strains to the initial dimensions. "True strain" has been defined, therefore, in the following manner:

$$\text{True strain} = \int_{l_0}^{l} \frac{dl}{l}$$
$$= \log_e \frac{l}{l_0}$$

where l_0 is the original length between two gage marks, l is the instantaneous length between these marks as the specimen is stretched, and e is the base of Naperian or natural logarithms. $\log_e$ is usually expressed simply as ln, so that the above expression may be written: True strain $= \ln \frac{l}{l_0}$. According to this definition, true strain is simply a summation of strain increments in which each increment is referred to the instantaneous specimen dimensions. A comparison of true strain to ordinary or nominal strain can be seen in Figure 50—15. This relationship is valid as long as the elongation is uniform within the gage length being used for the ordinary strain measurement; that is, the elongation values prior to the onset of necking. For most materials used in engineering structures, this limit rarely exceeds 0.25 to 0.30 millimetre per millimetre (0.25 to 0.30 inch per inch) nominal strain.

A stress-strain curve making use of true stresses and true strains provides a more fundamental description of the plastic behavior of a material than the ordinary nominal stress-strain curve and can, therefore, be expected to be of greater general significance. In practice, the method of determining true stress depends to some extent upon the shape of the specimen. For a cylindrical specimen, a convenient method is based on simultaneous measurements of load and minimum specimen diameter. The true strain can be calculated from area changes as well as from length changes, since the volume of the specimen does not change by a significant amount during plastic deformation. Therefore,

$$A_0 l_0 = Al$$
and
$$\frac{A_0}{A} = \frac{l}{l_0}$$

where A_0 and l_0 are the original cross-sectional area and length, A and l are the instantaneous area and length, respectively. Since true strain was defined as $\log_e \frac{l}{l_0}$, it follows that $\log_e \frac{A_0}{A}$ also represents true strain. Pointed micrometers should be used in the diameter measurements for the determination of true strain, especially after maximum load when the local constriction produced by necking down has become

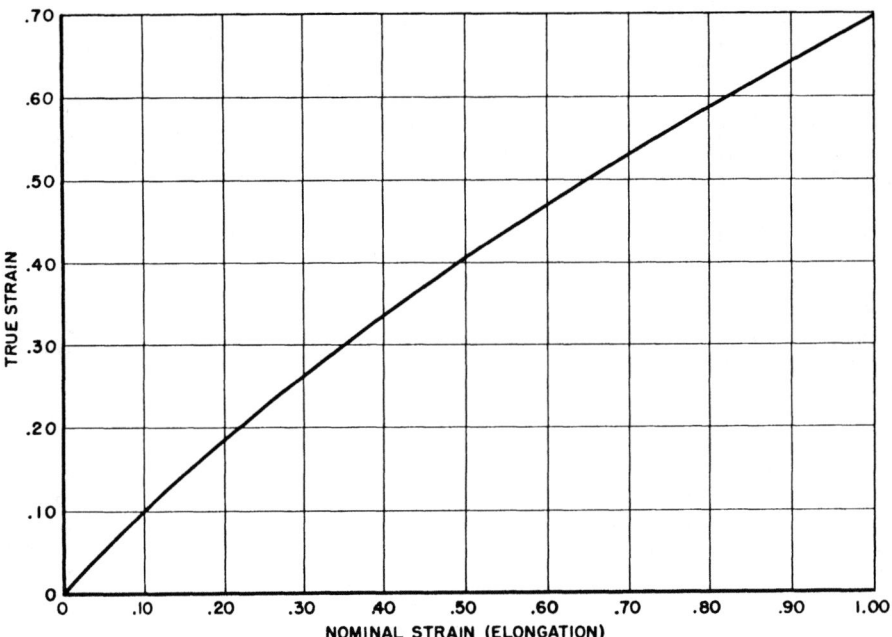

FIG. 50—15. Comparison of nominal and true strains, expressed in in. per in.

appreciable.

With thin sheet specimens, where area measurements are difficult to make, an extensometer can be used up to maximum load; i.e., during the period of uniform elongation, for obtaining length measurements to be used in calculating true strain. In order to obtain instantaneous area values for the calculation of true stress, the constancy of volume condition is used, thus permitting a computation of area changes from length changes. This procedure must not be used after maximum load because of the nonuniformity of the elongation along the gage length which arises as a result of necking down.

Using this procedure, the following equations are applicable:

$$\text{True strain} = \ln(1 + \epsilon)$$

and

$$\text{True stress} = \frac{P(1 + \epsilon)}{A_0}$$

where ϵ is the nominal strain, P is the corresponding load, and A_0 is the original area as previously defined.

An example of a true stress-strain curve is shown in Figure 50—16 along with an ordinary or nominal stress-strain curve obtained from a single test on a 9-mm (0.357-in.)-diameter specimen. The true stress-strain data were obtained from instantaneous diameter measurements, while the ordinary stress-strain data were obtained from a load-elongation diagram (the elongation was measured over a 35.6-mm (1.4-inch) gage length). It can be seen that the true stress rises continuously with increasing true strain. This is a consequence of strain hardening which is active during the entire test up to the instant of fracture. It is apparent that the ordinary load-extension diagram does not provide a clear-cut description of strain hardening. The strain-hardening characteristics of a material are of especial importance in sheet metal forming operations where the ability of the sheet to work harden and thereby pass the deformation along from element to element is critical.

The true stress-true strain curve is often represented mathematically by a power curve in the form of:

$$\sigma = K\epsilon^n$$

where K is the strength coefficient and n is the strain-hardening exponent. This relationship assumes that when the true-stress-true-strain curve of Figure 50—16 is plotted using logarithmic coordinates (either base 10 or base e), the entire curve or portions of it can be represented by a straight line or by a series of straight-line segments. The n value is useful for estimating the stretch formability of a material and is the rate of increase with increasing strain. A method for determining n value from a tension test is described in ASTM Method E 646. Other methods for determining n are in use. Generally, cold-rolled drawing quality special-killed steels exhibit a range of n values from 0.210 to 0.240. This is one example in which the true stress-strain curve, which is based on actual physically existent stresses and strains rather than fictitious ones, provides a more rational approach to the understanding of material behavior under loading.

In addition to the strain-hardening exponent, a value mentioned above, the plastic-strain ratio, r, can be determined from the tension test. Both of these quantities are used as estimates of the formability of sheet steel. The plastic-strain ratio is useful for estimating the deep-drawability of sheet steel; it is a measure of plastic anisotropy and is related to the preferred crystallographic orientations within a polycrystalline metal. As a

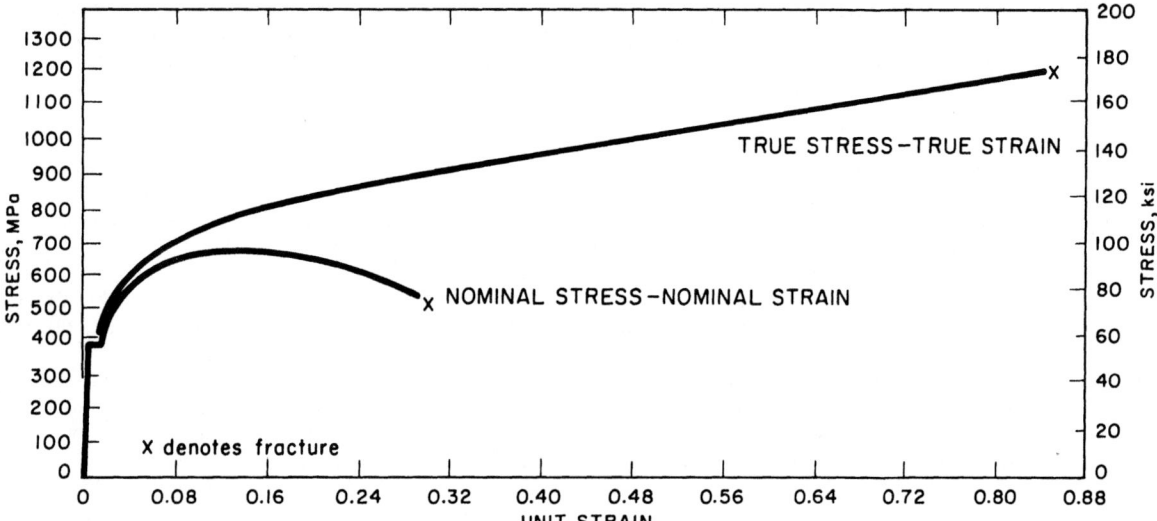

FIG. 50—16. True stress-true strain curve and nominal stress-nominal strain curve. (Prepared from data in "True Stress vs. Elongation Recorder" by D. E. Driscoll and T. S. DeSisto; Watertown Arsenal Report No. WAL 111/23, PR131104; July 15, 1955.)

measure of the drawability, it indicates the ability of a sheet to resist thinning or thickening when subjected to either the tension or compression forces in the plane of the sheet that occur during deep drawing.

The plastic-strain ratio, r (in sheet metal that has been strained by uniaxial tension sufficiently to induce uniform plastic flow but less than that at which local deformation—necking—begins) is the ratio of the true strain in the width of the tensile specimen (which is perpendicular to the direction of the applied stress) to the true strain in the thickness direction t. Thus, r is numerically equal to

$$r = \epsilon_w/\epsilon_t$$

where ϵ_w and ϵ_t are the true strains in the width and thickness directions, respectively. For some sheet steels, r has been found to vary with the amount of plastic flow. Usually, r is determined with the specimen axis oriented 0°, 45° and 90° with respect to the rolling direction of the sheet. Since r measured in a single direction generally has little significance, a weighted average value, R_m, is used as a measure of normal anisotropy (i.e., normal in the sense of orientation with respect to the sheet surface). This value is calculated from

$$R_m = (r_0 + 2r_{45} + r_{90})/4$$

where the numerical subscripts are the angular inclination of the longitudinal specimen axis with respect to the sheet rolling direction. Typical values for R_m are shown in Table 50—I. A measure of the planar anisotropy is Δr. This is a measure of whether a sheet will draw nonuniformly, that is, form ears on deep-drawn cylindrical parts. The value of Δr can be calculated from the following formula

$$\Delta r = (r_0 + r_{90} - 2r_{45})/2$$

Typical values for Δr are also shown in Table 50—I. In general, when Δr is positive, ears of approximately equal height form at 0° and 90° (with respect to the

Table 50—I. Approximate Values for Plastic-Strain Ratio of Sheet Steels.

Steel	R_m	Δr
Hot-Rolled Sheet	~1.0	~0.0
DQ Sheet	~1.2	~+0.4
DQ SK Sheet	~1.6	~+0.6
Tin-Mill Products	Generally less than cold-rolled sheet.	Generally less than cold-rolled sheet and approximately zero.

sheet rolling direction), while when Δr is negative, the ears form at 45°. As can be seen from Table 50—I, r_0 and r_{90} are usually greater than r_{45} for cold-rolled and annealed sheets, whereas, r_{45} is about the same as r_0 and r_{90} for hot-rolled sheet. A direct method for determining the plastic-strain ratio is described in ASTM Method E 517. Indirect methods have been developed that measure changes in crystallographic phenomena. One such method is Modul-r which relates changes in the resonant frequency of small specimens cut from a sheet at 0°, 45° and 90° to the rolling direction to the r values—this is described in the paper by Mould and Johnson in the bibliography at the end of this chapter.

The significance of the ductility properties determined in the tension test with respect to service is much more difficult to interpret. Gillette has presented a very capable discussion of this question in the American Society for Testing and Materials "Symposium on The Significance of the Tension Test of Metals in Relation to Design" (see bibliography). Tensile ductility values are rarely used in design, primarily because it is indeed seldom that the amount of ductility needed for a certain service application is known. Furthermore, even if such a value were known, it is doubtful if it could be translated into terms of ductility measured in a tension test. As already mentioned, ductility specifications are almost universally based on engineering judgment and experience. Because of the many uncertainties in interpretation with our present state of knowledge, however, it must be recognized that the significance of tensile ductility is somewhat limited.

Gillette mentions a number of categories of service which require that the materials used possess a certain amount of ductility. The first example is that of plastic forming, expecially where operations such as bending or deep drawing are involved, and the deformations may be sufficiently large to substantially exhaust the capacity of the material to deform. In attempting to apply tension-test ductility data to analyses of formability, it is important that care be taken to interpret the ductility data in the proper manner. Many attempts have been made to correlate the percentage elongation in 50 millimeters (2 inches) in the tension test with formability, and the general lack of success is well known. The total elongation in the tension test can only be considered as a rough indication of relative ductility to be expected in actual drawing operations. One obvious reason for the failure of the total elongation in the tension test to correlate with drawability lies in the fact that, as already pointed out, the total elongation value includes both the uniform and the local or necking elongation. In a stretching type of sheet metal forming operation, it is more logical to expect a correlation of uniform elongation with performance. Once a pronounced necking down or localization of deformation occurs in a forming operation, the useful limit of elongation has been reached, since the material at positions removed from the neck can no longer contribute to the total deformation needed to successfully form the part.

The problem of bending provides another example of the lack of fundamental significance of total elongation in the tension test. In bending, the local elongation at fracture is of critical importance, and there is some evidence that the reduction of area in the tension test, which is a measure of the ability to deform locally, correlates with the ability of a material to be bent. This correlation is based on the fact that the peak elongations at fracture in the tension fibers of a bend approach the values of the local elongation in the tension test for the same material, provided the comparison is made on a true strain basis.

The second example of service cited by Gillette, which involves ductility, is that in which the normal service calls for plastic extension. In many structural applications, a small amount of local plastic extension may be very important in relieving local stress concentrations, thereby possibly preventing rupture. Readjustment of local stresses by local plastic flow may also occur in cyclic loading applications. Although the availability of sufficient ductility to permit these readjustments is very important, it is extremely difficult to predict just what amount of ductility is adequate and how it should be measured; i.e., by elongation over some definite gage length or by reduction of area. Again, the engineer must call upon his past experience and best judgment.

Ductility is sometimes demanded as an "insurance factor," that is, extra protection in the event of accidents or overloads. The ability to deform, locally, especially as expressed by test values for the reduction of area, and the ability to absorb overloads without rupture may be desired, although again the necessary amount of ductility may be unpredictable. Actually, what is required in this class of application is the ability to absorb energy, or toughness, an attribute which depends not only on ductility but on strength as well. Toughness will be discussed in more detail in a subsequent section dealing with impact testing and notch toughness.

In considering the significance of ductility, it is important to recognize that ductility in the tension test, especially as measured by elongation or by reduction of area, is strongly dependent upon microstructure and is considered, therefore, to be a "structure sensitive" property. The tensile and yield strength, on the other hand, are less structure-sensitive. Tensile ductility, therefore, may provide a useful means of detecting the presence of undesired microstructures, particularly in heat-treated steels, which could result in inferior mechanical performance.

As will be discussed in the section on fatigue, attempts are being made to relate the tensile properties to the fatigue behavior of metals.

In summary, it can be said that the tension test, if properly conducted and interpreted, is an informative and versatile test, providing information on both the strength and ductility properties of materials. In addition to the direct application of some of the tensile properties in design, practical experience built up around the tension test makes it useful in specifying materials for particular applications as well as in the control of the uniformity of material supplied for those applications.

SECTION 3

HARDNESS TESTING

Hardness is a material characteristic which can perhaps best be defined in terms of resistance to deformation. The degree of hardness of a material can be manifested in a number of different ways depending upon the conditions to which the material is subjected. In metals, the most commonly used measure of hardness depends upon the resistance to penetration by a much harder body. Hardness may also be manifested as a resistance to abrasion or wear, as a resistance to cutting, as a resistance to crushing, as a resistance to deformation as in tension or compression, as a manifestation of resilience, i.e., rebound hardness, and others. In this discussion, attention will be focused on the type of hardness tests which measure the resistance to penetration under certain specified conditions, since these are by far the most widely used.

The extremely wide use of hardness tests, especially in conjunction with the making, shaping, and treating of steel, can be attributed not only to the extreme simplicity of sample preparation and test procedure, but also to the close relationship between hardness and other mechanical properties. The best example of the correlation between hardness and other mechanical properties is provided by quenched and tempered steels, where a hardness measurement permits a good estimate of most of the other mechanical properties. Hardness tests are especially well adapted to checks of uniformity of product, because of the great ease with which they can be made. If a process or treatment passes out of control, the departure from uniformity can frequently be detected in hardness changes in the product. It is primarily because of this usefulness in control of uniformity that hardness testing is used so extensively in the steel industry.

The two hardness tests used most widely are the Brinell test and the Rockwell test, each of these tests having been standardized by the American Society for Testing and Materials. Because of their universal usage, these two tests will be described in considerable detail. Other types of hardness tests which have certain specialized applications will also be discussed briefly.

THE BRINELL HARDNESS TEST

In the Brinell hardness test, which was proposed by Dr. J. A. Brinell of Sweden around 1900, a spherical ball, usually made of hardened steel, is forced into the specimen under a definite static load. The size of the resulting indentation provides a measurement of hardness as it is manifested under the particular conditions of the Brinell test.

A **Brinell hardness tester** consists of a device for applying a predetermined static load to the specimen through the indenter. One of the most commonly used types of Brinell machines is shown in Figure 50—17. In this machine, the load is applied hydraulically, and a weighted yoke is provided to prevent the maximum load desired from being exceeded. The yoke, which carries weight proportional to the desired load, acts on a small piston in the hydraulic system. As the pressure is increased by pumping, the load on the indenter and on the yoke piston will gradually increase until the yoke and weight float, indicating that the desired load has been attained. As long as the weights float, the load will remain constant. A Bourdon tube is usually provided for the purpose of giving an indication of the rate of loading and the approach to the desired testing load. If the parts of the hydraulic system are well made and the pistons accurately lapped, this type of machine will provide a very accurate load application and is much to be preferred to the use of a Bourdon gage alone.

The standard indenter for the Brinell test used in this country is a 10-millimetre (0.39-inch) spherical ball which is usually made of hardened steel. For tests on extremely hard materials, cemented carbide balls may be employed. According to ASTM Standards for the Brinell test (A 370 and E 10), a ball must not exhibit a permanent change in diameter greater than 0.01 mm (0.0004 in.) when pressed with a force of 3000 kgf (29 421 N or 6614 lbf) against the test specimen.

Once an indentation has been made, either its diameter or depth must be measured in order to obtain the Brinell hardness number. The diameter is the usual

FIG. 50—17. Brinell hardness testing machine, with sheet-metal covers removed to show details of weight yoke, pump, etc. (Courtesy, Tinius Olsen Testing Machine Company.)

measurement and is determined by a special measuring microscope that is fitted with a glass scale graduated in tenths of a millimetre. The depth of indentation can be measured by a special device attached to the indenter. Although measurements of indentation depth can be made very rapidly, it has been found that, in general, hardness values determined on the basis of depth measurements are less reliable than those determined from diameter measurements. For this reason, measurement of depth of indentation has never been approved by ASTM as a standard method. Unless great speed is desired, therefore, the diameter of the indentation should be used in determining the Brinell hardness number.

Brinell-Testing Technique—In selecting and preparing the specimen for use in the Brinell test, a number of precautions should be observed. First of all, the specimen must be thick enough that no **anvil effect** is encountered. After a test, the side of the specimen opposite the impression must show no effect from the loading such as a local bulging. In order to avoid such effects, which may lead to fictitious hardness values, the thickness of the specimen should never be less than ten times the depth of the impression. Care should also be taken that an indentation is not made too near the edge of the specimen. The distance from the specimen edge to the center of the impression must be not less than 2.5

times the diameter of the indentation to eliminate **edge effects.** The specimen should be flat, and its surface should be sufficiently smooth that the periphery of the indentation appears sharply defined under the measuring microscope. Another precaution which should be observed is concerned with spacing of multiple indentations. A spacing of at least two and a half indentation diameters should be maintained between indentations in order to avoid testing metal which has been disturbed by a previous indentation.

In the United States, the load for the Brinell test on iron and steel is customarily 3000 kilograms; however, loads of 1500 or 500 kilograms may be used. In conducting the test, the prepared specimen is placed on the anvil, which is raised until the specimen is in contact with the penetrator ball, and the load applied as smoothly as possible. The load is held for 10 to 15 seconds in the case of ferrous materials and for at least 30 seconds for softer metals. In order to be acceptable by ASTM Standard E 10, the load-measuring device should indicate actual loads within two per cent tolerance.

In measuring the diameter of the indentation, most satisfactory results may be obtained by measuring in at least two directions and using the average in determining the hardness number. If the indentation is not circular, a directional variation in hardness may be present in the material being tested. On the other hand, out of roundness of the indentation may also indicate deformation of the indenting ball.

The Brinell hardness number is given by the quotient of the applied load and the surface area of the indentation, i.e.,

$$\text{Brinell Hardness Number (HB)} = \frac{P}{A},$$

where P is the applied load in kilograms, and A is the area of the surface of the indentation expressed in square millimetres. It is important to note that the area referred to is the actual surface area and not the projected area of the indentation. If "D" is the diameter of the ball indenter, and "d" is the diameter of the indentation as measured with the Brinell microscope, the Brinell hardness number will be given by the relationship

$$HB = \frac{P}{\frac{\pi D}{2}(D - \sqrt{D^2 - d^2})}$$

where "P" is expressed in kilograms and "D" and "d" are expressed in millimetres. Tables have been calculated from this equation for the standard test conditions of 500-kilogram, 1500-kilogram and 3000-kilogram loads and the 10-millimetre ball. Hardness numbers are tabulated for a wide range of impression diameters, so that it is merely necessary to locate in the table the diameter measured in the test and to read the corresponding number.

The Brinell hardness number followed by the symbol HB without any suffix numbers denote the following test conditions:

Ball diameter: 10 mm
Load: 3000 kgf
Duration of Loading: 10 to 15s

For other conditions, the hardness number and the symbol HB is supplemented by numbers indicating the test conditions in the following order: diameter of ball, load, and the duration of loading. For example, 63HB10/500/30 indicates a Brinell hardness of 63 measured with a 10-mm ball, and a 500-kg load applied for 30 seconds.

It may sometimes be desirable to obtain Brinell hardness values on very small specimens or thin specimens in which the standard loads and indenter would be too large to obtain a satisfactory test from the point of view of anvil or edge effects. In such instances, it is possible to obtain an approximate Brinell hardness number by reducing the size of the ball indenter, and at the same time reducing the applied load in proportion to the square of the reduction in diameter of the ball. In other words, comparable test results should be obtained with different sizes of indenting balls, provided the applied loads are in the same ratios as the squares of the diameters of the indenting balls. For iron and steel, where the standard conditions call for a 3000-kilogram load and a 10-millimetre ball, the load "P" which should be used for a ball diameter "D" will be given by the equation:

$$\frac{P}{3000} = \frac{D^2}{10^2}$$

or

$$P = 30D^2$$

A test carried out under other than the standard conditions is not considered as standard by the ASTM, but is merely recommended as an alternate if the specimen size prohibits a standard test. From Table 50—II, however, it can be seen that it is possible to obtain consistent results under a wide range of test conditions.

In considering the Brinell test, it should be remembered that the deformation of the indenter under load, in addition to a certain amount of recovery of the indented metal, prevents a perfectly spherical surface from being formed. When the indenting ball is pressed into the specimen, it tends to flatten to an extent which depends on the magnitude of the applied load, and thus creates a larger diameter of indentation than would have resulted had no deformation of the indenter occurred. This circumstance, of course, causes some error in the hardness tables, particularly at higher hardness levels. One of the principal reasons for standardizing on the size and characteristics of the indenting ball and on the magnitude of the applied load has been to circumvent this difficulty; in reality, an attempt has been made to standardize the error. In the hardness-conversion tables for steel presented in the ASTM Standards for Methods of Testing of Metals (E 140), the effect of

Table 50—II. Results of Brinell Hardness Determinations Using Various Loads and Sizes of Indenters.

Diameter of Ball, mm	Load, kg	Diameter of Impression, mm	Brinell Hardness Number
10.00	3000.0	6.300	85
7.00	1470.0	4.400	85
5.00	750.0	3.130	87
1.19	42.5	0.748	86

varying amounts of indenter deformation can be seen. Brinell hardness numbers are shown for standard steel balls and carbide balls for a 3000-kilogram load. Up to 432 Brinell hardness, the values agree for both types of balls. Above 432HB, the carbide ball deforms least and indicates the greater hardness of the two indenters. When the indicated hardness with the carbide ball is 512, it is only 500 with a standard steel ball. Above 500HB no hardness values are given for the standard steel ball since in tests on harder materials it no longer conforms to the requirements for permanent set.

For testing articles too large or unwieldy to be tested in the type of Brinell testing machine just discussed, and for testing parts of structures or for testing under conditions where the indenting force must be applied in a direction other than vertical, portable Brinell testing units have been devised that are discussed later under "Portable Hardness-Testing Units."

THE ROCKWELL HARDNESS TEST

The Rockwell hardness test, like the Brinell test, measures that aspect of hardness which manifests itself as a resistance to penetration. Because of its simplicity, accuracy, and extreme versatility, the Rockwell test is more widely used today than any other type of hardness test. A wide variety of testing conditions is available, which permits testing over a wide range of hardnesses and also permits testing of very thin materials. In the Rockwell test, in contrast to the Brinell test, the hardness numbers do not bear a mathematical relation to diameter of indentation but are dial divisions, which indicate the depth of impression. Much of the inaccuracy associated with a measurement of total indentation depth is eliminated by the use of a differential depth measurement. The indenter is first seated by a minor load, after which a standardized major load is applied. It is the increment in indentation depth produced by the major load over that produced by the minor load which provides the basis of the Rockwell hardness. In this manner, the effects of small surface irregularities and surface disturbances caused by the indentation itself are eliminated and a very reproducible measurement is made possible.

The Rockwell hardness machine which is shown in Figure 50—18 is a precision-built apparatus which permits the application of accurate, predetermined loads to standardized indenters, as well as a device for measuring the depth of indentation produced. The load is applied through a system of weights and levers, and the rate of loading is controlled by a dashpot mechanism which provides a smooth, steady application of load. A dial gage which indicates depth directly during the test is provided for measurement of indentation depth, thus eliminating a separate operation such as required in the Brinell test. On the normal Rockwell machine, one dial division is equivalent to 0.002 millimetre penetration.

(Portable units for carrying out the Rockwell-type test have been developed for testing articles or structures that cannot be handled on the ordinary stationary Rockwell machine: these are discussed later under "Portable Hardness-Testing Units.")

The penetrators which are most frequently used are

FIG. 50—18. Rockwell hardness testing machine. (Courtesy, Wilson Mechanical Instrument Div., American Chain and Cable Co., Inc.)

the spheroconical diamond penetrator and the $\frac{1}{16}$-inch (1.59-mm) spherical steel ball, which are designated as the C-scale and B-scale penetrators respectively. The C-scale penetrator consists of a conical portion with a spherical tip lapped tangent to the cone. The angle of the cone is 120 degrees, and the radius of the spherical tip is 0.200 millimetre.

Specimen preparation for the Rockwell test is somewhat more critical than for the Brinell test, since the size of the indentation is much smaller. The surface of the specimen should be smooth, clean, and dry, and if a high degree of accuracy is desired, polishing through 2/0 or 3/0 metallographic paper is advisable. The surface to be tested should be free of scale and other foreign particles. The bottom surface should be reasonably flat, parallel to the test surface, and should also be free of scale or other matter which would tend to crush under the applied load. Pitted surfaces should be avoided. The presence of oil on the test surface will also tend to cause a low reading. The thickness of the specimen should be sufficient to prevent any anvil effect. It is desirable that no effect of the indentation be evident on the back side of the specimen, although if any such effect is not too pronounced, the hardness reading may not be greatly affected. It has been found in tests on specimens of commercially pure iron of Rockwell hardness B-35 that the readings were not affected by thickness down to a thickness of 0.040 inch (1 millimetre), although the impressions showed through the specimen at 0.060-inch (1.5-millimetre) thickness. ASTM has established tables of limiting thicknesses of various hardness levels for selected Rockwell scales using $\frac{1}{16}$-inch (1.59 millimetre) diameter ball and the diamond cone indenters. Nevertheless some products such as tin plate are tested with anvil effect, however, the hardness observed is a fraction of the hardness of the anvil—the harder the anvil, the higher the apparent hardness of the material.

Table 50—III. Rockwell Hardness Scales

Scale Symbol	Penetrator	Major Load, kgf*	Dial Figures
Group One			
B	¹⁄₁₆-in. (1.59-mm) ball	100	Red
C	Spheroconical diamond	150	Black
Group Two			
A	Spheroconical diamond	60	Black
D	Spheroconical diamond	100	Black
E	⅛-in. (3.2-mm) ball	100	Red
F	¹⁄₁₆-in. (1.59-mm) ball	60	Red
G	¹⁄₁₆-in. (1.59-mm) ball	150	Red
H	⅛-in. (3.2-mm) ball	60	Red
K	⅛-in. (3.2-mm) ball	150	Red
Group Three			
L	¼-in. (6.35-mm) ball	60	Red
M	¼-in. (6.35-mm) ball	100	Red
P	¼-in. (6.35-mm) ball	150	Red
R	½-in. (12.7-mm) ball	60	Red
S	½-in. (12.7-mm) ball	100	Red
V	½-in. (12.7-mm) ball	150	Red

*Minor load all scales = 10 kgf.

The principle of the Rockwell hardness test is shown in Figure 50—19. As can be seen, the test consists of three distinct steps after the specimen has been prepared. The first step is the application of the minor load and the zeroing of the dial. The second step is the application of the major load (which varies depending upon the scale being used—see Table 50—III). The third step is the removal of the major load and reading of the dial. The application of the minor and major loads and the removal of the major load must be done smoothly and without shock. The scale of the dial is numbered in such a way that low numbers correspond to deep impressions and thus relatively soft materials, whereas a high number corresponds to a shallow impression and therefore relatively hard material.

In Table 50—III, the various standard Rockwell scales are listed. As already indicated, the "B" and "C" scales are most widely used. The B-scale is used in what might be considered a medium hardness range, and is especially useful in testing low-carbon and medium-carbon steels in the annealed condition. The ¹⁄₁₆-inch (1.59 millimetre) diameter hardened steel ball, used as the penetrator in the B-scale test, is carried in a special chuck which permits rapid change of balls. The ball must not differ by over ± 0.0001 inch (0.0025 millimetre) from its nominal diameter, and must not show a variation of over ± 0.0002 inch (0.005 millimetre) in diameter within itself.

The C-scale is the scale used most frequently for hardness above 20 HRC. There is some overlapping in a number of the Rockwell scales, and for the sake of accuracy, it is desirable to select a scale such that the test value will fall in the middle of the scale range. In making B-scale tests, the steel ball tends to flatten at hardnesses of 100 and above, while at very low hardness near 0 HRB, the impression is so deep that the cap

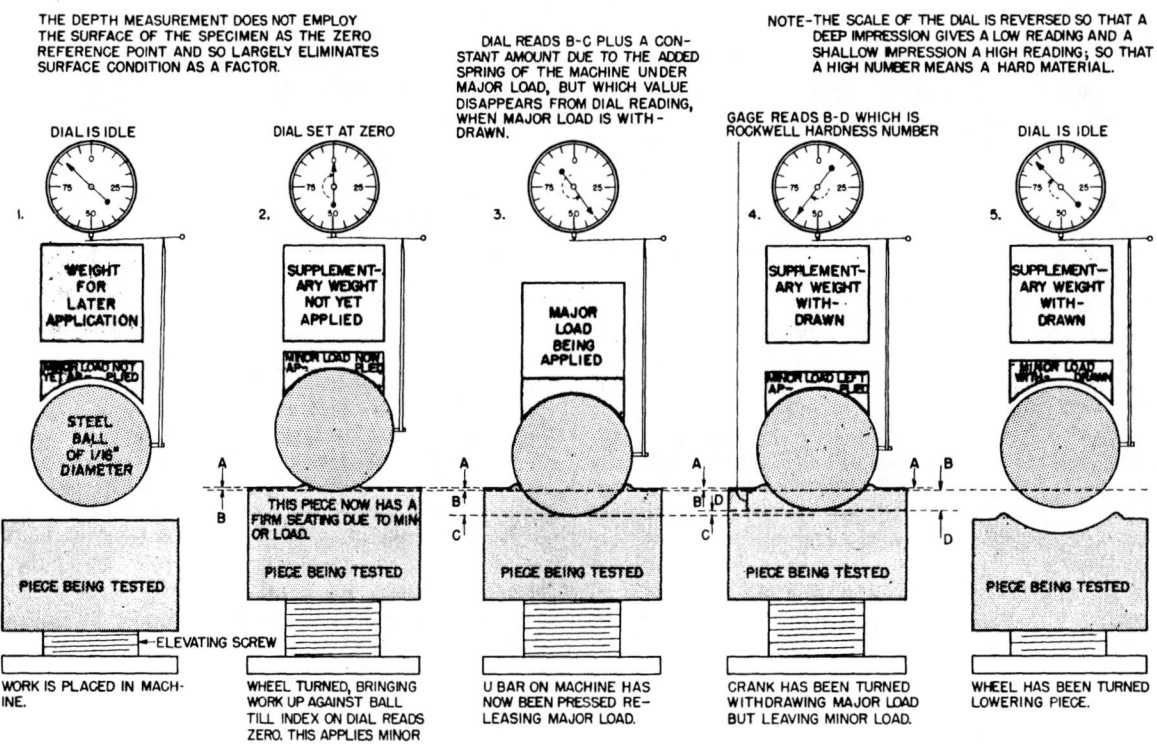

FIG. 50—19. Sketches illustrating principle of operation of Rockwell hardness testing machine. (Courtesy, Wilson Mechanical Instrument Div., American Chain and Cable Co., Inc.)

Table 50—IV. Rockwell Superficial Hardness Scales.

Major Load, kgf*	Prefix Symbols				
	N	T	W	X	Y
	Scale, Diamond	Scale, $\frac{1}{16}$-in. (1.59 mm) Ball	Scale, $\frac{1}{8}$-in. (3.18 mm) Ball	Scale, $\frac{1}{4}$-in. (6.35 mm) Ball	Scale, $\frac{1}{2}$-in. (12.7 mm) Ball
15.....	15 N	15 T	15 W	15 X	15 Y
30.....	30 N	30 T	30 W	30 X	30 Y
45.....	45 N	45 T	45 W	45 X	45 Y

*Minor Load = 3 kgf.

holding the ball may contact the specimen surface and affect the reading. Likewise, the C-scale should not be used for hardnesses below 20 HRC since the impression is so deep that inaccuracies which may exist in the upper portion of the cone may affect the readings.

Rockwell hardness numbers are always quoted with a scale symbol representing the penetrator, load, and dial used. The hardness number is followed by the symbol HR and the scale designation. For example, 64 HRC is a hardness number of 64 on the Rockwell C scale and 81 HR30N is a hardness of 81 on the Rockwell 30 N scale.

Rockwell Superficial Hardness Tests—It is frequently necessary to obtain a hardness value under conditions which prohibit the use of other than an extremely shallow indentation. The Rockwell superficial hardness tester has been developed for such applications. This test operates on the same principle as the regular Rockwell test, but utilizes much lighter loads and a more sensitive dial gage. The superficial test is particularly useful in hardness determinations on very thin strip, on nitrided or lightly carburized surfaces, and on very small parts or parts shaped in such a manner that they would collapse under the heavy load used in the standard test. The small indentation is also frequently useful in obtaining hardness readings very close to the edge of an object where an edge effect would affect the reading in a regular test. ASTM requirements for Rockwell superficial hardness tests are also found in the ASTM E 18 specification.

Two types of indenters are used in superficial hardness testing. One indenter is the same $\frac{1}{16}$-inch (1.59 millimetre) diameter ball used in the standard test, and is used in superficial tests on the softer metals, such as brasses, bronzes, and unhardened steel. The superficial hardness scales using this indenter are designated by "T". In superficial tests on harder materials, a diamond penetrator having the same configuration as the standard C-scale penetrator is used, the only difference being a more accurate finishing to final dimensions. This penetrator is designated as the "N diamond," and the corresponding hardness scales are referred to as "N" scales. Table **50**—IV shows the various superficial hardness scales, with the indenters and major and minor loads used in each. It will be noted that major loads of 15, 30, and 45 kilograms are used. The minor load in every instance is 3 kilograms. Each scale division on the hardness dial represents 0.001 millimetre penetration, as compared to 0.002 millimetre on the regular dial.

Specimen preparation for the superficial test is very critical, and polished surfaces are advisable. Specimens should be flat and free from dirt and foreign matter on both upper and lower surfaces. Test blocks are available for standardization as for the regular Rockwell test.

Superficial hardness numbers are prefixed by a number indicating the major load used and a letter designating the indenter. For example, if a reading of 42 were obtained in a test using the N-diamond and a 30-kilogram major load, the hardness would be designated by 42 HR30N. If the $\frac{1}{16}$-inch (1.59-mm) ball had been used, on the other hand, the hardness would be indicated as 42 HR30T.

THE VICKERS OR DIAMOND PYRAMID HARDNESS TEST

The Vickers hardness test is another of the class of tests which measures resistance to penetration. It is similar in principle to the Brinell test, but utilizes a different indenter and different magnitudes of loads. The indenter used in the Vickers test is a square-based diamond pyramid, and the hardness value obtained when using this penetrator is frequently referred to as the diamond pyramid hardness. The angle between opposite faces of the pyramid is 136 degrees, which was chosen so that the Vickers hardness scale would correspond approximately to the Brinell scale. This choice is based on the fact that it is recommended that loads be used in the Brinell test which result in indentations having diameters in the range of 0.25 to 0.50 times the ball diameter. The average of this range is 0.375 times the ball diameter, and if tangents are constructed to an impression of this size at the specimen surface, it will be found that the angle between the tangents is 136 degrees.

In making the Vickers test, the indenter is forced into the specimen and the diagonals of the square impression measured and averaged. From the known geometry of the indenter, the surface area of the indentation can be calculated once the diagonals have been measured. The diagonals are measured rather than the sides of the impression in order to obtain greater accuracy. The diamond pyramid hardness number is then calculated as the ratio of the applied load to the surface area of the impression. For the 136-degree square-based pyramid, the hardness can be calculated from the formula:

$$\text{Diamond pyramid hardness} = 1.854\ \frac{P}{D^2}$$

where "P" is the load in kilograms applied in making the indentation, and "D" is the average of the measured diagonals of the indentation expressed in millimetres. Vickers hardness tests are made using test loads ranging from 1 gf to 120 kgf; however, when loads between 1 gf and 1000 gf are used, the test is considered as a microhardness test and is discussed in the following subsection. The Vickers hardness number is followed by the symbol HV with a suffix number denoting the load in kilograms and a second suffix number indicating the duration of loading when the latter differs from the normal loading time of 10 to 15 seconds: for example, 402 HV50/20 indicates a hardness of 402 when a 50 kgf load is applied for 20 seconds.

The principal advantage of the diamond pyramid type of test is that geometrically similar indentations are always obtained regardless of the load applied. This useful characteristic of the impression geometry is not obtained with the spherical indenter, since the ratio of impression diameter to depth varies with the actual depth of the impression. In the discussion of the Brinell test, it was shown that, in order to make tests on thin material, it is not possible merely to decrease the load, but that it is also necessary to decrease the diameter of the indenter in proportion to the square root of the change in load in order to obtain a comparable hardness value. Since geometrically similar indentations are always obtained with the pyramidal indenter, decrease in load permits a satisfactory test on thin material, thus permitting the test to be applied over a wide range of thicknesses and over a wide range of hardnesses.

Specimen surface preparation is very important for the Vickers test, and for very light loads should approach a metallographic polish. It is also recommended that the thickness of the specimen be at least 1.5 times the diagonal of the indentation. The Vickers test is especially useful at high hardness levels because the diamond indenter deforms very little as compared to the balls used in the Brinell test.

For certain types of research work, the Vickers test has seen considerable use and has been standardized in ASTM Specifications E 92 for loads from 1 kgf to 120 kgf and E 384 for loads from 1 gf to 1000 gf (see also ASTM E110 for requirements covering portable diamond pyramid testers and the description of the Penetrascope under "Portable Hardness-Testing Units" below).

MICROHARDNESS TESTERS

Although the applications of microhardness testing are highly specialized, more and more applications are being found, and many special problems can be studied with this type of test. One of the primary uses of a microhardness tester is in fundamental studies of the hardness of various phases in the microstructure of a metal. By the use of very light loads, (usually in the range of 1 gf to 1000 gf) extremely small indentations can be placed in different phases in the microstructure and their differences in hardness determined. Extremely small scale variations in hardness such as variations across the diameter of very fine wire or across the thickness of very thin sheet can also be measured.

Several microhardness tests are commonly used. As mentioned previously, Vickers hardness tests are performed using loads as low as 1 gram force. The indenter and principles of this test are the same as those employed using the heavier loads; however, the machines are specially constructed. Knoop hardness tests are performed using loads in the range of 1 g to 1000 g also.

The Knoop indenter, a pyramidal type of indenter, was developed by the National Bureau of Standards. The indentation produced is a long, narrow, diamond-shaped impression. It is claimed that the advantage of this indenter lies in the fact that elastic recovery along the long axis of the indentation is very small, thus reducing variation from this source, which could be especially troublesome at very low loads. Hardness numbers

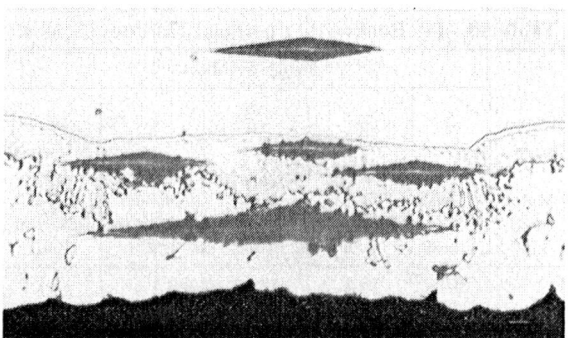

FIG. 50—20. Microhardness tests of various layers in galvanized sheet steel (Knoop indenter).

are based on the long dimension of the indentation and are calculated as the ratio of the indenting load to the projected area of the indentation. An application of the Knoop indenter to a study of the variation of hardness in the various layers of a galvanized steel sheet is shown in Figure 50—20. These tests have been standardized by the ASTM in Specification E 384.

Another principle of microhardness measurement is utilized in the **microcharacter**. The microcharacter is a small, cubically pointed diamond mounted under a very light load. A scratch is produced on a polished specimen by drawing the specimen under the diamond, and the width of the scratch is measured under a microscope. Variations in scratch width indicate variations in hardness, the narrower the scratch the higher the hardness.

PORTABLE HARDNESS-TESTING UNITS

Portable hardness testers are used principally for testing articles that are too large or unwieldy to be tested in the usual types of testing machines, for testing parts of fixed structures, or for testing under any conditions which require that the indenting force be applied in a direction other than vertical. There are two general types of portable hardness testers: (1) those that employ the indentation principle, and (2) those that operate on the rebound principle such as the Scleroscope. Among the first type are portable Brinell testers, Rockwell-type testers and diamond-pyramid testers, which are provided with various means for holding the indenter in contact with the surface to be tested; some of these means include chain clamps, "C"-clamps and magnets. However the tester is held to the piece being tested, there shall be no relative motion between the tester and the piece when the load is applied, and the tester must be mounted so that the axis of the indenter is normal to the surface of the test piece. The procedures for obtaining the indentation hardness of metallic materials with portable hardness testers are described in ASTM Method E 110.

Portable Brinell Testers—Portable Brinell testers generally apply the load by a hydraulic cylinder equipped with both a pressure gage and a spring-loaded relief valve. With this arrangement it is not possible to maintain the load at the point where the relief valve opens for any appreciable time. It is therefore necessary to bring up the load several times to the point where the pressure is released. For steel, when testing with a 3000-kg load, three load applications are equivalent to holding the load for 15 seconds as required in the standard method.

Portable Rockwell-Type Testers—Portable Rockwell type testers generally apply the load through a calibrated spring by a screw and are generally equipped with two indicators, one a dial gage that measures deflection of the spring to indicate the load, and the other a dial gage or micrometer screw to indicate the depth of penetration. The minor load is first applied as shown by the load indicator. The index on the depth indicator is set to the proper point. Then the major load is applied. The loading screw is then turned in the opposite direction until the minor load is again indicated on the load dial. The hardness is then read on the depth indicator as the difference between the readings at the minor load before and after application of the major load.

Portable Diamond-Pyramid Tester—An instrument designated as the Penetrascope is a portable hardness tester which is similar in principle to the Vickers test in that a square-based 136-degree pyramidal diamond is utilized as an indenter. In operation, a load is applied on the specimen through the indenter and the diagonals of the resulting square indentation are measured. The load is applied through a hydraulic thrust unit and can be varied through a range up to 40 kilograms. The diagonals are measured and averaged in the same manner as is done in the Vickers test.

The Shore Scleroscope Test—This test, which operates on the rebound principle, was introduced commercially about 1907, shortly after the advent of the Brinell test. It was used to a considerable extent for many years as a supplement to the Brinell test. The advantages claimed for the Scleroscope are its portability and the small size of indentation made. The portability feature permitted the testing of massive objects which otherwise could not be readily tested. A small smooth spot was prepared, the Scleroscope placed over it and a reading taken. The small impression made by the tester also made it suitable for testing finished articles on which a large indentation would have been undesirable.

The Scleroscope itself consists of a small diamond-tipped hammer enclosed in a glass tube which is provided with a suction bulb whereby the hammer may be raised to the top of the tube and dropped from a fixed and predetermined height. When the hammer is dropped on the object being tested, it rebounds to a height which is considered as a measure of hardness. If the impact were perfectly elastic, the hammer would rebound to its original height. A slight amount of energy will be dissipated in deforming the specimen, however, and this energy is not available for the rebound. A scale is provided on the instrument for measuring the height of the rebound, the units of the scale being obtained by dividing the average rebound from quenched high carbon steel into one hundred equal parts. In one model the height of rebound is determined by careful observation, while another model is equipped with an indicating dial.

In making the test, precautions must be taken that the specimen is solidly supported, the sound of the impact providing some indication of the solidity. The specimen surface must, of course, be smooth and flat. The tube of the Scleroscope must be vertical and the surface being tested must be horizontal so that true vertical impact and rebound can be obtained. For these reasons, considerable experience is required to obtain repeatable hardness values using a Scleroscope. The ASTM has prepared a standard recommended practice, E 448, covering the use of the Scleroscope.

The Scleroscope is not used as much as in the past, because the advantages once held over the standard Brinell test do not hold over newer types of tests. Furthermore, the nature of rebound hardness is not as well understood as indentation hardness. Perhaps the principal use of the Scleroscope test has been in testing the surface hardness of rolls for rolling mills.

It is of interest to note that the rebound principle has also been utilized in establishing a go/no-go gage for the hardness of steel balls. Balls are allowed to fall from an incline onto a hardened steel anvil. The balls rebound into two bins, one placed above the other, so that hard balls rebound into the top bin and softer balls rebound into the lower bin.

MISCELLANEOUS HARDNESS TESTS

File Hardness—Hardness testing with a file is an old, and crude method of measuring relative hardness, but one which can be useful in the hands of an experienced operator. A standard file is rubbed against the surface to be tested. If the file does not bite, the piece is designated as **file hard**. The test is, of course, greatly dependent upon the human factor and can only be considered useful where relatively large differences in hardness are of interest.

The Telebrineller (or Brinell Meter)—This instrument is a portable hardness tester consisting of an anvil containing a 10-mm Brinell ball that protrudes through the base of the anvil, and a 12.7-millimetre (½-inch) square steel bar, of known hardness, that is inserted into the anvil to back up the Brinell ball. To obtain a hardness number the tester is held in such a way that the ball is between the bar of known hardness and the specimen. The anvil is struck a sharp blow with a hammer and the diameters of the indentations made in the bar and the specimen are measured.

The Brinell hardness number of the specimen is then determined by multiplying the ratio of the diameter of impression in the test bar to the diameter of the impression in the specimen by the Brinell hardness number of the test bar. This calculation is done on a hardness calculator furnished with the instrument. The hardness of the bar should be approximately the same as the piece being investigated; that is, the bar should be varied so that the ratio is about 1.

The instrument is used in determining the hardness of large castings, railroad rails, pipe, etc.

Hardness Conversion Tables—It is evident from the descriptions of the various methods of hardness testing, that each test possesses certain inherent advantages. The choice of test for a particular hardness determination will depend on a number of factors, including the size of specimen and the hardness level. Frequently it may be desirable to convert a hardness reading obtained on one scale, say Rockwell "C", to some other scale, say Brinell hardness, for purposes of comparison with other data. Hardness conversion tables have been prepared for this purpose. The Society of Automotive Engineers, American Society for Metals, and American Society for Testing and Materials have jointly prepared a set of conversion data for steels harder than 220 HB, which is presented in Table **50—V**. Considerable discretion must be exercised in making hardness conversions, and particular care must be taken with regard to testing details if reliable conversions are to be made. Hardness conversion tables for other metals, including austenitic stainless steels, can be found in the ASM Handbook and in ASTM E 140.

When reporting converted hardness numbers, the measured values and the test scale shall be indicated in parentheses as in the following example: 353 HB (38 HRC).

Significance and General Utility of Hardness Tests—It has been seen that the most widely used hardness tests measure resistance to penetration under certain arbitrarily chosen conditions. In forcing a penetrator into a metal specimen, the metal in the vicinity of the penetration is plastically deformed, in order that the penetrator can be accommodated. The factors controlling the amount of distribution of this plastic deformation are very complex and no exact

quantitative analysis of the penetration of a metal by any shape of indenter has been developed. Qualitatively, however, it can be seen that resistance to penetration is a measure of resistance to plastic flow and should be related in some manner to the stress-strain curve of the material being tested, i.e., to the strength and strain-hardening properties of the material. In general for steels, a good correlation has been found to exist between hardness and tensile strength. A particularly close correlation exists in quenched and tempered steels where, in addition, yield strength and ductility can also be predicted reasonably well from hardness measurements. For this reason, the hardness test is of particular value in the field of heat-treatable steels.

In the making, shaping, and treating of steel products, the most extensive use of the hardness test is for inspection and control of uniformity. A good example of such use is provided by the extensive use of hardness tests in the production of deep-drawing sheets. It is usually known from experience that the hardness must be within a certain range if the sheet is to successfully form a certain part. Because of the great ease of making hardness tests on sheet products, very frequent tests can be made to assure that all the product is within the desired hardness range. Most uses of hardness tests are of a similar nature, depending upon a combination of ease of testing and correlation with other properties.

SECTION 4

FATIGUE TESTING

In a great many types of service applications, steel parts are required to withstand repeated or cyclic stressing; moving parts of machinery such as shafts, connecting rods, gears, etc., are examples of such applications. It has long been recognized that failures can occur in a machine part under repeated application of stresses well below those which the part is capable of withstanding under static load application. The failures which occur under repeated or cyclic stressing are referred to as **fatigue failures**. The importance of fatigue failures is well attested by the large percentage of failures in machine elements which are attributable to fatigue. It has been estimated that over 80 per cent of the failures in machines are a direct result of fatigue action.

Fatigue failures are progressive in nature, in that a crack is formed at some local spot or nucleus on the surface of the part after a certain number of load reversals, and is gradually propagated across the part. Finally, the remaining section becomes so small that it can no longer carry the applied load and complete failure ensues. A fatigue fracture generally appears brittle, even in metal which would be considered quite ductile in an ordinary tension test; the bright facets of the fracture led to the erroneous concept that the metal had "crystallized." The extremely localized nature of the fatigue failure is one of its most distinguishing characteristics and one which must be constantly kept in mind in considering the danger of fatigue.

Fatigue failures can almost always be traced to a nu-

cleus which is situated at some surface irregularity, such as a notch, a scratch, a flaw, an abrupt change in section, or a weld imperfection. It is evident, therefore, that the specific details of design and loading are of paramount importance in establishing the fatigue life of a particular part. The environment of the part is another aspect of the service conditions which is of great importance in establishing fatigue life in service. If the environment is at all corrosive, the fatigue resistance can be greatly impaired. It is obvious that the fatigue problem provides an excellent example of the importance of evaluating the actual service conditions to which a material is to be subjected, as discussed in the introduction to this chapter. Considerable data have now been accumulated which indicate that, if the actual loading conditions in the critical area of a part, are accurately determined, the fatigue life of a particular material in that part can be predicted from simple laboratory tests. This conclusion has been reached on the basis of a comparison of results of both full-scale fatigue tests on actual parts, and service performance, with results on simple laboratory specimens.

The fatigue tests which are to be discussed here are small-scale laboratory tests which are employed primarily to study materials, as opposed to tests of actual parts. Applicability of the results to service performance is subject to the considerations of the relationship between test and service conditions mentioned above. It is assumed that the tests to be discussed are at least capable of rating the relative fatigue properties of ma-

Table 50—V. Approximate Hardness Conversion Numbers for Non-Austenitic Steels (Rockwell C to Other Hardness Numbers) [1]

Rock-well C Hard-ness Num-ber	Vickers Hardness Number	Brinell Hardness Number [2]		Rockwell Hardness Number		Rockwell Superficial Hardness Number			Sclero-scope Hardness [3]	Rock-well C Hard-ness Num-ber
		10-mm Standard Ball, 3000-kgf Load	10-mm Carbide Ball, 3000-kgf Load	A Scale, 60-kgf Load, Diamond Penetra-tor	D Scale, 100-kgf Load, Diamond Penetra-tor	15-N Scale, 15-kgf Load, Superfi-cial Dia-mond Penetra-tor	30-N Scale, 30-kgf Load, Superfi-cial Dia-mond Penetra-tor	45-N Scale, 45-kgf Load, Superfi-cial Dia-mond Penetra-tor		
68	940	...	...	85.6	76.9	93.2	84.4	75.4	97.3	68
67	900	...	...	85.0	76.1	92.9	83.6	74.2	95.0	67
66	865	...	...	84.5	75.4	92.5	82.8	73.3	92.7	66
65	832	...	739	83.9	74.5	92.2	81.9	72.0	90.6	65
64	800	...	722	83.4	73.8	91.8	81.1	71.0	88.5	64
63	772	...	705	82.8	73.0	91.4	80.1	69.9	86.5	63
62	746	...	688	82.3	72.2	91.1	79.3	68.8	84.5	62
61	720	...	670	81.8	71.5	90.7	78.4	67.7	82.6	61
60	697	...	654	81.2	70.7	90.2	77.5	66.6	80.8	60
59	674	...	634	80.7	69.9	89.8	76.6	65.5	79.0	59
58	653	...	615	80.1	69.2	89.3	75.7	64.3	77.3	58
57	633	...	595	79.6	68.5	88.9	74.8	63.2	75.6	57
56	613	...	577	79.0	67.7	88.3	73.9	62.0	74.0	56
55	595	...	560	78.5	66.9	87.9	73.0	60.9	72.4	55
54	577	...	543	78.0	66.1	87.4	72.0	59.8	70.9	54
53	560	...	525	77.4	65.4	86.9	71.2	58.6	69.4	53
52	544	500	512	76.8	64.6	86.4	70.2	57.4	67.9	52
51	528	487	496	76.3	63.8	85.9	69.4	56.1	66.5	51
50	513	475	481	75.9	63.1	85.5	68.5	55.0	65.1	50
49	498	464	469	75.2	62.1	85.0	67.6	53.8	63.7	49
48	484	451	455	74.7	61.4	84.5	66.7	52.5	62.4	48
47	471	442	443	74.1	60.8	83.9	65.8	51.4	61.1	47
46	458	432	432	73.6	60.0	83.5	64.8	50.3	59.8	46
45	446	421	421	73.1	59.2	83.0	64.0	49.0	58.5	45
44	434	409	409	72.5	58.5	82.5	63.1	47.8	57.3	44
43	423	400	400	72.0	57.7	82.0	62.2	46.7	56.1	43
42	412	390	390	71.5	56.9	81.5	61.3	45.5	54.9	42
41	402	381	381	70.9	56.2	80.9	60.4	44.3	53.7	41
40	392	371	371	70.4	55.4	80.4	59.5	43.1	52.6	40
39	382	362	362	69.9	54.6	79.9	58.6	41.9	51.5	39
38	372	353	353	69.4	53.8	79.4	57.7	40.8	50.4	38
37	363	344	344	68.9	53.1	78.8	56.8	39.6	49.3	37
36	354	336	336	68.4	52.3	78.3	55.9	38.4	48.2	36
35	345	327	327	67.9	51.5	77.7	55.0	37.2	47.1	35
34	336	319	319	67.4	50.8	77.2	54.2	36.1	46.1	34
33	327	311	311	66.8	50.0	76.6	53.3	34.9	45.1	33
32	318	301	301	66.3	49.2	76.1	52.1	33.7	44.1	32
31	310	294	294	65.8	48.4	75.6	51.3	32.5	43.1	31
30	302	286	286	65.3	47.7	75.0	50.4	31.3	42.2	30
29	294	279	279	64.6	47.0	74.5	49.5	30.1	41.3	29
28	286	271	271	64.3	46.1	73.9	48.6	28.9	40.4	28
27	279	264	264	63.8	45.2	73.3	47.7	27.8	39.5	27
26	272	258	258	63.3	44.6	72.8	46.8	26.7	38.7	26
25	266	253	253	62.8	43.8	72.2	45.9	25.5	37.8	25
24	260	247	247	62.4	43.1	71.6	45.0	24.3	37.0	24
23	254	243	243	62.0	•42.1	71.0	44.0	23.1	36.3	23
22	248	237	237	61.5	41.6	70.5	43.2	22.0	35.5	22
21	243	231	231	61.0	40.9	69.9	42.3	20.7	34.8	21
20	238	226	226	60.5	40.1	69.4	41.5	19.6	34.2	20

[1] From Designation E 140 79, published by the American Society for Testing and Materials, 1916 Race Street, Philadelphia, Pa., 19103 which contains additional tables of hardness conversion numbers for other scales and other materials.

[2] The Brinell hardness numbers in boldface type are outside the range recommended for Brinell hardness testing in 3.2.2. of ASTM Method E 10, Test for Brinell Hardness of Metallic Materials.

[3] These Scleroscope hardness conversions are based on Vickers—Scleroscope hardness relationships developed from Vickers hardness data provided by the National Bureau of Standards for 13 steel reference blocks, Scleroscope hardness values obtained on these blocks by the Shore Instrument and Mfg. Co., Inc., the Roll Manufacturers Institute, and members of this institute, and also on hardness conversions, previously published by the American Society for Metals and the Roll Manufacturers Institute.

terials, and that, by properly taking into account particular service conditions, they may in many cases be directly applicable for design purposes.

Types of Fatigue Tests—Although fatigue tests formerly were known by the name of the inventor or the
manufacturer of the machine used to perform the test,
this is no longer the case. To adequately describe a
fatigue test, several of the following five test conditions
must be stated: (1) the variable being controlled (for
example, stress or load, strain, angle of bend, and so
on); (2) the relative number of cycles to failure (namely,
"low cycle" or "high cycle"); (3) whether the range of
the controlled variable is constant throughout the test
or whether it varies during the test; (4) the algebraic
ratio of two specified values of the controlled variable
during one cycle (that is, the ratio of the minimum and
maximum values); and (5) the type of loading (for example, axial, bending, torsion, biaxial, and so on). For example, a fatigue test may be performed in a machine
that always applies the same axial-load range (with the
maximum load tension and the minimum load compression and equal in magnitude to the tensile load)
throughout the test until failure occurs usually after 10^6
(one million) cycles; another test might be conducted in
a different machine that bends a specimen through a
different bend angle each cycle, with no constant relationship between the maximum and minimum angles,
but of different magnitude to cause failure to occur
within a few hundred cycles. The first test can be
termed a constant-amplitude, completely reversed,
axial-load-controlled fatigue test; the latter test can be
termed a random-amplitude, low-cycle, bending fatigue test. The algebraic ratio of the specified values of
the control variable is often called the stress ratio or the
strain ratio and is often defined as either

$$R = \frac{\text{Minimum value}}{\text{Maximum value}}, \text{ or}$$

$$A = \frac{\text{Alternating amplitude}}{\text{Mean value}}$$

Figure 50—21 defines these terms. Several specific machines are described in the following paragraphs.

A commonly performed fatigue test is the rotating-
beam test, in which the specimen is subjected to a constant bending moment while being rotated; thus, it is a
stress-controlled test. Any given fiber of the specimen
is thus subjected alternately to compression and tension stresses of equal magnitude (i.e., R = 1, A = infin-
ity). One widely used machine of this type is the R. R.
Moore machine shown in Figure 50—22. A schematic
diagram of the loading arrangement is shown in Figure
50—23A, from which it is evident that a uniform bending moment is applied over the length of the specimen.
Another type of rotating bending test utilizes a cantilever loading arrangement shown schematically in Figure 50—23B, from which it is evident that a nonuniform bending moment is applied over the length of the
specimen, although the moment is constant, and the
individual fibers are subjected to completely reversed
loading as with the R. R. Moore machine.

Another important fatigue test is the plane bending
or direct flexure test, in which the specimen is bent

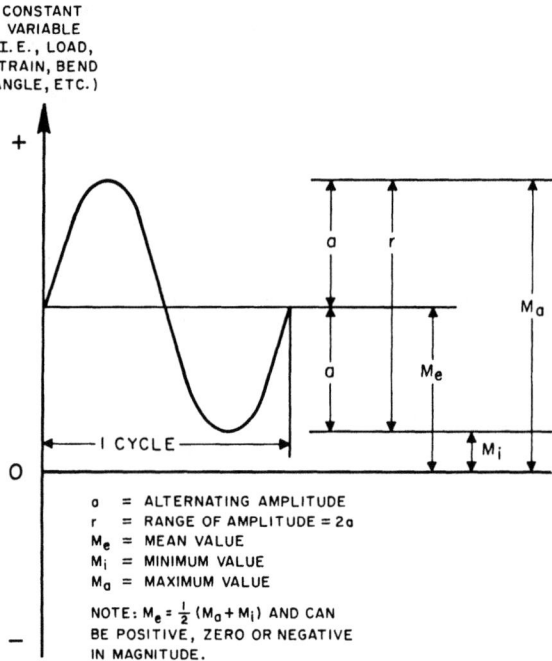

FIG. 50—21. Definition of terms for constant-amplitude test.

back and forth but not rotated. This type of test is particularly useful in the testing of specimens of flat rolled
products. The direct flexure test has the further advantage that surface preparation of the specimen is not
necessary, thus permitting the test to be made on
specimens having the actual surface to be exposed in
service. The mechanical type of machine shown in Figure 50—24 introduces the load into the specimen,
which is fixed at one end, through an adjustable crank;
since the deflection is of constant amplitude, this test is
a strain-controlled test. Generally, the mean strain is
zero, but these machines can often be adjusted so that
a nonzero mean strain can be obtained.

Direct flexure tests may also be of the resonant frequency type, in which the specimen is vibrated at its
fundamental frequency by some oscillating applied
force. Because of the characteristics of resonant
vibrations, very small forces applied at or near the resonance frequency of the specimen are capable of producing large amplitudes of vibration and correspondingly high stresses. Some resonant frequency machines
make use of an oscillating magnetic field tunable to the
resonant frequency of the specimen. The specimen is
supported at the nodes and vibrates as a free-free
beam. By taking advantage of resonance vibration in
this manner, specimens of a relatively large cross-
section can be tested which would require very large
machines if direct mechanical loading were employed.
In addition to magnetic oscillators other resonant-
frequency machines make use of mechanical
oscillators, or of hydraulic oscillators.

In recent years, considerable testing has been performed using completely-reversed (i.e., R = −1) axial
loads with strain as the controlled parameter. The
strain ranges investigated have resulted in fatigue lives
of less than 10^3 (one thousand) cycles to greater than

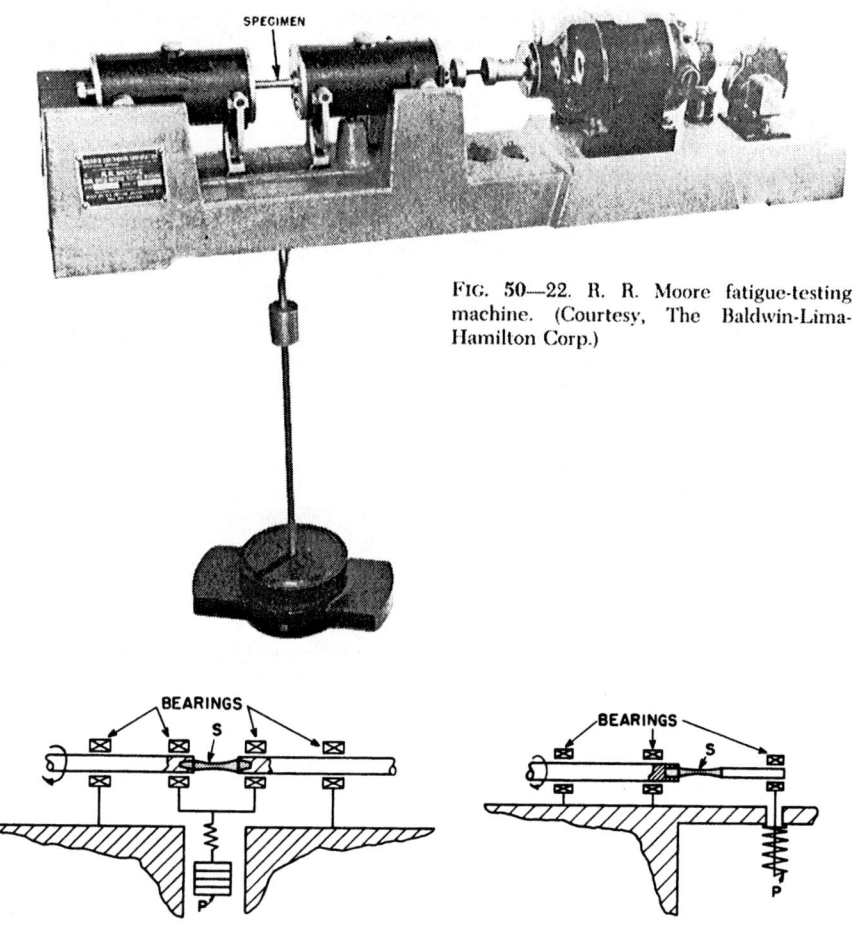

FIG. 50—22. R. R. Moore fatigue-testing machine. (Courtesy, The Baldwin-Lima-Hamilton Corp.)

FIG. 50—23. Schematic loading arrangements for: (Left) R. R. Moore and (Right) cantilever-type of fatigue-testing machines. "S" indicates specimen and "P" indicates load, in both cases. (From STP 91, "Manual on Fatigue Testing," published by American Society for Testing and Materials, 1949.)

FIG. 50—24. Krouse direct flexure fatigue-testing machine. (Courtesy, Krouse Testing Machine Company.)

10^6 (one million) cycles. This type of testing is described in the paper by Holt and Stewart and has been standardized by the ASTM as Method E 606. Axial-load tests are also performed using stress as the controlled parameter. At values of stress below the yield phenomenon of the material, the two tests are identical because the stress is proportional to the strain; however, above yield, proportionality is no longer the case and the choice of control parameters affects the results obtained.

One of the major considerations in the design of axial-load fatigue machines is the provision of a loading arrangement which will insure the application of a truly axial load. Eccentricity of loading introduces bending moments in the specimen and may have a very pronounced effect on the observed fatigue properties. Most axial-load machines are designed so that tests can be conducted in which the mean stress is not zero, i.e., in which the stress is not completely reversed, and therefore the stress ratio (or strain ratio), R, is other than −1. In many service applications, a part may not be subjected to alternate compressive and tensile stresses of equal magnitude, as is the case in the rotating beam type of test. For example, the stress may vary from zero to a maximum of 70 MPa (10 ksi) tensile stress, in which case the fatigue properties are quite different than if the loading produced a range of stresses from 70 MPa (10 ksi) in compression to 70 MPa (ksi) in tension. It should be pointed out that the direct flexure test also affords some possibility of variation in

mean stress. In some types of direct-flexure machines, it is possible to bend the specimen back and forth about an average position different from its equilibrium position, thus arriving at an average stress other than zero.

Fatigue results obtained from tests in which the specimens are bent should not be compared with results obtained from tests in which the specimens were axially loaded even if the mean stress in both cases was zero. This is illustrated in Figure 50—25. This figure also shows the effect of variation of the mean stress on fatigue results.

As already indicated, the environment to which a part is subjected can exert a profound influence on the observed fatigue behavior. If corrosive conditions are involved in a cyclic loading application, it is necessary to conduct the fatigue test under the same corrosive conditions. Corrosion fatigue is a field that is developing at the present time and it is suggested that the reader consult a reference, such as Craig, et al.

In the planning of corrosion fatigue tests, it is important to consider the relationship of testing frequency and rate of corrosion and to maintain a similar relationship to that existing in service.

Fatigue-Testing Specimen Preparation—A critical aspect of fatigue testing is the preparation of the test specimen. As already pointed out, fatigue behavior is very sensitive to surface conditions, and unless extreme care is taken in surface preparation, large scatter and unreliable results can be expected. Table 50—VI gives some indication of the effects of surface finish on ob-

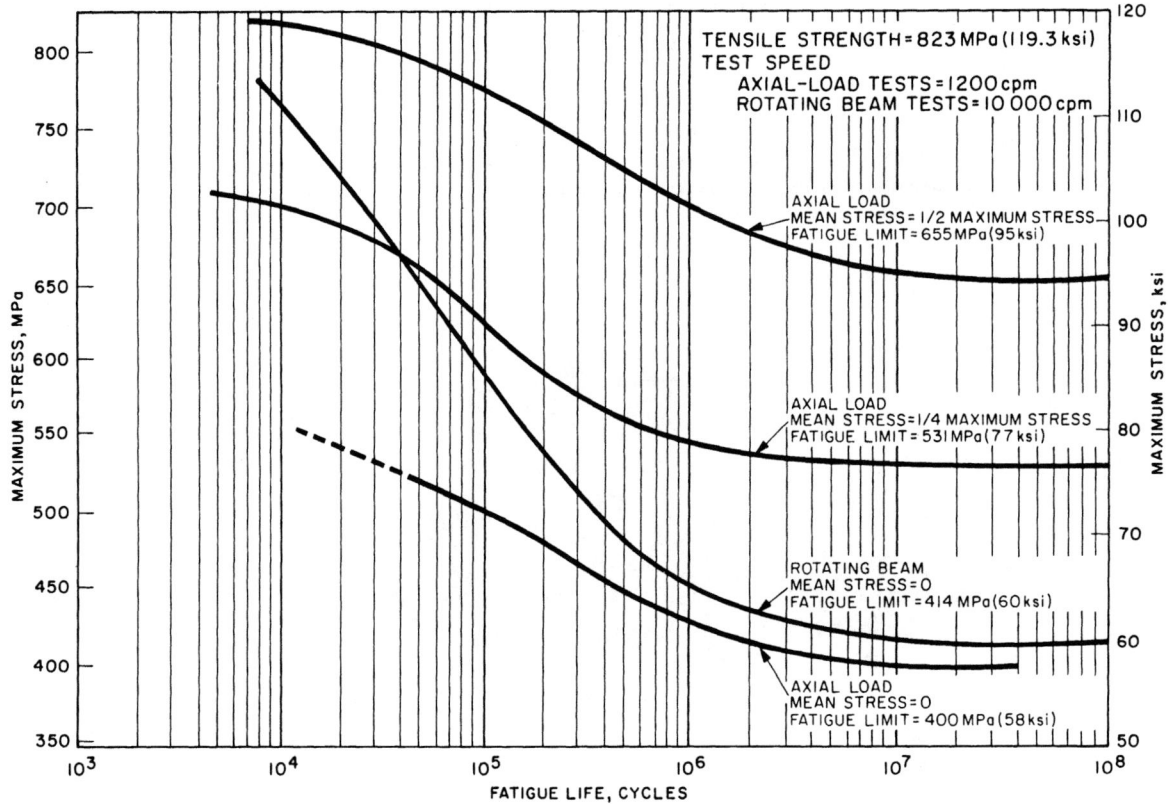

FIG. 50—25. Composite results of fatigue tests made on specimens of USS "T-1" steel under conditions indicated on the individual curves.

served fatigue limit for a steel.

In fatigue tests on notched specimens, great precautions must be taken in preparation of the root of the notch. The root of the notch should be very carefully lapped with a rotating, abrasive-bearing wire of the same radius as the notch root. Lapping should be continued until all circumferential scratches left by the machining operation have been removed.

Presentation of Fatigue-Test Data—Data on the fatigue strength of metals are generally presented in the form of so-called "S-N" diagrams, in which fatigue life is plotted as a function of the applied stress or strain. (Definitions relating to the terms used in fatigue testing may be found in ASTM E 206 and E 513.) Such a diagram has already been shown in Figure 50—25 for stress-life data; another diagram for strain-life data is shown in Figure 50—26. It is convenient to use a logarithmic scale in plotting the number of cycles, since the duration of tests at low stress levels may extend to hundreds of millions of cycles. The stress axis may be either an arithmetical or logarithmic scale although a logarithmic scale is usually used when plotting strain. Most steels exhibit what is known as a **fatigue limit**, i.e., a limiting stress below which an infinite number of stress cycles can be applied. The value of the fatigue limit is the fatigue characteristic most frequently reported for steels. In some applications, such as aircraft parts, however, steels may be used at stresses or strains above their fatigue limit in order to permit savings in weight. Such parts are removed from service after some period of use based on some selected number of cycles or based upon relationships between the fatigue crack growth rate (crack size versus number of cycles) and linear elastic fracture mechanics concepts (see discussion of fracture mechanics in Section 6 of this chapter). It is important to know, therefore, the highest value of stress or strain which the material can withstand without failure for a given number of cycles. This value is known as the **fatigue strength** (even when referring to

Table 50—VI. Effect of Surface Condition on Fatigue Limit.

Moore & Kommers, rotating beam

0.49% C quenched and tempered, 197 Brinell steel

Finish	Fatigue Limit, MPa	Fatigue Limit, psi
High polish (long.)	352	51 000
00 Emery	331	48 000
Ground	310	45 000
Smooth turned	296	43 000
Rough turned	290	42 000

Thomas, rotating cantilever

0.33% C steel

Finish	Fatigue Limit, MPa	Fatigue Limit, psi
High polish (long.)	286	41 500
FF Emery	279	40 500
No. 1 Emery	276	40 000
Coarse emery	269	39 000
Smooth file	265	38 500
Turned	252	36 500
Bastard file	245	35 500
Coarse file	228-234	33 000-34 000

strain) and is used not only for steels above the fatigue limit but also in describing the fatigue characteristics of materials which do not show a true fatigue limit, such as austenitic stainless steels. Another fatigue characteristic sometimes reported is the fatigue life at some given stress or strain level. For example, a part may be required to operate at a certain stress level. It then

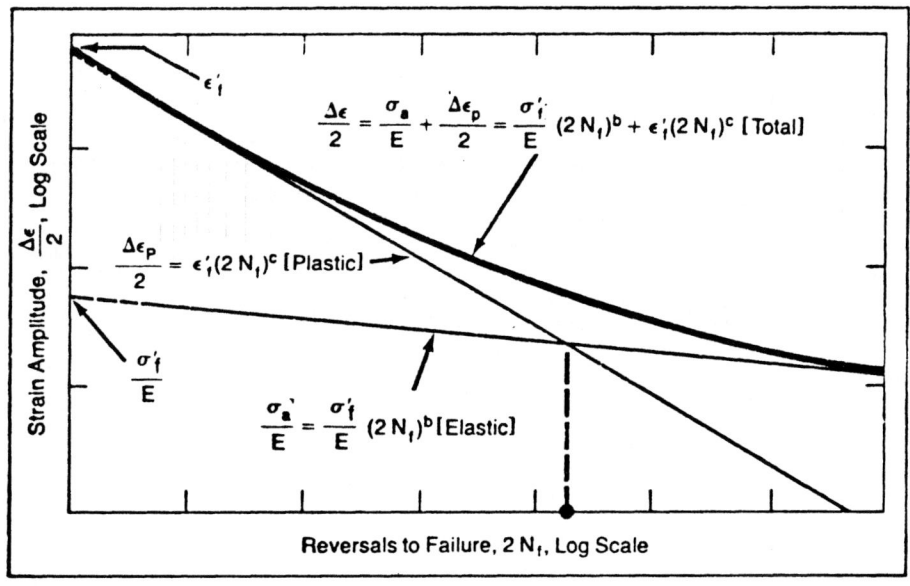

$$\frac{\Delta\epsilon}{2} = \frac{\sigma_a}{E} + \frac{\Delta\epsilon_p}{2} = \frac{\sigma_f'}{E}(2N_f)^b + \epsilon_f'(2N_f)^c \text{ [Total]}$$

$$\frac{\Delta\epsilon_p}{2} = \epsilon_f'(2N_f)^c \text{ [Plastic]}$$

$$\frac{\sigma_a'}{E} = \frac{\sigma_f'}{E}(2N_f)^b \text{ [Elastic]}$$

Strain Amplitude, $\frac{\Delta\epsilon}{2}$, Log Scale

Reversals to Failure, $2N_f$, Log Scale

Fig. 50—26. Strain-life plot.

becomes necessary to select a material which can endure this stress for the greatest number of cycles.

In experimental determination of the fatigue limit, it is customary to make the first test at a stress level well above the fatigue limit, and to gradually lower the stress level in subsequent tests until the fatigue limit is reached. (Generally, in tests near the fatigue limit, the stress in the specimen is elastic and, therefore, proportional to strain. At shorter lives, however, the stress may be above the yield strength—see Figure 50—25—and therefore the stress and strain are *not* proportional.) If the approximate value of the fatigue limit is not known beforehand, a tension test can be made, and the first fatigue test made at a stress level corresponding to about two-thirds of the tensile strength; this recommendation is valid only if the stress ratio is less than ¼ to ½ and if polished specimens are being tested. It is obviously important to make the first tests at high stress levels, since initial tests at stresses below the fatigue limit are of very little value, because the proximity of a selected stress to the value of the fatigue limit cannot be ascertained.

The S-N diagram for most steels consists of a sloping portion and a horizontal portion corresponding to the fatigue limit. For steels in a medium range of hardness, the knee of the curve will generally occur somewhere between one million and ten million cycles, so that a fatigue limit value based on ten million cycles is usually satisfactory. Fifty million cycles may be somewhat more reliable, however, and judgment based on experience with the shapes of S-N diagrams for various materials should be used in selecting the maximum number of cycles for the series of tests to determine the fatigue limit.

Strain-Life Fatigue—In recent years, design procedures utilizing the concept of strain life have been evolving. This concept is based on the fact that a zone of plastically deformed material exists as a result of almost all stress concentrations. Thus, in fatigue situations, although the structure is subjected to nominally elastic stresses, which cause elastic deformations, these elastic deformations subject the plastic zone to a strain-controlled condition. The fatigue behavior of this localized plastic zone is simulated by testing "smooth" (that is, without a notch) specimens under strain-controlled conditions (Figure 50—27).

The cyclic properties of a material are obtained by subjecting a smooth specimen to fully reversed cyclic strain amplitudes of constant magnitude. Such a fully reversed cycle results in a stress-strain hysteresis loop such as that shown in Figure 50—28.

The total strain range $\Delta\epsilon$, for a hysteresis loop is equal to twice the strain amplitude, ϵ_a (i.e., $\Delta\epsilon = 2\epsilon_a$), and the total stress range, $\Delta\sigma$, is equal to twice the stress amplitude σ_a (i.e., $\Delta\sigma = 2\sigma_a$). Moreover, the total strain amplitude can be represented as the sum of its elastic and plastic components, Figure 50—28, such that

$$\epsilon_a = \frac{\Delta\epsilon}{2} = \frac{\Delta\epsilon_e}{2} + \frac{\Delta\epsilon_p}{2} = \frac{\Delta\sigma}{2E} + \frac{\Delta\epsilon_p}{2} \qquad (1)$$

because $\Delta\epsilon_e = \dfrac{\Delta\sigma}{E}$, where E is Young's modulus.

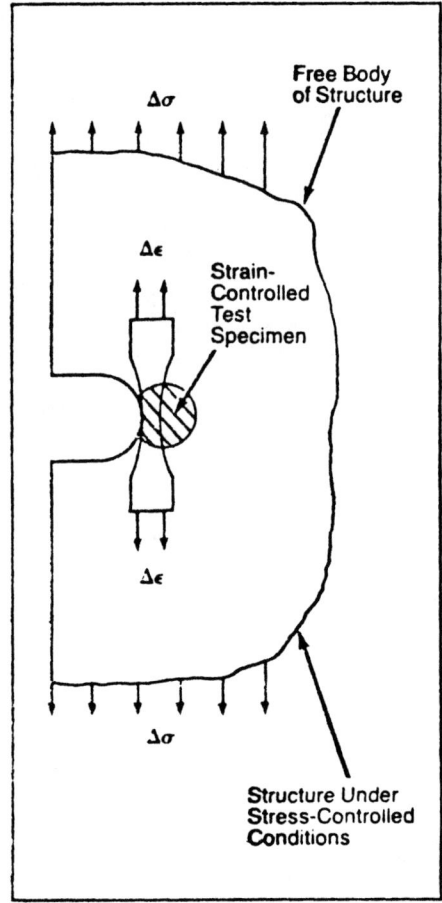

FIG. 50—27. Strain-controlled test specimen simulation for stress concentrations in structures.

A log-log plot of the stable plastic-strain amplitude $\Delta\epsilon_p/2$, versus the number of reversals to failure, $2N_f$, generally results in a straight-line relationship, Figure 50—26, given by the equation

$$\frac{\Delta\epsilon_p}{2} = \epsilon'_f (2N_f)^c \qquad (2)$$

where ϵ'_f = fatigue-ductility coefficient

        c = fatigue-ductility exponent

        N_f = number of cycles to failure; therefore $2N_f$ is equal to the number of reversals to failure.

Similarly, a log-log plot of the stable stress amplitude, $\Delta\sigma/2$, versus the number of reversals to failure, $2N_f$, results in a straight-line relationship given by the equation

$$\frac{\Delta\sigma}{2} = \sigma_a = \sigma'_f (2N_f)^b \qquad (3)$$

where

        σ_a = true stress amplitude

        σ'_f = fatigue-strength coefficient

        b = fatigue-strength exponent

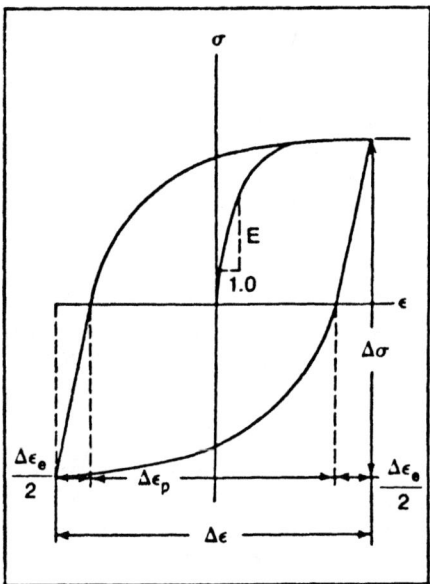

FIG. 50—28. Schematic of a stress-strain hysteresis loop.

Dividing this latter equation by the modulus of elasticity E, gives the elastic-strain amplitude in terms of the fatigue-strength coefficient, the fatigue-strength exponent and the fatigue life:

$$\frac{\Delta\epsilon_e}{2} = \frac{\sigma_a}{E} = \frac{\sigma'_t}{E}(2N_f)^b \qquad (4)$$

Combining Equations 1, 2 and 4 results in the strain-life relationship

$$\frac{\Delta\epsilon}{2} = \frac{\sigma'_t}{E}(2N_f)^b + \epsilon'_f(2N_f)^c \qquad (5)$$

which is represented as a heavy curve in Figure 50—26.

The transition-fatigue life, $2N_t$, obtained when the elastic and plastic components of the total strain are equal (i.e.,$\Delta\epsilon_p/\Delta\epsilon_e = 1$) is given by the relation

$$2N_t = \left[\frac{E\epsilon'_f}{\sigma'_t}\right]^{\frac{1}{b-c}} \qquad (6)$$

Equations 5 and 6, and Figure 50—26 show that total fatigue lives less than $2N_t$ are governed primarily by the plastic-strain amplitude (ductility), whereas the total fatigue lives greater than $2N_t$, are governed primarily by the elastic-strain amplitude (static strength). Moreover, it is claimed that the fatigue-ductility coefficient, ϵ'_f, and the fatigue-strength coefficient, σ'_t, determined from regression analysis of fatigue data, can be approximated by the true fracture ductility, ϵ_f, and the true fracture strength, σ_f, respectively, obtained from a monotonic tension test.

True stresses (σ) and true strains (ϵ) are used in the foregoing relationships; however, engineering stresses (S) and engineering strains (e) are obtained from the hysteresis loops and may be converted as follows:

$$\sigma = S(1+e)$$
$$\epsilon = \ln(1+e)$$

provided that uniform elongation is occurring. Since the engineering strains of the hysteresis loops are small ($\epsilon < 10^{-2}$) the error in assuming that $\sigma = S$ and $\epsilon = e$ is small and is therefore usually neglected.

Strain-life fatigue test terms and methods have been standardized in ASTM Definitions E 513 and Method E 606, respectively. The results of strain-life tests are discussed in the reference by Holt (1981, see bibliography at end of chapter) and the use of these data is described in AISI "Sheet-Steel Properties and Fatigue Design" and in "Fatigue Design Handbook" edited by Graham.

Significance of Small-Scale Fatigue Tests—There has been much discussion in recent years of the applicability of fatigue data derived from tests on small, highly polished bars to the prediction of fatigue life in service. One extreme point of view has been that such data are practically worthless and may even rate materials in the wrong order with respect to their service performance. This appears to be an unduly pessimistic attitude, especially in view of certain recent researches which indicate that if the service conditions are properly analyzed and if the state of stress at the critical point of a part is known, the fatigue limit of the part can be related to that determined in a simple polished-bar test. Fatigue tests on full-sized automobile parts have provided the basis for this conclusion. It should not be inferred from these remarks, however, that results of polished-bar tests should be applied indiscriminately to predictions of service behavior. Frequently, the service conditions will be so complex as to preclude the possibility of accurate analysis. In this event, the only alternative is a simulated service test.

In general, the fatigue limits of structural steels vary in a fairly regular way with tensile strength. An outstanding exception is provided by certain steels such as Bessemer steel or some of the high-strength low-alloy, phosphorus-bearing steels, in which the ratio of yield strength to tensile strength is higher than that normally found in most steels. In these steels, the fatigue limit is relatively high for a given tensile strength, which may indicate that fatigue limit is more closely related to the yield point than to tensile strength. The **fatigue ratio** is defined as the ratio of the fatigue limit to the static tensile strength and is in the neighborhood of 0.50 for most steels when determined using polished specimens tested at a stress ratio of −1 (completely reversed loading). Figure 50—29 is a schematic diagram for ordinary steels indicating the general relationship between fatigue limit and tensile strength for polished specimens, severely notched specimens, and corroding specimens.

In considering the effect of notches on fatigue, it is customary to use as a measure of the notch effect the **fatigue strength reduction factor,** which is usually designated as K_f, and is defined as the ratio of the fatigue strength of a member or specimen with no stress concentration to the fatigue strength in the presence of stress concentration. It is obvious that the strength reduction factor has no meaning except in terms of a specific notch geometry. A concept of notch sensitivity originated by Peterson (see bibliography) relates the strength reduction factor K_f to the theoretical stress concentration factor K_t resulting from the notch. K_t is the ratio of the maximum stress in a notched section to the nominal or average stress across the entire section computed from mathematical analyses or determined experimentally. Peterson defines notch sensitivity by a "q" factor which depends on the relative values of K_f and K_t in the following way:

$$q = \frac{K_f - 1}{K_t - 1}$$

Notch sensitivity according to this definition varies from zero (where $K_f = 1$) to unity (where $K_f = K_t$). This concept is useful in describing the reaction of a material to notches of varying degrees of severity. Since most actual parts subjected to cyclic loading actually contain stress concentrations of some type, it is important that the effect of notches on fatigue properties be considered.

In summary, it can be said that small-scale tests are very useful in comparing the behavior of different ma-

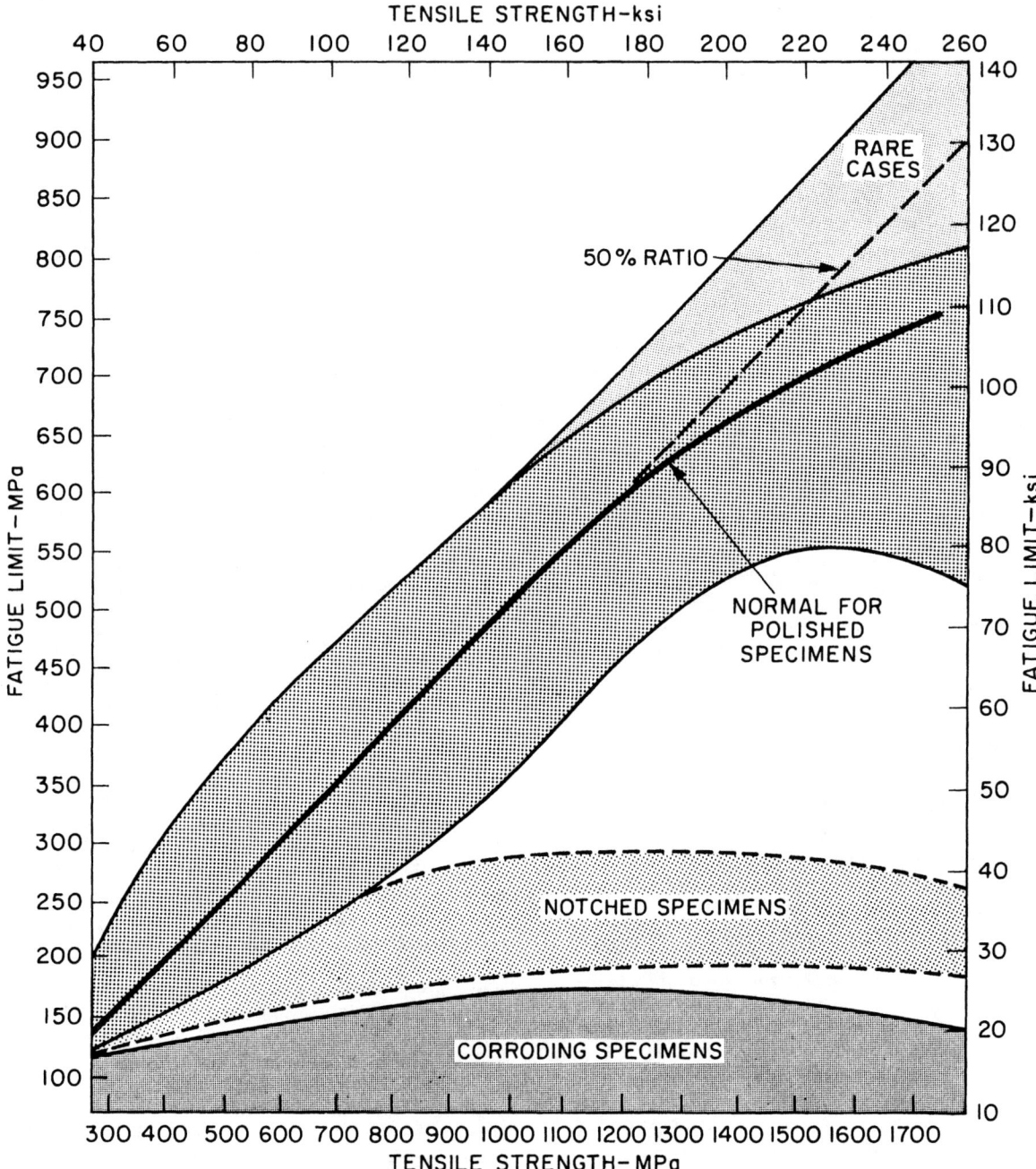

FIG. 50—29. Fatigue limit as a function of tensile strength for polished, notched and corroding specimens. (From "Prevention of Fatigue of Metals," by the staff of Battelle Memorial Institute. Published by John Wiley & Sons, Inc., New York, 1941.)

terials under repeated stress and can be used in studies of the effects of metallurgical variables on fatigue characteristics. Such tests not only form a useful guide in the development of new materials of improved fatigue strength, but also permit the evaluation of various surface treatments such as carburizing, nitriding, shot peening, cold rolling, etc. The effect of different environments on fatigue behavior also can be studied conveniently in laboratory tests.

Some aspects of statistical analysis of these data may be found in the latest edition of "A Guide for Fatigue Testing and the Statistical Analysis of Fatigue Data," ASTM STP (Special Technical Publication) 91-A.

As mentioned previously, considerable research is being expended in the area of the mechanical behavior of metals, especially in the area of fatigue. The reader is urged to peruse the current literature, so as to be aware of the changes that are occurring.

SECTION 5

FRACTURE TOUGHNESS TESTS

Beginning around 1940 the number of large monolithic structures, such as welded ships, pipelines, storage vessels, etc., increased very rapidly. The sudden and complete failure of several of these structures at stresses well below their yield strengths indicated that other considerations besides the conventional tensile properties must be included during the design of such structures. Analysis of these failures and reanalysis of previous failures indicated that the fracture usually initiated at notches. These notches were caused by (1) design features, e.g. two rigid members rigidly attached to one another, (2) fabrication procedures, e.g. weld-arc strikes, tool gouges, etc., or (3) flaws in the material, e.g. flakes, seams, weld porosity, etc. (For further information about service failures the reader is referred to Chapters XI and XII of Parker's book listed in the bibliography.) Almost all fabricated structures have notches of one type or another in them; the problem is to assess the severity of the various types of notches and the effect of a notch on the material. As a result of failures on large structures, considerable research has been and is being directed toward this problem and toward an understanding of fracture itself.

Visual observation of the fracture surfaces of a structure or member that failed in a sudden manner during service tends to show that there is little deformation, the surfaces are flat and at right angles to the plate surfaces, and they have a shiny, crystalline appearance. Similar observations made on fracture surfaces of a structure or member that failed in a ductile manner tend to have different characteristics; there are signs of yielding along the edges of the fracture, the fracture surfaces are at 45° angles with the plate surfaces, and they have a dull, fibrous appearance. Because of these observations, the sudden fractures at low stresses have tended to be known as **brittle fractures** or **low-energy fractures,** since only a very limited amount of energy is absorbed before fracture, mainly by elastic deformation. Brittle fracture literally means fracture without any plastic flow; while this condition may be closely approached in ferritic steels, minute amounts of plastic flow always occur. The appearance of the brittle, low energy fracture is caused by the individual grains separating by cleavage along definite crystallographic planes as occurs with the separation of layers of mica, or the cleavage of ordinary rock-salt crystals. Cleavage fracture in a ferritic steel occurs essentially along a single plane in each ferrite grain thus exposing a series of bright crystalline facets. The **ductile** or **shear fracture,** on the other hand, occurs after plastic deformation by a sliding action, that is, two portions of the grain separate (slip) by sliding one over the other. Because of the much lower energy requirements to cause cleavage, cleavage-type fracture, once started, can propagate over great distances at extremely high velocities of the order of a thousand or more metres (several thousand feet) per second. The energy necessary for fracture is provided by the release of elastic strain energy in the structure.

As is well known, the fracture behavior of ferritic steels is dependent upon many variables that affect the mechanical properties, including temperature, strain rate, section size, state of stress (which is directly related to section size), and notch acuity. As these variables are changed, there may be a transition from ductile to brittle behavior; i.e., a steel that is ductile and insensitive to a notch at one temperature may be brittle at a lower temperature; a steel that appears ductile and insensitive to a notch at a slow strain rate, may be brittle at a higher strain rate; and small notched specimens may be ductile while geometrically larger specimens may be brittle. The degree to which some of these mechanical-property variables influence the transition in behavior is dependent upon the strength of the steel; generally the transition behavior of ferritic steels with tensile strengths less than about 1000 MPa (150 ksi) show large effects due to variations in temperature and strain rate, while higher strength steels are relatively unaffected by these two variables. Steels of all strength levels generally are affected by the state of stress and notch acuity.

Effect of Temperature—There exists, for the lower strength steels, a temperature range in which the fracture appearance changes from the dull, oblique surface to a shiny, flat surface; at a high temperature, the entire fracture surface is ductile appearing, at a low temperature, the entire surface is brittle appearing, and at intermediate temperatures, the surface demonstrates characteristics of both types of behavior, as shown in Figure 50—30. This intermediate range is referred to as the transition-temperature range and usually some arbitrary criterion is used to select a specific temperature which is known as the **transition temperature.** Naturally, changes in the other mechanical-property

SHEAR FRACTURE, PER CENT
(measured in accordance with ASTM Method E436)

| 15 | 10 | 20 | 15 | 10 | 15 | 85 | 95 | 90 | 98 |

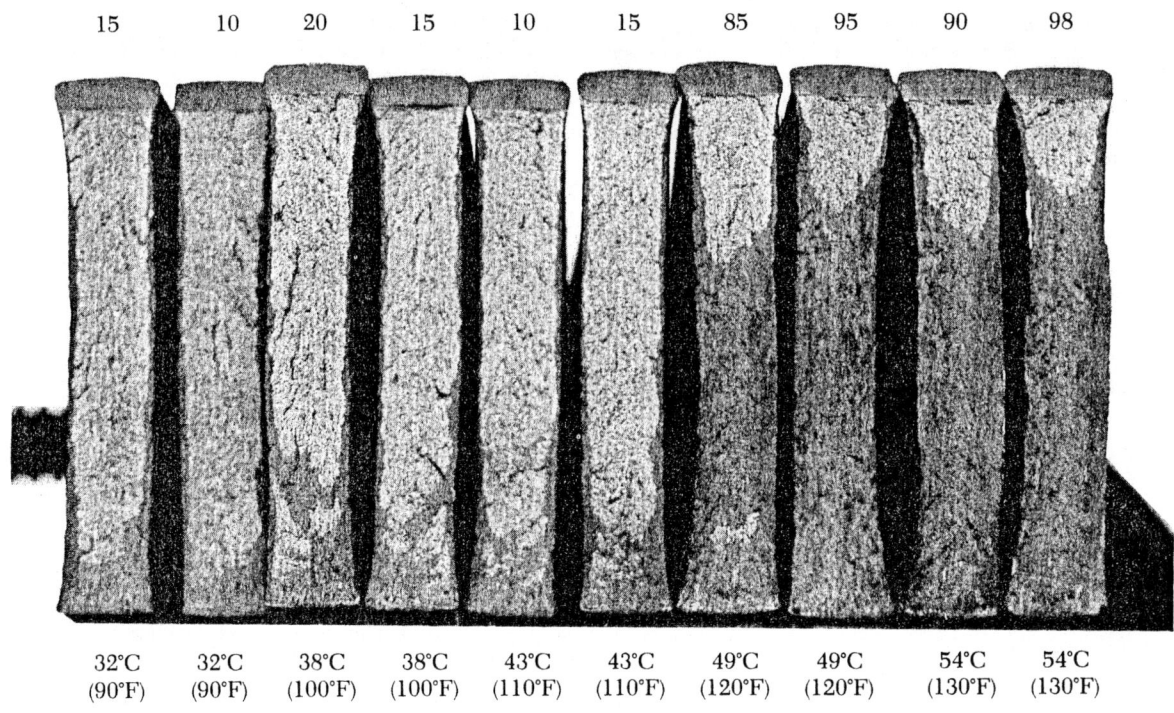

| 32°C | 32°C | 38°C | 38°C | 43°C | 43°C | 49°C | 49°C | 54°C | 54°C |
| (90°F) | (90°F) | (100°F) | (100°F) | (110°F) | (110°F) | (120°F) | (120°F) | (130°F) | (130°F) |

TEMPERATURE

FIG. 50—30. The variation in fracture appearance of drop-weight tear test specimens tested over a temperature range.

variants tend to shift the transition-temperature range to higher or lower temperatures. The effects of changing some of the mechanical-property variants are discussed below.

Effect of Strain Rate—Increasing the strain rate will increase the transition temperature of the lower strength steels since their yield points are strain-rate dependent and thus may be raised to a value equal to the fracture strength. In addition, the ductile-brittle transition becomes sharper at higher strain rates. The fracture behavior of steels with yield strengths above 1400 MPa (200 ksi) is relatively unaffected by variations in strain rate.

State of Stress—When a bar is pulled in tension so that the stress is uniform across any section, the state of stress, as previously discussed in this chapter in Section 2 on tension testing, is one of uniaxial tension and the principal stresses are σ_1, the tensile stress, and zero. The bar is free to undergo contraction, as governed by Poisson's ratio, in both the width and thickness directions. If the bar is notched, a much more complex stress system develops near the apex of the notch. The stress in the longitudinal direction of the base of the notch is raised to several times the net section stress because of the stress concentration factor; in addition, there are tensile stresses in both the width and thickness directions. This triaxial state of stress is a maximum at the midthickness of the member. The stress through the thickness is dependent upon the thickness of the mate-

rial, to the extent that no deformation occurs, and is greater for thick members than for thin members—this is the explanation for the dependency of the transition temperature on the section size. Various amounts of thinning, or contraction, are observed at or near the apex of the notch in fractured notched plates; this contraction is the result of stress relaxation and a reduction in the triaxial state of stress.

Notch Acuity—In the elastic range, the sharpness of the notch influences the elastic stress concentration factor; therefore, by increasing the notch acuity, the net-section stress at which yielding initiates at the notch tip is reduced. The notch acuity greatly affects the fracture behavior of all ferritic steels when the conditions are conducive to brittle behavior.

Fracture Behavior—Generally, when a monolithic structure fails in brittle manner, the crack initiates at some type of notch and may propagate a great distance through adjacent plates and other structural members. It is readily apparent that there are three modes of behavior—initiation, propagation, and arrest. Because the Charpy notched-bar impact test has long been standardized in this country, the Charpy test has been widely used to evaluate the transition temperatures of the plates, etc., taken from failed structures. From these studies, it was found that the notch toughness (by whatever measure) was least for the plates in which the fracture initiated. If the fractures propagated through other plates and stopped in yet others, the plates in which the fracture stopped had the greatest notch toughness. These results confirmed that once a crack started, it was quite easy for it to propagate. The higher strain rate of the dynamic crack, and the attendant decrease in notch toughness, are responsible for the ease of propagation and the need for a tougher steel to arrest the crack. This is somewhat analogous to a block on an inclined plane in that the static coefficient of friction is greater than the dynamic coefficient of friction; however, once the block starts to slide (initiation), its resistance to motion is decreased, and it continues down the plane (propagation) until it reaches the bottom (a free edge of the structure) or until it reaches a section of the plane that has a higher coefficient of friction (arrest). Because there are these three modes of behavior, tests have been developed and criteria have been formulated to evaluate these modes. It has been shown that the criteria of notch toughness may be divided roughly into two groups. Those in one group tend to evaluate fracture-propagation characteristics and thus depend upon the mode of fracture (shear or cleavage) and the fracture appearance. Those in the second group tend to evaluate the fracture-initiation characteristics and thus depend upon the amount of plastic deformation that occurs before fracture.

Several of the tests used to evaluate the notch toughness of steels are discussed individually in the following sections.

As mentioned previously, considerable research is being expended in the area of the mechanical behavior of metals, especially in the area of notch toughness. For this reason, new tests are being developed, old tests are being modified, and new interpretations of test results are being made. Thus the reader should refer to current literature so as to be aware of these changes.

DYNAMIC TESTS

Charpy Impact Tests and Other Notched-Bar Impact Tests—The Charpy and Izod notched-bar impact tests were developed early in this century. Originally it was assumed that the amount of energy absorbed by a material was proportional to its notch toughness, and that it was necessary to test only at room temperature. However, as the concept of the transition temperature developed, it was realized that tests had to be made over a range of temperatures. Because the Izod specimen is gripped by a massive vise at the base of the notch (Figure 50—31), it is quite difficult to perform this test at temperatures other than room temperature, since the vise acts as a heat sink or heat source. For this reason, today the Izod test is rarely used for the evaluation of steels. The Charpy test has become widely used in both assessing the behavior of a steel in the presence of a notch and in quality control during the steel-producing operations; it does not provide a number that can be used directly in engineering design calculations.

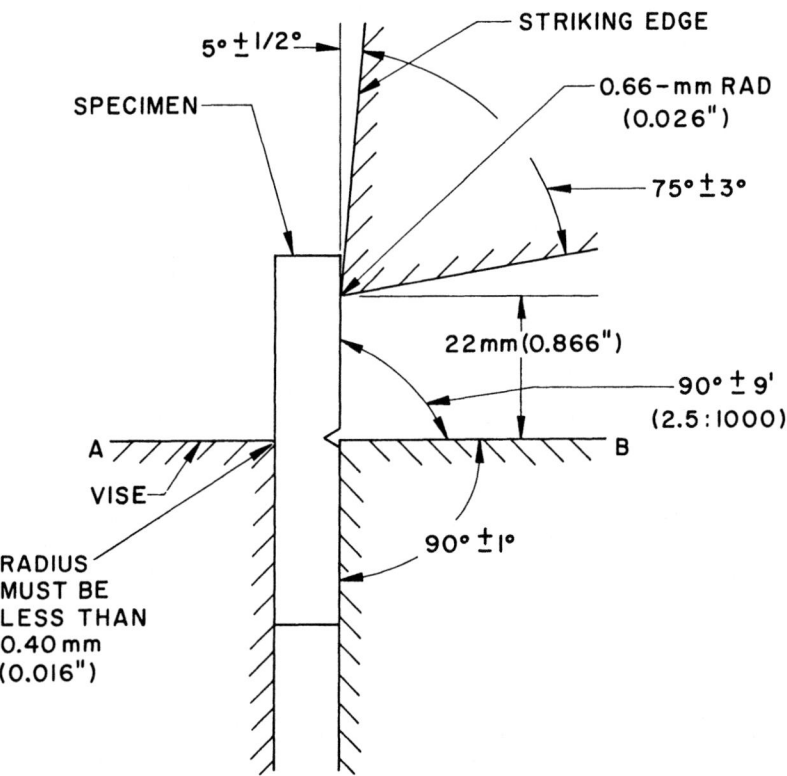

All dimensional tolerances shall be ±0.05 mm (0.002 in.) unless otherwise specified.
NOTE 1—The clamping surfaces of A and B shall be flat and parallel within 0.025 mm (0.001 in.).
NOTE 2—Finish on unmarked parts shall be 2 micrometres (63 microinch).

FIG. 50—31. Izod (cantilever-beam) impact test.

Although the keyhole-notch specimen was used in the past for impact-test specifications, it has been almost completely phased out in preference to the V-notch type that has a sharper notch and thus is a more severe test. In the Charpy test, a rectangular bar with a square cross section and a notch of specified geometry at the midlength (Figure 50—32) is supported near its ends and struck a single blow directly behind the notch sufficient to break the specimen generally by the swing of a weighted pendulum (Figure 50—33). The details of the test procedure are prescribed in the ASTM specifications designated as E 23, "Standard Methods for Notched-Bar Impact Testing of Metallic Materials," and A 370, "Standard Methods and Definitions for Mechanical Testing of Steel Products."

Interpretation of Charpy V-Notch Impact Test Results—Before attempting to evaluate the significance of any notched-bar test, it is of the utmost importance to remember that the behavior of a metal in such a test is strongly dependent on the geometry of the test piece, rate of loading, and temperature of testing chosen as the basis of the test. Because of the extremely complex interrelationships among these variables, results obtained in a test cannot be translated directly, at least with our present state of knowledge, to a service application in which the conditions of loading are altogether different. Even if the energy-absorbing capacity needed in structures for safe performance could be estimated, which is rarely the case, the energy absorption in an impact test could not be used directly as a design figure because of the lack of similarity of geometry and loading conditions. Impact specifications are, therefore, usually based on experience or engineering estimates, as has been seen to be the case in so many other mechanical-property specifications. Many im-

pact-strength specifications are currently written in such a way as to specify the minimum energy absorption required, as for example, 20 joules (15 foot-pounds) in the Charpy V-notch test, at the minimum service temperature to be encountered, although other requirements such as a fracture appearance of some specified amount of shear or so many micrometres or mils lateral expansion are increasing in importance. Even though a steel meeting these specifications would probably perform better in service than one which did not pass, it must be recognized that meeting a given requirement does not necessarily assure the desired service performance.

As discussed previously, the concept of transition temperature does much to provide a general basis of interpretation, not only for notched-bar impact tests, but for all other toughness tests. If a series of notched-bar impact tests is made on a ferritic steel at successively lower temperatures, it will be found that the energy absorption, the per cent shear and the amount of lateral expansion decrease to very low values over a range of temperature in the transition temperature range.

Presently there are three commonly used criteria for assessing the notch toughness of a steel by means of the Charpy test: (1) the per cent shear appearance of the fracture surface, (2) the energy absorbed, and (3) the amount of lateral deformation of the compression surface of the specimen behind the notch, lateral expansion. (It would be more desirable to measure the amount of contraction that occurs at the root of the notch due to the triaxial stresses developed; however,

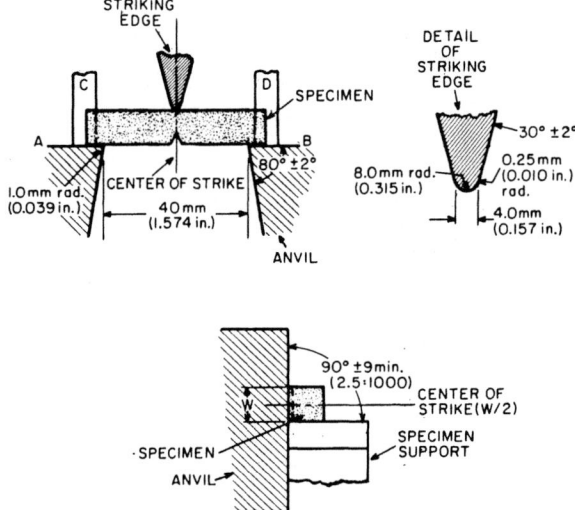

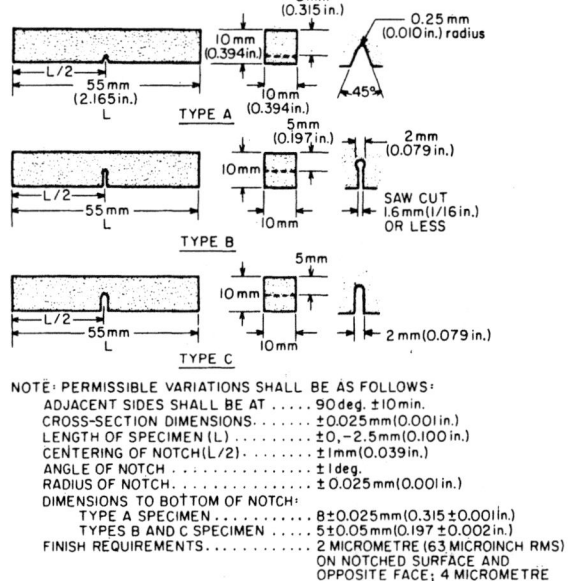

FIG. 50—32. Types of specimens for Charpy (simple beam) impact tests. (Based on information in ASTM Specification E 23.)

FIG. 50—33. Method of striking and supporting Charpy (simple beam) impact specimens. (Based on information in ASTM Specification E 23.)

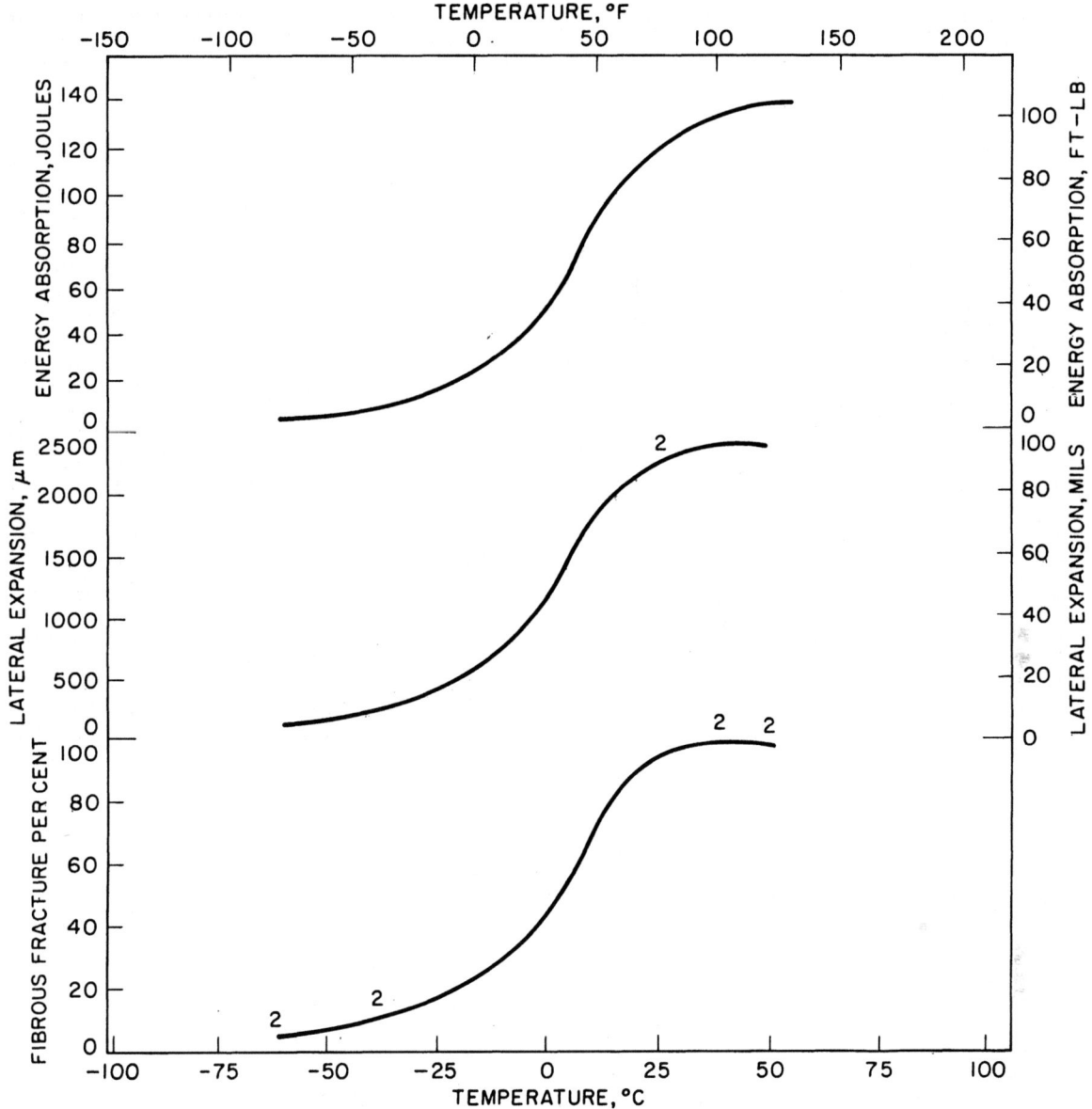

FIG. **50**—34. Charpy V-notch impact test results for ABS Class C Steel.

since for the Charpy specimen, the expansion has been shown to be related to the contraction and is easier to measure, the lateral expansion is the quantity customarily measured.) The variations of these three quantities with temperature for a low-carbon steel are shown in Figure **50**—34. Smooth curves have been fitted through the individual data points—it should be noted that values for all three criteria can be obtained from each specimen tested. The energy absorption measurements are obtained directly from the testing machine; the fracture appearance values are estimated from a visual observation of the fracture surfaces of the specimen; and the lateral expansion measured by a special jig described in ASTM Method A 370 (Figure **50**—35) and discussed in the paper by Holt.

It will be noted in Figure **50**—34 that the transition occurs over a temperature range. Consequently the criteria chosen have typically been a fixed number, e.g. 50 per cent shear, or 380 micrometres (15 mils) lateral expansion, for all grades and thicknesses of steel; however, papers by Gross have pointed out and the Boiler and Pressure Vessel Committee of the American Society of Mechanical Engineers have recognized that the values must be adjusted to comprehend the difference in the yield strengths of the many grades of steels available to structural designers and to comprehend the differences in the metallurgy and the state of stress of plates of different thickness of the same grade of steel. For a further discussion of these points see the papers by Gross listed in the bibliography.

In summary, in the Charpy V-notch test, the strain rate is fixed, the temperature is varied, and the notch

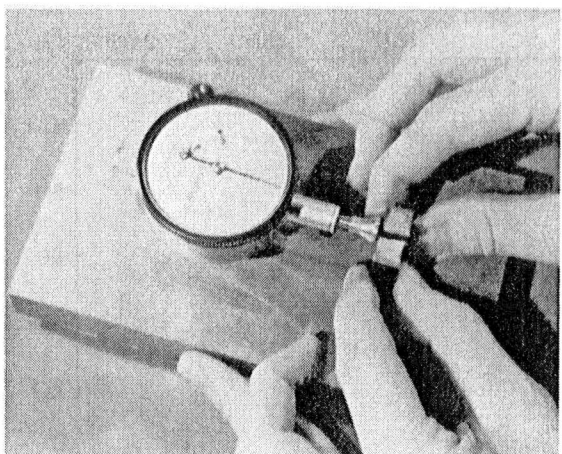

FIG. 50—35. Lateral expansion gage for Charpy impact specimens.

acuity is fixed. The Charpy V-notch test is primarily a measure of fracture initiation.

The Drop-Weight Test—The drop-weight test, developed at the U. S. Naval Research Laboratory is used extensively to investigate the conditions required for the initiation of brittle fracture in steels. The method has been standardized by the ASTM as Method E 208, "Standard Method for Conducting Drop-Weight Test for Determining Nil-Ductility Transition Temperature of Ferritic Steels," and the nil-ductility transition temperature obtained is frequently used in industrial specifications and codes.

The nil-ductility transition (NDT) temperature is defined as the maximum temperature at which a standard drop-weight specimen breaks when tested in accordance with Method E 208. The method uses simple beam specimens with a "crack starter" on the tensile surface that creates a crack very early during the test. The test is conducted by subjecting each of a series of specimens (usually four to eight) of the steel to a single impact load at a sequence of selected temperatures to determine the maximum temperature at which the triaxial stresses caused by the notch are sufficient to cause

fracture of the specimen at a fixed maximum surface strain. The impact load is provided by a guided, free-falling weight with an energy generally in the range of 340 to 1600 joules (250 to 1200 foot-pounds), depending upon the yield strength of the steel being tested. The specimens are prevented by a stop from deflecting more than about a millimetre or several hundredths of an inch (approximately five degrees) which limits the strains on the surface of the specimen to about the yield point strain.

The crack starter is a notched weld bead of a hard surfacing material applied to a plate surface—the opposite surface may be machined so as to make the thickness of the specimen conform to a standard size. The specimen is brought to the desired temperature, then placed upon the specimen supports with the notch centered and struck by the falling weight. The weld bead cracks after approximately a three degree bend occurs. If the test temperature is below the NDT temperature, the crack initiated in the plate by the weld-bead crack is sufficient to cause the specimen to fail with the additional small amount of strain that occurs in bending the specimen enough to contact the stops. If the temperature is above the NDT temperature, the crack will propagate some amount into the specimen, but the specimen will not fail. Figure 50—36 illustrates this test and Figure 50—37 shows the degree of cracking that occurs at temperatures above and below the NDT temperature. The drop-weight test is limited by the ASTM to plates thicker than about 16 millimeters (⅝ inch).

This test is a "go, no-go" test in that the specimen will either break or not break. Although some specimens of the series tested will not break at temperatures slightly below the NDT temperature, by definition, no specimens will break above the NDT temperature; it will be noted in Figure 50—37 that one of the two specimens tested at the NDT temperature did not break. The ASTM Test Method requires that duplicate specimens be tested and that both do not break at the "no break" temperature; this temperature is 5°C (10°F) higher than the NDT temperature, that is, the highest temperature at which one or more specimen break.

In summary, in the drop-weight test, the strain rate is fixed, the temperature is varied, and the notch acuity is fixed. The drop-weight test is a fracture-initiation test.

The Drop-Weight Tear Test—This test was developed as a method of measuring the fracture-propagation transition temperature of carbon and low-alloy steels used in line-pipe applications and has been standardized as Test Method E 436. The test procedure is somewhat similar to the Charpy test except that the specimens and testing machines are much larger. The specimens are 75 millimeters (3 inches) wide by the plate thickness (which may range from about 3 to 19 millimeters or ⅛ to ¾ inch by about 300 millimeters or 12 inches long). A notch is pressed (not machined) with a hardened tool into one of the plate thickness-by-300-millimetre (12-inch) sides at the midlength. The specimens, tested over a range of temperatures, are supported as simple beams and each is struck behind the notch by an impact blow sufficient to cause fracture. The fracture appearance is rated as to the amount of shear. Figure 50—30 shows a group of specimens

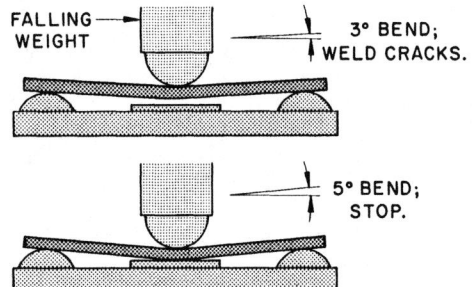

FALLING WEIGHT

3° BEND; WELD CRACKS.

5° BEND; STOP.

FRACTURE MUST BE DEVELOPED WITH 2° BEND AFTER SHARP NOTCH IS FORMED

FIG. 50—36. Schematic representation of the drop weight test used to determine the nil-ductility transition temperature by testing a series of specimens over a range of temperatures.

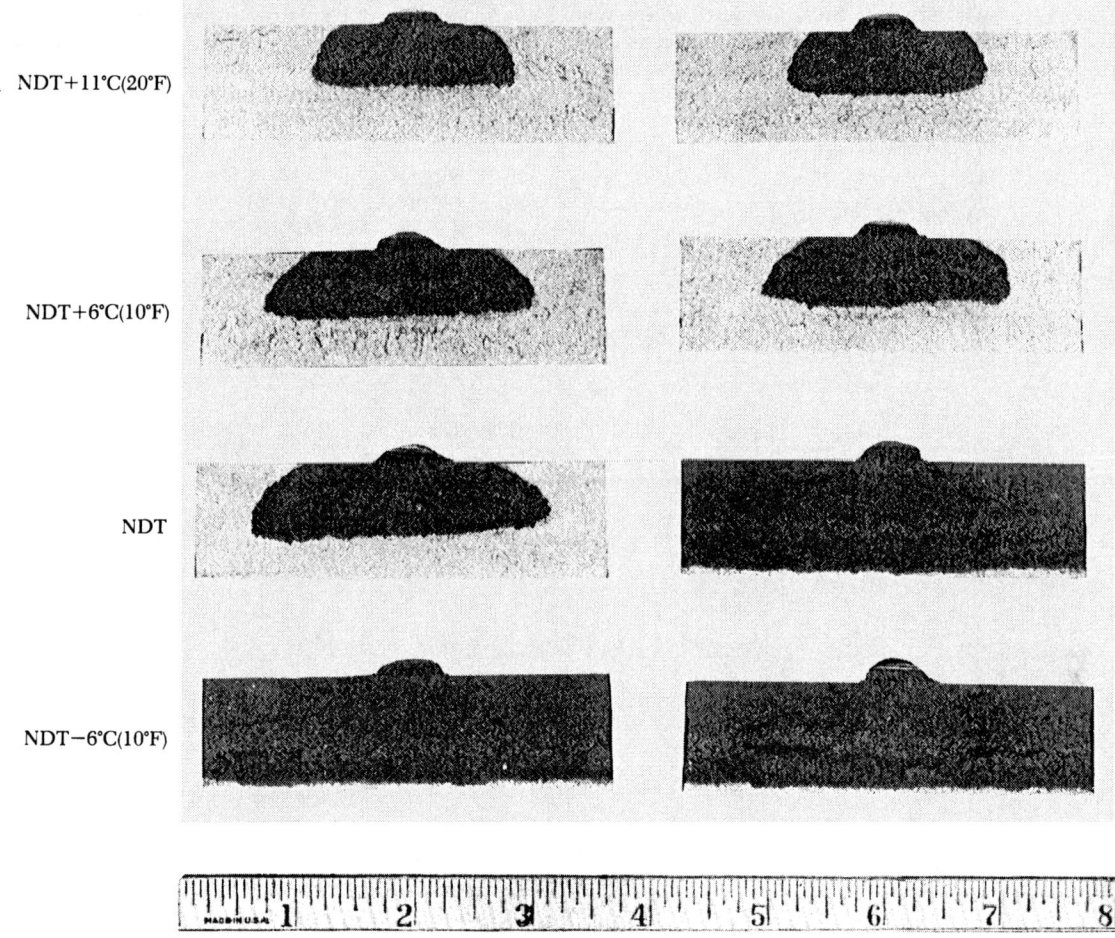

NDT+11°C(20°F)

NDT+6°C(10°F)

NDT

NDT−6°C(10°F)

FIG. 50—37. Photographs of drop-weight test specimens showing amount of cracking that occurs. The specimens were first tested in accordance with ASTM Method E 208; ink was flowed into the crack that developed, allowed to dry, and the specimens were then broken open by retesting at much lower temperatures. Thus, the original crack is defined by the black areas.

tested over a range of temperatures.

Considerable research has been conducted on the significance of the DWTT results. Correlation of the results of fullscale pipe tests with the results of the DWTT indicates that the transition in full-scale propagation appearance (fracture appearance remote from the initiation region) occurred at the same temperature as the transition in the DWTT per cent shear area.

In summary, in the drop-weight tear test, the strain rate is fixed, the temperature is varied, and the notch acuity is fixed. The drop-weight tear test is a test to measure the crack propagation characteristics of steel.

SECTION 6

FRACTURE MECHANICS AND FRACTURE CONTROL

Fracture mechanics technology provides a means of describing the life of a structural component subjected to repeated (cyclic) loading by separately treating the fatigue-crack initiation, fatigue-crack growth, and terminal fracture. Basically, fatigue initiation addresses the development of a microcrack in a structure while fatigue-crack growth addresses the macrocrack regions, and terminal fracture describes the role of material properties in the determination of the crack size when the component loses load-carrying capacity.

The fracture-mechanics approach is based on the fact that intensification of the elastic stress at the tip of a crack of length α can be completely described by the stress intensity factor, K_I, by the relationship

$$K_I = C\sigma\alpha^{1/2}$$

where σ is the applied stress and C is a geometric constant. If the stresses acting on a component and the appropriate material properties are known, fracture mechanics can be used to assess maximum flaw sizes which limit the component's design life. Solutions for evaluating K_I for numerous loading and crack geometries have been derived and are contained in the literature, for example, see Toda, et al.

FATIGUE-CRACK INITIATION

The number of cycles of a stress or strain range applied to a material that are necessary to generate a

crack determines the fatigue-crack-initiation behavior of the material. Classically, fatigue-initiation behavior is determined from smooth unnotched test specimens which are cyclically loaded to failure. A record of the number of cycles (n) to failure at various applied stress (s) or strain levels is developed for the particular material of interest. However, often a structure contains a stress riser which increases the cyclic stress locally. To evaluate these cases, test specimens are notched to develop S-N data in the presence of stress risers.

More recently, fracture mechanics methodologies

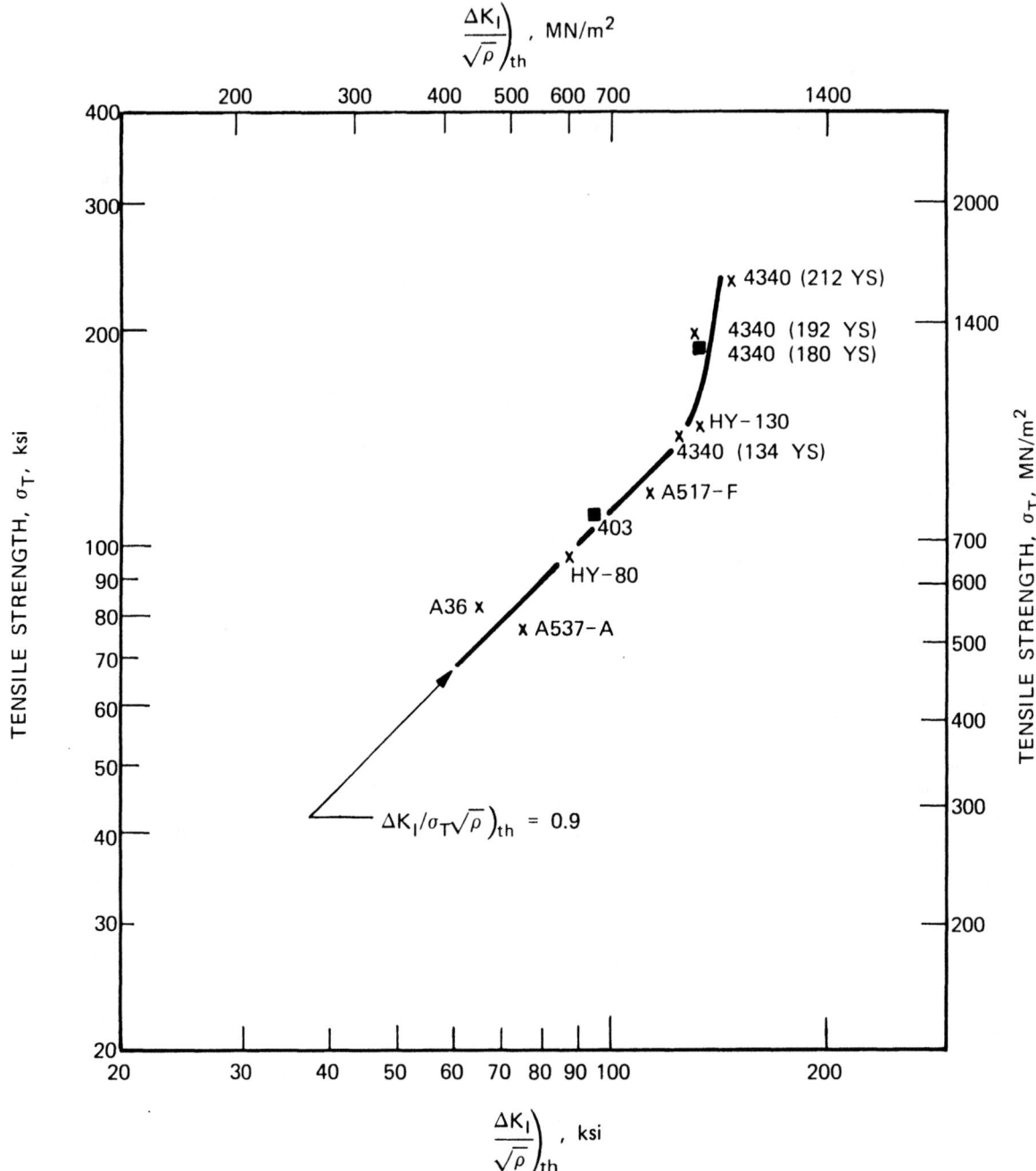

FIG. **50**—38. Dependence of fatigue-crack-initiation threshold on tensile strength.

have been used to quantify fatigue behavior. When testing notched specimens in fatigue, the alternating stress can be expressed in terms of an alternating K_I called ΔK_I. Fatigue-crack-initiation thresholds of various steels tested in this condition have been evaluated.

The radius, ρ, at the root of the notched specimen plays a significant role in the maximum stress at the notch tip. When the ratio $(\Delta K_I / \sqrt{\rho})_{th}$ is plotted against the tensile strength of the material, it appears that the fatigue-crack-initiation threshold (infinite life) increases with increasing tensile strength up to 1030 MPa

(150 ksi), Figure 50—38. At tensile strengths greater than 1030 MPa (150 ksi), the fatigue-crack initiation threshold is essentially independent of tensile strength.

For the type of specimen used to generate the fatigue-crack growth threshold data given in Figure 50—38, the maximum elastic stress, σ_{max}, at the root of the notch is described in the equation

$$\sigma_{max} = 2K_I / \sqrt{\pi \rho}$$

Comparing this equation to the linear region of Figure

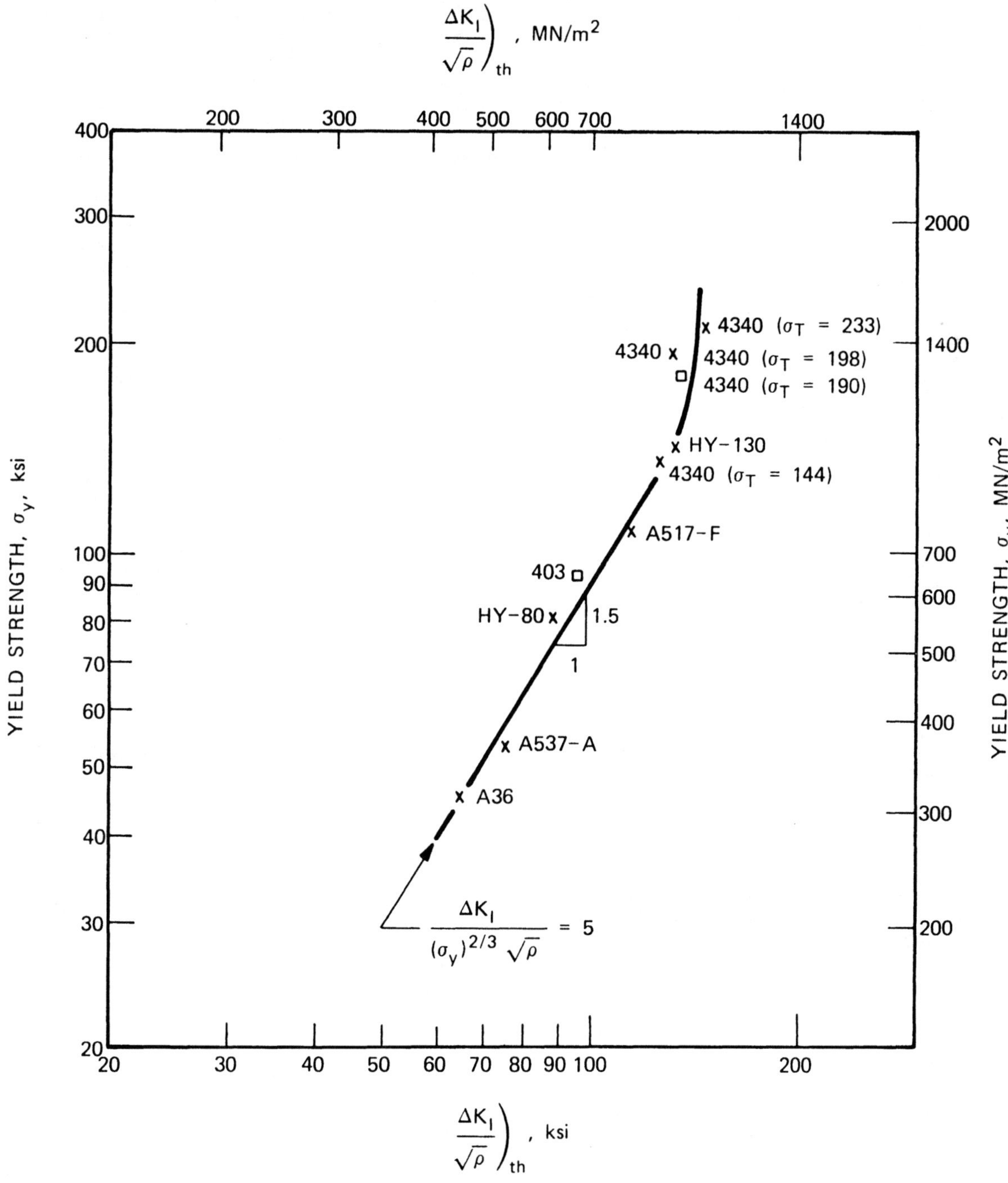

FIG. 50—39. Dependence of fatigue-crack-initiation threshold on yield strength.

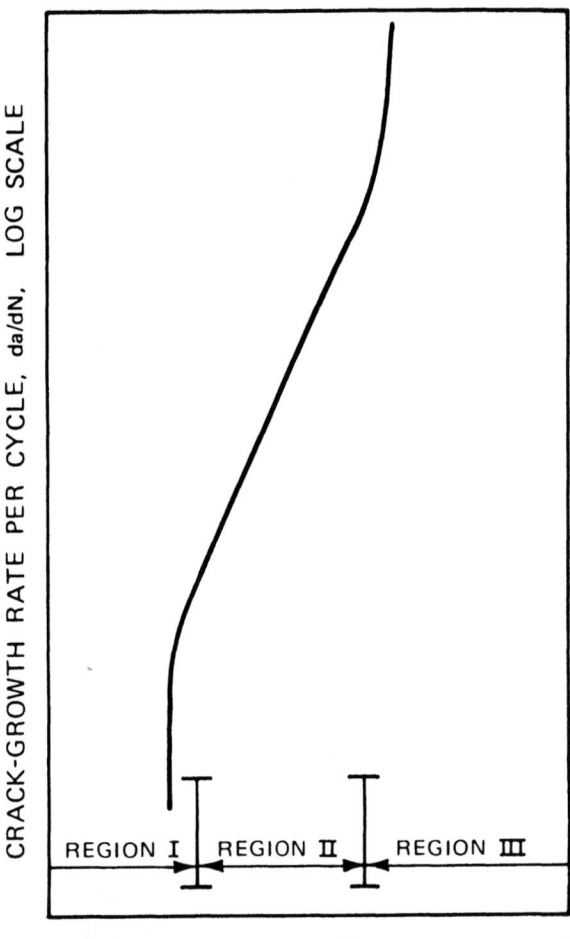

CRACK-GROWTH RATE PER CYCLE, da/dN, LOG SCALE

REGION I | REGION II | REGION III

STRESS-INTENSITY-FACTOR RANGE, ΔK_I LOG SCALE

FIG. 50—40. Schematic representation of fatigue-crack growth in steel.

50—38, it can be concluded that the fatigue-crack growth-initiation threshold corresponds to a maximum elastic stress in the notch root approximately equal to the tensile strength of the steel. When the threshold parameter $(\Delta K_I / \sqrt{\rho})_{th}$ is plotted against the yield strength of various steels as shown in Figure 50—39 a linear relation between 275 and 965 MPa (40 and 140 ksi) yield strength levels results.

FATIGUE-CRACK PROPAGATION

The growth of a fatigue crack is shown schematically in Figure 50—40. In Region I, the fatigue-crack-propagation threshold is described. Below a certain level of fluctuation of the stress-intensity factor, ΔK_I, a fatigue crack does not propagate. Region II describes stable-crack growth, in which the plot of da/dn and ΔK_I is linear (on a log-log scale). In Region III, crack growth is rapid and unstable because the ΔK_I is approaching a critical level, K_C, at which terminal fracture occurs. Terminal fracture will be discussed in the following section. Stable-crack growth (Region II) of steels has been investigated extensively and is described in Figure 50—41 for steels which exhibit yield strengths up to 550 MPa (80 ksi) with ferrite-pearlite microstructures, and in Figure 50—42 for steels with yield strengths above 550 MPa (80 ksi) and martensitic microstructures.

The data in Figures 50—38 and 50—39 lead to the conclusion that fatigue-crack-initiation behavior depends on the tensile properties of the various steels and are independent of their metallurgical properties. Figures 50—41 and 50—42 show that all steels exhibit essentially the same fatigue-crack-growth rate behavior. However, it will be shown in the following section that the conditions necessary for terminal fracture of steels depend strongly on their metallurgical nature.

SUDDEN UNSTABLE FRACTURE

Fracture mechanics equations such as those shown in

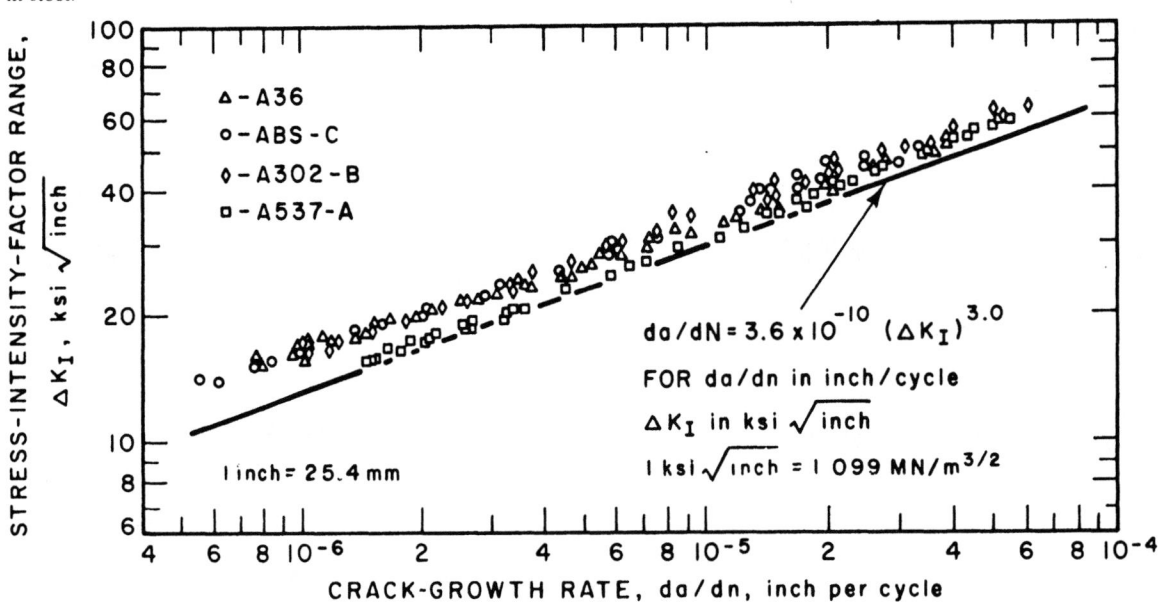

STRESS-INTENSITY-FACTOR RANGE, ΔK_I, ksi $\sqrt{inch}$

CRACK-GROWTH RATE, da/dn, inch per cycle

△ — A36
o — ABS-C
◇ — A302-B
□ — A537-A

$$da/dN = 3.6 \times 10^{-10} (\Delta K_I)^{3.0}$$
FOR da/dn in inch/cycle
ΔK_I in ksi $\sqrt{inch}$
1 ksi $\sqrt{inch}$ = 1 099 MN/m$^{3/2}$

1 inch = 25.4 mm

FIG. 50—41. Summary of fatigue-crack-growth data for ferrite-pearlite steels.

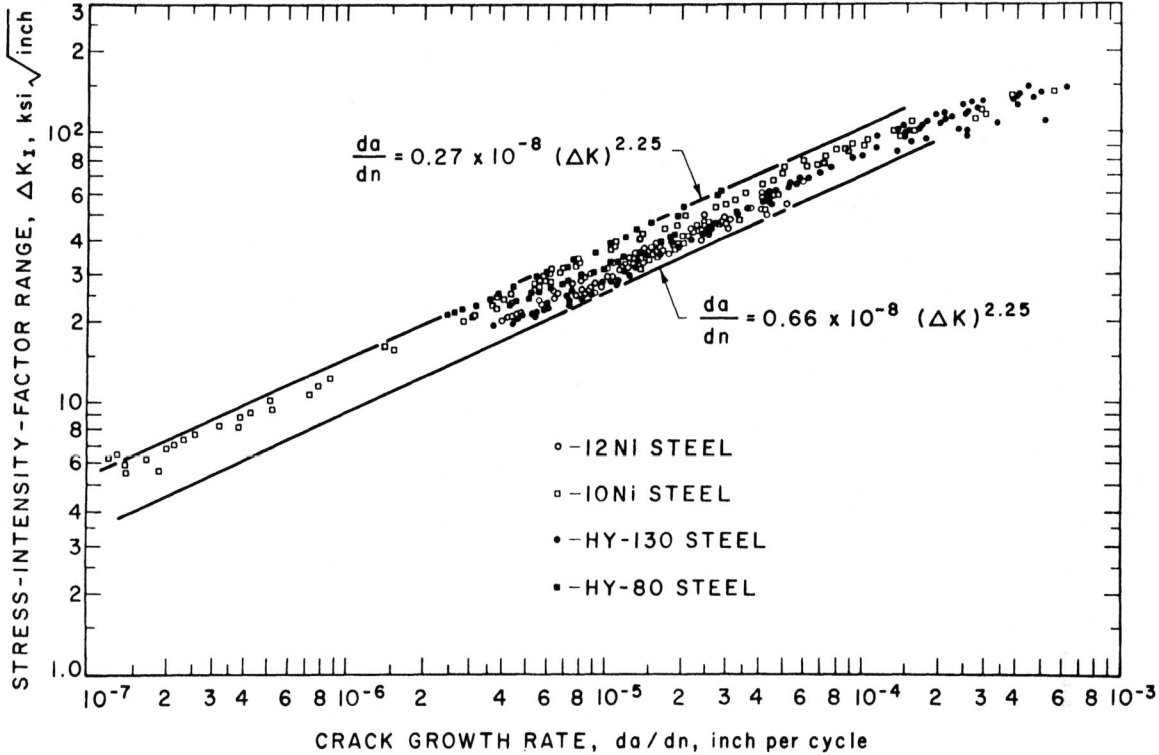

FIG. 50—42. Summary of fatigue-crack-propagation data for martensitic steels.

Figure 50—43 describe the interrelationship of stress, σ, crack size, a, and stress-intensity factor, K_I. If a critical value, K_{IC}, is exceeded by some combination of stress and crack size, sudden unstable brittle fracture takes place. K_{IC} is a material property which varies with temperature and strain rate. Thus, the K_{IC} of a material is a measure of its crack tolerance, or fracture toughness, under given conditions of temperature and loading rate. Extensive data are available on the K_{IC} levels exhibited by various steels. In contrast to the fatigue behavior described earlier, the fracture toughness of steels is highly dependent on the metallurgical nature and history of steels.

The dependence of K_{IC} on temperature and strain rate is shown in Figure 50—44 for a typical high-strength low-alloy steel. An equivalent description of the loading rates is that the dynamic, intermediate, and static (slow-bend) loads were applied in 0.001 second, 1 second, and 100 seconds, respectively. The temperature at which the K_{IC} value begins to rise sharply is a fracture-toughness transition temperature, and the difference between the static and dynamic loading rates is frequently called a "strain-rate shift." For a steel with a room temperature yield strength of 245 MPa (36 ksi), the shift is 71°C (160°F). As yield strength increases, the magnitude of the shift decreases, until at a yield-strength level of 965 MPa (140 ksi), the fracture-toughness transition is independent of loading rate. Thus, the common structural steels with room-temperature yield strengths 275 to 340 MPa (40 to 50 ksi) exhibit a higher level of fracture toughness (crack tolerance) at a given temperature when subjected to static loads than dynamic loads. It should be noted,

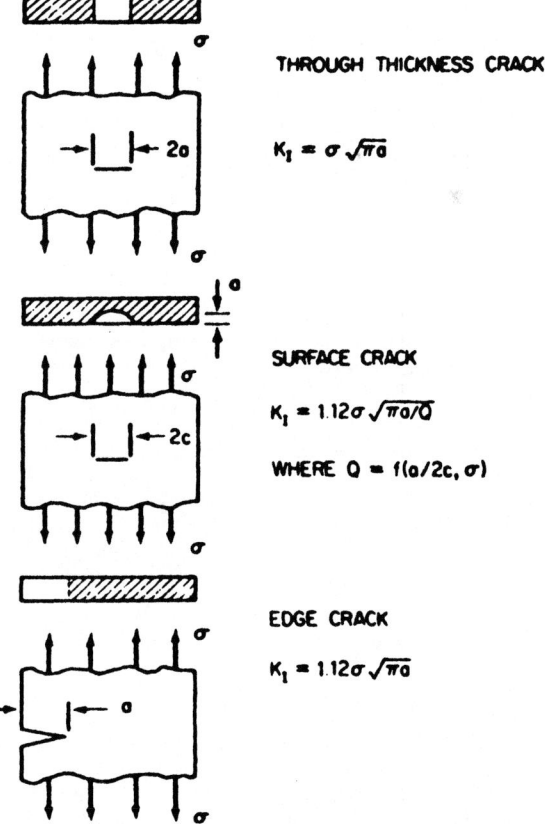

THROUGH THICKNESS CRACK

$$K_I = \sigma \sqrt{\pi a}$$

SURFACE CRACK

$$K_I = 1.12\sigma\sqrt{\pi a/Q}$$

WHERE $Q = f(a/2c, \sigma)$

EDGE CRACK

$$K_I = 1.12\sigma\sqrt{\pi a}$$

FIG. 50—43. K_I values for various crack geometries.

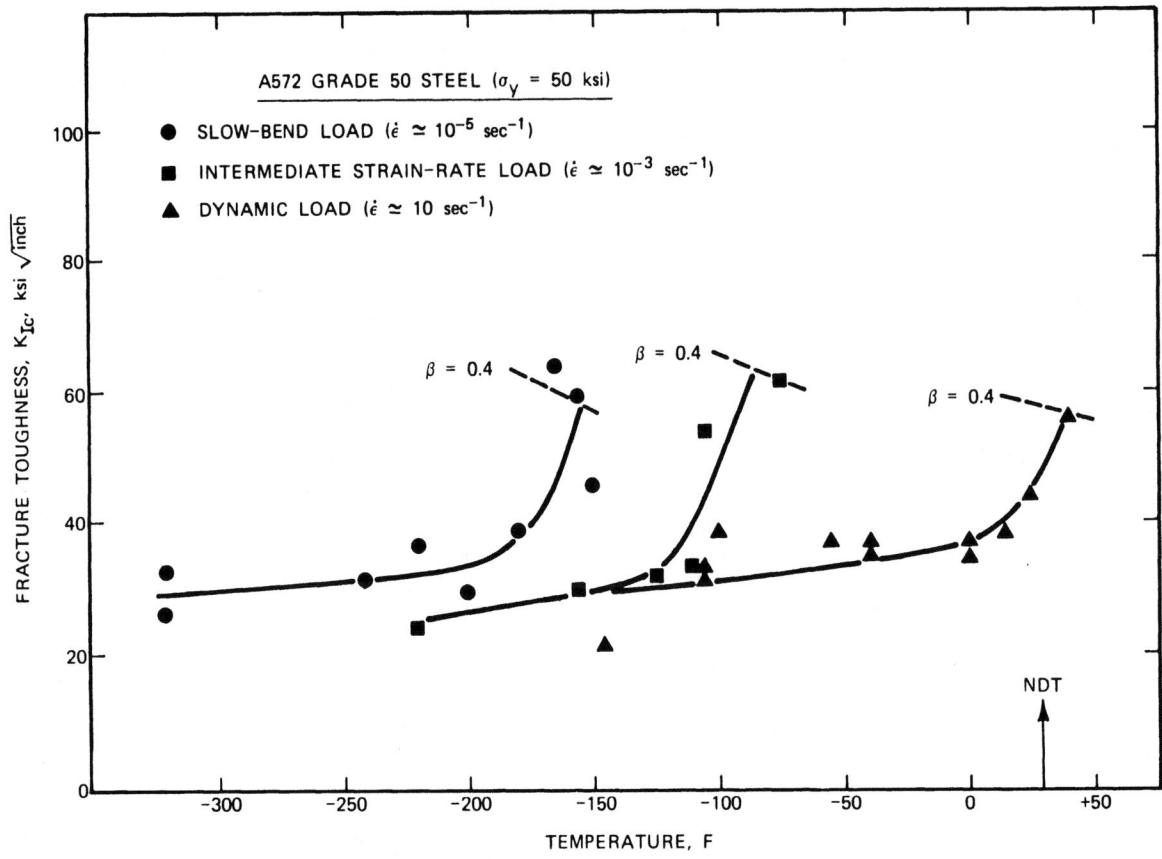

FIG. 50—44. Dependence of K_{IC} on temperature and strain rate.

however, that the strain-rate shift is effective only during the initiation of sudden unstable fracture. A propagating fracture relates more closely to the dynamic-loaded curve, because the strain rate at the tip of a propagating crack is very high.

Fracture-mechanics technology has extensive application in the field of failure analysis, material selection, and fracture-control plans. It is the first technology to relate stress history, crack size, component geometry, and material properties, thereby permitting a realistic assessment of the importance of cracks discovered in structural components in service. In failure analysis, measurement of critical crack size and material toughness permits calculations of the stress that was required to cause the fracture. Furthermore, a knowledge of crack tolerance in a given material leads to very practical inspection requirements.

A schematic diagram plotting flaw size and the number of cycles of fatigue loading is shown in Figure 50—45. Improvements in service life related to initial flaw size, stress level, and level of fracture toughness are included also. The fatigue-crack propagation rate, Figures 50—41 and 49—42, is very relevant to this diagram, because the greater the depth of a fatigue crack, the faster it grows. Thus, most of the life of a component is spent in initiating a crack and in growth of the crack when it is very small. Increasing the toughness and critical crack size of a material will lead to only minor increases in fatigue life. Decreasing the applied stress (or reducing stress concentrations by using generous fillet radius) can significantly increase the fatigue life of structural components.

SECTION 7

HIGH-TEMPERATURE TENSION, CREEP, AND RUPTURE TESTING

Introduction—The design of load-bearing structures for service at ambient temperature is based on the yield or tensile strengths of a material which are determined by the room-temperature tension test. However, at elevated temperatures the behavior is different. The purpose of this section is to outline briefly the procedures that are used to obtain the mechanical-property data which are pertinent to the design of structures for elevated-temperature service. As discussed in detail in the chapter dealing with steels for elevated-temperature service, the fact that metals generally deform more easily under load at elevated tem-

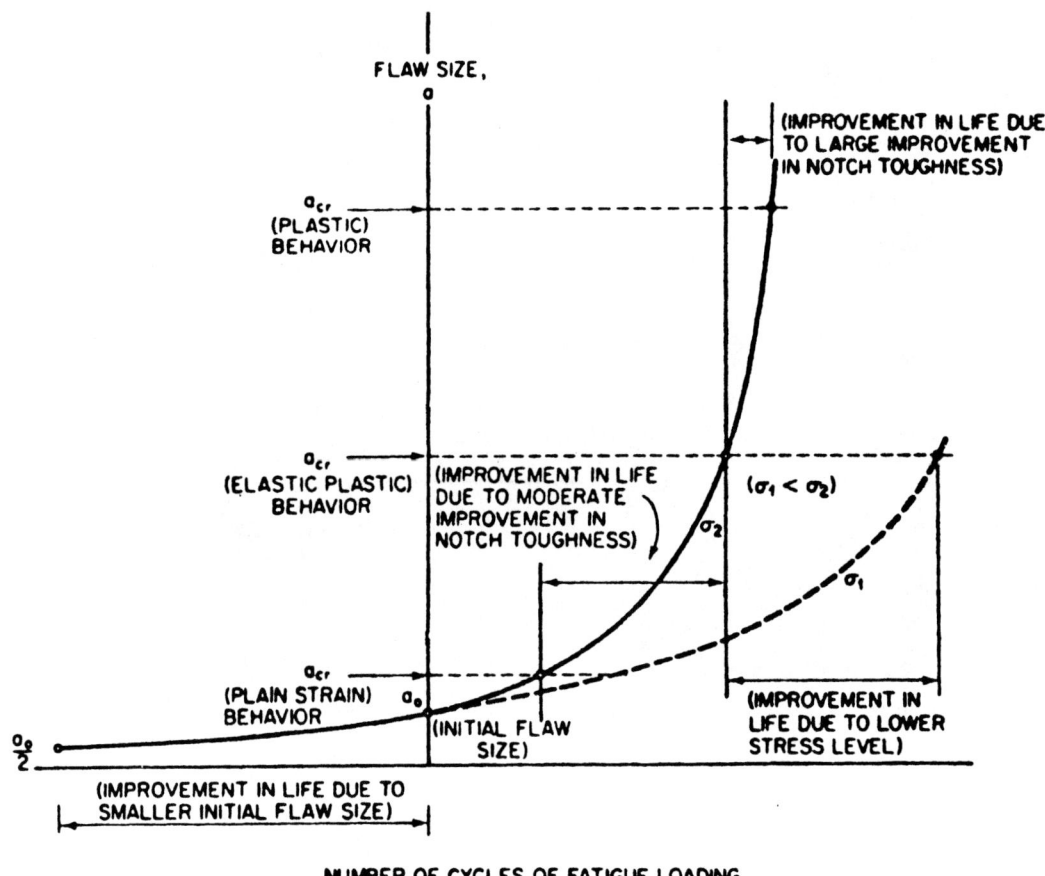

FIG. 50—45. Schematic showing effect of notch toughness, stress, and flaw size on improvement of life of a structure subjected to fatigue loading.

perature requires that the design stress be based on different principles than those for service at ambient temperature.

The design stress of a material, in the range from room-temperature to the temperature at which appreciable creep occurs, is based on the elevated-temperature tensile properties determined at the service temperature. The temperature at which creep becomes important depends on the particular metal. For carbon steel, creep becomes a design consideration at about 425°C (800°F), for alloy steels about 480°C (900°F), and for austenitic stainless steels about 535°C (1000°F). Above these temperatures, the design stresses are usually based on the creep and creep-rupture properties. However, for applications that involve relatively short-time service, the tensile properties may also be of interest in the creep range.

THE ELEVATED-TEMPERATURE TENSION TEST

The most common method for determining the strength of metals at elevated temperatures is the tension test, which provides design information up to the temperature at which appreciable creep is encountered. The elevated-temperature tension test gives a useful estimate of the static load-carrying capacity of metals under short-time tensile loading and a comparative measurement of the ductility that can be expected. The special techniques involved in performing the tension test at elevated temperatures are covered by ASTM Specifications E 21 and E 151. Elevated-temperature tension tests may be made in conventional testing machines. A typical arrangement used in a conventional machine is shown in Figure 50—46; the temperature controls are also shown. With this arrangement, the specimen is heated by a furnace supported on one column of the machine, the furnace being free to swing into and out of the axial position.

Tests may be made on either bar or flat stock. For bar tests, a 0.505-inch (12.83-mm) diameter specimen with threaded ends is usually used, although other sizes may be employed. Threaded-end specimens are customarily gripped with the furnace. For sheet tests, a regular ½-inch (12.7-mm) wide specimen is used. If sufficient material is available, the specimen may be gripped outside the furnace; however, it is customary to grip the specimen within the furnace by use of pins or wedges or a combination of both. Various other specimen configurations that are generally suitable for tests at elevated temperatures are also described in

FIG. 50—46. Elevated-temperature tension-test equipment showing specimen ready for testing in hydraulic tension test machine.

detail in ASTM Specification E 8.

The specimen extension (strain) is usually recorded autographically by an extensometer. ASTM Specification E 83 is recommended as a guide for selecting an extensometer with the required sensitivity and accuracy.

Tensile strength, yield strength, reduction of area, and elongation values obtained from short-time tension tests above the range 425° to 565°C (800° to 1050°F) are usually not used in design when long-time service is contemplated. For long-time service applications above these temperatures, design information is obtained from creep and creep-rupture tests. However, at temperatures above 815°C (1500°F), data obtained from short-time tension tests are often useful as a guide for hot-working operations.

CREEP AND CREEP-RUPTURE TESTS

Creep Testing—Creep is defined as the time-dependent deformation which occurs after the application of a load to a solid. The creep strength of steel is important above the range 425° to 535°C (800° to 1000°F) and is a primary factor in determining design stresses above these temperatures. The rate of creep is directly related to the applied stress and test temperature and the purpose of the creep test is to determine the creep rate as a function of applied stress and temperature.

The method of measuring creep resistance is simple enough in principle, but in practice it requires considerable laboratory apparatus, and great care and precision in its operation. Disregarding for the moment the exact type of apparatus, the following fundamental steps are almost universally employed. The specimen is held at a constant temperature in an electric-resistance heating furnace, and is subjected to a static tensile load. The load causes the specimen to elongate gradually, and the amount of elongation is measured periodically. The total elapsed time of each test may vary from a few hours to tens of thousands of hours. The recommended practice for conducting creep tests is covered by ASTM Specification E 139. To obtain satisfactory results, precise control of furnace temperature, specimen align-

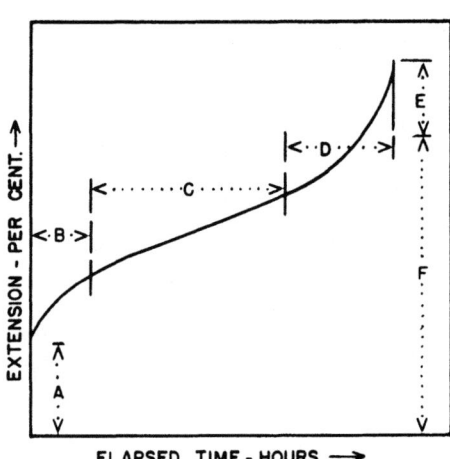

FIG. 50—47. Schematic creep curve. Extension plotted against elapsed time. (A) Elastic extension; (B) creep at decreasing rate; (C) creep at approximately constant rate; (D) creep at increasing rate; (E) elastic contraction; (F) permanent change of length.

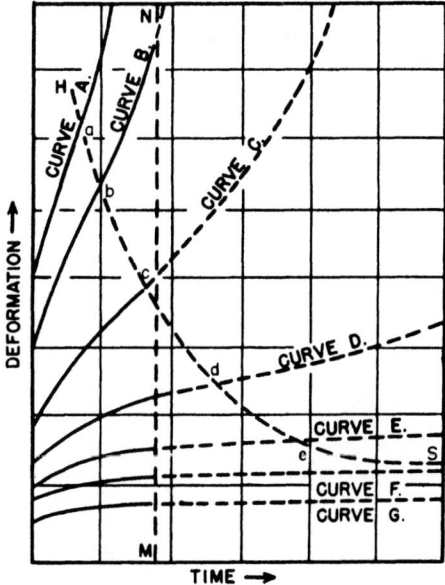

FIG. 50—48. Schematic plot of typical creep curves for seven specimens of the same steel; each specimen tested under a different stress ranging from a very high stress, curve A. to a very low stress, curve G. at a constant temperature. (Taken from "The Interpretation of Creep Tests" by P. G. McVetty, Proc. ASTM, Vol. 34, Part 2, 1934; pp. 105-116.)

ment, and applied stress are mandatory.

When the change of length taking place in a specimen over a period of time is plotted against the elapsed time, a creep curve is obtained, whose typical form is shown in Figure 50—47. When the load is first applied, an initial elongation (A) occurs. Then the specimen strains gradually, at a decreasing creep rate, during the primary stage of creep (B). The creep rate then becomes essentially constant for a period of time during the secondary or steady-state stage of creep (C). The slope of the creep curve in this secondary stage of creep is also referred to as the minimum-creep rate. Finally, if the time is long enough, the creep rate will increase (D), eventually leading to fracture of the specimen. At the end of the testing period, if fracture has not occurred, the load is removed and elastic contraction (E) occurs, corresponding approximately to the elastic extension found upon application of the load at

the start of the test. Thus, it is apparent that metals creeping under stress at high temperature can and do show both plastic and elastic properties simultaneously. The amount of permanent deformation is represented by (F).

Two standards of creep strength are commonly used: (a) the stress to produce a minimum-creep rate of 0.00001 per cent per hour (1 per cent per 100 000 hours) and (b) the stress to produce a total creep strain of 1 per cent in 100 000 hours. The use of minimum creep rate data as a basis for design is generally used in the United States whereas European practice is to use total creep strain data.

The stress for a selected creep rate for the material in question must be ascertained by experiment. This is done in the following manner: several creep tests are run under different stresses at a single temperature and the creep curves plotted on the same chart. A family of curves is shown in Figure 50—48. Five or six tests on any given material are usually sufficient to indicate its behavior at a given temperature. The creep rate during the second stage of creep is measured and plotted against the applied stress. On log-log coordinates, the points are found to normally exhibit a straight line as shown in Figure 50—49. The stress for a creep rate of 0.00001 per cent per hour must then be obtained by extrapolation or in some cases interpolation.

The Creep-Rupture Test—The creep-rupture test (sometimes called the stress-rupture test), is similar to the creep test. Generally the loads and consequently the creep rates are higher, and the test is carried to failure of the material. The apparatus for carrying out the rupture test is usually the same as that employed for the creep test, except that a less sensitive instrument may be used for measurement of the strain.

In reporting rupture data, it is convenient to plot the applied stress against the time for failure on log-log coordinates as shown in Figure 50—50, as this type of plot will usually yield a straight line. However, it is often observed that with increasing time, the slope of log-log plots of stress versus time to rupture exhibits a change in slope. Because of the change in slope, it is desirable to carry the rupture test out to a considerable time period. Rupture tests lasting in excess of 10 000 hours are now considered to be essential in determining, by extrapolation, the 100 000-hour rupture strength. Rupture values are usually reported as the stress for rupture in 100, 1000, 10 000 and 100 000 hours. In recent years there has been increased interest in using time-temperature parameters for determining the creep-rupture strength. An excellent review of par-

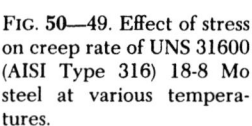

FIG. 50—49. Effect of stress on creep rate of UNS 31600 (AISI Type 316) 18-8 Mo steel at various temperatures.

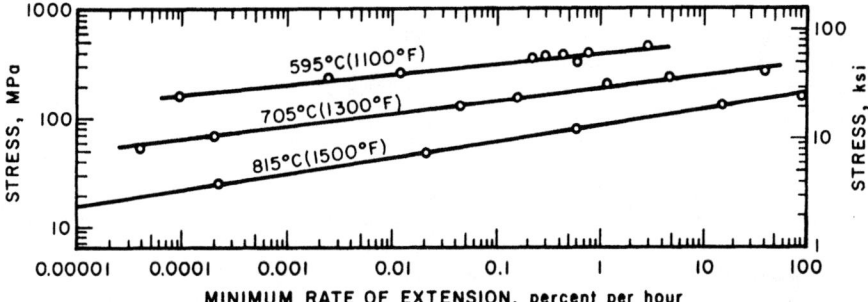

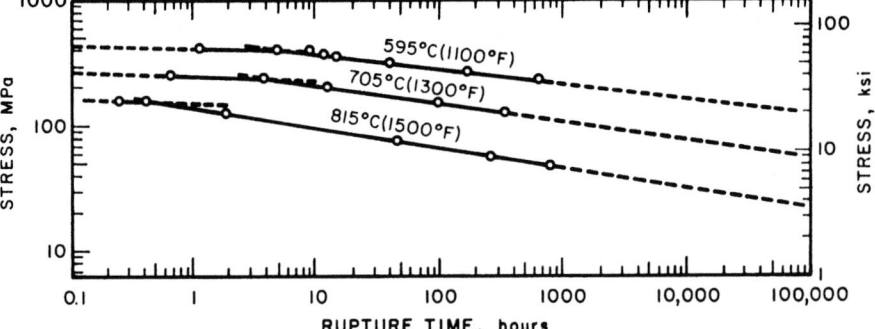

FIG. 50—50. Effect of stress on time for rupture of UNS S31600 (AISI Type 316) 18-8 Mo steel at various temperatures.

ametric methods is discussed in the paper by R. M. Goldhoff listed in the bibliography. The 100 000-hour rupture strength is one of the basic properties used in the establishment of design stresses.

In addition, rupture tests of notched specimens are sometimes required to provide an indication of the ability of the material to deform locally under multi-axial stress conditions. Tests of this type are important for high-strength materials which exhibit low ductility. Recommended practice for notched-specimen rupture tests is covered in ASTM Specification E 292.

In recent years, it has also become desirable to conduct creep and creep-rupture tests on metallic materials under conditions of rapid heating and short-time periods. Tests of this nature are covered by ASTM Specification E 150.

It should be noted that creep and creep-rupture tests are seldom carried out to the times corresponding to the intended service life. Thus, the stress for a minimum creep rate of 1 per cent in 100 000 hours or the 100 000-hour rupture strength are generally based on extrapolation of short-time tests to long times. In addition to extrapolation based on log-log plots of stress versus time to rupture and stress versus minimum creep rate, a number of parameters such as the Larson-Miller and Manson-Haferd, have been derived for extrapolation of creep-rupture data

Creep and Creep-Rupture Test Equipment—The design of equipment for research and development testing of steels for elevated-temperature service varies to permit: (a) constant load testing and (b) testing with constant stress. Figure **50—51** shows equipment of the United States Steel Corporation for carrying out tests of type (a).

As has been outlined above, a specimen in either the creep or creep-rupture test, enclosed within an electric-resistance heating furnace, normally is subjected to a tensile load imposed by dead weights and a lever. As has been stated above, creep and creep-rupture tests differ only in the magnitude of the stresses employed; in the rupture test greater stresses are used and the test is continued until fracture of the specimen occurs to determine the relation between the stress and the time to fracture. In either test, the progressive variation of extension with time is observed and correlated with the imposed stress, thus permitting an estimate of the minimum-creep rate that may be expected during elevated-temperature service.

The extension that is of interest in the creep test is

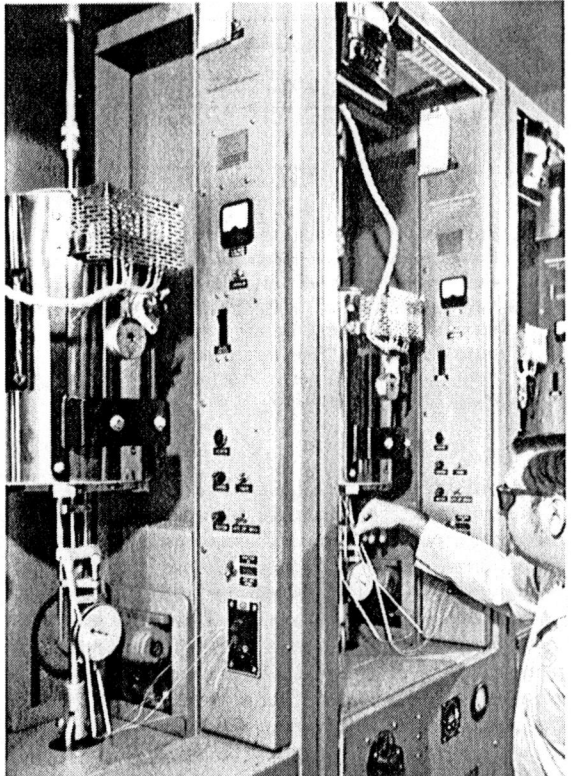

FIG. 50—51. Rupture-test equipment for constant-load testing up to 1205°C (2200°F) at United States Steel's Technical Center at Monroeville, Pa.

generally quite small, the rate being about 0.00001 per cent per hour (1 per cent per 100 000 hours) with the total elongation not exceeding one to two per cent. The extension occurring in the creep-rupture test is much greater, ranging up to 100 per cent per hour, with the total elongation ranging up to 100 per cent or more. Accordingly, different means are employed in the two types of tests for measuring the extension. Creep tests involve small amounts of strain; therefore, the strain measuring equipment must be appropriately sensitive to accurately measure the deformation that occurs. One very sensitive system measures the movement of reference marks engraved on polished platinum beads on arms extending from cylinders seated against the shoulders of the specimen. The measurement is made by a microscope sighted through a window in the furnace wall.

In creep tests requiring less sensitivity, the creep extension is measured with mechanical extensometers that are clamped to the shoulders of the test specimens. The extension is recorded autographically on a strip-type recorder with the use of a linear-differential transformer or recorded manually from a dial gage attached to the extensometer.

SECTION 8

MISCELLANEOUS MECHANICAL TESTS

Metals are subjected to a great variety of applications which may involve mechanical properties not directly measured in the more common mechanical tests already described. For this reason, a great many specialized tests have been developed which are usually aimed at a closer approximation to some important aspect of the actual service condition than the ordinary tests provide. A few of the more important miscellaneous mechanical tests are briefly described in this section, in order to provide some indication of the types of tests which have been developed.

A. Compression Testing—Frequently, in the design of structural members which are to be subjected to compressive working stresses, it is desirable to design on the basis of compressive yield strength rather than tensile yield strength, particularly if there is reason to believe that the compression properties of the material under consideration differ from the tension properties. The data obtainable from a compression test may include the yield strength or yield point, and in some cases "compressive strength." The term compressive strength has been defined by the American Society for Testing and Materials as the maximum compressive stress which a material is capable of developing. This strength figure has a definite value only for a material which fractures in compression. For other materials, arbitrary compression strength values are sometimes reported which are based on some degree of distortion which is regarded as indicating complete failure of the material.

An ASTM specification for the compression testing of metallic material is designated as E 9. It is recommended that standard specimens be in the form of circular cylinders, the important feature in specimen preparation being parallelism of the ends and perpendicularity of the planes of the ends to the specimen axis. As in the case of tension testing, axial loading is of great importance. In some instances, a special subpress is used in conjunction with the regular testing machine in order to facilitate truly axial application of the compression load.

Compression members are frequently fabricated from sheet material, particularly for use in aircraft. In the design of such members, it is necessary to know the compression properties of the sheet material. Obviously, edgewise compression tests are not simple on thin sheets because of buckling difficulties. Several methods of testing sheet specimens in compression have been proposed. One of these is the "pack" method, in which a composite specimen is built up of several layers of sheets. In this way, a specimen of sufficient thickness to avoid buckling is provided. Another type of test provides support against buckling by special jigs. One such jig consists of a number of rollers which rest against the faces of the specimen. Another type of fixture which was developed at the National Bureau of Standards simply uses flat tool-steel blocks lubricated with a high-pressure lubricant to support the specimen (Figure 50—52). The specimen is allowed to overhang the supports slightly on all sides for loading and attachment of extensometers. Still another type of sheet metal compression test is the cylinder method, in which the flat specimen is formed, by bending rolls, into a cylinder about 38 millimeters (1½ inches) in diameter and soldered along the longitudinal joint. This cylinder is very resistant to buckling and the ends can be accurately machined to insure axial loading. Another advantage of this method is the accessibility of the specimen for strain measurement. The principal disadvantage of the cylinder method is that a small amount of cold work is unavoidably introduced in forming the cylinder, and the inhomogeneity at the solder joint.

B. Bend Testing—The bend test, as the name implies, is intended to evaluate the ability of a material to undergo bending during forming operations to which it may be subjected. Generally, the bend test is conducted as a "go, no-go" test; i.e., either the specimen meets the desired bend requirement or fails by cracking. In some instances, however, a ductility value is

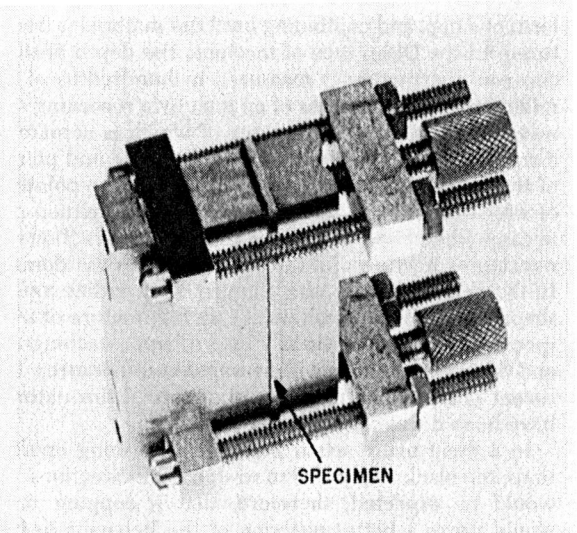

Fig. 50—52. National Bureau of Standards jig for compression tests on sheet metals. (Courtesy, U.S. Department of Commerce, National Bureau of Standards.)

derived from the bend test by placing gage marks on the outside or tension side of the bend and measuring the elongation after completing the bend. This procedure is covered by ASTM Specifications, E 190, E 290, A 720 and A 721 and by specifications of the American Welding Society. The first method was developed primarily for testing of welds.

Ordinarily the bend test is quite simple, merely involving a determination of whether or not a specified bend can be made satisfactorily. A typical method of stating a bend-test specification for a plate material, for example, is as follows: "The bend-test specimens shall stand being bent cold through 180 degrees without cracking on the outside of the bent portion to an inside diameter which shall have the following relation to thickness or gage of material." A set of bend diameters is then specified for various thickness ranges, the bend diameters increasing with increasing plate thickness. Edge conditions are very important and generally sufficient edge preparation is permitted to avoid an initial edge fracture. The method of bending is not specified and a large number of bending devices have been designed, the design usually being aimed at convenience so that large numbers of tests can be run in a relatively short time.

C. Cupping Tests—A number of so-called "cupping" tests have been developed for the purpose of measuring the ductility of sheet metal under conditions where the sheet is stretched in all directions simultaneously. Cupping tests are made on different machines, or testers, known as the Erichsen, the Olsen, the Guillery, the Wazau, etc., of which the first two are the most commonly used. While they differ in many respects, the Erichsen and the Olsen testers are similar in the manner of applying the test. A description of these tests is found in ASTM Method E 643. In both tests, the specimen of sheet or strip is clamped between two rings or dies, and a smooth ball, mounted upon or attached to a plunger, is forced against the flat surface of the specimen enclosed within the area of the ring, as shown in Figure 50—53, thus stretching it into the form of a cup, and continuing until the material is fractured. In the Olsen type of machine, the depth of the cup causing fracture is measured in hundredths of a millimetre or thousandths of an inch by a recording or measuring device, the indicator of which is actuated directly from the surface of the sample. The end point of the test is indicated by a pressure gage, the pointer of which drops back upon fracture of the specimen or in cases where there is no gage, by visual observation of necking or fracture of the test specimen in the dome. In the Erichsen tester, the plunger is somewhat cone-shaped with a smooth spherical end; the fracture of the specimen is detected visually by a mirror attachment, and the depth of the cup is measured in millimetres. In recent years, additional types of testers of this nature have been developed.

In a great many actual sheet-metal forming operations, the blank is required to stretch in all directions. It would be expected, therefore, that a cupping test would prove a better criterion of the behavior to be expected in such forming operations than would a simple tension test in which the material is stretched in only one direction. Actually, the correlation between

cupping tests and actual performance has in general been disappointing, except in cases where large differences in formability exist. It is thought by some that the reasons for the lack of correlation is that the test variables, primarily lubricating and ball size, are not representative of actual sheet-metal forming operations. Cupping tests are widely used for inspection purposes, however, since they provide a quick indication of ductility and some indication of the surface condition to be expected after drawing by the degree of roughness or coarseness developed on the cup during the cupping test.

D. Strain-Sensitivity and Strain-Aging Sensitivity Tests—Steel products are very frequently subject to cold-forming operations prior to or during fabrication for their final use and may go into service in the cold-worked condition. As discussed in the chapter on plain carbon steels, the properties of cold-worked steel may change progressively with time, this change being known as strain-aging. The question arises, therefore, as to how the changes in properties brought about by cold working and strain-aging will affect the performance of the material in service.

It is well known that straining and strain-aging exert a profound influence on the notch toughness characteristics of certain steels, tending to increase the susceptibility of these steels to brittle fracture. One of the most informative methods of evaluating the effect of straining and strain-aging of plate steels on notch toughness is the determination of the shift in transition temperature in the Charpy impact test. The interpretation of the relationship between test results and service behavior is, of course, subject to the same limitations as emphasized earlier in the general discussion of notch-toughness tests. It is possible, however, to obtain extremely useful comparisons among different steels and to provide relative measures of the extent to which the notch toughness is impaired by straining and strain-aging.

One testing procedure which has proved convenient and useful involves the cold rolling of oversize blanks for Charpy-type specimens. The degree of oversize is based on the desired amount of cold working; for example, if ten per cent reduction is desired, the blank is made about 1 millimetre (0.0394 inch) oversize and reduced to the standard dimension of 10 millimetres (0.394 inch). Two sets of specimens sufficient for the determination of transition temperatures are prepared in this manner. One set is tested as soon as possible after rolling and notching, while the other set is artificially aged for one hour at 285°C (550°F). This treatment is believed to produce the maximum effect of strain aging on notch toughness. The shifts of transition temperature caused first by straining, and second by straining and aging give an indication of the extent to which the ability to resist brittle fracture has been impaired. Examples of the types of transition behavior which may be obtained are shown in Figure 50—54.

Another test which is sometimes used to indicate the effects of straining and strain-aging on notch toughness is the Graham tapered-bar test. Varying amounts of cold work are produced by drawing a tapered circular bar through a die so as to produce a uniform cross section. The bar is then notched at various points along its length which correspond to various amounts of cold work and is broken as a cantilever specimen at each notch. The maximum amount of cold work is usually ten per cent. The cold-drawn bar can also be aged before testing. Although this test is relatively fast and simple, it has the disadvantage of not permitting the determination of transition temperatures, and is thus subject to the uncertainties which arise in impact testing at a single temperature.

Steels which exhibit pronounced strain aging show an increase in tensile strength when tested at temperatures in the neighborhood of 200 to 300°C (400 to 600°F) over that obtained in room-temperature tests. This increase in tensile strength is sometimes used as a measure of the effects of strain aging, but should not be substituted for a notch toughness test unless a correlation has been established.

Another aspect of strain aging, which is of great practical importance, is the return of the yield point in temper-rolled sheets intended for deep-drawing operations. Such sheets are normally temper rolled after annealing in order to eliminate the yield-point elongation and the accompanying tendency for the formation of stretcher strains or fluting during forming. In steels which are susceptible to strain aging, however, the

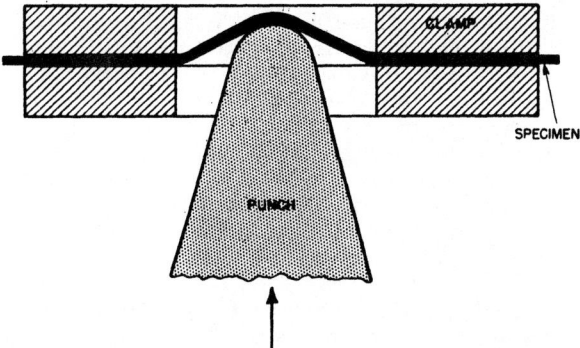

FIG. 50—53. Schematic representation of cupping test for sheet metals.

yield-point elongation and stretcher-strain tendency will return with time. Accelerated aging, in which the temper-rolled material is held at an elevated temperature such as 200°C (400°F) for some predetermined length of time, is frequently applied in order to indicate the aging tendency. The extent to which the yield point elongation returns in a tension test provides some indication of the manner in which the material can be expected to behave in a drawing operation after a con-

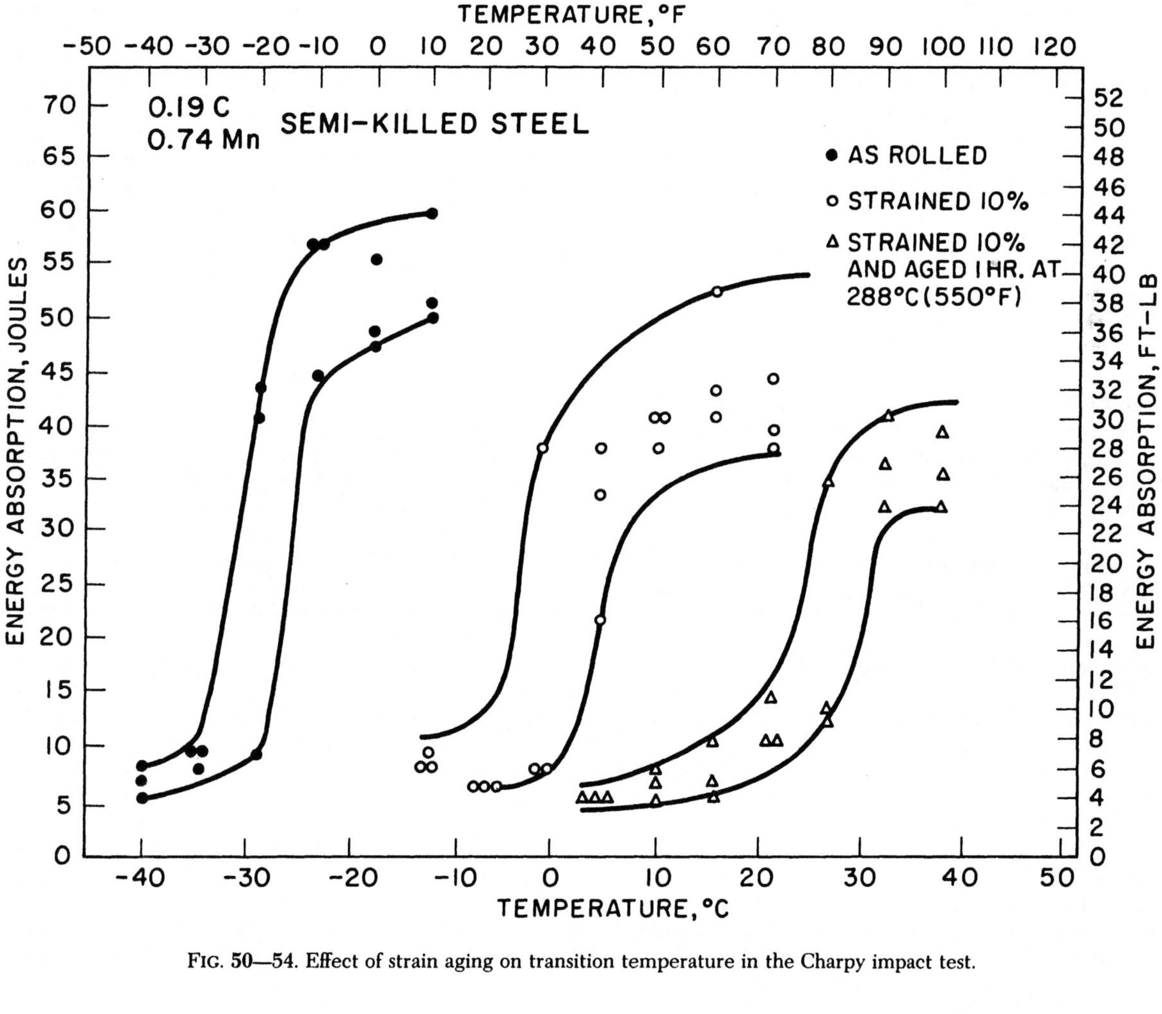

FIG. 50—54. Effect of strain aging on transition temperature in the Charpy impact test.

siderable lapse of time at atmospheric temperature.

E. Torsion Testing—In the torsion test, a specimen is subjected to twisting or torsional loads analogous to those encountered in drive shafts, crank shafts, etc. Information on the strength in torsion, particularly the yield point or yield strength, is important in designing for such applications. Torsion tests are not extensively used, since a satisfactory estimate of the yield point in torsion can generally be made from the yield point in tension as discussed previously in the section on "Significance of the Tension Test." In some instances, however, a direct measurement of torsion properties may be desirable and occasionally may be specified.

Torsion data are usually obtained in the form of a torque-twist curve, in which the applied torque is plotted against the angle of twist. Torsion produces a state of stress known as pure shear, and the shear stress at yielding can be calculated from the torque at yielding and the specimen dimensions. Actually, the stress varies from a maximum at the surface of the specimen to zero at the axis. In the elastic range, the variation is linear, and the maximum stress for a cylindrical specimen can be readily calculated from the following relation:

$$S = \frac{16T}{\pi d^3}$$

where:

S = maximum shear stress (MPa or psi)

T = torque in joules or inch-pounds

d = diameter of specimen in mm or inches.

In the plastic range, the calculation of the maximum shear stress is more complicated and the reader is referred to the method developed by A. Nadai, which considers the twisting of a cylindrical bar in the plastic range.

In the elastic range, the shear strain is proportional to the shear stress, the constant of proportionality being known as the **shear modulus** (modulus of rigidity). The shear modulus for steel is about 68 950 MPa (10 000 000 psi). The procedure for determining this value is given in ASTM method E 143.

Impact tests utilizing dynamic torsional loads are used to a considerable extent in testing brittle materials such as tool steels. Since the ductility in torsion is greater than in tension, a greater energy absorption is obtained in the torsion impact test than in a beam type of impact test. Sensitivity is thereby improved making separations possible in the torsion impact which are not possible in the notched-bar impact test. The energy absorbed by the specimen in the torsion impact test is measured by the loss in rotational energy of a flywheel which engages one end of the specimen and breaks it.

Torsion tests are also performed on wire products as a measure of ductility (see ASTM A 112, A 411, and E 558). A modification of the torsion test, the hot-twist test, is used as a measure of the hot-working characteristics of steels; that is, the optimum rolling temperature, forgeability, and so on.

F. Shear Testing—The term "shear testing" as used here refers to determinations of the resistance of metals to shearing in dies, i.e., cutting by shearing. "Shear test" is sometimes also used to refer to the torsion test, which, as indicated above, measures the resistance to deformation under shear stresses. The sense in which the terms "shear" or "shear strength" are used should always be clearly indicated in order to avoid any misinterpretation.

If a load-penetration diagram is determined while shearing a metal in dies, it will be found to be similar in general shape to the load-extension diagram in tension testing. The maximum load observed during the shearing operation divided by the area being sheared is taken as the shearing strength or shearing resistance of a material. It is necessary to state the exact testing conditions in reporting shearing resistance, since the value obtained will depend markedly on the die arrangement. The clearances, shear angles, and sharpness of the cutting edges will all affect the observed value of shearing resistance. The degree of penetration of the punch into the metal when the fracture begins is also usually reported. It has been generally observed that the shearing resistance of medium carbon steels is from two-thirds to three-fourths of the ultimate tensile strength.

G. Wear Testing—H. W. Gillette has defined the wear of a metal part as "its undesired gradual change in dimensions in service under frictional pressure." Wear generally involves two stages, in the first of which deformation occurs, and in the second of which removal of material may occur. Wear of metals may involve the contact of metal on metal, as in shafts and bearings, brakes and wheels, valves and seats; or it may involve the contact of non-metals on metals, as in the case of coke chutes, or in shovel buckets, etc. The phenomenon of wear is so complex that it is extremely difficult to interpret, and is one kind of service for which suitability can be reliably evaluated only in terms of actual service tests. Various wear tests have been used for specific purposes, but are only valid if the test method produces wear in the same manner in which it is produced in service. A more detailed discussion of wear and wear testing can be found in the publications by Haworth, and Gardner that are listed in the bibliography at the end of this chapter.

H. Damping Capacity Tests—Damping capacity is a measure of the rate at which a material dissipates energy of vibration, or in other words, a measure of the ability to damp out vibrations. Damping depends upon internal friction in the metal, which is manifested at stresses well below those at which gross yielding occurs. Internal friction probably arises from minute amounts of plastic flow on a submicroscopic scale, a process which results in heating and a loss of vibrational energy. The ability to damp vibrations is of importance in certain structures subjected to vibrations, where there may be a danger of resonant vibrations arising. Resonance can lead to large amplitudes of vibrations and excessively high stresses. A choice of material with relatively high damping capacity, which can also satisfy the ordinary mechanical requirements, may be of some benefit in avoiding resonance conditions. High damping capacity materials are also of value in supports for moving machinery, in that the transmission of vibrations to the supporting structure may be reduced.

A commonly used method of measuring damping capacity involves the measurement of the rate of decrease of amplitude of torsional vibrations. One end of a cylindrical specimen is clamped in a rigid base with the specimen in a vertical position. On the other end of the specimen, a heavy inertia bar is clamped. This inertia bar is rotated through an angle corresponding to the desired stress level in the specimen, usually by means of magnets, and then released. The specimen is thus set in torsional vibration, and the rate of decrease of vibration amplitude is measured by some suitable method. In a recently developed machine, a light beam is focused on a mirror on the inertia bar, and the beam is reflected onto a rotating drum carrying a strip of sensitive photographic paper. Measurements of the rate of decrease of vibration amplitude from the photographic record permit a calculation of the damping capacity. Damping capacity is usually expressed in terms of **specific damping capacity,** which is defined as the ratio of the energy loss per cycle to the elastic potential energy at the maximum amplitude of the cycle. Electrical-resistance strain gages and high-speed recording devices can also be used to determine damping capacity.

Bibliography

Am. Iron and Steel Institute, Sheet steel properties and fatigue design. The Institute, Washington, D. C., 1981.

Am. Society for Metals, Metals handbook; Cleveland, The Society, Ninth ed. 1985.

Am. Society for Testing Materials, Manual on fatigue testing (Special technical publication no. 91). Phila., The Society, 1949.

Am. Society for Testing Materials, A tentative guide for fatigue testing and the statistical analysis of fatigue date (Special technical publication no. 91A). Phila., The Society, 1958.

Am. Society for Testing and Materials, ASTM Standards, Part I, Ferrous metals. Phila., The Society, current edition. Section on testing methods.

Am. Society for Testing Materials, Symposium on wear of metals (Special technical publication no. 30). Phila., The Society, 1937.

Am. Society for Testing and Materials, Standard method of testing for plane-strain fracture toughness of metallic materials. ASTM Standard E399, Part 10, 1981.

Am. Society for Testing and Materials, Fracture toughness testing and its applications (Special Tech. Publication 381). Phila., The Society, 1965.

Am. Society for Testing and Materials, Correlation of short- and long-time elevated temperature test methods (Project no. 25); Appendix II of Report of Joint Research Committee on effect of temperature on properties of metals. Am. Society for Testing Materials Proc. 44: 186-215 (1944).

Barsom, J. M., Fatigue behavior of pressure-vessel steels. WRC Bulletin 194, Welding Research Council, New York, May 1974.

Barsom, J. M., Fatigue-crack propagation in steels of various yield strengths. Trans. ASME, Jour. of Engineering for Industry, Series B, 93, No. 4, November 1971.

Barsom, J. M., and Rolfe, S. T., Correlations between K_{IC} and Charpy V-notch test results with transition-temperature range. ASTM Special Technical Publication 466, ASTM, 1970.

Brown, W. F., Jr., and J. E. Srawley, Plane strain crack toughness testing of high strength metallic materials. ASTM Special Tech. Publication 410. Phila., The American Society for Testing and Materials, 1967.

Craig, H. L., Jr., Crooker, T. W., and Hoeppner, D. W., (eds.) Corrosion fatigue technology. Special Technical Publication 642, ASTM, 1978.

Fisher, J. W., Guide to 1974 AASHTO fatigue specifications.

American Institute of Steel Construction, 1974.

Foster, P. F., Mechanical testing of metals and alloys; 4th rev. ed. N. Y., Pitman, 1948.

Gardner, D., These machines grind, scratch, gouge to measure abrasive wear. Prod. Eng., Nov. 9, 1959.

Garofalo, F. and G. V. Smith, Effect of time and temperature on various mechanical properties during strain aging of normalized low carbon steels. Am. Society for Metals Trans. 47: 957-983 (1955).

Garofalo, F., P. R. Malenock and G. V. Smith, Hardness of various steels at elevated temperatures. Am. Society for Metals Trans. 45: 377-396 (1953).

Garofalo, R., P. R. Malenock and G. V. Smith, Influence of temperature on elastic constants of some commercial steels. Am. Society for Testing Materials, Symposium on determination of elastic constants (Special Technical Publication 129) Phila., the Society, 1952.

Goldhoff, R. M., Towards the standardization of time- temperature parameter usage in elevated-temperature data analysis. ASTM Jour. of Testing and Evaluation, Vol. 5, No. 5, Sept. 1974, p. 387.

Greham, J. A., (ed.), Fatigue design handbook. SAE Advances in Engineering, Vol. 4, Soc. of Automotive Engineers, Warrendale, Pa., 1968.

Griffith, A. A., Trans. Royal Society A, Vol. 221, page 163, 1920.

Gross, J. H., The effect of strength and thickness on notch ductility. The Welding Journal 48 (10) Research Supplement, 441s to 453s, 1969.

Gross, J. H., and R. D. Stout, Ductility and energy relations in Charpy tests of structural steels. The Welding Journal 37 (4), Research Supplement, 151s to 159s, 1958.

Haworth, R. D., The abrasion resistance of metals. Am. Society for Metals Trans. 41: 819-869 (1949).

Holt, J. M., Use and reproducibility of a gage to measure the lateral expansion of Charpy V-notch specimens. Special Technical Publication 608, American Society for Testing and Materials, 1976, p. 67-90.

Holt, J. M., Strain-controlled fatigue properties of hot-rolled DQSK steel compared with those of EX-TEN F50 and dual phase 80 steels. Paper 810434, Society of Automotive Engineers, Warrendale, Pa., 1981.

Holt, J. M., and Stewart, B. K., Strain-controlled fatigue properties of USS EX-TEN F50 steel. SAE Technical Paper No. 790460, SAE Congress and Exposition, 1979.

Irwin, G. R., Fracturing of metals (ASMS, 1947), Am. Society for Metals, Cleveland, Ohio, 1948.

Irwin, G. R., Fracture mechanics. Published in Structural Mechanics, Proceedings of the first Symposium on Naval Structural Mechanics. Pergamon Press, 1960, pages 577-594.

Kula, E. B., and Fahey, N. H., Effect of specimen geometry on determination of elongation in sheet tension specimens. Materials and Research Standards, August, 1961; pages 631-636.

Lysaght, V. E., Indentation hardness testing. N. Y., Reinhold, 1949.

McVetty, P. G., The interpretation of creep tests. Am. Society for Testing Materials Proc. 34 (Part 2): 105-116 (1934).

Miller, R. F., G. V. Smith and G. L. Kehl, Influence of strain rate on strength and type of failure of carbon-molybdenum steel at 850, 1000 and 1100 degrees Fahrenheit. Am. Society for Metals Trans. 31: 817-848 (1943).

Mould, P. R., and Johnson, T. E., Jr., Rapid assessments of drawability of cold-rolled low-carbon sheet steel. Sheet Metal Industries, June 1973.

Nadai, A., Plasticity. N. Y., McGraw-Hill, 1931.

Orowan, E., Fatigue and fracture of metals (MIT Symposium, 1950). New York, John Wiley and Sons, Inc.

Parker, E. R., Brittle behavior of engineering structures. New York, John Wiley and Sons, Inc., 1957.

Peterson, R. E., Stress concentration factors. John Wiley and Sons, New York, 1974.

Rolfe, S. T. and Barsom, J. M., Fracture and fatigue control in structures. Prentice-Hall, Englewood Cliffs, N. J., 1977.

Rolfe, S. T., Fracture-control guidelines for welded steel ship holds. Significance of Defects in Welded Structures, Proc. of the Japan-U.S. Seminar, 1973, Tokyo, edited by F. Kamozara and A. S. Kobanzaski. University of Tokyo Press, 1974.

Smith, G. V., Properties of metals at elevated temperatures. N. Y., McGraw-Hill, 1950.

Symposium on significance of the tension test of metals in relation to design. Am. Society for Testing Materials Proc. 40: 501-609 (1940).

Templin, R. L., The determination and significance of the proportional limit in the testing of metals. Am. Society for Testing Materials Proc. 29 (Part 2): 523-553 (1929).

Toda, H., Paris, P. C. and Ironin, G. R., The stress analysis of cracks handbook. Del Research Corp., 1973.

Willems, N., Easley, J. and Rolfe, S. T., Strength of materials. McGraw-Hill, New York, 1981 (specifically Chapter 13).

CHAPTER 51

Nondestructive Inspection of Steel

Introduction—Nondestructive inspection (NDI) encompasses all methods used to evaluate materials for mechanical and dimensional characteristics, for composition and for the presence of imperfections without impairing their serviceability in any way. NDI is much more than the measurement of dimensions and the visual assessment of product acceptability. It employs some form of controlled energy or excitation applied to the product and measures the response to that impetus, thereby permitting vital characteristics to be inferred or measured directly. For instance, responses to electro-magnetic fields, ultrasonic vibrations, or penetrating radiation such as X- or gamma rays are commonly used to measure thicknesses, indicate hardness, and detect imperfections without harm to the part—that is, nondestructively. It is a reliable and cost-effective way to make certain measurements and monitor parameters that would otherwise require destructive tests. In the steel industry, NDI is employed to detect and locate harmful discontinuities, to monitor dimensions such as thickness, width, and diameter, and to check for correct chemical composition and mechanical properties without destroying or affecting the serviceability of the product in any way. It is, furthermore, often specified for critical oil-country, pipeline, nuclear, and automotive products.

Inherent in NDI are parameters which permit relatively high-speed product assessment and have given rise to on-line monitoring of products for dimensional and mechanical parameters, and to provide early detection of gross imperfections and the generation of repetitive mechanical discontinuities. The vast increases in electronic capabilities for collecting and assessing large volumes of data now make automatic product-quality assessment and data tracking feasible. NDI applications in the steel industry now include real-time quality control and quality data-tracking systems as part of a plant quality-assurance program in addition to an array of reliable, portable individual instruments for specific manual or automatic functions. Through judicious selection of methods and systems design, steel products can be monitored from the caster or pouring floor through to the finished product. NDI permits timely feed-back and feed-forward of product-quality information, can divert or scrap unacceptable product, and can pinpoint production trouble spots for corrective action, thereby improving yield and reducing production costs and scrap losses. NDI, in concert with modern product-handling facilities and computerized data processing and control, offers opportunities to achieve the goal of high quality in conjunction with low production costs. This chapter reviews the several nondestructive methods commonly used in the steel industry with emphasis on those that are widely used and/or offer potential for use in cost-effective inspection systems. These include visual inspection and aids to visual inspection, radiography, ultrasonics, and magnetic and electromagnetic methods.

VISUAL INSPECTION

Visual inspection is the oldest and most commonly used nondestructive method. It requires inspectors with good eyesight and sharpened visual acuity to recognize salient features in a product and make judgments as to its acceptability. The combination of the human eye and brain provides high-resolution, full-color, automatic-focusing sensors capable of selective or detailed surface scanning coupled with an adaptive-learning system for feature recognition, classification, and disposition. Visual inspection may be employed on an almost limitless number of products under a wide variety of environmental conditions. Used with specific aids to visual inspection, both the scope and accuracy of the visual method can be increased significantly.

The visual inspector may be trained to recognize and assess certain features such as seams, laps, slivers, pits, and roll-ins on hot-rolled surfaces, to inspect assemblies for the presence of all working components, or to classify auto-body sheets for dinges, scratches, and discoloration. The inspector may be required to make rapid, repetitive judgments on high volumes of materials or to carefully assess welded joints in pressure vessels, bridges, or pipe. Visual inspection responds to the shape and color of the product, its surface texture, and contrast. How well these variables may be seen and assessed depends upon the illumination: its intensity, color, and the angle of incidence that might either highlight or mask indications or cause glare that would impair the inspector's efficiency. In addition to proper lighting, the surface of the product must be free from loose scale, dirt, oil, grease, or any other substance that would impair inspection.

Of major importance are the inspector's basic visual capabilities, training, and the development of visual acuity. No matter how good the inspector's eyesight, real performance is based upon the ability to rapidly detect, recognize, and assess specific parameters and/or discontinuities based upon training and written acceptance criteria. These may all be modified significantly by the inspector's physical and/or emotional condition and stress, as well as environmental and production-related variables.

The visual method is highly versatile, applying to nearly any size or shape of product; it provides direct and immediate information; it requires few, if any, aids; and it is easily learned. It is, however, limited to visible surfaces, is subject to vagaries in inspector performance and environment, and provides little, if any, information regarding the depth of imperfections.

Aids to Visual Inspection—The scope and accuracy of visual inspection can be increased significantly by the use of proper aids. Simple magnifiers and magnifiers with graduated reticles are very useful in assessing surface imperfections. Optical comparators are used routinely to measure compliance of machined surfaces with standards.

The rigid-tube borescope has been in use for years for the internal inspection of pipe, heat exchangers, and the like. The developments in fiber optics have resulted in flexible, high-resolution borescopes with diameters less than 12.7 mm (0.5 inch) and a wide selection of optics that permit visual inspection of hitherto inaccessible locations. These devices are often used with cameras and television systems to further enhance their utility and to provide permanent record capabilities.

High-resolution, high-sensitivity television systems are ideally suited for expanding visual-inspection capabilities. Systems are available that are sensitive to infrared or to visible light. In their simplest form they permit remote inspection of product in locations or under conditions that render conventional visual inspection impractical. Such installations are used to monitor the operation of various steelmaking and processing functions, permitting an operator to view various operations from one location, and are also used to inspect the surfaces of hot slabs, plate, strip, and cold-finished products by employing special television and illuminating techniques. For instance, by using stroboscopic techniques the surface of hot strip may be inspected during rolling. Such units provide stop-frame images which are held for sufficient time to permit evaluation. Furthermore, digital data-gathering and analysis methods are being used for the detection and identification of specific surface imperfections on hot slabs. It is expected that such pattern-recognition techniques can be used for real-time surface inspection of a wide variety of steel products under diverse operating conditions. Such developments portend automatic, in-line surface inspection and classification of steel products, and will allow automatic diversion of product for conditioning or for scrap. Acceptable product can continue on without the delays that were heretofore necessary for manual surface inspection and conditioning. It is evident that considerable savings may accrue, for instance, in the direct rolling of slabs to plate without the delays and reheating costs attendant with conventional methods where slabs are cooled, inspected, and conditioned prior to charging into the plate mill. Computer-controlled surface-inspection systems may be used on hot or cold products and will provide permanent records of the nature and location of surface imperfections; piece and heat number; date, turn, and order numbers; statistical classification of product quality; and pertinent data for plant data/tracking systems.

Other aids to visual inspection are classified as separate nondestructive methods. These include the liquid-dye-penetrant and the magnetic-particle methods that enhance the visibility of surface imperfections severalfold and penetrating-radiation or radiologic methods which allow inspection of the product interior as well as surfaces. These will be treated separately later on in this chapter.

The major requirements for effective visual inspection must be met whether it be in its simplest manual form or by computer-controlled pattern recognition and analysis. There must be adequate illumination of the proper type and incidence to provide for discrimination of changes in color, texture, and contour; the product surface must be relatively free of scale, dirt, grease, or other material that would adversely affect inspection; and the environmental conditions must be conducive to accurate and consistent product classification. Otherwise, the effectiveness of visual inspection will be reduced.

Visual inspection is the most widely used method in the steel industry and is used for the inspection of semi-finished as well as finished products to detect seams, laps, cracks, slivers, roll-ins, rolled-in scale, scum patches, and similar surface imperfections. It is used in the inspection of machined products, notably pipe, for bevel and thread geometry and finish, and for the inspection of welds for cracks, undercut, weld geometry, and the like. Visual inspection is highly versatile and effective when performed by properly trained inspectors under proper conditions. It is, however, less consistent than other methods that are less operator-dependent.

MAGNETIC-PARTICLE INSPECTION

The magnetic-particle method originated in the 1920's and is widely used throughout the steel industry. It is basically a surface-inspection technique and can be considered an aid to visual inspection. It is simple in concept and in application and relies on the physical principle that any discontinuity that cuts across the magnetic-flux lines in a magnetized ferromagnetic material forms local magnetic poles with a north pole being at one surface of the imperfection and a south pole being at the opposite surface. For a given discontinuity, the strength of the local poles is maximized when the major plane of the discontinuity is normal to the flux lines. As the angle of the plane of the discontinuity decreases toward zero, the pole strengths also decrease until there are no poles formed when the discontinuity is parallel to the flux lines. In practice, the minimum permissible angle is 45 degrees. Aside from orientation, the major contributors to localized pole strength are the nature and intensity of the applied field and magnetic properties of the material. The physical characteristics of the imperfection such as its depth, width, and length, and local decarburization around the imperfection all affect the pole strength. The effect of the N-S opposing poles across the discontinuity is to produce a magnetic leakage field or fringe flux diverted from its normal path within the material into the space immediately adjacent to the discontinuity. The shape and magnitude of the diverted magnetic field is, of course, affected by the factors mentioned above. It is reduced in amplitude and somewhat broadened by magnetic scale lying within or bridging across the discontinuity at its external surface.

It is this diverted magnetic field (leakage field or fringe flux) that attracts and holds finely divided magnetic particles sprinkled on the surface of the part. The configuration of the magnetic-particle indication and the amount of particle buildup give some indication of the nature and depth of the discontinuity. For the most part only those discontinuities that break the surface and that cut across the applied field (at angles equal to or greater than 45 degrees) will produce significant indications. In some cases subsurface discontinuities lying close to the surface will produce fuzzy but identifiable indications.

As in any nondestructive method a knowledge of its salient concepts is necessary for it to be successful. Powder selection is important. Usually, powders are dyed to contrast with the surface of the product being inspected, gray, red, and black being the most common colors. The powder grain size is usually a combination of relatively large, low-mobility and smaller, lighter particles. They are usually made of easily magnetized soft iron, and when they congregate the larger particles tend to attract some of the smaller particles to further build up the indications. The powders may be dry and applied with a shaker, blower, or squeeze bottle, or they may be in a liquid suspension to provide greater mobility. In addition to contrasting color dyes, some magnetic particles are covered with a fluorescent dye and the inspection conducted under ultraviolet light. The magnetic-particle-inspection method is commonly referred to as the Magnaflux method and the fluorescent particle, Magnaglo—trade names of the primary developer of the method. When applying the powders care must be exercised so that powder motion is gentle, allowing maximum powder capture at a discontinuity. In addition to the proper powder-application technique, the nature of the magnetizing field and the time at which powder is applied are all variables that must be understood and controlled. Powder mobility is essential so that all inspection surfaces receive an adequate "dusting" to achieve optimum sensitivity. This is not only a matter of application and powder characteristics, but is also directly affected by the strength and nature of the applied magnetic field. In general a pulsating direct-current magnetization is recommended for optimum mobility and flaw-depth discrimination. Furthermore, powders in liquid carriers are generally more sensitive than dry-powder methods.

Magnetization may be accomplished in a number of ways, but the most common are the direct, induction,

and yoke methods. In the direct method, current is passed through the part and the magnetic field thus generated is at right angles to the current path. For instance, if a current is passed lengthwise through a billet, bar, or pipe, a transverse field around the outside surface will be generated. If two probes are placed on the surface of an object and a current passed between them, the resulting magnetic fields would be transverse to the current paths between the probes. To magnetize an object longitudinally, therefore, several turns of a current-carrying conductor are wrapped around the part or the the part may be placed in a solenoid coil. In such cases, the orientation of the field is along the major axis of the coil or solenoid. Magnetic yokes generally take the general form of a "U" of solid or laminated magnetic-core material with a number of turns of copper wire wound around its base. The result is an electromagnet with poles at the open end of the "U". Yokes may be placed on or around objects to obtain magnetic fields of the correct orientation. The control of the type and strength of the field is a function of the power supply used (ac, dc, or pulsating dc) and the design of the magnetizing circuit. For direct magnetization, low-voltage supplies delivering several thousands of amperes are necessary. Yokes employ, in general, voltages from 100 to 200V and currents up to 5 or 10 amperes. Coils and solenoids may be designed for either type of supply.

The final, but by no means least, consideration is the time of powder application. Applying the powder during part magnetization provides the highest sensitivity but also may result in non-relevant indications. By carefully controlling the magnetizing fields and powder application during "active" magnetic-particle inspection, however, consistent and reliable results are obtainable. In many cases, where the inspection requirements permit, powder application after the magnetizing energy has been removed is preferable. This is known as the residual method and is commonly used in the inspection of billets and blooms.

In general, large, mechanized magnetic-particle inspection units are used for the inspection of billets, blooms, and certain pipe and casing grades. These normally employ fluorescent magnetic particles using the "residual" field technique. With three inspectors, an inspection rate of 2 to 2½ 12.2-metre (40 foot) 102- by 102-mm (4-inch by 4-inch) billets per minute can be expected. At 70 per cent utilization, this amounts to approximately 910 metric tons (1000 net tons) per turn. In such installations an inspector marks all rejectable indications by crayon or paint stick, and the billet is then conditioned by grinding, scarfing, or chipping.

Recently magnetic particles which can be "fixed" on the billet surface and that can withstand subsequent handling and storage without losing the indication have been developed. These will essentially eliminate the inspector/marker at the magnetic-particle-inspection station and will rely on the operator of the grinder to remove all indications. The average inspection speeds for a 12.2-metre (40 foot) long 152- by 152-mm (6 inch by 6 inch) billet is 2 per minute. This amounts to about 1490 metric tons (1640 net tons) per turn at 70 per cent utilization. The use of such systems will reduce manpower requirements and should increase inspection reliability.

The magnetic-particle method is simple in concept, and applications range from a simple, manual inspection to highly mechanized, self-marking systems. The inspection requirements are similar to those for visual inspection, requiring a product with a surface that is free from loose scale, grease, and dirt, and that can be mechanically manipulated for magnetization and inspection without jamming or cobbling. Finally, the choice of powder, method of application, and the control of magnetization are essential for successful magnetic-particle inspection.

EDDY-CURRENT METHODS

Eddy-current methods are used to measure physical and mechanical parameters and for the detection of surface imperfections in steel products. Eddy currents are so-named because their paths often resemble the circular eddies in water. The eddy-current method measures the electromagnetic interaction between a transducer or test coil and the part being inspected (Figure 51—1). By this interaction such physical parameters as the hardness, steel grade, and case depth can be inferred. In addition, cladding thickness, foil thicknesses, and the presence of surface and subsurface imperfections may be indicated.

The basic concepts in eddy-current inspection are well-known electromagnetic phenomena. When an alternating current is passed through a coil (the test probe or transducer) an alternating electromagnetic field of the same frequency is produced which surrounds the coil and emanates much like a radio wave into the surrounding space. Upon encountering another conductor this alternating field will induce a current in that conductor. The induced eddy current will be of such direction as to produce a second (reaction) field to oppose or reduce the field that produced it. If the test coil and test piece are far apart, the eddy current is feeble and the reaction field negligible. As the spacing is reduced, however, the effect becomes more pronounced and increases rapidly at close proximity. The exact values of this reaction are directly related to the size and electromagnetic properties of the test piece, the design of the test coil, the test frequency, and the orientation of the coil with respect to the test piece (conductor). This interaction is often referred to as "loading" of the test coil and can be detected directly by any measurement of the current through or voltage across the coil. The electrical property that changes the loading is known as impedance. As the spacing between the coil and part is increased, the coupling is decreased and the loading less pronounced. This is known as the "lift-off" effect. It is important, therefore, to maintain the coil-to-part spacing and the coil-excitation frequency and current constant if eddy-current measurements are to be consistent. The signals generated in eddy-current inspection reflect the changes in the coil-part interaction and are measured as voltages that are vector quantities, having both amplitude and direction (phase). By comparison and analysis of impedance changes, eddy currents are used to disclose imperfections and to measure product mechanical or physical parameters.

Eddy-current measurement of bulk properties such

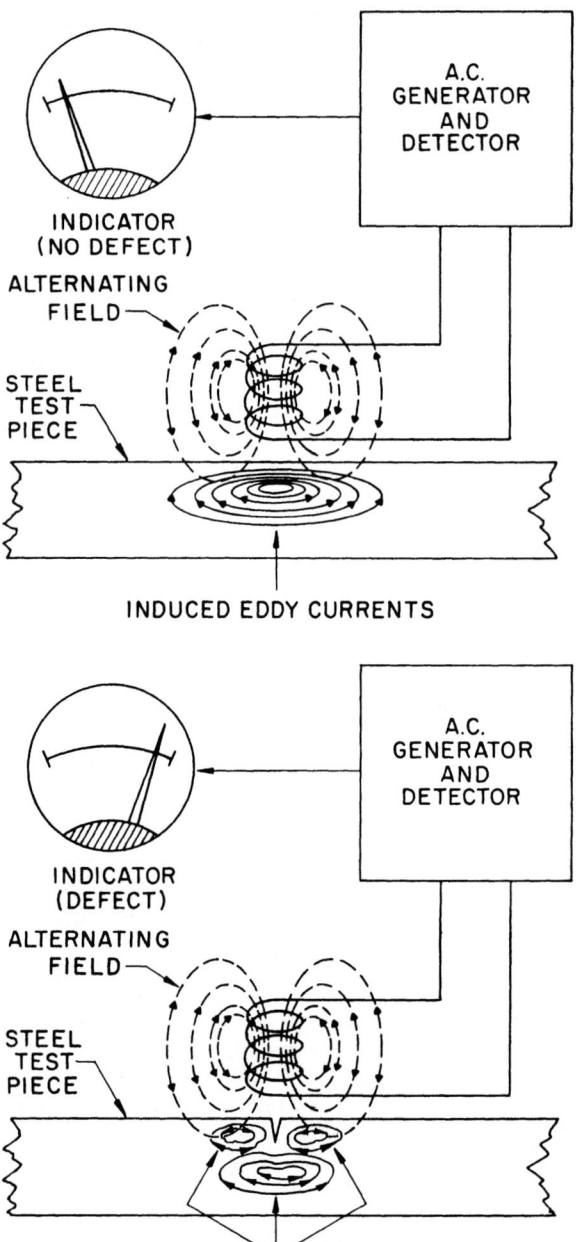

FIG. 51—1. Simplified illustration (above) of eddy-current generation in a uniform conducting test piece, and (below) eddy-current generation in a conducting test piece containing an imperfection. It should be noted that the eddy-current paths actually are confined to a shallower depth than this schematic sketch indicates.

as composition, hardness, and case depth involves analysis of the total signal generated in the transducer or test coil. High-strength, high-carbon, and alloy steel grades are magnetically "hard" and relatively difficult to magnetize (low magnetic permeability). Low-strength and low-carbon steels are magnetically "soft" and easily magnetized (high magnetic permeability).

Under similar test conditions, the magnetically soft material will result in signals with greater amplitude than magnetically hard materials. The preferrable configuration for bulk-properties tests, such as grade verification, is to use encircling coils wherein the test piece comprises the core. The test coil usually has two windings, and the secondary voltage is compared with a signal which is representative of a product with known parameters. If they are similar, the steel grades are judged to be similar. The degree and nature of any dissimilarity can be analyzed to indicate the properties of the unknown sample. In some instances the major difference is in the direction or phase of the signal, requiring phase as well as amplitude sensitivity in the measuring instrumentation. This is known as impedance analysis. Eddy-current instrumentation is used routinely in the automotive industry to verify incoming steel grade and to check heat-treated components such as bearings and fasteners for hardness and case depth automatically and at high production rates.

Coating or plating thicknesses of nonmetallics or nonferrous metals such as epoxy coatings and brass or copper plating on steel can be measured accurately using probe coils designed to respond to lift-off (probe-to-steel spacing). Coatings from several mils to claddings on the order of 2.54 mm (0.10 inch) can be directly indicated. Because lift-off is a major factor in eddy-current coating-thickness response, precise control of the probe-coil to part spacing is necessary. Plating and plastic-coating thicknesses ranging from 0.025 mm (0.001 inch) to 2.54 mm (0.10 inch) may be measured with accuracies on the order of ± 3 per cent. Once calibrated for specific base and coating materials, the instruments are direct reading. Typically, the frequencies used range from 100 kHz to 6 MHz.

The ability of the eddy-current method to detect surface imperfections at high speeds and without contact between the test coil and the test piece accounts for its popularity for automatic, in-line flaw detection. Imperfections which disrupt eddy-current paths in the proximity of the test coil or transducer will affect coil loading and thus be reflected in the signal response. In its simplest form a single coil is used, but most automatic systems employ the differential-coil method where one segment of the product is compared with adjacent segments. A typical coil configuration for rods, bars, and tubulars up to about 114 mm (4½ inches) in diameter consists of a solenoid primary winding, two or more diameters long, with a pair of identical, short, closely-spaced secondary coils concentric with the primary, and connected series-opposing so that under the same conditions their net signal output is zero. If a rod, bar, or tube of uniform physical dimensions, mechanical properties, and composition were passed through the test coil, the voltages developed in each of the secondary windings would be exactly equal in amplitude and direction (phase). By adding these voltages so that they cancel out (180 degrees out of phase through the series-opposing connection) the net output voltage will be zero and will remain so until something occurs to create an imbalance. Consequently, any imperfection which will change the "loading" or impedance in one of the coils and produce a change in signal amplitude, phase, or both, will result in an imbalance signal that is amplified and analyzed by appropriate instrumentation. As the imperfection moves out of one differential winding into the other, the imbalance shifts to the second coil and results in a signal with similar amplitude but with opposite sense. If the imperfection is so long as to extend through both coils, then during that period the resultant imbalance signal is reduced significantly and, in the case of a uniform longitudinal imperfection, the result may be zero or nearly so. By increasing the spacing between differential coils, this effect is reduced, but the increased spacing will also result in isolated imbalance signals for imperfections that are short compared to the spacing between coils.

The length and number of turns in the individual differential windings should be chosen on the basis of the length of the imperfections to be detected and throughput speeds. Coils that are short compared to the length of the average imperfection will sacrifice output signal in favor of resolution, and tend to produce imbalance signals of short duration. If the winding is long compared to the imperfection, the signal tends to flatten out with imbalance signals of comparatively long duration, and results in low resolution. These factors have a mitigating effect on the speed at which a product can be inspected by an eddy-current system. It is also necessary that an imperfection be in a test-coil winding for several cycles at any test frequency in order to produce consistently usable signals. The test frequency also directly affects the depth of penetration of eddy currents in the test piece, as given by the following formula:

$$S = k \sqrt{\frac{\rho}{\mu f}}$$

where:
S = depth of penetration
k = 50 290 for S in mm;
  = 1980 for S in in.
ρ = electrical resistivity of test piece (ohm·cm)
μ = magnetic permeability of test piece
f = test frequency in Hz

The frequency (which affects the depth discrimination of discontinuities) is multiplied by the magnetic permeability (μ) which for carbon and alloy steels may vary from values on the order of 50 to several hundred depending upon the magnetic circuits. Thus the depth of penetration will be considerably less for ferromagnetic materials than for nonferromagnetic materials with similar electrical resistivity, and can also vary greatly from grade to grade or even within a specific piece depending upon the effective permeability in the portion of the product under inspection. Localized permeability changes may result from patches of magnetic scale, burnished areas such as straightener rings, dings, decarburization, temperature, and segregation. These contribute to nonrelevant imbalance signals or "noise" that can obscure valid signals. Most eddy-current systems employ some means to suppress these noise signals.

The most commonly used technique for suppression of noise arising from random variations in magnetic permeability is magnetic saturation. By saturating the test piece using a strong, steady, magnetic field the effective permeability of the entire piece is reduced to approximately one, thus effectively suppressing the noise associated with permeability changes. In most cases, saturation is achieved by using yokes or encircling coils through which direct current is passed. The current is adjustable so that saturation may be achieved for various grades and sizes of steel components. This technique is used in eddy-current systems for the inspection of finished wire, rod, bar, and tubular products. Magnetic saturation not only reduces the noise but also increases the depth of penetration because the permeability multiplier in the depth-of-penetration-equation denominator is reduced to approximately one. This permits more accurate evaluation of deep imperfections. For instance, the depth of penetration for alloy steel at a test frequency of 10 kHz is approximately 1.5 mm (0.06 inch). Using magnetic saturation, the depth of penetration is about 4.1 mm (0.16 inch).

Magnetic permeability is also affected by product temperature, and can definitely affect test performance because it decreases at an increasing rate with increasing temperature until it approaches one at temperatures above the Curie point when carbon steels go into the austenitic phase. The Curie point is approximately 770°C (1418°F) for carbon steels. This characteristic is used to advantage for in-line inspection. In particular, water-cooled eddy-current encircling-coil transducers with appropriate instrumentation have been used for in-process inspection of hot rod and bar, as well as for continuously butt-welded and seamless tubular products. These inspection systems employ computer techniques for data analysis and are capable of detecting rejectable imperfections of both the random and repetitive natures. In the latter case, things like broken rolls and misaligned guides can be detected early through the repetitive flaws they generate, and corrective action taken before tons of rejectable product are produced. In rod and bar product, an analysis of the increased eddy-current background noise can be used to signal conditions where the potential for slug generation is high. Rod mill operators claim that eddy-current inspection and computer analysis of hot rod product provide assurance of cold-heading quality. These in-line systems are capable of handling diameters up to about 114 mm (4½ inches) at temperatures up to about 1093 to 1204°C (2000 to 2200°F), and at speeds in excess of 1830 m/min (6000 fpm). Surface imperfections such as seams, laps, roll-ins, slugs, pits, and slivers are detected by such systems. Furthermore, weld imperfections such as open welds, cave welds, incomplete welds, black spots, and cold welds (to some degree) are all detectable by in-line eddy-current inspection.

For flat surfaces and for round bars, tubes, and the like that are greater than 114 mm (4½ inches) in diameter, encircling coils are replaced with probe coils displaced around the periphery to provide complete coverage. Probe coils are also used for the inspection of longitudinal welds in electric resistance-welded and butt-welded pipe and with portable instruments for hand-probing of surface imperfections.

The inherent capabilities of the eddy-current method to reliably detect imperfections at high speeds and at elevated temperatures without contact with the product make it particularly attractive for in-plant inspection and quality-control applications. Furthermore, the ability of the eddy-current method to respond to steel grade, hardness, and heat treatment as well as its ability to measure coating thickness make it a most versatile tool. This very versatility, however, is the source of its greatest disadvantage. It is often difficult to detect and separate the parameter of concern from among the responses to all of the other variables. By selection of test frequency, coil design, control of product motion, and the appropriate data-collection and analysis programs, however, in-line eddy-current quality control and inspection is highly feasible.

LIQUID-PENETRANT INSPECTION

The liquid-penetrant method is an aid to visual inspection that is characterized by its simplicity and low cost. It is designed to enhance the visibility of surface imperfections in solid, nonporous materials. It relies upon the ability of certain substances (dyes) to flow uniformly over the surface of a material and penetrate into surface cavities.

In operation, the surface of the piece is thoroughly cleaned to remove all grease, oil, dirt, and loose scale. The liquid-dye penetrant is applied by spraying, brushing, or immersion and the piece is set aside for a period of time (dwell time) to permit full penetration by the dye. Typical dwell times range from five to thirty minutes. At the completion of the dwell time the part is thoroughly cleaned using a suitable solvent, sprays, clean cloths, or by immersion, and then dried. The surface of the cleaned, dried part must be free from excess penetrant on the surface. A developer is then sprayed on the clean, dry surface. The developer is usually a chalky powder in a solvent suspension that is applied uniformly over the entire surface using a spray can. Care must be exercised to cover the entire surface with a uniform coating. After a second waiting period (developing time) which is usually around 10 minutes, the part is ready for inspection. A surface imperfection is disclosed as a dye mark that bleeds out of a discontinuity and through the developer. It discloses the shape of the discontinuity on the developer. Upon completion of the inspection and recording of results, the part is thoroughly cleaned to remove all traces of the developer and dye indications.

The liquid-penetrant method may be divided into several general categories that pertain to the kinds of dyes, developers, and solvents used. There are more specific subdivisions which are beyond the scope of this section. In general, the penetrants are visible dyes and fluorescent materials. The solvents used may be aqueous or petroleum based. The resultant classifications are (1) the water-washable penetrant system, which may employ a fluorescent or visible dye, (2) the postemulsifiable system, which uses petroleum-based dyes for added sensitivity but adds an emulsification step in the cleaning process to allow the excess dye to be rinsed away with water, and (3) the solvent-removable system, which is generally restricted to small-lot applications where spray cans of dyes and solvents are usually employed. For most applications in the steel industry, the latter method is most commonly used. The first two techniques are amenable to large-lot, routine penetrant inspection where a mechanized inspection line is feasible. Interpretation of indications and marking are, however, manual.

The precautions in liquid-penetrant inspection are relatively few, but important. Cleanliness of the inspection surface is essential, but grinding and shot blasting are to be avoided because they tend to close up surface seams. Penetrant testing should be carried out in the 21° to 32°C (70° to 90°F) temperature range. It should never be allowed to exceed 50°C (122°F), and operation at 5°C (41°F) should be avoided. The recommended dwell and developer times must be adhered to in order to achieve consistent results. Finally, the choice of dyes, solvents, and developers may be dictated by the materials to be inspected and their intended uses, as well as safety considerations for inspection personnel. For instance, halogens and sulphur in dyes may be harmful to certain materials. Also, the local and OSHA restrictions on the uses of chemicals, particularly petroleum-based dyes and solvents, must be adhered to.

The major advantages of the liquid-penetrant method are quite similar to those of the visual method. It is inexpensive, direct, easily learned, portable, and may be used on components of almost any size. It discloses surface imperfections that would go undetected by unaided visual inspection. This is particularly true when fluorescent dyes and ultraviolet lights are used. It is, however, highly operator dependent and gives little indication of the severity of indications. Aside from mechanization of the various processing steps, little has been done to make liquid-penetrant inspection automatic. Consequently, aside from certain cast and forged products where liquid-penetrant inspection is specified, it is used occasionally where other methods are not feasible, to supplement other methods, and for maintenance inspection.

ULTRASONIC INSPECTION

The ultrasonic method is a versatile and highly reliable way to inspect steel products. It is widely used for the detection and location of imperfections and for thickness gaging. Systems range in complexity from simple, portable, manual flaw-detection or gaging units to highly mechanized, computer-controlled systems for automatic product inspection and classification. The method employs mechanical vibrations at frequencies above the upper limit of human hearing (approximately 20 kHz) and well into the megahertz (MHz) range. The frequencies used in the inspection of steel products range from 500 kHz to 25 MHz for most applications. The most commonly employed frequencies are 1.0, 2.25, and 5.0 MHz, and are used in the inspection of billets, blooms, and slabs for the detection of pipe and for the detection of non-metallic inclusions, seams, laps, slivers, laminations, and other internal and surface imperfections in forged and rolled products.

The ultrasonic method employs instruments that generate high-frequency mechanical vibrations through a transducer (variously known as a crystal or search-unit) which converts electrical impulses into mechanical vibrations of the prescribed mode. The transducer can also transform mechanical vibrations back into electrical impulses which are amplified and displayed by the instrument (Figure 51—2). In its usual

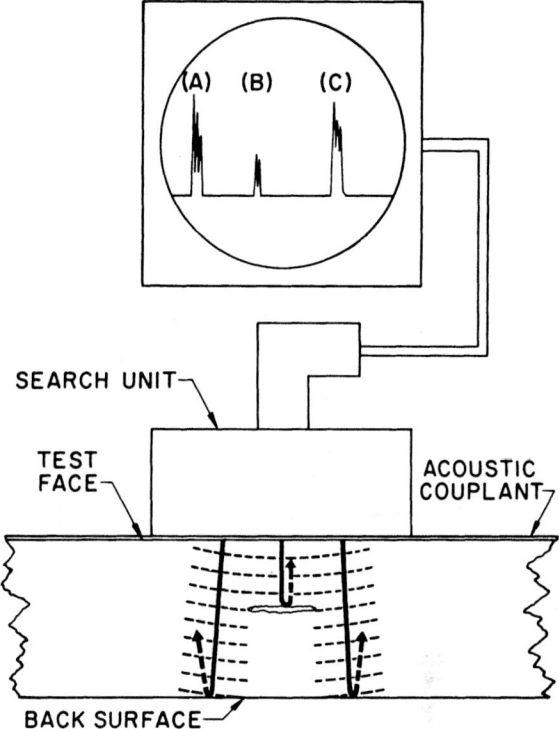

FIG. 51—2. Diagram illustrating principle of basic pulse-echo method of inspection. Idealized oscilloscope indications (A), (B) and (C) relate respectively to the signal received from the test face (initial pulse), a discontinuity and the back surface of the test piece.

form, ultrasonic inspection is analogous to radar or sonar where short bursts or impulses of energy are transmitted by an antenna or sonar transducer into the air or water and return echoes are received by the same antenna or transducer, converted into an electrical signal, amplified, and displayed on a cathode-ray-tube (CRT) screen. Through suitable timing and direction-sensing circuits, the range, azimuth, and elevation (depth) of targets can be determined. In more sophisticated units, the detailed nature of the target can be determined.

In ultrasonic inspection, there are three basic modes of ultrasonic energy that may exist in steel: (1) the longitudinal or compression wave (Figure 51—3), (2) the transverse or shear wave (Figure 51—4), and (3) the Rayleigh or surface wave. A special case of the latter which occurs in thin plate and sheet product is known as the Lamb or plate wave. Each has specific attributes making them useful for a variety of applications.

The longitudinal or compression wave is the same mode of sound vibration as occurs in hearing. In the longitudinal mode the molecules in the medium are alternately forced closer together (compression) or further apart (rarefaction) in the same direction that the ultrasonic wave is traveling. The stresses generated are compressional or tensional. The wave travels in straight lines in a material and is useful for the detection of laminations and non-metallics in rolled products and for the measurement of plate and pipe-wall thick-

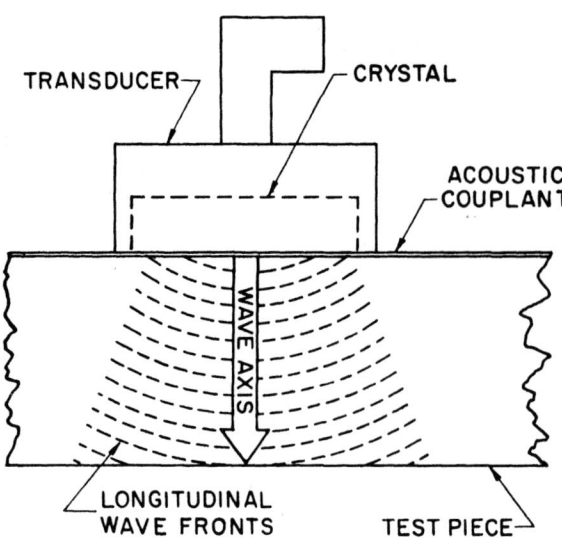

FIG. 51—3. Diagram of longitudinal-wave generation and wave pattern.

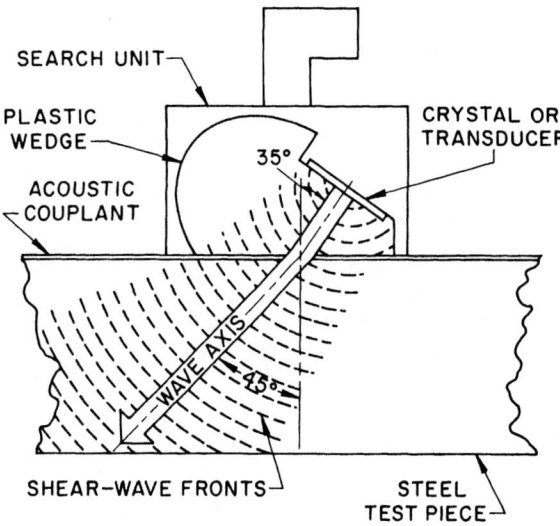

FIG. 51—4. Diagram of shear-wave generation and wave pattern.

nesses. The velocity of the compressional wave is a function of the physical constants in the transmitting medium and does not change with frequency. For annealed carbon steel, the longitudinal-wave velocity is 5.94 km/s (234 × 10³ in./s), for UNS S30200 (AISI Type 302) stainless steel it is 5.66 km/s (223 × 10³ in./s) and for cast iron 3.5 to 5.6 km/s (138 × 10³ to 221 × 10³ in./s). The longitudinal-wave velocity is altered by temperature and material physical properties such as density, hardness, and strength. For most installations, however, these parameters do not vary enough in any given product to be significant.

In the transverse or shear-wave mode, the particle displacement or vibration is at right angles (transverse) to the direction of the ultrasonic wave propagation,

and the stresses generated in the medium are shear stresses. The shear wave mode is not supported in air, water, oil, rubber, and waxes. Often known as the angle-beam mode, shear waves are generally used in the detection and location of seams, laps, slivers, and cracks located at or near either surface of rolled product. They may also be used for the detection of stringers and nonmetallics within the product. In steel, the shear-wave velocity is roughly one half that of the longitudinal wave: 3.24 km/s (126 × 10³ in./s) for carbon steel, 2.2 to 3.2 km/s (87 × 10³ to 126 × 10³ in./s) for cast iron, and 3.12 km/s (123 × 10³ in./s) for UNS S30200 (AISI Type 302) stainless steel. For many applications the shear-wave entry angle at the steel surface is 45 degrees, although 60- and 70-degree angles are used in weld inspection.

The Rayleigh or surface wave propagates along the surface of the part with little penetration. It has a peculiar wave motion in which the particles are displaced transversely and on a rotating, radial axis which describes an elongated spiral along the direction of wave propagation. In the case of Lamb or plate waves, similar wave modes are generated at both surfaces and sometimes down the center thickness of plate or strip. In any of these cases the velocity is approximately half of the longitudinal wave velocity. As with shear waves, surface waves are not propagated in water, oil, waxes, and the like. The surface wave velocity in carbon steel is 3.0 km/s (118 × 10³ in./s) and 3.12 km/s (123 × 10³ in./s) in UNS S30200 (AISI 302) stainless. No values are given for cast iron because of the high attenuation resulting from coarse grain structure.

The pulse-echo ultrasonic-inspection technique is the most commonly used. The following describes the basic performance parameters in checking a heavy (76-mm- or 3-inch-thick) steel plate using a longitudinal-wave 2.25 MHz transducer. The transducer is connected to the pulser/receiver of the ultrasonic instrument which is designed both to transmit and to receive ultrasonic impulses. The pulser is controlled by a "clock" which is adjustable to generate timing signals that are adjustable from, perhaps, 10 pulses per second (pps) to several thousand pps. The pulser circuit generates electrical impulses (600 to 1100 volts) of short duration (1 or 2 microseconds) for each pulse from the timer. When applied to the transducer this electrical spike is converted into an ultrasonic pulse at the selected frequency (say 2.25 MHz) and perhaps 3 microseconds in length. The frequency and pulse lengths are selectable by transducer selection and instrument control settings. The ultrasonic energy is transmitted through a couplant into the test piece. Ultrasonic energy is highly attenuated by air, and some sort of couplant that will fill the interface between the transducer and test piece and permit uniform sound transmission is essential. The couplant may be water, oil, grease, glycerine, water-soluble oil, or other material and must be free from air bubbles and foreign matter that will attenuate and scatter the sound. When the transducer is placed directly on the part, sound is reflected directly from the front surface of the part and back to the transducer and actually reverberates in the transducer/part interface. This results in a broadening of the initial or transmitted pulse and is analogous to ground clutter in radar and obscures imperfections located close to the front surface, which in this case may be as far as 10.2 to 13 mm (0.4 to 0.5 inch). For these sections, shorter pulse lengths and high-resolution transducer techniques are used to improve front-surface resolution.

Once the ultrasonic impulse enters the plate it travels in a straight line toward the opposite surface. Upon striking the rear surface it bounces back toward the front surface and the transducer, making the 152-mm (6-inch) round trip in about 26 microseconds. Upon reaching the front surface, a portion of the back-surface echo passes back through the front surface to the transducer and is seen as a strong impulse some distance to the right of the initial pulse on the CRT display. The bulk of the back-surface echo remains in the plate and makes a number of round trips until it is fully attenuated. By knowing the thickness of the material and the velocity of sound in that material, distance from the front surface can easily be estimated. At a pulse-repetition-frequency (prf) of 1000 pulses per second there would be 1000 microseconds between transmitted pulses. This would allow enough time for a part as thick as 3 metres (118 inches) to be inspected. The thickness limitations actually lie in the energy of the initial pulse and the sound-attenuation characteristics and geometry of the material being inspected. Inspection distances of 508 to 762 mm (20 to 30 inches) in rotor forgings are typical and 6.1 metres (20 feet) in shafts are possible.

The presence of a reflector in the form of pipe, or a lamination, or nonmetallic stringers, or flakes in the path of the sound beam will alter its path. In the case of the three-inch-thick plate, pipe- or center-lamination of any magnitude will probably intercept the sound beam entirely. As a consequence, an echo would be displayed at a distance equivalent to 13 microseconds round-trip travel time, midway between the initial pulse and original bulk surface indication. If the lamination is large and open, all of the sound beam will be intercepted and no back-surface echo will appear. It is important to note that in such cases where the lamination is at about mid-thickness, a sound reflection from it may appear at the normal location of the back echo, requiring additional analysis. If the lamination is tight or spotty, then some of the energy will pass through to the back surface. The relationships between lamination and back-surface-echo amplitudes provide the bases for numerous plate-quality rating procedures.

The foregoing discussion of the pulse-echo technique references primarily the longitudinal wave mode of operation. The principles described, however, are also applicable to the shear-wave and surface-wave modes. Whereas, in the example given, the longitudinal-wave mode interrogates only that plate material in the sound path beneath the transducer, the shear-wave sound path extends away from the transducer, alternately reflecting from each plate surface (Figure 51—5). If the sound-beam angle in the product is known, the location of surface and internal imperfections can be determined. Note that, because of the angular wave path, planar imperfections parallel to the plate surface are not readily detected. Similarly, surface or Rayleigh waves emanate from the transducer along a shallow path and do not penetrate to any extent. In any of these cases, however, the fundamentals of flaw size, type, and orientation, uniformity of couplant, and surface condition are all applicable. The mode selected is based upon which one is best suited to the application.

The echo amplitude received from any discontinuity is a function of several interrelated factors: (1) the nature, size, location, and orientation of imperfections, (2) the structure and internal cleanliness of the material being inspected, and (3) the transducer size, type, operating frequency, and sound-beam characteristics. These factors may take on various degrees of importance depending upon the inspection task and operating parameters. Nevertheless, they must all be considered in the design of an inspection system and in the interpreta-

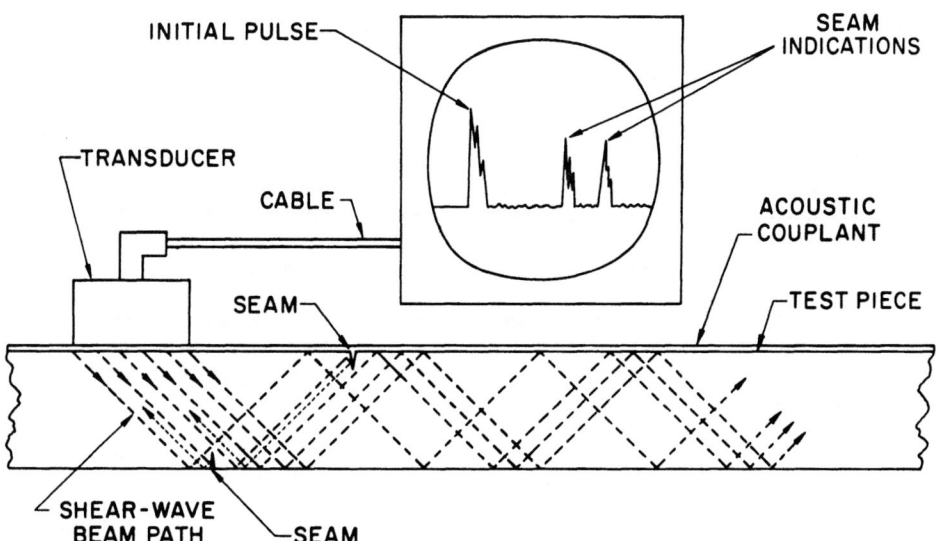

FIG. 51—5. Schematic diagram of shear-wave pattern in plate product. Dotted lines show paths of reflected sound impulses. (Actually, the transducer-beam pattern should show some divergence: this has been ignored for the sake of clarity and simplicity.)

tion of the results.

The nature of an imperfection has a direct bearing on the return echo. Smooth, planar reflectors reflect sound uniformly, the angle of reflected energy being equal to the angle of incidence throughout. Spherical, volume-type, or rough-surfaced imperfections will scatter, and mode conversion of the ultrasonic energy and the echo-return paths will vary in time resulting in a broadened, lower-amplitude response as compared to a smooth, planar discontinuity of the same cross-sectional area. The orientation of the discontinuity has a major effect on echo response. The maximum response is achieved where the reflecting surface is normal to the sound-beam path. Deviations from this normalcy of even a few degrees can result in a complete loss of the echo, depending upon the size of the imperfection and the sound-path distance. The presence of sound-transmitting material within discontinuities will reduce echo size as compared to that from the same discontinuity with no included material. Furthermore, if the cross-sectional area of a discontinuity is less than two or three times the ultrasonic wavelength, it may be difficult to detect. Similarly, planar reflectors which are approximately one-half wavelength thick are poor reflectors, but are good reflectors at a quarter wavelength thickness. At 2.25 MHz, the wavelength of longitudinal wave ultrasound in carbon steel is 2.64 mm (0.104 inch) and is 1.44 mm (0.056 inch) for the shear-wave mode. Reflectors less than about 0.254 mm (0.01-inch) thick tend to pass ultrasound as well as to reflect it. In fact, cracks under high compressive loads may not be detected by the ultrasonic method. Coarse-grained steels and those containing numerous inclusions and porosity are highly attenuative to ultrasonic energy because of scatter, absorption, and diffraction. Certain stainless and cast materials are characterized by high levels of noise resulting from coarse grain structure. The effect of grain size can be reduced by lowering the test frequency and also by grain-refinement practices. Surface roughness both at the entry and at reflecting surfaces also results in diffraction and scattering, and markedly decreases ultrasonic inspection responses.

The selection of the transducer, test frequency, and ancillary mechanisms is probably of the greatest importance in the design of an ultrasonic inspection system. It is essential that inspection-system parameters such as product shape and size, imperfections to be detected, throughput, product surface and internal condition, and operating environmental conditions be specified. By defining the size, nature, and locations of the imperfections to be detected, the size and frequency of the transducer can be estimated. The size of the sound beam and that of the smallest imperfection to be detected should be comparable. Otherwise overinspection and noisy operation are probable. The sound-beam size may be controlled by the choice of transducer and the use of focusing and collimating techniques. The location of imperfections and their nature will influence the selection of the ultrasonic wave mode and whether dual-transducers, ultrasonic delay lines, or focused transducers should be used. The design must also include suitable mechanisms to optimize and maintain transducer orientation with respect to the product. Otherwise soundbeam paths will be uncontrolled and the inspection inconsistent.

With all of the instrumentation, transducers, and manipulators at hand, successful ultrasonic inspection cannot be assured unless all of the foregoing factors are carefully analyzed for each installation and product, and the necessary ultrasonic, electronic, environmental, and mechanical parameters are specified and adhered to.

Applications—Portable ultrasonic instruments, some weighing less than 4.5 kg or 10 pounds, are routinely used in steel plants and in the field for confirmation checks of inspection results obtained with other equipment, maintenance inspection, weld inspection in pipe, structures and bridges (Figures 51—6 and 51—7), and inspection of large forgings. Typically these instruments use hand-held transducers, with oil, grease, or cellulose-gum and water mixtures for couplant. The instruments are calibrated against standards prescribed by such organizations as the American Welding Society, the American Society for Testing and Materials, and the American Society of Mechanical Engineers. The operation of these units requires considerable training and skill because the results are dependent upon the operator's interpretation of the CRT display, and minor changes in transducer location and orientation can cause major changes in the displayed echo. A skilled operator with knowledge of his product can make use of the subtleties of transducer positioning and pattern changes to arrive at accurate inspection results. He must constantly be aware of the salient parameters that affect manual ultrasonic inspection and make sure that the test surface is clean, that he has adequate couplant, and that he can position the transducer properly on the test piece. He must also be aware of the sources of irrelevant echoes, problems associated with the location and orientation of imperfections, and how the nature, size, location, and orientation affect the acceptability of the product. Because of these operator-related factors, many specifications require personnel who have been specially certified to conduct evaluations.

The ultrasonic method is more amenable to automatic operation and data analysis than other NDI methods. The precise time frames that are inherent with the pulse-echo technique are particularly useful with microprocessor controls and computer-aided data analysis. An automatic system can perform such operations as thickness gaging, flaw-detection recording and marking, and product classification at surface speeds up to about 91.5 metres per minute (60 inches per second). This speed is compatible with uniform ultrasonic coupling, low-noise operation, and consistent automatic product classification. It is achieved, however, only through careful control of the ultrasonic coupling and the mechanisms used to position the transducers with respect to the product. In most instances, product surface contour variations, differences in the amount of surface scale, and product straightness make it difficult to employ simple contact methods in an automatic system with any degree of consistency. The coupling medium for the majority of ultrasonic systems is water. The immersion technique, perhaps the oldest of these, is without peer in precision and versatility. It can be used on simple geometries such as billets and plate

FIG. **51**—6. Manual ultrasonic inspection of a large forging.

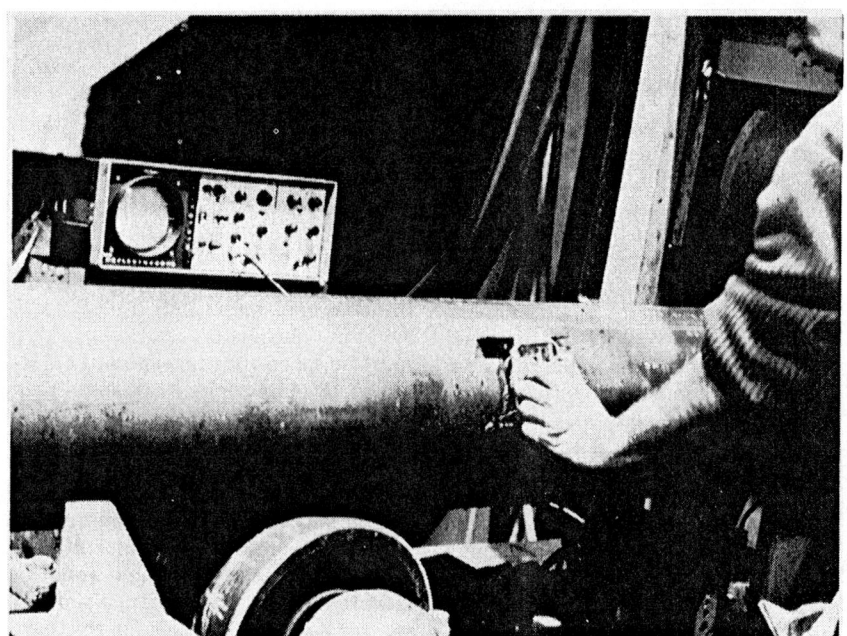

FIG. **51**—7. Manual ultrasonic shear-wave inspection of ERW pipe weld using contact transducer.

FIG. 51—8. Immersion tank for automatic ultrasonic inspection of plates.

(Figure 51—8) and can be programmed to track complex shapes by numerical control in the X-Y-Z directions. The immersion system is readily adapted to microprocessor control, digital, and/or chart recording of data for three-dimensional presentation and analysis. The immersion technique lends itself to focusing and shaping of the ultrasonic beam and because the transducer-to-part spacing is generally 25 mm (an inch) or more, the front surface dead zone resulting from reverberations is reduced significantly. The immersion technique is used routinely in the ultrasonic cleanliness rating of plate and billets, and in the inspection of complex structures such as wing sections for jet-liners. By manipulation of the transducer to part angle, varying degrees of ultrasonic entry angle in the part, from 0° longitudinal wave through angled longitudinal waves and shear waves to surface waves, are readily obtained. Surface wave inspection using the immersion technique is considered impractical, however, because of the high attenuation resulting from water along the surface of the sound path. The so-called squirters, bubblers, water columns, and wheel transducers are really extensions or modifications of the immersion technique. All of these require a sound path through the water that is free from bubbles and foreign material

that will absorb and scatter sound. In addition, they also require precision positioning and manipulation. For example, 1° of angle change in water from an incident angle of 0° to 1° will result in a 4° change in the sound beam in steel, and a shift from 20° to 21° in water will result in a shear-wave angle change of 3.1°. These facts must be taken into consideration in the design of any automatic ultrasonic inspection system, particularly where curved surfaces are involved. For instance, if the beam cross section were equal to one half of the diameter and all components of the incident beam were parallel, the incident angles at the beam edges would be 30° when the center component was normal (0°). If the beam cross section is one fourth of the diameter, the incident angles at the edges would be 7.2°. In either case the sound beam normally diverges resulting in even greater divergence in the part. Consequently, beam focusing is often necessary for control of the sound beam in the part. If the part has a complex contour, the incident angle of all sound beam components with the test piece must be analyzed to avoid unexpected responses. Aside from the immersion tank, water-column arrangements are used in the inspection of seamless and welded pipe and tubing, billets, bars, and rails. Each has its own set of inspection parameters which must be met and its own kind of transducer positioning and tracking.

Another commonly used method employs a Lucite shoe contoured to mate the component being inspected and containing a transducer or transducers located so as to provide sound beam entry at the prescribed location and entry angle in the part. These Lucite shoes do not ride directly on the part but are held away by means such as tungsten carbide buttons or strips and a gimballed shoe holder to maintain a precise gap between the transducer shoe face and the test piece. Such transducer arrangements are commonly called gap probes. The gap is generally of the order of 0.5 mm (0.02 inch) and contains a lamellar flow of water which assures uniform coupling of the ultrasonic impulses. In properly designed gap probes the water flow is uniform and free from turbulence, and the ultrasonic signals are quite consistent and free from coupling variations and noise.

Some hand-held transducers utilize a contained water column in a cylinder with the piezoelectric element at one end and a flexible membrane at the other. These units are useful on surfaces that are rough or have contours that are not amenable to inspection with the normal transducer. The contained water or couplant column is approximately 51 mm (two inches) long and manual positioning is critical because a slight shift in transducer orientation can affect response significantly. This disadvantage is overcome when using the wheel

transducer. The basic components are similar: a contained water column, a transducer affixed to the axle housing of the wheel, and a flexible membrane (Adiprene) tire that rolls along the surface of the test piece. Wheel transducers are available with longitudinal-wave, shear-wave, surface-wave, and variable-angle transducers. Although the longitudinal-wave wheel is supplied only with a downward radial sound path, the others can be supplied with a forward, sideways, or angled sound path with respect to wheel travel. As with all immersion or water-column units, precise orientation of the wheel transducer with the test surface is essential. In fact, with the shear-wave units the angle of the shear wave can be changed deliberately by appropriate angulation of the wheel. Only a thin film of couplant between the tire and test surface is necessary. Wheels are available in sizes from approximately 51 mm (two inches) to 191 mm (7.5 inches) in diameter and are used in the ultrasonic inspection of rails and the longitudinal weld in electric-resistance-welded (ERW) pipe.

The immersion (Figure 51—8), water-gap, and water-column coupling (Figure 51—9) techniques are particularly useful for automatic ultrasonic inspection systems. The nominal 300 fpm (91.4 m/min) ultrasonic inspection speed can be achieved with these techniques when the appropriate transducer positioning and product-tracking mechanisms are provided. This speed is more than adequate for the in-line ultrasonic

FIG. 51—9. Ultrasonic longitudinal-wave inspection of plate using water-column coupling.

FIG. 51—10. Automatic ultrasonic shear-wave inspection of ERW pipe weld using wheel transducer.

inspection of the longitudinal weld in ERW pipe (Figure 51—10) and for automatic rail inspection. For seamless pipe, however, where the entire pipe body must be inspected it is common practice to employ one or more transducer arrays that inspect the pipe in spiral strips by either passing the pipe without rotation through rotating ultrasonic heads or by rotating the pipe over stationary ultrasonic heads. Ultrasonic inspection systems for seamless pipe have been described that inspect the pipe with shear waves in both the clockwise and counter-clockwise directions and measure wall thickness as well. By using multiple heads, micro-processor control and data sorting, average throughputs of from 14.6 m/min (48 fpm) or 1.2 pieces per minute to 19.8 m/min (65 fpm) or 1.63 pieces/min of 140-mm (5½-inch) OD, 9.53-mm (0.375-inch) wall range III (12.2 metres or 40 feet) pipe are feasible.

Computer controlled and analyzed ultrasonic cleanliness rating of plate is conducted in an immersion tank on 127-mm by 229 mm (5-inch by 9-inch) plate samples. The cleanliness data are used for acceptance of plate for use in the production of submerged-arc (SA) welded line pipe, and a complete three-dimensional record is obtained in approximately five minutes. The scanning results may be transmitted from the plant to a central computer where it can be retrieved upon command for further study and evaluation.

These inspection capabilities and the ability of computers to collect, assess, and store large amounts of data form the bases for nearly all state-of-the-art inspection systems where NDI is an integral part of plant quality-tracking and quality-assurance programs.

RADIOGRAPHY

The commonly-used term radiography is used somewhat loosely in defining the nondestructive method wherein penetrating radiation is used to form images of a subject on some sensitive surface such as a film., fluorescent screen, or X-ray sensitive semiconductor. Radiology, which covers both film and real-time applications, is more apt. It is an extension of the visual method and discloses the nature, size, location, and distribution of imperfections either on a surface or internal. It can also indicate thickness and inspect mechanisms for the presence and location of imperfections.

Penetrating radiation is electromagnetic radiation, of very short wavelengths, that penetrates substances and form "radiographic" images of the substance on a sensitive surface. The penetrating radiation is usually in the forms of X-rays or gamma rays although thermal neutrons are used in a special form of radiography known as neutron radiography. X-rays and gamma rays are the same in nature, the difference being that X-rays are generated electronically when high-velocity electrons give up energy when striking a target anode such as tungsten, and gamma rays are produced as characteristic radiations produced during disintegration of radioisotopes such as Radium and Cobalt 60. Both the penetrating power (wavelength) and intensity of X-rays are controllable by adjusting the voltage applied and the current in the X-ray tube, respectively. When the X-ray tube is turned off, no X-rays are generated. Gamma rays, on the other hand, have wavelengths that are characteristic of a particular radioactive element, and thus the penetrating energy level is obtained by selecting the proper isotope. The intensity of the gamma radiation is selectable only by the specific activity of the source which is usually specified in curies. The specific activity is continually decreasing with the decay of the radioisotope. Each radioisotope has a half life which defines the period of time over which its specific activity drops to one half of its specified or nameplate value. For instance, the half life of Cobalt 60 is 5.3 years, in which time a source with a nameplate activity of 30 curies would drop to 15 curies.

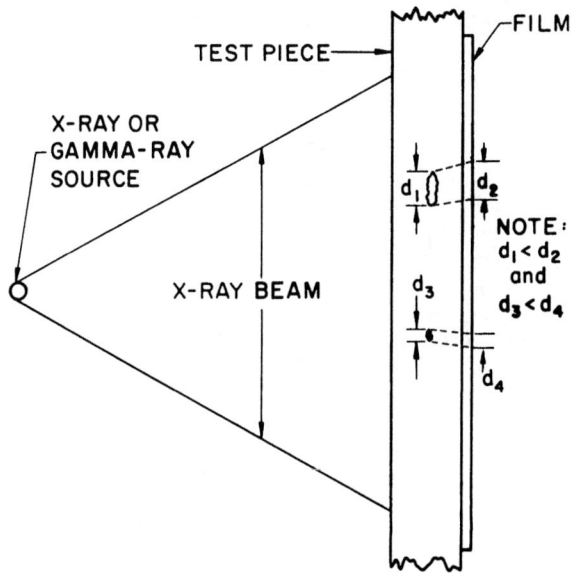

FILM

TEST PIECE

X-RAY OR
GAMMA-RAY
SOURCE

X-RAY BEAM

NOTE:
$d_1 < d_2$
and
$d_3 < d_4$

FIG. 51—11. Schematic diagram showing how a void and an inclusion in a steel test piece cast "shadows" that are recorded on a photographic film by affecting the intensity of the radiation that penetrates the test section. The shadow of the void will result in a dark spot on the film, while the inclusion will be indicated by a spot the appearance of which is a direct function of the density of the inclusion. Thus, inclusions more dense than the parent metal will appear as light spots, while less dense inclusions will appear as darker areas in relation to the uniform background on the film that represents the areas where the penetrating beam passed without disturbance. Divergence of the ray beam causes the "shadows" to be somewhat larger than the void or inclusion.

The half lives of sources used in industrial radiography vary from 75 days for Iridium 192 to 30 years for Cesium 137. It is necessary, therefore, to know how long a particular source has been in use. This is particularly true for Iridium 192 and Thulium 170, which have half lives of 75 and 130 days, respectively.

Penetrating radiation is absorbed and scattered by a material as a function of its density, structure, contour, and the presence of discontinuities. The thicker or more dense the material, the greater the absorption. On the other hand, the more voltage that is applied to the X-ray source, the shorter the wavelength and the greater its penetrating power. Also, by increasing the current in the X-ray tube or by selecting a radioisotope with a higher curie rating (specific activity), the amount of penetrating rays (intensity) at any wavelength will be increased.

In practice, the film or X-ray-sensitive detector is placed on or as close to one surface of the part being examined as possible (Figure 51—11). In film radiography the film holder is clamped up against the part. The X-ray source, on the other hand, is located two or three feet away from the opposite side of the part with the centerline of the source aimed directly through the part to the center of the film. Many of the principles governing the formation of a radiographic image are based on optics. The X-rays travel in straight lines, em-

anating from the source in a cone. Typically, the spot size of an X-ray tube ranges from 0.3 to 2.5 mm (0.012 to 0.098 inch), and the cone of radiation has a 40-degree included angle. To radiograph a 305-mm (12-inch) wide section, the minimum distance from source to object would be 419 mm (16.5 inches). There would be degradation at the edges of the image, however, coming from the periphery of the projected focal spot, and there is also a noticeable loss of X-ray intensity in the proximity of the edges. Consequently the source-to-object distance is increased to, perhaps, 610 mm (24 inches) to alleviate this condition. Secondly, the radiographic image is formed by "shadows" cast by changes of section, discontinuities, and inclusions in the part. Unless these are in direct contact with the film or image plane, some magnification of size and image unsharpness results. The magnification can be calculated directly by knowing the distances from the X-ray source to the object (source-object distance, or SOD) producing the image and from the source to the film (source-film-distance, or SFD). The magnification (M) is the ratio of SFD to SOD, and the unsharpness is the product of M and the focal spot size. It is generally advantageous to position the film as close to the back side of the object as possible. Even so, when radiographing thick sections, imperfections at or near the surface on the source side will be magnified and be less sharp than those at the film side. These considerations are of increased significance in fluoroscopic systems where the image plane is necessarily separated from the object by two or more inches.

In producing a radiograph it is common practice to use lead letters and numbers to identify the radiograph, to identify the product and the physical relationships of the area being X-rayed with adjacent areas, and to display one or more penetrameter images to establish the sensitivity of the radiograph. Penetrameters or image-quality indicators (IQIs) are made of the same material as that being radiographed, and their design is established by one or more of the standards-writing bodies. The most commonly used are the plaque type and the wire type. Typical for the plaque-type penetrameter is the American Society for Testing and Materials (ASTM) design. For specimen thicknesses up to 191 mm (7.5 inches) thick the penetrameter consists of a 12.7-mm (½ inch) by 38.1-mm (1.5-inch) piece of shim stock of the same basic material (e.g., carbon steel, stainless steel, aluminum) as the object being inspected. Its thickness may be one of ten values ranging from 0.127 mm (0.005 inch) to 3.810 mm (0.150 inch). The penetrameter thickness selected is the one closest to the nominal per cent sensitivity required, which for nearly all cases ranges from one per cent to four per cent. The most commonly specified value is two per cent, which means that the radiograph must display a penetrameter whose thickness is two per cent of the thickness of the material being radiographed. The penetrameter thickness (T) for radiographing 12.7 mm (0.5 inch) of carbon steel at a two per cent sensitivity would be 0.02 times 12.7 mm (0.500 inch) or 0.254 mm (0.01 inch). Penetrameter thickness is designated in hundredths of a millimetre (thousandths of an inch) in lead letters at one end of the penetrameter. In this case, (using English units) the penetrameter would have "10" in 0.25-inch (6.35-mm) high lead letters at one end to designate a thickness of 10 one-thousandths of an inch. The penetrameter material is designated by a notching system at the end opposite the lead numbers. An unnotched penetrameter denotes carbon steel and UNS S30400 (AISI Type 304) stainless. Note that the 10 penetrameter would be a one per cent penetrameter for 25.4-mm (1.0-inch) and four per cent for 6.35 mm (0.25-inch) section thicknesses. The penetrameter thickness is used to determine the contrast sensitivity of a radiograph but does little to establish its resolution. Consequently, each penetrameter contains three holes of different diameters drilled in it. These are equal to the penetrameter thickness T, 2T, and 4T. The sensitivity reading for any radiographic image is then given as the per cent contrast sensitivity plus the smallest penetrameter hole discernible. For a two per cent penetrameter, the required sensitivity is usually 2-2T (two per cent contrast and the 2T hole are displayed). (A 2-1T sensitivity is more sensitive than required, and a 2-4T reading would be less sensitive than required.) The plaque penetrameter is the most commonly used in the domestic steel industry. In Europe and Japan, however, the wire penetrameter or IQI is popular. These consist of a flexible, plastic strip with lead identification symbols for material and specimen thickness range and a number (usually seven) of thin wires of different thicknesses which must be resolved to achieve specified sensitivity. These are popular in weld radiography.

The procedures outlined thus far pertain to those geometric precautions that are necessary to produce radiographic images with acceptable image quality. In this discussion it was assumed that there was sufficient penetrating radiation energy and intensity to produce a radiologic image on a film or screen. In practice, there are a number of important factors that must be considered in the generation of the radiologic image.

Penetrating radiation is absorbed and scattered as a function of its wavelength and the density, contour, and soundness of the product being inspected. The thicker and more dense the material, the greater the attenuation or absorption of the rays and fewer that reach the film or screen. Consequently a radiologic image displays, through variations in X-ray beam intensity at the film or screen, a shadow image of sorts that reveals changes in section, density, and the presence of imperfections. In radiography an increase in intensity results in greater film blackening. In fluoroscopy, the increased radiation intensity results in a brightening of the image. Because the X-ray beam is scattered as well as absorbed, the X-ray image can be fogged by the scattered rays which contain no information. Fortunately, thin lead screens (about 0.076 mm or 0.003 inch thick) effectively absorb the longer-wavelength scattered radiation preventing fogging. In addition, the lead screens emit electrons when the higher-energy penetrating rays strike them, and these electrons contribute to film blackening thus intensifying the image. Thicker lead screens (about 0.127 mm or 0.005 inch) are used to back up the film and absorb back-scatter coming from behind the film. These lead screens are normally contained in a frame called a cassette which is opaque to light. The film is loaded in cassettes in a dark room to provide a convenient way to handle X-ray film on the job.

The degree of film blackening or image contrast is governed by the wavelength and intensity of the penetrating radiation in addition to the thickness and density of the object. Furthermore, X-ray films blacken in proportion to exposure time and film speed. A film density of 2 is a commonly accepted value for blackening. It is, by definition, the logarithm of the rates of the incident light to transmitted light through the radiograph. For a density of 2, this ratio is 100:1.

The amount of absorption for any material is given in half-value layers (HVL). The HVL changes with the energy or wavelength of the radiation and is defined as that thickness of material that reduces the intensity of the transmitted radiation to one half of the value of the incident radiation. One half-value layer would, therefore, reduce the transmitted radiation to one half of the incident radiation, two half values would reduce it to one fourth, three half values would reduce it to one eighth, and so on.

Technique charts are prepared that relate material thickness and relative exposures (X-ray intensity by time) for various X-ray energy levels or wavelengths as expressed in kilovolts. These charts are prepared for specific X-ray sources under controlled conditions: film speed, source-to-film distance, what screens are used (if any), and film density (usually 2.0). Curves are prepared showing how much exposure is required to produce a 2.0 density at various thicknesses of a given absorber: steel, aluminum, etc. Since the energy level from an X-ray source is a function of the voltage applied to the tube, these are expressed in kilovolts (kV). The intensity is controlled by the tube current which is usually in milliamperes (mA), and the exposure is given in milliampere-minutes or milliampere-seconds. These charts do not guarantee any specific sensitivity, but they do supply basic parameters from which an acceptable technique can be developed.

It is evident that by increasing the kilovoltage, greater penetration can be achieved and exposure times reduced. This reduces, however, the film density changes (contrast) between sections with small thickness changes. Thus, sensitivity can be adversely affected. It is advisable to use the lowest kilovoltage consistent with practical exposure times and required sensitivities. Note that at a selected kilovoltage, an increase in tube current can be used to improve contrast or to reduce exposure times. High currents, however, result in greater tube energy dissipation and higher operating temperatures. Care must be exercised not to exceed tube rating. (Many modern X-ray sources automatically control current so that the tube ratings are not exceeded.) Subject to the geometric restrictions cited earlier in this section, exposure times can also be altered significantly by changing the source-object-distance (SOD) because X-ray intensity varies inversely with the square of the SOD. By halving this distance, therefore, the intensity at the object is increased fourfold, whereas if the SOD is doubled that intensity is one quarter of its original value. These factors provide considerable flexibility in the determination of the technique for any specific radiographic problem, but care must be exercised to remain within the bounds of good practice such as the control of scatter, maintenance of film density (1.5 to 2.5 is recommended), and to observe equipment operating values and safety precautions. These considerations apply to the use of radioisotope as well, although the flexibility of X-ray sources in adjusting kV and mA is not present. Sources must be changed instead. In general, radioisotopes are not generally used in steel plants, except for special applications where their advantages of portability and high source energy are primary considerations.

Density and contrast are also a function of the film used and whether screens are used. Radiographic film is rated at three types by the American Society for Testing and Materials based on their response without screens: (1) low speed, very high contrast, very low graininess, (2) medium speed, high contrast, low graininess, and (3) high speed, medium contrast, high graininess. A fourth level is also listed, but applies only to film designed for use with fluorescent screens. Most industrial radiography employs Type 2 film. The relative exposure times from Type 1 to Type 3 are about 10 to 1, the Type 1 requiring the longest time. For steel thicknesses of 0.25 inch (6.4 mm) or more, lead screens provide improved image quality and reduce exposure times by a factor of one third. Fluorescent screens are plastic or paper sheets coated with a thin film of chemicals that fluoresce in the presence of X-rays. Calcium tungstate and barium lead sulphate are two of these. These screens will reduce exposure times by a factor of 0.01 or more, but introduce unsharpness through scatter and the graininess of the phosphor. The use of fluorescent screens is not recommended for critical applications such as weld inspection. The use of lead screens (front and back) should be standard practice.

The radiographic method provides direct and permanent indications of product contour, thickness changes, the nature and size of imperfections, the location and distribution of imperfections, and the presence and condition of internal mechanisms. It is, however, slow and costly as compared to other methods,

and often entails delays in final product acceptance and shipping while radiographs are being developed and evaluated. Real-time radiologic imaging provides the same basic advantages without the attendant delays. In fact, motion of the radiologic image during inspection often makes small imperfections easier to see.

Early real-time radiology used fluorescent screens. These had very low light output and required about one half hour dark adaptation for the inspector. Furthermore, special optical and safety precautions were necessary to protect the operator from X-rays. X-ray fluoroscopy was used sparingly in industry until the development of the X-ray image intensifier. Developed originally for medical applications, industrial versions of the image intensifier became available in the early 1950's. They had a field of view of 5 inches (127 mm) and an image brightness gain of 1200 to 1500. These early image intensifiers made it possible to perform 100 per cent inspection of the longitudinal weld in SA-welded line pipe. The pipe weld was fluoroscopically inspected to within 152 mm (6 inches) of either end, and about 356 to 406 mm (14 to 16 inches) of each end was inspected by radiography at the end X-ray station. These systems employed a 150 kVCP (constant potential) X-ray source and TV presentation of the fluoroscopic image (Figure 51—12). Two per cent static X-ray sensitivity was achieved using an American Petroleum Institute (API) penetrameter. The minimum hole size for the two per cent API penetrameters is 1.59 mm ($\frac{1}{16}$ inch) in the range of 6.4-mm (0.25-inch) to 19.1-mm (0.75-inch) pipe walls inspected. Inspection speeds of 6.1 metres per minute (20 fpm), the speed at which a 4 per cent penetrameter can be viewed "definitively," are used in this inspection, and a complete inspection cycle is approximately 2½ minutes per 12.2 metres (40 feet) of pipe. Present-day image intensifiers have fields of view up to 305 mm (12 inches) and brightness gains in excess of 10 000. Two problems with the image intensifier led to the development of alternative systems for weld inspection. First the geometry of the image

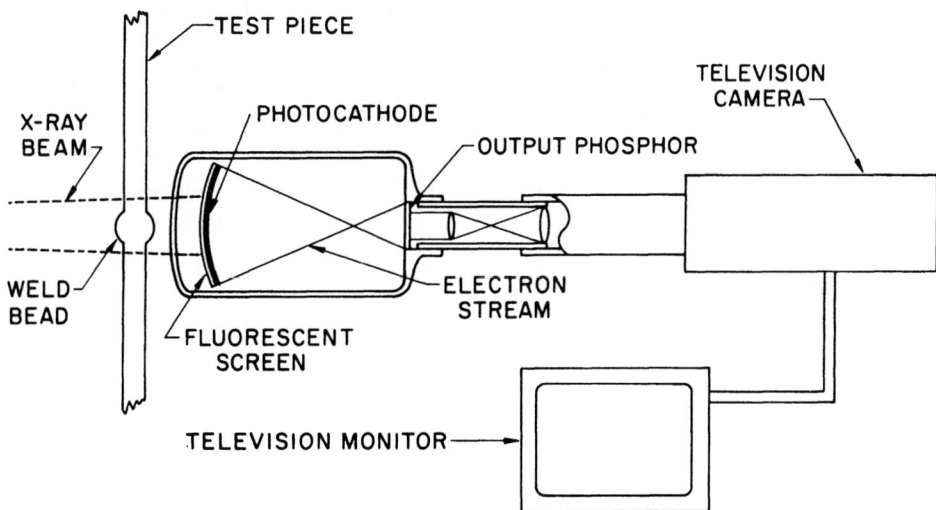

FIG. 51—12. Diagram showing principle of the use of a fluoroscopic image intensifier tube in conjunction with a television camera and monitor applied to the inspection of a weld bead.

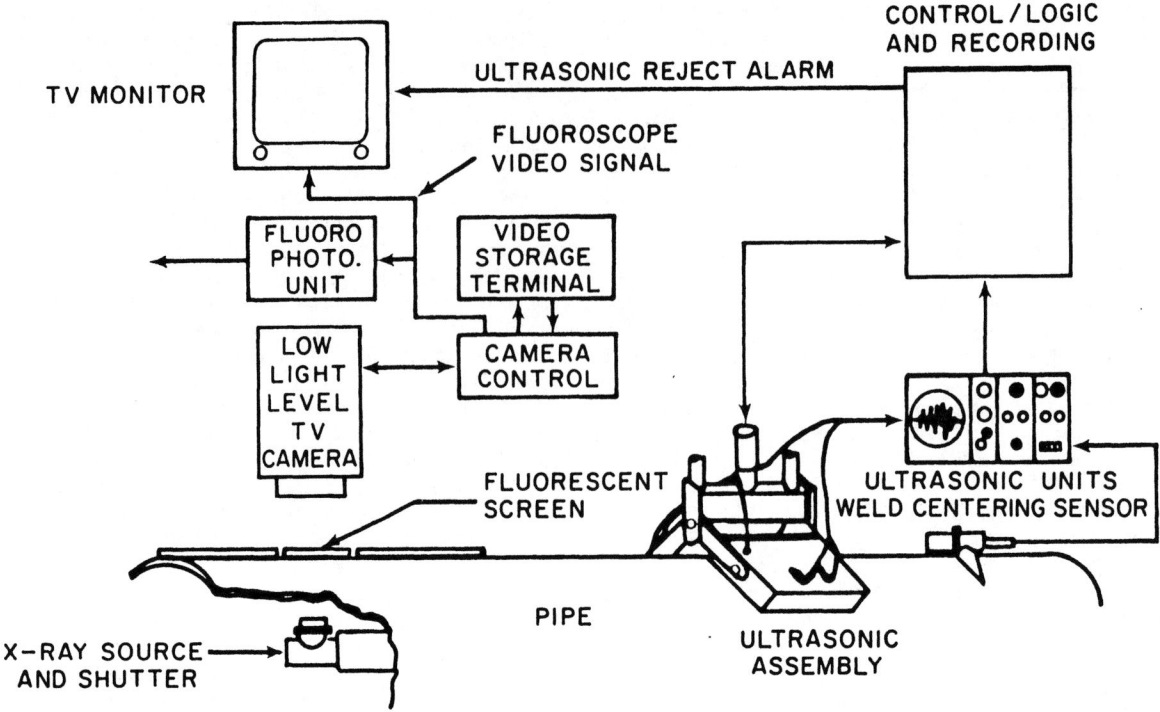

FIG. 51—13. Schematic assemblage of units to provide real-time imaging systems with the capability of making permanent records by means of photographic add-ons.

intensifier tube resulted in a pincushion effect in the image: the image was broadened and defocused somewhat at either end, and only about 60 per cent of the screen provided a consistently usable image. In addition, magnetic fields at either end of the pipe would often distort and defocus the image making it nearly unusable. The development of high-resolution scintillating screens and high-performance low-light-level television (LLLTV) systems provide system gains equal to or superior to that of an image intensifier. The screens do not exhibit the pincushioning effect and are not disturbed by magnetic fields. This system has been adopted in lieu of the image intensifier/TV combination and provides uniform image quality and sensitivity across the screen. The real-time imaging systems can be provided with photographic add-ons to provide permanent records similar to radiographs (Figure 51—13). Image enhancement and image analysis equipment is now being developed to further increase the utility of real-time systems. Real-time radiologic systems offer the basic advantages of radiography plus the advantages of rapid and timely evaluation of product quality compatible with pipe-mill production rates.

The radiographic method is one of the most widely accepted nondestructive methods because it provides direct information that requires little or no interpretation, and the sensitivity of the test and product identification are permanently recorded on the film. Safety is of primary importance in all radiologic installations, and strict adherence to established radiation-safety rules and regulations is required.

Bibliography

Annual Book of ASTM Standards, Section 3, Metals Test Methods and Analytical Procedures, Vol. 03.03—Metallography; Nondestructive Testing. American Society for Testing and Materials, Philadelphia, PA.

Betz, C. E., Principles of Magnetic Particle Testing, Magnaflux Corporation, Chicago, IL, 1967.

Betz, C. E., Principles of Penetrants, Magnaflux Corporation, Chicago, IL, 1963.

Krautkramer, Josef and Herbert, Ultrasonic Testing of Materials, Springer Verlag, Berlin, 1969.

McMaster, R.C., ed., Nondestructive Testing Handbook, Vols. I, II, Ronald Press, New York, NY, 1959.

Metals Handbook—Eighth Edition, Vol. 11, Nondestructive Inspection and Quality Control, Amer. Society for Metals, Metals Park, OH, 1976.

Radiography in Modern Industry, Eastman Kodak Company, Rochester, NY, 1981.

CHAPTER 52

Machinability of Carbon and Alloy Steels

Steels for Machining Applications—Machinability may be defined as that combination of properties in a material which affects its response to removal by a cutting tool. Many steels, commonly designated free-machining steels, contain special additives for improving their machining characteristics. The additives most commonly used for this purpose are sulphur and lead. In recent years, bismuth, selenium, and tellurium have been utilized to impart improved machinability to steels, often in combination with lead and sulphur. (Selenium has been utilized for many years to enhance the machining characteristics of certain stainless grades.) Many steels that do not contain the aforementioned elements are subjected to machining operations to produce a variety of parts. In some instances, the amount of metal to be removed by machining does not justify the extra cost of the additives. The effects of these additives on properties may preclude their use in certain steels, such as ultra high-strength steels, produced for applications in which optimum toughness is a prime consideration.

Machinability Evaluations—There is no universally accepted test method or criterion for rating the machinability of materials. Therefore, to aid in the development of new and improved steels for machining applications, steel producers have developed their own test methods, some of which are discussed in several of the references listed in the bibliography at the end of this chapter. Much of the machining information presented in this chapter relative to bar product was obtained in tests made on a special constant-pressure lathe at the United States Steel Technical Center. This lathe does not have a lead screw or fixed-feed mechanism but its carriage is mounted on ball bearings and loaded with a fixed weight. The feed obtained from tests conducted under controlled conditions is used as a measure of the relative ease with which a steel can be machined.

Additional information was obtained from tests con-

ducted on a multiple-spindle automatic screw machine producing a commercial part, Figure 52—1, from 15.88-mm- (⅝-inch) diameter cold-drawn bars. The cutting time for initial tool failure, or in some instances, the time in which tool wear was such that regrinding became desirable, was used as the criterion of performance of the steel.

Some of the information was obtained on a 31.75-mm- (1¼-inch) capacity single spindle No. 2 Brown and Sharpe automatic screw machine in the production of a test part, Figure 52—2, at the United States Steel Technical Center. The test part, produced from 23.81-mm- (¹⁵⁄₁₆-inch) diameter cold-drawn bars, involves the common machining operations—rough and finish forming, drilling, and cutting off. High-speed-steel tools are used to produce the part.

The objective of the screw-machine test is to establish the maximum production rate at which a steel can be machined into the standard part over an eight-hour period. During the test, specified limits of tool wear,

FIG. 52—1. Drum nut produced in automatic screw machine. (Magnification: 1½X)

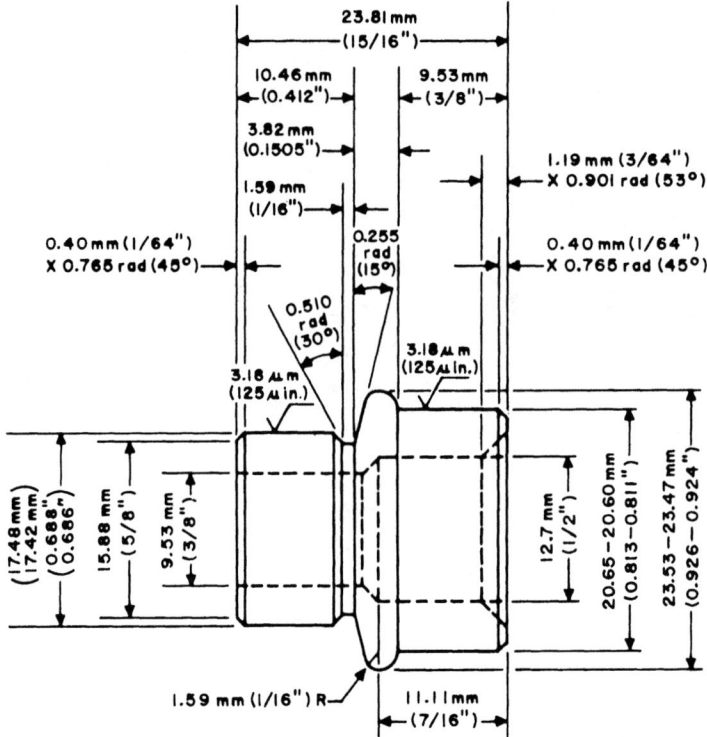

FIG. 52—2. The U.S.S. Technical Center machinability-test part.

surface finish, dimensional accuracy, and concentricity must not be exceeded. To establish the machining potential of each steel, the cutting speeds, tool feeds, and production rates are systematically varied until the maximum production rate is determined. This procedure, though expensive and time-consuming, is indispensable as the final discriminating evaluation in the development of steels with improved machinability.

In some instances, the machinability of steels was determined on the basis of the cutting speed that results in a 60-minute tool life in lathe-turning tests. These values of cutting speed are customarily termed V_{60} ratings. In other instances, the machinability of steels was determined by the amount of uniform wear on the flank of a carbide tool in lathe-turning tests at a constant cutting speed.

SECTION 2

FREE-MACHINING CARBON-STEEL BARS

The representative free-machining carbon-steel bars may be classified as shown in Table 52—I. These steels can provide substantial savings by increasing the production rate in high-speed machining operations. Over a period of many years, steel producers have conducted intensive research programs to develop steels having improved machinability. This has been particularly true for the low-carbon free-cutting steels—containing sulphur, or sulphur and lead, or sulphur, lead, and other additives—which are utilized extensively for the production of a wide variety of parts in automatic screw machines operating at high production rates.

The low-carbon free-machining steels may be classified into three categories according to the additives used, namely resulphurized steels, leaded resulphurized steels, and leaded resulphurized steels with special additives. Of these three, the resulphurized steels are

the most commonly used and were the first to be developed. The leaded resulphurized steels have also become widely accepted, since they provide appreciably superior machinability for a relatively small increase in price over the nonleaded steels. The leaded resulphurized steels with special additives are more recent and are designed for applications requiring a higher level of machinability than is attainable with steels containing only lead and sulphur as additives.

Resulphurized Rephosphorized Low-Carbon Steels —Although resulphurized, rephosphorized low-carbon steels nominally containing 0.1, 0.2, and 0.3 per cent sulphur are commonly used in machining applications, the most popular are those that contain about 0.3 per cent sulphur. For this reason, the 0.3 per cent sulphur steels, such as B1113, 1213, and 1215 grades, will be discussed in detail. Although the Bessemer process is

Table 52—I. Compositions of Free-Machining Carbon-Steel Bars.

Grade Designation		Composition, Per Cent					
UNS	SAE, AISI	C	Mn	P	S	Pb	Bi[1]
Low-Carbon Resulphurized Rephosphorized Steels							
G12110	1211	0.13 max.	0.60-0.90	0.07-0.12	0.10-0.15	—	—
G12120	1212	0.13 max.	0.70-1.00	0.07-0.12	0.16-0.23	—	—
G12130	1213	0.13 max.	0.70-1.00	0.07-0.12	0.24-0.33	—	—
G12150	1215	0.09 max.	0.75-1.05	0.04-0.09	0.26-0.35	—	—
	USS MX[2]	0.09 max.	0.75-1.05	0.07-0.12	0.26-0.35	—	—
G12134	12L13[3]	0.13 max.	0.70-1.00	0.07-0.12	0.24-0.33	0.15-0.35	—
G12144	12L14	0.15 max.	0.85-1.15	0.04-0.09	0.26-0.35	0.15-0.35	—
	Leaded B[2]	0.13 max.	1.05-1.35	0.04-0.09	0.40 min.	0.15-0.35	—
	USS MACH-5[2]	0.09 max.	0.85-1.15	0.04-0.09	0.26-0.35	0.15-0.35	0.05-0.15
Resulphurized Low-Phosphorus Steels							
G11080[3]	1108[3]	0.08-0.13	0.50-0.80	0.040 max.	0.08-0.13	—	—
G11100	1110	0.08-0.13	0.30-0.60	0.040 max.	0.08-0.13	—	—
G11170	1117	0.14-0.20	1.00-1.30	0.040 max.	0.08-0.13	—	—
G11180	1118	0.14-0.20	1.30-1.60	0.040 max.	0.08-0.13	—	—
G11370	1137	0.32-0.39	1.35-1.65	0.040 max.	0.08-0.13	—	—
G11390	1139	0.35-0.43	1.35-1.65	0.040 max.	0.13-0.20	—	—
G11400	1140	0.37-0.44	0.70-1.00	0.040 max.	0.08-0.13	—	—
G11410	1141	0.37-0.45	1.35-1.65	0.040 max.	0.08-0.13	—	—
G11440	1144	0.40-0.48	1.35-1.65	0.040 max.	0.24-0.33	—	—
G11460	1146	0.42-0.49	0.70-1.00	0.040 max.	0.08-0.13	—	—
G11510	1151	0.48-0.55	0.70-1.00	0.040 max.	0.08-0.13	—	—
Leaded Low-Phosphorus Steels[4]							
G10184	Leaded 1018	0.15-0.20	0.60-0.90	0.040 max.	0.050 max.	0.15-0.35	—
G11174	Leaded 1117	0.14-0.20	1.00-1.30	0.040 max.	0.08-0.13	0.15-0.35	—
G11374	Leaded 1137	0.32-0.39	1.35-1.65	0.040 max.	0.08-0.13	0.15-0.35	—
G11414	Leaded 1141	0.37-0.45	1.35-1.65	0.040 max.	0.08-0.13	0.15-0.35	—

[1] Bismuth content not specified—values shown represent usual levels of this element.
[2] Proprietary grade.
[3] SAE only.
[4] Lead may be added to any of the steels of the 1000 and 1100 series. The foregoing grades are presented as typical examples of leaded steels.

no longer being used to produce such steels, much information regarding their machining characteristics was developed over the years. Since data developed on Bessemer free-machining steels are applicable to basic oxygen or open-hearth free-machining steels, results of tests conducted on the Bessemer B1113 grade are included in this chapter.

Sulphur was the first additive used to impart "free-machining" characteristics to steel, and because of its effectiveness and low cost, it continues to be the most widely used additive. Steel is usually resulphurized by adding sulphur to the molten steel during tapping of the heat into the ladle.

Sulphur tends to make steel hot-short—that is, it is prone to develop surface cracks during rolling—as a result of the formation of iron-sulphide inclusions that form at the grain boundaries and are molten at rolling temperatures. Steelmakers minimize hot-shortness by adding sufficient manganese to ensure the formation of manganese-sulphide inclusions that are less fusible than iron-sulphide inclusions and do not form grain-boundary films. Thus, sulphur is present in steel primarily as manganese-sulphide inclusions.

Sulphur may have a beneficial effect on the quality of some steels in that it tends to suppress the evolution of carbon monoxide. Thus, sulphur acts somewhat as a "killing" agent and reduces the tendency for blowholes to form during solidification of the low-carbon free-machining steels. Fortunately, the steels that are utilized in applications requiring the highest levels of machinability are made to sulphur and manganese levels which are sufficiently high so that the steels are inherently semikilled without the addition of a deoxidizer such as silicon, which, as will be discussed, is very detrimental to machinability.

Whereas AISI and SAE specifications permit carbon contents as high as 0.13 per cent in resulphurized low-carbon phosphorus-bearing steels, most of these steels are currently made to carbon levels slightly below 0.10 per cent. The reason for this is evident from the curves in Figure 52—3, which show that in the constant-pressure test the highest machinability indexes are attained at the low-carbon (below about 0.10 per cent) levels. Furthermore, the data indicate that above and below a peak value of carbon the machinability index decreases.

In addition to increased machinability, maintaining low-carbon contents improves the performance of

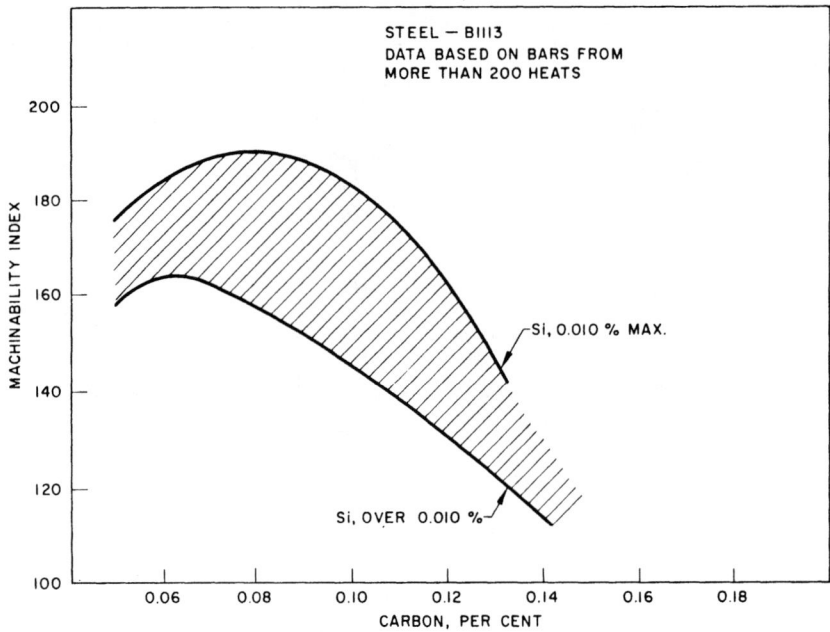

FIG. 52—3. Effect of carbon content on the machinability index of hot-rolled B1113 steel bars in the constant-pressure-lathe turning test.

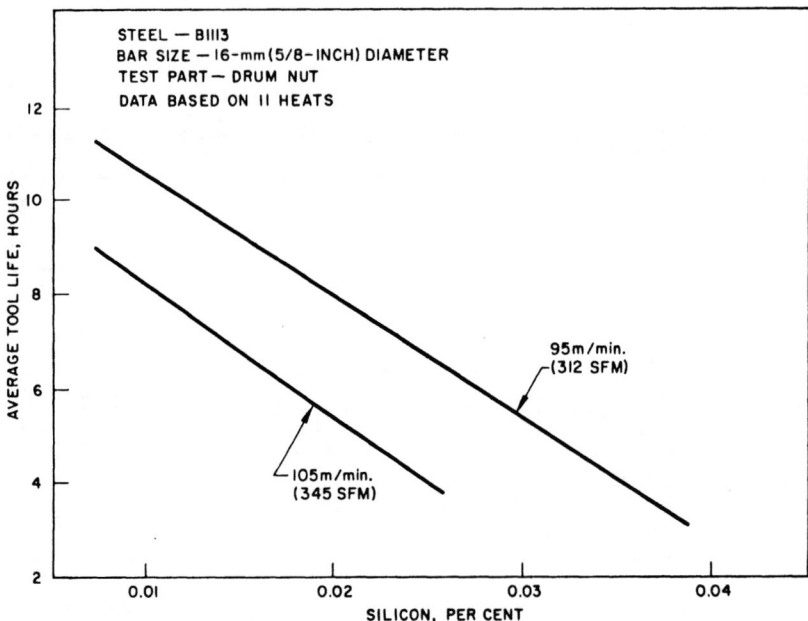

FIG. 52—4. Effect of silicon content on the performance of cold-drawn B1113 steel bars in the production of drum nuts.

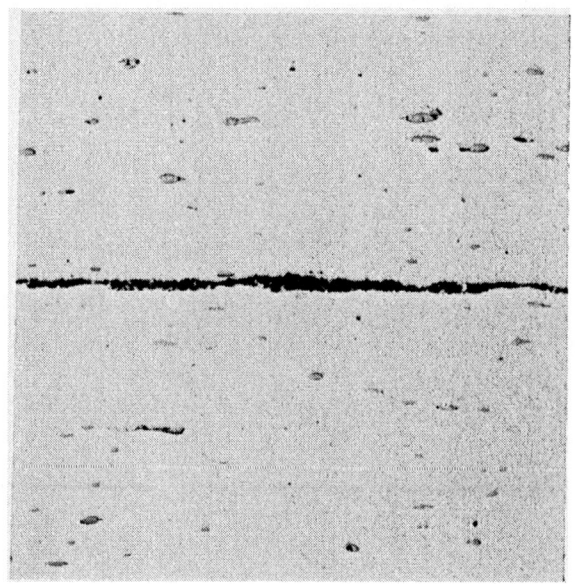

FIG. 52—5. Photomicrograph showing silicate stringer near the surface of a 24-mm (¹⁵⁄₁₆-inch) diameter cold-drawn bar. Unetched. Magnification: 100X.

parts subjected to cold-working operations such as crimping. In some applications, good cold-working performance is as important a requirement as good machining performance

As may be observed from Figure 52—3, silicon, even in minute amounts, has a very detrimental effect on the machinability of the steels under consideration. This effect is more specifically illustrated in Figure 52—4, which shows the existence of a linear relationship between the silicon content of cold-drawn B1113 steel bars and the average tool life obtained in the production of drum nuts in an automatic screw machine at cutting speeds of 105 and 95 surface metres per minute (m/min.), equivalent to 345 and 312 surface feet per minute (sfm).

Silicon impairs machinability in two ways. When it is present as abrasive silicate inclusions or stringers, Fig-

ure 52—5, tool life is drastically reduced. Silicon also influences the shape of manganese-sulphide inclusions —as the silicon content is increased, the proportion of desirable globular sulphides decreases. This effect of silicon content on the shape of sulphide inclusions and on the machinability index, as determined in the constant-pressure lathe, is shown in Figure 52—6, which compares the inclusion characteristics of G12130 (AISI or SAE 1213) steel bars containing 0.004 per cent silicon with the inclusions in bars from adjacent ingots of the same commercial heat to which ferrosilicon mold additions were made to provide silicon contents of 0.025 and 0.044 per cent.

The disproportionately large adverse effect of minute amounts of silicon on machinability and the need for low silicon and carbon contents in free-machining steel were initially recognized in a United States Steel-sponsored program conducted at the Battelle Memorial Institute. The results of this program led to the development of the practice followed in producing USS MX steel and other resulphurized low-carbon free-machining steels.

Because of the adverse effects of silicon, steel producers utilize steelmaking practices which ensure that the silicon level of the low-carbon free-machining steels is maintained uniformly at an absolute minimum. The improved practices include use of low-silicon ferromanganese products as the manganese addition agents for these steels, despite the higher cost of these addition agents compared to the cost of those normally used.

Sulphide-inclusion characteristics are affected by the oxygen level of the steel. Paliwoda (see bibliography) has reported that oxygen promotes the formation of sulphides exhibiting the desired globularity. He has shown that machinability increases as the oxygen-to-sulphur ratio increases (Figure 52—7).

The unfavorable effect of silicon (and other deoxidizers) on sulphide morphology is believed to be related to the deoxidizing effect of silicon. In fact, it has also been shown that deoxidation with carbon under vacuum affects the sulphide inclusions in a manner somewhat similar to that of a solid deoxidizer such as silicon. The effect of vacuum-carbon-deoxidation on sulphide morphology and the attendant machinability

Table 52—II. Effect of Manganese-Sulphide Inclusion Shape on Performance of G12150 (SAE or AISI 1215) Steel in Technical Center Automatic-Screw-Machine Tests.

Steel	Predominant Inclusion Shape	Time, h	Rough-Form Tool Feed, mmpr	Rough-Form Tool Feed, ipr	Maximum Flank Wear on Rough-Form Tool, mm	Maximum Flank Wear on Rough-Form Tool, in.	Surface Finish, μm	Surface Finish, μin.
Standard	Globular	8.0	0.089	0.0035	1.47	0.058	1.52-1.70	60-70
Vacuum Carbon Deoxidized	Linear	6.1	0.089	0.0035	1.85	0.073	2.79-2.92	110-115
Vacuum Carbon Deoxidized	Linear	8.0	0.069	0.0027	1.35	0.053	2.72-2.76	107-148

NOTE 1: Cutting speed 8.3 surface metres per minute (271 sfm); finish-form tool feed, 0.014 mm per revolution (0.00054 ipr); cycle time, 12 seconds.
NOTE 2: Sulphide-inclusion characteristics of these steels are shown in Figure 52—8.

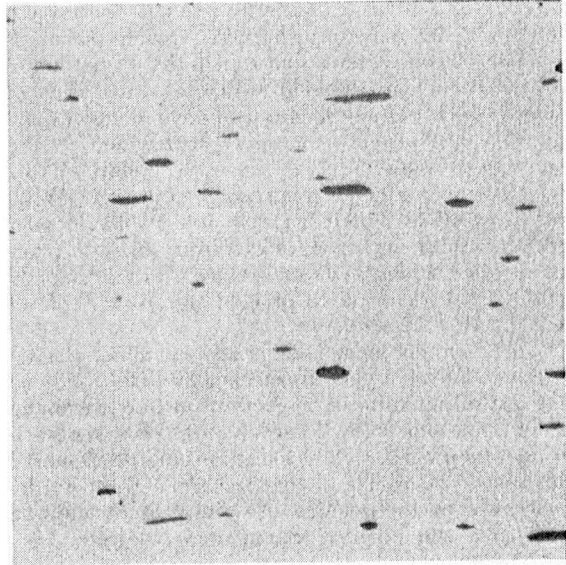

A. 0.004 per cent silicon; machinability index 177.

B. 0.025 per cent silicon; machinability index 136.

C. 0.044 per cent silicon; machinability index 101.

FIG. 52—6. Effect of silicon content on manganese-sulphide inclusion characteristics and machinability of 22-mm-diameter (⅞-inch-diameter) hot-rolled bars of USS MX open-hearth steel. Unetched. Magnification: 100X.

of low-carbon resulphurized steels is shown in Table 52—II which compares the performance of vacuum-carbon-deoxidized G12150 (AISI or SAE 1215) steel with standard G12150 steel in the production of the test part in the Technical Center's automatic screw machine. The inclusion characteristics of the steels are shown in Figure 52—8. As shown in Table 52—II, the steel with the globular inclusions exhibited appreciably less rough-form tool wear and a superior finish on the machined parts. Actually, for sustained machining, the rough-form feed had to be reduced for the steel with the linear inclusions.

The role that manganese sulphides play in improving machinability is related to their effect on chip formation. Several investigations have shown that an increase in sulphur content decreases the strain, that is, the amount of plastic deformation undergone by the chip. This means that the addition of sulphur decreases the work required in machining. Moreover, increasing the sulphur content of the steel has been shown to reduce the friction force between the chip and the cutting tool. A portion of this lower friction is a result of a reduction in the extent of contact between the chip and the tool. This latter effect is shown in Figure 52—9, which presents the results of lathe-turning tests on 0.08 per cent carbon-steel bars from the same heat at two

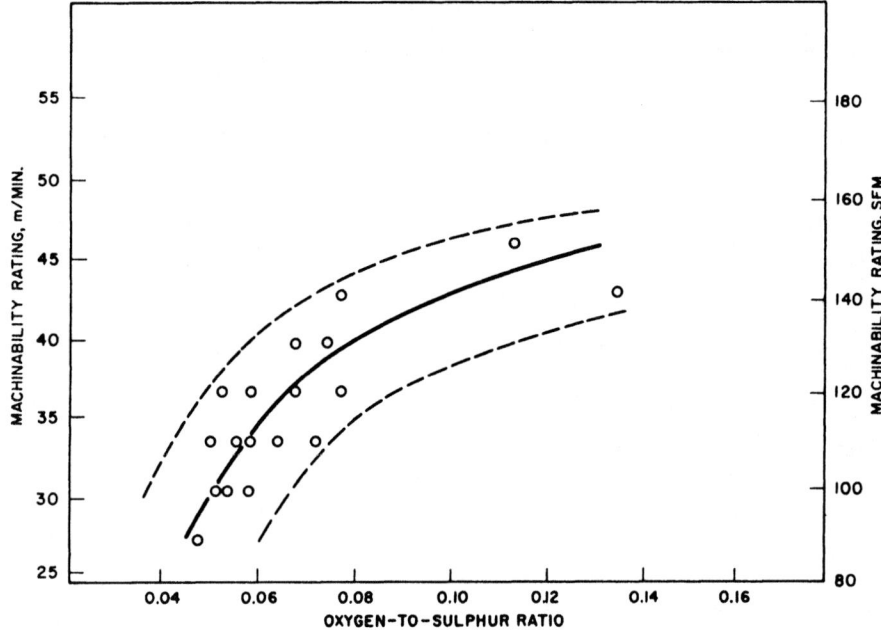

FIG. 52—7. Effect of ratio of oxygen to sulphur on the machinability rating of leaded 0.3 per cent sulphur steels. (After Paliwoda: see bibliography.)

levels of sulphur, the higher level being obtained by a mold addition of sulphur.

For many years, resulphurized low-carbon phosphorus-bearing steels were produced by the Bessemer process. These steels exhibited a high level of machinability and, consequently, were extremely popular in screw-machine shops. With the gradual decline in the Bessemer converter as a steelmaking facility, the steel industry undertook the development of open-hearth steels that would have machinability at least as good as that of the Bessemer grades. In contrast to open-hearth product, Bessemer steels inherently contain comparatively large quantities of phosphorus and nitrogen, two elements known to contribute to the machining performance of these steels. Consequently,

phosphorus and nitrogen, as well as sulphur, were added to the basic oxygen process steels and open-hearth steels produced for high-speed machining applications in which only Bessemer steel had previously been used.

As a result of continuing research, steel producers have been able to supply the screw-machine industry with basic oxygen process steels and open-hearth steels having machining characteristics even surpassing those of the best Bessemer steels. One advantage of the basic oxygen and open-hearth processes is that phosphorus and nitrogen contents can be controlled more readily, and restricted to lower levels when necessary, than in the Bessemer process. Machinability evaluations have disclosed the optimum phosphorus and nitrogen ranges

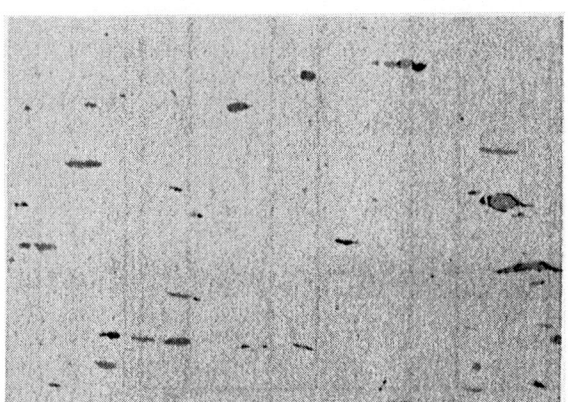 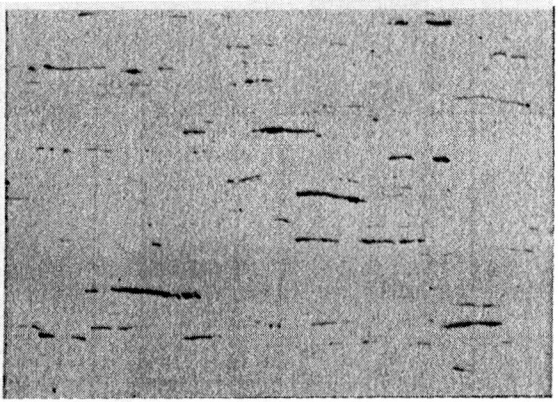

FIG. 52—8. Manganese-sulphide inclusions at mid-radius position of cold-drawn bars of (left) standard G12150 (AISI or SAE 1215) steel and (right) vacuum carbon deoxidized G12150 steel. Unetched. Magnification: 100X.

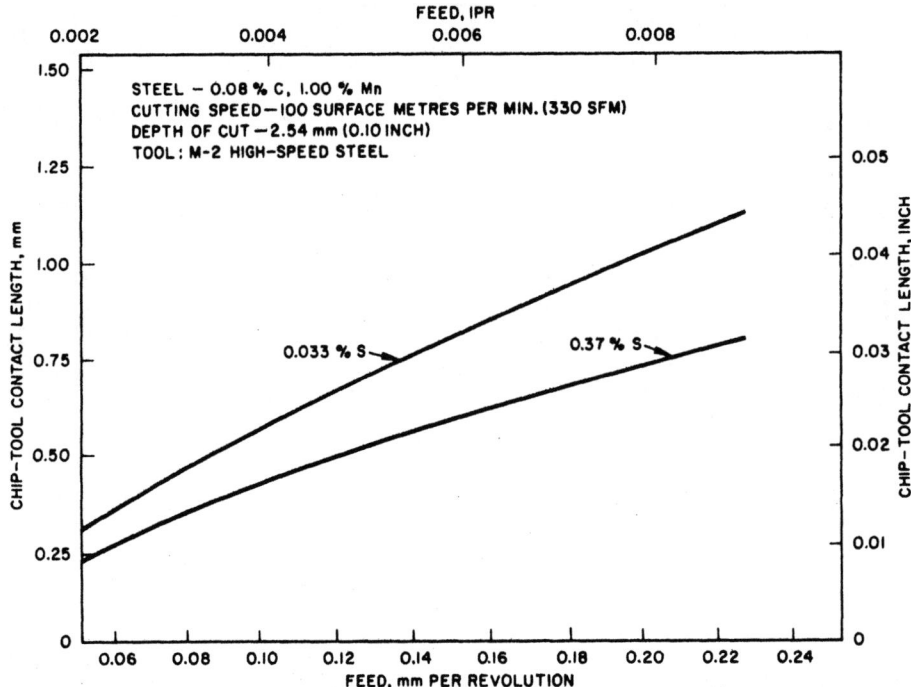

FIG. 52—9. Effect of sulphur content and feed rate on chip-tool contact length in turning.

for the low-carbon resulphurized steels to be about 0.05 to 0.09 per cent phosphorus and 0.007 to 0.012 per cent nitrogen.

The preceding findings resulted in the introduction, in 1962, of the free-machining steel now designated G12150 (AISI 1215). A comparison of the performance of typical G12150 and G12130 (AISI 1215 and 1213) steels in the Technical Center's automatic screw machine is included in Figure 52—10. It is to be noted that the G12150 steel could be machined satisfactorily at a production rate of 300 parts per hour. At this production rate, cutting tools had to be reground and re-

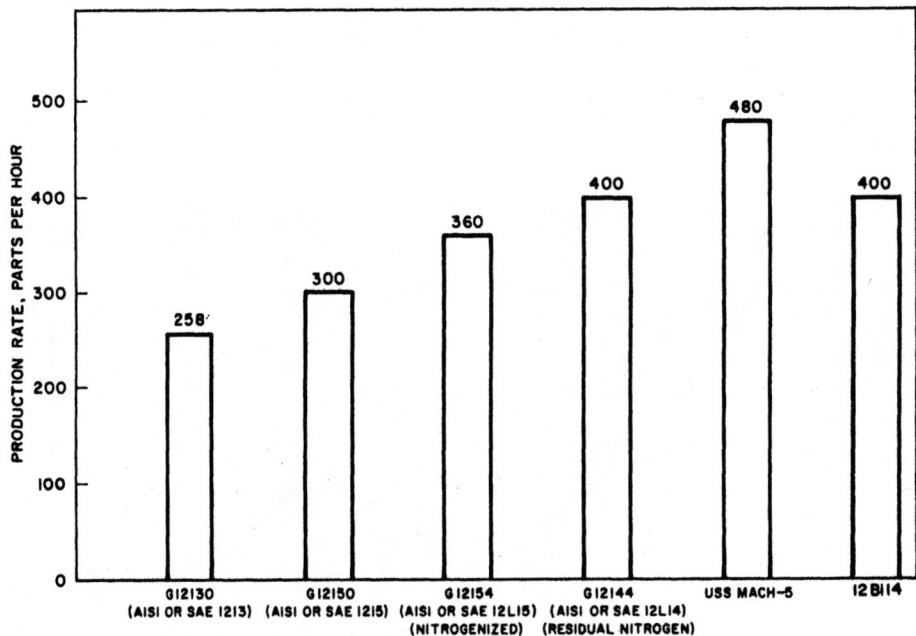

FIG. 52—10. Performance of low-carbon free-machining steels in the Technical Center's automatic screw machine.

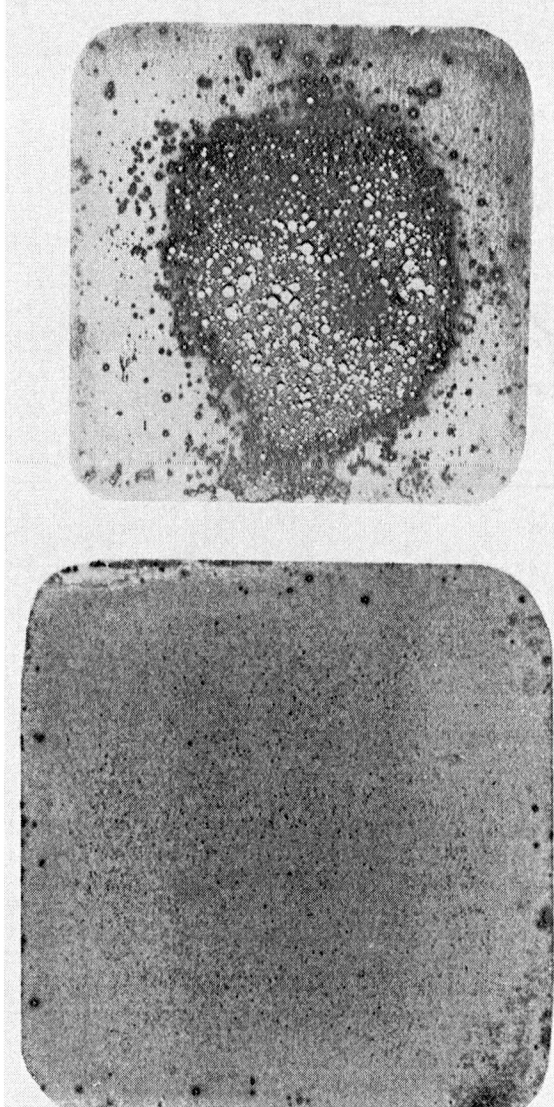

FIG. 52—11. Lead-exudation pattern on billet sections of G12144 (AISI or SAE 12L14) steel. (Above) A highly segregated billet. (Below) A slightly segregated billet. Light areas are lead. (Actual size).

Leaded Resulphurized Rephosphorized Low-Carbon Steels—Several versions of leaded resulphurized rephosphorized low-carbon steels are commercially available, Table 52—I, most of which contain about 0.30 per cent sulphur and 0.25 per cent lead. Although the different grades permit different maximum carbon contents, they are all normally made to carbon levels below about 0.10 per cent. These leaded steels are produced with or without added nitrogen, depending on customer preference.

Because lead has a limited solubility in molten steel and because it has a much greater density than steel, it tends to settle in the molten steel, and consequently, heavy lead segregation may occur at the bottom of ingots. Lead additions are made to individual ingots, and extreme care is taken to minimize lead segregation. To make the lead addition, a stream of high-purity, closely sized (–0.84 mm, + 0.42 mm or –20, + 40 mesh) lead shot is "shot" through a lead air-blast gun into the stream of molten steel as it is being teemed from the ladle to the ingot mold. The pressure in the air gun is carefully controlled so that the rate of addition is proportional to the rate of teeming. About 3 kilograms (6½ pounds) of lead shot are added per ton of molten steel so that the lead content in the finished product is in the range 0.15 to 0.35 per cent.

To determine whether a sufficient portion of the bottom of the ingot has been discarded to eliminate the aforementioned lead segregation, a test piece is cut immediately above the discarded portion and subjected to a lead exudation or "sweat" test. In this test, the bloom or billet test-piece is heated at 700°C (1290°F) for 10 to 20 minutes and then visually examined for beads of lead that "sweat" out of the steel at points of heavy lead segregation. Figure 52—11 shows the results of sweat tests on a billet section with little lead segregation and a billet section that is heavily segregated. In practice, the sweat-test specimens are com-

placed prematurely in machining the G12130 steel and the rate had to be lowered to 258 parts per hour to prevent undue wear of the cutting tools. This involved a reduced output of 14 per cent.

The lower sulphur G12120 or G12110 (AISI 1212 or 1211) grades shown in Table 52—I are used for those applications where limitations in attainable cutting speeds or other machining conditions would not permit realization of the benefits contributed by further increases in sulphur to the levels of the steels discussed above. These lower sulphur steels are also utilized for the production of certain parts in which additional sulphur would not be wanted because of possible undesirable effects on properties of the parts.

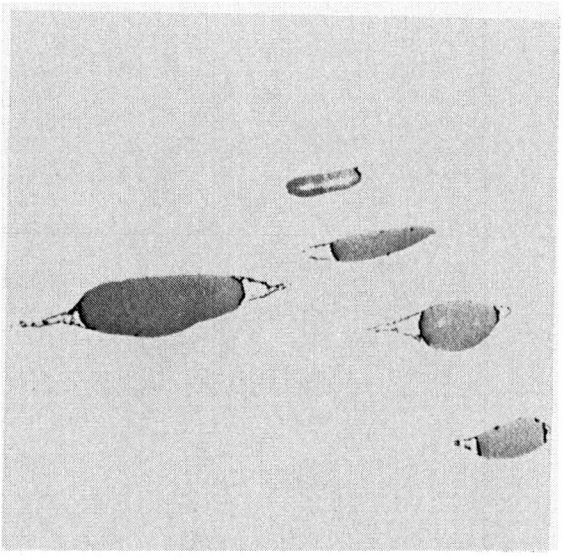

FIG. 52—12. Manganese-sulphide inclusions (gray) and lead inclusions (white) in leaded resulphurized steel. Unetched. Magnification: 500X.

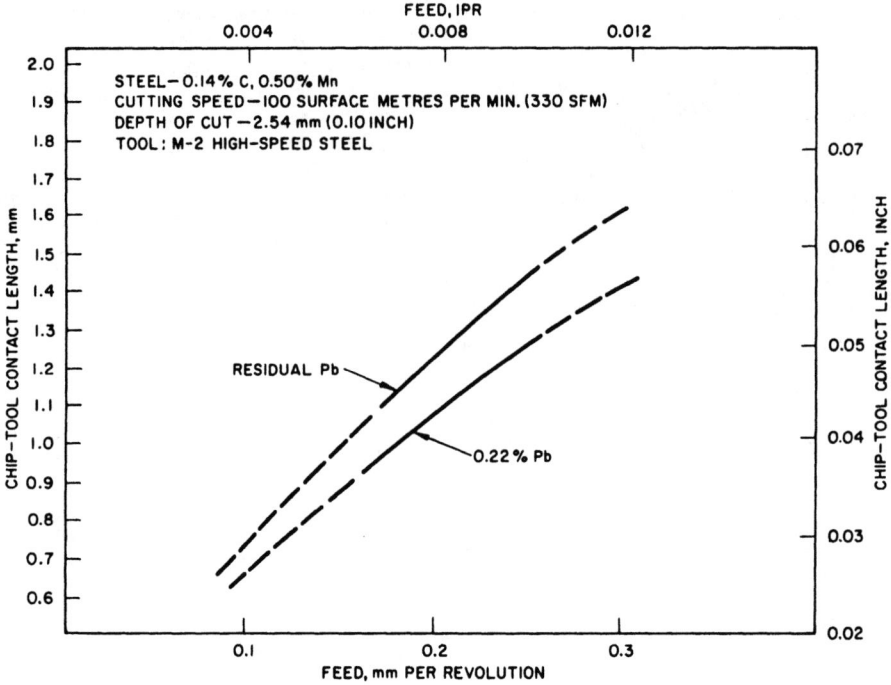

FIG. 52—13. Effect of lead content and feed rate on chip-tool contact length in turning.

pared with a rating chart to determine the degree of segregation. If the degree of segregation exceeds the minimum permitted by the exudation standard, the billet or bloom is cut back until the minimum standard is met before conversion to bar product.

More recently, techniques have been developed for making lead additions in a ladle properly equipped to exhaust lead fumes. The addition of lead in the ladle ensures a uniform lead content throughout the heat.

For many years, lead was believed to exist in steel as a submicroscopic dispersion. However, it is now well established that lead is present in elemental form as small inclusions, which are visible under the microscope and which are usually associated with manganese-sulphide inclusions in leaded resulphurized steels, Figure 52—12. .

The lead inclusions are soft and act as internal lubricants. Like sulphur, lead also reduces frictional forces

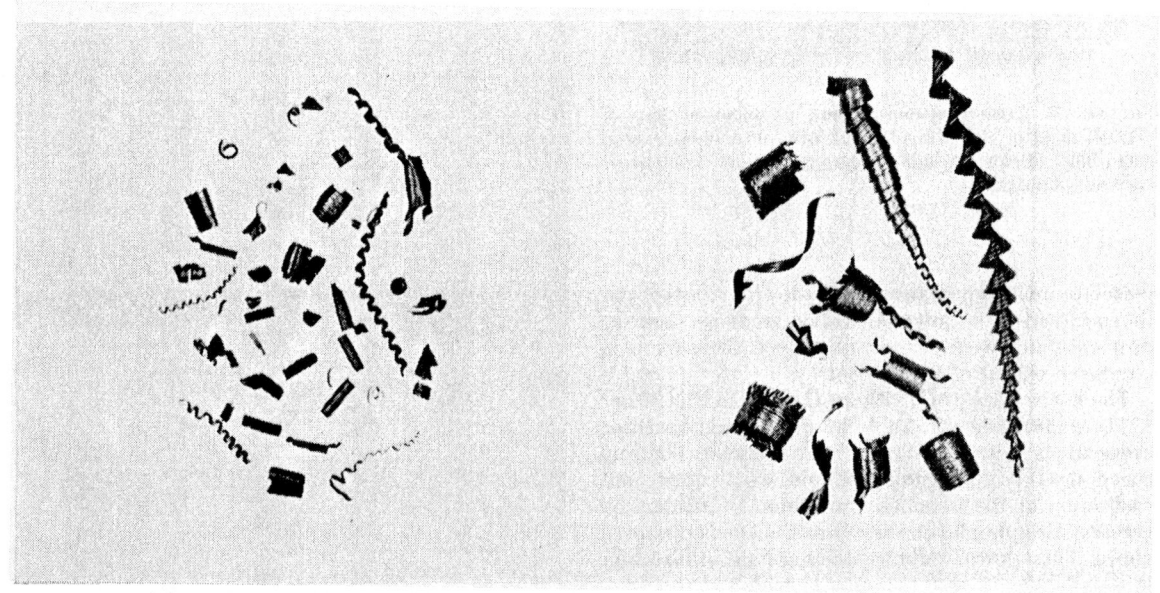

FIG. 52—14. Effect of lead on chips obtained in production of drum nuts from (left) G12144 (AISI or SAE 12L14) steel and (right) G12130 (AISI or SAE 1213) steel.

Table 52—III. Effect of Bismuth on Tool Performance in the Production of Drum Nuts from Cold-Drawn G12144 (AISI or SAE 12L14) Steel Bars.

Steel	Cutting Speed		Tool Life (Hours)	
	sm/min	sfm	Rough Form	9.1-mm ($^{23}\!/_{64}$-Inch) Drill
USS MACH 5 (12L14 + Bi)	129	422	6.6	7.2
G12144	129	422	4.4	5.7

NOTE: Rough-form feed 0.07 mm (0.0026 in.) per revolution; drill feed; 0.20 mm (0.0078 in.) per revolution.

by lowering the tool-chip contact length, Figure 52—13, and thus reduces the contact area for given cutting conditions. These reductions in tool-chip friction increase the useful life of cutting tools and improve the finish of machined surfaces. In addition, lead is very effective in promoting the formation of well-broken fine chips, Figure 52—14, which is a distinct advantage in screw-machine operations and in any machining process where chip handling or chip disposal may be a problem.

The effect of lead on the performance of steels in the Technical Center's automatic screw machine is included in Figure 52—10 in which a comparison can be made between G12150 and G12154 (AISI or SAE 1215 and 12L15) steels, which are similar except for the presence of lead in the G12154 grade. The G12154 steel performed satisfactorily at a production rate of 360 parts per hour, whereas the production rate had to be reduced to 300 parts per hour in machining the G12150 steel to provide satisfactory tool performance.

As indicated previously, leaded resulphurized low-carbon non-nitrogenized and nitrogenized steels are produced for applications in which the machinability of the leaded steels can be used to advantage. Customer preference varies with respect to which grade provides the better performance in machining applications. However, in those instances were severe cold-working is involved, the low-nitrogen steel is preferred and, indeed, often required, because it is much less prone to crack when subjected to high amounts of cold deformation. It is to be noted that in machining the laboratory-test part, Figure 52—10, the production rate for low-nitrogen G12144 (SAE or AISI 12L14) steel is 400 parts per hour whereas that for nitrogenized G12154 (AISI or SAE 12L15) steel is 10 per cent lower (360 parts per hour).

Leaded Resulphurized Steels With Special Additives —Several other additives—bismuth, selenium, and tellurium—are now being used in conjunction with lead and sulphur for improving the machinability of the low-carbon steels. These steels are used in applications where metal-removal rates faster than those attainable with the leaded resulphurized steels are feasible and more economical. The additions are usually made to individual ingots concurrently with the addition of the lead shot. These elements are present as inclusions in

combination with or associated with other elements. Of the foregoing elements, U.S. Steel is utilizing bismuth in the steels under consideration and, hence, this discussion will be on the role of bismuth in such free-machining steels.

In the production of the drum nut, the addition of bismuth to G12144 (AISI or SAE 12L14) steel (USS MACH-5 Steel) increased rough-form tool life by 50 per cent and drill life by about 25 per cent over that obtained with G12144 steel, Table 52—III. Actually, 10 300 parts were produced from the MACH-5 steel, and only 6 850 parts were produced from the G12144 steel before the rough-form tool was reground.

Tests on the Technical Center's automatic screw machine, Figure 52—10, have shown that in the production of the test part, MACH-5 steel can be machined satisfactorily at a production rate of 480 parts per hour, whereas the corresponding steel without bismuth (G12144 or AISI or SAE 12L14) did not perform satisfactorily at this production rate, but could be machined at a slower production rate of 400 parts per hour. The gross production rate for the bismuth-bearing steel was, therefore, 20 per cent higher than that for the standard G12144 steel.

An electron-microprobe scanning technique was used to study the inclusions in MACH-5 steel. The photographs in Figure 52—15 show that the characteristic $L\alpha$ radiation indicative of lead and bismuth emanates from the same inclusions, and thus disclose that the lead and bismuth, which are added separately, occur as an alloy in the steel bar. Furthermore, the results of quantitative point analyses for lead and bismuth in several lead-bismuth areas showed that the average composition of these regions is 65 per cent lead and 25 per cent bismuth. It is of interest to note that the lead-to-bismuth ratio determined from point analysis (2.6 to 1) is essentially identical with the ratio of the overall lead and bismuth contents of the steel (2.5 to 1). The improved machinability resulting from the bismuth addition is believed to be largely attributable to the alloying of the bismuth with the lead to provide a more effective internal lubricant.

As shown in Figure 52—15, the scanning images also disclosed that the gray inclusions, with which the lead-bismuth inclusions are associated, contain mainly manganese and sulphur, as indicated by the characteristic $K\alpha$ radiation of these two elements.

Bismuth-Containing Resulfurized Steels—The concern of steel producers over the toxicity of lead and the need to develop a suitable replacement for G12144 steel led to the development of 12Bi14 steel which, as illustrated in Figure 52—10, has machinability equivalent to that of G12144. The 12Bi14 steel is also equivalent to G12144 in tensile and impact properties, and can be case hardened without difficulty.

Free-Machining Steels for Case-Hardening—A wide variety of low-carbon free-machining-steel parts are subsequently case-hardened. To obtain a uniform case that is free of soft spots and of nonmartensitic bands, resulphurized high-phosphorus steels with or without additives such as lead or bismuth should be carburized at temperatures above about 955°C (1750°F). Typical data shown in Figure 52—16 disclose that the addition of lead or bismuth has a negligible effect on the case

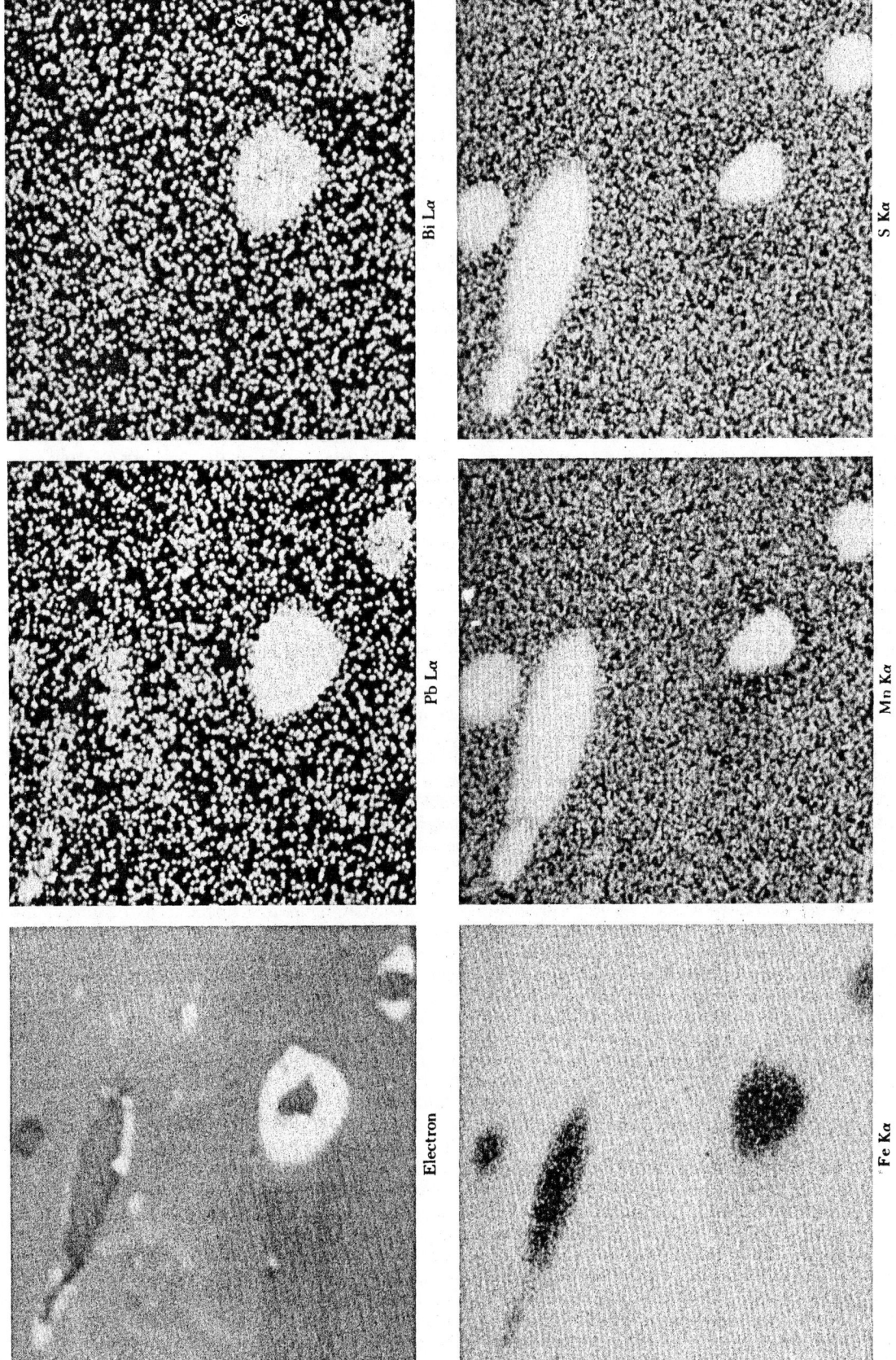

Fig. 52—15. Microscan images showing the distribution of lead, bismuth, iron, manganese and sulphur in the inclusions of a USS MACH-5 steel. Unetched. Magnification: 800X. (Courtesy, J. A. Gula.)

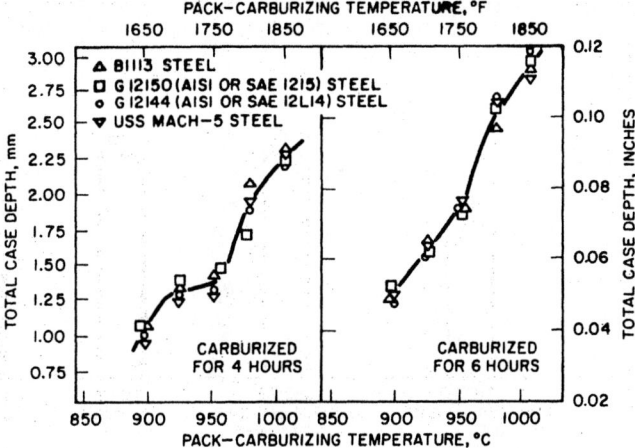

FIG. 52—16. Effect of carburizing time and temperature on total case depth produced in free-machining steels.

depths produced in the resulphurized free-machining steels. Gas carburizing and carbonitriding studies have also indicated that the foregoing additives have little influence on case depth.

Low-Phosphorus Free-Machining Steels—When factors other than machinability are also of major importance, the steel user has a choice of basic oxygen or open-hearth low-phosphorus resulphurized carbon steels (the G11000 or AISI or SAE 1100 series shown in Table 52—1) that embrace a wide variation in composition and therefore provide a broad range of attainable properties. These steels also possess good free-machining characteristics. Whereas carbon contents not greater than about 0.10 per cent are desirable for optimum machinability in the rephosphorized resulphurized low-carbon screw-stock steels, higher carbon contents provide the best machining characteristics in steels of the G11000 series. The effect of carbon content on bars of representative steels of the G11000 (AISI or SAE 1100) category is shown in Figure 52—17.

The lower carbon steels in this group, which are principally used for the production of cold-formed scrapless nuts, are resulphurized to facilitate tapping of the threads. In addition to production by the conventional method of adding sulphur to the ladle, steels are being made for subsequent scrapless-nut manufacture by the technique of inoculating low-sulphur ingots of rimmed steel with sulphur after they have rimmed sufficiently long to produce a satisfactory skin. Their low-sulphur rim and high-sulphur core provide a combination of excellent cold formability and tappability.

Steels containing more carbon (0.14 to 0.20 per cent C) are used in applications requiring a combination of good machinability and superior properties in the case-hardened condition compared to the properties attainable with the resulphurized high-phosphorus steels previously discussed. They also find use in nonhardened parts. The G11170 (AISI or SAE 1117) grade is the most popular of these steels but G11180 (AISI or SAE 1118) steels are used when increased

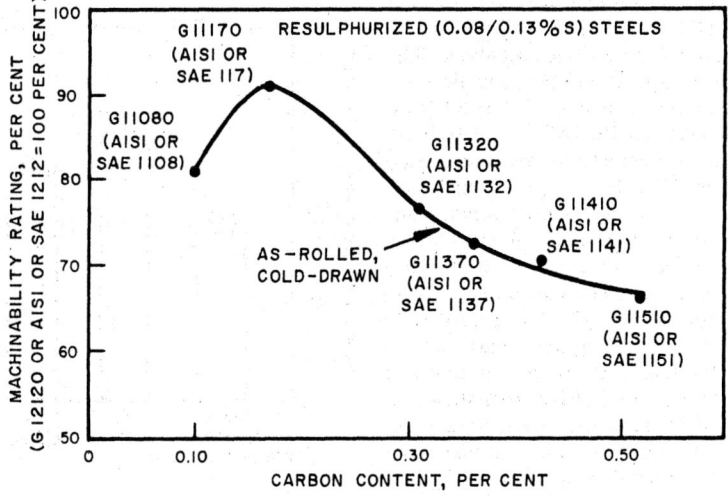

FIG. 52—17. Relationship between carbon content and the machinability of cold-drawn resulphurized steel bars of the G11000 (AISI or SAE 1100) series.

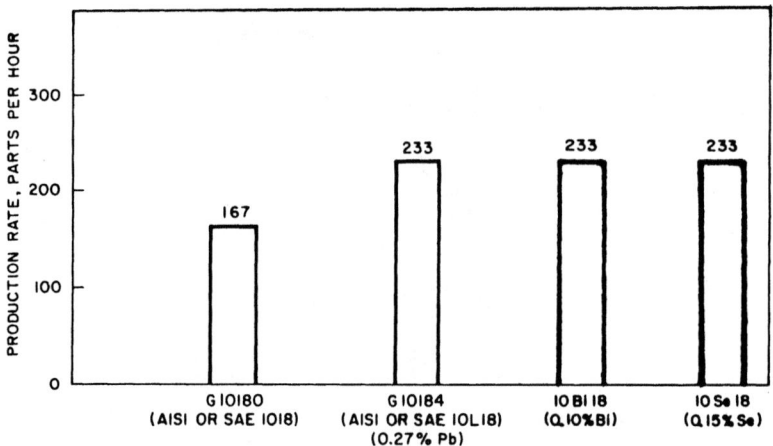

FIG. 52—18. Effect of additives on the performance of G10180 (AISI or SAE 1018) steel in the Technical Center's automatic screw machine.

hardenability, resulting from the higher manganese content, is required.

A number of medium-carbon free-machining steels such as the G11370, G11410, G11440, and G11510 (AISI or SAE 1137, 1141, 1144 and 1151) grades are available for producing parts requiring a relatively high strength. Depending on specific requirements, these steels are machined in either the hot-rolled, cold-drawn, or heat-treated condition. Bars of the G11440 (AISI or SAE 1144) grade are receiving very favorable acceptance as either cold-drawn and stress-relieved or warm-drawn product, in which conditions they provide desired mechanical properties and good machinability.

Leaded resulphurized steels are produced by adding lead to the steels representing the different carbon levels of the G11000 (AISI or SAE 1100) series. Typical steels in this category are shown in Table 52—I. Whether such a leaded steel should be selected for a given application is dependent on many factors; often the user can choose the proper steel only after comparative testing.

Lead is also added to non-resulphurized steels; that is, to steels of the G10000 (AISI or SAE 1000) series, as well as to the resulphurized steels discussed above. The advantage resulting from a lead addition can be illustrated by comparing the performance of G10184 (AISI or SAE 10L18) leaded and G10180 (AISI or SAE 1018) nonleaded steel in the Technical Center's automatic screw machine (Figure 52—18). As previously indicated, the presence of lead in resulphurized low-carbon steels results in the formation of small well-broken chips. This effect of lead is also very pronounced in machining the plain carbon steels. Because the chips from the leaded G10184 (AISI 10L18) steel more readily clear the tooling area, and also because lead reduces chip-tool friction, the addition of lead to the G10180 (AISI or SAE 1018) steel shown in Figure 52—18 resulted in a 40 per cent increase in production rate from 167 parts per hour to 233 parts per hour. Note that a similar increase in production rate was attained by the addition of 0.10 per cent bismuth and of 0.15 per cent selenium.

In some instances, fine-grain (aluminum-killed) free-machining resulphurized steels are specified. Because small aluminum additions to carbon steels adversely affect tool life, fine-grain free-machining carbon steels should be used only when necessary to provide the desired properties in a heat-treated product.

Selenium and tellurium are currently being used to enhance the machinability of carbon and alloy steels. The results of screw-machine tests conducted on medium-carbon G10400 (AISI or SAE 1040) steel with different additions are shown in Figure 52—19. The production rate attained with the standard G10400 steel was 136 parts per hour. The addition of 0.09 per cent selenium to this steel made it possible to increase the production rate to 193 parts per hour, an increase of 42 per cent. The addition of 0.09 per cent tellurium to the steel permitted a 25 per cent increase in production rate to 170 parts per hour. Figure 52—19 also shows the production rates attained with 0.10 per cent bismuth and with a combination of 0.19 per cent lead

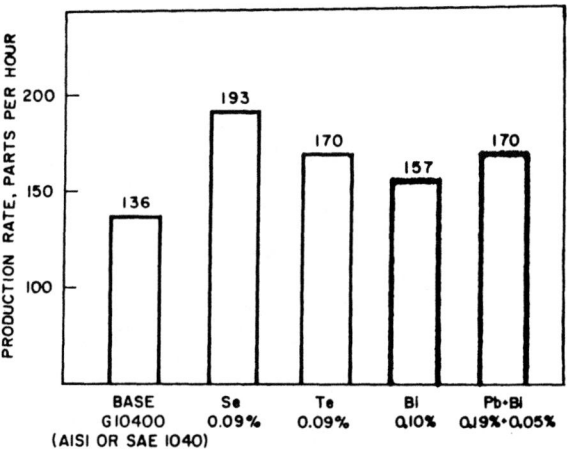

FIG. 52—19. Effect of additive on the performance of G10400 (AISI or SAE 1040) steels in the Technical Center's automatic screw machine.

and 0.05 per cent bismuth.

Screw-machine tests on resulphurized steels made by the basic-oxygen process have disclosed that they exhibit as high a level of machinability as correspond-

ing steels made by the open-hearth process. Therefore, the increasing utilization of the basic-oxygen steelmaking process will not adversely influence the performance of the free-machining steels.

SECTION 3

NON-FREE-MACHINING CARBON STEEL BARS

There are many applications involving machining for which steels containing no additive for improving machinability are utilized. The low-carbon steels in this category, which comprise the steels of the G10000 (AISI or SAE 1000) series are relatively soft and ductile and consequently exhibit "gummy" characteristics during machining with attendant formation of long stringy chips that are often objectionable and difficult to remove from the tool area.

In the plain-carbon steels of the G10000 (AISI or SAE 1000) series, machinability increases with increasing carbon up to 0.20 to 0.25 per cent, but decreases with

further increases in carbon as shown in Figure 52—20. Annealing improves the machining characteristics of the medium and higher carbon steels in this category, particularly at carbon levels of 0.45 to 0.50 per cent and higher. Pearlitic microstructures are generally preferable in the medium-carbon steels containing up to about 0.50 per cent carbon. Spheroidize-annealing treatments are definitely advantageous for enhancing the machinability of steels containing 0.60 per cent carbon and higher. For some applications, it is worthwhile to partly spheroidize steels of somewhat lower carbon levels.

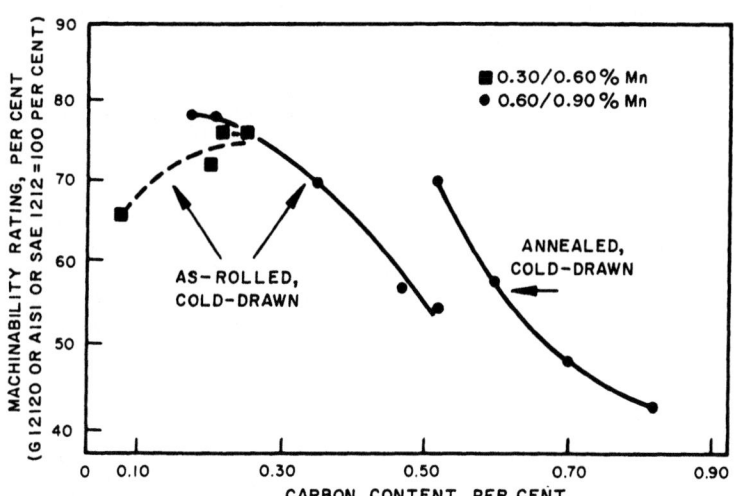

FIG. 52—20. Relationship between carbon content and the machinability of cold-drawn carbon-steel bars of the G10000 (AISI or SAE 1000) series.

SECTION 4

FREE-MACHINING ALLOY-STEEL BARS

Leaded alloy steels and resulphurized alloy steels are produced both in the low-carbon carburizing grades and in the medium-carbon grades. The compositions of typical steels in these categories are shown in Table 52—IV. Recently, selenium-containing alloy steels such as the G41400 (AISI or SAE 4140) steel with selenium, shown in Table 52—IV, have been developed and made commercially available. These alloy steel bars provide significant improvements in machine output and in the

surface finish of the machined part. Tellurium-containing alloy steels also have been marketed in recent years. Both selenium, which is present in steel as manganese-selenide inclusions, and tellurium, which is present as manganese-telluride inclusions, are more effective than either lead or sulphur for improving the machinability of medium-carbon steels.

The performance of annealed cold-drawn G41400 (AISI or SAE 4140) bars—containing either selenium,

Table 52—IV. Compositions of Alloy-Steel Bars with Sulphur, Lead or Selenium.

Grade		Composition, Per Cent							
UNS	AISI or SAE	C	Mn	Ni	Cr	Mo	Pb	S	Se(1)
G40240	4024	0.20-0.25	0.70-0.90	—	—	0.20-0.30	—	0.035-0.050	—
G40280	4028	0.25-0.30	0.70-0.90	—	—	0.20-0.30	—	0.035-0.050	—
G86204	Leaded 8620	0.18-0.23	0.70-0.90	0.40-0.70	0.40-0.60	0.15-0.25	0.15-0.35	—	—
G41404	Leaded 4140	0.38-0.43	0.75-1.00	—	0.80-1.10	0.15-0.25	0.15-0.35	—	—
	USS Carilloy FC(2)	0.47-0.55	0.95-1.30	—	0.60-0.90	0.15-0.25	—	0.06-0.10	—
Se-Containing									
G41404	4140	0.38-0.43	0.75-1.00	—	0.80-1.10	0.15-0.25	—	—	0.05-0.10

(1) Selenium content not specified—values shown represent usual level of this element.
(2) Proprietary grade.

tellurium, or lead—in the production of the standard test part at United States Steel's Technical Center is shown in Figure 52—21. The results indicate that G41400 bars contianing 0.11 per cent selenium had the best machinability of all the steels tested. The production rate of G41400 steel with 0.08 per cent selenium was the same as that of the G41404 steel with 0.19 per cent lead and the same as that of the G41400 steel containing 0.08 per cent tellurium. At the production rate shown in Figure 52—21, less tool wear was encountered in machining the steel containing 0.08 per cent selenium than in machining the leaded steel. Moreover, selenium does not cause the segregation problems encountered with lead additions and selenium steels do not exhibit the severe hot-shortness characteristics of tellurium-containing steels.

As shown in Figure 52—22, the effectiveness of selenium does not fade out as the tensile strength of the bars is increased from 550 MPa (80 ksi) to as much as 1240 MPa (180 ksi). This figure shows the previously mentioned data on G10400 and G41400 (AISI or SAE 1040 and 4140) steels and includes data on bars of G10180, G41420 and G41450 (AISI or SAE 1018, 4142 and 4145) steels. The bars G41420 (AISI or SAE 4142) steel contained about 0.08 per cent selenium and had been processed to tensile strength of 1090 and 1240 MPa (158 and 180 ksi). The bars of the G41450 (AISI or SAE 4145) steel contained about 0.10 per cent selenium, had been cold-drawn after quenching and tem-

pering, and had a tensile strength of about 1160 MPa (168 ksi). These data show that in the range of tensile strengths under consideration, the improvement in machinability imparted by selenium does not decrease as tensile strength (or hardness) increases. The upward pointing arrows on two of the data points signify that the actual increment in production rate should be greater than the values indicated; the difficulty in determining the increments more accurately was due to the difficulty in accurately establishing the maximum production rate for the corresponding control steels which exhibited very low levels of machinability.

The additives have no significant effect on the hardenability response of G41400 (AISI or SAE 4140) steel or on the tensile properties in the annealed condition or in the quenched and tempered condition up to at least 1380 MPa 4200 ksi). However, at about 316°C (600°F), leaded steels show a pronounced decrease in the reduction of area as shown in Figure 52—23. this phenomenon has been the subject of extensive studies. Also, all the additives affect the impact of fatigue properties of the steels in the quenched and tempered condition. For example, as shown in Table 52—V, in bars of the G41400 (AISI or SAE 4140) steels that were quenched and tempered to a tensile strength of about 1070 MPa (155 ksi), the data from Charpy V-notch impact tests and from rotating-beam fatigue tests show that the additives decrease the energy absorbed, and decrease the fatigue limit in notched specimens. It is

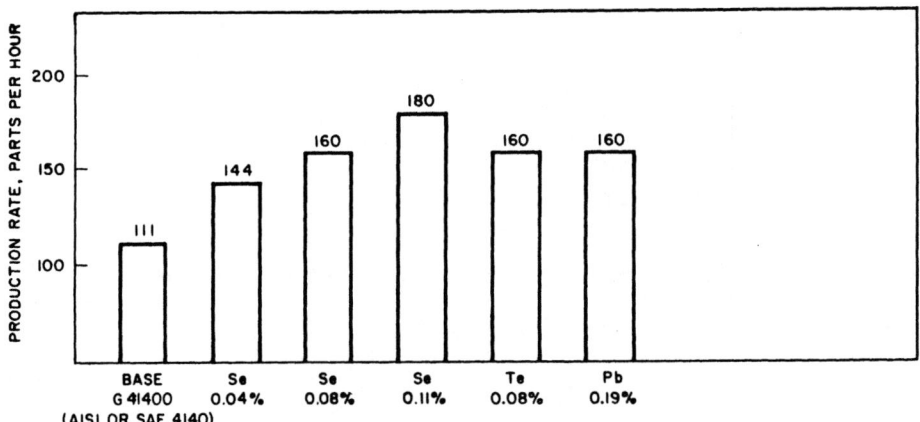

FIG. 52—21. Performance of G41400 (AISI or SAE 4140) steel with selected additions in the Technical Center's automatic screw machine.

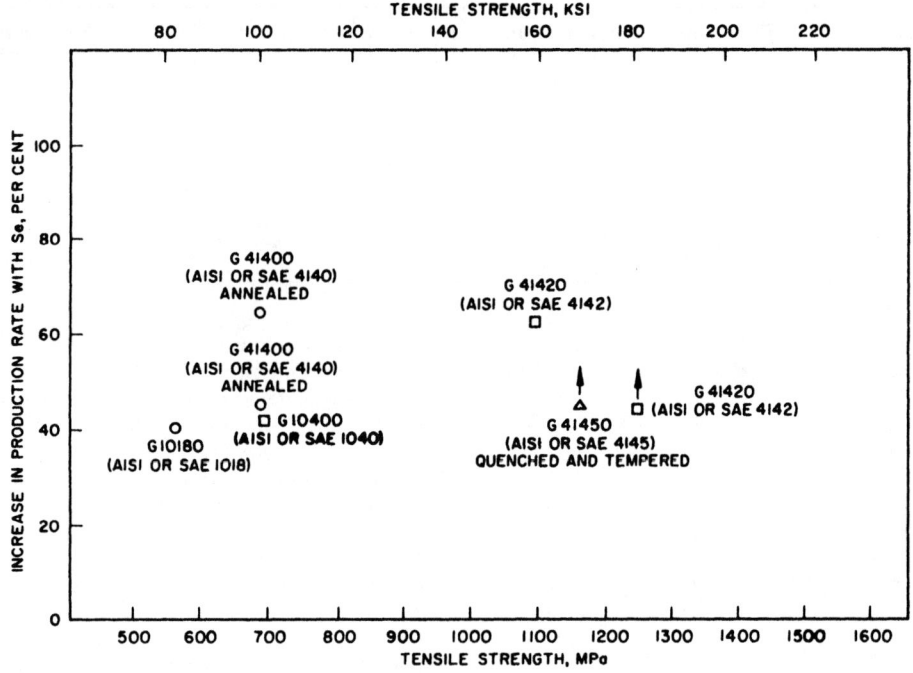

FIG. 52—22. Relationship between tensile strength and performance of various selenium-containing steels in the Technical Center's automatic screw machine.

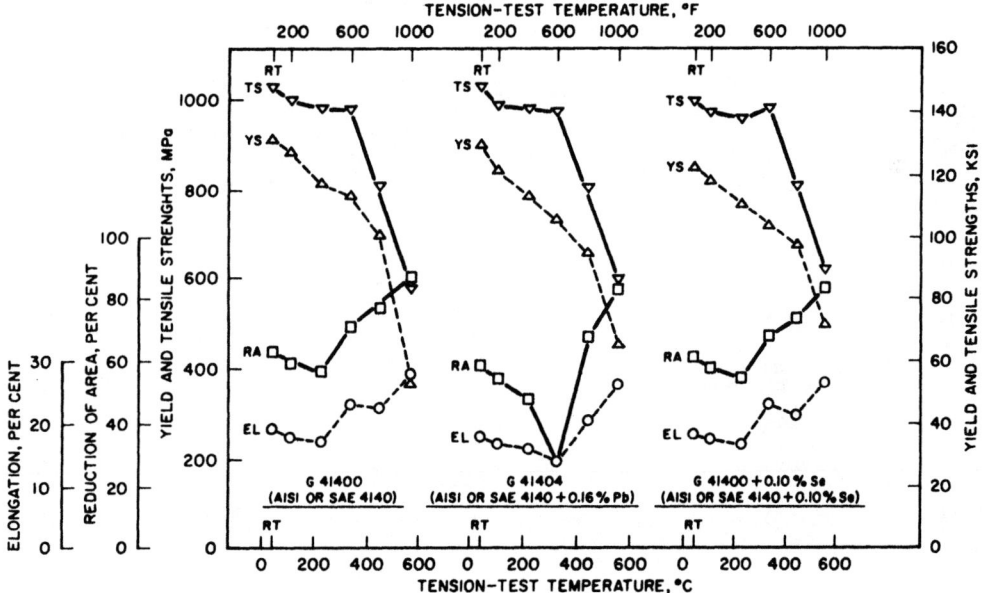

FIG. 52—23. Effect of tension-test temperature on the tensile properties of quenched and tempered bars of G41400 (AISI or SAE 4140) steels.

Table 52—V. Mechanical Properties of Quenched and Tempered G41400 (AISI or SAE 4140) Steel with Additives.

Additive	Tensile Strength		Energy Absorbed in Charpy V-Notch Test				Notched Fatigue Limit	
			-18°C	(0°F)	100% Shear			
	MPa	psi	joules	ft-lb	joules	ft-lb	MPa	psi
None	1082	157 000	83	61	104	77	448	64 000
0.08% Se	1069	155 000	56	41	68	50	379	55 000
0.08% Te	1082	157 000	50	37	57	42	365	53 000
0.19% Pb	1076	156 000	56	41	61	45	386	56 000

important to note, however, that the properties of the steels containing selenium, tellurium or lead are essentially similar.

Although the same amount of lead (0.15-0.35 per cent Pb) is added to the alloy grades as to the carbon

Table 52—VI. Effect of Sulphur on Tool Performance in Lathe Turning Tests.

Grade	Heat Treatment	Sulphur Content (Per Cent)	V_{60}*	
			Surface Metres per Min.	sfm
G86200 (AISI or SAE 8620)	Normalized	0.011	72.5	238
		0.039	77.1	253
		0.064	79.9	262
		0.087	90.2	296
G43400 (AISI or SAE 4340)	Annealed	0.017	57.9	190
		0.041	70.1	230
		0.066	76.2	250

*Cutting speed in surface metres or surface feet per minute for a 60-minute tool life.
NOTE: The tests were made with high-speed-steel tools at a feed of 0.05 mm (0.002 inch) per revolution and a cut depth of 1.27 mm (0.050 inch).

grades, the sulphur level of the resulphurized alloy steels is generally less than that of the resulphurized carbon grades. An indication of the effectiveness of sulphur additions on machinability is shown in Table 52—VI which shows the V_{60} values (cutting speed for a 60-minute tool life) obtained on normalized G86200 (AISI or SAE 8620) steel bars and on annealed G43400 (AISI or SAE 4340) steel bars, containing different amounts of sulphur. The various levels of sulphur were obtained by making mold additions of sulphur to the heats as they were being teemed. The data show that the tool performance was progressively improved as the sulphur content was increased. Tool life also substantially increased with the sulphur content of the G43400 (AISI or SAE 4340) bars when they were machined in the quenched and tempered condition at hardness levels as high as 363 BHN. The effectiveness of sulphur did not decrease as the hardness (and strength) of the bars was increased.

A resulphurized modification of G41500 (AISI or SAE 4150) steel known as USS Carilloy FC steel and containing 0.06 to 0.10 per cent sulphur has enabled users to increase tool life substantially (as much as 300 per cent) and increase machine output considerably in the machining of alloy parts at hardnesses well over 300 BHN. The improvement in output has been obtained with an attendant enhancement in surface fin-

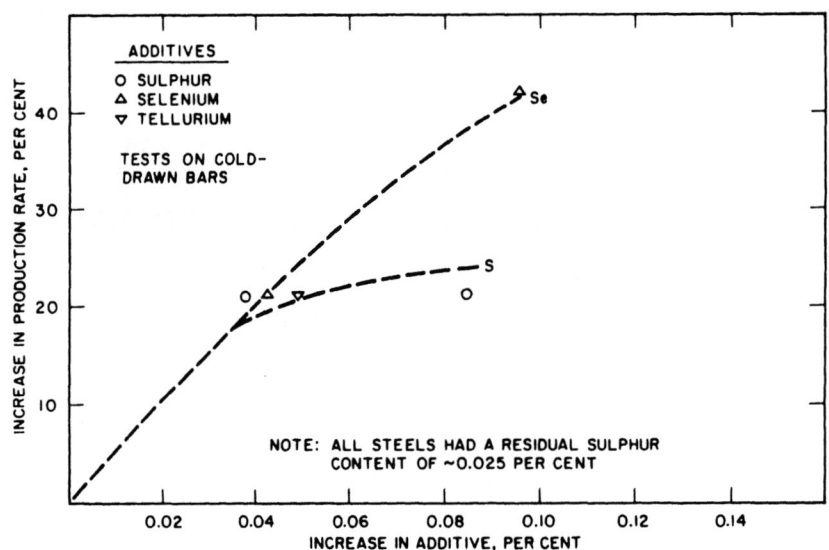

FIG. 52—24. Effect of additives in G86200 (AISI or SAE 8620) steel on increase in production rate.

Table 52—VII. Effect of Additives in G86200 (AISI or SAE 8620) Steel on Production Rate in Automatic-Screw-Machine Tests.

Additive (%)	Production Rate for Satisfactory Performance (pph)	Maximum Flank Wear on Rough-Form Tool		Test Results				Increase in Production Rate (%)
				Surface Roughness				
				Medium		Range		
		mm	in.	μm	μin.	μm	μin.	
None	138	0.508	0.020	2.67	105	2.26-3.10	89-122	—
0.063 S	167	0.711	0.028	2.49	98	1.91-2.90	75-114	21
0.11 S	167	0.406	0.016	2.57	101	2.08-2.82	82-111	21
0.043 Se	167	0.330	0.013	3.02	119	2.51-3.28	99-129	21
0.096 Se	195	0.457	0.018	2.97	117	2.06-3.40	81-134	41
0.049 Te	167	0.432	0.017	2.90	114	2.59-3.23	102-127	21
0.20 Pb	167	0.762	0.030	2.44	96	1.80-2.79	71-110	21
0.22 Pb	167	0.330	0.013	2.77	109	2.11-3.05	83-120	21

ish.

The effect of free-machining additives on the machinability of one-inch-diameter cold-drawn bars of G86200 (AISI or SAE 8620) steel is shown in Table 52—VII. The production rate in machining was at least 41 per cent higher in steel containing 0.096 per cent selenium and at least 21 per cent higher in steel containing 0.063 to 0.11 per cent sulphur, 0.043 per cent selenium, 0.049 per cent tellurium, or about 0.2 per cent lead. The effect of an increase in additive content on the increase in production rate is shown in Figure 52—24. Note that there is no difference in the increase in production rate imparted by small amounts up to about 0.04 per cent of S, Se and Te. Nevertheless because of the significant amounts of sulphur residuals in steels produced by conventional processes, better machinability would be consistently attained with steels containing Se and Te than with resulphurized steels containing an equivalent weight per cent of sulphur. Increasing the sulphur content of the steel to 0.063 and

0.11 per cent increased both the amount and the number of inclusions, with only a slight increase in the size of the inclusions. All the other additives increased the amount, number and size of inclusions in the steel. The effect of the amount of additive-related inclusion material on the machinability (maximum flank corner wear) is shown in Figure 52—25. Initially, small increases in inclusion area markedly reduce tool wear, and at high inclusion contents, major increases in the amount of inclusion materials provide relatively small reductions in tool wear.

Small amounts of sulphur, selenium and tellurium increased the grain-coarsening temperature (maximum temperature for austenite grain size number higher than ASTM No. 5) by about 110°C (200°F). These additives appear to have no significant influence on end-quench hardenability, or on the response of the steel to carburizing and carbonitriding.

The addition of lead to steel will not improve the life of cutting tools under all conditions. For example, un-

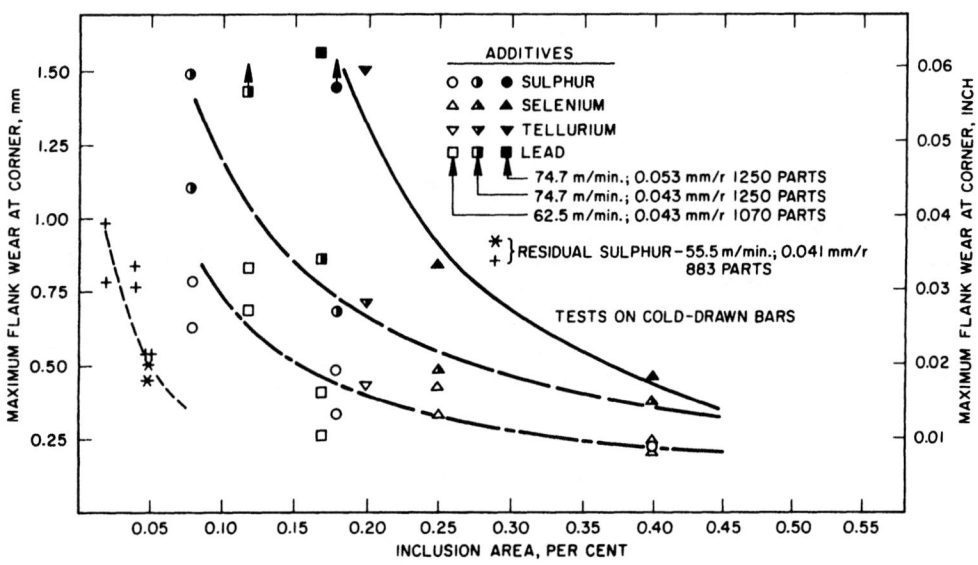

FIG. 52—25. Effect of inclusion area in G86200 (AISI or SAE 8620) steels in flank wear of rough form tools.

der certain cutting conditions, the life of carbide tools attained in machining alloy bars having high levels of hardness of the order of Rc 35 and above, has not been enhanced by the lead addition. On the other hand, the presence of lead has provided improved surface finish and improved chip characteristics even at the higher hardnesses.

In general, for a given level of carbon, an alloy steel will be more difficult to machine than a plain carbon steel. Machinability tends to decrease as the alloy content is increased. Alloy G40230 (AISI or SAE 4023) steel has about the same relative machinability as plain carbon G10220 (AISI or SAE 1022) steel. On the other hand, the machinability rating for the more highly alloyed G86200 (AISI or SAE 8620) grade is about the same as that for plain carbon G10400 (AISI or SAE 1040) steel.

Annealing improves the machinability of the medium-carbon and high-carbon alloy steels. For alloy steels containing about 0.40 per cent carbon, a coarse pearlitic microstructure is generally preferred to provide the optimum combination of cutting-tool performance and machined-surface finish. High-carbon alloy steels, such as the G15216 (AISI E52100 or SAE 52100) grade, require a spheroidize-annealing treatment to provide optimum machining characteristics. If a high level of machinability is a requisite, at least some degree of spheroidization is usually desirable in alloy steels containing more than about 0.50 per cent carbon.

SECTION 5

CALCIUM-CONTAINING MACHINING STEELS

In the last few years there has been a steady growth in the use of calcium-containing carbon and low-alloy steels to increase productivity in machining operations, particularly in operations that utilize carbide cutting tools. In the production of these steels the calcium addition agent is introduced into the steel in the ladle. The amount of calcium retained in the product is very small, generally less than 50 ppm, mainly in the form of an oxide phase and to a lesser extent as a sulfide phase. In the production of calcium-treated aluminum fine-grain steels, the steel refining practice is designed to expel alumina inclusions from the molten steel (because they are abrasive and accelerate tool wear) and

to promote the formation of calcium aluminates which are less abrasive than alumina. In the production of silicon-killed steels in which columbium and/or vanadium are used to inhibit austenite grain growth, the calcium, aluminum and silicon form low-melting-point silicates that coat the cutting tool and thus provide a protective barrier which reduces tool wear. Figure 52—26 shows the amount of wear generated by a standard G10450 (AISI or SAE 1045) aluminum fine-grain steel on a carbide tool in a lathe turning test, and the much lower amount of wear generated by an appropriately made calcium-containing G10450 aluminum fine grain steel.

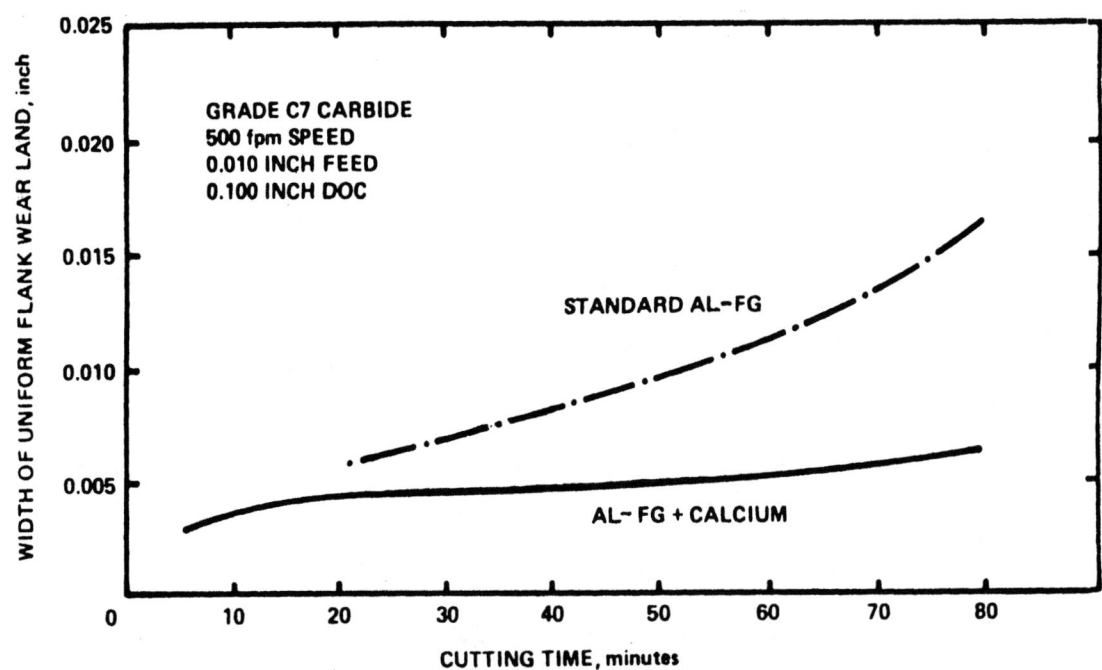

FIG. 52—26. Tool-wear tests on hot-rolled bars of G10450 (AISI or SAE 1045) steel using carbide tools.

SECTION 6

EFFECT OF BAR-FINISHING PRACTICE

It is generally recognized that cold-drawn bars are better adapted for certain machining operations, particularly for use in automatic screw machines, than bars in the hot-rolled condition. The cold-drawn bars are often preferable because they are straighter and have closer tolerances than the hot-rolled bars; moreover, the cold-drawn bars are free from scale and have a smooth surface finish. These attributes make cold-drawn bars ideal for the production of parts in screw machines. On the other hand, cold-drawn bars are costlier than hot-rolled bars and, consequently, some operators prefer to utilize the hot-rolled product.

In addition to the foregoing attributes of cold-drawn bars, cold drawing increases the strength and hardness of bars and decreases their ductility. These metallurgical factors influence the inherent machinability of steels. To determine the extent of the improvement in machinability obtained by cold drawing and to determine which of the aforementioned factors has the greater influence on machinability, two separate studies were conducted on bars of MX steel in the production of the standard drum nut.

In the first study, selected* pickled and machine-straightened hot-rolled bars (hardness of R_B 80) and cold-drawn bars (1.6 mm or $\frac{1}{16}$-inch draft, hardness of R_B 95) from the same heat of MX steel were compared in tests conducted at 95 and 105 surface metres per minute (312 and 345 sfm). As shown in Table 52—VIII, the cold-drawn bars exhibited an improvement in tool life of 43 per cent and 60 per cent at 95 and 105 metres per minute (312 and 345 sfm), respectively. The lower tool-life values exhibited by bars in this study as com-

pared with the values obtained for other heats of this grade is attributed to the high nitrogen content (0.016 per cent) of the steel.

In the second study, bars from four heats of MX steel were tested in the centerless-ground condition and in the cold-drawn condition. The 16-mm ($\frac{5}{8}$-inch) diameter centerless-ground bars were produced by grinding from 16.6-mm ($^{21}\!/_{32}$-inch) diameter hot-rolled bars and were not machined-straightened to avoid cold-working effects). As shown in Figure 52—27, the centerless-ground bars (hardness range of R_B 56 to 70) exhibited an average tool life of 8.6 hours, this time being slightly less than the average of 9.2 hours for bars cold-drawn with a 0.8-mm ($\frac{1}{32}$-inch) draft (hardness range of R_B 86 to 93) and the average value of 9.5 hours for bars cold

Table 52—VIII. Effect of Cold Drawing on the Performance of USS MX Steel Bars in the Production of Drum Nuts.

Cutting Speed		Average Tool Life, h		Improvement by Cold Drawing (Per Cent)
Surface Metres Per Minute	(sfm)	Hot Rolled*	Cold Drawn 1.6-mm ($\frac{1}{16}$-inch) Draft	
95	312	7.6	10.9	43
105	345	3.5	5.6	60

*Bars were selected to minimize variation in bar diameter within the lot. Only bars that were from 15.7 to 15.9 mm (0.618 to 0.627 inch) in diameter and from 0.05 to 0.18 mm (0.002 to 0.007 inch) out-of-round were used.

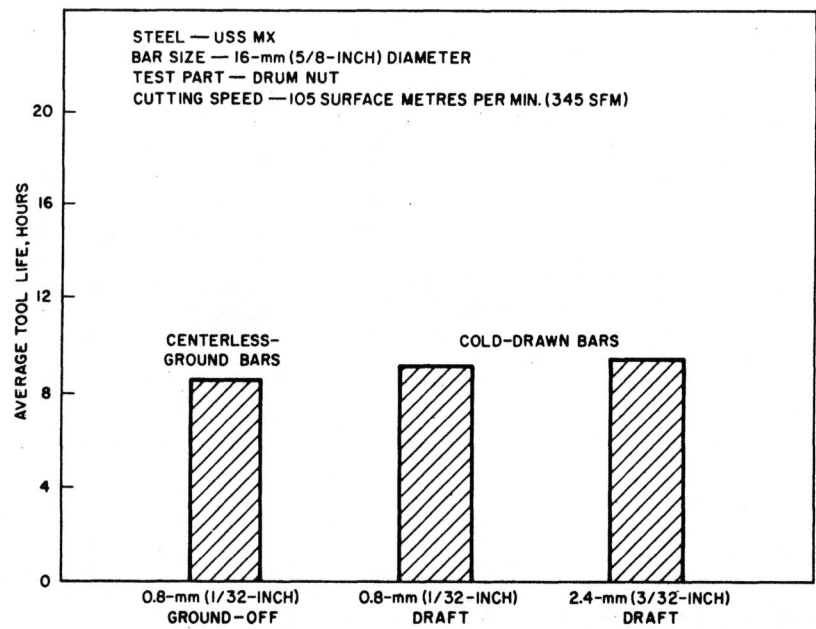

FIG. 52—27. Performance of centerless-ground bars and of cold-drawn bars in the production of drum nuts.

drawn with a 2.4-mm (³⁄₃₂-inch) draft (hardness range of R_B 95 to 98). Despite the considerable difference in hardness among the steels in this series of tests, there was no major effect of hardness on tool performance.

In general, the surface finish on the machined portion of parts produced from the cold-drawn bars in the preceding test was superior to that obtained on parts produced from the centerless-ground bars, as shown in Figure 52—28. The beneficial effect of cold drawing on machined-surface finish is corroborated by tests conducted on hot-rolled and cold-drawn MX bars in a United States Steel-sponsored project at the Massachusetts Institute of Technology. The results of tests conducted with a carbide tool are shown in Figure 52—29. For almost every cutting speed, the machined-surface finish of the cold-drawn bars is superior to that of the hot-rolled bars. The sudden impairment in surface finish at a critical value of cutting speed (for example about 18 surface metres per minute or 60 sfm for the cold-drawn bars) is associated with the initial development of a built-up edge. The projection of the built-up edge into the machined surface decreased as cutting speed increased with an attendant improvement in surface finish.

These tests show that bar-finishing practices have a marked influence on the performance of free-machining steels in automatic screw machines, and that for low-carbon free-machining steel the superior tool performance attained in machining cold-drawn bars (and centerless-ground bars) over that obtained on hot-rolled bars is probably mainly attributable to the better dimensional characteristics rather than to different mechanical properties. The difference in the surface finish of the parts, however, is probably related to differences in mechanical properties.

FIG. 52—28. Effect of cold drawing on the machined-surface finish of drum nuts produced from USS MX steel (Left) Centerless ground, with 0.8-mm (¹⁄₃₂-inch ground off; (Right) Cold-drawn, 0.8-mm (¹⁄₃₂-inch) draft. Magnification: 2X.

The annealing practices discussed previously improve the performance of cold-drawn high-carbon steel bars and certain medium-carbon steels in machining operations. However, the improvement in machinability that is attained with unalloyed medium-carbon bars does not always compensate for the added cost of the annealing.

The annealing operations are conducted prior to drawing. For some applications, it is a common practice to reheat the bars after drawing in the temperature range from 285° to 705°C (550° to 1300°F) either to obtain a desired combination of mechanical properties, to alleviate distortion after machining through the relief of residual stresses, or in some instances to improve cutting-tool performance.

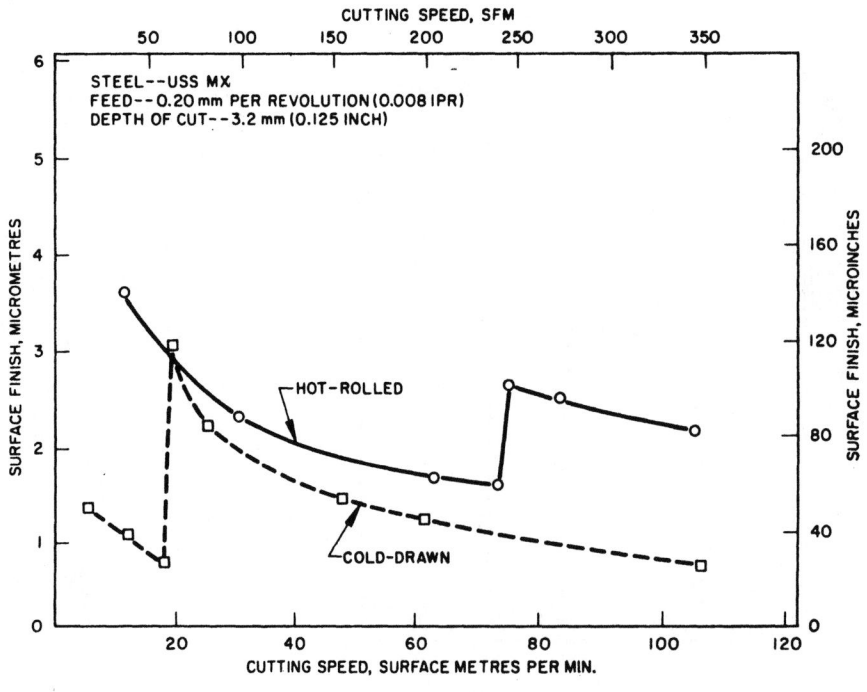

FIG. 52—29. Effect of cold drawing on the machined surface finish of USS MX steel at various cutting speeds in lathe tests.

SECTION 7

FREE-MACHINING PLATE STEELS

Resulphurized plate steels are supplied in both the low-carbon and medium-carbon categories for numerous applications. The two grades listed in Table 52—IX are examples of these products.

These steels can be machined at significantly higher cutting speeds and feeds than G10200 or G10450 (AISI or SAE 1020 or 1045) steels which they can replace in applications that require extensive machining. They are ideal for broaching and milling and can be readily drilled and tapped.

Constant-pressure drilling tests have shown that USS Fremax 45 steel has a drillability rating 25 per cent higher than that of G10450 (AISI or SAE 1045) steel. USS Fremax 45 steel is ideal for applications that require tough wear-resistant surfaces. It can be flame-hardened to a hardness of 62 R_C.

Typical applications for free-machining plate steels are molds, platens, sprockets, gears, bushings, cams, rings, fixtures, dies, die holders, and machining components.

Table 52—IX. Composition and Mechanical Properties of Two Free-Machining Plate Steels.

						Typical Tensile Strength			
						38mm (1½-inch) thick		76mm (3 inch) thick	
Grade	Composition (Per Cent)								
	C	Mn	P	S	Si	MPa	psi	MPa	psi
USS Fremax 15	0.20	0.90	0.06	0.20	0.10	446	64 700	419	60 800
	max.	1.30	0.12	0.30	0.30				
USS Fremax 45	0.40	0.90	0.040	0.20	0.15	633	91 800	616	89 400
	0.50	1.30	max.	0.30	max.				

Bibliography

Aborn, R. M., The role of metallurgy, particularly bismuth, selenium and tellurium in the machinability of steels. American Smelting and Refining Co., 1969.

Almand, E., Influence of low selenium additions on the transverse properties of structural steels. Revue de Metallurgie 66, No. 11, November 1969, page 749.

American Society for Metals, Metals handbook (Vol. 1, 1961, Properties and selection of metals; Section on gas nitriding in Vol. 2, Heat treating, cleaning and finishing (1964); and Vol. 3, 1967, Machining) Metals Park, Ohio, The Society.

Anon., Multiple spindle screw machines. Automatic Machining, April 1968, page 57.

Armour, J. D., Metallurgy and machinability of steels. Contribution to "Machining—Theory and Practice," published by American Society for Metals, Metals Park, Ohio, 1950.

Asada, Chiaki, Free cutting steels. U. S. Pat. No. 3 634 074, January 11, 1972.

Bagley, F. L., Jr. and R. Mennell, Machining characteristics of leaded steels. Jrl. Basic Engineering. Trans. ASME, June 1970, page 347.

Bhattacharya, D. and Schmitt, L. B., New Free-Machining Steels with Bismuth" Part I—Development Part II—Evaluation, Mechanical Working and Steel Processing—XVIII, Iron and Steel Society of AIME, 1980, p153-182.

Boulger, F. W., The effect of selenium on the machinability and tensile properties of five percent chromium steel. Trans. Am. Soc. for Metals, 52 (1960), page 698.

Boulger, F. W., Materials and machinability. Contribution to "Machining—Theory and Practice," published by American society for Metals, Metals Park, Ohio, 1950.

Boulger, F. W., H. E. Hartner, W. T. Lankford, Jr. and T. M. Garvey, Force relationships in the machining of low-carbon steels of different sulphur contents. Trans. Am. Soc. of Mech. Engrs. 79, pages 1155-1164, 1957.

Boulger, F. W., H. A. Moorhead, and T. M. Garvey, Superior machinability of MX explained. Iron Age 167, pages 90-95, May 17, 1951.

Boulger, F. W., H. L. Shaw and H. E. Johnston, Constant-pressure lathe test for measuring the machinability of free-cutting steels. Trans. Am. Soc. of Mech. Engrs. 71, pages 431-438, 1949.

Carney, D. J. and E. C. Rudolphy, Examination of high sulfur free-machining ingot, bloom and billet sections. Trans. Am. Inst. Min. and Met. Engrs. 197, pages 999-1008, 1953.

Case, S. L., U.S. Patent No. 2 485 358, October 18, 1949.

Chalfant, G. M., Revealing lead inclusions in leaded steel. Metal Progress 78, No. 3, September 1960, pages 77-79.

Cook, N. H., P. Jhaveri, and N. Nayak. The mechanism of chip curl and its importance in metal cutting. Trans. Am. Soc. of Mech. Engrs. Journal of Engineering for Industry 85, pages 374-380, 1963.

Dolan, T. J. and B. R. Price, Properites and machinability of a leaded steel. Metals and Alloys 11 (1940), page 20.

Douglas, A. W., W. E. Heitmann and E. S. Madrzyk, The effect of tellurium on the metallurgical and mechanical properties of steel. Trace Additives: Bismuth, Selenium and Tellurium in Iron and Steel. ASM Symposium. Materials Engineering Congress. Oct. 15, 1970.

Garvey, T. M. and H. J. Tata, Factors affecting the machinability of low-carbon free-machining steels. Mechanical working of steel 2, AIME Metallurgical Society Conferences 26. pages 99-132, 1965.

Garvey. T. M. and H. J. Tata. Machinability and metallurgy of resulphurized low-carbon free-machining steels. Am. Soc. of Tool and Mfg. Engrs. Preprint EM66-180, 1966.

Hugo, M. J. Ballot and J. Frey, Study of the hot-forming capacity of structural steels containing tellurium. Revue de

Metallurgie, June 1971, pages 397-409.

Ito, Tetsuro, and Goshi Kato, Free cutting stainless steels. U.S. Pat. No. 3 598 574, Aug. 10, 1971.

Ito, Tetsuro, and Goshi Kato, Free cutting stainless steels. U.S. Pat. No. 3 634 074, Jan. 11, 1972.

Kanter, J. J., Reducing creep in alloy steel bolting materials. Steel, 106, 1940.

Metcut Research Associates, Inc., Machining Data Handbook, Cincinnati, Ohio, 1966.

Miyashita, Yoshio, Katuhiko Nishikawa and Toshio Sata, Free cutting steels containing aluminum and calcium. U.S. Pat. No. 3 647 425, Mar. 7, 1972.

Mostovoy, S. and N. N. Breyer, The effect of lead on the mechanical properties of 4145 steel. Trans. Am. Soc. for Metals 61 (1968), page 219.

Murphy, D. W., Machining performance of steels. Iron and Steel Engineer 44, pages 123-129, April 1967.

Nachtman, E. S., Recent developments in improving the machinability of steel. International Conference of ASTME-Manufacturing Technology (1967), page 673.

Nead, J. E., C. E. Sims and O. E. Harder, Properties of some free-machining lead-bearing steels. Metals and Alloys 10, 1939, pages 68 and 109.

Odin, G., and E. Almand, Constructional steels with improved machinability; III, Influence of selenium and tellurium on the mechanical and physicochemical properties of different carbon and alloy steels. Materiaux et Techniques, March 1971.

Opitz, H., Unsolved problems associated with metal removal operations. Proceedings of the Second Buhl International Conference on materials Metal transformations (1966), page 272.

Opitz, H., "Tool Wear and Tool Life," Proccedings of the International Production Engineering Conference, AIME, 1963, pp107-113.

Paliwoda, E. J., The role of oxygen in free-cutting steels. Mechanical working of steel 2, AIME Metallurgical Society Conferences 26, pages 27-47, 1965.

Paliwoda, E. J., The influence of chemical composition on the machinability of rephosphorized open-hearth screw steel. Trans. Am. Soc. for Metals, 47, pages 680-691, 1955.

Paliwoda, E. J., The new free-cutting steels. Automatic Machining pages 61-64, May 1965.

Palmer, F. R., Free machining structural alloy. U.S. Pat. No. 2 009 716, July 30, 1935.

Schmitt, L. B., and E. S. Nachtman, Effect of tellurium on the machinability of 180 000 psi strength level alloy steel. Trace Additives. Bismuth, Selenium and Tellurium in Iron and Steel. Am. Soc. for Metals Symposium, Materials Engineering Congress. Oct. 15, 1970.

Schmitt, L. B., and E. S. Nachtman, An investigation of the machinability of elevated temperature drawn alloy steel bar stock. Mechanical Working of Steel 2, AIME Metallurgical Society Conferences 26, pages 151-179, 1965.

Shaw, M. C., E. Usui and P. A. Smith, Free-machining steel: III. Cutting forces, surface finish and chip formation. Trans. Am. Soc. of Mech. Engrs. 83(1) pages 181-193, 1961.

Shaw, M. C., P. A. Smith and N. H. Cook, Free-machining steel—tool wear characteristics of resulfurized steels. ASME Paper No. 60-PROD-2, May 1960.

Swinden, T., Leaded manganese-molybdenum steel. Paper No. 12/1943, Alloy Steels Research Committee, Jour. Iron and Steel Inst. 148, No. 2 (1943), page 441.

Tata, H. J., Phosphorus level—effect on machinability of low carbon steel bars. Automatic Machining, pages 43-45, January 1969.

Tata, H. J., The effect of special additives on the machinability of steel. Trace Additives: Bismuth. Selenium and Tellurium in Iron and Steel. Am. Soc. for Metals Symposium. Materials Engineering Congress, Oct. 15, 1970.

Tata, H. J. and R. E. Sampsell, Machinability of AISI 4140 Type steels maximized by selenium. Mechanical Working and Steel Processing—X. The Metallurgical Society of AIME (1972), page 378.

Tata, H. J. and R. E. Sampsell, Effect of additives on the machinability and properties of alloy-steel bars. SAE International Automotive Engineering Congress, 1973. No. 730114.

Tipnis, V. A. and S. W. Poole, The effect of selenium additions on the machinability of low-carbon resulfurized steels. Trace Additives: Bismuth. Selenium and Tellurium in Iron and Steel. Am. Soc. for Metals Symposium, Materials Engineering Congress, Oct. 15, 1970.

Tipnis, V. A. Joseph, R. A. and Dubrava, J. H., "Calcium Deoxidized Improved Machining Steels for Automotive Gears," SAE Paper No. 730115, Jan. 1973.

United States Air Force Machinability Report, Vol. 2, 1951, page 72.

Van Vlack, L. H. Correlation of machinability with inclusion characteristics in resulphurized Bessemer steels. Trans. Am. Soc. for Metals 45, pages 741-753, 1953.

Woolman, J. and A. Jacques, Influence of lead additions on the mechanical properties and machinability of some alloy steels. Jour. Iron and Steel Inst., July 1950, page 257.

Wragge, W. B. Discussion of T. Swinden paper listed above.

Yeo, R. B. G., The effect of oxygen in resulphurized steels. Journal of Metals; Part I, page 29-32, June 1967, Part II, pages 23-27, July 1967.

Zlatin, N., and J. V. Gould, The effect of a lead additive on the machinability of alloy steels. Trans. Am. Inst. Min. and Met. Engrs.—Journal of Engineering for Industry 81, No. 2, May 1959, pages 131-138.

CHAPTER 53

Gage Numbers

INTRODUCTION

In the metal industries, the word gage formerly was used in various systems, or scales, for expressing the thickness or weight per unit area of thin plates, sheet and strip, or the diameters of rods and wire. Specific diameters, thicknesses, or weights per square foot were denoted in gage systems by certain numerals followed by the word gage; for example, No. 12 gage, No. 20 gage, No. 30 gage, or simply 12 gage, 20 gage, and 30 gage. Gage numbers for flat-rolled products were used only in connection with thin materials; that is, usually when the thickness was not more than one-quarter inch or the weight per square foot was not more than 10 pounds, although most gage tables began at about one-half inch, or 20 pounds per square foot, and one table began at double these quantities. Heavier and thicker flat-rolled materials always have been designated by weight per unit area or length, or by thickness in English or metric units.

The danger of confusion in the use of gage numbers to indicate thicknesses and diameters must be emphasized. This danger was present in domestic as well as in foreign trade, and could be avoided by specifying thickness or diameter in inches, centimetres, or millimetres, or in weights per square foot or per square metre, or by giving other equivalents in absolute units. This latter method of specification in absolute units became general for most steel products, and, with the general acceptance of SI, weights and dimensions are now expressed either in absolute units of that system, or in English units. It is anticipated that eventually all weights and dimensions will be in absolute units of SI.

The American Society for Testing and Materials and other groups were active in setting up specifications aimed at eliminating the use of gages. Some of these specifications have been referred to in various chapters of this book dealing with specific steel products. Official publications of these groups should be consulted as a guide to modern ordering practices.

HISTORICAL DEVELOPMENT OF GAGES

Origin of Gages—The custom of indicating thickness and diameter by gage numbers originated in the early days of the metal industries, and the gage numbers were probably first employed to designate the different sizes that could be most readily produced by different stages, or steps, in the processes of manufacture. Inasmuch as these manufacturing processes sometimes varied considerably, not only for different commodities but often among different manufacturers of the same product, and as an individual system of measurement was often considered a trade advantage, a great number of gage systems came into existence. It has been said that at one time there were in use in this country and in England more than fifty different wire gages and several distinct gages for sheets. This condition alone would give rise to considerable confusion, but, as if confusion were to be sought rather than avoided, different names were often applied to the same or practically the same gage system. Also, the same names or symbols were frequently employed to designate different gage systems. All these systems were not only dissimilar with respect to each other in actual thickness denoted by the gage numbers, but the different numbers in the same table seldom bore any mathematical relation to each other.

Relation of Gage Number to Thickness—These gage systems had but one characteristic in common, namely, that the higher the gage number the thinner was the material. This relation of gage numbers to actual thicknesses was always maintained, and the association of high gage number with thin material or with small diameter became fixed, by long custom, in the minds of people associated with the metal industries. With but few exceptions, this relation of gage number to thickness or diameter persisted to very recent times. The exceptions were the Sheet Zinc Gage, Belgian Zinc Gage, Paris (French) Gage, and Music Wire Gage, in which the gage numbers increased with the thickness of the sheet or the diameter of the wire.

British Gages—Chief among the early gage systems were the Birmingham gages, one for sheets and another for wire, and the Stubs' gages. As Stubs was from Warrington, his most popular gage was often called the Warrington Wire Gage, but was also known as Peter Stubs' Gage. In this country, Peter Stubs' Gage and the Birmingham Wire Gage were considered to be identical. The first attempt at reform in gages was to list the equivalent of the gage numbers in decimals of an inch. This was done first individually by Stubs, and later by organized action of the British Board of Trade, in 1883. In that year, a gage was prepared which was intended as a standard for both wire and sheets but was later found to be unsuitable for sheets. This gage became the legal British standard gage on March 1, 1884, and became known as the British Imperial Standard Wire Gage, designated in the British Empire by the initials W.G. or B.W.G., and in the United States by the initials I.S.W.G. or S.W.G. When this gage was found to be unsuitable for sheets, a new gage, called the Birmingham Sheet and Hoop Iron Gage, was prepared by revising the old sheet gage. The new gage was used in England, merely by common consent until 1914, at which time it was established legally as the British Standard Gage for Iron and Steel Sheets and Hoops, represented by the symbols B.G. Custom gages for galvanized sheets usually had B.G. suffixed to the gage numbers, but the weights had no systematic relation to the weights of the British Standard Gage (B.G.) for uncoated sheets. The earlier English gages, Birmingham Wire Gage and Stubs' Warrington Wire Gage, fell into disuse except for telephone and telegraph wire. The relationships between the more common gages mentioned above and some others are given in Table 53—I.

United States Sheet Gages—The next attempt at standardizing gages was made in the United States, in 1892. On March 3rd of that year, the United States Standard Gage for Sheet and Plate Iron and Steel (Table 53—I) was established by an Act of Congress as the only standard gage for these materials after July 1, 1893. This gage is a weight gage based upon weights per square foot in pounds avoirdupois. The gage table as established by Congress began with 20 lb/ft², No. 7/0's gage, and ended with 0.25 lb/ft², No. 38 gage, but the light side of the table was extended by custom to 0.1875 lb/ft², or No. 44 gage. In this country the gage was standard for all uncoated iron and steel sheet and plate, and also was used for tin plate in the lighter gages.

Galvanized and long terne sheet had individual gages based on the U.S. Standard Gage with allowances made for the thickness of the coating in each table. Thus, for the same gage number the weight shown in the Galvanized Sheet Gage regardless of coating weight, was 0.1562 lb/ft² heavier than the weight shown in the U.S. Standard Gage (uncoated product). In the Long Terne Sheet Gage, each gage number may have had various weights depending upon the coating weight. With a commercial coating (6 lb per double base box) each gage number was 0.016 lb/ft² heavier than the weight shown for the corresponding U.S. Standard Gage number (see Table 53—II). Chapters 37 and 38 give the modern classifications of long terne sheet (now terne coated sheet) and galvanized sheet, respectively.

As stated in the preceding paragraph, the U.S. Standard Gage was based on weights per square foot in pounds avoirdupois. Table 53—I shows the approximate thickness for each gage number adopted by the originators of the gage, who based these thicknesses on the density of wrought iron, which is 480 lb/ft³ or 0.2778 lb/in.³. Since the adopted standard density for steel was about 2 per cent heavier than that of wrought iron (489.6 vs. 480 lb/ft³), the thickness equivalents for steel were slightly less than those listed in the U.S. Standard Gage. Although this change was acceptable because the governing factors in the gage schedule were weights and not thicknesses much confusion occurred in converting from weight to thickness for steel sheets. Consequently, the manufacturers of steel sheets in this country adopted a new gage, known as the Manufacturers' Standard Gage for Sheet Steel (Table 53—III). The gage numbers and corresponding weights in this gage were identical to those contained in the U.S. Standard Gage, but the equivalent thicknesses were less since they were based on the density of steel, not that of wrought iron. The conversion factor used in determining these thicknesses was actually greater than the density of steel by an amount necessary to allow for the fact that sheets were thicker in the center than at the edges where thickness is commonly and most conveniently measured, and other factors.

The factor commonly used in converting from weight to thickness of steel *sheets* was 41.82 lb/ft² per inch thick (see footnote, Table 53—I).

Effective January 1, 1970 domestic sheet producers adopted theoretical minimum weight billing, which required that sheets be specified to minimum decimal thickness. On June 1, 1970 the use of gage numbers for sheets was generally discontinued with all to be specified to the minimum decimal thickness. Weights for billing purposes were calculated on a density of 0.2833 lb/in.³ or 40.80 lb/ft² one inch thick.

Density of Iron and Steel—In the foregoing discussion, the density of steel was given as 489.6 lb/ft³ or 40.8 lb/ft² per inch of thickness, which figure was adopted as the standard density of steel of the grades and kinds generally used in plates and sheets. The actual density of steel varies slightly with composition and treatment, and thus may be at variance with the adopted standard density as can be seen from Table 53—IV, which presents data from various sources.

From these values, it is evident that the weight gage thickness equivalents cannot be applied with accuracy to many of the high alloy steels.

The Tin Plate Gage—The commonly accepted types of tin-mill products formerly were hot-dipped tin plate, short terne plate and black plate. Only black plate has survived until the present in the United States. For these products long custom had established the use of the Tin Plate Gage, which was practically the same for this country and England. This gage was expressed in pounds per base box, rather than in gage numbers; this practice continued in use for specifying electrolytic tin plate, tin-free steel and black plate, until the decision was made to specify these products in English or SI

Table 53—I. Relationship of Historical Gage Numbers

Gage No.	U.S.S.G. Uncoated Carbon Steel Sheets and Light Plates — Approx. Equiv. Thickness, Inch	U.S.S.G. lb/ft²	G.S.G. Galvanized Sheet Steel lb/ft²	G.S.G. oz/ft²	T.P.G. Tin Plate lb/bb	T.P.G. Symbol	Steel Wire Gage (W. & M. / U.S. Steel W.G.) Thickness, Inch	Music Wire Gage, M.W.G. Thickness, Inch	Brown & Sharpe, A.W.G. Thickness, Inch	Stubs' Iron Wire Gage, B.W.G. Thickness, Inch	Gage No.
7/0's	0.4902	20.0000					0.4900				7/0's
6/0's	.4596	18.7500					.4615	0.004	0.5800		6/0's
5/0's	.4289	17.5000					.4305	.005	.5165	0.500	5/0's
4/0's	.3983	16.2500					.3938	.006	.4600	.454	4/0's
3/0's	.3676	15.0000					.3625	.007	.4096	.425	3/0's
2/0's	.3370	13.7500					.3310	.008	.3648	.380	2/0's
0	.3064	12.5000					.3065	.009	.3249	.340	0
1	.2757	11.2500					.2830	.010	.2893	.300	1
2	.2604	10.6250					.2625	.011	.2576	.284	2
3	.2451	10.0000					.2437	.012	.2294	.259	3
4	.2298	9.3750					.2253	.013	.2043	.238	4
5	.2145	8.7500					.2070°	.014	.1819	.220	5
6	.1991	8.1250					.1920°	.016	.1620	.203	6
7	.1838	7.5000					.1770°	.018	.1443	.180	7
8	.1685	6.8750	7.0312	112.5			.1620°	.020	.1285	.165	8
9	.1532	6.2500	6.4062	102.5			.1483°	.022	.1144	.148	9
10	.1379	5.6250	5.7812	92.5			.1350°	.024	.1019	.134	10
11	.1225	5.0000	5.1562	82.5			.1205°	.026	.0907	.120	11
12	.1072	4.3750	4.5312	72.5			.1055°	.029	.0808	.109	12
13	.0919	3.7500	3.9062	62.5			.0915°	.031	.0720	.095	13
14	.0766	3.1250	3.2812	52.5			.0800°	.033	.0641	.083	14
15	.0689	2.8125	2.9687	47.5			.0720°	.035	.0571	.072	15
16	.0613	2.5000	2.6562	42.5			.0625°	.037	.0508	.065	16
17	.0551	2.2500	2.4062	38.5			.0540°	.039	.0453	.058	17
18	.0490	2.0000	2.1562	34.5			.0475°	.041	.0403	.049	18
19	.0429	1.7500	1.9062	30.5			.0410°	.043	.0359	.042	19
20	.0368	1.5000	1.6562	26.5			.0348	.045	.0320	.035	20
21	.0337	1.3750	1.5312	24.5			.0317	.047	.0285	.032	21
22	.0306	1.2500	1.4062	22.5			.0286	.049	.0253	.028	22

WEIGHT GAGES — THICKNESS GAGES

(Continued on next page)

Table 53—I. (Continued).

	WEIGHT GAGES							THICKNESS GAGES				
Name of Gage — Principal Use — Gage No.	United States Standard Gage, U.S.S.G. — Uncoated Carbon Steel Sheets and Light Plates — Approximate Equivalent Thickness, Inch	lb/ft²	Galvanized Sheet Gage, G.S.G. — Galvanized Sheet Steel — lb/ft²	oz/ft²	Tin Plate Gage, T.P.G. — Tin Plate — lb/ft²	lb/bb	Symbol	Steel Wire Gage Washburn & Moen or W. & M. Wire G. U.S. Steel W.G. Steel W.G. — Steel Wire, except Music Wire — Thickness, Inch	Music Wire Gage, M.W.G. — Steel Music Wire — Thickness, Inch	Brown & Sharpe Gage, B. & S.G. A.W.G. — Non-ferrous Sheets and Wire — Thickness, Inch	Stubs' Iron Wire Gage, W.W.G. B.W.G. — Flats, Plates and Wire — Thickness, Inch	Name of Gage — Principal Use — Gage No.
23	.0276	1.1250	1.2812	20.5	1.1250 1.079 1.047	235 228	6X 6XL	.0258	.051	.0226	.025	23
24	.0245	1.0000	1.1562	18.5	1.0000 .987 .964 .955 .895	215 210 208 195	5X D2X 5XL 4X	.0230	.055	.0201	.022	24
25	.0214	.8750	1.0312	16.5	.8750 .863 .827 .804 .771	188 180 175 168	4XL DX 3X 3XL	.0204	.059	.0179	.020	25
26	.0184	.7500	.9062	14.5	.7500 .748 .712	163 155	2X	.0181	.063	.0159	.018	26
27	.0169	.6875	.8437	13.5	.6875 .680 .657 .638	148 143 139	2XL DC	.0173	.067	.0142	.016	27
28	.0153	.6250	.7812	12.5	.6250 .620 .588 .574 .565	135 128 125 123	1X 1XL	.0162	.071	.0126	.014	28
29	.0138	.5625	.7187	11.5	.5625 .542 .514 .505	118 112 110		.0150	.075	.0113	.013	29

(Continued on next page)

Table 53—I. Relationship of Historical Gage Numbers (Concluded).

	WEIGHT GAGES							THICKNESS GAGES				
Name of Gage	United States Standard Gage, U.S.S.G.		Galvanized Sheet Gage, G.S.G.			Tin Plate Gage, T.P.G.		Steel Wire Gage Washburn & Moen or W. & M. Wire G. U.S. Steel W.G. Steel W.G.	Music Wire Gage, M.W.G.	Brown & Sharpe Gage, B. & S.G. A.W.G.	Stubs' Iron Wire Gage, W.W.G. B.W.G.	Name of Gage
Principal Use	Uncoated Carbon Steel Sheets and Light Plates		Galvanized Sheet Steel			Tin Plate		Steel Wire, except Music Wire	Steel Music Wire	Non-ferrous Sheets and Wire	Flats, Plates and Wire	Principal Use
Gage No.	Approximate Equivalent Thickness, Inch	lb/ft²	lb/ft²	oz/ft²	lb/ft²	lb/bb	Symbol	Thickness, Inch	Thickness, Inch	Thickness, Inch	Thickness, Inch	Gage No.
30	.0123	.5000	.6562	10.5	.5000 / .491 / .459	107 / 100	IC / ICL	.0140	.080	.0100	.012	30
31	.0107	.4375	.5937	9.5	.4375 / .436 / .413	95 / 90		.0132	.085	.0089	.010	31
32	.0100	.4062	.5625	9.0	.4062 / .390	85		.0128	.090	.0080	.009	32
33	.0092	.3750	.5312	8.5	.3750 / .367	80		.0118	.095	.0071	.008	33
34	.0084	.3437	.5000	8.0	.3437 / .321	75 / 70		.0104	.100	.0063	.007	34
35	.0077	.3125			.3125 / .298	65		.0095	.106	.0056	.005	35
36	.0069	.2812			.2812 / .276	60		.0090	.112	.0050	.004	36
37	.0065	.2656			.2656 / .253	55		.0085	.118	.0045		37
38	.0061	.2500			.2500			.0080	.124	.0040		38
39	.0057	.2344			.2343 / .2295	50		.0075	.130	.0035		39
40	.0054	.2187			.2187			.0070	.138	.0031		40
41	.0052	.2109			.2109 / .2066	45		.0066	.146	.0028		41
42	.0050	.2031			.2031			.0062	.154	.0025		42
43	.0048	.1953			.1953			.0060	.162	.0022		43
44	.0046	.1875			.1875 / .1836	40		.0058	.170	.0020		44

Three intermediate fractional gages sometimes used are omitted in this table.

Table above was based on the theoretical weight, which made the weight of a plate one foot square and one inch thick 40.8 pounds. Sheets and light plates were gaged on the edge, because spring in the rolls caused the centers to be slightly thicker than the edges. To have the estimated weights of sheets and light plates equal the actual weight, the average weight of a square foot one inch thick was taken as 41.82 pounds.

Table 53—II. Gage Weights Formerly Used for Long Terne Sheet of Various Coating Weights.

Gage Weights in Ounces and Pounds per Square Foot, for the Gages and Coatings Given

Long Terne Gage No.	Commercial		0.35 Ounce		0.45 Ounce		0.55 Ounce		0.75 Ounce		1.10 Ounce		1.45 Ounce	
	oz/ft²	lb/ft²	oz/ft²	lb/ft²	oz/ft²	lb/ft²	oz/ft²	lb/ft²	oz/ft²	lb/ft²	oz/ft²	lb/ft²	oz/ft²	lb/ft²
10	90.25	5.641												
11	80.25	5.016												
12	70.25	4.391												
13	60.25	3.766												
14	50.25	3.141												
15	45.25	2.828												
16	40.25	2.516	40.35	2.522										
17	36.25	2.266	36.35	2.272										
18	32.25	2.016	32.35	2.022	32.45	2.028								
19	28.25	1.766	28.35	1.772	28.45	1.778								
20	24.25	1.516	24.35	1.522	24.45	1.528	24.55	1.534	24.75	1.547				
21	22.25	1.391	22.35	1.397	22.45	1.403	22.55	1.409	22.75	1.422				
22	20.25	1.266	20.35	1.272	20.45	1.278	20.55	1.284	20.75	1.297	21.10	1.319	21.45	1.341
23	18.25	1.141	18.35	1.147	18.45	1.153	18.55	1.159	18.75	1.172	19.10	1.194	19.45	1.216
24	16.25	1.016	16.35	1.022	16.45	1.028	16.55	1.034	16.75	1.047	17.10	1.069	17.45	1.091
25	14.25	0.892	14.35	0.897	14.45	0.903	14.55	0.909	14.75	0.922	15.10	0.944	15.45	0.966
26	12.25	0.766	12.35	0.722	12.45	0.778	12.55	0.784	12.75	0.797	13.10	0.819	13.45	0.841
27	11.25	0.703	11.35	0.709	11.45	0.716	11.55	0.722	11.75	0.734	12.10	0.756	12.45	0.778
28	10.25	0.641	10.35	0.647	10.45	0.653	10.55	0.659	10.75	0.672	11.10	0.694	11.45	0.716
29	9.25	0.578	9.35	0.584	9.45	0.591	9.55	0.597	9.75	0.609	10.10	0.631	10.45	0.653
30	8.25	0.516	8.35	0.522	8.45	0.528	8.55	0.534	8.75	0.547	9.10	0.569	9.45	0.591

Nominal Coating Weights, pounds per double base box

6	9	12	15	20	30	40

Table 53—III. Manufacturers' Standard Gage for Sheet Steel.

Gage thickness equivalents were based on 0.0014945 in. per oz/ft²; 0.023912 in per lb/ft² (reciprocal of 41.820 lb/ft² per in. thick); 3.443329 in. per lb/in².

Manufacturers Standard Gage	oz/ft²	lb/in.²	lb/ft²	Inch Equivalent for Steel Sheet Thickness	Manufacturers Standard Gage No.	Manufacturers Standard Gage	oz/ft²	lb/in.²	lb/ft²	Inch Equivalent for Steel Sheet Thickness	Manufacturers Standard Gage No.
3	160	0.069444	10.0000	0.2391	3	21	22	.0095486	1.3750	.0329	21
4	150	.065104	9.3750	.2242	4	22	20	.0086806	1.2500	.0299	22
5	140	.060764	8.7500	.2092	5	23	18	.0078125	1.1250	.0269	23
6	130	.056424	8.1250	.1943	6	24	16	.0069444	1.0000	.0239	24
7	120	.052083	7.5000	.1793	7	25	14	.0060764	0.87500	.0209	25
8	110	.047743	6.8750	.1644	8	26	12	.0052083	.75000	.0179	26
9	100	.043403	6.2500	.1495	9	27	11	.0047743	.68750	.0164	27
10	90	.039062	5.6250	.1345	10	28	10	.0043403	.62500	.0149	28
						29	9	.0039062	.56250	.0135	29
11	80	.034722	5.0000	.1196	11	30	8	.0034722	.50000	.0120	30
12	70	.030382	4.3750	.1046	12						
13	60	.026042	3.7500	.0897	13	31	7	.0030382	.43750	.0105	31
14	50	.021701	3.1250	.0747	14	32	6.5	.0028212	.40625	.0097	32
15	45	.019531	2.8125	.0673	15	33	6	.0026042	.37500	.0090	33
16	40	.017361	2.5000	.0598	16	34	5.5	.0023872	.34375	.0082	34
17	36	.015625	2.2500	.0538	17	35	5	.0021701	.31250	.0075	35
18	32	.013889	2.0000	.0478	18	36	4.5	.0019531	.28125	.0067	36
19	28	.012153	1.7500	.0418	19	37	4.25	.0018446	.26562	.0064	37
20	24	.010417	1.5000	.0359	20	38	4	.0017361	.25000	.0060	38

Table 53—IV. Approximate Densities of Different Varieties of Iron and Steel.

Material (In Wrought Form)	Density (at 60°F.)		
	grams per cm³	lb/in.³	lb/ft³
Pure Iron (99.9% Fe)	7.86	0.284	491
Soft Steel (0.06% C) ..	7.87	0.284	491
Carbon Steel (0.40% C)	7.84	0.283	489
Tool Steel (0.90% C) ..	7.82	0.282	487
Wrought Iron	7.40-7.90	0.267-0.285	461-493
Stainless Steel (18% Cr, 8% Ni) ...	8.03	0.29	501
Stainless Steel (17% Cr, 0.12% C) .	7.75	0.28	484
Stainless Steel (27% Cr, 0.35% C) .	7.47	0.27	467
High Speed Tool Steel (18% W)	8.75	0.316	546

units as discussed in Chapter 36. By base box was meant 112 sheets, each 14 by 20 inches, or other combinations of number and size of sheets that would cover an area of 31 360 square inches. The gages of tin plate were formerly designated by symbols and names, as IC (Common), IX (X or Extra), DC (Double Common), 2X (two-X), ICL (Light), corresponding respectively to 107 lb, 135 lb, 139 lb, 155 lb, and 100 lb per base box. In these symbols, each X represented a specific additional weight and each L a specific decrease in weight. These symbols fell into disuse, giving way to the more logical method of designation of pounds per base box. Tin plate and black plate of 60-lb basis weight and lighter were usually referred to as "light basis weight plate" and were produced by the double cold-reduction method; that is, a second cold reduction given the plate after annealing, prior to coating. This type of product is still produced, but the designation "light basis weight plate" has been abandoned in favor of dimensional specification in English or SI units.

U.S. Wire Gages—The wire gages in general use were never standardized officially in the United States; the practice was for each of the steel wire manufacturers to adopt his own gage. Chief among the historical gages were the gage of the former American Steel and Wire Company, now an integral part of United States Steel, which adopted the Washburn and Moen gage, and the nearly identical Roebling gage. Therefore, upon recommendation of the Bureau of Standards, these manufacturers' gages were merged into one gage designated as the Steel Wire Gage (Stl. W.G.), or the United States Steel Wire Gage (U.S. Stl. W.G.), which was accepted as the standard gage for all steel wire other than music wire. This became the most commonly used steel-wire gage in this country. For all sheets and wire made of metals other than iron and steel, the Brown and Sharpe Gage (B. & S.G.), or American Wire Gage (A.W.G.), was recognized as the standard gage in the United States. It was prepared by Messrs. Brown and Sharpe of Providence, R.I., at the request of leading manufacturers of nonferrous wire in this country. Another gage, known as the Edison or Circular Mil gage, is still used by electrical engineers to simplify their calculations. This gage is based on the circular mil which is the area of a circle with a diameter of one mil (0.001 inch or 0.0254 mm). Other gages in use are the Trenton Iron Company's gages, and Stubs' Steel Wire Gage. None of the wire gages mentioned above, however, had any official authorization in this country. The only wire gage recognized in Acts of Congress was the Birmingham Wire Gage (B.W.G.), also known as Stubs' Iron Wire Gage, although it was not used to any extent by the wire manufacturers in the United States, except for telephone and telegraph wire (See ASTM A-111-80). This gage was sometimes used in designating the thickness of hoop and other strip-steel products, but the use of gages gradually is disappearing entirely and all thicknesses are generally specified in thousandths of an inch or in millimetres.

APPENDICES

Appendix A. Periodic Table of the Elements

LEGEND

ATOMIC NUMBER → 26 → Fe ← SYMBOL
Iron ← NAME
ATOMIC WEIGHT (Based on ^{12}C) → 55.847

* Starred elements have only one stable isotope.
○ Circled atomic numbers indicate elements synthetically prepared.
( ) indicates most stable or best-known isotope.

NOTES:

The elements in Group IA, (excluding hydrogen) also are known as "alkali metals."

The elements in Group IIA also are known as "alkaline-earth metals."

The elements in Group VIIA also are known as "halogens."

The lanthanide metals also are known as "rare-earth metals."

Period	Group IA	Group IIA	Group IIIB	Group IVB	Group VB	Group VIB	Group VIIB	Group VIII			Group IB	Group IIB	Group IIIA	Group IVA	Group VA	Group VIA	Group VIIA	Inert Gases
1	1 H Hydrogen 1.0079																	2 He Helium 4.00260
2	3 Li Lithium 6.941	4 Be* Beryllium 9.01218											5 B Boron 10.81	6 C Carbon 12.011	7 N Nitrogen 14.0067	8 O Oxygen 15.9994	9 F* Fluorine 18.9984	10 Ne Neon 20.179
3	11 Na* Sodium 22.98977	12 Mg Magnesium 24.305											13 Al* Aluminum 26.98154	14 Si Silicon 28.086	15 P* Phosphorus 30.97376	16 S Sulphur 32.06	17 Cl Chlorine 35.453	18 Ar Argon 39.948
4	19 K Potassium 39.098	20 Ca Calcium 40.08	21 Sc* Scandium 44.9559	22 Ti Titanium 47.90	23 V* Vanadium 50.9414	24 Cr Chromium 51.996	25 Mn* Manganese 54.9380	26 Fe Iron 55.847	27 Co* Cobalt 58.9332	28 Ni Nickel 58.71	29 Cu Copper 63.546	30 Zn Zinc 65.38	31 Ga Gallium 69.72	32 Ge Germanium 72.59	33 As* Arsenic 74.9216	34 Se Selenium 78.96	35 Br Bromine 79.904	36 Kr Krypton 83.80
5	37 Rb Rubidium 85.4678	38 Sr Strontium 87.62	39 Y* Yttrium 88.9059	40 Zr Zirconium 91.22	41 Nb*(a) Niobium 92.9064	42 Mo Molybdenum 95.94	43 Tc Technetium (97)	44 Ru Ruthenium 101.07	45 Rh* Rhodium 102.9055	46 Pd Palladium 106.4	47 Ag Silver 107.868	48 Cd Cadmium 112.40	49 In Indium 114.82	50 Sn Tin 118.69	51 Sb Antimony 121.75	52 Te Tellurium 127.60	53 I* Iodine 126.9045	54 Xe Xenon 131.30
6	55 Cs* Cesium 132.9054	56 Ba Barium 137.34	57–71 Lanthanide Metals (see below)	72 Hf Hafnium 178.49	73 Ta* Tantalum 180.9479	74 W(b) Tungsten 183.85	75 Re Rhenium 186.2	76 Os Osmium 190.2	77 Ir Iridium 192.22	78 Pt Platinum 195.09	79 Au* Gold 196.9665	80 Hg Mercury 200.59	81 Tl Thallium 204.37	82 Pb Lead 207.2	83 Bi* Bismuth 208.9804	84 Po Polonium (209)	85 At Astatine (210)	86 Rn Radon (222)
7	87 Fr Francium (223)	88 Ra Radium (226.0254)	89–103 Actinide Metals (see below)															

The LANTHANIDE METALS are:

57 La* Lanthanum 138.9055	58 Ce Cerium 140.12	59 Pr* Praseodymium 140.9077	60 Nd Neodymium 144.24	61 Pm Promethium (145)	62 Sm Samarium 150.4	63 Eu Europium 151.96	64 Gd Gadolinium 157.25	65 Tb* Terbium 158.9254	66 Dy Dysprosium 162.50	67 Ho* Holmium 164.9303	68 Er Erbium 167.26	69 Tm Thulium 168.9342	70 Yb Ytterbium 173.04	71 Lu Lutetium 174.97

The ACTINIDE METALS are:

89 Ac Actinium (227)	90 Th Thorium 232.0381	91 Pa Protactinium (231.0359)	92 U Uranium 238.029	93 Np Neptunium 237.0482	94 Pu Plutonium (244)	95 Am Americium (243)	96 Cm Curium (247)	97 Bk Berkelium (247)	98 Cf Californium (251)	99 Es Einsteinium (254)	100 Fm Fermium (257)	101 Md Mendelevium (258)	102 No Nobelium (255)	103 Lr Lawrencium (256)

(a) Known also as Columbium (symbol - Cb).

(b) Known also as Wolfram (symbol - W).

Atomic weights are those adopted by the International Union of Pure and Applied Chemistry (IUPAC) in 1971. See for reference "Metals Reference Book," Fifth Edition, C. J. Smithels, editor, published by Butterworths, London and Boston, 1976. IUPAC values are subject to certain limitations for which publications of IUPAC should be consulted.

Appendix B. Some Physical Properties of the Elements[w]

Element	Symbol	Melting Point[t] (°C)	Boiling Point[t] (°C)	Specific Heat[u] (kJ/kg·K)	Heat of Fusion[t] (kJ/kg)	Density at 20°C[t] (kg/m³)	Modulus of Elasticity in Tension[v] (GPa)	Structure at 20°C[s]
Actinium	Ac	1050	3330 (est.)	—	60	10 700	—	FCC
Aluminum	Al	660.4	2447	0.90	395	2698	60	FCC
Americium	Am	994	2600 (est.)	—	50	13 670	—	H
Antimony	Sb	630.7	1640	0.20	160	6684	78	R
Argon	Ar	-189.4	-185.87	0.52	29	1.782	—	g
Arsenic (gray)	As	817 (at 2837 kPa)[a]	613[a]	0.33	285	5720	—	R[f]
Astatine	At	302	334	—	113 (est.)	—	—	—
Barium	Ba	725.1	1849	0.28	58	3590	—	BCC[f]
Berkelium	Bk	—	—	—	—	—	—	—
Beryllium	Be	1277	2484	1.83	1300	1860	270 to 300	CPH[f]
Bismuth	Bi	271.4	1579	0.12	52	9800	32	R
Boron	B	2177	3658	1.20	2100	2460	—	R
Bromine	Br	-7.2	58.76	0.47	66	3103	—	l
Cadmium	Cd	321.1	767	0.23	54	8642	55[r]	CPH
Calcium	Ca	850	1487	0.62	213	1550	22 to 26[b]	FCC[t]
Californium	Cf	—	—	—	—	—	—	—
Carbon (graphite)	C	3652[a]	3930	0.69	8725	2267	5	H[f]
Cerium	Ce	795	3470	0.19	37	6770	40[d]	FCC[f]
Cesium	Cs	28.8	678.5	0.24	16	1879[b]	—	BCC
Chlorine	Cl	-101.0	-34.1	0.48	90	2.98	—	g
Chromium	Cr	1857	2682	0.44	395	7200	250	BCC[f]
Cobalt	Co	1494	2897	0.42	275	8900	200	CPH[f]
Columbium (see Niobium)	Cb	—	—	—	—	—	—	—
Copper	Cu	1084.5	2582	0.39	209	8920	110	FCC
Curium	Cm	—	—	—	—	13 510	—	CPH
Dysprosium	Dy	1500	2600	0.17	100 (est.)	8536	69 to 96[d]	CPH
Einsteinium	Es	—	—	—	—	—	—	—
Erbium	Er	1497	2900	0.17	100 (est.)	9051	110[d]	CPH
Europium	Eu	826	1440	0.18	69	5259	—	BCC
Fermium	Fm	—	—	—	—	—	—	—

(Continued on next page)

Appendix B (Continued)

Element	Symbol	Melting Point[t] (°C)	Boiling Point[t] (°C)	Specific Heat[u] (kJ/kg·K)	Heat of Fusion[t] (kJ/kg)	Density at 20°C[t] (kg/m³)	Modulus of Elasticity in Tension[v] (GPa)	Structure at 20°C[s]
Fluorine	F	-219.6	-188.2	0.83	42	1.580	—	g
Francium	Fr	—	—	—	—	—	—	—
Gadolinium	Gd	1306	3000	0.27	98	7895	55 to 96[d]	CPH
Gallium	Ga	29.78	1980	0.37	80	6972	—	O
Germanium	Ge	940	2852	0.32	508	5323	—	DC
Gold	Au	1064.42	2808	0.13	63	19 320	80	FCC
Hafnium	Hf	2222	4450	0.14	135	13 310	—	CPH[f]
Helium	He	-272.2	-268.935	5.19	3.4	0.1785	—	g
Holmium	Ho	1461	2600	0.17	104 (est.)	8803	75[d]	CPH
Hydrogen	H	-259.19	-252.87	14.3	58	0.0899	—	g
Indium	In	156.60	2070	0.23	28.4	7280	11	FCT
Iodine	I	113.7	184.2	0.22	61	4660	—	O
Iridium	Ir	2454	4389	0.13	137	22 650	520	FCC
Iron	Fe	1537	2872	0.44	272	7860	195	BCC[f]
Krypton	Kr	-157.2	-153.4	0.25	20	3.7	—	g
Lanthanum	La	920	3470	0.19	61	6174	70 to 75[d]	H[f]
Lawrencium	Lr	—	—	—	—	—	—	—
Lead	Pb	327.5	1751	0.13	23	11 340	14	FCC
Lithium	Li	180.5	1336	3.59	432	535	—	BCC
Lutetium	Lu	1652[q]	3330	0.64	110 (est.)	9842	—	CPH
Magnesium	Mg	650	1105	1.02	368	1740	44[e]	CPH
Manganese	Mn	1244	2120	0.48	270	7300	160	CC[f]
Mendelevium	Md	—	—	—	—	—	—	—
Mercury	Hg	-38.862	356.66	0.14	12	13 595	—	l
Molybdenum	Mo	2610	4646	0.25	290	10 220	320	BCC
Neodymium	Nd	1019	3111	0.19	49	7000	—	H[f]
Neon	Ne	-248.6	-246.05	—	17	0.90	—	g
Neptunium	Np	630	—	—	—	20 450	—	O
Nickel	Ni	1455	2920	0.44	300	8902	205[b]	FCC
Niobium	Nb	2477	4863	0.27	290	8570	—	BCC

(Continued on next page)

Appendix B (Continued)

Element	Symbol	Melting Point[t] (°C)	Boiling Point[t] (°C)	Specific Heat[u] (kJ/kg·K)	Heat of Fusion[t] (kJ/kg)	Density at 20°C[t] (kg/m³)	Modulus of Elasticity in Tension[v] (GPa)	Structure at 20°C[s]
Nitrogen	N	−209.97	−195.798	1.04	26	1.165	—	g
Nobelium	No	—	—	—	—	—	—	—
Osmium	Os	2727	5500	0.13	167	22 610	560	CPH
Oxygen	O	−218.787	−182.962	0.92	14	1.381	—	g
Palladium	Pd	1554	2940	0.24	160	12 023	112	FCCᵞ
Phosphorus (white)	P	44.2	280.3	0.74	21	1830	—	H[f]
Platinum	Pt	1772	3824	0.13	100	21 450	146[k]	FCC
Plutonium	Pu	640	3235	0.14	96	19 820	96	M[f]
Polonium	Po	254	962	0.13	59.8	9200	—	C
Potassium	K	63.7	765.5	0.76	58.6	870	—	BCC
Praseodymium	Pr	919	3130	0.19	80.2	6782	50 to 96[d]	H[f]
Promethium	Pm	1080	2460	0.18	86.5	—	—	H
Protactinium	Pa	1230 (est.)	4030 (est.)	0.12	63	15 400	—	—
Radium	Ra	700	1737	0.12	37	6000	—	—
Radon	Rn	−71	−62	0.09	13	10	—	g
Rhenium	Re	3180	5687	0.13	180	21 040	460[l]	CPH[f]
Rhodium	Rh	1963	3727	0.24	210	12 410	290[m]	FCC
Rubidium	Rb	39.0	694	0.36	26	1530	—	BCC
Ruthenium	Ru	2427	4119	0.23	256	12 450	400 (approx)	CPH[f]
Samarium	Sm	1072.1	1803	0.19	59	7536	55[d]	R[f]
Scandium	Sc	1539	2730	0.57	360	2990	—	CPH
Selenium	Se	221	685	0.32	69	4790	58	H[f]
Silicon	Si	1412	2680	0.71	1790	2330	112[c]	DC
Silver	Ag	961.93	2164	0.23	194	10 500	76	FCC
Sodium	Na	97.80	881.4	1.23	115	970	—	O[f]
Strontium	Sr	769	1381	0.30	94	2600	—	FCC[f]
Sulphur (yellow)	S	115.21	—	0.733	54	2070	—	O[f]
Tantalum	Ta	2985	5513	0.14	201	16 600	185[l]	BCC
Technetium	Tc	2250	4567	0.24	240	11 487	—	CPH
Tellurium	Te	449.6	1009	0.20	135	6240	40	H

(Continued on next page)

Appendix B (Continued)

Element	Symbol	Melting Point[t] (°C)	Boiling Point[t] (°C)	Specific Heat[u] (kJ/kg·K)	Heat of Fusion[t] (kJ/kg)	Density at 20°C[t] (kg/m³)	Modulus of Elasticity in Tension[v] (GPa)	Structure at 20°C[s]
Terbium	Tb	1356	2800	0.18	100 (est.)	8272	—	CPH
Thallium	Tl	303.1	1487	0.13	20	11 800	—	CPH
Thorium	Th	1750	4787	0.12	70	11 710	—	FCC
Thulium	Tm	1543	1730	0.16	110 (est.)	9332	—	CPH
Tin (white, β)	Sn	231.97	2623	0.22	58.9	7280	40 to 45[o]	T
Titanium (α)	Ti	1660	3318	0.42	390	4507	115	CPH[f]
Tungsten	W	3407	5663	0.13	216	19 350	340	BCC[f]
Uranium	U	1130	3927	0.12	54	19 050	165	O[f]
Vanadium	V	1917	3421	0.49	450	6100	120 to 140	BCC
Wolfram (see Tungsten)	W							
Xenon	Xe	−111.8	−108.1	0.16	17.5	5.896	—	g
Ytterbium	Yb	824	1430	0.15	53.22	6977	—	FCC[f]
Yttrium	Y	1530	3304	0.30	129	4478	120[d]	CPH[f]
Zinc	Zn	419.6	911	0.11	113	7100	(p)	CPH
Zirconium	Zr	1852	4504	0.28	230	6520	94	CPH[f]

(a) Sublimes.
(b) At 15°C.
(c) Face-centered cubic.
(d) Measured from stress-strain relationship on as-cast metal.
(e) Dynamic: static, 5.77; both for 99.98 per cent Mg.
(f) Ordinary form: other modifications known or probable.
(g) At 27.8°C.
(h) At 26.7°C.
(i) Rhombohedral.
(j) At 2634 kPa pressure.
(k) For small cyclic strains.
(l) At 20°C.
(m) For hard wire.
(n) Chill-cast specimen.
(o) Cast tin.
(p) Pure zinc has no clearly defined modulus of elasticity.
(q) Distilled metal.

(r) Sand cast.
(s) l = liquid; g = gas: for solid crystal structures, BCC = body-centered cubic; C = cubic; CC = complex cubic; CPH = close-packed hexagonal; DC = diamond cubic; FCC = face-centered cubic; FCT = face-centered tetragonal; H = hexagonal; M = monoclinic; O = orthorhombic; R = rhombohedral; T = tetragonal. From: "CRC Handbook of Chemistry and Physics," 59th edition, R. C. West, editor, published by CRC Press, West Palm Beach, Florida, 1978.
(t) From: "Lange's Handbook of Chemistry," Twelfth Edition, J. A. Dean, editor. McGraw-Hill Book Co., New York, 1979.
(u) From: "Thermophysical Properties of Matter," Y. S. Touloukian and E. H. Buyco, editors. IFI/Plenum, New York, 1970.
(v) From: "Metals Handbook," Vol. I, Eighth Edition, T. Lyman, editor. American Society for Metals, Metals Park, Ohio, 1961.
(w) Factors for converting the SI values in this table to fps units are given in Appendix C.

APPENDIX C. SI FACTORS FOR COMMONLY USED PHYSICAL UNITS*

To Convert from	to	Multiply by
ACCELERATION		
foot/second²	metre/second²	$3.048\,000* \times 10^{-1}$
inch/second²	metre/second²	$2.540\,000* \times 10^{-2}$
AREA		
foot²	metre²	$9.290\,304* \times 10^{-2}$
inch²	metre²	$6.451\,600* \times 10^{-4}$
MASS PER UNIT VOLUME **(DENSITY and MASS PER UNIT CAPACITY)**		
gram/centimetre³	kilogram/metre³	$1.000\,000 \times 10^{+3}$
pound-mass/foot³	kilogram/metre³	$1.601\,846 \times 10^{+1}$
pound-mass/inch³	kilogram/metre³	$2.767\,990 \times 10^{+4}$
slug/foot³	kilogram/metre³	$5.153\,788 \times 10^{+2}$
ELECTRICITY and MAGNETISM		
ampere (international of 1948)	ampere	$9.998\,35 \times 10^{-1}$
ampere-hour	coulomb	$3.600\,000* \times 10^{+3}$
coulomb (international of 1948)	coulomb	$9.998\,35 \times 10^{-1}$
farad (international of 1948)	farad	$9.995\,05 \times 10^{-1}$
faraday (physical)	coulomb	$9.652\,19 \times 10^{+4}$
gauss	tesla	$1.000\,000* \times 10^{-4}$
henry (international of 1948)	henry	$1.000\,495$
maxwell	weber	$1.000\,000* \times 10^{-8}$
ohm (international of 1948)	ohm	$1.000\,495$
volt (international of 1948)	volt	$1.000\,330$
ENERGY (INCLUDES WORK)		
British thermal unit (thermochemical)	joule	$1.054\,350 \times 10^{+3}$
calorie (mean)	joule	$4.190\,02$
calorie (thermochemical)	joule	$4.184\,000*$
electron volt	joule	$1.602\,19 \times 10^{-19}$
erg	joule	$1.000\,000* \times 10^{-7}$
foot-pound-force	joule	$1.355\,818$
foot-poundal	joule	$4.214\,011 \times 10^{-2}$
joule (international of 1948)	joule	$1.000\,165$
kilowatt-hour (international of 1948)	joule	$3.600\,59 \times 10^{+6}$
ton (nuclear equivalent of TNT)	joule	$4.184 \times 10^{+9}$
watt-hour	joule	$3.600\,000* \times 10^{+3}$
ENERGY PER UNIT AREA TIME		
Btu (thermochemical)/foot²-second	watt/metre²	$1.134\,893 \times 10^{+4}$
Btu (thermochemical)/foot²-minute	watt/metre²	$1.891\,489 \times 10^{+2}$
erg/centimetre²-second	watt/metre²	$1.000\,000* \times 10^{-3}$
watt/centimetre²	watt/metre²	$1.000\,000* \times 10^{+4}$
FORCE		
dyne	newton	$1.000\,000* \times 10^{-5}$
kilogram-force	newton	$9.806\,650*$
ounce-force (avoirdupois)	newton	$2.780\,139 \times 10^{-1}$
pound-force (lbf avoirdupois)	newton	$4.448\,222$
HEAT		
Btu (thermochemical) in./s ft² deg F (k, thermal conductivity)	watt/metre-kelvin	$5.188\,732 \times 10^{+2}$
Btu (thermochemical) in./h ft² deg F (k, thermal conductivity)	watt/metre-kelvin	$1.441\,314 \times 10^{-1}$
Btu (thermochemical)/ft²	joule/metre²	$1.134\,893 \times 10^{+4}$

(Continued on next page)

APPENDIX C (Continued)

To Convert from	to	Multiply by
HEAT (Continued)		
Btu (thermochemical)/h ft² deg F (C, thermal conductance)	watt/metre²-kelvin	5.674 466
Btu (thermochemical)/pound-mass	joule/kilogram	$2.324\ 444\ \times 10^{+3}$
Btu (thermochemical)/lbm deg F (c, heat capacity)	joule/kilogram-kelvin	$4.184\ 000\ \times 10^{+3}$
Btu (thermochemical)/s ft² deg F	watt/metre²-kelvin	$2.042\ 808\ \times 10^{+4}$
cal (thermochemical)/cm²	joule/metre²	$4.184\ 000^*\times 10^{+4}$
cal (thermochemical)/cm²s	watt/metre²	$4.184\ 000^*\times 10^{+4}$
cal (thermochemical)/cm s deg C	watt/metre-kelvin	$4.184\ 000^*\times 10^{+2}$
cal (thermochemical)/g	joule/kilogram	$4.184\ 000^*\times 10^{+3}$
cal (thermochemical)/g deg C	joule/kilogram-kelvin	$4.184\ 000^*\times 10^{+3}$
deg F h ft²/Btu (thermochemical) (R, thermal resistance)	kelvin-metre²/watt	$1.762\ 280\ \times 10^{-1}$
ft²/h (thermal diffusivity)	metre²/second	$2.580\ 640^*\times 10^{-5}$
LENGTH		
angstrom	metre	$1.000\ 000^*\times 10^{-10}$
foot	metre	$3.048\ 000^*\times 10^{-1}$
inch	metre	$2.540\ 000^*\times 10^{-2}$
micron	metre	$1.000\ 000^*\times 10^{-6}$
mil	metre	$2.540\ 000^*\times 10^{-5}$
mile (U.S. statute)	metre	$1.609\ 3\ \ \ \ \times 10^{+3}$
yard	metre	$9.144\ 000^*\times 10^{-1}$
MASS		
gram	kilogram	$1.000\ 000^*\times 10^{-3}$
kilogram-mass	kilogram	$1.000\ 000^*$
ounce-mass (avoirdupois)	kilogram	$2.834\ 952\ \times 10^{-2}$
ounce-mass (troy or apothecary)	kilogram	$3.110\ 348\ \times 10^{-2}$
pound-mass (lbm avoirdupois)	kilogram	$4.535\ 924\ \times 10^{-1}$
slug	kilogram	$1.459\ 390\ \times 10^{+1}$
ton (long, 2240 lbm)	kilogram	$1.016\ 047\ \times 10^{+3}$
POWER		
Btu (thermochemical)/second	watt	$1.054\ 350\ \times 10^{+3}$
Btu (thermochemical)/minute	watt	$1.757\ 250\ \times 10^{+1}$
calorie (thermochemical)/second	watt	$4.184\ 000^*$
Calorie (thermochemical)/minute	wa-t	$6.973\ 333\ \times 10^{-2}$
foot-pound-force/hour	watt	$3.766\ 161\ \times 10^{-4}$
foot-pound-force/minute	watt	$2.259\ 697\ \times 10^{-2}$
foot-pound-force/second	watt	$1.355\ 818$
horsepower (550 ft lbf/s)	watt	$7.456\ 999\ \times 10^{+2}$
horsepower (electric)	watt	$7.460\ 000^*\times 10^{+2}$
kilocalorie (thermochemical)/minute	watt	$6.973\ 333\ \times 10^{+1}$
kilocalorie (thermochemical)/second	watt	$4.184\ 000^*\times 10^{+3}$
watt (international of 1948)	watt	$1.000\ 165$
PRESSURE or STRESS (FORCE PER UNIT AREA)		
atmosphere (normal = 760 torr)	pascal	$1.013\ 250^*\times 10^{+5}$
bar	pascal	$1.000\ 000^*\times 10^{+5}$
centimetre of mercury (0°C)	pascal	$1.333\ 22\ \ \ \times 10^{+3}$
centimetre of water (4°C)	pascal	$9.806\ 38\ \ \ \times 10^{+1}$
dyne/centimetre²	pascal	$1.000\ 000^*\times 10^{-1}$
foot of water (3g.2°F)	pascal	$2.988\ 98\ \ \ \times 10^{+3}$
hectobar	pascal	$1.000\ 000^*\times 10^{+7}$
inch of mercury (60°F)	pascal	$3.376\ 85\ \ \ \times 10^{+3}$
kilobar	pascal	$1.000\ 000^*\times 10^{+8}$

(Continued on next page)

APPENDIX C (Continued)

To Convert from	to	Multiply by

PRESSURE or STRESS (FORCE PER UNIT AREA) (Continued)

kilogram-force/centimetre²	pascal	$9.806\ 650^* \times 10^{+4}$
pound-force/foot²	pascal	$4.788\ 026 \times 10^{+1}$
pound-force/inch² (psi)	pascal	$6.894\ 757 \times 10^{+3}$
torr (mm Hg, 0°C)	pascal	$1.333\ 22 \times 10^{+2}$

TEMPERATURE

degree Celsius	kelvin	$t_K = t_c + 273.15$
degree Fahrenheit	kelvin	$t_K = (t_F + 459.67)/1.8$
degree Rankine	kelvin	$t_K = t_R/1.8$
degree Fahrenheit	degree Celsius	$t_C = (t_F - 32)/1.8$
kelvin	degree Celsius	$t_C = t_K - 273.15$

VELOCITY (INCLUDES SPEED)

foot/hour	metre/second	$8.466\ 667 \times 10^{-5}$
foot/minute	metre/second	$5.080\ 000^* \times 10^{-3}$
foot/second	metre/second	$3.048\ 000^* \times 10^{-1}$
inch/second	metre/second	$2.540\ 000^* \times 10^{-2}$
kilometre/hour	metre/second	$2.777\ 778 \times 10^{-1}$
mile/minute (U. S. statute)	metre/second	$2.682\ 240^* \times 10^{+1}$
mile/second (U. S. statute)	metre/second	$1.609\ 344^* \times 10^{+3}$
mile/hour (U. S. statute)	kilometre/hour	$1.609\ 344^*$

VISCOSITY

centipoise	pascal•second	$1.000\ 000^* \times 10^{-3}$
foot²/second	metre²/second	$9.290\ 304^* \times 10^{-2}$
poise	pascal•second	$1.000\ 000^* \times 10^{-1}$
slug/foot-second	pascal•second	$4.788\ 026 \times 10^{+1}$

VOLUME (INCLUDES CAPACITY)

fluid ounce (U. S.)	metre³	$2.957\ 353 \times 10^{-5}$
foot³	metre³	$2.831\ 685 \times 10^{-2}$
gallon (U. S. liquid)	metre³	$3.785\ 412 \times 10^{-3}$
inch³	metre³	$1.638\ 706 \times 10^{-5}$
liter	metre³	$1.000\ 000^* \times 10^{-3}$
pint (U. S. liquid)	metre³	$4.731\ 765 \times 10^{-4}$
quart (U. S. liquid)	metre³	$9.463\ 529 \times 10^{-4}$
ton (register)	metre³	$2.831\ 685$

MULTIPLE AND SUBMULTIPLE UNITS

Multiplication Factors	Prefix	SI Symbol
$1\,000\,000\,000\,000 = 10^{12}$	tera	T
$1\,000\,000\,000 = 10^{9}$	giga	G
$1\,000\,000 = 10^{6}$	mega	M
$1\,000 = 10^{3}$	kilo	k
$100 = 10^{2}$	hecto	h
$10 = 10$	deka	da
$0.1 = 10^{-1}$	deci	d
$0.01 = 10^{-2}$	centi	c
$0.001 = 10^{-3}$	milli	m
$0.000001 = 10^{-6}$	micro	μ(mu)
$0.000000001 = 10^{-9}$	nano	n
$0.000000000001 = 10^{-12}$	pico	p
$0.000000000000001 = 10^{-15}$	femto	f
$0.000000000000000001 = 10^{-18}$	atto	a

*Values followed by asterisk are exact.

NOTE: The reader is referred to ASTM Designation E 380 (latest edition) entitled "Standard for Metric Practice", published by the American Society for Testing and Materials, 1916 Race St., Philadelphia, Pa. 19103; and "AISI Metric Practice Guide", published by the American Iron and Steel Institute, 1000 Sixteenth St., Washington, D.C., 20036; for further details.

Appendix E—Annual Production of Raw Steel in the United States (1930-1981)[a]

A. Thousands of Metric Tons

Year	Open Hearth Basic	Open Hearth Acid	Open Hearth Total[c]	Bessemer	Basic Oxygen Process	Electric	Crucible	Grand Total
1981			12 203		66 434	30 976		109 613
1982			11 842		61 339	28 273		101 455
1979			17 380		75 529	30 778		123 687
1978			19 332		75 735	29 245		124 312
1977			18 183		70 223	25 294		113 700
1976			21 292		72 500	22 328		116 120
1975			20 104		65 136	20 575		105 815
1974			32 204		73 983	26 008		132 195
1973			36 088		75 532	25 183		136 803
1972			31 693		67 662	21 519		120 874
1971			32 259		58 008	18 997		109 264
1970			43 565		57 451	18 291		119 307
1969			55 243		54 645	18 263		128 151
1968			59 724		44 281	15 253		119 260
1967	64 002	127	64 129	[d]	37 588	13 689		115 406
1966	76 933	200	77 133	252	30 779	13 490		121 654
1965	85 153	297	85 450	532	20 755	12 523		119 260
1964	88 590	402	88 992	778	14 009	11 502		115 281
1963	80 229	360	80 589	874	7 751	9 906		99 120
1962	74 914	344	75 258	730	5 038	8 176		89 202
1961	76 3Q2	357	76 660	799	3 599	7 860		88 917
1960	77 985	̣367	78 352	1 079	3 035	7 601		90 067
1959	73 686	403	74 089	1 252	1 691	7 741		84 773
1958	68 495	343	68 838	1 266	1 200	6 038		77 342
1957	91 651	572	92 223	2 245	554	7 231		102 253
1956	92 685	611	93 296	2 928	459	7 839		104 522
1955	95 076	503	95 579	3 012	279	7 303		106 173
1954	72 505	367	72 872	2 312		4 931		80 115
1953	90 563	586	91 149	3 498		6 604		101 251
1952	74 518	638	75 156	3 197		6 167		84 520
1951	83 813	707	84 520	4 437		6 479		95 436
1950	77 711	545	78 256	4 114		5 478		87 848
1949	63 268	460	63 728	3 580		3 432		70 740
1948	71 409	567	71 976	3 849		4 588		80 413
1947	69 137	603	69 740	3 839		3 436		77 015
1946	54 533	544	55 077	3 019		2 325		60 421
1945	64 474	789	65 263	3 905		3 136		72 304
1944	71 820	1 085	72 905	4 572		3 845		81 322
1943	70 042	1 283	71 325	5 104		4 163		80 592
1942	68 204	1 197	69 401	5 038		3 608		78 047
1941	66 508	977	67 485	5 060		2 603	2.1[b]	75 150
1940	55 232	626	55 858	3 365		1 542	0.9	60 766
1939	43 390	527	43 917	3 047		933	0.8	47 898
1938	26 103	277	26 380	1 911		513	0.006	28 804
1937	46 507	508	47 015	3 505		859	0.9	51 380
1936	43 806	428	44 234	3 514		785	0.8	48 534
1935	30 849	360	31 209	2 880		550	0.6	34 640
1934	23 630	279	23 909	2 197		367	0.5	26 474
1933	20 379	329	20 708	2 468		428	0.7	23 605
1932	11 931	167	12 098	1 587		245	0.6	13 901
1931	22 485	386	22 871	3 072		417	1.5	26 362
1930	34 818	794	35 612	5 117		613	2.3	41 345

(Continued)

Appendix D. Blast-Furnace Production of Pig Iron (1945-1981)[1] (Continued)

B. Thousands of Net Tons

Year	Pig Iron[2], [3], [4]							Ferro-alloys[3] and Silvery Pig Iron[4]
	Basic	Bessemer	Low Phos-phorus	Foundry	Malleable	All Others[2]	Total Pig Iron	
1981	71 166	56	132	1 221	929	66	73 570	
1980	66 964	39	72	713	799	134	68 721	
1979	84 047	12	44	1 591	1 003	306	87 003	
1978	84 497	295	110	1 856	471	450	87 679	
1977	78 198	254	122	1 610	690	454	81 328	
1976	82 894	1 150	112	1 ,489	800	425	86 870	
1975	75 911	1 087	140	1 321	814	650	79 923	
1974	91 193	1 100	106	1 528	1 382	600	95 909	
1973	96 202	1 242	101	1 581	1 352	359	100 837	371
1972	83 961	1 336	105	1 977	1 193	370	88 942	458
1971	77 341	1 215	145	1 214	1 336	48	81 299	393
1970	86 438	1 447	146	1 707	1 415	282	91 435	381
1969	89 888	1 306	96	1 850	1 456	421	95 017	463
1968	83 396	1 437	169	1 536	1 804	438	88 780	553
1967	81 344	1 722	166	1 550	1 830	372	86 984	663
1966	83 473	2 882	209	1 673	2 848	415	91 500	650
1965	80 431	2 716	220	1 599	2 806	413	88 185	674
1964	73 ,398	2 727	196	1 623	2 274	383	85 601	611
1963	64 950	2 769	188	1 533	2 186	218	71 ,844	631
1962	58 806	2 823	175	1 429	2 154	255	65 642	650
1961	58 150	2 601	176	1 362	2 103	239	64 631	664
1960	58 261	3 404	387	1 467	2 673	288	66 480	839
1959	52 114	3 056	374	1 881	2 488	280	60 193	635
1958	49 115	3 600	320	1 606	2 305	212	57 158	606
1957	65 378	6 344	580	2 279	3 459	335	78 375	964
1956	61 639	6 665	504	2 398	3 467	395	75 069	891
1955	62 485	7 436	263	2 755	3 531	387	76 857	932
1954	47 023	5 626	212	2 273	2 630	202	57 966	721
1953	59 883	8 111	297	2 501	3 784	326	74 902	955
1952	47 511	7 446	307	2 670	3 120	258	61 312	838
1951	54 213	9 046	315	3 051	3 363	287	70 275	953
1950	49 880	8 091	335	2 807	3 181	293	64 587	853
1949	40 905	7 059	302	2 504	2 409	233	53 412	763
1948	46 315	7 732	384	2 770	2 591	264	60 056	989
1947	44 805	7 182	331	2 953	2 875	183	58 329	985
1946	33 728	5 932	167	2 546	2 190	215	44 778	769
1945	39 867	8 256	314	2 249	2 350	188	53 224	943

[1] From American Iron and Steel Institute Annual Statistical Reports for corresponding years.

[2] Includes ferroalloys after 1973: also includes silvery pig iron after 1963; also includes direct castings.

[3] Ferroalloys included in "All Others" after 1973.

[4] Silvery pig iron included with total pig iron after 1963: included with ferroalloys in 1963 and previous years.

Appendix E—Annual Production of Raw Steel in the United States (1930-1981)[a]

A. Thousands of Metric Tons

Year	Open Hearth			Bessemer	Basic Oxygen Process	Electric	Crucible	Grand Total
	Basic	Acid	Total[c]					
1981			12 203		66 434	30 976		109 613
1982			11 842		61 339	28 273		101 455
1979			17 380		75 529	30 778		123 687
1978			19 332		75 735	29 245		124 312
1977			18 183		70 223	25 294		113 700
1976			21 292		72 500	22 328		116 120
1975			20 104		65 136	20 575		105 815
1974			32 204		73 983	26 008		132 195
1973			36 088		75 532	25 183		136 803
1972			31 693		67 662	21 519		120 874
1971			32 259		58 008	18 997		109 264
1970			43 565		57 451	18 291		119 307
1969			55 243		54 645	18 263		128 151
1968			59 724		44 281	15 253		119 260
1967	64 002	127	64 129	[d]	37 588	13 689		115 406
1966	76 933	200	77 133	252	30 779	13 490		121 654
1965	85 153	297	85 450	532	20 755	12 523		119 260
1964	88 590	402	88 992	778	14 009	11 502		115 281
1963	80 229	360	80 589	874	7 751	9 906		99 120
1962	74 914	344	75 258	730	5 038	8 176		89 202
1961	76 302	357	76 660	799	3 599	7 860		88 917
1960	77 985	367	78 352	1 079	3 035	7 601		90 067
1959	73 686	403	74 089	1 252	1 691	7 741		84 773
1958	68 495	343	68 838	1 266	1 200	6 038		77 342
1957	91 651	572	92 223	2 245	554	7 231		102 253
1956	92 685	611	93 296	2 928	459	7 839		104 522
1955	95 076	503	95 579	3 012	279	7 303		106 173
1954	72 505	367	72 872	2 312		4 931		80 115
1953	90 563	586	91 149	3 498		6 604		101 251
1952	74 518	638	75 156	3 197		6 167		84 520
1951	83 813	707	84 520	4 437		6 479		95 436
1950	77 711	545	78 256	4 114		5 478		87 848
1949	63 268	460	63 728	3 580		3 432		70 740
1948	71 409	567	71 976	3 849		4 588		80 413
1947	69 137	603	69 740	3 839		3 436		77 015
1946	54 533	544	55 077	3 019		2 325		60 421
1945	64 474	789	65 263	3 905		3 136		72 304
1944	71 820	1 085	72 905	4 572		3 845		81 322
1943	70 042	1 283	71 325	5 104		4 163		80 592
1942	68 204	1 197	69 401	5 038		3 608		78 047
1941	66 508	977	67 485	5 060		2 603	2.1[b]	75 150
1940	55 232	626	55 858	3 365		1 542	0.9	60 766
1939	43 390	527	43 917	3 047		933	0.8	47 898
1938	26 103	277	26 380	1 911		513	0.006	28 804
1937	46 507	508	47 015	3 505		859	0.9	51 380
1936	43 806	428	44 234	3 514		785	0.8	48 534
1935	30 849	360	31 209	2 880		550	0.6	34 640
1934	23 630	279	23 909	2 197		367	0.5	26 474
1933	20 379	329	20 708	2 468		428	0.7	23 605
1932	11 931	167	12 098	1 587		245	0.6	13 901
1931	22 485	386	22 871	3 072		417	1.5	26 362
1930	34 818	794	35 612	5 117		613	2.3	41 345

(Continued)

Appendix E—Annual Production of Raw Steel in the United States (1930-1981)[a] (Continued)

B. Thousands of Net Tons

Year	Open Hearth			Bessemer	Basic Oxygen Process	Electric	Crucible	Total
	Basic	Acid	Total[c]					
1981			13 452		73 231	34 145		120 828
1980			13 054		67 615	31 166		111 835
1979			19 158		83 256	33 927		136 341
1978			21 310		83 484	32 237		137 031
1977			20 043		77 408	27 882		125 333
1976			23 470		79 918	24 612		128 000
1975			22 161		71 800	22 680		116 642
1974			35 499		81 552	28 669		145 720
1973			39 780		83 260	27 759		150 799
1972			34 936		74 584	23 721		133 241
1971			35 559		63 943	20 941		120 443
1970			48 022		63 330	20 162		131 514
1969			60 894		60 236	20 132		141 262
1968			65 836		48 812	16 814		131 462
1967	70 550	140	70 690	[d]	41 434	15 089		127 213
1966	84 804	221	85 025	278	33 928	14 870		134 101
1965	93 866	327	94 193	586	22 879	13 804		131 462
1964	97 655	443	98 098	858	15 442	12 678		127 076
1963	88 437	397	88 834	963	8 544	10 920		109 261
1962	82 578	379	82 957	805	5 553	9 013		98 328
1961	84 108	394	84 502	881	3 967	8 664		98 014
1960	85 964	404	86 368	1 189	3 346	8 379		99 282
1959	81 225	444	81 669	1 380	1 864	8 533		93 446
1958	75 502	378	75 880	1 396	1 323	6 656		85 255
1957	101 028	630	101 658	2 475	611	7 971		112 715
1956	102 168	673	102 841	3 228	506	8 641		115 216
1955	104 804	555	105 359	3 320	307	8 050		117 036
1954	79 923	405	80 328	2 548		5 436		88 312
1953	99 828	646	100 474	3 856		7 280		111 610
1952	82 143	703	82 846	3 524		6 798		93 168
1951	92 388	779	93 167	4 891		7 142		105 200
1950	85 661	601	86 262	4 535		6 039		96 836
1949	69 742	507	70 249	3 946		3 783		77 978
1948	78 715	625	79 340	4 243		5 057		88 640
1947	76 209	665	76 874	4 232		3 788		84 894
1946	60 112	600	60 712	3 328		2 563		66 603
1945	71 070	870	71 940	4 305		3 457		79 702
1944	79 168	1 196	80 364	5 040		4 238		89 642
1943	77 208	1 414	78 622	5 626		4 589		88 837
1942	75 183	1 319	76 502	5 553		3 977		86 032
1941	73 313	1 077	74 390	5 578		2 869	2.3[b]	82 839
1940	60 883	690	61 573	3 709		1 700	1.0	66 983
1939	47 829	581	48 410	3 359		1 029	0.9	52 799
1938	28 775	305	29 080	2 106		566	0.007	31 752
1937	51 265	560	51 825	3 864		947	1.0	56 637
1936	48 289	472	48 761	3 873		865	0.9	53 500
1935	34 005	397	34 402	3 175		606	0.7	38 184
1934	26 047	308	26 355	2 422		405	0.6	29 183
1933	22 464	363	22 827	2 720		472	0.8	26 020
1932	13 152	184	13 336	1 716		270	0.7	15 323
1931	24 786	425	25 211	3 386		460	1.7	29 059
1930	38 381	875	39 256	5 640		676	2.5	45 575

[a] From: Annual Statistical Reports, American Iron and Steel Institute.

[b] Any crucible steel produced after 1941 is included with electric-furnace steel.

[c] Basic and acid open hearth combined after 1967. Shown separately prior to 1968.

[d] Included with open hearth.

APPENDIX F—Common Abbreviations Used in This Book

Unit or Term	Abbreviation
absolute	abs[1]
alternating current	ac[2]
American wire gage	AWG
ampere(s)	A
angström(s)	A
anno Domini	A.D.
annum	a
approximate(ly)	approx
argon-oxygen decarburizing	AOD
atmosphere(s)	atm
atomic weight(s)	at. wt
average	avg
barrel(s)	bbl
base box	bb
basic oxygen furnace (top blown)	BOF
basic oxygen process (top blown)	BOP
basic oxygen process (bottom blown)	Q-BOP
basic open-hearth (furnace or process	BOH
Baumé	Bé
before Christ	B.C.
Birmingham wire gage	BWG
body-centered cubic	bcc
boiling point	bp
Brinell hardness number	Bhn
British thermal unit(s)	Btu
Browne & Sharpe gage	B&SG
bushel(s)	bu
calorie(s)	cal
candela(s)	cd
centimetre(s)	cm
centimetre-gram-second (system)	cgs
centipoise(s)	cP
coulomb(s)	C
cubic centimetre(s)	cm³
cubic decimetre(s)	dm³
cubic foot (cubic feet)	ft³
cubic foot, standard (see "standard cubic foot")	
cubic foot (feet) per minute	ft³/min
cubic foot (feet) per second	ft³/s
cubic inch(es)	in.³
cubic metre(s)	m³
cubic metre(s) per second	m³/s
cubic millimetre(s)	mm³
cubic yard(s)	yd³
cycles per second (electrical)	Hz[3]
day(s)	d
decimetre(s)	dm
degree(s)	deg or °
degree(s) Celsius	°C
degree(s) Fahrenheit	°F
degree(s) Kelvin	K
degree(s) Rankine	°R
direct current	dc[4]
direct-reduced iron	DRI
Dortmund-Hörder	D-H
double-cold-reduced	2CR
electromagnetic unit(s)	emu
electroslag remelting	ESR
electrostatic unit(s)	esu
electromotive force	emf
estimated	est.
face-centered cubic	fcc
farad(s)	F
foot (feet)	ft

Unit or Term	Abbreviation
foot (feet) per minute	ft/min
foot (feet) per second	ft/s
foot (feet) per second per second	ft/s²
foot-pound(s)	ft-lb
foot-pound-second (system)	fps
gallon(s)	gal
gallon(s) per minute	gal/min
gallon(s) per second	gal/s
gas-stirred	GS
gauss(es)	G
gram(s)	g
gram-atom(s)	g-atom(s)
gram-calories	g-cal
henry (henries)	H
hertz	Hz
hexagonal close-packed	hcp
high-frequency	HF
horsepower	hp
hour(s)	h[5]
hydrogen-ion concentration	pH
inch(es)	in.
inch-pound(s)	in.-lb
inch(es) per minute	ipm
inch(es) per second	ips
inclusive	incl.
induction-stirred	IS
inside diameter	ID
International System of Units	SI[6]
joule(s)	J
kelvin(s)	K
kilocalorie(s)	kcal
kilogauss(es)	kG
kilogram(s)	kg
kilogram-calorie(s)	kg-cal
kilogram(s) per cubic metre	kg/m³
kilogram(s) per second	kg/s
kilojoule(s)	kJ
kilometre(s)	km
kilometre(s) per hour	km/h
kilopascal(s)	kPa
kilovolt(s)	kV
kilovolt-ampere(s)	kVA
kilowatt(s)	kW
kilowatt-hour(s)	kW·h
ladle-to-ladle	LL
ladle-to-mold	LM
logarithm (common)	log
logarithm (natural)	ln
maximum	max
megapascal(s)	MPa
melting point	mp
metre(s)	m
metre(s) per minute	m/min
metre(s) per second	m/s
metre(s) per second per second	m/s²
metric tons of hot metal	mtHM
micrometre(s)	μm
miles per hour	mph
millimetre(s)	mm
millimetre(s) per revolution	mmpr
minimum	min
minute(s)	min
mole(s)	mol(s)
molten-aluminum rim-killed	MA-RK
net tons of hot metal	NTHM
newton(s)	N
newton-metre(s)	N·m

Unit or Term	Abbreviation
newton(s) per square metre	N/m²
ohm(s)	Ω
outside diameter	OD
parts per billion	ppb
parts per million	ppm
pascal(s)	Pa
per cent	%
poise(s)	P
pound(s)	lb
pound-foot (pound-feet)	lb-ft
pound(s)-force	lbf
pound-inch(es)	lb-in.
pound(s)-mass	lbm
pound(s) per cubic foot	lb/ft³
pound(s) per square foot	psf
pound(s) per square inch	psi
pound(s) per square inch absolute	psia
pound(s) per square inch gage	psig
radian(s)	rad
radian(s) per second	rad/s
radian(s) per second per second	rad/s²
rare earth metals	REM
revolution(s) per minute	rpm
revolution(s) per second	rps
Rockwell hardness, B scale	R_B
Rockwell hardness, C scale	R_C
root mean square	rms
Ruhrstahl-Heraeus	R-H
Saybolt Furol seconds	SFS
Saybolt Universal seconds	SUS
second(s) (time)	s
Siemens	S
silicon-controlled rectifier	SCR
specific gravity	sp gr
specific heat	sp ht
square centimetre(s)	cm²
square foot (square feet)	ft²
square inch(es)	in.²
square kilometre(s)	km²
square metre(s)	m²
square metre(s) per second	m²/s
square millimetre(s)	mm²
square yard(s)	yd²
standard cubic foot	scf
steradian(s)	sr
surface feet per minute	sfm
Système International d'Unités, le	SI
Système International Tin Plate Area	SITA
tap degassing	TD
temperature	temp
tesla(s)	T
thousand pounds per square inch	ksi
ton(s) (metric)	t
vacuum arc remelting	VAR
vacuum oxygen decarburizing	VOD
Vickers hardness number	Vhn
volt(s)	V
volume (capacity)	vol
watt(s)	W
watt(s) per metre-kelvin	W/m·K
weber(s)	Wb
weight(s)	wt
weight per cent	wt %
yard(s)	yd
year(s)	a[7]

(Footnotes on next page)

APPENDIX F (Continued)

Footnotes

[1]   Or, occasionally a (as in psia, the abbreviation for "pounds per square inch absolute").

[2]   as adjective, a-c.

[3]   Abbreviation for "hertz."

[4]   As adjective, d-c.

[5]   Sometimes, hr(s) when fps units are used.

[6]   See "Système International d'Unitès, le".

[7]   Abbreviation for "annum."

INDEX

1515

INDEX